HANDBOOK OF

Microbiological Media

By

RONALD M. ATLAS

Edited by

LAWRENCE C. PARKS

[redacted]

CRC Press
Boca Raton Ann Arbor London Tokyo

Library of Congress Cataloging-in-Publication Data

Catalog record is available from the Library of Congress.

Direct all inquiries to CRC Press, Inc., 2000 Corporate Blvd., N.W., Boca Raton, Florida 33431.

International Standard Book Number 0-8493-2944-2
Printed in the United States of America 2 3 4 5 6 7 8 9 0
Printed on acid-free paper

Preface

The *Handbook of Microbiological Media* began as a result of my frustration with not being able easily to locate the most basic information I needed to culture microorganisms. Too often my library didn't have information on commercial products. Frequently I couldn't find descriptions of many media even in most major research libraries because the information was never published or was contained only in technical manuals. There was a clear need for a comprehensive reference on the media used to cultivate microorganisms.

The proliferation of media, the propensity of microbiologists to develop new modifications, variations and formulations of media, the use of differing names to describe equivalent media, the increased number of commercial manufacturers in diverse nations, proliferation of journals, the decreased availability of the *Difco Manual* as well as the limited distribution of similar manuals from other companies, and the loss of available resources that describe classical media, all led to loss of time and unnecessary difficulties in culturing microorganisms.

The *Handbook of Microbiological Media* compiles in one place in a consistent style the formulations, methods of preparation, and uses for several thousand different media. The *Handbook of Microbiological Media* includes descriptions of the media produced by major suppliers of dehydrated media—including Difco, BBL, and Oxoid, all the media used to cultivate the bacteria of the American Type Culture Collection, the media used for the testing of waters, wastewaters, and foods—including those recommended by the USEPA and FDA for the standard methods examination of waters, wastewaters, and foods. It also includes both classic and modern media used for the identification, cultivation, and maintenance of diverse bacteria. The compositions of various media can be compared so that alternate media can be used. Similar or identical media described by different names that actually have the same formulations can be readily identified. Using the *Handbook of Microbiological Media* should save time and effort for anyone cultivating microorganisms.

The *Handbook of Microbiological Media* is simple to use. The media are organized alphabetically. Synonyms for media are also listed and cross referenced within the manual. Each medium includes the composition, instructions for preparation, commercial sources, and uses. Using the *Handbook of Microbiological Media*, the information needed to prepare media for the cultivation of microorganisms can easily be found.

About the Author:

Ronald M. Atlas is Professor of Biology at the University of Louisville. He received his B.S. from the State University of New York at Stony Brook in 1968 and his M.S. and Ph.D. from Rutgers in 1970 and 1972, respectively. He quickly advanced through the academic ranks at the University of Louisville after spending a year at the Jet Propulsion Laboratory in Pasadena, CA. Dr. Atlas has received a number of honors including: The University of Louisville Excellence in Research Award, Johnson and Johnson Fellowship for Biology, and he is listed in *American Men and Women of Science*. In 1991 he received the ASM Award in Applied and Environmental Microbiology.

He teaches a variety of courses in microbiology at the University of Louisville and has authored several textbooks in general microbiology and microbial ecology. He has written well over 100 research papers. He has conducted studies on the fate of oil in the sea. As part of these studies, he has extensively characterized marine bacterial populations and examined the diversity of microorganisms. He pioneered the field of bioremediation for marine oil spills. His recent studies have focused on the application of molecular techniques to environmental problems. His studies have included the development of "suicide vectors" for the containment of genetically engineered microorganisms and the use of gene probes and the polymerase chain reaction for environmental monitoring, including the detection of pathogens and indicator bacteria for water quality monitoring.

He has served on the NIH Recombinant Advisory Committee (RAC), as well as on various advisory boards for the Environmental Protection Agency (EPA). He is Chairperson of the Environmental Committee of the Public and Scientific Affairs Board of the American Society for Microbiology. He has been a national lecturer for Sigma Xi, an American Society for Microbiology Foundation lecturer, and an Australian Society for Microbiology national lecturer. He has served on the editorial boards of *Applied and Environmental Microbiology, Binary, Advances in Microbial Ecology, BioScience, Biotechniques* and *Journal of Industrial Microbiology*. He is editor of *CRC Critical Reviews in Microbiology*.

About the Editor:

Lawrence C. Parks received his B.S. from the State University of New York at Stony Brook in 1971 and his Ph.D. from Rutgers in 1976. He did a postdoctoral fellowship at Temple University Medical School and then joined the faculty of the Department of Microbiology and Immunology of the University of Louisville in 1981. He currently teaches in the Department of Biology at the University of Louisville. His research interests are in bacterial physiology. He has studied photosynthetic bacteria, cell structure, and the physiological response of bacteria in the oral cavity to antimicrobics. He has taught general microbiology, microbial physiology, immunology, medical microbiology, and dental microbiology.

Acknowledgments

Many individuals contributed to the development and production of the *Handbook of Microbiological Media*. Rosalie Cote and Robert Gherna of the American Type Culture Collection provided information on the media employed by the ATCC for the cultivation of microorganisms in their collections. Keith Lampel provided information on the media used for the cultivation of microorganisms from foods. John Simlar provided technical information on media produced by Difco and Martin Cunningham on media produced by Oxoid. Deborah Atlas, Matthew Atlas, Anne Little, and Barbara Turgeon typed the manuscript. Michel Atlas, Betsy Tice, and D. J. Dorney proofread the manuscript. Janice Morey provided assistance in manuscript production. Harvey Kane of CRC Press supported the project from its inception to its completion.

Table of Contents

Introduction

Organization

The media described in the *Handbook of Microbiological Media* are organized alphabetically. Synonyms for media are listed and cross referenced. The description of each medium includes its name(s), composition, instructions for preparation, commercial sources, safety cautions where needed, and uses.

Names of Media

Media often have numerous names. In many cases media with identical compositions produced by different companies have different names. For example Trypticase Soy Agar produced by BBL Microbiology Systems, Tryptone Soy Agar produced by Oxoid Unipath, and Tryptic Soy Agar produced by Difco Laboratories have identical compositions. Many media also are known by acronyms. TSA, for example, is the common acronym for Trypticase Soy Agar. The *Handbook of Microbiological Media* gives the various synonymous names and directs the reader to see the entry where the information about that medium is given. In cases where modifications to a medium yield a new medium, such media generally are listed with the original medium name, followed by the term modified—for example, TSA, Modified rather than Modified TSA. Media that do not have formal names are listed according to the organism grown on that medium—for example, *Bacillus stearothermophilus* Broth.

Trademarks

The names of some media, components of media, and other terms are registered trademarks. The trademarked items referred to in the *Handbook of Microbiological Media* are listed below.

American Type Culture Collection® and ATCC® are trademarks of the American Type Culture Collection.

Bacto®, BiTek® and Difco® are trademarks of Difco Laboratories.

Oxoid® and Lab–Lemco® are trademarks of Unipath Ltd.

Acidase®, BBL®, Biosate®, CTA Medium®, DTA Medium®, DCLS Agar®, Desoxycholate®, Desoxycholate Agar®, Desoxycholate Citrate Agar®, Enterococcosel®, Eugonagar®, Eugonbroth®, GC-Lect®, Gelysate®, IsoVitaleX®, Mycobactosel®, Mycophil®, Mycosel®, Myosate®, Phytone®, Polypeptone®, Selenite-F Enrichment®, Thiotone®, Trichosel®, Trypticase®, TSA II®, and TSI Agar® are trademarks of Becton Dickinson and Co.

Composition of Media

Media for the cultivation of microorganisms contain the substances necessary to support the growth of microorganisms. Due to the diversity of microorganisms and their diverse metabolic pathways, there are numerous media. Even slight differences in the composition of a medium can result in dramatically different growth characteristics of microorganisms.

When methods for culturing microorganisms were first developed in the nineteenth century, largely by Robert Koch and his colleagues, animal and plant tissues were principally used as sources of nutrients used to support microbial growth. One of the major discoveries of Fanny Hesse in Koch's laboratory was that agar could be used to form solidified culture media on which microorganisms would grow. Extracts of plants and animal tissues were prepared as broths or mixed with agar to form a variety of culture media. Virtually any plant, animal, or animal organ was considered for use in preparing media. Infusions were prepared from beef heart, calf brains, and beef liver, as a few examples. These classic infusions still form the primary components of many media that are widely used today, such as Brain Heart Infusion Agar and Liver Broth.

The composition section of each medium describes the ingredients that make up the medium, their amounts, and the pH. It lists those ingredients in order of decreasing amount. Solids are listed first showing the weights to be added, followed by liquids showing the volumes to be included in the medium.

The composition uses generic terms where these are applicable. For example, pancreatic digest of casein is marketed by various manufacturers as trypticase, tryptone, and other commercial product names. While there may well be differences between these products, such differences are undefined. Variations also occur between batches of products produced as digests of animal tissues.

Media for the cultivation of microorganisms have a source of carbon for incorporation into biomass. For autotrophs the carbon source most often is carbon dioxide which may be supplied as bicarbonate within the medium. Carbohydrates, such as glucose, or other organic compounds, such as acetate, various lipids, proteins, hydrocarbons, and other organic compounds are included in media as sources of carbon for heterotrophs. These carbon sources may also serve as the supply of energy. Other compounds, such as ammonium ions, nitrite ions, elemental sulfur and reduced iron may be used as the sources of energy for the cultivation of autotrophs. Nitrogen also is required for microbial growth. It may be supplied as inorganic nitrogen compounds for the cultivation of some microorganisms but more commonly is supplied as proteins, peptones, or amino acids. Phosphates and metals—such as magnesium and iron, are also necessary components of microbiological media. Phosphates may also serve as buffers to maintain the pH of the medium within the growth tolerance limits of the microorganism being cultivated. Various additional growth factors may also be included in the media.

Agars

Agar is the most common solidifying agent used in microbiological media. Agar is a polysaccharide extract from marine algae. It melts at 84°C and solidifies at 38°C. Agar concentrations of 15.0g/L typically are used to form solid media. Lower concentrations of 7.5–10.0g/L are used to produce soft agars or semisolid media. Below are some agars used as solidifying agents in various media.

Agar Bacteriological (Agar No. 1)
: An agar with low calcium and magnesium. Available from Oxoid Unipath.

Agar, Bacto
: A purified agar with reduced pigmented compounds, salts, and extraneous matter. Available from Difco Laboratories.

Agar, BiTek™
: Agar prepared as a special technical grade. Available from Difco Laboratories.

Agar, Flake
: A technical grade agar. Available from Difco Laboratories.

Agar, Grade A
: A select grade agar containing minerals. Available from BBL Microbiology Systems.

Agar, Granulated
: A high grade granulated agar that has been filtered, decolorized, and purified. Available from BBL Microbiology Systems.

Agarose
: A low sulfate neutral gelling fraction of agar that is a complex galactose polysaccharide of near neutral charge.

Agar, Purified
: A very high grade agar that has been filtered, decolorized, and purified by washing and extraction of refined agars. It has reduced mineral content. Available from BBL Microbiology Systems.

Agar Technical (Agar No. 3)
: A technical grade agar. Available from Difco Laboratories and Oxoid Unipath.

Ionagar
: A purified agar. Available from Oxoid Unipath.

Noble Agar
: An agar that has been extensively washed and is essentially free of impurities. Available from Difco Laboratories.

Purified Agar
: An agar that has been extensively washed and extracted with water and organic solvent. Available from Difco Laboratories and Oxoid Unipath.

Peptones

Many complex media, that is, media in which not all the specific chemical components are known, contain peptones as the source of nitrogen. Peptones are hydrolyzed proteins formed by enzymatic or acidic digestion. Casein most often is used as the protein substrate for forming peptones, but other substances, such as soybean meal, also are commonly employed. Below is a list of some of the peptones that are used as ingredients in various media.

Acidase™ Peptone
: A hydrochloric acid hydrolysate of casein. It has a nitrogen content of 8% and is deficient in cystine and tryptophan. Available from BBL Microbiology Systems.

Bacto Casitone
: A pancreatic digest of casein. Available from Difco Laboratories.

Bacto Peptamin
: A peptic digest of animal tissues. Available from Difco Laboratories.

Bacto Peptone
: An enzymatic digest of animal tissues. It has a high concentration of low molecular weight peptones and amino acids. Available from Difco Laboratories.

Bacto Proteose Peptone
: An enzymatic digest of animal tissues. It has a high concentration of high molecular weight peptones. Available from Difco Laboratories.

Bacto Soytone
: A enzymatic hydrolysate of soybean meal. Available from Difco Laboratories.

Bacto Tryptone
: A pancreatic digest of casein. Available from Difco Laboratories.

Bacto Tryptose
: An enzymatic hydrolysate containing numerous peptides including those of higher molecular weights. Available from Difco Laboratories.

Biosate™ Peptone
A hydrolysate of plant and animal proteins. Available from BBL Microbiology Systems.

Casein Hydrolysate
A hydrolysate of casein prepared with hydrochloric acid digestion under pressure and neutralized with sodium hydroxide. It contains total nitrogen of 7.6% and NaCl of 28.3%. Available from Oxoid Unipath.

Gelatone
A pancreatic digest of gelatin. Available from Difco Laboratories.

Gelysate™ Peptone
A pancreatic digest of gelatin deficient in cystine and tryptophan and which has a low carbohydrate content. Available from Oxoid Unipath.

Lactoalbumin Hydrolysate
A pancreatic digest of lactoalbumin, a milk whey protein. It has high levels of amino acids. It contains total nitrogen of 11.9% and NaCl of 1.4%. Available from Difco Laboratories and Oxoid Unipath.

Liver Digest Neutralized
A papaic digest of liver that contains total nitrogen of 11.0% and NaCl of 1.6%. Available from Oxoid Unipath.

Mycological Peptone
A peptone that contains total nitrogen of 9.5% and NaCl of 1.1%. Available from Oxoid Unipath.

Myosate™ Peptone
A pancreatic digest of heart muscle. Available from BBL Microbiology Systems.

Neopeptone
An enzymatic digest of protein. Available from Difco Laboratories.

Peptone Bacteriological Neutralized
A mixed pancreatic and papaic digest of animal tissues. It contains total nitrogen of 14.0% and NaCl of 1.6%. Available from Difco Laboratories and Oxoid Unipath.

Peptone P
A peptic digest of fresh meat that has a high sulfur content and contains total nitrogen of 11.12% and NaCl of 9.3%. Available from Difco Laboratories and Oxoid Unipath.

Peptonized Milk
A pancreatic digest of high grade skim milk powder. It has a high carbohydrate and calcium concentration. It contains total nitrogen of 5.3% and NaCl of 1.6%. Available from Oxoid Unipath.

Phytone™ Peptone
A papaic digest of soybean meal. It has a high vitamin and a high carbohydrate content. Available from BBL Microbiology Systems.

Polypeptone™ Peptone
A mixture of peptones composed of equal parts of pancreatic digest of casein and peptic digest of animal tissue. Available from BBL Microbiology Systems.

Proteose Peptone
A specialized peptone prepared from a mixture of peptones that contains a wide variety of high molecular weight peptides. It contains total nitrogen of 12.7% and NaCl of 8.0%. Available from Difco Laboratories and Oxoid Unipath.

Proteose Peptone No. 2
An enzymatic digest of animal tissues with a high concentration of high molecular weight peptones. Available from Difco Laboratories.

Proteose Peptone No. 3
An enzymatic digest of animal tissues. It has a high concentration of high molecular weight peptones. Available from Difco Laboratories.

Soya Peptone
A papaic digest of soybean meal with a high carbohydrate concentration. It contains total nitrogen of 8.7% and NaCl of 0.4%. Available from Oxoid Unipath.

Soytone
A papaic digest of soybean meal. Available from Difco Laboratories and Oxoid Unipath.

Special Peptone
A mixture of peptones, including meat, plant and yeast digests. It contains a wide variety of peptides, nucleotides, and minerals. It contains total nitrogen of 11.7% and NaCl of 3.5%. Available from Oxoid Unipath.

Thiotone™ E Peptone
An enzymatic digest of animal tissue. Available from BBL Microbiology Systems.

Trypticase™ Peptone
A pancreatic digest of casein. It has a very low carbohydrate content and a relatively high tryptophan content. Available from BBL Microbiology Systems.

Tryptone
A pancreatic digest of casein. It contains total nitrogen of 12.7% and NaCl of 0.4%. Available from Oxoid Unipath.

Tryptone T

A pancreatic digest of casein with lower levels of calcium, magnesium, and iron than tryptone. It contains total nitrogen of 11.7% and NaCl of 4.9%. Available from Difco Laboratories and Oxoid Unipath.

Tryptose

An enzymatic hydrolysate containing high molecular weight peptides. It contains total nitrogen of 12.2% and NaCl of 5.7%. Available from Difco Laboratories and Oxoid Unipath.

Meat and Plant Extracts

Meat and plant infusions are aqueous extracts that are commonly used as sources of nutrients for the cultivation of microorganisms. Such infusions contain amino acids and low molecular weight peptides, carbohydrates, vitamins, minerals, and trace metals. Extracts of animal tissues contain relatively high concentrations of water soluble protein components and glycogen. Extracts of plant tissues contain relatively high concentrations of carbohydrates.

With regard to infusions, many media list as an ingredient infusion from beef heart or another animal tissue. This ingredient is prepared by boiling a given amount of the animal tissue, for example 500.0g, and then using the liquid or, more commonly, drying the broth and using the solids from the infusion. The actual weight of the dry solids extracted from the hot water used to create the infusion varies and so the ingredient typically is simply listed as 500.0g beef heart infusion although the actual weight of solids recovered from the infusion and used in the medium is far less. Brain heart infusion is prepared from calf brains and beef heart.

Below is a list of some of the meat and plant extracts that are used as ingredients in various media.

Bacto Beef

A desiccated powder of lean beef. Available from Difco Laboratories.

Bacto Beef Extract

An extract of beef (paste). Available from Difco Laboratories.

Bacto Beef Extract Desiccated

An extract of desiccated beef. Available from Difco Laboratories.

Bacto Beef Heart for Infusion

A desiccated powder of beef heart. Available from Difco Laboratories.

Bacto Liver

A desiccated powder of beef liver. Available from Difco Laboratories.

Lab-Lemco

A meat extract powder. Available from Oxoid Unipath.

Liver Desiccated

Dehydrated ox livers. Available from Oxoid Unipath.

Malt Extract

A water soluble extract from germinated grain dried by low temperature evaporation. It has a high carbohydrate content. It contains total nitrogen of 1.1% and NaCl of 0.1%.

Growth Factors

Many microorganisms have specific growth factor requirements that must be included in media for their successful cultivation. Vitamins, amino acids, fatty acids, trace metals, and blood components often must be added to media. In some cases specific defined components are used to meet the growth factor requirements. Incorporation of growth factors are used to enrich, that is, to increase the numbers of particular species of microorganisms. Most often mixtures of growth factors are used in microbiological media. Acid hydrolysates of casein commonly are used as sources of amino acids. Extracts of yeast cells also are employed as sources of amino acids and vitamins for the cultivation of microorganisms. Many media, particularly those employed in the clinical laboratory, contain blood or blood components that serve as essential nutrients for fastidious microorganisms. X factor (heme) and V factor (nicotinamide adenine dinucleotide) often are supplied by adding hemoglobin (BBL and Difco Laboratories), IsoVitaleX (BBL Microbiology Systems) and or Supplement VX (Difco Laboratories). Below is a list of some of the growth factors that are used as ingredients in various media.

Bacto Casamino Acids

A mixture of amino acids formed by acid hydrolysis of casein. Available from Difco Laboratories.

Bacto Vitamin Free Casamino Acids

A mixture of amino acids formed by acid hydrolysis of casein that is free of vitamins. Available from Difco Laboratories.

Bovine Albumin

Bovine albumin fraction V 0.2% in 0.85% saline solution. Available from BBL Microbiology Systems.

Bovine Blood, Citrated

Calf blood washed and treated with sodium citrate as an anticoagulant. Available from BBL Microbiology Systems.

Bovine Blood, Defibrinated

Calf blood treated to denature fibrinogen without causing cell lysis. Available from BBL Microbiology Systems.

Campylobacter Growth Supplement
Sodium pyruvate, sodium metabisulfite, and $FeSO_4$.

Castenholtz Salts
Agar, $NaNO_3$, Na_2HPO_4, KNO_3, nitrilotriacetic acid, $MgSO_4 \cdot 7H_2O$, $CaSO_4 \cdot 2H_2O$, NaCl, $FeCl_3$, $MnSO_4$, H_3BO_3, $ZnSO_4$, $CoCl_2 \cdot 6H_2O$, Na_2MoO_4, $CuSO_4$, and H_2SO_4.

CVA Enrichment
Glucose, L-cysteine·HCl·H_2O, vitamin B_{12}, L-glutamine, L-cystine·2HCl, adenine, nicotinamide adenine dinucleotide, cocarboxylase, guanine·HCl, $Fe(NO_3)_3$, *p*-aminobenzoic acid, and thiamine·HCl.

Cysteine Sulfide Reducing Agent
L-Cysteine·HCl·H_2O and $Na_2S \cdot 9H_2O$.

Dubos Medium Albumin
Albumin fraction V, glucose, and saline solution. Available from Difco Laboratories.

Dubos Oleic Albumin Complex
Alkalinized oleic acid, albumin fraction V, and saline solution. Available from Difco Laboratories.

Egg Yolk Emulsion
Chicken egg yolks and whole chicken egg. Available from Difco Laboratories and Oxoid Unipath.

Egg Yolk Emulsion, 50%
Chicken egg yolks, whole chicken egg, and saline solution. Available from Difco Laboratories.

EY Tellurite Enrichment
Egg yolk suspension with potassium tellurite. Available from Difco Laboratories and Oxoid Unipath.

Fresh Yeast Extract Solution
Live, pressed, starch-free, hydrolyzed Baker's yeast.

Fildes Enrichment
A peptic digest of sheep or horse blood that is a rich source of growth factors including hemin and nicotinamide adenine dinucleotide. Available from BBL Microbiology Systems, Difco Laboratories and Oxoid Unipath.

Hemin Solution
Hemin and NaOH.

Hemoglobin
Dried bovine hemoglobin. Used to provide hemin required by many fastidious microorganisms. Available from BBL Microbiology Systems and Difco Laboratories.

Hemoglobin Solution 2%
Provides hemin required by many fastidious microorganisms. Available from BBL Microbiology Systems and Difco Laboratories.

Hoagland Trace Element Solution, Modified
H_3BO_3, $MnCl_2 \cdot 4H_2O$, $AlCl_3$, $CoCl_2$, $CuCl_2$, KI, $NiCl_2$, $ZnCl_2$, $BaCl_2$, Na_2MoO_4, $SeCl_4$, $SnCl_2 \cdot 2H_2O$, $NaVO_3 \cdot H_2O$, KBr, and LiCl.

Horse Blood, Citrated
Horse blood washed and treated with sodium citrate used as an anticoagulant. Available from BBL Microbiology Systems.

Horse Blood, Defibrinated
Horse blood treated to denature fibrinogen without causing cell lysis. Available from BBL Microbiology Systems and Oxoid Unipath.

Horse Blood, Hemolysed
Horse blood treated to lyse cells. Available from Oxoid Unipath.

Horse Blood, Oxalated
Horse blood treated with potassium oxalate as an anticoagulant. Available from Oxoid Unipath.

Horse Serum
Horse blood is allowed to clot at 2°C–8°C so that the serum separates; the serum is filter sterilized. Serum usually is inactivated by heating to 56°C for 30 minutes to eliminate lipases that would cause degradation of lipids and inactivation of complement. Available from Difco Laboratories and Oxoid Unipath.

Hutner's Mineral Base
$MgSO_4 \cdot 7H_2O$, $CaCl_2 \cdot 2H_2O$, $FeSO_4 \cdot 7H_2O$, $(NH_4)_2MoO_4$, $FeSO_4 \cdot 7H_2O$, $ZnSO_4 \cdot 7H_2O$, EDTA, $MnSO_4 \cdot 7H_2O$, $Co(NO_3)_2 \cdot 6H_2O$, $CuSO_4 \cdot 5H_2O$, $Na_2B_4O_7 \cdot 10H_2O$, and nitrilotriacetic acid.

IsoVitaleX® Enrichment
Glucose, L-cysteine·HCl, L-glutamine, L-cystine, adenine, nicotinamide adenine dinucleotide, vitamin B_{12}, thiamine pyrophosphate, guanine·HCl, $Fe(NO_3)_3 \cdot 6H_2O$, *p*-aminobenzoic acid, and thiamine·HCl. Available from BBL Microbiology Systems.

Legionella Agar Enrichment
L-Cysteine and ferric pyrophosphate. Available from Difco Laboratories.

Legionella BCYE Growth Supplement
ACES buffer/KOH, ferric pyrophosphate, L-cysteine-HCl, and α-ketoglutarate. For the enrichment of *Legionella* species. Available from Oxoid Unipath.

Leptospira Enrichment
Lyophilized pooled rabbit serum containing hemoglobin that provides long chain fatty acids and B vitamins for growth of *Leptospira* species. Available from BBL Microbiology Systems and Difco Laboratories.

Metals "44"
$ZnSO_4 \cdot 7H_2O$, $FeSO_4 \cdot 7H_2O$, $MnSO_4 \cdot 7H_2O$, $CuSO_4 \cdot 5H_2O$, $Co(NO_3)_2 \cdot 6H_2O$, EDTA, and $Na_2B_4O_7 \cdot 10H_2O$.

Middlebrook ADC Enrichment
NaCl, bovine albumin fraction V, glucose and catalase. The albumin binds free fatty acids that may be toxic to mycobacteria. Available from BBL Microbiology Systems and Difco Laboratories.

Middlebrook OADC Enrichment
NaCl, bovine albumin, glucose, oleic acid, and catalase. The albumin binds free fatty acids that may be toxic to mycobacteria; the enrichment provides oleic acid used by *Mycobacterium tuberculosis* for growth. Available from BBL Microbiology Systems and Difco Laboratories.

Mycoplasma Enrichment without Penicillin
Horse serum, fresh autolysate of yeast—yeast extract, and thallium acetate. Provides cholesterol and nucleic acids for growth of *Mycoplasma* species. The thallium selectively inhibits other microorganisms. Available from BBL Microbiology Systems.

Mycoplasma Supplement
Yeast extract and horse serum. Available from Difco Laboratories.

Nitsch's Trace Elements
$MnSO_4$, H_3BO_3, $ZnSO_4$, Na_2MoO_4, $CuSO_4$, $CoCl_2 \cdot 6H_2O$, and H_2SO_4.

Oleic Albumin Complex
NaCl, bovine albumin fraction V, and oleic acid. The albumin binds free fatty acids that may be toxic to mycobacteria and the enrichment provides oleic acid that is used by *Mycobacterium tuberculosis* for growth. Available from BBL Microbiology Systems.

PPLO Serum Fraction
Serum fraction A. Available from Difco Laboratories.

Rabbit Blood, Citrated
Rabbit blood washed and treated with sodium citrate as an anticoagulant. Available from BBL Microbiology Systems.

Rabbit Blood, Defibrinated
Rabbit blood treated to denature fibrinogen without causing cell lysis. Available from BBL Microbiology Systems.

RPF Supplement
Fibrinogen, rabbit plasma, trypsin inhibitor, and potassium tellurite. For the selection and nutrient supplementation of *Staphylococcus aureus*. Available from Oxoid Unipath.

Sheep Blood, Citrated
Sheep blood washed and treated with sodium citrate as an anticoagulant. Available from BBL Microbiology Systems.

Sheep Blood, Defibrinated
Sheep blood treated to denature fibrinogen without causing cell lysis. Available from Oxoid Unipath.

SLA Trace Elements
$FeCl_2 \cdot 4H_2O$, H_3BO_3, $CoCl_2 \cdot 6H_2O$, $ZnCl_2$, $MnCl_2 \cdot 4H_2O$, $NiCl_2 \cdot 6H_2O$, $CuCl_2 \cdot 2H_2O$, $Na_2MoO_4 \cdot 2H_2O$, and $Na_2SeO_3 \cdot 5H_2O$.

Soil Extract
African Violet soil and Na_2CO_3.

Supplement A
Yeast concentrate with Crystal Violet. Available from Difco Laboratories.

Supplement B
Yeast concentrate, glutamine, coenzyme, cocarboxylase, hematin and growth factors. Available from Difco Laboratories.

Supplement C
Yeast concentrate. Available from Difco Laboratories.

Supplement VX
Essential growth factors V and X. Available from Difco Laboratories.

Trace Element Mixture
Ethylenediamine tetraacetic acid (EDTA), $ZnSO_4 \cdot 7H_2O$, $CaCl_2$, $MnCl_2 \cdot 4H_2O$, $FeSO_4 \cdot 7H_2O$, $CoCl_2 \cdot 6H_2O$, $CuSO_4 \cdot 5H_2O$, and $(NH_4)_6Mo_7O_{24} \cdot 4H_2O$.

Trace Element Solution HO-LE
H_3BO_3, $MnCl_2 \cdot 4H_2O$, sodium tartrate, $FeSO_4 \cdot 7H_2O$, $CoCl_2 \cdot 6H_2O$, $CuCl_2 \cdot 2H_2O$, $Na_2MoO_4 \cdot 2H_2O$, and $ZnCl_2$.

Trace Elements Solution SL-6
H_3BO_3, $CoCl_2 \cdot 6H_2O$, $ZnSO_4 \cdot 7H_2O$, $MnCl_2 \cdot 4H_2O$, $Na_2MoO_4 \cdot H_2O$, $NiCl_2 \cdot 6H_2O$, and $CuCl_2 \cdot 2H_2O$.

Trace Elements Solution SL-7
$FeCl_2 \cdot 4H_2O$, $CoCl_2 \cdot 6H_2O$, $MnCl_2 \cdot 4H_2O$, $ZnCl_2$, H_3BO_3, $Na_2MoO_4 \cdot 2H_2O$, $NiCl_2 \cdot 6H_2O$, $CuCl_2 \cdot 2H_2O$, and HCl.

Trace Element Solution SL-8
Disodium EDTA, $FeCl_2 \cdot 4H_2O$, $CoCl_2 \cdot 6H_2O$, $MnCl_2 \cdot 4H_2O$, $NiCl_2 \cdot 6H_20$, $ZnCl_2$, H_3BO_3, $NaMoO_4 \cdot 2H_2O$, and $CuCl_2 \cdot 2H_2O$.

Trace Elements Solution SL-10
$FeCl_2 \cdot 4H_2O$, $CoCl_2 \cdot 6H_2O$, $MnCl_2 \cdot 4H_2O$, $ZnCl_2$, $Na_2MoO_4 \cdot 2H_2O$, $NiCl_2 \cdot 6H_2O$, H_3BO_3, $CuCl_2 \cdot 2H_2O$ and HCl (25% solution).

Trace Metals A-5 Mix
H_3BO_3, $MnCl_2 \cdot 4H_2O$, $ZnSO_4 \cdot 7H_2O$, $CuSO_4 \cdot 5H_2O$, $Na_2MoO_4 \cdot 2H_2O$, and $Co(NO_3)_2 \cdot 6H_2O$.

VA Vitamin Solution
Nicotinamide, thiamine·HCl, *p*-aminobenzoic acid, biotin, calcium pantothenate, pyridoxine·2HCl, and cyanocobalamin.

Vitamin K_1 Solution
Vitamin K_1 and ethanol.

Vitox Supplement
Glucose, L-cysteine·HCl, L-glutamine, L-cystine, adenine sulfate, nicotinamide adenine dinucleotide, cocarboxylase, guanine·HCl, $Fe(NO_3)_3 \cdot 6H_2O$, *p*-aminobenzoic acid, vitamin B_{12}, and thiamine·HCl. Available from Oxoid Unipath.

Wolfe's Mineral Solution
$MgSO_4 \cdot 7H_2O$, nitriloacetic acid, NaCl, $MnSO_4 \cdot H_2O$, $FeSO_4 \cdot 7H_2O$, $CoCl_2 \cdot 6H_2O$, $CaCl_2$, $ZnSO_4 \cdot 7H_2O$, $CuSO_4 \cdot 5H_2O$, $AlK(SO_4)_2 \cdot 12H_2O$, H_3BO_3, and $Na_2MoO_4 \cdot 2H_2O$.

Wolfe's Vitamin Solution
Pyridoxine·HCl, thiamine·HCl, riboflavin, nicotinic acid, calcium pantothenate, *p*-aminobenzoic acid, thioctic acid, biotin, folic acid and cyanocobalamin.

Yeast Autolysate Growth Supplement
Yeast autolysate fractions, glucose, and $NaHCO_3$. Available from Oxoid Unipath.

Yeast Dialysate
Active, dried yeast.

Yeast Extract
A water soluble extract of autolyzed yeast cells. Available from BBL Microbiology Systems, Difco Laboratories, and Oxoid Unipath.

Yeastolate
A water soluble fraction of autolyzed yeast cells rich in vitamin B complex. Available from Difco Laboratories.

Selective Components

Many media contain selective components that inhibit the growth of nontarget microorganisms and favor the growth of specific organisms. Selective media are especially useful in the isolation of specific microorganisms from mixed populations. In many media for the study of microorganisms in nature, compounds are included in the media as sole sources of carbon or nitrogen so that only a few types of microorganisms can grow. Selective toxic compounds are also frequently used to select for the cultivation of particular microbial species. The isolation of a pathogen from a stool specimen, for example, where there is a high abundance of nonpathogenic normal microbiota, requires selective media. Often antimicrobics or other selectively toxic compounds are incorporated into media to suppress the growth of the background microbiota while permitting the cultivation of the target organism of interest. Bile salts, selenite, tetrathionate, tellurite, azide, phenylethanol, sodium lauryl sulfate, high sodium chloride concentrations, and various dyes—such as eosin, Crystal Violet, and Methylene Blue—are used as selective toxic chemicals. Antimicrobial agents used to suppress specific types of microorganisms include ampicillin, chloramphenicol, colistin, cycloheximide, gentamicin, kanamycin, nalidixic acid, sulfadiazine, and vancomycin. Various combinations of antimicrobics are effective in suppressing classes of microorganisms, such as enteric bacteria. Below are some of the selective agents, principally antimicrobic mixtures used for the selective isolation of pathogens, that are commonly used as selective agents in microbiological media.

Ampicillin Selective Supplement
Ampicillin. Used in media for the selection of *Aeromonas hydrophila*. Available from Oxoid Unipath.

Anaerobe Selective Supplement GN
Hemin, menadione, sodium succinate, nalidixic acid, and vancomycin. For the selection of Gram-negative anaerobes. Available from Oxoid Unipath.

Anaerobe Selective Supplement NS
Hemin, menadione, sodium pyruvate, and nalidixic acid. For the selection of nonsporulating anaerobes. Available from Oxoid Unipath.

Bacillus cereus Selective Supplement
Polymyxin B. For the selection of *Bacillus cereus*. Available from Oxoid Unipath.

Bordetella Selective Supplement
Cephalexin. For the selection of *Bordetella* species. Available from Oxoid Unipath.

Brucella Selective Supplement
Polymyxin B, bacitracin, cycloheximide, nalidixic acid, nystatin, and vancomycin.

For the selection of *Brucella* species. Available from Oxoid Unipath.

Campylobacter Selective Supplement Blaser-Wang
Vancomycin, polymyxin B, trimethoprim, amphotericin B, cephalothin. For the selection of *Campylobacter* species. Available from Oxoid Unipath.

Campylobacter Selective Supplement Butzler
Bacitracin, cycloheximide, colistin sulfate, sodium cephazolin, and novobiocin. For the selection of *Campylobacter* species. Available from Oxoid Unipath.

Campylobacter Selective Supplement Preston
Polymyxin B, rifampicin, trimethoprim, and cycloheximide. For the selection of *Campylobacter* species. Available from Oxoid Unipath.

Campylobacter Selective Supplement Skirrow
Vancomycin, trimethoprim, and polymyxin B. For the selection of *Campylobacter* species. Available from Oxoid Unipath.

CCDA Selective Supplement
Cefoperazone and amphotericin B. For the selection of *Campylobacter* species. Available from Oxoid Unipath.

Cefoperazone Selective Supplement
Cefoperazone. For the selection of *Campylobacter* species. Available from Oxoid Unipath.

CFC Selective Supplement
Cetrimide, fucidin, and cephaloridine. For the selection of pseudomonads. Available from Oxoid Unipath.

Chapman Tellurite Solution
Potassium tellurite 1% solution. Available from Difco Laboratories.

Chloramphenicol Selective Supplement
Chloramphenicol. For the selection of yeasts and filamentous fungi. Available from Oxoid Unipath.

Clostridium difficile Selective Supplement
D-Cycloserine and cefoxitin. For the selection of *Clostridium difficile*. Available from Difco Laboratories and Oxoid Unipath.

CN Inhibitor
Cesulodin and novobiocin. It inhibits enteric Gram-negative microorganisms. Available from BBL Microbiology Systems.

CNV Antimicrobic
Colistin sulfate, nystatin, and vancomycin. Available from Difco Laboratories.

CNVT Antimicrobic
Colistin sulfate, nystatin, vancomycin, and trimethoprim lactate. Available from Difco Laboratories.

Colbeck's Egg Broth
Egg emulsion and saline solution. Formerly available from Difco Laboratories—replaced with egg emulsion.

Fraser Supplement
Ferric ammonium sulfate, nalidixic acid, and Acriflavin hydrochloride. For the selection of *Listeria* species. Available from Oxoid Unipath.

Gardnerella vaginalis Selective Supplement
Gentamicin sulfate, nalidixic acid, and amphotericin B. For the selection of *Gardnerella vaginalis*. Available from Oxoid Unipath.

GC Selective Supplement
Yeast autolysate, glucose, Na_2HCO_3, vancomycin, colistin methane sulfonate, nystatin, and trimethoprim. For the selection of *Neisseria* species. Available from Oxoid Unipath.

Helicobacter pylori Selective Supplement Dent
Vancomycin, trimethoprim, cefulodin and amphotericin B. For the selection of *Helicobacter pylori*.

Kanamycin Sulfate Selective Supplement
Kanamycin sulfate. For the selection of enterococci. Available from Oxoid Unipath.

LCAT Selective Supplement
Lincomycin, colistin sulfate, amphotericin B, and trimethoprim. For the selection of *Neisseria* species. Available from Oxoid Unipath.

Legionella BMPA Selective Supplement
Cefamandole, polymyxin B, and anisomycin. For the selection of *Legionella* species. Available from Oxoid Unipath.

Legionella GVPC Selective Supplement
Glycine, vancomycin hydrochloride, polymixin B sulfate, and cycloheximide. For the selection of *Legionella* species. Available from Oxoid Unipath.

Legionella MWY Selective Supplement
Glycine, polymyxin B, anisomycin, vancomycin, Bromthymol B, and Bromcresol Purple. For the selection of *Legionella* species. Available from Oxoid Unipath.

Listeria Primary Selective Enrichment Supplement
Nalidixic acid and acriflavin. For the selection of *Listeria* species. Available from Oxoid Unipath.

Listeria Selective Enrichment Supplement
Nalidixic acid, cycloheximide and acriflavin. For the selection of *Listeria* species. Available from Oxoid Unipath.

Listeria Selective Supplement MOX
Colistin and moxalactam. For the selection of *Listeria monocytogenes*. Available from Oxoid Unipath.

Listeria Selective Supplement Oxford
Cycloheximide, colistin sulfate, acriflavin, cefotetan, and fosfomycin. For the selection of *Listeria* species. Available from Oxoid Unipath.

Modified Oxford Antimicrobic Supplement
Moxalactam and colistin sulfate. Available from Difco Laboratories.

MSRV Selective Supplement
Novobiocin. For the selection of *Salmonella*. Available from Oxoid Unipath.

Mycoplasma Supplement G
Horse serum, yeast extract, thallous acetate, and penicillin. For the selection of *Mycoplasma* species. Available from Oxoid Unipath.

Mycoplasma Supplement P
Horse serum, yeast extract, thallous acetate, glucose, Phenol Red, Methylene Blue, penicillin, and *Mycoplasma* broth base. For the selection of *Mycoplasma* species. Available from Oxoid Unipath.

Mycoplasma Supplement S
Yeast extract, horse serum, thallium acetate, and penicillin. Available from Difco Laboratories.

Oxford Antimicrobic Supplement
Cycloheximide, colistin sulfate, acriflavin, cefotetan, and fosfomycin. Available from Difco Laboratories.

Oxgall
Dehydrated fresh bile. For the selection of bile tolerant bacteria. Available from Difco Laboratories.

Oxytetracycline GYE Supplement
Oxytetracycline in a buffer. For the selection of yeasts and filamentous fungi. Available from Oxoid Unipath.

PALCAM Selective Supplement
Polymyxin B, acriflavin hydrochloride, and ceftazidime. For the selection of *Listeria monocytogenes*. Available from Oxoid Unipath.

Perfringens OPSP Selective Supplement A
Sodium sulfadiazine. For the selection of *Clostridium perfringens*. Available from Oxoid Unipath.

Perfringens SFP Selective Supplement A
Kanamycin sulfate and polymyxin B. For the selection of *Clostridium perfringens*. Available from Oxoid Unipath.

Perfringens TSC Selective Supplement A
D-Cycloserine. For the selection of *Clostridium perfringens*. Available from Oxoid Unipath.

Sodium Desoxycholate
Sodium salt of desoxycholic acid. Available from Difco Laboratories.

Sodium Taurocholate
Sodium salt of conjugated bile acid—75% sodium taurocholate and 25% bile salts. For the selection of bile tolerant bacteria. Available from Difco Laboratories.

STAA Selective Supplement
Streptomycin sulfate, cycloheximide, and thallous acetate. For the selection of *Brochothrix thermosphacta*. Available from Oxoid Unipath.

Staph/Strep Selective Supplement
Nalidixic acid and colistin sulfate. For the selection of *Staphylococcus* species and *Streptococcus* species. Available from Oxoid Unipath.

Streptococcus Selective Supplement COA
Colistin sulfate and oxolinic acid. For the selection of *Streptococcus* species. Available from Oxoid Unipath.

Sulfamandelate Supplement
Sodium sulfacetamide and sodium mandelate. For the selection of *Salmonella* species. Available from Oxoid Unipath.

Tellurite Solution
A solution containing potassium tellurite. Inhibits Gram-negative and most Gram-positive microorganisms. It is used for the isolation of *Corynebacterium* species, *Streptococcus* species, *Listeria* species, and *Candida albicans*. Available from BBL Microbiology Systems.

Tinsdale Supplement
Serum, potassium tellurite and sodium thiosulfate. For the selection of *Corynebacterium diphtheriae*. Available from Oxoid Unipath.

V C A Inhibitor
Vancomycin, colistin, anisomycin, and trimethoprim. Inhibits most Gram-negative and Gram-positive bacteria and yeasts. It is

used for the isolation of *Neisseria* species. Available from BBL Microbiology Systems.

V C A T Inhibitor

Vancomycin, colistin, anisomycin, and trimethoprim lactate. Inhibits most Gram-negative and Gram-positive bacteria and yeasts. It is used for the isolation of *Neisseria* species. Available from BBL Microbiology Systems and Oxoid Unipath.

V C N Inhibitor

Colistin, vancomycin, and nystatin. Inhibits most Gram-negative and Gram-positive bacteria and yeasts. It is used for the isolation of *Neisseria* species. Available from BBL Microbiology Systems and Oxoid Unipath.

V C N T Inhibitor

Colistin, vancomycin, nystatin, and trimethoprim lactate. Inhibits most Gram-negative and Gram-positive bacteria and yeasts. It is used for the isolation of *Neisseria* species. Available from BBL Microbiology Systems and Oxoid Unipath.

Yersinia Selective Supplement

Cefsulodin, irgasan, and novobiocin. For the selection of *Yersinia enterocolitica*. Available from Oxoid Unipath.

pH Buffers

Maintaining the pH of media usually is accomplished by the inclusion of suitable buffers. Since microorganisms grow optimally only within certain limits of a pH range, the pH generally is maintained within a few tenths of a pH unit. Phosphate buffers commonly are used. The pH is established by using varying volumes of equimolar concentrations of Na_2HPO_4 and NaH_2PO_4.

pH	Na_2HPO_4 (mL)	NaH_2PO_4 (mL)
5.4	3.0	97.0
5.6	5.0	95.0
5.8	7.8	92.2
6.0	12.0	88.0
6.2	18.5	81.5
6.4	26.5	73.5
6.6	37.5	62.5
6.8	50.0	50.0
7.0	61.1	38.9
7.2	71.5	28.5
7.4	80.4	19.6
7.6	86.8	13.2
7.8	91.4	8.6
8.0	94.5	5.5

Differential Components

The differentiation of many microorganisms is based upon the production of acid from various carbohydrates and other carbon sources or the decarboxylation of amino acids. Some media include indicators, particularly of pH that permit the visual detection of changes in pH resulting from such metabolic reactions. Below is a list of some commonly used pH indicators and their color reactions.

pH Indicator	pH Range	Acid Color	Alkaline Color
m-Cresol Purple	0.5–2.5	Red	Yellow
Thymol Blue	1.2–2.8	Red	Yellow
Bromphenol Blue	3.0–4.6	Yellow	Blue
Bromcresol Green	3.8–5.4	Yellow	Blue
Chlorcresol Green	4.0–5.6	Yellow	Blue
Methyl Red	4.2–6.3	Red	Yellow
Chlorphenol Red	5.0–6.6	Yellow	Red
Bromcresol Purple	5.2–6.8	Yellow	Purple
Bromthymol Blue	6.0–7.6	Yellow	Blue
Phenol Red	6.8–8.4	Yellow	Red
Cresol Red	7.2–8.8	Yellow	Red
m-Cresol Purple	7.4–9.0	Yellow	Purple
Thymol Blue	8.0–9.6	Yellow	Blue
Cresolphthalein	8.2–9.8	Colorless	Red
Phenolphthalein	8.3–10.0	Colorless	Red

Sources of Media

The *Handbook of Microbiological Media* includes the media produced by major suppliers of dehydrated media—including Difco, BBL, and Oxoid Unipath.

BBL Products

The media available from Beckton Dickinson Microbiology Systems, BBL Division, PO Box 243, 250 Schilling Circle, Cockeysville MD 21030 are listed below.

BBL Product Number	Medium Name
95891	A 1 Broth
21828	Acetamide Agar
21375	Acetate Differential Agar
10920	*Actinomyces* Broth
10912	AK Agar No. 2
21377	Amies Modified Transport Medium with Charcoal
97643	Amies Transport Medium without Charcoal
10926	Anaerobic Agar

BBL Product Number	Medium Name
97165	Anaerobic CNA Agar
97127	Andrade's Broth
10937	Antibiotic Medium 1
10943	Antibiotic Medium 2
10932	Antibiotic Medium 3
10949	Antibiotic Medium 4
10953	Antibiotic Medium 5
10965	Antibiotic Medium 8
10969	Antibiotic Medium 9
10973	Antibiotic Medium 10
10977	Antibiotic Medium 11
10986	Antibiotic Medium 13
10982	Antibiotic Medium 19
10914	AOAC Letheen Broth
10916	APT Agar
10918	APT Broth
11819	Arylsulfatase Agar
20827	ATS Medium
10996	Azide Blood Agar
11000	Azide Broth
11001	B12 Assay Medium
21836	*Bacteroides* Bile Esculin Agar
21379	BAGG Broth
11023	Baird–Parker Agar
21808	BCYE Agar
97881	BCYE Differential Agar
97878	BCYE Selective Agar with CCVC
97879	BCYE Selective Agar with PAC
97880	BCYE Selective Agar with PAV
97531	Beef Extract V
11027	BiGGY Agar
21838	Bile Esculin Agar
97350	Bile Esculin Agar with Kanamycin
11030	Bismuth Sulfite Agar
11037	Blood Agar Base
11049	Blood Agar with Low pH
11050	Bordet Gengou Agar
11057	Brain Heart CC Agar
11059	Brain Heart Infusion
11065	Brain Heart Infusion Agar
11069	Brain Heart Infusion with PABA
11071	Brain Heart Infusion with PABA and Agar
11073	Brilliant Green Agar
12150	Brilliant Green Agar with Sulfadiazine
11078	Brilliant Green Bile Agar
11079	Brilliant Green Bile Broth
98083	Brilliant Green Bile Broth with MUG
11086	*Brucella* Agar
97402	*Brucella* Blood Agar with Hemin and Vitamin K1
11088	*Brucella* Broth
12367	Buffered Peptone Water
21727	*Campylobacter* Agar with 5 Antimicrobics and 10% Sheep Blood
21747	*Campylobacter* Thioglycollate Medium with 5 Antimicrobics
11102	Cary and Blair Transport Medium
11106	Casman Agar Base
21733	CDC Anaerobe Blood Agar
21735	CDC Anaerobe Blood Agar with Kanamycin and Vancomycin
21733	CDC Anaerobe Blood Agar with Phenylethyl Alcohol
21846	CDC Anaerobe Laked Blood Agar with Kanamycin and Vancomycin
11533	Cetrimide Agar, USP
11108	Chapman Stone Agar
21169	Chocolate II Agar with Hemoglobin and IsoVitaleX®
12309	CIN Agar
12218	CLED Agar
21852	*Clostridium difficile* Agar
11114	*Clostridium* Selective Agar
11116	Coagulase Mannitol Agar
11124	Columbia Agar
12161	Columbia Broth
12104	Columbia CNA Agar
99933	Columbia CNA Agar, Modified with Sheep Blood
11127	Cooked Meat Medium
95982	Cooked Meat Medium with Glucose, Hemin and Vitamin K
11132	Corn Meal Agar
11094	CTA Agar
11096	CTA Medium
11144	DCLS Agar
11150	Deoxycholate Agar
11154	Deoxycholate Citrate Agar
11160	Deoxycholate Lactose Agar
12330	Dermatophyte Test Medium Base
11165	Dextrose Agar
11167	Dextrose Broth
11175	Dextrose Tryptone Agar
11178	DNase Test Agar
11180	Dubos Broth
98111	E Agar
11187	EC Broth

BBL Product Number	Medium Name
12332	EC Broth with MUG
97873	Egg Yolk Agar, Modified
11195	Emerson Agar
11199	Endo Agar
11203	Endo Agar, LES
11119	Endo Broth
12204	Enterococcosel™ Agar
12207	Enterococcosel™ Broth
11213	*Enterococcus* Agar
11221	Eosin Methylene Blue Agar, Levine
11215	Eosin Methylene Blue Agar, Modified
98118	Esculin Iron Agar
11226	Ethyl Violet Azide Broth
11230	Eugon Agar
11235	Eugon Broth
12338	FC Agar
11364	FC Broth
11245	FDA Broth
11258	Flo Agar
11260	Fluid Thioglycollate Medium, without Glucose or Eh Indicator
11267	Folic Acid Assay Medium
98158	GBNA Medium
11275	GC Agar Base
12207	GC II Agar Base
97715	GC-Lect™ Agar
11279	GN Broth, Hajna
11286	Green Yeast and Mold Broth
21779	Group A Selective Strep Agar with Sheep Blood
97884	HBT Bilayer Medium
97110	HC Agar Base
12210	Hektoen Enteric Agar
11298	Indole Nitrite Medium
11303	Infusion Broth
11310	Inhibitory Mold Agar
21889	Jordan's Tartrate Agar
11313	KF *Streptococcus* Agar
11315	KF *Streptococcus* Broth
11317	Kligler Iron Agar
11333	Lactose Broth
11338	Lauryl Sulfate Broth
98076	Lauryl Sulfate Broth with MUG
11327	LBS™ Agar
11331	LBS™ Broth
21858	Lecithin Lactose Agar
97920	*Legionella* Selective Agar
11341	Letheen Agar

BBL Product Number	Medium Name
12333	*Listeria* Enrichment Broth
11343	Litmus Milk
11355	Loeffler Medium
11359	Lowenstein–Jensen Medium
21896	Lowenstein–Jensen Medium with NaCl
12336	LPM Agar
98125	Lysine Agar, Selective
11362	Lysine Iron Agar
98126	M Broth
98148	M17 Broth
11387	MacConkey Agar
12306	MacConkey Agar No. 2
11393	MacConkey Agar without Crystal Violet
11397	MacConkey Broth
11398	Malonate Broth, Ewing Modified
11401	Malt Agar
11403	Malt Extract Agar
11405	Malt Extract Broth
11407	Mannitol Salt Agar
21557	Martin–Lewis Agar
98149	McBride *Listeria* Agar
11417	Middlebrook 7H9 Broth Base
11422	Middlebrook and Cohn 7H10 Agar
11429	Møller Decarboxylase Broth
21665	Møller KCN Broth Base
21517	Motility Indole Ornithine Medium
11436	Motility Test Medium
11381	MPH Agar
95617	MRS Agar
98156	MRS Broth
11383	MRVP Broth
11443	Mueller Hinton Broth
11438	Mueller Hinton II Agar
21411	Mycobactosel™ Agar
21413	Mycobactosel™ L–J Medium
11445	Mycophil™ Agar
11450	Mycophil™ Agar with Low pH
11452	Mycophil™ Broth
11456	*Mycoplasma* Agar Base
11458	*Mycoplasma* Broth Base
12346	*Mycoplasma* Broth Base, Frey
11461	Mycosel™ Agar
21791	Neomycin Blood Agar
21839	Nitrate Broth
11472	Nutrient Agar
11476	Nutrient Agar, 1.5%
11478	Nutrient Broth
11481	Nutrient Gelatin

BBL Product Number	Medium Name
11486	Orange Serum Agar
11484	Oxidation–Fermentation Medium
98153	PA C Agar
11492	Pantothenate Assay Medium
20976	Petragnani Medium
11538	Phenethyl Alcohol Agar
11502	Phenol Red Agar
11506	Phenol Red Broth
11514	Phenol Red Glucose Broth
11519	Phenol Red Lactose Broth
11527	Phenol Red Mannitol Broth
11533	Phenol Red Sucrose Broth
11537	Phenylalanine Agar
11546	Phytone™ Yeast Extract Agar
11488	PKU Test Agar
11549	Potato Dextrose Agar
98142	Presence–Absence Broth
96349	*Pseudomonas* Isolation Agar
11558	Purple Broth
98122	R2A Agar
12339	R3A Agar
21890	Rapid Fermentation Medium
98123	Regan–Lowe Charcoal Agar
11564	Reinforced Clostridial Agar
11567	Rice Extract Agar
97802	SABHI Agar
11584	Sabouraud Glucose Agar
11589	Sabouraud Glucose Agar, Emmons
97815	Sabouraud Maltose Agar
11596	*Salmonella–Shigella* Agar
21820	Salt Broth, Modified
12189	Schaedler Broth
11606	Selenite Cystine Broth
11607	Selenite F Broth
11612	Sellers Agar
21183	Serum Tellurite Agar
21868	Seven H11 Agar
11576	SF Broth
11578	SIM Medium
11619	Simmons' Citrate Agar
11624	Snyder Agar
21618	Sodium Hippurate Broth
11636	Spirolate Broth
11580	SPS Agar
11638	Standard Methods Agar
11643	Standard Methods Agar with Lecithin and Polysorbate 80
11369	Standard Methods Broth

BBL Product Number	Medium Name
11647	*Staphylococcus* Agar No. 110
11651	Sterility Test Broth
11664	Streptosel™ Agar
11666	Streptosel™ Broth
21809	SXT Blood Agar
99011	T7 Agar Base
12201	TAT Broth Base
11685	TCBS Agar
11694	Tech Agar
11698	Tellurite Glycine Agar
11702	Tergitol 7 Agar
11704	Tergitol 7 Broth
11706	Tetrathionate Broth
21184	Thayer–Martin Selective Agar
11712	Thiogel ® Medium
11716	Thioglycollate Medium, Brewer Modified
11718	Thioglycollate Medium without Glucose
11720	Thioglycollate Medium without Indicator–135C
97885	TOC Agar
11735	Todd–Hewitt Broth
11739	Tomato Juice Agar
11688	TPEY Agar
21417	Transgrow Medium with Trimethoprim
21361	Transgrow Medium without Trimethoprim
11743	Transport Medium
11747	Trichosel™ Broth, Modified
11749	Triple Sugar Iron Agar
11756	Trypticase Agar Base
11760	Trypticase Glucose Extract Agar
11043	Trypticase Soy Agar
11763	Trypticase Soy Agar with Lecithin and Polysorbate 80
12305	Trypticase Soy Agar, Modified
97457	Trypticase Soy Agar with Sheep Blood and Gentamicin
11768	Trypticase Soy Broth
11777	Trypticase Soy Broth with 0.1% Agar
11774	Trypticase Soy Broth without Glucose
11696	Trypticase Tellurite Agar Base
11920	Tryptophan 1% Solution
12365	Tryptose Broth
12342	Tryptose Phosphate Broth, Modified
11690	TSN Agar
12243	Universal Beer Agar
11795	Urea Agar
11797	Urease Test Broth
12348	UVM Modified *Listeria* Enrichment Broth

BBL Product Number	Medium Name
21874	V Agar
11801	Veal Infusion Agar
11803	Veal Infusion Broth
11807	Violet Red Bile Agar
98081	Violet Red Bile Agar with MUG
11812	Vogel and Johnson Agar
11816	WL Differential Agar
11818	WL Nutrient Agar
11826	Wort Agar
11836	XL Agar Base
11837	XLD Agar
97886	Yeast Extract Phosphate Agar
97299	Yeast Fermentation Broth
12101	Yeast Nitrogen Base

Difco Products

The media available from Difco Laboratories, Detroit MI 28401 are listed below.

Difco Product Number	Medium Name
1823	A 1 Broth
0316	AC Agar
0317	AC Broth
0742	Acetate Differential Agar
0840	*Actinomyces* Broth
0957	Actinomycete Isolation Agar
0744	Amies Transport Medium with Charcoal
0832	Amies Transport Medium without Charcoal
0536	Anaerobic Agar
4040	Anaerobic LKV Blood Agar
0263	Antibiotic Medium 1
0270	Antibiotic Medium 2
0243	Antibiotic Medium 3
0244	Antibiotic Medium 4
0277	Antibiotic Medium 5
0660	Antibiotic Medium 6
0667	Antibiotic Medium 8
0462	Antibiotic Medium 9
0463	Antibiotic Medium 10
0593	Antibiotic Medium 11
0669	Antibiotic Medium 12
0043	Antibiotic Medium 19
0654	APT Agar
0655	APT Broth
1019	ATS Medium

Difco Product Number	Medium Name
0409	Azide Blood Agar
0387	Azide Broth
0457	B12 Assay Medium
0541	B12 Culture Agar, USP
0542	B12 Inoculum Broth, USP
0442	BAGG Broth
0768	Baird–Parker Agar
0487	BCP D Agar
0717	BG Sulfa Agar
0635	BiGGY Agar
0879	Bile Esculin Agar
0878	Bile Esculin Agar Base
0525	Bile Esculin Azide Agar
0419	Biotin Assay Medium
0073	Bismuth Sulfite Agar
0073	Bismuth Sulfite Broth
0045	Blood Agar Base
0696	Blood Agar No. 2
0048	Bordet–Gengou Agar
0483	Brain Heart CC Agar
0037	Brain Heart Infusion
0418	Brain Heart Infusion Agar
0498	Brain Heart Infusion with PABA
0499	Brain Heart Infusion with PABA and Agar
0279	Brewer Anaerobic Agar
0236	Brewer Thioglycollate Medium
0237	Brewer Thioglycollate Medium, Modified
0285	Brilliant Green Agar
0014	Brilliant Green Bile Agar
0007	Brilliant Green Bile Broth
0494	Brilliant Green Broth
0964	*Brucella* Agar
0495	*Brucella* Broth
1810	Buffered Peptone Water
0578	Bushnell-Haas Broth
1820	*Campylobacter* Agar Base
3279	*Campylobacter* Agar, Blaser's
3280	*Campylobacter* Agar, Skirrows
0835	*Candida* BCG Agar Base
0290	Casman Agar Base
0854	Cetrimide Agar, USP
0456	CF Assay Medium
0313	Chapman Stone Agar
0894	Charcoal Agar
0513	Chlamydospore Agar
4140	Chocolate Agar
1002	Chocolate Agar, Enriched

Difco Product Number	Medium Name
0460	Choline Assay Medium
0695	Christensen Agar
0971	CLED Agar
0501	Coagulase Agar Base
0792	Columbia Blood Agar
0944	Columbia Broth
0867	Columbia CNA Agar
0083	Conradi Drigalski Agar
0703	Cooke Rose Bengal Agar
0267	Cooked Meat Medium
0386	Corn Meal Agar
0114	Corn Meal Agar with Dextrose
0077	Crystal Violet Agar
0047	Cystine Heart Agar
0523	Cystine Tryptic Agar
0338	Czapek Dox Broth
0339	Czapek Solution Agar
0686	D/E Neutralizing Agar
0819	D/E Neutralizing Broth
0823	D/E Neutralizing Broth Base
0759	DCLS Agar
0890	Decarboxylase Base, Møller
0872	Decarboxylase Medium Base, Falkow
0273	Deoxycholate Agar
0274	Deoxycholate Citrate Agar
0420	Deoxycholate Lactose Agar
0067	Dextrose Agar
0063	Dextrose Broth
0068	Dextrose Proteose No. 3 Agar
0066	Dextrose Starch Agar
0080	Dextrose Tryptone Agar
0706	Dextrose Tryptone Broth
0351	Disinfectant Test Broth AOAC
0632	DNase Test Agar
0220	DNase Test Agar with Methyl Green
1006	Dorset Egg Medium
0942	DTM Agar
0385	Dubos Broth
0373	Dubos Oleic Agar
0333	E Agar
0314	EC Medium
0566	EE Broth, Mossel
0017	Eijkman Lactose Medium
0974	Elliker Broth
0076	EMB Agar
0511	EMB Agar Base
0587	Emerson Agar
0739	Emerson Yp Ss Agar

Difco Product Number	Medium Name
0006	Endo Agar
0736	Endo Agar, LES
0749	Endo Broth
0053	*Endamoeba* Medium
1828	Enteric Fermentation Base
0301	Enterococci Confirmatory Agar
0302	Enterococci Confirmatory Broth
0300	Enterococci Presumptive Broth
0746	*Enterococcus* Agar
0606	Ethyl Violet Azide Broth
0532	Euglena B12 Medium
0589	Eugon Agar
0590	Eugon Broth
0675	F35M Hajna Broth
0677	FC Agar
0883	FC Broth
0841	Fermentation Broth
0349	Fildes Enrichment Agar
0987	Fletcher Medium
0642	Fluid Sabouraud Medium
0256	Fluid Thioglycollate Medium
0697	Fluid Thioglycollate Medium with Beef Extract
0607	Fluid Thioglycollate Medium with K Agar
0318	Folic Acid Assay Medium
0822	Folic Acid Casei Medium
0967	Folic AOAC Medium
0289	GC Agar
1809	Giolitti-Cantoni Broth
0486	GN Broth, Hajna
0451	H Broth
0044	Heart Infusion Agar
0038	Heart Infusion Broth
0853	Hektoen Enteric Agar
0909	Herellea Agar
0995	Inositol Assay Medium
0771	ISP Medium 3
0772	ISP Medium 4
0647	KCN Broth
0496	KF *Streptococcus* Agar
0997	KF *Streptococcus* Broth
0985	K–L Virulence Agar
0086	Kligler Iron Agar
0015	Koser Citrate Medium
0910	Kupferberg Trichomonas Base
0911	Kupferberg Trichomonas Broth
0900	Lactobacilli Agar, AOAC

Difco Product Number	Medium Name
0901	Lactobacilli Broth, AOAC
0881	Lactobacilli MRS Broth
0004	Lactose Broth
1061	Lash Serum Medium
0241	Lauryl Sulfate Broth
0088	Lead Acetate Agar
1830	*Legionella* Agar Base
0794	*Leptospira* Medium, EMJH
0680	Letheen Agar
0681	Letheen Broth
0005	Levine EMB Agar
1806	LICNR Broth
0117	Lima Bean Agar
0294	Littman Oxgall Agar
0052	Liver Infusion Agar
0269	Liver Infusion Broth
0059	Liver Veal Agar
0070	Loeffler Blood Serum Medium 389
0444	Lowenstein–Jensen Medium
0215	Lysine Decarboxylase Broth
0849	Lysine Iron Agar
0940	M Broth
0075	MacConkey Agar
0470	MacConkey Agar without Crystal Violet
1818	MacConkey Agar, CS
0331	MacConkey Agar without Salt
0020	MacConkey Broth
0395	Malonate Broth
0569	Malonate Broth, Ewing Modified
0024	Malt Agar
0112	Malt Extract Agar
0113	Malt Extract Broth
0306	Mannitol Salt Agar
0926	Mannitol Salt Broth
0979	Marine Agar 2216
0791	Marine Broth 2216
0922	McBride *Listeria* Agar
0941	McClung Toabe Agar
0319	Micro Assay Culture Agar
0320	Micro Inoculum Broth
0553	Microbial Content Test Agar
0627	Middlebrook 7H10 Agar with Middlebrook OADC Enrichment
0627	Middlebrook 7H10 Agar with Middlebrook OADC Enrichment and Hemin
0627	Middlebrook 7H10 Agar with Streptomycin
0838	Middlebrook 7H11 Agar with Middlebrook ADC Enrichment

Difco Product Number	Medium Name
0838	Middlebrook 7H11 Agar with Middlebrook OADC Enrichment
0838	Middlebrook 7H11 Agar with Middlebrook OADC Enrichment and Triton WR 1339
0714	Middlebrook ADC Enrichment
0801	Middlebrook OADC Enrichment with Triton WR 1339
0722	Middlebrook OADC Enrichment
1804	MIL Medium
0554	Minimal Agar, Davis
0756	Minimal Broth, Davis
0298	Mitis–Salivarius Agar
0869	Motility GI Medium
0735	Motility Indole Ornithine Medium
0761	Motility Medium S
0450	Motility Sulfide Medium
0105	Motility Test Medium
0016	MRVP Broth
0252	Mueller Hinton Agar
0757	Mueller Hinton Broth
0264	Mueller Tellurite Medium
0689	Mycobiotic Agar
0405	Mycological Agar
0305	Mycological Agar with Low pH
0406	Mycological Broth
0304	Mycological Broth with Low pH
0321	*Neurospora* Culture Agar
0817	*Neurospora* Minimal Medium
0322	Niacin Assay Medium
0258	NIH Agar
0257	NIH Thioglycollate Broth
0106	Nitrate Agar
0268	Nitrate Broth
0001	Nutrient Agar
0069	Nutrient Agar, 1.5%
0634	Nutrient Agar, pH 6.0
0003	Nutrient Broth
0011	Nutrient Gelatin
0552	Oatmeal Agar
0731	Oatmeal Agar
0521	Orange Serum Agar
0518	Orange Serum Broth Concentrate 10X
0688	Oxidative Fermentative Medium
1811	Oxytetracycline Glucose Yeast Extract Agar
0141	Pagano–Levin Agar
0994	Panthenol Assay Medium

Difco Product Number	Medium Name
0604	Pantothenate Assay Medium
0816	Pantothenate Medium, AOAC USP
0881	*Pediococcus* Medium
0089	Peptone Iron Agar
1807	Peptone Water
1010	Petragnani Medium
0098	Phenol Red Agar
0100	Phenol Red Lactose Agar
0103	Phenol Red Mannitol Agar
0090	Phenol Red Tartrate Agar
0745	Phenylalanine Agar
0806	Phenylalanine Malonate Broth
0504	Phenylethanol Agar
0417	*Photobacterium* Broth
0898	Pike Streptococcal Broth
0980	PKU Test Agar
0479	Plate Count Agar
1812	Plate Count Agar, Special
0751	Plate Count Broth
1800	PM Indicator Agar
0670	Porcine Heart Agar
0013	Potato Dextrose Agar
0549	Potato Dextrose Broth
0051	Potato Infusion Agar
0625	Potato Malt Agar
0412	PPLO Agar
0410	PPLO Broth with Crystal Violet
0554	PPLO Broth without Crystal Violet
0065	Proteose No. 3 Agar
0448	*Pseudomonas* Agar F
0449	*Pseudomonas* Agar P
0927	*Pseudomonas* Isolation Agar
0228	Purple Agar
0227	Purple Broth
0082	Purple Lactose Agar
0324	Pyridoxine Assay Medium
0951	Pyridoxine Y Medium
1808	Reinforced Clostridial Agar
1808	Reinforced Clostridial Medium
0899	Rice Extract Agar
0480	Rogosa SL Agar
0478	Rogosa SL Broth
1831	Rose Bengal Chloramphenicol Agar
0084	Russell Double Sugar Agar
0797	SABHI Agar
0797	SABHI Blood Agar
0747	Sabouraud Agar, Modified
0109	Sabouraud Glucose Agar
0382	Sabouraud Glucose Broth
0110	Sabouraud Maltose Agar
0429	Sabouraud Maltose Broth
0074	*Salmonella–Shigella* Agar
0661	SBG Enrichment Broth
0715	SBG Sulfa Enrichment
0408	Schaedler Agar
0534	Schaedler Broth
0275	Selenite Broth
0687	Selenite Cystine Broth
0895	Sellers Agar
0315	SF Broth
0811	SFP Agar
0271	SIM Medium
0091	Simmons' Citrate Agar
0247	Snyder Test Agar
0079	Sorbitol MacConkey Agar
0950	Spirit Blue Agar
0845	SPS Agar
1816	Standard II Nutrient Agar
0649	*Staphylococcus* Broth
0297	*Staphylococcus* Medium No. 110
0072	Starch Agar
0054	Stock Culture Agar
0988	Stuart Medium Base
0500	Sulfate API Broth
0972	Sulfite Agar
0352	Synthetic Broth, AOAC
0984	TAT Broth Base
0650	TCBS Agar
0617	Tellurite Glycine Agar
0455	Tergitol 7 Agar
0800	Tergitol 7 Agar H
0912	Tergitol 7 Broth
0104	Tetrathionate Broth
0580	Tetrathionate Broth
0750	TGE Broth
1247	Thayer–Martin Medium
0303	Thermoacidurans Agar
0326	Thiamine Assay Medium
0808	Thiamine Assay Medium LV
0530	Thioglycollate Gelatin Medium
0363	Thioglycollate Medium without Glucose
0432	Thioglycollate Medium without Glucose and Indicator
0430	Thioglycollate Medium without Indicator
0434	Thiol Broth
0307	Thiol Medium

Difco Product Number	Medium Name
0786	Tinsdale Agar
5790	Tissue Culture Amino Acids, HeLa 100X
5785	Tissue Culture Dulbecco Solution
5351	Tissue Culture Earle Solution
5508	Tissue Culture Hanks Solution
5477	Tissue Culture Medium 199
5068	Tissue Culture Medium Eagle with Earle Balanced Salt Solution
5071	Tissue Culture Medium Eagle with Hanks Balanced Salt Solution
5651	Tissue Culture Medium Eagle, HeLa
5825	Tissue Culture Medium Ham F10
5926	Tissue Culture Medium NCTC 109
5087	Tissue Culture Medium RPMI #1640
5675	Tissue Culture Minimal Medium Eagle with Earle Balanced Salts Solution
5839	Tissue Culture Minimal Medium Eagle Spinner Modified
5556	Tissue Culture Tyrode Solution
5833	Tissue Culture Vitamins Minimal Eagle, 100X
0492	Todd–Hewitt Broth
0031	Tomato Juice Agar
0389	Tomato Juice Agar Special
0517	Tomato Juice Broth
0556	TPEY Agar
1410	Transgrow Medium
0877	*Trichophyton* Agar 1
0874	*Trichophyton* Agar 2
0965	*Trichophyton* Agar 3
0197	*Trichophyton* Agar 4
0411	*Trichophyton Agar 5*
0524	*Trichophyton* Agar 6
0955	*Trichophyton* Agar 7
0265	Triple Sugar Iron Agar
0364	Tryptic Agar Base
1829	Tryptic Digest Broth
0367	Tryptic Nitrate Medium
0369	Tryptic Soy Agar
4490	Tryptic Soy Agar with 0.6% Yeast Extract
0370	Tryptic Soy Broth
0862	Tryptic Soy Broth without Glucose
0002	Tryptone Glucose Beef Extract Agar
0327	Tryptophan Assay Medium
0064	Tryptose Agar
0633	Tryptose Agar with Thiamine
0662	Tryptose Blood Agar Base with Yeast Extract
0062	Tryptose Broth

Difco Product Number	Medium Name
0623	Tryptose Broth with Thiamine
0060	Tryptose Phosphate Broth
0491	TT Broth Base, Hajna
0856	UBA Medium
0283	Urea Agar
0272	Urea R Broth
0343	Veal Infusion Agar
0344	Veal Infusion Broth
0917	*Veillonella* Agar
0012	Violet Red Bile Agar
0360	Vitamin B_{12} Assay Medium
0562	Vogel and Johnson Agar
1805	Wilkins–Chalgren Agar
0312	Wilson–Blair Base
0425	WL Differential Medium
0424	WL Nutrient Agar
0471	WL Nutrient Broth
0924	Worfel–Ferguson Agar
0111	Wort Agar
0555	XL Agar Base
0788	XLD Agar
0391	Yeast Carbon Base, 10X
0712	Yeast Malt Extract Agar
0770	Yeast Malt Extract Agar
0711	Yeast Malt Extract Broth
0393	Yeast Morphology Agar
0392	Yeast Nitrogen Base
1817	*Yersinia* Selective Agar Base

Oxoid Unipath Products

The media available from Unipath Ltd., Oxoid Unipath Division 9200 Rumsey Rd., Columbia MD 21045 and Unipath Ltd., Oxoid Unipath Division Wade Rd., Basingstoke, Hampshire, RG24 OPW, England are listed below.

Oxoid Unipath Product Number	Medium Name
PM1a	Acid Egg Medium
PM118	Actidione® Agar
CM833	*Aeromonas* Medium
CM731	AFPA
CM425	Amies Modified Transport Medium with Charcoal
CM335	Antibiotic Medium 2

Oxoid Unipath Product Number	Medium Name
CM287	Antibiotic Medium 3
CM259	Azide Blood Agar
CM868	Azide Broth, Rothe
CM617	*Bacillus cereus* Selective Agar Base
CM275	Baird–Parker Agar
CM589	BiGGY Agar
CM888	Bile Esculin Agar
CM201	Bismuth Sulfite Agar
CM55	Blood Agar Base
CM854	Blood Agar Base Sheep
CM271	Blood Agar No. 2
CM655	BMPA-α Medium
CM267	*Bordetella pertussis* Selective Medium with Bordet–Gengou Agar Base
CM119	*Bordetella pertussis* Selective Medium with Charcoal Agar Base
CM267	Bordet Gengou Agar
CM225	Brain Heart Infusion
CM375	Brain Heart Infusion Agar
CM263	Brilliant Green Agar
CM329	Brilliant Green Agar, Modified
CM31	Brilliant Green Bile Broth
CM691	*Brucella* Agar
CM908	*Campylobacter* Selective Medium Karmali
CM519	Cary and Blair Transport Medium
CM119	Charcoal Agar
CM209	China Blue Lactose Agar
CM333	Cholera Medium TCBS
CM353	Clausen Medium
CM301	CLED Agar
CM423	CLED Agar with Andrade Indicator
CM601	*Clostridium difficile* Agar
CM543	*Clostridium perfringens* Agar, OPSP
CM331	Columbia Blood Agar
CM81	Cooked Meat Medium
CM103	Corn Meal Agar
CM213	Crossley Milk Medium
CM97	Czapek Dox Agar, Modified
CM95	Czapek Dox Liquid Medium, Modified
CM393	DCLS Agar
CM163	Deoxycholate Agar
CM35	Deoxycholate Citrate Agar
CM227	Deoxycholate Citrate Agar, Hynes
CM539	Dermasel Agar Base
CM175	Dextrose Broth
CM75	Dextrose Tryptone Agar

Oxoid Unipath Product Number	Medium Name
CM73	Dextrose Tryptone Broth
CM261	Diagnostic Sensitivity Test Agar
CM729	Dichloran Glycerol Agar
CM321	DNase Agar
CM727	DRBC Agar
CM27	Edwards Medium, Modified
CM317	EE Broth
CM69	EMB Agar, Modified
CM37	Endo Agar
CM479	Endo Agar, LES
CM869	Ethyl Violet Azide Broth
CM895	Fraser Broth
CM331	*Gardnerella vaginalis* Selective Medium
CM755	GBS Agar Base, Islam
CM841	GBS Medium, Rapid
CM523	Giolitti-Cantoni Broth
CM898	*Haemophilus* Test Medium
CM419	Hektoen Enteric Agar
CM83	Hoyle Medium Base
CM867	Iron Agar, Lyngby
CM79	Iron Sulfite Agar
CM471	Iso–Sensitest Agar
CM473	Iso–Sensitest Broth
CM591	Kanamycin Esculin Azide Agar
CM771	Kanamycin Esculin Azide Broth
CM701	KF *Streptococcus* Agar
CM33	Kligler Iron Agar
CM441	Kranep Agar Base
CM17	Lab–Lemco Agar
CM15	Lab–Lemco Broth
CM137	Lactose Broth
CM451	Lauryl Sulfate Broth
CM831	Lauryl Tryptose Mannitol Broth with Tryptophan
CM862	*Listeria* Enrichment Broth I, USDA FSIS
CM863	*Listeria* Enrichment Broth II, USDA FSIS
CM45	Litmus Milk
CM77	Liver Broth
PM1	Lowenstein–Jensen Medium
CM495	LS Differential Medium
CM308	Lysine Decarboxylase Broth, Taylor Modification
CM381	Lysine Iron Agar
CM191	Lysine Medium
CM785	M17 Agar
CM817	M17 Broth
CM7	MacConkey Agar

Oxoid Unipath Product Number	Medium Name
CM109	MacConkey Agar No. 2
CM115	MacConkey Agar No. 3
CM7b	MacConkey Agar without Salt
CM5	MacConkey Broth
CM5a	MacConkey Broth, Purple
CM59	Malt Extract Agar
CM57	Malt Extract Broth
CM85	Mannitol Salt Agar
CM733	Maximum Recovery Diluent
CM21	Milk Agar
CM607	Minerals Modified Medium
CM783	MLCB Agar
CM910	Modified Semi-Solid Rappaport–Vassiliadis Medium
CM361	MRS Agar
CM359	MRS Broth
CM43	MRVP Medium
CM337	Mueller–Hinton Agar
CM405	Mueller–Hinton Broth
CM401	*Mycoplasma* Agar with Supplement G
CM401	*Mycoplasma* Agar with Supplement P
CM403	*Mycoplasma* Broth with Supplement G
CM403	*Mycoplasma* Broth with Supplement P
CM3	Nutrient Agar
CM1	Nutrient Broth
CM67	Nutrient Broth No. 2
CM135a	Nutrient Gelatin
CM657	Orange Serum Agar
CM856	Oxford Agar
CM883	Oxidative Fermentative Medium
CM545	Oxytetracycline Glucose Yeast Extract Agar
CM877	PALCAM Agar
CM509	Peptone Water
CM61	Peptone Water with Andrade's Indicator
CM325	Plate Count Agar
CM463	Plate Count Agar
CM681	Plate Count Agar with Antibiotic-Free Skim Milk
CM139	Potato Dextrose Agar
CM559	*Pseudomonas* CFC Agar
CM559	*Pseudomonas* CN Agar
PM2A	Pyruvic Acid Egg Medium
CM777	Raka–Ray Agar
CM669	Rappaport–Vassiliadis Enrichment Broth
CM866	Rappaport–Vassiliadis Soya Peptone Broth

Oxoid Unipath Product Number	Medium Name
CM151	Reinforced Clostridial Agar
CM149	Reinforced Clostridial Medium
CM627	Rogosa Agar
CM549	Rose Bengal Chloramphenicol Agar
CM41	Sabouraud Glucose Agar
CM41a	Sabouraud Maltose Agar
CM147	Sabouraud Medium, Fluid
CM99	*Salmonella–Shigella* Agar
CM533	*Salmonella–Shigella* Agar, Modified
CM94	Salt Meat Broth
CM437	Schaedler Agar
CM497	Schaedler Broth
R39	Selenite Broth
CM399	Selenite Broth Base, Mannitol
CM699	Selenite Cystine Broth
CM409	Sensitest Agar
CM435	SIM Medium
CM155	Simmons' Citrate Agar
CM377	Slanetz and Bartley Medium
CM813	Sorbitol MacConkey Agar
CM881	STAA Agar Base
CM145	*Staphylococcus* Agar No. 110
CM331	*Streptococcus* Selective Medium
CM111	Stuart Transport Medium, Modified
CM793	Tergitol 7 Agar
CM29	Tetrathionate Broth
CM671	Tetrathionate Broth, USA
CM391	Thioglycollate Broth USP, Alternative
CM23	Thioglycollate Medium, Brewer
CM415	Thioglycollate Medium without Indicator
CM173	Thioglycollate Medium, USP
CM487	Tinsdale Agar
CM189	Todd–Hewitt Broth
CM113	Tomato Juice Agar
PM4	Tributyrin Agar
CM161	*Trichomonas* Medium
R27	*Trichomonas* Medium No. 2
CM277	Triple Sugar Iron Agar
CM595	Tryptone Bile Agar
CM127	Tryptone Glucose Beef Extract Agar
CM131	Tryptone Soya Agar
TCM129	Tryptone Soya Broth
CM87	Tryptone Water Broth
CM233	Tryptose Blood Agar
CM283	Tryptose Phosphate Broth
CM587	Tryptose Sulfite Cycloserine Agar

Oxoid Unipath Product Number	Medium Name
CM651	UBA Medium
CM53	Urea Agar Base
CM71	Urea Broth Base
CM107	Violet Red Bile Agar
CM485	Violet Red Bile Glucose Agar
CM641	Vogel and Johnson Agar
CM619	Wilkins–Chalgren Anaerobe Agar
CM619	Wilkins–Chalgren Anaerobe Agar with G–N Supplement
CM643	Wilkins–Chalgren Anaerobe Agar with N–S Supplement
CM643	Wilkins–Chalgren Anaerobe Broth
CM309	WL Nutrient Agar
CM501	WL Nutrient Broth
CM247	Wort Agar
CM469	XLD Agar
CM19	Yeast Extract Agar
CM653	*Yersinia* Selective Agar Base

Preparation of Media

The ingredients in a medium are usually dissolved and the medium is then sterilized. When agar is used as a solidifying agent the medium must be heated gently, usually to boiling, to dissolve the agar. In some cases where interactions of components, such as metals, would cause precipitates, solutions must be prepared and occasionally sterilized separately before mixing the various solutions to prepare the complete medium. The pH often is adjusted prior to sterilization, but in some cases sterile acid or base is used to adjust the pH of the medium following sterilization. Many media are sterilized by exposure to elevated temperatures. The most common method is to autoclave the medium. Different sterilization procedures are employed when heat-labile compounds are included in the formulation of the medium.

Autoclaving

Autoclaving uses exposure to steam, generally under pressure, to kill microorganisms. Exposure for 15 min to steam at 115 psi—121°C is most commonly used. Such exposure kills vegetative bacterial cells and bacterial endospores. However, some substances do not tolerate such exposures and lower temperatures and different exposure times are sometimes employed. Media containing carbohydrates often are sterilized at 116°C–118°C in order to prevent the decomposition of the carbohydrate and the formation of toxic compounds that would inhibit microbial growth. Below is a list of pressure–temperature relationships.

Pressure—psi	Temperature—°C
0	100
1	101.9
2	103.6
3	105.3
4	106.9
5	108.4
6	109.8
7	111.3
8	112.6
9	113.9
10	115.2
11	116.4
12	117.6
13	118.8
14	119.9
15	121.0
16	122.0
17	123.0
18	124.0
19	125.0
20	126.0
21	126.9
22	127.8
23	128.7
24	129.6
25	130.4

Tyndallization

Exposure to steam at 100°C for 30 minutes will kill vegetative bacterial cells but not endospores. Such exposure can be achieved using flowing steam in an Arnold sterilizer. By allowing the medium to cool and incubate under conditions where endospore germination will occur and by repeating the 100°C–30 minute exposure on three successive days the medium can be sterilized because all the endospores will have germinated and the heat exposure will have killed all the vegetative cells. This process of repetitive exposure to 100°C is called tyndallization, after its discoverer John Tyndall.

Inspissation

Inspissation is a heat exposure method that is employed with high protein materials, such as egg-containing media, that cannot withstand the high temperatures used in autoclaving. This process causes coagulation of the protein without greatly altering its chemical properties. Several different protocols can be followed for inspissation. Using an Arnold sterilizer or a specialized inspissator, the medium is exposed to 75°C–80°C for 2 hours on each of three successive days. Inspissation using an autoclave employs exposure to 85°C–90°C for 10 minutes achieved by having a mixture of air and steam in the chamber, followed by 15 minutes exposure during which the temperature is raised to 121°C using only steam under pressure in the chamber; the temperature then is slowly lowered to less than 60°C.

Filtration

Filtration is commonly used to sterilize media containing heat-labile compounds. Liquid media are passed through sintered glass or membranes, typically made of cellulose acetate or nitrocellulose, with small pore sizes. A membrane with a pore size of 0.2mm will trap bacterial cells and, therefore, sometimes is called a bacteriological filter. By preventing the passage of microorganisms, filtration renders fluids free of bacteria and eukaryotic microorganisms, that is, free of living organisms, and hence sterile. Many carbohydrate solutions, antibiotic solutions, and vitamin solutions are filter sterilized and added to media that have been cooled to temperatures below 50°C.

Caution about Hazardous Components

Some media contain components that are toxic or carcinogenic. Appropriate safety precautions must be taken when using media with such components. Basic fuchsin and acid fuchsin are carcinogens and caution must be used in handling media with these compounds to avoid dangerous exposure that could lead to the development of malignancies. Thallium salts, sodium azide, sodium biselenite, and cyanide are among the toxic components found in some media. These compounds are poisonous and steps must be taken to avoid ingestion, inhalation, and skin contact. Azides also react with many metals, especially copper, to form explosive metal azides. The disposal of azides must avoid contact with copper or achieve sufficient dilution to avoid the formation of such hazardous explosive compounds. Media with sulfur-containing compounds may result in the formation of hydrogen sulfide which is a toxic gas. Care must be used to ensure proper ventilation. Media with human blood or human blood components must be handled with great caution to avoid exposure to human immunodeficiency virus and other pathogens that contaminate some blood supplies. Proper handling and disposal procedures must be followed with blood-containing as well as other media that are used to cultivate microorganisms.

Uses of Media

ATCC Media

The *Handbook of Microbiological Media* contains all the media used to cultivate the bacteria of the American Type Culture Collection. The ATCC media numbers and the equivalent media names used in the *Handbook of Microbiological Media* are listed below.

ATCC Medium Number	Medium Name
1	Mannitol Agar
2	Marine Agar 2216
3	Nutrient Agar
4	Rabbit Blood Agar
5	Sporulation Agar
6	Brain, Liver, Heart Semi-Solid
7	*Actinomyces* Broth
8	Nutrient Agar with 3% NaCl
9	Nutrient Agar with Yeast Extract
10	Nutrient Agar with Soil Extract
11	*Azotobacter* Supplement
12	*Azotobacter* Supplement
13	*Azotobacter* Supplement
14	*Azotobacter* Medium
15	*Azotobacter* Supplement
17	Tomato Juice Yeast Extract Milk Medium
18	Trypticase Soy Agar
20	JB Medium with Glucose
21	*Bacillus* Medium
22	*Bacillus* Agar, 1/4 Strength
22	*Bacillus* Broth, 1/4 Strength
24	Nutrient Agar with Potato Starch
25	Yeast Extract Glucose Medium
26	Nutrient Agar with Phytone
27	Glycerol Agar
28	Emmon's Modification of Sabouraud's Agar
29	Brain Heart Infusion Agar with Yeast Extract
30	*Bacterium* Medium
31	Peptone Broth
32	Green Top Agar
33	Tomato Juice Agar
34	Neomycin Agar
35	Bordet-Gengou Medium
36	*Caulobacter* Medium
37	*Chromatium* Medium
38	Beef Liver Medium for Anaerobes
39	*Clostridium* Medium

ATCC Medium Number	Medium Name
40	*Clostridium* Medium
42	*Desulfovibrio* Medium
43	*Clostridium* Medium
44	Brain Heart Infusion Agar
45	Nutrient Agar with Maltose
46	Fluid Thioglycollate Medium with Rabbit Serum
47	*Desulfovibrio* Medium with NaCl
49	Thiol Medium
50	Proteose No. 3 Agar
51	AC Broth
52	*Escherichia* Medium
53	*Escherichia* Medium
54	Davis Supplemented Minimal Medium
55	Nutrient Agar with Dihydrostreptomycin
56	Dorset Egg Medium
57	*Escherichia* Medium
58	Nutrient Agar with Uracil
59	*Pseudomonas* Medium
60	*Escherichia* Medium
62	*Escherichia* Medium
64	*Thiobacillus* Medium
65	*Flavobacterium* Medium
66	*Acetobacter* Medium
68	Brain Heart Infusion with Chicken Serum
69	Fish Peptone Agar
69	Fish Peptone Broth
70	Casman's Agar Base with Rabbit Blood
71	Trypticase Soy Agar with Glycerol
72	Heterotrophic Medium for *Hydrogenomonas*
73	YGC Medium
76	*Lactobacillus bifidus* Medium
77	Trypticase Soy Agar
78	*Lactobacillus* Medium
80	Lactic Bacteria Broth
84	*Betabacterium* Medium
85	Czapek's Dox Agar with 3% Glucose
86	Nutrient Agar with Glucose
87	Nutrient Agar with 10% NaCl
89	*Micrococcus-Sarcina* Medium
90	Lowenstein-Jensen Medium
91	Lowenstein-Jensen Medium with Streptomycin
92	Steenken and Smith Agar
93	*Mycobacterium* Medium
95	GC Medium with Supplement B
96	*Nitrobacter* Medium B
97	Potato Glucose Agar

ATCC Medium Number	Medium Name
101	*Photobacterium* Broth
102	*Pseudomonas denitrificans* Medium
103	Soybean Agar
104	Potato Dextrose Agar and Yeast Medium
105	Nutrient Agar with 1.5% NaCl
106	*Pseudomonas saccharophila* Medium
107	Nutrient Agar with Ethylene Glycol
108	Nutrient Agar with Ethanolamine
109	Malt Extract Agar
110	Nutrient Agar with V-8™ Juice
111	Rhizobium X Medium
112	Yeast Agar (Van Niel's)
114	Mineral Medium S with 1% Sucrose
115	Mineral Medium A
116	*Micrococcus* Medium, FDA
117	*Micrococcus* Medium
122	Horse Serum Agar
122	Horse Serum Broth
123	TGY Medium
124	*Streptococcus* Agar
125	*Thiobacillus* Medium
129	Nutrient Agar with 0.5% NaCl
132	*Zymobacterium* Agar
132	*Zymobacterium* Broth
134	Czapek's Dox Agar
136	Tryptose Blood Agar
137	Tris YP Agar
137	Tris YP Broth
138	*Beggiatoa* Medium
140	Antibiotic Medium 2
141	Tryptose Agar
142	*Lactobacillus* Sake Medium
144	PGY Agar
145	Potato Carrot Medium
146	N-Z Amine A Medium
147	CTA Agar
148	Sorangium Medium
150	Stock Culture Agar with L-asparagine
151	Oatmeal Agar
152	*Thiobacillus* Medium
153	*Achromobacter* Choline Medium
155	Feodorov Medium
158	RGCA Medium
159	*Brevibacterium* Medium
160	Sea Water Yeast Extract Agar
163	*Clostridium* Medium
164	*Cytophaga* Medium
164	Spirochete Medium

ATCC Medium Number	Medium Name
165	*Derxia* Medium
166	Tryptone Agar
167	L Medium
169	*Lactobacillus* Medium
171	XSM Agar
172	N-Z Amine A with Starch and Glucose
173	Middlebrook 7H10 Agar with Middlebrook OADC Enrichment
174	Bennett's Medium
175	Bushnell-Haas Medium
176	Trypticase Soy Agar with 3% NaCl
177	Fluid Thioglycollate Medium
179	*Pseudomonas* Medium
180	Nutrient Agar with Erythromycin
181	*Serratia* Medium
182	FDA Broth
184	Glucose Asparagine Agar
185	Bennett's Modified Agar Medium
186	*Pseudomonas* Medium
187	*Pseudomonas* Medium I
188	*Veillonella* Medium
189	*Zoogloea* Medium
190	*Azotobacter* Basal Agar
190	*Azotobacter* Basal Broth
192	Cystine Heart Agar with Rabbit Blood
195	YEPP Medium
196	Yeast Malt Extract Agar
197	*Alcaligenes* Medium
198	*Actinoplanes* Medium
199	Emerson Agar
200	Yeast Malt Extract Agar
200	Yeast Malt Extract Broth
201	TDC Medium
202	Furoate Agar
203	Mycophil Agar
204	CM Agar
207	Starkey's Medium C, Modified
209	Marine *Cytophaga* Agar
210	Starkey's Medium C, Modified with Salt
211	Antibiotic Medium 1
212	Yeast Tryptone Medium
213	*Halobacterium* Medium
214	*Leptotrichia buccalis* Medium
215	YGC Medium with Cysteine
216	YGC Medium
217	Yeast Agar, Van Niel's with 25% NaCl
218	Heart Infusion Agar with Yeast Extract
219	Lowenstein-Jensen Medium without Glycerol

ATCC Medium Number	Medium Name
220	Casitone Agar
221	*Nitrosomonas* Medium
224	*Peptococcus glycinophilus* Medium
225	Beef Extract Agar
225	Beef Extract Broth
226	*Pseudomonas* Medium
227	Glycerol Beef Extract Medium
228	Marine *Pseudomonas* Medium
229	Quinolinic Acid Medium
230	Antibiotic Medium 1 with Tetracycline, Streptomycin and Chloramphenicol
231	SWMTY Marine Medium
233	*Spirillum gracile* Medium
234	Peptone Succinate Agar
235	Todd-Hewitt Broth
236	Krainsky's Asparagine Agar
237	Trace Element Solution HO-LE
238	*Thiobacillus* Medium B
239	Swampy Medium
240	*Azotobacter* Medium
241	Waksman's Glucose Agar
243	Heart Infusion Broth with Inactivated Horse Serum and Fresh Yeast Extract
245	Heart Infusion Broth with Human Serum and Yeast Extract
247	PPLO Broth without Crystal Violet with Horse Serum and Fresh Yeast Extract
248	PPLO Broth without Crystal Violet with Horse Serum
249	Trypticase Soy Broth with Inactivated Horse Serum
250	Trypticase Soy Agar with NaCl, Horse Serum and Penicillin
251	Veal Infusion Broth with Rabbit Serum
252	Brain Heart Infusion with Agar, Yeast Extract, Sucrose, Horse Serum and Penicillin
253	Mueller-Hinton Chocolate Agar
254	Heart Infusion Agar
255	Heart Infusion Agar with Inactivated Horse Serum, NaCl and Penicillin
257	Brain Heart Infusion with Sucrose and Horse Serum
258	Heart Infusion Agar with Inactivated Horse Serum
259	Brain Heart Infusion with Inactivated Horse Serum, NaCl, Agar, Yeast Extract and Penicillin
260	Trypticase Soy Agar with Sheep Blood
261	Brain Heart Infusion

ATCC Medium Number	Medium Name
262	Brain Heart Infusion with Inactivated Horse Serum, Agar, Yeast Extract, Sucrose and Penicillin
265	Enriched Nutrient Broth
266	Trypticase Broth, Supplemented
267	PGP Broth
268	Peptone Sucrose Broth
269	Antibiotic Medium 3
270	Nutrient Agar with 0.5% NaCl and Sodium Citrate
271	*Escherichia* Medium
272	Nutrient Yeast Glucose Medium
273	Nutrient Broth Salts Medium
274	Tryptone Broth
275	Tryptone Broth with $CaCl_2$
276	*Achromobacter pestifer* Medium
278	*Pseudomonas bathycetes* Medium
279	Manganese Medium for *Pseudomonas* species
280	*Leptotrichia* Medium
283	Imidazole Utilization Medium
284	K101 *Flexibacter* Medium
287	Serum Glucose Agar
289	Medium for *Prosthecomicrobium* and *Ancalomicrobium*
290	S6 Medium for Thiobacilli
291	Acidic Tomato Medium for *Leuconostoc*
292	Baar's Medium for Sulfate Reducers
293	*Thermoactinopolyspora* Medium
294	Tryptone with NaCl Broth
295	S-8 Medium For Thiobacilli
298	*Nocardia histidans* Medium
299	Dextrose Starch Agar
312	Czapek Agar
329	Mineral Salts Agar
336	Potato Dextrose Agar
337	Potato Dextrose Yeast Agar
341	Chu's No. 10 Medium
365	Starch Agar Medium for *Pseudomonas*
368	Blood Agar Base
373	Maleate Medium for *Pseudomonas fluorescens*
375	Soil Extract Medium
377	Skim Milk Agar
394	Mycobactin Medium
396	Methylamine Salts Medium
399	*Streptomyces* Medium
404	Loeffler Medium
405	BSL for *Corynebacterium*
410	Nocardia Medium
412	Methanol Medium for *Achromobacter*
414	Beijerinck's *Thiobacillus* Medium
415	Potato Extract Agar
416	Lactobacilli MRS Broth
418	HY Medium for *Flavobacterium*
420	*Cytophaga* Medium
421	Potato Infusion Agar
422	Medium 2508-85-1 with Amino Acids
423	Spizizen Potato Agar
424	*Arthrobacter* Medium
425	*Alcaligenes* N5 Medium
426	*Thiobacillus* Medium
429	*Leucothrix* Medium
430	*Leucothrix* Medium
432	SP Medium
433	Tomato Juice Broth
434	Heart Infusion Agar with Rabbit Blood
435	Yeast Extract Peptone Starch Agar
436	Methanol Medium
437	Diaminopimelic Acid Medium
438	*Nitrosolobus* Medium
439	Mist Agar
440	Emerson Agar, Half Strength
444	PYS Agar
445	Yeast Tryptone Starch Medium
447	CP Medium
450	T2 Medium For *Thiobacillus*
452	Beef Extract with NaCl
454	*Staphylococcus* Medium
455	*Bacillus* Medium
456	Pyrrolidone Agar
457	*Achromobacter* Medium
459	YGC Medium
461	Castenholz TYE Medium
462	*Prosthecobacter* Medium
463	Folic Acid Agar
464	*Cellulomonas* PTYG Medium
465	*Alcaligenes* NB YE Medium
468	Tryptone in Sea Water Agar
469	Burke's Modified Nitrogen-Free Medium
470	Potato Infusion with Inorganic Salts
472	*Thiobacillus* A2 Agar
472	*Thiobacillus* A2 Broth
474	*Eubacterium* Medium
480	*Nitrobacter* Medium 203
481	*Nitrobacter* Medium 204
484	PPLO Broth with Additives for *Mycoplasma*

ATCC Medium Number	Medium Name
485	PPLO Broth, pH 7.6, with Additives for *Mycoplasma*
486	Chalquist's Antigen Medium, Modified
488	Heart Infusion Broth with Additives for *Streptobacillus*
489	*Brucella* Broth with Additives
490	*Brucella* Broth with Additives
492	Heart Infusion Broth with Additives for *Staphylococcus*
493	Heart Infusion Broth with Inactivated Horse Serum
497	Davis and Mingioli Medium with B1 and Asparagine
498	Chu's No. 10 Medium, Modified
501	Davis and Mingioli Medium with Proline
502	ATS Medium
504	*Leuconostoc* Medium
505	Brain Heart Infusion with Casein
507	Tryptone Glucose Beef Extract Agar with Yeast Extract
508	Mineral Salts Medium with Methanol
509	Medium for *Prosthecomicrobium* and *Ancalomicrobium* with Nicotinamide
510	*Acidaminococcus* Medium VR
511	*Clostridium* Medium
517	Oatmeal Soy Peptone Medium
518	MPY Agar
519	Uric Acid Agar for Clostridia
519	Uric Acid Broth for Clostridia
520	Thiamine Salts Medium
521	Veal Infusion Agar
521	Veal Infusion Broth
522	Czapek Agar with Peptone
525	Eugon Broth
526	Peptone Yeast Extract Agar
527	ISP Medium 4
528	*Thiobacillus* Medium
531	Gauze's Medium No. 1
535	PYGM Medium
538	Ionic Medium with Pipecolate
540	Brain Heart Infusion with Thiamine
542	Peptone Starch Carbonate Medium
543	*Rhodopseudomonas* Medium
544	Heart Infusion Broth with Glucose
545	Heart Infusion Broth with Horse Serum, Fresh Yeast Extract and Penicillin
546	Tryptose Blood Agar
548	Soil Extract Potato Extract Medium
549	Trypticase Phytone Glucose Medium
550	R8AH Medium
551	Oatmeal Agar
552	*Bacillus* Medium
553	OZR Medium
554	Yeast Extract Glycerol Medium
555	*Mycoplasma* Agar
555	*Mycoplasma* Broth
557	Casamino Peptone Czapek Medium
559	*Mycoplasma* Medium
561	*Enterobacter* Medium
562	*Pediococcus* Medium
563	*Pediococcus* Medium with Mevalonic Acid
564	GYPT Medium
566	*Megasphaera* Medium
568	*Clostridium* Medium
569	*Thermoplasma* Agar
569	*Thermoplasma* Broth
573	*Bacillus* Agar
573	*Bacillus* Broth
577	*Clostridium* Alginate Medium
580	Tryptone Glucose Beef Extract Agar
582	MPY Agar
583	EG NaCl Medium No. 7
586	*Bifidobacterium* Medium
587	YPSC Medium
588	AMS Medium
589	*Achromobacter* Medium
590	Casamino Acids Medium
591	*Clostridium* Medium
593	Chopped Meat Medium
594	Brain Heart Infusion with Rabbit Serum
595	TNT Medium
596	L Medium For *Salmonella*
598	GPY Salts Medium
599	*Treponema* Medium
602	E Medium for Anaerobes
603	TYG Medium
605	Maleate Medium for *Pseudomonas fluorescens* with Glucose and Phenol Red
607	Sodium Chloride Sucrose Medium 900
608	Sodium Chloride Sucrose Medium 900 with Penicillin G
609	*Pseudomonas* Medium
610	Veal Infusion Broth with Horse Serum
611	PGT Medium
613	Lactate, Sea Water Minimal Medium
615	*Salmonella* Medium

ATCC Medium Number	Medium Name
616	BG-11 Agar
616	BG-11 Broth
617	BG-11 Marine Agar
617	BG-11 Marine Broth
620	Nutrient Agar with 1% Methanol
621	PYEX Glucose Salt Medium
623	Basal Synthetic Medium
625	Gorham's Medium for Algae
628	Nutrient Agar with Methanol
634	Brain Heart Infusion with Rabbit Serum and Yeast Extract
635	*Thermoactinomyces* Medium
639	Heart Infusion with Porcine Serum and Fresh Yeast Extract
640	Beef Extract Peptone Serum Medium
641	PGLE Medium
642	TPGY Medium
646	Modified Trypticase Soy
647	*Flavobacterium* Medium
648	Penassay Broth with Magnesium
649	Soil Extract Glucose Yeast Extract Agar
650	*Rhodopseudomonas* Medium
651	Asparaginate Glycerol Agar
652	Lactobacilli MRS Broth with Ethanol
653	Trypticase Soy Agar, Modified with Horse Serum
654	Soil Extract Peptone Beef Extract Medium
655	*Thermobacterium* Medium
656	*Hyphomicrobium* Medium
657	Soy Bean Extract, M-1
658	TYGS Medium
661	Alkaline *Bacillus* Medium
662	Sea Water Medium
663	PYG Medium
664	Trypticase Phytone Glucose Medium
665	YTN Medium
666	Wakimoto Medium, Modified
668	SMC Medium
673	TYES Medium
674	B/1t 7A Medium
675	*Mycoplasma* Medium
676	Brain Heart Infusion with 3% Sodium Chloride
677	*Brevibacterium* Medium
679	TGYM Medium
681	*Brevibacterium* Medium
682	Nutrient Agar with Horse Serum

ATCC Medium Number	Medium Name
683	Koch's K1 Medium
688	Nutrient Agar, pH 8.0
689	*Aplanobacterium* Medium
691	*Pseudomonas* Phage Medium
692	ISP Medium 4 with Yeast Extract
693	GTYE Medium
694	Sour Dough Medium
697	*Thermus* Medium
698	Peptone Succinate Agar in Sea Water
704	Methanol Medium with 1% Peptone
713	Medium A for Producing Lysates
715	Trypticase Phytone Glucose Medium with Tween™ 80
716	PMY Medium
717	PFS Medium
723	Castenholz TYE Medium with 2% Trypticase Yeast Extract
724	CML Medium
725	Liver Infusion Sake Medium
730	*Microbacterium* Medium
731	Luminous Medium
733	SCY Medium
735	Chopped Meat Medium with 1% Glucose
737	Chopped Meat Medium with 1% Tween™ 80
738	Glucose Agar, 9K
738	Glucose Broth, 9K
739	Horse Blood Agar
741	TYG Medium
742	Folic Acid Casei Medium
743	Alkvisco Medium
744	Heart Infusion Broth with Glucose
746	Lactobacilli MRS Broth with Cysteine
747	Corn Milk Medium
748	SYA Medium
751	Medium for *Prosthecomicrobium* and *Ancalomicrobium*, Modified
752	ISP Medium 4 with Glucose
753	Yeast Milk Medium
754	*Flexibacter* Medium
756	TYN
757	YB Medium
759	Neomycin Agar, Modified
761	Chase's Medium SP
763	NBY Medium
764	*Streptococcus* Medium
765	*Escherichia* Medium
766	*Desulfovibrio* Medium with Lactate
769	*Halobius* Medium

ATCC Medium Number	Medium Name
771	Mysorens Medium
773	*Mycoplasma* Liquid Medium
775	*Pseudomonas* Medium
776	Mineral Salts Medium with Methanol and Yeast Extract
778	Davis and Mingioli Glucose Minimal Medium
779	G Medium
782	Synthetic Sea Water Medium
783	ML Medium
784	AMS Agar
785	Trypticase Soy Broth with Fetal Calf Serum
787	SYC Medium
788	Benzene Sulfonate Medium
791	*Myxococcus xanthus* Medium
792	PY 1% Medium
793	*Cytophaga* Agarase Agar
794	PP Starch Medium
795	Amino-butyric Acid Medium
796	Isoleucine Hydroxamate Medium
797	*Flavobacterium* M1 Agar
798	Plant *Mycoplasma* Agar
798	Plant *Mycoplasma* Broth
800	Salt Medium
804	Sea Water Yeast Extract Broth, Modified
805	CM plus YE Medium
806	*Thiobacillus* Heterotrophic Medium
810	Myxobacteria Medium
812	Nutrient Agar with Cysteine
814	GC Agar
815	NBY Medium
818	Erythritol Agar
818	Erythritol Broth
819	Blue-Green Nitrogen-Fixing Agar
819	Blue-Green Nitrogen-Fixing Broth
820	Glutamate Medium
822	C 3N *Spiroplasma* Medium
824	ASN-III Agar
824	ASN-III Broth
825	*Xanthomonas* Medium
826	YGC Medium with Glutamic Acid
828	Peptone Yeast Extract Medium
829	Alcal Mannose Medium
833	Mueller-Hinton Medium with Garden Soil
834	*Sarcina ventriculi* Growth Medium
835	Cyclohexanone Medium
837	Yeast Extract Medium

ATCC Medium Number	Medium Name
838	*Spirillum* Nitrogen-Fixing Medium
840	Wilbrinck's Agar for *Xanthomonas albilineans*
841	Bouillon Medium
842	PPLO Broth without Crystal Violet with Horse Serum, Fresh Yeast Extract and Glucose
843	PPLO Broth without Crystal Violet with Calf Serum, Fresh Yeast Extract and Sodium Acetate
844	*Halodurans* Medium
845	*Zymomonas* Medium
846	Glucose Yeast Extract Medium
852	*Serratia* Hd-MHr
854	Blue-Green Agar
854	Blue-Green Broth
871	ESA Medium
872	Nitrate Broth
873	Sorbitol Agar
876	Casein Medium
877	Glucose Broth
878	Maintenance of L Antigen in *Neisseria*
879	YEPB Medium
880	Pept Carb Soluble Starch Agar
883	Rabbit Blood Agar
884	Guanosine Medium
887	Vitamin Medium for *Microbacterium*
892	Uric Acid Medium
894	TEP Uric Acid Medium
895	Uric Acid Utilization
896	Trypticase Soy Agar with Defibrinated Human Blood
896	Trypticase Soy Broth with Defibrinated Human Blood
905	Petragnani Medium
909	*Arthrobacter* YCWD
910	Phosphate Mineral Salts Medium with Octane
920	*Chloroflexus* Agar
920	*Chloroflexus* Broth
924	Medium AS4
925	HP101 Halophile Medium
926	Phthalic Acid Medium
928	*Nitrosococcus* Medium
929	*Nitrosolobus* Medium
936	Corn Steep Liquor Medium
937	Inorganic Salts Maltose Medium
939	M56 Medium

ATCC Medium Number	Medium Name
941	L Medium with Methanol
944	MS Agar
948	*Zymomonas* Medium
949	L and F Basal Salts, Modified with Heptadecane
953	Mueller-Hinton Agar
953	Mueller-Hinton Broth
954	C/10 Medium Reichenbach
955	Dibenzothiophene Mineral Medium
956	*Microcyclus* Medium
957	MN Marine Medium
958	Middlebrook 7H10 Agar with Middlebrook OADC Enrichment and Hemin
959	Peptone Succinate Salts in Sea Water
963	*Meniscus glaucopis* Broth
964	Winogradsky's N-Free Medium
965	Tomato Dextrin Yeast Medium
966	Lactose Distillers Solubles Medium
968	Nutrient Broth NaCl Thymine Medium
969	Ribose Production Medium
970	Antibiotic Medium 3 Plus
971	*Chromobacterium* Medium
972	Lactate Agar
972	Lactate Broth
973	Camphor Minimal Medium
974	*Halobacterium* Medium
975	Skim Milk Acetate Medium
976	Proskauer-Beck Medium for *Mycobacterium*
977	Sodium Acetate Medium I
978	Medium for Lactobacilli
980	Medium for Lactobacilli
982	E Medium for Anaerobes with 0.3% Phloroglucinol
983	E Medium for Anaerobes with 0.2% Rutin
985	Glucose Yeast Extract Medium
988	SP4 Medium
989	Chopped Meat Medium with Formate and Fumarate
990	*Mycobacterium* Yeast Extract Medium
991	TSYES Medium
993	TSY Medium
995	Tryptose Phosphate Broth
998	Trypticase Soy Broth with 1.5% NaCl
1004	Sweet E Broth For Anaerobes
1006	*Lactobacillus* Medium
1013	Malt and Peptone Medium
1015	Chopped Meat Carbohydrate Medium
1016	Chopped Meat Carbohydrate Medium with Rumen Fluid
1017	Chopped Meat Glucose Medium
1018	Mineral Medium with Glucose
1019	*Acetobacterium* Medium
1020	LB Broth, Modified
1022	Yeast Glucose Broth
1024	MED IIa
1027	Acidophilic *Bacillus stearothermophilus* Agar
1027	Acidophilic *Bacillus stearothermophilus* Broth
1030	Heart Infusion Broth with Horse Serum and Fresh Yeast Extract
1031	Chopped Meat Medium with Menadione
1032	Peptone Medium
1036	*Thiomicrospira* Medium
1037	Meat Extract with Peptone and 1.5% Salt
1038	Glucose Yeast Medium with Calcium Carbonate
1039	*Acholeplasma* Medium
1040	*Mycoplasma* Broth with 10% Swine Serum
1042	JD-3 Medium
1043	*Methanosarcina* Medium
1044	Waxy Maize Starch Medium
1045	Methanobacteria Medium
1047	MN Marine Medium with Vitamin B_{12}
1048	Plate Count Agar
1050	*Aquaspirillum* Autotrophic Agar
1050	*Aquaspirillum* Autotrophic Broth
1053	Reinforced Clostridial Medium
1057	*Methylococcus* Medium
1058	Trypticase Soy Agar with Yeast Extract and Glucose
1059	PYG Medium for *Spirillum*
1060	*Frankia* Medium
1061	Trypticase Soy Broth with Tween™ 80
1063	Yeast Tryptone Medium with Streptomycin
1065	LB Medium
1066	LB Medium with Ampicillin
1073	*Caryophanon* Medium
1074	GC Medium with Defined Supplements
1077	Nitrogen-Fixing Marine Medium
1080	Schaedler Broth
1082	LB Medium
1083	LB Medium for *X*1776
1084	Brain Heart Infusion Soil Extract Medium
1086	*Thermus* PMY Agar

ATCC Medium Number	Medium Name
1086	*Thermus* PMY Broth
1088	CYE Agar
1090	Marine Methanol Medium
1092	B Broth
1093	Peptone Yeast Extract Agar
1094	Davis and Mingioli Medium, Modified
1095	Heart Infusion Broth with Glucose and Antibiotics
1096	Methanol Medium
1097	*Halomonas* Medium
1098	RF Medium
1099	CYE Agar, Buffered
1100	Sucrose Peptone Agar
1101	*Sphaericus* Spore Medium
1102	Chopped Meat Carbohydrate Medium with Tween™ 80
1103	*Sphaerotilus* CGYA Medium
1104	Penassay G-THY Medium
1105	CPC Medium
1106	PPLO Broth with Bovine Serum
1107	*Thermoanaerobium brockii* Medium
1108	*Renibacterium* KDM2 Medium
1109	*Flexiligladius* Medium
1111	Double Strength Crude Medium for *Lactobacillus*
1113	TS Soil Extract
1115	*Brucella* Broth
1116	*Brucella* Albimi Broth with 0.16% Agar
1118	Methanobacteria Medium with Glucose and Yeast Extract
1120	*Clostridium kluyveri* Medium
1122	LB Medium with YPTG
1123	Mineral Salts Medium
1124	Medium D
1126	Potato Medium
1127	Stanier's Basal Medium with Pyridoxine and Yeast Extract
1128	Peptone Recovery Broth
1129	Peptone Sodium Cholate
1130	Peptone Cholic Acid Recovery
1133	L Medium with DAP and THY
1135	Methanobacteria Medium with Xylose, Yeast Extract and Tryptone
1136	Methanobacteria Medium with Yeast Extract, Sodium Acetate and Methanol
1137	Modified E Medium
1138	*Sphingobacterium* Medium
1139	Defined Medium for *Rhodopseudomonas*

ATCC Medium Number	Medium Name
1142	Allen and Arnon Medium with Nitrate
1144	Xylose YP Agar
1144	Xylose YP Broth
1145	LBE Medium
1146	Nutrient Agar with Tetracycline
1147	*Brucella* Albimi Broth with Agar and 1.5% NaCl
1148	Glucose Yeast Broth with NaCl
1150	Peptone Yeast Medium with $MgSO_4$
1151	Marine Peptone Yeast Medium with Magnesium Sulfate
1154	L Medium
1155	L Medium with Ampicillin
1159	*Xanthomonas* TYG Agar
1160	Dubos Agar with Filter Paper
1161	Heart Infusion Broth with Inactivated Horse Serum, Fresh Yeast Extract and Sucrose
1163	*Photobacterium* MPY Medium
1167	*Pseudomonas solanacearum* Medium
1168	LHET2 Medium
1169	Glucose Tetrazolium Medium
1170	*Rhodopseudomonas blastica* Medium
1176	*Halobacterium* Medium
1177	*Spiroplasma* Medium
1179	V Y Medium
1181	L Medium with DAP, THY and AMP
1182	L Medium with Tetracycline
1188	*Spiroplasma* Agar MID
1188	*Spiroplasma* Broth MID
1189	Trypticase Soy Broth with 10mM Glucose
1190	*Thermoanaerobacter ethanolicus* Medium
1191	*Clostridium thermocellum* Medium
1193	*Beggiatoa* Medium
1198	AT5N Medium
1199	Lactose Minimal Medium
1200	Defined Glucose Medium EMSY-1
1201	ISP 5
1203	CM4 Medium
1204	C 3G *Spiroplasma* Medium
1205	Yeast Mannitol Agar
1206	Tryptose Broth with Thiamine
1207	BC Medium
1209	NSMP, Modified
1210	Avian *Mycoplasma* Agar
1210	Avian *Mycoplasma* Broth
1213	King's Medium B
1214	NBY Medium

ATCC Medium Number	Medium Name
1215	*Acholeplasma* Medium
1216	Modified SMC
1217	Nutrient Agar, pH 5.0
1218	*Halobacterium* Starch Medium
1223	Medium for *Treponema pectinovorum*
1225	LB Medium for *X*1776 with Tetracycline and Ampicillin
1226	LB Medium with Tetracycline and Ampicillin
1227	LB Medium with Ampicillin
1228	Chopped Meat Medium with 0.025% Tween™ 80
1230	*Haloarcola* Medium
1231	Trichlorophenol Medium
1232	Thermophilic *Bacillus* Medium
1233	Anthranilic Acid Medium, Revised
1234	BG 11 Uracil Agar
1234	BG 11 Uracil Broth
1235	LB Medium with Tetracycline and Ampicillin
1236	LB Medium with Kanamycin
1237	Peptone Yeast Extract Glucose Medium, Modified
1238	Chopped Meat with 10% Reduced Filtered Rumen Fluid
1239	Creatinine Medium
1240	*Clostridium* Cellulose Medium
1241	CT Medium
1243	Yeast Agar, Van Niel's with Succinate
1244	Yeast Agar, Van Niel's with Glutamate
1246	Mineral Medium for Hydrogen Bacteria
1247	Nutrient Broth with Bovine Serum
1249	Baar's Medium for Sulfate Reducers, Modified
1250	Baar's Medium for Sulfate Reducers, Modified with 2.5% NaCl
1252	Reinforced Clostridial Medium with Sodium Lactate
1255	*Thiomicrospira denitrificans* Medium
1256	*Sulfolobus* Medium
1257	ETSA Medium
1260	Thermus BP Medium
1261	GC Agar with Ampicillin and Tetracycline
1265	GC Agar with Chloramphenicol, Ampicillin, and Tetracycline
1267	Marine Glucose Trypticase Yeast Extract Agar
1267	Marine Glucose Trypticase Yeast Extract Broth
1270	*Halobacterium* Medium

ATCC Medium Number	Medium Name
1271	Benzoate Medium
1272	Hydroxybenzoate Medium
1273	LB Medium with Tetracycline
1274	Benzoate Medium II
1275	*Haloanaerobium* Medium
1276	*Thermodesulfotobacterium* Agar
1276	*Thermodesulfotobacterium* Broth
1277	*Propionispira* Medium
1279	*Haloanaerobium praevalens* Medium
1281	M9 Medium with Casamino Acids
1282	Medium for Sulfate Reducers
1283	Medium for Sulfate Reducers
1284	GC Agar with Ampicillin and Gentamicin
1285	GC Agar with Ampicillin
1286	Sea Water Complete Medium
1287	Kerosene Mineral Salts Medium
1290	*Microcyclus-Spirosoma* Medium
1291	SC Agar
1291	SC Broth
1293	Brain Heart Infusion, Supplemented
1294	Medium A
1296	*Spirochaeta zuelzereae* Medium
1297	Nutrient Agar with Sucrose
1298	Brackish *Prosthecomicrobium* Medium
1299	*Cytophaga* Medium
1300	*Clostridium acidiurici* Medium
1301	Marine Peptone Succinate Salts Medium
1302	Halophilic *Clostridium* Agar
1302	Halophilic *Clostridium* Broth
1304	*Sulfolobus solfataricus* Medium
1306	Nitrate Mineral Salts Medium
1308	*Rhodospirillum* Medium
1311	TBAB 298 Medium
1312	*Azospirillum amazonense* Medium
1315	LB Medium with Ampicillin
1316	BSK Medium
1317	PGS Agar
1320	YT Medium
1321	IFO Agar
1321	IFO Broth
1322	Chopped Meat Glucose Medium with NaCl
1327	Medium VTY
1331	Urea Broth 10B for *Ureaplasma urealyticum*
1332	Differential Agar Medium A8 for *Ureaplasma urealyticum*
1333	Trypticase Soy Glucose Medium
1334	TPl Medium

ATCC Medium Number	Medium Name
1339	R2YE Medium
1340	MS Medium for Methanogens
1341	RmM Medium
1342	*Halobacteroides* Medium
1343	General Salts Medium for Estuarine Methanogens
1344	*Clostridium noterae* Medium
1345	Anaerobic Cellulolytic Medium
1349	Halophile Medium
1350	Trypticase Soy Broth with Neomycin
1351	GC Agar
1352	Oxalate Maintenance Medium
1354	Nitrate Mineral Salts Medium with Methanol
1355	*Methanosarcina acetovorans* Medium
1356	Defined Medium with Povidone Iodine
1357	NOS Spirochete Medium
1358	Nutrient Broth, Half Strength
1360	Tomato Paste Oatmeal Agar
1364	LB Medium with Ampicillin
1365	E Medium for Anaerobes with 0.1% Cellobiose
1366	Peptone Yeast Extract Medium
1367	OTI Medium, Modified
1368	*Clostridium* Cellulolytic Medium
1369	Chopped Meat Medium with 10% Fetal Calf Serum
1370	Yeast Agar, Van Niel's with NaCl
1372	Glutamate Medium
1374	Cholesterol Medium
1375	Dubos Broth Base with Horse Serum
1377	*Bacillus pasteurii* NH_4 YE Medium
1377	*Haemophilus ducreyi* Medium
1378	GC Agar with Penicillin G
1379	Marine *Rhodococcus* Medium
1380	Trypticase Soy Broth with Calcium Chloride
1381	PPYG Medium
1383	TYE HES Medium
1385	Marine *Rhodopseudomonas* Medium
1386	Trypticase Soy Agar, Modified
1389	PY Salt Medium
1390	CHCA Salts Medium
1394	*Halobacterium pharaonis* Medium
1396	*Thiosphaera* Agar
1396	*Thiosphaera* Broth
1398	Low Phosphate Buffered Basal Medium, Modified
1399	*Serratia* Medium

ATCC Medium Number	Medium Name
1400	Thiocyanate Utilization Medium
1402	MVL Medium
1403	*Methylophaga* Agar
1403	*Methylophaga* Broth
1407	*Rhodopila globiformis* Medium
1408	*Rhodospirillum* Medium
1409	Imhoff's Medium, Modified
1410	*Ectothiorhodospira* Medium
1411	Cellulolytic Agar for Thermophiles
1411	Cellulolytic Broth for Thermophiles
1412	Cellulolytic Agar with Sea Salts
1412	Cellulolytic Broth with Sea Salts
1413	Phenol Nutrient Supplemented Agar
1414	*Spiroplasma* Medium with 25 mg/L Phenol Red
1415	LB Medium with Rifampicin
1416	Hydroxybenzoic Acid Medium
1417	Low Iron YC Agar
1417	Low Iron YC Broth
1419	SI Agar
1420	Trypticase Soy Agar with Sheep Blood, Sucrose and Tetracycline
1421	LHET2 Medium with Yeast Extract or Yeast Autolysate
1422	*Thiomicrospira* Medium
1423	*Brucella* Broth, Modified
1425	*Sporomusa* Medium
1426	Cyclohexanecarboxylic Acid Medium
1428	*Halomethanococcus* Medium
1430	*Melissococcus pluton* Medium
1431	*Blastobacter* Medium
1432	Potato P-YE *Thermus* Medium
1433	Chlorohydroxybenzoic Acid Medium
1434	*Halobacterium denitrificans* Medium
1435	*Mycoplasma* Agar
1436	Marine Spirochaete Medium
1438	*Rhodobacter veldkampii* Medium
1439	*Methanogenium* Medium
1441	Cornstarch Soluble Medium
1443	*Nannocystis* Agar
1445	Nutrient Agar with Streptomycin
1446	Penassay Broth with Chloramphenicol
1448	*Ectothiorhodospira* Medium, Modified
1449	*Chromatium* Medium
1452	Mueller-Hinton Medium with Rabbit Serum
1455	Alkaline Yeast Extract Malt Medium
1460	Mineral Medium with Dichlorobenzoate

ATCC Medium Number	Medium Name
1463	Hayflick Medium, Modified
1464	*Dictyoglomus* Medium
1465	Basal Thermophile Medium
1467	*Ilyobacter* Agar
1467	*Ilyobacter* Broth
1468	LB Medium with Kanamycin
1470	*Leptospira* Medium, Modified
1471	Sludge Medium for Methanobacteria
1473	LPBM Acido-Thermophile Medium
1474	Methanol Mineral Salts Medium
1475	Glucose Yeast Chalk Agar
1476	TYEG Medium
1478	Tryptone Glucose Beef Extract Agar with Sucrose
1479	Mineral Medium with Chloridazon
1480	Mineral Medium with Antipyrin
1481	Trypticase Soy Agar, Modified
1483	Dichloromethane Medium for *Hyphomicrobium*
1484	Citrate Phosphate Buffered Glucose Medium
1485	*Actinopolyspora* Medium
1486	Oatmeal Nitrate Agar
1487	Marine Salts Medium for *Sporosarcina halophila*
1489	Hickey-Tresner Agar
1490	Chopped Meat Medium, Modified
1492	Mineral Base E for Autotrophic Growth
1493	Buffered Marine Yeast Medium
1494	NOS Medium, Modified
1499	*Clostridium aminobutyricum* Medium
1500	Cornstarch Soluble Medium without *n*-Butanol
1501	Brackish Acetate
1502	Medium E for *Bacillus*
1503	*Leptothrix* 2X PGY Medium
1506	Peptone Succinate Salts Medium
1507	Middlebrook 7H9 Broth, Supplemented
1508	*Rhizobium* X Medium with Thiram
1509	Nutrient Broth, Diluted 1:100
1510	*Thiobacillus tepidarius* Medium
1511	Vitamin B_6 Blood Agar
1512	Burke's Modified Nitrogen-Free Medium with Benzoate
1513	PY CMC Medium
1514	Oxalate Medium, Modified
1515	PY Inositol Medium
1516	PPLO Broth without Crystal Violet with Horse Serum and Yeast Extract

ATCC Medium Number	Medium Name
1518	*Methanolobus* Medium
1519	MVTY Medium
1521	PYGV Medium
1522	Trypticase Soy Agar with Tobramycin
1522	Trypticase Soy Broth with Tobramycin
1524	Peptone Yeast Extract Medium
1526	Rila Marine Medium
1527	PY Medium with Glucose
1528	PY Medium with Fructose
1530	Thermophilic Maintenance Medium
1531	*Thiocapsa* Medium
1533	GC Agar with Ampicillin
1535	Nitrate Methanol Medium
1536	*Anaerospirillum* Medium
1538	*Thermoproteus* Medium
1539	Chopped Meat Medium with Formate and Fumarate
1540	*Brucella* Albimi Broth with Formate and Fumarate
1541	Singh's Medium, Modified
1543	Mineral Medium with Santonin
1545	Methanol Ammonium Salts Medium
1547	Anaerobic D-Gluconate
1548	Anaerobic Glucuronic Acid Medium
1549	Benzoate Minimal Salts Medium
1550	Glycocholate Mineral Medium
1551	*Methanomicrobium* Medium
1552	Van Niel's Yeast Medium with Pyruvate, Modified
1553	Alginate Utilization Medium
1554	Mineral Salts for Thermophiles
1555	*Desulfobulbus* Medium
1556	*Clostridium aerotolerans* Medium
1558	*Desulfurococcus* Medium
1559	*Flexibacter* Medium
1560	Cellulolytic Medium with Rumen Fluid
1561	Middlebrook 7H10 Agar with Streptomycin
1562	Standard Agar with Methanol and Yeast Extract
1565	Casitone Yeast Extract Agar
1567	*Pedomicrobium* PSM Medium with Ribose
1570	Folic Acid Casei Medium with Chloramphenicol
1572	Sea Water Yeast Extract Peptone Medium
1573	*Nitrosomonas europaea* Medium
1575	Marine Agar with κ- and λ-Carrageenan Broth
1576	Marine Agar with ι-Carrageenan Broth

ATCC Medium Number	Medium Name
1581	Mineral Salts with Butane
1582	Aolpha Medium
1585	*Centenum* Medium
1590	Natronobacteria Medium
1591	*Bacillus schlegelii* Medium
1595	KPL Medium
1596	Cholic Acid Medium
1600	S Salts
1601	*Desulfonema limicola* Medium
1602	*Desulfonema magnum* Medium
1603	DNB Medium
1604	Nutrient Agar with Yeast Extract
1607	Gliding Medium
1611	U4 Medium
1612	*Acetobacterium* Medium
1613	*Clostridium thermolacticum* Medium
1616	*Methanobacterium alcalophilum* Medium
1617	Sludge Medium for Methanobacteria, pH 7.9
1618	Heterotrophic Medium for Hydrogen-Oxidizing Bacteria
1620	CH 1 Medium
1621	Proteose Yeast Extract Medium
1622	LB Medium with Glucose
1626	*Clostridium aminovalericum* Medium
1627	*Desulfobacterium* Medium
1628	*Desulfobacterium phenolicum* Medium, Modified
1629	Heart Infusion Medium with Fetal Bovine Serum
1645	Trypticase Soy Agar with Sheep Blood, Formate and Fumarate
1646	*Desulfovibrio sulfodismutans* Medium
1647	*Desulfobacterium indolicum* Medium
1648	*Desulfobacterium* Medium, Modified
1649	*Selenomonas adiaminophila* Medium
1653	Magnetic *Spirillum* Growth Medium, Revised
1655	BAM Agar
1655	BAM Broth
1656	BAM SM Agar
1656	BAM SM Broth
1657	M14 Medium
1658	MMS Medium for *Thermotoga neapolitana*
1659	YA Halophile Medium
1660	MH Salts
1661	M13 *Verrucomicrobium* Medium
1662	*Eubacterium acidaminophilum* Medium
1663	Dilute Potato Medium
1664	Anaerobic Trypticase Soy Agar with Calf Blood
1669	Acetogen Medium
1673	Burke's Modified Nitrogen-Free Medium
1675	LB Medium with Chloramphenicol
1676	Van Niel's Medium, Modified
1678	YGCP Medium
1679	SOT Medium
1682	Artificial Deep Lake Medium
1683	AMS Agar without Methanol
1686	Sorbitol Medium
1687	*Flavobacterium* Medium
1689	MH Medium
1690	*Desulfomonile tiedjei* Medium
1691	*Agrobacterium*-Mannitol Medium
1693	VY2 Agar
1694	Stanier's Basal Medium with Trichlorophenoxyacetate
1695	*Bacillus* Pullulan Salts
1696	*Bacillus* Xylose Salts
1699	*Mycoplasma* Medium, Revised
1700	*Rhizomonas* Medium
1701	*Agrobacterium* Mannitol Medium
1702	Hydroxybenzoate Agar
1702	Hydroxybenzoate Broth
1703	Chopped Meat Carbohydrate Medium with Rumen Fluid
1705	*Brucella* Albimi Broth with Sheep Blood
1710	Carnitine Chloride Medium
1712	Spirochete Medium
1713	*Azotobacter* Agar, Modified I
1713	*Azotobacter* Broth, Modified I
1714	*Azotobacter* Agar, Modified II
1714	*Azotobacter* Broth, Modified II
1715	E Medium for Anaerobes with Filtered Rumen Fluid and 0.1% Cellobiose
1716	Corn Steep Liquor Medium
1717	Buffered S & H Medium
1718	R70-2 Agar, Modified with Fructose Broth
1719	R70-2 Agar, Modified with Glucose Broth
1722	*Dichotomicrobium* Medium
1723	*Sulfolobus* Medium, Revised
1724	Ducreyi Medium, Revised
1725	Nutrient Broth with 6% NaCl
1729	*Ruminococcus pasteuri* Medium
1731	*Zymomonas* Sucrose Medium
1732	PYGV Marine Medium

ATCC Medium Number	Medium Name
1734	Cellulolytic Medium with Rumen Fluid and Soluble Starch
1735	*Dermabacter* Medium
1741	Mineral Medium with Phenol
1742	Glucose Yeast Extract Medium
1745	LB Medium with TMP
1746	LB Medium with TPP
1748	Siderophore Mineral Medium
1750	Anacker and Ordal Medium
1751	Anacker and Ordal Medium, Enriched
1758	Chopped Meat Medium, Modified with Tween™ 80
1760	*Veillonella* Medium, DSM
1761	Chopped Meat Medium, Modified with Arginine
1763	Yeast Malt Extract Catalase Agar
1764	Mineral Medium
1765	*Acetobacter xylinum* Medium
1766	Xylan Medium
1771	*Azotobacter* Medium
1775	*Archaeoglobus* Medium
1776	Tyrosine Agar
1777	*Fervidobacterium* Medium
1782	*Eubacterium angustum* Medium
1786	*Ilyobacter* Medium
1787	KC Bottom Agar
1787	KC Broth
1787	KC Top Agar
1788	Cinnamate Medium
1790	*Haliscomenobacter* Medium
1792	*Formivibrio citricus* Medium

Standard Methods Media for the Examination of Water and Wastewater

The *Handbook of Microbiological Media* contains the media used for the testing of waters and wastewaters according to standard methods.

A-1 Broth
Acetamide Broth
Asparagine Broth
Aureomycin® Bengal Glucose Peptone Agar
Azide Dextrose Broth
Baird–Parker Agar Base
Bile Esculin Agar
Brain-Heart Infusion Agar
Brain-Heart Infusion Broth
Brilliant Green Lactose Bile Broth
Buffered Charcoal Yeast Extract Alpha Base
Casitone–Glycerol Yeast Autolysate Broth (CGY)
CCVC Medium (BCYE Selective Agar with CCVC)
Czapek (or Czapek-Dox) Agar
Diamalt Agar
EC Medium (Fecal Coliform Test)
EIA Substrate
Extracted Hay Medium
Ferrous Sulfide Agar
GPVA Medium (BCYE Selective Agar with GPVA)
Iron Oxidizing Medium
Isolation Medium (Iron Bacteria Isolation Medium)
Lauryl Tryptose Broth
LES Endo Agar
Lipovitellin Salt Mannitol Agar
M–7 h FC Agar
MacConkey Agar
mE Agar
M-Endo Medium
m-*Enterococcus* Agar
M-FC Agar, Modified
M-FC Medium
m-HPC Agar
Milk Agar
M-KLEB Agar
Mn Agar No. 1
Mn Agar No. 2
MP Agar
M-PA Agar
M-PA Agar, Modified
M-ST Holding Medium
M-*Staphylococcus* Broth
MSV AcS Agar
MSV Agar
MSV Broth
MSV GS Agar
MSV I Agar
MSV LT Agar
MSV S Agar
MSV SS Agar
MSV SUC Agar
m-T7 Agar

mTEC Agar
MY Agar
Neopeptone Glucose Rose Bengal Aureomycin ®Agar
Nutrient Agar
NWRI Agar (HPCA)
Pfizer Selective Enterococcus (PSE) Agar
Plate Count Agar
Presence-Absence (P-A) Broth
R2A Agar
SCY Medium
Starch Casein Agar
Streptomycin Terramycin® Malt Extract Agar
Sulfate-Reducing Medium
Sulfur Medium
Thiosulfate Oxidizing Medium
Thiothrix Medium
Tryptic Soy Agar
Trypticase Soy Agar (TSA)
Trypticase Soy Broth (TSB)
Urea Substrate
Yeast Extract Malt Extract Glucose Agar
Yeast Nitrogen Base Glucose Broth

FDA Recommended Media for the Testing of Foods

The *Handbook of Microbiological Media* contains the media used for the testing of foods—including those recommended by the FDA.

FDA No.	Medium Name
M1	A 1 Broth
M2	Acetamide Agar
M2	Acetamide Broth
M3	Acetate Differential Agar
M4	Acid Broth
M5	AE Sporulation Medium, Modified
M6	Agar Medium P
M7	AKI Medium
M8	Akaline Peptone Agar
M9	Alkaline Peptone Salt Broth
M10	Alkaline Peptone Water
M11	Anaerobe Agar
M12	Anaerobic Egg Yolk Agar
M13	Andrade's Broth
M14	Antibiotic Medium 1
M15	Antibiotic Medium 4
M16	Arginine Glucose Slants
M17	Baird–Parker Agar, Supplemented
M18	Bile Esculin Agar
M19	Bismuth Sulfite Agar
M20	Blood Agar
M20a	Blood Agar Base
M21	Blood Agar Base
M22	Blood Agar No. 2
M23	Brain Heart Infusion Agar, 0.7%
M24	Brain Heart Infusion Agar
M24	Brain Heart Infusion Broth
M25	Brilliant Green Lactose Bile Broth
M26	Bromcresol Purple Broth
M27	Bromcresol Purple Dextrose Broth
M28	*Brucella* Semisolid Medium with Cysteine
M28	*Brucella* Semisolid Medium with Glycine
M28	*Brucella* Semisolid Medium with NaCl
M28	*Brucella* Semisolid Medium with Nitrate
M29	*Campylobacter* Enrichment Broth
M30	*Campylobacter* Isolation Agar A
M30	*Campylobacter* Isolation Agar B
M31	Cary and Blair Transport Medium
M32	Casamino Acids Yeast Extract Broth (CYE Broth)
M34	Casamino Acids Yeast Extract Salts Broth, Gorbach (CA YE Broth)
M35	CIN Agar
M36	Cell Growth Medium
M37	Cetrimide Agar, USP
M38	Chopped Liver Broth
M39	Christensen Citrate Sulfide Medium
M40	Urea Agar
M41	Congo Red BHI Agarose Medium
M42	Cooked Meat Medium
M43	Cooked Meat Medium, Modified
M44	Decarboxylase Medium Base, Falkow
M45	Duncan–Strong Sporulation Medium, Modified
M46	Eagle Medium
M47	Earle's Balanced Salts, Phenol Red–Free
M48	EB Motility Medium
M49	*EC* Broth
M50	*EC* Broth with MUG
M52	Enrichment Broth, pH 7.3 with Pyruvate
M53	Esculin Agar
M54	Gelatin Agar
M55	Gelatin Salt Agar
M58	Ham's F–10 Medium
M59	Heart Infusion Agar
M60	Heart Infusion Agar
M60	Heart Infusion Broth

FDA No.	Medium Name
M61	Hektoen Enteric Agar
M62	Hemorrhagic coli Agar
M63	Hugh–Leifson's Glucose Broth
M64	Indole Medium
M65	Indole Medium
M66	Indole Nitrate Medium
M67	Irgasan® Ticarcillin Chlorate Broth
M68	Iron Milk Medium, Modified
M69	King's Medium B
M70	Oxidation Fermentation Medium, King's
M71	Kliger Iron Agar
M72	Koser Citrate Medium
M73	L15 Medium, Modified Leibovitz
M74	Lactose Broth
M75	Lactose Gelatin Medium
M76	Lauryl Sulfate Broth
M77	Lauryl Sulfate Broth with MUG
M78	Letheen Agar, Modified
M79	Letheen Broth, Modified
M80	Levine EMB Agar
M81	LPM Agar
M82	LPM Agar with Esculin and Ferric Iron
M83	Liver Veal Agar
M84	Liver Veal Egg Yolk Agar
M85	Long–Term Preservation Medium
M86	Lysine Arginine Iron Agar
M87	Lysine Decarboxylase Broth, Falkow
M88	Lysine Decarboxylase Medium
M89	Lysine Iron Agar
M90	Lysozyme Broth
M91	MacConkey Agar
M92	Malonate Broth, Ewing Modified
M93	Malt Extract Agar
M94	Malt Extract Broth
M95	Mannitol Egg Yolk Polymyxin Agar
M96	Mannitol Maltose Agar
M97	Mannitol Salt Agar
M98	Cellobiose Polymyxin B Colistin Agar, Modified
M99	Motility Indole Ornithine Medium
M100	Motility Medium
M101	Motility Nitrate Medium
M102	Motility Nitrate Medium, Buffered
M103	Motility Test Medium
M104	MRVP Broth
M105	Mucate Broth
M106	Mucate Control Broth
M107	Mueller–Hinton Agar
M108	Nitrate Broth
M109	Nitrate Broth, Enriched

FDA No.	Medium Name
M110	Nitrate Reduction Broth
M111	Nonfat Dry Milk, Reconstituted
M112	Nutrient Agar
M113	Nutrient Agar
M114	Nutrient Broth
M115	Nutrient Gelatin
M116	Oxidative–Fermentative Glucose Medium, Semisolid
M116	Oxidative–Fermentative Glucose Medium, Semisolid with NaCl
M117	Oxidative–Fermentative Test Medium
M118	Oxford Agar
M120	Peptone Sorbitol Bile Broth
M121	Phenol Red Broth
M122	Phenol Red Glucose Broth
M123	Phenylalanine Agar
M124	Plate Count Agar
M125	PMP Broth
M126	Potassium Cyanide Broth
M127	Potato Dextrose Agar
M128	*Pseudomonas* Agar F
M129	*Pseudomonas* Agar P
M130	Purple Broth
M130a	Purple Broth
M131	Pyrazinamidase Agar
M132	Rappaport–Vassiliadis Enrichment Broth
M133	Sabouraud's Glucose Agar
M133	Sabouraud's Glucose Broth
M134	Selenite Cystine Broth
M135	Blood Agar No. 2
M136	*Shigella* Broth
M137	SIM Medium
M138	Simmons Citrate Agar
M139	Sorbitol MacConkey Agar
M140	Sporulation Broth
M141	Spray's Fermentation Medium
M142	*Staphylococcus* Agar No. 110
M143	Starch Agar
M144	T_1N_1 Medium
M145	Tetrathionate Broth
M146	Fluid Thioglycollate Medium
M147	TCBS Agar
M148	Toluidine Blue DNA Agar
M149	Triple Sugar Iron Agar
M150	Trypticase Novobiocin Broth
M151	Trypticase Peptone Glucose Yeast Extract Broth
M151a	Trypticase Peptone Glucose Yeast Extract Broth with Trypsin
M152	Trypticase Soy Agar

FDA No.	Medium Name
M153	Trypticase Soy Agar Yeast Extract
M154	Trypticase Soy Broth
M154a	Trypticase Soy Broth with Sodium Chloride and Sodium Pyruvate
M155	Trypticase Soy Broth with Glycerol
M156	Trypticase Soy Broth, Modified
M157	Trypticase Soy Broth with Yeast Extract
M158	Trypticase Soy Polymyxin Broth
M159	Trypticase Soy Agar with Sheep Blood
M160	Trypticase Soy Tryptose Broth
M161	T_1N_0 Broth
M161	T_1N_1 Broth
M161	T_1N_3 Medium
M161	T_1N_6 Broth
M161	T_1N_8 Broth
M161	T_1N_{10} Broth
M162	Tryptone Phosphate Broth
M163	T_1N_1 Agar
M163	T_1N_2 Agar
M164	Tryptone Broth
M165	Tryptone Yeast Extract Agar
M166	Tryptone Blood Agar Base
M167	Tryptose Agar
M167	Tryptose Broth
M168	Tryptose Phosphate Broth
M169	Tryptose Sulfite Cycloserine Agar
M170	Tyrosine Agar
M171	Urease Test Broth
M172	Urea R Broth
M173	Veal Infusion Agar
M173	Veal Infusion Broth
M174	Violet Red Bile Agar
M175	Violet Red Bile Agar with MUG
M176	Vogel and Johnson Agar
M177	VP Medium
M178	Wagatsuma Agar
M179	XLD Agar
M180	Y 1 Adrenal Cell Growth Medium
M181	Yeast Extract Agar

Media for the Isolation and Identification of Microorganisms from Clinical Specimens

The *Handbook of Microbiological Media* includes both classic and modern media used for the identification, cultivation, and maintenance of diverse bacteria described in *The Manual of Clinical Microbiology* for medically important microorganisms. Below are some of the primary media used in clinical microbiology laboratories for the isolation of pathogens.

Medium	Use
Anaerobic Blood Agar	Cultivation of a wide variety of anaerobic pathogens
Anaerobic Kanamycin Vancomycin Laked Blood Agar	Cultivation of *Bacteroides* species
Azide Blood Agar	Cultivation of *Streptococcus pyogenes* which forms small colonies and demonstrates beta hemolysis
Bacteroides Bile Esculin Agar	Cultivation of *Bacteroides* species
Bile Esculin Agar	Cultivation of group D streptococci including *Enterococcus faecalis*
Bismuth Sulfite Agar	Cultivation of *Salmonella* including *S. typhi* from faecal specimens
Blood Agar	Cultivation of many pathogens
Blood Agar with Colistin and Nalixic Acid	Selective cultivation of Gram-positive bacteria
Brain Heart Infusion Medium	Cultivation of a wide variety of aerobic and anaerobic human pathogens
Brilliant Green Agar	Selective cultivation of *Salmonella* species
Campylobacter Agar	Selective cultivation of *Campylobacter* species
Chocolate Agar	Cultivation of a wide variety of human pathogens including *Haemophilus* and *Neisseria* species
Chopped Meat Glucose Medium	Cultivation of a wide variety of anaerobes
Deoxycholate Agar	Selective cultivation of Gram-negative enteric bacteria
Deoxycholate Citrate Agar	Selective cultivation of *Salmonella* and some *Shigella* species
Enterococcus Agar	Selective cultivation of enterococci
Eosin Methylene Blue Agar	Selective cultivation of Gram-negative enteric bacteria
GN Broth	Enrichment for *Salmonella* and *Shigella* species
Hektoen Enteric Agar	Selective cultivation of enteric bacteria including *Shigella* species
MacConkey Agar	Selective cultivation of Gram-negative enteric bacteria
Mannitol Salt Agar	Selective cultivation of *Staphylococcus* species
Martin–Lewis Agar	Selective cultivation of *Neisseria* species
New York City Agar	Selective cultivation of *Neisseria* and *Mycoplasma* species

Medium	Use
Phenylethyl Alcohol Medium	Selective cultivation of *Staphylococcus* and *Streptococcus* species.
Selenite F Broth	Enrichment for *Salmonella* species
Tetrathionate Broth	Enrichment for *Salmonella* and some *Shigella* species
Thayer Martin Agar	Selective cultivation of *Neisseria* species
Thioglycollate Broth	Cultivation of a wide variety of anaerobes
Xylose Lysine Deoxycholate Agar	Selective cultivation of *Salmonella* and some *Shigella* species

References

Below is a list of references that can be consulted for further information about media used for the isolation, cultivation, and differentiation of microorganisms.

A Compilation of Culture Media for the Cultivation of Microorganisms. 1930. M. Levine and H. W. Schoenlein. Williams and Wilkins Co., Baltimore.

ATCC Catalogue of Bacteria and Bacteriophages. 1992. R. Gherna, P. Pienta, and R. Cote, eds. American Type Culture Collection, Rockville MD.

ATCC Catalogue of Filamentous Fungi. 1991. S. C. Jong and M. J. Edwards, eds. American Type Culture Collection, Rockville MD.

ATCC Catalogue of Recombinant DNA Materials. 1991. D. R. Maglott and W. C. Nierman, eds. American Type Culture Collection, Rockville MD.

Color Atlas and Textbook of Diagnostic Microbiology. 1992. E. W. Koneman, S. D. Allen, W. M. Janda, P. C. Schreckenberger, and W. C. Winn, Jr., eds. J. B. Lippincott Co., Philadelphia PA.

Compendium of Methods for the Microbiological Examination of Foods. 1992. C. Vanderzant and D. F. Splittstoessser, eds. American Public Health Association,Washington DC.

CRC Handbook Series in Nutrition and Food Section G: Diets, Culture Media, Food Supplements; Volume III: Culture Media for Microorganisms and Plants. 1978. M. Rechcigl, Jr. ed. CRC Press, Boca Raton FL.

Diagnostic Microbiology. 1990. S. M. Finegold and W. J. Martin. C. V. Mosby Co., St. Louis.

Difco Manual: Dehydrated Culture Media and Reagents for Microbiology. Difco Laboratories, Detroit MI.

Food and Drug Administration Bacteriological Analytical Manual. 1992. AOAC International, Arlington VA.

Manual of BBl Products and Laboratory Procedures. 1988. D. A. Power and P. J. McCuen, eds. Beckton Dickinson and Company, Cockeysville, MD.

Manual of Clinical Microbiology. 1991. A. Ballows, W. J. Hausler, K. L. Herman, H. D. Isenberg, and H. J. Shadomy, eds. American Society for Microbiology, Washington DC.

Media for Isolation–Cultivation–Identification–Maintenance of Medical Bacteria. 1985. J. F. McFaddin. Williams and Wilkins, Baltimore MD.

The Oxoid Manual. 1990. E. Y. Bridson, ed. Unipath Ltd. Basingstoke, Hampshire England.

Practical Microbiology. 1969. G. Sirockin and S. Cullimore. McGraw-Hill, London.

Standard Methods for the Examination of Water and Wastewater. 1992. A. E. Greenberg, L. S. Clesceri, and A. D. Eaton, eds. American Public Health Association, Washington DC.

A 1 Broth

Composition per liter:

Pancreatic digest of casein	20.0g
Lactose	5.0g
NaCl	5.0g
Salicin	0.5g
Triton™ X-100	1.0mL

pH 6.9 ± 0.1 at 25°C

Source: This medium is available as a premixed powder from Difco Laboratories and BBL Microbiology Systems.

Preparation of Medium: Add components to distilled/deionized water and bring volume to 1.0L. Mix thoroughly. Gently heat and bring to boiling. Distribute into test tubes containing an inverted Durham tube. Autoclave for 10 min at 15 psi pressure–121°C.

Use: For the detection of fecal coliforms in foods, treated wastewater and seawater by a most-probable-number (MPN) method. Multiple dilutions of samples (3, 5, or 10 replicates per dilution) are added to tubes containing A 1 Broth. After incubation test tubes with gas accumulation in the Durham tubes are scored positive and those with no gas as negative. A MPN table is consulted to determine the most probable number of fecal coliforms.

A1 Minimal Medium

Composition per liter:

L-Asparagine	5.0g
$(NH_4)_2SO_4$	5.0g
Sodium pyruvate	5.0g
$MgSO_4 \cdot 7H_2O$	2.0g
Spermadine·3HCl	0.125g
L-Asparagine	0.10g
L-Isoleucine	0.10g
L-Methionine	0.10g
L-Phenylalanine	0.10g
L-Valine	0.10g
L-Leucine	0.05g
KH_2PO_4	0.013g
$FeCl_3 \cdot 6H_2O$	2.7mg
$CaCl_2$	1.1mg
Cyanocobalamin	1.0mg
Tris(hydroxymethyl)aminomethane buffer (0.01M solution, pH 7.6)	1.0L

pH 7.6 ± 0.2 at 25°C

Preparation of Medium: Add solid components to 1.0L of Tris buffer. Mix thoroughly. Filter sterilize. Aseptically distribute into tubes or flasks.

Use: For the cultivation of *Myxococcus xanthus*.

A 3 Agar

Composition per 202.4mL:

Agar base	140.0mL
Supplement solution	62.4mL

pH 6.0 ± 0.2 at 25°C

Agar Base:
Composition per liter:

Pancreatic digest of casein	17.0g
Ionagar No. 2, Oxoid Unipath	7.5g
NaCl	5.0g
Papaic digest of soybean meal	3.0g
K_2HPO_4	2.5g
Glucose	2.5g

Preparation of Agar Base: Add components, except agar, to distilled/deionized water and bring volume to 1.0L. Adjust pH to 5.5. Add agar. Mix thoroughly. Gently heat and bring to boiling. Distribute into screw-capped bottles in 140.0mL volumes. Autoclave for 15 min at 15 psi pressure–121°C. Cool to 45°–50°C.

Supplement Solution:
Composition per 62.4mL:

Horse serum-urea solution	40.0mL
Fresh yeast extract solution	20.0mL
Penicillin solution	2.0mL
Phenol Red solution	0.4mL

Preparation of Supplement Solution: Aseptically combine components. Mix thoroughly.

Horse Serum-Urea Solution:
Composition per 40mL:

Urea	0.2g
Horse serum, unheated	40.0mL

Preparation of Horse Serum-Urea Solution: Add urea to 40.0mL of horse serum. Mix thoroughly. Filter sterilize.

Fresh Yeast Extract Solution:
Composition per 100mL:

Live, pressed, starch-free, Baker's yeast	25.0g

Penicillin Solution:
Composition per 10mL:

Penicillin G	1,000,000U

Preparation of Penicillin Solution: Add penicillin to distilled/deionized water and bring volume to 10.0mL. Mix thoroughly. Filter sterilize.

Phenol Red Solution:
Composition per 10mL:

Phenol Red	0.1g

Preparation of Phenol Red Solution: Add Phenol Red to distilled/deionized water and bring volume to 10.0mL. Mix thoroughly. Filter sterilize.

Preparation of Medium: Aseptically combine 140.0mL of cooled, sterile agar base and 62.4mL of sterile supplement solution. Mix thoroughly. Pour into sterile Petri dishes or distribute into sterile tubes.

Use: For the cultivation of *Ureaplasma urealyticum* from urine. Also used for the cultivation of other *Ureaplasma* species.

A 3B Agar

Composition per 101.5mL:

Agar base	80.0mL
Supplement solution	21.5mL

pH 6.0 ± 0.2 at 25°C

Agar Base:
Composition per liter:

Pancreatic digest of casein	17.0g
Ionagar No. 2, Oxoid Unipath	7.5g
NaCl	5.0g
Papaic digest of soybean meal	3.0g
K_2HPO_4	2.5g
Glucose	2.5g

Preparation of Agar Base: Add components, except agar, to distilled/deionized water and bring volume to 1.0L. Adjust pH to 5.5. Add agar. Mix thoroughly. Gently heat and bring to boiling. Distribute into screw-capped bottles in 80.0mL volumes. Autoclave for 15 min at 15 psi pressure–121°C. Cool to 45°–50°C.

Supplement Solution:
Composition per 21.5mL:

Horse serum-urea solution	20.0mL
Penicillin solution	1.0mL
Cysteine·HCl·H_2O solution	0.5mL

Preparation of Supplement Solution: Aseptically combine components. Mix thoroughly.

Horse Serum-Urea Solution:
Composition per 40mL:

Urea	0.2g
Horse serum, unheated	40.0mL

Preparation of Horse Serum-Urea Solution: Add urea to 40.0mL of horse serum. Mix thoroughly. Filter sterilize.

Penicillin Solution:
Composition per 10mL:

Penicillin G	1,000,000U

Preparation of Penicillin Solution: Add penicillin to distilled/deionized water and bring volume to 10.0mL. Mix thoroughly. Filter sterilize.

Cysteine·HCl·H_2O Solution:
Composition per 10mL:

L-Cysteine·HCl·H_2O	0.2g

Preparation of Cysteine·HCl·H_2O Solution: Add L-Cysteine·HCl·H_2O to distilled/deionized water and bring volume to 10.0mL. Mix thoroughly. Filter sterilize.

Preparation of Medium: Aseptically combine 80.0mL of cooled, sterile agar base and 21.5mL of sterile supplement solution. Mix thoroughly.

Use: For the cultivation of *Ureaplasma urealyticum* from urine. Also used for the cultivation of other *Ureaplasma* species.

A 7 Agar (Shepard's Differential Agar)

Composition per 205.7mL:

Agar base	160.0mL
Supplement solution	45.7mL

pH 6.0 ± 0.2 at 25°C

Agar Base:
Composition per 165mL:

Pancreatic digest of casein	2.72g
Agar	2.1g
NaCl	0.8g
Papaic digest of soybean meal	0.48g
K_2HPO_4	0.4g
Glucose	0.4g
$MnSO_4$·H_2O	0.15g

Preparation of Agar Base: Add components, except agar, to distilled/deionized water and bring volume to 165.0mL. Adjust pH to 5.5. Add agar. Mix thoroughly. Autoclave for 15 min at 15 psi pressure–121°C. Cool to 45°–50°C.

Supplement Solution:
Composition per 45.72mL:

Horse serum, unheated	40.0mL
Fresh yeast extract solution	2.0mL
Penicillin solution	2.0mL
CVA enrichment	1.0mL
Cysteine·HCl·H_2O solution	0.5mL
Urea solution	0.22mL

Preparation of Supplement Solution: Aseptically combine components. Mix thoroughly.

Fresh Yeast Extract Solution:
Composition per 100mL:

Live, pressed, starch-free, Baker's yeast	25.0g

Preparation of Fresh Yeast Extract Solution: Add the live Baker's yeast to 100.0mL of distilled/deionized water. Autoclave for 90 min at 15 psi pressure–121°C. Allow to stand. Remove supernatant solution. Adjust pH to 6.6–6.8. Filter sterilize.

Penicillin Solution:

Composition per 10mL:

Penicillin G 1,000,000U

Preparation of Penicillin Solution: Add penicillin to distilled/deionized water and bring volume to 10.0mL. Mix thoroughly. Filter sterilize.

CVA Enrichment:

Composition per liter:

Glucose 100.0g
L-Cysteine·HCl·H_2O 25.9g
L-Glutamine 10.0g
Vitamin B_{12} 0.01g
L-Cystine·2HCl 1.0g
Adenine 1.0g
Nicotinamide adenine dinucleotide 0.25g
Cocarboxylase 0.1g
Guanine·HCl 0.03g
$Fe(NO_3)_3$ 0.02g
p-Aminobenzoic acid 0.013g
Thiamine·HCl 3.0mg

Preparation of CVA Enrichment: Add components to distilled/deionized water and bring volume to 1.0L. Mix thoroughly. Filter sterilize.

Cysteine·HCl·H_2O Solution:

Composition per 10mL:

L-Cysteine·HCl·H_2O 0.4g

Preparation of Cysteine·HCl·H_2O Solution: Add L-Cysteine·HCl·H_2O solution to distilled/deionized water and bring volume to 10.0mL. Mix thoroughly. Filter sterilize.

Urea Solution:

Composition per 10mL:

Urea, ultrapure 1.0g

Preparation of Urea Solution: Add urea to distilled/deionized water and bring volume to 10.0mL. Mix thoroughly. Filter sterilize.

Preparation of Medium: Aseptically combine 160.0mL of cooled, sterile agar base and 45.9mL of sterile supplement solution. Mix thoroughly. Pour into sterile Petri dishes or distribute into sterile tubes.

Use: For the cultivation and differentiation of *Ureaplasma urealyticum* from urine based on their ability to produce ammonia from urea. Bacteria that produce ammonia appear as golden to dark brown colonies. Also used for the cultivation of other *Ureaplasma* species.

A 7 Agar, Modified

Composition per 205.7mL:

Agar base 160.0mL
Supplement solution 45.7mL

pH 6.0 ± 0.2 at 25°C

Agar Base:

Composition per 165mL:

Agar 10.0g
Pancreatic digest of casein 2.72g
NaCl 0.8g
Papaic digest of soybean meal 0.48g
K_2HPO_4 .... 0.4g
Glucose 0.4g
$MnSO_4 \cdot H_2O$ 0.15g

Preparation of Agar Base: Add components, except agar, to distilled/deionized water and bring volume to 165.0mL. Adjust pH to 5.5. Add agar. Mix thoroughly. Autoclave for 15 min at 15 psi pressure–121°C. Cool to 45°–50°C.

Supplement Solution:

Composition per 45.72mL:

Horse serum, unheated 40.0mL
Fresh yeast extract solution 2.0mL
Penicillin solution 2.0mL
CVA enrichment 1.0mL
Cysteine·HCl·H_2O solution 0.5mL
Urea solution 0.22mL

Preparation of Supplement Solution: Aseptically combine components. Mix thoroughly.

Fresh Yeast Extract Solution:

Composition per 100mL:

Live, pressed, starch-free, Baker's yeast 25.0g

Preparation of Fresh Yeast Extract Solution: Add the live Baker's yeast to 100.0mL of distilled/deionized water. Autoclave for 90 min at 15 psi pressure–121°C. Allow to stand. Remove supernatant solution. Adjust pH to 6.6–6.8. Filter sterilize.

Penicillin Solution:

Composition per 10mL:

Penicillin G 1,000,000U

Preparation of Penicillin Solution: Add penicillin to distilled/deionized water and bring volume to 10.0mL. Mix thoroughly. Filter sterilize.

CVA Enrichment:

Composition per liter:

Glucose 100.0g
L-Cysteine·HCl·H_2O 25.9g
L-Glutamine 10.0g
Vitamin B_{12} 0.01g
L-Cystine·2HCl 1.0g
Adenine 1.0g
Nicotinamide adenine dinucleotide 0.25g
Cocarboxylase 0.1g
Guanine·HCl 0.03g
$Fe(NO_3)_3$ 0.02g
p-Aminobenzoic acid 0.013g
Thiamine·HCl 3.0mg

Preparation of CVA Enrichment: Add components to distilled/deionized water and bring volume to 1.0L. Mix thoroughly. Filter sterilize.

Cysteine·HCl·H_2O Solution:
Composition per 10mL:

L-Cysteine·HCl·H_2O	0.4g

Preparation of Cysteine·HCl·H_2O Solution: Add L-Cysteine·HCl·H_2O solution to distilled/deionized water and bring volume to 10.0mL. Mix thoroughly. Filter sterilize.

Urea Solution:
Composition per 10mL:

Urea, ultrapure	1.0g

Preparation of Urea Solution: Add urea to distilled/deionized water and bring volume to 10.0mL. Mix thoroughly. Filter sterilize.

Preparation of Medium: Aseptically combine 160.0mL of cooled, sterile agar base and 45.9mL of sterile supplement solution. Mix thoroughly. Pour into sterile Petri dishes or distribute into sterile tubes.

Use: For the cultivation and differentiation of *Ureaplasma urealyticum* from urine based on their ability to produce ammonia from urea. Bacteria that produce ammonia appear as golden to dark brown colonies. Also used for the cultivation of other *Ureaplasma* species.

A 7B Agar

Composition per 205.7mL:

Agar base	160.0mL
Supplement solution	45.7mL

pH 6.0 ± 0.2 at 25°C

Agar Base:
Composition per 165mL:

Pancreatic digest of casein	2.72g
Agar	2.1g
NaCl	0.8g
Papaic digest of soybean meal	0.48g
K_2HPO_4	0.4g
Glucose	0.4g
Putrescine·2HCl	0.33g
$MnSO_4$·H_2O	0.15g

Preparation of Agar Base: Add components, except agar, to distilled/deionized water and bring volume to 165.0mL. Adjust pH to 5.5. Add agar. Mix thoroughly. Autoclave for 15 min at 15 psi pressure–121°C. Cool to 45°–50°C.

Supplement Solution:
Composition per 45.72mL:

Horse serum, unheated	40.0mL
Fresh yeast extract solution	2.0mL
Penicillin solution	2.0mL
CVA enrichment	1.0mL
Cysteine·HCl·H_2O solution	0.5mL
Urea solution	0.22mL

Preparation of Supplement Solution: Aseptically combine components. Mix thoroughly.

Fresh Yeast Extract Solution:
Composition per 100mL:

Live, pressed, starch-free, Baker's yeast	25.0g

Preparation of Fresh Yeast Extract Solution: Add the live Baker's yeast to 100.0mL of distilled/deionized water. Autoclave for 90 min at 15 psi pressure–121°C. Allow to stand. Remove supernatant solution. Adjust pH to 6.6–6.8. Filter sterilize.

Penicillin Solution:
Composition per 10mL:

Penicillin G	1,000,000U

Preparation of Penicillin Solution: Add penicillin to distilled/deionized water and bring volume to 10.0mL. Mix thoroughly. Filter sterilize.

CVA Enrichment:
Composition per liter:

Glucose	100.0g
L-Cysteine·HCl·H_2O	25.9g
L-Glutamine	10.0g
Vitamin B_{12}	0.01g
L-Cystine·2HCl	1.0g
Adenine	1.0g
Nicotinamide adenine dinucleotide	0.25g
Cocarboxylase	0.1g
Guanine·HCl	0.03g
$Fe(NO_3)_3$	0.02g
p-Aminobenzoic acid	0.013g
Thiamine·HCl	3.0mg

Preparation of CVA Enrichment: Add components to distilled/deionized water and bring volume to 1.0L. Mix thoroughly. Filter sterilize.

Cysteine·HCl·H_2O Solution:
Composition per 10mL:

L-Cysteine·HCl·H_2O	0.4g

Preparation of Cysteine·HCl·H_2O Solution: Add L-Cysteine·HCl·H_2O solution to distilled/deionized water and bring volume to 10.0mL. Mix thoroughly. Filter sterilize.

Urea Solution:
Composition per 10mL:

Urea, ultrapure	1.0g

Preparation of Urea Solution: Add urea to distilled/deionized water and bring volume to 10.0mL. Mix thoroughly. Filter sterilize.

Preparation of Medium: Aseptically combine 160.0mL of cooled, sterile agar base and 45.9mL of sterile supplement solution. Mix thoroughly. Pour into sterile Petri dishes or distribute into sterile tubes.

Use: For the cultivation and differentiation of *Ureaplasma urealyticum* from urine based on their ability to produce ammonia from urea. Bacteria that produce ammonia appear as golden to dark brown colonies. Also used for the cultivation of other *Ureaplasma* species.

A 8B Agar

Composition per 84.6mL:

Agar base	80.0mL
Supplement solution	4.6mL

pH 6.0 ± 0.2 at 25°C

Agar Base:

Composition per 165mL:

Pancreatic digest of casein	2.72g
Agar	2.1g
NaCl	0.8g
Papaic digest of soybean meal	0.48g
K_2HPO_4	0.4g
Glucose	0.4g
$MnSO_4 \cdot H_2O$	0.15g
$CaCl_2 \cdot 2H_2O$	0.03g
Putrescine·2HCl	34g

Preparation of Agar Base: Add components, except agar, to distilled/deionized water and bring volume to 165.0mL. Adjust pH to 5.5. Add agar. Mix thoroughly. Autoclave for 15 min at 15 psi pressure–121°C. Cool to 45°–50°C.

Supplement Solution:

Composition per 4.6mL:

Horse serum, unheated	1.0mL
Fresh yeast extract solution	1.0mL
Penicillin solution	1.0mL
Urea solution	1.0mL
$Cysteine \cdot HCl \cdot H_2O$ solution	0.5mL
GHL tripeptide solution	0.1mL

Preparation of Supplement Solution: Aseptically combine components. Mix thoroughly.

Fresh Yeast Extract Solution:

Composition per 100mL:

Live, pressed, starch-free, Baker's yeast	25.0g

Preparation of Fresh Yeast Extract Solution: Add the live Baker's yeast to 100.0mL of distilled/deionized water. Autoclave for 90 min at 15 psi pressure–121°C. Allow to stand. Remove supernatant solution. Adjust pH to 6.6–6.8. Filter sterilize.

Penicillin Solution:

Composition per 10mL:

Penicillin G	1,000,000U

Preparation of Penicillin Solution: Add penicillin to distilled/deionized water and bring volume to 10.0mL. Mix thoroughly. Filter sterilize.

GHL Tripeptide Solution:

Composition per 10mL:

GHL (Glycyl-L-histidyl-L-lysine acetate) tripeptide	0.2g

Preparation of GHL Tripeptide Solution: Add components to distilled/deionized water and bring volume to 10.0mL. Mix thoroughly. Filter sterilize.

$Cysteine \cdot HCl \cdot H_2O$ Solution:

Composition per 10mL:

L-$Cysteine \cdot HCl \cdot H_2O$	0.4g

Preparation of $Cysteine \cdot HCl \cdot H_2O$ Solution: Add L-$Cysteine \cdot HCl \cdot H_2O$ solution to distilled/deionized water and bring volume to 10.0mL. Mix thoroughly. Filter sterilize.

Urea Solution:

Composition per 10mL:

Urea, ultrapure	1.0g

Preparation of Urea Solution: Add urea to distilled/deionized water and bring volume to 10.0mL. Mix thoroughly. Filter sterilize.

Preparation of Medium: Aseptically combine 80.0mL of cooled, sterile agar base and 4.6mL of sterile supplement solution. Mix thoroughly. Pour into sterile Petri dishes or distribute into sterile tubes.

Use: For the cultivation of *Ureaplasma urealyticum* from urine. Also used for the cultivation of other *Ureaplasma* species.

A Medium, 5X

Composition per liter:

K_2HPO_4	52.5g
KH_2PO_4	22.5g
$(NH_4)_2SO_4$	5.0g
Sodium citrate·$2H_2O$	2.5g
Carbon source solution	10.0mL
$MgSO_4 \cdot 7H_2O$ solution	1.0mL

pH 7.0 ± 0.2 at 25°C

Carbon Source Solution:

Composition per 100mL:

Carbon source	20.0g

Preparation of Carbon Source Solution: Add glycerol or glucose to distilled/deionized water and bring volume to 100.0mL. Mix thoroughly. Filter sterilize.

$MgSO_4 \cdot 7H_2O$ Solution:

Composition per 100mL:

$MgSO_4 \cdot 7H_2O$	24.65g

Preparation of $MgSO_4 \cdot 7H_2O$ Solution: Add $MgSO_4 \cdot 7H_2O$ to distilled/deionized water and bring volume to 100.0mL. Mix thoroughly. Filter sterilize.

Preparation of Medium: Add components, except carbon source solution and $MgSO_4 \cdot 7H_2O$ solution, to distilled/deionized water and bring volume to 1.0L. Mix thoroughly. Gently heat and bring to boiling. Autoclave for 15 min at 15 psi pressure–121°C. Cool to 45°–50°C. To prepare medium for use (1×) aseptically dilute 200.0mL of 5× stock solution with 789.0mL of sterile distilled/deionized water. Aseptically add 10.0mL of sterile carbon source solution and 1.0mL of sterile $MgSO_4 \cdot 7H_2O$ solution. Mix thoroughly. Aseptically distribute into sterile tubes or flasks.

Use: For the cultivation of *Escherichia coli.*

AATCC Bacteriostasis Agar (American Association of Textile Chemists and Colorists Bacteriostasis Agar)

Composition per liter:

Agar	15.0g
Peptone	10.0g
Beef extract	5.0g
NaCl	5.0g

pH 7.2 ± 0.2 at 25°C

Preparation of Medium: Add components to distilled/deionized water and bring volume to 1.0L. Mix thoroughly. Gently heat and bring to boiling. Distribute into tubes or flasks. Autoclave for 15 min at 15 psi pressure–121°C.

Use: For the maintenance of cultures of *Escherichia coli* and *Staphylococcus aureus*. For the detection of antibacterial activity of fabrics. Test cultures of *Escherichia coli* or *Staphylococcus aureus* are inoculated onto an agar plate and a sample of sterile fabric is placed on the surface. Lack of bacterial growth indicates the fabric has antibacterial activity.

AATCC Bacteriostasis Agar *See:* FDA Agar

AATCC Bacteriostasis Broth *See:* FDA Broth

AATCC Mineral Salts Iron Agar (American Association of Textile Chemists and Colorists Mineral Salts Iron Agar)

Composition per liter:

Agar	20.0g
$(NH_4)_2NO_3$	3.0g
KH_2PO_4	2.5g
K_2HPO_4	2.0g
$MgSO_4 \cdot 7H_2O$	0.2g
$FeSO_4 \cdot 7H_2O$	0.1g

pH 5.6 ± 0.2 at 25°C

Preparation of Medium: Add components to distilled/deionized water and bring volume to 1.0L. Mix thoroughly. Gently heat and bring to boiling. Distribute into tubes or flasks. Autoclave for 15 min at 15 psi pressure–121°C.

Use: For testing the resistance of textiles to fungi that cause mildew and rot. It is also used to test the effectiveness of fungicides used on textiles for preventing growth of fungi. Cultures of *Chaetomium globosum* or *Aspergillus niger* are inoculated onto the plate and a sample of fabric is placed on top. Lack of growth of these fungi on the textile is indicative of resistance to mildew.

ABY Agar (Acid Bismuth Yeast Agar)

Composition per liter:

Agar	20.0g
Glucose	20.0g
$Bi_2(SO_3)_2$	8.0g
$(NH_4)_2SO_4$	3.0g
KH_2PO_4	3.0g
$MgSO_4 \cdot 7H_2O$	0.25g
$CaCl_2 \cdot 2H_2O$	0.25g
Biotin	10.0μg

pH 7.2 ± 0.2 at 25°C

Preparation of Medium: Add components to distilled/deionized water and bring volume to 1.0L. Mix thoroughly. Gently heat and bring to boiling. Distribute into tubes or flasks. Autoclave for 15 min at 15 psi pressure–121°C. Cool tubes in a slanted position.

Use: For selective isolation and differentiation of *Candida albicans* from other *Candida* species. *C. albicans* and *C. tropicalis* colonies appear as smooth brownish-black round colonies. Other *Candida* species are differentially pigmented or produce diffusible pigments. Usually used in conjunction with BiGGY agar to differentiate further *Candida*; on BiGGY agar *C. albicans* appears as brown to black colonies with no pigment diffusion and no sheen, whereas *C. tropicalis* appears as dark brown colonies with black centers, black pigment diffusion and a sheen.

AC Agar (AC Medium)

Composition per liter:

Proteose peptone No. 3	20.0g
Glucose	5.0g
Beef extract	3.0g
Yeast extract	3.0g
Malt extract	3.0g
Ascorbic acid	0.2g
Agar	1.0g

pH 7.2 ± 0.2 at 25°C

Source: This medium is available as a premixed powder from Difco Laboratories.

Preparation of Medium: Add components to distilled/deionized water and bring volume to 1.0L. Mix thoroughly. Gently heat and bring to boiling. Distribute into tubes or flasks. Autoclave for 15 min at 15 psi pressure–121°C.

Use: For the cultivation and isolation of anaerobes, microaerophiles and aerobes. It is recommended for sterility testing of solutions and other materials not containing mercurial preservatives.

AC Broth

Composition per liter:

Proteose peptone No. 3	20.0g
Glucose	5.0g
Beef extract	3.0g
Yeast extract	3.0g
Malt extract	3.0g
Ascorbic acid	0.2g

pH 7.2 ± 0.2 at 25°C

Source: This medium is available as a premixed powder from Difco Laboratories.

Preparation of Medium: Add components to distilled/deionized water and bring volume to 1.0L. Mix thoroughly. Gently heat and bring to boiling. Distribute into tubes or flasks. Autoclave for 15 min at 15 psi pressure–121°C.

Use: For the cultivation and isolation of a wide variety of microorganisms including anaerobes, microaerophiles and aerobes. It is recommended for sterility testing of solutions and other materials not containing mercurial preservatives.

AC Medium
See: **AC Agar**

ACC Medium

Composition per liter:

Proteose peptone	20.0g
Agar	12.0g
Glycerol	1.5g
K_2SO_4	1.5g
$MgSO_4 \cdot 7H_2O$	1.5g
Antibiotic solution	10.0mL

pH 7.2 ± 0.2 at 25°C.

Antibiotic Solution:

Composition per 10.0mL:

Cycloheximide	0.075g
Ampicillin	0.050g
Chloramphenicol	0.0125g

Preparation of Antibiotic Solution: Add components to distilled/deionized water and bring volume to 10.0mL. Mix thoroughly. Filter sterilize.

Preparation of Medium: Add components, except antibiotic solution, to distilled/deionized water and bring volume to 990.0mL. Mix thoroughly. Gently heat and bring to boiling. Autoclave for 15 min at 15 psi pressure–121°C. Cool to 45°–50°C. Aseptically add sterile antibiotic solution. Mix thoroughly. Pour into sterile Petri dishes or distribute into sterile tubes.

Use: For the selective isolation and cultivation of fluorescent *Pseudomonas* species.

Acetamide Agar

Composition per liter:

Agar	15.0g
Acetamide	10.0g
NaCl	5.0g
K_2HPO_4	1.0g
$NH_4H_2PO_4$	1.0g
$MgSO_4 \cdot 7H_2O$	0.2g
Bromthymol Blue	0.08g

pH 6.9 ± 0.2 at 25°C

Preparation of Medium: Add components to distilled/deionized water and bring volume to 1.0L. Mix thoroughly. Gently heat and bring to boiling. Adjust pH. Distribute into tubes or flasks. Autoclave for 15 min at 15 psi pressure–121°C. Cool tubes in a slanted position to produce a long slant.

Use: For the differentiation of nonfermentative Gram-negative bacteria, especially *Pseudomonas aeruginosa.* Can be used as a confirmatory test for water analysis. Bacteria that deamidate acetamide turn the medium blue.

Acetamide Agar

Composition per liter:

Agar	15.0g
Acetamide	10.0g
NaCl	5.0g
K_2HPO_4	1.39g
KH_2PO_4	0.73g
$MgSO_4 \cdot 7H_2O$	0.5g
Phenol Red	0.012g

pH 6.9 ± 0.2 at 25°C

Source: This medium is available as a premixed powder from BBL Microbiology Systems.

Preparation of Medium: Add components to distilled/deionized water and bring volume to 1.0L. Mix thoroughly. Gently heat and bring to boiling. Adjust pH. Distribute into tubes or flasks. Autoclave for 15 min at 15 psi pressure–121°C. Cool tubes in a slanted position to produce a long slant.

Use: For the differentiation of nonfermentative Gram-negative bacteria, especially *Pseudomonas aeruginosa*. Can be used as a confirmatory test for water analysis. Bacteria that deamidate acetamide turn the medium blue.

Acetamide Broth

Composition per liter:

Acetamide	10.0g
NaCl	5.0g
K_2HPO_4	1.39g
KH_2PO_4	0.73g
$MgSO_4 \cdot 7H_2O$	0.5g
Phenol Red	0.012g

pH 6.9 ± 0.2 at 25°C

Preparation of Medium: Add components to distilled/deionized water and bring volume to 1.0L. Mix thoroughly. Adjust pH. Autoclave for 15 min at 15 psi pressure–121°C.

Use: For the differentiation of nonfermentative Gram-negative bacteria, especially *Pseudomonas aeruginosa*. Can be used as a confirmatory test for water analysis. Bacteria that deamidate acetamide turn the broth purplish red.

Acetamide Cetrimide Glycerol Mannitol Selective Medium

Composition per liter:

Agar	15.0g
K_2SO_4	10.0g
D-Mannitol	5.0g
$MgCl_2 \cdot 6H_2O$	1.4g
Cetrimide	0.3g
Peptone	0.2g
Acetamide solution	100.0mL
Glycerol	5.0mL

pH 7.0 ± 0.2 at 25°C

Acetamide Solution:
Composition per 100mL:

Acetamide	10.0g
Phenol Red	0.012g

Preparation of Acetamide Solution: Add components to distilled/deionized water and bring volume to 100.0mL. Mix thoroughly. Filter sterilize.

Preparation of Medium: Add components, except acetamide solution, to distilled/deionized water and bring volume to 900.0mL. Mix thoroughly. Adjust pH to 7.0. Gently heat and bring to boiling. Autoclave for 20 min at 15 psi pressure–121°C. Cool to 45°–50°C. Aseptically add sterile acetamide solution. Mix thoroughly. Pour into sterile Petri dishes.

Use: For the cultivation of *Pseudomonas aeruginosa, P. fluorescens, P. putida, P. alcaligenes, P. cepacia,* and *P. pseudoalcaligenes.*

Acetate Agar

Composition per liter:

Meat extract	50.0g
Glucose	10.0g
Peptone	5.0g
Yeast extract	5.0g
Sodium acetate buffer	100.0mL
Tween™ 80	0.5mL

pH 5.4 ± 0.2 at 25°C

Sodium Acetate Buffer:
Composition per liter:

Sodium acetate·$3H_2O$	272.2g

Preparation of Sodium Acetate Buffer: Add sodium acetate to distilled/deionized water and bring volume to 1.0L. Mix thoroughly. Adjust pH to 5.4 with glacial acetic acid. Filter sterilize.

Preparation of Medium: Add components, except sodium acetate buffer, to distilled/deionized water and bring volume to 900.0mL. Mix thoroughly. Gently heat and bring to boiling. Adjust pH to 5.4. Autoclave for 15 min at 15 psi pressure–121°C. Cool to 45°–50°C. Aseptically add 100.0mL of sterile sodium acetate buffer. Mix thoroughly. Aseptically distribute into sterile tubes or flasks.

Use: For the isolation and cultivation of *Leuconostoc* species and *Pediococcus* species.

Acetate Differential Agar (Sodium Acetate Agar) (Simmons' Citrate Agar, Modified)

Composition per liter:

Agar	20.0g
NaCl	5.0g
Sodium acetate	2.0g
$(NH_4)H_2PO_4$	1.0g
K_2HPO_4	1.0g
$MgSO_4 \cdot 7H_2O$	0.2g
Bromthymol Blue	0.08g

pH 6.8 ± 0.2 at 25°C

Source: This medium is available as a premixed powder from Difco Laboratories and BBL Microbiology Systems.

Preparation of Medium: Add components to cold distilled/deionized water and bring volume to 1.0L. Mix thoroughly. Gently heat and bring to boiling. Distribute into tubes to produce a 1 cm butt and 30 cm slant. Autoclave for 15 min at 15 psi pressure–121°C. Cool tubes in a slanted position.

Use: For the differentiation of *Shigella* species from *Escherichia coli* and also for the differentiation of nonfermenting Gram-negative bacteria. Bacteria that can utilize acetate as the sole carbon source turn the medium blue.

Acetobacter Medium

Composition per liter:

Agar	15.0g
Autolyzed yeast	10.0g
$CaCO_3$	10.0g
Glucose	3.0g

pH 7.0 ± 0.2 at 25°C

Preparation of Medium: Add components to distilled/deionized water and bring volume to 1.0L. Mix thoroughly. Gently heat and bring to boiling. Distribute into tubes to produce a 1 cm butt and 30 cm slant. Autoclave for 15 min at 15 psi pressure–121°C. Agitate tubes to mix $CaCO_3$. Cool tubes rapidly in a slanted position to keep the $CaCO_3$ in suspension.

Use: For the cultivation and maintenance of *Acetobacter* species and *Gluconobacter* species.

Acetobacter xylinum Medium

Composition per liter:

Glucose	20.0g
Peptone	5.0g
Yeast extract	5.0g
Na_2HPO_4	2.7g
Citric acid	1.5g

Preparation of Medium: Add components to distilled/deionized water and bring volume to 1.0L. Mix thoroughly. Gently heat and bring to boiling. Distribute into tubes or flasks. Autoclave for 15 min at 15 psi pressure–121°C.

Use: For the cultivation and maintenance of *Acetobacter xylinum*.

Acetobacterium Medium (ATCC Medium 1612)

Composition per liter:

Fructose	10.0g
$NaHCO_3$	10.0g
Yeast extract	2.0g
NH_4Cl	1.0g
$Cysteine \cdot HCl \cdot H_2O$	0.5g
$Na_2S \cdot 9H_2O$	0.5g
K_2HPO_4	0.45g
KH_2PO_4	0.33g
$MgSO_4 \cdot 7H_2O$	0.1g
Resazurin	1.0mg
Wolfe's mineral solution	20.0mL
Wolfe's vitamin solution	20.0mL

pH 7.4 ± 0.2 at 25°C

Wolfe's Mineral Solution:

Composition per liter:

$MgSO_4 \cdot 7H_2O$	3.0g
Nitrilotriacetic acid	1.5g
NaCl	1.0g
$MnSO_4 \cdot H_2O$	0.5g
$FeSO_4 \cdot 7H_2O$	0.1g
$CoCl_2 \cdot 6H_2O$	0.1g
$CaCl_2$	0.1g
$ZnSO_4 \cdot 7H_2O$	0.1g
$CuSO_4 \cdot 5H_2O$	0.01g
$AlK(SO_4)_2 \cdot 12H_2O$	0.01g
H_3BO_3	0.01g
$Na_2MoO_4 \cdot 2H_2O$	0.01g

Preparation of Wolfe's Mineral Solution: Add nitrilotriacetic acid to 500.0mL of distilled/deionized water. Dissolve by adjusting pH to 6.5 with KOH. Add distilled/deionized water to 1.0L. Add remaining components.

Wolfe's Vitamin Solution:

Composition per liter:

Pyridoxine·HCl	10.0mg
Thiamine·HCl	5.0mg
Riboflavin	5.0mg
Nicotinic acid	5.0mg
Calcium pantothenate	5.0mg
p-Aminobenzoic acid	5.0mg
Thioctic acid	5.0mg

Biotin	2.0mg
Folic acid	2.0mg
Cyanocobalamin	100.0μg

Preparation of Wolfe's Vitamin Solution: Add components to distilled/deionized water and bring volume to 1.0L. Mix thoroughly.

Preparation of Medium: Add all components, except fructose, and bring volume to 1.0L with distilled/deionized water. Mix thoroughly. Equilibrate to pH 7.4 by gassing with 80% N_2 + 20% CO_2. Distribute into test tubes. Autoclave for 15 min at 15 psi pressure–121°C. Add sterile anaerobic Na_2CO_3 (0.25mL of 5% Na_2CO_3 per 10.0mL medium) to bring the pH to 8.2. Add sterile fructose solution to give a final concentration of 1%. If autotrophic growth is desired omit fructose and gas with 80% H_2 + 20% CO_2.

Use: For the cultivation and maintenance of *Acetobacterium* species, *Clostridium aceticum* and other bacteria which can ferment fructose to acetic acid.

Acetobacterium Medium (ATCC Medium 1019)

Composition per liter:

$NaHCO_3$	3.0g
Yeast extract	1.0g
NH_4Cl	1.0g
KH_2PO_4	0.4g
K_2HPO_4	0.4g
$MgSO_4 \cdot 7H_2O$	0.1g
Fructose (20% solution)	25.0mL
Wolfe's vitamin solution	10.0mL
Wolfe's mineral solution	10.0mL
Resazurin (0.01% solution)	1.0mL

pH 6.7 ± 0.2 at 25°C

Wolfe's Vitamin Solution:
Composition per liter:

Pyridoxine·HCl	10.0mg
Thiamine·HCl	5.0mg
Riboflavin	5.0mg
Nicotinic acid	5.0mg
Calcium pantothenate	5.0mg
p-Aminobenzoic acid	5.0mg
Thioctic acid	5.0mg
Biotin	2.0mg
Folic acid	2.0mg
Cyanocobalamin	100.0μg

Preparation of Wolfe's Vitamin Solution: Add components to distilled/deionized water and bring volume to 1.0L. Mix thoroughly.

Wolfe's Mineral Solution:
Composition per liter:

$MgSO_4 \cdot 7H_2O$	3.0g
Nitrilotriacetic acid	1.5g
$MnSO_4 \cdot H_2O$	0.5g
NaCl	1.0g
$FeSO_4 \cdot 7H_2O$	0.1g
$CoCl_2 \cdot 6H_2O$	0.1g
$CaCl_2$	0.1g
$ZnSO_4 \cdot 7H_2O$	0.1g
$CuSO_4 \cdot 5H_2O$	0.01g
$AlK(SO_4)_2 \cdot 12H_2O$	0.01g
H_3BO_3	0.01g
$Na_2MoO_4 \cdot 2H_2O$	0.01g

Preparation of Wolfe's Mineral Solution: Add nitrilotriacetic acid to 500.0mL of distilled/deionized water. Dissolve by adjusting pH to 6.5 with KOH. Add distilled/deionized water to 1.0L. Add remaining components.

Preparation of Medium: Add all components, except fructose, to distilled/deionized water and bring volume to 975.0mL. Boil to remove dissolved O_2. Add 40.0mL of a solution containing 1.25% L-cysteine·HCl·H_2O and 1.25% $Na_2S \cdot 9H_2O$. Autoclave for 15 min at 15 psi pressure–121°C. Immediately gas with 90% N_2 + 10% CO_2 to maintain anaerobiosis until cooled to 50°C. Add 25.0mL of a filter-sterilized 20% fructose solution. If necessary, adjust pH to 6.7. Aseptically distribute into tubes under anaerobic conditions. Cap with rubber stoppers.

Use: For the cultivation and maintenance of *Acetobacterium* species.

Acetogen Medium

Composition per 421.8:

$NaHCO_3$	2.4g
NH_4Cl	0.2g
Yeast extract	0.2g
Stock salts solution #1	40.0mL
Potassium phosphate buffer	20.0mL
Clarified rumen fluid	20.0mL
Stock salts solution #2	4.0mL
Trace minerals	4.0mL
Vitamin solution	4.0mL
Reducing agent	4.0mL
Tungstate solution	0.4mL
Resazurin (0.1% solution)	0.4mL

Potassium Phosphate Buffer:
Composition per 830mL:

K_2HPO_4	15.68g
KH_2PO_4	4.72g

Preparation of Potassium Phosphate Buffer: Dissolve K_2HPO_4 in 600.0mL distilled/deionized water and KH_2PO_4 in 230.0mL distilled/deionized water. Mix the two solutions together and use.

Stock Salts Solution #1:
Composition per liter:

KCl	1.6g
NaCl	1.4g
$MgSO_4 \cdot 7H_2O$	0.2g

Preparation of Stock Salts Solution #1: Add components to distilled/deionized water and bring volume to 1.0L. Mix thoroughly.

Stock Salts Solution #2:
Composition per liter:

$CaCl_2 \cdot 2H_2O$	0.1g

Preparation of Stock Salts Solution #2: Add components to distilled/deionized water and bring volume to 1.0L. Mix thoroughly.

Trace Minerals:
Composition per liter:

Nitrilotriacetic acid	1.5g
$MgSO_4 \cdot 7H_2O$	3.0g
$MnSO_4 \cdot H_2O$	0.5g
NaCl	1.0g
$NiCl_2 \cdot 6H_2O$	0.1g
$FeSO_4 \cdot 7H_2O$	0.1g
$CoCl_2 \cdot 6H_2O$	0.1g
$CaCl_2$	0.1g
$ZnSO_4 \cdot 7H_2O$	0.1g
$Na_2SeO_3 \cdot 5H_2O$	0.01g
$CuSO_4 \cdot 5H_2O$	0.01g
$AlK(SO_4)_2 \cdot 12H_2O$	0.01g
H_3BO_3	0.01g
$Na_2MoO_4 \cdot 2H_2O$	0.01g

Preparation of Trace Minerals: Add nitrilotriacetic acid to 500.0mL of distilled/deionized water. Dissolve by adjusting pH to 6.5 with KOH. Bring volume to 1.0L with distilled/deionized water. Add remaining components. Mix thoroughly.

Vitamin Solution:
Composition per liter:

Pyridoxine·HCl	10.0mg
Ascorbic acid	5.0mg
Calcium pantothenate	5.0mg
Choline chloride	5.0mg
Lipoic acid	5.0mg
i-Inositol	5.0mg
Niacinamide	5.0mg
Nicotinic acid	5.0mg
p-Aminobenzoic acid	5.0mg
Pyridoxal·HCl	5.0mg
Riboflavin	5.0mg
Thiamine·HCl	5.0mg
Biotin	2.0mg
Folic acid	2.0mg
Vitamin B_{12}	0.1mg

Preparation of Vitamin Solution: Add components to distilled/deionized water and bring volume to 1.0L. Mix thoroughly. Store frozen.

Tungstate Solution:
Composition per liter:

$Na_2WO_4 \cdot 2H_2O$	99.0mg

Preparation of Tungstate Solution: Add components to distilled/deionized water and bring volume to 1.0L. Mix thoroughly.

Reducing Agent:
Composition per 110mL:

Cysteine·HCl·H_2O	2.5g
$Na_2S \cdot 9H_2O$	2.5g

Preparation of Reducing Agent: Add 110.0mL distilled/deionized water to a 250.0mL round bottom flask. Boil under N_2 gas for 1 minute. Cool to room temperature. Add cysteine·HCl and dissolve. Adjust to pH 9 with 5*N* NaOH. Add washed $Na_2S \cdot 9H_2O$ and dissolve. Distribute in amounts needed into tubes or flasks. Autoclave for 10 min at 15 psi pressure–121°C.

Preparation of Medium: Add components, except $NaHCO_3$ and reducing agent, to distilled/deionized water and bring volume to 417.8mL. Mix thoroughly. Gently heat and bring to boiling under 80% N_2 + 20% CO_2. Cool to 45°–50°C. Add $NaHCO_3$ and reducing agent. Distribute into tubes or flasks under 80% N_2 + 20% CO_2. Autoclave for 15 min at 15 psi pressure–121°C. After inoculation, exchange headspace with 80% H_2 + 20% CO_2.

Use: For the cultivation and maintenance of acetogenic anaerobes such as some *Clostridium* species.

Acholeplasma Medium (ATCC Medium 1039)

Composition per liter:

Papaic digest of soybean meal	10.0g
Agar	3.0g
PPLO broth without Crystal Violet	900.0mL
Fresh yeast extract solution	100.0mL

pH 7.8 ± 0.2 at 25°C

PPLO Broth without Crystal Violet:
Composition per 900mL:

Beef heart, infusion from	225.0g
Peptone	9.0g
NaCl	4.5g

Source: PPLO broth without Crystal Violet is available as a premixed powder from Difco Laboratories.

Preparation of PPLO Broth without Crystal Violet: Add components to distilled/deionized water and bring volume to 900.0mL. Mix thoroughly. Autoclave for 15 min at 15 psi pressure–121°C. Cool to room temperature.

Fresh Yeast Extract Solution:
Composition per 100mL:

Live, pressed, starch-free, Baker's yeast	25.0g

Preparation of Fresh Yeast Extract Solution: Add the live Baker's yeast to 100.0mL of distilled/deionized water. Mix thoroughly. Autoclave for 90 min at 15 psi pressure–121°C. Allow to stand. Remove supernatant solution. Adjust pH to 6.6–6.8.

Preparation of Medium: Add components to distilled/deionized water and bring volume to 1.0L. Mix thoroughly. Gently heat and bring to boiling. Distribute into test tubes or flasks. Autoclave for 10 min at 15 psi pressure–121°C.

Use: For the cultivation and maintenance of *Acholeplasma* species.

Acholeplasma Medium (ATCC Medium 1215)

Composition per 1020mL:

PPLO Broth without Crystal Violet	700.0mL
Fetal bovine serum, heat-inactivated	100.0mL
Fresh yeast extract solution	100.0mL
Tween™-glucose-BSA solution	100.0mL
Phenol Red (0.1% solution)	20.0mL

PPLO Broth without Crystal Violet:
Composition per 700mL:

Beef heart, infusion from	175.0g
Peptone	7.0g
NaCl	3.5g

Source: PPLO broth without Crystal Violet is available as a premixed powder from Difco Laboratories.

Preparation of PPLO Broth without Crystal Violet: Add components to distilled/deionized water and bring volume to 700.0mL. Autoclave for 15 min at 15 psi pressure–121°C. Cool to room temperature.

Fresh Yeast Extract Solution:
Composition per 100mL:

Live, pressed, starch-free, Baker's yeast	25.0g

Preparation of Fresh Yeast Extract Solution: Add the live Baker's yeast to 100.0mL of distilled/deionized water. Autoclave for 90 min at 15 psi pressure–121°C. Allow to stand. Remove supernatant solution. Adjust pH to 6.6–6.8.

Tween™-Glucose-BSA Solution:

Glucose	2.0g
Tween™ 80	0.1g
Bovine serum albumin, Fraction V (1% solution)	100.0mL

Preparation of Tween™-Glucose-BSA Solution: Add glucose and Tween™ 80 to 100.0mL of bovine serum albumin solution and mix thoroughly. Filter sterilize solution through a 0.2μm membrane filter.

Preparation of Medium: Aseptically mix components. Distribute into sterile tubes or flasks.

Use: For the cultivation and maintenance of *Acholeplasma* species.

Achromobacter Choline Medium

Composition per liter:

NaCl	30.0g
Agar	18.0g
Choline chloride	5.0g
K_2HPO_4	1.0g
$MgSO_4 \cdot 7H_2O$	0.5g
$FeSO_4 \cdot 7H_2O$	0.01g

Preparation of Medium: Add components to distilled/deionized water and bring volume to 1.0L. Mix well and warm gently until dissolved. Autoclave for 15 min at 15 psi pressure–121°C. Pour into sterile Petri dishes.

Use: For the cultivation and maintenance of *Achromobacter cholinophagum* and other bacteria that can utilize choline as a carbon source.

Achromobacter Medium (ATCC Medium 457)

Composition per liter:

K_2HPO_4	7.32g
Ammonium tartrate	4.6g
KH_2PO_4	1.09g
$MgSO_4 \cdot 7H_2O$	0.04g
$FeSO_4 \cdot 7H_2O$	0.04g
$CaCl_2 \cdot 2H_2O$	0.014g
$MgSO_4 \cdot 7H_2O$	0.002g

pH 7.5 ± 0.2 at 25°C

Preparation of Medium: Add components to distilled/deionized water and bring volume to 1.0L. Mix well and warm gently until dissolved. Distribute into test tubes or flasks. Autoclave for 15 min at 15 psi pressure–121°C.

Use: For the cultivation and maintenance of *Achromobacter* species and *Alcaligenes* species.

Achromobacter Medium (ATCC Medium 589)

Composition per liter:

Agar	20.0g
K_2HPO_4	7.0g
Methionine	5.0g
KH_2PO_4	2.0g
$(NH_4)_2SO_4$	1.0g
Sodium citrate	0.4g
$MgSO_4 \cdot 7H_2O$	0.1g

Preparation of Medium: Add components to distilled/deionized water and bring volume to 1.0L. Mix thoroughly. Gently heat and bring to boiling. Autoclave for 15 min at 15 psi pressure–121°C. Pour into sterile Petri dishes.

Use: For the cultivation and maintenance of *Achromobacter* species.

Achromobacter pestifer Medium

Composition per liter:

Agar	15.0g
Yeast extract	12.5g
Beef extract	10.0g
Peptone	10.0g
NaCl	5.0g

pH 7.2 ± 0.2 at 25°C

Preparation of Medium: Add components to distilled/deionized water and bring volume to1.0L. Mix thoroughly. Gently heat and bring to boiling. Distribute into tubes or flasks. Autoclave for 15 min at 15 psi pressure–121°C.

Use: For the cultivation and maintenance of *Achromobacter pestifer.*

Acid Bismuth Yeast Agar
See: **ABY Agar**

Acid Broth

Composition per liter:

Glucose	5.0g
Proteose peptone	5.0g
Yeast extract	5.0g
K_2HPO_4	4.0g

pH 5.0 ± 0.2 at 25°C

Preparation of Medium: Add components to distilled/deionized water and bring volume to 1.0L. Mix thoroughly. Distribute into tubes or flasks. Autoclave for 15 min at 15 psi pressure–121°C.

Use: For the isolation of bacteria from canned foods.

Acid Egg Medium

Composition per 1640mL:

Potato starch	30.0g
KH_2PO_4	12.3g
Malachite Green	0.4g
$MgSO_4 \cdot 7H_2O$	0.3g
Penicillin G	100,000IU
Fresh egg mixture	1000.0mL
Glycerol	12.0mL

Source: Available as a prepared medium from Oxoid Unipath.

Preparation of Medium: Add components to 1.0L of fresh egg mixture. Mix thoroughly. Gently heat and bring to boiling. Bring volume to 1640.0mL with distilled/deionized water. Distribute into tubes or flasks. Autoclave for 15 min at 15 psi pressure–121°C with tubes in an upright position.

Use: For the cultivation and maintenance of *Mycobacterium tuberculosis.*

Acid Products Test Broth

Composition per liter:

Invert sugar	10.0g
Peptone	10.0g
Yeast extract	7.5g

pH 4.0 ± 0.2 at 25°C

Preparation of Medium: Add components to distilled/deionized water and bring volume to 1.0L. Mix thoroughly. Gently heat while stirring and bring to boiling. Cool to 25°C. Adjust pH to 4.0 with 25% tartaric acid solution. Distribute into screw-capped flasks in 300mL volumes. Autoclave for 15 min at 15 psi pressure–121°C.

Use: For the cultivation of acid tolerant microorganisms from foods. For the sterility testing of canned foods.

Acidaminococcus Medium VR

Composition per liter:

Acid hydrolyzed casein (vitamin and salt free)	20.0g
Glucose	5.0g
L-Cysteine·HCl·H_2O	0.35g
DL-Tryptophan	0.1g
Guanine	0.01g
Uracil	0.01g
Hypoxanthine	0.01g
Pyridoxal	1.0mg
Calcium pantothenate	1.0mg
Thiamine	50.0μg
Niacin	50.0μg

Riboflavin....................50.0μg
p-Aminobenzoic acid....................10.0μg
Biotin....................2.0μg
Folic acid....................1.0μg
Vitamin B_{12}....................1.0μg
VR Salts A....................30.0mL
VR Salts B....................4.0mL
pH 7.0 ± 0.2 at 25°C

VR Salts A:
Composition per 500mL:
Na_2HPO_4....................37.5g
KH_2PO_4....................12.5g

Preparation of VR Salts A: Add components to distilled/deionized water and bring volume to 500mL. Mix thoroughly.

VR Salts B:
Composition per liter:
$MgSO_4 \cdot 7H_2O$....................24.0g
$CaCl_2 \cdot 2H_2O$....................0.5g
$FeSO_4 \cdot 7H_2O$....................0.5g
$ZnSO_4 \cdot 7H_2O$....................0.25g
$MnSO_4 \cdot H_2O$....................0.25g
$CoCl_2 \cdot 6H_2O$....................0.25g
$VSO_4 \cdot 7H_2O$....................0.25g
$Na_2MoO_4 \cdot 2H_2O$....................0.25g
$CuSO_4 \cdot 5H_2O$....................0.125g

Preparation of VR Salts B: Add components to distilled/deionized water and bring volume to 700.0mL. Add 2.0mL of concentrated HCl and heat until dissolved. Add 5.0g of nitrilotriacetic acid to 300.0mL distilled/deionized water. Adjust pH with 10*N* NaOH to 7.0. Stir vigorously and slowly add the nitrilotriacetic acid solution to the larger volume of salt solutions. Continue stirring until dissolved. Add distilled/deionized water and bring volume to 1.0L. Filter through paper. Store in a cool place.

Preparation of Medium: Filter-sterilize vitamins as separate solution. Add aseptically to sterile basal medium. If necessary, adjust pH with solid K_2CO_3 to 7.0. Prepare and distribute medium anaerobically using Hungate techniques with 100% N_2 gas.

Use: For the cultivation and maintenance of *Acidaminococcus fermentans*.

Acidic Tomato Medium for *Leuconostoc*

Composition per liter:
Agar (if needed)....................15.0g
Glucose....................10.0g
Peptone....................10.0g
Yeast extract....................5.0g
$MgSO_4 \cdot 7H_2O$....................0.20g
$MnSO_4 \cdot 4H_2O$....................0.05g
Tomato juice....................250.0mL
pH 4.8 ± 0.2 at 25°C

Preparation of Medium: Add solid components to 750.0mL distilled/deionized water. Add tomato juice. Mix well and warm gently until dissolved. Autoclave for 15 min at 15 psi pressure–121°C. Pour into sterile Petri dishes.

Use: For the cultivation and maintenance of *Leuconostoc oenos* and other *Leuconostoc* species.

Acidophilic *Bacillus stearothermophilus* Agar

Composition per liter:
Part A....................400.0mL
Part B....................600.0mL
pH 5.0 ± 0.2 at 25°C

Part A:
Composition per 400mL:
Soluble starch....................10.0g
Pancreatic digest of casein....................5.0g
Yeast extract....................5.0g
KH_2PO_4....................1.0g
$CaCl_2 \cdot 2H_2O$....................0.5g
$MnCl_2 \cdot 4H_2O$....................0.5g

Preparation of Part A: Add components to distilled/deionized water and bring volume to 400.0mL. Mix thoroughly. Gently heat and bring to boiling. Adjust pH to 4.7. Autoclave for 15 min at 15 psi pressure–121°C. Cool to 50°C.

Part B:
Composition per 600mL:
Agar....................20.0g

Preparation of Part B: Add agar to distilled/deionized water and bring volume to 600.0mL. Autoclave for 15 min at 15 psi pressure–121°C. Cool to 50°C.

Preparation of Medium: Aseptically combine solution A and solution B. Mix thoroughly. Adjust pH to 5.0. Pour into sterile Petri dishes.

Use: For the cultivation and maintenance of *Bacillus stearothermophilus* and other acidophilic *Bacillus* species.

Acidophilic *Bacillus stearothermophilus* Broth

Composition per liter:
Soluble starch....................10.0g
Pancreatic digest of casein....................5.0g

Yeast extract..5.0g
KH_2PO_4...1.0g
$CaCl_2 \cdot 2H_2O$..0.5g
$MnCl_2 \cdot 4H_2O$..0.5g
pH 5.0 ± 0.2 at 25°C

Preparation of Medium: Dissolve all components except agar in 1.0L distilled/deionized water. Mix thoroughly. Gently heat and bring to boiling. Adjust to pH 5.0. Autoclave for 15 min at 15 psi pressure–121°C. Precipitate will dissolve after cooling and mixing.

Use: For the cultivation and maintenance of *Bacillus stearothermophilus* and other acidophilic *Bacillus* species.

Actinobacillus lignieresii Medium

Composition per 1010mL:

Agar...10.0g
Hartley's digest broth...............................900.0mL
Filde's enrichment.....................................100.0mL
Antibiotic solution 10.0mL
pH 7.5 ± 0.2 at 25°C

Hartley's Digest Broth:
Composition per 10L:

Ox heart...3000.0g
Pancreatin...50.0g
Na_2CO_3, anhydrous (0.8% solution)..................5.0L
HCl, concentrated 80.0mL

Preparation of Hartley's Digest Broth: Finely mince the ox heart. Add the meat to 5.0L of distilled/deionized water. Gently heat and bring to 80°C. Add Na_2CO_3 solution. Cool to 45°C. Add pancreatin and maintain at 45°C for 4 hr while stirring. Add the HCl and steam at 100°C for 30 min. Cool to room temperature. Adjust pH to 8.0 with 1*N* NaOH. Gently heat and bring to boiling. Continue boiling for 25 min. Filter while hot through Whatman #1 filter paper. Cool to room temperature. Adjust pH to 7.5.

Fildes Enrichment Solution:
Composition 206mL:

Pepsin...1.0g
NaCl (0.85% solution)...............................150.0mL
Sheep blood, defibrinated............................50.0mL
HCl...6.0mL

Source: Fildes enrichment solution is available as a premixed powder from Difco Laboratories and Oxoid Unipath.

Preparation of Fildes Enrichment Solution: Combine components. Mix thoroughly. Incubate at 56°C for 4 hr. Bring pH to 7.0 with 20% NaOH. Adjust pH to 7.2 with HCl. Do not autoclave. Add 0.25 mL of chloroform and store at 4°C. Before use heat to 56°C to remove chloroform.

Antibiotic Solution:
Composition per 10mL:

Oleandomycin phosphate................................0.02g
Neomycin sulfate 1.5mg

Preparation of Antibiotic Solution: Add components to distilled/deionized water and bring volume to 10.0mL. Mix thoroughly. Filter sterilize.

Preparation of Medium: Add agar to 900.0mL of Hartley's digest broth. Mix thoroughly. Gently heat and bring to boiling. Autoclave for 15 min at 15 psi pressure–121°C. Cool to 45°–50°C. Aseptically add 100.0mL of Filde's enrichment and 10.0mL of antibiotic solution. Mix thoroughly. Pour into sterile Petri dishes or distribute into sterile tubes.

Use: For the isolation and cultivation of *Actinobacillus lignieresii.*

Actidione® Agar (Cycloheximide Agar)

Composition per liter:

Glucose 50.0g
Agar 15.0g
Pancreatic digest of casein................................5.0g
Yeast extract...4.0g
KH_2PO_4...0.55g
KCl...0.425g
$CaCl_2 \cdot 2H_2O$..0.125g
$MgSO_4 \cdot 7H_2O$ 0.125g
Bromocresol Green 22.0mg
Actidione® (cycloheximide) 10.0mg
$FeCl_3$ 2.5mg
pH 5.5 ± 0.2 at 25°C

Source: Available as a prepared medium from Oxoid Unipath.

Preparation of Medium: Add components to distilled/deionized water and bring volume to 1.0L. Mix thoroughly. Gently heat and bring to boiling. Distribute into tubes or flasks. Autoclave for 15 min at 15 psi pressure–121°C. Pour into sterile Petri dishes or leave in tubes.

Use: For the enumeration and detection of bacteria in specimens containing large numbers of yeasts and molds.

Actinomyces Agar

Composition per liter:

Agar...20.0g
K_2HPO_4...13.0g
Heart muscle, solids from infusion..................10.0g
Peptic digest of animal tissue...........................10.0g
Glucose 5.0g

Yeast extract 5.0g
NaCl 5.0g
Pancreatic digest of casein 4.0g
KH_2PO_4 .... 2.0g
$(NH_4)_2SO_4$ 1.0g
L-Cysteine·HCl·H_2O 1.0g
Soluble starch 1.0g
$MgSO_4 \cdot 7H_2O$ 0.2g
$CaCl_2 \cdot 2H_2O$ 0.01g

pH 6.9 ± 0.2 at 25°C

Preparation of Medium: Add components to distilled/deionized water and bring volume to 1.0L. If a semisolid medium is desired, add 7.0g of agar instead of 20.0g. Mix thoroughly. Gently heat and bring to boiling. Distribute into tubes or flasks. Autoclave for 10 min at 15 psi pressure–121°C. Pour into sterile Petri dishes or leave in tubes.

Use: For the maintenance or cultivation of a variety of anaerobic bacteria including *Actinomyces* species, *Eubacterium* species, *Fusobacterium* species, *Propionibacterium* species and others.

Actinomyces Broth

Composition per liter:

K_2HPO_4 .... 13.0g
Heart muscle, solids from infusion 10.0g
Peptic digest of animal tissue 10.0g
Glucose 5.0g
Yeast extract 5.0g
NaCl 5.0g
Pancreatic digest of casein 4.0g
KH_2PO_4 .... 2.0g
$(NH_4)_2SO_4$ 1.0g
L-Cysteine·HCl·H_2O 1.0g
Soluble starch 1.0g
$MgSO_4 \cdot 7H_2O$ 0.2g
$CaCl_2 \cdot 2H_2O$ 0.01g

pH 6.9 ± 0.2 at 25°C

Source: This medium is available as a premixed powder from BBL Microbiology Systems.

Preparation of Medium: Add components to distilled/deionized water and bring volume to 1.0L. Mix thoroughly. Distribute into tubes or flasks. Autoclave for 10 min at 15 psi pressure–121°C.

Use: For the maintenance or cultivation of a variety of anaerobic bacteria including *Actinomyces* species, *Eubacterium* species, *Fusobacterium* species, *Propionibacterium* species and others.

Actinomyces Broth

Composition per liter:

Beef heart, infusion from 500.0g
KH_2PO_4 .... 15.0g
Peptic digest of animal tissue 10.0g
Glucose 5.0g
Yeast extract 5.0g
NaCl 5.0g
Pancreatic digest of casein 4.0g
KH_2PO_4 .... 2.0g
$(NH_4)_2SO_4$ 1.0g
L-Cysteine·HCl·H_2O 1.0g
Soluble starch 1.0g
$MgSO_4 \cdot 7H_2O$ 0.2g
$CaCl_2 \cdot 2H_2O$ 0.02g

pH 7.2 ± 0.2 at 25°C

Source: This medium is available as a premixed powder from Difco Laboratories.

Preparation of Medium: Add components to distilled/deionized water and bring volume to 1.0L. Mix thoroughly. Distribute into tubes or flasks. Autoclave for 10 min at 15 psi pressure–121°C.

Use: For the maintenance or cultivation of a variety of anaerobic bacteria including *Actinomyces* species, *Eubacterium* species, *Fusobacterium* species, *Propionibacterium* species and others.

Actinomyces Isolation Agar

Composition per liter:

Agar 15.0g
Glycerol 5.0g
Sodium propionate 4.0g
Sodium caseinate 2.0g
K_2HPO_4 .... 0.5g
Asparagine 0.1g
$MgSO_4 \cdot 7H_2O$ 0.1g
$FeSO_4 \cdot 7H_2O$ 0.001g

Preparation of Medium: Add components to distilled/deionized water and bring volume to 1.0L. Mix thoroughly. Gently heat and bring to boiling. Distribute into tubes or flasks. Autoclave for 15 min at 15 psi pressure–121°C. Pour into sterile Petri dishes or leave in tubes.

Use: For the isolation and cultivation of *Actinomyces* species.

Actinomycete Growth Medium

Composition per liter:

Succinic acid 1.18g
L-Glutamine 0.29g
$CaCl_2 \cdot 2H_2O$ 0.2g
KH_2PO_4 .... 0.2g
$MgSO_4 \cdot 7H_2O$ 0.2g
NaCl 0.1g
m-Inositol 0.090g

Ferric EDTA....0.037g
$MnSO_4·H_2O$ 4.5mg
H_3BO_3 .... 1.5mg
$ZnSO_4·7H_2O$.... 1.5mg
Nicotonic acid 0.5mg
Pyridoxine-HCl 0.5mg
Thiamine-HCl 0.1mg
$CuSO_4·5H_2O$ 0.04mg
$Na_2MoO_4·2H_2O$.... 0.025mg
pH 6.4 ± 0.2 at 25°C

Preparation of Medium: Add components to distilled/deionized water and bring volume to 1.0L. Mix thoroughly. Distribute into tubes or flasks. Autoclave for 15 min at 15 psi pressure–121°C.

Use: For the cultivation of actimomycetes.

Actinomycete Isolation Agar

Composition per liter:

Agar....15.0g
Sodium propionate....4.0g
Sodium caseinate 2.0g
K_2HPO_4....0.5g
Asparagine 0.1g
$MgSO_4·7H_2O$ 0.1g
$FeSO_4·7H_2O$.... 1.0mg
pH 8.1± 0.2 at 25°C

Source: This medium is available as a premixed powder from Difco Laboratories.

Preparation of Medium: Add components to distilled/deionized water and bring volume to 1.0L. Mix thoroughly. Gently heat and bring to boiling. Add 5.0g of glycerol. Distribute into tubes or flasks. Autoclave for 15 min at 15 psi pressure–121°C.

Use: For the isolation and cultivation of aerobic *Actinomyces* from soil and water.

Actinoplanes Medium

Composition per liter:

Oatmeal, baby cereal....60.0g
Yeast 2.5g
K_2HPO_4....1.0g
KCl....0.5g
$MgSO_4·7H_2O$ 0.5g
$FeSO_4·7H_2O$....0.01g

Preparation of Medium: Add components to distilled/deionized water and bring volume to 1.0L. Mix thoroughly. Distribute into tubes or flasks. Autoclave for 15 min at 15 psi pressure–121°C.

Use: For the cultivation and maintenance of *Actinoplanes* species.

Actinopolyspora Medium

Composition per liter:

Agar....20.0g
Maltose....10.0g
N-Z-Amine A 2.0g
Yeast extract....1.0g
Beef extract....1.0g
pH 7.3 ± 0.2 at 25°C

Preparation of Medium: Add components to distilled/deionized water and bring volume to 1.0L. Mix thoroughly. Gently heat and bring to boiling. Distribute into tubes or flasks. Autoclave for 15 min at 15 psi pressure–121°C. Pour into sterile Petri dishes or leave in tubes.

Use: For the cultivation and maintenance of *Actinopolyspora thermovinacea.*

AE Sporulation Medium, Modified

Composition per 1079.2mL:

Polypeptone™ 10.0g
Yeast extract....10.0g
Na_2HPO_4....4.36g
Ammonium acetate 1.5g
KH_2PO_4....0.25g
$MgSO_4·7H_2O$ 0.2g
Raffinose solution 39.6mL
Na_2CO_3 solution....13.2mL
$CoCl_2·6H_2O$ solution 13.2mL
Sodium ascorbate solution 13.2mL
pH 7.8 ± 0.1 at 25°C

Raffinose Solution:
Composition per 100mL:

Raffinose 10.0g

Preparation of Raffinose Solution: Add raffinose to distilled/deionized water and bring volume to 100.0mL. Mix thoroughly. Filter sterilize.

Na_2CO_3 Solution:
Composition per 100mL:

Na_2CO_3 ....7.0g

Preparation of Na_2CO_3 Solution: Add Na_2CO_3 to distilled/deionized water and bring volume to 100.0mL. Mix thoroughly. Filter sterilize.

$CoCl_2$ Solution:
Composition per 100mL:

$CoCl_2·6H_2O$ 0.32g

Preparation of $CoCl_2$ Solution: Add $CoCl_2·6H_2O$ to distilled/deionized water and bring volume to 100.0mL. Mix thoroughly. Filter sterilize.

Sodium Ascorbate Solution:

Composition per 100mL:

Sodium ascorbate 1.5g

Preparation of Sodium Ascorbate Solution: Add sodium ascorbate to distilled/deionized water and bring volume to 100.0mL. Mix thoroughly. Filter sterilize. Use freshly prepared solution.

Preparation of Medium: Add components—except raffinose solution, Na_2CO_3 solution, $CoCl_2$ solution, and sodium ascorbate solution—to distilled/deionized water and bring volume to 1.0L. Mix thoroughly. Adjust pH to 7.5 using 2M sodium carbonate solution. Distribute into tubes in 15.0mL volumes. Autoclave for 15 min at 15 psi pressure–121°C. Aseptically add 0.6mL of sterile raffinose solution, 0.2mL of sterile Na_2CO_3 solution, and 0.2mL of sterile $CoCl_2$ solution to each tube. Mix thoroughly. Prior to inoculation, steam medium for 10 min. Cool to 25°C. Aseptically add 0.2mL of sterile sodium ascorbate solution to each tube.

Use: For the cultivation and sporulation of *Clostridium perfringens*.

Aerobic Low Peptone Basal Medium *See*: ALP Basal Medium

Aeromonas Differential Agar (Dextrin Fuchsin Sulfite Agar)

Composition per liter:

Dextrin 15.0g
Agar 13.0g
Pancreatic digest of casein 10.0g
Na_2HPO_4 .... 7.75g
NaCl 5.0g
Beef extract 3.0g
Na_2SO_3 .... 1.6g
Acid Fuchsin solution 50.0mL
pH 7.5 ± 0.2 at 25°C

Acid Fuchsin Solution:

Composition per 50mL:

Acid Fuchsin 0.25g

Preparation of Acid Fuchsin Solution: Add Acid Fuchsin to 50.0mL of 5% aqueous dioxan. Mix well to dissolve.

Caution: Acid Fuchsin is a potential carcinogen and care must be taken to avoid inhalation of the powdered dye and contamination of the skin.

Preparation of Medium: Add components to distilled/deionized water and bring volume to 1.0L. Mix thoroughly. Gently heat while stirring and bring to boiling. Distribute into tubes or flasks. Autoclave for 15 min at 15 psi pressure–121°C. Pour into sterile Petri dishes or leave in tubes.

Use: For the isolation and differentiation of *Aeromonas* species from other Gram-negative rods such as *Pseudomonas* and Enterobacteriaceae. Specimens with low numbers of *Aeromonas* may first be enriched by growth in Starch Broth for 4–9 days. After 24 hours of growth on this agar, colonies are sprayed with Nadi reagent (1% solution of N,N,N´,N´-tetramethyl-*p*-phenylene-diammonium dichloride). A positive Nadi reaction (dextrin degradation) is indicated by a purple color at the periphery of the colony. Dextrin fermentation is also indicated by red colonies. *Aeromonas* species appear as large, convex dark red colonies with purple periphery.

Aeromonas hydrophila Medium

Composition per liter:

Inositol 10.0g
Pancreatic digest of casein 10.0g
L-Ornithine·HCl 5.0g
Proteose peptone 5.0g
Agar 3.0g
Yeast extract 3.0g
Mannitol 1.0g
Ferric ammonium citrate 0.5g
$Na_2S_2O_3 \cdot 5H_2O$ 0.4g
Bromcresol Purple 0.02g
pH 6.7 ± 0.2 at 25°C

Preparation of Medium: Add components to distilled/deionized water and bring volume to 1.0L. Mix thoroughly. Gently heat until dissolved. Adjust pH to 6.7. Distribute into tubes in 5.0mL volumes. Autoclave for 12 min at 15 psi pressure–121°C.

Use: For the isolation and cultivation of *Aeromonas hydrophila*.

Aeromonas Medium (Ryan's *Aeromonas* Medium)

Composition per liter:

Agar 12.5g
$Na_2S_2O_3$ 10.67g
Proteose peptone 5.0g
NaCl 5.0g
Xylose 3.75g
L-Lysine·HCl 3.5g
Yeast extract 3.0g
Sorbitol 3.0g
Bile salts No, 3 .... 3.0g
Inositol 2.5g
L-Arginine·HCl 2.0g

Lactose 1.5g
Ferric ammonium citrate 0.8g
Bromthymol Blue 0.04g
Thymol Blue 0.04g
pH 8.0 ± 0.1 at 25°C

Source: Available as a dehydrated powder from Oxoid Unipath.

Preparation of Medium: Add components to distilled/deionized water and bring volume to 1.0L. Mix thoroughly. Gently heat and bring to boiling. Do not autoclave. Cool to 50°C and aseptically add 5.0mg ampicillin. Pour into sterile Petri dishes.

Use: For the isolation and selective differentiation of *Aeromonas hydrophila* and other *Aeromonas* species from clinical and non-clinical specimens. *Aeromonas* species appear as small (0.5-1.5mm), dark green colonies with darker centers.

AFPA (*Aspergillus flavus/parasiticus* Agar)

Composition per liter:
Yeast extract 20.0g
Agar 15.0g
Peptone 10.0g
Ferric ammonium citrate 0.5g
Dichloran (botran®) 2.0mg
pH 6.3 ± 0.2 at 25°C

Source: Available as a dehydrated powder from Oxoid Unipath.

Preparation of Medium: Add components to distilled/deionized water and bring volume to 1.0L. Mix thoroughly. Gently heat while stirring and bring to boiling. Add 100.0mg chloramphenicol. Autoclave for 15 min at 15 psi pressure–121°C. Pour into sterile Petri dishes.

Use: For the selective isolation and enumeration of *Aspergillus flavus* and *A. parasiticus*. Colonies of these fungi appear with dark yellow-orange color on the reverse side.

Agar Medium A
See: **Antibiotic Medium 1**

Agar Medium C
See: **Antibiotic Medium 4**

Agar Medium for Differential Enumeration of Lactic Streptococci

Composition per 1170mL:
Agar 15.0g
Carboxymethylcellulose 15.0g
Calcium citrate 10.0g
Pancreatic digest of casein 5.0g
Yeast extract 5.0g
L-Arginine·HCl 5.0g
Casamino acids 2.5g
K_2HPO_4 .... 1.25g
Calcium carbonate solution 100.0mL
Nonfat milk solution 50.0mL
Bromcresol Purple solution 20.0mL
pH 5.9 ± 0.2 at 25°C

Calcium Carbonate Solution:
Composition per 100mL:
$CaCO_3$ 3.0g

Preparation of Calcium Carbonate Solution: Add $CaCO_3$ to distilled/deionized water and bring volume to 100.0mL. Mix thoroughly. Autoclave for 15 min at 15 psi pressure–121°C.

Nonfat Milk Solution:
Composition per 100mL:
Nonfat milk 11.0g

Preparation of Nonfat Milk Solution: Add nonfat milk to distilled/deionized water and bring volume to 100.0mL. Mix thoroughly. Autoclave for 15 min at 15 psi pressure–121°C.

Bromcresol Purple Solution:
Composition per 20mL:
Bromcresol Purple 0.02g

Preparation of Bromcresol Purple Solution: Add Bromcresol Purple to distilled/deionized water and bring volume to 20.0mL. Mix thoroughly. Filter sterilize.

Preparation of Medium: Add agar to 500.0mL of distilled/deionized water. Gently heat and bring to boiling. In a separate flask add carboxymethylcellulose and calcium citrate to 500.0mL of distilled/deionized water. Gently heat while stirring until a white, turbid suspension is formed. Combine the two solutions. Add the pancreatic digest of casein, yeast extract, K_2HPO_4, casamino acids and arginine. Mix thoroughly. Gently heat and bring to boiling. Adjust pH to 5.6 with 6*N* HCl. Distribute into screw-capped bottles in 100.0mL volumes. Autoclave for 15 min at 15 psi pressure–121°C. Cool to 45°–50°C. Immediately prior to pouring plates, aseptically add 5.0mL of sterile nonfat milk solution, 10.0mL of sterile calcium carbonate solution, and 2.0mL of sterile Bromcresol Purple solution to each screw-capped bottle. Mix thoroughly. The pH should be 5.9. Pour into cold, sterile Petri dishes.

Use: For the cultivation, differentiation and enumeration of *Lactobacillus lactis, L. lactis* subspecies *cremoris* and *L. lactis* subspecies *diacetylactis*. Lactose-fermenting bacteria such as *L. lactis* subspecies *cremoris* appear as yellow colonies. Arginine-utilizing bacteria such as *L. lactis* and *L. lactis* subspecies *diacetylactis* appear as purple colonies. Citrate utilizing bacteria such as *L. lactis* subspecies *diacetylactis* appear as colonies surrounded by a clear zone.

Agar Medium P (PM Indicator Agar)

Composition per liter:

Agar	15.0g
Glucose	5.25g
Peptone	5.0g
Beef extract	3.0g
Pancreatic digest of casein	1.7g
Tween™ 80	1.0g
NaCl	0.5g
Papaic digest of soybean meal	0.3g
K_2HPO_4	0.25g
Bromcresol Purple	0.06g

pH 7.8 ± 0.2 at 25°C

Preparation of Medium: Add components to distilled/deionized water and bring volume to 1.0L. Mix thoroughly. Gently heat and bring to boiling. Distribute into tubes or flasks. Autoclave for 15 min at 15 psi pressure–121°C. Pour into sterile Petri dishes or leave in tubes.

Use: For the cultivation of *Bacillus stearothermophilus* for the detection of penicillin in milk.

Agrobacterium Mannitol Medium

Composition per liter:

Mannitol	10.0g
L-Glutamate	2.0g
KH_2PO_4	0.5g
Yeast extract	0.3g
$MgSO_4 \cdot 7H_2O$	0.2g
NaCl	0.2g

pH 7.0 ± 0.2 at 25°C

Preparation of Medium: Add components to distilled/deionized water and bring volume to 1.0L. Mix thoroughly. Adjust pH to 7.0. Autoclave for 15 min at 15 psi pressure–121°C.

Use: For the cultivation of *Agrobacterium rhizogenes.*

Agrobacterium Medium

Composition per liter:

Agar	18.0g
Erythritol	5.0g
$NaNO_3$	2.5g
$CaCl_2$	0.2g
$MgSO_4 \cdot 7H_2O$	0.2g
NaCl	0.2g
KH_2PO_4	0.1g
Ferric EDTA	1.3mg
Biotin	2μg
Supplement	10.0mL

pH 7.0 ± 0.2 at 25°C

Supplement:

Composition per liter:

Cycloheximide	0.25g
Bacitracin	0.1g
Na_2SeO_3	0.1g
Tyrothricin	1.0mg

Preparation of Supplement: Add components to distilled/deionized water and bring volume to 10.0mL. Mix thoroughly. Filter sterilize.

Preparation of Medium: Add components, except supplement, to distilled/deionized water and bring volume to 990.0mL. Mix thoroughly. Adjust pH to 7.0 with 1*N* NaOH. Gently heat and bring to boiling. Autoclave for 15 min at 15 psi pressure–121°C. Cool to 45°–50°C. Aseptically add sterile supplement. Mix thoroughly. Pour into sterile Petri dishes or distribute into sterile tubes.

Use: For the selective isolation and cultivation of *Agrobacterium* species biotype 2.

Agrobacterium Medium

Composition per liter:

Agar	20.0g
Mannitol	10.0g
$NaNO_3$	4.0g
$MgCl_2$	2.0g
Calcium propionate	1.2g
$Mg_3(PO_4)_2$	0.2g
$MgSO_4$	0.1g
$MgCO_3$	0.075g
$NaHCO_3$	0.075g
Supplement	100.0mL

pH 7.1 ± 0.2 at 25°C

Supplement:

Composition per 100mL:

Berberine	0.275g
Cycloheximide	0.2g
Bacitracin	0.1g
Na_2SeO_3	0.1g

Penicillin G 0.06g
Streptomycin sulfate 0.03g
Tyrothricin.. 1.0mg

Preparation of Supplement: Add components to distilled/deionized water and bring volume to 100.0mL. Mix thoroughly. Filter sterilize.

Preparation of Medium: Add components, except supplement, to distilled/deionized water and bring volume to 900.0mL. Mix thoroughly. Gently heat and bring to boiling. Autoclave for 15 min at 15 psi pressure–121°C. Cool to 45°–50°C. Aseptically add 100.0mL of sterile supplement. Mix thoroughly. Pour into sterile Petri dishes or distribute into sterile tubes.

Use: For the selective isolation and cultivation of *Agrobacterium* species.

Agrobacterium Medium

Composition per liter:

Agar..12.0g
Lactose 5.0g
Na_2HPO_4 ....1.8g
KNO_3...1.0g
$MgSO_4 \cdot 7H_2O$ 0.1g
Supplement 100.0mL

pH 6.8 ± 0.2 at 25°C

Supplement:

Composition per 100mL:

$MnSO_4 \cdot 4H_2O$ 3.35g
Ferric EDTA... 2.5mg

Preparation of Supplement: Add components to distilled/deionized water and bring volume to 100.0mL. Mix thoroughly. Filter sterilize.

Preparation of Medium: Add components, except supplement, to distilled/deionized water and bring volume to 900.0mL. Mix thoroughly. Gently heat and bring to boiling. Autoclave for 1 min at 25 psi pressure–130°C. Cool to 45°–50°C. Aseptically add sterile supplement. Mix thoroughly. Pour into sterile Petri dishes or distribute into sterile tubes.

Use: For the selective isolation and cultivation of *Agrobacterium* species.

Agrobacterium Medium D1

Composition per liter:

Agar..15.0g
Mannitol...15.0g
LiCl 6.0g
$NaNO_3$...5.0g
K_2HPO_4 ....2.0g
$MgSO_4 \cdot 7H_2O$ 0.2g
Bromthymol Blue...0.1g
$Ca(NO_3)_2 \cdot 4H_2O$ 0.02g

pH 7.2 ± 0.2 at 25°C

Preparation of Medium: Add components to distilled/deionized water and bring volume to 1.0L. Mix thoroughly. Gently heat and bring to boiling. Distribute into tubes or flasks. Autoclave for 15 min at 15 psi pressure–121°C. Cool to 45°–50°C. Adjust pH to 7.2. Pour into sterile Petri dishes or leave in tubes.

Use: For the selective isolation and cultivation of *Agrobacterium* species.

AGS

See: **Arginine Glucose Slant**

AH5 Medium

Composition per 205.9mL:

Agar base 160.0mL
Supplement solution.......................................45.9mL

pH 6.0 ± 0.2 at 25°C

Agar Base:

Composition per 165mL:

Pancreatic digest of casein 2.72g
Agar...2.1g
NaCl 0.8g
Papaic digest of soybean meal 0.48g
K_2HPO_4...0.4g
Glucose 0.4g

Preparation of Agar Base: Add components, except agar, to distilled/deionized water and bring volume to 165.0mL. Adjust pH to 5.5. Add agar. Mix thoroughly. Autoclave for 15 min at 15 psi pressure–121°C. Cool to 45°–50°C.

Supplement Solution:

Composition per 45.9mL:

Horse serum, unheated...................................40.0mL
Fresh yeast extract solution..............................2.0mL
Penicillin solution 2.0mL
CVA enrichment 1.0mL
$Cysteine \cdot HCl \cdot H_2O$ solution 0.5mL
Urea solution..0.4mL

Preparation of Supplement Solution: Aseptically combine components. Mix thoroughly.

Fresh Yeast Extract Solution:

Composition per 100mL:

Live, pressed, starch-free, Baker's yeast...........25.0g

Preparation of Fresh Yeast Extract Solution: Add the live Baker's yeast to 100.0mL of distilled/deionized water. Autoclave for 90 min at 15 psi pressure–121°C. Allow to stand. Remove supernatant solution. Adjust pH to 6.6–6.8.

Penicillin Solution:

Composition per 10mL:

Penicillin G 1,000,000U

Preparation of Penicillin Solution: Add penicillin to distilled/deionized water and bring volume to 10.0mL. Mix thoroughly. Filter sterilize.

CVA Enrichment:

Composition per liter:

Glucose 100.0g
L-Cysteine·HCl·H_2O 25.9g
L-Glutamine 10.0g
Vitamin B_{12} 0.01g
L-Cystine·2HCl 1.0g
Adenine 1.0g
Nicotinamide adenine dinucleotide 0.25g
Cocarboxylase 0.1g
Guanine·HCl 0.03g
$Fe(NO_3)_3$ 0.02g
p-Aminobenzoic acid 0.013g
Thiamine·HCl 3.0mg

Preparation of CVA Enrichment: Add components to distilled/deionized water and bring volume to 1.0L. Mix thoroughly. Filter sterilize.

Cysteine·HCl·H_2O Solution:

Composition per 10mL:

L-Cysteine·HCl·H_2O 0.4g

Preparation of Cysteine·HCl·H_2O Solution: Add L-Cysteine·HCl·H_2O solution to distilled/deionized water and bring volume to 10.0mL. Mix thoroughly. Filter sterilize.

Urea Solution:

Composition per 10mL:

Urea 1.0g

Preparation of Urea Solution: Add urea to distilled/deionized water and bring volume to 10.0mL. Mix thoroughly. Filter sterilize.

Preparation of Medium: Aseptically combine 160.0mL of cooled, sterile agar base and 45.9mL of sterile supplement solution. Mix thoroughly. Pour into sterile Petri dishes or distribute into sterile tubes.

Use: For the cultivation of *Ureaplasma urealyticum* from urine and exudates and for the cultivation of other *Ureaplasma* species.

AJYE Medium
See: **Apple Juice Yeast Extract Medium**

AK Agar No. 2 (Sporulating Agar)

Composition per liter:

Agar 15.0g
Pancreatic digest of gelatin 6.0g
Pancreatic digest of casein 4.0g
Yeast extract 3.0g
Beef extract 1.5g
Glucose 1.0g
$MnSO_4 \cdot 7H_2O$ 0.3g

pH 6.6 ± 0.2 at 25°C

Source: This medium is available as a premixed powder from BBL Microbiology Systems.

Preparation of Medium: Add components to distilled/deionized water and bring volume to 1.0L. Mix thoroughly. Gently heat while stirring and bring to boiling. Distribute into tubes or flasks. Autoclave for 20 min at 15 psi pressure–121°C. Make sure medium is dissolved before autoclaving.

Use: For the preparation of spore suspensions used to detect antibiotic residues in milk and dairy products.

AKI Medium

Composition per liter:

Peptone 15.0g
NaCl 5.0g
Yeast extract 4.0g
Sodium bicarbonate solution 30.0mL

pH 7.2 ± 0.2 at 25°C

Sodium Bicarbonate Solution:

Composition per 100mL:

$NaHCO_3$ 10.0g

Preparation of Sodium Bicarbonate Solution: Add sodium bicarbonate to distilled/deionized water and bring volume to 100.0mL. Mix thoroughly. Filter sterilize. Use freshly prepared solution.

Preparation of Medium: Add components, except sodium bicarbonate solution, to distilled/deionized water and bring volume to 970.0mL. Mix thoroughly. Autoclave for 15 min at 15 psi pressure–121°C. Cool to 45°–50°C. Aseptically add sterile sodium bicarbonate solution. Mix thoroughly. Aseptically distribute into sterile tubes or flasks. Prepare medium freshly.

Use: For the cultivation of *Vibrio cholerae* and other *Vibrio* species.

Albumin Fatty Acid Broth, *Leptospira* Medium
See: Bovine Albumin Tween™ 80 Medium, Ellinghausen and McCullough, Modified

Albumin Fatty Acid Semisolid Medium, Modified
See: Bovine Albumin Tween™ 80 Semisolid Medium, Ellinghausen and McCullough, Modified

Alcal Mannose Medium

Composition per liter:

K_2HPO_4	15.1g
KH_2PO_4	5.6g
Mannose	1.0g
Yeast extract	1.0g
Casamino acids	0.5g
$MgSO_4 \cdot 7H_2O$	0.4g
$CaCl_2 \cdot 2H_2O$	50.0mg
$FeSO_4 \cdot 7H_2O$	10.0mg

Preparation of Medium: Add components to distilled/deionized water and bring volume to 1.0L. Mix thoroughly. Distribute into tubes or flasks. Autoclave for 15 min at 15 psi pressure–121°C.

Use: For the cultivation of *Bacillus circulans*.

Alcaligenes Agar

Composition per liter:

Agar	10.0g
Peptone	5.0g
Ammonium lactate	3.0g
Meat extract	3.0g
Ferric citrate	0.2g

pH 7.0 ± 0.2 at 25°C

Preparation of Medium: Add ferric citrate to distilled/deionized water and bring volume to 100.0mL. In a separate flask, add remaining components to distilled/deionized water and bring volume to 900.0mL. Mix thoroughly. Adjust pH to 7.0. Steam the two solutions for 20 min on three consecutive days. Aseptically combine the two solutions. Pour into sterile Petri dishes or distribute into sterile tubes.

Use: For the cultivation of *Alcaligenes* species.

Alcaligenes Medium

Composition per liter:

Peptone	5.0g
Beef extract	3.0g
Ferric citrate	0.2g
Ammonium lactate solution	3.0mL

pH 7.0 ± 0.2 at 25°C

Ammonium Lactate Solution:
Composition per 100mL:

Lactic acid	60.0g

Preparation of Ammonium Lactate Solution: Dissolve lactic acid in 100.0mL distilled/deionized water. Neutralize with NH_4OH to pH 7.0.

Preparation of Medium: Add peptone, beef extract, and ammonium lactate to distilled/deionized water and bring volume to 1.0L. Mix thoroughly. Gently heat and bring to boiling. Autoclave for 15 min at 15 psi pressure–121°C. Add ferric citrate aseptically. Mix thoroughly. Aseptically distribute into tubes or flasks.

Use: For the cultivation of *Alcaligenes tolerans*.

Alcaligenes N5 Medium

Composition per liter:

Sodium succinate·$2H_2O$	5.0g
KH_2PO_4	0.75g
NH_4Cl	0.67g
K_2HPO_4	0.61g
$MgSO_4 \cdot 7H_2O$	0.2g
$CaCl_2 \cdot 2H_2O$	0.03g
$MnCl_2 \cdot 4H_2O$	3.0mg
$FeCl_3$	2.4mg
$Na_2MoO_4 \cdot 2H_2O$	1.0mg

Preparation of Medium: Add components to distilled/deionized water and bring volume to 1.0L. Mix thoroughly. Gently heat while stirring and bring to boiling. Distribute into tubes or flasks. Autoclave for 15 min at 15 psi pressure–121°C.

Use: For the cultivation and maintenance of *Alcaligenes faecalis*.

Alcaligenes NA YE Medium (*Alcaligenes* Nutrient Agar Yeast Extract Medium)

Composition per liter:

Agar	15.0g
Pancreatic digest of gelatin	5.0g
Yeast extract	5.0g
Beef extract	3.0g

pH 7.0 ± 0.2 at 25°C

Preparation of Medium: Add components to distilled/deionized water and bring volume to 1.0L. Mix thoroughly. Gently heat while stirring and bring to boiling. Distribute into tubes or flasks. Autoclave for 15 min at 15 psi pressure–121°C. Pour into sterile Petri dishes or leave in tubes.

Use: For the cultivation and maintenance of *Alcaligenes* species.

Alcaligenes NB YE Agar (*Alcaligenes* Nutrient Broth Yeast Extract Agar)

Composition per liter:

Agar	15.0g
Pancreatic digest of gelatin	5.0g
Yeast extract	5.0g
Beef extract	3.0g

Preparation of Medium: Add components to distilled/deionized water and bring volume to 1.0L. Mix thoroughly. Gently heat while stirring and bring to boiling. Distribute into tubes or flasks. Autoclave for 15 min at 15 psi pressure–121°C. Pour into sterile Petri dishes or leave in tubes.

Use: For the cultivation and maintenance of *Alcaligenes faecalis*.

Alcaligenes NB YE Broth (Alcaligenes Nutrient Broth Yeast Extract Broth)

Composition per liter:

Pancreatic digest of gelatin	5.0g
Yeast extract	5.0g
Beef extract	3.0g

Preparation of Medium: Add components to distilled/deionized water and bring volume to 1.0L. Mix thoroughly. Gently heat while stirring and bring to boiling. Distribute into tubes or flasks. Autoclave for 15 min at 15 psi pressure–121°C.

Use: For the cultivation of *Alcaligenes faecalis*.

Alcaligenes NB YE Medium (*Alcaligenes* Nutrient Broth Yeast Extract Medium)

Composition per liter:

Pancreatic digest of gelatin	5.0g
Yeast extract	5.0g
Beef extract	3.0g

pH 7.0 ± 0.2 at 25°C

Preparation of Medium: Add components to distilled/deionized water and bring volume to 1.0L. Mix thoroughly. Distribute into tubes or flasks. Autoclave for 15 min at 15 psi pressure–121°C.

Use: For the cultivation and maintenance of *Alcaligenes* species.

Alcaligenes Nutrient Agar Yeast Extract Medium

See: Alcaligenes NA YE Medium

Alcaligenes Nutrient Broth Yeast Extract Agar

See: Alcaligenes NB YE Agar

Alcaligenes Nutrient Broth Yeast Extract Broth

See: Alcaligenes NB YE Broth

Alcaligenes Nutrient Broth Yeast Extract Medium

See: Alcaligenes NB YE Medium

Alginate Utilization Medium

Composition per liter:

Solution B	500.0mL
Solution A	400.0mL
Solution C	100.0mL

Solution A:

Composition per 400mL:

Marine salts	38.0g

Preparation of Solution A: Add marine salts to distilled/deionized water and bring volume to 400.0mL. Mix thoroughly. Autoclave for 15 min at 15 psi pressure–121°C.

Solution B:

Composition per 500mL:

Agar	20.0g
Sodium alginate	10.0g

Preparation of Solution B: Add components to distilled/deionized water and bring volume to 500.0mL. Mix thoroughly. Autoclave for 15 min at 15 psi pressure–121°C.

Solution C:

Composition per 100mL:

Tris·HCl buffer	0.067g

$NaNO_3$ 0.047g
Ferric EDTA 66.5mg
Sodium glycerophosphate 6.67mg
Thiamine·HCl 67.0μg
Vitamin B_{12} 1.3μg
Biotin 0.67μg

Preparation of Solution C: Add components to distilled/deionized water and bring volume to 100.0mL. Mix thoroughly. Filter sterilize.

Preparation of Medium: Aseptically combine Solutions A, B and C. For liquid medium, omit agar from Solution B.

Use: For the cultivation of microorganisms that can utilize alginate as a carbon source. Growth on alginate (production of alginase) is a diagnostic test used in the differentiation of *Vibrio* species.

Alkaline *Bacillus* Medium

Composition per liter:

Agar 15.0g
Peptone 10.0g
Glucose 10.0g
Yeast extract 5.0g
K_2HPO_4 .... 1.0g
Na_2CO_3 solution 100.0mL
pH 8.5–11.0 at 25°C

Na_2CO_3 Solution:
Composition per 100mL:

Na_2CO_3 .... 10.0g

Preparation of Na_2CO_3 Solution: Add Na_2CO_3 to distilled/deionized water and bring volume to 100.0mL. Mix thoroughly. Filter sterilize.

Preparation of Medium: Add components, except Na_2CO_3 solution, to distilled/deionized water and bring volume to 900.0mL. Gently heat while stirring and bring to boiling. Autoclave for 15 min at 10 psi pressure–115°C. Cool to 45°–50°C. Aseptically add sterile Na_2CO_3 solution. Mix thoroughly. Pour into sterile Petri dishes or distribute into sterile tubes.

Use: For the cultivation and maintenance of alkalophilic microorganisms such as *Bacillus alcalophilus*, *Bacillus circulans* and other *Bacillus* species.

Alkaline Peptone Agar

Composition per liter:

NaCl 20.0g
Agar 15.0g
Peptone 10.0g
pH 8.5 ± 0.2 at 25°C

Preparation of Medium: Add components to distilled/deionized water and bring volume to 1.0L. Mix thoroughly. Gently heat and bring to boiling. Adjust pH to 8.5. Distribute into tubes. Autoclave for 15 min at 15 psi pressure–121°C. Allow tubes to cool in a slanted position.

Use: For the cultivation of *Vibrio cholerae* and other *Vibrio* species.

Alkaline Peptone Salt Broth (APS Broth)

Composition per liter:

NaCl 30.0g
Peptone 10.0g

Preparation of Medium: Add components to distilled/deionized water and bring volume to 1.0L. Mix thoroughly. Adjust pH to 8.5. Distribute into tubes in 10.0mL volumes. Autoclave for 10 min at 15 psi pressure–121°C.

Use: For the cultivation of *Vibrio cholerae* and other *Vibrio* species from foods.

Alkaline Peptone Water

Composition per liter:

NaCl 10.0g
Peptone 10.0g
pH 8.5 ± 0.2 at 25°C

Preparation of Medium: Add components to distilled/deionized water and bring volume to 1.0L. Mix thoroughly. Adjust pH to 8.5. Distribute into tubes or flasks. Autoclave for 10 min at 15 psi pressure–121°C.

Use: For the cultivation and transport of *Vibrio cholerae* and other *Vibrio* species from foods.

Alkaline Peptone Water

Composition per liter:

Peptone 10.0g
NaCl 5.0g
pH 9.0 ± 0.2 at 25°C

Preparation of Medium: Add components to distilled/deionized water and bring volume to 1.0L. Mix thoroughly. Adjust pH to 9.0. Distribute into tubes or flasks. Autoclave for 20 min at 15 psi pressure–121°C.

Use: For the cultivation of a variety of alkalophilic microorganisms, especially *Vibrio* species.

Alkaline Peptone Water

Composition per liter:

Peptone	10.0g
NaCl	5.0g

pH 8.4 ± 0.2 at 25°C

Preparation of Medium: Add components to distilled/deionized water and bring volume to 1.0L. Mix thoroughly. Adjust pH to 8.4. Distribute into tubes or flasks. Autoclave for 20 min at 15 psi pressure–121°C.

Use: For the cultivation of a variety of alkalophilic microorganisms.

Alkaline Yeast Extract Malt Medium

Composition per liter:

Malt extract	10.0g
Yeast extract	4.0g
Glucose	4.0g
Na_2CO_3 (10% solution)	100.0mL

pH 8.5–11.0 at 25°C

Preparation of Medium: Add components except Na_2CO_3 to distilled/deionized water and bring volume to 900.0mL. Mix thoroughly. Gently heat and bring to boiling. Autoclave for 15 min at 15 psi pressure–121°C. Separately sterilize a 10% Na_2CO_3 solution and aseptically add 100.0mL. Adjust pH, if necessary, to 8.5–11.0.

Use: For the cultivation of *Nocardiopsis dassonvillei*.

Alkvisco Medium

Composition per liter:

Agar	15.0g
Beef extract	10.0g
Peptone	10.0g
NaCl	5.0g
Acrylonitrile	0.5g
KCN	10.0mg

pH 6.5-8.0 at 25°C

Caution: Cyanide is toxic.

Preparation of Medium: Add components, except acrylonitrile, to distilled/deionized water and bring volume to 980.0mL. Mix thoroughly. Gently heat and bring to boiling. Autoclave for 10 min at 15 psi pressure–121°C. Add acrylonitrile to 20.0mL of distilled/deionized water and filter-sterilize. Add aseptically to the sterile basal medium.

Caution: Acrylonitrile is a carcinogen; use appropriate precautions.

Use: For the cultivation and maintenance of *Bacillus subtilis* and *Corynebacterium* species.

Allen and Arnon Medium with Nitrate

Composition per 1000.25mL:

Noble agar	10.0g
KNO_3	0.253g
$NaNO_3$	0.212g
Solution A	25.0mL
Solution B	6.25mL

Solution A:

Composition per 2 liters:

$MgSO_4 \cdot 7H_2O$ (4% solution)	500.0mL
$CaCl_2 \cdot 2H_2O$ (1.2% solution)	500.0mL
NaCl (3.8% solution)	500.0mL
Microelements stock solution	500.0mL

Preparation of Solution A: Prepare individual solutions and combine.

Microelements Stock Solution:

Composition per 1090mL:

H_3BO_3	572.0mg
$MnCl_2 \cdot 4H_2O$	360.0mg
$ZnSO_4 \cdot 7H_2O$	44.0mg
MoO_3	36.0mg
$CuSO_4 \cdot 5H_2O$	15.8mg
$CoCl_2 \cdot 6H_2O$	8.0mg
NH_4VO_3	4.6mg
A & A FeEDTA solution	160.0mL

Preparation of Microelements Stock Solution: Add components to distilled/deionized water and bring volume to 1090.0mL. Mix well.

A & A FeEDTA Solution:

Composition per 550mL:

Disodium EDTA·$2H_2O$	8.0mg
$FeSO_4 \cdot 7H_2O$	4.6mg

Preparation of A & A FeEDTA Solution: Dissolve 5.2g KOH in 186.0mL distilled/deionized water. Add 20.4g disodium EDTA·$2H_2O$. Add 13.7g $FeSO_4 \cdot 7H_2O$ to 364.0mL distilled/deionized water. Combine the EDTA solution with the $FeSO_4$ solution. Sparge solution with filtered air until color changes. The pH of ferrous EDTA solution is about 3.5.

Solution B:

Composition per 500mL:

K_2HPO_4	28.0g

Preparation of Solution B: Add K_2HPO_4 to distilled/deionized water and bring volume to 500.0mL.

Preparation of Medium: Add agar, KNO_3, and $NaNO_3$ to distilled/deionized water and bring volume to 969.0mL. Mix thoroughly. Gently heat and bring to boiling. Add 25.0mL Solution A. Autoclave for 15 min at 15 psi pressure–121°C. Add 6.25mL Solution B aseptically after sterilization.

Use: For the cultivation and maintenance of *Anabaena* species and *Nostoc* species.

ALP Basal Medium (Aerobic Low Peptone Basal Medium)

Composition per liter:

Agar	15.0g
$(NH_4)_2SO_4$	1.0g
Pancreatic digest of casein	0.5g
Yeast extract	0.5g
$MgSO_4·7H_2O$	0.2g
KCl	0.2g
Phenol Red	0.02g
Substrate solution	50.0mL

pH 7.8 ± 0.2 at 25°C

Substrate Solution:
Composition per 50mL:

Substrate	0.1g

Preparation of Substrate Solution: Add substrate to distilled/deionized water and bring volume to 50.0mL. Use sugars, carbohydrates, *n*-butanol, other alcohols, or any acidogenic carbon source. Mix thoroughly. Filter sterilize.

Preparation of Medium: Add components, except substrate solution, to distilled/deionized water and bring volume to 950.0mL. Mix thoroughly. Gently heat and bring to boiling. Adjust pH to 7.8. Distribute into screw-capped tubes in 3.0mL volumes. Autoclave for 15 min at 15 psi pressure–121°C. Cool to 45°–50°C. Aseptically add 0.15mL of sterile substrate solution to each tube. Mix thoroughly. Allow tubes to cool in a slanted position.

Use: For the cultivation and differentiation of microorganisms based on their ability to utilize a variety of carbon sources such as carbohydrates, alcohols and other acidogenic substrates.

ALP Basal Medium (Aerobic Low Peptone Basal Medium)

Composition per liter:

Agar	15.0g
$(NH_4)_2SO_4$	1.0g
Pancreatic digest of casein	0.5g
Yeast extract	0.5g
Glucose	0.2g
$MgSO_4·7H_2O$	0.2g
KCl	0.2g
Phenol Red	0.02g
Substrate solution	50.0mL

pH 6.5 ± 0.2 at 25°C

Substrate Solution:
Composition per 50mL:

Substrate	0.1g

Preparation of Substrate Solution: Add substrate to distilled/deionized water and bring volume to 50.0mL. Use gelatin, aliphatic acids or any alkalogenic carbon source. Mix thoroughly. Filter sterilize.

Preparation of Medium: Add components, except substrate solution, to distilled/deionized water and bring volume to 950.0mL. Mix thoroughly. Gently heat and bring to boiling. Adjust pH to 6.5. Distribute into screw-capped tubes in 3.0mL volumes. Autoclave for 15 min at 15 psi pressure–121°C. Cool to 45°–50°C. Aseptically add 0.15mL of sterile substrate solution to each tube. Mix thoroughly. Allow tubes to cool in a slanted position.

Use: For the cultivation and differentiation of microorganisms based on their ability to utilize a variety of carbon sources such as gelatin, aliphatic acids and other alkalophilic substrates.

AMB Agar

Composition per liter:

Agar	15.0g
Starch, soluble	5.0g
Pancreatic digest of casein	2.5g
$MgSO_4·7H_2O$	0.5g
K_2HPO_4	0.25g

Preparation of Medium: Add components to distilled/deionized water and bring volume to 1.0L. Mix thoroughly. Gently heat and bring to boiling. Distribute into tubes or flasks. Autoclave for 15 min at 15 psi pressure–121°C. Pour into sterile Petri dishes or leave in tubes.

Use: For the cultivation of myxobacteria.

AMB Broth

Composition per liter:

Starch, soluble	5.0g
Pancreatic digest of casein	2.5g
$MgSO_4·7H_2O$	0.5g
K_2HPO_4	0.25g

Preparation of Medium: Add components to distilled/deionized water and bring volume to 1.0L. Mix thoroughly. Distribute into tubes or flasks. Autoclave for 15 min at 15 psi pressure–121°C.

Use: For the cultivation of myxobacteria.

American Association of Textile Chemists and Colorists Bacteriostasis Agar
See: **AATCC Bacteriostasis Agar**

American Association of Textile Chemists and Colorists Bacteriostasis Broth
See: **FDA Broth**

American Association of Textile Chemists and Colorists Mineral Salts Iron Agar
See: **AATCC Mineral Salts Iron Agar**

American Society for Testing and Materials Nutrient Salts Agar
See: **ASTM Nutrient Salts Agar**

American Trudeau Society Medium
See: **ATS Medium**

Amies Modified Transport Medium with Charcoal

Composition per liter:

Charcoal	10.0g
Agar	4.0g
NaCl	3.0g
Na_2HPO_4	1.15g
Sodium thioglycollate	1.0g
KCl	0.2g
$CaCl_2 \cdot 2H_2O$	0.1g
$MgCl_2 \cdot 6H_2O$	0.1g
KH_2PO_4	0.2g

pH 7.2 ± 0.2 at 25°C

Source: This medium is available as a premixed powder from Difco Laboratories.

Preparation of Medium: Add components to distilled/deionized water and bring volume to 1.0L. Mix thoroughly. Gently heat and bring to boiling. Distribute into flasks or tubes. Autoclave for 20 min at 15 psi pressure–121°C. While cooling, turn tubes to uniformly suspend charcoal.

Use: For the transport of swab specimens to prolong the survival of microorganisms, especially *Neisseria gonorrhoeae,* between collection and culturing. Addition of charcoal to this medium neutralizes metabolic products which may be toxic to *Neisseria gonorrhoeae.*

Amies Modified Transport Medium with Charcoal

Composition per liter:

Charcoal	10.0g
NaCl	8.0g
Agar	3.6g
Na_2HPO_4	1.15g
Sodium thioglycollate	1.0g
KCl	0.2g
$CaCl_2 \cdot 2H_2O$	0.1g
$MgCl_2 \cdot 6H_2O$	0.1g
KH_2PO_4	0.2g

pH 7.2 ± 0.2 at 25°C

Source: This medium is available as a premixed powder from BBL Microbiology Systems and Oxoid Unipath.

Preparation of Medium: Add components to distilled/deionized water and bring volume to 1.0L. Mix thoroughly. Gently heat and bring to boiling. Distribute into flasks or tubes. Autoclave for 20 min at 15 psi pressure–121°C. While cooling, turn tubes to uniformly suspend charcoal.

Use: For the transport of swab specimens to prolong the survival of microorganisms, especially *Neisseria gonorrhoeae,* between collection and culturing. Addition of charcoal to this medium neutralizes metabolic products which may be toxic to *Neisseria gonorrhoeae.*

Amies Transport Medium without Charcoal

Composition per liter:

Agar	4.0g
NaCl	3.0g
Na_2HPO_4	1.15g
Sodium thioglycollate	1.0g
KCl	0.2g
$CaCl_2 \cdot 2H_2O$	0.1g
$MgCl_2 \cdot 6H_2O$	0.1g
KH_2PO_4	0.2g

pH 7.2 ± 0.2 at 25°C

Source: This medium is available as a premixed powder from Difco Laboratories.

Preparation of Medium: Add components to distilled/deionized water and bring volume to 1.0L. Mix

thoroughly. Gently heat and bring to boiling. Distribute into flasks or tubes. Autoclave for 20 min at 15 psi pressure–121°C.

Use: For the transport of swab specimens to prolong the survival of microorganisms, especially *Neisseria gonorrhoeae,* between collection and culturing.

Amies Transport Medium without Charcoal

Composition per liter:

NaCl	8.0g
Agar	3.6g
Na_2HPO_4	1.15g
Sodium thioglycollate	1.0g
KCl	0.2g
$CaCl_2 \cdot 2H_2O$	0.1g
$MgCl_2 \cdot 6H_2O$	0.1g
KH_2PO_4	0.2g

pH 7.2 ± 0.2 at 25°C

Source: This medium is available as a premixed powder from BBL Microbiology Systems.

Preparation of Medium: Add components to distilled/deionized water and bring volume to 1.0L. Mix thoroughly. Gently heat and bring to boiling. Distribute into flasks or tubes. Autoclave for 20 min at 15 psi pressure–121°C.

Use: For the transport of swab specimens to prolong the survival of microorganisms, especially *Neisseria gonorrhoeae,* between collection and culturing.

Amino Acid Assay Medium

Composition per liter:

Glucose	50.0g
Sodium acetate	40.0g
NH_4Cl	6.0g
KH_2PO_4	1.2g
K_2HPO_4	1.2g
Asparagine	0.8g
L-Glutamic acid	0.6g
Pyridoxamine·HCl	0.6g
Pyridoxal·HCl	0.6g
DL-Valine	0.5g
L-Lysine·HCl	0.5g
DL-Isoleucine	0.5g
DL-Leucine	0.5g
L-Arginine·HCl	0.5g
DL-Threonine	0.4g
$MgSO_4 \cdot 7H_2O$	0.4g
DL-Alanine	0.4g
DL-Phenylalanine	0.2g
L-Tyrosine	0.2g
Lysine	0.2g
L-Aspartic acid	0.2g
DL-Methionine	0.2g
L-Proline	0.2g
L-Histidine·HCl	0.12g
DL-Serine	0.1g
L-Cystine	0.1g
DL-Tryptophan	0.08g
$MnSO_4 \cdot 7H_2O$	0.04g
$FeSO_4$	0.02g
NaCl	0.02g
Adenine sulfate	0.02g
Guanine·HCl	0.02g
Uracil	0.02g
Xanthine	0.02g
Nicotinic acid	2.0mg
Pyridoxine·HCl	2.0mg
Thiamine·HCl	1.0mg
Calcium pantothenate	1.0mg
Riboflavin	1.0mg
p-Aminobenzoic acid	0.2mg
Folic acid	0.02mg
Biotin	2.0μg

pH 6.7 ± 0.2 at 25°C

Source: This medium is available, without cystine, lysine or methionine, as a premixed powder from Difco Laboratories.

Preparation of Medium: Add components to 1.0L distilled water, omitting the specific amino acid to be assayed for in the procedure. Heat to boiling for 2–3 min. Distribute to tubes. Autoclave for 10 min at 15 psi pressure–121°C.

Use: For the microbiological assay for amino acids. *Pediococcus acidilactici* ATCC 8042 and *Enterococcus hirae* ATCC 8043 are used as test microorganisms.

Amino-butyric Acid Medium

Composition per liter:

Agar	15.0g
DL-Amino-butyric acid	10.0g
K_2HPO_4	7.0g
Glucose	5.0g
KH_2PO_4	3.0g
$(NH_4)_2SO_4$	1.0g
$MgSO_4 \cdot 7H_2O$	0.5g

pH 7.0 ± 0.2 at 25°C

Preparation of Medium: Add components to distilled/deionized water and bring volume to 1.0L. Mix thoroughly. Gently heat and bring to boiling. Autoclave for 15 minutes at 15 psi–121°C. Pour into sterile Petri dishes.

Use: For the cultivation and maintenance of *Serratia marcescens* and other microorganisms which can utilize amino-butyric acid as a carbon source.

Ammonium Mineral Salts Agar
See: **AMS Agar**

Ammonium Mineral Salts Agar without Methanol
See: **AMS Agar without Methanol**

Ammonium Yeast Extract Medium
See: ***Bacillus pasteurii* NH_4 YE Medium**

AMS Agar (Ammonium Mineral Salts Agar)

Composition per liter:

Agar	15.0g
$MgSO_4 \cdot 7H_2O$	1.0g
K_2HPO_4	0.7g
KH_2PO_4	0.54g
NH_4Cl	0.5g
$CaCl_2 \cdot 2H_2O$	0.2g
$FeSO_4 \cdot 7H_2O$	4.0mg
H_3BO_4	0.3mg
$CoCl_2 \cdot 6H_2O$	0.2mg
$ZnSO_4 \cdot 7H_2O$	0.1mg
$Na_2MoO_4 \cdot 2H_2O$	0.06mg
$MnCl_2 \cdot 4H_2O$	0.03mg
$NiCl_2 \cdot 6H_2O$	0.02mg
$CuCl_2 \cdot 2H_2O$	0.01mg

pH 6.8 ± 0.2 at 25°C

Preparation of Medium: Add components to distilled/deionized water and bring volume to 1.0L. Mix thoroughly. Gently heat and bring to boiling. Autoclave for 15 minutes at 15 psi–121°C. Add sterile methanol to a concentration of 0.5% aseptically to cooled basal medium.

Use: For the cultivation and maintenance of bacteria which can utilize methanol as a carbon source such as *Methylobacterium* species, *Methylomonas* species, and *Methylophilus* species.

AMS Agar without Methanol (Ammonium Mineral Salts Agar without Methanol)

Composition per liter:

Agar	15.0g
$MgSO_4 \cdot 7H_2O$	1.0g
K_2HPO_4	0.7g
KH_2PO_4	0.54g
NH_4Cl	0.5g
$CaCl_2 \cdot 2H_2O$	0.2g
$FeSO_4 \cdot 7H_2O$	4.0mg
H_3BO_4	0.3mg
$CoCl_2 \cdot 6H_2O$	0.2mg
$ZnSO_4 \cdot 7H_2O$	0.1mg
$Na_2MoO_4 \cdot 2H_2O$	0.06mg
$MnCl_2 \cdot 4H_2O$	0.03mg
$NiCl_2 \cdot 6H_2O$	0.02mg
$CuCl_2 \cdot 2H_2O$	0.01mg

pH 6.8 ± 0.2 at 25°C

Preparation of Medium: Add components to distilled/deionized water and bring volume to 1.0L. Mix thoroughly. Gently heat and bring to boiling. Autoclave for 15 minutes at 15 psi–121°C.

Use: For the cultivation and maintenance of *Methylosinus trichosporium* and other methane-oxidizing bacteria. Cultures are grown under an atmosphere of 50% methane.

AMS Medium

Composition per liter:

NaCl	26.0g
$MgSO_4 \cdot 7H_2O$	12.0g
Peptone	5.0g
Beef extract	3.0g
$CaCl_2 \cdot 2H_2O$	1.5g
KCl	0.7g

Preparation of Medium: Add components to distilled/deionized water and bring volume to 1.0L. Mix thoroughly. Gently heat and bring to boiling. Distribute into tubes or flasks. Autoclave for 15 minutes at 15 psi–121°C.

Use: For the cultivation of *Alteromonas espejiana.*

Amygdalin Medium

Composition per liter:

Peptone	10.0g
Beef extract	5.0g
NaCl	5.0g
Agar	3.0g
Amygdalin solution	200.0mL
Bromthymol Blue (0.05% solution)	5.0mL

pH 7.0 ± 0.2 at 25°C

Amygdalin Solution:
Composition per 200mL:

Amygdalin	10.0g

Preparation of Amygdalin Solution: Add amygdalin to distilled/deionized water and bring volume to 200.0mL. Mix thoroughly. Filter sterilize.

Preparation of Medium: Add components, except amygdalin solution, to distilled/deionized water and bring volume to 800.0mL. Mix thoroughly. Gently heat and bring to boiling. Adjust pH to 7.0. Autoclave for 20 min at 15 psi pressure–121°C. Cool to 45°–50°C. Aseptically add sterile amygdalin solution. Mix thoroughly. Aseptically distribute into sterile tubes with cotton plugs. Allow tubes to cool in a slanted position forming a short slant.

Use: For the cultivation and differentiation of *Serratia* species based on their ability to produce acid and HCN from amygdalin.

Anacker and Ordal Medium

Composition per liter:

Agar	10.0g
Pancreatic digest of casein	0.5g
Yeast extract	0.5g
Sodium acetate	0.2g
Beef extract	0.2g

pH 7.3 ± 0.1 at 25°C

Preparation of Medium: Add components to distilled/deionized water and bring volume to 1.0L. Mix thoroughly. Gently heat and bring to boiling. Distribute into tubes or flasks. Autoclave for 15 minutes at 15 psi–121°C. Pour into sterile Petri dishes or leave in tubes.

Use: For the cultivation and maintenance of *Flexibacter columnaris*.

Anacker and Ordal Medium, Enriched

Composition per liter:

Agar	10.0g
Pancreatic digest of casein	5.0g
Yeast extract	0.5g
Sodium acetate	0.2g
Beef extract	0.2g

pH 7.3 ± 0.1 at 25°C

Preparation of Medium: Add components to distilled/deionized water and bring volume to 1.0L. Mix thoroughly. Gently heat and bring to boiling. Distribute into tubes or flasks. Autoclave for 15 minutes at 15 psi–121°C. Pour into sterile Petri dishes or leave in tubes.

Use: For the cultivation and maintenance of *Flexibacter psychrophilus*.

Anaerobe Agar

Composition per 1001.5mL:

Agar	20.0g
Pancreatic digest of casein	17.0g
NaCl	5.0g
Yeast extract	5.0g
Papaic digest of soybean meal	3.0g
K_2HPO_4	2.5g
Glucose	2.5g
Cystine solution	5.0mL
Vitamin K_1 solution	1.0mL
Hemin solution	0.5mL

pH 7.5 ± 0.2 at 25°C

Cystine Solution:
Composition per 5mL:

L-Cystine	0.4g
NaOH (1N solution)	5.0mL

Preparation of Cystine Solution: Add L-Cystine to 5.0mL of NaOH solution. Mix thoroughly.

Vitamin K_1 Solution:
Composition per 100mL:

Vitamin K_1	1.0g
Ethanol	99.0mL

Preparation of Vitamin K_1 Solution: Add vitamin K_1 to 99.0mL of absolute ethanol. Mix thoroughly. Filter sterilize.

Hemin Solution:
Composition per 100mL:

Hemin	1.0g
NaOH (1*N* solution)	20.0mL

Preparation of Hemin Solution: Add hemin to 20.0mL of 1*N* NaOH solution. Mix thoroughly. Bring volume to 100.0mL with distilled/deionized water. Autoclave for 15 min at 15 psi pressure–121°C. Cool to 45°–50°C.

Preparation of Medium: Add components, except vitamin K_1 solution and hemin solution, to distilled/deionized water and bring volume to 1.0L. Mix thoroughly. Gently heat while stirring and bring to boiling. Adjust pH to 7.5. Autoclave for 15 min at 15 psi pressure–121°C. Cool to 45°–50°C. Aseptically add 1.0mL of sterile vitamin K_1 solution and 0.5mL of sterile hemin solution. Mix thoroughly. Pour into sterile Petri dishes or distribute into sterile tubes.

Use: For the cultivation of anaerobes from cosmetic products.

Anaerobic Agar

Composition per liter:

Pancreatic digest of casein	20.0g
Agar	15.0g
NaCl	5.0g
Sodium thioglycollate	2.0g
Sodium formaldehyde sulfoxylate	1.0g

pH 7.2 ± 0.2 at 25°C

Preparation of Medium: Add components to distilled/deionized water and bring volume to 1.0L. Mix thoroughly. Gently heat and bring to boiling. Adjust ph to 7.2. Distribute into tubes until medium is 3 inches deep. Autoclave for 20 min at 15 psi pressure–121°C.

Use: For the anaerobic cultivation of *Bacillus* species and *Sporolactobacillus* species.

Anaerobic Agar

Composition per liter:

Pancreatic digest of casein	20.0g
Agar	15.0g
Yeast extract	15.0g
NaCl	5.0g
Sodium thioglycollate	2.0g
Sodium formaldehyde sulfoxylate	1.0g

pH 7.2 ± 0.2 at 25°C

Preparation of Medium: Add components to distilled/deionized water and bring volume to 1.0L. Mix thoroughly. Gently heat and bring to boiling. Adjust pH to 7.2. Distribute into tubes until medium is 3 inches deep. Autoclave for 20 min at 15 psi pressure–121°C.

Use: For the anaerobic cultivation of *Bacillus* species, especially *B. larvae*, *B. popilliae*, and *B. lentimorbus*.

Anaerobic Agar

Composition per liter:

Agar	20.0g
Pancreatic digest of casein	20.0g
Glucose	10.0g
NaCl	5.0g
Sodium thioglycollate	2.0g
Sodium formaldehyde sulfoxylate	1.0g
Methylene Blue	2.0mg

pH 7.2 ± 0.2 at 25°C

Source: This medium is available as a premixed powder from Difco Laboratories.

Preparation of Medium: Add components to distilled/deionized water and bring volume to 1.0L. Mix thoroughly. Gently heat and bring to boiling. Adjust pH to 7.2. Distribute into tubes until medium is 3 inches deep. Autoclave for 15 min at 15 psi pressure–121°C.

Use: For the cultivation of a variety of anaerobic microorganisms, especially *Clostridium* species.

Anaerobic Agar

Composition per liter:

Pancreatic digest of casein	17.5g
Agar	15.0g
Glucose	10.0g
Papaic digest of soybean meal	2.5g
NaCl	2.5g
Sodium thioglycollate	2.0g
Sodium formaldehyde sulfoxylate	1.0g
L-Cystine	0.4g
Methylene Blue	2.0mg

pH 7.2 ± 0.2 at 25°C

Source: This medium is available as a premixed powder from BBL Microbiology Systems.

Preparation of Medium: Add components to distilled/deionized water and bring volume to 1.0L. Mix thoroughly. Gently heat and bring to boiling. Autoclave for 15 minutes at 15 psi–121°C. Use with Brewer anaerobic Petri dishes or in tubes or ordinary plates and incubate in anaerobic jars.

Use: For the cultivation of *Clostridium* species and for anaerobic microorganisms.

Anaerobic Broth

Composition per liter:

Pancreatic digest of casein	17.5g
Glucose	10.0g
NaCl	2.5g
Papaic digest of soybean meal	2.5g
Sodium thioglycollate	2.0g
Sodium formaldehyde sulfoxylate	1.0g
L-Cystine	0.4g
Methylene Blue	2.0mg

pH 7.2 ± 0.2 at 25°C

Preparation of Medium: Add components to distilled/deionized water and bring volume to 1.0L. Mix thoroughly. Gently heat and bring to boiling. Distribute into tubes or flasks. Autoclave for 15 min at 15 psi pressure–121°C.

Use: For the cultivation of a variety of anaerobic and microaerophilic microorganisms.

Anaerobic Cellulolytic Medium

Composition per liter:

NH_4Cl	1.0g
Cellobiose	1.0g
Yeast extract	1.0g

$MgSO_4$	0.5g
KCl	0.5g
L-Cysteine·HCl·H_2O	0.5g
K_2HPO_4	0.4g
Resazurin	1.0mg
Wolfe's mineral solution	20.0mL
Na_2CO_3 solution	10.0mL
$Na_2S \cdot 9H_2O$ solution	10.0mL

pH 6.9 ± 0.1 at 25°C

Wolfe's Mineral Solution:
Composition per liter

$MgSO_4 \cdot 7H_2O$	3.0g
Nitrilotriacetic acid	1.5g
NaCl	1.0g
$MnSO_4 \cdot H_2O$	0.5g
$FeSO_4 \cdot 7H_2O$	0.1g
$CoCl_2 \cdot 6H_2O$	0.1g
$CaCl_2$	0.1g
$ZnSO_4 \cdot 7H_2O$	0.1g
$CuSO_4 \cdot 5H_2O$	0.01g
$AlK(SO_4)_2 \cdot 12H_2O$	0.01g
H_3BO_3	0.01g
$Na_2MoO_4 \cdot 2H_2O$	0.01g

Preparation of Wolfe's Mineral Solution: Add nitrilotriacetic acid to approximately 500.0mL of distilled/deionized water and adjust pH to 6.5 with KOH to dissolve. Bring volume to 1.0L with distilled/deionized water. Add remaining compounds one at a time. Mix thoroughly.

Na_2CO_3 Solution:
Composition per 100mL:

Na_2CO_3	10.0g

Preparation of Na_2CO_3 Solution: Add Na_2CO_3 to distilled/deionized water and bring volume to 100.0mL. Mix thoroughly. Filter sterilize.

$Na_2S \cdot 9H_2O$ Solution:
Composition per 100mL:

$Na_2S \cdot 9H_2O$	15.0g

Preparation of $Na_2S \cdot 9H_2O$ Solution: Add $Na_2S \cdot 9H_2O$ to distilled/deionized water and bring volume to 100.0mL. Mix thoroughly. Filter sterilize.

Preparation of Medium: Add components, except Na_2CO_3 solution and $Na_2S \cdot 9H_2O$ solution, to distilled/deionized water and bring volume to 980.0mL. Boil medium under a stream of 80% N_2 + 10% CO_2 + 10% H_2 until the resazurin indicator is colorless. Cool medium and distribute anaerobically into test tubes in 10.0mL volumes using 80% N_2 + 10% CO_2 + 10% H_2. Stopper the tubes anaerobically. Autoclave for 15 min at 15 psi pressure–121°C. Cool medium to room temperature. Aseptically add 0.1mL of sterile Na_2CO_3 solution and 0.1mL of sterile $Na_2S \cdot 9H_2O$ solution to each tube. Mix thoroughly.

Use: For the cultivation and maintenance of microorganisms which can utilize cellobiose as sole carbon source such as *Clostridium cellulovorans*.

Anaerobic CNA Agar (Anaerobic Colistin Nalidixic Acid Agar)

Composition per liter:

Agar	13.0g
Pancreatic digest of casein	12.0g
Peptic digest of animal tissue	5.0g
Yeast extract	3.0g
Beef extract	3.0g
NaCl	5.0g
Cornstarch	1.0g
Glucose	1.0g
L-Cysteine·HCl·H_2O	0.5g
Vitamin K_1	10.0mg
Hemin	10.0mg
Colistin	10.0mg
Nalidixic acid	10.0mg
Sheep blood, defibrinated	50.0mL

Source: This medium is available as a premixed powder from BBL Microbiology Systems.

Preparation of Medium: Add components, except sheep blood, to distilled/deionized water and bring volume to 950.0mL. Mix thoroughly. Gently heat and bring to boiling. Autoclave for 15 minutes at 15 psi–121°C. Cool to 45–50°C. Aseptically add 50.0mL of sterile, defibrinated sheep blood. Mix thoroughly. Pour into sterile Petri dishes.

Use: For the selective isolation of anaerobic streptococci.

Anaerobic Egg Yolk Agar

Composition per 1080mL:

Agar	20.0g
Proteose peptone	20.0g
NaCl	5.0g
Pancreatic digest of casein	5.0g
Yeast extract	5.0g
Egg yolk emulsion, 50%	80.0mL

pH 7.0 ± 0.2 at 25°C

Egg Yolk Emulsion, 50%:
Composition per 100mL:

Chicken egg yolks	11
Whole chicken egg	1
NaCl (0.9% solution)	50.0mL

Preparation of Egg Yolk Emulsion, 50%: Soak eggs with 1:100 dilution of saturated mercuric chloride solution for 1 min. Crack 11 eggs and separate yolks from whites. Mix egg yolks with 1 chicken egg. Measure 50.0mL of egg yolk emulsion and add to 50.0mL of 0.9% NaCl solution. Mix thoroughly. Filter sterilize. Warm to 45°–50°C.

Preparation of Medium: Add components, except egg yolk emulsion, 50%, to distilled/deionized water and bring volume to 1.0L. Mix thoroughly. Gently heat and bring to boiling. Autoclave for 15 min at 15 psi pressure–121°C. Cool to 45°–50°C. Aseptically add 80.0mL of sterile egg yolk emulsion, 50%. Mix thoroughly. Pour into sterile Petri dishes or distribute into sterile tubes. Allow plates to dry at 35°C for 24 hr.

Use: For the cultivation of *Clostridium* species.

Anaerobic Egg Yolk Agar

Composition per liter:

Agar	20.0g
Proteose peptone	20.0g
Pancreatic digest of casein	5.0g
NaCl	5.0g
Yeast extract	5.0g
Egg yolk emulsion, 50%	20.0mL

pH 7.0 ± 0.2 at 25°C

Egg Yolk Emulsion, 50%:
Composition per 100mL:

Chicken egg yolks	2
NaCl (0.9% solution)	10.0mL

Preparation of Egg Yolk Emulsion, 50%: Soak eggs with 1:100 dilution of saturated mercuric chloride solution for 1 min. Crack eggs and separate yolks from whites. Measure 10.0mL of egg yolk emulsion and add to 10.0mL of 0.9% NaCl solution. Mix thoroughly. Filter sterilize. Warm to 45°–50°C.

Preparation of Medium: Add components, except egg yolk emulsion, to distilled/deionized water and bring volume to 980.0mL. Mix thoroughly. Gently heat and bring to boiling. Autoclave for 15 min at 15 psi pressure–121°C. Cool to 45°–50°C. Aseptically add sterile egg yolk emulsion. Mix thoroughly. Pour into sterile Petri dishes. Allow plates to dry at 35°C for 24 hr.

Use: For the cultivation of *Yersinia enterocolitica*.

Anaerobic D-Gluconate Medium

Composition per liter:

Agar	15.0g
Pancreatic digest of casein	10.0g
Yeast extract	5.0g
D-Gluconate	4.0g
$MgSO_4 \cdot 7H_2O$	2.5g
$(NH_4)_2SO_4$	1.4g
L-Cysteine·HCl·H_2O	1.0g
$CaCl_2 \cdot 2H_2O$	0.15g
$FeSO_4 \cdot 7H_2O$	0.02g
Resazurin	1.0mg
$NaHCO_3$ solution	10.0mL

pH 7.1 ± 0.2 at 25°C

$NaHCO_3$ Solution:
Composition per 100mL:

$NaHCO_3$	10.0g

Preparation of $NaHCO_3$ Solution: Add $NaHCO_3$ to distilled/deionized water and bring volume to 100.0mL. Mix thoroughly. Filter sterilize.

Preparation of Medium: Add components, except $NaHCO_3$ solution, to distilled/deionized water and bring volume to 990.0mL. Prepare anaerobically under 100% N_2. Autoclave for 15 min at 15 psi pressure–121°C. Aseptically add 10.0mL of the sterile $NaHCO_3$ solution. Mix thoroughly. Adjust pH to 7.1.

Use: For the cultivation and maintenance of microorganisms which can utilize D-gluconate as a carbon source such as *Bacteroides pectinophilus*.

Anaerobic Glucuronic Acid Medium

Composition per liter:

Agar	15.0g
Pancreatic digest of casein	10.0g
Yeast extract	5.0g
Glucuronic acid	4.0g
$MgSO_4 \cdot 7H_2O$	2.5g
$(NH_4)_2SO_4$	1.4g
L-Cysteine·HCl·H_2O	1.0g
$CaCl_2 \cdot 2H_2O$	0.15g
$FeSO_4 \cdot 7H_2O$	0.02g
Resazurin	1.0mg
$NaHCO_3$ solution	10.0mL

pH 7.1 ± 0.2 at 25°C

$NaHCO_3$ Solution:
Composition per 100mL:

$NaHCO_3$	10.0g

Preparation of $NaHCO_3$ Solution: Add $NaHCO_3$ to distilled/deionized water and bring volume to 100.0mL. Mix thoroughly. Filter sterilize.

Preparation of Medium: Add components, except $NaHCO_3$ solution, to distilled/deionized water and bring volume to 990.0mL. Prepare anaerobically under 100% N_2. Autoclave for 15 min at 15 psi pressure–121°C. Aseptically add 10.0mL of the sterile $NaHCO_3$ solution. Mix thoroughly. Adjust pH to 7.1.

Use: For the cultivation and maintenance of microorganisms which can utilize D-glucuronate as a carbon source such as *Bacteroides galacturonicus*.

Anaerobic LKV Blood Agar

Composition per liter:

Agar	15.0g
Pancreatic digest of casein	13.0g
Peptic digest of animal tissue	10.0g
NaCl	5.0g
Yeast extract	2.0g
Glucose	1.0g
$NaHSO_3$	0.1g
Sheep blood, laked	50.0mL
Antibiotic solution	10.0mL
Hemin	1.0mL
Vitamin K_1 solution	1.0mL

pH 7.1–7.8 at 25°C

Source: This medium is available as a premixed powder from Difco Laboratories.

Antibiotic Solution:
Composition per 10mL:

Kanamycin	0.075g
Vancomycin	7.5mg

Preparation of Antibiotic Solution: Add components to distilled/deionized water and bring volume to 10.0mL. Mix thoroughly. Filter sterilize.

Vitamin K_1 Solution:
Composition per 100mL:

Vitamin K_1	0.1g
Ethanol	99.0mL

Preparation of Vitamin K_1 Solution: Add vitamin K_1 to 99.0mL of absolute ethanol. Mix thoroughly.

Hemin Solution:
Composition per 100mL:

Hemin	0.01g
NaOH (1*N* solution)	20.0mL

Preparation of Hemin Solution: Add hemin to 20.0mL of 1*N* NaOH solution. Mix thoroughly. Bring volume to 100.0mL with distilled/deionized water.

Preparation of Medium: Add components—except sheep blood, antibiotic solution, and vitamin K_1 solution—to distilled/deionized water and bring volume to 939.0mL. Mix thoroughly. Gently heat and bring to boiling. Autoclave for 15 min at 15 psi pressure–121°C. Cool to 45°–50°C. Aseptically add 50.0mL of sterile sheep blood, 10.0mL of sterile antibiotic solution, and 1.0mL of sterile vitamin K_1 solution. Mix thoroughly. Pour into sterile Petri dishes or distribute into sterile tubes.

Use: For the isolation and cultivation of anaerobic Gram-negative microorganisms, expecially *Bacteroides* species.

Anaerobic Trypticase™ Soy Medium with Calf Blood

Composition per liter:

Pancreatic digest of casein	15.0g
Agar	15.0g
Papaic digest of soybean meal	5.0g
NaCl	5.0g
Calf blood, defibrinated	100.0mL

pH 7.3 ± 0.2 at 25°C

Preparation of Medium: Add components, except calf blood, to distilled/deionized water and bring volume to 900.0mL. Mix thoroughly. Prepare medium anaerobically with 80% N_2 + 10% CO_2 + 10% H_2. Gently heat while stirring and bring to boiling for 1 min. Autoclave for 15 min at 15 psi pressure–121°C. Do not overheat. Cool to 45°–50°C. Aseptically add 100.0mL sterile, defibrinated calf blood. Pour into sterile Petri dishes.

Use: For the isolation and cultivation of fastidious as well as nonfastidious microorganisms. For the differentiation of *Haemophilus* species.

Anaerobic TVLS Medium

Composition per liter:

Pancreatic digest of casein	17.0g
Beef extract	7.5g
Glucose	6.0g
Enzymatic hydrolysate of soybean meal	3.0g
Liver hydrolysate	3.0g
NaCl	2.5g
Na_2SO_3	0.7g
Sodium thioglycollate	0.5g
L-Cysteine·HCl·H_2O	0.25g
Agar	0.1g
Bovine serum	100.0mL

pH 7.3 ± 0.2 at 25°C

Preparation of Medium: Add components, except bovine serum, to distilled/deionized water and bring volume to 900.0mL. Mix thoroughly. Gently heat and bring to boiling. Autoclave for 15 minutes at 15 psi–121°C. Cool to 45–50°C. Aseptically add 100.0mL bovine serum. Distribute into sterile tubes.

Use: For the isolation and cultivation of anaerobic microorganisms.

Anaerospirillum Medium

Composition per liter:

Polypeptone™	10.0g
Glucose	10.0g
Yeast extract	5.0g
Na_2CO_3	3.0g
NaCl	2.0g
K_2HPO_4	1.0g
$MgSO_4{\cdot}7H_2O$	0.2g

pH 6.5 ± 0.2 at 25°C

Preparation of Medium: Add components to distilled/deionized water and bring volume to 1.0L. Mix thoroughly. Gently heat and bring to boiling. Autoclave for 15 minutes at 15 psi–121°C. Distribute into tubes or flasks using anaerobic techniques and 100% CO_2 as gas phase.

Use: For the cultivation of *Anaerospirillum succiniciproducens.*

Andersen's Pork Pea Agar

Composition per 1685mL:

Agar	16.0g
Peptone	5.0g
Pancreatic digest of casein	1.6g
K_2HPO_4	1.25g
Soluble starch	1.0g
Sodium thioglycollate	0.5g
Pork infusion	800.0mL
Thioglycollate agar	660.0mL
Pea infusion	200.0mL
$NaHCO_3$ solution	25.0mL

pH 7.2 ± 0.2 at 25°C

Pork Infusion:

Composition per liter:

Pork, fresh lean ground	454.0g

Preparation of Pork Infusion: Add ground pork to distilled/deionized water and bring volume to 1.0L. Autoclave for 60 min at 0 psi pressure–100°C. Filter through two layers of cheesecloth. Cool to 4°C. Skim fat from surface. Warm to 25°C. Centrifuge at 5,000 rpm for 10 min. Discard pellet.

Pea Infusion:

Composition per 450mL:

Green peas, fresh or frozen	454.0g
Diatomaceous earth (celite)	10.0g

Preparation of Pea Infusion: Add green peas to 450.0mL of distilled/deionized water. Blend until smooth. Autoclave for 60 min at 0 psi pressure–100°C. Centrifuge at 5,000 rpm for 10 min. Discard pellet. Clarify supernatant solution with diatomaceous earth (celite). Filter through Whatman #4 filter paper. Use filtrate solution.

Thioglycollate Agar:

Composition per liter:

Agar	20.75g
Pancreatic digest of casein	15.0g
Glucose	5.5g
Yeast extract	5.0g
NaCl	2.5g
L-Cystine	0.5g
Sodium thioglycollate	0.5g
Resazurin	1.0mg

pH 7.1 ± 0.2 at 25°C

Preparation of Thioglycollate Agar: Add components to distilled/deionized water and bring volume to 1.0L. Mix thoroughly. Gently heat and bring to boiling. Autoclave for 15 min at 15 psi pressure–121°C. Cool to 45°–50°C.

$NaHCO_3$ Solution:

Composition per 100mL:

$NaHCO_3$	5.0g

Preparation of $NaHCO_3$ Solution: Add $NaHCO_3$ to distilled/deionized water and bring volume to 100.0mL. Mix thoroughly. Filter sterilize.

Preparation of Medium: Combine components, except $NaHCO_3$ solution and thioglycollate agar. Mix thoroughly. Adjust pH to 7.2. Autoclave for 5 min at 15 psi pressure–121°C. While medium is still hot add 25.0g of celite. Filter through Whatman #4 filter paper with suction. Autoclave for 12 min at 15 psi pressure–121°C. Cool to 45°–50°C. Aseptically add 25.0mL of sterile $NaHCO_3$ solution. Mix thoroughly. Pour into sterile Petri dishes in 15.0mL volumes. Allow agar to solidify. Cover agar with 10.0mL of sterile, cooled thioglycollate agar.

Use: For the cultivation of mesophilic *Clostridium* species. For the recovery of endospores from foods following heat treatments.

Andrade's Broth

Composition per liter:

Pancreatic digest of gelatin	10.0g
NaCl	5.0g
Beef extract	3.0g
Andrade's indicator	10.0mL
Carbohydrate solution	50.0mL

pH 7.4 ± 0.2 at 25°C

Source: Available as a prepared medium from BBL Microbiology Systems, in tubes containing adonitol, arabinose, cellobiose, glucose, dulcitol, fructose, galactose, inositol, lactose, maltose, mannitol, raffinose, rhamnose, salicin, sorbitol, sucrose, trehalose, or xylose.

Andrade's Indicator
Composition per 100mL:

NaOH (1*N* solution)	16.0mL
Acid Fuchsin	0.1 g

Preparation of Andrade's Indicator: Add Acid Fuchsin to NaOH solution and bring volume to 100.0mL with distilled/deionized water.

Carbohydrate Solution:
Composition per 100mL:

Carbohydrate	10.0g

Preparation of Carbohydrate Solution: Add carbohydrate to distilled/deionized water and bring volume to 100.0mL. Adonitol, arabinose, cellobiose, glucose, dulcitol, fructose, galactose, inositol, lactose, maltose, mannitol, raffinose, rhamnose, salicin, sorbitol, sucrose, trehalose, xylose, or other carbohydrates may be used. Mix thoroughly. Filter sterilize.

Preparation of Medium: Add components, except carbohydrate solution, to distilled/deionized water and bring volume to 1.0L. Mix thoroughly. Gently heat and bring to boiling. Distribute in 10.0mL volumes into test tubes containing inverted Durham tubes. Autoclave for 15 minutes at 15 psi–121°C. Cool to 25°C. Add 0.5mL of sterile carbohydrate solution to each tube.

Caution: Acid Fuchsin is a potential carcinogen and care must be taken to avoid inhalation of the powdered dye and contact with the skin.

Use: For the determination of carbohydrate fermentation reactions of microorganisms, particularly members of the Enterobacteriaceae. A Durham tube is used to collect gas produced during the fermentation reaction. Acid production is indicated by a pink color.

Anthranilic Acid Medium, Revised

Composition per 1040mL:

Na_2HPO_4	6.0g
KH_2PO_4	3.0g
NH_4Cl	1.0g
NaCl	0.5g
Glucose solution	25.0mL
$CaCl_2$ solution	10.0mL
$MgSO_4$ solution	10.0mL
Anthranilic acid solution	5.0mL

Glucose Solution:
Composition per 100mL:

D-Glucose	20.0g

Preparation of Glucose Solution: Add glucose to distilled/deionized water and bring volume to 100.0mL. Mix thoroughly. Filter sterilize.

$CaCl_2$ Solution:
Composition per 100mL:

$CaCl_2 \cdot 2H_2O$	0.147g

Preparation of $CaCl_2$ Solution: Add $CaCl_2$ to distilled/deionized water and bring volume to 100.0mL. Mix thoroughly. Filter sterilize.

$MgSO_4$ Solution:
Composition per 10mL:

$MgSO_4 \cdot 7H_2O$	2.47g

Preparation of $MgSO_4$ Solution: Add $MgSO_4$ to distilled/deionized water and bring volume to 100.0mL. Mix thoroughly. Filter sterilize.

Anthranilic Acid Solution:
Composition per 100mL:

Anthranilic acid	1.0g
Ethanol (95% solution)	100.0mL

Preparation of Anthranilic Acid Solution: Add anthranilic acid to 100.0mL of ethanol. Mix thoroughly. Filter sterilize.

Preparation of Medium: Add components—except glucose solution, $CaCl_2$ solution, $MgSO_4$ solution, and anthranilic acid solution—to distilled/deionized water and bring volume to 990.0mL. Mix thoroughly. Gently heat and bring to boiling. Autoclave for 15 min at 15 psi pressure–121°C. Cool to 45°–50°C. Aseptically add 25.0mL of sterile glucose solution, 10.0mL of sterile $CaCl_2$ solution, 10.0mL of sterile $MgSO_4$ solution, and 5.0mL of sterile anthranilic acid solution. Mix thoroughly. Pour into sterile Petri dishes or distribute into sterile tubes.

Use: For the cultivation of *Escherichia coli*.

Antibiotic Medium 1 (Penassay Seed Agar) (Seed Agar) (Agar Medium A)

Composition per liter:

Agar	15.0g
Pancreatic digest of gelatin	6.0g
Pancreatic digest of casein	4.0g
Yeast extract	3.0g
Beef extract	1.5g
Glucose	1.0g

pH 6.6 ± 0.1 at 25°C

Source: This medium is available as a premixed powder from BBL Microbiology Systems and Difco Laboratories.

Preparation of Medium: Add components to distilled/deionized water and bring volume to 1.0L. Mix thoroughly. Gently heat and bring to boiling. Distrib-

ute into tubes or flasks. Autoclave for 15 minutes at 15 psi–121°C. Pour into sterile Petri dishes or leave in tubes.

Use: For antibiotic assay testing, detection of antibiotics in milk and determination of the antimicrobial effectiveness of antibiotics.

Antibiotic Medium 1 with Tetracycline

Composition per liter:

Agar	15.0g
Pancreatic digest of gelatin	6.0g
Pancreatic digest of casein	4.0g
Yeast extract	3.0g
Beef extract	1.5g
Glucose	1.0g
Tetracycline solution	10.0mL

pH 6.6 ± 0.1 at 25°C

Tetracycline Solution:

Composition per 10mL:

Tetracycline	0.02g

Preparation of Tetracycline Solution: Add tetracycline to distilled/deionized water and bring volume to 10.0mL. Mix thoroughly. Filter sterilize.

Preparation of Medium: Add components, except tetracycline solution, to distilled/deionized water and bring volume to 1.0L. Mix thoroughly. Gently heat and bring to boiling. Autoclave for 15 minutes at 15 psi–121°C. Cool to 45°–50°C. Aseptically add sterile tetracycline solution. Mix thoroughly. Pour into sterile Petri dishes or distribute into sterile tubes.

Use: For selective cultivation and maintenance of *Salmonella choleraesuis*.

Antibiotic Medium 1 with Tetracycline, Streptomycin and Chloramphenicol

Composition per liter:

Agar	15.0g
Pancreatic digest of gelatin	6.0g
Pancreatic digest of casein	4.0g
Yeast extract	3.0g
Beef extract	1.5g
Glucose	1.0g
Antibiotic solution	10.0mL

pH 6.6 ± 0.1 at 25°C

Antibiotic Solution:

Composition per 10mL:

Tetracycline	0.02g
Streptomycin	0.02g
Chloramphenicol	0.02g

Preparation of Antibiotic Solution: Add components to distilled/deionized water and bring volume to 10.0mL. Mix thoroughly. Filter sterilize.

Preparation of Medium: Add components, except tetracycline solution, to distilled/deionized water and bring volume to 1.0L. Mix thoroughly. Gently heat and bring to boiling. Autoclave for 15 minutes at 15 psi–121°C. Cool to 45°–50°C. Aseptically add sterile antibiotic solution. Mix thoroughly. Pour into sterile Petri dishes or distribute into sterile tubes.

Use: For selective cultivation and maintenance of *Salmonella choleraesuis*.

Antibiotic Medium 2 (Penassay Base Agar) (Base Agar)

Composition per liter:

Agar	15.0g
Pancreatic digest of gelatin	6.0g
Yeast extract	3.0g
Beef extract	1.5g

pH 6.6 ± 0.1 at 25°C

Source: This medium is available as a premixed powder from Difco Laboratories, Oxoid Unipath, and BBL Microbiology Systems.

Preparation of Medium: Add components to distilled/deionized water and bring volume to 1.0L. Mix thoroughly. Gently heat and bring to boiling. Distribute into tubes or flasks. Autoclave for 15 minutes at 15 psi–121°C. Pour into sterile Petri dishes.

Use: For use as a base layer in antibiotic assay testing. Especially useful for the plate assay of bacitracin and penicillin G.

Antibiotic Medium 3 (Penassay Broth)

Composition per liter:

Pancreatic digest of gelatin	5.0g
NaCl	3.5g
Yeast extract	1.5g
Beef extract	1.5g
Glucose	1.0g
K_2HPO_4	3.68g
KH_2PO_4	1.32g

pH 7.0 ± 0.05 at 25°C

Source: This medium is available as a premixed powder from Difco Laboratories, BBL Microbiology Systems, and Oxoid Unipath.

Preparation of Medium: Add components to distilled/deionized water and bring volume to 1.0L. Mix

thoroughly. Gently heat and bring to boiling. Distribute into tubes or flasks. Autoclave for 15 minutes at 15 psi–121°C.

Use: For antibiotic assay testing. Used for the serial dilution assay of penicillins and other antibiotics. Used in the turbidimetric assay of penicillin and tetracycline with *Staphylococcus aureus*. For the cultivation and maintenance of *Bacillus subtilis, Salmonella choleraesuis,* and *Staphylococcus aureus*.

Antibiotic Medium 3 Plus

Composition per liter:

Agar	15.0g
Peptone	5.0g
K_2HPO_4	3.68g
NaCl	3.5g
Yeast extract	2.5g
Glucose	1.75g
Beef extract	1.50g
KH_2PO_4	1.32g

pH 7.0 ± 0.05 at 25°C

Preparation of Medium: Add components to distilled/deionized water and bring volume to 1.0L. Mix thoroughly. Gently heat and bring to boiling. Distribute into tubes or flasks. Autoclave for 15 minutes at 15 psi–121°C. Pour into sterile Petri dishes or leave in tubes.

Use: For antibiotic assay testing and for the cultivation of *Escherichia coli*.

Antibiotic Medium 4 (Yeast Beef Agar) (Agar Medium C)

Composition per liter:

Agar	15.0g
Pancreatic digest of gelatin	6.0g
Yeast extract	3.0g
Beef extract	1.5g
Glucose	1.0g

pH 6.6 ± 0.05 at 25°C

Source: This medium is available as a premixed powder from BBL Microbiology Systems and Difco Laboratories.

Preparation of Medium: Add components to distilled/deionized water and bring volume to 1.0L. Mix thoroughly. Gently heat and bring to boiling. Distribute into tubes or flasks. Autoclave for 15 minutes at 15 psi–121°C. Pour into sterile Petri dishes or leave in tubes.

Use: For antibiotic assay testing.

Antibiotic Medium 5 (Streptomycin Assay Agar with Yeast Extract)

Composition per liter:

Agar	15.0g
Pancreatic digest of gelatin	6.0g
Yeast extract	3.0g
Beef extract	1.5g

pH 7.9 ± 0.1 at 25°C

Source: This medium is available as a premixed powder from BBL Microbiology Systems, Difco Laboratories and Oxoid Unipath.

Preparation of Medium: Add components to distilled/deionized water and bring volume to 1.0L. Mix thoroughly. Gently heat and bring to boiling. Distribute into tubes or flasks. Autoclave for 15 minutes at 15 psi–121°C. Pour into sterile Petri dishes.

Use: For antibiotic assay testing. For the streptomycin assay using the cylinder plate technique and *Bacillus subtilis* as test organism.

Antibiotic Medium 6

Composition per liter:

Pancreatic digest of casein	17.0g
NaCl	5.0g
Papaic digest of soybean meal	3.0g
Glucose	2.5g
K_2HPO_4	2.5g
$MnSO_4 \cdot H_2O$	0.03g

pH 7.0 ± 0.1 at 25°C

Source: This medium is available as a premixed powder from Difco Laboratories.

Preparation of Medium: Add components to distilled/deionized water and bring volume to 1.0L. Mix thoroughly. Gently heat and bring to boiling. Distribute into tubes or flasks. Autoclave for 15 minutes at 15 psi–121°C. Pour into sterile Petri dishes.

Use: For antibiotic assay testing.

Antibiotic Medium 7

Composition per liter:

Agar	15.0g
Pancreatic digest of gelatin	6.0g
Yeast extract	3.0g
Beef extract	1.5g

pH 7.0 ± 0.1 at 25°C

Preparation of Medium: Add components to distilled/deionized water and bring volume to 1.0L. Mix thoroughly. Gently heat and bring to boiling.Adjust pH to 7.0. Distribute into tubes or flasks. Autoclave for 15 minutes at 15 psi–121°C. Pour into sterile Petri dishes.

Use: For use as a base layer in antibiotic assay testing. Especially useful for the plate assay of bacitracin and penicillin G.

Antibiotic Medium 8 (Base Agar with Low pH)

Composition per liter:

Agar	15.0g
Pancreatic digest of gelatin	6.0g
Yeast extract	3.0g
Beef extract	1.5g

pH 5.9 ± 0.1 at 25°C

Source: This medium is available as a premixed powder from Difco Laboratories, and BBL Microbiology Systems.

Preparation of Medium: Add components to distilled/deionized water and bring volume to 1.0L. Mix thoroughly. Gently heat and bring to boiling. Distribute into tubes or flasks. Autoclave for 15 minutes at 15 psi–121°C. Pour into sterile Petri dishes.

Use: For antibiotic assay testing. For use as the base agar and the seed agar in the plate assay of tetracycline. For use as the seed agar in the plate assay of vancomycin, mitomycin and mithramycin.

Antibiotic Medium 9 (Polymyxin Base Agar)

Composition per liter:

Agar	20.0g
Pancreatic digest of casein	17.0g
NaCl	5.0g
Papaic digest of soybean meal	3.0g
K_2HPO_4	2.5g
Glucose	2.5g

pH 7.2 ± 0.1 at 25°C

Source: This medium is available as a premixed powder from BBL Microbiology Systems and Difco Laboratories.

Preparation of Medium: Add components to distilled/deionized water and bring volume to 1.0L. Mix thoroughly. Gently heat and bring to boiling. Distribute into tubes or flasks. Autoclave for 15 minutes at 15 psi–121°C. Pour into sterile Petri dishes.

Use: For antibiotic assay testing. For base agar for the plate assay of carbenicillin, colistimethate, and polymyxin B.

Antibiotic Medium 10 (Polymyxin Seed Agar)

Composition per liter:

Pancreatic digest of casein	17.0g
Agar	12.0g
Polysorbate 80	10.0g
NaCl	5.0g
Papaic digest of soybean meal	3.0g
K_2HPO_4	2.5g
Glucose	2.5g

pH 7.3 ± 0.2 at 25°C

Source: This medium is available as a premixed powder from BBL Microbiology Systems and Difco Laboratories.

Preparation of Medium: Add components to distilled/deionized water and bring volume to 1.0L. Mix thoroughly. Gently heat and bring to boiling. Distribute into tubes or flasks. Autoclave for 15 minutes at 15 psi–121°C. Pour into sterile Petri dishes.

Use: For antibiotic assay testing. For seed agar for the plate assay of carbenicillin, colistimethate, and polymyxin B.

Antibiotic Medium 11 (Neomycin Assay Agar)

Composition per liter:

Agar	15.0g
Pancreatic digest of gelatin	6.0g
Pancreatic digest of casein	4.0g
Yeast extract	3.0g
Beef extract	1.5g
Glucose	1.0g

pH 8.0 ± 0.1 at 25°C

Source: This medium is available as a premixed powder from BBL Microbiology Systems, Difco Laboratories and Oxoid Unipath.

Preparation of Medium: Add components to distilled/deionized water and bring volume to 1.0L. Mix thoroughly. Gently heat and bring to boiling. Distribute into tubes or flasks. Autoclave for 15 minutes at 15 psi–121°C. Pour into sterile Petri dishes.

Use: For antibiotic assay testing. For base agar and seed agar for the plate assay to test the effectiveness of neomycin sulfate, amoxicillin, ampicillin, clindamycin, cyclacillin, erythromycin, gentamycin, neomycin, oleandomycin, and sisomycin.

Antibiotic Medium 12

Composition per liter:

Agar	25.0g
Peptone	10.0g

Glucose 10.0g
NaCl 10.0g
Yeast extract...5.0g
Beef extract 2.5g

pH 6.0 ± 0.1 at 25°C

Source: This medium is available as a premixed powder from Difco Laboratories.

Preparation of Medium: Add components to distilled/deionized water and bring volume to 1.0L. Mix thoroughly. Gently heat and bring to boiling. Distribute into tubes or flasks. Autoclave for 15 minutes at 15 psi–121°C. Pour into sterile Petri dishes.

Use: For antibiotic assay effectiveness testing.

Antibiotic Medium 13 (Sabouraud Liquid Broth, Modified) (Fluid Saboraud Medium)

Composition per liter:

Glucose 20.0g
Pancreatic digest of casein 5.0g
Peptic digest of animal tissue..............................5.0g

pH 5.7 ± 0.1 at 25°C

Source: This medium is available as a premixed powder from BBL Microbiology Systems and Difco Laboratories.

Preparation of Medium: Add components to distilled/deionized water and bring volume to 1.0L. Mix thoroughly. Gently heat and bring to boiling. Distribute into tubes or flasks. Autoclave for 15 minutes at 15 psi–121°C. Pour into sterile Petri dishes.

Use: For testing the effectivness of antibiotics on yeast and molds.

Antibiotic Medium 19 (Nystatin Assay Agar)

Composition per liter:

Agar..23.5g
Glucose 10.0g
NaCl 10.0g
Pancreatic digest of gelatin 9.4g
Yeast extract...4.7g
Beef extract 2.4g

pH 6.1 ± 0.2 at 25°C

Source: This medium is available as a premixed powder from BBL Microbiology Systems and also from Difco Laboratories.

Preparation of Medium: Add components to distilled/deionized water and bring volume to 1.0L. Mix thoroughly. Gently heat and bring to boiling. Distribute into tubes or flasks. Autoclave for 15 minutes at 15 psi–121°C. Pour into sterile Petri dishes.

Use: For assaying the mycostatic activity of pharmaceutical preparations. For seed agar for the plate assay to test the effectiveness of nystatin, amphotericin B, and natamycin.

Antibiotic Medium 20

Composition per liter:

Glucose 11.0g
Pancreatic digest of casein 10.0g
Yeast extract...6.5g
Pancreatic digest of gelatin 5.0g
K_2HPO_4 ....3.68g
NaCl 3.5g
Beef extract 1.5g
KH_2PO_4 ....1.32g

pH 6.6 ± 0.2 at 25°C

Preparation of Medium: Add components to distilled/deionized water and bring volume to 1.0L. Mix thoroughly. Gently heat and bring to boiling. Distribute into tubes or flasks. Autoclave for 15 minutes at 15 psi–121°C. Pour into sterile Petri dishes.

Use: For assaying the mycostatic activity of pharmaceutical preparations.

Antibiotic Medium 21

Composition per liter:

Glucose 11.0g
Pancreatic digest of gelatin 5.0g
K_2HPO_4 ....3.68g
NaCl 3.5g
Yeast extract...1.5g
Beef extract 1.5g
KH_2PO_4 ....1.32g

pH 6.6 ± 0.2 at 25°C

Preparation of Medium: Add components to distilled/deionized water and bring volume to 1.0L. Mix thoroughly. Gently heat and bring to boiling. Distribute into tubes or flasks. Autoclave for 15 minutes at 15 psi–121°C. Pour into sterile Petri dishes.

Use: For assaying the mycostatic activity of pharmaceutical preparations.

Antibiotic Sulfonamide Sensitivity Test Agar (ASS Agar)

Composition per liter:

Agar..12.0g
Proteose peptone 10.0g

Beef extract	10.0g
NaCl	3.0g
Glucose	2.0g
Na_2HPO_4	2.0g
Sodium acetate	1.0g
Adenine	0.01g
Guanine	0.01g
Uracil	0.01g
Xanthine	0.01g

pH 7.4 ± 0.2 at 25°C

Preparation of Medium: Add components to distilled/deionized water and bring volume to 1.0L. Mix thoroughly. Gently heat and bring to boiling. Distribute into tubes or flasks. Autoclave for 15 minutes at 15 psi–121°C. Pour into sterile Petri dishes or leave in tubes.

Use: For testing the antimicrobial effectiveness of antibiotics and sulfonamides. Also, for detecting the presence of antimicrobial substances in milk, urine and other fluids.

AO Agar

Composition per liter:

Agar	11.0g
Sodium acetate	0.5g
Pancreatic digest of casein	0.5g
Yeast extract	0.5g
Beef extract	0.2g

pH 7.2 ± 0.2 at 25°C

Preparation of Medium: Add components to distilled/deionized water and bring volume to 1.0L. Mix thoroughly. Gently heat and bring to boiling. Distribute into tubes or flasks. Autoclave for 15 min at 15 psi pressure–121°C. Pour into sterile Petri dishes or leave in tubes.

Use: For the isolation and cultivation of *Cytophaga* species, *Herpetosiphon* species, *Saprospira* species, and *Flexithrix* species.

AO Agar

Composition per liter:

Agar	4.0g
Sodium acetate	0.5g
Pancreatic digest of casein	0.5g
Yeast extract	0.5g
Beef extract	0.2g

pH 7.2 ± 0.2 at 25°C

Preparation of Medium: Add components to distilled/deionized water and bring volume to 1.0L. Mix thoroughly. Gently heat and bring to boiling. Distribute into tubes or flasks. Autoclave for 15 min at 15 psi pressure–121°C. Pour into sterile Petri dishes or leave in tubes.

Use: For the maintenance of *Cytophaga* species, *Herpetosiphon* species, *Saprospira* species, and *Flexithrix* species.

AOAC Letheen Broth (Association of Official Analytical Chemists Letheen Broth)

Composition per liter:

Peptic digest of animal tissue	10.0g
Polysorbate 80	5.0g
NaCl	5.0g
Beef extract	5.0g
Lecithin	0.7g

pH 7.0 ± 0.2 at 25°C

Source: This medium is available as a premixed powder from Difco Laboratories and BBL Microbiology Systems.

Preparation of Medium: Add components to distilled/deionized water and bring volume to 1.0L. Mix thoroughly. Gently heat and bring to boiling. Distribute into tubes in 10.0mL volumes. Autoclave for 15 minutes at 15 psi–121°C.

Use: For the determination of phenol coefficients of disinfectant products containing cationic surface-active materials. Use according to *Official Methods of Analysis* of the *Association of Official Analytical Chemists* (AOAC).

Aolpha Medium

Composition per 1041mL:

NaCl	100.0g
Agar	15.0g
$MgSO_4 \cdot 7H_2O$	9.5g
$MgCl_2 \cdot 6H_2O$	5.0g
KCl	5.0g
Peptone	5.0g
Yeast extract	1.0g
$CaCl_2 \cdot 2H_2O$	0.2g
$(NH_4)_2SO_4$	0.1g
KNO_3	0.1g
Metals solution	20.0mL
Phosphate solution	20.0mL
Vitamin solution	1.0mL

pH 7.0 ± 0.2 at 25°C

Metals Solution:

Composition per liter:

$MgSO_4 \cdot 7H_2O$	29.7g
Nitrilotriacetic acid	10.0g
$CaCl_2 \cdot 2H_2O$	3.3g

$FeSO_4 \cdot 7H_2O$ 99.0mg
$Na_2MoO_4 \cdot 2H_2O$ 12.7mg
Metals "44" 50.0mL

Preparation of Metals Solution: Solubilize nitrilotriacetic acid with KOH. Dissolve remaining ingredients. Adjust pH to 7.2 with KOH or H_2SO_4. Autoclave for 15 minutes at 15 psi–121°C. Add aseptically to sterile basal medium.

Metals "44":

Composition per 100mL:

$ZnSO_4 \cdot 7H_2O$ 1.1g
$FeSO_4 \cdot 7H_2O$ 0.5g
EDTA 0.25g
$MnSO_4 \cdot 7H_2O$ 0.154g
$CuSO_4 \cdot 5H_2O$ 0.04g
$Co(NO_3)_2 \cdot 6H_2O$ 0.025g
$Na_2B_4O_7 \cdot 10H_2O$ 0.018g

Preparation of Metals "44": Add components to distilled/deionized water and bring volume to 100.0mL. Mix thoroughly. Autoclave for 15 minutes at 15 psi–121°C. Add aseptically to sterile basal medium.

Phosphate Solution:

Composition per liter:

K_2HPO_4 .... 2.5g
KH_2PO_4 .... 2.5g

Preparation of Phosphate Solution: Add components to distilled/deionized water and bring volume to 1.0L. Mix thoroughly. Autoclave for 15 minutes at 15 psi–121°C. Add aseptically to sterile basal medium.

Vitamin Solution:

Composition per liter:

Pyridoxine·HCl 10.0mg
Calcium pantothenate 5.0mg
Nicotinamide 5.0mg
Riboflavin 5.0mg
Thiamine·HCl 5.0mg
Biotin 2.0mg
Folic acid 2.0mg
Cyanocobalamin 0.1mg

Preparation of Vitamin Solution: Add components to distilled/deionized water and bring volume to 1.0L. Mix thoroughly. Filter sterilize and add aseptically to sterile basal medium.

Preparation of Medium: Add components—except metals "44", phosphate solution and vitamin solution—to distilled/deionized water and bring volume to 1.0L. Mix thoroughly. Gently heat and bring to boiling. Adjust pH of basal medium to 7.0. Autoclave for 15 minutes at 15 psi–121°C. Cool to 50°C and aseptically add the metals "44", phosphate and vitamin solutions.

Use: For the cultivation and maintenance of *Halomonas meridiana* and other *Halomonas* species.

Aphanomyces Synthetic Medium

Composition per liter:

D-Glucose 5.0g
KH_2PO_4 .... 2.0g
L-Asparagine 0.75g
$MgCl_2$ 0.05g
$FeCl_3$ 5.0mg
$MnCl_2$ 5.0mg
$ZnCl_2$ 5.0mg
L-Methionine 0.02mg

pH 5.5 ± 0.2 at 25°C

Preparation of Medium: Add components to distilled/deionized water and bring volume to 1.0L. Mix thoroughly. Autoclave for 15 min at 15 psi pressure–121°C. Adjust pH to 5.5. Aseptically distribute into sterile tubes or flasks.

Use: For the cultivation of *Aphanomyces* species.

Aplanobacterium Medium

Composition per liter:

Agar 20.0g
Glucose 10.0g
Peptone 5.0g
Yeast extract 5.0g

pH 7.2 ± 0.2 at 25°C

Preparation of Medium: Add components to distilled/deionized water and bring volume to 1.0L. Mix thoroughly. Gently heat and bring to boiling. Autoclave for 15 minutes at 15 psi–121°C. Pour into sterile Petri dishes or leave in tubes.

Use: For the cultivation and maintenance of *Xanthomonas* species.

Apple Juice Yeast Extract Medium (AJYE Medium)

Composition per 1200mL:

Agar 30.0g
Yeast extract 10.0g
Apple juice 1.0L

pH 4.8 ± 0.2 at 25°C

Preparation of Medium: Add yeast extract to 1.0L of apple juice. Mix thoroughly. Adjust pH to 4.8. Autoclave for 10 min at 9 psi pressure–114°C. Cool to 45°–50°C. In a separate flask, add agar to

200.0mL of distilled/deionized water and bring volume to 1.0L. Mix thoroughly. Gently heat and bring to boiling. Autoclave for 15 min at 15 psi pressure–121°C. Cool to 45°–50°C. Aseptically combine the sterile apple juice solution with the sterile agar solution. Mix thoroughly. Pour into sterile Petri dishes.

Use: For the cultivation of *Zymomonas* species.

APT Agar

Composition per liter:

Agar	15.0g
Pancreatic digest of casein	12.5g
Glucose	10.0g
Yeast extract	7.5g
NaCl	5.0g
K_2HPO_4	5.0g
Sodium citrate	5.0g
Na_2CO_3	1.25g
$MnCl_2 \cdot 4H_2O$	0.14g
$MgSO_4 \cdot 7H_2O$	0.8g
Polysorbate 80	0.2g
$FeSO_4 \cdot 7H_2O$	0.04g
Thiamine·HCl	1.0mg

pH 6.7 ± 0.2 at 25°C

Source: This medium is available as a premixed powder from Difco Laboratories

Preparation of Medium: Add components to distilled/deionized water and bring volume to 1.0L. Mix thoroughly. Gently heat and bring to boiling. Distribute into tubes or flasks. Autoclave for 15 minutes at 13 psi–118°–121°C. Pour into sterile Petri dishes or leave in tubes.

Use: For the cultivation and enumeration of bacteria, especially heterofermentative lactobacilli, from meat and other foods. Also, for the cultivation of streptococci.

APT Agar

Composition per liter:

Agar	13.5g
Pancreatic digest of casein	10.0g
Glucose	10.0g
Yeast extract	7.5g
NaCl	5.0g
KH_2PO_4	5.0g
Sodium citrate	5.0g
Na_2CO_3	1.25g
$MnCl_2 \cdot 4H_2O$	0.14g
$MgSO_4 \cdot 7H_2O$	0.8g
Polysorbate 80	0.2g
$FeSO_4 \cdot 7H_2O$	0.04g

pH 6.7 ± 0.2 at 25°C

Source: This medium is available as a premixed powder from BBL Microbiology Systems.

Preparation of Medium: Add components to distilled/deionized water and bring volume to 1.0L. Mix thoroughly. Gently heat and bring to boiling. Distribute into tubes or flasks. Autoclave for 15 minutes at 13 psi–118°–121°C. Pour into sterile Petri dishes or leave in tubes.

Use: For the cultivation and enumeration of bacteria, especially heterofermentative lactobacilli, from meat and other foods. Also, for the cultivation of streptococci.

APT Broth

Composition per liter:

Pancreatic digest of casein	12.5g
Glucose	10.0g
Yeast extract	7.5g
NaCl	5.0g
K_2HPO_4	5.0g
Sodium citrate	5.0g
$MgSO_4 \cdot 7H_2O$	0.8g
$MnCl_2 \cdot 4H_2O$	0.14g
$FeSO_4 \cdot 7H_2O$	0.04g
Polysorbate 80	0.2g
Na_2CO_3	1.25g
Thiamine·HCl	1.0mg

pH 7.7 ± 0.2 at 25°C

Source: This medium is available as a premixed powder from Difco Laboratories.

Preparation of Medium: Add components to distilled/deionized water and bring volume to 1.0L. Mix thoroughly. Gently heat and bring to boiling. Distribute into tubes or flasks. Autoclave for 15 minutes at 15 psi–118°–121°C.

Use: For the cultivation of lactic acid bacteria. Use for the cultivation of heterofermentative lactobacilli from meat and other foods. The American Public Health Association recommends adding 100μg/L of thiamine to this medium.

APT Broth

Composition per liter:

Pancreatic digest of casein	10.0g
Glucose	10.0g
Yeast extract	7.5g
NaCl	5.0g
KH_2PO_4	5.0g
Sodium citrate	5.0g
$MgSO_4 \cdot 7H_2O$	0.8g
$MnCl_2 \cdot 4H_2O$	0.14g

$FeSO_4 \cdot 7H_2O$ 0.04g
Polysorbate 80 .... 0.2g
Na_2CO_3 .... 1.25g
pH 7.7 ± 0.2 at 25°C

Source: This medium is available as a premixed powder from BBL Microbiology Systems.

Preparation of Medium: Add components to distilled/deionized water and bring volume to 1.0L. Mix thoroughly. Gently heat and bring to boiling. Distribute into tubes or flasks. Autoclave for 15 minutes at 15 psi–118°–121°C.

Use: For the cultivation of lactic acid bacteria. Use for the cultivation of heterofermentative lactobacilli from meat and other foods. The American Public Health Association recommends adding 100μg/L of thiamine to this medium.

Aquaspirillum Autotrophic Agar

Composition per liter:
Noble agar 15.0g
$Na_2HPO_4 \cdot 12H_2O$ 9.0g
KH_2PO_4 .... 1.5g
NH_4Cl 1.0g
$MgSO_4 \cdot 7H_2O$ 0.2g
$CaCl_2 \cdot 2H_2O$ 0.01g
Ferric ammonium citrate 5.0mg
$NaHCO_3$ solution 10.0mL
Trace elements solution 3.0mL
pH 7.1 ± 0.2 at 25°C

$NaHCO_3$ Solution:
Composition per 10mL:
$NaHCO_3$ 0.5g

Preparation of $NaHCO_3$ Solution: Add the $NaHCO_3$ to distilled/deionized water and bring volume to 10.0mL. Mix thoroughly. Filter sterilize.

Trace Elements Solution:
Composition per liter:
H_3BO_3 .... 30.0mg
$CoCl_2 \cdot 6H_2O$ 20.0mg
$ZnSO_4 \cdot 7H_2O$ 10.0mg
$MnCl_2 \cdot 4H_2O$ 3.0mg
$Na_2MoO_4 \cdot 2H_2O$ 3.0mg
$NiCl_2 \cdot 6H_2O$ 2.0mg
$CuCl_2 \cdot 2H_2O$ 1.0mg

Preparation of Trace Elements Solution: Add components to distilled/deionized water and bring volume to 1.0L. Mix thoroughly.

Preparation of Medium: Add components, except $NaHCO_3$ solution, to double distilled water and bring volume to 990.0mL. Mix thoroughly. Gently heat and bring to boiling. Autoclave for 15 minutes at 15 psi–121°C. Cool to 45°–50°C. Aseptically add sterile $NaHCO_3$ solution. Mix thoroughly. Pour into sterile Petri dishes or distribute into sterile tubes. For autotrophic growth, incubate under 85% H_2 + 10% CO_2 + 5% O_2.

Use: For the autotrophic cultivation and maintenance of *Aquaspirillum autotrophicum.*

Aquaspirillum Autotrophic Broth

Composition per liter:
$Na_2HPO_4 \cdot 12H_2O$ 9.0g
KH_2PO_4 .... 1.5g
NH_4Cl 1.0g
$MgSO_4 \cdot 7H_2O$ 0.2g
$CaCl_2 \cdot 2H_2O$ 0.01g
Ferric ammonium citrate 5.0mg
$NaHCO_3$ solution 10.0mL
Trace elements solution 3.0mL
pH 7.1 ± 0.2 at 25°C

$NaHCO_3$ Solution:
Composition per 10mL:
$NaHCO_3$ 0.5g

Preparation of $NaHCO_3$ Solution: Add the $NaHCO_3$ to distilled/deionized water and bring volume to 10.0mL. Mix thoroughly. Filter sterilize.

Trace Elements Solution:
Composition per liter:
H_3BO_3 .... 30.0mg
$CoCl_2 \cdot 6H_2O$ 20.0mg
$ZnSO_4 \cdot 7H_2O$ 10.0mg
$MnCl_2 \cdot 4H_2O$ 3.0mg
$Na_2MoO_4 \cdot 2H_2O$ 3.0mg
$NiCl_2 \cdot 6H_2O$ 2.0mg
$CuCl_2 \cdot 2H_2O$ 1.0mg

Preparation of Trace Elements Solution: Add components to distilled/deionized water and bring volume to 1.0L. Mix thoroughly.

Preparation of Medium: Add components, except $NaHCO_3$ solution, to double distilled water and bring volume to 990.0mL. Mix thoroughly. Gently heat and bring to boiling. Autoclave for 15 minutes at 15 psi–121°C. Cool to 45°–50°C. Aseptically add sterile $NaHCO_3$ solution. Mix thoroughly. Aseptically distribute into sterile tubes or flasks. To grow autotrophically, incubate under 85% H_2 + 10% CO_2 + 5% O_2.

Use: For the autotrophic cultivation of *Aquaspirillum autotrophicum.*

Aquaspirillum Heterotrophic Agar

Composition per liter:
Noble agar 15.0g
$Na_2HPO_4 \cdot 12H_2O$ 9.0g

KH_2PO_4	1.5g
NH_4Cl	1.0g
Sodium succinate	1.0g
$MgSO_4 \cdot 7H_2O$	0.2g
$CaCl_2 \cdot 2H_2O$	0.01g
Ferric ammonium citrate	5.0mg
Trace elements solution	3.0mL

pH 7.1 ± 0.2 at 25°C

Trace Elements Solution:
Composition per liter:

H_3BO_3	30.0mg
$CoCl_2 \cdot 6H_2O$	20.0mg
$ZnSO_4 \cdot 7H_2O$	10.0mg
$MnCl_2 \cdot 4H_2O$	3.0mg
$Na_2MoO_4 \cdot 2H_2O$	3.0mg
$NiCl_2 \cdot 6H_2O$	2.0mg
$CuCl_2 \cdot 2H_2O$	1.0mg

Preparation of Trace Elements Solution: Add components to distilled/deionized water and bring volume to 1.0L. Mix thoroughly.

Preparation of Medium: Add components to double distilled water and bring volume to 1.0L. Mix thoroughly. Gently heat and bring to boiling. Autoclave for 15 minutes at 15 psi–121°C. Pour into sterile Petri dishes or distribute into sterile tubes.

Use: For the heterotrophic cultivation and maintenance of *Aquaspirillum autotrophicum.*

Aquaspirillum Heterotrophic Broth

Composition per liter:

$Na_2HPO_4 \cdot 12H_2O$	9.0g
KH_2PO_4	1.5g
NH_4Cl	1.0g
Sodium succinate	1.0g
$MgSO_4 \cdot 7H_2O$	0.2g
$CaCl_2 \cdot 2H_2O$	0.01g
Ferric ammonium citrate	5.0mg
Trace elements solution	3.0mL

pH 7.1 ± 0.2 at 25°C

Trace Elements Solution:
Composition per liter:

H_3BO_3	30.0mg
$CoCl_2 \cdot 6H_2O$	20.0mg
$ZnSO_4 \cdot 7H_2O$	10.0mg
$MnCl_2 \cdot 4H_2O$	3.0mg
$Na_2MoO_4 \cdot 2H_2O$	3.0mg
$NiCl_2 \cdot 6H_2O$	2.0mg
$CuCl_2 \cdot 2H_2O$	1.0mg

Preparation of Trace Elements Solution: Add components to distilled/deionized water and bring volume to 1.0L. Mix thoroughly.

Preparation of Medium: Add components to double distilled water and bring volume to 1.0L. Mix thoroughly. Gently heat and bring to boiling. Autoclave for 15 minutes at 15 psi–121°C. Aseptically distribute into sterile tubes or flasks.

Use: For the heterotrophic cultivation of *Aquaspirillum autotrophicum.*

Archaeoglobus Medium

Composition per liter:

NaCl	18.0g
$NaHCO_3$	5.0g
$MgCl_2 \cdot 6H_2O$	4.0g
$MgSO_4 \cdot 7H_2O$	3.45g
Sodium L-lactate	1.5g
Yeast extract	0.5g
KCl	0.34g
NH_4Cl	0.25g
$CaCl_2 \cdot 2H_2O$	0.14g
K_2HPO_4	0.14g
$Fe(NH_4)_2(SO_4)_2 \cdot 7H_2O$	2.0mg
Resazurin	1.0mg
$Na_2S \cdot 9H_2O$ solution	25.0mL
Trace elements solution	10.0mL

pH 6.9 ± 0.2 at 25°C

Trace Elements Solution:
Composition per liter:

$MgSO_4 \cdot 7H_2O$	3.0g
Nitrilotriacetic acid	1.5g
NaCl	1.0g
$MnSO_4 \cdot 2H_2O$	0.5g
$CoSO_4 \cdot 7H_2O$	0.18g
$ZnSO_4 \cdot 7H_2O$	0.18g
$FeSO_4 \cdot 7H_2O$	0.1g
$CaCl_2 \cdot 2H_2O$	0.1g
$NiCl_2 \cdot 6H_2O$	0.025g
$KAl(SO_4)_2 \cdot 12H_2O$	0.02g
$CuSO_4 \cdot 5H_2O$	0.01g
H_3BO_3	0.01g
$Na_2MoO_4 \cdot 2H_2O$	0.01g
$Na_2SeO_3 \cdot 5H_2O$	0.3mg

$Na_2S \cdot 9H_2O$ Solution:
Composition per 50mL:

$Na_2S \cdot 9H_2O$	1.0g

Preparation of $Na_2S \cdot 9H_2O$ Solution: Prepare and dispense solution anaerobically under with 80% N_2 + 20% CO_2. Add $Na_2S \cdot 9H_2O$ to distilled/deionized water and bring volume to 50.0mL. Mix thoroughly. Adjust pH to 7.0. Autoclave for 15 minutes at 15 psi–121°C.

Preparation of Trace Elements Solution: Add nitrilotriacetic acid to approximately 500.0mL distilled/deionized water. Dissolve by adding KOH and adjust pH to 6.5. Add remaining components. Bring volume to 1.0L with additional distilled/deionized water. Adjust pH to 7.0 with KOH.

Preparation of Medium: Add components, except $NaHCO_3$ and $Na_2S{\cdot}9H_2O$ solution to distilled/deionized water and bring volume to 1.0L. Mix well and heat to boiling for a few minutes. Cool rapidly to room temperature while gassing with 80% N_2 + 20% CO_2. Add $NaHCO_3$ and adjust pH to 6.9. Distribute anaerobically under 80% N_2 + 20% CO_2 and pressurize sealed containers up to 2 bar pressure. Autoclave for 15 minutes at 15 psi–121°C. Prior to inoculation of cultures, add 0.25mL of sterile $Na_2S{\cdot}9H_2O$ solution to each tube containing 9.75mL of sterile basal medium.

Use: For the cultivation and maintenance of *Archaeoglobus fulgidus*.

Archangium violaceum Medium

Composition per liter:

Monosodium glutamate	1.0g
L-Leucine	0.50g
L-Tyrosine	0.50g
L-Isoleucine	0.30g
L-Proline	0.25g
$MgSO_4{\cdot}7H_2O$	0.20g
L-Lysine	0.15g
L-Arginine	0.10g
L-Asparagine	0.10g
L-Serine	0.10g
L-Threonine	0.10g
L-Valine	0.10g
L-Alanine	0.05g
L-Glycine	0.05g
L-Histidine	0.05g
L-Methionine	0.05g
$Ca_3(PO_4)_2$	0.02g
KCl	0.02g
Tris(hydroxymethyl)aminomethane buffer (0.02M solution, pH 7.5)	1.0L

pH 7.5 ± 0.2 at 25°C

Preparation of Medium: Add solid components to 1.0L of Tris buffer. Mix thoroughly. Filter sterilize. Aseptically distribute into tubes or flasks.

Use: For the cultivation of *Archangium violaceum*.

Arenavirus Plaquing Medium

Composition per liter:

Eagle basal medium	1.0L
Agarose solution	1.0L
Fetal calf serum, inactivated	100.0mL

pH 7.0 ± 0.2 at 25°C

Eagle's Basal Medium:
Composition per liter:

HEPES (N-2-Hydroxyethylpiperazine-N′-2-ethanesulfonic acid) buffer	9.53g
NaCl	6.8g
$NaHCO_3$	2.2g
Glucose	1.0g
KCl	0.4g
$CaCl_2{\cdot}2H_2O$	0.2g
NaH_2PO_4	0.125g
$MgSO_4{\cdot}7H_2O$	0.1g
L-Isoleucine	0.026g
L-Leucine	0.026g
L-Lysine	0.026g
L-Threonine	0.024g
L-Valine	0.0235g
L-Tyrosine	0.018g
L-Arginine	0.0174g
L-Phenylalanine	0.0165g
L-Cystine	0.012g
L-Histidine	8.0mg
L-Methionine	7.5mg
L-Tryptophan	4.0mg
Inositol	1.8mg
Biotin	1.0mg
Calcium pantothenate	1.0mg
Choline chloride	1.0mg
Folic acid	1.0mg
Nicotinamide	1.0mg
Pyridoxal·HCl	1.0mg
Thiamine·HCl	1.0mg
Riboflavin	0.1mg

pH 7.2–7.4 at 25°C

Source: This medium is available as a premixed powder from Difco Laboratories.

Preparation of Eagle's Basal Medium: Add components to distilled/deionized water and bring volume to 1.0L. Mix thoroughly. Adjust pH to 7.0 with NaOH. Filter sterilize.

Agarose Solution:
Composition per liter:

Agarose	20.0g

Preparation of Agarose Solution: Add agarose to distilled/deionized water and bring volume to 1.0L. Mix thoroughly. Autoclave for 15 min at 15 psi pressure–121°C. Cool to 45°–50°C.

Preparation of Medium: To 1.0L of sterile Eagle basal medium, aseptically add 1.0L of sterile, cooled agarose solution and 100.0mL of fetal calf serum. Mix thoroughly. Pour into sterile Petri dishes.

Use: For the cultivation of animal tissue culture cells used for the growth of arenaviruses.

Arginine Assay Medium
See: **Amino Acid Assay Medium**

Arginine Glucose Slants (AGS)

Composition per liter:

NaCl	20.0g
Agar	13.5g
Pancreatic digest of casein	10.0g
L-Arginine·HCl	5.0g
Peptone	5.0g
Yeast extract	3.0g
Glucose	1.0g
Ferric ammonium citrate	0.5g
$Na_2S_2O_3 \cdot 5H_2O$	0.3g
Bromcresol Purple	0.02g

pH 6.8-7.0 at 25°C

Preparation of Medium: Add components to distilled/deionized water and bring volume to 1.0L. Mix thoroughly. Gently heat and bring to boiling. Distribute into tubes. Autoclave for 12 min at 15 psi pressure–121°C. Allow tubes to cool in a slanted position.

Use: For the cultivation and differentation of *Vibrio* species.

Armstrong *Fusarium* Medium

Composition per liter:

Glucose	20.0g
$Ca(NO_3)_2 \cdot 4H_2O$	8.4g
KH_2PO_4	1.09g
KCl	0.22g
$FeCl_3$	0.2μg
$MnSO_4$	0.2μg
$ZnSO_4$	0.2μg

Preparation of Medium: Add components to distilled/deionized water and bring volume to 1.0L. Mix thoroughly. Filter sterilize.

Use: For the cultivation of *Fusarium* species.

Arthrobacter Medium

Composition per liter:

Agar	10.0g
Casein	1.0g
Glucose	1.0g
K_2HPO_4	1.0g
Yeast extract	0.7g
$MgSO_4 \cdot 7H_2O$	0.25g
$(NH_4)_2SO_4$	0.25g

pH 6.9–7.0 at 25°C

Preparation of Medium: Add components to tap water and bring volume to 1.0L. Mix thoroughly. Gently heat and bring to boiling. Autoclave for 15 min at 15 psi pressure–121°C. Pour into sterile Petri dishes.

Use: For the isolation, cultivation and enumeration of *Arthrobacter* species from soil.

Arthrobacter Medium

Composition per liter:

Agar	15.0g
Peptone	10.0g
Yeast extract	10.0g
K_2HPO_4	2.0g
Rhodotorulic acid (δ-*N*-acetyl-L-ornithine) or desferal	20.0μg

pH 7.4 ± 0.2 at 25°C

Preparation of Medium: Add components to distilled/deionized water and bring volume to 1.0L. Mix thoroughly. Gently heat and bring to boiling. Distribute into tubes or flasks. Autoclave for 15 minutes at 15 psi–121°C. Pour into sterile Petri dishes or leave in tubes.

Use: For the cultivation and maintenance of *Aureobacterium flavescens.*

Arthrobacter YCWD

Composition per liter:

Pancreatic digest of casein	10.0g
Yeast extract	1.0g

pH 7.2 ± 0.2 at 25°C

Preparation of Medium: Add components to distilled/deionized water and bring volume to 1.0L. Mix thoroughly. Gently heat and bring to boiling. Distribute into tubes or flasks. Autoclave for 15 minutes at 15 psi–121°C.

Use: For the cultivation and maintenance of *Arthrobacter* species.

Artificial Deep Lake Medium

Composition per liter:

NaCl	180.0g
$MgCl_2 \cdot 6H_2O$	75.0g
Noble agar	15.0g
Sodium succinate	10.0g
$MgSO_4 \cdot 7H_2O$	7.4g
KCl	7.4g
$CaCl_2 \cdot 2H_2O$	1.0g
Yeast extract	1.0g
Vitamin solution	10.0mL

pH 7.4 ± 0.2 at 25°C

Vitamin Solution:

Composition per liter:

Biotin	30.0mg
Cyanocobalamin	20.0mg
Thiamine·HCl	10.0mg

Preparation of Vitamin Solution: Add components to distilled/deionized water and bring volume to 1.0L. Mix thoroughly. Filter sterilize and add aseptically to sterile basal medium.

Preparation of Medium: Add components, except vitamin solution, to distilled/deionized water and bring volume to 990.0mL. Mix thoroughly. Gently heat and bring to boiling. Adjust medium to pH 7.4. Autoclave for 15 minutes at 15 psi–121°C. Cool to 50°C. Aseptically add 10.0mL of vitamin solution. Pour into sterile Petri dishes or leave in tubes.

Use: For the cultivation and maintenance *Halobacterium lacusprofundi*.

Arylsulfatase Agar (Wayne Sulfatase Agar)

Composition per liter:

Agar	15.0g
Na_2HPO_4	2.5g
L-Asparagine	1.0g
KH_2PO_4	1.0g
K_2HPO_4	1.0g
Trisodium phenolphthalein sulfate	0.65g
Pancreatic digest of casein	0.5g
Ferric ammonium citrate	0.05g
$MgSO_4 \cdot 7H_2O$	0.01g
$CaCl_2 \cdot 2H_2O$	0.5mg
$ZnSO_4 \cdot 7H_2O$	0.1mg
$CuSO_4$	0.1mg
Glycerol	10.0mL

pH 7.0 ± 0.2 at 25°C

Source: This medium is available as a premixed powder from BBL Microbiology Systems.

Preparation of Medium: Add glycerol to approximately 800.0mL of distilled/deionized water. Mix thoroughly. Add remaining components and bring volume to 1.0L with distilled/deionized water. Mix thoroughly. Gently heat and bring to boiling. Distribute into tubes. Autoclave for 15 minutes at 15 psi–121°C. Cool tubes in an upright position.

Use: For the biochemical differentiation of species of *Mycobacterium*. Inoculate tubes with *Mycobacterium* cultures and incubate aerobically at 35°C for 3–14 days. Add 0.5–1.0mL of 2*N* Na_2CO_3 to each tube and observe color change within 30 minutes. Development of a pink color is indicative of *M. fortuitum* or *M. chelonae*. *M. tuberculosis* gives a negative reaction.

Ashbey's Nitrogen–Free Agar

Composition per liter:

Agar	15.0g
Mannitol	15.0g
$CaCl_2 \cdot 2H_2O$	0.2g
K_2HPO_4	0.2g
$MgSO_4 \cdot 7H_2O$	0.2g
MoO_3 (10% solution)	0.1mL
$FeCl_3$ (10% solution	0.05mL

pH 7.2 ± 0.2 at 25°C

Preparation of Medium: Add components to distilled/deionized water and bring volume to 1.0L. Mix thoroughly. Gently heat and bring to boiling. Distribute into tubes or flasks. Autoclave for 15 min at 15 psi pressure–121°C. Pour into sterile Petri dishes or leave in tubes.

Use: For the isolation and cultivation of bacteria, such as *Azotobacter* species and cyanobacteria, that can utilize atmospheric N_2 as sole nitrogen source.

Ascospore Agar

Composition per liter :

Potassium acetate	30.0g
Yeast extract	2.5g
Glucose	1.0g

pH 6.4 ± 0.2 at 25°C

Preparation of Medium: Add components to distilled/deionized water and bring volume to 1.0L. Mix thoroughly. Gently heat and bring to boiling. Distribute into tubes or flasks. Autoclave for 15 minutes at 15 psi–121°C. Pour into sterile Petri dishes or leave in tubes.

Use: For the enrichment of ascosporogenous yeasts and their production of ascospores.

ASLA Agar

Composition per liter:

Sodium lactate	20.0g
Davis agar	10.0g
$(NH_4)_2SO_4$	3.0g
Na_2HPO_4	1.2g
Cysteine·HCl	0.5g
$MgSO_4 \cdot 7H_2O$	0.2g
$MnSO_4 \cdot 4H_2O$	0.05g
$FeSO_4 \cdot 7H_2O$	0.04g
Vitamin solution	10.0mL

pH 6.5 ± 0.2 at 25°C

Vitamin Solution:

Composition per liter:

Biotin	0.1g
Calcium pantothenate	0.1g

p-Aminobenzoic acid 0.1g
Thiamine 0.1g

Preparation of Vitamin Solution: Add components to distilled/deionized water and bring volume to 1.0L. Mix thoroughly. Filter sterilize. Store solution at –20°C.

Preparation of Medium: Add components, except vitamin solution, to distilled/deionized water and bring volume to 990.0mL. Mix thoroughly. Gently heat and bring to boiling. Autoclave for 15 min at 15 psi pressure–121°C. Cool to 45°–50°C. Aseptically add 10.0mL of sterile vitamin solution. Mix thoroughly. Pour into sterile Petri dishes or distribute into sterile tubes.

Use: For the selective isolation and cultivation of most *Propionibacterium* species from foods.

ASN-III Agar

Composition per liter:

NaCl 25.0g
$MgSO_4 \cdot 7H_2O$ 3.5g
$MgCl_2 \cdot 6H_2O$ 2.0g
$NaNO_3$ 0.75g
$K_2HPO_4 \cdot 3H_2O$ 0.75g
$CaCl_2 \cdot 2H_2O$ 0.5g
KCl 0.5g
Na_2CO_3 .... 0.02g
Citric acid 3.0mg
Ferric ammonium citrate 3.0mg
Magnesium EDTA 0.5mg
Vitamin B_{12} 10.0μg
Agar solution 100.0mL
A-5 Trace metals 1.0mL

pH 7.3 ± 0.2 at 25°C

Agar Solution:
Composition per 100mL:

Noble agar 10.0g

Preparation of Agar Solution: Add agar to glass distilled water and bring volume to 100.0mL. Mix thoroughly. Gently heat and bring to boiling. Autoclave for 15 min at 15 psi pressure–121°C. Cool to 45°–50°C.

A-5 Trace Metals:
Composition per liter:

H_3BO_3 .... 2.86g
$MnCl_2 \cdot 4H_2O$ 1.81g
$ZnSO_4 \cdot 7H_2O$ 0.222g
$CuSO_4 \cdot 5H_2O$ 0.079g
$Co(NO_3)_2 \cdot 6H_2O$ 0.049g
$Na_2MoO_4 \cdot 2H_2O$ 0.039g

Preparation of A-5 Trace Metals: Add components to distilled/deionized water and bring volume to 1.0L. Mix thoroughly.

Preparation of Medium: Add components, except agar solution, to glass distilled water and bring volume to 900.0mL. Mix well and heat gently until dissolved. Filter sterilize. Warm to 45°–50°C. Aseptically add agar solution. Mix thoroughly. Pour into sterile Petri dishes or distribute into sterile tubes.

Use: For the cultivation of *Xenococcus* species. Also, for the isolation of cyanobacteria from marine habitats.

ASN-III Broth

Composition per liter:

NaCl 25.0g
$MgSO_4 \cdot 7H_2O$ 3.5g
$MgCl_2 \cdot 6H_2O$ 2.0g
$NaNO_3$ 0.75g
$K_2HPO_4 \cdot 3H_2O$ 0.75g
$CaCl_2 \cdot 2H_2O$ 0.5g
KCl 0.5g
Na_2CO_3 .... 0.02g
Citric acid 3.0mg
Ferric ammonium citrate 3.0mg
Magnesium EDTA 0.5mg
Vitamin B_{12} 10.0μg
A-5 Trace metals 1.0mL

pH 7.3 ± 0.2 at 25°C

A-5 Trace Metals:
Composition per liter:

H_3BO_3 .... 2.86g
$MnCl_2 \cdot 4H_2O$ 1.81g
$ZnSO_4 \cdot 7H_2O$ 0.222g
$CuSO_4 \cdot 5H_2O$ 0.079g
$Co(NO_3)_2 \cdot 6H_2O$ 0.049g
$Na_2MoO_4 \cdot 2H_2O$ 0.039g

Preparation of A-5 Trace Metals: Add components to distilled/deionized water and bring volume to 1.0L. Mix thoroughly.

Preparation of Medium: Add components to glass distilled water and bring volume to 1.0L. Mix well and heat gently until dissolved. Filter sterilize.

Use: For the cultivation of *Xenococcus* species. Also, for the isolation of cyanobacteria from marine habitats.

Asparaginate Glycerol Agar

Composition per liter:

Agar 15.0g
Sodium asparaginate 1.0g
K_2HPO_4 .... 1.0g
Glycerol 10.0mL

pH 7.0 ± 0.2 at 25°C

Preparation of Medium: Add components to distilled/deionized water and bring volume to 1.0L. Mix thoroughly. Gently heat and bring to boiling. Distribute into tubes or flasks. Autoclave for 15 minutes at 15 psi–121°C. Pour into sterile Petri dishes or leave in tubes.

Use: For the cultivation and maintenance of *Nocardia transvalensis*.

Asparagine Broth

Composition per liter:

DL-Asparagine	30g
K_2HPO_4	1.0g
$MgSO_4 \cdot 7H_2O$	0.5g

pH 6.9–7.2 at 25°C

Preparation of Medium: Add components to distilled/deionized water and bring volume to 1.0L. Mix well until dissolved. Adjust pH to between 6.9 and 7.2. Distribute into tubes or flasks. Autoclave for 15 min at 15 psi pressure–121°C.

Use: For a presumptive test medium in the differentiation of nonfermentative Gram-negative bacteria, especially *Pseudomonas aeruginosa*. For use in the multiple tube technique in the microbiological analysis of recreational waters.

Aspergillus Differential Medium

Composition per liter:

Agar	15.0g
Pancreatic digest of casein	15.0g
Yeast extract	10.0g
Ferric citrate	0.5g

Preparation of Medium: Add components to distilled/deionized water and bring volume to 1.0L. Mix thoroughly. Gently heat and bring to boiling. Distribute into tubes in 7.0mL volumes. Autoclave for 15 min at 15 psi pressure–121°C. Allow tubes to cool in a slanted position.

Use: For the cultivation and differentiation of *Aspergillus flavus*. *A. flavus* appears as bright orange colonies.

Aspergillus flavus/parasiticus Agar Base

See: **AFPA Base**

ASS Agar

See: **Antibiotic Sulfonamide Sensitivity Test Agar**

Association of Official Analytical Chemists Letheen Broth

See: **AOAC Letheen Broth**

Asticcacaulis Medium

Composition per liter:

Agar	15.0g
Pancreatic digest of casein	0.5g
Yeast extract	0.5g
Sodium acetate	0.2g

pH 6.8 ± 0.2 at 25°C

Preparation of Medium: Add components to distilled/deionized water and bring volume to 1.0L. Mix thoroughly. Gently heat and bring to boiling. Distribute into tubes or flasks. Autoclave for 15 min at 15 psi pressure–121°C. Pour into sterile Petri dishes or leave in tubes.

Use: For the isolation and cultivation of *Asticcacaulis* species.

ASTM Nutrient Salts Agar (American Society for Testing and Materials Nutrient Salts Agar)

Composition per liter:

Agar	15.0g
KH_2PO_4	0.7g
K_2HPO_4	0.7g
$MgSO_4 \cdot 7H_2O$	0.7g
NH_4NO_3	1.0g
NaCl	5.0mg
$FeSO_4 \cdot 7H_2O$	2.0mg
$ZnSO_4$	2.0mg
$MnSO_4 \cdot H_2O$	1.0mg

pH 6.5 ± 0.2 at 25°C

Preparation of Medium: Add components to tap water and bring volume to 1.0L. Mix thoroughly. Gently heat and bring to boiling. Distribute into tubes or flasks. Autoclave for 15 min at 15 psi pressure–121°C. Pour into sterile Petri dishes or leave in tubes.

Use: For determination of susceptibility of plastics to fungal degradation.

AT5N Medium

Composition per liter:

$CaCO_3$	10.0g
$(NH_4)_2SO_4$	1.5g
K_2HPO_4	0.5g
$MgSO_4$	50.0mg

$KHCO_3$ 30.0mg
$CaCl_2 \cdot 2H_2O$ 20.0mg

Preparation of Medium: Add components to tap water and bring volume to 1.0L. Mix thoroughly. Gently heat and bring to boiling. Distribute into tubes or flasks. Autoclave for 15 min at 15 psi pressure–121°C.

Use: For the cultivation of bacteria that oxidize ammonia, especially those from wastewater.

Atlas Oil Agar

Composition per liter:

Bushnell-Haas Agar 990.0mL
Oil 10.0mL
pH 7.0 ± 0.2 at 25°C

Bushnell-Haas Agar:
Composition per 990mL:

Agar 15.0g
KH_2PO_4 .... 1.0g
K_2HPO_4 .... 1.0g
NH_4NO_3 .... 1.0g
$MgSO_4 \cdot 7H_2O$ 0.2g
$FeCl_3$ 0.05g
$CaCl_2 \cdot 2H_2O$ 0.02g

Preparation of Bushnell-Haas Agar: Add components to distilled/deionized water and bring volume to 990.0mL. Mix thoroughly. Gently heat and bring to boiling. Autoclave for 15 min at 15 psi pressure–121°C. Cool to 60°C.

Preparation of Medium: Filter sterilize oil. Aseptically add 10.0mL of sterile oil to 990.0mL of cooled, sterile Bushnell-Haas Agar. Put mixture into a sterile blender container. Blend on low speed to minimize the incorporation of air into the medium. Pour into sterile Petri dishes.

Use: For the cultivation and enumeration of hydrocarbon-utilizing bacteria by direct plating of water and sediment samples.

ATS Medium (American Trudeau Society Medium)

Composition per liter:

Potato 20.0g
Malachite green 0.2g
Egg yolk emulsion 500.0mL
Glycerol 10.0mL
pH 6.5–7.0 at 25°C

Source: Available as a prepared medium from Difco Laboratories and BBL Microbiology Systems.

Egg Yolk Emulsion:
Composition:

Chicken egg yolks 11
Whole chicken egg 1

Preparation of Egg Yolk Emulsion: Soak eggs with 1:100 dilution of saturated mercuric chloride solution for 1 min. Crack eggs and separate yolks from whites. Mix egg yolks with 1 chicken egg.

Preparation of Medium: Add components to distilled/deionized water and bring volume to 1.0L. Distribute into tubes. Autoclave for 15 min at 15 psi pressure–121°C in a slanted position.

Use: For the isolation and cultivation of *Mycobacterium* species other than *M. leprae*. Especially useful for detection of *Mycobacterium tuberculosis,* from clinical specimens such as cerebrospinal fluid, pleural fluid and tissues.

Aureomycin® Rose Bengal Glucose Peptone Agar

Composition per liter:

Agar 20.0g
Glucose 10.0g
Peptone 5.0g
KH_2PO_4 .... 1.0g
$MgSO_4 \cdot 7H_2O$ 0.5g
Rose Bengal 0.035g
Aureomycin solution 200.0mL
pH 5.4 ± 0.2 at 25°C

Aureomycin Solution:
Composition per 200mL:

Aureomycin·HCl 0.07g

Preparation of Aureomycin Solution: Add aureomycin·HCl to distilled/deionized water and bring volume to 200.0mL. Mix thoroughly. Filter sterilize.

Preparation of Medium: Add components, except aureomycin solution, to distilled/deionized water and bring volume to 800.0mL. Mix thoroughly. Gently heat and bring to boiling. Autoclave for 15 min at 15 psi pressure–121°C. Cool to 45°–50°C. Aseptically add 200.0mL of sterile aureomycin solution. Mix thoroughly. Pour into sterile Petri dishes or distribute into sterile tubes.

Use: For the cultivation and enumeration of fungi isolated from sewage and polluted waters.

Auxanographic Agar Medium

See: **Carbon Assimilation Medium**

AV Agar with Vitamins

Composition per liter:

Agar	15.0g
Glucose	1.0g
Glycerol	1.0g
L-Arginine	0.3g
K_2HPO_4	0.3g
NaCl	0.3g
$MgSO_4 \cdot 7H_2O$	0.2g
Vitamin solution	100.0mL
Trace salts solution	1.0mL

Vitamin Solution:

Composition per 100mL:

p-Aminobenzoic acid	0.5mg
Calcium pantothenate	0.5mg
HCl	0.5mg
Inositol	0.5mg
Niacin	0.5mg
Pyridoxine	0.5mg
Ribovlavin	0.5mg
Thiamine·HCl	0.5mg
Biotin	0.25mg

Preparation of Vitamin Solution: Add components to distilled/deionized water and bring volume to 1.0L. Mix thoroughly. Filter sterilize.

Trace Salts Solution:

Composition per liter:

$FeSO_4 \cdot 7H_2O$	10.0g
$CuSO_4 \cdot 5H_2O$	1.0g
$MnSO_4 \cdot 7H_2O$	1.0g
$ZnSO_4 \cdot 7H_2O$	1.0g

Preparation of Trace Salts Solution: Add components to distilled/deionized water and bring volume to 1.0L. Mix thoroughly.

Preparation of Medium: Add components, except vitamin solution, to distilled/deionized water and bring volume to 900.0mL. Mix thoroughly. Gently heat and bring to boiling. Autoclave for 15 min at 15 psi pressure–121°C. Cool to 45°–50°C. Aseptically add 100.0mL of sterile vitamin solution. Mix thoroughly. Pour into sterile Petri dishes or distribute into sterile tubes.

Use: For the isolation and cultivation of *Actinomadura* species, *Actinopolyspora* species, *Excellospora* species and *Microspora* species.

Avian *Mycoplasma* Agar

Composition per liter:

Agar, not inhibitory to mycoplasmas	10.0g
PPLO broth without Crystal Violet	700.0mL
Swine or horse serum, heat-inactivated at 56°C for 30 min.	150.0mL
Fresh yeast extract solution	100.0mL
Phenol Red solution	20.0mL
Glucose solution	10.0mL
Arginine solution	10.0mL
NAD	10.0mL

PPLO Broth without Crystal Violet:

Composition per 700mL:

Beef heart, infusion from	175.0g
Peptone	7.0g
NaCl	3.5g

Source: PPLO Broth without Crystal Violet is available as a premixed powder from Difco Laboratories.

Preparation of PPLO Broth without Crystal Violet: Add components to distilled/deionized water and bring volume to 700.0mL. Autoclave for 15 min at 15 psi pressure–121°C. Cool to 25°C. Beef heart for infusion may be substituted; 100g of beef heart for infusion are equivalent to 500g of fresh heart tissue.

Fresh Yeast Extract Solution:

Composition per 100mL:

Live, pressed, starch-free, Baker's yeast	25.0g

Preparation of Fresh Yeast Extract Solution: Add the live Baker's yeast to 100.0mL of distilled/deionized water. Autoclave for 90 min at 15 psi pressure–121°C. Allow to stand. Remove supernatant solution. Adjust pH to 6.6–6.8.

Phenol Red Solution:

Composition per 20mL:

Phenol Red	0.02g

Preparation of Phenol Red Solution: Add Phenol Red to distilled/deionized water and bring volume to 20.0mL. Mix thoroughly. Filter sterilize.

Glucose Solution:

Composition per 10mL:

Glucose	1.0g

Preparation of Glucose Solution: Add glucose to distilled/deionized water and bring volume to 10.0mL. Mix thoroughly. Filter sterilize.

Arginine Solution:

Composition per 10mL:

Arginine	1.0g

Preparation of Arginine Solution: Add arginine to distilled/deionized water and bring volume to 10.0mL. Mix thoroughly. Filter sterilize.

NAD Solution:

Composition per 10mL:

NAD	0.1g

Preparation of NAD Solution: Add NAD to distilled/deionized water and bring volume to 10.0mL. Mix thoroughly. Filter sterilize.

Preparation of Medium: Add 10.0g agar to 700.0mL PPLO broth without Crystal Violet. Gently heat to boiling with frequent mixing. Autoclave for 15 min at 15 psi pressure–121°C. Cool to 50–55°C. Warm other components to 50–55°C using a water bath. Aseptically combine all components. Mix thoroughly. Pour into sterile Petri dishes or into sterile tubes.

Use: For the cultivation and maintenance of *Mycoplasma* species.

Avian *Mycoplasma* Broth

Composition per liter:

PPLO broth without Crystal Violet	700.0mL
Swine or horse serum, heat-inactivated at 56°C for 30 min.	150.0mL
Fresh yeast extract solution	100.0mL
Phenol Red solution	20.0mL
Glucose solution	10.0mL
Arginine solution	10.0mL
NAD	10.0mL

PPLO Broth without Crystal Violet:
Composition per 700mL:

Beef heart, infusion from	175.0g
Peptone	7.0g
NaCl	3.5g

Source: PPLO Broth without Crystal Violet is available as a premixed powder from Difco Laboratories.

Preparation of PPLO Broth without Crystal Violet: Add components to distilled/deionized water and bring volume to 700.0mL. Autoclave for 15 min at 15 psi pressure–121°C. Cool to 25°C. Beef heart for infusion may be substituted; 100g of beef heart for infusion are equivalent to 500g of fresh heart tissue.

Fresh Yeast Extract Solution:
Composition per 100mL:

Live, pressed, starch-free, Baker's yeast	25.0g

Preparation of Fresh Yeast Extract Solution: Add the live Baker's yeast to 100.0mL of distilled/deionized water. Autoclave for 90 min at 15 psi pressure–121°C. Allow to stand. Remove supernatant solution. Adjust pH to 6.6–6.8.

Phenol Red Solution:
Composition per 20mL:

Phenol Red	0.02g

Preparation of Phenol Red Solution: Add Phenol Red to distilled/deionized water and bring volume to 20.0mL. Mix thoroughly. Filter sterilize.

Glucose Solution:
Composition per 10mL:

Glucose	1.0g

Preparation of Glucose Solution: Add glucose to distilled/deionized water and bring volume to 10.0mL. Mix thoroughly. Filter sterilize.

Arginine Solution:
Composition per 10mL:

Arginine	1.0g

Preparation of Arginine Solution: Add arginine to distilled/deionized water and bring volume to 10.0mL. Mix thoroughly. Filter sterilize.

NAD Solution:
Composition per 10mL:

NAD	0.1g

Preparation of NAD Solution: Add NAD to distilled/deionized water and bring volume to 10.0mL. Mix thoroughly. Filter sterilize.

Preparation of Medium: Aseptically combine components. Distribute into sterile tubes or flasks.

Use: For the cultivation and maintenance of *Mycoplasma* species.

Azide Agar
See: ***Enterococcus*** **Agar**

Azide Blood Agar

Composition per liter:

Agar	15.0g
Pancreatic digest of casein	5.0g
Peptic digest of animal tissue	5.0g
NaCl	5.0g
Beef extract	3.0g
NaN_3	0.2g
Sheep blood, defibrinated	50.0mL

pH 7.2 ± 0.2 at 25°C

Source: This medium is available as a premixed powder from Difco Laboratories, BBL Microbiology Systems and Oxoid Unipath.

Caution: Sodium azide is toxic. Azides also react with metals and disposal must be highly diluted.

Preparation of Medium: Add components, except sheep blood, to distilled/deionized water and bring volume to 950.0mL. Mix thoroughly. Gently heat and bring to boiling. Autoclave for 15 min at 15 psi pressure–121°C. Cool to 45–50°C. Aseptically add 50.0mL of sterile defibrinated sheep blood. Pour into sterile Petri dishes or distribute into sterile tubes. Allow tubes to cool in a slanted position.

Use: For the isolation and differentiation of streptococci and staphylococci from specimens containing mixed flora and from nonclinical specimens such as water and sewage.

Azide Blood Agar with Crystal Violet (Packer's Agar)

Composition per liter:

Agar	15.0g
Pancreatic digest of casein	5.0g
Peptic digest of animal tissue	5.0g
NaCl	5.0g
Beef extract	3.0g
NaN_3	0.9g
Crystal Violet	2.0mg
Sheep blood, defibrinated	50.0mL

pH 7.2 ± 0.2 at 25°C

Caution: Sodium azide is toxic. Azides also react with metals and disposal must be highly diluted.

Preparation of Medium: Add components, except sheep blood, to distilled/deionized water and bring volume to 950.0mL. Mix thoroughly. Gently heat and bring to boiling. Autoclave for 15 min at 15 psi pressure–121°C. Cool to 45–50°C. Aseptically add 50.0mL of sterile defibrinated sheep blood. Pour into sterile Petri dishes or distribute into sterile tubes. Allow tubes to cool in a slanted position.

Use: For the isolation and enumeration of fecal streptococci from nonclinical specimens such as water and food. Also used for the isolation of *Streptococcus pneumoniae* and *Erysipelothrix rhusiopathiae*.

Azide Broth (Azide Glucose Broth) (Azide Dextrose Broth)

Composition per liter:

Pancreatic digest of casein	15.0g
Glucose	7.5g
Beef extract	4.5g
NaCl	7.5g
NaN_3	0.2g

pH 7.2 ± 0.2 at 25°C

Source: This medium is available as a premixed powder from Difco Laboratories and BBL Microbiology Systems.

Caution: Sodium azide is toxic. Azides also react with metals and disposal must be highly diluted.

Preparation of Medium: Add components to distilled/deionized water and bring volume to 1.0L. Mix thoroughly. Gently heat and bring to boiling. Distribute into tubes or flasks. Autoclave for 15 min at 15 psi pressure–121°C. Prepare double strength broth for samples larger than 1.0mL.

Use: For the detection and enrichment of fecal streptococci in water and sewage. Also used in the Multiple-Tube Technique as a presumptive test for the presence of fecal streptococci.

Azide Broth, Rothe (Azide Glucose Broth, Rothe) (Azide Dextrose Broth, Rothe)

Composition per liter:

Peptone	20.0g
Glucose	5.0g
NaCl	5.0g
K_2HPO_4	2.7g
KH_2PO_4	2.7g
NaN_3	0.2g

pH 6.8 ± 0.2 at 25°C

Source: This medium is available as a premixed powder from Oxoid Unipath.

Caution: Sodium azide is toxic. Azides also react with metals and disposal must be highly diluted.

Preparation of Medium: Add components to distilled/deionized water and bring volume to 1.0L. Mix thoroughly. Gently heat and bring to boiling. Distribute into tubes or flasks. Autoclave for 15 min at 15 psi pressure–121°C. Prepare double strength broth for samples larger than 1.0mL.

Use: For the detection of enterococci in water and sewage.

Azide Citrate Broth

Composition per liter:

Pancreatic digest of casein	20.0g
Sodium citrate	10.0g
Yeast extract	5.0g
Glucose	5.0g
NaCl	5.0g
K_2HPO_4	4.0g
KH_2PO_4	1.5g
NaN_3	0.25g

pH 7.0 ± 0.2 at 25°C

Caution: Sodium azide is toxic. Azides also react with metals and disposal must be highly diluted.

Preparation of Medium: Add components to distilled/deionized water and bring volume to 1.0L. Mix thoroughly. Gently heat and bring to boiling. Distribute into tubes or flasks. Autoclave for 15 min at 15 psi

pressure–118°C. Prepare double strength broth for samples larger than 1.0mL.

Use: For the detection and enrichment of fecal streptococci in water and sewage.

Azide Dextrose Broth
See: **Azide Broth**

Azide Dextrose Broth, Rothe
See: **Azide Broth, Rothe**

Azide Glucose Broth
See: **Azide Broth**

Azide Glucose Broth, Rothe
See: **Azide Broth, Rothe**

Azide Medium

Composition per liter:

Peptone	10.0g
K_2HPO_4	5.0g
Glucose	5.0g
NaCl	5.0g
Yeast extract	3.0g
KH_2PO_4	2.0g
NaN_3	0.25g
Bromcresol Purple solution	2.0mL

pH 7.2 ± 0.2 at 25°C

Bromcresol Purple Solution:

Composition per 10mL:

Bromcresol Purple	0.16g
Ethanol	10.0mL

Preparation of Bromcresol Purple Solution: Add bromcresol purple to ethanol and bring volume to 10.0mL. Mix thoroughly.

Caution: Sodium azide is toxic. Azides also react with metals and disposal must be highly diluted.

Preparation of Medium: Add components to distilled/deionized water and bring volume to 1.0L. Mix thoroughly. Distribute into tubes or flasks. Autoclave for 15 min at 15 psi pressure–121°C.

Use: For cultivation of *Streptococcus* species and *Staphylococcus* species from clinical and nonclinical specimens.

Azospirillum amazonense Medium (LGI Medium)

Composition per liter:

Sucrose	5.0g
Agar	1.75g
KH_2PO_4	0.6g
K_2HPO_4	0.2g
$MgSO_4 \cdot 7H_2O$	0.2g
$CaCl_2 \cdot 2H_2O$	0.02g
$FeCl_3$	0.01g
$Na_2MoO_4 \cdot 2H_2O$	2.0mg
Bromothymol Blue (0.5% in 0.2 *N* KOH)	5.0mL

pH 6.0 ± 0.2 at 25°C

Preparation of Medium: Add components to distilled/deionized water and bring volume to 1.0L. Mix thoroughly. Gently heat and bring to boiling. Distribute into tubes or flasks. Autoclave for 15 min at 15 psi pressure–121°C. Pour into sterile Petri dishes or leave in tubes.

Use: For the cultivation and maintenance of *Azospirillum amazonense*.

Azospirillum lipoferum Agar Medium

Composition per liter:

Glucose	20.0g
Agar	15.0g
K_2HPO_4	0.8g
$MgSO_4 \cdot 7H_2O$	0.5g
KH_2PO_4	0.2g
$FeCl_3 \cdot 6H_2O$	0.1g
Yeast extract	0.1g
$CaCl_2 \cdot 2H_2O$	0.02g
$Na_2MoO_4 \cdot 2H_2O$	0.02g

pH 6.9 ± 0.2 at 25°C

Preparation of Medium: Add components to distilled/deionized water and bring volume to 1.0L. Mix thoroughly. Gently heat and bring to boiling. Distribute into tubes or flasks. Autoclave for 15 min at 15 psi pressure–121°C. Pour into sterile Petri dishes or leave in tubes.

Use: For the cultivation of *Azospirillum lipoferum*.

Azospirillum lipoferum Agar Medium

Composition per liter:

Agar	15.0g
Calcium malate	10.0g
K_2HPO_4	0.8g
$MgSO_4 \cdot 7H_2O$	0.5g

KH_2PO_4	0.2g
$FeCl_3 \cdot 6H_2O$	0.1g
Yeast extract	0.1g
$CaCl_2 \cdot 2H_2O$	0.02g
$Na_2MoO_4 \cdot 2H_2O$	0.02g

pH 6.9 ± 0.2 at 25°C

Preparation of Medium: Add components to distilled/deionized water and bring volume to 1.0L. Mix thoroughly. Gently heat and bring to boiling. Distribute into tubes or flasks. Autoclave for 15 min at 15 psi pressure–121°C. Pour into sterile Petri dishes or leave in tubes.

Use: For the cultivation of *Azospirillum lipoferum*.

Azospirillum lipoferum Medium

Composition per liter:

Calcium malate	10.0g
K_2HPO_4	1.0g
$MgSO_4 \cdot 7H_2O$	0.5g
$CaCl_2 \cdot 2H_2O$	0.02g

pH 6.5 ± 0.2 at 25°C

Preparation of Medium: Add components to distilled/deionized water and bring volume to 1.0L. Mix thoroughly. Distribute into tubes or flasks. Autoclave for 15 min at 15 psi pressure–121°C.

Use: For the isolation and cultivation of *Azospirillum lipoferum*.

Azospirillum Medium

Composition per liter:

Sodium malate	5.0g
Agar	1.75g
KH_2PO_4	0.4g
$MgSO_4 \cdot 7H_2O$	0.2g
K_2HPO_4	0.1g
NaCl	0.1g
$FeCl_3$	0.01g
$CaCl_2 \cdot 2H_2O$	0.02g
$Na_2MoO_4 \cdot 2H_2O$	2.0mg
Bromthymol Blue solution	5.0mL

pH 6.8 ± 0.2 at 25°C

Bromthymol Blue Solution:
Composition per 10mL:

Bromthymol Blue	0.5g
Ethanol	10.0mL

Preparation of Bromthymol Blue Solution: Add Bromthymol Blue to 10.0mL of ethanol. Mix thoroughly.

Preparation of Medium: Add components to distilled/deionized water and bring volume to 1.0L. Mix thoroughly. Distribute into tubes or flasks. Autoclave for 15 min at 15 psi pressure–121°C.

Use: For the cultivation of *Azospirillum* species isolated from roots.

Azotobacter Agar, Modified I

Composition per liter:

Agar	15.0g
Sucrose	10.0g
Glucose	10.0g
$MgSO_4 \cdot 7H_2O$	0.2g
KH_2PO_4	0.15g
$CaSO_4 \cdot 2H_2O$	0.1g
K_2HPO_4	0.05g
$CaCl_2$	0.02g
Na_2MoO_4	2.0mg
$FeCl_3$	1.0mg
$Na_2MoO_4 \cdot 2H_2O$	1.0mg

pH 7.2 ± 0.2 at 25°C

Preparation of Medium: Add components to distilled/deionized water and bring volume to 1.0L. Mix thoroughly. Gently heat and bring to boiling. Adjust pH to 7.2. Distribute into tubes or flasks. Autoclave for 15 min at 15 psi pressure–121°C. Pour into sterile Petri dishes or leave in tubes.

Use: For the cultivation and maintenance of *Azotobacter* species.

Azotobacter Agar, Modified II

Composition per liter:

Sucrose	20.0g
Agar	15.0g
KH_2PO_4	0.15g
$MgSO_4 \cdot 7H_2O$	0.2g
K_2HPO_4	0.05g
$CaCl_2$	0.02g
Na_2MoO_4	2.0mg
$FeCl_3$	1.0mg
$Na_2MoO_4 \cdot 2H_2O$	1.0mg

pH 6.2 ± 0.2 at 25°C

Preparation of Medium: Add components to distilled/deionized water and bring volume to 1.0L. Mix thoroughly. Gently heat and bring to boiling. Adjust pH to 6.2. Distribute into tubes or flasks. Autoclave for 15 min at 15 psi pressure–121°C. Pour into sterile Petri dishes or leave in tubes.

Use: For the cultivation and maintenance of *Azotobacter* species *and Beijerinckia derxii*.

Azotobacter Basal Agar

Composition per liter:

Agar	15.0g
K_2HPO_4	1.0g

$MgSO_4 \cdot 7H_2O$ 0.2g
NaCl 0.2g
$FeSO_4 \cdot 7H_2O$ 5.0mg
Soil extract 100.0mL
pH 7.2 ± 0.2 at 25°C

Soil Extract:
Composition per 200mL:
African Violet soil 0.5g
Na_2CO_3 .... 0.5g

Preparation of Soil Extract: Add components to tap water and bring volume to 200.0mL. Autoclave for 60 min at 15 psi pressure–121°C. Filter through Whatman filter paper.

Preparation of Medium: Add components including filtered soil extract to tap water and bring volume to 1.0L. Mix thoroughly. Gently heat and bring to boiling. Distribute into tubes or flasks. Autoclave for 15 min at 15 psi pressure–121°C. Pour into sterile Petri dishes or leave in tubes.

Use: For the cultivation of a variety of bacteria including *Azomonas* species, *Azotobacter* species, and others when a carbon source is added.

Azotobacter Basal Broth

Composition per liter:
K_2HPO_4 .... 1.0g
$MgSO_4 \cdot 7H_2O$ 0.2g
NaCl 0.2g
$FeSO_4 \cdot 7H_2O$ 5.0mg
Soil extract 100.0mL
pH 7.2 ± 0.2 at 25°C

Soil Extract:
Composition per 200mL:
African Violet soil 0.5g
Na_2CO_3 .... 0.5g

Preparation of Soil Extract: Add components to tap water and bring volume to 200.0mL. Autoclave for 60 min at 15 psi pressure–121°C. Filter through Whatman filter paper.

Preparation of Medium: Add components including filtered soil extract to tap water and bring volume to 1.0L. Mix thoroughly. Distribute into tubes or flasks. Autoclave for 15 min at 15 psi pressure–121°C.

Use: For the cultivation of a variety of bacteria including *Azomonas* species, *Azotobacter* species, and others when a carbon source is added.

Azotobacter Broth, Modified I

Composition per liter:
Sucrose 10.0g
Glucose 10.0g
$MgSO_4 \cdot 7H_2O$ 0.2g
KH_2PO_4 .... 0.15g
$CaSO_4 \cdot 2H_2O$ 0.1g
K_2HPO_4 .... 0.05g
$CaCl_2$ 0.02g
Na_2MoO_4 .... 2.0mg
$FeCl_3$ 1.0mg
$Na_2MoO_4 \cdot 2H_2O$ 1.0mg
pH 7.2 ± 0.2 at 25°C

Preparation of Medium: Add components to distilled/deionized water and bring volume to 1.0L. Mix thoroughly. Gently heat and bring to boiling. Adjust pH to 7.2. Distribute into tubes or flasks. Autoclave for 15 min at 15 psi pressure–121°C.

Use: For the cultivation of *Azotobacter* species.

Azotobacter Broth, Modified II

Composition per liter:
Sucrose 20.0g
KH_2PO_4 .... 0.15g
$MgSO_4 \cdot 7H_2O$ 0.2g
K_2HPO_4 .... 0.05g
$CaCl_2$ 0.02g
Na_2MoO_4 .... 2.0mg
$FeCl_3$ 1.0mg
$Na_2MoO_4 \cdot 2H_2O$ 1.0mg
pH 6.2 ± 0.2 at 25°C

Preparation of Medium: Add components to distilled/deionized water and bring volume to 1.0L. Mix thoroughly. Gently heat and bring to boiling. Adjust pH to 6.2. Distribute into tubes or flasks. Autoclave for 15 min at 15 psi pressure–121°C.

Use: For the cultivation of *Azotobacter* species and *Beijerinckia derxii*.

Azotobacter chroococcum Agar

Composition per liter:
Agar 20.0g
$CaCO_3$ 20.0g
Glucose 20.0g
K_2HPO_4 .... 0.8g
$MgSO_4 \cdot 7H_2O$ 0.5g
KH_2PO_4 .... 0.2g
$FeCl_3 \cdot 6H_2O$ 0.1g
$Na_2MoO_4 \cdot 2H_2O$ 0.05g
pH 7.4–7.6 at 25°C

Preparation of Medium: Add components to distilled/deionized water and bring volume to 1.0L. Mix thoroughly. Gently heat and bring to boiling. Distribute into tubes or flasks. Autoclave for 15 min at 15 psi

pressure–121°C. Pour into sterile Petri dishes or leave in tubes.

Use: For the cultivation and maintenance of *Azotobacter chroococcum.*

Azotobacter chroococcum Agar

Composition per liter:

Agar	20.0g
Glucose	20.0g
K_2HPO_4	0.8g
$MgSO_4 \cdot 7H_2O$	0.5g
KH_2PO_4	0.2g
$FeCl_3 \cdot 6H_2O$	0.1g
$CaCl_2 \cdot 2H_2O$	0.05g
$Na_2MoO_4 \cdot 2H_2O$	0.05g

pH 7.4-7.6 ± 0.2 at 25°C

Preparation of Medium: Add components to distilled/deionized water and bring volume to 1.0L. Mix thoroughly. Gently heat and bring to boiling. Distribute into tubes or flasks. Autoclave for 15 min at 15 psi pressure–121°C. Pour into sterile Petri dishes or leave in tubes.

Use: For the cultivation and maintenance of *Azotobacter chroococcum.*

Azotobacter chroococcum Medium

Composition per liter:

$CaCO_3$	20.0g
Glucose	20.0g
K_2HPO_4	1.0g
$MgSO_4 \cdot 7H_2O$	0.5g

Preparation of Medium: Add components to distilled/deionized water and bring volume to 1.0L. Mix thoroughly. Distribute into tubes or flasks. Autoclave for 15 min at 15 psi pressure–121°C.

Use: For the cultivation of *Azotobacter chroococcum.*

Azotobacter Medium (ATCC Medium 14)

Composition per liter:

Sucrose	20.0g
Agar	15.0g
K_2HPO_4	0.8g
Yeast extract	0.5g
KH_2PO_4	0.2g
$MgSO_4 \cdot 7H_2O$	0.2g
$CaSO_4 \cdot 2H_2O$	0.1g
$FeCl_3$	1.0mg
$Na_2MoO_4 \cdot 2H_2O$	1.0mg

pH 7.2 ± 0.2 at 25°C

Preparation of Medium: Add components to distilled/deionized water and bring volume to 1.0L. Mix thoroughly. Gently heat and bring to boiling. Distribute into tubes or flasks. Autoclave for 15 min at 15 psi pressure–121°C. Pour into sterile Petri dishes or leave in tubes.

Use: For the cultivation of a variety of bacteria including *Azomonas* species, *Azotobacter* species, *Beijerinckia derxii, Pseudomonas azotocolligans,* and *Rhodococcus erythropolis.*

Azotobacter Medium (ATCC Medium 240)

Composition per liter:

Agar	15.0g
$MgSO_4 \cdot 7H_2O$	0.2g
KH_2PO_4	0.15g
K_2HPO_4	0.05g
$CaCl_2$	0.02g
$Na_2MoO_4 \cdot 2H_2O$	2.0mg
$FeCl_3$	1.0mg

pH 7.2 ± 0.2 at 25°C

Preparation of Medium: Add components to distilled/deionized water and bring volume to 1.0L. If required, add sucrose to a concentration of 1%. Mix thoroughly. Gently heat and bring to boiling. Distribute into tubes or flasks. Autoclave for 15 min at 15 psi pressure–121°C. Pour agar medium into sterile Petri dishes or leave in tubes.

Use: For the cultivation and maintenance of a variety of bacteria including *Azotobacter* species.

Azotobacter Medium (ATCC Medium 1771)

Composition per liter:

Agar	15.0g
Glucose	10.0g
KH_2PO_4	0.22g
$CaSO_4 \cdot 2H_2O$	0.1g
$MgSO_4 \cdot 7H_2O$	0.098g
NaCl	0.058g
K_2HPO_4	0.058g
$FeSO_4 \cdot 7H_2O$	5.0mg
$Na_2MoO_4 \cdot 2H_2O$	0.2mg

pH 7.2 ± 0.2 at 25°C

Preparation of Medium: Add components to distilled/deionized water and bring volume to 1.0L. Mix thoroughly. Gently heat and bring to boiling. Distribute into tubes or flasks. Autoclave for 15 min at 15 psi pressure–121°C. Pour into sterile Petri dishes or leave in tubes.

Use: For the cultivation and maintenance of a variety of bacteria including *Azotobacter* species.

Azotobacter paspali Medium

Composition per liter:

Agar	20.0g
Sucrose	20.0g
$CaCO_3$	1.0g
$MgSO_4 \cdot 7H_2O$	0.20g
KH_2PO_4	0.15g
K_2HPO_4	0.05g
$CaCl_2$	0.02g
$Na_2MoO_4 \cdot 2H_20$	2.0mg
Bromthymol Blue solution	10.0mL
$FeCl_3$ (10% solution)	0.1mL

pH 7.0 ± 0.2 at 25°C

Bromthymol Blue Solution:

Composition per 10mL:

Bromthymol Blue	0.5g
Ethanol	10.0mL

Preparation of Bromthymol Blue Solution: Add Bromthymol Blue to 10.0mL of ethanol. Mix thoroughly.

Preparation of Medium: Add components to distilled/deionized water and bring volume to 1.0L. Mix thoroughly. Gently heat and bring to boiling. Distribute into tubes or flasks. Autoclave for 15 min at 15 psi pressure–121°C. Pour into sterile Petri dishes or leave in tubes.

Use: For the cultivation and maintenance of *Azotobacter paspali*.

Azotobacter Supplement (ATCC Medium 11)

Composition per liter:

Agar	15.0g
K_2HPO_4	1.0g
$MgSO_4 \cdot 7H_2O$	0.2g
NaCl	0.2g
$FeSO_4 \cdot 7H_2O$	5.0mg
Soil extract	100.0mL
Glucose solution	100.0mL

pH 7.6 ± 0.2 at 25°C

Soil Extract:

Composition per 200mL:

African Violet soil	0.5g
Na_2CO_3	0.5g

Preparation of Soil Extract: Add components to tap water and bring volume to 200.0mL. Autoclave for 60 min at 15 psi pressure–121°C. Filter through Whatman filter paper.

Glucose Solution:

Composition per 100mL:

Glucose	20.0g

Preparation of Glucose Solution: Add glucose to distilled/deionized water and bring volume to 100.0mL. Mix thoroughly. Filter sterilize.

Preparation of Medium: Add components, except glucose solution, to tap water and bring volume to 900.0mL. Mix thoroughly. Adjust pH to 7.6. Autoclave for 15 min at 15 psi pressure–121°C. Cool to 50–55°C. Aseptically add 100.0mL sterile glucose solution. Mix thoroughly. Pour into sterile Petri dishes or leave in tubes.

Use: For the cultivation of *Azomonas agilis*, and *Azotobacter chroococcum*.

Azotobacter Supplement (ATCC Medium 12)

Composition per liter:

Agar	15.0g
K_2HPO_4	1.0g
$MgSO_4 \cdot 7H_2O$	0.2g
NaCl	0.2g
$FeSO_4 \cdot 7H_2O$	5.0mg
Soil extract	100.0mL
Mannitol solution	100.0mL

pH 7.6 ± 0.2 at 25°C

Soil Extract:

Composition per 200mL:

African Violet soil	0.5g
Na_2CO_3	0.5g

Preparation of Soil Extract: Add components to distilled/deionized water and bring volume to 200.0mL. Autoclave for 60 min at 15 psi pressure–121°C. Filter through Whatman filter paper.

Mannitol Solution:

Composition per 100mL:

Mannitol	20.0g

Preparation of Mannitol Solution: Add mannitol to distilled/deionized water and bring volume to 100.0mL. Mix thoroughly. Filter sterilize.

Preparation of Medium: Add components, except mannitol solution, to tap water and bring volume to 900.0mL. Mix thoroughly. Adjust pH to7.6. Autoclave for 15 min at 15 psi pressure–121°C. Cool to 50–55°C. Aseptically add 100.0mL sterile mannitol solution. Mix thoroughly. Pour into sterile Petri dishes or leave in tubes.

Use: For the cultivation of *Azotobacter* species and *Azomonas* species.

Azotobacter Supplement (ATCC Medium 13)

Composition per liter:

Agar	15.0g
K_2HPO_4	1.0g
$MgSO_4 \cdot 7H_2O$	0.2g
NaCl	0.2g
$FeSO_4 \cdot 7H_2O$	5.0mg
Soil extract	100.0mL
Glucose solution	100.0mL

pH 6.0 ± 0.2 at 25°C

Soil Extract:

Composition per 200mL:

African Violet soil	0.5g
Na_2CO_3	0.5g

Preparation of Soil Extract: Add components to tap water and bring volume to 200.0mL. Autoclave for 60 min at 15 psi pressure–121°C. Filter through Whatman filter paper.

Glucose Solution:

Composition per 100mL:

Glucose	20.0g

Preparation of Glucose Solution: Add glucose to distilled/deionized water and bring volume to 100.0mL. Mix thoroughly. Filter sterilize.

Preparation of Medium: Add components, except glucose solution, to tap water and bring volume to 900.0mL. Mix thoroughly. Adjust pH to 6.0. Autoclave for 15 min at 15 psi pressure–121°C. Cool to 50–55°C. Aseptically add 100.0mL sterile glucose solution. Mix thorougly. Pour into sterile Petri dishes or leave in tubes.

Use: For the cultivation of *Beijerinckia* species.

Azotobacter Supplement (ATCC Medium 15)

Composition per liter:

Agar	15.0g
K_2HPO_4	1.0g
$MgSO_4 \cdot 7H_2O$	0.2g
NaCl	0.2g
$FeSO_4 \cdot 7H_2O$	5.0mg
Soil extract	100.0mL
Mannitol solution	100.0mL

pH 6.0 ± 0.2 at 25°C

Soil Extract:

Composition per 200mL:

African Violet soil	0.5g
Na_2CO_3	0.5g

Preparation of Soil Extract: Add components to distilled/deionized water and bring volume to 200.0mL. Autoclave for 60 min at 15 psi pressure–121°C. Filter through Whatman filter paper.

Mannitol Solution:

Composition per 100mL:

Mannitol	20.0g

Preparation of Mannitol Solution: Add mannitol to distilled/deionized water and bring volume to 100.0mL. Mix thoroughly. Filter sterilize.

Preparation of Medium: Add components, except mannitol solution, to tap water and bring volume to 900.0mL. Mix thoroughly. Adjust pH to 6.0. Autoclave for 15 min at 15 psi pressure–121°C. Cool to 50–55°C. Aseptically add 100.0mL sterile mannitol solution. Mix thoroughly. Pour into sterile Petri dishes or leave in tubes.

Use: For the cultivation of *Azomonas macrocytogenes*.

Azotobacter vinelandii Medium

Composition per liter:

Sodium benzoate	1.0g
K_2HPO_4	0.5g
Mannitol	0.5g

Preparation of Medium: Add components to distilled/deionized water and bring volume to 1.0L. Mix thoroughly. Distribute into tubes or flasks. Autoclave for 15 min at 15 psi pressure–121°C.

Use: For the cultivation of *Azotobacter vinelandii* from water samples.

Azotobacter vinelandii Medium

Composition per liter:

Sodium benzoate	1.0g
K_2HPO_4	0.5g
Ethanol	1.0mL

Preparation of Medium: Add components, except ethanol, to distilled/deionized water and bring volume to 999.0mL. Mix thoroughly. Autoclave for 15 min at 15 psi pressure–121°C. Cool to 45°–50°C. Aseptically add 1.0mL of filter-sterilized ethanol. Mix thoroughly. Aseptically distribute into sterile tubes or flasks.

Use: For the cultivation of *Azotobacter vinelandii* from soil.

B Broth (Medium for *Ureaplasma*)

Composition per 100.25mL:

Yeast extract	0.1g
GHL (Glycyl-L-histidyl-L-lysine)	2.0μg
PPLO broth without Crystal Violet	50.0mL
Horse serum, not inactivated	10.0mL
Bromothymol Blue (0.4% solution)	1.0mL
Urea solution	0.25mL

pH 6.0 ± 0.2 at 25°C

PPLO Broth without Crystal Violet:

Composition per 50mL:

Beef heart, infusion from	1.62g
Peptone	0.32g
NaCl	0.16g

Source: PPLO broth without Crystal Violet is available as a premixed powder from Difco Laboratories.

Preparation of PPLO Broth without Crystal Violet: Add components to distilled/deionized water and bring volume to 50.0mL. Mix thoroughly.

Urea Solution:

Composition per 10mL:

Urea	1.0g

Preparation of Urea Solution: Add urea to distilled/deionized water and bring volume to 10.0mL. Mix thoroughly. Filter sterilize.

Preparation of Medium: Add components—except GHL, urea solution and horse serum—to double glass-distilled water and bring volume to 90.0mL. Mix thoroughly. Gently heat and bring to boiling. Autoclave for 15 min at 15 psi pressure–121°C. Cool to 50–55°C. To 90.0mL of the sterile medium, aseptically add 2.0μg GHL, 10.0mL horse serum and 0.25mL of sterile urea solution. Mix thoroughly. Aseptically distribute into tubes or flasks.

Use: For the cultivation and maintenance of *Ureaplasma urealyticum* and other *Ureaplasma* species.

B/1t 7 A Medium

Composition per liter:

Agar	20.0g
K_2HPO_4	7.0g
KH_2PO_4	3.0g
Glucose	2.0g
$(NH_4)_2SO_4$	1.0g
$MgSO_4 \cdot 7H_2O$	0.1g
$CaCl_2 \cdot 2H_2O$	0.01g
Indole	0.01g
$FeSO_4 \cdot 7H_2O$	0.5mg

Preparation of Medium: Add components to distilled/deionized water and bring volume to 1.0L. Mix thoroughly. Gently heat and bring to boiling. Distribute into tubes or flasks. Autoclave for 15 min at 15 psi pressure–121°C. Pour into sterile Petri dishes or leave in tubes.

Use: For the cultivation and maintenance of *Escherichia coli* and other bacteria.

B_{12} Assay Medium

Composition per liter:

Glucose	20.5g
Lactose	20.0g
Amino acids, vitamin-free casamino acids	15.0g
Sodium acetate	10.0g
K_2HPO_4	2.5g
Polysorbate 80	2.0g
Ascorbic acid	1.0g
L-Arginine	0.5g
L-Histidine	0.25g
L-Phenylalanine	0.25g
L-Valine	0.25g
L-Asparagine	0.2g
$MgSO_4 \cdot 7H_2O$	0.2g
Mercaptoacetic acid	0.13g
Calcium pantothenate	0.1g
L-Tryptophan	0.1g
$MnSO_4$	0.08g
Adenine	0.04g
Guanine	0.04g
Thymine	0.04g
Uracil	0.04g
$(NH_4)_2SO_4 \cdot FeSO_4 \cdot 6H_2O$	0.03g
KCN	5.0mg
Pyridoxal·HCl	1.0mg
Niacin	1.0mg
Riboflavin	1.0mg
Thiamine·HCl	0.5mg
p-Aminobenzoic acid	0.5mg
Folic acid	0.05mg

pH 6.0 ± 0.2 at 25°C

Source: This medium is available as a premixed powder from Difco Laboratories and BBL Microbiology Systems.

Caution: Cyanide is toxic.

Preparation of Medium: Add components to distilled/deionized water and bring volume to 1.0L. Mix thoroughly. Gently heat and bring to boiling. Continue boiling 2–3 min. Allow precipitate to settle out. Distribute supernatant into tubes in 5.0mL volumes. Add standard solution or test solutions to each tube. Adjust the volume of each tube to 10.0mL with distilled/deionized water. Autoclave for 15 min at 15 psi pressure–121°C.

Use: For the determination of the vitamin B_{12} content of pharmaceutical products and other materials. *Lactobacillus leichmanii* ATCC 7830 is used as a test organism. A standard curve can be generated by adding known concentrations of cyanocobalamin and measuring the growth response turbidimetrically at 530 nm.

B_{12} Culture Agar, USP

Composition per liter:

Agar	15.0g
Glucose	10.0g
Proteose peptone No. 3	7.5g
Yeast extract	7.5g
KH_2PO_4	2.0g
Polysorbate 80	0.1g
Tomato juice	100.0mL

pH 6.8 ± 0.1 at 25°C

Source: This medium is available as a premixed powder from Difco Laboratories.

Preparation of Medium: Add components to distilled/deionized water and bring volume to 1.0L. Mix thoroughly. Gently heat and bring to boiling. Distribute into tubes in 10.0mL volumes. Autoclave for 15 min at 15 psi pressure–121°C. Cool tubes in an upright position.

Use: For the cultivation and maintenance of *Lactobacillus leichmannii* ATCC 7830 to be used as the test organism in the Vitamin B_{12} assay according to the USP.

B_{12} Inoculum Broth, USP

Composition per liter:

Glucose	10.0g
Proteose peptone No. 3	7.5g
Yeast extract	7.5g
K_2HPO_4	2.0g
Polysorbate 80	0.1g
Tomato juice	100.0mL

pH 6.8 ± 0.1 at 25°C

Source: This medium is available as a premixed powder from Difco Laboratories.

Preparation of Medium: Add components to distilled/deionized water and bring volume to 1.0L. Mix thoroughly. Gently heat and bring to boiling. Distribute into tubes in 10.0mL volumes. Autoclave for 15 min at 15 psi pressure–121°C.

Use: For preparation of inoculum cultures of *Lactobacillus leichmanii* ATCC 7830 to be used as the test organism in the Vitamin B_{12} assay according to the USP.

Baar's Medium for Sulfate Reducers

Composition per liter:

Sodium lactate	3.5g
$MgSO_4 \cdot 7H_2O$	2.0g
K_2HPO_4	1.0g
$CaSO_4$	1.0g
NH_4Cl	0.5g
Ferrous ammonium sulfate solution	10.0mL
Yeast extract solution	10.0mL

pH 7.5 ± 0.2 at 25°C

Ferrous Ammonium Sulfate Solution:
Composition per 10mL:

$Fe(NH_4)_2(SO_4)_2$	0.5g

Preparation of Ferrous Ammonium Sulfate Solution: Add $Fe(NH_4)_2(SO_4)_2$ to distilled/deionized water and bring volume to 10.0mL. Mix thoroughly. Autoclave for 15 min at 15 psi pressure–121°C.

Yeast Extract Solution:
Composition per 10mL:

Yeast extract	1.0g

Preparation of Yeast Extract Solution: Add yeast extract to distilled/deionized water and bring volume to 10.0mL. Mix thoroughly. Autoclave for 15 min at 15 psi pressure–121°C.

Preparation of Medium: Add components, except ferrous ammonium sulfate solution and yeast extract solution, to tap water and bring volume to 980.0mL. Mix thoroughly. Gently heat and bring to boiling. Autoclave for 15 min at 15 psi pressure–121°C. Cool to 45°–50°C. Aseptically add 10.0mL of sterile ferrous ammonium sulfate solution and sterile yeast extract solution. Mix thoroughly. Aseptically distribute into tubes or flasks.

Use: For the cultivation and maintenance of *Desulfotomaculum nigrificans*.

Baar's Medium for Sulfate Reducers, Modified

Composition per 1020mL:

Component I	400.0mL
Component II	200.0mL
Component III	400.0mL
Ferrous ammonium sulfate solution	20.0mL

pH 7.5 ± 0.2 at 25°C

Component I:
Composition per 400mL:

Sodium citrate	5.0g
$MgSO_4$	2.0g

$CaSO_4$	1.0g
NH_4Cl	1.0g

Preparation of Component I: Add components to distilled/deionized water and bring volume to 400.0mL. Mix thoroughly. Adjust pH to 7.5. Autoclave for 15 min at 15 psi pressure–121°C.

Component II:

Composition per 200mL:

K_2HPO_4	0.5g

Preparation of Component II: Add K_2HPO_4 to distilled/deionized water and bring volume to 200.0mL. Mix thoroughly. Adjust pH to 7.5. Autoclave for 15 min at 15 psi pressure–121°C.

Component III:

Composition per 400mL:

Sodium lactate	3.5g
Yeast extract	1.0g

Preparation of Component III: Add components to distilled/deionized water and bring volume to 400.0mL. Mix thoroughly. Adjust pH to 7.5. Autoclave for 15 min at 15 psi pressure–121°C.

Ferrous Ammonium Sulfate Solution:

Composition per 20mL:

$Fe(NH_4)_2(SO_4)_2$	1.0g

Preparation of Ferrous Ammonium Sulfate Solution: Add $Fe(NH_4)_2(SO_4)_2$ to distilled/deionized water and bring volume to 20.0mL. Mix thoroughly. Filter sterilize.

Preparation of Medium: Aseptically combine component I, component II and component III. Mix thoroughly. Distribute 5.0mL volumes into tubes under 97% N_2 + 3% H_2. Add medium to tubes while still warm to exclude as much O_2 as possible. Aseptically add 0.1mL of sterile ferrous ammonium sulfate solution to 5.0mL of medium immediately prior to inoculation.

Use: For the cultivation and maintenance of *Desulfovibrio, Desulfobulbus, Desulfotomaculum,* and *Thermodesulfobacterium* species.

Baar's Medium for Sulfate Reducers, Modified with 2.5% NaCl

Composition per 1020mL:

Component I	400.0mL
Component II	200.0mL
Component III	400.0mL
Ferrous ammonium sulfate solution	20.0mL

pH 7.5 ± 0.2 at 25°C

Component I:

Composition per 400mL:

NaCl	25.0g
Sodium citrate	5.0g
$MgSO_4$	2.0g
$CaSO_4$	1.0g
NH_4Cl	1.0g

Preparation of Component I: Add components to distilled/deionized water and bring volume to 400.0mL. Mix thoroughly. Adjust pH to 7.5. Autoclave for 15 min at 15 psi pressure–121°C.

Component II:

Composition per 200mL:

K_2HPO_4	0.5g

Preparation of Component II: Add K_2HPO_4 to distilled/deionized water and bring volume to 200.0mL. Mix thoroughly. Adjust pH to 7.5. Autoclave for 15 min at 15 psi pressure–121°C.

Component III:

Composition per 400mL:

Sodium lactate	3.5g
Yeast extract	1.0g

Preparation of Component III: Add components to distilled/deionized water and bring volume to 400.0mL. Mix thoroughly. Adjust pH to 7.5. Autoclave for 15 min at 15 psi pressure–121°C.

Ferrous Ammonium Sulfate Solution:

Composition per 20mL:

$Fe(NH_4)_2(SO_4)_2$	1.0g

Preparation of Ferrous Ammonium Sulfate Solution: Add $Fe(NH_4)_2(SO_4)_2$ to distilled/deionized water and bring volume to 20.0mL. Mix thoroughly. Filter sterilize.

Preparation of Medium: Aseptically combine component I, component II and component III. Mix thoroughly. Distribute 5.0mL volumes into tubes under 97% N_2 + 3% H_2. Add medium to tubes while still warm to exclude as much O_2 as possible. Aseptically add 0.1mL of sterile ferrous ammonium sulfate solution to 5.0mL of medium immediately prior to inoculation.

Use: For the cultivation of *Desulfovibrio africanus* and other *Desulfovibrio* species which prefer 2.5% NaCl.

Bacillus Agar

Composition per liter:

Agar	20.0g
$(NH_4)_2SO_4$	1.3g
Glucose	1.0g
Yeast extract	1.0g

KH_2PO_4	0.37g
$MgSO_4 \cdot 7H_2O$	0.25g
$CaCl_2 \cdot 2H_2O$	0.07g
$FeCl_3$	0.02g

pH 4.0 ± 0.2 at 25°C

Preparation of Medium: Add components to distilled/deionized water and bring volume to 500.0mL. Mix thoroughly. Gently heat and bring to boiling. Adjust pH to 3.5. Prepare a separate agar solution by adding 20.0g/500.0mL of distilled/deionized water. Autoclave solutions separately for 15 min at 15 psi pressure–121°C. Cool to 50–55°C. Aseptically combine both solutions. This procedure avoids acid hydrolysis of the agar. Pour into sterile Petri dishes or leave in tubes.

Use: For the cultivation of acidophilic *Bacillus* species such as *Bacillus acidocaldarius*.

Bacillus Agar, 1/4 Strength

Composition per liter:

Agar	18.0g
Yeast extract	2.5g
Pancreatic digest of casein	1.0g

pH 7.2 ± 0.2 at 25°C

Preparation of Medium: Add components to distilled/deionized water and bring volume to 1.0L. Mix thoroughly. Gently heat and bring to boiling. Distribute into tubes or flasks. Autoclave for 15 min at 15 psi pressure–121°C. Pour into sterile Petri dishes or leave in tubes.

Use: For the cultivation and maintenance of *Bacillus megaterium*.

Bacillus Broth

Composition per liter:

$(NH_4)_2SO_4$	1.3g
Glucose	1.0g
Yeast extract	1.0g
KH_2PO_4	0.37g
$MgSO_4 \cdot 7H_2O$	0.25g
$CaCl_2 \cdot 2H_2O$	0.07g
$FeCl_3$	0.02g

pH 4.0 ± 0.2 at 25°C

Preparation of Medium: Add components to distilled/deionized water and bring volume to 1.0L. Mix thoroughly. Gently heat and bring to boiling. Adjust pH to 4.0 with 10*N* H_2SO_4. Distribute into tubes or flasks. Autoclave for 15 min at 15 psi pressure–121°C.

Use: For the cultivation of acidophilic *Bacillus* species such as *Bacillus acidocaldarius*.

Bacillus Broth, 1/4 Strength

Composition per liter:

Yeast extract	2.5g
Pancreatic digest of casein	1.0g

pH 7.2 ± 0.2 at 25°C

Preparation of Medium: Add components to distilled/deionized water and bring volume to 1.0L. Mix thoroughly. Distribute into tubes or flasks. Autoclave for 15 min at 15 psi pressure–121°C.

Use: For the cultivation of *Bacillus megaterium*.

Bacillus cereus Medium (BCM)

Composition per 110mL:

Agar	2.0g
D-Mannitol	1.0g
$(NH_4)_2PO_4$	0.1g
KCl	0.02g
$MgSO_4 \cdot 7H_2O$	0.02g
Yeast extract	0.02g
Bromcresol Purple	4.0mg
Egg yolk emulsion, 20%	10.0mL

pH 7.0 ± 0.2 at 25°C

Egg Yolk Emulsion, 20%:

Composition per 100mL:

Chicken egg yolks	11
Whole chicken egg	1
NaCl (0.9% solution)	80.0mL

Preparation of Egg Yolk Emulsion, 20%: Soak eggs with 1:100 dilution of saturated mercuric chloride solution for 1 min. Crack eggs and separate yolks from whites. Mix egg yolks with 1 chicken egg. Measure 20.0mL of egg yolk emulsion and add to 80.0mL of 0.9% NaCl solution. Mix thoroughly. Filter sterilize. Warm to 45°–50°C.

Preparation of Medium: Add components—except egg yolk emulsion, 20%—to distilled/deionized water and bring volume to 100.0mL. Mix thoroughly. Gently heat and bring to boiling. Autoclave for 15 min at 15 psi pressure–121°C. Cool to 45°–50°C. Aseptically add 10.0mL of sterile egg yolk emulsion, 20%. Mix thoroughly. Pour into sterile Petri dishes or distribute into sterile tubes.

Use: For the cultivation of *Bacillus cereus*.

Bacillus cereus Motility Medium

See: **BC Motility Medium**

Bacillus cereus Selective Agar Base

Composition per liter:

Agar	15.0g
Sodium pyruvate	10.0g
Mannitol	10.0g
Na_2HPO_4	2.5g
NaCl	2.0g
Peptone	1.0g
KH_2PO_4	0.25g
Bromthymol Blue	0.12g
$MgSO_4·7H_2O$	0.1g
Egg yolk emulsion	25.0mL
Polymyxin B solution	10.0mL

pH 7.2 ± 0.2 at 25°C

Source: This medium is available as a premixed powder from Oxoid Unipath.

Egg Yolk Emulsion:
Composition:

Chicken egg yolks	11
Whole chicken egg	1

Preparation of Egg Yolk Emulsion: Soak eggs with 1:100 dilution of saturated mercuric chloride solution for 1 min. Crack eggs and separate yolks from whites. Mix egg yolks with 1 chicken egg.

Polymyxin B Solution:
Composition per 10mL:

Polymyxin B	100,000U

Preparation of Polymyxin B Solution: Add Polymyxin B to distilled/deionized water and bring volume to 10.0mL. Mix thoroughly. Filter sterilize.

Preparation of Medium: Add components, except egg yolk emulsion and plymyxin B solution to distilled/deionized water and bring volume to 965.0mL. Gently heat and bring to boiling. Distribute into tubes or flasks. Autoclave for 15 min at 15 psi pressure–121°C. Cool to 50°C. Aseptically add sterile polymyxin B and 25.0mL of sterile egg yolk emulsion. Mix thoroughly. Pour into sterile Petri dishes or leave in tubes.

Use: For the selection and presumptive identification of *Bacillus cereus*. Also, for the isolation and enumeration of these bacteria. *B. cereus* grows as moderate-sized (5mm) crenated colonies which are turquoise surrounded by a precipitate of egg yolk which is also turquoise.

Bacillus coagulans Medium

Composition per liter:

Agar	20.0g
Glucose	5.0g
Proteose peptone	5.0g
Yeast extract	5.0g
K_2HPO_4	4.0g
$MnSO_4·4H_2O$ solution	10.0mL
$CaCl_2$ solution	10.0mL

pH 5.0 ± 0.2 at 25°C

$MnSO_4·4H_2O$ Solution:
Composition per 10mL:

$MnSO_4·4H_2O$	0.05mg

Preparation of $MnSO_4·4H_2O$ Solution: Add $MnSO_4·4H_2O$ to distilled/deionized water and bring volume to 10.0mL. Mix thoroughly. Filter sterilize.

$CaCl_2$ Solution:
Composition per 10mL:

$CaCl_2$	0.045mg

Preparation of $CaCl_2$ Solution: Add $CaCl_2$ to distilled/deionized water and bring volume to 10.0mL. Mix thoroughly. Filter sterilize.

Preparation of Medium: Add components, except $MnSO_4·4H_2O$ solution and $CaCl_2$ solution, to distilled/deionized water and bring volume to 980.0mL. Mix thoroughly. Gently heat and bring to boiling. Autoclave for 15 min at 15 psi pressure–121°C. Avoid overheating. Cool to 45°–50°C. Aseptically add sterile $MnSO_4·4H_2O$ solution and $CaCl_2$ solution. Mix thoroughly. Pour into sterile Petri dishes or distribute into sterile tubes.

Use: For the cultivation of *Bacillus coagulans*.

Bacillus fastidiosus Medium

Composition per liter:

Agar	15.0g
$Na_2HPO_4·12H_2O$	6.0g
Yeast extract	2.5g
Uric acid	1.0g
Mineral solution	100.0mL

pH 7.0 ± 0.2 at 25°C

Mineral Solution:
Composition per 100mL:

KH_2PO_4	0.1g
$MgSO_4·7H_2O$	0.03g
$CaCl_2$	0.01g
NaCl	0.01g
$FeCl_3·6H_2O$	1.0mg

Preparation of Mineral Solution: Add components to distilled/deionized water and bring volume to 100.0mL. Mix thoroughly.

Preparation of Medium: Add components to distilled/deionized water and bring volume to 1.0L. Mix thoroughly. Gently heat and bring to boiling. Distribute into tubes or flasks. Autoclave for 15 min at 15 psi

pressure–121°C. Pour into sterile Petri dishes or leave in tubes.

Use: For the cultivation of *Bacillus fastidiosus*.

Bacillus Medium

Composition per liter:

Agar	25.0g
Peptone	6.0g
Pancreatic digest of casein	3.0g
Yeast extract	3.0g
Beef extract	1.5g
$MnSO_4 \cdot 4H_2O$	1.0μg

pH 7.0 ± 0.2 at 25°C

Preparation of Medium: Add components to distilled/deionized water and bring volume to 1.0L. Mix thoroughly. Gently heat and bring to boiling. Distribute into tubes or flasks. Autoclave for 15 min at 15 psi pressure–121°C. Pour into sterile Petri dishes or leave in tubes.

Use: For the cultivation of *Bacillus* species.

Bacillus Medium (ATCC Medium 455)

Composition per liter:

Soluble starch	30.0g
Agar	20.0g
Polypeptone™	5.0g
Yeast extract	5.0g

Preparation of Medium: Add components to distilled/deionized water and bring volume to 1.0L. Mix thoroughly. Gently heat and bring to boiling. Distribute into tubes or flasks. Autoclave for 15 min at 15 psi pressure–121°C. Swirl medium to resuspend starch. Pour into sterile Petri dishes or leave in tubes.

Use: For the cultivation and maintenance of *Bacillus subtilis*. Also used to detect amylase producing microorganisms.

Bacillus Medium

Composition per liter:

$(NH_4)_2HPO_4$	1.0g
$MgSO_4 \cdot 7H_2O$	0.2g
KCl	0.2g
Yeast extract	0.2g
Glucose solution	50.0mL
Bromcresol Purple solution	15.0mL

pH 7.0 ± 0.2 at 25°C

Glucose Solution:

Composition per 100mL:

Glucose	10.0g

Preparation of Glucose Solution: Add glucose to distilled/deionized water and bring volume to 100.0mL. Mix thoroughly. Filter sterilize.

Preparation of Medium: Add components, except glucose solution, to distilled/deionized water and bring volume to 1.0L. Mix thoroughly. Gently heat and bring to boiling. Distribute 9.5mL volumes into test tubes that contain an inverted Durham tube. Autoclave for 20 min at 15 psi pressure–121°C. Cool to 25°C. Aseptically add 0.5mL of sterile glucose to each tube. Mix thoroughly.

Use: For cultivation and differentiation of *Bacillus* species based on acid and gas production from glucose.

Bacillus Medium (ATCC Medium 552)

Composition per liter:

Peptone	10.0g
Lactose	5.0g
NaCl	5.0g
Beef extract	3.0g
K_2HPO_4	2.0g

pH 7.2 ± 0.2 at 25°C

Preparation of Medium: Add components to distilled/deionized water and bring volume to 1.0L. Mix thoroughly. Gently heat and bring to boiling. Distribute into tubes or flasks. Autoclave for 15 min at 15 psi pressure–121°C.

Use: For the cultivation and maintenance of *Bacillus* species.

Bacillus Medium (ATCC Medium 21)

Composition per liter:

Glycerol	20.0g
L-Glutamic acid	4.0g
Citric acid	2.0g
K_2HPO_4	0.5g
Ferric ammonium citrate	0.5g
$MgSO_4$	0.5g

pH 7.4 ± 0.2 at 25°C

Preparation of Medium: Add components to tap water and bring volume to 1.0L. Mix thoroughly. Gently heat and bring to boiling. Distribute into tubes or flasks. Autoclave for 15 min at 15 psi pressure–121°C.

Use: For the cultivation of *Bacillus licheniformis*.

Bacillus pasteurii NH_4 YE Medium (Ammonium Yeast Extract Medium)

Composition per liter:

Yeast extract	20.0g
Agar	20.0g
$(NH_4)_2SO_4$	10.0g

pH 9.0 ± 0.2 at 25°C

Preparation of Medium: Add each component to a separate flask and bring volume of each to 333.0mL with 0.13M Tris buffer, pH 9.0. Autoclave ingredients separately for 15 min at 15 psi pressure–121°C. No growth occurs if components are sterilized together. Cool to 50–55°C and aseptically combine solutions. Pour into sterile Petri dishes.

Use: For the cultivation and maintenance of *Bacillus pasteurii.*

Bacillus popilliae Maintenance Medium

Composition per liter:

Agar	20.0g
Yeast extract	15.0g
Pancreatic digest of casein	5.0g
K_2HPO_4	3.0g
Glucose solution	10.0mL

pH 7.2 ± 0.2 at 25°C

Glucose Solution:

Composition per 10mL:

Glucose	2.0g

Preparation of Glucose Solution: Add glucose to distilled/deionized water and bring volume to 10.0mL. Mix thoroughly. Filter sterilize.

Preparation of Medium: Add components, except glucose solution, to distilled/deionized water and bring volume to 990.0mL. Mix thoroughly. Gently heat and bring to boiling. Autoclave for 15 min at 15 psi pressure–121°C. Cool to 45°–50°C. Aseptically add sterile glucose solution. Mix thoroughly. Pour into sterile Petri dishes or distribute into sterile tubes.

Use: For the cultivation and maintenance of *Bacillus popilliae*.

Bacillus popilliae Medium

Composition per liter:

Yeast extract	10.0g
Acid hydrolysate of casein	7.95g
K_2HPO_4	3.0g
Beef extract	1.36g
Trehalose	1.0g
Starch	0.68g

pH 7.3 ± 0.1 at 25°C

Preparation of Medium: Add components to distilled/deionized water and bring to 1.0L. Mix thoroughly. Gently heat until dissolved. Do not overheat. Filter sterilize. Aseptically distribute into sterile tubes or flasks.

Use: For the cultivation of *Bacillus popilliae.*

Bacillus popilliae Medium

Composition per liter:

Yeast extract	15.0g
K_2HPO_4	3.0g

pH 7.2 ± 0.2 at 25°C

Preparation of Medium: Add components to distilled/deionized water and bring volume to 1.0L. Mix thoroughly. Distribute into tubes or flasks. Autoclave for 15 min at 15 psi pressure–121°C.

Use: For the cultivation of *Bacillus popilliae.*

Bacillus Pullulan Salts

Composition per liter:

Pullulan	2.50g
NaCl	1.00g
NH_4Cl	1.00g
KH_2PO_4	0.50g
$MgSO_4·7H_2O$	0.50g
Yeast extract	0.10g
$CaCl_2·2H_2O$	0.05g
Trace mineral solution	10.0mL
Vitamin solution	10.0mL

pH 6.0 ± 0.2 at 25°C

Trace Mineral Solution:

Composition per liter:

$CoCl_2·6H_2O$	0.2g
$FeSO_4·7H_2O$	0.13g
$ZnCl_2·2H_2O$	0.1g
$MnCl_2·4H_2O$	0.1g
$CaCl_2·2H_2O$	20.0mg
Na_2SeO_3	20.0mg
$Na_2WO_4·2H_2O$	20.0mg
$NaMoO_4·2H_2O$	1.0mg
H_3BO_3	0.5mg
$CuSO_4·5H_2O$	0.4mg
KI	0.1mg

Preparation of Trace Mineral Solution: Add components to distilled/deionized water and bring volume to 1.0L. Mix thoroughly.

Vitamin Solution:

Composition per liter:

Pyridoxine·HCl	10.0mg
Thiamine·HCl	5.0mg
Riboflavin	5.0mg
Nicotinic acid	5.0mg
Calcium pantothenate	5.0mg
p-Aminobenzoic acid	5.0mg
Thioctic acid	5.0mg
Biotin	2.0mg
Folic acid	2.0mg
Cyanocobalamin	0.1mg

Preparation of Vitamin Solution: Add components to distilled/deionized water and bring volume to 1.0L. Mix thoroughly. Filter sterilize.

Preparation of Medium: Add components, except vitamin solution, to distilled/deionized water and bring volume to 990.0L. Mix thoroughly. Gently heat and bring to boiling. Adjust pH to 6.0. Autoclave for 15 min at 15 psi pressure–121°C. Cool to 25°C. Aseptically add sterile vitamin solution. Mix thoroughly. Aseptically distribute into sterile tubes or flasks.

Use: For the cultivation and maintenance of *Bacillus* species which can degrade pullulan.

Bacillus schlegelii Agar

Composition per liter:

Noble agar	30.0g
$Na_2HPO_4 \cdot 2H_2O$	4.5g
KH_2PO_4	1.5g
Sodium pyruvate	1.5g
NH_4Cl	1.0g
$MgSO_4 \cdot 7H_2O$	0.2g
$MnSO_4 \cdot H_2O$	0.01g
$CaCl_2 \cdot 2H_2O$	0.01g
Ferric ammonium citrate	5.0mg
Agar solution	200.0mL
Sodium pyruvate	100.0mL
SL-6 trace elements	3.0mL

pH 7.1 ± 0.2 at 25°C

Agar Solution:

Composition per 200mL:

Noble agar	30.0g

Preparation of Agar Solution: Add agar to distilled/deionized water and bring volume to 200.0mL. Mix thoroughly. Gently heat and bring to boiling. Autoclave for 15 min at 15 psi pressure–121°C. Cool to 45°–50°C.

Pyruvate Solution:

Composition per 100mL:

Sodium pyruvate	1.5g

Preparation of Pyruvate Solution: Add sodium pyruvate to distilled/deionized water and bring volume to 100.0mL. Mix thoroughly. Filter sterilize. Warm to 45°–50°C.

SL-6 Trace Elements Solution:

Composition per liter:

H_3BO_3	0.3g
$CoCl_2 \cdot 6H_2O$	0.2g
$ZnSO_4 \cdot 7H_2O$	0.10g
$MnCl_2 \cdot 4H_2O$	0.03g
$Na_2MoO_4 \cdot H_2O$	0.03g
$NiCl_2 \cdot 6H_2O$	0.02g
$CuCl_2 \cdot 2H_2O$	0.01g

Preparation of SL-6 Trace Elements Solution: Add components to distilled/deionized water and bring volume to 1.0L. Mix thoroughly. Adjust pH to 3.4.

Preparation of Medium: Add components, except sodium pyruvate solution and agar solution, to distilled/deionized water and bring volume to 700.0mL. Mix thoroughly. Gently heat and bring to boiling. Adjust pH to 7.1. Autoclave for 15 min at 15 psi pressure–121°C. Cool to 50°C. Add sodium pyruvate solution and agar solution. Mix thoroughly. Pour into sterile Petri dishes or distribute into sterile tubes.

Use: For the cultivation and maintenance of *Bacillus schlegelii*.

Bacillus schlegelii Broth

Composition per liter:

$Na_2HPO_4 \cdot 2H_2O$	4.5g
KH_2PO_4	1.5g
Sodium pyruvate	1.5g
NH_4Cl	1.0g
$MgSO_4 \cdot 7H_2O$	0.2g
$MnSO_4 \cdot H_2O$	0.01g
$CaCl_2 \cdot 2H_2O$	0.01g
Ferric ammonium citrate	5.0mg
Sodium pyruvate	100.0mL
SL-6 trace elements	3.0mL

pH 7.1 ± 0.2 at 25°C

Pyruvate Solution:

Composition per 100mL:

Sodium pyruvate	1.5g

Preparation of Pyruvate Solution: Add sodium pyruvate to distilled/deionized water and bring volume to 100.0mL. Mix thoroughly. Filter sterilize.

SL-6 Trace Elements Solution:

Composition per liter:

H_3BO_3	0.3g
$CoCl_2 \cdot 6H_2O$	0.2g

$ZnSO_4 \cdot 7H_2O$ 0.10g
$MnCl_2 \cdot 4H_2O$ 0.03g
$Na_2MoO_4 \cdot H_2O$ 0.03g
$NiCl_2 \cdot 6H_2O$ 0.02g
$CuCl_2 \cdot 2H_2O$ 0.01g

Preparation of SL-6 Trace Elements Solution: Add components to distilled/deionized water and bring volume to 1.0L. Mix thoroughly. Adjust pH to 3.4.

Preparation of Medium: Add components, except sodium pyruvate solution, to distilled/deionized water and bring volume to 900.0mL. Mix thoroughly. Gently heat and bring to boiling. Adjust pH to 7.1. Autoclave for 15 min at 15 psi pressure–121°C. Cool to 50°C. Aseptically add sodium pyruvate solution. Aseptically distribute into sterile tubes or flasks.

Use: For the cultivation and maintenance of *Bacillus schlegelii.*

Bacillus stearothermophilus Broth

Composition per liter:

Pancreatic digest of casein 10.0g
Yeast extract 5.0g
K_2HPO_4 ..... 2.0g

pH 7.2 ± 0.2 at 25°C

Preparation of Medium: Add components to distilled/deionized water and bring volume to 1.0L. Mix thoroughly. Distribute into tubes or flasks. Autoclave for 15 min at 15 psi pressure–121°C.

Use: For the cultivation of *Bacillus stearothermophilus.*

Bacillus stearothermophilus Defined Broth

Composition per 100mL:

Mineral salts solution 10.0mL
Potassium phosphate buffer 5.0mL
L-Glutamate·HCl (1% solution) 4.0mL
L-Leucine (1% solution) 1.64mL
L-Lysine·HCl (1% solution) 1.40mL
L-Serine (1% solution) 1.40mL
L-Aspartate (1% solution) 1.30mL
L-Valine (1% solution) 1.26mL
Biotin (0.01% solution) 1.0mL
Glucose (20% solution) 1.0mL
L-Isoleucine (1% solution) 1.0mL
L-Proline (1% solution) 1.0mL
Nicotinic acid (0.01% solution) 1.0mL
Thiamine·HCl (0.01% solution) 1.0mL
L-Phenylalanine (1% solution) 0.86mL
L-Alanine (1% solution) 0.84mL
L-Threonine (1% solution) 0.84mL
L-Arginine·HCl (1% solution) 0.64mL
L-Tyrosine (1% solution) 0.56mL
L-Methionine (1% solution) 0.52mL
Glycine (1% solution) 0.50mL
L-Asparagine·H_2O (1% solution) 0.50mL
L-Cystine (1% solution) 0.50mL
L-Glutamine (1% solution) 0.50mL
L-Histidine·HCl·H_2O (1% solution) 0.42mL
L-Tryptophan (1% solution) 0.30mL
$CaCl_2$ (5% solution) 0.01mL
$FeCl_3 \cdot 6H_2O$ (0.05% solution) 0.01mL
$MnCl_2$ (10mM solution) 0.01mL
$ZnSO_4 \cdot 7H_2O$ (5% solution) 0.01mL

pH 7.3 ± 0.2 at 25°C

Mineral Salts Solution:

Composition per liter:

NaCl 10.0g
NH_4Cl 10.0g
$MgSO_4$ 4.0g

Preparation of Mineral Salts Solution: Add components to distilled/deionized water and bring volume to 1.0L. Mix thoroughly.

Potassium Phosphate Buffer:

Composition per 500mL:

K_2HPO_4 ..... 125.0g
KH_2PO_4 ..... 30.0g

Preparation of Potassium Phosphate Buffer: Add components to distilled/deionized water and bring volume to 500.0mL. Mix thoroughly.

Preparation of Medium: Add components to distilled/deionized water and bring volume to 100.0mL. Mix thoroughly. Filter sterilize.

Use: For the cultivation of *Bacillus stearothermophilus* in a chemically defined medium.

Bacillus stearothermophilus Sporulation Broth

Composition per liter:

Agar 20.0g
Pancreatic digest of gelatin 5.0g
Yeast extract 4.0g
Beef extract 3.0g
$MnCl_2 \cdot 4H_2O$ 10.0μg

pH 7.2 ± 0.2 at 25°C

Preparation of Medium: Add components to distilled/deionized water and bring volume to 1.0L. Mix thoroughly. Gently heat and bring to boiling. Distribute into tubes or flasks. Autoclave for 15 min at 15 psi pressure–121°C. Pour into sterile Petri dishes or leave in tubes.

Use: For the cultivation and sporulation of *Bacillus stearothermophilus*.

Bacillus thuringiensis Medium

Composition per liter:

Glucose	3.0g
$(NH_4)_2SO_4$	2.0g
Yeast extract	2.0g
$K_2HPO_4 \cdot 3H_2O$	0.5g
$MgSO_4 \cdot 7H_2O$	0.2g
$CaCl_2 \cdot 2H_2O$	0.08g
$MnSO_4 \cdot 4H_2O$	0.05g

pH 7.3 ± 0.2 at 25°C

Preparation of Medium: Add components to distilled/deionized water and bring volume to 1.0L. Mix thoroughly. Adjust pH to 7.3. Distribute into tubes or flasks. Autoclave for 15 min at 15 psi pressure–121°C.

Use: For the cultivation of *Bacillus thuringiensis*.

Bacillus Xylose Salts

Composition per liter:

Yeast extract	5.0g
Xylose	5.0g
NaCl	1.0g
NH_4Cl	1.0g
KH_2PO_4	0.50g
$MgSO_4 \cdot 7H_2O$	0.50g
$CaCl_2 \cdot 2H_2O$	0.05g
Trace mineral solution	10.0mL
Vitamin solution	10.0mL

pH 4.0 ± 0.2 at 25°C

Trace Mineral Solution:

Composition per liter:

$CoCl_2 \cdot 6H_2O$	0.2g
$FeSO_4 \cdot 7H_2O$	0.13g
$ZnCl_2 \cdot 2H_2O$	0.1g
$MnCl_2 \cdot 4H_2O$	0.1g
$CaCl_2 \cdot 2H_2O$	20.0mg
Na_2SeO_3	20.0mg
$Na_2WO_4 \cdot 2H_2O$	20.0mg
$NaMoO_4 \cdot 2H_2O$	1.0mg
H_3BO_3	0.5mg
$CuSO_4 \cdot 5H_2O$	0.4mg
KI	0.1mg

Preparation of Trace Mineral Solution: Add components to distilled/deionized water and bring volume to 1.0L. Mix thoroughly.

Vitamin Solution:

Composition per liter:

Pyridoxine·HCl	10.0mg
Thiamine·HCl	5.0mg
Riboflavin	5.0mg
Nicotinic acid	5.0mg
Calcium pantothenate	5.0mg
p-Aminobenzoic acid	5.0mg
Thioctic acid	5.0mg
Biotin	2.0mg
Folic acid	2.0mg
Cyanocobalamin	0.1mg

Preparation of Vitamin Solution: Add components to distilled/deionized water and bring volume to 1.0L. Mix thoroughly.

Preparation of Medium: Add components to distilled/deionized water and bring volume to 1.0L. Mix thoroughly. Gently heat and bring to boiling. Adjust pH of medium to 4.0. Distribute into tubes or flasks. Autoclave for 15 min at 15 psi pressure–121°C.

Use: For the cultivation and maintenance of *Bacillus* species which can utilize xylose as a carbon source.

Bacterial Cell Agar (BCA)

Composition per liter:

Tryptose	17.36g
Agar	15.0g
NaCl	8.68g
Beef extract	5,2g
Yeast extract	1.7g

pH 7.3 ± 0.2 at 25°C

Preparation of Medium: Add components, except agar, to distilled/deionized water and bring volume to 1.0L. Mix thoroughly. Autoclave for 15 min at 15 psi pressure–121°C. Cool to 30°C. Inoculate with a culture of *Aeromonas hydrophila*. Incubate with shaking at 30°C for 72 hr. Centrifuge culture in 40.0mL volumes at 10,000 × *g* for 10 min. Wash the cells four times in sterile 0.85% saline. Resuspend the cell pellet in 25.0mL of distilled/deionized water. Autoclave for 15 min at 15 psi pressure–121°C. Cool to 45°–50°C. In a separate flask, add 15.0g of agar to 1.0L of distilled/deionized water. Mix thoroughly. Gently heat and bring to boiling. Autoclave for 15 min at 15 psi pressure–121°C. Cool to 45°–50°C. Aseptically combine 25.0mL of washed cells and 250.0mL of cooled, sterile agar solution. Mix thoroughly. Pour into sterile Petri dishes.

Use: For the cultivation of freshwater *Myxobacterium* species.

Bacterium Medium

Composition per liter:

Agar	20.0g

Peptone	6.0g
Yeast extract	3.0g
Beef extract	1.5g
Glucose	1.0g

Preparation of Medium: Add components to distilled/deionized water and bring volume to 1.0L. Mix thoroughly. Gently heat and bring to boiling. Distribute into tubes or flasks. Autoclave for 15 min at 15 psi pressure–121°C. Pour into sterile Petri dishes or leave in tubes.

Use: An archaic medium used for the cultivation and growth of bacteria originally classified in the genus *Bacterium* but now classified in the genera *Brevibacterium* and *Kurthia*.

Bacteroides Bile Esculin Agar (BBE Agar)

Composition per liter:

Oxgall	20.0g
Pancreatic digest of casein	15.0g
Agar	15.0g
Papaic digest of soybean meal	5.0g
NaCl	5.0g
Esculin	1.0g
Ferric ammonium citrate	0.5g
Gentamicin solution	2.5mL
Hemin solution	2.5mL
Vitamin K_1 solution	1.0mL

pH 7.0 ± 0.2 at 25°C

Source: This medium is available as a premixed powder from BBL Microbiology Systems.

Gentamicin Solution:

Composition per 10mL:

Gentamicin	0.4mg

Preparation of Gentamicin Solution: Add gentamicin to 10.0mL of distilled/deionized water. Mix thoroughly. Filter sterilize.

Hemin Solution:

Composition per 100mL:

Hemin	0.5g
NaOH (1*N* solution)	10.0mL

Preparation of Hemin Solution: Add components to 100.0mL of distilled/deionized water. Mix thoroughly. Autoclave for 15 min at 15 psi pressure–121°C. Cool to 45°–50°C.

Vitamin K_1 Solution:

Composition per 100mL:

Vitamin K_1	1.0g
Ethanol	99.0mL

Preparation of Vitamin K_1 Solution: Add vitamin K_1 to 99.0mL of absolute ethanol. Mix thoroughly. Filter sterilize.

Preparation of Medium: Add components, except hemin solution, gentamicin solution and vitamin K_1 solution to distilled/deionized water and bring volume to 994.0mL. Mix thoroughly. Gently heat and bring to boiling. Autoclave for 15 min at 15 psi pressure–121°C. Cool to 45–50°C. Aseptically add 2.5mL sterile hemin solution, 2.5mL sterilie gentamicin solution and 1.0mL sterile vitamin K_1 solution.

Use: For the selection and presumptive identification of the *Bacteriodes fragilis* group. Also, for the differentiation of *Bacteroides* species based on hydrolysis of esculin and presence of catalase. After incubation for 48 hours, bacteria of the *B. fragilis* group appear as gray, circular, raised colonies larger than 1mm. Esculin hydrolysis is indicated by the presence of a blackened zone around the colonies.

Bacteroides Medium

Composition per liter:

Pancreatic digest of casein	27.0g
Yeast extract	3.0g
K_2HPO_4	2.5g
K_2CO_3	2.0g
NaCl	2.0g
Hemin solution	10.0mL
Vitamin K_1 solution	0.2mL

Hemin Solution:

Composition per 100mL:

Hemin	1.0g
NaOH (1*N* solution)	20.0mL

Preparation of Hemin Solution: Add hemin to 20.0mL of 1*N* NaOH solution. Mix thoroughly. Bring volume to 100.0mL with distilled/deionized water.

Vitamin K_1 Solution:

Composition per 100mL:

Vitamin K_1	1.0g
Ethanol	99.0mL

Preparation of Vitamin K_1 Solution: Add vitamin K_1 to 99.0mL of absolute ethanol. Mix thoroughly.

Preparation of Medium: Add components to distilled/deionized water and bring volume to 1.0L. Mix thoroughly. Distribute into tubes or flasks. Autoclave for 15 min at 15 psi pressure–121°C.

Use: For the cultivation of *Bacteroides asaccharolyticus* and *Bacteroides melaninogenicus*.

BAGG Broth (Buffered Azide Glucose Glycerol Broth)

Composition per liter:

Pancreatic digest of casein	10.0g
Peptic digest of animal tissue	10.0g
Glucose	5.0g
NaCl	5.0g
K_2HPO_4	4.0g
KH_2PO_4	1.5g
NaN_3	0.5g
Bromcresol Purple	0.015g
Glycerol	5.0mL

pH 6.9 ± 0.2 at 25°C

Source: Available as a premixed powder from Difco Laboratories and BBL Microbiology Systems.

Caution: Sodium azide is toxic. Azides also react with metals and disposal must be highly diluted.

Preparation of Medium: Add 5.0mL glycerol to 900.0mL of distilled/deionized water. Add remaining components and bring volume to 1.0L. Mix thoroughly. Gently heat and bring to boiling. Distribute into tubes in 10.0mL volumes. Autoclave for 15 min at 10 psi pressure–116°C.

Use: For the cultivation of fecal streptococci from a variety of clinical and nonclinical specimens. It is recommended for qualitative presumptive and confirmatory tests for fecal streptococci.

Baird–Parker Agar

Composition per liter:

Agar	17.0g
Glycine	12.0g
Sodium pyruvate	10.0g
Pancreatic digest of casein	10.0g
Beef extract	5.0g
LiCl	5.0g
Yeast extract	1.0g

pH 7.0 ± 0.2 at 25°C

Source: This medium is available as a premixed powder from Difco Laboratories, Oxoid Unipath, and BBL Microbiology Systems.

Preparation of Medium: Add components to distilled/deionized water and bring volume to 1.0L. Mix thoroughly. Gently heat and bring to boiling. Autoclave for 15 min at 15 psi pressure–121°C. Cool to 45–50°C. Pour into sterile Petri dishes.

Use: Used as a base for the preparation of Egg-Tellurite-Glycine-Pyruvate Agar for the selective isolation and enumeration of coagulase-positive staphylococci from food, skin, soil, air and other materials.

Baird–Parker Agar

Composition per liter:

Agar	17.0g
Glycine	12.0g
Sodium pyruvate	10.0g
Pancreatic digest of casein	10.0g
Beef extract	5.0g
LiCl	5.0g
Yeast extract	1.0g
Sulfamethazine solution	10.0mL

pH 7.0 ± 0.2 at 25°C

Sulfamethazine Solution:
Composition per 10mL:

Sulfamethazine	0.05g

Preparation of Sulfamethazine Solution: Add sulfamethazine to distilled/deionized water and bring volume to 10.0mL. Mix thoroughly. Filter sterilize.

Preparation of Medium: Add components, except sulfamethazine solution, to distilled/deionized water and bring volume to 990.0mL. Mix thoroughly. Gently heat and bring to boiling. Autoclave for 15 min at 15 psi pressure–121°C. Cool to 45°–50°C. Aseptically add sterile sulfamethazine solution. Mix thoroughly. Pour into sterile Petri dishes or distribute into sterile tubes.

Use: Used as a base for the preparation of Egg-Tellurite-Glycine-Pyruvate Agar for the selective isolation and enumeration of coagulase-positive staphylococci from food, skin, soil, air and other materials.

Baird–Parker Agar, Supplemented

Composition per liter:

Agar	17.0g
Glycine	12.0g
Sodium pyruvate	10.0g
Pancreatic digest of casein	10.0g
Beef extract	5.0g
LiCl	5.0g
Yeast extract	1.0g
RPF supplement	100.0mL

pH 7.0 ± 0.2 at 25°C

RPF Supplement:
Composition per 100mL:

Bovine fibrinogen	3.75g
Trypsin inhibitor	25.0mg
K_2TeO_3	25.0mg
Rabbit plasma	25.0mL

Caution: Potassium tellurite is toxic.

Preparation of RPF Supplement: Add components to distilled/deionized water and bring volume to 100.0mL. Mix thoroughly. Filter sterilize.

Preparation of Medium: Add components, except RPF supplement, to distilled/deionized water and bring volume to 900.0mL. Mix thoroughly. Gently heat and bring to boiling. Autoclave for 15 min at 15 psi pressure–121°C. Cool to 45–50°C. Aseptically add 100.0mL of filter-sterilized RPF supplement. Mix thoroughly but gently. Pour into sterile Petri dishes.

Use: For the selective isolation and enumeration of coagulase-positive staphylococci from food, skin, soil, air and other materials. Also, for differentiation and identification of staphylococci on the basis of their ability to coagulate plasma. Colonies surrounded by an opaque zone of coagulated plasma are diagnostic for *Staphylococcus aureus*.

Balamuth Medium

Composition per 200mL:

Dehydrated egg yolk	36.0g
Dried liver concentrate	1.0g
Rice starch	0.2g
Potassium phosphate buffer, pH 7.5	125.0mL
NaCl solution	125.0mL

pH 7.3 ± 0.2 at 25°C

NaCl Solution

Composition per 200mL:

NaCl	1.6g

Preparation of NaCl Solution: Add NaCl to distilled/deionized water and bring volume to 200.0mL. Mix thoroughly.

Potassium Phosphate Buffer, 0.067M

Composition per 200mL:

K_2HPO_4 (1M solution)	8.6mL
KH_2PO_4 (1M solution)	4.66mL

Preparation of Potassium Phosphate Buffer: Combine the K_2HPO_4 and KH_2PO_4 solutions. Bring volume to 200.0mL with distilled/deionized water. Adjust pH to 7.5.

Preparation of Medium: Add dehydrated egg yolk to 36.0mL of distilled/deionized water. Add 125.0mL of 0.8% NaCl. Mix thoroughly in a blender. Heat in a covered, double boiler until infusion reaches 80°C and maintain at this temperature for 20 min. Add 20.0mL of distilled/deionized H_2O. Filter through a layer of cheesecloth. To 90–100.0mL of filtrate add 0.8% NaCl solution to bring volume to 125.0mL. Autoclave for 20 min at 15 psi pressure–121°C. Cool to 4°C. Filter. To filtrate add an equal volume of 0.067M potassium phosphate buffer, pH 7.5. Add1.0g of dried liver concentrate. Mix thoroughly. Distribute into tubes or flasks in 10.0mL volumes. Autoclave for 20 min at 15 psi pressure–121°C. Prior to inoculation, add 0.01g of rice starch to each tube.

Use: For the cultivation and maintenance of *Entamoeba histolytica* and other intestinal protozoa.

BAM Agar

Composition per liter:

Agar	30.0g
Glucose	5.0g
KH_2PO_4	3.0g
Yeast extract	1.0g
$MgSO_4 \cdot 7H_2O$	0.5g
$CaCl_2 \cdot 2H_2O$	0.25g
$(NH_4)_2SO_4$	0.2g
Trace elements	1.0mL

pH 4.0 ± 0.2 at 25°C

Trace Elements:

Composition per liter:

$CaCl_2 \cdot 2H_2O$	0.66g
$Na_2MoO_4 \cdot 2H_2O$	0.30g
$ZnSO_4 \cdot 7H_2O$	0.18g
$CoCl_2 \cdot 6H_2O$	0.18g
$CuSO_4 \cdot 5H_2O$	0.16g
$MnSO_4 \cdot 4H_2O$	0.15g
H_3BO_3	0.10g

Preparation of Trace Elements: Add components to 1.0L of distilled/deionized water. Mix thoroughly.

Preparation of Medium: Add components, except agar to distilled/deionized water and bring volume to 800.0mL. Mix thoroughly. Gently heat and bring to boiling. Adjust medium to pH 4.0 with H_2SO_4. Add agar to 200.0mL distilled/deionized water. Autoclave agar separately to avoid acid hydrolysis. Autoclave for 15 min at 15 psi pressure–121°C. Mix two solutions together. Pour into sterile Petri dishes or distribute into sterile tubes.

Use: For the cultivation and maintenance of *Bacillus acidoterrestris*.

BAM Broth

Composition per liter:

Glucose	5.0g
KH_2PO_4	3.0g
Yeast extract	1.0g
$MgSO_4 \cdot 7H_2O$	0.5g
$CaCl_2 \cdot 2H_2O$	0.25g

$(NH_4)_2SO_4$	0.2g
Trace elements	1.0mL

pH 4.0 ± 0.2 at 25°C

Trace Elements:
Composition per liter:

$CaCl_2 \cdot 2H_2O$	0.66g
$Na_2MoO_4 \cdot 2H_2O$	0.30g
$ZnSO_4 \cdot 7H_2O$	0.18g
$CoCl_2 \cdot 6H_2O$	0.18g
$CuSO_4 \cdot 5H_2O$	0.16g
$MnSO_4 \cdot 4H_2O$	0.15g
H_3BO_3	0.10g

Preparation of Trace Elements: Add components to 1.0L of distilled/deionized water. Mix thoroughly.

Preparation of Medium: Add components to distilled/deionized water and bring volume to 1.0L. Mix thoroughly. Gently heat and bring to boiling. Adjust medium to pH 4.0 with H_2SO_4. Distribute into tubes or flasks. Autoclave for 15 min at 15 psi pressure–121°C.

Use: For the cultivation and maintenance of *Bacillus acidoterrestris*.

BAM SM Agar

Composition per liter:

Agar	20.0g
Glucose	5.0g
KH_2PO_4	3.0g
Yeast extract	6.0g
$MgSO_4 \cdot 7H_2O$	0.5g
$CaCl_2 \cdot 2H_2O$	0.25g
$(NH_4)_2SO_4$	0.2g
Trace elements	1.0mL

pH 4.0 ± 0.2 at 25°C

Trace Elements:
Composition per liter:

$CaCl_2 \cdot 2H_2O$	0.66g
$Na_2MoO_4 \cdot 2H_2O$	0.30g
$ZnSO_4 \cdot 7H_2O$	0.18g
$CoCl_2 \cdot 6H_2O$	0.18g
$CuSO_4 \cdot 5H_2O$	0.16g
$MnSO_4 \cdot 4H_2O$	0.15g
H_3BO_3	0.10g

Preparation of Trace Elements: Add components to 1.0L of distilled/deionized water. Mix thoroughly.

Preparation of Medium: Add components, except agar to distilled/deionized water and bring volume to 800.0mL. Mix thoroughly. Gently heat and bring to boiling. Adjust medium to pH 4.0 with H_2SO_4. Add agar to 200.0mL distilled/deionized water. Autoclave agar separately to avoid acid hydrolysis. Autoclave for 15 min at 15 psi pressure–121°C. Mix two solutions together. Pour into sterile Petri dishes or distribute into sterile tubes.

Use: For the cultivation and maintenance of *Bacillus cycloheptanicus*.

BAM SM Broth

Composition per liter:

Glucose	5.0g
KH_2PO_4	3.0g
Yeast extract	6.0g
$MgSO_4 \cdot 7H_2O$	0.5g
$CaCl_2 \cdot 2H_2O$	0.25g
$(NH_4)_2SO_4$	0.2g
Trace elements	1.0mL

pH 4.0 ± 0.2 at 25°C

Trace Elements:
Composition per liter:

$CaCl_2 \cdot 2H_2O$	0.66g
$Na_2MoO_4 \cdot 2H_2O$	0.30g
$ZnSO_4 \cdot 7H_2O$	0.18g
$CoCl_2 \cdot 6H_2O$	0.18g
$CuSO_4 \cdot 5H_2O$	0.16g
$MnSO_4 \cdot 4H_2O$	0.15g
H_3BO_3	0.10g

Preparation of Trace Elements: Add components to 1.0L of distilled/deionized water. Mix thoroughly.

Preparation of Medium: Add components to distilled/deionized water and bring volume to 1.0L. Mix thoroughly. Gently heat and bring to boiling. Adjust medium to pH 4.0 with H_2SO_4. Distribute into tubes or flasks. Autoclave for 15 min at 15 psi pressure–121°C.

Use: For the cultivation and maintenance of *Bacillus cycloheptanicus*.

Basal Mineral Medium

Composition per liter:

NH_4Cl	0.80g
K_2HPO_4	0.70g
$MgSO_4 \cdot 7H_2O$	0.01g
Disodium EDTA	9.2mg
$FeSO_4 \cdot 7H_2O$	7.0mg
$CaSO4 \cdot 2H_2O$	2.0mg
H_3BO_3	0.1mg
$ZnSO_4 \cdot 7H_2O$	0.1mg
$MnSO_4 \cdot 4H_2O$	0.02mg
$Co(NO_3)_2$	0.01mg
$NaMoO_4 \cdot 2H_2O$	0.01mg
$CuSO_4 \cdot 5H_2O$	0.5µg

Preparation of Medium: Add components to distilled/deionized water and bring volume to 1.0L. Mix thoroughly. Filter sterilize.

Use: For the cultivation of *Beggiatoa* species.

Basal Synthetic Medium

Composition per liter:

L-Glutamic acid	20.0g
$(NH_4)_2SO_4$	4.0g
K_2HPO_4	1.88g
KH_2PO_4	0.57g
$MgSO_4 \cdot 7H_2O$	0.2g
Salt solution	10.0mL

Salt Solution:

Composition per liter:

$FeCl_3 \cdot 6H_2O$	0.6g
$MnCl_2 \cdot 4H_2O$	0.6g
$ZnCl_2$	0.6g
$CuSO_4 \cdot 5H_2O$	0.6g
$CaCl_2 \cdot 2H_2O$	0.6g
NaCl	0.6g

Preparation of Salt Solution: Add components to 1.0L of distilled/deionized water. Mix thoroughly.

Preparation of Medium: Add components to distilled/deionized water and bring volume to 1.0L. Mix thoroughly. Gently heat and bring to boiling. Distribute into tubes or flasks. Autoclave for 15 min at 15 psi pressure–121°C.

Use: For the cultivation and maintenance of *Acinetobacter lwoffii.*

Basal Thermophile Medium

Composition per liter:

Solution 1	850.0mL
Solution 2	100.0mL
Solution 3	50.0mL

Solution 1:

Composition per 850mL:

Pancreatic digest of casein	10.0g
K_2HPO_4	1.5g
NH_4Cl	0.9g
KH_2PO_4	0.75g
$MgCl_2 \cdot 6H_2O$	0.2g
Trace element solution	9.0mL
Vitamin solution	5.0mL
Resazurin (0.2% solution)	1.0mL
$FeSO_4 \cdot 7H_2O$ (10% solution)	0.03mL

Preparation of Solution 1: Add components to distilled/deionized water and bring volume to 850.0mL. Mix thoroughly. Autoclave for 45 min at 15 psi pressure–121°C. Cool to 45°–50°C.

Solution 2:

Composition per 100mL:

Yeast extract	3.0g

Preparation of Solution 2: Add yeast extract to distilled/deionized water and bring volume to 100.0mL. Mix thoroughly. Autoclave for 45 min at 15 psi pressure–121°C. Cool to 45°–50°C.

Solution 3:

Composition per 50mL:

Glucose	5.0g

Preparation of Solution 3: Add glucose to distilled/deionized water and bring volume to 50.0mL. Mix thoroughly. Autoclave for 45 min at 15 psi pressure–121°C. Cool to 45°–50°C.

Trace Element Solution:

Composition per liter:

Nitrilotriacetic acid	12.5g
NaCl	1.0g
$FeCl_3 \cdot 4H_2O$	0.2g
$MnCl_2 \cdot 4H_2O$	0.1g
$CaCl_2 \cdot 2H_2O$	0.1g
$ZnCl_2$	0.1g
$CuCl_2$	0.02g
Na_2SeO_3	0.02g
$CoCl_2 \cdot 6H_2O$	0.017g
H_3BO_3	0.01g
$Na_2MoO_4 \cdot 2H_2O$	0.01g

Preparation of Trace Element Solution: Add nitrilotriacetic acid to 100.0mL of distilled/deionized water. Adjust pH to 6.5 with KOH. Add remaining components and bring volume to 1.0L. Mix thoroughly.

Wolfe's Vitamin Solution:

Composition per liter:

Pyridoxine·HCl	10.0mg
Thiamine·HCl	5.0mg
Riboflavin	5.0mg
Nicotinic acid	5.0mg
Calcium pantothenate	5.0mg
p-Aminobenzoic acid	5.0mg
Thioctic acid	5.0mg
Biotin	2.0mg
Folic acid	2.0mg
Cyanocobalamin	0.1mg

Preparation of Wolfe's Vitamin Solution: Add components to distilled/deionized water and bring volume to 1.0L. Mix thoroughly.

$Na_2S \cdot 9H_2O$ Solution:

Composition per 100mL:

$Na_2S \cdot 9H_2O$	10.0g

Preparation of $Na_2S \cdot 9H_2O$ Solution: Add $Na_2S \cdot 9H_2O$ to distilled/deionized water and bring volume to 100.0mL. Mix thoroughly. Autoclave for 15 min at 15 psi pressure–121°C.

Preparation of Medium: Aseptically combine solution 1, solution 2 and solution 3 under 100% N_2. Distribute into tubes in 10.0mL volumes under 100% N_2. Immediately prior to inoculation, aseptically add 0.1mL of sterile $Na_2S \cdot 9H_2O$ solution to each tube.

Use: For the cultivation and maintenance of *Clostridium* species, *Fervidobacterium nodosum* and *Thermoanaerobium brockii*.

Base Agar
See: **Antibiotic Medium 2**

Base Agar with Low pH
See: **Antibiotic Medium 8**

Base Layer Agar with Nutrient Overlay Agar

Composition per 2.5L:

Fat substrate	50.0g
Nutrient agar	1.5L
Basal medium	1.0L

Fat Substrate:
Composition:

Fat	50.0g

Preparation of Fat Substrate: Tributyrin, corn oil, soybean oil, any cooking oil, lard, tallow, or triglycerides that do not contain antioxidants or other inhibitory substances may be used. Remove free fatty acids in the fat substrate by dissolving 50.0g of fat substrate in 500.0mL of petroleum ether. Pass the solution through an activated alumina column. Remove the petroleum ether by evaporation on a steam table under 100% N_2. Autoclave for 30 min at 15 psi pressure–121°C. Cool to 50°C.

Nutrient Agar:
Composition per liter:

Agar	15.0g
Pancreatic digest of gelatin	5.0g
Beef extract	3.0g

Preparation of Nutrient Agar: Add components to distilled/deionized water and bring volume to 1.0L. Mix thoroughly. Gently heat while stirring and bring to boiling. Distribute into tubes or flasks. Autoclave for 15 min at 15 psi pressure–121°C. Cool to 45°–50°C.

Source: Available as a premixed powder from Difco Laboratories and BBL Microbiology Systems.

Basal Medium:
Composition per liter:

Agar	15.0g
Victoria Blue B solution	200.0mL

Preparation of Basal Medium: Add agar to 800.0mL of distilled/deionized water. If tributyrin is used as the fat substrate, add agar to 1.0L of distilled/deionized water. Autoclave for 15 min at 15 psi pressure–121°C. Cool to 50°C. If tributyrin is not used as the fat substrate, aseptically add 200.0mL of Victoria Blue B solution. Mix thoroughly.

Victoria Blue B Solution:
Composition per 200mL:

Victoria Blue B	0.12g

Preparation of Victoria Blue B Solution: Add the Victoria Blue B to 200.0mL of distilled/deionized water. Mix thoroughly. Filter sterilize. Warm to 50°C.

Preparation of Medium: Aseptically combine 1.0L of sterile basal medium with 50.0g of sterile fat substrate in a warm sterile blender container. Blend for 1 min until homogenized. Rapidly pour into sterile Petri dishes in 7.0mL volumes. Dry the surface of the plates by partially opening the lids in a laminar flow hood for 15 min. Add dilution of food samples to be tested. When the inoculum is dry, pour nutrient agar as an overlay onto each plate. Use 10–12mL of nutrient agar per plate.

Use: For the isolation, cultivation and identification of lipolytic microorganisms from food.

BBE Agar
See: ***Bacteroides* Bile Esculin Agar**

BBGS Agar
See: **Bile Salts Brilliant Green Starch Agar**

BCM
See: ***Bacillus cereus* Medium**

BC Medium (Medium for *Acetivibrio cellulolyticus*)

Composition per liter:

Cellulose powder	3.0g
$NaHCO_3$	2.0g
Mineral solution 1	75.0mL

Mineral solution 2 ....75.0mL
Cysteine-sulfide reducing solution 12.8mL
$FeSO_4 \cdot 7H_2O$ solution 10.0mL
Vitamin mixture 10.0mL
Wolfe's mineral solution 10.0mL
Resazurin (0.1% solution)..............................1.0mL
pH 7.6 ± 0.2 at 25°C

Caution: This medium contains sodium sulfide and may produce toxic H_2S gas. Prepare in a chemical fume hood.

Mineral Solution 1:
Composition per liter:
K_2HPO_4 ....3.9g

Preparation of Mineral Solution 1: Add K_2HPO_4 to distilled/deionized water and bring volume to 1.0L. Mix thoroughly.

Mineral Solution 2:
Composition per liter:
NH_4Cl..12.0g
Na_2SO_4 ....2.5g
KH_2PO_4..2.4g
$MgSO_4 \cdot 7H_2O$ 1.2g
$CaCl_2 \cdot 2H_2O$..0.8g

Preparation of Mineral Solution 2: Add components to distilled/deionized water and bring volume to 1.0L. Mix thoroughly.

$FeSO_4 \cdot 7H_2O$ Solution:
Composition per 100mL:
$FeSO_4 \cdot 7H_2O$..0.2g

Preparation of $FeSO_4 \cdot 7H_2O$ Solution: Dissolve $FeSO_4 \cdot 7H_2O$ in 100.0mL distilled/deionized water. Add 3 drops concentrated HCl. Mix thoroughly.

Vitamin Mixture:
Composition per liter:
Pyridoxine·HCl 10.0mg
Thiamine·HCl.. 5.0mg
Cyanocobalamin 5.0mg
Lipoic acid (thioctic acid) 5.0mg
Biotin 2.0mg
p-Aminobenzoic acid 0.5mg

Preparation of Vitamin Mixture: Add components to distilled/deionized water and bring volume to 1.0L. Store below –20°C.

Wolfe's Mineral Solution:
Composition per liter
$MgSO_4 \cdot 7H_2O$ 3.0g
Nitriloacetic acid..1.5g
$MnSO_4 \cdot H_2O$ 0.5g
NaCl 1.0g
$FeSO_4 \cdot 7H_2O$..0.1g
$CoCl_2 \cdot 6H_2O$..0.1g
$CaCl_2$ 0.1g
$ZnSO_4 \cdot 7H_2O$..0.1g
$CuSO_4 \cdot 5H_2O$ 0.01g
$AlK(SO_4)_2 \cdot 12H_2O$ 0.01g
H_3BO_3 ....0.01g
$Na_2MoO_4 \cdot 2H_2O$..0.01g

Preparation of Wolfe's Mineral Solution: Add nitrilotriacetic acid to 500.0mL of distilled/deionized water and adjust to pH 6.5 with KOH to dissolve. Bring volume to 1.0L with distilled/deionized water. Add remaining components one at a time.

Cysteine-Sulfide Reducing Solution:
Composition per 200mL:
L-Cysteine·HCl·H_2O..2.5g
$Na_2S \cdot 9H_2O$ 2.5g

Preparation of Cysteine-Sulfide Reducing Solution: Add L-Cysteine·HCl·H_2O to 50.0mL distilled/deionized water. Quickly adjust pH to 10 with fresh 3*N* NaOH and flush under 100% N_2. Add $Na_2S \cdot 9H_2O$. Bring volume to 200.0mL with distilled/deionized water. Boil under 100% N_2. Transfer anaerobically to tubes or flasks and stopper. Autoclave for 15 min at 15 psi pressure–121°C.

Preparation of Medium: Add cellulose and $NaHCO_3$ to distilled/deionized water and bring volume to 800.0mL. Add all other components except cysteine-sulfide reducing solution. Heat and boil under 90% N_2 + 10% CO_2. Cool and continue flushing under 90% N_2 + 10% CO_2. The pH should be 7.6 at room temperature; do not adjust. Add 8.0mL cysteine-sulfide reducing solution. Add 4.8mL more of cysteine-sulfide reducing solution. Distribute anaerobically into tubes in 7.0mL volumes and cap.

Use: For the cultivation and maintenance of *Acetivibrio cellulolyticus*, *Acetivibrio cellulosolvens*, *Bacteroides cellulosolvens* and other cellulose degrading microorganisms.

BC Motility Medium (*Bacillus cereus* Motility Medium)

Composition per liter:
Pancreatic digest of casein..............................10.0g
Glucose 5.0g
Agar..3.0g
Na_2HPO_4 ....2.5g
Yeast extract 2.5g
pH 7.4 ± 0.2 at 25°C

Preparation of Medium: Add components to distilled/deionized water and bring volume to 1.0L. Mix thoroughly. Gently heat and bring to boiling. Distribute into tubes in 2.0mL volumes. Autoclave for 15 min at 15 psi pressure–121°C.

Use: For the cultivation and observation of motility of *Bacillus cereus*.

BCA
See: Bacterial Cell Agar

BCP Azide Broth (Bromcresol Purple Azide Broth)

Composition per liter:

Casein peptone	10.0g
Yeast extract	10.0g
D-Glucose	5.0g
NaCl	5.0g
K_2HPO_4	2.7g
KH_2PO_4	2.7g
NaN_3	0.5g
Bromcresol purple	0.032g

pH 6.9 ± 0.2 at 25°C

Caution: Sodium azide is toxic. Azides also react with metals and disposal must be highly diluted.

Preparation of Medium: Add components, except cysteine, to 900.0mL distilled/deionized water. Mix thoroughly. Gently heat to boiling. Distribute into tubes or flasks. Autoclave for 15 min at 15 psi pressure–121°C.

Use: For use in the confirmation test for the presence of fecal streptococci in water and wastewater.

BCP Broth
See: Bromcresol Purple Dextrose Broth

BCP D Agar (Bromcresol Purple Deoxycholate Agar)

Composition per liter:

Agar	25.0g
Lactose	10.0g
Sucrose	10.0g
Pancreatic digest of casein	7.5g
Thiopeptone	7.5g
NaCl	5.0g
Yeast extract	2.0g
Sodium citrate	2.0g
Sodium deoxycholate	1.0g
Bromcresol Purple	0.02g

pH 7.2 ± 0.2 at 25°C

Preparation of Medium: Add components to distilled/deionized water and bring volume to 1.0L. Mix thoroughly. Gently heat and bring to boiling. Pour into sterile Petri dishes without sterilization. Do not autoclave. Use the same day.

Use: For the isolation, cultivation, and differentiation of Gram-negative enteric bacilli from clinical and nonclinical specimens. For the isolation, cultivation and identification of microorganisms from fecal specimens. For the isolation and cultivation of *Salmonella, Shigella* and other nonlactose and nonsucrose fermenting microorganisms. Nonlactose/nonsucrose fermenting microorganisms appear as colorless or blue colonies. Lactose/sucrose fermenting microorganisms, such as coliform bacteria, appear as yellow-opaque white colonies surrounded by a zone of precipitated deoxycholate.

BCP DCLS Agar (Bromcresol Purple Deoxycholate Citrate Lactose Sucrose Agar)

Composition per liter:

Agar	14.0g
Sodium citrate	10.0g
Lactose	7.5g
Sucrose	7.5g
Pancreatic digest of casein	7.5g
Peptone	7.5g
NaCl	5.0g
$Na_2S_2O_3 \cdot 5H_2O$	5.0g
Yeast extract	3.0g
Meat extract	3.0g
Sodium deoxycholate	2.5g
Bromcresol Purple	0.02g

pH 7.2 ± 0.2 at 25°C

Preparation of Medium: Add components to distilled/deionized water and bring volume to 1.0L. Mix thoroughly. Gently heat and bring to boiling. Pour into sterile Petri dishes without sterilization. Do not autoclave. Use the same day.

Use: For the differential isolation of Gram-negative enteric bacilli from clinical and nonclinical specimens. For the isolation and identification of microorganisms from fecal specimens. For the isolation of *Salmonella*, *Shigella* and other nonlactose and nonsucrose fermenting microorganisms. Nonlactose/nonsucrose fermenting microorganisms appear as colorless or blue colonies. Lactose/sucrose fermenting microorganisms, such as coliform bacteria, appear as yellow-opaque white colonies surrounded by a zone of precipitated deoxycholate.

BCP MS G Agar
See: **Bromocresol Purple Milk Solids Glucose Agar**

BCYE Agar (BCYE Alpha Base) (Buffered Charcoal Yeast Extract Agar)

Composition per liter:

Agar 15.0g
Yeast extract 10.0g
ACES buffer (2-[(2-Amino-2-oxoethyl)-amino]-ethane sulfonic acid) 10.0g
Charcoal, activated 2.0g
α-Ketogluatrate 1.0g
L-Cysteine·HCl·H_2O 0.4g
$Fe_4(P_2O_7)_3 \cdot 9H_2O$ 0.25g
pH 6.9 ± 0.2 at 25°C

Source: This medium is available as a premixed powder from BBL Microbiology Systems.

Preparation of Medium: Add components, except cysteine, to distilled/deionized water and bring volume to 1.0L. Mix thoroughly. Adjust medium to pH 6.9 with 1*N* KOH. Heat gently and bring to boil for 1 minute. Autoclave for 15 min at 15 psi pressure–121°C. Cool to 50–55°C. Add 4.0mL of a 10% solution of L-cysteine·HCl·H_2O which has been filter-sterilized. Mix thoroughly. Pour into sterile Petri dishes with constant agitation to keep charcoal in suspension.

Use: For the isolation, cultivation, and maintenance of *Legionella pneumophila* and other *Legionella* species from environmental and clinical specimens.

BCYE α Agar, Modified
See: ***Legionella*** **Agar Base**

BCYEα with Alb (Buffered Charcoal Yeast Extract Agar with Albumin)

Composition per liter:

Agar 15.0g
Yeast extract 10.0g
ACES buffer (2-[(2-Amino-2-oxoethyl)-amino]-ethane sulfonic acid) 10.0g
Charcoal, activated 2.0g
α-Ketogluatrate 1.0g
Bovine serum albumin solution 10.0mL
Cysteine·HCl·H_2O solution 10.0mL
$Fe_4(P_2O_7)_3 \cdot 9H_2O$ solution 10.0mL
pH 6.9 ± 0.2 at 25°C

Bovine Serum Albumin Solution:
Composition per 10mL:

Bovine serum albumin 0.1g

Preparation of Bovine Serum Albumin Solution: Add bovine serum albumin to distilled/deionized water and bring volume to 10.0mL. Mix thoroughly. Filter sterilize.

Cysteine·HCl·H_2O Solution:
Composition per 10mL:

L-Cysteine·HCl·H_2O 0.4g

Preparation of Cysteine·HCl·H_2O Solution: Add cysteine·HCl·H_2O to distilled/deionized water and bring volume to 10.0mL. Mix thoroughly. Filter sterilize.

$Fe_4(P_2O_7)_3 \cdot 9H_2O$ Solution:
Composition per 10mL:

$Fe_4(P_2O_7)_3 \cdot 9H_2O$ 0.25g

Preparation of $Fe_4(P_2O_7)_3 \cdot 9H_2O$ Solution: Add $Fe_4(P_2O_7)_3 \cdot 9H_2O$ to distilled/deionized water and bring volume to 10.0mL. Mix thoroughly. Filter sterilize.

Preparation of Medium: Add components—except cysteine·HCl·H_2O solution, $Fe_4(P_2O_7)_3 \cdot 9H_2O$ solution, and bovine serum albumin solution—to distilled/deionized water and bring volume to 970.0mL. Mix thoroughly. Adjust medium to pH 6.9 with 1*N* KOH. Heat gently and bring to boil for 1 minute. Autoclave for 15 min at 15 psi pressure–121°C. Cool to 50–55°C. Aseptically add the cysteine·HCl·H_2O solution, $Fe_4(P_2O_7)_3 \cdot 9H_2O$ solution and 10.0mL of sterile bovine serum albumin solution. Mix thoroughly. Pour into sterile Petri dishes with constant agitation to keep charcoal in suspension.

Use: For the isolation, cultivation, and maintenance of *Legionella pneumophila* and other *Legionella* species from environmental and clinical specimens.

BCYEα without L-Cysteine (Buffered Charcoal Yeast Extract Agar without L-Cysteine)

Composition per liter:

Agar 15.0g
Yeast extract 10.0g
ACES buffer (2-[(2-Amino-2-oxoethyl)-amino]-ethane sulfonic acid) 10.0g

Charcoal, activated 2.0g
α-Ketogluatrate 1.0g
$Fe_4(P_2O_7)_3 \cdot 9H_2O$ solution 10.0mL
pH 6.9 ± 0.2 at 25°C

$Fe_4(P_2O_7)_3 \cdot 9H_2O$ Solution:
Composition per 10mL:
$Fe_4(P_2O_7)_3 \cdot 9H_2O$ 0.25g

Preparation of $Fe_4(P_2O_7)_3 \cdot 9H_2O$ Solution: Add $Fe_4(P_2O_7)_3 \cdot 9H_2O$ to distilled/deionized water and bring volume to 10.0mL. Mix thoroughly. Filter sterilize.

Preparation of Medium: Add components, except $Fe_4(P_2O_7)_3 \cdot 9H_2O$ solution, to distilled/deionized water and bring volume to 990.0mL. Mix thoroughly. Adjust medium to pH 6.9 with 1*N* KOH. Heat gently and bring to boil for 1 minute. Autoclave for 15 min at 15 psi pressure–121°C. Cool to 50–55°C. Aseptically add 10.0mL of sterile $Fe_4(P_2O_7)_3 \cdot 9H_2O$ solution. Mix thoroughly. Pour into sterile Petri dishes with constant agitation to keep charcoal in suspension.

Use: For the isolation, cultivation, and maintenance of *Legionella pneumophila* and other *Legionella* species from environmental and clinical specimens.

BCYE Differential Agar (Buffered Charcoal Yeast Extract Differential Agar)

Composition per liter:
Agar 15.0g
Yeast extract 10.0g
ACES buffer (2-[(2-Amino-2-oxoethyl)-amino]-ethane sulfonic acid) 10.0g
Charcoal, activated 2.0g
α-Ketogluatrate 1.0g
L-Cysteine·HCl·H_2O 0.4g
$Fe_4(P_2O_7)_3 \cdot 9H_2O$ 0.25g
Bromcresol Purple 0.01g
Bromthymol Blue 0.01g
pH 6.9 ± 0.2 at 25°C

Source: This medium is available as a premixed powder from BBL Microbiology Systems.

Preparation of Medium: Add components, except cysteine, to distilled/deionized water and bring volume to 1.0L. Mix thoroughly. Adjust medium to pH 6.9 with 1*N* KOH. Heat gently and bring to boil for 1 minute. Autoclave for 15 min at 15 psi pressure–121°C. Cool to 50–55°C. Add 4.0mL of a 10% solution of L-cysteine·HCl·H_2O which has been filter-sterilized. Mix thoroughly. Pour into sterile Petri dishes with constant agitation to keep charcoal in suspension.

Use: For the isolation, cultivation, and maintenance of *Legionella pneumophila* and other *Legionella* species from environmental and clinical specimens. For the presumptive differential identification of *Legionella* species based on colony color and morphology. *Legionella pneumophila* appear as light blue/green colonies. *Legionella micdadei* appear as blue/gray or dark blue colonies.

BCYE Medium, Diphasic Blood Culture (Buffered Charcoal Yeast Extract Medium, Diphasic Blood Culture)

Composition per liter:
Agar phase 1.0L
Broth phase 1.0L
pH 6.9 ± 0.2 at 25°C

Agar Phase:
Composition per liter:
Agar 20.0g
ACES buffer (2-[(2-Amino-2-oxoethyl)-amino]-ethane sulfonic acid) 10.0g
Yeast extract 10.0g
Charcoal, activated, acid washed 4.0g
KOH 2.80g
α-Ketoglutarate 1.0g
Cysteine·HCl·H_2O solution 10.0mL
$Fe_4(P_2O_7)_3 \cdot 9H_2O$ solution 10.0mL

Cysteine·HCl·H_2O Solution:
Composition per 10mL:
L-Cysteine·HCl·H_2O 0.4g

Preparation of Cysteine·HCl·H_2O Solution: Add cysteine·HCl·H_2O to distilled/deionized water and bring volume to 10.0mL. Mix thoroughly. Filter sterilize.

$Fe_4(P_2O_7)_3 \cdot 9H_2O$ Solution:
Composition per 10mL:
$Fe_4(P_2O_7)_3 \cdot 9H_2O$ 0.25g

Preparation of $Fe_4(P_2O_7)_3 \cdot 9H_2O$ Solution: Add $Fe_4(P_2O_7)_3 \cdot 9H_2O$ to distilled/deionized water and bring volume to 10.0mL. Mix thoroughly. Filter sterilize.

Preparation of Agar Phase: Add components, except cysteine·HCl·H_2O solution and $Fe_4(P_2O_7)_3$ solution, to distilled/deionized water and bring volume to 980.0mL. Mix thoroughly. Adjust medium to pH 6.9 with 1*N* KOH. Heat gently and bring to boiling for 1 minute. Autoclave for 15 min at 15 psi pressure–121°C. Cool to 50–55°C. Aseptically add the cysteine·HCl·H_2O solution and $Fe_4(P_2O_7)_3 \cdot 9H_2O$ solution. Mix thoroughly.

Broth Phase:

Composition per liter:

ACES buffer (2-[(2-Amino-2-oxoethyl)-amino]-ethane sulfonic acid) 10.0g
Yeast extract 10.0g
Charcoal, activated, acid washed 4.0g
KOH 2.40g
α-Ketoglutarate 1.0g
Sodium polyaneolsulfonate 0.30g
Cysteine·HCl·H_2O solution 10.0mL
$Fe_4(P_2O_7)_3$·$9H_2O$ solution 10.0mL

Cysteine·HCl·H_2O Solution:

Composition per 10mL:

L-Cysteine·HCl·H_2O 0.4g

Preparation of Cysteine·HCl·H_2O Solution: Add cysteine·HCl·H_2O to distilled/deionized water and bring volume to 10.0mL. Mix thoroughly. Filter sterilize.

$Fe_4(P_2O_7)_3$·$9H_2O$ Solution:

Composition per 10mL:

$Fe_4(P_2O_7)_3$·$9H_2O$ 0.25g

Preparation of $Fe_4(P_2O_7)_3$·$9H_2O$ Solution: Add $Fe_4(P_2O_7)_3$·$9H_2O$ to distilled/deionized water and bring volume to 10.0mL. Mix thoroughly. Filter sterilize.

Preparation of Broth Phase: Add components, except cysteine·HCl·H_2O solution and $Fe_4(P_2O_7)_3$ solution, to distilled/deionized water and bring volume to 980.0mL. Mix thoroughly. Adjust medium to pH 6.9 with 1*N* KOH. Heat gently and bring to boiling for 1 minute. Autoclave for 15 min at 15 psi pressure–121°C. Cool to 50–55°C. Aseptically add the cysteine·HCl·H_2O solution and $Fe_4(P_2O_7)_3$·$9H_2O$ solution. Mix thoroughly.

Preparation of Medium: Aseptically distribute cooled sterile agar phase into sterile blood culture bottles in 100.0mL volumes. Allow bottles to cool in a slanted position. Aseptically add 50.0mL of sterile broth phase to each blood culture bottle.

Use: For the isolation and cultivation of *Legionella pneumophila* and other *Legionella* species from blood samples.

BCYE Selective Agar with CCVC (Buffered Charcoal Yeast Extract Selective Agar with Cephalothin, Colistin, Vancomycin and Cycloheximide)

Composition per 1014mL:

Agar 15.0g
Yeast extract 10.0g
ACES buffer (2-[(2-Amino-2-oxoethyl)-amino]-ethane sulfonic acid) 10.0g
Charcoal, activated 2.0g
α-Ketogluatrate 1.0g
$Fe_4(P_2O_7)_3$·$9H_2O$ 0.25g
Antibiotic solution 10.0mL
Cysteine·HCl·H_2O solution 4.0mL

pH 6.9 ± 0.2 at 25°C

Source: This medium is available as a premixed powder from BBL Microbiology Systems.

Cysteine·HCl·H_2O Solution:

Composition per 10mL:

L-Cysteine·HCl·H_2O 1.0g

Preparation of Cysteine·HCl·H_2O Solution: Add L-Cysteine·HCl·H_2O to distilled/deionized water and bring volume to 10.0mL. Mix thoroughly. Filter sterilize.

Antibiotic Solution:

Composition per 10mL:

Cycloheximide 80.0mg
Colistin 16.0mg
Cephalothin 4.0mg
Vancomycin 0.5mg

Preparation of Antibiotic Solution: Add components to distilled/deionized water and bring volume to 10.0mL. Mix thoroughly. Filter sterilize.

Preparation of Medium: Add components, except cysteine and antibiotic solution, to distilled/deionized water and bring volume to 1.0L. Mix thoroughly. Adjust medium to pH 6.9 with 1*N* KOH. Heat gently and bring to boil for 1 minute. Autoclave for 15 min at 15 psi pressure–121°C. Cool to 50–55°C. Add 4.0mL of L-cysteine·HCl·H_2O solution and 10.0mL of sterile antibiotic solution. Mix thoroughly. Pour into sterile Petri dishes with constant agitation to keep charcoal in suspension.

Use: For the isolation, cultivation, and maintenance of *Legionella pneumophila* and other *Legionella* species from environmental and clinical specimens. Used for the selective recovery of *Legionella pneumophila* while reducing contaminating microorganisms from environmental water samples.

BCYE Selective Agar with GPVA (Buffered Charcoal Yeast Extract Selective Agar with Glycine, Polymyxin B, Vancomycin, and Anisomycin)

Composition per 1014mL:

Agar 15.0g

Yeast extract	10.0g
ACES buffer (2-[(2-Amino-2-oxoethyl)-amino]-ethane sulfonic acid)	10.0g
Charcoal, activated	2.0g
α-Ketogluatrate	1.0g
$Fe_4(P_2O_7)_3 \cdot 9H_2O$	0.25g
Antibiotic solution	10.0mL
Cysteine·HCl·H_2O solution	4.0mL

pH 6.9 ± 0.2 at 25°C

Cysteine·HCl·H_2O Solution:
Composition per 10mL:

L-Cysteine·HCl·H_2O	1.0g

Preparation of Cysteine·HCl·H_2O Solution: Add L-Cysteine·HCl·H_2O to distilled/deionized water and bring volume to 10.0mL. Mix thoroughly. Filter sterilize.

Antibiotic Solution:
Composition per 10mL:

Glycine	3.0g
Anisomycin	0.08g
Vancomycin	5.0mg
Polymyxin B	100,000U

Preparation of Antibiotic Solution: Add components to distilled/deionized water and bring volume to 10.0mL. Mix thoroughly. Filter sterilize.

Preparation of Medium: Add components, except cysteine solution and antibiotic solution, to distilled/deionized water and bring volume to 1.0L. Mix thoroughly. Adjust medium to pH 6.9 with 1*N* KOH. Heat gently and bring to boil for 1 minute. Autoclave for 15 min at 15 psi pressure–121°C. Cool to 50–55°C. Add 4.0mL of L-cysteine·HCl·H_2O solution and 10.0mL of sterile antibiotic solution. Mix thoroughly. Pour into sterile Petri dishes with constant agitation to keep charcoal in suspension.

Use: For the isolation, cultivation, and maintenance of *Legionella pneumophila* and other *Legionella* species from environmental and clinical specimens. Used for the selective recovery of *Legionella pneumophila* while reducing contaminating microorganisms from potable water samples.

BCYE Selective Agar with GVPC (Buffered Charcoal Yeast Extract Selective Agar with Glycine, Vancomycin, Polymyxin B, and Cycloheximide)

Composition per 1014mL:

Agar	15.0g
Yeast extract	10.0g
ACES buffer (2-[(2-Amino-2-oxoethyl)-amino]-ethane sulfonic acid)	10.0g
Charcoal, activated	2.0g
α-Ketogluatrate	1.0g
$Fe_4(P_2O_7)_3 \cdot 9H_2O$	0.25g
Antibiotic solution	10.0mL
Cysteine·HCl·H_2O solution	4.0mL

pH 6.9 ± 0.2 at 25°C

Source: This medium is available as a premixed powder from Oxoid Unipath.

Cysteine·HCl·H_2O Solution:
Composition per 10mL:

L-Cysteine·HCl·H_2O	1.0g

Preparation of Cysteine·HCl·H_2O Solution: Add L-Cysteine·HCl·H_2O to distilled/deionized water and bring volume to 10.0mL. Mix thoroughly. Filter sterilize.

Antibiotic Solution:
Composition per 10mL:

Glycine	3.0g
Cycloheximide	0.08g
Vancomycin	1.0mg
Polymyxin B	79,200U

Preparation of Antibiotic Solution: Add components to distilled/deionized water and bring volume to 10.0mL. Mix thoroughly. Filter sterilize.

Preparation of Medium: Add components, except cysteine solution and antibiotic solution, to distilled/deionized water and bring volume to 1.0L. Mix thoroughly. Adjust medium to pH 6.9 with 1*N* KOH. Heat gently and bring to boil for 1 minute. Autoclave for 15 min at 15 psi pressure–121°C. Cool to 50–55°C. Add 4.0mL of L-cysteine·HCl·H_2O solution and 10.0mL of sterile antibiotic solution. Mix thoroughly. Pour into sterile Petri dishes with constant agitation to keep charcoal in suspension.

Use: For the isolation, cultivation, and maintenance of *Legionella pneumophila* and other *Legionella* species from environmental and clinical specimens. Used for the selective recovery of *Legionella pneumophila* while reducing contaminating microorganisms from potable water samples.

BCYE Selective Agar with PAC (Buffered Charcoal Yeast Extract Selective Agar with Polymyxin B, Anisomycin and Cefamandole)

Composition per 1014mL:

Agar	15.0g
Yeast extract	10.0g

ACES buffer (2-[(2-Amino-2-oxoethyl)-amino]-ethane sulfonic acid) 10.0g
Charcoal, activated 2.0g
α-Ketogluatrate 1.0g
$Fe_4(P_2O_7)_3 \cdot 9H_2O$ 0.25g
Antibiotic solution 10.0mL
Cysteine·HCl·H_2O solution 4.0mL

pH 6.9 ± 0.2 at 25°C

Source: This medium is available as a premixed powder from BBL Microbiology Systems.

Cysteine·HCl·H_2O Solution:
Composition per 10mL:
L-Cysteine·HCl·H_2O 1.0g

Preparation of Cysteine·HCl·H_2O Solution: Add L-Cysteine·HCl·H_2O to distilled/deionized water and bring volume to 10.0mL. Mix thoroughly. Filter sterilize.

Antibiotic Solution:
Composition per 10mL:
Polymyxin B ....80,000 units
Anisomycin 80.0mg
Cefamandole 2.0mg

Preparation of Antibiotic Solution: Add components to distilled/deionized water and bring volume to 10.0mL. Mix thoroughly. Filter sterilize.

Preparation of Medium: Add components, except cysteine and antibiotic solution, to distilled/deionized water and bring volume to 1.0L. Mix thoroughly. Adjust medium to pH 6.9 with 1*N* KOH. Heat gently and bring to boil for 1 minute. Autoclave for 15 min at 15 psi pressure–121°C. Cool to 50–55°C. Add 4.0mL of L-cysteine·HCl·H_2O solution and 10.0mL of sterile antibiotic solution. Mix thoroughly. Pour into sterile Petri dishes with constant agitation to keep charcoal in suspension.

Use: For the isolation, cultivation, and maintenance of *Legionella pneumophila* and other *Legionella* species from environmental and clinical specimens. Used for the selective recovery of *Legionella pneumophila* while reducing contaminating microorganisms from potable water samples.

BCYE Selective Agar with PAV (Buffered Charcoal Yeast Extract Selective Agar with Polymyxin B, Anisomicin and Vancomycin) (Wadowsky–Yee Medium)

Composition per 1014mL:
Agar 15.0g
Yeast extract 10.0g
ACES buffer (2-[(2-Amino-2-oxoethyl)-amino]-ethane sulfonic acid) 10.0g
Charcoal, activated 2.0g
α-Ketogluatrate 1.0g
$Fe_4(P_2O_7)_3 \cdot 9H_2O$ 0.25g
Antibiotic solution 10.0mL
Cysteine·HCl·H_2O solution 4.0mL

pH 6.9 ± 0.2 at 25°C

Source: This medium is available as a premixed powder from BBL Microbiology Systems.

Cysteine·HCl·H_2O Solution:
Composition per 10mL:
L-Cysteine·HCl·H_2O 1.0g

Preparation of Cysteine·HCl·H_2O Solution: Add L-Cysteine·HCl·H_2O to distilled/deionized water and bring volume to 10.0mL. Mix thoroughly. Filter sterilize.

Antibiotic Solution:
Composition per 10mL:
Polymyxin B ....40,000 units
Anisomycin 80.0mg
Vancomycin 0.5mg

Preparation of Antibiotic Solution: Add components to distilled/deionized water and bring volume to 10.0mL. Mix thoroughly. Filter sterilize.

Preparation of Medium: Add components, except cysteine and antibiotic solution, to distilled/deionized water and bring volume to 1.0L. Mix thoroughly. Adjust medium to pH 6.9 with 1*N* KOH. Heat gently and bring to boil for 1 minute. Autoclave for 15 min at 15 psi pressure–121°C. Cool to 50–55°C. Add 4.0mL of L-cysteine·HCl·H_2O solution and 10.0mL of sterile antibiotic solution. Mix thoroughly. Pour into sterile Petri dishes with constant agitation to keep charcoal in suspension.

Use: For the isolation, cultivation, and maintenance of *Legionella pneumophila* and other *Legionella* species from environmental and clinical specimens. Used for the selective recovery of *Legionella pneumophila* while reducing contaminating microorganisms from potable water samples.

Beef Extract Agar

Composition per liter:
Agar 15.0g
Peptone 5.0g
Beef extract 3.0g

pH 7.4 ± 0.2 at 25°C

Preparation of Medium: Add components to distilled/deionized water and bring volume to 1.0L. Mix thoroughly. Heat gently and bring to boiling. Distribute into tubes or flasks. Autoclave for 15 min at 15 psi pressure–121°C. Pour into Petri dishes or leave in tubes.

Use: For the cultivation and maintenance of a wide variety of microorganisms and recommended for culture of microorganisms from milk and water.

Beef Extract Agar (ATCC Medium 225)

Composition per liter:

Agar	25.0g
Beef extract	10.0g
Peptone	10.0g
NaCl	5.0g

pH 7.2 ± 0.2 at 25°C

Preparation of Medium: Add components to distilled/deionized water and bring volume to 1.0L. Mix thoroughly. Heat gently and bring to boiling. Distribute into tubes or flasks. Autoclave for 15 min at 15 psi pressure–121°C. Pour into Petri dishes or leave in tubes.

Use: For the cultivation and maintenance of a wide variety of microorganisms including *Alcaligenes* species, *Pseudomonas aeruginosa* and *Bacillus sphaericus*.

Beef Extract Broth

Composition per liter:

Peptone	5.0g
Beef extract	3.0g

pH 7.4 ± 0.2 at 25°C

Preparation of Medium: Add components to distilled/deionized water and bring volume to 1.0L. Mix thoroughly. Heat gently and bring to boiling. Distribute into tubes or flasks. Autoclave for 15 min at 15 psi pressure–121°C.

Use: For the cultivation and maintenance of a wide variety of microorganisms and recommended for culture of microorganisms from milk and water.

Beef Extract Broth (ATCC Medium 225)

Composition per liter:

Beef extract	10.0g
Peptone	10.0g
NaCl	5.0g

pH 7.2 ± 0.2 at 25°C

Preparation of Medium: Add components to distilled/deionized water and bring volume to 1.0L. Mix thoroughly. Heat gently and bring to boiling. Distribute into tubes or flasks. Autoclave for 15 min at 15 psi pressure–121°C.

Use: For the cultivation of a wide variety of microorganisms including *Alcaligenes* species, *Pseudomonas aeruginosa* and *Bacillus sphaericus*.

Beef Extract Peptone Serum Medium

Composition per liter:

Agar	25.0g
Beef extract	10.0g
Peptone	10.0g
NaCl	1.0g
Bovine serum	50.0mL

pH 8.5 ± 0.2 at 25°C

Preparation of Medium: Add components, except bovine serum, to distilled/deionized water and bring volume to 950.0mL. Mix thoroughly. Adjust pH to 8.5. Heat gently and bring to boiling. Autoclave for 15 min at 15 psi pressure–121°C. Cool to 50°–55°C. Aseptically add 50.0mL of sterile bovine serum. Pour into sterile Petri dishes or leave in tubes.

Use: For the cultivation and maintenance of *Serratia marcescens*.

Beef Extract V

Composition per liter:

Beef extract	24 g

pH 9.0 at 25°C

Source: This medium is available as a premixed powder from BBL Microbiology Systems.

Preparation of Medium: Add component to distilled/deionized water and bring volume to 1.0L. Mix thoroughly. Adjust pH to 9.0 with NaOH. Autoclave for 15 min at 15 psi pressure–118°–121°C.

Use: For use in the elution of viruses which have been adsorbed onto filters during filtration of water and wastewater samples.

Beef Extract with NaCl

Composition per liter:

Beef extract	10.0g
NaCl	5.0g

Preparation of Medium: Add components to distilled/deionized water and bring volume to 1.0L. Mix thoroughly. Distribute into tubes or flasks. Autoclave for 15 min at 15 psi pressure–121°C.

Use: For the cultivation of *Bacillus megaterium*.

Beef Infusion Agar

Composition per liter:

Ground defatted beef	453.6g
Agar	20.0g
Peptone	10.0g
NaCl	5.0g

pH 7.6 ± 0.2 at 25°C

Preparation of Medium: Add ground beef to 1.0L of distilled/deionized water. Let stand overnight at 4°C. Gently heat and bring to 80°–90°C for 60 min. Let stand for 2 hr. Filter through muslin. To filtrate add peptone and salt. Mix thoroughly. Adjust pH to 7.6 with 4% NaOH. Filter through Whatman #1 filter paper. Bring volume of filtrate to 1.0L. Add agar. Gently heat and bring to boiling. Distribute into tubes or flasks. Autoclave for 15 min at 15 psi pressure–121°C. Pour into sterile Petri dishes or leave in tubes.

Use: For the cultivation of a variety of microorganisms.

Beef Infusion Broth

Composition per liter:

Ground defatted beef	453.6g
Peptone	10.0g
NaCl	5.0g

pH 7.6 ± 0.2 at 25°C

Preparation of Medium: Add ground beef to 1.0L of distilled/deionized water. Let stand overnight at 4°C. Gently heat and bring to 80°–90°C for 60 min. Let stand for 2 hr. Filter through muslin. To filtrate add peptone and salt. Mix thoroughly. Adjust pH to 7.6 with 4% NaOH. Filter through Whatman #1 filter paper. Bring volume of filtrate to 1.0L. Add agar. Gently heat and bring to boiling. Distribute into tubes or flasks. Autoclave for 15 min at 15 psi pressure–121°C.

Use: For the cultivation of a variety of microorganisms.

Beef Liver Medium for Anaerobes

Composition per liter:

Beef liver, minced	500.0g
Peptone	10.0g
K_2HPO_4	1.0g

pH 8.0 ± 0.2 at 25°C

Preparation of Medium: Add beef liver to 1.0L of tap water. Soak for 12–24 hours at 4°C. Skim fat off top. Autoclave for 10 min at 15 psi pressure–121°C. Filter through cheesecloth. Save meat. To flitrate, add peptone and K_2HPO_4. Adjust pH to 8.0. Filter through paper. Add tap water and bring volume to 1.0L. Add a small amount of $CaCO_3$ to a flask or test tube. Add 0.5 inch of reserved liver. Cover meat with 2 inches of broth. Cap tubes and autoclave for 15 min at 15 psi pressure–121°C.

Use: For the cultivation and maintenance of a variety of *Clostridium* species.

Beggiotoa and *Thiothrix* Medium

Composition per liter:

$CaSO_4 \cdot 2H_2O$ (saturated solution)	20.0mL
NH_4Cl (4% solution)	5.0mL
Trace elements	5.0mL
K_2HPO_4 (1% solution)	1.0mL
$MgSO_4 \cdot 7H_2O$ (1% solution)	1.0mL

Trace Elements:

Composition per liter:

EDTA solution	20.0mL
$Co(NO_3)_2$ (0.01% solution)	10.0mL
$CuSO_4 \cdot 5H_2O$ (0.00005% solution)	10.0mL
H_3BO_3 (0.1% solution)	10.0mL
$MnSO_4 \cdot 4H_2O$ (0.02% solution)	10.0mL
$Na_2MoO_4 \cdot 2H_2O$ (0.01% solution)	10.0mL
$ZnSO_4 \cdot 7H_2O$ (0.1% solution)	10.0mL

Preparation of Trace Elements:: Add components to distilled/deionized water and bring volume to 1.0L. Mix thoroughly.

EDTA Solution:

Composition per 100mL:

$FeSO_4$	7.0g
EDTA	2.0g
HCl, concentrated	1.0mL

Preparation of EDTA Solution: Add EDTA and $FeSO_4$ to concentrated HCl. Mix thoroughly. Carefully add to distilled/deionized water and bring volume to 100.0mL.

Preparation of Medium: Add components to distilled/deionized water and bring volume to 1.0L. Mix thoroughly. Distribute into tubes or flasks. Autoclave for 15 min at 15 psi pressure–121°C.

Use: For the cultivation of *Beggiatoa* species and myxotrophic *Thiothrix* species.

Beggiatoa Medium (ATCC Medium 1193)

Composition per liter:

Sodium sulfide	0.5g
Sodium acetate	0.01g
Yeast extract	0.01g
Nutrient broth	0.01g

pH 7.5 ± 0.2 at 25°C

Preparation of Medium: Add components to distilled/deionized water and bring volume to 1.0L. Mix thoroughly. Autoclave for 15 min at 15 psi pressure–121°C. Distribute into tubes or flasks.

Use: For the cultivation of *Beggiatoa alba*.

Beggiatoa Medium (ATCC Medium 138)

Composition per liter:

Yeast extract	2.0g
Agar	2.0g
Sodium acetate	0.5g
$CaCl_2$	0.1g
Catalase	10,000U

pH 7.2 ± 0.2 at 25°C

Preparation of Medium: Add components, except catalase, to tap water and bring volume to 1.0L. Mix thoroughly. Autoclave for 15 min at 15 psi pressure–121°C. Cool to 45°–50°C. Aspetically add 10,000 units of sterile catalase.

Use: For the cultivation and maintenance of *Beggiatoa alba* and *Vitreoscilla* species.

Beijerinckia Medium

Composition per liter:

Glucose	20.0g
KH_2PO_4	1.0g
$MgSO_4 \cdot 7H_2O$	0.5g

pH 5.0 ± 0.2 at 25°C

Preparation of Medium: Add components to distilled/deionized water and bring volume to 1.0L. Mix thoroughly. Distribute into tubes or flasks. Autoclave for 15 min at 15 psi pressure–121°C.

Use: For the cultivation of *Beijerinckia* species.

Beijerinckia Medium

Composition per liter:

Glucose	20.0g
K_2HPO_4	0.8g
$MgSO_4 \cdot 7H_2O$	0.5g
KH_2PO_4	0.2g
$CaCl_2$	0.05g
$FeCl_3 \cdot 6H_2O$	0.025g
$Na_2MoO_4 \cdot 2H_2O$	5.0mg

pH 6.9 ± 0.2 at 25°C

Preparation of Medium: Add components to distilled/deionized water and bring volume to 1.0L. Mix thoroughly. Distribute into tubes or flasks. Autoclave for 15 min at 15 psi pressure–121°C.

Use: For the isolation and cultivation of *Beijerinckia* species.

Beijerinck's *Thiobacillus* Medium

Composition per liter:

Noble agar	20.0g
Na_2HPO_4	0.2g
$MgCl_2$	0.1g
NH_4Cl	0.1g
$Na_2S_2O_3$ solution	100.0mL
$NaHCO_3$ solution	10.0mL

pH 7.0–7.2 at 25°C

$Na_2S_2O_3$ Solution:
Composition per 100mL:

$Na_2S_2O_3$	5.0g

Preparation of $Na_2S_2O_3$ Solution: Add $Na_2S_2O_3$ to distilled/deionized water and bring volume to 100.0mL. Mix thoroughly. Filter sterilize.

$NaHCO_3$ Solution:
Composition per 10mL:

$NaHCO_3$	1.0g

Preparation of $NaHCO_3$ Solution: Add $NaHCO_3$ to distilled/deionized water and bring volume to 10.0mL. Mix thoroughly. Filter sterilize.

Preparation of Medium: Add components, except $Na_2S_2O_3$ solution and $NaHCO_3$ solution, to distilled/deionized water and bring volume to 890.0mL. Mix thoroughly. Autoclave for 15 min at 15 psi pressure–121°C. Aseptically add 100.0mL of sterile $Na_2S_2O_3$ solution and 10.0mL of sterile $NaHCO_3$ solution. Mix thoroughly. Pour into sterile Petri dishes or leave in tubes.

Use: For the cultivation and maintenance of *Thiobacillus thermophilica*.

Bennett's Medium

Composition per liter:

Agar	15.0g
Glucose	10.0g
Pancreatic digest of casein	2.0g
Yeast extract	1.0g
Beef extract	1.0g

pH 7.0 ± 0.2 at 25°C

Preparation of Medium: Add components to distilled/deionized water and bring volume to 1.0L. Mix thoroughly. Heat gently to boiling. Distribute into tubes or flasks. Autoclave for 15 min at 15 psi pressure–121°C. Pour into sterile Petri dishes or leave in tubes.

Use: For the cultivation and maintenance of a variety of soil microorganisms such as *Streptomyces* species, *Nocardia* species, *Flexibacter* species, *Micromonospora* species and others.

Bennett's Modified Agar Medium

Composition per liter:

Meer agar (washed agar)	20.0g
Dextrin	10.0g
Pancreatic digest of casein	2.0g
Yeast extract	1.0g
Beef extract	1.0g
$CoCl_2 \cdot 6H_2O$	0.01g

pH 7.0 ± 0.2 at 25°C

Preparation of Medium: Add components to distilled/deionized water and bring volume to 1.0L. Mix thoroughly. Heat gently to boiling. Distribute into tubes or flasks. Autoclave for 15 min at 15 psi pressure–121°C. Pour into sterile Petri dishes or leave in tubes.

Use: For the cultivation and maintenance of *Streptomyces* species.

Benzene Sulfonate Medium

Composition per liter:

Agar	15.0g
Sodium benzene sulfonate	1.0g
$(NH_4)_2SO_4$	1.0g
K_2HPO_4	0.7g
KH_2PO_4	0.3g
$MgSO_4 \cdot 7H_2O$	0.2g
$CaCl_2$	10.0mg
$FeSO_4 \cdot 7H_2O$	5.0mg
$ZnSO_4 \cdot 7H_2O$	70.0μg
$CuSO_4$	50.0μg
H_3BO_3	10.0μg
$MoO_3 \cdot 2H_2O$	10.0μg
$MnSO_4 \cdot 5H_2O$	2.0μg

Preparation of Medium: Add components to distilled/deionized water and bring volume to 1.0L. Mix thoroughly. Heat gently to boiling. Distribute into tubes or flasks. Autoclave for 15 min at 15 psi pressure–121°C. Pour into sterile Petri dishes or leave in tubes.

Use: For the cultivation and maintenance of *Comamonas testosteroni*.

Benzoate Medium

Composition per liter:

Noble agar	20.0g
NaCl	5.0g
$(NH_4)_2HPO_4$	3.0g
Sodium benzoate	3.0g
KH_2PO_4	1.2g
Yeast extract	0.5g
$MgSO_4 \cdot 7H_2O$	0.2g
Benzoate solution	25.0mL

Benzoate Solution:
Composition 25mL:

Sodium benzoate	3.0g

Preparation of Benzoate Solution: Add sodium benzoate to distilled/deionized water and bring volume to 25.0mL. Mix thoroughly. Filter sterilize.

Preparation of Medium: Add components except benzoate solution to distilled/deionized water and bring volume to 975.0mL. Mix thoroughly. Heat gently to boiling. Autoclave for 15 min at 15 psi pressure–121°C. Cool to 45°–50°C. Aseptically add 25.0mL sterile benzoate solution. Mix thoroughly and pour into sterile Petri dishes or leave in tubes.

Use: For the cultivation of *Pseudomonas putida* and other microorganisms which can utilize benzoate as a carbon source.

Benzoate Medium II

Composition per 1.5 liters:

Noble agar	30.0g
$(NH_4)_2HPO_4$	3.0g
NaCl	1.67g
KH_2PO_4	1.2g
Yeast extract	0.5g
$MgSO_4 \cdot 7H_2O$	0.2g
$FeSO_4 \cdot 7H_2O$	0.1g

Benzoate Solution:
Composition 25mL:

Sodium benzoate	1.0g

Preparation of Benzoate Solution: Add sodium benzoate to distilled/deionized water and bring volume to 25.0mL. Mix thoroughly. Filter sterilize.

Preparation of Medium: Add components, except agar and sodium benzoate, to distilled/deionized water and bring volume to 600.0mL. Mix thoroughly. Autoclave for 15 min at 15 psi pressure–121°C. Cool to 45°–50°C. In a separate flask add agar to distilled/deionized water and bring volume to 375.0mL. Mix thoroughly. Gently heat and bring to boiling. Autoclave for 15 min at 15 psi pressure–121°C. Cool to 45°–50°C. Aseptically combine the two autoclave sterilized solutions. Mix thoroughly. Aseptically add the sterile benzoate solution. Mix thoroughly. Pour into sterile Petri dishes or leave in tubes.

Use: For the cultivation of *Pseudomonas putida* and other microorganisms which can utilize benzoate as a carbon source.

Benzoate Minimal Salts Medium

Composition per liter:

K_2HPO_4	10.0g
$NaNH_4HPO_4 \cdot 4H_2O$	3.5g
$MgSO_4 \cdot 7H_2O$	0.2g
Citric acid (anhydrous)	0.2g
Benzoate solution	25.0mL

pH 7.0 ± 0.2 at 25°C

Benzoate Solution:

Composition 25mL:

Sodium benzoate	2.5g

Preparation of Benzoate Solution: Add sodium benzoate to distilled/deionized water and bring volume to 25.0mL. Mix thoroughly. Filter sterilize.

Preparation of Medium: Add components to distilled/deionized water and bring volume to 950.0mL. Mix thoroughly. Adjust pH to 7.0. Autoclave for 15 min at 15 psi pressure–121°C. Cool to 45°C. Aseptically add 25.0mL sterile benzoate solution. Mix thoroughly. Aseptically distribute into sterile tubes or flasks.

Use: For the cultivation of microorganisms which can utilize benzoate as a carbon source.

Betabacterium Medium

Composition per liter:

Pancreatic digest of casein	10.0g
Agar	10.0g
Yeast extract	5.0g
Glucose	5.0g
K_2HPO_4	2.0g
Liver extract	100.0mL

pH 7.2 ± 0.2 at 25°C

Preparation of Medium: Add 1 pound of finely ground beef liver to 2.0L of distilled/deionized water. Autoclave for 2.5–3 hours at 15 psi pressure–121°C under flowing steam. The liquid should become fluorescent yellow. Filter through sterile cheesecloth. Save solids and dry at 50°C. Add a few pieces of the dried liver to sterile test tubes or flasks. Prepare basal medium by adding components to distilled/deionized water and bring volume to 1.0L. Autoclave for 15 min at 15 psi pressure–121°C. Aseptically add sterile basal medium to each test tube or flask containing liver. Commercial liver extract may be used at a concentration of 0.1%.

Use: For growth and maintenance of *Lactobacillus* species. *Betabacterium* is an archaic name which was used to describe several bacteria as a subgenus of the *Lactobacillus* group.

BG 11 Agar (Medium BG 11 for Cyanobacteria)

Composition per liter:

Agar	10.0g
$NaNO_3$	1.5g
$MgSO_4 \cdot 7H_2O$	0.075g
K_2HPO_4	0.04g
$CaCl_2 \cdot 2H_2O$	0.036g
Na_2CO_3	0.02g
Citric acid	6.0mg
Ferric ammonium citrate	6.0mg
Disodium EDTA	1.0mg
Trace metal mix A5	1.0mL

pH 7.1 ± 0.2 at 25°C

Trace Metal Mix A5:

Composition per liter:

H_3BO_3	2.86g
$MnCl_2 \cdot 4H_2O$	1.81g
$Na_2MoO_4 \cdot 2H_2O$	0.39g
$ZnSO_4 \cdot 7H_2O$	0.222g
$CuSO_4 \cdot 5H_2O$	0.079g
$Co(NO_3)_2 \cdot 6H_2O$	0.049g

Preparation of Trace Metal Mix A5: Add components to distilled/deionized water and bring volume to 1.0L. Mix thoroughly.

Preparation of Medium: Add components to distilled/deionized water and bring volume to 1.0L. Mix thoroughly. Heat gently to boiling. Distribute into tubes or flasks. Autoclave for 15 min at 15 psi pressure–121°C. For solid medium, pour into sterile Petri dishes or leave in tubes.

Use: For the cultivation and maintenance of a variety of cyanobacteria including *Anabaena* species, *Calothrix* species, *Chaemisiphon* species, *Chorogloeopsis* species, *Chroococcidiopsis* species, *Cylindrospermum* species, *Dermocarpa* species, *Fischerella* species, *Gloebacter* species, *Gloeocapsa* species, *Gloeothece* species, *Nostoc* species, *Oscillatoria* species, *Phormidium* species, *Pleurocapsa* species *Pseudanabaena* species, *Scytonema* species, *Spirulina* species, *Synechococcus* species, *Synechocystis* species, and others.

BG 11 Marine Agar (Medium BG 11 for Marine Cyanobacteria)

Composition per liter:

Agar	10.0g
NaCl	10.0g
$NaNO_3$	1.5g

$MgSO_4 \cdot 7H_2O$ 0.075g
K_2HPO_4 .... 0.04g
$CaCl_2 \cdot 2H_2O$ 0.036g
Na_2CO_3 .... 0.02g
Citric acid 6.0mg
Ferric ammonium citrate 6.0mg
EDTA disodium salt 1.0mg
Vitamin B_{12} solution 100.0mL
Trace metal mix A5 .... 1.0mL
pH 7.1 ± 0.2 at 25°C

Trace Metal Mix A5:
Composition per liter:
H_3BO_3 .... 2.86g
$MnCl_2 \cdot 4H_2O$ 1.81g
$Na_2MoO_4 \cdot 2H_2O$ 0.39g
$ZnSO_4 \cdot 7H_2O$ 0.222g
$CuSO_4 \cdot 5H_2O$ 0.079g
$Co(NO_3)_2 \cdot 6H_2O$ 0.049g

Preparation of Trace Metal Mix A5: Add components to distilled/deionized water and bring volume to 1.0L. Mix thoroughly.

Vitamin B_{12} Solution:
Composition per 100mL:
Vitamin B_{12} 1.0μg

Preparation of Vitamin B_{12} Solution: Add vitamin B_{12} to distilled/deionized water and bring volume to 100.0mL. Mix thoroughly. Filter sterilize.

Preparation of Medium: Add components, except vitamin B_{12} solution, to distilled/deionized water and bring volume to 900.0mL. Mix thoroughly. Heat gently to boiling. Autoclave for 15 min at 15 psi pressure–121°C. Aseptically add 100.0mL of sterile vitamin B_{12} solution. Mix thoroughly. Pour into sterile Petri dishes or leave in tubes.

Use: For the cultivation and maintenance of *Synechococcus* species. Also, used for the isolation of cyanobacteria from freshwater habitats.

BG 11 Marine Broth (Medium BG 11 for Marine Cyanobacteria)

Composition per liter:
NaCl 10.0g
$NaNO_3$ 1.5g
$MgSO_4 \cdot 7H_2O$ 0.075g
K_2HPO_4 .... 0.04g
$CaCl_2 \cdot 2H_2O$ 0.036g
Na_2CO_3 .... 0.02g
Citric acid 6.0mg
Ferric ammonium citrate 6.0mg
EDTA disodium salt 1.0mg
Vitamin B_{12} solution 100.0mL
Trace metal mix A5 .... 1.0mL
pH 7.1 ± 0.2 at 25°C

Trace Metal Mix A5:
Composition per liter:
H_3BO_3 .... 2.86g
$MnCl_2 \cdot 4H_2O$ 1.81g
$Na_2MoO_4 \cdot 2H_2O$ 0.39g
$ZnSO_4 \cdot 7H_2O$ 0.222g
$CuSO_4 \cdot 5H_2O$ 0.079g
$Co(NO_3)_2 \cdot 6H_2O$ 0.049g

Preparation of Trace metal mix A5: Add components to distilled/deionized water and bring volume to 1.0L. Mix thoroughly.

Vitamin B_{12} Solution:
Composition per 100mL:
Vitamin B_{12} 1.0μg

Preparation of Vitamin B_{12} Solution: Add vitamin B_{12} to distilled/deionized water and bring volume to 100.0mL. Mix thoroughly. Filter sterilize.

Preparation of Medium: Add components, except vitamin B_{12} solution, to distilled/deionized water and bring volume to 900.0mL. Mix thoroughly. Heat gently to boiling. Autoclave for 15 min at 15 psi pressure–121°C. Aseptically add 100.0mL of sterile vitamin B_{12} solution. Mix thoroughly. Distribute into sterile tubes or flasks.

Use: For the cultivation and maintenance of *Synechococcus* species. Also, used for the isolation of cyanobacteria from freshwater habitats.

BG 11 Medium (Medium BG 11 for Cyanobacteria)

Composition per liter:
Agar 10.0g
$NaNO_3$ 1.5g
$MgSO_4 \cdot 7H_2O$ 0.075g
K_2HPO_4 .... 0.04g
$CaCl_2 \cdot 2H_2O$ 0.036g
Na_2CO_3 .... 0.02g
Citric acid 6.0mg
Ferric ammonium citrate 6.0mg
EDTA disodium salt 1.0mg
Trace metal mix A5 .... 1.0mL
pH 7.1 ± 0.2 at 25°C

Trace Metal Mix A5:
Composition per liter:
H_3BO_3 .... 2.86g
$MnCl_2 \cdot 4H_2O$ 1.81g
$Na_2MoO_4 \cdot 2H_2O$ 0.39g

$ZnSO_4 \cdot 7H_2O$ 0.222g
$CuSO_4 \cdot 5H_2O$ 0.079g
$Co(NO_3)_2 \cdot 6H_2O$ 0.049g

Preparation of Trace Metal Mix A5: Add components to distilled/deionized water and bring volume to 1.0L. Mix thoroughly.

Preparation of Medium: Add components to distilled/deionized water and bring volume to 1.0L. Mix thoroughly. Gently heat and bring to boiling. Distribute into tubes or flasks. Autoclave for 15 min at 15 psi pressure–121°C. Pour into sterile Petri dishes or leave in tubes.

Use: For the cultivation and maintenance of *Anabaena* species, *Calothrix* species, *Chaemisiphon* species, *Chorogloeopsis* species, *Chroococcidiopsis* species, *Crinalium epipsammum*, *Cylindrospermum* species, *Dermocarpa* species, *Fischerella* species, *Gloebacter violaceus*, *Gloeocapsa* species, *Gloeothece* species, *Hapalosiphon fontinalis*, *Nostoc* species, *Oscillatoria* species, *Phormidium* species, *Pleurocapsa* species, *Pseudanabaena* species, *Scytonema* species, *Spirulina* species, *Synechococcus* species, *Synechocystis* species, and *Tolypothrix tenuis*.

BG 11 Uracil Agar

Composition per liter:

Agar 10.0g
Uracil 2.8g
$NaNO_3$ 1.5g
$MgSO_4 \cdot 7H_2O$ 0.075g
K_2HPO_4 .... 0.04g
$CaCl_2 \cdot 2H_2O$ 0.036g
Na_2CO_3 .... 0.02g
Citric acid 6.0mg
Ferric ammonium citrate 6.0mg
EDTA disodium salt 1.0mg
Trace metal mix A5 .... 1.0mL

pH 7.1 ± 0.2 at 25°C

Trace Metal Mix A5:
Composition per liter:

H_3BO_3 .... 2.86g
$MnCl_2 \cdot 4H_2O$ 1.81g
$Na_2MoO_4 \cdot 2H_2O$ 0.39g
$ZnSO_4 \cdot 7H_2O$ 0.222g
$CuSO_4 \cdot 5H_2O$ 0.079g
$Co(NO_3)_2 \cdot 6H_2O$ 0.049g

Preparation of Trace Metal Mix A5: Add components to distilled/deionized water and bring volume to 1.0L. Mix thoroughly.

Preparation of Medium: Add components to distilled/deionized water and bring volume to 1.0L. Mix thoroughly. Heat gently to boiling. Distribute into tubes or flasks. Autoclave for 15 min at 15 psi pressure–121°C. Pour into sterile Petri dishes.

Use: For the cultivation and maintenance of *Anabaena variabilis*.

BG 11 Uracil Broth

Composition per liter:

Uracil 2.8g
$NaNO_3$ 1.5g
$MgSO_4 \cdot 7H_2O$ 0.075g
K_2HPO_4 .... 0.04g
$CaCl_2 \cdot 2H_2O$ 0.036g
Na_2CO_3 .... 0.02g
Citric acid 6.0mg
Ferric ammonium citrate 6.0mg
EDTA disodium salt 1.0mg
Trace metal mix A5 .... 1.0mL

pH 7.1 ± 0.2 at 25°C

Trace Metal Mix A5:
Composition per liter:

H_3BO_3 .... 2.86g
$MnCl_2 \cdot 4H_2O$ 1.81g
$Na_2MoO_4 \cdot 2H_2O$ 0.39g
$ZnSO_4 \cdot 7H_2O$ 0.222g
$CuSO_4 \cdot 5H_2O$ 0.079g
$Co(NO_3)_2 \cdot 6H_2O$ 0.049g

Preparation of Trace Metal Mix A5: Add components to distilled/deionized water and bring volume to 1.0L. Mix thoroughly.

Preparation of Medium: Add components to distilled/deionized water and bring volume to 1.0L. Mix thoroughly. Heat gently to boiling. Distribute into tubes or flasks. Autoclave for 15 min at 15 psi pressure–121°C.

Use: For the cultivation and maintenance of *Anabena variabilis*.

BG Sulfa Agar (Brilliant Green Sulfapyridine Agar)

Composition per liter:

Agar 20.0g
Proteose peptone No. 3 .... 10.0g
Lactose 10.0g
Sucrose 10.0g
NaCl 5.0g
Yeast extract 3.0g
Sodium sulfapyridine 1.0g
Brilliant Green 0.125g

pH 6.9 ± 0.2 at 25°C

Source: This medium is available as a premixed powder from Difco Laboratories.

Preparation of Medium: Add components to distilled/deionized water and bring volume to 1.0L. Mix thoroughly. Heat gently to boiling. Distribute into tubes or flasks. Autoclave for no longer than 15 min at 15 psi pressure–121°C. Pour into sterile Petri dishes if desired.

Use: For the selective isolation of *Salmonella* species other than *S. typhi* from food, dairy products, eggs and egg products and feed. *Salmonella* appear as red, pink, or white colonies surrounded by zones of bright red.

BHI
See: Brain Heart Infusion

BHI Agar
See: Brain Heart Infusion Agar

BHI Broth
See: Brain Heart Infusion Broth

BHIS
See: Brain Heart Infusion, Supplemented

Bicarbonate Agar

Composition per 100mL:

Soybean-casein digest agar	90.0mL
Sodium bicarbonate solution	10.0mL

pH 7.3 ± 0.2 at 25°C

Soybean-Casein Digest Agar :
Composition per liter:

Agar	15.0g
Pancreatic digest of casein	15.0g
Papaic digest of soybean meal	5.0g
NaCl	5.0g

Preparation of Soybean-Casein Digest Agar: Add components to distilled/deionized water and bring volume to 1.0L. Mix thoroughly. Gently heat and bring to boiling. Autoclave for 15 min at 15 psi pressure–121°C.

Sodium Bicarbonate Solution:
Composition per 10mL:

$NaHCO_3$	0.7g

Preparation of Sodium Bicarbonate Solution: Add $NaHCO_3$ to distilled/deionized water and bring volume to 10.0mL. Mix thoroughly. Filter sterilize. Use freshly prepared solution.

Preparation of Medium: To 90.0mL of cooled, sterile soybean-casein digest agar, aseptically add 10.0mL of sterile sodium bicarbonate solution. Mix thoroughly. Pour into sterile Petri dishes.

Use: For the cultivation of *Vibrio* species from foods.

Bifidobacterium Medium

Composition per liter:

Glucose	20.0g
Pancreatic digest of casein	20.0g
Yeast extract	10.0g
Peptone	10.0g
Tomato juice	333.0mL
Tween™ 80	2.0mL

pH 6.8 ± 0.2 at 25°C

Preparation of Medium: Combine 333.0mL of tomato juice with 666.0mL of distilled/deionized water. Bring to boiling. Filter through paper. Add remaining components to filtrate. Mix thoroughly. Bring volume to 1.0L with distilled/deionized water. Distribute into tubes or flasks. Autoclave for 30 min at 15 psi pressure–110°C.

Use: For the cultivation of *Bifidobacterium infantis*.

BiGGY Agar (Bismuth Sulfite Glucose Glycerin Yeast Extract Agar) (Nickerson Medium)

Composition per liter:

Agar	16.0g
Glucose	10.0g
Glycine	10.0g
Bismuth ammonium citrate	5.0g
Na_2SO_3	3.0g
Yeast extract	1.0g

pH 6.8 ± 0.2 at 25°

Source: This medium is available as a premixed powder from Difco Laboratories, Oxoid Unipath and BBL Microbiology Systems.

Preparation of Medium: Add components to distilled/deionized water and bring volume to 1.0L. Mix thoroughly and heat with frequent agitation until boiling. Distribute into tubes or flasks. Do not autoclave. Cool to approximately 45°–50°C. If desired, add 2mg/L neomycin sulfate. Swirl to disperse the insoluble material and pour into sterile Petri dishes.

Use: For the detection, isolation and presumptive

identification of *Candida* species. Addition of neomycin helps inhibit bacterial species. *C. albicans* appears as brown to black colonies with no pigment diffusion and no sheen. *C. tropicalis* appears as dark brown colonies with black centers, black pigment diffusion and a sheen. *C. krusei* appears as shiny, wrinkled, brown to black colonies with yellow pigment diffusion. *C. pseudotropicalis* appears as flat, shiny red to brown colonies with no pigment diffusion. *C. parakrusei* appears as flat, shiny, wrinkled dark reddish-brown colonies with light reddish-brown peripheries and a yellow fringe. *C. stellatoidea* appears as flat dark brown colonies with a light fringe.

Bile Esculin Agar

Composition per liter:

Oxgall	20.0g
Agar	15.0g
Pancreatic digest of gelatin	5.0g
Beef extract	3.0g
Esculin	1.0g
Ferric citrate	0.5g
Horse serum	50.0mL

pH 6.8 ± 0.2 at 25°

Source: This medium is available as a premixed powder from Oxoid Unipath and BBL Microbiology Systems.

Preparation of Medium: Add components, except horse serum, to distilled/deionized water and bring volume to 950.0L. Mix thoroughly and heat with frequent agitation until boiling. Autoclave for 15 min at 15 psi pressure–121°C. Cool to 45–50°C. Aseptically add 50.0mL of filter sterilized horse serum. Distribute into sterile Petri dishes or test tubes. Cool tubes in a slanted position.

Use: For differentiation between group D streptococci and non-group D streptococci. Also, to differentiate members of the Enterobacteriaceae, particularly *Klebsiella, Enterobacter* and *Serratia* from other enteric bacteria. Also, to differentiate *Listeria monocytogenes*. Bile tolerance and esculin hydrolysis (seen as a dark brown to black complex) are presumptive for enterococci (Group D streptococci).

Bile Esculin Agar

Composition per liter:

Esculin	1.0g
Bile esculin agar base	1.0L

pH 6.6 ± 0.2 at 25°C

Bile Esculin Agar Base:
Composition per liter:

Oxgall	40.0g
Agar	15.0g
Peptone	5.0g
Beef extract	3.0g
Ferric citrate	0.5g

Source: This medium is available as a premixed powder from Difco Laboratories.

Preparation of Bile Esculin Agar Base: Add components to distilled/deionized water and bring volume to 1.0L. Mix thoroughly.

Preparation of Medium: Add desired amount of esculin—typically 1.0g—to bile esculin agar base. Mix thoroughly and heat with frequent agitation until boiling. Autoclave for 15 min at 15 psi pressure–121°C. Cool to 45–50°C. Distribute into sterile Petri dishes or test tubes. Cool tubes in a slanted position.

Use: For the isolation and presumptive identification of group D streptococci.

Bile Esculin Agar with Kanamycin

Composition per liter:

Oxgall	20.0g
Agar	15.0g
Beef extract	3.0g
Esculin	1.0g
Ferric citrate	0.5g
Hemin	10.0mg
Vitamin K_1	10.0mg
Horse serum	50.0mL
Kanamycin solution	10.0mL

pH 7.1 ± 0.2 at 25°

Source: This medium is available as a premixed powder from BBL Microbiology Systems.

Kanamycin Solution:
Composition per 10mL:

Kanamycin	1.0g

Preparation of Kanamycin Solution: Add Kanamycin to distilled/deionized water and bring volume to 10.0mL. Mix thoroughly. Filter sterilize.

Preparation of Medium: Add components to distilled/deionized water and bring volume to 1.0L. Mix thoroughly and heat with frequent agitation until boiling. Autoclave for 15 min at 15 psi pressure–121°C. Cool to 45–50°C. Aseptically add 50.0mL of 5% filter sterilized horse serum and 10.0mL sterile kanamycin solution. Distribute into test tubes or flasks. Cool tubes in a slanted position.

Use: For the selective isolation and/or presumptive identification of bacteria of the *Bacteroides fragilis* group from specimens containing mixed flora. Examine colonies with a long-wavelength UV light. Pigmented colonies of the *Bacteroides* group will fluoresce red-orange. Growth on this medium with blackening of the medium is presumptive for *B. fragilis*.

Bile Esculin Azide Agar

Composition per liter:

Pancreatic digest of casein	17.0g
Agar	15.0g
Oxgall	10.0g
NaCl	5.0g
Yeast extract	5.0g
Proteose peptone No. 3	3.0g
Esculin	1.0g
Ferric ammonium citrate	0.5g
NaN_3	0.15g

pH 7.1 ± 0.2 at 25°C

Source: This medium is available as a premixed powder from Difco Laboratories.

Caution: Sodium azide is toxic. Azides also react with metals and disposal must be highly diluted.

Preparation: Add components to distilled/deionized water and bring volume to 1.0L. Mix thoroughly and heat with frequent agitation until boiling. Distribute into tubes or flasks. Autoclave for 15 min at 15 psi pressure–121°C. Cool to 45–50°C. Pour into sterile Petri dishes or leave in tubes. Cool tubes in a slanted position.

Use: For the isolation and presumptive identification of group D streptococci.

Bile Oxalate Sorbose Broth (BOS Broth)

Composition per liter:

Na_2HPO_4	9.14g
Sodium oxalate	5.0g
Bile salts	2.0g
NaCl	1.0g
$CaCl_2 \cdot 2H_2O$	0.01g
$MgSO_4 \cdot 7H_2O$	0.01g
Asparagine solution	100.0mL
Methionine solution	100.0mL
Sorbose solution	100.0mL
Yeast extract solution	10.0mL
Sodium pyruvate solution	10.0mL
Metanil Yellow solution	10.0mL
Sodium nitrofurantoin solution	10.0mL
Irgasan® solution	1.0mL

pH 7.6 ± 0.2 at 25°C

Asparagine Solution:
Composition per 100mL:

Asparagine	1.0g

Preparation of Asparagine Solution: Add asparagine to distilled/deionized water and bring volume to 100.0mL. Mix thoroughly. Filter sterilize.

Methionine Solution:
Composition per 100mL:

Methionine	1.0g

Preparation of Methionine Solution: Add methionine to distilled/deionized water and bring volume to 100.0mL. Mix thoroughly. Filter sterilize.

Sorbose Solution:
Composition per 100mL:

Sorbose	10.0g

Preparation of Sorbose Solution: Add sorbose to distilled/deionized water and bring volume to 100.0mL. Mix thoroughly. Filter sterilize.

Yeast Extract Solution:
Composition per 10mL:

Yeast extract	0.025g

Preparation of Yeast Extract Solution: Add yeast extract to distilled/deionized water and bring volume to 10.0mL. Mix thoroughly. Filter sterilize.

Sodium Pyruvate Solution:
Composition per 10mL:

Sodium pyruvate	0.05g

Preparation of Sodium Pyruvate Solution: Add sodium pyruvate to distilled/deionized water and bring volume to 10.0mL. Mix thoroughly. Filter sterilize.

Metanil Yellow Solution:
Composition per 10mL:

Metanil Yellow	0.025g

Preparation of Metanil Yellow Solution: Add Metanil Yellow to distilled/deionized water and bring volume to 10.0mL. Mix thoroughly. Filter sterilize.

Sodium Nitrofurantoin Solution:
Composition per 10mL:

Sodium nitrofurantoin	0.010g

Preparation of Sodium Nitrofurantoin Solution: Add sodium nitrofurantoin to distilled/deionized water and bring volume to 10.0mL. Mix thoroughly. Filter sterilize.

Irgasan® Solution:
Composition per 10mL:

Irgasan	0.04g
Ethanol (95% solution)	10.0mL

Preparation of Irgasan Solution: Add Irgasan to 10.0mL of ethanol. Mix thoroughly. Filter sterilize.

Preparation of Medium: Add components—except asparagine solution, methionine solution, sorbose solution, yeast extract solution, sodium pyruvate solution, Metanil Yellow solution, sodium nitrofurantoin solution, and Irgasan solution—to distilled/deionized water and bring volume to 659.0mL. Mix thoroughly. Gently heat and bring to boiling. Autoclave for 15 min at 15 psi pressure–121°C. Cool to 45°–50°C. Aseptically add 100.0mL of sterile asparagine solution, 100.0mL of sterile methionine solution, 100.0mL of sterile sorbose solution, 10.0mL of sterile yeast extract solution, 10.0mL of sterile sodium pyruvate solution, 10.0mL of sterile Metanil Yellow solution, 10.0mL of sterile sodium nitrofurantoin solution, and 1.0mL of sterile Irgasan solution. Mix thoroughly. Pour into sterile Petri dishes or distribute into sterile tubes.

Use: For the isolation and cultivation of *Yersinia enterocolitica* from foods.

Bile Salts Brilliant Green Starch Agar (BBGS Agar)

Composition per liter:

Agar	15.0g
Soluble starch	10.0g
Proteose peptone	10.0g
Beef extract	5.0g
Bile salts	5.0g
Brilliant Green (0.05% solution)	1.0mL

pH 7.2 ± 0.2 at 25°C

Preparation of Medium: Add components to distilled/deionized water and bring volume to 1.0L. Mix thoroughly. Gently heat while stirring and bring to boiling. Distribute into tubes or flasks. Autoclave for 15 min at 15 psi pressure–121°C. Pour into sterile Petri dishes or leave in tubes.

Preparation of Medium: Suspend the ingredients in distilled water and dissolve by heating while stirring. Autoclave for 15 min at 121°C.

Use: For the isolation and cultivation of *Aeromonas hydrophila* from foods.

Bile Salt Gelatin Agar

Composition per 100mL:

Gelatin	3.0g
Agar	1.5g
Pancreatic digest of casein	1.0g
NaCl	1.0g
Sodium taurocholate	0.5g
Na_2CO_3	0.1g
Water	100.0mL

pH 8.5 ± 0.2 at 25°C

Preparation of Medium: Add components to distilled/deionized water and bring volume to 1.0L. Mix thoroughly. Gently heat and bring to boiling. Distribute into tubes or flasks. Autoclave for 15 min at 15 psi pressure–121°C. Pour into sterile Petri dishes or leave in tubes.

Use: For the cultivation of *Vibrio cholerae*.

Biotin Assay Medium

Composition per liter:

Glucose	40.0g
Sodium acetate	20.0g
Vitamin assay casamino acids	12.0g
K_2HPO_4	1.0g
KH_2PO_4	1.0g
$MgSO_4 \cdot 7H_2O$	0.4g
DL-Tryptophane	0.2g
L-Cystine	0.2g
Adenine sulfate	0.02g
$FeSO_4$	0.02g
Guanine·HCl	0.02g
$MgSO_4 \cdot 7H_2O$	0.02g
NaCl	0.02g
Uracil	0.02g
Calcium pantothenate	2.0mg
Niacin	2.0mg
Pyridoxine·HCl	2.0mg
Riboflavin	2.0mg
Thiamine·HCl	2.0mg
p-Aminobenzoic acid	0.2mg

pH 6.7 ± 0.2 at 25°C

Source: This medium is available as a premixed powder from Difco Laboratories.

Preparation of Medium: Add components to distilled/deionized water and bring volume to 1.0L. Mix thoroughly. Gently heat and bring to boiling. Continue boiling 2–3 min. Distribute into tubes in 5.0mL volumes. Add standard solution or test solutions to each tube. Adjust the volume of each tube to 10.0mL with distilled/deionized water. Autoclave for 15 min at 15 psi pressure–121°C.

Use: For use in the microbiological assay of biotin using *Lactobacillus plantarum* as the test microorganism.

Biphasic Medium for *Neisseria*

Composition per liter:

Glucose starch agar	1.0L
Glucose starch broth	1.0L

pH 7.3 ± 0.2 at 25°

Glucose Starch Agar:
Composition per liter:

Agar	20.0g
Gelatin	20.0g
Proteose peptone No. 3	15.0g
Soluble starch	10.0g
NaCl	5.0g
Glucose	2.0g
Na_2HPO_4	3.0g

Preparation of Glucose Starch Agar: Add components to distilled/deionized water and bring volume to 1.0L. Mix thoroughly. Gently heat and bring to boiling. Autoclave for 15 min at 15 psi pressure–121°C. Cool to 50°C.

Glucose Starch Broth:
Composition per liter:

Gelatin	20.0g
Proteose peptone No. 3	15.0g
Soluble starch	10.0g
NaCl	5.0g
Glucose	2.0g
Na_2HPO_4	3.0g

Preparation of Glucose Starch Broth: Add components to distilled/deionized water and bring volume to 1.0L. Mix thoroughly. Gently heat and bring to boiling. Autoclave for 15 min at 15 psi pressure–121°C. Cool to 25°C.

Preparation of Medium: Aseptically distibute glucose starch agar into flasks in 100–125mL volumes. Allow agar to solidify. Overlay agar with 25.0mL of sterile glucose starch broth.

Use: For selective isolation and cultivation of *Neisseria* species.

Birdseed Agar (*Guizotia abyssinica* Creatinine Agar) (Niger seed Agar) (Staib Agar)

Composition per liter:

Agar	15.0g
Glucose	15.0g
Creatinine	5.0g
KH_2PO_4	3.0g
Biphenyl	1.0g
Chloramphenicol	0.5g
Guizotia abyssinica seed (niger seed) extract	1000.0mL

pH 6.7 ± 0.2 at 25°

Preparation of Medium: Prepare seed extract by grinding 50.0g of *Guizotia abyssinica* seed in 1.0L of distilled/deionized water. Boil for 30 min. Filter through cheesecloth and filter paper. Add remaining components to seed filtrate. Mix thoroughly and heat with frequent agitation until boiling. Distribute into flasks or tubes. Autoclave for 25 min at 15 psi pressure–110°C.

Use: For selective isolation and differentiation between *Cryptococcus neoformans* from other *Cryptococcus* species and other yeasts.

Bismuth Sulfite Agar

Composition per liter:

Agar	20.0g
$Bi_2(SO_3)_3$	8.0g
Pancreatic digest of casein	5.0g
Peptic digest of animal tissue	5.0g
Beef extract	5.0g
Glucose	5.0g
Na_2HPO_4	4.0g
$FeSO_4 \cdot 7H_2O$	0.3g

pH 7.5 ± 0.2 at 25°C

Source: This medium is available as a premixed powder from Difco Laboratories, Oxoid Unipath and BBL Microbiology Systems.

Preparation of Medium: Add components to distilled/deionized water and bring volume to 1.0L. Mix thoroughly and heat with frequent agitation until boiling. Boil for 1 min. Do not autoclave. Cool to 45°–50°C. Pour into sterile Petri dishes while gently shaking flask to disperse precipitate. Use plates the same day as prepared.

Use: For the selective isolation and identification of *Salmonella typhi* and other enteric bacilli. *Salmonella typhi* appear as flat, black, "rabbit-eye" colonies surrounded by a zone of black with a metallic sheen.

Bismuth Sulfite Broth (m-Bismuth Sulfite Broth)

Composition per liter:

$Bi_2(SO_3)_3$	16.0g
Pancreatic digest of casein	10.0g
Peptic digest of animal tissue	10.0g
Beef extract	10.0g
Glucose	10.0g
Na_2HPO_4	8.0g
$FeSO_4 \cdot 7H_2O$	0.6g

pH 7.7 ± 0.2 at 25°C

Preparation of Medium: Add components to distilled/deionized water and bring volume to 1.0L. Mix thoroughly and heat with frequent agitation until boiling. Boil for 1 min. Do not autoclave. Cool to

45°–50°C. Mix to disperse the precipitate and aseptically distribute into sterile tubes or flasks. Use 2.0–2.2mL of medium for each membrane filter.

Use: For the selective isolation of *Salmonella typhi* and other enteric bacilli and for the detection of *Salmonella* by the membrane filter method.

Bismuth Sulfite Glucose Glycerin Yeast Extract Agar
See: **BiGGY Agar**

Blaser's Agar
See: ***Campylobacter* Selective Medium, Blaser-Wang**

Blaser's *Campylobacter* Agar
See: ***Campylobacter* Agar, Blaser's**

Blaser-Wang *Campylobacter* Medium
See: ***Campylobacter* Selective Medium, Blaser-Wang**

Blaser-Wang *Campylobacter* Medium
See: **Blaser-Wang**

Blastobacter Enrichment Medium

Composition per liter:

Agar	18.0g
Peptone	0.5g
$MgSO_4 \cdot 7H_2O$	0.13g
$KH_2PO_4 \cdot 3H_2O$	0.13g

pH 7.2 ± 0.2 at 25°C

Preparation of Medium: Add components to tap water and bring volume to 1.0L. Mix thoroughly. Gently heat and bring to boiling. Distribute into tubes or flasks. Autoclave for 15 min at 15 psi pressure–121°C. Pour into sterile Petri dishes or leave in tubes.

Use: For the enrichment and cultivation of *Blastobacter* species.

Blastobacter Medium

Composition per liter:

Agar	15.0g
Peptone	10.0g
Yeast extract	10.0g
NaCl	5.0g

pH 7.2 ± 0.2 at 25°C

Preparation of Medium: Add components to distilled/deionized water and bring volume to 1.0L. Mix thoroughly and heat with frequent agitation until boiling. Autoclave for 15 min at 15 psi pressure–121°C.

Use: For the cultivation and maintenance of *Blastobacter natatorius* and other *Blastobacter* species.

Blood Agar

Composition per liter:

Agar	15.0g
Pancreatic digest of casein	15.0g
Papaic digest of soybean meal	5.0g
NaCl	5.0g
Sheep blood, defibrinated	50.0mL

pH 7.6 ± 0.2 at 25°C

Preparation of Medium: Add components, except sheep blood, to distilled/deionized water and bring volume to 950.0mL. Mix thoroughly. Gently heat and bring to boiling. Autoclave for 15 min at 15 psi pressure–121°C. Cool to 45°–50°C. Aseptically add 50.0mL of sterile sheep blood. Mix thoroughly. Pour into sterile Petri dishes in 20.0mL volumes.

Use: For the cultivation of fastidious microorganisms.

Blood Agar Base (Infusion Agar) (FDA Medium M21)

Composition per liter:

Heart muscle, infusion from	375.0g
Agar	15.0g
Thiotone	10.0g
NaCl	5.0g

pH 7.3 ± 0.2 at 25°C

Preparation of Medium: Add components to distilled/deionized water and bring volume to 1.0L. Mix thoroughly. Gently heat and bring to boiling. Distribute into tubes or flasks. Autoclave for 20 min at 15 psi pressure–121°C. Pour into sterile Petri dishes or leave in tubes.

Use: For the cultivation of a variety of microorganisms. For the preparation of blood agar by the addition of sterile blood.

Blood Agar Base (ATCC Medium 368)

Composition per liter:

Beef heart, infusion from	500.0g
Agar	15.0g
Tryptose	10.0g
NaCl	5.0g

pH 6.8 ± 0.2 at 25°C

Source: This medium is available as a premixed powder from Difco Laboratories.

Preparation of Medium: Add components to distilled/deionized water and bring volume to 1.0L. Mix thoroughly. Heat with frequent agitation and boil for 1 min to completely dissolve. Autoclave for 15 min at 15 psi–121°C. Cool the basal medium to 45–50°C. Aseptically add sterile, defibrinated blood to a final concentration of 5%. Mix thoroughly and pour into sterile Petri dishes.

Use: For the isolation, cultivation and detection of hemolytic activity of staphylococci, streptococci and other fastidious microorganisms.

Blood Agar Base (Infusion Agar)

Composition per liter:

Agar	15.0g
Pancreatic digest of casein	13.0g
NaCl	5.0g
Yeast extract	5.0g
Heart muscle, solids from infusion	2.0g
Sheep blood, defibrinated	50.0mL

pH 7.3 ± 0.2 at 25°C

Source: This medium is available as a premixed powder from BBL Microbiology Systems.

Preparation of Medium: Add components, except sheep blood, to distilled/deionized water and bring volume to 950.0mL. Mix thoroughly. Heat with frequent agitation and boil for 1 min to completely dissolve. Autoclave for 15 min at 15 psi–121°C. Cool to 45–50°C. Aseptically add 50.0mL of sterile, defibrinated sheep blood. Mix thoroughly and pour into sterile Petri dishes.

Use: For the isolation, cultivation and detection of hemolytic activity of streptococci and other fastidious microorganisms.

Blood Agar Base

Composition per liter:

Agar	15.0g
Beef extract	10.0g
NaCl	5.0g
Peptone	10.0g
Sheep blood, defibrinated	50.0mL

pH 7.3 ± 0.2 at 25°C

Source: This medium is available as a premixed powder from Oxoid Unipath.

Preparation of Medium: Add components, except sheep blood, to distilled/deionized water and bring volume to 950.0mL. Mix thoroughly. Heat with frequent agitation and boil for 1 min to completely dissolve. Autoclave for 15 min at 15 psi–121°C. Cool to 45–50°C. Aseptically add 50.0mL of sterile, defibrinated sheep blood. Mix thoroughly and pour into sterile Petri dishes.

Use: For the isolation, cultivation and detection of hemolytic activity of streptococci and other fastidious microorganisms.

Blood Agar Base Sheep

Composition per liter:

Pancreatic digest of casein	14.0g
Agar	12.5g
NaCl	5.0g
Peptone	4.5g
Yeast extract	4.5g
Sheep blood, defibrinated	70.0mL

ph 7.3 ± 0.2 at 25°C

Source: This medium is available as a premixed powder from Oxoid Unipath.

Preparation: Add components to distilled/deionized water and bring volume to 1.0L. Mix thoroughly. Autoclave for 15 min at 15 psi–121°C. Cool the basal medium to 45–50°C. Aseptically add 70.0mL sterile, defibrinated sheep blood. Pour into sterile Petri dishes.

Use: For giving improved hemolytic reactions with sheep blood.

Blood Agar, Diphasic

Composition per 800mL:

Lean beef, desiccated	25.0g
Agar	10.0g
Neopeptone	10.0g
NaCl	2.5g
Locke solution	200.0mL
Rabbit blood, defibrinated	100.0mL

pH 7.2–7.4 at 25°C

Locke Solution:

Composition per liter:

NaCl	8.0g
Glucose	2.5g

KH_2PO_4	0.3g
KCl	0.2g
$CaCl_2 \cdot 2H_2O$	0.2g

Preparation of Locke Solution: Add components to distilled/deionized water and bring volume to 1.0L. Mix thoroughly. Filter sterilize.

Preparation of Medium: Add beef to 500.0mL of distilled/deionized water. Let stand for 60 min. Gently heat and bring to 80°C for 5 min. Filter through Whatman #1 filter paper. To filtrate add remaining components, except Locke solution and rabbit blood. Mix thoroughly. Adjust pH to 7.2–7.4 with NaOH. Autoclave for 20 min at 15 psi pressure–121°C. Cool to 45°–50°C. Aseptically add sterile rabbit blood. Mix thoroughly. Aseptically distribute into sterile tubes in 5.0mL volumes. Allow tubes to cool in a slanted position. Immediately prior to inoculation, overlay agar in each tube with 2.0mL of sterile Locke solution.

Use: For the cultivation of *Trypanosoma* species and *Leishmania* species.

Blood Agar No. 2

Composition per liter:

Proteose peptone	15.0g
Agar	12.0g
NaCl	5.0g
Yeast extract	5.0g
Liver digest	2.5g

pH 7.4 ± 0.2 at 25°C

Source: This medium is available as a premixed powder from Difco Laboratories and Oxoid Unipath.

Preparation of Medium: Add components to distilled/deionized water and bring volume to 1.0L. Mix thoroughly. Heat with frequent agitation and boil for 1 min to completely dissolve. Autoclave for 15 min at 15 psi–121°C. Cool the basal medium to 45–50°C. Aseptically add sterile, defibrinated blood to a final concentration of 7%. Pour into sterile Petri dishes.

Use: For the isolation, cultivation and detection of hemolytic activity of streptococci, pneumococci and other particularly fastidious microorganisms.

Blood Agar with Low pH

Composition per liter:

Beef heart, solids from infusion	500.0g
Agar	15.0g
Tryptose	10.0g
NaCl	5.0g
Sheep blood, defibrinated	50.0mL

pH 6. 8 ± 0.2 at 25°C

Source: This medium is available as a premixed powder from BBL Microbiology Systems.

Preparation of Medium: Add components, except sheep blood, to distilled/deionized water and bring volume to 950.0mL. Mix thoroughly. Heat with frequent agitation and boil for 1 min to completely dissolve. Autoclave for 15 min at 15 psi–121°C. Cool to 45–50°C. Aseptically add 50.0mL of sterile, defibrinated sheep blood. Mix thoroughly and pour into sterile Petri dishes.

Use: For the isolation and growth of a wide variety of microorganisms and for the detection of the hemolytic reactions of streptococci and other fastidious microorganisms. The slightly acid acid pH of this medium enhances distinct hemolytic reactions.

Blood Glucose Cystine Agar

Composition per 100mL:

Nutrient agar	85.0mL
Glucose cystine solution	10.0mL
Human blood, fresh	5.0mL

pH 6.8 ± 0.2 at 25°C

Nutrient Agar:

Composition per liter:

Agar	15.0g
Pancreatic digest of gelatin	5.0g
Beef extract	3.0g

Source: Nutrient agar is available as a premixed powder from BBL Microbiology Systems and Difco Laboratories.

Preparation of Nutrient Agar: Add components to distilled/deionized water and bring volume to 1.0L. Mix thoroughly. Gently heat while stirring and bring to boiling. Distribute into tubes or flasks. Autoclave for 15 min at 15 psi pressure–121°C. Cool to 45°–50°C.

Glucose Cystine Solution:

Composition per 50mL:

Glucose	12.5g
Cystine·HCl	0.5g

Preparation of Glucose Cystine Solution: Add components to distilled/deionized water and bring volume to 50.0mL. Mix thoroughly. Filter sterilize.

Preparation of Medium: To 85.0mL of cooled, sterile agar solution, aseptically add 10.0mL of sterile glucose cystine solution and 5.0mL of human blood. Mix thoroughly. Pour into sterile Petri dishes or distribute into sterile tubes.

Use: For the cultivation of *Francisella tularensis*.

Blue–Green Agar

Composition per liter:

Agar (if needed)	10.0g
$NaNO_3$	1.5g
$MgSO_4 \cdot 7H_2O$	0.075g
K_2HPO_4	0.04g
$CaCl_2 \cdot 2H_2O$	0.036g
Na_2CO_3	0.02g
Citric acid	6.0mg
Ferric ammonium citrate	6.0mg
EDTA disodium salt	1.0mg
Vitamin B_{12} solution	50.0mL
Trace metal mix A5	1.0mL

pH 7.1 ± 0.2 at 25°C

Trace Metal Mix A5:

Composition per liter:

H_3BO_3	2.86g
$MnCl_2 \cdot 4H_2O$	1.81g
$Na_2MoO_4 \cdot 2H_2O$	0.39g
$ZnSO_4 \cdot 7H_2O$	0.222g
$CuSO_4 \cdot 5H_2O$	0.079g
$Co(NO_3)_2 \cdot 6H_2O$	0.049g

Preparation of Trace Metal Mix A5: Add components to distilled/deionized water and bring volume to 1.0L. Mix thoroughly.

Vitamin B_{12} Solution:

Composition per 50mL:

Vitamin B_{12}	0.01g

Preparation of Vitamin B_{12} Solution: Add vitamin B_{12} to distilled/deionized water and bring volume to 50.0mL. Mix thoroughly. Filter sterilize.

Preparation of Medium: Add components, except vitamin B_{12} solution, to glass distilled water and bring volume to 950.0mL. Mix thoroughly. Heat gently and bring to boiling. Autoclave for 15 min at 15 psi–121°C. Cool the basal medium to 45–50°C. Add vitamin B_{12} solution. Mix thoroughly. Pour into sterile Petri dishes or distribute into sterile tubes.

Use: For the cultivation and maintenance of *Synechococcus* species.

Blue–Green Broth

Composition per liter:

$NaNO_3$	1.5g
$MgSO_4 \cdot 7H_2O$	0.075g
K_2HPO_4	0.04g
$CaCl_2 \cdot 2H_2O$	0.036g
Na_2CO_3	0.02g
Citric acid	6.0mg
Ferric ammonium citrate	6.0mg
EDTA disodium salt	1.0mg
Vitamin B_{12} solution	50.0mL
Trace metal mix A5	1.0mL

pH 7.1 ± 0.2 at 25°C

Trace Metal Mix A5:

Composition per liter:

H_3BO_3	2.86g
$MnCl_2 \cdot 4H_2O$	1.81g
$Na_2MoO_4 \cdot 2H_2O$	0.39g
$ZnSO_4 \cdot 7H_2O$	0.222g
$CuSO_4 \cdot 5H_2O$	0.079g
$Co(NO_3)_2 \cdot 6H_2O$	0.049g

Preparation of Trace Metal Mix A5: Add components to distilled/deionized water and bring volume to 1.0L. Mix thoroughly.

Vitamin B_{12} Solution:

Composition per 50mL:

Vitamin B_{12}	0.01g

Preparation of Vitamin B_{12} Solution: Add vitamin B_{12} to distilled/deionized water and bring volume to 50.0mL. Mix thoroughly. Filter sterilize.

Preparation of Medium: Add components, except vitamin B_{12}, to glass distilled water and bring volume to 950.0mL. Mix thoroughly. Heat gently and bring to boiling. Autoclave for 15 min at 15 psi–121°C. Cool the basal medium to 45–50°C. Add vitamin B_{12} solution. Mix thoroughly. Distribute into sterile tubes or flasks.

Use: For the cultivation and maintenance of *Synechococcus* species.

Blue–Green Nitrogen-Fixing Agar

Composition per liter:

Noble agar	10.0g
$MgSO_4 \cdot 7H_2O$	0.075g
K_2HPO_4	0.04g
$CaCl_2 \cdot 2H_2O$	0.036g
Na_2CO_3	0.02g
Citric acid	6.0mg
Ferric ammonium citrate	6.0mg
EDTA disodium salt	1.0mg
Trace metal mix A5	1.0mL

pH 7.1 ± 0.2 at 25°C

Trace Metal Mix A5:

Composition per liter:

H_3BO_3	2.86g
$MnCl_2 \cdot 4H_2O$	1.81g
$Na_2MoO_4 \cdot 2H_2O$	0.39g
$ZnSO_4 \cdot 7H_2O$	0.222g
$CuSO_4 \cdot 5H_2O$	0.079g
$Co(NO_3)_2 \cdot 6H_2O$	0.049g

Preparation of Trace Metal Mix A5: Add components to distilled/deionized water and bring volume to 1.0L. Mix thoroughly.

Preparation of Medium: Add components to glass distilled water and bring volume to 1.0L. Mix thoroughly. Heat gently and bring to boiling. Autoclave for 15 min at 15 psi–121°C. Check pH after autoclaving and readjust if necessary. Pour into sterile Petri dishes or distribute into sterile tubes.

Use: For the cultivation and maintenance of *Calothrix*, *Fischerella* and *Nostoc* species.

Blue–Green Nitrogen-Fixing Broth

Composition per liter:

$MgSO_4 \cdot 7H_2O$	0.075g
K_2HPO_4	0.04g
$CaCl_2 \cdot 2H_2O$	0.036g
Na_2CO_3	0.02g
Citric acid	6.0mg
Ferric ammonium citrate	6.0mg
EDTA disodium salt	1.0mg
Trace metal mix A5	1.0mL

pH 7.1 ± 0.2 at 25°C

Trace Metal Mix A5:

Composition per liter:

H_3BO_3	2.86g
$MnCl_2 \cdot 4H_2O$	1.81g
$Na_2MoO_4 \cdot 2H_2O$	0.39g
$ZnSO_4 \cdot 7H_2O$	0.222g
$CuSO_4 \cdot 5H_2O$	0.079g
$Co(NO_3)_2 \cdot 6H_2O$	0.049g

Preparation of Trace Metal Mix A5: Add components to distilled/deionized water and bring volume to 1.0L. Mix thoroughly.

Preparation of Medium: Add components to glass distilled water and bring volume to 1.0L. Mix thoroughly. Heat gently and bring to boiling. Autoclave for 15 min at 15 psi–121°C. Check pH after autoclaving and readjust if necessary. Aseptically distribute into sterile tubes or flasks.

Use: For the cultivation and maintenance of *Calothrix*, *Fischerella* and *Nostoc* species.

BMPA–α Medium (Semiselective Medium for *Legionella pneumophila*)

Composition per liter:

Agar	15.0g
Yeast extract	10.0g
ACES buffer (2-[(2-Amino-2-oxoethyl)-amino]-ethane sulfonic acid)	10.0g
Charcoal, activated	2.0g
α-Ketogluatrate	1.0g
$Fe_4(P_2O_7)_3 \cdot 9H_2O$	0.25g
Antibiotic inhibitor	10.0mL
L-Cysteine·HCl·H_2O solution	10.0mL

pH 6.9 ± 0.2 at 25°C

Antibiotic Inhibitor:

Composition per 10mL:

Anisomycin	0.08g
Cefamandole	4.0mg
Polymyxin B	80,000U

Preparation of Antibiotic Inhibitor: Add components to distilled/deionized water and bring volume to 10.0mL. Mix thoroughly. Filter sterilize.

L-Cysteine·HCl·H_2O Solution:

Composition per 10mL:

L-Cysteine·HCl·H_2O	0.4g

Preparation of L-Cysteine·HCl·H_2O Solution: Add L-Cysteine·HCl·H_2O to distilled/deionized water and bring volume to 10.0mL. Mix thoroughly. Filter sterilize.

Preparation of Medium: Add components, except cysteine and antibiotic inhibitor, to distilled/deionized water and bring volume to 980.0mL. Mix thoroughly. Adjust medium to pH 6.9 with 1*N* KOH. Heat gently and bring to boiling for 1 min. Autoclave for 15 min at 15 psi pressure–121°C. Cool to 50°–55°C. Add 10.0mL of the sterile L-cysteine·HCl·H_2O solution and 10.0mL of the sterile antibiotic solution. Mix thoroughly. Pour into sterile Petri dishes with constant agitation to keep charcoal in suspension.

Use: For the selective isolation and cultivation of *Legionella pneumophila* and other *Legionella* species.

BMPA–α Medium (Edelstein BMPA–α Medium)

Composition per liter:

Agar	13.0g
Yeast extract	10.0g
ACES buffer (2-[(2-Amino-2-oxoethyl)-amino]-ethane sulfonic acid)	2.0g
Charcoal, activated	2.0g
α-Ketogluatrate	0.2g
$Fe_4(P_2O_7)_3 \cdot 9H_2O$	0.05g
Antibiotic inhibitor	10.0mL
L-Cysteine·HCl·H_2O solution	10.0mL

pH 6.9 ± 0.2 at 25°C

Source: This medium is available as premixed vials from Oxoid Unipath.

Antibiotic Inhibitor:
Composition per 10mL:

Anisomycin	0.08g
Cefamandole	4.0mg
Polymyxin B	80,000U

Preparation of Antibiotic Inhibitor: Add components to distilled/deionized water and bring volume to 10.0mL. Mix thoroughly. Filter sterilize.

L-Cysteine·HCl·H_2O Solution:
Composition per 10mL:

L-Cysteine·HCl·H_2O	0.08g

Preparation of L-Cysteine·HCl·H_2O Solution: Add L-Cysteine·HCl·H_2O to distilled/deionized water and bring volume to 10.0mL. Mix thoroughly. Filter sterilize.

Preparation of Medium: Add components, except cysteine and antibiotic inhibitor, to distilled/deionized water and bring volume to 980.0mL. Mix thoroughly. Adjust medium to pH 6.9 with 1*N* KOH. Heat gently and bring to boiling for 1 min. Autoclave for 15 min at 15 psi pressure–121°C. Cool to 50°–55°C. Add 10.0mL of the sterile L-cysteine·HCl·H_2O solution and 10.0mL of the sterile antibiotic solution. Mix thoroughly. Pour into sterile Petri dishes with constant agitation to keep charcoal in suspension.

Use: For the selective isolation and cultivation of *Legionella pneumophila* and other *Legionella* species.

Bonner–Addicott Medium

Composition per liter:

Agar	25.0g
Glucose	20.0g
$Ca(NO_3)_2 \cdot 4H_2O$	0.236g
KNO_3	0.081g
KCl	0.065g
$MgSO_4 \cdot 7H_2O$	0.036g
KH_2PO_4	0.012g
Ferric tartrate	1.0mg

Preparation of Medium: Add components to distilled/deionized water and bring volume to 1.0L. Mix thoroughly. Gently heat and bring to boiling. Distribute into tubes or flasks. Autoclave for 15 min at 15 psi pressure–121°C. Pour into sterile Petri dishes or leave in tubes.

Use: For the cultivation of a variety of fungi.

Bordetella pertussis Selective Medium with Bordet-Gengou Agar Base

Composition 1210mL:

Bordet-Gengou agar base	1.0L
Defibrinated horse blood	200.0mL
Cephalexin solution	10.0mL

pH 6.7± 0.2 at 25°C

Source: This medium is available as a premixed powder from Oxoid Unipath.

Bordet-Gengou Agar Base:
Composition per liter:

Agar	20.0g
NaCl	5.5g
Pancreatic digest of casein	5.0g
Peptic digest of animal tissue	5.0g

Preparation of Bordet-Gengou Agar Base: Add components of Bordet-Gengou Agar Base to 1.0L of 1% glycerol solution. Autoclave for 15 min at 15 psi pressure–121°C. Cool to 50°C.

Cephalexin Solution:
Composition per 10mL:

Cephalexin	0.04g

Preparation of Cephalexin Solution: Add cephalexin to distilled/deionized water and bring volume to 10.0mL. Mix thoroughly. Filter sterilize.

Preparation of Medium: Aseptically add 10.0mL sterile cephalexin solution and 200.0mL of defibrinated horse blood to 1.0L Bordet-Gengou Agar Base. Mix thoroughly and pour into sterile Petri dishes.

Use: For selective isolation and presumptive identification of *Bordetella pertussis* and *Bordetella parapertussis*. *B. pertussis* appears as small, nearly transparent, "bisected pearl-like" colonies.

Bordetella pertussis Selective Medium with Charcoal Agar Base

Composition 1110mL:

Charcoal agar base	1.0L
Horse blood, defibrinated	100.0mL
Cephalexin solution	10.0mL

pH 6.7± 0.2 at 25°C

Source: This medium is available as a premixed powder from Oxoid Unipath.

Charcoal Agar Base:
Composition per liter:

Agar	12.0g
Beef extract	10.0g
Starch	10.0g
NaCl	5.0g
Pancreatic digest of casein	5.0g
Peptic digest of animal tissue	5.0g
Charcoal	4.0g
Nicotinic acid	1.0mg

Preparation of Charcoal Agar Base: Add components of Charcoal Agar Base to distilled/deionized water and bring volume to 1.0L. Autoclave for 15 min at 15 psi pressure–121°C. Cool to 50°C.

Cephalexin Solution:
Composition per 10mL:

Cephalexin	0.04g

Preparation of Cephalexin Solution: Add cephalexin to distilled/deionized water and bring volume to 10.0mL. Mix thoroughly. Filter sterilize.

Preparation of Medium: Aseptically add 10.0mL sterile cephalexin solution and 100.0mL defibrinated horse blood to charcoal agar base. Mix thoroughly and pour into sterile Petri dishes.

Use: For selective isolation and presumptive identification of *Bordetella pertussis* and *Bordetella parapertussis*. *B. pertussis* appears as small, pale, shiny colonies.

Bordet–Gengou Agar

Composition per liter:

Agar	20.0g
Glycerol	10.0g
NaCl	5.5g
Pancreatic digest of casein	5.0g
Peptic digest of animal tissue	5.0g
Potato, solids from infusion	4.5g
Rabbit blood	200.0mL

pH 6.7± 0.2 at 25°C

Source: This medium is available as a premixed powder from Difco Laboratories, Oxoid Unipath and BBL Microbiology Systems.

Preparation of Medium: Add 10.0g glycerol to 980.0mL distilled/deionized water. Add other components, except rabbit blood, to the glycerol solution. Mix thoroughly. Heat with occasional agitation of the medium. Boil for 1 min. Autoclave for 15 min at 15 psi pressure–121°C. Cool medium to 50°C. Aseptically add 200.0mL rabbit blood (prewarmed to 35°C) or rabbit blood to a concentration of 15%–30%. 150.0–200.0mL of sterile, defibrinated horse blood may be used in place of rabbit blood. Mix thoroughly and pour plates or prepare slants.

Use: For the detection and isolation of *Bordetella pertussis* and *Bordetella parapertussis* from clinical specimens. The medium is rendered selective by the addition of methicillin. *Bordetella pertussis* appears as small (<1mm), smooth, pearl-like colonies surrounded by a narrow zone of hemolysis. *Bordetella parapertussis* appears as brown, non-shiny colonies with a green-black coloration on the reverse side. *Bordetella bronchiseptica* appears as brown, non-shiny, moderately sized colonies with a roughly pitted surface.

Bordet–Gengou Medium (ATCC Medium 35)

Composition per liter:

Agar	20.0g
Glycerol	10.0g
Proteose peptone	10.0g
NaCl	5.5g
Pancreatic digest of casein	5.0g
Peptic digest of animal tissue	5.0g
Potato, solids from infusion	4.5g
Rabbit blood	150.0mL

pH 6.7± 0.2 at 25°C

Source: This medium is available as a premixed powder from Difco Laboratories, Oxoid Unipath and BBL Microbiology Systems.

Preparation of Medium: Add 10.0g glycerol to 980.0mL distilled/deionized water. Add other components, except rabbit blood, to the glycerol solution. Mix thoroughly. Heat with occasional agitation of the medium. Boil for 1 min. Autoclave for 15 min at 15 psi pressure–121°C. Cool medium to 50°C. Aseptically add 150.0mL rabbit blood (prewarmed to 35°C) Mix thoroughly. Pour into sterile Petri dishes or distribute into sterile tubes. Allow tubes to cool in a slanted position.

Use: For the detection and isolation of *Bordetella pertussis* and *Bordetella parapertussis* from clinical specimens. The medium is rendered selective by the addition of methicillin. *Bordetella pertussis* appears as small (<1mm), smooth, pearl-like colonies surrounded by a narrow zone of hemolysis. *Bordetella parapertussis* appears as brown, non-shiny colonies with a green-black coloration on the reverse side. *Bordetella bronchiseptica* appears as brown, non-shiny, moderately sized, colonies with a roughly pitted surface.

Borrelia Medium

Composition per 370mL:

Solution 4	240.0mL
Solution 1	80.0mL
Solution 2	34.0mL
Rabbit serum, sterile	10.0mL
Solution 3	4.0mL
Solution 5	0.7mL

Solution 1:
Composition per liter:

$Na_2HPO_4 \cdot 7H_2O$	26.52g
Glucose	12.75g

Proteose peptone No.2	5.95g
Pancreatic digest of casein	2.55g
NaCl	1.20g
Sodium pyruvate	1.06g
$NaH_2PO_4 \cdot H_2O$	1.03g
KCl	0.85g
$MgCl_2 \cdot 6H_2O$	0.68g
N-acetylglucosamine	0.53g
Sodium citrate·$2H_2O$	0.47g

Preparation of Solution 1: Add components to distilled/deionized water and bring volume to 1.0L. Mix thoroughly. Store at -20° C.

Solution 2:

Composition per 100mL:

Bovine albumin fraction V	10.0g

Preparation of Solution 2: Add bovine albumin to distilled/deionized water and bring volume to 100.0mL. Mix thoroughly. Adjust pH to 7.8 with NaOH. Store at –20°C.

Solution 3:

Composition per 100mL:

$NaHCO_3$	4.5g

Preparation of Solution 3: Add $NaHCO_3$ to distilled/deionized water and bring volume to 100.0mL. Mix thoroughly. Prepare solution freshly.

Solution 4:

Composition per 100mL:

Gelatin	7.0g

Preparation of Solution 4: Add gelatin to distilled/deionized water and bring volume to 100.0mL. Mix thoroughly. Autoclave for 15 min at 10 psi pressure–115°C. Store at 4°C.

Solution 5:

Composition per 100mL:

Phenol Red	0.5g

Preparation of Solution 5: Add Phenol Red to distilled/deionized water and bring volume to 100.0mL. Mix thoroughly. Store at 4°C.

Preparation of Medium: Combine 80.0mL of solution 1, 34.0mL of solution 2, 4.0mL of solution 3, 0.7mL of solution 5 and 1.3mL of distilled/deionized water. Mix thoroughly. Filter sterilize under pressure. Aseptically distribute into sterile borosilicate screw-capped tubes in 6.0mL volumes. Melt solution 4 by immersing tube in warm water. Add 2.0mL of solution 4 to each screw-capped tube. Add 0.5mL of sterile rabbit serum to each screw-capped tube.

Use: For the cultivation of *Borrelia hermsii*, *B. turicatae*, and *B. parkeri*.

BOS Broth
See: Bile Oxalate Sorbose Broth

Botrytis Separation Agar

Composition per liter:

Agar	20.0g
Pancreatic digest of casein	5.0g
Glycerol	5.0g
$NaNO_3$	3.0g
Yeast extract	3.0g
Sorbose	2.5g
KCl	1.0g
$MgSO_4$	0.5g
KH_2PO_4	0.15g

Preparation of Medium: Add components to distilled/deionized water and bring volume to 1.0L. Mix thoroughly. Gently heat and bring to boiling. Distribute into tubes or flasks. Autoclave for 15 min at 15 psi pressure–121°C. Pour into sterile Petri dishes or leave in tubes.

Use: For the cultivation and differentiation of *Botrytis* species. *B. cinerea* will grow equally well with and without sorbose. *B. alli* is inhibited by sorbose.

Bouillon Medium

Composition per liter:

Peptone	15.0g
Meat extract	5.0g
NaCl	5.0g
K_2HPO_4	5.0g

pH 7.0 ± 0.2 at 25°C

Preparation of Medium: Add components to distilled/deionized water and bring volume to 1.0L. Mix thoroughly. Heat with frequent agitation and bring to boiling. Distribute into tubes or flasks. Autoclave for 15 min at 15 psi–121°C.

Use: For general cultivation of heterotrophic microorganisms.

Bovine Albumin Tween™ 80 Medium, Ellinghausen and McCullough, Modified (Albumin Fatty Acid Broth, *Leptospira* Medium)

Composition per liter:

Basal medium	900.0mL
Albumin fatty acid supplement	100.0mL

Basal Medium:

Composition per liter:

Na_2HPO_4, anhydrous	1.0g
NaCl	1.0g
KH_2PO_4, anhydrous	0.3g
NH_4Cl (25% solution)	1.0mL
Glycerol (10% solution)	1.0mL
Sodium pyruvate (10% solution)	1.0mL
Thiamine·HCl (0.5% solution)	1.0mL

pH 7.4 ± 0.2 at 25°C

Preparation of Basal Medium: Add components to distilled/deionized water and bring volume to 1.0L. Mix thoroughly. Adjust pH to 7.4. Gently heat and bring to boiling. Autoclave for 15 min at 15 psi pressure–121°C. Cool to 25°C.

Albumin Fatty Acid Supplement:

Composition per 200mL:

Bovine albumin fraction V	20.0g
Polysorbate (Tween™) 80 (10% solution)	25.0mL
$FeSO_4 \cdot 7H_2O$ (0.5% solution)	20.0mL
$CaCl_2 \cdot 2H_2O$ (1.5% solution)	2.0mL
$MgCl_2 \cdot 2H_2O$ (1.5% solution)	2.0mL
Vitamin B_{12} (0.2% solution)	2.0mL
$ZnSO_4 \cdot 7H_2O$ (0.4% solution)	2.0mL
$CuSO_4 \cdot 5H_2O$ (0.3% solution)	0.2mL

Preparation of Albumin Fatty Acid Supplement: Add bovine albumin to 100.0mL of distilled/deionized water. Mix thoroughly. Add remaining components while stirring. Adjust pH to 7.4. Bring volume to 200.0mL with distilled/deionized water. Filter sterilize. Store at −20°C.

Preparation of Medium: Aseptically combine 100.0mL of sterile albumin fatty acid supplement and 900.0mL of sterile basal medium. Mix thoroughly. Aseptically distribute into sterile tubes or flasks.

Use: For the cultivation of *Leptospira* species.

Bovine Albumin Tween™ 80 Semisolid Medium, Ellinghausen and McCullough, Modified (Albumin Fatty Acid Semisolid Medium, Modified)

Composition per liter:

Basal medium	900.0mL
Albumin fatty acid supplement	100.0mL

Basal Medium:

Composition per liter:

Agar	2.2g
Na_2HPO_4, anhydrous	1.0g
NaCl	1.0g
KH_2PO_4, anhydrous	0.3g
NH_4Cl (25% solution)	1.0mL
Glycerol (10% solution)	1.0mL
Sodium pyruvate (10% solution)	1.0mL
Thiamine·HCl (0.5% solution)	1.0mL

pH 7.4 ± 0.2 at 25°C

Preparation of Basal Medium: Add components to distilled/deionized water and bring volume to 1.0L. Mix thoroughly. Adjust pH to 7.4. Gently heat and bring to boiling. Autoclave for 15 min at 15 psi pressure–121°C. Cool to 25°C.

Albumin Fatty Acid Supplement:

Composition per 200mL:

Bovine albumin fraction V	20.0g
Polysorbate (Tween™) 80 (10% solution)	25.0mL
$FeSO_4 \cdot 7H_2O$ (0.5% solution)	20.0mL
$CaCl_2 \cdot 2H_2O$ (1.5% solution)	2.0mL
$MgCl_2 \cdot 2H_2O$ (1.5% solution)	2.0mL
Vitamin B_{12} (0.2% solution)	2.0mL
$ZnSO_4 \cdot 7H_2O$ (0.4% solution)	2.0mL
$CuSO_4 \cdot 5H_2O$ (0.3% solution)	0.2mL

Preparation of Albumin Fatty Acid Supplement: Add bovine albumin to 100.0mL of distilled/deionized water. Mix thoroughly. Add remaining components while stirring. Adjust pH to 7.4. Bring volume to 200.0mL with distilled/deionized water. Filter sterilize. Store at −20°C.

Preparation of Medium: Aseptically combine 100.0mL of sterile albumin fatty acid supplement and 900.0mL of sterile basal medium. Mix thoroughly. Aseptically distribute into sterile tubes or flasks.

Use: For the cultivation of *Leptospira* species.

Bovine Serum Albumin Tween™ 80 Agar (BSA Tween™ 80 Agar)

Composition per liter:

Basal medium	900.0mL
Albumin supplement	100.0mL

Basal Medium:

Composition per liter:

Agar	11.0g
Na_2HPO_4	1.0g
NaCl	1.0g
KH_2PO_4	0.3g
Glycerol (10% solution)	1.0mL
NH_4Cl (25% solution)	1.0mL
Sodium pyruvate (10% solution)	1.0mL
Thiamine (0.5% solution	1.0mL

Preparation of Basal Medium: Add components to distilled/deionized water and bring volume to 1.0L. Mix thoroughly. Adjust pH to 7.4. Autoclave for 15 min at 15 psi pressure–121°C. Cool to 25°C.

Albumin Supplement:

Composition per 100mL:

Bovine albumin	10.0g
Tween™ 80 (10% solution)	12.5mL
$FeSO_4$ (0.5% solution)	10.0mL
$MgCl_2$-$CaCl_2$ solution	1.0mL
Cyanocobalamin (0.02% solution)	1.0mL
$ZnSO_4$ (0.4% solution)	1.0mL

Preparation of Albumin Supplement: Add components to distilled/deionized water and bring volume to 100.0mL. Mix thoroughly. Adjust pH to 7.4. Filter sterilize.

$MgCl_2$–$CaCl_2$ Solution:

Composition per 100mL:

$CaCl_2 \cdot 2H_2O$	1.5g
$MgCl_2 \cdot 6H_2O$	1.5g

Preparation of $MgCl_2$–$CaCl_2$ Solution: Add components to distilled/deionized water and bring volume to 100.0mL. Mix thoroughly.

Preparation of Medium: To 900.0mL of cooled, sterile basal medium, aseptically add 100.0mL of sterile albumin supplement. Mix thoroughly. Aseptically distribute into sterile tubes or flasks.

Use: For the cultivation and maintenance of *Leptospira* species.

Bovine Serum Albumin Tween™ 80 Broth (BSA Tween™ 80 Broth)

Composition per liter:

Basal medium	900.0mL
Albumin supplement	100.0mL

pH 7.4 ± 0.2 at 25°C

Basal Medium:

Composition per liter:

Na_2HPO_4	1.0g
NaCl	1.0g
KH_2PO_4	0.3g
Glycerol (10% solution)	1.0mL
NH_4Cl (25% solution)	1.0mL
Sodium pyruvate (10% solution)	1.0mL
Thiamine (0.5% solution	1.0mL

Preparation of Basal Medium: Add components to distilled/deionized water and bring volume to 1.0L. Mix thoroughly. Adjust pH to 7.4. Autoclave for 15 min at 15 psi pressure–121°C. Cool to 25°C.

Albumin Supplement:

Composition per 100mL:

Bovine albumin	10.0g
Tween™ 80 (10% solution)	12.5mL
$FeSO_4$ (0.5% solution)	10.0mL
$MgCl_2$-$CaCl_2$ solution	1.0mL
Cyanocobalamin (0.02% solution)	1.0mL
$ZnSO_4$ (0.4% solution)	1.0mL

Preparation of Albumin Supplement: Add components to distilled/deionized water and bring volume to 100.0mL. Mix thoroughly. Adjust pH to 7.4. Filter sterilize.

$MgCl_2$–$CaCl_2$ Solution:

Composition per 100mL:

$CaCl_2 \cdot 2H_2O$	1.5g
$MgCl_2 \cdot 6H_2O$	1.5g

Preparation of $MgCl_2$–$CaCl_2$ Solution: Add components to distilled/deionized water and bring volume to 100.0mL. Mix thoroughly.

Preparation of Medium: To 900.0mL of cooled, sterile basal medium, aseptically add 100.0mL of sterile albumin supplement. Mix thoroughly. Aseptically distribute into sterile tubes or flasks.

Use: For the isolation and cultivation of *Leptospira* species.

Bovine Serum Albumin Tween™ 80 Soft Agar (BSA Tween™ 80 Soft Agar) (Semisolid BSA Tween™ 80 Medium)

Composition per liter:

Basal medium	900.0mL
Albumin supplement	100.0mL

Basal Medium:

Composition per liter:

Agar	2.0g
Na_2HPO_4	1.0g
NaCl	1.0g
KH_2PO_4	0.3g
Glycerol (10% solution)	1.0mL
NH_4Cl (25% solution)	1.0mL
Sodium pyruvate (10% solution)	1.0mL
Thiamine (0.5% solution	1.0mL

Preparation of Basal Medium: Add components to distilled/deionized water and bring volume to 1.0L. Mix thoroughly. Adjust pH to 7.4. Autoclave for 15 min at 15 psi pressure–121°C. Cool to 25°C.

Albumin Supplement:

Composition per 100mL:

Bovine albumin	10.0g
Tween™ 80 (10% solution)	12.5mL
$FeSO_4$ (0.5% solution)	10.0mL
$CaCl_2$-$MgCl_2$ solution	1.0mL
Cyanocobalamin (0.02% solution)	1.0mL
$ZnSO_4$ (0.4% solution)	1.0mL

Preparation of Albumin Supplement: Add components to distilled/deionized water and bring volume to 100.0mL. Mix thoroughly. Adjust pH to 7.4. Filter sterilize.

$MgCl_2$–$CaCl_2$ Solution:

Composition per 100mL:

$CaCl_2 \cdot 2H_2O$	1.5g
$MgCl_2 \cdot 6H_2O$	1.5g

Preparation of $MgCl_2$–$CaCl_2$ Solution: Add components to distilled/deionized water and bring volume to 100.0mL. Mix thoroughly.

Preparation of Medium: To 900.0mL of cooled, sterile basal medium aseptically add 100.0mL of sterile albumin supplement. Mix thoroughly. Aseptically distribute into sterile tubes or flasks.

Use: For the cultivation of *Leptospira* species.

Brackish Acetate

Composition per liter:

Sodium acetate	1.0g
KNO_3	1.0g
$NaH_2PO_4 \cdot 2H_2O$	0.05g
Artifical seawater	250.0mL
Modified Hutner's basal salts	20.0mL
Vitamin solution	10.0mL

pH 7.2 ± 0.2 at 25°C

Artificial Seawater:

Composition per liter:

NaCl	23.5g
$MgCl_2$	5.0g
Na_2SO_4	3.9g
$CaCl_2$	1.1g
KCl	0.66g
$NaHCO_3$	0.19g
KBr	0.1g
H_3BO_3	0.026g
$SrCl_2$	0.024g
NaF	3.0mg

Preparation of Artificial Seawater: Add components to distilled/deionized water and bring volume to 100.0mL. Mix thoroughly.

Metals "44":

Composition per 100mL:

$ZnSO_4 \cdot 7H_2O$	1.1g
$FeSO_4 \cdot 7H_2O$	0.5g
EDTA	0.25g
$MnSO_4 \cdot 7H_2O$	0.154g
$CuSO_4 \cdot 5H_2O$	0.04g
$Co(NO_3)_2 \cdot 6H_2O$	0.025g
$Na_2B_4O_7 \cdot 10H_2O$	0.018g

Preparation of Metals "44": Add components to distilled/deionized water and bring volume to 100.0mL. Mix thoroughly. Autoclave for 15 minutes at 15 psi–121°C. Add aseptically to sterile basal medium.

Modified Hutner's Basal Salts:

Composition per liter:

$MgSO_4 \cdot 7H_2O$	29.7g
Nitrilotriacetic acid	10.0g
$CaCl_2 \cdot 2H_2O$	3.34g
$FeSO_4 \cdot 7H_2O$	0.1g
$(NH_4)_2MoO_4$	9.25mg
Metals "44"	50.0mL

Preparation of Modified Hutner's Basal Salts: Dissolve the nitrilotriacetic acid first and neutralize the solution with KOH. Add other components and adjust the pH to 7.2 with KOH or H_2SO_4. There may be a slight precipitate. Store at 5°C.

Vitamin Solution:

Composition per liter:

Thiamine·HCl	5.0mg
D-Calcium pantothenate	5.0mg
Riboflavin	5.0mg
Biotin	2.0mg
Folic acid	2.0mg
Vitamin B_{12}	0.1mg

Preparation of Vitamin Solution: Add components to distilled/deionized water and bring volume to 1.0L. Mix thoroughly. Filter-sterilize and add aseptically to sterile basal medium.

Preparation of Medium: Add a few drops of H_2SO_4 to the distilled water to retard precipitation of the metal salts. Add components, except for Metals "44" and vitamin solutions, to 250.0mL of artificial sea water and 720.0mL of distilled/deionized water. Adjust pH to 7.2. Distribute into tubes or flasks. Sterilize by autoclaving for 15 minutes at 15 lbs. pressure–121°C. Aseptically add Metals "44" and vitamin solutions. Mix thoroughly. Aseptically distribute into sterile tubes or flasks.

Use: For the cultivation of *Filomicrobium fusiforme*.

Brackish *Prosthecomicrobium* Medium

Composition per liter:

Agar	15.0g
Peptone	0.25g
Yeast extract	0.25g
Glucose	0.25g
Artificial seawater	250.0mL
Modified Hutner's basal salts	20.0mL
Vitamins	10.0mL

pH 7.2 ± 0.2 at 25°C

Artificial Seawater:

Composition per liter:

NaCl	23.477g
$MgCl_2$	4.981g
Na_2SO_4	3.917g
$CaCl_2$	1.102g
KCl	0.664g
$NaHCO_3$	0.192g
KBr	0.096g
H_3BO_3	0.026g
$SrCl_2$	0.024g
NaF	3.0mg

Preparation of Artificial Seawater: Add components to distilled/deionized water and bring volume to 100.0mL. Mix thoroughly.

Modified Hutner's Basal Salts:

Composition per liter:

$MgSO_4.7H_2O$	29.7g
Nitrilotriacetic acid	10.0g
$CaCl_2 \cdot 2H_2O$	3.34g
$FeSO_4 \cdot 7H_2O$	0.1g
$(NH_4)_2MoO_4$	9.25mg
Metals "44"	50.0mL

Preparation of Modified Hutner's Basal Salts: Dissolve the nitrilotriacetic acid first and neutralize the solution with KOH. Add other ingredients and readjust the pH with KOH and/or H_2SO_4 to 7.2. There may be a slight precipitate. Store at 5°C.

Metals "44":

Composition per 100mL:

$ZnSO_4 \cdot 7H_2O$	1.1g
$FeSO_4 \cdot 7H_2O$	0.5g
EDTA	0.25g
$MnSO_4 \cdot 7H_2O$	0.154g
$CuSO_4 \cdot 5H_2O$	0.04g
$Co(NO_3)_2 \cdot 6H_2O$	0.025g
$Na_2B_4O_7 \cdot 10H_2O$	0.018g

Preparation of Metals "44": Add components to distilled/deionized water and bring volume to 100.0mL. Mix thoroughly. Autoclave for 15 minutes at 15 psi–121°C. Add aseptically to sterile basal medium.

Vitamin Solution:

Composition per liter:

Thiamine·HCl	5.0mg
D-Calcium pantothenate	5.0mg
Riboflavin	5.0mg
Biotin	2.0mg
Folic acid	2.0mg
Vitamin B_{12}	0.1mg

Preparation of Vitamin Solution: Add components to distilled/deionized water and bring volume to 1.0L. Mix thoroughly. Filter-sterilize and add aseptically to sterile basal medium.

Preparation of Medium: Add a few drops of H_2SO_4 to the distilled water to retard precipitation of the metal salts. Add components, except for metals"44" and vitamin solutions, to 250.0mL of artificial sea water and 720.0mL of distilled/deionized water. Adjust pH to 7.2. Distribute into tubes or flasks. Sterilize by autoclaving for 15 minutes at 15 lbs. pressure–121°C. Aseptically add Metals "44" and vitamin solutions. Mix thoroughly. Aseptically distribute into sterile tubes or flasks.

Use: For the cultivation of *Prosthecomicrobium litoralum.*

Brain Heart CC Agar (Brain Heart Cycloheximide Chloramphenicol Agar)

Composition per liter:

Pancreatic digest of casein	16.0g
Agar	13.5g
Brain heart, solids from infusion	8.0g
Peptic digest of animal tissue	5.0g
NaCl	5.0g
Na_2HPO_4	2.5g
Glucose	2.0g
Cycloheximide	0.5g
Chloramphenicol	0.05g

pH 7.4 ± 0.2 at 25°C

Source: This medium is available as a premixed powder from Difco Laboratories and BBL Microbiology Systems.

Preparation of Medium: Add components to distilled/deionized water and bring volume to 1.0L. Mix thoroughly. Distribute into tubes or flasks while shaking to distribute precipitate. Autoclave for 15 min at 15 psi–118°C.

Use: For the selective isolation of fastidious pathogenic fungi such as *Histoplasma capsulatum* and *Blastomyces dermatitidis* from specimens heavily contaminated with bacteria and other fungi. It may also be used as a base supplemented with sheep blood and gentamicin for enrichment and additional selectivity.

Brain Heart Infusion (BHI)

Composition per liter:

Pancreatic digest of gelatin	14.5g
Brain heart, solids from infusion	6.0g
Peptic digest of animal tissue	6.0g
NaCl	5.0g
Glucose	3.0g
Na_2HPO_4	2.5g

pH 7.4 ± 0.2 at 25°C

Source: This medium is available as a premixed powder from Difco Laboratories, Oxoid Unipath and BBL Microbiology Systems.

Preparation of Medium: Add components to distilled/deionized water and bring volume to 1.0L. Mix thoroughly. Distribute into tubes or flasks. Autoclave for 15 min at 15 psi–121°C.

Use: For the cultivation of fastidious and nonfastidious microorganisms, including aerobic and anaerobic bacteria, from a variety of clinical and nonclinical specimens. It is particularly useful for culturing streptococci, pneumococci and meningococci. It is also used for the preparation of inocula for use in antimicrobial susceptibility tests and as a base for blood culture.

Brain Heart Infusion Agar (BHI Agar)

Composition per liter:

Beef heart infusion	250.0g
Calf brain infusion	200.0g
Proteose peptone	10.0g
NaCl	5.0g
$Na_2HPO_4 \cdot 12H_2O$	2.5g
Glucose	2.0g

pH 7.4 ± 0.2 at 25°C

Preparation of Medium: Add components to distilled/deionized water and bring volume to 1.0L. Mix thoroughly. Gently heat and bring to boiling. Distribute into tubes or flasks. Autoclave for 15 min at 15 psi pressure–121°C. Pour into sterile Petri dishes or leave in tubes.

Use: For the cultivation of a variety of fastidious and nonfastidious, aerobic and anaerobic microorganisms.

Brain Heart Infusion Agar

Composition per liter:

Pancreatic digest of casein	16.0g
Agar	13.5g
Brain heart, solids from infusion	8.0g
Peptic digest of animal tissue	5.0g
NaCl	5.0g
Glucose	2.0g
Na_2HPO_4	2.5g

pH 7.4 ± 0.2 at 25°C

Source: This medium is available as a premixed powder from Difco Laboratories, Oxoid Unipath and BBL Microbiology Systems.

Preparation of Medium: Add components to distilled/deionized water and bring volume to 1.0L. Mix thoroughly. Distribute into tubes or flasks while shaking to distribute precipitate. Autoclave for 15 min at 15 psi–121°C.

Use: For the cultivation of a wide variety of fastidious microorganisms, including bacteria, yeast and molds. With the addition of 10% sheep blood, it is used for the isolation and cultivation of many fungal species, including systemic fungi, from clinical and nonclinical specimens. The addition of gentamicin and chloramphenicol with 10% sheep blood produces a selective medium used for the isolation of pathogenic fungi from specimens heavily contaminated with bacteria and saprophtic fungi. It is recommended for the isolation of *Histoplasma capsulatum* and other pathogenic fungi including *Coccidioides immitis*.

Brain Heart Infusion Agar with Yeast Extract

Composition per liter:

Yeast extract	20.0g
Pancreatic digest of casein	16.0g
Agar	13.5g
Brain heart, solids from infusion	8.0g
Peptic digest of animal tissue	5.0g
NaCl	5.0g
Glucose	2.0g
Na_2HPO_4	2.5g

pH 7.4 ± 0.2 at 25°C

Source: This medium is available as a premixed powder from Difco Laboratories, Oxoid Unipath and BBL Microbiology Systems.

Preparation of Medium: Add components to distilled/deionized water and bring volume to 1.0L. Mix thoroughly. Distribute into tubes or flasks while shaking to distribute precipitate. Autoclave for 15 min at 15 psi–121°C.

Use: For the cultivation of *Mycoplasma equirhinis*.

Brain Heart Infusion Broth (BHI Broth)

Composition per liter:

Beef heart infusion 250.0g
Calf brain infusion 200.0g
Proteose peptone 10.0g
NaCl 5.0g
$Na_2HPO_4 \cdot 12H_2O$ 2.5g
Glucose 2.0g

pH 7.4 ± 0.2 at 25°C

Preparation of Medium: Add components to distilled/deionized water and bring volume to 1.0L. Mix thoroughly. Distribute into tubes or flasks. Autoclave for 15 min at 15 psi pressure–121°C.

Use: For the cultivation of a variety of fastidious and nonfastidious, aerobic and anaerobic microorganisms.

Brain Heart Infusion Broth

Composition per liter:

Calf brain infusion 200.0g
Beef heart infusion 250.0g
Proteose p eptone 10.0g
Glucose 2.0g
NaCl 5.0g
Na_2HPO_4 .... 2.5g

pH 7.4 ± 0.2 at 25°C

Preparation of Medium: Add components to distilled/deionized water and bring volume to 1.0L. Mix thoroughly. Distribute into tubes or flasks while shaking to distribute precipitate. Autoclave for 15 min at 15 psi–121°C.

Use: For the cultivation of a wide variety of microorganisms, including bacteria, yeast and molds, especially fastidious species.

Brain Heart Infusion with 0.7% Agar

Composition per liter:

Beef heart infusion 250.0g
Calf brain infusion 200.0g
Proteose peptone 10.0g
Agar 7.0g
NaCl 5.0g
$Na_2HPO_4 \cdot 12H_2O$ 2.5g
Glucose 2.0g

pH 5.3 ± 0.2 at 25°C

Preparation of Medium: Add components to distilled/deionized water and bring volume to 1.0L. Mix thoroughly. Gently heat and bring to boiling. Adjust pH to 5.3 with 1*N* HCl. Distribute into tubes in 25.0mL volumes. Autoclave for 10 min at 15 psi pressure–121°C.

Use: For the cultivation of *Staphylococcal* species for the production of enterotoxin.

Brain Heart Infusion Agar with Chloramphenicol

Composition per liter:

Pancreatic digest of casein 16.0g
Agar 13.5g
Brain/heart, solids from infusion 8.0g
Peptic digest of animal tissue 5.0g
NaCl 5.0g
Glucose 2.0g
Na_2HPO_4 .... 2.5g
Sheep blood, defibrinated 50.0mL
Chloramphenicol solution 10.0mL

pH 7.4 ± 0.2 at 25°C

Chloramphenicol Solution:
Composition per 10mL:

Chloramphenicol 0.05g

Preparation of Chloramphenicol Solution: Add chloramphenicol to distilled/deionized water and bring volume to 10.0mL. Mix thoroughly. Filter sterilize.

Preparation of Medium: Add components, except chloramphenicol solution and sheep blood, to distilled/deionized water and bring volume to 940.0mL. Mix thoroughly. Gently heat and bring to boiling. Autoclave for 15 min at 15 psi pressure–121°C. Cool to 45°–50°C. Aseptically add sterile chloramphenicol solution and sheep blood. Mix thoroughly. Pour into sterile Petri dishes or distribute into sterile tubes.

Use: For the isolation and cultivation of a wide variety of fungal species, especially systemic fungi, from clinical and nonclinical specimens. Also used for the selective isolation of pathogenic fungi from specimens heavily contaminated with bacteria and saprophtic fungi. For maintenance of fungal species on slant cultures.

Brain Heart Infusion Agar with Penicillin and Streptomycin

Composition per liter:

Pancreatic digest of casein 16.0g
Agar 13.5g
Brain/heart, solids from infusion 8.0g
Peptic digest of animal tissue 5.0g
NaCl 5.0g

Glucose	2.0g
Na_2HPO_4	2.5g
Streptomycin	40.0mg
Penicillin	20,000U
Sheep blood, defibrinated	50.0mL

pH 7.4 ± 0.2 at 25°C

Preparation of Medium: Add components, except sheep blood, to distilled/deionized water and bring volume to 950.0mL. Mix thoroughly and while stirring bring to a boil for 1 min to completely dissolve. Autoclave for 15 min at 15 psi pressure–121°C. Cool to 50°C. Aseptically add 50.0mL defibrinated sheep blood. Mix thoroughly. Pour into sterile Petri dishes while agitating gently to distribute the precipitate through the medium.

Use: For the isolation and cultivation of a wide variety of fungal species, especially systemic fungi, from clinical and nonclinical specimens. Also used for the selective isolation of pathogenic fungi from specimens heavily contaminated with bacteria and saprophtic fungi. For maintenance of fungal species on slant cultures.

Brain Heart Infusion Agar with 10% Sheep Blood, Gentamicin and Chloramphenicol

Composition per liter:

Pancreatic digest of casein	16.0g
Agar	13.5g
Brain/heart, solids from infusion	8.0g
Peptic digest of animal tissue	5.0g
NaCl	5.0g
Glucose	2.0g
Na_2HPO_4	2.5g
Sheep blood, defibrinated	100.0mL
Antibiotic solution	10.0mL

pH 7.4 ± 0.2 at 25°C

Antibiotic Solution:

Composition per 10mL:

Chloramphenicol	0.05g
Gentamicin	0.05g

Preparation of Antibiotic Solution: Add components to distilled/deionized water and bring volume to 10.0mL. Mix thoroughly. Filter sterilize.

Preparation of Medium: Add components, except antibiotic solution and sheep blood, to distilled/deionized water and bring volume to 940.0mL. Mix thoroughly. Gently heat and bring to boiling. Autoclave for 15 min at 15 psi pressure–121°C. Cool to 45°–50°C. Aseptically add sterile antibiotic solution and sheep blood. Mix thoroughly. Pour into sterile Petri dishes or distribute into sterile tubes.

Use: For the isolation and cultivation of a wide variety of fungal species, especially systemic fungi, from clinical and nonclinical specimens. Also used for the selective isolation of pathogenic fungi from specimens heavily contaminated with bacteria and saprophytic fungi. For maintenance of fungal species on slant cultures.

Brain Heart Infusion Soil Extract Medium

Composition per liter:

Yeast extract	20.0g
Pancreatic digest of casein	16.0g
Brain/heart, solids from infusion	8.0g
Peptic digest of animal tissue	5.0g
NaCl	5.0g
Glucose	2.0g
Na_2HPO_4	2.5g
Soil extract	250.0mL
Vitamin B_{12} solution	1.0mL

pH 7.2 ± 0.2 at 25°C

Soil Extract:

Composition per 400mL:

African Violet soil	1.0g
Na_2CO_3	1.0g

Preparation of Soil Extract: Autoclave for 60 min at 15 psi pressure–121°C. Filter through paper before using in medium.

Vitamin B_{12} Solution:

Composition per 1mL:

Vitamin B_{12}	2.0μg

Preparation of Vitamin B_{12} Solution: Add vitamin B_{12} to distilled/deionized water and bring volume to 1.0mL. Mix thoroughly. Filter sterilize.

Preparation of Medium: Add components, except glucose, yeast extract, and vitamin B_{12} solution, to tap water and bring volume to 799.0mL. Mix thoroughly. Autoclave for 15 min at 15 psi–121°C. Add yeast extract and glucose to 200.0mL of tap water. Filter sterilize and add aseptically to cooled sterile basal medium. Aseptically add 1.0mL of vitamin B_{12} solution. Mix thoroughly. Aseptically distribute into sterile tubes or flasks.

Use: For the cultivation of a wide variety of microorganisms, including bacteria, yeast and molds, especially fastidious species from soil. It is useful for the isolation of *Histoplasma capsulatum* and other pathogenic fungi including *Coccidioides immitis*.

Brain Heart Infusion, Supplemented (BHIS)

Composition per liter:

Brain heart infusion broth, dehydrated	37.0g
Yeast extract	5.0g
Supplement solution	10.2mL
Resazurin (0.25% solution)	4.0mL

pH 7.4 ± 0.2 at 25°C

Supplement Solution:

Composition per 10.2mL:

Cysteine·HCl·H_2O	0.5g
Hemin solution	10.0mL
Vitamin K_1 solution	0.2mL

Preparation of Supplement Solution: Add cysteine·HCl·H_2O and vitamin K_1 to 10.0mL hemin solution. Mix thoroughly.

Hemin Solution:

Composition per 100mL:

Hemin	50.0mg
1*N* NaOH	1.0mL

Preparation of Hemin Solution: Add components to 100.0mL distilled/deionized water and bring volume to 1.0L. Mix thoroughly.

Vitamin K_1 Solution:

Composition per 30mL:

Ethanol (95% solution)	30.0mL
Vitamin K_1	0.15mL

Preparation of Vitamin K_1 Solution: Add vitamin K_1 to 30.0mL 95% ethanol. Mix thoroughly.

Preparation of Medium: Add components (Brain heart infusion broth powder, yeast extract and resazurin) to distilled/deionized water and bring volume to 990.0mL. Mix thoroughly and while stirring bring to boiling for 1 min. Cool to 45–50°C. Add supplement solution containing hemin solution, vitamin K_1 solution and cysteine. Autoclave for 15 min at 15 psi pressure–121°C. Pour into sterile Petri dishes while agitating gently to distribute the precipitate through the medium.

Use: For the cultivation of *Centipeda periodontii.*

Brain Heart Infusion with Casein

Composition per liter:

Pancreatic digest of gelatin	14.5g
Brain/heart, solids from infusion	6.0g
Peptic digest of animal tissue	6.0g
NaCl	5.0g
Casein	5.0g
Glucose	3.0g
Na_2HPO_4	2.5g

pH 7.4 ± 0.2 at 25°C

Preparation of Medium: Add components to distilled/deionized water and bring volume to 1.0L. Mix thoroughly. Distribute into tubes or flasks. Autoclave for 15 min at 15 psi–121°C.

Use: For the cultivation of *Serratia marcescens.*

Brain Heart Infusion with Chicken Serum

Composition per liter:

Yeast extract	20.0g
Pancreatic digest of casein	16.0g
Brain/heart, solids from infusion	8.0g
Peptic digest of animal tissue	5.0g
NaCl	5.0g
Glucose	2.0g
Na_2HPO_4	2.5g
Chicken serum, heat inactivated	50.0mL
Nicotinamide adenine dinucleotide solution	10.0mL

pH 7.2 ± 0.2 at 25°C

Nicotinamide Adenine Dinucleotide Solution:

Composition per 10mL:

Nicotinamide adenine dinucleotide	0.1g

Preparation of Nicotinamide Adenine Dinucleotide Solution: Add 0.1g nicotinamide adenine dinucleotide to distilled/deinonized water and bring volume to 10.0mL Filter sterilize.

Preparation of Medium: Add components, except chicken serum and nicotinamide adenine dinucleotide solution, to distilled/deionized water and bring volume to 940.0mL. Autoclave for 15 min at 15 psi–121°C. Cool to 50°–55°C. Aspetically add 10.0mL of nicotinamide adenine dinucleotide solution and 50.0mL of heat inactivated chicken serum. Mix thoroughly. Aseptically distribute aseptically into sterile tubes or flasks.

Use: For cultlivation of *Haemophilus paragallinarum* and *Pasteurella avium.*

Brain Heart Infusion with PABA (Brain Heart Infusion with *p*-Aminobenzoic Acid)

Composition per liter:

Pancreatic digest of gelatin	14.5g
Brain/heart, solids from infusion	6.0g
Peptic digest of animal tissue	6.0g

NaCl 5.0g
Glucose 3.0g
Na_2HPO_4 ....2.5g
p-Aminobenzoic acid 0.05g

pH 7.4 ± 0.2 at 25°C

Source: This medium is available as a premixed powder from BBL Microbiology Systems and Difco Laboratories.

Preparation of Medium: Add components to distilled/deionized water and bring volume to 1.0L. Mix thoroughly. The addition of 1.0g agar to the medium enhances the growth of anaerobic and microaerophilic microorganisms. Heat with frequent agitation and boil for 1 min to dissolve. Distribute into tubes or flasks. Autoclave for 15 min at 15 psi–121°C.

Use: For the detection of microorganisms in the blood of patients who have received sulfonamide therapy.

Brain Heart Infusion with Agar, Yeast Extract, Sucrose, Horse Serum and Penicillin

Composition per liter:

Sucrose 100.0g
Pancreatic digest of gelatin 14.5g
Agar..12.0g
Yeast extract 5.0g
Brain heart, solids from infusion 6.0g
Peptic digest of animal tissue................................6.0g
NaCl 5.0g
Glucose 3.0g
Na_2HPO_4 ....2.5g
Horse serum 100.0mL
Penicillin solution 10.0mL

pH 7.4 ± 0.2 at 25°C

Penicillin Solution:

Composition per 10mL:

Penicillin 100,000U

Preparation of Penicillin Solution: Add penicillin to distilled/deionized water and bring volume to 10.0mL. Mix thoroughly. Filter sterilize.

Preparation of Medium: Add components, except penicillin solution and horse serum, to distilled/deionized water and bring volume to 890.0mL. Mix thoroughly. Gently heat and bring to boiling. Autoclave for 15 min at 15 psi pressure–121°C. Cool to 45°–50°C. Aseptically add sterile penicillin solution and horse serum. Mix thoroughly. Pour into sterile Petri dishes or distribute into sterile tubes.

Use: For the cultivation of fastidious fungi.

Brain Heart Infusion with Sucrose and Horse Serum

Composition per liter:

Sucrose 171.0g
Pancreatic digest of gelatin 14.5g
Brain heart, solids from infusion 6.0g
Peptic digest of animal tissue................................6.0g
NaCl 5.0g
Glucose 3.0g
Na_2HPO_4 ....2.5g
Horse serum 100.0mL

pH 7.4 ± 0.2 at 25°C

Preparation of Medium: Add components, except horse serum, to distilled/deionized water and bring volume to 900.0mL. Mix thoroughly. Gently heat and bring to boiling. Autoclave for 15 min at 15 psi pressure–121°C. Cool to 45°–50°C. Aseptically add sterile horse serum. Mix thoroughly. Pour into sterile Petri dishes or distribute into sterile tubes.

Use: For the cultivation of fastidious fungi.

Brain Heart Infusion with Agar, Yeast Extract, Sucrose, Inactivated Horse Serum and Penicillin

Composition per liter:

Agar..12.0g
Yeast extract 5.0g
Sucrose 100.0g
Pancreatic digest of gelatin 14.5g
Brain heart, solids from infusion 6.0g
Peptic digest of animal tissue................................6.0g
NaCl 5.0g
Glucose 3.0g
Na_2HPO_4 ....2.5g
Horse serum, inactivated..............................100.0mL
Penicillin solution 10.0mL

pH 7.4 ± 0.2 at 25°C

Penicillin Solution:

Composition per 10mL:

Penicillin 1,000,000U

Preparation of Penicillin Solution: Add penicillin to distilled/deionized water and bring volume to 10.0mL. Mix thoroughly. Filter sterilize.

Preparation of Medium: Add components, except penicillin solution and inactivated horse serum, to distilled/deionized water and bring volume to 890.0mL. Mix thoroughly. Gently heat and bring to boiling. Autoclave for 15 min at 15 psi pressure–121°C. Cool to 45°–50°C. Aseptically add sterile penicillin solution and horse serum. Mix thoroughly. Pour into sterile Petri dishes or distribute into sterile tubes.

Use: For the cultivation of fastidious fungi.

Brain Heart Infusion with Agar, Yeast Extract, NaCl, Inactivated Horse Serum and Penicillin

Composition per liter:

Agar	12.0g
Yeast extract	5.0g
NaCl	20.0g
Pancreatic digest of gelatin	14.5g
Brain heart, solids from infusion	6.0g
Peptic digest of animal tissue	6.0g
NaCl	5.0g
Glucose	3.0g
Na_2HPO_4	2.5g
Horse serum, inactivated	100.0mL
Penicillin solution	10.0mL

pH 7.4 ± 0.2 at 25°C

Penicillin Solution:

Composition per 10mL:

Penicillin	1,000,000U

Preparation of Penicillin Solution: Add penicillin to distilled/deionized water and bring volume to 10.0mL. Mix thoroughly. Filter sterilize.

Preparation of Medium: Add components, except penicillin solution and inactivated horse serum, to distilled/deionized water and bring volume to 890.0mL. Mix thoroughly. Gently heat and bring to boiling. Autoclave for 15 min at 15 psi pressure–121°C. Cool to 45°–50°C. Aseptically add sterile penicillin solution and horse serum. Mix thoroughly. Pour into sterile Petri dishes or distribute into sterile tubes.

Use: For the cultivation of fastidious fungi.

Brain Heart Infusion with PABA and Agar (Brain Heart Infusion with *p*-Aminobenzoic Acid and Agar)

Composition per liter:

Pancreatic digest of gelatin	14.5g
Brain/heart, solids from infusion	6.0g
Peptic digest of animal tissue	6.0g
NaCl	5.0g
Glucose	3.0g
Na_2HPO_4	2.5g
Agar	1.0g
p-Aminobenzoic acid	0.05g

pH 7.4 ± 0.2 at 25°C

Source: This medium is available as a premixed powder from BBL Microbiology Systems and Difco Laboratories.

Preparation of Medium: Add components to distilled/deionized water and bring volume to 1.0L. Mix thoroughly. The addition of 1.0g agar to the medium enhances the growth of anaerobic and microaerophilic microorganisms. Heat with frequent agitation and boil for 1 min to dissolve. Distribute into tubes or flasks. Autoclave for 15 min at 15 psi–121°C.

Use: For the detection of microorganisms in the blood of patients who have received sulfonamide therapy.

Brain Heart Infusion with Rabbit Serum

Composition per liter:

Yeast extract	20.0g
Pancreatic digest of casein	16.0g
Brain/heart, solids from infusion	8.0g
Peptic digest of animal tissue	5.0g
NaCl	5.0g
Glucose	2.0g
Na_2HPO_4	2.5g
Rabbit serum, heat inactivated	50.0mL
Nicotinamide adenine dinucleotide solution	10.0mL

pH 7.2 ± 0.2 at 25°C

Nicotinamide Adenine Dinucleotide Solution:

Composition per 10mL:

Nicotinamide adenine dinucleotide	0.1g

Preparation of Nicotinamide Adenine Dinucleotide Solution: Add 0.1g nicotinamide adenine dinucleotide to distilled/deinonized water and bring volume to 10.0mL. Filter sterilize.

Preparation of Medium: Add components, except nicotinamide adenine dinucleotide solution and rabbit serum, to distilled/deionized water and bring volume to 940.0mL. Autoclave for 15 min at 15 psi–121°C. Cool to 50°–55°C. Aspetically add 10.0mL of a 0.1% filter-sterilized solution of nicotinamide adenine dinucleotide. Aseptically add 50.0mL of heat inactivated rabbit serum. Distribute aseptically into tubes or flasks.

Use: For the cultivation and maintenance of *Actinobacillus* species.

Brain Heart Infusion with Rabbit Serum and Yeast Extract

Composition per liter:

Pancreatic digest of casein	16.0g
Brain/heart, solids from infusion	8.0g
Peptic digest of animal tissue	5.0g
NaCl	5.0g
Glucose	2.0g
Na_2HPO_4	2.5g
Rabbit serum, heat inactivated	200.0mL
Fresh yeast extract (25% solution)	100.0mL

pH 7.2 ± 0.2 at 25°C

Fresh Yeast Extract Solution:

Composition per 100mL:

Live, pressed, starch-free, Baker's yeast	25.0g

Preparation of Fresh Yeast Extract Solution: Add the live Baker's yeast to 100.0mL of distilled/deionized water. Autoclave for 90 min at 15 psi pressure–121°C. Allow to stand. Remove supernatant solution. Adjust pH to 6.6–6.8.

Preparation of Medium: Add components, except rabbit serum, to distilled/deionized water and bring volume to 800.0mL. Autoclave for 15 min at 15 psi–121°C. Cool to 50°–55°C. Aseptically add 200.0mL of sterile heat inactivated rabbit serum. Distribute aseptically into tubes or flasks.

Use: For the cultivation and maintenance of *Mycoplasma equirhinis*.

Brain Heart Infusion with 3% Sodium Chloride

Composition per liter:

NaCl	30.0g
Pancreatic digest of gelatin	14.5g
Brain/heart, solids from infusion	6.0g
Peptic digest of animal tissue	6.0g
Glucose	3.0g
Na_2HPO_4	2.5g

pH 7.4 ± 0.2 at 25°C

Preparation of Medium: Add components to distilled/deionized water and bring volume to 1.0L. Mix thoroughly. Distribute into tubes or flasks. Autoclave for 15 min at 15 psi–121°C.

Use: For the cultivation of *Vibrio parahaemolyticus*.

Brain Heart Infusion with Thiamine

Composition per liter:

NaCl	30.0g
Pancreatic digest of gelatin	14.5g
Brain/heart, solids from infusion	6.0g
Peptic digest of animal tissue	6.0g
Glucose	3.0g
Na_2HPO_4	2.5g
Thiamine·HCl	1.0mg

pH 7.4 ± 0.2 at 25°C

Preparation of Medium: Add components to distilled/deionized water and bring volume to 1.0L. Mix thoroughly. Distribute into tubes or flasks. Autoclave for 15 min at 15 psi–121°C.

Use: For the cultivation *Bacillus larvae*.

Brain Liver Heart Semisolid Medium

Composition per liter:

Beef heart, infusion from	250.0g
Calf brains, infusion from	200.0g
Liver, infusion from	50.0g
Proteose peptone	10.0g
NaCl	5.0g
Neopeptone	3.25g
Pancreatic digest of casein	3.25g
Na_2HPO_4	2.5g
Glucose	2.0g
Agar	1.75g

pH 7.3 ± 0.2 at 25°C

Preparation of Medium: Add components to distilled/deionized water and bring volume to 1.0L. Mix thoroughly. Gently heat and bring to boiling. Distribute into tubes or flasks. Autoclave for 15 min at 15 psi pressure–121°C. Leave in tubes.

Use: For the cultivation of fastidious microorganisms. For the cultivation of *Actinomyces bovis*, *Actinomyces israelii*, *Actinomyces naeslundi*, *Clostridium symbiosum* and *Eubacterium limosum*.

Brevibacterium Medium (ATCC Medium 677)

Composition per liter:

Agar	30.0g
KH_2PO_4	2.0g
Na_2HPO_4	2.0g
$(NH_4)_2SO_4$	2.0g
Yeast extract	2.0g
Tween™ 60	2.0g
$MgSO_4·7H_2O$	0.2g
$FeSO_4·7H_2O$	0.1g
$MnSO_4$	0.01g
n-Hexadecane	50.0mL

pH 7.0 ± 0.2 at 25°C

Preparation of Medium: Add components to distilled/deionized water and bring volume to 1.0L. Mix thoroughly. Blend for 30 minutes in a blender to disperse the *n*-hexadecane. Distribute into tubes or flasks. Autoclave for 20 min at 15 psi–121°C.

Use: For the cultivation and maintenance of *Brevibacterium alkanophilum* and other microorganisms which can utilize hexadecane as a carbon source.

Brevibacterium Medium (ATCC Medium 681)

Composition per liter:

Glucose	10.0g
Peptone	5.0g
Yeast extract	5.0g

pH 5.0–6.0 at 25°C

Preparation of Medium: Add components to distilled/deionized water and bring volume to 1.0L. Mix thoroughly. Distribute into tubes or flasks. Autoclave for 15 min at 15 psi–121°C.

Use: For the cultivation and maintenance of *Brevibacterium* spp. and *Enterobacter cloacae*.

Brevibacterium Medium (ATCC Medium 159)

Composition per liter:

Agar	25.0g
Glucose	20.0g
$CaCO_3$	20.0g
Yeast extract	10.0g

Preparation of Medium: Add components to distilled/deionized water and bring volume to 1.0L. Mix thoroughly. Gently heat to boiling. Distribute into tubes or flasks. Autoclave for 15 min at 15 psi–121°C. Pour into sterile Petri dishes or leave in tubes.

Use: For the cultivation and maintenance of *Brevibacterium* species.

Brewer Anaerobic Agar

Composition per liter:

Agar	20.0g
Proteose peptone No. 3	10.0g
Glucose	10.0g
Pancreatic digest of casein	5.0g
Yeast extract	5.0g
NaCl	5.0g
Sodium thioglycollate	2.0g
Sodium formaldehyde sulfoxylate	1.0g
Resazurin	2.0mg

pH 7.2 ± 0.2 at 25°C

Source: This medium is available as a premixed powder from Difco Laboratories.

Preparation of Medium: Add components to distilled/deionized water and bring volume to 1.0L. Mix thoroughly. Distribute into tubes or flasks. Autoclave for 15 min at 15 psi–121°C.

Use: For the cultivation and maintenance of anaerobic and microaerophilic microorganisms.

Brewer Thioglycollate Medium

Composition per liter:

Beef, infusion from	500.0g
Proteose peptone	10.0g
NaCl	5.0g
Glucose	5.0g
K_2HPO_4	2.0g
Sodium thioglycollate	0.5g
Agar	0.5g
Methylene Blue	2.0mg

pH 7.2 ± 0.2 at 25°C

Source: This medium is available as a premixed powder from Difco Laboratories.

Preparation of Medium: Add components to distilled/deionized water and bring volume to 1.0L. Mix thoroughly. Gently heat to boiling. Distribute into tubes or flasks. Autoclave for 15 min at 15 psi–121°C.

Use: For the cultivation and maintenance of anaerobic and microaerophilic microorganisms. Also used for testing the sterility of biological products and materials.

Brewer Thioglycollate Medium, Modified

Composition per liter:

Tryptic digest of casein	17.0g
Glucose	10.0g
NaCl	5.0g
Enzymatic hydrolysate of soybean meal	3.0g
K_2HPO_4	2.0g
Sodium thioglycollate	1.0g
Agar	0.5g
Methylene Blue	2.0mg

pH 7.2 ± 0.2 at 25°C

Source: This medium is available as a premixed powder from Difco Laboratories.

Preparation of Medium: Add components to distilled/deionized water and bring volume to 1.0L. Mix thoroughly. Gently heat to boiling. Distribute into tubes or flasks. Autoclave for 15 min at 15 psi–121°C.

Use: For the cultivation and maintenance of anaerobic and microaerophilic microorganisms. Also used for testing the sterility of biological products and materials.

Brilliant Green Agar

Composition per liter:

Agar	20.0g
Lactose	10.0g
Sucrose	10.0g
Peptic digest of animal tissue	5.0g
Pancreatic digest of casein	5.0g
NaCl	5.0g
Phenol Red	0.08g
Brilliant Green	0.0125g

pH 6.9 ± 0.2 at 25°C

Source: This medium is available as a premixed powder from Difco Laboratories, Oxoid Unipath and BBL Microbiology Systems.

Preparation of Medium: Add components to distilled/deionized water and bring volume to1.0L. Mix thoroughly. Gently heat and bring to boiling. Distribute into tubes or flasks. Autoclave for 15 min at 15 psi pressure–121°C. Pour into sterile Petri dishes.

Use: For the selective isolation of *Salmonella* other than *S. typhi* from feces and other specimens, and food and dairy products. *Salmonella* other than *S. typhi* appear as red/pink/white colonies surrounded by a zone of red in the agar indicating nonlactose/sucrose fermentation. *Proteus* or *Pseudomonas* species may appear as small red colonies. Lactose- or sucrose-fermenting bacteria appear as yellow-green colonies surrounded by a zone of yellow-green in the agar.

Brilliant Green Agar with Sulfadiazine

Composition per liter:

Agar	20.0g
Lactose	10.0g
Sucrose	10.0g
Pancreatic digest of casein	5.0g
Peptic digest of animal tissue	5.0g
NaCl	5.0g
Yeast extract	3.0g
Phenol Red	0.08g
Sulfadiazine	0.08g
Brilliant Green	0.0125g

pH 6.9 ± 0.2 at 25°C

Source: This medium is available as a premixed powder from BBL Microbiology Systems.

Preparation of Medium: Add components to distilled/deionized water and bring volume to1.0L. Mix thoroughly. Gently heat and bring to boiling. Distribute into tubes or flasks. Autoclave for 15 min at 15 psi pressure–121°C. Pour into sterile Petri dishes.

Use: For the selective detection of *Salmonella* in foods, especially from egg products. *Salmonella* other than *S. typhi* appear as red/pink colonies surrounded by a zone of red in the agar indicating nonlactose/sucrose fermentation. *Proteus* or *Pseudomonas* species may appear as small red colonies. Lactose- or sucrose-fermenting bacteria appear as yellow-green colonies surrounded by a zone of yellow-green in the agar.

Brilliant Green Agar, Modified

Composition per liter:

Agar	12.0g
Lactose	10.0g
Sucrose	10.0g
Beef extract	5.0g
Peptone	5.0g
NaCl	5.0g
NaCl	5.0g
Yeast extract	3.0g
Na_2HPO_4	1.0g
NaH_2PO_4	0.6g
Phenol Red	0.09g
Brilliant Green	4.7mg

pH 6.9 ± 0.2 at 25°C

Source: This medium is available as a premixed powder from Oxoid Unipath.

Preparation of Medium: Add components to distilled/deionized water and bring volume to1.0L. Mix thoroughly. Gently heat and bring to boiling. Do not autoclave. Cool to 45–50°C. Addition of 1.0g sodium sulfacetamide and 250.0mg sodium mandelate enhances inhibition of contaminating microorganisms. Pour into sterile Petri dishes.

Use: For the selective isolation of *Salmonella* other than *S. typhi* from feces and other specimens, and food and dairy products. *Salmonella* other than *S. typhi* appear as red/pink/white colonies surrounded by a zone of red in the agar indicating nonlactose/sucrose fermentation. *Proteus* or *Pseudomonas* species may appear as small red colonies. Lactose- or sucrose-fermenting bacteria appear as yellow-green colonies surrounded by a zone of yellow-green in the agar.

Brilliant Green Bile Agar

Composition per liter:

Noble agar	10.15g
Pancreatic digest of gelatin	8.25g

Lactose	1.9g
Na_2SO_3	0.205g
$FeCl_3$	0.0295g
Basic Fuchsin	0.078g
Erioglaucine	0.065g
KH_2PO_4	0.015g
Oxgall, dehydrated	2.95mg
Brilliant Green	0.03mg

pH 6.9 ± 0.2 at 25°C

Source: This medium is available as a premixed powder from Difco Laboratories and BBL Microbiology Systems.

Caution: Basic Fuchsin is a potential carcinogen and care must be taken to avoid inhalation of the powdered dye and contamination of the skin.

Preparation of Medium: Add components to distilled/deionized water and bring volume to1.0L. For plating 10.0mL samples, prepare the medium double strength. Mix thoroughly. Gently heat and bring to boiling. Distribute into tubes or flasks. Autoclave for 15 min at 15 psi pressure–121°C. Pour into sterile Petri dishes. Care should be taken to avoid exposure of the prepared medium to light.

Use: For the detection and enumeration of coliform bacteria in materials of sanitary importance such as water, sewage and foods. *Escherichia coli* appears as dark red colonies with a pink halo. *Enterobacter* species appear as pink colonies.

Brilliant Green Bile Broth (Brilliant Green Lactose Bile Broth)

Composition per liter:

Oxgall, dehydrated	20.0g
Lactose	10.0g
Pancreatic digest of gelatin	10.0g
Brilliant Green	0.013g

pH 7.2 ± 0.2 at 25°C

Source: This medium is available as a premixed powder from Difco Laboratories, BBL Microbiology Systems and Oxoid Unipath.

Preparation of Medium: Add components to distilled/deionized water and bring volume to1.0L. Mix thoroughly. Distribute into tubes containing inverted Durham tubes, in 10.0mL amounts for testing 1.0mL or less of sample. Autoclave for 12 min (not longer than 15 min) at 15 psi pressure–121°C. After sterilization, cool the broth rapidly. Medium is sensitive to light.

Use: For the detection of coliform microorganisms in foods, dairy products, water and wastewater as well as in other materials of sanitary importance. Turbidity in the broth and gas in the Durham tube are positive indications of *Escherichia coli*.

Brilliant Green Bile Broth with MUG

Composition per liter:

Oxgall, dehydrated	20.0g
Lactose	10.0g
Pancreatic digest of gelatin	10.0g
MUG (4-methyl umbelliferyl-β-D-glucuronide)	0.05g
Brilliant Green	0.013g

pH 7.2 ± 0.2 at 25°C

Source: This medium is available as a premixed powder from BBL Microbiology Systems.

Preparation of Medium: Add components to distilled/deionized water and bring volume to1.0L. Mix thoroughly. Distribute into tubes containing inverted Durham tubes, in 10.0mL amounts for testing 1.0mL or less of sample. Autoclave for 12 min (not longer than 15 min) at 15 psi pressure–121°C. After sterilization, cool the broth rapidly.

Use: For the detection of coliform microorganisms in foods, dairy products, water and wastewater as well as in other materials of sanitary importance. The presence of *E. coli* and other coliforms is determined by the presence of fluorescence in the tube.

Brilliant Green Broth (m–Brilliant Green Broth)

Composition per liter:

Proteose peptone No.3	20.0g
Lactose	20.0g
Sucrose	20.0g
NaCl	10.0g
Yeast extract	6.0g
Phenol Red	0.16g
Brilliant Green	0.025g

pH 6.9 ± 0.2 at 25°C

Source: This medium is available as a premixed powder from Difco Laboratories.

Preparation of Medium: Add components to distilled/deionized water and bring volume to 1.0L. Mix thoroughly. Gently heat with frequent mixing. Boil for 1 min. Cool to 25°C. Add 2.0mL to each sterile absorbent filter used.

Use: For the selective isolation and differentiation of *Salmonella* from polluted water by the membrane filter method.

Brilliant Green Lactose Bile Broth
See: **Brilliant Green Bile Broth**

Brilliant Green Phenol Red Agar

Composition per liter:

Agar	15.0g
Lactose	15.0g
Peptone	10.0g
Meat extract	5.0g
NaCl	5.0g
Phenol Red	0.08g
Brilliant Green	0.0125g

pH 6.9 ± 0.2 at 25°C

Preparation of Medium: Add components to distilled/deionized water and bring volume to 1.0L. Mix thoroughly. Gently heat and bring to boiling. Distribute into tubes or flasks. Autoclave for 15 min at 15 psi pressure–121°C. Pour into sterile Petri dishes or leave in tubes.

Use: For the cultivation of *Salmonella* species.

Brilliant Green Sulfapyridine Agar
See: **BG Sulfa Agar**

Brochothrix thermosphacta Medium

Composition per liter:

Peptone	20.0g
Glycerol	15.0g
Agar	13.0g
Yeast extract	2.0g
K_2HPO_4	1.0g
$MgSO_4 \cdot 7H_2O$	1.0g

pH 7.0 ± 0.2 at 25°C

Preparation of Medium: Add components to distilled/deionized water and bring volume to 1.0L. Mix thoroughly. Gently heat and bring to boiling. Distribute into tubes or flasks. Autoclave for 15 min at 15 psi pressure–121°C. Pour into sterile Petri dishes or leave in tubes.

Use: For the isolation and cultivation of *Brochothrix thermosphacta* from meats and meat products.

Brolacin Agar
See: **CLED Agar**

Bromcresol Purple Azide Broth
See: **BCP Azide Broth**

Bromcresol Purple Broth

Composition per liter:

Peptone	10.0g
NaCl	5.0g
Beef extract	3.0g
Bromcresol Purple	0.04g
Carbohydrate solution	10.0mL

pH 7.0 ± 0.2 at 25°C

Carbohydrate Solution:
Composition per 10mL:

Carbohydrate	5.0g

Preparation of Carbohydrate Solution: Add carbohydrate to distilled/deionized water and bring volume to 10.0mL. Mix thoroughly. Filter sterilize.

Preparation of Medium: Add components to distilled/deionized water and bring volume to 1.0L. Mix thoroughly. Gently heat and bring to boiling. Distribute into test tubes that contain an inverted Durham tube. Autoclave for 10 min at 15 psi pressure–121°C.

Use: For the differentiation of a variety of microorganisms based on their fermentation of specific carbohydrates. Bacteria that ferment the specific carbohydrate turn the medium yellow. When bacteria produce gas, the gas is trapped in the Durham tube.

Bromcresol Purple Deoxycholate Agar
See: **BCP D Agar**

Bromcresol Purple Deoxycholate Citrate Lactose Sucrose Agar
See: **BCP DCLS Agar**

Bromcresol Purple Dextrose Broth (BCP Broth)

Composition per liter:

Glucose	10.0g
Peptone	5.0g
Beef extract	3.0g
Bromcresol Purple solution	2.0mL

pH 7.0 ± 0.2 at 25°C

Bromcresol Purple Solution:
Composition per 10mL:

Bromcresol Purple	0.16g
Ethanol (95% solution)	10.0mL

Preparation of Bromcresol Purple Solution: Add Bromcresol Purple to 10.0mL of ethanol. Mix thoroughly.

Preparation of Medium: Add components to distilled/deionized water and bring volume to 1.0L. Mix thoroughly. Distribute into tubes in 12–15mL volumes. Autoclave for 15 min at 15 psi pressure–121°C.

Use: For the cultivation and differentiation of bacteria based on their ability to ferment glucose. Bacteria that ferment glucose turn the medium yellow.

Bromocresol Purple Milk Solids Glucose Agar (BCP MS G Agar)

Composition per 2L:

Skim milk powder	80.0g
Glucose	40.0g
Agar	30.0g
Bromcresol Purple solution	2.0mL

pH 6.6 ± 0.2 at 25°C

Bromcresol Purple Solution:
Composition per 10mL:

Bromcresol Purple	0.16g
Ethanol (95% solution)	10.0mL

Preparation of Bromcresol Purple Solution: Add Bromcresol Purple to 10.0mL of ethanol. Mix thoroughly.

Preparation of Medium: Add skim milk powder and Bromcresol Purple solution to distilled/deionized water and bring volume to 1.0L. Mix thoroughly. Autoclave for 8 min at 11 psi pressure–116°C. Cool to 45°–50°C. In a separate flask add glucose to distilled/deionized water and bring volume to 200.0mL. Mix thoroughly. Autoclave for 8 min at 11 psi pressure–116°C. Cool to 45°–50°C. In a third flask add agar to distilled/deionized water and bring volume to 800.0mL. Mix thoroughly. Gently heat and bring to boiling. Autoclave for 15 min at 15 psi pressure–121°C. Cool to 45°–50°C. Aseptically combine the three sterile solutions. Mix thoroughly. Aseptically adjust the pH to 6.6 with sterile 1*N* HCl. Aseptically distribute into sterile tubes. Allow tubes to cool in a slanted position.

Use: For the cultivation and differentiation of *Trychophyton mentagrophytes, Trychophyton rubrum,* and *Microsporum persicolor.*

Bromcresol Purple Milk Yeast Extract with CCG

Composition per liter:

Milk solution	1.0L
Agar solution	900.0mL
Yeast extract solution	40.0mL
Chloramphenicol solution	10.0mL
Cycloheximide solution	10.0mL
Gentamicin solution	0.8mL

Milk Solution:
Composition per liter:

Skim milk powder	80.0g
Bromcresol Purple solution	2.0mL

Preparation of Milk Solution: Add components to distilled/deionized water and bring volume to 1.0L. Mix thoroughly. Autoclave for 8 min at 11 psi pressure–116°C. Cool to 45°–50°C.

Bromcresol Purple Solution:
Composition per 10mL:

Bromcresol Purple	0.16g
Ethanol (95% solution)	10.0mL

Preparation of Bromcresol Purple Solution: Add Bromcresol Purple to 10.0mL of ethanol. Mix thoroughly.

Agar Solution:
Composition per 900mL:

Agar	30.0g

Preparation of Agar Solution: Add agar to distilled/deionized water and bring volume to 900.0mL. Mix thoroughly. Gently heat and bring to boiling. Autoclave for 15 min at 15 psi pressure–121°C. Cool to 45°–50°C.

Yeast Extract Solution:
Composition per 100mL:

Yeast extract	40.0g

Preparation of Yeast Extract Solution: Add yeast extract to distilled/deionized water and bring volume to 100.0mL. Mix thoroughly. Filter sterilize.

Chloramphenicol Solution:
Composition per 10mL:

Chloramphenicol	0.1g

Preparation of Chloramphenicol Solution: Add chloramphenicol to distilled/deionized water and bring volume to 10.0mL. Mix thoroughly. Filter sterilize.

Cycloheximide Solution:
Composition per 10mL:

Cycloheximide	0.2g

Preparation of Cycloheximide Solution: Add cycloheximide to distilled/deionized water and bring volume to 10.0mL. Mix thoroughly. Filter sterilize.

Gentamicin Solution:
Composition per 10mL:

Gentamicin	0.5g

Preparation of Gentamicin Solution: Add gentamicin to distilled/deionized water and bring volume to 10.0mL. Mix thoroughly. Filter sterilize.

Preparation of Medium: Aseptically combine the sterile milk solution and sterile agar solution. Aseptically add 40.0mL of sterile yeast extract solution, 10.0mL of sterile chloramphenicol solution, 10.0mL of sterile cycloheximide solution and 0.8mL of sterile gentamicin solution. Mix thoroughly. Aseptically distribute into sterile tubes. Allow tubes to cool in a slanted position.

Use: For the isolation, cultivation and differentiation of *Trichophyton verrucosum* and *Trichophyton schoenleinii*.

Bromthymol Blue Agar

Composition per liter:

Agar	11.0g
Peptone	10.0g
NaCl	5.0g
Yeast extract	5.0g
Lactose (33% solution)	27.0mL
Bromthymol Blue (1% solution)	10.0mL
Sodium thiosulfate (50% solution)	2.0mL
Glucose (33% solution)	1.2mL
Maranil solution (5% solution)	1.0mL

pH 7.7–7.8 at 25°C

Preparation of Medium: Add agar, peptone, NaCl, and yeast extract to distilled/deionized water and bring volume to 1.0L. Mix thoroughly. Adjust pH to 8.0. Autoclave for 20 min at 15 psi pressure–121°C. Cool to 45°–50°C. Filter sterilize separately the lactose solution, Bromthymol Blue solution, sodium thiosulfate solution, glucose solution and maranil solution. To the cooled, sterile agar solution aseptically add 27.0mL of sterile lactose solution, 10.0mL of sterile Bromthymol Blue solution, 2.0mL of sterile sodium thiosulfate solution, 1.2mL of sterile glucose solution and1.0mL of sterile maranil solution. Mix thoroughly. Adjust pH to 7.7–7.8. Pour into sterile Petri dishes or distribute into sterile tubes.

Use: For the selective isolation and cultivation of members of the Enterobacteriaceae.

Bromthymol Blue Broth

Composition per 101.45mL:

Pancreatic digest of casein	0.7g
NaCl	0.5g
Beef extract	0.3g
Yeast extract	0.3g
Beef heart, solids from infusion	0.2g
Yeast extract	0.1g
Horse serum	10.0mL
Bromthymol Blue solution	1.0mL
Ampicillin solution	1.0mL
Urea solution	0.25mL
Nystatin solution	0.1mL
Tripeptide solution	0.1mL

pH 6.0 ± 0.2 at 25°C

Bromothymol Blue Solution:
Composition per 50mL:

Bromothymol Blue	0.2g
NaOH (0.01*N* solution)	32.0mL

Preparation of Bromothymol Blue Solution: Add Bromthymol Blue to NaOH solution. Mix thoroughly. Bring volume to 50.0mL with distilled/deionized water. Autoclave for 15 min at 15 psi pressure–121°C. Store at 25°C.

Ampicillin Solution:
Composition per 10mL:

Ampicillin	1.0g

Preparation of Ampicillin Solution: Add ampicillin to distilled/deionized water and bring volume to 10.0mL. Mix thoroughly. Filter sterilize.

Urea Solution:
Composition per 100mL:

Urea	10.0g

Preparation of Urea Solution: Add urea to distilled/deionized water and bring volume to 100.0mL. Filter sterilize. Store at –20°C.

Nystatin Solution:
Composition per 1mL:

Nystatin	50,000U

Preparation of Nystatin Solution: Add nystatin to distilled/deionized water and bring volume to 1.0mL. Filter sterilize.

Tripeptide Solution:
Composition per 10mL:

Glycyl-L-histidyl-L-lysine acetate	0.2mg

Preparation of Tripeptide Solution: Add glycyl-L-histidyl-L-lysine acetate to distilled/deionized water and bring volume to 10.0mL. Mix thoroughly. Filter sterilize. Store at –20°C.

Preparation of Medium: Add components—except horse serum, ampicillin solution, urea solution, nystatin solution and tripeptide solution—to distilled/deionized water and bring volume to 90.0mL. Mix thoroughly. Gently heat and bring to boiling. Autoclave for 15 min at 15 psi pressure–121°C. Cool to 45°–50°C. Aseptically add 10.0mL of sterile horse serum, 1.0mL of sterile ampicillin solution, 0.25mL of sterile urea solution, 0.1mL of sterile nystatin solution and 0.1mL of sterile tripeptide solution. Mix thoroughly. Pour into sterile Petri dishes.

Use: For the cultivation of *Ureaplasma* species from clinical specimens.

Bromthymol Blue Lactose Agar
See: BTB Lactose Agar

Brucella Agar

Composition per liter:

Agar	15.0g
Pancreatic digest of casein	10.0g
Peptic digest of animal tissue	10.0g
NaCl	5.0g
Yeast extract	2.0g
Glucose	1.0g
$NaHSO_3$	0.1g
Horse blood, defibrinated	100.0mL

pH 7.0 ± 0.2 at 25°C

Source: This medium is available as a premixed powder from Difco Laboratories and BBL Microbiology Systems and Oxoid Unipath.

Preparation of Medium: Add components to distilled/deionized water and bring volume to 900.0mL. Mix thoroughly. Heat gently with frequent mixing. Boil for 1 min. Autoclave for 15 min at 15 psi pressure–121°C. Cool to 45°–50°C. Add 100.0mL sterile defibrinated horse blood. Mix gently and pour into sterile Petri dishes.

Use: For the cultivation and maintenance of *Brucella* species. Also used for the isolation and cultivation of nonfastidious and fastidious microorganisms from a variety of clinical and nonclinical specimens.

Brucella Agar Base *Campylobacter* Medium

Composition per 1100mL:

Cycloheximide (actidione)	0.05g
Sodium cephazolin	0.015g
Novobiocin	5.0mg
Bacitracin	25,000U
Colistin sulfate	10,000U
Brucella agar base	1.0L
Horse blood, defibrinated	100.0mL

Brucella Agar Base

Composition per liter:

Agar	15.0g
Pancreatic digest of casein	10.0g
Peptic digest of animal tissue	10.0g
NaCl	5.0g
Yeast extract	2.0g
Glucose	1.0g
$NaHSO_3$	0.1g

Preparation of Brucella Agar Base: Add components to distilled/deionized water and bring volume to 1.0L. Mix thoroughly.

Optional Supplement:

Composition per 10mL:

Sodium pyruvate	0.25g
$NaHSO_3$	0.25g
$FeSO_4 \cdot 7H_2O$	0.25g

Preparation of Optional Supplement: Add components to distilled/deionized water and bring volume to 10.0mL. Filter sterilize.

Preparation of Medium: Add components to 1.0L of prepared Brucella Agar Base. Mix thoroughly. Autoclave for 15 min at 15 psi pressure–121°C. Cool to 45°–50°C. Add 100.0mL sterile, defibrinated horse blood. Addition of 10.0mL of optional supplement will improve growth. Mix thoroughly. Pour into sterile Petri dishes.

Use: For the selective isolation and cultivation of *Campylobacter jejuni* from fecal specimens or rectal swabs.

Brucella Agar with Vitamin K_1

Composition per liter:

Agar	17.5g
Pancreatic digest of casein	10.0g
Peptic digest of animal tissue	10.0g
NaCl	5.0g
Yeast extract	2.0g
Glucose	1.0g
$NaHSO_3$	0.1g
Sheep blood, defibrinated	50.0mL
Vitamin K_1 solution	1.0mL

pH 7.0 ± 0.2 at 25°C

Vitamin K_1 Solution:

Composition per 20mL:

Vitamin K_1	0.2g
Ethanol, absolute	20.0mL

Preparation of Vitamin K_1 Solution: Add vitamin K_1 to 20.0mL of ethanol. Mix thoroughly. Filter sterilize.

Preparation of Medium: Add components, except sheep blood and vitamin K_1 solution, to distilled/deionized water and bring volume to 949.0mL. Mix thoroughly. Gently heat and bring to boiling. Autoclave for 15 min at 15 psi pressure–121°C. Cool to 45°–50°C. Aseptically add 50.0mL of sterile sheep blood and 1.0mL of sterile vitamin K_1 solution. Mix thoroughly. Pour into sterile Petri dishes or distribute into sterile tubes.

Use: For the cultivation of *Brucella* species.

Brucella Albimi Broth
See: *Brucella* Broth

Brucella Albimi Broth with 0.16% Agar

Composition per liter:

Pancreatic digest of casein	10.0g
Peptic digest of animal tissue	10.0g
NaCl	5.0g
Yeast extract	2.0g
Agar	1.6g
Glucose	1.0g
$NaHSO_3$	0.1g
Horse blood, defibrinated	100.0mL

pH 7.0 ± 0.2 at 25°C

Preparation of Medium: Add components, except horse blood, to distilled/deionized water and bring volume to 900.0mL. Mix thoroughly. Heat gently with frequent mixing. Boil for 1 min. Autoclave for 15 min at 15 psi pressure–121°C. Cool to 45°–50°C. Aseptically add 100.0mL of sterile defibrinated horse blood. Mix thoroughly. Aseptically distribute into sterile tubes or flasks.

Use: For the cultivation and maintenance of *Campylobacter* species.

Brucella Albimi Broth with Agar and 1.5% NaCl

Composition per liter:

NaCl	15.0g
Pancreatic digest of casein	10.0g
Peptic digest of animal tissue	10.0g
Yeast extract	2.0g
Agar	1.6g
Glucose	1.0g
$NaHSO_3$	0.1g
Horse blood, defibrinated	100.0mL

pH 7.0 ± 0.2 at 25°C

Preparation of Medium: Add components, except horse blood, to distilled/deionized water and bring volume to 900.0mL. Mix thoroughly. Heat gently with frequent mixing. Boil for 1 min. Autoclave for 15 min at 15 psi pressure–121°C. Cool to 45°–50°C. Aseptically add 100.0mL of sterile defibrinated horse blood. Mix thoroughly. Aseptically distribute into sterile tubes or flasks.

Use: For the cultivation and maintenance of *Campylobacter nitrofigilis*.

Brucella Albimi Broth with Formate and Fumarate

Composition per 1050mL:

Pancreatic digest of casein	10.0g
Peptic digest of animal tissue	10.0g
NaCl	5.0g
Yeast extract	2.0g
Glucose	1.0g
$NaHSO_3$	0.1g
Horse blood, defibrinated	100.0mL
Formate-fumarate solution	50.0mL

pH 7.0 ± 0.2 at 25°C

Formate-Fumarate Solution:
Composition per 100mL:

Sodium formate	6.0g
Fumaric acid	6.0g

Preparation of Formate-Fumarate Solution: Add components to distilled/deionized water and bring volume to 100.0mL. Mix thoroughly. Adjust pH to 7.0. Filter sterilize.

Preparation of Medium: Add components, except formate-fumarate solution and horse blood, to distilled/deionized water and bring volume to 900.0mL. Mix thoroughly. Heat gently with frequent mixing. Boil for 1 min. Autoclave for 15 min at 15 psi pressure–121°C. Cool to 45°–50°C. Add 100.0mL sterile defibrinated horse blood. Mix gently and aseptically distribute into sterile tubes in 5.0mL volumes. Aseptically add 0.25mL of formate/fumarate solution to each tube containing 5.0mL of medium immediately prior to inoculation.

Use: For the cultivation and maintenance of *Campylobacter mucosalis*.

Brucella Albimi Broth with Sheep Blood

Composition per liter:

Pancreatic digest of casein	10.0g
Peptic digest of animal tissue	10.0g
NaCl	5.0g

Yeast extract	2.0g
Glucose	1.0g
$NaHSO_3$	0.1g
Sheep blood, defibrinated	100.0mL

pH 7.0 ± 0.2 at 25°C

Preparation of Medium: Add components, except sheep blood, to distilled/deionized water and bring volume to 900.0mL. Mix thoroughly. Heat gently with frequent mixing. Boil for 1 min. Autoclave for 15 min at 15 psi pressure–121°C. Cool to 45°–50°C. Aseptically add 100.0mL of sterile defibrinated sheep blood. Mix thoroughly. Aseptically distribute into sterile tubes or flasks.

Use: For the cultivation and maintenance of *Helicobacter nemestrinae*.

Brucella Albimi Medium, Semisolid

Composition per liter:

Pancreatic digest of casein	10.0g
Peptic digest of animal tissue	10.0g
Glycine	10.0g
NaCl	8.5g
Yeast extract	2.0g
Agar	1.6g
Glucose	1.0g
L-Cysteine·HCl·H_2O	0.2g
$NaHSO_3$	0.1g

pH 7.0 ± 0.2 at 25°C

Preparation of Medium: Add components to distilled/deionized water and bring volume to 1.0L. Mix thoroughly. Adjust pH to 7.0. Gently heat and bring to boiling. Distribute into tubes in 10.0mL volumes. Autoclave for 15 min at 15 psi pressure–121°C. Allow tubes to cool in an upright position.

Use: For the cultivation and identification of *Campylobacter* species.

Brucella Anaerobic Blood Agar

Composition per liter:

Vitamin K_1	0.01g
Anaerobic agar base	1000.0mL
Sheep blood, sterile, defibrinated	50.0mL

Anaerobic Agar Base

Composition per liter:

Pancreatic digest of casein	17.5g
Agar	15.0g
Glucose	10.0g
Papaic digest of soybean meal	2.5g
NaCl	2.5g
Sodium thioglycollate	2.0g
Sodium formaldehyde sulfoxylate	1.0g
L-Cystine·HCl·H_2O	0.4g
Methylene Blue	0.002g

pH 7.0 ± 0.2 at 25°C

Preparation of Anaerobic Agar Base: Add components to distilled/deionized water and bring volume to 1.0L. Mix thoroughly. Autoclave for 15 min at 15 psi pressure–121°C. Cool to 45°–50°C.

Preparation of Medium: To 950.0mL of cooled, sterile anaerobic agar base, aseptically add 10.0mg of vitamin K_1 and 50.0mL sterile, defibrinated sheep blood.

Use: For the isolation of anaerobes.

Brucella Blood Agar with Hemin and Vitamin K_1

Composition per liter:

Agar	15.0g
Pancreatic digest of casein	10.0g
Peptic digest of animal tissue	10.0g
NaCl	5.0g
Yeast extract	2.0g
Glucose	1.0g
$NaHSO_3$	0.1g
Vitamin K_1	1.0mL
Hemin	1.0mL
Sheep blood, defibrinated	50.0mL

Source: Available as a prepared medium from BBL Microbiology Systems.

Vitamin K_1 Solution:

Composition per 100mL:

Vitamin K_1	1.0g
Ethanol	99.0mL

Preparation of Vitamin K_1 Solution: Add vitamin K_1 to 99.0mL of absolute ethanol. Mix thoroughly. Filter sterilize.

Hemin Solution:

Composition per 100mL:

Hemin	1.0g
NaOH (1*N* solution)	20.0mL

Preparation of Hemin Solution: Add hemin to 20.0mL of 1*N* NaOH solution. Mix thoroughly. Bring volume to 100.0mL with distilled/deionized water.

Preparation of Medium: Add components, except vitamin K_1 solution and sheep blood, to distilled/deionized water and bring volume to 949.0mL. Mix thoroughly. Gently heat and bring to boiling. Autoclave for 15 min at 15 psi pressure–121°C. Cool to 45°–50°C. Aseptically add 1.0mL sterile vitamin

K_1 solution and 50.0mL of sterile defibrinated sheep blood. Mix gently and pour into sterile Petri dishes.

Use: For the isolation and cultivation of anaerobic microorganisms from clinical and nonclinical specimens. After growth on agar plates colonies should be examined under a dissecting microscope under long-wave UV light. Members of the pigmented *Bacteroides* group appear as red/orange fluorescent colonies.

Brucella Blood Culture Broth

Composition per liter:

Sucrose	100.0g
Hemin	0.5g
Sodium polyanetholsulfonate (SPS)	0.25g
Brucella broth base	1000.0mL
Vitamin K_1 solution	1.0mL

pH 7.0 ± 0.2 at 25°C

***Brucella* Broth Base:**

Composition per liter:

Pancreatic digest of casein	10.0g
Peptic digest of animal tissue	10.0g
NaCl	5.0g
Yeast extract	2.0g
Glucose	1.0g
$NaHSO_3$	0.1g

Preparation of *Brucella* Broth Base: Add components to distilled/deionized water and bring volume to 1.0L. Mix thoroughly.

Vitamin K_1 Solution:

Composition per 100mL:

Vitamin K_1	1.09g

Preparation of Vitamin K_1 Solution: Add vitamin K_1 to 99.0mL of absolute ethanol. Store in the dark at 4°C.

Preparation of Medium: Add components, except Vitamin K_1 solution, to prepared *Brucella* Broth Base. Autoclave for 15 min at 15 psi pressure–121°C. Cool to 45°–50°C. Aseptically add 1.0mL of vitamin K_1 solution. Distribute into sterile tubes or flasks.

Use: For the isolation and cultivation of microorganisms from blood. Especially useful for the cultivation of anaerobes.

Brucella Broth (*Brucella* Albimi Broth)

Composition per liter:

Pancreatic digest of casein	10.0g
Peptic digest of animal tissue	10.0g
NaCl	5.0g
Yeast extract	2.0g
Glucose	1.0g
$NaHSO_3$	0.1g
Horse blood, defibrinated	50.0mL

pH 7.0 ± 0.2 at 25°C

Source: This medium is available as a premixed powder from Difco Laboratories and BBL Microbiology Systems.

Preparation of Medium: Add components, except horse blood, to distilled/deionized water and bring volume to 950.0mL. Mix thoroughly. Heat gently with frequent mixing. Boil for 1 min. Autoclave for 15 min at 15 psi pressure–121°C. Cool to 45°–50°C. Aseptically add 50.0mL of sterile horse blood. Mix thoroughly. Aseptically distribute into sterile tubes or flasks.

Use: For the cultivation and maintenance of *Campylobacter coli, Campylobacter fecalis*, and *Brucella* species. Also used for the isolation and cultivation of a wide variety of fastidious and nonfastidious microorganisms.

Brucella Broth Base *Campylobacter* Medium

Composition per liter:

Cycloheximide (actidione®)	50.0mg
Sodium cephazolin	15.0mg
Novobiocin	5.0mg
Bacitracin	25,000U
Colistin sulfate	10,000U
Brucella broth base	900.0mL
Horse blood, defibrinated	100.0mL

pH 7.0 ± 0.2 at 25°C

***Brucella* Broth Base:**

Composition per liter:

Pancreatic digest of casein	10.0g
Peptic digest of animal tissue	10.0g
NaCl	5.0g
Yeast extract	2.0g
Glucose	1.0g
$NaHSO_3$	0.1g

Preparation of *Brucella* Broth Base: Add components to distilled/deionized water and bring volume to 1.0L. Mix thoroughly.

Optional Supplement:

Composition per 10mL:

Sodium pyruvate	0.25g
$NaHSO_3$	0.25g
$FeSO_4 \cdot 7H_2O$	0.25g

Preparation of Optional Supplement: Add components to distilled/deionized water and bring volume to 10.0mL. Filter sterilize.

Preparation of Medium: Add components, except horse blood, to 900.0mL of prepared *Brucella* broth base. Mix thoroughly. Autoclave for 15 min at 15 psi pressure–121°C. Cool to 45°–50°C. Aseptically add 100.0mL sterile, defibrinated horse blood. Addition of 10.0mL of optional supplement will improve growth. Mix thoroughly. Pour into sterile Petri dishes.

Use: For the selective isolation and cultivation of *Campylobacter jejuni* from fecal specimens or rectal swabs. Addition of the optional supplement improves growth.

Brucella Broth with Additives (ATCC Medium 489)

Composition per liter:

Pancreatic digest of casein	10.0g
Peptic digest of animal tissue	10.0g
NaCl	3.5g
Yeast extract	2.0g
Glucose	1.0g
$NaHSO_3$	0.1g
Horse serum, inactivated	100.0mL
Fresh yeast extract solution	50.0mL

pH 7.0 ± 0.2 at 25°C

Fresh Yeast Extract Solution:

Composition per 100mL:

Live, pressed, starch-free, Baker's yeast	25.0g

Preparation of Fresh Yeast Extract Solution: Add the live Baker's yeast to 100.0mL of distilled/deionized water. Autoclave for 90 min at 15 psi pressure–121°C. Allow to stand. Remove supernatant solution. Adjust pH to 6.6–6.8.

Preparation of Medium: Add components, except horse serum and fresh yeast extract solution, to distilled/deionized water and bring volume to 850.0mL. Mix thoroughly. Gently heat and bring to boiling. Autoclave for 15 min at 15 psi pressure–121°C. Cool to 45°–50°C. Aseptically add 100.0mL of sterile horse serum and 50.0mL of sterile fresh yeast extract solution. Mix thoroughly. Aseptically distribute into sterile tubes or flasks.

Use: For the cultivation of *Corynebacterium* species.

Brucella Broth with Additives (ATCC Medium 490)

Composition per liter:

NaCl	30.0g
Pancreatic digest of casein	10.0g
Peptic digest of animal tissue	10.0g
Yeast extract	2.0g
Glucose	1.0g
$NaHSO_3$	0.1g
Horse serum, inactivated	100.0mL
Fresh yeast extract solution	50.0mL

pH 7.0 ± 0.2 at 25°C

Fresh Yeast Extract Solution:

Composition per 100mL:

Live, pressed, starch-free, Baker's yeast	25.0g

Preparation of Fresh Yeast Extract Solution: Add the live Baker's yeast to 100.0mL of distilled/deionized water. Autoclave for 90 min at 15 psi pressure–121°C. Allow to stand. Remove supernatant solution. Adjust pH to 6.6–6.8.

Preparation of Medium: Add components, except horse serum and fresh yeast extract solution, to distilled/deionized water and bring volume to 850.0mL. Mix thoroughly. Gently heat and bring to boiling. Autoclave for 15 min at 15 psi pressure–121°C. Cool to 45°–50°C. Aseptically add 100.0mL of horse serum and 50.0mL of sterile fresh yeast extract solution. Mix thoroughly. Aseptically distribute into sterile tubes or flasks.

Use: For the cultivation of salt-tolerant *Corynebacterium* species.

Brucella Broth, Modified

Composition per liter:

Pancreatic digest of casein	10.0g
Peptic digest of animal tissue	10.0g
NaCl	5.0g
$MgSO_4 \cdot 7H_2O$	2.46g
Yeast extract	2.0g
$CaCl_2$	1.1g
Glucose	1.0g
$NaHSO_3$	0.1g
Horse blood, defibrinated	100.0mL

pH 7.0 ± 0.2 at 25°C

Preparation of Medium: Add components, except horse blood, to distilled/deionized water and bring volume to 900.0mL. Mix thoroughly. Gently heat and bring to boiling. Autoclave for 15 min at 15 psi pressure–121°C. Cool to 45°–50°C. Aseptically add sterile horse blood. Mix thoroughly. Aseptically distribute into sterile tubes or flasks.

Use: For the cultivation and maintenance of *Campylobacter coli* and *Campylobacter fecalis*.

Brucella FBP Agar

Composition per liter:

Agar	15.0g
Pancreatic digest of casein	10.0g
Peptic digest of animal tissue	10.0g
NaCl	5.0g
Yeast extract	2.0g
Glucose	1.0g
$NaHSO_3$	0.1g
FBP solution	30.0mL

pH 7.0 ± 0.2 at 25°C

FBP Solution:

Composition per 30mL:

$FeSO_4$	0.25g
$NaHSO_3$	0.25g
Sodium pyruvate	0.25g

Preparation of FBP Solution: Add components to distilled/deionized water and bring volume to 30.0mL. Mix thoroughly. Filter sterilize.

Preparation of Medium: Add components, except FBP solution, to distilled/deionized water and bring volume to 970.0mL. Mix thoroughly. Gently heat and bring to boiling. Autoclave for 15 min at 15 psi pressure–121°C. Cool to 45°–50°C. Aseptically add 30.0mL of sterile FBP solution. Mix thoroughly. Pour into sterile Petri dishes or distribute into sterile tubes.

Use: For the cultivation of *Brucella* species.

Brucella FBP Broth

Composition per liter:

Pancreatic digest of casein	10.0g
Peptic digest of animal tissue	10.0g
NaCl	5.0g
Yeast extract	2.0g
Glucose	1.0g
$NaHSO_3$	0.1g
FBP solution	30.0mL

pH 7.0 ± 0.2 at 25°C

FBP Solution:

Composition per 30mL:

$FeSO_4$	0.25g
Sodium metabisulfite, anhydrous	0.25g
Sodium pyruvate, anhydrous	0.25g

Preparation of FBP Solution: Add components to distilled/deionized water and bring volume to 30.0mL. Mix thoroughly. Filter sterilize.

Preparation of Medium: Add components, except FBP solution, to distilled/deionized water and bring volume to 970.0mL. Mix thoroughly. Gently heat and bring to boiling. Autoclave for 15 min at 15 psi pressure–121°C. Cool to 45°–50°C. Aseptically add 30.0mL of sterile FBP solution. Mix thoroughly. Aseptically distribute into sterile tubes.

Use: For the cultivation of *Brucella* species.

Brucella Medium Base

Composition per liter:

Agar	15.0g
Glucose	10.0g
Peptone	10.0g
Beef extract	5.0g
NaCl	5.0g

pH 7.5 ± 0.2 at 25°C

Preparation: Add components to distilled/deionized water and bring volume to 1.0L. Mix thoroughly. Heat gently and bring to boiling. Distribute into tubes or flasks. Autoclave for 15 min at 15 psi pressure–121°C. Pour into sterile Petri dishes or leave in tubes.

Use: For the isolation of *Campylobacter* species.

Brucella Medium, Selective

Composition per liter:

Agar	15.0g
Pancreatic digest of casein	10.0g
Peptic digest of animal tissue	10.0g
NaCl	5.0g
Yeast extract	2.0g
Glucose	1.0g
$NaHSO_3$	0.1g
Horse serum	100.0mL
VCNF antibiotic solution	10.0mL

pH 7.0 ± 0.2 at 25°C

Preparation of VCNF Antibiotic Solution: Add components to distilled/deionized water and bring volume to 10.0mL. Mix thoroughly. Filter sterilize.

Preparation of Medium: Add components, except horse serum and VCNF antibiotic solution, to distilled/deionized water and bring volume to 890.0mL. Mix thoroughly. Gently heat and bring to boiling. Autoclave for 15 min at 15 psi pressure–121°C. Cool to 45°–50°C. Aseptically add 100.0mL of sterile horse serum and 10.0mL of VCNF antibiotic solution. Mix thoroughly. Pour into sterile Petri dishes or distribute into sterile tubes.

Use: For the selective isolation, cultivation and maintenance of *Brucella* species.

Brucella Selective Medium

Composition per liter:

Beef heart, infusion from	500.0g
Agar	15.0g
Tryptose	10.0g
NaCl	5.0g
Glucose	2.5g
Gelatin	1.0g
Sheep blood	100.0mL
Antibiotic solution	10.0mL

pH 7.4 ± 0.2 at 25°C

Antibiotic Solution:

Composition per 10mL:

Cycloheximide	1.0g
Bacitracin	250,000U
Circulin	250,000U
Polymyxin B	100,000U

Preparation of Antibiotic Solution: Add components to distilled/deionized water and bring volume to 10.0mL. Mix thoroughly. Filter sterilize.

Preparation of Medium: Add components, except sheep blood and antibiotic solution, to distilled/deionized water and bring volume to 890.0mL. Mix thoroughly. Gently heat and bring to boiling. Autoclave for 15 min at 15 psi pressure–121°C. Cool to 45°–50°C. Aseptically add 100.0mL of sterile sheep blood and 10.0mL of sterile antibiotic solution. Mix thoroughly. Pour into sterile Petri dishes or distribute into sterile tubes.

Use: For the selective isolation and cultivation of *Brucella* species.

Brucella Semisolid Medium with Cysteine

Composition per liter:

Peptamin	10.0g
Pancreatic digest of casein	10.0g
Glycine	10.0g
NaCl	5.0g
Yeast extract	2.0g
Agar	1.8g
Glucose	1.0g
Cysteine·HCl·H_2O	0.2g
$NaHSO_3$	0.1g
Sodium citrate	0.1g
Neutral Red solution	10.0mL

pH 7.0 ± 0.2 at 25°C

Neutral Red Solution:

Composition per 100mL:

Neutral Red	0.2g
Ethanol	10.0mL

Preparation of Neutral Red Solution: Add Neutral Red to 10.0mL of ethanol. Bring volume to 100.0mL.

Preparation of Medium: Add components to distilled/deionized water and bring volume to 1.0L. Mix thoroughly. Gently heat and bring to boiling. Distribute into tubes in 10.0mL volumes. Autoclave for 15 min at 15 psi pressure–121°C.

Use: For the cultivation and differentiation of *Campylobacter* species based on H_2S production from cysteine.

Brucella Semisolid Medium with Glycine

Composition per liter:

Peptamine	10.0g
Pancreatic digest of casein	10.0g
Glycine	10.0g
NaCl	5.0g
Yeast extract	2.0g
Agar	1.8g
Glucose	1.0g
$NaHSO_3$	0.1g
Sodium citrate	0.1g
Neutral Red solution	10.0mL

pH 7.0 ± 0.2 at 25°C

Neutral Red Solution:

Composition per 100mL:

Neutral Red	0.2g
Ethanol	10.0mL

Preparation of Neutral Red Solution: Add Neutral Red to 10.0mL of ethanol. Bring volume to 100.0mL.

Preparation of Medium: Add components to distilled/deionized water and bring volume to 1.0L. Mix thoroughly. Gently heat and bring to boiling. Distribute into tubes in 10.0mL volumes. Autoclave for 15 min at 15 psi pressure–121°C.

Use: For the cultivation and differentiation of *Campylobacter* species based on glycine utilization.

Brucella Semisolid Medium with NaCl

Composition per liter:

NaCl	35.0g
Peptamin	10.0g
Pancreatic digest of casein	10.0g
Yeast extract	2.0g
Agar	1.8g

Glucose 1.0g
$NaHSO_3$ 0.1g
Sodium citrate 0.1g
Neutral Red solution 10.0mL

pH 7.0 ± 0.2 at 25°C

Neutral Red Solution:
Composition per 100mL:

Neutral Red 0.2g
Ethanol 10.0mL

Preparation of Neutral Red Solution: Add Neutral Red to 10.0mL of ethanol. Bring volume to 100.0mL.

Preparation of Medium: Add components to distilled/deionized water and bring volume to 1.0L. Mix thoroughly. Gently heat and bring to boiling. Distribute into tubes in 10.0mL volumes. Autoclave for 15 min at 15 psi pressure–121°C.

Use: For the cultivation and differentiation of *Campylobacter* species based on glycine utilization.

Brucella Semisolid Medium with Nitrate

Composition per liter:

Peptamin 10.0g
Pancreatic digest of casein 10.0g
Glycine 10.0g
KNO_3 .... 10.0g
NaCl 5.0g
Yeast extract 2.0g
Agar 1.8g
Glucose 1.0g
$NaHSO_3$ 0.1g
Sodium citrate 0.1g

pH 7.0 ± 0.2 at 25°C

Preparation of Medium: Add components to distilled/deionized water and bring volume to 1.0L. Mix thoroughly. Gently heat and bring to boiling. Distribute into tubes in 10.0mL volumes. Autoclave for 15 min at 15 psi pressure–121°C.

Use: For the cultivation and differentiation of *Campylobacter* species based on nitrate reduction.

Bryant-Robinson Medium

Composition per 1010mL:

Glucose, cellobiose, or maltose 5.0g
L-Methionine 0.08g
Mineral solution 50.0mL
Na_2CO_3 solution 50.0mL
Hemin solution 10.0mL
Cysteine·HCl–Na_2S solution 10.0mL
Vitamin solution 5.0mL
VFA solution 4.5mL
Resazurin 1.0mL

pH 6.5 ± 0.2 at 25°C

Mineral Solution:
Composition per liter:

KH_2PO_4 .... 18.0g
NaCl 18.0g
$(NH_4)_2SO_4$ 8.0g
$CaCl_2·6H_2O$ 0.53g
$MgCl_2·6H_2O$ 0.4g
$CoCl_2·6H_2O$ 0.2g
$MnCl_2·4H_2O$ 0.2g
$FeSO_4·7H_2O$ 0.08g

Preparation of Mineral Solution: Add components to distilled/deionized water and bring volume to 1.0L. Mix thoroughly.

Na_2CO_3 Solution:
Composition per 100mL:

Na_2CO_3 .... 8.0g

Preparation of Na_2CO_3 Solution: Add Na_2CO_3 to O_2-free distilled/deionized water. Mix thoroughly. Gas with 100% CO_2 for 15 min. Autoclave for 15 min at 15 psi pressure–121°C.

Hemin Solution:
Composition per 100mL:

Hemin 0.01g
NaOH (0.002% solution) 100.0mL

Preparation of Hemin Solution: Add hemin to 100.0mL of NaOH solution. Mix thoroughly.

Cysteine·HCl–Na_2S Solution:
Composition per 100mL:

Cysteine·HCl 2.5g
$Na_2S·9H_2O$ 2.5g

Preparation of Cysteine·HCl–Na_2S Solution: Add cysteine·HCl to distilled/deionized water and bring volume to 80.0mL. Mix thoroughly. Adjust pH to 11 with NaOH. Add $Na_2S·9H_2O$. Mix thoroughly. Bring volume to 100.0mL with distilled/deionized water. Gently heat and bring to boiling under 100% N_2. Cool to 25°C under 100% N_2. Autoclave for 15 min at 15 psi pressure–121°C.

Vitamin Solution:
Composition per 100mL:

Calcium pantothenate 0.02g
Nicotinamide 0.02g
Pyridoxine·HCl 0.02g
Riboflavin 0.02g
Thiamine·HCl 0.02g
p-Aminobenzoic acid 1.0mg
Biotin 0.25mg
Folic acid 0.25mg
Vitamin B_{12} 0.1mg

Preparation of Vitamin Solution: Add components to distilled/deionized water and bring volume to 100.0mL. Mix thoroughly. Filter sterilize.

Volatile Fatty Acid Solution:
Composition per liter:

Acetic acid	36.0mL
DL-α-Methylbutyric acid	2.0mL
Isovaleric acid	2.0mL
n-Valeric acid	2.0mL
Isobutyric acid	1.8mL

Preparation of Vitamin Solution: Add components to distilled/deionized water and bring volume to 1.0L. Mix thoroughly.

Preparation of Medium: Add components, except cysteine·HCl–Na_2S solution, to distilled/deionized water and bring volume to 1.0L. Mix thoroughly. Gently heat and bring to boiling. Continue boiling until resazurin turns colorless indicating reduction. Anaerobically distribute into tubes in 10.0mL volumes. Cap with butyl rubber stoppers. Place tubes in a press. Autoclave for 15 min at 15 psi pressure–121°C. Immediately prior to inoculation aseptically and anaerobically add 0.1mL of cysteine·HCl–Na_2S solution per tube.

Use: For the cultivation of *Bacteroides* species from rumens.

BSA Tween™ 80 Agar
See: **Bovine Serum Albumin Tween™ 80 Agar**

BSA Tween™ 80 Broth
See: **Bovine Serum Albumin Tween™ 80 Broth**

BSA Tween™ 80 Soft Agar
See: **Bovine Serum Albumin Tween™ 80 Soft Agar**

BSK Medium

Composition per 1260mL:

Bovine albumin Fraction V	50.0g
HEPES (N-[2-Hydroxyethyl]piperazine-N´-2-ethanesulfonic acid) buffer	6.0g
Neopeptone	5.0g
Glucose	5.0g
$NaHCO_3$	2.2g
Sodium pyruvate	0.8g
Sodium citrate	0.7g
N-Acetyl glucosamine	0.4g
Gelatin solution	200.0mL
CMRL 1066, without glutamine, without bicarbonate, 10X	100.0mL
Rabbit serum	72.0mL

pH 7.6-7.65 at 25°C

Gelatin Solution:
Composition per 200mL:

Gelatin	14.0g

Preparation of Gelatin Solution: Add gelatin to distilled/deionized water and bring volume to 200.0mL. Heat gently to boiling. Mix thoroughly. Filter sterilize.

CMRL 1066 Medium without Glutamine, without bicarbonate, 10X:
Composition per liter:

NaCl	6.8g
D-Glucose	1.0g
KCl	0.4g
L-Cysteine·HCl·H_2O	0.26g
$CaCl_2$, anhydrous	0.2g
$MgSO_4 \cdot 7H_2O$	0.2g
$NaH_2PO_4 \cdot H_2O$	0.14g
Sodium acetate·$3H_2O$	0.083g
L-Glutamic acid	0.075g
L-Arginine·HCl	0.070g
L-Lysine·HCl	0.070g
L-Leucine	0.060g
Glycine	0.050g
Ascorbic acid	0.050g
L-Proline	0.040g
L-Tyrosine	0.040g
L-Aspartic acid	0.030g
L-Threonine	0.030g
L-Alanine	0.025g
L-Phenylalanine	0.025g
L-Serine	0.025g
L-Valine	0.025g
L-Cystine	0.020g
L-Histidine·HCl·H_2O	0.020g
L-Isoleucine	0.020g
Phenol red	0.020g
L-Methionine	0.015g
Deoxyadenosine	0.010g
Deoxycytidine	0.010g
Deoxyguanosine	0.010g
Glutathione, reduced	0.010g
Thymidine	0.010g
Hydroxy-L-proline	0.010g
L-Tryptophan	0.010g
Nicotinamide adenine dinucleotide	7.0mg
Tween™ 80	5.0mg
Sodium glucoronate·H_2O	4.2mg

Coenzyme A	2.5mg
Cocarboxylase	1.0mg
Flavin adenine dinucleotide	1.0mg
Nicotinamide adenine dinucleotide phosphate	1.0mg
Uridine triphosphate	1.0mg
Choline chloride	0.50mg
Cholesterol	0.20mg
5-Methyldeoxycytidine	0.10mg
Inositol	0.05mg
p-Aminobenzoic acid	0.05mg
Niacin	0.025mg
Niacinamide	0.025mg
Pyridoxine	0.025mg
Pyridoxal·HCl	0.025mg
Biotin	0.01mg
D-Calcium pantothenate	0.01mg
Folic acid	0.01mg
Riboflavin	0.01mg
Thiamine·HCl	0.01mg

pH 7.2 ± 0.2 at 25°C

Preparation of CMRL 1066 Medium without Glutamine, without bicarbonate, 10X: Add components to distilled/deionized water and bring volume to 1.0L. Mix thoroughly. Adjust pH to 7.2. Filter sterilize.

Preparation of Medium: Add components, except gelatin solution and rabbit serum, to 628.0mL of glass distilled water. Mix thoroughly. Adjust pH to 7.6–7.65. Add 200.0mL 7% aqueous gelatin solution. Filter-sterilize entire medium. Aseptically add 72.0mL of sterile rabbit serum.

Use: For the cultivation of a wide variety of microorganisms in a chemically defined medium. For the cultivation of *Borrelia* and *Spirochaeta* species.

BSL for *Corynebacterium* (Buffered Soy Lactose for *Corynebacterium*)

Composition per liter:

Agar	15.0g
Papaic digest of soybean meal	10.0g
Na_2HPO_4	6.0g
KH_2PO_4	3.0g
NH_4Cl	1.0g
$MgSO_4.7H_2O$	0.2g
Lactose solution	100.0mL

pH 6.8-7.2 at 25°C

Lactose Solution:

Composition per 100mL:

Lactose	10.0g

Preparation of Lactose Solution: Add lactose to distilled/deionized water and bring volume to 100.0mL. Mix thoroughly. Autoclave for 15 min at 15 psi pressure–121°C. Cool to 45°–50°C.

Preparation of Medium: Add components, except lactose solution, to distilled/deionized water and bring volume to 900.0mL. Mix thoroughly. Heat gently with frequent mixing. Adjust pH to 6.8–7.2. Autoclave for 15 min at 15 psi pressure–121°C. Cool to 45°–50°C. Aseptically add sterile lactose solution. Mix thoroughly. Pour into sterile Petri dishes or distribute into sterile tubes.

Use: For the cultivation and maintenance of *Curtobacterium flaccumfaciens.*

BSR Medium

Composition per liter:

Beef heart, solids from infusion	500.0g
Sorbitol	70.0g
Sucrose	10.0g
Tryptose	10.0g
NaCl	5.0g
Fructose	1.0g
Glucose	1.0g
Phenol Red	0.020g
Horse serum	100.0mL

pH 7.6 ± 0.2 at 25°C

Preparation of Medium: Add components, except horse serum, to distilled/deionized water and bring volume to 900.0mL. Mix thoroughly. Autoclave for 15 min at 15 psi pressure–121°C. Cool to 45°–50°C. Aseptically add 100.0mL of horse serum. Mix thoroughly. Aseptically distribute into sterile tubes or flasks.

Use: For the isolation and cultivation of *Spiroplasma citri*.

BSTSY Agar

Composition per liter:

Pancreatic digest of casein	17.0g
Agar	15.0g
NaCl	5.0g
Yeast extract	4.0g
Papaic digest of soybean meal	3.0g
K_2HPO_4	2.5g
Glucose	2.5g
Bovine serum	100.0mL

pH 7.3 ± 0.2 at 25°C

Preparation of Medium: Add components, except bovine serum, to distilled/deionized water and bring volume to 900.0mL. Mix thoroughly. Gently heat and bring to boiling. Autoclave for 15 min at 15 psi pressure–121°C. Cool to 45°–50°C. Aseptically add sterile bovine serum. Mix thoroughly. Pour into sterile Petri dishes or distribute into sterile tubes.

Use: For the isolation and cultivation of *Simonsiella* species and *Alysiella* species.

BTB Lactose Agar (Bromthymol Blue Lactose Agar)

Composition per liter:

Agar	15.0g
Lactose	10.0g
Proteose peptone	5.0g
Beef extract	3.0g
Bromthymol Blue	0.17g

pH 8.7–7.2 at 25°C

Preparation of Medium: Add components to distilled/deionized water and bring volume to 1.0L. Mix thoroughly. Heat gently with frequent mixing. Bring to boiling. Distribute into tubes or flasks. Autoclave for 15 min at 15 psi pressure–121°C. Pour into sterile Petri dishes if desired.

Use: For the isolation and cultivation of pathogenic staphylococci.

BTB Teepol® Agar

Composition per liter:

NaCl	20.0g
Agar	15.0g
Peptone	10.0g
Sucrose	10.0g
Beef extract	5.0g
Bromthymol Blue	0.08g
Teepol	2.0mL

pH 7.8 ± 0.2 at 25°C

Preparation of Medium: Add components to distilled/deionized water and bring volume to 1.0L. Teepol may be substituted by 0.1mL of Tergitol™ 7. Mix thoroughly. Gently heat and bring to boiling. Adjust pH to 7.8. Autoclave for 15 min at 15 psi pressure–121°C. Pour into sterile Petri dishes.

Use: For the isolation and cultivation of *Vibrio anguillarum.*

Buffered Azide Glucose Glycerol Broth
See: **BAGG Broth**

Buffered Charcoal Yeast Extract Agar
See: **BCYE Agar**

Buffered Charcoal Yeast Extract Agar with Albumin
See: **BCYEα with Alb**

Buffered Charcoal Yeast Extract Agar without L-Cysteine
See: **BCYEα without L-Cysteine**

Buffered Charcoal Yeast Extract Differential Agar (DIFF/BCYE)

Composition per 1014mL:

Agar	17.0g
ACES (2-[(2-Amino-2-oxoethyl)-amino]-ethane sulfonic acid) buffer	10.0g
Yeast extract	10.0g
Charcoal, activated	1.5g
$Fe_4(P_2O_7)_3 \cdot 9H_2O$	0.25g
Bromcresol Purple	0.01g
Bromthymol Blue	0.01g
Antibiotic solution	10.0mL
Cysteine·HCl·H_2O solution	4.0mL

pH 6.9 ± 0.2 at 25°C

Antibiotic Solution:
Composition per 10mL:

Vancomycin	1.0mg
Polymyxin B	50,000U

Preparation of Antibiotic Solution: Add components to distilled/deionized water and bring volume to 10.0mL. Mix thoroughly. Filter sterilize.

Cysteine·HCl·H_2O Solution:
Composition per 10mL:

L-Cysteine·HCl·H_2O	1.0g

Preparation of Cysteine·HCl·H_2O Solution: Add L-Cysteine·HCl·H_2O to distilled/deionized water and bring volume to 10.0mL. Mix thoroughly. Filter sterilize.

Preparation of Medium: Add components, except cysteine solution and antibiotic solution, to distilled/deionized water and bring volume to 1.0L. Mix thoroughly. Adjust medium to pH 6.9 with 1*N* KOH. Heat gently and bring to boil for 1 minute. Autoclave for 15 min at 15 psi pressure–121°C. Cool to 50–55°C. Add 4.0mL of sterile L-cysteine·HCl·H_2O

solution and 10.0mL of sterile antibiotic solution. Mix thoroughly. Pour into sterile Petri dishes with constant agitation to keep charcoal in suspension.

Use: For the isolation, cultivation, and maintenance of *Legionella pneumophila* and other *Legionella* species from environmental and clinical specimens. Used for the selective recovery of *Legionella pneumophila* while reducing contaminating microorganisms from environmental water samples.

Buffered Charcoal Yeast Extract Medium, Diphasic Blood Culture
See: **BCYE Medium, Diphasic Blood Culture**

Buffered Charcoal Yeast Extract Selective Agar with Cephalothin, Colistin, Vancomycin and Cycloheximide
See: **BCYE Selective Agar with CCVC**

Buffered Charcoal Yeast Extract Selective Agar with Glycine, Polymyxin B, Vancomycin, and Anisomycin
See: **BCYE Selective Agar with GPVA**

Buffered Charcoal Yeast Extract Selective Agar with Glycine, Vancomycin, Polymyxin B, and Cycloheximide
See: **BCYE Selective Agar with GVPC**

Buffered Charcoal Yeast Extract Selective Agar with Polymyxin B, Anisomycin and Cefamandole
See: **BCYE Selective Agar with PAC**

Buffered Charcoal Yeast Extract Selective Agar with Polymyxin B, Anisomicin and Vancomycin
See: **BCYE Selective Agar with PAV**

Buffered Marine Yeast Medium

Composition per liter:

NaCl	24.0g
Agar	20.0g
Yeast extract	5.0g
1M Phosphate buffer, pH 6.8	20.0mL
Hutner's mineral base	20.0mL
KOH (1*N*)	7.0mL

pH 6.8 ± 0.2 at 25°C

1M Phosphate Buffer, pH 6.8:
Composition per liter:

$K_2H_2PO_4$	85.4g
$NaH_2PO_4 \cdot H_2O$	70.4g

Preparation of Phosphate Buffer: Add components to distilled/deionized water and bring volume to 1.0L. Mix thoroughly. Adjust pH to 6.8.

Hutner's Mineral Base:
Composition per liter:

$MgSO_4.7H_2O$	29.7g
Nitrilotriacetic acid	10.0g
$CaCl_2 \cdot 2H_2O$	3.34g
$FeSO_4 \cdot 7H_2O$	0.01g
$(NH_4)_2MoO_4$	9.25mg
Metals "44"	50.0mL

Preparation of Hutner's Mineral Base: Initially add a few drops of H_2SO_4 to the distilled water to retard precipitation. Dissolve the nitrilotriacetic acid first and neutralize the solution with KOH. Add other ingredients and adjust the pH to 7.2 with KOH and/or H_2SO_4. There may be a slight precipitate. Store at 5°C.

Metals "44":
Composition per 100mL:

$ZnSO_4 \cdot 7H_2O$	1.1g
$FeSO_4 \cdot 7H_2O$	0.5g
EDTA	0.25g
$MnSO_4 \cdot 7H_2O$	0.154g
$CuSO_4 \cdot 5H_2O$	0.04g
$Co(NO_3)_2 \cdot 6H_2O$	0.025g
$Na_2B_4O_7 \cdot 10H_2O$	0.018g

Preparation of Metals "44":

Preparation of Medium: Add components to distilled/deionized water and bring volume to 1.0L. Mix thoroughly. Distribute into tubes or flasks. Autoclave for 15 min at 15 psi pressure–121°C.

Use: For the cultivation and maintenance of *Pseudomonas* species.

Buffered Peptone Water

Composition per liter:

Pancreatic digest of gelatin	10.0g
NaCl	5.0g
Na_2HPO_4	3.5g
KH_2PO_4	1.5g

pH 7.2 ± 0.2 at 25°C

Source: This medium is available as a premixed powder from BBL Laboratories, Difco Laboratories, and Oxoid Unipath.

Preparation of Medium: Add components to distilled/deionized water and bring volume to 1.0L. Mix thoroughly. Distribute into tubes or flasks. Autoclave for 15 min at 15 psi pressure–121°C.

Use: Use as a pre-enrichment medium for the isolation of *Salmonella*, especially injured microorganisms, from various food sources.

Buffered S & H Agar

Composition per liter:

Agar	15.0g
Peptone	5.0g
Yeast extract	5.0g
Na_2HPO_4	2.7g
Citric acid·H_2O	1.15g
Glucose solution	40.0mL

pH 5.0 ± 0.2 at 25°C

Glucose Solution:

Composition per 50mL:

Glucose	25.0g

Preparation of Glucose Solution: Add 25.0g of glucose to 50.0mL of distilled/deionized water. Mix thoroughly and gently heat to dissolve. Filter sterilize.

Preparation of Medium: Add components, except glucose solution,to distilled/deionized water and bring volume to 960.0mL. Mix thoroughly. Gently heat to boiling. Adjust pH to 5.0 with HCl. Autoclave for 15 min at 15 psi pressure–121°C. Cool to 50°C. Aseptically add 40.0mL of sterile glucose solution to sterile basal medium. Pour into sterile Petri dishes or distribute into sterile tubes.

Use: For the cultivation and maintenance of *Acetobacter xylinum*.

Buffered S & H Broth

Composition per liter:

Peptone	5.0g
Yeast extract	5.0g
Na_2HPO_4	2.7g
Citric acid·H_2O	1.15g
Glucose solution	40.0mL

pH 5.0 ± 0.2 at 25°C

Glucose Solution:

Composition per 50mL:

Glucose	25.0g

Preparation of Glucose Solution: Add glucose to 50.0mL of distilled/deionized water. Mix thoroughly and gently heat to dissolve. Filter sterilize.

Preparation of Medium: Add components, except glucose solution, to distilled/deionized water and bring volume to 960.0mL. Mix thoroughly. Gently heat to boiling. Adjust pH to 5.0 with HCl. Autoclave for 15 min at 15 psi pressure–121°C. Cool to 50°C. Aseptically add 40.0mL of sterile glucose solution to sterile basal medium. Aseptically distribute into sterile tubes or flasks.

Use: For the cultivation of *Acetobacter xylinum*.

Buffered Soy Lactose for *Corynebacterium*

See: BSL for *Corynebacterium*

Buffered Yeast Extract Broth

See: **BYEB**

Burke's Modified Nitrogen-Free Medium

Composition per liter:

$MgSO_4·7H_2O$	0.2g
Na_2HPO_4	0.19g
$NaHCO_3$	0.05g
$CaSO_4·2H_2O$	0.02g
KH_2PO_4	0.011g
$SrCl_2·6H_2O$	0.01g
NaCl	0.01g
Adenine	0.01g
$FeSO_4.7H_2O$	6.0mg
Na_2MoO_3	0.5mg

pH 7.8 ± 0.2 at 25°C

Preparation of Medium: Add components to distilled/deionized water and bring volume to 1.0L. Mix thoroughly. Distribute into tubes or flasks. Autoclave for 15 min at 15 psi pressure–121°C.

Use: For the cultivation of *Azotobacter vinelandii*.

Burke's Modified Nitrogen-Free Medium

Composition per liter:

Noble agar	15.0g
Glucose	10.0g
Cellulose	10.0g
K_2HPO_4	1.0g
$CaCl_2 \cdot 2H_2O$	0.1g
$MgSO_4 \cdot 7H_2O$	0.02g
$FeSO_4 \cdot 7H_2O$	50.0mg
$Na_2MoO_4 \cdot 2H_2O$	25.0mg
Vitamin B_{12}	0.1mg
Vitamin solution	1.0mL

pH 7.2–7.3 ± 0.2 at 25°C

Vitamin Solution:

Composition per 50mL:

Thiamine·HCl	843.3mg
Pantothenic acid	595.8mg
Nicotinic acid	307.8mg
p-Aminobenzoic acid	68.6mg
Pyridoxamine·2HCl	60.3mg
Biotin	50.0mg
Folic acid	11.0mg
Vitamin B_{12}	3.5mg

Preparation of Vitamin Solution: Add components to distilled/deionized water and bring volume to 50.0mL. Mix thoroughly. Filter sterilize.

Preparation of Medium: Add components, except glucose, cellulose, K_2HPO_4, and vitamin solution to distilled/deionized water and bring volume to 850.0mL. Mix thoroughly. Adjust pH to 7.2–7.3. In three separate flasks add glucose, cellulose and K_2HPO_4 to 50.0mL of distilled/deionized water. Filter sterilize the vitamin solution. Autoclave the other solutions separately for 15 min at 15 psi pressure–121°C. Cool to 25°C. Aseptically combine all the solutions and mix thoroughly. Distribute into sterile tubes or flasks or pour into sterile Petri dishes.

Use: For the cultivation and maintenance of *Streptomyces* species.

Burke's Modified Nitrogen-Free Medium with Benzoate

Composition per liter:

Sodium benzoate	0.72g
$MgSO_4 \cdot 7H_2O$	0.2g
Na_2HPO_4	0.189g
$NaHCO_3$	0.05g
$CaSO_4 \cdot 2H_2O$	0.02g
KH_2PO_4	0.011g
$SrCl_2 \cdot 6H_2O$	0.01g
NaCl	0.01g
Adenine	0.01g
$FeSO_4 \cdot 7H_2O$	6.0mg
Na_2MoO_3	0.5mg

pH 7.8 ± 0.2 at 25°C

Preparation of Medium: Add components to distilled/deionized water and bring volume to 1.0L. Mix thoroughly. Distribute into tubes or flasks. Autoclave for 15 min at 15 psi pressure–121°C.

Use: For the cultivation of *Pseudomonas* species and other microorganisms which can utilize benzoate as sole carbon source.

Bushnell-Haas Agar

Composition per liter:

Agar	15.0g
KH_2PO_4	1.0g
K_2HPO_4	1.0g
NH_4NO_3	1.0g
$MgSO_4 \cdot 7H_2O$	0.2g
$FeCl_3$	0.05g
$CaCl_2 \cdot 2H_2O$	0.02g

pH 7.0 ± 0.2 at 25°C

Preparation of Medium: Add components to distilled/deionized water and bring volume to 1.0L. Mix thoroughly. Gently heat and bring to boiling. Distribute into tubes or flasks. Autoclave for 15 min at 15 psi pressure–121°C. Pour into sterile Petri dishes or leave in tubes. Layer hydrocarbon on agar surface or add aseptically add sterile hydrocarbon to cooled agar prior to pouring plates.

Use: For examining fuels for microbial contamination and for studying the hydrocarbon utilization by microorganisms. Also for the cultivation of *Nocardia* species.

Bushnell-Haas Broth

Composition per liter:

KH_2PO_4	1.0g
K_2HPO_4	1.0g
NH_4NO_3	1.0g
$MgSO_4 \cdot 7H_2O$	0.2g
$FeCl_3$	0.05g
$CaCl_2 \cdot 2H_2O$	0.02g

pH 7.0 ± 0.2 at 25°C

Source: This medium is available as a premixed powder from Difco Laboratories.

Preparation of Medium: Add components to distilled/deionized water and bring volume to 1.0L. Mix thoroughly. Distribute into tubes or flasks. Autoclave for 15 min at 15 psi pressure–121°C. Layer hydrocarbon on broth surface or add directly to broth.

Use: For examining fuels for microbial contamination and for studying the hydrocarbon utilization by microorganisms. Also for the cultivation of *Nocardia* species.

Bushnell-Haas Medium

Composition per liter:

KH_2PO_4	1.0g
K_2HPO_4	1.0g
NH_4NO_3	1.0g
Cholesterol	0.3g
$MgSO_4 \cdot 7H_2O$	0.2g
$FeCl_3$	0.05g
$CaCl_2 \cdot 2H_2O$	0.02g

pH 7.0 ± 0.2 at 25°C

Source: This medium is available as a premixed powder from Difco Laboratories.

Preparation of Medium: Add components to distilled/deionized water and bring volume to 1.0L. Mix thoroughly. Distribute into tubes or flasks. Autoclave for 15 min at 15 psi pressure–121°C. Layer hydrocarbon on broth surface or add directly to broth.

Use: For the cultivation of *Nocardia* species.

Butzler's *Campylobacter* Medium
See: Campylobacter Selective Medium, Butzler's

Butzler Medium
See: *Campylobacter* Selective Medium, Butzler's

BYE Agar

Composition per liter:

Pancreatic digest of casein	16.0g
Agar	13.5g
Brain heart, solids from infusion	8.0g
Peptic digest of animal tissue	5.0g
NaCl	5.0g
Glucose	2.0g
Na_2HPO_4	2.5g
Yeast extract	2.0g
Blood, human or animal, sterile	150.0mL

pH 7.8–8.0 ± 0.2 at 25°C

Preparation of Medium: Add components except blood to distilled/deionized water and bring volume to 850.0mL. Mix thoroughly. Autoclave for 15 min at 15 psi–121°C. Cool to 45°–50°C. Aseptically add 150.0mL of sterile blood. Outdated, citrated or heparinized blood bank blood is acceptable. Pour into sterile Petri dishes.

Use: For the isolation and cultivation of *Mycoplasma* species and L-forms of bacteria. Used for the detection of *Mycoplasma* species in tissue culture and cell lines.

BYEB
(Buffered Yeast Extract Broth)

Composition per liter:

ACES buffer (2-[(2-Amino-2-oxoethyl)-amino]-ethane sulfonic acid)	10.0g
Yeast extract	10.0g
α-Ketogluatrate	1.0g
L-Cysteine·HCl·H_2O	0.4g
$Fe_4(P_2O_7)_3 \cdot 9H_2O$	0.25g

pH 6.9 ± 0.2 at 25°C

Preparation of Medium: Add components to distilled/deionized water and bring volume to 1.0L. Mix thoroughly. Adjust pH to 6.9. Filter sterilize. Aseptically distribute into sterile tubes or flasks.

Use: For the cultivation of *Legionella pneumophila*.

C 3G *Spiroplasma* Medium

Composition per liter:

Sucrose	100.0g
Phenol Red	10.0mg
PPLO broth without Crystal Violet	500.0mL
Horse serum	150.0mL
Fresh yeast extract solution	50.0mL
CMRL-1066 medium	5.0mL

pH 7.5 ± 0.2 at 25°C

PPLO Broth without Crystal Violet:
Composition per 500mL:

Beef heart, infusion from	11.52g
Peptone	2.32g
NaCl	1.15g

Source: PPLO broth without Crystal Violet is available as a premixed powder from Difco Laboratories.

Preparation of PPLO Broth without Crystal Violet: Add components to distilled/deionized water and bring volume to 500.0mL. Mix thoroughly.

Fresh Yeast Extract Solution:
Composition per 100mL:

Baker's yeast, live, pressed, starch-free	25.0g

Preparation of Fresh Yeast Extract Solution: Add the live Baker's yeast to 100.0mL of distilled/deionized water. Autoclave for 90 min at 15 psi pressure–121°C. Allow to stand. Remove supernatant solution. Adjust pH to 6.6–6.8. Filter sterilize.

CMRL-1066 Medium:
Composition per liter:

NaCl	6.8g
$NaHCO_3$	2.2g
D-Glucose	1.0g
KCl	0.4g
L-Cysteine·HCl·H_2O	0.26g
$CaCl_2$, anhydrous	0.2g
$MgSO_4 \cdot 7H_2O$	0.2g
$NaH_2PO_4 \cdot H_2O$	0.14g
L-Glutamine	0.1g
Sodium acetate·$3H_2O$	0.083g
L-Glutamic acid	0.075g
L-Arginine·HCl	0.070g
L-Lysine·HCl	0.070g
L-Leucine	0.060g
Glycine	0.050g
Ascorbic acid	0.050g
L-Proline	0.040g
L-Tyrosine	0.040g
L-Aspartic acid	0.030g
L-Threonine	0.030g
L-Alanine	0.025g
L-Phenylalanine	0.025g
L-Serine	0.025g
L-Valine	0.025g
L-Cystine	0.020g
L-Histidine·HCl·H_2O	0.020g
L-Isoleucine	0.020g
Phenol Red	0.020g
L-Methionine	0.015g
Deoxyadenosine	0.010g
Deoxycytidine	0.010g
Deoxyguanosine	0.010g
Glutathione, reduced	0.010g
Thymidine	0.010g
Hydroxy-L-proline	0.010g
L-Tryptophan	0.010g
Nicotinamide adenine dinucleotide	7.0mg
Tween™ 80	5.0mg
Sodium glucoronate·H_2O	4.2mg
Coenzyme A	2.5mg
Cocarboxylase	1.0mg
Flavin adenine dinucleotide	1.0mg
Nicotinamide adenine dinucleotide phosphate	1.0mg
Uridine triphosphate	1.0mg
Choline chloride	0.50mg
Cholesterol	0.20mg
5-Methyldeoxycytidine	0.10mg
Inositol	0.05mg
p-Aminobenzoic acid	0.05mg
Niacin	0.025mg
Niacinamide	0.025mg
Pyridoxine	0.025mg
Pyridoxal·HCl	0.025mg
Biotin	0.01mg
D-Calcium pantothenate	0.01mg
Folic acid	0.01mg
Riboflavin	0.01mg
Thiamine·HCl	0.01mg

Source: CMRL-1066 medium is available as a premixed powder from GIBCO BRL.

Preparation of CMRL-1066 Medium: Add components to distilled/deionized water and bring volume to 1.0L. Mix thoroughly. Adjust pH to 7.2. Filter sterilize.

Preparation of Medium: Add components—except horse serum, fresh yeast extract, and CMRL medium—to distilled/deionized water and bring volume to 795.0mL. Adjust pH to 7.5. Autoclave for 15 min at 15 psi pressure–121°C. Aseptically add 150.0mL of sterile horse serum, 50.0mL of sterile fresh yeast extract solution and 5.0mL of sterile CMRL medium. Distribute into sterile tubes or flasks.

Use: For cultivation and maintenance of *Spiroplasma* species.

C 3N *Spiroplasma* Medium

Composition per 100mL:

Sucrose	12.0g
Phenol Red	10.0mg
PPLO Broth without Crystal Violet	50.0mL
Horse serum	20.0mL
Fresh yeast extract solution	5.0mL
CMRL-1066 medium	0.5mL

pH 7.5 ± 0.2 at 25°C

PPLO Broth without Crystal Violet:

Composition per 500mL:

Beef heart, infusion from	11.52g
Peptone	2.32g
NaCl	1.15g

Source: PPLO broth without Crystal Violet is available as a premixed powder from Difco Laboratories.

Preparation of PPLO Broth without Crystal Violet: Add components to distilled/deionized water and bring volume to 500.0mL. Mix thoroughly.

Fresh Yeast Extract Solution:

Composition per 100mL:

Baker's yeast, live, pressed, starch-free	25.0g

Preparation of Fresh Yeast Extract Solution: Add the live Baker's yeast to 100.0mL of distilled/deionized water. Autoclave for 90 min at 15 psi pressure–121°C. Allow to stand. Remove supernatant solution. Adjust pH to 6.6–6.8. Filter sterilize.

CMRL-1066 Medium:

Composition per liter:

NaCl	6.8g
$NaHCO_3$	2.2g
D-Glucose	1.0g
KCl	0.4g
L-Cysteine·HCl·H_2O	0.26g
$CaCl_2$, anhydrous	0.2g
$MgSO_4 \cdot 7H_2O$	0.2g
$NaH_2PO_4 \cdot H_2O$	0.14g
L-Glutamine	0.1g
Sodium acetate·$3H_2O$	0.083g
L-Glutamic acid	0.075g
L-Arginine·HCl	0.070g
L-Lysine·HCl	0.070g
L-Leucine	0.060g
Glycine	0.050g
Ascorbic acid	0.050g
L-Proline	0.040g
L-Tyrosine	0.040g
L-Aspartic acid	0.030g
L-Threonine	0.030g
L-Alanine	0.025g
L-Phenylalanine	0.025g
L-Serine	0.025g
L-Valine	0.025g
L-Cystine	0.020g
L-Histidine·HCl·H_2O	0.020g
L-Isoleucine	0.020g
Phenol Red	0.020g
L-Methionine	0.015g
Deoxyadenosine	0.010g
Deoxycytidine	0.010g
Deoxyguanosine	0.010g
Glutathione, reduced	0.010g
Thymidine	0.010g
Hydroxy-L-proline	0.010g
L-Tryptophan	0.010g
Nicotinamide adenine dinucleotide	7.0mg
Tween™ 80	5.0mg
Sodium glucoronate·H_2O	4.2mg
Coenzyme A	2.5mg
Cocarboxylase	1.0mg
Flavin adenine dinucleotide	1.0mg
Nicotinamide adenine dinucleotide phosphate	1.0mg
Uridine triphosphate	1.0mg
Choline chloride	0.50mg
Cholesterol	0.20mg
5-Methyldeoxycytidine	0.10mg
Inositol	0.05mg
p-Aminobenzoic acid	0.05mg
Niacin	0.025mg
Niacinamide	0.025mg
Pyridoxine	0.025mg
Pyridoxal·HCl	0.025mg
Biotin	0.01mg
D-Calcium pantothenate	0.01mg
Folic acid	0.01mg
Riboflavin	0.01mg
Thiamine·HCl	0.01mg

Source: CMRL-1066 medium is available as a premixed powder from GIBCO BRL.

Preparation of CMRL-1066 Medium: Add components to distilled/deionized water and bring volume to 1.0L. Mix thoroughly. Adjust pH to 7.2. Filter sterilize.

Preparation of Medium: Add components—except horse serum, fresh yeast extract, and CMRL medium—to distilled/deionized water and bring volume to 75.0mL. Adjust pH to 7.5. Autoclave for 15 min at 15 psi pressure–121°C. Aseptically add 20.0mL of sterile horse serum, 5.0mL of sterile yeast extract and 0.5mL of sterile CMRL medium. Distribute into sterile tubes or flasks.

Use: For cultivation and maintenance of *Spiroplasma kunkelii.*

C/10 Medium Reichenbach

Composition per liter:

Agar	15.0g
Pancreatic digest of casein	3.0g
$CaCl_2$	1.0g

pH 7.2 ± 0.2 at 25°C

Preparation of Medium: Add components, except agar, to distilled/deionized water and bring volume to 1.0L. Adjust pH to 7.2. Add agar. Mix thoroughly. Gently heat to boiling. Distribute into tubes or flasks. Autoclave for 15 min at 15 psi pressure–121°C. Pour into sterile Petri dishes or leave in tubes.

Use: For cultivation and maintenance of *Flexibacter filiformis*.

CA YE Broth
See: **Casamino Acids Yeast Extract Salts Broth, Gorbach**

Cadmium Fluoride Acriflavin Tellurite Medium
See: **CFAT Medium**

Caffeic Acid Ferric Citrate Test Medium (CAFC Test Medium) (Caffeic Acid Agar)

Composition per liter:

Agar	20.0g
$(NH_4)_2SO_4$	5.0g
Glucose	5.0g
Yeast extract	2.0g
K_2HPO_4	0.8g
$MgSO_4·3H_2O$	0.7g
Caffeic acid·$1/2H_2O$	0.18g
Chloramphenicol	0.05g
Ferric citrate solution	4.0mL

pH 6.5 ± 0.2 at 25°C

Ferric Citrate Solution:
Composition per 20mL:

Agar	100.0mg

Preparation of Ferric Citrate Solution: Add ferric citrate to 20.0mL of distilled/deionized water. Mix thoroughly.

Preparation of Medium: Add components, except chloramphenicol, to distilled/deionized water and bring volume to 1.0L. Mix thoroughly. Heat to boiling. Autoclave for 15 min at 15 psi pressure–121°C. Cool to 45°–50°C. Aseptically add 0.05g of chloramphenicol. Mix thoroughly. Pour into sterile Petri dishes.

Use: For the isolation and presumptive identification of *Cryptococcus neoformans*. *C. neoformans* appears as dark brown colonies. All other *Cryptococcus* species appear as light brown or nonpigmented colonies.

CAGV Medium
See: **Casamino Acid Glucose Medium**

CAL Agar (Cellobiose Arginine Lysine Agar) (Yersinia Isolation Agar)

Composition per liter:

Agar	20.0g
L-Arginine·HCl	6.5g
L-Lysine·HCl	6.5g
NaCl	5.0g
Cellobiose	3.5g
Yeast extract	3.0g
Sodium deoxycholate	1.5g
Neutral Red	0.03g

Preparation of Medium: Add components to distilled/deionized water and bring volume to 1.0L. Mix thoroughly. Heat to boiling. Do not autoclave. Pour into sterile Petri dishes.

Use: For isolation and characterization of *Yersinia enterocolitica* from fecal specimens and enumeration of *Y. enterocolitica* from water and other liquid specimens.

CAL Broth (Cellobiose Arginine Lysine Broth)

Composition per liter:

L-Arginine·HCl	6.5g
L-Lysine·HCl	6.5g
NaCl	5.0g
Cellobiose	3.5g
Yeast extract	3.0g
Sodium deoxycholate	1.5g
Neutral Red	0.03g

Preparation of Medium: Add components to distilled/deionized water and bring volume to 1.0L. Mix

thoroughly. Heat to boiling. Do not autoclave. Distribute into sterile tubes in 6.0–8.0mL volumes.

Use: For isolation and characterization of *Yersinia enterocolitica* from fecal specimens and enumeration of *Y. enterocolitica* from water and other liquid specimens.

Calymmatobacterium granulomatis Semidefined Medium

Composition per liter:

Papaic digest of soybean meal	20.0g
NaCl	2.5g
K_2HPO_4	1.5g
Sodium thioglycollate	0.6g
L-Cystine	0.4g

pH 7.2 ± 0.2 at 25°C

Preparation of Medium: Add components to distilled/deionized water and bring volume to 1.0L. Mix thoroughly. Adjust pH to 7.2. Distribute into screw-capped tubes in 20–22mL volumes. Autoclave for 15 min at 15 psi pressure–121°C. Tighten screw caps.

Use: For the cultivation of *Calymmatobacterium granulomatis*.

Camphor Minimal Medium

Composition per liter:

Agar	20.0g
K_2HPO_4	4.4g
NH_4Cl	2.1g
KH_2PO_4	1.7g
100× salt solution	10.0mL

100X Salt Solution:

Composition per liter:

$MgSO_4$	19.5g
$FeSO_4 \cdot 7H_2O$	5.0g
$MnSO_4 \cdot H_2O$	5.0g
Ascorbic acid	1.0g
$CaCl_2 \cdot 2H_2O$	0.3g

Preparation of 100X Salt Solution: Add components to distilled/deionized water and bring volume to 1.0L. Mix thoroughly.

Preparation of Medium: Add components to distilled/deionized water and bring volume to 1.0L. Gently heat and bring to boiling. Autoclave for 15 min at 15 psi pressure–121°C. Pour into sterile Petri dishes. Allow to cool to room temperature. Invert Petri dishes. Spread 0.2mL of 2M D-(+) camphor solution in methylene chloride (CH_2Cl_2) on the inside cover of each plate.

Use: For cultivation and maintenance of *Pseudomonas putida*.

Campy BAP Medium

See: Campylobacter **Selective Medium, Blaser-Wang**

Campy Cefex Agar

See: Campylobacter **Isolation Agar B**

Campy THIO Medium

Composition per liter:

Pancreatic digest of casein	20.0g
Agar	15.0g
NaCl	2.5g
K_2HPO_4	1.5g
Sodium thioglycollate	0.6g
L-Cystine	0.4g
Na_2SO_3	0.2g
Antibiotic supplement	10.0mL

Antibiotic Supplement:

Composition per 10mL:

Cephalothin	15.0mg
Vancomycin	10.0mg
Trimethoprim	5.0mg
Amphotericin B	2.0mg
Polymyxin B	2,500U

Preparation of Antibiotic Supplement: Add components to 10.0mL of distilled/deionized water. Filter sterilize.

Preparation of Medium: Add components, except antibiotic solution, to distilled/deionized water and bring volume to 990.0mL. Mix thoroughly. Gently heat and bring to boiling. Autoclave for 15 min at 15 psi pressure–121°C. Cool to 45°–50°C. Aseptically add 10.0mL of sterile antibiotic solution. Mix thoroughly. Aseptcially distribute into sterile screw-capped tubes in 3.0mL volumes for 1.5-cm swabs or 5.0mL volumes for 3.0-cm swabs.

Use: For the maintenance—as a holding or transport medium—of *Campylobacter* species isolated from clinical specimens on swabs.

Campylobacter Agar with 5 Antimicrobics and 10% Sheep Blood

Composition per liter:

Agar	15.0g
Pancreatic digest of casein	10.0g
Peptic digest of animal tissue	10.0g
NaCl	5.0g

Yeast extract 2.0g
Glucose 1.0g
$NaHSO_3$ 0.1g
Sheep blood, defibrinated 100.0mL
Antibiotic supplement 10.0mL
pH 7.2 ± 0.2 at 25°C

Antibiotic Supplement:
Composition per 10mL:
Cephalothin 0.015g
Vancomycin 0.01g
Trimethoprim 5.0mg
Amphotericin B 2.0mg
Polymyxin B 2,500U

Preparation of Antibiotic Supplement: Add components to 10.0mL of distilled/deionized water. Filter sterilize.

Source: This medium is available as a prepared medium from BBL Microbiology Systems.

Preparation of Medium: Add components, except sheep blood and antibiotic solution, to distilled/deionized water and bring volume to 890.0mL. Mix thoroughly. Gently heat and bring to boiling. Autoclave for 15 min at 15 psi pressure–121°C. Cool to 45°–50°C. Aseptically add 100.0mL of sterile sheep blood and 10.0mL of sterile antibiotic solution. Mix thoroughly. Pour into sterile Petri dishes or distribute into sterile tubes.

Use: For the primary selective isolation and cultivation of *Campylobacter jejuni* from human fecal specimens.

Campylobacter Agar, Blaser's (Blaser's *Campylobacter* Agar)

Composition per liter:
Campylobacter agar base 990.0mL
Supplement B 10.0mL
pH 7.4 ± 0.2 at 25°C

Campylobacter Agar Base:
Composition per liter:
Proteose peptone 15.0g
Agar 12.0g
NaCl 5.0g
Yeast extract 5.0g
Liver digest 2.5g

Source: *Campylobacter* agar base and *Campylobacter* antimicrobic supplement B are available as a premixed powder from Difco Laboratories.

Preparation of *Campylobacter* Agar Base: Add components to distilled/deionized water and bring volume to 990.0mL. Mix thoroughly. Gently heat and bring to boiling. Autoclave for 15 min at 15 psi pressure–121°C. Cool to 45°–50°C.

Supplement B:
Composition per 10mL:
Cephalothin 15mg
Vancomycin 10mg
Trimethoprim 5mg
Amphotericin B 2mg
Polymyxin B 2,500units

Preparation of Supplement B: Add components to 10.0mL of distilled/deionized water. Filter sterilize.

Preparation of Medium: Prepare 990.0mL of *Campylobacter* Agar Base. Autoclave and cool to 45°–50°C. Aseptically add 10.0mL of sterile Supplement B. Mix thoroughly. Pour into sterile Petri dishes.

Use: For the selective isolation of *Campylobacter jejuni* from fecal specimens, food and environmental specimens.

Campylobacter Agar, Skirrow's (Skirrow's *Campylobacter* Agar)

Composition per liter:
Campylobacter agar base 990.0mL
Supplement S 10.0mL
pH 7.4 ± 0.2 at 25°C

***Campylobacter* Agar Base:**
Composition per liter:
Proteose peptone 15.0g
Agar 12.0g
NaCl 5.0g
Yeast extract 5.0g
Liver digest 2.5g

Source: *Campylobacter* agar base and *Campylobacter* antimicrobic supplement S are available as a premixed powder from Difco Laboratories.

Preparation of *Campylobacter* Agar Base: Add components to distilled/deionized water and bring volume to 990.0mL. Mix thoroughly. Gently heat and bring to boiling. Autoclave for 15 min at 15 psi pressure–121°C. Cool to 45°–50°C.

Supplement S:
Composition per 10mL:
Vancomycin 10mg
Trimethoprim 5mg
Polymyxin B 2,500units

Preparation of Supplement S: Add components to 10.0mL of distilled/deionized water. Filter sterilize.

Preparation of Medium: Prepare 990.0mL of *Campylobacter* Agar Base. Autoclave and cool to 45°–50°C. Aseptically add 10.0mL of sterile Supplement S. Mix thoroughly. Pour into sterile Petri dishes.

Use: For the selective isolation of *Campylobacter jejuni* from fecal specimens, food and environmental specimens.

Campylobacter Blood-Free Selective Agar

Composition per liter:

Agar	12.0g
Beef extract	10.0g
Peptone	10.0g
Charcoal	4.0g
Casein hydrolysate	3.0g
Sodium deoxycholate	1.0g
$Fe_2SO_4 \cdot H_2O$	0.25g
Sodium pyruvate	0.25g
Cefoperazone solution	10.0mL

pH 7.4 ± 0.2 at 25°C

Cefoperazone Solution:
Composition per 10mL:

Sodium cefoperazone	0.032g

Preparation of Cefoperazone Solution: Add sodium cefoperazone to distilled/deionized water and bring volume to 10.0mL. Mix thoroughly. Filter sterilize.

Preparation of Medium: Add components, except cefoperazone solution, to distilled/deionized water and bring volume to 990.0mL. Mix throughly. Heat with frequent agitation and boil for 1 min to completely dissolve. Autoclave for 15 min at 15 psi–121°C. Cool to 50–55°C. Add 10.0mL of sterile cefoperazone solution. Addition of 10mg/L amphotericin B improves the selectivity of the medium. Mix thoroughly. Pour into sterile Petri dishes.

Use: For the selective isolation of *Campylobacter* species, especially *C. jejuni, C. coli* and *C. laridis.*

Campylobacter Charcoal Differential Agar (CCDA) (Preston Blood–Free Medium)

Composition per liter:

Agar	12.0g
Beef extract	10.0g
Peptone	10.0g
NaCl	5.0g
Charcoal	4.0g
Casein hydrolysate	3.0g
Sodium deoxycholate	1.0g
$FeSO_4$	0.25g
Sodium pyruvate	0.25g
Cefoperazone solution	10.0mL

pH 7.5 ± 0.2 at 25°C

Cefoperazone Solution:
Composition per 10mL:

Sodium cefoperazone	0.032g

Preparation of Cefoperazone Solution: Add sodium cefoperazone to distilled/deionized water and bring volume to 10.0mL. Mix thoroughly. Filter sterilize.

Preparation of Medium: Add components, except cefoperazone solution, to distilled/deionized water and bring volume to 990.0mL. Mix thoroughly. Gently heat and bring to boiling. Autoclave for 15 min at 15 psi pressure–121°C. Cool to 45°–50°C. Aseptically add 10.0mL of sterile cefoperazone solution. Mix thoroughly. Pour into sterile Petri dishes or distribute into sterile tubes.

Use: For the cultivation of *Campylobacter* species.

Campylobacter Enrichment Broth

Composition per liter:

Beef extract	10.0g
Peptone	10.0g
Yeast extract	6.0g
NaCl	5.0g
Horse blood, laked	50.0mL
FBP solution	4.0mL
Antibiotic solution	4.0mL

pH 7.5 ± 0.2 at 25·C

Horse Blood, Laked:
Composition per 50mL:

Horse blood, fresh	50.0mL

Preparation of Horse Blood, Laked: Add blood to a sterile polypropylene bottle. Freeze overnight at –20°C. Thaw at 8°C. Refreeze at –20°C. Thaw again at 8°C.

FBP Solution:
Composition per 100mL:

$FeSO_4$	6.25g
$NaHSO_3$	6.25g
Sodium pyruvate	6.25g

Preparation of FBP Solution: Add components to distilled/deionized water and bring volume to 100.0mL. Mix thoroughly. Filter sterilize.

Antibiotic Solution:
Composition per 10mL:

Cycloheximide	0.1g
Sodium cefoperazone	0.030g
Trimethoprim lactate	0.0125g
Rifampicin	0.010g

Preparation of Antibiotic Solution: Add components to distilled/deionized water and bring volume to 10.0mL. Mix thoroughly. Filter sterilize.

Preparation of Medium: Add components—except laked horse blood, FBP solution, and antibiotic solution—to distilled/deionized water and bring volume to 942.0mL. Mix thoroughly. Gently heat and bring to boiling. Autoclave for 15 min at 15 psi pressure–121°C. Cool to 45°–50°C. Aseptically add 50.0mL of sterile laked horse blood, 4.0mL of FBP solution, and 4.0mL antibiotic solution. Mix thoroughly. Pour into sterile Petri dishes or distribute into sterile tubes.

Use: For the isolation and cultivation of *Campylobacter* species from dairy products.

Campylobacter Enrichment Broth (FDA Medium M29)

Composition per 1024mL:

Basal medium	950.0mL
Horse blood, lysed	50.0mL
Cefoperazone solution	8.0mL
FBP solution	4.0mL
Trimethoprim lactate solution	4.0mL
Vancomycin solution	4.0mL
Cycloheximide solution	4.0mL

pH 7.5 ± 0.2 at 25°C

Basal Medium:
Composition per 950mL:

Beef extract	10.0g
Peptone	10.0g
Yeast extract	6.0g
NaCl	5.0g

Preparation of Basal Medium: Add components to distilled/deionized water and bring volume to 950.0mL. Mix thoroughly. Autoclave for 15 min at 15 psi pressure–121°C. Cool to 45°–50°C.

FBP Solution:
Composition per 100mL:

$FeSO_4 \cdot 7H_2O$	6.25g
$Na_2S_2O_5$	6.25g
Sodium pyruvate	6.25g

Preparation of FBP: Add components to distilled/deionized water and bring volume to 100.0mL. Mix thoroughly. Filter sterilize.

Cefoperazone Solution:
Composition per 10mL:

Cefoperazone	0.037g

Preparation of Cefoperazone Solution: Add cefoperazone to distilled/deionized water and bring volume to 10.0mL. Mix thoroughly. Filter sterilize.

Trimethoprim Lactate Solution:
Composition per 10mL:

Trimethoprim lactate	0.031g

Preparation of Trimethoprim Lactate Solution: Add trimethoprim lactate to distilled/deionized water and bring volume to 10.0mL. Mix thoroughly. Filter sterilize.

Vancomycin Solution:
Composition per 10mL:

Vancomycin	0.025g

Preparation of Vancomycin Solution: Add vancomycin to distilled/deionized water and bring volume to 10.0mL. Mix thoroughly. Filter sterilize.

Cycloheximide Solution:
Composition per 10mL:

Cycloheximide	0.025g

Preparation of Cycloheximide Solution: Add cycloheximide to distilled/deionized water and bring volume to 10.0mL. Mix thoroughly. Filter sterilize.

Preparation of Medium: To 950.0mL of cooled sterile basal medium, aseptically add 50.0mL of lysed (fresh, frozen and thawed) horse blood, 4.0mL of sterile FBP solution, 8.0mL of sterile cefoperazone solution, 4.0mL of sterile trimethoprim lactate solution, 4.0mL of sterile vancomycin solution, and 4.0mL of sterile cycloheximide solution. Mix thoroughly. Aseptically distribute into sterile screw-capped tubes or bottles. Close caps tightly to reduce O_2 absorbtion. Use within two weeks.

Use: For the selective isolation and cultivation of *Campylobacter* species.

Campylobacter Enrichment Broth (FDA Medium M29)

Composition per 1020mL:

Basal medium	950.0mL
Horse blood, lysed	50.0mL
FBP solution	4.0mL
Cefoperazone solution	4.0mL
Trimethoprim lactate solution	4.0mL
Vancomycin solution	4.0mL
Cycloheximide solution	4.0mL

pH 7.5 ± 0.2 at 25°C

Basal Medium:
Composition per 950mL:

Beef extract	10.0g
Peptone	10.0g
Yeast extract	6.0g
NaCl	5.0g

Preparation of Basal Medium: Add components to distilled/deionized water and bring volume to 950.0mL. Mix thoroughly. Autoclave for 15 min at 15 psi pressure–121°C. Cool to 45°–50°C.

FBP Solution:

Composition per 100mL:

$FeSO_4 \cdot 7H_2O$	6.25g
$Na_2S_2O_5$	6.25g
Sodium pyruvate	6.25g

Preparation of FBP: Add components to distilled/deionized water and bring volume to 100.0mL. Mix thoroughly. Filter sterilize.

Cefoperazone Solution:

Composition per 10mL:

Cefoperazone	0.037g

Preparation of Cefoperazone Solution: Add cefoperazone to distilled/deionized water and bring volume to 10.0mL. Mix thoroughly. Filter sterilize.

Trimethoprim Lactate Solution:

Composition per 10mL:

Trimethoprim lactate	0.031g

Preparation of Trimethoprim Lactate Solution: Add trimethoprim lactate to distilled/deionized water and bring volume to 10.0mL. Mix thoroughly. Filter sterilize.

Vancomycin Solution:

Composition per 10mL:

Vancomycin	0.025g

Preparation of Vancomycin Solution: Add vancomycin to distilled/deionized water and bring volume to 10.0mL. Mix thoroughly. Filter sterilize.

Cycloheximide Solution:

Composition per 10mL:

Cycloheximide	0.025g

Preparation of Cycloheximide Solution: Add cycloheximide to distilled/deionized water and bring volume to 10.0mL. Mix thoroughly. Filter sterilize.

Preparation of Medium: To 950.0mL of cooled sterile basal medium, aseptically add 50.0mL of lysed (fresh, frozen and thawed) horse blood, 4.0mL of sterile FBP solution, 4.0mL of sterile cefoperazone solution, 4.0mL of sterile trimethoprim lactate solution, 4.0mL of sterile vancomycin solution, and 4.0mL of sterile cycloheximide solution. Mix thoroughly. Aseptically distribute into sterile screw-capped tubes or bottles. Close caps tightly to reduce O_2 absorbtion. Use within two weeks.

Use: For the selective isolation and cultivation of *Campylobacter* species.

Campylobacter Enrichment Broth (FDA Medium M29)

Composition per 1020mL:

Basal medium	950.0mL
Horse blood, lysed	50.0mL
FBP solution	4.0mL
Cefoperazone solution	4.0mL
Trimethoprim lactate solution	4.0mL
Rifampicin solution	4.0mL
Cycloheximide solution	4.0mL

pH 7.5 ± 0.2 at 25°C

Basal Medium:

Composition per 950mL:

Beef extract	10.0g
Peptone	10.0g
Yeast extract	6.0g
NaCl	5.0g

Preparation of Basal Medium: Add components to distilled/deionized water and bring volume to 950.0mL. Mix thoroughly. Autoclave for 15 min at 15 psi pressure–121°C. Cool to 45°–50°C.

FBP Solution:

Composition per 100mL:

$FeSO_4 \cdot 7H_2O$	6.25g
$Na_2S_2O_5$	6.25g
Sodium pyruvate	6.25g

Preparation of FBP: Add components to distilled/deionized water and bring volume to 100.0mL. Mix thoroughly. Filter sterilize.

Cefoperazone Solution:

Composition per 10mL:

Cefoperazone	0.037g

Preparation of Cefoperazone Solution: Add cefoperazone to distilled/deionized water and bring volume to 10.0mL. Mix thoroughly. Filter sterilize.

Trimethoprim Lactate Solution:

Composition per 10mL:

Trimethoprim lactate	0.031g

Preparation of Trimethoprim Lactate Solution: Add trimethoprim lactate to distilled/deionized water and bring volume to 10.0mL. Mix thoroughly. Filter sterilize.

Rifampicin Solution:

Composition per 100mL:

Rifampicin	0.25g
Ethanol, absolute	50.0mL

Preparation of Rifampicin Solution: Add rifampicin to 50.0mL of ethanol. Mix thoroughly. Bring volume to 100.0mL with distilled/deionized water. Filter sterilize.

Cycloheximide Solution:
Composition per 10mL:
Cycloheximide 0.025g

Preparation of Cycloheximide Solution: Add cycloheximide to distilled/deionized water and bring volume to 10.0mL. Mix thoroughly. Filter sterilize.

Preparation of Medium: To 950.0mL of cooled sterile basal medium, aseptically add 50.0mL of lysed (fresh, frozen and thawed) horse blood, 4.0mL of sterile FBP solution, 4.0mL of sterile cefoperazone solution, 4.0mL of sterile trimethoprim lactate solution, 4.0mL of sterile rifampicin solution, and 4.0mL of sterile cycloheximide solution. Mix thoroughly. Aseptically distribute into sterile screw-capped tubes or bottles. Close caps tightly to reduce O_2 absorbtion. Use within two weeks.

Use: For the selective isolation and cultivation of *Campylobacter* species from dairy products.

Campylobacter Enrichment Broth

Composition per liter:
Beef extract 10.0g
Peptone 10.0g
Yeast extract 6.0g
NaCl 5.0g
Horse blood, laked 50.0mL
FBP solution 4.0mL
Antibiotic solution 4.0mL
pH 7.5 ± 0.2 at 25·C

Horse Blood, Laked:
Composition per 50mL:
Horse blood, fresh 50.0mL

Preparation of Horse Blood, Laked: Add blood to a sterile polypropylene bottle. Freeze overnight at –20°C. Thaw at 8°C. Refreeze at –20°C. Thaw again at 8°C.

FBP Solution:
Composition per 100mL:
$FeSO_4$ 6.25g
$NaHSO_3$ 6.25g
Sodium pyruvate 6.25g

Preparation of FBP Solution: Add components to distilled/deionized water and bring volume to 100.0mL. Mix thoroughly. Filter sterilize.

Antibiotic Solution:
Composition per 10mL:
Cycloheximide 0.1g
Sodium cefoperazone 0.030g
Trimethoprim lactate 0.0125g
Vancomycin 0.010g

Preparation of Antibiotic Solution: Add components to distilled/deionized water and bring volume to 10.0mL. Mix thoroughly. Filter sterilize.

Preparation of Medium: Add components—except horse blood, FBP solution, and antibiotic solution—to distilled/deionized water and bring volume to 942.0mL. Mix thoroughly. Gently heat and bring to boiling. Autoclave for 15 min at 15 psi pressure–121°C. Cool to 45°–50°C. Aseptically add 50.0mL of sterile horse blood, 4.0mL of FBP solution, and 4.0mL antibiotic solution. Mix thoroughly. Pour into sterile Petri dishes or distribute into sterile tubes.

Use: For the isolation and cultivation of *Campylobacter* species from foods.

Campylobacter fecalis Medium

Composition per 1133.6mL:
Starch, soluble 2.4g
Yeast extract 2.4g
Blood agar base 1.0L
Bovine blood 120.0mL
Antibiotic solution 10.0mL
Sodium lactate (60% syrup) 3.6mL
pH 6.8 ± 0.2 at 25°C

Blood Agar Base:
Composition per liter:
Beef heart, solids from infusion 500.0g
Agar 15.0g
Tryptose 10.0g
NaCl 5.0g

Preparation of Blood Agar Base: Add components to distilled/deionized water and bring volume to 1.0L. Mix thoroughly. Gently heat while stirring and bring to boiling.

Antibiotic Solution:
Composition per 10mL:
Cycloheximide 0.12g
Albamycin 6.0mg
Bacitracin 6,000U

Preparation of Antibiotic Solution: Add components to distilled/deionized water and bring volume to 10.0mL. Mix thoroughly. Filter sterilize.

Preparation of Medium: To 1.0L of blood agar base, add soluble starch, yeast extract, and sodium lactate. Mix thoroughly. Autoclave for 15 min at 15 psi pressure–121°C. Cool to 45°–50°C. Aseptically add sterile bovine blood and antibiotic solution. Mix thoroughly. Pour into sterile Petri dishes or distribute into sterile tubes.

Use: For the cultivation and isolation of *Campylobacter fecalis*.

Campylobacter fetus Medium

Composition per liter:

Proteose peptone	10.0g
NaCl	5.0g
Beef extract	3.0g
Bovine blood	50.0mL
Antibiotic solution	10.0mL

pH 7.2–7.4 at 25°C

Antibiotic Solution:
Composition per 10mL:

Novobiocin	2.0mg
Bacitracin	2,000U

Preparation of Antibiotic Solution: Add components to distilled/deionized water and bring volume to 10.0mL. Mix thoroughly. Filter sterilize.

Preparation of Medium: Add components, except bovine blood and antibiotic solution, to distilled/deionized water and bring volume to 940.0mL. Mix thoroughly. Gently heat and bring to boiling. Autoclave for 15 min at 15 psi pressure–121°C. Cool to 45°–50°C. Aseptically add sterile bovine blood and antibiotic solution. Mix thoroughly. Aseptically distribute into sterile tubes or flasks.

Use: For the isolation and cultivation of *Campylobacter fetus*.

Campylobacter fetus Medium

Composition per 1160mL:

Fluid thioglycollate agar	1.0L
Sheep blood, defibrinated	150.0mL
Antibiotic solution	10.0mL

pH 7.1 ± 0.2 at 25°C

Fluid Thioglycollate Agar:
Composition per liter:

Agar	15.0g
Pancreatic digest of casein	15.0g
Glucose	5.5g
Yeast extract	5.0g
NaCl	2.5g
Agar	0.75g
L-Cystine	0.5g
Sodium thioglycollate	0.5g
Resazurin	1.0mg

Preparation of Fluid Thioglycollate Agar: Add components to distilled/deionized water and bring volume to 1.0L. Mix thoroughly. Gently heat and bring to boiling. Autoclave for 15 min at 15 psi pressure–121°C. Cool to 25°C.

Antibiotic Solution:
Composition per 10mL:

Cycloheximide	0.05g
Novobiocin	5.0mg
Bacitracin	25,000U
Polymyxin B sulfate	10,000U

Preparation of Antibiotic Solution: Add components to distilled/deionized water and bring volume to 10.0mL. Mix thoroughly. Filter sterilize.

Preparation of Medium: To 1.0L of cooled sterile fluid thioglycollate agar, aseptically add 150.0mL of sterile sheep blood and 10.0mL of sterile antibiotic solution.

Use: For the isolation and cultivation of *Campylobacter fetus* from human specimens.

Campylobacter fetus Selective Medium

Composition per liter:

Fluid thioglycollate agar	1.0L
Sheep blood, defibrinated	150.0mL
Antibiotic solution	10.0mL

Fluid Thioglycollate Agar:
Composition per liter:

Agar	15.0g
Pancreatic digest of casein	15.0g
Glucose	5.5g
Yeast extract	5.0g
NaCl	2.5g
Agar	0.75g
L-Cystine	0.5g
Sodium thioglycollate	0.5g
Resazurin	1.0mg

Preparation of Fluid Thioglycollate Agar: Add components to distilled/deionized water and bring volume to 1.0L. Mix thoroughly. Gently heat and bring to boiling. Autoclave for 15 min at 15 psi pressure–121°C. Cool to 25°C.

Antibiotic Solution:
Composition per 10mL:

Cycloheximide	0.05g
Cephalothin	0.02g
Novobiocin	5.0mg
Bacitracin	25,000U
Colistin	10,000U

Preparation of Antibiotic Solution: Add components to distilled/deionized water and bring volume to 10.0mL. Mix thoroughly. Filter sterilize.

Preparation of Medium: To 1.0L of cooled sterile fluid thioglycollate agar, aseptically add 150.0mL of sterile sheep blood and 10.0mL of sterile antibiotic solution.

Use: For the isolation and cultivation of *Campylobacter fetus*.

Campylobacter Isolation Agar A

Composition per liter:

Agar	12.0g
Beef extract	10.0g
Peptone	10.0g
NaCl	5.0g
Charcoal	4.0g
Casein hydrolysate	3.0g
Yeast extract	2.0g
Sodium deoxycholate	1.0g
$FeSO_4$	0.25g
Sodium pyruvate	0.25g
Antibiotic solution	10.0mL

pH 7.4 ± 0.2 at 25°C

Antibiotic Solution:
Composition per 10mL:

Cycloheximide	0.1g
Sodium cefoperazone	0.030g

Preparation of Antibiotic Solution: Add components to distilled/deionized water and bring volume to 10.0mL. Mix thoroughly. Filter sterilize.

Preparation of Medium: Add components, except antibiotic solution, to distilled/deionized water and bring volume to 990.0mL. Mix thoroughly. Gently heat and bring to boiling. Autoclave for 15 min at 15 psi pressure–121°C. Cool to 45°–50°C. Aseptically add sterile antibiotic solution. Mix thoroughly. Pour into sterile Petri dishes. Swirl flask while pouring to distribute charcoal.

Use: For the isolation and cultivation of *Campylobacter* species.

Campylobacter Isolation Agar B (*Campy* Cefex Agar)

Composition per liter:

Agar	15.0g
Pancreatic digest of casein	10.0g
Peptic digest of animal tissue	10.0g
NaCl	5.0g
Yeast extract	2.0g
Glucose	1.0g
$FeSO_4$	0.5g
Sodium pyruvate	0.5g
$NaHSO_3$	0.35g
Horse blood, laked	50.0mL
Antibiotic solution	10.0mL

pH 7.0 ± 0.2 at 25°C

Horse Blood, Laked:
Composition per 50mL:

Horse blood, fresh	50.0mL

Preparation of Horse Blood, Laked: Add blood to a sterile polypropylene bottle. Freeze overnight at –20°C. Thaw at 8°C. Refreeze at –20°C. Thaw again at 8°C.

Antibiotic Solution:
Composition per 10mL:

Cycloheximide	0.1g
Sodium cefoperazone	0.033g

Preparation of Antibiotic Solution: Add components to distilled/deionized water and bring volume to 10.0mL. Mix thoroughly. Filter sterilize.

Preparation of Medium: Add components, except horse blood and antibiotic solution, to distilled/deionized water and bring volume to 940.0mL. Mix thoroughly. Gently heat and bring to boiling. Autoclave for 15 min at 15 psi pressure–121°C. Cool to 45°–50°C. Aseptically add sterile horse blood and antibiotic solution. Mix thoroughly. Pour into sterile Petri dishes or distribute into sterile tubes.

Use: For the isolation and cultivation of *Campylobacter* species.

Campylobacter Selective Medium, Blaser-Wang (Blaser-Wang *Campylobacter* Medium) (Blaser's Agar) (Campy BAP Medium)

Composition per liter:

Brucella Agar Base	890.0mL
Sheep blood	100.0mL
Antibiotic supplement	10.0mL

***Brucella* Agar Base:**
Composition per 890mL:

Agar	15.0g
Glucose	10.0g
Pancreatic digest of casein	10.0g
NaCl	5.0g
Peptic digest of animal tissue	5.0g

pH 7.5 ± 0.2 at 25°C

Preparation of *Brucella* Agar Base: Add components to distilled/deionized water and bring volume to 890.0mL. Mix thoroughly. Gently heat and bring to boiling. Autoclave for 15 min at 15 psi pressure–121°C. Cool to 45°–50°C.

Antibiotic Supplement:
Composition per 10mL:

Cephalothin	15.0mg
Vancomycin	10.0mg
Trimethoprim	5.0mg
Amphotericin B	2.0mg
Polymyxin B	2,500U

Preparation of Antibiotic Supplement: Add components to 10.0mL of distilled/deionized water. Filter sterilize.

Preparation of Medium: Prepare 890.0mL of *Brucella* agar base. Sterilize as directed. Cool to 50–55°C and add 100.0mL sheep blood or 50.0–70.0mL laked horse blood. Laked blood is prepared by freezing whole blood overnight and thawing to room temperature. Aseptically add 10.0mL of sterile antibiotic supplement. Mix thoroughly. Pour into sterile Petri dishes.

Use: For the selective isolation of *Campylobacter* species.

Campylobacter Selective Medium, Blaser-Wang (Blaser-Wang *Campylobacter* Medium)

Composition per liter:

Columbia agar base	890.0mL
Sheep blood	100.0mL
Antibiotic supplement	10.0mL

pH 7.3 ± 0.2 at 25°C

Columbia Agar Base:
Composition per liter:

Special peptone	25.0g
Agar	10.0g
NaCl	5.0g
Starch	1.0g

Preparation of Columbia Agar Base: Add components to distilled/deionized water and bring volume to 890.0mL. Mix thoroughly. Gently heat and bring to boiling. Autoclave for 15 min at 15 psi pressure–121°C. Cool to 45°–50°C.

Antibiotic Supplement:
Composition per 10mL:

Cephalothin	15.0mg
Vancomycin	10.0mg
Trimethoprim	5.0mg
Amphotericin B	2.0mg
Polymyxin B	2,500U

Preparation of Antibiotic Supplement: Add components to 10.0mL of distilled/deionized water. Filter sterilize.

Preparation of Medium: To 890.0mL of cooled, sterile Columbia agar base, aseptically add 100.0mL sheep blood or 50.0–70.0mL laked horse blood. Laked blood is prepared by freezing whole blood overnight and thawing to room temperature. Aseptically add 10.0mL of sterile antibiotic supplement. Mix thoroughly. Pour into sterile Petri dishes.

Use: For the selective isolation of *Campylobacter* species.

Campylobacter Selective Medium, Butzler's (Butzler's *Campylobacter* Medium)

Composition per liter:

Brucella Agar Base	940.0mL
Sheep or horse blood, defibrinated	50.0mL
Antibiotic supplement	10.0mL

pH 7.5 ± 0.2 at 25°C

***Brucella* Agar Base:**
Composition per liter:

Agar	15.0g
Glucose	10.0g
Pancreatic digest of casein	10.0g
NaCl	5.0g
Peptic digest of animal tissue	5.0g

Preparation of *Brucella* Agar Base: Add components to distilled/deionized water and bring volume to 940.0mL. Mix thoroughly. Gently heat and bring to boiling. Autoclave for 15 min at 15 psi pressure–121°C. Cool to 45°–50°C.

Antibiotic Supplement:
Composition per 10mL:

Cycloheximide	50.0mg
Cephazolin	15.0mg
Novobiocin	5.0mg
Bacitracin	25,000U
Colistin sulfate	10,000U

Preparation of Antibiotic Supplement: Add components to 10.0mL of distilled/deionized water. Filter sterilize.

Preparation of Medium: To 940.0mL of cooled sterile *Brucella* agar base, aseptically add 50.0mL of defibrinated sheep or horse blood and 10.0mL of sterile antibiotic supplement. Mix thoroughly. For enhanced growth, medium may also be supplemented with 0.25g $Fe_2SO_4 \cdot H_2O$, 0.25g sodium metabisulfite and 0.25g sodium pyruvate. Pour into sterile Petri dishes.

Use: For the selective isolation of *Campylobacter* species.

Campylobacter Selective Medium, Butzler's (Butzler's *Campylobacter* Medium)

Composition per liter:

Columbia agar base	940.0mL
Blood, horse or sheep	50.0mL
Antibiotic supplement	10.0mL

pH 7.3 ± 0.2 at 25°C

Columbia Agar Base:
Composition per liter:

Peptone	25.0g
Agar	10.0g
NaCl	5.0g
Starch	1.0g

Preparation of Columbia Agar Base: Add components to distilled/deionized water and bring volume to 940.0mL. Mix thoroughly. Gently heat and bring to boiling. Autoclave for 15 min at 15 psi pressure–121°C. Cool to 45°–50°C.

Antibiotic Supplement:
Composition per 10mL:

Cycloheximide	50.0mg
Cephazolin	15.0mg
Novobiocin	5.0mg
Bacitracin	25,000U
Colistin sulfate	10,000U

Preparation of Antibiotic Supplement: Add components to 10.0mL of distilled/deionized water. Filter sterilize.

Preparation of Medium: To 940.0mL of cooled, sterile Columbia agar base, aseptically add 50.0mL of defibrinated sheep or horse blood and 10.0mL of sterile antibiotic supplement. Mix thoroughly. The medium may also be supplemented with 0.25g $Fe_2SO_4 \cdot H_2O$, 0.25g sodium metabisulfite and 0.25g sodium pyruvate. Pour into sterile Petri dishes.

Use: For the selective isolation of *Campylobacter* species.

Campylobacter Selective Medium, Karmali's (Karmali's *Campylobacter* Medium)

Composition per liter:

Activated charcoal	4.0g
Columbia agar base	990.0mL
Antibiotic supplement	10.0mL

pH 7.4 ± 0.2 at 25°C

Source: This medium is available as a premixed powder from Oxoid.

Columbia Agar Base:
Composition per 990mL:

Peptone	25.0g
Agar	10.0g
NaCl	5.0g
Starch	1.0g

Preparation of Columbia Agar Base: Add components to distilled/deionized water and bring volume to 990.0mL. Mix thoroughly. Gently heat and bring to boiling. Autoclave for 15 min at 15 psi pressure–121°C. Cool to 45°–50°C.

Antibiotic Supplement:
Composition per 10mL:

Sodium pyruvate	0.05g
Cycloheximide	0.05g
Cefoperazone	0.016g
Hemin	0.016g
Vancomycin	0.01g

Preparation of Antibiotic Supplement: Add components to 10.0mL of distilled/deionized water. Filter sterilize.

Preparation of Medium: Prepare 990.0mL of Columbia agar base. Sterilize as directed. Cool to 50–55°C. Add defibrinated sheep or horse blood to a final concentration of 5–7%. Add 10.0mL of sterile antibiotic supplement. Mix thoroughly. For enhanced growth, medium may also be supplemented with 0.25g $Fe_2SO_4 \cdot H_2O$, 0.25g sodium metabisulfite and 0.25g sodium pyruvate. Pour into sterile Petri dishes. Swirl while pouring to keep charcoal in suspension.

Use: For the selective isolation of *Campylobacter* species.

Campylobacter Selective Medium, Preston's (Preston's *Campylobacter* Medium)

Composition per liter:

Campylobacter agar base	940.0mL
Horse blood, lysed	50.0mL
Antibiotic supplement	10.0mL

pH 7.5 ± 0.2 at 25°C

***Campylobacter* Agar Base:**
Composition per liter:

Agar	12.0g
Beef extract	10.0g
Peptone	10.0g
NaCl	5.0g

Preparation of *Campylobacter* Agar Base: Add components to distilled/deionized water and bring volume to 940.0mL. Mix thoroughly. Gently heat and bring to boiling. Autoclave for 15 min at 15 psi pressure–121°C. Cool to 45°–50°C.

Antibiotic Supplement:
Composition per 10mL:

Cycloheximide	0.1g
Rifampicin	0.01g
Trimethoprim lactate	0.01g
Polmyxin B	5,000U

Preparation of Antibiotic Supplement: Add components to 10.0mL of 50:50 acetone:distilled/deionized water. Filter sterilize.

Preparation of Medium: To 940.0mL of cooled, sterile *Campylobacter* agar base, aseptically add 50.0mL of lysed horse blood and 10.0mL of sterile antibiotic supplement. Mix thoroughly. Pour into sterile Petri dishes.

Use: For the selective isolation of *Campylobacter* species.

Campylobacter sputorum subspecies *bubulus* Medium

Composition per liter:

Agar	1.5g
Brilliant Green	0.01g
Ethyl Violet	1.25mg
Brucella broth base	1.0L
Antibiotic solution	10.0mL

pH 7.0 ± 0.2 at 25°C

***Brucella* Broth Base:**
Composition per liter:

Pancreatic digest of casein	10.0g
Peptic digest of animal tissue	10.0g
NaCl	5.0g
Yeast extract	2.0g
Glucose	1.0g
$NaHSO_3$	0.1g

Preparation of *Brucella* Broth Base: Add components to distilled/deionized water and bring volume to 1.0L. Mix thoroughly.

Antibiotic Solution:
Composition per 10mL:

Cycloheximide	0.1g
Bacitracin	20,000U

Preparation of Antibiotic Solution: Add components to distilled/deionized water and bring volume to 10.0mL. Mix thoroughly. Filter sterilize.

Preparation of Medium: To 1.0L of *Brucella* broth base, add agar, Brilliant Green, and Ethyl Violet. Mix thoroughly. Autoclave for 15 min at 15 psi pressure–121°C. Cool to 45°–50°C. Aseptically add sterile antibiotic solution. Mix thoroughly. Aseptically distribute into sterile tubes or flasks.

Use: For the cultivation and isolation of *Campylobacter sputorum* subspecies *bubulus*.

Campylobacter sputorum subspecies *mucosalis* Medium

Composition per liter:

Yeast extract	2.8g
KNO_3	1.0g
Fluid thioglycollate broth without glucose	1.0L

Fluid Thioglycollate Broth without Glucose:
Composition per liter:

Pancreatic digest of casein	15.0g
Yeast extract	5.0g
NaCl	2.5g
Agar	0.75g
L-Cystine	0.5g
Sodium thioglycollate	0.5g
Resazurin	1.0mg

Preparation of Fluid Thioglycollate Broth without Glucose: Add components to distilled/deionized water and bring volume to 1.0L. Mix thoroughly.

Preparation of Medium: Combine components. Mix thoroughly. Gently heat and bring to boiling. Autoclave for 15 min at 15 psi pressure–121°C. Cool to 25°C. Aseptically distribute into sterile tubes or flasks.

Use: For the cultivation and isolation of *Campylobacter sputorum* subspecies *mucosalis*.

Campylobacter Thioglycollate Medium with 5 Antimicrobics

Composition per liter:

Pancreatic digest of casein	17.0g
Glucose	6.0g
Papaic digest of soybean meal	3.0g
NaCl	2.5g
Agar	1.6g
Sodium thioglycollate	0.5g
Na_2SO_3	0.1g
Antibiotic supplement solution	10.0mL

pH 7.0 ± 0.2 at 25°C

Source: This medium is available as a premixed powder from BBL Microbiology Systems.

Antibiotic Supplement Solution:
Composition per 10mL:

Cephalothin	0.015g
Vancomycin	0.01g
Trimethoprim	5.0mg
Amphotericin B	2.0mg
Polymyxin B	2,500U

Preparation of Antibiotic Supplement Solution: Add componenets to 10.0mL of distilled/deionized water. Mix thoroughly. Filter sterilize.

Preparation of Medium: Add components, except cephalothin, vancomycin, trimethoprim, amphotericin, and polymyxin B, to distilled deionized water and bring volume to 990.0mL. Mix thoroughly. Gently heat and bring to boiling. Autoclave for 15 min at 15 psi pressure–121°C. Cool to 45°–50°C. Add 10.0mL of sterile antibiotic supplement. Mix thoroughly. Pour into sterile Petri dishes.

Use: For maintenenace–as a holding medium or transport medium– of fecal specimens or swabs suspected of containing *Campylobacter jejuni* or other *Campylobacter* species when immediate inoculation of *Campylobacter* growth medium is unavailable.

Candida BCG Agar Base (*Candida* Bromcresol Green Agar Base)

Composition per liter:

Glucose	40.0g
Agar	15.0g
Peptone	10.0g
Yeast extract	1.0g
Bromcresol Green	0.02g
Neomycin solution	10.0mL

pH 6.1 ± 0.1 at 25°C

Source: This medium is available as a premixed powder from Difco Laboratories.

Neomycin Solution:
Composition per 10mL:

Neomycin	0.5g

Preparation of Neomycin Solution: Add neomycin to distilled/deionized water and bring volume to 10.0mL. Mix thoroughly. Filter sterilize.

Preparation of Medium: Add components, except neomycin solution, to distilled/deionized water and bring volume to 1.0L. Mix thoroughly and heat gently until boiling. Autoclave for 15 min at 15 psi pressure–121°C. Cool to 50–55°C. Aseptically add 10.0mL sterile neomycin solution. Mix thoroughly. Pour into sterile Petri dishes or leave in tubes.

Use: For the selective isolation and identification of *Candida* species. It is a highly differential medium that is used for demonstrating morphological and biochemical reactions characterizing different *Candida* species. *C. albicans* appears as blunt conical colonies with smooth edges and yellow to blue-green color. *C. stellatoidea* appears as convex colonies with smooth edges and yellow to green color. *C. tropicalis* appears as convex colonies with wavy edges and yellow-green to green color with a dark blue-green base. *C. pseudotropicalis* appears as convex, shiny colonies with smooth edges and green color with a light green edge. *C. krusei* appears as low conical colonies with spreading edges and blue-green color. *C. stellatoidea* appears as convex colonies with smooth edges and yellow to green color.

Candida Bromcresol Green Agar Base
See: Candida **BCG Agar Base**

Candida Isolation Agar

Composition per liter:

Agar	20.0g
Glucose	10.0g
Peptone	5.0g
Yeast extract	3.0g
Malt extract	3.0g
Aniline Blue	0.1g

pH 5.9 ± 0.5 at 25°C

Source: This medium is as available as a premixed powder from Difco Laboratories.

Preparation of Medium: Add components to distilled/deionized water and bring volume to 1.0L. Mix thoroughly. Gently heat and bring to boiling. Distribute into tubes or flasks. Autoclave for 15 min at 15 psi pressure–121°C. Pour into sterile Petri dishes or leave in tubes.

Use: For the isolation and differentiation of *Candida albicans*. *C. albicans* turns the medium blue.

Caprylate Thallous Agar
See: **CT Agar**

Carbohydrate Fermentation Broth

Composition per liter:

Peptone	10.0g
NaCl	5.0g

Meat extract	3.0g
Carbohydrate solution	50.0mL
Andrade's indicator	10.0mL

pH 7.1 ± 0.2 at 25°C

Andrade's Indicator:
Composition per 100mL:

Acid Fuchsin	0.1 g
NaOH (1*N* solution)	16.0mL

Preparation of Andrade's Indicator: Add components to distilled/deionized water and bring volume to 100.0mL. Mix thoroughly.

Carbohydrate Solution:
Composition per 100mL:

Carbohydrate	10.0g

Preparation of Carbohydrate Solution: Add carbohydrate to distilled/deionized water and bring volume to 100.0mL. Adonitol, arabinose, cellobiose, glucose, dulcitol, fructose, galactose, inositol, lactose, maltose, mannitol, raffinose, rhamnose, salicin, sorbitol, sucrose, trehalose, xylose, or other carbohydrates may be used. Mix thoroughly. Filter sterilize.

Caution: Acid Fuchsin is a potential carcinogen and care must be taken to avoid inhalation of the powdered dye and contact with the skin.

Preparation of Medium: Add components, except carbohydrate solution, to distilled/deionized water and bring volume to 1.0L. Mix thoroughly. Gently heat and bring to boiling. Distribute in 10.0mL volumes into test tubes containing inverted Durham tubes. Autoclave for 15 minutes at 15 psi–121°C. Cool to 25°C. Add 0.5mL of sterile carbohydrate solution to each tube.

Use: For the determination of carbohydrate fermentation reactions of microorganisms, particularly members of the Enterobacteriaceae. A Durham tube is used to collect gas produced during the fermentation reaction. Acid production is indicated by a pink reaction.

Carbohydrate Medium Base
See: **CHO Medium Base**

Carbon Assimilation Medium

Composition per liter:

Agar solution	500.0mL
Mineral base medium	500.0mL

pH 6.5 ± 0.1 at 25°C

Agar Solution:
Composition per liter:

Agar	32.0g

Preparation of Agar Solution: Add agar to distilled/deionized water and bring volume to 1.0L. Mix thoroughly. Gently heat and bring to boiling. Autoclave for 15 min at 15 psi pressure–121°C. Cool to 45°–50°C.

Mineral Base Medium:
Composition per 500mL:

Carbohydrate	10.0g
NaCl	5.0g
NH_4HPO_4	1.0g
K_2HPO_4	1.0g
$MgSO_4·7H_2O$, anhydrous	0.1g

Preparation of Mineral Base Medium: Add components to distilled/deionized water and bring volume to 500.0mL. Mix thoroughly. Gently heat until dissolved. Filter sterilize. Warm to 45°–50°C.

Preparation of Medium: Combine 500.0mL of cooled, sterile agar solution and 500.0mL of sterile mineral base medium. Mix thoroughly. Aseptically distribute into sterile tubes. Allow tubes to cool in a slanted position.

Use: For the cultivation and differentiation of microorganisms based on their ability to utilize a particular carbon source.

Carbon Assimilation Medium, Auxanographic Method for Yeast Identification

Composition per liter:

Noble agar	20.00g
$(NH_4)_2SO_4$	0.5g
KH_2PO_4	0.1g
$MgSO_4·7H_2O$	0.05g
NaCl	0.01g
$CaCl_2·2H_2O$	0.01g
DL-Methionine	2.0mg
DL-Tryptophan	2.0mg
L-Histidine·HCl	1.0mg
Inositol	0.2mg
KI	0.01mg
H_3BO_3	0.05mg
$ZnSO_4·7H_2O$	0.04mg
$MnSO_4·4H_2O$	0.04mg
Thiamine·HCl	0.04mg
Pyroxidine·HCl	0.04mg
Niacin	0.04mg
Calcium pantothenate	0.04mg
p-Aminobenzoic acid	0.02mg
Riboflavin	0.02mg
$FeCl_3$	0.02mg
$Na_2MoO_4·4H_2O$	0.02mg
$CuSO_4·5H_2O$	4.0μg

Folic acid.. 0.2μg
Biotin 0.2μg

pH 4.5 ± 0.2 at 25°C

Preparation of Medium: Add components to distilled/deionized water and bring volume to 1.0L. Mix thoroughly. Gently heat and bring to boiling. Distribute into screw-capped tubes in 20.0mL volumes. Autoclave for 15 min at 15 psi pressure–121°C.

Use: For carbohydrate assimilation tests by the auxanographic method for identification of yeasts.

Carbon Utilization Test

Composition per liter:

Ionagar 10.0g
NH_4Cl..1.0g
$MgSO_4 \cdot 7H_2O$ 0.5g
Ferric ammonium citrate..................................0.05g
$CaCl_2$ 0.5mg
Sodium potassium phosphate
buffer (0.33M solution, pH 6.8) 1.0L
Carbon source 10.0mL

pH 6.8 ± 0.2 at 25°C

Carbon Source:

Composition per 10mL:

Carbon source 1.0g

Preparation of Carbon Source: Add carbon source to distilled/deionized water and bring volume to 10.0mL. Mix thoroughly. Filter sterilize.

Preparation of Medium: Add components, except carbon source, to distilled/deionized water and bring volume to 990.0mL. Mix thoroughly. Gently heat and bring to boiling. Autoclave for 15 min at 15 psi pressure–121°C. Cool to 45°–50°C. Aseptically add sterile carbon source. Mix thoroughly. Pour into sterile Petri dishes or distribute into sterile tubes.

Use: For the cultivation and differentiation of *Pseudomonas* species based on their ability to utilize a specific carbon source.

Cardiobacterium hominis Medium

Composition per liter:

Glucose 5.0g
Leucine...0.43g
Threonine 0.28g
Glutamic acid..0.2g
Valine 0.19g
Glycine..0.18g
Arginine 0.16g
Histidine..0.13g
Proline 0.1g
Tyrosine...0.04g
Buffered salts solution 100.0mL
Vitamin solution... 10.0mL

pH 7.0 ± 0.2 at 25°C

Buffered Salts Solution:

Composition per liter:

Na_2PHO_4 ....284.0.g
KH_2PO_4 ....272.0.g
NaCl 5.0g
$FeSO_4 \cdot 7H_2O$ 4.0g
$MgSO_4 \cdot 7H_2O$ 4.0g
$ZnSO_4 \cdot 7H_2O$..0.4g
$MnSO_4 \cdot H_2O$ 0.3g
$CuSO_4 \cdot 5H_2O$ 0.05g

Preparation of Buffered Salts Solution: Add components to distilled/deionized water and bring volume to 1.0L. Mix thoroughly.

Vitamin Solution:

Composition per liter:

Pyridoxine·HCl 2.0mg
Calcium pantothenate.................................... 1.0mg
Nicotinamide.. 1.0mg
Thiamine·HCl... 1.0mg
Biotin 0.1mg

Preparation of Vitamin Solution: Add components to distilled/deionized water and bring volume to 1.0L. Mix thoroughly.

Preparation of Medium: Add components to distilled/deionized water and bring volume to 1.0L. Mix thoroughly. Adjust pH to 7.0. Filter sterilize.

Use: For the isolation and cultivation of *Cardiobacterium hominis*.

Cardiobacterium hominis Medium

Composition per liter:

K_2HPO_4 ....7.0g
Yeast extract 5.0g
KH_2PO_4 ....3.0g
$(NH_4)_2SO_4$ 0.1g
$MgSO_4 \cdot 7H_2O$ 0.01g

pH 7.0 ± 0.2 at 25°C

Preparation of Medium: Add components to distilled/deionized water and bring volume to 1.0L. Mix thoroughly. Distribute into tubes or flasks. Autoclave for 15 min at 15 psi pressure–121°C.

Use: For the cultivation of *Cardiobacterium hominis*.

Carnitine Chloride Medium

Composition per liter:

Noble agar...15.0g
DL-Carnitine chloride 10.0g

Na_2HPO_4 .... 10.0g
KH_2PO_4 .... 5.5g
$(NH_4)_2HPO_4$ 2.0g
$NH_4H_2PO_4$ 1.5g
$MgSO_4 \cdot 7H_2O$ 0.2g
Yeast extract 0.05g
$CaCl_2$ 0.015g
$Fe_2(SO_4)_3$ 0.6mg
$CuSO_4 \cdot 5H_2O$ 0.2mg
$MnSO_4 \cdot H_2O$ 0.2mg
$ZnSO_4 \cdot 7H_2O$ 0.2mg

pH 7.0 ± 0.1 at 25°C

Preparation of Medium: Add components to distilled/deionized water and bring volume to 1.0L. Adjust pH to 7.0 with NaOH. Mix thoroughly. Heat gently until boiling. Distribute into tubes or flasks. Autoclave for 15 min at 15 psi pressure–121°C. Pour into sterile Petri dishes or leave in tubes.

Use: For the cultivation and maintenance of bacteria that can use carnitine as a carbon source.

Cary and Blair Transport Medium

Composition per liter:

Agar 5.0g
NaCl 5.0g
Sodium thioglycollate 1.5g
Na_2HPO_4 .... 1.1g
$CaCl_2$ solution 9.0mL

pH 8.0 ± 0.5 at 25°C

Source: This medium is available as a premixed powder from BBL Microbiology Systems, Difco Laboratories and Oxoid Unipath.

$CaCl_2$ Solution:
Composition per 10mL:

$CaCl_2$ 0.1g

Preparation of $CaCl_2$ Solution: Add $CaCl_2$ to distilled/deionized water and bring volume to 10.0mL. Mix thoroughly. Filter sterilize.

Preparation of Medium: Add components to distilled/deionized water and bring volume to 1.0L. Mix thoroughly and heat gently until boiling. Cool to 50°C. Add 9.0mL of a 1% $CaCl_2$ solution. Adjust the pH to 8.4. Distribute into screw-capped tubes in 7.0mL volumes. Sterilize under flowing steam for 15 min. After sterilization, tighten screw-caps.

Use: For the maintenance—as a holding medium or transport medium—of clinical specimens during collection or shipment.

Cary and Blair Transport Medium, Modified

Composition per liter:

Agar 5.0g
NaCl 5.0g
Sodium thioglycollate 1.5g
L-Cysteine·HCl·H_2O 0.5g
$CaCl_2 \cdot 2H_2O$ 0.1g
Na_2HPO_4 .... 0.1g
$NaHSO_3$ 0.1g
Resazurin solution 4.0mL

pH 8.4 ± 0.2 at 25°C

Resazurin Solution:
Composition per 380mL:

Resazurin 0.05g
Ethanol (95% solution) 200.0mL

Preparation of Resazurin Solution: Add resazurin to 200.0mL of ethanol. Mix thoroughly. Bring volume to 380.0mL with distilled/deionized water.

Preparation of Medium: Add components, except L-Cysteine·HCl·H_2O, to distilled/deionized water and bring volume to 1.0L. Mix thoroughly. Gas the solution with 100% CO_2 for 10–15 min. Add the L-Cysteine·HCl·H_2O. Mix thoroughly. Adjust pH to 8.4. Anaerobically distribute into tubes under 100% N_2. Cap tubes with butyl rubber stoppers. Autoclave for 15 min at 0 psi pressure–100°C on three consecutive days.

Use: For the maintenance—as a holding medium—of clinical specimens during collection or shipment.

Caryophanon Medium

Composition per liter:

Agar 15.0g
Yeast extract 2.0g
Sodium acetate 1.0g
Pancreatic digest of casein 1.0g

pH 7.5 ± 0.2 at 25°C

Preparation of Medium: Add components to distilled/deionized water and bring volume to 1.0L. Mix thoroughly. Gently heat and bring to boiling. Distribute into tubes or flasks. Autoclave for 15 min at 15 psi pressure–121°C. Pour into sterile Petri dishes or leave in tubes.

Use: For cultivation and maintenance of *Caryophanon tenue* and other *Caryophanon* species.

CAS Medium

Composition per liter:

Pancreatic digest of casein 10.0g

$MgSO_4·7H_2O$ 1.0g
K_2HPO_4 .... 0.25g
pH 6.8 ± 0.2 at 25°C

Preparation of Medium: Add components to distilled/deionized water and bring volume to 1.0L. Mix thoroughly. Distribute into tubes or flasks. Autoclave for 15 min at 15 psi pressure–121°C.

Use: For the cultivation of myxobacteria.

Casamino Acids Glucose Medium (CAGV Medium)

Composition per liter:

Agar 20.0g
Glucose 1.0g
Vitamin-free casamino acids 1.0g
Mineral salts solution A 20.0mL
Vitamin solution 10.0mL
pH 7.2 ± 0.2 at 25°C

Mineral Salts Solution A:

Composition per liter:

$MgSO_4·7H_2O$ 29.7g
$NaMoO_4·2H_2O$ 12.67g
Nitrilotriacetic acid 10.0g
$CaCl_2·2H_20$ 3.34g
$FeSO_4·7H_2O$ 0.10g
Metallic salts solution B 50.0mL

Preparation of Mineral Salts Solution A: Add nitrilotriacetic acid to 500.0mL of distilled/deionized water. Dissolve by adjusting pH to 6.5 with KOH. Add remaining components. Readjust pH to 7.2 with H_2SO_4 or KOH. Add distilled/deionized water to 1.0L.

Metallic Salts Solution B:

Composition per 100mL:

$ZnSO_4·7H_2O$ 1.1g
$FeSO_4·7H_20$ 0.5g
Ethylenediaminetetraacetic acid 0.3g
$MnSO_4·H_2O$ 0.3g
$CuSO_4·5H_2O$ 0.04g
$CoCL_2·6H_20$ 0.02g
$Na_2B_4O_7·10H_20$ 0.02g

Preparation of Metallic Salts Solution B: Add a few drops of H_2SO_4 to distilled/deionized water to inhibit precipitate formation. Add components to acidified distilled/deionized water and bring volume to 100.0mL. Mix thoroughly.

Vitamin Solution:

Pyridoxine·HCL 0.01g
Calcium pantothenate 5.0mg
Nicotinamide 5.0mg
Riboflavin 5.0mg
Thiamine·HCl 5.0mg
Biotin 2.0mg
Folic acid 2.0mg
Vitamin B_{12} 0.1mg

Preparation of Vitamin Solution: Add components to distilled/deionized water and bring volume to 1.0L. Mix thoroughly. Filter sterilize.

Preparation of Medium: Add components to distilled/deionized water and bring volume to 1.0L. Mix thoroughly. Gently heat and bring to boiling. Distribute into tubes or flasks. Autoclave for 15 min at 15 psi pressure–121°C. Pour into sterile Petri dishes or leave in tubes.

Use: For the cultivation and maintenance of *Microcyclus aquaticus.*

Casamino Acids Medium

Composition per liter:

Casamino acids 1.0g
Glucose 1.0g
Biotin 0.02mg
Modified Hutner's basal salts 20.0mL

Modified Hutner's Basal Salts:

Composition per liter:

$MgSO_4·7H_2O$ 29.7g
Nitrilotriacetic acid 10.0g
$CaCl_2·2H_2O$ 3.34g
$FeSO_4·7H_2O$ 0.1g
$(NH_4)_2MoO_4$ 9.25mg
Metals "44" 50.0mL

Preparation of Modified Hutner's Basal Salts: Add nitrilotriacetic acid to 500.0mL of distilled/deionized water. Dissolve by adjusting pH to 6.5 with KOH. Add remaining components. Add distilled/deionized water to 1.0L.

Metals "44":

Composition per 100mL:

$ZnSO_4·7H_2O$ 1.1g
$FeSO_4·7H_2O$ 0.5g
EDTA 0.25g
$MnSO_4·7H_2O$ 0.154g
$CuSO_4·5H_2O$ 0.04g
$Co(NO_3)_2·6H_2O$ 0.025g
$Na_2B_4O_7·10H_2O$ 0.018g

Preparation of Metals "44": Acidify distilled/deionized water with a drop of H_2SO_4 to retard precipitation of salts. Add components to distilled/deionized water and bring volume to 100.0mL.

Preparation of Medium: Add components to distilled/deionized water and bring volume to 1.0L. Mix

thoroughly. Distribute into tubes or flasks. Autoclave for 15 min at 15 psi pressure–121°C.

Use: For cultivation of *Ancylobacter aquaticus* and *Enhydrobacter aerosaccus*.

Casamino Acids Peptone Czapek's Agar

Composition per liter:

Sucrose	30.0g
Agar	15.0g
Peptone	2.0g
Casamino acids	1.0g
K_2HPO_4	1.0g
KCl	0.5g
$MgSO_4 \cdot 7H_2O$	0.5g
$FeSO_4 \cdot 7H_2O$	0.01g

Preparation of Medium: Add components to distilled/deionized water and bring volume to 1.0L. Mix thoroughly. Gently heat and bring to boiling. Distribute into tubes or flasks. Autoclave for 15 min at 15 psi pressure–121°C. Pour into sterile Petri dishes or leave in tubes.

Use: For the isolation and cultivation of *Actinomadura* species, *Actinopolyspora* species, *Excellospora* species and *Microspora* species.

Casamino Acids Yeast Extract Broth (CYE Broth)

Composition per liter:

Casamino acids	30.0g
Yeast extract	4.0g
K_2HPO_4	0.5g

pH 7.4 ± 0.2 at 25°C.

Preparation of Medium: Add components to distilled/deionized water and bring volume to 1.0L. Mix thoroughly. Distribute into tubes or flasks. Autoclave for 15 min at 15 psi pressure–121°C.

Use: For the cultivation of *Vibrio* species from foods.

Casamino Acids Yeast Extract Lincomycin Medium

Composition per liter:

Casamino acids	20.0g
K_2HPO_4	8.71g
Yeast extract	6.0g
NaCl	2.5g
Lincomycin solution	5.0mL
Trace salts solution	1.0mL

pH 8.5 ± 0.2 at 25°C.

Lincomycin Solution:
Composition per 5mL:

Lincomycin	45.0mg

Preparation of Lincomycin Solution: Add lincomycin to distilled/deionized water and bring volume to 5.0mL. Mix thoroughly. Filter sterilize.

Trace Salts Solution:
Composition per liter:

$MgSO_4 \cdot 7H_2O$	50.0g
$MnCl_2 \cdot 4H_2O$	5.0g
$FeCl_2$	5.0g

Preparation of Trace Salts Solution: Add components to distilled/deionized water and bring volume to 1.0L. Add sufficient 0.1*N* H_2SO_4 to dissolve components. Mix thoroughly. Filter sterilize.

Preparation of Medium: Add components, except trace salts solution and lincomycin solution, to distilled/deionized water and bring volume to 994.0mL. Mix thoroughly. Adjust pH to 8.5. Autoclave for 15 min at 15 psi pressure–121°C. Cool to 25°C. Aseptically add 1.0mL sterile trace salts solution and 5.0mL of sterile lincomycin solution. Mix thoroughly. Aseptically distribute into sterile tubes or flasks.

Use: For the cultivation of heat-labile toxin-producing enterotoxigenic *Escherichia coli*.

Casamino Acids Yeast Extract Salts Broth, Gorbach (CA YE Broth)

Composition per liter:

Casamino acids	20.0g
K_2HPO_4	8.71g
Yeast extract	6.0g
NaCl	2.5g
Trace salts solution	1.0mL

pH 8.5 ± 0.2 at 25°C.

Trace Salts Solution:
Composition per liter:

$MgSO_4 \cdot 7H_2O$	50.0g
$MnCl_2 \cdot 4H_2O$	5.0g
$FeCl_2$	5.0g

Preparation of Trace Salts Solution: Add components to distilled/deionized water and bring volume to 1.0L. Add sufficient 0.1*N* H_2SO_4 to dissolve components. Mix thoroughly. Filter sterilize.

Preparation of Medium: Add components, except trace salts solution, to distilled/deionized water and bring volume to 999.0mL. Mix thoroughly. Adjust pH to 8.5. Autoclave for 15 min at 15 psi pressure–121°C. Cool to 25°C. Aseptically add 1.0mL sterile trace salts solution. Mix thoroughly. Aseptically distribute into sterile tubes or flasks.

Use: For the cultivation of heat-labile-toxin producing enterotoxigenic *Escherichia coli*.

Casamino Peptone Czapek Medium

Composition per liter:

Sucrose	30.0g
Agar	15.0g
Peptone	2.0g
Casamino acids	1.0g
K_2HPO_4	1.0g
KCl	0.5g
$MgSO_4 \cdot 7H_2O$	0.5g
$FeSO_4 \cdot 7H_2O$	0.01g

Preparation of Medium: Add components to distilled/deionized water and bring volume to 1.0L. Mix thoroughly. Gently heat to boiling. Distribute into tubes or flasks. Autoclave for 15 min at 15 psi pressure–121°C. Pour into sterile Petri dishes or leave in tubes.

Use: For cultivation and maintenance of *Actinoplanes* species, *Pseudonocardia compacta*, and *Streptomyces* species.

Casein Agar

Composition per liter:

Agar	10.0g
Skim milk	50.0mL

Preparation of Medium: Add components to distilled/deionized water and bring volume to 1.0L. Mix thoroughly. Gently heat and bring to boiling. Distribute into tubes or flasks. Autoclave for 15 min at 15 psi pressure–121°C. Pour into sterile Petri dishes or leave in tubes.

Use: For the cultivation and differentiation of aerobic actinomycetes based on casein utilization. Bacteria that utilize casein, such as *Streptomyces* and *Actinomadura* species, appear as colonies surrounded by a clear zone. *Nocardia asteroides, N. caviae,* and *Mycobacterium fortuitum* do not utilize casein.

Casein Medium

Composition per liter:

NaCl	250.0g
Agar	20.0g
$MgCl_2 \cdot 6H_2O$	20.0g
Casein hydrolysate	5.0g
Yeast extract	5.0g
KCl	2.0g
$CaCl_2 \cdot 2H_2O$	0.2g

pH 7.4 ± 0.2 at 25°C

Preparation of Medium: Add components to distilled/deionized water and bring volume to 950.0mL. Mix thoroughly. Gently heat to boiling. Adjust pH to 7.4. Bring volume to 1.0L with distilled/deionized water. Distribute into tubes or flasks. Autoclave for 15 min at 15 psi pressure–121°C. Pour into sterile Petri dishes or leave in tubes.

Use: For cultivation and maintenance of *Halobacterium* species and other halophilic bacteria.

Casein Yeast Extract Glucose Agar (CYG Agar)

Composition per liter:

Agar	20.0g
Glucose	5.0g
Casein hydrolysate	5.0g
Yeast extract	5.0g

pH 7.0 ± 0.2 at 25°C

Preparation of Medium: Add components to distilled/deionized water and bring volume to 1.0L. Mix thoroughly. Gently heat and bring to boiling. Distribute into tubes or flasks. Autoclave for 15 min at 15 psi pressure–121°C.

Use: For agar dilution susceptibility tests with imidazole antifungal agents.

Casein Yeast Extract Glucose Broth (CYG Broth)

Composition per liter:

Casein hydrolysate	5.0g
Glucose	5.0g
Yeast extract	5.0g

pH 7.0 ± 0.2 at 25°C

Preparation of Medium: Add components to distilled/deionized water and bring volume to 1.0L. Mix thoroughly. Gently heat and bring to boiling. Distribute into tubes or flasks. Autoclave for 15 min at 15 psi pressure–121°C.

Use: For agar dilution susceptibility tests with imidazole antifungal agents.

Casitone Agar

Composition per liter:

Pancreatic digest of casein 20.0g
Agar 15.0g
$MgSO_4 \cdot 7H_2O$ 1.0g
Potassium phosphate buffer, pH 7.2 .... 1.0L
pH 7.2 ± 0.2 at 25°C

Preparation of Medium: Combine components. Mix thoroughly. Gently heat to boiling. Distribute into tubes or flasks. Autoclave for 15 min at 15 psi pressure–121°C. Pour into sterile Petri dishes or leave in tubes.

Use: For cultivation and maintenance of *Myxococcus* species.

Casitone Glycerol Yeast Autolysate Broth (CGY Autolysate Broth)

Composition per liter:

Pancreatic digest of casein 5.0g
Yeast autolysate 1.0g
Glycerol 10.0mL

Preparation of Medium: Add components to distilled/deionized water and bring volume to 1.0L. Mix thoroughly. Distribute into tubes or flasks. Autoclave for 15 min at 15 psi pressure–121°C.

Use: For the isolation, cultivation and enumeration, of iron and sulfur bacteria from the *Sphaerotilus* group.

Casitone Yeast Extract Agar

Composition per liter:

Agar 15.0g
Pancreatic digest of casein 5.0g
Yeast extract 3.0g
$MgSO_4 \cdot 7H_2O$ 1.0g

Preparation of Medium: Add components to distilled/deionized water and bring volume to 1.0L. Mix thoroughly. Gently heat to boiling. Distribute into tubes or flasks. Autoclave for 15 min at 15 psi pressure–121°C. Pour into sterile Petri dishes or leave in tubes.

Use: For cultivation and maintenance of *Chitinophaga pinensis.*

Casman Agar Base

Composition per liter:

Noble agar 14.0g
Proteose Peptone No. 3 .... 10.0g
Tryptose 10.0g
NaCl 5.0g
Beef extract 3.0g
Cornstarch 1.0g
Glucose 0.5g
p-Aminobenzoic acid 0.05g
Nicotinamide 0.05g
Blood 50.0mL
Water-lysed blood solution 1.5mL
pH 7.3 ± 0.2 at 25°C

Source: This medium is available as a premixed powder from Difco Laboratories and BBL Microbiology Systems.

Water—Lysed Blood Solution:
Composition per 8mL:

Blood 2.0mL

Preparation of Water—Lysed Blood Solution: Add blood to distilled/deionized water and bring volume to 8.0mL. Mix thoroughly. Filter sterilize.

Preparation of Medium: Add components, except blood and water-lysed blood solution, to distilled/deionized water and bring volume to 948.5mL. Mix thoroughly. Gently heat to boiling. Autoclave for 15 min at 15 psi pressure–121°C. Cool to 50°C. Aseptically add 50.0mL sterile blood and 1.5mL sterile water-lysed blood solution (one part blood to three parts water). Water-lysed blood may be omitted if sterile blood is partially lysed due to storage. Mix thoroughly. Pour into sterile Petri dishes or distribute into sterile tubes.

Use: For the isolation of fastidious bacteria from clinical specimens. For the cultivation under reduced oxygen tension of fastidious microorganisms such as *Haemophilus influenzae*, *Neisseria meningitidis* and *Neisseria gonorrhoeae*.

Casman Agar Base with Rabbit Blood

Composition per liter:

Noble agar 14.0g
Proteose Peptone No. 3 .... 10.0g
Tryptose 10.0g
NaCl 5.0g
Beef extract 3.0g
Cornstarch 1.0g
Glucose 0.5g
p-Aminobenzoic acid 0.05g
Nicotinamide 0.05g
Rabbit blood 50.0mL
Water-lysed blood solution 1.5mL
pH 7.3 ± 0.2 at 25°C

Source: Casman Agar Base is available as a premixed powder from BBL Microbiology Systemsand Difco Laboratories.

Water—Lysed Blood Solution:

Composition per 8mL:

Rabbit blood 2.0mL

Preparation of Water—Lysed Blood Solution: Add blood to distilled/deionized water and bring volume to 8.0mL. Mix thoroughly. Filter sterilize.

Preparation of Medium: Add components, except rabbit blood and water-lysed blood solution, to distilled/deionized water and bring volume to 950.0L. Mix thoroughly. Gently heat to boiling. Autoclave for 15 min at 15 psi pressure–121°C. Cool to 50°C. Aseptically add 50.0mL sterile rabbit blood and 1.5mL sterile water-lysed blood solution. Water-lysed blood may be omitted if sterile blood is partially lysed due to storage. Mix thoroughly. Pour into sterile Petri dishes or distribute into sterile tubes.

Use: For the cultivation and maintenance of *Gardnerella vaginalis*.

Castenholtz D Medium (Medium D)

Composition per liter:

$NaNO_3$ 0.7g
Na_2HPO_4 .... 0.11g
KNO_3 .... 0.10g
$MgSO_4 \cdot 7H_2O$ 0.10g
Nitrilotriacetic acid 0.10g
$CaSO_4 \cdot 2H_2O$ 0.06g
NaCl 8.0mg
$FeCl_3$ solution 1.0mL
Micronutrient solution 0.5mL

pH 7.5 ± 0.2 at 25°C

$FeCl_3$ Solution:

Composition per liter:

$FeCl_3 \cdot 6H_2O$ 2.28g

Preparation of $FeCl_3$ Solution: Add $FeCl_3 \cdot 6H_2O$ to distilled/deionized water and bring volume to 1.0L. Mix thoroughly.

Micronutrient Solution:

Composition per liter:

$MnSO_4 \cdot H_2O$ 2.28g
H_3BO_3 .... 0.5g
$ZnSO_4 \cdot 7H_2O$ 0.5g
$CoCl_2 \cdot 6H_2O$ 0.025g
$CuSO_4 \cdot 5H_2O$ 0.025g
$Na_2MoO_4 \cdot 2H_2O$ 0.025g
H_2SO_4 .... 0.5mL

Preparation of Micronutrient Solution: Add components to distilled/deionized water and bring volume to 1.0L. Mix thoroughly.

Preparation of Medium: Add nitrilotriacetic acid to 500.0mL of distilled/deionized water. Dissolve by adjusting pH to 6.5 with KOH. Add remaining components. Mix thoroughly. Readjust pH to 7.5. Bring volume to 1.0L with distilled/deionized water. Mix thoroughly. Distribute into tubes or flasks. Autoclave for 15 min at 15 psi pressure–121°C.

Use: For the isolation of cyanobacteria, including thermophilic species. For the cultivation of *Chloroflexus* species and *Fischerella* species.

Castenholtz D Medium, Modified (Medium D, Modified)

Composition per liter:

NaCl 160.0g
$NaNO_3$ 0.69g
Na_2HPO_4 .... 0.111g
KNO_3 .... 0.103g
$MgSO_4 \cdot 7H_2O$ 0.1g
Nitrilotriacetic acid 0.1g
$CaSO_4 \cdot 2H_2O$ 0.06g
$FeCl_3$ 0.3mg
Trace metal solution Castenholz 1.0mL

pH 7.5 ± 0.2 at 25°C

Trace Metal Solution Castenholz:

Composition per liter:

$MnSO_4 \cdot H_2O$ 2.28g
H_3BO_3 .... 0.5g
$ZnSO_4 \cdot 7H_2O$ 0.5g
$Co(NO_3)_2 \cdot 6H_2O$ 0.025g
$CuSO_4 \cdot 5H_2O$ 0.025g
$Na_2MoO_4 \cdot 2H_2O$ 0.025g
H_2SO_4 .... 0.5mL

Preparation of Trace Metal Solution Castenholz: Add components to distilled/deionized water and bring volume to 1.0L. Mix thoroughly.

Preparation of Medium: Add nitrilotriacetic acid to 500.0mL of distilled/deionized water. Dissolve by adjusting pH to 6.5 with KOH. Add remaining components. Mix thoroughly. Readjust pH to 7.5. Bring volume to 1.0L with distilled/deionized water. Mix thoroughly. Distribute into screw-capped tubes or flasks. Autoclave for 15 min at 15 psi pressure–121°C.

Use: For the isolation of halophilic cyanobacteria.

Castenholtz DG Medium (Medium DG)

Composition per liter:

Glycyl-glycine buffer	0.8g
$NaNO_3$	0.7g
Na_2HPO_4	0.11g
KNO_3	0.10g
$MgSO_4 \cdot 7H_2O$	0.10g
Nitrilotriacetic acid	0.10g
$CaSO_4 \cdot 2H_2O$	0.06g
NaCl	8.0mg
$FeCl_3$ solution	1.0mL
Micronutrient solution	0.5mL

pH 7.5 ± 0.2 at 25°C

$FeCl_3$ Solution:

Composition per liter:

$FeCl_3 \cdot 6H_2O$	2.28g

Preparation of $FeCl_3$ Solution: Add $FeCl_3 \cdot 6H_2O$ to distilled/deionized water and bring volume to 1.0L. Mix thoroughly.

Micronutrient Solution:

Composition per liter:

$MnSO_4 \cdot H_2O$	2.28g
H_3BO_3	0.5g
$ZnSO_4 \cdot 7H_2O$	0.5g
$CoCl_2 \cdot 6H_2O$	0.025g
$CuSO_4 \cdot 5H_2O$	0.025g
$Na_2MoO_4 \cdot 2H_2O$	0.025g
H_2SO_4	0.5mL

Preparation of Micronutrient Solution: Add components to distilled/deionized water and bring volume to 1.0L. Mix thoroughly.

Preparation of Medium: Add nitrilotriacetic acid to 500.0mL of distilled/deionized water. Dissolve by adjusting pH to 6.5 with KOH. Add remaining components. Mix thoroughly. Readjust pH to 8.1. Bring volume to 1.0L with distilled/deionized water. Mix thoroughly. Distribute into tubes or flasks. Autoclave for 15 min at 15 psi pressure–121°C.

Use: For the isolation of cyanobacteria, including thermophilic species.

Castenholtz DGN Medium (Medium DGN)

Composition per liter:

Glycyl-glycine buffer	0.8g
$NaNO_3$	0.7g
NH_4Cl	0.2g
Na_2HPO_4	0.11g
KNO_3	0.10g
$MgSO_4 \cdot 7H_2O$	0.10g
Nitrilotriacetic acid	0.10g
$CaSO_4 \cdot 2H_2O$	0.06g
NaCl	8.0mg
$FeCl_3$ solution	1.0mL
Micronutrient solution	0.5mL

pH 7.5 ± 0.2 at 25°C

$FeCl_3$ Solution:

Composition per liter:

$FeCl_3 \cdot 6H_2O$	2.28g

Preparation of $FeCl_3$ Solution: Add $FeCl_3 \cdot 6H_2O$ to distilled/deionized water and bring volume to 1.0L. Mix thoroughly.

Micronutrient Solution:

Composition per liter:

$MnSO_4 \cdot H_2O$	2.28g
H_3BO_3	0.5g
$ZnSO_4 \cdot 7H_2O$	0.5g
$CoCl_2 \cdot 6H_2O$	0.025g
$CuSO_4 \cdot 5H_2O$	0.025g
$Na_2MoO_4 \cdot 2H_2O$	0.025g
H_2SO_4	0.5mL

Preparation of Micronutrient Solution: Add components to distilled/deionized water and bring volume to 1.0L. Mix thoroughly.

Preparation of Medium: Add nitrilotriacetic acid to 500.0mL of distilled/deionized water. Dissolve by adjusting pH to 6.5 with KOH. Add remaining components. Mix thoroughly. Readjust pH to 8.2. Bring volume to 1.0L with distilled/deionized water. Mix thoroughly. Distribute into tubes or flasks. Autoclave for 15 min at 15 psi pressure–121°C.

Use: For the isolation of cyanobacteria, including thermophilic species.

Castenholtz ND Medium (Medium ND)

Composition per liter:

Na_2HPO_4	0.11g
$MgSO_4 \cdot 7H_2O$	0.10g
Nitrilotriacetic acid	0.10g
$CaSO_4 \cdot 2H_2O$	0.06g
NaCl	8.0mg
$FeCl_3$ solution	1.0mL
Micronutrient solution	0.5mL

pH 7.5 ± 0.2 at 25°C

$FeCl_3$ Solution:

Composition per liter:

$FeCl_3 \cdot 6H_2O$	2.28g

Preparation of $FeCl_3$ Solution: Add $FeCl_3 \cdot 6H_2O$ to distilled/deionized water and bring volume to 1.0L. Mix thoroughly.

Micronutrient Solution:

Composition per liter:

$MnSO_4 \cdot H_2O$	2.28g
H_3BO_3	0.5g
$ZnSO_4 \cdot 7H_2O$	0.5g
$CoCl_2 \cdot 6H_2O$	0.025g
$CuSO_4 \cdot 5H_2O$	0.025g
$Na_2MoO_4 \cdot 2H_2O$	0.025g
H_2SO_4	0.5mL

Preparation of Micronutrient Solution: Add components to distilled/deionized water and bring volume to 1.0L. Mix thoroughly.

Preparation of Medium: Add nitrilotriacetic acid to 500.0mL of distilled/deionized water. Dissolve by adjusting pH to 6.5 with KOH. Add remaining components. Mix thoroughly. Readjust pH to 8.2. Bring volume to 1.0L with distilled/deionized water. Mix thoroughly. Distribute into tubes or flasks. Autoclave for 15 min at 15 psi pressure–121°C.

Use: For the isolation of cyanobacteria, including thermophilic species, that require reduced nitrogen concentrations.

Castenholz TYE Medium (Castenholz Trypticase™ Yeast Extract Medium)

Composition per liter:

Castenholz salts, 2X	500.0mL
1% TYE	100.0mL

pH 7.6 ± 0.2 at 25°C

Castenholz Salts, 2X:

Composition per liter:

Agar	30.0g
$NaNO_3$	1.4g
Na_2HPO_4	0.22g
KNO_3	0.21g
Nitrilotriacetic acid	0.2g
$MgSO_4 \cdot 7H_2O$	0.2g
$CaSO_4 \cdot 2H_2O$	0.12g
NaCl	0.016g
$FeCl_3$ (0.03% solution)	2.0mL
Nitsch's Trace Elements	2.0mL

Preparation of Castenholz Salts, 2X: Add components to distilled/deionized water and bring volume to 1.0L. Mix thoroughly. Gently heat and bring to boiling. Adjust pH to 8.2. Autoclave for 15 min at 15 psi pressure–121°C.

Nitsch's Trace Elements:

Composition per liter:

$MnSO_4$	2.2g
H_3BO_3	0.5g
$ZnSO_4$	0.5g
$CoCl_2 \cdot 6H_2O$	0.046g
Na_2MoO_4	0.025g
$CuSO_4$	0.016g
H_2SO_4	0.5mL

Preparation of Nitsch's Trace Elements: Add components to distilled/deionized water and bring volume to 1.0L. Mix thoroughly.

1% TYE

Composition per liter:

Pancreatic digest of casein	10.0g
Yeast extract	10.0g

Preparation of 1% TYE: Add components to distilled/deionized water and bring volume to 1.0L. Mix thoroughly. Autoclave for 15 min at 15 psi pressure–121°C.

Preparation of Medium: Aseptically combine 500.0mL sterile Castenholz salts, 2X, 100.0mL sterile 1% TYE and 400.0mL sterile distilled/deionized water. Adjust pH to 7.6.

Use: For the cultivation and maintenance of *Thermonema lapsum* and *Thermus* species.

Castenholz TYE Medium with 2% Trypticase™ Yeast Extract

Composition per liter:

Castenholz salts, 2X	500.0mL
2% TYE	100.0mL

pH 7.6 ± 0.2 at 25°C

Castenholz Salts, 2X:

Composition per liter:

Agar	30.0g
$NaNO_3$	1.4g
Na_2HPO_4	0.22g
KNO_3	0.21g
$MgSO_4 \cdot 7H_2O$	0.2g
Nitrilotriacetic acid	0.2g
$CaSO_4 \cdot 2H_2O$	0.12g
NaCl	0.016g
$FeCl_3$ solution (0.03% solution)	2.0mL
Nitsch's Trace Elements	2.0mL

Preparation of Castenholz Salts, 2X: Add components to distilled/deionized water and bring volume to 1.0L. Mix thoroughly. Gently heat and bring to boiling. Adjust pH to 8.2. Autoclave for 15 min at 15 psi pressure–121°C.

Nitsch's Trace Elements:

Composition per liter:

$MnSO_4$	2.2g
H_3BO_3	0.5g

$ZnSO_4$	0.5g
$CoCl_2 \cdot 6H_2O$	0.046g
Na_2MoO_4	0.025g
$CuSO_4$	0.016g
H_2SO_4	0.5mL

Preparation of Nitsch's Trace Elements: Add components to distilled/deionized water and bring volume to 1.0L. Mix thoroughly.

2% TYE

Composition per liter:

Pancreatic digest of casein	20.0g
Yeast extract	20.0g

Preparation of 2% TYE: Add components to distilled/deionized water and bring volume to 1.0L. Mix thoroughly. Autoclave for 15 min at 15 psi pressure–121°C.

Preparation of Medium: Aseptically combine 500.0mL sterile Castenholz salts, 2X, 100.0mL sterile 2% TYE and 400.0mL sterile distilled/deionized water. Adjust pH to 7.6.

Use: For the cultivation and maintenance of *Thermus* species.

Caulobacter Medium

Composition per liter:

Agar	10.0g
Peptone	2.0g
Yeast extract	1.0g
$MgSO_4 \cdot 7H_2O$	0.2g
Riboflavin	1.0mg

pH 7.0 ± 0.2 at 25°C

Preparation of Medium: Add components to tap water and bring volume to 1.0L. Mix thoroughly. Gently heat and bring to boiling. Distribute into tubes or flasks. Autoclave for 15 min at 15 psi pressure–121°C. Pour into sterile Petri dishes or leave in tubes.

Use: For the cultivation of *Caulobacter* species from fresh water.

Caulobacter Medium

Composition per liter:

Agar	10.0g
Peptone	0.5g
Sea water, filtered	1.0L

pH 7.0 ± 0.2 at 25°C

Preparation of Medium: Combine components. Mix thoroughly. Gently heat and bring to boiling. Distribute into tubes or flasks. Autoclave for 15 min at 15 psi pressure–121°C. Pour into sterile Petri dishes or leave in tubes.

Use: For the cultivation of *Caulobacter* species from marine isolates.

Caulobacter Medium

Composition per liter:

Glucose	1.0g
Peptone	1.0g
Yeast extract	1.0g
Salt solution	100.0mL

Salt Solution:

Composition per 100mL:

EDTA	0.1g
KNO_3	0.1g
K_2HPO_4	0.066g
$MgSO_4$	0.033g
$FeSO_4 \cdot 7H_2O$	9.3mg
$NaBO_3 \cdot 4H_2O$	2.63mg
$MgCl_2 \cdot 4H_2O$	1.81mg
$CaCl_2$	1.2mg
$(NH_4)_6Mo_7O_{24} \cdot 7H_2O$	1.0mg
$ZnSO_4 \cdot 7H_2O$	0.22mg
$CuSO_4 \cdot 5H_2O$	0.079mg
$Co(NO_3)_2 \cdot H_2O$	0.02mg

Preparation of Salt Solution: Add components to distilled/deionized water and bring volume to 100.0mL. Mix thoroughly.

Preparation of Medium: Add components to distilled/deionized water and bring volume to 1.0L. Mix thoroughly. Distribute into tubes or flasks. Autoclave for 15 min at 15 psi pressure–121°C.

Use: For the enrichment of *Stella* species from polluted waters.

Caulobacter Medium

Composition per liter:

Agar	10.0g
Peptone	2.0g
Yeast extract	1.0g
$MgSO_4 \cdot 7H_2O$	0.2g

Preparation of Medium: Add components to tap water and bring volume to 1.0L. Mix throughly. Gently heat to boiling. Distribute into tubes or flasks. Autoclave for 15 min at 15 psi–121°C. Pour into sterile Petri dishes or leave in tubes.

Use: For cultivation and maintenance of *Asticcacaulis excentricus, Caulobacter* species, *Labrys monachus, Pedomicrobium* species, *Pirellula staleyi, Pseudomonas carboxydohydrogena* and *Stella* species.

CBI Agar
See: ***Clostridium botulinum*** **Isolation Agar**

CC Medium

Composition per liter:

Agar	20.0g
KH_2PO_4	4.0g
Potato starch	0.5g
Solution 3	100.0mL
Solution 1	10.0mL

pH 7.3 ± 0.2 at 25°C

Solution 1:

Composition per liter:

$MgSO_4 \cdot 7H_2O$	20.0g
$CaCl_2 \cdot 2H_2O$	2.0g
$FeSO_4 \cdot 7H_2O$	0.4g
H_3BO_3	0.02g
$MnSO_4 \cdot 2H_2O$	0.015g
$NaMoO_4 \cdot 2H_2O$	0.015g
KJ	0.010g
$ZnSO_4$	4.0mg
$CoCl_2 \cdot 4H_2O$	0.4mg
$CuSO_4 \cdot 5H_2O$	0.4mg

Preparation of Solution 1: Add components to distilled/deionized water and bring volume to 1.0L. Mix thoroughly. Adjust pH with 10.0mL of 10% HCl solution.

Solution 3:

Composition per 100mL:

Pancreatic digest of casein	12.0g
Yeast extract	12.0g
L-Cysteine·HCl	0.5g
L-Asparagine	0.03g
DL-Tryptophan	0.02g
Solution 2	12.0mL

Preparation of Solution 3: Add components to distilled/deionized water and bring volume to 100.0mL. Mix thoroughly. Filter sterilize.

Solution 2:

Composition per 100mL:

p-Aminobenzoic acid	0.02g
Calcium pantothenate	0.02g
m-Inositol	0.02g
Pyridoxine·HCl	0.02g
Thiamine·HCl	0.02g
Nicotinamide	0.01g
Nicotinic acid	0.01g
Folic acid	5.0mg
Biotin	1.0mg
Vitamin B_{12}	1.0mL

Preparation of Solution 2: Add components to distilled/deionized water and bring volume to 100.0mL. Mix thoroughly.

Preparation of Medium: Add KH_2PO_4, to distilled/deionized water and bring volume to 250.0mL. Mix thoroughly. Adjust pH to 7.6 with NaOH. Add 10.0mL of solution 1. In a separate flask, add potato starch to 70.0mL of boiling distilled/deionized water. Add potato starch solution to other solution. Add agar. Bring volume to 900.0mL of distilled/deionized water. Autoclave for 15 min at 15 psi pressure–121°C. Cool to 45°–50°C. Aseptically add 100.0mL of sterile solution 3. Mix thoroughly. Pour into sterile Petri dishes or distribute into sterile tubes.

Use: For the isolation and cultivation of *Actinomycetes* species.

CCDA
See: ***Campylobacter*** **Charcoal Differential Agar**

CCFA
See: ***Clostridium*** **Difficile Agar**

CCVC Medium (Cephalothin Cycloheximide Vancomycin Colistin Medium

Composition per liter:

BCYE-alpha base	990.0mL
Antibiotic supplement	10.0mL

pH 6.9 ± 0.2 at 25°C

Source: This medium is available as a premixed powder from BBL Microbiology Systems.

BCYE—Alpha Base:

Composition per liter:

Agar	15.0g
Yeast extract	10.0g
ACES buffer (2-[(2-Amino-2-oxoethyl)-amino]-ethane sulfonic acid)	10.0g
Charcoal, activated	2.0g
α-Ketogluatrate	1.0g
$Fe_4(P_2O_7)_3 \cdot 9H_2O$	0.25g
Cysteine·HCl·H_2O solution	10.0mL

Cysteine·HCl·H_2O Solution:

Composition per 10mL:

L-Cysteine·HCl·H_2O	0.4g

Preparation of Cysteine·HCl·H_2O Solution: Add cysteine·HCl·H_2O to distilled/deionized water and bring volume to 10.0mL. Mix thoroughly. Filter sterilize.

Preparation of BCYE—Alpha Base: Add components, except cysteine solution, to distilled/deionized water and bring volume to 990.0mL. Mix thoroughly. Adjust medium to pH 6.9 with 1*N* KOH. Heat gently and bring to boiling for 1 min. Autoclave for 15 min at 15 psi pressure–121°C. Cool to 50–55°C. Add 4.0mL of L-cysteine·HCl·H_2O solution . Mix thoroughly.

Antibiotic Supplement Solution:

Composition per 10.0mL:

Cycloheximide	80.0mg
Colistin	16.0mg
Cephalothin	4.0mg
Vancomycin	0.5mg

Preparation of Antibiotic Supplement Solution: Add components to 10.0mL of distilled/deionized water. Filter sterilize.

Preparation of Medium: To cooled BCYE-alpha base add 10.0mL sterile antibiotic supplement. Mix thoroughly. Adjust pH to 6.9 with sterile 1*N* KOH. Pour into sterile Petri dishes with constant agitation to keep charcoal in suspension.

Use: For the selective isolation and cultivation of *Legionella* species from environmental samples.

CDC Anaerobe Blood Agar

Composition per liter:

Agar	20.0g
Pancreatic digest of casein	15.0g
Papaic digest of soybean meal	5.0g
NaCl	5.0g
Yeast extract	5.0g
L-Cystine	0.4g
Sheep blood, defibrinated	50.0mL
Vitamin K_1 solution	1.0mL
Hemin solution	0.5mL

pH 7.5 ± 0.2 at 25°C

Source: This medium is available as a prepared medium from BBL Microbiology Systems.

Vitamin K_1 Solution:

Composition per 100mL:

Vitamin K_1	1.0g
Ethanol	99.0mL

Preparation of Vitamin K_1 Solution: Add vitamin K_1 to 99.0mL of absolute ethanol. Mix thoroughly. Filter sterilize.

Hemin Solution:

Composition per 100mL:

Hemin	1.0g
NaOH (1*N* solution)	20.0mL

Preparation of Hemin Solution: Add hemin to 20.0mL of 1*N* NaOH solution. Mix thoroughly. Bring volume to 100.0mL with distilled/deionized water.

Preparation of Medium: Add components, except Vitamin K_1 and sheep blood, to distilled/deionized water and bring volume to 949.0mL. Mix thoroughly. Heat gently and bring to boiling for 1 min. Autoclave for 15 min at 15 psi pressure–121°C. Cool to 50–55°C. Aseptically add 1.0mL vitamin K_1 solution and 50.0mL sterile, defibrinated sheep blood. Mix thoroughly. Pour into sterile Petri dishes.

Use: For the isolation and cultivation of fastidious and slow-growing, obligate anaerobic bacteria from a variety of clinical and nonclinical specimens. For the isolation and cultivation of *Actinomyces israelii, Bacteroides melaninogenicus, Bacteroides thetaiotaomicron, Clostridium haemolyticum,* and *Fusobacterium necrophorum.*

CDC Anaerobe Blood Agar with Kanamycin and Vancomycin

Composition per liter:

Agar	20.0g
Pancreatic digest of casein	15.0g
NaCl	5.0g
Papaic digest of soybean meal	5.0g
Yeast extract	5.0g
L-Cystine	0.4g
Sheep blood, defibrinated	50.0mL
Antibiotic solution	10.0mL
Vitamin K_1 solution	1.0mL
Hemin solution	0.5mL

pH 7.5 ± 0.2 at 25°C

Source: This medium is available as a prepared medium from BBL Microbiology Systems.

Antibiotic Solution:

Composition per 10mL:

Kanamycin	0.1g
Vancomycin	7.5mg

Preparation of Antibiotic Solution: Add components to distilled/deionized water and bring volume to 10.0mL. Mix thoroughly. Filter sterilize.

Vitamin K_1 Solution:

Composition per 100mL:

Vitamin K_1	1.0g
Ethanol	99.0mL

Preparation of Vitamin K_1 Solution: Add vitamin K_1 to 99.0mL of absolute ethanol. Mix thoroughly. Filter sterilize.

Hemin Solution:
Composition per 100mL:

Hemin	1.0g
NaOH (1*N* solution)	20.0mL

Preparation of Hemin Solution: Add hemin to 20.0mL of 1*N* NaOH solution. Mix thoroughly. Bring volume to 100.0mL with distilled/deionized water.

Preparation of Medium: Add components, except vitamin K_1 solution and sheep blood, to distilled/deionized water and bring volume to 949.0mL. Mix thoroughly. Heat gently and bring to boiling for 1 min. Autoclave for 15 min at 15 psi pressure–121°C. Cool to 50–55°C. Aseptically add 1.0mL sterile vitamin K_1 solution and 50.0mL sterile, defibrinated sheep blood. Mix thoroughly. Pour into sterile Petri dishes.

Use: For the selective isolation of fastidious and slow-growing, obligate anaerobic Gram-negative bacteria, especially *Bacteroides* species, from a variety of clinical and nonclinical specimens.

CDC Anaerobe Blood Agar with Phenylethyl Alcohol (CDC Anaerobe Blood Agar with PEA)

Composition per liter:

Agar	20.0g
Pancreatic digest of casein	15.0g
NaCl	5.0g
Papaic digest of soybean meal	5.0g
Yeast extract	5.0g
L-Cystine	0.4g
Sheep blood, defibrinated	50.0mL
Vitamin K_1 solution	10.0mL
Hemin solution	0.5mL

pH 7.5 ± 0.2 at 25°C

Source: This medium is available as a prepared medium from BBL Microbiology Systems.

Vitamin K_1 Solution:
Composition per 100mL:

Vitamin K_1	0.1g
Phenylethyl alcohol	25.0g
Ethanol	74.0mL

Preparation of Vitamin K_1 Solution: Add components to 74.0mL of absolute ethanol. Mix thoroughly. Filter sterilize.

Hemin Solution:
Composition per 100mL:

Hemin	1.0g
NaOH (1*N* solution)	20.0mL

Preparation of Hemin Solution: Add hemin to 20.0mL of 1*N* NaOH solution. Mix thoroughly. Bring volume to 100.0mL with distilled/deionized water.

Preparation of Medium: Add components, except vitamin K_1 solution and sheep blood, to distilled/deionized water and bring volume to 940.0mL. Mix thoroughly. Heat gently and bring to boiling for 1 min. Autoclave for 15 min at 15 psi pressure–121°C. Cool to 50–55°C. Aseptically add 1.0mL vitamin K_1 solution add 50.0mL sterile, defibrinated sheep blood. Mix thoroughly. Pour into sterile Petri dishes.

Use: For the selective isolation of fastidious and slow-growing obligate anaerobic bacteria from a variety of clinical and nonclinical specimens.

CDC Anaerobe Laked Blood Agar with Kanamycin And Vancomycin (CDC Anaerobe Laked Blood Agar with KV)

Composition per liter:

Agar	20.0g
Pancreatic digest of casein	15.0g
Papaic digest of soybean meal	5.0g
Yeast extract	5.0g
L-Cystine	0.4g
Sheep blood, defibrinated, laked	50.0mL
Antibiotic solution	10.0mL
Vitamin K_1 solution	1.0mL
Hemin solution	0.5mL

pH 7.5 ± 0.2 at 25°C

Source: This medium is available as a prepared medium from BBL Microbiology Systems.

Antibiotic Solution:
Composition per 10mL:

Kanamycin	0.1g
Vancomycin	7.5mg

Preparation of Antibiotic Solution: Addcomponents to distilled/deionized water and bring volume to 10.0mL. Mix thoroughly. Filter sterilize.

Vitamin K_1 Solution:
Composition per 100mL:

Vitamin K_1	1.0g
Ethanol	99.0mL

Preparation of Vitamin K_1 Solution: Add vitamin K_1 to 99.0mL of absolute ethanol. Mix thoroughly. Filter sterilize.

Hemin Solution:

Composition per 100mL:

Hemin	1.0g
NaOH (1*N* solution)	20.0mL

Preparation of Hemin Solution: Add hemin to 20.0mL of 1*N* NaOH solution. Mix thoroughly. Bring volume to 100.0mL with distilled/deionized water.

Preparation of Medium: Add components, except antibiotic solution, vitamin K_1 and laked sheep blood, to distilled/deionized water and bring volume to 939.0mL. Mix thoroughly. Heat gently and bring to boiling for 1 min. Autoclave for 15 min at 15 psi pressure–121°C. Cool to 50–55°C. Aseptically add the 1.0mL sterile vitamin K_1 solution and 10.0mL sterile antibiotic solution. Mix thoroughly. Aseptically add 50.0mL sterile, defibrinated, laked sheep blood. Laked blood is prepared by freezing whole blood overnight and thawing to room temperature. Mix thoroughly. Pour into sterile Petri dishes.

Use: For the selective isolation of fastidious and slow-growing, obligate anaerobic bacteria from a variety of clinical and nonclinical specimens.

CDC Modified McClung–Toabe Egg Yolk Agar

See: **McClung–Toabe Egg Yolk Agar, CDC Modified**

Cefoperazone Vancomycin Amphotericin Medium

See: **CVA Medium**

Cefsulodin Irgasan® Novobiocin Agar

See: **CIN Agar**

Cell Growth Medium

Composition per 2250mL:

Eagle's minimal esential medium with Hanks' salts (MEMH)	1.0L
L 15 medium, modified Leibovitz	1.0L
Fetal calf serum	200.0mL
$NaHCO_3$ solution	50.0mL

pH 7.5 at 25°C

MEMH:

Composition per liter:

NaCl	8.0g
Glucose	1.0g
KCl	0.4g
$CaCl_2 \cdot 2H_2O$	0.14g
$MgSO_4 \cdot 7H_2O$	0.1g
KH_2PO_4	0.06g
Na_2HPO_4	0.05g
L-Isoleucine	0.026g
L-Leucine	0.026g
L-Lysine	0.026g
L-Threonine	0.024g
L-Valine	0.0235g
L-Tyrosine	0.018g
L-Arginine	0.0174g
L-Phenylalanine	0.0165g
L-Cystine	0.012g
L-Histidine	8.0mg
L-Methionine	7.5mg
Phenol Red	5.0mg
L-Tryptophan	4.0mg
Inositol	1.8mg
Biotin	1.0mg
Folic acid	1.0mg
Calcium pantothenate	1.0mg
Choline chloride	1.0mg
Nicotinamide	1.0mg
Pyridoxal·HCl	1.0mg
Thiamine·HCl	1.0mg
Riboflavin	0.1mg

Preparation of MEMH: Add components to distilled/deionized water and bring volume to 1.0L. Mix thoroughly. Filter sterilize.

L 15 Medium, Modified Leibovitz

Composition per liter:

NaCl	8.0g
DL-Threonine	0.6g
Sodium pyruvate	0.6g
DL-Alanine	0.5g
L-Arginine, free base	0.5g
KCl	0.4g
L-Asparagine·H_2O	0.3g
L-Histidine, free base	0.3g
L-Glutamine	0.3g
L-Isoleucine	0.3g
L-Phenylalanine	0.3g
L-Tyrosine	0.3g
DL-Methionine	0.2g
DL-Valine	0.2g
Glycine	0.2g
L-Serine	0.2g
Na_2HPO_4, anhydrous	0.2g
$CaCl_2$, anhydrous	0.1g

L-Cysteine, free base	0.1g
L-Leucine·HCl	0.1g
Streptomycin	0.1g
$MgCl_2$, anhydrous	0.094g
D-Galactose	0.090g
KH_2PO_4	0.060g
Gentamicin	0.050g
L-Tryptophan	0.020g
Phenol Red	0.010g
i-Inositol	2.0mg
Choline chloride	1.0mg
D-Calcium pantothenate	1.0mg
Folic acid	1.0mg
Nicotinamide	1.0mg
Pyridoxine·HCl	1.0mg
Thiamine monophosphate·$2H_2O$	1.0mg
Riboflavin-5-phosphate	0.1mg
Penicillin G	100,000U

pH 7.5 ± 0.2 at 25°C

Preparation of L 15 Medium, Modified Leibovitz: Add components to distilled/deionized water and bring volume to 1.0L. Mix thoroughly. Filter sterilize. Store at 5°C.

$NaHCO_3$ Solution:
Composition per 100mL:

$NaHCO_3$	7.5g

Preparation of $NaHCO_3$ Solution: Add $NaHCO_3$ to distilled/deionized water and bring volume to 100.0mL. Mix thoroughly. Filter sterilize.

Preparation of Medium: Aseptically combine 1.0L of sterile Eagle's minimal esential medium with Hanks' salts (MEMH) and 1.0L of sterile L 15 medium, modified Leibovitz. Mix thoroughly. Immediately prior to use, aseptically add 200.0mL of fetal calf serum and 50.0mL of $NaHCO_3$ solution. Mix thoroughly. Aseptically distribute into sterile containers.

Use: For the cultivation of mammalian HeLa or Vero tissue culture cells to test the cytopathic effects of *Escherichia coli.*

Cellobiose Arginine Lysine Agar
See: CAL Agar/Broth

Cellobiose Polymyxin Colistin Agar (CPC Agar)

Composition per liter:

Solution A	900.0mL
Solution B	100.0mL

pH 7.6 ± 0.2 at 25°C.

Solution A:
Composition per 900mL:

NaCl	20.0g
Agar	15.0g
Peptone	10.0g
Beef extract	5.0g
Bromthymol Blue	0.04g
Cresol Red	0.04g

Preparation of Solution A: Add components to distilled/deionized water and bring volume to 900.0mL. Mix thoroughly. Adjust pH to 7.6. Gently heat and bring to boiling. Autoclave for 15 min at 15 psi pressure–121°C. Cool to 50°–55°C.

Solution B:
Composition per 100mL:

Cellobiose	15.0g
Colistin	1,360,000U
Polymyxin B	100,000U

Preparation of Solution B: Add components to distilled/deionized water and bring volume to 100.0mL. Mix thoroughly. Filter sterilize.

Preparation of Medium: Aseptically combine 900.0mL of cooled, sterile solution A and 100.0mL of sterile solution B. Mix thoroughly. Pour into sterile Petri dishes. Use within 7 days.

Use: For the cultivation and identification of *Vibrio* species from foods.

Cellobiose Polymyxin B Colistin Agar, Modified

Composition per liter:

Solution 1	900.0mL
Solution 2	100.0mL

pH 7.6 ± 0.2 at 25°C

Solution 1:
Composition per 900mL:

NaCl	20.0g
Agar	15.0g
Peptone	10.0g
Beef extract	5.0g
1000× dye stock solution	1.0mL

Preparation of Solution 1: Add components to distilled/deionized water and bring volume to 900.0mL. Mix thoroughly. Adjust pH to 7.6. Gently heat and bring to boiling. Do not autoclave. Cool to 48°–55°C.

1000X Dye Stock Solution:

Bromthymol Blue	4.0g
Cresol Red	4.0g
Ethanol (95% solution)	100.0mL

Preparation of 1000X Dye Stock Solution: Add Bromthymol Blue and Cresol Red to 100.0mL of ethanol. Mix thoroughly.

Solution 2:

Composition per 100mL:

Cellobiose	10.0g
Colistin	400,000U
Polymyxin B	100,000U

Preparation of Solution 2: Add cellobiose to distilled/deionized water and bring volume to 100.0mL. Mix thoroughly. Gently heat until dissolved. Cool to 25°C. Add colistin and polymyxin B. Mix thoroughly.

Preparation of Medium: Combine cooled solution 1 and solution 2. Mix thoroughly. Do not autoclave. Pour into sterile Petri dishes.

Use: For the cultivation of *Vibrio* species from foods.

Cellulolytic Agar for Thermophiles

Composition per liter:

Agar	30.0g
K_2HPO_4	1.65g
NH_4SO_4	1.6g
Yeast extract	1.0g
NaCl	0.96g
Cysteine·HCl·H_2O	0.5g
$CaCl_2$	0.096g
$MgSO_4$	0.096g
Cellulose suspension	200.0mL
Resazurin (0.1% solution)	1.0mL

pH 7.2 ± 0.2 at 25°C

Cellulose Suspension:

Composition per 200mL:

Cellulose powder, Whatman CF11	8.0g

Preparation of Cellulose Suspension: Add cellulose powder to 200.0mL of distilled/deionized water and mix thoroughly.

Preparation of Medium: Prepare and dispense medium anaerobically in 100% N_2. Add components to distilled/deionized water and bring volume to 1.0L. Mix thoroughly. Adjust pH to 7.2 with 5M NaOH. Distribute into tubes or flasks. Autoclave for 15 min at 15 psi pressure–121°C.

Use: For cultivation of *Clostridium stercorarium* and other bacteria which can utilize cellulose as a carbon source.

Cellulolytic Agar with Sea Salts

Composition per liter:

Agar	20.0g
NH_4Cl	2.0g
K_2HPO_4	1.65g
Yeast extract	1.2g
Cysteine·HCl·H_2O	0.5g
Cellulose suspension	200.0mL
Filtered sea water	200.0mL
Mineral solution	150.0mL
Resazurin (0.1% solution)	1.0mL

pH 7.2 ± 0.2 at 25°C

Cellulose Suspension:

Composition per 200mL:

Cellulose powder, Whatman CF11	8.0g

Preparation of Cellulose Suspension: Add cellulose powder to 200.0mL of distilled/deionized water and mix thoroughly.

Mineral Solution:

Composition per liter:

NaCl	6.0g
$(NH_4)_2SO_4$	6.0g
$CaCl_2$	0.6g
$MgSO_4$	0.6g

Preparation of Mineral Solution: Add components to distilled/deionized water and bring volume to 1.0L. Mix thoroughly.

Preparation of Medium: Prepare and dispense medium anaerobically under 100% N_2. Add components to distilled/deionized water and bring volume to 1.0L. Mix thoroughly. Adjust pH to 7.2 with 5M NaOH. Distribute into tubes or flasks. Autoclave for 15 min at 15 psi pressure–121°C.

Use: For cultivation and maintenance of *Clostridium papyrosolvens* and other marine bacteria which can utilize cellulose as a carbon source.

Cellulolytic Broth for Thermophiles

Composition per liter:

K_2HPO_4	1.65g
NH_4SO_4	1.6g
Yeast extract	1.0g
NaCl	0.96g
Cysteine·HCl·H_2O	0.5g
$CaCl_2$	0.096g
$MgSO_4$	0.096g
Resazurin (0.1% solution)	1.0mL

pH 7.2 ± 0.2 at 25°C

Preparation of Medium: Prepare and dispense medium anaerobically in 100% N_2. Add components to distilled/deionized water and bring volume to 1.0L. Mix thoroughly. Adjust pH to 7.2 with 5M NaOH. Distribute into tubes or flasks which contain cellulose as a strip (4.5cm × 1.0cm) of Whatman No. 1 filter paper. Autoclave for 15 min at 15 psi pressure–121°C.

Use: For cultivation of *Clostridium stercorarium* and other bacteria which can utilize cellulose as a carbon source.

Cellulolytic Broth with Sea Salts

Composition per liter:

K_2HPO_4	1.65g
NH_4Cl	1.0g
Yeast extract	0.6g
Cysteine·HCl·H_2O	0.5g
Filtered sea water	200.0mL
Mineral solution	150.0mL
Resazurin (0.1% solution)	1.0mL

pH 7.2 ± 0.2 at 25°C

Mineral Solution:

Composition per liter:

NaCl	6.0g
$(NH_4)_2SO_4$	6.0g
$CaCl_2$	0.6g
$MgSO_4$	0.6g

Preparation of Mineral Solution: Add components to distilled/deionized water and bring volume to 1.0L. Mix thoroughly.

Preparation of Medium: Prepare and dispense medium anaerobically in 100% N_2 atmosphere. Add components to distilled/deionized water and bring volume to 1.0L. Adjust pH to 7.2 with 5 M NaOH. Distribute into tubes or flasks which contain cellulose as a strip (4.5cm × 1.0cm) of Whatman No. 1 filter paper. Autoclave for 15 min at 15 psi pressure–121°C.

Use: For cultivation and maintenance of *Clostridium papyrosolvens* and other marine bacteria which can utilize cellulose as a carbon source.

Cellulolytic Clostridia Medium

Composition per liter:

Cellulose	20.0g
$CaCO_3$	2.0g
K_2HPO_4	1.0g
$(NH_4)_2SO_4$	1.0g
$MgSO_4·7H_2O$	0.5g
NaCl	0.5g
Resazurin	1.0mg

pH 7.1 ± 0.2 at 25°C

Preparation of Medium: Add components to distilled/deionized water and bring volume to 1.0L. Mix thoroughly. Distribute into tubes or flasks. Autoclave for 15 min at 15 psi pressure–121°C.

Use: For the isolation, cultivation and enrichment of cellulolytic *Clostridium* species.

Cellulolytic Medium with Rumen Fluid

Composition per liter:

Basal medium	975.0mL
Alkaline solution	25.0mL

pH 6.8 ± 0.2 at 25°C

Basal Medium:

Composition per 975mL:

Agar	15.0g
$NaHCO_3$	6.37g
Pancreatic digest of casein	5.0g
Cellobiose	5.0g
NaCl	0.90g
$(NH_4)_2SO_4$	0.90g
K_2HPO_4	0.45g
KH_2PO_4	0.45g
$MgSO_4·7H_2O$	0.18g
$CaCl_2$	0.09g
Resazurin	1.0mg
Rumen fluid, clarified	400.0mL

Preparation of Basal Medium: Add components to distilled/deionized water and bring volume to 975.0mL. Mix thoroughly. Gently heat and bring to boiling under a gas phase of 98% CO_2 + 2% H_2. Cool slightly.

Alkaline Solution:

Composition per 25mL:

Cysteine·HCl·H_2O	0.25g
$Na_2S·9H_2O$	0.25g

Preparation of Alkaline Solution: Add components to 25.0mL of distilled/deionized water. Mix thoroughly. Prepare freshly.

Preparation of Medium: Prepare 975.0mL of basal medium. Heat to boiling and cool as directed. Add 25.0mL of freshly prepared alkaline solution. Distribute into tubes using anaerobic techniques under a gas phase of 98% CO_2 + 2% H_2. Autoclave for 15 min at 15 psi pressure–121°C. Adjust pH to 6.8.

Use: For cultivation and maintenance of *Clostridium polysaccharolyticum.*

Cellulolytic Medium with Rumen Fluid and Soluble Starch

Composition per liter:

Basal medium	975.0mL
Alkaline solution	25.0mL

pH 6.8 ± 0.2 at 25°C

Basal Medium:

Composition per 975mL:

Agar	15.0g
$NaHCO_3$	6.37g
Pancreatic digest of casein	5.0g
Cellobiose	5.0g
Soluble starch	5.0g
NaCl	0.90g
$(NH_4)_2SO_4$	0.90g
K_2HPO_4	0.45g
KH_2PO_4	0.45g
$MgSO_4 \cdot 7H_2O$	0.18g
$CaCl_2$	0.09g
Resazurin	1.0mg

Preparation of Basal Medium: Add components to distilled/deionized water and bring volume to 975.0mL. Mix thoroughly. Gently heat and bring to boiling under a gas phase of 98% CO_2 + 2% H_2. Cool slightly.

Alkaline Solution:

Composition per 25mL:

Cysteine·HCl·H_2O	0.25g
$Na_2S \cdot 9H_2O$	0.25g

Preparation of Alkaline Solution: Add components to 25.0mL of distilled/deionized water. Mix thoroughly. Prepare freshly.

Preparation of Medium: Prepare 975.0mL of basal medium. Heat to boiling and cool as directed. Add 25.0mL of freshly prepared alkaline solution. Distribute into tubes using anaerobic techniques under a gas phase of 98% CO_2 + 2% H_2. Autoclave for 15 min at 15 psi pressure–121°C. Adjust pH to 6.8.

Use: For cultivation of *Selenomonas ruminantium* and *Succinimonas amylolytica.*

Cellulomonas PTYG Medium (*Cellulomonas* Peptone Tryptone Yeast Extract Glucose Medium)

Composition per liter:

Agar	15.0g
Glucose	5.0g
Peptone	5.0g
Pancreatic digest of casein	5.0g
Yeast extract	5.0g

Preparation of Medium: Add components to distilled/deionized water and bring volume to 1.0L. Mix thoroughly. Gently heat and bring to boiling. Distribute into tubes or flasks. Autoclave for 15 min at 15 psi pressure–121°C. Pour into sterile Petri dishes or leave in tubes.

Use: For cultivation and maintenance of *Cellulomonas* species.

Cellulose Broth

Composition per liter:

Cellulose, powdered	1.0g
K_2HPO_4	1.0g
$(NH_4)_2SO_4$	1.0g
$MgSO_4 \cdot 7H_2O$	0.2g
$CaCl_2 \cdot 2H_2O$	0.1g
$FeCl_3$	0.02g

pH 7.0–7.5 at 25°C

Preparation of Medium: Add cellulose to 100.0mL of distilled/deionized water. Mix thoroughly. In a separate flask, add remaining components to distilled/deionized water and bring volume to 900.0mL. Mix thoroughly. Autoclave both solutions separately for 15 min at 15 psi pressure–121°C. Cool to 45°–50°C. Aseptically combine the two sterile solutions. Mix thoroughly. Aseptically distribute into sterile tubes or flasks.

Use: For the isolation and cultivation of *Cytophaga* species, *Herpetosiphon* species, *Saprospira* species, and *Flexithrix* species.

Cellulose Overlay Agar

Composition per plate:

Stan 5 agar	15.0mL
Cellulose overlay agar	5.0mL

Stan 5 Agar:

Composition per liter:

Solution B	650.0mL
Solution A	350.0mL

Solution A:

Composition per 350mL:

$CaCl_2 \cdot 2H_2O$	1.0g
$(NH_4)_2SO_4$	1.0g
$MgSO_4 \cdot 7H_2O$	1.0g
Trace element solution	1.0mL

Preparation of Solution A: Add components to distilled/deionized water and bring volume to 350.0mL. Mix thoroughly. Gently heat and bring to boiling. Autoclave for 15 min at 15 psi pressure–121°C. Cool to 45°–50°C.

Trace Element Solution:

Composition per liter:

EDTA	8.0g
$MnCl_2 \cdot 4H_2O$	0.1g
$CoCl_2$	0.02g
KBr	0.02g
$ZnCl_2$	0.02g
$CuSO_4$	0.01g
H_3BO_3	0.01g
$NaMoO_4 \cdot 2H_2O$	0.01g
$BaCl_2$	5.0mg
LiCl	5.0mg
$SnCl_2 \cdot 2H_2O$	5.0mg

Preparation of Trace Element Solution: Add components to distilled/deionized water and bring volume to 1.0L. Mix thoroughly.

Solution B:

Composition per 650mL:

Agar	10.0g
K_2HPO_4	1.0g

Preparation of Solution B: Add components to distilled/deionized water and bring volume to 650.0mL. Mix thoroughly. Gently heat and bring to boiling. Autoclave for 15 min at 15 psi pressure–121°C. Cool to 45°–50°C.

Preparation of Stan 5 Agar: Aseptically combine 350.0mL of cooled, sterile solution A and 650.0mL of cooled sterile solution B. Mix thoroughly.

Cellulose Overlay Agar:

Composition per liter:

Solution A	350.0mL
Solution B	650.0mL

Solution A:

Composition per 350mL:

$CaCl_2 \cdot 2H_2O$	1.0g
$(NH_4)_2SO_4$	1.0g
$MgSO_4 \cdot 7H_2O$	1.0g
Trace element solution	1.0mL

Preparation of Solution A: Add components to distilled/deionized water and bring volume to 350.0mL. Mix thoroughly. Gently heat and bring to boiling. Autoclave for 15 min at 15 psi pressure–121°C. Cool to 45°–50°C.

Trace Element Solution:

Composition per liter:

EDTA	8.0g
$MnCl_2 \cdot 4H_2O$	0.1g
$CoCl_2$	0.02g
KBr	0.02g
$ZnCl_2$	0.02g
$CuSO_4$	0.01g
H_3BO_3	0.01g
$NaMoO_4 \cdot 2H_2O$	0.01g
$BaCl_2$	5.0mg
LiCl	5.0mg
$SnCl_2 \cdot 2H_2O$	5.0mg

Preparation of Trace Element Solution: Add components to distilled/deionized water and bring volume to 1.0L. Mix thoroughly.

Solution B:

Composition per 650mL:

Agar	10.0g
K_2HPO_4	1.0g

Preparation of Solution B: Add components to distilled/deionized water and bring volume to 650.0mL. Mix thoroughly. Gently heat and bring to boiling. Autoclave for 15 min at 15 psi pressure–121°C. Cool to 45°–50°C.

Preparation of Cellulose Overlay Agar: Aseptically combine 350.0mL of cooled, sterile solution A and 650.0mL of cooled sterile solution B. Mix thoroughly.

Preparation of Medium: Pour cooled sterile Stan 5 agar into sterile Petri dishes in 15.0mL volumes. Allow agar to solidify. Overlay each plate with 5.0mL of cellulose overlay agar.

Use: For the cultivation of myxobacteria.

Centenum Medium

Composition per liter of tap water:

Agar	20.0g
Yeast extract	10.0g
Sodium pyruvate	2.2g
K_2HPO_4	1.0g
$MgSO_4$	0.5g
Vitamin B_{12}	0.02mg

pH 7.0–7.2 at 25°C

Preparation of Medium: Add components to distilled/deionized water and bring volume to 1.0L. Mix thoroughly. Gently heat and bring to boiling. Distribute into tubes or flasks. Autoclave for 15 min at 15 psi pressure–121°C. Pour into sterile Petri dishes or leave in tubes.

Use: For cultivation and maintenance of *Rhodospirillum* species.

Cephalothin Cycloheximide Vancomycin Colistin Medium

See: **CCVC Medium**

Cereal Agar

Composition per liter:

Cereal, precooked mixed	100.0g
Agar	15.0g

Preparation of Medium: Add components to distilled/deionized water and bring volume to 1.0L. Mix thoroughly. Gently heat and bring to boiling. Autoclave for 15 min at 15 psi pressure–121°C. Pour into sterile Petri dishes or distribute into sterile tubes. Allow tubes to cool in a slanted position.

Use: For the cultivation and sporulation of fungi.

Cetrimide Agar, Non-USP

Composition per liter:

Beef heart, solids from infusion	500.0g
Agar	15.0g
Tryptose	10.0g
NaCl	5.0g
Cetrimide	0.9g

pH 7.2 ± 0.2 at 25°C

Preparation of Medium: Add components to distilled/deionized water and bring volume to 1.0L. Mix thoroughly. Gently heat and bring to boiling. Distribute into tubes or flasks. Autoclave for 15 min at 13 psi pressure–118°C. Pour into sterile Petri dishes or leave in tubes.

Use: For the selective isolation, cultivation and identification of *Pseudomonas aeruginosa* and other Gram-negative nonfermentative bacteria.

Cetrimide Agar, USP (Pseudosel® Agar)

Composition per liter:

Pancreatic digest of gelatin	20.0g
Agar	13.6g
K_2SO_4	10.0g
$MgCl_2$	1.4g
Cetrimide	0.3g
Glycerol	10.0mL

pH 7.2 ± 0.2 at 25°C

Source: This medium is available as a premixed powder from Difco Laboratories and BBL Microbiology Systems.

Preparation of Medium: Add components to distilled/deionized water and bring volume to 1.0L. Mix thoroughly. Gently heat and bring to boiling. Distribute into tubes or flasks. Autoclave for 15 min at 13 psi pressure–118°C. Pour into sterile Petri dishes or leave in tubes.

Use: For the selective isolation, cultivation and identification of *Pseudomonas aeruginosa* and other Gram-negative nonfermentative bacteria.

CF Assay Medium (Citrovorum Factor Assay Medium)

Composition per liter:

Glucose	50.0g
Sodium acetate	40.0g
Vitamin assay casamino acids	10.0g
NH_4Cl	6.0g
K_2HPO_4	1.2g
KH_2PO_4	1.2g
$MgSO_4·7H_2O$	0.4g
DL-Alanine	0.2g
DL-Tryptophan	0.2g
L-Cystine	0.2g
L-Cystine·HCl	0.2g
$MgSO_4·7H_2O$	0.04g
Adenine sulfate	0.02g
$FeSO_4$	0.02g
Glycine	0.02g
Guanine·HCl	0.02g
NaCl	0.02g
Uracil	0.02g
Xanthine	0.02g
Pyridoxamine·HCl	6.0mg
Nicotinic acid	2.0mg
Pyridoxine·HCl	2.0mg
Calcium pantothenate	1.0mg
Riboflavin	1.0mg
Thiamine·HCl	1.0mg
Pyridoxal·HCl	600.0μg
p-Aminobenzoic acid	200.0μg
Folic acid	20.0μg
Biotin	2.0μg

pH 6.7 ± 0.2 at 25°C

Preparation of Medium: Add components to distilled/deionized water and bring volume to 1.0L. Mix thoroughly. Gently heat and bring to boiling. Continue boiling 2–3 min. Allow precipitate to settle out. Distribute supernatant into tubes in 5.0mL volumes. Add standard solution or test solutions to each tube. Adjust the volume of each tube to 10.0mL with distilled/deionized water. Autoclave for 10 min at 15 psi pressure–121°C.

Use: For the microbiological assay of citrovorum factor using *Pediococcus acidilactici*.

CFAT Medium (Cadmium Fluoride Acriflavin Tellurite Medium)

Composition per liter:

Pancreatic digest of casein	17.0g
Agar	15.0g
Glucose	7.5g
NaCl	5.0g
Papaic digest of soybean meal	3.0g
K_2HPO_4	2.5g
NaF	0.8g
$CdSO_4$	0.013g
K_2TeO_3	2.5mg
Neutral acriflavin	1.2mg
Basic Fuchsin	0.25mg
Sheep blood, defibrinated	50.0mL

Caution: Potassium tellurite is toxic.

Preparation of Medium: Add components, except sheep blood, to distilled/deionized water and bring volume to 950.0mL. Mix thoroughly. Gently heat and bring to boiling. Autoclave for 15 min at 15 psi pressure–121°C. Cool to 45°–50°C. Add 50.0mL sterile, defibrinated sheep blood. Mix thoroughly. Pour into sterile Petri dishes or leave in tubes.

Use: For isolation, cultivation and enumeration of *Actinomyces viscosus* and *Actinomyces naeslundii* from clinical specimens, especially dental plaque.

CGY Autolysate Broth *See:* Casitone Glycerol Yeast Autolysate Broth

CH 1 Medium

Composition per liter:

NaCl	250.0g
Tris	12.0g
Glycerol	10.0g
Hy-Case SF	5.0g
Yeast extract	5.0g
Solution 1	50.0mL

Solution 1:

Composition per liter:

$MgCl_2 \cdot 6H_2O$	40.0g
KCl	4.0g
$CaCl_2 \cdot 2H_2O$	0.4g

pH 7.4 ± 0.2 at 25°C

Preparation of Solution 1: Add components to distilled/deionized water and bring volume to 1.0L. Mix thoroughly.

Preparation of Medium: Add components to distilled/deionized water and bring volume to 1.0L. Mix thoroughly. Adjust pH to 7.4. Distribute into tubes or flasks. Autoclave for 15 min at 15 psi pressure–121°C.

Use: For the cultivation of *Haloarcula vallismortis*.

Chalquist's Antigen Medium, Modified

Composition:

Soluble starch	0.5g
Pancreatic digest of casein	0.05g
Cysteine·HCl·H_2O	0.01g
NAD (nicotinamide adeninedinucleotide)	0.01g
PPLO broth without Crystal Violet	90.0mL
Swine serum, inactivated	10.0mL
Phenol Red (1% solution)	0.25mL

pH 7.6 ± 0.2 at 25°C

PPLO Broth without Crystal Violet:

Composition per 500mL:

Beef heart, infusion from	11.52g
Peptone	2.32g
NaCl	1.15g

Source: PPLO broth without Crystal Violet is available as a premixed powder from Difco Laboratories.

Preparation of PPLO Broth without Crystal Violet: Add components to distilled/deionized water and bring volume to 500.0mL. Mix thoroughly.

Preparation of Medium: Add components to distilled/deionized water and bring volume to 1.0L. Mix thoroughly. Distribute into tubes or flasks. Autoclave for 15 min at 15 psi pressure–121°C.

Use: For the cultivation of *Mycoplasma synoviae*.

Chapman Stone Agar

Composition per liter:

$(NH_4)_2SO_4$	75.0g
NaCl	55.0g
Gelatin	30.0g
Agar	15.0g
D-Mannitol	10.0g
Pancreatic digest of casein	10.0g
K_2HPO_4	5.0g
Yeast extract	2.0g

pH 7.0 ± 0.2 at 25°C

Source: This medium is available as a premixed powder from BBL Microbiology Systems and Difco Laboratories.

Preparation of Medium: Add components to distilled/deionized water and bring volume to 1.0L. Mix

thoroughly. Autoclave for 10 min at 15 psi pressure–121°C. Pour into sterile Petri dishes while the medium is still hot. Add 25.0mL of medium per Petri dish.

Use: For the isolation of staphylococci from a variety of specimens.

Charcoal Agar

Composition per liter:

Beef heart, solids from infusion	500.0g
Agar	18.0g
Peptone	10.0g
Soluble starch	10.0g
NaCl	5.0g
Charcoal, activated, acid-washed	4.0g
Yeast extract	3.5g

pH 7.3 ± 0.2 at 25°C

Source: This medium is available as a premixed powder from Difco Laboratories.

Preparation of Medium: Add components to distilled/deionized water and bring volume to 1.0L. Mix thoroughly. Gently heat and bring to boiling with frequent stirring. Autoclave for 15 min at 15 psi pressure–121°C. Cool to 45°–50°C. Pour into sterile Petri dishes or leave in tubes. Shake flask while dispensing to keep charcoal in suspension. Allow tubes to cool in a slanted position.

Use: For the cultivation and maintenance of fastidious microorganisms, especially *Bordetella pertussis*, for production of vaccines.

Charcoal Agar

Composition per liter:

Agar	12.0g
Beef extract	10.0g
Peptone	10.0g
Starch	10.0g
NaCl	5.0g
Charcoal	4.0g
Nicotinic acid	0.001g

pH 7.4 ± 0.2 at 25°C

Source: This medium is available as a premixed powder from Oxoid Unipath.

Preparation of Medium: Add components to distilled/deionized water and bring volume to 1.0L. Mix thoroughly. Gently heat and bring to boiling with frequent stirring. Autoclave for 15 min at 15 psi pressure–121°C. Cool to 45°–50°C. This medium may be enriched by the addition of blood. Pour into sterile Petri dishes or distribute into tubes. Shake flask while dispensing to keep charcoal in suspension.

Use: For the cultivation and isolation of various bacteria; with the addition of blood for the cultivation of fastidious bacteria.

Charcoal Agar Slants
See: **Diphasic Medium for Amoeba**

Charcoal Agar with Horse Blood

Composition per liter:

Agar	12.0g
Beef extract	10.0g
Peptone	10.0g
Starch	10.0g
NaCl	5.0g
Charcoal, bacteriological	4.0g
Nicotinic acid	1.0mg
Horse blood, defibrinated	100.0mL

pH 7.4 ± 0.2 at 25°C

Preparation of Medium: Add components to distilled/deionized water and bring volume to 900.0L. Mix thoroughly. Gently heat and bring to boiling with frequent stirring. Autoclave for 15 min at 15 psi pressure–121°C. Cool to 80°C. Aseptically add 100.0mL of sterile, defibrinated horse blood. Maintain at 80°C for 10 min to form chocolate agar. Pour into sterile Petri dishes or distribute into tubes. Shake flask while dispensing to keep charcoal in suspension.

Use: For the cultivation and isolation of *Haemophilus influenzae*.

Charcoal Agar with Horse Blood and Cephalexin

Composition per liter:

Agar	12.0g
Beef extract	10.0g
Peptone	10.0g
Starch	10.0g
NaCl	5.0g
Charcoal	4.0g
Nicotinic acid	1.0mg
Horse blood, defibrinated	100.0mL
Cephalexin solution	10.0mL

pH 7.4 ± 0.2 at 25°C

Cephalexin Solution:
Composition per 10mL:

Cephalexin	0.04g

Preparation of Cephalexin Solution: Add cephalexin to distilled/deionized water and bring volume to 10.0mL. Mix thoroughly. Filter sterilize.

Preparation of Medium: Add components, except cephalexin solution and horse blood, to distilled/deionized water and bring volume to 890.0L. Mix thoroughly. Gently heat and bring to boiling with frequent stirring. Autoclave for 15 min at 15 psi pressure–121°C. Cool to 45°–50°C. Aseptically add 100.0mL of sterile, defibrinated horse blood and 10.0mL sterile cephalexin solution. Pour into sterile Petri dishes or distribute into tubes. Shake flask while dispensing to keep charcoal in suspension.

Use: For the cultivation and isolation of *Bordetella pertussis*.

Charcoal Blood Medium

Composition per liter:

Beef heart, solids from infusion	500.0g
Agar	18.0g
Peptone	10.0g
Soluble starch	10.0g
NaCl	5.0g
Charcoal, activated, acid-washed	4.0g
Yeast extract	3.5g
Horse or sheep blood, defibrinated	100.0mL
Cephalexin solution	10.0mL

pH 7.4 ± 0.2 at 25°C

Cephalexin Solution:
Composition per 10mL:

Cephalexin	0.04g

Preparation of Cephalexin Solution: Add cephalexin to distilled/deionized water and bring volume to 10.0mL. Mix thoroughly. Filter sterilize.

Preparation of Medium: Add components, except blood and cephalexin solution, to distilled/deionized water and bring volume to 890.0mL. Mix thoroughly. Gently heat and bring to boiling. Autoclave for 15 min at 15 psi pressure–121°C. Cool to 45°–50°C. Aseptically add sterile blood and cephalexin solution. Mix thoroughly. Pour into sterile Petri dishes or distribute into sterile tubes.

Use: For the cultivation of *Haemophilus influenzae*.

Charcoal Yeast Extract Agar
See: **CYE Agar**

Charcoal Yeast Extract Agar, Buffered
See: **CYE Agar, Buffered**

Charcoal Yeast Extract Diphasic Blood Culture Medium
*See: **Legionella pneumophila*** **Medium**

Chase's Medium SP

Composition per liter:

Agar	10.0g
Proteose peptone	10.0g
K_2HPO_4	2.0g
$(NH_4)_2SO_4$	1.0g
Sucrose solution	100.0mL

pH 6.5 ± 0.2 at 25°C

Sucrose Solution:
Composition per 100mL:

Sucrose	10.0g

Preparation of Sucrose Solution: Add sucrose to distilled/deionized water and bring volume to 100.0mL. Mix thoroughly. Filter sterilize.

Preparation of Medium: Add components, except sucrose solution, to tap water and bring volume to 900.0mL. Mix thoroughly. Gently heat and bring to boiling. Autoclave for 15 min at 15 psi pressure–121°C. Cool to 45°–50°C. Aseptically add 100.0mL of sterile sucrose solution. Mix thoroughly. Pour into sterile Petri dishes or distribute into sterile tubes.

Use: For the cultivation and maintenance of ATCC strain 13949.

CHCA Salts Medium (Cyclohexane Carboxylic Acid Salts Medium)

Composition per liter:

K_2HPO_4	3.5g
KH_2PO_4	1.5g
Cyclohexane carboxylic acid	1.0g
NH_4NO_3	1.0g
$MgSO_4 \cdot 7H_2O$	0.5g
$FeSO_4 \cdot 7H_2O$	0.1g
Yeast extract	0.1g
$CaCl_2 \cdot 2H_2O$	0.01g
$Na_2MoO_2 \cdot 2H_2O$	0.01g
$ZnSO_4 \cdot 7H_2O$	0.01g

pH 7.0 ± 0.2 at 25°C

Preparation of Medium: Add components to distilled/deionized water and bring volume to 1.0L. Mix thoroughly. Adjust pH to 7.0. Gently heat and bring to boiling. Distribute into tubes or flasks. Autoclave for 15 min at 15 psi pressure–121°C. Pour into sterile Petri dishes or leave in tubes.

Use: For the cultivation and maintenance of bacteria that can utilize cyclohexane carboxylic acid as a carbon source. For the cultivation and maintenance of *Arthrobacter globiformis*.

Cheese Agar

Composition per liter:

Cheese, ripened	100.0g
NaCl	50.0g
Agar	15.0g
Peptone	10.0g
Potassium citrate	10.0g
Sodium oxalate	2.0g

pH 7.4 ± 0.2 at 25°C

Preparation of Medium: Add the cheese and potassium citrate to distilled/deionized water and bring volume to 300.0mL. Gently heat and bring to 50°C to separate the fat. Discard the fat. In a separate flask add the remaining components to distilled/deionized water and bring volume to 700.0mL. Gently heat and bring to boiling. Add the 300.0mL of aqueous suspension of cheese solids. Adjust pH to 7.4. Autoclave for 25 min at 15 psi pressure–121°C. Pour into sterile Petri dishes or distribute into sterile tubes.

Use: For the isolation and cultivation of *Brevibacterium linens*.

Chicken Soup Broth

Composition per 5L:

Chicken	2.5kg
Peppercorns	6
Cloves	3
Bay leaf	2
Celery, stalks including leaves	2
Onion, large	1
Carrot	1
Dill, fresh	1/4 cup
NaCl	0.1g

Preparation of Medium: Add a nice, whole chicken to a large pot. Add enough tap water to cover the chicken by about 1 in. Stud the whole, peeled onion with the three cloves. Add the onion and remaining ingredients to the pot. Rapidly heat and bring to boiling. Lower heat to a simmer and cook for 1 to 1.5 hr. Remove the chicken and vegetables from the broth. Remove skin and bones from the chicken. Cut up the meat into 1 inch pieces. Return the meat to the broth. If desired, slice the carrot and celery and return them to the broth.

Use: For the growth and nutrition of microbiologists.

China Blue Lactose Agar

Composition per liter:

Agar	12.0g
Lactose	10.0g
Peptone	5.0g
NaCl	5.0g
Beef extract	3.0g
China Blue	0.375g

pH 7.0 ± 0.2 at 25°C

Source: This medium is available as a premixed powder from Oxoid Unipath.

Preparation of Medium: Add components to distilled/deionized water and bring volume to 1.0L. Mix thoroughly. Gently heat and bring to boiling. Distribute into tubes or flasks. Autoclave for 15 min at 15 psi pressure–121°C. Pour into sterile Petri dishes or leave in tubes.

Use: For the cultivation, differentiation and enumeration of bacteria from dairy products. Lactose-fermenting bacteria appear as blue colonies. Non-lactose fermenting bacteria appear as colorless colonies.

Chitin Agar

Composition per liter:

Agar	15.0g
Chitin, precipitated	3.0g
$(NH_4)_2SO_4$	2.0g
Na_2HPO_4	1.1g
KH_2PO_4	0.7g
$MgSO_4 \cdot 7H_2O$	0.2g
$FeSO_4$	1.0mg
$MnSO_4$	1.0mg

Chitin, Precipitated:

Composition:

Chitin	40.0g
HCl, concentrated	400.0mL

Preparation of Chitin, Precipitated: Add chitin to 400.0mL of cold concentrated HCl. Add this solution to 2.0L of distilled/deionized water at 5°C. Filter the solution through Whatman #1 filter paper. Dialyze the precipitated chitin against tap water for 12 hr. Adjust the pH to 7.0 with KOH.

Preparation of Medium: Add components to distilled/deionized water and bring volume to 1.0L. Mix thoroughly. Gently heat and bring to boiling. Distribute into tubes or flasks. Autoclave for 15 min at 15 psi pressure–121°C. Pour into sterile Petri dishes or leave in tubes.

Use: For the isolation and cultivation of *Cytophaga* species, *Herpetosiphon* species, *Saprospira* species, and *Flexithrix* species.

Chitin Agar

Composition per liter:

Agar	20.0g
Chitin	4.0g
K_2HPO_4	0.7g
$MgSO_4 \cdot 7H_2O$	0.5g
KH_2PO_4	0.3g
$FeSO_4 \cdot 7H_2O$	0.01g
$MnCl_2 \cdot 4H_2O$	0.001g
$ZnSO_4 \cdot 7H_2O$	0.001g

pH 8.0 ± 0.2 at 25°C

Preparation of Medium: Add components to distilled/deionized water and bring volume to 1.0L. Mix thoroughly. Gently heat and bring to boiling. Distribute into tubes or flasks. Autoclave for 15 min at 15 psi pressure–121°C. Pour into sterile Petri dishes or leave in tubes.

Use: For the selective isolation and cultivation of streptomycetes.

Chlamydia Growth Medium

Composition per 500mL:

Eagle minimum essential medium with Earle salts, 10X	50.0mL
Fetal calf serum	50.0mL
L-Glutamine solution	5.0mL

pH 7.4 ± 0.2 at 25°C

Eagle Minimum Essential Medium with Earle Salts, 10X:

Composition per liter:

NaCl	6.8g
Glucose	1.0g
KCl	0.4g
$CaCl_2 \cdot 2H_2O$	0.2g
$MgCl_2 \cdot 6H_2O$	0.2g
NaH_2PO_4	0.15g
L-Arginine	0.10g
L-Lysine	0.06g
L-Isoleucine	0.05g
L-Leucine	0.05g
L-Threonine	0.05g
L-Valine	0.05g
L-Tyrosine	0.04g
L-Phenylalanine	0.03g
L-Histidine	0.03g
L-Cystine	0.02g
L-Methionine	0.02g
L-Tryptophan	0.01g
i-Inositol	2.0mg
Calcium pantothenate	1.0mg
Choline chloride	1.0mg
Folic acid	1.0mg
Nicotinamide	1.0mg
Pyridoxal	1.0mg
Thiamine·HCl	1.0mg
Riboflavin	0.1mg

Preparation of Eagle Minimum Essential Medium With Earle Salts, 10X: Add components to distilled/deionized water and bring volume to 1.0L. Mix thoroughly. Adjust pH to 7.4 with 7.5% Na_2CO_3 solution. Filter sterilize.

Glutamine Solution:

Composition per 100mL:

L-Glutamine	2.92g
NaCl (0.85% solution)	100.0mL

Preparation of Glutamine Solution: Add the glutamine to the 0.85% NaCl solution. Mix thoroughly. Filter sterilize.

Preparation of Medium: Aseptically combine 50.0mL of sterile Eagle minimum essential medium with Earle salts, 10X, 50.0mL of fetal calf serum and 5.0mL of sterile glutamine solution. Bring volume to 500.0mL with sterile distilled/deionized water. Mix thoroughly. Aseptically distribute into sterile tubes or flasks.

Use: For the cultivation of *Chlamydia* species.

Chlamydia Isolation Medium

Composition per 500mL:

Eagle minimum essential medium with Earle salts, 10X	50.0mL
Fetal calf serum	50.0mL
Selective supplement	10.0mL
L-Glutamine solution	5.0mL

pH 7.4 ± 0.2 at 25°C

Eagle Minimum Essential Medium with Earle Salts, 10X:

Composition per liter:

NaCl	6.8g
Glucose	1.0g
KCl	0.4g
$CaCl_2 \cdot 2H_2O$	0.2g
$MgCl_2 \cdot 6H_2O$	0.2g
NaH_2PO_4	0.15g
L-Arginine	0.10g
L-Lysine	0.06g
L-Isoleucine	0.05g
L-Leucine	0.05g
L-Threonine	0.05g
L-Valine	0.05g
L-Tyrosine	0.04g
L-Phenylalanine	0.03g
L-Histidine	0.03g
L-Cystine	0.02g
L-Methionine	0.02g

L-Tryptophan	0.01g
i-Inositol	2.0mg
Calcium pantothenate	1.0mg
Choline chloride	1.0mg
Folic acid	1.0mg
Nicotinamide	1.0mg
Pyridoxal	1.0mg
Thiamine·HCl	1.0mg
Riboflavin	0.1mg

Preparation of Eagle Minimum Essential Medium With Earle Salts, 10X: Add components to distilled/deionized water and bring volume to 1.0L. Mix thoroughly. Adjust pH to 7.4 with 7.5% Na_2CO_3 solution. Filter sterilize.

Selective Supplement:
Composition per 10mL:

Glucose	0.594g
Vancomycin	0.050g
Gentamicin	0.010g
Amphotericin B	2.0mg
Cycloheximide	2.0mg

Preparation of Selective Supplement: Add components to distilled/deionized water and bring volume to 10.0mL. Mix thoroughly. Filter sterilize.

Glutamine Solution:
Composition per 100mL:

L-Glutamine	2.92g
NaCl (0.85% solution)	100.0mL

Preparation of Glutamine Solution: Add the glutamine to the 0.85% NaCl solution. Mix thoroughly. Filter sterilize.

Preparation of Medium: Aseptically combine 50.0mL of sterile Eagle minimum essential medium with Earle salts, 10X, 50.0mL of fetal calf serum, 10.0mL of selective supplement and 5.0mL of sterile glutamine solution. Bring volume to 500.0mL with sterile distilled/deionized water. Mix thoroughly. Aseptically distribute into sterile tubes or flasks.

Use: For the isolation and cultivation of *Chlamydia* species.

Chlamydospore Agar

Composition per liter:

Purified polysaccharide	20.0g
Agar	15.0g
KH_2PO_4	1.0g
$(NH_4)_2SO_4$	1.0g
Trypan Blue	0.1g
Biotin	5.0μg

pH 5.1 ± 0.2 at 25°C

Source: This medium is available as a premixed powder from Difco Laboratories.

Preparation of Medium: Add components to distilled/deionized water and bring volume to 1.0L. Mix thoroughly. Gently heat and bring to boiling. Distribute into tubes or flasks. Autoclave for 15 min at 15 psi pressure–121°C. Pour into sterile Petri dishes or leave in tubes.

Use: For differentiating *Candida albicans* from other *Candida* species of on the basis of chlamydospore formation.

Chlorobiaceae Medium 1

Composition per 4990mL:

Solution 1	4.0L
O_2-free water	860.0mL
$NaHCO_3$ solution	100.0mL
$Na_2S{\cdot}9H_2O$ solution	20.0mL
Trace element solution	5.0mL
Vitamin B_{12} solution	5.0mL

pH 6.8 ± 0.2 at 25°C

Solution 1:
Composition per 4L:

$MgSO_4{\cdot}7H_2O$	2.5g
KCl	1.7g
KH_2PO_4	1.7g
NH_4Cl	1.7g
$CaCl_2{\cdot}2H_2O$	1.25g

Preparation of Solution 1: Add components to distilled/deionized water and bring volume to 4.0L. Mix thoroughly. Autoclave for 45 min at 15 psi pressure–121°C. Cool to 25°C under 100% N_2. Saturate with CO_2 by stirring under 100% CO_2 for 30 min.

O_2-Free Water:
Composition per 860mL:

H_2O	860.0mL

Preparation of O_2-Free Water: Autoclave H_2O for 15 min at 15 psi pressure–121°C. Cool to 25°C under 100% N_2.

$NaHCO_3$ Solution:
Composition per 100mL:

$NaHCO_3$	7.5g

Preparation of $NaHCO_3$ Solution: Add the $NaHCO_3$ to distilled/deionized water and bring volume to 100.0mL. Mix thoroughly. Gas with 100% CO_2 for 20 min. Filter sterilize with positive CO_2 pressure.

$Na_2S{\cdot}9H_2O$ Solution:
Composition per 100mL:

$Na_2S{\cdot}9H_2O$	10.0g

Preparation of $Na_2S{\cdot}9H_2O$ Solution: Add $Na_2S{\cdot}9H_2O$ to distilled/deionized water. Mix thoroughly. Gas with 100% N_2 for 15 min in a screw-capped bottle. Tightly close cap. Autoclave for 15 min at 15 psi pressure–121°C. Cool to 25°C.

Trace Element Solution:

Composition per liter:

$FeCl_2 \cdot 4H_2O$	1.5g
$CoCl_2 \cdot 6H_2O$	0.19g
$MnCl_2 \cdot 4H_2O$	0.10g
$ZnCl_2$	0.07g
H_3BO_3	0.06g
$NaMoO_4 \cdot 2H_2O$	0.04g
$CuCl_2 \cdot 2H_2O$	0.02g
$NiCl_2 \cdot 6H_20$	0.02g
HCl (25% solution)	6.5mL

Preparation of Trace Element Solution: Add components to distilled/deionized water and bring volume to 1.0L. Mix thoroughly. Autoclave for 15 min at 15 psi pressure–121°C. Cool to 25°C.

Vitamin B_{12} Solution:

Composition per 100mL:

Vitamin B_{12}	2.0mg

Preparation of Vitamin B_{12} Solution: Add vitamin B_{12} to distilled/deionized water and bring volume to 100.0mL. Mix thoroughly. Filter sterilize.

Preparation of Medium: To 4.0L of sterile, CO_2-saturated solution 1, aseptically add the remaining components. Mix thoroughly. Adjust pH to 6.8. Aseptically distribute into sterile 100.0mL bottles using positive pressure of 95% N_2 + 5% CO_2. Completely fill bottles with medium except for a pea-sized air bubble.

Use: For the isolation and cultivation of members of the Chlorobiaceae.

Chlorobiaceae Medium 2

Composition per 1051mL:

Solution 1	950.0mL
$Na_2S \cdot 9H_2O$ solution	60.0mL
$NaHCO_3$ solution	40.0mL
Vitamin B_{12} solution	1.0mL

pH 6.8 ± 0.2 at 25°C

Solution 1:

Composition per 950mL:

KH_2PO_4	1.0g
NH_4Cl	0.5g
$MgSO_4 \cdot 7H_2O$	0.4g
$CaCl_2 \cdot 2H_2O$	0.05g
Trace element solution SL-8	1.0mL

Preparation of Solution 1: Add components to distilled/deionized water and bring volume to 950.0mL. Mix thoroughly. Autoclave for 15 min at 15 psi pressure–121°C. Cool to 45°–50°C.

Trace Element Solution SL-8:

Composition per liter:

Disodium EDTA	5.2g
$FeCl_2 \cdot 4H_2O$	1.5g
$CoCl_2 \cdot 6H_2O$	0.19g
$MnCl_2 \cdot 4H_2O$	0.10g
$ZnCl_2$	0.07g
H_3BO_3	0.06g
$NaMoO_4 \cdot 2H_2O$	0.04g
$CuCl_2 \cdot 2H_2O$	0.02g
$NiCl_2 \cdot 6H_20$	0.02g

Preparation of Trace Element Solution SL-8: Add components to distilled/deionized water and bring volume to 1.0L. Mix thoroughly.

$Na_2S \cdot 9H_2O$ Solution:

Composition per 100mL:

$Na_2S \cdot 9H_2O$	5.0g

Preparation of $Na_2S \cdot 9H_2O$ Solution: Add $Na_2S \cdot 9H_2O$ to distilled/deionized water and bring volume to 100.0mL. Autoclave for 15 min at 15 psi pressure–121°C. Cool to 45°–50°C.

$NaHCO_3$ Solution:

Composition per 100mL:

$NaHCO_3$	5.0g

Preparation of $NaHCO_3$ Solution: Add $NaHCO_3$ to distilled/deionized water and bring volume to 100.0mL. Mix thoroughly. Filter sterilize.

Vitamin B_{12} Solution:

Composition per 100mL:

Vitamin B_{12}	2.0mg

Preparation of Vitamin B_{12} Solution: Add vitamin B_{12} to distilled/deionized water and bring volume to 100.0mL. Mix thoroughly. Filter sterilize.

Preparation of Medium: To 950.0mL of cooled, sterile solution 1, aseptically add 60.0mL of sterile $Na_2S \cdot 9H_2O$ solution, 40.0mL of sterile $NaHCO_3$ solution, and 1.0mL of sterile vitamin B_{12} solution. Mix thoroughly. Adjust pH to 6.8 with sterile H_2SO_4 or Na_2CO_3. Aseptically distribute into sterile 50.0mL or 100.0mL bottles with metal screw-caps and rubber seals. Completely fill bottles with medium except for a pea-sized air bubble.

Use: For the isolation and cultivation of freshwater and soil members of the Chlorobiaceae.

Chloroflexus Agar

Composition per liter:

Agar	15.0g
Glycyl-glycine	0.5g
Yeast extract	0.5g

Na_2S	0.5g
NH_4Cl	0.2g
$MgSO_4 \cdot 7H_2O$	0.1g
Nitrilotriacetic acid	0.1g
$NaNO_3$	0.689g
Na_2HPO_4	0.111g
KNO_3	0.103g
$CaSO_4 \cdot 2H_2O$	0.06g
NaCl	8.0mg
$FeCl_3$ solution	1.0mL
Micronutrient solution	1.0mL

pH 8.2–8.4 at 25°C

$FeCl_3$ Solution:

Composition per liter:

$FeCl_3$	0.29g

Preparation of $FeCl_3$ Solution: Add $FeCl_3$ to distilled/deionized water and bring volume to 1.0L. Mix thoroughly.

Micronutrient Solution:

Composition per liter:

$MnSO_4 \cdot 7H_2O$	2.28g
H_3BO_3	0.50g
$ZnSO_4 \cdot 7H_2O$	0.50g
$CoCl_2 \cdot 6H_2O$	0.045g
$CuSO_4 \cdot 2H_2O$	0.025g
$Na_2MoO_4 \cdot 2H_2O$	0.025g
H_2SO_4,concentrated	0.5mL

Preparation of Micronutrient Solution: Add components to distilled/deionized water and bring volume to 1.0L. Mix thoroughly.

Preparation of Medium: Add components, except Na_2S, to distilled/deionized water and bring volume to 1.0L. Mix thoroughly. Adjust pH to 8.2–8.4. Add Na_2S. Readjust pH to 8.2–8.4. Gently heat and bring to boiling. Distribute into tubes or flasks. Autoclave for 15 min at 15 psi pressure–121°C. Pour into sterile Petri dishes or leave in tubes.

Use: For the cultivation of *Chloroflexus aurantiacus.*

Chloroflexus Broth

Composition per liter:

Glycyl-glycine	0.5g
Yeast extract	0.5g
NH_4Cl	0.2g
$NaNO_3$	0.689g
Na_2S	0.5g
Na_2HPO_4	0.111g
KNO_3	0.103g
$MgSO_4 \cdot 7H_2O$	0.1g
Nitrilotriacetic acid	0.1g
$CaSO_4 \cdot 2H_2O$	0.06g
NaCl	8.0mg
$FeCl_3$ solution	1.0mL
Micronutrient solution	1.0mL

pH 8.2–8.4 at 25°C

$FeCl_3$ Solution:

Composition per liter:

$FeCl_3$	0.29g

Preparation of $FeCl_3$ Solution: Add $FeCl_3$ to distilled/deionized water and bring volume to 1.0L. Mix thoroughly.

Micronutrient Solution:

Composition per liter:

$MnSO_4 \cdot 7H_2O$	2.28g
H_3BO_3	0.50g
$ZnSO_4 \cdot 7H_2O$	0.50g
$CoCl_2 \cdot 6H_2O$	0.045g
$CuSO_4 \cdot 2H_2O$	0.025g
$Na_2MoO_4 \cdot 2H_2O$	0.025g
H_2SO_4 (concentrated)	0.5mL

Preparation of Micronutrient Solution: Add components to distilled/deionized water and bring volume to 1.0L. Mix thoroughly.

Preparation of Medium: Add components, except Na_2S, to distilled/deionized water and bring volume to 1.0L. Mix thoroughly. Adjust pH to 8.2–8.4. Add Na_2S. Readjust pH to 8.2–8.4. Filter sterilize. Distribute into sterile tubes or flasks.

Use: For the cultivation of *Chloroflexus aurantiacus.*

Chlorohydroxybenzoic Acid Medium

Composition per liter:

$K_2HPO_4 \cdot 3H_2O$	4.25g
NH_4Cl	2.0g
$NaH_2PO_4 \cdot H_2O$	1.0g
5-Chloro-2-hydroxybenzoic acid	0.5g
$MgSO_4 \cdot 7H_2O$	0.2g
Nitrilotriacetic acid	0.1g
$FeSO_4 \cdot 7H_2O$	0.012g
$MnSO_4 \cdot H_2O$	3.0mg
$ZnSO_4 \cdot 7H_2O$	3.0mg
$CoSO_4$	1.0mg

pH 7.0-7.4 at 25°C

Preparation of Medium: Add 5-chloro-2-hydroxybenzoic acid to 800.0mL of distilled/deionized water. Adjust pH to 7.0 with NaOH. Add remaining components and bring volume to 1.0L. Distribute into tubes or flasks. Autoclave for 15 min at 15 psi pressure–121°C.

Use: For the cultivation of bacteria that can utilize 5-chloro-hydroxybenzoic acid. For the cultivation of ATCC strain 35944.

CHO Medium
See: **Fermentation Broth**

CHO Medium Base (Carbohydrate Medium Base)

Composition per liter:

Pancreatic digest of casein	15.0g
Yeast extract	7.0g
NaCl	2.5g
Agar	0.75g
Sodium thioglycollate	0.5g
L-Cystine	0.25g
Ascorbic acid	0.1g
Bromthymol Blue	0.01g

pH 7.0 ± 0.2 at 25°C

Preparation of Medium: Add components to distilled/deionized water and bring volume to 1.0L. Mix thoroughly. Gently heat and bring to boiling. Distribute into tubes or flasks. Autoclave for 15 min at 15 psi pressure–121°C. Cool to 45–50°C.

Use: Used as a basal medium to which carbohydrates are added for fermentation studies of anaerobic bacteria. Generally, 6.25mL of 10% filter-sterilized solution of carbohydrate is added to the sterile basal medium.

Chocolate Agar

Composition per liter:

Agar	15.0g
Pantone	10.0g
Bitone	10.0g
NaCl	5.0g
Tryptic digest of beef heart	3.0g
Cornstarch	1.0g
Sheep blood, defibrinated	100.0mL
Supplement B (Difco)	10.0mL

pH 7.3 ± 0.2 at 25°C

Supplement B:

Composition per 10mL:

Supplement B contains yeast concentrate, glutamine, coenzyme, cocarboxylase, hematin and growth factors.

Preparation of Supplement B: Add components to distilled/deionized water and bring volume to 10.0mL. Mix thoroughly. Filter sterilize.

Preparation of Medium: Add components, except hemoglobin solution and sheep blood, to distilled/deionized water and bring volume to 890.0mL. Mix thoroughly. Gently heat until boiling. Autoclave for 15 min at 15 psi pressure–121°C. Cool to 45°–50°C. Aseptically add 100.0mL of sterile defibrinated sheep blood. Gently heat while stirring and bring to 85°C for 5–10 min. Cool to 50°C. Aseptically add 10.0mL of sterile supplement B. Mix thoroughly. Pour into sterile Petri dishes or distribute into sterile tubes.

Use: For the isolation and cultivation of a variety of fastidious microorganisms.

Chocolate Agar

Composition per liter:

Proteose peptone No. 3	15.0g
Agar	10.0g
NaCl	5.0g
K_2HPO_4	4.0g
Cornstarch	1.0g
KH_2PO_4	1.0g
Hemoglobin solution	100.0mL
Supplement B	10.0mL

pH 7.0 ± 0.2 at 25°C

Source: Available from Difco Laboratories.

Supplement B:

Composition per 10mL:

Supplement B contains yeast concentrate, glutamine, coenzyme, cocarboxylase, hematin and growth factors.

Preparation of Supplement B: Add components to distilled/deionized water and bring volume to 10.0mL. Mix thoroughly. Filter sterilize.

Hemoglobin Solution:

Composition per 100mL:

Hemoglobin	10.0g

Preparation of Hemoglobin Solution: Add hemoglobin to distilled/deionized water and bring volume to 100.0mL. Mix thoroughly. Filter sterilize.

Preparation of Medium: Add components, except hemoglobin solution and supplement B, to distilled/deionized water and bring volume to 990.0mL. Mix thoroughly. Gently heat and bring to boiling. Autoclave for 15 min at 15 psi pressure–121°C. Cool to 45°–50°C. Aseptically add 100.0mL of sterile hemoglobin solution. Gently heat while stirring and bring to 85°C for 5–10 min. Cool to 50°C. Aseptically add 10.0mL of sterile supplement B. Mix thoroughly. Pour into sterile Petri dishes or distribute into sterile tubes.

Use: For the isolation and cultivation of fastidious microorganisms.

Chocolate Agar, Enriched

Composition per liter:

GC medium base	740.0mL
Hemoglobin solution	250.0mL
Supplement B	10.0mL

pH 7.3 ± 0.2 at 25°C

Source: Available from Difco Laboratories.

GC Medium Base:

Composition per 740mL:

Proteose peptone No. 3	15.0g
Agar	20.0g
NaCl	5.0g
K_2HPO_4	4.0g
Glucose	1.5g
Cornstarch	1.0g
KH_2PO_4	1.0g

pH 7.2 ± 0.2 at 25°C

Preparation of GC Medium Base: Add components to distilled/deionized water and bring volume to 740.0mL. Mix thoroughly. Gently heat until boiling. Autoclave for 15 min at 15 psi pressure–121°C. Cool to 45°–50°C.

Hemoglobin Solution:

Composition per 250mL:

Hemoglobin	10.0g

Preparation of Hemoglobin Solution: Add hemoglobin to distilled/deionized water and bring volume to 250.0mL. Mix thoroughly. Autoclave for 15 min at 15 psi pressure–121°C. Cool to 45°–50°C.

Supplement B:

Composition per 10mL:

Supplement B contains yeast concentrate, glutamine, coenzyme, cocarboxylase, hematin and growth factors.

Preparation of Supplement B: Add components to distilled/deionized water and bring volume to 10.0mL. Mix thoroughly. Filter sterilize.

Preparation of Medium: To 740.0mL of cooled sterile GC medium base, aseptically add 250.0mL of sterile hemoglobin solution, and 10.0mL of sterile supplement B. Mix thoroughly. Pour into sterile Petri dishes or distribute into sterile tubes.

Use: For the cultivation of fastidious microorganisms, especially *Neisseria* species.

Chocolate Agar, Enriched

Composition per liter:

GC medium base	740.0mL
Hemoglobin solution	250.0mL
Supplement VX	10.0mL

pH 7.3 ± 0.2 at 25°C

Source: Available Difco Laboratories.

GC Medium Base:

Composition per 740mL:

Proteose peptone No. 3	15.0g
Agar	20.0g
NaCl	5.0g
K_2HPO_4	4.0g
Glucose	1.5g
Cornstarch	1.0g
KH_2PO_4	1.0g

pH 7.2 ± 0.2 at 25°C

Preparation of GC Medium Base: Add components to distilled/deionized water and bring volume to 740.0mL. Mix thoroughly. Gently heat until boiling. Autoclave for 15 min at 15 psi pressure–121°C. Cool to 45°–50°C.

Hemoglobin Solution:

Composition per 250mL:

Hemoglobin	10.0g

Preparation of Hemoglobin Solution: Add hemoglobin to distilled/deionized water and bring volume to 250.0mL. Mix thoroughly. Autoclave for 15 min at 15 psi pressure–121°C. Cool to 45°–50°C.

Supplement VX:

Composition per 10mL:

Supplement VX contains essential growth factors.

Preparation of Supplement VX: Add components to distilled/deionized water and bring volume to 10.0mL. Mix thoroughly. Filter sterilize.

Preparation of Medium: To 740.0mL of cooled sterile GC medium base, aseptically add 250.0mL of sterile hemoglobin solution, and 10.0mL of sterile supplement B. Mix thoroughly. Pour into sterile Petri dishes or distribute into sterile tubes.

Use: For the cultivation of fastidious microorganisms, especially *Neisseria* species.

Chocolate II Agar

Composition per liter:

Agar	12.0g
Hemoglobin	10.0g
Pancreatic digest of casein	7.5g
Selected meat peptone	7.5g
NaCl	5.0g
K_2HPO_4	4.0g
Cornstarch	1.0g
KH_2PO_4	1.0g

Preparation of Medium: Add components to distilled/deionized water and bring volume to 1.0L. Mix thoroughly. Gently heat to boiling. Autoclave for 15 min at 15 psi pressure–121°C. Pour into sterile Petri dishes or leave in tubes.

Use: For the isolation and cultivation of fastidious microorganisms.

Chocolate II Agar with Hemoglobin and IsoVitaleX® (GCII Agar with Hemoglobin and IsoVitaleX®)

Composition per liter:

GCII agar base	990.0mL
IsoVitaleX enrichment	10.0mL

pH 7.3 ± 0.2 at 25°C

Source: Available as a prepared medium from BBL Microbiology Systems.

GCII Agar Base:

Composition per liter:

Agar	12.0g
Hemoglobin	10.0g
Pancreatic digest of casein	7.5g
Selected meat peptone	7.5g
NaCl	5.0g
K_2HPO_4	4.0g
Cornstarch	1.0g
KH_2PO_4	1.0g

Preparation of GCII Agar Base: Add components to distilled/deionized water and bring volume to 1.0L. Mix thoroughly. Gently heat to boiling. Autoclave for 15 min at 15 psi pressure–121°C. Cool to 45°–50°C.

IsoVitaleX® Enrichment:

Composition per liter:

Glucose	100.0g
L-Cysteine·HCl	25.9g
L-Glutamine	10.0g
L-Cystine	1.1g
Adenine	1.0g
Nicotinamide adenine dinucleotide	0.25g
Vitamin B_{12}	0.1g
Thiamine pyrophosphate	0.1g
Guanine·HCl	0.03g
$Fe(NO_3)_3 \cdot 6H_2O$	0.02g
p-Aminobenzoic acid	0.013g
Thiamine·HCl	3.0mg

Preparation of IsoVitaleX®: Add components to distilled/deionized water and bring volume to 1.0L. Mix thoroughly. Filter sterilize.

Preparation of Medium: Aseptically add 10.0mL of sterile IsoVitaleX® enrichment to 990.0L of sterile, cooled GCII agar base. Mix thoroughly. Pour into sterile Petri dishes or distribute into sterile tubes.

Use: For the isolation and cultivation of fastidious microorganisms, especially *Neisseria* and *Haemophilus* species, from a variety of clinical specimens.

Chocolate Tellurite Agar (Tellurite Blood Agar)

Composition per liter:

Agar	10.0g
Casein/meat (50/50) peptone	10.0g
Hemoglobin	10.0g
NaCl	5.0g
K_2HPO_4	4.0g
Cornstarch	1.0g
KH_2PO_4	1.0g
K_2TeO_3	0.1g
Bio-X enrichment	10.0mL

Bio-X Enrichment:

Composition:

Glucose	100.0g
Cysteine·HCl	25.9g
L-Glutamate	10.0g
L-Cystine	1.1g
Adenine	1.0g
Cocarboxylase	0.1g
NAD (nicotinamide adenine dinucleotide)	250.0mg
Guanine·HCl	0.03g
$FeNO_3$	0.02g
p-Aminobenzoic acid	0.013g
Vitamin B_{12}	0.01g
Thiamine·HCl	3.0mg

pH 7.2 ± 0.2 at 25°C

Preparation of Bio-X Enrichment: Add components to distilled/deionized water and bring volume to 1.0L. Mix thoroughly. Filter sterilize.

Caution: Potassium tellurite is toxic.

Preparation of Medium: Add components, except Bio-X enrichment to distilled/deionized water and bring volume to 990.0mL. Mix thoroughly. Gently heat and bring to boiling. Autoclave for 15 min at 15 psi pressure–121°C. Cool to 45°–50°C. Aseptically add filter-sterilized Bio-X enrichment. Mix thoroughly. Pour into sterile Petri dishes or distribute into sterile tubes.

Use: For the selective isolation and cultivation of *Corynebacterium* species. *C. diphtheriae* appears as gray-black colonies.

Cholera Medium TCBS

Composition per liter:

Sucrose	20.0g
Agar	14.0g

Peptone	10.0g
NaCl	10.0g
Sodium citrate	10.0g
$Na_2S_2O_3 \cdot 5H_2O$	10.0g
Ox bile	8.0 g
Yeast extract	5.0g
Ferric citrate	1.0g
Bromothymol Blue	0.04g
Thymol Blue	0.04g

pH 8.6 ± 0.2 at 25°C

Source: This medium is available as a premixed powder from Oxoid Unipath.

Preparation of Medium: Add components to distilled/deionized water and bring volume to 1.0L. Mix thoroughly. Gently heat and bring to boiling. Do not autoclave. Pour into sterile Petri dishes. Dry agar plates before using.

Use: For the growth of *Vibrio cholerae*, *V. parahaemolyticus*, and other *Vibrio* species.

Cholesterol Medium

Composition per 1030mL:

Solution A	500.0mL
Solution B	500.0mL
Amino acid solution	20.0mL
Vitamin solution	10.0mL

pH 6.8 ± 0.2 at 25°C

Solution A:

Composition per liter:

$(NH_4)_2SO_4$	5.0g
KH_2PO_4	1.0g
$MgSO_4 \cdot 7H_2O$	0.5g
$CaCl_2 \cdot 2H_2O$	0.1g
NaCl	0.1g
Wolfe's mineral solution	10.0mL

Preparation of Solution A: Add components to distilled/deionized water and bring volume to 1.0L. Mix thoroughly. Autoclave for 15 min at 15 psi pressure–121°C. Cool to 45°–50°C.

Wolfe's Mineral Solution:

Composition per liter:

$MgSO_4 \cdot 7H_2O$	3.0g
Nitrilotriacetic acid	1.5g
NaCl	1.0g
$MnSO_4 \cdot H_2O$	0.5g
$CaCl_2$	0.1g
$CoCl_2 \cdot 6H_2O$	0.1g
$FeSO_4 \cdot 7H_2O$	0.1g
$ZnSO_4 \cdot 7H_2O$	0.1g
$AlK(SO_4)_2 \cdot 12H_2O$	0.01g
$CuSO_4 \cdot 5H_2O$	0.01g
H_3BO_3	0.01g
$Na_2MoO_4 \cdot 2H_2O$	0.01g

Preparation of Wolfe's Mineral Solution: Add nitrilotriacetic acid to 500.0mL of distilled/deionized water. Dissolve by adjusting pH to 6.5 with KOH. Add distilled/deionized water to 1.0L. Add remaining components.

Solution B:

Composition:

Noble agar	15.0g
Cholesterol	2.0g
Tween™ 80	1.0g
Yeast extract	0.5g

Preparation of Solution B: Add components to distilled/deionized water and bring volume to 1.0L. Mix thoroughly. Gently heat to boiling. Autoclave for 15 min at 15 psi pressure–121°C. Cool to 45°–50°C.

Amino Acid Solution:

Composition per 100mL:

L-Histidine	0.5g
DL-Methionine	0.1g
DL-Tryptophan	0.1g

Preparation of Amino Acid Solution: Add components to distilled/deionized water and bring to 100.0mL. Filter sterilize.

Vitamin Solution:

Composition per liter:

myo-Inositol	200.0mg
Calcium pantothenate	40.0mg
Niacin	40.0mg
Pyridoxine·HCl	40.0mg
Thiamine	40.0mg
p-Aminobenzoic acid	20.0mg
Riboflavin	20.0mg
Biotin	200.0μg
Folic acid	200.0μg

Preparation of Vitamin Solution: Add components to distilled/deionized water and bring to 1.0L. Filter sterilize.

Preparation of Medium: Combine cooled sterile solution A and cooled sterile solution B.Aseptically add filter-sterilized amino acid solution and vitamin solution. Adjust pH to 6.8. Pour into sterile Petri dishes or distribute into sterile tubes.

Use: For the cultivation of ATCC strain 31384.

Cholic Acid Medium

Composition per liter:

Noble agar	15.0g
K_2HPO_4	3.5g

Cholic acid 2.0g
$(NH_4)_2SO_4$ 2.0g
KH_2PO_4 .... 1.5g
$MgSO_4 \cdot 7H_2O$ 0.1g
$CaCl_2 \cdot 2H_2O$ 0.01g
$FeSO_4 \cdot 7H_2O$ 0.5mg

pH 7.0 ± 0.2 at 25°C

Preparation of Medium: Add components to distilled/deionized water and bring volume to 1.0L. Mix thoroughly. Adjust pH to 7.0. Gently heat and bring to boiling. Autoclave for 15 min at 15 psi pressure–121°C. Pour into sterile Petri dishes or distribute into sterile tubes.

Use: For for the cultivation and maintenance of *Nocardia* species and other bacteria which can utilize cholic acid as a carbon source.

Choline Assay Medium

Composition per liter:

Sucrose 40.0g
Potassium sodium tartrate 11.4g
$(NH_4)_2NO_3$ 2.0g
KH_2PO_4 .... 2.0g
$MgSO_4 \cdot 7H_2O$ 1.0g
$CaCl_2 \cdot 2H_2O$ 0.2g
NaCl 0.2g
$ZnSO_4 \cdot 7H_2O$ 0.02g
$FeSO_4$ 1.1mg
Na_3BO_3 .... 0.7mg
$(NH_4)_2MoO_3$ 0.5mg
CuCl 0.3mg
$MgSO_4 \cdot 7H_2O$ 0.11mg
Biotin 0.010mg

pH5.5 ± 0.2 at 25°C

Source: This medium is available as a premixed powder from Difco Laboratories.

Preparation of Medium: Add components to distilled/deionized water and bring volume to 1.0L. Mix thoroughly. Gently heat and bring to boiling. Continue boiling 2–3 min. Allow precipitate to settle out. Distribute supernatant into 125mL flasks in 10.0mL volumes. Add standard solution or test solutions to each flask. Adjust the volume of each flask to 20.0mL with distilled/deionized water. Autoclave for 10 min at 15 psi pressure–121°C.

Use: For the microbiological assay of choline using *Neurospora crassa* as the test microorganism.

Chopped Liver Broth

Composition per liter:

Fresh beef liver 500.0g
Peptone 10.0g
K_2HPO_4 .... 1.0g
Soluble starch 1.0g

pH 7.0 ± 0.2 at 25°C

Preparation of Medium: Grind fresh beef liver. Add to 1.0L of distilled/deionized water. Gently heat and bring to boiling. Continue boiling for 60 min. Cool to 25°C. Adjust pH to 7.0. Gently heat and bring to boiling. Continue boiling for 10 min. Filter through cheesecloth. Save chopped liver particles. To filtrate, add remaining components. Bring volume to 1.0L with distilled/deionized water. Adjust pH to 7.0. Filter through Whatman #1 filter paper. Add chopped liver particles to test tubes to a depth of 1.2–2.5 cm. Add 10.0mL of broth to each tube. Autoclave for 15 min at 15 psi pressure–121°C.

Use: For the isolation and cultivation of *Clostridium botulinum, Clostridium perfringens* and other anaerobic bacteria from foods.

Chopped Meat Carbohydrate Medium

Composition per 1240mL:

Peptone 30.0g
K_2HPO_4 .... 5.0g
Yeast extract 5.0g
Cellobiose 1.0g
Maltose 1.0g
Starch 1.0g
L-Cysteine·HCl·H_2O 0.5g
Chopped meat extract filtrate 1.0L
Chopped meat extract solids 200.0mL
Resazurin (0.025% solution) 4.0mL

pH 7.0 ± 0.2 at 25°C

Chopped Meat Extract:

Composition per liter:

Beef or horse meat 500.0g
NaOH (1*N* solution) 25.0mL

Preparation of Chopped Meat Extract: Use lean beef or horse meat. Remove fat and connective tissue. Grind. Add meat and NaOH to distilled/deionized water and bring volume to 1.0L. Gently heat and bring to boiling while stirring. Cool to 25°C. Remove fat from surface. Filter. Reserve ground meat particles and filtrate. Add distilled/deionized water to filtrate and bring volume to 1.0L.

Preparation of Medium: To 1.0L of chopped meat extract filtrate, add the remaining components, except cysteine and chopped meat solids. Mix thoroughly. Gently heat to boiling. Cool to room temperature. Add the cysteine. Adjust pH to 7.0. Distribute 1 part chopped meat solids (by volume) and 5 parts of liquid (by volume) into tubes under O_2-free

97% N_2 + 3% H_2. Cap with rubber stoppers and place tubes in a press. Autoclave for 15 min at 15 psi pressure–121°C with fast exhaust.

Use: For the cultivation of anaerobic bacteria including *Clostridium* species, *Eubacterium* species and *Gemmiger formicilis*.

Chopped Meat Carbohydrate Medium with Rumen Fluid (ATCC Medium 1016)

Composition per 1390mL:

Peptone	30.0g
K_2HPO_4	5.0g
Yeast extract	5.0g
Cellobiose	1.0g
Maltose	1.0g
Starch	1.0g
L-Cysteine·HCl·H_2O	0.5g
Chopped meat extract filtrate	1.0L
Chopped meat extract solids	200.0mL
Rumen fluid	150.0mL
Resazurin (0.025% solution)	4.0mL

pH 7.0 ± 0.2 at 25°C

Chopped Meat Extract:

Composition per liter:

Beef or horse meat	500.0g
NaOH (1*N* solution)	25.0mL

Preparation of Chopped Meat Extract: Use lean beef or horse meat. Remove fat and connective tissue. Grind. Add meat and NaOH to distilled/deionized water and bring volume to 1.0L. Gently heat and bring to boiling while stirring. Cool to 25°C. Remove fat from surface. Filter. Reserve ground meat particles and filtrate. Add distilled/deionized water to filtrate and bring volume to 1.0L.

Preparation of Medium: To 1.0L of chopped meat extract filtrate, add the remaining components, except cysteine and chopped meat solids. Mix thoroughly. Gently heat to boiling. Cool to room temperature. Add the L-cysteine. Adjust pH to 7.0. Distribute 1 part chopped meat solids (by volume) and 5 parts of liquid (by volume) into tubes under O_2-free 97% N_2 + 3% H_2. Cap with rubber stoppers and place tubes in a press. Autoclave for 15 min at 15 psi pressure–121°C with fast exhaust.

Use: For the cultivation of anaerobic bacteria including *Butyrivibrio crossotus*, *Eubacterium* species, and *Ruminococcus* species.

Chopped Meat Carbohydrate Medium with Rumen Fluid

Composition per 1390mL:

Peptone	30.0g
K_2HPO_4	5.0g
Yeast extract	5.0g
Glucose	4.0g
Cellobiose	1.0g
Maltose	1.0g
Starch	1.0g
L-Cysteine·HCl·H_2O	0.5g
Chopped meat extract filtrate	1.0L
Chopped meat extract solids	200.0mL
Rumen fluid	150.0mL
Resazurin (0.025% solution)	4.0mL

pH 7.0 ± 0.2 at 25°C

Chopped Meat Extract:

Composition per liter:

Beef or horse meat	500.0g
NaOH (1*N* solution)	25.0mL

Preparation of Chopped Meat Extract: Use lean beef or horse meat. Remove fat and connective tissue. Grind. Add meat and NaOH to distilled/deionized water and bring volume to 1.0L. Gently heat and bring to boiling while stirring. Cool to 25°C. Remove fat from surface. Filter. Reserve ground meat particles and filtrate. Add distilled/deionized water to filtrate and bring volume to 1.0L.

Preparation of Medium: To 1.0L of chopped meat extract filtrate, add the remaining components, except cysteine and chopped meat solids. Mix thoroughly. Gently heat to boiling. Cool to room temperature. Add the L-cysteine. Adjust pH to 7.0. Distribute 1 part chopped meat solids (by volume) and 5 parts of liquid (by volume) into tubes under O_2-free 97% N_2 + 3% H_2. Cap with rubber stoppers and place tubes in a press. Autoclave for 15 min at 15 psi pressure–121°C with fast exhaust.

Use: For the cultivation of anaerobic bacteria including *Fusobacterium prausnitzii*, *Eubacterium* species, and *Prevotella ruminicola*.

Chopped Meat Carbohydrate Medium with Tween™ 80

Composition per 1240mL:

Peptone	30.0g
K_2HPO_4	5.0g
Yeast extract	5.0g
Cellobiose	1.0g
Maltose	1.0g
Starch	1.0g

Tween™ 80 .... 1.0g
L-Cysteine·HCl·H_2O 0.5g
Chopped meat extract filtrate 1.0L
Chopped meat extract solids 200.0mL
Resazurin (0.025% solution) 4.0mL
pH 7.0 ± 0.2 at 25°C

Chopped Meat Extract:
Composition per liter:
Beef or horse meat 500.0g
NaOH (1*N* solution) 25.0mL

Preparation of Chopped Meat Extract: Use lean beef or horse meat. Remove fat and connective tissue. Grind. Add meat and NaOH to distilled/deionized water and bring volume to 1.0L. Gently heat and bring to boiling while stirring. Cool to 25°C. Remove fat from surface. Filter. Reserve ground meat particles and filtrate. Add distilled/deionized water to filtrate and bring volume to 1.0L.

Preparation of Medium: To 1.0L of chopped meat extract filtrate, add the remaining components, except cysteine and chopped meat solids. Mix thoroughly. Gently heat to boiling. Cool to room temperature. Add the L-cysteine. Adjust pH to 7.0. Distribute 1 part chopped meat solids (by volume) and 5 parts of liquid (by volume) into tubes under O_2-free 97% N_2 + 3% H_2. Cap with rubber stoppers and place tubes in a press. Autoclave for 15 min at 15 psi pressure–121°C with fast exhaust.

Use: For the cultivation of *Coprococcus* species and *Peptostreptococcus micros.*

Chopped Meat Glucose Medium

Composition per 1240mL:
Peptone 30.0g
K_2HPO_4 .... 5.0g
Yeast extract 5.0g
Glucose 5.0g
L-Cysteine·HCl·H_2O 0.5g
Chopped meat extract filtrate 1.0L
Chopped meat extract solids 200.0mL
Resazurin (0.025% solution) 4.0mL
pH 7.0 ± 0.2 at 25°C

Chopped Meat Extract:
Composition per liter:
Beef or horse meat 500.0g
NaOH (1*N* solution) 25.0mL

Preparation of Chopped Meat Extract: Use lean beef or horse meat. Remove fat and connective tissue. Grind. Add meat and NaOH to distilled/deionized water and bring volume to 1.0L. Gently heat and bring to boiling while stirring. Cool to 25°C. Remove fat from surface. Filter. Reserve ground meat particles and filtrate. Add distilled/deionized water to filtrate and bring volume to 1.0L.

Preparation of Medium: To 1.0L of chopped meat extract filtrate, add the remaining components, except cysteine and chopped meat solids. Mix thoroughly. Gently heat to boiling. Cool to room temperature. Add the cysteine. Adjust pH to 7.0. Distribute 1 part chopped meat solids (by volume) and 5 parts of liquid (by volume) into tubes under O_2-free 97% N_2 + 3% H_2. Cap with rubber stoppers and place tubes in a press. Autoclave for 15 min at 15 psi pressure–121°C with fast exhaust.

Use: For the cultivation of *Clostridium* species and *Selenomonas noxia.*

Chopped Meat Glucose Medium with NaCl

Composition per 1205mL:
NaCl 30.0g
Peptone 30.0g
K_2HPO_4 .... 5.0g
Yeast extract 5.0g
Glucose 5.0g
L-Cysteine·HCl·H_2O 0.5g
Chopped meat extract filtrate 1.0L
Chopped meat extract solids 200.0mL
Resazurin (0.025% solution) 4.0mL
pH 7.0 ± 0.2 at 25°C

Chopped Meat Extract:
Composition per liter:
Beef or horse meat 500.0g
NaOH (1*N* solution) 25.0mL

Preparation of Chopped Meat Extract: Use lean beef or horse meat. Remove fat and connective tissue. Grind. Add meat and NaOH to distilled/deionized water and bring volume to 1.0L. Gently heat and bring to boiling while stirring. Cool to 25°C. Remove fat from surface. Filter. Reserve ground meat particles and filtrate. Add distilled/deionized water to filtrate and bring volume to 1.0L.

Preparation of Medium: To 1.0L of chopped meat extract filtrate, add the remaining components, except cysteine and chopped meat solids. Mix thoroughly. Gently heat to boiling. Cool to room temperature. Add the L-cysteine. Adjust pH to 7.0. Distribute 1 part chopped meat solids (by volume) and 5 parts of liquid (by volume) into tubes under O_2-free 97% N_2 + 3% H_2. Cap with rubber stoppers and place tubes in a press. Autoclave for 15 min at 15 psi pressure–121°C with fast exhaust.

Use: For the cultivation and maintenance of anaerobic halophilic bacteria.

Chopped Meat Medium

Composition per 1205mL:

Peptone	30.0g
K_2HPO_4	5.0g
Yeast extract	5.0g
L-Cysteine·HCl·H_2O	0.5g
Chopped meat extract filtrate	1.0L
Chopped meat extract solids	200.0mL
Resazurin (0.025% solution)	4.0mL

pH 7.0 ± 0.2 at 25°C

Chopped Meat Extract:

Composition per liter:

Beef or horse meat	500.0g
NaOH (1*N* solution)	25.0mL

Preparation of Chopped Meat Extract: Use lean beef or horse meat. Remove fat and connective tissue. Grind. Add meat and NaOH to distilled/deionized water and bring volume to 1.0L. Gently heat and bring to boiling while stirring. Cool to 25°C. Remove fat from surface. Filter. Reserve ground meat particles and filtrate. Add distilled/deionized water to filtrate and bring volume to 1.0L.

Preparation of Medium: To 1.0L of chopped meat extract filtrate, add the remaining components, except cysteine and chopped meat solids. Mix thoroughly. Gently heat to boiling. Cool to room temperature. Add the cysteine. Adjust pH to 7.0. Distribute 1 part chopped meat solids (by volume) and 5 parts of liquid (by volume) into tubes under O_2-free 97% N_2 + 3% H_2. Cap with rubber stoppers and place tubes in a press. Autoclave for 15 min at 15 psi pressure–121°C with fast exhaust.

Use: For cultivation and maintenance of a variety of anaerobic bacteria including *Bacteroides* species, *Bifidobacterium* species, *Capnocytophaga* species, *Clostridium* species, *Eubacterium* species, *Fusobacterium* species, *Peptostreptococcus* species, *Prevotella* species, *Propionibacterium* species, *Ruminococcus* species and others.

Chopped Meat Medium with 10% Fetal Calf Serum

Composition per 1230mL:

Peptone	30.0g
K_2HPO_4	5.0g
Yeast extract	5.0g
L-Cysteine·HCl·H_2O	0.5g
Chopped meat extract filtrate	1.0L
Chopped meat extract solids	200.0mL
Fetal calf serum	100.0mL
Resazurin (0.025% solution)	4.0mL

pH 7.0 ± 0.2 at 25°C

Chopped Meat Extract:

Composition per liter:

Beef or horse meat	500.0g
NaOH (1*N* solution)	25.0mL

Preparation of Chopped Meat Extract: Use lean beef or horse meat. Remove fat and connective tissue. Grind. Add meat and NaOH to distilled/deionized water and bring volume to 1.0L. Gently heat and bring to boiling while stirring. Cool to 25°C. Remove fat from surface. Filter. Reserve ground meat particles and filtrate. Add distilled/deionized water to filtrate and bring volume to 1.0L.

Preparation of Medium: To 1.0L of chopped meat extract filtrate, add the remaining components, except L-cysteine and chopped meat solids. Mix thoroughly. Gently heat to boiling. Cool to room temperature. Add the L-cysteine. Adjust pH to 7.0. Distribute 1 part chopped meat solids (by volume) and 5 parts of liquid (by volume) into tubes under O_2-free 97% N_2 + 3% H_2. Cap with rubber stoppers and place tubes in a press. Autoclave for 15 min at 15 psi pressure–121°C with fast exhaust.

Use: For the cultivation and maintenance of *Actinomyces hordeovulneris*.

Chopped Meat Medium with Formate and Fumarate

Composition per 1230mL:

Peptone	30.0g
K_2HPO_4	5.0g
Yeast extract	5.0g
L-Cysteine·HCl·H_2O	0.5g
Chopped meat extract filtrate	1.0L
Chopped meat extract solids	200.0mL
Resazurin (0.025% solution)	4.0mL
Formate-fumarate solution	0.05mL

pH 7.0 ± 0.2 at 25°C

Chopped Meat Extract:

Composition per liter:

Beef or horse meat	500.0g
NaOH (1*N* solution)	25.0mL

Preparation of Chopped Meat Extract: Use lean beef or horse meat. Remove fat and connective tissue. Grind. Add meat and NaOH to distilled/deionized water and bring volume to 1.0L. Gently heat and bring to boiling while stirring. Cool to 25°C. Remove fat from surface. Filter. Reserve ground meat particles and filtrate. Add distilled/deionized water to filtrate and bring volume to 1.0L.

Formate-Fumarate Solution:

Composition per 100mL:

Sodium formate	6.0g
Fumaric acid	6.0g

Preparation of Formate-Fumarate Solution: Add components to distilled/deionized water and bring volume to 100.0mL. Adjust pH to 7.0. Filter sterilize.

Preparation of Medium: To 1.0L of chopped meat extract filtrate, add the remaining components, except L-cysteine, formate-fumarate solution and chopped meat solids. Mix thoroughly. Gently heat to boiling. Cool to room temperature. Add the L-cysteine. Adjust pH to 7.0. Distribute 1 part chopped meat solids (by volume) and 5 parts of liquid (by volume) into tubes under O_2-free 97% N_2 + 3% H_2. Cap with rubber stoppers and place tubes in a press. Autoclave for 15 min at 15 psi pressure–121°C with fast exhaust. Prior to inoculation, add 0.05mL of formate-fumarate solution to each tube containing approximately 6.5mL of Chopped Meat Medium.

Use: For the cultivation and maintenance of *Bacteroides ureolyticus* and *Wolinella* species.

Chopped Meat Medium with 1% Glucose

Composition per 1230mL:

Peptone	30.0g
Glucose	10.0g
K_2HPO_4	5.0g
Yeast extract	5.0g
L-Cysteine·HCl·H_2O	0.5g
Chopped meat extract filtrate	1.0L
Chopped meat extract solids	200.0mL
Resazurin (0.025% solution)	4.0mL

pH 7.0 ± 0.2 at 25°C

Chopped Meat Extract:

Composition per liter:

Beef or horse meat	500.0g
NaOH (1*N* solution)	25.0mL

Preparation of Chopped Meat Extract: Use lean beef or horse meat. Remove fat and connective tissue. Grind. Add meat and NaOH to distilled/deionized water and bring volume to 1.0L. Gently heat and bring to boiling while stirring. Cool to 25°C. Remove fat from surface. Filter. Reserve ground meat particles and filtrate. Add distilled/deionized water to filtrate and bring volume to 1.0L.

Preparation of Medium: To 1.0L of chopped meat extract filtrate, add the remaining components, except L-cysteine and chopped meat solids. Mix thoroughly. Gently heat to boiling. Cool to room temperature. Add the L-cysteine. Adjust pH to 7.0. Distribute 1 part chopped meat solids (by volume) and 5 parts of liquid (by volume) into tubes under O_2-free 97% N_2 + 3% H_2. Cap with rubber stoppers and place tubes in a press. Autoclave for 15 min at 15 psi pressure–121°C with fast exhaust.

Use: For the cultivation and maintenance of anaerobic bacteria including *Bacteroides disiens, Coprococcus eutastus, Eubacterium formicigenerans, Prevotella disiens, Ruminococcus torques,* and *Streptococcus hansenii.*

Chopped Meat Medium with Menadione

Composition per 1230mL:

Peptone	30.0g
K_2HPO_4	5.0g
Yeast extract	5.0g
L-Cysteine·HCl·H_2O	0.5g
Chopped meat extract filtrate	1.0L
Chopped meat extract solids	200.0mL
Resazurin (0.025% solution)	4.0mL
Menadione solution	0.25mL

pH 7.0 ± 0.2 at 25°C

Chopped Meat Extract:

Composition per liter:

Beef or horse meat	500.0g
NaOH (1*N* solution)	25.0mL

Preparation of Chopped Meat Extract: Use lean beef or horse meat. Remove fat and connective tissue. Grind. Add meat and NaOH to distilled/deionized water and bring volume to 1.0L. Gently heat and bring to boiling while stirring. Cool to 25°C. Remove fat from surface. Filter. Reserve ground meat particles and filtrate. Add distilled/deionized water to filtrate and bring volume to 1.0L.

Menadione Solution:

Composition per liter:

Menadione (Vitamin K_3)	50.0μg
Ethanol (20% solution)	25.0mL

Preparation of Menadione Solution: Dissolve menadione in ethanol. Filter sterilize.

Preparation of Medium: To 1.0L of chopped meat extract filtrate, add the remaining components, except L-cysteine, menadione solution and chopped meat solids. Mix thoroughly. Gently heat to boiling. Cool to room temperature. Add the L-cysteine. Adjust pH to 7.0. Distribute 1 part chopped meat solids (by volume) and 5 parts of liquid (by volume) into tubes under O_2-free 97% N_2 + 3% H_2. Cap with rubber stoppers and place tubes in a press. Autoclave for 15 min at 15 psi pressure–121°C with fast exhaust. Prior to inoculation, add 0.25mL of menadione solution to each tube containing approximately 5.0mL of Chopped Meat Medium.

Use: For the cultivation and maintenance of *Bacteroides gingivalis, Bacteroides macacae* and *Porphyromonas gingivalis.*

Chopped Meat Medium with 1% Tween™ 80 (ATCC Medium 737)

Composition per 1230mL:

Peptone	30.0g
Tween™ 80	10.0g
K_2HPO_4	5.0g
Yeast extract	5.0g
L-Cysteine·HCl·H_2O	0.5g
Chopped meat extract filtrate	1.0L
Chopped meat extract solids	200.0mL
Resazurin (0.025% solution)	4.0mL

pH 7.0 ± 0.2 at 25°C

Chopped Meat Extract:

Composition per liter:

Beef or horse meat	500.0g
NaOH (1*N* solution)	25.0mL

Preparation of Chopped Meat Extract: Use lean beef or horse meat. Remove fat and connective tissue. Grind. Add meat and NaOH to distilled/deionized water and bring volume to 1.0L. Gently heat and bring to boiling while stirring. Cool to 25°C. Remove fat from surface. Filter. Reserve ground meat particles and filtrate. Add distilled/deionized water to filtrate and bring volume to 1.0L.

Preparation of Medium: To 1.0L of chopped meat extract filtrate, add the remaining components, except L-cysteine and chopped meat solids. Mix thoroughly. Gently heat to boiling. Cool to room temperature. Add the L-cysteine. Adjust pH to 7.0. Distribute 1 part chopped meat solids (by volume) and 5 parts of liquid (by volume) into tubes under O_2-free 97% N_2 + 3% H_2. Cap with rubber stoppers and place tubes in a press. Autoclave for 15 min at 15 psi pressure–121°C with fast exhaust.

Use: For the cultivation and maintenance of *Eubacterium biforme* and *Lactobacillus* species.

Chopped Meat Medium with 0.025% Tween™ 80 (ATCC Medium 1228)

Composition per 1230mL:

Peptone	30.0g
K_2HPO_4	5.0g
Yeast extract	5.0g
L-Cysteine·HCl·H_2O	0.5g
Tween™ 80	0.25g
Chopped meat extract filtrate	1.0L
Chopped meat extract solids	200.0mL
Resazurin (0.025% solution)	4.0mL

pH 7.0 ± 0.2 at 25°C

Chopped Meat Extract:

Composition per liter:

Beef or horse meat	500.0g
NaOH (1*N* solution)	25.0mL

Preparation of Chopped Meat Extract: Use lean beef or horse meat. Remove fat and connective tissue. Grind. Add meat and NaOH to distilled/deionized water and bring volume to 1.0L. Gently heat and bring to boiling while stirring. Cool to 25°C. Remove fat from surface. Filter. Reserve ground meat particles and filtrate. Add distilled/deionized water to filtrate and bring volume to 1.0L.

Preparation of Medium: To 1.0L of chopped meat extract filtrate, add the remaining components, except L-cysteine and chopped meat solids. Mix thoroughly. Gently heat to boiling. Cool to room temperature. Add the L-cysteine. Adjust pH to 7.0. Distribute 1 part chopped meat solids (by volume) and 5 parts of liquid (by volume) into tubes under O_2-free 97% N_2 + 3% H_2. Cap with rubber stoppers and place tubes in a press. Autoclave for 15 min at 15 psi pressure–121°C with fast exhaust.

Use: For the cultivation and maintenance of *Eubacterium* and *Lactobacillus* species.

Chopped Meat Medium with 10% Reduced Filtered Rumen Fluid

Composition per 1330mL:

Peptone	30.0g
K_2HPO_4	5.0g
Yeast extract	5.0g
L-Cysteine·HCl·H_2O	0.5g
Chopped meat extract filtrate	1.0L
Chopped meat extract solids	200.0mL
Rumen fluid, reduced and filtered	100.0mL
Resazurin (0.025% solution)	4.0mL

pH 7.0 ± 0.2 at 25°C

Chopped Meat Extract:

Composition per liter:

Beef or horse meat	500.0g
NaOH (1*N* solution)	25.0mL

Preparation of Chopped Meat Extract: Use lean beef or horse meat. Remove fat and connective tissue. Grind. Add meat and NaOH to distilled/deionized water and bring volume to 1.0L. Gently heat and bring to boiling while stirring. Cool to 25°C. Remove fat from surface. Filter. Reserve ground meat particles and filtrate. Add distilled/deionized water to filtrate and bring volume to 1.0L.

Preparation of Medium: To 1.0L of chopped meat extract filtrate, add the remaining components, except L-cysteine and chopped meat solids. Mix thoroughly. Gently heat to boiling. Cool to room temperature. Add the L-cysteine. Adjust pH to 7.0. Distribute 1 part chopped meat solids (by volume) and 5 parts of liquid (by volume) into tubes under O_2-free 97% N_2 + 3% H_2. Cap with rubber stoppers and place tubes in a press. Autoclave for 15 min at 15 psi pressure–121°C

Use: For cultivation and maintenance of *Eubacterium hallii.*

Chopped Meat Medium, Modified

Composition per 1230mL:

Pancreatic digest of casein	30.0g
Peptone	30.0g
Agar	20.0g
K_2HPO_4	5.0g
Yeast extract	5.0g
L-Cysteine·HCl·H_2O	0.5g
Chopped meat extract filtrate	1.0L
Chopped meat extract solids	200.0mL
Hemin solution	10.0mL
Resazurin (0.025% solution)	4.0mL
Vitamin K_1 solution	0.2mL

pH 7.0 ± 0.2 at 25°C

Chopped Meat Extract:

Composition per liter:

Beef or horse meat	500.0g
NaOH (1*N* solution)	25.0mL

Preparation of Chopped Meat Extract: Use lean beef or horse meat. Remove fat and connective tissue. Grind. Add meat and NaOH to distilled/deionized water and bring volume to 1.0L. Gently heat and bring to boiling while stirring. Cool to 25°C. Remove fat from surface. Filter. Reserve ground meat particles and filtrate. Add distilled/deionized water to filtrate and bring volume to 1.0L.

Hemin Solution:

Composition per 100mL:

Hemin	0.05g
NaOH (1*N* solution)	1.0mL

Preparation of Hemin Solution: Add components to distilled/deionized water and bring volume to 100.0mL. Mix thoroughly.

Vitamin K_1 Solution:

Composition per 30mL:

Ethanol (95% solution)	30.0mL
Vitamin K_1	0.15mL

Preparation of Vitamin K_1 Solution: Mix components. Store solution protected from light at 5°C. Discard after one month.

Preparation of Medium: To 1.0L of chopped meat extract filtrate, add the remaining components, except cysteine, hemin solution, vitamin K_1 solution and chopped meat solids. Mix thoroughly. Gently heat to boiling. Cool to room temperature. Add the cysteine, hemin solution and vitamin K_1 solution. Adjust pH to 7.0. Distribute 1 part chopped meat solids (by volume) and 5 parts of liquid (by volume) into tubes under O_2-free 97% N_2 + 3% H_2. Cap with rubber stoppers and place tubes in a press. Autoclave for 15 min at 15 psi pressure–121°C with fast exhaust.

Use: For cultivation and maintenance of a variety of anaerobic bacteria including *Actinomyces* species, *Bacteroides* species, *Clostridium* species, *Eubacterium* species, *Fusobacterium* species, *Peptostreptococcus* species, *Porphyromonas* species, *Prevotella* species, *Propionibacterium* species, *Selenomonas* species and others.

Chopped Meat Medium, Modified with Arginine

Composition per 1230mL:

Pancreatic digest of casein	30.0g
Peptone	30.0g
Agar	20.0g
Arginine	5.0g
K_2HPO_4	5.0g
Yeast extract	5.0g
L-Cysteine·HCl·H_2O	0.5g
Chopped meat extract filtrate	1.0L
Chopped meat extract solids	200.0mL
Hemin solution	10.0mL
Resazurin (0.025% solution)	4.0mL
Vitamin K_1 solution	0.2mL

pH 7.0 ± 0.2 at 25°C

Chopped Meat Extract:

Composition per liter:

Beef or horse meat	500.0g

NaOH (1*N* solution) 25.0mL
Tween™ 80 .... 25.0mL

Preparation of Chopped Meat Extract: Use lean beef or horse meat. Remove fat and connective tissue. Grind. Add meat and NaOH to distilled/deionized water and bring volume to 1.0L. Gently heat and bring to boiling while stirring. Cool to 25°C. Remove fat from surface. Filter. Reserve ground meat particles and filtrate. Add distilled/deionized water to filtrate and bring volume to 1.0L.

Hemin Solution:
Composition per 100mL:
Hemin 0.05g
NaOH (1*N* solution) 1.0mL

Preparation of Hemin Solution: Add components to distilled/deionized water and bring volume to 100.0mL. Mix thoroughly.

Vitamin K_1 Solution:
Composition per 30mL:
Ethanol (95% solution) 30.0mL
Vitamin K_1 .... 0.15mL

Preparation of Vitamin K_1 Solution: Mix components. Store solution protected from light at 5°C. Discard after one month.

Preparation of Medium: To 1.0L of chopped meat extract filtrate, add the remaining components, except L-cysteine, hemin solution, vitamin K_1 solution and chopped meat solids. Mix thoroughly. Gently heat to boiling. Cool to room temperature. Add the L-cysteine, hemin solution and vitamin K_1 solution. Adjust pH to 7.0. Distribute 1 part chopped meat solids (by volume) and 5 parts of liquid (by volume) into tubes under O_2-free 97% N_2 + 3% H_2. Cap with rubber stoppers and place tubes in a press. Autoclave for 15 min at 15 psi pressure–121°C with fast exhaust.

Use: For the cultivation and maintenance of *Eubacterium lentum*.

Chopped Meat Medium, Modified with Formate and Fumarate

Composition per 1230mL:
Pancreatic digest of casein 30.0g
Peptone 30.0g
Agar 20.0g
K_2HPO_4 .... 5.0g
Yeast extract 5.0g
L-Cysteine·HCl·H_2O 0.5g
Chopped meat extract filtrate 1.0L
Chopped meat extract solids 200.0mL
Hemin solution 10.0mL
Resazurin (0.025% solution) 4.0mL
Formate-fumarate solution 0.25mL
Vitamin K_1 solution 0.2mL
pH 7.0 ± 0.2 at 25°C

Chopped Meat Extract:
Composition per liter:
Beef or horse meat 500.0g
NaOH (1*N* solution) 25.0mL

Preparation of Chopped Meat Extract: Use lean beef or horse meat. Remove fat and connective tissue. Grind. Add meat and NaOH to distilled/deionized water and bring volume to 1.0L. Gently heat and bring to boiling while stirring. Cool to 25°C. Remove fat from surface. Filter. Reserve ground meat particles and filtrate. Add distilled/deionized water to filtrate and bring volume to 1.0L.

Hemin Solution:
Composition per 100mL:
Hemin 0.05g
NaOH (1*N* solution) 1.0mL

Preparation of Hemin Solution: Add components to distilled/deionized water and bring volume to 100.0mL. Mix thoroughly.

Formate-Fumarate Solution:
Composition per 100mL:
Sodium formate 6.0g
Fumaric acid 6.0g

Preparation of Formate-Fumarate Solution: Add components to distilled/deionized water and bring volume to 100.0mL. Adjust pH to 7.0. Filter sterilize.

Vitamin K_1 Solution:
Composition per 30mL:
Ethanol (95% solution) 30.0mL
Vitamin K_1 .... 0.15mL

Preparation of Vitamin K_1 Solution: Mix components. Store solution protected from light at 5°C. Discard after one month.

Preparation of Medium: To 1.0L of chopped meat extract filtrate, add the remaining components, except L-cysteine, hemin solution, vitamin K_1 solution, formate-fumarate solution and chopped meat solids. Mix thoroughly. Gently heat to boiling. Cool to room temperature. Add the L-cysteine, hemin solution and vitamin K_1 solution. Adjust pH to 7.0. Distribute 1 part chopped meat solids (by volume) and 5 parts of liquid (by volume) into tubes under O_2-free 97% N_2 + 3% H_2. Cap with rubber stoppers and place tubes in a press. Autoclave for 15 min at 15 psi pressure–121°C with fast exhaust. Prior to inoculation, add 0.25mL of formate-fumarate solution to each tube containing approximately 5.0mL of Chopped Meat Medium, Modified.

Use: For the cultivation and maintenance of *Bacteroides gracilis, Bacteroides ureolyticus, Campylobacter mucosalis* and *Wolinella succinogenes.*

Chopped Meat Medium, Modified with Tween™ 80

Composition per 1230mL:

Pancreatic digest of casein	30.0g
Peptone	30.0g
Agar	20.0g
K_2HPO_4	5.0g
Yeast extract	5.0g
L-Cysteine·HCl·H_2O	0.5g
Chopped meat extract filtrate	1.0L
Chopped meat extract solids	200.0mL
Hemin solution	10.0mL
Resazurin (0.025% solution)	4.0mL
Vitamin K_1 solution	0.2mL

pH 7.0 ± 0.2 at 25°C

Chopped Meat Extract:

Composition per liter:

Beef or horse meat	500.0g
NaOH (1*N* solution)	25.0mL
Tween™ 80	25.0mL

Preparation of Chopped Meat Extract: Use lean beef or horse meat. Remove fat and connective tissue. Grind. Add meat and NaOH to distilled/deionized water and bring volume to 1.0L. Gently heat and bring to boiling while stirring. Cool to 25°C. Remove fat from surface. Filter. Reserve ground meat particles and filtrate. Add distilled/deionized water to filtrate and bring volume to 1.0L.

Hemin Solution:

Composition per 100mL:

Hemin	0.05g
NaOH (1*N* solution)	1.0mL

Preparation of Hemin Solution: Add components to distilled/deionized water and bring volume to 100.0mL. Mix thoroughly.

Vitamin K_1 Solution:

Composition per 30mL:

Ethanol (95% solution)	30.0mL
Vitamin K_1	0.15mL

Preparation of Vitamin K_1 Solution: Mix components. Store solution protected from light at 5°C. Discard after one month.

Preparation of Medium: To 1.0L of chopped meat extract filtrate, add the remaining components, except L-cysteine, hemin solution, vitamin K_1 solution and chopped meat solids. Mix thoroughly. Gently heat to boiling. Cool to room temperature. Add the L-cysteine, hemin solution and vitamin K_1 solution. Adjust pH to 7.0. Distribute 1 part chopped meat solids (by volume) and 5 parts of liquid (by volume) into tubes under O_2-free 97% N_2 + 3% H_2. Cap with rubber stoppers and place tubes in a press. Autoclave for 15 min at 15 psi pressure–121°C with fast exhaust.

Use: For the cultivation and maintenance of *Lactobacillus* species and *Eubacterium biforme.*

Christensen Agar

Composition per liter:

Agar	15.0g
NaCl	5.0g
Sodium citrate	3.0g
KH_2PO_4	1.0g
Cysteine·HCl·H_2O	0.1g
Phenol Red	12.0mg

pH 6.9 ± 0.2 at 25°C

Preparation of Medium: Add components to distilled/deionized water and bring volume to 1.0L. Mix thoroughly. Gently heat and bring to boiling. Dispense into tubes or flasks. Autoclave for 15 min at 15 psi pressure–121°C. Pour into sterile Petri dishes or leave in tubes. Allow tubes to cool in a slanted position.

Use: For the differentiation of enteric pathogens, especially members of the Enterobacteriaceae, and coliforms based on their ability to utilize citrate as a carbon source. Bacteria that can utilize citrate turn the medium pink-red.

Christensen Agar

Composition per liter:

Agar	15.0g
NaCl	5.0g
Sodium citrate	3.0g
KH_2PO_4	1.0g
Yeast extract	0.5g
Glucose	0.2g
Cysteine·HCl·H_2O	0.1g
Phenol Red	12.0mg

pH 6.9 ± 0.2 at 25°C

Source: This medium is available as a premixed powder from Difco Laboratories.

Preparation of Medium: Add components to distilled/deionized water and bring volume to 1.0L. Mix thoroughly. Gently heat and bring to boiling. Dispense into tubes or flasks. Autoclave for 15 min at 15 psi pressure–121°C. Pour into sterile Petri dishes or leave in tubes. Allow tubes to cool in a slanted position.

Use: For the differentiation of enteric pathogens, especially members of the Enterobacteriaceae, and coliforms based on their ability to utilize citrate as a carbon source. Bacteria that can utilize citrate turn the medium pink-red.

Christensen Citrate Agar, Modified (Citrate Agar)

Agar	12.0g
NaCl	5.0g
Sodium citrate	3.8g
KH_2PO_4	1.0g
Yeast extract	0.5g
Glucose	0.2g
Cysteine·HCl·H_2O	0.1g
Phenol Red	0.02g

pH 6.7 ± 0.2 at 25°C

Preparation of Medium: Add components to distilled/deionized water and bring volume to 1.0L. Mix thoroughly. Gently heat and bring to boiling. Dispense into tubes or flasks. Autoclave for 15 min at 15 psi pressure–121°C. Pour into sterile Petri dishes or leave in tubes. Allow tubes to cool in a slanted position.

Use: For the differentiation of enteric pathogens, especially members of the Enterobacteriaceae, and coliforms based on their ability to utilize citrate as a carbon source. Bacteria that can utilize citrate turn the medium pink-red.

Christensen Citrate Sulfide Medium

Composition per liter:

Agar	15.0g
NaCl	5.0g
Sodium citrate·2H_2O	3.0g
KH_2HPO_4	1.0g
Yeast extract	0.5g
Ferric citrate	0.2g
Ammonium citrate	0.2g
Glucose	0.2g
Cysteine·HCl·H_2O	0.1g
$Na_2S_2O_3$·5H_2O	0.08g
Phenol Red	0.012g

pH 6.7± 0.2 at 25°C

Preparation of Medium: Add components to distilled/deionized water and bring volume to 1.0L. Mix thoroughly. Gently heat and bring to boiling. Dispense into tubes or flasks. Autoclave for 15 min at 15 psi pressure–121°C. Pour into sterile Petri dishes or leave in tubes. Allow tubes to cool in a slanted position.

Use: For the differentiation of enteric pathogens, especially members of the Enterobacteriaceae, and coliforms based on their ability to utilize citrate as a carbon source and production of H_2S. Bacteria that can utilize citrate turn the medium pink-red. H_2S production appears as a blackening of the butt of the tube.

Christensen Urea Agar Base
See: **Urea Agar**

Christensen's Urea Broth
See: **Urea Broth**

Chromatiaceae Medium 1

Composition per 4990mL:

Solution 1	4.0L
O_2-free water	860.0mL
$NaHCO_3$ solution	100.0mL
Na_2S·9H_2O solution	20.0mL
Trace element solution	5.0mL
Vitamin B_{12} solution	5.0mL

pH 7.3 ± 0.2 at 25°C

Solution 1:

Composition per 4L:

$MgSO_4$·7H_2O	2.5g
KCl	1.7g
KH_2PO_4	1.7g
NH_4Cl	1.7g
$CaCl_2$·2H_2O	1.25g

Preparation of Solution 1: Add components to distilled/deionized water and bring volume to 4.0L. Mix thoroughly. Autoclave for 45 min at 15 psi pressure–121°C. Cool to 25°C under 100% N_2. Saturate with CO_2 by stirring under 100% CO_2 for 30 min.

O_2-Free Water:

Composition per 860mL:

H_2O	860.0mL

Preparation of O_2-Free Water: Autoclave H_2O-for 15 min at 15 psi pressure–121°C. Cool to 25°C under 100% N_2.

$NaHCO_3$ Solution:

Composition per 100mL:

$NaHCO_3$	7.5g

Preparation of $NaHCO_3$ Solution: Add the $NaHCO_3$ to distilled/deionized water and bring volume to 100.0mL. Mix thoroughly. Gas with 100% CO_2 for 20 min. Filter sterilize with positive CO_2 pressure.

$Na_2S \cdot 9H_2O$ Solution:

Composition per 100mL:

$Na_2S \cdot 9H_2O$ 10.0g

Preparation of $Na_2S \cdot 9H_2O$ Solution: Add $Na_2S \cdot 9H_2O$ to distilled/deionized water. Mix thoroughly. Gas with 100% N_2 for 15 min in a screw-capped bottle. Tightly close cap. Autoclave for 15 min at 15 psi pressure–121°C. Cool to 25°C.

Trace Element Solution:

Composition per liter:

$FeCl_2 \cdot 4H_2O$ 1.5g
$CoCl_2 \cdot 6H_2O$ 0.19g
$MnCl_2 \cdot 4H_2O$ 0.10g
$ZnCl_2$ 0.07g
H_3BO_3 .... 0.06g
$NaMoO_4 \cdot 2H_2O$ 0.04g
$CuCl_2 \cdot 2H_2O$ 0.02g
$NiCl_2 \cdot 6H_20$ 0.02g
HCl (25% solution) 6.5mL

Preparation of Trace Element Solution: Add components to distilled/deionized water and bring volume to 1.0L. Mix thoroughly. Autoclave for 15 min at 15 psi pressure–121°C. Cool to 25°C.

Vitamin B_{12} Solution:

Composition per 100mL:

Vitamin B_{12} 2.0mg

Preparation of Vitamin B_{12} Solution: Add components to distilled/deionized water and bring volume to 100.0mL. Mix thoroughly. Filter sterilize.

Preparation of Medium: To 4.0L of sterile, CO_2-saturated solution 1, aseptically add the remaining components. Mix thoroughly. Adjust pH to 7.3. Aseptically distribute into sterile 100.0mL bottles using positive pressure of 95% N_2 + 5% CO_2. Completely fill bottles with medium except for a pea-sized air bubble.

Use: For the isolation and cultivation of members of the Chlorobiaceae.

Chromatiaceae Medium 2

Composition per 1051mL:

Solution 1 .... 950.0mL
$Na_2S \cdot 9H_2O$ solution 60.0mL
$NaHCO_3$ solution 40.0mL
Vitamin B_{12} solution 1.0mL

pH 7.3 ± 0.2 at 25°C

Solution 1:

Composition per 950mL:

KH_2PO_4 .... 1.0g
NH_4Cl 0.5g
$MgSO_4 \cdot 7H_2O$ 0.4g
$CaCl_2 \cdot 2H_2O$ 0.05g
Trace element solution SL-8 .... 1.0mL

Preparation of Solution 1: Add components to distilled/deionized water and bring volume to 950.0mL. Mix thoroughly. Autoclave for 15 min at 15 psi pressure–121°C. Cool to 45°–50°C.

Trace Element Solution SL-8:

Composition per liter:

Disodium EDTA 5.2g
$FeCl_2 \cdot 4H_2O$ 1.5g
$CoCl_2 \cdot 6H_2O$ 0.19g
$MnCl_2 \cdot 4H_2O$ 0.10g
$ZnCl_2$ 0.07g
H_3BO_3 .... 0.06g
$NaMoO_4 \cdot 2H_2O$ 0.04g
$CuCl_2 \cdot 2H_2O$ 0.02g
$NiCl_2 \cdot 6H_20$ 0.02g

Preparation of Trace Element Solution SL-8: Add components to distilled/deionized water and bring volume to 1.0L. Mix thoroughly.

$Na_2S \cdot 9H_2O$ Solution:

Composition per 100mL:

$Na_2S \cdot 9H_2O$ 5.0g

Preparation of $Na_2S \cdot 9H_2O$ Solution: Add $Na_2S \cdot 9H_2O$ to distilled/deionized water and bring volume to 100.0mL. Autoclave for 15 min at 15 psi pressure–121°C. Cool to 45°–50°C.

$NaHCO_3$ Solution:

Composition per 100mL:

$NaHCO_3$ 5.0g

Preparation of $NaHCO_3$ Solution: Add $NaHCO_3$ to distilled/deionized water and bring volume to 100.0mL. Mix thoroughly. Filter sterilize.

Vitamin B_{12} Solution:

Composition per 100mL:

Vitamin B_{12} 2.0mg

Preparation of Vitamin B_{12} Solution: Add vitamin B_{12} to distilled/deionized water and bring volume to 100.0mL. Mix thoroughly. Filter sterilize.

Preparation of Medium: To 950.0mL of cooled, sterile solution 1, aseptically add 60.0mL of sterile $Na_2S \cdot 9H_2O$ solution, 40.0mL of sterile $NaHCO_3$ solution, and 1.0mL of sterile vitamin B_{12} solution. Mix thoroughly. Adjust pH to 7.3 with sterile H_2SO_4 or Na_2CO_3. Aseptically distribute into sterile 50.0mL or 100.0mL bottles with metal screw-caps and rubber seals. Completely fill bottles with medium except for a pea-sized air bubble.

Use: For the isolation and cultivation of freshwater and soil members of Chromatiaceae.

Chromatium Medium (ATCC Medium 1449)

Composition per liter:

KH_2PO_4	0.5g
NH_4Cl	0.4g
$MgSO_4 \cdot 7H_2O$	0.2g
$CaCl_2 \cdot 2H_2O$	0.05g
Disodium EDTA	0.01g
Trace elements	1.0mL
$NaHCO_3$ solution	50.0mL
$Na_2S \cdot 9H_2O$ solution	50.0mL
Sodium pyruvate solution	50.0mL

pH 7.0 ± 0.2 at 25°C

Trace Elements:

Composition per liter:

Disodium EDTA	5.2g
$FeCl_2 \cdot 4H_2O$	1.5g
$CoCl_2 \cdot 6H_2O$	0.19g
$Na_2MoO_4 \cdot 2H_2O$	0.188g
$MnCl_2 \cdot 4H_2O$	0.1g
$ZnCl_2$	0.07g
$VOSO_4 \cdot 2H_2O$	0.03g
$NiCl_2 \cdot 6H_2O$	0.025g
H_3BO_3	6.0mg
$CuCl_2 \cdot 2H_2O$	2.0mg
Na_2SeO_3	2.0mg

Preparation of Trace Elements: Add components to distilled/deionized water and bring volume to 1.0L. Mix thoroughly.

$NaHCO_3$ Solution:

Composition per 50mL:

$NaHCO_3$	2.0g

Preparation of $NaHCO_3$ Solution: Add $NaHCO_3$ to distilled/deionized water and bring volume to 50.0mL. Filter sterilize. Use freshly prepared solution.

$Na_2S \cdot 9H_2O$ Solution:

Composition per 50mL:

$Na_2S \cdot 9H_2O$	1.0g

Preparation of $Na_2S \cdot 9H_2O$ Solution: Add $Na_2S \cdot 9H_2O$ to distilled/deionized water and bring volume to 50.0mL. Autoclave for 15 min at 15 psi pressure–121°C. Use freshly prepared solution.

Sodium Pyruvate Solution:

Composition per 50mL:

Sodium pyruvate	0.5g

Preparation of Sodium Pyruvate Solution: Add $NaHCO_3$ to distilled/deionized water and bring volume to 50.0mL. Filter sterilize. Use freshly prepared solution. Sodium acetate may be substituted for the sodium pyruvate.

Preparation of Medium: Add components, except $NaHCO_3$ solution, $Na_2S \cdot 9H_2O$ solution, and sodium pyruvate solution, to distilled deionized water and bring volume to 850.0mL. Autoclave for 15 min at 15 psi pressure–121°C. Cool to room temperature. Add the sterile $NaHCO_3$ solution, the sterile $Na_2S \cdot 9H_2O$ solution, and the sterile sodium pyruvate solution, in that order. Adjust the pH to 7.0. Distribute into screw-capped tubes or flasks. Fill to capacity.

Use: For cultivation and maintenance of *Chromatium* species.

Chromatium Medium (ATCC Medium 37)

Composition per 127mL:

Solution 1	76.2mL
Solution 2 + Solution 3	44.8mL
Solution 4	6.0mL

Solution 1:

Composition per 2.5 liters:

$CaCl_2$	2.0g

Preparation of Solution 1: Add $CaCl_2$ to distilled/deionized water and bring volume to 2.5L. Distribute in 80.0mL volumes into 127.0mL screw-capped bottles. Autoclave for 15 min at 15 psi pressure–121°C.

Solution 2:

Composition per 100mL:

Sodium ascorbate	2.4g
KCl	1.0g
KH_2PO_4	1.0g
$MgCl_2 \cdot 6H_2O$	0.8g
NH_4Cl	0.8g
Heavy metal solution	50.0mL
Vitamin solution	15.0mL
Vitamin B_{12} solution	3.0mL

Preparation of Solution 2: Add components to distilled/deionized water and bring volume to 100.0mL. Mix thoroughly.

Solution 3:

Composition per 900mL:

$NaHCO_3$	4.5g

Preparation of Solution 3: Add $NaHCO_3$ to distilled/deionized water and bring volume to 900.0mL. Mix thoroughly. Bubble 100% CO_2 through the solution for 30 min. After CO_2 saturation of Solution 3, add Solution 2 and immediately filter the mixture through a Seitz filter (or a Millipore) using positive CO_2 pressure to push the liquid through.

Solution 4:

Composition per 200mL:

$Na_2S \cdot 9H_2O$	3.0g

Preparation of Solution 4: Add $Na_2S \cdot 9H_2O$ to distilled/deionized water and bring volume to 200.0mL. Add a magnetic stir bar to the flask. Autoclave for 15 min at 15 psi pressure–121°C. On a magnetic stirrer, slowly add 2.0mL of sterile 2M H_2SO_4. This partially neutralizes the solution. The solution should turn yellow. H_2S gas will be liberated—neutralization and distribution of the solution should be done as rapidly as possible under adequate ventilation.

Heavy Metal Solution:
Composition per liter:

Ethylenediamine tetraacetate (EDTA)	1.5g
$FeSO_4 \cdot 7H_2O$	0.2g
$ZnSO_4 \cdot 7H_2O$	0.1g
$MnCl_2 \cdot 4H_2O$	0.02g
Modified Hoagland trace element solution	6.0mL

Preparation of Heavy Metal Solution: Dissolve EDTA in approximately 800.0mL of distilled/deionized water. Add remaining components. Bring volume to 1.0L with distilled/deionized water. Mix thoroughly.

Vitamin B_{12} Solution:
Composition per 100mL:

Vitamin B_{12} (cyanocobalamin)	2.0mg

Preparation of Vitamin B_{12} Solution: Add vitamin B_{12} to distilled/deionized water and bring volume to 100.0mL. Mix thoroughly.

Vitamin Solution:
Composition per 100mL:

Pyridoxamine·2HCl	5.0mg
Nicotinic acid	2.0mg
Thiamine	1.0mg
Pantothenic acid	0.5mg
Biotin	0.2mg
p-Aminobenzoic acid	0.1mg

Preparation of Vitamin Solution: Add components to distilled/deionized water and bring volume to 100.0mL. Mix thoroughly.

Modified Hoagland Trace Element Solution:
Composition per 3.6L:

H_3BO_3	11.0g
$MnCl_2 \cdot 4H_2O$	7.0g
$AlCl_3$	1.0g
$CoCl_2$	1.0g
$CuCl_2$	1.0g
KI	1.0g
$NiCl_2$	1.0g
$ZnCl_2$	1.0g
$BaCl_2$	0.5g
KBr	0.5g
LiCl	0.5g
Na_2MoO_4	0.5g
$SeCl_4$	0.5g
$SnCl_2 \cdot 2H_2O$	0.5g
$NaVO_3 \cdot H_2O$	0.1g

Preparation of Modified Hoagland Trace Element Solution: Prepare each component as a separate solution. Dissolve each salt in approximately 100.0mL of distilled/deionized water. Adjust the pH of each solution to below 7.0. Combine all the salt solutions and bring the volume to 3.6L with distilled/deionized water. Adjust the pH to 3–4. A yellow precipitate may form after mixing. After a few days, it will turn into a fine white precipitate. Mix the solution thoroughly before using.

Preparation of Medium: To the 80.0mL of sterile solution 1 in screw-capped bottles, add combined solutions 2 and 3 immediately after filtration and fill bottles to capacity. Mix thoroughly. Aseptically remove 6.0mL of the medium from the bottles and replace it with 6.0mL of neutralized solution 4. Let stand for 24 hr. The medium should form a fine white precipitate before using. To inoculate, remove 6.0mL of the completed medium from the bottles and replace it with 6.0mL of inoculum.

Use: For the cultivation and maintenance of *Chromatium tepidum*.

Chromobacterium Medium

Composition per liter:

NaCl	30.0g
$MgCl_2$	10.8g
$MgSO_4$	5.4g
Peptone	5.0g
$CaCl_2$	1.0g
KCl	0.7g

pH 7.0 ± 0.2 at 25°C

Preparation of Medium: Add components to distilled/deionized water and bring volume to 1.0L. Mix thoroughly. Distribute into tubes or flasks. Autoclave for 15 min at 15 psi pressure–121°C.

Use: For the cultivation and maintenance of *Chromobacterium* species and *Alteromonas luteoviolacea*.

Chromogenic Substrate Broth

Composition per liter:

NaCl	10.0g
HEPES (N-[2-Hydroxyethyl] piperazine-N´-[2-ethane-sulfonic acid]) buffer	6.9g
$(NH_4)_2SO_4$	5.0g
o-Nitrophenyl-β-D-galactopyranoside	0.50g

Solanium 0.50g
$MgSO_4$...0.10g
4-Methylumbelliferyl-β-D-glucuronide...........0.075g
$CaCl_2$...0.05g
Na_2SO_3...0.04g
Amphotericin B.. 1.0mg
$MnSO_4$.. 0.5mg
$ZnSO_4$.. 0.5mg

Preparation of Medium: Add components to distilled/deionized water and bring volume to 1.0L. Mix thoroughly. Distribute into tubes or flasks. Autoclave for 15 min at 15 psi pressure–121°C.

Use: For the detection of coliform bacteria based on their hydrolysis of chromogenic substrates by production of β-D-galactopyranosidase. Bacteria that produce β-D-galactopyranosidase turn the medium yellow.

Chu's No. 10 Medium

Composition per liter:

Agar...15.0g
$Ca(NO_3)_2·4H_2O$...0.232g
$Na_2SiO_3·5H_2O$...0.044g
$MgSO_4·7H_2O$ 0.025g
Na_2CO_3..0.02g
K_2HPO_4..0.01g
Citric acid.. 3.5mg
Ferric citrate.. 3.5mg

Preparation of Medium: Add components to distilled/deionized water and bring volume to 1.0L. Mix thoroughly. Gently heat to boiling. Distribute into tubes or flasks. Autoclave for 15 min at 15 psi pressure–121°C. Pour into sterile Petri dishes or leave in tubes.

Use: For the cultivation and maintenance of *Anabaena* species and *Plectomena boryanum.*

Chu's No. 10 Medium, Modified

Composition per liter:

Agar...15.0g
$Ca(NO_3)_2·4H_2O$...0.232g
$Na_2SiO_3·5H_2O$...0.044g
$MgSO_4·7H_2O$ 0.025g
Na_2CO_3..0.02g
K_2HPO_4..0.01g
Citric acid.. 3.5mg
Ferric citrate.. 3.5mg
Metal solution 1.0mL

Metal Solution:

Composition per liter:

H_3BO_3 ....2.4g
$MnCl_2·4H_2O$...1.4g
$ZnCl_2$..0.4g
$CoCl_2·6H_2O$...0.02g
$CuCl_2·2H_2O$... 0.1mg

Preparation of Metal Solution: Add components to distilled/deionized water and bring volume to 1.0L. Mix thoroughly.

Preparation of Medium: Add components to distilled/deionized water and bring volume to 1.0L. Mix thoroughly. Gently heat to boiling. Distribute into tubes or flasks. Autoclave for 15 min at 15 psi pressure–121°C. Pour into sterile Petri dishes or leave in tubes.

Use: For the cultivation and maintenance of *Anabaena* species and *Plectomena boryanum.*

Chu's No. 11 Medium, Modified

Composition per liter:

$NaNO_3$...1.5g
$MgSO_4·7H_2O$ 0.08g
$Na_2SiO_3·9H_2O$...0.06g
$CaCl_2·2H_2O$...0.04g
$K_2HPO_4·3H_2O$...0.04g
Na_2CO_3..0.02g
Citric acid.. 6.0mg
Ferric ammonium citrate............................... 6.0mg
EDTA 1.0mg
Trace metal solution A5 with cobalt 1.0mL

pH 7.5 ± 0.2 at 25°C

Trace Metal Solution A5 With Cobalt:

Composition per liter:

H_3BO_3 ....2.86g
$MnCl_2·4H_2O$...1.81g
$Na_2MoO_4·2H_2O$...0.390g
$ZnSO_4·7H_2O$..0.222g
$CuSO_4·H_2O$ 0.079g
$Co(NO_3)_2·6H_2O$...0.049g

Preparation of Trace Metal Solution A5 With Cobalt: Add components to distilled/deionized water and bring volume to 1.0L. Mix thoroughly.

Preparation of Medium: Add components to seawater and bring volume to 1.0L. Mix thoroughly. Gently heat and bring to boiling. Distribute into tubes or flasks. Autoclave for 15 min at 15 psi pressure–121°C.

Use: For the isolation and cultivation of cyanobacteria from marine habitats.

CIN Agar (*Yersinia* Selective Agar) (Cefsulodin Irgasan® Novobiocin Agar)

Composition per liter:

Mannitol	20.0g
Agar	12.0g
Pancreatic digest of gelatin	10.0g
Beef extract	5.0g
Peptic digest of animal tissue	5.0g
Sodium pyruvate	2.0g
Yeast extract	2.0g
NaCl	1.0g
Sodium deoxycholate	0.5g
Neutral Red	0.03g
Cefsulodin	0.015g
Irgasan®(triclosan)	4.0mg
Novobiocin	2.5mg
Crystal Violet	1.0mg

pH 7.4 ± 0.2 at 25°C

Source: This medium is available as a premixed powder from BBL Microbiology Systems.

Preparation of Medium: Add components, except cefsulodin and novobiocin, to distilled/deionized water and bring volume to 1.0L. Heat, mixing continuously, until boiling. Do not autoclave. Cool to 45°–50°C. Aseptically add cefsulodin and novobiocin. Mix thoroughly. Pour into sterile Petri dishes or distribute into sterile tubes.

Use: For the selective isolation and differentiation of *Yersinia enterocolitica* from a variety of clinical and nonclinical specimens based on mannitol fermentation. *Yersinia enterocolitica* appears as "bull's eye" colonies with deep red centers surrounded by a transparent periphery.

Cinnamate Medium

Composition per liter:

$NaHCO_3$	2.5g
$MgCl_2 \cdot 6H_2O$	2.03g
Cinnamic acid	1.48g
KH_2PO_4	1.36g
NH_4Cl	0.53g
$Na_2S \cdot 9H_2O$	0.24g
$CaCl_2 \cdot 2H_2O$	0.15g
Yeast extract	0.05g
Modified Wolfe's metals	10.0mL
Wolfe's vitamin solution	10.0mL

pH 7.5–7.7 at 25°C

Modified Wolfe's Metals:

Composition per liter:

Na_2SeO_3	10.0mg
$NaWO_4 \cdot 2H_2O$	10.0mg
$NiCl_2 \cdot 6H_2O$	10.0mg
Wolfe's metals solution	1.0L

Preparation of Modified Wolfe's Metals: Combine the components. Mix thoroughly.

Wolfe's Metals Solution:

Composition per liter:

$MgSO_4 \cdot 7H_2O$	3.0g
Nitriloacetic acid	1.5g
NaCl	1.0g
$MnSO_4 \cdot H_2O$	0.5g
$CaCl_2$	0.1g
$CoCl_2 \cdot 6H_2O$	0.1g
$FeSO_4 \cdot 7H_2O$	0.1g
$ZnSO_4 \cdot 7H_2O$	0.1g
$AlK(SO_4)_2 \cdot 12H_2O$	0.01g
$CuSO_4 \cdot 5H_2O$	0.01g
H_3BO_3	0.01g
$Na_2MoO_4 \cdot 2H_2O$	0.01g

Preparation of Wolfe's Metals Solution: Add nitrilotriacetic acid to 500.0mL of distilled/deionized water. Adjust pH to 6.5 with KOH. Add distilled/deionized water to 1.0L. Add remaining components. Mix thoroughly.

Wolfe's Vitamin Solution:

Composition per liter:

Pyridoxine·HCl	0.01g
p-Aminobenzoic acid	5.0mg
Calcium pantothenate	5.0mg
Nicotinic acid	5.0mg
Riboflavin	5.0mg
Thiamine·HCl	5.0mg
Thioctic acid	5.0mg
Biotin	2.0mg
Folic acid	2.0mg
Cyanocobalamin	100.0μg

Preparation of Wolfe's Vitamin Solution: Add components to distilled/deionized water and bring volume to 1.0L. Mix thoroughly.

Preparation of Medium: Add all components, except $NaHCO_3$ and $Na_2S \cdot 9H_2O$, to distilled/deionized water and bring volume to 1.0L. Gently heat and bring to boiling under 90% N_2 + 10% CO_2. Cool medium to room temperature while continuing to gas with 90%N_2 + 10%CO_2. Add $NaHCO_3$ and $Na_2S \cdot 9H_2O$. Adjust pH to 7.5–7.7. Distribute into tubes under 90% N_2 + 10% CO_2 using anaerobic techniques. Autoclave for 15 min at 15 psi pressure–121°C.

Use: For the cultivation of anaerobic bacteria that can utilize cinnamic acid as a carbon source. As a basal medium for the cultivation of *Formivibrio citricus*.

Citrate Agar
See: Simmons Citrate Agar
See: Christensen Citrate Agar, Modified

Citrate Medium, Koser's Modified

Composition per liter:

NaCl	5.0g
Citric acid	2.0g
$(NH_4)H_2PO_4$	1.0g
K_2HPO_4	1.0g
$MgSO_4 \cdot 7H_2O$	0.2g

pH 6.8 ± 0.2 at 25°C

Preparation of Medium: Add components to distilled/deionized water and bring volume to 1.0L. Mix thoroughly. Adjust pH to 6.8. Distribute into tubes in 5.0mL volumes. Autoclave for 15 min at 15 psi pressure–121°C.

Use: For the cultivation and differentiation of bacteria based on their ability to utilize citrate as carbon source.

Citrate Phosphate Buffered Glucose Medium

Composition per liter:

Solution A	750.0mL
Solution C	200.0mL
Solution B	50.0mL

pH 3.5 ± 0.2 at 25°C

Solution A:

Composition per 750mL:

$(NH_4)_2SO_4$	3.0g
Citric acid, anhydrous	1.92g
Na_2HPO_4	1.23g
$MgSO_4 \cdot 7H_2O$	0.5g
KCl	0.1g
$Ca(NO_3)_2 \cdot 4H_2O$	0.02g
$FeSO_4 \cdot 7H_2O$	0.01g

Preparation of Solution A: Add components to distilled/deionized water and bring volume to 750.0mL. Mix thoroughly.

Solution B:

Composition per 50mL:

Glucose	10.0g
Yeast extract	1.0g

Preparation of Solution B: Add components to distilled/deionized water and bring volume to 50.0mL. Mix thoroughly.

Solution C:

Composition per 200mL:

Agarose (electrophoresis grade)	6.0g

Preparation of Solution C: Add agarose to distilled/deionized water and bring volume to 200.0mL. Mix thoroughly.

Preparation of Medium: Prepare Solutions A, B, and C. Autoclave solutions separately for 15 min at 15 psi pressure–121°C. Cool to 50°–55°C. Combine Solutions A, B, and C. Mix thoroughly. Immediately distribute into sterile tubes or flasks.

Use: For the cultivation and maintenance of *Acidiphilium organovorum.*

Citrobacter diversus Medium

Composition per liter:

Component 1	500.0mL
Component 2	250.0mL
Component 3	250.0mL

Component 1:

Composition per 500mL:

Agar	7.5g
Pancreatic digest of casein	7.5g
Papaic digest of soybean meal	2.5g

Preparation of Component 1: Add components to distilled/deionized water and bring volume to 500.0mL. Mix thoroughly. Gently heat and bring to boiling. Autoclave for 15 min at 15 psi pressure–121°C. Cool to 45°–50°C.

Component 2:

Composition per 250mL:

Agar	7.5g
Pancreatic digest of casein	7.5g
Papaic digest of soybean meal	2.5g
L-Tyrosine	1.0g

Preparation of Component 2: Add components to distilled/deionized water and bring volume to 250.0mL. Mix thoroughly. Gently heat and bring to boiling. Autoclave for 15 min at 15 psi pressure–121°C. Cool to 45°–50°C.

Component 3:

Composition per 250mL:

L-Tyrosine	1.0g

Preparation of Component 3: Add components to distilled/deionized water and bring volume to 250.0mL. Mix thoroughly. Gently heat and bring to boiling. Autoclave for 15 min at 15 psi pressure–121°C. Rapidly cool in ice water (0°C) for 20 min. A fine white crystalline precipitate should form.

Preparation of Medium: Pour sterile component 1 into sterile Petri dishes in 18.0mL volumes. Allow agar to solidify. Warm component 3 to 50°C for 20 min. Aseptically combine component 2 and component 3. Mix thoroughly. Pour 7.0mL of mixture over the solidified component 1 agar.

Use: For the isolation and cultivation of *Citrobacter diversus*.

Citrovorum Factor Assay Medium
See: **CF Assay Medium**

CK Agar

Composition per liter:

Agar	15.0g
Glucose	5.0g
KNO_3	2.0g
$CaCl_2$	1.5g
$MgSO_4 \cdot 7H_2O$	1.5g
K_2HPO_4	0.25g
Ferric citrate	0.02g

Preparation of Medium: Add components to distilled/deionized water and bring volume to 1.0L. Mix thoroughly. Gently heat and bring to boiling. Distribute into tubes or flasks. Autoclave for 15 min at 15 psi pressure–121°C. Pour into sterile Petri dishes or leave in tubes.

Use: For the cultivation of myxobacteria.

CK1 Medium

Composition per liter:

$MgSO_4 \cdot 7H_2O$	3.0g
KNO_3	2.0g
$CaCl_2$	1.4g
Ferric citrate	0.02g
Glucose solution	100.0mL
K_2HPO_4 solution	10.0mL

Glucose Solution:

Composition per 100mL:

D-Glucose	10.0g

Preparation of Glucose Solution: Add D-glucose to distilled/deionized water and bring volume to 100.0mL. Mix thoroughly. Autoclave for 15 min at 15 psi pressure–121°C. Cool to 25°C.

K_2HPO_4 Solution:

Composition per 10mL:

K_2HPO_4	2.5mg

Preparation of K_2HPO_4 Solution: Add K_2HPO_4 to distilled/deionized water and bring volume to 10.0mL. Mix thoroughly. Autoclave for 15 min at 15 psi pressure–121°C. Cool to 25°C.

Preparation of Medium: Add components, except glucose solution and K_2HPO_4 solution, to distilled/deionized water and bring volume to 890.0mL. Mix thoroughly. Autoclave for 15 min at 15 psi pressure–121°C. Cool to 25°C. Aseptically add sterile glucose solution and K_2HPO_4 solution. Mix thoroughly. Aseptically distribute into sterile tubes or flasks.

Use: For the cultivation of myxobacteria.

Clausen Medium (Dithionite Thioglycollate, HS T, Broth)

Composition per liter:

Pancreatic digest of casein	15.0g
Glucose	6.0g
Yeast extract	6.0g
Glycerol	5.0g
Papaic digest of soybean meal	3.0g
Tween™ 80	3.0g
NaCl	2.5g
K_2HPO_4	2.0g
L-Asparagine	1.25g
Sodium citrate	1.0g
Agar	0.75g
L-Cystine	0.5g
Sodium thioglycollate	0.5g
$MgSO_4 \cdot 7H_2O$	0.4g
Sodium dithionite	0.4g
Lecithin	0.3g
$CaCl_2 \cdot 2H_2O$	4.0mg
$MnCl_2 \cdot 4H_2O$	2.0mg
$CoSO_4 \cdot 7H_2O$	1.0mg
$CuSO_4 \cdot 5H_2O$	1.0mg
$FeSO_4 \cdot 7H_2O$	1.0mg
$ZnSO_4 \cdot 7H_2O$	1.0mg
Resazurin	1.0mg

pH 7.1 ± 0.2 at 25°C

Source: This medium is available as a premixed powder from Oxoid Unipath.

Preparation of Medium: Add Tween™ 80 and glycerol to distilled/deionized water and bring volume to 1.0L. Gently heat and bring to boiling. Distribute into tubes or flasks. Autoclave for 15 min at 15 psi pressure–121°C. The medium must not be resterilized.

Use: For sterility testing by the membrane filter method or the tube dilution method to determine the presence of microbial contamination in a variety of specimens.

CLED Agar (Cystine Lactose Electrolyte Deficient Agar) (Brolacin Agar)

Composition per liter:

Agar	15.0g
Lactose	10.0g
Pancreatic digest of casein	4.0g
Pancreatic digest of gelatin	4.0g
Beef extract	3.0g
L-Cystine	0.128g
Bromthymol Blue	0.02g

pH 7.3 ± 0.2 at 25°C

Source: This medium is available as a premixed powder from BBL Microbiology Systems, Difco Laboratories and Oxoid Unipath.

Preparation of Medium: Add components to distilled/deionized water and bring volume to 1.0L. Mix thoroughly. Gently heat while stirring and bring to boiling. Autoclave for 15 min at 15 psi pressure–121°C. Cool to 50°–55°C. Pour into sterile Petri dishes or distribute into sterile tubes.

Use: For the isolation, enumeration and presumptive identification of microorganisms from urine.

CLED Agar with Andrade Indicator (Cystine Lactose Electrolyte Deficient Agar with Andrade Indicator)

Composition per liter:

Agar	15.0g
Pancreatic digest of casein	10.0g
Peptone	4.0g
Beef extract	3.0g
Cystine	0.128g
Bromthymol Blue	0.02g
Andrade indicator	10.0mL

pH 7.5 ± 0.2

Source: This medium is available as a premixed powder from Oxoid Unipath.

Caution: Acid Fuchsin is a potential carcinogen and care must be taken to avoid inhalation of the powdered dye and contamination of the skin.

Andrade's Indicator:

Composition per 100mL:

NaOH (1*N* solution)	16.0mL
Acid Fuchsin	0.1g

Preparation of Andrade's Indicator: Add Acid Fuchsin to NaOH solution and bring volume to 100.0mL with distilled/deionized water.

Preparation of Medium: Add components to distilled/deionized water and bring volume to 1.0L. Mix thoroughly. Gently heat while stirring and bring to boiling. Autoclave for 15 min at 15 psi pressure–121°C. Cool to 50°–55°C. Pour into sterile Petri dishes or distribute into sterile tubes.

Use: For the differentiation of microorganisms based on colony characteristics.

Clostridia Medium

Composition per liter:

Sodium L-lactate	10.0g
Sodium acetate	8.0g
K_2HPO_4	0.5g
$(NH_4)_2 \cdot 7H_2O$	0.5g
Sodium thioglycollate	0.5g
Yeast extract	0.5g
$MgSO_4 \cdot 7H_2O$	0.1g
$FeSO_4 \cdot 7H_2O$	0.02g
p-Aminobenzoate	100.0μg
Biotin	0.1μg

pH 6.0–7.0 at 25°C

Preparation of Medium: Add components to distilled/deionized water and bring volume to 1.0L. Mix thoroughly. Adjust pH to 6.0–7.0. Distribute into tubes or flasks. Autoclave for 20 min at 15 psi pressure–121°C.

Use: For the isolation and cultivation of *Clostridium* species that ferment lactate and acetate.

Clostridium acidurici Medium

Composition per liter:

Uric acid	2.0g
Yeast extract	1.0g
K_2HPO_4	0.91g
KOH	0.67g
Sodium thioglycollate	0.5g
$MgSO_4 \cdot 7H_2O$	0.25g
$CaCl_2 \cdot 2H_2O$	0.015g
$FeSO_4 \cdot 7H_2O$	6.0mg
$NaHCO_3$ solution	25.0mL
Sodium thioglycollate solution	25.0mL

pH 7.0–7.5 at 25°C

$NaHCO_3$ Solution:
Composition per 25mL:

$NaHCO_3$	5.0g

Preparation of $NaHCO_3$ Solution: Add $NaHCO_3$ to distilled/deionized water and bring volume to 25.0mL. Mix thoroughly. Filter sterilize.

Sodium Thioglycollate Solution:
Composition per 25mL:

Sodium thioglycollate	0.5g

Preparation of Sodium Thioglycollate Solution: Add sodium thioglycollate to distilled/deionized water and bring volume to 25.0mL. Mix thoroughly. Autoclave solution separately for 15 min at 15 psi pressure–121°C.

Preparation of Medium: Add K_2HPO_4 and KOH to distilled/deionized water and bring volume to 1.0L. Add uric acid. Gently heat until boiling. Add the remaining components, except $NaHCO_3$ and sodium thioglycollate. Mix thoroughly. Adjust pH to 7.0–7.5. Distribute into tubes or flasks. Autoclave for 15 min at 15 psi pressure–121°C. Add 0.25mL of sterile $NaHCO_3$ solution and 0.25mL of sterile sodium thioglycollate solution for each 10.0mL of sterile basal medium.

Use: For the cultivation and maintenance of *Clostridium acidurici, Clostridium purinolyticum* and other bacteria that can utilize uric acid as a carbon source.

Clostridium aerotolerans Medium

Composition per liter:

Agar	15.0g
Xylan	5.0g
Yeast extract	5.0g
Na_2CO_3	4.0g
NaCl	0.45g
$(NH_4)_2SO_4$	0.45g
K_2HPO_4	0.225g
KH_2PO_4	0.225g
Cysteine·HCl·H_2O	0.125g
$Na_2S·9H_2O$	0.125g
$MgSO_4·7H_2O$	0.09g
$CaCl_2$	0.045g

pH 7.0 ± 0.2 at 25°C

Preparation of Medium: Prepare medium anaerobically under 100% CO_2. Add components to distilled/deionized water and bring volume to 1.0L. Mix thoroughly. Gently heat while stirring and bring to boiling. Autoclave for 15 min at 15 psi pressure–121°C. Cool to 50°–55°C. Adjust pH to 7.0. Pour into sterile Petri dishes or distribute into sterile tubes.

Use: For the cultivation and maintenance of *Clostridium aerotolerans*.

Clostridium Alginate Medium

Composition per liter of sea water:

Agar	15.0g
Sodium alginate	10.0g
K_2HPO_4	2.0g
Peptone	1.0g
Yeast extract	1.0g
Sea water	1.0L

pH 7.0–7.5 at 25°C

Preparation of Medium: Add K_2HPO_4 to 1.0L of sea water. Mix thoroughly. Gently heat while stirring to dissolve. Filter solution twice. Add remaining components. Mix thoroughly. Adjust pH to 7.0–7.5. Autoclave for 15 min at 15 psi pressure–121°C. Pour into sterile Petri dishes or distribute into sterile tubes.

Use: For the cultivation and maintenance of *Clostridium alginolyticum* and other bacteria which can utilize alginate as a carbon source.

Clostridium aminobutyricum Medium

Composition per liter:

K_2HPO_4	7.05g
Yeast extract	3.0g
KH_2PO_4	1.29g
$MgCl_2·6H_2O$	0.2g
$CaCl_2·2H_2O$	0.01g
$FeCl_3·6H_2O$	0.01g
Methylene Blue	2.0mg
$MnSO_4·H_2O$	1.0mg
$Na_2MoO_4·2H_2O$	1.0mg
γ-Aminobutyrate solution	100.0mL
Na_2CO_3 solution	100.0mL
$Na_2S·9H_2O$ solution	50.0mL

pH 7.4–7.7 at 25°C

γ-Aminobutyrate Solution:
Composition per 100mL:

γ-Aminobutyrate	5.0g

Preparation of γ-Aminobutyrate Solution: Add γ-aminobutyrate to distilled/deionized water and bring volume to 100.0mL. Mix thoroughly. Filter sterilize.

Na_2CO_3 Solution:
Composition per 100mL:

Na_2CO_3	2.0g

Preparation of Na_2CO_3 Solution: Add Na_2CO_3 to distilled/deionized water and bring volume to 100.0mL. Mix thoroughly. Autoclave for 15 min at 15 psi pressure–121°C.

$Na_2S·9H_2O$ Solution:
Composition per 50mL:

$Na_2S·9H_2O$	0.3g

Preparation of $Na_2S·9H_2O$ Solution: Add $Na_2S·9H_2O$ to distilled/deionized water and bring volume to 50.0mL. Mix thoroughly. Autoclave for 15 min at 15 psi pressure–121°C.

Preparation of Medium: Add components, except γ-aminobutyrate solution, Na_2CO_3 solution, and $Na_2S·9H_2O$ solution, to distilled/deionized water and bring volume to 750.0mL. Autoclave for 15 min at 15 psi pressure–121°C. Cool under 80% N_2 + 10% CO_2 + 10% H_2. Aseptically add the sterile γ-aminobutyrate solution, Na_2CO_3 solution, and $Na_2S·9H_2O$ solution. Adjust pH to 7.4–7.7. Distribute using anaerobic technique into tubes or flasks.

Use: For the cultivation and maintenance of *Clostridium aminobutyricum* and other bacteria which can utilize aminobutyric acid as a carbon source.

Clostridium aminovalericum Medium

Composition per liter:

K_2HPO_4	9.77g
5-Aminovaleric acid	6.0g
Yeast extract	5.0g
Mannitol	1.0g
KH_2PO_4	0.54g
Sodium thioglycollate	0.5g
$MgSO_4$	0.06g
$CaSO_4·2H_2O$	0.034g
Trace elements SL-6	1.0mL

pH 7.9 ± 0.2 at 25°C

Trace Elements Solution SL-6:
Composition per liter:

H_3BO_3	0.3g
$CoCl_2·6H_2O$	0.2g
$ZnSO_4·7H_2O$	0.10g
$MnCl_2·4H_2O$	0.03g
$Na_2MoO_4·H_2O$	0.03g
$NiCl_2·6H_2O$	0.02g
$CuCl_2·2H_2O$	0.01g

Preparation of Trace Elements Solution SL-6: Add components to distilled deionized water and bring volume to 1.0L. Mix thoroughly. Adjust pH to 3.4.

Preparation of Medium: Prepare medium under 100% N_2. Add components to distilled/deionized water and bring volume to 1.0L. Autoclave for 15 min at 15 psi pressure–121°C. Adjust pH to 7.9. Distribute into tubes or flasks under 100% N_2.

Use: For the cultivation and maintenance of *Clostridium aminovalericum* and other bacteria that can utilize aminovaleric acid as a carbon source.

Clostridium botulinum Isolation Agar (CBI Agar)

Composition per 1033mL:

Egg yolk agar base	900.0mL
Egg yolk emulsion, 50%	100.0mL
Cycloserine solution	25.0mL
Sulfamethoxazole solution	4.0mL
Trimethoprim solution	4.0mL

pH7.4 ± 0.2 at 25°C

Egg Yolk Agar Base:
Composition per 900mL:

Pancreatic digest of casein	40.0g
Agar	20.0g
Na_2HPO_4	5.0g
Yeast extract	5.0g
Glucose	2.0g
NaCl	2.0g
$MgSO_4·7H_2O$ solution	0.2mL

Preparation of Egg Yolk Agar Base: Add components to distilled/deionized water and bring volume to 900.0mL. Mix thoroughly. Gently heat to boiling. Autoclave for 15 min at 15 psi pressure–121°C. Cool to 45°–50°C.

$MgSO_4·7H_2O$ Solution:
Composition per 100mL:

$MgSO_4·7H_2O$	5.0g

Preparation of $MgSO_4·7H_2O$ Solution: Add $MgSO_4·7H_2O$ to distilled/deionized water and bring volume to 100.0mL. Mix thoroughly.

Cycloserine Solution:
Composition per 100mL:

Cycloserine	1.0g

Preparation of Cycloserine Solution: Add cycloserine to distilled/deionized water and bring volume to 100.0mL. Mix thoroughly. Filter sterilize.

Sulfamethoxazole Solution:
Composition per 100mL:

Sulfamethoxazole	1.9g

Preparation of Sulfamethoxazole Solution: Add sulfamethoxazole to distilled/deionized water and bring volume to 50.0mL. Add sufficient 10% NaOH to dissolve. Bring volume to 100.0mL with distilled/deionized water. Mix thoroughly. Filter sterilize.

Trimethoprim Solution:
Composition per 100mL:

Trimethoprim	0.1g

Preparation of Trimethoprim Solution: Add trimethoprim to distilled/deionized water and bring volume to 50.0mL. Gently heat to 55°C. Add sufficient 0.05*N* HCl to dissolve. Bring volume to 100.0mL with distilled/deionized water. Mix thoroughly. Filter sterilize.

Egg Yolk Emulsion, 50%:
Composition per 100mL:

Chicken egg yolks	11
Whole chicken egg	1
NaCl (0.9% solution)	50.0mL

Preparation of Egg Yolk Emulsion, 50%: Soak eggs with 1:100 dilution of saturated mercuric chloride solution for 1 min. Crack eggs and separate yolks from whites. Mix egg yolks with 1 chicken egg. Measure 50.0mL of egg yolk emulsion and add to 50.0mL of 0.9% NaCl solution. Mix thoroughly. Filter sterilize. Warm to 45°–50°C.

Preparation of Medium: Aseptically add warmed, sterile egg yolk emulsion, 50%, and sterile cycloserine solution, sterile sulfamethoxazole solution, and sterile trimethoprim solution to cooled, sterile egg yolk agar base. Mix thoroughly. Pour into sterile Petri dishes.

Use: For isolation, cultivation and differentiation based on lipase activity of *Clostridium botulinum*, types A, B, and F from fecal specimens associated with food-borne and infant botulism. *C. botulinum* types A, B, and F appear as raised colonies surrounded by an opaque zone. Other *Clostridium* species and *C. botulinum* type G appear as pinpoint colonies with no opaque zone.

Clostridium Cellulolytic Medium

Composition per liter:

Agar	20.0g
Cellulose	7.5g
$K_2HPO_4 \cdot 3H_2O$	2.9g
Yeast extract	2.0g
KH_2PO_4	1.5g
$(NH_4)_2SO_4$	1.3g
$FeSO_4$	1.25g
Cysteine·HCl·H_2O	1.0g
$MgCl_2 \cdot 6H_2O$	1.0g
$CaCl_2 \cdot 2H_2O$	0.15g
Resazurin	2.0mg

pH 7.5 ± 0.2 at 25°C

Preparation of Medium: Add components, except cysteine·HCl·H_2O, to distilled/deionized water and bring volume to 1.0L. Mix thoroughly. Heat to boiling. Adjust pH to 7.5. Prereduce under 100% N_2. Add cysteine·HCl·H_2O. Distribute into tubes under 100% N_2. Cap tubes with rubber stoppers. Autoclave for 15 min at 15 psi pressure–121°C.

Use: For the cultivation and maintenance of *Clostridium cellulolyticum* and other bacteria which can degrade cellulose.

Clostridium Cellulose Medium

Composition per liter:

Agar	20.0g
Filter paper (or 5.0g Avicel)	10.0g
$CaCO_3$	5.0g
Polypeptone™	5.0g
$Na_2CO_3 \cdot 10H_2O$	4.0g
K_2HPO_4	2.2g
Yeast extract	2.0g
KH_2PO_4	1.5g
$(NH_4)_2SO_4$	1.3g
$MgCl_2 \cdot 6H_2O$	1.0g
Cysteine·HCl·H_2O	0.5g
$CaCl_2$	0.15g
$FeSO_4 \cdot 7H_2O$	6.0mg

pH 7.0 ± 0.2 at 25°C

Preparation of Medium: Add components to distilled/deionized water and bring volume to 1.0L. Mix thoroughly. Heat to boiling. Autoclave for 15 min at 15 psi pressure–121°C. Pour into sterile Petri dishes or distribute into sterile tubes.

Use: For the cultivation and maintenance of *Clostridium cellulolyticum* and other bacteria which can degrade cellulose.

Clostridium chauvoei Blood Agar

Composition per 100mL:

Liver extract	3.0g
Agar	1.6g
Glucose	1.0g
VL broth base	94.0mL
Sheep blood, defibrinated	5.0mL

pH 7.2–7.4 at 25°C

VL Broth Base:
Composition per liter:

Pancreatic digest of casein	10.0g
NaCl	5.0g
Yeast extract	5.0g
Meat extract	2.0g
Agar	0.6g
Cysteine·HCl·H_2O	0.4g

Preparation of VL Broth Base: Add components to distilled/deionized water and bring volume to 1.0L. Mix thoroughly. Gently heat until dissolved. Adjust pH to 7.2–7.4.

Preparation of Medium: Add liver extract, glucose and agar to 94.0mL of VL broth base. Mix thoroughly. Gently heat and bring to boiling. Autoclave for 15 min at 15 psi pressure–121°C. Cool to 45°–50°C. Aseptically add sterile sheep blood. Mix thoroughly. Pour into sterile Petri dishes or distribute into sterile tubes

Use: For the isolation and cultivation of *Clostridium chauvoei*.

Clostridium difficile Agar

Composition per liter:

Clostridum difficile agar base	920.0mL
Clostridium difficile selective supplement	10.0mL
Horse blood, defibrinated	70.0mL

pH 7.4 ± 0.2 at 25°C

Source: This medium is available as a premixed powder from Oxoid Unipath.

***Clostridum difficile* Agar Base:**
Composition per 920mL:

Proteose peptone	40.0g
Agar	15.0g
Fructose	6.0g
Na_2HPO_4	5.0g
NaCl	2.0g
KH_2PO_4	1.0g
$MgSO_4 \cdot 7H_2O$	0.1g

Preparation of *Clostridum difficile* Agar Base: Add components to distilled/deionized water and bring volume to 920.0mL. Mix thoroughly. Gently heat to boiling. Autoclave for 15 min at 15 psi pressure–121°C. Cool to 45°–50°C.

***Clostridium difficile* Selective Supplement:**
Composition per 10mL:

D-Cycloserine	500.0mg
Cefoxitin	16.0mg

Preparation of *Clostridium difficile* Selective Supplement: Add components to distilled/deionized water and bring volume to 10.0mL. Mix thoroughly. Filter sterilize.

Preparation of Medium: Add 10.0mL of sterile *Clostridium difficile* selective supplement and 70.0mL of sterile, defibrinated horse blood to 920.0mL of cooled, sterile *Clostridum difficile* Agar Base. Mix thoroughly. Pour into sterile Petri dishes or distribute into sterile tubes.

Use: For the selective isolation and cultivation of *Clostridium difficile* from clinical and nonclinical specimens.

Clostridium difficile Agar (Cycloserine Cefoxitin Fructose Agar) (CCFA)

Composition per liter:

Peptic digest of animal tissue	32.0g
Agar	20.0g
Fructose	6.0g
Na_2HPO_4	5.0g
NaCl	2.0g
KH_2PO_4	1.0g
Cycloserine	0.25g
$MgSO_4$	0.1g
Neutral Red	0.03g
Cefoxitin solution	10.0mL

pH 7.2 ± 0.2 at 25°C

Source: This medium is available as a premixed powder from BBL Microbiology Systems.

Cefoxitin Solution:
Composition per 10mL:

Cefoxitin	16.0mg

Preparation of Cefoxitin Solution: Add cefoxitin to distilled/deionized water and bring volume to 10.0mL. Mix thoroughly. Filter sterilize.

Preparation of Medium: Add components to distilled/deionized water and bring volume to 990.0mL. Mix thoroughly. Gently heat to boiling. Autoclave for 15 min at 15 psi pressure–121°C. Cool to 45°–50°C. Aseptically add 10.0mL of sterile cefoxitin solution. Mix thoroughly. Pour into sterile Petri dishes or distribute into sterile tubes.

Use: For the selective isolation and cultivation of *Clostridium difficile* from clinical and nonclinical specimens.

Clostridium kluyveri Medium

Composition per liter:

Potassium acetate	5.0g
Sodium thioglycollate	0.5g
K_2HPO_4	0.3g
NH_4Cl	0.25g
KH_2PO_4	0.2g
$MgSO_4 \cdot 7H_2O$	0.2g
$CaCl_2 \cdot 2H_2O$	0.01g
$FeSO_4 \cdot 7H_2O$	5.0mg
$MnSO_4 \cdot 4H_2O$	2.0mg
$Na_2MoO_4 \cdot 2H_2O$	2.0mg
p-Aminobenzoate	0.2mg
Biotin	0.01mg
Ethanol	20.0mL
Acetic acid, glacial	2.5mL

pH 7.0 ± 0.2 at 25°C

Preparation of Medium: Add components, except sodium thioglycollate to distilled/deionized water and bring volume to 1.0L. Gently heat and bring to boiling. Mix thoroughly. Add sodium thioglycollate immediately prior to sterilization. Mix thoroughly. Autoclave for 15 min at 15 psi pressure–121°C. Adjust pH to 7.0 with sterile 60% K_2CO_3 solution.

Use: For the isolation and cultivation of *Clostridium kluyveri*.

Clostridium kluyveri Medium

Composition per liter:

Part A	965.0mL
Part B	35.0mL

pH 7.0 ± 0.2 at 25°C

Part A:

Composition per 965mL:

Sodium acetate·$3H_2O$	7.5g
$(NH_4)_2SO_4$	2.65g
Agar	2.0g
Yeast extract	2.0g
Sodium thioglycollate	0.5g
p-Aminobenzoic acid	0.1mg
Biotin	5.0μg
Potassium phosphate buffer (2M, pH 7.0)	10.0mL
Salt solution	10.0mL

Preparation of Part A: Add components to distilled/deionized water and bring volume to 965.0mL. Mix thoroughly. Gently heat and bring to boiling. Autoclave for 15 min at 15 psi pressure–121°C. Cool to room temperature.

Salt Solution:

Composition per 100mL:

$MgSO_4 \cdot H_2O$	2.5g
$CaCl_2$	0.15g
$FeSO_4 \cdot 7H_2O$	0.15g
$MnSO_4 \cdot 2H_2O$	0.02g
$Na_2MoO_4 \cdot 2H_2O$	0.02g

Preparation of Salt Solution: Add components to distilled/deionized water and bring volume to 100.0mL. Mix thoroughly.

Part B:

Composition:

K_2CO_3 (1 M solution)	20.0mL
Ethanol (95% solution)	15.0mL

Preparation of Part B: Prepare a 1M solution of K_2CO_3 and filter sterilize. Filter sterilize 25.0mL of 95% ethanol solution. Aseptically combine 20.0mL of sterile K_2CO_3 solution and 15.0mL of sterile ethanol.

Preparation of Medium: Add 35.0mL of sterile Part B to 965.0mL of sterile, cooled Part A. Adjust pH to 7.0 with HCl. Aseptically distribute into tubes under 97% N_2 + 3% H_2. Cap with rubber stoppers.

Use: For the cultivation and maintenance of *Clostridium kluyveri*.

Clostridium Medium (ATCC Medium 39)

Composition per liter:

K_2HPO_4	7.0g
γ-Aminobutyric acid	5.0g
Yeast extract	3.0g
Agar	1.5g
KH_2PO_4	1.3g
$MgCl_2 \cdot 6H_2O$	0.2g
$CaCl_2 \cdot 2H_2O$	0.01g
$FeCl_3 \cdot 6H_2O$	0.01g
Methylene Blue	2.0mg
$MnSO_4$	1.0mg
Na_2MoO_4	1.0mg
$Na_2S \cdot 9H_2O$ solution	10.0mL

$Na_2S \cdot 9H_2O$ solution:

Composition per 20mL:

$Na_2S \cdot 9H_2O$	0.6g

Preparation of $Na_2S \cdot 9H_2O$ Solution: Add $Na_2S \cdot 9H_2O$ to distilled/deionized water and bring volume to 20.0mL. Autoclave for 15 min at 15 psi pressure–121°C. Use freshly prepared solution.

Preparation of Medium: Add components, except $Na_2S \cdot 9H_2O$, to distilled/deionized water and bring volume to 1.0L. Mix thoroughly. Gently heat to boiling. Autoclave for 15 min at 15 psi pressure–121°C. Cool to 45°–50°C. Distribute anaerobically into sterile tubes. Aseptically add 0.1mL of sterile 1.5% $Na_2S \cdot 9H_2O$ solution to each 5.0mL of the medium. Cap with rubber stoppers.

Use: For the cultivation and maintenance of a variety of *Clostridium* species.

Clostridium Medium

Composition per liter:

Sodium L-glutamate	10.0g
Sodium thioglycollate	0.5g
Yeast extract	0.5g
K_2HPO_4	0.2g
$MgSO_4 \cdot 7H_2O$	0.1g

pH 7.6 ± 0.2 at 25°C

Preparation of Medium: Add components to distilled/deionized water and bring volume to 1.0L. Mix thoroughly. Distribute into tubes or flasks. Autoclave for 15 min at 15 psi pressure–121°C.

Use: For the enrichment and isolation of glutamate-fermenting *Clostridium* species.

Clostridium Medium

Composition per liter:

Uric acid	2.0g
Yeast extract	1.2g
$MgSO_4 \cdot 7H_2O$	0.05g
$CaCl_2 \cdot 2H_2O$	5.0mg
$FeSO_4 \cdot 7H_2O$	2.0mg
Resazurin	1.0mg
KOH (10*N* solution)	3.0mL
$K_2HPO_4 \cdot 3H_2O$ (70% solution)	1.5mL
Mercaptoacetic acid	1.5mL

pH 7.2 ± 0.2 at 25°C

Preparation of Medium: Add KOH solution and $K_2HPO_4 \cdot 3H_2O$ solution to distilled/deionized water and bring volume to 500.0mL. Gently heat and bring to boiling. Mix thoroughly. Add uric acid slowly. Cool to 45°–50°C. Add remaining components. Add mercaptoacetic acid immediately prior to sterilization. Bring volume to 1.0L with distilled/deionized water. Mix thoroughly. Autoclave for 15 min at 15 psi pressure–121°C. Adjust pH to 7.2 with sterile 60% K_2CO_3 solution.

Use: For the isolation and cultivation of purine-fermenting *Clostridium* species.

Clostridium Medium (ATCC Medium 40)

Composition per liter:

K_2HPO_4	7.0g
δ-Aminovaleric acid·HCl (neutralized)	5.0g
Agar	1.5g
KH_2PO_4	1.3g
Yeast extract	1.0g
$MgCl_2 \cdot 6H_2O$	0.2g
$CaCl_2 \cdot 2H_2O$	0.01g
$FeCl_3 \cdot 6H_2O$	0.01g
Methylene Blue	2.0mg
$MnSO_4$	1.0mg
Na_2MoO_4	1.0mg
$Na_2S \cdot 9H_2O$ solution	20.0mL

$Na_2S \cdot 9H_2O$ Solution:

Composition per 100mL:

$Na_2S \cdot 9H_2O$	1.5g

Preparation of $Na_2S \cdot 9H_2O$ Solution: Add $Na_2S \cdot 9H_2O$ to distilled/deionized water and bring volume to 100.0mL. Autoclave for 15 min at 15 psi pressure–121°C. Use freshly prepared solution.

Preparation of Medium: Add components, except $Na_2S \cdot 9H_2O$ solution, to distilled/deionized water and bring volume to 1.0L. Mix thoroughly. Gently heat to boiling. Autoclave for 15 min at 15 psi pressure–121°C. Cool to 45°–50°C. Distribute anaerobically into sterile tubes. Aseptically add 0.1mL of sterile $Na_2S \cdot 9H_2O$ solution to each 5.0mL of the medium. Cap with rubber stoppers.

Use: For the cultivation and maintenance of a variety of *Clostridium* species.

Clostridium Medium (ATCC Medium 43)

Composition per liter:

Agar	15.0g
Yeast extract	5.0g
L-Arginine·HCl	2.0g
L-Lysine·HCl	2.0g
NH_4Cl	2.0g
Sodium formate	2.0g
K_2HPO_4	1.75g
$MgSO_4 \cdot 7H_2O$	0.2g
$CaCl_2 \cdot 2H_20$	0.01g
$FeSO_4 \cdot 7H_2O$	0.01g
Methylene Blue	2.0mg
$Na_2S \cdot 9H_2O$ solution	30.0mL

$Na_2S \cdot 9H_2O$ Solution:

Composition per 100mL:

$Na_2S \cdot 9H_2O$	1.0g

Preparation of $Na_2S \cdot 9H_2O$ Solution: Add $Na_2S \cdot 9H_2O$ to distilled/deionized water and bring volume to 100.0mL. Autoclave for 15 min at 15 psi pressure–121°C. Use freshly prepared solution.

Preparation of Medium: Add components, except $Na_2S \cdot 9H_2O$ solution, to tap water and bring volume to 1.0L. Mix thoroughly. Gently heat to boiling. Autoclave for 15 min at 15 psi pressure–121°C. Cool to 45°–50°C. Distribute anaerobically into sterile tubes. Aseptically add 0.15mL of sterile $Na_2S \cdot 9H_2O$ solution to each 5.0mL of the medium. Cap with rubber stoppers.

Use: For the cultivation and maintenance of a variety of *Clostridium* species.

Clostridium Medium (ATCC Medium 163)

Composition per liter:

Agar	20.0g
Sodium glutamate	17.0g
Yeast extract	6.0g
Sodium thioglycollate	0.5g

Phosphate buffer (1.0 M, pH 7.4)	40.0mL
$MgSO_4$ (2.0M solution)	0.5mL
$FeSO_4$ (0.2M solution)	0.2mL
$CaCl_2$ (1.0M solution)	0.1mL
$CoCl_2$ (0.1M solution)	0.1mL
$MnCl_2$ (0.1M solution)	0.1mL
Na_2MoO_4 (0.1M solution)	0.1mL

Preparation of Medium: Add components to distilled/deionized water and bring volume to 1.0L. Mix thoroughly. Gently heat to boiling. Autoclave for 15 min at 15 psi pressure–121°C. Pour into sterile Petri dishes or distribute into sterile tubes.

Use: For the cultivation and maintenance of a variety of *Clostridium* species.

Clostridium Medium (ATCC Medium 511)

Composition per liter:

Yeast extract	4.0g
Alanine	3.0g
Peptone	3.0g
Cysteine	0.2g
$MgSO_4$	0.05g
$FeSO_4$	0.01g
Potassium phosphate buffer (1M, pH 7.1)	5.0mL
$CaSO_4$ (saturated solution)	2.5mL

Preparation of Medium: Add components to distilled/deionized water and bring volume to 1.0L. Mix thoroughly. Distribute into tubes or flasks. Autoclave for 15 min at 15 psi pressure–121°C.

Use: For the cultivation of a variety of *Clostridium* species.

Clostridium Medium (ATCC Medium 568)

Composition per liter:

Na_2CO_3	10.0g
Fructose	3.0g
K_2HPO_4	2.0g
Yeast extract	2.0g
$(NH_4)_2SO_4$	1.0g
$MgSO_4 \cdot 7H_2O$	0.5g
Sodium thioglycollate	0.05g
$CaSO_4$	0.015g
$FeSO_4 \cdot 7H_2O$	2.5mg
$MnSO_4 \cdot H_2O$	0.5mg
$Na_2MoO_4 \cdot 2H_2O$	0.5mg

pH 7.8 ± 0.2 at 25°C

Preparation of Medium: Add components to distilled/deionized water and bring volume to 1.0L. Mix thoroughly. Distribute into tubes or flasks. Autoclave for 15 min at 15 psi pressure–121°C.

Use: For the cultivation of a variety of *Clostridium* species.

Clostridium Medium (ATCC Medium 591)

Composition per liter:

Solution 1	600.0mL
Solution 2	400.0mL

pH 8.0 ± 0.2 at 25°C

Solution 1:

Composition per 600mL:

Peptone	5.0g

Preparation of Solution 1: Add component to distilled/deionized water and bring volume to 600.0mL. Mix thoroughly. Autoclave for 15 min at 15 psi pressure–121°C.

Solution 2:

Composition per 400mL:

$NaHCO_3$	20.0g
Fructose	10.0g
K_2HPO_4	10.0g
Sodium thioglycollate	0.75g
Vitamin solution	14.0mL
Trace elements solution	10.0mL

Preparation of Solution 2: Add components, except sodium thioglycollate, to distilled/deionized water and bring volume to 400.0mL. Mix thoroughly. Gas with 100% CO_2. Add sodium thioglycollate. Adjust pH to 8.0. Filter sterilize.

Vitamin Solution:

Composition per 100mL:

Thiamine	0.1g
Nicotinic acid	0.05g
Pyridoxine	0.05g
Pantothenic acid	0.025g
p-Aminobenzoic acid	5.0mg
Vitamin B_{12}	2.0mg
Biotin	1.0mg

Preparation of Vitamin Solution: Add components to distilled/deionized water and bring volume to 100.0mL. Mix thoroughly.

Trace Elements Solution:

Composition per liter:

EDTA	0.5g
$FeSO_4 \cdot 7H_2O$	0.2g
H_3BO_3	0.03g
$CoCl_2 \cdot 6H_2O$	0.02g
$ZnSO_4 \cdot 7H_2O$	0.01g
$MnCl_2 \cdot 4H_2O$	3.0mg

$Na_2MoO_4 \cdot 2H_2O$	3.0mg
$NiCl_2 \cdot 6H_2O$	2.0mg
$CuCl_2 \cdot 2H_2O$	1.0mg

Preparation of Trace Elements Solution: Add components to distilled/deionized water and bring volume to 1.0L. Mix thoroughly.

Preparation of Medium: Aseptically combine 600.0mL of sterile solution 1 and 400.0mL of sterile solution 2. Distribute into sterile tubes or flasks.

Use: For the cultivation and maintenance of a variety of *Clostridium* species.

Clostridium noterae Medium

Composition per liter:

Yeast extract	2.0g
NH_4Cl	1.0g
NaCl	0.45g
$K_2HPO_4 \cdot 3H_2O$	0.4g
Cysteine·HCl·H_2O	0.15g
Na_2CO_3 solution	30.0mL
Trace metals solution	10.0mL
$Na_2S \cdot 9H_2O$ solution	10.0mL

pH 7.9 ± 0.1 at 25°C

Na_2CO_3 Solution:
Composition per 50mL:

Na_2CO_3	5.0g

Preparation of Na_2CO_3 Solution: Add Na_2CO_3 to distilled/deionized water and bring volume to 50.0mL. Autoclave for 15 min at 15 psi pressure–121°C. Use freshly prepared solution.

Trace Metals Solution:
Composition per liter:

$Na_2EDTA \cdot 2H_2O$	0.50g
$CoCl_2 \cdot 6H_2O$	0.15g
$FeSO_4 \cdot 7H_2O$	0.10g
$MnCl_2 \cdot 4H_2O$	0.10g
$ZnCl_2$	0.10g
$AlCl_3 \cdot 6H_2O$	0.04g
$CuCl_2 \cdot 2H_2O$	0.02g
$NiSO_4 \cdot 6H_2O$	0.02g
H_2SeO_3	0.01g
H_3BO_3	0.01g
$Na_2MoO_4 \cdot 2H_2O$	0.01g

Preparation of Trace Metals Solution: Add components to distilled/deionized water and bring volume to 1.0L. Mix thoroughly.

$Na_2S \cdot 9H_2O$ Solution:
Composition per 10mL:

$Na_2S \cdot 9H_2O$	0.15g

Preparation of $Na_2S \cdot 9H_2O$ Solution: Add $Na_2S \cdot 9H_2O$ to distilled/deionized water and bring volume to 10.0mL. Autoclave for 15 min at 15 psi pressure–121°C. Use freshly prepared solution.

Preparation of Medium: Add components, except Na_2CO_3 solution and $Na_2S \cdot 9H_2O$ solution, to distilled/deionized water and bring volume to 1.0L. Mix thoroughly. Adjust pH to 7.0 with 10M NaOH. Gently heat to boiling. Distribute under O_2-free 100% N_2 gas into tubes in 5.0mL volumes. Cap with rubber stoppers. Autoclave for 15 min at 15 psi pressure–121°C. Prior to inoculation, add to each tube 0.15mL Na_2CO_3 solution and 0.05mL $Na_2S \cdot 9H_2O$ solution. Incubate under 80% H_2 + 20% CO_2 to provide conditions for H_2-fixation.

Use: For the cultivation and maintenance of *Clostridium noterae*.

Clostridium novyi Blood Agar

Composition per 100mL:

Agar	2.0g
Glucose	1.0g
Neopeptone	1.0g
Proteolyzed liver	0.5g
Yeast extract	0.5g
Horse blood, defibrinated	10.0mL
Reducing solution	0.75mL
Salts solution	0.5mL

pH 7.6–7.8 at 25°C

Salts Solution:
Composition per 100mL:

$MgSO_4 \cdot 7H_2O$	4.0g
$MnSO_4 \cdot 4H_2O$	0.2g
HCl	0.05g
$FeCl_3$	0.04g

Preparation of Salts Solution: Add components to distilled/deionized water and bring volume to 100.0mL. Mix thoroughly.

Reducing Solution:
Composition per 10mL:

Cysteine·HCl·H_2O	0.12g
Dithiothreitol	0.12g
Glutamine	0.06g

Preparation of Reducing Solution: Add components to distilled/deionized water and bring volume to 10.0mL. Mix thoroughly. Adjust pH to 7.6–7.8. Filter sterilize.

Preparation of Medium: Add agar to distilled/deionized water and bring volume to 50.0mL. Mix thoroughly. Gently heat and bring to boiling. In another flask add neopeptone, yeast extract, liver extract and salts solution to distilled/deionized water and bring volume to 50.0mL. Mix thoroughly. Gently heat until dissolved. Combine the two solutions. Distibute into screw-capped bottles in 18.0mL volumes. Autoclave for 10 min at 10 psi pressure–115°C. Cool to 45°–50°C. Medium may be stored at 4°C at this point. Immediately prior to inoculation, aseptically add 2.0mL of horse blood and 0.15mL of sterile reducing solution to each tube of melted agar at 50°C. Mix thoroughly. Pour the contents of each tube into a sterile Petri dish.

Use: For the cultivation of *Clostridium novyi*.

Clostridium perfringens Agar, OPSP (Perfringens Agar, OPSP)

Composition per liter:

Pancreatic digest of casein	15.0g
Agar	10.0g
Liver extract	7.0g
Papaic digest of soybean meal	5.0g
Yeast extract	5.0g
Tris(hydroxymethyl)aminomethane buffer	1.5g
Ferric ammonium citrate	1.0g
$Na_2S_2O_5$	1.0g
Antibiotic inhibitor	10.0mL

pH 7.3 ± 0.2 at 25°C

Source: This medium is available as a premixed powder from Oxoid Unipath.

Antibiotic Inhibitor:

Composition per 10mL:

Sodium sulfadiazine	0.1g
Oleandomycin phosphate	0.5mg
Polymyxin B	10,000U

Preparation of Antibiotic Inhibitor: Add components to distilled/deionized water and bring volume to 10.0mL. Mix thoroughly. Filter sterilize.

Preparation of Medium: Add components, except antibiotic inhibitor, to distilled/deionized water and bring volume to 990.0mL. Mix thoroughly. Gently heat and bring to boiling. Autoclave for 15 min at 15 psi pressure–121°C. Cool to 45°–50°C. Aseptically add sterile antibiotic inhibitor. Mix thoroughly. Pour into sterile Petri dishes or distribute into sterile tubes.

Use: For the presumptive identification and enumeration of *Clostridium perfringens* in foods.

Clostridium Selective Agar (Clostrisel Agar)

Composition per liter:

Pancreatic digest of casein	17.0g
Agar	14.0g
Glucose	6.0g
Papaic digest of soybean meal	3.0g
NaCl	2.5g
Sodium thioglycollate	1.8g
Sodium formaldehyde sulfoxylate	1.0g
L-Cystine	0.25g
NaN_3	0.15g
Neomycin sulfate	0.15g

pH 7.0 ± 0.2 at 25ºC

Source: This medium is available as a premixed powder from BBL Microbiology Systems.

Preparation of Medium: Add components to distilled/deionized water and bring volume to 1.0L. Mix thoroughly. Gently heat while stirring and bring to boiling. Distribute into tubes or flasks. Autoclave for 15 min at 15 psi pressure–118°C. Pour into sterile Petri dishes or leave in tubes.

Caution: Sodium azide is toxic. Azides also react with metals and disposal must be highly diluted.

Use: For the selective isolation of pathogenic *Clostridium* species from specimens containing mixed flora, e.g., from wounds, fecal specimens, soil, and other specimens.

Clostridium sphenoides Medium

Composition per liter:

Agar	15.0g
Trisodium citrate·$2H_2O$	14.7g
Yeast extract	4.0g
KH_2PO_4	3.4g
K_2HPO_4	2.0g
Peptone	2.0g
NaCl	0.6g
L-Cysteine·HCl	0.3g
$(NH_4)_2SO_4$	0.3g
$MgSO_4 \cdot 7H_2O$	0.2g
$CaCl_2 \cdot 2H_2O$	0.06g
Resazurin	1.0mg

pH 6.7–7.0 ± 0.2 at 25°C

Preparation of Medium: Add components, except L-cysteine·HCl, to distilled/deionized water and bring volume to 1.0L. Mix thoroughly. Gently heat and bring to boiling. Add L-cysteine·HCl. Distribute anaerobically into tubes in 5.0mL volumes. Autoclave for 20 min at 15 psi pressure–121°C. Cool to 45°–50°C. Inoculate with serial dilution of mud specimens before agar solidifies.

Use: For the isolation of *Clostridium sphenoides* from mud.

Clostridium thermocellum Medium

Composition per liter:

Filter paper	18.75g
$Na_2HPO_4 \cdot 12H_2O$	4.2g
Yeast extract	2.0g
KH_2PO_4	1.5g
NH_4Cl	0.5g
$MgCl_2 \cdot 6H_2O$	0.18g
Reducing solution	40.0mL
Wolfe's modified mineral elixir	5.0mL
Resazurin (0.1% solution)	1.0mL
Vitamin solution	0.5mL

Caution: This medium contains Na_2S, and H_2S production will occur, especially upon prolonged boiling. H_2S is hazardous and preparation of this medium should be done in a chemical fume hood.

Reducing Solution:

Composition per 200mL:

Cysteine·HCl·H_2O	2.5g
$Na_2S \cdot 9H_2O$	2.5g
NaOH (0.2*N* solution)	200.0mL

Preparation of Reducing Solution: Gently heat the NaOH solution and bring to boiling. Gas with 95% N_2 + 5% H_2. Cool to room temperature. Add the cysteine·HCl·H_2O and $Na_2S \cdot 9H_2O$. Anaerobically distribute into tubes. Cap with rubber stoppers. Autoclave for 15 min at 15 psi pressure–121°C.

Vitamin Solution:

Composition per 500mL:

Pyridoxine HCl	0.1g
p-Aminobenzoic acid	0.05g
Calcium pantothenate	0.05g
Nicotinic acid	0.05g
Thioctic acid	0.05g
Biotin	0.02g
Folic acid	0.02g
Riboflavin	5.0mg
Thiamine·HCl	5.0mg
Vitamin B_{12}	1.0mg

Preparation of Vitamin Solution: Add components to distilled/deionized water and bring volume to 500.0mL. Mix thoroughly. Store solution in the dark at –10°C.

Wolfe's Modified Mineral Elixir:

Composition per liter:

$MgSO_4 \cdot 7H_2O$	3.0g
Nitrilotriacetic acid	1.5g
NaCl	1.0g
$MnSO_4 \cdot H_2O$	0.5g
$CaCl_2$, anhydrous	0.1g
$Co(NO_3)_2 \cdot 6H_2O$	0.1g
$FeSO_4 \cdot 7H_2O$	0.1g
$ZnSO_4 \cdot 7H_2O$	0.1g
$AlK(SO_4)_2$, anhydrous	0.01g
$CuSO_4 \cdot 5H_2O$	0.01g
H_3BO_3	0.01g
$Na_2MoO_4 \cdot 2H_2O$	0.01g
Na_2SeO_3, anhydrous	1.0mg

Preparation of Wolfe's Modified Mineral Elixir: Add nitrilotriacetic acid to 500.0mL of distilled/deionized water. Dissolve by adjusting pH to 6.5 with KOH. Add remaining components. Add distilled/deionized water to 1.0L.

Preparation of Medium: Add components, except reducing solution, to distilled/deionized water and bring volume to 1.0L. If medium is to be distributed into tubes, omit bulk filter paper and substitute one Whatman #1 filter paper strip (8mm × 70mm) per tube of broth. Gently heat and bring to boiling under 95% N_2 + 5% H_2. Continue boiling until color changes from blue to pink. Add the reducing solution. The pink color will disappear, indicating that the solution has been reduced. Distribute into tubes or flasks under 95% N_2 + 5% H_2 using anerobic techniques. If tubes are used, remember to add Whatman #1 filter paper strips, prior to the addition of broth. Cap tubes with rubber stoppers. Autoclave for 15 min at 15 psi pressure–121°C.

Use: For the cultivation and maintenance of *Clostridium thermocellum*.

Clostridium thermolacticum Medium

Composition per liter:

$KHCO_3$	4.5g
NaCl	2.25g
Sucrose	2.0g
Yeast extract	2.0g
$MgSO_4 \cdot 7H_2O$	0.5g
NH_4Cl	0.5g
K_2HPO_4	0.348g
Cysteine·HCl·H_2O solution	0.3g
$Na_2S \cdot 9H_2O$ solution	0.3g
$CaCl_2 \cdot 2H_2O$	0.25g
KH_2PO_4	0.227g
$FeSO_4 \cdot 7H_2O$	2.0mg
Resazurin	1.0mg
Cysteine·HCl·H_2O solution	25.0mL
$Na_2S \cdot 9H_2O$ solution	25.0mL
Wolfe's vitamin solution	10.0mL
Trace element solution SL-6	3.0mL

pH 7.0–7.2 at 25°C

Wolfe's Vitamin Solution:

Composition per liter:

Pyridoxine·HCl	10.0mg
p-Aminobenzoic acid	5.0mg
Calcium pantothenate	5.0mg
Nicotinic acid	5.0mg
Riboflavin	5.0mg
Thiamine·HCl	5.0mg
Thioctic acid	5.0mg
Biotin	2.0mg
Folic acid	2.0mg
Cyanocobalamin	100.0μg

Preparation of Medium: Add components to distilled/deionized water and bring volume to 1.0L. Mix thoroughly.

Trace Elements Solution SL-6:

Composition per liter:

H_3BO_3	0.3g
$CoCl_2 \cdot 6H_2O$	0.2g
$ZnSO_4 \cdot 7H_2O$	0.10g
$MnCl_2 \cdot 4H_2O$	0.03g
$Na_2MoO_4 \cdot H_2O$	0.03g
$NiCl_2 \cdot 6H_2O$	0.02g
$CuCl_2 \cdot 2H_2O$	0.01g

Preparation of Trace Elements Solution SL-6: Add components to distilled/deionized water and bring volume to 1.0L. Mix thoroughly. Adjust pH to 3.4.

Cysteine·HCl·H_2O Solution:

Composition per 10mL:

Cysteine·HCl·H_2O	0.12g

Preparation of Cysteine·HCl·H_2O Solution: Add cysteine·HCl·H_2O to distilled/deionized water and bring volume to 10.0mL. Gas tubes under 100% N_2 and tightly seal. Autoclave for 15 min at 15 psi pressure–121°C. Use freshly prepared solution.

$Na_2S \cdot 9H_2O$ Solution:

Composition per 10mL:

$Na_2S \cdot 9H_2O$	0.12g

Preparation of $Na_2S \cdot 9H_2O$ Solution: Add $Na_2S \cdot 9H_2O$ to distilled/deionized water and bring volume to 10.0mL. Gas tube under 100% N_2 and tightly seal. Autoclave for 15 min at 15 psi pressure–121°C. Use freshly prepared solution.

Preparation of Medium: Prepare anaerobically under 80% N_2 + 20% CO_2. Add components to distilled/deionized water and bring volume to 1.0L. Mix thoroughly. Distribute into tubes using anaerobic techniques. Autoclave for 15 min at 15 psi pressure–121°C. Prior to inoculation of cultures, inject 0.25mL of sterile cysteine·HCl·H_2O solution and 0.25mL of sterile $Na_2S \cdot 9H_2O$ solution per 10.0mL of medium.

Use: For the cultivation and maintenance of *Clostridium thermolacticum*.

Clostrisel Agar
See: ***Clostridium*** **Selective Agar**

CM
See: Coliform Medium

CM Agar

Composition per liter:

Agar	20.0g
Polypeptone™	10.0g
Yeast extract	10.0g
NaCl	5.0g

pH 7.0 ± 0.2 at 25°C

Preparation of Medium: Add components to distilled/deionized water and bring volume to 1.0L. Mix thoroughly. Gently heat until boiling. Distribute into tubes or flasks. Autoclave for 15 min at 15 psi pressure–121°C. Pour into sterile Petri dishes or leave in tubes.

Use: For the cultivation and maintenance of *Bacillus subtilis*.

CM plus YE Medium

Composition per liter:

NaCl	200.0g
$MgSO_4 \cdot 7H_2O$	20.0g
Yeast extract	10.0g
Casamino acids (vitamin free)	7.5g
Sodium citrate	3.0g
KCl	2.0g
$FeSO_4 \cdot 7H_2O$ (4.98% solution)	1.0mL

pH 7.4 ± 0.2 at 25°C

Preparation of Medium: Add components to distilled/deionized water and bring volume to 1.0L. Mix thoroughly. Gently heat until boiling. Adjust pH to 7.4 with NaOH. Distribute into tubes or flasks. Autoclave for 10 min at 15 psi pressure–121°C.

Use: For the cultivation and maintenance of *Actinopolyspora halophila*.

CM4 Medium

Composition per liter:

Cellobiose	6.0g
Yeast extract	5.0g
K_2HPO_4	2.9g
KH_2PO_4	1.5g
$(NH_4)_2SO_4$	1.3g
NaCl	1.0g

$MgCl_2$ 0.75g
Sodium thioglycollate 0.5g
$CaCl_2$ 0.0132g
Resazurin (1.0% solution) 0.2mL
$FeSO_4$ (1.25% solution) 0.1mL

Preparation of Medium: Add components to distilled/deionized water and bring volume to 1.0L. Mix thoroughly. Gently heat until boiling. Boil until color changes from red to colorless, indicating reduced state. Cool. Distribute into tubes or flasks under 97% N_2 + 3% H_2. Cap with rubber stoppers. Autoclave for 15 min at 15 psi pressure–121°C. Pour into sterile Petri dishes or leave in tubes.

Use: For the cultivation and maintenance of *Clostridium* species and other bacteria that can utilize cellobiose as a carbon source.

CML Medium (Cooked Meat Liver Medium)

Composition per liter:

Cooked meat 57.0g
Glucose 10.0g
Tryptose 10.0g
Liver infusion broth 10.0mL

pH 6.9 ± 0.2 at 25°C

Liver Infusion Broth:

Composition per liter:

Beef liver, infusion from 500.0g
Proteose peptone 10.0g
NaCl 5.0g

Preparation of Liver Infusion Broth: Add components to distilled/deionized water and bring volume to 1.0L. Mix thoroughly.

Preparation of Medium: Add cooked meat to distilled/deionized water and bring volume to 1.0L. Chill to 4°C until liquid is clear. Filter through cheesecloth. Add remaining components to filtrate. Distribute into tubes or flasks. Autoclave for 10 min at 15 psi pressure–121°C.

Use: For the cultivation and maintenance of *Fusobacterium varium*.

CMRL 1066 Medium with Glutamine, 10X (Connaught Medical Research Laboratories Medium with Glutamine, 10X)

Composition per liter:

NaCl 6.8g
$NaHCO_3$ 2.2g
D-Glucose 1.0g
KCl 0.4g
L-Cysteine·HCl·H_2O 0.26g
$CaCl_2$, anhydrous 0.2g
$MgSO_4$·$7H_2O$ 0.2g
NaH_2PO_4·H_2O 0.14g
L-Glutamine 0.1g
Sodium acetate·$3H_2O$ 0.083g
L-Glutamic acid 0.075g
L-Arginine·HCl 0.070g
L-Lysine·HCl 0.070g
L-Leucine 0.060g
Glycine 0.050g
Ascorbic acid 0.050g
L-Proline 0.040g
L-Tyrosine 0.040g
L-Aspartic acid 0.030g
L-Threonine 0.030g
L-Alanine 0.025g
L-Phenylalanine 0.025g
L-Serine 0.025g
L-Valine 0.025g
L-Cystine 0.020g
L-Histidine·HCl·H_2O 0.020g
L-Isoleucine 0.020g
Phenol Red 0.020g
L-Methionine 0.015g
Deoxyadenosine 0.010g
Deoxycytidine 0.010g
Deoxyguanosine 0.010g
Glutathione, reduced 0.010g
Thymidine 0.010g
Hydroxy-L-proline 0.010g
L-Tryptophan 0.010g
Nicotinamide adenine dinucleotide 7.0mg
Tween™ 80 .... 5.0mg
Sodium glucoronate·H_2O 4.2mg
Coenzyme A 2.5mg
Cocarboxylase 1.0mg
Flavin adenine dinucleotide 1.0mg
Nicotinamide adenine dinucleotide phosphate 1.0mg
Uridine triphosphate 1.0mg
Choline chloride 0.50mg
Cholesterol 0.20mg
5-Methyldeoxycytidine 0.10mg
Inositol 0.05mg
p-Aminobenzoic acid 0.05mg
Niacin 0.025mg
Niacinamide 0.025mg
Pyridoxine 0.025mg
Pyridoxal·HCl 0.025mg
Biotin 0.01mg
D-Calcium pantothenate 0.01mg
Folic acid 0.01mg

Riboflavin..0.01mg
Thiamine·HCl..0.01mg

pH 7.2 ± 0.2 at 25°C

Source: This medium is available as a premixed powder from GIBCO BRL.

Preparation of Medium: Add components to distilled/deionized water and bring volume to 1.0L. Mix thoroughly. Adjust pH to 7.2. Filter sterilize.

Use: For the cultivation of a wide variety of microorganisms in a chemically defined basal medium.

CN Screen Medium (*Cryptococcus neoformans* Screen Medium)

Composition per liter:

Agar	15.0g
Asparagine	1.0g
Glutamine	1.0g
Glycine	1.0g
Thiamine·HCl	1.0g
Tryptophan	1.0g
K_2HPO_4	4.0g
$MgSO_4 \cdot 7H_2O$	2.5g
Glucose	1.25g
EDTA	0.6g
Biotin	0.51g
Dihydroxyphenylalanine (Dopa)	0.2g
Phenol Red	0.2g

pH 5.5–5.6 ± 0.2 at 25°C

Preparation of Medium: Add components to distilled/deionized water and bring volume to 1.0L. Mix thoroughly. Gently heat until boiling. Distribute into tubes or flasks. Autoclave for 15 min at 15 psi pressure–121°C.

Use: For the screening of yeast isolates for the presumptive identification of *Cryptococcus neoformans*. *Cryptococcus neoformans* form black colonies.

CNS Agar

Composition per liter:

Agar	15.0g
LiCl	10.0g
Pancreatic digest of gelatin	5.0g
Beef extract	3.0g
K_2HPO_4	2.0g
Yeast extract	2.0g
KH_2PO_4	0.5g
Glucose solution	50.0mL
$MgSO_4 \cdot 7H_2O$ solution	10.0mL
Antibiotic solution	10.0mL
Bravo 500	0.082mL

pH 6.9 ± 0.2 at 25°C

Glucose Solution:

Composition per 50mL:

Glucose	5.0g

Preparation of Glucose Solution: Add glucose to distilled/deionized water and bring volume to 50.0mL. Mix thoroughly. Filter sterilize.

$MgSO_4 \cdot 7H_2O$ Solution:

Composition per 10mL:

$MgSO_4 \cdot 7H_2O$	0.25g

Preparation of $MgSO_4 \cdot 7H_2O$ Solution: Add $MgSO_4 \cdot 7H_2O$ to distilled/deionized water and bring volume to 10.0mL. Mix thoroughly. Filter sterilize.

Antibiotic Solution:

Composition per 10mL:

Cycloheximide	0.04g
Polymyxin B sulfate	0.032g
Nalidixic acid	0.025g

Preparation of Antibiotic Solution: Add components to distilled/deionized water and bring volume to 10.0mL. Mix thoroughly. Filter sterilize.

Preparation of Medium: Add components—except glucose solution, $MgSO_4 \cdot 7H_2O$ solution and antibiotic solution—to distilled/deionized water and bring volume to 930.0mL. Mix thoroughly. Gently heat and bring to boiling. Autoclave for 15 min at 15 psi pressure–121°C. Cool to 45°–50°C. Aseptically add 50.0mL of sterile glucose solution, 10.0mL of sterile $MgSO_4 \cdot 7H_2O$ solution and 10.0mL of sterile antibiotic solution. Mix thoroughly. Pour into sterile Petri dishes or distribute into sterile tubes.

Use: For the isolation and cultivation of *Corynbacterium nebraskense*.

Coagulase Agar Base

Composition per liter:

Agar	25.0g
Brain heart infusion	10.5g
Pancreatic digest of casein	10.5g
D-Mannitol	10.0g
Brain heart infusion	5.0g
NaCl	3.5g
Papaic digest of soybean meal	3.5g
Bromcresol Purple	0.02g
Rabbit plasma	100.0mL

pH 7.4 ± 0.2 at 25°C

Preparation of Medium: Add components, except rabbit plasma, to distilled/deionized water and bring volume to 1.0L. Mix thoroughly. Gently heat while stirring until boiling. Distribute into tubes or flasks. Autoclave for 15 min at 15 psi pressure–121°C. Cool to 45°–50°C. Add rabbit plasma to a final concentration of 7–15%. Mix thoroughly. Pour into sterile Petri dishes in 18.0mL volume per plate.

Use: For the cultivation and differentiation of *Staphylococcus aureus* from other *Staphylococcus* species based on coagulase production.

Coagulase Mannitol Agar

Composition per liter:

Agar	14.5g
Pancreatic digest of casein	10.5g
D-Mannitol	10.0g
Brain heart infusion	5.0g
NaCl	3.5g
Papaic digest of soybean meal	3.5g
Bromcresol Purple	0.02g
Rabbit plasma with 0.15% EDTA	100.0mL

pH 7.3 ± 0.2 at 25°C

Source: This medium is available as a premixed powder from BBL Microbiology Systems.

Preparation of Medium: Add components, except rabbit plasma, to distilled/deionized water and bring volume to 1.0L. Mix thoroughly. Gently heat while stirring until boiling. Distribute into tubes or flasks. Autoclave for 15 min at 15 psi pressure–121°C. Cool to 45°–50°C. Add rabbit plasma with 0.15% EDTA to a final concentration of 7–15%. Mix thoroughly. Pour into sterile Petri dishes in 18.0mL volume per plate.

Use: For the cultivation and differentiation of *Staphylococcus aureus* from other *Staphylococcus* species based on coagulase production and mannitol fermentation.

COBA (Colistin Oxolinic Acid Blood Agar)

Composition per liter:

Columbia agar base	930.0mL
Horse blood, defibrinated, sterile	50.0mL
Colistin sulfate solution	10.0mL
Oxolinic acid solution	10.0mL

pH 7.3 ± 0.2 at 25°C

Columbia Agar Base:
Composition per 930mL:

Agar	13.5g
Pancreatic digest of casein	10.0g
Peptic digest of animal tissue	10.0g
NaCl	5.0g
Beef extract	3.0g
Yeast extract	3.0g
Cornstarch	1.0g

Preparation of Columbia Agar Base: Add components to distilled/deionized water and bring volume to 930.0mL. Mix thoroughly. Gently heat until boiling. Autoclave for 15 min at 15 psi pressure–121°C. Cool to 45°–50°C.

Colistin Sulfate Solution:
Composition per 10mL:

Colistin sulfate	10.0mg

Preparation of Colistin Sulfate Solution: Add colistin sulfate to distilled/deionized water and bring volume to 10.0mL. Mix thoroughly. Filter sterilize.

Oxolinic Acid Solution:
Composition per 10mL:

Oxolinic acid	5.0–10.0mg

Preparation of Oxolinic Acid Solution: Add Oxolinic acid to distilled/deionized water and bring volume to 10.0mL. Mix thoroughly. Filter sterilize.

Preparation of Medium: To 930.0mL of sterile, cooled Columbia Agar Base, add sterile colistin sulfate, sterile oxolinic acid and sterile, defibrinated horse blood. Mix thoroughly. Pour into sterile Petri dishes.

Use: For the isolation and cultivation of *streptococci* in pure culture from mixed flora in clinical specimens.

Coliform Medium (CM)

Composition per liter:

Bile salts No. 3	20.0g
Lactose	20.0g
Proteose peptone No. 3	10.0g
Yeast extract	6.0g
Sodium lauryl sulfate	1.0g
Sodium deoxycholate	0.1g
Bromcresol Purple solution	10.0mL

pH 7.0 ± 0.2 at 25°C

Bromcresol Purple Solution:
Composition per 100mL:

Bromcresol Purple	0.35g
NaOH (0.1*N* solution)	2.0mL

Preparation of Bromcresol Purple Solution: Combine Bromcresol Purple and NaOH solution. Mix thoroughly. Bring volume to 100.0mL with distilled/deionized water. Filter sterilize.

Preparation of Medium: Add components, except Bromcresol Purple solution, to distilled/deionized water and bring volume to 990.0mL. Mix thoroughly. Add 10.0mL of Bromcresol Purple solution. Mix thoroughly. Adjust pH to 7.0 with 1*N* HCl.

Distribute into flasks in 95.0mL volumes. Autoclave for 15 min at 15 psi pressure–121°C.

Use: For the isolation and cultivation of coliform microorganisms from cream.

Coliform Medium (CM)

Composition per liter:

Bile salts No. 3	20.0g
Lactose	20.0g
Proteose peptone No. 3	10.0g
Yeast extract	6.0g
Sodium lauryl sulfate	1.0g
Sodium deoxycholate	0.1g
Bromcresol Purple solution	10.0mL

pH 6.8 ± 0.2 at 25°C

Bromcresol Purple Solution:
Composition per 100mL:

Bromcresol Purple	0.35g
NaOH (0.1*N* solution)	2.0mL

Preparation of Bromcresol Purple Solution: Combine Bromcresol Purple and NaOH solution. Mix thoroughly. Bring volume to 100.0mL with distilled/deionized water. Filter sterilize.

Preparation of Medium: Add components, except Bromcresol Purple solution, to distilled/deionized water and bring volume to 990.0mL. Mix thoroughly. Add 10.0mL of Bromcresol Purple solution. Mix thoroughly. Adjust pH to 6.8 with 1*N* NH_4OH. Distribute into flasks in 95.0mL volumes. Autoclave for 25 min at 15 psi pressure–121°C.

Use: For the isolation and cultivation of coliform microorganisms from yogurt and raw milk.

Coliform Medium, Modified (MCM)

Composition per liter:

Lactose	20.0g
Tris(hydroxymethyl)aminomethane buffer	12.1g
Proteose peptone No. 3	10.0g
Yeast extract	6.0g
Bile salts No. 3	1.0g
Sodium lauryl sulfate	1.0g
Sodium deoxycholate	0.1g
Bromcresol Purple solution	10.0mL

pH 7.0 ± 0.2 at 25°C

Bromcresol Purple Solution:
Composition per 100mL:

Bromcresol Purple	0.35g
NaOH (0.1*N* solution)	2.0mL

Preparation of Bromcresol Purple Solution: Combine Bromcresol Purple and NaOH solution. Mix thoroughly. Bring volume to 100.0mL with distilled/deionized water. Filter sterilize.

Preparation of Medium: Add components, except Bromcresol Purple solution, to distilled/deionized water and bring volume to 990.0mL. Mix thoroughly. Add 10.0mL of Bromcresol Purple solution. Mix thoroughly. Adjust pH to 7.0 with 1*N* HCl. Distribute into flasks in 95.0mL volumes. Autoclave for 15 min at 15 psi pressure–121°C.

Use: For the isolation and cultivation of coliform microorganisms from cream.

Coliform Medium, Modified (MCM)

Composition per liter:

Lactose	20.0g
Tris(hydroxymethyl)aminomethane buffer	12.1g
Proteose peptone No. 3	10.0g
Yeast extract	6.0g
Bile salts No. 3	1.0g
Sodium lauryl sulfate	1.0g
Sodium deoxycholate	0.1g
Bromcresol Purple solution	10.0mL

pH 6.8 ± 0.2 at 25°C

Bromcresol Purple Solution:
Composition per 100mL:

Bromcresol Purple	0.35g
NaOH (0.1*N* solution)	2.0mL

Preparation of Bromcresol Purple Solution: Combine Bromcresol Purple and NaOH solution. Mix thoroughly. Bring volume to 100.0mL with distilled/deionized water. Filter sterilize.

Preparation of Medium: Add components, except Bromcresol Purple solution, to distilled/deionized water and bring volume to 990.0mL. Mix thoroughly. Add 10.0mL of Bromcresol Purple solution. Mix thoroughly. Adjust pH to 6.8 with 1*N* NH_4OH. Distribute into flasks in 95.0mL volumes. Autoclave for 25 min at 15 psi pressure–121°C.

Use: For the isolation and cultivation of coliform microorganisms from yogurt.

Colloidal Chitin Agar

Composition per liter:

Agar	20.0g
Chitin, colloidal	4.0g
K_2HPO_4	0.7g
$MgSO_4 \cdot 5H_2O$	0.5g
KH_2PO_4	0.3g

$FeSO_4 \cdot 7H_2O$ 0.01g
$MnCl_2$ 1.0mg
$ZnSO_4$ 1.0mg
pH 7.0 ± 0.2 at 25°C

Preparation of Medium: Add components to distilled/deionized water and bring volume to 1.0L. Mix thoroughly. Gently heat and bring to boiling. Distribute into tubes or flasks. Autoclave for 15 min at 15 psi pressure–121°C. Pour into sterile Petri dishes or leave in tubes.

Use: For the isolation and cultivation of *Micromonospora* species from water, soil or sediment. For the germination of spores of *Micromonospora* species.

Colonization Medium

Composition per 700mL:
Mannose 1.0g
Pancreatic digest of gelatin 0.2g
Brain heart, solids from infusion 0.08g
Peptic digest of animal tissue 0.08g
NaCl 0.07g
Glucose 0.04g
Na_2HPO_4 .... 0.03g
Bile salts No. 3 .... 0.1g
Dulbecco's phosphate-buffered saline 700.0mL
pH 7.4 ± 0.2 at 25°C

Dulbecco's Phosphate-Buffered Saline :
Composition per liter:
NaCl 8.0g
$Na_2HPO_4 \cdot 7H_2O$ 2.16g
KCl 0.2g
KH_2PO_4 .... 0.2g
$CaCl_2$ 0.1g
$MnCl_2 \cdot 6H_2O$ 0.1g

Preparation of Dulbecco's Phosphate-Buffered Saline: Add components to distilled/deionized water and bring volume to 1.0L. Mix thoroughly.

Preparation of Medium: Combine components. Mix thoroughly. Filter sterilize. Aseptically distribute into sterile tubes or flasks.

Use: For the differentiation of enterotoxigenic *Escherichia coli* from foods, based on the HeLa cell test for colonization.

Columbia Agar

Composition per liter:
Columbia agar base 950.0mL
Sheep blood 50.0mL
pH 7.3 ± 0.2 at 25°C

Source: This medium is available as a premixed powder from BBL Microbiology Systems.

Columbia Agar Base:
Composition per liter:
Agar 13.5g
Pancreatic digest of casein 12.0g
NaCl 5.0g
Peptic digest of animal tissue 5.0g
Beef extract 3.0g
Yeast extract 3.0g
Cornstarch 1.0g

Preparation of Columbia Agar Base: Add components to distilled/deionized water and bring volume to 1.0L. Mix thoroughly. Gently heat until boiling. Autoclave for 15 min at 15 psi pressure–121°C. Cool to 45°–50°C.

Preparation of Medium: To 950.0mL of cooled, sterile Columbia Agar Base, aseptically add 50.0mL sterile, defibrinated sheep blood. Mix thoroughly. Pour into sterile Petri dishes or Distribute into sterile tubes.

Use: For the isolation and cultivation of nonfastidious and fastidious microorganisms from a variety of clinical and nonclinical specimens.

Columbia Blood Agar

Composition per liter:
Columbia blood agar base 950.0mL
Sheep blood 50.0mL
pH 7.3 ± 0.2 at 25°C

Columbia Blood Agar Base:
Composition per liter:
Agar 15.0g
Pantone 10.0g
Bitone 10.0g
NaCl 5.0g
Tryptic digest of beef heart 3.0g
Cornstarch 1.0g

Source: Columbia blood agar base is available as a premixed powder from Difco Laboratories.

Preparation of Columbia Blood Agar Base: Add components to distilled/deionized water and bring volume to 1.0L. Mix thoroughly. Gently heat until boiling. Autoclave for 15 min at 15 psi pressure–121°C. Cool to 45°–50°C.

Preparation of Medium: To 950.0mL of cooled, sterile Columbia Blood Agar Base, aseptically add 50.0mL sterile, defibrinated sheep blood. Mix thoroughly. Pour into sterile Petri dishes or Distribute into sterile tubes.

Use: With the addition of blood or other enrichments, used for the isolation and cultivation of fastidious microorganisms.

Columbia Blood Agar

Composition per liter:

Columbia blood agar base	950.0mL
Sheep blood	50.0mL

pH 7.3 ± 0.2 at 25°C

Source: This medium is available as a premixed powder from Oxoid Unipath.

Columbia Blood Agar Base:

Composition per liter:

Special peptone	23.0g
Agar	10.0g
NaCl	5.0g
Starch	1.0g

Preparation of Columbia Blood Agar Base: Add components to distilled/deionized water and bring volume to 1.0L. Mix thoroughly. Gently heat until boiling. Autoclave for 15 min at 15 psi pressure–121°C. Cool to 45°–50°C.

Preparation of Medium: To 950.0mL of cooled, sterile Columbia blood agar base, aseptically add 50.0mL sterile, defibrinated sheep blood. Mix thoroughly. Pour into sterile Petri dishes or distribute into sterile tubes.

Use: For the cultivation of a variety of fastidious microorganisms.

Columbia Broth

Composition per liter:

Bitone	10.0g
Pancreatic digest of casein	5.0g
Peptic digest of animal tissue	5.0g
NaCl	5.0g
Tryptic digest of beef heart	3.0g
Tris(hydroxymethyl)aminomethane·HCl	2.86g
Glucose	2.5g
Tris(hydroxymethyl)aminomethane	0.83g
Na_2CO_3	0.6g
L-Cysteine·HCl	0.1g
$MgSO_4$, anhydrous	0.1g
$FeSO_4$	0.02g

pH 7.5 ± 0.2 at 25°C

Source: This medium is available as a premixed powder from Difco Laboratories.

Preparation of Medium: Add components to distilled/deionized water and bring volume to 1.0L. Mix thoroughly. Gently heat until boiling. Distribute into tubes or flasks. Autoclave for 15 min at 15 psi pressure–121°C.

Use: For the cultivation and isolation of fastidious bacteria from clinical specimens or as a general purpose broth.

Columbia Broth

Composition per liter:

Pancreatic digest of casein	10.0g
Peptic digest of animal tissue	8.0g
NaCl	5.0g
Yeast extract	5.0g
Tris(hydroxymethyl) aminomethane·HCl buffer	2.86g
Glucose	2.5g
Tris(hydroxymethyl) aminomethane buffer	0.83g
Cysteine·HCl·H_2O	0.1g
$MgSO_4 \cdot 7H_2O$	0.05g
$FeSO_4$	0.012g

pH 7.4 ± 0.2 at 25°C

Source: This medium is available as a premixed powder from BBL Microbiology Systems.

Preparation of Medium: Add components to distilled/deionized water and bring volume to 1.0L. Mix thoroughly. Gently heat until boiling. Distribute into tubes or flasks. Autoclave for 15 min at 15 psi pressure–121°C.

Use: For the cultivation of a wide variety of microorganisms. Use as a general purpose medium.

Columbia CNA Agar (Columbia Colistin Nalidixic Acid Agar)

Composition per liter:

Columbia blood agar base	950.0L
Sheep blood	50.0mL

pH 7.3 ± 0.2 at 25°C

Source: This medium is available as a premixed powder from BBL Microbiology Systems and Difco Laboratories.

Columbia Blood Agar Base:

Composition per liter:

Agar	13.5g
Pancreatic digest of casein	12.0g
NaCl	5.0g
Peptic digest of animal tissue	5.0g
Beef extract	3.0g
Yeast extract	3.0g
Cornstarch	1.0g
Nalidixic acid	15.0mg
Colistin	10.0mg

Preparation of Columbia Blood Agar Base: Add components to distilled/deionized water and bring volume to 1.0L. Mix thoroughly. Gently heat until boiling. Autoclave for 15 min at 15 psi pressure–121°C. Cool to 45°–50°C.

Preparation of Medium: To 950.0mL of cooled, sterile Columbia blood agar base, aseptically add 50.0mL of sterile, defibrinated sheep blood. Mix thoroughly. Pour into sterile Petri dishes or distribute into sterile tubes.

Use: For the selective isolation, cultivation and differentiation of Gram-positive cocci from clinical and nonclinical specimens.

Columbia CNA Agar, Modified with Sheep Blood

Composition per liter:

Columbia blood agar base	950.0L
Sheep blood, defribinated	50.0mL

pH 7.3 ± 0.2 at 25°C

Source: This medium is available as a premixed powder from BBL Microbiology Systems.

Columbia Blood Agar Base:

Composition per liter:

Agar	13.5g
Pancreatic digest of casein	12.0g
NaCl	5.0g
Peptic digest of animal tissue	5.0g
Beef extract	3.0g
Yeast extract	3.0g
Cornstarch	1.0g
Nalidixic acid	5.0mg
Colistin	10.0mg

Preparation of Columbia Blood Agar Base: Add components to distilled/deionized water and bring volume to 1.0L. Mix thoroughly. Gently heat until boiling. Autoclave for 15 min at 15 psi pressure–121°C. Cool to 45°–50°C.

Preparation of Medium: To 950.0L of cooled, sterile Columbia blood agar base, aseptically add 50.0mL of sterile, defibrinated sheep blood. Mix thoroughly. Pour into sterile Petri dishes or distribute into sterile tubes.

Use: For the selective isolation, cultivation and differentiation of Gram-positive cocci from clinical and nonclinical materials.

Columbia Colistin Nalidixic Acid Agar *See:* Columbia CNA Agar

Complex Medium

Composition per liter:

NaCl	250.0g
$MgSO_4 \cdot 7H_2O$	20.0g
Yeast extract	10.0g
Casamino acids	7.5g
Trisodium citrate	3.0g
KCl	2.0g

pH 7.5–7.8 at 25°C

Preparation of Medium: Add components to distilled/deionized water and bring volume to 1.0L. Mix thoroughly. Autoclave for 5 min at 15 psi pressure–121°C. Filter through Whatman #1 filter paper. Adjust pH of filtrate to 7.4. Distribute into tubes or flasks. Autoclave for 15 min at 15 psi pressure–121°C.

Use: For the isolation and cultivation of *Actinomadura* species, *Actinopolyspora* species, *Excellospora* species and *Microspora* species.

Congo Red Acid Morpholinepropanesulfonic Acid Pigmentation Agar (CRAMP Agar)

Composition per liter:

Agarose	14.0g
Morpholinepropanesulfonic acid	8.4g
NaCl	2.9g
Casamino acids	2.0g
Galactose	2.0g
Tricine (N-Tris-hydroxymethyl-methylglycine) buffer	1.8g
$Na_2S_2O_3 \cdot 5H_2O$	0.6g
NH_4Cl	0.5g
K_2HPO_4	0.24g
$MgSO_4 \cdot 7H_2O$	0.10g
Congo Red	5.0mg

pH 5.3 ± 0.2 at 25°C

Preparation of Medium: Add components to distilled/deionized water and bring volume to 1.0L. Mix thoroughly. Gently heat and bring to boiling. Adjust pH to 5.3. Distribute into tubes or flasks. Autoclave for 15 min at 15 psi pressure–121°C. Pour into sterile Petri dishes or leave in tubes.

Use: For the cultivation of *Yersinia* species with plasmids.

Congo Red Agar (CR Agar)

Composition per liter:

GC agar base	890.0mL
Hemoglobin solution	100.0mL
Supplement solution	10.0mL
Congo Red (0.01% solution)	0.1mL

pH 7.2 ± 0.2 at 25°C

GC Agar Base:
Composition per 890mL:

Agar	10.0g
Pancreatic digest of casein	7.5g
Peptic digest of animal tissue	7.5g
NaCl	5.0g
K_2HPO_4	4.0g
Cornstarch	1.0g
KH_2PO_4	1.0g

Preparation of GC Agar Base: Add components to distilled/deionized water and bring volume to 890.0mL. Mix thoroughly. Gently heat until boiling. Autoclave for 15 min at 15 psi pressure–121°C. Cool to 45°–50°C.

Hemoglobin Solution:
Composition per 100mL:

Hemoglobin	2.0g

Preparation of Hemoglobin Solution: Add hemoglobin to distilled/deionized water and bring volume to 100.0mL. Mix thoroughly. Autoclave for 15 min at 15 psi pressure–121°C. Cool to 50°C.

Congo Red Solution:
Composition per 100mL:

Congo Red	0.01g

Preparation of Congo Red Solution: Add Congo Red to 100.0mL of distilled/deionized water. Mix thoroughly. Autoclave for 15 min at 15 psi pressure–121°C.

Supplement Solution:
Composition per liter:

Glucose	100.0g
L-Cysteine·HCl	25.9g
L-Glutamine	10.0g
L-Cystine	1.1g
Adenine	1.0g
Nicotinamide adenine dinucleotide	0.25g
Vitamin B_{12}	0.1g
Thiamine pyrophosphate	0.1g
Guanine·HCl	0.03g
$Fe(NO_3)_3 \cdot 6H_2O$	0.02g
p-Aminobenzoic acid	0.013g
Thiamine·HCl	3.0mg

Preparation of Supplement Solution: Add components to distilled/deionized water and bring volume to 1.0L. Mix thoroughly. Filter sterilize.

Source: The supplement solution IsoVitaleX® enrichment is available from BBL Microbiology Laboratories. This enrichment may be replaced by supplement VX from Difco Laboratories.

Preparation of Medium: To 890.0mL of sterile, cooled GC Agar Base aseptically add 100.0mL of sterile cooled hemoglobin solution, 10.0mL of sterile supplement solution, and 0.1mL of sterile Congo Red solution. Mix thoroughly. Pour into sterile Petri dishes.

Use: For the isolation and differentiation of virulent and avirulent strains of *Shigella, Vibrio cholerae, Escherichia coli,* and *Neisseria meningitidis*. Use for the detection and differentiation of "iron-responsive" avirulent mutants. Used in the preparation of live vaccines. Used for the differentiation of sensitive *Neisseria gonorrhoeae* (no growth) from other *Neisseria* species (growth) that are resistant to Congo Red.

Congo Red Agar (CR Agar)

Composition per liter:

Soybean-casein digest agar	890.0mL
Hemoglobin solution	100.0mL
Supplement solution	10.0mL
Congo Red (0.01% solution)	0.1mL

pH 7.3 ± 0.2 at 25°C

Soybean-Casein Digest Agar:
Composition per 890mL:

Pancreatic digest of casein	17.0g
Agar	15.0g
NaCl	5.0g
Papaic digest of soybean meal	3.0g
Glucose	2.5g
K_2HPO_4	2.5g

Preparation of Soybean-Casein Digest Agar: Add components to distilled/deionized water and bring volume to 890.0mL. Mix thoroughly. Gently heat until boiling. Autoclave for 15 min at 15 psi pressure–121°C. Cool to 45°–50°C.

Hemoglobin Solution:
Composition per 100mL:

Hemoglobin	2.0g

Preparation of Hemoglobin Solution: Add hemoglobin to distilled/deionized water and bring volume to 100.0mL. Mix thoroughly. Autoclave for 15 min at 15 psi pressure–121°C. Cool to 50°C.

Congo Red Solution:
Composition per 100mL:

Congo Red	0.01g

Preparation of Congo Red Solution: Add Congo Red to 100.0mL of distilled/deionized water. Mix thoroughly. Autoclave for 15 min at 15 psi pressure–121°C.

Supplement Solution:
Composition per liter:

Glucose	100.0g
L-Cysteine·HCl	25.9g

L-Glutamine 10.0g
L-Cystine 1.1g
Adenine 1.0g
Nicotinamide adenine dinucleotide 0.25g
Vitamin B_{12} 0.1g
Thiamine pyrophosphate 0.1g
Guanine·HCl 0.03g
$Fe(NO_3)_3·6H_2O$ 0.02g
p-Aminobenzoic acid 0.013g
Thiamine·HCl 3.0mg

Preparation of Supplement Solution: Add components to distilled/deionized water and bring volume to 1.0L. Mix thoroughly. Filter sterilize.

Preparation of Medium: To 890.0mL of sterile, cooled soybean-casein digest agar aseptically add 100.0mL of sterile cooled hemoglobin solution, 10.0mL of sterile supplement solution, and 0.1mL of sterile Congo Red solution. Mix thoroughly. Pour into sterile Petri dishes.

Source: The supplement solution IsoVitaleX® enrichment is available from BBL Microbiology Laboratories. This enrichment may be replaced by supplement VX from Difco Laboratories.

Use: For the isolation and differentiation of virulent and avirulent strains of *Shigella, Vibrio cholerae, Escherichia coli,* and *Neisseria meningitidis.* Use for the detection and differentiation of "iron-responsive" avirulent mutants. Used in the preparation of live vaccines. Used for the differentiation of sensitive *Neisseria gonorrhoeae* (no growth) from other *Neisseria* species (growth) that are resistant to Congo Red.

Congo Red BHI Agarose Medium

Composition per liter:

Agarose 15.0g
Pancreatic digest of gelatin 14.5g
Brain heart, solids from infusion 6.0g
Peptic digest of animal tissue 6.0g
NaCl 5.0g
Glucose 3.0g
Na_2HPO_4 .... 2.5g
Congo Red 0.075g

pH 7.4 ± 0.2 at 25°C

Preparation of Medium: Add components to distilled/deionized water and bring volume to 1.0L. Mix thoroughly. Gently heat and bring to boiling. Distribute into tubes or flasks. Autoclave for 15 min at 15 psi pressure–121°C. Pour into sterile Petri dishes in 20.0mL volumes.

Use: For the isolation, cultivation and detection of virulent strains of *Yersinia enterocolitica.*

Congo Red Magnesium Oxalate Agar (CRMOX Agar)

Composition per liter:

Solution 1 .... 825.0mL
Solution 2 .... 80.0mL
Solution 3 .... 80.0mL
Solution 4 .... 10.0mL
Solution 5 .... 5.0mL

pH 7.3 ± 0.2 at 25°C

Solution 1:

Composition per 825mL:

Pancreatic digest of casein 15.0g
Agar 15.0g
Papaic digest of soybean meal 5.0g
NaCl 5.0g

pH 7.3 ± 0.2 at 25°C

Preparation of Solution 1: Add components to distilled/deionized water and bring volume to 825.0mL. Mix thoroughly. Gently heat and bring to boiling. Autoclave for 15 min at 15 psi pressure–121°C. Do not overheat.

Solution 2:

Composition per liter:

$MgCl_2·6H_2O$ 50.8g

Preparation of Solution 2: Add $MgCl_2·6H_2O$ to distilled/deionized water and bring volume to 1.0L. Mix thoroughly. Autoclave for 15 min at 15 psi pressure–121°C.

Solution 3:

Composition per liter:

Sodium oxalate 33.2g

Preparation of Solution 3: Add sodium oxalate to distilled/deionized water and bring volume to 1.0L. Mix thoroughly. Autoclave for 15 min at 15 psi pressure–121°C.

Solution 4:

Composition per 100mL:

D-Galactose 20.0g

Preparation of Solution 4: Add D-galactose to distilled/deionized water and bring volume to 100.0mL. Mix thoroughly. Filter sterilize.

Solution 5:

Composition per 10.0mL:

Congo Red 0.1g

Preparation of Solution 5: Add Congo Red to distilled/deionized water and bring volume to 10.0mL. Mix thoroughly. Autoclave for 15 min at 15 psi pressure–121°C.

Preparation of Medium: Aseptically combine 80.0mL of sterile solution 2, 80.0mL of sterile solution 3, 10.0mL of sterile solution 4 and 5.0mL of sterile solution 5. Mix thoroughly. Warm to 50°C. Add this mixture to 825.0mL of cooled, sterile solution 1. Mix thoroughly. Pour into sterile Petri dishes.

Use: For the cultivation and identification of pathogenic serotypes of *Yersinia enterocolitica*. For the determination of whether *Yersinia* strains contain the *Yersinia* virulence plasmid.

Connaught Medical Research Laboratories Medium with Glutamine, 10X
See: CMRL 1066 Medium with Glutamine, 10X

Conradi Drigalski Agar

Composition per liter:

Agar	15.0g
Casein	10.0g
Lactose	10.0g
Peptone	10.0g
NaCl	5.0g
Bromcresol Purple	0.03g
Crystal Violet	4.0mg

pH 6.8 ± 0.2 at 25°C

Preparation of Medium: Add components to distilled/deionized water and bring volume to 1.0L. Mix thoroughly. Gently heat until boiling. Distribute into tubes or flasks. Autoclave for 15 min at 15 psi pressure–121°C. Pour into sterile Petri dishes or leave in tubes.

Use: For the isolation and cultivation of Gram-negative enteric bacilli.

Converse Liquid Medium, Levine Modification

Composition per liter:

Ionagar No. 2 or Noble agar	10.0g
Glucose	4.0g
Ammonium acetate	1.23g
K_2HPO_4	0.52g
Tamol	0.5g
$MgSO_4 \cdot 7H_2O$	0.4g
KH_2PO_4	0.4g
NaCl	0.014g
Na_2CO_3	0.012g
$CaCl_2 \cdot 2H_2O$	0.002g
$ZnSO_4 \cdot 7H_2O$	0.002g

Preparation of Medium: Add components to distilled/deionized water and bring volume to 1.0L. Mix thoroughly. Gently heat and bring to boiling. Autoclave for 15 min at 15 psi pressure–121°C. Pour into sterile Petri dishes in 15.0mL volumes.

Use: For the cultivation and induction of spherules of *Coccidioides immitis*.

Cooke Rose Bengal Agar

Composition per liter:

Agar	20.0g
Glucose	10.0g
Enzymatic hydrolysate of soybean meal	5.0g
KH_2PO_4	1.0g
$MgSO_4 \cdot 7H_2O$	0.5g
Rose Bengal	35.0mg

pH 6.0 ± 0.2 at 25°C

Source: This medium is available as a premixed powder from Difco Laboratories.

Preparation of Medium: Add components to distilled/deionized water and bring volume to 1.0L. Mix thoroughly. Gently heat until boiling. Distribute into tubes or flasks. Autoclave for 15 min at 15 psi pressure–121°C. Pour into sterile Petri dishes or leave in tubes.

Use: For the isolation of fungi.

Cooked Meat Liver Medium
See: CML Medium

Cooked Meat Medium

Composition per liter:

Beef heart	454.0g
Proteose peptone	20.0g
NaCl	5.0g
Glucose	2.0g

pH 7.2 ± 0.2 at 25°C

Source: This medium is available as a premixed powder from Difco Laboratories.

Preparation of Medium: Finely chop beef heart. Add approximately 1.5g of heart particles to test tubes. Add remaining components to distilled/deionized water and bring volume to 1.0L. Mix thoroughly. Distribute into tubes in 10.0mL volumes. Autoclave for 15 min at 15 psi pressure–121°C. Slowly cool tubes to prevent expulsion of meat particles.

Use: For cultivation and maintenance of anaerobic microorganisms.

Cooked Meat Medium

Composition per liter:

Heart muscle	454.0g
Beef extract	10.0g
Peptone	10.0g
NaCl	5.0g
Glucose	2.0g

pH 7.2 ± 0.2 at 25°C

Source: This medium is available as a premixed powder from Oxoid Unipath.

Preparation of Medium: Finely chop beef heart. Add approximately 1.5g of heart particles to test tubes. Add remaining components to distilled/deionized water and bring volume to 1.0L. Mix thoroughly. Distribute into tubes in 10.0mL volumes. Autoclave for 15 min at 15 psi pressure–121°C. Slowly cool tubes to prevent expulsion of meat particles.

Use: For the cultivation and maintenance of aerobic and anaerobic microorganisms. Used for the cultivation of anaerobes, especially pathogenic *Clostridia*.

Cooked Meat Medium

Composition per liter:

Heart tissue granules	98.0g
Peptic digest of animal tissue	20.0g
NaCl	5.0g
Glucose	2.0g

pH 7.2 ± 0.2 at 25°C

Source: This medium is available as a premixed powder from BBL Microbiology Systems.

Preparation of Medium: Add approximately 1.0g of heart tissue granules to test tubes. Add remaining components to distilled/deionized water and bring volume to 1.0L. Mix thoroughly. Distribute into tubes in 10.0mL volumes. Autoclave for 15 min at 15 psi pressure–121°C. Slowly cool tubes to prevent expulsion of meat particles.

Use: For the cultivation of anaerobes, especially pathogenic *Clostridia*.

Cooked Meat Medium, Modified

Composition per liter:

Cooked meat medium	66.0g
Solution A	1.0L

pH 6.8 ± 0.2 at 25°C

Solution A:

Composition per liter:

Pancreatic digest of casein	10.0g
Glucose	2.0g
Soluble starch	1.0g
Sodium thioglycollate	1.0g
Neutral Red (1% aqueous)	5.0mL

Preparation of Solution A: Add components to distilled/deionized water and bring volume to 1.0L. Mix thoroughly. Gently heat until dissolved.

Preparation of Medium: Add 1.0g of cooked meat medium to each of 66 test tubes. Add 15.0mL of solution A to each test tube. Allow meat particles to rehydrate. Autoclave for 15 min at 15 psi pressure–121°C.

Use: For the cultivation of a variety of anaerobic microorganisms.

Cooked Meat Medium with Glucose, Hemin and Vitamin K

Composition per liter:

Heart tissue granules	98.0g
Peptic digest of animal tissue	20.0g
NaCl	5.0g
Glucose	5.0g
Yeast extract	5.0g
Hemin	5.0mg
Vitamin K	1.0mg

pH 7.2 ± 0.2 at 25°C

Source: This medium is available as a premixed powder from BBL Microbiology Systems.

Preparation of Medium: Add approximately 1.0g of heart tissue granules to test tubes. Add remaining components to distilled/deionized water and bring volume to 1.0L. Mix thoroughly. Distribute into tubes in 10.0mL volumes. Autoclave for 15 min at 15 psi pressure–121°C. Slowly cool tubes to prevent expulsion of meat particles.

Use: For the cultivation of anaerobes, especially pathogenic *Clostridia*.

Corn Meal Agar (Corn Meal Agar with Polysorbate 80)

Composition per liter:

Agar	15.0g
Corn meal, solids from infusion	2.0g
Tween™ 80	1.0g

pH 5.6–6.0 at 25°C

Source: This medium is available as a premixed powder from Difco Laboratories, BBL Microbiology Systems and Oxoid Unipath.

Preparation of Medium: Add components to distilled/deionized water and bring volume to 1.0L. Mix thoroughly. Gently heat until boiling. Distribute into tubes or flasks. Autoclave for 15 min at 15 psi pressure–121°C. Pour into sterile Petri dishes or leave in tubes.

Use: For the cultivation and maintenance of fungi. For the production of chlamydospores by *Candida albicans* and for the cultivation of phytopathological fungi.

Corn Meal Agar with Dextrose

Composition per liter:

Agar	15.0g
Corn meal, solids from infusion	2.0g
Glucose	2.0g
Tween™ 80	1.0g

pH 5.6–6.0 at 25°C

Source: This medium is available as a premixed powder from Difco Laboratories.

Preparation of Medium: Add components to distilled/deionized water and bring volume to 1.0L. Mix thoroughly. Gently heat until boiling. Distribute into tubes or flasks. Autoclave for 15 min at 15 psi pressure–121°C. Pour into sterile Petri dishes or leave in tubes.

Use: For the cultivation of phytopathological and other fungi.

Corn Meal Agar with Polysorbate 80

See: **Corn Meal Agar**

Corn Meal *Phytophthora* Isolation Medium No. 1

Composition per liter:

Agar	15.0g
Corn meal, solids from infusion	2.0g
Vancomycin	0.2g
Pentachloronitrobenzene (PCNB)	0.1g
Pimaricin	0.01g

pH 5.6–6.0 at 25°C

Preparation of Medium: Add components, except pimaricin and vancomycin, to distilled/deionized water and bring volume to 1.0L. Mix thoroughly. Gently heat until boiling. Autoclave for 15 min at 15 psi pressure–121°C. Aseptically add pimaricin and vancomycin. Mix thoroughly. Pour into sterile Petri dishes.

Use: For the cultivation of *Phytophthora* species.

Corn Meal *Phytophthora* Isolation Medium No. 2

Composition per liter:

Agar	15.0g
Corn meal, solids from infusion	2.0g
Vancomycin	0.3g
Pentachloronitrobenzene (PCNB)	0.025g
Pimaricin	5.0mg

pH 5.6–6.0 at 25°C

Preparation of Medium: Add components, except pimaricin and vancomycin, to distilled/deionized water and bring volume to 1.0L. Mix thoroughly. Gently heat until boiling. Autoclave for 15 min at 15 psi pressure–121°C. Aseptically add pimaricin and vancomycin. Mix thoroughly. Pour into sterile Petri dishes.

Use: For the cultivation of *Phytophthora* species.

Corn Milk Medium

Composition per liter:

Skim milk	20.0g
Agar	15.0g
Yeast extract	12.5g
Peptone	10.0g
Beef extract	5.0g
K_2HPO_4	5.0g
NaCl	5.0g
$MgSO_4 \cdot 7H_2O$	1.0g
Corn steep liquor	7.0mL

pH 7.0 ± 0.2 at 25°C

Preparation of Medium: Add components to distilled/deionized water and bring volume to 1.0L. Mix thoroughly. Gently heat until boiling. Distribute into tubes or flasks. Autoclave for 15 min at 15 psi pressure–121°C. Pour into sterile Petri dishes or leave in tubes.

Use: For the cultivation and maintenance of *Bacillus subtilis*.

Corn Steep Liquor Medium

Composition per liter:

Glucose	60.0g
Corn steep liquor	40.0g
Urea	8.0g
KH_2PO_4	5.0g
Fumaric acid	1.0g
$MgSO_4 \cdot 7H_2O$	0.5g
Hutner's mineral base	20.0mL

pH 7.0 ± 0.2 at 25°C

Hutner's Mineral Base:

Composition per liter:

$MgSO_4 \cdot 7H_2O$	29.7g
Nitrilotriacetic acid	10.0g
$CaCl_2 \cdot 2H_2O$	3.34g
$FeSO_4 \cdot 7H_2O$	99.0mg
$(NH_4)_2MoO_4$	9.25mg
Metals "44"	50.0mL

Preparation of Hutner's Mineral Base: Add nitrilotriacetic acid to 500.0mL of distilled/deionized water. Dissolve by adjusting pH to 6.5 with KOH. Add remaining components. Add distilled/deionized water to 1.0L.

Metals "44":

Composition per 100mL:

$ZnSO_4 \cdot 7H_2O$	1.1g
$FeSO_4 \cdot 7H_2O$	0.5g
EDTA	0.25g
$MnSO_4 \cdot 7H_2O$	0.154g
$CuSO_4 \cdot 5H_2O$	0.04g
$Co(NO_3)_2 \cdot 6H_2O$	0.025g
$Na_2B_4O_7 \cdot 10H_2O$	0.018g

Preparation of Metals "44": Add components to distilled/deionized water and bring volume to 100.0mL. Mix thoroughly.

Preparation of Medium: Add components to distilled/deionized water and bring volume to 1.0L. Mix thoroughly. Distribute into tubes or flasks. Autoclave for 15 min at 15 psi pressure–121°C.

Use: For the cultivation of *Pseudomonas* species.

Cornstarch Soluble Medium (CSSM)

Composition per liter:

Cornstarch	42.0g
n-Butanol	18.0g
Yeast extract	10.0g
Asparagine·H_2O	2.0g
$(NH_4)_2SO_4$	2.0g
NaCl	1.0g
KH_2PO_4	0.75g
K_2HPO_4	0.75g
Cysteine·HCl·H_2O	0.5g
$MgSO_4$	0.02g
$FeSO_4 \cdot 7H_2O$	0.01g
$MnSO_4 \cdot H_2O$	0.01g

Preparation of Medium: Add components to distilled/deionized water and bring volume to 1.0L. Mix thoroughly. Gently heat until boiling. Boil and cool under 80% N_2 + 10% H_2 + 10% CO_2. Distribute anaerobically into tubes under the same gas mixture. Cap with butyl rubber stoppers. Autoclave for 15 min at 15 psi pressure–121°C.

Use: For the cultivation and maintenance of *Clostridium thermoamylolyticum.*

Cornstarch Soluble Medium (CSSM) (ATCC Medium 1500)

Composition per liter:

Cornstarch	42.0g
Yeast extract	10.0g
Asparagine·H_2O	2.0g
$(NH_4)_2SO_4$	2.0g
NaCl	1.0g
KH_2PO_4	0.75g
K_2HPO_4	0.75g
Cysteine·HCl·H_2O	0.5g
$MgSO_4$	0.02g
$FeSO_4 \cdot 7H_2O$	0.01g
$MnSO_4 \cdot H_2O$	0.01g

Preparation of Medium: Add components to distilled/deionized water and bring volume to 1.0L. Mix thoroughly. Gently heat until boiling. Boil and cool under 80% N_2 + 10% H_2 + 10% CO_2. Distribute anaerobically into tubes under the same gas mixture. Cap with butyl rubber stoppers. Autoclave for 15 min at 15 psi pressure–121°C.

Use: For the cultivation and maintenance of *Clostridium thermoamylolyticum.*

Corynebacterium diphtheriae Virulence Test Medium

See: **K–L Virulence Agar**

Corynebacterium Liquid Enrichment Medium

Composition per 2000mL:

Fosfomycin	0.15g
Glucose 6-phosphate	0.03g
Solution A	985.0mL
Bovine serum	100.0mL
Egg yolk suspension	10eggs
Nystatin solution	1.15mL
L-Cystine (1% solution)	1.0mL

pH 7.4 ± 0.2 at 25°C

Solution A:

Composition per liter:

Meat extract	9.0g
Proteose peptone no. 3	9.0g

NaCl 2.7g
Glucose 1.8g
$Na_2HPO_4 \cdot 12H_2O$ 1.8g
K_2TeO_3 (2% solution) 75.0mL
L-Cystine (1% solution) 10.0mL

Preparation of Solution A: Add components to distilled/deionized water and bring volume to 985.0mL. Mix thoroughly. Filter sterilize.

Egg Yolk Emulsion:
Composition:
Chicken egg yolks 9
Whole chicken egg 1

Preparation of Egg Yolk Emulsion: Soak eggs with 1:100 dilution of saturated mercuric chloride solution for 1 min. Crack eggs and separate yolks from whites. Mix egg yolks with 1 chicken egg. Filter sterilize.

Nystatin Solution:
Composition per 10mL:
Nystatin 10,000U

Preparation of Nystatin Solution: Add nystatin to distilled/deionized water and bring volume to 10.0mL. Mix thoroughly. Filter sterilize.

Cystine Solution:
Composition per 10mL:
L-Cystine 0.1g

Preparation of Cystine Solution: Add cystine to distilled/deionized water and bring volume to 10.0mL. Mix thoroughly. Filter sterilize.

Caution: Potassium tellurite is toxic.

Preparation of Medium: To 985.0mL of sterile solution A, aseptically add the remaining components. Mix thoroughly. Aseptically distribute into sterile tubes in 2.0–3.0mL volumes.

Use: For the isolation and cultivation of *Corynebacterium diphtheriae.*

CP Medium

Composition per liter:
Peptone 2.5g
Starch 2.0g
$NaNO_3$ 0.38g
Tris buffer 0.25g
K_2HPO_4 .... 0.038g
$MgSO_4 \cdot 7H_2O$ 0.038g
$CaCl_2 \cdot 2H_2O$ 0.017g
NaCl 0.013g
TC vitamins minimal eagle, 100X 5.0mL
Solution 1 .... 1.0mL
Solution 2 .... 1.0mL
Solution 3 .... 1.0mL
Solution 4 .... 1.0mL
Vitamin B_{12} solution 0.2mL
pH 8.7 ± 0.2 at 25°C

TC Vitamins Minimal Eagle 100X:
Composition per liter:
Inositol 2.0mg
Choline chloride 1.0mg
Folic acid 1.0mg
Nicotinamide 1.0mg
Calcium pantothenate 1.0mg
Pyridoxal 1.0mg
Thiamine·HCl 1.0mg
Riboflavin 0.1mg

Preparation of TC Vitamins Minimal Eagle, 100X: Add components to distilled/deionized water and bring volume to 1.0L. Mix thoroughly. Filter sterilize.

Solution 1:
Composition per 100mL:
EDTA 5.0g
KOH 3.1g

Preparation of Solution 1: Add components to distilled/deionized water and bring volume to 100.0mL. Mix thoroughly.

Solution 2:
Composition per liter:
$FeSO_4 \cdot 7H_2O$ 4.98g

Preparation of Solution 2: Add components to distilled/deionized water acidified with 1.0mL of H_2SO_4. Bring volume to 1.0L. Mix thoroughly.

Solution 3:
Composition per 100mL:
H_3BO_3 .... 1.14g

Preparation of Solution 3: Add components to distilled/deionized water and bring volume to 100.0mL. Mix thoroughly.

Solution 4:
Composition per 100mL:
$ZnSO_4 \cdot 7H_2O$ 0.88g
$MnCl_2 \cdot 4H_2O$ 0.144g
MoO_3 .... 0.071g
$CoNO_3 \cdot 6H_2O$ 0.049g
$CuSO_4 \cdot 5H_2O$ 0.016g

Preparation of Solution 4: Add components to distilled/deionized water and bring volume to 100.0mL. Mix thoroughly.

Vitamin B_{12} Solution
Composition per 10mL:
Vitamin B_{12} 10.0mg

Preparation of Vitamin B_{12} Solution: Add Vitamin B_{12} to distilled/deionized water and bring volume to 10.0mL. Mix thoroughly. Filter sterilize.

Preparation of Medium: Add components except for vitamin solutions to distilled/deionized water and bring volume to 995.0mL. Mix thoroughly. Distribute into tubes or flasks. Autoclave for 15 min at 15 psi pressure–121°C. Aseptically add vitamin solutions. Mix thoroughly.

Use: For the cultivation of *Lysobacter* species.

CPC Agar
See: Cellobiose Polymyxin Colistin Agar

CPC Medium

Composition per liter:

Sucrose	30.0g
Peptone	2.0g
Casein hydrolysate	1.0g
$K_2HPO_4 \cdot 3H_2O$	1.0g
KCl	0.5g
$MgSO_4 \cdot 7H_2O$	0.5g
$FeSO_4 \cdot 7H_2O$	0.1g

pH 7.2 ± 0.2 at 25°C

Preparation of Medium: Add components to tap water and bring volume to 1.0L. Mix thoroughly. Distribute into tubes or flasks. Autoclave for 15 min at 15 psi pressure–121°C.

Use: For the cultivation of *Actinoplanes* species.

CR Agar
See: Congo Red Agar

CRAMP Agar
See: Congo Red Acid Morpholinepropanesulfonic Acid Pigmentation Agar

Creatinine Medium

Composition per liter:

Creatinine	5.0g
Agar	2.0g
Fumaric acid	2.0g
K_2HPO_4	2.0g
Yeast extract	1.0g
Salt solution	10.0mL

pH 6.8 ± 0.2 at 25°C

Salt Solution:

Composition per liter:

$MgSO_4$	12.2g
$FeSO_4 \cdot 7H_2O$	2.8g
$MnSO_4 \cdot H_2O$	1.7g
$CaCl_2 \cdot 2H_2O$	0.76g
NaCl	0.6g
$Na_2MoO_4 \cdot 2H_2O$	0.1g
$ZnSO_4 \cdot 7H_2O$	0.06g
HCl (0.1*N* solution)	1.0L

Preparation of Salt Solution: Dissolve salts in 1.0L of 0.1*N* HCl solution. Mix thoroughly.

Preparation of Medium: Add components to distilled/deionized water and bring volume to 1.0L. Mix thoroughly. Adjust pH to 6.8 with NaOH or KOH. Gently heat until boiling. Distribute into tubes or flasks. Autoclave for 15 min at 15 psi pressure–121°C.

Use: For the cultivation and maintenance of *Pseudomonas* species.

CRMOX Agar
See: Congo Red–Magnesium Oxalate Agar

Crossley Milk Medium

Composition per liter:

Skim milk powder	100.0g
Peptone	10.0g
Bromocresol Purple	0.1g

pH 5.8 ± 0.2 at 25°C

Source: This medium is available as a premixed powder from Oxoid Unipath.

Preparation of Medium: Add components to a very small volume of distilled/deionized water and mix to a paste. Gradually add more distilled/deionized water and bring volume to 1.0L. Distribute in 10.0mL volumes into tubes. Autoclave for 5 min at 15 psi pressure–121°C.

Use: For the routine examination of canned food samples for anaerobic bacteria.

Cryptococcus neoformans Screen Medium
See: CN Screen Medium

Crystal Violet Agar

Composition per liter:

Agar	15.0g
Lactose	10.0g
Proteose peptone	5.0g
Beef extract	3.0g
Crystal Violet	3.3mg

pH 6.8 ± 0.1 at 25°C

Preparation of Medium: Add components to distilled/deionized water and bring volume to 1.0L. Mix thoroughly. Gently heat until boiling. Distribute into tubes or flasks. Autoclave for 15 min at 15 psi pressure–121°C. Pour into sterile Petri dishes or leave in tubes.

Use: For the differentiation of pathogenic from nonpathogenic staphylococci. Hemolytic and coagulating strains of *Staphylococcus aureus* appear as purple or yellow colonies. Nonhemolytic and noncoagulating strains of *Staphylococcus* species appear as white colonies.

Crystal Violet Azide Esculin Agar

Composition per liter:

Agar	15.0g
Glucose	5.0g
NaCl	5.0g
Proteose peptone	5.0g
Pancreatic digest of casein	5.0g
Meat extract	3.0g
Esculin	1.0g
NaN_3	1.0g
Crystal Violet	0.1g
Bovine blood, citrated	100.0mL

pH 7.5 ± 0.2 at 25°C

Preparation of Medium: Add components, except citrated bovine blood, to distilled/deionized water and bring volume to 900.0mL. Mix thoroughly. Gently heat and bring to boiling. Autoclave for 15 min at 15 psi pressure–121°C. Cool to 45°–50°C. Aseptically add sterile citrated bovine blood. Mix thoroughly. Pour into sterile Petri dishes or distribute into sterile tubes.

Caution: Sodium azide is toxic. Azides also react with metals and disposal must be highly diluted.

Use: For the cultivation of *Erysipelothrix rhusiopathiae*.

Crystal Violet Esculin Agar

Composition per liter:

Agar	15.0g
Glucose	5.0g
NaCl	5.0g
Proteose peptone	5.0g
Pancreatic digest of casein	5.0g
Meat extract	3.0g
Esculin	1.0g
Crystal Violet	2.0mg
Blood, citrated	100.0mL

pH 7.5 ± 0.2 at 25°C

Preparation of Medium: Add components, except citrated blood, to distilled/deionized water and bring volume to 900.0mL. Mix thoroughly. Gently heat and bring to boiling. Autoclave for 15 min at 15 psi pressure–121°C. Cool to 45°–50°C. Aseptically add citrated blood. Mix thoroughly. Pour into sterile Petri dishes or distribute into sterile tubes.

Use: For the cultivation of *Erysipelothrix rhusiopathiae*.

Crystal Violet Pectate Medium (CVP Medium)

Composition per liter:

Sodium polypectate	9.0g
Agar	2.0g
$NaNO_3$	1.0g
NaOH (1*N* solution)	4.5mL
$CaCl_2 \cdot H_2O$ (10% solution)	3.0mL
Crystal Violet (0.075% solution)	1.0mL
Sodium lauryl sulfate (10% solution)	0.5mL

pH 7.2 ± 0.2 at 25°C

Preparation of Medium: In a preheated blender, add 500.0mL of boiling distilled/deionized water and the components, except sodium polypectate and sodium lauryl sulfate solution. Blend at high speed for 15 sec. Continue blending at low speed and slowly add 9.0g of sodium polypectate. Pour the incomplete medium into a 2-L flask and add 0.5mL of sodium lauryl sulfate solution. Mix thoroughly. Cap flask with an aluminum foil seal rather than cotton. Autoclave for 25 min at 15 psi pressure–121°C. Pour medium quickly into sterile Petri dishes. Allow plates to dry at 25°C for 48 hr before use.

Use: For the cultivation pectinolytic microorganisms, such as *Erwinia* species, from foods.

CS Vitamin B_{12} Agar

Composition per liter:

Glucose	20.0g
K_2SO_4	20.0g
Agar	15.0g
Sodium acetate	12.0g
Vitamin assay casamino acids	10.0g
Papaic digest of soybean meal	5.0g

Sodium thioglycollate 1.7g
K_2HPO_4 .... 1.0g
KH_2PO_4 .... 1.0g
Ribonucleic acid 1.0g
Sorbitan monooleate complex 1.0g
$MgSO_4 \cdot 7H_2O$ 0.4g
DL-Tryptophan 0.2g
L-Cystine 0.2g
$FeSO_4$ 0.02g
$MgSO_4 \cdot 7H_2O$ 0.02g
NaCl 0.02g
Adenine sulfate 0.018g
Guanine·HCl 0.012g
Uracil 0.01g
Xanthine 0.01g
Pyridoxal 4.0mg
Pyridoxine 4.0mg
Calcium pentothenate 2.0mg
Niacin 2.0mg
Riboflavin 2.0mg
Thiamine·HCl 2.0mg
Folic acid 1.0mg
Biotin 1.0μg
Lactobacillus leichmannii suspension 1.0mL

pH 6.2 ± 0.1 at 25°C

Preparation of Medium: Add components, except *Lactobacillus leichmannii* suspension, to distilled/deionized water and bring volume to 1.0L. Mix thoroughly. Gently heat and bring to boiling. Autoclave for 15 min at 15 psi pressure–121°C. Cool to 45°–50°C. Inoculate medium with 1.0mL of *Lactobacillus leichmannii* suspension. Mix thoroughly. Pour into sterile 150 mm Petri dishes in 50mL volumes. Allow agar surface to dry before using.

Use: For the microbiological assay of vitamin B_{12} by the cup plate or disk method using *Lactobacillus leichmannii* as the test microorganism.

CSSM

See: **Cornstarch Soluble Medium**

CT Agar (Caprylate Thallous Agar)

Composition per liter:

Solution A 500.0mL
Solution B 500.0mL

pH 7.2 ± 0.2 at 25°C

Solution A:

Composition per 500mL:

K_2HPO_4 .... 2.61g
KH_2PO_4 .... 0.68g
Thallous sulfate 0.25g
$MgSO_4 \cdot 7H_2O$ 0.12g
$CaCl_2 \cdot 2H_2O$ 0.016g
Trace element solution 10.0mL
Yeast extract 2.0mL
Caprylic acid 1.1mL

Preparation of Solution A: Add components to distilled/deionized water and bring volume to 500.0mL. Mix thoroughly. Adjust pH to 7.2 with NaOH. Autoclave for 20 min at 10 psi pressure–115°C.

Trace Element Solution:

Composition per liter:

H_3PO_4 .... 1.96g
$FeSO_4 \cdot 7H_2O$ 0.056g
$ZnSO_4 \cdot 4H_2O$ 0.029g
$CuSO_4 \cdot 5H_2O$ 0.025g
$MnSO_4 \cdot 4H_2O$ 0.022g
H_3BO_3 .... 6.2mg
$Co(NO_3)_2 \cdot 6H_2O$ 3.0mg

Preparation of Trace Element Solution: Add components to distilled/deionized water and bring volume to 1.0L. Mix thoroughly. Store at 4°C.

Solution B:

Composition per liter:

Agar 15.0g
NaCl 7.0g
$(NH_4)_2SO_4$ 1.0g

Preparation of Solution B: Add components to distilled/deionized water and bring volume to 500.0mL. Mix thoroughly. Gently heat and bring to boiling. Adjust pH to 7.2. Autoclave for 20 min at 10 psi pressure–115°C.

Preparation of Medium: Aseptically combine 500.0mL of sterile solution A and 500.0mL of sterile solution B. Mix thoroughly. Pour into sterile Petri dishes in 25.0–30.0mL volumes.

Use: For the isolation and cultivation of the *Serratia* species.

CT Agar

Composition per liter:

Agar 20.0g
Pancreatic digest of casein 20.0g
$MgSO_4 \cdot 7H_2O$ 2.0g
Potassium phosphate
buffer (0.02M solution, pH 7.6) 500.0mL

pH 7.6 ± 0.2 at 25°C

Preparation of Medium: Add agar, pancreatic digest of casein, and $MgSO_4 \cdot 7H_2O$ to distilled/deionized water and bring volume to 500.0mL. Mix thoroughly. Gently heat and bring to boiling. Autoclave agar–pancreatic digest of casein–$MgSO_4 \cdot 7H_2O$ solution and potassium phosphate buffer solution separately for 15 min at 15 psi pressure–121°C. Cool to 25°C. Aseptically combine the two solutions. Aseptically add sterile components. Mix thoroughly. Pour into sterile Petri dishes or distribute into sterile tubes.

Use: For the cultivation of myxobacteria.

CT Broth

Composition per liter:

Pancreatic digest of casein	20.0g
$MgSO_4 \cdot 7H_2O$	2.0g
Potassium phosphate buffer (0.02M solution, pH 7.6)	500.0mL

pH 7.6 ± 0.2 at 25°C

Preparation of Medium: Add pancreatic digest of casein and $MgSO_4 \cdot 7H_2O$ to distilled/deionized water and bring volume to 500.0mL. Mix thoroughly. Autoclave pancreatic digest of casein-$MgSO_4 \cdot 7H_2O$ solution and potassium phosphate buffer solution separately for 15 min at 15 psi pressure–121°C. Cool to 25°C. Aseptically combine the two solutions. Aseptically distribute into sterile tubes or flasks.

Use: For the cultivation of myxobacteria.

CT Medium

Composition per liter:

Agar	15.0g
Pancreatic digest of casein	10.0g
Yeast extract	3.5g
$MgSO_4$	0.96g

Preparation of Medium: Add components to distilled/deionized water and bring volume to 1.0L. Mix thoroughly. Gently heat until boiling. Distribute into tubes or flasks. Autoclave for 15 min at 15 psi pressure–121°C. Pour into sterile Petri dishes or leave in tubes.

Use: For the cultivation and maintenance of *Stigmatella aurantiaca.*

CTA Agar (Cystine Trypticase™ Agar)

Composition per liter:

Pancreatic digest of casein	20.0g
Agar	14.0g
NaCl	5.0g
L-Cystine	0.5g
Na_2SO_3	0.5g
Phenol Red	0.017g

pH 7.3 ± 0.2 at 25°C

Preparation of Medium: Add components to distilled/deionized water and bring volume to 1.0L. Mix thoroughly. Gently heat until boiling. Distribute into tubes or flasks. Autoclave for 15 min at 15 psi pressure–118°C. Pour into sterile Petri dishes or leave in tubes. Two drops of sterile rabbit serum added per tube prior to solidification enhance the recovery of *Corynebacterium diphtheriae.*

Use: For cultivation and maintenance of a variety of fastidious microorganisms. Used for carbohydrate fermentation tests in the differentiation of *Neisseria* species.

CTA Medium (Cystine Trypticase™ Agar Medium) (Cystine Tryptic Agar)

Composition per liter:

Pancreatic digest of casein	20.0g
NaCl	5.0g
Carbohydrate	5.0g
Agar	2.5g
L-Cystine	0.5g
Na_2SO_3	0.5g
Phenol Red	0.017g

pH 7.3 ± 0.2 at 25°C

Source: Available as a premixed powder from BBL Microbiology Systems and Difco Laboratories.

Preparation of Medium: Add components to distilled/deionized water and bring volume to 1.0L. Mix thoroughly. Adjust pH to 7.3. Gently heat until boiling. Distribute into tubes or flasks. Autoclave for 15 min at 15 psi pressure–118°C. Cool tubes in an upright position. Store at room temperature.

Use: For cultivation and maintenance of a variety of fastidious microorganisms. Used for the detection of bacterial motility. Used, with added specific carbohydrate, for fermentation reactions of fastidious microorganisms, especially, *Neisseria* species, pneumococci, streptococci and non-sporeforming anaerobes.

CTA Medium with Yeast Extract and Rabbit Serum (Cystine Trypticase™ Agar Medium with Yeast Extract and Rabbit Serum)

Composition per liter:

Yeast extract	50.0g
Pancreatic digest of casein	20.0g

NaCl....5.0g
Carbohydrate....5.0g
Agar....2.5g
L-Cystine....0.5g
Na_2SO_3....0.5g
Phenol Red....0.017g
Rabbit serum....250.0mL

pH 7.3 ± 0.2 at 25ºC

Preparation of Medium: Add components, except rabbit serum to distilled/deionized water and bring volume to 750.0mL. Mix thoroughly. Adjust pH to 7.3. Gently heat until boiling. Autoclave for 15 min at 15 psi pressure–118°C. Cool to 50°C. Aseptically add sterile rabbit serum. Mix thoroughly. Distribute into sterile tubes. Store at room temperature—do not refrigerate.

Use: For the cultivation and maintenance of fastidious microorganisms, especially mycoplasmas and related microorganisms.

CTT Medium

Composition per liter:

Agar....15.0g
Pancreatic digest of casein....10.0g
Tris(hydroxymethyl)aminomethane buffer....1.21g
Potassium phosphate
buffer (1 mM, pH 7.6)....1.0L
Magnesium sulfate solution....10.0mL

pH 7.6 ± 0.2 at 25°C

Magnesium Sulfate Solution:

Composition per 10mL:

$MgSO_4 \cdot 7H_2O$....2.0g

Preparation of Magnesium Sulfate Solution: Add components to 10.0mL of distilled/deionized water. Mix thoroughly.

Preparation of Medium: Combine components. Mix thoroughly. Adjust pH to 7.6. Gently heat and bring to boiling. Distribute into tubes or flasks. Autoclave for 15 min at 15 psi pressure–121°C. Pour into sterile Petri dishes or leave in tubes.

Use: For the cultivation of myxobacteria.

CVA Medium (Cefoperazone Vancomycin Amphotericin Medium)

Composition per liter:

Agar....15.0g
Casein peptone....10.0g
Meat peptone....10.0g
NaCl....5.0g
Yeast autolysate....2.0g
Glucose....1.0g
$NaHSO_3$....0.1g
Sheep blood, defibrinated....50.0mL
CVA antibiotic solution....10.0mL

pH 7.0 ± 0.2 at 25°C

CVA Antibiotic Solution:

Composition per 10mL:

Cefoperazone....20.0mg
Vancomycin....10.0mg
Amphotericin B....2.0mg

Preparation of CVA Antibiotic Solution: Add components to distilled/deionized water and bring volume to 10.0mL. Mix thoroughly. Filter sterilize.

Preparation of Medium: Add components, except CVA antibiotic solution and sheep blood, to distilled/deionized water and bring volume to 940.0mL. Mix thoroughly. Gently heat until boiling. Autoclave for 15 min at 15 psi pressure–121°C. Cool to 45°–50°C. Aseptically add sterile CVA antibiotic solution and sterile, defibrinated sheep blood. Mix thoroughly. Pour into sterile Petri dishes.

Use: For the isolation and cultivation of *Campylobacter* species from clinical specimens.

CVP Medium

See: **Crystal Violet Pectate Medium**

CY Agar

Composition per liter:

Agar....15.0g
Pancreatic digest of casein....3.0g
$CaCl_2 \cdot 2H_2O$....1.0g
Yeast extract....1.0g
Cyanocobalamin....0.5mg

pH 7.2 ± 0.2 at 25°C

Preparation of Medium: Add components to distilled/deionized water and bring volume to 1.0L. Mix thoroughly. Gently heat and bring to boiling. Distribute into tubes or flasks. Autoclave for 15 min at 15 psi pressure–121°C. Pour into sterile Petri dishes or leave in tubes.

Use: For the cultivation of myxobacteria.

CYA Agar

See: **Czapek Yeast Autolysate Agar**

CYC Medium

Composition per liter:

Sucrose	30.0g
Casamino acids, vitamin-free	6.0g
$NaNO_3$	3.0g
Yeast extract	2.0g
K_2HPO_4	1.0g
$MgSO_4 \cdot 7H_2O$	0.5g
KCl	0.5g
$FeSO_4 \cdot 7H_2O$	0.01g
Antibiotic solution	10.0mL

pH 7.2 ± 0.2 at 25°C

Preparation of Medium: Add components to distilled/deionized water and bring volume to 1.0L. Mix thoroughly. Distribute into tubes or flasks. Autoclave for 15 min at 15 psi pressure–121°C.

Antibiotic Solution:

Composition per 10mL:

Cycloheximide	0.050g
Novobiocin	0.025g

Preparation of Antibiotic Solution: Add components to distilled/deionized water and bring volume to 10.0mL. Mix thoroughly. Filter sterilize.

Preparation of Medium: Add components, except antibiotic solution, to distilled/deionized water and bring volume to 990.0mL. Mix thoroughly. Gently heat and bring to boiling. Autoclave for 15 min at 15 psi pressure–121°C. Cool to 45°–50°C. Aseptically add sterile antibiotic solution. Mix thoroughly. Aseptically distribute into sterile tubes.

Use: For the isolation and cultivation of *Thermoactinomyces* species.

Cyclohexanecarboxylic Acid Medium

Composition per liter:

K_2HPO_4	3.5g
Cyclohexanecarboxylic acid	2.0g
KH_2PO_4	1.5g
NH_4NO_3	1.0g
$MgSO_4 \cdot 7H_2O$	0.5g
Yeast extract	0.1g
$CaCl_2 \cdot 2H_2O$	0.01g
$FeCl_3 \cdot 6H_2O$	0.01g
$NaMoO_4 \cdot 7H_2O$	0.01g
$ZnSO_4 \cdot 7H_2O$	0.01g

pH 7.0 ± 0.2 at 25°C

Preparation of Medium: Add components to distilled/deionized water and bring volume to 1.0L. Mix thoroughly. Distribute into tubes or flasks. Autoclave for 15 min at 15 psi pressure–121°C.

Use: For the cultivation and maintenance of *Alcaligenes faecalis* and other bacteria that can utilize cyclohexanecarboxylic acid as a carbon source.

Cyclohexanecarboxylic Acid Salts Medium
See: CHCA Salts Medium

Cyclohexanone Medium

Composition per liter:

NH_4NO_3	3.0g
K_2HPO_4	0.25g
$MgSO_4 \cdot 7H_2O$	0.20g
$CaCl_2 \cdot 2H_2O$	0.01g
$FeCl_3 \cdot 6H_2O$	1.0mg
Cyclohexanone	1.0mL

Preparation of Medium: Add components, except cyclohexanone, to distilled/deionized water and bring volume to 999.0mL. Mix thoroughly. Distribute into tubes or flasks. Autoclave for 20 min at 15 psi pressure–121°C. Filter sterilize cyclohexanone. Aseptically add 1.0mL cyclohexanone. Mix thoroughly.

Use: For the cultivation and maintenance of *Nocardia* species and other bacteria that can utilize cyclohexanone as a carbon source.

Cycloheximide Agar
See: Actidione® Agar

Cycloheximide Chloramphenicol Agar
See: Mycosel™ Agar
See: Mycobiotic Agar

Cycloserine Cefoxitin Egg Yolk Fructose Agar

Composition per liter:

Proteose peptone No. 2	40.0g
Agar	25.0g
Fructose	6.0g
Na_2HPO_4	5.0g
NaCl	2.0g
KH_2PO_4	1.0g
$MgSO_4 \cdot 7H_2O$	0.1g
Egg yolk emulsion	100.0mL
Antibiotic solution	10.0mL
Neutral Red solution	3.0mL
Hemin solution	1.0mL

Egg Yolk Emulsion:
Composition:

Chicken egg yolks	11
Whole chicken egg	1

Preparation of Egg Yolk Emulsion: Soak eggs with 1:100 dilution of saturated mercuric chloride solution for 1 min. Crack eggs. Separate yolks from whites for 11 eggs. Mix egg yolks with 1 chicken egg.

Antibiotic Solution:
Composition per 10mL:

Cycloserine	0.5g
Cefoxitin	0.016g

Preparation of Antibiotic Solution: Add components to distilled/deionized water and bring volume to 10.0mL. Mix thoroughly. Filter sterilize.

Neutral Red Solution:
Composition per 10mL:

Neutral Red	0.1g
Ethanol	10.0mL

Preparation of Neutral Red Solution: Add Neutral Red to 10.0mL of ethanol. Mix thoroughly.

Hemin Solution:
Composition per 100mL:

Hemin	0.5g
NaOH (1*N* solution)	10.0mL

Preparation of Hemin Solution: Add hemin to 10.0mL of 1*N* NaOH solution. Mix thoroughly. Bring volume to 100.0mL with distilled/deionized water.

Preparation of Medium: Add components, except egg yolk emulsion and antibiotic solution, to distilled/deionized water and bring volume to 890.0mL. Mix thoroughly. Gently heat and bring to boiling. Autoclave for 15 min at 15 psi pressure–121°C. Cool to 45°–50°C. Aseptically add sterile egg yolk emulsion and antibiotic solution. Mix thoroughly. Pour into sterile Petri dishes.

Use: For the selective isolation and cultivation of *Clostridium difficile* from feces.

Cycloserine Cefoxitin Fructose Agar

See: ***Clostridium difficile*** **Agar**

CYE Agar (Charcoal Yeast Extract Agar)

Composition per liter:

Agar	17.0g
Yeast extract	10.0g
Charcoal, activated, acid-washed	2.0g
L-Cysteine·HCl·H_2O solution	10.0mL
$Fe_4(P_2O_7)_3$ solution	10.0mL

pH 6.9 ± .05 at 50°C

L-Cysteine·HCl·H_2O Solution:
Composition per 10mL:

L-Cysteine·HCl·H_2O	0.4g

Preparation of L-Cysteine·HCl·H_2O solution: Add L-Cysteine·HCl·H_2O to distilled/deionized water and bring volume to 10.0mL. Mix thoroughly. Filter sterilize.

$Fe_4(P_2O_7)_3$ Solution:
Composition per liter:

$Fe_4(P_2O_7)_3$	0.25g

Preparation of $Fe_4(P_2O_7)_3$ Solution: Add soluble $Fe_4(P_2O_7)_3$ to distilled/deionized water and bring volume to 10.0mL. Mix thoroughly. Filter sterilize. The soluble $Fe_4(P_2O_7)_3$ must be kept dry and in the dark. Do not use if brown or yellow. Prepare solutions freshly. Do not heat over 60°C to dissolve. The mixture dissolves readily in a 50°C water bath.

Preparation of Medium: Add components, except L-cysteine·HCl·H_2O solution and $Fe_4(P_2O_7)_3$ solution to distilled/deionized water and bring volume to 980.0mL. Mix thoroughly. Gently heat to boiling. Autoclave for 15 min at 15 psi pressure–121°C. Cool to 50° C. Add 10.0mL of sterile L-cysteine·HCl·H_2O solution. Add 10.0mL of sterile $Fe_4(P_2O_7)_3$ solution. Adjust pH to 6.9 at 50°C by adding 4.0–4.5mL of 1.0 *N* KOH. This is a critical step. Mix thoroughly. Pour in 20.0mL volumes into sterile Petri dishes. Swirl medium while pouring to keep charcoal in suspension.

Use: For the cultivation and maintenance of *Legionella* species and *Tatlockia micdadei.*

CYE Agar, Buffered (Charcoal Yeast Extract Agar, Buffered)

Composition per liter:

Agar	17.0g
ACES buffer (N-2-acetamido-2-aminoethane sulfonic acid)	10.0g
Yeast extract	10.0g
Charcoal, activated, acid-washed	2.0g
L-Cysteine·HCl·H_2O solution	10.0mL
$Fe_4(P_2O_7)_3$ solution	10.0mL

pH 6.9 ± .05 at 50°C

L-Cysteine·HCl·H_2O Solution:
Composition per 10mL:

L-Cysteine·HCl·H_2O	0.4g

Preparation of L-Cysteine·HCl·H_2O solution: Add L-Cysteine·HCl·H_2O to distilled/deionized water and bring volume to 10.0mL. Mix thoroughly. Filter sterilize.

$Fe_4(P_2O_7)_3$ solution:
Composition per liter:

$Fe_4(P_2O_7)_3$ 0.25g

Preparation of $Fe_4(P_2O_7)_3$ solution: Add soluble $Fe_4(P_2O_7)_3$ to distilled/deionized water and bring volume to 10.0mL. Mix thoroughly. Filter sterilize. The soluble $Fe_4(P_2O_7)_3$ must be kept dry and in the dark. Do not use if brown or yellow. Prepare solutions freshly. Do not heat over 60°C to dissolve. The mixture dissolves readily in a 50°C water bath.

Preparation of Medium: Add components, except L-cysteine·HCl·H_2O solution and $Fe_4(P_2O_7)_3$ solution to distilled/deionized water and bring volume to 980.0mL. Mix thoroughly. Gently heat to boiling. Autoclave for 15 min at 15 psi pressure–121°C. Cool to 50° C. Add 10.0mL of sterile L-cysteine·HCl·H_2O solution. Add 10.0mL of sterile $Fe_4(P_2O_7)_3$ solution. Adjust pH to 6.9 at 50°C by adding 4.0–4.5mL of 1.0 *N* KOH. This is a critical step. Mix thoroughly. Pour in 20.0mL volumes into sterile Petri dishes. Swirl medium while pouring to keep charcoal in suspension.

Use: For the cultivation and maintenance of *Legionella* species and *Xylella fastidiosa*.

CYE Broth
See: Casamino Acids Yeast Extract Broth

CYE DBCM
See: Legionella pneumophila Medium Charcoal Yeast Extract Diphasic Blood Culture Medium

CYG Agar
See: Casein Yeast Extract Glucose Agar

CYG Agar

Composition per liter:

Agar 15.0g
Pancreatic digest of casein 3.0g
$CaCl_2 \cdot 2H_2O$ 1.0g
Yeast extract 1.0g
Cyanocobalamin 0.5mg
Glucose solution 100.0mL

pH 7.2 ± 0.2 at 25°C

Glucose Solution:
Composition per 100mL:

D-Glucose 5.0g

Preparation of Glucose Solution: Add D-glucose to distilled/deionized water and bring volume to 100.0mL. Mix thoroughly. Autoclave for 15 min at 15 psi pressure–121°C. Cool to 25°C.

Preparation of Medium: Add components, except glucose solution, to distilled/deionized water and bring volume to 900.0mL. Mix thoroughly. Gently heat and bring to boiling. Autoclave for 15 min at 15 psi pressure–121°C. Cool to 45°–50°C. Aseptically add sterile glucose solution. Mix thoroughly. Pour into sterile Petri dishes or distribute into sterile tubes.

Use: For the isolation and cultivation of *Cytophaga* species, *Herpetosiphon* species, *Saprospira* species, and *Flexithrix* species.

Cylindrocladium Isolation Medium

Composition per liter:

Agar 20.0g
Glucose 15.0g
KH_2PO_4 .... 1.0g
KNO_3 .... 0.5g
$MgSO_4 \cdot 7H_2O$ 0.5g
Yeast extract 0.5g
Chloramphenicol solution 10.0mL
Chlortetracycline solution 10.0mL
Thiabendazole solution 10.0mL
Tergitol NPX® (Union Carbide) 1.0mL

Chloramphenicol Solution:
Composition per 10mL:

Chloramphenicol 0.1g
Ethanol (95% solution) 10.0mL

Preparation of Chloramphenicol Solution: Add chloramphenicol to 10.0mL of ethanol. Mix thoroughly. Filter sterilize.

Chlortetracycline Solution:
Composition per 10mL:

Chlortetracycline 0.04g
Ethanol, absolute 5.0mL

Preparation of Chlortetracycline Solution: Add chlortetracycline to 5.0mL of ethanol. Mix thoroughly. Bring volume to 10.0mL with distilled/deionized water. Filter sterilize.

Thiabendazole Solution:
Composition per 10mL:

Thiabendazole	1.0mg

Preparation of Thiabendazole Solution: Add thiabendazole to distilled/deionized water and bring volume to 10.0mL. Mix thoroughly. Filter sterilize.

Preparation of Medium: Filter sterilize tergitol NPX. Add components—except tergitol NPX, thiabendazole solution, chloramphenicol solution and chlortetracycline solution—to distilled/deionized water and bring volume to 969.0mL. Mix thoroughly. Gently heat and bring to boiling. Autoclave for 15 min at 15 psi pressure–121°C. Cool to 45°–50°C. Aseptically add sterile tergitol NPX, thiabendazole solution, chloramphenicol solution and chlortetracycline solution. Mix thoroughly. Pour into sterile Petri dishes or distribute into sterile tubes.

Use: For the isolation and cultivation of *Cylindrocladium* species.

Cystine Heart Agar

Composition per liter:

Beef heart, solids from infusion	500.0g
Agar	15.0g
Glucose	10.0g
Proteose peptone	10.0g
NaCl	5.0g
L-Cystine	1.0g
Hemoglobin solution	100.0mL

pH 6.8 ± 0.2 at 25°C

Source: This medium is available as a premixed powder from Difco Laboratories.

Hemoglobin Solution:
Composition per 100mL:

Hemoglobin	2.0g

Preparation of Hemoglobin Solution: Add hemoglobin to cold distilled/deionized water and bring volume to 100.0mL. Mix thoroughly by shaking for 10–15 min. Autoclave for 15 min at 15 psi pressure–121°C. Cool to 50–60°C.

Preparation of Medium: Add components, except hemoglobin solution, to distilled/deionized water and bring volume to 900.0mL. Mix thoroughly. Gently heat until boiling. Autoclave for 15 min at 15 psi pressure–121°C. Cool to 50–60°C. Aseptically add 100.0mL of sterile cooled hemoglobin solution. Mix thoroughly. Pour into sterile Petri dishes or distribute into sterile tubes.

Use: For the cultivation and maintenance of *Francisella tularensis* and *Francisella philomiragia.* Without the hemoglobin enrichment, it supports excellent growth of Gram-negative cocci and other pathogenic microorganisms.

Cystine Heart Agar with Rabbit Blood

Composition per liter:

Beef heart, solids from infusion	500.0g
Agar	15.0g
Glucose	10.0g
Proteose peptone	10.0g
NaCl	5.0g
L-Cystine	1.0g
Rabbit blood, defibrinated	50.0mL

pH 6.8 ± 0.2 at 25°C

Source: This medium is available as a premixed powder from Difco Laboratories.

Preparation of Medium: Add components, except rabbit blood, to distilled/deionized water and bring volume to 950.0mL. Mix thoroughly. Gently heat until boiling. Autoclave for 15 min at 15 psi pressure–121°C. Cool to 50–60°C. Aseptically add 50.0mL of sterile, defibrinated rabbit blood. Mix thoroughly. Pour into sterile Petri dishes or distribute into sterile tubes.

Use: For the cultivation and maintenance of *Francisella tularensis* and *Francisella philomiragia.* Without the hemoglobin enrichment, it supports excellent growth of Gram-negative cocci and other pathogenic microorganisms.

Cystine Lactose Electrolyte Deficient Agar

See: **CLED Agar**

Cystine Lactose Electrolyte Deficient Agar with Andrade Indicator

See: **CLED Agar with Andrade Indicator**

Cystine Tellurite Blood Agar

Composition per 120mL:

Heart infusion agar	100.0mL
K_2TeO_3 solution	15.0mL
Sheep blood	5.0mL
L-Cystine	5.0mg

pH 7.4 ± 0.2 at 25°C

Heart Infusion Agar:

Composition per liter:

Beef heart, infusion from	500.0g
Agar	20.0g
Tryptose	10.0g
Yeast extract	5.0g
NaCl	5.0g

Preparation of Heart Infusion Agar: Add components to distilled/deionized water and bring volume to 1.0L. Mix thoroughly. Autoclave for 15 min at 15 psi pressure–121°C. Cool to 45°–50°C.

K_2TeO_3 Solution:

Composition per 100mL:

K_2TeO_3	0.3g

Preparation of K_2TeO_3 Solution: Add K_2TeO_3 to distilled/deionized water and bring volume to 100.0mL. Mix thoroughly. Autoclave for 15 min at 15 psi pressure–121°C.

Caution: Potassium tellurite is toxic.

Preparation of Medium: Add sterile K_2TeO_3 solution, sterile defibrinated sheep blood and sterile solid L-cystine to sterile, cooled Heart Infusion Agar. Mix thoroughly. Pour into sterile Petri dishes or distribute into sterile tubes.

Use: For the isolation, differentiation and cultivation of *Corynebacterium diphtheriae*. *C. diphtheriae* appear as dark gray to black colnies.

Cystine Tellurite Blood Agar

Composition per liter:

Heart infusion agar	900.0mL
K_2TeO_3 solution	75.0mL
Rabbit blood	25.0mL
L-Cystine	22.0mg

pH 7.4 ± 0.2 at 25°C

Heart Infusion Agar:

Composition per 900mL:

Beef heart, solids from infusion	500.0g
Agar	20.0g
Tryptose	10.0g
Yeast extract	5.0g
NaCl	5.0g

Preparation of Heart Infusion Agar: Add components to distilled/deionized water and bring volume to 900.0mL. Mix thoroughly. Autoclave for 15 min at 15 psi pressure–121°C. Cool to 45°–50°C.

K_2TeO_3 Solution:

Composition per 100mL:

K_2TeO_3	0.3g

Preparation of K_2TeO_3 Solution: Add K_2TeO_3 to distilled/deionized water and bring volume to 100.0mL. Mix thoroughly. Autoclave for 15 min at 15 psi pressure–121°C.

Caution: Potassium tellurite is toxic.

Preparation of Medium: Add sterile K_2TeO_3 solution, sterile rabbit blood and sterile solid L-cystine to sterile, cooled Heart Infusion Agar. Mix thoroughly. Pour into sterile Petri dishes or distribute into sterile tubes.

Use: For the isolation, differentiation and cultivation of *Corynebacterium diphtheriae*. *C. diphtheriae* appear as dark gray to black colnies.

Cystine Tryptic Agar
See: **CTA Agar**

Cystine Trypticase Agar
See: **CTA Agar**

Cystine Trypticase Agar Medium
See: **CTA Medium**

Cystine Trypticase Agar Medium with Yeast Extract and Rabbit Serum
See: **CTA Medium with Yeast Extract and Rabbit Serum**

CYT Agar

Composition per liter:

Agar	15.0g
Pancreatic digest of casein	1.0g
$CaCl_2 \cdot 2H_2O$	0.5g
$MgSO_4 \cdot 7H_2O$	0.5g
Yeast extract	0.5g

pH 7.2 ± 0.2 at 25°C

Preparation of Medium: Add components to distilled/deionized water and bring volume to 1.0L. Mix thoroughly. Gently heat and bring to boiling. Distribute into tubes or flasks. Autoclave for 15 min at 15 psi pressure–121°C. Pour into sterile Petri dishes or leave in tubes.

Use: For the isolation and cultivation of *Cytophaga* species, *Herpetosiphon* species, *Saprospira* species, and *Flexithrix* species.

Cytophaga Agarase Agar (ATCC Medium 793)

Composition per liter:

Agar	15.0g
KH_2PO_4	1.0g
$MgSO_4 \cdot 7H_2O$	0.5g
NH_4Cl	0.5g
$CaCl_2 \cdot H_2O$	0.02g
Vishniac and Santer trace element mixture	0.2mL

pH 7.2 ± 0.2 at 25°C

Vishniac and Santer Trace Element Mixture:
Composition per liter:

Ethylenediamine tetraacetic acid (EDTA)	50.0g
$ZnSO_4 \cdot 7H_2O$	22.0g
$CaCl_2$	5.54g
$MnCl_2 \cdot 4H_2O$	5.06g
$FeSO_4 \cdot 7H_2O$	4.99g
$CoCl_2 \cdot 6H_2O$	1.61g
$CuSO_4 \cdot 5H_2O$	1.57g
$(NH_4)_6Mo_7O_{24} \cdot 4H_2O$	1.10g

Preparation of Vishniac and Santer Trace Element Mixture: Add components to distilled/deionized water and bring volume to 1.0L. Adjust pH to 6.0 with KOH. Mix thoroughly.

Preparation of Medium: Add components to distilled/deionized water and bring volume to 1.0L. Adjust pH to 7.2. Mix thoroughly. Distribute into tubes or flasks. Autoclave for 15 min at 15 psi pressure–121°C. Pour into sterile Petri dishes or leave in tubes.

Use: For the cultivation and maintenance of *Cytophaga flevensis*.

Cytophaga Agarase Broth

Composition per liter:

Agar	1.0g
KH_2PO_4	1.0g
$MgSO_4 \cdot 7H_2O$	0.5g
NH_4Cl	0.5g
$CaCl_2$·H2O	0.02g
Vishniac and Santer trace element mixture	0.2mL

pH 7.2 ± 0.2 at 25°C

Vishniac and Santer Trace Element Mixture:
Composition per liter:

Ethylenediamine tetraacetic acid (EDTA)	50.0g
$ZnSO_4 \cdot 7H_2O$	22.0g
$CaCl_2$	5.54g
$MnCl_2 \cdot 4H_2O$	5.06g
$FeSO_4 \cdot 7H_2O$	4.99g
$CoCl_2 \cdot 6H_2O$	1.61g
$CuSO_4 \cdot 5H_2O$	1.57g
$(NH_4)_6Mo_7O_{24} \cdot 4H_2O$	1.10g

Preparation of Vishniac and Santer Trace Element Mixture: Add components to distilled/deionized water and bring volume to 1.0L. Adjust pH to 6.0 with KOH. Mix thoroughly.

Preparation of Medium: Add components to distilled/deionized water and bring volume to 1.0L. Adjust pH to 7.2. Mix thoroughly. Distribute into tubes or flasks. Autoclave for 15 min at 15 psi pressure–121°C.

Use: For the cultivation of *Cytophaga flevensis*.

Cytophaga fermentans Medium

Composition per liter:

NaCl	30.0g
Agar	5.0g
$NaHCO_3$	5.0g
KH_2PO_4	1.0g
NH_4Cl	1.0g
$MgCl_2 \cdot 6H_2O$	0.5g
Yeast extract	0.3g
$Na_2S \cdot 9H_2O$	0.1g
$CaCl_2$	0.04g
Ferric citrate (4mM solution)	5.0mL
Trace element solution	2.0mL

pH 7.0 ± 0.2 at 25°C

Trace Element Solution:
Composition per 100mL:

H_3BO_3	0.28g
$MnSO_4 \cdot 6H_2O$	0.21g
$Na_2MoO_4 \cdot 2H_2O$	0.075g
$Zn(NO_3)_2 \cdot 6H_2O$	0.025g
$CoCl_2 \cdot 6H_2O$	0.02g
$Cu(NO_3)_2 \cdot 3H_2O$	0.02g

Preparation of Trace Element Solution: Add components to distilled/deionized water and bring volume to 100.0mL. Mix thoroughly.

Preparation of Medium: Add components to distilled/deionized water and bring volume to 1.0L. Mix thoroughly. Gently heat and bring to boiling. Distribute into tubes or flasks. Autoclave for 15 min at 15 psi pressure–121°C.

Use: For the cultivation of agar-digesting *Cytophaga fermentans*.

Cytophaga Medium (ATCC Medium 420)

Composition per liter:

NaCl	20.0g
Yeast extract	10.0g
Agar	3.0g
$MgSO_4 \cdot 7H_2O$	1.0g

NH_4Cl 1.0g
K_2HPO_4 .... 0.2g
$FeCl_3$ 1.0μg
pH 7.5 ± 0.2 at 25°C

Preparation of Medium: Add components to tap water and bring volume to 1.0L. Adjust pH to 7.5. Mix thoroughly. Distribute into tubes or flasks. Autoclave for 15 min at 15 psi pressure–121°C.

Use: For the cultivation of *Cytophaga fermentens.*

Cytophaga Medium (ATCC Medium 1299)

Composition per liter:
Agar 11.0g
Pancreatic digest of casein 0.5g
Yeast extract 0.5g
Beef extract 0.2g
Sodium acetate 0.2g
pH 7.2 ± 0.2 at 25°C

Preparation of Medium: Add components to distilled/deionized water and bring volume to 1.0L. Mix thoroughly. Distribute into tubes or flasks. Autoclave for 15 min at 15 psi pressure–121°C. Pour into sterile Petri dishes or leave in tubes.

Use: For the cultivation and maintenance of *Cytophaga* species and *Flavobacterium branchiophilum.*

Cytophaga Medium

Composition per liter:
NaCl 30.0g
Agar 15.0g
KH_2PO_4 .... 1.0g
NH_4Cl 1.0g
Yeast extract 1.0g
$MgSO_4$ 0.5g
$CaCl_2$ 0.04g
$FeCl_3·6H_2O$ 1.25mg
$NaHCO_3$ solution 100.0mL
Glucose solution 10.0mL
$Na_2S·9H_2O$ solution 1.0mL

Glucose Solution:
Composition per 100mL:
Glucose 10.0g

Preparation of Glucose Solution: Add glucose to distilled/deionized water and bring volume to 100.0mL. Mix thoroughly. Autoclave for 15 min at 15 psi pressure–121°C.

$NaHCO_3$ Solution:
Composition per 100mL:
$NaHCO_3$ 5.0g

Preparation of $NaHCO_3$ Solution: Add the $NaHCO_3$ to distilled/deionized water and bring volume to 100.0mL. Mix thoroughly. Filter sterilize.

$Na_2S·9H_2O$ Solution:
Composition per 100mL:
$Na_2S·9H_2O$ 10.0g

Preparation of $Na_2S·9H_2O$ Solution: Add $Na_2S·9H_2O$ to distilled/deionized water and bring volume to 100.0mL. Mix thoroughly. Autoclave for 15 min at 15 psi pressure–121°C.

Preparation of Medium: Add components, except $NaHCO_3$ solution, glucose solution and $Na_2S·9H_2O$ solution to distilled/deionized water and bring volume to 889.0mL. Autoclave for 15 min at 15 psi pressure–121°C. Cool to 50°C. Aseptically add sterile $NaHCO_3$ solution, sterile glucose solution and sterile $Na_2S·9H_2O$ solution. Mix thoroughly. Pour into sterile Petri dishes or distribute into sterile tubes.

Use: The cultivation of *Cytophaga agarovorans.*

Cytophaga Spirochete Medium

See: ***Cytophaga*** **Medium**
See: **Spirochete Medium**

Czapek Agar (ATCC Medium 312)

Composition per liter:
Sucrose 30.0g
Agar 15.0g
$NaNO_3$ 3.0g
K_2HPO_4 .... 1.0g
KCl 0.5g
$MgSO_4·7H_20$ 0.5g
$FeSO_4·7H_2O$ 0.01g
pH 7.3 ± 0.2 at 25°C

Preparation of Medium: Add components, except sucrose, to distilled/deionized water and bring volume to 900.0mL. Mix thoroughly. Distribute into tubes or flasks. In a separate flask, add sucrose to distilled/deionized water and bring volume to 100.0mL. Mix thoroughly. Autoclave both solutions separately for 15 min at 15 psi pressure–121°C. Cool to 50°C. Combine the sterile solutions. Mix thoroughly. Pour into sterile Petri dishes or distribute into sterile tubes.

Use: For the cultivation and maintenance of *Streptomyces* species. Also, for the cultivation of Actinoplanaceae.

Czapek Agar with Peptone (ATCC Medium 522)

Composition per liter:

Sucrose	30.0g
Agar	15.0g
Peptone	5.0g
$NaNO_3$	3.0g
K_2HPO_4	1.0g
KCl	0.5g
$MgSO_4 \cdot 7H_20$	0.5g
$FeSO_4 \cdot 7H_2O$	0.01g

pH 7.3 ± 0.2 at 25°C

Preparation of Medium: Add components, except sucrose, to distilled/deionized water and bring volume to 900.0mL. Mix thoroughly. Distribute into tubes or flasks. In a separate flask, add sucrose to distilled/deionized water and bring volume to 100.0mL. Mix thoroughly. Autoclave both solutions separately for 15 min at 15 psi pressure–121°C. Cool to 50°C. Combine the sterile solutions. Mix thoroughly. Pour into sterile Petri dishes or distribute into sterile tubes.

Use: For the cultivation and maintenance of *Streptomyces* species. Also, for the cultivation of Actinoplanaceae.

Czapek Agar
See: Czapek Yeast Autolysate Agar

Czapek Dox Agar

Composition per liter:

Sucrose	30.0g
Agar	15.0g
$NaNO_3$	3.0g
K_2HPO_4	1.0g
$MgSO_4 \cdot 7H_2O$	0.5g
KCl	0.5g
$FeSO_4 \cdot 7H_2O$	0.01g

pH 7.3 ± 0.2 at 25°C

Preparation of Medium: Add components to distilled/deionized water and bring volume to 1.0L. Mix thoroughly. Distribute into tubes or flasks. Autoclave for 15 min at 15 psi pressure–121°C. Pour into sterile Petri dishes or leave in tubes.

Use: For the cultivation and maintenance of *Actinoplanes* species, *Amorphosporangium auranticolor*, *Ampullariella* species, *Spirillospora albida* and *Streptomyces armeniacus*.

Czapek Dox Agar with 3% Glucose

Composition per liter:

Glucose	30.0g
Agar	15.0g
$NaNO_3$	3.0g
K_2HPO_4	1.0g
KCl	0.5g
$MgSO_4 \cdot 7H_2O$	0.5g
$FeSO_4 \cdot 7H_2O$	0.01g

pH 7.3 ± 0.2 at 25°C

Preparation of Medium: Sterilize by autoclaving for 15 minutes at 15 lbs. pressure (121°C). Pour into sterile Petri dishes or leave in tubes.

Use: For the cultivation and maintenance of *Microbispora rosea* and *Streptomyces* species.

Czapek Dox Agar, Modified

Composition per liter:

Sucrose	30.0g
Agar	12.0g
$NaNO_3$	2.0g
Magnesium glycerophosphate	0.5g
KCl	0.5g
K_2SO_4	0.35g
$FeSO_4$	0.01g

pH 6.8 ± 0.2 at 25°C

Source: This medium is available as a premixed powder from Oxoid Unipath.

Preparation of Medium: Add components to distilled/deionized water and bring volume to 1.0L. Mix thoroughly. Distribute into tubes or flasks. Autoclave for 15 min at 15 psi pressure–121°C. Pour into sterile Petri dishes or leave in tubes.

Use: For the cultivation and maintenance of numerous fungal species. For chlamydospore production by *Candida albicans*.

Czapek Dox Broth

Composition per liter:

Sucrose	30.0g
$NaNO_3$	3.0g
K_2HPO_4	1.0g
$MgSO_4 \cdot 7H_2O$	0.5g
KCl	0.5g
$FeSO_4 \cdot 7H_2O$	0.01g

pH 7.3 ± 0.2 at 25°C

Source: This medium is available as a premixed powder from Difco Laboratories.

Preparation of Medium: Add components to distilled/deionized water and bring volume to 1.0L. Mix thoroughly. Distribute into tubes or flasks. Autoclave for 15 min at 15 psi pressure–121°C.

Use: For the cultivation and maintenance of a variety of fungal and bacterial species that can use nitrate as sole nitrogen source.

Czapek Dox Liquid Medium, Modified

Composition per liter:

Sucrose	30.0g
$NaNO_3$	2.0g
Magnesium glycerophosphate	0.5g
KCl	0.5g
K_2SO_4	0.35g
$FeSO_4$	0.01g

pH 6.8 ± 0.2 at 25°C

Source: This medium is available as a premixed powder from Oxoid Unipath.

Preparation of Medium: Add components to distilled/deionized water and bring volume to 1.0L. Mix thoroughly. Distribute into tubes or flasks. Autoclave for 15 min at 15 psi pressure–121°C.

Use: For the cultivation of fungi and bacteria capable of utilizing sodium nitrate as the sole source of nitrogen.

Czapek Solution Agar

Composition per liter:

Sucrose	30.0g
Agar	15.0g
$NaNO_3$	2.0g
K_2HPO_4	1.0g
KCl	0.5g
$MgSO_4 \cdot 7H_20$	0.5g
$FeSO_4 \cdot 7H_2O$	0.01g

pH 7.3 ± 0.2 at 25°C

Source: This medium is available as a premixed powder from Difco Laboratories.

Preparation of Medium: Add components to distilled/deionized water and bring volume to 1.0L. Mix thoroughly. Distribute into tubes or flasks. Autoclave for 15 min at 15 psi pressure–121°C. Pour into sterile Petri dishes or leave in tubes.

Use: For the cultivation of *Aspergillus, Penicillium* and other fungi. For the cultivation and maintenance of microorganisms that can utilize nitrate as sole nitrogen source.

Czapek Yeast Autolysate Agar (CYA Agar) (Czapek Agar)

Composition per liter:

Sucrose	30.0g
Agar	15.0g
Yeast extract	5.0g
$NaNO_3$	3.0g
K_2HPO_4	1.0g
KCl	0.5g
$MgSO_4 \cdot 7H_2O$	0.5g
$FeSO_4 \cdot 7H_2O$	0.01g

pH 7.3 ± 0.2 at 25°C

Sucrose Solution:

Composition per 100mL:

Sucrose	30.0g

Preparation of Sucrose Solution: Add sucrose to distilled/deionized water and bring volume to 100.0mL. Mix thoroughly. Autoclave for 15 min at 15 psi pressure–121°C. Cool to 50°C.

Preparation of Medium: Add components, except sucrose solution, to distilled/deionized water and bring volume to 900.0mL. Mix thoroughly. Autoclave for 15 min at 15 psi pressure–121°C. Cool to 50°C. Aseptically add sterile sucrose solution. Mix thoroughly. Pour into sterile Petri dishes or leave in tubes.

Use: For the isolation and cultivation of heat-resistant filamentous fungi (molds) from foods.

Davis and Mingioli Glucose Minimal Medium

Composition per liter:

Agar	15.0g
K_2HPO_4	7.0g
KH_2PO_4	3.0g
$(NH_4)_2SO_4$	1.0g
Sodium citrate·$3H_2O$	0.5g
$MgSO_4 \cdot 7H_2O$	0.1g
L-Arginine	0.02g
L-Tryptophan	0.02g
Glucose solution	10.0mL

pH 7.0 ± 0.2 at 25°C

Glucose Solution:

Composition per 10mL:

Glucose	2.0g

Preparation of Glucose Solution: Add glucose to distilled/deionized water and bring volume to 10.0mL. Mix thoroughly. Filter sterilize.

Preparation of Medium: Add components, except glucose solution, to distilled/deionized water and bring volume to 990.0mL. Gently heat and bring to boiling. Autoclave for 15 min at 15 psi pressure–121°C. Cool to 45°–50°C. Aseptically add sterile glucose solution. Mix thoroughly. Pour into sterile Petri dishes or distribute into sterile tubes.

Use: For the cultivation and maintenance of *Escherichia coli.*

Davis and Mingioli Medium, Modified

Composition per liter:

K_2HPO_4	7.0g
KH_2PO_4	2.0g
$(NH_4)_2SO_4$	1.0g
Na citrate·$2H_2O$	0.5g
$MgSO_4 \cdot 7H_2O$	0.1g
Glucose solution	10.0mL
Streptomycin solution	10.0mL
Additives solution	10.0mL

pH 7.2 ± 0.2 at 25°C

Glucose Solution:

Composition per 10mL:

Glucose	4.0g

Preparation of Glucose Solution: Add glucose to distilled/deionized water and bring volume to 10.0mL. Mix thoroughly. Filter sterilize.

Streptomycin Solution:

Composition per 10mL:

Streptomycin	4.0g

Preparation of Streptomycin Solution: Add streptomycin to distilled/deionized water and bring volume to 10.0mL. Mix well. Filter sterilize.

Additives Solution:

Composition per 10mL:

DL-Threonine	0.1g
DL- or L-Leucine	0.1g
Thiamine·HCl	0.5mg

Preparation of Additives Solution: Add components to distilled/deionized water and bring volume to 10.0mL. Mix thoroughly. Filter sterilize.

Preparation of Medium: Add components—except glucose solution, streptomycin solution and additives solution—to distilled/deionized water and bring volume to 970.0mL. Mix thoroughly. Adjust pH to 7.2. Gently heat and bring to boiling. Autoclave for 25 min at 15 psi pressure–121°C. Cool to 45°–50°C. Aseptically add sterile glucose solution, streptomycin solution and additives solution. Mix thoroughly. Aseptically distribute into sterile tubes or flasks.

Use: For the cultivation and maintenance of *Escherichia coli.*

Davis and Mingioli Medium with B_1 and Asparagine

Composition per liter:

K_2HPO_4	7.0g
KH_2PO_4	3.0g
$(NH_4)_2SO_4$	1.0g
Sodium citrate·$3H_2O$	0.5g
L-Asparagine	0.4g
$MgSO_4 \cdot 7H_2O$	0.1g
Vitamin B_1	0.1mg
Glucose solution	1.0mL

pH 7.0 ± 0.2 at 25°C

Glucose Solution:

Composition per 100mL:

Glucose	20.0g

Preparation of Glucose Solution: Add glucose to distilled/deionized water and bring volume to 100.0mL. Mix thoroughly. Filter sterilize.

Preparation of Medium: Add components, except glucose solution, to distilled/deionized water and bring volume to 999.0mL. Gently heat and bring to boiling. Autoclave for 15 min at 15 psi pressure–121°C. Cool to 25°C. Aseptically add 1.0mL of sterile glucose solution. Mix thoroughly. Aseptically distribute into sterile tubes or flasks.

Use: For the cultivation and maintenance of *Escherichia coli.*

Davis and Mingioli Medium with Proline

Composition per liter:

K_2HPO_4	7.0g
KH_2PO_4	3.0g
Glucose	2.0g
$(NH_4)_2SO_4$	1.0g
L-Proline	0.5g
Sodium citrate·$3H_2O$	0.5g
$MgSO_4 \cdot 7H_2O$	0.1g
Glucose solution	1.0mL

pH 7.0 ± 0.2 at 25°C

Glucose Solution:

Composition per 100mL:

Glucose	20.0g

Preparation of Glucose Solution: Add glucose to distilled/deionized water and bring volume to 100.0mL. Mix thoroughly. Filter sterilize.

Preparation of Medium: Add components, except glucose solution, to distilled/deionized water and bring volume to 999.0mL. Gently heat and bring to boiling. Autoclave for 15 min at 15 psi pressure–121°C. Cool to 25°C. Aseptically add 1.0mL of sterile glucose solution. Mix thoroughly. Aseptically distribute into sterile tubes or flasks.

Use: For the cultivation and maintenance of *Escherichia coli.*

Davis Supplemented Minimal Medium

Composition per liter:

Agar	15.0g
K_2HPO_4	7.0g
KH_2PO_4	3.0g
Casein hydrolysate	2.0g
Yeast extract	2.0g
$(NH_4)_2SO_4$	1.0g
Sodium citrate·$3H_2O$	0.5g
$MgSO_4 \cdot 7H_2O$	0.1g
Glucose solution	20.0mL

pH 7.0 ± 0.2 at 25°C

Glucose Solution

Composition per 100mL:

Glucose	10.0g

Preparation of Glucose Solution: Add glucose to distilled/deionized water and bring volume to 100.0mL. Mix thoroughly. Filter sterilize.

Preparation of Medium: Add components, except glucose solution, to distilled/deionized water and bring volume to 980.0mL. Gently heat and bring to boiling. Autoclave for 15 min at 15 psi pressure–121°C. Cool to 45°–50°C. Aseptically add 20.0mL of sterile glucose solution. Mix thoroughly. Pour into sterile Petri dishes or distribute into sterile tubes.

Use: For the cultivation and maintenance of *Escherichia coli.*

DCLS Agar (Deoxycholate Citrate Lactose Sucrose Agar)

Composition per liter:

Agar	12.0g
Sodium citrate·$3H_2O$	10.5g
Lactose	5.0g
$Na_2S_2O_3$	5.0g
Sucrose	5.0g
Pancreatic digest of casein	3.5g
Peptic digest of animal tissue	3.5g
Beef extract	3.0g
Sodium deoxycholate	2.5g
Neutral Red	0.03g

pH 7.2 ± 0.1 at 25°C

Source: This medium is available as a premixed powder from BBL Microbiology Systems, Difco Laboratories and Oxoid Unipath Unipath.

Preparation of Medium: Add components to distilled/deionized water and bring volume to 1.0L. Mix thoroughly. Gently heat while stirring and bring to boiling. Do not overheat. Do not autoclave. Pour into sterile Petri dishes in 20.0mL volumes.

Use: For the selective isolation of *Salmonella* species, *Shigella* species and *Vibrio* species from fecal specimens.

Decarboxylase Base, Møller

Composition per liter:

Amino acid	10.0g
Beef extract	5.0g
Peptone	5.0g
Glucose	0.5g
Bromcresol Purple	0.01g
Cresol Red	5.0mg
Pyridoxal	5.0mg
Mineral oil	200.0mL

pH 6.0 ± 0.2 at 25°C

Source: This medium is available as a premixed powder Difco Laboratories.

Preparation of Medium: Add components, except mineral oil, to distilled/deionized water and bring volume to 1.0L. For amino acid, use L-arginine, L-lysine or L-ornithine. Mix thoroughly. Distribute into screw-capped tubes in 5.0mL volumes. Autoclave medium and mineral oil separately for 15 min at 15 psi pressure–121°C. After inoculation, overlay medium with 1.0mL of sterile mineral oil per tube.

Use: For the cultivation and differentiation of bacteria based on their ability to decarboxylate the amino acid. Bacteria that decarboxylate arginine, lysine or ornithine turn the medium turbid purple.

Decarboxylase Medium Base, Falkow

Composition per liter:

Amino acid	5.0g
Peptone	5.0g
Yeast extract	3.0g
Glucose	1.0g
Bromcresol Purple	0.02g
Mineral oil	200.0mL

pH 6.8 ± 0.2 at 25°C

Source: This medium is available as a premixed powder from Difco Laboratories.

Preparation of Medium: Add components, except mineral oil, to distilled/deionized water and bring volume to 1.0L. For amino acid, use L-arginine, L-lysine or L-ornithine. Mix thoroughly. Distribute into screw-capped tubes in 5.0mL volumes. Autoclave medium and mineral oil separately for 15 min at 15 psi pressure–121°C. After inoculation, overlay medium with 1.0mL of sterile mineral oil per tube.

Use: For the cultivation and differentiation of bacteria based on their ability to decarboxylate the amino acid. Bacteria that decarboxylate arginine, lysine or ornithine turn the medium turbid purple.

Decarboxylase Medium, Ornithine Modified

Composition per liter:

L-Ornithine	10.0g
Meat peptone	5.0g
Yeast extract	3.0g
Bromcresol Purple solution	5.0mL

pH 5.5 ± 0.2 at 25°C

Bromcresol Purple Solution:
Composition per 100mL:

Bromcresol Purple	0.2g
Ethanol	50.0mL

Preparation of Bromcresol Purple Solution: Add bromcresol purple to ethanol. Mix thoroughly. Bring volume to 100.0mL with distilled/deionized water. Mix thoroughly. Filter sterilize.

Preparation of Medium: Add components to distilled/deionized water and bring volume to 1.0L. Mix thoroughly. Gently heat until dissolved. Adjust pH to 5.5 with HCl or NaOH. Distribute into screw-capped tubes. Autoclave for 15 min at 15 psi pressure–121°C.

Use: For the cultivation and differentiation of bacteria based on their ability to decarboxylate ornithine. Bacteria that decarboxylate ornithine turn the medium turbid purple.

D/E Neutralizing Agar

Composition per liter:

Agar	15.0g
Glucose	10.0g
Soybean lecithin	7.0g
$Na_2S_2O_3 \cdot 5H_2O$	6.0g
Polysorbate 80	5.0g
Pancreatic digest of casein	5.0g
$NaHSO_3$	2.5g
Yeast extract	2.5g
Sodium thioglycollate	1.0g
Bromcresol Purple	0.02g

pH 7.6 ± 0.2 at 25°C

Source: This medium is available as a premixed powder from Difco Laboratories.

Preparation of Medium: Add components to distilled/deionized water and bring volume to 1.0L. Mix thoroughly. Gently heat and bring to boiling. Distribute into flasks in 9.0mL volumes. Autoclave for 15 min at 15 psi pressure–121°C.

Use: For the neutralization and testing of antiseptics and disinfectants.

D/E Neutralizing Broth (Dey/Engley Neutralizing Broth)

Composition per liter:

Glucose	10.0g
Soybean lecithin	7.0g
$Na_2S_2O_3 \cdot 5H_2O$	6.0g
Tween™ 80	5.0g
Pancreatic digest of casein	5.0g
$NaHSO_3$	2.5g
Yeast extract	2.5g
Sodium thioglycollate	1.0g
Bromcresol Purple	0.02g

pH 7.6± 0.2 at 25°C

Source: This medium is available as a premixed powder from Difco Laboratories.

Preparation of Medium: Add components to distilled/deionized water and bring volume to 1.0L. Mix

thoroughly. Distribute into tubes in 9.0mL volumes. Autoclave for 15 min at 15 psi pressure–121°C.

Use: For the neutralization and testing of antiseptics and disinfectants.

D/E Neutralizing Broth Base (Dey/Engley Neutralizing Broth Base)

Composition per liter:

Glucose	10.0g
Pancreatic digest of casein	5.0g
Yeast extract	2.5g
Bromcresol Purple	0.02g

pH 7.6± 0.2 at 25°C

Preparation of Medium: Add components to distilled/deionized water and bring volume to 1.0L. Mix thoroughly. Distribute into tubes in 9.0mL volumes. Autoclave for 15 min at 15 psi pressure–121°C.

Use: For the neutralization and testing of antiseptics and disinfectants.

Defined Glucose Medium EMSY-1

Composition per liter:

Na_2HPO_4	1.79g
KH_2PO_4	1.7g
Citric acid	0.5g
NH_4Cl	0.43g
$MgSO_4·7H_2O$	0.41g
$CaCl_2·2H_2O$	0.04g
NaCl	0.03g
$FeCl_3·6H_2O$	4.84mg
Glucose solution	100.0mL
Yeast extract solution	10.0mL
TK6-3 solution	1.0mL

pH 7.2 ± 0.2 at 25°C

Glucose Solution:

Composition per 100mL:

Glucose	10.0g

Preparation of Glucose Solution: Add glucose to distilled/deionized water and bring volume to 100.0mL. Mix thoroughly. Filter sterilize.

Yeast Extract Solution:

Composition per 10mL:

Yeast extract	0.4g

Preparation of Yeast Extract Solution: Add yeast extract to distilled/deionized water and bring volume to 10.0mL. Mix thoroughly. Filter sterilize.

TK6-3 Solution:

Composition per liter:

$ZnSO_4·7H_2O$	1.45g
$CuSO_4·5H_2O$	0.76g
$MnSO_4·H_2O$	0.31g
H_3BO_3	0.19g
$Na_2MoO_4·2H_2O$	0.17g
KI	0.04g
H_2SO_4 (1*N* solution)	1.0mL

Preparation of TK6-3 Solution: Add components to distilled/deionized water and bring volume to 1.0L. Mix thoroughly.

Preparation of Medium: Add components, except glucose solution and yeast extract solution, to distilled/deionized water and bring volume to 890.0mL. Mix thoroughly. Gently heat and bring to boiling. Autoclave for 15 min at 15 psi pressure–121°C. Cool rapidly to 25°C. Aseptically add 100.0mL of sterile glucose solution and 10.0mL of sterile yeast extract solution. Mix thoroughly. Aseptically distribute into sterile tubes or flasks.

Use: For the cultivation and maintenance of *Xanthomonas campestris*.

Defined Medium for *Rhodopseudomonas*

Composition per liter:

Malic acid	4.0g
$(NH_4)_2SO_4$	1.0g
K_2HPO_4	0.9g
KH_2PO_4	0.6g
$MgSO_4·7H_2O$	0.2g
$CaCl_2·2H_2O$	0.075g
EDTA	0.020g
$FeSO_4·7H_2O$	0.012g
Thiamine	1.0mg
Biotin	0.015mg
Trace elements	1.0mL

pH 6.8 ± 0.2 at 25°C

Trace Elements:

Composition per 250mL:

H_3BO_3	0.7g
$MnSO_4·H_2O$	0.4g
$Na_2MoO_4·2H_2O$	0.19g
$ZnSO_4·7H_2O$	0.060g
$CoCl_2·6H_2O$	0.050g
$Cu(NO_3)_2·3H_2O$	0.010g

Preparation of Trace Elements: Add components to distilled/deionized water and bring volume to 250.0mL. Mix thoroughly.

Preparation of Medium: Add components to distilled/deionized water and bring volume to 1.0L. Mix

thoroughly. Adjust pH to 6.8. Distribute into tubes or flasks. Autoclave for 15 min at 15 psi pressure–121°C.

Use: For the cultivation and maintenance of *Rhodobacter capsulatus.*

Defined Medium with Povidone Iodine

Composition per 1025mL:

Basal solution	1.0L
Solution B	10.0mL
Solution C	10.0mL
Solution A	5.0mL

Basal Solution:

Composition per liter:

Agar	20.0g
Na_2HPO_4	4.8g
KH_2PO_4	4.4g
NH_4Cl	1.0g
$MgSO_4·7H_2O$	0.5g

Preparation of Basal Solution: Add components to distilled/deionized water and bring volume to 1.0L. Mix thoroughly. Gently heat and bring to boiling. Autoclave for 15 min at 15 psi pressure–121°C. Cool to 45°–50°C.

Solution A:

Composition per 100mL:

Ferric ammonium citrate	1.0g
$CaCl_2·2H_2O$	0.1g

Preparation of Solution A: Add components to distilled/deionized water and bring volume to 100.0mL. Mix thoroughly. Filter sterilize.

Solution B:

Composition per 100mL:

D-Glucose	10.0g

Preparation of Solution B: Add glucose to distilled/deionized water and bring volume to 100.0mL. Mix thoroughly. Filter sterilize.

Solution C:

Composition per 100mL:

Povidone-iodine	0.1g

Preparation of Solution C: Add povidone-iodine to distilled/deionized water and bring volume to 100.0mL. Mix thoroughly. Filter sterilize.

Preparation of Medium: To 1.0L of cooled, sterile basal solution aseptically add 5.0mL of sterile solution A, 10.0mL of sterile solution B, and 10.0mL of sterile solution C. Mix thoroughly. Pour into sterile Petri dishes or distribute into sterile tubes.

Use: For the cultivation and maintenance of *Pseudomonas aeruginosa* and *Pseudomonas cepacia.*

DeMan, Rogosa, Sharpe Agar
See: **MRS Agar**

DeMan, Rogosa, Sharpe Broth
See: **MRS Broth**

Deoxycholate Agar

Composition per liter:

Agar	16.0g
Lactose	10.0g
NaCl	5.0g
Pancreatic digest of casein	5.0g
Peptic digest of animal tissue	5.0g
K_2HPO_4	2.0g
Ferric citrate	1.0g
Sodium citrate	1.0g
Sodium deoxycholate	1.0g
Neutral Red	0.033g

pH 7.3 ± 0.2 at 25°C

Source: This medium is available as a premixed powder from BBL Microbiology Systems.

Preparation of Medium: Add components to distilled/deionized water and bring volume to 1.0L. Mix thoroughly. Gently heat and bring to boiling. Do not autoclave. Cool to 45°–50°C. Pour into sterile Petri dishes.

Use: For the selective isolation, cultivation, enumeration and differentiation of Gram-negative enteric microorganisms from a variety of clinical and nonclinical specimens. *Escherichia coli* appears as large, flat rose-red colonies. *Enterobacter* and *Klebsiella* species appear as large mucoid pale colonies with a pink center. *Proteus* and *Salmonella* species appear as large, colorless to tan colonies. *Shigella* species appear as colorless to pink colonies. *Pseudomonas* species appear as irregular colorless to brown colonies.

Deoxycholate Agar (Desoxycholate Agar)

Composition per liter:

Agar	15.0g
Lactose	10.0g
Peptone	10.0g
NaCl	5.0g
K_2HPO_4	2.0g
Ferric citrate	1.0g
Sodium citrate	1.0g
Sodium deoxycholate	1.0g
Neutral Red	0.03g

pH 7.3 ± 0.2 at 25°C

Source: This medium is available as a premixed powder from Oxoid Unipath and Difco Laboratories.

Preparation of Medium: Add components to distilled/deionized water and bring volume to 1.0L. Mix thoroughly. Gently heat and bring to boiling. Do not autoclave. Cool to 50°C. Pour into sterile Petri dishes.

Use: For the selective isolation, cultivation, enumeration and differentiation of Gram-negative enteric microorganisms from a variety of clinical and nonclinical specimens. *Escherichia coli* appears as large, flat rose-red colonies. *Enterobacter* and *Klebsiella* species appear as large mucoid pale colonies with a pink center. *Proteus* and *Salmonella* species appear as large, colorless to tan colonies. *Shigella* species appear as colorless to pink colonies. *Pseudomonas* species appear as irregular colorless to brown colonies.

Deoxycholate Citrate Agar

Composition per liter:

Sodium citrate	50.0g
Agar	15.0g
Lactose	10.0g
Beef extract	5.0g
Peptone	5.0g
$Na_2S_2O_3 \cdot 5H_2O$	5.0g
Sodium deoxycholate	2.5g
Ferric citrate	1.0g
Neutral Red	0.025g

pH 7.3 ± 0.2 at 25°C

Source: This medium is available as a premixed powder from Oxoid Unipath.

Preparation of Medium: Add components to distilled/deionized water and bring volume to 1.0L. Mix thoroughly. Gently heat and bring to boiling. Do not autoclave. Cool to 45°–50°C. Pour into sterile Petri dishes. Dry the agar surface before use.

Use: For the selective isolation and cultivation of enteric pathogens, especially *Salmonella* and *Shigella* species.

Deoxycholate Citrate Agar (Desoxycholate Citrate Agar)

Composition per liter:

Pork infusion	330.0g
Sodium citrate	20.0g
Agar	13.5g
Lactose	10.0g
Proteose peptone No. 3	10.0g
Sodium deoxycholate	5.0g
Ferric ammonium citrate	2.0g
Neutral Red	0.02g

pH 7.5 ± 0.2 at 25°C

Source: This medium is available as a premixed powder from Difco Laboratories.

Preparation of Medium: Add components to distilled/deionized water and bring volume to 1.0L. Mix thoroughly. Gently heat and bring to boiling. Do not autoclave. Cool to 45°–50°C. Pour into sterile Petri dishes. Dry the agar surface before use.

Use: For the selective isolation and cultivation of enteric pathogens, especially *Salmonella* and *Shigella* species.

Deoxycholate Citrate Agar

Composition per liter:

Sodium citrate	20.0g
Agar	17.0g
Lactose	10.0g
Meat, solids from infusion	10.0g
Peptic digest of animal tissue	10.0g
Sodium deoxycholate	5.0g
Ferric citrate	1.0g
Neutral Red	0.02g

pH 7.3 ± 0.2 at 25°C

Source: This medium is available as a premixed powder from BBL Microbiology Systems.

Preparation of Medium: Add components to distilled/deionized water and bring volume to 1.0L. Mix thoroughly. Gently heat and bring to boiling. Do not autoclave. Cool to 45°–50°C. Pour into sterile Petri dishes. Dry the agar surface before use.

Use: For the selective isolation and cultivation of enteric pathogens, especially *Salmonella* and *Shigella* species.

Deoxycholate Citrate Agar, Hynes

Composition per liter:

Agar	12.0g
Lactose	10.0g
Sodium citrate	8.5g
$Na_2S_2O_3 \cdot 5H_2O$	5.4g
Beef extract powder	5.0g
Peptone	5.0g
Sodium deoxycholate	5.0g
Ferric citrate	1.0g
Neutral Red	0.02g

pH 7.3 ± 0.2 at 25°C

Source: This medium is available as a premixed powder from Oxoid Unipath.

Preparation of Medium: Add components to distilled/deionized water and bring volume to 1.0L. Mix

thoroughly. Gently heat and bring to boiling. Do not autoclave. Cool to 45°–50°C. Pour into sterile Petri dishes. Dry the agar surface before use.

Use: For the selective isolation, cultivation, and differentiation of enteric pathogens, especially *Salmonella* and *Shigella* species. Lactose-fermenting bacteria appear as pink colonies that may or may not be surrounded by a zone of precipitated deoxycholate. Nonlactose-fermenting bacteria appear as colorless colonies that are surrounded by a clear orange-yellow zone.

Deoxycholate Citrate Lactose Sucrose Agar *See:* DCLS Agar

Deoxycholate Lactose Agar

Composition per liter:

Agar	15.0g
Lactose	10.0g
NaCl	5.0g
Pancreatic digest of casein	5.0g
Peptic digest of animal tissue	5.0g
Sodium citrate	2.0g
Sodium deoxycholate	0.5g
Neutral Red	0.033g

pH 7.1 ± 0.2 at 25°C

Source: This medium is available as a premixed powder from BBL Microbiology Systems and Difco Laboratories.

Preparation of Medium: Add components to distilled/deionized water and bring volume to 1.0L. Mix thoroughly. Gently heat and bring to boiling. Do not autoclave. Cool to 45°–50°C. Pour into sterile Petri dishes. Dry the agar surface before use.

Use: For the selective isolation, cultivation, and differentiation of enteric pathogens, especially *Salmonella* and *Shigella* species. Lactose-fermenting bacteria appear as pink colonies that may or may not be surrounded by a zone of precipitated deoxycholate. Nonlactose-fermenting bacteria appear as colorless colonies that are surrounded by a clear orange-yellow zone. Also used for the enumeration of coliform bacteria from water, milk and dairy products.

Deoxycholate Lactose Sucrose Sorbitol Agar

Composition per liter:

Sodium citrate	20.0g
Agar	15.0g
D-Sorbitol	10.0g
Lactose	10.0g
Sucrose	5.0g
Pancreatic digest of casein	5.0g
Yeast extract	5.0g
Sodium deoxycholate	2.5g
Ferric citrate	1.0g
Neutral Red	0.02g

pH 7.4 ± 0.2 at 25°C

Preparation of Medium: Add components to distilled/deionized water and bring volume to 1.0L. Mix thoroughly. Gently heat and bring to boiling. Do not overheat. Adjust pH to 7.4. Do not autoclave. Pour into sterile Petri dishes or distribute into sterile tubes.

Use: For the isolation and cultivation of the *Hafnia* species.

Dermabacter Medium

Composition per liter:

Pancreatic digest of casein	10.0g
Glucose	5.0g
NaCl	5.0g
Yeast extract	5.0g

pH 7.4 ± 0.2 at 25°C

Preparation of Medium: Add components to distilled/deionized water and bring volume to 1.0L. Mix thoroughly. Distribute into tubes or flasks. Autoclave for 15 min at 15 psi pressure–121°C.

Use: For the cultivation and maintenance of *Dermabacter hominis*.

Dermasel Agar Base

Composition per liter:

Glucose	20.0g
Agar	14.5g
Papaic digest of soybean meal	10.0g
Antibiotic inhibitor	10.0mL

pH 6.8–7.0 at 25°C

Source: This medium is available as a premixed powder from Oxoid Unipath.

Antibiotic Inhibitor:

Composition per 10mL:

Cycloheximide	0.4g
Chloramphenicol	0.05g
Acetone	10.0mL

Preparation of Antibiotic Inhibitor: Add cycloheximide and chloramphenicol to 10.0mL of acetone. Mix thoroughly.

Preparation of Medium: Add components to distilled/deionized water and bring volume to 990.0mL.

Mix thoroughly. Gently heat and bring to boiling. Do not overheat. Add antibiotic inhibitor. Mix thoroughly. Autoclave for 10 min at 15 psi pressure–121°C. Pour into sterile Petri dishes.

Use: For the isolation and cultivation of dermatophytic fungi isolated from hair, nails or skin scrapings.

Dermatophyte Test Medium Agar
See: DTM Agar

Dermatophyte Test Medium Base

Composition per liter:

Agar	20.0g
Glucose	10.0g
Papaic digest of soybean meal	10.0g
Cycloheximide	0.5g
Phenol Red	0.2g
Gentamycin sulfate	0.1g
Chlortetracycline	0.1g

pH 5.5 ± 0.2 at 25°C

Source: This medium is available as a premixed powder from BBL Microbiology Systems.

Preparation of Medium: Add components, except gentamycin sulfate and chlortetracycline, to distilled/deionized water and bring volume to 1.0L. Mix thoroughly. Gently heat while stirring and bring to boiling. Autoclave for 15 min at 15 psi pressure–121°C. Cool to 45°–50°C. Aseptically add gentamycin sulfate and chlortetracycline. Mix thoroughly. Pour into sterile Petri dishes.

Use: For the selective isolation and cultivation of pathogenic fungi from cutaneous sources.

Derxia gummosa Medium

Composition per liter:

Agar	20.0g
Starch	20.0g
$MgSO_4 \cdot 7H_2O$	0.2g
KH_2PO_4	0.15g
$NaHCO_3$	0.1g
K_2HPO_4	0.05g
$CaCl_2$	0.02g
$Na_2MoO_4 \cdot 2H_2O$	2.0mg
Bromthymol Blue solution	5.0mL
$FeCl_3 \cdot 6H_2O$ (10% solution)	0.1mL

pH 6.9 ± 0.2 at 25°C

Bromthymol Blue Solution:

Composition per 10mL:

Bromthymol Blue	0.5g
Ethanol	10.0mL

Preparation of Bromthymol Blue Solution: Add Bromthymol Blue to 10.0mL of ethanol. Mix thoroughly.

Preparation of Medium: Add components to distilled/deionized water and bring volume to 1.0L. Mix thoroughly. Distribute into tubes or flasks. Autoclave for 15 min at 15 psi pressure–121°C.

Use: For the cultivation of *Derxia gummosa.*

Derxia Medium

Composition per liter:

Agar	20.0g
Glucose	20.0g
NH_4Cl	2.0g
K_2HPO_4	1.0g
$MgSO_4 \cdot 7H_2O$	0.2g
$CaSO_4$	5.0mg
$FeSO_4 \cdot 7H_2O$	5.0mg
$Na_2MoO_4 \cdot 2H_2O$	0.5mg

pH 6.7 ± 0.2 at 25°C

Preparation of Medium: Add components to distilled/deionized water and bring volume to 1.0L. Mix thoroughly. Gently heat and bring to boiling. Adjust pH to 6.7. Distribute into tubes or flasks. Autoclave for 15 min at 15 psi pressure–121°C. Pour into sterile Petri dishes or leave in tubes.

Use: For the cultivation and maintenance of *Derxia gummosa.*

Desoxycholate Agar
See: Deoxycholate Agar

Desoxycholate Citrate Agar
See: Deoxycholate Citrate Agar

Desulfobacterium indolicum Medium

Composition per 1002.4mL:

Solution A	900.0mL
Solution C	50.0mL
Solution D	30.0mL
Solution E—Wolfe's vitamin solution	10.0mL
Solution G	10.0mL
Solution B—Trace elements solution SL-10	1.0mL
Solution F	1.0mL
Solution H	0.4mL

pH 7.6 ± 0.2 at 25°C

Solution A:

Composition per 900mL:

NaCl	21.0g
$MgCl_2·6H_2O$	3.0g
Na_2SO_4	3.0g
KCl	0.5g
NH_4Cl	0.3g
KH_2PO_4	0.2g
$CaCl_2·2H_2O$	0.15g
Resazurin	1.0mg

Preparation of Solution A: Prepare and dispense solution anaerobically under 80% N_2 + 20% CO_2. Add components to distilled/deionized water and bring volume to 900.0mL. Mix thoroughly. Gently heat and bring to boiling. Continue boiling until resazurin turns colorless indicating reduction. Cap with rubber stoppers. Autoclave for 15 min at 15 psi pressure–121°C. Cool to 45°–50°C.

Solution B—Trace Elements Solution SL-10:

Composition per liter:

$FeCl_2·4H_2O$	1.5g
$CoCl_2·6H_2O$	0.19g
$MnCl_2·4H_2O$	0.10g
$ZnCl_2$	0.070g
$Na_2MoO_4·2H_2O$	0.036g
$NiCl_2·6H_2O$	0.024g
H_3BO_3	6.0mg
$CuCl_2·2H_2O$	2.0mg
HCl (25% solution)	10.0mL

Preparation of Trace Elements Solution SL-10: Add the $FeCl_2·4H_2O$ to 10.0mL of HCl solution. Mix thoroughly. Bring volume to approximately 900.0mL with distilled/deionized water. Mix thoroughly. Adjust pH to 6.0 with NaOH. Bring volume to 1.0L with distilled/deionized water. Filter sterilize. Aseptically gas under 100% N_2 for 20 min.

Solution C:

Composition per 50mL:

$NaHCO_3$	2.5g

Preparation of Solution C: Add $NaHCO_3$ to distilled/deionized water and bring volume to 50.0mL. Mix thoroughly. Filter sterilize. Aseptically gas under 80% N_2 + 20% CO_2 for 20 min.

Solution D:

Composition per 107.7mL:

Indole	0.3g
NaCl (30% solution)	7.0mL
$MgCl_2·6H_2O$ (40% solution)	0.7mL

Preparation of Solution D: Prepare and dispense all solutions anaerobically under 100% N_2. Add indole to distilled/deionized water and bring volume to 100.0mL. Mix thoroughly. Gently heat while stirring until dissolved. Prepare the NaCl solution and the $MgCl_2·6H_2O$ solution separately. Autoclave the three solutions separately for 15 min at 15 psi pressure–121°C. Cool to 45°–50°C. To 100.0mL of sterile indole solution, aseptically and anaerobically add 7.0mL of sterile NaCl solution and 0.7mL of sterile $MgCl_2·6H_2O$ solution. Mix thoroughly.

Solution E—Wolfe's Vitamin Solution:

Composition per liter:

Pyridoxine·HCl	0.01g
Thiamine·HCl	5.0mg
Riboflavin	5.0mg
Nicotinic acid	5.0mg
Calcium pantothenate	5.0mg
p-Aminobenzoic acid	5.0mg
Thioctic acid	5.0mg
Biotin	2.0mg
Folic acid	2.0mg
Cyanocobalamin	0.1mg

Preparation of Wolfe's Vitamin Solution: Add components to distilled/deionized water and bring volume to 1.0L. Mix thoroughly. Filter sterilize. Aseptically gas under 100% N_2 for 20 min.

Solution F:

Composition per liter:

$Na_2SeO_3·5H_2O$	3.0mg
NaOH (0.01M solution)	1.0L

Preparation of Solution F: Add $Na_2SeO_3·5H_2O$ to 1.0L of NaOH solution. Mix thoroughly. Filter sterilize. Aseptically gas under 100% N_2 for 20 min.

Solution G:

Composition per 10mL:

$Na_2S·9H_2O$	0.4g

Preparation of Solution G: Add $Na_2S·9H_2O$ to distilled/deionized water and bring volume to 10.0mL. Gas under 100% N_2 for 20 min. Cap with a rubber stopper. Autoclave for 15 min at 15 psi pressure–121°C. Cool to 25°C.

Solution H:

Composition per 10mL:

$Na_2S_2O_4$	0.5g

Preparation of Solution H: Add $Na_2S_2O_4$ to distilled/deionized water and bring volume to 10.0mL. Mix thoroughly. Filter sterilize. Aseptically gas under 100% N_2 for 20 min. Prepare solution freshly.

Preparation of Medium: To 900.0mL of cooled, sterile solution A, aseptically and anaerobically add in the following order: 1.0mL of sterile solution B, 50.0mL of sterile solution C, 10.0mL of sterile solution E, 1.0mL of sterile solution F, and 10.0mL of sterile solution G. Mix thoroughly. Immediately prior to inoculation aseptically and anaerobically add

30.0mL of sterile solution D and 0.4mL of sterile solution H. Mix thoroughly. Aseptically and anaerobically distribute into sterile tubes or flasks.

Use: For the cultivation and maintenance of *Desulfobacterium indolicum*.

Desulfobacterium Medium

Composition per 1002.4mL:

Solution A	930.0mL
Solution C	50.0mL
Solution D—Wolfe's vitamin solution	10.0mL
Solution F	10.0mL
Solution B—Trace elements solution SL-10	1.0mL
Solution E	1.0mL
Solution G	0.4mL

pH 7.0 ± 0.2 at 25°C

Solution A:

Composition per 930mL:

NaCl	21.0g
$MgCl_2 \cdot 6H_2O$	3.0g
Na_2SO_4	3.0g
KCl	0.5g
NH_4Cl	0.3g
KH_2PO_4	0.2g
$CaCl_2 \cdot 2H_2O$	0.15g
Resazurin	1.0mg

Preparation of Solution A: Prepare and dispense solution anaerobically under 80% N_2 + 20% CO_2. Add components to distilled/deionized water and bring volume to 930.0mL. Mix thoroughly. Gently heat and bring to boiling. Continue boiling until resazurin turns colorless indicating reduction. Cap with rubber stoppers. Autoclave for 15 min at 15 psi pressure–121°C. Cool to 45°–50°C.

Solution B—Trace Elements Solution SL-10:

Composition per liter:

$FeCl_2 \cdot 4H_2O$	1.5g
$CoCl_2 \cdot 6H_2O$	0.19g
$MnCl_2 \cdot 4H_2O$	0.10g
$ZnCl_2$	0.070g
$Na_2MoO_4 \cdot 2H_2O$	0.036g
$NiCl_2 \cdot 6H_2O$	0.024g
H_3BO_3	6.0mg
$CuCl_2 \cdot 2H_2O$	2.0mg
HCl (25% solution)	10.0mL

Preparation of Trace Elements Solution SL-10: Add the $FeCl_2 \cdot 4H_2O$ to 10.0mL of HCl solution. Mix thoroughly. Bring volume to approximately 900.0mL with distilled/deionized water. Mix thoroughly. Adjust pH to 6.0 with NaOH. Bring volume to 1.0L with distilled/deionized water. Filter sterilize. Aseptically gas under 100% N_2 for 20 min.

Solution C:

Composition per 50mL:

$NaHCO_3$	2.5g

Preparation of Solution C: Add $NaHCO_3$ to distilled/deionized water and bring volume to 50.0mL. Mix thoroughly. Filter sterilize. Aseptically gas under 80% N_2 + 20% CO_2 for 20 min.

Solution D—Wolfe's Vitamin Solution:

Composition per liter:

Pyridoxine·HCl	0.01g
Thiamine·HCl	5.0mg
Riboflavin	5.0mg
Nicotinic acid	5.0mg
Calcium pantothenate	5.0mg
p-Aminobenzoic acid	5.0mg
Thioctic acid	5.0mg
Biotin	2.0mg
Folic acid	2.0mg
Cyanocobalamin	0.1mg

Preparation of Wolfe's Vitamin Solution: Add components to distilled/deionized water and bring volume to 1.0L. Mix thoroughly. Filter sterilize. Aseptically gas under 100% N_2 for 20 min.

Solution E:

Composition per liter:

$Na_2SeO_3 \cdot 5H_2O$	3.0mg
NaOH (0.01M solution)	1.0L

Preparation of Solution E: Add $Na_2SeO_3 \cdot 5H_2O$ to 1.0L of NaOH solution. Mix thoroughly. Filter sterilize. Aseptically gas under 100% N_2 for 20 min.

Solution F:

Composition per 10mL:

$Na_2S \cdot 9H_2O$	0.4g

Preparation of Solution F: Add $Na_2S \cdot 9H_2O$ to distilled/deionized water and bring volume to 10.0mL. Gas under 100% N_2 for 20 min. Cap with a rubber stopper. Autoclave for 15 min at 15 psi pressure–121°C. Cool to 25°C.

Solution G:

Composition per 10mL:

$Na_2S_2O_4$	0.5g

Preparation of Solution G: Add $Na_2S_2O_4$ to distilled/deionized water and bring volume to 10.0mL. Mix thoroughly. Filter sterilize. Aseptically gas under 100% N_2 for 20 min. Prepare solution freshly.

Preparation of Medium: To 900.0mL of cooled, sterile solution A, aseptically and anaerobically add in the following order: 1.0mL of sterile solution B, 50.0mL of sterile solution C, 10.0mL of sterile solution D, 1.0mL of sterile solution E, and 10.0mL of sterile solution F. Mix thoroughly. Immediately prior to inoculation aseptically and anaerobically add 0.4mL of sterile solution G. Mix thoroughly. Aseptically and anaerobically distribute into sterile tubes or flasks.

Use: For the cultivation and maintenance of *Desulfobacterium autotrophicum*.

Desulfobacterium Medium, Modified

Composition per 1002.4mL:

Solution A	920.0mL
Solution C	50.0mL
Solution D	10.0mL
Solution E—Wolfe's vitamin solution	10.0mL
Solution G	10.0mL
Solution B—Trace elements solution SL-10	1.0mL
Solution F	1.0mL
Solution H	0.4mL

pH 7.0 ± 0.2 at 25°C

Solution A:

Composition per 920mL:

NaCl	21.0g
$MgCl_2 \cdot 6H_2O$	3.0g
Na_2SO_4	3.0g
KCl	0.5g
NH_4Cl	0.3g
KH_2PO_4	0.2g
$CaCl_2 \cdot 2H_2O$	0.15g
Resazurin	1.0mg

Preparation of Solution A: Prepare and dispense solution anaerobically under 80% N_2 + 20% CO_2. Add components to distilled/deionized water and bring volume to 920.0mL. Mix thoroughly. Gently heat and bring to boiling. Continue boiling until resazurin turns colorless indicating reduction. Cap with rubber stoppers. Autoclave for 15 min at 15 psi pressure–121°C. Cool to 45°–50°C.

Solution B—Trace Elements Solution SL-10:

Composition per liter:

$FeCl_2 \cdot 4H_2O$	1.5g
$CoCl_2 \cdot 6H_2O$	0.19g
$MnCl_2 \cdot 4H_2O$	0.10g
$ZnCl_2$	0.070g
$Na_2MoO_4 \cdot 2H_2O$	0.036g
$NiCl_2 \cdot 6H_2O$	0.024g
H_3BO_3	6.0mg
$CuCl_2 \cdot 2H_2O$	2.0mg
HCl (25% solution)	10.0mL

Preparation of Trace Elements Solution SL-10: Add the $FeCl_2 \cdot 4H_2O$ to 10.0mL of HCl solution. Mix thoroughly. Bring volume to approximately 900.0mL with distilled/deionized water. Mix thoroughly. Adjust pH to 6.0 with NaOH. Bring volume to 1.0L with distilled/deionized water. Filter sterilize. Aseptically gas under 100% N_2 for 20 min.

Solution C:

Composition per 50mL:

$NaHCO_3$	2.5g

Preparation of Solution C: Add $NaHCO_3$ to distilled/deionized water and bring volume to 50.0mL. Mix thoroughly. Filter sterilize. Aseptically gas under 80% N_2 + 20% CO_2 for 20 min.

Solution D:

Composition per 10mL:

Sodium acetate·$3H_2O$	2.5g

Preparation of Solution D: Prepare and dispense solution anaerobically under 80% N_2 + 20% CO_2. Add sodium acetate to distilled/deionized water and bring volume to 10.0mL. Mix thoroughly. Cap with rubber stopper. Autoclave for 15 min at 15 psi pressure–121°C. Cool to 45°–50°C.

Solution E—Wolfe's Vitamin Solution:

Composition per liter:

Pyridoxine·HCl	0.01g
Thiamine·HCl	5.0mg
Riboflavin	5.0mg
Nicotinic acid	5.0mg
Calcium pantothenate	5.0mg
p-Aminobenzoic acid	5.0mg
Thioctic acid	5.0mg
Biotin	2.0mg
Folic acid	2.0mg
Cyanocobalamin	0.1mg

Preparation of Wolfe's Vitamin Solution: Add components to distilled/deionized water and bring volume to 1.0L. Mix thoroughly. Filter sterilize. Aseptically gas under 100% N_2 for 20 min.

Solution F:

Composition per liter:

$Na_2SeO_3 \cdot 5H_2O$	3.0mg
NaOH (0.01M solution)	1.0L

Preparation of Solution F: Add $Na_2SeO_3 \cdot 5H_2O$ to 1.0L of NaOH solution. Mix thoroughly. Filter sterilize. Aseptically gas under 100% N_2 for 20 min.

Solution G:

Composition per 10mL:

$Na_2S \cdot 9H_2O$	0.4g

Preparation of Solution G: Add $Na_2S \cdot 9H_2O$ to distilled/deionized water and bring volume to 10.0mL. Gas under 100% N_2 for 20 min. Cap with a rubber stopper. Autoclave for 15 min at 15 psi pressure–121°C. Cool to 25°C.

Solution H:

Composition per 10mL:

$Na_2S_2O_4$	0.5g

Preparation of Solution H: Add $Na_2S_2O_4$ to distilled/deionized water and bring volume to 10.0mL. Mix thoroughly. Filter sterilize. Aseptically gas under 100% N_2 for 20 min. Prepare solution freshly.

Preparation of Medium: To 920.0mL of cooled, sterile solution A, aseptically and anaerobically add in the following order: 1.0mL of sterile solution B, 50.0mL of sterile solution C, 10.0mL of sterile solution D, 10.0mL of sterile solution E, 1.0mL of sterile solution F, and 10.0mL of sterile solution G. Mix thoroughly. Immediately prior to inoculation aseptically and anaerobically add 0.4mL of sterile solution H. Mix thoroughly. Aseptically and anaerobically distribute into sterile tubes or flasks.

Use: For the cultivation and maintenance of *Desulfobacter curvatus* and *Desulfobacter latus*.

Desulfobacterium phenolicum Medium

Composition per 1002.4mL:

Solution A	930.0mL
Solution C	50.0mL
Solution E—Wolfe's vitamin solution	10.0mL
Solution G	10.0mL
Solution D	4.0mL
Solution B—Trace elements solution SL-10	1.0mL
Solution F	1.0mL
Solution H	0.4mL

pH 7.0 ± 0.2 at 25°C

Solution A:

Composition per 920mL:

NaCl	21.0g
$MgCl_2 \cdot 6H_2O$	3.0g
Na_2SO_4	3.0g
KCl	0.5g
NH_4Cl	0.3g
KH_2PO_4	0.2g
$CaCl_2 \cdot 2H_2O$	0.15g
Resazurin	1.0mg

Preparation of Solution A: Prepare and dispense solution anaerobically under 80% N_2 + 20% CO_2. Add components to distilled/deionized water and bring volume to 920.0mL. Mix thoroughly. Gently heat and bring to boiling. Continue boiling until resazurin turns colorless indicating reduction. Cap with rubber stoppers. Autoclave for 15 min at 15 psi pressure–121°C. Cool to 45°–50°C.

Solution B—Trace Elements Solution SL-10:

Composition per liter:

$FeCl_2 \cdot 4H_2O$	1.5g
$CoCl_2 \cdot 6H_2O$	0.19g
$MnCl_2 \cdot 4H_2O$	0.10g
$ZnCl_2$	0.070g
$Na_2MoO_4 \cdot 2H_2O$	0.036g
$NiCl_2 \cdot 6H_2O$	0.024g
H_3BO_3	6.0mg
$CuCl_2 \cdot 2H_2O$	2.0mg
HCl (25% solution)	10.0mL

Preparation of Trace Elements Solution SL-10: Add $FeCl_2 \cdot 4H_2O$ to 10.0mL of HCl solution. Mix thoroughly. Bring volume to 900.0mL with distilled/deionized water. Mix thoroughly. Adjust pH to 6.0 with NaOH. Bring volume to 1.0L with distilled/deionized water. Filter sterilize. Aseptically gas under 100% N_2 for 20 min.

Solution C:

Composition per 50mL:

$NaHCO_3$	2.5g

Preparation of Solution C: Add $NaHCO_3$ to distilled/deionized water and bring volume to 50.0mL. Mix thoroughly. Filter sterilize. Aseptically gas under 80% N_2 + 20% CO_2 for 20 min.

Solution D:

Composition per 10mL:

Sodium benzoate	1.0g
Phenol	0.1g

Preparation of Solution D: Add components to distilled/deionized water and bring volume to 10.0mL. Mix thoroughly. Filter sterilize. Aseptically gas under 100% N_2 for 20 min.

Solution E—Wolfe's Vitamin Solution:

Composition per liter:

Pyridoxine·HCl	0.01g
Thiamine·HCl	5.0mg
Riboflavin	5.0mg
Nicotinic acid	5.0mg
Calcium pantothenate	5.0mg
p-Aminobenzoic acid	5.0mg
Thioctic acid	5.0mg
Biotin	2.0mg
Folic acid	2.0mg
Cyanocobalamin	0.1mg

Preparation of Wolfe's Vitamin Solution: Add components to distilled/deionized water and bring volume to 1.0L. Mix thoroughly. Filter sterilize. Aseptically gas under 100% N_2 for 20 min.

Solution F:

Composition per liter:

$Na_2SeO_3 \cdot 5H_2O$	3.0mg
NaOH (0.01M solution)	1.0L

Preparation of Solution F: Add $Na_2SeO_3 \cdot 5H_2O$ to 1.0L of NaOH solution. Mix thoroughly. Filter sterilize. Aseptically gas under 100% N_2 for 20 min.

Solution G:

Composition per 10mL:

$Na_2S \cdot 9H_2O$	0.4g

Preparation of Solution G: Add $Na_2S \cdot 9H_2O$ to distilled/deionized water and bring volume to 10.0mL. Gas under 100% N_2 for 20 min. Cap with a rubber stopper. Autoclave for 15 min at 15 psi pressure–121°C. Cool to 25°C.

Solution H:

Composition per 10mL:

$Na_2S_2O_4$	0.5g

Preparation of Solution H: Add $Na_2S_2O_4$ to distilled/deionized water and bring volume to 10.0mL. Mix thoroughly. Filter sterilize. Aseptically gas under 100% N_2 for 20 min. Prepare solution freshly.

Preparation of Medium: To 920.0mL of cooled, sterile solution A, aseptically and anaerobically add in the following order: 1.0mL of sterile solution B, 50.0mL of sterile solution C, 10.0mL of sterile solution D, 10.0mL of sterile solution E, 1.0mL of sterile solution F, and 10.0mL of sterile solution G. Mix thoroughly. Immediately prior to inoculation aseptically and anaerobically add 0.4mL of sterile solution H. Mix thoroughly. Aseptically and anaerobically distribute into sterile tubes or flasks.

Use: For the cultivation and maintenance of *Desulfobacterium phenolicum*.

Desulfobulbus Medium

Composition per liter:

Sodium propionate	1.85g
Na_2SO_4	1.5g
$(NH_4)_2SO_4$	1.24g
NaCl	0.6g
Cysteine·HCl·H_2O	0.5g
KH_2PO_4	0.3g
$MgSO_4 \cdot 7H_2O$	0.12g
$CaCl_2 \cdot 2H_2O$	0.08g
Trace minerals	10.0mL
Vitamin solution	10.0mL

pH 7.0 ± 0.2 at 25°C

Trace Minerals:

Composition per liter:

Nitrilotriacetic acid	12.8g
$FeSO_4 \cdot 7H_2O$	0.3g
$CoCl_2 \cdot 2H_2O$	0.1g
$MnCl_2 \cdot 4H_2O$	0.1g
$ZnCl_2$	0.1g
$CuCl_2$	0.02g
H_3BO_3	0.01g
Na_2MoO_4	0.01g
$NiSO_4 \cdot 6H_2O$	2.6mg
Na_2SeO_3	1.7mg

Preparation of Trace Minerals: Add nitrilotriacetic acid to 500.0mL of distilled/deionized water. Dissolve by adjusting pH to 6.5 with KOH. Add remaining components. Add distilled/deionized water to 1.0L.

Wolfe's Vitamin Solution:

Composition per liter:

Pyridoxine·HCl	0.01g
Thiamine·HCl	5.0mg
Riboflavin	5.0mg
Nicotinic acid	5.0mg
Calcium pantothenate	5.0mg
p-Aminobenzoic acid	5.0mg
Thioctic acid	5.0mg
Biotin	2.0mg
Folic acid	2.0mg
Cyanocobalamin	0.1mg

Preparation of Wolfe's Vitamin Solution: Add components to distilled/deionized water and bring volume to 1.0L. Mix thoroughly.

Preparation of Medium: Add components, except cysteine·HCl·H_2O, to distilled/deionized water and bring volume to 1.0L. Mix thoroughly. Gently heat and bring to boiling. Cool to 25°C under 85% N_2 + 15% CO_2. Add cysteine·HCl·H_2O. Mix thoroughly. Adjust pH to 7.2 with $KHCO_3$. Anaerobically distribute into tubes or flasks under 85% N_2 + 15% CO_2. Autoclave for 15 min at 15 psi pressure–121°C. Adjust pH to 7.0, if necessary.

Use: For the cultivation and maintenance of *Desulfobulbus elongatus*.

Desulfomonile tiedjei Medium

Composition per liter:

$NaHCO_3$	3.0g
PIPES (Piperazine-N,N´-bis-2-ethanesulfonic acid) buffer	1.5g
Na_2SO_4	1.42g
Yeast extract	1.0g
Mineral solution	20.0mL
Trace metal solution	10.0mL
$Na_2S_2O_4$ solution	10.0mL
Vitamin solution	10.0mL
Sodium pyruvate solution	10.0mL
Resazurin (0.1% solution)	1.0mL

pH 7.3 ± 0.2 at 25°C

Mineral Solution:

Composition per liter:

NH_4Cl	50.0g
NaCl	40.0g
$MgCl_2 \cdot 6H_2O$	8.3g
KCl	5.0g
KH_2PO_4	5.0g
$CaCl_2 \cdot 2H_2O$	1.0g

Preparation of Mineral Solution: Add components to distilled/deionized water and bring volume to 1.0L. Mix thoroughly.

Trace Metal Solution:

Composition per liter:

Nitrilotriacetic acid	2.0g
$MnSO_4 \cdot H_2O$	1.0g
$Fe(NH_4)_2(SO_4)_2 \cdot 6H_2O$	0.8g
$CoCl_2 \cdot 6H_2O$	0.2g
$ZnSO_4 \cdot 7H_2O$	0.2g
$CuCl_2 \cdot 2H_2O$	0.02g
$Na_2MoO_4 \cdot H_2O$	0.02g
Na_2SeO_4	0.02g
Na_2WO_4	0.02g
$NiCl_2 \cdot 6H_2O$	0.02g

Preparation of Trace Metal Solution: Add nitrilotriacetic acid to 500.0mL of distilled/deionized water. Dissolve by adjusting pH to 6.5 with KOH. Add remaining components. Add distilled/deionized water to 1.0L.

Vitamin Solution:

Composition per liter:

Nicotinamide	0.050g
1,4-Naphthoquinone	0.020g
p-Aminobenzoic acid	5.0mg
Biotin	5.0mg
Calcium pantothenate	5.0mg
Cyanocobalamin	5.0mg
Folic acid	5.0mg
Hemin	5.0mg
Pyridoxine·HCl	5.0mg
Riboflavin	5.0mg
Thioctic acid	5.0mg
NaOH (0.1*N* solution)	5.0mL

Preparation of Vitamin Solution: Add thioctic acid, 1,4-naphthoquinone, and hemin to 5.0mL of 0.1*N* NaOH solution. Mix thoroughly. Bring volume to 1.0L with distilled/deionized water. Add remaining components. Mix thoroughly.

$Na_2S_2O_4$ Solution:

Composition per 10mL:

$Na_2S_2O_4$	0.087g

Preparation of $Na_2S_2O_4$ Solution: Add $Na_2S_2O_4$ to distilled/deionized water and bring volume to 10.0mL. Filter sterilize. Prepare freshly.

Sodium Pyruvate Solution:

Composition per 10mL:

Sodium pyruvate	4.4g

Preparation of Sodium Pyruvate Solution: Add sodium pyruvate to distilled/deionized water and bring volume to 10.0mL. Filter sterilize.

Preparation of Medium: Add PIPES buffer, Na_2SO_4, yeast extract, mineral solution and trace metal solution to distilled/deionized water and bring volume to 970.0mL. Mix thoroughly. Adjust pH to 7.3 with HCl. Add $NaHCO_3$ and resazurin. Gently heat and bring to boiling under 80% N_2 + 20% CO_2. Replace headspace with 2 atmospheres pressure of the same gas phase. Autoclave for 15 min at 15 psi pressure–121°C. Cool to 25°C. Anaerobically and aseptically add sterile vitamin solution, sodium pyruvate solution and $Na_2S_2O_4$ solution. Mix thoroughly.

Use: For the cultivation and maintenance of *Desulfomonile tiedjei*.

Desulfonema limicola Medium

Composition per 1009mL:

Solution A	850.0mL
Solution C	100.0mL
Solution H	20.0mL
Solution D	10.0mL
Solution F—Wolfe's vitamin solution	10.0mL
Solution I	10.0mL
Solution G	6.6mL
Solution B—Trace elements solution SL-10	1.0mL
Solution E	1.0mL
Solution J	0.4mL

pH 7.6 ± 0.2 at 25°C

Solution A:

Composition per 920mL:

NaCl	13.0g
$MgCl_2 \cdot 6H_2O$	2.2g
Na_2SO_4	3.0g
KCl	0.5g
NH_4Cl	0.3g
KH_2PO_4	0.2g
$CaCl_2 \cdot 2H_2O$	0.15g
Resazurin	0.5mg

Preparation of Solution A: Prepare and dispense solution anaerobically under 80% N_2 + 20% CO_2. Add components to distilled/deionized water and bring volume to 920.0mL. Mix thoroughly. Gently heat and bring to boiling. Continue boiling until resazurin turns colorless indicating reduction and a pH of 6.0 is reached. Cap with rubber stoppers. Autoclave for 15 min at 15 psi pressure–121°C. Cool to 25°C.

Solution B—Trace Elements Solution SL-10:
Composition per liter:

$FeCl_2 \cdot 4H_2O$	1.5g
$CoCl_2 \cdot 6H_2O$	0.19g
$MnCl_2 \cdot 4H_2O$	0.10g
$ZnCl_2$	0.070g
$Na_2MoO_4 \cdot 2H_2O$	0.036g
$NiCl_2 \cdot 6H_2O$	0.024g
H_3BO_3	6.0mg
$CuCl_2 \cdot 2H_2O$	2.0mg
HCl (25% solution)	10.0mL

Preparation of Trace Elements Solution SL-10: Add the $FeCl_2 \cdot 4H_2O$ to 10.0mL of HCl solution. Mix thoroughly. Bring volume to approximately 900.0mL with distilled/deionized water. Mix thoroughly. Adjust pH to 6.0 with NaOH. Bring volume to 1.0L with distilled/deionized water. Filter sterilize. Aseptically gas under 100% N_2 for 20 min.

Solution C:
Composition per 100mL:

$NaHCO_3$	5.0g

Preparation of Solution C: Add $NaHCO_3$ to distilled/deionized water and bring volume to 100.0mL. Mix thoroughly. Filter sterilize. Aseptically gas under 80% N_2 + 20% CO_2 for 20 min.

Solution D:
Composition per 10mL:

Sodium acetate·$3H_2O$	2.5g

Preparation of Solution D: Prepare and dispense solution anaerobically under 80% N_2 + 20% CO_2. Add sodium acetate to distilled/deionized water and bring volume to 10.0mL. Mix thoroughly. Cap with rubber stopper. Autoclave for 15 min at 15 psi pressure–121°C. Cool to 25°C.

Solution E:
Composition per 1mL:

Disodium succinate	0.1g

Preparation of Solution E: Add disodium succinate to distilled/deionized water and bring volume to 1.0mL. Mix thoroughly. Gas with 80% N_2 + 20% CO_2. Cap with rubber stopper. Autoclave for 15 min at 15 psi pressure–121°C. Cool to 25°C.

Solution F—Wolfe's Vitamin Solution:
Composition per liter:

Pyridoxine·HCl	0.01g
Thiamine·HCl	5.0mg
Riboflavin	5.0mg
Nicotinic acid	5.0mg
Calcium pantothenate	5.0mg
p-Aminobenzoic acid	5.0mg
Thioctic acid	5.0mg
Biotin	2.0mg
Folic acid	2.0mg
Cyanocobalamin	0.1mg

Preparation of Wolfe's Vitamin Solution: Add components to distilled/deionized water and bring volume to 1.0L. Mix thoroughly. Filter sterilize. Aseptically gas under 100% N_2 for 20 min.

Solution G:
Composition per 6.6mL:

$AlCl_3 \cdot 6H_2O$ (4.9% solution)	5.0mL
Na_2CO_3 (10.6% solution)	1.6mL

Preparation of Solution G: Combine both solutions. Mix thoroughly. Gas with 100% N_2. Cap with a rubber stopper. Autoclave for 15 min at 15 psi pressure–121°C. Cool to 25°C.

Solution H:
Composition per 10mL:

Rumen fluid, clarified	20.0mL

Preparation of Solution H: Gas rumen fluid under 100% N_2 for 20 min. Cap with a rubber stopper. Autoclave for 15 min at 15 psi pressure–121°C. Cool to 25°C.

Solution I:
Composition per 10mL:

$Na_2S \cdot 9H_2O$	0.4g

Preparation of Solution I: Add $Na_2S \cdot 9H_2O$ to distilled/deionized water and bring volume to 10.0mL. Gas under 100% N_2 for 20 min. Cap with a rubber stopper. Autoclave for 15 min at 15 psi pressure–121°C. Cool to 25°C.

Solution J:
Composition per 10mL:

$Na_2S_2O_4$	0.5g

Preparation of Solution J: Add $Na_2S_2O_4$ to distilled/deionized water and bring volume to 10.0mL. Mix thoroughly. Filter sterilize. Aseptically gas under 100% N_2 for 20 min. Prepare solution freshly.

Preparation of Medium: To 850.0mL of cooled, sterile solution A, aseptically and anaerobically add in the following order: 1.0mL of sterile solution B, 1000.0mL of sterile solution C, 10.0mL of sterile solution D, 1.0mL of sterile solution E, 10.0mL of sterile solution F, and 6.6mL of sterile solution G, 20.0mL of sterile solution H and 10.0mL of sterile solution I. Mix thoroughly. Immediately prior to inoculation aseptically and anaerobically add 0.4mL of sterile solution J. Mix thoroughly. Aseptically and anaerobically distribute into sterile tubes or flasks.

Use: For the cultivation and maintenance of *Desulfonema limicola*.

Desulfonema magnum Medium

Composition per 1001mL:

Solution A	890.0mL
Solution C	50.0mL
Solution J	20.0mL
Solution D	10.0mL
Solution G—Wolfe's vitamin solution	10.0mL
Solution K	10.0mL
Solution I	6.6mL
Solution B—Trace elements solution SL-10	1.0mL
Solution E	1.0mL
Solution F	1.0mL
Solution H	1.0mL
Solution L	0.4mL

pH 6.9 ± 0.2 at 25°C

Solution A:

Composition per 890mL:

NaCl	21.0g
$MgCl_2 \cdot 6H_2O$	5.5g
Na_2SO_4	3.0g
$CaCl_2 \cdot 2H_2O$	1.35g
KCl	0.5g
NH_4Cl	0.3g
KH_2PO_4	0.2g
Resazurin	0.5mg

Preparation of Solution A: Prepare and dispense solution anaerobically under 80% N_2 + 20% CO_2. Add components to distilled/deionized water and bring volume to 890.0mL. Mix thoroughly. Gently heat and bring to boiling. Continue boiling until resazurin turns colorless indicating reduction and a pH of 6.0 is reached. Cap with rubber stoppers. Autoclave for 15 min at 15 psi pressure–121°C. Cool to 25°C.

Solution B—Trace Elements Solution SL-10:

Composition per liter:

$FeCl_2 \cdot 4H_2O$	1.5g
$CoCl_2 \cdot 6H_2O$	0.19g
$MnCl_2 \cdot 4H_2O$	0.10g
$ZnCl_2$	0.070g
$Na_2MoO_4 \cdot 2H_2O$	0.036g
$NiCl_2 \cdot 6H_2O$	0.024g
H_3BO_3	6.0mg
$CuCl_2 \cdot 2H_2O$	2.0mg
HCl (25% solution)	10.0mL

Preparation of Trace Elements Solution SL-10: Add the $FeCl_2 \cdot 4H_2O$ to 10.0mL of HCl solution. Mix thoroughly. Bring volume to approximately 900.0mL with distilled/deionized water. Mix thoroughly. Adjust pH to 6.0 with NaOH. Bring volume to 1.0L with distilled/deionized water. Filter sterilize. Aseptically gas under 100% N_2 for 20 min.

Solution C:

Composition per 50mL:

$NaHCO_3$	2.5g

Preparation of Solution C: Add $NaHCO_3$ to distilled/deionized water and bring volume to 50.0mL. Mix thoroughly. Filter sterilize. Aseptically gas under 80% N_2 + 20% CO_2 for 20 min.

Solution D:

Composition per 10mL:

Sodium benzoate	0.6g

Preparation of Solution D: Add sodium benzoate to distilled/deionized water and bring volume to 10.0mL. Mix thoroughly. Gas with 100% N_2 for 10 min. Cap with a rubber stopper. Autoclave for 15 min at 15 psi pressure–121°C. Cool to 25°C.

Solution E:

Composition per 1mL:

$Na_2SeO_3 \cdot 5H_2O$	3.5μg

Preparation of Solution E: Add $Na_2SeO_3 \cdot 5H_2O$ to distilled/deionized water and bring volume to 1.0mL. Mix thoroughly. Gas with 100% N_2 for 10 min. Cap with a rubber stopper. Autoclave for 15 min at 15 psi pressure–121°C. Cool to 25°C.

Solution F:

Composition per 1mL:

Disodium succinate	0.1g

Preparation of Solution F: Add disodium succinate to distilled/deionized water and bring volume to 1.0mL. Mix thoroughly. Gas with 100% N_2. Cap with a rubber stopper. Autoclave for 15 min at 15 psi pressure–121°C. Cool to 25°C.

Solution G—Wolfe's Vitamin Solution:

Composition per liter:

Pyridoxine·HCl	0.01g
Thiamine·HCl	5.0mg
Riboflavin	5.0mg
Nicotinic acid	5.0mg
Calcium pantothenate	5.0mg
p-Aminobenzoic acid	5.0mg
Thioctic acid	5.0mg
Biotin	2.0mg
Folic acid	2.0mg
Cyanocobalamin	0.1mg

Preparation of Wolfe's Vitamin Solution: Add components to distilled/deionized water and bring volume to 1.0L. Mix thoroughly. Filter sterilize. Aseptically gas under 100% N_2 for 20 min.

Solution H:

Composition per 1mL:

Vitamin B_{12}	0.05mg

Preparation of Solution H: Add disodium succinate to distilled/deionized water and bring volume to 1.0mL. Mix thoroughly. Gas with 100% N_2. Cap with a rubber stopper. Autoclave for 15 min at 15 psi pressure–121°C. Cool to 25°C.

Solution I:

Composition per 6.6mL:

$AlCl_3 \cdot 6H_2O$ (4.9% solution)	5.0mL
Na_2CO_3 (10.6% solution)	1.6mL

Preparation of Solution I: Combine both solutions. Mix thoroughly. Gas with 100% N_2. Cap with a rubber stopper. Autoclave for 15 min at 15 psi pressure–121°C. Cool to 25°C.

Solution J:

Composition per 10mL:

Rumen fluid, clarified	20.0mL

Preparation of Solution J: Gas rumen fluid under 100% N_2 for 20 min. Cap with a rubber stopper. Autoclave for 15 min at 15 psi pressure–121°C. Cool to 25°C.

Solution K:

Composition per 10mL:

$Na_2S \cdot 9H_2O$	0.4g

Preparation of Solution K: Add $Na_2S \cdot 9H_2O$ to distilled/deionized water and bring volume to 10.0mL. Gas under 100% N_2 for 20 min. Cap with a rubber stopper. Autoclave for 15 min at 15 psi pressure–121°C. Cool to 25°C.

Solution L:

Composition per 10mL:

$Na_2S_2O_4$	0.5g

Preparation of Solution L: Add $Na_2S_2O_4$ to distilled/deionized water and bring volume to 10.0mL. Mix thoroughly. Filter sterilize. Aseptically gas under 100% N_2 for 20 min. Use freshly prepared solution.

Preparation of Medium: To 890.0mL of cooled, sterile solution A, aseptically and anaerobically add in the following order: 1.0mL of sterile solution B, 50.0mL of sterile solution C, 10.0mL of sterile solution D, 1.0mL of sterile solution E, 1.0mL of sterile solution F, and 10.0mL of sterile solution G, 1.0mL of sterile solution H and 6.6mL of sterile solution I, 20.0mL of sterile solution J, and 10.0mL of sterile solution K. Mix thoroughly. Immediately prior to inoculation aseptically and anaerobically add 0.4mL of sterile solution L. Mix thoroughly. Aseptically and anaerobically distribute into sterile tubes or flasks.

Use: For the cultivation and maintenance of *Desulfonema magnum.*

Desulfovibrio Medium

Composition per 1056.5mL:

$(NH_4)_2SO_4$	5.3g
Sodium acetate	2.0g
NaCl	1.0g
KH_2PO_4	0.5g
$MgSO_4 \cdot 7H_2O$	0.2g
$CaCl_2 \cdot 2H_2O$	0.1g
Na_2CO_3 solution	50.0mL
Solution 1	10.0mL
Solution 2	1.0mL

pH 7.2 ± 0.2 at 25°C

Solution 1:

Composition per liter:

Nitrilotriacetic acid	12.8g
$FeCl_2 \cdot 4H_2O$	0.3g
$CoCl_2 \cdot 6H_2O$	0.17g
$MnCl_2 \cdot 4H_2O$	0.1g
$ZnCl_2$	0.1g
$CuCl_2$	0.02g
H_3BO_3	0.01g
$Na_2MoO_4 \cdot 2H_2O$	0.01g

Preparation of Solution 1: Add nitrilotriacetic acid to 500.0mL of distilled/deionized water. Dissolve by adjusting pH to 6.5 with NaOH. Add remaining components. Readjust pH to 7.2 with H_2SO_4 or NaOH. Add distilled/deionized water to 1.0L.

Solution 2:

Composition per 100mL:

Resazurin	0.2g

Preparation of Solution 2: Add resazurin to distilled/deionized water and bring volume to 100.0mL. Mix thoroughly.

Na_2CO_3 Solution:

Composition per 100mL:

Na_2CO_3	8.0g

Preparation of Na_2CO_3 Solution Solution: Add Na_2CO_3 to distilled/deionized water and bring volume to 100.0mL. Mix thoroughly. Filter sterilize. Gas with 100% N_2 for 20 min.

HCl Solution:

Composition per 100mL:

HCl	25.0mL

Preparation of HCl Solution: Add HCl to distilled/deionized water and bring volume to 100.0mL. Mix thoroughly. Autoclave for 15 min at 15 psi pressure–121°C. Cool to 25°C. Gas with 100% N_2 for 20 min.

$Na_2S_2O_4$ Solution:

Composition per 100mL:

$Na_2S_2O_4$	8.7g

Preparation of $Na_2S_2O_4$ Solution: Add $Na_2S_2O_4$ to distilled/deionized water and bring volume to 100.0mL. Mix thoroughly. Autoclave for 15 min at 15 psi pressure–121°C. Cool to 25°C. Gas with 100% N_2 for 20 min.

Preparation of Medium: Add components—except Na_2CO_3 solution, HCl solution, and $Na_2S_2O_4$ solution—to distilled/deionized water and bring volume to 1.0L. Mix thoroughly. Gently heat and bring to boiling. Autoclave for 15 min at 15 psi pressure–121°C. Cool to 45°–50°C. Anaerobically and aseptically add 50.0mL of sterile Na_2CO_3 solution, 5.5mL of sterile HCl solution, and 1.0mL of sterile $Na_2S_2O_4$ solution. Mix thoroughly. Anaerobically and aseptically distribute into sterile tubes or flasks.

Use; For the isolation, cultivation and enrichment of *Desulfovibrio* species.

Desulfovibrio Medium

Composition per liter of tap water:

Agar	15.0g
Glucose	5.0g
Peptone	5.0g
Beef extract	3.0g
$MgSO_4$	1.5g
Na_2SO_4	1.5g
Yeast extract	0.2g
$Fe(NH_4)_2(SO_4)_2$	0.1g

pH 7.0 ± 0.2 at 25°C

Preparation of Medium: Sterilize by autoclaving for 15 minutes at 15 lbs. pressure (121°C).

Use: For the cultivation and maintenance of *Desulfomaculum nigrificans, Desulfovibiro desulfuricans,* and *Desolfovibrio gigas.*

Desulfovibrio Medium with Lactate

Composition per liter:

Agar	15.0g
Lactate	10.0g
Glucose	5.0g
Peptone	5.0g
Beef extract	3.0g
$MgSO_4$	1.5g
Na_2SO_4	1.5g
Yeast extract	0.2g
$Fe(NH_4)_2(SO_4)_2$	0.1g

pH 7.0 ± 0.2 at 25°C

Preparation of Medium: Add components to tap water and bring volume to 1.0L. Mix thoroughly. Gently heat and bring to boiling. Distribute into tubes or flasks. Autoclave for 15 min at 15 psi pressure–121°C. Pour into sterile Petri dishes or leave in tubes.

Use: For the cultivation and maintenance of *Desulfovibrio desulfuricans.*

Desulfovibrio Medium with NaCl

Composition per liter:

NaCl	30.0g
Agar	15.0g
Glucose	5.0g
Peptone	5.0g
Beef extract	3.0g
$MgSO_4$	1.5g
Na_2SO_4	1.5g
Yeast extract	0.2g
$Fe(NH_4)_2(SO_4)_2$	0.1g

pH 7.0 ± 0.2 at 25°C

Preparation of Medium: Add components to tap water and bring volume to 1.0L. Mix thoroughly. Gently heat and bring to boiling. Distribute into tubes or flasks. Autoclave for 15 min at 15 psi pressure–121°C. Pour into sterile Petri dishes or leave in tubes.

Use: For the cultivation and maintenance of *Desulfovibrio desulfuricans* and *D. salexigens.*

Desulfovibrio sulfodismutans Medium

Composition per 1002mL:

Solution A	920.0mL
Solution C	50.0mL
Solution D	10.0mL
Solution F	10.0mL
Solution G	10.0mL
Solution B—	
Trace elements solution SL-10	1.0mL
Solution E	1.0mL

pH 7.1–7.4 at 25°C

Solution A:

Composition per 920mL:

NaCl	1.0g
KCl	0.5g
$MgCl_2·6H_2O$	0.4g
NH_4Cl	0.3g
KH_2PO_4	0.2g
$CaCl_2·2H_2O$	0.15g

Preparation of Solution A: Prepare and dispense solution anaerobically under 80% N_2 + 20% CO_2. Add components to distilled/deionized water and bring volume to 920.0mL. Mix thoroughly. Gently heat and bring to boiling. Continue boiling until resazurin turns colorless indicating reduction. Cap with rubber stoppers. Autoclave for 15 min at 15 psi pressure–121°C. Cool to 45°–50°C.

Solution B—Trace Elements Solution SL-10:
Composition per liter:

$FeCl_2 \cdot 4H_2O$	1.5g
$CoCl_2 \cdot 6H_2O$	0.19g
$MnCl_2 \cdot 4H_2O$	0.10g
$ZnCl_2$	0.070g
$Na_2MoO_4 \cdot 2H_2O$	0.036g
$NiCl_2 \cdot 6H_2O$	0.024g
H_3BO_3	6.0mg
$CuCl_2 \cdot 2H_2O$	2.0mg
HCl (25% solution)	10.0mL

Preparation of Trace Elements Solution SL-10: Add the $FeCl_2 \cdot 4H_2O$ to 10.0mL of HCl solution. Mix thoroughly. Bring volume to approximately 900.0mL with distilled/deionized water. Mix thoroughly. Adjust pH to 6.0 with NaOH. Bring volume to 1.0L with distilled/deionized water. Filter sterilize. Aseptically gas under 100% N_2 for 20 min.

Solution C:
Composition per 50mL:

$NaKCO_3$	2.5g

Preparation of Solution C: Add $NaHCO_3$ to distilled/deionized water and bring volume to 50.0mL. Mix thoroughly. Filter sterilize. Aseptically gas under 80% N_2 + 20% CO_2 for 20 min.

Solution D:
Composition per 10mL:

Sodium acetate·$3H_2O$	0.3g

Preparation of Solution D: Prepare and dispense solution anaerobically under 80% N_2 + 20% CO_2. Add sodium acetate to distilled/deionized water and bring volume to 10.0mL. Mix thoroughly. Cap with rubber stopper. Autoclave for 15 min at 15 psi pressure–121°C. Cool to 45°–50°C.

Solution E:
Composition per liter:

Calcium pantothenate	0.050mg
Biotin	0.010mg

Preparation of Solution E: Add components to distilled/deionized water and bring volume to 1.0L. Mix thoroughly. Filter sterilize. Aseptically gas under 100% N_2 for 20 min.

Solution F:
Composition per 10mL:

$Na_2S \cdot 9H_2O$	0.4g

Preparation of Solution F: Add $Na_2S \cdot 9H_2O$ to distilled/deionized water and bring volume to 10.0mL. Gas under 100% N_2 for 20 min. Cap with a rubber stopper. Autoclave for 15 min at 15 psi pressure–121°C. Cool to 25°C.

Solution G:
Composition per 10mL:

$Na_2S_2O_5$	1.05g

Preparation of Solution G: Add $Na_2S_2O_5$ to distilled/deionized water and bring volume to 10.0mL. Mix thoroughly. Filter sterilize. Aseptically gas under 100% N_2 for 20 min. Prepare solution freshly.

Preparation of Medium: To 920.0mL of cooled, sterile solution A, aseptically and anaerobically add in the following order: 1.0mL of sterile solution B, 50.0mL of sterile solution C, 10.0mL of sterile solution D, 1.0mL of sterile solution E, 10.0mL of sterile solution F, and 10.0mL of sterile solution G. Mix thoroughly. Aseptically and anaerobically distribute into sterile tubes or flasks.

Use: For the cultivation and maintenance of *Desulfovibrio sulfodismutans*.

Desulfurococcus Medium

Composition per 1300mL:

Solution C	500.0mL
Sloution B	450.0mL
Solution A	300.0mL
Solution D	50.0mL

Solution A:
Composition per 300mL:

$(NH_4)_2SO_4$	1.3g
KH_2PO_4	0.28g
$MgSO_4 \cdot 7H_2O$	0.25g
$CaCl_2 \cdot 2H_2O$	0.07g
$FeSO_4 \cdot 7H_2O$	0.028g
$Na_2B_4O_7 \cdot 10H_2O$	4.5mg
$MnCl_2 \cdot 4H_2O$	1.8mg
$ZnSO_4 \cdot 7H_2O$	0.220mg
$CuCl_2 \cdot 2H_2O$	0.050mg
$Na_2MoO_4 \cdot 2H_2O$	0.030mg
$VOSO_4 \cdot 2H_2O$	0.030mg
$CoSO_4 \cdot 7H_2O$	0.010mg

Preparation of Solution A: Add components to distilled/deionized water and bring volume to 300.0mL. Mix thoroughly. Gently heat and bring to boiling. Autoclave for 15 min at 15 psi pressure–121°C. Cool to 25°C. Gas under 100% N_2 for 20 min.

Solution B:
Composition per 450mL:

Sulfur	5.0g

Preparation of Solution B: Add sulfur to distilled/deionized water and bring volume to 450.0mL. Autoclave for 30 min at 0 psi pressure–100°C on three consecutive days. Gas under 100% N_2 for 20 min.

Solution C:

Composition per 500mL:

Pancreatic digest of casein	2.0g
Yeast extract	2.0g
Resazurin	1.0mg

Preparation of Solution C: Add components to distilled/deionized water and bring volume to 500.0mL. Mix thoroughly. Gently heat and bring to boiling. Autoclave for 15 min at 15 psi pressure–121°C. Cool to 25°C. Gas under 100% N_2 for 20 min.

Solution D:

$Na_2S \cdot 9H_2O$	0.5g

Preparation of Solution D: Add components to distilled/deionized water and bring volume to 50.0mL. Mix thoroughly. Autoclave for 15 min at 15 psi pressure–121°C. Cool to 25°C. Gas under 100% N_2 for 20 min.

Preparation of Medium: Aseptically combine solutions A–D under nitrogen gas. Seal containers with butyl rubber stoppers.

Use: For the cultivation and maintenance of *Desulfurococcus mobilit* and *Desulfurococcus mucosus*.

Desulfuromonas Medium

Composition per 1051mL:

Elemental sulfur slurry	10.0g
Solution 1	1.0L
Solution 3	40.0mL
Solution 4	6.0mL
Solution 5	5.0mL
Solution 2	1.0mL

pH 7.2 ± 0.2 at 25°C

Solution 1:

Composition per liter:

NaCl	20.0g
$MgCl_2 \cdot 6H_2O$	3.0g
KH_2PO_4	1.0g
NH_4Cl	0.3g
$CaCl_2 \cdot 2H_2O$	0.1g
HCl (2*N* solution)	4.0mL

Preparation of Solution 1: Add components to distilled/deionized water and bring volume to 1.0L. Mix thoroughly. Autoclave for 15 min at 15 psi pressure–121°C. Cool to 25°C.

Solution 2:

Composition per liter:

Disodium EDTA	5.2g
$CoCl_2 \cdot 6H_2O$	1.9g
$FeCl_2 \cdot 4H_2O$	1.5g
$MnCl_2 \cdot 4H_2O$	1.0g
$ZnCl_2$	0.7g
H_3BO_3	0.62g
$Na_2MoO_4 \cdot 2H_2O$	0.36g
$NiCl_2 \cdot 6H_2O$	0.24g
$CuCl_2 \cdot 2H_2O$	0.17g

pH 6.5 ± 0.2 at 25°C

Preparation of Solution 2: Add components to distilled/deionized water and bring volume to 1.0L. Mix thoroughly. Adjust pH to 6.5. Autoclave for 15 min at 15 psi pressure–121°C. Cool to 25°C.

Solution 3:

Composition per 100mL:

$NaHCO_3$	10.0g

Preparation of Solution 3: Add $NaHCO_3$ to distilled/deionized water and bring volume to 100.0mL. Mix thoroughly. Autoclave for 15 min at 15 psi pressure–121°C. Cool to 25°C.

Solution 4:

Composition per 100mL:

$Na_2S \cdot 9H_2O$	5.0g

Preparation of Solution 4: Add $Na_2S \cdot 9H_2O$ to distilled/deionized water and bring volume to 100.0mL. Mix thoroughly. Autoclave for 15 min at 15 psi pressure–121°C. Cool to 25°C.

Solution 5:

Composition per 200mL:

Pyridoxamine·HCl	0.01g
Nicotinic acid	4.0mg
p-Aminobenzoic acid	2.0mg
Thiamine	2.0mg
Cyanocobalamin	1.0mg
Pantothenic acid	1.0mg
Biotin	0.5mg

Preparation of Solution 5: Add components to distilled/deionized water and bring volume to 200.0mL. Mix thoroughly. Filter sterilize.

Elemental Sulfur Slurry:

Composition per 10g:

Sulfur flowers	10.0g

Preparation of Elemental Sulfur Slurry: Add highly purified sulfur flowers to a mortar and grind to a fine powder. Add sufficient distilled/deionized water to produce a slurry. Distribute into 100-mL screw-capped bottles in 20.0mL volumes. Autoclave for 30 min at 10 psi pressure–115°C. Decant supernatant solution. Reserve sulfur slurry.

Preparation of Medium: To 1.0L of cooled, sterile solution 1, aseptically add 1.0mL of sterile solution 2, 40.0mL of sterile solution 3, 6.0mL of sterile solution 4 and 5.0mL of sterile solution 5. Mix thoroughly. Adjust pH to 7.2. Aseptically distribute into sterile 50.0mL screw-capped bottles. Fill bottles completely with medium except for a pea-sized air bubble. Aseptically add a pea-sized piece of sulfur slurry to each 50.0mL of medium.

Use: For the isolation and cultivation of marine *Desulfuromonas* species.

Desulfuromonas Medium

Composition per 1031mL:

Elemental sulfur slurry	10.0g
Solution 1	1.0L
Solution 3	20.0mL
Solution 4	6.0mL
Solution 5	5.0mL
Solution 2	1.0mL

pH 7.2 ± 0.2 at 25°C

Solution 1:

Composition per liter:

KH_2PO_4	1.0g
$MgCl_2 \cdot 6H_2O$	0.4g
NH_4Cl	0.3g
$CaCl_2 \cdot 2H_2O$	0.1g
HCl (2*N* solution)	4.0mL

Preparation of Solution 1: Add components to distilled/deionized water and bring volume to 1.0L. Mix thoroughly. Autoclave for 15 min at 15 psi pressure–121°C. Cool to 25°C.

Solution 2:

Composition per liter:

Disodium EDTA	5.2g
$CoCl_2 \cdot 6H_2O$	1.9g
$FeCl_2 \cdot 4H_2O$	1.5g
$MnCl_2 \cdot 4H_2O$	1.0g
$ZnCl_2$	0.7g
H_3BO_3	0.62g
$Na_2MoO_4 \cdot 2H_2O$	0.36g
$NiCl_2 \cdot 6H_2O$	0.24g
$CuCl_2 \cdot 2H_2O$	0.17g

Preparation of Solution 2: Add components to distilled/deionized water and bring volume to 1.0L. Mix thoroughly. Adjust pH to 6.5. Autoclave for 15 min at 15 psi pressure–121°C. Cool to 25°C.

Solution 3:

Composition per 100mL:

$NaHCO_3$	10.0g

Preparation of Solution 3: Add $NaHCO_3$ to distilled/deionized water and bring volume to 100.0mL. Mix thoroughly. Autoclave for 15 min at 15 psi pressure–121°C. Cool to 25°C.

Solution 4:

Composition per 100mL:

$Na_2S \cdot 9H_2O$	5.0g

Preparation of Solution 4: Add $Na_2S \cdot 9H_2O$ to distilled/deionized water and bring volume to 100.0mL. Mix thoroughly. Autoclave for 15 min at 15 psi pressure–121°C. Cool to 25°C.

Solution 5:

Composition per 200mL:

Pyridoxamine·HCl	0.01g
Nicotinic acid	4.0mg
p-Aminobenzoic acid	2.0mg
Thiamine	2.0mg
Cyanocobalamin	1.0mg
Pantothenic acid	1.0mg
Biotin	0.5mg

Preparation of Solution 5: Add components to distilled/deionized water and bring volume to 200.0mL. Mix thoroughly. Filter sterilize.

Elemental Sulfur Slurry:

Composition per 10g:

Sulfur flowers	10.0g

Preparation of Elemental Sulfur Slurry: Add highly purified sulfur flowers to a mortar and grind to a fine powder. Add sufficient distilled/deionized water to produce a slurry. Distribute into 100.0mL screw-capped bottles in 20.0mL volumes. Autoclave for 30 min at 10 psi pressure–115°C. Decant supernatant solution. Reserve sulfur slurry.

Preparation of Medium: To 1.0L of cooled, sterile solution 1, aseptically add 1.0mL of sterile solution 2, 40.0mL of sterile solution 3, 6.0mL of sterile solution 4 and 5.0mL of sterile solution 5. Mix thoroughly. Adjust pH to 7.2. Aseptically distribute into sterile 50.0mL screw-capped bottles. Fill bottles completely with medium except for a pea-sized air bubble. Aseptically add a pea-sized piece of sulfur slurry to each 50.0mL of medium.

Use: For the isolation and cultivation of freshwater *Desulfuromonas* species.

Dextrin Fuchsin Sulfite Agar

See: ***Aeromonas*** **Differential Agar**

Dextrose Agar

Composition per liter:

Agar	15.0g
Glucose	10.0g
NaCl	5.0g
Pancreatic digest of casein	5.0g
Peptic digest of animal tissue	5.0g
Beef extract	3.0g

pH 6.9 ± 0.2 at 25°C

Source: This medium is available as a premixed powder from BBL Microbiology Systems .

Preparation of Medium: Add components to distilled/deionized water and bring volume to 1.0L. Mix thoroughly. Gently heat and bring to boiling. Distribute into tubes or flasks. Autoclave for 15 min at 15 psi pressure–121°C. Pour into sterile Petri dishes or leave in tubes.

Use: For the cultivation and enumeration of microorganisms from foods. For use as a base for the preparation of blood agar and for general laboratory procedures.

Dextrose Agar

Composition per liter:

Agar	15.0g
Glucose	10.0g
Tryptose	10.0g
NaCl	5.0g
Beef extract	3.0g

pH 7.3 ± 0.2 at 25°C

Source: This medium is available as a premixed powder from Difco Laboratories.

Preparation of Medium: Add components to distilled/deionized water and bring volume to 1.0L. Mix thoroughly. Gently heat and bring to boiling. Distribute into tubes or flasks. Autoclave for 15 min at 15 psi pressure–121°C. Pour into sterile Petri dishes or leave in tubes.

Use: For the cultivation of a wide variety of microorganisms. For use as a base for the preparation of blood agar and for general laboratory procedures.

Dextrose Ascitic Fluid Semisolid Agar

Composition per liter:

Pancreatic digest of casein	2.66g
NaCl	1.33g
Agar	0.5g
Phenol Red	4.8mg
Ascitic fluid	50.0mL
Glucose solution	15.0mL

pH 7.4 ± 0.2 at 25°C

Glucose Solution:

Composition per 15mL:

Carbohydrate	3.0g

Preparation of Glucose Solution: Add glucose to distilled/deionized water and bring volume to 15.0mL. Mix thoroughly. Filter sterilize.

Preparation of Medium: Add components, except ascitic fluid and glucose solution, to distilled/deionized water and bring volume to 935.0mL. Mix thoroughly. Gently heat and bring to boiling. Autoclave for 15 min at 15 psi pressure–121°C. Cool to 45°–50°C. Aseptically add sterile ascitic fluid and glucose solution. Mix thoroughly. Aseptically distribute into sterile tubes.

Use: For the isolation and cultivation of microorganisms from spinal fluid.

Dextrose Broth

Composition per liter:

Tryptose	10.0g
Glucose	5.0g
NaCl	5.0g
Beef extract	3.0g

pH 7.2 ± 0.2 at 25°C

Source: This medium is available as a premixed powder from Difco Laboratories and Oxoid Unipath.

Preparation of Medium: Add components to distilled/deionized water and bring volume to 1.0L. Mix thoroughly. Distribute into tubes or flasks. Autoclave for 15 min at 15 psi pressure–121°C.

Use: For the isolation and enrichment of fastidious or damaged microorganisms.

Dextrose Broth

Composition per liter:

Pancreatic digest of casein	10.0g
Glucose	5.0g
NaCl	5.0g

pH 7.3 ± 0.2

Source: This medium is available as a premixed powder from BBL Microbiology Systems.

Preparation of Medium: Add components to distilled/deionized water and bring volume to 1.0L. Mix thoroughly. Distribute into tubes or flasks. Autoclave for 15 min at 15 psi pressure–121°C.

Use: For the cultivation and differentiation of microorganisms based on their ability to ferment glucose. If desired, a Durham tube may be added to the test tubes to determine gas production.

Dextrose Proteose No. 3 Agar

Composition per liter:

Proteose peptone No. 3	20.0g
Agar	13.0g
NaCl	5.0g

Glucose 2.0g
Tellurite blood solution 50.0mL
pH 7.4 ± 0.2 at 25°C

Tellurite Blood Solution:
Composition per 60mL:
Sheep blood, defibrinated.............................50.0mL
Chapman tellurite solution...........................10.0mL

Preparation of Tellurite Blood Solution: Aseptically combine 10.0mL of Chapman tellurite solution with 50.0mL of sterile, defibrinated sheep blood. Mix thoroughly.

Chapman Tellurite Solution:
Composition per 100mL:
K_2TeO_3 ....1.0g

Preparation of Chapman Tellurite Solution: Add K_2TeO_3 to distilled/deionized water and bring volume to 100.0mL. Mix thoroughly. Filter sterilize.

Caution: Potassium tellurite is toxic.

Preparation of Medium: Add components, except tellurite blood solution, to distilled/deionized water and bring volume to 940.0mL. Mix thoroughly. Gently heat and bring to boiling. Autoclave for 15 min at 15 psi pressure–121°C. Cool to 75°–80°C. Aseptically add 50.0mL of sterile tellurite blood solution. Mix thoroughly. Maintain at 75°–80°C for 10–15 min or until the agar becomes chocolatized. Cool slowly to 50°C. Pour into sterile Petri dishes or distribute into sterile tubes.

Use: For propagating pure cultures of *Neisseria gonorrhoeae* and other fastidious microorganisms.

Dextrose Starch Agar

Composition per liter:
Gelatin..20.0g
Proteose peptone 15.0g
Agar...10.0g
Starch 10.0g
Glucose 5.0g
NaCl 5.0g
Na_2HPO_4 ....3.0g
pH 7.3 ± 0.2 at 25°C

Source: This medium is available as a premixed powder from Difco Laboratories.

Preparation of Medium: Add components to distilled/deionized water and bring volume to 1.0L. Mix thoroughly. Gently heat while stirring and bring to boiling. Distribute into tubes or flasks. Autoclave for 15 min at 15 psi pressure–121°C. Pour into sterile Petri dishes or leave in tubes.

Use: For the cultivation and maintenance of *Neisseria gonorrhoeae*, *Neisseria animalis*, and other fastidious microorganisms.

Dextrose Tryptone Agar

Composition per liter:
Agar...15.0g
Pancreatic digest of casein 10.0g
Glucose 5.0g
Bromcresol Purple 0.04g
pH 6.9 ± 0.2 at 25°C

Source: This medium is available as a premixed powder from BBL Microbiology Systems and Difco Laboratories.

Preparation of Medium: Add components to distilled/deionized water and bring volume to 1.0L. Mix thoroughly. Gently heat and bring to boiling. Distribute into tubes or flasks. Autoclave for 15 min at 15 psi pressure–121°C. Pour into sterile Petri dishes or leave in tubes.

Use: For the isolation and cultivation of mesophilic and thermophilic aerobic microorganisms in food.

Dextrose Tryptone Agar

Composition per liter:
Agar...12.0g
Pancreatic digest of casein 10.0g
Glucose 5.0g
Bromcresol Purple 0.04g
pH 6.9 ± 0.2 at 25°C

Source: This medium is available as a premixed powder from Oxoid Unipath.

Preparation of Medium: Add components to distilled/deionized water and bring volume to 1.0L. Mix thoroughly. Gently heat and bring to boiling. Distribute into tubes or flasks. Autoclave for 15 min at 15 psi pressure–121°C. Pour into sterile Petri dishes or leave in tubes.

Use: For the isolation, cultivation and enumeration of "flat-sour" thermophiles and mesophiles in food. Acid-producing microorganisms such as "flat-sour" hermophiles appear as yellow colonies surrounded by a yellow zone.

Dextrose Tryptone Broth (m–Dextrose Tryptone Broth)

Composition per liter:
Pancreatic digest of casein 20.0g
Glucose 10.0g
Bromcresol Purple 0.04g
pH 6.7 ± 0.2 at 25°C

Preparation of Medium: Add components to distilled/deionized water and bring volume to 1.0L. Mix thoroughly. Distribute into tubes or flasks. Autoclave for 15 min at 15 psi pressure–121°C.

Use: For the isolation, cultivation and enumeration of "flat-sour" thermophiles and mesophiles in food by the membrane filter technique. Acid-producing microorganisms such as "flat-sour" thermophiles turn the medium yellow.

Dextrose Tryptone Broth

Composition per liter:

Pancreatic digest of casein	10.0g
Glucose	5.0g
Bromcresol Purple	0.04g

pH 6.9 ± 0.2 at 25°C

Source: This medium is available as a premixed powder from Oxoid Unipath.

Preparation of Medium: Add components to distilled/deionized water and bring volume to 1.0L. Mix thoroughly. Distribute into tubes or flasks. Autoclave for 15 min at 15 psi pressure–121°C.

Use: For the isolation and cultivation of "flat-sour" thermophiles and mesophiles in food. Acid-producing microorganisms such as "flat-sour" thermophiles turn the medium yellow.

Dey/Engley Neutralizing Broth
See: **D/E Neutralizing Broth**

DG18 Agar
See: **Dichloran Glycerol Agar**

Diagnostic Sensitivity Test Agar (DST Agar)

Composition per liter:

Agar	12.0g
Proteose peptone	10.0g
Veal infusion solids	10.0g
NaCl	3.0g
Na_2HPO_4	2.0g
Glucose	2.0g
Sodium acetate	1.0g
Adenine sulfate	0.01g
Guanine·HCl	0.01g
Uracil	0.01g
Xanthine	0.01g
Thiamine	0.02mg
Horse blood, defibrinated	70.0mL

pH 7.4 ± 0.2 at 25°C

Source: This medium is available as a premixed powder from Oxoid Unipath.

Preparation of Medium: Add components, except horse blood, to distilled/deionized water and bring volume to 930.0mL. Mix thoroughly. Gently heat and bring to boiling. Autoclave for 15 min at 15 psi pressure–121°C. Cool to 45°–50°C. Aseptically add sterile horse blood. Mix thoroughly. Pour into sterile Petri dishes or distribute into sterile tubes.

Use: For the cultivation of microorganisms for antimicrobial sensitivity testing.

Diamalt Agar

Composition per liter:

Diamalt	150.0g
Agar	20.0g

Preparation of Medium: Add components to distilled/deionized water and bring volume to 1.0L. Mix thoroughly. Gently heat and bring to boiling. Distribute into tubes or flasks. Autoclave for 15 min at 15 psi pressure–121°C. Pour into sterile Petri dishes or leave in tubes.

Use: For the isolation and cultivation of yeasts.

Diaminopimelic Acid Medium

Composition per liter:

Pancreatic digest of gelatin	5.0g
Beef extract	3.0g
Diaminopimelic acid	0.050g

pH 6.9 ± 0.2 at 25°C

Preparation of Medium: Add components to distilled/deionized water and bring volume to 1.0L. Mix thoroughly. Distribute into tubes or flasks. Autoclave for 15 min at 15 psi pressure–121°C.

Use: For the cultivation and maintenance of *Bacillus megaterium.*

Diamonds Medium, Modified

Composition per liter:

Pancreatic digest of casein	20.0g
Yeast extract	1.0g
L-Cysteine·HCl·H_2O	0.5g
Maltose	0.5g
L-Ascorbic acid	0.02g
Horse serum, inactivated	100.0mL
Antibiotic inhibitor	10.0mL

pH 6.5 ± 0.2 at 25°C

Antibiotic Inhibitor:

Composition per 10mL:

Streptomycin sulfate	0.15g

Amphotericin B	0.2mg
Penicillin G	100,000U

Preparation of Antibiotic Inhibitor: Add components to distilled/deionized water and bring volume to 10.0mL. Mix thoroughly. Filter sterilize.

Preparation of Medium: Add components, except antibiotic inhibitor and horse serum, to distilled/deionized water and bring volume to 890.0mL. Mix thoroughly. Gently heat and bring to boiling. Autoclave for 15 min at 15 psi pressure–121°C. Cool to 25°C. Aseptically add sterile antibiotic inhibitor and horse serum. Mix thoroughly. Aseptically distribute into sterile tubes in 5.0mL volumes.

Use: For the cultivation of *Trichomonas* species.

Dibenzothiophene Mineral Medium

Composition per liter:

Beef extract	10.0g
Na_2HPO_4	3.0g
KH_2PO_4	2.0g
NH_4Cl	2.0g
Dibenzothiophene	0.5g
$MgCl_2 \cdot 6H_2O$	0.2g
$FeCl_3 \cdot 6H_2O$	0.028g

Preparation of Medium: Add components to distilled/deionized water and bring volume to 1.0L. Mix thoroughly. Distribute into tubes or flasks. Autoclave for 15 min at 15 psi pressure–121°C.

Use: For the cultivation of bacteria that can metabolize dibenzothiophene.

Dichloran Glycerol Agar (DG18 Agar)

Composition per liter:

Agar	15.0g
Glucose	10.0g
Peptone	5.0g
KH_2PO_4	1.0g
$MgSO_4 \cdot 7H_2O$	0.5g
Dichloran	2.0mg
Chloramphenicol solution	10.0mL

pH 5.6 ± 0.2 at 25°C

Source: This medium is available as a premixed powder from Oxoid Unipath.

Chloramphenicol Solution:

Chloramphenicol	0.1g

Preparation of Chloramphenicol Solution: Add chloramphenicol to distilled/deionized water and bring volume to 10.0mL. Mix thoroughly. Filter sterilize.

Preparation of Medium: Add components, except chloramphenicol solution, to distilled/deionized water and bring volume to 990.0mL. Mix thoroughly. Gently heat and bring to boiling. Autoclave for 15 min at 15 psi pressure–121°C. Cool to 45°–50°C. Aseptically add sterile chloramphenicol solution. Mix thoroughly. Pour into sterile Petri dishes or distribute into sterile tubes.

Use: For the enumeration and isolation of xerophilic molds from dried and semi-dried foods.

Dichloran Rose Bengal Chloramphenicol Agar

See: **DRBC Agar**

Dichloromethane Medium for *Hyphomicrobium*

Composition per liter:

$K_2HPO_4 \cdot 3H_2O$	4.1g
KH_2PO_4	1.4g
$MgSO_4 \cdot 7H_2O$	0.2g
$(NH_4)_2SO_4$	0.2g
Dichloromethane (methylene chloride)	1.0mL
Trace elements solution	1.0mL

pH 7.2 ± 0.2 at 25°C

Trace Elements Solution:

Composition per liter:

$Ca(NO_3)_2$	25.0g
$FeSO_4 \cdot 7H_2O$	1.0g
H_3BO_3	1.0g
$MnSO_4 \cdot H_2O$	1.0g
$Co(NO_3)_2 \cdot 6H_2O$	0.25g
$CuCl_2 \cdot 2H_2O$	0.25g
$(NH_4)_6Mo_7O_{24} \cdot 4H_2O$	0.25g
$ZnCl_2$	0.25g
NH_4VO_3	0.1g

Preparation of Medium: Filter sterilize dichloromethane. Add components, except dichloromethane, to distilled/deionized water and bring volume to 999.0mL. Mix thoroughly. Gently heat and bring to boiling. Adjust pH to 7.2. Autoclave for 15 min at 15 psi pressure–121°C. Cool to 45°–50°C. Aseptically add sterile dichloromethane. Mix thoroughly. Aseptically distribute into sterile tubes or flasks.

Use: For the cultivation and maintenance of *Hyphomicrobium* species

Dichotomicrobium Medium

Composition per liter:

Agar	18.0g
Sodium DL-malate	1.0g
Yeast extract	1.0g
Articial sea water, 3X solution	980.0mL
Hutner's mineral base	20.0mL

pH 7.1 ± 0.1 at 25°C

Articial Sea Water, 3X Solution:

Composition per 5 liters:

NaCl	352.14g
$MgCl_2 \cdot 6H_2O$	159.30g
Na_2SO_4	58.76g
$CaCl_2 \cdot 2H_2O$	21.75g
KCl	9.96g
$NaHCO_3$	2.88g
KBr	1.44g
H_3BO_3	0.39g

Preparation of Articial Seawater, 3X Solution: Add components to distilled/deionized water and bring volume to 5.0L. Mix thoroughly.

Hutner's Mineral Base:

Composition per liter:

$MgSO_4 \cdot 7H_2O$	29.7g
Nitrilotriacetic acid	10.0g
$CaCl_2 \cdot 2H_2O$	3.34g
$FeSO_4 \cdot 7H_2O$	0.099g
$(NH_4)_2MoO_4$	9.25mg
Metals "44"	50.0mL

Preparation of Hutner's Mineral Base: Add nitrilotriacetic acid to 500.0mL of distilled/deionized water. Dissolve by adjusting pH to 6.5 with KOH. Add remaining components. Readjust pH to 7.2 with H_2SO_4 or KOH. Add distilled/deionized water to 1.0L. Store at 5°C.

Metals "44":

Composition per 100mL:

$ZnSO_4 \cdot 7H_2O$	1.1g
$FeSO_4 \cdot 7H_2O$	0.5g
EDTA	0.25g
$MnSO_4 \cdot 7H_2O$	0.154g
$CuSO_4 \cdot 5H_2O$	0.04g
$Co(NO_3)_2 \cdot 6H_2O$	0.025g
$Na_2B_4O_7 \cdot 10H_2O$	0.018g

Preparation of Metals "44": Add a few drops of H_2SO_4 to distilled/deionized water to inhibit precipitate formation. Add components to acidified distilled/deionized water and bring volume to 100.0mL. Mix thoroughly.

Preparation of Medium: Combine components. Mix thoroughly. Gently heat and bring to boiling. Adjust pH to 7.1. Distribute into tubes or flasks. Autoclave for 15 min at 15 psi pressure–121°C. Pour into sterile Petri dishes or leave in tubes.

Use: For the cultivation and maintenance of *Dichotomicrobium thermohalophilium.*

Dictyoglomus Medium

Composition per liter:

Soluble starch	5.0g
$Na_2HPO_4 \cdot 12H_2O$	4.2g
Polypeptone™	2.0g
Yeast extract	2.0g
KH_2PO_4	1.5g
Cysteine·HCl·H_2O	1.0g
Na_2CO_3	1.0g
NH_4Cl	0.5g
$MgCl_2 \cdot 6H_2O$	0.38g
$CaCl_2$	0.05g
$Fe(NH_4)_2(SO_4)_2 \cdot 6H_2O$	0.039g
Resazurin	2.0mg
Trace metals	10.0mL
Wolfe's vitamin solution	10.0mL

pH 7.2 ± 0.2 at 25°C

Trace Metals:

Composition per liter:

$CoCl_2 \cdot 6H_2O$	0.29g
$ZnSO_4 \cdot 7H_2O$	0.28g
$Na_2MoO_4 \cdot 2H_2O$	0.24g
$MnCl_2 \cdot 4H_2O$	0.20g
Na_2SeO_3	0.017g

Preparation of Trace Metals: Add components to distilled/deionized water and bring volume to 1.0L. Adjust pH to 6.0 with KOH. Mix thoroughly.

Wolfe's Vitamin Solution:

Composition per liter:

Pyridoxine·HCl	0.01g
Thiamine·HCl	5.0mg
Riboflavin	5.0mg
Nicotinic acid	5.0mg
Calcium pantothenate	5.0mg
p-Aminobenzoic acid	5.0mg
Thioctic acid	5.0mg
Biotin	2.0mg
Folic acid	2.0mg
Cyanocobalamin	0.1mg

Preparation of Wolfe's Vitamin Solution: Add components to distilled/deionized water and bring volume to 1.0L. Mix thoroughly. Filter sterilize.

Preparation of Medium: Prepare and dispense medium under 100% N_2. Add components, except Wolfe's vitamin solution, to distilled/deionized water and bring volume to 990.0mL. Mix thoroughly. Gently heat and bring to boiling. Continue boiling until resazurin turns colorless indicating reduction. Autoclave for 15 min at 15 psi pressure–121°C. Cool to 25°C under 100% N_2. Aseptically add sterile Wolfe's vitamin solution. Mix thoroughly. Adjust pH to 7.2 if necessary. Aseptically and anaerobically distribute into sterile tubes or flasks.

Use: For the cultivation and maintenance of *Dictyoglomus thermophilum.*

DIFF/BCYE
See: Buffered Charcoal Yeast Extract Differential Agar

Differential Agar Medium A8 for *Ureaplasma urealyticum*

Composition per 103.1mL:

Basal agar	80.0mL
Horse serum, unheated	20.0mL
Fresh yeast extract solution	1.0mL
Urea solution	1.0mL
CVA enrichment	0.5mL
L-Cysteine·HCl·H_2O solution	0.5mL
GHL tripeptide solution	0.1mL

pH 5.5 ± 0.2 at 25°C

Basal Agar:
Composition per 80mL:

Tryptic soy broth	2.4g
Noble agar	1.05g
Putrescine·2HCl	0.17g
$CaCl_2·2H_2O$	0.015g

Preparation of Basal Agar: Add components to distilled/deionized water and bring volume to 80.0mL. Mix thoroughly. Adjust pH to 5.5 with 2*N* HCl. Gently heat and bring to boiling. Autoclave for 15 min at 15 psi pressure–121°C. Cool to 50°–55°C.

Fresh Yeast Extract Solution:
Composition per 100mL:

Baker's yeast, live, pressed, starch-free	25.0g

Preparation of Fresh Yeast Extract Solution: Add the live Baker's yeast to 100.0mL of distilled/deionized water. Autoclave for 90 min at 15 psi pressure–121°C. Allow to stand. Remove supernatant solution. Adjust pH to 6.6–6.8. Filter sterilize.

Urea Solution:
Composition per 30mL:

Urea	3.0g

Preparation of Urea Solution: Add urea to distilled/deionized water and bring volume to 30.0mL. Mix thoroughly. Filter sterilize.

CVA Enrichment:
Composition per liter:

Glucose	100.0g
L-Cysteine·HCl·H_2O	25.9g
L-Glutamine	10.0g
Adenine	1.0g
L-Cystine·2HCl	1.0g
Nicotinamide adenine dinucleotide	0.25g
Cocarboxylase	0.1g
Guanine·HCl	0.03g
$Fe(NO_3)_3$	0.02g
Vitamin B_{12}	0.01g
p-Aminobenzoic acid	0.013g
Thiamine·HCl	3.0mg

Preparation of CVA Enrichment: Add components to distilled/deionized water and bring volume to 1.0L. Mix thoroughly. Filter sterilize.

Cysteine·HCl·H_2O Solution:
Composition per 50mL:

Cysteine·HCl·H_2O	1.0g

Preparation of Cysteine·HCl·H_2O Solution: Add cysteine·HCl·H_2O to distilled/deionized water and bring volume to 50.0mL. Mix thoroughly. Filter sterilize.

GHL Tripeptide Solution:
Composition per 10mL:

GHL tripeptide	0.2mg

Preparation of GHL Tripeptide Solution: Add GHL tripeptide (glycyl-L-histidyl-L-lysine acetate) to distilled/deionized water and bring volume to 10.0mL. Mix thoroughly. Filter sterilize.

Preparation of Medium: To 80.0mL of cooled, sterile basal agar, aseptically add 20.0mL of sterile horse serum, 1.0mL of sterile fresh yeast extract solution, 1.0mL of sterile urea solution, 0.5mL of sterile CVA enrichment, 0.5mL of sterile cysteine·HCl·H_2O solution and 0.1mL of sterile GHL tripeptide solution. Mix thoroughly. Pour into sterile Petri dishes in 20.0mL volumes.

Use: For the cultivation and maintenance of *Ureaplasma urealyticum.*

Differential Broth for Lactic Streptococci

Composition per liter:

Sodium citrate	20.0g
Arginine	5.0g
Pancreatic digest of casein	5.0g
Yeast extract	5.0g

K_2HPO_4	1.0g
Bromcresol Purple	0.02g
Skim milk (11% solution)	35.0mL

pH 6.2 ± 0.2 at 25°C

Preparation of Medium: Add components, except skim milk solution, to distilled/deionized water and bring volume to 800.0mL. Mix thoroughly. Add 35.0mL of skim milk solution. Bring volume to 1.0L with distilled/deionized water. Place medium in a steam bath for 15 min. Cool to 25°C. Adjust pH to 6.2. Distribute 7.0mL volumes into screw-capped tubes that contain an inverted Durham tube. Autoclave for 15 min at 15 psi pressure–121°C. Allow autoclave to cool below 70°C before opening door.

Use: For the cultivation and differentiation of *Lactobacillus lactis, L. lactis* subspecies *cremoris* and *L. lactis* subspecies *diacetylactis*. Lactose-fermenting bacteria such as *L. lactis* subspecies *cremoris* turn the medium yellow. Arginine-utilizing bacteria such as *L. lactis* initially turn the medium yellow but then turn it back to violet. Citrate-utilizing bacteria such as *L. lactis* subspecies *diacetylactis* turn the medium violet and produce CO_2 which is trapped as a bubble in the Durham tube.

Dilute Potato Medium

Composition per 1090mL:

Glucose	1.0g
Na_2HPO_4	0.12g
$Ca(NO_3)_2 \cdot 4H_2O)$	0.05g
Peptone	0.05g
Potato decoction	100.0mL

pH 6.8 ± 0.2 at 25°C

Potato Decoction:

Composition per liter:

Potato	20.0g

Preparation of Potato Decoction: Peel and dice potato. Add to 1.0L of distilled/deionized water. Gently heat and bring to boiling. Continue boiling for 30 min. Filter through Whatman #1 filter paper. Bring volume of filtrate to 1.0L with distilled/deionized water.

Preparation of Medium: Add components to distilled/deionized water and bring volume to 1090.0mL. Mix thoroughly. Adjust pH to 6.8. Distribute into tubes or flasks. Autoclave for 15 min at 15 psi pressure–121°C.

Use: For the cultivation and maintenance of *Rhizobacter daucus*.

Diphasic Blood Culture Buffered Charcoal Yeast Extract Medium

See: ***Legionella pneumophila* Medium Charcoal Yeast Extract Diphasic Blood Culture Medium**

Diphasic Medium for Amoeba (Charcoal Agar Slants)

Composition per liter:

Agar slants	1.0L
Buffered saline overlay	1.0L

pH 7.4 ± 0.2 at 25°C

Agar Slants:

Composition per liter:

Agar	10.0g
Charcoal, activated	10.0g
Pancreatic digest of casein	5.0g
KH_2PO_4	4.0g
Na_2HPO_4	3.0g
Asparagine	2.0g
Sodium citrate	1.0g
Ferric ammonium citrate	0.1g
$MgSO_4 \cdot 7H_2O$	0.1g
Cholesterol solution	25.0mL
Glycerol	10.0mL

Cholesterol Solution:

Composition per 25mL:

Cholesterol	0.25g
Acetone	25.0mL

Preparation of Cholesterol Solution: Add cholesterol to 25.0mL of acetone. Mix thoroughly.

Preparation of Agar Slants: Add components, except agar, charcoal and cholesterol solution, to distilled/deionized water and bring volume to 1.0L. Mix thoroughly. Gently heat to dissolve. Do not boil. Add agar, charcoal and cholesterol solution. Mix thoroughly. Gently heat and bring to boiling. Distribute into tubes in 3.0mL volumes. Autoclave for 15 min at 15 psi pressure–121°C. Resuspend charcoal. Allow tubes to cool in a slanted position with short butts or no butts.

Buffered Saline Overlay:

Composition per liter:

NaCl	5.0g
Solution B	810.0mL
Solution A	190.0mL

Solution A:

Composition per liter:

KH_2PO_4, anhydrous	9.07g

Preparation of Medium: Add KH_2PO_4 to distilled/deionized water and bring volume to 1.0L. Mix thoroughly.

Solution B:

Composition per liter:

Na_2HPO_4, anhydrous	9.46g

Preparation of Solution B: Add Na_2HPO_4 to distilled/deionized water and bring volume to 1.0L. Mix thoroughly.

Preparation of Buffered Saline Overlay: Combine 810.0mL of solution A and 190.0mL of solution B. Add the NaCl. Mix thoroughly. Autoclave for 15 min at 15 psi pressure–121°C. Cool to 25°C. Store at 4°C.

Preparation of Medium: To each agar slant, aseptically add 3.0mL of sterile buffered saline overlay.

Use: For the cultivation and maintenance of *Amoebae* species.

Diphosphothiamine Medium

Composition per liter:

Proteose peptone	20.0g
Glucose	10.0g
NaCl	5.0g
Tween™ 40	0.05g
Diphosphothiamine	1.0mg

pH 7.3 ± 0.2 at 25°C

Preparation of Medium: Add components, except diphosphothiamine, to distilled/deionized water and bring volume to 1.0L. Mix thoroughly. Gently heat until dissolved. Autoclave for 15 min at 15 psi pressure–121°C. Cool to 45°–50°C. Aseptically add 1.0mg of diphosphothiamine. Mix thoroughly. Aseptically distribute into sterile tubes or flasks.

Use: For the cultivation of *Haemophilus piscium.*

Disinfectant Test Broth AOAC

Composition per liter:

Peptic digest of animal tissue	10.0g
Beef extract	5.0g
NaCl	5.0g

pH 6.8 ± 0.2 at 25°C

Source: This medium is available as a premixed powder from Difco Laboratories.

Preparation of Medium: Add components to distilled/deionized water and bring volume to 1.0L. Mix thoroughly. Distribute into tubes or flasks. Autoclave for 15 min at 13 psi pressure–118°C.

Use: For the determination of phenol coefficients of disinfectants.

Dithionite Thioglycollate, HS T, Broth

See: **Clausen Medium**

DM Medium

Composition per liter:

Starch, soluble	5.0g
$MgSO_4 \cdot 7H_2O$	0.50g
K_2HPO_4	0.25g

Preparation of Medium: Add components to distilled/deionized water and bring volume to 1.0L. Mix thoroughly. Distribute into tubes or flasks. Autoclave for 15 min at 15 psi pressure–121°C.

Use: For the cultivation of myxobacteria.

DNase Agar

Composition per liter:

Tryptose	20.0g
Agar	12.0g
NaCl	5.0g
Deoxyribonucleic acid	2.0g

pH 7.3 ± 0.2 at 25°C

Source: This medium is available as a premixed powder from Oxoid Unipath.

Preparation of Medium: Add components to distilled/deionized water and bring volume to 1.0L. Mix thoroughly. Gently heat and bring to boiling. Distribute into tubes or flasks. Autoclave for 15 min at 15 psi pressure–121°C. Pour into sterile Petri dishes or leave in tubes.

Use: For the differentiation of microorganisms, especially *Staphylococcus* species and *Serratia marcescens*, based on their production of deoxyribonuclease.

DNase Medium

Composition per liter:

Agar	15.0g
Pancreatic digest of casein	10.0g
Peptic digest of animal tissue	10.0g
L-Arabinose	10.0g
NaCl	5.0g
Deoxyribonucleic acid	2.0g
Methyl Green	0.09g
Phenol Red	0.05g
Antibiotic solution	10.0mL

pH 7.3 ± 0.2 at 25°C

Antibiotic Solution:

Composition per 10mL:

Cephalothin	0.01g
Ampicillin	5.0mg

Colistimethate	5.0mg
Amphotericin B	2.5mg

Preparation of Antibiotic Solution: Add components to distilled/deionized water and bring volume to 10.0mL. Mix thoroughly. Filter sterilize.

Preparation of Medium: Add components, except antibiotic solution, to distilled/deionized water and bring volume to 990.0mL. Mix thoroughly. Gently heat and bring to boiling. Autoclave for 15 min at 15 psi pressure–121°C. Cool to 45°–50°C. Aseptically add sterile components. Mix thoroughly. Pour into sterile Petri dishes or distribute into sterile tubes.

Use: For the isolation and cultivation of *Serratia marcescens*.

DNase Test Agar

Composition per liter:

Agar	15.0g
Pancreatic digest of casein	15.0g
NaCl	5.0g
Papaic digest of soybean meal	5.0g
Deoxyribonucleic acid	2.0g

pH 7.3 ± 0.2 at 25°C

Source: This medium is available as a premixed powder from BBL Microbiology Systems and Difco Laboratories.

Preparation of Medium: Add components to distilled/deionized water and bring volume to 1.0L. Mix thoroughly. Gently heat while stirring and bring to boiling. Distribute into tubes or flasks. Autoclave for 15 min at 13 psi pressure–118°C. Pour into sterile Petri dishes or leave in tubes.

Use: For the differentiation of microorganisms, especially *Staphylococcus* species and *Serratia marcescens*, based on their production of deoxyribonuclease.

DNase Test Agar with Methyl Green

Composition per liter:

Agar	15.0g
Pancreatic digest of casein	10.0g
Peptic digest of animal tissue	10.0g
NaCl	5.0g
Deoxyribonucleic acid	2.0g
Methyl Green	0.05g

pH 7.3 ± 0.2 at 25°C

Source: This medium is available as a premixed powder from Difco Laboratories.

Preparation of Medium: Add components to distilled/deionized water and bring volume to 1.0L. Mix thoroughly. Gently heat while stirring and bring to boiling. Distribute into tubes or flasks. Autoclave for 15 min at 13 psi pressure–118°C. Pour into sterile Petri dishes or leave in tubes.

Use: For the differentiation of microorganisms, especially *Staphylococcus* species and *Serratia marcescens*, based on their production of deoxyribonuclease.

DNase Test Agar with Toluidine Blue

Composition per liter:

Agar	15.0g
Pancreatic digest of casein	10.0g
Peptic digest of animal tissue	10.0g
NaCl	5.0g
Deoxyribonucleic acid	2.0g
Toluidine Blue	0.1g

pH 7.3 ± 0.2

Preparation of Medium: Add components to distilled/deionized water and bring volume to 1.0L. Mix thoroughly. Gently heat while stirring and bring to boiling. Distribute into tubes or flasks. Autoclave for 15 min at 13 psi pressure–118°C. Pour into sterile Petri dishes or leave in tubes.

Use: For the differentiation of microorganisms, especially *Staphylococcus* species and *Serratia marcescens*, based on their production of deoxyribonuclease.

DNB Medium

Composition per liter:

Nutrient broth	2.4g
Yeast extract	1.5g

Preparation of Medium: Add components to distilled/deionized water and bring volume to 1.0L. Mix thoroughly. Distribute into tubes or flasks. Autoclave for 15 min at 15 psi pressure–121°C.

Use: For the cultivation of *Bdellovibrio bacteriovorus* and ATCC strain 43826.

Dorset Egg Medium

Composition per liter:

Homogenized whole egg	950.0mL
Glycerol	50.0mL

pH 6.8–7.4 at 25°C

Source: This medium is available as a prepared medium from Difco Laboratories.

Homogenized Whole Egg:
Composition per liter:

Whole eggs	18–24

Preparation of Homogenized Whole Egg: Use fresh eggs, less than 1 week old. Scrub the shells with soap. Let stand in a soap solution for 30 min. Rinse in running water. Soak eggs in 70% ethanol for 15 min. Break the eggs into a sterile container. Homogenize by shaking. Filter through four layers of sterile cheesecloth into a sterile graduated cylinder. Measure out 1.0L.

Preparation of Medium: Filter sterilize glycerol. Combine glycerol and homogenized whole egg. Mix thoroughly. Distribute into sterile screw-capped tubes. Place tubes in a slanted position. Inspissate at 85°C (moist heat) for 45 min.

Use: For the maintenance of *Mycobacterium* species.

Double–Strength Crude Medium for *Lactobacillus*

Composition per liter:

Sucrose	20.0g
Yeast extract	20.0g
Casein hydrolysate	15.0g
Potassium acetate	3.0g
Histidine·HCl	2.0g
Ascorbic acid	1.0g
Pyridoxamine·HCl	33.0μg
Salts A	20.0mL
Salts B	5.0mL

pH 5.4 ± 0.2 at 25°C

Salts A:
Composition per 100mL:

$KH_2PO_4 \cdot H_2O$	16.5g
$K_2HPO_4 \cdot 3H_2O$	16.5g

Preparation of Salts A: Add components to distilled/deionized water and bring volume to 100.0mL. Mix thoroughly.

Salts B:
Composition per 100mL:

$MgSO_4 \cdot 7H_2O$	8.0g
$FeSO_4 \cdot 7H_2O$	0.4g
$MnSO_4 \cdot H2O$	0.4g
NaCl	0.4g
HCl, concentrated	0.1mL

Preparation of Salts B: Add components to distilled/deionized water and bring volume to 100.0mL. Mix thoroughly.

Preparation of Medium: Add components to distilled/deionized water and bring volume to 1.0L. Mix thoroughly. Gently heat and bring to boiling. Adjust pH to 5.4 with acetic acid. Distribute into tubes or flasks. Autoclave for 15 min at 15 psi pressure–121°C.

Use: For the cultivation and maintenance of *Lactobacillus* species.

Doyle and Roman Enrichment Medium

Composition per liter:

Pancreatic digest of casein	10.0g
Peptic digest of animal tissue	10.0g
NaCl	5.0g
Sodium succinate	3.0g
Yeast extract	2.0g
Glucose	1.0g
$NaHSO_3$	0.1g
Cysteine·HCl·H_2O	0.1g
Horse blood, lysed	70.0mL
Antibiotic solution	10.0mL

pH 7.0 ± 0.2 at 25°C

Antibiotic Solution:
Composition per 10mL:

Cycloheximide	0.05g
Vancomycin	0.015g
Trimethoprim lactate	5.0mg
Polymyxin B	200,000U

Preparation of Antibiotic Solution: Add components to distilled/deionized water and bring volume to 10.0mL. Mix thoroughly. Filter sterilize.

Preparation of Medium: Add components, except antibiotic solution and horse blood, to distilled/deionized water and bring volume to 920.0mL. Mix thoroughly. Gently heat and bring to boiling. Autoclave for 15 min at 15 psi pressure–121°C. Cool to 45°–50°C. Aseptically add sterile antibiotic solution and horse blood. Mix thoroughly. Aseptically distribute into sterile flasks in 90.0–100.00mL volumes.

Use: For the cultivation and enrichment of *Campylobacter* species from foods.

DRBC Agar (Dichloran Rose Bengal Chloramphenicol Agar)

Composition per liter:

Agar	15.0g
Glucose	10.0g
Peptone	5.0g
KH_2PO_4	1.0g
$MgSO_4 \cdot 7H_2O$	0.5g
Rose Bengal	0.025g

Dichloran...0.002g
Chloramphenicol solution............................10.0mL

pH 5.6 ± 0.2 at 25°C

Source: This medium is available as a premixed powder from Oxoid Unipath.

Chloramphenicol Solution:
Composition per 10mL:

Chloramphenicol...0.1g

Preparation of Chloramphenicol Solution: Add chloramphenicol to distilled/deionized water and bring volume to 10.0mL. Mix thoroughly. Filter sterilize.

Preparation of Medium: Add components, except chloramphenicol solution, to distilled/deionized water and bring volume to 990.0mL. Mix thoroughly. Gently heat and bring to boiling. Autoclave for 15 min at 15 psi pressure–121°C. Cool to 45°–50°C. Aseptically add sterile chloramphenicol solution. Mix thoroughly. Pour into sterile Petri dishes or distribute into sterile tubes.

Use: For the isolation, cultivation and enumeration of yeasts and molds associated with food spoilage.

DS Sporulation Medium, Modified
See: Duncan–Strong Sporulation Medium, Modified

DSM 134
See: ***Haliscomenobacter*** Medium

DST Agar
See: Diagnostic Sensitivity Test Agar

DTC Agar

Composition per liter:

Agar...20.0g
Pancreatic digest of casein..............................15.0g
NaCl..5.0g
Papaic digest of soybean meal..........................5.0g
Deoxyribonucleic acid.....................................2.0g
Toluidine Blue O...0.1g
Cephalothin solution...................................10.0mL

Cephalothin Solution:
Composition per 10mL:

Cephalothin...1.0g

Preparation of Cephalothin Solution: Add cephalothin to distilled/deionized water and bring volume to 10.0mL. Mix thoroughly. Filter sterilize.

Preparation of Medium: Add components, except cephalothin solution, to distilled/deionized water and bring volume to 990.0mL. Mix thoroughly. Gently heat and bring to boiling. Autoclave for 15 min at 15 psi pressure–121°C. Cool to 45°–50°C. Aseptically add sterile cephalothin solution. Mix thoroughly. Pour into sterile Petri dishes.

Use: For the isolation and cultivation of *Serratia* species. *Serratia* appear as colonies with red halos.

DTM Agar (Dermatophyte Test Medium Agar)

Composition per liter:

Agar...20.0g
Enzymatic digest of soybean meal....................10.0g
Glucose..10.0g
Cycloheximide...0.5g
Phenol Red...0.2g
Chlortetracycline...0.1g
Gentamicin...0.1g

pH 7.3 ± 0.2 at 25°C

Source: Available as a prepared medium from Difco Laboratories.

Preparation of Medium: Add components to distilled/deionized water and bring volume to 1.0L. Mix thoroughly. Gently heat and bring to boiling. Distribute into tubes or flasks. Autoclave for 15 min at 15 psi pressure–121°C. Pour into sterile Petri dishes or leave in tubes.

Use: For the isolation and cultivation of dermatophytic fungi.

Dubos Agar with Filter Paper

Composition per liter:

Agar...15.0g
K_2HPO_4...1.0g
KCl...0.5g
$MgSO_4·7H_2O$...0.5g
$NaNO_3$...0.5g
$FeSO_4·7H_2O$...0.01g

pH 7.2 ± 0.2 at 25°C

Preparation of Medium: Add components to distilled/deionized water and bring volume to 1.0L. Mix

thoroughly. Gently heat and bring to boiling. Adjust pH to 7.2. Autoclave for 15 min at 15 psi pressure–121°C. Pour into sterile Petri dishes. Lay sterile strips of Whatman #1 filter paper on the surface of the agar.

Use: For the cultivation and maintenance of *Cytophaga hutchinsonii*.

Dubos Broth

Composition per liter:

Na_2HPO_4	2.5g
L-Asparagine	2.0g
KH_2PO_4	1.0g
Pancreatic digest of casein	0.5g
Tween™ 80	0.2g
$CaCl_2 \cdot 2H_2O$	0.5mg
$CuSO_4$	0.1mg
$ZnSO_4 \cdot 7H_2O$	0.1mg
Ferric ammonium citrate	0.05g
$MgSO_4 \cdot 7H_2O$	0.01g
Bovine serum albumin or bovine serum	20.0mL

pH 6.5 ± 0.2

Source: This medium is available as a premixed powder from BBL Microbiology Systems and Difco Laboratories.

Preparation of Medium: Add components, except bovine serum or bovine serum albumin, to distilled/deionized water and bring volume to 980.0mL. Mix thoroughly. Gently heat and bring to boiling. Autoclave for 15 min at 15 psi pressure–121°C. Cool to 45°–50°C. Aseptically add sterile bovine serum or bovine serum albumin. Mix thoroughly. Aseptically distribute into sterile tubes.

Use: For the cultivation of *Mycobacterium tuberculosis* and other *Mycobacterium* species.

Dubos Broth with Horse Serum

Composition per liter:

Na_2HPO_4	2.5g
L-Asparagine	2.0g
KH_2PO_4	1.0g
Pancreatic digest of casein	0.5g
Tween™ 80	0.2g
$CaCl_2 \cdot 2H_2O$	0.5mg
$CuSO_4$	0.1mg
$ZnSO_4 \cdot 7H_2O$	0.1mg
Ferric ammonium citrate	0.05g
$MgSO_4 \cdot 7H_2O$	0.01g
Horse serum	50.0mL

pH 6.5 ± 0.2 at 25°C

Preparation of Medium: Add components, except horse serum, to distilled/deionized water and bring volume to 950.0mL. Mix thoroughly. Gently heat and bring to boiling. Autoclave for 15 min at 15 psi pressure–121°C. Cool to 45°–50°C. Aseptically add sterile horse serum. Mix thoroughly. Aseptically distribute into sterile tubes.

Use: For the cultivation and maintenance of the *Corynebacterium* species.

Dubos Mineral Medium

Composition per liter:

K_2HPO_4	1.0g
KCl	0.5g
$MgSO_4 \cdot 7H_2O$	0.5g
$NaNO_3$	0.5g
$FeSO_4 \cdot 7H_2O$	0.01g

pH 7.2 ± 0.2 at 25°C

Preparation of Medium: Add components to distilled/deionized water and bring volume to 1.0L. Mix thoroughly. Distribute into tubes or flasks. Autoclave for 15 min at 15 psi pressure–121°C.

Use: For the isolation and cultivation of *Cytophaga* species, *Herpetosiphon* species, *Saprospira* species, and *Flexithrix* species.

Dubos Oleic Agar

Composition per liter:

Agar	15.0g
Na_2HPO_4	2.5g
KH_2PO_4	1.0g
L-Asparagine	1.0g
Pancreatic digest of casein	0.5g
Ferric ammonium citrate	0.05g
$MgSO_4 \cdot 7H_2O$	0.01g
$CaCl_2 \cdot 2H_2O$	0.5mg
$CuSO_4$	0.1mg
$ZnSO_4 \cdot 7H_2O$	0.1mg
Dubos oleic albumin complex	20.0mL
Penicillin solution	10.0mL

pH 6.6 ± 0.2 at 25°C

Source: This medium is available as a premixed powder from Difco Laboratories.

Dubos Oleic Albumin Complex:

Composition per 100mL:

Bovine serum albumin, fraction V	5.0g
Oleic acid, sodium salt	0.05g
NaCl (0.85% solution)	100.0mL

Preparation of Dubos Oleic Albumin Complex: Add bovine serum albumin and oleic acid to 100.0mL of NaCl solution. Mix thoroughly. Filter sterilize.

Penicillin Solution:
Composition per 10mL:

Penicillin	10,000U

Preparation of Penicillin Solution: Add penicillin to distilled/deionized water and bring volume to 10.0mL. Mix thoroughly. Filter sterilize.

Preparation of Medium: Add components, except Dubos oleic albumin complex and penicillin solution, to distilled/deionized water and bring volume to 970.0mL. Mix thoroughly. Gently heat and bring to boiling. Autoclave for 15 min at 15 psi pressure–121°C. Cool to 45°–50°C. Aseptically add sterile Dubos oleic albumin complex and penicillin solution. Mix thoroughly. Pour into sterile Petri dishes or distribute into sterile tubes. Allow tubes to cool in a slanted position.

Use: For the isolation of *Mycobacterium tuberculosis* and determining its sensitivity to chemotherapeutic agents.

Ducreyi Medium, Revised
See: ***Haemophilus ducreyi* Medium, Revised**

Dulaney Slants

Composition per liter:

Egg yolks	50.0mL
Locke solution	50.0mL

Locke Solution:
Composition per 100.0mL:

NaCl	0.9g
Glucose	0.25g
$CaCl_2 \cdot 2H_2O$	0.024g
KCl	0.042g
Na_2CO_3	0.020g

Preparation of Locke Solution: Add components to distilled/deionized water and bring volume to 100.0mL. Mix thoroughly. Filter sterilize.

Preparation of Medium: Aseptically remove the yolks from 5–8 day old hen egg embryos. Add an equal volume of sterile Locke solution containing sterile glass beads. Mix thoroughly to homogenize. Aseptically distribute into sterile tubes. Inspissate tubes in a slanted position at 80°C (moist heat) for 15 min.

Use: For the cultivation of *Calymmatobacter granulomatis* from clinical specimens.

Duncan–Strong Sporulation Medium, Modified (DS Sporulation Medium, Modified) (Sporulation Medium, Modified)

Composition per liter:

Proteose peptone	15.0g
$Na_2HPO_4 \cdot 7H_2O$	10.0g
Raffinose	4.0g
Yeast extract	4.0g
Sodium thioglycollate	1.0g

pH 7.8 ± 0.2 at 25°C

Preparation of Medium: Add components to distilled/deionized water and bring volume to 1.0L. Mix thoroughly. Gently heat and bring to boiling. Distribute into tubes or flasks. Autoclave for 15 min at 15 psi pressure–121°C. Adjust pH to 7.8 with filter-sterilized 0.66M Na_2CO_3. Pour into sterile Petri dishes or leave in tubes.

Use: For the cultivation and induction of sporulation of *Clostridium perfringens*.

Dunkelberg Agar
See: **Peptone Starch Dextrose Agar**

Dunkelberg Carbohydrate Medium, Modified

Composition per 100mL:

Proteose peptone No. 3	1.5g
Carbohydrate	1.0g
$Na_2HPO_4 \cdot 2H_2O$	0.207g
Phenol Red	0.055g
$NaH_2PO_4 \cdot H_2O$	0.038g
Horse serum	5.0mL

pH 7.4 ± 0.2 at 25°C

Preparation of Medium: Add components, except horse serum, to distilled/deionized water and bring volue to 95.0mL. For carbohydrate use glucose, maltose or starch. Mix thoroughly. Filter sterilize. Aseptically add sterile horse serum. Mix thoroughly. Aseptically distribute into sterile tubes or flasks.

Use: For the cultivation and differentiation of *Gardnerella vaginalis* based on their ability to ferment glucose, maltose or starch.

Dunkelberg Maintenance Medium

Composition per liter:

Proteose peptone No. 3	20.0g
Soluble starch	10.0g
Agar	8.0g
Glucose	2.0g
Na_2HPO_4	1.0g
NaH_2PO_4	1.0g

pH 6.8 ± 0.2 at 25°C

Preparation of Medium: Add starch to approximately 100.0mL of cold distilled/deionized water. Mix thoroughly. Add starch solution to 400.0mL of boiling distilled/deionized water. Add remaining components. Mix thoroughly. Bring volume to 1.0L with distilled/deionized water. Distribute into screw-capped tubes. Autoclave for 12 min at 8 psi pressure–112°C.

Use: For the cultivation and maintenance of *Gardnerella vaginalis*.

Dunkelberg Semisolid Carbohydrate Fermentation Medium

Composition per liter:

Proteose peptone No. 3	20.0g
Agar	5.0g
Carbohydrate	10.0g
Bromcresol Purple solution	1.0mL

pH 7.4 ± 0.2 at 25°C

Bromcresol Purple Solution:
Composition per 10mL:

Bromcresol Purple	0.16g
Ethanol (95% solution)	10.0mL

Preparation of Bromcresol Purple Solution: Add bromcresol purple to 10.0mL of ethanol. Mix thoroughly. Filter sterilize.

Preparation of Medium: Add components to distilled/deionized water and bring volue to 1.0L. For carbohydrate use glucose, maltose or starch. Mix thoroughly. Gently heat and bring to boiling. Filter sterilize. Aseptically distribute into sterile tubes or flasks.

Use: For the cultivation and differentiation of *Gardnerella vaginalis* based on their ability to ferment glucose, maltose or starch.

E Agar
(m–E Agar)

Composition per liter:

Yeast extract	30.0g
Agar	15.0g
NaCl	15.0g
Pancreatic digest of gelatin	10.0g
Esculin	1.0g
Nalidixic acid	0.25g
NaN_3	0.15g
Cycloheximide	0.05g
TTC solution	15.0mL

pH 7.1 ± 0.2 at 25°C

Source: This medium is available as a premixed powder from BBL Microbiology Systems and Difco Laboratories.

TTC Solution:

Composition per 15mL:

2,3,5-Triphenyltetrazolium chloride	0.15g

Preparation of TTC Solution: Add triphenyltetrazolium chloride to distilled/deionized water and bring volume to 15.0mL. Mix thoroughly. Filter sterilize.

Preparation of Medium: Add components, except TTC solution, to distilled/deionized water and bring volume to 1.0L. Mix thoroughly. Gently heat and bring to boiling. Autoclave for 15 min at 15 psi pressure–121°C. Cool to 45°–50°C. Aseptically add sterile TTC solution. Mix thoroughly. Pour into sterile Petri dishes or distribute into sterile tubes.

Caution: Sodium azide is toxic. Azides also react with metals and disposal must be highly diluted.

Use: For the isolation, cultivation and enumeration of enterococci in water by the membrane filter method. It is used in conjunction with Esculin Iron Agar.

E Medium for Anaerobes

Composition per 100mL:

Glucose	0.05g
L-Cysteine·HCl·H_2O	0.05g
Maltose	0.05g
$(NH_4)_2SO_4$	0.05g
Peptone	0.05g
Soluble starch	0.05g
Yeast extract	0.05g
Salts solution	50.0mL
Rumen fluid	30.0mL
Resazurin solution	0.4mL

pH 7.0 ± 0.2 at 25°C

Salts Solution:

Composition per liter:

$NaHCO_3$	10.0g
NaCl	2.0g
K_2HPO_4	1.0g
KH_2PO_4	1.0g
$CaCl_2$, anhydrous	0.2g
$MgSO_4$	0.2g

Preparation of Salts Solution: Add $CaCl_2$ and $MgSO_4$ to approximately 300.0mL of distilled/deionized water. Mix thoroughly. Bring volume to 800.0mL with distilled/deionized water. Add remaining components. Mix thoroughly. Bring volume to 1.0L with distilled/deionized water. Mix thoroughly. Store at 4°C.

Rumen Fluid:

Composition per 100mL:

Rumen fluid	100.0mL

Preparation of Rumen Fluid: Obtain the rumen contents from a cow that has been fed an alfalfa-hay ration. Filter rumen contents through two layers of cheesecloth. Store under 100% CO_2 in the refrigerator. The particulate material will settle out. Use only the supernatant liquid.

Resazurin Solution:

Composition per 44mL:

Resazurin	0.011g

Preparation of Resazurin Solution: Add resazurin to distilled/deionized water and bring volume to 44.0mL. Mix thoroughly.

Preparation of Medium: Add components, except cysteine·HCl·H_2O, to distilled/deionized water and bring volume to 100.0mL. Mix thoroughly. Gently heat and bring to boiling. Continue boiling until resazurin turns colorless indicating reduction. Cool in an ice-water bath under 100% CO_2. Add the cysteine·HCl·H_2O. Adjust pH to 7.0 with 8*N* NaOH or 5*N* HCl. Anaerobically distribute into tubes under O_2-free 100% N_2. Cap tubes with butyl rubber stoppers. Place tubes in a press. Autoclave for 12 min at 15 psi pressure–121°C with fast exhaust.

Use: For the cultivation and maintenance of *Bacteroides ruminicola, Bacteroides succinogenes, Butyrivibrio fibrisolvens, Clostridium methylpentosum, Eubacterium ruminantium, Lachnospira multipara, Micromonospora ruminantium, Prevotella ruminicola, Propionibacterium acidipropionici, Selenomonas ruminantium, Selenomonas suis, Succinivibrio dextrinosolvens, Treponema bryantii,* and *Treponema succinifaciens,*

E Medium for Anaerobes with 0.1% Cellobiose

Composition per 100mL:

Cellobiose	0.1g
Glucose	0.05g
L-Cysteine·HCl·H_2O	0.05g
Maltose	0.05g
$(NH_4)_2SO_4$	0.05g
Peptone	0.05g
Soluble starch	0.05g
Yeast extract	0.05g
Salts solution	50.0mL
Rumen fluid	30.0mL
Resazurin solution	0.4mL

pH 7.0 ± 0.2 at 25°C

Salts Solution:

Composition per liter:

$NaHCO_3$	10.0g
NaCl	2.0g
K_2HPO_4	1.0g
KH_2PO_4	1.0g
$CaCl_2$, anhydrous	0.2g
$MgSO_4$	0.2g

Preparation of Salts Solution: Add $CaCl_2$ and $MgSO_4$ to approximately 300.0mL of distilled/deionized water. Mix thoroughly. Bring volume to 800.0mL with distilled/deionized water. Add remaining components. Mix thoroughly. Bring volume to 1.0L with distilled/deionized water. Mix thoroughly. Store at 4°C.

Rumen Fluid:

Composition per 100mL:

Rumen fluid	100.0mL

Preparation of Rumen Fluid: Obtain the rumen contents from a cow that has been fed an alfalfa-hay ration. Filter rumen contents through two layers of cheesecloth. Store under 100% CO_2 in the refrigerator. The particulate material will settle out. Use only the supernatant liquid.

Resazurin Solution:

Composition per 44mL:

Resazurin	0.011g

Preparation of Resazurin Solution: Add resazurin to distilled/deionized water and bring volume to 44.0mL. Mix thoroughly.

Preparation of Medium: Add components, except cysteine·HCl·H_2O, to distilled/deionized water and bring volume to 100.0mL. Mix thoroughly. Gently heat and bring to boiling. Continue boiling until resazurin turns colorless indicating reduction. Cool in an ice-water bath under 100% CO_2. Add the cysteine·HCl·H_2O. Adjust pH to 7.0 with 8*N* NaOH or 5*N* HCl. Anaerobically distribute into tubes under O_2-free 100% N_2. Cap tubes with butyl rubber stoppers. Place tubes in a press. Autoclave for 12 min at 15 psi pressure–121°C with fast exhaust.

Use: For the cultivation and maintenance of *Eubacterium cellulosolvens* and *Fibrobacter inyrdyinslid.*

E Medium for Anaerobes with Filtered Rumen Fluid and 0.1% Cellobiose

Composition per 100mL:

Cellobiose	0.1g
Glucose	0.05g
L-Cysteine·HCl·H_2O	0.05g
Maltose	0.05g
$(NH_4)_2SO_4$	0.05g
Peptone	0.05g
Soluble starch	0.05g
Yeast extract	0.05g
Salts solution	50.0mL
Rumen fluid, filtered	30.0mL
Resazurin solution	0.4mL

pH 7.0 ± 0.2 at 25°C

Salts Solution:

Composition per liter:

$NaHCO_3$	10.0g
NaCl	2.0g
K_2HPO_4	1.0g
KH_2PO_4	1.0g
$CaCl_2$ (anhydrous)	0.2g
$MgSO_4$	0.2g

Preparation of Salts Solution: Add $CaCl_2$ and $MgSO_4$ to approximately 300.0mL of distilled/deionized water. Mix thoroughly. Bring volume to 800.0mL with distilled/deionized water. Add remaining components. Mix thoroughly. Bring volume to 1.0L with distilled/deionized water. Mix thoroughly. Store at 4°C.

Rumen Fluid:

Composition per 100mL:

Rumen fluid	100.0mL

Preparation of Rumen Fluid: Obtain the rumen contents from a cow that has been fed an alfalfa-hay ration. Filter rumen contents through two layers of cheesecloth. Store under 100% CO_2 in the refrigerator. The particulate material will settle out. Use only the supernatant liquid. Filter through a 0.20 μm filter.

Resazurin Solution:

Composition per 44mL:

Resazurin	0.011g

Preparation of Resazurin Solution: Add resazurin to distilled/deionized water and bring volume to 44.0mL. Mix thoroughly.

Preparation of Medium: Add components, except cysteine·HCl·H_2O, to distilled/deionized water and bring volume to 100.0mL. Mix thoroughly. Gently heat and bring to boiling. Continue boiling until resazurin turns colorless indicating reduction. Cool in an ice-water bath under 100% CO_2. Add the cysteine·HCl·H_2O. Adjust pH to 7.0 with 8*N* NaOH or 5*N* HCl. Anaerobically distribute into tubes under O_2-free 100% N_2. Cap tubes with butyl rubber stoppers. Place tubes in a press. Autoclave for 12 min at 15 psi pressure–121°C with fast exhaust.

Use: For the cultivation and maintenance of the *Fibrobacter* species.

E Medium for Anaerobes with 0.3% Phloroglucinol

Composition per 110.4mL:

$(NH_4)_2SO_4$	0.5g
L-Cysteine·HCl·H_2O	0.05g
Soluble starch	0.05g
Salts solution	50.0mL
Rumen fluid	30.0mL
Phloroglucinol solution	30.0mL
Resazurin solution	0.4mL

pH 6.6 ± 0.2 at 25°C

Salts Solution:
Composition per liter:

$NaHCO_3$	10.0g
NaCl	2.0g
K_2HPO_4	1.0g
KH_2PO_4	1.0g
$CaCl_2$, anhydrous	0.2g
$MgSO_4$	0.2g

Preparation of Salts Solution: Add $CaCl_2$ and $MgSO_4$ to approximately 300.0mL of distilled/deionized water. Mix thoroughly. Bring volume to 800.0mL with distilled/deionized water. Add remaining components. Mix thoroughly. Bring volume to 1.0L with distilled/deionized water. Mix thoroughly. Store at 4°C.

Rumen Fluid:
Composition per 100mL:

Rumen fluid	100.0mL

Preparation of Rumen Fluid: Obtain the rumen contents from a cow that has been fed an alfalfa-hay ration. Filter rumen contents through two layers of cheesecloth. Store under 100% CO_2 in the refrigerator. The particulate material will settle out. Use only the supernatant liquid.

Phloroglucinol Solution:
Composition per 100mL:

Phloroglucinol	1.0g

Preparation of Phloroglucinol Solution: Add phloroglucinol to distilled/deionized water and bring volume to 100.0mL. Mix thoroughly. Filter sterilize. Keep away from light.

Resazurin Solution:
Composition per 44mL:

Resazurin	0.011g

Preparation of Resazurin Solution: Add resazurin to distilled/deionized water and bring volume to 44.0mL. Mix thoroughly.

Preparation of Medium: Add components, except cysteine·HCl·H_2O, to distilled/deionized water and bring volume to 100.0mL. Mix thoroughly. Gently heat and bring to boiling. Continue boiling until resazurin turns colorless indicating reduction. Cool in an ice-water bath under 100% CO_2. Add the cysteine·HCl·H_2O. Adjust pH to 6.6 with 8*N* NaOH or 5*N* HCl. Anaerobically distribute into tubes under O_2-free 100% N_2. Cap tubes with butyl rubber stoppers. Place tubes in a press. Autoclave for 12 min at 15 psi pressure–121°C with fast exhaust.

Use: For the cultivation and maintenance of *Coprococcus* species.

E Medium for Anaerobes with 0.2% Rutin

Composition per 110.4mL:

$(NH_4)_2SO_4$	0.5g
L-Cysteine·HCl·H_2O	0.05g
Soluble starch	0.05g
Salts solution	50.0mL
Rumen fluid	30.0mL
Rutin solution	30.0mL
Resazurin solution	0.4mL

pH 6.6 ± 0.2 at 25°C

Salts Solution:
Composition per liter:

$NaHCO_3$	10.0g
NaCl	2.0g
K_2HPO_4	1.0g
KH_2PO_4	1.0g
$CaCl_2$, anhydrous	0.2g
$MgSO_4$	0.2g

Preparation of Salts Solution: Add $CaCl_2$ and $MgSO_4$ to approximately 300.0mL of distilled/deionized water. Mix thoroughly. Bring volume to 800.0mL with distilled/deionized water. Add remaining components. Mix thoroughly. Bring volume to

1.0L with distilled/deionized water. Mix thoroughly. Store at 4°C.

Rumen Fluid:
Composition per 100mL:

Rumen fluid	100.0mL

Preparation of Rumen Fluid: Obtain the rumen contents from a cow that has been fed an alfalfa-hay ration. Filter rumen contents through two layers of cheesecloth. Store under 100% CO_2 at 4°C. The particulate material will settle out. Use the liquid.

Rutin Solution:
Composition per 100mL:

Rutin	0.2g

Preparation of Rutin Solution: Add rutin to distilled/deionized water and bring volume to 100.0mL. Mix thoroughly. Filter sterilize.

Resazurin Solution:
Composition per 44mL:

Resazurin	0.011g

Preparation of Resazurin Solution: Add resazurin to distilled/deionized water and bring volume to 44.0mL. Mix thoroughly.

Preparation of Medium: Add components, except cysteine·HCl·H_2O, to distilled/deionized water and bring volume to 100.0mL. Mix thoroughly. Gently heat and bring to boiling. Continue boiling until resazurin turns colorless indicating reduction. Cool in an ice-water bath under 100% CO_2. Add the cysteine·HCl·H_2O. Adjust pH to 6.6 with 8*N* NaOH or 5*N* HCl. Anaerobically distribute into tubes under O_2-free 100% N_2. Cap tubes with butyl rubber stoppers. Place tubes in a press. Autoclave for 12 min at 15 psi pressure–121°C with fast exhaust.

Use: For the cultivation of *Butyrivibrio* species.

E Medium for Anaerobes, Modified

Composition per 103.6mL:

L-Cysteine·HCl·H_2O	0.05g
$(NH_4)_2SO_4$	0.05g
Peptone	0.05g
Yeast extract	0.05g
Salts solution	50.0mL
Rumen fluid	30.0mL
Potassium phosphate buffer (1M, pH 6.5)	2.8mL
Hemin solution	1.0mL
Glucose-maltose solution	1.4mL
Starch solution	1.4mL
Resazurin (0.025% solution)	0.4mL
Vitamin K_3 solution	0.2mL

pH 6.5 ± 0.2 at 25°C

Salts Solution:
Composition per liter:

$NaHCO_3$	10.0g
NaCl	2.0g
K_2HPO_4	1.0g
KH_2PO_4	1.0g
$CaCl_2$, anhydrous	0.2g
$MgSO_4$	0.2g

Preparation of Salts Solution: Add $CaCl_2$ and $MgSO_4$ to approximately 300.0mL of distilled/deionized water. Mix thoroughly. Bring volume to 800.0mL with distilled/deionized water. Add remaining components. Mix thoroughly. Bring volume to 1.0L with distilled/deionized water. Mix thoroughly. Store at 4°C.

Rumen Fluid:
Composition per 100mL:

Rumen fluid	100.0mL

Preparation of Rumen Fluid: Obtain the rumen contents from a cow that has been fed an alfalfa-hay ration. Filter rumen contents through two layers of cheesecloth. Store under 100% CO_2 in the refrigerator. The particulate material will settle out. Use only the supernatant liquid.

Hemin Solution:
Composition per 100mL:

Hemin	0.050g
NaOH (1*N* solution)	1.0mL

Preparation of Hemin Solution: Add hemin to 1.0mL of 1*N* NaOH solution. Mix thoroughly. Bring volume to 100.0mL with distilled/deionized water. Autoclave for 15 min at 15 psi pressure–121°C. Cool to 45°–50°C.

Glucose-Maltose Solution:
Composition per 10mL:

Glucose	0.5g
Maltose	0.5g

Preparation of Glucose-Maltose Solution: Add components to distilled/deionized water and bring volume to 10.0mL. Mix thoroughly. Filter sterilize.

Starch Solution:
Composition per 10mL:

Starch, soluble	0.5g

Preparation of Starch Solution: Add starch to distilled/deionized water and bring volume to 10.0mL. Mix thoroughly. Autoclave for 15 min at 15 psi pressure–121°C.

Resazurin Solution:
Composition per 44mL:

Resazurin	0.011g

Preparation of Resazurin Solution: Add resazurin to distilled/deionized water and bring volume to 44.0mL. Mix thoroughly.

Vitamin K_3 Solution:
Composition per 25mL:

Vitamin K_3 (menadione)	0.0125g
Ethanol, absolute	25.0mL

Preparation of Vitamin K_1 Solution: Add vitamin K_3 to 99.0mL of ethanol. Mix thoroughly.

Preparation of Medium: Filter sterilize potassium phosphate buffer. Add components—except cysteine·HCl·H_2O, vitamin K_3 solution, potassium phosphate buffer, glucose-maltose solution, and starch solution—to distilled/deionized water and bring volume to 98.0mL. Mix thoroughly. Gently heat and bring to boiling. Continue boiling until resazurin turns colorless indicating reduction. Cool in an ice-water bath under O_2-free 97% N_2 + 3% H_2. Add the cysteine·HCl·H_2O and vitamin K_3 solution. Mix thoroughly. Adjust pH to 6.5 with 8*N* NaOH or 5*N* HCl. Anaerobically distribute into tubes under O_2-free 97% N_2 + 3% H_2 in 7.0mL volumes. Cap tubes with butyl rubber stoppers. Place tubes in a press. Autoclave for 12 min at 15 psi pressure–121°C with fast exhaust. Immediately prior to inoculation, aseptically add 0.2mL of filter-sterilized potassium phosphate buffer, 0.1mL of sterile glucose-maltose solution, and 0.1mL of sterile starch solution to each tube. Mix thoroughly.

Use: For the cultivation and maintenance of *Bacteroides* species, *Butyrivibrio fibrisolvens, Clostridium methylpentosum, Eubacterium ruminantium, Lachnospira multipara, Micromonospora ruminantium, Prevotella ruminicola, Propionibacterium acidipropionici, Selenomonas* species, *Succinivibrio dextrinosolvens,* and *Treponema* species.

Eagle Medium

Composition per 99.1mL:

Eagle MEM in Hanks BSS	87.0mL
Fetal bovine serum	10.0mL
$NaHCO_3$ (7.5% solution)	1.0mL
Penicillin-streptomycin solution	1.0mL
Amphotericin B solution	0.1mL

pH 7.2–7.4 at 25°C

Eagle MEM in Hanks BSS:
Composition per liter:

NaCl	8.0g
Glucose	1.0g
KCl	0.4g
$CaCl_2 \cdot 2H_2O$	0.14g
$MgSO_4 \cdot 7H_2O$	0.1g
KH_2PO_4	0.06g
Na_2HPO_4	0.05g
L-Isoleucine	0.026g
L-Leucine	0.026g
L-Lysine	0.026g
L-Threonine	0.024g
L-Valine	0.0235g
L-Tyrosine	0.018g
L-Arginine	0.0174g
L-Phenylalanine	0.0165g
L-Cystine	0.012g
L-Histidine	8.0mg
L-Methionine	7.5mg
Phenol Red	5.0mg
L-Tryptophan	4.0mg
Inositol	1.8mg
Biotin	1.0mg
Folic acid	1.0mg
Calcium pantothenate	1.0mg
Choline chloride	1.0mg
Nicotinamide	1.0mg
Pyridoxal·HCl	1.0mg
Thiamine·HCl	1.0mg
Riboflavin	0.1mg

Preparation of Eagle MEM in Hanks BSS: Add components to distilled/deionized water and bring volume to 1.0L. Mix thoroughly.

Penicillin-Streptomycin Solution:
Composition per 1mL:

Streptomycin	0.01g
Penicillin	10,000U

Preparation of Penicillin-Streptomycin Solution: Add components to distilled/deionized water and bring volume to 1.0mL. Mix thoroughly.

Amphotericin B Solution:
Composition per 1mL:

Amphotericin B	1.0mg

Preparation of Amphotericin B Solution: Add amphotericin B to distilled/deionized water and bring volume to 1.0mL. Mix thoroughly.

Preparation of Medium: Combine components. Mix thoroughly. Filter sterilize.

Use: For the cultivation of animal tissue culture cell lines.

Eagle Medium

Composition per liter:

Hanks balanced salt solution (10X)	100.0mL
Calf serum	50.0mL
$NaHCO_3$ (7.5% solution)	29.6mL
Tissue culture amino acids (50X)	20.0mL
Tissue culture vitamins (100X)	10.0mL

Glutamine solution 10.0mL
Phenol Red (0.5% solution) 4.0mL
Penicillin solution 1.0mL
Streptomycin solution 0.4mL
pH 7.0 ± 0.2 at 25°C

Hanks Balanced Salt Solution (10X):
Composition per 100mL:
NaCl 8.0g
Glucose 1.0g
KCl 0.4g
$NaHCO_3$ 0.35g
$CaCl_2 \cdot 2H_2O$ 0.14g
$MgCl_2 \cdot 6H_2O$ 0.1g
$MgSO_4 \cdot 7H_2O$ 0.1g
Na_2HPO_4 .... 0.06g
KH_2PO_4 .... 0.06g
Phenol Red 0.02g

Preparation of Hanks Balanced Salt Solution (10X): Add components to distilled/deionized water and bring volume to 100.0mL. Mix thoroughly.

Tissue Culture Amino Acids (50X):
Composition per liter:
L-Arginine 0.1g
L-Lysine 0.058g
L-Isoleucine 0.052g
L-Leucine 0.052g
L-Threonine 0.048g
L-Valine 0.046g
L-Tyrosine 0.036g
L-Phenylalanine 0.032g
L-Histidine 0.031g
L-Cystine 0.024g
L-Methionine 0.015g
L-Tryptophan 0.010g

Preparation of Tissue Culture Amino Acids, Minimal Eagle 50X: Add components to distilled/deionized water and bring volume to 1.0L. Mix thoroughly.

Tissue Culture Vitamins (100X):
Composition per liter:
Inositol 2.0mg
Calcium pantothenate 1.0mg
Choline chloride 1.0mg
Folic acid 1.0mg
Nicotinamide 1.0mg
Pyridoxal 1.0mg
Thiamine·HCl 1.0mg
Riboflavin 0.1mg

Preparation of TC Vitamins, Minimal Eagle 100X: Add components to distilled/deionized water and bring volume to 1.0L. Mix thoroughly.

Glutamine Solution:
Composition per 100mL:
L-Glutamine 2.9g

Preparation of Glutamine Solution: Add glutamine to distilled/deionized water and bring volume to 100.0mL. Mix thoroughly.

Penicillin Solution:
Composition per 1mL:
Penicillin 200,000U

Preparation of Penicillin Solution: Add penicillin to distilled/deionized water and bring volume to 1.0mL. Mix thoroughly.

Streptomycin Solution:
Composition per 1mL:
Streptomycin 0.5g

Preparation of Streptomycin Solution: Add streptomycin to distilled/deionized water and bring volume to 1.0mL. Mix thoroughly.

Preparation of Medium: Combine components. Mix thoroughly. Adjust pH to 7.0 with 1*N* NaOH. Filter sterilize.

Use: For the cultivation of animal tissue culture cell lines especially for use with rhinoviruses.

Eagle Medium

Composition per 100.1mL:
Eagle MEM in Earle BSS 94.0mL
$NaHCO_3$ (7.5% solution) 3.0mL
Fetal bovine serum, inactivated 2.0mL
Penicillin-streptomycin solution 1.0mL
Amphotericin B solution 0.1mL
pH 7.2–7.4 at 25°C

Eagle MEM in Earle BSS:
Composition per liter:
NaCl 6.8g
Glucose 1.0g
KCl 0.4g
$CaCl_2 \cdot 2H_2O$ 0.2g
$MgCl_2 \cdot 6H_2O$ 0.2g
NaH_2PO_4 .... 0.15g
L-Arginine 0.10g
L-Lysine 0.06g
L-Isoleucine 0.05g
L-Leucine 0.05g
L-Threonine 0.05g
L-Valine 0.05g
L-Tyrosine 0.04g
L-Phenylalanine 0.03g
L-Histidine 0.03g
L-Cystine 0.02g
L-Methionine 0.02g

L-Tryptophan 0.01g
i-Inositol 2.0mg
Calcium pantothenate 1.0mg
Choline chloride 1.0mg
Folic acid 1.0mg
Nicotinamide 1.0mg
Pyridoxal 1.0mg
Thiamine·HCl 1.0mg
Riboflavin 0.1mg

Preparation of Eagle MEM in Earle BSS: Add components to distilled/deionized water and bring volume to 1.0L. Mix thoroughly.

Penicillin-Streptomycin Solution:
Composition per 1mL:
Streptomycin 0.01g
Penicillin 10,000U

Preparation of Penicillin-Streptomycin Solution: Add components to distilled/deionized water and bring volume to 1.0mL. Mix thoroughly.

Amphotericin B Solution:
Composition per 1mL:
Amphotericin B 1.0mg

Preparation of Amphotericin B Solution: Add amphotericin B to distilled/deionized water and bring volume to 1.0mL. Mix thoroughly.

Preparation of Medium: Combine components. Mix thoroughly. Filter sterilize.

Use: For the cultivation of animal tissue culture cell lines.

Eagle Medium, Modified

Composition per liter:
Eagle MEM (10X) 100.0mL
Fetal bovine serum 100.0mL
Glucose solution 20.0mL
HEPES (N-2-Hydroxyethyl piperazine-N´-2-ethanesulfonic acid) buffer, 1M, pH 7.2 .... 20.0mL
Glutamine solution 10.0mL
$NaHCO_3$ (7.5% solution) 7.5mL
Gentamicin sulfate solution 0.2mL
pH 7.2 ± 0.2 at 25°C

Eagle MEM (10X):
Composition per 100mL:
Sterile salt solution 97.0mL
TC amino acids, minimal Eagle 50X 2.0mL
TC vitamins, minimal Eagle 100X 1.0mL

Preparation of Eagle MEM (10X): Combine components. Mix thoroughly. Filter sterilize.

Sterile Salt Solution:
Composition per 100mL:
NaCl 6.8g
Glucose 1.0g
KCl 0.4g
$CaCl_2$ 0.2g
$MgCl_2$ 0.2g
NaH_2PO_4 .... 0.15g

Preparation of Sterile Salt Solution: Add components to distilled/deionized water and bring volume to 100.0mL. Mix thoroughly. Filter sterilize.

Tissue Culture Amino Acids, Minimal Eagle 50X:
Composition per liter:
L-Arginine 0.1g
L-Lysine 0.06g
L-Isoleucine 0.05g
L-Leucine 0.05g
L-Threonine 0.05g
L-Valine 0.05g
L-Tyrosine 0.04g
L-Phenylalanine 0.03g
L-Histidine 0.03g
L-Cystine 0.02g
L-Methionine 0.02g
L-Tryptophan 0.01g

Preparation of Tissue Culture Amino Acids, Minimal Eagle 50X: Add components to distilled/deionized water and bring volume to 1.0L. Mix thoroughly. Adjust pH to 7.2–7.4. Filter sterilize.

TC Vitamins, Minimal Eagle 100X:
Composition per liter:
Inositol 2.0mg
Calcium pantothenate 1.0mg
Choline chloride 1.0mg
Folic acid 1.0mg
Nicotinamide 1.0mg
Pyridoxal 1.0mg
Thiamine·HCl 1.0mg
Riboflavin 0.1mg

Preparation of TC Vitamins, Minimal Eagle 100X: Add components to distilled/deionized water and bring volume to 1.0L. Mix thoroughly. Filter sterilize.

Glucose Solution:
Composition per 100mL:
Glucose 27.0g

Preparation of Glucose Solution: Add glucose to distilled/deionized water and bring volume to 100.0mL. Mix thoroughly. Filter sterilize.

Glutamine Solution:

Composition per 10mL:

L-Glutamine	5.0g

Preparation of Glutamine Solution: Add glutamine to distilled/deionized water and bring volume to 10.0mL. Mix thoroughly. Filter sterilize.

Gentamicin Solution:

Composition per 1mL:

Gentamicin sulfate	0.05g

Preparation of Gentamicin Solution: Add gentamicin sulfate to distilled/deionized water and bring volume to 1.0mL. Mix thoroughly. Filter sterilize.

Preparation of Medium: Combine components. Mix thoroughly. Filter sterilize.

Use: For the cultivation of animal tissue culture cell lines, especially for McCoy cells.

Earle's Balanced Salts, Phenol Red–Free

Composition per liter:

NaCl	6.8g
$NaHCO_3$	2.2g
Glucose	1.0g
KCl	0.40g
$CaCl_2 \cdot 2H_2O$	0.265g
$MgSO_4 \cdot 7H_2O$	0.20g
$NaH_2PO_4 \cdot H_2O$	0.14g

pH 7.2 ± 0.2 at 25°C

Preparation of Medium: Add components to distilled/deionized water and bring volume to 1.0L. Mix thoroughly. Filter sterilize.

Use: For the preparation of tissue culture media where Phenol Red is not desired.

EB Motility Medium

Composition per liter:

Peptone or gelysate	10.0g
NaCl	5.0g
Agar	4.0g
Beef extract	3.0g

pH 7.4 ± 0.2 at 25°C

Preparation of Medium: Add components to distilled/deionized water and bring volume to 1.0L. Mix thoroughly. Gently heat and bring to boiling. Distribute into tubes in 8.0mL volumes. Autoclave for 15 min at 15 psi pressure–121°C.

Use: For the cultivation and differentiation of bacteria based on motility.

EC Broth (*Escherichia coli* Broth) (EC Medium)

Composition per liter:

Pancreatic digest of casein	20.0g
Lactose	5.0g
NaCl	5.0g
K_2HPO_4	4.0g
Bile salts mixture	1.5g
KH_2PO_4	1.5g

pH 6.9 ± 0.2 at 25°C

Source: This medium is available as a premixed powder from BBL Microbiology Systems and Difco Laboratories.

Preparation of Medium: Add components to distilled/deionized water and bring volume to 1.0L. Mix thoroughly. Distribute into test tubes that contain an inverted Durham tube. Autoclave for 12 min at 15 psi pressure–121°C. Cool broth as quickly as possible.

Use: For the cultivation and differentiation of coliform bacteria at 37°C and of *Escherichia coli* at 45.5°C.

EC Broth with MUG

Composition per liter:

Pancreatic digest of casein	20.0g
Lactose	5.0g
NaCl	5.0g
K_2HPO_4	4.0g
Bile salts mixture	1.5g
KH_2PO_4	1.5g
4-Methylumbeliferyl-β-D-glucuronide (MUG)	0.05g

pH 6.9 ± 0.2 at 25°C

Source: This medium is available as a premixed powder from BBL Microbiology Systems.

Preparation of Medium: Add components to distilled/deionized water and bring volume to 1.0L. Mix thoroughly. Distribute into test tubes that contain an inverted Durham tube in 10.0mL volumes. Autoclave for 15 min at 15 psi pressure–121°C.

Use: For the detection of *Escherichia coli* in water and food samples by a fluorogenic procedure.

ECM Agar

Composition per liter:

Agar	15.0g
NaCl	6.0g
Escherichia coli cells, washed	1.0g
$MgSO_4 \cdot 7H_2O$	0.50g

Preparation of Medium: Add components to distilled/deionized water and bring volume to 1.0L. Mix thoroughly. Gently heat and bring to boiling. Distribute into tubes or flasks. Autoclave for 15 min at 15 psi pressure–121°C. Pour into sterile Petri dishes or leave in tubes.

Use: For the cultivation of myxobacteria.

Ectothiorhodospira halochloris Medium

Composition per liter:

NaCl	180.0g
Na_2SO_4	20.0g
$NaHCO_3$	14.0g
Na_2CO_3	6.0g
$Na_2S\cdot 9H_2O$	1.0g
Sodium succinate	1.0g
NH_4Cl	0.8g
KH_2PO_4	0.5g
Yeast extract	0.5g
$MgCl_2\cdot 6H_2O$	0.10g
$CaCl_2\cdot 2H_2O$	0.05g
Vitamin solution VA	1.0mL
Trace element solution SLA	1.0mL

pH 8.5–8.7 at 25°C

Vitamin Solution VA:

Composition per liter:

Nicotinamide	0.04g
Thiamine dichloride	0.03g
p-Aminobenzoic acid	0.02g
Biotin	0.01g
Calcium pantothenate	0.01g
Pyridoxal chloride	0.01g
Vitamin B_{12}	5.0mg

Preparation of Vitamin Solution VA: Add components to distilled/deionized water and bring volume to 1.0L. Mix thoroughly.

Trace Element Solution SLA:

Composition per liter:

$FeCl_2\cdot 4H_2O$	1.8g
H_3BO_3	0.5g
$CoCl_2\cdot 6H_2O$	0.25g
$ZnCl_2$	0.10g
$MnCl_2\cdot 4H_2O$	0.07g
$NaMoO_4\cdot 2H_2O$	0.03g
$CuCl_2\cdot 2H_2O$	0.01g
Na_2SeO_3	0.01g
$NiCl_2\cdot 6H_20$	0.01g

Preparation of Trace Element Solution SLA: Add components to distilled/deionized water and bring volume to 1.0L. Mix thoroughly. Adjust pH to 3.0 with 2*N* HCl.

Preparation of Medium: Add components, except trace element solution SLA, to distilled/deionized water and bring volume to 999.0mL. Mix thoroughly. Filter sterilize. Aseptically add 1.0mL of sterile trace element solution SLA. Mix thoroughly. Aseptically distribute into flasks or bottles. Completely fill bottles with medium except for a pea-sized air bubble.

Use: For the enrichment and isolation of *Ectothiorhodospira halochloris*.

Ectothiorhodospira halophila Medium

Composition per liter:

NaCl	200.0g
NH_4Cl	0.4g
$(NH_4)_2SO_4$	0.1g
Na_2CO_3 solution	100.0mL
Tris buffer (1M solution, pH 7.5)	30.0mL
Solution C	5.0mL
Potassium phosphate buffer (1M solution, pH 7.5)	3.0mL
Additional solution	2.5mL

pH 7.4–8.0 at 25°C

Na_2CO_3 Solution:

Composition per 100mL:

Na_2CO_3	10.0g

Preparation of Na_2CO_3 Solution: Add Na_2CO_3 to distilled/deionized water and bring volume to 100.0mL. Mix thoroughly. Autoclave for 15 min at 15 psi pressure–121°C. Cool to 25°C.

Solution C:

Composition per liter:

$MgCl_2\cdot 6H_2O$	24.0g
$CaCl_2\cdot 2H_2O$	3.3g
$FeCl_3\cdot 4H_2O$	1.1g
$(NH_4)_6Mo_7O_{24}\cdot 4H_2O$	0.1g
Nitrilotriacetic acid	10mg
Trace elements solution	50.0mL

Preparation of Solution C: Add nitrilotriacetic acid to 500.0mL of distilled/deionized water. Dissolve by adjusting pH to 6.5 with KOH. Add remaining components. Readjust pH to 7.2 with H_2SO_4 or KOH. Add distilled/deionized water to 1.0L.

Trace Elements Solution:

Composition per 100mL:

$ZnCl_2$	0.52g
EDTA	0.25g
$MnCl_2\cdot 4H_2O$	0.08g
$FeCl_3\cdot 4H_2O$	0.03g
$Co(NO_3)_2\cdot 6H_2O$	0.02g

$CuCl_2 \cdot 2H_2O$ 0.02g
H_3BO_3 .... 0.01g

Preparation of Trace Elements Solution: Add comonents to distilled/deionized water and bring volume to 1.0L. Mix thoroughly. Adjust pH to 3.0 with 2*N* HCl.

Additional Solution:

Composition per 50mL:

$NaS_2O_3 \cdot 6H_2O$ 6.0g
Sodium succinate 5.0g
Sodium ascorbate 1.0g

Preparation of Additional Solution: Add components to distilled/deionized water and bring volume to 50.0mL. Mix thoroughly. Filter sterilize.

Preparation of Medium: Add components, except Na_2CO_3 solution and additional solution, to distilled/deionized water and bring volume to 900.0mL. Autoclave for 15 min at 15 psi pressure–121°C. Cool to 45°–50°C. Aseptically adjust pH to 7.4–7.8 with filter-sterilized HCl. Aseptically distribute into 50.0mL screw-capped bottles. Fill each bottle almost to the top, leaving a space of 2.8mL in the neck. Aseptically add 2.5mL of sterile additional solution to each bottle. Mix thoroughly.

Use: For the isolation and cultivation of *Ectothiorhodospira halophila*.

Ectothiorhodospira halophila Medium

Composition per liter:

NaCl 220.0g
Potassium succinate 1.0g
$Na_2S \cdot 9H_2O$ 0.1g
K_2HPO_4 solution 20.0mL
$NaHCO_3$ solution 20.0mL
Solution C 20.0mL
$(NH_4)_2SO_4$ solution 5.0mL
Vitamin solution 0.5mL

pH 7.4–8.0 at 25°C

K_2HPO_4 Solution:

Composition per liter:

K_2HPO_4 .... 125.0g

Preparation of K_2HPO_4 Solution: Add K_2HPO_4 to distilled/deionized water and bring volume to 1.0L. Mix thoroughly.

$NaHCO_3$ Solution:

Composition per liter:

$NaHCO_3$ 100.0g

Preparation of $NaHCO_3$ Solution: Add $NaHCO_3$ to distilled/deionized water and bring volume to 1.0L. Mix thoroughly.

Solution C:

Composition per liter:

$MgCl_2 \cdot 6H_2O$ 24.0g
$CaCl_2 \cdot 2H_2O$ 3.3g
$FeCl_3 \cdot 4H_2O$ 1.1g
$(NH_4)_6Mo_7O_{24} \cdot 4H_2O$ 0.1g
Nitrilotriacetic acid 10.0mg
Trace elements solution 50.0mL

Preparation of Solution C: Add components to distilled/deionized water and bring volume to 1.0L. Mix thoroughly.

Trace Elements Solution:

Composition per 100mL:

$ZnCl_2$ 0.52g
EDTA 0.25g
$MnCl_2 \cdot 4H_2O$ 0.08g
$FeCl_3 \cdot 4H_2O$ 0.03g
$Co(NO_3)_2 \cdot 6H_2O$ 0.02g
$CuCl_2 \cdot 2H_2O$ 0.02g
H_3BO_3 .... 0.01g

Preparation of Trace Elements Solution: Add comonents to distilled/deionized water and bring volume to 1.0L. Mix thoroughly. Adjust pH to 3.0 with 2*N* HCl.

$(NH_4)_2SO_4$ Solution:

Composition per liter:

$(NH_4)_2SO_4$ 100.0g

Preparation of $(NH_4)_2SO_4$ Solution: Add components to distilled/deionized water and bring volume to 1.0L. Mix thoroughly.

Vitamin Solution:

Composition per liter:

Nicotinic acid 2.0mg
Thiamine 1.0mg
p-Aminobenzoic acid 0.2mg
Biotin 0.02mg
Vitamin B_{12} 1.0μg

Preparation of Vitamin Solution: Add components to distilled/deionized water and bring volume to 1.0L. Mix thoroughly.

Preparation of Medium: Add components to distilled/deionized water and bring volume to 1.0L. Mix thoroughly. Adjust pH to 7.4–8.0. Filter sterilize. Aseptically distribute into flasks or bottles. Completely fill bottles with medium except for a pea-sized air bubble.

Use: For the isolation and cultivation of *Ectothiorhodospira halophila*.

Ectothiorhodospira Medium

Composition per liter:

NaCl	180.0g
Na_2SO_4	20.0g
$NaHCO_3$	14.0g
Na_2CO_3	6.0g
Sodium succinate	1.0g
NH_4Cl	0.8g
KH_2PO_4	0.5g
$MgCl_2 \cdot 6H_2O$	0.10g
$CaCl_2 \cdot 2H_2O$	0.05g
Feeding solution	10.0mL
Trace element solution SLA	1.0mL
Vitamin solution VA	1.0mL

pH 8.5–8.7 at 25°C

Feeding Solution:

Composition per 100mL:

NaCl	10.0g
$NaHCO_3$	10.0g
$Na_2S \cdot 9H_2O$	5.0g

Preparation of Feeding Solution: Add components to distilled/deionized water and bring volume to 100.0mL. Mix thoroughly. Filter sterilize.

Trace Element Solution SLA:

Composition per liter:

$FeCl_2 \cdot 4H_2O$	1.8g
H_3BO_3	0.5g
$CoCl_2 \cdot 6H_2O$	0.25g
$ZnCl_2$	0.1g
$MnCl_2 \cdot 4H_2O$	0.07g
$NaMoO_4 \cdot 2H_2O$	0.03g
$CuCl_2 \cdot 2H_2O$	0.01g
Na_2SeO_3	0.01g
$NiCl_2 \cdot 6H_2O$	0.01g

Preparation of Trace Element Solution SLA: Add components to distilled/deionized water and bring volume to 1.0L. Mix thoroughly. Adjust pH to 3.0 with 2*N* HCl.

Vitamin Solution VA:

Composition per liter:

Nicotinamide	0.04g
Thiamine dichloride	0.03g
p-Aminobenzoic acid	0.02g
Biotin	0.01g
Calcium pantothenate	0.01g
Pyridoxal chloride	0.01g
Vitamin B_{12}	5.0mg

Preparation of Vitamin Solution VA: Add components to distilled/deionized water and bring volume to 1.0L. Mix thoroughly.

Preparation of Medium: Add components, except trace element solution SLA and feeding solution, to distilled/deionized water and bring volume to 999.0mL. Mix thoroughly. Filter sterilize. Aseptically add 1.0mL of sterile trace element solution SLA. Mix thoroughly. Aseptically distribute into flasks or bottles. Completely fill bottles with medium except for a pea-sized air bubble. Prior to inoculation aseptically remove sufficient amount of medium to permit the addition of feeding medium. Add 1.0mL of feeding solution per each 100.0mL of medium.

Use: For the isolation and cultivation of *Ectothiorhodospira halochloris* and *Ectothiorhodospira halophila*.

Ectothiorhodospira Medium

Composition per liter:

NaCl	130.0g
Na_2SO_4	10.0g
Sodium acetate	2.0g
KH_2PO_4	0.8g
Sodium carbonate buffer, (1M, pH 9.0)	200.0mL
$MgCl_2 \cdot 6H_2O$ solution	10.0mL
$Na_2S \cdot 9H_2O$ solution	6.0mL
$CaCl_2 \cdot 2H_2O$ solution	5.0mL
NH_4Cl solution	4.0mL
SLA trace elements	1.0mL
VA vitamin solution	1.0mL

pH 9.0 ± 0.2 at 25°C

$MgCl_2 \cdot 6H_2O$ Solution:

Composition per 10mL:

$MgCl_2 \cdot 6H_2O$	0.1g

Preparation of $MgCl_2 \cdot 6H_2O$ Solution: Add $MgCl_2 \cdot 6H_2O$ to distilled/deionized water and bring volume to 10.0mL. Mix thoroughly. Filter sterilize.

$Na_2S \cdot 9H_2O$ Solution:

Composition per 10mL:

$Na_2S \cdot 9H_2O$	0.5g

Preparation of $Na_2S \cdot 9H_2O$ Solution: Add $Na_2S \cdot 9H_2O$ to distilled/deionized water and bring volume to 10.0mL. Mix thoroughly. Filter sterilize. Use freshly prepared solution.

$CaCl_2 \cdot 2H_2O$ Solution:

Composition per 10mL:

$CaCl_2 \cdot 2H_2O$	0.1g

Preparation of $CaCl_2 \cdot 2H_2O$ Solution: Add $CaCl_2 \cdot 2H_2O$ to distilled/deionized water and bring volume to 10.0mL. Mix thoroughly. Filter sterilize.

NH_4Cl Solution:

Composition per 10mL:

NH_4Cl	2.0g

Preparation of NH_4Cl Solution: Add NH_4Cl to distilled/deionized water and bring volume to 10.0mL. Mix thoroughly. Filter sterilize.

SLA Trace Elements:

Composition per liter:

$FeCl_2 \cdot 4H_2O$	1.8g
H_3BO_3	0.5g
$CoCl_2 \cdot 6H_2O$	0.25g
$ZnCl_2$	0.1g
$MnCl_2 \cdot 4H_2O$	0.07g
$Na_2MoO_4 \cdot 2H_2O$	0.03g
$CuCl_2 \cdot 2H_2O$	0.01g
$Na_2SeO_3 \cdot 5H_2O$	0.01g
$NiCl_2 \cdot 6H_2O$	0.01g

Preparation of SLA Trace Elements: Add components to distilled/deionized water and bring volume to 1.0L. Mix thoroughly. Adjust pH to 2–3.

VA Vitamin Solution:

Composition per 500mL:

Nicotinamide	0.175g
Thiamine·HCl	0.15g
p-Aminobenzoic acid	0.10g
Biotin	0.05g
Calcium pantothenate	0.05g
Pyridoxine·2HCl	0.05g
Cyanocobalamin	0.025g

Preparation of VA Vitamin Solution: Add components to distilled/deionized water and bring volume to 500.0mL. Mix thoroughly.

Preparation of Medium: Add components—except $MgCl_2 \cdot 6H_2O$ solution, $Na_2S \cdot 9H_2O$ solution, $CaCl_2 \cdot 2H_2O$ solution, and NH_4Cl solution—to distilled/deionized water and bring volume to 975.0mL. Mix thoroughly. Gently heat and bring to boiling. Autoclave for 15 min at 15 psi pressure–121°C. Cool to 25°C. Aseptically add 10.0mL of sterile $MgCl_2 \cdot 6H_2O$ solution, 6.0mL of sterile $Na_2S \cdot 9H_2O$ solution, 5.0mL of sterile $CaCl_2 \cdot 2H_2O$ solution, and 4.0mL of sterile NH_4Cl solution. Mix thoroughly. Aseptically distribute into culture bottles. Incubate for 2 days before inoculation.

Use: For the cultivation and maintenance of *Ectothiorhodospira abdelmalekii* and *Ectothiorhodospira halochloris*.

Ectothiorhodospira Medium, Modified

Composition per liter:

NaCl	30.0g
Na_2SO_4	10.0g
Sodium acetate	2.0g
KH_2PO_4	0.8g
Sodium carbonate buffer (1M, pH 9.0)	200.0mL
$MgCl_2 \cdot 6H_2O$ solution	10.0mL
$Na_2S \cdot 9H_2O$ solution	6.0mL
$CaCl_2 \cdot 2H_2O$ solution	5.0mL
NH_4Cl solution	4.0mL
SLA trace elements	1.0mL
VA vitamin solution	1.0mL

pH 9.0 ± 0.2 at 25°C

$MgCl_2 \cdot 6H_2O$ Solution:

Composition per 10mL:

$MgCl_2 \cdot 6H_2O$	0.1g

Preparation of $MgCl_2 \cdot 6H_2O$ Solution: Add $MgCl_2 \cdot 6H_2O$ to distilled/deionized water and bring volume to 10.0mL. Mix thoroughly. Filter sterilize.

$Na_2S \cdot 9H_2O$ Solution:

Composition per 10mL:

$Na_2S \cdot 9H_2O$	0.5g

Preparation of $Na_2S \cdot 9H_2O$ Solution: Add $Na_2S \cdot 9H_2O$ to distilled/deionized water and bring volume to 10.0mL. Mix thoroughly. Filter sterilize. Use freshly prepared solution.

$CaCl_2 \cdot 2H_2O$ Solution:

Composition per 10mL:

$CaCl_2 \cdot 2H_2O$	0.1g

Preparation of $CaCl_2 \cdot 2H_2O$ Solution: Add $CaCl_2 \cdot 2H_2O$ to distilled/deionized water and bring volume to 10.0mL. Mix thoroughly. Filter sterilize.

NH_4Cl Solution:

Composition per 10mL:

NH_4Cl	2.0g

Preparation of NH_4Cl Solution: Add NH_4Cl to distilled/deionized water and bring volume to 10.0mL. Mix thoroughly. Filter sterilize.

SLA Trace Elements:

Composition per liter:

$FeCl_2 \cdot 4H_2O$	1.8g
H_3BO_3	0.5g
$CoCl_2 \cdot 6H_2O$	0.25g
$ZnCl_2$	0.1g
$MnCl_2 \cdot 4H_2O$	0.07g
$Na_2MoO_4 \cdot 2H_2O$	0.03g
$CuCl_2 \cdot 2H_2O$	0.01g
$Na_2SeO_3 \cdot 5H_2O$	0.01g
$NiCl_2 \cdot 6H_2O$	0.01g

Preparation of SLA Trace Elements: Add components to distilled/deionized water and bring volume to 1.0L. Mix thoroughly. Adjust pH to 2–3.

VA Vitamin Solution:

Composition per 500mL:

Nicotinamide	0.175g
Thiamine·HCl	0.15g
p-Aminobenzoic acid	0.1g
Biotin	0.05g
Calcium pantothenate	0.05g

Pyridoxine·2HCl 0.05g
Cyanocobalamin 0.025g

Preparation of VA Vitamin Solution: Add components to distilled/deionized water and bring volume to 500.0mL. Mix thoroughly.

Preparation of Medium: Add components—except $MgCl_2 \cdot 6H_2O$ solution, $Na_2S \cdot 9H_2O$ solution, $CaCl_2 \cdot 2H_2O$ solution, and NH_4Cl solution—to distilled/deionized water and bring volume to 975.0mL. Mix thoroughly. Gently heat and bring to boiling. Autoclave for 15 min at 15 psi pressure–121°C. Cool to 25°C. Aseptically add 10.0mL of sterile $MgCl_2 \cdot 6H_2O$ solution, 6.0mL of sterile $Na_2S \cdot 9H_2O$ solution, 5.0mL of sterile $CaCl_2 \cdot 2H_2O$ solution, and 4.0mL of sterile NH_4Cl solution. Mix thoroughly. Aseptically distribute into culture bottles. Incubate for 2 days before inoculation.

Use: For the cultivation and maintenance of *Ectothiorhodospira vacuotalta.*

Edelstein BMPA–α Medium *See:* BMPA–α Medium

Edwards Medium, Modified

Composition per liter:

Agar 15.0g
Beef extract 10.0g
Peptone 10.0g
NaCl 5.0g
Esculin 1.0g
Tl_2SO_4 .... 0.33g
Crystal Violet 1.3mg
Bovine or sheep blood 50.0mL

pH 7.4 ± 0.2 at 25°C

Source: This medium is available as a premixed powder from Oxoid.

Preparation of Medium: Add components, except blood, to distilled/deionized water and bring volume to 950.0mL. Mix thoroughly. Gently heat and bring to boiling. Autoclave for 20 min at 10 psi pressure–115°C. Cool to 45°–50°C. Aseptically add 50.0mL of sterile bovine blood or sheep blood. Mix thoroughly. Pour into sterile Petri dishes.

Use: For the selective isolation and cultivation of *Streptococcus agalactiae* and other streptococci involved in bovine mastitis.

EE Broth (Enterobacteriaceae Enrichment Broth)

Composition per liter:

Ox Bile 20.0g
Peptone 10.0g
Na_2HPO_4 .... 6.45g
Glucose 5.0g
KH_2PO_4 .... 2.0g
Brilliant Green 0.0135g

pH 7.2 ± 0.2 at 25°C

Source: This medium is available as a premixed powder from Oxoid.

Preparation of Medium: Add components to distilled/deionized water and bring volume to 1.0L. Mix thoroughly. Distribute into flasks in 100.0mL volumes. Gently heat at 100°C for 30 min. Do not autoclave. Cool rapidly to 25°C.

Use: For the cultivation and enrichment of members of the Enterobateriaceae in the examination of foods and animal feed. Used in conjunction with tryptone soy broth. Bacteria belonging to the Enterobacteriaceae turn this medium turbid and yellow-green.

EE Broth, Mossel (Enterobacteriaceae Enrichment Broth, Mossel)

Composition per liter:

Enzymatic hydrolysate of protein 10.0g
Na_2HPO_4 .... 8.0g
Glucose 5.0g
KH_2PO_4 .... 2.0g
Oxgall 0.1g
Brilliant Green 0.0135g

pH 7.2 ± 0.2 at 25°C

Source: This medium is available as a premixed powder from Difco Laboratories.

Preparation of Medium: Add components to distilled/deionized water and bring volume to 1.0L. Mix thoroughly. Distribute into flasks in 120.0mL volumes. Gently heat at 100°C for 30 min. Do not autoclave. Cool rapidly to 25°C.

Use: For the cultivation and enrichment of members of the Enterobateriaceae in the examination of foods and animal feed. Used in conjunction with tryptone soy broth. Bacteria belonging to the Enterobacteriaceae turn this medium turbid and yellow-green.

EG NaCl Medium No. 7 (Ethylene Glycol NaCl Medium No. 7)

Composition per liter:

NaCl	70.0g
Agar	15.0g
K_2HPO_4	7.5g
KH_2PO_4	1.0g
$(NH_4)_2SO_4$	0.8g
$MgSO_4·7H_2O$	0.1g
$FeSO_4·7H_2O$	0.01g
Ethylene glycol	10.0mL
Salt solution	1.0mL

Salt Solution:
Composition per liter:

$CaCl_2·2H_2O$	6.0g
$ZnSO_4·7H_2O$	4.4g
$MnSO_4·H_2O$	3.0g
$(NH_4)_6Mo_7O_{24}·4H_2O$	1.82g
$CuCl_2·2H_2O$	0.2g

Preparation of Salt Solution: Add components to distilled/deionized water and bring volume to 1.0L. Mix thoroughly.

Preparation of Medium: Add components to distilled/deionized water and bring volume to 1.0L. Mix thoroughly. Gently heat and bring to boiling. Distribute into tubes or flasks. Autoclave for 15 min at 15 psi pressure–121°C. Pour into sterile Petri dishes or leave in tubes.

Use: For the cultivation of ATCC strain 27042.

Egg Tellurite Glycine Pyruvate Agar
See: **ETGPA**

Egg Yolk Agar

Composition per liter:

Proteose peptone No. 2	40.0g
Agar	25.0g
Na_2HPO_4	5.0g
Glucose	2.0g
NaCl	2.0g
KH_2PO_4	1.0g
$MgSO_4·7H_2O$	0.1g
Egg yolk emulsion	100.0mL
Hemin solution	1.0mL

pH 7.6 ± 0.2 at 25°C

Hemin Solution:
Composition per 100mL:

Hemin	0.5g
NaOH (1*N* solution)	20.0mL

Preparation of Hemin Solution: Add hemin to 20.0mL of 1*N* NaOH solution. Mix thoroughly. Bring volume to 100.0mL with distilled/deionized water.

Egg Yolk Emulsion:
Composition:

Chicken egg yolks	11
Whole chicken egg	1

Preparation of Egg Yolk Emulsion: Soak eggs with 1:100 dilution of saturated mercuric chloride solution for 1 min. Crack eggs and separate yolks from whites. Mix egg yolks with 1 chicken egg.

Preparation of Medium: Add components, except egg yolk emulsion, to distilled/deionized water and bring volume to 900.0mL. Mix thoroughly. Gently heat and bring to boiling. Autoclave for 15 min at 15 psi pressure–121°C. Cool to 45°–50°C. Aseptically add sterile egg yolk emulsion. Mix thoroughly. Pour into sterile Petri dishes.

Use: For the isolation, cultivation and differentiation of *Clostridium* species and some other anaerobic bacteria.

Egg Yolk Agar, Lombard–Dowell
See: **Lombard–Dowell Egg Yolk Agar**

Egg Yolk Agar, Modified

Composition per liter:

Agar	20.0g
Pancreatic digest of casein	15.0g
Vitamin K_1	10.0g
NaCl	5.0g
Papaic digest of soybean meal	5.0g
Yeast extract	5.0g
L-Cystine	0.4g
Hemin	5.0mg
Egg yolk emulsion	100.0mL

Source: Available as a prepared medium from BBL Microbiology Systems.

Egg Yolk Emulsion:
Composition:

Chicken egg yolks	11
Whole chicken egg	1

Preparation of Egg Yolk Emulsion: Soak eggs with 1:100 dilution of saturated mercuric chloride solution for 1 min. Crack eggs and separate yolks from whites. Mix egg yolks with 1 chicken egg.

Preparation of Medium: Add components, except egg yolk emulsion, to distilled/deionized water and bring volume to 900.0mL. Mix thoroughly. Gently heat and bring to boiling. Autoclave for 15 min at 15 psi pressure–121°C. Cool to 45°–50°C. Aseptically add sterile egg yolk emulsion. Mix thoroughly. Pour into sterile Petri dishes.

Use: For the isolation, cultivation and differentiation of *Clostridium* species and some other anaerobic bacteria.

Egg Yolk Agar with Neomycin *See:* Lombard–Dowell Neomycin Agar

EIA Substrate

Composition per liter:

Agar	15.0g
Esculin	1.0g
Ferric citrate	0.5g

pH 7.1 ± 0.1 at 25°C

Source: This medium is available as a premixed powder from Difco Laboratories.

Preparation of Medium: Add components to distilled/deionized water and bring volume to 1.0L. Mix thoroughly. Gently heat and bring to boiling. Adjust pH to 7.1. Distribute into tubes or flasks. Autoclave for 15 min at 15 psi pressure–121°C. Pour into sterile Petri dishes.

Use: For the cultivation and enumeration of marine enterococci by the membrane filter method.

Eijkman Lactose Medium

Composition per liter:

Pancreatic digest of casein	15.0g
K_2HPO_4	10.0g
KH_2PO_4	4.0g
Lactose	3.0g
NaCl	2.5g

pH 6.8 ± 0.1 at 25°C

Preparation of Medium: Add components to distilled/deionized water and bring volume to 1.0L. Mix thoroughly. Distribute into test tubes that contain an inverted Durham tube. Autoclave for 15 min at 15 psi pressure–121°C.

Use: For the cultivation and differentiation of *Escherichia coli* from other coliform organisms based on their ability to ferment lactose and produce gas.

Eijkman Lactose Medium

Composition per liter:

Tryptose	15.0g
NaCl	5.0g
K_2HPO_4	4.0g
Lactose	3.0g
KH_2PO_4	1.5g

pH 6.8 ± 0.1 at 25°C

Preparation of Medium: Add components to distilled/deionized water and bring volume to 1.0L. Mix thoroughly. Distribute into test tubes that contain an inverted Durham tube. Autoclave for 15 min at 15 psi pressure–121°C.

Use: For the cultivation and differentiation of *Escherichia coli* from other coliform organisms based on their ability to ferment lactose and produce gas.

Elek Agar *See*: K–L Virulence Agar

Elliker Agar

Composition per liter:

Pancreatic digest of casein	20.0g
Agar	15.0g
Glucose	5.0g
Lactose	5.0g
Sucrose	5.0g
Yeast extract	5.0g
NaCl	4.0g
Gelatin	2.5g
Sodium acetate	1.5g
Ascorbic acid	0.5g

pH 6.8 ± 0.2 at 25°C

Preparation of Medium: Add components to distilled/deionized water and bring volume to 1.0L. Mix thoroughly. Gently heat and bring to boiling. Distribute into tubes or flasks. Autoclave for 15 min at 15 psi pressure–121°C. Pour into sterile Petri dishes or leave in tubes.

Use: For the isolation and cultivation of lactic streptococci.

Elliker Agar

Composition per liter:

Pancreatic digest of casein	20.0g
Agar	15.0g
Yeast extract	10.0g
Gelatin	4.0g
Glucose	3.0g
Ascorbic acid	2.5g

Lactose	2.5g
NaCl	2.5g
Sodium acetate	2.5g
Sucrose	2.5g

pH 6.8 ± 0.1 at 25°C

Preparation of Medium: Add components to distilled/deionized water and bring volume to 1.0L. Mix thoroughly. Gently heat and bring to boiling. Autoclave for 15 min at 15 psi pressure–121°C. Pour into sterile Petri dishes or distribute into sterile tubes.

Use: For the cultivation of streptococci and lactobacilli of importance in the dairy industry.

Elliker Broth

Composition per liter:

Pancreatic digest of casein	20.0g
Yeast extract	10.0g
Gelatin	4.0g
Glucose	3.0g
Ascorbic acid	2.5g
Lactose	2.5g
NaCl	2.5g
Sodium acetate	2.5g
Sucrose	2.5g

pH 6.8 ± 0.1 at 25°C

Preparation of Medium: Add components to distilled/deionized water and bring volume to 1.0L. Mix thoroughly. Distribute into test tubes that contain an inverted Durham tube. Autoclave for 15 min at 15 psi pressure–121°C.

Use: For the cultivation of streptococci and lactobacilli of importance in the dairy industry.

Elliker Broth

Composition per liter:

Pancreatic digest of casein	20.0g
Dextrose	5.0g
Lactose	5.0g
Sucrose	5.0g
Yeast extract	5.0g
NaCl	4.0g
Gelatin	2.5g
Sodium acetate	1.5g
Ascorbic acid	0.5g

pH 6.8 ± 0.2 at 25°C

Source: This medium is available as a premixed powder from Difco Laboratories.

Preparation of Medium: Add components to distilled/deionized water and bring volume to 1.0L. Mix thoroughly. Distribute into test tubes that contain an inverted Durham tube. Autoclave for 15 min at 15 psi pressure–121°C.

Use: For the cultivation of streptococci and lactobacilli of importance in the dairy industry.

EMB Agar (Eosin Methylene Blue Agar)

Composition per liter:

Agar	13.5g
Pancreatic digest of casein	10.0g
Lactose	5.0g
Sucrose	5.0g
K_2HPO_4	2.0g
Eosin Y	0.4g
Methylene Blue	0.065g

pH 7.2 ± 0.2 at 25°C

Source: This medium is available as a premixed powder from BBL Microbiology Systems and Difco Laboratories.

Preparation of Medium: Add components to distilled/deionized water and bring volume to 1.0L. Mix thoroughly. Gently heat and bring to boiling. Distribute into tubes or flasks. Autoclave for 15 min at 15 psi pressure–121°C. Pour into sterile Petri dishes.

Use: For the isolation, cultivation and differentiation of Gram-negative enteric bacteria based on lactose fermentation. Bacteria that ferment lactose, especially the coliform bacterium *Escherichia coli*, appear as colonies with a green metallic sheen or blue-black to brown color. Bacteria that do not ferment lactose appear as colorless or transparent light purple colonies.

EMB Agar Base

Composition per liter:

Agar	15.0g
Peptone	10.0g
K_2HPO_4	2.0g
Eosin Y	0.4g
Methylene Blue	0.065g

pH 7.3 ± 0.2 at 25°C

Preparation of Medium: Add components to distilled/deionized water and bring volume to 1.0L. Mix thoroughly. Gently heat and bring to boiling. Distribute into tubes or flasks. Autoclave for 15 min at 15 psi pressure–121°C. Pour into sterile Petri dishes.

Use: For the isolation, cultivation and differentiation of Gram-negative enteric bacteria based on lactose fermentation. Bacteria that ferment lactose, especially the coliform bacterium *Escherichia coli*, appear as colonies with a green metallic sheen or blue-black to brown color. Bacteria that do not ferment lactose appear as colorless or transparent light purple colonies.

EMB Agar, Modified (Eosin Methylene Blue Agar, Modified)

Composition per liter:

Agar	15.0g
Lactose	10.0g
Pancreatic digest of gelatin	10.0g
K_2HPO_4	2.0g
Eosin Y	0.4g
Methylene Blue	0.065g

pH 6.8 ± 0.2 at 25°C

Source: This medium is available as a premixed powder from Oxoid.

Preparation of Medium: Add components to distilled/deionized water and bring volume to 1.0L. Mix thoroughly. Gently heat and bring to boiling. Distribute into tubes or flasks. Autoclave for 15 min at 15 psi pressure–121°C. Cool to 60°C. Shake medium to oxidize methylene blue. Pour into sterile Petri dishes. Swirl flask while pouring plates to distribute precipitate.

Use: For the isolation, cultivation and differentiation of Gram-negative enteric bacteria based on lactose fermentation. Bacteria that ferment lactose, especially the coliform bacterium *Escherichia coli*, appear as colonies with a green metallic sheen or blue-black to brown color. Bacteria that do not ferment lactose appear as colorless or transparent light purple colonies.

Emerson Agar

Composition per liter:

Agar	20.0g
Glucose	10.0g
Beef extract	4.0g
Pancreatic digest of gelatin	4.0g
NaCl	2.5g
Yeast extract	1.0g

pH 7.0 ± 0.2 at 25°C

Source: This medium is available as a premixed powder from BBL Microbiology Systems.

Preparation of Medium: Add components to distilled/deionized water and bring volume to 1.0L. Mix thoroughly. Gently heat and bring to boiling. Distribute into tubes or flasks. Autoclave for 15 min at 13 psi pressure–118°C. Pour into sterile Petri dishes or leave in tubes.

Use: For the isolation, cultivation, and maintenance of members of the Actinomycetaceae, Streptomycetaceae and molds. For the cultivation and maintenance of *Arthrobacter* species, *Microbispora rosea*, *Micromonospora coerulea*, *Mycobacterium* species, *Nocardia asteroides*, *Nocardiopsis dassonvillei*, *Pseudonocardia thermophila*, *Staphylococcus epidermidis*, *Streptomyces flaveus*, *Streptomyces olivaceus*, *Streptomyces thermoviolaceus*, *Streptomyces thermovulgaris*, and *Streptomyces vendargensis*.

Emerson Agar, Half Strength

Composition per liter:

Agar	20.0g
Soluble starch	7.5g
Yeast extract	2.0g
K_2HPO_4	0.5g
$MgSO_4 \cdot 7H_2O$	0.25g

pH 7.0 ± 0.2 at 25°C

Preparation of Medium: Add components to distilled/deionized water and bring volume to 1.0L. Mix thoroughly. Gently heat and bring to boiling. Distribute into tubes or flasks. Autoclave for 15 min at 15 psi pressure–121°C. Pour into sterile Petri dishes or leave in tubes.

Use: For the cultivation and maintenance of *Streptosporangium longisporum*.

Emerson Yp Ss Agar

Composition per liter:

Agar	20.0g
Soluble starch	15.0g
Yeast extract	4.0g
K_2HPO_4	1.0g
$MgSO_4 \cdot 7H_2O$	0.5g

pH 7.0 ± 0.2 at 25°C

Source: This medium is available as a premixed powder from Difco Laboratories.

Preparation of Medium: Add components to distilled/deionized water and bring volume to 1.0L. Mix thoroughly. Gently heat and bring to boiling. Distribute into tubes or flasks. Autoclave for 15 min at 13 psi pressure–118°C. Pour into sterile Petri dishes or leave in tubes.

Use: For the cultivation and maintenance of *Allomyces* and other fungi.

Emmon's Modification of Sabouraud's Agar (Sabouraud's Agar, Modified)

Composition per liter:

Agar	20.0g
Glucose	20.0g
Pancreatic digest of casein	5.0g
Peptic digest of animal tissue	5.0g

pH 6.8–7.0 at 25°C

Source: This medium is available as a premixed powder from Difco Laboratories.

Preparation of Medium: Add components to distilled/deionized water and bring volume to 1.0L. Mix thoroughly. Gently heat and bring to boiling. Distribute into tubes or flasks. Autoclave for 15 min at 15 psi pressure–121°C. Pour into sterile Petri dishes or leave in tubes.

Use: For the cultivation and maintenance of *Bacillus subtilis, Nocardia* species, and *Streptomyces albus.*

Endamoeba Medium
See: Entamoeba Medium

Endo Agar

Composition per liter:

Agar	15.0g
Lactose	10.0g
Peptic digest of animal tissue	10.0g
K_2HPO_4	3.5g
Na_2SO_3	2.5g
Basic Fuchsin	0.5g

pH 7.4 ± 0.2 at 25°C

Source: This medium is available as a premixed powder from BBL Microbiology Systems and Difco Laboratories.

Caution: Basic Fuchsin is a potential carcinogen and care must be taken to avoid inhalation of the powdered dye and contact with the skin.

Preparation of Medium: Add components to distilled/deionized water and bring volume to 1.0L. Mix thoroughly. Gently heat and bring to boiling. Autoclave for 15 min at 15 psi pressure–121°C. Cool to 45°–50°C. Pour into sterile Petri dishes. Swirl flask while pouring plates to keep precipitate in suspension. Protect from the light.

Use: For the selective isolation, cultivation and differentiation of coliform and other enteric microorganisms based on their ability to ferment lactose. Lactose-fermenting bacteria appear as dark red colonies with a gold metallic sheen. Lactose-nonfermenting bacteria appear as colorless or translucent colonies.

Endo Agar

Composition per liter:

Agar	10.0g
Lactose	10.0g
Peptic digest of animal tissue	10.0g
K_2HPO_4	3.5g
Na_2SO_3	2.5g
Basic Fuchsin solution	4.0mL

pH 7.5 ± 0.2 at 25°C

Source: This medium is available as a premixed powder from Oxoid.

Caution: Basic Fuchsin is a potential carcinogen and care must be taken to avoid inhalation of the powdered dye and contact with the skin.

Basic Fuchsin Solution:
Composition per 10mL:

Basic Fuchsin	1.0g
Ethanol (95% solution)	10.0mL

Preparation of Basic Fuchsin Solution: Add Basic Fuchsin to 10.0mL of ethanol. Mix thoroughly.

Preparation of Medium: Add components to distilled/deionized water and bring volume to 1.0L. Mix thoroughly. Gently heat and bring to boiling. Autoclave for 15 min at 15 psi pressure–121°C. Cool to 45°–50°C. Pour into sterile Petri dishes. Swirl flask while pouring plates to keep precipitate in suspension. Protect from the light.

Use: For the selective isolation, cultivation and differentiation of coliform and other enteric microorganisms based on their ability to ferment lactose. Lactose-fermenting bacteria appear as dark red colonies with a gold metallic sheen. Lactose-nonfermenting bacteria appear as colorless or translucent colonies.

Endo Agar, LES (Endo Agar, Laurance Experimental Station) (m-Endo Agar, LES) (m–LES, Endo Agar)

Composition per liter:

Agar	14.0g
Lactose	9.4g
Peptones (pancreatic digest of casein 65% and yeast extract 35%)	7.5g
NaCl	3.7g
Pancreatic digest of casein	3.7g
Peptic digest of animal tissue	3.7g
K_2HPO_4	3.3g
Na_2SO_3	1.6g
Yeast extract	1.2g
KH_2PO_4	1.0g
Basic Fuchsin	0.8g
Sodium lauryl sulfate	0.05g
Ethanol	20.0mL

pH 7.2 ± 0.2 at 25°C

Source: This medium is available as a premixed powder from BBL Microbiology Systems and Difco Laboratories.

Caution: Basic Fuchsin is a potential carcinogen and care must be taken to avoid inhalation of the powdered dye and contact with the skin.

Preparation of Medium: Add ethanol to approximately 900.0mL of distilled/deionized water. Add remaining components. Bring volume to 1.0L with distilled/deionized water. Mix thoroughly. Gently heat and bring to boiling. Autoclave for 15 min at 15 psi pressure–121°C. Pour into sterile 60-mm Petri dishes in 4.0mL volumes. Protect from the light.

Use: For the cultivation and enumeration of coliform bacteria by the membrane filter method.

Endo Agar, LES (m-Endo Agar, LES)

Composition per liter:

Agar	10.0g
Lactose	9.4g
Tryptose	7.5g
NaCl	3.7g
Peptone	3.7g
Pancreatic digest of casein	3.7g
K_2HPO_4	3.3g
Na_2SO_3	1.6g
Yeast extract	1.2g
KH_2PO_4	1.0g
Sodium deoxycholate	0.1g
Sodium lauryl sulfate	0.05g
Basic Fuchsin solution	8.0mL

pH 7.2 ± 0.2 at 25°C

Caution: Basic Fuchsin is a potential carcinogen and care must be taken to avoid inhalation of the powdered dye and contact with the skin.

Basic Fuchsin Solution:

Composition per 10mL:

Basic Fuchsin	1.0g
Ethanol (95% solution)	10.0mL

Preparation of Basic Fuchsin Solution: Add Basic Fuchsin to 10.0mL of ethanol. Mix thoroughly.

Preparation of Medium: Add components to distilled/deionized water and bring volume to 1.0L. Mix thoroughly. Gently heat and bring to boiling. Autoclave for 15 min at 15 psi pressure–121°C. Cool to 45°–50°C. Pour into sterile Petri dishes. Swirl flask while pouring plates to keep precipitate in suspension. Protect from the light.

Use: For the cultivation and enumeration of coliform bacteria from water by the membrane filter method.

Endo Broth (m-Endo Broth)

Composition per liter:

Lactose	12.5g
Peptone	10.0g
NaCl	5.0g
Pancreatic digest of casein	5.0g
Peptic digest of animal tissue	5.0g
K_2HPO_4	4.375g
Na_2SO_3	2.1g
Yeast extract	1.5g
KH_2PO_4	1.375g
Basic Fuchsin	1.05g
Sodium deoxycholate	0.1g
Ethanol (95% solution)	20.0mL

pH 7.2 ± 0.1 at 25°C

Source: This medium is available as a premixed powder from BBL Microbiology Systems and Difco Laboratories.

Caution: Basic Fuchsin is a potential carcinogen and care must be taken to avoid inhalation of the powdered dye and contact with the skin.

Preparation of Medium: Add ethanol to approximately 900.0mL of distilled/deionized water. Add remaining components. Bring volume to 1.0L with distilled/deionized water. Mix thoroughly. Gently heat and bring to boiling. Rapidly cool broth below 45°C. Do not autoclave. Use 1.8–2.0mL for each filter pad. Protect from the light. Prepare broth freshly.

Use: For the cultivation and enumeration of coliform bacteria from water by the membrane filter method.

Enriched Nutrient Broth

Composition per liter:

Beef heart, infusion from	250.0g
Tryptose	5.0g
Pancreatic digest of gelatin	3.38g
NaCl	2.5g
Yeast extract	2.5g
Beef extract	2.02g

pH 7.2 ± 0.1 at 25°C

Preparation of Medium: Add components to distilled/deionized water and bring volume to 1.0L. Mix thoroughly. Gently heat and bring to boiling. Distribute into tubes or flasks. Autoclave for 15 min at 15 psi pressure–121°C. Pour into sterile Petri dishes or leave in tubes.

Use: For the cultivation and maintenance of *Bacillus licheniformis, Bacillus polymyxa, Bacillus subtilis, Escherichia coli, Listonella anguillarum, Micrococcus* *luteus, Pseudomonas aeruginosa, Salmonella choleraesuis, Staphylococcus aureus, Staphylococcus epidermidis, Streptococcus* species, and *Vibrio cholerae.*

Enrichment Broth, pH 7.3

Composition per liter:

Pancreatic digest of casein	17.0g
NaCl	5.0g
Papaic digest of soybean meal	3.0g
Glucose	2.5g
K_2HPO_4	2.5g
Yeast extract	6.0g
Nalidixic acid solution	8.0mL
Cycloheximide solution	5.1mL
Acriflavin·HCl solution	3.0mL

pH 7.3 ± 0.2 at 25°C

Cycloheximide Solution:

Composition per 10mL:

Cycloheximide	0.1g
Ethanol, absolute	4.0mL

Preparation of Cycloheximide Solution: Add components to distilled/deionized water and bring volume to 10.0mL. Mix thoroughly. Filter sterilize.

Nalidixic Acid Solution:

Composition per 10mL:

Nalidixic acid	0.05g

Preparation of Nalidixic Acid Solution: Add nalidixic acid to distilled/deionized water and bring volume to 10.0mL. Mix thoroughly. Filter sterilize.

Acriflavin·HCl Solution:

Composition per 10mL:

Acriflavin·HCl	0.05g

Preparation of Acriflavin·HCl Solution: Add acriflavin·HCl to distilled/deionized water and bring volume to 10.0mL. Mix thoroughly. Filter sterilize.

Preparation of Medium: Add components—except cycloheximide solution, nalidixic acid solution, and acriflavin·HCl solution—to distilled/deionized water and bring volume to 983.9mL. Mix thoroughly. Gently heat and bring to boiling. Autoclave for 15 min at 15 psi pressure–121°C. Cool to 45°–50°C. Aseptically add 5.1mL of sterile cycloheximide solution, 8.0mL of sterile nalidixic acid solution, and 3.0mL of sterile acriflavin·HCl solution. Mix thoroughly. Aseptically distribute into sterile tubes.

Use: For the isolation, cultivation, and enrichment of a variety of microorganisms from nondairy foods.

Enrichment Broth, pH 7.3

Composition per liter:

Pancreatic digest of casein	17.0g
NaCl	5.0g
Papaic digest of soybean meal	3.0g
Glucose	2.5g
K_2HPO_4	2.5g
Yeast extract	6.0g
Nalidixic acid solution	8.0mL
Cycloheximide solution	5.1mL
Acriflavin·HCl solution	2.0mL

pH 7.3 ± 0.2 at 25°C

Cycloheximide Solution:

Composition per 10mL:

Cycloheximide	0.1g
Ethanol, absolute	4.0mL

Preparation of Cycloheximide Solution: Add components to distilled/deionized water and bring volume to 10.0mL. Mix thoroughly. Filter sterilize.

Nalidixic Acid Solution:

Composition per 10mL:

Nalidixic acid	0.05g

Preparation of Nalidixic Acid Solution: Add nalidixic acid to distilled/deionized water and bring volume to 10.0mL. Mix thoroughly. Filter sterilize.

Acriflavin·HCl Solution:

Composition per 10mL:

Acriflavin·HCl	0.05g

Preparation of Acriflavin·HCl Solution: Add acriflavin·HCl to distilled/deionized water and bring volume to 10.0mL. Mix thoroughly. Filter sterilize.

Preparation of Medium: Add components—except cycloheximide solution, nalidixic acid solution, and acriflavin·HCl solution—to distilled/deionized water and bring volume to 984.9mL. Mix thoroughly. Gently heat and bring to boiling. Autoclave for 15 min at 15 psi pressure–121°C. Cool to 45°–50°C. Aseptically add 5.1mL of cycloheximide solution, 8.0mL of nalidixic acid solution, and 2.0mL of acriflavin·HCl solution. Mix thoroughly. Aseptically distribute into sterile tubes.

Use: For the isolation, cultivation, and enrichment of a variety of microorganisms from milk and dairy products.

Enrichment Broth, pH 7.3 with Pyruvate

Composition per liter:

Pancreatic digest of casein	17.0g
NaCl	5.0g
Papaic digest of soybean meal	3.0g

Glucose 2.5g
K_2HPO_4 .... 2.5g
Yeast extract 6.0g
Pyruvate solution 11.1mL
Nalidixic acid solution 8.0mL
Cycloheximide solution 5.1mL
Acriflavin·HCl solution 3.0mL

pH 7.3 ± 0.2 at 25°C

Pyruvate Solution:

Composition per 20mL:

Sodium pyruvate 2.0g

Preparation of Pyruvate Solution: Add sodium pyruvate to distilled/deionized water and bring volume to 20.0mL. Mix thoroughly. Filter sterilize.

Cycloheximide Solution:

Composition per 10mL:

Cycloheximide 0.1g
Ethanol, absolute 4.0mL

Preparation of Cycloheximide Solution: Add components to distilled/deionized water and bring volume to 10.0mL. Mix thoroughly. Filter sterilize.

Nalidixic Acid Solution:

Composition per 10mL:

Nalidixic acid 0.05g

Preparation of Nalidixic Acid Solution: Add nalidixic acid to distilled/deionized water and bring volume to 10.0mL. Mix thoroughly. Filter sterilize.

Acriflavin·HCl Solution:

Composition per 10mL:

Acriflavin·HCl 0.05g

Preparation of Acriflavin·HCl Solution: Add acriflavin·HCl to distilled/deionized water and bring volume to 10.0mL. Mix thoroughly. Filter sterilize.

Preparation of Medium: Add components—except pyruvate solution, cycloheximide solution, nalidixic acid solution, and acriflavin·HCl solution—to distilled/deionized water and bring volume to 972.8mL. Mix thoroughly. Gently heat and bring to boiling. Autoclave for 15 min at 15 psi pressure–121°C. Cool to 45°–50°C. Aseptically add 11.1mL of sterile pyruvate solution. Mix thoroughly. Inoculate medium and incubate at 30°C for 6 hr. Aseptically add 5.1mL of sterile cycloheximide solution, 8.0mL of sterile nalidixic acid solution, and 3.0mL of sterile acriflavin·HCl solution. Mix thoroughly.

Use: For the isolation, cultivation, and enrichment of a *Listeria* species from nondairy foods.

Enrichment Broth, pH 7.3 with Pyruvate

Composition per liter:

Pancreatic digest of casein 17.0g
NaCl 5.0g
Papaic digest of soybean meal 3.0g
Glucose 2.5g
K_2HPO_4 .... 2.5g
Yeast extract 6.0g
Pyruvate solution 11.1mL
Nalidixic acid solution 8.0mL
Cycloheximide solution 5.1mL
Acriflavin·HCl solution 2.0mL

pH 7.3 ± 0.2 at 25°C

Pyruvate Solution:

Composition per 20mL:

Sodium pyruvate 2.0g

Preparation of Pyruvate Solution: Add sodium pyruvate to distilled/deionized water and bring volume to 20.0mL. Mix thoroughly. Filter sterilize.

Cycloheximide Solution:

Composition per 10mL:

Cycloheximide 0.1g
Ethanol, absolute 4.0mL

Preparation of Cycloheximide Solution: Add components to distilled/deionized water and bring volume to 10.0mL. Mix thoroughly. Filter sterilize.

Nalidixic Acid Solution:

Composition per 10mL:

Nalidixic acid 0.05g

Preparation of Nalidixic Acid Solution: Add nalidixic acid to distilled/deionized water and bring volume to 10.0mL. Mix thoroughly. Filter sterilize.

Acriflavin·HCl Solution:

Composition per 10mL:

Acriflavin·HCl 0.05g

Preparation of Acriflavin·HCl Solution: Add acriflavin·HCl to distilled/deionized water and bring volume to 10.0mL. Mix thoroughly. Filter sterilize.

Preparation of Medium: Add components—except pyruvate solution, cycloheximide solution, nalidixic acid solution, and acriflavin·HCl solution—to distilled/deionized water and bring volume to 973.8mL. Mix thoroughly. Gently heat and bring to boiling. Autoclave for 15 min at 15 psi pressure–121°C. Cool to 45°–50°C. Aseptically add 11.1mL of sterile pyruvate solution. Mix thoroughly. Inoculate medium and incubate at 30°C for 6 hr. Aseptically add 5.1mL of sterile cycloheximide solution, 8.0mL of sterile nalidixic acid solution, and 2.0mL of sterile acriflavin·HCl solution. Mix thoroughly.

Use: For the isolation, cultivation, and enrichment of a *Listeria* species from milk and dairy products.

Enrichment Broth for *Aeromonas hydrophila*

Composition per liter:

NaCl	5.0g
Maltose	3.5g
Yeast extract	3.0g
Bile salts No. 3	1.0g
L-Cysteine·HCl·H_2O	0.3g
Bromthymol Blue	0.03g
Novobiocin	5.0mg

pH 7.0 ± 0.2 at 25°C

Preparation of Medium: Add components to distilled/deionized water and bring volume to 1.0L. Mix thoroughly. Distribute into tubes or flasks. Autoclave for 15 min at 15 psi pressure–121°C.

Use: For the cultivation and enrichment of *Aeromonas hydrophila*.

Entamoeba Medium (Endamoeba Medium)

Composition per liter:

Liver infusion	272.0g
Rice powder	14.2g
Agar	11.0g
Proteose peptone	5.5g
Sodium glycerophosphate	3.0g
NaCl	2.7g
Horse serum	50.0mL

pH 7.0 ± 0.2 at 25°C

Source: This medium is available as a premixed powder from Difco Laboratories.

Rice Powder:
Composition per 15g:

Rice powder	15.0g

Preparation of Rice Powder: Sterilize rice powder at 160°C for 60 min. Do not overheat or rice powder will scorch.

Preparation of Medium: Add components, except horse serum and rice powder, to distilled/deionized water and bring volume to 994.0mL. Mix thoroughly. Gently heat and bring to boiling. Distribute into tubes in 7.0mL volumes. Autoclave for 15 min at 15 psi pressure–121°C. Allow tubes to cool in a slanted position. Aseptically add enough sterile horse serumm to each tube to cover about half the slant. Aseptically add 0.1g of sterile rice powder to each tube.

Use: For the cultivation of *Entamoeba histolytica*.

Enteric Fermentation Base (Fermentation Base for *Campylobacter*)

Composition per liter:

Peptic digest of animal tissue	10.0g
NaCl	5.0g
Beef extract	3.0g
Carbohydrate solution	100.0mL
Andrade's indicator	10.0mL

pH 7.2 ± 0.1 at 25°C

Source: This medium is available as a premixed powder from Difco Laboratories.

Carbohydrate Solution:
Composition per 100mL:

Carbohydrate	10.0g

Preparation of Carbohydrate Solution: Add carbohydrate to distilled/deionized water and bring volume to 100.0mL. Mix thoroughly. Filter sterilize. Glucose, lactose, mannitol, sucrose, adonitol, arabinose, cellobiose, dulcitol, glycerol, inositol, salicin, xylose or other carbohydrates may be used. For the preparation of expensive carbohydrate solutions (adonitol, arabinose, cellobiose, dulcitol, glycerol, inositol, salicin or xylose), 5.0g of carbohydrate per 100.0mL of distilled/deionized water may be used.

Andrade's Indicator:
Composition per 100mL:

NaOH (1*N* solution)	16.0mL
Acid Fuchsin	0.1 g

Caution: Acid Fuchsin is a potential carcinogen and care must be taken to avoid inhalation of the powdered dye and contact with the skin.

Preparation of Andrade's Indicator: Add components to distilled/deionized water and bring volume to 100.0mL. Mix thoroughly.

Preparation of Medium: Add components, except carbohydrate solution, to distilled/deionized water and bring volume to 900.0mL. Mix thoroughly. Gently heat and bring to boiling. Distribute into tubes that contain an inverted Durham tube in 9.0mL volumes. Autoclave for 15 min at 15 psi pressure–121°C. Cool to 25°C. Aseptically add 1.0mL of sterile carbohydrate solution per tube. Mix thoroughly.

Use: For the cultivation and differentiation of a variety of bacteria based on their ability to ferment different carbohydrates. Bacteria that produce acid from carbohydrate fermentation turn the medium dark pink to red. Bacteria that produce gas have a bubble trapped in the Durham tube.

Enterobacter Medium

Composition per 800mL:

Casein hydrolysate	2.0g
K_2HPO_4	1.4g
K_2SO_4	1.0g
Yeast extract	1.0g
KH_2PO_4	0.6g
$MgSO_4$	0.5g
Glycerol	20.0mL

Preparation of Medium: Add components to distilled/deionized water and bring volume to 800.0mL. Mix thoroughly. Distribute into tubes or flasks. Autoclave for 15 min at 15 psi pressure–121°C.

Use: For the cultivation and maintenance of *Enterobacter* species and *Klebsiella pneumoniae*.

Enterobacteriaceae Enrichment Broth
See: **EE Broth**

Enterobacteriaceae Enrichment Broth, Mossel
See: **EE Broth, Mossel**

Enterococci Confirmatory Agar

Composition per liter:

Agar	15.0g
Glucose	5.0g
Pancreatic digest of casein	5.0g
Yeast extract	5.0g
NaN_3	0.4g
Methylene Blue	10.0mg

pH 8.0 ± 0.2 at 25°C

Source: This medium is available as a premixed powder from Difco Laboratories.

Caution: Sodium azide is toxic. Azides also react with metals and disposal must be highly diluted.

Preparation of Medium: Add components to distilled/deionized water and bring volume to 1.0L. Mix thoroughly. Gently heat and bring to boiling. Distribute into tubes. Autoclave for 15 min at 15 psi pressure–121°C. Allow tubes to cool in a slanted position. Add sufficient amount of Enterococci confirmatory broth to cover half the slant.

Use: For the identification of enterococci from water by the confirmatory test.

Enterococci Confirmatory Broth

Composition per liter:

NaCl	65.0g
Glucose	5.0g
Pancreatic digest of casein	5.0g
Yeast extract	5.0g
NaN_3	0.4g
Methylene Blue	10.0mg
Penicillin	650U

pH 8.0 ± 0.2 at 25°C

Source: This medium is available as a premixed powder from Difco Laboratories.

Caution: Sodium azide is toxic. Azides also react with metals and disposal must be highly diluted.

Preparation of Medium: Add components, except penicillin, to distilled/deionized water and bring volume to 1.0L. Mix thoroughly. Gently heat and bring to boiling. Autoclave for 15 min at 15 psi pressure–121°C. Cool to 25°C. Aseptically add penicillin. Mix thoroughly. Add sufficient amount of Enterococci confirmatory broth to cover half the slant of Enterococci confirmatory agar.

Use: For the identification of enterococci from water by the confirmatory test.

Enterococci Presumptive Broth

Composition per liter:

Glucose	5.0g
Pancreatic digest of casein	5.0g
Yeast extract	5.0g
NaN_3	0.4g
Bromthymol Blue	32.0mg

pH 8.4 ± 0.2 at 25°C

Source: This medium is available as a premixed powder from Difco Laboratories.

Caution: Sodium azide is toxic. Azides also react with metals and disposal must be highly diluted.

Preparation of Medium: Add components to distilled/deionized water and bring volume to 1.0L. Mix thoroughly. Distribute into tubes or flasks. Autoclave for 15 min at 15 psi pressure–121°C.

Use: For the isolation and identification of enterococci from water by the presumptive test. Bacteria that produce acid and turn the medium yellow and turbid after incubation at 45°C are presumptive enterococci.

Enterococcosel™ Agar

Composition per liter:

Pancreatic digest of casein	17.0g
Agar	13.5g

Oxgall	10.0g
NaCl	5.0g
Yeast extract	5.0g
Peptic digest of animal tissue	3.0g
Esculin	1.0g
Sodium citrate	1.0g
Ferric ammonium citrate	0.5g
NaN_3	0.25g

pH 7.1 ± 0.2 at 25°C

Source: This medium is available as a premixed powder from BBL Microbiology Systems.

Caution: Sodium azide is toxic. Azides also react with metals and disposal must be highly diluted.

Preparation of Medium: Add components to distilled/deionized water and bring volume to 1.0L. Mix thoroughly. Gently heat while stirring and bring to boiling. Distribute into tubes or flasks. Autoclave for 15 min at 15 psi pressure–121°C. Pour into sterile Petri dishes or leave in tubes.

Use: For the rapid, selective isolation, cultivation and enumeration of fecal Group D streptococci (enterococci). For the cultivation of staphylococci and *Listeria monocytogenes*.

Enterococcosel™ Broth

Composition per liter:

Pancreatic digest of casein	17.0g
Oxgall	10.0g
NaCl	5.0g
Yeast extract	5.0g
Peptic digest of animal tissue	3.0g
Esculin	1.0g
Sodium citrate	1.0g
Ferric ammonium citrate	0.5g
NaN_3	0.25g

pH 7.1 ± 0.2 at 25°C

Source: This medium is available as a premixed powder from BBL Microbiology Systems.

Caution: Sodium azide is toxic. Azides also react with metals and disposal must be highly diluted.

Preparation of Medium: Add components to distilled/deionized water and bring volume to 1.0L. Mix thoroughly. Gently heat while stirring until dissolved. Distribute into tubes or flasks. Autoclave for 15 min at 15 psi pressure–121°C.

Use: For the cultivation and differentiation of Group D streptococci (enterococci).

Enterococcus Agar (m-*Enterococcus* Agar) (Azide Agar)

Composition per liter:

Pancreatic digest of casein	15.0g
Agar	10.0g
Papaic digest of soybean meal	5.0g
Yeast extract	5.0g
KH_2PO_4	4.0g
Glucose	2.0g
NaN_3	0.4g
Triphenyltetrazolium chloride	0.1g

pH 7.2 ± 0.2 at 25°C

Source: This medium is available as a premixed powder from BBL Microbiology Systems and Difco Laboratories.

Caution: Sodium azide is toxic. Azides also react with metals and disposal must be highly diluted.

Preparation of Medium: Add components to distilled/deionized water and bring volume to 1.0L. Mix thoroughly. Gently heat and bring to boiling. Cool to 45°–50°C. Do not autoclave. Pour into sterile Petri dishes.

Use: For the isolation, cultivation and enumeration of entercocci in water, sewage and feces by the membrane filter method. Also used for the direct plating of specimens for detection and enumeration of fecal streptococci.

Eosin Methylene Blue Agar
See: **EMB Agar**

Eosin Methylene Blue Agar, Levine
See: **Levine EMB Agar**

Eosin Methylene Blue Agar, Modified
See: **EMB Agar, Modified**

Epoxysuccinic Acid Medium
See: **ESA Medium**

Erwinia amylovora Selective Medium

Composition per liter:

Agar	20.0g
Mannitol	10.0g
L-Asparagine	3.0g
Sodium taurocholate	2.5g
K_2HPO_4	2.0g
Nicotinic acid	0.5g
$MgSO_4 \cdot 7H_2O$	0.2g
Nitrilotriacetic acid	10.0mL
Actidione (cycloheximide) solution	10.0mL
Bromthymol Blue	9.0mL
Neutral Red	2.5mL
$TlNO_3$ solution	1.75mL
Tergitol™ 7	0.1mL

pH 7.2–7.3 at 25°C

Cycloheximide Solution:

Composition per 10mL:

Cycloheximide	0.05g

Preparation of Cycloheximide Solution: Add cycloheximide to distilled/deionized water and bring volume to 10.0mL. Mix thoroughly. Filter sterilize.

$TlNO_3$ Solution:

Composition per 10mL:

$TlNO_3$	0.1g

Preparation of $TlNO_3$ Solution: Add $TlNO_3$ to distilled/deionized water and bring volume to 10.0mL. Mix thoroughly. Filter sterilize.

Preparation of Medium: Add components, except $TlNO_3$ solution and cycloheximide solution, to distilled/deionized water and bring volume to 988.25mL. Mix thoroughly. Adjust pH to 7.2–7.3. Gently heat and bring to boiling. Autoclave for 15 min at 15 psi pressure–121°C. Cool to 45°–50°C. Aseptically add 1.75mL of sterile $TlNO_3$ solution and 10.0mL of sterile cycloheximide solution. Mix thoroughly. Pour into sterile Petri dishes or distribute into sterile tubes.

Use: For the isolation and cultivation of *Erwinia amylovora*.

Erwinia Medium D3

Composition per liter:

Agar	15.0g
Arabinose	10.0g
Sucrose	10.0g
LiCl	7.0g
Casein hydrolysate	5.0g
NaCl	5.0g
$MgSO_4 \cdot 7H_2O$	0.3g
Acid Fuchsin	0.1g
Bromthymol Blue	0.06g
Sodium dodecyl sulfate	0.05g

pH 7.0 ± 0.2 at 25°C

Caution: Acid Fuchsin is a potential carcinogen and care must be taken to avoid inhalation of the powdered dye and contact with the skin.

Preparation of Medium: Add components to distilled/deionized water and bring volume to 1.0L. Mix thoroughly. Gently heat and bring to boiling. Adjust pH to 8.2. Autoclave for 15 min at 15 psi pressure–121°C. Cool to 45°–50°C. The pH after autoclaving should be 7.0. Pour into sterile Petri dishes or distribute into sterile tubes.

Use: For the isolation and cultivation of *Erwinia* species.

Erwinia Selective Medium

Composition per liter:

Agar	15.0g
$(NH_4)_2SO_4$	5.0g
K_2HPO_4	2.0g
Eosin Y	0.4g
Methylene Blue	0.065g
Glycerol	10.0mL
Antibiotic solution	10.0mL

Antibiotic Solution:

Composition per 10mL:

Cycloheximide	0.25g
Novobiocin	0.04g
Neomycin sulfate	0.04g

Preparation of Antibiotic Solution: Add components to distilled/deionized water and bring volume to 10.0mL. Mix thoroughly. Filter sterilize.

Preparation of Medium: Add components, except antibiotic solution, to distilled/deionized water and bring volume to 990.0mL. Mix thoroughly. Gently heat and bring to boiling. Autoclave for 15 min at 15 psi pressure–121°C. Cool to 45°–50°C. Aseptically add sterile antibiotic solution. Mix thoroughly. Pour into sterile Petri dishes or distribute into sterile tubes.

Use: For the selective isolation and cultivation of *Erwinia* species.

Erysipelothrix Medium

Composition per liter:

$Na_2HPO_4 \cdot 12H_2O$	18.0g
Glucose	6.0g
Peptone S	5.0g
Yeast extract	5.0g

L-Arginine·HCl....0.5g
Tween™ 80....0.5mL

pH 7.8–8.0 at 25°C

Preparation of Medium: Add components to distilled/deionized water and bring volume to 1.0L. Mix thoroughly. Distribute into tubes or flasks. Autoclave for 15 min at 15 psi pressure–121°C.

Use: For the cultivation of *Erysipelothrix* species.

Erythritol Agar

Composition per liter:

Agar....15.0g
Erythritol....2.0g
K_2HPO_4....1.15g
NH_4NO_3....1.0g
KH_2PO_4....0.625g
$MgSO_4 \cdot 7H_2O$....0.02g

Preparation of Medium: Add components to distilled/deionized water and bring volume to 1.0L. Mix thoroughly. Gently heat and bring to boiling. Distribute into tubes or flasks. Autoclave for 15 min at 15 psi pressure–121°C. Pour into sterile Petri dishes or leave in tubes.

Use: For the cultivation and maintenance of *Klebsiella pneumoniae*.

Erythritol Broth

Composition per liter:

Erythritol....2.0g
K_2HPO_4....1.15g
NH_4NO_3....1.0g
Yeast extract....1.0g
KH_2PO_4....0.625g
$MgSO_4 \cdot 7H_2O$....0.02g

Preparation of Medium: Add components to distilled/deionized water and bring volume to 1.0L. Mix thoroughly. Distribute into tubes or flasks. Autoclave for 15 min at 15 psi pressure–121°C.

Use: For the cultivation of *Klebsiella pneumoniae*.

ESA Medium (Epoxysuccinic Acid Medium)

Composition per liter:

Agar....20.0g
$(NH_4)_2SO_4$....3.0g
KH_2PO_4....1.5g
Na_2HPO_4....1.5g
$MgSO_4 \cdot 7H_2O$....0.5g
Yeast extract....0.5g
$CaCl_2 \cdot 2H_2O$....0.01g
$FeSO_4 \cdot 7H_2O$....0.01g
$MnSO_4 \cdot 4H_2O$....0.001g
Epoxysuccinate solution....100.0mL

pH 7.0 ± 0.2 at 25°C

Epoxysuccinate Solution:
Composition per 100mL:

Sodium *cis*-epoxysuccinate....10.0g

Preparation of Epoxysuccinate Solution: Add sodium *cis*-epoxysuccinate to distilled/deionized water and bring volume to 100.0mL. Mix thoroughly. Filter sterilize.

Preparation of Medium: Add components, except epoxysuccinate solution, to distilled/deionized water and bring volume to 900.0mL. Mix thoroughly. Gently heat and bring to boiling. Autoclave for 15 min at 15 psi pressure–121°C. Cool to 50°–60°C. Aseptically add sterile epoxysuccinate solution. Mix thoroughly. Pour into sterile Petri dishes or distribute into sterile tubes.

Use: For the cultivation and maintenance of *Nocardia tartaricans*.

Escherichia coli Broth
See: EC Broth

Escherichia Medium (ATCC Medium 765)

Composition per liter:

K_2HPO_4....11.67g
Casamino acids....11.0g
KH_2PO_4....4.49g
$(NH_4)_2SO_4$....1.98g
$MgSO_4 \cdot 7H_2O$....0.25g
$FeSO_4 \cdot 7H_2O$....0.5mg
Glycerol....20.25mL

Preparation of Medium: Add components to distilled/deionized water and bring volume to 1.0L. Mix thoroughly. Distribute into tubes or flasks. Autoclave for 15 min at 15 psi pressure–121°C.

Use: For the cultivation and maintenance of *Escherichia coli*.

Escherichia Medium (ATCC Medium 52)

Composition per liter:

Glucose....40.0g
Agar....30.0g
Peptone....10.0g

Preparation of Medium: Add components to distilled/deionized water and bring volume to 1.0L. Mix thoroughly. Gently heat and bring to boiling. Distribute into tubes or flasks. Autoclave for 15 min at 15 psi pressure–121°C. Pour into sterile Petri dishes or leave in tubes.

Use: For the cultivation and maintenance of *Escherichia coli.*

Escherichia Medium (ATCC Medium 53)

Composition per liter:

Agar	15.0g
Casein hydrolysate	6.0g
K_2HPO_4	0.2g
$MgSO_4 \cdot 7H_2O$	0.2g
Asparagine	0.15g
Glycerol	2.0mL
$FeSO_4 \cdot 7H_2O$	1.0μg
Vitamin B_{12} solution	10.0mL

pH 7.0 ± 0.2 at 25°C

Vitamin B_{12} Solution:

Composition per 10mL:

Vitamin B_{12}	0.04g

Preparation of Vitamin B_{12} Solution: Add vitamin B_{12} to distilled/deionized water and bring volume to 10.0mL. Mix thoroughly.

Preparation of Medium: Add components—except glycerol, agar and vitamin B_{12} solution—to distilled/deionized water and bring volume to 900.0mL. Mix thoroughly. Gently heat and bring to boiling. Adjust pH to 7.0. Filter through Whatman #1 filter paper. Add glycerol and agar. Mix thoroughly. Gently heat and bring to boiling. Add vitamin B_{12} solution. Distribute into tubes or flasks. Autoclave for 15 min at 15 psi pressure–121°C. Pour into sterile Petri dishes or leave in tubes.

Use: For the cultivation and maintenance of *Escherichia coli.*

Escherichia Medium (ATCC Medium 57)

Composition per liter:

Agar	15.0g
K_2HPO_4	7.0g
Glycerol	5.0g
KH_2PO_4	3.0g
$(NH_4)_2SO_4$	1.5g
L-Lysine	0.1g
$MgSO_4$	0.1g
$CaCl_2$	0.01g
$FeSO_4 \cdot 7H_2O$	0.5mg

pH 7.1 ± 0.2 at 25°C

Preparation of Medium: Add components to distilled/deionized water and bring volume to 1.0L. Mix thoroughly. Gently heat and bring to boiling. Distribute into tubes or flasks. Autoclave for 15 min at 15 psi pressure–121°C. Pour into sterile Petri dishes or leave in tubes.

Use: For the cultivation and maintenance of *Escherichia coli.*

Escherichia Medium (ATCC Medium 60)

Composition per liter:

Agar	15.0g
K_2HPO_4	7.0g
KH_2PO_4	3.0g
Casein hydrolysate	2.0g
Sodium citrate, anhydrous	0.4g
$MgSO_4 \cdot 7H_2O$	0.1g
$(NH_4)_2SO_4$	0.1g
Glycerol	2.0mL

Preparation of Medium: Add components to distilled/deionized water and bring volume to 1.0L. Mix thoroughly. Gently heat and bring to boiling. Distribute into tubes or flasks. Autoclave for 15 min at 15 psi pressure–121°C. Pour into sterile Petri dishes or leave in tubes.

Use: For the cultivation and maintenance of *Escherichia coli.*

Escherichia Medium (ATCC Medium 62)

Composition per 700mL:

Agar	12.5g
Casein hydrolysate	6.0g
Glycerol	2.0g
K_2HPO_4	0.2g
$MgSO_4 \cdot 7H_2O$	0.2g
L-Asparagine	0.15g
$FeSO_4 \cdot 7H_2O$	5.0mg
Vitamin B_{12}	0.4mg

pH 7.2 ± 0.2 at 25°C

Preparation of Medium: Add components, except glycerol, agar and vitamin B_{12}, to distilled/deionized water and bring volume to 500.0mL. Adjust pH to 7.2. Mix thoroughly. Gently heat and bring to boiling. Filter through Whatman #1 filter paper. Cool to 25°C. Add agar and glycerol. Bring volume to 1.0L with distilled/deionized water. Mix thoroughly. Gently heat and bring to boiling. Readjust pH to 7.2. Add vitamin B_{12}. Mix thoroughly. Distribute into tubes in 10.0mL volumes. Autoclave for 15 min at 15 psi pressure–121°C. Allow tubes to cool in a slanted position.

Use: For the cultivation and maintenance of *Escherichia coli.*

Escherichia Medium (ATCC Medium 271)

Composition per liter:

Pancreatic digest of casein	10.0g
NaCl	8.0g
Yeast extract	1.0g

Preparation of Medium: Add components to distilled/deionized water and bring volume to 1.0L. Mix thoroughly. Distribute into tubes or flasks. Autoclave for 15 min at 15 psi pressure–121°C.

Use: For the cultivation of *Escherichia coli.*

Esculin Agar

Composition per liter:

Agar	15.0g
Pancreatic digest of casein	13.0g
NaCl	5.0g
Yeast extract	5.0g
Heart muscle, solids from infusion	2.0g
Esculin	1.0g
Ferric citrate	0.5g

pH 7.3 ± 0.2 at 25°C

Preparation of Medium: Add components to distilled/deionized water and bring volume to 1.0L. Mix thoroughly. Gently heat and bring to boiling. Distribute into screw-capped tubes in 3.0mL volumes. Autoclave for 15 min at 15 psi pressure–121°C. Allow tubes to cool in a slanted position.

Use: For the cultivation and differentiation of bacteria based on their ability to hydrolyze esculin and produce H_2S. Bacteria that hydrolyze esculin appear as colonies surrounded by a reddish-brown to dark brown zone. Bacteria that produce H_2S appear as black colonies.

Esculin Agar, Lombard–Dowell

See: **Lombard–Dowell Esculin Agar**

Esculin Broth

Composition per liter:

Beef heart, solids from infusion	500.0g
Tryptose	10.0g
NaCl	5.0g
Agar	1.0g
Esculin	1.0g

pH 7.0 ± 0.2 at 25°C

Preparation of Medium: Add components to distilled/deionized water and bring volume to 1.0L. Mix thoroughly. Gently heat and bring to boiling. Distribute into screw-capped tubes in 7.0mL volumes. Autoclave for 15 min at 15 psi pressure–121°C.

Use: For the cultivation and differentiation of bacteria based on their ability to hydrolyze esculin. Bacteria that hydrolyze esculin turn the medium brown-black to black.

Esculin Iron Agar

Composition per liter:

Agar	15.0g
Esculin	1.0g
Ferric ammonium citrate	0.5g

pH 7.1 ± 0.2 at 25°C

Source: This medium is available as a premixed powder from BBL Microbiology Systems.

Preparation of Medium: Add components to distilled/deionized water and bring volume to 1.0L. Mix thoroughly. Gently heat and bring to boiling. Distribute into tubes or flasks. Autoclave for 15 min at 15 psi pressure–121°C. Pour into sterile Petri dishes.

Use: For the cultivation and identification of enterococci based on their ability to hydrolyze esculin. Used in conjunction with E agar and the membrane filter method.

Esculin Mannitol Agar

Composition per liter:

Agar	13.5g
Polypeptone™	10.0g
D-Mannitol	10.0g
Pancreatic digest of casein	5.0g
Yeast extract	5.0g
NaCl	5.0g
Heart peptone	3.0g
Cornstarch	1.0g
Esculin	1.0g
Ferric ammonium citrate	0.5g
Phenol Red	0.025g
Nalidixic acid solution	10.0mL
Colistin solution	10.0mL

pH 7.3 ± 0.2 at 25°C

Nalidixic Acid Solution:

Composition per 10mL:

Nalidixic acid 0.015g

Preparation of Nalidixic Acid Solution: Add nalidixic acid to distilled/deionized water and bring volume to 10.0mL. Mix thoroughly. Filter sterilize.

Colistin Solution:

Composition per 10mL:

Colistin 0.01g

Preparation of Colistin Solution: Add colistin to distilled/deionized water and bring volume to 10.0mL. Mix thoroughly. Filter sterilize.

Preparation of Medium: Add components, except nalidixic acid solution and colistin solution, to distilled/deionized water and bring volume to 980.0mL. Mix thoroughly. Gently heat and bring to boiling. Autoclave for 15 min at 15 psi pressure–121°C. Cool to 45°–50°C. Aseptically add sterile nalidixic acid solution and colistin solution. Mix thoroughly. Pour into sterile Petri dishes or distribute into sterile tubes.

Use: For the selective isolation, cultivation and differentiation of *Staphylococcus aureus* and Group D streptococci based on mannitol fermentation and hydrolysis of esculin. Bacteria that ferment mannitol appear as yellow colonies surrounded by a yellow zone. Bacteria that hydrolyze esculin appear as dark brown to black colonies surrounded by a dark brown to black zone.

Esculin Thallium Medium

Composition per liter:

Agar 15.0g
Beef extract 10.0g
Peptone 10.0g
NaCl 5.0g
Esculin 1.0g
Thallous sulfate 0.33g
Crystal Violet 1.3mg
Blood, bovine or sheep 50.0mL

pH 7.4 ± 0.2 at 25°C

Preparation of Medium: Add components, except blood, to distilled/deionized water and bring volume to 950.0mL. Mix thoroughly. Gently heat and bring to boiling. Autoclave for 15 min at 15 psi pressure–121°C. Cool to 45°–50°C. Aseptically add sterile blood. Mix thoroughly. Pour into sterile Petri dishes or distribute into sterile tubes.

Caution: Thallium salts are toxic.

Use: For the cultivation of *Streptococcus* species that cause bovine mastitis.

ETGPA (Egg Tellurite Glycine Pyruvate Agar)

Composition per liter:

Agar 17.0g
Glycine 12.0g
Sodium pyruvate 10.0g
Pancreatic digest of casein 10.0g
Beef extract 5.0g
LiCl 5.0g
Yeast extract 1.0g
Egg yolk emulsion 50.0mL
K_2TeO_3 solution 10.0mL

pH 7.0 ± 0.2 at 25°C

Source: This medium is available as a premixed powder from BBL Microbiology Systems.

Egg Yolk Emulsion:

Composition:

Chicken egg yolks 11
Whole chicken egg 1

Preparation of Egg Yolk Emulsion: Soak egg with 1:100 dilution of saturated mercuric chloride solution for 1 min. Crack eggs and separate yolks from whites. Mix egg yolks with 1 chicken egg.

K_2TeO_3 Solution:

Composition per 100mL:

K_2TeO_3 ..... 1.0g

Preparation of K_2TeO_3 Solution: Add K_2TeO_3 to distilled/deionized water and bring volume to 100.0mL. Mix thoroughly. Filter sterilize.

Caution: Potassium tellurite is toxic.

Preparation of Medium: Add components to distilled/deionized water and bring volume to 940.0mL. Mix thoroughly. Gently heat and bring to boiling. Autoclave for 15 min at 15 psi pressure–121°C. Cool to 45–50°C. Add 10.0mL of sterile 1% tellurite solution and 50.0mL of sterile egg yolk emulsion. If desired, add sulphamethazine to a final concentration of 50mg/mL. Mix thoroughly but gently and pour into sterile Petri dishes.

Use: For the selective isolation and enumeration of coagulase-positive staphylococci from food, skin, soil, air and other materials. For the differentiation and identification of staphylococci on the basis of their ability to clear egg yolk. Addition of sulphamethazine inhibits the growth of *Proteus*. Gray-black colonies surrounded by a clear zone are diagnostic for *Staphylococcus aureus*.

Ethyl Violet Azide Broth (EVA Broth)

Composition per liter:

Pancreatic digest of casein	13.5g
Yeast extract	6.5g
Glucose	5.0g
NaCl	5.0g
K_2HPO_4	2.7g
KH_2PO_4	2.7g
NaN_3	0.4g
Ethyl Violet	0.83mg

pH 7.0 ± 0.2 at 25°C

Source: This medium is available as a premixed powder from BBL Microbiology Systems.

Preparation of Medium: Add components to distilled/deionized water and bring volume to 1.0L. Mix thoroughly. Gently heat and bring to boiling. Distribute into tubes in 10.0mL volumes. Autoclave for 15 min at 15 psi pressure–121°C.

Caution: Sodium azide is toxic. Azides also react with metals and disposal must be highly diluted.

Use: For the isolation, cultivation and enumeration of enterococci from water and other specimens. Fecal enterococci turn the medium turbid with a purple sediment on the bottom of the tube.

Ethyl Violet Azide Broth (EVA Broth)

Composition per liter:

Tryptose	20.0g
Glucose	5.0g
NaCl	5.0g
K_2HPO_4	2.7g
KH_2PO_4	2.7g
NaN_3	0.4g
Ethyl Violet	0.83mg

pH 7.0 ± 0.2 at 25°C

Source: This medium is available as a premixed powder from Difco Laboratories.

Preparation of Medium: Add components to distilled/deionized water and bring volume to 1.0L. Mix thoroughly. Gently heat and bring to boiling. Distribute into tubes in 10.0mL volumes. Autoclave for 15 min at 15 psi pressure–121°C.

Caution: Sodium azide is toxic. Azides also react with metals and disposal must be highly diluted.

Use: For the isolation, cultivation and enumeration of enterococci from water and other specimens. Fecal enterococci turn the medium turbid with a purple sediment on the bottom of the tube.

Ethyl Violet Azide Broth (EVA Broth)

Composition per liter:

Tryptose	20.0g
Glucose	5.0g
NaCl	5.0g
K_2HPO_4	2.7g
KH_2PO_4	2.7g
NaN_3	0.3g
Ethyl Violet	0.5mg

pH 6.8 ± 0.2 at 25°C

Source: This medium is available as a premixed powder from Oxoid.

Preparation of Medium: Add components to distilled/deionized water and bring volume to 1.0L. Mix thoroughly. Gently heat and bring to boiling. Distribute into tubes in 10.0mL volumes. Autoclave for 15 min at 15 psi pressure–121°C.

Caution: Sodium azide is toxic. Azides also react with metals and disposal must be highly diluted.

Use: For the isolation, cultivation and enumeration of enterococci from water and other specimens. Fecal enterococci turn the medium turbid with a purple sediment on the bottom of the tube.

Ethylene Glycol NaCl Medium No. 7

See: **EG NaCl Medium No. 7**

ETSA Medium

Composition per 998mL:

Agar	19.0g
Pancreatic digest of casein	15.0g
Papaic digest of soybean meal	5.0g
NaCl	5.0g
Yeast extract	1.0g
KNO_3	0.5g
Sodium formate	0.5g
Sodium succinate	0.5g
Sheep blood, defibrinated	30.0mL
Cysteine·HCl·H_2O solution	10.0mL
Dithiothreitol solution	10.0mL
Glucose solution	10.0mL
Na_2CO_3 solution	10.0mL
Menadione solution	2.0mL
Sodium fumarate solution	2.0mL
Sodium lactate (60% syrup)	1.3mL
Hemin solution	1.0mL

pH 7.3 ± 0.2 at 25°C

Hemin Solution:
Composition per 200mL:

KOH	1.12g
Hemin	0.2g
Ethanol (95% solution)	100.0mL

Preparation of Hemin Solution: Add KOH to distilled/deionized water and bring volume to 100.0mL. Mix thoroughly. Add ethanol. Mix thoroughly. Add hemin. Mix thoroughly.

Menadione Solution:
Composition per 100mL:

Menadione (vitamin K_3)	1.0g
Ethanol (95% solution)	50.0mL

Preparation of Menadione Solution: Add menadione to 50.0mL of ethanol. Mix thoroughly. Bring volume to 100.0mL with distilled/deionized water. Filter sterilize.

Cysteine·HCl·H_2O Solution:
Composition per 10mL:

L-Cysteine·HCl·H_2O	0.4g

Preparation of Cysteine·HCl·H_2O Solution: Add cysteine·HCl·H_2O to distilled/deionized water and bring volume to 10.0mL. Mix thoroughly. Filter sterilize.

Dithiothreitol Solution:
Composition per 10mL:

Dithiothreitol	0.05g

Preparation of Dithiothreitol Solution: Add dithiothreitol to distilled/deionized water and bring volume to 10.0mL. Mix thoroughly. Filter sterilize.

Glucose Solution:
Composition per 10mL:

D-Glucose	1.0g

Preparation of Glucose Solution: Add glucose to distilled/deionized water and bring volume to 10.0mL. Mix thoroughly. Filter sterilize.

Sodium Fumarate Solution:
Composition per 10mL:

Sodium fumarate	0.1g

Preparation of Sodium Fumarate Solution: Add sodium fumarate to distilled/deionized water and bring volume to 10.0mL. Mix thoroughly. Filter sterilize.

Na_2CO_3 Solution:
Composition per 10mL:

Na_2CO_3	0.4g

Preparation of Na_2CO_3 Solution: Add Na_2CO_3 to distilled/deionized water and bring volume to 10.0mL. Mix thoroughly. Filter sterilize.

Preparation of Medium: Add components—except menadione solution, cysteine·HCl·H_2O solution, dithiothreitol solution, glucose solution, sodium fumarate solution, Na_2CO_3 solution, and sheep blood—to distilled/deionized water and bring volume to 926.0mL. Mix thoroughly. Gently heat and bring to boiling. Autoclave for 15 min at 15 psi pressure–121°C. Cool to 50°–60°C. Aseptically add in the following order: 2.0mL of sterile menadione solution, 10.0mL of sterile cysteine·HCl·H_2O solution, 10.0mL of sterile dithiothreitol solution, 10.0mL of sterile glucose solution, 2.0mL of sterile sodium fumarate solution, 10.0mL of sterile Na_2CO_3 solution, and 30.0mL of sterile sheep blood. Mix thoroughly. Aseptically and anaerobically distribute into tubes under 80% N_2 + 10% CO_2 + 10% H_2. Cap tubes with butyl rubber stoppers. Place tubes in a press. Autoclave for 15 min at 15 psi pressure–121°C with fast exhaust.

Use: For the cultivation and maintenance of *Bacteroides pneumosintes, Falcivibrio grandis, Falcivibrio vaginalis, Mobiluncus curtisii, Mobiluncus mulieris, Serpula hyodysenteriae,* and *Treponema* species.

Eubacterium acidaminophilum Medium

Composition per liter:

KH_2PO_4	1.0g
NH_4Cl	0.5g
$MgCl_2·6H_2O$	0.4g
$CaCl_2·2H_2O$	0.1g
$NaHCO_3$ solution	20.0mL
Sodium selenite solution	10.0mL
$Na_2S·9H_2O$ solution	3.0mL
Vitamin solution	1.0mL
Trace element solution	1.0mL

pH 7.2–7.3 at 25°C

Sodium Selenite Solution:
Composition per liter:

NaOH	0.5g
$Na_2SeO_3·5H_2O$	0.03g

Preparation of Sodium Selenite Solution: Add components to distilled/deionized water and bring volume to 1.0L. Mix thoroughly. Filter sterilize. Gas with 100% N_2 for 20 min.

$Na_2S·9H_2O$ Solution:
Composition per 10mL:

$Na_2S·9H_2O$	1.0g

Preparation of $Na_2S·9H_2O$ Solution: Add $Na_2S·9H_2O$ to distilled/deionized water and bring volume to 10.0mL. Mix thoroughly. Gas with 100%

N_2 for 20 min. Autoclave for 15 min at 15 psi pressure–121°C. Cool to 45°–50°C.

Vitamin Solution:

Composition per 100mL:

p-Aminobenzoic acid	4.0mg
Biotin	1.0mg

Preparation of Vitamin Solution: Add components to distilled/deionized water and bring volume to 100.0mL. Mix thoroughly. Filter sterilize.

Trace Elements Solution SL-7:

Composition per liter:

$FeCl_2·4H_2O$	1.5g
$CoCl_2·6H_2O$	0.19g
$MnCl_2·4H_2O$	0.1g
$ZnCl_2$	0.07g
$Na_2MoO_4·2H_2O$	0.036g
$NiCl_2·6H_2O$	0.024g
H_3BO_3	6.0mg
$CuCl_2·2H_2O$	2.0mg
HCl (25% solution)	10.0mL

Preparation of Trace Elements Solution SL-7: Add the $FeCl_2·4H_2O$ to the HCl. Add distilled/deionized water and bring volume to 1.0L. Add remaining components. Mix thoroughly. Autoclave for 15 min at 15 psi pressure–121°C under 100% N_2. Cool to 25°C.

Preparation of Medium: Add components—except $NaHCO_3$ solution, $Na_2S·9H_2O$ solution, vitamin solution, and sodium selenite solution—to distilled/deionized water and bring volume to 966.0mL. Mix thoroughly. Gently heat and bring to boiling. Autoclave for 15 min at 15 psi pressure–121°C. Cool to 45°–50°C under 80% N_2 and 20% CO_2. Aseptically add 20.0mL of sterile $NaHCO_3$ solution, 3.0mL of sterile $Na_2S·9H_2O$ solution, 1.0mL of sterile vitamin solution, and 10.0mL of sterile sodium selenite solution. Mix thoroughly. Adjust pH to 7.2–7.3 with dilute sterile HCl or Na_2CO_3, if necessary. Aseptically and anaerobically distribute into sterile tubes under 80% N_2 and 20% CO_2. Cap tubes with rubber stoppers.

Use: For the cultivation and maintenance of *Eubacterium acidaminophilum.*

Eubacterium angustum Medium

Composition per liter:

$NaHCO_3$	5.0g
Tris(hydroxymethyl)aminomethane buffer	3.0g
Uric acid	3.0g
Yeast extract	1.0g
Cysteine·HCl·H_2O	0.5g
NH_4Cl	0.5g
K_2HPO_4	0.1g
$MgSO_4·7H_2O$	0.05g
$Na_2S·9H_2O$	0.05g
Resazurin	1.0mg
Wolfe's mineral solution	10.0mL
Selenium solution	1.0mL

pH 7.9 ± 0.2 at 25°C

Selenium Solution:

Composition per liter:

$Na_2SeO_3·5H_2O$	3.0mg
NaOH (0.01M solution)	1.0L

Preparation of Selenium Solution: Add $Na_2SeO_3·5H_2O$ to 1.0L of NaOH solution. Mix thoroughly. Filter sterilize. Aseptically gas under 100% N_2 for 20 min.

Wolfe's Mineral Solution:

Composition per liter

$MgSO_4·7H_2O$	3.0g
Nitriloacetic acid	1.5g
NaCl	1.0g
$MnSO_4·H_2O$	0.5g
$FeSO_4·7H_2O$	0.1g
$CoCl_2·6H_2O$	0.1g
$CaCl_2$	0.1g
$ZnSO_4·7H_2O$	0.1g
$CuSO_4·5H_2O$	0.01g
$AlK(SO_4)_2·12H_2O$	0.01g
H_3BO_3	0.01g
$Na_2MoO_4·2H_2O$	0.01g

Preparation of Wolfe's Mineral Solution: Add nitrilotriacetic acid to 500.0mL of distilled/deionized water. Dissolve by adjusting pH to 6.5 with KOH. Add remaining components. Add distilled/deionized water to 1.0L.

Preparation of Medium: Add the uric acid and the tris(hydroxymethyl)aminomethane buffer to 900.0mL of distilled/deionized water. Mix thoroughly. Gently heat while stirring until dissolved. Add remaining components, except cysteine·HCl·H_2O, $NaHCO_3$, and $Na_2S·9H_2O$. Gently heat and bring to boiling. Cool to 25°C under 80% N_2 + 10% CO_2 + 10% H_2. Add cysteine·HCl·H_2O, $NaHCO_3$, and $Na_2S·9H_2O$. Mix thoroughly. Anaerobically distribute into tubes under 80% N_2 + 10% CO_2 + 10% H_2. Cap tubes with rubber stoppers. Place tubes in a press. Autoclave for 15 min at 15 psi pressure–121°C with fast exhaust.

Use: the cultivation and maintenance of *Eubacterium angustum.*

Eubacterium Medium

Composition per liter:

Pancreatic digest of casein	20.0g
Agar	15.0g
Meat extract	15.0g

Glucose 5.0g
$Na_2HPO_4·12H_2O$ 4.0g
Cysteine·HCl..0.5g

pH 7.4 ± 0.2 at 25°C

Preparation of Medium: Add components to distilled/deionized water and bring volume to 1.0L. Mix thoroughly. Gently heat and bring to boiling. Distribute into tubes or flasks. Autoclave for 15 min at 15 psi pressure–121°C. Pour into sterile Petri dishes or leave in tubes. Use freshly prepared medium.

Use: For the cultivation of *Eubacterium* species.

Eubacterium Medium

Composition per liter:

Beef brain powder...33.33g
Pancreatic digest of casein...............................15.0g
Yeast extract...10.0g
Glucose 5.5g
Yeast extract...5.0g
NaCl..2.5g
Sodium thioglycollate.......................................1.8g
L-Cystine 0.5g

pH 7.0 ± 0.2 at 25°C

Preparation of Medium: Add components, except beef brain powder, to distilled/deionized water and bring volume to 1.0L. Mix thoroughly. Gently heat and bring to boiling under 97% N_2 + 3% H_2. Continue boiling for 15–20 min. Adjust pH to 7.0. Cool to 25°C under 97% N_2 + 3% H_2. Anaerobically distribute into tubes in 9.0mL volumes. Add 0.3g of beef brain powder to each tube. Cap tubes with rubber stoppers. Place tubes in a press. Autoclave for 15 min at 15 psi pressure–121°C with fast exhaust.

Use: For the cultivation of *Eubacterium* species.

Euglena B_{12} Medium

Composition per liter:

Sucrose..30.0g
L-Glutamic acid..6.0g
Glycine..5.0g
DL-Aspartic acid 4.0g
DL-Malic acid 2.0g
Succinic acid..1.04g
$MgSO_4·7H_2O$ 0.8g
$(NH_4)_2CO_3$...0.72g
KH_2PO_4..0.6g
$CaCO_3$ 0.16g
$FeCl_3$ 0.060g
$ZnSO_4·7H_2O$...0.040g
Thiamine·HCl...0.012g
$MnSO_4·H_2O$ 6.0mg
$CoSO_4$ 5.0mg
$(NH_4)_2MoO_3$.. 1.34mg
H_3BO_3 .... 1.14mg
$CuSO_4·5H_2O$ 0.62mg

pH 3.5 ± 0.1 at 25°C

Preparation of Medium: Add components to distilled/deionized water and bring volume to 1.0L. Mix thoroughly. Gently heat until dissolved. Distribute into tubes in 2.0mL volumes. Add standard solution or test solution to each tube. Bring volume of each tube to 4.0mL with distilled/deionized water. Autoclave for 15 min at 0 psi pressure–100°C.

Use: For the assay of vitamin B_{12} using *Euglena gracillis* as the test organism.

Eugon Agar

Composition per liter:

Agar...15.0g
Pancreatic digest of casein...............................15.0g
Glucose 5.5g
Papaic digest of soybean meal 5.0g
NaCl..4.0g
L-Cystine 0.3g
Na_2SO_3..0.2g

pH 7.0 ± 2.0 at 25°C

Source: This medium is available as a premixed powder fromand Difco Laboratories.

Preparation of Medium: Add components to distilled/deionized water and bring volume to 1.0L. Mix thoroughly. Gently heat and bring to boiling. Distribute into tubes or flasks. Autoclave for 15 min at 13 psi pressure–118°C. Pour into sterile Petri dishes.

Use: For the cultivation and maintenance of a variety of fastidious microorganisms. For the cultivation and maintenance of *Bifidobacterium* species.

Eugon Blood Agar (Eugonagar™)

Composition per liter:

Agar...15.0g
Pancreatic digest of casein...............................15.0g
Glucose 5.5g
Papaic digest of soybean meal 5.0g
NaCl..4.0g
L-Cystine 0.3g
Na_2SO_3..0.2g
Sheep blood, defibrinated........................... 100.0mL

pH 7.0 ± 2.0 at 25°C

Source: This medium is available as a premixed powder from BBL Microbiology Systems.

Preparation of Medium: Add components, except sheep blood, to distilled/deionized water and bring volume to 900.0mL. Mix thoroughly. Gently heat and bring to boiling. Autoclave for 15 min at 13 psi pressure–118°C. Cool to 45°–50°C. Aseptically add sterile sheep blood. Mix thoroughly. Pour into sterile Petri dishes or distribute into sterile tubes. If desired, medium may be chocolatized by maintaining at 80°–85°C for 20 min after the addition of sheep blood.

Use: For the cultivation and maintenance of fastidious microorganisms. For the cultivation and maintenance of *Bifidobacterium* species.

Eugon Broth (Eugonbroth™)

Composition per liter:

Pancreatic digest of casein	15.0g
Glucose	5.5g
Papaic digest of soybean meal	5.0g
NaCl	4.0g
L-Cystine	0.3g
Na_2SO_3	0.2g

pH 7.0 ± 0.2 at 25°C

Source: This medium is available as a premixed powder from BBL Microbiology Systems and Difco Laboratories.

Preparation of Medium: Add components to distilled/deionized water and bring volume to 1.0L. Mix thoroughly. Gently heat while stirring and bring to boiling. Distribute into tubes or flasks. Autoclave for 15 min at 13 psi pressure–118°C.

Use: For the cultivation and maintenance of a variety of fastidious microorganisms.

EVA Broth
See: **Ethyl Violet Azide Broth**

Extracted Hay Medium

Composition:

Hay or grass	50.0g

Preparation of Medium: Add hay or grass to 1.0L of distilled/deionized water. Gently heat and bring to boiling. Continue boiling for 30 min. Rinse with cold water twice. Add 1.0L of of distilled/deionized water, boil 30 min and rinse. Repeat this process at least five times. Dry the extracted hay or grass. Add 10–30 blades of extracted hay or grass to a large test tube. Autoclave for 15 min at 15 psi pressure–121°C.

Use: For the isolation and cultivation of *Beggiatoa* species and myxotrophic *Thiothrix* species.

FAA Alternative Selective
See: **Fastidious Anaerobe Agar, Alternative Selective**

FAA Alternative Selective with Neomycin, Vancomycin, and Josamycin
See: **Fastidious Anaerobe Agar, Alternative Selective with Neomycin, Vancomycin, and Josamycin**

FAA Selective with Neomycin and Vancomycin
See: **Fastidious Anaerobe Agar, Selective with Neomycin and Vancomycin**

Fastidious Anaerobe Agar (FAA)

Composition per liter:

Peptone	23.0g
Agar	12.0g
NaCl	5.0g
Glucose	1.0g
L-Arginine	1.0g
Sodium pyruvate	1.0g
Soluble starch	1.0g
Cysteine·HCl·H_2O	0.5g
Sodium succinate	0.5g
$NaHCO_3$	0.4g
$Na_4P_2O_7 \cdot 10H_2O$	0.25g
Sheep blood, defibrinated	50.0mL
Hemin solution	1.0mL
Vitamin K_1	0.1mL

pH 7.2 ± 0.2 at 25°C

Vitamin K_1 Solution:
Composition per 100mL:

Vitamin K_1	1.0g
Ethanol	99.0mL

Preparation of Vitamin K_1 Solution: Add vitamin K_1 to 99.0mL of absolute ethanol. Mix thoroughly.

Hemin Solution:
Composition per 100mL:

Hemin	1.0g
NaOH (1*N* solution)	20.0mL

Preparation of Hemin Solution: Add hemin to 20.0mL of 1*N* NaOH solution. Mix thoroughly. Bring volume to 100.0mL with distilled/deionized water.

Preparation of Medium: Add components, except defibrinated sheep blood, to distilled/deionized water and bring volume to 950.0mL. Mix thoroughly. Gently heat and bring to boiling. Autoclave for 15 min at 15 psi pressure–121°C. Cool to 45°–50°C. Aseptically add 50.0mL sterile defibrinated sheep blood. Mix thoroughly. Pour into sterile Petri dishes or distribute into sterile tubes.

Use: For the cultivation of a variety of fastidious anaerobes from clinical and nonclinical specimens.

Fastidious Anaerobe Agar, Alternative Selective (FAA Alternative Selective)

Composition per liter:

Peptone	23.0g
Agar	12.0g
NaCl	5.0g
Glucose	1.0g
L-Arginine	1.0g
Sodium pyruvate	1.0g
Soluble starch	1.0g
Cysteine·HCl·H_2O	0.5g
Sodium succinate	0.5g
$NaHCO_3$	0.4g
$Na_4P_2O_7 \cdot 10H_2O$	0.25g
Sheep blood, defibrinated	50.0mL
Hemin solution	1.0mL
Vitamin K_1	0.1mL

pH 7.2 ± 0.2 at 25°C

Vitamin K_1 Solution:
Composition per 100mL:

Vitamin K_1	1.0g
Ethanol	99.0mL

Preparation of Vitamin K_1 Solution: Add vitamin K_1 to 99.0mL of absolute ethanol. Mix thoroughly.

Hemin Solution:
Composition per 100mL:

Hemin	1.0g
NaOH (1*N* solution)	20.0mL

Preparation of Hemin Solution: Add hemin to 20.0mL of 1*N* NaOH solution. Mix thoroughly. Bring volume to 100.0mL with distilled/deionized water.

Preparation of Medium: Add components, except defibrinated sheep blood, to distilled/deionized water and bring volume to 950.0mL. Mix thoroughly. Gently heat and bring to boiling. Autoclave for 15 min at 15 psi pressure–121°C. Cool to 45°–50°C. Aseptically add 50.0mL sterile defibrinated sheep blood. Mix thoroughly. Pour into sterile Petri dishes or distribute into sterile tubes.

Use: For the cultivation of a variety of fastidious anaerobes from clinical andnon clinical specimens.

Fastidious Anaerobe Agar, Alternative Selective with Neomycin, Vancomycin, and Josamycin (FAA Alternative Selective with Neomycin, Vancomycin, and Josamycin)

Composition per liter:

Peptone	23.0g
Agar	12.0g
NaCl	5.0g
Glucose	1.0g
L-Arginine	1.0g
Sodium pyruvate	1.0g
Soluble starch	1.0g
Cysteine·HCl·H_2O	0.5g
Sodium succinate	0.5g
$NaHCO_3$	0.4g
$Na_4P_2O_7 \cdot 10H_2O$	0.25g
Neomycin	0.1g
Sheep blood, defibrinated	50.0mL
Vancomycin solution	10.0mL
Josamycin	10.0mL
Hemin solution	1.0mL
Vitamin K_1	0.1mL

pH 7.2 ± 0.2 at 25°C

Vitamin K_1 Solution:
Composition per 100mL:

Vitamin K_1	1.0g
Ethanol	99.0mL

Preparation of Vitamin K_1 Solution: Add vitamin K_1 to 99.0mL of absolute ethanol. Mix thoroughly.

Hemin Solution:
Composition per 100mL:

Hemin	1.0g
NaOH (1*N* solution)	20.0mL

Preparation of Hemin Solution: Add hemin to 20.0mL of 1*N* NaOH solution. Mix thoroughly. Bring volume to 100.0mL with distilled/deionized water.

Vancomycin Solution:
Composition per 10mL:

Vancomycin	5.0mg

Preparation of Vancomycin Solution: Add vancomycin to distilled/deionized water and bring volume to 10.0mL. Mix thoroughly. Filter sterilize.

Josamycin Solution:
Composition per 10mL:

Josamycin	3.0mg

Preparation of Josamycin Solution: Add josamycin to distilled/deionized water and bring volume to 10.0mL. Mix thoroughly. Filter sterilize.

Preparation of Medium: Add components, except defibrinated sheep blood, vancomycin solution, and josamycin solution, to distilled/deionized water and bring volume to 930.0mL. Mix thoroughly. Gently heat and bring to boiling. Autoclave for 15 min at 15 psi pressure–121°C. Cool to 45°–50°C. Aseptically add 50.0mL sterile defibrinated sheep blood, 10.0mL vancomycin solution, and 10.0mL josamycin solution. Mix thoroughly. Pour into sterile Petri dishes or distribute into sterile tubes.

Use: For the selective cultivation of *Fusobacterium* species from clinical and nonclinical specimens.

Fastidious Anaerobe Agar, Selective (FAA Selective)

Composition per liter:

Peptone	23.0g
Agar	12.0g
NaCl	5.0g
Glucose	1.0g
L-Arginine	1.0g
Sodium pyruvate	1.0g
Soluble starch	1.0g
Cysteine·HCl·H_2O	0.5g
Sodium succinate	0.5g
$NaHCO_3$	0.4g
$Na_4P_2O_7 \cdot 10H_2O$	0.25g
Sheep blood, defibrinated	50.0mL
Hemin solution	1.0mL
Vitamin K_1	0.1mL

pH 7.2 ± 0.2 at 25°C

Vitamin K_1 Solution:
Composition per 100mL:

Vitamin K_1	1.0g
Ethanol	99.0mL

Preparation of Vitamin K_1 Solution: Add vitamin K_1 to 99.0mL of absolute ethanol. Mix thoroughly.

Hemin Solution:
Composition per 100mL:

Hemin	1.0g
NaOH (1*N* solution)	20.0mL

Preparation of Hemin Solution: Add hemin to 20.0mL of 1*N* NaOH solution. Mix thoroughly. Bring volume to 100.0mL with distilled/deionized water.

Preparation of Medium: Add components, except defibrinated sheep blood, to distilled/deionized water and bring volume to 950.0mL. Mix thoroughly. Gently heat and bring to boiling. Autoclave for 15 min at 15 psi pressure–121°C. Cool to 45°–50°C. Aseptically add 50.0mL sterile defibrinated sheep blood. Mix thoroughly. Pour into sterile Petri dishes or distribute into sterile tubes.

Use: For the cultivation of a variety of fastidious anaerobes from clinical andnon clinical specimens.

Fastidious Anaerobe Agar, Selective with Neomycin and Vancomycin (FAA Selective with Neomycin and Vancomycin)

Composition per liter:

Peptone	23.0g
Agar	12.0g
NaCl	5.0g
Glucose	1.0g
L-Arginine	1.0g
Sodium pyruvate	1.0g
Soluble starch	1.0g
Cysteine·HCl·H_2O	0.5g
Sodium succinate	0.5g
$NaHCO_3$	0.4g
$Na_4P_2O_7 \cdot 10H_2O$	0.25g
Neomycin	0.1g
Sheep blood, defibrinated	50.0mL
Vancomycin solution	10.0mL
Hemin solution	1.0mL
Vitamin K_1	0.1mL

pH 7.2 ± 0.2 at 25°C

Vitamin K_1 Solution:
Composition per 100mL:

Vitamin K_1	1.0g
Ethanol	99.0mL

Preparation of Vitamin K_1 Solution: Add vitamin K_1 to 99.0mL of absolute ethanol. Mix thoroughly.

Hemin Solution:
Composition per 100mL:

Hemin	1.0g
NaOH (1*N* solution)	20.0mL

Preparation of Hemin Solution: Add hemin to 20.0mL of 1*N* NaOH solution. Mix thoroughly. Bring volume to 100.0mL with distilled/deionized water.

Vancomycin Solution:
Composition per 10mL:

Vancomycin	7.5mg

Preparation of Vancomycin Solution: Add vancomycin to distilled/deionized water and bring volume to 10.0mL. Mix thoroughly. Filter sterilize.

Preparation of Medium: Add components, except defibrinated sheep blood and vancomycin solution, to distilled/deionized water and bring volume to 940.0mL. Mix thoroughly. Gently heat and bring to boiling. Autoclave for 15 min at 15 psi pressure–121°C. Cool to 45°–50°C. Aseptically add 50.0mL sterile defibrinated sheep blood and 10.0mL vancomycin solution. Mix thoroughly. Pour into sterile Petri dishes or distribute into sterile tubes.

Use: For the selective cultivation of *Fusobacterium* species from clinical and nonclinical specimens.

Fay and Barry Medium

Composition per liter:

Amino acid	10.0g
Peptone	5.0g
Yeast extract	3.0g
Bromcresol Purple solution	5.0mL

pH 5.5 ± 0.2 at 25°C

Bromcresol Purple Solution:
Composition per 100mL:

Bromcresol Purple	0.2g
Ethanol	50.0mL

Preparation of Bromcresol Purple Solution: Add Bromcresol Purple to 50.0mL of absolute ethanol. Add distilled/deionized water and bring volume to 100.0mL. Mix thoroughly.

Preparation of Medium: Add components to distilled/deionized water and bring volume to 1.0L. The amino acid may be L-arginine, L-ornithine, or L-lysine, depending on which amino acid decarboxylase activity is being measured. Mix thoroughly. Distribute into tubes or flasks. Autoclave for 15 min at 15 psi pressure–121°C.

Use: For determination of decarboxylase activities of *Aeromonas* species.

FC Agar (Fecal Coliform Agar) (m-FC Agar) (m–Fecal Coliform Agar)

Composition per liter:

Agar	15.0g
Lactose	12.5g

NaCl 5.0g
Proteose peptone No. 3 ....5.0g
Yeast extract..3.0g
Bile salts..1.5g
Aniline Blue 0.1g
Rosolic acid solution.................................... 10.0mL

pH 7.4 ± 0.2 at 25°C

Source: This medium is available as a premixed powder from BBL Microbiology Systems and Difco Laboratories.

Rosolic Acid Solution:
Composition per 100mL:

Rosolic acid..1.0g

Preparation of Rosolic Acid Solution: Add rosolic acid to 0.2*N* NaOH and bring volume to 100.0L. Mix thoroughly.

Preparation of Medium: Add 10.0mL rosolic acid solution to 950.0mL distilled/deionized water. Mix thoroughly. Add other components and bring volume to 1.0L with distilled/deionized water. Mix thoroughly. Gently heat and bring to boiling with frequent mixing. Do not autoclave. Pour into sterile Petri dishes or leave in tubes.

Use: For the cultivation of fecal coliform bacteria from waters and the enumeration of coliform bacteria using the membrane filtration method.

FC Agar (Fecal Coliform Agar) (m-FC Agar) (m–Fecal Coliform Agar)

Composition per liter:

Agar...15.0g
Lactose 12.5g
Tryptose 10.0g
NaCl 5.0g
Proteose peptone No. 3 ....5.0g
Yeast extract..3.0g
Bile salts..1.5g
Aniline Blue 0.1g
Rosolic acid solution.................................... 10.0mL

pH 7.4 ± 0.2 at 25°C

Rosolic Acid Solution:
Composition per 100mL:

Rosolic acid..1.0g

Preparation of Rosolic Acid Solution: Add rosolic acid to 0.2*N* NaOH and bring volume to 100.0L. Mix thoroughly.

Preparation of Medium: Add 10.0mL rosolic acid solution to 950.0mL distilled/deionized water. Mix thoroughly. Add other components and bring volume to 1.0L with distilled/deionized water. Mix thoroughly. Gently heat and bring to boiling with frequent mixing. Do not autoclave. Pour into sterile Petri dishes or leave in tubes.

Use: For the cultivation of fecal coliform bacteria from waters and the enumeration of coliform bacteria using the membrane filtration method.

FC Broth (Fecal Coliform Broth) (m-FC Broth) (m–Fecal Coliform Broth)

Composition per liter:

Lactose 12.5g
Tryptose 10.0g
NaCl 5.0g
Proteose peptone No. 3 ....5.0g
Yeast extract..3.0g
Bile salts..1.5g
Aniline Blue 0.1g
Rosolic acid solution.................................... 10.0mL

pH 7.4 ± 0.2 at 25°C

Rosolic Acid Solution:
Composition per 100mL:

Rosolic acid..1.0g

Preparation of Rosolic Acid Solution: Add rosolic acid to 0.2*N* NaOH and bring volume to 100.0L. Mix thoroughly.

Preparation of Medium: Add 10.0mL rosolic acid solution to 950.0mL distilled/deionized water. Mix thoroughly. Add other components and bring volume to 1.0L with distilled/deionized water. Mix thoroughly. Gently heat and bring to boiling with frequent mixing. Do not autoclave. Pour into sterile Petri dishes or leave in tubes.

Use: For the cultivation of fecal coliform bacteria from waters and the enumeration of coliform bacteria using the membrane filtration method.

FC Broth (Fecal Coliform Broth) (m-FC Broth) (m–Fecal Coliform Broth)

Composition per liter:

Lactose 12.5g
NaCl 5.0g
Proteose peptone No. 3 ....5.0g
Yeast extract..3.0g

Bile salts ... 1.5g
Aniline Blue ... 0.1g
Rosolic acid solution ... 10.0mL
pH 7.4 ± 0.2 at 25°C

Source: This medium is available as a premixed powder from BBL Microbiology Systems and Difco Laboratories.

Rosolic Acid Solution:
Composition per 100mL:
Rosolic acid ... 1.0g

Preparation of Rosolic Acid Solution: Add rosolic acid to 0.2*N* NaOH and bring volume to 100.0L. Mix thoroughly.

Preparation of Medium: Add 10.0mL rosolic acid solution to 950.0mL distilled/deionized water. Mix thoroughly. Add other components and bring volume to 1.0L with distilled/deionized water. Mix thoroughly. Gently heat and bring to boiling with frequent mixing. Do not autoclave. Pour into sterile Petri dishes or leave in tubes.

Use: For the cultivation of fecal coliform bacteria from waters and the enumeration of coliform bacteria using the membrane filtration method.

(FCIC)
See: **Fecal Coliform Agar, Modified**

FDA Agar (ATCC Medium 182) (AATCC Bacteriostasis Agar) (American Association of Textile Chemists and Colorists Bacteriostasis Agar)

Composition per liter:
Agar ... 15.0g
Peptic digest of animal tissue ... 10.0g
Beef extract ... 5.0g
NaCl ... 5.0g
pH 6.8 ± 0.1 at 25°C

Source: This medium is available as a premixed powder from BBL Microbiology Systems.

Preparation of Medium: Add components to distilled/deionized water and bring volume to 1.0L. Mix thoroughly. Gently heat and bring to boiling. Distribute into tubes or flasks. Autoclave for 15 min at 15 psi pressure–121°C. Pour into sterile Petri dishes or leave in tubes.

Use: For testing the antibacterial activities of antiseptics and disinfectants.

FDA Broth (AATCC Bacteriostasis Broth) (American Association of Textile Chemists and Colorists Bacteriostasis Broth)

Composition per liter:
Peptic digest of animal tissue ... 10.0g
Beef extract ... 5.0g
NaCl ... 5.0g
pH 6.8 ± 0.1 at 25°C

Source: This medium is available as a premixed powder from BBL Microbiology Systems.

Preparation of Medium: Add components to distilled/deionized water and bring volume to 1.0L. Mix thoroughly. Distribute into tubes or flasks. Autoclave for 15 min at 15 psi pressure–121°C.

Use: For testing the antibacterial activities of antiseptics and disinfectants.

Fecal Coliform Agar
See: **FC Agar**

Fecal Coliform Agar, Modified (m–Fecal Coliform Agar, Modified) (FCIC)

Composition per liter:
Agar ... 15.0g
Inositol ... 10.0g
Tryptose ... 10.0g
Proteose peptone No. 3 ... 5.0g
NaCl ... 5.0g
Yeast extract ... 3.0g
Bile salts No. 3 ... 1.5g
Aniline Blue ... 0.1g
pH 7.4 ± 0.2 at 25°C

Preparation of Medium: Add components and bring volume to 1.0L. Mix thoroughly. Gently heat and bring to boiling. Do not autoclave. Cool to 50°C. Adjust pH to 7.4. Pour into sterile Petri dishes in 20.0mL volumes. Allow surface of plates to dry before using.

Use: For the isolation, cultivation and enumeration of *Klebsiella* species using the membrane filter method.

Fecal Coliform Agar, Modified

Composition per liter:
Agar ... 15.0g

Lactose 12.5g
Tryptose 10.0g
Proteose peptone No. 3 ....5.0g
NaCl 5.0g
Yeast extract 3.0g
Bile salts No. 3 ....1.5g
Aniline Blue 0.1g

pH 7.4 ± 0.2 at 25°C

Preparation of Medium: Add components and bring volume to 1.0L. Mix thoroughly. Gently heat and bring to boiling. Do not autoclave. Cool to 50°C. Adjust pH to 7.4. Pour into sterile Petri dishes in 20.0mL volumes. Allow surface of plates to dry before using.

Use: For the isolation, cultivation and identification of stressed fecal coliform microorganisms based on their ability to ferment lactose. Lactose-fermenting bacteria turn the medium blue.

Fecal Coliform Broth

See: **FC Broth**

Feeley–Gorman Agar

See: **F–G Agar**

Feeley–Gorman Agar with Selenium

See: **F–G Agar with Selenium**

Feeley–Gorman Broth

See: **F–G Broth**

Feodorov Medium

Composition per liter:

Mannitol or glucose 20.0g
Marine salts mixture 18.0g
$CaCO_3$ 0.5g
K_2HPO_4 ....0.3g
$MgSO_4$ 0.3g
$CaHPO_4$ 0.2g
K_2SO_4 ....0.2g
$FeCl_3$ 0.1g
Trace elements solution 1.0mL

Trace Elements Solution:

Composition per 100mL:

H_3BO_3 ....0.5g
$(NH_4)_6Mo_7O_{24}\cdot 4H_2O$ 0.5g
KI 0.05g
NaBr 0.05g
$Al_2(SO_4)_3\cdot 18H_2O$ 0.03g
$ZnSO_4$ 0.02g

Preparation of Trace Elements Solution: Add components to distilled/deionized water and bring volume to 100.0mL. Mix thoroughly.

Preparation of Medium: Add components to distilled/deionized water and bring volume to 1.0L. Mix thoroughly. Distribute into tubes or flasks. Autoclave for 15 min at 15 psi pressure–121°C.

Use: For the cultivation and maintenance of *Azotobacter vinelandii*.

Fermentation Basal Medium

Composition per liter:

Agar 15.0g
$(NH_4)_2HPO_4$ 1.0g
$MgSO_4\cdot 7H_2O$ 0.2g
KCl 0.02g
Carbohydrate solution 100.0mL
Bromcresol Purple solution 20.0mL

pH 7.0 ± 0.2 at 25°C

Carbohydrate Solution:

Composition per 100mL:

Carbohydrate 10.0g

Preparation of Carbohydrate Solution: Add carbohydrate to distilled/deionized water and bring volume to 100.0mL. Mix thoroughly. Filter sterilize.

Bromcresol Purple Solution:

Composition per 100mL:

Bromcresol Purple 0.04g
Ethanol 50.0mL

Preparation of Bromcresol Purple Solution: Add Bromcresol Purple to 50.0mL of absolute ethanol. Add distilled/deionized water and bring volume to 100.0mL. Mix thoroughly.

Preparation of Medium: Add components, except carbohydrate solution, to distilled/deionized water and bring volume to 900.0mL. Mix thoroughly. Gently heat and bring to boiling. Autoclave for 15 min at 15 psi pressure–121°C. Cool to 45°–50°C. Aseptically add 100.0mL sterile carbohydrate solution. Various carbohydrates are used for different fermentation tests. Mix thoroughly. Pour into sterile Petri dishes or distribute into sterile tubes.

Use: For differentiation of aerobic actinomycetes based upon carbohydrate fermentation. Actinomycetes that produce acid from carbohydrates turn the medium yellow.

Fermentation Base for *Campylobacter*
See: **Enteric Fermentation Base**

Fermentation Broth (CHO Medium)

Composition per liter:

Pancreatic digest of casein	15.0g
Yeast extract	7.0g
NaCl	2.5g
Agar	0.75g
Sodium thioglycollate	0.5g
L-Cystine	0.25g
Ascorbic acid	0.1g
Bromthymol Blue	0.01g
Carbohydrate or starch solution	100.0mL

pH 7.0 ± 0.1 at 25°C

Source: This medium is available as a premixed powder from Difco Laboratories.

Carbohydrate Solution:
Composition per 100mL:

Carbohydrate	6.0g

Preparation of Carbohydrate Solution: Add carbohydrate to distilled/deionized water and bring volume to 10.0mL. Mix thoroughly. Filter sterilize.

Starch Solution:
Composition per 100mL:

Starch	2.5g

Preparation of Starch Solution: Add starch to distilled/deionized water and bring volume to 100.0mL. Mix thoroughly. Filter sterilize.

Preparation of Medium: Add components, except carbohydrate solution, to distilled/deionized water and bring volume to 900.0mL. Mix thoroughly. Distribute into tubes or flasks. Autoclave for 15 min at 15 psi pressure–121°C. Cool to 45–50°C. Aseptically add 100.0mL sterile carbohydrate solution. Mix thoroughly. Aseptically distribute into sterile tubes or flasks. Loosen caps on tubes. Place in an anaerobic chamber under an atmosphere of 85% N_2, 10% H_2, and 5% CO_2. Fasten the caps securely or maintain in an anaerobic chamber.

Use: For differentiation of anaerobic bacteria based upon carbohydrate fermentation. Bacteria that ferment carbohydrates turn the medium yellow.

Fermentation Medium

Composition per liter:

Glucose or mannitol	10.0g
Pancreatic digest of casein	10.0g
Agar	2.2g
Yeast extract	1.0
Bromcresol Purple	0.04g

pH 7.0 ± 0.2 at 25°C

Preparation of Medium: Add components to distilled/deionized water and bring volume to 1.0L. Mix thoroughly. Gently heat and bring to boiling. Distribute into tubes or flasks. Autoclave for 10 min at 15 psi pressure–121°C. Pour into sterile Petri dishes or leave in tubes.

Use: For differentiating *Staphylococcus* and *Micrococcus* species based upon the fermentation of glucose and mannitol.

Ferrous Sulfide Agar

Composition per 1200mL:

Agar layer	1.0L
Liquid overlay	200.0mL

Agar Layer:
Composition per liter:

Agar	30.0g
FeS washed precipitate supension	500.0mL

Preparation of Agar Layer: Add agar to distilled/deionized water and bring volume to 500.0mL. Mix thoroughly. Gently heat and bring to boiling. Autoclave for 15 min at 15 psi pressure–121°C. Cool to 45°–50°C. Heat FeS washed precipitate suspension to 45°–50°C. Mix thoroughly. Aseptically add 500.0mL sterile FeS washed precipitate supension to 500.0mL sterile agar at 45°–50°C. Mix thoroughly.

FeS Washed Precipitate Suspension:
Composition per 500mL:

$Fe(NH_4)_2(SO_4)_2 \cdot 6H_2O$	78.4g
$Na_2S \cdot 9H_2O$	15.6g

Preparation of FeS Washed Precipitate Suspension: Add $Na_2S \cdot 9H_2O$ and $Fe(NH_4)_2(SO_4)_2$ to 500.0mL boiling distilled/deionized water. Let precipitate settle from the hot solution in a completely filled and stoppered bottle. Wash precipitate four times by decanting supernatant and replacing each time with 500.0mL boiling distilled/deionizedwater. Store FeS washed precipitate suspension in a completely filled 500.0mL glass stoppered bottle.

Liquid Overlay:
Composition per liter:

$(NH_4)_2Cl$	1.0g
K_2HPO_4	0.5g
$MgSO_4 \cdot 7H_2O$	0.2g
$CaCl_2$	0.1g

Preparation of Liquid Overlay: Add components to distilled/deionized water and bring volume to 1.0L. Mix thoroughly. Autoclave for 15 min at 15 psi pressure–121°C. Cool to 25°C. Aseptically bubble 100% CO_2 for 15 seconds.

Preparation of Medium: Aseptically distribute agar layer into sterile tubes in 10.0mL volumes. Allow tubes to cool in a slanted poistion. Aseptically add 2.0mL of sterile liquid overlay to each tube.

Use: For enumeration, enrichment and isolation of iron and sulfur bacteria, including *Gallionella ferruginea*.

Fervidobacterium Medium

Composition per liter:

Pancreatic digest of casein	10.0g
Glucose	5.0g
Yeast extract	3.0g
K_2HPO_4	1.5g
NH_4Cl	0.9g
KH_2PO_4	0.75g
$MgCl_2 \cdot 6H_2O$	0.2g
$Na_2S \cdot 9H_2O$ solution	10.0mL
Trace elements solution	9.0mL
Wolfe's vitamin solution	5.0mL
Resazurin (0.2% solution)	1.0mL
$FeSO_4 \cdot 7H_2O$ (10% solution)	0.03mL

pH 7.0 ± 0.1 at 25°C

Trace Elements Solution:
Composition per liter:

Nitrilotriacetic acid	12.5g
NaCl	1.0g
$FeCl_3 \cdot 4H_2O$	0.2g
$CaCl_2 \cdot 2H_2O$	0.1g
$MnCl_2 \cdot 4H_2O$	0.1g
$ZnCl_2$	0.1g
$CuCl_2$	0.02g
Na_2SeO_3	0.02g
$CoCl_2 \cdot 6H_2O$	0.017g
H_3BO_3	0.01g
$Na_2MoO_4 \cdot 2H_2O$	0.01g

Preparation of Trace Elements Solution: Add nitrilotriacetic acid to 500.0mL of distilled/deionized water. Dissolve by adjusting pH to 6.5 with KOH. Add remaining components. Add distilled/deionized water to 1.0L. Filter sterilize. Maintain under an atmosphere of 100% N_2.

Wolfe's Vitamin Solution:
Composition per liter:

Pyridoxine·HCl	0.01g
Thiamine·HCl	5.0mg
Riboflavin	5.0mg
Nicotinic acid	5.0mg
Calcium pantothenate	5.0mg
p-Aminobenzoic acid	5.0mg
Thioctic acid	5.0mg
Biotin	2.0mg
Folic acid	2.0mg
Cyanocobalamin	0.1mg

Preparation of Wolfe's Vitamin Solution: Add components to distilled/deionized water and bring volume to 1.0L. Mix thoroughly. Filter sterilize. Maintain under an atmosphere of 100% N_2.

$Na_2S \cdot 9H_2O$ Solution:
Composition per 10mL:

$Na_2S \cdot 9H_2O$	0.5g

Preparation of $Na_2S \cdot 9H_2O$ Solution: Add $Na_2S \cdot 9H_2O$ to distilled/deionized water and bring volume to 10.0mL. Mix thoroughly. Filter sterilize. Maintain under an atmosphere of 100% N_2.

Preparation of Medium: Add components, except sodium sulfide solution, trace elements solution, and Wolfe's vitamin solution, to distilled/deionized water and bring volume to 976.0mL. Mix thoroughly. Autoclave for 15 min at 15 psi pressure–121°C. Cool under an atmosphere of 100% N_2. Aseptically add 9.0mL of trace elements solution and 5.0mL of Wolfe's vitamin solution under an atmosphere of 100% N_2. Mix thoroughly. Aseptically distribute into sterile tubes or flasks under an atmosphere of 100% N_2. Add $Na_2S \cdot 9H_2O$ solution just prior to use to a concentration 0.1 percent.

Use: For cultivation and maintenance of *Clostridium* species, *Fervidobacterium nodosum, F. islandicum* and *Thermoanaerobium brockii*.

F–G Agar (Feeley–Gorman Agar)

Composition per liter:

Casein, acid hydrolyzed	17.5g
Agar	17.0g
Beef extract	3.0g
Starch	1.5g
Cysteine solution	10.0mL
$Fe_4(P_2O_7)_3$ solution	10.0mL

pH 6.9 ± 0.05 at 25°C

Cysteine Solution:
Composition per 10mL:

L-Cysteine HCl·H_2O	0.4g

Preparation of Cysteine Solution: Add L-Cysteine HCl·H_2O to distilled/deionized water and bring volume to 10.0mL. Mix thoroughly. Filter sterilize.

$Fe_4(P_2O_7)_3$ Solution:
Composition per 10mL:

$Fe_4(P_2O_7)_3$	0.25g

Preparation of $Fe_4(P_2O_7)_3$ Solution: Add $Fe_4(P_2O_7)_3$ to distilled/deionized water and bring volume to 10.0mL. Mix thoroughly. Filter sterilize.

Preparation of Medium: Add components, except cysteine solution and $Fe_4(P_2O_7)_3$ solution, to distilled/deionized water and bring volume to 980.0mL. Mix thoroughly. Gently heat and bring to boiling. Autoclave for 15 min at 15 psi pressure–121°C. Cool to 45°–50°C. Aseptically add 10.0mL cysteine solution. Mix thoroughly. Aseptically add 10.0mL $Fe_4(P_2O_7)_3$ solution. Mix thoroughly. Adjust pH to 6.9. Pour into sterile Petri dishes or distribute into sterile tubes.

Use: For the isolation and cultivation of *Legionella pneumophila*.

F–G Agar with Selenium (Feeley–Gorman Agar with Selenium)

Composition per liter:

Casein, acid hydrolyzed	17.5g
Agar	17.0g
Beef extract	3.0g
Starch	1.5g
Cysteine solution	10.0mL
$Fe_4(P_2O_7)_3$ solution	10.0mL
$Na_2SeO_3 \cdot 5H_2O$ solution	10.0mL

pH 6.9 ± 0.05 at 25°C

Cysteine Solution:

Composition per 10mL:

L-Cysteine HCl·H_2O	0.4g

Preparation of Cysteine Solution: Add L-Cysteine HCl·H_2O to distilled/deionized water and bring volume to 10.0mL. Mix thoroughly. Filter sterilize.

$Fe_4(P_2O_7)_3$ Solution:

Composition per 10mL:

$Fe_4(P_2O_7)_3$	0.25g

Preparation of $Fe_4(P_2O_7)_3$ Solution: Add $Fe_4(P_2O_7)_3$ to distilled/deionized water and bring volume to 10.0mL. Mix thoroughly. Filter sterilize.

$Na_2SeO_3 \cdot 5H_2O$ Solution:

Composition per 10mL:

$Na_2SeO_3 \cdot 5H_2O$	0.010g

Preparation of $Na_2SeO_3 \cdot 5H_2O$ Solution: Add $Na_2SeO_3 \cdot 5H_2O$ to distilled/deionized water and bring volume to 10.0mL. Mix thoroughly. Filter sterilize.

Preparation of Medium: Add components—except cysteine solution, $Fe_4(P_2O_7)_3$ solution, and $Na_2SeO_3 \cdot 5H_2O$ solution—to distilled/deionized water and bring volume to 970.0mL. Mix thoroughly. Gently heat and bring to boiling. Autoclave for 15 min at 15 psi pressure–121°C. Cool to 45°–50°C. Aseptically add 10.0mL of sterile cysteine solution. Mix thoroughly. Aseptically add 10.0mL of sterile $Fe_4(P_2O_7)_3$ solution and 10.0mL of sterile $Na_2SeO_3 \cdot 5H_2O$ solution. Mix thoroughly. Adjust pH to 6.9. Pour into sterile Petri dishes or distribute into sterile tubes.

Use: For the isolation and cultivation of *Legionella pneumophila*.

F–G Broth (Feeley–Gorman Broth)

Composition per liter:

Casein, acid hydrolyzed	17.5g
Beef extract	3.0g
Starch	1.5g
Cysteine solution	10.0mL
$Fe_4(P_2O_7)_3$ solution	10.0mL

pH 6.9 ± 0.05 at 25°C

Cysteine Solution:

Composition per 10mL:

L-Cysteine HCl·H_2O	0.4g

Preparation of Cysteine Solution: Add L-Cysteine HCl·H_2O to distilled/deionized water and bring volume to 10.0mL. Mix thoroughly. Filter sterilize.

$Fe_4(P_2O_7)_3$ Solution:

Composition per 10mL:

$Fe_4(P_2O_7)_3$	0.25g

Preparation of $Fe_4(P_2O_7)_3$ Solution: Add $Fe_4(P_2O_7)_3$ to distilled/deionized water and bring volume to 10.0mL. Mix thoroughly. Filter sterilize.

Preparation of Medium: Add components, except cysteine solution and $Fe_4(P_2O_7)_3$ solution, to distilled/deionized water and bring volume to 980.0mL. Mix thoroughly. Gently heat and bring to boiling. Autoclave for 15 min at 15 psi pressure–121°C. Cool to 45°–50°C. Aseptically add 10.0mL cysteine solution. Mix thoroughly. Aseptically add 10.0mL $Fe_4(P_2O_7)_3$ solution. Mix thoroughly. Adjust pH to 6.9. Aseptically distribute into sterile tubes or flasks.

Use: For the cultivation of *Legionella pneumophila*.

fGTC Agar

Composition per liter:

Pancreatic digest of casein	15.0g
Agar	15.0g
Papaic digest of soybean meal	5.0g
NaCl	5.0g
KH_2PO_4	5.0g

Amylose Azure..3.0g
Galactose..1.0g
Thallous acetate 0.5g
MUG (4-Methylumbelliferyl-
α-D-galactoside 0.1g
$NaHCO_3$ solution 20.0mL
Gentamicin solution.......................................2.5mL
Tween™ 80 ....0.75mL
pH 7.3 ± 0.2 at 25°C

Gentamicin Solution:
Composition per 10mL:
Gentamicin..0.01g

Preparation of Gentamicin Solution: Add gentamicin to distilled/deionized water and bring volume to 10.0mL. Mix thoroughly.

$NaHCO_3$ Solution:
Composition per 20mL:
$NaHCO_3$ 2.0g

Preparation of $NaHCO_3$ Solution: Add the $NaHCO_3$ to distilled/deionized water and bring volume to 20.0mL. Mix thoroughly. Filter sterilize. Use freshly prepared solution.

Preparation of Medium: Add components, except $NaHCO_3$ solution, to distilled/deionized water and bring volume to 980.0mL. Mix thoroughly. Gently heat and bring to boiling. Autoclave for 15 min at 15 psi pressure–121°C. Cool to 50°C. Aseptically add sterile $NaHCO_3$ solution. Mix thoroughly. Pour into sterile Petri dishes.

Use: For the cultivation, differentiation and enumeration of *Enterococcus* species based on starch hydrolysis and production of fluorescence. Bacteria that hydrolyze starch, such as *Streptococcus bovis*, appear as colonies surrounded by a clear zone. Bacteria that produce fluorescence, such as *Streptococcus bovis* and *Enterococcus faecium*, appear as colonies surrounded by a zone of of bright bluish fluorescence when viewed under a long-wave UV lamp. Other bacteria, such as *Enterococcus faecalis*, *Enterococcus avium*, or *Streptococcus equinus*, do not hydrolyze starch or produce fluorescence.

Fildes Enrichment Agar

Composition per liter:
Agar..15.0g
Peptone...5.0g
Beef extract 3.0g
Fildes enrichment solution...........................50.0mL

Fildes Enrichment Solution:
Composition 206mL:
Pepsin...1.0g
NaCl (0.85% solution) 150.0mL
Sheep blood, defibrinated.............................50.0mL
HCl 6.0mL
pH 7.0–7.2 at 25°C

Source: Fildes enrichment solution is available from Difco Laboratories.

Preparation of Fildes Enrichment Solution: Combine components. Mix thoroughly. Incubate at 56°C for 4 h. Bring pH to 7.0 with 20% NaOH. Adjust pH to 7.2 with HCl. Do not autoclave. Add 0.25 mL of chloroform and store at 4°C. Before use heat to 56°C to remove chloroform.

Preparation of Medium: Add components, except Fildes enrichment solution, to distilled/deionized water and bring volume to 950.0mL. Mix thoroughly. Gently heat and bring to boiling. Autoclave for 15 min at 15 psi pressure–121°C. Cool to 56°C. Aseptically add 50.0mL sterile Fildes enrichment solution. Mix thoroughly. Pour into sterile Petri dishes or distribute into sterile tubes.

Use: For the isolation and cultivation of *Haemophilus influenzae*.

Fish Peptone Agar

Composition per liter:
Agar...5.0g
Maltose...5.0g
NaCl 5.0g
Peptone 5.0g
Pancreatic digest of casein...............................5.0g
Yeast extract..5.0g
Trout tissue extract solution.........................50.0mL
pH 7.0 ± 0.2 at 25°C

Trout Tissue Extract Solution:
Composition per liter:
Fish (brook trout) 500.0g
Pepsin..1.0g
HCl, concentrated 15.0mL

Preparation of Trout Tissue Extract: Add 1.0L distilled/deionized water to brook trout and blend for 20–30 min. Add 1.0g pepsin and 15.0mL concentrated HCl to digest the trout proteins. Incubate for 12 h at 45°C. Adjust pH to 7.0. Allow solids to settle. Filter sterilize. Do not autoclave. Store at 5°C.

Preparation of Medium: Add components, except trout tissue extract, to distilled/deionized water and bring volume to 950.0L. Mix thoroughly. Gently heat and bring to boiling. Autoclave for 15 min at 13 psi pressure–118°C. Cool to 45°–50°C. Aseptically add 50.0mL sterile trout tissue extract. Mix thoroughly. Pour into sterile Petri dishes or distribute into sterile tubes.

Use: For the cultivation and maintenance of *Aeromonas salmonicida*.

Fish Peptone Broth

Composition per liter:

Maltose	5.0g
NaCl	5.0g
Peptone	5.0g
Pancreatic digest of casein	5.0g
Yeast extract	5.0g
Trout tissue extract solution	50.0mL

pH 7.0 ± 0.2 at 25°C

Trout Tissue Extract Solution:
Composition per liter:

Fish (brook trout)	500.0g
Pepsin	1.0g
HCl, concentrated	15.0mL

Preparation of Trout Tissue Extract: Add 1.0L distilled/deionized water to brook trout and blend for 20–30 min. Add 1.0 g pepsin and 15.0mL concentrated HCl to digest the trout proteins. Incubate for 12 h at 45°C. Adjust pH to 7.0. Allow solids to settle. Filter sterilize. Do not autoclave. Store at 5°C.

Preparation of Medium: Add components, except trout tissue extract, to distilled/deionized water and bring volume to 950.0L. Mix thoroughly. Gently heat and bring to boiling. Autoclave for 15 min at 10 psi pressure–118°C. Cool to 45°–50°C. Aseptically add 50.0mL sterile trout tissue extract. Mix thoroughly. Aseptically distribute into sterile tubes or flasks.

Use: For the cultivation of *Aeromonas salmonicida*.

Flagella Broth

Composition per liter:

Tryptose or biosate	10.0g
NaCl	2.5g
K_2HPO_4	1.0g

pH 7.0 ± 0.1 at 25°C

Preparation of Medium: Add components to distilled/deionized water and bring volume to 1.0L. Mix thoroughly. Gently heat and bring to boiling. Distribute into tubes or flasks. Autoclave for 15 min at 15 psi pressure–121°C. Pour into sterile Petri dishes or leave in tubes.

Use: For the cultivation of flagella-producing bacteria.

Flavobacterium M1 Agar

Composition per liter:

Agar	15.0g
Proteose peptone	5.0g
NaCl	3.0g
Beef extract	2.0g
Yeast extract	1.0g

pH 7.0–7.2 at 25°C

Preparation of Medium: Add components to distilled/deionized water and bring volume to 1.0L. Mix thoroughly. Gently heat and bring to boiling. Distribute into tubes or flasks. Autoclave for 15 min at 15 psi pressure–121°C. Pour into sterile Petri dishes or leave in tubes.

Use: For the cultivation and maintenance of *Flavobacterium indolthelicum*.

Flavobacterium Medium (ATCC Medium 647)

Composition per liter:

Agar	12.0g
Beef extract	10.0g
Peptone	10.0g
NaCl	5.0g

pH 7.2–7.3 at 25°C

Preparation of Medium: Add components to distilled/deionized water and bring volume to 1.0L. Mix thoroughly. Gently heat and bring to boiling. Distribute into tubes or flasks. Autoclave for 15 min at 15 psi pressure–121°C. Pour into sterile Petri dishes or leave in tubes.

Use: For the isolation and cultivation of *Flavobacterium* species from food and food-processing equipment.

Flavobacterium Medium

Composition per liter:

Na_2SO_4	1.0g
Pancreatic digest of casein	1.0g
Yeast extract	1.0g

pH 6.0 ± 0.2 at 25°C

Preparation of Medium: Add components to distilled/deionized water and bring volume to 1.0L. Mix thoroughly. Adjust pH to 6.0 with H_2SO_4. Distribute into tubes or flasks. Autoclave for 15 min at 15 psi pressure–121°C.

Use: For the cultivation and maintenance of *Flavobacterium acidurans*.

Flavobacterium Medium (ATCC Medium 65)

Composition per liter:

Agar	12.0g
Sodium caseinate	2.0g
Peptone	1.0g
K_2HPO_4	0.5g
Yeast extract	0.5g

pH 7.4 ± 0.2 at 25°C

Preparation of Medium: Add components to distilled/deionized water and bring volume to 1.0L. Mix thoroughly. Gently heat and bring to boiling. Distribute into tubes or flasks. Autoclave for 15 min at 15 psi pressure–121°C. Pour into sterile Petri dishes or leave in tubes.

Use: For the cultivation and maintenance of *Flavobacterium aquatile*.

Flavobacterium Medium (ATCC Medium 1687)

Composition per liter:

Sodium glutamate	4.0g
K_2HPO_4	0.65g
$NaNO_3$	0.5g
KH_2PO_4	0.19g
$MgSO_4 \cdot 7H_2O$	0.1g
$FeSO_4$ solution	2.0mL

pH 7.4 ± 0.2 at 25°C

$FeSO_4$ Solution:

Composition per 10mL:

$FeSO_4 \cdot 7H_2O$	0.03g

Preparation of $FeSO_4$ Solution: Add $FeSO_4$ to distilled/deionized water and bring volume to 10.0mL. Mix thoroughly. Filter sterilize.

Preparation of Medium: Add components, except $FeSO_4$ solution, to distilled/deionized water and bring volume to 998.0mL. Mix thoroughly. Autoclave for 15 min at 15 psi pressure–121°C. Cool to 25°C. Aseptically add 2.0mL sterile $FeSO_4$ solution. Mix thoroughly. Adjust pH to 7.4. Aseptically distribute into sterile tubes or flasks.

Use: For the cultivation of *Flavobacterium* species.

Flavobacterium Medium M1

Composition per liter:

Agar	12.0g
Proteose peptone	5.0g
NaCl	3.0g
Beef extract	2.0g
Yeast extract	0.2g

pH 7.2–7.4 at 25°C

Preparation of Medium: Add components to distilled/deionized water and bring volume to 1.0L. Mix thoroughly. Gently heat and bring to boiling. Distribute into tubes or flasks. Autoclave for 15 min at 15 psi pressure–121°C. Pour into sterile Petri dishes or leave in tubes.

Use: For the isolation and cultivation of *Flavobacterium* species.

Fletcher Medium

Composition per liter:

Agar	1.5g
NaCl	0.5g
Peptone	0.3g
Beef extract	0.2g
Rabbit serum	50.0mL

pH 7.9 ± 0.1 at 25°C

Source: This medium is available as a premixed powder from Difco Laboratories.

Preparation of Medium: Add components, except rabbit serum, to distilled/deionized water and bring volume to 950.0mL. Mix thoroughly. Gently heat and bring to boiling. Autoclave for 15 min at 15 psi pressure–121°C. Cool to 50°–55°C. Aseptically add 50.0mL sterile rabbit serum. Mix thoroughly. Aseptically distribute into sterile tubes or flasks.

Use: For the isolation, cultivation and maintenance of cultures of *Leptospira* species.

Fletcher Medium with Fluorouracil (Flurouracil Leptospira Medium)

Composition per liter:

Agar	1.5g
NaCl	0.5g
Peptone	0.3g
Beef extract	0.2g
Rabbit serum	50.0mL
Fluorouracil solution	20.0mL

pH 7.9 ± 0.1 at 25°C

Fluorouracil Solution:

Composition per 100mL:

Fluorouracil	10.0g

Preparation of Fluorouracil Solution: Add fluorouracil to 50.0mL distilled/deionized water. Add 1.0mL 2*N* NaOH and bring volume to 100.0mL. Gently heat to 56°C for 2 hr. Adjust pH to 7.4–7.6 with NaOH. Mix thoroughly. Filter sterilize.

Preparation of Medium: Add components, except rabbit serum and fluorouracil solution, to distilled/deionized water and bring volume to 930.0mL. Mix thoroughly. Gently heat and bring to boiling. Autoclave for 15 min at 15 psi pressure–121°C. Cool to 50°–55°C. Aseptically add 80.0mL sterile rabbit serum. Mix thoroughly. Aseptically distribute into sterile tubes or flasks. Immediately prior to use add 0.1mL fluorouracil solution per 5.0mL medium.

Use: For the isolation, cultivation and maintenance of cultures of *Leptospira* species.

Fletcher's Semisolid Medium

Composition per 2120mL:

Agar	1.5g
NaCl	0.5g
Peptone	0.3g
Beef extract	0.2g
Rabbit serum	240.0mL

pH 7.9 ± 0.1 at 25°C

Preparation of Medium: Add components, except rabbit serum, to distilled/deionized water and bring volume to 1880.0mL. Mix thoroughly. Gently heat and bring to boiling. Autoclave for 15 min at 15 psi pressure–121°C. Cool to 50°–55°C. Aseptically add 240.0mL sterile rabbit serum. Mix thoroughly. Aseptically distribute into sterile tubes or flasks.

Use: For the isolation, cultivation and maintenance of cultures of *Leptospira* species.

Flexibacter Agar

Composition per liter:

Agar	15.0g
Monosodium glutamate	5.0g
Pancreatic digest of casein	1.0g
Vitamin-free casamino acids	1.0g
Sodium glycerophosphate	0.1g
Vitamin B_{12}	1.0μg
Trace element solution HO-LE	1.0mL

Trace Element Solution HO-LE:
Composition per liter:

H_3BO_3	2.85g
$MnCl_2 \cdot 4H_2O$	1.8g
Sodium tartrate	1.77g
$FeSO_4 \cdot 7H_2O$	1.36g
$CoCl_2 \cdot 6H_2O$	0.04g
$CuCl_2 . 2H_2O$	0.027g
$Na_2MoO_4 \cdot 2H_2O$	0.025g
$ZnCl_2$	0.020g

Preparation of Trace Element Solution HO-LE: Add components to distilled/deionized water and bring volume to 1.0L. Mix thoroughly. Filter sterilize.

Preparation of Medium: Add components to filtered sea water and bring volume to 999.0mL. Mix thoroughly. Gently heat and bring to boiling. Autoclave for 15 min at 15 psi pressure–121°C. Cool to 45°–50°C. Aseptically add 1.0mL sterile trace element solution HO-LE. Mix thoroughly. Pour into sterile Petri dishes or distribute into sterile tubes.

Use: For the cultivation and maintenance of *Flexibacter polymorphus.*

Flexibacter Broth

Composition per liter:

Monosodium glutamate	5.0g
Pancreatic digest of casein	1.0g
Vitamin-free casamino acids	1.0g
Sodium glycerophosphate	0.1g
Vitamin B_{12}	1.0μg
Trace element solution HO-LE	1.0mL

Trace Element Solution HO-LE:
Composition per liter:

H_3BO_3	2.85g
$MnCl_2 \cdot 4H_2O$	1.8g
Sodium tartrate	1.77g
$FeSO_4 \cdot 7H_2O$	1.36g
$CoCl_2 \cdot 6H_2O$	0.04g
$CuCl_2 . 2H_2O$	0.027g
$Na_2MoO_4 \cdot 2H_2O$	0.025g
$ZnCl_2$	0.020g

Preparation of Trace Element Solution HO-LE: Add components to distilled/deionized water and bring volume to 1.0L. Mix thoroughly. Filter sterilize.

Preparation of Medium: Add components to filtered seawater and bring volume to 999.0mL. Mix thoroughly. Autoclave for 15 min at 15 psi pressure–121°C. Cool to 45°–50°C. Aseptically add 1.0mL sterile trace element solution HO-LE. Mix thoroughly. Aseptically distribute into sterile tubes or flasks.

Use: For the cultivation of *Flexibacter polymorphus.*

Flexibacter Medium (ATCC Medium 1559)

Composition per liter:

Solution B	700.0mL
Solution A	300.0mL

Solution A:
Composition per 300 mL:

Pancreatic digest of casein	0.5g
Yeast extract	0.5g
Beef extract	0.2g
Sodium acetate	0.2g

Preparation of Solution A: Add components to distilled/deionized water and bring volume to 1.0L. Mix thoroughly. Autoclave for 15 min at 15 psi pressure–121°C. Cool to 45°–50°C.

Solution B:

Aged sea water	700.0 mL

Preparation of Solution B: Allow sea water to sit for 7 days. Autoclave for 15 min at 15 psi pressure–121°C. Cool to 45°–50°C.

Preparation of Medium: Aseptically add 300.0mL of sterile solution A to 700.0mL of sterile solution B at 45°–50°C. Mix thoroughly. Aseptically distribute into sterile tubes or flasks.

Use: For the cultivation of *Flexibacter maritimus*.

Flexibacterium Medium

Composition per 1060mL:

Yeast extract	1.0g
$Ca(NO_3)_2 \cdot 4H_2O$	0.1g
K_2HPO_4	0.02g
Sea water, filtered	1.0L
Glucose solution	50.0mL
Trace elements	10.0mL

pH 7.0 ± 0.2 at 25°C

Glucose Solution:

Composition per 50mL:

Glucose	1.0g

Preparation of Glucose Solution: Add glucose to distilled/deionized water and bring volume to 50.0mL. Mix thoroughly. Autoclave for 15 min at 15 psi pressure–121°C. Cool to 25°C.

Trace Elements:

Composition per liter:

$FeSO_4 \cdot 7H_2O$	0.5mg
$ZnSO_4 \cdot 7H_2O$	0.3mg
H_3BO_3	0.1mg
$CoCl_2 \cdot 6H_2O$	0.1mg
$CuSO_4 \cdot 5H_2O$	0.1mg
$MnSO_4 \cdot 4H_2O$	0.1mg
$Na_2MoO_4 \cdot 2H_2O$	0.1mg

Preparation of Trace Elements: Add components to distilled/deionized water and bring volume to 1.0L. Mix thoroughly.

Preparation of Medium: Combine components, except glucose solution. Mix thoroughly. Adjust pH to 7.2. Gently heat and bring to boiling. Autoclave for 15 min at 15 psi pressure–121°C. Cool to 45°–50°C. Aseptically add sterile glucose solution. Mix thoroughly. Aseptically distribute into sterile tubes or bottles.

Use: For the cultivation of *Flexibacter litorale* and *F. marinum*.

Flexibacterium Medium

Composition per 1050mL:

Tris(hydroxymethyl)aminomethane buffer	1.0g
Yeast extract	1.0g
$CaCl_2 \cdot 2H_2O$	0.1g
KCl	0.1g
$MgSO_4 \cdot 7H_2O$	0.1g
Sodium glycerophosphate	0.1g
$NaNO_3$	0.1g
Cobalamin	1.0μg
Glucose solution	50.0mL
Trace elements	10.0mL

pH 7.5 ± 0.2 at 25°C

Glucose Solution:

Composition per 50mL:

Glucose	1.0g

Preparation of Glucose Solution: Add glucose to distilled/deionized water and bring volume to 50.0mL. Mix thoroughly. Autoclave for 15 min at 15 psi pressure–121°C. Cool to 25°C.

Trace Elements:

Composition per liter:

$FeSO_4 \cdot 7H_2O$	0.5mg
$ZnSO_4 \cdot 7H_2O$	0.3mg
H_3BO_3	0.1mg
$CoCl_2 \cdot 6H_2O$	0.1mg
$CuSO_4 \cdot 5H_2O$	0.1mg
$MnSO_4 \cdot 4H_2O$	0.1mg
$Na_2MoO_4 \cdot 2H_2O$	0.1mg

Preparation of Trace Elements: Add components to distilled/deionized water and bring volume to 1.0L. Mix thoroughly.

Preparation of Medium: Add components, except glucose solution, to distilled/deionized water and bring volume to 1.0L. Mix thoroughly. Adjust pH to 7.5. Autoclave for 15 min at 15 psi pressure–121°C. Cool to 45°–50°C. Aseptically add sterile glucose solution. Mix thoroughly. Aseptically distribute into sterile tubes or flasks.

Use: For the cultivation of *Saprospira thermalis*, *Flexibacter elegans* and *Flexibacter rubrum*.

Flexibacterium Medium

Composition per 1050mL:

Tris(hydroxymethyl)aminomethane buffer	1.0g
Yeast extract	1.0g
Glycerol	1.0g
$CaCl_2 \cdot 2H_2O$	0.1g
KCl	0.1g
$MgSO_4 \cdot 7H_2O$	0.1g
Sodium glycerophosphate	0.1g
$NaNO_3$	0.1g
Cobalamin	1.0μg
Trace elements	10.0mL

pH 7.5 ± 0.2 at 25°C

Trace Elements:

Composition per liter:

$FeSO_4 \cdot 7H_2O$	0.5mg
$ZnSO_4 \cdot 7H_2O$	0.3mg

H_3BO_3 .... 0.1mg
$CoCl_2 \cdot 6H_2O$ 0.1mg
$CuSO_4 \cdot 5H_2O$ 0.1mg
$MnSO_4 \cdot 4H_2O$ 0.1mg
$Na_2MoO_4 \cdot 2H_2O$ 0.1mg

Preparation of Trace Elements: Add components to distilled/deionized water and bring volume to 1.0L. Mix thoroughly.

Preparation of Medium: Add components to distilled/deionized water and bring volume to 1.0L. Mix thoroughly. Adjust pH to 7.5. Distribute into sterile tubes or flasks. Autoclave for 15 min at 15 psi pressure–121°C.

Use: For the cultivation of *Saprospira albida*.

Flexiligladius Medium

Composition per liter:
Beef heart, solids from infusion 50.0g
Agar 15.0g
Tryptose 1.0g
NaCl 0.5g
pH 7.4 ± 0.2 at 25°C

Preparation of Medium: Add components to distilled/deionized water and bring volume to 1.0L. Mix thoroughly. Gently heat and bring to boiling. Distribute into tubes or flasks. Autoclave for 15 min at 15 psi pressure–121°C. Pour into sterile Petri dishes or leave in tubes.

Use: For the cultivation and maintenance of *Ensifer adhaerens*.

FLN Medium (Fluorescence Lactose Nitrate Medium)

Composition per liter:
Lactose 20.0g
Agar 15.0g
Proteose peptone No. 3 .... 10.0g
KNO_3 .... 2.0g
K_2HPO_4 .... 1.5g
$MgSO_4 \cdot 7H_2O$ 1.5g
$NaNO_2$ 0.5g
Phenol Red. 0.02g
pH 7.2 ±0.2 at 25°C

Preparation of Medium: Add components to distilled/deionized water and bring volume to 1.0L. Mix thoroughly. Gently heat and bring to boiling. Distribute into tubes or flasks. Autoclave for 15 min at 15 psi pressure–121°C. Pour into sterile Petri dishes or leave in tubes.

Use: For the differentiation of pseudomonads from other non–fermentative bacilli. Lactose fermentation is indicated by the medium turning yellow. *Pseudomonas cepacia* often produces acid from lactose. Denitrification from nitrate or nitrite is indicated by the formation of gas bubbles in the solid medium. P. *aeruginosa*, *P. mendocina*, and *P. denitrificans* are positive for denitrification. Fluorescein production is indicated by fluorescence under UV light. *P. aeruginosa* is positive for fluorescein production; *P. denitrificans* does not produce fluorescein.

Flo Agar

Composition per liter:
Agar 14.0g
Pancreatic digest of casein 10.0g
Peptic digest of animal tissue 10.0g
K_2HPO_4 .... 1.5g
$MgSO_4 \cdot 7H_2O$ 1.5g
pH 7.2 ± 0.2 at 25°C

Source: This medium is available as a premixed powder from BBL Microbiology Systems.

Preparation of Medium: Add components to distilled/deionized water and bring volume to 1.0L. Mix thoroughly. Gently heat and bring to boiling. Distribute into tubes or flasks. Autoclave for 15 min at 15 psi pressure–121°C. Pour into sterile Petri dishes or leave in tubes.

Use: For cultivation of fluorescent *Pseudomonas* species.

Fluid Sabouraud Medium
See: **Antibiotic Medium 13**

Fluid Sabouraud Medium

Composition per liter:
Glucose 20.0g
Pancreatic digest of casein 5.0g
Peptic digest of animal tissue 5.0g
pH5.7 ± 0.2 at 25°C

Source: This medium is available as a premixed powder from Difco Laboratories.

Preparation of Medium: Add components to distilled/deionized water and bring volume to 1.0L. Mix thoroughly. Distribute into tubes or flasks. Autoclave for 15 min at 15 psi pressure–121°C.

Use: For cultivation of yeasts, molds and aciduric microorganisms.

Fluid Thioglycollate Agar

Composition per liter:

Pancreatic digest of casein	15.0g
Glucose	5.0g
Yeast extract	5.0g
NaCl	2.5g
Agar	0.75g
L-Cystine	0.5g
Sodium thioglycollate	0.5g
Resazurin	1.0mg

pH 7.1 ±0.2 at 25°C

Preparation of Medium: Add components to distilled/deionized water and bring volume to 1.0L. Mix thoroughly. Gently heat and bring to boiling. Distribute into tubes or flasks. Autoclave for 15 min at 15 psi pressure–121°C. Pour into sterile Petri dishes or leave in tubes.

Use: For cultivation of anaerobic bacteria. For the cultivation of *Campylobacter fetus, C. jejuni, Leptotrichia buccalis,* and *Streptococcus* species.

Fluid Thioglycollate Agar with Calcium Carbonate

Composition per liter:

Agar	75.0g
Pancreatic digest of casein	15.0g
$CaCO_3$	10.0g
Glucose	5.0g
Yeast extract	5.0g
NaCl	2.5g
L-Cystine	0.5g
Sodium thioglycollate	0.5g
Resazurin	0.001g

pH 7.1 ±0.2 at 25°C

Preparation of Medium: Add components, except $CaCO_3$, to distilled/deionized water and bring volume to 1.0L. Mix thoroughly. Gently heat and bring to boiling. Distribute into tubes or flasks. Autoclave for 15 min at 15 psi pressure–121°C. Sterilize $CaCO_3$, by autoclaving for 15 min at 15 psi pressure–121°C. Aseptically add sterile $CaCO_3$, to sterile tubes or plates—0.1g sterile $CaCO_3$, per 10.0mL medium to be added. Pour medium into sterile Petri dishes or distribute into sterile tubes.

Use: For rapid cultivation of anaerobic bacteria. For the cultivation and maintenance of *Campylobacter fetus, C. jejuni, Leptotrichia buccalis,* and *Streptococcus* species.

Fluid Thioglycollate Medium

Composition per liter:

Pancreatic digest of casein	15.0g
Glucose	5.5g
Yeast extract	5.0g
NaCl	2.5g
Agar	0.75g
L-Cystine	0.5g
Sodium thioglycollate	0.5g
Resazurin	1.0mg

pH 7.1 ± 0.2 at 25°C

Source: This medium is available as a premixed powder from Difco Laboratories.

Preparation of Medium: Add components to distilled/deionized water and bring volume to 1.0L. Mix thoroughly. Gently heat and bring to boiling. Distribute into tubes or flasks. Autoclave for 15 min at 15 psi pressure–121°C. If medium becomes oxidized before use (resazurin turns red) heat in a boiling water bath to expel absorbed O_2. Cool to 25°C.

Use: For the cultivation of anaerobic, microaerophilic and aerobic microorganisms. For use in sterility testing of a variety of specimens.

Fluid Thioglycollate Medium with Beef Extract

Composition per liter:

Pancreatic digest of casein	15.0g
Glucose	5.5g
Yeast extract	5.0g
Beef extract	5.0g
NaCl	2.5g
Agar	0.75g
L-Cystine	0.5g
Sodium thioglycollate	0.5g
Resazurin	1.0mg

pH 7.2 ± 0.2 at 25°C

Source: This medium is available as a premixed powder from Difco Laboratories.

Preparation of Medium: Add components to distilled/deionized water and bring volume to 1.0L. Mix thoroughly. Gently heat and bring to boiling. Distribute into tubes or flasks. Autoclave for 15 min at 15 psi pressure–121°C. If medium becomes oxidized before use (resazurin turns red) heat in a boiling water bath to expel absorbed O_2. Cool to 25°C.

Use: For the cultivation of anaerobic, microaerophilic and aerobic microorganisms. For use in sterility testing of a variety of specimens.

Fluid Thioglycollate Medium with Rabbit Serum

Composition per liter:

Pancreatic digest of casein	15.0g
Glucose	5.5g
Yeast extract	5.0g
NaCl	2.5g
Agar	0.75g
L-Cystine	0.5g
Sodium thioglycollate	0.5g
Resazurin	1.0mg
Rabbit serum	100.0mL

pH 7.1 ± 0.2 at 25°C

Preparation of Medium: Add components, except rabbit serum, to distilled/deionized water and bring volume to 900.0mL. Mix thoroughly. Gently heat and bring to boiling. Distribute into tubes in 9.0mL volumes. Autoclave for 15 min at 15 psi pressure–121°C. If medium becomes oxidized before use (resazurin turns red) heat in a boiling water bath to expel absorbed O_2. Cool to 25°C. Immediately prior to inoculation, aseptically and anerobically add 1.0mL of sterile rabbit serum to each tube. Mix thoroughly.

Use: For the cultivation of anaerobic, microaerophilic and aerobic microorganisms. For use in sterility testing of a variety of specimens.

Fluid Thioglycollate Medium without Glucose or E_h Indicator

Composition per liter:

Pancreatic digest of casein	20.0g
NaCl	2.5g
Agar	0.75g
L-Cystine	0.5g
Sodium thioglycollate	0.5g

pH 7.1 ± 0.2 at 25°C

Source: This medium is available as a premixed powder from BBL Microbiology Systems.

Preparation of Medium: Add components to distilled/deionized water and bring volume to 1.0L. Mix thoroughly. Gently heat and bring to boiling. Distribute into tubes or flasks. Autoclave for 15 min at 15 psi pressure–121°C. Aseptically distribute into sterile screw cap tubes or flasks.

Use: For cultivation of anaerobic bacteria. For use as a basal medium in carbohydrate fermentation tests for differentiating anaerobic bacteria. For carbohydrate fermentation tests 1.0 mL of a 10% filter sterilized carbohydrate solution is aseptically added to 9.0mL of fluid thioglycollate medium.

Fluid Thioglycollate Medium with K Agar

Composition per liter:

Pancreatic digest of casein	15.0g
Glucose	5.0g
Yeast extract	5.0g
KCl	2.5g
K agar	0.45g
L-Cystine	0.5g
Resazurin	1.0mg
Thioglycollic acid	0.3mL

pH 7.2 ± 0.2 at 25°C

Source: This medium is available as a premixed powder from Difco Laboratories.

Preparation of Medium: Add components to distilled/deionized water and bring volume to 1.0L. Mix thoroughly. Gently heat and bring to boiling. Distribute into tubes or flasks. Autoclave for 15 min at 15 psi pressure–121°C. If medium becomes oxidized before use (resazurin turns red) heat in a boiling water bath to expel absorbed O_2. Cool to 25°C.

Use: For the cultivation of anaerobic, microaerophilic and aerobic microorganisms. For use in sterility testing of a variety of specimens.

Fluorescence Denitrification Medium

See: **FN Medium**

Fluorescence Lactose Nitrate Medium

See: **FLN Medium**

Fluorescent Pectolytic Agar (FPA Medium)

Composition per liter:

Proteose peptone No. 3	20.0g
Agar	15.0g
Pectin	5.0g
K_2HPO_4	1.5g
$MgSO_4 \cdot 7H_2O$	0.73g
Antibiotic solution	10.0mL

Antibiotic Solution:

Composition per 10mL:

Cycloheximide	0.075g
Novobiocin	0.045g
Penicillin G	75,000U
Ethanol	1.0mL

Preparation of Antibiotic Solution: Add components to 1.0mL of ethanol. Mix thoroughly. Let stand for 30 min. Bring volume to 10.0mL with distilled/deionized water. Mix thoroughly. Filter sterilize.

Preparation of Medium: Add components, except antibiotic solution, to distilled/deionized water and bring volume to 990.0mL. Mix thoroughly. Gently heat and bring to boiling. Autoclave for 15 min at 15 psi pressure–121°C. Cool to 45°–50°C. Aseptically add sterile antibiotic solution. Mix thoroughly. Pour into sterile Petri dishes.

Use: For the cultivation of fluorescent *Pseudomonas* species that are pectinolytic.

Flurouracil *Leptospira* Medium *See:* Fletcher Medium with Fluoruracil

FM Medium

Composition per liter:

Agar	15.0g
Pancreatic digest of casein	15.0g
Glucose	5.0g
NaCl	5.0g
Yeast extract	5.0g
L-Cysteine·HCl·H_2O	0.75g
Crystal Violet	0.01g
Horse serum	50.0mL
Streptomycin solution	10.0mL

pH 7.2 ± 0.2 at 25°C

Streptomycin Solution:

Composition per 10mL:

Streptomycin	0.01g

Preparation of Streptomycin Solution: Add streptomycin to distilled/deionized water and bring volume to 10.0mL. Mix thoroughly. Filter sterilize.

Preparation of Medium: Add components, except horse serum and streptomycin solution, to distilled/deionized water and bring volume to 940.0mL. Mix thoroughly. Gently heat and bring to boiling. Autoclave for 15 min at 15 psi pressure–121°C. Cool to 45°–50°C. Aseptically add sterile horse serum and streptomycin solution. Mix thoroughly. Pour into sterile Petri dishes or distribute into sterile tubes.

Use: For the selective isolation and cultivation of *Fusobacterium* species.

FN Medium (Fluorescence Denitrification Medium)

Composition per liter:

Agar	15.0g
Proteose peptone No. 3	10.0g
KNO_3	2.0g
K_2HPO_4	1.5g
$MgSO_4 \cdot 7H_2O$	1.5g
$NaNO_2$	0.5g

pH 7.2 ±0.2 at 25°C

Preparation of Medium: Add components to distilled/deionized water and bring volume to 1.0L. Mix thoroughly. Gently heat and bring to boiling. Distribute into tubes or flasks. Autoclave for 15 min at 15 psi pressure–121°C. Pour into sterile Petri dishes or leave in tubes.

Use: For the differentiation of pseudomonads from other non–fermentative bacilli. Denitrification from nitrate or nitrite is indicated by the formation of gas bubbles in the solid medium. *P. aeruginosa, P. mendocina,* and *P. denitrificans* are positive for denitrification. Fluorescein production is indicated by fluorescence under UV light. *P. aeruginosa* is positive for fluorescein production; *P. denitrificans* does not produce fluorescein.

Folic Acid Agar

Composition per liter:

Noble agar	15.0g
$K_2HPO_4 \cdot 3H_2O$	1.2g
Folic acid	1.0g
KH_2PO_4	0.5g
Salts A	5.0mL
Salts B	1.5mL

Salts A:

Composition per 100mL:

$MgSO_4 \cdot 7H_2O$	1.0g
$CaCl_2 \cdot 2H_2O$	0.1g
$FeSO_4 \cdot 7H_2O$	0.1g

Preparation of Salts A Solution: Add components to distilled/deionized water and bring volume to 100.0mL. Mix thoroughly. Maintain for 3 days at 25°C to dissolve. Filter sterilize the supernatant.

Salts B:

Composition per 100mL:

$MnSO_4$	0.1g
Na_2MoO_4	0.1g

Preparation of Salts B Solution: Add components to distilled/deionized water and bring volume to 100.0mL. Mix thoroughly. Filter sterilize.

Preparation of Medium: Add components, except salts A solution and salts B solution, to distilled/deionized water and bring volume to 994.5mL. Mix thoroughly. Gently heat and bring to boiling. Autoclave for 15 min at 15 psi pressure–121°C. Cool to 45°–50°C. Aseptically add 5.0mL sterile salts A solution and 1.5mL sterile salts B solution. Mix thoroughly. Pour into sterile Petri dishes or distribute into sterile tubes.

Use: For the cultivation and maintenance of *Pseudomonas* species.

Folic Acid Assay Medium

Composition per liter:

Glucose	40.0g
Sodium citrate	20.0g
Vitamin assay casamino acids	12.0g
K_2HPO_4	1.0g
KH_2PO_4	1.0g
$MgSO_4 \cdot 7H_2O$	0.4g
L-Cystine	0.2g
DL-Tryptophan	0.2g
Adenine sulfate	0.02g
$FeSO_4 \cdot 7H_2O$	0.02g
Guanine·HCl	0.02g
$MnSO_4 \cdot 7H_2O$	0.02g
NaCl	0.02g
Uracil	0.02g
Pyridoxine·HCl	4.0mg
Niacin	2.0mg
Riboflavin	2.0mg
Thiamine·HCl	2.0mg
Calcium pantothenate	0.4mg
p-Aminobenzoic acid	0.2mg
Biotin	0.8μg

pH 6.8 ± 0.2 at 25°C

Source: This medium is available as a premixed powder from Difco Laboratories.

Preparation of Medium: Add components to distilled/deionized water and bring volume to 1.0L. Mix thoroughly. Gently heat and bring to boiling with frequent mixing. Cool to 45°–50°C. Mix thoroughly. Distribute into tubes or flasks. Autoclave for 10 min at 15 psi pressure–121°C.

Use: For the microbiological assaying of folic acid using *Enterococcus hirae* as the test organism.

Folic Acid Assay Medium

Composition per liter:

Glucose	20.3g
Lactose	20.0g
Sodium acetate	10.0g
Vitamin assay casamino acids	5.0g
K_2HPO_4	2.5g
Polysorbate 80 (Tween™ 80)	2.0g
L-Alanine	0.4g
L-Asparagine	0.2g
$MgSO_4 \cdot 7H_2O$	0.2g
Mercaptoacetic acid	0.125g
L-Tryptophan	0.1g
$MnSO_4 \cdot H_2O$	0.08g
Adenine	0.04g
Guanine	0.04g
Uracil	0.04g
$(NH_4)_2SO_4 \cdot FeSO_4 \cdot 6H_2O$	0.03g
Glutathione	5.0mg
Pyridoxine·HCl	4.0mg
p-Aminobenzoic acid	2.0mg
Riboflavin	1.0mg
Calcium pantothenate	0.8mg
Niacin	0.8mg
Thiamine·HCl	0.4mg
Biotin	0.02mg

Source: This medium is available as a premixed powder from BBL Microbiology Systems.

Preparation of Medium: Add components to distilled/deionized water and bring volume to 1.0L. Mix thoroughly. Gently heat and bring to boiling. Distribute into tubes or flasks. Autoclave for 15 min at 0 psi pressure–100°C.

Use: For the determination of the folic acid content of pharmaceutical products and other materials based upon the extent of growth of the folic acid requiring *Lactobacillus* test species.

Folic Acid Casei Medium

Composition per liter:

Glucose	40.0g
Sodium acetate	40.0g
Pancreatic digest of casein (charcoal treated)	10.0g
K_2HPO_4	1.0g
KH_2PO_4	1.0g
L-Asparagine	0.6g
L-Cysteine·HCl·H_2O	0.5g
$MgSO_4 \cdot 7H_2O$	0.4g
L-Tryptophan	0.2g
Sorbitan monooleate complex	0.1g
$FeSO_4 \cdot 7H_2O$	0.02g
NaCl	0.02g
Xanthine	0.02g
$MnSO_4$	0.015g
Adenine sulfate	0.01g
Guanine·HCl	0.01g
Uracil	0.01g
Glutathione, reduced	5.0mg
Pyridoxine·HCl	4.0mg
p-Aminobenzoic acid	2.0mg
Riboflavin	1.0mg
Calcium pantothenate	0.8mg
Niacin	0.8mg

Thiamine·HCl 0.4mg
Biotin 0.02mg

pH 6.7 ± 0.1 at 25°C

Source: This medium is available as a premixed powder from Difco Laboratories.

Preparation of Medium: Add components to distilled/deionized water and bring volume to 1.0L. Mix thoroughly. Gently heat and bring to boiling with frequent mixing. Cool to 45°–50°C. Mix thoroughly. Distribute into tubes or flasks. Autoclave for 10 min at 15 psi pressure–121°C.

Use: For assaying concentrations of folic acid, especially folic acid in blood serum, using *Lactobacillus casei* subspecies *rhamnosus* as the test organism. The extent of growth of *L. casei* is proportional to the amount of folic acid in the test substance.

Folic Acid Casei Medium with Chloramphenicol

Composition per liter:

Glucose	40.0g
Sodium acetate	40.0g
Pancreatic digest of casein, charcoal treated	10.0g
K_2HPO_4	1.0g
KH_2PO_4	1.0g
L-Asparagine	0.6g
L-Cysteine·HCl·H_2O	0.5g
$MgSO_4 \cdot 7H_2O$	0.4g
L-Tryptophan	0.2g
Sorbitan monooleate complex	0.1g
$FeSO_4 \cdot 7H_2O$	0.02g
NaCl	0.02g
Xanthine	0.02g
$MnSO_4 \cdot 7H_2O$	0.015g
Adenine sulfate	0.01g
Guanine·HCl	0.01g
Uracil	0.01g
Glutathione, reduced	5.0mg
Pyridoxine·HCl	4.0mg
p-Aminobenzoic acid	2.0mg
Riboflavin	1.0mg
Calcium pantothenate	0.8mg
Niacin	0.8mg
Thiamine·HCl	0.4mg
Biotin	0.02mg
Chloramphenicol solution	10.0mL
Folic acid solution	10.0mL

pH 6.7 ± 0.1 at 25°C

Chloramphenicol Solution:
Composition per 10mL:

Chloramphenicol 0.02g

Preparation of Chloramphenicol Solution: Add chloramphenicol to distilled/deionized water and bring volume to 10.0mL. Mix thoroughly. Filter sterilize.

Folic Acid Solution:
Composition per 10mL:

Folic acid 0.1μg

Preparation of Folic Acid Solution: Add folic acid to distilled/deionized water and bring volume to 10.0mL. Mix thoroughly. Filter sterilize.

Preparation of Medium: Add components, except chloramphenicol solution and folic acid solution, to distilled/deionized water and bring volume to 980.0mL. Mix thoroughly. Gently heat and bring to boiling. Autoclave for 15 min at 15 psi pressure–121°C. Cool to 45°–50°C. Aseptically add 10.0mL sterile chloramphenicol solution and 10.0mL sterile folic acid solution. Mix thoroughly. Aseptically distribute into sterile tubes or flasks.

Use: For the cultivation and maintenance of *Lactobacillus casei.*

Folic AOAC Medium

Composition per liter:

Sodium citrate	52.0g
Glucose	40.0g
Vitamin assay casamino acids	10.0g
K_2HPO_4	6.4g
L-Cysteine·HCl	0.76g
L-Asparagine	0.6g
$MgSO_4 \cdot 7H_2O$	0.4g
L-Tryptophan	0.2g
Sorbitan monooleate complex	0.1g
$FeSO_4 \cdot 7H_2O$	0.02g
$MnSO_4 \cdot 7H_2O$	0.02g
NaCl	0.02g
Xanthine	0.02g
Adenine sulfate	0.01g
Guanine·HCl	0.01g
Uracil	0.01g
Glutathione	5.2mg
Pyridoxine·HCl	4.0mg
p-Aminobenzoic acid	1.0mg
Riboflavin	1.0mg
Calcium pantothenate	0.8mg
Nicotinic acid	0.8mg
Thiamine·HCl	0.4mg
Biotin	0.02mg

pH 6.7 ± 0.1 at 25°C

Source: This medium is available as a premixed powder from Difco Laboratories.

Preparation of Medium: Add components to distilled/deionized water and bring volume to 1.0L. Mix thoroughly. Gently heat and bring to boiling with frequent mixing. Cool to 45°–50°C. Mix thoroughly. Distribute into tubes or flasks. Autoclave for 5 min at 15 psi pressure–121°C.

Use: For the microbiological assaying of folic acid using *Enterococcus hirae* as the test organism.

Fomes annosus Isolation Medium No. 1

Composition per liter:

Agar	20.0g
Peptone	5.0g
KH_2PO_4	0.5g
$MgSO_4$	0.25g
Pentachloronitrobenzene (PCNB)	0.19mg
Streptomycin	0.1mg
Ethanol (95% solution)	20.0mL
Lactic acid (50% solution)	2.0mL

Preparation of Medium: Filter sterilize ethanol and lactic acid. Add components, except ethanol and lactic acid, to distilled/deionized water and bring volume to 978.0mL. Mix thoroughly. Gently heat and bring to boiling. Autoclave for 15 min at 15 psi pressure–121°C. Cool to 45°–50°C. Aseptically add sterile ethanol and lactic acid. Mix thoroughly. Pour into sterile Petri dishes or distribute into sterile tubes

Use: For the cultivation of *Fomes annosus*.

Fomes annosus Isolation Medium No. 2

Composition per liter:

Malt extract	20.0g
Agar	17.0g
Streptomycin sulfate solution	10.0mL
Phenylphenol solution	2.5mL
Lactic acid (50% solution)	1.0mL

Streptomycin Sulfate Solution:
Composition per 10mL:

Streptomycin sulfate	0.1g

Preparation of Streptomycin Sulfate Solution: Add streptomycin sulfate to distilled/deionized water and bring volume to 10.0mL. Mix thoroughly. Filter sterilize.

Phenylphenol Solution:
Composition per 20mL:

o-Phenylphenol	0.48g
Ethanol (95% solution)	20.0mL

Preparation of Phenylphenol Solution: Add *o*-phenylphenol to 20.0mL of ethanol. Mix thoroughly. Filter sterilize.

Preparation of Medium: Filter sterilize lactic acid. Add components—except lactic acid, phenylphenol solution, and streptomycin sulfate solution—to distilled/deionized water and bring volume to 986.5mL. Mix thoroughly. Gently heat and bring to boiling. Autoclave for 15 min at 15 psi pressure–121°C. Cool to 45°–50°C. Aseptically add sterile lactic acid, phenylphenol solution, and streptomycin sulfate solution. Mix thoroughly. Pour into sterile Petri dishes or distribute into sterile tubes.

Use: For the cultivation of *Fomes annosus*.

Formivibrio citricus Medium

Composition per liter:

Trisodium citrate	2.94g
$NaHCO_3$	2.5g
$MgCl_2·6H_2O$	2.03g
KH_2PO_4	1.36g
NH_4Cl	0.53g
$Na_2S·9H_2O$	0.24g
$CaCl_2·2H_2O$	0.15g
Modified Wolfe's metals solution	10.0mL
Wolfe's vitamins solution	10.0mL

pH 7.5–7.7 at 25°C

Modified Wolfe's Metals Solution:
Composition per liter

$NaWO_4·2H_2O$	10.01g
Na_2SeO_3	0.01g
$NiCl_2·6H_2O$	0.01g
Wolfe's mineral solution	1.0L

Preparation of Modified Wolfe's Metals Solution: Add components to Wolfe's mineral solution and bring volume to 10.0mL. Mix thoroughly.

Wolfe's Mineral Solution:
Composition per liter

$MgSO_4·7H_2O$	3.0g
Nitriloacetic acid	1.5g
NaCl	1.0g
$MnSO_4·H_2O$	0.5g
$FeSO_4·7H_2O$	0.1g
$CoCl_2·6H_2O$	0.1g
$CaCl_2$	0.1g
$ZnSO_4·7H_2O$	0.1g
$CuSO_4·5H_2O$	0.01g
$AlK(SO_4)_2·12H_2O$	0.01g
H_3BO_3	0.01g
$Na_2MoO_4·2H_2O$	0.01g

Preparation of Wolfe's Mineral Solution: Add nitrilotriacetic acid to 500.0mL of distilled/deionized water. Dissolve by adjusting pH to 6.5 with KOH. Add remaining components. Add distilled/deionized water to 1.0L.

Wolfe's Vitamin Solution:

Composition per liter:

Pyridoxine·HCl	0.01g
Thiamine·HCl	5.0mg
Riboflavin	5.0mg
Nicotinic acid	5.0mg
Calcium pantothenate	5.0mg
p-Aminobenzoic acid	5.0mg
Thioctic acid	5.0mg
Biotin	2.0mg
Folic acid	2.0mg
Cyanocobalamin	0.1mg

Preparation of Wolfe's Vitamin Solution: Add components to distilled/deionized water and bring volume to 1.0L. Mix thoroughly.

Preparation of Medium: Add components, except $NaHCO_3$ and sodium sulfide, to distilled/deionized water and bring volume to 1.0L. Mix thoroughly. Gently heat and bring to boiling. Cool under an atmosphere of 90% N_2 + 10% CO_2. Add bicarbonate and sodium sulfide. Mix thoroughly. Adjust pH to 7.5–7.7. Anaerobically distribute into tubes under 90% N_2 + 10% CO_2. Cap tubes with rubber stoppers. Place tubes in a press. Autoclave at 121°C for 15 minutes with fast exhaust.

Use: For the cultivation and maintenance of *Formivibrio citricus*.

Fowells Acetate Agar

Composition per liter:

Agar	20.0g
Sodium acetate trihydrate	5.0g

pH 6.5–7.0 at 25°C

Preparation of Medium: Add sodium acetate trihydrate to distilled/deionized water and bring volume to 1.0L. Adjust pH to 6.5–7.0. Add agar. Mix thoroughly. Gently heat and bring to boiling. Distribute into tubes or flasks. Autoclave for 15 min at 15 psi pressure–121°C. Pour into sterile Petri dishes or leave in tubes.

Use: For the cultivation of fungi with the production of ascospores.

FPA Medium

See: **Fluorescent Pectolytic Agar**

FPA Medium

Composition per liter:

Proteose peptone No. 3	20.0g
Agar	15.0g
Pectin, citrus	5.0g
K_2HPO_4	1.5g
$MgSO_4 \cdot 7H_2O$	1.5g
Antibiotic solution	10.0mL

pH 7.0 ± 0.2 at 25°C

Antibiotic Solution:

Composition per 10.0mL:

Cycloheximide	0.075g
Novobiocin	0.045g
Penicillin G	75,000U

Preparation of Antibiotic Solution: Add components to distilled/deionized water and bring volume to 10.0mL. Mix thoroughly. Filter sterilize.

Preparation of Medium: Add components, except agar and antibiotic solution, to distilled/deionized water and bring volume to 990.0mL. Mix thoroughly. Adjust pH to 7.0. Add agar. Mix thoroughly. Gently heat and bring to boiling. Autoclave for 15 min at 15 psi pressure–121°C. Cool to 45°–50°C. Aseptically add sterile antibiotic solution. Mix thoroughly. Pour into sterile Petri dishes or distribute into sterile tubes.

Use: For the isolation and cultivation of *Pseudomonas* species that cause soft-rot.

FRAG Agar (Fragilis Agar)

Composition per 1025mL:

L-Cysteine·HCl·H_2O	0.5g
Basal solution	995.0mL
Glucuronic acid solution	25.0mL
Gentamicin solution	1.0mL
Hemin-vitamin K_1 solution	1.0mL
Ferric sulfate solution	1.0mL
Mineral solution	1.0mL
Phenol Red (1% solution)	1.0mL
Vitamin B_{12} solution	0.05mL

pH 7.0 ± 0.1 at 25°C

Basal Solution:

Composition per 995mL:

Oxgall	20.0g
Agar	15.4g
K_2HPO_4	2.26g
Yeast extract	2.0g
Pancreatic digest of casein	1.4g
$(NH_4)_2HPO_4$	1.0g
K_2HPO_4	0.9g
Papaic digest of soybean meal	0.12g
NaCl	0.12g

Preparation of Basal Solution: Add components to distilled/deionized water and bring volume to 995.0mL. Mix thoroughly. Gently heat and bring to boiling. Autoclave for 15 min at 15 psi pressure–121°C. Cool to 45°–50°C.

Glucuronic Acid Solution:
Composition per 100mL:

D-Glucuronic acid	40.0g

Preparation of Glucuronic Acid Solution: Add glucuronic acid to distilled/deionized water and bring volume to 100.0mL. Mix thoroughly. Filter sterilize.

Gentamicin Solution:
Composition per 10mL:

Gentamicin	0.1mg

Preparation of Gentamicin Solution: Add gentamicin to distilled/deionized water and bring volume to 10.0mL. Mix thoroughly. Filter sterilize.

Vitamin K_1–Hemin Solution:
Composition per liter:

Vitamin K_1 solution	10.0mL
Hemin solution	10.0mL

Preparation of Vitamin K_1–Hemin Solution: Add components to distilled/deionized water and bring volume to 1.0L. Mix thoroughly. Filter sterilize.

Vitamin K_1 Solution:
Composition per 100mL:

Vitamin K_1	1.0g
Ethanol	99.0mL

Preparation of Vitamin K_1 Solution: Add vitamin K_1 to 99.0mL of absolute ethanol. Mix thoroughly.

Hemin Solution:
Composition per 10mL:

Hemin	0.5g
NaOH	0.4g

Preparation of Hemin Solution: Add hemin and NaOH to distilled/deionized water and bring volume to 10.0mL. Mix thoroughly.

Ferric Sulfate Solution:
Composition per 100mL:

$FeSO_4 \cdot 9H_2O$	0.04g

Preparation of Ferric Sulfate Solution: Add $FeSO_4 \cdot 9H_2O$ to distilled/deionized water and bring volume to 100.0mL. Mix thoroughly. Filter sterilize.

Mineral Solution:
Composition per 100mL:

NaCl	9.0g
$CaCl_2 \cdot 2H_2O$	0.27g
$MgCl_2 \cdot 6H_2O$	0.2g
$CoCl_2 \cdot 6H_2O$	0.1g
$MnCl_2 \cdot 4H_2O$	0.1g

Preparation of Mineral Solution: Add components to distilled/deionized water and bring volume to 100.0mL. Mix thoroughly. Filter sterilize.

Vitamin B_{12} Solution:
Composition per 10mL:

Vitamin B_{12}	0.1mg

Preparation of Vitamin B_{12} Solution: Add vitamin B_{12} to distilled/deionized water and bring volume to 10.0mL. Mix thoroughly. Filter sterilize.

Preparation of Medium: To 995.0mL of cooled, sterile basal solution aseptically add, 0.5g of cysteine·HCl·H_2O, 25.0mL of sterile glucuronic acid solution, 1.0mL of sterile gentamicin solution, 1.0mL of sterile hemin-vitamin K_1 solution,1.0mL of sterile ferric sulfate solution, 1.0mL of sterile mineral solution, 1.0mL of Phenol Red solution, and 0.05mL of sterile vitamin B_{12} solution. Mix thoroughly. Pour into sterile Petri dishes or distribute into sterile tubes.

Use: For the isolation, cultivation and differentiation of the *Bacteroides fragilis* Group (*B. fragilis, B. thetaiotamicron, B. vulgatus, B. distasonis, B. ovatus,* and *B. uniformis*) from clinical specimens.

Fragilis Agar
See: **FRAG Agar**

Francisella tularensis Isolation Medium

Composition per liter:

Agar	10.0g
Glucose	10.0g
Pancreatic digest of casein	10.0g
Peptic digest of animal tissue	10.0g
Cysteine·HCl·H_2O	5.0g
NaCl	5.0g
Sodium thioglycollate	2.0g
Glucose	1.0g
Thiamine·HCl	5.0mg

pH 7.2 ± 0.2 at 25°C

Preparation of Medium: Add components, except agar, to distilled/deionized water and bring volume to 1.0L. Mix thoroughly. Adjust pH to 7.2. Add agar. Gently heat and bring to boiling. Autoclave for 20 min at 15 psi pressure–121°C. Cool to 45°–50°C. Pour into sterile Petri dishes.

Use: For the isolation and cultivation of *Francisella tularensis*.

Frankia Isolation Medium

Composition per liter:

Sucrose	40.0g
$Ca(NO_3)_2 \cdot 4H_2O$	0.242g
KNO_3	0.085g
KCl	0.061g
$MgSO_4 \cdot 7H_2O$	0.042g
KH_2PO_4	0.020g
$MnSO_4 \cdot H_2O$	4.5mg
$FeCl_3 \cdot 6H_2O$	2.5mg
H_3BO_3	1.5mg
$ZnSO_4 \cdot 7H_2O$	1.5mg
Nicotinic acid	0.5mg
Pyridoxine·HCl	0.5mg
$Na_2MoO_4 \cdot 2H_2O$	0.25mg
Thiamine·HCl	0.1mg
$CuSO_4 \cdot 5H_2O$	0.04mg
Mannitol solution	10.0mL
Supplement solution	10.0mL

pH 5.5 ± 0.2 at 25°C

Mannitol Solution:

Composition per 100mL:

Mannitol	11.84g

Preparation of Mannitol Solution: Add mannitol to distilled/deionized water and bring volume to 100.0mL. Mix thoroughly. Filter sterilize.

Supplement Solution:

Composition per 10mL:

L-Glutamic acid	0.185g
L-Arginine	0.174g
L-Glutamine	0.146g
L-Aspartic acid	0.133g
L-Asparagine	0.132g
Glycine	0.075g
Urea	0.060g
Naphthaloneacetic acid	2.0mg
Zeatin	1.0μg

Preparation of Supplement Solution: Add components to distilled/deionized water and bring volume to 10.0mL. Mix thoroughly. Filter sterilize.

Preparation of Medium: Add components, except mannitol solution and supplement solution, to distilled/deionized water and bring volume to 980.0mL. Mix thoroughly. Adjust pH to 5.5. Gently heat and bring to boiling. Autoclave for 15 min at 15 psi pressure–121°C. Cool to 45°–50°C. Aseptically add 10.0mL of sterile mannitol solution and 10.0mL of sterile supplement solution. Mix thoroughly. Aseptically distribute into sterile tubes or flasks.

Use: For the isolation and cultivation of *Frankia* species from root nodules.

Frankia Medium

Composition per liter:

Sucrose	20.0g
Agar	10.0g
Edamin	1.0g
Mannitol	1.0g
$CaCO_3$	0.5g
K_2HPO_4	0.5g
Yeast extract	0.5g
$MgCl_2 \cdot 7H_2O$	0.2g
NaCl	0.1g
H_3BO_3	0.1g
$MnSO_4 \cdot H_2O$	0.025g
$ZnSO_4 \cdot 7H_2O$	0.010g
Nicotinic acid	0.5mg
$Na_2MoO_4 \cdot 2H_2O$	0.25mg
Pyridoxine·HCl	0.1mg
Thiamine·HCl	0.1mg
$CuSO_4 \cdot 5H_2O$	0.025mg

pH 7.0 ± 0.2 at 25°C

Preparation of Medium: Add components to distilled/deionized water and bring volume to 1.0L. Mix thoroughly. Gently heat and bring to boiling. Adjust pH to 7.0. Distribute into tubes or flasks. Autoclave for 15 min at 15 psi pressure–121°C. Pour into sterile Petri dishes or leave in tubes.

Use: For the cultivation of *Frankia* species from root nodules.

Frankia Medium

Composition per 1200mL:

Na_2HPO_4	1.0g
NaCl	1.0g
KH_2PO_4	0.3g
Glycerol	1.0mL
NH_4Cl	1.0mL
Thiamine	1.0mL
Albumin fatty acid supplement	200.0mL

pH 7.4 ± 0.2 at 25°C

Albumin Fatty Acid Supplement:

Composition per 200mL:

Bovine albumin fraction V	20.0g
Tween™ 80	25.0mL
$FeSO_4 \cdot 7H_2O$	20.0mL
$CaCl_2 \cdot 2H_2O$	2.0mL
$MgCl_2 \cdot 6H_2O$	2.0mL
Vitamin B_{12}	2.0mL
$ZnSO_4 \cdot 7H_2O$	2.0mL
$CuSO_4 \cdot 5H_2O$	0.2mL

Preparation of Albumin Fatty Acid Supplement: Add components to distilled/deionized water and bring volume to 200.0mL. Mix thoroughly. Filter sterilize.

Preparation of Medium: Add components, except albumin fatty acid supplement, to distilled/deionized water and bring volume to 800.0mL. Mix thoroughly. Gently heat and bring to boiling. Autoclave for 15 min at 15 psi pressure–121°C. Cool to 45°–50°C. Aseptically add 200.0mL of sterile albumin fatty acid supplement. Mix thoroughly. Aseptically distribute into sterile tubes or flasks.

Use: For the cultivation and maintenance of *Frankia* species.

Frankia Medium

Composition per liter:

Basal medium	900.0mL
Albumin fatty acid supplement	100.0mL

pH 7.4 ± 0.2 at 25°C

Basal Medium:

Composition per liter:

Na_2HPO_4	1.0g
NaCl	1.0g
KH_2PO_4	0.3g
Glycerol (10% solution)	1.0mL
NH_4Cl (25% solution)	1.0mL
Thiamine (0.5% solution)	1.0mL

Preparation of Medium: Add components to distilled/deionized water and bring volume to 1.0L. Mix thoroughly. Adjust pH to 7.4. Distribute into tubes or flasks. Autoclave for 15 min at 15 psi pressure–121°C.

Albumin Fatty Acid Supplement:

Composition per 200mL:

Bovine albumin, fraction V	20.0g
Tween™ 80 (10% solution)	25.0mL
$FeSO_4·7H_2O$ (0.5% solution)	20.0mL
$CaCl_2·2H_2O$ (1% solution)	2.0mL
$MgCl_2·6H_2O$ (1% solution)	2.0mL
Vitamin B_{12} (0.2% solution)	2.0mL
$ZnSO_4·7H_2O$ (0.4% solution)	2.0mL
$CuSO_4·5H_2O$ (0.3% solution)	0.2mL

Preparation of Albumin Fatty Acid Supplement: Add albumin to 100.0mL distilled/deionized water. Mix thoroughly. Slowly add Tween™ 80 and then other components. Adjust pH to 7.4. Add distilled/deionized water and bring volume to 200.0mL. Mix thoroughly. Autoclave for 15 min at 15 psi pressure–121°C.

Preparation of Medium: Aseptically combine 900.0mL sterile basal medium and 100.0mL sterile albumin fatty acid supplement. Mix thoroughly. Aseptically distribute into sterile tubes or flasks.

Use: For the cultivation and maintenance of *Frankia* species.

Fraser Broth

Composition per liter:

NaCl	20.0g
Na_2HPO_4	12.0g
Beef Extract	5.0g
Proteose peptone	5.0g
Pancreatic digest of casein	5.0g
Yeast extract	5.0g
LiCl	3.0g
KH_2PO_4	1.35g
Aesculin	1.0g
Fraser supplement solution	10.0mL

pH 7.2 ± 0.2 at 25°C

Source: This medium is available as a premixed powder from Difco Laboratories and Oxoid Unipath.

Fraser Supplement Solution:

Composition per 10mL:

Ferric ammonium citrate	0.5g
Acriflavine·HCl	0.25g
Nalidixic acid	0.1g
Ethanol	5.0mL

Preparation of Fraser Supplement Solution Solution: Add components to distilled/deionized water and bring volume to 10.0mL. Mix thoroughly. Filter sterilize.

Preparation of Medium: Add components, except Fraser supplement solution, to distilled/deionized water and bring volume to 990.0mL. Mix thoroughly. Gently heat and bring to boiling. Autoclave for 15 min at 15 psi pressure–121°C. Cool to 45°–50°C. Aseptically add sterile Fraser supplement solution. Mix thoroughly. Aseptically distribute into sterile tubes or flasks.

Use: For the isolation of *Listeria* species from food and environmental species.

Fraser Secondary Enrichment Broth

Composition per liter:

NaCl	20.0g
Na_2HPO_4	12.0g
Beef extract	5.0g
Proteose peptone	5.0g
Pancreatic digest of casein	5.0g
Yeast extract	5.0g
LiCl	3.0g
KH_2PO_4	1.35g
Esculin	1.0g
Acriflavin solution	10.0mL
Ferric ammonium citrate solution	10.0mL
Nalidixic acid solution	1.0mL

Ferric Ammonium Citrate Solution:
Composition per 10mL:

Ferric ammonium citrate	0.5g

Preparation of Ferric Ammonium Citrate Solution: Add ferric ammonium citrate to distilled/deionized water and bring volume to 10.0mL. Mix thoroughly. Filter sterilize.

Acriflavin Solution:
Composition per 10mL:

Acriflavin	0.025g

Preparation of Acriflavin Solution: Add acriflavin to distilled/deionized water and bring volume to 10.0mL. Mix thoroughly. Filter sterilize.

Nalidixic Acid Solution:
Composition per 10mL:

Nalidixic acid	0.04g
NaOH (0.1*N* solution)	10.0mL

Preparation of Nalidixic Acid Solution: Add nalidixic acid to 10.0mL of NaOH solution. Mix thoroughly. Filter sterilize.

Preparation of Medium: Add components, except acriflavin solution and ferric ammonium citrate solution, to distilled/deionized water and bring volume to 980.0mL. Mix thoroughly. Gently heat and bring to boiling. Distribute into tubes in 10.0mL volumes. Autoclave for 12 min at 15 psi pressure–121°C. Cool rapidly to 25°C. Immediately prior to inoculation, aseptically add 0.1mL of sterile acriflavin solution and 0.1mL of ferric ammonium citrate solution to each tube. Mix thoroughly.

Use: For the isolation, cultivation and enrichment of *Listeria monocytogenes* from foods and environmental specimens based on esculin hydrolysis. Bacteria that hydrolyze esculin appear as black colonies.

FSM Selective Medium

Composition per liter:

Agar	18.0g
Peptone	10.0g
Glucose	4.0g
$(NH_4)_2SO_4$	1.32g
K_2HPO_4	1.18g
Casamino acids	1.0g
Yeast extract	1.0g
KH_2PO_4	0.44g
$MgSO_4 \cdot 7H_2O$	0.2g
$FeC_6H_5O_7 \cdot 5H_2O$	3.0mg
Citric acid	1.9mg
$ZnSO_4 \cdot 7H_2O$	1.6mg
$MnSO_4 \cdot H_2O$	1.5mg
2,3,5-Triphenyltetrazolium·HCl solution	10.0mL
Benomyl solution	10.0mL
Polymyxin B solution	10.0mL
Chloroneb solution	10.0mL
Dichloran solution	10.0mL
Bacitracin solution	10.0mL
Cycloheximide solution	10.0mL
Pentachloronitrobenzene solution	10.0mL
Pimaricin solution	10.0mL
Tyrothricin solution	10.0mL
Vancomycin solution	10.0mL
Chloromycetin solution	10.0mL
Penicillin G solution	10.0mL

2,3,5-Triphenyltetrazolium·HCl Solution:
Composition per 10mL:

2,3,5-Triphenyltetrazolium·HCl	0.5mg

Preparation of 2,3,5-Triphenyltetrazolium-HCl Solution: Add 2,3,5-triphenyltetrazolium·HCl to distilled/deionized water and bring volume to 10.0mL. Mix thoroughly. Autoclave for 7 min at 15 psi pressure–121°C.

Benomyl Solution:
Composition per 10mL:

Benomyl	0.5mg

Preparation of Benomyl Solution: Add benomyl to distilled/deionized water and bring volume to 10.0mL. Mix thoroughly. Filter sterilize.

Polymyxin B Solution:
Composition per 10mL:

Polymyxin B	0.1mg

Preparation of Polymyxin B Solution: Add polymyxin B to distilled/deionized water and bring volume to 10.0mL. Mix thoroughly. Filter sterilize.

Chloroneb Solution:
Composition per 10mL:

Chloroneb	0.1mg

Preparation of Chloroneb Solution: Add chloroneb to distilled/deionized water and bring volume to 10.0mL. Mix thoroughly. Filter sterilize.

Dichloran Solution:
Composition per 10mL:

Dichloran	0.1mg

Preparation of Dichloran Solution: Add dichloran to distilled/deionized water and bring volume to 10.0mL. Mix thoroughly. Filter sterilize.

Bacitracin Solution:
Composition per 10mL:

Bacitracin	0.05mg

Preparation of Bacitracin Solution: Add bacitracin to distilled/deionized water and bring volume to 10.0mL. Mix thoroughly. Filter sterilize.

Cycloheximide Solution:

Composition per 10mL:

Cycloheximide 0.05mg

Preparation of Cycloheximide Solution: Add cycloheximide to distilled/deionized water and bring volume to 10.0mL. Mix thoroughly. Filter sterilize.

Pentachloronitrobenzene Solution:

Composition per 10mL:

Pentachloronitrobenzene.............................. 0.03mg

Preparation of Pentachloronitrobenzene Solution: Add pentachloronitrobenzene to distilled/deionized water and bring volume to 10.0mL. Mix thoroughly. Filter sterilize.

Pimaricin Solution:

Composition per 10mL:

Pimaricin.. 0.02mg

Preparation of Pimaricin Solution: Add pimaricin to distilled/deionized water and bring volume to 10.0mL. Mix thoroughly. Filter sterilize.

Tyrothricin Solution:

Composition per 10mL:

Tyrothricin.. 0.02mg

Preparation of Tyrothricin Solution: Add tyrothricin to distilled/deionized water and bring volume to 10.0mL. Mix thoroughly. Filter sterilize.

Vancomycin Solution:

Composition per 10mL:

Vancomycin.. 0.01mg

Preparation of Vancomycin Solution: Add vancomycin to distilled/deionized water and bring volume to 10.0mL. Mix thoroughly. Filter sterilize.

Chloromycetin Solution:

Composition per 10mL:

Chloromycetin.. 5.0μg

Preparation of Chloromycetin Solution: Add chloromycetin to distilled/deionized water and bring volume to 10.0mL. Mix thoroughly. Filter sterilize.

Penicillin G Solution:

Composition per 10mL:

Penicillin G 1.0μg

Preparation of Penicillin G Solution: Add penicillin G to distilled/deionized water and bring volume to 10.0mL. Mix thoroughly. Filter sterilize.

Preparation of Medium: Add components—except 2,3,5-triphenyltetrazolium chloride solution, benomyl solution, polymyxin B solution, chloroneb solution, dichloran solution, bacitracin solution, cycloheximide solution, pentachloronitrobenzene solution, pimaricin solution, tyrothricin solution, vancomycin solution, chloromycetin solution, and Penicillin G solution—to distilled/deionized water and bring volume to 870.0mL. Mix thoroughly. Gently heat and bring to boiling. Autoclave for 15 min at 15 psi pressure–121°C. Cool to 45°–50°C. Aseptically add 10.0mL each of sterile 2,3,5-triphenyltetrazolium chloride solution, benomyl solution, polymyxin B solution, chloroneb solution, dichloran solution, bacitracin solution, cycloheximide solution, pentachloronitrobenzene solution, pimaricin solution, tyrothricin solution, vancomycin solution, chloromycetin solution, and Penicillin G solution. Mix thoroughly. Pour into sterile Petri dishes. Dry plates for 24 hr at 30°C.

Use: For the isolation and cultivation of *Pseudomonas solanacearum* from soil.

FTX Broth

Composition per 1001mL:

Sodium glutamate 10.0g
Glucose 2.0g
Tris 2.0g
Sodium glycerophosphate 0.1g
Artificial seawater 1.0L
Trace element solution 1.0mL

pH 8.0 ± 0.2 at 25°C

Artificial Seawater:

Composition per liter:

NaCl 24.7g
$MgSO_4 \cdot 7H_2O$ 6.3g
$MgCl_2 \cdot 6H_2O$ 4.6g
$CaCl_2$ 1.0g
KCl 0.7g
$NaHCO_3$ 0.2g

Preparation of Artificial Seawater: Add components to distilled/deionized water and bring volume to 1.0L. Mix thoroughly.

Trace Element Solution:

Composition per liter:

Disodium EDTA 8.0g
$MnCl_2 \cdot 4H_2O$ 0.1g
$CoCl_2 \cdot 6H_2O$ 0.02g
KBr 0.02g
KI 0.02g
$ZnCl_2$ 0.02g
$CuSO_4$ 0.01g
H_3BO_3 .... 0.01g
$Na_2MoO_4 \cdot 2H_2O$ 0.01g
LiCl 5.0mg
$SnCl_2 \cdot 2H_2O$ 5.0mg

Preparation of Trace Element Solution: Add components to distilled/deionized water and bring volume to 1.0L. Mix thoroughly.

Preparation of Medium: Add components to 1.0L of artificial seawater. Mix thoroughly. Gently heat and bring to boiling. Distribute into tubes or flasks. Autoclave for 15 min at 15 psi pressure–121°C.

Use: For the isolation and cultivation of *Cytophaga* species, *Herpetosiphon* species, *Saprospira* species, and *Flexithrix* species.

Furoate Agar

Composition per liter:

Agar	20.0g
2-Furoic acid	2.0g
K_2HPO_4	1.0g
NH_4Cl	1.0g
$MgSO_4 \cdot 7H_2O$	0.1g

pH 7.0 ± 0.2 at 25°C

Preparation of Medium: Add components to distilled/deionized water and bring volume to 1.0L. Mix thoroughly. Gently heat and bring to boiling. Distribute into tubes or flasks. Autoclave for 15 min at 15 psi pressure–121°C. Pour into sterile Petri dishes or leave in tubes.

Use: For the cultivation and maintenance of *Bacillus megaterium* and *Pseudomonas* species.

Fusobacterium Medium

Composition per liter:

Agar	15.0g
Pancreatic digest of casein	15.0g
Glucose	5.0g
NaCl	5.0g
Yeast extract	5.0g
L-Cysteine	0.75g
Crystal Violet	0.01g
Bovine serum	50.0mL
Streptomycin solution	10.0mL

pH 7.2 ± 0.2 at 25°C

Streptomycin Solution:
Composition per 10mL:

Streptomycin	0.01g

Preparation of Streptomycin Solution: Add streptomycin to distilled/deionized water and bring volume to 10.0mL. Mix thoroughly. Filter sterilize.

Preparation of Medium: Add components, except bovine serum and streptomycin solution, to distilled/deionized water and bring volume to 940.0mL. Mix thoroughly. Gently heat and bring to boiling. Autoclave for 15 min at 15 psi pressure–121°C. Cool to 45°–50°C. Aseptically add 50.0mL of sterile bovine serum and 10.0mL of sterile streptomycin solution. Mix thoroughly. Pour into sterile Petri dishes or distribute into sterile tubes.

Use: For the cultivation of *Fusobacterium* species.

Fusobacterium necrophorum Medium

Composition per 500mL:

Pancreatic digest of casein	16.0g
Agar	8.3g
Biosate	4.0g
$MgSO_4$	2.5g
Na_2HPO_4	2.5g
Thiotone	2.0g
Glucose	0.5g
Egg yolk emulsion, 50%	45.0mL
Crystal Violet solution	25.0mL
Phenylethyl alcohol solution	1.35mL

pH 7.3 ± 0.2 at 25°C

Egg Yolk Emulsion, 50%:
Composition per 100mL:

Chicken egg yolks	11
Whole chicken egg	1
NaCl (0.9% solution)	50.0mL

Preparation of Egg Yolk Emulsion, 50%: Soak eggs with 1:100 dilution of saturated mercuric chloride solution for 1 min. Crack 11 eggs separating yolks from whites. Mix egg yolks with 1 chicken egg. Measure 50.0mL of egg yolk emulsion and add to 50.0mL of 0.9% NaCl solution. Mix thoroughly.

Crystal Violet Solution:
Composition per 25mL:

Crystal Violet	0.0115g

Preparation of Crystal Violet Solution: Aseptically add Crystal Violet to sterile distilled/deionized water and bring volume to 25.0mL. Mix thoroughly.

Phenylethyl Alcohol Solution:
Composition per 100mL:

Phenylethyl alcohol	0.27g

Preparation of Phenylethyl Alcohol Solution: Aseptically add phenylethyl alcohol to sterile distilled/deionized water and bring volume to 100.0mL. Mix thoroughly.

Preparation of Medium: Add components—except egg yolk emulsion, 50%, Crystal Violet solution, and phenylethyl alcohol solution—to distilled/deionized water and bring volume to 428.65mL. Mix thoroughly. Gently heat and bring to boiling. Autoclave for 15 min at 15 psi pressure–121°C. Cool to 45°–50°C. Aseptically add 45.0mL of sterile egg yolk emulsion, 50%, 25.0mL of Crystal Violet solution, and 1.35mL of sterile phenylethyl alcohol solution.

Mix thoroughly. Pour into sterile Petri dishes or distribute into sterile tubes.

Use: For the isolation and cultivation of *Fusobacterium necrophorum.*

FX A Broth

Composition per liter:

Pancreatic digest of casein	10.0g
Yeast extract	2.0g
$MgSO_4 \cdot 7H_2O$	1.0g

pH 7.0 ± 0.2 at 25°C

Preparation of Medium: Add components to distilled/deionized water and bring volume to 1.0L. Mix thoroughly. Distribute into tubes or flasks. Autoclave for 15 min at 15 psi pressure–121°C.

Use: For the isolation and cultivation of *Cytophaga* species, *Herpetosiphon* species, *Saprospira* species, and *Flexithrix* species.

FX AG Broth

Composition per liter:

Pancreatic digest of casein	10.0g
Yeast extract	2.0g
$MgSO_4 \cdot 7H_2O$	1.0g
Glucose solution	100.0mL

pH 7.0 ± 0.2 at 25°C

Glucose Solution:

Composition per 100mL:

D-Glucose	2.0g

Preparation of Glucose Solution: Add D-glucose to distilled/deionized water and bring volume to 100.0mL. Mix thoroughly. Autoclave for 15 min at 15 psi pressure–121°C. Cool to 25°C.

Preparation of Medium: Add components, except glucose solution, to distilled/deionized water and bring volume to 900.0mL. Mix thoroughly. Gently heat and bring to boiling. Autoclave for 15 min at 15 psi pressure–121°C. Cool to 45°–50°C. Aseptically add sterile glucose solution. Mix thoroughly. Pour into sterile Petri dishes or distribute into sterile tubes.

Use: For the isolation and cultivation of *Cytophaga* species, *Herpetosiphon* species, *Saprospira* species, and *Flexithrix* species.

G Medium

Composition per liter:

$(NH_4)_2SO_4$	2.0g
Yeast extract	2.0g
Glucose	1.0g
K_2HPO_4	0.6g
KH_2PO_4	0.4g
$MgSO_4 \cdot 7H_2O$	0.2g
$CaCl_2$	0.08g
$MnSO_4 \cdot H_2O$	0.05g
$CuSO_4 \cdot 5H_2O$	5.0mg
$ZnSO_4 \cdot 7H_2O$	5.0mg
$FeSO_4 \cdot 7H_2O$	0.5mg

pH 7.8 ± 0.2 at 25°C

Preparation of Medium: Add components to distilled/deionized water and bring volume to 1.0L. Mix thoroughly. Distribute into tubes or flasks. Autoclave for 15 min at 15 psi pressure–121°C.

Use: For the cultivation and maintenance of *Bacillus cereus*.

GA Medium
See: Gelatin Agar

Gaizotia abyssinica Creatinine Agar
See: Birdseed Agar

Gardnerella vaginalis Selective Medium

Composition per liter:

Columbia blood agar base	940.0mL
Rabbit or horse serum	50.0mL
Antibiotic inhibitor solution	10.0mL

pH 7.2 ± 0.2 at 25°C

Source: This medium is available as a premixed powder from Oxoid Unipath.

Columbia Blood Agar Base:
Composition per liter:

Special peptone	23.0g
Agar	10.0g
NaCl	5.0g
Starch	1.0g

Source: Columbia blood agar base is available as a premixed powder from Oxoid Unipath.

Preparation of Columbia Blood Agar Base: Add components to distilled/deionized water and bring volume to 1.0L. Mix thoroughly. Gently heat until boiling. Autoclave for 15 min at 15 psi pressure–121°C. Cool to 45°–50°C.

Antibiotic Inhibitor Solution:
Composition per 10mL:

Nalidixic acid	0.035g
Gentamicin sulFate	4.0mg
Amphotericin B	2.0mg
Ethanol	4.0mL

Preparation of Antibiotic Inhibitor Solution: Add components to distilled/deionized water and bring volume to 10.0mL. Mix thoroughly. Filter sterilize.

Preparation of Medium: To 940.0mL of cooled sterile Columbia blood agar base, aseptically add 50.0mL rabbit or horse blood serum and 10.0mL sterile antibiotic inhibitor solution. Pour into sterile Petri dishes or distribute into sterile tubes.

Use: For the selective isolation, cultivation and differentiation of *Gardnerella vaginalis* from clinical specimens, such as the vaginal discharge of patients with vaginitis. *G. vaginalis* exhibits β hemolysis on this medium.

Gauze's Medium No. 1

Composition per liter:

Agar	30.0g
Soluble starch	20.0g
KNO_3	1.0g
K_2HPO_4	0.5g
$MgSO_4$	0.5g
NaCl	0.5g
$FeSO_4$	0.01g

Preparation of Medium: Add components to tap water and bring volume to 1.0L. Mix thoroughly. Gently heat and bring to boiling. Distribute into tubes or flasks. Autoclave for 15 min at 15 psi pressure–121°C. Pour into sterile Petri dishes or leave in tubes.

Use: For the cultivation and maintenance of *Streptomyces phaeopurpureus*.

GBNA Medium (Gum Base Nalidixic Acid Medium)

Composition per liter:

Gellan gum	8.0g
Pancreatic digest of casein	5.7g
NaCl	1.7g
Papaic digest of soybean meal	1.0g
Glucose	0.83g
K_2HPO_4	0.83g
$MgCl_2 \cdot 6H_2O$	0.33g
Nalidixic acid	0.05g

pH 7.2 ± 0.2 at 25°C

Source: This medium is available as a premixed powder from BBL Microbiology Systems.

Preparation of Medium: Add components to tap water and bring volume to 1.0L. Mix thoroughly. Gently heat and bring to boiling. Distribute into tubes or flasks. Autoclave for 15 min at 15 psi pressure–121°C. Pour into sterile Petri dishes or leave in tubes.

Use: For the isolation and cultivation of *Listeria monocytogenes* from clinical and nonclinical specimens.

GBS Agar Base, Islam (Group B Streptococci Agar) (Islam GBS Agar)

Composition per liter:

Proteose peptone	23.0g
Agar	10.0g
Na_2HPO_4	5.75g
Soluble starch	5.0g
NaH_2PO_4	1.5g
Horse serum, heat inactivated	50.0mL

pH 7.5 ± 0.1 at 25°C

Source: This medium is available as a premixed powder from Oxoid Unipath.

Preparation of Medium: Add components, except horse serum, to distilled/deionized water and bring volume to 950.0mL. Mix thoroughly. Gently heat and bring to boiling. Autoclave for 15 min at 15 psi pressure–121°C. Cool to 45°–50°C. Aseptically add 50.0mL sterile inactivated horse serum. Mix thoroughly. Pour into sterile Petri dishes or distribute into sterile tubes.

Use: For the isolation and detection of Group B streptococci from clinical specimens. Group B streptococci produce orange pigmented colonies when incubated under anaerobic conditions.

GBS Medium, Rapid (Group B Streptococci Medium)

Starch	80.0g
Proteose peptone	23.0g
Na_2HPO_4	5.75g
NaH_2PO_4	1.5g
Horse serum, inactivated	50.0mL
Antibiotic inhibitor solution	10.0 mL

pH 7.5 ± 0.2 at 25°C

Source: This medium is available as a premixed powder from Oxoid Unipath.

Antibiotic Inhibitor Solution:
Composition per 10mL:

Metronidazole	10.0mg
Gentamicin	2.0mg

Preparation of Antibiotic Inhibitor Solution: Add components to distilled/deionized water and bring volume to 10.0mL. Mix thoroughly. Filter sterilize.

Preparation of Medium: Add components, except horse serum and antibiotic inhibitor solution, to distilled/deionized water and bring volume to 940.0mL. Mix thoroughly. Gently heat and bring to boiling. Autoclave for 15 min at 15 psi pressure–121°C. Cool to 45°–50°C. Aseptically add 50.0mL of sterile heat inactivated horse serum and 10.0mL sterile antibiotic inhibitor solution. Mix thoroughly. Aseptically distribute into sterile tubes or flasks. Cool to 5°C and hold at that temperature for 12 hr prior to use.

Use: For the rapid isolation and cultivation of Group B streptococci from clinical specimens.

GC Agar (ATCC Medium 814)

Composition per 1010mL:

GC agar base, 2×	500.0mL
Hemoglobin solution	500.0mL
Supplement solution	10.0mL

pH 7.2 ± 0.2 at 25°C

GC Agar Base, 2X:
Composition 500mL:

Agar	10.0g
Pancreatic digest of casein	7.5g
Peptic digest of animal tissue	7.5g
NaCl	5.0g
K_2HPO_4	4.0g
Cornstarch	1.0g
KH_2PO_4	1.0g

Source: GC agar base is available as a premixed powder from BBL Microbiology Systems. This base may be replaced by GC medium base available from Difco Laboratories.

Preparation of GC Agar Base, 2X: Add components to distilled/deionized water and bring volume to 500.0mL. Mix thoroughly. Gently heat until boiling. Autoclave for 15 min at 15 psi pressure–121°C. Cool to 45°–50°C.

Hemoglobin Solution:
Composition 500mL:

Bovine hemoglobin	10.0g

Preparation of Hemoglobin Solution: Add bovine hemoglobin to distilled/deionized water and bring volume to 500.0mL. Mix thoroughly. Autoclave for 15 min at 15 psi pressure–121°C. Cool to 45°–50°C.

Supplement Solution:

Composition per liter:

Glucose	100.0g
L-Cysteine·HCl	25.9g
L-Glutamine	10.0g
L-Cystine	1.1g
Adenine	1.0g
Nicotinamide adenine dinucleotide	0.25g
Vitamin B_{12}	0.1g
Thiamine pyrophosphate	0.1g
Guanine·HCl	0.03g
$Fe(NO_3)_3 \cdot 6H_2O$	0.02g
p-Aminobenzoic acid	0.013g
Thiamine·HCl	3.0mg

Source: The supplement solution IsoVitaleX® enrichment is available from BBL Microbiology Laboratories. This enrichment may be replaced by supplement VX from Difco Laboratories.

Preparation of Supplement Solution: Add components to distilled/deionized water and bring volume to 1.0L. Mix thoroughly. Filter sterilize.

Preparation of Medium: To 500.0mL of sterile GC agar base aseptically add 500.0mL of sterile hemoglobin solution at 45°–50°C. Mix thoroughly. Aseptically add 10.0mL sterile supplement solution. Mix thoroughly. Pour into sterile Petri dishes or distribute into sterile tubes.

Use: For the isolation and cultivation of fastidious bacteria, especially *Neisseria* and *Haemophilus* species. For the cultivation and maintenance of *Branhamella catarrhalis, Campylocbacter pylori, Eikenella corrodens, Helicobacter pylori, Moraxella nonliquefaciens, Morococcus cerebrosis, Oligella ureolytica. O. urethralis, Pasteurella volantium, Proteus mirabilis,* and *Taylorella equigenitalis.*

GC Agar (GC Medium) (ATCC Medium 1351)

Composition per 1.0L:

GC agar base	950.0mL
Blood, defibrinated	50.0mL

pH 7.2 ± 0.2 at 25°C

GC Agar Base:

Composition per liter:

Agar	10.0g
Pancreatic digest of casein	7.5g
Peptic digest of animal tissue	7.5g
NaCl	5.0g
K_2HPO_4	4.0g
Cornstarch	1.0g
KH_2PO_4	1.0g

Source: GC agar base is available as a premixed powder from BBL Microbiology Systems. This base may be replaced by GC medium base available from Difco Laboratories.

Preparation of GC Agar Base: Add components to distilled/deionized water and bring volume to 1.0L. Mix thoroughly. Gently heat until boiling. Autoclave for 15 min at 15 psi pressure–121°C. Cool to 75°–80°C.

Preparation of Medium: To 950.0mL of sterile GC agar base aseptically add 50.0mL sterile defibrinated blood with thorough mixing and maintain at 75°–80°C for 15–20 min until the medium is chocolatized. Pour into sterile Petri dishes or distribute into sterile tubes.

Use: For the isolation and cultivation of fastidious bacteria, especially *Neisseria* and *Haemophilus* species. For the cultivation and maintenance of *Branhamella catarrhalis, Campylocbacter pylori, Eikenella corrodens, Helicobacter pylori, Moraxella nonliquefaciens, Morococcus cerebrosis, Oligella ureolytica. O. urethralis, Pasteurella volantium, Proteus mirabilis,* and *Taylorella equigenitalis.*

GC Agar with Ampicillin

Composition per 1020mL:

GC agar base, 2×	500.0mL
Hemoglobin solution	500.0mL
Supplement solution	10.0mL
Ampicillin solution	10.0mL

pH 7.2 ± 0.2 at 25°C

GC Agar Base, 2X:

Composition 500mL:

Agar	10.0g
Pancreatic digest of casein	7.5g
Peptic digest of animal tissue	7.5g
NaCl	5.0g
K_2HPO_4	4.0g
Cornstarch	1.0g
KH_2PO_4	1.0g

Source: GC agar base is available as a premixed powder from BBL Microbiology Systems. This base may be replaced by GC medium base available from Difco Laboratories.

Preparation of GC Agar Base, 2X: Add components to distilled/deionized water and bring volume to 500.0mL. Mix thoroughly. Gently heat until boiling. Autoclave for 15 min at 15 psi pressure–121°C. Cool to 45°–50°C.

Hemoglobin Solution:

Composition 500mL:

Bovine hemoglobin	10.0g

Preparation of Hemoglobin Solution: Add bovine hemoglobin to distilled/deionized water and bring volume to 500.0mL. Mix thoroughly. Autoclave for 15 min at 15 psi pressure–121°C. Cool to 45°–50°C.

Supplement Solution:
Composition per liter:

Glucose	100.0g
L-Cysteine·HCl	25.9g
L-Glutamine	10.0g
L-Cystine	1.1g
Adenine	1.0g
Nicotinamide adenine dinucleotide	0.25g
Vitamin B_{12}	0.1g
Thiamine pyrophosphate	0.1g
Guanine·HCl	0.03g
$Fe(NO_3)_3 \cdot 6H_2O$	0.02g
p-Aminobenzoic acid	0.013g
Thiamine·HCl	3.0mg

Source: The supplement solution IsoVitaleX® enrichment is available from BBL Microbiology Laboratories. This enrichment may be replaced by supplement VX from Difco Laboratories.

Preparation of Supplement Solution: Add components to distilled/deionized water and bring volume to 1.0L. Mix thoroughly. Filter sterilize.

Ampicillin Solution:
Composition per 10mL:

Ampicillin	0.02g

Preparation of Ampicillin Solution: Add ampicillin to distilled/deionized water and bring volume to 10.0mL. Mix thoroughly. Filter sterilize.

Preparation of Medium: To 500.0mL of sterile GC agar base aseptically add 500.0mL of sterile hemoglobin solution at 45°–50°C. Mix thoroughly. Aseptically add 10.0mL of sterile supplement solution and 10.0mL of sterile ampicillin solution. Mix thoroughly. Pour into sterile Petri dishes or distribute into sterile tubes.

Use: For the cultivation and maintenance of *Branhamella catarrhalis, Haemophilus influenzae,* and *H. parainfluenzae.*

GC Agar with Ampicillin and Gentamicin

Composition per 1030mL:

GC agar base, 2×	500.0mL
Hemoglobin solution	500.0mL
Supplement solution	10.0mL
Ampicillin solution	10.0mL
Gentamicin solution	10.0mL

pH 7.2 ± 0.2 at 25°C

GC Agar Base, 2X:
Composition 500mL:

Agar	10.0g
Pancreatic digest of casein	7.5g
Peptic digest of animal tissue	7.5g
NaCl	5.0g
K_2HPO_4	4.0g
Cornstarch	1.0g
KH_2PO_4	1.0g

Source: GC agar base is available as a premixed powder from BBL Microbiology Systems. This base may be replaced by GC medium base available from Difco Laboratories.

Preparation of GC Agar Base, 2X: Add components to distilled/deionized water and bring volume to 500.0mL. Mix thoroughly. Gently heat until boiling. Autoclave for 15 min at 15 psi pressure–121°C. Cool to 45°–50°C.

Hemoglobin Solution:
Composition 500mL:

Bovine hemoglobin	10.0g

Preparation of Hemoglobin Solution: Add bovine hemoglobin to distilled/deionized water and bring volume to 500.0mL. Mix thoroughly. Autoclave for 15 min at 15 psi pressure–121°C. Cool to 45°–50°C.

Supplement Solution:
Composition per liter:

Glucose	100.0g
L-Cysteine·HCl	25.9g
L-Glutamine	10.0g
L-Cystine	1.1g
Adenine	1.0g
Nicotinamide adenine dinucleotide	0.25g
Vitamin B_{12}	0.1g
Thiamine pyrophosphate	0.1g
Guanine·HCl	0.03g
$Fe(NO_3)_3 \cdot 6H_2O$	0.02g
p-Aminobenzoic acid	0.013g
Thiamine·HCl	3.0mg

Source: The supplement solution IsoVitaleX® enrichment is available from BBL Microbiology Laboratories. This enrichment may be replaced by supplement VX from Difco Laboratories.

Preparation of Supplement Solution: Add components to distilled/deionized water and bring volume to 1.0L. Mix thoroughly. Filter sterilize.

Ampicillin Solution:
Composition per 10mL:

Ampicillin	0.01g

Preparation of Ampicillin Solution: Add ampicillin to distilled/deionized water and bring volume to 10.0mL. Mix thoroughly. Filter sterilize.

Gentamicin Solution:
Composition per 10mL:

Gentamicin	2.0mg

Preparation of Gentamicin Solution: Add gentamicin to distilled/deionized water and bring volume to 10.0mL. Mix thoroughly. Filter sterilize.

Preparation of Medium: To 500.0mL of sterile GC agar base aseptically add 500.0mL of sterile hemoglobin solution at 45°–50°C. Mix thoroughly. Aseptically add 10.0mL sterile supplement solution, 10.0mL sterile ampicillin solution, and 10.0mL sterile gentamicin solution. Mix thoroughly. Pour into sterile Petri dishes or distribute into sterile tubes.

Use: For the cultivation and maintenance of *Haemophilus parainfluenzae*.

GC Agar with Ampicillin and Tetracycline

Composition per 1030mL:

GC agar base, 2×	500.0mL
Hemoglobin solution	500.0mL
Supplement solution	10.0mL
Ampicillin solution	10.0mL
Tetracycline solution	10.0mL

pH 7.2 ± 0.2 at 25°C

GC Agar Base, 2X:
Composition 500mL:

Agar	10.0g
Pancreatic digest of casein	7.5g
Peptic digest of animal tissue	7.5g
NaCl	5.0g
K_2HPO_4	4.0g
Cornstarch	1.0g
KH_2PO_4	1.0g

Source: GC agar base is available as a premixed powder from BBL Microbiology Systems. This base may be replaced by GC medium base available from Difco Laboratories.

Preparation of GC Agar Base, 2X: Add components to distilled/deionized water and bring volume to 500.0mL. Mix thoroughly. Gently heat until boiling. Autoclave for 15 min at 15 psi pressure–121°C. Cool to 45°–50°C.

Hemoglobin Solution:
Composition 500mL:

Bovine hemoglobin	10.0g

Preparation of Hemoglobin Solution: Add bovine hemoglobin to distilled/deionized water and bring volume to 500.0mL. Mix thoroughly. Autoclave for 15 min at 15 psi pressure–121°C. Cool to 45°–50°C.

Supplement Solution:
Composition per liter:

Glucose	100.0g
L-Cysteine·HCl	25.9g
L-Glutamine	10.0g
L-Cystine	1.1g
Adenine	1.0g
Nicotinamide adenine dinucleotide	0.25g
Vitamin B_{12}	0.1g
Thiamine pyrophosphate	0.1g
Guanine·HCl	0.03g
$Fe(NO_3)_3·6H_2O$	0.02g
p-Aminobenzoic acid	0.013g
Thiamine·HCl	3.0mg

Source: The supplement solution IsoVitaleX® enrichment is available from BBL Microbiology Laboratories. This enrichment may be replaced by supplement VX from Difco Laboratories.

Preparation of Supplement Solution: Add components to distilled/deionized water and bring volume to 1.0L. Mix thoroughly. Filter sterilize.

Ampicillin Solution:
Composition per 10mL:

Ampicillin	0.01g

Preparation of Ampicillin Solution: Add ampicillin to distilled/deionized water and bring volume to 10.0mL. Mix thoroughly. Filter sterilize.

Tetracycline Solution:
Composition per 10mL:

Tetracycline	5.0mg

Preparation of Tetracycline Solution: Add tetracycline to distilled/deionized water and bring volume to 10.0mL. Mix thoroughly. Filter sterilize.

Preparation of Medium: To 500.0mL of sterile GC agar base aseptically add 500.0mL of sterile hemoglobin solution at 45°–50°C. Mix thoroughly. Aseptically add 10.0mL sterile supplement solution, 10.0mL sterile ampicillin solution, and 10.0mL sterile tetracycline solution. Mix thoroughly. Pour into sterile Petri dishes or distribute into sterile tubes.

Use: For the cultivation and maintenance of *Haemophilus parainfluenzae*.

GC Agar with Chloramphenicol, Tetracycline, and Ampicillin

Composition per 1040mL:

GC agar base, 2×	500.0mL
Hemoglobin solution	500.0mL
Supplement solution	10.0mL
Ampicillin solution	10.0mL

Tetracycline solution 10.0mL
Chloramphenicol solution 10.0mL
pH 7.2 ± 0.2 at 25°C

GC Agar Base, 2X:
Composition per 500mL:
Agar 10.0g
Pancreatic digest of casein 7.5g
Peptic digest of animal tissue 7.5g
NaCl 5.0g
K_2HPO_4 .... 4.0g
Cornstarch 1.0g
KH_2PO_4 .... 1.0g

Source: GC agar base is available as a premixed powder from BBL Microbiology Systems. This base may be replaced by GC medium base available from Difco Laboratories.

Preparation of GC Agar Base, 2X: Add components to distilled/deionized water and bring volume to 500.0mL. Mix thoroughly. Gently heat until boiling. Autoclave for 15 min at 15 psi pressure–121°C. Cool to 45°–50°C.

Hemoglobin Solution:
Composition per500mL:
Bovine hemoglobin 10.0g

Preparation of Hemoglobin Solution: Add bovine hemoglobin to distilled/deionized water and bring volume to 500.0mL. Mix thoroughly. Autoclave for 15 min at 15 psi pressure–121°C. Cool to 45°–50°C.

Supplement Solution:
Composition per liter:
Glucose 100.0g
L-Cysteine·HCl 25.9g
L-Glutamine 10.0g
L-Cystine 1.1g
Adenine 1.0g
Nicotinamide adenine dinucleotide 0.25g
Vitamin B_{12} 0.1g
Thiamine pyrophosphate 0.1g
Guanine·HCl 0.03g
$Fe(NO_3)_3 \cdot 6H_2O$ 0.02g
p-Aminobenzoic acid 0.013g
Thiamine·HCl 3.0mg

Source: The supplement solution IsoVitaleX® enrichment is available from BBL Microbiology Laboratories. This enrichment may be replaced by supplement VX from Difco Laboratories.

Preparation of Supplement Solution: Add components to distilled/deionized water and bring volume to 1.0L. Mix thoroughly. Filter sterilize.

Ampicillin Solution:
Composition per 10mL:
Ampicillin 0.01g

Preparation of Ampicillin Solution: Add ampicillin to distilled/deionized water and bring volume to 10.0mL. Mix thoroughly. Filter sterilize.

Tetracycline Solution:
Composition per 10mL:
Tetracycline 5.0mg

Preparation of Tetracycline Solution: Add tetracycline to distilled/deionized water and bring volume to 10.0mL. Mix thoroughly. Filter sterilize.

Chloramphenicol Solution:
Composition per 10mL:
Chloramphenicol 5.0mg

Preparation of Chloramphenicol Solution: Add chloramphenicol to distilled/deionized water and bring volume to 10.0mL. Mix thoroughly. Filter sterilize.

Preparation of Medium: To 500.0mL of sterile GC agar base aseptically add 500.0mL of sterile hemoglobin solution at 45°–50°C. Mix thoroughly. Aseptically add 10.0mL sterile supplement solution, 10.0mL sterile ampicillin solution, 10.0mL sterile chloramphenicol, and 10.0mL sterile tetracycline solution. Mix thoroughly. Pour into sterile Petri dishes or distribute into sterile tubes.

Use: For the cultivation and maintenance of *Haemophilus parainfluenzae*.

GC Agar with Defined Supplements

GC agar base 990.0mL
Defined supplements solution 10.0mL
pH 7.2 ± 0.2 at 25°C

GC Agar Base:
Composition per liter:
Agar 10.0g
Pancreatic digest of casein 7.5g
Peptic digest of animal tissue 7.5g
NaCl 5.0g
K_2HPO_4 .... 4.0g
Cornstarch 1.0g
KH_2PO_4 .... 1.0g

Source: GC agar base is available as a premixed powder from BBL Microbiology Systems. This base may be replaced by GC medium base available from Difco Laboratories.

Preparation of GC Agar Base: Add components to distilled/deionized water and bring volume to 1.0L. Mix thoroughly. Gently heat until boiling. Autoclave for 15 min at 15 psi pressure–121°C. Cool to 45°–50°C.

Defined Supplements Solution:
Composition per 100mL:

Glucose	40.0g
Glutamine	1.0g
$Fe(NO_3)_3 \cdot 6H_2O$	0.05g
Cocarboxylase	2.0mg

Preparation of Defined Supplements Solution: Add components to distilled/deionized water and bring volume to 100.0mL. Mix thoroughly. Filter sterilize.

Preparation of Medium: To 990.0mL of sterile GC agar base aseptically add 10.0mL sterile defined supplements solution. Mix thoroughly. Pour into sterile Petri dishes or distribute into sterile tubes.

Use: For the cultivation and maintenance of *Neisseria gonorrhoeae*.

GC Agar with Penicillin G

Composition per 1020mL:

GC medium base, 2×	500.0mL
Hemoglobin solution	500.0mL
Supplement solution	10.0mL
Penicillin G solution	10.0mL

pH 7.2 ± 0.2 at 25°C

GC Medium Base, 2X:
Composition 500mL:

Proteose peptone No. 3	15.0g
Agar	10.0g
NaCl	5.0g
K_2HPO_4	4.0g
Cornstarch	1.0g
KH_2PO_4	1.0g

Source: GC medium base is available as a premixed powder from Difco Laboratories.

Preparation of GC Medium Base, 2X: Add components to distilled/deionized water and bring volume to 500.0mL. Mix thoroughly. Gently heat until boiling. Autoclave for 15 min at 15 psi pressure–121°C. Cool to 45°–50°C.

Hemoglobin Solution:
Composition 500mL:

Bovine hemoglobin	10.0g

Preparation of Hemoglobin Solution: Add bovine hemoglobin to distilled/deionized water and bring volume to 500.0mL. Mix thoroughly. Autoclave for 15 min at 15 psi pressure–121°C. Cool to 45°–50°C.

Supplement Solution:
Composition per liter:

Glucose	100.0g
L-Cysteine·HCl	25.9g
L-Glutamine	10.0g
L-Cystine	1.1g
Adenine	1.0g
Nicotinamide adenine dinucleotide	0.25g
Vitamin B_{12}	0.1g
Thiamine pyrophosphate	0.1g
Guanine·HCl	0.03g
$Fe(NO_3)_3 \cdot 6H_2O$	0.02g
p-Aminobenzoic acid	0.013g
Thiamine·HCl	3.0mg

Source: The supplement solution IsoVitaleX® enrichment is available from BBL Microbiology Laboratories. This enrichment may be replaced by supplement VX from Difco Laboratories.

Preparation of Supplement Solution: Add components to distilled/deionized water and bring volume to 1.0L. Mix thoroughly. Filter sterilize.

Penicillin G Solution:
Composition per 10mL:

Penicillin G	0.05g

Preparation of Penicillin G Solution: Add penicillin G to distilled/deionized water and bring volume to 10.0mL. Mix thoroughly. Filter sterilize.

Preparation of Medium: To 500.0mL of sterile GC medium base aseptically add 500.0mL of sterile hemoglobin solution at 45°–50°C. Mix thoroughly. Aseptically add 10.0mL of sterile supplement solutionand 10.0mL of sterile penicillin G solution. Mix thoroughly. Pour into sterile Petri dishes or distribute into sterile tubes.

Use: For the cultivation and maintenance of *Neisseria gonorrhoea*e.

GC Agar with Supplement A

Composition per 1020mL:

GC medium base, 2×	500.0mL
Hemoglobin solution	500.0mL
Supplement solution	10.0mL
Supplement A, Difco	10.0mL

pH 7.2 ± 0.2 at 25°C

GC Medium Base, 2X:
Composition 500mL:

Proteose peptone No. 3	15.0g
Agar	10.0g
NaCl	5.0g
K_2HPO_4	4.0g
Cornstarch	1.0g
KH_2PO_4	1.0g

Source: GC medium base is available as a premixed powder from Difco Laboratories.

Preparation of GC Medium Base, 2X: Add components to distilled/deionized water and bring volume to 500.0mL. Mix thoroughly. Gently heat until boiling. Autoclave for 15 min at 15 psi pressure–121°C. Cool to 45°–50°C.

Hemoglobin Solution:
Composition 500mL:

Bovine hemoglobin	10.0g

Preparation of Hemoglobin Solution: Add bovine hemoglobin to distilled/deionized water and bring volume to 500.0mL. Mix thoroughly. Autoclave for 15 min at 15 psi pressure–121°C. Cool to 45°–50°C.

Supplement Solution:
Composition per liter:

Glucose	100.0g
L-Cysteine·HCl	25.9g
L-Glutamine	10.0g
L-Cystine	1.1g
Adenine	1.0g
Nicotinamide adenine dinucleotide	0.25g
Vitamin B_{12}	0.1g
Thiamine pyrophosphate	0.1g
Guanine·HCl	0.03g
$Fe(NO_3)_3 \cdot 6H_2O$	0.02g
p-Aminobenzoic acid	0.013g
Thiamine·HCl	3.0mg

Source: The supplement solution IsoVitaleX® enrichment is available from BBL Microbiology Laboratories. This enrichment may be replaced by supplement VX from Difco Laboratories.

Preparation of Supplement Solution: Add components to distilled/deionized water and bring volume to 1.0L. Mix thoroughly. Filter sterilize.

Supplement A:
Composition per 10mL:

Supplement A contains yeast concentrate with Crystal Violet.

Preparation of Supplement A: Add components to distilled/deionized water and bring volume to 10.0mL. Mix thoroughly. Filter sterilize.

Preparation of Medium: To 500.0mL of sterile GC medium base aseptically add 500.0mL of sterile hemoglobin solution at 45°–50°C. Mix thoroughly. Aseptically add 10.0mL of sterile supplement solution and 10.0mL of sterile supplement A solution. Mix thoroughly. Pour into sterile Petri dishes or distribute into sterile tubes.

Use: For the cultivation of *Neisseria gonorrhoeae*, other *Neisseria* species and *Haemophilus* species.

GC Agar with Supplement A and with VCN Inhibitor

Composition per 1030mL:

GC medium base, 2×	500.0mL
Hemoglobin solution	500.0mL
Supplement solution	10.0mL
Supplement A, Difco	10.0mL
VCN inhibitor	10.0mL

pH 7.2 ± 0.2 at 25°C

GC Medium Base, 2X:
Composition 500mL:

Proteose peptone No. 3	15.0g
Agar	10.0g
NaCl	5.0g
K_2HPO_4	4.0g
Cornstarch	1.0g
KH_2PO_4	1.0g

Source: GC medium base is available as a premixed powder from Difco Laboraotries.

Preparation of GC Medium Base, 2X: Add components to distilled/deionized water and bring volume to 500.0mL. Mix thoroughly. Gently heat until boiling. Autoclave for 15 min at 15 psi pressure–121°C. Cool to 45°–50°C.

Hemoglobin Solution:
Composition 500mL:

Bovine hemoglobin	10.0g

Preparation of Hemoglobin Solution: Add bovine hemoglobin to distilled/deionized water and bring volume to 500.0mL. Mix thoroughly. Autoclave for 15 min at 15 psi pressure–121°C. Cool to 45°–50°C.

Supplement Solution:
Composition per liter:

Glucose	100.0g
L-Cysteine·HCl	25.9g
L-Glutamine	10.0g
L-Cystine	1.1g
Adenine	1.0g
Nicotinamide adenine dinucleotide	0.25g
Vitamin B_{12}	0.1g
Thiamine pyrophosphate	0.1g
Guanine·HCl	0.03g
$Fe(NO_3)_3 \cdot 6H_2O$	0.02g
p-Aminobenzoic acid	0.013g
Thiamine·HCl	3.0mg

Source: The supplement solution IsoVitaleX® enrichment is available from BBL Microbiology Laboratories. This enrichment may be replaced by supplement VX from Difco Laboratories.

Preparation of Supplement Solution: Add components to distilled/deionized water and bring volume to 1.0L. Mix thoroughly. Filter sterilize.

Supplement A:
Composition per 10mL:
Supplement A contains yeast concentrate with Crystal Violet.

Preparation of Supplement A: Add components to distilled/deionized water and bring volume to 10.0mL. Mix thoroughly. Filter sterilize.

VCN Inhibitor:
Composition per 10mL:

Colistin	7.5mg
Vancomycin	3.0mg
Nystatin	12,500U

Preparation of VCN Inhibitor: Add components to distilled/deionized water and bring volume to 10.0mL. Mix thoroughly. Filter sterilize.

Preparation of Medium: To 500.0mL of sterile GC medium base aseptically add 500.0mL of sterile hemoglobin solution at 45°–50°C. Mix thoroughly. Aseptically add 10.0mL of sterile supplement solution, 10.0mL of sterile supplement A solution and 10.0mL sterile VCN Inhibitor. Mix thoroughly. Pour into sterile Petri dishes or distribute into sterile tubes.

Use: For the cultivation of *Neisseria gonorrhoeae*, other *Neisseria* species and *Haemophilus* species.

GC Agar with Supplement A and with VCTN Inhibitor

Composition per 1030mL:

GC medium base, 2×	500.0mL
Hemoglobin solution	500.0mL
Supplement solution	10.0mL
Supplement A, Difco	10.0mL
VCTN inhibitor	10.0mL

pH 7.2 ± 0.2 at 25°C

GC Medium Base, 2X:
Composition 500mL:

Proteose peptone No. 3	15.0g
Agar	10.0g
NaCl	5.0g
K_2HPO_4	4.0g
Cornstarch	1.0g
KH_2PO_4	1.0g

Source: GC medium base is available as a premixed powder from Difco Laboraotries.

Preparation of GC Medium Base, 2X: Add components to distilled/deionized water and bring volume to 500.0mL. Mix thoroughly. Gently heat until boiling. Autoclave for 15 min at 15 psi pressure–121°C. Cool to 45°–50°C.

Hemoglobin Solution:
Composition 500mL:

Bovine hemoglobin	10.0g

Preparation of Hemoglobin Solution: Add bovine hemoglobin to distilled/deionized water and bring volume to 500.0mL. Mix thoroughly. Autoclave for 15 min at 15 psi pressure–121°C. Cool to 45°–50°C.

Supplement Solution:
Composition per liter:

Glucose	100.0g
L-Cysteine·HCl	25.9g
L-Glutamine	10.0g
L-Cystine	1.1g
Adenine	1.0g
Nicotinamide adenine dinucleotide	0.25g
Vitamin B_{12}	0.1g
Thiamine pyrophosphate	0.1g
Guanine·HCl	0.03g
$Fe(NO_3)_3·6H_2O$	0.02g
p-Aminobenzoic acid	0.013g
Thiamine·HCl	3.0mg

Source: The supplement solution IsoVitaleX® enrichment is available from BBL Microbiology Laboratories. This enrichment may be replaced by supplement VX from Difco Laboratories.

Preparation of Supplement Solution: Add components to distilled/deionized water and bring volume to 1.0L. Mix thoroughly. Filter sterilize.

Supplement A:
Composition per 10mL:
Supplement A contains yeast concentrate with Crystal Violet.

Preparation of Supplement A: Add components to distilled/deionized water and bring volume to 10.0mL. Mix thoroughly. Filter sterilize.

VCTN Inhibitor:
Composition per liter:

Vancomycin	4.0mg
Colistin	7.5mg
Trimethoprim lactate	5.0mg
Nystatin	12,500U

Preparation of VCTN Inhibitor: Add components to distilled/deionized water and bring volume to 10.0mL. Mix thoroughly. Filter sterilize.

Preparation of Medium: To 500.0mL of sterile GC medium base aseptically add 500.0mL of sterile hemoglobin solution at 45°–50°C. Mix thoroughly. Aseptically add 10.0mL of sterile supplement solution, 10.0mL of sterile supplement A, and 10.0mL sterile VCTN inhibitor. Mix thoroughly. Pour into sterile Petri dishes or distribute into sterile tubes.

Use: For the cultivation of *Neisseria gonorrhoeae*, other *Neisseria* species and *Haemophilus* species.

GC Agar with Supplement B

GC agar base	990.0mL
Supplement B	10.0mL

pH 7.2 ± 0.2 at 25°C

GC Agar Base:

Composition per liter:

Agar	10.0g
Pancreatic digest of casein	7.5g
Peptic digest of animal tissue	7.5g
NaCl	5.0g
K_2HPO_4	4.0g
Cornstarch	1.0g
KH_2PO_4	1.0g

Source: GC agar base is available as a premixed powder from BBL Microbiology Systems. This base may be replaced by GC medium base available from Difco Laboratories.

Preparation of GC Agar Base: Add components to distilled/deionized water and bring volume to 1.0L. Mix thoroughly. Gently heat until boiling. Autoclave for 15 min at 15 psi pressure–121°C. Cool to 45°–50°C.

Supplement B:

Composition per 10mL:

Supplement B contains yeast concentrate, glutamine, coenzyme, cocarboxylase, hematin and growth factors.

Preparation of Supplement B: Add components to distilled/deionized water and bring volume to 10.0mL. Mix thoroughly. Filter sterilize.

Preparation of Medium: To 990.0mL of sterile GC agar base aseptically add 10.0mL sterile defined supplement B. Mix thoroughly. Pour into sterile Petri dishes or distribute into sterile tubes.

Use: For the cultivation and maintenance of *Neisseria gonorrhoeae*.

GC Medium, New York City Formulation

Composition per liter:

GC agar base	850.0mL
Horse blood, lysed	100.0mL
Yeast autolysate supplement	30.0mL
LCAT antibiotic solution	20.0mL

pH 7.3 ± 0.2 at 25°C

GC Agar Base:

Composition per 850mL:

Special peptone	15.0g
Agar	10.0g
NaCl	5.0g
K_2HPO_4	4.0g
Cornstarch	1.0g
KH_2PO_4	1.0g

pH 7.2 ± 0.2 at 25°C

Preparation of GC Agar Base: Add components of GC medium base and the hemoglobin to distilled/deionized water and bring volume to 850.0mL. Mix thoroughly. Gently heat until boiling. Autoclave for 15 min at 15 psi pressure–121°C. Cool to 45°–50°C.

Horse Blood, Lysed:

Composition per 100mL:

Saponin	0.5g
Horse blood, defibrinated	100.0mL

Preparation of Horse Blood, Lysed: Add saponin to defibrinated horse blood. Mix thoroughly. Allow blood to lyse.

Yeast Autolysate Supplement:

Composition per 30.0mL:

Yeast autolysate	10.0g
Glucose	1.0g
$NaHCO_3$	0.15g

Preparation of Yeast Autolysate Supplement: Add components to distilled/deionized water and bring volume to 30.0mL. Mix thoroughly. Filter sterilize.

LCAT Antibiotic Solution:

Composition per 20mL:

Colistin	6.0mg
Trimethoprim lactate	5.0mg
Lincomycin	1.0mg
Amphotericin B	1.0mg

Preparation of LCAT Antibiotic Solution: Add components to distilled/deionized water and bring volume to 20.0mL. Mix thoroughly. Filter sterilize.

Preparation of Medium: To 850.0mL of cooled sterile GC agar base, aseptically add 100.0mL of sterile lysed horse blood, 30.0mL of sterile yeast autolysate supplement, and 20.0mL of LCAT antibiotic solution. Mix thoroughly. Pour into sterile Petri dishes or distribute into sterile tubes.

Use: For the selective isolation and cultivation of fastidious microorganisms, especially *Neisseria* species.

GC Medium

See: **GC Agar**

GCII Agar

Composition per liter:

GCII agar base, 2×	490.0mL
Hemoglobin solution	490.0mL
Supplement solution	10.0mL

pH 7.2 ± 0.2 at 25°C

GCII Agar Base, 2X:

Composition per liter:

Agar	10.0g
Pancreatic digest of casein	7.5g
Selected meat peptone	7.5g
NaCl	5.0g
K_2HPO_4	4.0g
Cornstarch	1.0g
KH_2PO_4	1.0g

Source: GCII agar base is available as a premixed powder from BBL Microbiology Systems.

Preparation of GCII Agar Base, 2X: Add components to distilled/deionized water and bring volume to 500.0mL. Mix thoroughly. Gently heat until boiling. Autoclave for 15 min at 15 psi pressure–121°C. Cool to 45°–50°C.

Hemoglobin Solution:

Composition 500mL:

Hemoglobin	10.0g

Preparation of Hemoglobin Solution: Add hemoglobin to distilled/deionized water and bring volume to 500.0mL. Mix thoroughly. Autoclave for 15 min at 15 psi pressure–121°C. Cool to 45°–50°C.

Supplement Solution:

Composition per liter:

Glucose	100.0g
L-Cysteine·HCl	25.9g
L-Glutamine	10.0g
L-Cystine	1.1g
Adenine	1.0g
Nicotinamide adenine dinucleotide	0.25g
Vitamin B_{12}	0.1g
Thiamine pyrophosphate	0.1g
Guanine·HCl	0.03g
$Fe(NO_3)_3 \cdot 6H_2O$	0.02g
p-Aminobenzoic acid	0.013g
Thiamine·HCl	3.0mg

Preparation of Supplement Solution: Add components to distilled/deionized water and bring volume to 1.0L. Mix thoroughly. Filter sterilize.

Preparation of Medium: To 490.0mL of sterile GC II agar base aseptically add 490.0mL of sterile hemoglobin solution at 45°–50°C. Mix thoroughly. Aseptically add 10.0mL sterile supplement solution. Mix thoroughly. Pour into sterile Petri dishes or distribute into sterile tubes.

Use: For the isolation and cultivation of fastidious microorganisms, especially *Neisseria* and *Haemophilus* species, from clinical specimens.

GCII Agar

Composition per liter:

GCII agar base	950.0mL
Blood, defibrinated	50.0mL

pH 7.2 ± 0.2 at 25°C

GCII Agar Base with Extra Agar:

Composition per liter:

Agar	12.0g
Pancreatic digest of casein	7.5g
Selected meat peptone	7.5g
NaCl	5.0g
K_2HPO_4	4.0g
Cornstarch	1.0g
KH_2PO_4	1.0g

Source: GCII agar base is available as a premixed powder from BBL Microbiology Systems.

Preparation of GCII Agar Base Extra Agar: Add components to distilled/deionized water and bring volume to 1.0L. Mix thoroughly. Gently heat until boiling. Autoclave for 15 min at 15 psi pressure–121°C. Cool to 45°–50°C.

Preparation of Medium: To 950.0mL of sterile GC agar base aseptically add 50.0mL sterile defibrinated blood with thorough mixing and maintain at 75°–80°C for 15–20 min until the medium is chocolatized. Pour into sterile Petri dishes or distribute into sterile tubes.

Use: For the isolation and cultivation of fastidious microorganisms, especially *Neisseria* and *Haemophilus* species, from clinical specimens.

GCII Agar with Hemoglogin and IsoVitaleX®

See: **Chocolate II Agar with Hemoglogin and IsoVitaleX®**

GCA Agar with Thiamine

Composition per liter:

Glucose	25.0g
Agar	14.0g
Papaic digest of soybean meal	10.0g
NaCl	5.0g
Pancreatic digest of heart muscle	3.0g
Cysteine·HCl·H_2O	1.0g
Thiamine	0.05mg
Rabbit blood, defibrinated	50.0mL

pH 6.8 ± 0.2 at 25°C

Preparation of Medium: Add components, except rabbit blood, to distilled/deionized water and bring volume to 950.0mL. Mix thoroughly. Gently heat and bring to boiling. Autoclave for 15 min at 15 psi pressure–121°C. Cool to 45°–50°C. Aseptically add sterile rabbit blood. Mix thoroughly. Pour into sterile Petri dishes or distribute into sterile tubes.

Use: For the isolation and cultivation of *Francisella tularensis*.

GC–Lect™ Agar

Composition per liter:

GCII agar base, 2×	500.0mL
Hemoglobin solution	500.0mL
Supplement solution	10.0mL
Selective agent solution	10.0mL

pH 7.2 ± 0.2 at 25°C

Source: This medium is available as a prepared medium from BBL Microbiology Systems.

GCII Agar Base, 2X with Extra Agar:
Composition per liter:

Agar	12.0g
Pancreatic digest of casein	7.5g
Selected meat peptone	7.5g
NaCl	5.0g
K_2HPO_4	4.0g
Cornstarch	1.0g
KH_2PO_4	1.0g

Source: GCII agar base is available as a premixed powder from BBL Microbiology Systems.

Preparation of GCII Agar Base, 2X with Extra Agar: Add components to distilled/deionized water and bring volume to 500.0mL. Mix thoroughly. Gently heat until boiling. Autoclave for 15 min at 15 psi pressure–121°C. Cool to 45°–50°C.

Hemoglobin Solution:
Composition 500mL:

Hemoglobin	10.0g

Preparation of Hemoglobin Solution: Add hemoglobin to distilled/deionized water and bring volume to 500.0mL. Mix thoroughly. Autoclave for 15 min at 15 psi pressure–121°C. Cool to 45°–50°C.

Supplement Solution:
Composition per liter:

Glucose	100.0g
L-Cysteine·HCl	25.9g
L-Glutamine	10.0g
L-Cystine	1.1g
Adenine	1.0g
Nicotinamide adenine dinucleotide	0.25g
Vitamin B_{12}	0.1g
Thiamine pyrophosphate	0.1g
Guanine·HCl	0.03g
$Fe(NO_3)_3 \cdot 6H_2O$	0.02g
p-Aminobenzoic acid	0.013g
Thiamine·HCl	3.0mg

Source: The supplement solution IsoVitaleX® enrichment is available from BBL Microbiology Laboratories. This enrichment may be replaced by supplement VX from Difco Laboratories.

Preparation of Supplement Solution: Add components to distilled/deionized water and bring volume to 1.0L. Mix thoroughly. Filter sterilize.

Selective Agents:
Composition per 10mL:

Selective agents	0.017g

Preparation of Selective Agents: Add components to distilled/deionized water and bring volume to 10.0mL. Mix thoroughly. Filter sterilize.

Preparation of Medium: To 500.0mL of sterile GC II agar base aseptically add 500.0mL of sterile hemoglobin solution at 45°–50°C. Mix thoroughly. Aseptically add 10.0mL sterile supplement solution and 10.0mL selective agents solution. Mix thoroughly. Pour into sterile Petri dishes or distribute into sterile tubes.

Use: For the isolation and cultivation of *Neisseria gonorrhoeae* from clinical specimens.

Gelatin Agar

Composition per liter:

Gelatin	30.0g
Agar	15.0g
Pancreatic digest of casein	10.0g
NaCl	10.0g

pH 7.2 ± 0.2 at 25°C

Preparation of Medium: Add components to distilled/deionized water and bring volume to 1.0L. Mix thoroughly. Gently heat and bring to boiling. Distribute into tubes or flasks. Autoclave for 15 min at 15 psi pressure–121°C.

Use: For the cultivation of bacteria isolated from foods and their differentiation based on proteolytic activity.

Gelatin Agar (GA Medium)

Composition per liter:

Solution 1	950.0mL
Solution 2	50.0mL

pH 7.2 ± 0.2 at 25°C

Solution 1:

Composition per 950mL:

Gelatin	30.0g
Agar	15.0g
Pancreatic digest of casein	10.0g
NaCl	2.0g
D-Mannitol	1.0g
Glucose	1.0g
KNO_3	1.0g
Sodium acetate	1.0g
Sodium formate	1.0g
Sodium succinate	1.0g
Yeast extract	1.0g
Sodium lactate (60% solution)	5.0mL

Preparation of Solution 1: Add components to distilled/deionized water and bring volume to 950.0mL. Mix thoroughly. Gently heat and bring to boiling. Autoclave for 15 min at 15 psi pressure–121°C.

Solution 2:

Composition per 50mL:

Na_2HPO_4	1.0g
L-Cysteine·HCl·H_2O	0.5g
$Na_2CO_3 \cdot H_2O$	0.5g
Sucrose	0.5g
Dithiothreitol	0.1g
Menadione solution	2.0mL

Preparation of Solution 2: Add components to distilled/deionized water and bring volume to 50.0mL. Mix thoroughly. Filter sterilize.

Menadione Solution:

Composition per 100mL:

Menadione (vitamin K_3)	0.05g
Ethanol	99.0mL

Preparation of Menadione Solution: Add menadione to 99.0mL of absolute ethanol. Mix thoroughly.

Preparation of Medium: Aseptically combine sterile solution 1 with sterile solution 2. Mix thoroughly. Pour into sterile Petri dishes.

Use: For the cultivation and differentiation of microorganisms from dental plaque based on their ability to produce gelatinase. For the differentiation of aerobic, anaerobic and facultative microorganisms of clinical significance.

Gelatin Agar

Composition per liter:

Agar	15.0g
Gelatin	15.0g
Peptone	4.0g
Yeast extract	1.0g

pH 7.2 ± 0.2 at 25°C

Preparation of Medium: Add components to distilled/deionized water and bring volume to 1.0L. Mix thoroughly. Gently heat and bring to boiling. Distribute into tubes or flasks. Autoclave for 15 min at 15 psi pressure–121°C. Pour into sterile Petri dishes or leave in tubes.

Use: For the cultivation of a variety of heterotrophic bacteria based upon their utilization of gelatin.

Gelatin Infusion Broth

Composition per liter:

Beef heart, solids from infusion	500.0g
Gelatin	40.0g
Tryptose	10.0g
NaCl	5.0g

pH 7.4 ± 0.2 at 25°C

Preparation of Medium: Add components to distilled/deionized water and bring volume to 1.0L. Mix thoroughly. Gently heat and bring to boiling. Distribute into tubes or flasks. Autoclave for 15 min at 15 psi pressure–121°C. Pour into sterile Petri dishes or leave in tubes.

Use: For the cultivation and differentiation of a variety of heterotrophic bacteria based upon their production of gelatinase. The gelatinase liquifies the medium.

Gelatin Medium

Composition per liter:

Gelatin	4.0g

pH 7.0 ± 0.2 at 25°C

Preparation of Medium: Add gelatin to distilled/deionized water and bring volume to 1.0L. Mix thoroughly. Gently heat and bring to boiling. Distribute into tubes. Autoclave for 15 min at 15 psi pressure–121°C.

Use: For the cultivation and differentiation of *Nocardia* and *Streptomyces* species based on utilization of gelatin. *Nocardia asteroides* usually exhibits no growth. *Nocardia brasiliensis* shows good growth and round compact colonies. *Streptomyces* species show varying degrees of growth.

Gelatin Medium

Composition per liter:

Gelatin	120.0g
Pancreatic digest of casein	13.0g
Sodium chloride	5.0g
Yeast extract	5.0g
Heart muscle, solids from infusion	2.0g
Sodium thioglycollate	0.5g

pH 7.0 ± 0.2 at 25°C

Preparation of Medium: Add components to distilled/deionized water and bring volume to 1.0L. Mix thoroughly. Gently heat and bring to boiling. Distribute into tubes. Autoclave for 15 min at 15 psi pressure–121°C. Pour into sterile Petri dishes or leave in tubes.

Use: For the cultivation of gelatin-utilizing *Clostridium* species.

Gelatin Metronidazole Cadmium Medium (GMC Medium)

Composition per liter:

Solution 1	950.0mL
Solution 2	50.0mL

pH 7.2 ± 0.2 at 25°C

Solution 1:

Composition per 950mL:

Gelatin	30.0g
Agar	15.0g
Pancreatic digest of casein	10.0g
NaCl	2.0g
D-Mannitol	1.0g
Glucose	1.0g
KNO_3	1.0g
Sodium acetate	1.0g
Sodium formate	1.0g
Sodium succinate	1.0g
Yeast extract	1.0g
$CdSO_4 \cdot 8H_2O$	0.020g
Metronidazole	0.010g
Sodium lactate (60% solution)	5.0mL

Preparation of Solution 1: Add components to distilled/deionized water and bring volume to 950.0mL. Mix thoroughly. Gently heat and bring to boiling. Autoclave for 15 min at 15 psi pressure–121°C.

Solution 2:

Composition per 50mL:

Na_2HPO_4	1.0g
L-Cysteine·HCl·H_2O	0.5g
$Na_2CO_3 \cdot H_2O$	0.5g
Sucrose	0.5g
Dithiothreitol	0.1g
Menadione solution	2.0mL

Preparation of Solution 2: Add components to distilled/deionized water and bring volume to 50.0mL. Mix thoroughly. Filter sterilize.

Menadione Solution:

Composition per 100mL:

Menadione (vitamin K_3)	0.05g
Ethanol	99.0mL

Preparation of Menadione Solution: Add menadione to 99.0mL of absolute ethanol. Mix thoroughly.

Preparation of Medium: Aseptically combine sterile solution 1 with sterile solution 2. Mix thoroughly. Pour into sterile Petri dishes.

Use: For the cultivation and differentiation of microorganisms from dental plaque based on their ability to produce gelatinase. For the differentiation of aerobic, anaerobic and facultative microorganisms of clinical significance.

Gelatin Phosphate Salt Agar (GPS Agar)

Composition per liter:

Agar	15.0g
Gelatin	10.0g
NaCl	10.0g
K_2HPO_4	5.0g

pH 7.2 ± 0.2 at 25°C

Preparation of Medium: Add components to distilled/deionized water and bring volume to 1.0L. Mix thoroughly. Gently heat and bring to boiling. Distribute into tubes or flasks. Autoclave for 15 min at 15 psi pressure–121°C. Pour into sterile Petri dishes.

Use: For the cultivation and differentiation of *Vibrio* species from foods.

Gelatin Phosphate Salt Broth (GPS Broth)

Composition per liter:

Gelatin	10.0g
NaCl	10.0g
K_2HPO_4	5.0g

pH 7.2 ± 0.2 at 25°C

Preparation of Medium: Add components to distilled/deionized water and bring volume to 1.0L. Mix thoroughly. Gently heat and bring to boiling. Distribute into tubes or flasks. Autoclave for 15 min at 15 psi pressure–121°C.

Use: For the cultivation of *Vibrio* species from foods.

Gelatin Salt Agar

Composition per liter:

NaCl	30.0g
Agar	15.0g
Gelatin	15.0g
Peptone	4.0g
Yeast extract	1.0g

pH 7.2 ± 0.2 at 25°C

Preparation of Medium: Add components to distilled/deionized water and bring volume to 1.0L. Mix thoroughly. Gently heat and bring to boiling. Distribute into tubes or flasks. Autoclave for 15 min at 15 psi pressure–121°C. Pour into sterile Petri dishes or leave in tubes.

Use: For the cultivation and differentiation of *Vibrio* species from foods.

Gelatinase Test Medium

Composition per liter:

Gelatin	3.0g
ACES buffer	1.0g
Yeast extract	1.0g
Charcoal, activated	0.15g
a-Ketoglutarate monopotassium salt	0.1g
L-Cysteine·HCl·H_2O (4% solution)	1.0mL
KOH (85% solution)	1.0mL
$Fe_4(P_2O_7)_3$ solution	1.0mL

pH 6.9 ± 0.2 at 25°C

L-Cysteine·HCl·H_2O Solution:
Composition per 10mL:

L-Cysteine·HCl·H_2O	0.4g

Preparation of L-Cysteine·HCl·H_2O Solution: Add L-cysteine·HCl·H_2O to distilled/deionized water and bring volume to 10.0mL. Mix thoroughly. Filter sterilize.

$Fe_4(P_2O_7)_3$ Solution:
Composition per 10mL:

$Fe_4(P_2O_7)_3$	0.15g

Preparation of $Fe_4(P_2O_7)_3$ Solution: Add $Fe_4(P_2O_7)_3$ to distilled/deionized water and bring volume to 10.0mL. Mix thoroughly. Filter sterilize.

Preparation of Medium: Add ACES buffer to distilled/deionized water and bring volume to 899.0mL. Mix thoroughly. Gently heat to 50°C. Add 1.0mL KOH solution. Mix thoroughly. Add charcoal, yeast extract and a-ketoglutarate. Add 80.0mL distilled/deionized water to wash sides of flask. Mix thoroughly. Autoclave for 15 min at 15 psi pressure–121°C. Cool to 50°C. Aseptically add 10.0mL sterile cysteine solution and 10.0mL sterile $Fe_4(P_2O_7)_3$ solution. Mix thoroughly. Adjust pH to 6.9. Aseptically distribute into sterile screw cap tubes.

Use: For the cultivation and differentiation of gelatinase producing bacteria.

General Salts Medium for Estuarine Methanogens

Composition per 410.8mL:

Agar	8.0g
NaCl	3.6g
$NaHCO_3$	2.0g
Complete salts solution	200.0mL
Cysteine-sulfide reducing agent	16.0mL
Wolfe's mineral solution	4.0mL
Vitamin solution	4.0mL
Yeast extract–trypticase solution	4.0mL
Sodium acetate (25% solution)	2.0mL
$Fe(NH_4)_2SO_4$ (0.2% solution)	0.4mL
Resazurin (0.1% solution)	0.4mL

pH 7.0 ± 0.2 at 25°C

Complete Salts Solution:
Composition per liter:

$MgSO_4 \cdot 7H_2O$	6.9g
$MgCl_2 \cdot 6H_2O$	5.5g
KCl	0.67g
NH_4Cl	0.5g
$CaCl_2 \cdot 2H_2O$	0.28g
K_2HPO_4	0.28g

Preparation of Complete Salts Solution: Add components to distilled/deionized water and bring volume to 1.0L. Mix thoroughly.

Cysteine-Sulfide Reducing Agent:
Composition per 400mL:

L-Cysteine·HCl·H_2O	5.0g
Na_2S (12.5% solution)	40.0mL
NaOH (1*N* solution)	30.0mL

Preparation of Cysteine-Sulfide Reducing Agent: Add distilled/deionized water to a 500.0mL round bottom flask. Add freshly prepared NaOH solution. Gently heat and bring to boiling under 100% N_2. Remove gassing probe. Add cysteine·HCl·H_2O. Add freshly prepared Na_2S solution. Renew gassing for several minutes. Cap with rubber stoppers. Distribute into 8.0mL/18mm Hungate tubes.

Yeast Extract Trypticase Solution:
Composition per 100mL:

Yeast extract	20.0g
Pancreatic digest of casein	20.0g

Preparation of Yeast Extract Trypticase Solution: Add components to distilled/deionized water and bring volume to 100.0mL. Mix thoroughly.

Wolfe's Mineral Solution:
Composition per liter:

$MgSO_4 \cdot 7H_2O$	3.0g
Nitriloacetic acid	1.5g
NaCl	1.0g
$MnSO_4 \cdot H_2O$	0.5g
$FeSO_4 \cdot 7H_2O$	0.1g
$CoCl_2 \cdot 6H_2O$	0.1g
$CaCl_2$	0.1g
$ZnSO_4 \cdot 7H_2O$	0.1g

$CuSO_4 \cdot 5H_2O$	0.01g
$AlK(SO_4)_2 \cdot 12H_2O$	0.01g
H_3BO_3	0.01g
$Na_2MoO_4 \cdot 2H_2O$	0.01g

Preparation of Wolfe's Mineral Solution: Add nitrilotriacetic acid to 500.0mL of distilled/deionized water. Dissolve by adjusting pH to 6.5 with KOH. Add remaining components. Add distilled/deionized water to 1.0L.

Preparation of Medium: Add components, except cysteine-sulfide reducing agent, to distilled/deionized water and bring volume to 410.8mL. Mix thoroughly. Adjust pH to 7.0. Gently heat and bring to boiling under 80% N_2 + 20% CO_2. Add cysteine-sulfide reducing agent. Continue boiling until resazurin turns colorless indicating reduction. Distribute anaerobically into culture tubes with aluminum crimp seals. Autoclave for 15 min at 15 psi pressure–121°C.

Use: For the cultivation and maintenance of *Methanococcus deltae, Methanococcus vannielii, Methanococcus voltae, Methanogenium cariaci, Methanogenium marisnigri,* and *Methanogenium olentangyi.*

Geodermatophilus obscurus Medium

Composition per liter:

Agar	20.0g
Malt extract, purified solids	15.0g
Starch, soluble	10.0g
Sucrose	10.0g
Yeast extract	5.0g
$CaCO_3$	2.0g

Preparation of Medium: Add components to distilled/deionized water and bring volume to 1.0L. Mix thoroughly. Gently heat and bring to boiling. Distribute into tubes or flasks. Autoclave for 15 min at 15 psi pressure–121°C. Pour into sterile Petri dishes or leave in tubes.

Use: For the isolation and cultivation of *Geodermatophilus obscurus.*

George's Medium, Modified

Composition per liter:

Agar	15.0g
Peptone	1.0g
KNO_3	0.2g
K_2HPO_4	0.02g
$MgSO_4 \cdot 7H_2O$	0.02g
Ferric citrate	0.035mg

Preparation of Medium: Add components to tap water and bring volume to 1.0L. Mix thoroughly. Gently heat and bring to boiling. Distribute into tubes. Autoclave for 15 min at 15 psi pressure–121°C. Pour into sterile Petri dishes or leave in tubes.

Use: For the cultivation of a variety of algae.

Gilardi Motility Test and Maintenance Medium

See: Motility Test and Maintenance Medium, Gilardi

Giolitti-Cantoni Broth (Giolitti-Cantoni Broth Base with Tellurite)

Composition per 1030mL:

Mannitol	20.0g
Pancreatic digest of casein	10.0g
Beef extract	5.0g
LiCl	5.0g
NaCl	5.0g
Yeast extract	5.0g
Sodium pyruvate	3.0g
Glycine	1.2g
Chapman tellurite solution	0.3mL

pH 6.9 ± 0.2 at 25°C

Source: This medium is available as a premixed powder from Oxoid Unipath and Difco Laboratories.

Chapman Tellurite Solution:

Composition per 100mL:

K_2TeO_3	3.5g

Preparation of Chapman Tellurite Solution: Add K_2TeO_3 to distilled/deionized water and bring volume to 100.0mL. Mix thoroughly. Filter sterilize.

Caution: Potassium tellurite is toxic.

Preparation of Medium: Add components, except Chapman tellurite solution, to distilled/deionized water and bring volume to 1.0L. Gently heat and bring to boiling. Mix thoroughly. Distribute into tubes in 10.0mL volumes. Autoclave for 15 min at 15 psi pressure–121°C. Cool rapidly to 25°C. Immediately prior to use incubate tubes at 100°C to expel oxygen. Aseptically add 0.3mL sterile Chapman tellurite solution to each tube. Mix thoroughly. Cool to 25°C.

Use: For cultivation and enrichment of *Staphylococcus aureus* from foods.

Giolitti-Cantoni Broth Base with Tellurite

See: **Giolitti-Cantoni Broth**

Gliding Medium

Composition per liter:

Pancreatic digest of casein	0.5g
Yeast extract	0.5g

pH 7.0 ± 0.2 at 25°C

Preparation of Medium: Add components to distilled/deionized water and bring volume to 1.0L. Mix thoroughly. Distribute into tubes or flasks. Autoclave for 15 min at 15 psi pressure–121°C.

Use: For the cultivation and maintenance of the gliding bacteria *Cytophaga* species and *Flexibacter columnaris*.

Gluconate Peptone Broth

Composition per liter:

Potassium gluconate	40.0g
Casein peptone	1.5g
K_2HPO_4	1.0g
Yeast extract	1.0g

pH 7.0 ± 0.2 at 25°C

Preparation of Medium: Add components to distilled/deionized water and bring volume to 1.0L. Mix thoroughly. Distribute into tubes or flasks. Autoclave for 15 min at 15 psi pressure–121°C.

Use: For the cultivation and differentiation of Gram-negative bacteria based on their ability to oxidize gluconate to 2-ketogluconate. For the differentiation of fluorescent *Pseudomonas* species. After inoculation with bacteria and 48 hr of growth in this medium, Benedict's reagent is added. Bacteria that produce the reducing sugar, 2-ketogluconate turn the reagent yellow-orange to orange-red.

Glucose Agar, 9K

Composition per liter:

$(NH_4)_2SO_4$	3.0g
KH_2PO_4	0.5g
$MgSO_4 \cdot 7H_2O$	0.5g
KCl	0.1g
$Ca(NO_3)_2$	0.0125g
$FeSO_4 \cdot 7H_2O$	0.01mg
Agar solution	500.0mL
Glucose solution	100.0mL

pH 4.5 ± 0.2 at 25°C

Agar Solution:

Composition per 500mL:

Agar	15.0g

Preparation of Agar Solution: Add agar to distilled/deionized water and bring volume to 500.0mL. Mix thoroughly. Autoclave for 15 min at 15 psi pressure–121°C. Cool to 55°C.

Glucose Solution:

Composition per 100mL:

Glucose	10.0g

Preparation of Glucose Solution: Add glucose to distilled/deionized water and bring volume to 100.0mL. Mix thoroughly. Filter sterilize.

Preparation of Medium: Add components, except agar solution and glucose solution, to distilled/deionized water and bring volume to 400.0mL. Mix thoroughly. Adjust pH to 4.5 with H_2SO_4. Autoclave for 15 min at 15 psi pressure–121°C. Cool to 55°C. Aseptically add 500.0mL sterile agar solution and 100.0mL sterile glucose solution. Mix thoroughly. Pour into sterile Petri dishes or distribute into sterile tubes.

Use: For the cultivation of *Thiobacillus acidophilus*.

Glucose Asparagine Agar

Composition per liter:

Agar	15.0g
Glucose	10.0g
Asparagine	0.5g
K_2HPO_4	0.5g

pH 7.0 ± 0.2 at 25°C

Preparation of Medium: Add components to distilled/deionized water and bring volume to 1.0L. Mix thoroughly. Gently heat and bring to boiling. Distribute into tubes or flasks. Autoclave for 15 min at 15 psi pressure–121°C. Pour into sterile Petri dishes or leave in tubes.

Use: For the cultivation and maintenance of *Actinoplanes violaceus, Ampullariella kummingensis, Streptomyces avermitilis, S. calvus, S. hygroscopicus, S. tosaensis,* and *Streptoverticillium morookaense*.

Glucose Broth

Composition per 800 mL:

Agar	10.0g
Beef extract	10.0g
Peptone	10.0g
NaCl	5.0g
Glucose solution	200.0mL

pH 7.0 ± 0.2 at 25°C

Glucose Solution:
Composition per 200mL:
Glucose 20.0g

Preparation of Glucose Solution: Add glucose to distilled/deionized water and bring volume to 200.0mL. Mix thoroughly. Filter sterilize.

Preparation of Medium: Add components, except glucose solution, to distilled/deionized water and bring volume to 800.0mL. Mix thoroughly. Gently heat and bring to boiling. Adjust pH to 7.0. Autoclave for 15 min at 15 psi pressure–121°C. Cool to 45°C. Aseptically add 200.0mL sterile glucose solution. Mix thoroughly. Aseptically distribute into sterile tubes or flasks.

Use: For the cultivation of *Pseudomonas* species.

Glucose Broth, 9K

Composition per liter:
Glucose 10.0g
$(NH_4)_2SO_4$ 3.0g
KH_2PO_4 .... 0.5g
$MgSO_4 \cdot 7H_2O$ 0.5g
KCl 0.1g
$Ca(NO_3)_2$ 0.0125g
$FeSO_4 \cdot 7H_2O$ 0.01mg
Glucose solution 100.0mL
pH 3.5 ± 0.2 at 25°C

Glucose Solution:
Composition per 100mL:
Glucose 10.0g

Preparation of Glucose Solution: Add glucose to distilled/deionized water and bring volume to 100.0mL. Mix thoroughly. Filter sterilize.

Preparation of Medium: Add components, except glucose solution, to distilled/deionized water and bring volume to 900.0mL. Mix thoroughly. Adjust pH to 3.5 with H_2SO_4. Autoclave for 15 min at 15 psi pressure–121°C. Cool to 25°C. Aseptically add 100.0mL sterile glucose solution. Mix thoroughly. Aseptically distribute into sterile tubes or flasks.

Use: For the cultivation of *Thiobacillus acidophilus*.

Glucose Nitrogen–Free Salt Agar

Composition per liter:
Agar 15.0g
$CaCO_3$ 1.0g
K_2HPO_4 .... 1.0g
$MgSO_4 \cdot 7H_2O$ 0.2g
NaCl 0.2g
$FeSO_4 \cdot 7H_2O$ 0.1g
$Na_2MoO_4 \cdot 2H_2O$ 5.0mg
Glucose solution 100.0mL
pH 7.0 ± 0.2 at 25°C

Glucose Solution:
Composition per 100mL:
Glucose 10.0g

Preparation of Glucose Solution: Add glucose to distilled/deionized water and bring volume to 100.0mL. Mix thoroughly. Filter sterilize.

Preparation of Medium: Add components, except glucose solution, to distilled/deionized water and bring volume to 900.0mL. Mix thoroughly. Gently heat and bring to boiling. Autoclave for 15 min at 15 psi pressure–121°C. Cool to 45°–50°C. Aseptically add 100.0mL of sterile glucose solution. Mix thoroughly. Pour into sterile Petri dishes or distribute into sterile tubes.

Use: For the cultivation of *Azotobacter* species.

Glucose Nitrogen–Free Salt Solution

Composition per liter:
$CaCO_3$ 1.0g
K_2HPO_4 .... 1.0g
$MgSO_4 \cdot 7H_2O$ 0.2g
NaCl 0.2g
$FeSO_4 \cdot 7H_2O$ 0.1g
$Na_2MoO_4 \cdot 2H_2O$ 5.0mg
Glucose solution 100.0mL
pH 7.0 ± 0.2 at 25°C

Glucose Solution:
Composition per 100mL:
Glucose 10.0g

Preparation of Glucose Solution: Add glucose to distilled/deionized water and bring volume to 100.0mL. Mix thoroughly. Filter sterilize.

Preparation of Medium: Add components, except glucose solution, to distilled/deionized water and bring volume to 900.0mL. Mix thoroughly. Gently heat and bring to boiling. Autoclave for 15 min at 15 psi pressure–121°C. Cool to 45°–50°C. Aseptically add 100.0mL of sterile glucose solution. Mix thoroughly. Aseptically distribute into sterile tubes or flasks.

Use: For the cultivation of *Azotobacter* species.

Glucose Nutrient Agar

Composition per liter:
Agar 15.0g
Pancreatic digest of casein 10.0g

Glucose	5.0g
K_2HPO_4	5.0g
NaCl	5.0g
Yeast extract	5.0g

Preparation of Medium: Add components to distilled/deionized water and bring volume to 1.0L. Mix thoroughly. Gently heat and bring to boiling. Distribute into tubes or flasks. Autoclave for 15 min at 15 psi pressure–121°C. Pour into sterile Petri dishes or leave in tubes.

Use: For the isolation and cultivation of *Brochothrix thermosphacta*.

Glucose Peptone Agar

Composition per liter:

Peptone	20.0g
Agar	15.0g
Glucose	10.0g
NaCl	5.0g

pH 7.2 ± 0.2 at 25°C

Preparation of Medium: Add components to distilled/deionized water and bring volume to 1.0L. Mix thoroughly. Gently heat and bring to boiling. Distribute into tubes or flasks. Autoclave for 15 min at 15 psi pressure–121°C. Pour into sterile Petri dishes or leave in tubes.

Use: For the cultivation of *Agrobacterium* species.

Glucose Peptone Yeast Extract Salts Medium *See:* GPY Salts Medium

Glucose Phosphate Broth

Composition per liter:

Peptone	10.0g
K_2HPO_4	5.0g
Glucose	5.0g

pH 7.5 ± 0.2 at 25°C

Preparation of Medium: Add components, except glucose, to distilled/deionized water and bring volume to 1.0L. Mix thoroughly. Gently heat and bring to boiling. Filter while hot through Whatman filter paper. Cool to 25°C. Adjust pH to 7.5. Add 5.0g glucose. Mix thoroughly. Distribute into sterile tubes or flasks. Autoclave for 10 min at 10 psi pressure–115°C.

Use: For the cultivation of a variety of nonfastidious heterotrophic microorganisms.

Glucose Salt Teepol Broth (GSTB)

Composition per liter:

NaCl	30.0g
Peptone	10.0g
Glucose	5.0g
Beef extract	3.0g
Methyl Violet	2.0mg
Sodium lauryl sulfate (Teepol—0.1% solution)	4.0mL

pH 8.8 ± 0.2 at 25°C

Preparation of Medium: Add components to distilled/deionized water and bring volume to 1.0L. Mix thoroughly. Adjust pH to 8.8. Distribute into tubes or flasks. Autoclave for 15 min at 15 psi pressure–121°C.

Use: For the cultivation of *Vibrio* species from foods.

Glucose Tetrazolium Medium

Composition per liter:

Agar	15.0g
Pancreatic digest of gelatin	6.0g
Yeast extract	3.0g
Beef extract	1.5g
1,3,5-Triphenyl tetrazolium chloride	0.05g
Glucose solution	100.0mL

pH 6.6 ± 0.1 at 25°C

Glucose Solution:

Composition per 100mL:

Glucose	10.0g

Preparation of Glucose Solution: Add glucose to distilled/deionized water and bring volume to 100.0mL. Mix thoroughly. Filter sterilize.

Preparation of Medium: Add components, except glucose solution, to distilled/deionized water and bring volume to 900.0mL. Mix thoroughly. Gently heat and bring to boiling. Autoclave for 15 min at 15 psi pressure–121°C. Cool to 45°–50°C. Aseptically add glucose solution. Mix thoroughly. Pour into sterile Petri dishes or distribute into sterile tubes.

Use: For the cultivation and maintenance of *Streptococcus mutans*.

Glucose Tryptone Yeast Extract Medium *See:* GTYE Medium

Glucose Yeast Broth with NaCl

Composition per liter:

NaCl	50.0g
Agar	15.0g
Yeast extract	3.0g
Glucose	1.0g

pH 7.0 ± 0.2 at 25°C

Preparation of Medium: Add components to distilled/deionized water and bring volume to 1.0L. Mix thoroughly. Gently heat and bring to boiling. Distribute into tubes or flasks. Autoclave for 15 min at 15 psi pressure–121°C. Pour into sterile Petri dishes or leave in tubes.

Use: For the cultivation and maintenance of *Pediococcus halophilus*.

Glucose Yeast Chalk Agar

Composition per liter:

Chalk	40.0g
Agar	15.0g
Glucose	5.0g
Yeast extract	5.0g

Preparation of Medium: Add components to distilled/deionized water and bring volume to 1.0L. Mix thoroughly. Gently heat and bring to boiling. Distribute into tubes or flasks. Autoclave for 15 min at 15 psi pressure–121°C. Pour into sterile Petri dishes or leave in tubes.

Use: For the cultivation and maintenance of *Xanthomonas* species.

Glucose Yeast Extract Agar

Composition per liter:

Agar	15.0g
Glucose	5.0g
Meat extract	5.0g
Peptone	5.0g
Yeast extract	5.0g

pH 6.5 ± 0.2 at 25°C

Preparation of Medium: Add components to distilled/deionized water and bring volume to 1.0L. Mix thoroughly. Gently heat and bring to boiling. Distribute into tubes or flasks. Autoclave for 15 min at 15 psi pressure–121°C. Pour into sterile Petri dishes or leave in tubes.

Use: For the isolation and cultivation of *Leuconostoc* species and *Pediococcus* species.

Glucose Yeast Extract Medium (ATCC Medium 846)

Composition per liter:

Noble agar	13.0g
Yeast extract	5.0g
KH_2PO_4	1.0g
NaCl	1.0g
Peptone	1.0g

pH 6.8 ± 0.2 at 25°C

Preparation of Medium: Add components to distilled/deionized water and bring volume to 1.0L. Mix thoroughly. Gently heat and bring to boiling. Distribute into tubes or flasks. Autoclave for 15 min at 15 psi pressure–121°C. Pour into sterile Petri dishes or leave in tubes.

Use: For the cultivation and maintenance of *Pseudomonas glathei*.

Glucose Yeast Extract Medium (ATCC Medium 985)

Composition per liter:

Agar	15.0g
Yeast extract	3.0g
Glucose	1.0g

pH 7.0 ± 0.2 at 25°C

Preparation of Medium: Add components to distilled/deionized water and bring volume to 1.0L. Mix thoroughly. Gently heat and bring to boiling. Distribute into tubes or flasks. Autoclave for 15 min at 15 psi pressure–121°C. Pour into sterile Petri dishes or leave in tubes.

Use: For the cultivation and maintenance of *Acinetobacter tartarogenes*, *Agrobacterium viscosum*, and *Pseudomonas* species.

Glucose Yeast Extract Medium (ATCC Medium 1742)

Composition per liter:

Agar	15.0g
Glucose	5.0g
Yeast extract	3.5g

Preparation of Medium: Add components to distilled/deionized water and bring volume to 1.0L. Mix thoroughly. Gently heat and bring to boiling. Distribute into tubes or flasks. Autoclave for 15 min at 15 psi pressure–121°C. Pour into sterile Petri dishes or leave in tubes.

Use: For the cultivation and maintenance of *Xanthobacter* species.

Glucose Yeast Extract Peptone Medium (GYP Medium)

Composition per liter:

Glucose	20.0g
Agar	10.0g
Peptone	5.0g
Yeast extract	5.0g
$CaCO_3$	0.1g

Preparation of Medium: Add components to distilled/deionized water and bring volume to 1.0L. Mix thoroughly. Gently heat and bring to boiling. Distribute into tubes or flasks. Autoclave for 15 min at 15 psi pressure–121°C. Pour into sterile Petri dishes or leave in tubes.

Use: For the isolation and cultivation of *Sporolactobacillus* species.

Glucose Yeast Extract Peptone Thioglycollate Medium *See:* GYPT Medium

Glucose Yeast Medium with Calcium Carbonate

Composition per liter:

Agar	15.0g
$CaCO_3$	7.5g
Peptone	5.0g
Yeast extract	5.0g
Glucose	3.0g

pH 7.0 ± 0.2 at 25°C

Preparation of Medium: Add components to distilled/deionized water and bring volume to 1.0L. Mix thoroughly. Gently heat and bring to boiling. Distribute into tubes or flasks. Adjust pH to 6.3. Autoclave for 15 min at 15 psi pressure–121°C. Pour into sterile Petri dishes or leave in tubes.

Use: For the cultivation and maintenance of *Erwinia herbicola*.

Glucose Yeast Peptone Medium

Composition per liter:

Agar	20.0g
Glucose	5.0g
Peptone	5.0g
Yeast extract	3.0g

pH 7.0 ± 0.2 at 25°C

Preparation of Medium: Add components to distilled/deionized water and bring volume to 1.0L. Mix thoroughly. Gently heat and bring to boiling. Distribute into tubes or flasks. Autoclave for 15 min at 15 psi pressure–121°C. Pour into sterile Petri dishes or leave in tubes.

Use: For the cultivation and maintenance of a variety of heterotrophic microorganisms.

Glutamate Medium (ATCC Medium 1372)

Composition per liter:

K_2HPO_4	6.0g
Sodium glutamate	5.0g
Peptone	4.0g
Yeast extract	4.0g

Preparation of Medium: Add components to distilled/deionized water and bring volume to 1.0L. Mix thoroughly. Gently heat and bring to boiling. Distribute into tubes or flasks. Autoclave for 15 min at 15 psi pressure–121°C. Pour into sterile Petri dishes or leave in tubes.

Use: For the cultivation and maintenance of *Escherichia coli*.

Glutamate Medium

Composition per liter:

Solution A	500.0mL
Solution B	250.0mL
Solution C	250.0mL

Solution A:

Composition per 500mL:

Mannitol	10.0g
K_2HPO_4	0.22g

Preparation of Solution A: Add components to distilled/deionized water and bring volume to 500.0mL. Mix thoroughly. Autoclave for 15 min at 15 psi pressure–121°C. Cool to 25°C.

Solution B:

Composition per 250mL:

$MgSO_4 \cdot 7H_2O$	0.1g
$CaCl_2 \cdot 6H_2O$	0.08g
$FeCl_3 \cdot 6H_2O$	0.05g

Preparation of Solution B: Add components to distilled/deionized water and bring volume to 250.0mL. Mix thoroughly. Autoclave for 15 min at 15 psi pressure–121°C. Cool to 25°C.

Solution C:

Composition per 250mL:

Sodium glutamate	1.1g

Calcium pantothenate 0.5mg
Thiamine·HCl 0.1mg
Biotin 0.5μg

Preparation of Solution C: Add components to distilled/deionized water and bring volume to 250.0mL. Mix thoroughly. Filter sterilize.

Preparation of Medium: Aseptically combine 500.0mL of cooled, sterile solution A, 250.0mL of cooled, sterile solution B, and 250.0mL of sterile solution C. Mix thoroughly. Aseptically distribute into sterile tubes or flasks.

Use: For the isolation of *Rhizobium* species.

Glutamate Medium (ATCC Medium 820)

Composition per liter:

Agar	15.0g
Sodium glutamate	5.0g
KH_2PO_4	1.0g
$MgSO_4·7H_2O$	0.2g
KCl	0.1g
Glucose solution	100.0mL

pH 6.5 ± 0.2 at 25°C

Glucose Solution:

Composition per 100mL:

Glucose	10.0g

Preparation of Glucose Solution: Add glucose to distilled/deionized water and bring volume to 100.0mL. Mix thoroughly. Filter sterilize.

Preparation of Medium: Add components, except glucose solution, to distilled/deionized water and bring volume to 900.0mL. Mix thoroughly. Gently heat and bring to boiling. Autoclave for 15 min at 15 psi pressure–121°C. Cool to 45°–50°C. Aseptically add 100.0mL sterile glucose solution. Mix thoroughly. Pour into sterile Petri dishes or distribute into sterile tubes.

Use: For the cultivation and maintenance of *Pseudomonas* species.

Glycerol Agar

Composition per 1070mL:

Agar	15.0g
Peptone	5.0g
Beef extract	3.0g
Soil extract	1.0L
Glycerol	70.0mL

pH 7.0 ± 0.2 at 25°C

Soil Extract:

Composition per liter:

Soil, air-dried	1.0Kg

Preparation of Soil Extract: Sift soil through a No. 9 mesh screen. Add to 2.4L of tap water. Mix thoroughly. Autoclave for 60 min at 15 psi pressure–121°C. Cool to 25°C. Filter through Whatman #1 filter paper. Bring volume to 1.0L with tap water.

Preparation of Medium: Combine components. Mix thoroughly. Gently heat and bring to boiling. Autoclave for 15 min at 15 psi pressure–121°C. Pour into sterile Petri dishes or distribute into sterile tubes.

Use: For the selective isolation and cultivation of *Nocardia* species and *Rhodococcus* species.

Glycerol Agar

Composition per liter:

Beef heart, solids from infusion	250.0g
Glycerol	60.0g
Agar	15.0g
Pancreatic digest of gelatin	5.0g
Beef extract	3.0g
Tryptose	5.0g
NaCl	2.5g

pH 7.3 ± 0.2 at 25°C

Preparation of Medium: Add components to distilled/deionized water and bring volume to 1.0L. Mix thoroughly. Gently heat and bring to boiling. Distribute into tubes or flasks. Autoclave for 15 min at 15 psi pressure–121°C. Pour into sterile Petri dishes or leave in tubes.

Use: For the cultivation and maintenance of *Bacillus subtilis, Enterococcus faecalis, Erwinia chrysanthemi, Gordona rubropertinctus, Mycobacterium* species, *Nocardia brevicatena, Rhodococcus equi,* and *R. rhodochrous.*

Glycerol Arginine Agar

Composition per liter:

Agar	15.0g
Glycerol	12.5g
Arginine	1.0g
K_2HPO_4	1.0g
NaCl	1.0g
$MgSO_4·7H_2O$	0.5g
$Fe_2(SO_4)_3·6H_2O$	0.01g
$CuSO_4·5H_2O$	1.0mg
$MnSO_4·H_2O$	1.0mg
$ZnSO_4·7H_2O$	1.0mg

Preparation of Medium: Add components to distilled/deionized water and bring volume to 1.0L. Mix thoroughly. Gently heat and bring to boiling. Distribute into tubes or flasks. Autoclave for 15 min at 15 psi pressure–121°C. Pour into sterile Petri dishes or leave in tubes.

Use: For the selective isolation and cultivation of streptomycetes.

Glycerol Asparagine Agar
See: **ISP 5**

Glycerol Beef Extract Medium

Composition per liter:

Agar	15.0g
Peptone	10.0g
NaCl	5.0g
Beef extract	3.0g
Glycerol	40.0mL

Preparation of Medium: Add components to distilled/deionized water and bring volume to 1.0L. Mix thoroughly. Gently heat and bring to boiling. Distribute into tubes or flasks. Autoclave for 15 min at 15 psi pressure–121°C. Pour into sterile Petri dishes or leave in tubes.

Use: For the cultivation and maintenance of *Corynebactrium alkanolyticum, Pseudomonas mallei, P. pseudomallei,* and *Rhodococcus* species.

Glycerol Chalk Agar

Composition per liter:

NaCl	30.0g
Agar	15.0g
Glycerol	10.0g
$CaCO_3$	5.0g
Peptone	5.0g
Yeast extract	3.0g

Preparation of Medium: Add components to distilled/deionized water and bring volume to 1.0L. Mix thoroughly. Gently heat and bring to boiling. Autoclave for 15 min at 15 psi pressure–121°C. Pour into sterile Petri dishes. Swirl flask while dispensing medium to keep $CaCO_3$ in suspension.

Use: For the cultivation of *Photobacterium* species and *Lucibacterium* species.

Glycerol Enriched Medium

Composition per liter:

Glycerol	30.0g
Peptone	20.0g
Yeast extract	10.0g

Preparation of Medium: Add components to distilled/deionized water and bring volume to 1.0L. Mix thoroughly. Gently heat and bring to boiling. Distribute into tubes or flasks. Autoclave for 15 min at 15 psi pressure–121°C. Pour into sterile Petri dishes or leave in tubes.

Use: For the cultivation of a variety of heterotrophic microorganisms.

Glycerol–Free Medium

Composition per liter:

L-Asparagine	4.0g
L-Glutamine	4.0g
Monosodium glutamate	4.0g
Na_2HPO_4	2.5g
Pancreatic digest of casein	1.0g
KH_2PO_4	1.0g
Triton®WR 1339	0.25g
Ferric ammonium citrate	0.05g
$CaCl_2$	1.0mg
$CuSO_4$	0.5mg
$ZnSO_4$	0.5mg

Preparation of Medium: Add components to distilled/deionized water and bring volume to 1.0L. Mix thoroughly. Distribute into tubes or flasks. Autoclave for 15 min at 15 psi pressure–121°C.

Use: For the cultivation of *Mycobacterium tuberculosis* Bacille Calmette-Guèrin (BCG) for vaccine production.

Glycerol Glycine Agar

Composition per liter:

Agar	15.0g
Glycine	2.5g
K_2HPO_4	1.0g
NaCl	1.0g
$CaCO_3$	0.1g
$FeSO_4$	0.1g
$MgSO_4$	0.1g
Glycerol	20.0mL

pH 7.0 ± 0.2 at 25°C

Preparation of Medium: Add components, except glycerol, to distilled/deionized water and bring volume to 1.0L. Mix thoroughly. Gently heat and bring to boiling. Add glycerol. Mix thoroughly. Distribute into tubes or flasks. Autoclave for 15 min at 15 psi pressure–121°C. Pour into sterile Petri dishes or leave in tubes.

Use: For the cultivation of *Streptomyces* species.

Glycine Cycloheximide Phenol Red Agar

Composition per liter:

Solution B	800.0mL
Solution A	200.0mL

Solution A:

Composition per 200mL:

Glycine	10.0g
$(NH_4)_2SO_4$	5.0g
KH_2PO_4	1.0g
$MgSO_4 \cdot 7H_2O$	0.5g
NaCl	0.1g
$CaCl_2 \cdot 2H_2O$	0.1g
DL-Methionine	0.02g
DL-Tryptophan	0.02g
L-Histidine·HCl	0.01g
Inositol	2.0mg
H_3BO_3	0.5mg
$ZnSO_4 \cdot 7H_2O$	0.4mg
$MnSO_4 \cdot 4H_2O$	0.4mg
Thiamine·HCl	0.4mg
Pyroxidine·HCl	0.4mg
Niacin	0.4mg
Calcium pantothenate	0.4mg
p-Aminobenzoic acid	0.2mg
Riboflavin	0.2mg
$FeCl_3$	0.2mg
$Na_2MoO_4 \cdot 4H_2O$	0.2mg
KI	0.1mg
$CuSO_4 \cdot 5H_2O$	0.04mg
Folic acid	2.0μg
Biotin	2.0μg
Cycloheximide solution	1.6mL

Preparation of Solution A: Add components to distilled/deionized water and bring volume to 200.0mL. Mix thoroughly. Filter sterilize.

Cycloheximide Solution:

Composition per 100mL:

Cycloheximide	0.1g

Preparation of Cycloheximide Solution: Add cylcohexamide to distilled/deionized water and bring volume to 100.0mL. Mix thoroughly. Filter sterilize.

Solution B:

Composition per 800mL:

Agar	20.0g
Phenol Red solution	30.0mL

Preparation of Solution B: Add components to distilled/deionized water and bring volume to 800.0mL. Mix thoroughly. Gently heat and bring to boiling. Autoclave for 15 min at 15 psi pressure–121°C. Cool to 50°–55°C.

Phenol Red Solution:

Composition per 100mL:

Phenol Red	0.5g

Preparation of Phenol Red Solution: Add phenol Red to distilled/deionized water and bring volume to 100.0mL. Mix thoroughly.

Preparation of Medium: To 800.0mL sterile solution B, aseptically add add 200.0mL sterile solution A at 55°C. Mix thoroughly. Pour into sterile Petri dishes or distribute into sterile tubes.

Use: For the selective cultivation and differentiation of fungi from clinical and nonclinical specimens. *Cryptococcus neoformans* turns the medium bright red.

Glycocholate Mineral Medium

Composition per liter:

Agar	15.0g
K_2HPO_4	3.5g
$(NH_4)_2SO_4$	2.0g
Sodium glycocholate	2.0g
KH_2PO_4	1.5g
$MgSO_4 \cdot 7H_2O$	0.1g
Yeast extract	0.1g
$CaCl_2 \cdot 2H_2O$	0.01g
$FeSO_4 \cdot 7H_2O$	0.5mg

pH 7.0 ± 0.2 at 25°C

Preparation of Medium: Add components to distilled/deionized water and bring volume to 1.0L. Mix thoroughly. Gently heat and bring to boiling. Distribute into tubes or flasks. Autoclave for 15 min at 15 psi pressure–121°C. Pour into sterile Petri dishes or leave in tubes.

Use: For the cultivation and maintenance of *Pseudomonas putida* and *Pseudomonas* species.

GMC Medium
See: **Gelatin Metronidazole Cadmium Medium**

GN Broth, Hajna

Composition per liter:

Pancreatic digest of casein	10.0g
Peptic digest of animal tissue	10.0g
NaCl	5.0g
Sodium citrate	5.0g
K_2HPO_4	4.0g
D-Mannitol	2.0g
KH_2PO_4	1.5g
Glucose	1.0g
Sodium deoxycholate	0.5g

pH 7.0 ± 0.2 at 25°C

Source: This medium is available as a premixed powder from BBL Microbiology Systems and Difco Laboratories.

Preparation of Medium: Add components to distilled/deionized water and bring volume to 1.0L. Mix

thoroughly. Gently heat and bring to boiling. Distribute into tubes or flasks. Autoclave for 15 min at 13 psi pressure–118°C. Pour into sterile Petri dishes or leave in tubes.

Use: For the selective cultivation of *Salmonella* and *Shigella* species.

Gonococcus Medium

Composition per 623mL:

Part II	500.0mL
Part I	123.0mL

pH 7.4 ± 0.2 at 25°C

Part I:

Composition per 123mL:

K_2HPO_4	10.5g
Glucose	7.5g
NaCl	5.25g
KH_2PO_4	4.5g
Sodium acetate	1.5g
L-Cysteine·HCl·H_2O	1.2g
Sodium citrate	1.13g
$NaHCO_3$	1.0g
K_2SO_4	0.9g
Na_2SO_4	0.75g
$MgCl_2 \cdot 6H_2O$	0.45g
KCl	0.3g
NH_4Cl	0.3g
L-Arginine·HCl	0.25g
L-Proline	0.25g
Oxaloacetate	0.25g
L-Glutamic acid	0.19g
L-Methionine	0.19g
L-Asparagine·H_2O	0.13g
L-Isoleucine	0.13g
L-Serine	0.13g
L-Cystine	0.05g
Calcium pantothenate	0.02g
Thiamine·HCl	0.02g
Thiamine pyrophosphate chloride	0.02g
Nicotinamide adenine dinucleotide	0.01g
$CaCl_2 \cdot 2H_2O$	5.0mg
$Fe(NO_3)_3 \cdot 9H_2O$	5.0mg
Uracil	5.0mg
Biotin	4.0mg
Hypoxanthine	2.5mg
Sodium thioglycollate	0.025mg

Preparation of Part I: Add components to distilled/deionized water and bring volume to 123.0mL. Mix thoroughly. Adjust pH to 7.2 with 6*N* NaOH. Warm to 50°C for 45 min. Filter sterilize.

Part II:

Composition per 500mL:

Agar	10.0g
Soluble starch	7.5g

Preparation of Part II: Add components to distilled/deionized water and bring volume to 500.0mL. Mix thoroughly. Gently heat and bring to boiling. Autoclave for 15 min at 15 psi pressure–121°C. Cool to 45°–50°C.

Preparation of Medium: Aseptically combine 123.0mL of sterile part I and 500.0mL of cooled, sterile part II. Mix thoroughly. Pour into sterile Petri dishes.

Use: For the cultivation of gonococci.

Gorham's Medium for Algae

Composition per liter:

$NaNO_3$	4.96g
$MgSO_4 \cdot 7H_2O$	0.075g
$Na_2SiO_3 \cdot 9H_2O$	0.058g
K_2HPO_4	0.039g
$CaCl_2 \cdot 2H_2O$	0.036g
Na_2CO_3	0.020g
Citric acid	6.0mg
EDTA	6.0mg
Ferric citrate	6.0mg

pH 7.5 ± 0.5 at 25°C

Preparation of Medium: Add components to distilled/deionized water and bring volume to 1.0L. Mix thoroughly. Distribute into tubes or flasks. Autoclave for 15 min at 15 psi pressure–121°C.

Use: For the cultivation and maintenance of *Anabaena flos-aquae* and *Microcystis aeruginosa*.

GPS Agar
See: **Gelatin Phosphate Salt Agar**

GPS Broth
See: **Gelatin Phosphate Salt Broth**

GPVA Medium

Composition per liter:

Agar	15.0g
Yeast extract	10.0g
ACES buffer (2-[(2-Amino-2-oxoethyl)-amino]-ethane sulfonic acid)	10.0g
Glycine	3.0g
Charcoal, activated	2.0g
α-Ketogluatrate	1.0g
$Fe_4(P_2O_7)_3 \cdot 9H_2O$	0.25g
Antibiotic inhibitor solution	10.0mL

pH 6.9 ± 0.2 at 25°C

Antibiotic Inhibitor Solution:

Composition per 10mL:

Anisomycin	0.08g
Vancomycin	5.0mg
Polymyxin B	100,000U

Preparation of Antibiotic Inhibitor Solution: Add components to distilled/deionized water and bring volume to 10.0mL. Mix thoroughly. Filter sterilize.

Preparation of Medium: Add components, except antibiotic inhibitor solution, to distilled/deionized water and bring volume to 990.0mL. Mix thoroughly. Gently heat and bring to boiling. Autoclave for 15 min at 15 psi pressure–121°C. Cool to 45°–50°C. Adjust pH to 6.9. Aseptically add 10.0mL sterile antibiotic inhibitor solution. Mix thoroughly. Pour into sterile Petri dishes or distribute into sterile tubes.

Use: For the isolation and cultivation of *Legionella* species from environmental waters.

GPY Salts Medium (Glucose Peptone Yeast Extract Salts Medium)

Composition per liter:

Glucose	1.0g
Peptone	0.5g
Yeast extract	0.1g
Modified Hutner's mineral base	20.0mL

Hutner's Mineral Base:

Composition per liter:

$MgSO_4 \cdot 7H_2O$	29.7g
Nitrilotriacetic acid	10.0g
$CaCl_2 \cdot 2H_2O$	3.34g
$FeSO_4 \cdot 7H_2O$	99.0mg
$(NH_4)_2MoO_4$	9.25mg
Metals "44"	50.0mL

Preparation of Hutner's Mineral Base: Add nitrilotriacetic acid to 500.0mL of distilled/deionized water. Dissolve by adjusting pH to 6.5 with KOH. Add remaining components. Readjust pH to 7.2 with H_2SO_4 or KOH. Add distilled/deionized water to 1.0L. Store at 5°C.

Metals "44":

Composition per 100mL:

$ZnSO_4 \cdot 7H_2O$	1.1g
$FeSO_4 \cdot 7H_2O$	0.5g
EDTA	0.25g
$MnSO_4 \cdot 7H_2O$	0.154g
$CuSO_4 \cdot 5H_2O$	0.04g
$Co(NO_3)_2 \cdot 6H_2O$	0.025g
$Na_2B_4O_7 \cdot 10H_2O$	0.018g

Preparation of Metals "44": Add a few drops of H_2SO_4 to distilled/deionized water to inhibit precipitate formation. Add components to acidified distilled/deionized water and bring volume to 100.0mL. Mix thoroughly.

Preparation of Medium: Add components to distilled/deionized water and bring volume to 1.0L. Mix thoroughly. Distribute into tubes or flasks. Autoclave for 15 min at 15 psi pressure–121°C.

Use: For the cultivation and maintenance of *Prosthecobacter fusiformis*.

Grahamella Medium

Composition per liter:

Agar	20.0g
Saline (0.9% solution)	800.0mL
Rabbit serum, filter-sterilized	90.0mL
Rabbit hemoglobin (2% solution)	45.0mL

pH 7.2 ± 0.2 at 25°C

Preparation of Medium: Add components, except rabbit serum and rabbit hemoglobin, to distilled/deionized water and bring volume to 900.0mL. Mix thoroughly. Gently heat and bring to boiling. Distribute into small tubes in 2.0mL volumes. Autoclave for 15 min at 15 psi pressure–121°C. Cool to 60°C. Aseptically add 0.2mL of sterile rabbit serum and 0.1mL of rabbit hemoglobin solution to each tube. Mix thoroughly. Autoclave for 30 min with a mixture of steam and air at 0 psi pressure–56°C.

Use: For the cultivation of *Grahamella* species.

Granada Medium

Composition per liter:

Starch, soluble	150.0g
Proteose peptone No. 3	38.0g
NaCl	3.0g
Trimethoprim lactate	0.015g
Sodium phosphate buffer (0.06M, pH 7.4)	900.0mL
Horse serum, coagulated	100.0mL

pH 7.4 ± 0.2 at 25°C

Preparation of Medium: Add proteose peptone No. 3 and NaCl to 200.0mL of sodium phosphate buffer and bring to boiling. Add 400.0mL of cold sodium phosphate buffer, starch and trimethoprim lactate. Mix thoroughly. Bring volume to 900.0mL with sodium phosphate buffer. Gently heat while stirring in a boiling water bath for exactly 20 min. Do not autoclave. Cool to 90–95°C. Add horse serum. Mix thoroughly. Cool to 60–65°C while stirring. Pour into sterile Petri dishes. Medium will solidify in 2–3 hr.

Use: For the early selective isolation and cultivation of Group B streptococci from clinical specimens.

Green Top Agar

Composition per liter:

Agar	15.0g
Yeast extract	2.0g
Sodium acetate	1.0g
Pancreatic digest of casein	1.0g
Soil extract	50.0mL

pH 7.4 ± 0.2 at 25°C

Soil Extract:
Composition per 200mL:

African Violet soil	0.5g
Na_2CO_3	0.5g

Preparation of Soil Extract: Add components to tap water and bring volume to 200.0mL. Autoclave for 60 min at 15 psi pressure–121°C. Filter through Whatman filter paper.

Preparation of Medium: Add components to distilled/deionized water and bring volume to 1.0L. Mix thoroughly. Gently heat and bring to boiling. Distribute into tubes or flasks. Autoclave for 15 min at 15 psi pressure–121°C. Pour into sterile Petri dishes or leave in tubes.

Use: For the cultivation and maintenance of *Bacillus macroides, Caryophanon latum, Lampropedia lyalina,* and *Vittreoscilla stercoraria.*

Green Yeast and Mold Broth (m–Green Yeast and Mold Broth)

Composition per liter:

Glucose	50.0g
Yeast extract	9.0g
Pancreatic digest of casein	5.0g
Peptic digest of animal tissue	5.0g
$MgSO_4 \cdot 7H_2O$	2.1g
K_2HPO_4	2.0g
α-Amylase	0.05g
Thiamine	0.05g
Bromcresol Green	0.026g

pH 4.6 ± 0.2 at 25°C

Source: This medium is available as a premixed powder from BBL Microbiology Systems.

Preparation of Medium: Add components to distilled/deionized water and bring volume to 1.0L. Mix thoroughly. Distribute into tubes or flasks. Autoclave for 10 min at 15 psi pressure–121°C.

Use: For the cultivation of fungi from beverages.

Group A Selective Strep Agar with Sheep Blood

Composition per liter:

Pancreatic digest of casein	14.5g
Agar	14.0g
NaCl	5.0g
Papaic digest of soybean meal	5.0g
Sheep blood	50.0mL
Growth factor solution	10.0mL
Selective agents solution	10.0mL

pH 7.4 ± 0.2 at 25°C

Source: This medium is available as a prepared medium from BBL Microbiology Systems.

Growth Factor Solution:
Composition per 10mL:

Growth factors, BBL	1.5g

Preparation of Growth Factor Solution: Add growth factors to distilled/deionized water and bring volume to 10.0mL. Mix thoroughly. Filter sterilize.

Selective Agent Solution:
Composition per 10mL:

Selective agents	0.042g

Preparation of Selective Agent Solution: Add selective agents to distilled/deionized water and bring volume to 10.0mL. Mix thoroughly. Filter sterilize.

Preparation of Medium: Add components, except sheep blood, growth factor solution, and selective agent solution, to distilled/deionized water and bring volume to 930.0mL. Mix thoroughly. Gently heat and bring to boiling. Autoclave for 15 min at 15 psi pressure–121°C. Cool to 45°–50°C. Aseptically add 50.0mL sheep blood, 10.0mL sterile growth factor solution, and 10.0mL sterile selective agent components. Mix thoroughly. Pour into sterile Petri dishes or distribute into sterile tubes.

Use: For the selective cultivation and primary isolation of group A streptococci, especially *Streptococcus pyogenes* from clinical specimens.

Group B Streptococci Agar
See: **GBS Agar Base, Islam**

Group B Streptococci Medium
See: **GBS Medium, Rapid**

GSP Agar

Composition per liter:

Starch, soluble	20.0g
Agar	15.0g

Sodium glutamate	10.0g
K_2HPO_4	2.0g
$MgSO_4 \cdot 7H_2O$	0.5g
Phenol Red	0.36g

pH 7.2–7.4 at 25°C

Preparation of Medium: Add components to distilled/deionized water and bring volume to 1.0L. Mix thoroughly. Distribute into tubes or flasks. Autoclave for 15 min at 15 psi pressure–121°C.

Use: For the selective isolation and cultivation of *Pseudomonas* species.

GSTB
See: Glucose Salt Teepol Broth

GTYE Medium (Glucose Tryptone Yeast Extract Medium)

Composition per liter:

NaCl	30.0g
Agar	15.0g
Pancreatic digest of casein	10.0g
Yeast extract	10.0g
Glucose	5.0g
$MgSO_4 \cdot 7H_2O$	5.0g

Preparation of Medium: Add components to tap water and bring volume to 1.0L. Mix thoroughly. Gently heat and bring to boiling. Distribute into tubes or flasks. Autoclave for 15 min at 15 psi pressure–121°C. Pour into sterile Petri dishes or leave in tubes.

Use: For the cultivation of ATCC strain 21081.

Guanosine Medium

Composition per liter:

Glucose	20.0g
Peptone	10.0g
Yeast extract	10.0g
Guanosine	0.02g

Preparation of Medium: Add components to distilled/deionized water and bring volume to 1.0L. Mix thoroughly. Distribute into tubes or flasks. Autoclave for 15 min at 15 psi pressure–121°C.

Use: For the cultivation and maintenance of *Salmonella choleraesuis.*

Guizotia abyssinica Creatinine Agar
See: Birdseed Agar

Gum Base Nalidixic Acid Medium
See: GBNA Medium

Gum Tragacanth Gum Arabic Medium

Composition per 100mL:

Gum tragacanth	2.0g
Gum arabic	1.0g

Preparation of Medium: Add components to distilled/deionized water and bring volume to 100.0mL. Mix thoroughly. Gently heat and bring to boiling. Autoclave for 15 min at 15 psi pressure–121°C. Pour into sterile Petri dishes.

Use: For the cultivation of aciduric flat sour sporeformers from foods. For the isolation and cultivation of *Desulfotomaculum nigrificans* from foods.

GYP Medium
See: Glucose Yeast Extract Peptone Medium

GYPT Medium (Glucose Yeast Extract Peptone Thioglycollate Medium)

Composition per liter:

Agar	8.0g
Yeast extract	6.0g
Glucose	5.0g
Peptone	2.0g
Sodium thioglycollate	0.5g

pH 7.4 ± 0.2 at 25°C

Preparation of Medium: Add components to distilled/deionized water and bring volume to 1.0L. Mix thoroughly. Adjust pH to 7.4. Gently heat and bring to boiling. Anaerobically distribute into tubes under 97% N_2 + 3% H_2. Cap tubes with rubber stoppers. Place tubes in a press. Autoclave for 15 min at 15 psi pressure–121°C with fast exhaust.

Use: the cultivation and maintenance of *Spirochaeta stenostrepta*.

H Agar

Composition per liter:

Agar	15.0g
Pancreatic digest of casein	10.0g
NaCl	8.0g

Preparation of Medium: Add components to distilled/deionized water and bring volume to 1.0L. Mix thoroughly. Gently heat and bring to boiling. Autoclave for 15 min at 15 psi pressure–121°C. Pour into sterile Petri dishes.

Preparation of Medium: Add components to distilled/deionized water and bring volume to 1.0L. Mix thoroughly. Distribute into tubes or flasks. Autoclave for 15 min at 15 psi pressure–121°C.

Use: For the cultivation of *Escherichia coli* and a variety of other bacteria.

H Agar (Hominis Agar)

Composition per 98mL:

Base agar	65.0mL
Horse serum	20.0mL
Yeast dialysate	10.0mL
Penicillin solution	2.0mL
Thallium acetate solution	1.0mL

pH 7.3 ± 0.2 at 25°C

Base Agar:
Composition per liter:

Papaic digest of soybean meal	20.0g
Agarose	10.0g
NaCl	5.0g
Phenol Red (2% solution)	1.0mL

Preparation of Base Agar: Add components to distilled/deionized water and bring volume to 1.0L. Mix thoroughly. Gently heat and bring to boiling. Adjust pH to 7.3. Autoclave for 15 min at 15 psi pressure–121°C. Cool to 45°–50°C.

Yeast Dialysate:
Composition per 10mL:

Active, dried yeast	450.0g

Preparation of Yeast Dialysate: Add active, dried yeast to distilled/deionized water and bring volume to 1250.0mL. Gently heat and bring to 40°C. Autoclave for 15 min at 15 psi pressure–121°C. Put into dialysis tubing. Dialyze against 1.0L of distilled/deionized water for 2 days at 4°C. Discard tubing and its contents. Autoclave dialysate for 15 min at 15 psi pressure–121°C. Store at –20°C.

Penicillin Solution:
Composition per 10mL:

Penicillin	100,000U

Preparation of Penicillin Solution: Add penicillin to distilled/deionized water and bring volume to 10.0mL. Mix thoroughly. Filter sterilize.

Thallium Acetate Solution:
Composition per 10mL:

Thallium acetate	0.33g

Preparation of Thallium Acetate Solution: Add thallium acetate to distilled/deionized water and bring volume to 10.0mL. Mix thoroughly. Filter sterilize.

Caution: Thallium salts are toxic.

Preparation of Medium: To 65.0mL of cooled, sterile base agar, aseptically add 10.0mL of sterile yeast dialysate, 20.0mL of horse serum, 2.0mL of sterile penicillin solution and 1.0mL of sterile thallium acetate solution. Mix thoroughly. Pour into 10mm × 35mm Petri dishes in 5.0mL volumes. Allow plates to stand overnight at 25°C to remove excess surface moisture.

Use: For isolation of *Mycoplasma pneumoniae* and *M. hominis*.

H Broth

Composition per liter:

NaCl	5.0g
Pancreatic digest of casein	5.0g
Peptone	5.0g
Beef extract	3.0g
K_2HPO_4	2.5g
Glucose	1.0g

pH 7.2 ± 0.2 at 25°C

Preparation of Medium: Add components to distilled/deionized water and bring volume to 1.0L. Mix thoroughly. Distribute into tubes in 4.0mL volumes. Autoclave for 15 min at 10 psi pressure–115°C.

Use: For the preparation of the H agglutination antigen used in the differentiation and identification of *Salmonella* species types and subtypes.

H Broth (Hominis Broth)

Composition per 99mL:

Base broth	65.0mL
Horse serum	20.0mL
Yeast dialysate	10.0mL
Penicillin solution	2.0mL
Glucose solution	1.0mL
Thallium acetate solution	1.0mL

pH 7.3 ± 0.2 at 25°C

Base Broth:
Composition per liter:

Papaic digest of soybean meal	20.0g

NaCl	5.0g
Phenol Red (2% solution)	1.0mL

Preparation of Base Broth: Add components to distilled/deionized water and bring volume to 1.0L. Mix thoroughly. Gently heat and bring to boiling. Adjust pH to 7.3. Autoclave for 15 min at 15 psi pressure–121°C. Cool to 25°C.

Yeast Dialysate:

Composition per 10mL:

Active, dried yeast	450.0g

Preparation of Yeast Dialysate: Add active, dried yeast to distilled/deionized water and bring volume to 1250.0mL. Gently heat and bring to 40°C. Autoclave for 15 min at 15 psi pressure–121°C. Put into dialysis tubing. Dialyze against 1.0L of distilled/deionized water for 2 days at 4°C. Discard tubing and its contents. Autoclave dialysate for 15 min at 15 psi pressure–121°C. Store at –20°C.

Penicillin Solution:

Composition per 10mL:

Penicillin	100,000U

Preparation of Penicillin Solution: Add penicillin to distilled/deionized water and bring volume to 10.0mL. Mix thoroughly. Filter sterilize.

Glucose Solution:

Composition per 10mL:

D-Glucose	1.8g

Preparation of Glucose Solution: Add D-glucose to distilled/deionized water and bring volume to 10.0mL. Mix thoroughly. Filter sterilize.

Thallium Acetate Solution:

Composition per 10mL:

Thallium acetate	0.33g

Preparation of Thallium Acetate Solution: Add thallium acetate to distilled/deionized water and bring volume to 10.0mL. Mix thoroughly. Filter sterilize.

Caution: Thallium salts are toxic.

Preparation of Medium: To 65.0mL of cooled, sterile base broth, aseptically add 10.0mL of sterile yeast dialysate, 20.0mL of horse serum, 2.0mL of sterile penicillin solution, 1.0mL of sterile glucose solution and 1.0mL of sterile thallium acetate solution. Mix thoroughly. Aseptically distribute into sterile screw-capped tubes in 5.0mL volumes. Screw caps down tightly.

Use: For isolation and cultivation of *Mycoplasma pneumoniae*.

H Broth (Hominis Broth)

Composition per 100mL:

Base broth	65.0mL
Horse serum	20.0mL
Yeast dialysate	10.0mL
Penicillin solution	2.0mL
Arginine solution	2.0mL
Thallium acetate solution	1.0mL

pH 7.3 ± 0.2 at 25°C

Base Broth:

Composition per liter:

Papaic digest of soybean meal	20.0g
NaCl	5.0g
Phenol Red (2% solution)	1.0mL

Preparation of Base Broth: Add components to distilled/deionized water and bring volume to 1.0L. Mix thoroughly. Gently heat and bring to boiling. Adjust pH to 7.3. Autoclave for 15 min at 15 psi pressure–121°C. Cool to 25°C.

Yeast Dialysate:

Composition per 10mL:

Active, dried yeast	450.0g

Preparation of Yeast Dialysate: Add active, dried yeast to distilled/deionized water and bring volume to 1250.0mL. Gently heat and bring to 40°C. Autoclave for 15 min at 15 psi pressure–121°C. Put into dialysis tubing. Dialyze against 1.0L of distilled/deionized water for 2 days at 4°C. Discard tubing and its contents. Autoclave dialysate for 15 min at 15 psi pressure–121°C. Store at –20°C.

Penicillin Solution:

Composition per 10mL:

Penicillin	100,000U

Preparation of Penicillin Solution: Add penicillin to distilled/deionized water and bring volume to 10.0mL. Mix thoroughly. Filter sterilize.

Arginine Solution:

Composition per 10mL:

L-Arginine	1.74g

Preparation of Arginine Solution: Add L-arginine to distilled/deionized water and bring volume to 10.0mL. Mix thoroughly. Filter sterilize.

Thallium Acetate Solution:

Composition per 10mL:

Thallium acetate	0.33g

Preparation of Thallium Acetate Solution: Add thallium acetate to distilled/deionized water and bring volume to 10.0mL. Mix thoroughly. Filter sterilize.

Caution: Thallium salts are toxic.

Preparation of Medium: To 65.0mL of cooled, sterile base broth, aseptically add 10.0mL of sterile yeast dialysate, 20.0mL of horse serum, 2.0mL of sterile penicillin solution, 1.0mL of sterile glucose solution and 1.0mL of sterile thallium acetate solution. Mix thoroughly. Aseptically distribute into sterile screw-capped tubes in 5.0mL volumes. Screw caps down tightly.

Use: For isolation and cultivation of *Mycoplasma hominis*.

H Diphasic Medium

Composition per 197mL:

Base agar	65.0mL
Base broth	65.0mL
Horse serum	40.0mL
Yeast dialysate	20.0mL
Penicillin solution	4.0mL
Glucose solution	1.0mL
Thallium acetate solution	2.0mL

pH 7.3 ± 0.2 at 25°C

Base Agar:

Composition per liter:

Papaic digest of soybean meal	20.0g
Agarose	10.0g
NaCl	5.0g
Phenol Red (2% solution)	1.0mL

Preparation of Base Agar: Add components to distilled/deionized water and bring volume to 1.0L. Mix thoroughly. Gently heat and bring to boiling. Adjust pH to 7.3. Autoclave for 15 min at 15 psi pressure–121°C. Cool to 45°–50°C.

Base Broth:

Composition per liter:

Papaic digest of soybean meal	20.0g
NaCl	5.0g
Phenol Red (2% solution)	1.0mL

Preparation of Base Broth: Add components to distilled/deionized water and bring volume to 1.0L. Mix thoroughly. Gently heat and bring to boiling. Adjust pH to 7.3. Autoclave for 15 min at 15 psi pressure–121°C. Cool to 25°C.

Yeast Dialysate:

Composition per 10mL:

Dried yeast, active	450.0g

Preparation of Yeast Dialysate: Add active, dried yeast to distilled/deionized water and bring volume to 1250.0mL. Gently heat and bring to 40°C. Autoclave for 15 min at 15 psi pressure–121°C. Put into dialysis tubing. Dialyze against 1.0L of distilled/deionized water for 2 days at 4°C. Discard tubing and its contents. Autoclave dialysate for 15 min at 15 psi pressure–121°C. Store at –20°C.

Penicillin Solution:

Composition per 10mL:

Penicillin	100,000U

Preparation of Penicillin Solution: Add penicillin to distilled/deionized water and bring volume to 10.0mL. Mix thoroughly. Filter sterilize.

Glucose Solution:

Composition per 10mL:

D-Glucose	1.8g

Preparation of Glucose Solution: Add D-glucose to distilled/deionized water and bring volume to 10.0mL. Mix thoroughly. Filter sterilize.

Thallium Acetate Solution:

Composition per 10mL:

Thallium acetate	0.33g

Preparation of Thallium Acetate Solution: Add thallium acetate to distilled/deionized water and bring volume to 10.0mL. Mix thoroughly. Filter sterilize.

Caution: Thallium salts are toxic.

Preparation of Medium: To 65.0mL of cooled, sterile base agar, aseptically add 10.0mL of sterile yeast dialysate, 20.0mL of horse serum, 2.0mL of sterile penicillin solution and 1.0mL of sterile thallium acetate solution. Mix thoroughly. Aseptically distribute into screw-capped tubes in 3.0mL volumes. Allow agar to solidify. To 65.0mL of cooled, sterile base broth, aseptically add 10.0mL of sterile yeast dialysate, 20.0mL of horse serum, 2.0mL of sterile penicillin solution, 1.0mL of sterile glucose solution and 1.0mL of sterile thallium acetate solution. Mix thoroughly. Aseptically distribute 3.0mL of broth solution on top of the 3.0mL of solidified base agar in each tube. Screw caps down tightly.

Use: For the isolation and cultivation of *Mycoplasma pneumoniae*.

H Medium

Composition per liter:

Pancreatic digest of casein	10.0g
NaCl	8.0g

Preparation of Medium: Add components to distilled/deionized water and bring volume to 1.0L. Mix thoroughly. Distribute into tubes or flasks. Autoclave for 15 min at 15 psi pressure–121°C.

Use: For the cultivation of *Escherichia coli* and a variety of other bacteria.

H Top Agar

Composition per liter:

Pancreatic digest of casein	10.0g
NaCl	8.0g
Agar	7.0g

Preparation of Medium: Add components to distilled/deionized water and bring volume to 1.0L. Mix thoroughly. Gently heat and bring to boiling. Autoclave for 15 min at 15 psi pressure–121°C. Pour into sterile Petri dishes that contain H agar.

Use: For the cultivation of *Escherichia coli* and a variety of other bacteria.

HA
See: **Halophilic Agar**

HAEB
See: **Horie Arabinose Ethyl Violet Broth**

Haemophilus ducreyi Medium

Composition per liter:

Columbia blood agar base	675.0mL
Rabbit blood	300.0mL
Fresh yeast extract solution	25.0mL

pH 6.5–7.0 at 25°C

Columbia Blood Agar Base:
Composition per 675mL:

Agar	15.0g
Pantone	10.0g
Bitone	10.0g
NaCl	5.0g
Tryptic digest of beef heart	3.0g
Cornstarch	1.0g

Preparation of Columbia Blood Agar Base: Add components to distilled/deionized water and bring volume to 675.0mL. Mix thoroughly. Gently heat until boiling. Autoclave for 15 min at 15 psi pressure–121°C. Cool to 45°–50°C.

Fresh Yeast Extract Solution:
Composition per 100mL:

Baker's yeast live, pressed, starch-free	25.0g

Preparation of Fresh Yeast Extract Solution: Add the live Baker's yeast to 100.0mL of distilled/deionized water. Autoclave for 90 min at 15 psi pressure–121°C. Allow to stand. Remove supernatant solution. Adjust pH to 6.6–6.8. Filter sterilize.

Preparation of Medium: To 675.0mL of cooled, sterile Columbia blood agar base, aseptically add rabbit blood and sterile fresh yeast extract solution. Aseptically adjust pH to 6.5–7.0.

Use: For the cultivation and maintenance of *Haemophilus ducreyi*.

Haemophilus ducreyi Medium, Revised (Ducreyi Medium, Revised)

Composition per 1010mL:

Solution B	500.0mL
Solution A	400.0mL
Solution C	110.0mL

pH 7.4 ± 0.2 at 25°C

Solution A:
Composition per 400mL:

Beef heart, infusion from	500.0g
Agar	15.0g
Tryptose	10.0g
NaCl	5.0g

Preparation of Solution A: Add components to distilled/deionized water and bring volume to 400.0L. Mix thoroughly. Gently heat and bring to boiling. Autoclave for 15 min at 15 psi pressure–121°C. Cool to 45°–50°C.

Solution B:
Composition per 500mL:

Hemoglobin	10.0g

Preparation of Solution B: Add components to distilled/deionized water and bring volume to 500.0L. Mix thoroughly. Gently heat and bring to boiling. Autoclave for 15 min at 15 psi pressure–121°C. Cool to 45°–50°C.

Solution C:
Composition per 110mL:

Fetal bovine serum	100.0mL
Supplement solution	10.0mL

Supplement Solution:
Composition per liter:

Glucose	100.0g
L-Cysteine·HCl	25.9g
L-Glutamine	10.0g
L-Cystine	1.1g
Adenine	1.0g
Nicotinamide adenine dinucleotide	0.25g
Vitamin B_{12}	0.1g
Thiamine pyrophosphate	0.1g
Guanine·HCl	0.03g
$Fe(NO_3)_3 \cdot 6H_2O$	0.02g
p-Aminobenzoic acid	0.013g
Thiamine·HCl	3.0mg

Source: The supplement solution IsoVitaleX® enrichment is available from BBL Microbiology Laboratories. This enrichment may be replaced by supplement VX from Difco Laboratories.

Preparation of Supplement Solution: Add components to distilled/deionized water and bring volume to 1.0L. Mix thoroughly. Filter sterilize.

Preparation of Solution C: Combine components. Mix thoroughly. Filter sterilize. Warm to 45°–50°C.

Preparation of Medium: Aseptically combine solution A, solution B and solution C. Mix thoroughly. Pour into sterile Petri dishes or distribute into sterile tubes.

Use: For the cultivation and maintenance of *Haemophilus ducreyi.*

Haemophilus influenzae Defined Medium MI

Composition per liter:

NaCl	5.8g
K_2HPO_4	3.5g
Glycerol	3.0g
KH_2PO_4	2.7g
Inosine	2.0g
L-Glutamic acid	1.3g
K_2SO_4	1.0g
Sodium lactate	0.8g
L-Aspartic acid	0.5g
Nitrilotriethanol	0.4g
L-Arginine	0.3g
L-Leucine	0.3g
L-Cystine	0.2g
$MgCl_2$	0.2g
L-Tyrosine	0.2g
L-Methionine	0.1g
L-Serine	0.1g
Uracil	0.1g
L-Lysine	0.05g
Glycine	0.03g
$CaCl_2$	0.022g
Hypoxanthine	0.02g
Polyvinyl alcohol	0.02g
Tween™ 80	0.02g
Hemin	0.01g
L-Histidine	0.01g
Calcium pantothenate	4.0mg
Ethylenediaminetetraacetate	4.0mg
Nicotinamide adenine dinucleotide	4.0mg
Thiamine	4.0mg

Preparation of Medium: Add components to distilled/deionized water and bring volume to 1.0L. Mix thoroughly. Filter sterilize.

Use: For the cultivation of *Haemophilus influenza* in a chemically defined medium.

Haemophilus influenzae Defined Medium MI–Cit

Composition per liter:

NaCl	5.8g
K_2HPO_4	3.5g
Glycerol	3.0g
KH_2PO_4	2.7g
Inosine	2.0g
L-Glutamic acid	1.3g
K_2SO_4	1.0g
Sodium lactate	0.8g
L-Aspartic acid	0.5g
Nitrilotriethanol	0.4g
L-Leucine	0.3g
L-Cystine	0.2g
$MgCl_2$	0.2g
L-Tyrosine	0.2g
Citrulline	0.15g
L-Methionine	0.1g
L-Serine	0.1g
L-Lysine	0.05g
Glycine	0.03g
$CaCl_2$	0.022g
Hypoxanthine	0.02g
Polyvinyl alcohol	0.02g
Tween™ 80	0.02g
Hemin	0.01g
L-Histidine	0.01g
Calcium pantothenate	4.0mg
Ethylenediaminetetraacetate	4.0mg
Nicotinamide adenine dinucleotide	4.0mg
Thiamine	4.0mg

Preparation of Medium: Add components to distilled/deionized water and bring volume to 1.0L. Mix thoroughly. Filter sterilize.

Use: For the cultivation of *Haemophilus influenza* in a chemically defined medium.

Haemophilus Test Medium (HTM)

Composition per liter:

Beef infusion	300.0g
Acid hydrolysate of casein	17.5g
Agar	17.0g
Starch	1.5g
Yeast extract	5.0g
HTM supplement	10.0mL

pH 7.4 ± 0.2 at 25°C

Source: This medium is available as a premixed powder from Oxoid Unipath.

HTM Supplement:
Composition per 10mL:

Nicotinamide adenine dinucleotide	0.03g
Hematin	0.03g

Preparation of HTM Supplement: Add components to distilled/deionized water and bring volume to 10.0mL. Mix thoroughly. Filter sterilize.

Preparation of Medium: Add components, except HTM supplement, to distilled/deionized water and bring volume to 990.0mL. Mix thoroughly. Gently heat and bring to boiling. Autoclave for 15 min at 15 psi pressure–121°C. Cool to 45°–50°C. Aseptically add 10.0mL of sterile HTM supplement. Mix thoroughly. Pour into sterile Petri dishes or distribute into sterile tubes.

Use: For antimicrobial susceptibility testing of *Haemophilus influenzae*.

Hagedorn and Holt Selective Medium

Composition per liter:

NaCl	20.0g
Agar	15.0g
Yeast extract	2.0g
Pancreatic digest of casein	1.7g
Agar	1.5g
NaCl	0.5g
Papaic digest of soybean meal	0.3g
K_2HPO_4	0.25g
Glucose	0.25g
Cycloheximide	0.1g
Methyl Red	0.15mg

pH 7.3 ± 0.2 at 25°C

Preparation of Medium: Add components to distilled/deionized water and bring volume to 1.0L. Mix thoroughly. Gently heat and bring to boiling. Distribute into tubes or flasks. Autoclave for 15 min at 15 psi pressure–121°C. Pour into sterile Petri dishes or leave in tubes.

Use: For the selective isolation of *Arthrobacter* species in soil.

Haliscomenobacter Medium

Composition per liter:

Agar	10.0g
$(NH_4)_2SO_4$	0.5g
Glucose	0.15g
$CaCO_3$	0.1g
KCl	0.05g
K_2HPO_4	0.05g
$MgSO_4·7H_2O$	0.05g
$Ca(NO_3)_2$	0.01g
Vitamin solution	10.0mL

Vitamin Solution:
Composition per 10mL:

Thiamine	0.4mg
Vitamin B_{12}	0.05mg

Preparation of Vitamin Solution: Add components to distilled/deionized water and bring volume to 10.0mL. Mix thoroughly. Filter sterilize.

Preparation of Medium: Add components, except vitamin solution, to distilled/deionized water and bring volume to 990.0mL. Mix thoroughly. Gently heat and bring to boiling. Autoclave for 15 min at 15 psi pressure–121°C. Cool to 45°–50°C. Aseptically add sterile vitamin solution. Mix thoroughly. Pour into sterile Petri dishes or distribute into sterile tubes.

Use: For the isolation of *Haliscomenobacter* species from activated sludge.

Haliscomenobacter Medium (DSM 134)

Glutamic acid	1.31g
$MgSO_4·7H_2O$	0.075g
$CaCl_2·2H_2O$	0.050g
K_2HPO_4	0.040g
$Na_2HPO_4·2H_2O$	0.040g
KH_2PO_4	0.027g
$FeCl_3·6H_2O$	5.0mg
$MnSO_4·H_2O$	3.0mg
Pancreatic digest of casein	1.7mg
NaCl	0.5mg
Papaic digest of soybean meal	0.3mg
K_2HPO_4	0.25mg
Vitamin solution	10.0mL
Glucose solution	5.0mL
Trace element solution SL-6	1.0mL

pH 7.5 ± 0.2 at 25°C

Vitamin Solution:
Composition per 10mL:

Thiamine	0.4mg
Vitamin B_{12}	0.01mg

Preparation of Vitamin Solution: Add components to distilled/deionized water and bring volume to 10.0mL. Mix thoroughly. Filter sterilize.

Glucose Solution:
Composition per 5mL:

D-Glucose	2.0g

Preparation of Glucose Solution: Add glucose to distilled/deionized water and bring volume to 5.0mL. Mix thoroughly. Autoclave for 15 min at 15 psi pressure–121°C.

Trace Elements Solution SL-6:

Composition per liter:

H_3BO_3	0.3g
$CoCl_2 \cdot 6H_2O$	0.2g
$ZnSO_4 \cdot 7H_2O$	0.1g
$MnCl_2 \cdot 4H_2O$	0.03g
$Na_2MoO_4 \cdot H_2O$	0.03g
$NiCl_2 \cdot 6H_2O$	0.02g
$CuCl_2 \cdot 2H_2O$	0.01g

Preparation of Trace Elements Solution SL-6: Add components to distilled/deionized water and bring volume to 1.0L. Mix thoroughly. Adjust pH to 3.4.

Preparation of Medium: Add components, except vitamin solution and glucose solution, to distilled/deionized water and bring volume to 985.0mL. Mix thoroughly. Adjust pH to 7.5. Gently heat and bring to boiling. Autoclave for 15 min at 15 psi pressure–121°C. Cool to 25°C. Aseptically add sterile vitamin solution and glucose solution. Mix thoroughly. Aseptically distribute into sterile tubes or flasks.

Use: For the cultivation and maintenance of *Haliscomenobacter hydrossis*.

Haloanaerobium Medium

Composition per 1066mL:

NaCl	130.0g
Pancreatic digest of casein	10.0g
Yeast extract	10.0g
$MgSO_4 \cdot H_2O$	5.0g
KCl	1.0g
Thioglycollate-ascorbate reducing agent	30.9mL
Glucose solution	25.75mL
NaOH solution	10.3mL
Wolfe's vitamin solution	10.0mL
Wolfe's mineral solution	10.0mL

pH 7.0 ± 0.2 at 25°C

Glucose Solution:

Composition per 30mL:

D-Glucose	3.0g

Preparation of Glucose Solution: Add D-glucose to distilled/deionized water and bring volume to 30.0mL. Mix thoroughly. Filter sterilize.

NaOH Solution:

Composition per 20mL:

NaOH	1.6g

Preparation of NaOH Solution: Add NaOH to distilled/deionized water and bring volume to 20.0mL. Mix thoroughly. Autoclave for 15 min at 15 psi pressure–121°C.

Wolfe's Vitamin Solution:

Composition per liter:

Pyridoxine·HCl	0.01g
Thiamine·HCl	5.0mg
Riboflavin	5.0mg
Nicotinic acid	5.0mg
Calcium pantothenate	5.0mg
p-Aminobenzoic acid	5.0mg
Thioctic acid	5.0mg
Biotin	2.0mg
Folic acid	2.0mg
Cyanocobalamin	0.1mg

Preparation of Wolfe's Vitamin Solution: Add components to distilled/deionized water and bring volume to 1.0L. Mix thoroughly.

Wolfe's Mineral Solution:

Composition per liter

$MgSO_4 \cdot 7H_2O$	3.0g
Nitriloacetic acid	1.5g
NaCl	1.0g
$MnSO_4 \cdot H_2O$	0.5g
$FeSO_4 \cdot 7H_2O$	0.1g
$CoCl_2 \cdot 6H_2O$	0.1g
$CaCl_2$	0.1g
$ZnSO_4 \cdot 7H_2O$	0.1g
$CuSO_4 \cdot 5H_2O$	0.01g
$AlK(SO_4)_2 \cdot 12H_2O$	0.01g
H_3BO_3	0.01g
$Na_2MoO_4 \cdot 2H_2O$	0.01g

Preparation of Wolfe's Mineral Solution: Add nitrilotriacetic acid to 500.0mL of distilled/deionized water. Dissolve by adjusting pH to 6.5 with KOH. Add remaining components. Add distilled/deionized water to 1.0L.

Thioglycollate-Ascorbate Reducing Agent:

Composition per 100mL:

Ascorbic acid	1.0g
Sodium thioglycollate	1.0g

Preparation of Thioglycollate-Ascorbate Reducing Agent: Add components to distilled/deionized water and bring volume to 100.0mL. Mix thoroughly. Adjust pH to 7.0. Filter sterilze.

Preparation of Medium: Add components, except thioglycollate-ascorbate reducing agent, glucose and NaOH solutions, to distilled/deionized water and bring volume to 990.0mL. Mix thoroughly. Gently heat and bring to boiling. Anaerobically distribute into tubes under 97% N_2 + 3% H_2 in 9.7mL volumes. Cap tubes with rubber stoppers. Autoclave for 15 min at 15 psi pressure–121°C. Cool to 25°C. Immediately prior to inoculation, aseptically add 0.3mL of sterile thioglycollate-ascorbate reducing agent, 0.25mL of sterile glucose solution and 0.1mL of sterile NaOH solution to each tube.

Use: For the cultivation and maintenance of *Haloanaerobium praevalens.*

Haloanaerobium praevalens Medium

Composition per liter:

NaCl	130.0g
Agar	20.0g
Yeast extract	2.00g
Pancreatic digest of casein	2.00g
NH_4Cl	0.50g
$MgSO_4 \cdot 7H_2O$	0.50g
K_2HPO_4	0.35g
$CaCl_2 \cdot 2H_2O$	0.25g
KH_2PO_4	0.23g
$FeSO_4 \cdot 7H_2O$	2.0mg
$NaHCO_3$ solution	20.0mL
Cysteine-sulfide reducing agent	20.0mL
Wolfe's vitamin solution	10.0mL
Methanol	10.0mL
Resazurin (0.025% solution)	4.0mL
Trace elements SL-6	3.0mL

pH 6.8 ± 0.2 at 25°C

$NaHCO_3$ Solution:

Composition per 20mL:

$NaHCO_3$	850.0mg

Preparation of $NaHCO_3$ Solution: Add $NaHCO_3$ to distilled/deionized water and bring volume to 20.0mL. Mix thoroughly. Filter sterilize. Gas with 100% CO_2 for 20 min.

Cysteine-Sulfide Reducing Agent:

Composition per 20mL:

L-Cysteine·HCl·H_2O	0.3g
$Na_2S \cdot 9H_2O$	0.3g

Preparation of Cysteine-Sulfide Reducing Agent: Add L-Cysteine·HCl·H_2O to 10.0mL of distilled/deionized water. Mix thoroughly. In a separate tube, add $Na_2S \cdot 9H_2O$ to 10.0mL of distilled/deionized water. Mix thoroughly. Gas both solutions with 100% N_2 and cap tubes. Autoclave both solutions for 15 min at 15 psi pressure–121°C using fast exhaust. Cool to 50°C. Aseptically combine the two solutions under 100% N_2.

Wolfe's Vitamin Solution:

Composition per liter:

Pyridoxine·HCl	10.0mg
Thiamine·HCl	5.0mg
Riboflavin	5.0mg
Nicotinic acid	5.0mg
Calcium pantothenate	5.0mg
p-Aminobenzoic acid	5.0mg
Thioctic acid	5.0mg
Biotin	2.0mg
Folic acid	2.0mg
Cyanocobalamin	100.0μg

Preparation of Wolfe's Vitamin Solution: Add components to distilled/deionized water and bring volume to 1.0L. Mix thoroughly. Filter sterilize.

Trace Elements Solution SL-6:

Composition per liter:

H_3BO_3	0.3g
$CoCl_2 \cdot 6H_2O$	0.2g
$ZnSO_4 \cdot 7H_2O$	0.1g
$MnCl_2 \cdot 4H_2O$	0.03g
$Na_2MoO_4 \cdot H_2O$	0.03g
$NiCl_2 \cdot 6H_2O$	0.02g
$CuCl_2 \cdot 2H_2O$	0.01g

Preparation of Trace Elements Solution SL-6: Add components to distilled/deionized water and bring volume to 1.0L. Mix thoroughly. Adjust pH to 3.4.

Preparation of Medium: Add components, except $NaHCO_3$ solution, cysteine-sulfide reducing agent, Wolfe's vitamin solution and methanol, to distilled/deionized water and bring volume to 940.0mL. Mix thoroughly. Autoclave for 15 min at 15 psi pressure–121°C. Cool under 80% N_2 + 20% CO_2. Aseptically and anaerobically add the sterile $NaHCO_3$ solution, the sterile cysteine-sulfide reducing agent, the sterile Wolfe's vitamin solution and filter-sterilized methanol. Mix thoroughly. Adjust pH to 6.8. Aseptically and anaerobically distribute into sterile tubes or flasks.

Use: For the cultivation and maintenance of *Haloanaerobium praevalens.*

Haloarcula Medium

Composition per 1001mL:

NaCl	250.0g
$MgSO_4 \cdot 7H_2O$	20.0g
Agar	15.0g
Sodium citrate	3.0g
KCl	2.0g
$CaCl_2$	0.2g
Peptone solution	100.0mL
Trace elements solution	1.0mL

pH 7.4 ± 0.1 at 25°C

Peptone Solution:

Composition per 100mL:

Bacteriological peptone, Oxoid Unipath	10.0g

Preparation of Peptone Solution: Add peptone to distilled/deionized water and bring volume to 100.0mL. Mix thoroughly. Filter sterilize.

Trace Elements Solution:
Composition per 100mL:

$FeCl_2 \cdot 4H_2O$	0.36g
$MnCl_2 \cdot 4H_2O$	0.022g

Preparation of Trace Elements Solution: Add components to distilled/deionized water and bring volume to 100.0mL. Mix thoroughly. Filter sterilize.

Preparation of Medium: Add components, except peptone solution and trace elements solution, to distilled/deionized water and bring volume to 900.0mL. Mix thoroughly. Gently heat and bring to boiling. Adjust pH to 7.4 with NaOH. Autoclave for 15 min at 15 psi pressure–121°C. Cool to 45°–50°C. Aseptically add 100.0mL of sterile peptone solution and 1.0mL of sterile trace elements solution. Mix thoroughly. Pour into sterile Petri dishes or distribute into sterile tubes.

Use: For the cultivation and maintenance of *Haloanaerobium praevalens.*

Halobacteria Medium

Composition per liter:

NaCl	220.0g
Agar	10.0g
$MgSO_4 \cdot 7H_2O$	10.0g
Casein hydrolysate	5.0g
KCl	5.0g
Disodium citrate	3.0g
KNO_3	1.0g
Yeast extract	1.0g
$CaCl_2 \cdot 6H_2O$	0.2g

pH 7.2–7.4 at 25°C

Preparation of Medium: Add components to distilled/deionized water and bring volume to 1.0L. Mix thoroughly. Gently heat until dissolved. Adjust pH to 7.2–7.4. Distribute into tubes or flasks. Autoclave for 15 min at 15 psi pressure–121°C.

Use: For the cultivation and enumeration of halobacteria.

Halobacteriaceae Medium 1

Composition per liter:

Salt, crude solar	250.0g
$MgSO_4 \cdot 7H_2O$	20.0g
KCl	5.0g
Pancreatic digest of casein	5.0g
Yeast extract	5.0g
$CaCl_2 \cdot 6H_2O$	0.2g

pH 7.0 ± 0.2 at 25°C

Preparation of Medium: Add components to distilled/deionized water and bring volume to 1.0L. Mix thoroughly. Gently heat until dissolved. Adjust pH to 7.0. Distribute into tubes or flasks. Autoclave for 15 min at 15 psi pressure–121°C.

Use: For axenic cultivation of members of the Halobacteriacea.

Halobacteriaceae Medium 2

Composition per liter:

NaCl	250.0g
$MgSO_4 \cdot 7H_2O$	20.0g
Yeast extract	10.0g
Casamino acids	7.5g
Trisodium citrate	3.0g
KCl	2.0g
$FeCl_2$	2.3mg

pH 7.5–7.8 at 25°C

Preparation of Medium: Add components to distilled/deionized water and bring volume to 1.0L. Mix thoroughly. Gently heat until dissolved. Adjust pH to 7.5–7.8. Distribute into tubes or flasks. Autoclave for 15 min at 15 psi pressure–121°C.

Use: For the axenic cultivation of halobacteria and halococci.

Halobacteriaceae Medium 3

Composition per liter:

NaCl	240.0g
L-Glutamine	15.0g
KCl	5.0g
K_2SO_4	5.0g
$MgCl_2 \cdot 6H_2O$	5.0g
$MgSO_4$, anhydrous	5.0g
NH_4Cl	5.0g
Pancreatic digest of casein	5.0g
Yeast extract	5.0g
K_2HPO_4	0.5g
L-Arginine	0.5g
L-Isoleucine	0.25g
L-Leucine	0.25g
L-Lysine	0.25g
L-Proline	0.25g
L-Valine	0.25g
Cytidylic acid	0.2g
$CaCl_2 \cdot 2H_2O$	0.1g
L-Methionine	0.1g
L-Tyrosine	0.1g
L-Phenylalanine	0.05g
$FeCl_2 \cdot 6H_2O$	5.0mg

pH 6.8 ± 0.2 at 25°C

Preparation of Medium: Add components to distilled/deionized water and bring volume to 1.0L. Mix thoroughly. Gently heat until dissolved. Adjust pH to

6.8. Distribute into tubes or flasks. Autoclave for 15 min at 15 psi pressure–121°C.

Use: For the cultivation of some halobacteria and halococci.

Halobacteriaceae Medium 4

Composition per liter:

NaCl	250.0g
$MgSO_4 \cdot 7H_2O$	20.0g
NH_4Cl	5.0g
L-Glutamic acid	1.3g
DL-Valine	1.0g
Glycerol	1.0g
L-Lysine	0.85g
L-Leucine	0.80g
DL-Serine	0.61g
DL-Threonine	0.50g
DL-Isoleucine	0.44g
DL-Alanine	0.43g
L-Arginine	0.40g
DL-Methionine	0.37g
DL-Phenylalanine	0.26g
L-Tyrosine	0.20g
Adenylic acid	0.1g
KNO_3	0.1g
Uridylic acid	0.1g
Glycine	0.06g
KH_2PO_4	0.05g
K_2HPO_4	0.05g
L-Cysteine	0.05g
L-Proline	0.05g
Sodium citrate	0.05g
$FeCl_2$	2.3mg
$CaCl_2 \cdot 2H_2O$	0.7mg
$ZnSO_4 \cdot 7H_2O$	0.44mg
$MnSO_4 \cdot H_2O$	0.3mg
$CuSO_4 \cdot 5H_2O$	0.05mg

pH 6.2 ± 0.2 at 25°C

Preparation of Medium: Add components to distilled/deionized water and bring volume to 1.0L. Mix thoroughly. Gently heat until dissolved. Adjust pH to 6.2. Distribute into tubes or flasks. Autoclave for 15 min at 15 psi pressure–121°C.

Use: For the cultivation of members of the Halobacteriaceae.

Halobacterium denitrificans Medium

Composition per liter:

NaCl	176.0g
Agar	20.0g
$MgCl_2 \cdot 6H_2O$	20.0g
HEPES (N-2-Hydroxyethylpiperazine-N´-2-ethanesulfonic acid) buffer	11.9g
Yeast extract	5.0g
Hy-Case SF (Humko-Sheffield)	2.0g
KCl	2.0g
$CaCl_2 \cdot 2H_2O$	0.1g

pH 6.7 ± 0.2 at 25°C

Preparation of Medium: Add components to distilled/deionized water and bring volume to 1.0L. Mix thoroughly. Gently heat and bring to boiling. Adjust pH to 6.7. Distribute into tubes or flasks. Autoclave for 15 min at 15 psi pressure–121°C. Pour into sterile Petri dishes or leave in tubes.

Use: For the aerobic cultivation and maintenance of *Haloferax (Halobacterium) denitrificans.*

Halobacterium denitrificans Medium

Composition per liter:

NaCl	176.0g
Agar	20.0g
$MgCl_2 \cdot 6H_2O$	20.0g
HEPES (N-2-Hydroxyethylpiperazine-N´-2-ethanesulfonic acid) buffer	11.9g
KNO_3	5.0g
Yeast extract	5.0g
Hy-Case SF (Humko-Sheffield)	2.0g
KCl	2.0g
$CaCl_2 \cdot 2H_2O$	0.1g

pH 6.7 ± 0.2 at 25°C

Preparation of Medium: Add components to distilled/deionized water and bring volume to 1.0L. Mix thoroughly. Gently heat and bring to boiling. Adjust pH to 6.7. Distribute into tubes or flasks. Autoclave for 15 min at 15 psi pressure–121°C. Pour into sterile Petri dishes or leave in tubes.

Use: For the anerobic cultivation and maintenance of *Haloferax (Halobacterium) denitrificans.*

Halobacterium Medium (ATCC Medium 974)

Composition per 100mL:

Solution 1	75.0mL
Solution 2	25.0mL

pH 6.8 ± 0.2 at 25°C

Solution 1:

Composition per 75mL:

NaCl	12.5g
$MgCl_2 \cdot 6H_2O$	5.0g
K_2SO_4	0.5g
$CaCl_2 \cdot 6H_2O$	0.02g

Preparation of Solution 1: Add components to distilled/deionized water and bring volume to 75.0mL. Mix thoroughly. Adjust pH to 6.8. Autoclave for 15 min at 15 psi pressure–121°C. Cool to 45°–50°C.

Solution 2:

Composition per 25mL:

Agar	2.0g
Pancreatic digest of casein	0.5g
Yeast extract	0.5g

Preparation of Solution 2: Add components to distilled/deionized water and bring volume to 25.0mL. Mix thoroughly. Adjust pH to 6.8. Autoclave for 15 min at 15 psi pressure–121°C. Cool to 45°–50°C.

Preparation of Medium: Aseptically combine sterile solution 1 and sterile solution 2. Mix thoroughly. Pour into sterile Petri dishes or distribute into sterile tubes.

Use: For the cultivation and maintenance of *Haloferax volcanii.*

Halobacterium Medium (ATCC Medium 1270)

Composition per liter:

NaCl	194.0g
$MgSO_4$	24.0g
$MgCl_2$	16.0g
KCl	5.0g
Yeast extract	5.0g
$CaCl_2$	1.0g
NaBr	0.5g
$NaHCO_3$	0.2g

pH 7.3 ± 0.2 at 25°C

Preparation ofPreparation of Medium: Add components to distilled/deionized water and bring volume to 1.0L. Mix thoroughly. Distribute into tubes or flasks. Autoclave for 15 min at 15 psi pressure–121°C.

Use: For the cultivation of *Haloarcula hispanica* and *Haloferax gibbonsii.*

Halobacterium Medium (ATCC Medium 213)

Composition per liter:

Solution 1	500.0mL
Solution 2	500.0mL

pH 7.0 ± 0.2 at 25°C

Solution 1:

Composition per 500mL:

Yeast extract	10.0g
Pancreatic digest of casein	2.5g

Preparation of Solution 1: Add components to distilled/deionized water and bring volume to 500.0mL. Mix thoroughly. Gently heat and bring to boiling. Adjust pH to 7.0. Autoclave for 15 min at 15 psi pressure–121°C. Cool to 45°–50°C.

Solution 2:

Composition per 500mL:

NaCl	250.0g
Agar	20.0g
$MgSO_4 \cdot 7H_2O$	10.0g
KCl	5.0g
$CaCl_2 \cdot 6H_2O$	0.2g

Preparation of Solution 2: Add components to distilled/deionized water and bring volume to 500.0mL. Mix thoroughly. Gently heat and bring to boiling. Autoclave for 15 min at 15 psi pressure–121°C. Cool to 45°–50°C.

Preparation of Medium: Aseptically combine sterile solution 1 and sterile solution 2. Mix thoroughly. Aseptically distribute into sterile tubes or flasks.

Use: For the cultivation of *Halobacterium salinarium.*

Halobacterium Medium

Composition per liter:

Solution 1	500.0mL
Solution 2	500.0mL

pH 7.0 ± 0.2 at 25°C

Solution 1:

Composition per 500mL:

Yeast extract	10.0g
Pancreatic digest of casein	2.5g

Preparation of Solution 1: Add components to distilled/deionized water and bring volume to 500.0mL. Mix thoroughly. Gently heat and bring to boiling. Adjust pH to 7.0. Autoclave for 15 min at 15 psi pressure–121°C. Cool to 45°–50°C.

Solution 2:

Composition per 500mL:

NaCl	250.0g
$MgSO_4 \cdot 7H_2O$	10.0g
KCl	5.0g
$CaCl_2 \cdot 6H_2O$	0.2g

Preparation of Solution 2: Add components to distilled/deionized water and bring volume to 500.0mL. Mix thoroughly. Gently heat and bring to boiling. Autoclave for 15 min at 15 psi pressure–121°C. Cool to 45°–50°C.

Preparation of Medium: Aseptically combine sterile solution 1 and sterile solution 2. Mix thoroughly. Aseptically distribute into sterile tubes or flasks.

Use: For the cultivation of *Halobacterium salinarium.*

Halobacterium Medium (ATCC Medium 1176)

Composition per liter:

NaCl	156.0g
$MgSO_4 \cdot 7H_2O$	20.0g
$MgCl_2 \cdot 6H_2O$	13.0g
Yeast extract	5.0g
KCl	4.0g
$CaCl_2 \cdot 6H_2O$	1.0g
Glucose	1.0g
NaBr	0.5g
$NaHCO_3$	0.2g

pH 7.0 ± 0.2 at 25°C

Preparation of Medium: Add components to distilled/deionized water and bring volume to 1.0L. Mix thoroughly. Distribute into tubes or flasks. Autoclave for 15 min at 15 psi pressure–121°C.

Use: For the cultivation of *Haloferax mediterranei,*

Halobacterium pharaonis Medium

Composition per liter:

NaCl	250.0g
Agar	20.0g
Casamino acids	15.0g
Trisodium citrate·$2H_2O$	3.0g
Glutamic acid	2.5g
$MgSO_4 \cdot 7H_2O$	2.5g
KCl	2.0g

pH 8.5 ± 0.2 at 25°C

Preparation of Medium: Add components to distilled/deionized water and bring volume to 1.0L. Mix thoroughly. Gently heat and bring to boiling. Adjust pH to 6.0. Autoclave for 15 min at 15 psi pressure–121°C. Cool to 50°C. Readjust pH to 8.5. Pour into sterile Petri dishes or distribute into sterile tubes.

Use: For the cultivation and maintenance of *Natronobacterium (Halobacterium) pharaonis.*

Halobacterium Starch Medium

Composition per liter:

$MgCl_2 \cdot 6H_2O$	160.0g
NaCl	125.0g
K_2SO_4	5.0g
Soluble starch	2.0g
Peptone	1.0g
Yeast extract	1.0g
$CaCl_2 \cdot 2H_2O$	0.13g

pH 7.0 ± 0.2 at 25°C

Preparation of Medium: Add components to distilled/deionized water and bring volume to 1.0L. Mix thoroughly. Distribute into tubes or flasks. Autoclave for 15 min at 15 psi pressure–121°C.

Use: For the cultivation and maintenance of *Haloarcula marismortui* and *Halobacterium sodomense.*

Halobacterium volcanii Medium

Composition per 100mL:

NaCl	25.0g
$MgSO_4 \cdot 7H_2O$	1.0g
KCl	0.5g
Glycine	0.2g
$CaCl_2 \cdot 6H_2O$	0.02g
Yeast autolysate	1.0mL

pH 7.0 ± 0.2 at 25°C

Preparation of Medium: Add components to distilled/deionized water and bring volume to 100.0mL. Mix thoroughly. Adjust pH to 7.0. Distribute into tubes or flasks. Autoclave for 15 min at 15 psi pressure–121°C.

Use: For the specific enrichment of *Halobacterium volcanii.*

Halobacterium volcanii Medium

Composition per liter:

NaCl	125.0g
$MgCl_2 \cdot 6H_2O$	50.0g
K_2SO_4	5.0g
Pancreatic digest of casein	5.0g
Yeast extract	5.0g
$CaCl_2 \cdot 6H_2O$	0.2g

pH 6.8 ± 0.2 at 25°C

Preparation of Medium: Add components to distilled/deionized water and bring volume to 1.0L. Mix thoroughly. Gently heat until dissolved. Adjust pH to 6.8. Distribute into tubes or flasks. Autoclave for 15 min at 15 psi pressure–121°C.

Use: For the cultivation of *Halobacterium volcanii.*

Halobacteroides Medium

Composition per 990mL:

NaCl	88.0g
$MgCl_2 \cdot 6H_2O$	20.0g
$CaCl_2 \cdot 2H_2O$	7.4g
Yeast extract	5.0g
KCl	3.7g
L-Cysteine·HCl·H_2O	0.5g
Resazurin	1.0mg
Glucose (10% solution)	50.0mL
Sodium PIPES (Piperazine-N,N′-bis-2-ethane sulfonate buffer, 1M, pH 6.8–7.0)	40.0mL

pH 6.8–7.0 at 25°C

Preparation of Medium: Filter sterilize glucose solution and PIPES buffer solution separately. Add remaining components—except glucose solution, PIPES buffer solution, and cysteine·HCl·H_2O—to distilled/deionized water and bring volume to 900.0mL. Mix thoroughly. Gently heat and bring to boiling under 100% N_2. Add cysteine·HCl·H_2O. Mix thoroughly. Autoclave for 15 min at 15 psi pressure–121°C. Cool to 45°–50°C. Aseptically add 50.0mL of sterile glucose solution and 40.0mL of sterile PIPES buffer solution. Mix thoroughly. Aseptically and anaerobically distribute into sterile tubes or flasks.

Use: For the cultivation and maintenance of *Halobacteroides halobius* and *Sporohalobacter marismortui*.

Halobius Medium

Composition per liter:

NaCl	116.0g
Agar	20.0g
$MgSO_4 \cdot 7H_2O$	20.0g
Yeast extract	10.0g
Vitamin-free casamino acids	7.5g
Sodium citrate	3.0g
KCl	2.0g
$FeCl_2$	0.023g

pH 6.2 ± 0.2 at 25°C

Preparation of Medium: Add components to distilled/deionized water and bring volume to 1.0L. Mix thoroughly. Gently heat and bring to boiling. Distribute into tubes or flasks. Autoclave for 15 min at 15 psi pressure–121°C. Pour into sterile Petri dishes or leave in tubes.

Use: For the cultivation and maintenance of *Micrococcus halobius*.

Halodurans Medium

Composition per liter:

NaCl	150.0g
Agar	20.0g
$MgSO_4 \cdot 7H_2O$	20.0g
Yeast extract	10.0g
Casamino acids	7.5g
Sodium citrate	3.0g
KCl	2.0g
$FeCl_2$	0.023g

pH 7.0 ± 0.2 at 25°C

Preparation of Medium: Add components to distilled/deionized water and bring volume to 1.0L. Mix thoroughly. Gently heat and bring to boiling. Distribute into tubes or flasks. Autoclave for 15 min at 15 psi pressure–121°C. Pour into sterile Petri dishes or leave in tubes.

Use: For the cultivation and maintenance of *Micrococcus varians*.

Halomethanococcus Medium (*Methanohalophilus* Medium)

Composition per 1030mL:

Trimethylamine·HCl	2.5g
Na_2CO_3	2.0g
$NaHCO_3$	2.0g
Casamino acids	0.5g
L-Cysteine·HCl·H_2O	0.5g
$Na_2S \cdot 9H_2O$ solution	0.5g
NH_4Cl	0.5g
Pancreatic digest of casein	0.5g
Yeast extract	0.5g
K_2HPO_4	0.2g
Ammonium-2-mercaptoethanesulfonate	1.0mg
Artificial brine	1.0L
Wolfe's vitamin solution	10.0mL
Wolfe's mineral solution	10.0mL
Volatile acids solution	10.0mL

pH 7.1 ± 0.2 at 25°C

$Na_2S \cdot 9H_2O$ Solution:

Composition per 10mL:

$Na_2S \cdot 9H_2O$	2.0g

Preparation of $Na_2S \cdot 9H_2O$ Solution: Add $Na_2S \cdot 9H_2O$ to distilled/deionized water and bring volume to 10.0mL. Mix thoroughly. Filter sterilize.

Artificial Brine:

Composition per liter:

NaCl	80.7g
$MgCl_2 \cdot 6H_2O$	35.1g
Na_2SO_4	12.9g
KCl	5.7g
$CaCl_2$	0.55g
$LiCl_2$	0.13g
H_3BO_3	0.12g

Preparation of Artificial Brine: Add components to distilled/deionized water and bring volume to 1.0L. Mix thoroughly.

Wolfe's Vitamin Solution:

Composition per liter:

Pyridoxine·HCl	0.01g
Thiamine·HCl	5.0mg
Riboflavin	5.0mg
Nicotinic acid	5.0mg
Calcium pantothenate	5.0mg
p-Aminobenzoic acid	5.0mg
Thioctic acid	5.0mg
Biotin	2.0mg
Folic acid	2.0mg
Cyanocobalamin	0.1mg

Preparation of Wolfe's Vitamin Solution: Add components to distilled/deionized water and bring volume to 1.0L. Mix thoroughly. Filter sterilize. Store at 4°C.

Wolfe's Mineral Solution:

Composition per liter:

$MgSO_4 \cdot 7H_2O$	3.0g
Nitriloacetic acid	1.5g
NaCl	1.0g
$MnSO_4 \cdot H_2O$	0.5g
$FeSO_4 \cdot 7H_2O$	0.1g
$CoCl_2 \cdot 6H_2O$	0.1g
$CaCl_2$	0.1g
$ZnSO_4 \cdot 7H_2O$	0.1g
$CuSO_4 \cdot 5H_2O$	0.01g
$AlK(SO_4)_2 \cdot 12H_2O$	0.01g
H_3BO_3	0.01g
$Na_2MoO_4 \cdot 2H_2O$	0.01g

Preparation of Wolfe's Mineral Solution: Add nitrilotriacetic acid to 500.0mL of distilled/deionized water. Dissolve by adjusting pH to 6.5 with KOH. Add remaining components. Add distilled/deionized water to 1.0L.

Volatile Acids Solution:

Composition per liter:

α-Methylbutyric acid	0.5mL
Isobutyric acid	0.5mL
Isovaleric acid	0.5mL
Valeric acid	0.5mL

Preparation of Volatile Acids Solution: Add components to distilled/deionized water and bring volume to 1.0L. Mix thoroughly.

Preparation of Medium: Combine components, except the cysteine·HCl·H_2O, trimethylamine·HCl, and $Na_2S \cdot 9H_2O$ solution. Mix thoroughly. Gently heat and bring to boiling. Add cysteine·HCl·H_2O. Mix thoroughly. Cool in an ice water bath under 80% N_2 + 20% CO_2. Add trimethylamine·HCl. Mix thoroughly. Adjust pH to 7.1. Aseptically and anaerobically distribute into tubes in 10.0mL volumes under 80% N_2 + 20% CO_2. Autoclave for 15 min at 15 psi pressure–121°C. Immediately prior to inoculation, aseptically add 0.25mL of sterile $Na_2S \cdot 9H_2O$ solution to each tube. Mix thoroughly.

Use: For the cultivation and maintenance of *Methanohalophilus mahii.*

Halomonas Medium

Composition per liter:

NaCl	80.0g
$MgSO_4 \cdot 7H_2O$	20.0g
Casamino acids with vitamins	7.5g
Proteose peptone No. 3	5.0g
Sodium citrate	3.0g
Yeast extract	1.0g
K_2HPO_4	0.5g
$Fe(NH_4)_2(SO_4)_2 \cdot 6H_2O$	0.05g

pH 7.0 ± 0.2 at 25°C

Preparation of Medium: Add components to distilled/deionized water and bring volume to 1.0L. Mix thoroughly. Adjust pH to 7.0 with KOH. Distribute into tubes or flasks. Autoclave for 15 min at 15 psi pressure–121°C.

Use: For the cultivation and maintenance of *Halomonas elongata.*

Halophile Medium

Composition per liter:

NaCl	100.0g
KCl	5.0g
$MgCl_2 \cdot 6H_2O$	5.0g
$MgSO_4 \cdot 7H_2O$	5.0g
NH_4Cl	5.0g
Peptone solution (15% solution)	30.0mL
Yeast extract solution (15% solution)	30.0mL
Ferric citrate solution (1% solution)	10.0mL
Trace element solution	5.0mL

Trace Element Solution:

Composition per liter:

$ZnSO_4 \cdot 7H_2O$	0.22g
$MgCl_2 \cdot 4H_2O$	0.18g
$CoCl_2 \cdot 6H_2O$	0.01g
$Na_2MoO_4 \cdot H_2O$	6.3mg
$CuSO_4 \cdot 5H_2O$	1.0mg

Preparation of Trace Element Solution: Add components to distilled/deionized water and bring volume to 1.0L. Mix thoroughly.

Preparation of Medium: Add components to distilled/deionized water and bring volume to 1.0L. Mix thoroughly. Gently heat and bring to boiling. Distribute into tubes or flasks. Autoclave for 15 min at 15 psi pressure–121°C.

Use: For the cultivation of *Rhodospirillum salinarum.*

Halophilic Agar (HA)

Composition per liter:

NaCl	250.0g
$MgSO_4 \cdot 7H_2O$	25.0g
Agar	20.0g
Casamino acids	10.0g
Yeast extract	10.0g

Proteose peptone 5.0g
Trisodium citrate 3.0g
KCl 2.0g

pH 7.2 ± 0.2 at 25°C

Preparation of Medium: Combine the ingredients with distilled water and heat to boiling to dissolve completely. Autoclave at 121°C for 15 min.

Use: For the isolation and cultivation of halophilic microorganisms from foods, such as *Pseudomonas* species and *Flavobacterium* species from fish and salted foods.

Halophilic Broth (HB)

Composition per liter:

NaCl 250.0g
$MgSO_4 \cdot 7H_2O$ 25.0g
Casamino acids 10.0g
Yeast extract 10.0g
Proteose peptone 5.0g
Trisodium citrate 3.0g
KCl 2.0g

pH 7.2 ± 0.2 at 25°C

Preparation of Medium: Add components to distilled/deionized water and bring volume to 1.0L. Mix thoroughly. Gently heat and bring to boiling. Distribute into tubes or flasks. Autoclave for 15 min at 15 psi pressure–121°C.

Use: For the isolation and cultivation of halophilic microorganisms from foods, such as *Pseudomonas* species and *Flavobacterium* species from fish and salted foods.

Halophilic *Clostridium* Agar

Composition per liter:

L-Cysteine·HCl·H_2O 0.5g
Solution 1 ....1.0L
Solution 2 ....100.0mL

pH 6.2–7.0 at 25°C

Solution 1:

Composition per liter:

NaCl 105.0g
Agar 20.0g
KCl 7.5g
$CaCO_3$ 5.0g
L-Glutamic acid 4.0g
Soluble starch 2.0g
Casamino acids 2.0g
Nutrient broth 2.0g
Yeast extract 2.0g
$FeSO_4 \cdot 7H_2O$ 2.0mg
Resazurin 1.0mg
NaOH (2.5 *N* solution) 12.5mL
Wolfe's vitamin solution 10.0mL
Wolfe's mineral solution 10.0mL

Preparation of Solution 1: Add components, except $CaCO_3$, to distilled/deionized water and bring volume to 1.0L. Mix thoroughly. Gently heat and bring to boiling. When all components have dissolved, add the $CaCO_3$. Mix thoroughly.

Wolfe's Vitamin Solution:

Composition per liter:

Pyridoxine·HCl 0.01g
Thiamine·HCl 5.0mg
Riboflavin 5.0mg
Nicotinic acid 5.0mg
Calcium pantothenate 5.0mg
p-Aminobenzoic acid 5.0mg
Thioctic acid 5.0mg
Biotin 2.0mg
Folic acid 2.0mg
Cyanocobalamin 0.1mg

Preparation of Wolfe's Vitamin Solution: Add components to distilled/deionized water and bring volume to 1.0L. Mix thoroughly.

Wolfe's Mineral Solution:

Composition per liter

$MgSO_4 \cdot 7H_2O$ 3.0g
Nitriloacetic acid 1.5g
NaCl 1.0g
$MnSO_4 \cdot H_2O$ 0.5g
$FeSO_4 \cdot 7H_2O$ 0.1g
$CoCl_2 \cdot 6H_2O$ 0.1g
$CaCl_2$ 0.1g
$ZnSO_4 \cdot 7H_2O$ 0.1g
$CuSO_4 \cdot 5H_2O$ 0.01g
$AlK(SO_4)_2 \cdot 12H_2O$ 0.01g
H_3BO_3 ....0.01g
$Na_2MoO_4 \cdot 2H_2O$ 0.01g

Preparation of Wolfe's Mineral Solution: Add nitrilotriacetic acid to 500.0mL of distilled/deionized water. Dissolve by adjusting pH to 6.5 with KOH. Add remaining components. Add distilled/deionized water to 1.0L.

Solution 2:

Composition per 100mL:

$MgCl_2 \cdot 6H_2O$ 20.3g
$CaCl_2 \cdot 2H_2O$ 7.35g

Preparation of Solution 2: Add components to distilled/deionized water and bring volume to 100.0mL. Mix thoroughly. Gas with 100% N_2 for 20 min. Autoclave for 15 min at 15 psi pressure–121°C. Cool to 45°–50°C.

Preparation of Medium: Gently heat 1.0L of solution 1 and bring to boiling under 100% N_2. Add the cysteine·HCl·H_2O. Continue boiling until resazurin turns colorless indicating reduction. Volume of solution 1 should be about 900.0mL. Anaerobically distribute into tubes in 9.0mL volumes under 100% N_2. Cap tubes with rubber stoppers. Place tubes in a press. Autoclave for 15 min at 15 psi pressure–121°C with fast exhaust. Cool to 50°C. Aseptically add 1.0mL of sterile solution 2 to each tube. In the presence of $CaCO_3$, the pH may be higher than 7.0. Do not adjust pH.

Use: For the cultivation and maintenance of *Sporohalobacter lortetii.*

Halophilic *Clostridium* Broth

Composition per liter:

L-Cysteine·HCl·H_2O	0.5g
Solution 1	1.0L
Solution 2	100.0mL

pH 6.2–7.0 at 25°C

Solution 1:

Composition per liter:

NaCl	105.0g
KCl	7.5g
L-Glutamic acid	4.0g
Casamino acids	2.0g
Nutrient broth	2.0g
Yeast extract	2.0g
$FeSO_4 \cdot 7H_2O$	2.0mg
Resazurin	1.0mg
NaOH (2.5 *N* solution)	12.5mL
Wolfe's vitamin solution	10.0mL
Wolfe's elements solution	10.0mL

Preparation of Solution 1: Add components to distilled/deionized water and bring volume to 1.0L. Mix thoroughly.

Wolfe's Vitamin Solution:

Composition per liter:

Pyridoxine·HCl	0.01g
Thiamine·HCl	5.0mg
Riboflavin	5.0mg
Nicotinic acid	5.0mg
Calcium pantothenate	5.0mg
p-Aminobenzoic acid	5.0mg
Thioctic acid	5.0mg
Biotin	2.0mg
Folic acid	2.0mg
Cyanocobalamin	0.1mg

Preparation of Wolfe's Vitamin Solution: Add components to distilled/deionized water and bring volume to 1.0L. Mix thoroughly.

Wolfe's Mineral Solution:

Composition per liter

$MgSO_4 \cdot 7H_2O$	3.0g
Nitriloacetic acid	1.5g
NaCl	1.0g
$MnSO_4 \cdot H_2O$	0.5g
$FeSO_4 \cdot 7H_2O$	0.1g
$CoCl_2 \cdot 6H_2O$	0.1g
$CaCl_2$	0.1g
$ZnSO_4 \cdot 7H_2O$	0.1g
$CuSO_4 \cdot 5H_2O$	0.01g
$AlK(SO_4)_2 \cdot 12H_2O$	0.01g
H_3BO_3	0.01g
$Na_2MoO_4 \cdot 2H_2O$	0.01g

Preparation of Wolfe's Mineral Solution: Add nitrilotriacetic acid to 500.0mL of distilled/deionized water. Dissolve by adjusting pH to 6.5 with KOH. Add remaining components. Add distilled/deionized water to 1.0L.

Solution 2:

Composition per 100mL:

$MgCl_2 \cdot 6H_2O$	20.3g
$CaCl_2 \cdot 2H_2O$	7.35g

Preparation of Solution 2: Add components to distilled/deionized water and bring volume to 100.0mL. Mix thoroughly. Gas with 100% N_2 for 20 min. Autoclave for 15 min at 15 psi pressure–121°C.

Preparation of Medium: Gently heat 1.0L of solution 1 and bring to boiling under 100% N_2. Add the cysteine·HCl·H_2O. Continue boiling until resazurin turns colorless indicating reduction. Volume of solution 1 should be about 900.0mL. Anaerobically distribute into tubes in 9.0mL volumes under 100% N_2. Cap tubes with rubber stoppers. Place tubes in a press. Autoclave for 15 min at 15 psi pressure–121°C with fast exhaust. Cool to 25°C. Aseptically add 1.0mL of sterile solution 2 to each tube. Adjust pH to 6.2–7.0 if necessary with sterile O_2-free NaOH or HCl.

Use: For the cultivation and maintenance of *Sporohalobacter lortetii.*

Halophilic *Halobacterium* Medium

Composition per liter:

NaCl	200.0g
$MgSO_4 \cdot 7H_2O$	37.0g
$CaCl_2 \cdot 2H_2O$	0.7g
KCl	0.5g
$MnCl_2 \cdot 4H_2O$	0.05g
Yeast extract	100.0mL

pH 7.0 ± 0.2 at 25°C

Preparation of Medium: Add components to distilled/deionized water and bring volume to 1.0L. Mix thoroughly. Gently heat until dissolved. Adjust pH to 7.0. Distribute into tubes or flasks. Autoclave for 15 min at 15 psi pressure–121°C.

Use: For the cultivation of extremely halophilic *Halobacterium* species.

Halophilic Synthetic Medium

Composition per liter:

Glucose	0.1g
KNO_3	0.05g
$FePO_4$	0.01g
Artificial sea water	100.0mL

Artificial Seawater:

Composition per 100mL:

NaCl	2.4g
$MgCl_2 \cdot 6H_2O$	1.1g
Na_2SO_4	0.4g
$CaCl_2 \cdot 6H_2O$	0.2g
KCl	0.07g
$NaHCO_3$	0.02g
KBr	0.01g
$SrCl_2 \cdot 6H_2O$	4.0mg
H_3BO_3	3.0mg
$Na_2SiO_3 \cdot 9H_2O$	0.5mg
NaF	0.3mg

Preparation of Artificial Seawater: Add components to distilled/deionized water and bring volume to 100.0mL. Mix thoroughly.

Preparation of Medium: Add components to distilled/deionized water and bring volume to 1.0L. Mix thoroughly. Distribute into tubes or flasks. Autoclave for 15 min at 15 psi pressure–121°C.

Use: For the cultivation of halophilic bacteria.

Ham's F–10 Medium

Composition per liter:

NaCl	7.4g
$NaHCO_3$	1.2g
Glucose	1.1g
$NaH_2PO_4 \cdot H_2O$	0.29g
KCl	0.28g
L-Arginine·HCl	0.21g
L-Glutamine	0.15g
$MgSO_4 \cdot 7H_2O$	0.15g
Sodium pyruvate	0.11g
KH_2PO_4	0.08g
$CaCl_2 \cdot 2H_2O$	0.04g
L-Cystine·2HCl	0.04g
L-Histidine·HCl·H_2O	0.02g
L-Lysine·HCl	0.02g
L-Asparagine-H_2O	0.01g
L-Aspartic Acid	0.01g
L-Glutamic acid	0.01g
L-Leucine	0.01g
L-Proline	0.01g
L-Serine	0.01g
L-Alanine	8.9mg
Glycine	7.5mg
D-Phenylalanine	5.0mg
L-Methionine	4.5mg
Hypoxanthine	4.1mg
L-Threonine	3.6mg
L-Valine	3.5mg
L-Isoleucine	2.6mg
L-Tyrosine	1.8mg
Vitamin B_{12}	1.4mg
Folic acid	1.3mg
Phenol red	1.2mg
Thiamine·HCl	1.0mg
$FeSO_4 \cdot 7H_2O$	0.8mg
Choline chloride	0.7mg
D-Calcium pantothenate	0.7mg
Thymidine	0.7mg
Niacinamide	0.6mg
L-Tryptophan	0.6mg
Isoinositol	0.5mg
Riboflavin	0.4mg
Lipoic acid	0.2mg
Pyridoxine·HCl	0.2mg
$ZnSO_4 \cdot 7H_2O$	0.03mg
Biotin	0.02mg
$CuSO_4 \cdot 5H_2O$	3.0μg

pH 7.0 ± 0.2 at 25°C

Preparation of Medium: Add components to distilled/deionized water and bring volume to 1.0L. Mix thoroughly. Filter sterilize.

Use: For the growth of Y-1 cell cultures used in the mouse adrenal assay for heat-labile toxin of enterotoxigenic *Escherichia coli* and *Vibrio* species.

Hartley's Digest Broth

Composition per 10L:

Ox heart	3000.0g
Pancreatin	50.0g
Na_2CO_3, anhydrous (0.8% solution)	5.0L
HCl, concentrated	80.0mL

pH 7.5 ± 0.2 at 25°C

Preparation of Medium: Finely mince the ox heart. Add the meat to 5.0L of distilled/deionized water. Gently heat and bring to 80°C. Add the 5.0L of Na_2CO_3 solution. Cool to 45°C. Add pancreatin and maintain at 45°C for 4 hr while stirring. Add the HCl and steam at 100°C for 30 min. Cool to room temperature. Adjust pH to 8.0 with 1*N* NaOH. Gently heat and bring to boiling. Continue boiling for 25 min. Filter while hot through Whatman #1 filter paper. Cool to room temperature. Adjust pH to 7.5. Autoclave for 15 min at 15 psi pressure–121°C.

Use: For the isolation and cultivation of *Actinobacillus lignieresii* from cattle and *Actinobacillus lignieresii*.

Hay Extract Medium
See: HE Medium

Hayflick Medium

Composition per 107.5mL:

Mycoplasma broth base	70.0mL
Horse serum	20.0mL
Fresh yeast extract solution	10.0mL
Penicillin solution	5.0mL
Thallous acetate solution	2.5mL

pH 7.8 ± 0.2 at 25°C

Mycoplasma Broth Base:

Pancreatic digest of casein	7.0g
NaCl	5.0g
Beef extract	3.0g
Yeast extract	3.0g
Beef heart, solids from infusion	2.0g

pH 7.8 ± 0.2 at 25°C

Preparation of Mycoplasma Broth Base: Add components to distilled/deionized water and bring volume to 1.0L. Gently heat and bring to boiling. Mix thoroughly. Distribute into tubes or flasks. Autoclave for 15 min at 15 psi pressure–121°C.

Fresh Yeast Extract Solution:
Composition per 100mL:

Baker's yeast, live, pressed, starch-free	25.0g

Preparation of Fresh Yeast Extract Solution: Add live Baker's yeast to 100.0mL of distilled/deionized water. Autoclave for 90 min at 15 psi pressure–121°C. Allow to stand. Remove supernatant solution. Adjust pH to 6.6–6.8. Filter sterilize.

Penicillin Solution:
Composition per 5mL:

Penicillin	20,000U

Preparation of Penicillin Solution: Add penicillin to distilled/deionized water and bring volume to 5.0mL. Mix thoroughly. Filter sterilize.

Thallous Acetate Solution:
Composition per 10mL:

Thallous acetate	0.1g

Preparation of Thallous Acetate Solution: Add thallous acetate to distilled/deionized water and bring volume to 10.0mL. Mix thoroughly. Filter sterilize.

Use: For the cultivation of *Mycoplasma* species.

Hayflick Medium, Modified

Composition per 1212mL:

Beef heart, infusion from	500.0g
Tryptose	10.0g
Noble agar	9.6g
NaCl	5.0g
Horse serum, normal	200.0mL
Fresh yeast extract solution	100.0mL
Calf thymus DNA solution	12.0mL

pH 7.8 ± 0.2 at 25°C

Fresh Yeast Extract Solution:
Composition per 100mL:

Baker's yeast live, pressed, starch-free	25.0g

Preparation of Fresh Yeast Extract Solution: Add the live Baker's yeast to 100.0mL of distilled/deionized water. Autoclave for 90 min at 15 psi pressure–121°C. Allow to stand. Remove supernatant solution. Adjust pH to 6.6–6.8. Filter sterilize.

Calf Thymus DNA Solution:
Composition per 20mL:

Calf thymus DNA	0.04g

Preparation of Calf Thymus DNA Solution: Add calf thymus DNA to distilled/deionized water and bring volume to 20.0mL. Mix thoroughly. Filter sterilize.

Preparation of Medium: Add components, except horse serum, fresh yeast extract solution, and calf thymus DNA solution, to distilled/deionized water and bring volume to 900.0mL. Mix thoroughly. Gently heat and bring to boiling. Autoclave for 15 min at 15 psi pressure–121°C. Cool to 45°–50°C. Aseptically add 200.0mL of sterile horse serum, 100.0mL of sterile fresh yeast extract solution, and 12.0mL of sterile calf thymus DNA solution. Mix thoroughly. Aseptically distribute into sterile tubes.

Use: For the cultivation and maintenance of *Mycoplasma mustelae*.

HB
See: Halophilic Broth

HBT Bilayer Medium (Human Blood Tween™ Bilayer Medium)

Composition per 1062.5mL:

Agar	13.5g

Pancreatic digest of casein	12.0g
Casein/meat peptone	10.0g
NaCl	5.0g
Peptic digest of animal tissue	5.0g
Beef extract	3.0g
Yeast extract	3.0g
Cornstarch	1.0g
Human blood, anticoagulated	25.0mL
Colistin solution	10.0mL
Nalidixic acid solution	10.0mL
Amphotericin B solution	10.0mL
Polysorbate 80 (Tween™ 80) solution	7.5mL

pH 7.3 ± 0.2 at 25°C

Source: This medium is available as a premixed powder from BBL Microbiology Systems.

Colistin Solution:
Composition per liter:

Colistin	0.010g

Preparation of Colistin: Add colistin to distilled/deionized water and bring volume to 10.0mL. Mix thoroughly. Filter sterilize.

Nalidixic Acid Solution:
Composition per liter:

Nalidixic acid	0.020g

Preparation of Nalidixic Acid Solution: Add nalidixic acid to distilled/deionized water and bring volume to 10.0mL. Mix thoroughly. Filter sterilize.

Amphotericin B Solution:
Composition per liter:

Amphotericin B	3.0mg

Preparation of Amphotericin B Solution: Add Amphotericin B to distilled/deionized water and bring volume to 10.0mL. Mix thoroughly. Filter sterilize.

Tween™ 80 Solution:
Composition per 100mL:

Tween™ 80	1.0mL

Preparation of Tween™ 80 Solution: Add Tween™ 80 to distilled/deionized water and bring volume to 100.0mL. Mix thoroughly. Adjust pH to 7.3. Filter sterilize.

Preparation of Medium: Add components, except amphotericin B, Tween™ 80 and human blood, to distilled/deionized water and bring volume to 1.0L. Mix thoroughly. Gently heat and bring to boiling. Divide the medium into two 500.0mL fractions. Autoclave both flasks of media for 15 min at 15 psi pressure–121°C. Cool to 45°–50°C. To one flask aseptically add 5.0mL of sterile colistin solution, 5.0mL of sterile nalidixic acid solution, 5.0mL of sterile amphotericin B solution, and 3.75mL of Tween™ 80 solution. Mix thoroughly. Pour into sterile Petri dishes in 7.0mL volumes. Allow agar to harden. To remaining flask aseptically add 5.0mL of sterile colistin solution, 5.0mL of sterile nalidixic acid solution, 5.0mL of sterile amphotericin B solution, 3.75mL of sterile Tween™ 80 solution, and 25.0mL of sterile human blood. Mix thoroughly. Pour into the same Petri dishes that each contain 7.0mL of the agar medium without blood. The top layer should be approximately 14.0mL per plate.

Use: For the selective isolaltion, cultivation and differentiation of *Gardnerella vaginalis* from clinical specimens.

HC Agar
See: **Hemorrhagic coli Agar**

HC Agar Base

Composition per liter:

Glucose	20.0g
Agar	15.0g
Yeast extract	5.0g
Na_2HPO_4	3.5g
KH_2PO_4	3.4g
Pancreatic digest of casein	2.5g
Peptic digest of animal tissue	2.5g
NH_4Cl	1.4g
Na_2CO_3	1.0g
Chloramphenicol	0.1g
$MgSO_4 \cdot 7H_2O$	0.06g
Polysorbate 80 (Tween™ 80) solution	20.0mL

pH 7.0 ± 0.2 at 25°C

Source: This medium is available as a premixed powder from BBL Microbiology Systems.

Preparation of Medium: Add components, except Tween™ 80, to distilled/deionized water and bring volume to 980.0mL. Mix thoroughly. Gently heat and bring to boiling. Add Tween™ 80. Mix thoroughly. Autoclave for 15 min at 15 psi pressure–121°C. Pour into sterile Petri dishes.

Use: For the cultivation and enumeration of molds in cosmetics and toiletries.

HE Medium
(Hay Extract Medium)

Composition per liter:

Agar	10.0g
Peptone	1.0g
Yeast extract	1.0g
Hay extract solution	500.0mL

pH 6.5 ± 0.2 at 25°C

Hay Extract Solution:
Composition per liter:
Hay, dried..50.0g

Preparation of Hay Extract Solution: Add dried barn hay to distilled/deionized water and bring volume to 1.0L. Mix thoroughly. Gently heat and bring to boiling. Filter through Whatman #40 filter paper.

Preparation of Medium: Add components to distilled/deionized water and bring volume to 1.0L. Mix thoroughly. Gently heat and bring to boiling. Distribute into tubes or flasks. Autoclave for 15 min at 15 psi pressure–121°C. Pour into sterile Petri dishes or leave in tubes.

Use: For the isolation and cultivation of *Spirochaeta aurantia*.

Heart Infusion Agar

Composition per liter:
Beef heart, infusion from..............................500.0g
Agar..15.0g
Tryptose..10.0g
NaCl..5.0g
pH 7.4 ± 0.2 at 25°C

Source: This medium is available as a premixed powder from Difco Laboratories.

Preparation of Medium: Add components to distilled/deionized water and bring volume to 1.0L. Mix thoroughly. Gently heat and bring to boiling. Distribute into tubes or flasks. Autoclave for 15 min at 15 psi pressure–121°C. Pour into sterile Petri dishes or leave in tubes.

Use: For the isolation and cultivation of a wide variety of fastidious microorganisms. It can also be used as a base for the preparation of blood agar in determining hemolytic reactions. For the cultivation and maintenance of *Bacillus anthracis, Bacillus cereus, Bacillus mycoides, Serratia rubidaea, Staphylococcus aureus, Tsatumella ptyseos,* and *Vibrio vulnificus*.

Heart Infusion Agar with Glucose

Composition per liter:
Beef heart, infusion from..............................500.0g
Agar..15.0g
Tryptose..10.0g
NaCl..5.0g
Glucose...1.0g
pH 7.4 ± 0.2 at 25°C

Preparation of Medium: Add components to distilled/deionized water and bring volume to 1.0L. Mix thoroughly. Gently heat and bring to boiling. Distribute into tubes or flasks. Autoclave for 15 min at 15 psi pressure–121°C. Pour into sterile Petri dishes or leave in tubes.

Use: For the cultivation and maintenance of the *Bacillus* species and *Pseudomonas* species.

Heart Infusion Agar with Horse Serum and Fresh Yeast Extract

Composition per 930mL:
Heart infusion agar....................................720.0mL
Horse serum, unheated...............................200.0mL
Fresh yeast extract solution..........................10.0mL
pH 7.4 ± 0.2 at 25°C

Heart Infusion Agar:
Composition per liter:
Beef heart, infusion from..............................500.0g
Agar..15.0g
Tryptose..10.0g
NaCl..5.0g

Preparation of Heart Infusion Agar: Add components to distilled/deionized water and bring volume to 1.0L. Mix thoroughly. Gently heat and bring to boiling. Autoclave for 15 min at 15 psi pressure–121°C. Cool to 45°–50°C.

Fresh Yeast Extract Solution:
Composition per 100mL:
Baker's yeast live, pressed, starch-free............25.0g

Preparation of Fresh Yeast Extract Solution: Add the live Baker's yeast to 100.0mL of distilled/deionized water. Autoclave for 90 min at 15 psi pressure–121°C. Allow to stand. Remove supernatant solution. Adjust pH to 6.6–6.8.

Preparation of Medium: To 720.0mL of sterile cooled heart infusion broth, aseptically add 200.0mL of horse serum and 10.0mL of fresh yeast exctract solution. Mix thoroughly. Pour into sterile Petri dishes or distribute into sterile tubes.

Use: For the cultivation and maintenance of *Mycoplasma equigenitalium* and *Mycoplasma subdolum*.

Heart Infusion Agar with Inactivated Horse Serum

Composition per liter:
Beef heart, infusion from..............................500.0g
Agar..15.0g
Tryptose..10.0g

NaCl..5.0g
Horse serum, inactivated.............................100.0mL
pH 7.4 ± 0.2 at 25°C

Preparation of Medium: Add components, except horse serum, to distilled/deionized water and bring volume to 900.0mL. Mix thoroughly. Gently heat and bring to boiling. Autoclave for 15 min at 15 psi pressure–121°C. Cool to 45°–50°C. Aseptically add sterile horse serum. Mix thoroughly. Pour into sterile Petri dishes or distribute into sterile tubes.

Use: For the cultivation and maintenance of the *Corynebacterium* species.

Heart Infusion Agar with Inactivated Horse Serum, NaCl and Penicillin

Composition per liter:
Beef heart, infusion from 500.0g
NaCl..35.0g
Agar...15.0g
Tryptose 10.0g
Horse serum, inactivated.............................100.0mL
Penicillin solution 10.0mL
pH 7.4 ± 0.2 at 25°C

Penicillin Solution:
Composition per 10mL:
Penicillin 1,000,000U

Preparation of Penicillin Solution: Add penicillin to distilled/deionized water and bring volume to 10.0mL. Mix thoroughly. Filter sterilize.

Preparation of Medium: Add components, except penicillin solution and horse serum, to distilled/deionized water and bring volume to 890.0mL. Mix thoroughly. Gently heat and bring to boiling. Autoclave for 15 min at 15 psi pressure–121°C. Cool to 45°–50°C. Aseptically add 10.0mL of sterile penicillin solution and 100.0mL of sterile horse serum. Mix thoroughly. Pour into sterile Petri dishes or distribute into sterile tubes.

Use: For the cultivation and maintenance of the Corynebacterium species.

Heart Infusion Agar with Rabbit Blood

Composition per liter:
Beef heart, infusion from 500.0g
Agar...15.0g
Tryptose 10.0g
NaCl..5.0g
Rabbit blood...50.0mL
pH 7.4 ± 0.2 at 25°C

Preparation of Medium: Add components, except rabbit blood, to distilled/deionized water and bring volume to 950.0mL. Mix thoroughly. Gently heat and bring to boiling. Autoclave for 15 min at 15 psi pressure–121°C. Cool to 45°–50°C. Aseptically add sterile rabbit blood. Mix thoroughly. Pour into sterile Petri dishes or distribute into sterile tubes.

Use: For the cultivation and maintenance of *Neisseria lactamica*.

Heart Infusion Agar with Yeast Extract

Composition per liter:
Beef heart, infusion from 500.0g
Agar...20.0g
Tryptose 10.0g
NaCl..5.0g
Yeast extract 5.0g
pH 7.4 ± 0.2 at 25°C

Preparation of Medium: Add components to distilled/deionized water and bring volume to 1.0L. Mix thoroughly. Gently heat and bring to boiling. Distribute into tubes or flasks. Autoclave for 15 min at 15 psi pressure–121°C. Pour into sterile Petri dishes or leave in tubes.

Use: For the cultivation and maintenance of *Moraxella nonliquefaciens*.

Heart Infusion Broth

Composition per liter:
Beef heart, infusion from 500.0g
Tryptose 10.0g
NaCl..5.0g
pH 7.4 ± 0.2 at 25°C

Source: This medium is available as a premixed powder from Difco Laboratories.

Preparation of Medium: Add components to distilled/deionized water and bring volume to 1.0L. Mix thoroughly. Distribute into tubes or flasks. Autoclave for 15 min at 15 psi pressure–121°C.

Use: For the isolation and cultivation of a wide variety of fastidious microorganisms.

Heart Infusion Broth with Additives for *Staphylococcus*

Composition per liter:
Beef heart, infusion from 500.0g
NaCl..30.0 g
Tryptose 10.0g

Horse serum, inactivated 100.0mL
Penicillin solution 10.0mL
Fresh yeast extract solution............................ 5.0mL
pH 7.4 ± 0.2 at 25°C

Penicillin Solution:
Composition per 10mL:
Penicillin 1,000,000U

Preparation of Penicillin Solution: Add penicillin to distilled/deionized water and bring volume to 10.0mL. Mix thoroughly. Filter sterilize.

Fresh Yeast Extract Solution:
Composition per 100mL:
Baker's yeast live, pressed, starch-free 10.0g

Preparation of Fresh Yeast Extract Solution: Add the live Baker's yeast to 100.0mL of distilled/deionized water. Autoclave for 90 min at 15 psi pressure–121°C. Allow to stand. Remove supernatant solution. Adjust pH to 6.6–6.8.

Preparation of Medium: Add components—except horse serum, fresh yeast extract solution, and penicillin solution—to distilled/deionized water and bring volume to 800.0mL. Mix thoroughly. Gently heat and bring to boiling. Autoclave for 15 min at 15 psi pressure–121°C. Cool to 45°–50°C. Aseptically add sterile horse serum. Mix thoroughly. Aseptically distribute into sterile tubes or flasks.

Use: For the cultivation of *Staphylococcus* species.

Heart Infusion Broth with Additives for *Streptobacillus*

Composition per liter:
Beef heart, infusion from 500.0g
Tryptose 10.0g
Peptone.. 10.0g
NaCl 5.0g
Glucose 0.5g
Horse serum, inactivated........................... 200.0mL
pH 7.5 ± 0.2 at 25°C

Preparation of Medium: Add components, except horse serum, to distilled/deionized water and bring volume to 800.0mL. Mix thoroughly. Gently heat and bring to boiling. Autoclave for 15 min at 15 psi pressure–121°C. Cool to 45°–50°C. Aseptically add sterile horse serum. Mix thoroughly. Aseptically distribute into sterile tubes or flasks.

Use: For the cultivation and maintenance of *Streptobacillus moniliformis.*

Heart Infusion Broth with Glucose

Composition per liter:
Beef heart, infusion from 500.0g
Tryptose 10.0g
NaCl 5.0g
Glucose 1.0g
pH 7.4 ± 0.2 at 25°C

Preparation of Medium: Add components to distilled/deionized water and bring volume to 1.0L. Mix thoroughly. Distribute into tubes or flasks. Autoclave for 15 min at 15 psi pressure–121°C.

Use: For the cultivation and maintenance of *Arthrobacter* species, *Bacillus* species, and *Pseudomonas* species.

Heart Infusion Broth with Glucose and Antibiotics

Composition per liter:
Beef heart, infusion from 500.0g
Tryptose 10.0g
NaCl 5.0g
Antibiotic inhibitor solution......................... 10.0mL
Glucose solution.. 10.0mL
pH 7.4 ± 0.2 at 25°C

Antibiotic Inhibitor Solution:
Composition per 10mL:
Streptomycin sulfate 0.10g
Tetracycline·HCl 0.025g

Preparation of Antibiotic Inhibitor Solution: Add components to distilled/deionized water and bring volume to 10.0mL. Mix thoroughly. Filter sterilize.

Glucose Solution:
Composition per 10mL:
D-Glucose 1.0g

Preparation of Glucose Solution: Add D-glucose to distilled/deionized water and bring volume to 10.0mL. Mix thoroughly. Filter sterilize.

Preparation of Medium: Add components, except antibiotic inhibitor solution and glucose solution, to distilled/deionized water and bring volume to 980.0mL. Mix thoroughly. Gently heat and bring to boiling. Autoclave for 15 min at 15 psi pressure–121°C. Cool to 45°–50°C. Aseptically add sterile antibiotic inhibitor solution and glucose solution. Mix thoroughly. Aseptically distribute into sterile tubes or flasks.

Use: For the cultivation of *Escherichia coli.*

Heart Infusion Broth with Horse Serum and Fresh Yeast Extract

Composition per 930mL:

Heart infusion broth	720.0mL
Horse serum, unheated	200.0mL
Fresh yeast extract solution	10.0mL

pH 7.4 ± 0.2 at 25°C

Heart Infusion Broth:
Composition per liter:

Beef heart, infusion from	500.0g
Tryptose	10.0g
NaCl	5.0g

Preparation of Heart Infusion Broth: Add components to distilled/deionized water and bring volume to 1.0L. Mix thoroughly. Autoclave for 15 min at 15 psi pressure–121°C. Cool to 45°–50°C.

Fresh Yeast Extract Solution:
Composition per 100mL:

Baker's yeast, live, pressed, starch-free	25.0g

Preparation of Fresh Yeast Extract Solution: Add the live Baker's yeast to 100.0mL of distilled/deionized water. Autoclave for 90 min at 15 psi pressure–121°C. Allow to stand. Remove supernatant solution. Adjust pH to 6.6–6.8.

Preparation of Medium: To 720.0mL of sterile cooled heart infusion broth, aseptically add 200.0mL of horse serum and 10.0mL of fresh yeast extract solution. Mix thoroughly. Aseptically distribute into sterile tubes or flasks.

Use: For the cultivation and maintenance of *Mycoplasma equigenitalium* and *Mycoplasma subdolum.*

Heart Infusion Broth with Horse Serum, Fresh Yeast Extract and Penicillin

Composition per 940mL:

Heart infusion broth	720.0mL
Horse serum, unheated	200.0mL
Fresh yeast extract solution	10.0mL
Penicillin solution	10.0mL

pH 7.4 ± 0.2 at 25°C

Heart Infusion Broth:
Composition per liter:

Beef heart, infusion from	500.0g
Tryptose	10.0g
NaCl	5.0g

Preparation of Heart Infusion Broth: Add components to distilled/deionized water and bring volume to 1.0L. Mix thoroughly. Autoclave for 15 min at 15 psi pressure–121°C. Cool to 45°–50°C.

Fresh Yeast Extract Solution:
Composition per 100mL:

Baker's yeast, live, pressed, starch-free	25.0g

Preparation of Fresh Yeast Extract Solution: Add the live Baker's yeast to 100.0mL of distilled/deionized water. Autoclave for 90 min at 15 psi pressure–121°C. Allow to stand. Remove supernatant solution. Adjust pH to 6.6–6.8.

Penicillin Solution:
Composition per 10mL:

Penicillin	100,000U

Preparation of Penicillin Solution: Add penicillin to distilled/deionized water and bring volume to 10.0mL. Mix thoroughly. Filter sterilize.

Preparation of Medium: To 720.0mL of sterile cooled heart infusion broth, aseptically add 200.0mL of horse serum, 10.0mL of fresh yeast extract solution, and 10.0mL of sterile penicillin solution. Mix thoroughly. Aseptically distribute into sterile tubes or flasks.

Use: For the cultivation and maintenance of *Mycoplasma equigenitalium* and *Mycoplasma subdolum.*

Heart Infusion Broth with Human Serum and Fresh Yeast Extract

Composition per 930mL:

Heart infusion broth	720.0mL
Human serum, unheated	200.0mL
Fresh yeast extract solution	10.0mL

pH 7.4 ± 0.2 at 25°C

Heart Infusion Broth:
Composition per liter:

Beef heart, infusion from	500.0g
Tryptose	10.0g
NaCl	5.0g

Preparation of Heart Infusion Broth: Add components to distilled/deionized water and bring volume to 1.0L. Mix thoroughly. Autoclave for 15 min at 15 psi pressure–121°C. Cool to 45°–50°C.

Fresh Yeast Extract Solution:
Composition per 100mL:

Baker's yeast, live, pressed, starch-free	25.0g

Preparation of Fresh Yeast Extract Solution: Add the live Baker's yeast to 100.0mL of distilled/deionized water. Autoclave for 90 min at 15 psi pressure–121°C. Allow to stand. Remove supernatant solution. Adjust pH to 6.6–6.8.

Preparation of Medium: To 720.0mL of sterile cooled heart infusion broth, aseptically add 200.0mL of human serum and 10.0mL of fresh yeast extract solution. Mix thoroughly. Aseptically distribute into sterile tubes or flasks.

Use: For the cultivation and maintenance of *Mycoplasma equigenitalium* and *Mycoplasma subdolum*.

Heart Infusion Broth with Inactivated Horse Serum

Composition per liter:

Beef heart, infusion from	500.0g
Tryptose	10.0g
NaCl	5.0g
Horse serum, inactivated	100.0mL

pH 7.6 ± 0.2 at 25°C

Preparation of Medium: Add components, except horse serum, to distilled/deionized water and bring volume to 900.0mL. Mix thoroughly. Gently heat and bring to boiling. Autoclave for 15 min at 15 psi pressure–121°C. Cool to 45°–50°C. Aseptically add sterile horse serum. Mix thoroughly. Aseptically distribute into sterile tubes or flasks.

Use: For the cultivation and maintenance of *Enterococcus faecium*.

Heart Infusion Broth with Inactivated Horse Serum and Fresh Yeast Extract

Composition per liter:

Beef heart, infusion from	500.0g
Tryptose	10.0g
NaCl	5.0g
Horse serum, inactivated	200.0mL
Fresh yeast extract solution	100.0mL

pH 7.5 ± 0.2 at 25°C

Source: This medium is available as a premixed powder from Difco Laboratories.

Fresh Yeast Extract Solution:
Composition per 100mL:

Baker's yeast, live, pressed, starch-free	25.0g

Preparation of Fresh Yeast Extract Solution: Add the live Baker's yeast to 100.0mL of distilled/deionized water. Autoclave for 90 min at 15 psi pressure–121°C. Allow to stand. Remove supernatant solution. Adjust pH to 6.6–6.8.

Preparation of Medium: Add components, except horse serum and fresh yeast extract solution, to distilled/deionized water and bring volume to 700.0mL. Mix thoroughly. Gently heat and bring to boiling. Autoclave for 15 min at 15 psi pressure–121°C. Cool to 45°–50°C. Aseptically add sterile horse serum and fresh yeast extract solution. Mix thoroughly. Aseptically distribute into sterile tubes or flasks.

Use: For the cultivation and maintenance of *Acholeplasma* species, *Mycoplasma* species, and *Streptobacillus* species.

Heart Infusion Broth with Inactivated Horse Serum, Fresh Yeast Extract, and Sucrose

Composition per liter:

Beef heart, infusion from	500.0g
Sucrose	40.0g
Tryptose	10.0g
NaCl	5.0g
Horse serum, inactivated	200.0mL
Fresh yeast extract solution	100.0mL

pH 7.5 ± 0.2 at 25°C

Source: This medium is available as a premixed powder from Difco Laboratories.

Fresh Yeast Extract Solution:
Composition per 100mL:

Baker's yeast, live, pressed, starch-free	25.0g

Preparation of Fresh Yeast Extract Solution: Add the live Baker's yeast to 100.0mL of distilled/deionized water. Autoclave for 90 min at 15 psi pressure–121°C. Allow to stand. Remove supernatant solution. Adjust pH to 6.6–6.8.

Preparation of Medium: Add components, except horse serum and fresh yeast extract solution, to distilled/deionized water and bring volume to 700.0mL. Mix thoroughly. Gently heat and bring to boiling. Autoclave for 15 min at 15 psi pressure–121°C. Cool to 45°–50°C. Aseptically add sterile horse serum and fresh yeast extract solution. Mix thoroughly. Aseptically distribute into sterile tubes or flasks.

Use: For the cultivation and maintenance of *Acholeplasma* species, *Mycoplasma* species, and *Streptobacillus* species.

Heart Infusion Broth with Porcine Serum and Fresh Yeast Extract

Composition per liter:

Beef heart, infusion from	500.0g
Tryptose	10.0g
NaCl	5.0g

Porcine serum, inactivated 200.0mL
Fresh yeast extract (25% w/v solution) 100.0mL
pH 7.5 ± 0.2 at 25°C

Porcine Serum:
Composition per 200mL:
Porcine serum 200.0mL

Preparation of Porcine Serum: Adjust the pH of 200.0mL of porcine serum to 4.4 with sterile 1*N* HCl. Do not overshoot pH below 4.2. Allow serum to stand at 4°C for 18–20 hr. Adjust pH to 7.0 with sterile NaOH. Aseptically centrifuge at 9,000 rpm for 20 min. Discard sediment. Filter supernatant solution through 0.85μm filter. Filter sterilize through a 0.22μm filter. Store at –70°C.

Fresh Yeast Extract Solution:
Composition per 100mL:
Baker's yeast, live, pressed, starch-free 25.0g

Preparation of Fresh Yeast Extract Solution: Add the live Baker's yeast to 100.0mL of distilled/deionized water. Autoclave for 90 min at 15 psi pressure–121°C. Allow to stand. Remove supernatant solution. Adjust pH to 6.6–6.8.

Preparation of Medium: Add components, except porcine serum and fresh yeast extract solution, to distilled/deionized water and bring volume to 700.0mL. Mix thoroughly. Gently heat and bring to boiling. Autoclave for 15 min at 15 psi pressure–121°C. Cool to 45°–50°C. Aseptically add sterile porcine serum and fresh yeast extract solution. Mix thoroughly. Aseptically distribute into sterile tubes or flasks.

Use: For the cultivation of *Acholeplasma* species.

Heart Infusion Medium with Fetal Bovine Serum

Composition per liter:
Beef heart, infusion from 500.0g
Agar 15.0g
Tryptose 10.0g
NaCl 5.0g
Fetal bovine serum 100.0mL
pH 7.4 ± 0.2 at 25°C

Preparation of Medium: Add components, except fetal bovine serum, to distilled/deionized water and bring volume to 900.0mL. Mix thoroughly. Gently heat and bring to boiling. Autoclave for 15 min at 15 psi pressure–121°C. Cool to 45°–50°C. Aseptically add sterile fetal bovine serum. Mix thoroughly. Aseptically distribute into sterile tubes or flasks.

Use: For the cultivation of *Haemophilus ducreyi*.

Heart Infusion Tyrosine Agar

Composition per liter:
Beef heart, infusion from 500.0g
Agar 15.0g
Tryptose 10.0g
NaCl 5.0g
L-Tyrosine 1.0g
pH 7.4 ± 0.2 at 25°C

Preparation of Medium: Add components to distilled/deionized water and bring volume to 1.0L. Mix thoroughly. Gently heat and bring to boiling. Distribute into tubes. Autoclave for 15 min at 15 psi pressure–121°C. Allow tubes to cool in a slanted position.

Use: For the cultivation and differentiation of *Bordetella parapertussis* based on browning of blood-free medium.

Hektoen Enteric Agar

Composition per liter:
Agar 13.5g
Lactose 12.0g
Peptic digest of animal tissue 12.0g
Sucrose 12.0g
Bile salts 9.0g
NaCl 5.0g
$Na_2S_2O_3$ 5.0g
Yeast extract 3.0g
Salicin 2.0g
Ferric ammonium citrate 1.5g
Acid Fuchsin 0.1g
Bromthymol Blue 0.064g
pH 7.6 ± 0.2 at 25°C

Source: This medium is available as a premixed powder from BBL Microbiology Systems, Difco Laboratories and Oxoid Unipath.

Caution: Acid Fuchsin is a potential carcinogen and care must be taken to avoid inhalation of the powdered dye and contact with the skin.

Preparation of Medium: Add components to distilled/deionized water and bring volume to 1.0L. Mix thoroughly. Gently heat while stirring until components are dissolved. Do not autoclave. Pour into sterile Petri dishes. Allow agar to solidify with the Petri dish covers partially off.

Use: For the isolation and cultivation of Gram-negative enteric microorganisms from a variety of clinical and nonclinical specimens based on lactose or sucrose fermentation and H_2S production. It is used for the isolation and differentiation of *Salmonella* and *Shigella*. Bacteria that ferment lactose or sucrose appear as yellow to orange colonies. Bacteria that produce H_2S appear as colonies with black centers.

Helicobacter pylori Isolation Agar

Composition per liter:

Agar	15.0g
Bitone	10.0g
Pancreatic digest of casein	5.0g
NaCl	5.0g
Peptic digest of animal tissue	5.0g
Tryptic digest of beef heart	3.0g
Cornstarch	1.0g
Horse blood, laked	35.0mL
Antibiotic inhibitor solution	10.0mL

pH 7.3 ± 0.2 at 25°C

Antibiotic Inhibitor Solution:
Composition per 10mL:

Vancomycin	0.010g
Amphotericin B	5.0mg
Cefsulodin	5.0mg
Trimethoprim lactate	5.0mg

Preparation of Antibiotic Inhibitor Solution: Add components to distilled/deionized water and bring volume to 10.0mL. Mix thoroughly. Filter sterilize.

Preparation of Medium: Add components, except horse blood and antibiotic inhibitor solution, to distilled/deionized water and bring volume to 955.0mL. Mix thoroughly. Gently heat and bring to boiling. Autoclave for 15 min at 15 psi pressure–121°C. Cool to 45°–50°C. Aseptically add sterile horse blood and sterile antibiotic inhibitor solution. Mix thoroughly. Pour into sterile Petri dishes or distribute into sterile tubes.

Use: For the isolation and cultivation of *Helicobacter pylori* from clinical specimens.

Hemin Medium for *Mycobacterium*

See: **Middlebrook 7H10 Agar with Middlebrook OADC Enrichment and Hemin**

Hemo ID Quad Plate with Growth Factors (*Hemophilus* Identification Quadrant Plate with Growth Factors)

Composition per plate:

Quadrant I	5.0mL
Quadrant II	5.0mL
Quadrant III	5.0mL
Quadrant IV	5.0mL

Quadrant I:
Composition per 5mL:

Hemin	0.1mg
Brain heart infusion agar	5.0mL

Quadrant II:
Composition per 5mL:

Brain heart infusion agar	5.0mL
Supplement solution	0.05mL

Quadrant III:
Composition per 5mL:

Hemin	0.1mg
Brain heart infusion agar	5.0mL
Supplement solution	0.05mL

Quadrant IV:
Composition per 5mL:

Hemin	0.1mg
Brain heart infusion agar	5.0mL
Horse blood	0.25mL
Supplement solution	0.05mL

Source: The supplement solution IsoVitaleX® enrichment is available from BBL Microbiology Laboratories. This enrichment may be replaced by supplement VX from Difco Laboratories.

Use: For the differentiation and presumptive identification of *Haemophilus* species. The Hemo ID Quad Plate is a four-sectored plate each containing a different medium.

Hemorrhagic coli Agar (HC Agar)

Composition per liter:

Sorbitol	20.0g
Pancreatic digest of casein	20.0g
Agar	15.0g
NaCl	5.0g
Bile salts No. 3	1.12g
Bromcresol Purple	0.015g

pH 7.2 ± 0.2 at 25°C

Preparation of Medium: Add components to distilled/deionized water and bring volume to 1.0L. Mix thoroughly. Gently heat and bring to boiling. Distribute into tubes or flasks. Autoclave for 15 min at 15 psi pressure–121°C. Pour into sterile Petri dishes.

Use: For the isolation and cultivation of enterohemorraghic *Escherichia coli* from food.

Heparin Medium

Composition per liter:

Agar	15.0g
Pancreatic digest of casein	3.5g

NaCl	1.0g
Pancreatic digest of soybean meal	0.6g
Glucose	0.5g
K_2HPO_4	0.5g
Heparin solution	10.0mL

pH 6.5 ± 0.2 at 25°C

Heparin Solution:

Composition per 10mL:

Heparin	0.02g

Preparation of Heparin Solution: Add heparin to distilled/deionized water and bring volume to 10.0mL. Mix thoroughly. Filter sterilize.

Preparation of Medium: Add components, except heparin solution, to distilled/deionized water and bring volume to 990.0mL. Mix thoroughly. Gently heat and bring to boiling. Autoclave for 15 min at 15 psi pressure–121°C. Cool to 45°–50°C. Aseptically add sterile heparin solution. Mix thoroughly. Pour into sterile Petri dishes or distribute into sterile tubes.

Use: For the cultivation of *Flavobacterium leparinum.*

Herellea Agar

Composition per liter:

Agar	16.0g
Pancreatic digest of casein	15.0g
Lactose	10.0g
Maltose	10.0g
Enzymatic digest of soybean meal	5.0g
NaCl	5.0g
Bile salts	1.25g
Bromcresol Purple	0.020g

pH 6.8 ± 0.2 at 25°C

Source: This medium is available as a premixed powder from Difco Laboratories.

Preparation of Medium: Add components to distilled/deionized water and bring volume to 1.0L. Mix thoroughly. Gently heat and bring to boiling. Distribute into tubes or flasks. Autoclave for 15 min at 15 psi pressure–121°C. Pour into sterile Petri dishes or leave in tubes.

Use: For the isolation, cultivation and differentiation of Gram-negative nonfermentative and fermentative bacteria. It is especially recommended for the differentiation of *Acinetobacter (Herellea)* species from *Neisseria gonorrhoeae* in urethral or vaginal specimens. Fermentative bacteria appear as yellow colonies surrounded by yellow zones. Nonfermentative bacteria, such as *Acinetobacter* species appear as pale lavender colonies.

Hershey's Tris–Buffered Salts Medium

Composition per liter:

Tris(hydroxymethyl)aminomethane buffer (0.1M solution)	12.1g
NaCl	5.4g
KCl	3.0g
NH_4Cl	1.1g
$MgCl_2$	0.095g
KH_2PO_4	0.087g
Na_2SO_4	0.023g
$CaCl_2$	0.011g
$FeCl_3$	0.16mg
Glucose solution	100.0mL

pH 7.4 ± 0.2 at 25°C

Glucose Solution:

Composition per 100mL:

Glucose	2.0g

Preparation of Glucose Solution: Add glucose to distilled/deionized water and bring volume to 100.0mL. Mix thoroughly. Autoclave for 15 min at 15 psi pressure–121°C. Cool to 25°C.

Preparation of Medium: Add components, except glucose solution, to distilled/deionized water and bring volume to 900.0mL. Mix thoroughly. Gently heat and bring to boiling. Autoclave for 15 min at 15 psi pressure–121°C. Cool to 25°C. Aseptically add sterile glucose solution. Mix thoroughly. Aseptically distribute into sterile tubes or flasks.

Use: For the cultivation of a variety of heterotrophic microorganisms.

Heterotrophic Medium for *Hydrogenomonas*

Composition per liter:

Agar	15.0g
Tryptose	5.0g
Cornstarch	2.0g
Sodium succinate·$6H_2O$	2.0g
Sodium glutamate	1.0g
Yeast extract	1.0g
Sodium citrate·$2H_2O$	0.5g
Sodium acetate·$3H_2O$	0.3g
KH_2PO_4	0.2g
$MgSO_4$	0.1g

pH 6.8–7.2 at 25°C

Preparation of Medium: Add components to distilled/deionized water and bring volume to 1.0L. Mix thoroughly. Gently heat and bring to boiling. Distribute into tubes or flasks. Autoclave for 15 min at 15 psi pressure–121°C. Pour into sterile Petri dishes or leave in tubes.

Use: For the heterotrophic cultivation of *Hydrogenomonas* species.

Heterotrophic Medium for Hydrogen–Oxidizing Bacteria

Composition per 1010mL:

Solution A	900.0mL
Solution C	100.0mL
Solution B	10.0mL

Solution A:

Composition per 900mL:

Noble agar	17.0g
$Na_2HPO_4 \cdot 12H_2O$	9.0g
KH_2PO_4	1.5g
NH_4Cl	1.0g
$MgSO_4 \cdot 7H_2O$	0.2g
Trace elements solution SL-6	1.0mL

Preparation of Solution A: Add components to distilled/deionized water and bring volume to 900.0mL. Mix thoroughly. Autoclave for 15 min at 15 psi pressure–121°C. Cool to 45°–50°C.

Trace Elements Solution SL-6:

Composition per liter:

H_3BO_3	0.3g
$CoCl_2 \cdot 6H_2O$	0.2g
$ZnSO_4 \cdot 7H_2O$	0.1g
$MnCl_2 \cdot 4H_2O$	0.03g
$Na_2MoO_4 \cdot H_2O$	0.03g
$NiCl_2 \cdot 6H_2O$	0.02g
$CuCl_2 .2H_2O$	0.01g

Preparation of Trace Elements Solution SL-6: Add components to distilled/deionized water and bring volume to 1.0L. Mix thoroughly. Adjust pH to 3.4.

Solution B:

Composition per 10mL:

$CaCl_2 \cdot 2H_2O$	0.01g
Ferric ammonium citrate	5.0mg

Preparation of Solution B: Add components to distilled/deionized water and bring volume to 10.0mL. Mix thoroughly. Autoclave for 15 min at 15 psi pressure–121°C. Cool to 45°–50°C.

Solution C:

Composition per 100mL:

Sodium 3-hydroxybutyrate	2.0g

Preparation of Solution C: Add sodium 3-hydroxybutyrate to distilled/deionized water and bring volume to 100.0mL. Mix thoroughly. Filter sterilize. Warm to 45°–50°C.

Preparation of Medium: Aseptically combine 900.0mL of sterile solution A, 10.0mL of sterile solution B and 100.0mL of sterile solution C. Mix thoroughly. Pour into sterile Petri dishes or distribute into sterile tubes.

Use: For the heterotrophic cultivation and maintenance of *Xanthobacter agilis*.

Heterotrophic Plate Count
See: **HPC Agar**

HHD Medium

Composition per liter:

Agar	20.0g
Pancreatic digest of casein	10.0g
Casamino acids	3.0g
Fructose	2.5g
KH_2PO_4	2.5g
Papaic digest of soybean meal	1.5g
Tween™ 80	1.0g
Yeast extract	1.0g
Bromcresol Green solution	20.0mL

pH 7.0 ± 0.2 at 25°C

Bromcresol Green Solution:

Composition per 30mL:

Bromcresol Green	0.1g
NaOH (0.01*N* solution)	30.0mL

Preparation of Bromcresol Green Solution: Add Bromcresol Green to 30.0mL of NaOH solution. Mix thoroughly. Filter sterilize.

Preparation of Medium: Add components to distilled/deionized water and bring volume to 1.0L. Mix thoroughly. Gently heat and bring to boiling. Adjust pH to 7.0. Distribute into tubes or flasks. Autoclave for 15 min at 15 psi pressure–121°C. Pour into sterile Petri dishes or leave in tubes.

Use: For the cultivation of *Salmonella* species from foods.

Hickey–Tresner Agar

Composition per liter:

Agar	15.0g
Dextrin	10.0g
Pancreatic digest of casein	2.0g
Meat extract	1.0g
Yeast extract	1.0g
$CaCl_2$	2.0mg

pH 7.2 ± 0.2 at 25°C

Preparation of Medium: Add components to distilled/deionized water and bring volume to 1.0L. Mix thoroughly. Gently heat and bring to boiling. Distribute into tubes or flasks. Autoclave for 15 min at 15 psi

pressure–121°C. Pour into sterile Petri dishes or leave in tubes.

Use: For the cultivation and maintenance of *Thermomonospora curvata*.

Hippurate Broth
See: Sodium Hippurate Broth

Histoplasma capsulatum Agar

Composition per liter:

Agar	12.5g
Glucose	10.0g
Citric acid	10.0g
Potato starch	2.0g
α-Ketoglutaric acid	1.0g
L-Cystine·HCl·H_2O	1.0g
Glutathione, reduced	0.5g
L-Asparagine	0.1g
L-Tryptophan	0.02g
Solution 1	250.0mL
Solution 3	40.0mL
Solution 2	10.0mL
Solution 4	10.0mL
Solution 8	10.0mL
Solution 5	1.0mL
Solution 6	0.1mL
Solution 7	0.1mL

pH 6.5 ± 0.2 at 25°C

Solution 1:

Composition per liter:

KH_2PO_4	8.0g
$(NH_4)_2SO_4$	8.0g
$MgSO_4·7H_2O$	0.86g
$CaCl_2$, anhydrous	0.08g
$ZnSO_4·7H_2O$	0.05g

Preparation of Solution 1: Add components to distilled/deionized water and bring volume to 500.0mL. Mix thoroughly. Bring volume to 1.0L with distilled/deionized water. Store at 5°C.

Solution 2:

Composition per liter:

$FeSO_4·7H_2O$	5.70g
$MnCl_2·6H_2O$	0.80g
$NaMoO_4·2H_2O$	0.15g
HCl, concentrated	1.0mL

Preparation for Solution 2: Add 1.0mL of concentrated HCl to 100.0mL of distilled water in a 1.0L volumetric flask. Dissolve each component completely in the sequence given. Bring volume to 1.0L with distilled/deionized water. Store at 5°C. Discard if red color or red precipitate appears.

Solution 3:

Composition per 100mL:

Casein, acid-hydrolyzed, vitamin-free	10.0g

Preparation for Solution 3: Add casein to distilled/deionized water and bring volume to 100.0mL. Do not use enzymatically-digested casein.

Solution 4:

Composition per liter:

Calcium pantothenate	0.2g
Inositol	0.2g
Riboflavin	0.2g
Thiamine·HCl	0.2g
Nicotinamide	0.1g
Biotin	0.01g

Preparation for Solution 4: Add components to distilled/deionized water and bring volume to 1.0L. Mix thoroughly. Store at –20°C.

Solution 5:

Composition per 100mL:

Hemin	0.2g
NH_4OH, concentrated	0.3mL

Preparation for Solution 5: Add hemin to approximately 30.0mL of distilled/deionized water. Add NH_4OH. Mix thoroughly until dissolved. Bring volume to 100.0mL with distilled/deionized water. Store at 5° C.

Solution 6:

Composition per 10mL:

DL-Thioctic acid	0.01g
Ethanol (95% solution)	10.0mL

Preparation for Solution 6: Add DL-thioctic acid to 10.0mL of ethanol. Mix thoroughly. Store at –20°C.

Solution 7:

Composition per 10mL:

Coenzyme A	0.01g
$Na_2S·5H_2O$ (0.05% solution)	0.2mL

Preparation for Solution 7: Prepare $Na_2S·5H_2O$ solution in freshly boiled distilled/deionized water. Add coenzyme A to 9.8mL of distilled/deionized water. Mix thoroughly. Add freshly prepared $Na_2S·5H_2O$ solution. Mix thoroughly. Store the solution at –20°C.

Solution 8:

Composition per 100mL:

Oleic acid	0.10g

Preparation for Solution 8: Add oleic acid to 50.0mL of distilled/deionized water. Adjust pH to 9.0 with NaOH. Gently heat until dissolved. Bring volume to 100.0mL with distilled/deionized water. Store at 5°C.

Preparation of Medium: Add components—except agar, potato starch, and solution 8—to distilled/deionized water and bring volume to 400.0mL. Mix thoroughly. Adjust pH to 6.5 with 20% KOH solution. Filter sterilize. In a separate flask add potato starch to 50.0mL of distilled/deionized water. Add the starch solution to 450.0mL of boiling distilled/deionized water. Add 10.0mL of solution 8 and the agar. Mix thoroughly. Autoclave for 15 min at 15 psi pressure–121°C. Cool to 70°C. Aseptically combine the two sterile solutions. Pour into sterile Petri dishes or distribute into sterile tubes.

Use: For cultivation and maintenance of *Histoplasma capsulatum* in the yeast phase. Also for the cultivation of *Histoplasma duboisii, Blastomyces dermatitidis* and *Sprotrichum schenckii.*

Histoplasma capsulatum Agar

Composition per liter:

Agar	15.0g
Glucose	10.0g
Potato starch	2.0g
α-Ketoglutaric acid	1.0g
L-Cystine·HCl·H_2O	1.0g
Glutathione, reduced	0.5g
L-Asparagine	0.1g
L-Tryptophan	0.02g
Solution 1	250.0mL
Solution 3	40.0mL
Solution 2	10.0mL
Solution 4	10.0mL
Solution 8	10.0mL
Solution 5	1.0mL
Solution 6	0.1mL
Solution 7	0.1mL

pH 6.5 ± 0.2 at 25°C

Solution 1:

Composition per liter:

KH_2PO_4	8.0g
$(NH_4)_2SO_4$	8.0g
$MgSO_4 \cdot 7H_2O$	0.86g
$CaCl_2$, anhydrous	0.08g

Preparation of Solution 1: Add components to distilled/deionized water and bring volume to 500.0mL. Mix thoroughly. Bring volume to 1.0L with distilled/deionized water. Store at 5°C.

Solution 2:

Composition per liter:

$FeSO_4 \cdot 7H_2O$	5.70g
$MnCl_2 \cdot 6H_2O$	0.80g
$NaMoO_4 \cdot 2H_2O$	0.15g
HCl, concentrated	1.0mL

Preparation for Solution 2: Add the 1.0mL of concentrated HCl to 100.0mL of distilled water in a 1.0L volumetric flask. Dissolve each component completely in the sequence given. Bring volume to 1.0L with distilled/deionized water. Store at 5°C. Discard if red color or red precipitate appears.

Solution 3:

Composition per 100mL:

Casein, acid-hydrolyzed, vitamin-free	10.0g

Preparation for Solution 3: Add casein to distilled/deionized water and bring volume to 100.0mL. Do not use enzymatically-digested casein.

Solution 4:

Composition per liter:

Calcium pantothenate	0.2g
Inositol	0.2g
Riboflavin	0.2g
Thiamine·HCl	0.2g
Nicotinamide	0.1g
Biotin	0.01g

Preparation for Solution 4: Add components to distilled/deionized water and bring volume to 1.0L. Mix thoroughly. Store at –20°C.

Solution 5:

Composition per 100mL:

Hemin	0.2g
NH_4OH, concentrated	0.3mL

Preparation for Solution 5: Add hemin to approximately 30.0mL of distilled/deionized water. Add NH_4OH. Mix thoroughly until dissolved. Bring volume to 100.0mL with distilled/deionized water. Store at 5° C.

Solution 6:

Composition per 10mL:

DL-Thioctic acid	0.01g
Ethanol (95% solution)	10.0mL

Preparation for Solution 6: Add DL-thioctic acid to 10.0mL of ethanol. Mix thoroughly. Store at –20°C.

Solution 7:

Composition per 10mL:

Coenzyme A	0.01g
$Na_2S \cdot 5H_2O$ (0.05% solution)	0.2mL

Preparation for Solution 7: Prepare $Na_2S \cdot 5H_2O$ solution in freshly boiled distilled/deionized water. Add coenzyme A to 9.8mL of distilled/deionized water. Mix thoroughly. Add freshly prepared $Na_2S \cdot 5H_2O$ solution. Mix thoroughly. Store the solution at –20°C.

Solution 8:

Composition per 100mL:

Oleic acid	0.10g

Preparation for Solution 8: Add oleic acid to 50.0mL of distilled/deionized water. Adjust pH to 9.0 with NaOH. Gently heat until dissolved. Bring volume to 100.0mL with distilled/deionized water. Store at 5°C.

Preparation of Medium: Add components—except agar, potato starch, and solution 8—to distilled/deionized water and bring volume to 400.0mL. Mix thoroughly. Adjust pH to 6.5 with 20% KOH solution. Filter sterilize. In a separate flask add potato starch to 50.0mL of distilled/deionized water. Add the starch solution to 450.0mL of boiling distilled/deionized water. Add 10.0mL of solution 8 and the agar. Mix thoroughly. Autoclave for 15 min at 15 psi pressure–121°C. Cool to 70°C. Aseptically combine the two sterile solutions. Pour into sterile Petri dishes or distribute into sterile tubes.

Use: For cultivation and maintenance of *Histoplasma capsulatum* in the mycelial phase.

Histoplasma capsulatum Broth

Composition per liter:

Glucose	10.0g
Citric acid	10.0g
α-Ketoglutaric acid	1.0g
L-Cystine·HCl·H_2O	1.0g
Potato starch	0.5g
Glutathione, reduced	0.5g
L-Asparagine	0.1g
L-Tryptophan	0.02g
Solution 1	250.0mL
Solution 3	40.0mL
Solution 2	10.0mL
Solution 4	10.0mL
Solution 5	1.0mL
Solution 8	1.0mL
Solution 6	0.1mL
Solution 7	0.1mL

pH 6.5 ± 0.2 at 25°C

Solution 1:

Composition per liter:

KH_2PO_4	8.0g
$(NH_4)_2SO_4$	8.0g
$MgSO_4 \cdot 7H_2O$	0.86g
$CaCl_2$, anhydrous	0.08g
$ZnSO_4 \cdot 7H_2O$	0.05g

Preparation of Solution 1: Add components to distilled/deionized water and bring volume to 500.0mL. Mix thoroughly. Bring volume to 1.0L with distilled/deionized water. Store at 5°C.

Solution 2:

Composition per liter:

$FeSO_4 \cdot 7H_2O$	5.70g
$MnCl_2 \cdot 6H_2O$	0.80g
$NaMoO_4 \cdot 2H_2O$	0.15g
HCl, concentrated	1.0mL

Preparation for Solution 2: Add 1.0mL of concentrated HCl to 100mL of distilled water in a 1.0L volumetric flask. Dissolve each component completely in the sequence given. Bring volume to 1.0L with distilled/deionized water. Store at 5°C. Discard if red color or red precipitate appears.

Solution 3:

Composition per 100mL:

Casein, acid-hydrolyzed, vitamin-free	10.0g

Preparation for Solution 3: Add casein to distilled/deionized water and bring volume to 100.0mL. Do not use enzymatically-digested casein.

Solution 4:

Composition per liter:

Calcium pantothenate	0.2g
Inositol	0.2g
Riboflavin	0.2g
Thiamine·HCl	0.2g
Nicotinamide	0.1g
Biotin	0.01g

Preparation for Solution 4: Add components to distilled/deionized water and bring volume to 1.0L. Mix thoroughly. Store at –20°C.

Solution 5:

Composition per 100mL:

Hemin	0.2g
NH_4OH, concentrated	0.3mL

Preparation for Solution 5: Add hemin to approximately 30.0mL of distilled/deionized water. Add NH_4OH. Mix thoroughly until dissolved. Bring volume to 100.0mL with distilled/deionized water. Store at 5° C.

Solution 6:

Composition per 10mL:

DL-Thioctic acid	0.01g
Ethanol (95% solution)	10.0mL

Preparation for Solution 6: Add DL-thioctic acid to 10.0mL of ethanol. Mix thoroughly. Store at –20°C.

Solution 7:

Composition per 10mL:

Coenzyme A	0.01g
$Na_2S \cdot 5H_2O$ (0.05% solution)	0.2mL

Preparation for Solution 7: Prepare $Na_2S \cdot 5H_2O$ solution in freshly boiled distilled/deionized water. Add coenzyme A to 9.8mL of distilled/deionized water. Mix thoroughly. Add freshly prepared $Na_2S \cdot 5H_2O$ solution. Mix thoroughly. Store the solution at –20°C.

Solution 8:
Composition per 100mL:
Oleic acid 0.10g

Preparation for Solution 8: Add oleic acid to 50.0mL of distilled/deionized water. Adjust pH to 9.0 with NaOH. Gently heat until dissolved. Bring volume to 100.0mL with distilled/deionized water. Store at 5°C.

Preparation of Medium: Add components—except potato starch, and solution 8—to distilled/deionized water and bring volume to 400.0mL. Mix thoroughly. Adjust pH to 6.5 with 20% KOH solution. Filter sterilize. In a separate flask add potato starch to 50.0mL of distilled/deionized water. Add the starch solution to 450.0mL of boiling distilled/deionized water. Add 1.0mL of solution 8. Mix thoroughly. Autoclave for 15 min at 15 psi pressure–121°C. Cool to 70°C. Aseptically combine the two sterile solutions. Pour into sterile Petri dishes or distribute into sterile tubes.

Use: For cultivation of *Histoplasma capsulatum* in the yeast phase. Also for the cultivation of *Histoplasma duboisii, Blastomyces dermatitidis* and *Sprotrichum schenckii.*

HL Agar

Composition per plate:
Columbia agar base 10.0mL
Columbia blood top agar 5.0mL
pH 7.3 ± 0.2 at 25°C

Columbia Agar Base:
Composition per liter:
Agar 13.5g
Pancreatic digest of casein 12.0g
NaCl 5.0g
Peptic digest of animal tissue 5.0g
Beef extract 3.0g
Yeast extract 3.0g
Cornstarch 1.0g

Preparation of Columbia Agar Base: Add components to distilled/deionized water and bring volume to 1.0L. Mix thoroughly. Gently heat until boiling. Autoclave for 15 min at 15 psi pressure–121°C. Cool to 45°–50°C.

Columbia Blood Top Agar:
Composition per liter:
Agar 13.5g
Pancreatic digest of casein 12.0g
NaCl 5.0g
Peptic digest of animal tissue 5.0g
Beef extract 3.0g
Yeast extract 3.0g
Cornstarch 1.0g
Horse blood, defibrinated 50.0mL

Preparation of Columbia Blood Top Agar: Add components, except horse blood, to distilled/deionized water and bring volume to 950.0mL. Mix thoroughly. Gently heat until boiling. Autoclave for 15 min at 15 psi pressure–121°C. Cool to 45°–50°C. Aseptically add sterile horse blood. Mix thoroughly.

Preparation of Medium: Pour cooled, sterile Columbia agar base into sterile Petri dishes in 10.0mL volumes. Allow agar to solidify. Pour 5.0mL of cooled, sterile Columbia blood top agar over Columbia agar base which has solidified but is still warm.

Use: For the cultivation of *Listeria monocytogenes* from foods.

Hohn's Medium, Modified
See: **Steenken and Smith Agar**

HO-LE Trace Element Solution

Composition per liter:
H_3BO_3 .... 2.85g
$MnCl_2 \cdot 4H_2O$ 1.8g
Sodium tartrate 1.77g
$FeSO_4$ 1.36g
$CoCl_2 \cdot 6H_2O$ 0.04g
$CuCl_2 \cdot 2H_2O$ 0.026g
$Na_2MoO_4 \cdot 2H_2O$ 0.025g
$ZnCl_2$ 0.021g

Preparation of Medium: Add components to distilled/deionized water and bring volume to 1.0L. Mix thoroughly. Distribute into tubes or flasks. Autoclave for 15 min at 15 psi pressure–121°C.

Use: For use as an enrichment to other media that require trace minerals.

Hominis Agar
See: **H Agar**

Hominis Broth
See: **H Broth**

Horie Arabinose Ethyl Violet Broth (HAEB)

Composition per liter:
NaCl 30.0g
Peptone 5.0g

Beef extract 3.0g
Bromthymol Blue.. 0.03g
Ethyl Violet 1.0mg
Arabinose solution 100.0mL

pH 9.0 ± 0.2 at 25°C

Arabinose Solution:

Composition per 100mL:

Arabinose 5.0g

Preparation of Arabinose Solution: Add arabinose to distilled/deionized water and bring volume to 100.0mL. Mix thoroughly. Filter sterilize.

Preparation of Medium: Add components, except arabinose solution, to distilled/deionized water and bring volume to 900.0mL. Mix thoroughly. Gently heat and bring to boiling. Adjust pH to 9.0. Autoclave for 15 min at 15 psi pressure–121°C. Cool to 45°–50°C. Aseptically add sterile arabinose solution. Mix thoroughly. Aseptically distribute into sterile tubes or flasks.

Preparation of Medium: Dissolve all ingredients, except arabinose in 900.0mL of distilled water. Adjust the pH to 9.0 ± 0.2 and sterilize at 121°C for 15 min. Cool and aseptically add 100.0mL of a filter-sterilized 5% arabinose solution in distilled water. After mixing, aseptically dispense the medium into strile tubes or bottles.

Use: For the cultivation of *Vibrio* species from foods.

Horse Blood Agar

Composition per liter:

Beef heart, infusion from 500.0g
Agar.. 15.0g
Tryptose 10.0g
NaCl 5.0g
Horse blood, defibrinated............................. 50.0mL

pH 6.8 ± 0.2 at 25°C

Preparation of Medium: Add components, except horse blood, to distilled/deionized water and bring volume to 950.0mL. Mix thoroughly. Gently heat and bring to boiling. Autoclave for 15 min at 15 psi pressure–121°C. Cool to 45°–50°C. Aseptically add sterile horse blood. Mix thoroughly. Pour into sterile Petri dishes or distribute into sterile tubes.

Use: For the cultivation and maintenance of *Yersinia pseudotuberculosis*.

Horse Serum Agar

Composition per liter:

Agar.. 15.0g
Pancreatic digest of gelatin 5.0g
Beef extract 3.0g
Horse serum 200.0mL

pH 6.8 ± 0.2 at 25°C

Preparation of Medium: Add components, except horse serum, to distilled/deionized water and bring volume to 800.0mL. Mix thoroughly. Gently heat and bring to boiling. Autoclave for 15 min at 15 psi pressure–121°C. Cool to 45°–50°C. Aseptically add sterile horse serum. Mix thoroughly. Pour into sterile Petri dishes or distribute into sterile tubes.

Use: For the cultivation and maintenance of *Pseudomonas aeruginosa* and *Streptobacillus moniliformis*.

Horse Serum Broth

Composition per liter:

Pancreatic digest of gelatin 5.0g
Beef extract 3.0g
Horse serum 200.0mL

pH 6.8 ± 0.2 at 25°C

Source: This medium is available as a premixed powder from Difco Laboratories.

Preparation of Medium: Add components, except horse serum, to distilled/deionized water and bring volume to 800.0mL. Mix thoroughly. Gently heat and bring to boiling. Autoclave for 15 min at 15 psi pressure–121°C. Cool to 45°–50°C. Aseptically add sterile horse serum. Mix thoroughly. Aseptically distribute into sterile tubes or flasks.

Use: For the cultivation and maintenance of *Pseudomonas aeruginosa* and *Streptobacillus moniliformis*.

Hoyer's Medium

Composition per liter:

$(NH_4)_2SO_4$ 1.0g
KH_2PO_4 .... 0.9g
$MgSO_4 \cdot 7H_2O$ 0.25g
K_2HPO_4 .... 0.1g
$FeCl_3 \cdot 6H_2O$ 0.02g
Ethanol solution 200.0mL

Ethanol Solution:

Composition per 200mL:

Ethanol 30.0mL

Preparation of Ethanol Solution: Add ethanol to distilled/deionized water and bring volume to 200.0mL. Mix thoroughly. Filter sterilize.

Preparation of Medium: Add components, except ethanol solution, to distilled/deionized water and bring volume to 800.0mL. Mix thoroughly. Distribute into tubes in 4.0mL volumes. Autoclave for 15 min at 15 psi pressure–121°C. Cool to 25°C. Aseptically add 1.0mL of sterile ethanol solution to each tube. Mix thoroughly.

Use: For the cultivation of *Acetobacter* species.

Hoyle Medium Base

Composition per liter:

Agar	15.0g
Beef extract	10.0g
Peptone	10.0g
NaCl	5.0g
Blood, laked	50.0mL
Tellurite solution	10.0mL

pH 7.8 ± 0.2 at 25°C

Source: This medium is available as a premixed powder from Oxoid Unipath.

Tellurite Solution:

Composition per 100mL:

K_2TeO_3	3.5g

Preparation of Tellurite Solution: Add K_2TeO_3 to distilled/deionized water and bring volume to 100.0mL. Mix thoroughly. Filter sterilize.

Caution: Potassium tellurite is toxic.

Preparation of Medium: Add components, except laked blood, to distilled/deionized water and bring volume to 940.0mL. Mix thoroughly. Gently heat and bring to boiling. Autoclave for 15 min at 15 psi pressure–121°C. Cool to 45°–50°C. Aseptically add sterile components. Mix thoroughly. Pour into sterile Petri dishes or distribute into sterile tubes.

Use: For the isolation and differentiation of *Corynebacterium diphtheriae*.

HP 6 Agar

Composition per liter:

Agar	15.0g
Sodium glutaminate	10.0g
$MgSO_4 \cdot 7H_2O$	1.0g
Yeast extract	1.0g
Cyanocobalamin	0.5mg
Glucose solution	100.0mL

pH 7.2 ± 0.2 at 25°C

Glucose Solution:

Composition per 100mL:

D-Glucose	5.0g

Preparation of Glucose Solution: Add D-glucose to distilled/deionized water and bring volume to 100.0mL. Mix thoroughly. Autoclave for 15 min at 15 psi pressure–121°C. Cool to 25°C.

Preparation of Medium: Add components, except glucose solution, to distilled/deionized water and bring volume to 900.0mL. Mix thoroughly. Gently heat and bring to boiling. Autoclave for 15 min at 15 psi pressure–121°C. Cool to 45°–50°C. Aseptically add sterile glucose solution. Mix thoroughly. Pour into sterile Petri dishes or distribute into sterile tubes.

Use: For the isolation and cultivation of *Cytophaga* species, *Herpetosiphon* species, *Saprospira* species, and *Flexithrix* species.

HP 74 Broth

Composition per liter:

Sodium glutaminate	10.0g
$MgSO_4 \cdot 7H_2O$	2.0g
Yeast extract	2.0g
Glucose solution	100.0mL
Phosphate buffer solution	20.0mL

pH 6.5 ± 0.2 at 25°C

Glucose Solution:

Composition per 100mL:

D-Glucose	10.0g

Preparation of Glucose Solution: Add D-glucose to distilled/deionized water and bring volume to 100.0mL. Mix thoroughly. Autoclave for 15 min at 15 psi pressure–121°C. Cool to 25°C.

Phosphate Buffer Solution:

Composition per 100mL:

K_2HPO_4	6.81g

Preparation of Phosphate Buffer Solution: Add K_2HPO_4 to distilled/deionized water and bring volume to 100.0mL. Mix thoroughly. Adjust pH to 6.5. Autoclave for 15 min at 15 psi pressure–121°C. Cool to 25°C.

Preparation of Medium: Add components, except glucose solution and phosphate buffer solution, to distilled/deionized water and bring volume to 880.0mL. Mix thoroughly. Gently heat and bring to boiling. Autoclave for 15 min at 15 psi pressure–121°C. Cool to 45°–50°C. Aseptically add 100.0mL of sterile glucose solution and 20.0mL of sterile phosphate buffer solution. Mix thoroughly. Aseptically distribute into sterile tubes or flasks.

Use: For the isolation and cultivation of *Cytophaga* species, *Herpetosiphon* species, *Saprospira* species, and *Flexithrix* species.

HP 101 Halophile Medium

Composition per liter:

NaCl	100.0g
Agar	20.0g

Peptone	10.0g
$MgSO_4 \cdot 7H_2O$	4.3g
$NaNO_3$	2.0g
Yeast extract	1.0g

pH 7.2 ± 0.2 at 25°C

Preparation of Medium: Add components to distilled/deionized water and bring volume to 1.0L. Mix thoroughly. Gently heat and bring to boiling. Distribute into tubes or flasks. Autoclave for 15 min at 15 psi pressure–121°C. Pour into sterile Petri dishes or leave in tubes.

Use: For the cultivation and maintenance of *Pseudomonas* species.

HP Medium

Composition per liter:

Pancreatic digest of soybean meal	20.0g
Beef extract	10.0g
Yeast extract	6.0g
Ammonium citrate	5.0g
Tween™ 80	0.5g
$MgSO_4 \cdot 7H_2O$	0.2g
$MnSO_4 \cdot 4H_2O$	0.05g
$FeSO_4 \cdot 7H_2O$	0.04g
Glucose solution	10.0mL
Tetracycline solution	10.0mL

Glucose Solution:

Composition per 100mL:

Glucose	10.0g

Preparation of Glucose Solution: Add glucose to distilled/deionized water and bring volume to 100.0mL. Mix thoroughly. Filter sterilize.

Tetracycline Solution:

Composition per 100mL:

Tetracycline	10.0g

Preparation of Tetracycline Solution: Add tetracycline to distilled/deionized water and bring volume to 100.0mL. Mix thoroughly. Filter sterilize.

Preparation of Medium: Add components, except glucose solution and tetracycline solution, to distilled/deionized water and bring volume to 990.0mL. Mix thoroughly. Gently heat and bring to boiling. Autoclave for 15 min at 15 psi pressure–121°C. Cool to 45°–50°C. Aseptically add sterile glucose solution and tetracycline solution. Mix thoroughly. Aseptically distribute into sterile tubes or flasks.

Use: For the cultivation and enumeration of *Leuconostoc* species.

HPC Agar
See: **NWRI Agar**

HPC Agar (Heterotrophic Plate Count Agar) (m–HPC Agar)

Composition per liter:

Gelatin	25.0g
Pancreatic digest of gelatin	20.0g
Agar	15.0g
Glycerol	10.0mL

pH 7.1 ± 0.2 at 25°C

Source: Available from Difco Laboratories.

Preparation of Medium: Add components, except glycerol, to distilled/deionized water and bring volume to 990.0mL. Mix thoroughly. Gently heat and bring to boiling. Add glycerol. Mix thoroughly. Autoclave for 15 min at 15 psi pressure–121°C. Cool to 45°–50°C. Pour into sterile Petri dishes.

Use: For the the cultivation and enumeration of microorganisms from potable water sources, swimming pools and other water specimens, by the membrane filter method and heterotrophic plate count technique.

HR Antifungal Assay Medium Buffered with MOPS

Composition per liter:

MOPS (3-N-Morpholino-propanesulfonic acid) buffer	34.53g
Glucose	10.0g
$(NH_4)_2SO_4$	2.5g
KH_2PO_4	1.0g
$NaHCO_3$	1.0g
Glutamine	0.58g
$MgSO_4 \cdot 7H_2O$	0.5g
$CaCl_2 \cdot 2H_2O$	0.1g
NaCl	0.1g
L-Lysine	0.07g
L-Isoleucine	0.05g
L-Leucine	0.05g
L-Threonine	0.05g
L-Valine	0.05g
L-Arginine	0.04g
L-Histidine	0.02g
L-Methionine	0.01g
L-Tryptophan	8.2mg
DL-Methionine	2.0mg
DL-Tryptophan	2.0mg
Inositol	2.0mg
L-Histidine·HCl	1.0mg
H_3BO_3	0.5mg

Calciun pantothenate	0.4mg
$MnSO_4 \cdot H_2O$	0.4mg
Niacin	0.4mg
Pyridoxine	0.4mg
Thiamine·HCl	0.4mg
$ZnSO_4 \cdot 7H_2O$	0.4mg
p-Aminobenzoic acid	0.2mg
$FeCl_3$	0.2mg
Riboflavin	0.2mg
Na_2MoO_3	0.2mg
KI	0.1mg
$CuSO_4 \cdot 5H_2O$	0.04mg
Biotin	2.0μg
Folic acid	2.0μg

pH 7.0 ± 0.2 at 25°C

Preparation of Medium: Add components, except $NaHCO_3$ and MOPS buffer, to distilled/deionized water and bring volume to 900.0mL. Mix thoroughly. Add $NaHCO_3$ and MOPS buffer. Mix thoroughly. Adjust pH to 7.0. Bring volume to 1.0L with distilled/deionized water. Filter sterilize.

Use: For the testing of the effectiveness of antifungal agents against clinical fungal isolates using the broth dilution susceptibility testing method.

HTM
See: ***Haemophilus*** **Test Medium**

Hugh–Leifson's Glucose Broth

Composition per liter:

NaCl	30.0g
Glucose	10.0g
Agar	3.0g
Peptone	2.0g
Yeast extract	0.5g
Bromcresol Purple	0.015g

pH 7.4 ± 0.2 at 25°C

Preparation of Medium: Add components to distilled/deionized water and bring volume to 1.0L. Mix thoroughly. Gently heat while stirring and bring to boiling. Adjust pH to 7.4. Distibute into tubes or flasks. Autoclave for 15 min at 15 psi pressure–121°C.

Use: For the cultivation and differentiation of bacteria based on their ability to ferment glucose. Bacteria that ferment glucose turn the medium yellow.

Hugh–Leifson's Oxidation-Fermentation Medium
See: **Oxidation-Fermentation Medium, Hugh–Leifson**

Human Blood Tween™ Bilayer Medium
See: **HBT Bilayer Medium**

Hungate's Habitat–Simulating Medium

Composition per 1140.2mL:

Rumen fluid	333.0mL
Mineral solution A	167.0mL
Mineral solution B	167.0mL
$NaHCO_3$ solution	53.0mL
Cysteine·HCl solution	10.6mL
Substrate solution	10.6mL
Resazurin solution	1.0mL

Mineral Solution A:

Composition per liter:

NaCl	6.0g
KH_2PO_4	3.0g
$(NH_4)_2SO_4$	3.0g
$CaCl_2$	0.6g
$MgSO_4$	0.6g

Preparation of Solution A: Add components to distilled/deionized water and bring volume to 1.0L. Mix thoroughly.

Mineral Solution B:

K_2HPO_4	3.0

Preparation of Solution B: Add K_2HPO_4 to distilled/deionized water and bring volume to 1.0L. Mix thoroughly.

Resazurin Solution:

Composition per 100mL:

Resazurin	0.1g

Preparation of Resazurin Solution: Add resazurin to distilled/deionized water and bring volume to 100.0L. Mix thoroughly.

Cysteine·HCl Solution:

Composition per 100mL:

Cysteine·HCl	3.0g

Preparation of Cysteine·HCl Solution: Add Cysteine·HCl to O_2-free distilled/deionized water and bring volume to 100.0L. Mix thoroughly. Gently heat and bring to boiling. Continue boiling for 2 min. Cool to 25°C under 100% N_2. Seal tube with a stopper that is wired in place. Autoclave for 15 min at 15 psi pressure–121°C. Cool to 25°.

$NaHCO_3$ Solution:

Composition per 10mL:

$NaHCO_3$	1.0g

Preparation of $NaHCO_3$ Solution: Add $NaHCO_3$ to O_2-free distilled/deionized water and bring volume to 10.0mL. Mix thoroughly. Filter sterilize. Gas with 100% CO_2 for 15 min.

Substrate Solution:
Composition per 100mL:

Sugar 10.0g

Preparation of Substrate Solution: Add sugar to O_2-free distilled/deionized water. Mix thoroughly. Gas with 100% N_2 for 15 min. Autoclave for 15 min at 15 psi pressure–121°C. Cool to 45°–50°C.

Preparation of Medium: Add 167.0mL of solution A, 167.0mL of solution B, and 1.0mL of resazurin solution to distilled/deionized water and bring volume to 733.0mL. Mix thoroughly. Gently heat and bring to boiling. Continue boiling until resazurin turns colorless indicating reduction. Bring volume back to 733.0mL (some evaporation will have occurred) with O_2-free distilled/deionized water. Cool to 45°–50°C under O_2-free 100% CO_2. Anaerobically add rumen fluid. Anaerobically distribute into tubes in 10.0mL volumes. Cap with butyl rubber stoppers. Place tubes in a press. Autoclave for 15 min at 15 psi pressure–121°C. Cool to 25°C. Immediately prior to inoculation, aseptically and anaerobically add 0.1mL of sterile cysteine·HCl solution, 0.5mL of sterile $NaHCO_3$ solution and 0.1mL of substrate solution per 10.0mL of medium in each tube.

Use: For the cultivation of *Bacteroides* species from rumens.

HY Agar for *Flavobacterium*

Composition per liter:

Agar 8.0g
Glutamic acid 5.0g
K_2HPO_4 .... 0.1g
$MgSO_4·7H_2O$ 0.1g

pH 7.3 ± 0.2 at 25°C

Preparation of Medium: Add components to distilled/deionized water and bring volume to 1.0L. Glutamic acid may be replaced by 1.0g folic acid if desired. Mix thoroughly. Gently heat and bring to boiling. Distribute into tubes or flasks. Autoclave for 15 min at 15 psi pressure–121°C.

Use: For the cultivation of *Flavobacterium* species.

HY Medium for *Flavobacterium*

Composition per liter:

Glutamic acid 5.0g
K_2HPO_4 .... 0.1g
$MgSO_4·7H_2O$ 0.1g

pH 7.3 ± 0.2 at 25°C

Preparation of Medium: Add components to distilled/deionized water and bring volume to 1.0L. Glutamic acid may be replaced by 1.0g folic acid if desired. Mix thoroughly. Gently heat and bring to boiling. Distribute into tubes or flasks. Autoclave for 15 min at 15 psi pressure–121°C.

Use: For the cultivation of *Flavobacterium* species.

HYA Agar

Composition per liter:

Agar 15.0g
Proteose peptone No. 3 .... 10.0g
Beef extract 1.0g
Lactose solution 10.0mL
Galactose solution 10.0mL
Glucose solution 10.0mL

pH 6.8 ± 0.2 at 25°C

Lactose Solution:
Composition per 10mL:

Lactose 5.0g

Preparation of Lactose Solution: Add lactose to distilled/deionized water and bring volume to 10.0mL. Mix thoroughly. Filter sterilize.

Galactose Solution:
Composition per 10mL:

Galactose 2.5g

Preparation of Galactose Solution: Add galactose to distilled/deionized water and bring volume to 10.0mL. Mix thoroughly. Filter sterilize.

Glucose Solution:
Composition per 10mL:

Glucose 2.5g

Preparation of Glucose Solution: Add glucose to distilled/deionized water and bring volume to 10.0mL. Mix thoroughly. Filter sterilize.

Preparation of Medium: Add components—except lactose solution, galactose solution, and glucose solution—to distilled/deionized water and bring volume to 970.0mL. Mix thoroughly. Gently heat and bring to boiling. Adjust pH to 6.8, Autoclave for 15 min at 15 psi pressure–121°C. Cool to 45°–50°C. Aseptically add sterile lactose solution, galactose solution, and glucose solution. Mix thoroughly. Pour into sterile Petri dishes.

Use: For the cultivation of acidogenic microorganisms, especially *Lactobacillus bulgaricus* and *Streptococcus thermophilus,* from foods.

Hydrogen–Oxidizing Bacteria Medium

Composition per 1020mL:

Solution I 1.0L
Solution II 10.0mL
Solution III 10.0mL

Solution I:

Composition per liter:

$Na_2HPO_4 \cdot 12H_2O$ 9.0g
KH_2PO_4 .... 1.5g
NH_4Cl 1.0g
$MgSO_4 \cdot 7H_2O$ 0.2g
Trace elements solution 1.0mL

Preparation of Solution I: Add components to distilled/deionized water and bring volume to 1.0L. Mix thoroughly. Gently heat until dissolved. Autoclave for 15 min at 15 psi pressure–121°C. Cool to 25°C.

Trace Elements Solution:

Composition per liter:

H_3BO_3 .... 0.3g
$CoCl_2 \cdot 6H_2O$ 0.2g
$ZnSO_4 \cdot 7H_2O$ 0.1g
$MnCl_2 \cdot 4H_2O$ 0.03g
$NaMoO_4 \cdot 2H_2O$ 0.03g
$NiCl_2 \cdot 6H_2O$ 0.02g
$CuCl_2 \cdot 2H_2O$ 0.01g

Preparation of Trace Elements Solution: Add components to distilled/deionized water and bring volume to 1.0L. Mix thoroughly.

Solution II:

Composition per 100mL:

$CaCl_2 \cdot 2H_2O$ 0.1g
Ferric ammonium citrate 0.05g

Preparation of Solution II: Add components to distilled/deionized water and bring volume to 100.0mL. Mix thoroughly. Autoclave for 15 min at 15 psi pressure–121°C. Cool to 25°C.

Solution III:

Composition per 100mL:

$NaHCO_3$ 5.0g

Preparation of Solution III: Add $NaHCO_3$ to distilled/deionized water and bring volume to 100.0mL. Mix thoroughly. Filter sterilize.

Preparation of Medium: Aseptically combine 1.0L of cooled, sterile solution I, 10.0mL of cooled, sterile solution II and 10.0mL of sterile solution III. Mix thoroughly. Aseptically distribute into sterile tubes or flasks.

Use: For the cultivation of hydrogen–oxidizing bacteria.

Hydroxybenzoate Agar

Composition per 1001mL:

Solution A 990.0mL
Solution D 500.0mL
Solution B 10.0mL
Solution C 1.0mL

pH 7.0 ± 0.2 at 25°C

Solution A:

Composition per 990mL:

4-Hydroxybenzoic acid 3.0g
$(NH_4)_2SO_4$ 3.0g
NaCl 2.5g
K_2HPO_4 .... 1.6g
Yeast extract 0.5g

Preparation of Solution A: Add components to distilled/deionized water and bring volume to 990.0mL. Mix thoroughly. Gently heat and bring to boiling. Autoclave for 15 min at 15 psi pressure–121°C. Cool to 45°–50°C.

Solution B:

Composition per 10mL:

$MgSO_4 \cdot 7H_2O$ 0.27g

Preparation of Solution B: Add $MgSO_4 \cdot 7H_2O$ to distilled/deionized water and bring volume to 10.0mL. Mix thoroughly. Autoclave for 15 min at 15 psi pressure–121°C. Cool to 45°–50°C.

Solution C:

Composition per 1mL:

$Fe(NH_4)_2(SO_4)_2 \cdot 6H_2O$ 0.05g

Preparation of Solution C: Add component to distilled/deionized water and bring volume to 1.0mL. Mix thoroughly. Filter sterilize. Prepare solution immediately before adding to solutions A and B.

Solution D:

Composition per 500mL:

Agar 14.0g

Preparation of Solution D: Add agar to distilled/deionized water and bring volume to 500.0mL. Mix thoroughly. Autoclave for 15 min at 15 psi pressure–121°C. Cool to 45°–50°C.

Preparation of Medium: Aseptically combine cooled sterile solution A, cooled sterile solution B and cooled sterile solution D. Immediately add 1.0mL freshly prepared sterile solution C. Adjust pH to 7.0 with 6*N* NaOH. Mix thoroughly. Pour into sterile Petri dishes or distribute into sterile tubes.

Use: For the cultivation and maintenance of *Comamonas testosteroni*.

Hydroxybenzoate Broth

Composition per 1001mL:

Solution A	990.0mL
Solution B	10.0mL
Solution C	1.0mL

pH 7.0 ± 0.2 at 25°C

Solution A:

Composition per 990mL:

4-Hydroxybenzoic acid	3.0g
$(NH_4)_2SO_4$	3.0g
NaCl	2.5g
K_2HPO_4	1.6g
Yeast extract	0.5g

Preparation of Solution A: Add components to distilled/deionized water and bring volume to 990.0mL. Mix thoroughly. Gently heat and bring to boiling. Autoclave for 15 min at 15 psi pressure–121°C. Cool to 45°–50°C.

Solution B:

Composition per 10mL:

$MgSO_4 \cdot 7H_2O$	0.27g

Preparation of Solution B: Add $MgSO_4 \cdot 7H_2O$ to distilled/deionized water and bring volume to 10.0mL. Mix thoroughly. Autoclave for 15 min at 15 psi pressure–121°C. Cool to 45°–50°C.

Solution C:

Composition per 1mL:

$Fe(NH_4)_2(SO_4)_2 \cdot 6H_2O$	0.05g

Preparation of Solution C: Add component to distilled/deionized water and bring volume to 1.0mL. Mix thoroughly. Filter sterilize. Prepare solution immediately before adding to solutions A and B.

Preparation of Medium: Aseptically combine cooled sterile solution A and cooled sterile solution B. Immediately add 1.0mL freshly prepared sterile solution C. Adjust pH to 7.0 with 6*N* NaOH. Mix thoroughly. Aseptically distribute into sterile tubes or flasks.

Use: For the cultivation and maintenance of *Comamonas testosteroni.*

Hydroxybenzoate Medium

Composition per liter:

Noble agar	20.0g
$(NH_4)_2HPO_4$	3.0g
K_2HPO_4	1.2g
NaCl	0.5g
$MgSO_4 \cdot 7H_2O$	0.2g
$FeSO_4 \cdot 7H_2O$	0.1g
p-Hydroxybenzoic acid solution	50.0mL

pH 7.0 ± 0.2 at 25°C

***p*-Hydroxybenzoic Acid Solution:**

Composition per 50mL:

p-Hydroxybenzoic acid	3.0g

Preparation of *p*-Hydroxybenzoic Acid Solution: Add *p*-hydroxybenzoic acid to distilled/deionized water and bring volume to 50.0mL. Mix thoroughly. Filter sterilize.

Preparation of Medium: Add components, except *p*-hydroxybenzoic acid solution, to distilled/deionized water and bring volume to 950.0mL. Mix thoroughly. Gently heat and bring to boiling. Adjust pH to 7.0 with 5*N* NaOH. Autoclave for 15 min at 15 psi pressure–121°C. Cool to 45°–50°C. Aseptically add sterile *p*-hydroxybenzoic acid solution. Mix thoroughly. Pour into sterile Petri dishes or distribute into sterile tubes.

Use: For the cultivation and maintenance of *Pseudomonas putida.*

Hydroxybenzoic Acid Medium

Composition per liter:

Agar	15.0g
$K_2HPO_4 \cdot 3H_2O$	4.25g
NH_4Cl	2.0g
4-Hydroxybenzoic acid	1.0g
$NaH_2PO_4 \cdot H_2O$	1.0g
$MgSO_4 \cdot 7H_2O$	0.2g
Nitrilotriacetic acid	0.1g
$FeSO_4 \cdot 7H_2O$	0.012g
$MnSO_4 \cdot H_2O$	3.0mg
$ZnSO_4 \cdot 7H_2O$	3.0mg
$CoSO_4$	1.0mg

pH 7.2 ± 0.2 at 25°C

Preparation of Medium: Add 4-hydroxybenzoic acid and nitrilotriacetic acid to approximately 600.0mL of distilled/deionized water. Adjust pH to 8.0 with concentrated NaOH. Add remaining components. Mix thoroughly. Readjust pH to 7.2. Bring volume to 1.0L with distilled/deionized water. Distribute into tubes or flasks. Autoclave for 15 min at 15 psi pressure–121°C.

Use: For the cultivation and maintenance of *Bacillus* species.

Hyphomicrobium Enrichment Medium

Composition per 100mL:

KNO_3	0.04g
$Na_2HPO_4 \cdot 7H_2O$	0.02g
$MgSO_4 \cdot 7H_2O$	0.48mg
$FeCl_3 \cdot 7H_2O$	0.02mg
$MnCl_2 \cdot 4H_2O$	0.01mg

pH 7.2 ± 0.2 at 25°C

Preparation of Medium: Add components to distilled/deionized water and bring volume to 1.0L. Mix thoroughly. Adjust pH to 7.2. Distribute into tubes or flasks. Autoclave for 15 min at 15 psi pressure–121°C.

Use: For the cultivation and enrichment of *Hyphomicrobium* species.

Hyphomicrobium Medium

Composition per liter:

Agar	15.0g
Na_2HPO_4	2.13g
KH_2PO_4	1.36g
$MgSO_4 \cdot 7H_2O$	0.2g
$CaCl_2 \cdot 2H_2O$	9.95mg
$FeSO_4 \cdot 7H_2O$	5.0mg
$MnSO_4 \cdot 4H_2O$	2.5mg
$Na_2MoO_4 \cdot 2H_2O$	2.5mg
Urea solution	30.0mL
Methanol	4.0mL

Urea Solution:

Composition per 100mL:

Urea	20.0g

Preparation of Urea Solution: Add urea to distilled/deionized water and bring volume to 100.0mL. Mix thoroughly. Filter sterilize.

Preparation of Medium: Filter sterilize methanol. Add components, except urea solution and methanol, to distilled/deionized water and bring volume to 966.0mL. Mix thoroughly. Gently heat and bring to boiling. Autoclave for 15 min at 15 psi pressure–121°C. Cool to 45°–50°C. Aseptically add sterile urea solution and sterile methanol. Mix thoroughly. Aseptically distribute into sterile tubes or bottles.

Use: For the cultivation of *Hyphomicrobium* species.

Hyphomicrobium Medium

Composition per liter:

Noble agar	18.0g
Na_2HPO_4	2.15g
KH_2PO_4	1.36g
$(NH_4)_2SO_4$	0.5g
$MgSO_4 \cdot 7H_2O$	0.2g
Trace element solution	5.0mL
Methylamine·HCl solution	20.0mL

pH 7.1 ± 0.1 at 25°C

Trace Element Solution:

Composition per 100mL:

$CuCl_2$	0.15g
$FeSO_4 \cdot 7H_2O$	0.1g
$Na_2MoO_4 \cdot 2H_2O$	0.05g
$MnSO_4 \cdot H_2O$	0.035g

Preparation of Trace Element Solution: Add components to distilled/deionized water and bring volume to 100.0mL. Mix thoroughly.

Methylamine·HCl Solution:

Composition per 20mL:

Methylamine·HCl	3.38g

Preparation of Methylamine·HCl Solution: Add methylamine·HCl to distilled/deionized water and bring volume to 20.0mL. Mix thoroughly. Filter sterilize.

Preparation of Medium: Add components, except methylamine·HCl solution, to distilled/deionized water and bring volume to 980.0mL. Mix thoroughly. Gently heat and bring to boiling. Autoclave for 15 min at 15 psi pressure–121°C. Cool to 45°–50°C. Aseptically add sterile methylamine·HCl solution. Mix thoroughly. Adjust pH to 7.1, if necessary. Pour into sterile Petri dishes or distribute into sterile tubes.

Use: For the cultivation and maintenance of *Hyphomicrobium aestuarii, Hyphomicrobium facilis, Hyphomicrobium hollandicum, Hyphomicrobium vulgare,* and *Hyphomicrobium zavarzinii.*

Hyphomicrobium Medium 337a

Composition per liter:

KH_2PO_4	1.3g
Na_2HPO_4	1.13g
$(NH_4)_2SO_4$	0.50g
$MgSO_4 \cdot 7H_2O$	0.20g
$CaCl_2 \cdot 2H_2O$	3.09mg
$FeSO_4 \cdot 7H_2O$	2.0mg
$Na_2MoO_4 \cdot 2H_2O$	1.0mg
$MnSO_4 \cdot 4H_2O$	0.88mg

pH 7.2–7.5 at 25°C

Preparation of Medium: Add components to distilled/deionized water and bring volume to 1.0L. Mix thoroughly. Distribute into tubes or flasks. Autoclave for 15 min at 15 psi pressure–121°C.

Use: For the enrichment and cultivation of *Hyphomicrobium* species.

Hyphomonas Enrichment Medium

Composition per liter:

Peptone	0.05g
Yeast extract	0.05g

Preparation of Medium: Add components to distilled/deionized water and bring volume to 1.0L. Mix thoroughly. Distribute into tubes or flasks. Autoclave for 15 min at 15 psi pressure–121°C.

Use: For the isolation and cultivation of *Hyphomonas* species.

Hypoxanthine Agar

Composition per 1100mL:

Agar	15.0g
Beef extract	3.0g
Peptone	5.0g
Hypoxanthine solution	5.0g

pH 7.0 ± 0.1 at 25°C

Hypoxanthine Solution:
Composition per 100mL:

Hypoxanthine	5.0g

Preparation of Hypoxanthine Solution: Add hypoxanthine to distilled/deionized water and bring volume to 100.0mL. Mix thoroughly. Filter sterilize.

Preparation of Medium: Add components, except hypoxanthine solution, to distilled/deionized water and bring volume to 900.0mL. Mix thoroughly. Autoclave for 15 min at 15 psi pressure–121°C. Cool to 45°–50°C. Aseptically add 100.0mL of sterile hypoxanthine solution. Mix thoroughly. Pour into sterile 15 mm × 100 mm Petri dishes in 25.0 mL volumes.

Use: For the cultivation and differentiation of bacteria based on hypoxanthine hydrolysis. Bacteria that hydrolyze hypoxanthine, such as *Streptomyces griseus*, appear with a clear zone under and around the colonies. *Nocardia asteroides* does not hydrolyze hypoxanthine.

IBB Agar
See: **Inositol Brilliant Green Bile Salts Agar**

IFO Agar

Composition per liter:

Agar	20.0g
$(NH_4)_2HPO_4$	3.0g
NaCl	1.0g
$MgSO_4 \cdot 7H_2O$	0.2g
$FeSO_4 \cdot 7H_2O$	10.0mg
$MnSO_4 \cdot 4\text{-}6H_2O$	5.0mg
Riboflavin	0.02mg
Calcium pantothenate	0.02mg
Pyridoxine·HCl	0.02mg
Nicotinic acid	0.02mg
p-Aminobenzoic acid	0.01mg
Thiamine·HCl	0.01mg
Biotin	1.0μg
Methanol	10.0mL

pH 7.0 ± 0.2 at 25°C

Preparation of Medium: Add components, except agar and methanol, to distilled/deionized water and bring volume to 490.0mL. Mix thoroughly. Autoclave for 15 min at 15 psi pressure–121°C. Cool to 45°–50°C. In a separate flask, add agar to distilled/deionized water and bring volume to 500.0mL. Mix thoroughly. Gently heat and bring to boiling. Autoclave for 15 min at 15 psi pressure–121°C. Cool to 45°–50°C. Aseptically combine the two sterile solutions. Aseptically add 10.0mL filter sterilized methanol. Mix thoroughly. Adjust pH to 7.0. Pour into sterile Petri dishes or distribute into sterile tubes.

Use: For the cultivation and maintenance of *Hyphomicrobium methylovorum.*

IFO Broth

Composition per liter:

$(NH_4)_2HPO_4$	3.0g
NaCl	1.0g
$MgSO_4 \cdot 7H_2O$	0.2g
$FeSO_4 \cdot 7H_2O$	10.0mg
$MnSO_4 \cdot 4\text{-}6H_2O$	5.0mg
Riboflavin	20.0μg
Calcium pantothenate	20.0μg
Pyridoxine·HCl	20.0μg
Nicotinic acid	20.0μg
p-Aminobenzoic acid	10.0μg
Thiamine·HCl	10.0μg
Biotin	1.0μg
Methanol	10.0mL

pH 7.0 ± 0.2 at 25°C

Preparation of Medium: Add components, except methanol, to distilled/deionized water and bring volume to 990.0mL. Mix thoroughly. Autoclave for 15 min at 15 psi pressure–121°C. Aseptically add 10.0mL filter sterilized methanol. Mix thoroughly. Adjust pH to 7.0. Aseptically distribute into sterile tubes or flasks.

Use: For the cultivation and maintenance of *Hyphomicrobium methylovorum.*

IGP Medium
See: **Intracellular Growth Phase Medium**

Ilyobacter Agar

Composition per liter:

NaCl	20.0g
Agar	15.0g
$MgCl_2 \cdot 6H_2O$	3.0g
KCl	0.5g
NH_4Cl	0.25g
KH_2PO_4	0.2g
$CaCl_2 \cdot 2H_2O$	0.15g
Resazurin	1.0mg
Sodium sulfide solution	10.0mL
Sodium L-tartrate solution	10.0mL
$NaHCO_3$ solution	10.0mL
Trace element solution SL-7	1.0mL

pH 7.2 ± 0.2 at 25°C

Sodium Sulfide Solution:
Composition per 100mL:

$Na_2S \cdot 9H_2O$	3.6g

Preparation of Sodium Sulfide Solution: Add $Na_2S \cdot 9H_2O$ to distilled/deionized water and bring volume to 100.0mL. Mix thoroughly. Autoclave for 15 min at 15 psi pressure–121°C under N_2. Maintain under 100% N_2.

Sodium L-Tartrate Solution:
Composition per 10mL:

Sodium L-tartrate	2.0g

Preparation of Sodium L-Tartrate Solution: Add sodium L-tartrate to distilled/deionized water and bring volume to 10.0mL. Mix thoroughly. Filter sterilize. Maintain under 80% N_2 + 20% CO_2.

$NaHCO_3$ Solution:
Composition per 10mL:

$NaHCO_3$	2.5g

Preparation of $NaHCO_3$ Solution: Add $NaHCO_3$ to distilled/deionized water and bring volume to 10.0mL. Mix thoroughly. Filter sterilize. Maintain under 80% N_2 + 20% CO_2.

Trace Elements Solution SL-7:

Composition per liter:

$FeCl_2 \cdot 4H_2O$	1.5g
$CoCl_2 \cdot 6H_2O$	0.19g
$MnCl_2 \cdot 4H_2O$	0.1g
$ZnCl_2$	0.07g
H_3BO_3	0.062g
$Na_2MoO_4 \cdot 2H_2O$	0.036g
$NiCl_2 \cdot 6H_2O$	0.024g
$CuCl_2 \cdot 2H_2O$	0.017g
HCl (25% solution)	10.0mL

Preparation of Trace Elements Solution SL-7: Add the $FeCl_2 \cdot 4H_2O$ to the HCl. Add distilled/deionized water and bring volume to 1.0L. Add remaining components. Mix thoroughly. Filter sterilize. Maintain under 80% N_2 + 20% CO_2.

Preparation of Medium: Add components—except agar, sodium sulfide solution, sodium L-tartrate solution, $NaHCO_3$ solution, and trace element solution SL-7 solution—to distilled/deionized water and bring volume to 469.0mL. Mix thoroughly. Gently heat and bring to boiling. Autoclave for 15 min at 15 psi pressure–121°C. Cool to 45°–50°C. Maintain under 80% N_2 + 20% CO_2. Aseptically add 10.0mL L-tartrate solution, 10.0mL $NaHCO_3$ solution, and 1.0mL trace element solution SL-7 solution under 80% N_2 + 20% CO_2. Mix thoroughly. In a separate flask add agar to distilled/deionized water and bring volume to 500.0mL. Mix thoroughly. Gently heat and bring to boiling. Autoclave for 15 min at 15 psi pressure–121°C. Cool to 45°–50°C. Combine sterile agar and sterile basal medium. Adjust pH to 7.2. Aseptically add 10.0mL sodium sulfide solution. Pour into sterile Petri dishes or distribute into sterile tubes. Maintain under 80% N_2 + 20% CO_2.

Use: For the cultivation and maintenance of *Ilyobacter tartaricus.*

Ilyobacter Broth

Composition per liter:

NaCl	20.0g
$MgCl_2 \cdot 6H_2O$	3.0g
KCl	0.5g
NH_4Cl	0.25g
KH_2PO_4	0.2g
$CaCl_2 \cdot 2H_2O$	0.15g
Resazurin	1.0mg
Sodium sulfide solution	10.0mL
Sodium L-tartrate solution	10.0mL
$NaHCO_3$ solution	10.0mL
Trace element solution SL-7	1.0mL

pH 7.2 ± 0.2 at 25°C

Sodium Sulfide Solution:

Composition per 100mL:

$Na_2S \cdot 9H_2O$	3.6g

Preparation of Sodium Sulfide Solution: Add $Na_2S \cdot 9H_2O$ to distilled/deionized water and bring volume to 100.0mL. Mix thoroughly. Autoclave for 15 min at 15 psi pressure–121°C under 100% N_2. Maintain under 100% N_2.

Sodium L-Tartrate Solution:

Composition per 10mL:

Sodium L-tartrate	2.0g

Preparation of Sodium L-Tartrate Solution: Add sodium L-tartrate to distilled/deionized water and bring volume to 10.0mL. Mix thoroughly. Filter sterilize. Maintain under 80% N_2 + 20% CO_2.

$NaHCO_3$ Solution:

Composition per 10mL:

$NaHCO_3$	2.5g

Preparation of $NaHCO_3$ Solution: Add $NaHCO_3$ to distilled/deionized water and bring volume to 10.0mL. Mix thoroughly. Filter sterilize. Maintain under 80% N_2 + 20% CO_2.

Trace Elements Solution SL-7:

Composition per liter:

$FeCl_2 \cdot 4H_2O$	1.5g
$CoCl_2 \cdot 6H_2O$	0.19g
$MnCl_2 \cdot 4H_2O$	0.1g
$ZnCl_2$	0.07g
H_3BO_3	0.062g
$Na_2MoO_4 \cdot 2H_2O$	0.036g
$NiCl_2 \cdot 6H_2O$	0.024g
$CuCl_2 \cdot 2H_2O$	0.017g
HCl (25% solution)	10.0mL

Preparation of Trace Elements Solution SL-7: Add the $FeCl_2 \cdot 4H_2O$ to the HCl. Add distilled/deionized water and bring volume to 1.0L. Add remaining components. Mix thoroughly. Filter sterilize. Maintain under 80% N_2 + 20% CO_2.

Preparation of Medium: Add components—except sodium sulfide solution, sodium L-tartrate solution, $NaHCO_3$ solution, and trace element solution SL-7 solution—to distilled/deionized water and bring volume to 969.0mL. Mix thoroughly. Gently heat and bring to boiling. Autoclave for 15 min at 15 psi pressure–121°C. Cool to 45°–50°C. Maintain under 80% N_2 + 20% CO_2. Aseptically add 10.0mL L-tartrate solution, 10.0mL $NaHCO_3$ solution, and 1.0mL trace element solution SL-7 solution under 80% N_2 + 20% CO_2. Mix thoroughly. Aseptically distribute into sterile tubes or flasks under 80% N_2 + 20% CO_2. Adjust pH to 7.2. At time of inoculation add sodium sulfide solution to a final concentration of 0.1%.

Use: For the cultivation and maintenance of *Ilyobacter tartaricus.*

Ilyobacter Medium

Composition per liter:

Crotonic acid	1.7g
NaCl	1.0g
Yeast extract	1.0g
$Na_2HPO_4·12H_2O$	0.7g
KCl	0.5g
$MgCl_2·6H_2O$	0.4g
NH_4Cl	0.3g
Na_2SO_4	0.1g
Sodium sulfide solution	10.0mL
$CaCl_2·2H_2O$ (1.0%)	1.0mL
$FeCl_3$ (0.5%)	1.0mL
Modified SL-7 trace elements solution	1.0mL
Resazurin (0.1%)	1.0mL
Selenite/tungstate solution	1.0mL

pH 6.8–7.2 at 25°C

Sodium Sulfide Solution:

Composition per 100mL:

$Na_2S·9H_2O$	3.6g

Preparation of Sodium Sulfide Solution: Add $Na_2S·9H_2O$ to distilled/deionized water and bring volume to 100.0mL. Mix thoroughly. Autoclave for 15 min at 15 psi pressure–121°C under N_2. Maintain under 100% N_2.

Modified Sl-7 Trace Elements Solution:

Composition per liter:

$CoCl_2·6H_2O$	0.2g
$MnCl_2·4H_2O$	0.1g
$ZnCl_2$	0.07g
H_3BO_3	0.06g
$Na_2MoO_4·2H_2O$	0.04g
$CuCl_2·2H_2O$	0.02g
$NiCl_2·6H_2O$	0.02g
HCl (1*N*)	3.0mL

Preparation of Modified SL-7 Trace Elements Solution: Add components to distilled/deionized water and bring volume to 1.0L. Mix thoroughly. Filter sterilize. Maintain under 80% N_2 + 20% CO_2.

Selenite/Tungstate Solution:

Composition per liter:

NaOH	0.5g
$Na_2WO_4·2H_2O$	4.0mg
$Na_2SeO_3·5HO$	3.0mg

Preparation of Selenite/Tungstate Solution: Add components to distilled/deionized water and bring volume to 1.0L. Mix thoroughly. Filter sterilize. Maintain under 80% N_2 + 20% CO_2.

Preparation of Medium: Add components—except sodium sulfide solution, modified SL-7 trace elements solution, and selenite/tungstate solution—to distilled/deionized water and bring volume to 969.0mL. Mix thoroughly. Gently heat and bring to boiling. Autoclave for 15 min at 15 psi pressure–121°C. Adjust pH to 5.5. Cool to 45°–50°C under 100% N_2. Maintain under 100% N_2. Aseptically add 1.0mL sterile modified SL-7 trace elements solution and 1.0mL sterile selenite/tungstate solution under 100% N_2. Mix thoroughly. Aseptically distribute into sterile tubes or flasks under 100% N_2. At time of inoculation add sodium sulfide solution to a final concentration of 1.0% sodium sulfide solution. Maintain under 100% N_2.

Use: For the cultivation and maintenance of *Ilyobacter delafieldii.*

IM
See: Infection Medium

Imhoff's Medium, Modified

Composition per liter:

NaCl	30.0g
$NaHCO_3$	3.0g
KH_2PO_4	1.0g
NH_4Cl	1.0g
Sodium acetate	1.0g
Na_2SO_4	0.7g
$MgCl_2·6H_2O$	0.5g
Sodium ascorbate	0.5g
$CaCl_2·2H_2O$	0.1g
Yeast extract	0.1g
Sodium sulfide solution	10.0mL
SLA trace elements solution	1.0mL
VA vitamin solution	1.0mL

pH 6.9–7.0 at 25°C

Sodium Sulfide Solution:

Composition per 100mL:

$Na_2S·9H_2O$	2.0g

Preparation of Sodium Sulfide Solution: Add $Na_2S·9H_2O$ to distilled/deionized water and bring volume to 100.0mL. Mix thoroughly. Autoclave for 15 min at 15 psi pressure–121°C under N_2. Maintain under 100% N_2.

SLA Trace Elements Solution:

Composition per liter:

$FeCl_2·4H_2O$	1.8g
H_3BO_3	0.5g
$CoCl_2·6H_2O$	0.25g
$ZnCl_2$	0.1g

$MnCl_2 \cdot 4H_2O$ 0.07g
$Na_2MoO_4 \cdot 2H_2O$ 0.03g
$CuCl_2 \cdot 2H_2O$ 0.01g
$Na_2SeO_3 \cdot 5H_2O$ 0.01g
$NiCl_2 \cdot 6H_2O$ 0.01g

Preparation of SLA Trace Elements Solution: Add components to distilled/deionized water and bring volume to 1.0L. Mix thoroughly. Adjust pH to pH 2–3. Filter sterilize.

VA Vitamin Solution:
Composition per 500 mL:
Nicotinamide 0.17g
Thiamine·HCl 0.15g
p-Aminobenzoic acid 0.1g
Biotin 0.05g
Calcium pantothenate 0.05g
Pyridoxine·2HCl 0.05g
Cyanocobalamin 0.02g

Preparation of VA Vitamin Solution: Add components to distilled/deionized water and bring volume to 500.0mL. Mix thoroughly. Filter sterilize.

Preparation of Medium: Add components—except sodium sulfide solution, SLA trace elements solution, and VA vitamin solution—to distilled/deionized water and bring volume to 988.0mL. Mix thoroughly. Autoclave for 15 min at 15 psi pressure–121°C. Cool to 25°C. Aseptically add 1.0mL sterile SLA trace elements solution and 1.0mL sterile VA vitamin solution. Aseptically add 10.0mL sterile sodium sulfide solution. Mix thoroughly. Aseptically distribute into sterile tubes or flasks.

Use: For the cultivation and maintenance of *Rhodobacter adriaticus* and *Rhodobacter sulfidophilus*.

Imidazole Utilization Medium

Composition per liter:
Imidazole 5.0g
KH_2PO_4 .... 0.5g
$MgSO_4 \cdot 7H_2O$ 0.5g
$CaCl_2$ 3.0mg
$FeSO_4 \cdot 7H_2O$ 3.0mg
Molybdenum solution 1.0mL
Trace element solution 1.0mL
pH 6.0 ± 0.2 at 25°C

Molybdenum Solution:
Composition per 18mL:
$Na_2MoO_4 \cdot 2H_2O$ 0.5mg

Preparation of Molybdenum Solution: Add components to distilled/deionized water and bring volume to 18.0mL. Mix thoroughly. Filter sterilize.

Trace Element Solution:
Composition per 18mL:
H_3BO_3 .... 11.0mg
$MnCl_2 \cdot 4H_2O$ 7.0mg
$Al_2(SO_4)_3 \cdot 18\ H_2O$ 1.94mg
$Co(NO_3)_2 \cdot 6H_2O$ 1.0mg
$CuSO_4 \cdot 5H_2O$ 1.0mg
$NiSO_4 \cdot 6H_2O$ 1.0mg
$ZnSO_4 \cdot H_2O$ 0.62mg
KBr 0.5mg
KI 0.5mg
LiCl 0.5mg
$SnCl_2 \cdot 2H_2O$ 0.5mg

Preparation of Trace Element Solution: Add components to distilled/deionized water and bring volume to 18.0mL. Mix thoroughly. Filter sterilize.

Preparation of Medium: Add components, except molybdenum solution and trace elements solution, to distilled/deionized water and bring volume to 998.0mL. Mix thoroughly. Distribute into tubes or flasks. Autoclave for 15 min at 15 psi pressure–121°C. Cool to 25°C. Aseptically add 1.0mL molybdenum solution and 1.0mL trace element solution. Mix thoroughly. Adjust pH to 6.0 with phosphoric acid. Mix thoroughly. Aseptically distribute into sterile tubes or flasks.

Use: For the cultivation and maintenance of *Pseudomonas* species.

Indole Medium

Composition per 200mL:
K_2HPO_4 .... 3.13g
L-Tryptophan 1.0g
NaCl 1.0g
KH_2PO_4 .... 0.27g
pH 7.2 ± 0.2 at 25°C

Preparation of Medium: Add components to distilled/deionized water and bring volume to 200.0mL. Mix thoroughly. Distribute into tubes or flasks. Autoclave for 15 min at 15 psi pressure–121°C.

Use: For the differentiation of microorganisms by means of the indole production from tryptophan test.

Indole Medium

Composition per liter:
Pancreatic digest of casein 20.0g
pH 7.3 ± 0.2 at 25°C

Preparation of Medium: Add pancreatic digest of casein to distilled/deionized water and bring volume to 1.0L. Mix thoroughly. Distribute into tubes or flasks. Autoclave for 15 min at 15 psi pressure–121°C.

Use: For the differentiation of microorganisms by means of the indole test.

Indole Nitrite Medium (Trypticase™ Nitrate Broth)

Composition per liter:

Pancreatic digest of casein	20.0g
Na_2HPO_4	2.0g
Agar	1.0g
Glucose	1.0g
KNO_3	1.0g

pH 7.2 ± 0.2 at 25°C

Source: This medium is available as a premixed powder from BBL Microbiology Systems.

Preparation of Medium: Add components to distilled/deionized water and bring volume to 1.0L. Mix thoroughly. Gently heat and bring to boiling with frequent agitation. Distribute into tubes or flasks. Autoclave for 15 min at 15 psi pressure–121°C.

Use: For the identification of microorganisms by means of the nitrate reduction and indole tests.

Infection Medium (IM)

Composition per 100mL:

Pancreatic digest of gelatin	0.05g
Bile salts No. 3	0.05g
Brain heart, solids from infusion	0.02g
Peptic digest of animal tissue	0.02g
NaCl	0.017g
Glucose	0.01g
Na_2HPO_4	8.0mg
Earle's balanced salts solution	80.0mL
Fetal bovine serum, heat-inactivated (2 hr at 55°C)	20.0mL

pH 7.4 ± 0.2 at 25°C

Earle's Balanced Salts Solution:

Composition per liter:

NaCl	6.8g
$NaHCO_3$	2.2g
Glucose	1.0g
KCl	0.40g
$CaCl_2 \cdot 2H_2O$	0.265g
$MgSO_4 \cdot 7H_2O$	0.20g
$NaH_2PO_4 \cdot H_2O$	0.14g

Preparation of Earle's Balanced Salts Solution: Add components to distilled/deionized water and bring volume to 1.0L. Mix thoroughly. Filter sterilize.

Preparation of Medium: Combine components. Mix thoroughly. Filter sterilize. Store at 4°–10°C.

Use: For screening of *Escherichia coli* for pathogenicity using the HeLa cell test for invasiveness.

Infusion Agar

See: **Blood Agar Base**

Infusion Broth

Composition per liter:

Pancreatic digest of casein	13.0g
NaCl	5.0g
Yeast extract	5.0g
Heart muscle, solids from infusion	2.0g

pH 7.4 ± 0.2 at 25°C

Source: This medium is available as a premixed powder from BBL Microbiology Systems.

Preparation of Medium: Add components to distilled/deionized water and bring volume to 1.0L. Mix thoroughly. Distribute into tubes or flasks. Autoclave for 15 min at 15 psi pressure–121°C.

Use: For the cultivation of a wide variety of microorganisms.

Inhibitory Mold Agar

Composition per liter:

Agar	15.0g
Glucose	5.0g
Yeast extract	5.0g
Pancreatic digest of casein	3.0g
Na_2HPO_4	2.0g
Peptic digest of animal tissue	2.0g
Starch	2.0g
Dextrin	1.0g
$MgSO_4 \cdot 7H_2O$	0.8g
Chloramphenicol	0.125g
$FeSO_4$	0.04g
NaCl	0.04g
$MnSO_4$	0.16g

pH 6.7 ± 0.2 at 25°C

Source: This medium is available as a premixed powder from BBL Microbiology Systems.

Preparation of Medium: Add components to distilled/deionized water and bring volume to 1.0L. Mix thoroughly. Gently heat and bring to boiling with frequent agitation. Distribute into tubes or flasks. Autoclave for 15 min at 15 psi pressure–121°C. Pour into sterile Petri dishes or leave in tubes.

Use: For the isolation of pathogenic fungi.

Inorganic Salts Maltose Medium

Composition per liter:

Yeast extract	4.0g
Peptone	2.0g
Inorganic salts solution	980.0mL
Maltose solution	20.0mL

pH 7.5 ± 0.2 at 25°C

Inorganic Salts Solution:

Composition per liter:

$MgSO_4 \cdot 7H_2O$	49.37g
NaCl	43.8g
$CaCl_2 \cdot 2H_2O$	1.29g

Preparation of Inorganic Salts Solution: Add NaCl first and then other components to distilled/deionized water and bring volume to 1.0L. Mix thoroughly.

Maltose Solution:

Composition per 100mL:

Maltose	25.0g

Preparation of Maltose Solution: Add maltose to distilled/deionized water and bring volume to 100.0mL. Mix thoroughly. Filter sterilize.

Preparation of Medium: Add components, except maltose solution, to inorganic salts solution and bring volume to 980.0mL. Mix thoroughly. Adjust pH to 7.5 with KOH. Autoclave for 15 min at 15 psi pressure–121°C. Cool to 25°C. Aseptically add 20.0mL sterile maltose solution. Mix thoroughly. Aseptically distribute into sterile tubes or flasks.

Use: For the cultivation and maintenance of *Spirochaeta halophila*.

Inorganic Salts Starch Agar

See: **ISP Medium 4**

Inositol Assay Medium

Composition per liter:

Glucose	100.0g
Potassium citrate	10.0g
Citric acid	2.0g
KH_2PO_4	1.1g
KCl	0.85g
L-Asparagine	0.8g
L-Glutamic acid	0.6g
L-Isoleucine	0.5g
L-Leucine	0.5g
L-Lysine	0.5g
L-Valine	0.5g
L-Arginine	0.48g
DL-Alanine	0.4g
DL-Threonine	0.4g
$CaCl_2$	0.25g
$MgSO_4 \cdot 7H_2O$	0.25g
DL-Aspartic acid	0.2g
DL-Phenylalanine	0.2g
Glycine	0.2g
L-Methionine	0.2g
L-Tyrosine	0.2g
L-Proline	0.2g
L-Histidine	0.124g
DL-Serine	0.1g
L-Cystine	0.1g
DL-Tryptophan	0.08g
$FeCl_3$	0.05g
$MnSO_4 \cdot 7H_2O$	0.05g
Calcium pantothenate	5.0mg
Pyridoxine·HCl	1.0mg
Thiamine·HCl	0.5mg
Biotin	0.01mg

pH 5.2 ± 0.2 at 25°C

Source: This medium is available as a premixed powder from Difco Laboratories.

Preparation of Medium: Add components to distilled/deionized water and bring volume to 1.0L. Mix thoroughly. Distribute into tubes or flasks. Autoclave for 5 min at 15 psi pressure–121°C.

Use: For the microbiological assaying of inositol using *Saccharomyces uvarum* as the test organism.

Inositol Assay Medium KB

Composition per liter:

Glucose	40.0g
$(NH_4)_2SO_4$	4.0g
DL-Asparagine	4.0g
KH_2PO_4	3.0g
$MgSO_4 \cdot 7H_2O$	1.0g
$CaCl_2$	0.98g
$FeSO_4 \cdot 7H_2O$	0.5mg
Calcium pantothenate	0.4mg
Nicotinic acid	0.4mg
Pyridoxine	0.4mg
Thiamine	0.4mg
H_3BO_3	0.2mg
KI	0.2mg
$CuSO_4 \cdot 5H_2O$	0.09mg
$MnSO_4 \cdot 7H_2O$	0.08mg
$ZnSO_4 \cdot 7H_2O$	0.08mg
$(NH_4)_6Mo_7O_{24} \cdot 4H_2O$	0.04mg
Riboflavin	0.02mg
Biotin	0.4μg

pH 5.0 ± 0.2 at 25°C

Preparation of Medium: Add components to distilled/deionized water and bring volume to 1.0L. Mix

thoroughly. Distribute into tubes or flasks. Autoclave for 5 min at 15 psi pressure–121°C.

Use: For the microbiological assaying of inositol using *Kloeckera apiculata* as the test organism.

Inositol Brilliant Green Bile Salts Agar (IBB Agar) (*Pleisomonas* Differential Agar)

Composition per liter:

Agar	15.0g
meso-Inositol	10.0g
Proteose peptone	10.0g
Bile salts no. 3	8.5g
Meat extract	5.0g
NaCL	5.0g
Neutral Red (2% solution)	1.25mL
Brilliant Green (0.1% solution)	0.33mL

pH 7.2 ± 0.1 at 25°C

Preparation of Medium: Add components to distilled/deionized water and bring volume to 1.0L. Mix thoroughly. Gently heat and bring to boiling. Distribute into tubes or flasks. Autoclave for 15 min at 15 psi pressure–121°C. Pour into sterile Petri dishes or leave in tubes.

Use: For isolation of *Aeromonas* and *Plesiomonas* species.

Inositol Gelatin Deeps

Composition per liter:

Gelatin	120.0g
Inositol	10.0g
Na_2HPO_4	5.0g
Yeast extract	5.0g
Phenol Red	0.05g

pH 7.4 ± 0.2 at 25°C

Preparation of Medium: Add components to distilled/deionized water and bring volume to 1.0L. Mix thoroughly. Gently heat and bring to boiling. Adjust pH to 7.4. Distribute into tubes in 5.0mL volumes. Autoclave for 15 min at 10 psi pressure–115°C.

Use: For the cultivation of *Pleisomonas shigelloides* from foods.

Inositol Urea Caffeic Acid Medium

Composition per liter:

Agar solution	900.0mL
Base solution	100.0mL

Agar Solution:

Composition per 900mL:

Agar	15.0g

Preparation of Agar Solution: Add agar to distilled/deionized water and bring volume to 900.0mL. Mix thoroughly. Gently heat and bring to boiling. Autoclave for 15 min at 15 psi pressure–121°C. Cool to 45°–50°C.

Base Solution:

Composition per 100mL:

Inositol	10.0g
Urea	5.0g
KH_2PO_4	1.0g
$MgSO_4 \cdot 7H_2O$	0.5g
Caffeic acid	0.2g
NaCl	0.1g
$CaCl_2 \cdot 2H_2O$	0.1g
Gentamicin sulfate	0.04g
H_3BO_3	0.5mg
$ZnSO_4 \cdot 7H_2O$	0.4mg
$MnSO_4 \cdot 4H_2O$	0.4mg
Thiamine·HCl	0.4mg
Pyroxidine·HCl	0.4mg
Niacin	0.4mg
Calcium pantothenate	0.4mg
p-Aminobenzoic acid	0.2mg
Riboflavin	0.2mg
$FeCl_3$	0.2mg
$Na_2MoO_4 \cdot 4H_2O$	0.2mg
KI	0.1mg
$CuSO_4 \cdot 5H_2O$	0.04mg
Folic acid	2.0μg
Biotin	2.0μg
Ferric citrate solution (1% solution)	1.0mL

Preparation of Base Solution: Add components, except urea, to distilled/deionized water and bring volume to 100.0mL. Mix thoroughly. Gently heat just until components are dissolved. Cool to 75°–80°C. Add urea. Mix thoroughly. Do not heat after addition of urea. Do not autoclave. Filter sterilize.

Preparation of Medium: Aseptically combine the cooled, sterile agar solution with the sterile base solution. Mix thoroughly. Pour into sterile Petri dishes.

Use: For the selective isolation and differentiation of *Cryptococcus* species based on inositol and urea utilization and pigment production from caffeic acid. On this medium, only *Cryptococcus* species utilize inositol as sole carbon source and urea as sole nitrogen source. *Cryptococcus neoformans* appears as dark brown colonies. Other *Cryptococcus* species are unpigmented.

International *Streptomyces* Project Medium 1
See: **ISP Medium 1**

International *Streptomyces* Project Medium 2
See: **ISP Medium 2**

International *Streptomyces* Project Medium 3
See: **ISP Medium 3**

International *Streptomyces* Project Medium 4
See: **ISP Medium 4**

International *Streptomyces* Project Medium 4 with Glucose
See: **ISP Medium 4 with Glucose**

International *Streptomyces* Project Medium 4 with Yeast Extract
See: **ISP Medium 4 with Yeast Extract**

International *Streptomyces* Project Medium 5
See: **ISP Medium 5**

International *Streptomyces* Project Medium 6
See: **ISP Medium 6**

International *Streptomyces* Project Medium 7
See: **Tyrosine Agar**

International *Streptomyces* Project Medium 8
See: **Nitrate Broth**

International *Streptomyces* Project Medium 9
See: **ISP Medium 9**

Intracellular Growth Phase Medium (IGP Medium)

Composition per 100mL:

Gentamicin sulfate	500.0mg
Lysozyme	30.0mg
Eagle MEM	72.0mL
Dulbecco's phosphate-buffered saline	20.0mL
Fetal bovine serum	8.0mL

pH 7.2–7.4 at 25°C

Eagle MEM:
Composition per liter:

NaCl	8.0g
Glucose	1.0g
KCl	0.4g
$CaCl_2 \cdot 2H_2O$	0.14g
$MgSO_4 \cdot 7H_2O$	0.1g
KH_2PO_4	0.06g
Na_2HPO_4	0.05g
L-Isoleucine	0.026g
L-Leucine	0.026g
L-Lysine	0.026g
L-Threonine	0.024g
L-Valine	0.0235g
L-Tyrosine	0.018g
L-Arginine	0.0174g
L-Phenylalanine	0.0165g
L-Cystine	0.012g
L-Histidine	8.0mg
L-Methionine	7.5mg
Phenol Red	5.0mg
L-Tryptophan	4.0mg
Inositol	1.8mg
Biotin	1.0mg
Folic acid	1.0mg
Calcium pantothenate	1.0mg
Choline chloride	1.0mg
Nicotinamide	1.0mg
Pyridoxal·HCl	1.0mg
Thiamine·HCl	1.0mg
Riboflavin	0.1mg

Preparation of Eagle MEM: Add components to distilled/deionized water and bring volume to 1.0L. Mix thoroughly.

Dulbecco's Phosphate-Buffered Saline :
Composition per liter:

NaCl	8.0g

$Na_2HPO_4 \cdot 7H_2O$ 2.16g
KCl 0.2g
KH_2PO_4 .... 0.2g
$CaCl_2$ 0.1g
$MnCl_2 \cdot 6H_2O$ 0.1g

Preparation of Dulbecco's Phosphate-Buffered Saline: Add components to distilled/deionized water and bring volume to 1.0L. Mix thoroughly.

Preparation of Medium: Combine components. Mix thoroughly. Filter sterilize. Aseptically distribute into sterile tubes or flasks.

Use: For screening of *Escherichia coli* for pathogenicity using the HeLa cell test for invasiveness.

Ion Agar for *Ureaplasma*

Composition per 101.45mL:
HEPES (N-[2-Hydroxyethyl] piperazine-N´-[2-ethane-sulfonic acid]) buffer 1.19g
Ionagar No. 2 .... 0.75g
Pancreatic digest of casein 0.7g
NaCl 0.5g
Beef extract 0.3g
Yeast extract 0.3g
Beef heart, solids from infusion 0.2g
Yeast extract 0.1g
Horse serum, normal sterile 10.0mL
Ampicillin solution 1.0mL
Urea solution 0.25mL
Nystatin solution 0.1mL
Tripeptide solution 0.1mL
pH 7.2 ± 0.2 at 25°C

Ampicillin Solution:
Composition per 10mL:
Ampicillin 1.0g

Preparation of Ampicillin Solution: Add ampicillin to distilled/deionized water and bring volume to 10.0mL. Mix thoroughly. Filter sterilize.

Urea Solution:
Composition per 100mL:
Urea 10.0g

Preparation of Urea solution: Add urea to distilled/deionized water and bring volume to 100.0mL. Filter sterilize. Store at –20°C.

Nystatin Solution:
Composition per 1mL:
Nystatin 50,000U

Preparation of Nystatin Solution: Add nystatin to distilled/deionized water and bring volume to 1.0mL. Filter sterilize.

Tripeptide Solution:
Composition per 10mL:
Glycyl-L-histidyl-L-lysine acetate 0.2mg

Preparation of Tripeptide Solution: Add glycyl-L-histidyl-L-lysine acetate to distilled/deionized water and bring volume to 10.0mL. Mix thoroughly. Filter sterilize. Store at –20°C.

Preparation of Medium: Add components—except horse serum, ampicillin solution, urea solution, nystatin solution and tripeptide solution—to distilled/deionized water and bring volume to 90.0mL. Mix thoroughly. Gently heat and bring to boiling. Autoclave for 15 min at 15 psi pressure–121°C. Cool to 45°–50°C. Aseptically add 10.0mL of sterile horse serum, 1.0mL of sterile ampicillin solution, 0.25mL of sterile urea solution, 0.1mL of sterile nystatin solution and 0.1mL of sterile tripeptide solution. Mix thoroughly. Pour into sterile Petri dishes.

Use: For the cultivation of *Ureaplasma* species from clinical specimens.

Ionic Medium with Pipecolate

Composition per liter:
Agar 30.0g
Ionic medium 950.0mL
Pipecolic acid·HCl solution 50.0mL

Ionic Medium:
Composition per liter:
K_2HPO_4 .... 4.10g
Na_2HPO_4 .... 3.34g
KH_2PO_4 .... 2.26g
NaH_2PO_4 .... 2.24g
Salt solution 10.0mL

Preparation of Ionic Medium: Add components to distilled/deionized water and bring volume to 1.0L. Mix thoroughly.

Salt Solution:
Composition per liter:
$MgSO_4 \cdot 7H_2O$ 14.8g
$FeSO_4 \cdot 7H_2O$ 0.55g
$MnSO_4$ 0.045g
H_2SO_4, concentrated 0.2mL

Preparation of Salt Solution: Add components to distilled/deionized water and bring volume to 1.0L. Mix thoroughly.

Pipecolic Acid·HCl Solution:
Composition 100mL:
Pipecolic acid·HCl 4.14g

Preparation of Pipecolic acid·HCl Solution: Add pipecolic acid·HCl to distilled/deionized water and bring volume to 100.0mL. Mix thoroughly. Adjust pH to 7.0.

Preparation of Medium: Add agar to 950.0mL ionic medium. Mix thoroughly. Gently heat and bring to boiling. Add 50.0mL pipecolic acid·HCl. Mix thoroughly. Distribute into tubes or flasks. Autoclave for 15 min at 15 psi pressure–121°C. Pour into sterile Petri dishes or leave in tubes.

Use: For the cultivation and maintenance of *Pseudomonas putida*.

Irgasan® Ticarcillin Chlorate Broth (ITC Broth)

Composition per liter:

$MgCl_2 \cdot 6H_20$	60.0g
Pancreatic digest of casein	10.0g
NaCl	5.0g
$KClO_4$	1.0g
Yeast extract	1.0g
Malachite Green (0.2% solution)	5.0mL
Irgasan solution	1.0mL
Ticarcillin solution	1.0mL

pH 7.6 ± 0.2 at 25°C

Irgasan Solution:

Composition per 10mL:

Irgasan (triclosan)	1.0mg

Preparation of Irgasan Solution: Add Irgasan to distilled/deionized water and bring volume to 10.0mL. Mix thoroughly. Filter sterilize.

Ticarcillin Solution:

Composition per 10mL:

Ticarcillin	1.0 mg

Preparation of Ticarcillin Solution: Add ticarcillin to distilled/deionized water and bring volume to 10.0mL. Mix thoroughly. Filter sterilize.

Preparation of Medium: Add components, except Irgasan solution and ticarcillin solution, to distilled/deionized water and bring volume to 998.0mL. Mix thoroughly. Autoclave for 15 min at 15 psi pressure–121°C. Cool to 45°–50°C. Adjust to pH 7.6. Aseptically add 1.0mL Irgasan solution and 1.0mL ticarcillin solution. Mix thoroughly. Aseptically distribute into sterile tubes or flasks.

Use: For the selective isolation and cultivation of *Yersinia* species.

Iron Agar, Lyngby (Lyngby Iron Agar)

Composition per liter:

Peptone	20.0g
Agar	12.0g
NaCl	5.0g
Beef extract	3.0g
Yeast extract	3.0g
L-Cysteine	0.6g
Ferric citrate	0.3g
$Na_2S_2O_3$	0.3g

pH 7.4 ± 0.2 at 25°C

Source: This medium is available as a premixed powder from Oxoid Unipath.

Preparation of Medium: Add components to distilled/deionized water and bring volume to 1.0L. Mix thoroughly. Gently heat and bring to boiling. Distribute into tubes or flasks. Autoclave for 15 min at 15 psi pressure–121°C. Pour into sterile Petri dishes or leave in tubes.

Use: For the cultivation and enumeration of H_2S-producing bacteria and total counts of heterotrophic bacteria from fish and fish products.

Iron Bacteria Isolation Medium

Composition per liter:

Agar	10.0g
$(NH_4)_2SO_4$	0.5g
Glucose	0.15g
$CaCO_3$	0.1g
K_2HPO_4	0.05g
$MgSO_4 \cdot 7H_2O$	0.05g
KCl	0.05g
$Ca(NO_3)_2$	0.01g
Vitamin solution	10.0mL

Vitamin Solution:

Composition per 10mL:

Thiamine	0.4mg
Cyanocobalamin	0.01mg

Preparation of Vitamin Solution: Add components to distilled/deionized water and bring volume to 10.0mL. Mix thoroughly. Filter sterilize.

Preparation of Medium: Add components, except vitamin solution, to distilled/deionized water and bring volume to 990.0mL. Mix thoroughly. Gently heat and bring to boiling. Autoclave for 15 min at 15 psi pressure–121°C. Cool to 45°–50°C. Aseptically add 10.0mL vitamin solution. Mix thoroughly. Pour into sterile Petri dishes or distribute into sterile tubes.

Use: For the isolation of iron bacteria.

Iron Milk Medium

Composition per liter:

Iron filings	1.0g
Whole milk	1.0L

pH 6.8 ± 0.2 at 25°C

Preparation of Medium: Add iron filings, which may be small balls of steel wool, to whole milk and bring volume to 1.0L. Mix thoroughly. Distribute into tubes or flasks. Autoclave for 15 min at 15 psi pressure–121°C.

Use: For the cultivation of lactic acid bacteria. For the cultivation and differentiation of *Clostridium* species. The medium turns black if H_2S is produced. The medium turns red if acid is produced from milk carbohydrate fermentation. Acid and gas production is characteristic of *C. perfringens* and *C. butyricum.*

Iron Milk Medium, Modified

Composition per 1050mL:

Whole milk, fresh	1.0L
$FeSO_4·7H_2O$	1.0g

Preparation of Medium: Add $FeSO_4·7H_2O$ to distilled/deionized water and bring volume to 50.0mL. Add slowly to 1.0L of whole milk. Mix thoroughly. Distribute into tubes in 11.0mL volumes. Autoclave for 12 min at 13 psi pressure–118°C.

Use: For the cultivation and enumeration of *Clostridium perfringens* in foods.

Iron–Oxidizing Medium

Composition per liter:

$(NH_4)_2SO_4$	3.0g
K_2HPO_4	0.50g
$MgSO_4·7H_2O$	0.50g
KCl	0.10g
$Ca(NO_3)_2$	0.01g
$FeSO_4·7H_2O$ solution	300.0mL
H_2SO_4 (10*N*)	1.0mL

pH 3.0–3.6 at 25°C

$FeSO_4·7H_2O$ Solution:

Composition per 300mL:

$FeSO_4·7H_2O$	44.22g

Preparation of Medium: Add $FeSO_4·7H_2O$ to distilled/deionized water and bring volume to 300.0mL. Mix thoroughly. Autoclave for 15 min at 15 psi pressure–121°C. Cool to 25°C.

Preparation of Medium: Add components, except $FeSO_4·7H_2O$ solution, to distilled/deionized water and bring volume to 700.0mL. Mix thoroughly. Gently heat and bring to boiling. Autoclave for 15 min at 15 psi pressure–121°C. Cool to 25°C. Aseptically add 300.0mL sterile $FeSO_4·7H_2O$ solution. Mix thoroughly. Aseptically distribute into sterile tubes or flasks.

Use: For enumeration, isolation, and cultivation of iron and sulfur bacteria, such as *Thiobacillus ferrooxidans.*

Iron Sulfite Agar

Composition per liter:

Agar	12.0g
Pancreatic digest of casein	10.0g
Ferric citrate	0.5g
$Na_2S·9H_2O$	0.5g

pH 7.1 ± 0.2 at 25°C

Source: This medium is available as a premixed powder from Oxoid Unipath.

Preparation of Medium: Add components to distilled/deionized water and bring volume to 1.0L. Mix thoroughly. Gently heat and bring to boiling. Distribute into tubes or flasks. Autoclave for 15 min at 15 psi pressure–121°C. Mix thoroughly. Pour into sterile Petri dishes or leave in tubes.

Use: For the detection of thermophilic anaerobic organisms.

Islam GBS Agar

See: **GBS Agar Base, Islam**

ISM Agar

Composition per liter:

$MgSO_4·7H_2O$	49.2g
NaCl	43.5g
Agar	7.5g
Yeast exract	4.0g
Peptone	2.0g
$CaCl_2·2H_2O$	1.5g
Maltose solution	100.0mL

Maltose Solution:

Composition per 100mL:

Maltose	5.0g

Preparation of Maltose Solution: Add maltose to distilled/deionized water and bring volume to 100.0mL. Mix thoroughly. Filter sterilize.

Preparation of Medium: Add components, except maltose solution, to distilled/deionized water and bring volume to 900.0mL. Mix thoroughly. Gently heat and bring to boiling. Autoclave for 15 min at 15 psi pressure–121°C. Cool to 45°–50°C. Aseptically add sterile maltose solution. Mix thoroughly. Pour into sterile Petri dishes or distribute into sterile tubes.

Use: For the cultivation and maintenance of *Spirochaeta halophila.*

ISM Broth

Composition per liter:

$MgSO_4 \cdot 7H_2O$	49.2g
NaCl	43.5g
Yeast exract	4.0g
Peptone	2.0g
$CaCl_2 \cdot 2H_2O$	1.5g
Maltose solution	100.0mL

Maltose Solution:

Composition per 100mL:

Maltose	5.0g

Preparation of Maltose Solution: Add maltose to distilled/deionized water and bring volume to 100.0mL. Mix thoroughly. Filter sterilize.

Preparation of Medium: Add components, except maltose solution, to distilled/deionized water and bring volume to 900.0mL. Mix thoroughly. Gently heat and bring to boiling. Autoclave for 15 min at 15 psi pressure–121°C. Cool to 45°–50°C. Aseptically add sterile maltose solution. Mix thoroughly. Aseptically distribute into sterile tubes or flasks.

Use: For the cultivation of *Spirochaeta halophila*.

Isoleucine Hydroxamate Medium

Composition per liter:

Agar	15.0g
K_2HPO_4	7.0g
Glucose	5.0g
KH_2PO_4	3.0g
L-Isoleucine hydroxamate	1.0g
$(NH_4)_2SO_4$	1.0g

pH 7.0 ± 0.2 at 25°C

Preparation of Medium: Add components to distilled/deionized water and bring volume to 1.0L. Mix thoroughly. Gently heat and bring to boiling. Distribute into tubes or flasks. Autoclave for 15 min at 15 psi pressure–121°C. Pour into sterile Petri dishes or leave in tubes.

Use: For the cultivation and maintenance of *Serratia marcescens*.

Iso–Sensitest Agar

Composition per liter:

Casein, hydrolyzed	11.0g
Agar	8.0g
Peptones	3.0g
NaCl	3.0g
Na_2HPO_4	2.0g
Glucose	2.0g
Sodium acetate	1.0g
Soluble starch	1.0g
Magnesium glycerophosphate	0.2g
Calcium gluconate	0.1g
L-Cysteine·HCl	0.02g
L-Tryptophan	0.02g
Adenine	0.01g
Guanine	0.01g
Xanthine	0.01g
Uracil	0.01g
Nicotinamide	3.0mg
Pantothenate	3.0mg
Pyridoxine	3.0mg
$MnCl_2 \cdot 4H_2O$	2.0mg
$CoSO_4$	1.0mg
$CuSO_4 \cdot 5H_2O$	1.0mg
$FeSO_4 \cdot 7H_2O$	1.0mg
Menadione	1.0mg
Cyanocobalamin	1.0mg
$ZnSO_4 \cdot 7H_2O$	1.0mg
Biotin	0.3mg
Thiamine	.04mg

pH 7.4 ± 0.2 at 25°C

Source: This medium is available as a premixed powder from Oxoid Unipath.

Preparation of Medium: Add components to distilled/deionized water and bring volume to 1.0L. Mix thoroughly. Gently heat and bring to boiling. Distribute into tubes or flasks. Autoclave for 15 min at 15 psi pressure–121°C. Pour into sterile Petri dishes or leave in tubes.

Use: For antimicrobial susceptibility testing.

Iso–Sensitest Broth

Composition per liter:

Casein, hydrolysed	11.0g
Peptones	3.0g
NaCl	3.0g
Glucose	2.0g
Na_2HPO_4	2.0g
Sodium acetate	1.0g
Soluble starch	1.0g
Magnesium glycerophosphate	0.2g
Calcium gluconate	0.1g
L-Cysteine·HCl	0.02g
L-Tryptophan	0.02g
Adenine	0.01g
Guanine	0.01g
Xanthine	0.01g
Uracil	0.01g
Nicotinamide	3.0mg
Pantothenate	3.0mg
Pyridoxine	3.0mg
$MnCl_2 \cdot 4H_2O$	2.0mg
$CoSO_4$	1.0mg
$CuSO_4 \cdot 5H_2O$	1.0mg

$FeSO_4 \cdot 7H_2O$	1.0mg
Menadione	1.0mg
Cyanocobalamin	1.0mg
$ZnSO_4 \cdot 7H_2O$	1.0mg
Biotin	0.3mg
Thiamine	0.04mg

pH 7.4 ± 0.2 at 25°C

Source: This medium is available as a premixed powder from Oxoid Unipath.

Preparation of Medium: Add components to distilled/deionized water and bring volume to 1.0L. Mix thoroughly. Gently heat and bring to boiling. Distribute into tubes or flasks. Autoclave for 15 min at 15 psi pressure–121°C. Pour into sterile Petri dishes or leave in tubes.

Use: For antimicrobial susceptibility testing.

IsoVitaleX® Enrichment

Composition per liter:

Glucose	100.0g
L-Cysteine·HCl	25.9g
L-Glutamine	10.0g
Adenine	1.0g
Thiamine pyrophosphate	0.1g
Vitamin B_{12}	0.1g
Guanine·HCl	0.03g
$Fe(NO_3)_3 \cdot 9H_2O$	0.02g
p-Aminobenzoic acid	0.013g
Thiamine·HCl	0.003g

Preparation of IsoVitaleX® Enrichment: Add components to distilled/deionized water and bring volume to 1.0L. Mix thoroughly. Filter sterilize.

Use: As a nutrient supplement for the isolation and cultivation of fastidious microorganisms.

ISP Medium 1 (International *Streptomyces* Project Medium 1) (Tryptone Yeast Extract Broth)

Composition per liter:

Pancreatic digest of casein	5.0g
Yeast extract	3.0g

pH 7.0–7.2 at 25°C

Preparation of Medium: Add components to distilled/deionized water and bring volume to 1.0L. Mix thoroughly. Distribute into tubes or flasks. Autoclave for 15 min at 15 psi pressure–121°C.

Use: For cultivation of *Streptomyces* species according to the International *Streptomyces* Project.

ISP Medium 2 (International *Streptomyces* Project Medium 2) (Yeast Extract Malt Extract Agar)

Composition per liter:

Agar	20.0g
Malt extract	10.0g
Yeast extract	4.0g
Glucose	4.0g

pH 7.3 ± 0.2 at 25°C

Preparation of Medium: Add components to distilled/deionized water and bring volume to 1.0L. Mix thoroughly. Gently heat and bring to boiling. Distribute into tubes or flasks. Autoclave for 15 min at 15 psi pressure–121°C. Pour into sterile Petri dishes or leave in tubes.

Use: For cultivation of *Streptomyces* species according to the International *Streptomyces* Project.

ISP Medium 3 (International *Streptomyces* Project Medium 3) (Oatmeal Agar)

Composition per liter:

Oatmeal	20.0g
Agar	18.0g
Trace salts solution	1.0mL

Source: This medium is available as a premixed powder from Difco Laboratories.

Trace Salts Solution:
Composition per 100mL:

$FeSO_4 \cdot 7H_2O$	0.1g
$MnCl_2 \cdot 4H_2O$	0.1g
$ZnSO_4 \cdot 7H_2O$	0.1g

Preparation of Trace Salts Solution: Add components to distilled/deionized water and bring the volume to 100.0mL. Mix thoroughly. Filter sterilize.

Preparation of Medium: Add oatmeal to distilled/deionized water and bring volume to 1.0L. Mix thoroughly. Gently heat and bring to boiling. Steam for 20 min. Filter through cheesecloth. Add agar. Add sufficient distilled/deionized water to bring volume to 999.0mL. Gently heat and bring to boiling. Mix thoroughly. Distribute into tubes or flasks. Autoclave for 15 min at 15 psi pressure–121°C. Cool to 45°–50°C. Aseptically add 1.0mL sterile trace salts solution. Mix thoroughly. Pour into sterile Petri dishes or distribute into sterile tubes.

Use: For cultivation of *Streptomyces* species according to the International *Streptomyces* Project.

ISP Medium 4 (International *Streptomyces* Project Medium 4) (Inorganic Salts Starch Agar)

Composition per liter:

Agar	20.0g
Soluble starch	10.0g
$CaCO_3$	2.0g
$(NH_4)_2SO_4$	2.0g
K_2HPO_4	1.0g
$MgSO_4 \cdot 7H_2O$	1.0g
NaCl	1.0g
$FeSO_4 \cdot 7H_2O$	1.0mg
$MnCl_2 \cdot 7H_2O$	1.0mg
$ZnSO_4 \cdot 7H_2O$	1.0mg

pH 7.2 ± 0.2 at 25°C

Source: This medium is available as a premixed powder from Difco Laboratories.

Preparation of Medium: Add components to distilled/deionized water and bring volume to 1.0L. Mix thoroughly. Gently heat and bring to boiling with frequent agitation. Distribute into tubes or flasks. Autoclave for 15 min at 15 psi pressure–121°C. Pour into sterile Petri dishes or leave in tubes.

Use: For characterizing *Streptomyces* species. For the cultivation and maintenance of *Actinomadura fastidiosa, Actinomadura roseoviolacea, Actinomadura* species, *Actinoplanes* species, *Amycolatopsis mediterranei, Kitasatosporia grisea, Kitasatosporia papulosa, Saccharomonospora internatus, Saccharopolyspora hirsuta, Streptomyces* species, *Streptosporangium* species, and *Streptoverticillium* species.

ISP Medium 4 with Glucose (International *Streptomyces* Project Medium 4 with Glucose)

Composition per liter:

Agar	20.0g
Glucose	20.0g
Soluble starch	10.0g
$CaCO_3$	2.0g
$(NH_4)_2SO_4$	2.0g
K_2HPO_4	1.0g
$MgSO_4 \cdot 7H_2O$	1.0g
NaCl	1.0g
$FeSO_4 \cdot 7H_2O$	1.0mg
$MnCl_2 \cdot 7H_2O$	1.0mg
$ZnSO_4 \cdot 7H_2O$	1.0mg

pH 7.2 ± 0.2 at 25°C

Preparation of Medium: Add components to distilled/deionized water and bring volume to 1.0L. Mix thoroughly. Gently heat and bring to boiling with frequent agitation. Distribute into tubes or flasks. Autoclave for 15 min at 15 psi pressure–121°C. Pour into sterile Petri dishes with swirling or leave in tubes.

Use: For the cultivation and maintenance of *Streptomyces purpureus.*

ISP Medium 4 with Yeast Extract (International *Streptomyces* Project Medium 4 with Yeast Extract)

Composition per liter:

Agar	20.0g
Soluble starch	10.0g
$CaCO_3$	2.0g
$(NH_4)_2SO_4$	2.0g
K_2HPO_4	1.0g
$MgSO_4 \cdot 7H_2O$	1.0g
NaCl	1.0g
Yeast extract	1.0g
$FeSO_4 \cdot 7H_2O$	1.0mg
$MnCl_2 \cdot 7H_2O$	1.0mg
$ZnSO_4 \cdot 7H_2O$	1.0mg

pH 7.2 ± 0.2 at 25°C

Preparation of Medium: Add components to distilled/deionized water and bring volume to 1.0L. Mix thoroughly. Gently heat and bring to boiling with frequent agitation. Distribute into tubes or flasks. Autoclave for 15 min at 15 psi pressure–121°C. Pour into sterile Petri dishes with swirling or leave in tubes.

Use: For the cultivation and maintenance of *Thermomonospora mesouviformis.*

ISP Medium 5 (International *Streptomyces* Project Medium 5) (Glycerol Asparagine Agar)

Composition per liter:

Agar	20.0g
Glycerol	10.0g
L-asparagine	1.0g
K_2HPO_4	1.0g
Trace salts solution	1.0mL

pH 7.4 ± 0.2 at 25°C

Trace Salts Solution:

Composition per 100mL:

$FeSO_4 \cdot 7H_2O$	0.1g

$MnCl_2 \cdot 4H_2O$ 0.1g
$ZnSO_4 \cdot 7H_2O$ 0.1g

Preparation of Trace Salts Solution: Add components to distilled/deionized water and bring the volume to 100.0mL. Mix thoroughly. Filter sterilize.

Preparation of Medium: Add components, except trace salts solution, to distilled/deionized water and bring volume to 999.0mL. Mix thoroughly. Gently heat and bring to boiling. Autoclave for 15 min at 15 psi pressure–121°C. Cool to 45°–50°C. Aseptically add 1.0mL sterile trace salts solution. Mix thoroughly. Pour into sterile Petri dishes or distribute into sterile tubes.

Use: For the cultivation and maintenance of the *Pseudonocardia* species and *Streptomyces peucetius*.

ISP Medium 6 (International *Streptomyces* Project Medium 6) (Peptone Yeast Extract Iron Agar)

Composition per liter:

Agar 15.0g
Peptone 15.0g
Proteose peptone 5.0g
K_2HPO_4 .... 1.0g
Yeast extract 1.0g
Ferric ammonium citrate 0.5g
$Na_2S_2O_3$ 0.08g

Preparation of Medium: Add components to distilled/deionized water and bring volume to 1.0L. Mix thoroughly. Gently heat and bring to boiling. Distribute into tubes or flasks. Autoclave for 15 min at 15 psi pressure–121°C. Pour into sterile Petri dishes or leave in tubes.

Use: For the cultivation and maintenance of *Streptomyces* species.

ISP Medium 7
See: **Tyrosine Agar**

ISP Medium 8
See: **Nitrate Broth**

ISP Medium 9 (International Streptomyces Project Medium 9)

Composition per liter:

$K_2HPO_4 \cdot 3H_2O$ 5.65g
$(NH_4)_2SO_4$ 2.64g
KH_2PO_4 .... 2.38g
$MgSO_4 \cdot 7H_2O$ 1.0g
Carbohydrate solution 100.0mL
Pridham and Gottlieb trace salts 1.0mL
pH 6.8–7.0 at 25°C

Carbohydrate Solution:
Composition per 100mL:

Carbohydrate 10.0g

Preparation of Carbohydrate Solution: Add carbohydrate to distilled/deionized water and bring volume to 100.0mL.Use glucose, arabinose, sucrose, xylose, inositol, mannitol, fructose, rhamnose, raffinose or cellulose. Mix thoroughly. Filter sterilize.

Pridham And Gottlieb Trace Salts:
Composition per 100mL:

$MnCl_2 \cdot 7H_2O$ 0.79g
$CuSO_4 \cdot 5H_2O$ 0.64g
$ZnSO_4 \cdot 7H_2O$ 0.15g
$FeSO_4 \cdot 7H_2O$ 0.11g

Preparation of Pridham And Gottlieb Trace Salts: Add components to distilled/deionized water and bring volume to 100.0mL. Mix thoroughly.

Preparation of Medium: Add components, except carbohydrate solution, to distilled/deionized water and bring volume to 900.0mL. Mix thoroughly. Gently heat and bring to boiling with frequent agitation. Autoclave for 15 min at 15 psi pressure–121°C. Cool to 45°–50°C. Aseptically add sterile carbohydrate solution. Mix thoroughly. Aseptically distribute into sterile tubes or flasks.

Use: For the cultivation and differentiation of *Streptomyces purpureus* and other *Streptomyces* species based on carbohydrate utilization.

ITC Broth
See: **Irgasan® Ticarcillin Chlorate Broth**

J Agar

Composition per liter:

Agar	20.0g
Yeast extract	15.0g
Pancreatic digest of casein	5.0g
K_2HPO_4	3.0g
Glucose solution	10.0mL

pH 7.3–7.5 at 25°C

Glucose Solution:
Composition per 10mL:

Glucose	2.0g

Preparation of Glucose Solution: Add glucose to distilled/deionized water and bring volume to 10.0mL. Mix thoroughly. Filter sterilize.

Preparation of Medium: Add components, except glucose solution, to distilled/deionized water and bring volume to 990.0mL. Mix thoroughly. Gently heat and bring to boiling. Autoclave for 15 min at 15 psi pressure–121°C. Cool to 45°–50°C. Aseptically add sterile glucose solution. Mix thoroughly. Pour into sterile Petri dishes.

Use: For the cultivation of *Bacillus* species and *Sporolactobacillus* species.

J Broth

Composition per liter:

Yeast extract	15.0g
Pancreatic digest of casein	5.0g

pH 7.3–7.5 at 25°C

Preparation of Medium: Add components to distilled/deionized water and bring volume to 1.0L. Mix thoroughly. Adjust pH to 7.3–7.5. Distribute into tubes or flasks. Autoclave for 20 min at 15 psi pressure–121°C.

Use: For the cultivation of *Bacillus* species and *Sporolactobacillus* species for performing the Voges-Proskauer test.

JB Medium with Glucose

Composition per liter:

Yeast extract	15.0g
Pancreatic digest of casein	5.0g
K_2HPO_4	3.0g
Glucose	2.0g

pH 7.3–7.5 at 25°C

Preparation of Medium: Add components to distilled/deionized water and bring volume to 1.0L. Mix thoroughly. Gently heat and bring to boiling. Distribute into tubes or flasks. Autoclave for 15 min at 15 psi pressure–121°C. Pour into sterile Petri dishes or leave in tubes.

Use: For the cultivation and maintenance of *Bacillus popilliae*.

JD1 Medium

Composition per liter:

Beef heart, solids from infusion	25.0g
Agar	15.0g
Peptone	5.0g
NaCl	2.5g
Bovine albumin	0.5g
Hemin chloride	0.04g

pH 7.8 ± 0.2 at 25°C

Preparation of Medium: Add components to distilled/deionized water and bring volume to 1.0L. Mix thoroughly. Gently heat and bring to boiling. Distribute into tubes or flasks. Autoclave for 15 min at 15 psi pressure–121°C. Pour into sterile Petri dishes or leave in tubes.

Use: For the isolation and cultivation of PD-ALS (Pierce's disease-almond leaf scorch) bacteria.

JD3 Medium

Composition per liter:

Pancreatic digest of casein	4.0g
Papaic digest of soybean meal	2.0g
Trisodium citrate	2.0g
K_2HPO_4	1.5g
KH_2PO_4	1.0g
$MgSO_4 \cdot 7H_2O$	1.0g
Disodium succinate	0.01g
Bovine serum albumin solution	100.0mL
Hemin chloride solution	10.0mL

pH 7.0 ± 0.2 at 25°C

Bovine Serum Albumin Solution:
Composition per 100mL:

Bovine serum albumin Fraction V	2.0g

Preparation of Bovine Serum Albumin Solution: Add 2.0g bovine serum albumin Fraction V to 100.0 mL distilled/deionized water. Mix thoroughly. Filter sterilize.

Hemin Chloride Solution:
Composition per 10mL:

Hemin chloride	0.01g

Preparation of Hemin Chloride Solution: Add 0.01g hemin chloride to to 10.0 mL 0.5*N* NaOH. Mix thoroughly.

Preparation of Medium: Add components, except bovine serum albumin solution, to distilled/

deionized water and bring volume to 900.0mL. Mix thoroughly. Adjust to pH 7.0. Autoclave for 20 min at 15 psi pressure–121°C. Cool to 25°C. Aseptically add 100.0mL sterile bovine serum albumin solution. Mix thoroughly. Aseptically distribute into sterile tubes or flasks.

Use: For the cultivation of ATCC strain 33107.

Jones–Kendrick Pertussis Transport Medium

Composition per liter:

Beef heart, solids from infusion	500.0g
Tryptose	10.0g
NaCl	5.0g
Agar	20.0g
Soluble starch	10.0g
Charcoal powder, activated	4.0g
Yeast extract	3.5g
Penicillin solution	10.0mL

pH 7.4 ± 0.2 at 25°C

Penicillin Solution:
Composition per 10mL:

Penicillin	300U

Preparation of Penicillin Solution: Add penicillin to distilled/deionized water and bring volume to 10.0mL. Mix thoroughly. Filter sterilize.

Preparation: Add components, except penicillin solution, starch, yeast extract, heart infusion, and agar to water. Boil to dissolve. Add charcoal, mix well and autoclave. Cool to 50°C, add penicillin and dispense into small bottles as slants. Cool and seal tightly. Store at 5°C. Stable for 2 to 3 months.

Use: For the cultivation and transport of *Bordetella pertussis* between clinical isolation and laboratory cultivation.

Jordan's Tartrate Agar

Composition per liter:

Agar	15.0g
Pancreatic digest of casein	10.0g
Sodium potassium tartrate	10.0g
NaCl	5.0g
Phenol Red	0.024g

pH 7.7 ± 0.3 at 25°C

Source: This medium is available as a prepared medium in tubes from BBL Microbiology Systems.

Preparation of Medium: Add components to distilled/deionized water and bring volume to 1.0L. Mix thoroughly. Gently heat and bring to boiling. Adjust pH to 7.7. Distribute into tubes. Autoclave for 15 min at 15 psi pressure–121°C.

Use: For differentiation and identification of members of the Enterobacteriaceae, especially *Salmonella* species, based upon the ability to utilize tartrate. Utilization of tartrate turns the medium yellow. *Salmonella enteritidis* utilizes tartrate. *S. paratyphi* A does not utilize tartrate.

K101 *Flexibacter* Medium

Composition per liter:

Agar	10.0g
Casamino acids	1.0g
Glucose	1.0g
Tris(hydroxymethyl)aminomethane buffer	1.0g
$CaCl_2$	0.1g
KNO_3	0.1g
$MgSO_4 \cdot 7H_2O$	0.1g
Sodium glycerophosphate	0.1g
Thiamine·HCl	1.0mg
Cyanocobalamin	1.0μg
Trace elements solution HO-LE	1.0mL

pH 7.5 ± 0.2 at 25°C

Trace Element Solution HO-LE:
Composition per liter:

H_3BO_3	2.85g
$MnCl_2 \cdot 4H_2O$	1.8g
Sodium tartrate	1.77g
$FeSO_4 \cdot 7H_2O$	1.36g
$CoCl_2 \cdot 6H_2O$	0.04g
$CuCl_2 .2H_2O$	0.027g
$Na_2MoO_4 \cdot 2H_2O$	0.025g
$ZnCl_2$	0.020g

Preparation of Trace Element Solution HO-LE: Add components to distilled/deionized water and bring volume to 1.0L. Mix thoroughly. Filter sterilize.

Preparation of Medium: Add components, except trace elements solution HO-LE, to distilled/deionized water and bring volume to 999.0mL. Mix thoroughly. Gently heat and bring to boiling. Autoclave for 15 min at 15 psi pressure–121°C. Cool to 45°–50°C. Aseptically add 1.0mL trace elements solution HO-LE. Mix thoroughly. Pour into sterile Petri dishes or distribute into sterile tubes.

Use: For the cultivation and maintenance of *Cytophaga* species, *Flexibacter* species, *Herpetosiphon geysericola,* and *Myxococcus fulvus*.

Kanamycin Esculin Azide Agar

Composition per liter:

Pancreatic digest of casein	20.0g
Agar	10.0g
NaCl	5.0g
Yeast extract	5.0g
Esculin	1.0g
Sodium citrate	1.0g
Ferric ammonium citrate	0.5g
NaN_3	0.15g
Kanamycin sulfate solution	10.0mL

pH 7.0 ± 0.2 at 25°C

Source: This medium is available as a premixed powder from Oxoid Unipath.

Kanamycin Sulfate Solution:
Composition per 10mL:

Kanamycin sulfate	20.0mg

Preparation of Kanamycin Sulfate Solution: Add kanamycin sulfate to distilled/deionized water and bring volume to 10.0mL. Mix thoroughly.

Caution: Sodium azide is toxic. Azides also react with metals and disposal must be highly diluted.

Preparation of Medium: Add components to distilled/deionized water and bring volume to 1.0L. Mix thoroughly. Gently heat and bring to boiling. Distribute into tubes or flasks. Autoclave for 15 min at 15 psi pressure–121°C. Pour into sterile Petri dishes or leave in tubes.

Use: For the isolation of enterococci from foods.

Kanamycin Esculin Azide Broth

Composition per liter:

Pancreatic digest of casein	20.0g
NaCl	5.0g
Yeast extract	5.0g
Esculin	1.0g
Sodium citrate	1.0g
Ferric ammonium citrate	0.5g
NaN_3	0.15g
Kanamycin sulfate solution	10.0mL

pH 7.0 ± 0.2 at 25°C

Source: This medium is available as a premixed powder from Oxoid Unipath.

Kanamycin Sulfate Solution:
Composition per 10mL:

Kanamycin sulfate	0.02g

Preparation of Kanamycin Sulfate Solution: Add kanamycin sulfate to distilled/deionized water and bring volume to 10.0mL. Mix thoroughly.

Caution: Sodium azide is toxic. Azides also react with metals and disposal must be highly diluted.

Preparation of Medium: Add components to distilled/deionized water and bring volume to 1.0L. Mix thoroughly. Distribute into tubes or flasks. Autoclave for 15 min at 15 psi pressure–121°C.

Use: For the isolation of enterococci from foods.

Kanamycin Vancomycin Blood Agar (KVBA)

Composition per liter:

Agar	17.5g
Pancreatic digest of casein	15.0g
Papaic digest of soybean meal	5.0g

NaCl	5.0g
Kanamycin	0.1g
Sheep blood, defibrinated	50.0mL
Vancomycin solution	10.0mL
Vitamin K_1 solution	1.0mL

Vancomycin Solution:
Composition per 10mL:

Vancomycin	7.5mg

Preparation of Vancomycin Solution: Add vancomycin to distilled/deionized water and bring volume to 10.0mL. Mix thoroughly. Filter sterilize.

Vitamin K_1 Solution:
Composition per 100mL:

Vitamin K_1	1.0g

Preparation of Vitamin K_1 Solution: Add vitamin K_1 to 99.0mL of absolute ethanol. Mix thoroughly. Filter sterilize.

Preparation of Medium: Add components, except sheep blood, vancomycin and vitamin K_1 solution, to distilled/deionized water and bring volume to 939.0mL. Mix thoroughly. Gently heat and bring to boiling. Autoclave for 15 min at 15 psi pressure–121°C. Cool to 45°–50°C. Aseptically add sheep blood, vancomycin solution and 1.0mL vitamin K_1 solution. Mix thoroughly. Pour into sterile Petri dishes or distribute into sterile tubes.

Use: For the selective isolation of anaerobes, particularly *Bacteroides*, from clinical specimens.

Kanamycin Vancomycin Laked Blood Agar

Composition per liter:

Agar	17.5g
Pancreatic digest of casein	15.0g
Papaic digest of soybean meal	5.0g
NaCl	5.0g
Kanamycin	0.075g
Sheep blood, laked	50.0mL
Vancomycin solution	10.0mL
Vitamin K_1 solution	1.0mL

Vancomycin Solution:
Composition per 10mL:

Vancomycin	7.5mg

Preparation of Vancomycin Solution: Add vancomycin to distilled/deionized water and bring volume to 10.0mL. Mix thoroughly. Filter sterilize.

Vitamin K_1 Solution:
Composition per 100mL:

Vitamin K_1	1.0g

Preparation of Vitamin K_1 Solution: Add vitamin K_1 to 99.0mL of absolute ethanol. Mix thoroughly. Filter sterilize.

Preparation of Medium: The blood is laked (hemolyzed) by freezing whole blood overnight and then thawing. Add components, except sheep blood, vancomycin and vitamin K_1 solution, to distilled/deionized water and bring volume to 939.0mL. Mix thoroughly. Gently heat and bring to boiling. Autoclave for 15 min at 15 psi pressure–121°C. Cool to 45°–50°C. Aseptically add sheep blood, vancomycin solution and 1.0mL vitamin K_1 solution. Mix thoroughly. Pour into sterile Petri dishes or distribute into sterile tubes.

Use: For the isolation of *Bacteroides melaninogenicus* group.

Karmali's *Campylobacter* Medium
See: Campylobacter Selective Medium, Karmali's

Kasai Medium

Composition per liter:

Pancreatic digest of casein	20.0g
Soluble starch	20.0g
Cysteine·HCl·H_2O	5.0g
K_2HPO_4	5.0g
NaCl	5.0g
Yeast extract	2.0g

Preparation of Medium: Add components to distilled/deionized water and bring volume to 1.0L. Mix thoroughly. Gently heat and bring to boiling. Distribute into tubes or flasks. Autoclave for 15 min at 15 psi pressure–121°C.

Use: For the isolation and cultivation of *Leptotrichia buccalis* from saliva and plaque.

KC Bottom Agar

Composition per liter:

Agar	10.0g
Pancreatic digest of casein	10.0g
KCl	2.5g
NaCl	2.5g
$CaCl_2$ solution	1.0mL

$CaCl_2$ Solution:
Composition per 10mL:

$CaCl_2 \cdot 2H_2O$	1.47g

Preparation of $CaCl_2$ Solution: Add $CaCl_2 \cdot 2H_2O$ to distilled/deionized water and bring volume to 10.0mL. Mix thoroughly. Filter sterilize.

Preparation of Medium: Add components, except $CaCl_2$ solution, to distilled/deionized water and bring volume to 999.0mL. Mix thoroughly. Gently heat and bring to boiling. Autoclave for 15 min at 15 psi pressure–121°C. Cool to 45°–50°C. Aseptically add 1.0mL $CaCl_2$ solution. Mix thoroughly. Pour into sterile Petri dishes or distribute into sterile tubes.

Use: For the cultivation and maintenance of *Escherichia coli.*

KC Broth

Composition per liter:

Pancreatic digest of casein	10.0g
KCl	5.0g
$CaCl_2$ solution	0.5mL

$CaCl_2$ Solution:
Composition per 10mL:

$CaCl_2 \cdot 2H_2O$	1.47g

Preparation of $CaCl_2$ Solution: Add $CaCl_2 \cdot 2H_2O$ to distilled/deionized water and bring volume to 10.0mL. Mix thoroughly. Filter sterilize.

Preparation of Medium: Add components, except $CaCl_2$ solution, to distilled/deionized water and bring volume to 999.5mL. Mix thoroughly. Distribute into tubes or flasks. Autoclave for 15 min at 15 psi pressure–121°C. Cool to 25°C. Aseptically add 0.5mL $CaCl_2$ solution. Mix thoroughly. Aseptically distribute into sterile tubes.

Use: For the cultivation of *Escherichia coli.*

KC Top Agar

Composition per liter:

Pancreatic digest of casein	10.0g
Agar	8.0g
NaCl	5.0g

Preparation of Medium: Add components to distilled/deionized water and bring volume to 1.0L. Mix thoroughly. Gently heat and bring to boiling. Distribute into tubes or flasks. Autoclave for 15 min at 15 psi pressure–121°C. Pour into sterile Petri dishes or leave in tubes.

Use: For the cultivation and maintenance of *Escherichia coli.*

KCN Broth

Composition per liter:

Na_2HPO_4	5.64g
NaCl	5.0g
Peptone	3.0g
KH_2PO_4	0.225g
KCN (0.5% solution)	15.0mL

pH 7.6 ± 0.2 at 25°C

Caution: Cyanide is toxic.

Preparation of Medium: Add components, except KCN solution, to distilled/deionized water and bring volume to 985.0mL. Mix thoroughly. Autoclave for 15 min at 15 psi pressure–121°C. Cool to 25°C. Aseptically add KCN solution. Mix thoroughly. Aseptically distribute into sterile tubes. Stopper immediately.

Use: For the differentiation of Enterobacteriaceae based upon growth in the presence of potassium cyanide.

Kelly Medium, Nonselective Modified

Composition per 1430mL:

HEPES buffer (*N*-2-Hydroxyethylpiperazine-*N*-2-ethanesulfonic acid)	6.0g
Proteose peptone No. 2	5.0g
D-Glucose	3.0g
$NaHCO_3$	2.2g
Pancreatic digest of casein	1.0g
Yeast, autolyzed	1.0g
Sodium pyruvate	0.8g
Sodium citrate	0.7g
N-Acetylglucosamine	0.4g
$MgCl_2 \cdot 6H_2O$	0.3g
Gelatin solution	200.0mL
Bovine serum albumin	143.0mL
CMRL-1066 medium with glutamine, 10X	100.0mL
Rabbit serum, heat inactivated	86.0mL
Hemin solution	1.0mL

pH 7.2 ± 0.2 at 25°C

CMRL-1066 Medium with Glutamine, 10X:
Composition per liter:

NaCl	6.8g
$NaHCO_3$	2.2g
D-Glucose	1.0g
KCl	0.4g
L-Cysteine·HCl·H_2O	0.26g
$CaCl_2$, anhydrous	0.2g
$MgSO_4 \cdot 7H_2O$	0.2g
$NaH_2PO_4 \cdot H_2O$	0.14g
L-Glutamine	0.1g
Sodium acetate·$3H_2O$	0.083g
L-Glutamic acid	0.075g
L-Arginine·HCl	0.070g
L-Lysine·HCl	0.070g
L-Leucine	0.060g
Glycine	0.050g

Ascorbic acid	0.050g
L-Proline	0.040g
L-Tyrosine	0.040g
L-Aspartic acid	0.030g
L-Threonine	0.030g
L-Alanine	0.025g
L-Phenylalanine	0.025g
L-Serine	0.025g
L-Valine	0.025g
L-Cystine	0.020g
L-Histidine·HCl·H_2O	0.020g
L-Isoleucine	0.020g
Phenol Red	0.020g
L-Methionine	0.015g
Deoxyadenosine	0.010g
Deoxycytidine	0.010g
Deoxyguanosine	0.010g
Glutathione, reduced	0.010g
Thymidine	0.010g
Hydroxy-L-proline	0.010g
L-Tryptophan	0.010g
Nicotinamide adenine dinucleotide	7.0mg
Tween™ 80	5.0mg
Sodium glucuronate·H_2O	4.2mg
Coenzyme A	2.5mg
Cocarboxylase	1.0mg
Flavin adenine dinucleotide	1.0mg
Nicotinamide adenine dinucleotide phosphate	1.0mg
Uridine triphosphate	1.0mg
Choline chloride	0.50mg
Cholesterol	0.20mg
5-Methyldeoxycytidine	0.10mg
Inositol	0.05mg
p-Aminobenzoic acid	0.05mg
Niacin	0.025mg
Niacinamide	0.025mg
Pyridoxine	0.025mg
Pyridoxal·HCl	0.025mg
Biotin	0.01mg
D-Calcium pantothenate	0.01mg
Folic acid	0.01mg
Riboflavin	0.01mg
Thiamine·HCl	0.01mg

pH 7.2 ± 0.2 at 25°C

Source: This solution is available as a premixed powder from GIBCO BRL.

Preparation of CMRL-1066 Medium with Glutamine, 10X: Add components to distilled/deionized water and bring volume to 1.0L. Mix thoroughly. Adjust pH to 7.2. Filter sterilize.

Gelatin Solution:

Composition per 200mL:

Gelatin	14.0g

Preparation of Gelatin Solution: Add gelatin to distilled/deionized water and bring volume to 1.0L. Mix thoroughly. Gently heat and bring to boiling. Autoclave for 15 min at 15 psi pressure–121°C. Cool to 50°C.

Hemin Solution:

Composition per 100mL:

Hemin	1.0g
NaOH (1*N* solution)	20.0mL

Preparation of Hemin Solution: Add hemin to 20.0mL of 1*N* NaOH solution. Mix thoroughly. Bring volume to 100.0mL with distilled/deionized water.

Bovine Serum Albumin Solution:

Composition per 200mL:

Bovine serum albumin	70.0g

Preparation of Bovine Serum Albumin Solution: Add bovine serum albumin to distilled/deionized water and bring volume to 200.0mL. Filter sterilize.

Preparation of Medium: Add components, except gelatin solution, bovine serum albumin solution, and rabbit serum to distilled/deionized water and bring volume to 1001.0mL. Mix thoroughly. Bring pH to 7.6 with 5 *N* NaOH. Filter sterilize. Aseptically add 200.0mL sterile gelatin solution, 143.0mL sterile bovine serum albumin and 86.0mL sterile heat-inactivated rabbit serum. Mix thoroughly. Aseptically dispense into sterile tubes or flasks.

Use: For isolation of *Borrelia burgdorferi* and other spirochetes.

Kelly Medium, Selective Modified

Composition per liter:

Bovine serum albumin fraction V	50.0g
HEPES buffer (*N*-2-Hydroxyethylpiperazine-*N*-2-ethanesulfonic acid)	6.0g
Glucose	5.0g
Neopeptone	5.0g
$NaHCO_3$	2.2g
Sodium pyruvate	0.8g
Sodium citrate	0.7g
N-Acetylglucosamine	0.4g
Kanamycin	8.0mg
5-Fluorouracil	2.3mg
Gelatin solution	200.0mL
CMRL-1066 medium with glutamine, 10X	100.0mL
Rabbit serum, partially hemolyzed	70.0mL

pH 7.7 ± 0.2 at 25°C

Gelatin Solution:

Composition per 200mL:

Gelatin 14.0g

Preparation of Gelatin Solution: Add gelatin to distilled/deionized water and bring volume to 1.0L. Mix thoroughly. Gently heat and bring to boiling. Autoclave for 15 min at 15 psi pressure–121°C. Cool to 50°C.

CMRL-1066 Medium with Glutamine, 10X:

Composition per liter:

NaCl 6.8g
$NaHCO_3$ 2.2g
D-Glucose 1.0g
KCl 0.4g
L-Cysteine·HCl·H_2O 0.26g
$CaCl_2$, anhydrous 0.2g
$MgSO_4 \cdot 7H_2O$ 0.2g
$NaH_2PO_4 \cdot H_2O$ 0.14g
L-Glutamine 0.1g
Sodium acetate·$3H_2O$ 0.083g
L-Glutamic acid 0.075g
L-Arginine·HCl 0.070g
L-Lysine·HCl 0.070g
L-Leucine 0.060g
Glycine 0.050g
Ascorbic acid 0.050g
L-Proline 0.040g
L-Tyrosine 0.040g
L-Aspartic acid 0.030g
L-Threonine 0.030g
L-Alanine 0.025g
L-Phenylalanine 0.025g
L-Serine 0.025g
L-Valine 0.025g
L-Cystine 0.020g
L-Histidine·HCl·H_2O 0.020g
L-Isoleucine 0.020g
Phenol Red 0.020g
L-Methionine 0.015g
Deoxyadenosine 0.010g
Deoxycytidine 0.010g
Deoxyguanosine 0.010g
Glutathione, reduced 0.010g
Thymidine 0.010g
Hydroxy-L-proline 0.010g
L-Tryptophan 0.010g
Nicotinamide adenine dinucleotide 7.0mg
Tween™ 80 .... 5.0mg
Sodium glucuronate·H_2O 4.2mg
Coenzyme A 2.5mg
Cocarboxylase 1.0mg
Flavin adenine dinucleotide 1.0mg
Nicotinamide adenine dinucleotide phosphate 1.0mg
Uridine triphosphate 1.0mg
Choline chloride 0.50mg
Cholesterol 0.20mg
5-Methyldeoxycytidine 0.10mg
Inositol 0.05mg
p-Aminobenzoic acid 0.05mg
Niacin 0.025mg
Niacinamide 0.025mg
Pyridoxine 0.025mg
Pyridoxal·HCl 0.025mg
Biotin 0.01mg
D-Calcium pantothenate 0.01mg
Folic acid 0.01mg
Riboflavin 0.01mg
Thiamine·HCl 0.01mg

pH 7.2 ± 0.2 at 25°C

Source: This solution is available as a premixed powder from GIBCO BRL.

Preparation of CMRL-1066 Medium with Glutamine, 10X: Add components to distilled/deionized water and bring volume to 1.0L. Mix thoroughly. Adjust pH to 7.2. Filter sterilize.

Preparation of Medium: Add components, except gelatin solution, partially hemolyzed rabbit serum solution, kanamycin, and 5-fluorouracil to distilled/deionized water and bring volume to 1.0mL. Mix thoroughly. Bring pH to 7.6 with 5 *N* NaOH. Filter sterilize. Aseptically add 200.0mL sterile gelatin solution, 70.0mL of partially hemolyzed rabbit serum, 8.0mg kanamycin, and 230.0mg 5-fluorouracil. Mix thoroughly. Aseptically distribute into sterile tubes or flasks.

Use: For the isolation of *Borrelia burgdorferi.*

Kempler–McKay Agar
See: **KM Agar**

Kerosene Mineral Salts Medium

Composition per liter:

KH_2PO_4 .... 1.0g
K_2HPO_4 .... 1.0g
NH_4NO_3 .... 1.0g
$MgSO_4 \cdot 7H_2O$ 0.2g
$CaCl_2$ 0.02g
$FeCl_3$ 0.05g
Kerosene 20.0mL

pH 6.9–7.0 at 25°C

Preparation of Medium: Add components, except kerosene, to distilled/deionized water and bring volume to 1.0L. Mix thoroughly. Adjust pH to 6.9–7.0 with dilute NaOH. Distribute into tubes in

10.0mL volumes or flasks in 100.0mL volumes. Autoclave for 15 min at 15 psi pressure–121°C. Overlay tubes with 0.2mL of kerosene per tube. Overlay flasks with 2.0mL of kerosene per flask.

Use: For the cultivation and maintenance of *Pseudomonas aeruginosa.*

Ketogluconate Broth

Composition per liter:

Potassium gluconate	20.0g
KH_2PO_4	5.4g
KNO_3	2.0g

pH 6.5 ± 0.2 at 25°C

Preparation of Medium: Add components to distilled/deionized water and bring volume to 1.0L. Mix thoroughly. Filter sterilize. Aseptically distribute into sterile tubes.

Use: For use in identifying bacteria that can utilize 2-ketogluconate.

Ketolactonate Broth

Composition per liter:

Agar	20.0g
Lactose	10.0g
Yeast extract	10.0g

Preparation of Medium: Add components to distilled/deionized water and bring volume to 1.0L. Mix thoroughly. Gently heat and bring to boiling. Distribute into tubes or flasks. Autoclave for 15 min at 15 psi pressure–121°C. Pour into sterile Petri dishes or leave in tubes.

Use: For use in identification of agrobacteria and other bacteria based upon utilization of 3-ketogluconate. After incubation Benedicts solution is added to the plates. Yellow zones around colonies indicate positive utilization of 3-ketogluconate.

KF *Streptococcus* Agar

Composition per liter:

Agar	20.0g
Maltose	20.0g
Proteose peptone	10.0g
Sodium glycerophosphate	10.0g
Yeast extract	10.0g
NaCl	5.5g
Lactose	1.0g
NaN_3	0.4g
Bromcresol Purple	0.015g
2,3,5-Triphenyltetrazolium chloride solution	10.0mL

pH 7.2 ± 0.2 at 25°C

Source: This medium is available as a premixed powder from Difco Laboratories and Oxoid Unipath.

2,3,5-Triphenyltetrazolium Chloride Solution: Composition per 10mL:

2,3,5-Triphenyltetrazolium chloride	0.1g

Preparation of 2,3,5-Triphenyltetrazolium Chloride Solution: Add 2,3,5-triphenyltetrazolium chloride to distilled/deionized water and bring volume to 10.0mL. Mix thoroughly. Filter sterilize.

Caution: Sodium azide is toxic. Azides also react with metals and disposal must be highly diluted.

Preparation of Medium: Add components, except 2,3,5-triphenyltetrazolium chloride solution, to distilled/deionized water and bring volume to 990.0mL. Mix thoroughly. Gently heat and bring to boiling. Autoclave for 15 min at 15 psi pressure–121°C. Cool to 45°–50°C. Aseptically add 2,3,5-triphenyltetrazolium chloride solution. Mix thoroughly. Pour into sterile Petri dishes or distribute into sterile tubes.

Use: For the isolation and enumeration of enterococci.

KF *Streptococcus* Agar

Composition per liter:

Agar	20.0g
Maltose	20.0g
Sodium glycerophosphate	10.0g
Yeast extract	10.0g
NaCl	5.0g
Pancreatic digest of casein	5.0g
Peptic digest of animal tissue	5.0g
Lactose	1.0g
NaN_3	0.4g
2,3,5-Triphenyltetrazolium chloride solution	10.0mL

pH 7.2 ± 0.2 at 25°C

Source: This medium is available as a premixed powder from BBL Microbiology Systems.

2,3,5-Triphenyltetrazolium Chloride Solution: Composition per 10mL:

2,3,5-Triphenyltetrazolium chloride	0.1g

Preparation of 2,3,5-Triphenyltetrazolium Chloride Solution: Add 2,3,5-triphenyltetrazolium chloride to distilled/deionized water and bring volume to 10.0mL. Mix thoroughly. Filter sterilize.

Caution: Sodium azide is toxic. Azides also react with metals and disposal must be highly diluted.

Preparation of Medium: Add components, except 2,3,5-triphenyltetrazolium chloride solution, to distilled/deionized water and bring volume to

990.0mL. Mix thoroughly. Gently heat and bring to boiling. Autoclave for 15 min at 15 psi pressure–121°C. Cool to 45°–50°C. Aseptically add 2,3,5-triphenyltetrazolium chloride solution. Mix thoroughly. Pour into sterile Petri dishes or distribute into sterile tubes.

Use: For the selective cultivation and enumeration of fecal streptococci.

KF *Streptococcus* Broth

Composition per liter:

Maltose	20.0g
Sodium glycerophosphate	10.0g
Yeast extract	10.0g
NaCl	5.0g
Pancreatic digest of casein	5.0g
Peptic digest of animal tissue	5.0g
Lactose	1.0g
Na_2CO_3	0.636g
NaN_3	0.4g
Phenol Red	0.018g
2,3,5-Triphenyltetrazolium chloride solution	10.0mL

pH 7.2 ± 0.2 at 25°C

Source: This medium is available as a premixed powder from BBL Microbiology Systems and Difco Laboratories.

Caution: Sodium azide is toxic. Azides also react with metals and disposal must be highly diluted.

Preparation of Medium: Add components, except 2,3,5-triphenyltetrazolium chloride solution, to distilled/deionized water and bring volume to 990.0mL. Mix thoroughly. Gently heat and bring to boiling. Autoclave for 15 min at 15 psi pressure–121°C. Cool to 45°–50°C. Aseptically add 2,3,5-triphenyltetrazolium chloride solution. Mix thoroughly. Aseptically distribute into sterile tubes or flasks.

Use: For the selective cultivation of fecal streptococci.

KG Agar
See: **Kim–Goepfert Agar**

Kim–Goepfert Agar (KG Agar)

Composition per 1001mL:

Solution A	900.0mL
Egg yolk emulsion, 50%	100.0mL
Polymyxin B solution	1.0mL

pH 6.8 ± 0.2 at 25°C

Source: This medium is available as a premixed powder from BBL Microbiology Systems and Difco Laboratories.

Solution A:

Composition per 900mL:

Agar	18.0g
Peptone	1.0g
Yeast extract	0.5g
Phenol Red	0.025g

Preparation of Solution A: Add components to distilled/deionized water and bring volume to 900.0mL. Mix thoroughly. Adjust pH to 6.8. Autoclave for 15 min at 15 psi pressure–121°C. Cool to 45°–50°C.

Egg Yolk Emulsion, 50%:

Composition per 100mL:

Chicken egg yolks	11
Whole chicken egg	1
NaCl (0.9% solution)	50.0mL

Preparation of Egg Yolk Emulsion, 50%: Soak eggs with 1:100 dilution of saturated mercuric chloride solution for 1 min. Crack eggs and separate yolks from whites. Mix egg yolks with 1 chicken egg. Measure 50.0mL of egg yolk emulsion and add to 50.0mL of 0.9% NaCl solution. Mix thoroughly. Filter sterilize. Warm to 45°–50°C.

Polymyxin B Solution:

Composition per 5mL:

Polymyxin B sulfate	500,000U

Preparation of Polymyxin B Solution: Add polymyxin B to distilled/deionized water and bring volume to 5.0mL. Mix thoroughly. Filter sterilize.

Preparation of Medium: To 900.0mL of cooled, sterile solution A, aseptically add 100.0mL of sterile egg yolk emulsion, 50% and 1.0mL of sterile polymyxin B solution. Mix thoroughly. Pour into sterile Petri dishes. Allow plates to dry in the dark at 30°C for 24 hr before using.

Use: For the cultivation and differentiation of *Bacillus cereus*.

Kimmig's Agar

Composition per liter:

Agar	15.0g
Glucose	10.0g
Pancreatic digest of gelatin	9.5g
Beef extract	5.5g
NaCl	5.0g
Peptone	5.0g
Glycerol	5.0mL

pH 6.9 ± 0.2 at 35°C

Preparation of Medium: Add glycerol and then other components to distilled/deionized water and bring volume to 1.0L. Mix thoroughly. Gently heat and bring to boiling. Distribute into tubes or flasks. Autoclave for 15 min at 15 psi pressure–121°C. Pour into sterile Petri dishes or leave in tubes.

Use: For the assay of fungistatic agents. For agar dilution test of antifungal agents. For cultivation and preservation of various fungi.

King's Medium A

Composition per liter:

Proteose peptone	20.0g
Agar	15.0g
Glycerol	10.0g
K_2SO_4	10.0g
$MgCl_2 \cdot 6H_2O$	3.5g

pH 7.2–7.4 ± 0.2 at 25°C

Preparation of Medium: Add components to distilled/deionized water and bring volume to 1.0L. Mix thoroughly. Gently heat and bring to boiling. Distribute into tubes or flasks. Autoclave for 15 min at 15 psi pressure–121°C. Pour into sterile Petri dishes or leave in tubes.

Use: For the nonselective isolation, cultivation and pigment production of *Pseudomonas*.

King's Medium B

Composition per liter:

Agar	20.0g
Proteose peptone No. 3	20.0g
K_2HPO_4, anhydrous	1.5g
$MgSO_4 \cdot 7H_2O$	1.5g
Glycerol	15.0mL

pH 7.2 ± 0.2 at 25°C

Preparation of Medium: Add components to distilled/deionized water and bring volume to 1.0L. Mix thoroughly. Gently heat and bring to boiling. Distribute into tubes or flasks. Autoclave for 15 min at 15 psi pressure–121°C. Pour into sterile Petri dishes or leave in tubes.

Use: For the nonselective isolation, cultivation and pigment production of *Pseudomonas* species.

King's Medium B

Composition per liter:

Proteose peptone No.3	20.0g
Agar	15.0g
K_2HPO_4	1.5g
$MgSO_4 \cdot 7H_2O$	1.5g
Glycerol	10.0mL

pH 7.2 ± 0.2 at 25°C

Preparation of Medium: Add components to distilled/deionized water and bring volume to 1.0L. Mix thoroughly. Gently heat and bring to boiling. Distribute into tubes or flasks. Autoclave for 15 min at 15 psi pressure–121°C. Pour into sterile Petri dishes or leave in tubes.

Use: For the cultivation and maintenance of *Pseudomonas glumae*.

King's O–F Medium
See: **Oxidation–Fermentation Medium, King's**

Kirchner's Enrichment Medium

Composition per liter:

$Na_2HPO_4 \cdot 12H_2O$	19.0g
Asparagine	5.0g
KH_2PO_4	2.5g
Sodium citrate	2.5g
$MgSO_4$	0.6g
Serum	100.0mL
Glycerol	20.0mL
Penicillin solution	10.0mL
Phenol Red (0.4% solution)	3.0mL

pH 7.4–7.6 at 25°C

Penicillin Solution:

Composition per 10mL:

Penicillin	100,000U

Preparation of Penicillin Solution: Add penicillin to distilled/deionized water and bring volume to 10.0mL. Mix thoroughly. Filter sterilize.

Preparation of Medium: Add components, except serum and penicillin solution, to distilled/deionized water and bring volume to 890.0mL. Mix thoroughly. Gently heat and bring to boiling. Autoclave for 15 min at 15 psi pressure–121°C. Cool to 45°–50°C. Aseptically add sterile serum and penicillin solution. Mix thoroughly. Aseptically distribute into sterile tubes or flasks.

Use: For the cultivation and enrichment of *Mycobacterium* species.

K–L Virulence Agar (Klebs–Loeffler Virulence Agar) (Elek Agar) (*Corynebacterium diphtheriae* Virulence Test Medium)

Composition per 1300mL:

K-L agar base	1.0L
Rabbit serum	200.0mL
K_2TeO_3 solution	100.0mL
K-L filter strips	100

pH 7.8 ± 0.2 at 25°C

Source: This medium is available as a premixed powder from Difco Laboratories.

K-L Agar Base:

Composition per liter:

Meat peptone	20.0g
Agar	15.0g
NaCl	2.5g

Preparation of K-L Agar Base: Add components to distilled/deionized water and bring volume to 1.0L. Mix thoroughly. Gently heat and bring to boiling. Autoclave for 15 min at 15 psi pressure–121°C. Cool to 50°C.

K_2TeO_3 Solution:

Composition per 100mL:

K_2TeO_3	0.3g

Preparation of K_2TeO_3 Solution: Add K_2TeO_3 to distilled/deionized water and bring volume to 100.0mL. Mix thoroughly. Filter sterilize.

K-L Filter Strips:

Composition:

Whatman No. 3 filter paper	as needed
Diphtheria toxin solution	10.0mL

Preparation of K-L Strips: Cut Whatman No. 3 filter paper into 1.5cm × 7cm strips. Autoclave for 15 min at 15 psi pressure–121°C. Aseptically dip each strip into a sterile solution containing 1000U of purified diphtheria toxin/mL. Drain off excess liquid. Store each strip in a sterile bottle.

Caution: Potassium tellurite is toxic.

Preparation of Medium: Filter sterilize rabbit serum. To 1.0L of cooled, sterile K-L agar base, aseptically add sterile rabbit serum and sterile K_2TeO_3 solution. Mix thoroughly. Pour into sterile Petri dishes in 13.0mL volumes. Before the agar solidifies, aseptically add one K-L filter strip across the diameter of the plate. Allow the filter strip to sink to the bottom of the plate or press it down with sterile forceps. Allow the agar to solidify. Dry the surface of the plates by incubating at 35°C with lid of plate ajar for 2 hr.

Use: For *in vitro* toxigenicity testing of *Corynebacterium diphtheriae* by the agar diffusion technique. *C. diphtheriae* that produce toxin form white precipitin lines at approximately 45° angles from the culture streak line.

K–L Virulence Agar (Klebs–Loeffler Virulence Agar)

Composition per 1250mL:

K-L agar base	1.0L
K-L enrichment	200.0mL
K_2TeO_3 solution	50.0mL
K-L filter strips	100

pH 7.8 ± 0.2 at 25°C

Source: This medium is available as a premixed powder from Difco Laboratories.

K-L Agar Base:

Composition per liter:

Meat peptone	20.0g
Agar	15.0g
NaCl	2.5g

Preparation of K-L Agar Base: Add components to distilled/deionized water and bring volume to 1.0L. Mix thoroughly. Gently heat and bring to boiling. Autoclave for 15 min at 15 psi pressure–121°C. Cool to 50°C.

K-L Enrichment:

Composition per 200mL:

Casamino acids	4.0g
Glycerol	100.0mL
Tween™ 80	100.0mL

Preparation of K-L Enrichment: Combine components. Mix thoroughly. Filter sterilize.

K_2TeO_3 Solution:

Composition per 100mL:

K_2TeO_3	1.0g

Preparation of K_2TeO_3 Solution: Add K_2TeO_3 to distilled/deionized water and bring volume to 100.0mL. Mix thoroughly. Filter sterilize.

K-L Filter Strips:

Composition:

Whatman No. 3 filter paper	as needed
Diphtheria toxin solution	as needed

Preparation of K-L Strips: Cut Whatman No. 3 filter paper into 1.5cm × 7cm strips. Autoclave for 15 min at 15 psi pressure–121°C. Aseptically dip each strip into a sterile solution containing 1000U of purified diphtheria toxin/mL. Drain off excess liquid. Store each strip in a sterile bottle.

Caution: Potassium tellurite is toxic.

Preparation of Medium: To 1.0L of cooled, sterile K-L agar base, aseptically add sterile K-L enrichment and sterile K_2TeO_3 solution. Mix thoroughly. Pour into sterile Petri dishes in 13.0mL volumes. Before the agar solidifies, aseptically add one K-L filter strip across the diameter of the plate. Allow the filter strip to sink to the bottom of the plate or press it down with sterile forceps. Allow the agar to solidify. Dry the surface of the plates by incubating at 35°C with lid of plate ajar for 2 hr.

Use: For *in vitro* toxigenicity testing of *Corynebacterium diphtheriae* by the agar diffusion technique. *C. diphtheriae* that produce toxin form white precipitin lines at approximately 45° angles from the culture streak line.

Kleb Agar (m-Kleb Agar)

Composition per liter:

Agar	15.0g
Proteose peptone No. 3	10.0g
NaCl	5.0g
Adonitol	5.0g
Beef extract	1.0g
Aniline Blue	0.1g
Sodium lauryl sulfate	0.1g
Phenol Red	0.025g
Ethanol (95% solution)	20.0mL
Carbenicillin solution	10.0mL

pH 7.4 ± 0.2 at 25°C

Carbenicillin Solution:
Composition per 10mL:

Carbenicillin	0.05g

Preparation of Carbenicillin Solution: Add carbenicillin to distilled/deionized water and bring volume to 10.0mL. Mix thoroughly. Filter sterilize.

Preparation of Medium: Add components, except ethanol and carbenicillin solution, to distilled/deionized water and bring volume to 970.0mL. Mix thoroughly. Gently heat and bring to boiling. Autoclave for 15 min at 15 psi pressure–121°C. Cool to 45°–50°C. Aseptically add 20.0mL ethanol and 10.0mL carbenicillin solution. Mix thoroughly. Pour into sterile Petri dishes or distribute into sterile tubes.

Use: For the enumeration of bacteria from waters.

Klebsiella Medium (m-*Klebsiella* Medium)

Composition per 1041mL:

Agar	15.0g
Adonitol	4.0g
2× salt solution	500.0mL
Uric acid solution	200.0mL
Phenol Red solution	10.0mL
Sodium taurocholate solution	30.0mL
Carbenicillin solution	1.0mL

2X Salt Solution:
Composition per liter:

KCl	8.0g
K_2HPO_4	3.0g
NaCl	2.0g
KH_2PO_4	1.0g
$MgSO_4 \cdot 7H_2O$	0.2g

Preparation of 2X Salt Solution: Add components to distilled/deionized water and bring volume to 1.0L. Mix thoroughly.

Uric Acid Solution:
Composition per 200mL:

Uric acid	0.3g

Preparation of Uric Acid Solution: Dissolve uric acid in a small volume of 1*N* NaOH. Bring volume to 200.0mL with distilled/deionized water. Adjust pH to 7.1 with 1*N* HCl. Filter sterilize.

Phenol Red Solution:
Composition per 10mL:

Phenol Red	0.1g

Preparation of Phenol Red Solution: Add Phenol Red to sterile distilled/deionized water and bring volume to 10.0mL. Mix thoroughly.

Sodium Taurocholate Solution:
Composition per 30mL:

Sodium taurocholate	0.4g

Preparation of Sodium Taurocholate Solution: Add sodium taurocholate to sterile distilled/deionized water and bring volume to 30.0mL. Mix thoroughly.

Carbenicillin Solution:
Composition per 1mL:

Carbenicillin	5.0mg

Preparation of Carbenicillin Solution: Add carbenicillin to distilled/deionized water and bring volume to 1.0mL. Mix thoroughly. Filter sterilize.

Preparation of Medium: Add adonitol and agar to 500.0mL of 2× salt solution. Bring volume to 800.0mL with distilled/deionized water. Mix thoroughly. Gently heat and bring to boiling. Autoclave for 15 min at 15 psi pressure–121°C. Cool to 45°–50°C. Aseptically add 200.0mL of uric acid solution, 30.0mL of sodium taurocholate solution, 10.0mL of Phenol Red solution, and 1.0mL of carbenicillin solution. Mix thoroughly. Pour into sterile Petri dishes or distribute into sterile tubes.

Use: For the enumeration of *Klebsiella* species by the membrane filter method.

Klebsiella Selective Agar

Composition per liter:

Agar	26.0g
DL–Phenylalanine	10.0g
L-Ornithine·HCl	10.0g
Raffinose	7.0g
Pancreatic digest of casein	2.5g
Yeast extract	2.5g
K_2HPO_4	2.0g
Phenol Red solution	10.0mL
Carbenicillin solution	10.0mL

pH 5.6 ± 0.2 at 25°C

Phenol Red Solution:

Composition per 10mL:

Phenol Red	0.5g

Preparation of Phenol Red Solution: Add Phenol Red to 50% ethanol and bring volume to 10.0mL. Mix thoroughly.

Preparation of Medium: Add components, except carbenicillin solution, to distilled/deionized water and bring volume to 990.0mL. Mix thoroughly. Gently heat and bring to boiling. Autoclave for 15 min at 15 psi pressure–121°C. Cool to 45°–50°C. Aseptically add 10.0mL carbenicillin solution. Mix thoroughly. Adjust pH to 5.6–5.7 with sterile 1 *N* HCl. Pour into sterile Petri dishes or distribute into sterile tubes.

Use: For the isolation and identification of *Klebsiella pneumoniae* from clinical specimens.

Klebs-Loeffler Virulence Agar
See: K-L Virulence Agar

Kligler Iron Agar

Composition per liter:

Peptone	20.0g
Agar	12.0g
Lactose	10.0g
NaCl	5.0g
Beef extract	3.0g
Yeast extract	3.0g
Glucose	1.0g
Ferric citrate	0.3g
$Na_2S_2O_3$	0.3g
Phenol Red	0.05g

pH 7.4 ± 0.2 at 25°C

Source: This medium is available as a premixed powder from Difco Laboratories and Oxoid Unipath.

Preparation of Medium: Add components to distilled/deionized water and bring volume to 1.0L. Mix thoroughly. Gently heat and bring to boiling. Distribute into tubes. Autoclave for 15 min at 15 psi pressure–121°C. Pour into sterile Petri dishes or leave in tubes.

Use: For the differentiation and identification of Enterobacteriaceae based upon sugar fermentation and hydrogen sulfide production. Sugar fermentation is indicated by the medium turning yellow. H_2S production results in the medium turning black.

Kligler Iron Agar

Composition per liter:

Agar	15.0g
Lactose	10.0g
Pancreatic digest of casein	10.0g
Peptic digest of animal tissue	10.0g
NaCl	5.0g
Glucose	1.0g
Ferric ammonium citrate	0.5g
$Na_2S_2O_3$	0.5g
Phenol Red	0.025g

pH 7.4 ± 0.2 at 25°C

Source: This medium is available as a premixed powder from BBL Microbiology Systems.

Preparation of Medium: Add components to distilled/deionized water and bring volume to 1.0L. Mix thoroughly. Gently heat and bring to boiling. Distribute into tubes or flasks. Autoclave for 15 min at 15 psi pressure–121°C. Pour into sterile Petri dishes or leave in tubes.

Use: For the differentiation and identification of Enterobacteriaceae based upon sugar fermentation and hydrogen sulfide production. Sugar fermentation is indicated by the medium turning yellow. H_2S production results in the medium turning black.

Kligler Iron Agar (FDA M71)

Composition per liter:

Agar	15.0g
Lactose	20.0g
Pancreatic digest of casein	10.0g
Peptic digest of animal tissue	10.0g
NaCl	5.0g
Glucose	1.0g
Ferric ammonium citrate	0.5g
$Na_2S_2O_3$	0.5g
Phenol Red	0.025g

pH 7.4 ± 0.2 at 25°C

Preparation of Medium: Add components to distilled/deionized water and bring volume to 1.0L. Mix thoroughly. Gently heat and bring to boiling. Distribute into tubes or flasks. Autoclave for 15 min at 15 psi pressure–121°C. Pour into sterile Petri dishes or leave in tubes.

Use: For the differentiation and identification of Enterobacteriaceae based upon sugar fermentation and hydrogen sulfide production. Sugar fermentation is indicated by the medium turning yellow. H_2S production results in the medium turning black.

KM Agar (Kempler–McKay Agar)

Composition per liter:

Agar	15.0g
Milk, nonfat	10.0g
Glucose	5.0g
Milk protein hydrolysate	2.5g
Solution 1	10.0mL
Solution 2	10.0mL

pH 6.6 ± 0.2 at 25°C

Solution 1:
Composition per 100mL:

$K_3Fe(CN)_6$	10.0g

Preparation of Solution 1: Add components to distilled/deionized water and bring volume to 100.0mL. Mix thoroughly. Filter sterilize.

Solution 2:
Composition per 40mL:

Ferric citrate	1.0g
Sodium citrate	1.0g

Preparation of Solution 2: Add components to distilled/deionized water and bring volume to 40.0mL. Mix thoroughly. Filter sterilize.

Caution: Cyanide is toxic.

Preparation of Medium: Add components, except solution 1 and solution 2, to distilled/deionized water and bring volume to 980.0mL. Mix thoroughly. Gently heat and bring to boiling. Adjust pH to 6.6. Autoclave for 12 min at 10 psi pressure–115°C. Cool to 45°–50°C. Aseptically add sterile solution 1 and solution 2. Mix thoroughly. Pour into sterile Petri dishes. Allow plates to dry in the dark at 30°C for 24 hr before using.

Use: For the isolation and cultivation of acidogenic microorganisms from foods. For the differentiation of citrate-fermenting lactic bacteria, such as *Lactobacillus lactis* subspecies *diacetylactis*, from the noncitrate-fermenting *Lactobacillus lactis* subspecies *cremoris*.

Knisely Medium for *Bacillus anthracis*

Composition per liter:

Beef heart, solids from infusion	500.0g
Agar	15.0g
Pancreatic digest of casein	10.0g
NaCl	5.0g
EDTA	200.0mg
Lysozyme	40.0mg
Thallous acetate	40.0mg
Polymyxin	30,000U

Preparation of Medium: Add components, except EDTA, lysozyme, thallous acetate, and polymyxin, to distilled/deionized water and bring volume to 1.0mL. Mix thoroughly. Gently heat and bring to boiling. Adjust pH to 7.3. Autoclave for 15 min at 15 psi pressure–121°C. Cool to 45°–50°C. Aseptically add sterile EDTA, lysozyme, thallous acetate, and polymyxin. Mix thoroughly. Pour into sterile Petri dishes or distribute into sterile tubes.

Use: For the cultivation and maintenance of *Bacillus anthracis*.

Koch's K1 Medium

Composition per liter:

Glucose	1.8g
Peptone	0.6g
Yeast extract	0.4g

Preparation of Medium: Add components to distilled/deionized water and bring volume to 1.0L. Mix thoroughly. Distribute into tubes or flasks. Autoclave for 15 min at 15 psi pressure–121°C.

Use: For the cultivation of a variety of fungi.

Korthof Medium

Composition per 1088mL:

NaCl	1.4g
$Na_2HPO_4 \cdot 2H_2O$	0.88g
Peptone	0.8g
KH_2PO_4	0.24g
$CaCl_2$	0.04g
KCl	0.04g
$NaHCO_3$	0.02g
Rabbit serum, inactivated	80.0mL
Rabbit hemoglobin solution	8.0mL

pH 7.2 ± 0.2 at 25°C

Rabbit Hemoglobin Solution:
Composition per 20mL:

Rabbit blood clot	10.0mL

Preparation of Rabbit Hemoglobin Solution: Add rabbit blood clot to 10.0mL of distilled/deionized water. Lyse the clot by freezing and thawing.

Preparation of Medium: Add components, except rabbit serum and rabbit hemoglobin solution, to distilled/deionized water and bring volume to 1.0L. Mix thoroughly. Gently heat and bring to boiling. Cool to 25°C. Filter through Whatman #1 filter paper. Distribute into flasks in 100.0mL volumes. Autoclave for 15 min at 15 psi pressure–121°C. Cool to 45°–50°C. Aseptically add 8.0mL of rabbit serum and 0.8mL of rabbit hemoglobin solution to each flask. Mix thoroughly.

Use: For the cultivation of *Leptospira* species.

Korthof Medium, Modified

Composition per liter:

NaCl	1.40g
$Na_2HPO_4 \cdot 2H_2O$	0.88g
Peptone	0.80g
KH_2PO_4	0.24g
$CaCl_2$	0.04g
KCl	0.04g
$NaHCO_3$	0.02g
Rabbit serum, heat inactivated at 56°C	100.0mL

pH 7.2–7.6 at 25°C

Preparation of Medium: Add components, except rabbit serum, to distilled/deionized water and bring volume to 900.0L. Mix thoroughly. Gently heat and bring to boiling. Boil for 20 min. Cool overnight at 4°C. Filter through Whatman No. 2 filter paper. Distribute into tubes or flasks. Autoclave for 15 min at 15 psi pressure–121°C. Cool to50°–56°C. Aseptically add 100.0mL rabbit serum. Mix thoroughly.

Use: For the cultivation of *Leptospira* species.

Koser Citrate Medium

Composition per liter:

Sodium citrate	3.0g
$NaNH_4HPO_4 \cdot 4H_2O$	1.5g
KH_2PO_4	1.0g
$MgSO_4 \cdot 7H_2O$	0.2g

pH 6.7 ± 0.2 at 25°C

Source: This medium is available as a premixed powder from Difco Laboratories.

Preparation of Medium: Add components to distilled/deionized water and bring volume to 1.0L. Mix thoroughly. Gently heat and bring to boiling. Distribute into tubes or flasks. Autoclave for 15 min at 15 psi pressure–121°C. Pour into sterile Petri dishes or leave in tubes.

Use: For the differentiation of *Escherichia coli* and *Enterobacter aerogenes* based oncitrate utilization.

Kosmachev's Medium

Composition per liter:

Agar	15.0g
$CaCO_3$	4.0g
KNO_3	1.0g
$(NH_4)_2SO_4$	1.0g
Na_2HPO_4	1.0g
$MgSO_4 \cdot 7H_2O$	0.5g
$FeSO_4 \cdot 7H_2O$	0.01g
Yeast autolysate (30% solution)	15.0mL

Preparation of Medium: Add components to distilled/deionized water and bring volume to 1.0L. Mix thoroughly. Gently heat and bring to boiling. Distribute into tubes or flasks. Autoclave for 15 min at 15 psi pressure–121°C. Pour into sterile Petri dishes or leave in tubes.

Use: For the isolation and cultivation of *Actinomadura* species, *Actinopolyspora* species, *Excellospora* species and *Microspora* species.

KPL Medium

Composition per 1141mL:

Agar	15.0g
Galactose	10.0g
Glucose	10.0g
Lactic acid whey	1.0L
White table wine	140.0mL
Tween™ 80	1.0mL

pH 5.5 ± 0.2 at 25°C

Lactic Acid Whey:

Composition per liter:

Skim milk (10% solution)	1.0L

Preparation of Lactic Acid Whey: Adjust the pH of the skim milk to 5.5 with lactic acid. Gently heat and bring to boiling. Continue boiling for 30 min. Filter through Whatman #1 filter paper.

Preparation of Medium: Combine components, except white table wine. Mix thoroughly. Gently heat and bring to boiling. Adjust pH to 5.5. Autoclave for 15 min at 15 psi pressure–121°C. Cool to 45°–50°C. Filter sterilize white table wine. To cooled sterile basal medium, aseptically add sterile white table wine. Mix thoroughly. Pour into sterile Petri dishes or distribute into sterile tubes.

Use: For the cultivation and maintenance of *Lactobacillus kefiranofaciens*.

Krainsky's Asparagine Agar

Composition per liter:

Agar	15.0g
Glucose	10.0g
K_2HPO_4	0.5g
L-Asparagine	0.5g

pH 7.0 ± 0.2 at 25°C

Preparation of Medium: Add components to distilled/deionized water and bring volume to 1.0L. Mix thoroughly. Gently heat and bring to boiling. Distribute into tubes or flasks. Autoclave for 15 min at 15 psi pressure–121°C. Pour into sterile Petri dishes or leave in tubes.

Use: For the cultivation and maintenance of *Streptomyces fragmentans.*

Kranep Agar Base

Composition per liter:

KSCN	25.5g
Agar	18.3g
Sodium pyruvate	8.2g
Pancreatic digest of gelatin	6.1g
LiCl	5.1g
Mannitol	5.1g
Beef extract	3.7g
NaN_3	0.05g
Cycloheximide	0.041g
Egg yolk emulsion	100.0mL

pH 6.8 ± 0.2 at 25°C

Source: This medium is available as a premixed powder from Oxoid Unipath.

Egg Yolk Emulsion:

Composition:

Chicken egg yolks	11
Whole chicken egg	1

Preparation of Egg Yolk Emulsion: Soak eggs with 1:100 dilution of saturated mercuric chloride solution for 1 min. Crack eggs and separate yolks from whites. Mix egg yolks with 1 chicken egg.

Caution: Sodium azide is toxic. Azides also react with metals and disposal must be highly diluted.

Preparation of Medium: Add components, except egg yolk emulsion, to distilled/deionized water and bring volume to 900.0mL. Mix thoroughly. Gently heat and bring to boiling. Autoclave for 15 min at 15 psi pressure–121°C. Cool to 45°–50°C. Aseptically add 100.0mL egg yolk emulsion. Mix thoroughly. Pour into sterile Petri dishes or distribute into sterile tubes.

Use: For the isolation and enumeration of staphylococci from foods.

Kupferberg *Trichomonas* Base

Composition per liter:

Pancreatic digest of casein	20.0g
Cysteine·HCl·H_2O	1.5g
Agar	1.0g
Maltose	1.0g
Methylene Blue	3.0mg
Bovine serum	50.0mL

pH 6.0 ± 0.2 at 25°C

Source: This medium is available as a premixed powder from Difco Laboratories.

Preparation of Medium: Add components, except bovine serum, to distilled/deionized water and bring volume to 950.0mL. Mix thoroughly. Gently heat and bring to boiling. Autoclave for 15 min at 15 psi pressure–121°C. Cool to 45°–50°C. Aseptically add 50.0mL bovine serum. If desired, additional selectivity can be obtained by aseptically adding 250,000U penicillin and 1.0g streptomycin or 1.0g chloramphenicol. Mix thoroughly. Pour into sterile Petri dishes or distribute into sterile tubes.

Use: For the cultivation of the *Trichomonas* species from clinical specimens.

Kupferberg *Trichomonas* Broth

Composition per liter:

Enzymatic digest of protein	20.0g
Cysteine·HCl·H_2O	1.5g
Agar	1.0g
Maltose	1.0g
Chloramphenicol	0.1g
Methylene Blue	3.0mg
Bovine serum	50.0mL

pH 6.0 ± 0.2 at 25°C

Source: This medium is available as a premixed powder from Difco Laboratories.

Preparation of Medium: Add components, except bovine serum, to distilled/deionized water and bring volume to 950.0mL. Mix thoroughly. Gently heat and bring to boiling. Autoclave for 15 min at 15 psi pressure–121°C. Cool to 45°–50°C. Aseptically add bovine serum. If desired, additional selectivity can be obtained by aseptically adding 250,000U penicillin and 1.0g streptomycin or 1.0g chloramphenicol. Mix thoroughly. Pour into sterile Petri dishes or distribute into sterile tubes.

Use: For the cultivation of the *Trichomonas* species from clinical specimens.

KVBA

See: **Kanamycin Vancomycin Blood Agar**

L and F Basal Salts, Modified with Heptadecane

Composition per liter:

NH_4Cl	2.0g
Na_2HPO_4	0.21g
$MgSO_4 \cdot 7H_2O$	0.2g
NaH_2PO_4	0.09g
KCl	0.04g
$CaCl_2$	0.015g
$FeSO_4 \cdot 7H_2O$	1.0mg
$ZnSO_4 \cdot 7H_2O$	0.070mg
H_3BO_3	0.010mg
$MnSO_4 \cdot 5H_2O$	0.010mg
MoO_3	0.010mg
$CuSO_4 \cdot 5H_2O$	5.0μg
Heptadecane	2.0mL

pH 7.2 ± 0.2 at 25°C

Preparation of Medium: Add components, except heptadecane, to distilled/deionized water and bring volume to 1.0L. Mix thoroughly. Gently heat and bring to boiling. Distribute equally into four 250.0mL volumes. Autoclave for 15 min at 15 psi pressure–121°C. Cool to 60°C. Filter sterilize heptadecane. To one 250.0mL fraction of basal salts, aseptically add 0.5mL of sterile heptadecane. Pour mixture into a sterile blender. Homogenize slowly to mix heptadecane with basal salts and not to create excess bubbles. Rapidly distribute medium to sterile screw-capped tubes. Chill tubes quickly in an ice pack or in the refrigerator. Allow tubes to solidify in a slanted position.

Use: For the cultivation and maintenance of *Thermoleophilum album* and *Thermoleophilum minutum.*

L Medium (ATCC Medium 1154)

Composition per liter:

Pancreatic digest of casein	10.0g
NaCl	5.0g
Yeast extract	5.0g

pH 7.0 ± 0.2 at 25°C

Preparation of Medium: Add components to distilled/deionized water and bring volume to 1.0L. Mix thoroughly. Adjust pH to 7.0. Distribute into tubes or flasks. Autoclave for 25 min at 15 psi pressure–121°C.

Use: For the cultivation and maintenance of *Escherichia coli.*

L Medium (ATCC Medium 167)

Composition per liter:

Agar	20.0g
$NaNO_3$	2.0g
Na_2HPO_4	0.21g
$MgSO_4 \cdot 7H_2O$	0.2g
NaH_2PO_4	0.09g
KCl	0.04g
$CaCl_2$	0.015g
$FeSO_4 \cdot 7H_2O$	1.0mg
Salts solution	1.0mL

Salts Solution:

Composition per 100mL:

$ZnSO_4 \cdot 7H_2O$	7.0mg
H_3BO_3	1.0mg
$MnSO_4 \cdot 5H_2O$	1.0mg
MoO_3	1.0mg
$CuSO_4 \cdot 5H_2O$	0.5mg

Preparation of Salts Solution: Add components to distilled/deionized water and bring volume to 100.0mL. Mix thoroughly.

Preparation of Medium: Add components to distilled/deionized water and bring volume to 1.0L. Mix thoroughly. Gently heat and bring to boiling. Distribute into tubes or flasks. Autoclave for 15 min at 15 psi pressure–121°C. Pour into sterile Petri dishes or leave in tubes.

Use: For the cultivation and maintenance of *Methylococcus capsulatus* and *Pseudomonas methanica.*

L Medium for *Salmonella*

Composition per liter:

Pancreatic digest of casein	10.0g
NaCl	5.0g
Yeast extract	5.0g
Glucose	1.0g

pH 7.2 ± 0.2 at 25°C

Preparation of Medium: Add components to distilled/deionized water and bring volume to 1.0L. Mix thoroughly. Adjust pH to 7.2. Distribute into tubes or flasks. Autoclave for 15 min at 15 psi pressure–121°C.

Use: For the cultivation and maintenance of *Salmonella choleraesuis.*

L Medium with Ampicillin

Composition per liter:

Pancreatic digest of casein	10.0g
NaCl	5.0g

Yeast extract 5.0g
Ampicillin solution 20.0mg

pH 7.0 ± 0.2 at 25°C

Ampicillin Solution:
Composition per 10mL:
Ampicillin 0.02g

Preparation of Ampicillin Solution: Add ampicillin to distilled/deionized water and bring volume to 10.0mL. Mix thoroughly. Filter sterilize.

Preparation of Medium: Add components to distilled/deionized water and bring volume to 1.0L. Mix thoroughly. Adjust pH to 7.0. Distribute into tubes or flasks. Autoclave for 15 min at 15 psi pressure–121°C.

Use: For the cultivation and maintenance of *Escherichia coli.*

L Medium with DAP and THY (L Medium with Diaminopimelic Acid and Thymidine)

Composition per liter:
Pancreatic digest of casein 10.0g
NaCl 5.0g
Yeast extract 5.0g
Glucose solution... 10.0mL
Diaminopimelic acid solution 10.0mL
Thymidine solution 10.0mL

pH 7.0 ± 0.2 at 25°C

Glucose Solution:
Composition per 10mL:
D-Glucose 1.0g

Preparation of Glucose Solution: Add D-glucose to distilled/deionized water and bring volume to 10.0mL. Mix thoroughly. Filter sterilize.

Diaminopimelic Acid Solution:
Composition per 10mL:
DL-Diaminopimelic acid.....................................0.1g

Preparation of Diaminopimelic Acid Solution: Add diaminopimelic acid to distilled/deionized water and bring volume to 10.0mL. Mix thoroughly. Filter sterilize.

Thymidine Solution:
Composition per 10mL:
Thymidine 0.02g

Preparation of Thymidine Solution: Add thymidine to distilled/deionized water and bring volume to 10.0mL. Mix thoroughly. Filter sterilize.

Preparation of Medium: Add components—except glucose solution, diaminopimelic acid solution, and thymidine solution—to distilled/deionized water and bring volume to 970.0mL. Mix thoroughly. Gently heat and bring to boiling. Adjust pH to 7.0. Autoclave for 15 min at 15 psi pressure–121°C. Cool to 45°–50°C. Aseptically add sterile glucose solution, diaminopimelic acid solution, and thymidine solution. Mix thoroughly. Aseptically distribute into sterile tubes or flasks.

Use: For the cultivation and maintenance of *Escherichia coli.*

L Medium with DAP, THY and AMP (L Medium with Diaminopimelic Acid, Thymidine and Ampicillin)

Composition per liter:
Pancreatic digest of casein 10.0g
NaCl 5.0g
Yeast extract 5.0g
Diaminopimelic acid solution 10.0mL
Thymidine solution 10.0mL
Ampicillin solution 10.0mL

pH 7.0 ± 0.2 at 25°C

Diaminopimelic Acid Solution:
Composition per 10mL:
DL-Diaminopimelic acid.....................................0.1g

Preparation of Diaminopimelic Acid Solution: Add DL-diaminopimelic acid to distilled/deionized water and bring volume to 10.0mL. Mix thoroughly. Filter sterilize.

Thymidine Solution:
Composition per 10mL:
Thymidine 0.04g

Preparation of Thymidine Solution: Add thymidine to distilled/deionized water and bring volume to 10.0mL. Mix thoroughly. Filter sterilize.

Ampicillin Solution:
Composition per 10mL:
Ampicillin 0.02mg

Preparation of Ampicillin Solution: Add ampicillin to distilled/deionized water and bring volume to 10.0mL. Mix thoroughly. Filter sterilize.

Preparation of Medium: Add components—except ampicillin solution, diaminopimelic acid solution, and thymidine solution—to distilled/deionized water and bring volume to 970.0mL. Mix thoroughly. Gently heat and bring to boiling. Adjust pH to 7.0. Autoclave for 15 min at 15 psi pressure–121°C. Cool to 45°–50°C. Aseptically add sterile ampicillin solution, diaminopimelic acid solution, and thymidine solution. Mix thoroughly. Aseptically distribute into sterile tubes or flasks.

Use: For the cultivation and maintenance of *Escherichia coli.*

L Medium with Diaminopimelic Acid and Thymidine

See: **L Medium with DAP and THY**

L Medium with Diaminopimelic Acid, Thymidine and Ampicillin

See: **L Medium with DAP, THY and AMP**

L Medium with Methanol

Composition per liter:

Agar	20.0g
$NaNO_3$	2.0g
Na_2HPO_4	0.21g
$MgSO_4·7H_2O$	0.2g
NaH_2PO_4	0.09g
KCl	0.04g
$CaCl_2$	0.015g
$FeSO_4·7H_2O$	1.0mg
Methanol	20.0mL
Salts solution	1.0mL

Salts Solution:

Composition per 100mL:

$ZnSO_4·7H_2O$	7.0mg
H_3BO_3	1.0mg
$MnSO_4·5H_2O$	1.0mg
MoO_3	1.0mg
$CuSO_4·5H_2O$	0.5mg

Preparation of Salts Solution: Add components to distilled/deionized water and bring volume to 100.0mL. Mix thoroughly.

Preparation of Medium: Add components, except methanol, to distilled/deionized water and bring volume to 980.0mL. Mix thoroughly. Gently heat and bring to boiling. Autoclave for 15 min at 15 psi pressure–121°C. Cool to 45°–50°C. Filter sterilize methanol. Aseptically add 20.0mL of sterile methanol to cooled sterile basal medium. Mix thoroughly. Pour into sterile Petri dishes or distribute into sterile tubes.

Use: For the cultivation and maintenance of *Methylobacillus glycogenes.*

L Medium with Tetracycline

Composition per liter:

Pancreatic digest of casein	10.0g
NaCl	5.0g
Yeast extract	5.0g
Tetracycline solution	10.0mL

pH 7.0 ± 0.2 at 25°C

Tetracycline Solution:

Composition per 10mL:

Tetracycline	0.02mg

Preparation of Tetracycline Solution: Add tetracycline to distilled/deionized water and bring volume to 10.0mL. Mix thoroughly. Filter sterilize.

Preparation of Medium: Add components, except tetracycline solution, to distilled/deionized water and bring volume to 990.0mL. Mix thoroughly. Gently heat and bring to boiling. Adjust pH to 7.0. Autoclave for 15 min at 15 psi pressure–121°C. Cool to 45°–50°C. Aseptically add sterile tetracycline solution. Mix thoroughly. Aseptically distribute into sterile tubes or flasks.

Use: For the cultivation and maintenance of *Escherichia coli.*

L Salts for Thermophiles

See: **Mineral Salts for Thermophiles**

L15 Medium, Modified Leibovitz

Composition per liter:

NaCl	8.0g
DL-Threonine	0.6g
Sodium pyruvate	0.6g
DL-Alanine	0.5g
L-Arginine, free base	0.5g
KCl	0.4g
L-Asparagine·H_2O	0.3g
L-Histidine, free base	0.3g
L-Glutamine	0.3g
L-Isoleucine	0.3g
L-Phenylalanine	0.3g
L-Tyrosine	0.3g
DL-Methionine	0.2g
DL-Valine	0.2g
Glycine	0.2g
L-Serine	0.2g
Na_2HPO_4, anhydrous	0.2g
$CaCl_2$, anhydrous	0.1g
L-Cysteine, free base	0.1g
L-Leucine·HCl	0.1g

$MgCl_2$, anhydrous 0.094g
D-Galactose 0.090g
KH_2PO_4 .... 0.060g
L-Tryptophan 0.020g
Phenol Red 0.010g
i-Inositol 2.0mg
Choline chloride 1.0mg
D-Calcium pantothenate 1.0mg
Folic acid 1.0mg
Nicotinamide 1.0mg
Pyridoxine·HCl 1.0mg
Thiamine monophosphate·$2H_2O$ 1.0mg
Riboflavin-5-phosphate 0.1mg

pH 7.5 ± 0.2 at 25°C

Preparation of Medium: Add components to distilled/deionized water and bring volume to 1.0L. Mix thoroughly. Filter sterilize. Store at 5°C.

Use: For the cultivation of oysters used for the growth of enteroviruses.

Lab-Lemco Agar

Composition per liter:

Agar 15.0g
Peptone 5.0g
Lab-lemco meat extract 3.0g

pH 7.4 ± 0.2 at 25°C

Preparation of Medium: Add components to distilled/deionized water and bring volume to 1.0L. Mix thoroughly. Gently heat and bring to boiling. Distribute into tubes or flasks. Autoclave for 15 min at 15 psi pressure–121°C. Pour into sterile Petri dishes or leave in tubes.

Use: For the cultivation and maintenance of a variety of heterotrophic microorganisms.

Lab-Lemco Broth

Composition per liter:

Peptone 5.0g
Lab-lemco meat extract 3.0g

pH 7.4 ± 0.2 at 25°C

Preparation of Medium: Add components to distilled/deionized water and bring volume to 1.0L. Mix thoroughly. Distribute into tubes or flasks. Autoclave for 15 min at 15 psi pressure–121°C.

Use: For the cultivation of a variety of heterotrophic microorganisms, including microorganisms from wastewater.

Lachica's Medium

See: **SA Agar, Modified**

Lactate Agar

Composition per liter:

Yeast extract 3.0g
K_2HPO_4 .... 2.80g
Agar 2.0g
Peptone 2.0g
KH_2PO_4 .... 0.52g
Sodium lactate (60% solution) 10.0mL

pH 7.2 ± 0.2 at 25°C

Preparation of Medium: Add components to distilled/deionized water and bring volume to 1.0L. Mix thoroughly. Gently heat and bring to boiling. Adjust pH to 7.2. Distribute into tubes or flasks. Autoclave for 15 min at 15 psi pressure–121°C. Pour into sterile Petri dishes or leave in tubes.

Use: For the cultivation and maintenance of *Serpens flexibilis*.

Lactate Broth

Composition per liter:

Yeast extract 3.0g
K_2HPO_4 .... 2.80g
Peptone 2.0g
KH_2PO_4 .... 0.52g
Sodium lactate (60% solution) 10.0mL

pH 7.2 ± 0.2 at 25°C

Preparation of Medium: Add components to distilled/deionized water and bring volume to 1.0L. Mix thoroughly. Gently heat and bring to boiling. Adjust pH to 7.2. Distribute into tubes or flasks. Autoclave for 15 min at 15 psi pressure–121°C.

Use: For the cultivation and maintenance of *Serpens flexibilis*.

Lactate Sea Water Minimal Medium

Composition per liter:

Tris(hydroxymethyl)aminomethane·HCl 7.88g
Sodium or potassium lactate 2.0g
NH_4Cl 1.0g
$K_2HPO_4·3H_2O$ 0.075g
$FeSO_4·7H_2O$ 0.028g
Artificial sea water 500.0mL

pH 7.5 ± 0.2 at 25°C

Artificial Sea Water:

Composition per liter:

$MgSO_4·7H_2O$ 24.7g
NaCl 23.4g
$CaCl_2·2H_2O$ 2.9g
KCl 1.5g

Preparation of Artificial Sea Water: Add components to distilled/deionized water and bring volume to 1.0L. Mix thoroughly.

Preparation of Medium: Add components to distilled/deionized water and bring volume to 1.0L. Mix thoroughly.Adjust pH to 7.5. Distribute into tubes or flasks. Autoclave for 15 min at 15 psi pressure–121°C.

Use: For the cultivation of ATCC strain s 27134 and 27136.

Lactic Acid Bacteria Medium

Composition per liter:

Agar	15.0g
Glucose	10.0g
KH_2PO_4	5.0g
Peptone	5.0g
Sodium acetate·$3H_2O$	5.0g
Yeast extract	5.0g
Diammonium hydrogen citrate	2.0g
Tween™ 80	1.0g
Sorbic acid	0.5g
$MgSO_4·7H_2O$	0.5g
$MnSO_4·4H_2O$	0.2g
$FeSO_4·7H_2O$	0.05g

pH 5.3–5.4 at 25°C

Preparation of Medium: Add components to distilled/deionized water and bring volume to 1.0L. Mix thoroughly. Adjust pH to 5.3–5.4. Gently heat and bring to boiling. Distribute into tubes or flasks. Autoclave for 15 min at 15 psi pressure–121°C. Pour into sterile Petri dishes or leave in tubes.

Use: For the isolation and cultivation of lactic acid bacteria from wine.

Lactic Agar for Yogurt Bacteria, Modified

Composition per liter:

Elliker agar	1.0L
Milk solution	70.0mL

pH 6.8 ± 0.1 at 25°C

Elliker Agar:

Composition per liter:

Pancreatic digest of casein	20.0g
Agar	15.0g
Yeast extract	10.0g
Gelatin	4.0g
Glucose	3.0g
Ascorbic acid	2.5g
Lactose	2.5g
NaCl	2.5g
Sodium acetate	2.5g
Sucrose	2.5g

Preparation of Elliker Agar: Add components to distilled/deionized water and bring volume to 1.0L. Mix thoroughly. Gently heat and bring to boiling. Autoclave for 15 min at 15 psi pressure–121°C. Cool to 45°–50°C.

Milk Solution:

Composition per 100mL:

Nonfat dry milk solids	11.0g

Preparation of Milk Solution: Add nonfat dry milk solids to distilled/deionized water and bring volume to 100.0mL. Mix thoroughly. Autoclave for 12 min at 15 psi pressure–121°C. Cool to 45°–50°C.

Preparation of Medium: Add 70.0mL of sterile milk solution to 1.0L of cooled, sterile Elliker agar. Mix thoroughly. Pour into sterile Petri dishes. Allow plates to dry at 28°–30°C for 18–24 hr.

Use: For the cultivation of acidogenic microorganisms, especially *Lactobacillus* species and lactic streptococci, from foods.

Lactic Bacteria Broth

Composition per liter:

Sodium acetate	12.0g
Glucose	10.0g
Pancreatic digest of casein	10.0g
Yeast autolysate	5.0g
Solution A	5.0mL
Solution B	5.0mL

pH 5.1–5.3 at 25°C

Solution A:

Composition per 100mL:

K_2HPO_4	10.0g
KH_2PO_4	10.0g

Preparation of Solution A: Add components to distilled/deionized water and bring volume to 100.0mL. Mix thoroughly.

Solution B:

Composition per 100mL:

$MgSO_4·7H_2O$	4.0g
$FeSO_4$	0.2g
$MnSO_4·5H_2O$	0.2g
NaCl	0.2g

Preparation of Solution B: Add components to distilled/deionized water and bring volume to 100.0mL. Mix thoroughly.

Preparation of Medium: Add components to distilled/deionized water and bring volume to 1.0L. Mix thoroughly. Distribute into tubes or flasks. Autoclave for 15 min at 15 psi pressure–121°C.

Use: For the cultivation and maintenance of *Lactobacillus buchneri, Lactobacillus delbrueckii,* and *Pediococcus damnosus.*

Lactic Streak Agar

Composition per liter:

Agar	15.0g
Sodium carboxymethylcellulose	10.0g
Calcium citrate	10.0g
Beef extract	5.0g
Papaic digest of soybean meal	5.0g
Polypeptone™	5.0g
Yeast extract	5.0g
Lactose	1.5g
L-Arginine·HCl	1.5g
Bromcresol Purple (0.1% solution)	20.0mL

pH 6.0 ± 0.1 at 25°C

Preparation of Medium: Add components—except Bromcresol Purple solution, calcium citrate, and sodium carboxymethylcellulose—to distilled/deionized water and bring volume to 800.0mL. Mix thoroughly. Gently heat and bring to boiling. In a blender, add the calcium citrate and sodium carboxymethylcellulose to 200.0mL of distilled/deionized water. Blend until mixed. Add the 200.0mL of citrate/carboxymethylcellulose solution to the hot agar solution. Mix thoroughly. Adjust pH to 6.0. Distribute into flasks in 100.0mL volumes. Autoclave for 10 min at 10 psi pressure–115°C. Cool to 45°–50°C. Aseptically add 2.0mL of sterile Bromcresol Purple to each flask. Mix thoroughly. Pour into sterile Petri dishes. Allow plates to dry at 37°C for 1 hr before use.

Use: For the differentiation of lactic streptococci. Bacteria that produce acid from lactose appear as yellow colonies.

Lactobacilli Agar, AOAC (Lactobacilli Agar, Association of Official Analytical Chemists)

Composition per liter:

Milk, peptonized	15.0g
Agar	10.0g
Glucose	10.0g
Yeast extract	5.0g
KH_2PO_4	2.0g
Sorbitan monooleate complex	1.0g
Tomato juice	100.0mL

pH 6.8 ± 0.2 at 25°C

Source: This medium is available as a premixed powder from Difco Laboratories.

Preparation of Medium: Add components to distilled/deionized water and bring volume to 1.0L. Mix thoroughly. Gently heat and bring to boiling. Continue boiling for 2–3 min. Distribute into tubes in 10.0mL volumes. Autoclave for 15 min at 15 psi pressure–121°C. Allow tubes to cool in an upright position.

Use: For the cultivation and maintenance of stock cultures of *Lactobacillus casei* ATCC 7469, *Lactobacillus fermentum* ATCC 9338, *Lactobacillus leichmannii* ATCC 4797, and *Lactobacillus viridescens* ATCC 12706 used in the microbiological assay of vitamins.

Lactobacilli Broth, AOAC (Lactobacilli Broth, Association of Official Analytical Chemists)

Composition per liter:

Milk, peptonized	15.0g
Glucose	10.0g
Yeast extract	5.0g
KH_2PO_4	2.0g
Sorbitan monooleate complex	1.0g
Tomato juice	100.0mL

pH 6.8 ± 0.2 at 25°C

Source: This medium is available as a premixed powder from Difco Laboratories.

Preparation of Medium: Add components to distilled/deionized water and bring volume to 1.0L. Mix thoroughly. Gently heat and bring to boiling. Continue boiling for 2–3 min. Distribute into tubes in 10.0mL volumes. Autoclave for 15 min at 15 psi pressure–121°C.

Use: For the cultivation and preparation of inocula of stock cultures of *Lactobacillus casei* ATCC 7469, *Lactobacillus fermentum* ATCC 9338, *Lactobacillus leichmannii* ATCC 4797, and *Lactobacillus viridescens* ATCC 12706 used in the microbiological assay of vitamins.

Lactobacilli deMan-Rogosa-Sharpe Broth

See: **Lactobacilli MRS Broth**

Lactobacilli MRS Broth (Lactobacilli deMan-Rogosa-Sharpe Broth)

Composition per liter:

Glucose	20.0g

Beef extract	10.0g
Peptone	10.0g
Sodium acetate	5.0g
Yeast extract	5.0g
Ammonium citrate	2.0g
Na_2HPO_4	2.0g
Tween™ 80	1.0g
$MgSO_4 \cdot 7H_2O$	0.1g
$MnSO_4 \cdot 5H_2O$	0.05g

pH 6.5 ± 0.2 at 25°C

Source: This medium is available as a premixed powder from Difco Laboratories.

Preparation of Medium: Add components to distilled/deionized water and bring volume to 1.0L. Mix thoroughly. Gently heat and bring to boiling. Distribute into tubes or flasks. Autoclave for 15 min at 15 psi pressure–121°C.

Use: For the cultivation of *Lactobacillus* species. Also used for the cultivation and maintenance of *Aerococcus viridans, Bifidobacterium coryneforme, Lactococcus plantarum, Leuconostoc* species, *Pectinatus cerevisiiphilus, Pediococcus* species, and *Sporolactobacillus inulinus.*

Lactobacilli MRS Broth with Melvalonic Acid
See: Pediococcus Medium with Mevalonic Acid

Lactobacilli MRS Broth
See: Pediococcus Medium

Lactobacilli MRS Broth with Cysteine

Composition per liter:

Glucose	20.0g
Beef extract	10.0g
Peptone	10.0g
Sodium acetate	5.0g
Yeast extract	5.0g
Cysteine	2.0g
Ammonium citrate	2.0g
Na_2HPO_4	2.0g
Tween™ 80	1.0g
$MgSO_4 \cdot 7H_2O$	0.1g
$MnSO_4 \cdot 5H_2O$	0.05g

pH 6.5 ± 0.2 at 25°C

Preparation of Medium: Add components to distilled/deionized water and bring volume to 1.0L. Mix thoroughly. Gently heat and bring to boiling. Distribute into tubes or flasks. Autoclave for 15 min at 15 psi pressure–121°C.

Use: For the cultivation and maintenance of *Lactobacillus* species.

Lactobacilli MRS Broth with Ethanol

Composition per liter:

Glucose	20.0g
Beef extract	10.0g
Peptone	10.0g
Sodium acetate	5.0g
Yeast extract	5.0g
Cysteine	2.0g
Ammonium citrate	2.0g
Na_2HPO_4	2.0g
Tween™ 80	1.0g
$MgSO_4 \cdot 7H_2O$	0.1g
$MnSO_4 \cdot 5H_2O$	0.05g
Ethanol (95% solution)	100.0mL

pH5.0 ± 0.2 at 25°C

Preparation of Medium: Add components, except ethanol, to distilled/deionized water and bring volume to 990.0mL. Mix thoroughly. Gently heat and bring to boiling. Autoclave for 15 min at 15 psi pressure–121°C. Cool to 25°C. Filter sterilize ethanol. Aseptically add 100.0mL of sterile ethanol. Mix thoroughly. Aseptically distribute into sterile tubes or flasks.

Use: For the cultivation and maintenance of *Lactobacillus* species.

Lactobacillus bifidus Medium

Composition for one liter:

Lactose	70.0g
Sodium acetate, anhydrous	50.0g
Pancreatic digest of casein	10.0g
K_2HPO_4	5.0g
Tween™ 80	1.0g
Alanine	0.4g
Cystine	0.4g
Tryptophan	0.4g
Asparagine	0.2g
Adenine	0.02g
Guanine	0.02g
Uracil	0.02g
Xanthine	0.02g
Pyridoxine·HCl	2.4mg
Nicotinic acid	1.2mg
Calcium pantothenate	0.8mg
Riboflavin	0.4mg
Thiamine·HCl	0.4mg

p-Aminobenzoic acid	0.020mg
Folic acid	0.020mg
Biotin	8.0μg
Ascorbic acid	100.0mL
Human milk, skimmed	20.0mL
Salts B	10.0mL

pH 6.8 ± 0.2 at 25°C

Salts B:
Composition per 250mL:

$MgSO_4 \cdot 7H_2O$	10.0g
$FeSO_4 \cdot 7H_2O$	0.5g
NaCl	0.5g
$MnSO_4 \cdot 2H_2O$	0.337g

Preparation of Salts B: Add components to distilled/deionized water and bring volume to 250.0mL. Mix thoroughly.

Ascorbic Acid Solution:
Composition per 100mL:

Ascorbic acid	1.0g

Preparation of Ascorbic Acid Solution: Add ascorbic acid to distilled/deionized water and bring volume to 100.0mL. Mix thoroughly. Adjust pH to 6.5. Filter sterilize.

Preparation of Medium: Add components, except ascorbic acid solution and human milk, to distilled/deionized water and bring volume to 900.0mL. Mix thoroughly. Gently heat and bring to boiling. Autoclave for 15 min at 15 psi pressure–121°C. Cool to 45°–50°C. Aseptically add 100.0mL of sterile ascorbic acid solution. Mix thoroughly. This constitutes a double-strength medium. To prepare a single-strength medium, aseptically combine 500.0mL of sterile double-strength medium, 480.0mL of sterile distilled/deionized water and 20.0mL of sterile human milk. Aseptically distribute into sterile tubes or flasks.

Use: For the cultivation and maintenance of *Bifidobacterium (Lactobacillus) bifidum.*

Lactobacillus bulgaricus Agar (LB Agar)

Composition per 900mL:

Agar	20.0g
Glucose	20.0g
Beef extract	10.0g
Pancreatic digest of casein	10.0g
Yeast extract	5.0g
K_2HPO_4	2.0g
Tween™ 80	1.0g
Acetate buffer	80.0mL
Tomato juice, filtered	40.0mL

pH 6.8 ± 0.2 at 25°C

Acetate Buffer:
Composition per liter:

Sodium acetate	113.55g
Acetic acid	10.0mL

Preparation of Acetate Buffer: Add components to distilled/deionized water and bring volume to 1.0L. Mix thoroughly.

Preparation of Medium: Add components, except acetate buffer, to distilled/deionized water and bring volume to 820.0mL. Mix thoroughly. Gently heat and bring to boiling. Adjust pH to 6.8. Add 80.0mL of acetate buffer. Mix thoroughly. Autoclave for 15 min at 15 psi pressure–121°C. Pour into sterile Petri dishes.

Use: For the isolation, cultivation and enumeration of *Lactobacillus bulgaricus* from foods.

Lactobacillus Heteroferm Screen Agar (MRS, Modified)

Composition per liter:

Glucose	20.0g
Agar	15.0g
Proteose peptone No. 3	10.0g
Sodium acetate	5.0g
Yeast extract	5.0g
2-Phenylethyl alcohol	3.0g
Ammonium citrate	2.0g
K_2HPO_4	2.0g
$MgSO_4 \cdot 7H_2O$	0.1g
$MnSO_4 \cdot H_2O$	0.05g
Bromcresol Green	0.04g
Cycloheximide	4.0mg
Tween™ 80	1.0mL

pH 5.5 ± 0.01 at 25°C

Preparation of Medium: Add components to distilled/deionized water and bring volume to 1.0L. Mix thoroughly. Autoclave for 15 min at 15 psi pressure–121°C. Cool to 50°C. Adjust pH to 5.5 with glacial acetic acid. Pour into sterile Petri dishes.

Use: For the isolation and cultivation of *Lactobacillus* species from salad dressings.

Lactobacillus Heteroferm Screen Broth (MRS, Modified)

Composition per liter:

Glucose	20.0g
Proteose peptone No. 3	10.0g
Sodium acetate	5.0g

Yeast extract 5.0g
2-Phenylethyl alcohol 3.0g
Ammonium citrate 2.0g
K_2HPO_4 .... 2.0g
$MgSO_4 \cdot 7H_2O$ 0.1g
$MnSO_4 \cdot H_2O$ 0.05g
Bromcresol Green 0.04g
Cycloheximide 4.0mg
Tween™ 80 .... 1.0mL

pH 4.3 ± 0.01 at 25°C

Preparation of Medium: Add components to distilled/deionized water and bring volume to 1.0L. Mix thoroughly. Adjust pH to 4.3 with concentrated HCl. Distribute into test tubes that contain an inverted Durham tube. Autoclave for 15 min at 15 psi pressure–121°C.

Use: For the isolation and cultivation of *Lactobacillus* species from foods.

Lactobacillus Medium (ATCC Medium 1006)

Composition per liter:

Agar 20.0g
Pancreatic digest of casein 10.0g
Glucose 5.0g
Yeast extract 5.0g
KH_2PO_4 .... 3.0g
K_2HPO_4 .... 3.0g
Tryptose 3.0g
Sodium acetate 1.0g
L-Cysteine·HCl·H_2O 0.2g
Salt solution R 5.0mL
Tween™ 80 .... 1.0mL

pH 6.3 ± 0.2 at 25°C

Salt Solution R:

Composition per 100mL:

$MgSO_4 \cdot 7H_2O$ 11.5g
$MnSO_4 \cdot 2H_2O$ 2.4g
$FeSO_4 \cdot 7H_2O$ 0.68g

Preparation of Salt Solution R: Add components to distilled/deionized water and bring volume to 100.0mL. Mix thoroughly. Store at 4°C.

Preparation of Medium: Add components to distilled/deionized water and bring volume to 1.0L. Mix thoroughly. Gently heat and bring to boiling. Distribute into tubes or flasks. Autoclave for 15 min at 15 psi pressure–121°C. Pour into sterile Petri dishes or leave in tubes.

Use: For the cultivation and maintenance of *Lactobacillus delbrueckii* and *Lactobacillus jensenii*.

Lactobacillus Medium (ATCC Medium 78)

Composition per liter:

Pancreatic digest of casein 20.0g
Agar 15.0g
Tryptose 5.0g
Yeast extract 5.0g
Glucose 3.0g
Lactose 2.0g
Liver extract concentrate 1.0g
Tween™ 80 .... 0.05g
Tomato juice, filtered 200.0mL

pH 6.5 ± 0.2 at 25°C

Preparation of Medium: Add components to distilled/deionized water and bring volume to 1.0L. Mix thoroughly. Gently heat and bring to boiling. Adjust pH to 6.5. Distribute into tubes or flasks. Autoclave for 15 min at 15 psi pressure–121°C. Pour into sterile Petri dishes or leave in tubes.

Use: For the cultivation and maintenance of *Lactobacillus fermentum* and *Lactobacillus salivarius*.

Lactobacillus Medium (ATCC Medium 169)

Composition per liter:

Glucose 10.0g
Proteose peptone 7.5g
Yeast extract 7.5g
KH_2PO_4 .... 2.0g
L-Cysteine·HCl·H_2O 1.0g
Tween™ 80 .... 0.1g
Tomato juice, filtered 100.0mL

pH 7.0 ± 0.2 at 25°C

Preparation of Medium: Add components to distilled/deionized water and bring volume to 1.0L. Mix thoroughly. Gently heat and bring to boiling. Adjust pH to 7.0. Distribute into tubes or flasks. Autoclave for 15 min at 15 psi pressure–121°C. Pour into sterile Petri dishes or leave in tubes.

Use: For the cultivation and maintenance of *Lactobacillus casei*.

Lactobacillus Sake Medium

Composition per liter:

Agar 13.0g
Yeast extract 5.0g
Liver extract concentrate 0.2g
Sake 700.0mL

Preparation of Medium: Add components to distilled/deionized water and bring volume to 1.0L. Mix

thoroughly. Gently heat and bring to boiling. Distribute into tubes or flasks. Autoclave for 15 min at 15 psi pressure–121°C. Pour into sterile Petri dishes or leave in tubes.

Use: For the cultivation and maintenance of *Lactobacillus fructivorans and Lactobacillus homohiochi.*

Lactobacillus Selection Agar
See: **LBS™ Agar**

Lactobacillus Selection Broth
See: **LBS™ Broth**

Lactobacillus Selection Oxgall Agar (LBS™ Oxgall Agar)

Composition per liter:

Sodium acetate·$3H_2O$	25.0g
Glucose	20.0g
Agar	15.0g
Pancreatic digest of casein	10.0g
KH_2PO_4	6.0g
Yeast extract	5.0g
Ammonium citrate	2.0g
Oxgall	1.5g
Polysorbate 80	1.0g
$MgSO_4$	0.575g
$FeSO_4$	0.034g
$MnSO_4$	0.12g
Acetic acid, glacial	1.32mL

pH 5.5 ± 0.2 at 25°C

Preparation of Medium: Add components, except acetic acid, to distilled/deionized water and bring volume to 998.7mL. Mix thoroughly. Gently heat and bring to boiling. Add glacial acetic acid. Mix thoroughly. Gently heat while stirring and bring to 90°–100°C for 2–3 min. Do not autoclave. Pour into sterile Petri dishes or distribute into sterile tubes.

Use: For the selective isolation, cultivation and enumeration of lactobacilli.

Lactobacillus–Streptococcus Differential Medium
See: **L-S Differential Medium**

Lactose Broth

Composition per liter:

Lactose	5.0g
Pancreatic digest of gelatin	5.0g
Beef extract	3.0g

pH 6.9 ± 0.2 at 25°C

Source: This medium is available as a premixed powder from BBL Microbiology Systems, Difco Laboratories, and Oxoid Unipath.

Preparation of Medium: Add components to distilled/deionized water and bring volume to 1.0L. Mix thoroughly. Distribute into tubes containing an inverted Durham tube in 10.0mL volumes. Autoclave for 12 min at 15 psi pressure–121°C. Cool broth quickly to 25°C. For testing water samples with 10.0mL volumes, prepare medium double strength.

Use: For detection of lactose-fermenting, Gram-negative coliforms, as a pre-enrichment broth for *Salmonella* species and in the study of lactose fermentation of bacteria in general.

Lactose Distillers Solubles Medium

Composition per liter:

Lactose	20.0g
Distillers solubles	15.0g
Yeast, autolyzed	5.0g

pH 7.0 ± 0.2 at 25°C

Preparation of Medium: Add components to distilled/deionized water and bring volume to 1.0L. Mix thoroughly. Gently heat and bring to boiling. Distribute into tubes or flasks. Autoclave for 15 min at 15 psi pressure–121°C. Pour into sterile Petri dishes or leave in tubes.

Use: For the cultivation and maintenance of *Streptomyces avermitilis.*

Lactose Egg Yolk Milk Agar

Composition per 1206mL:

Lactose	12.0g
Agar	1.0g
Columbia blood agar base	800.0mL
Skim milk	150.0mL
Egg yolk emulsion, 50%	36.0mL
Inhibitor solution	20.0mL
Neutral Red (1% solution)	3.25mL

pH 7.0 ± 0.2 at 25°C

Columbia Blood Agar Base:
Composition per 800mL:

Agar	15.0g
Pantone	10.0g
Bitone	10.0g
NaCl	5.0g

Tryptic digest of beef heart 3.0g
Cornstarch 1.0g

Preparation of Columbia Blood Agar Base: Add components to distilled/deionized water and bring volume to 1.0L. Mix thoroughly. Gently heat until boiling.

Egg Yolk Emulsion, 50%:
Composition per 100mL:

Chicken egg yolks 11
Whole chicken egg 1
NaCl (0.9% solution) 50.0mL

Preparation of Egg Yolk Emulsion, 50%: Soak eggs with 1:100 dilution of saturated mercuric chloride solution for 1 min. Crack eggs and separate yolks from whites. Mix egg yolks with 1 chicken egg. Measure 50.0mL of egg yolk emulsion and add to 50.0mL of 0.9% NaCl solution. Mix thoroughly. Filter sterilize. Warm to 45°–50°C.

Inhibitor Solution:
Composition per 20mL:

Neomycin sulfate 0.18g
NaN_3 .... 0.24g

Preparation of Inhibitor Solution: Add neomycin sulfate and NaN_3 to distilled/deionized water and bring volume to 20.0mL. Mix thoroughly. Filter sterilize.

Caution: Sodium azide is toxic. Azides also react with metals and disposal must be highly diluted.

Preparation of Medium: Combine Columbia blood agar base, lactose, agar and Neutral Red and bring volume to 1.0L. Adjust pH to 7.0. Autoclave for 15 min at 15 psi pressure–121°C. Cool to 55°C. Filter sterilize skim milk. To 1.0L of cooled sterile agar mixture, aseptically add 150.0mL of sterile skim milk, 36.0mL of sterile egg yolk emulsion, 50%, and 20.0mL of sterile inhibitor solution. Mix thoroughly. Pour into sterile Petri dishes or distribute into sterile tubes.

Use: For the cultivation and maintenance of *Clostridium* species.

Lactose Gelatin Medium

Composition per liter:

Gelatin 120.0g
Tryptose 15.0g
Lactose 10.0g
Yeast extract 10.0g
Phenol Red (0.5% solution) 10.0mL

pH 7.5 ± 0.2 at 25°C

Preparation of Medium: Add tryptose, yeast extract and lactose to distilled/deionized water and bring volume to 400.0mL. Mix thoroughly. Add gelatin to distilled/deionized water and bring volume to 590.0mL Gently heat gelatin solution while stirring and bring to 50–60°C. Add Phenol Red. Mix the two solutions together. Distribute into tubes in 10.0mL volumes. Autoclave for 10 min at 15 psi pressure–121°C. If medium is not used in 8 hr, deoxygenate by heating to 50°–70°C for 2–3 hr prior to inoculation.

Use: For the cultivation of *Clostridium perfringens*.

Lactose Minimal Medium

Composition per liter:

Agar 20.0g
Lactose 15.0g
K_2HPO_4 .... 5.0g
NH_4Cl 2.0g
NaCl 1.0g
$MgSO_4$ 0.1g
Yeast extract 0.1g

Preparation of Medium: Add components to distilled/deionized water and bring volume to 1.0L. Mix thoroughly. Gently heat and bring to boiling. Distribute into tubes or flasks. Autoclave for 15 min at 15 psi pressure–121°C. Pour into sterile Petri dishes or leave in tubes.

Use: For the cultivation and maintenance of *Xanthomonas campestris*.

Lactose Ricinoleate Broth

Composition per liter:

Lactose 10.0g
Peptone 5.0g
Sodium ricinoleate 1.0g

pH 7.6 ± 0.2 at 25°C

Preparation of Medium: Add components to distilled/deionized water and bring volume to 1.0L. Mix thoroughly. Distribute into tubes or flasks. Autoclave for 15 min at 15 psi pressure–121°C.

Use: For the selective cultivation of members of the Enterobacteriaceae.

Lambda Broth

Composition per liter:

Pancreatic digest of casein 10.0g
NaCl 2.5g

Preparation of Medium: Add components to distilled/deionized water and bring volume to 1.0L. Mix thoroughly. Distribute into tubes or flasks. Autoclave for 25 min at 15 psi pressure–121°C.

Use: For the cultivation of *Escherichia coli* in the preparation of bacteriophage lysates.

Lambda Plates

Composition per liter:

Agar	10.0g
Pancreatic digest of casein	10.0g
NaCl	2.5g

Preparation of Medium: Add components to distilled/deionized water and bring volume to 1.0L. Mix thoroughly. Gently heat and bring to boiling. Distribute into tubes or flasks. Autoclave for 15 min at 15 psi pressure–121°C. Pour into sterile Petri dishes in 45.0mL volumes per plate.

Use: For use as a base agar to support the cultivation of *Escherichia coli* in the preparation of bacteriophages.

Lambda Top Agar

Composition per liter:

Pancreatic digest of casein	10.0g
Agar	7.0g
NaCl	2.5g

Preparation of Medium: Add components to distilled/deionized water and bring volume to 1.0L. Mix thoroughly. Gently heat and bring to boiling. Autoclave for 15 min at 15 psi pressure–121°C. Cool to 50°C. Distribute into flasks in 100.0mL volumes. Reautoclave for 15 min at 15 psi pressure–121°C. Store at 25°C.

Use: For use as a top agar for the distribution of bacteriophage or *Escherichia coli*.

Lash Serum Medium

Composition per liter:

Casamino acids	14.0g
NaCl	6.0g
Glucose	2.0g
Maltose	1.5g
Sodium lactate (60% solution)	0.5g
KCl	0.1g
$CaCl_2 \cdot 2H_2O$	0.1g
Serum solution	500.0mL

pH 5.8 ± 0.2 at 25°C

Serum Solution:

Composition per500mL:

$NaHCO_3$	0.1g
Bovine serum	200.0mL

Preparation of Serum Solution: Add components to distilled/deionized water and bring volume to 500.0mL. Mix thoroughly. Filter sterilize.

Preparation of Medium: Add components, except serum solution, to distilled/deionized water and bring volume to 500.0mL. Mix thoroughly. Distribute into tubes in 5.0mL volumes. Autoclave for 15 min at 15 psi pressure–121°C. Cool to 25°C. Aseptically add 5.0mL of sterile serum solution to each tube. Mix thoroughly.

Use: For the cultivation of *Trichomonas vaginalis* from clinical specimens.

Lauryl Sulfate Broth (m-Lauryl Sulfate Broth)

Composition per liter:

Peptone	39.0g
Lactose	30.0g
Yeast extract	6.0g
Sodium lauryl sulfate	1.0g
Phenol Red	0.2g

pH 7.4 ± 0.2 at 25°C

Source: This medium is available as a premixed powder from Oxoid Unipath.

Preparation of Medium: Add components to distilled/deionized water and bring volume to 1.0L. Mix thoroughly. Distribute into bottles or flasks. Autoclave for 15 min at 15 psi pressure–121°C.

Use: For the cultivation and enumeration of coliform bacteria, especially *Escherichia coli*, in water by the membrane filter method.

Lauryl Sulfate Broth (Lauryl Tryptose Broth)

Composition per liter:

Pancreatic digest of casein	20.0g
Lactose	5.0g
NaCl	5.0g
K_2HPO_4	2.75g
KH_2PO_4	2.75g
Sodium lauryl sulfate	0.1g

pH 6.8 ± 0.2 at 25°C

Source: This medium is available as a premixed powder from BBL Microbiology Systems and Difco Laboratories.

Preparation of Medium: Add components to distilled/deionized water and bring volume to 1.0L. Mix thoroughly. Distribute into tubes containing an inverted Durham tube in 10.0mL volumes. Autoclave for 12 min at 15 psi pressure–121°C. Cool broth quickly to 25°C. For testing water samples with 10mL volumes, prepare medium double strength.

Use: For the detection of coliform bacteria in a variety of specimens. Also, for the enumeration of coliform bacteria by the multiple-tube fermentation technique.

Lauryl Sulfate Broth with MUG

Composition per liter:

Pancreatic digest of casein	20.0g
Lactose	5.0g
NaCl	5.0g
K_2HPO_4	2.75g
KH_2PO_4	2.75g
Sodium lauryl sulfate	0.1g
4-Methylumbellferyl-β-D-glucuronide (MUG)	0.05g

pH 6.8 ± 0.2 at 25°C

Source: This medium is available as a premixed powder from BBL Microbiology Systems.

Preparation of Medium: Add components to distilled/deionized water and bring volume to 1.0L. Mix thoroughly. Distribute into tubes containing an inverted Durham tube in 10.0mL volumes. Autoclave for 12 min at 15 psi pressure–121°C. Cool broth quickly to 25°C. For testing water samples with 10.0mL volumes, prepare medium double strength.

Use: For the detection of *Escherichia coli* in water and food samples by a fluorogenic procedure.

Lauryl Tryptose Broth

See: **Lauryl Sulfate Broth**

Lauryl Tryptose Mannitol Broth with Tryptophan

Composition per liter:

Pancreatic digest of casein	20.0g
Lactose	5.0g
NaCl	5.0g
K_2HPO_4	2.75g
KH_2PO_4	2.75g
Sodium lauryl sulfate	0.1g
L-Tryptophan	0.2g

pH 6.8 ± 0.2 at 25°C

Source: This medium is available as a premixed powder from Oxoid Unipath.

Preparation of Medium: Add components to distilled/deionized water and bring volume to 1.0L. Mix thoroughly. Distribute into tubes containing an inverted Durham tube in 10.0mL volumes. Autoclave for 10 min at 10 psi pressure–115°C. Cool broth quickly to 25°C.

Use: For the detection of *Escherichia coli* in water samples.

LAVMm2 Medium

Composition per liter:

Lactalbumin hydrolysate	10.0g
Sodium acetate	5.0g
$MgCl_2 \cdot 6H_2O$	20.3mg
Nitrilotriacetic acid	19.1mg
$CaCl_2$	11.1mg
$FeSO_4$	0.152mg
Thiamine·HCl	0.05mg
Cupric acetate	0.04mg
Biotin	0.02mg

pH 8.0–8.1 at 25°C

Preparation of Medium: Add components to distilled/deionized water and bring volume to 1.0L. Mix thoroughly. Adjust pH to 7.5 with Na_2CO_3. Distribute into tubes or flasks. Autoclave for 15 min at 15 psi pressure–121°C. The pH should be 8.0–8.1 after autoclaving.

Use: For the cultivation of *Caryophanon latum.*

LB Agar

Composition per liter:

Agar	15.0g
Pancreatic digest of casein	10.0g
NaCl	5.0g
Yeast extract	5.0g
1*N* NaOH	1.0mL

pH 7.0 ± 0.2 at 25°C

Preparation of Medium: Add components to distilled/deionized water and bring volume to 1.0L. Mix thoroughly. Gently heat and bring to boiling. Adjust pH to 7.0. Distribute into tubes or flasks. Autoclave for 25 min at 15 psi pressure–121°C. Pour into sterile Petri dishes in 35–40mL volumes.

Use: For the cultivation of *Escherichia coli.*

LB Agar

See: ***Lactobacillus bulgaricus* Agar**

LB Broth, Modified

Composition per liter:

Pancreatic digest of casein	10.0g
NaCl	5.8g
Yeast extract	5.0g
NaCl solution	16.8mL
Glucose solution	10.0mL
$CaCl_2 \cdot 2H_2O$ solution	2.0mL
$MgCl_2$ solution	1.6mL

pH 7.0 ± 0.2 at 25°C

NaCl Solution:
Composition per 100mL:

NaCl	25.0g

Preparation of NaCl Solution: Add NaCl to distilled/deionized water and bring volume to 100.0mL. Mix thoroughly. Filter sterilize.

Glucose Solution:
Composition per 100mL:

D-Glucose	40.0g

Preparation of Glucose Solution: Add D-glucose to distilled/deionized water and bring volume to 100.0mL. Mix thoroughly. Filter sterilize.

$CaCl_2{\cdot}2H_2O$ Solution:
Composition per 10mL:

$CaCl_2{\cdot}2H_2O$	0.735g

Preparation of $CaCl_2{\cdot}2H_2O$ Solution: Add $CaCl_2{\cdot}2H_2O$ to distilled/deionized water and bring volume to 10.0mL. Mix thoroughly. Filter sterilize.

$MgCl_2$ Solution:
Composition per 10mL:

$MgCl_2$	0.95g

Preparation of $MgCl_2$ Solution: Add $MgCl_2$ to distilled/deionized water and bring volume to 10.0mL. Mix thoroughly. Filter sterilize.

Preparation of Medium: Add components—except NaCl solution, glucose solution, $CaCl_2{\cdot}2H_2O$ solution, and $MgCl_2$ solution—to distilled/deionized water and bring volume to 969.6mL. Mix thoroughly. Adjust pH to 7.0. Autoclave for 30 min at 15 psi pressure–121°C. Cool to 45°–50°C. Aseptically add 16.8mL of sterile NaCl solution, 10.0mL of sterile glucose solution, 2.0mL of sterile $CaCl_2{\cdot}2H_2O$ solution, and 1.6mL of sterile $MgCl_2$ solution. Mix thoroughly. Aseptically distribute into sterile tubes or flasks.

Use: For the cultivation and maintenance of *Escherichia coli.*

LB Medium (LB Broth, Miller) (ATCC Medium 1065) (ATCC Medium 1082)

Composition per liter:

Pancreatic digest of casein	10.0g
NaCl	10.0g
Yeast extract	5.0g

Source: Available from Difco Laboratories.

Preparation of Medium: Add components to distilled/deionized water and bring volume to 1.0L. Mix thoroughly. Distribute into tubes or flasks. Autoclave for 15 min at 15 psi pressure–121°C.

Use: For the cultivation and maintenance of *Bacillus subtilis, Corynebacterium glutamicum, Enterobacter cloacae, Erwinia uredovora, Escherichia coli, Klebsiella oxytoca,* and *Salmonella choleraesuis.*

LB Medium (Luria Broth) (Lenox Broth)

Composition per liter:

Pancreatic digest of casein	10.0g
NaCl	5.0g
Yeast extract	5.0g

pH 7.0 ± 0.2 at 25°C

Preparation of Medium: Add components to distilled/deionized water and bring volume to 1.0L. Mix thoroughly. Adjust pH to 7.0. Distribute into tubes or flasks. Autoclave for 25 min at 15 psi pressure–121°C.

Use: For the cultivation of *Escherichia coli.*

LB Medium for *X*1776

Composition per liter:

NaCl	10.0g
Pancreatic digest of casein	10.0g
Yeast extract	5.0g
Glucose solution	10.0mL
Diaminopimelic acid solution	10.0mL
Thymidine solution	10.0mL

pH 7.0 ± 0.2 at 25°C

Glucose Solution:
Composition per 10mL:

D-Glucose	0.8g

Preparation of Glucose Solution: Add D-glucose to distilled/deionized water and bring volume to 10.0mL. Mix thoroughly. Filter sterilize.

Diaminopimelic Acid Solution:
Composition per 10mL:

DL-Diaminopimelic acid	0.1g

Preparation of Diaminopimelic Acid Solution: Add diaminopimelic acid to distilled/deionized water and bring volume to 10.0mL. Mix thoroughly. Filter sterilize.

Thymidine Solution:
Composition per 10mL:

Thymidine	0.02g

Preparation of Thymidine Solution: Add thymidine to distilled/deionized water and bring volume to 10.0mL. Mix thoroughly. Filter sterilize.

Preparation of Medium: Add components—except glucose solution, diaminopimelic acid solution, and thymidine solution—to distilled/deionized water and bring volume to 970.0mL. Mix thoroughly. Gently heat and bring to boiling. Autoclave for 15 min at 15 psi pressure–121°C. Cool to 45°–50°C. Aseptically add sterile glucose solution, diaminopimelic acid solution, and thymidine solution. Mix thoroughly. Aseptically distribute into sterile tubes or flasks.

Use: For the cultivation and maintenance of *Bacillus subtilis* and *Escherichia coli.*

LB Medium for X1776 with Tetracycline and Ampicillin

Composition per liter:

NaCl	10.0g
Pancreatic digest of casein	10.0g
Yeast extract	5.0g
Antibiotic solution	10.0mL
Glucose solution	10.0mL
Diaminopimelic acid solution	10.0mL
Thymidine solution	10.0mL

pH 7.0 ± 0.2 at 25°C

Antibiotic Solution:
Composition per 10mL:

Ampicillin	0.01g
Tetracycline	0.01g

Preparation of Antibiotic Solution: Add components to distilled/deionized water and bring volume to 10.0mL. Mix thoroughly. Filter sterilize.

Glucose Solution:
Composition per 10mL:

D-Glucose	0.8g

Preparation of Glucose Solution: Add D-glucose to distilled/deionized water and bring volume to 10.0mL. Mix thoroughly. Filter sterilize.

Diaminopimelic Acid Solution:
Composition per 10mL:

DL-Diaminopimelic acid	0.1g

Preparation of Diaminopimelic Acid Solution: Add diaminopimelic acid to distilled/deionized water and bring volume to 10.0mL. Mix thoroughly. Filter sterilize.

Thymidine Solution:
Composition per 10mL:

Thymidine	0.02g

Preparation of Thymidine Solution: Add thymidine to distilled/deionized water and bring volume to 10.0mL. Mix thoroughly. Filter sterilize.

Preparation of Medium: Add components—except glucose solution, diaminopimelic acid solution, and thymidine solution—to distilled/deionized water and bring volume to 960.0mL. Mix thoroughly. Gently heat and bring to boiling. Autoclave for 15 min at 15 psi pressure–121°C. Cool to 45°–50°C. Aseptically add sterile antibiotic solution, glucose solution, diaminopimelic acid solution, and thymidine solution. Mix thoroughly. Aseptically distribute into sterile tubes or flasks.

Use: For the cultivation and maintenance of *Bacillus subtilis* and *Escherichia coli.*

LB Medium with Ampicillin (ATCC Medium 1315)

Composition per liter:

NaCl	10.0g
Pancreatic digest of casein	10.0g
Yeast extract	5.0g
Ampicillin solution	10.0mL

pH 7.0 ± 0.2 at 25°C

Ampicillin Solution:
Composition per 10mL:

Ampicillin	0.1mg

Preparation of Ampicillin Solution: Add ampicillin to distilled/deionized water and bring volume to 10.0mL. Mix thoroughly. Filter sterilize.

Preparation of Medium: Add components, except ampicillin solution, to distilled/deionized water and bring volume to 990.0mL. Mix thoroughly. Adjust pH to 7.0. Autoclave for 15 min at 15 psi pressure–121°C. Cool to 45°–50°C. Aseptically add sterile ampicillin solution. Mix thoroughly. Aseptically distribute into sterile tubes or flasks.

Use: For the cultivation and maintenance of *Escherichia coli.*

LB Medium with Ampicillin (ATCC Medium 1364)

Composition per liter:

NaCl	10.0g
Pancreatic digest of casein	10.0g
Yeast extract	5.0g
Ampicillin solution	10.0mL

pH 7.0 ± 0.2 at 25°C

Ampicillin Solution:
Composition per 10mL:

Ampicillin	0.02mg

Preparation of Ampicillin Solution: Add ampicillin to distilled/deionized water and bring volume to 10.0mL. Mix thoroughly. Filter sterilize.

Preparation of Medium: Add components, except ampicillin solution, to distilled/deionized water and bring volume to 990.0mL. Mix thoroughly. Adjust pH to 7.0. Autoclave for 15 min at 15 psi pressure–121°C. Cool to 45°–50°C. Aseptically add sterile ampicillin solution. Mix thoroughly. Aseptically distribute into sterile tubes or flasks.

Use: For the cultivation and maintenance of *Escherichia coli.*

LB Medium with Chloramphenicol

Composition per liter:

NaCl	10.0g
Pancreatic digest of casein	10.0g
Yeast extract	5.0g
Chloramphenicol	0.01g

pH 7.0 ± 0.2 at 25°C

Preparation of Medium: Add components to distilled/deionized water and bring volume to 1.0L. Mix thoroughly. Adjust pH to 7.0. Distribute into tubes or flasks. Autoclave for 15 min at 15 psi pressure–121°C.

Use: For the cultivation and maintenance of *Escherichia coli.*

LB Medium with Glucose

Composition per liter:

NaCl	10.0g
Pancreatic digest of casein	10.0g
Glucose	5.0g
Yeast extract	5.0g

pH 7.0 ± 0.2 at 25°C

Preparation of Medium: Add components to distilled/deionized water and bring volume to 1.0L. Mix thoroughly. Adjust pH to 7.0. Distribute into tubes or flasks. Autoclave for 15 min at 15 psi pressure–121°C.

Use: For the cultivation and maintenance of *Escherichia coli.*

LB Medium with IPTG Medium

Composition per liter:

NaCl	10.0g
Pancreatic digest of casein	10.0g
Yeast extract	5.0g
IPTG solution	10.0mL

pH 7.0 ± 0.2 at 25°C

IPTG Solution:

Composition per 10mL:

IPTG (Isopropylthio-β-galactoside)	0.24g

Preparation of IPTG Solution: Add IPTG to distilled/deionized water and bring volume to 10.0mL. Mix thoroughly. Filter sterilize.

Preparation of Medium: Add components, except IPTG solution, to distilled/deionized water and bring volume to 990.0mL. Mix thoroughly. Adjust pH to 7.0. Autoclave for 15 min at 15 psi pressure–121°C. Aseptically add sterile IPTG solution. Mix thoroughly. Aseptically distribute into sterile tubes or flasks.

Use: For the cultivation and maintenance of *Escherichia coli.*

LB Medium with 25mg Kanamycin (ATCC Medium 1236)

Composition per liter:

NaCl	10.0g
Pancreatic digest of casein	10.0g
Yeast extract	5.0g
Kanamycin	0.025g

pH 7.0 ± 0.2 at 25°C

Preparation of Medium: Add components to distilled/deionized water and bring volume to 1.0L. Mix thoroughly. Adjust pH to 7.0. Distribute into tubes or flasks. Autoclave for 15 min at 15 psi pressure–121°C.

Use: For the cultivation and maintnenace of *Escherichia coli.*

LB Medium with 100mg Kanamycin (ATCC Medium 1468)

Composition per liter:

NaCl	10.0g
Pancreatic digest of casein	10.0g
Yeast extract	5.0g
Kanamycin	0.1g

pH 7.0 ± 0.2 at 25°C

Preparation of Medium: Add components to distilled/deionized water and bring volume to 1.0L. Mix thoroughly. Adjust pH to 7.0. Distribute into tubes or flasks. Autoclave for 15 min at 15 psi pressure–121°C.

Use: For the cultivation and maintnenace of *Erwinia uredovora*.

LB Medium with Rifampicin

Composition per liter:

NaCl	10.0g
Pancreatic digest of casein	10.0g
Yeast extract	5.0g
Rifampicin	0.1g

pH 7.0 ± 0.2 at 25°C

Preparation of Medium: Add components to distilled/deionized water and bring volume to 1.0L. Mix thoroughly. Adjust pH to 7.0. Distribute into tubes or flasks. Autoclave for 15 min at 15 psi pressure–121°C.

Use: For the cultivation and maintenance of *Enterobacter cloacae*.

LB Medium with Tetracycline

Composition per liter:

NaCl	10.0g
Pancreatic digest of casein	10.0g
Yeast extract	5.0g
Tetracycline	0.020g

pH 7.0 ± 0.2 at 25°C

Preparation of Medium: Add components to distilled/deionized water and bring volume to 1.0L. Mix thoroughly. Adjust pH to 7.0. Distribute into tubes or flasks. Autoclave for 15 min at 15 psi pressure–121°C.

Use: For the cultivation and maintenance of *Escherichia coli*.

LB Medium with Tetracycline and Ampicillin (ATCC Medium 1226)

Composition per liter:

NaCl	10.0g
Pancreatic digest of casein	10.0g
Yeast extract	5.0g
Antibiotic solution	10.0mL

pH 7.0 ± 0.2 at 25°C

Antibiotic Solution:

Composition per 10mL:

Ampicillin	0.01g
Tetracycline	0.01g

Preparation of Antibiotic Solution: Add components to distilled/deionized water and bring volume to 10.0mL. Mix thoroughly. Filter sterilize.

Preparation of Medium: Add components, except antibiotic solution, to distilled/deionized water and bring volume to 990.0mL. Mix thoroughly. Adjust pH to 7.0. Autoclave for 15 min at 15 psi pressure–121°C. Cool to 45°–50°C. Aseptically add sterile antibiotic solution. Mix thoroughly. Aseptically distribute into sterile tubes or flasks.

Use: For the cultivation and maintenance of *Escherichia coli*.

LB Medium with Tetracycline and Ampicillin (ATCC Medium 1235)

Composition per liter:

NaCl	10.0g
Pancreatic digest of casein	10.0g
Yeast extract	5.0g
Antibiotic solution	10.0mL

pH 7.0 ± 0.2 at 25°C

Antibiotic Solution:

Composition per 10mL:

Ampicillin	0.01g
Tetracycline	5.0mg

Preparation of Antibiotic Solution: Add components to distilled/deionized water and bring volume to 10.0mL. Mix thoroughly. Filter sterilize.

Preparation of Medium: Add components, except antibiotic solution, to distilled/deionized water and bring volume to 990.0mL. Mix thoroughly. Adjust pH to 7.0. Autoclave for 15 min at 15 psi pressure–121°C. Cool to 45°–50°C. Aseptically add sterile antibiotic solution. Mix thoroughly. Aseptically distribute into sterile tubes or flasks.

Use: For the cultivation and maintenance of *Escherichia coli*.

LB Medium with Thiamine Monophosphate

See: **LB Medium with TMP**

LB Medium with Thiamine Pyrophosphate

See: **LB Medium with TPP**

LB Medium with TMP (LB Medium with Thiamine Monophosphate)

Composition per liter:

NaCl	10.0g
Pancreatic digest of casein	10.0g
Yeast extract	5.0g
Thiamine monophosphate	0.038mg

pH 7.0 ± 0.2 at 25°C

Preparation of Medium: Add components to distilled/deionized water and bring volume to 1.0L. Mix thoroughly. Adjust pH to 7.0. Distribute into tubes or flasks. Autoclave for 15 min at 15 psi pressure–121°C.

Use: For the cultivation and maintenance of *Escherichia coli.*

LB Medium with TPP (LB Medium with Thiamine Pyrophosphate)

Composition per liter:

NaCl	10.0g
Pancreatic digest of casein	10.0g
Yeast extract	5.0g
Thiamine pyrophosphate	0.046mg

pH 7.0 ± 0.2 at 25°C

Preparation of Medium: Add components to distilled/deionized water and bring volume to 1.0L. Mix thoroughly. Adjust pH to 7.0. Distribute into tubes or flasks. Autoclave for 15 min at 15 psi pressure–121°C.

Use: For the cultivation and maintenance of *Escherichia coli.*

LB Top Agar

Composition per liter:

Pancreatic digest of casein	10.0g
Agar	7.0g
NaCl	5.0g
Yeast extract	5.0g

pH 7.0 ± 0.2 at 25°C

Preparation of Medium: Add components to distilled/deionized water and bring volume to 1.0L. Mix thoroughly. Gently heat and bring to boiling. Adjust pH to 7.0. Autoclave for 15 min at 15 psi pressure–121°C. Cool to 50°C. Distribute into flasks in 100.0mL volumes. Reautoclave for 15 min at 15 psi pressure–121°C. Store at 25°C.

Use: For use as a top agar for the distribution of bacteriophage or *Escherichia coli.*

LBE Medium

Composition per liter:

NaCl	10.0g
Pancreatic digest of casein	10.0g
Yeast extract	5.0g
Glucose solution	10.0mL
50X medium E	4.0mL

pH 7.0 ± 0.2 at 25°C

Glucose Solution:
Composition per 100mL:

D-Glucose	20.0g

Preparation of Glucose Solution: Add D-glucose to distilled/deionized water and bring volume to 100.0mL. Mix thoroughly. Filter sterilize.

50X Medium E:
Composition per liter:

K_2HPO_4, anhydrous	500.0g
$Na(NH_4)HPO_4 \cdot 4H_2O$	175.0g
Citric acid·H_2O	100.0g
$MgSO_4 \cdot 7H_2O$	10.0g

Preparation of 50X Medium E: Add components to 670.0mL of distilled/deionized water in the following order: $MgSO_4 \cdot 7H_2O$, citric acid·H_2O, K_2HPO_4, and $Na(NH_4)HPO_4 \cdot 4H_2O$. Mix thoroughly. Bring volume to 1.0L with distilled/deionized water.

Preparation of Medium: Add components—except glucose solution and 50X medium E—to distilled/deionized water and bring volume to 986.0mL. Mix thoroughly. Autoclave for 15 min at 15 psi pressure–121°C. Cool to 45°–50°C. Aseptically add 10.0mL of sterile glucose solution and 4.0mL of sterile 50X medium E. Mix thoroughly. Aseptically distribute into sterile tubes or flasks.

Use: For the cultivation and maintenance of *Escherichia coli.*

LBS™ Agar (*Lactobacillus* Selection Agar)

Composition per liter:

Sodium acetate·$3H_2O$	25.0g
Glucose	20.0g
Agar	15.0g
Pancreatic digest of casein	10.0g
KH_2PO_4	6.0g
Yeast extract	5.0g
Ammonium citrate	2.0g
Polysorbate 80	1.0g
$MgSO_4$	0.575g
$FeSO_4$	0.034g
$MnSO_4$	0.12g
Acetic acid, glacial	1.32mL

pH 5.5 ± 0.2 at 25°C

Source: This medium is available as a premixed powder from BBL Microbiology Systems.

Preparation of Medium: Add components, except acetic acid, to distilled/deionized water and bring volume to 998.7mL. Mix thoroughly. Gently heat and bring to boiling. Add glacial acetic acid. Mix thoroughly. Gently heat while stirring and bring to 90°–100°C for 2–3 min. Do not autoclave. Pour into sterile Petri dishes or distribute into sterile tubes.

Use: For the selective isolation, cultivation and enumeration of lactobacilli.

LBS™ Broth (*Lactobacillus* Selection Broth)

Composition per liter:

Sodium acetate·$3H_2O$	25.0g
Glucose	20.0g
Pancreatic digest of casein	10.0g
KH_2PO_4	6.0g
Yeast extract	6.0g
Ammonium citrate	2.0g
Polysorbate 80	1.0g
$MgSO_4$	0.575g
$FeSO_4$	0.034g
$MnSO_4$	0.12g
Acetic acid, glacial	1.32mL

pH 5.4 ± 0.2 at 25°C

Source: This medium is available as a premixed powder from BBL Microbiology Systems.

Preparation of Medium: Add components, except acetic acid, to distilled/deionized water and bring volume to 998.7mL. Mix thoroughly. Gently heat and bring to boiling. Add glacial acetic acid. Mix thoroughly. Gently heat while stirring and bring to 90°–100°C for 2–3 min. Do not autoclave. Aseptically distribute into sterile tubes.

Use: For the selective isolation and cultivation of lactobacilli.

LBS Oxgall Agar
See: Lactobacillus **Selection Oxgall Agar**

LD Agar
See: **Lombard–Dowell Agar**

LD Bile Agar
See: **Lombard–Dowell Bile Agar**

LD Broth
See: **Lombard–Dowell Broth**

LD Egg Yolk Agar
See: **Lombard–Dowell Egg Yolk Agar**

LD Esculin Agar
See: **Lombard–Dowell Esculin Agar**

LD Gelatin Agar
See: **Lombard–Dowell Gelatin Agar**

Lead Acetate Agar

Composition per liter:

Agar	15.0g
Peptone	15.0g
Proteose peptone	5.0g
Glucose	1.0g
Lead acetate	0.2g
$Na_2S_2O_3$	0.08g

pH 6.6 ± 0.2 at 25°C

Preparation of Medium: Add components to distilled/deionized water and bring volume to 1.0L. Mix thoroughly. Gently heat and bring to boiling. Distribute into tubes or flasks. Autoclave for 15 min at 15 psi pressure–121°C. Pour into sterile Petri dishes or leave in tubes. Allow tubes to cool in a slanted position.

Use: For the cultivation and differentiation of Gram-negative coliform bacteria based on H_2S production. Bacteria that produce H_2S turn the medium brown.

LEB, FDA
See: Listeria **Enrichment Broth, FDA**

Lecithin Lactose Agar

Composition per liter:

Agar	15.0g
Pancreatic digest of casein	12.7g
Lactose	10.0g
NaCl	5.5g

Peptic digest of animal tissue	5.5g
Yeast extract	3.9g
Pancreatic digest of heart muscle	3.3g
Cornstarch	1.1g
Egg lecithin	0.66g
L-Cysteine·HCl·H_2O	0.5g
NaN_3	0.2g
Neomycin sulfate	0.15g
$CaCl_2$	0.05g
Bromcresol Purple	0.02g

pH 6.8 ± 0.2 at 25°C

Source: Available as a prepared medium from BBL Microbiology Systems.

Caution: Sodium azide is toxic. Azides also react with metals and disposal must be highly diluted.

Preparation of Medium: Add components to distilled/deionized water and bring volume to 1.0L. Mix thoroughly. Gently heat and bring to boiling. Distribute into tubes or flasks. Autoclave for 15 min at 15 psi pressure–121°C. Pour into sterile Petri dishes.

Use: For the isolation, cultivation and differentiation of histolytic clostridia from clinical specimens based on lecithinase production and lactose fermentation. It is especially useful for the differentiation of *Clostridium perfringens, C. sordelli, C. novyi, C. septicum* and *C. histolyticum*. Bacteria that produce lecithinase appear as colonies surrounded by an opalescent zone. Bacteria that ferment lactose appear as colonies surrounded by a yellow zone.

Lecithin Lipase Anaerobic Agar

Composition per liter:

Pancreatic digest of casein	40.0g
Agar	25.0g
Yeast extract	5.0g
Glucose	2.0g
NaCl	2.0g
KH_2PO_4	1.0g
$Na_2HPO_4 \cdot 12H_2O$	5.0g
$MgSO_4 \cdot 7H_2O$	0.1g
Egg yolk emulsion	100.0mL

pH 7.6 ± 0.2 at 25°C

Egg Yolk Emulsion:

Composition:

Chicken egg yolks	11
Whole chicken egg	1

Preparation of Egg Yolk Emulsion: Soak eggs with 1:100 dilution of saturated mercuric chloride solution for 1 min. Crack eggs and separate yolks from whites. Mix egg yolks with 1 chicken egg. Filter sterilize.

Preparation of Medium: Add components, except egg yolk emulsion, to distilled/deionized water and bring volume to 900.0mL. Mix thoroughly. Gently heat and bring to boiling. Autoclave for 15 min at 15 psi pressure–121°C. Cool to 45°–50°C. Aseptically add sterile egg yolk emulsion. Mix thoroughly. Pour into sterile Petri dishes or distribute into sterile tubes.

Use: For the isolation, cultivation and differentiation of *Clostridium* species based on lecithinase production and lipase production. Bacteria that produce lecithinase appear as colonies surrounded by a zone of insoluble precipitate. Bacteria that produce lipase appear as colonies with a pearly iridescent sheen.

Lecithin Tween™ Medium (LT Medium)

Composition per liter:

Tween™ 80	30.0g
Agar	15.0g
Pancreatic digest of casein	10.0g
Peptic digest of animal tissue	10.0g
NaCl	5.0g
Glucose	1.0g
Lecithin	5.0g
$Na_2S_2O_3 \cdot 5H_2O$	5.0g
Glycerol	3.0g
Histidine, free base	1.0g

pH 7.5 ± 0.2 at 25°C

Antibiotic Solution:

Composition per 10mL:

5–Fluorocytosine	0.2g
Fosfomicin	0.1g
Ticarcillin	0.1g

Preparation of Antibiotic Solution: Add components to distilled/deionized water and bring volume to 10.0mL. Mix thoroughly. Filter sterilize.

Preparation of Medium: Add components, except antibiotic solution, to distilled/deionized water and bring volume to 990.0mL. Mix thoroughly. Gently heat and bring to boiling. Autoclave for 15 min at 15 psi pressure–121°C. Cool to 45°–50°C. Aseptically add sterile antibiotic solution. Mix thoroughly. Pour into sterile Petri dishes in 20.0mL volumes.

Use: For the isolation and cultivation of multiresistant lipophilic *Corynebacterium* species, especially *Corynebacterium* Group JK found primarily in infections in immunocompromised hosts and patients with prosthetic valve endocarditis.

Lee's Agar

Composition per liter:

Agar	18.0g
Pancreatic digest of casein	10.0g
Yeast extract	10.0g
Lactose	5.0g
Sucrose	5.0g
$CaCO_3$	3.0g
K_2HPO_4	0.5g
Bromcresol Purple (0.2% solution)	10.0mL

pH 7.0 ± 0.2 at 25°C

Bromcresol Purple Solution:
Composition per 10mL:

Bromcresol Purple	0.02g

Preparation of Bromcresol Purple Solution: Add Bromcresol Purple to distilled/deionized water and bring volume to 10.0mL. Mix thoroughly. Filter sterilize.

Preparation of Medium: Add components, except Bromcresol Purple solution, to distilled/deionized water and bring volume to 990.0mL. Mix thoroughly. Adjust pH to 7.0. Gently heat and bring to boiling. Autoclave for 20 min at 15 psi pressure–121°C. Cool to 45°–50°C. Aseptically add sterile Bromcresol Purple solution. Mix thoroughly. Pour into sterile, chilled Petri dishes in 20–25mL volumes. Swirl flask while dispensing to evenly suspend $CaCO_3$. Dry plates at 30°C for 18–24 hr before use.

Use: For the isolation, cultivation and enumeration of *Lactobacillus bulgaricus* from yogurt.

Legionella Agar Base (*Legionella* Medium) (BCYEα Agar, Modified)

Composition per liter:

Agar	17.0g
Yeast extract	10.0g
ACES buffer (N-2-acetamido-2-aminoethane sulfonic acid)	6.0g
Charcoal, activated	1.5g
KOH	1.5g
α-Ketoglutarate	1.0g

pH 6.85–7.0 at 25°C

Source: This medium is available as a prepared medium from Difco Laboratories.

***Legionella* Agar Enrichment:**
Composition per 10mL:

L-Cysteine·HCl·H_2O	0.4g
$Fe_4(P_2O_7)_3$	0.25g

Preparation of *Legionella* Agar Enrichment: Add components to distilled/deionized water and bring volume to 10.0mL. Mix thoroughly. Filter sterilize.

Preparation of Medium: Add components, except *Legionella* agar enrichment, to distilled/deionized water and bring volume to 990.0mL. Mix thoroughly. Gently heat to boiling. Autoclave for 15 min at 15 psi pressure–121°C. Cool to 50° C. Add 10.0mL of sterile *Legionella* agar enrichment. Adjust pH to 6.9 at 50°C by adding 4.0–4.5mL of 1.0*N* KOH—this is a critical step. Mix thoroughly. Pour into sterile Petri dishes in 20.0mL volumes. Swirl medium while pouring to keep charcoal in suspension.

Use: For the preparation of *Legionella* agars. Also, for the isolation and cultivation of *Legionella* species from clinical and nonclinical materials.

Legionella Medium
See: ***Legionella* Agar Base**

Legionella pneumophila Medium (Charcoal Yeast Extract Diphasic Blood Culture Medium) (Diphasic Blood Culture Buffered Charcoal Yeast Extract Medium) (CYE–DBCM)

Composition per liter:

Agar phase	500.0mL
Broth phase	500.0mL

pH 6.9–7.0 at 25°C

Agar Phase:
Composition per 500mL:

Agar	17.0g
Charcoal, activated	2.0g

Preparation of Agar Phase: Add components to distilled/deionized water and bring volume to 500.0mL. Mix thoroughly. Gently heat and bring to boiling. Distribute in 20.0mL volumes into 125.0mL serum bottles with aluminum crimp seals and rubber stoppers. Autoclave for 20 min at 15 psi pressure–121°C. Cool to 50°C. Swirl medium to put charcoal in suspension. Allow agar to solidify so that a slant with a 6.0cm height is formed.

Broth Phase:
Composition per 500mL:

Yeast extract	20.0g
L-Cysteine·HCl·H_2O solution	0.40g
$Fe(NO_3)_3·9H_2O$ solution	0.10g

Preparation of Broth Phase: Add yeast extract to distilled/deionized water and bring volume to 480.0mL. Mix thoroughly. Autoclave for 15 min at 15 psi pressure–121°C. Cool to 25°C. Aseptically add sterile cysteine·HCl·H_2O solution and $Fe(NO_3)_3 \cdot 9H_2O$ solution. Mix thoroughly. Adjust pH to 6.9 with 6.0mL of sterile 1*N* KOH.

L-Cysteine·HCl·H_2O Solution:
Composition per 10mL:
L-Cysteine·HCl·H_2O 0.04g

Preparation of L-Cysteine·HCl·H_2O Solution: Add cysteine·HCl·H_2O to distilled/deionized water and bring volume to 10.0mL. Mix thoroughly. Filter sterilize.

$Fe(NO_3)_3 \cdot 9H_2O$ Solution:
Composition per 10mL:
$Fe(NO_3)_3 \cdot 9H_2O$ 0.04g

Preparation of $Fe(NO_3)_3 \cdot 9H_2O$ Solution: Add $Fe(NO_3)_3 \cdot 9H_2O$ to distilled/deionized water and bring volume to 10.0mL. Mix thoroughly. Filter sterilize.

Preparation of Medium: Add 20.0mL of sterile broth phase to 125.0mL serum bottles containing 20.0mL of solidified agar phase. Seal bottles by crimping metal caps over rubber stoppers.

Use: For the isolation and cultivation of *Legionella pneumophila* from blood cultures.

Legionella Selective Agar

Composition per liter:
Agar 15.0g
ACES (2-[(2-amino-2-oxoethyl)-amino]ethane sulfonic acid) buffer 10.0g
Yeast extract 10.0g
Charcoal, activated 2.0g
α-Ketoglutarate 1.0g
L-Cysteine·HCl·H_2O solution 10.0mL
$Fe_4(P_2O_7)_3$ solution 10.0mL
Antibiotic solution 10.0mL
pH 6.85–7.0 at 25°C

Source: This medium is available as a prepared medium from BBL Microbiology Systems.

Cysteine·HCl·H_2O Solution:
Composition per 10mL:
L-Cysteine·HCl·H_2O 0.4g

Preparation of Cysteine·HCl·H_2O Solution: Add cysteine·HCl·H_2O to distilled/deionized water and bring volume to 10.0mL. Mix thoroughly. Filter sterilize.

$Fe_4(P_2O_7)_3$ Solution:
Composition per 10mL:
$Fe_4(P_2O_7)_3$ 0.25g

Preparation of $Fe_4(P_2O_7)_3$ Solution: Add $Fe_4(P_2O_7)_3$ to distilled/deionized water and bring volume to 10.0mL. Mix thoroughly. Filter sterilize.

Antibiotic Solution:
Composition per 10mL:
Anisomycin 10.0mg
Colistin 3.75mg
Vancomycin 2.0mg

Preparation of Antibiotic Solution: Add components to distilled/deionized water and bring volume to 10.0mL. Mix thoroughly. Filter sterilize.

Preparation of Medium: Add components—except cysteine·HCl·H_2O, $Fe_4(P_2O_7)_3$, and antibiotic solutions—to distilled/deionized water and bring volume to 970.0mL. Mix thoroughly. Gently heat and bring to boiling. Autoclave for 15 min at 15 psi pressure–121°C. Cool to 45°–50°C. Aseptically add sterile cysteine·HCl·H_2O, $Fe_4(P_2O_7)_3$, and antibiotic solutions. Mix thoroughly. Pour into sterile Petri dishes. Swirl medium while pouring to keep charcoal in suspension.

Use: *Legionella* Selective Agar is used in qualitative procedures for the isolation of *Legionella* species from clinical and nonclinical specimens.

Legume Extract Agar

Composition per liter:
Alfalfa roots 35.0g
Agar 20.0g
Soybean meal 10.0g
Sucrose 10.0g
$CaCO_3$ 5.0g
Glucose 5.0g
K_2HPO_4 ....1.0g
$MgSO_4 \cdot 7H_2O$ 0.2g
$CaCl_2$ 0.1g
NaCl 0.1g
$FeCl_3$ 1.0mg

Preparation of Medium: Wash the alfalfa roots well and cut them up. Add 10.0g of soybean meal. Add three times the volume of distilled/deionized water. Steam for 1 hr. Let stand at 25°C overnight. Bring volume to 1.0L with distilled/deionized water. Filter through paper pulp. To the filtrate, add the K_2HPO_4, $CaCl_2$, $MgSO_4 \cdot 7H_2O$, NaCl, $FeCl_3$, and agar. Autoclave for 20 min at 15 psi pressure–121°C. Cool to 45°–50°C. Add the $CaCO_3$, sucrose and glucose. Mix thoroughly. Distribute into tubes or flasks. Autoclave for 20 min at 10 psi pressure–115°C.

Use: For the cultivation of *Rhizobium* species.

Leifson Medium

Composition per liter:

Agar	15.0g
Pancreatic digest of casein	2.0g
$MgCl_2$	1.0g
Yeast extract	1.0g

pH 8.0 ± 0.2 at 25°C

Preparation of Medium: Add components to distilled/deionized water and bring volume to 1.0L. Mix thoroughly. Gently heat and bring to boiling. Adjust pH to 8.0. Distribute into tubes or flasks. Autoclave for 15 min at 15 psi pressure–121°C. Pour into sterile Petri dishes or leave in tubes.

Use: For the direct isolation and routine culturing of *Hyphomonas* species.

LEMB Agar
See: Levine EMB Agar

Lenox Broth
See: LB Medium

Leptospira Medium

Composition per liter:

$(NH_4)_2Fe(SO_4)_2 \cdot 6H_2O$	6.0g
NaH_2PO_4	0.53g
L-Asparagine	0.5g
Glycerol	0.2g
Tween™ 60	0.2g
$MgSO_4 \cdot 7H_2O$	0.15g
KH_2PO_4	0.069g
Tween™ 80	0.05g
EDTA	0.01g
$CaCO_3$	4.0mg
Thiamine·HCl	1.0mg
Vitamin B_{12}	1.0μg

pH 7.4–7.6 at 25°C

Preparation of Medium: Add components, except thiamine·HCl to distilled/deionized water and bring volume to 990.0mL. Mix thoroughly. Gently heat and bring to boiling. Autoclave for 15 min at 15 psi pressure–121°C. Aseptically add 1.0mg of thiamine·HCl. Aseptically distribute into sterile tubes or flasks.

Use: For the cultivation of *Leptospira* species.

Leptospira Medium, EMJH (*Leptospira* Medium, Ellinghausen–McCullough/ Johnson–Harris)

Composition per liter:

Na_2HPO_4	1.0g
NaCl	1.0g
KH_2PO_4	0.3g
NH_4Cl	0.25g
Thiamine	5.0mg
Rabbit serum	100.0mL

pH 7.5 ± 0.2 at 25°C

Source: This medium is available as a premixed powder from Difco Laboratories.

Preparation of Medium: Add components, except rabbit serum, to distilled/deionized water and bring volume to 900.0mL. Mix thoroughly. Gently heat and bring to boiling. Autoclave for 15 min at 15 psi pressure–121°C. Cool to 25°C. Aseptically add sterile rabbit serum. Mix thoroughly. Aseptically distribute into sterile tubes or flasks.

Use: For the cultivation and maintenance of *Leptospira* species.

Leptospira Medium, Modified

Composition per liter:

Agar	1.5g
NaCl	0.5g
Peptone	0.3g
Beef extract	0.2g
Hemin solution	2.5mL
Sterile rabbit serum	100.0mL

pH 7.3 ± 0.1 at 25°C

Hemin Solution:
Composition per 100mL:

Hemin	0.05g
NaOH (1*N* solution)	1.0mL

Preparation of Hemin Solution: Add hemin to 1.0mL of 1*N* NaOH solution. Mix thoroughly. Bring volume to 100.0mL with distilled/deionized water. Autoclave for 15 min at 15 psi pressure–121°C. Cool to 45°–50°C.

Preparation of Medium: Add components, except hemin solution and rabbit serum, to distilled/ deionized water and bring volume to 897.5mL. Mix thoroughly. Gently heat and bring to boiling. Adjust pH to 7.4. Autoclave for 15 min at 15 psi pressure–121°C. Cool to 45°–50°C. Aseptically add 2.5mL of sterile hemin solution and 100.0mL of sterile rabbit serum. Mix thoroughly. The pH of the medium should be 7.3. Store at 4°C for 24 hr. Inactivate medium at 56°C for 60 min. Aseptically distribute into sterile tubes or flasks.

Use: For the cultivation and maintenance of *Leptospira biflexa, Leptospira borgpetersenii, Leptospira interrogans, Leptospira meyeri, Leptospira noguchii, Leptospira santarosai,* and *Leptospira weili.*

Leptospira Protein–Free Medium (*Leptospira* PF Medium)

Composition per liter:

TES (*N*-tris[hydroxymethyl]methyl-2-aminoethane sulfonic acid) buffer 1.2g
NaCl 0.9g
Sodium pyruvate 0.2g
CT-Tween™ 60 .... 12.0mL
CT-Tween™ 40 .... 3.0mL
$MgCl_2$-$CaCl_2$ solution 1.0mL
Cyanocobalamin (0.02% solution) 1.0mL
Glycerol (10% solution) 1.0mL
KH_2PO_4 (1% solution) 1.0mL
$MnSO_4 \cdot H_2O$ (0.1% solution) 1.0mL
$ZnSO_4$ (0.4% solution) 0.1mL

pH 7.6 ± 0.2 at 25°C

CT-Tween™ 60:

Composition per 200mL:

Charcoal, Norit A 40.0g
Tween™ 60 .... 20.0g

Preparation of CT-Tween™ 60: Add Tween™ 60 to 200.0mL of distilled/deionized water. Mix thoroughly. While stirring, add charcoal. Stir mixture for 18 hr at 25°C. Allow charcoal to settle out of suspension for 18 hr at 4°C. Carefully decant the Tween™ solution off the sediment. Centrifuge the Tween™ solution at 10,000 × g for 1 hr. Decant supernatant solution. Pass Tween™ solution through a thin channel ultrafiltration XM 100 membrane. Store stock solution at –20°C.

CT-Tween™ 40:

Composition per 200mL:

Charcoal, Norit A 40.0g
Tween™ 40 .... 20.0g

Preparation of CT-Tween™ 40: Add Tween™ 40 to 200.0mL of distilled/deionized water. Mix thoroughly. While stirring, add charcoal. Stir mixture for 18 hr at 25°C. Allow charcoal to settle out of suspension for 18 hr at 4°C. Carefully decant the Tween™ solution off the sediment. Centrifuge the Tween™ solution at 10,000 × g for 1 hr. Decant supernatant solution. Pass Tween™ solution through a thin channel ultrafiltration XM 100 membrane. Store stock solution at –20°C.

$MgCl_2$–$CaCl_2$ Solution:

Composition per 100mL:

$CaCl_2 \cdot 2H_2O$ 1.5g
$MgCl_2 \cdot 6H_2O$ 1.5g

Preparation of $MgCl_2$–$CaCl_2$ Solution: Add components to distilled/deionized water and bring volume to 100.0mL. Mix thoroughly.

Preparation of Medium: Add components to distilled/deionized water and bring volume to 1.0L. Mix thoroughly. Filter sterilize. Aseptically distribute into sterile tubes or flasks.

Use: For the cultivation of *Leptospira* species.

Leptothrix 2X PYG Medium

Composition per liter:

HEPES (N-2-hydroxyethyl piperazine-N′-2-ethanesulfonic acid) buffer 3.57g
$MgSO_4 \cdot 7H_2O$ 0.6g
Glucose 0.5g
Peptone 0.5g
Yeast extract 0.5g
$CaCl_2 \cdot 2H_2O$ 0.07g
$MnSO_4 \cdot H_2O$ 0.017g

pH 7.3 ± 0.2 at 25°C

Preparation of Medium: Add components to distilled/deionized water and bring volume to 1.0L. Mix thoroughly. Adjust pH to 7.3. Distribute into tubes or flasks. Autoclave for 15 min at 15 psi pressure–121°C.

Use: For the cultivation and maintenance of *Leptothrix discophora.*

Leptothrix ochracea Medium

Composition per liter:

Agar 10.0g
Manganous acetate 0.1g
Manganese bicarbonate solution 100.0mL

Manganese Bicarbonate Solution:

Composition per 100mL:

$MnCO_3$ 2.0g

Preparation of Manganese Bicarbonate Solution: Add $MnCO_3$ to distilled/deionized water and bring volume to 100.0mL. Mix thoroughly. Gas with 100% CO_2 for 20 min. Filter through Whatman #1 filter paper.

Preparation of Medium: Add components to distilled/deionized water and bring volume to 1.0L. Mix thoroughly. Gently heat and bring to boiling. Distribute into tubes or flasks. Autoclave for 15 min at 15 psi pressure–121°C. Pour into sterile Petri dishes or leave in tubes.

Use: For the cultivation of *Leptothrix ochracea.*

Leptothrix Strains Medium

Composition per liter:

Agar	7.5g
$MnCO_2$	2.0g
Beef extract	1.0g
$Fe(NH_4)_2(SO_4)_2$	0.15g
Sodium citrate	0.15g
Yeast extract	0.075g
Vitamin B_{12}	5.0μg

Preparation of Medium: Add components to distilled/deionized water and bring volume to 1.0L. Mix thoroughly. Distribute into tubes or flasks. Autoclave for 15 min at 15 psi pressure–121°C.

Use: For the isolation of *Leptothrix* species.

Leptothrix Strains Medium

Composition per liter:

Agar	12.0g
Peptone	5.0g
$MgSO_4 \cdot 7H_2O$	0.20g
Ferric ammonium citrate	0.15g
$CaCl_2$	0.05g
$FeCl_3 \cdot 6H_2O$	0.01g
$MnSO_4 \cdot H_2O$	0.01g

Preparation of Medium: Add components to distilled/deionized water and bring volume to 1.0L. Mix thoroughly. Gently heat and bring to boiling. Distribute into tubes or flasks. Autoclave for 15 min at 15 psi pressure–121°C. Pour into sterile Petri dishes or leave in tubes.

Use: For the isolation and cultivation of *Leptothrix* species.

Leptotrichia buccalis Medium

Composition per liter:

Agar	15.0g
Nutrient broth	8.0g
Yeast extract	2.0g
Glucose solution	10.0mL
L-Cysteine·HCl·H_2O solution	10.0mL
Hemin solution	4.0mL

pH 7.2–7.6 at 25°C

Glucose Solution:

Composition per 10mL:

D-Glucose	2.0g

Preparation of Glucose Solution: Add glucose to distilled/deionized water and bring volume to 10.0mL. Mix thoroughly. Filter sterilize.

Cysteine·HCl·H_2O Solution:

Composition per 10mL:

L-Cysteine·HCl·H_2O	1.0g

Preparation of Cysteine·HCl·H_2O Solution: Add cysteine·HCl·H_2O to distilled/deionized water and bring volume to 10.0mL. Mix thoroughly. Filter sterilize.

Hemin Solution:

Composition per 10mL:

Hemin	2.5mg
Triethanolamine (50% solution)	10.0mL

Preparation of Hemin Solution: Add hemin to 10.0mL of triethanolamine solution. Mix thoroughly.

Preparation of Medium: Add components, except glucose solution, to distilled/deionized water and bring volume to 990.0mL. Mix thoroughly. Gently heat and bring to boiling. Autoclave for 15 min at 15 psi pressure–121°C. Cool to 45°–50°C. Aseptically add sterile glucose solution. Mix thoroughly. Pour into sterile Petri dishes or distribute into sterile tubes.

Use: For the cultivation and maintenance of *Leptotrichia buccalis.*

Leptotrichia Medium

Composition per liter:

Pancreatic digest of casein	10.0g
NaCl	5.0g
Peptone	5.0g
Yeast extract	3.0g
Na_2HPO_4	2.5g
L-Cysteine·HCl·H_2O	0.5g
Horse serum	100.0mL
Tomato decoction	50.0mL

pH 7.2–7.4 at 25°C

Tomato Decoction:

Composition per 100mL:

Tomatoes	50.0mL

Preparation of Tomato Decoction: Mince fresh tomatoes and measure 50.0mL. Add 50.0mL of tap water. Mix thoroughly. Gently heat and bring to boiling. Continue boiling for 10 min. Filter through Whatman #1 filter paper. Autoclave filtrate for 15 min at 15 psi pressure–121°C.

Preparation of Medium: Add components, except horse serum and tomato decoction, to distilled/deionized water and bring volume to 850.0mL. Mix thoroughly. Gently heat and bring to boiling. Adjust pH to 7.2–7.4. Autoclave for 15 min at 15 psi pressure–121°C. Cool to 25°C. Aseptically add sterile horse serum and tomato decoction. Mix thoroughly. Aseptically distribute into sterile tubes or flasks.

Use: For the cultivation and maintenance of *Leptotrichia buccalis.*

LES Endo Agar
See: Endo Agar, LES

Letheen Agar

Composition per liter:

Agar	15.0g
Tween™ 80	7.0g
Pancreatic digest of casein	5.0g
Beef extract	3.0g
Glucose	1.0g
Lecithin	1.0g

pH 7.0 ± 0.2 at 25°C

Source: This medium is available as a premixed powder from BBL Microbiology Systems and Difco Laboratories.

Preparation of Medium: Add components to distilled/deionized water and bring volume to 1.0L. Mix thoroughly. Gently heat and bring to boiling. Distribute into tubes or flasks. Autoclave for 15 min at 15 psi pressure–121°C. Pour into sterile Petri dishes or leave in tubes.

Use: For the determination of the antimicrobial activity of quaternary ammonium compounds.

Letheen Agar, Modified

Composition per liter:

Agar	20.0g
Thiotone	10.0g
Pancreatic digest of casein	10.0g
Tween™ 80	7.0g
NaCl	5.0g
Beef extract	3.0g
Yeast extract	2.0g
Glucose	1.0g
Lecithin	1.0g
$NaHSO_3$	0.1g

pH 7.2 ± 0.2 at 25°C

Preparation of Medium: Add components to distilled/deionized water and bring volume to 1.0L. Mix thoroughly. Gently heat and bring to boiling. Distribute into tubes or flasks. Autoclave for 15 min at 15 psi pressure–121°C. Pour into sterile Petri dishes in 20.0mL.

Use: For the determination of the antimicrobial activity of quaternary ammonium compounds.

Letheen Broth

Composition per liter:

Peptic digest of animal tissue	10.0g
Beef extract	5.0g
NaCl	5.0g
Tween™ 80	5.0g
Lecithin	0.7g

pH 7.0 ± 0.2 at 25°C

Source: This medium is available as a premixed powder from Difco Laboratories.

Preparation of Medium: Add components to distilled/deionized water and bring volume to 1.0L. Mix thoroughly. Distribute into tubes or flasks. Autoclave for 15 min at 15 psi pressure–121°C.

Use: For the determination of the antimicrobial activity of quaternary ammonium compounds.

Letheen Broth, Modified

Composition per liter:

Peptic digest of animal tissue	10.0g
Thiotone peptone	10.0g
Beef extract	5.0g
NaCl	5.0g
Tween™ 80	5.0g
Pancreatic digest of casein	5.0g
Yeast extract	2.0g
Lecithin	0.7g
$NaHSO_3$	0.1g

pH 7.2 ± 0.2 at 25°C

Preparation of Medium: Add components to distilled/deionized water and bring volume to 1.0L. Mix thoroughly. Distribute into screw-capped bottles in 90.0mL volumes. Autoclave for 15 min at 15 psi pressure–121°C.

Use: For the determination of the antimicrobial activity of quaternary ammonium compounds.

Leuconostoc Medium

Composition per liter:

$CaCO_3$	50.0g
Malt extract	50.0g
Agar	15.0g
NaCl	2.5g
Beef extract	1.0g
Polypeptone™	1.0g

Preparation of Medium: Add components to distilled/deionized water and bring volume to 1.0L. Mix thoroughly. Gently heat and bring to boiling. Distribute into tubes or flasks. Autoclave for 10 min at 15 psi pressure–121°C. Pour into sterile Petri dishes or leave in tubes.

Use: For the cultivation and maintenance of *Leuconostoc mesenteroides.*

Leuconostoc oenos Medium

Composition per liter:

Glucose	10.0g
Peptone	10.0g
Yeast extract	5.0g
$MnSO_4 \cdot 4H_2O$	0.1g
Tomato juice	250.0mL
Cysteine·HCl solution	10.0mL

pH 4.8 ± 0.2 at 25°C

Cysteine·HCl Solution:
Composition per 10mL:

Cysteine·HCl	0.5g

Preparation of Cysteine·HCl Solution: Add cysteine·HCl to distilled/deionized water and bring volume to 10.0mL. Mix thoroughly. Filter sterilize.

Preparation of Medium: Add components, except cysteine·HCl solution, to distilled/deionized water and bring volume to 990.0mL. Mix thoroughly. Gently heat and bring to boiling. Autoclave for 15 min at 15 psi pressure–121°C. Cool to 25°C. Aseptically add sterile cysteine·HCl solution. Mix thoroughly. Aseptically distribute into sterile tubes or flasks.

Use: For the cultivation of *Leuconostoc oenos*.

Leucothrix Medium (ATCC Medium 429)

Composition per liter:

NaCl	11.7g
Monosodium glutamate	10.0g
$MgCl_2 \cdot 6H_2O$	5.35g
Na_2SO_4	2.0g
$CaCl_2 \cdot 2H_2O$	0.75g
Tris(hydroxymethyl)aminomethane buffer	0.5g
KCl	0.35g
Na_2HPO_4	0.05g

pH 7.6 ± 0.2 at 25°C

Preparation of Medium: Add components to distilled/deionized water and bring volume to 1.0L. Mix thoroughly. Distribute into tubes or flasks. Autoclave for 15 min at 15 psi pressure–121°C.

Use: For the cultivation and maintenance of *Leucothrix mucor*.

Leucothrix Medium (ATCC Medium 430)

Composition per liter:

NaCl	11.7g
$MgCl_2 \cdot 6H_2O$	5.35g
Na_2SO_4	2.0g
$CaCl_2 \cdot 2H_2O$	0.75g
Pancreatic digest of casein	0.5g
Yeast extract	0.5g
Tris(hydroxymethyl)aminomethane buffer	0.5g
KCl	0.35g
Na_2HPO_4	0.05g

pH 7.6 ± 0.2 at 25°C

Preparation of Medium: Add components to distilled/deionized water and bring volume to 1.0L. Mix thoroughly. Distribute into tubes or flasks. Autoclave for 15 min at 15 psi pressure–121°C.

Use: For the cultivation and maintenance of *Leucothrix mucor*.

Levine EMB Agar (Levine Eosin Methylene Blue Agar) (Eosin Methylene Blue Agar, Levine) (LEMB Agar)

Composition per liter:

Agar	15.0g
Lactose	10.0g
Peptone	10.0g
K_2HPO_4	2.0g
Eosin Y	0.4g
Methylene Blue	0.065mg

pH 7.1 ± 0.2 at 25°C

Source: This medium is available as a premixed powder from BBL Microbiology Systems and Difco Laboratories.

Preparation of Medium: Add components to distilled/deionized water and bring volume to 1.0L. Mix thoroughly. Gently heat and bring to boiling. Distribute into tubes or flasks. Autoclave for 15 min at 15 psi pressure–121°C. Pour into sterile Petri dishes or leave in tubes.

Use: For the isolation, cultivation and differentiation of Gram-negative enteric bacteria based on lactose fermentation. Bacteria that ferment lactose, especially the coliform bacterium *Escherichia coli*, appear as colonies with a green metallic sheen or blue-black to brown color. Bacteria that do not ferment lactose appear as colorless or transparent light purple colonies.

Levinthal's Agar

Composition per 105mL:

Nutrient agar, sterile	100.0mL
Rabbit blood or human blood, sterile	5.0mL

pH 6.8 ± 0.2 at 25°C

Nutrient Agar:
Composition per liter:

Agar	15.0g
Pancreatic digest of gelatin	5.0g
Beef extract	3.0g

Source: Nutrient agar is available as a premixed powder from BBL Microbiology Systems and Difco Laboratories.

Preparation of Nutrient Agar: Add components to distilled/deionized water and bring volume to 1.0L. Mix thoroughly. Gently heat while stirring and bring to boiling. Distribute into tubes or flasks. Autoclave for 15 min at 15 psi pressure–121°C. Cool to 45°–50°C.

Preparation of Medium: To 100.0mL of cooled, sterile nutrient agar, aseptically add 5.0mL of human blood or rabbit blood. Mix thoroughly. Heat in a boiling water bath for 5 min. Allow the deposit to settle out of suspension. Pour the clear supernatant solution into sterile Petri dishes or distribute into sterile tubes.

Use: For the cultivation of *Haemophilus* species.

LGI Medium

See: ***Azospirillum amazonense*** **Medium**

LHET2 Medium

Composition per liter:

Solution A	500.0mL
Solution B	500.0mL

pH 2.5–3.0 at 25°C

Solution A:
Composition per 500mL:

$(NH_4)_2SO_4$	2.0g
K_2HPO_4	0.51g
$MgSO_4 \cdot 7H_2O$	0.5g
KCl	0.1g
Pancreatic digest of casein	0.06g
NaCl	0.02g
Papaic digest of soybean meal	0.01g

Preparation of Solution A: Add components to distilled/deionized water and bring volume to 500.0mL. Mix thoroughly. Gently heat and bring to boiling. Adjust pH to 2.5–3.0 with 1*N* H_2SO_4. Autoclave for 15 min at 15 psi pressure–121°C. Cool to 45°–50°C.

Solution B:
Composition per 500mL:

Agar	12.0g
Glucose	1.0g

Preparation of Solution B: Add components to distilled/deionized water and bring volume to 500.0mL. Mix thoroughly. Gently heat and bring to boiling. Autoclave for 15 min at 15 psi pressure–121°C. Cool to 45°–50°C.

Preparation of Medium: Aseptically combine sterile solution A and sterile solution B. Mix thoroughly. Pour into sterile Petri dishes or distribute into sterile tubes.

Use: For the cultivation and maintenance of *Acidiphilium cryptum*.

LHET2 Medium with Yeast Extract or Yeast Autolysate

Composition per liter:

Solution A	500.0mL
Solution B	500.0mL

pH 2.5–3.0 at 25°C

Solution A:
Composition per 500mL:

$(NH_4)_2SO_4$	2.0g
K_2HPO_4	0.51g
$MgSO_4 \cdot 7H_2O$	0.5g
KCl	0.1g
Yeast extract or yeast autolysate	0.1g
Pancreatic digest of casein	0.06g
NaCl	0.02g
Papaic digest of soybean meal	0.01g

Preparation of Solution A: Add components to distilled/deionized water and bring volume to 500.0mL. Mix thoroughly. Gently heat and bring to boiling. Adjust pH to 2.5–3.0 with 1*N* H_2SO_4. Autoclave for 15 min at 15 psi pressure–121°C. Cool to 45°–50°C.

Solution B:
Composition per 500mL:

Agar	12.0g
Glucose	1.0g

Preparation of Solution B: Add components to distilled/deionized water and bring volume to 500.0mL. Mix thoroughly. Gently heat and bring to boiling. Autoclave for 15 min at 15 psi pressure–121°C. Cool to 45°–50°C.

Preparation of Medium: Aseptically combine sterile solution A and sterile solution B. Mix thoroughly. Pour into sterile Petri dishes or distribute into sterile tubes.

Use: For the cultivation and maintenance of *Acidiphilium angustum, Acidiphilium facilis,* and *Acidiphilium rubrum.*

LICNR Broth (Lysine Iron Cystine Neutral Red Broth)

Composition per 500mL:

L-Lysine·HCl	10.0g
Mannitol	5.0g
Pancreatic digest of casein	5.0g
Yeast extract	3.0g
Glucose	1.0g
Salicin	1.0g
Ferric ammonium citrate	0.5g
L-Cystine	0.1g
$Na_2S_2O_3$	0.1g
Neutral Red	0.025g
Novobiocin solution	10.0mL

pH 6.2 ± 0.2 at 25°C

Novobiocin Solution:

Composition per 10mL:

Novobiocin	0.015g

Preparation of Novobiocin Solution: Add novobiocin to distilled/deionized water and bring volume to 10.0mL. Mix thoroughly. Filter sterilize.

Preparation of Medium: Add components, except novobiocin solution, to distilled/deionized water and bring volume to 990.0mL. Mix thoroughly. Gently heat and bring to boiling. Continue boiling for 2–3 min. Distribute into tubes in 10.0mL volumes. Autoclave for 15 min at 15 psi pressure–121°C. Cool to 45°–50°C. Aseptically add 0.1mL of sterile novobiocin solution to each tube. Mix thoroughly.

Use: For the rapid presumptive detection of *Salmonella* in foods, food ingredients and feed materials.

Lima Bean Agar

Composition per liter:

Lima beans, solids from infusion	62.5g
Agar	15.0g

pH 5.6 ± 0.2 at 25°C

Source: This medium is available as a premixed powder from Difco Laboratories.

Preparation of Medium: Add components to distilled/deionized water and bring volume to 1.0L. Mix thoroughly. Gently heat and bring to boiling. Distribute into tubes or flasks. Autoclave for 15 min at 15 psi pressure–121°C. Pour into sterile Petri dishes or leave in tubes.

Use: For the cultivation of a variety of phytopathological fungi and other fungi.

Lipovitellin Salt Mannitol Agar

Composition per liter:

NaCl	75.0g
Egg yolk	20.0g
Agar	15.0g
D-Mannitol	10.0g
Polypeptone™	10.0g
Beef extract	1.0g
Phenol Red	0.025g

Preparation of Medium: Add components to distilled/deionized water and bring volume to 1.0L. Mix thoroughly. Gently heat and bring to boiling. Distribute into tubes or flasks. Autoclave for 15 min at 15 psi pressure–121°C. Pour into sterile Petri dishes or leave in tubes.

Use: For the detection of *Staphylococcus aureus* in swimming pool water based on lipovitellin-lipase activity and mannitol fermentation. *Staphylococcus aureus* and other bacteria with lipovitellin-lipase activity attack the egg yolk and appear as colonies surrounded by an opaque zone. Bacteria that ferment mannitol appear as colonies surrounded by a yellow zone.

Listeria Enrichment Broth

Composition per liter:

Pancreatic digest of casein	17.0g
Yeast extract	6.0g
NaCl	5.0g
Papaic digest of soybean meal	3.0g
Glucose	2.5g
K_2HPO_4	2.5g
Cycloheximide	0.05g
Nalidixic acid	0.04g
Acriflavine·HCl	0.015g

pH 7.3 ± 0.2 at 25°C

Source: This medium is available as a premixed powder from BBL Microbiology Systems.

Preparation of Medium: Add components to distilled/deionized water and bring volume to 1.0L. Mix thoroughly. Gently heat and bring to boiling. Distribute into tubes or flasks. Autoclave for 15 min at 15 psi pressure–121°C.

Use: For the isolation and cultivation of *Listeria monocytogenes* from food, milk and dairy products according to the FDA formula.

Listeria Enrichment Broth, FDA (LEB, FDA)

Composition per liter:

Soybean casein digest broth yeast extract	1.0L

Nalidixic acid solution 8.0mL
Cycloheximide solution 5.1mL
Acriflavin·HCl solution.................................. 3.0mL
pH 7.3 ± 0.2 at 25°C

Soybean Casein Digest Broth Yeast Extract:
Composition per liter:

Pancreatic digest of casein 17.0g
Yeast extract 6.0g
NaCl 5.0g
Papaic digest of soybean meal 3.0g
K_2HPO_4 .... 2.5g
Glucose 2.5g

Source: This medium is available as a premixed powder from BBL Microbiology Systems and Difco Laboratories.

Preparation of Soybean Casein Digest Broth Yeast Extract: Add components to distilled/deionized water and bring volume to 1.0L. Mix thoroughly. Autoclave for 15 min at 15 psi pressure–121°C.

Nalidixic Acid Solution:
Composition per 100mL:

Nalidixic acid 0.5g

Preparation of Nalidixic Acid Solution: Add nalidixic acid to distilled/deionized water and bring volume to 100.0mL. Mix thoroughly. Filter sterilize.

Cycloheximide Solution:
Composition per 100mL:

Cycloheximide 1.5g
Ethanol 40.0mL

Preparation of Cycloheximide Solution: Add cycloheximide to 40.0mL of ethanol. Mix thoroughly. Bring volume to 100.0mL with distilled/deionized water. Filter sterilize.

Acriflavin·HCl Solution:
Composition per 100mL:

Acriflavin·HCl solution...................................... 0.5g

Preparation of Acriflavin·HCl Solution: Add acriflavin·HCl solution to distilled/deionized water and bring volume to 100.0mL. Mix thoroughly. Filter sterilize.

Preparation of Medium: Add components—except nalidixic acid solution, acriflavin solution, and cycloheximide solution—to distilled/deionized water and bring volume to 990.0mL. Mix thoroughly. Gently heat and bring to boiling. Autoclave for 15 min at 15 psi pressure–121°C. Cool to 45°–50°C. Aseptically add 8.0mL of sterile nalidixic acid solution, 5.1mL of sterile cycloheximide solution, and 3.0mL of sterile acriflavin solution. Mix thoroughly. Pour into sterile Petri dishes or distribute into sterile tubes.

Use: For the isolation and enrichment of *Listeria monocytogenes* from foods.

Listeria Enrichment Broth I, USDA FSIS (*Listeria* Primary Selective Enrichment Broth, UVM I) (University of Vermont I *Listeria* Primary Selective Enrichment Broth)

Composition per liter:

NaCl 20.0g
Na_2HPO_4 .... 12.0g
Beef extract 5.0g
Proteose peptone 5.0g
Pancreatic digest of casein 5.0g
Yeast extract 5.0g
KH_2PO_4 .... 1.35g
Esculin... 1.0g
Nalidixic acid solution 1.0mL
Acriflavine solution... 1.0mL
pH 7.4 ± 0.2 at 25°C

Source: This medium is available as a premixed powder from Oxoid Unipath.

Nalidixic Acid Solution:
Composition per 10mL:

Nalidixic acid 0.2g
NaOH (0.1M solution) 10.0mL

Preparation of Nalidixic Acid Solution: Add nalidixic acid to 10.0mL of NaOH solution. Mix thoroughly. Filter sterilize.

Acriflavine Solution:
Composition per 10mL:

Acriflavine.. 0.12g

Preparation of Acriflavine Solution: Add acriflavine to distilled/deionized water and bring volume to 10.0mL. Mix thoroughly. Filter sterilize. Use freshly prepared solution.

Preparation of Medium: Add components, except nalidixic acid solution and acriflavine solution, to distilled/deionized water and bring volume to 998.0mL. Mix thoroughly. Gently heat and bring to boiling. Autoclave for 15 min at 15 psi pressure–121°C. Cool to 45°–50°C. Aseptically add sterile nalidixic acid solution. Mix thoroughly. Store at 4°C. Immediately prior to use, aseptically add 1.0mL of sterile acriflavine solution. Mix thoroughly. Aseptically distribute into sterile tubes or flasks.

Use: For the selective isolation, cultivation and enrichment of *Listeria monocytogenes* from food, milk and dairy products.

Listeria Enrichment Broth II, USDA FSIS (*Listeria* Primary Selective Enrichment Broth, UVM II) (University of Vermont II *Listeria* Primary Selective enrichment Broth

Composition per liter:

NaCl	20.0g
Na_2HPO_4	12.0g
Beef extract	5.0g
Protease peptone	5.0g
Pancreatic digest of casein	5.0g
Yeast extract	5.0g
KH_2PO_4	1.35g
Esculin	1.0g
Nalidixic acid solution	1.0mL
Acriflavine solution	1.0mL

pH 7.4 ± 0.2 at 25°C

Source: This medium is available as a premixed powder from Oxoid Unipath.

Nalidixic Acid Solution:
Composition per 10mL:

Nalidixic acid	0.2g
NaOH (0.1M solution)	10.0mL

Preparation of Nalidixic Acid Solution: Add nalidixic acid to 10.0mL of NaOH solution. Mix thoroughly. Filter sterilize.

Acriflavine Solution:
Composition per 10mL:

Acriflavine	0.25g

Preparation of Acriflavine Solution: Add acriflavine to distilled/deionized water and bring volume to 10.0mL. Mix thoroughly. Filter sterilize. Use freshly prepared solution.

Preparation of Medium: Add components, except nalidixic acid solution and acriflavine solution, to distilled/deionized water and bring volume to 998.0mL. Mix thoroughly. Gently heat and bring to boiling. Autoclave for 15 min at 15 psi pressure–121°C. Cool to 45°–50°C. Aseptically add sterile nalidixic acid solution. Mix thoroughly. Store at 4°C. Immediately prior to use, aseptically add 1.0mL of sterile acriflavine solution. Mix thoroughly. Aseptically distribute into sterile tubes or flasks.

Use: For the selective isolation, cultivation and enrichment of *Listeria monocytogenes* from food, milk and dairy products.

Listeria Fermentation Broth

Composition per liter:

Proteose peptone No. 3	10.0g
NaCl	5.0g
Beef extract	1.0g
Bromcresol Purple	0.1g
Carbohydrate solution	10.0mL

Carbohydrate Solution:
Composition per 10mL:

Carbohydrate	5.0g

Preparation of Carbohydrate Solution: Add carbohydrate to distilled/deionized water and bring volume to 10.0mL. Mix thoroughly. Filter sterilize. Use glucose, salicin, rhamnose, dulcitol or raffinose.

Preparation of Medium: Add components, except carbohydrate solution, to distilled/deionized water and bring volume to 990.0mL. Mix thoroughly. Gently heat and bring to boiling. Autoclave for 15 min at 15 psi pressure–121°C. Cool to 45°–50°C. Aseptically add sterile carbohydrate solution. Mix thoroughly. Aseptically distribute into sterile tubes or flasks.

Use: For the cultivation and differentiation of *Listeria* species based on fermentation of glucose, salicin, rhamnose, dulcitol and raffinose.

Listeria Primary Selective Enrichment Broth, UVM I
See: ***Listeria*** **Enrichment Broth I, USDA FSIS**

Listeria Primary Selective Enrichment Broth, UVM II
See: ***Listeria*** **Enrichment Broth II, USDA FSIS**

Listeria Selective Agar, Modified Oxford
See: **Oxford Agar, Modified**

Listeria Selective Agar, Oxford
See: **Oxford Agar**

Listeria Transport Enrichment Medium

Composition per liter:

Sodium glycerophosphate	10.0g
Agar	2.0g
Sodium thioglycollate	1.0g
$CaCl_2$	0.1g
Nalidixic acid	0.04g
Acridine solution	2.0mL

pH 7.4 ± 0.2 at 25°C

Acridine Solution:
Composition per 10mL:

Acridine	0.04g

Preparation of Acridine Solution: Add acridine to distilled/deionized water and bring volume to 10.0mL. Mix thoroughly. Autoclave for 15 min at 15 psi pressure–121°C. Cool to 45°–50°C.

Preparation of Medium: Add components, except acridine solution, to distilled/deionized water and bring volume to 998.0mL. Mix thoroughly. Gently heat and bring to boiling. Autoclave for 15 min at 15 psi pressure–121°C. Cool to 45°–50°C. Aseptically add acridine solution. Mix thoroughly. Aseptically distribute into sterile tubes in 10.0mL volumes or fill bottles 4/5's full.

Use: For the maintenance—as a transport medium—and enrichment of *Listeria* species.

Lithium Chloride Phenylethanol Moxalactam Plating Agar

See: **LPM Agar**

Litmus Lactose Agar (LL Agar)

Composition per liter:

Agar	10.0g
Lactose	10.0g
Meat peptone	5.0g
Beef extract	3.0g
Litmus	1.0g

pH 7.0 ± 0.2 at 25°C

Preparation of Medium: Add components to distilled/deionized water and bring volume to 1.0L. Mix thoroughly. Gently heat and bring to boiling. Distribute into tubes or flasks. Autoclave for 15 min at 15 psi pressure–121°C. Pour into sterile Petri dishes or leave in tubes.

Use: For the maintenance of lactic acid bacteria and for the differentiation of several bacteria on the basis of lactose fermentation. Bacteria that ferment lactose appear as red colonies. Bacteria that do not ferment lactose appear as dark blue-purple colonies.

Litmus Lactose Agar with Crystal Violet (LLK Agar)

Composition per liter:

Agar	10.0g
Lactose	10.0g
Meat peptone	5.0g
Beef extract	3.0g
Litmus	1.0g
Crystal Violet	5.0mg

pH 7.0 ± 0.2 at 25°C

Preparation of Medium: Add components to distilled/deionized water and bring volume to 1.0L. Mix thoroughly. Gently heat and bring to boiling. Distribute into tubes or flasks. Autoclave for 15 min at 15 psi pressure–121°C. Pour into sterile Petri dishes or leave in tubes.

Use: For the maintenance of lactic acid bacteria and for the differentiation of several bacteria on the basis of lactose fermentation. Bacteria that ferment lactose appear as red colonies. Bacteria that do not ferment lactose appear as dark blue-purple colonies.

Litmus Milk

Composition per liter:

Skim milk	100.0g
Azolitmin	0.5g
Na_2SO_3	0.5g

pH 6.5 ± 0.2 at 25°C

Source: This medium is available as a premixed powder from BBL Microbiology Systems, Difco Laboratories, and Oxoid Unipath.

Preparation of Medium: Add components to distilled/deionized water and bring volume to 1.0L. Mix thoroughly. Gently heat and bring to boiling. Distribute into tubes or flasks. Autoclave for 20 min at 10 psi pressure–115°C.

Use: For the maintenance of lactic acid bacteria and for the differentiation of several bacteria, especially *Clostridium* species, based on their action on milk. Bacteria that do not ferment carbohydrates, such as *Proteus vulgaris* or *Moraxella lacunata*, show no change in the litmus indicator. Bacteria that ferment lactose or glucose with the production of gas, such as *Clostridium perfringens*, turn the medium pink and frothy. Bacteria that proteolytically degrade lactalbumin turn the medium blue. Bacteria that coagulate casein form a curd or clot. Bacteria that peptonize casein, such as *Pseudomonas aeruginosa*, show a dissolution of the clot.

Littman Oxgall Agar

Composition per liter:

Agar....................20.0g
Oxgall....................15.0g
Glucose....................10.0g
Peptone....................10.0g
Crystal Violet....................0.01g
Streptomycin solution....................10.0mL

pH 6.5 ± 0.2 at 25°C

Source: This medium is available as a premixed powder from Difco Laboratories.

Streptomycin Solution:
Composition per 10mL:

Streptomycin....................0.03g

Preparation of Streptomycin Solution: Add streptomycin to distilled/deionized water and bring volume to 10.0mL. Mix thoroughly. Filter sterilize.

Preparation of Medium: Add components, except streptomycin solution, to distilled/deionized water and bring volume to 990.0mL. Mix thoroughly. Gently heat and bring to boiling. Autoclave for 15 min at 15 psi pressure–121°C. Cool to 45°–50°C. Aseptically add sterile streptomycin solution. Mix thoroughly. Pour into sterile Petri dishes or distribute into sterile tubes. Allow tubes to cool in a slanted position.

Use: For the selective isolation and cultivation of fungi, especially dermatophytes.

Liver Broth

Composition per liter:

Extracted liver tissue, minced....................30.0g
Infusion from fresh liver....................23.0g
Peptone....................10.0g
K_2HPO_4....................1.0g
Agar overlay solution....................200.0mL

pH 6.8 ± 0.2 at 25°C

Source: This medium is available as a premixed powder from Oxoid Unipath.

Agar Overlay Solution:
Composition per 200mL:

Agar, Oxoid Unipath No. 3....................4.0g

Preparation of Agar Overlay Solution: Add agar to distilled/deionized water and bring volume to 200.0mL. Mix thoroughly. Gently heat and bring to boiling. Autoclave for 15 min at 15 psi pressure–121°C. Cool to 45°–50°C.

Preparation of Medium: Add components, except agar, to distilled/deionized water and bring volume to 1.0L. Mix thoroughly. Gently heat and bring to boiling. Distribute into tubes to a depth of 50mm. Make sure that some liver particles are transferred to each tube. Autoclave for 20 min at 10 psi pressure–115°C. After inoculation, aseptically overlayer the broth with 2.0mL of sterile, cooled agar solution per tube.

Use: For the isolation and cultivation of saccharolytic or putrefactive mesophilic and thermophilic anaerobic bacteria from foods.

Liver Broth

Composition per liter:

Beef liver, fresh....................453.0g
Pancreatic digest of casein....................10.0g
K_2HPO_4....................1.0g
Soluble starch....................1.0g

pH 7.6 ± 0.2 at 25°C

Preparation of Medium: Remove the fat from fresh beef liver. Grind the liver. Add 1.0L of distilled/deionized water. Gently heat and bring to boiling. Continue boiling for 60 min. Adjust pH to 7.6. Filter through cheesecloth. Reserve meat. To filtrate, add pancreatic digest of casein, K_2HPO_4, and soluble starch. Bring volume to 1.0L with distilled/deionized water. Refilter solution. Add meat particles to test tubes to a depth of approximately 2 cm. Distribute broth into tubes with meat particles in 15.0mL volumes. Autoclave for 20 min at 15 psi pressure–121°C.

Use: For the isolation and cultivation of anaerobic microorganisms, especially *Clostridium botulinum*, from foods.

Liver Broth with NaCl

Composition per liter:

Beef liver, fresh....................453.0g
NaCl....................150.0g
Pancreatic digest of casein....................10.0g
K_2HPO_4....................1.0g
Soluble starch....................1.0g

pH 7.6 ± 0.2 at 25°C

Preparation of Medium: Remove the fat from fresh beef liver. Grind the liver. Add 1.0L of distilled/deionized water. Gently heat and bring to boiling. Continue boiling for 60 min. Adjust pH to 7.6. Filter through cheesecloth. Reserve meat. To filtrate, add pancreatic digest of casein, K_2HPO_4, NaCl and soluble starch. Bring volume to 1.0L with distilled/deionized water. Refilter solution. Add meat particles to test tubes to a depth of approximately 2 cm. Distribute broth into tubes with meat particles in 15.0mL volumes. Autoclave for 20 min at 15 psi pressure–121°C. After inoculation, overlay each tube with sterile, melted petroleum jelly.

Use: For the cultivation of obligate halophiles from brined and dry-salted vegetables.

Liver Infusion Agar

Composition per liter:

Beef liver, infusion from	500.0g
Agar	20.0g
Proteose peptone	10.0g
NaCl	5.0g

pH 6.9 ± 0.2 at 25°C

Source: This medium is available as a premixed powder from Difco Laboratories.

Preparation of Medium: Add components to distilled/deionized water and bring volume to 1.0L. Mix thoroughly. Gently heat and bring to boiling. Distribute into tubes or flasks. Autoclave for 15 min at 15 psi pressure–121°C. Pour into sterile Petri dishes or leave in tubes.

Use: For the cultivation of *Brucella* species and other fastidious pathogenic bacteria.

Liver Infusion Broth

Composition per liter:

Beef liver, infusion from	500.0g
Proteose peptone	10.0g
NaCl	5.0g

pH 6.9 ± 0.2 at 25°C

Source: This medium is available as a premixed powder from Difco Laboratories.

Preparation of Medium: Add components to distilled/deionized water and bring volume to 1.0L. Mix thoroughly. Gently heat and bring to boiling. Distribute into tubes or flasks. Autoclave for 15 min at 15 psi pressure–121°C.

Use: For the cultivation of *Brucella* species and other fastidious pathogenic bacteria.

Liver Infusion Sake Medium

Composition per liter:

Liver, fresh	400.0g
Agar	15.0g
Sake	400.0mL

Preparation of Medium: Finely mince liver and add to 600.0mL of distilled/deionized water. Gently heat and bring to boiling. Continue boiling for 5–10 min. Filter through Whatman #2 filter paper. To filtrate, add agar and sake. Bring volume to 1.0L with distilled/deionized water. Mix thoroughly. Gently heat and bring to boiling. Distribute into tubes or flasks. Autoclave for 25 min at 15 psi pressure–121°C. Pour into sterile Petri dishes or leave in tubes.

Use: For the cultivation and maintenance of *Lactobacillus fructivorans* and *Lactobacillus homohiochi*.

Liver Veal Agar

Composition per liter:

Veal, infusion from	500.0g
Beef liver, infusion from	50.0g
Gelatin	20.0g
Proteose peptone	20.0g
Agar	15.0g
Soluble starch	10.0g
Glucose	5.0g
NaCl	5.0g
Casein	2.0g
$NaNO_3$	2.0g
Enzymatic digest of protein	1.3g
Pancreatic digest of casein	1.3g

pH 7.3 ± 0.2 at 25°C

Source: This medium is available as a premixed powder from Difco Laboratories.

Preparation of Medium: Add components to distilled/deionized water and bring volume to 1.0L. Mix thoroughly. Gently heat and bring to boiling. Distribute into tubes or flasks. Make sure that some liver and veal particles are transferred to each tube. Autoclave for 15 min at 15 psi pressure–121°C. Pour into sterile Petri dishes or leave in tubes.

Use: For the cultivation of a variety of anaerobic organisms.

Liver Veal Egg Yolk Agar

Composition per 1080mL:

Liver veal agar	1.0L
Egg yolk emulsion, 50%	80.0mL

pH 7.3 ± 0.2 at 25°C

Liver Veal Agar:

Composition per liter:

Veal, infusion from	500.0g
Beef liver, infusion from	50.0g
Gelatin	20.0g
Proteose peptone	20.0g
Agar	15.0g

Soluble starch 10.0g
Glucose 5.0g
NaCl 5.0g
Casein 2.0g
$NaNO_3$ 2.0g
Enzymatic digest of protein 1.3g
Pancreatic digest of casein 1.3g

Source: This medium is available as a premixed powder from Difco Laboratories.

Preparation of Liver Veal Agar: Add components to distilled/deionized water and bring volume to 1.0L. Mix thoroughly. Gently heat and bring to boiling. Distribute into tubes or flasks. Make sure that some liver and veal particles are transferred to each tube. Autoclave for 15 min at 15 psi pressure–121°C. Cool to 50°C.

Egg Yolk Emulsion, 50%:
Composition per 100mL:
Chicken egg yolks 11
Whole chicken egg 1
NaCl (0.9% solution) 50.0mL

Preparation of Egg Yolk Emulsion, 50%: Soak eggs with 1:100 dilution of saturated mercuric chloride solution for 1 min. Crack eggs and separate yolks from whites. Mix egg yolks with 1 chicken egg. Measure 50.0mL of egg yolk emulsion and add to 50.0mL of 0.9% NaCl solution. Mix thoroughly. Filter sterilize. Warm to 45°–50°C.

Preparation of Medium: To 1.0L of cooled sterile liver veal agar, aseptically add 80.0mL of sterile egg yolk emulsion, 50%. Mix thoroughly. Pour into sterile Petri dishes. Dry plates at 35°C for 24 hr.

Use: For the cultivation of a variety of anaerobic organisms.

LL Agar
See: **Litmus Lactose Agar**

LLK Agar
See: **Litmus Lactose Agar with Crystal Violet**

Loeffler Blood Serum Medium

Composition per liter:
Beef blood serum 750.0mL
Dextrose broth 250.0mL
pH 7.1 ± 0.2 at 25°C

Source: This medium is available as a premixed powder from Difco Laboratories.

Dextrose Broth:
Composition per liter:
Tryptose 10.0g
Glucose 5.0g
Sodium chloride 5.0g
Beef extract 3.0g

Preparation of Dextrose Broth: Add components to distilled/deionized water and bring volume to 1.0L. Mix thoroughly.

Preparation of Medium: Combine 750.0mL of beef blood serum with 250.0mL of dextrose broth. Mix thoroughly. Distribute into screw-capped tubes. Slant tubes in the autoclave. Close the autoclave door loosely. Autoclave for 10 min at 0 psi pressure–100°C. Close the autoclave door tightly. Autoclave for 15 min at 15 psi pressure–121°C.

Use: For the cultivation of *Corynebacterium diphtheriae*. Also used for the demonstration of pigment production and proteolysis by *Corynebacterium diphtheriae*.

Loeffler Blood Serum Medium

Composition per liter:
Beef blood serum 750.0mL
Dextrose broth 250.0mL
pH 7.1 ± 0.2 at 25°C

Dextrose Broth:
Composition per liter:
Enzymatic digest of protein 2.50g
Glucose 1.25g
NaCl 1.25g
Beef extract 0.75g

Preparation of Dextrose Broth: Add components to distilled/deionized water and bring volume to 1.0L. Mix thoroughly.

Preparation of Medium: Combine 750.0mL of beef blood serum with 250.0mL of dextrose broth. Mix thoroughly. Distribute into screw-capped tubes. Slant tubes in the autoclave. Close the autoclave door loosely. Autoclave for 10 min at 0 psi pressure–100°C. Close the autoclave door tightly. Autoclave for 15 min at 15 psi pressure–121°C.

Use: For the cultivation of *Corynebacterium diphtheriae*. Also used for the demonstration of pigment production and proteolysis by *Corynebacterium diphtheriae*.

Loeffler Medium

Composition per liter:
Beef serum 70.0g

Egg, dried 7.5g
Heart muscle, solids from infusion 0.72g
Glucose 0.71g
Peptic digest of animal tissue 0.71g
NaCl 0.36g

pH 7.6 ± 0.2 at 25°C

Source: This medium is available as a premixed powder from BBL Microbiology Systems.

Preparation of Medium: Add components to distilled/deionized water and bring volume to 1.0L. Mix thoroughly. Distribute into screw-capped tubes. Slant tubes in the autoclave. Close the autoclave door loosely. Autoclave for 10 min at 0 psi pressure–100°C. Close the autoclave door tightly. Autoclave for 15 min at 15 psi pressure–121°C.

Use: For the cultivation of *Corynebacterium diphtheriae*. For the demonstration of pigment production and proteolysis by *Corynebacterium diphtheriae*. Also used for the cultivation and maintenance of *Moraxella lacunata*.

Loeffler Slant

Composition per liter:

Tryptose 5.0g
Glucose 1.0g
Beef serum 750.0mL

Preparation of Medium: Add components to distilled/deionized water and bring volume to 1.0L. Mix thoroughly. Distribute into screw-capped tubes. Slant tubes in the autoclave. Close the autoclave door loosely. Autoclave for 10 min at 0 psi pressure–100°C. Close the autoclave door tightly. Autoclave for 15 min at 15 psi pressure–121°C.

Use: For the cultivation of *Corynebacterium diphtheriae*. For the demonstration of pigment production and proteolysis by *Corynebacterium diphtheriae*. Also used for the cultivation and maintenance of *Moraxella lacunata*.

Loeffler Slant, Modified

Composition per liter:

Peptone 0.5g
Glucose 1.0g
Beef serum 300.0mL

pH 7.6 ± 0.2 at 25°C

Preparation of Medium: Add peptone and glucose to distilled/deionized water and bring volume to 100.0mL. Mix thoroughly. Add beef serum. Mix thoroughly. Adjust pH to 7.6. Distribute into screw-capped tubes in 3.0mL volumes. Slant tubes in the autoclave. Autoclave for 30 min at 0 psi pressure–100°C.

Use: For the cultivation of *Corynebacterium diphtheriae*. For the demonstration of pigment production and proteolysis by *Corynebacterium diphtheriae*. Also used for the cultivation and maintenance of *Moraxella lacunata*.

LOM Agar
See: Lysine Ornithine Mannitol Agar

Lombard–Dowell Agar (LD Agar)

Agar 20.0g
Pancreatic digest of casein 5.0g
Yeast extract 5.0g
NaCl 2.5g
L-Cystine 0.4g
L-Tryptophan 0.2g
Na_2SO_3 .... 0.1g
Hemin 10.0mg
NaOH (1*N* NaOH) 5.0mL
Vitamin K_1 solution 1.0mL

pH 7.5 ± 0.2 at 25°C

Vitamin K_1 Solution:

Composition per 100mL:

Vitamin K_1 .... 1.0g
Ethanol 99.0mL

Preparation of Vitamin K_1 Solution: Add vitamin K_1 to 99.0mL of absolute ethanol. Mix thoroughly.

Preparation of Medium: Add hemin and cystine to 5.0mL of NaOH. Mix thoroughly. Add remaining components. Bring volume to 1.0L with distilled/deionized water. Mix thoroughly. Gently heat and bring to boiling. Distribute into tubes or flasks. Autoclave for 15 min at 15 psi pressure–121°C. Pour into sterile Petri dishes.

Use: For the cultivation and identification of a variety of obligate anaerobic bacteria. For the cultivation of *Bacteroides species, Fusobacterium species, Clostridium* species, and non–spore–forming Gram–positive anaerobes.

Lombard–Dowell Bile Agar (LD Bile Agar)

Composition per liter:

Agar 20.0g
Oxgall 20.0g
Pancreatic digest of casein 5.0g
Yeast extract 5.0g

NaCl	2.5g
D-Glucose	1.0g
L-Cystine	0.4g
L-Tryptophan	0.2g
Na_2SO_3	0.1g
Hemin	10.0mg
NaOH (1*N* NaOH)	5.0mL
Vitamin K_1 solution	1.0mL

pH 7.5 ± 0.2 at 25°C

Vitamin K_1 Solution:
Composition per 100mL:

Vitamin K_1	1.0g
Ethanol	99.0mL

Preparation of Vitamin K_1 Solution: Add vitamin K_1 to 99.0mL of absolute ethanol. Mix thoroughly.

Preparation of Medium: Add hemin and cystine to 5.0mL of NaOH. Mix thoroughly. Add remaining components. Bring volume to 1.0L with distilled/deionized water. Mix thoroughly. Gently heat and bring to boiling. Distribute into tubes or flasks. Autoclave for 15 min at 15 psi pressure–121°C. Pour into sterile Petri dishes.

Use: For the cultivation and identification of a variety of obligate anaerobic bacteria in the presence of 20% bile.

Lombard–Dowell Broth (LD Broth)

Composition per liter:

Pancreatic digest of casein	5.0g
Yeast extract	5.0g
Agar	0.7g
NaCl	2.5g
L-Tryptophan	0.2g
Na_2SO_3	0.1g
NaOH (1*N* NaOH)	5.0mL
Hemin solution	1.0mL
Vitamin K_1 solution	1.0mL

pH 7.5 ± 0.2 at 25°C

Hemin Solution:
Composition per 100mL:

Hemin	1.0g
NaOH (1*N* solution)	20.0mL

Preparation of Hemin Solution: Add hemin to 20.0mL of 1*N* NaOH solution. Mix thoroughly. Bring volume to 100.0mL with distilled/deionized water.

Vitamin K_1 Solution:
Composition per 100mL:

Vitamin K_1	1.0g
Ethanol	99.0mL

Preparation of Vitamin K_1 Solution: Add vitamin K_1 to 99.0mL of absolute ethanol. Mix thoroughly.

Preparation of Medium: Add tryptophan to 5.0mL of NaOH. Mix thoroughly. Add remaining components. Bring volume to 1.0L with distilled/deionized water. Mix thoroughly. Gently heat and bring to boiling. Adjust pH to 7.5. Distribute into screw-capped tubes in 7.0mL volumes. Autoclave for 15 min at 15 psi pressure–121°C. Cool tubes, with caps loose, under 85% N_2 + 10% H_2 + 5% CO_2. Tighten caps.

Use: For the cultivation of a wide variety of anaerobic bacteria.

Lombard–Dowell Egg Yolk Agar (LD Egg Yolk Agar) (Egg Yolk Agar, Lombard–Dowell)

Composition per 9100mL:

$Na_2HPO_4 \cdot 12H_2O$	5.0g
Glucose	2.0g
LD Agar	9000.0mL
Egg yolk emulsion	100.0mL
$MgSO_4 \cdot 7H_20$ (5% solution)	0.2mL

pH 7.5 ± 0.2 at 25°C

LD Agar:
Composition per liter:

Agar	20.0g
Pancreatic digest of casein	5.0g
Yeast extract	5.0g
NaCl	2.5g
L-Cystine	0.4g
L-Tryptophan	0.2g
Na_2SO_3	0.1g
Hemin	10.0mg
NaOH (1*N* NaOH)	5.0mL
Vitamin K_1 solution	1.0mL

Preparation of LD Agar: Add hemin and cystine to 5.0mL of NaOH. Mix thoroughly. Add remaining components. Mix thoroughly. Gently heat and bring to boiling.

Vitamin K_1 Solution:
Composition per 100mL:

Vitamin K_1	1.0g
Ethanol	99.0mL

Preparation of Vitamin K_1 Solution: Add vitamin K_1 to 99.0mL of absolute ethanol. Mix thoroughly.

Egg Yolk Emulsion:
Composition:

Chicken egg yolks	11
Whole chicken egg	1

Preparation of Egg Yolk Emulsion: Soak eggs with 1:100 dilution of saturated mercuric chloride solution for 1 min. Crack eggs and separate yolks from whites. Mix egg yolks with 1 chicken egg.

Preparation of Medium: Combine components, except egg yolk emulsion. Mix thoroughly. Autoclave for 15 min at 15 psi pressure–121°C. Cool to 45°–50°C. Aseptically add 100.0mL of egg yolk emulsion. Mix thoroughly. Pour into sterile Petri dishes.

Use: For the cultivation of a wide variety of anaerobic bacteria. For the differentiation of anaerobic bacteria based on lecithinase production, lipase production and proteolytic ability. Bacteria that produce lecithinase appear as colonies surrounded by a zone of insoluble precipitate. Bacteria that produce lipase appear as colonies with a pearly iridescent sheen. Bacteria that produce proteolytic activity appear as colonies surrounded by a clear zone.

Lombard–Dowell Esculin Agar (LD Esculin Agar) (Esculin Agar, Lombard–Dowell)

Composition per liter:

Agar	20.0g
Pancreatic digest of casein	5.0g
Yeast extract	5.0g
NaCl	2.5g
Esculin	1.0g
Ferric citrate	0.5g
L-Cystine	0.4g
L-Tryptophan	0.2g
Hemin	10.0mg
NaOH (1*N* NaOH)	5.0mL
Vitamin K_1 solution	1.0mL

pH 7.5 ± 0.2 at 25°C

Vitamin K_1 Solution:

Composition per 100mL:

Vitamin K_1	1.0g
Ethanol	99.0mL

Preparation of Vitamin K_1 Solution: Add vitamin K_1 to 99.0mL of absolute ethanol. Mix thoroughly.

Preparation of Medium: Add hemin and cystine to 5.0mL of NaOH. Mix thoroughly. Add remaining components. Bring volume to 1.0L with distilled/deionized water. Mix thoroughly. Gently heat and bring to boiling. Distribute into tubes or flasks. Autoclave for 15 min at 15 psi pressure–121°C. Pour into sterile Petri dishes.

Use: For the cultivation of a wide variety of anaerobic bacteria. For the differentiation of anaerobic bacteria based on esculin hydrolysis, H_2S production and catalase production. Bacteria that hydrolyze esculin appear as colonies surrounded by a red-brown to dark brown zone. Bacteria that produce H_2S appear as black colonies.

Lombard–Dowell Gelatin Agar (LD Gelatin Agar)

Composition per liter:

Agar	20.0g
Pancreatic digest of casein	5.0g
Yeast extract	5.0g
Gelatin	4.0g
NaCl	2.5g
Glucose	1.0g
L-Cystine	0.4g
L-Tryptophan	0.2g
Na_2SO_3	0.1g
Hemin	10.0mg
NaOH (1*N* NaOH)	5.0mL
Vitamin K_1 solution	1.0mL

pH 7.5 ± 0.2 at 25°C

Vitamin K_1 Solution:

Composition per 100mL:

Vitamin K_1	1.0g
Ethanol	99.0mL

Preparation of Vitamin K_1 Solution: Add vitamin K_1 to 99.0mL of absolute ethanol. Mix thoroughly.

Preparation of Medium: Add hemin and cystine to 5.0mL of NaOH. Mix thoroughly. Add remaining components, except agar and gelatin. Bring volume to 750.0mL with distilled/deionized water. Mix thoroughly. Gently heat and bring to boiling. In a separate flask, add gelatin to 100.0mL of cold distilled/deionized water. Gently heat and bring to 70°C. Add gelatin solution to the 750.0mL of basal medium. Mix thoroughly. Add agar. Bring volume to 1.0L with distilled/deionized water. Autoclave for 15 min at 15 psi pressure–121°C. Pour into sterile Petri dishes.

Use: For the cultivation of a wide variety of anaerobic bacteria. For the differentiation of anaerobic bacteria based on gelatinase production. After incubation of plates, gelatinase activity is determined by addition of Frazier's reagent. Bacteria that hydrolyze gelatin appear as colonies surrounded by a clear zone.

Lombard–Dowell Neomycin Agar (Egg Yolk Agar with Neomycin)

Composition per 9100mL:

$Na_2HPO_4 \cdot 12H_2O$	5.0g

Glucose	2.0g
Neomycin sulfate	0.1g
LD Agar	9000.0mL
Egg yolk emulsion	100.0mL
$MgSO_4 \cdot 7H_20$ (5% solution)	0.2mL

pH 7.5 ± 0.2 at 25°C

LD Agar:
Composition per liter:

Agar	20.0g
Pancreatic digest of casein	5.0g
Yeast extract	5.0g
NaCl	2.5g
L-Cystine	0.4g
L-Tryptophan	0.2g
Na_2SO_3	0.1g
Hemin	10.0mg
NaOH (1*N* NaOH)	5.0mL
Vitamin K_1 solution	1.0mL

Preparation of LD Agar: Add hemin and cystine to 5.0mL of NaOH. Mix thoroughly. Add remaining components. Mix thoroughly. Gently heat and bring to boiling.

Vitamin K_1 Solution:
Composition per 100mL:

Vitamin K_1	1.0g
Ethanol	99.0mL

Preparation of Vitamin K_1 Solution: Add vitamin K_1 to 99.0mL of absolute ethanol. Mix thoroughly.

Egg Yolk Emulsion:
Composition:

Chicken egg yolks	11
Whole chicken egg	1

Preparation of Egg Yolk Emulsion: Soak eggs with 1:100 dilution of saturated mercuric chloride solution for 1 min. Crack eggs and separate yolks from whites. Mix egg yolks with 1 chicken egg.

Preparation of Medium: Combine components, except egg yolk emulsion and neomycin sulfate. Mix thoroughly. Autoclave for 15 min at 15 psi pressure–121°C. Cool to 45°–50°C. Aseptically add 100.0mL of egg yolk emulsion and neomycin sulfate. Mix thoroughly. Pour into sterile Petri dishes.

Use: For the selective cultivation of a wide variety of anaerobic bacteria. For the differentiation of anaerobic bacteria based on lecithinase production, lipase production and proteolytic ability. Bacteria that produce lecithinase appear as colonies surrounded by a zone of insoluble precipitate. Bacteria that produce lipase appear as colonies with a pearly iridescent sheen. Bacteria that produce proteolytic activity appear as colonies surrounded by a clear zone.

Long–Term Preservation Medium

Composition per liter:

NaCl	30.0g
Peptone	10.0g
Agar	3.0g
Yeast extract	3.0g

Preparation of Medium: Add components to distilled/deionized water and bring volume to 1.0L. Mix thoroughly. Gently heat and bring to boiling. Distribute into screw-capped tubes in 4.0mL volumes. Autoclave for 15 min at 15 psi pressure–121°C.

Use: For the cultivation and maintenance of a wide variety of bacteria.

Low Iron YC Agar

Composition per 1033mL:

Solution 1	1.0L
Solution 4	30.0mL
Solution 2	2.0mL
Solution 3	1.0mL

pH 7.4 ± 0.2 at 25°C

Solution 1:
Composition per liter:

Yeast extract	20.0g
Noble agar	10.0g
Casamino acids	10.0g
KH_2PO_4	5.0g
$CaCl_2$	1.0g
Tryptophan	0.05g

Preparation of Solution 1: Add components to distilled/deionized water and bring volume to 1.0L. Mix thoroughly. Adjust pH to 7.4. Gently heat and bring to boiling. Filter through #40 ashless filter paper.

Solution 2:
Composition per 100mL:

$MgSO_4 \cdot 7H_2O$	22.5g
$CuSO_4 \cdot 5H_2O$	0.5g
$ZnSO_4 \cdot 5H_2O$	0.2g
β-Alanine	0.115g
Nicotinic Acid	0.115g
$MnCl_2 \cdot 4H_2O$	0.075g
Pimelic acid	7.5mg
HCl, conc.	3.0mL

Preparation of Solution 2: Add components to distilled/deionized water and bring volume to 100.0mL. Mix thoroughly.

Solution 3:
Composition per 100mL:

L-Cystine	20.0g
HCl, concentrated	20.0mL

Preparation of Solution 3: Add components to distilled/deionized water and bring volume to 100.0mL. Mix thoroughly.

Solution 4:

Composition per 100mL:

Maltose	50.0g
$CaCl_2 \cdot 2H_2O$	0.5g

Preparation of Solution 4: Add components to distilled/deionized water and bring volume to 100.0mL. Mix thoroughly. Autoclave for 10 min at 11 psi pressure–116°C. Cool to 45°–50°C.

Preparation of Medium: To 1.0L of solution 1, add 2.0mL of solution 2 and 1.0mL of solution 3. Mix thoroughly. Autoclave for 15 min at 15 psi pressure–121°C. Cool to 45°–50°C. Aseptically add 30.0mL of sterile solution 4. Mix thoroughly. Pour into sterile Petri dishes or distribute into sterile tubes.

Use: For the cultivation and maintenance of *Corynebacterium diphtheriae*.

Low Iron YC Broth

Composition per 1033mL:

Solution 1	1.0L
Solution 4	30.0mL
Solution 2	2.0mL
Solution 3	1.0mL

pH 7.4 ± 0.2 at 25°C

Solution 1:

Composition per liter:

Yeast extract	20.0g
Casamino acids	10.0g
KH_2PO_4	5.0g
$CaCl_2 \cdot 2H_2O$	1.0g
Tryptophan	0.05g

Preparation of Solution 1: Add components to distilled/deionized water and bring volume to 1.0L. Mix thoroughly. Adjust pH to 7.4. Gently heat and bring to boiling. Filter through #40 ashless filter paper.

Solution 2:

Composition per 100mL:

$MgSO_4 \cdot 7H_2O$	22.5g
$CuSO_4 \cdot 5H_2O$	0.5g
$ZnSO_4 \cdot 5H_2O$	0.2g
β-Alanine	0.115g
Nicotinic Acid	0.115g
$MnCl_2 \cdot 4H_2O$	0.075g
Pimelic acid	7.5mg
HCl, conc.	3.0mL

Preparation of Solution 2: Add components to distilled/deionized water and bring volume to 100.0mL. Mix thoroughly.

Solution 3:

Composition per 100mL:

L-Cystine	20.0g
HCl, concentrated	20.0mL

Preparation of Solution 3: Add components to distilled/deionized water and bring volume to 100.0mL. Mix thoroughly.

Solution 4:

Composition per 100mL:

Maltose	50.0g
$CaCl_2 \cdot 2H_2O$	0.5g

Preparation of Solution 4: Add components to distilled/deionized water and bring volume to 100.0mL. Mix thoroughly. Autoclave for 10 min at 11 psi pressure–116°C. Cool to 25°C.

Preparation of Medium: To 1.0L of solution 1, add 2.0mL of solution 2 and 1.0mL of solution 3. Mix thoroughly. Autoclave for 15 min at 15 psi pressure–121°C. Cool to 25°C. Aseptically add 30.0mL of sterile solution 4. Mix thoroughly. Aseptically distribute into sterile tubes or flasks.

Use: For the cultivation and maintenance of *Corynebacterium diphtheriae*.

Low Phosphate Buffered Basal Medium, Modified

Composition per 1030mL:

Pectin	4.0g
NH_4Cl	1.0g
Na_2HPO_4	0.72g
KH_2PO_4	0.3g
$MgCl_2 \cdot 6H_2O$	0.2g
Reducing agent	20.0mL
Yeast extract solution	10.0mL
Trace minerals	10.0mL
Vitamins	5.0mL
Resazurin (0.2% solution)	1.0mL
$FeSO_4 \cdot 7H_2O$ (2.5% solution)	0.03mL

pH 7.3 ± 0.1 at 25°C

Reducing Agent:

Composition per 20mL:

$Na_2S \cdot 9H_2O$	0.5g

Preparation of Reducing Agent: Add $Na_2S \cdot 9H_2O$ to distilled/deionized water and bring volume to 20.0mL. Mix thoroughly. Gas with 100% N_2 for 20 min. Cap with a rubber stopper. Autoclave for 45 min at 15 psi pressure–121°C. Use freshly prepared solution.

Yeast Extract Solution:

Composition per 10mL:

Yeast extract	1.0g

Preparation of Yeast Extract Solution: Add yeast extract to distilled/deionized water and bring volume to 10.0mL. Mix thoroughly. Autoclave for 45 min at 15 psi pressure–121°C. Cool to 25°C.

Trace Minerals:

Composition per liter:

Nitrilotriacetic acid	12.8g
NaCl	1.0g
$CoCl_2 \cdot 6H_2O$	0.16g
$CaCl_2 \cdot 2H_2O$	0.1g
$FeSO_4 \cdot 7H_2O$	0.1g
$MnCl_2 \cdot 4H_2O$	0.1g
$ZnCl_2$	0.1g
$CuCl_2$	0.02g
Na_2SeO_3	0.02g
H_3BO_3	0.01g
$Na_2MoO_4 \cdot 2H_2O$	0.01g
$NiSO_4 \cdot 6H_2O$	0.026g

Preparation of Trace Minerals: Add nitrilotriacetic acid to 500.0mL of distilled/deionized water. Dissolve by adjusting pH to 6.5 with KOH. Add remaining components. Add distilled/deionized water to 1.0L.

Wolfe's Vitamin Solution:

Composition per liter:

Pyridoxine·HCl	0.01g
Thiamine·HCl	5.0mg
Riboflavin	5.0mg
Nicotinic acid	5.0mg
Calcium pantothenate	5.0mg
p-Aminobenzoic acid	5.0mg
Thioctic acid	5.0mg
Biotin	2.0mg
Folic acid	2.0mg
Cyanocobalamin	0.1mg

Preparation of Wolfe's Vitamin Solution: Add components to distilled/deionized water and bring volume to 1.0L. Mix thoroughly.

Preparation of Medium: Add components, except yeast extract solution and reducing agent, to distilled/deionized water and bring volume to 1.0L. Mix thoroughly. Gently heat and bring to boiling. Cool under 90% N_2 + 10% CO_2. Anaerobically distribute into tubes in 6.0mL volumes. Autoclave for 45 min at 15 psi pressure–121°C. Aseptically add 0.06mL of sterile yeast extract solution to each tube. Mix thoroughly. Immediately prior to inoculation aseptically add 0.12mL of sterile reducing agent to each tube. Mix thoroughly.

Use: For the cultivation and maintenance of *Clostridium thermosulfurogenes.*

Lowenstein–Gruft Medium

Composition per 1600mL:

Potato starch	30.0g
Asparagine	3.6g
KH_2PO_4	2.4g
Magnesium citrate	0.6g
Malachite Green	0.4g
$MgSO_4 \cdot 7H_2O$	0.24g
Nalidixic acid	0.056g
Ribonucleic acid	0.08mg
Homogenized whole egg	1.0L
Glycerol	12.0mL
Penicillin	80,000U

Homogenized Whole Egg:

Composition per liter:

Whole eggs	18–24

Preparation of Homogenized Whole Egg: Use fresh eggs, less than 1 week old. Scrub the shells with soap. Let stand in a soap solution for 30 min. Rinse in running water. Soak eggs in 70% ethanol for 15 min. Break the eggs into a sterile container. Homogenize by shaking. Filter through four layers of sterile cheesecloth into a sterile graduated cylinder. Measure out 1.0L.

Preparation of Medium: Add glycerol to 600.0mL of distilled/deionized water. Mix thoroughly. Add remaining components, except fresh egg mixture. Mix thoroughly. Gently heat while stirring and bring to boiling. Autoclave for 15 min at 15 psi pressure–121°C. Cool to 50°C. Aseptically add 1.0L of homogenized whole egg. Mix thoroughly. Distribute into sterile screw-capped tubes. Place tubes in a slanted position. Inspissate at 85°C (moist heat) for 45 min.

Use: For the cultivation and differentiation of *Mycobacterium* species. *M. tuberculosis* appears as granular, rough, dry colonies. *M. kansasii* appears as smooth to rough photochromogenic colonies. *M. gordonae* appears as smooth yellow-orange colonies. *M. avium* appears as smooth, colorless colonies. *M. smegmatis* appears as wrinkled, creamy white colonies.

Lowenstein–Jensen Medium

Composition per 1600mL:

Potato starch	30.0g
Asparagine	3.6g
KH_2PO_4	2.4g
Magnesium citrate	0.6g
Malachite Green	0.4g
$MgSO_4 \cdot 7H_2O$	0.24g
Homogenized whole egg	1.0L
Glycerol	12.0mL

Source: Available as a prepared medium from BBL Microbiology Systems, Difco Laboratories and Oxoid Unipath.

Homogenized Whole Egg:
Composition per liter:

Whole eggs	18-24

Preparation of Homogenized Whole Egg: Use fresh eggs, less than 1 week old. Scrub the shells with soap. Let stand in a soap solution for 30 min. Rinse in running water. Soak eggs in 70% ethanol for 15 min. Break the eggs into a sterile container. Homogenize by shaking. Filter through four layers of sterile cheesecloth into a sterile graduated cylinder. Measure out 1.0L.

Preparation of Medium: Add glycerol to 600.0mL of distilled/deionized water. Mix thoroughly. Add remaining components, except fresh egg mixture. Mix thoroughly. Gently heat while stirring and bring to boiling. Autoclave for 15 min at 15 psi pressure–121°C. Cool to 50°C. Aseptically add 1.0L of homogenized whole egg. Mix thoroughly. Distribute into sterile screw-capped tubes. Place tubes in a slanted position. Inspissate at 85°C (moist heat) for 45 min.

Use: For the cultivation and differentiation of *Mycobacteirum* species. *M. tuberculosis* appears as granular, rough, dry colonies. *M. kansasii* appears as smooth to rough photochromogenic colonies. *M. gordonae* appears as smooth yellow-orange colonies. *M. avium* appears as smooth, colorless colonies. *M. smegmatis* appears as wrinkled, creamy white colonies. Also used for the cultivation and maintenance of *Gordona* species, *Nocardia* species, *Rhodococcus* species, and *Tsukamurella paurometabolum*.

Lowenstein–Jensen Medium with NaCl

Composition per 1600mL:

NaCl	80.0g
Potato starch	30.0g
Asparagine	3.6g
KH_2PO_4	2.4g
Magnesium citrate	0.6g
Malachite Green	0.4g
$MgSO_4 \cdot 7H_2O$	0.24g
Homogenized whole egg	1.0L
Glycerol	12.0mL

Homogenized Whole Egg:
Composition per liter:

Whole eggs	18-24

Preparation of Homogenized Whole Egg: Use fresh eggs, less than 1 week old. Scrub the shells with soap. Let stand in a soap solution for 30 min. Rinse in running water. Soak eggs in 70% ethanol for 15 min. Break the eggs into a sterile container. Homogenize by shaking. Filter through four layers of sterile cheesecloth into a sterile graduated cylinder. Measure out 1.0L.

Preparation of Medium: Add glycerol to 600.0mL of distilled/deionized water. Mix thoroughly. Add remaining components, except fresh egg mixture. Mix thoroughly. Gently heat while stirring and bring to boiling. Autoclave for 15 min at 15 psi pressure–121°C. Cool to 50°C. Aseptically add 1.0L of homogenized whole egg. Mix thoroughly. Distribute into sterile screw-capped tubes. Place tubes in a slanted position. Inspissate at 85°C (moist heat) for 45 min.

Use: For the cultivation of *Mycobacterum smegmatis* and other salt-tolerant *Mycobacterum* species.

Lowenstein–Jensen Medium without Glycerol

Composition per liter:

Potato starch	30.0g
Asparagine	3.6g
KH_2PO_4	2.4g
Magnesium citrate	0.6g
Malachite green	0.4g
$MgSO_4 \cdot 7H_2O$	0.24g
Homogenized whole egg	1.0L

Homogenized Whole Egg:
Composition per liter:

Whole eggs	18–24

Preparation of Homogenized Whole Egg: Use fresh eggs, less than 1 week old. Scrub the shells with soap. Let stand in a soap solution for 30 min. Rinse in running water. Soak eggs in 70% ethanol for 15 min. Break the eggs into a sterile container. Homogenize by shaking. Filter through four layers of sterile cheesecloth into a sterile graduated cylinder. Measure out 1.0L.

Preparation of Medium: Add components, except fresh egg mixture, to 600.0mL of distilled/deionized water. Mix thoroughly. Gently heat while stirring and bring to boiling. Autoclave for 15 min at 15 psi pressure–121°C. Cool to 50°C. Aseptically add 1.0L of homogenized whole egg. Mix thoroughly. Distribute into sterile screw-capped tubes. Place tubes in a slanted position. Inspissate at 85°C (moist heat) for 45 min.

Use: For the cultivation and maintenance of *Mycobacterium* species, especially *M. bovis* and other species that are sensitive to glycerol.

Lowenstein–Jensen Medium with Streptomycin

Composition per 161..0mL:

Potato starch	30.0g
Asparagine	3.6g
KH_2PO_4	2.4g
Magnesium citrate	0.6g
Malachite Green	0.4g
$MgSO_4 \cdot 7H_2O$	0.24g
Homogenized whole egg	1.0L
Glycerol	12.0mL
Streptomycin solution	10.0mL

Homogenized Whole Egg:

Composition per liter:

Whole eggs	18-24

Preparation of Homogenized Whole Egg: Use fresh eggs, less than 1 week old. Scrub the shells with soap. Let stand in a soap solution for 30 min. Rinse in running water. Soak eggs in 70% ethanol for 15 min. Break the eggs into a sterile container. Homogenize by shaking. Filter through four layers of sterile cheesecloth into a sterile graduated cylinder. Measure out 1.0L.

Streptomycin Solution:

Composition per 10mL:

Streptomycin	0.1mg

Preparation of Streptomycin Solution: Add streptomycin to distilled/deionized water and bring volume to 10.0mL. Mix thoroughly. Filter sterilize.

Preparation of Medium: Add glycerol to 600.0mL of distilled/deionized water. Mix thoroughly. Add remaining components, except fresh egg mixture. Mix thoroughly. Gently heat while stirring and bring to boiling. Autoclave for 15 min at 15 psi pressure–121°C. Cool to 50°C. Aseptically add 1.0L of homogenized whole egg and 10.0mL of sterile streptomycin solution. Mix thoroughly. Distribute into sterile screw-capped tubes. Place tubes in a slanted position. Inspissate at 85°C (moist heat) for 45 min.

Use: For the cultivation and differentiation of *Mycobacterium* species. *M. tuberculosis* appears as granular, rough, dry colonies. *M. kansasii* appears as smooth to rough photochromogenic colonies. *M. gordonae* appears as smooth yellow-orange colonies. *M. avium* appears as smooth, colorless colonies. *M. smegmatis* appears as wrinkled, creamy white colonies. Also used for the cultivation and maintenance of *Gordona* species, *Nocardia* species, *Rhodococcus* species, and *Tsukamurella paurometabolum.*

LPBM Acido–Thermophile Medium

Composition per liter:

Agar	20.0g
Cellulose	5.0g
KH_2PO_4	1.0g
NH_4Cl	1.0g
Yeast extract	1.0g
Cellobiose	0.5g
$MgSO_4 \cdot 7H_2O$	0.2g
$Na_2HPO_4 \cdot 7H_2O$	0.1g
$CaCl_2 \cdot 2H_2O$	0.02g

pH 5.2 ± 0.2 at 25°C

Preparation of Medium: Add components, except cellulose and cellobiose, to distilled/deionized water and bring volume to 1.0L. Mix thoroughly. Adjust pH to 5.2 with H_3PO_4. Add cellulose and cellobiose. Mix thoroughly. Gently heat and bring to boiling. Distribute into tubes or flasks. Autoclave for 15 min at 15 psi pressure–121°C. Pour into sterile Petri dishes or leave in tubes.

Use: For the cultivation and maintenance of *Acidothermus cellulolyticus.*

LPM Agar (Lithium Chloride Phenylethanol Moxalactam Plating Agar)

Composition per liter:

Agar	15.0g
Glycine anhydride	10.0g
LiCl	5.0g
NaCl	5.0g
Pancreatic digest of casein	5.0g
Peptic digest of animal tissue	5.0g
Beef extract	3.0g
Phenylethyl alcohol	2.5g
Moxalactam solution	2.0mL

pH 7.3 ± 0.2 at 25°C

Source: This medium is available as a premixed powder from BBL Microbiology Systems and Difco Laboratories.

Moxalactam Solution:

Composition per 10mL:

Moxalactam	0.1g

Preparation of Moxalactam Solution: Add moxalactam to distilled/deionized water and bring volume to 10.0mL. Mix thoroughly. Filter sterilize.

Preparation of Medium: Add components, except moxalactam solution, to distilled/deionized water and bring volume to 998.0mL. Mix thoroughly.

Gently heat while stirring and bring to boiling. Autoclave for 12 min at 15 psi pressure–121°C. Cool to 45°–50°C. Aseptically add 2.0mL of sterile moxalactam solution. Mix thoroughly. Pour into sterile Petri dishes or distribute into sterile tubes.

Use: For the isolation and cultivation of *Listeria monocytogenes*.

LPM Agar with Esculin and Ferric Iron

Composition per liter:

Agar	15.0g
Glycine anhydride	10.0g
LiCl	5.0g
NaCl	5.0g
Pancreatic digest of casein	5.0g
Peptic digest of animal tissue	5.0g
Beef extract	3.0g
Phenylethyl alcohol	2.5g
Esculin	1.0g
Ferric ammonium citrate	0.5g
Moxalactam solution	2.0mL

pH 7.3 ± 0.2 at 25°C

Moxalactam Solution:
Composition per 10mL:

Moxalactam	0.1g

Preparation of Moxalactam Solution: Add moxalactam to distilled/deionized water and bring volume to 10.0mL. Mix thoroughly. Filter sterilize.

Preparation of Medium: Add components, except moxalactam solution, to distilled/deionized water and bring volume to 998.0mL. Mix thoroughly. Gently heat while stirring and bring to boiling. Autoclave for 12 min at 15 psi pressure–121°C. Cool to 45°–50°C. Aseptically add 2.0mL of sterile moxalactam solution. Mix thoroughly. Pour into sterile Petri dishes or distribute into sterile tubes.

Use: For the isolation and cultivation of *Listeria monocytogenes*.

LS Differential Medium (*Lactobacillus Streptococcus* Differential Medium)

Composition per liter:

Glucose	20.0g
Agar	13.0g
Pancreatic digest of casein	10.0g
Beef extract	5.0g
NaCl	5.0g
Papaic digest of soybean meal	5.0g
Yeast extract	5.0g
L-cysteine·HCl·H_2O	0.3g
Skim milk solution	100.0mL
Triphenyltetrazolium chloride solution	10.0mL

pH 6.1 ± 0.2 at 25°C

Source: This medium is available as a premixed powder from Oxoid Unipath.

Skim Milk Solution:
Composition per 100mL:

Skim milk, antibiotic free	10.0g

Preparation of Skim Milk Solution: Add skim milk to distilled/deionized water and bring volume to 100.0mL. Mix thoroughly. Autoclave for 5 min at 15 psi pressure–121°C. Cool to 50°C.

Triphenyltetrazolium Chloride Solution:
Composition per 10mL:

2,3,5-Triphenyltetrazolium chloride	0.2g

Preparation of Triphenyltetrazolium Chloride Solution: Add triphenyltetrazolium chloride to distilled/deionized water and bring volume to 10.0mL. Mix thoroughly. Filter sterilize. Warm to 50°C.

Preparation of Medium: Add components, except skim milk solution and triphenyltetrazolium chloride solution, to distilled/deionized water and bring volume to 890.0mL. Mix thoroughly. Gently heat and bring to boiling. Autoclave for 15 min at 15 psi pressure–121°C. Cool to 45°–50°C. Aseptically add sterile skim milk solution and sterile triphenyltetrazolium chloride solution. Mix thoroughly. Pour into sterile Petri dishes or distribute into sterile tubes.

Use: For the differentiation and enumeration of lactobacilli and streptococci in yogurt. *Lactobacillus* species appear as irregular, red colonies surrounded by a white, opaque zone. *Streptococcus* species appear as round, red colonies surrounded by a clear zone.

LT Medium
See: **Lecithin Tween™ Medium**

Luminous Medium

Composition per liter:

NaCl	30.0g
Agar	20.0g
NH_4Cl	5.0g
Pancreatic digest of casein	5.0g
Yeast extract	5.0g
K_2HPO_4	3.9g

KH_2PO_4 .... 2.1g
$CaCO_3$ 1.0g
$MgSO_4 \cdot 7H_2O$ 1.0g
KCl 0.75g
Tris buffer (1M solution, pH 7.5) 50.0mL
Glycerol 3.0mL

pH 7.2 ± 0.2 at 25°C

Preparation of Medium: Add components to distilled/deionized water and bring volume to 1.0L. Mix thoroughly. Gently heat and bring to boiling. Distribute into tubes or flasks. Autoclave for 15 min at 15 psi pressure–121°C. Pour into sterile Petri dishes or leave in tubes.

Use: For the cultivation and maintenance of *Alteromonas hanedai, Photobacterium* species, *Shewanella hanedai,* and *Vibrio* species.

Luria Agar Base, Miller

Composition per liter:

Agar 15.0g
Tryptone 10.0g
Yeast extract 5.0g
NaCl 0.5g

pH 7.0 ± 0.2 at 25°C

Preparation of Medium: Add components to distilled/deionized water and bring volume to 1.0L. Mix thoroughly. Gently heat and bring to boiling. Distribute into tubes or flasks. Autoclave for 15 min at 15 psi pressure–121°C. Pour into sterile Petri dishes or leave in tubes.

Use: For the cultivation and maintenance of bacteria for genetic and molecular studies.

Luria Bertani Agar, Miller (LB Agar, Miller)

Composition per liter:

Agar 15.0g
Tryptone 10.0g
NaCl 10.0g
Yeast extract 5.0g

pH 7.5 ± 0.2 at 25°C

Preparation of Medium: Add components to distilled/deionized water and bring volume to 1.0L. Mix thoroughly. Gently heat and bring to boiling. Distribute into tubes or flasks. Autoclave for 15 min at 15 psi pressure–121°C. Pour into sterile Petri dishes or leave in tubes.

Use: For the cultivation and maintenance of bacteria for genetic and molecular studies.

Luria Broth
See: **LB Medium**

Lyngby Iron Agar
See: **Iron Agar, Lyngby**

Lysine Agar, Selective

Composition per liter:

Agar 15.0g
L-Lysine 10.0g
Peptone 5.0g
Glucose 3.5g
Yeast extract 3.0g
Bile salts mixture 1.5g
Sulfapyridine 0.3g
Bromcresol Purple 0.03g
Crystal Violet 0.001g

pH 6.8 ± 0.1 at 25°C

Source: This medium is available as a premixed powder from BBL Microbiology Systems.

Preparation of Medium: Add components to distilled/deionized water and bring volume to 1.0L. Mix thoroughly. Gently heat and bring to boiling. Distribute into tubes or flasks. Autoclave for 15 min at 15 psi pressure–121°C. Pour into sterile Petri dishes or leave in tubes.

Use: For the selective isolation and cultivation of *Salmonella* species from food by the hydrophobic grid membrane filter method.

Lysine Arginine Iron Agar

Composition per liter:

Agar 15.0g
L-Arginine 10.0g
L-Lysine 10.0g
Peptone 5.0g
Yeast extract 3.0g
Glucose 1.0g
Ferric ammonium citrate 0.5g
Sodium thiosulfate 0.04g
Bromcresol Purple 0.02g

pH 6.8 ± 0.2 at 25°C

Preparation of Medium: Add components to distilled/deionized water and bring volume to 1.0L. Mix thoroughly. Gently heat and bring to boiling. Adjust pH to 6.8. Distribute into screw-capped tubes in 5.0mL volumes. Autoclave for 12 min at 15 psi pressure–121°C. Allow tubes to cool in a slanted position.

Use: For the cultivation and differentiation of bacteria based on their ability to decarboxylate lysine, decarboxylate arginine and produce H_2S. Bacteria that decarboxylate lysine or arginine turn the medium purple. Bacteria that produce H_2S appear as black colonies.

Lysine Decarboxylase Broth, Falkow

Composition per liter:

Peptone	5.0g
L-Lysine	5.0g
Yeast extract	3.0g
Glucose	1.0g
Bromcresol Purple	0.02g

pH 6.5–6.8 at 25°C

Preparation of Medium: Add components to distilled/deionized water and bring volume to 1.0L. Mix thoroughly. Gently heat and bring to boiling. Adjust pH to 6.5–6.8. Distribute into tubes in 5.0mL volumes. Autoclave for 15 min at 15 psi pressure–121°C.

Use: For the cultivation and differentiation of bacteria, especially *Salmonella*, based on their ability to decarboxylate lysine. Bacteria that decarboxylate lysine turn the medium turbid purple.

Lysine Decarboxylase Broth, Taylor Modification

Composition per liter:

L-Lysine	5.0g
Yeast extract	3.0g
Glucose	1.0g
Bromcresol Purple	0.02g

pH 6.1 ± 0.2 at 25°C

Source: This medium is available as a premixed powder from Oxoid Unipath.

Preparation of Medium: Add components to distilled/deionized water and bring volume to 1.0L. Mix thoroughly. Gently heat and bring to boiling. Adjust pH to 6.1. Distribute into tubes in 5.0mL volumes. Autoclave for 15 min at 15 psi pressure–121°C.

Use: For the cultivation and differentiation of bacteria, especially *Salmonella*, based on their ability to decarboxylate lysine. Bacteria that decarboxylate lysine turn the medium turbid purple.

Lysine Decarboxylase Broth, Taylor Modification (Lysine Decarboxylase Broth)

Composition per liter:

L-Lysine	5.0g
Peptone	5.0g
Yeast extract	3.0g
Glucose	1.0g
Bromcresol Purple	0.02g

pH 6.8 ± 0.2 at 25°C

Source: This medium is available as a premixed powder from Difco Laboratories.

Preparation of Medium: Add components to distilled/deionized water and bring volume to 1.0L. Mix thoroughly. Gently heat and bring to boiling. Adjust pH to 6.1. Distribute into tubes in 5.0mL volumes. Autoclave for 15 min at 15 psi pressure–121°C.

Use: For the cultivation and differentiation of bacteria, especially *Salmonella*, based on their ability to decarboxylate lysine. Bacteria that decarboxylate lysine turn the medium turbid purple.

Lysine Decarboxylase Medium

Composition per liter:

Glucose	0.5g
KH_2PO_4	0.5g
L-Lysine·HCl	0.5g

pH 4.6 ± 0.2 at 25°C

Preparation of Medium: Add components to distilled/deionized water and bring volume to 1.0L. Mix thoroughly. Gently heat and bring to boiling. Adjust pH to 4.6. Autoclave for 15 min at 15 psi pressure–121°C. Aseptically distribute into sterile tubes in 1.0mL volumes.

Use: For the cultivation and differentiation of Gram-negative nonfermentative bacteria based on their ability to decarboxylate lysine. Bacteria that decarboxylate lysine turn the medium turbid purple.

Lysine Iron Agar

Composition per liter:

Agar	13.5g
L-Lysine	10.0g
Pancreatic digest of gelatin	5.0g
Yeast extract	3.0g
Glucose	1.0g
Ferric ammonium citrate	0.5g
$Na_2S_2O_3 \cdot 5H_2O$	0.04g
Bromcresol Purple	0.02g

pH 6.7 ± 0.2 at 25°C

Source: This medium is available as a premixed powder from BBL Microbiology Systems, Difco Laboratories and Oxoid Unipath.

Preparation of Medium: Add components to distilled/deionized water and bring volume to 1.0L. Mix thoroughly. Gently heat while stirring and bring to boiling. Distribute into tubes in 10.0mL volumes.

Autoclave for 12 min at 15 psi pressure–121°C. Allow tubes to cool in a slanted position.

Use: For the cultivation and differentiation of members of the Enterobacteriaceae based on their ability to decarboxylate lysine and to form H_2S. Bacteria that decarboxylate lysine turn the medium purple. Bacteria that produce H_2S appear as black colonies.

Lysine Iron Cystine Neutral Red Broth *See:* LICNR Broth

Lysine Medium

Composition per liter:

Glucose	44.5g
Agar	17.8g
KH_2PO_4	1.78g
Lysine	1.0g
$MgSO_4 \cdot 7H_2O$	0.89g
$CaCl_2 \cdot 2H_2O$	0.178g
NaCl	0.089g
Inositol	0.02g
Calcium pantothenate	2.0mg
Adenine	1.78mg
DL-Methionine	0.89mg
L-Histidine	0.89mg
DL-Tryptophan	0.89mg
Thiamine·HCl	0.4mg
Pyridoxine	0.4mg
Nicotinic acid	0.4mg
$FeSO_4 \cdot 7H_2O$	0.22mg
p-Aminobenzoic acid	0.2mg
Riboflavin	0.2mg
$MnSO_4 \cdot H_2O$	0.035mg
$ZnSO_4 \cdot 7H_2O$	0.035mg
$(NH_4)_2MoO_4 \cdot 4H_2O$	0.018mg
H_3BO_3	8.9μg
Biotin	2.0μg
Folic acid	1.0μg
Potassium lactate (50% solution)	10.0mL
Lactic acid	1.0mL

pH 4.8 ± 0.2 at 25°C

Source: This medium is available as a premixed powder from Oxoid Unipath.

Preparation of Medium: Add potassium lactate to distilled/deionized water and bring volume to 900.0mL. Mix thoroughly. Add remaining components, except lactic acid. Mix thoroughly. Gently heat while stirring and bring to boiling. Do not autoclave. Cool to 50°C. Adjust pH to 4.8 by adding 1.0mL of lactic acid. Pour into sterile Petri dishes. Dry the surface of the plates by incubation at 37°C for 24 hr.

Use: For the isolation, cultivation and enumeration of wild yeasts encountered in brewing.

Lysine Ornithine Mannitol Agar (LOM Agar)

Composition per liter:

Agar	13.5g
L–Ornithine·HCl	6.5g
D–Mannitol	5.25g
L–Lysine·HCl	5.0g
NaCl	5.0g
Yeast extract	3.0g
Bromthymol Blue	0.3g
Vancomycin solution	10.0mL

pH 6.5 ± 0.2 at 25°C

Vancomycin Solution:
Composition per 10mL:

Vancomycin·HCl	0.03g

Preparation of Vancomycin Solution: Add vancomycin to distilled/deionized water and bring volume to 10.0mL. Mix thoroughly. Filter sterilize.

Preparation of Medium: Add components, except vancomycin solution, to distilled/deionized water and bring volume to 990.0mL. Mix thoroughly. Gently heat and bring to boiling. Autoclave for 15 min at 15 psi pressure–121°C. Cool to 45°–50°C. Aseptically add sterile vancomycin solution. Mix thoroughly. Pour into sterile Petri dishes or distribute into sterile tubes.

Use: For the cultivation and differentiation of Gram-negative bacilli based on their ability to decarboxylate lysine or ornithine and mannitol fermentation. Especially useful for the identification of *Enterobacter agglomerans*. Bacteria that ferment mannitol appear as dark yellow colonies. Bacteria that decarboxylate lysine or ornithine appear as green-yellow colonies.

Lysozyme Broth

Composition per 1005mL:

Basal glycerol broth	1.0L
Lysozyme solution	5.0mL

Basal Glycerol Broth:
Composition per liter:

Peptone	5.0g
Beef extract	3.0g
Glycerol	70.0mL

Preparation of Basal Glycerol Broth: Add components to distilled/deionized water and bring volume to 1.0L. Mix thoroughly. Distribute 500.0mL of the broth into screw-capped tubes in 5.0mL volumes. Autoclave the tubes and the flask with the remaining broth for 15 min at 15 psi pressure–121°C. Cool to 25°C.

Lysozyme Solution:

Composition per 100mL:

Lysozyme	0.1g
HCl (0.01*N* solution)	100.0mL

Preparation of Lysozyme Solution: Add lysozyme to 100.0mL of HCl solution. Mix thoroughly. Filter sterilize. Store for up to 1 week at 4°C.

Preparation of Medium: Add 5.0mL of the sterile lysozyme solution to 95.0mL of the cooled, sterile basal glycerol broth. Mix thoroughly. Aseptically distribute into sterile screw-capped tubes in 5.0mL volumes.

Use: For the cultivation and differentiation of *Nocardia asteroides, Streptomyces griseus* and *Actinomadura madurae* based on sensitivity to lysozyme. *Nocardia asteroides* grows well in both the basal glycerol broth and the lysozyme broth. *Actinomadura madurae* and *Streptomyces griseus* grow well in the basal glycerol broth but not in the lysozyme broth.

Lysozyme Broth

Composition per 1010mL:

Nutrient broth	1.0L
Lysozyme solution	10.0mL

pH 6.9 ± 0.2 at 25°C

Nutrient Broth:

Composition per liter:

Pancreatic digest of gelatin	5.0g
Beef extract	3.0g

Source: Nutrient broth is available as a premixed powder from BBL Microbiology Systems and Difco Laboratories.

Preparation of Nutrient Broth: Add components to distilled/deionized water and bring volume to 1.0L. Mix thoroughly. Distribute into bottles in 99.0mL volumes. Autoclave for 15 min at 15 psi pressure–121°C. Cool to 25°C.

Lysozyme Solution:

Composition per 100mL:

Lysozyme	0.1g

Preparation of Lysozyme Solution: Add lysozyme to distilled/deionized water and bring volume to 100.0mL. Mix thoroughly. Filter sterilize.

Preparation of Medium: Add 1.0mL of sterile lysozyme solution to 99.0mL of cooled, sterile nutrient broth. Mix thoroughly. Aseptically distribute into sterile tubes in 2.5mL volumes.

Use: For the cultivation and differentiation of *Bacillus cereus* in foods. *B. cereus* is resistant to lysozyme and will grow in this medium.

M Broth

Composition per liter:

Pancreatic digest of casein	12.5g
K_2HPO_4	5.0g
NaCl	5.0g
Sodium citrate	5.0g
Yeast extract	5.0g
Mannose	2.0g
$MgSO_4 \cdot 7H_2O$	0.8g
Polysorbate 80	0.75g
$FeSO_4$	0.04g

pH 7.0 ± 0.22 at 25°C

Source: Available as a premixed powder from BBL Microbiology Systems and Difco Laboratories.

Preparation of Medium: Add components to distilled/deionized water and bring volume to 1.0L. Mix thoroughly. Distribute into tubes or flasks. Autoclave for 15 min at 15 psi pressure–121°C.

Use: For the detection of *Salmonella* in dried foods and feeds.

M1 Medium

Composition per liter:

L-Leucine	2.0g
L-Alanine	1.0g
L-Isoleucine	1.0g
L-Phenylalanine	1.0g
L-Proline	1.0g
L-Tryptophane	1.0g
L-Asparagine	0.50g
L-Lysine	0.50g
L-Methionine	0.50g
L-Tyrosine	0.40g
L-Valine	0.20g
L-Serine	0.20g
$MgSO_4 \cdot 7H_2O$	0.20g
NaCl	0.20g
KH_2PO_4	0.14g
L-Arginine	0.10g
L-Cysteine	0.10g
L-Glycine	0.10g
L-Histidine	0.10g
L-Threonine	0.10g
$CaCl_2$	2.0mg
$FeCl_3 \cdot 6H_2O$	2.0mg
Tris(hydroxymethyl)aminomethane buffer (0.01M solution, pH 7.6)	1.0L

pH 7.6 ± 0.2 at 25°C

Preparation of Medium: Add solid components to 1.0L of Tris buffer. Mix thoroughly. Filter sterilize. Aseptically distribute into tubes or flasks.

Use: For the cultivation of *Myxococcus xanthus.*

M1A Medium

Composition per 1001mL:

Sorbitol	23.3g
Peptone	6.0g
Sucrose	3.3g
Pancreatic digest of casein	3.3g
Beef heart infusion	2.0g
Glucose	1.3g
Yeast extract	1.0g
Fructose	0.3g
Phenol Red	20.0mg
Schneider's *Drosophila* medium	533.0mL
Fetal calf serum, heat inactivated	167.0mL
Fresh yeast extract solution	33.0mL
Penicillin solution	8.0mL

Schneider's *Drosophila* Medium:

Composition per liter:

$MgSO_4 \cdot 7H_2O$	3.7g
NaCl	2.1g
Yeast extract	2.0g
Trehalose	2.0g
D-Glucose	2.0g
L-Glutamine	1.8g
L-Lysine·HCl	1.7g
L-Proline	1.7g
KCl	1.6g
$Na_2HPO_4 \cdot 7H_2O$	1.3g
L-Glutamic acid	0.8g
L-Methionine	0.8g
$CaCl_2$, anhydrous	0.6g
KH_2PO_4	0.5g
β-Alanine	0.5g
L-Tyrosine	0.5g
L-Arginine	0.4g
L-Aspartic acid	0.4g
L-Histidine	0.4g
L-Threonine	0.4g
$NaHCO_3$	0.4g
Glycine	0.3g
L-Serine	0.3g
L-Valine	0.3g
L-Isoleucine	0.2g
L-Leucine	0.2g
L-Phenylalanine	0.2g
α-Ketoglutaric acid	0.2g
Fumaric acid	0.1g
Malic acid	0.1g
Succinic acid	0.1g
L-Cystine	0.1g
L-Tryptophan	0.1g
L-Cysteine	0.06g

Preparation of Schneider's *Drosophila* Medium: Add components to distilled/deionized water and bring volume to 1.0L. Mix thoroughly. Filter sterilize.

Penicillin Solution:
Composition per 10mL:
Penicillin 2,500,000U

Preparation of Penicillin Solution: Add penicillin to distilled/deionized water and bring volume to 10.0mL. Filter sterilize.

Fresh Yeast Extract Solution:
Composition per 100mL:
Baker's yeast, live, pressed, starch-free 25.0g

Preparation of Fresh Yeast Extract Solution: Add the live Baker's yeast to 100.0mL of distilled/deionized water. Autoclave for 90 min at 15 psi pressure–121°C. Allow to stand. Remove supernatant solution. Adjust pH to 6.6–6.8. Filter sterilize.

Preparation of Medium: Add components—except Schneider's *Drosophila* medium, fetal calf serum, fresh yeast extract solution, and penicillin solution— to distilled/deionized water and bring volume to 260.0mL. Mix thoroughly. Gently heat and bring to boiling. Autoclave for 15 min at 15 psi pressure–121°C. Cool to 45°–50°C. Aseptically add 533.0mL of sterile Schneider's *Drosophila* medium, 167.0mL of sterile fetal calf serum, 33.0mL of sterile fresh yeast extract solution, and 8.0mL of sterile penicillin solution. Mix thoroughly. Pour into sterile Petri dishes or distribute into sterile tubes.

Use: For the isolation and cultivation of *Spiroplasma* species that cause corn stunt.

M3 Agar

Composition per 1020mL:
Agar 18.0g
Na_2HPO_4 .... 0.732g
KH_2PO_4 .... 0.466g
NaCl 0.29g
Sodium propionate 0.2g
$MgSO_4 \cdot 7H_2O$ 0.1g
$CaCO_3$ 0.02g
KNO_3 .... 0.01g
$FeSO_4 \cdot 7H_2O$ 0.2mg
$ZnSO_4 \cdot 7H_2O$ 0.18mg
$MnSO_4 \cdot 4H_2O$ 0.020mg
Cycloheximide solution 10.0mL
Thiamine·HCl solution 10.0mL
pH 7.0 ± 0.2 at 25°C

Cycloheximide Solution:
Composition per 10mL:
Cycloheximide 0.05g

Preparation of Cycloheximide Solution: Add cycloheximide to distilled/deionized water and bring volume to 10.0mL. Mix thoroughly. Filter sterilize.

Thiamine·HCl Solution:
Composition per 10mL:
Thiamine·HCl 4.0mg

Preparation of Thiamine·HCl Solution: Add thiamine·HCl to distilled/deionized water and bring volume to 10.0mL. Mix thoroughly. Filter sterilize.

Preparation of Medium: Add components, except cycloheximide solution and thiamine·HCl solution, to distilled/deionized water and bring volume to 980.0mL. Mix thoroughly. Gently heat and bring to boiling. Autoclave for 15 min at 15 psi pressure–121°C. Cool to 45°–50°C. Aseptically add 10.0mL of sterile cycloheximide solution and 10.0mL of thiamine·HCl solution. Mix thoroughly. Pour into sterile Petri dishes or distribute into sterile tubes.

Use: For the selective isolation and cultivation of *Nocardia* species and *Rhodococcus* species.

M3 Agar Medium

Composition per liter:
Agar 18.0g
Na_2HPO_4 .... 0.732g
KH_2PO_4 .... 0.466g
NaCl 0.29g
Sodium propionate 0.20g
KNO_3 .... 0.10g
$MgSO_4 \cdot 7H_2O$ 0.10g
$CaCO_3$ 0.02g
Thiamine·HCl 4.0mg
$FeSO_4 \cdot 7H_2O$ 0.2mg
$ZnSO_4 \cdot 7H_2O$ 0.18mg
$MnSO_4 \cdot 4H_2O$ 0.02mg
Cycloheximide solution 10.0mL
Thiamine·HCl solution 10.0mL
pH 7.0 ± 0.2 at 25°C

Cycloheximide Solution:
Composition per 10mL:
Cycloheximide 0.04g

Preparation of Cycloheximide Solution: Add cycloheximide to distilled/deionized water and bring volume to 10.0mL. Mix thoroughly. Filter sterilize.

Thiamine·HCl Solution:
Composition per 10mL:
Thiamine·HCl 0.04g

Preparation of Thiamine·HCl Solution: Add thiamine·HCl to distilled/deionized water and bring volume to 10.0mL. Mix thoroughly. Filter sterilize.

Preparation of Medium: Add components, except cycloheximide solution and thiamine·HCl solution, to distilled/deionized water and bring volume to 980.0mL. Mix thoroughly. Gently heat and bring to boiling. Autoclave for 15 min at 15 psi pressure–121°C. Cool to 45°–50°C. Aseptically add sterile cycloheximide solution and thiamine·HCl solution. Mix thoroughly. Pour into sterile Petri dishes or distribute into sterile tubes.

Use: For the cultivation of *Micromonospora* species.

M9 Medium

Composition per liter:

Na_2HPO_4	6.0g
KH_2PO_4	3.0g
NH_4Cl	1.0g
NaCl	0.5g
Glucose solution	10.0mL
$MgSO_4 \cdot 7H_2O$ solution	1.0mL
Thiamine·HCl solution	1.0mL
$CaCl_2$ solution	1.0mL

pH 7.0 ± 0.2 at 25°C

Glucose Solution:
Composition per 100mL:

D-Glucose	20.0g

Preparation of Glucose Solution: Add glucose to distilled/deionized water and bring volume to 1.0L. Mix thoroughly. Autoclave for 15 min at 15 psi pressure–121°C.

$MgSO_4 \cdot 7H_2O$ Solution:
Composition per liter:

$MgSO_4 \cdot 7H_2O$	246.5g

Preparation of $MgSO_4 \cdot 7H_2O$ Solution: Add $MgSO_4 \cdot 7H_2O$ to distilled/deionized water and bring volume to 1.0L. Mix thoroughly. Autoclave for 15 min at 15 psi pressure–121°C.

Thiamine·HCl Solution:
Composition per 10mL:

Thiamine·HCl	10.0mg

Preparation of Thiamine·HCl Solution: Add thiamine·HCl to distilled/deionized water and bring volume to 1.0L. Mix thoroughly. Filter sterilize.

$CaCl_2$ Solution:
Composition per liter:

$CaCl_2$ solution	14.7g

Preparation of $CaCl_2$ Solution: Add $CaCl_2$ solution to distilled/deionized water and bring volume to 1.0L. Mix thoroughly. Autoclave for 15 min at 15 psi pressure–121°C.

Preparation of Medium: Add components, except $MgSO_4 \cdot 7H_2O$ solution, glucose solution, thiamine·HCl solution, and $CaCl_2$ solution, to distilled/deionized water and bring volume to 987.0mL. Mix thoroughly. Adjust pH to 7.0. Autoclave for 15 min at 15 psi pressure–121°C. Cool to room temperature. Aseptically add sterile $MgSO_4 \cdot 7H_2O$ solution, sterile glucose solution, sterile thiamine·HCl solution, and sterile $CaCl_2$ solution. Mix thoroughly. Distribute into tubes or flasks.

Use: For the cultivation and maintenance of *Escherichia coli* and a variety of other bacteria.

M9 Medium with Casamino Acids

Composition per liter:

Na_2HPO_4	6.0g
Casamino acids	5.0g
KH_2PO_4	3.0g
NH_4Cl	1.0g
NaCl	0.5g
Glucose solution	10.0mL
$MgSO_4 \cdot 7H_2O$ solution	1.0mL
Thiamine·HCl solution	1.0mL
$CaCl_2$ solution	1.0mL

pH 7.0 ± 0.2 at 25°C

Glucose solution:
Composition per 100mL:

D-Glucose	20.0g

Preparation of Glucose Solution: Add glucose to distilled/deionized water and bring volume to 1.0L. Mix thoroughly. Autoclave for 15 min at 15 psi pressure–121°C.

$MgSO_4 \cdot 7H_2O$ Solution:
Composition per liter:

$MgSO_4 \cdot 7H_2O$	246.5g

Preparation of $MgSO_4 \cdot 7H_2O$ Solution: Add $MgSO_4 \cdot 7H_2O$ to distilled/deionized water and bring volume to 1.0L. Mix thoroughly. Autoclave for 15 min at 15 psi pressure–121°C.

Thiamine·HCl Solution:
Composition per 10mL:

Thiamine·HCl	10.0mg

Preparation of Thiamine·HCl Solution: Add thiamine·HCl to distilled/deionized water and bring volume to 1.0L. Mix thoroughly. Filter sterilize.

$CaCl_2$ Solution:
Composition per liter:

$CaCl_2$ solution	14.7g

Preparation of $CaCl_2$ Solution: Add $CaCl_2$ solution to distilled/deionized water and bring volume to 1.0L. Mix thoroughly. Autoclave for 15 min at 15 psi pressure–121°C.

Preparation of Medium: Add components, except $MgSO_4 \cdot 7H_2O$ solution, glucose solution, thiamine·HCl solution, and $CaCl_2$ solution, to distilled/deionized water and bring volume to 987.0mL. Mix thoroughly. Adjust pH to 7.0. Autoclave for 15 min at 15 psi pressure–121°C. Cool to room temperature. Aseptically add sterile $MgSO_4·7H_2O$ solution, sterile glucose solution, sterile thiamine·HCl solution, and sterile $CaCl_2$ solution. Mix thoroughly. Distribute into tubes or flasks.

Use: For the cultivation and maintenance of *Flavobacterium meningosepticum.*

M13 *Verrucomicrobium* Medium

Composition per liter:

Glucose	0.25g
Peptone	0.25g
Yeast extract	0.25g
Distilled water	670.0mL
Artificial seawater	250.0mL
Tris-HCl buffer, (0.1M solution, pH 7.5)	50.0mL
Modified Huntner's basal salts	20.0mL
Vitamin solution	10.0mL

pH 7.5 ± 0.2 at 25°C

Artificial Seawater:
Composition per liter:

NaCl	23.48g
$MgCl_2$	4.98g
Na_2SO_4	3.92g
$CaCl_2$	1.10g
KCl	0.66g
$NaHCO_3$	0.19g
H_3BO_3	0.026g
$SrCl_2$	0.024g
KBr	6.0mg
NaF	3.0mg

Preparation of Artificial Seawater: Add components to distilled/deionized water and bring volume to 1.0L. Mix thoroughly.

Modified Hutner's Basal Salts:
Composition per liter:

$MgSO_4·7H_2O$	29.7g
Nitrilotriacetic acid	10.0g
$CaCl_2·2H_2O$	3.34g
$FeSO_4·7H_2O$	99.0mg
$(NH_4)_2MoO_4$	9.25mg
Metals "44"	50.0mL

Preparation of Modified Hutner's Basal Salts: Add nitrilotriacetic acid to 500.0mL of distilled/deionized water. Dissolve by adjusting pH to 6.5 with KOH. Add remaining components. Add distilled/deionized water to 1.0L.

Metals "44":
Composition per 100mL:

$ZnSO_4·7H_2O$	1.1g
$FeSO_4·7H_2O$	0.5g
EDTA	0.25g
$MnSO_4·7H_2O$	0.154g
$CuSO_4·5H_2O$	0.04g
$Co(NO_3)_2·6H_2O$	0.025g
$Na_2B_4O_7·10H_2O$	0.018g

Preparation of Metals "44": Add components to distilled/deionized water and bring volume to 100.0mL. Mix thoroughly.

Vitamin Solution:
Composition per liter:

D-Calcium pantothenate	5.0mg
Riboflavin	5.0mg
Thiamine·HCl	5.0mg
Biotin	2.0mg
Folic acid	2.0mg
Vitamin B_{12}	0.1mg

Preparation of Vitamin Solution: Add components to distilled/deionized water and bring volume to 1.0L. Mix thoroughly. Filter sterilize.

Preparation of Medium: Add components, except Modified Hutner's Basal Salts, to distilled/deionized water and bring volume to 980.0mL. Mix thoroughly. Autoclave for 15 min at 15 psi pressure–121°C. Cool to room temperature. Aseptically add 20.0mL of sterile Modified Hutner's Basal Salts. Mix thoroughly. Aseptically distribute into sterile tubes or flasks.

Use: For the cultivation and maintenance of *Verrucomicrobium spinosum.*

M14 Medium

Composition per liter:

Yeast extract	1.0g
D-Glucose	1.0g
Tris(hydroxymethyl)aminomethane	0.753g
Artificial seawater	250.0mL
Modified Hutner's basal salts	20.0mL

pH 7.5 ± 0.2 at 25°C

Artificial Seawater:
Composition per liter:

NaCl	23.48g
$MgCl_2$	4.98g
Na_2SO_4	3.92g
$CaCl_2$	1.10g
KCl	0.66g
$NaHCO_3$	0.19g
H_3BO_3	0.026g
$SrCl_2$	0.024g

KBr	6.0mg
NaF	3.0mg

Preparation of Artificial Seawater: Add components to distilled/deionized water and bring volume to 1.0L. Mix thoroughly.

Modified Hutner's Basal Salts:

Composition per liter:

$MgSO_4 \cdot 7H_2O$	29.7g
Nitrilotriacetic acid	10.0g
$CaCl_2 \cdot 2H_2O$	3.34g
$FeSO_4 \cdot 7H_2O$	99.0mg
$(NH_4)_2MoO_4$	9.25mg
Metals "44"	50.0mL

Preparation of Modified Hutner's Basal Salts: Add nitrilotriacetic acid to 500.0mL of distilled/deionized water. Dissolve by adjusting pH to 6.5 with KOH. Add remaining components. Add distilled/deionized water to 1.0L.

Metals "44":

Composition per 100mL:

$ZnSO_4 \cdot 7H_2O$	1.1g
$FeSO_4 \cdot 7H_2O$	0.5g
EDTA	0.25g
$MnSO_4 \cdot 7H_2O$	0.154g
$CuSO_4 \cdot 5H_2O$	0.04g
$Co(NO_3)_2 \cdot 6H_2O$	0.025g
$Na_2B_4O_7 \cdot 10H_2O$	0.018g

Preparation of Metals "44": Add components to distilled/deionized water and bring volume to 100.0mL. Mix thoroughly.

Preparation of Medium: Add components, except Modified Hutner's Basal Salts, to distilled/deionized water and bring volume to 980.0mL. Mix thoroughly. Autoclave for 15 min at 15 psi pressure–121°C. Cool to room temperature. Aseptically add 20.0mL of sterile Modified Hutner's Basal Salts. Mix thoroughly. Aseptically distribute into sterile tubes or flasks.

Use: For the cultivation and maintenance of *Pirellula marina*.

M16 Agar

Composition per liter:

Agar	10.0g
Beef extract	5.0g
Pancreatic digest of soybean meal	5.0g
Polypeptone™	5.0g
Sodium acetate·$3H_2O$	3.0g
Yeast extract	2.5g
Ascorbic acid	0.5g
Carbohydrate solution	50.0mL

pH 7.2 ± 0.2 at 25°C

Carbohydrate Solution:

Composition per 50mL:

Carbohydrate	5.0g

Preparation of Carbohydrate Solution: Add carbohydrate to distilled/deionized water and bring volume to 50.0mL. Lactose or glucose may be used. Mix thoroughly. Filter sterilize.

Preparation of Medium: Add components, except carbohydrate solution, to distilled/deionized water and bring volume to 950.0mL. Mix thoroughly. Gently heat and bring to boiling. Adjust pH to 7.2 with 2*N* NaOH. Autoclave for 15 min at 15 psi pressure–121°C. Cool to 45°–50°C. Aseptically add 50.0mL of sterile carbohydrate solution. Mix thoroughly. Pour into sterile Petri dishes or distribute into sterile tubes.

Use: For the cultivation of lactobacilli from cheddar cheese.

M17 Agar

Composition per liter:

Disodium β-glycerophosphate	19.0g
Agar	11.0g
Polypeptone™	5.0g
Beef extract	5.0g
Papaic digest of soybean meal	5.0g
Yeast extract	2.5g
Ascorbic acid	0.5g
Lactose solution	50.0mL
$MgSO_4 \cdot 7H_2O$ (1M solution)	1.0mL

pH 6.9 ± 0.2 at 25°C

Lactose Solution:

Composition per 100mL:

Lactose	10.0g

Preparation of Lactose Solution: Add lactose to distilled/deionized water and bring volume to 100.0mL. Mix thoroughly. Autoclave for 15 min at 15 psi pressure–121°C.

Preparation of Medium: Add components, except lactose solution, to distilled/deionized water and bring volume to 950.0mL. Mix thoroughly. Gently heat until boiling. Autoclave for 15 min at 15 psi pressure–121°C. Cool to 45°–50°C. Aseptically add 50.0mL of sterile lactose solution. Mix thoroughly. Pour into sterile Petri dishes or distribute into sterile tubes.

Use: For the cultivation and maintenance of streptococci and their bacteriophages. Also used for cultivation and maintenance of starter cultures for cheese and yogurt manufacture as well as detecting streptococcal mutants which are unable to ferment lactose.

M17 Agar

Composition per liter:

Disodium β-glycerophosphate	19.0g
Agar	11.0g
Beef extract	5.0g
Papaic digest of soybean meal	5.0g
Yeast extract	2.5g
Ascorbic acid	0.5g
$MgSO_4 \cdot 7H_2O$	0.25g
Lactose solution	50.0mL

pH 6.9 ± 0.2 at 25°C

Source: This medium is available as a premixed powder from Oxoid Unipath.

Lactose Solution:
Composition per 100mL:

Lactose	10.0g

Preparation of Lactose Solution: Add lactose to distilled/deionized water and bring volume to 100.0mL. Mix thoroughly. Autoclave for 15 min at 15 psi pressure–121°C.

Preparation of Medium: Add components, except lactose solution, to distilled/deionized water and bring volume to 950.0mL. Mix thoroughly. Gently heat until boiling. Autoclave for 15 min at 15 psi pressure–121°C. Cool to 45°–50°C. Aseptically add 50.0mL of sterile lactose solution. Mix thoroughly. Pour into sterile Petri dishes or distribute into sterile tubes.

Use: For the cultivation and maintenance of streptococci and their bacteriophages. Also used for cultivation and maintenance of starter cultures for cheese and yogurt manufacture as well as detecting streptococcal mutants which are unable to ferment lactose.

M17 Broth

Composition per liter:

Disodium β-glycerophosphate	19.0g
Beef extract	5.0g
Lactose	5.0g
Papaic digest of soybean meal	5.0g
Pancreatic digest of casein	2.5g
Peptic digest of animal tissue	2.5g
Yeast extract	2.5g
Ascorbic acid	0.5g
$MgSO_4 \cdot 7H_2O$	0.25g

pH 7.15 ± 0.05 at 25°C

Source: This medium is available as a premixed powder from BBL Microbiology Systems and Oxoid Unipath.

Preparation of Medium: Add components to distilled/deionized water and bring volume to 1.0L. Mix thoroughly. Distribute into tubes or flasks. Autoclave for 15 min at 15 psi pressure–121°C.

Use: For the cultivation and maintenance of streptococci and their bacteriophages. Also used for cultivation and maintenance of starter cultures for cheese and yogurt manufacture as well as detecting streptococcal mutants which are unable to ferment lactose.

M56 Agar

Composition per liter:

Agar	15.0g
Na_2HPO_4	8.7g
KH_2PO_4	5.3g
D-Glucose	4.0g
$(NH_4)_2SO_4$	2.0g
$MgSO_4 \cdot 7H_2O$	0.10g
L-Histidine	0.05g
L-Leucine	0.05g
Uracil	0.03g
$Ca(NO_3)_2 \cdot 4H_2O$	5.0mg
$FeSO_4 \cdot 7H_2O$	5.0mg
$ZnSO_4 \cdot 7H_2O$	5.0mg

pH 7.0 ± 0.2 at 25°C

Preparation of Medium: Add components to distilled/deionized water and bring volume to 1.0L. Mix thoroughly. Gently heat until boiling. Autoclave for 15 min at 15 psi pressure–121°C. Pour into sterile Petri dishes or distribute into sterile tubes.

Use: For the cultivation and maintenance of *Escherichia coli*.

M56 Medium

Composition per liter:

Na_2HPO_4	8.7g
KH_2PO_4	5.3g
D-Glucose	4.0g
$(NH_4)_2SO_4$	2.0g
$MgSO_4 \cdot 7H_2O$	0.10g
L-Histidine	0.05g
L-Leucine	0.05g
Uracil	0.03g
$Ca(NO_3)_2 \cdot 4H_2O$	5.0mg
$FeSO_4 \cdot 7H_2O$	5.0mg
$ZnSO_4 \cdot 7H_2O$	5.0mg

pH 7.0 ± 0.2 at 25°C

Preparation of Medium: Add components to distilled/deionized water and bring volume to 1.0L. Mix thoroughly. Distribute into tubes or flasks. Autoclave for 15 min at 15 psi pressure–121°C.

Use: For the cultivation and maintenance of *Escherichia coli*.

M63 Medium, 5X

Composition per liter:

KH_2PO_4	68.0g
$(NH_4)_2SO_4$	10.0g
$FeSO_4 \cdot 7H_2O$	2.5mg
Carbohydrate solution	10.0mL
$MgSO_4 \cdot 7H_2O$ solution	1.0mL

pH 7.0 ± 0.2 at 25°C

Carbohydrate Solution:
Composition per 100mL:

Carbohydrate	20.0g

Preparation of Carbohydrate Solution: Add carbohydrate to distilled/deionized water and bring volume to 100.0mL. Glucose of glycerol may be used. Mix thoroughly. Filter sterilize.

$MgSO_4 \cdot 7H_2O$ Solution:
Composition per 100mL:

$MgSO_4 \cdot 7H_2O$	24.65g

Preparation of $MgSO_4 \cdot 7H_2O$ Solution: Add $MgSO_4 \cdot 7H_2O$ to distilled/deionized water and bring volume to 100.0mL. Mix thoroughly. Filter sterilize.

Preparation of Medium: Add components, except carbohydrate solution and $MgSO_4 \cdot 7H_2O$ solution, to distilled/deionized water and bring volume to 1.0L. Mix thoroughly. Gently heat and bring to boiling. Autoclave for 15 min at 15 psi pressure–121°C. Cool to 45°–50°C. To prepare medium for use (1×) aseptically dilute 200.0mL of 5× stock solution with 789.0mL of sterile distilled/deionized water. Aseptically add 10.0mL of sterile carbohydrate solution and 1.0mL of sterile $MgSO_4 \cdot 7H_2O$ solution. Mix thoroughly. Aseptically distribute into sterile tubes or flasks.

Use: For the cultivation of *Escherichia coli*.

MacConkey Agar

Composition per liter:

Pancreatic digest of gelatin	17.0g
Agar	13.5g
Lactose	10.0g
NaCl	5.0g
Bile salts	1.5g
Pancreatic digest of casein	1.5g
Peptic digest of animal tissue	1.5g
Neutral Red	0.03g
Crystal Violet	1.0mg

pH 7.1 ± 0.2 at 25°C

Source: This medium is available as a premixed powder from BBL Microbiology Systems and Difco Laboratories.

Preparation of Medium: Add components to distilled/deionized water and bring volume to 1.0L. Mix thoroughly. Gently heat while stirring until boiling. Autoclave for 15 min at 15 psi pressure–121°C. Pour into sterile Petri dishes or distribute into sterile tubes.

Use: For the selective isolation, cultivation and differentiation of coliforms and enteric pathogens based on the ability to ferment lactose. Lactose-fermenting organisms appear as red to pink colonies. Lactose-nonfermenting organisms appear as colorless or transparent colonies.

MacConkey Agar

Composition per liter:

Peptone	20.0g
Agar	12.0g
Lactose	10.0g
Bile salts	5.0g
NaCl	5.0g
Neutral Red	0.075g

pH 7.4 ± 0.2 at 25°C

Source: This medium is available as a premixed powder from Oxoid Unipath.

Preparation of Medium: Add components to distilled/deionized water and bring volume to 1.0L. Mix thoroughly. Gently heat while stirring until boiling. Autoclave for 15 min at 15 psi pressure–121°C. Pour into sterile Petri dishes or distribute into sterile tubes.

Use: For the selective isolation, cultivation and differentiation of coliforms and enteric pathogens based on the ability to ferment lactose. Lactose-fermenting organisms appear as red to pink colonies. Lactose-nonfermenting organisms appear as colorless or transparent colonies.

MacConkey Agar No. 2 (MacConkey II Agar)

Composition per liter:

Peptone	20.0g
Agar	15.0g
Lactose	10.0g
NaCl	5.0g
Bile salts No.2	1.5g
Neutral Red	0.05g
Crystal Violet	1.0mg

pH 7.2 ± 0.2 at 25°C

Source: This medium is available as a premixed powder from BBL Microbiology Systems and Oxoid Unipath.

Preparation of Medium: Add components to distilled/deionized water and bring volume to 1.0L. Mix thoroughly. Gently heat while stirring until boiling. Autoclave for 15 min at 15 psi pressure–121°C. Pour into sterile Petri dishes or distribute into sterile tubes.

Use: For the selective isolation, cultivation and differentiation of enteric pathogens, especially enterococci, in clinical specimens and in materials of sanitary importance.

MacConkey Agar No. 3

Composition per liter:

Peptone	20.0g
Agar	15.0g
Lactose	10.0g
NaCl	5.0g
Bile salts No.3	1.5g
Neutral Red	0.03g
Crystal Violet	0.001g

pH 7.1 ± 0.2 at 25°C

Source: This medium is available as a premixed powder from Oxoid Unipath.

Preparation of Medium: Add components to distilled/deionized water and bring volume to 1.0L. Mix thoroughly. Gently heat while stirring until boiling. Autoclave for 15 min at 15 psi pressure–121°C. Pour into sterile Petri dishes or distribute into sterile tubes.

Use: For the selective isolation, cultivation and differentiation of enteric pathogens, especially *Salmonella* and *Shigella*, in clinical specimens and in foods.

MacConkey Agar with Sorbitol
See: Sorbitol MacConkey Agar

MacConkey Agar without Crystal Violet

Composition per liter:

Agar	12.0g
Lactose	10.0g
Pancreatic digest of casein	10.0g
Peptic digest of animal tissue	10.0g
Bile salts	5.0g
NaCl	5.0g
Neutral Red	0.05g

pH 7.4 ± 0.2 at 25°C

Source: This medium is available as a premixed powder from BBL Microbiology Systems and Difco Laboratories.

Preparation of Medium: Add components to distilled/deionized water and bring volume to 1.0L. Mix thoroughly. Gently heat while stirring until boiling. Autoclave for 15 min at 15 psi pressure–121°C. Pour into sterile Petri dishes or distribute into sterile tubes.

Use: For the detection of members of the *Enterobacteriaceae* and enterococci as well as some staphylococci. Used for the isolation and detection of coliforms and enteric pathogens from water and wastewater.

MacConkey Agar without Salt

Composition per liter:

Peptone	20.0g
Agar	12.0g
Lactose	10.0g
Bile salts	5.0g
Neutral Red	0.075g

pH 7.4 ± 0.2 at 25°C

Source: This medium is available as a premixed powder from Difco Laboratories and Oxoid Unipath.

Preparation of Medium: Add components to distilled/deionized water and bring volume to 1.0L. Mix thoroughly. Gently heat while stirring until boiling. Autoclave for 15 min at 15 psi pressure–121°C. Pour into sterile Petri dishes or distribute into sterile tubes. Dry the surface of plates before inoculation.

Use: Used for the isolation and detection of coliforms and enteric pathogens from urine. Provides a low electrolyte medium on which most *Proteus* species will not swarm and therefore avoids overgrow of the plate.

MacConkey Agar, CS

Composition per liter:

Peptone	17.0g
Agar	13.5g
Lactose	10.0g
NaCl	5.0g
Proteose peptone	3.0g
Bile salts	1.5g
Neutral Red	0.03g
Crystal Violet	1.0mg

pH 7.1 ± 0.2 at 25°C

Source: Available as a prepared medium from Difco Laboratories.

Preparation of Medium: Add components to distilled/deionized water and bring volume to 1.0L. Mix thoroughly. Gently heat while stirring until boiling. Autoclave for 15 min at 15 psi pressure–121°C. Pour into sterile Petri dishes or distribute into sterile tubes.

Use: For the cultivation and differentiation of lactose-fermenting and nonfermenting Gram-negative bacteria while also controlling the swarming of *Proteus* species, if present. Lactose-fermenting organisms appear as red to pink colonies. Lactose-nonfermenting organisms appear as colorless or transparent colonies.

MacConkey Broth

Composition per liter:

Pancreatic digest of gelatin	20.0g
Lactose	10.0g
Oxgall	5.0g
Bromcresol Purple	0.02g

pH 7.3 ± 0.2 at 25°C

Source: This medium is available as a premixed powder from BBL Microbiology Systems and Difco Laboratories.

Preparation of Medium: Add components to distilled/deionized water and bring volume to 1.0L. If testing 10.0mL samples, prepare medium double strength. Mix thoroughly. Gently heat while stirring until boiling. Distribute into test tubes containing inverted Durham tubes. Autoclave for 15 min at 15 psi pressure–121°C.

Use: For the selective isolation and cultivation of coliforms in milk and water.

MacConkey Broth

Composition per liter:

Peptone	20.0g
Lactose	10.0g
Bile salts	5.0g
NaCl	5.0g
Neutral Red	0.075g

pH 7.4 ± 0.2 at 25°C

Source: This medium is available as a premixed powder from Oxoid Unipath.

Preparation of Medium: Add components to distilled/deionized water and bring volume to 1.0L. If testing 10.0mL samples, prepare medium double strength. Mix thoroughly. Gently heat while stirring until boiling. Distribute into test tubes containing inverted Durham tubes. Autoclave for 15 min at 15 psi pressure–121°C.

Use: For the selective isolation and cultivation of coliforms in milk and water.

MacConkey Broth, Purple

Composition per liter:

Peptone	20.0g
Lactose	10.0g
Bile salts	5.0g
NaCl	5.0g
Bromcresol Purple	0.01g

pH 7.4 ± 0.2 at 25°C

Source: This medium is available as a premixed powder or tablets from Oxoid Unipath.

Preparation of Medium: Add components to distilled/deionized water and bring volume to 1.0L. If testing 10.0mL samples, prepare medium double strength. Mix thoroughly. Gently heat while stirring until boiling. Distribute into test tubes containing inverted Durham tubes. Autoclave for 15 min at 15 psi pressure–121°C.

Use: For the selective isolation and cultivation of coliforms in milk and water.

MacConkey II Agar
See: **MacConkey Agar No. 2**

Magnesium Oxalate Agar (MOX Agar)

Composition per liter:

Pancreatic digest of casein	15.0g
Agar	15.0g
Papaic digest of soybean meal	5.0g
NaCl	5.0g
$MgCl_2 \cdot 6H_2O$	4.1g
Sodium oxalate	2.68g

pH 7.4–7.6 at 25°C

Preparation of Medium: Add components to distilled/deionized water and bring volume to 1.0L. Mix thoroughly. Gently heat and bring to boiling. Distribute into tubes or flasks. Autoclave for 15 min at 15 psi pressure–121°C. Pour into sterile Petri dishes or leave in tubes.

Use: For the cultivation of *Yersinia enterocolitica* from foods.

Magnetic *Spirillum* Growth Medium, Revised (MSGM, Revised)

Composition per liter:

Agar	1.3g
KH_2PO_4	0.68g
Tartaric acid	0.37g
Succinic acid	0.37g
$NaNO_3$	0.12g
Sodium acetate	0.05g
Ascorbic acid	0.035g
Wolfe's vitamin solution	10.0mL
Wolfe's mineral solution	5.0mL
Ferric quinate solution	2.0mL
Resazurin (0.1% solution)	0.45mL

pH 6.75 ± 0.2 at 25°C

Wolfe's Vitamin Solution:

Composition per liter:

Pyridoxine·HCl	10.0mg
Thiamine·HCl	5.0mg
Riboflavin	5.0mg
Nicotinic acid	5.0mg
Calcium pantothenate	5.0mg
p-Aminobenzoic acid	5.0mg
Thioctic acid	5.0mg
Biotin	2.0mg
Folic acid	2.0mg
Cyanocobalamin	100.0μg

Preparation of Wolfe's Vitamin Solution: Add components to distilled/deionized water and bring volume to 1.0L. Mix thoroughly.

Wolfe's Mineral Solution:

Composition per liter:

$MgSO_4 \cdot 7H_2O$	3.0g
Nitriloacetic acid	1.5g
NaCl	1.0g
$MnSO_4 \cdot H_2O$	0.5g
$FeSO_4 \cdot 7H_2O$	0.1g
$CoCl_2 \cdot 6H_2O$	0.1g
$CaCl_2$	0.1g
$ZnSO_4 \cdot 7H_2O$	0.1g
$CuSO_4 \cdot 5H_2O$	0.01g
$AlK(SO_4)_2 \cdot 12H_2O$	0.01g
H_3BO_3	0.01g
$Na_2MoO_4 \cdot 2H_2O$	0.01g

Preparation of Wolfe's Mineral Solution: Add nitrilotriacetic acid to 500.0mL of distilled/deionized water. Dissolve by adjusting pH to 6.5 with KOH. Add remaining components. Add distilled/deionized water to 1.0L.

Ferric Quinate Solution:

Composition per 100mL:

$FeCl_3$	0.27g
Quinic acid	0.19g

Preparation of Ferric Quinate Solution: Add components to distilled/deionized water and bring volume to 1.0L. Mix thoroughly. Autoclave for 15 min at 15 psi pressure–121°C.

Preparation of Medium: To 1.0L of distilled/deionized water add components in the following order: Wolfe's Vitamin solution, Wolfe's Mineral solution, ferric quinate solution, resazurin, KH_2PO_4, $NaNO_3$, ascorbic acid, tartaric acid, succinic acid, sodium acetate and agar. Mix thoroughly after each addition. Adjust pH to 6.75 with NaOH. Autoclave for 15 min at 15 psi pressure–121°C. Aseptically distribute into sterile screw-capped tubes. Fill tubes to capacity with medium. Use a heavy inoculum in each tube and do not introduce a head space of air. Screw down caps tightly.

Use: For the cultivation and maintenance of *Aquaspirillum magnetotacticum*.

Maintenance of L Antigen in *Neisseria*

Composition per liter:

Proteose peptone No. 3	20.0g
Agar	15.0g
Na_2HPO_4	5.0g
NaCl	5.0g
Glucose	0.5g
Rabbit blood, defibrinated	100.0mL

pH 7.4–7.6 at 25°C

Preparation of Medium: Add components, except rabbit blood, to distilled/deionized water and bring volume to 900.0mL. Mix thoroughly. Gently heat while stirring until boiling. Autoclave for 20 min at 15 psi pressure–121°C. Cool to 60°C. Aseptically add 100.0mL of sterile, defibrinated rabbit blood. Maintain at 75°C while shaking for 30 min. Pour into sterile Petri dishes or distribute into sterile tubes.

Use: For the cultivation and maintenance of *Neisseria gonorrhoeae*.

Maintenance SCY Medium
See: **SCY Medium**

Malachite Green Broth

Composition per liter:

Peptone	15.0g
Beef extract	9.0g
Malachite Green	0.01mg

pH 7.3 ± 0.2 at 25°C

Preparation of Medium: Add components to distilled/deionized water and bring volume to 1.0L. Mix thoroughly. Distribute into tubes or flasks. Autoclave for 15 min at 15 psi pressure–121°C.

Use: For the cultivation of *Pseudomonas aeruginosa*.

Maleate Medium for *Pseudomonas fluorescens*

Composition per liter:

Agar	15.0g
K_2HPO_4	1.13g
NH_4NO_3	1.0g
KH_2PO_4	0.48g
$MgSO_4 \cdot 7H_2O$	0.2g
Potassium maleate solution	8.61mL

pH 7.0 ± 0.2 at 25°C

Potassium Maleate Solution:
Composition per liter:

Maleic acid	116.07g
KOH (10*N* solution)	200.0mL

Preparation of Potassium Maleate Solution: Add maleic acid to distilled/deionized water and bring volume to 600.0mL. Slowly add KOH solution (generates heat). Bring volume to 1.0L with distilled/deionized water. Adjust pH to 7.0. Filter sterilize.

Preparation of Medium: Add components, except potassium maleate solution, to distilled/deionized water and bring volume to 991.4mL. Mix thoroughly. Gently heat while stirring until boiling. Autoclave for 20 min at 15 psi pressure–121°C. Cool to 50°C. Asptically add 8.61mL of the potassium maleate solution. Mix thoroughly. Pour into sterile Petri dishes or distribute into sterile tubes.

Use: For the cultivation and maintenance of *Pseudomonas fluorescens* and *Mycoplasma hyopneumoniae*.

Maleate Medium for *Pseudomonas fluorescens* with Glucose and Phenol Red

Composition per liter:

Agar	15.0g
Glucose	10.0g
K_2HPO_4	1.13g
NH_4NO_3	1.0g
KH_2PO_4	0.48g
$MgSO_4{\cdot}7H_2O$	0.2g
Phenol Red	0.04g
Potassium maleate solution	8.61mL

pH 7.0 ± 0.2 at 25°C

Potassium Maleate Solution:
Composition per liter:

Maleic acid	116.07g
KOH (10*N* solution)	200.0mL

Preparation of Potassium Maleate Solution: Add maleic acid to distilled/deionized water and bring volume to 600.0mL. Slowly add KOH solution (generates heat). Bring volume to 1.0L with distilled/deionized water. Adjust pH to 7.0. Filter sterilize.

Preparation of Medium: Add components, except potassium maleate solution, to distilled/deionized water and bring volume to 991.4mL. Mix thoroughly. Gently heat while stirring until boiling. Autoclave for 20 min at 15 psi pressure–121°C. Cool to 50°C. Aseptically add 8.61mL of the potassium maleate solution. Mix thoroughly. Pour into sterile Petri dishes or distribute into sterile tubes.

Use: For the cultivation and maintenance of *Pseudomonas fluorescens* and *Mycoplasma hyopneumoniae*.

Malonate Broth

Composition per liter:

Sodium malonate	3.0g
NaCl	2.0g
$(NH_4)_2SO_4$	2.0g
K_2HPO_4	0.6g
KH_2PO_4	0.4g
Bromthymol Blue	0.025g

pH 6.7 ± 0.2 at 25°C

Source: This medium is available as a premixed powder from Difco Laboratories.

Preparation of Medium: Add components to distilled/deionized water and bring volume to 1.0L. Mix thoroughly. Distribute into tubes or flasks. Autoclave for 15 min at 15 psi pressure–121°C. Avoid introduction of carbon and nitrogen from other sources.

Use: For the cultivation and differentiation of coliforms and other enteric organisms, particularly *Enterobacter* and *Escherichia* based on their ability to utilize malonate as the sole carbon source and ammonium sulfate as the sole nitrogen source. Malonate-utilizing organisms turn the medium blue.

Malonate Broth, Ewing Modified

Composition per liter:

Sodium malonate	3.0g
NaCl	2.0g
$(NH_4)_2SO_4$	2.0g
Yeast extract	1.0g
Glucose	0.25g
K_2HPO_4	0.6g
KH_2PO_4	0.4g
Bromthymol Blue	0.025g

pH 6.7 ± 0.2 at 25°C

Source: This medium is available as a premixed powder from BBL Microbiology Systems and Difco Laboratories.

Preparation of Medium: Add components to distilled/deionized water and bring volume to 1.0L. Mix thoroughly. Distribute into tubes or flasks. Autoclave for 15 min at 15 psi pressure–121°C.

Use: For the cultivation and differentiation of coliforms and other enteric organisms, particularly *Enterobacter* and *Escherichia* based on their ability to utilize malonate as a carbon source and ammonium sulfate as a nitrogen source. The small amount of yeast extract and glucose encourages the growth of some organisms that may be distressed or fail to respond. Malonate-utilizing organisms turn the medium blue.

Malt Agar

Composition per liter:

Malt extract	30.0g
Agar	15.0g

pH 5.5 ± 0.2 at 25°C

Source: This medium is available as a premixed powder from BBL Microbiology Systems and Difco Laboratories.

Preparation of Medium: Add components to distilled/deionized water and bring volume to 1.0L. Mix thoroughly. Gently heat while stirring until boiling. Distribute into tubes or flasks. Autoclave for 15 min at 15 psi pressure–118°C. Do not overheat or agar will not harden. Pour into sterile Petri dishes or distribute into sterile tubes.

Use: For the cultivation of yeasts and molds.

Malt and Peptone Medium

Composition per liter:

Agar	15.0g
Malt extract	10.0g
Peptone	5.0g
NaCl	1.0g

pH 6.5 ± 0.2 at 25°C

Preparation of Medium: Add components to distilled/deionized water and bring volume to 1.0L. Mix thoroughly. Distribute into tubes or flasks. Adjust pH to 6.5. Autoclave for 15 min at 15 psi pressure–121°C.

Use: For the cultivation and maintenance of *Flavobacterium* species.

Malt Extract Agar

Composition per liter:

Malt extract	30.0g
Agar	15.0g
Peptone	5.0g

pH 7.0 ± 0.2 at 25°C

Preparation of Medium: Add components to distilled/deionized water and bring volume to 1.0L. Mix thoroughly. Gently heat and bring to boiling. Distribute into tubes or flasks. Autoclave for 20 min at 10 psi pressure–115°C. Pour into sterile Petri dishes or leave in tubes.

Use: For the cultivation of *Xanthomonas* species.

Malt Extract Agar (MEA)

Composition per liter:

Agar	20.0g
Glucose	20.0g
Malt extract	20.0g
Peptone	1.0g

pH 5.5 ± 0.2 at 25°C

Preparation of Medium: Add components to distilled/deionized water and bring volume to 1.0L. Mix thoroughly. Gently heat and bring to boiling. Distribute into tubes or flasks. Autoclave for 15 min at 15 psi pressure–121°C. Pour into sterile Petri dishes or leave in tubes.

Use: For the isolation, cultivation and identification of heat-resistant filamentous fungi (molds) from foods.

Malt Extract Agar

Composition per liter:

Malt extract	30.0g
Agar	15.0g
Mycological peptone	5.0g

pH 5.4 ± 0.2 at 25°C

Source: This medium is available as a premixed powder from Oxoid Unipath.

Preparation of Medium: Add components to distilled/deionized water and bring volume to 1.0L. Mix thoroughly. Gently heat while stirring until boiling. Distribute into tubes or flasks. Autoclave for 10 min at 15 psi pressure–115°C. Do not overheat or agar will not harden. If a lower pH (3.5) is desired, cool medium to 55°C and aseptically add 100.0mL of sterile lactic acid. Pour into sterile Petri dishes or distribute into sterile tubes.

Use: For the detection, isolation and enumeration of yeasts and molds. The addition of lactic acid suppresses bacterial growth.

Malt Extract Agar (ATCC Medium 109)

Composition per liter:

Agar	15.0g
Maltose	12.75g
Dextrin	2.75g
Glycerol	2.35g
Pancreatic digest of gelatin	0.78g

pH 4.6 ± 0.2 at 25°C

Source: This medium is available as a premixed powder from BBL Microbiology Systems and Difco Laboratories.

Preparation of Medium: Add components to distilled/deionized water and bring volume to 1.0L. Mix thoroughly. Gently heat while stirring until boiling. Distribute into tubes or flasks. Autoclave for 15 min at 15 psi pressure–118°C. Do not overheat or agar will not harden. Pour into sterile Petri dishes or distribute into sterile tubes.

Use: For the cultivation and maintenance of yeasts, molds and *Flavobacterium lucecoloratum.*

Malt Extract Agar

Composition per liter:

Malt extract	30.0g
Agar	20.0g
Chlortetracycline solution	10.0mL

pH 5.5 ± 0.2 at 25°C

Chlortetracycline Solution:

Composition per 10mL:

Chlortetracycline	0.04mg

Preparation of Chlortetracycline Solution: Add chlortetracycline to distilled/deionized water and bring volume to 10.0mL. Mix thoroughly. Filter sterilize.

Preparation of Medium: Add components, except chlortetracycline solution, to distilled/deionized water and bring volume to 990.0mL. Mix thoroughly. Gently heat and bring to boiling. Autoclave for 15 min at 15 psi pressure–121°C. Cool to 45°–50°C. Aseptically add sterile chlortetracycline solution. Mix thoroughly. Pour into sterile Petri dishes in 20.0mL volumes.

Use: For the cultivation of yeasts and filamentous fungi (molds) from cosmetics.

Malt Extract Broth

Composition per liter:

Malt extract	6.0g
Maltose	1.8g
Glucose	6.0g
Yeast extract	1.2g

pH 4.7 ± 0.2 at 25°C

Source: This medium is available as a premixed powder from Difco Laboratories.

Preparation of Medium: Add components to distilled/deionized water and bring volume to 1.0L. Mix thoroughly. Distribute into tubes or flasks. Autoclave for 15 min at 15 psi pressure–121°C. Do not overheat—this results in darkening of the broth.

Use: For the isolation, cultivation and enumeration of yeast and filamentous fungi (mold).

Malt Extract Broth

Composition per liter:

Malt extract	17.0g
Mycological peptone	3.0g

pH 5.4 ± 0.2 at 25°C

Source: This medium is available as a premixed powder from Oxoid Unipath.

Preparation of Medium: Add components to distilled/deionized water and bring volume to 1.0L. Mix thoroughly. Distribute into tubes or flasks. Autoclave for 10 min at 15 psi pressure–115°C.

Use: For the cultivation of molds and yeasts, especially for sterility testing.

Malt Extract Broth

Composition per liter:

Malt extract, purified solids	15.0g

pH 4.7 ± 0.2 at 25°C

Source: This medium is available as a premixed powder from BBL Microbiology Systems.

Preparation of Medium: Add components to distilled/deionized water and bring volume to 1.0L. Mix thoroughly. Distribute into tubes or flasks. Autoclave for 15 min at 15 psi pressure–118°C. Do not overheat.

Use: For the cultivation of yeasts and molds.

Malt Extract Yeast Extract 40% Glucose Agar (MY40G)

Composition per liter:

Glucose	400.0g
Agar	12.0g
Malt extract powder	12.0g
Yeast extract	3.0g

pH 5.5 ± 0.2 at 25°C

Preparation of Medium: Add components, except glucose, to 550.0mL of distilled/deionized water. Mix thoroughly. Gently heat and bring to boiling. Bring volume to 600.0mL with distilled/deionized water. While the solution is still hot, add the glucose all at once while stirring to avoid formation of lumps. Autoclave for 30 min at 0 psi pressure–100°C.

Use: For the isolation and cultivation of osmotolerant microorganisms from foods.

Malt Peptone Yeast Extract Agar
See: **MPY Agar**

Maltose Peptone Yeast Extract Agar
See: **MPY Agar**

Maltose Peptone Yeast Extract Broth
See: **MPY Broth**

Maltose Peptone Yeast Extract Medium
See: **MPY Agar**

Manganese Agar No. 1 (Mn Agar No. 1)

Composition per liter:

Agar	10.0g
$MnCO_3$	2.0g
Beef extract	1.0g
$Fe(NH_4)_2(SO_4)_2$	0.15g
Sodium citrate	0.15g
Yeast extract	0.075g
Cyanocobalamin solution	10.0mL

Cyanocobalamin Solution:

Composition per 10mL:

Cyanocobalamin	0.005mg

Preparation of Cyanocobalamin Solution: Add cyanocobalamin to distilled/deionized water and bring volume to 10.0mL. Mix thoroughly. Filter sterilize.

Preparation of Medium: Add components, except cyanocobalamin, to distilled/deionized water and bring volume to 990.0mL. Mix thoroughly. Autoclave for 15 min at 15 psi pressure–121°C. Cool to 45°–50°C. Aseptically add 10.0mL of the sterile cyanocobalamin solution. Mix thoroughly. Pour into sterile Petri dishes or distribute into sterile tubes.

Use: For the isolation and cultivation of iron and sulfur bacteria. Also used to differentiate *Leptothrix* (*Sphaerotilus*) *discophorus* from *Sphaerotilus natans*.

Manganese Agar No. 2 (Mn Agar No. 2)

Composition per liter:

Agar	15.0g
$MnSO_4 \cdot H_2O$	10.0mg

Preparation of Medium: Add components to distilled/deionized water and bring volume to 1.0L. Mix thoroughly. Gently heat and bring to boiling. Distribute into tubes or flasks. Autoclave for 15 min at 15 psi pressure–121°C. Pour into sterile Petri dishes or leave in tubes. Use freshly prepared solution.

Use: For the enumeration, enrichment and isolation of iron and sulfur bacteria. For the isolation and cultivation of *Leptothrix* species from water.

Manganese Medium for *Pseudomonas* species

Composition per liter:

Noble agar	10.0g
$MnCO_3$	1.0g
$Fe(NH_4)_2(SO_4)_2 \cdot 6H_2O$	0.15g
Sodium citrate	0.15g
Yeast extract	0.075g
$Na_4P_2O_7 \cdot 10H_2O$	0.05g

pH 6.8 ± 0.2 at 25°C

Preparation of Medium: Add components to distilled/deionized water and bring volume to 1.0L. Mix thoroughly. Gently heat and bring to boiling. Distribute into tubes or flasks. Autoclave for 15 min at 15 psi pressure–121°C. Pour into sterile Petri dishes or leave in tubes.

Use: For the cultivation and maintenance of *Pseudomonas putida* and other *Pseudomonas* species.

Manganese Nutrient Agar

Composition per liter:

Agar	15.0g
Peptone	5.0g
Meat extract	3.0g
$MnSO_4 \cdot H_2O$	5.0mg

Preparation of Medium: Add components to distilled/deionized water and bring volume to 1.0L. Mix thoroughly. Gently heat and bring to boiling. Distribute into tubes or flasks. Autoclave for 15 min at 15 psi pressure–121°C. Pour into sterile Petri dishes or leave in tubes.

Use: For the cultivation and obtaining sporulation of *Bacillus* species.

Mannitol Agar

Composition per liter:

Mannitol	25.0g
Agar	15.0g
Yeast extract	5.0g
Peptone	3.0g

Preparation of Medium: Add components to distilled/deionized water and bring volume to 1.0L. Mix thoroughly. Gently heat and bring to boiling. Distribute into tubes or flasks. Autoclave for 15 min at 15 psi pressure–121°C. Pour into sterile Petri dishes or leave in tubes.

Use: For the cultivation and maintenance of *Acetobacter aceti, Acetobacter hansenii, Acetobacter pasteurianus, Frateuria aurantia, Gluconobacter asaii, Gluconobacter cerinus, Gluconobacter oxydans* and other bacteria that can utilize mannitol as a carbon source.

Mannitol Egg Yolk Polymyxin Agar

Composition per liter:

Agar	15.0g
D-Mannitol	10.0g
Peptone	10.0g
NaCl	10.0g
Beef extract	1.0g
Phenol Red	0.025g
Egg yolk emulsion, 50%	50.0mL
Polymyxin B	10.0mL

pH 7.1 ± 0.1 at 25°C

Source: Available as a prepared medium from Difco Laboratories.

Egg Yolk Emulsion, 50%:
Composition per 100mL:

Chicken egg yolks	11
Whole chicken egg	1
NaCl (0.9% solution)	80.0mL

Preparation of Egg Yolk Emulsion, 50%: Soak eggs with 1:100 dilution of saturated mercuric chloride solution for 1 min. Crack eggs and separate yolks from whites. Mix egg yolks with 1 chicken egg. Measure 50.0mL of egg yolk emulsion and add to 50.0mL of 0.9% NaCl solution. Mix thoroughly. Filter sterilize. Warm to 45°–50°C.

Polymyxin B Solution:
Composition per 10mL:

Polymyxin B	100,000U

Preparation of Polymyxin B Solution: Add polymyxin B to distilled/deionized water and bring volume to 10.0mL. Mix thoroughly. Filter sterilize.

Preparation of Medium: Add components—except egg yolk emulsion, 20% and polymyxin B solution—to distilled/deionized water and bring volume to 940.0mL. Mix thoroughly. Gently heat and bring to boiling. Autoclave for 15 min at 15 psi pressure–121°C. Cool to 45°–50°C. Aseptically add 50.0mL of sterile egg yolk emulsion, 50% and 10.0mL of sterile polymyxin B solution. Mix thoroughly. Pour into sterile Petri dishes.

Use: For the cultivation and enumeration of *Bacillus cereus* from foods.

Mannitol Lysine Crystal Violet Brilliant Green Agar
See: MLCB Agar

Mannitol Maltose Agar

Composition per liter:

NaCl	20.0g
Agar	13.0g
D-Mannitol	10.0g
Maltose	10.0g
Beef extract	5.0g
Papaic digest of soybean meal	5.0g
Polypeptone™	5.0g
Dye stock solution, 1000×	1.0mL

pH 7.8 ± 0.2 at 25°C

Dye Stock Solution, 1000X :

Bromthymol Blue	4.0g
Cresol Red	4.0g
Ethanol, 95%	100.0mL

Preparation of Dye Stock Solution, 1000X: Add Bromthymol Blue and Cresol Red to 100.0mL of ethanol. Mix thoroughly.

Preparation of Medium: Add components to distilled/deionized water and bring volume to 1.0L. Mix thoroughly. Adjust to pH 7.8. Gently heat and bring to boiling. Distribute into tubes or flasks. Autoclave for 15 min at 15 psi pressure–121°C. Pour into sterile Petri dishes or leave in tubes.

Use: For the cultivation of *Vibrio* species from foods.

Mannitol Salt Agar

Composition per liter:

NaCl	75.0g
Agar	15.0g
D-Mannitol	10.0g
Pancreatic digest of casein	5.0g
Peptic digest of animal tissue	5.0g

Beef extract 1.0g
Phenol Red 0.025g

pH 7.4 ± 0.2 at 25°C

Source: This medium is available as a premixed powder from BBL Microbiology Systems and Difco Laboratories and Oxoid Unipath.

Preparation of Medium: Add components to distilled/deionized water and bring volume to 1.0L. Mix thoroughly. Gently heat while stirring and bring to boiling. Distribute into tubes or flasks. Autoclave for 15 min at 15 psi pressure–121°C. Pour into sterile Petri dishes or leave in tubes.

Use: For the selective isolation, cultivation and enumeration of staphylococci from clinical and nonclinical specimens. Mannitol-utilizing organisms turn the medium yellow.

Mannitol Salt Broth

Composition per liter:

NaCl 100.0g
D-Mannitol...2.5g
Pancreatic digest casein 17.0g
Papaic digest soy bean meal 3.0g
K_2HPO_4 ....2.5g
Phenol Red 0.025g

pH 7.4 ± 0.2 at 25°C

Source: This medium is available as a premixed powder from Difco Laboratories.

Preparation of Medium: Add components to distilled/deionized water and bring volume to 1.0L. Mix thoroughly. Distribute into tubes or flasks. Autoclave for 15 min at 15 psi pressure–121°C.

Use: For the selective isolation and cultivation of staphylococci from foods and nonclinical specimens. Mannitol-utilizing organisms turn the medium yellow.

Mannitol Yolk Polymyxin Agar (MYP Agar)

Composition per 110mL:

Agar..1.5g
NaCl 1.0g
Peptone..1.0g
D-Mannitol..1.0g
$(NH_4)_2PO_4$ 0.1g
Meat extract 0.1g
Phenol Red 2.5mg
Egg yolk emulsion, 20% 10.0mL
Polymyxin B solution 1.0mL

pH 7.1 ± 0.2 at 25°C

Egg Yolk Emulsion, 20%:
Composition per 100mL:

Chicken egg yolks 11
Whole chicken egg..1
NaCl (0.9% solution) 80.0mL

Preparation of Egg Yolk Emulsion, 20%: Soak eggs with 1:100 dilution of saturated mercuric chloride solution for 1 min. Crack eggs and separate yolks from whites. Mix egg yolks with 1 chicken egg. Measure 20.0mL of egg yolk emulsion and add to 80.0mL of 0.9% NaCl solution. Mix thoroughly. Filter sterilize. Warm to 45°–50°C.

Polymyxin B Solution:
Composition per 1mL:

Polymyxin B 1.0mg

Preparation of Polymyxin B Solution: Add polymyxin B to distilled/deionized water and bring volume to 1.0mL. Mix thoroughly. Filter sterilize.

Preparation of Medium: Add components—except egg yolk emulsion, 20% and polymyxin B solution—to distilled/deionized water and bring volume to 100.0mL. Mix thoroughly. Gently heat and bring to boiling. Autoclave for 15 min at 15 psi pressure–121°C. Cool to 45°–50°C. Aseptically add 10.0mL of sterile egg yolk emulsion, 20% and 1.0mL of sterile polymyxin B solution. Mix thoroughly. Pour into sterile Petri dishes.

Use: For the cultivation and maintenance of *Bacillus cereus*.

Marine Agar 2216

Composition per liter:

NaCl 19.45g
Agar...15.0g
$MgCl_2$ 8.8g
Peptone..5.0g
Na_2SO_3 ....3.24g
$CaCl_2$ 1.8g
Yeast extract 1.0g
KCl 0.55g
$NaHCO_3$ 0.16g
Ferric citrate 0.1g
KBr..0.08g
$SrCl_2$ 0.03g
H_3BO_3 ....0.02g
Na_2HPO_4 .... 8.0mg
Na_2SiO_3 .... 4.0mg
NaF... 2.4mg
NH_4NO_3 .... 1.6mg

pH 7.6 ± 0.2 at 25°C

Source: This medium is available as a premixed powder from Difco Laboratories.

Preparation of Medium: Add components to distilled/deionized water and bring volume to 1.0L. Mix thoroughly. Gently heat while stirring and bring to boiling. Distribute into tubes or flasks. Autoclave for 15 min at 15 psi pressure–121°C. Pour into sterile Petri dishes or leave in tubes.

Use: For the isolation, cultivation and maintenance of a wide variety of heterotrophic marine bacteria.

Marine Agar with ι-Carrageenan

Composition per 1070mL:

Solution A	1.0L
Solution B	60.0mL
Solution C	10.0mL

pH 7.2 ± 0.2 at 25°C

Solution A:

Composition per liter:

NaCl	25.0g
Agar	15.0g
$MgSO_4 \cdot 7H_2O$	5.0g
Casamino acids	2.5g
$NaNO_3$	2.0g
ι-Carrageenan	2.5g
$CaCl_2 \cdot 2H_2O$	0.2g
KCl	0.1g

Preparation of Solution A: Add components to distilled/deionized water and bring volume to 1.0L. Mix thoroughly. Gently heat and bring to boiling. Autoclave for 15 min at 15 psi pressure–121°C.

Solution B:

Composition per 100mL:

$Na_2HPO_4 \cdot 2H_2O$	3.56g

Preparation of Solution B: Add component to distilled/deionized water and bring volume to 100.0mL. Mix thoroughly. Autoclave for 15 min at 15 psi pressure–121°C.

Solution C:

Composition per 100mL:

$FeSO_4 \cdot 7H_2O$	0.3g

Preparation of Solution C: Add component to distilled/deionized water and bring volume to 100.0mL. Mix thoroughly. Autoclave for 15 min at 15 psi pressure–121°C.

Preparation of Medium: Aseptically add 60.0mL of sterile solution B and 10.0mL of sterile solution C to 1.0L of sterile solution A. Mix thoroughly. Pour into sterile Petri dishes or distribute into sterile tubes.

Use: For the cultivation and maintenance of ATCC strain 43554.

Marine Agar with κ- and λ-Carrageenan

Composition per 1070mL:

Solution A	1.0L
Solution B	60.0mL
Solution C	10.0mL

pH 7.2 ± 0.2 at 25°C

Solution A:

Composition per liter:

NaCl	25.0g
Agar	15.0g
$MgSO_4 \cdot 7H_2O$	5.0g
Casamino acids	2.5g
$NaNO_3$	2.0g
κ-Carrageenan	1.25g
λ-Carrageenan	1.25g
$CaCl_2 \cdot 2H_2O$	0.2g
KCl	0.1g

Preparation of Solution A: Add components to distilled/deionized water and bring volume to 1.0L. Mix thoroughly. Gently heat and bring to boiling. Autoclave for 15 min at 15 psi pressure–121°C.

Solution B:

Composition per 100mL:

$Na_2HPO_4 \cdot 2H_2O$	3.56g

Preparation of Solution B: Add component to distilled/deionized water and bring volume to 100.0mL. Mix thoroughly. Autoclave for 15 min at 15 psi pressure–121°C.

Solution C:

Composition per 100mL:

$FeSO_4 \cdot 7H_2O$	0.3g

Preparation of Solution C: Add component to distilled/deionized water and bring volume to 100.0mL. Mix thoroughly. Autoclave for 15 min at 15 psi pressure–121°C.

Preparation of Medium: Aseptically add 60.0mL of sterile solution B and 10.0mL of sterile solution C to 1.0L of sterile solution A. Mix thoroughly. Pour into sterile Petri dishes or distribute into sterile tubes.

Use: For the cultivation and maintenance of *Pseudomonas carrageenovora*.

Marine Broth 2216

Composition per liter:

NaCl	19.45g
$MgCl_2$	8.8g
Peptone	5.0g
Na_2SO_3	3.24g

$CaCl_2$ 1.8g
Yeast extract 1.0g
KCl 0.55g
$NaHCO_3$ 0.16g
Ferric citrate 0.1g
KBr 0.08g
$SrCl_2$ 0.03g
H_3BO_3 .... 0.02g
Na_2HPO_4 .... 8.0mg
Na_2SiO_3 .... 4.0mg
NaF 2.4mg
NH_4NO_3 .... 1.6mg
pH 7.6 ± 0.2 at 25°C

Source: This medium is available as a premixed powder from Difco Laboratories.

Preparation of Medium: Add components to distilled/deionized water and bring volume to 1.0L. Mix thoroughly. Gently heat while stirring and bring to boiling. Distribute into tubes or flasks. Autoclave for 15 min at 15 psi pressure–121°C.

Use: For the isolation, cultivation and maintenance of a wide variety of heterotrophic marine bacteria.

Marine Broth with ι-Carrageenan

Composition per 1070mL:
Solution A 1.0L
Solution B 60.0mL
Solution C 10.0mL
pH 7.2 ± 0.2 at 25°C

Solution A:
Composition per liter:
NaCl 25.0g
$MgSO_4 \cdot 7H_2O$ 5.0g
Casamino acids 2.5g
$NaNO_3$ 2.0g
ι-Carrageenan 2.5g
$CaCl_2 \cdot 2H_2O$ 0.2g
KCl 0.1g

Preparation of Solution A: Add components to distilled/deionized water and bring volume to 1.0L. Mix thoroughly. Gently heat and bring to boiling. Autoclave for 15 min at 15 psi pressure–121°C.

Solution B:
Composition per 100mL:
$Na_2HPO_4 \cdot 2H_2O$ 3.56g

Preparation of Solution B: Add components to distilled/deionized water and bring volume to 100.0mL. Mix thoroughly. Autoclave for 15 min at 15 psi pressure–121°C.

Solution C:
Composition per 100mL:
$FeSO_4 \cdot 7H_2O$ 0.3g

Preparation of Solution C: Add components to distilled/deionized water and bring volume to 100.0mL. Mix thoroughly. Autoclave for 15 min at 15 psi pressure–121°C.

Preparation of Medium: Aseptically add 60.0mL of sterile solution B and 10.0mL of sterile solution C to 1.0L of sterile solution A. Mix thoroughly. Pour into sterile Petri dishes or distribute into sterile tubes.

Use: For the cultivation of ATCC strain 43554.

Marine Broth with κ- and λ-Carrageenan

Composition per 1070mL:
Solution A 1.0L
Solution B 60.0mL
Solution C 10.0mL
pH 7.2 ± 0.2 at 25°C

Solution A:
Composition per liter:
NaCl 25.0g
$MgSO_4 \cdot 7H_2O$ 5.0g
Casamino acids 2.5g
$NaNO_3$ 2.0g
κ-Carrageenan 1.25g
λ-Carrageenan 1.25g
$CaCl_2 \cdot 2H_2O$ 0.2g
KCl 0.1g

Preparation of Solution A: Add components to distilled/deionized water and bring volume to 1.0L. Mix thoroughly. Gently heat and bring to boiling. Autoclave for 15 min at 15 psi pressure–121°C.

Solution B:
Composition per 100mL:
$Na_2HPO_4 \cdot 2H_2O$ 3.56g

Preparation of Solution B: Add components to distilled/deionized water and bring volume to 100.0mL. Mix thoroughly. Autoclave for 15 min at 15 psi pressure–121°C.

Solution C:
Composition per 100mL:
$FeSO_4 \cdot 7H_2O$ 0.3g

Preparation of Solution C: Add components to distilled/deionized water and bring volume to 100.0mL. Mix thoroughly. Autoclave for 15 min at 15 psi pressure–121°C.

Preparation of Medium: Aseptically add 60.0mL of sterile solution B and 10.0mL of sterile solution C to 1.0L of sterile solution A. Mix thoroughly. Distribute into sterile tubes or flasks.

Use: For the cultivation and maintenance of *Pseudomonas carrageenovora*.

Marine Chlorobiaceae Medium 2

Composition per 1051mL:

Solution 1	950.0mL
$Na_2S·9H_2O$ solution	60.0mL
$NaHCO_3$ solution	40.0mL
Vitamin B_{12} solution	1.0mL

pH 6.8 ± 0.2 at 25°C

Solution 1:

Composition per 950mL:

NaCl	20.0g
$MgSO_4·7H_2O$	3.0g
KH_2PO_4	1.0g
NH_4Cl	0.5g
$CaCl_2·2H_2O$	0.05g
Trace element solution SL-8	1.0mL

Preparation of Solution 1: Add components to distilled/deionized water and bring volume to 950.0mL. Mix thoroughly. Autoclave for 15 min at 15 psi pressure–121°C. Cool to 45°–50°C.

Trace Element Solution SL-8:

Composition per liter:

Disodium EDTA	5.2g
$FeCl_2·4H_2O$	1.5g
$CoCl_2·6H_2O$	0.19g
$MnCl_2·4H_2O$	0.10g
$ZnCl_2$	0.07g
H_3BO_3	0.06g
$NaMoO_4·2H_2O$	0.04g
$CuCl_2·2H_2O$	0.02g
$NiCl_2·6H_20$	0.02g

Preparation of Trace Element Solution SL-8: Add components to distilled/deionized water and bring volume to 1.0L. Mix thoroughly.

$Na_2S·9H_2O$ Solution:

Composition per 100mL:

$Na_2S·9H_2O$	5.0g

Preparation of $Na_2S·9H_2O$ Solution: Add $Na_2S·9H_2O$ to distilled/deionized water and bring volume to 100.0mL. Autoclave for 15 min at 15 psi pressure–121°C. Cool to 45°–50°C.

$NaHCO_3$ Solution:

Composition per 100mL:

$NaHCO_3$	5.0g

Preparation of $NaHCO_3$ Solution: Add $NaHCO_3$ to distilled/deionized water and bring volume to 100.0mL. Mix thoroughly. Filter sterilize.

Vitamin B_{12} Solution:

Composition per 100mL:

Vitamin B_{12}	2.0mg

Preparation of Vitamin B_{12} Solution: Add vitamin B_{12} to distilled/deionized water and bring volume to 100.0mL. Mix thoroughly. Filter sterilize.

Preparation of Medium: To 950.0mL of cooled, sterile solution 1, aseptically add 60.0mL of sterile $Na_2S·9H_2O$ solution, 40.0mL of sterile $NaHCO_3$ solution, and 1.0mL of sterile vitamin B_{12} solution. Mix thoroughly. Adjust pH to 6.8 with sterile H_2SO_4 or Na_2CO_3. Aseptically distribute into sterile 50.0mL or 100.0mL bottles with metal screw-caps and rubber seals. Completely fill bottles with medium except for a pea-sized air bubble.

Use: For the isolation and cultivation of marine members of the Chlorobiaceae.

Marine Chromatiaceae Medium 2

Composition per 1051mL:

Solution 1	950.0mL
$Na_2S·9H_2O$ solution	60.0mL
$NaHCO_3$ solution	40.0mL
Vitamin B_{12} solution	1.0mL

pH 7.3± 0.2 at 25°C

Solution 1:

Composition per 950mL:

NaCl	20.0g
$MgSO_4·7H_2O$	3.0g
KH_2PO_4	1.0g
NH_4Cl	0.5g
$CaCl_2·2H_2O$	0.05g
Trace element solution SL-8	1.0mL

Preparation of Solution 1: Add components to distilled/deionized water and bring volume to 950.0mL. Mix thoroughly. Autoclave for 15 min at 15 psi pressure–121°C. Cool to 45°–50°C.

Trace Element Solution SL-8:

Composition per liter:

Disodium EDTA	5.2g
$FeCl_2·4H_2O$	1.5g
$CoCl_2·6H_2O$	0.19g
$MnCl_2·4H_2O$	0.10g
$ZnCl_2$	0.07g
H_3BO_3	0.06g
$NaMoO_4·2H_2O$	0.04g
$CuCl_2·2H_2O$	0.02g
$NiCl_2·6H_20$	0.02g

Preparation of Trace Element Solution SL-8: Add components to distilled/deionized water and bring volume to 1.0L. Mix thoroughly.

$Na_2S·9H_2O$ Solution:

Composition per 100mL:

$Na_2S·9H_2O$	5.0g

Preparation of $Na_2S·9H_2O$ Solution: Add $Na_2S·9H_2O$ to distilled/deionized water and bring volume to 100.0mL. Autoclave for 15 min at 15 psi pressure–121°C. Cool to 45°–50°C.

$NaHCO_3$ Solution:

Composition per 100mL:

$NaHCO_3$	5.0g

Preparation of $NaHCO_3$ Solution: Add $NaHCO_3$ to distilled/deionized water and bring volume to 100.0mL. Mix thoroughly. Filter sterilize.

Vitamin B_{12} Solution:

Composition per 100mL:

Vitamin B_{12}	2.0mg

Preparation of Vitamin B_{12} Solution: Add vitamin B_{12} to distilled/deionized water and bring volume to 100.0mL. Mix thoroughly. Filter sterilize.

Preparation of Medium: To 950.0mL of cooled, sterile solution 1, aseptically add 60.0mL of sterile $Na_2S·9H_2O$ solution, 40.0mL of sterile $NaHCO_3$ solution, and 1.0mL of sterile vitamin B_{12} solution. Mix thoroughly. Adjust pH to 7.3 with sterile H_2SO_4 or Na_2CO_3. Aseptically distribute into sterile 50.0mL or 100.0mL bottles with metal screw-caps and rubber seals. Completely fill bottles with medium except for a pea-sized air bubble.

Use: For the isolation and cultivation of marine members of the Chromatiaceae.

Marine *Cytophaga* Agar

Composition per liter:

Agar	15.0g
Nutrient broth	8.0g
Yeast extract	5.0g
Salt solution	1.0L

Salt Solution:

Composition per liter:

NaCl	12.86g
$MgCl_2$	2.48g
KCl	0.75g
$CaCl_2$	0.56g
$Fe(SO_4)_2(NH_4)_2$	0.048g

Preparation of Salt Solution: Add components to distilled/deionized water and bring volume to 1.0L. Mix thoroughly.

Preparation of Medium: Add solid components to 1.0L of salt solution. Mix thoroughly. Gently heat while stirring and bring to boiling. Distribute into tubes or flasks. Autoclave for 15 min at 15 psi pressure–121°C. Pour into sterile Petri dishes or leave in tubes.

Use: For the cultivation and maintenance of *Cytophaga* species.

Marine Glucose Trypticase™ Yeast Extract Agar (MGTY Agar)

Composition per liter:

Agar	8.0g
Glucose	2.0g
Pancreatic digest of casein	1.0g
Yeast extract	1.0g
L-Cysteine·HCl·H_2O	0.5g
Seawater	750.0mL
Tris-HCl buffer (5.0 mM, pH 7.5)	50.0mL
Resazurin (0.1% solution)	1.0mL

pH 7.5 ± 0.2 at 25°C

Preparation of Medium: Add components to distilled/deionized water and bring volume to 1.0L. Mix thoroughly. Gently heat while stirring and bring to boiling. Distribute into tubes or flasks under 97% N_2 + 3% H_2. Cap with rubber stoppers and place tubes in a press. Autoclave for 15 min at 15 psi pressure–121°C with fast exhaust.

Use: For the cultivation and maintenance of *Spirochaeta isovalerica*.

Marine Glucose Trypticase™ Yeast Extract Broth (MGTY Broth)

Composition per liter:

Glucose	2.0g
Pancreatic digest of casein	1.0g
Yeast extract	1.0g
L-Cysteine·HCl·H_2O	0.5g
Seawater	750.0mL
Tris-HCl buffer (5.0 mM, pH 7.5)	50.0mL
Resazurin (0.1% solution)	1.0mL

pH 7.5 ± 0.2 at 25°C

Preparation of Medium: Add components to distilled/deionized water and bring volume to 1.0L. Mix thoroughly. Gently heat while stirring and bring to boiling. Distribute into tubes or flasks under 97% N_2 + 3% H_2. Cap with rubber stoppers and place tubes in a press. Autoclave for 15 min at 15 psi pressure–121°C with fast exhaust.

Use: For the cultivation and maintenance of *Spirochaeta isovalerica.*

Marine Methanol Medium

Composition per liter:

NaCl	20.0g
$(NH_4)_2SO_4$	2.0g
K_2HPO_4	2.0g
KH_2PO_4	1.0g
$MgSO_4 \cdot 7H_2O$	0.3g
Methanol	10.0mL
Vitamin B_{12} solution	10.0mL
Trace metals solution	1.0mL

pH 7.0 ± 0.2 at 25°C

Vitamin B_{12} Solution:

Composition per 100mL:

Vitamin B_{12}	10.0μg

Preparation of Vitamin B_{12} Solution: Add the vitamin B_{12} to distilled/deionized water and bring volume to 100.0mL. Adjust pH to 5. Autoclave for 15 min at 15 psi pressure–121°C.

Trace Metals Solution:

Composition per liter:

$ZnSO_4 \cdot 7H_2O$	1.4g
$MnSO_4 \cdot H_2O$	0.84g
$FeSO_4 \cdot 7H_2O$	0.28g
$CuSO_4 \cdot 5H_2O$	0.25g
$Na_2MoO_4 \cdot 2H_2O$	0.24g
$CoCl_2 \cdot 6H_2O$	0.24g
$CaCl_2 \cdot 2H_2O$	0.15g

Preparation of Trace Metals Solution: Add components to distilled/deionized water and bring volume to 1.0L. Mix thoroughly.

Preparation of Medium: Add components, except vitamin B_{12} solution and methanol to distilled/deionized water and bring volume to 980.0mL. Adjust pH to 7.0 with NaOH. Autoclave for 15 min at 15 psi pressure–121°C. Filter sterilize methanol. Aseptically add sterile vitamin B_{12} solution and filter-sterilized methanol. Distibute into sterile tubes or flasks.

Use: For the cultivation and maintenance of *Methylophaga thalassica.*

Marine Peptone Succinate Salts Medium (PSS Medium)

Composition per liter:

Peptone	10.0g
Succinic acid	1.0g
$(NH_4)_2SO_4$	1.0g
$MgSO_4 \cdot 7H_2O$	1.0g
$FeCl_3 \cdot 6H_2O$	2.0mg
$MnSO_4 \cdot H_2O$	2.0mg
Synthetic sea water	1.0L

pH 6.8 ± 0.2 at 25°C

Synthetic Sea Water:

Composition per liter:

NaCl	27.5g
$MgCl_2$	5.0g
$MgSO_4 \cdot 7H_2O$	2.0g
KCl	1.0g
$CaCl_2$	0.5g
$FeSO_4$	1.0mg

Preparation of Synthetic Sea Water: Add components to distilled/deionized water and bring volume to 1.0L. Mix thoroughly.

Preparation of Medium: Add components to 1.0L of synthetic sea water. Mix thoroughly. Gently heat while stirring and bring to boiling. Adjust pH to 6.8 with KOH. Distribute into tubes or flasks. Autoclave for 15 min at 15 psi pressure–121°C.

Use: For the cultivation and maintenance of *Oceanospirillum beijerinckii* and *Oceanospirillum multiglobuliferum.*

Marine Peptone Yeast Medium with Magnesium Sulfate

Composition per liter:

NaCl	20.0g
Peptone	10.0g
$MgSO_4 \cdot 7H_2O$	2.0g
$(NH_4)_2SO_4$	2.0g
Yeast extract	1.0g

pH 7.0 ± 0.2 at 25°C

Preparation of Medium: Add components to distilled/deionized water and bring volume to 1.0L. Mix thoroughly. Distribute into tubes or flasks. Autoclave for 15 min at 15 psi pressure–121°C.

Use: For the cultivation and maintenance of *Oceanospirillum pusillum.*

Marine *Pseudomonas* Medium

Composition per liter:

Agar	15.0g
Nutrient broth	8.0g
Yeast extract	5.0g
Salt solution	1.0L

Salt Solution:

Composition per liter:

NaCl	12.86g
$MgCl_2$	2.48g

KCl	0.75g
$CaCl_2$	0.56g
$Fe(SO_4)_2(NH_4)_2$	0.048g

Preparation of Salt Solution: Add components to distilled/deionized water and bring volume to 1.0L. Mix thoroughly.

Preparation of Medium: Add components to 1.0L of salt solution. Mix thoroughly. Gently heat and bring to boiling. Distribute into tubes or flasks. Autoclave for 15 min at 15 psi pressure–121°C. Pour into sterile Petri dishes or leave in tubes.

Use: For the cultivation and maintenance of *Alteromonas haloplanktis.*

Marine *Rhodococcus* Medium

Composition per liter:

Yeast extract	10.0g
Malt extract	4.0g
Glucose	4.0g
Sea water	750.0mL

Preparation of Medium: Add components to distilled/deionized water and bring volume to 1.0L. Mix thoroughly. Gently heat while stirring and bring to boiling. Distribute into tubes or flasks. Autoclave for 15 min at 15 psi pressure–121°C.

Use: For the cultivation and maintenance of *Rhodococcus marinonascens.*

Marine *Rhodopseudomonas* Medium

Composition per liter:

NaCl	30.4g
Yeast extract	1.0g
Disodium succinate	1.0g
KH_2PO_4	0.5g
$MgSO_4·7H_2O$	0.4g
NH_4Cl	0.4g
$CaCl_2·2H_2O$	0.05g
Ferric citrate (0.1% solution)	5.0mL
Trace elements SL-6	1.0mL
Ethanol	0.5mL

pH 6.8 ± 0.2 at 25°C

Trace Elements Solution SL-6:
Composition per liter:

H_3BO_3	0.3g
$CoCl_2·6H_2O$	0.2g
$ZnSO_4·7H_2O$	0.10g
$MnCl_2·4H_2O$	0.03g
$Na_2MoO_4·H_2O$	0.03g
$NiCl_2·6H_2O$	0.02g
$CuCl_2·2H_2O$	0.01g

Preparation of Trace Elements Solution SL-6: Add components to distilled/deionized water and bring volume to 1.0L. Mix thoroughly. Adjust pH to 3.4.

Preparation of Medium: Add components to distilled/deionized water and bring volume to 1.0L. Mix thoroughly. Gently heat while stirring and bring to boiling. Distribute into tubes or flasks. Autoclave for 15 min at 15 psi pressure–121°C.

Use: For the cultivation and maintenance of *Rhodopseudomonas marina.*

Marine Salts Medium for *Sporosarcina halophila*

Composition per liter:

Marine salts mix	100.0g
Agar	20.0g
Yeast extract	10.0g
Proteose peptone No. 3	5.0g
Glucose	1.0g

pH 7.0 ± 0.2 at 25°C

Preparation of Medium: Add components to distilled/deionized water and bring volume to 1.0L. Mix thoroughly. Gently heat while stirring and bring to boiling. Distribute into tubes or flasks. Autoclave for 15 min at 15 psi pressure–121°C. Pour into sterile Petri dishes or leave in tubes.

Use: For the cultivation and maintenance of *Sporosarcina halophila.*

Marine Spirochete Medium

Composition per liter:

Cellobiose	2.0g
Peptone	2.0g
Yeast extract	1.0g
Sodium thioglycollate	1.0g
Sea water, charcoal-filtered	800.0mL

pH 7.5 ± 0.2 at 25°C

Preparation of Medium: Add components, except sodium thioglycollate, to glass distilled water and bring volume to 1.0L. Mix thoroughly. Bubble 100% N_2 into medium for 1.5 min. Add sodium thioglycollate. Adjust pH to 7.5 with 10*N* KOH. Distribute into tubes or flasks. Autoclave for 15 min at 15 psi pressure–121°C.

Use: For the cultivation and maintenance of *Spirochaeta bajacaliforniensis.*

Martin–Lewis Agar

Composition per liter:

Agar	12.0g
Hemoglobin	10.0g

Pancreatic digest of casein	7.5g
Selected meat peptone	7.5g
NaCl	5.0g
K_2HPO_4	4.0g
Cornstarch	1.0g
KH_2PO_4	1.0g
Supplement solution	10.0mL
VCAT inhibitor	10.0mL

pH 7.2 ± 0.22 at 25°C

Source: Available as a prepared medium from BBL Microbiology Systems.

Supplement Solution:
Composition per liter:

Glucose	100.0g
L-Cysteine·HCl	25.9g
L-Glutamine	10.0g
L-Cystine	1.1g
Adenine	1.0g
Nicotinamide adenine dinucleotide	0.25g
Vitamin B_{12}	0.1g
Thiamine pyrophosphate	0.1g
Guanine·HCl	0.03g
$Fe(NO_3)_3 \cdot 6H_2O$	0.02g
p-Aminobenzoic acid	0.013g
Thiamine·HCl	3.0mg

Source: The supplement solution IsoVitaleX® enrichment is available from BBL Microbiology Laboratories. This enrichment may be replaced by supplement VX from Difco Laboratories.

Preparation of Supplement Solution: Add components to distilled/deionized water and bring volume to 1.0L. Mix thoroughly. Filter sterilize.

VCAT Inhibitor:
Composition per 10mL:

Vancomycin	4.0mg
Colistin	7.5mg
Anisomycin	0.02g
Trimethoprim lactate	5.0mg

Preparation of VCAT Inhibitor: Add components to distilled/deionized water and bring volume to 10.0mL. Mix thoroughly. Filter sterilize.

Preparation of Medium: Add components, except supplement solution enrichment and VCAT inhibitor, to distilled/deionized water and bring volume to 980.0mL. Gently heat while stirring and bring to boiling. Autoclave for 15 min at 15 psi pressure–121°C. Cool to 45°–50°C. Aseptically add sterile supplement solution enrichment and sterile VCAT inhibitor. Mix thoroughly. Pour into sterile Petri dishes.

Use: For the isolation and cultivation of pathogenic *Neisseria* from specimens containing mixed flora of bacteria and fungi.

Martin–Lewis Agar, Enriched

Composition per liter:

Agar	12.0g
Pancreatic digest of casein	7.5g
Selected meat peptone	7.5g
NaCl	5.0g
K_2HPO_4	4.0g
Cornstarch	1.0g
KH_2PO_4	1.0g
Sarcina lutea suspension	20.0mL
Horse serum, inactivated	20.0mL
Supplement solution	10.0mL
PCAT inhibitor	10.0mL

pH 7.2 ± 0.22 at 25°C

***Sarcina lutea* Suspension:**
Composition per 20mL:

Sarcina lutea FDA 1001	10^6–10^7 cells

Preparation of *Sarcina lutea* Suspension: Aseptically wash the growth of 24-hr cultures of *Sarcina lutea* FDA 1001 cells from Thayer-Martin plates with sterile soybean casein digest broth. Standardize the suspension by adding additional sterile tryptic soy broth to yield 40% light transmission at 530nm wavelength.

Soybean Casein Digest Broth:
Composition per liter:

Pancreatic digest of casein	17.0g
NaCl	5.0g
Papaic digest of soybean meal	3.0g
K_2HPO_4	2.5g
Glucose	2.5g

pH 7.3 ± 0.2 at 25°C

Preparation of Soybean Casein Digest Broth: Add components to distilled/deionized water and bring volume to 1.0L. Mix thoroughly. Distribute into tubes or flasks. Autoclave for 15 min at 15 psi pressure–121°C.

Supplement Solution:
Composition per liter:

Glucose	100.0g
L-Cysteine·HCl	25.9g
L-Glutamine	10.0g
L-Cystine	1.1g
Adenine	1.0g
Nicotinamide adenine dinucleotide	0.25g
Vitamin B_{12}	0.1g
Thiamine pyrophosphate	0.1g
Guanine·HCl	0.03g
$Fe(NO_3)_3 \cdot 6H_2O$	0.02g
p-Aminobenzoic acid	0.013g
Thiamine·HCl	3.0mg

Source: The supplement solution IsoVitaleX® enrichment is available from BBL Microbiology Laboratories. This enrichment may be replaced by supplement VX from Difco Laboratories.

Preparation of Supplement Solution: Add components to distilled/deionized water and bring volume to 1.0L. Mix thoroughly. Filter sterilize.

PCAT Inhibitor:
Composition per 10mL:

Anisomycin	0.02g
Colistin	7.5mg
Trimethoprim lactate	5.0mg
Penicillin G	25,000U

Preparation of PCAT Inhibitor: Add components to distilled/deionized water and bring volume to 10.0mL. Mix thoroughly. Filter sterilize.

Preparation of Medium: Add components—except *Sarcina lutea* suspension, horse serum, supplement solution, and PCAT inhibitor—to distilled/deionized water and bring volume to 940.0mL. Gently heat while stirring and bring to boiling. Autoclave for 15 min at 15 psi pressure–121°C. Cool to 45°–50°C. Aseptically add 20.0mL of sterile *Sarcina lutea* suspension, 20.0mL of sterile horse serum, 10.0mL of supplement solution, and 10.0mL of sterile PCAT inhibitor. Mix thoroughly. Pour into sterile Petri dishes.

Use: For the isolation and cultivation of pathogenic *Neisseria*, especially penicillinase-producing strains, from specimens containing mixed flora of bacteria and fungi.

Maximum Recovery Diluent

Composition per liter:

NaCl	8.5g
Peptone	1.0g

pH 7.0 ± 0.2 at 25°C

Source: This medium is available as a premixed powder from Oxoid Unipath.

Preparation of Medium: Add components to distilled/deionized water and bring volume to 1.0L. Mix thoroughly. Distribute into tubes or flasks. Autoclave for 15 min at 15 psi pressure–121°C.

Use: This diluent is a physiologically isotonic and protective medium for maximal recovery of microorganisms from a variety of sources.

m-Bismuth Sulfite Broth
See: **Bismuth Sulfite Broth**

MBM Medium
See: **Methylene Blue Milk Medium**

MBM Acetate Medium (Mineral Base Medium with Acetate)

Composition per liter:

Agar	16.0g
NaCl	5.0g
K_2HPO_4	1.0g
$NH_4H_2PO_4$	1.0g
Sodium acetate·$3H_2O$	1.0g
$MgSO_4 \cdot 7H_2O$	0.1g
Bromothymol Blue	0.01g

pH 6.5 ± 0.2 at 25°C

Preparation of Medium: Add components to distilled/deionized water and bring volume to 1.0L. Mix thoroughly. Adjust pH to 6.5. Gently heat and bring to boiling. Distribute into screw-capped tubes in 3.0mL volumes. Autoclave for 15 min at 15 psi pressure–121°C. Allow tubes to cool in a slanted position.

Use: For determining the nutritional independence of bacteria. Bacteria that are nutritionally independent turn the medium blue.

m-Brilliant Green Broth
See: **Brilliant Green Broth**

McBride Agar, Modified

Composition per liter:

Agar	15.0g
Glycine anhydride	10.0g
Tryptose	10.0g
NaCl	5.0g
Beef extract	3.0g
Phenylethanol	2.5g
LiCl	0.5g
Cycloheximide solution	10.0mL

pH 7.3 ± 0.2 at 25°C

Cycloheximide Solution:
Composition per 10mL:

Cycloheximide	0.2g

Preparation of Cycloheximide Solution: Add cycloheximide to distilled/deionized water and bring volume to 10.0mL. Mix thoroughly. Filter sterilize.

Preparation of Medium: Add components, except cycloheximide solution, to distilled/deionized water and bring volume to 990.0mL. Mix thoroughly. Gently heat and bring to boiling. Autoclave for 15 min at 15 psi pressure–121°C. Cool to 45°–50°C. Aseptically add sterile cycloheximide solution. Mix thoroughly. Pour into sterile Petri dishes or distribute into sterile tubes.

Use: For the isolation of *Listeria monocytogenes* from dairy products.

McBride *Listeria* Agar

Composition per liter:

Agar	15.0g
Glycine	10.0g
Pancreatic digest of casein	5.0g
Peptic digest of animal tissue	5.0g
NaCl	5.0g
Beef extract	3.0g
Phenylethyl alcohol	2.5g
LiCl	0.5g

pH 7.3 ± 0.22 at 25°C

Source: This medium is available as a premixed powder from BBL Microbiology Systems and Difco Laboratories.

Preparation of Medium: Add components to distilled/deionized water and bring volume to 1.0L. Mix thoroughly. Gently heat while stirring and bring to boiling. Distribute into tubes or flasks. Autoclave for 15 min at 15 psi pressure–121°C. Pour into sterile Petri dishes or leave in tubes.

Use: For the selective isolation of *Listeria monocytogenes* from clinical and nonclinical specimens containing mixed flora.

McCarthy Agar

Composition per liter:

Cornstarch	10.0g
Naladixic acid	0.015g
Colistin	0.010g
GC Agar base	1.0L

pH 7.2 ± 0.2 at 25°C

GC Agar Base:
Composition per liter:

Agar	10.0g
Pancreatic digest of casein	7.5g
Peptic digest of animal tissue	7.5g
NaCl	5.0g
K_2HPO_4	4.0g
Cornstarch	1.0g
KH_2PO_4	1.0g

Preparation of GC Agar Base: Add components to distilled/deionized water and bring volume to 1.0L. Mix thoroughly.

Preparation of Medium: To 1.0L of GC Agar base add the cornstarch. Gently heat while stirring to dissolve. Add the naladixic acid and colistin. Mix thoroughly. Distribute into tubes or flasks. Autoclave for 15 min at 15 psi pressure–121°C. Pour into sterile Petri dishes or leave in tubes.

Use: For the isolation and differentiation of *Gardnerella vaginalis (Haemophilus vaginalis, Corynebacterium vaginale*) from genitourinary specimens. Bacteria which can utilize starch appear as colonies surrounded by a clear zone.

McClung Toabe Agar

Composition per liter:

Proteose peptone	40.0g
Agar	25.0g
Na_2HPO_4	5.0g
Glucose	2.0g
NaCl	2.0g
KH_2PO_4	1.0g
$MgSO_4 \cdot 7H_2O$	0.1g
Egg yolk emulsion, 50%	100.0mL

pH 7.3 ± 0.2 at 25°C

Source: This medium is available as a premixed powder from Difco Laboratories.

Egg Yolk Emulsion, 50%:
Composition per 100mL:

Chicken egg yolks	11
Whole chicken egg	1
NaCl (0.9% solution)	50.0mL

Preparation of Egg Yolk Emulsion, 50%: Soak eggs with 1:100 dilution of saturated mercuric chloride solution for 1 min. Crack eggs and separate yolks from whites. Mix egg yolks with 1 chicken egg. Measure 50.0mL of egg yolk emulsion and add to 50.0mL of 0.9% NaCl solution. Mix thoroughly. Filter sterilize. Warm to 45°–50°C.

Preparation of Medium: Add components, except egg yolk emulsion, 50%, to distilled/deionized water and bring volume to 900.0mL. Mix thoroughly. Gently heat while stirring and bring to boiling. Autoclave for 15 min at 15 psi pressure–121°C. Cool to 50°–55°C. Aseptically add 100.0mL of sterile egg yolk emulsion, 50%. Mix thoroughly. Pour into sterile Petri dishes in 15.0mL volumes.

Use: For the isolation and cultivation of *Clostridium perfringens* in foods.

McClung–Toabe Agar, Modified

Composition per liter:

Proteose peptone No. 2	40.0g
Agar	20.0g
Na_2HPO_4	5.0g
Glucose	2.0g
NaCl	2.0g
KH_2PO_4	1.0g
$MgSO_4 \cdot 7H_2O$	0.1g

Egg yolk emulsion, 50%	100.0mL
Hemin solution	1.0mL

pH 7.6 ± 0.2 at 25°C

Egg Yolk Emulsion, 50%:
Composition per 100mL:

Chicken egg yolks	11
Whole chicken egg	1
NaCl (0.9% solution)	50.0mL

Preparation of Egg Yolk Emulsion, 50%: Soak eggs with 1:100 dilution of saturated mercuric chloride solution for 1 min. Crack eggs and separate yolks from whites. Mix egg yolks with 1 chicken egg. Measure 50.0mL of egg yolk emulsion and add to 50.0mL of 0.9% NaCl solution. Mix thoroughly. Filter sterilize. Warm to 45°–50°C.

Hemin Solution:
Composition per 100mL:

Hemin	0.5g
NaOH (1*N* solution)	20.0mL

Preparation of Hemin Solution: Add hemin to 20.0mL of 1*N* NaOH solution. Mix thoroughly. Bring volume to 100.0mL with distilled/deionized water.

Preparation of Medium: Add components, except egg yolk emulsion, 50%, to distilled/deionized water and bring volume to 900.0mL. Mix thoroughly. Gently heat while stirring and bring to boiling. Autoclave for 15 min at 15 psi pressure–121°C. Cool to 50°–55°C. Aseptically add 100.0mL of sterile egg yolk emulsion, 50%. Mix thoroughly. Pour into sterile Petri dishes in 20.0mL volumes.

Use: For the cultivation of a wide variety of anaerobic bacteria. For the differentiation of anaerobic bacteria based on lecithinase production and lipase production. Bacteria that produce lecithinase appear as colonies surrounded by a zone of insoluble precipitate. Bacteria that produce lipase appear as colonies with a pearly iridescent sheen.

McClung Carbon–Free Broth

Composition per liter:

$NaNO_3$	2.0g
K_2HPO_4	0.8g
$MgSO_4 \cdot 7H_2O$	0.5g
$FeCl_3$	0.010g
$MnCl_2 \cdot 4H_2O$	8.0mg
$ZnSO_4$	2.0mg

pH 7.2 ± 0.2 at 25°C

Preparation of Medium: Add components to distilled/deionized water and bring volume to 1.0L. Mix thoroughly. Gently heat without boiling until salts dissolve. Cool to 25°C. Adjust pH to 7.2. Filter sterilize.

Use: For use as a basal medium in determining the carbon assimilation capabilities of microorganisms.

McClung–Toabe Agar, Modified

Composition per liter:

Proteose peptone No. 2	40.0g
Agar	20.0g
Na_2HPO_4	5.0g
Glucose	2.0g
NaCl	2.0g
KH_2PO_4	1.0g
$MgSO_4 \cdot 7H_2O$	0.1g
Neomycin	0.1g
Egg yolk emulsion, 50%	100.0mL
Hemin solution	1.0mL

pH 7.6 ± 0.2 at 25°C

Egg Yolk Emulsion, 50%:
Composition per 100mL:

Chicken egg yolks	11
Whole chicken egg	1
NaCl (0.9% solution)	50.0mL

Preparation of Egg Yolk Emulsion, 50%: Soak eggs with 1:100 dilution of saturated mercuric chloride solution for 1 min. Crack eggs and separate yolks from whites. Mix egg yolks with 1 chicken egg. Measure 50.0mL of egg yolk emulsion and add to 50.0mL of 0.9% NaCl solution. Mix thoroughly. Filter sterilize. Warm to 45°–50°C.

Hemin Solution:
Composition per 100mL:

Hemin	0.5g
NaOH (1*N* solution)	20.0mL

Preparation of Hemin Solution: Add hemin to 20.0mL of 1*N* NaOH solution. Mix thoroughly. Bring volume to 100.0mL with distilled/deionized water.

Preparation of Medium: Add components, except egg yolk emulsion, 50%, to distilled/deionized water and bring volume to 900.0mL. Mix thoroughly. Gently heat while stirring and bring to boiling. Autoclave for 15 min at 15 psi pressure–121°C. Cool to 50°–55°C. Aseptically add 100.0mL of sterile egg yolk emulsion, 50%. Mix thoroughly. Pour into sterile Petri dishes in 20.0mL volumes.

Use: For the cultivation of *Clostridium* species. For the differentiation of *Clostridium* species based on lecithinase production and lipase production. Bacteria that produce lecithinase appear as colonies surrounded by a zone of insoluble precipitate. Bacteria that produce lipase appear as colonies with a pearly iridescent sheen.

McClung–Toabe Agar, Modified

Composition per liter:

Proteose peptone No. 2	20.0g
Agar	20.0g
Yeast extract	5.0g
Pancreatic digest of casein	5.0g
NaCl	5.0g
Sodium thioglycollate	1.0g
Egg yolk emulsion, 50%	80.0mL

pH 7.6 ± 0.2 at 25°C

Egg Yolk Emulsion, 50%:
Composition per 100mL:

Chicken egg yolks	11
Whole chicken egg	1
NaCl (0.9% solution)	50.0mL

Preparation of Egg Yolk Emulsion, 50%: Soak eggs with 1:100 dilution of saturated mercuric chloride solution for 1 min. Crack eggs and separate yolks from whites. Mix egg yolks with 1 chicken egg. Measure 50.0mL of egg yolk emulsion and add to 50.0mL of 0.9% NaCl solution. Mix thoroughly. Filter sterilize. Warm to 45°–50°C.

Preparation of Medium: Add components, except egg yolk emulsion, 50%, to distilled/deionized water and bring volume to 920.0mL. Mix thoroughly. Gently heat while stirring and bring to boiling. Autoclave for 15 min at 15 psi pressure–121°C. Cool to 50°–55°C. Aseptically add 80.0mL of sterile egg yolk emulsion, 50%. Mix thoroughly. Pour into sterile Petri dishes in 20.0mL volumes.

Use: For the cultivation of *Clostridium botulinum.*

McClung–Toabe Egg Yolk Agar, CDC Modified (CDC Modified McClung–Toabe Egg Yolk Agar)

Composition per liter:

Pancreatic digest of casein	40.0g
Agar	25.0g
$NaHPO_4$	5.0g
Yeast extract	5.0g
D-Glucose	2.0g
NaCl	2.0g
Egg yolk emulsion	100.0mL
$MgSO_4$ (5% solution)	0.2mL

pH 7.4 ± 0.2 at 25°C

Egg Yolk Emulsion:
Composition:

Chicken egg yolks	11
Whole chicken egg	1

Preparation of Egg Yolk Emulsion: Soak eggs with 1:100 dilution of saturated mercuric chloride solution for 1 min. Crack eggs. Separate yolks from whites for 11 eggs. Mix egg yolks with 1 chicken egg.

Preparation of Medium: Add components, except egg yolk emulsion, to distilled/deionized water and bring volume to 900.0mL. Mix thoroughly. Gently heat while stirring and bring to boiling. Autoclave for 15 min at 15 psi pressure–121°C. Cool to 60°C. Aseptically add 100.0mL of sterile egg yolk emulsion. Mix thoroughly. Pour into sterile Petri dishes in 20.0mL volumes.

Use: For the isolation, cultivation, and differentiation of anaerobic bacteria from foods. Bacteria that produce lecithinase appear as colonies surrounded by an insoluble opaque precipitate. Bacteria that produce lipase activity appear as colonies with a sheen or "pearly" surface. Bacteria that possess proteolytic activity appear as colonies surrounded by a clear zone.

McClung–Toabe Egg Yolk Agar, CDC Modified (CDC Modified McClung–Toabe Egg Yolk Agar)

Composition per liter:

Pancreatic digest of casein	40.0g
Agar	25.0g
Na_2HPO_4	5.0g
Yeast extract	5.0g
Glucose	2.0g
NaCl	2.0g
KH_2PO_4	1.0g
Egg yolk emulsion, 50%	100.0mL
$MgSO_4 \cdot 7H_2O$ (5% solution)	0.2mL

pH 7.3 ± 0.2 at 25°C

Egg Yolk Emulsion, 50%:
Composition per 100mL:

Chicken egg yolks	11
Whole chicken egg	1
NaCl (0.9% solution)	50.0mL

Preparation of Egg Yolk Emulsion, 50%: Soak eggs with 1:100 dilution of saturated mercuric chloride solution for 1 min. Crack eggs and separate yolks from whites. Mix egg yolks with 1 chicken egg. Measure 50.0mL of egg yolk emulsion and add to 50.0mL of 0.9% NaCl solution. Mix thoroughly. Filter sterilize. Warm to 45°–50°C.

Preparation of Medium: Add components—except egg yolk emulsion, 50%—to distilled/deionized water and bring volume to 900.0mL. Mix thoroughly.

Gently heat while stirring and bring to boiling. Autoclave for 15 min at 15 psi pressure–121°C. Cool to 50°–55°C. Aseptically add 100.0mL of sterile egg yolk emulsion, 50%. Mix thoroughly. Pour into sterile Petri dishes in 15.0mL volumes.

Use: For the cultivation of a wide variety of anaerobic bacteria. For the differentiation of anaerobic bacteria based on lecithinase production, lipase production and proteolytic ability. Bacteria that produce lecithinase appear as colonies surrounded by a zone of insoluble precipitate. Bacteria that produce lipase appear as colonies with a pearly iridescent sheen. Bacteria that produce proteolytic activity appear as colonies surrounded by a clear zone.

MD 1 Medium

Composition per liter:

Pancreatic digest of casein	3.0g
$MgSO_4 \cdot 7H_2O$	2.0g
$CaCl_2$	0.5g
Trace element solution	1.0mL
Vitamin B_{12} solution	1.0mL

Trace Element Solution:
Composition per liter:

EDTA	8.0g
$MnCl_2 \cdot 4H_2O$	0.1g
$CoCl_2$	0.02g
KBr	0.02g
$ZnCl_2$	0.02g
$CuSO_4$	0.01g
H_3BO_3	0.01g
$NaMoO_4 \cdot 2H_2O$	0.01g
$BaCl_2$	5.0mg
LiCl	5.0mg
$SnCl_2 \cdot 2H_2O$	5.0mg

Preparation of Trace Element Solution: Add components to distilled/deionized water and bring volume to 1.0L. Mix thoroughly.

Vitamin B_{12} Solution:
Composition per 10mL:

Vitamin B_{12}	5.0mg

Preparation of Vitamin B_{12} Solution: Add vitamin B_{12} to distilled/deionized water and bring volume to 10.0mL. Mix thoroughly.

Preparation of Medium: Add components to distilled/deionized water and bring volume to 1.0L. Mix thoroughly. Distribute into tubes or flasks. Autoclave for 15 min at 15 psi pressure–121°C.

Use: For the cultivation of myxobacteria.

MD Medium

Composition per liter:

Agar	20.0g
L-Malic acid	20.0g
Pancreatic digest of casein	10.0g
D-Glucose	5.0g
Casamino acids	3.0g
Pancreatic digest of soybean meal	1.5g
Tween™ 80	1.0g
Yeast extract	1.0g
Bromcresol Green solution	20.0mL

pH 7.0 ± 0.2 at 25°C

Bromcresol Green Solution:
Composition per 30mL:

Bromcresol Green	0.1g
NaOH (0.01*N* solution)	30.0mL

Preparation of Bromcresol Green Solution: Add Bromcresol Green to 30.0mL of NaOH solution. Mix thoroughly. Filter sterilize.

Preparation of Medium: Add components to distilled/deionized water and bring volume to 1.0L. Mix thoroughly. Gently heat and bring to boiling. Distribute into tubes or flasks. Adjust pH to 7.0 with 10*N* KOH. Autoclave for 15 min at 15 psi pressure–121°C. Pour into sterile Petri dishes or leave in tubes.

Use: For the isolation and cultivation of *Salmonella* species from foods.

m–Dextrose Tryptone Broth
See: **Dextrose Tryptone Broth**

m–E Agar
See: **E Agar**

MEA
See: **Malt Extract Agar**

Meat Extract with Peptone and 1.5% Salt

Composition per liter:

NaCl	15.0g
Peptone	10.0g
Meat extract	3.0g

Preparation of Medium: Add components to distilled/deionized water and bring volume to 1.0L. Mix thoroughly. Distribute into tubes or flasks. Autoclave for 15 min at 15 psi pressure–121°C.

Use: For the cultivation and maintenance of *Alcaligenes* species.

Medium 2A

Composition per liter:

Arginine	10.0g
NaCl	5.0g
Agar	4.0g
Peptone	1.0g
$K_2HPO_4 \cdot 3H_2O$	0.3g
Phenol Red	0.01g

pH 7.2–7.4 at 25°C

Preparation of Medium: Add components to distilled/deionized water and bring volume to 1.0L. Mix thoroughly. Gently heat and bring to boiling. Distribute into tubes. Autoclave for 15 min at 15 psi pressure–121°C.

Use: For the cultivation and differentiation of *Pseudomonas* species based on their production of arginine dihydrolase activity.

MED IIa

Composition per liter:

Tris buffer stock solution	10.0mL
$CaCl_2$ (5.0% solution)	10.0mL
$MgSO_4 \cdot 7H_2O$ (3.33% solution)	1.0mL

pH 7.2 ± 0.2 at 25°C

Tris Buffer Stock Solution:

Composition per 500mL:

Tris(hydroxymethyl)aminomethane·HCl	35.01g
Tris(hydroxymethyl)aminomethane	3.35g

Preparation of Tris Buffer Stock Solution: Add components to distilled/deionized water and bring volume to 500.0mL. Mix thoroughly. Adjust pH to 7.2.

Preparation of Medium: Add components to distilled/deionized water and bring volume to 1.0L. Mix thoroughly. Distribute into tubes or flasks. Autoclave for 20 min at 15 psi pressure–121°C.

Use: For the cultivation and maintenance of *Vampirovibrio chlorellavorus.*

Medium 4 m 1

Composition per liter:

Agar	15.0g
Peptone	3.0g
Pancreatic digest of casein	3.0g
Yeast extract	3.0g
Maltose	2.0g
Lactose	1.0g
Sodium dichromate solution	100.0mL

pH 7.0 ± 0.2 at 25°C

Sodium Dichromate Solution:

Composition per 100mL:

Sodium dichromate	0.05g

Preparation of Sodium Dichromate Solution: Add sodium dichromate to distilled/deionized water and bring volume to 100.0mL. Mix thoroughly. Autoclave for 15 min at 15 psi pressure–121°C. Cool to 45°–50°C.

Preparation of Medium: Add components, except sodium dichromate solution, to distilled/deionized water and bring volume to 900.0mL. Mix thoroughly. Gently heat and bring to boiling. Adjust pH to 7.2. Autoclave for 15 min at 15 psi pressure–121°C. Cool to 45°–50°C. Aseptically add sterile sodium dichromate solution. Mix thoroughly. Pour into sterile Petri dishes or distribute into sterile tubes.

Use: For the isolation and cultivation of *Corynebacterium sepedonicum.*

Medium 2508-85-1 with Amino Acids

Composition per liter:

Agar	20.0g
Nutrient broth	8.0g
D-Glucose	5.0g
Polypeptone™	5.0g
Yeast extract	5.0g
L-Lysine	0.1g
L-Methionine	0.05g
Diaminopimelic acid	0.05g

Preparation of Medium: Add components to distilled/deionized water and bring volume to 1.0L. Mix thoroughly. Gently heat and bring to boiling. Distribute into tubes or flasks. Autoclave for 30 min at 15 psi pressure–121°C. Pour into sterile Petri dishes or leave in tubes.

Use: For the cultivation and maintenance of *Escherichia coli.*

Medium A

Composition per liter:

D-Glucose	20.0g
Agar	20.0g
Yeast extract	10.0g
Biotin	1.0mg
Calcium pantothenate	1.0mg

pH 7.3 ± 0.2 at 25°C

Preparation of Medium: Add components, except biotin and calcium pantothenate, to distilled/deionized water and bring volume to 990.0mL. Mix thoroughly. Gently heat and bring to boiling. Autoclave for 15 min at 15 psi pressure–121°C. Cool to 45°–50°C. Add biotin and calcium pantothenate to distilled deionized water and bring volume to 10.0mL. Mix thoroughly. Filter sterilize. Aseptically add the sterile biotin and calcium pantothenate solution to the cooled sterile basal medium. Mix thoroughly. Pour into sterile Petri dishes or distribute into sterile tubes.

Use: For the cultivation and maintenance of *Zymomonas mobilis*.

Medium A for Producing Lysates

Composition per liter:

Nutrient broth	8.0g
KCl	1.0g
$MgSO_4.7H_2O$	0.25g
$MnCl_2$	1.25mg
$FeSO_4$ (1.0mM solution)	1.0mL
$Ca(NO_3)_2$ (1.0M solution)	1.0mL

pH 7.0–7.2 at 25°C

Preparation of Medium: Add components, except $FeSO_4$ and $Ca(NO_3)_2$, to distilled/deionized water and bring volume to 998.0mL. Mix thoroughly. Adjust pH to 7.0–7.2. Autoclave for 30 min at 15 psi pressure–115°C. Cool to 45°–50°C. Prepare 1.0mM $FeSO_4$ solution and 1.0M $Ca(NO_3)_2$ solution separately. Filter sterilize both solutions. Aseptically add the sterile $FeSO_4$ solution and sterile $Ca(NO_3)_2$ solution to the cooled sterile basal medium. Mix thoroughly. Distribute into sterile tubes or flasks.

Use: For the cultivation of microorganisms to be lysed.

Medium for *Acetivibrio cellulolyticus*
See: BC Medium

Medium AS4

Composition per liter:

Sucrose	80.0g
PPLO broth without Crystal Violet	500.0mL
Horse serum	200.0mL
Phenol Red (0.5% solution)	5.0mL

pH 7.2 ± 0.2 at 25°C

PPLO Broth without Crystal Violet:
Composition per 500mL:

Beef heart (solids from infusion)	11.53g
Peptone	2.33g
NaCl	1.15g

Source: PPLO broth without Crystal Violet is available as a premixed powder from Difco Laboratories.

Preparation of PPLO Broth without Crystal Violet: Add components to distilled/deionized water and bring volume to 500.0mL.Mix thoroughly. Beef heart for infusion may be substituted; 100g of beef heart for infusion are equivalent to 500g of fresh heart tissue.

Preparation of Medium: Add components, except horse serum, to distilled/deionized water and bring volume to 800.0mL. Mix thoroughly. Adjust pH to 7.2. Autoclave for 10 min at 15 psi pressure–121°C. Cool to 45°–50°C. Aseptically add 200.0mL of non-inactivated, sterile horse serum. Mix thoroughly. Aseptically distribute into sterile tubes or flasks.

Use: For the cultivation and maintenance of *Spiroplasma melliferum.*

Medium B for Sulfate Reducers (Postgate's Medium B for Sulfate Reducers)

Composition per liter:

Sodium lactate	3.5g
$MgSO_4{\cdot}7H_2O$	2.0g
NH_4Cl	1.0g
$CaSO_4$	1.0g
Yeast extract	1.0g
KH_2PO_4	0.5g
$FeSO_4{\cdot}7H_2O$	0.5g
Ascorbic acid	0.1g
Thioglycollic acid	0.1g

pH 7.0–7.5 at 25°C

Preparation of Medium: Add components, except ascorbic acid and thioglycollic acid, to tap water and bring volume to 1.0L. For marine bacteria, NaCl may be added or sea water used in place of tap water. Mix thoroughly. Adjust pH to 7.0–7.5. Thioglycollate and ascorbate should be added immediately prior to sterilization. Distribute into tubes or flasks. Autoclave for 15 min at 15 psi pressure–121°C.

Use: For isolation, cultivation and maintenance of *Desulfovibrio* species and *Desulfotomaculum* species. This medium turns black as a result of H_2S production due to bacterial growth.

Medium BG11 for Cyanobacteria
See: BG11 Agar and BG11 Medium

Medium BG11 for Marine Cyanobacteria
See: **BG11 Marine Agar and BG11 Marine Broth**

Medium C for Sulfate Reducers (Postgate's Medium C for Sulfate Reducers)

Composition per liter:

Sodium lactate	6.0g
Na_2SO_4	4.5g
NH_4Cl	1.0g
Yeast extract	1.0g
KH_2PO_4	0.5g
Sodium citrate·$2H_2O$	0.3g
$CaCl_2·6H_2O$	0.06g
$MgSO_4·7H_2O$	0.06g
$FeSO_4·7H_2O$	0.004g

pH 7.5 ± 0.2 at 25°C

Preparation of Medium: Add components to distilled/deionized water and bring volume to 1.0L. For marine bacteria, NaCl may be added or sea water used in place of distilled/deionized water. Mix thoroughly. Adjust pH to 7.5. Distribute into tubes or flasks. Autoclave for 15 min at 15 psi pressure–121°C.

Use: For detection, culturing and storage of *Desulfovibrio* species and many *Desulfotomaculum* species. This medium should be used when a clear culture medium is desired such as for chemostat culture. This medium may be cloudy after sterilization but usually clears on cooling. It turns black as a result of H_2S production due to bacterial growth.

Medium D
See: **Castenholtz D Medium**

Medium D2

Composition per liter:

Agar	15.0g
Glucose	10.0g
LiCl	5.0g
Pancreatic digest of casein	4.0g
Yeast extract	2.0g
Tris(hydroxymethyl)amino-methane·HCl buffer	1.2g
NH_4Cl	1.0g
$MgSO_4·7H_2O$	0.3g
Polymyxin sulfate solution	10.0mL
NaN_3 solution	10.0mL

pH 6.9 ± 0.2 at 25°C

Polymyxin Sulfate Solution:
Composition per 10mL:

Polymyxin sulfate	0.04g

Preparation of Polymyxin Sulfate Solution: Add polymyxin sulfate to distilled/deionized water and bring volume to 10.0mL. Mix thoroughly. Filter sterilize. Use freshly prepared solution.

NaN_3 Solution:
Composition per 10mL:

NaN_3	2.0mg

Preparation of NaN_3 Solution: Add NaN_3 to distilled/deionized water and bring volume to 10.0mL. Mix thoroughly. Filter sterilize. Use freshly prepared solution.

Caution: Sodium azide is toxic. Azides also react with metals and disposal must be highly diluted.

Preparation of Medium: Add components, except polymyxin sulfate solution and NaN_3 solution, to distilled/deionized water and bring volume to 980.0mL. Mix thoroughly. Gently heat and bring to boiling. Autoclave for 15 min at 15 psi pressure–121°C. Cool to 45°–50°C. Aseptically add sterile polymyxin sulfate solution and NaN_3 solution. Mix thoroughly. Pour into sterile Petri dishes or distribute into sterile tubes.

Use: For the selective isolation and cultivation of *Corynebacterium* species.

Medium D4

Composition per liter:

Agar	15.0g
Sucrose	10.0g
NH_4Cl	5.0g
Na_2HPO_4, anhydrous	2.3g
Pancreatic digest of casein	1.0g
Sodium dodecyl sulfate	0.6g
Glycerol	10.0mL

Preparation of Medium: Add components to distilled/deionized water and bring volume to 1.0L. Mix thoroughly. Gently heat and bring to boiling. Distribute into tubes or flasks. Autoclave for 15 min at 15 psi pressure–121°C. Pour into sterile Petri dishes or leave in tubes.

Use: For the selective isolation and cultivation of *Pseudomonas syringae*.

Medium D, Modified
See: **Castenholtz D Medium, Modified**

Medium D for Sulfate Reducers (Postgate's Medium D for Sulfate Reducers)

Composition per liter:

Sodium pyruvate	3.5g
$MgCl_2 \cdot 6H_2O$	1.6g
NH_4Cl	1.0g
Yeast extract	1.0g
KH_2PO_4	0.5g
$CaCl_2 \cdot 2H_2O$	0.1g
$FeSO_4 \cdot 7H_2O$	0.004g

pH 7.5 ± 0.2 at 25°C

Preparation of Medium: Add components to distilled/deionized water and bring volume to 1.0L. Malate or fumarate may also be used as a carbon source. For marine bacteria, NaCl may be added or sea water used in place of distilled/deionized water. Mix thoroughly. Adjust pH to 7.5. Filter sterilize. Aseptically distribute into sterile tubes or flasks.

Use: For cultivation of *Desulfovibrio* species and *Desulfotomaculum* species that can grow in the absence of sulfate.

Medium D for Sulfate Reducers (Postgate's Medium D for Sulfate Reducers)

Composition per liter:

$MgCl_2 \cdot 6H_2O$	1.6g
Choline chloride	1.0g
NH_4Cl	1.0g
Yeast extract	1.0g
KH_2PO_4	0.5g
$CaCl_2 \cdot 2H_2O$	0.1g
$FeSO_4 \cdot 7H_2O$	0.004g

pH 7.5 ± 0.2 at 25°C

Preparation of Medium: Add components to distilled/deionized water and bring volume to 1.0L. Malate or fumarate may also be used as a carbon source. For marine bacteria, NaCl may be added or sea water used in place of distilled/deionized water. Mix thoroughly. Adjust pH to 7.5. Filter sterilize. Aseptically distribute into sterile tubes or flasks.

Use: For cultivation of *Desulfovibrio* species and *Desulfotomaculum* species that can grow in the absence of sulfate.

Medium D for *Thermus*

Composition per liter:

Pancreatic digest of casein	1.0g
Yeast extract	1.0g
$NaNO_3$	0.7g
KNO_3	0.1g
$MgSO_4 \cdot 7H_2O$	0.1g
Na_2HPO_4	0.1g
Nitrilotriacetic acid	0.1g
$CaSO_4 \cdot 2H_2O$	0.06g
NaCl	8.0mg
$MnSO_4 \cdot H_2O$	2.2mg
$ZnSO_4 \cdot 7H_2O$	0.5mg
H_3BO_3	0.5mg
$FeCl_3$	0.28mg
$Na_2MoO_4 \cdot 2H_2O$	0.03mg
$CuSO_4$	0.02mg

pH 8.2 ± 0.2 at 25°C

Preparation of Medium: Add nitrilotriacetic acid to 500.0mL of distilled/deionized water. Dissolve by adjusting pH to 6.5 with KOH. Add remaining components. Readjust pH to 8.2 with H_2SO_4 or KOH. Add distilled/deionized water to 1.0L. Distribute into tubes or flasks. Autoclave for 15 min at 15 psi pressure–121°C.

Use: For the cultivation of *Thermus* species.

Medium DG
See: **Castenholtz DG Medium**

Medium DGN
See: **Castenholtz DGN Medium**

Medium E for *Bacillus*

Composition per liter:

NaCl	50.0g
K_2HPO_4	10.6g
Sucrose	10.0g
KH_2PO_4	5.3g
$(NH_4)_2SO_4$	1.0g
$MgSO_4$	0.25g
Trace salts solution	10.0mL

Trace Salts Solution:

Composition per liter:

$MnSO_4 \cdot H_2O$	3.0g
Disodium EDTA	1.0g
$FeSO_4 \cdot 7H_2O$	0.1g
$CaCl_2 \cdot 2H_2O$	0.1g
$CoCl_2 \cdot 6H_2O$	0.1g
$ZnSO_4 \cdot 7H_2O$	0.1g
$CuSO_4 \cdot 5H_2O$	0.01g

$AlK(SO_4)_2 \cdot 12H_2O$ 0.01g
H_3BO_3 .... 0.01g
$Na_2MoO_4 \cdot 2H_2O$ 0.01g

Preparation of Trace Salts Solution: Add components to distilled/deionized water and bring volume to 1.0L. Mix thoroughly.

Preparation of Medium: Add components to distilled/deionized water and bring volume to 1.0L. Mix thoroughly. Autoclave for 15 min at 15 psi pressure–121°C. Aseptically distribute into sterile tubes or flasks.

Use: For the cultivation and maintenance of *Bacillus* species.

Medium E for Sulfate Reducers (Postgate's Medium E for Sulfate Reducers)

Composition per liter:

Agar 15.0g
Sodium lactate 3.5g
$MgCl_2 \cdot 6H_2O$ 2.0g
NH_4Cl 1.0g
Na_2SO_4 .... 1.0g
$CaCl_2 \cdot 2H_2O$ 1.0g
Yeast extract 1.0g
KH_2PO_4 .... 0.5g
Ascorbic acid 0.1g
Thioglycollic acid 0.1g
$FeSO_4 \cdot 7H_2O$ 0.004g

pH 7.6 ± 0.2 at 25°C

Preparation of Medium: Add components, except ascorbic acid and thioglycollic acid, to tap water and bring volume to 1.0L. For marine bacteria, NaCl may be added or sea water used in place of tap water. Mix thoroughly. Gently heat and bring to boiling. Adjust pH to 7.6. Thioglycollate and ascorbate should be added immediately prior to sterilization. Distribute into screw-capped tubes or flasks. Autoclave for 15 min at 15 psi pressure–121°C.

Use: For the cultivation and enumeration of *Desulfovibrio* species and *Desulfotomaculum* species as black colonies in deep agar cultures. Also used for isolation of pure cultures of of *Desulfovibrio* species and *Desulfotomaculum* species.

Medium F

Composition per liter:

$MgSO_4 \cdot 7H_2O$ 0.5g
$(NH_4)_2SO_4$ 0.15g
KCl 0.05g
KH_2PO_4 .... 0.05g
$Ca(NO_3)_2$ 0.01g
$FeSO_4 \cdot 7H_2O$ solution 10.0mL

pH 3.5 ± 0.2 at 25°C

$FeSO_4 \cdot 7H_2O$ Solution:

Composition per 10mL:

$FeSO_4 \cdot 7H_2O$ 1.0g

Preparation of $FeSO_4 \cdot 7H_2O$ Solution: Add the $FeSO_4 \cdot 7H_2O$ to distilled/deionized water and bring volume to 10.0mL. Mix thoroughly. Filter sterilize.

Preparation of Medium: Add components, except $FeSO_4 \cdot 7H_2O$ solution, to tap water and bring volume to 990.0mL. Mix thoroughly. Gently heat until dissolved. Adjust pH to 3.5. Autoclave for 15 min at 15 psi pressure–121°C. Cool to 45°–50°C. Aseptically add 10.0mL of sterile $FeSO_4 \cdot 7H_2O$ solution. Mix thoroughly. Aseptically distribute into sterile tubes or flasks.

Use: For the cultivation of *Thiobacillus* species.

Medium F for Sulfate Reducers (Postgate's Medium F for Sulfate Reducers)

Composition per liter:

Agar 12.0g
Pancreatic digest of casein 10.0g
Sodium lactate 3.5g
Ferrous citrate 0.5g
Na_2SO_3 .... 0.5g
$MgSO_4 \cdot 7H_2O$ 0.2g

pH 7.1 ± 0.2 at 25°C

Preparation of Medium: Add components, except ascorbic acid and thioglycollic acid, to tap water and bring volume to 1.0L. For marine bacteria, NaCl may be added or sea water used in place of tap water. Mix thoroughly. Gently heat and bring to boiling. Adjust pH to 7.1. Thioglycollate and ascorbate should be added immediately prior to sterilization. Distribute into screw-capped tubes or flasks. Autoclave for 15 min at 15 psi pressure–121°C.

Use: For isolation and cultivation of *Desulfotomaculum nigrificans, Desulfovibrio* species and other *Desulfotomaculum* species especially in food. These bacteria form black colonies in deep agar cultures.

Medium G for Sulfate Reducers (Postgate's Medium G for Sulfate Reducers)

Composition per 1015.2mL:

Solution 1 .... 970.0mL
Solution 4 .... 30.0mL

Solution 8A, 8B, 8C, 8D or 8E 10.0mL
Solution 5 .... 3.0mL
Solution 2 .... 1.0mL
Solution 3 .... 1.0mL
Solution 6 .... 0.1mL
Solution 7 .... 0.1mL

pH 7.2 ± 0.2 at 25°C

Solution 1:

Composition per 970mL:

Na_2SO_4 .... 3.0g
NaCl 1.2
$MgCl_2 \cdot 6H_2O$ 0.4g
KCl 0.3g
NH_4Cl 0.3g
KH_2PO_4 .... 0.2g
$CaCl_2 \cdot 2H_2O$ 0.15g

Preparation of Solution 1: Add components to distilled/deionized water and bring volume to 970.0mL. Mix thoroughly. Adjust pH to 7.2 with 2*N* HCl. Autoclave for 15 min at 15 psi pressure–121°C. Cool to 25°C.

Solution 2:

Composition per 10mL:

NaOH 5.0mg
Na_2SeO_3 .... 0.03mg

Preparation of Solution 2: Add NaOH and Na_2SeO_3 to distilled/deionized water and bring volume to 10.0mL. Mix thoroughly. Autoclave for 15 min at 15 psi pressure–121°C. Cool to 25°C.

Solution 3:

Composition per liter:

$FeCl_2 \cdot 4H_2O$ 1.5g
$CoCl_2 \cdot 6H_2O$ 0.12g
$MnCl_2 \cdot 4H_2O$ 0.1g
$ZnCl_2$ 0.070g
H_3BO_3 .... 0.060g
$NiCl_2 \cdot 6H_2O$ 0.025g
$NaMoO_4 \cdot 2H_2O$ 0.025g
$CuCl_2 \cdot 2H_2O$ 0.015g

Preparation of Solution 3: Add components to distilled/deionized water and bring volume to 1.0L. Mix thoroughly. Autoclave for 15 min at 15 psi pressure–121°C. Cool to 25°C.

Solution 4:

Composition per 30mL:

$NaHCO_3$ 2.55g

Preparation of Solution 4: Add $NaHCO_3$ to distilled/deionized water and bring volume to 30.0mL. Mix thoroughly. Gas with 100% CO_2 for 10–15 min. Filter sterilize.

Solution 5:

Composition per 3mL:

$Na_2S \cdot 9H_2O$ 0.36g

Preparation of Solution 5: Add $Na_2S \cdot 9H_2O$ to distilled/deionized water and bring volume to 3.0mL. Mix thoroughly. Gas with 100% N_2 for 5–10 min. Cap tube with a rubber stopper. Autoclave for 15 min at 15 psi pressure–121°C. Cool to 25°C.

Solution 6:

Composition per 100mL:

Thiamine·HCl 0.010g
Cyanocobalamin 5.0mg
p-Aminobenzoic acid 5.0mg
Biotin 1.0mg

Preparation of Solution 6: Add components to distilled/deionized water and bring volume to 100.0mL. Mix thoroughly. Filter sterilize.

Solution 7:

Composition per 100mL:

Succinic acid 0.6g
Isobutyric acid 0.5g
Valeric acid 0.5g
2-Methylbutyric acid 0.5g
3-Methylbutyric acid 0.5g
Caproic acid 0.2g

Preparation of Solution 7: Add components to distilled/deionized water and bring volume to 100.0mL. Mix thoroughly. Adjust pH to 9.0 with NaOH. Autoclave for 15 min at 15 psi pressure–121°C. Cool to 25°C.

Solution 8A:

Composition per 100mL:

Sodium acetate·$3H_2O$ 20.0g

Solution 8B:

Composition per 100mL:

Propionic acid 7.0g

Solution 8C:

Composition per 100mL:

n-Butyric acid 8.0g

Solution 8D:

Composition per 100mL:

Benzoic acid 5.0g

Solution 8E:

Composition per 100mL:

n-Palmitic acid 5.0g

Preparation of Solutions 8A-E: Add the appropriate amount of component to distilled/deionized water and bring volume to 100.0mL. Mix thoroughly. Adjust pH to 9.0 with NaOH. Autoclave for 15 min at 15 psi pressure–121°C. Cool to 25°C.

Preparation of Medium: To 970.0mL of cooled, sterile solution 1 aseptically add 1.0mL of sterile solution 2, 1.0mL of sterile solution 3, 30.0mL of sterile solution 4, 3.0mL of sterile solution 5, 0.1mL of sterile solution 6, 0.1mL of sterile solution 7, and 10.0mL of sterile solution 8A, 8B, 8C, 8D or 8E. Mix thoroughly. Aseptically distribute into sterile tubes or flasks.

Use: For the isolation and cultivation of *Desulfovibrio baarsii, Desulfovibrio sapovorans, Desulfobacter* species, *Desulfonema* species, *Desulfobulbus* species, and *Desulfotomaculum acetoxidans*.

Medium for Ammonia Oxidizers

Composition per liter:

$MgSO_4 \cdot 7H_2O$	0.20g
$(NH_4)_2SO_4$	0.13g
K_2HPO_4	0.09g
$CaCl_2 \cdot 2H_2O$	0.02g
Chelated iron	1.0mg
$MnCl_2 \cdot 4H_2O$	0.2mg
$Na_2MoO_4 \cdot 2H_2O$	0.1mg
$ZnSO_4 \cdot 7H_2O$	0.1mg
$CuSO_4 \cdot 5H_2O$	0.02mg
$CoCl_2 \cdot 6H_2O$	2.0μg

Preparation of Medium: Add components to distilled/deionized water and bring volume to 1.0L. Mix thoroughly. Distribute into tubes or flasks. Autoclave for 15 min at 15 psi pressure–121°C.

Use: For the isolation, cultivation and enrichment of ammonia-oxidizing bacteria from soil.

Medium for Ammonia Oxidizers

Composition per liter:

$(NH_4)_2SO_4$	2.0g
$MgSO_4 \cdot 7H_2O$	0.2g
$CaCl_2 \cdot 2H_2O$	0.02g
K_2HPO_4	0.02g
Chelated iron	1.0mg
$MnCl_2 \cdot 4H_2O$	0.2mg
$Na_2MoO_4 \cdot 2H_2O$	0.1mg
$ZnSO_4 \cdot 7H_2O$	0.1mg
$CuSO_4 \cdot 5H_2O$	0.02mg
$CoCl_2 \cdot 6H_2O$	2.0μg

Preparation of Medium: Add components to distilled/deionized water and bring volume to 1.0L. Mix thoroughly. Distribute into tubes or flasks. Autoclave for 15 min at 15 psi pressure–121°C.

Use: For the isolation, cultivation and enrichment of ammonia-oxidizing bacteria from soil.

Medium for Ammonia Oxidizers

Composition per liter:

K_2HPO_4	0.5g
$(NH_4)_2SO_4$	0.5g
Phenol Red	0.5g
$MgSO_4 \cdot 7H_2O$	0.05g
$CaCl_2 \cdot 2H_2O$	0.02g
NaCl	0.02g
$Na_2MoO_4 \cdot 2H_2O$	2.4μg
Metals "44"	1.0mL

Metals "44":

Composition per 100mL:

$ZnSO_4 \cdot 7H_2O$	1.1g
$FeSO_4 \cdot 7H_2O$	0.5g
EDTA	0.25g
$MnSO_4 \cdot 7H_2O$	0.154g
$CuSO_4 \cdot 5H_2O$	0.04g
$Co(NO_3)_2 \cdot 6H_2O$	0.025g
$Na_2B_4O_7 \cdot 10H_2O$	0.018g

Preparation of Metals "44": Add a few drops of H_2SO_4 to distilled/deionized water to inhibit precipitate formation. Add components to acidified distilled/deionized water and bring volume to 100.0mL. Mix thoroughly.

Preparation of Medium: Add components to distilled/deionized water and bring volume to 1.0L. Mix thoroughly. Distribute into tubes or flasks. Autoclave for 15 min at 15 psi pressure–121°C.

Use: For the isolation, cultivation and enrichment of ammonia-oxidizing bacteria from soil.

Medium for Ammonia Oxidizers

Composition per liter:

$(NH_4)_2SO_4$	0.5g
KH_2PO_4	0.2g
$CaCl_2 \cdot 2H_2O$	0.04g
$MgSO_4 \cdot 7H_2O$	0.04g
Ferric citrate	0.5mg
Phenol Red	0.5mg

Preparation of Medium: Add components to distilled/deionized water and bring volume to 1.0L. Mix thoroughly. Distribute into tubes or flasks. Autoclave for 15 min at 15 psi pressure–121°C.

Use: For the isolation, cultivation and enrichment of ammonia-oxidizing bacteria from soil.

Medium for Ammonia Oxidizers, Brackish

Composition per liter:

$CaCO_3$	5.0g
NH_4Cl	0.5g

K_2HPO_4 .... 0.05g
Sea water 400.0mL

Preparation of Medium: Add components to distilled/deionized water and bring volume to 1.0L. Mix thoroughly. Distribute into tubes or flasks. Autoclave for 15 min at 15 psi pressure–121°C.

Use: For the isolation, cultivation and enrichment of ammonia-oxidizing bacteria from brackish specimens.

Medium for Ammonia Oxidizers, Marine

Composition per liter:

$(NH_4)_2SO_4$ 1.32g
$MgSO_4 \cdot 7H_2O$ 0.20g
Chelated iron 0.13g
K_2HPO_4 .... 0.11g
$CaCl_2 \cdot 2H_2O$ 0.02g
$ZnSO_4 \cdot 7H_2O$ 0.1mg
$CuSO_4 \cdot 5H_2O$ 0.02mg
$CoCl_2 \cdot 6H_2O$ 2.0μg
$MnCl_2 \cdot 4H_2O$ 2.0μg
$Na_2MoO_4 \cdot 2H_2O$ 1.0μg
Sea water 1.0L

Preparation of Medium: Combine components. Mix thoroughly. Distribute into tubes or flasks. Autoclave for 15 min at 15 psi pressure–121°C.

Use: For the isolation, cultivation and enrichment of marine ammonia-oxidizing bacteria.

Medium for Hydrocarbon–Degrading Bacteria

Composition per 1020mL:

NH_4Cl 0.5g
$MgSO_4 \cdot 7H_2O$ 0.5g
NaCl 0.4g
Hydrocarbon 20.0mL
KH_2PO_4 solution 0.5mL
$Na_2HPO_4 \cdot H_2O$ solution 0.5mL

KH_2PO_4 Solution:
Composition per 100mL:

KH_2PO_4 .... 10.0g

Preparation of KH_2PO_4 Solution: Add KH_2PO_4 to distilled/deionized water and bring volume to 100.0mL. Mix thoroughly. Autoclave for 15 min at 15 psi pressure–121°C. Cool to 25°C.

$Na_2HPO_4 \cdot H_2O$ Solution:
Composition per 100mL:

$Na_2HPO_4 \cdot H_2O$ 10.0g

Preparation of $Na_2HPO_4 \cdot H_2O$ Solution: Add $Na_2HPO_4 \cdot H_2O$ to distilled/deionized water and bring volume to 100.0mL. Mix thoroughly. Autoclave for 15 min at 15 psi pressure–121°C. Cool to 25°C.

Preparation of Medium: Add components—except hydrocarbon, KH_2PO_4 solution and $Na_2HPO_4 \cdot H_2O$ solution—to distilled/deionized water and bring volume to 999.0mL. Mix thoroughly. Gently heat and bring to boiling. Autoclave for 15 min at 15 psi pressure–121°C. Cool to 45°–50°C. Aseptically add 0.5mL of sterile KH_2PO_4 solution and 0.5mL of the sterile $Na_2HPO_4 \cdot H_2O$ solution. Mix thoroughly. Aseptically distribute into sterile tubes in 10.0mL volumes. Add 0.2mL of sterile hydrocarbon to each tube.

Use: For the cultivation and enumeration of hydrocarbon-degrading bacteria in fresh water.

Medium for Hydrocarbon–Degrading Bacteria (Naphthalene Mineral Salts Medium)

Composition per liter:

K_2HPO_4 .... 1.0g
$(NH_4)_2SO_4$ 1.0g
$MgSO_4 \cdot 7H_2O$ 0.3g
$CaCl_2$ 0.1g
$FeSO_4 \cdot 7H_2O$ 0.02g
Naphthalene 2.0mL

pH 7.0 ± 0.2 at 25°C

Preparation of Medium: Add components, except naphthalene, to distilled/deionized water and bring volume to 998.0mL. Mix thoroughly. Gently heat and bring to boiling. Autoclave for 15 min at 15 psi pressure–121°C. Cool to 45°–50°C. Aseptically add 2.0mL of sterile naphthalene to 20.0mL of sterile basal salts. Ultrasonically homogenize the solution. Add the naphthalene–basal salts homogenate back to the remainder of the sterile basal salts medium. Mix thoroughly. Aseptically distribute into sterile tubes or flasks.

Use: For the cultivation and enrichment of hydrocarbon-degrading bacteria.

Medium for Lactobacilli (ATCC Medium 980)

Composition per liter:

Agar 20.0g
Peptone 12.5g
Glucose 11.0g
Sodium acetate 10.0g
Yeast extract 5.5g

KH_2PO_4	0.25g
K_2HPO_4	0.25g
$MgSO_4$	0.1g
$MnSO_4 \cdot 4H_2O$	0.05g
$FeSO_4 \cdot 7H_2O$	0.05g

pH 6.8 ± 0.2 at 25°C

Preparation of Medium: Add components to distilled/deionized water and bring volume to 1.0L. Mix thoroughly. Adjust pH to 6.8. Autoclave for 10 min at 15 psi pressure–120°C. Aseptically distribute into sterile tubes or flasks.

Use: For the cultivation and maintenance of *Pediococcus acidilactici* and *Bacillus* species.

Medium for Nitrite Oxidizers

Composition per liter:

$KHCO_3$	1.5g
KH_2PO_4	0.5g
K_2HPO_4	0.5g
KNO_2	0.3g
$MgSO_4 \cdot 7H_2O$	0.2g
NaCl	0.2g
$CaCl_2 \cdot 2H_2O$	0.01g
$FeSO_4 \cdot 7H_2O$	0.01g

Preparation of Medium: Add components to distilled/deionized water and bring volume to 1.0L. Mix thoroughly. Distribute into tubes or flasks. Autoclave for 15 min at 15 psi pressure–121°C.

Use: For the isolation, cultivation and enrichment of nitrate-oxidizing bacteria.

Medium for Nitrite Oxidizers, Marine

Composition per liter:

$MgSO_4 \cdot 7H_2O$	0.1g
$NaNO_2$	0.07g
$CaCl_2 \cdot 2H_2O$	6.0mg
K_2HPO_4	1.74mg
Chelated iron	1.0mg
$MnCl_2 \cdot 4H_2O$	66.0μg
$Na_2MoO_4 \cdot 2H_2O$	30.0μg
$ZnSO_4 \cdot 7H_2O$	30.0μg
$CuSO_4 \cdot 5H_2O$	6.0μg
$CoCl_2 \cdot 6H_2O$	0.6μg
Sea water	700.0mL

Preparation of Medium: Add components to distilled/deionized water and bring volume to 1.0L. Mix thoroughly. Distribute into tubes or flasks. Autoclave for 15 min at 15 psi pressure–121°C.

Use: For the isolation, cultivation and enrichment of marine nitrate-oxidizing bacteria.

Medium for *Prosthecomicrobium* and *Ancalomicrobium*

Composition per liter:

Agar	15.0g
Peptone	0.1g
Hutner's modified salts solution	20.0mL
Vitamin solution	10.0mL

Hutner's Mineral Base:

Composition per liter:

$MgSO_4 \cdot 7H_2O$	29.7g
Nitrilotriacetic acid	10.0g
$CaCl_2 \cdot 2H_2O$	3.34g
$FeSO_4 \cdot 7H_2O$	0.1g
$(NH_4)_2MoO_4$	9.25mg
Metals "44"	50.0mL

Preparation of Hutner's Mineral Base: Add nitrilotriacetic acid to 500.0mL of distilled/deionized water. Dissolve by adjusting pH to 6.5 with KOH. Add remaining components. Add distilled/deionized water to 1.0L.

Metals "44":

Composition per 100mL:

$ZnSO_4 \cdot 7H_2O$	1.1g
$FeSO_4 \cdot 7H_2O$	0.5g
EDTA	0.25g
$MnSO_4 \cdot 7H_2O$	0.154g
$CuSO_4 \cdot 5H_2O$	0.04g
$Co(NO_3)_2 \cdot 6H_2O$	0.025g
$Na_2B_4O_7 \cdot 10H_2O$	0.018g

Preparation of Metals "44": Add components to distilled/deionized water and bring volume to 100.0mL. Mix thoroughly.

Vitamin Solution:

Composition per liter:

Pyridoxine·HCl	0.01g
Calcium pantothenate	5.0mg
Nicotinamide	5.0mg
Riboflavin	5.0mg
Thiamine HCl	5.0mg
Biotin	2.0mg
Folic acid	2.0mg
Vitamin B_{12}	0.1mg

Preparation of Vitamin Solution: Add components to distilled/deionized water and bring volume to 1.0L. Mix thoroughly. Filter sterilize.

Preparation of Medium: Add components, except vitamin solution, to distilled/deionized water and bring volume to 990.0mL. Mix thoroughly. Gently heat and bring to boiling. Autoclave for 15 min at 15 psi pressure–121°C. Cool to 45°–50°C. Aseptically add sterile vitamin solution. Mix thoroughly. Pour into sterile Petri dishes or distribute into sterile tubes.

Use: For the isolation of *Prosthecomicrobium* species and *Ancalomicrobium* species.

Medium for *Prosthecomicrobium* and *Ancalomicrobium*

Composition per liter:

$(NH_4)_2SO_4$	0.25g
Glucose	0.25g
Na_2HPO_4	0.071g
Modified Hutner's basal salts	20.0mL
Vitamin solution	10.0mL

pH 7.2 ± 0.2 at 25°C

Modified Hutner's Basal Salts:

Composition per liter:

$MgSO_4 \cdot 7H_2O$	29.7g
Nitrilotriacetic acid	10.0g
$CaCl_2 \cdot 2H_2O$	3.34g
$FeSO_4 \cdot 7H_2O$	0.10g
$(NH_4)_2MoO_4$	9.25mg
Metals "44"	50.0mL

Preparation of Modified Hutner's Basal Salts: Add nitrilotriacetic acid to 500.0mL of distilled/deionized water. Dissolve by adjusting pH to 6.5 with KOH. Add remaining components. Readjust pH to 7.2 with H_2SO_4 or KOH. Add distilled/deionized water to 1.0L. Store at 5°C.

Metals "44":

Composition per 100mL:

$ZnSO_4 \cdot 7H_2O$	1.1g
$FeSO_4 \cdot 7H_2O$	0.5g
EDTA	0.25g
$MnSO_4 \cdot 7H_2O$	0.154g
$CuSO_4 \cdot 5H_2O$	0.04g
$Co(NO_3)_2 \cdot 6H_2O$	0.025g
$Na_2B_4O_7 \cdot 10H_2O$	0.018g

Preparation of Metals "44": Add a few drops of H_2SO_4 to distilled/deionized water to inhibit precipitate formation. Add components to acidified distilled/deionized water and bring volume to 100.0mL. Mix thoroughly.

Vitamin Solution:

Composition per liter:

Thiamine·HCl	5.0mg
D-Calcium pantothenate	5.0mg
Riboflavin	5.0mg
Biotin	2.0mg
Folic acid	2.0mg
Vitamin B_{12}	0.1mg

Preparation of Vitamin Solution: Add components to distilled/deionized water and bring volume to 1.0L. Mix thoroughly. Filter sterilize.

Preparation of Medium: Add components, except vitamin solution. to distilled deionized water and bring volume to 990.0mL. Mix thoroughly. Autoclave for 15 min at 15 psi pressure–121°C. Cool to room temperature. Aseptically add 10.0mL of sterile vitamin solution. Mix thoroughly. Aseptically distribute into sterile tubes or flasks.

Use: For the cultivation and maintenance of *Prosthecomicrobium enhydrum*, *Prosthecomicrobium pneumaticum*, and *Ancalomicrobium* species.

Medium for *Prosthecomicrobium* and *Ancalomicrobium* with Nicotinamide

Composition per liter:

$(NH_4)_2SO_4$	0.25g
Glucose	0.25g
Na_2HPO_4	0.071g
Modified Hutner's basal salts	20.0mL
Vitamin solution	10.0mL

pH 7.2 ± 0.2 at 25°C

Modified Hutner's Basal Salts:

Composition per liter:

$MgSO_4 \cdot 7H_2O$	29.7g
Nitrilotriacetic acid	10.0g
$CaCl_2 \cdot 2H_2O$	3.34g
$FeSO_4 \cdot 7H_2O$	0.10g
$(NH_4)_2MoO_4$	9.25mg
Metals "44"	50.0mL

Preparation of Modified Hutner's Basal Salts: Add nitrilotriacetic acid to 500.0mL of distilled/deionized water. Dissolve by adjusting pH to 6.5 with KOH. Add remaining components. Readjust pH to 7.2 with H_2SO_4 or KOH. Add distilled/deionized water to 1.0L. Store at 5°C.

Metals "44":

Composition per 100mL:

$ZnSO_4 \cdot 7H_2O$	1.1g
$FeSO_4 \cdot 7H_2O$	0.5g
EDTA	0.25g
$MnSO_4 \cdot 7H_2O$	0.154g
$CuSO_4 \cdot 5H_2O$	0.04g
$Co(NO_3)_2 \cdot 6H_2O$	0.025g
$Na_2B_4O_7 \cdot 10H_2O$	0.018g

Preparation of Metals "44": Add a few drops of H_2SO_4 to distilled/deionized water to inhibit precipitate formation. Add components to acidified distilled/deionized water and bring volume to 100.0mL. Mix thoroughly.

Vitamin Solution:

Composition per liter:

Thiamine·HCl	5.0mg
D-Calcium pantothenate	5.0mg
Riboflavin	5.0mg
Nicotinamide	5.0mg
Biotin	2.0mg
Folic acid	2.0mg
Vitamin B_{12}	0.1mg

Preparation of Vitamin Solution: Add components to distilled/deionized water and bring volume to 1.0L. Mix thoroughly. Filter sterilize.

Preparation of Medium: Add components, except vitamin solution, to distilled/deionized water and bring volume to 990.0mL. Mix thoroughly. Autoclave for 15 min at 15 psi pressure–121°C. Cool to room temperature. Aseptically add 10.0mL of sterile vitamin solution. Mix thoroughly. Aseptically distribute into sterile tubes or flasks.

Use: For the cultivation and maintenance of *Ancalomicrobium adetum* and *Prosthecomicrobium* species.

Medium for *Prosthecomicrobium* and *Ancalomicrobium*, Modified

Composition per liter:

Agar	15.0g
Glucose	1.0g
$(NH_4)_2SO_4$	0.25g
Peptone	0.15g
Yeast extract	0.15g
Modified Hutner's basal salts	20.0mL
Vitamin solution	10.0mL

Modified Hutner's Basal Salts:

Composition per liter:

$MgSO_4 \cdot 7H_2O$	29.7g
Nitrilotriacetic acid	10.0g
$CaCl_2 \cdot 2H_2O$	3.34g
$FeSO_4 \cdot 7H_2O$	0.10g
$(NH_4)_2MoO_4$	9.25mg
Metals "44"	50.0mL

Preparation of Modified Hutner's Basal Salts: Add nitrilotriacetic acid to 500.0mL of distilled/deionized water. Dissolve by adjusting pH to 6.5 with KOH. Add remaining components. Readjust pH to 7.2 with H_2SO_4 or KOH. Add distilled/deionized water to 1.0L. Store at 5°C.

Metals "44":

Composition per 100mL:

$ZnSO_4 \cdot 7H_2O$	1.1g
$FeSO_4 \cdot 7H_2O$	0.5g
EDTA	0.25g
$MnSO_4 \cdot 7H_2O$	0.154g
$CuSO_4 \cdot 5H_2O$	0.04g
$Co(NO_3)_2 \cdot 6H_2O$	0.025g
$Na_2B_4O_7 \cdot 10H_2O$	0.018g

Preparation of Metals "44": Add a few drops of H_2SO_4 to distilled/deionized water to inhibit precipitate formation. Add components to acidified distilled/deionized water and bring volume to 100.0mL. Mix thoroughly.

Vitamin Solution:

Composition per liter:

Thiamine·HCl	5.0mg
D-Calcium pantothenate	5.0mg
Riboflavin	5.0mg
Biotin	2.0mg
Folic acid	2.0mg
Vitamin B_{12}	0.1mg

Preparation of Vitamin Solution: Add components to distilled/deionized water and bring volume to 1.0L. Mix thoroughly. Filter sterilize.

Preparation of Medium: Add components, except vitamin solution, to distilled deionized water and bring volume to 990.0mL. Mix thoroughly. Autoclave for 15 min at 15 psi pressure–121°C. Cool to room temperature. Aseptically add 10.0mL of sterile vitamin solution. Mix thoroughly. Aseptically distribute into sterile tubes or flasks.

Use: For the cultivation and maintenance of *Ancalomicrobium adetum*, *Prosthecomicrobium hirschii*, and *Prosthecomicrobium* species.

Medium for Sulfate Reducers (ATCC Medium 1282)

Composition per 1050mL:

Modified Baar's medium for sulfate reducers	1020.0mL
Organic acid solution	10.0mL
Vitamin solution	10.0mL
Wolfe's mineral solution	10.0mL

pH 7.5 ± 0.2 at 25°C

Modified Baar's Medium for Sulfate Reducers:

Composition per 1020mL:

Component I	400.0mL
Component III	400.0mL
Component II	200.0mL
$Fe(NH_4)_2(SO_4)_2$ (5% solution)	20.0mL

Component I:

Composition per 400mL:

Sodium citrate	5.0g
$MgSO_4$	2.0g

$CaSO_4$ 1.0g
NH_4Cl 1.0g

Preparation of Component I: Add components to distilled/deionized water and bring volume to 400.0mL. Mix thoroughly. Adjust pH to 7.5. Autoclave for 15 min at 15 psi pressure–121°C.

Component II:
Composition per 200mL:
K_2HPO_4 .... 0.5g

Preparation of Component II: Add K_2HPO_4 to distilled/deionized water and bring volume to 200.0mL. Mix thoroughly. Adjust pH to 7.5. Autoclave for 15 min at 15 psi pressure–121°C.

Component III:
Composition per 400mL:
Sodium lactate 3.5g
Yeast extract 1.0g

Preparation of Component III: Add components to distilled/deionized water and bring volume to 400.0mL. Mix thoroughly. Adjust pH to 7.5. Autoclave for 15 min at 15 psi pressure–121°C.

Preparation of Modified Baar's Medium for Sulfate Reducers: Aseptically combine the three sterile solutions, except the $Fe(NH_4)_2(SO_4)_2$ solution. Mix thoroughly. Distribute 5.0mL volumes into tubes under 97% N_2 + 3% H_2. Add medium to tubes while still warm to exclude as much O_2 as possible. Prepare a 5% solution of ferrous ammonium sulfate, $Fe(NH_4)_2(SO_4)_2$. Sterilize by filtration. Add 0.2mL of sterile $Fe(NH_4)_2(SO_4)_2$ solution to 10.0mL of medium immediately prior to inoculation.

Organic Acid Solution:
Composition per 100mL:
Butyric acid 5.18mL
Caproic acid 2.4mL
Octanoic acid 1.25mL

Preparation of Organic Acid Solution: Add components to distilled/deionized water and bring volume to 75.0mL. Adjust pH to 7.0 with 5*N* NaOH. Bring volume to 100.0mL with distilled/deionized water. Filter sterilize.

Wolfe's Vitamin Solution:
Composition per liter:
Pyridoxine·HCl 10.0mg
Thiamine·HCl 5.0mg
Riboflavin 5.0mg
Nicotinic acid 5.0mg
Calcium pantothenate 5.0mg
p-Aminobenzoic acid 5.0mg
Thioctic acid 5.0mg
Biotin 2.0mg
Folic acid 2.0mg
Cyanocobalamin 100.0μg

Preparation of Wolfe's Vitamin Solution: Add components to distilled/deionized water and bring volume to 1.0L. Mix thoroughly. Filter sterilize.

Wolfe's Mineral Solution:
Composition per liter
$MgSO_4 \cdot 7H_2O$ 3.0g
Nitriloacetic acid 1.5g
NaCl 1.0g
$MnSO_4 \cdot H_2O$ 0.5g
$FeSO_4 \cdot 7H_2O$ 0.1g
$CoCl_2 \cdot 6H_2O$ 0.1g
$CaCl_2$ 0.1g
$ZnSO_4 \cdot 7H_2O$ 0.1g
$CuSO_4 \cdot 5H_2O$ 0.01g
$AlK(SO_4)_2 \cdot 12H_2O$ 0.01g
H_3BO_3 .... 0.01g
$Na_2MoO_4 \cdot 2H_2O$ 0.01g

Preparation of Wolfe's Mineral Solution: Add nitrilotriacetic acid to 500.0mL of distilled/deionized water. Dissolve by adjusting pH to 6.5 with KOH. Add remaining components. Add distilled/deionized water to 1.0L. Filter sterilize.

Preparation of Medium: To each test tube containing 10.0mL of modified Baar's medium for sulfate reducers aseptically add 0.1mL of sterile organic acid solution, 0.1mL of sterile Wolfe's vitamin solution and 0.1mL of sterile Wolfe's mineral solution immediately prior to inoculation.

Use: For the cultivation and maintenance of *Desulfotomaculum thermobenzoicum* and *Desulfovibrio sapovorans.*

Medium for Sulfate Reducers (Postgate's Medium for Sulfate Reducers) (ATCC Medium 1283)

Composition per liter:
Part A 869.0mL
Part C 100.0mL
Part D 10.0mL
Part E 10.0mL
Part F 10.0mL
Part B, trace element solution SL-7 .... 1.0mL
pH 7.7 ± 0.2 at 25°C

Part A:
Composition per 869mL:
Na_2SO_4 .... 3.0g
NaCl 1.0g
KCl 0.5g

$MgCl_2 \cdot 6H_2O$ 0.4g
NH_4Cl 0.3g
KH_2PO_4 .... 0.2g
$CaCl_2 \cdot 2H_2O$ 0.15g

Preparation of Part A: Add components to distilled/deionized water and bring volume to 869.0mL. Mix thoroughly. Prepare and autoclave part A under 90% N_2 + 10% CO_2. Autoclave for 15 min at 15 psi pressure–121°C. Cool to room temperature.

Part B, Trace element solution SL-7:
Composition per liter:
$FeCl_2 \cdot 4H_2O$ 1.5g
$CoCl_2 \cdot 6H_2O$ 0.19g
$MnCl_2 \cdot 4H_2O$ 0.10g
$ZnCl_2$ 0.07g
H_3BO_3 .... 0.06g
$Na_2MoO_4 \cdot 2H_2O$ 0.04g
$NiCl_2 \cdot 6H_2O$ 0.02g
$CuCl_2 \cdot 2H_2O$ 0.02g
HCl, 25% 10.0mL

Preparation of Part B: Add the $FeCl_2 \cdot 4H_2O$ to the HCl. Add distilled/deionized water and bring volume to 1.0L. Add remaining components. Mix thoroughly. Autoclave under 100% N_2 for 15 min at 15 psi pressure–121°C. Cool to room temperature.

Part C:
Composition per 100mL:
$NaHCO_3$ 5.0g

Preparation of Part C: Add the $NaHCO_3$ to distilled/deionized water and bring volume to 100.0mL. Mix thoroughly. Filter sterilize. Gas with 90% N_2 + 10% CO_2 to remove residual O_2.

Part D:
Composition per 10mL:
Sodium butyrate 0.7g
Sodium caproate 0.3g
Sodium octanoate 0.15g

Preparation of Part D: Add components to distilled/deionized water and bring volume to 10.0mL. Mix thoroughly. Autoclave under 100% N_2 for 15 min at 15 psi pressure–121°C. Cool to room temperature.

Part E:
Composition per 10mL:
Yeast extract 1.0g
Thiamine·HCl 100.0μg
p-Aminobenzoic acid 40.0μg
D(+)-Biotin 10.0μg

Preparation of Part E: Add components to distilled/deionized water and bring volume to 10.0mL. Mix thoroughly. Autoclave under 100% N_2 for 15 min at 15 psi pressure–121°C. Cool to room temperature.

Part F:
Composition per 10mL:
$Na_2S \cdot 9H_2O$ 0.4g

Preparation of Part F: Add $Na_2S \cdot 9H_2O$ to distilled/deionized water and bring volume to 10.0mL. Mix thoroughly. Autoclave under 100% N_2 for 15 min at 15 psi pressure–121°C. Cool to room temperature.

Preparation of Medium: To 869.0mL of sterile cooled Part A, aseptically add the remaining sterile solutions in the following order: Part B, Part C, Part D, Part E, and Part F. Mix thoroughly. Adjust pH to 7.7. Anaerobically distribute under 80% N_2 + 20% CO_2 into sterile tubes or flasks.

Use: For the cultivation and maintenance of *Desulfovibrio baarsii* and *Desulfovibrio sapovorans*.

Medium for *Treponema pectinovorum*

Composition per liter:
Polypeptone™ 5.0g
Heart infusion broth 5.0g
Yeast extract 5.0g
NaCl 5.0g
K_2HPO_4 .... 2.0g
$(NH_4)_2SO_4$ 2.0g
Agar 1.0g
Pectin 0.8g
L-Cysteine·HCl·H_2O 0.68g
Rumen fluid 500.0mL
Resazurin (25.0 mg/100.0mL water) 4.0mL
pH 7.0–7.2 at 25°C

Preparation of Medium: Add components to distilled/deionized water and bring volume to 1.0L. Prepare and distribute anaerobically under 90% N_2 + 10% CO_2. Mix thoroughly. Adjust pH to 7.0–7.2. Distribute into screw-capped tubes or flasks. Autoclave for 15 min at 15 psi pressure–121°C.

Use: For the cultivation and maintenance of *Treponema pectinovorum*.

Medium for *Ureaplasma*
See: **B Broth**

Medium M71

Composition per liter:
Agar 20.0g
Peptone 10.0g
Glucose 5.0g
H_3BO_3 .... 1.0g

Pancreatic digest of casein 1.0g
Cycloheximide 0.05g
2,3,5-Triphenyltetrazolium·HCl solution...... 10.0mL

2,3,5-Triphenyltetrazolium·HCl Solution:
Composition per 10mL:
2,3,5-Triphenyltetrazolium·HCl........................ 0.05g

Preparation of 2,3,5-Triphenyltetrazolium-HCl Solution: Add 2,3,5-triphenyltetrazolium·HCl to distilled/deionized water and bring volume to 10.0mL. Mix thoroughly. Autoclave for 15 min at 15 psi pressure–121°C.

Preparation of Medium: Add components, except 2,3,5-triphenyltetrazolium·HCl solution, to distilled/deionized water and bring volume to 990.0mL. Mix thoroughly. Gently heat and bring to boiling. Autoclave for 15 min at 15 psi pressure–121°C. Cool to 45°–50°C. Aseptically add 10.0mL of sterile 2,3,5-triphenyltetrazolium·HCl solution. Mix thoroughly. Pour into sterile Petri dishes.

Use: For the selective isolation and cultivation of *Pseudomonas syringae*.

Medium N for Sulfate Reducers (Postgate's Medium N for Sulfate Reducers)

Composition per liter:
$(NH_4)_2SO_4$ 7.0g
Sodium lactate.. 6.0g
NH_4Cl.. 1.0g
Yeast extract 1.0g
KH_2PO_4 .... 0.5g
Sodium citrate·$2H_2O$ 0.3g
$FeSO_4 \cdot 7H_2O$ 0.1g
$CaCl_2 \cdot 6H_2O$ 0.06g
$MgSO_4 \cdot 7H_2O$ 0.06g
pH 7.5 ± 0.2 at 25°C

Preparation of Medium: Add components to distilled/deionized water and bring volume to 1.0L. For marine bacteria, NaCl may be added or sea water used in place of distilled/deionized water. Mix thoroughly. Adjust pH to 7.5. Distribute into tubes or flasks. Autoclave for 15 min at 15 psi pressure–121°C.

Use: For detection, culturing and storage of *Desulfovibrio* species and many *Desulfotomaculum* species. This medium should be used when a clear culture medium is desired such as for chemostat culture. This medium may be cloudy after sterilization but usually clears on cooling. It turns black as a result of H_2S production due to bacterial growth.

Medium ND
See: **Castenholtz ND Medium**

Medium R

Composition per liter:
$Na_2S_2O_3 \cdot 5H_2O$.. 5.0g
KNO_3.. 2.0g
$MgCl_2 \cdot 6H_2O$... 0.5g
NH_4Cl.. 0.5g
KH_2PO_4 solution 10.0mL
$NaHCO_3$ solution 10.0mL
$FeSO_4 \cdot 7H_2O$ solution 10.0mL
pH 7.0 ± 0.2 at 25°C

KH_2PO_4 Solution:
Composition per 10mL:
KH_2PO_4 .... 2.0g

Preparation of KH_2PO_4 Solution: Add KH_2PO_4 to distilled/deionized water and bring volume to 10.0mL. Mix thoroughly. Filter sterilize.

$NaHCO_3$ Solution:
Composition per 10mL:
$NaHCO_3$ 1.0g

Preparation of $NaHCO_3$ Solution: Add the $NaHCO_3$ to distilled/deionized water and bring volume to 10.0mL. Mix thoroughly. Filter sterilize.

$FeSO_4 \cdot 7H_2O$ Solution:
Composition per 10mL:
$FeSO_4 \cdot 7H_2O$ 10.0mg

Preparation of $FeSO_4 \cdot 7H_2O$ Solution: Add the $FeSO_4 \cdot 7H_2O$ to distilled/deionized water and bring volume to 10.0mL. Mix thoroughly. Filter sterilize.

Preparation of Medium: Add components—except KH_2PO_4 solution, $NaHCO_3$ solution, and $FeSO_4 \cdot 7H_2O$ solution—to tap water and bring volume to 970.0mL. Mix thoroughly. Gently heat until dissolved. Adjust pH to 7.0. Autoclave for 15 min at 15 psi pressure–121°C. Cool to 45°–50°C. Aseptically add 10.0mL of sterile KH_2PO_4 solution, 10.0mL of $NaHCO_3$ solution, and 10.0mL of $FeSO_4 \cdot 7H_2O$ solution. Mix thoroughly. Aseptically distribute into sterile tubes or flasks.

Use: For the cultivation of *Thiobacillus denitrificans*.

Medium S

Composition per liter:
$Na_2S_2O_3 \cdot 5H_2O$.. 5.0g
$(NH_4)_2SO_4$ 4.0g
KH_2PO_4 .... 4.0g
$MgSO_4$.. 0.5g

$CaCl_2$ ... 0.25g
$FeSO_4$ ... 0.01g

Preparation of Medium: Add components to distilled/deionized water and bring volume to 1.0L. Mix thoroughly. Distribute into tubes or flasks. Autoclave for 15 min at 15 psi pressure–121°C.

Use: For the cultivation of *Thiobacillus* species.

Medium SP 4

Composition per liter:

Pancreatic digest of casein	11.0g
Peptone	5.3g
Glucose	5.0g
NaCl	0.875g
Beef extract	0.525g
Yeast extract	0.525g
Beef heart, solids from infusion	0.35g
Fetal bovine serum, heat inactivated	170.0mL
Yeast extract solution	100.0mL
CMRL 1066, 10× solution	50.0mL
Fresh yeast extract solution	35.0mL
Phenol Red solution	20.0mL
Penicillin solution	10.0mL

pH 7.6 ± 0.2 at 25°C

Yeast Extract Solution:

Composition per 100mL:

Yeast extract ... 2.0g

Preparation of Yeast Extract Solution: Add yeast extract to distilled/deionized water and bring volume to 100.0mL. Mix thoroughly. Autoclave for 15 min at 15 psi pressure–121°C.

CMRL 1066, 10X Solution:

Composition per liter:

NaCl	6.8g
$NaHCO_3$	2.2g
D-Glucose	1.0g
KCl	0.4g
L-Cysteine·HCl·H_2O	0.26g
$CaCl_2$, anhydrous	0.2g
$MgSO_4·7H_2O$	0.2g
$NaH_2PO_4·H_2O$	0.14g
L-Glutamine	0.1g
Sodium acetate·$3H_2O$	0.083g
L-Glutamic acid	0.075g
L-Arginine·HCl	0.070g
L-Lysine·HCl	0.070g
L-Leucine	0.060g
Glycine	0.050g
Ascorbic acid	0.050g
L-Proline	0.040g
L-Tyrosine	0.040g
L-Aspartic acid	0.030g
L-Threonine	0.030g
L-Alanine	0.025g
L-Phenylalanine	0.025g
L-Serine	0.025g
L-Valine	0.025g
L-Cystine	0.020g
L-Histidine·HCl·H_2O	0.020g
L-Isoleucine	0.020g
Phenol Red	0.020g
L-Methionine	0.015g
Deoxyadenosine	0.010g
Deoxycytidine	0.010g
Deoxyguanosine	0.010g
Glutathione, reduced	0.010g
Thymidine	0.010g
Hydroxy-L-proline	0.010g
L-Tryptophan	0.010g
Nicotinamide adenine dinucleotide	7.0mg
Tween™ 80	5.0mg
Sodium glucoronate·H_2O	4.2mg
Coenzyme A	2.5mg
Cocarboxylase	1.0mg
Flavin adenine dinucleotide	1.0mg
Nicotinamide adenine dinucleotide phosphate	1.0mg
Uridine triphosphate	1.0mg
Choline chloride	0.50mg
Cholesterol	0.20mg
5-Methyldeoxycytidine	0.10mg
Inositol	0.05mg
p-Aminobenzoic acid	0.05mg
Niacin	0.025mg
Niacinamide	0.025mg
Pyridoxine	0.025mg
Pyridoxal·HCl	0.025mg
Biotin	0.01mg
D-Calcium pantothenate	0.01mg
Folic acid	0.01mg
Riboflavin	0.01mg
Thiamine·HCl	0.01mg

Source: CMRL 1066, 10× medium is available as a premixed powder from GIBCO BRL.

Preparation of CMRL 1066, 10X Solution: Add components to distilled/deionized water and bring volume to 1.0L. Mix thoroughly. Adjust pH to 7.2. Filter sterilize.

Fresh Yeast Extract Solution:

Composition per 100mL:

Baker's yeast, live, pressed, starch-free ... 25.0g

Preparation of Fresh Yeast Extract Solution: Add the live Baker's yeast to 100.0mL of distilled/deionized water. Autoclave for 90 min at 15 psi pressure–121°C. Allow to stand. Remove supernatant solution. Adjust pH to 6.6–6.8. Filter sterilize.

Phenol Red Solution:

Composition per 100mL:

Phenol Red	0.01g

Preparation of Phenol Red Solution: Add Phenol Red to distilled/deionized water and bring volume to 10.0mL. Mix thoroughly. Filter sterilize.

Penicillin Solution:

Composition per 10mL:

Penicillin	1,000,000U

Preparation of Penicillin Solution: Add penicillin to distilled/deionized water and bring volume to 10.0mL. Filter sterilize.

Preparation of Medium: Add components—except fetal bovine serum, yeast extract solution, CMRL 1066, 10× solution, fresh yeast extract solution, Phenol Red solution, and penicillin solution—to distilled/deionized water and bring volume to 615.0mL. Mix thoroughly. Gently heat and bring to boiling. Autoclave for 15 min at 15 psi pressure–121°C. Cool to 45°–50°C. Aseptically add 170.0mL of sterile fetal bovine serum, 100.0mL of sterile yeast extract solution, 50.0mL of sterile CMRL 1066, 10× solution, 35.0mL of sterile fresh yeast extract solution, 20.0mL of sterile Phenol Red solution, and 10.0mL of sterile penicillin solution. Mix thoroughly. Aseptically distribute into sterile tubes or flasks.

Use: For the isolation and cultivation of *Spiroplasma* species from ticks.

Medium VTY

Composition per 100mL:

Peptone	1.0g
Noble agar	0.7g
Yeast extract	0.5g
L-Cysteine·HCl·H_2O	0.1g
Salts A	20.0mL
Salts B	20.0mL
Glucose (10% solution)	5.0mL
$NaHCO_3$ (5% solution)	1.0mL
Hemin solution	1.0mL
Volatile fatty acid solution	0.31mL
Resazurin (0.1% solution)	0.1mL

pH 7.2 ± 0.2 at 25°C

Salts A:

Composition per liter:

$CaCl_2 \cdot 2H_2O$	0.6g
$MgSO_4$	0.45g

Preparation of Salts A: Add components to distilled/deionized water and bring volume to 1.0L. Mix thoroughly.

Salts B:

Composition per liter:

NaCl	4.5g
$(NH_4)_2SO_4$	4.5g
Potassium phosphate buffer (0.05M, pH 7.4)	1.0L

Preparation of Salts B: Add NaCl and $(NH_4)_2SO_4$ to 1.0L of 0.05M potassium phosphate buffer, pH 7.4. Mix thoroughly.

Hemin Solution:

Composition per liter:

Hemin	0.5g
NaOH (0.01*N* solution)	1.0mL

Preparation of Hemin Solution: Add hemin to 1.0mL of 0.01*N* NaOH solution. Mix thoroughly.

Volatile Fatty Acid Solution:

Composition per 31mL:

Acetic acid	17.0mL
Propionic acid	6.0mL
n-Butyric acid	4.0mL
n-Valeric acid	1.0mL
Isovaleric acid	1.0mL
Isobutyric acid	1.0mL
DL-α-methylbutyric acid	1.0mL

Preparation of Volatile Fatty Acid Solution: Combine components. Mix thoroughly.

Preparation of Medium: Add components, except glucose and $NaHCO_3$, to distilled/deionized water and bring volume to 94.0mL. Mix thoroughly. Adjust pH to 7.2. Gently heat and gas with 95% N_2 + 5% CO_2 until reduced. Anaerobically distribute into tubes or flasks. Cap with rubber stoppers. Autoclave for 20 min at 15 psi pressure–121°C. Cool to 50°C. Filter sterilize glucose solution and $NaHCO_3$ solution separately. Aseptically and anaerobically add sterile glucose solution and sterile $NaHCO_3$ solution to cooled, sterile basal medium.

Use: For the cultivation and maintenance of *Roseburia cecicola*.

Megasphaera Medium

Composition per liter:

Yeast extract	4.0g
K_2HPO_4	3.2g
KH_2PO_4	1.6g
Agar	1.0g
NH_4Cl	0.5g
Sodium thioglycollate	0.45g
$CaCl_2$	0.2g
$MgCl_2$	0.2g
Sodium lactate (60% solution)	16.0mL

pH 7.0 ± 0.2 at 25°C

Preparation of Medium: Add components to distilled/deionized water and bring volume to 1.0L. Mix thoroughly. Gently heat and bring to boiling. Distribute into tubes or flasks. Autoclave for 15 min at 15 psi pressure–121°C.

Use: For the cultivation and maintenance of *Megasphaera elsdenii.*

Melissococcus pluton Medium

Composition per liter:

Glucose	10.0g
Neopeptone	5.0g
Peptone	2.5g
Yeast extract	2.5g
Soluble starch	2.0g
Pancreatic digest of casein	2.0g
L-Cysteine·HCl·H_2O	0.25g
Phosphate buffer (1M, pH 6.7)	50.0mL

pH 7.2 ± 0.2 at 25°C

Preparation of Medium: Add components to distilled/deionized water and bring volume to 1.0L. Mix thoroughly. Adjust pH to 7.2. Gently heat and bring to boiling. Distribute into tubes or flasks that have been flushed with 90% N_2 + 10% CO_2. Cap with butyl rubber stoppers. Place tubes in a press. Autoclave for 15 min at 15 psi pressure–121°C.

Use: For the cultivation and maintenance of *Melisococcus pluton.*

Membrane Lauryl Sulfate Broth

Composition per liter:

Peptone	39.0g
Lactose	30.0g
Yeast extract	6.0g
Sodium lauryl sulfate	1.0g
Phenol Red	0.2g

pH 7.4 ± 0.2 at 25°C

Preparation of Medium: Add components to distilled/deionized water and bring volume to 1.0L. Mix thoroughly. Distribute into tubes or flasks. Autoclave for 15 min at 15 psi pressure–121°C.

Use: For the enumeration of coliform organisms and *Escherichia coli* in water.

m-Endo Agar, LES
See: **Endo Agar, LES**

m-Endo Broth
See: **Endo Broth**

Meniscus glaucopis Agar

Composition per liter:

Agar	15.0g
$CaCO_3$	10.0g
Maltose	5.0g
Yeast extract	1.0g
KH_2PO_4	0.5g
NaCl	0.4g
NH_4Cl	0.4g
Sodium thioglycollate	0.3g
$MgSO_4·7H_2O$	0.2g
$CaCl_2·H_2O$	0.01g
$FeSO_4·7H_2O$	1.0mg
Resazurin (0.025% solution)	4.0mL
Trace elements solution SL-6	1.0mL
Vitamin solution	10.0mL

pH 7.3 ± 0.2 at 25°C

Trace Elements Solution SL-6:
Composition per liter:

H_3BO_3	0.30g
$CoCl_2·6H_2O$	0.20g
$ZnSO_4·7H_2O$	0.10g
$MnCl_2·4H_2O$	0.03g
$Na_2MoO_4·H_2O$	0.03g
$NiCl_2·6H_2O$	0.02g
$CuCl_2·2H_2O$	0.01g

Preparation of Trace Elements Solution SL-6: Add components to distilled/deionized water and bring volume to 1.0L. Mix thoroughly. Adjust pH to 3.4.

Vitamin Solution:
Composition per liter:

Vitamin B_{12}	2.8mg
Thiamine·HCl	0.28mg

Preparation of Vitamin Solution: Add components to distilled/deionized water and bring volume to 10.0mL. Mix thoroughly. Filter sterilize.

Preparation of Medium: Add components, except vitamin solution, to distilled/deionized water and bring volume to 990.0mL. Mix thoroughly. Adjust pH to 7.3 with 10% Na_2CO_3. Gently heat and bring to boiling. Continue boiling until resazurin changes color. Cool to 50°C. Distribute into tubes in 7.0mL volumes under O_2-free 97% N_2 + 3% H_2. Cap with rubber stoppers under O_2-free 97% N_2 + 3% H_2. Place tubes in a press. Autoclave for 15 min at 15 psi pressure–121°C using fast exhaust. Cool to 50°C. Aseptically add 0.25mL of sterile vitamin solution to each tube.

Use: For the cultivation and maintenance of *Meniscus glaucopis.*

Meniscus glaucopis Broth

Composition per liter:

Maltose	5.0g
Agar	3.0g
Yeast extract	1.0g
KH_2PO_4	0.5g
NaCl	0.4g
NH_4Cl	0.4g
Sodium thioglycollate	0.3g
$MgSO_4 \cdot 7H_2O$	0.2g
$CaCl_2 \cdot H_2O$	0.01g
$FeSO_4 \cdot 7H_2O$	1.0mg
Resazurin (0.025% solution)	4.0mL
Trace elements solution SL-6	1.0mL
Vitamin solution	10.0mL

pH 7.3 ± 0.2 at 25°C

Trace Elements Solution SL-6:

Composition per liter:

H_3BO_3	0.30g
$CoCl_2 \cdot 6H_2O$	0.20g
$ZnSO_4 \cdot 7H_2O$	0.10g
$MnCl_2 \cdot 4H_2O$	0.03g
$Na_2MoO_4 \cdot H_2O$	0.03g
$NiCl_2 \cdot 6H_2O$	0.02g
$CuCl_2 \cdot 2H_2O$	0.01g

Preparation of Trace Elements Solution SL-6: Add components to distilled/deionized water and bring volume to 1.0L. Mix thoroughly. Adjust pH to 3.4.

Vitamin Solution:

Composition per liter:

Vitamin B_{12}	2.8mg
Thiamine·HCl	0.28mg

Preparation of Vitamin Solution: Add components to distilled/deionized water and bring volume to 10.0mL. Mix thoroughly. Filter sterilize.

Preparation of Medium: Add components, except vitamin solution, to distilled/deionized water and bring volume to 990.0mL. Mix thoroughly. Adjust pH to 7.3 with 10% Na_2CO_3. Gently heat and bring to boiling. Continue boiling until resazurin changes color. Cool to 50°C. Distribute into tubes in 7.0mL volumes under O_2-free 97% N_2 + 3% H_2. Cap with rubber stoppers under O_2-free 97% N_2 + 3% H_2. Place tubes in a press. Autoclave for 15 min at 15 psi pressure–121°C using fast exhaust. Cool to 50°C. Aseptically add 0.25mL of sterile vitamin solution to each tube.

Use: For the cultivation and maintenance of *Meniscus glaucopis*.

m-Enterococcus Agar
See: **Enterococcus Agar**

MES Agar
See: **U Agar Plates**

Metallogenium Cultivation Broth

Composition per liter:

Gum arabic	20.0g
$MnCO_3$	0.5g

$MnCO_3$:

Composition per 100mL:

$MnCl_2$	20.0g
$NaHCO_3$ (25% solution)	25.0mL

Preparation of $MnCO_3$: Add $MnCl_2$ to distilled/deionized water and bring volume to 100.0mL. Mix thoroughly. Add $NaHCO_3$ solution. Filter through Whatman #1 filter paper. Save the $MnCO_3$ precipitate. Wash and store under distilled/deionized water.

Preparation of Medium: Add components to distilled/deionized water and bring volume to 1.0L. Mix thoroughly. Distribute into tubes or flasks. Autoclave for 15 min at 15 psi pressure–121°C.

Use: For the cultivation of *Metallogenium* species.

Metallogenium Cultivation Broth

Composition per liter:

Starch, hydrolyzed	20.0g
$MnCO_3$	0.5g

$MnCO_3$:

Composition per 100mL:

$MnCl_2$	20.0g
$NaHCO_3$ (25% solution)	25.0mL

Preparation of $MnCO_3$: Add $MnCl_2$ to distilled/deionized water and bring volume to 100.0mL. Mix thoroughly. Add $NaHCO_3$ solution. Filter through Whatman #1 filter paper. Save the $MnCO_3$ precipitate. Wash and store under distilled/deionized water.

Preparation of Medium: Hydrolyze starch with HCl. Add components to distilled/deionized water and bring volume to 1.0L. Mix thoroughly. Distribute into tubes or flasks. Autoclave for 15 min at 15 psi pressure–121°C.

Use: For the cultivation of *Metallogenium* species.

Metallogenium Isolation Agar

Composition per liter:

Agar	15.0g
Manganese acetate	0.1g

Preparation of Medium: Add components to distilled/deionized water and bring volume to 1.0L. Mix

thoroughly. Gently heat and bring to boiling. Distribute into tubes or flasks. Autoclave for 15 min at 15 psi pressure–121°C. Pour into sterile Petri dishes or leave in tubes.

Use: For the isolation and cultivation of *Metallogenium* species.

Metallogenium Medium

Composition per liter:

$MnCO_3$	2.0g
Starch, hydrolyzed	1.0g
DNA	0.01g
Catalase	5.0mg
Mycoplasma broth base	100.0mL
Yeast extract, ultrafiltrate	100.0mL
Horse serum	10.0mL

Mycoplasma Broth Base:

Composition per liter:

Pancreatic digest of casein	7.0g
NaCl	5.0g
Beef extract	3.0g
Yeast extract	3.0g
Beef heart, solids from infusion	2.0g

Preparation of _Mycoplasma_ Broth Base: Add components to distilled/deionized water and bring volume to 1.0L. Mix thoroughly. Autoclave for 15 min at 15 psi pressure–121°C. Cool to 25°C.

$MnCO_3$:

Composition per 100mL:

$MnCl_2$	20.0g
$NaHCO_3$ (25% solution)	25.0mL

Preparation of $MnCO_3$: Add $MnCl_2$ to distilled/deionized water and bring volume to 100.0mL. Mix thoroughly. Add $NaHCO_3$ solution. Filter through Whatman #1 filter paper. Save the $MnCO_3$ precipitate. Wash and store under distilled/deionized water.

Preparation of Medium: Add $MnCO_3$, hydrolyzed starch and DNA to 25.0mL of distilled/deionized water. Mix thoroughly. Autoclave for 15 min at 15 psi pressure–121°C. Cool to 45°–50°C. Aseptically add 100.0mL of sterile *Mycoplasma* broth base, 100.0mL of ultrafiltrate of yeast extract, 10.0mL of horse serum and 5.0mg of catalase. Mix thoroughly. Aseptically distribute into sterile tubes or flasks.

Use: For the cultivation of *Metallogenium* species.

Methanobacillus Medium

Composition per liter:

KH_2PO_4	9.0g
K_2HPO_4	6.0g
NH_4Cl	5.0g
$MgCl_2$	1.0g
$CaCl_2$	0.01g
$FeSO_4 \cdot 7H_2O$	0.01g
Ethanol	10.0mL

pH 7.4 ± 0.2 at 25°C

Preparation of Medium: Filter sterilize ethanol. Add components, except ethanol, to tap water and bring volume to 990.0mL. Mix thoroughly. Gently heat until dissolved. Autoclave for 20 min at 10psi pressure–115°C. Cool to 45°–50°C. Aseptically add sterile ethanol. Mix thoroughly. Aseptically distribute into sterile tubes or flasks.

Use: For the selective isolation and cultivation of *Methanobacillus* species from mixed cultures.

Methanobacteria Medium

Composition per liter:

Mineral Solution 2	50.0mL
Sodium carbonate solution	50.0mL
Mineral Solution 1	25.0mL
Cysteine-sulfide reducing agent	20.0mL
Wolfe's mineral solution	10.0mL
Vitamin solution	10.0mL
Resazurin (0.025% solution)	4.0mL

pH 7.2 ± 0.2 at 25°C

Mineral Solution 1:

Composition per liter:

K_2HPO_4	6.0g

Preparation of Medium: Add K_2HPO_4 to distilled/deionized water and bring volume to 1.0L. Mix thoroughly.

Mineral Solution 2:

Composition per liter:

NaCl	12.0g
KH_2PO_4	6.0g
$(NH_4)_2SO_4$	6.0g
$MgSO_4 \cdot 7H_2O$	2.4g
$CaCl_2 \cdot 2H_2O$	1.6g

Preparation of Mineral Solution 2: Add components to distilled/deionized water and bring volume to 1.0L. Mix thoroughly.

Sodium Carbonate Solution:

Composition per 100mL:

Na_2CO_3	8.0g

Preparation of Sodium Carbonate Solution: Add Na_2CO_3 to distilled/deionized water and bring volume to 100.0mL. Mix thoroughly.

Cysteine-Sulfide Reducing Agent:

Composition per 20mL:

L-Cysteine·HCl·H_2O	0.3g
$Na_2S \cdot 9H_2O$	0.3g

Preparation of Cysteine-Sulfide Reducing Agent: Add L-Cysteine·HCl·H_2O to 10.0mL of distilled/deionized water. Mix thoroughly. In a separate tube, add $Na_2S·9H_2O$ to 10.0mL of distilled/deionized water. Mix thoroughly. Gas both solutions with 100% N_2 and cap tubes. Autoclave both solutions for 15 min at 15 psi pressure–121°C using fast exhaust. Cool to 50°C. Aseptically combine the two solutions under 100% N_2.

Wolfe's Mineral Solution:

Composition per liter

$MgSO_4·7H_2O$	3.0g
Nitriloacetic acid	1.5g
NaCl	1.0g
$MnSO_4·H_2O$	0.5g
$FeSO_4·7H_2O$	0.1g
$CoCl_2·6H_2O$	0.1g
$CaCl_2$	0.1g
$ZnSO_4·7H_2O$	0.1g
$CuSO_4·5H_2O$	0.01g
$AlK(SO_4)_2·12H_2O$	0.01g
H_3BO_3	0.01g
$Na_2MoO_4·2H_2O$	0.01g

Preparation of Wolfe's Mineral Solution: Add nitrilotriacetic acid to 500.0mL of distilled/deionized water. Dissolve by adjusting pH to 6.5 with KOH. Add remaining components. Add distilled/deionized water and bring volume to 1.0L.

Wolfe's Vitamin Solution:

Composition per liter:

Pyridoxine·HCl	10.0mg
Thiamine·HCl	5.0mg
Riboflavin	5.0mg
Nicotinic acid	5.0mg
Calcium pantothenate	5.0mg
p-Aminobenzoic acid	5.0mg
Thioctic acid	5.0mg
Biotin	2.0mg
Folic acid	2.0mg
Cyanocobalamin	100.0μg

Preparation of Wolfe's Vitamin Solution: Add components to distilled/deionized water and bring volume to 1.0L. Mix thoroughly. Filter sterilize.

Preparation of Medium: Add components, except vitamin solution and cysteine-sulfide reducing agent, to distilled/deionized water and bring volume to 970.0mL. Mix thoroughly. Autoclave for 15 min at 15 psi pressure–121°C. Cool under 80% N_2 + 20% CO_2. Aseptically add the sterile vitamin solution and then the sterile cysteine-sulfide reducing agent. Adjust the pH to 7.2. Distribute aseptically and anaerobically into sterile tubes.

Use: For the cultivation and maintenance of *Acetogenium kivui, Methanobacterium formicicum, Methanobacterium thermoautotrophicum,* and *Methanobrevibacter arboriphilicus.*

Methanobacteria Medium with Glucose and Yeast Extract

Composition per liter:

Glucose	5.0g
Yeast extract	2.0g
Mineral solution 2	50.0mL
Sodium carbonate solution	50.0mL
Mineral solution 1	25.0mL
Cysteine-sulfide reducing agent	20.0mL
Wolfe's mineral solution	10.0mL
Vitamin solution	10.0mL
Resazurin (0.025% solution)	4.0mL

pH 7.2 ± 0.2 at 25°C

Mineral Solution 1:

Composition per liter:

K_2HPO_4	6.0g

Preparation of Mineral Solution 1: Add K_2HPO_4 to distilled/deionized water and bring volume to 1.0L. Mix thoroughly.

Mineral Solution 2:

Composition per liter:

NaCl	12.0g
KH_2PO_4	6.0g
$(NH_4)_2SO_4$	6.0g
$MgSO_4·7H_2O$	2.4g
$CaCl_2·2H_2O$	1.6g

Preparation of Mineral Solution 2: Add components to distilled/deionized water and bring volume to 1.0L. Mix thoroughly.

Sodium Carbonate Solution:

Composition per 100mL:

Na_2CO_3	8.0g

Preparation of Sodium Carbonate Solution: Add Na_2CO_3 to distilled/deionized water and bring volume to 100.0mL. Mix thoroughly.

Cysteine-Sulfide Reducing Agent:

Composition per 20mL:

L-Cysteine·HCl·H_2O	0.3g
$Na_2S·9H_2O$	0.3g

Preparation of Cysteine-Sulfide Reducing Agent: Add L-Cysteine·HCl·H_2O to 10.0mL of distilled/deionized water. Mix thoroughly. In a separate tube, add $Na_2S·9H_2O$ to 10.0mL of distilled/deionized water. Mix thoroughly. Gas both solutions with 100% N_2 and cap tubes. Autoclave both solutions for

15 min at 15 psi pressure–121°C using fast exhaust. Cool to 50°C. Aseptically combine the two solutions under 100% N_2.

Wolfe's Mineral Solution:

Composition per liter

$MgSO_4 \cdot 7H_2O$	3.0g
Nitriloacetic acid	1.5g
NaCl	1.0g
$MnSO_4 \cdot H_2O$	0.5g
$FeSO_4 \cdot 7H_2O$	0.1g
$CoCl_2 \cdot 6H_2O$	0.1g
$CaCl_2$	0.1g
$ZnSO_4 \cdot 7H_2O$	0.1g
$CuSO_4 \cdot 5H_2O$	0.01g
$AlK(SO_4)_2 \cdot 12H_2O$	0.01g
H_3BO_3	0.01g
$Na_2MoO_4 \cdot 2H_2O$	0.01g

Preparation of Wolfe's Mineral Solution: Add nitrilotriacetic acid to 500.0mL of distilled/deionized water. Dissolve by adjusting pH to 6.5 with KOH. Add remaining components. Add distilled/deionized water to 1.0L.

Wolfe's Vitamin Solution:

Composition per liter:

Pyridoxine·HCl	10.0mg
Thiamine·HCl	5.0mg
Riboflavin	5.0mg
Nicotinic acid	5.0mg
Calcium pantothenate	5.0mg
p-Aminobenzoic acid	5.0mg
Thioctic acid	5.0mg
Biotin	2.0mg
Folic acid	2.0mg
Cyanocobalamin	100.0μg

Preparation of Wolfe's Vitamin Solution: Add components to distilled/deionized water and bring volume to 1.0L. Mix thoroughly. Filter sterilize.

Preparation of Medium: Add components, except vitamin solution and cysteine-sulfide reducing agent, to distilled/deionized water and bring volume to 970.0mL. Mix thoroughly. Autoclave for 15 min at 15 psi pressure–121°C. Cool under 80% N_2 + 20% CO_2. Aseptically add the sterile vitamin solution and then the sterile cysteine-sulfide reducing agent. Adjust the pH to 7.2. Distribute aseptically and anaerobically into sterile tubes.

Use: For the cultivation and maintenance of *Clostridium saccharolyticum, Clostridium thermoaceticum,* and *Clostridium thermohydrosulfuricum.*

Methanobacteria Medium with Xylose, Yeast Extract and Tryptone

Composition per liter:

Pancreatic digest of casein	10.0g
Xylose	5.0g
Yeast extract	3.0g
Mineral solution 2	50.0mL
Sodium carbonate solution	50.0mL
Mineral solution 1	25.0mL
Cysteine-sulfide reducing agent	20.0mL
Wolfe's mineral solution	10.0mL
Vitamin solution	10.0mL
Resazurin (0.025% solution)	4.0mL

pH 7.2 ± 0.2 at 25°C

Mineral Solution 1:

Composition per liter:

K_2HPO_4	6.0g

Preparation of Medium: Add K_2HPO_4 to distilled/deionized water and bring volume to 1.0L. Mix thoroughly.

Mineral Solution 2:

Composition per liter:

NaCl	12.0g
KH_2PO_4	6.0g
$(NH_4)_2SO_4$	6.0g
$MgSO_4 \cdot 7H_2O$	2.4g
$CaCl_2 \cdot 2H_2O$	1.6g

Preparation of Mineral Solution 2: Add components to distilled/deionized water and bring volume to 1.0L. Mix thoroughly.

Sodium Carbonate Solution:

Composition per 100mL:

Na_2CO_3	8.0g

Preparation of Sodium Carbonate Solution: Add Na_2CO_3 to distilled/deionized water and bring volume to 100.0mL. Mix thoroughly.

Cysteine-Sulfide Reducing Agent:

Composition per 20mL:

L-Cysteine·HCl·H_2O	0.3g
$Na_2S \cdot 9H_2O$	0.3g

Preparation of Cysteine-Sulfide Reducing Agent: Add L-Cysteine·HCl·H_2O to 10.0mL of distilled/deionized water. Mix thoroughly. In a separate tube, add $Na_2S \cdot 9H_2O$ to 10.0mL of distilled/deionized water. Mix thoroughly. Gas both solutions with 100% N_2 and cap tubes. Autoclave both solutions for 15 min at 15 psi pressure–121°C using fast exhaust. Cool to 50°C. Aseptically combine the two solutions under 100% N_2.

Wolfe's Mineral Solution:
Composition per liter:

$MgSO_4 \cdot 7H_2O$	3.0g
Nitriloacetic acid	1.5g
NaCl	1.0g
$MnSO_4 \cdot H_2O$	0.5g
$FeSO_4 \cdot 7H_2O$	0.1g
$CoCl_2 \cdot 6H_2O$	0.1g
$CaCl_2$	0.1g
$ZnSO_4 \cdot 7H_2O$	0.1g
$CuSO_4 \cdot 5H_2O$	0.01g
$AlK(SO_4)_2 \cdot 12H_2O$	0.01g
H_3BO_3	0.01g
$Na_2MoO_4 \cdot 2H_2O$	0.01g

Preparation of Wolfe's Mineral Solution: Add nitrilotriacetic acid to 500.0mL of distilled/deionized water. Dissolve by adjusting pH to 6.5 with KOH. Add remaining components. Add distilled/deionized water to 1.0L.

Wolfe's Vitamin Solution:
Composition per liter:

Pyridoxine·HCl	10.0mg
Thiamine·HCl	5.0mg
Riboflavin	5.0mg
Nicotinic acid	5.0mg
Calcium pantothenate	5.0mg
p-Aminobenzoic acid	5.0mg
Thioctic acid	5.0mg
Biotin	2.0mg
Folic acid	2.0mg
Cyanocobalamin	100.0μg

Preparation of Wolfe's Vitamin Solution: Add components to distilled/deionized water and bring volume to 1.0L. Mix thoroughly. Filter sterilize.

Preparation of Medium: Add components, except vitamin solution and cysteine-sulfide reducing agent, to distilled/deionized water and bring volume to 970.0mL. Mix thoroughly. Autoclave for 15 min at 15 psi pressure–121°C. Cool under 80% N_2 + 20% CO_2. Aseptically add the sterile vitamin solution and then the sterile cysteine-sulfide reducing agent. Adjust the pH to 7.2. Distribute aseptically and anaerobically into sterile tubes.

Use: For the cultivation and maintenance of *Thermobacteroides acetoethylicus.*

Methanobacteria Medium with Yeast Extract, Sodium Acetate and Methanol

Composition per liter:

Glucose	5.0g
Sodium acetate	4.1g
Yeast extract	2.0g
Mineral solution 2	50.0mL
Sodium carbonate solution	50.0mL
Mineral solution 1	25.0mL
Cysteine-sulfide reducing agent	20.0mL
Wolfe's mineral solution	10.0mL
Vitamin solution	10.0mL
Methanol	4.0mL
Resazurin (0.025% solution)	4.0mL

pH 7.2 ± 0.2 at 25°C

Mineral Solution 1:
Composition per liter:

K_2HPO_4	6.0g

Preparation of Mineral Solution 1: Add K_2HPO_4 to distilled/deionized water and bring volume to 1.0L. Mix thoroughly.

Mineral Solution 2:
Composition per liter:

NaCl	12.0g
KH_2PO_4	6.0g
$(NH_4)_2SO_4$	6.0g
$MgSO_4 \cdot 7H_2O$	2.4g
$CaCl_2 \cdot 2H_2O$	1.6g

Preparation of Mineral Solution 2: Add components to distilled/deionized water and bring volume to 1.0L. Mix thoroughly.

Sodium Carbonate Solution:
Composition per 100mL:

Na_2CO_3	8.0g

Preparation of Sodium Carbonate Solution: Add Na_2CO_3 to distilled/deionized water and bring volume to 100.0mL. Mix thoroughly.

Cysteine-Sulfide Reducing Agent:
Composition per 20mL:

L-Cysteine·HCl·H_2O	300.0mg
$Na_2S \cdot 9H_2O$	300.0mg

Preparation of Cysteine-Sulfide Reducing Agent: Add L-Cysteine·HCl·H_2O to 10.0mL of distilled/deionized water. Mix thoroughly. In a separate tube, add $Na_2S \cdot 9H_2O$ to 10.0mL of distilled/deionized water. Mix thoroughly. Gas both solutions with 100% N_2 and cap tubes. Autoclave both solutions for 15 min at 15 psi pressure–121°C using fast exhaust. Cool to 50°C. Aseptically combine the two solutions under 100% N_2.

Wolfe's Mineral Solution:
Composition per liter:

$MgSO_4 \cdot 7H_2O$	3.0g
Nitriloacetic acid	1.5g
NaCl	1.0g
$MnSO_4 \cdot H_2O$	0.5g

$FeSO_4 \cdot 7H_2O$	0.1g
$CoCl_2 \cdot 6H_2O$	0.1g
$CaCl_2$	0.1g
$ZnSO_4 \cdot 7H_2O$	0.1g
$CuSO_4 \cdot 5H_2O$	0.01g
$AlK(SO_4)_2 \cdot 12H_2O$	0.01g
H_3BO_3	0.01g
$Na_2MoO_4 \cdot 2H_2O$	0.01g

Preparation of Wolfe's Mineral Solution: Add nitrilotriacetic acid to 500.0mL of distilled/deionized water. Dissolve by adjusting pH to 6.5 with KOH. Add remaining components. Add distilled/deionized water to 1.0L.

Wolfe's Vitamin Solution:

Composition per liter:

Pyridoxine·HCl	10.0mg
Thiamine·HCl	5.0mg
Riboflavin	5.0mg
Nicotinic acid	5.0mg
Calcium pantothenate	5.0mg
p-Aminobenzoic acid	5.0mg
Thioctic acid	5.0mg
Biotin	2.0mg
Folic acid	2.0mg
Cyanocobalamin	100.0μg

Preparation of Wolfe's Vitamin Solution: Add components to distilled/deionized water and bring volume to 1.0L. Mix thoroughly. Filter sterilize.

Preparation of Medium: Add components, except vitamin solution, cysteine-sulfide reducing agent and methanol, to distilled/deionized water and bring volume to 970.0mL. Mix thoroughly. Autoclave for 15 min at 15 psi pressure–121°C. Cool under 80% N_2 + 20% CO_2. Filter sterilize methanol. Aseptically add 4.0mL of sterile methanol to cooled, sterile basal medium. Aseptically add the sterile vitamin solution and then the sterile cysteine-sulfide reducing agent. Adjust the pH to 7.2. Distribute aseptically and anaerobically into sterile tubes.

Use: For the cultivation and maintenance of *Butyribacterium methylotrophicum.*

Methanobacterium alcaliphilum Medium

Composition per liter:

$NaHCO_3$	10.0g
Yeast extract	2.0g
Peptone	2.0g
NH_4Cl	1.0g
L-Cysteine·HCl·H_2O	0.5g
K_2HPO_4	0.4g
$MgCl_2 \cdot 6H_2O$	0.1g
$CaCl_2$	0.02g
Resazurin	1.0mg
Salt solution	5.0mL

pH 8.4 ± 0.2 at 25°C

Salt Solution:

Composition per 100mL:

Sodium EDTA·$2H_2O$	0.10g
$CoCl_2 \cdot 6H_2O$	0.03g
$MnCl_2 \cdot 4H_2O$	0.02g
$ZnCl_2$	0.02g
$AlCl_3 \cdot 6H_2O$	8.0mg
$CuCl_2 \cdot 2H_2O$	4.0mg
$NiSO_4 \cdot 6H_2O$	4.0mg
Na_2SeO_3	2.7mg
$FeSO_4 \cdot 7H_2O$	2.0mg
H_3BO_3	2.0mg
$NaMoO_4 \cdot 2H_2O$	2.0mg

Preparation of Salt Solution: Add components to distilled/deionized water and bring volume to 100.0mL. Mix thoroughly.

Preparation of Medium: Add components, except $NaHCO_3$, yeast extract, peptone and L-cysteine·-HCl·H_2O, to distilled/deionized water and bring volume to 990.0mL. Gently heat and bring to boiling under O_2-free 100% N_2. Continue boiling until resazurin becomes pale, indicating partial reduction. Add the yeast extract, peptone and L-cysteine·HCl·H_2O and continue boiling under O_2-free 100% N_2 until resazurin becomes colorless, indicating complete reduction. Cool to room temperature under O_2-free 100% N_2. Add $NaHCO_3$ to 10.0mL of distilled/deionized water. Mix thoroughly. Gas with O_2-free 100% N_2 in a sealed tube. Add reduced $NaHCO_3$ solution to cooled reduced medium. Distribute anaerobically into tubes. Cap with butyl rubber stoppers and secure with closures. Autoclave for 15 min at 15 psi pressure–121°C with fast exhaust.

Use: For cultivation and maintenance of *Methanobacterium alcaliphilum.*

Methanobacterium Enrichment Medium

Composition per liter:

$CaCO_3$	100.0g
K_2HPO_4	5.0g
$(NH_4)_2SO_4$	0.3g
$MgSO_4 7H_2O$	0.1g
$FeSO_4 \cdot 7H_2O$	0.02g
Na_2CO_3 solution	10.0mL
$Na_2S \cdot 9H_2O$ solution	10.0mL
Ethanol	10.0mL
Yeast autolysate	5.0mL

pH 7.2 ± 0.2 at 25°C

Na_2CO_3 Solution:
Composition per 10mL:

$NaHCO_3$	0.5g

Preparation of Na_2CO_3 Solution: Add Na_2CO_3 to distilled/deionized water and bring volume to 10.0mL. Mix thoroughly. Filter sterilize.

$Na_2S{\cdot}9H_2O$ Solution:
Composition per 10mL:

$Na_2S{\cdot}9H_2O$	0.1g

Preparation of $Na_2S{\cdot}9H_2O$ Solution: Add $Na_2S{\cdot}9H_2O$ to distilled/deionized water and bring volume to 10.0mL. Mix thoroughly. Filter sterilize.

Preparation of Medium: Filter sterilize ethanol. Add components—except ethanol, Na_2CO_3 solution, and $Na_2S{\cdot}9H_2O$ solution—to distilled/deionized water and bring volume to 970.0mL. Mix thoroughly. Gently heat and bring to boiling. Autoclave for 15 min at 15 psi pressure–121°C. Cool to 45°–50°C. Aseptically add sterile ethanol, Na_2CO_3 solution, and $Na_2S{\cdot}9H_2O$ solution. Mix thoroughly. Aseptically distribute into sterile tubes or flasks.

Use: For the cultivation and enrichment of *Methanobacterium* species.

Methanobacterium ruminantium Medium

Composition per liter:

$NaHCO_3$	6.0g
NaCl	2.0g
Cysteine·HCl·H_2O	1.0g
$K_2HPO_4{\cdot}3H_2O$	1.0g
KH_2PO_4	1.0g
NH_4Cl	1.0g
$CaCl_2{\cdot}2H_2O$	0.1g
$MgSO_4{\cdot}7H_2O$	0.1g
Resazurin	1.0mg
Rumen fluid	300.0mL
$Na_2S{\cdot}9H_2O$ solution	10.0mL

6.8 ± 0.2 at 25°C

$Na_2S{\cdot}9H_2O$ Solution:
Composition per 10mL:

$Na_2S{\cdot}9H_2O$	0.25g

Preparation of $Na_2S{\cdot}9H_2O$ Solution: Add $Na_2S{\cdot}9H_2O$ to distilled/deionized water and bring volume to 10.0mL. Mix thoroughly. Autoclave for 15 min at 15 psi pressure–121°C. Cool to 25°C.

Preparation of Medium: Prepare and distribute medium anaerobically under 80% H_2 + 20% CO_2. Add components, except rumen fluid and $Na_2S{\cdot}9H_2O$ solution, to distilled/deionized water and bring volume to 690.0mL. Mix thoroughly. Gently heat and bring to boiling. Continue boiling until resazurin turns colorless, indicating reduction. Autoclave for 15 min at 15 psi pressure–121°C. Cool to 25°C. Aseptically add 10.0mL of sterile $Na_2S{\cdot}9H_2O$ solution and 300.0mL of sterile rumen fluid. Mix thoroughly. Aseptically and anaerobically distribute into sterile tubes or flasks.

Use: For the cultivation of *Methanobacterium ruminantium.*

Methanobacterium thermoautotrophicum Medium, Taylor and Pirt

Composition per liter:

Na_2CO_3	4.0g
$(NH_4)_2SO_4$	3.0g
NaCl	1.2g
KH_2PO_4	0.6g
Cysteine·HCl·H_2O	0.5g
K_2HPO_4	0.3g
Nitrilotriacetic acid	0.03g
$CoCl_2$	0.02g
$CaCl_2$	0.01g
$FeSO_4$	0.01g
$MgSO_4$	0.01g
$MnSO_4$	0.01g
$ZnSO_4$	2.0mg
Resazurin	1.0mg
$AlK(SO_4)_2$	0.2mg
$CuSO_4$	0.2mg
H_3BO_3	0.2mg
Na_2MoO_4	0.2mg
$Na_2S{\cdot}9H_2O$ solution	10.0mL

pH 7.2 ± 0.2 at 25°C

$Na_2S{\cdot}9H_2O$ Solution:
Composition per 10mL:

$Na_2S{\cdot}9H_2O$	0.5g

Preparation of $Na_2S{\cdot}9H_2O$ Solution: Add $Na_2S{\cdot}9H_2O$ to distilled/deionized water and bring volume to 10.0mL. Mix thoroughly. Autoclave for 15 min at 15 psi pressure–121°C. Cool to 25°C.

Preparation of Medium: Prepare and distribute medium anaerobically under 80% H_2 + 20% CO_2. Add components, except $Na_2S{\cdot}9H_2O$ solution, to distilled/deionized water and bring volume to 990.0mL. Mix thoroughly. Gently heat and bring to boiling. Continue boiling until resazurin turns colorless, indicating reduction. Autoclave for 15 min at 15 psi pressure–121°C. Cool to 25°C. Aseptically add 10.0mL of sterile $Na_2S{\cdot}9H_2O$ solution. Mix thoroughly. Aseptically and anaerobically distribute into sterile tubes or flasks.

Use: For the cultivation of *Methanobacterium thermoautotrophicum*.

Methanococcus vannielii Medium

Composition per 1020mL:

Solution A	500.0mL
Inorganic salts solution	500.0mL
$Na_2S·9H_2O$ solution	10.0mL
Na_2CO_3 solution	10.0mL

Solution A:

Composition per 500mL:

Sodium formate	10.0g
Phenol Red	3.0mg
Methylene Blue	2.0mg

Preparation of Solution A: Add components to distilled/deionized water and bring volume to 500.0mL. Mix thoroughly. Autoclave for 15 min at 15 psi pressure–121°C. Cool to 25°C.

Inorganic Salts Solution:

Composition per 500mL:

$K_2HPO_4·3H_2O$	1.45g
NH_4Cl	1.0g
KH_2PO_4	0.75g
$MgCl_2·6H_2O$	0.2g
Nitrilotriacetic acid	0.04g
$CaCl_2·2H_2O$	0.02g
$FeCl_2·4H_2O$	3.6mg
$CoCl_2·6H_2O$	1.5mg
MnCl2·4H2O	0.9mg
$ZnCl_2$	0.9mg
H_3BO_2	0.17mg
$Na_2MoO_4·2H_2O$	.09mg

Preparation of Inorganic Salts Solution: Add nitrilotriacetic acid to 250.0mL of distilled/deionized water. Dissolve by adjusting pH to 6.5 with KOH. Add remaining components. Readjust pH to 7.2 with H_2SO_4 or KOH. Add distilled/deionized water to 500.0mL. Filter sterilize.

$Na_2S·9H_2O$ Solution:

Composition per 10mL:

$Na_2S·9H_2O$	0.3g

Preparation of $Na_2S·9H_2O$ Solution: Add $Na_2S·9H_2O$ to distilled/deionized water and bring volume to 10.0mL. Mix thoroughly. Autoclave for 15 min at 15 psi pressure–121°C. Cool to 25°C.

Na_2CO_3 Solution:

Composition per 10mL:

Na_2CO_3	2.5g

Preparation of Na_2CO_3 Solution: Add Na_2CO_3 to distilled/deionized water and bring volume to 10.0mL. Mix thoroughly. Autoclave for 15 min at 15 psi pressure–121°C. Cool to 25°C.

Preparation of Medium: Prepare and distribute medium anaerobically under 80% N_2 + 20% CO_2. Aseptically and anaerobically combine 500.0mL of sterile inorganic salts solution, 500.0mL of sterile solution A, 10.0mL of sterile $Na_2S·9H_2O$ solution, and 10.0mL of sterile Na_2CO_3 solution. Mix thoroughly. Aseptically and anaerobically distribute into sterile tubes or flasks.

Use: For the isolation and cultivation of *Methanococcus vannielii* from marine mud.

Methanococcus vannielii Medium

Composition per liter:

Sodium formate	15.0g
K_2HPO_4	3.48g
$CoCl_2·6H_2O$	2.38g
NH_4Cl	1.0g
Cysteine·HCl·H_2O	0.3g
$MgSO_4·7H_2O$	0.2g
$CaCl_2·2H_2O$	0.01g
$FeSO_4·7H_2O$	0.01g
$MnSO_4·H_2O$	7.5mg
$Na_2MoO_4·2H_2O$	7.5mg
Na_2SeO_3	1.7mg
$Na_2S·9H_2O$	10.0mL

$Na_2S·9H_2O$ Solution:

Composition per 10mL:

$Na_2S·9H_2O$	0.15g

Preparation of $Na_2S·9H_2O$ Solution: Add $Na_2S·9H_2O$ to distilled/deionized water and bring volume to 10.0mL. Mix thoroughly. Autoclave for 15 min at 15 psi pressure–121°C. Cool to 25°C.

Preparation of Medium: Prepare and distribute medium anaerobically under 100% N_2. Add components, except $Na_2S·9H_2O$ solution, to distilled/deionized water and bring volume to 990.0mL. Mix thoroughly. Gently heat and bring to boiling. Continue boiling until resazurin turns colorless, indicating reduction. Autoclave for 15 min at 15 psi pressure–121°C. Cool to 25°C. Aseptically add 10.0mL of sterile $Na_2S·9H_2O$ solution. Mix thoroughly. Aseptically and anaerobically distribute into sterile tubes or flasks.

Use: For the cultivation of *Methanococcus vannielii*.

Methanogen Enrichment Medium, Barker

Composition per liter:

$CaCO_3$	20.0g

NH_4Cl	1.0g
$K_2HPO_4 \cdot 3H_2O$	0.4g
$MgCl_2 \cdot 6H_2O$	0.1g
Methanol	20.0mL

pH 7.0 ± 0.2 at 25°C

Preparation of Medium: Add components, except methanol and $CaCO_3$, to distilled/deionized water and bring volume to 1.0L. Mix thoroughly. Gently heat and bring to boiling. Autoclave for 15 min at 15 psi pressure–121°C. Cool to 25°C. Aseptically add filter-sterilized methanol solution. Mix thoroughly. Add 1.0g of $CaCO_3$ to each of 50.0mL screw-capped bottles. Autoclave for 15 min at 15 psi pressure–121°C. Cool to 25°C. Fill each bottle to capacity with enrichment medium.

Use: For the cultivation of methanogenic bacteria.

Methanogen Medium

Composition per 106mL:

$CaCO_3$	10.0g
Calcium acetate	2.0g
NH_4Cl	0.1g
$K_2HPO_4 \cdot 3H_2O$	0.04g
$MgCl_2 \cdot 6H_2O$	0.01g
$Na_2S \cdot 9H_2O$ solution	3.0mL
Na_2CO_3 solution	3.0mL

$Na_2S \cdot 9H_2O$ Solution:
Composition per 10mL:

$Na_2S \cdot 9H_2O$	0.1g

Preparation of $Na_2S \cdot 9H_2O$ Solution: Add $Na_2S \cdot 9H_2O$ to distilled/deionized water and bring volume to 10.0mL. Mix thoroughly. Autoclave for 15 min at 15 psi pressure–121°C. Cool to 25°C.

Na_2CO_3 Solution:
Composition per 10mL:

Na_2CO_3	0.5g

Preparation of Na_2CO_3 Solution: Add Na_2CO_3 to distilled/deionized water and bring volume to 10.0mL. Mix thoroughly. Autoclave for 15 min at 15 psi pressure–121°C. Cool to 25°C.

Preparation of Medium: Prepare and distribute medium anaerobically under 100% N_2. Add components, except $Na_2S \cdot 9H_2O$ solution and Na_2CO_3 solution, to distilled/deionized water and bring volume to 100.0mL. Mix thoroughly. Autoclave for 15 min at 15 psi pressure–121°C. Cool to 25°C. Aseptically add 3.0mL of sterile $Na_2S \cdot 9H_2O$ solution and 3.0mL of sterile Na_2CO_3 solution. Mix thoroughly. Aseptically and anaerobically distribute into sterile tubes or flasks.

Use: For the cultivation and enrichment of acetate-utilizing methanogenic bacteria.

Methanogen Medium, Zeikus

Composition per 1010mL:

Inorganic salts solution	500.0mL
Vitamin solution	500.0mL
$Na_2S \cdot 9H_2O$ solution	10.0mL

pH 7.0 ± 0.2 at 25°C

Inorganic Salts Solution:
Composition per 500mL:

$K_2HPO_4 \cdot 3H_2O$	1.45g
NH_4Cl	1.0g
KH_2PO_4	0.75g
$MgCl_2 \cdot 6H_2O$	0.2g
Nitrilotriacetic acid	0.04g
$CaCl_2 \cdot 2H_2O$	0.02g
$FeCl_2 \cdot 4H_2O$	3.6mg
$CoCl_2 \cdot 6H_2O$	1.5mg
MnCl2·4H2O	0.9mg
$ZnCl_2$	0.9mg
H_3BO_2	0.17mg
$Na_2MoO_4 \cdot 2H_2O$	.09mg

Preparation of Inorganic Salts Solution: Add nitrilotriacetic acid to 250.0mL of distilled/deionized water. Dissolve by adjusting pH to 6.5 with KOH. Add remaining components. Readjust pH to 7.2 with H_2SO_4 or KOH. Add distilled/deionized water to 500.0mL. Filter sterilize.

Vitamin Solution:
Composition per 500mL:

Pyridoxine·HCl	1.0mg
p-Aminobenzoic acid	0.5mg
Ca-D-pantothenate	0.5mg
Nicotinic acid	0.5mg
Riboflavin	0.5mg
Thiamine·HCl	0.5mg
Thioctic acid	0.5mg
Biotin	0.2mg
Folic acid	0.2mg
Vitamin B_{12}	0.01mg

Preparation of Vitamin Solution: Add components to distilled/deionized water and bring volume to 500.0mL. Mix thoroughly. Filter sterilize.

$Na_2S \cdot 9H_2O$ Solution:
Composition per 10mL:

$Na_2S \cdot 9H_2O$	0.3g

Preparation of $Na_2S \cdot 9H_2O$ Solution: Add $Na_2S \cdot 9H_2O$ to distilled/deionized water and bring volume to 10.0mL. Mix thoroughly. Autoclave for 15 min at 15 psi pressure–121°C. Cool to 25°C.

Preparation of Medium: Prepare and distribute medium anaerobically under 95% N_2 + 5% CO_2. Aseptically and anaerobically combine 500.0mL of sterile inorganic salts solution, 500.0mL of sterile vitamin solution and 10.0mL of sterile $Na_2S \cdot 9H_2O$ solution. Mix thoroughly. Aseptically and anaerobically distribute into sterile tubes or flasks.

Use: For the cultivation of methanogenic bacteria.

Methanogenium Medium

Composition per liter:

NaCl	18.0g
$NaHCO_3$	5.0g
$MgSO_4 \cdot 7H_2O$	3.45g
$MgCl_2 \cdot 7H_2O$	2.75g
Yeast extract	2.0g
Pancreatic digest of casein	2.0g
Sodium acetate	1.0g
Resazurin	1.0g
L-Cysteine·HCl·H_2O	0.5g
$Na_2S \cdot 9H_2O$	0.5g
KCl	0.335g
NH_4Cl	0.25g
$CaCl_2 \cdot 2H_2O$	0.14g
K_2HPO_4	0.14g
$Fe(NH_4)_2(SO_4)_2 \cdot 7H_2O$	2.0mg
Trace elements solution SL-6	10.0mL
Wolfe's vitamin solution	10.0mL

pH 6.8 ± 0.2 at 25°C

Trace Elements Solution SL-6:

Composition per liter:

H_3BO_3	0.30g
$CoCl_2 \cdot 6H_2O$	0.20g
$ZnSO_4 \cdot 7H_2O$	0.10g
$MnCl_2 \cdot 4H_2O$	0.03g
$Na_2MoO_4 \cdot H_2O$	0.03g
$NiCl_2 \cdot 6H_2O$	0.02g
$CuCl_2 \cdot 2H_2O$	0.01g

Preparation of Trace Elements Solution SL-6: Add components to distilled/deionized water and bring volume to 1.0L. Mix thoroughly. Adjust pH to 3.4.

Wolfe's Vitamin Solution:

Composition per liter:

Pyridoxine·HCl	10.0mg
Thiamine·HCl	5.0mg
Riboflavin	5.0mg
Nicotinic acid	5.0mg
Calcium pantothenate	5.0mg
p-Aminobenzoic acid	5.0mg
Thioctic acid	5.0mg
Biotin	2.0mg
Folic acid	2.0mg
Cyanocobalamin	100.0μg

Preparation of Wolfe's Vitamin Solution: Add components to distilled/deionized water and bring volume to 1.0L. Mix thoroughly. Filter sterilize.

Preparation of Medium: Prepare and dispense medium under 80% N_2 + 20% CO_2. Add components, except Wolfe's vitamin solution, to distilled/deionized water and bring volume to 990.0mL. Mix thoroughly. Adjust pH to 6.8. Autoclave for 15 min at 15 psi pressure–121°C. Cool under 80% N_2 + 20% CO_2. Aseptically add sterile Wolfe's vitamin solution. Aseptically and anaerobically distribute into sterile tubes or flasks.

Use: For the cultivation and maintenance of *Methanococcus frisius, Methanococcus maripaludis, Methanococcus thermolithotrophicus,* and *Methanoplanus limicola.*

Methanohalophilus Medium
See: Halomethanococcus Medium

Methanol Ammonium Salts Medium

Composition per liter:

$MgSO_4 \cdot 7H_2O$	1.0g
NH_4Cl	0.5g
Na_2HPO_4	0.33g
KH_2PO_4	0.26g
$CaCl_2$	0.2g
Ferrous EDTA	5.0mg
$Na_2MoO_4 \cdot 2H_2O$	2.0mg
$FeSO_4 \cdot 7H_2O$	500.0μg
$ZnSO_4 \cdot 7H_2O$	400.0μg
EDTA	250.0μg
$CoCl_2 \cdot 6H_2O$	50.0μg
$MnCl_2 \cdot 4H_2O$	20.0μg
H_3BO_4	15.0μg
$NiCl_2 \cdot 6H_2O$	10.0μg
Methanol	5.0mL

pH 6.8 ± 0.2 at 25°C

Preparation of Medium: Add Na_2HPO_4 and KH_2PO_4 to distilled/deionized water and bring volume to 100.0mL. Mix thoroughly. In a separate container, add remaining components, except methanol, to distilled/deionized water and bring volume to 895.0mL. Mix thoroughly. Autoclave both solutions for 15 min at 15 psi pressure–121°C. Cool to room temperature. Filter sterilize methanol. Aseptically add the sterile phosphate solution and the sterile methanol to the cooled, sterile basal medium. Mix thoroughly. Aseptically distribute into sterile tubes or flasks.

Use: For the maintenance and cultivation of *Methylomonas methylotrophus.*

Methanol Medium (ATCC Medium 436)

Composition per liter:

Agar	15.0g
K_2HPO_4	7.0g
$(NH_4)_2SO_4$	3.0g
KH_2PO_4	2.0g
$MgSO_4 \cdot 7H_2O$	0.5g
Yeast extract	0.2g
$FeSO_4 \cdot 7H_2O$	0.01g
$MnSO_4 \cdot H_2O$	8.0mg
Biotin	0.2μg
Thiamine·HCl	0.2μg
Methanol	10.0mL

pH 7.0 ± 0.2 at 25°C

Preparation of Medium: Add components, except methanol, to distilled/deionized water and bring volume to 990.0mL. Mix thoroughly. Gently heat and bring to boiling. Autoclave for 15 min at 15 psi pressure–121°C. Cool to 50°–55°C. Filter sterilize methanol. Aseptically add the sterile methanol to the cooled, sterile basal medium. Mix thoroughly. Aseptically distribute into sterile tubes or flasks.

Use: For the cultivation and maintenance of *Ancylobacter* species, *Methanomonas methylovora*, and *Methylobacterium* species.

Methanol Medium (ATCC Medium 1096)

Composition per liter:

NH_4NO_3	0.75g
$FeCl_3$	0.743g
Methanol	0.45g
$MgSO_4$	0.09g
KH_2PO_4	0.044g
$Na_2MoO_4 \cdot 2H_2O$	0.1mg

pH 7.0 ± 0.2 at 25°C

Preparation of Medium: Prepare and dispense medium under 97% N_2 + 3% H_2. Add components, except methanol, to distilled/deionized water and bring volume to 999.0mL. Mix thoroughly. Gently heat and bring to boiling. Autoclave for 15 min at 15 psi pressure–121°C. Cool to 50°–55°C. Filter sterilize methanol. Aseptically add the sterile methanol to the cooled, sterile basal medium. Mix thoroughly. Aseptically distribute into sterile tubes or flasks.

Use: For the cultivation and maintenance of *Ancylobacter, Methylobacterium* species and *Methanomonas methylovora.*

Methanol Medium for *Achromobacter*

Composition per liter:

NH_4Cl	5.0g
KH_2PO_4	2.0g
NaCl	0.5g
$MgSO_4$	0.2g
Yeast extract	0.2g
$FeSO_4$	2.0mg
$MnCl_2$	2.0mg
Methanol	20.0mL

pH 7.0 ± 0.2 at 25°C

Preparation of Medium: Add components, except methanol, to distilled/deionized water and bring volume to 980.0mL. Mix thoroughly. Autoclave for 15 min at 15 psi pressure–121°C. Cool to 50°–55°C. Filter sterilize methanol. Aseptically add the sterile methanol to the cooled, sterile basal medium. Mix thoroughly. Aseptically distribute into sterile tubes or flasks.

Use: For the cultivation and maintenance of *Achromobacter methanolophila, Methylobacterium rhodesianum, Pseudomonas insueta,* and *Pseudomonas polysaccharogenes.*

Methanol Medium with 1% Peptone

Composition per liter:

Agar	15.0g
Peptone	10.0g
K_2HPO_4	7.0g
$(NH_4)_2SO_4$	3.0g
KH_2PO_4	2.0g
$MgSO_4 \cdot 7H_2O$	0.5g
Yeast extract	0.2g
$FeSO_4 \cdot 7H_2O$	0.01g
$MnSO_4 \cdot H_2O$	8.0mg
Biotin	0.2μg
Thiamine·HCl	0.2μg
Methanol	10.0mL

pH 7.0 ± 0.2 at 25°C

Preparation of Medium: Add components, except methanol, to distilled/deionized water and bring volume to 990.0mL. Mix thoroughly. Autoclave for 15 min at 15 psi pressure–121°C. Cool to 50°–55°C. Filter sterilize methanol. Aseptically add the sterile methanol to the cooled, sterile basal medium. Mix thoroughly. Aseptically distribute into sterile tubes or flasks.

Use: For the cultivation and maintenance of *Methylobacterium* species.

Methanol Mineral Salts Medium

Composition per liter:

Agar	20.0g
$(NH_4)_2SO_4$	2.0g
NH_4Cl	2.0g
$(NH_4)_2HPO_4$	2.0g
Yeast extract	2.0g
KH_2PO_4	1.0g
K_2HPO_4	1.0g
$MgSO_4 \cdot 7H_2O$	0.5g
$Fe_2SO_4 \cdot 7H_2O$	0.01g
$CaCl_2 \cdot 2H_2O$	0.01g
Methanol	10.0mL

pH 7.0 ± 0.2 at 25°C

Preparation of Medium: Add components, except methanol, to distilled/deionized water and bring volume to 990.0mL. Mix thoroughly. Gently heat and bring to boiling. Autoclave for 15 min at 15 psi pressure–121°C. Cool to 50°–55°C. Filter sterilize methanol. Aseptically add the sterile methanol to the cooled, sterile basal medium. Mix thoroughly. Aseptically distribute into sterile Petri dishes or sterile tubes.

Use: For the cultivation and maintenance of *Pseudomonas viscogena.*

Methanolobus Medium

Composition per liter:

NaCl	18.0g
$NaHCO_3$	5.0g
$MgSO_4 \cdot 7H_2O$	3.45g
$MgCl_2 \cdot 6H_2O$	2.75g
L-Cysteine·HCl·H_2O	0.5g
$Na_2S \cdot 9H_2O$	0.5g
KCl	0.335g
NH_4Cl	0.25g
$CaCl_2 \cdot 2H_2O$	0.14g
K_2HPO_4	0.14g
$Fe(NH_4)_2(SO_4)_2 \cdot 6H_2O$	2.0mg
Resazurin	1.0mg
Wolfe's mineral solution	10.0mL
Wolfe's vitamin solution	10.0mL
Methanol	5.0mL

pH 6.5 ± 0.2 at 25°C

Wolfe's Mineral Solution:

Composition per liter:

$MgSO_4 \cdot 7H_2O$	3.0g
Nitriloacetic acid	1.5g
NaCl	1.0g
$MnSO_4 \cdot H_2O$	0.5g
$FeSO_4 \cdot 7H_2O$	0.1g
$CoCl_2 \cdot 6H_2O$	0.1g
$CaCl_2$	0.1g
$ZnSO_4 \cdot 7H_2O$	0.1g
$CuSO_4 \cdot 5H_2O$	0.01g
$AlK(SO_4)_2 \cdot 12H_2O$	0.01g
H_3BO_3	0.01g
$Na_2MoO_4 \cdot 2H_2O$	0.01g

Preparation of Wolfe's Mineral Solution: Add nitrilotriacetic acid to 500.0mL of distilled/deionized water. Dissolve by adjusting pH to 6.5 with KOH. Add remaining components. Add distilled/deionized water to 1.0L.

Wolfe's Vitamin Solution:

Composition per liter:

Pyridoxine·HCl	10.0mg
Thiamine·HCl	5.0mg
Riboflavin	5.0mg
Nicotinic acid	5.0mg
Calcium pantothenate	5.0mg
p-Aminobenzoic acid	5.0mg
Thioctic acid	5.0mg
Biotin	2.0mg
Folic acid	2.0mg
Cyanocobalamin	100.0μg

Preparation of Wolfe's Vitamin Solution: Add components to distilled/deionized water and bring volume to 1.0L. Mix thoroughly. Filter sterilize.

Preparation of Medium: Prepare and dispense medium under 80% N_2 + 20% CO_2. Add components, except methanol and Wolfe's vitamin solution, to distilled/deionized water and bring volume to 985.0mL. Mix thoroughly. Autoclave for 15 min at 15 psi pressure–121°C. Cool under 80% N_2 + 20% CO_2. Aseptically add sterile Wolfe's vitamin solution and sterile methanol. Adjust pH to 6.5. Aseptically and anaerobically distribute into sterile tubes or flasks.

Use: For the cultivation and maintenance of *Methanolobus tindarius.*

Methanomicrobium Medium

Composition per liter:

$NaHCO_3$	2.0g
Yeast extract	1.0g
Pancreatic digest of casein	1.0g
NaCl	0.6g
L-Cysteine·HCl·H_2O	0.5g
$Na_2S \cdot 9H_2O$	0.5g
K_2HPO_4	0.3g
KH_2PO_4	0.3g
$(NH_4)_2SO_4$	0.3g
$MgSO_4 \cdot 7H_2O$	0.13g
$CaCl_2 \cdot 2H_2O$	8.0mg
$FeSO_4 \cdot 7H_2O$	2.0mg
Rumen fluid, clarified	300.0mL
Fatty acid mixture	20.0mL

Wolfe's mineral solution 10.0mL
Wolfe's vitamin solution 10.0mL
Resazurin (0.1% solution).............................. 1.0mL
pH 6.5 ± 0.2 at 25°C

Fatty Acid Mixture:
Composition per liter:
Valeric acid.. 0.7mL
Isovaleric acid 0.7mL
α-Methylbutyric acid 0.5mL
Isobutyric acid.. 0.5mL

Preparation of Fatty Acid Mixture: Add components to distilled/deionized water and bring volume to 1.0L. Mix thoroughly.

Wolfe's Mineral Solution:
Composition per liter:
$MgSO_4 \cdot 7H_2O$ 3.0g
Nitriloacetic acid...1.5g
NaCl 1.0g
$MnSO_4 \cdot H_2O$ 0.5g
$FeSO_4 \cdot 7H_2O$..0.1g
$CoCl_2 \cdot 6H_2O$...0.1g
$CaCl_2$ 0.1g
$ZnSO_4 \cdot 7H_2O$...0.1g
$CuSO_4 \cdot 5H_2O$ 0.01g
$AlK(SO_4)_2 \cdot 12H_2O$ 0.01g
H_3BO_3 ....0.01g
$Na_2MoO_4 \cdot 2H_2O$...0.01g

Preparation of Wolfe's Mineral Solution: Add nitrilotriacetic acid to 500.0mL of distilled/deionized water. Dissolve by adjusting pH to 6.5 with KOH. Add remaining components. Add distilled/deionized water to 1.0L.

Wolfe's Vitamin Solution:
Composition per liter:
Pyridoxine·HCl 10.0mg
Thiamine·HCl... 5.0mg
Riboflavin.. 5.0mg
Nicotinic acid 5.0mg
Calcium pantothenate.................................. 5.0mg
p-Aminobenzoic acid 5.0mg
Thioctic acid.. 5.0mg
Biotin 2.0mg
Folic acid.. 2.0mg
Cyanocobalamin 100.0μg

Preparation of Wolfe's Vitamin Solution: Add components to distilled/deionized water and bring volume to 1.0L. Mix thoroughly.

Preparation of Medium: Prepare and dispense medium under 80% N_2 + 20% CO_2. Add components to distilled/deionized water and bring volume to 1.0L. Mix thoroughly. Adjust pH to 6.5. Distribute into tubes or flasks under 80% N_2 + 20% CO_2. Cap with rubber stoppers. Autoclave for 15 min at 15 psi pressure–121°C.

Use: For the cultivation and maintenance of *Methanomicrobium mobile*.

Methanomonas Autotrophic Medium

Composition per liter:
$NaNO_3$...2.0g
Na_2HPO_4 ....0.21g
$MgSO_4 \cdot 7H_2O$ 0.2g
NaH_2PO_4 ....0.09g
KCl...0.04g
$CaCl_2$ 0.015g
$FeSO_4 \cdot 7H_2O$ 1.0mg
$ZnSO_4 \cdot 7H_2O$... 0.3mg
$CuSO_4 \cdot 5H_2O$ 0.2mg
H_3BO_3 .... 0.06mg
$MnSO_4 \cdot H_2O$ 0.03mg
MoO_3... 0.015mg

Preparation of Medium: Add components to distilled/deionized water and bring volume to 1.0L. Mix thoroughly. Gently heat until dissolved. Distribute into tubes or flasks. Autoclave for 15 min at 15 psi pressure–121°C.

Use: For the autotrophic cultivation of *Methanomonas* species.

Methanosarcina acetovorans Medium

Composition per liter:
NaCl 23.4g
Agar...10.0g
$MgSO_4$...6.3g
Na_2CO_3..5.0g
Trimethylamine·HCl 3.0g
Yeast extract 1.0g
NH_4Cl...1.0g
KCl..0.8g
Na_2HPO_4 ....0.6g
L-Cysteine·HCl·H_2O 0.25g
$Na_2S \cdot 9H_2O$ 0.25g
$CaCl_2 \cdot 2H_2O$...0.14g
Resazurin.. 1.0mg
Wolfe's mineral solution 10.0mL
pH 7.2 ± 0.2 at 25°C

Wolfe's Mineral Solution:
Composition per liter:
$MgSO_4 \cdot 7H_2O$ 3.0g
Nitriloacetic acid...1.5g
NaCl 1.0g

$MnSO_4 \cdot H_2O$	0.5g
$FeSO_4 \cdot 7H_2O$	0.1g
$CoCl_2 \cdot 6H_2O$	0.1g
$CaCl_2$	0.1g
$ZnSO_4 \cdot 7H_2O$	0.1g
$CuSO_4 \cdot 5H_2O$	0.01g
$AlK(SO_4)_2 \cdot 12H_2O$	0.01g
H_3BO_3	0.01g
$Na_2MoO_4 \cdot 2H_2O$	0.01g

Preparation of Wolfe's Mineral Solution: Add nitrilotriacetic acid to 500.0mL of distilled/deionized water. Dissolve by adjusting pH to 6.5 with KOH. Add remaining components. Add distilled/deionized water to 1.0L.

Preparation of Medium: Add components, except $Na_2S \cdot 9H_2O$, to glass distilled water and bring volume to 990.0mL. Mix thoroughly. Methanol or methylamine·HCl may be substituted for the trimethylamine·HCl at a concentration of 50 mM. Heat gently and bring to boiling. Adjust pH to 7.2 with 6*N* HCl. Autoclave for 5 min at 10 psi pressure–115°C. Cool to 50°C under 80% N_2 + 20% CO_2. If a large precipitate is present add a small amount of HCl and mix thoroughly. Add $Na_2S \cdot 9H_2O$. Mix thoroughly. Distribute into tubes under 80% N_2 + 20% CO_2. Cap with butyl rubber stoppers. Autoclave for 15 min at 15 psi pressure–121°C. A precipitate will form but resolubilizes as the medium cools. Invert tubes as they are cooling to facilitate resolubilization. Allow tubes to cool in a slanted position.

Use: For the cultivation and maintenance of *Methanococcoides methylutens* and *Methanosarcina acetivorans.*

Methanosarcina Medium

Composition per liter:

Agar	20.0g
NaCl	2.25g
Yeast extract	2.00g
Pancreatic digest of casein	2.00g
NH_4Cl	0.50g
$MgSO_4 \cdot 7H_2O$	0.50g
K_2HPO_4	0.35g
$CaCl_2 \cdot 2H_2O$	0.25g
KH_2PO_4	0.23g
$FeSO_4 \cdot 7H_2O$	2.0mg
$NaHCO_3$ solution	20.0mL
Cysteine-sulfide reducing agent	20.0mL
Wolfe's vitamin solution	10.0mL
Methanol	10.0mL
Resazurin (0.025% solution)	4.0mL
Trace elements SL-6	3.0mL

pH 6.8 ± 0.2 at 25°C

$NaHCO_3$ Solution:

Composition per 20mL:

$NaHCO_3$	850.0mg

Preparation of $NaHCO_3$ Solution: Add $NaHCO_3$ to distille/deionized water and bring volume to 20.0mL. Mix thoroughly. Filter sterilize. Gas with 100% CO_2 for 20 min.

Cysteine-Sulfide Reducing Agent:

Composition per 20mL:

L-Cysteine·HCl·H_2O	0.3g
$Na_2S \cdot 9H_2O$	0.3g

Preparation of Cysteine-Sulfide Reducing Agent: Add L-Cysteine·HCl·H_2O to 10.0mL of distilled/deionized water. Mix thoroughly. In a separate tube, add $Na_2S \cdot 9H_2O$ to 10.0mL of distilled/deionized water. Mix thoroughly. Gas both solutions with 100% N_2 and cap tubes. Autoclave both solutions for 15 min at 15 psi pressure–121°C using fast exhaust. Cool to 50°C. Aseptically combine the two solutions under 100% N_2.

Wolfe's Vitamin Solution:

Composition per liter:

Pyridoxine·HCl	10.0mg
Thiamine·HCl	5.0mg
Riboflavin	5.0mg
Nicotinic acid	5.0mg
Calcium pantothenate	5.0mg
p-Aminobenzoic acid	5.0mg
Thioctic acid	5.0mg
Biotin	2.0mg
Folic acid	2.0mg
Cyanocobalamin	100.0μg

Preparation of Wolfe's Vitamin Solution: Add components to distilled/deionized water and bring volume to 1.0L. Mix thoroughly. Filter sterilize.

Trace Elements Solution SL-6:

Composition per liter:

H_3BO_3	0.3g
$CoCl_2 \cdot 6H_2O$	0.2g
$ZnSO_4 \cdot 7H_2O$	0.10g
$MnCl_2 \cdot 4H_2O$	0.03g
$Na_2MoO_4 \cdot H_2O$	0.03g
$NiCl_2 \cdot 6H_2O$	0.02g
$CuCl_2 \cdot 2H_2O$	0.01g

Preparation of Trace Elements Solution SL-6: Add components to distilled/deionized water and bring volume to 1.0L. Mix thoroughly. Adjust pH to 3.4.

Preparation of Medium: Add components, except $NaHCO_3$ solution, cysteine-sulfide reducing agent, Wolfe's vitamin solution and methanol, to distilled/deionized water and bring volume to 940.0mL. Mix thoroughly. Autoclave for 15 min at 15 psi pressure–121°C. Cool under 80% N_2 + 20% CO_2. Aseptically and anaerobically add the sterile $NaHCO_3$ solution, the sterile cysteine-sulfide reducing agent, the sterile Wolfe's vitamin solution and filter-sterilized methanol. Mix thoroughly. Adjust pH to 6.8. Aseptically and anaerobically distribute into sterile tubes or flasks.

Use: For the cultivation and maintenance of *Bifidobacterium asteroides, Methanosarcina barkeri, Methanosarcina* species, and *Methanosarcina vacuolata.*

Methanospirillum hungatei Medium

Composition per 100mL:

Na_2CO_3	0.4g
Sodium formate	0.2g
Pancreatic digest of casein	0.2g
Yeast extract	0.2g
NaCl	0.05g
L-Cysteine·HCl·H_2O	0.03g
K_2HPO_4	0.02g
KH_2PO_4	0.02g
$(NH_4)_2SO_4$	0.02g
$MgSO_4 \cdot 7H_2O$	9.0mg
$CaCl_2 \cdot 2H_2O$	6.0mg
Resazurin	0.1mg
$Na_2S \cdot 9H_2O$ solution	10.0mL
Vitamin solution	1.0mL
Trace metal solution	1.0mL

pH 7.0 ± 0.2 at 25°C

$Na_2S \cdot 9H_2O$ Solution:

Composition per 10mL:

$Na_2S \cdot 9H_2O$	0.03g

Preparation of $Na_2S \cdot 9H_2O$ Solution: Add $Na_2S \cdot 9H_2O$ to distilled/deionized water and bring volume to 10.0mL. Mix thoroughly. Autoclave for 15 min at 15 psi pressure–121°C. Cool to 25°C.

Vitamin Solution:

Composition per 100mL:

Pyridoxine·HCl	1.0mg
p-Aminobenzoic acid	0.5mg
Calcium-D-pantothenate	0.5mg
Nicotinic acid	0.5mg
Riboflavin	0.5mg
Thiamine·HCl	0.5mg
Thioctic acid	0.5mg
Biotin	0.2mg
Folic acid	0.2mg
Vitamin B_{12}	0.01mg

Preparation of Vitamin Solution: Add components to distilled/deionized water and bring volume to 1.0L. Mix thoroughly. Filter sterilize.

Trace Metal Solution:

Composition per liter:

$K_2HPO_4 \cdot 3H_2O$	9.0g
K_2HPO_4	6.0g
NH_4Cl	5.0g
$MgCl_2 \cdot 6H_2O$	1.0g
$CaCl_2 \cdot 2H_2O$	0.01g

Preparation of Trace Metal Solution: Add components to distilled/deionized water and bring volume to 1.0L. Mix thoroughly.

Preparation of Medium: Prepare and distribute medium anaerobically under 80% H_2 + 20% CO_2. Add components, except $Na_2S \cdot 9H_2O$ solution, to distilled/deionized water and bring volume to 90.0mL. Mix thoroughly. Gently heat and bring to boiling. Continue boiling until resazurin turns colorless, indicating reduction. Autoclave for 15 min at 15 psi pressure–121°C. Cool to 25°C. Aseptically add 10.0mL of sterile $Na_2S \cdot 9H_2O$ solution. Mix thoroughly. Aseptically and anaerobically distribute into sterile tubes or flasks.

Use: For the cultivation of *Methanospirillum hungatei.*

Methyl Red Voges-Proskauer Broth
See: **MRVP Broth**

Methyl Red Voges–Proskauer Medium
See: **MRVP Medium**

Methylamine Salts Medium

Composition per liter:

Agar	15.0g
Methylamine·HCl	6.75g
K_2HPO_4	2.12g
KH_2PO_4	1.0g
Solution A	5.0mL
Solution B	1.0mL

pH 7.0 ± 0.2 at 25°C

Solution A:

Composition per 100mL:

$MgSO_4 \cdot 7H_2O$	2.0g
$CaCl_2 \cdot 2H_2O$	0.2g
$FeSO_4 \cdot 7H_2O$	0.2g

Preparation of Solution A: Add components to distilled/deionized water and bring volume to 100.0mL. Mix thoroughly.

Solution B:

Composition per 100mL:

$MnSO_4 \cdot 7H_2O$	0.05g
$Na_2MoO_4 \cdot 2H_2O$	0.05g

Preparation of Solution B: Add components to distilled/deionized water and bring volume to 100.0mL. Mix thoroughly.

Preparation of Medium: Add components to distilled/deionized water and bring volume to 1.0L. Mix thoroughly. Gently heat and bring to boiling. Distribute into tubes or flasks. Autoclave for 15 min at 15 psi pressure–121°C. Pour into sterile Petri dishes or leave in tubes.

Use: For the cultivation and maintenance of *Methylobacterium extorquens* and *Pseudomonas* species.

Methylene Blue Milk Medium (MBM Medium)

Composition per liter:

Skim milk, dehydrated	100.0g
Methylene Blue	10.0g

pH 6.4 ± 0.2 at 25°C

Preparation of Medium: Add components to distilled/deionized water and bring volume to 1.0L. Mix thoroughly. Distribute into tubes or flasks. Autoclave for 20 min at 10 psi pressure–115°C.

Use: For cultivation and differentiation of Group D streptococci (enterococci) from other *Streptococcus* species.

Methylococcus Medium

Composition per liter:

Agar	8.0g
$NaNO_3$ (20% solution)	10.0mL
L-F Salts solution	10.0mL
Sodium-potassium phosphate buffer	6.5mL

pH 7.1 ± 0.2 at 25°C

Sodium-Potassium Phosphate Buffer:

Composition per liter:

KH_2PO_4	136.0g
NaOH	28.8g

Preparation of Sodium-Potassium Phosphate Buffer: Add components to distilled/deionized water and bring volume to 1.0L. Mix thoroughly. Adjust pH to 7.1.

L-F Salts Solution:

Composition per liter:

$MgSO_4 \cdot 7H_2O$ (10% solution)	200.0mL
$CaCl_2 \cdot 2H_2O$ (10% solution)	20.0mL
$FeSO_4$ (10% solution)	10.0mL
$ZnSO_4 \cdot 7H_2O$ (1% solution)	4.9mL
H_3BO_3 (1% solution)	0.6mL
$MnSO_4 \cdot H_2O$ (1% solution)	0.27mL
$CuSO_4 \cdot 5H_2O$ (1% solution)	0.2mL

Preparation of L-F Salts Solution: Filter sterilize $FeSO_4$ solution immediately prior to use. Add all components to distilled/deionized water and bring volume to 1.0L. Mix thoroughly.

Preparation of Medium: Add components to distilled/deionized water and bring volume to 1.0L. Mix thoroughly. Adjust pH to 7.1. Autoclave for 15 min at 15 psi pressure–121°C. Pour into sterile Petri dishes or leave in tubes.

Use: For cultivation and maintenance of *Methylococcus* species.

Methylophaga Agar

Composition per 103mL:

Agar solution	50.0mL
Mineral base, 2X	50.0mL
Solution T	2.0mL
Vitamin B_{12} solution	1.0mL
Methanol	0.3mL

pH 7.3 ± 0.2 at 25°C

Agar Solution:

Composition per 500mL:

Agar	15.0g

Preparation of Agar Solution: Add agar to distilled/deionized water and bring volume to 500.0mL. Mix thoroughly. Autoclave for 15 min at 15 psi pressure–121°C. Cool to 50°C.

Mineral Base, 2X:

Composition per 500mL:

NaCl	24.0g
$MgCl_2 \cdot 6H_2O$	3.0g
$MgSO_4 \cdot 7H_2O$	2.0g
$CaCl_2 \cdot 2H_2O$	1.0g
KCl	0.5g
Bis-Tris buffer (bis[2-Hydroxyethyl]imino-tris[hydroxymethyl]-methane)	0.5g
Wolfe's mineral solution	10.0mL

Preparation of Mineral Base, 2X: Add components to distilled/deionized water and bring volume to 500.0mL. Mix thoroughly. Adjust pH to 7.3. Autoclave for 15 min at 15 psi pressure–121°C. Cool to 50°C.

Wolfe's Mineral Solution:
Composition per liter:

$MgSO_4 \cdot 7H_2O$	3.0g
Nitriloacetic acid	1.5g
NaCl	1.0g
$MnSO_4 \cdot H_2O$	0.5g
$FeSO_4 \cdot 7H_2O$	0.1g
$CoCl_2 \cdot 6H_2O$	0.1g
$CaCl_2$	0.1g
$ZnSO_4 \cdot 7H_2O$	0.1g
$CuSO_4 \cdot 5H_2O$	0.01g
$AlK(SO_4)_2 \cdot 12H_2O$	0.01g
H_3BO_3	0.01g
$Na_2MoO_4 \cdot 2H_2O$	0.01g

Preparation of Wolfe's Mineral Solution: Add nitrilotriacetic acid to 500.0mL of distilled/deionized water. Dissolve by adjusting pH to 6.5 with KOH. Add remaining components. Add distilled/deionized water to 1.0L.

Solution T:
Composition per 100mL:

NH_4Cl	10.0g
Bis-Tris buffer (bis[2-Hydroxyethyl]imino-tris[hydroxymethyl]-methane)	10.0g
KH_2PO_4	0.7g
Ferric ammonium citrate	0.3g

Preparation of Solution T: Add components to distilled/deionized water and bring volume to 100.0mL. Mix thoroughly. Adjust pH to 7.3. Autoclave for 15 min at 15 psi pressure–121°C.

Vitamin B_{12} Solution:
Composition per 10mL:

Vitamin B_{12}	1.0μg

Preparation of Vitamin B_{12} Solution: Add Vitamin B_{12} to 10.0mL of distilled/deionized water. Mix thoroughly. Filter sterilize.

Preparation of Medium: Aseptically mix 50.0mL of the sterile agar solution with 50.0mL of the sterile mineral base, 2X. Aseptically combine the sterile solution T and sterile vitamin B_{12} solution with the sterile mineral base. Filter sterilize methanol and add to basal medium. Pour into sterile Petri dishes or distribute into sterile tubes.

Use: For the cultivation and maintenance of *Methylophaga marina.*

Methylophaga Broth

Composition per 103mL:

Mineral base	100.0mL
Solution T	2.0mL
Vitamin B_{12} solution	1.0mL
Methanol	0.3mL

pH 7.3 ± 0.2 at 25°C

Mineral Base:
Composition per liter:

NaCl	24.0g
$MgCl_2 \cdot 6H_2O$	3.0g
$MgSO_4 \cdot 7H_2O$	2.0g
$CaCl_2 \cdot 2H_2O$	1.0g
KCl	0.5g
Bis-Tris buffer (bis[2-Hydroxyethyl]imino-tris[hydroxymethyl]-methane)	0.5g
Wolfe's mineral solution	10.0mL

Preparation of Mineral Base: Add components to distilled/deionized water and bring volume to 1.0L. Mix thoroughly. Adjust pH to 7.3. Autoclave for 15 min at 15 psi pressure–121°C.

Wolfe's Mineral Solution:
Composition per liter:

$MgSO_4 \cdot 7H_2O$	3.0g
Nitriloacetic acid	1.5g
NaCl	1.0g
$MnSO_4 \cdot H_2O$	0.5g
$FeSO_4 \cdot 7H_2O$	0.1g
$CoCl_2 \cdot 6H_2O$	0.1g
$CaCl_2$	0.1g
$ZnSO_4 \cdot 7H_2O$	0.1g
$CuSO_4 \cdot 5H_2O$	0.01g
$AlK(SO_4)_2 \cdot 12H_2O$	0.01g
H_3BO_3	0.01g
$Na_2MoO_4 \cdot 2H_2O$	0.01g

Preparation of Wolfe's Mineral Solution: Add nitrilotriacetic acid to 500.0mL of distilled/deionized water. Dissolve by adjusting pH to 6.5 with KOH. Add remaining components. Add distilled/deionized water to 1.0L.

Solution T:
Composition per 100mL:

NH_4Cl	10.0g
Bis-Tris buffer (bis[2-Hydroxyethyl]imino-tris[hydroxymethyl]-methane)	10.0g
KH_2PO_4	0.7g
Ferric ammonium citrate	0.3g

Preparation of Solution T: Add components to distilled/deionized water and bring volume to 100.0mL. Mix thoroughly. Adjust pH to 7.3. Autoclave for 15 min at 15 psi pressure–121°C.

Vitamin B_{12} Solution:
Composition per 10mL:

Vitamin B_{12}	1.0μg

Preparation of Vitamin B_{12} Solution: Add Vitamin B_{12} to 10.0mL of distilled/deionized water. Mix thoroughly. Filter sterilize.

Preparation of Medium: Aseptically combine the sterile solution T and sterile vitamin B_{12} solution with the sterile mineral base. Filter sterilize methanol and add to basal medium. Aseptically distribute into tubes or flasks.

Use: For the cultivation and maintenance of *Methylophaga marina*.

m–FC Agar
See: **FC Agar**

m–FC Broth
See: **FC Broth**

m–Fecal Coliform Agar
See: **FC Agar**

m–Fecal Coliform Agar, Modified
See: **Fecal Coliform Agar, Modified**

m–Fecal Coliform Broth
See: **FC Broth**

MGA Agar

Composition per liter:

Agar	20.0g
Glucose	2.0g
L-Asparagine	1.0g
K_2HPO_4	0.5g
$MgSO_4·7H_2O$	0.5g
Trace salts solution	1.0mL

pH 7.4 ± 0.2 at 25°C

Trace Salts Solution:
Composition per liter:

$FeSO_4·7H_2O$	10.0g
$CuSO_4·5H_2O$	1.0g
$MnSO_4·7H_2O$	1.0g
$ZnSO_4·7H_2O$	1.0g

Preparation of Trace Salts Solution: Add components to distilled/deionized water and bring volume to 1.0L. Mix thoroughly. Adjust pH to 8.0.

Preparation of Medium: Add components to distilled/deionized water and bring volume to 1.0L. Mix thoroughly. Gently heat and bring to boiling. Distribute into tubes or flasks. Autoclave for 15 min at 15 psi pressure–121°C. Pour into sterile Petri dishes or leave in tubes.

Use: For the isolation and cultivation of *Actinomadura* species, *Actinopolyspora* species, *Excellospora* species and *Microspora* species.

m–Green Yeast and Mold Broth
See: **Green Yeast and Mold Broth**

MGTY Agar
See: **Marine Glucose Trypticase Yeast Extract Agar**

MGTY Broth
See: **Marine Glucose Trypticase Yeast Extract Broth**

MH IH Agar

Composition per liter:

Solution A	490.0mL
Solution B	490.0mL
Supplement solution	20.0mL

pH 6.9 ± 0.2 at 25°C

Solution A:
Composition per 490mL:

Beef infusion	300.0g
Acid hydrolysate of casein	17.5g
Agar	17.0g
Starch	1.5g

Preparation of Solution A: Add components to distilled/deionized water and bring volume to 490.0mL. Mix thoroughly. Gently heat and bring to boiling. Autoclave for 15 min at 15 psi pressure–121°C. Cool to 45°–50°C.

Solution B:
Composition per 490mL:

Hemoglobin	10.0g

Preparation of Solution B: Add hemoglobin to distilled/deionized water and bring volume to 490.0mL. Mix thoroughly. Gently heat and bring to boiling. Autoclave for 15 min at 15 psi pressure–121°C. Cool to 45°–50°C.

Supplement Solution:
Composition per liter:

Glucose	100.0g
L-Cysteine·HCl	25.9g
L-Glutamine	10.0g
L-Cystine	1.1g
Adenine	1.0g
Nicotinamide adenine dinucleotide	0.25g

Vitamin B_{12}	0.1g
Thiamine pyrophosphate	0.1g
Guanine·HCl	0.03g
$Fe(NO_3)_3 \cdot 6H_2O$	0.02g
p-Aminobenzoic acid	0.013g
Thiamine·HCl	3.0mg

Source: The supplement solution IsoVitaleX® enrichment is available from BBL Microbiology Laboratories. This enrichment may be replaced by supplement VX from Difco Laboratories.

Preparation of Supplement Solution: Add components to distilled/deionized water and bring volume to 1.0L. Mix thoroughly. Filter sterilize.

Preparation of Medium: Aseptically combine cooled, sterile solution A and cooled, sterile solution B. Mix thoroughly. Adjust pH to 6.9 with sterile 1*N* HCl or sterile 1*N* KOH. Aseptically add 20.0mL of sterile supplement solution. Pour into sterile Petri dishes or distribute into sterile tubes.

Use: For the cultivation and differentiation of *Legionella* species.

MH Medium

Composition per liter:

NaCl	60.7g
Agar	20.0g
$MgCl_2 \cdot 6H_2O$	15.0g
Yeast extract	10.0g
$MgSO_4 \cdot 7H_2O$	7.4g
Proteose peptone No. 3	5.0g
KCl	1.5g
Glucose	1.0g
$CaCl_2$	0.27g
$NaHCO_3$	0.45g
NaBr	0.19g

Preparation of Medium: Add components to distilled/deionized water and bring volume to 1.0L. Mix thoroughly. Gently heat and bring to boiling. Distribute into tubes or flasks. Autoclave for 15 min at 15 psi pressure–121°C. Pour into sterile Petri dishes or leave in tubes.

Use: For the cultivation and maintenance of *Deleya salina* and *Volcaniella eurihalina*.

MH Salts

Composition per liter:

NaCl	120.5g
$MgCl_2 \cdot 6H_2O$	22.4g
Agar	20.0g
$MgSO_4$	14.4g
Yeast extract	10.0g
Proteose peptone No. 3	5.0g
KCl	3.0g
Glucose	1.0g
$CaCl_2$	0.54g
$NaHCO_3$	0.09g
NaBr	0.039g

pH 7.5 ± 0.2 at 25°C

Preparation of Medium: Add components to distilled/deionized water and bring volume to 1.0L. Mix thoroughly. Adjust pH to 7.5. Gently heat and bring to boiling. Distribute into tubes or flasks. Autoclave for 15 min at 15 psi pressure–121°C. Pour into sterile Petri dishes or leave in tubes.

Use: For the cultivation and maintenance of *Bacillus halophilus*.

m–HPC Agar
See: **HPC Agar**

Micro Assay Culture Agar

Composition per liter:

Yeast extract	20.0g
Agar	10.0g
Glucose	10.0g
Proteose peptone No. 3	5.0g
KH_2PO_4	2.0g
Sorbitan monooleate complex	0.1g

pH 6.7 ± 0.2 at 25°C

Source: This medium is available as a premixed powder from Difco Laboratories.

Preparation of Medium: Add components to distilled/deionized water and bring volume to 1.0L. Mix thoroughly. Gently heat and bring to boiling. Distribute into tubes in 10.0mL volumes. Autoclave for 15 min at 15 psi pressure–121°C. Just prior to solidification of the agar, disperse precipitate by gently twirling tube.

Use: For carrying stock cultures of lactobacilli and other test microorganisms used in microbiological assays. It is also used for general cultivation of lactobacilli.

Micro Inoculum Broth

Composition per liter:

Yeast extract	20.0g
Glucose	10.0g
Proteose peptone No. 3	5.0g
KH_2PO_4	2.0g
Sorbitan monooleate complex	0.1g

pH 6.7 ± 0.2 at 25°C

Source: This medium is available as a premixed powder from Difco Laboratories.

Preparation of Medium: Add components to distilled/deionized water and bring volume to 1.0L. Mix thoroughly. Distribute into tubes in 10.0mL volumes. Autoclave for 15 min at 15 psi pressure–121°C.

Use: For the cultivation of lactobacilli used in microbiological assays. It is of particular value in the preparation of the inoculum for these tests.

Microbacterium Medium

Composition per liter:

Glucose	10.0g
KH_2PO_4	5.0g
K_2HPO_4	5.0g
Potassium aspartate	5.0g
$(NH_4)_2SO_4$	2.0g
$MgSO_4·7H_2O$	0.5g
Calcium pantothenate	0.2g
β-Mercaptopurine	0.1g
$FeSO_4·7H_2O$	0.01g
Thiamine·HCl	0.01g
Biotin	0.1mg

pH 7.0 ± 0.2 at 25°C

Preparation of Medium: Add components to distilled/deionized water and bring volume to 1.0L. Mix thoroughly. Distribute into tubes or flasks. Autoclave for 10 min at 15 psi pressure–121°C.

Use: For the cultivation and maintenance of *Microbacterium* species.

Microbial Content Test Agar

Composition per liter:

Agar	15.0g
Pancreatic digest of casein	15.0g
NaCl	5.0g
Tween™ 80	5.0g
Enzymatic hydrolysate of soybean meal	5.0g
Lecithin	0.7g

pH 7.3 ± 0.2 at 25°C

Source: This medium is available as a premixed powder from Difco Laboratories.

Preparation of Medium: Add components to distilled/deionized water and bring volume to 1.0L. Mix thoroughly. Gently heat and bring to boiling. Boil for 1–2 min. Distribute into tubes or flasks. Autoclave for 15 min at 15 psi pressure–121°C. Pour into sterile Petri dishes or leave in tubes.

Use: For use in the microbial content test of water soluble cosmetic products. Also used for determining the efficiency of sanitization of containers, equipment and environmental surfaces.

Micrococcus Medium

Composition per liter:

Agar	15.0g
Peptone	5.0g
Yeast extract	3.0g
Beef extract	1.5g
Glucose	1.0g

pH 7.4 ± 0.2 at 25°C

Preparation of Medium: Add components to distilled/deionized water and bring volume to 1.0L. Mix thoroughly. Gently heat and bring to boiling. Distribute into tubes or flasks. Autoclave for 15 min at 15 psi pressure–121°C. Pour into sterile Petri dishes or leave in tubes.

Use: For the cultivation and maintenance of *Staphylococcus aureus* and *Micrococcus* species.

Micrococcus Medium, FDA

Composition per liter:

Agar	15.0g
Proteose peptone	10.0g
Beef extract	5.0g
NaCl	5.0g

pH 7.2 ± 0.2 at 25°C

Preparation of Medium: Add components to distilled/deionized water and bring volume to 1.0L. Mix thoroughly. Gently heat and bring to boiling. Distribute into tubes or flasks. Autoclave for 15 min at 15 psi pressure–121°C. Pour into sterile Petri dishes or leave in tubes.

Use: For the cultivation and maintenance of *Staphylococcus aureus* and *Micrococcus* species.

Micrococcus–Sarcina Medium

Composition per liter:

Agar	16.0g
Pancreatic digest of casein	5.0g
Sodium succinate·$6H_2O$	2.0g
Starch	2.0g
Yeast autolysate	1.0g
Sodium citrate·$2H_2O$	0.5g
Sodium acetate·$3H_2O$	0.3g
K_2HPO_4	0.2g

pH 7.0 ± 0.2 at 25°C

Preparation of Medium: Add components to distilled/deionized water and bring volume to 1.0L. Mix thoroughly. Gently heat and bring to boiling. Distribute into tubes or flasks. Autoclave for 15 min at 15 psi pressure–121°C. Pour into sterile Petri dishes or leave in tubes.

Use: For the cultivation and maintenance of *Micrococcus luteus* and *Sarcina* species.

Microcyclus eburneus Medium

Composition per liter:

K_2HPO_4	7.0g
$(NH_4)SO_4$	3.0g
KH_2PO_4	2.0g
$MgSO_4 \cdot 7H_2O$	0.5g
Yeast extract	0.2g
Thiamine·HCl	0.2mg
Biotin	0.02mg
$FeSO_4 \cdot 7H_2O$	2.0μg
$MnSO_4 \cdot 4H_2O$	2.0μg

Preparation of Medium: Add components to distilled/deionized water and bring volume to 1.0L. Mix thoroughly. Distribute into tubes or flasks. Autoclave for 15 min at 15 psi pressure–121°C.

Use: For the cultivation of *Microcyclus eburneus.*

Microcylus major Medium

Composition per liter:

Glucose	1.0g
Peptone	1.0g
KNO_3	0.1g
K_2HPO_4	0.07g
$MgSO_4 \cdot 7H_2O$	0.03g
Trace element solution	1.0mL

Trace Element Solution:

Composition per liter:

Disodium EDTA	10.0g
$FeSO_4 \cdot 7H_20$	9.3g
$NaBO_3 \cdot 4H_2O$	2.6g
$MnCl_2 \cdot 4H_2O$	1.8g
$CaCl_2$	1.2g
$(NH_4)_6Mo_7O_{24} \cdot 4H_2O$	1.0g
$ZnSO_4 \cdot 7H_2O$	0.2g
$CuSO_4 \cdot 5H_2O$	0.08g
$Co(NO_3)_2 \cdot H_2O$	0.02g

Preparation of Trace Element Solution: Add components to distilled/deionized water and bring volume to 1.0L. Mix thoroughly.

Preparation of Medium: Add components to distilled/deionized water and bring volume to 1.0L. Mix thoroughly. Distribute into tubes or flasks. Autoclave for 15 min at 15 psi pressure–121°C.

Use: For the cultivation of *Microcyclus major.*

Microcyclus marinus Medium

Composition per liter:

NaCl	23.5g
$MgCl_2$	5.0g
Na_2SO_4	4.0g
$CaCl_2 \cdot 2H_2O$	1.5g
KCl	0.7g
$NaHCO_3$	0.2g

Preparation of Medium: Add components to distilled/deionized water and bring volume to 1.0L. Mix thoroughly. Distribute into tubes or flasks. Autoclave for 15 min at 15 psi pressure–121°C.

Use: For the cultivation of *Microcyclus marinus.*

Microcyclus Medium

Composition per liter:

Agar	15.0g
Glucose	5.0g
Peptone	5.0g
Yeast extract	5.0g

pH 6.8 ± 0.2 at 25°C

Preparation of Medium: Add components to distilled/deionized water and bring volume to 1.0L. Mix thoroughly. Gently heat and bring to boiling. Distribute into tubes or flasks. Autoclave for 15 min at 15 psi pressure–121°C. Pour into sterile Petri dishes or leave in tubes.

Use: For the cultivation and maintenance of *Flectobacillus major* and *Microcyclus* species.

Microcyclus–Spirosoma Medium

Composition per liter:

Agar	15.0g
Glucose	1.0g
Peptone	1.0g
Yeast extract	1.0g

pH 6.8–7.0 at 25°C

Preparation of Medium: Add components to distilled/deionized water and bring volume to 1.0L. Mix thoroughly. Gently heat and bring to boiling. Distribute into tubes or flasks. Autoclave for 15 min at 15 psi pressure–121°C. Pour into sterile Petri dishes or leave in tubes.

Use: For the cultivation and maintenance of *Spirosoma linguale* and *Microcyclus* species.

Middlebrook 7H9 Broth, Supplemented

Composition per liter:

Na_2HPO_4	2.5g
KH_2PO_4	1.0g
Monosodium glutamate	0.5g
$(NH_4)_2SO_4$	0.5g
Tween™ 80	0.5g
Sodium citrate	0.1g

$MgSO_4 \cdot 7H_2O$	0.05g
Ferric ammonium citrate	0.04g
Mycobactin J, Allied Laboratories, Inc.	2.0mg
$CuSO_4 \cdot 5H_2O$	1.0mg
Pyridoxine	1.0mg
$ZnSO_4 \cdot 7H_2O$	1.0mg
Biotin	0.5mg
$CaCl_2 \cdot 2H_2O$	0.5mg
Dubos oleic albumin complex	100.0mL
Glycerol	2.0mL

pH 6.6 ± 0.2 at 25°C

Dubos Oleic Albumin Complex:

Composition per 100mL:

Bovine serum albumin, fraction V	5.0g
Oleic acid, sodium salt	0.05g
NaCl (0.85% solution)	100.0mL

Preparation of Dubos Oleic Albumin Complex: Add bovine serum albumin and oleic acid to 100.0mL of NaCl solution. Mix thoroughly. Filter sterilize.

Preparation of Medium: Add components, except Dubos oleic albumin complex, to distilled/deionized water and bring volume to 900.0mL. Mix thoroughly. Gently heat and bring to boiling. Autoclave for 15 min at 15 psi pressure–121°C. Cool to 45°–50°C. Aseptically add sterile Dubos oleic albumin complex. Mix thoroughly. Pour into sterile Petri dishes or distribute into sterile tubes.

Use: For the cultivation and maintenance of *Mycobacterium avium.*

Middlebrook 7H9 Broth with Middlebrook ADC Enrichment

Composition per liter:

Na_2HPO_4	2.5g
KH_2PO_4	1.0g
Monosodium glutamate	0.5g
$(NH_4)_2SO_4$	0.5g
Sodium citrate	0.1g
$MgSO_4 \cdot 7H_2O$	0.05g
Ferric ammonium citrate	0.04g
$CuSO_4 \cdot 5H_2O$	1.0mg
Pyridoxine	1.0mg
$ZnSO_4 \cdot 7H_2O$	1.0mg
Biotin	0.5mg
$CaCl_2 \cdot 2H_2O$	0.5mg
Middlebrook ADC enrichment	100.0mL
Glycerol	2.0mL

pH 6.6 ± 0.2 at 25°C

Source: This medium is available as a premixed powder from BBL Microbiology Systems and Difco Laboratories.

Middlebrook ADC Enrichment:

Composition per 100mL:

Bovine albumin fraction V	5.0g
Glucose	2.0g
Catalase	3.0mg

Source: This enrichment is available as a prepared enrichment from Difco Laboratories.

Preparation of Middlebrook ADC Enrichment: Add components to distilled/deionized water and bring volume to 100.0mL. Mix thoroughly. Filter sterilize.

Preparation of Medium: Add glycerol to 900.0mL of distilled/deionized water and add remaining components, except Middlebrook ADC enrichment. Mix thoroughly. Gently heat and bring to boiling. Autoclave for 15 min at 15 psi pressure–121°C. Cool to 50°–55°C. Aseptically add 100.0mL of sterile Middlebrook ADC enrichment. Mix thoroughly. Distribute into sterile tubes or flasks.

Use: For the isolation, cultivation and maintenance of *Mycobacterium* species, including *M. tuberculosis*. Also used for determining the antimicrobial susceptibility of mycobacteria.

Middlebrook 7H9 Broth with Middlebrook OADC Enrichment

Composition per liter:

Na_2HPO_4	2.5g
KH_2PO_4	1.0g
Monosodium glutamate	0.5g
$(NH_4)_2SO_4$	0.5g
Sodium citrate	0.1g
$MgSO_4 \cdot 7H_2O$	0.05g
Ferric ammonium citrate	0.04g
$CuSO_4 \cdot 5H_2O$	1.0mg
Pyridoxine	1.0mg
$ZnSO_4 \cdot 7H_2O$	1.0mg
Biotin	0.5mg
$CaCl_2 \cdot 2H_2O$	0.5mg
Middlebrook OADC enrichment	100.0mL
Glycerol	2.0mL

pH 6.6 ± 0.2 at 25°C

Source: This medium is available as a premixed powder from BBL Microbiology Systems and Difco Laboratories.

Middlebrook OADC Enrichment:

Composition per 100mL:

Bovine albumin fraction V	5.0g
Glucose	2.0g
NaCl	0.85g
Oleic acid	0.05g
Catalase	4.0mg

Source: Available as a prepared enrichment from Difco Laboratories.

Preparation of Middlebrook OADC Enrichment: Add components to distilled/deionized water and bring volume to 100.0mL. Mix thoroughly. Filter sterilize.

Preparation of Medium: Add glycerol to 900.0mL of distilled/deionized water and add remaining components, except Middlebrook OADC enrichment. Mix thoroughly. Gently heat and bring to boiling. Autoclave for 15 min at 15 psi pressure–121°C. Cool to 50°–55°C. Aseptically add 100.0mL of sterile Middlebrook OADC enrichment. Mix thoroughly. Distribute into sterile tubes or flasks.

Use: For the isolation, cultivation and maintenance of *Mycobacterium* species, including *M. tuberculosis*. Also used for determining the antimicrobial susceptibility of mycobacteria.

Middlebrook 7H9 Broth with Middlebrook OADC Enrichment and Triton WR 1339

Composition per liter:

Na_2HPO_4	2.5g
KH_2PO_4	1.0g
Monosodium glutamate	0.5g
$(NH_4)_2SO_4$	0.5g
Sodium citrate	0.1g
$MgSO_4 \cdot 7H_2O$	0.05g
Ferric ammonium citrate	0.04g
$CuSO_4 \cdot 5H_2O$	1.0mg
Pyridoxine	1.0mg
$ZnSO_4 \cdot 7H_2O$	1.0mg
Biotin	0.5mg
$CaCl_2 \cdot 2H_2O$	0.5mg
Middlebrook OADC enrichment with Triton WR 1339	100.0mL
Glycerol	2.0mL

pH 6.6 ± 0.2 at 25°C

Source: This medium is available as a premixed powder from BBL Microbiology Systems.

Middlebrook OADC Enrichment with Triton WR 1339:

Composition per 100mL:

Bovine albumin fraction V	5.0g
Glucose	2.0g
NaCl	0.85g
Triton WR-1339	0.25g
Oleic acid	0.05g
Catalase	4.0mg

Source: Available as a prepared enrichment from Difco Laboratories.

Preparation of Middlebrook OADC Enrichment with Triton WR 1339: Add components to distilled/deionized water and bring volume to 100.0mL. Mix thoroughly. Filter sterilize.

Preparation of Medium: Add glycerol to 900.0mL of distilled/deionized water and add remaining components, except Middlebrook OADC enrichment with Triton WR-1339. Mix thoroughly. Gently heat and bring to boiling. Autoclave for 15 min at 15 psi pressure–121°C. Cool to 50°–55°C. Aseptically add 100.0mL of sterile Middlebrook OADC enrichment with Triton WR-1339. Mix thoroughly. Distribute into sterile tubes or flasks.

Use: For the isolation, cultivation and maintenance of *Mycobacterium* species, including *M. tuberculosis*. Also used for determining the antimicrobial susceptibility of mycobacteria.

Middlebrook 7H10 Agar with Middlebrook ADC Enrichment

Composition per liter:

Agar	15.0g
Na_2HPO_4	1.5g
KH_2PO_4	1.5g
$(NH_4)_2SO_4$	0.5g
L-Glutamic Acid	0.5g
Sodium citrate	0.4g
Ferric ammonium citrate	0.04g
$MgSO_4 \cdot 7H_2O$	0.025g
$ZnSO_4 \cdot 7H_2O$	1.0mg
$CuSO_4 \cdot 5H_2O$	1.0mg
Pyridoxine	1.0mg
Biotin	0.5mg
$CaCl_2 \cdot 2H_2O$	0.5mg
Malachite Green	.0.25mg
Middlebrook ADC enrichment	100.0mL
Glycerol	5.0mL

pH 6.6 ± 0.2 at 25°C

Source: Available as a premixed powder from BBL Microbiology Systems and Difco Laboratories.

Middlebrook ADC Enrichment:

Composition per 100mL:

Bovine albumin fraction V	5.0g
Glucose	2.0g
Catalase	0.003g
Distilled water	100.0mL

Source: Available as a prepared enrichment from Difco Laboratories.

Preparation of Middlebrook ADC Enrichment: Add components to distilled/deionized water and bring volume to 100.0mL. Mix thoroughly. Filter sterilize.

Preparation of Medium: Add glycerol to 900.0mL of distilled/deionized water and add remaining components, except Middlebrook ADC enrichment. Mix thoroughly. Gently heat and bring to boiling. Autoclave for 15 min at 15 psi pressure–121°C. Cool to 50°–55°C. Aseptically add 100.0mL of sterile Middlebrook ADC enrichment. Mix thoroughly. Pour into sterile Petri dishes or distribute into sterile tubes.

Use: For the isolation, cultivation and maintenance of *Mycobacterium* species, including *M. tuberculosis*. Also used for determining the antimicrobial susceptibility of mycobacteria.

Middlebrook 7H10 Agar with Middlebrook OADC Enrichment (Middlebrook and Cohn 7H10 Agar)

Composition per liter:

Agar	15.0g
Na_2HPO_4	1.5g
KH_2PO_4	1.5g
$(NH_4)_2SO_4$	0.5g
L-Glutamic Acid	0.5g
Sodium citrate	0.4g
Ferric ammonium citrate	0.04g
$MgSO_4 \cdot 7H_2O$	0.025g
$ZnSO_4 \cdot 7H_2O$	1.0mg
$CuSO_4 \cdot 5H_2O$	1.0mg
Pyridoxine	1.0mg
Biotin	0.5mg
$CaCl_2 \cdot 2H_2O$	0.5mg
Malachite Green	.0.25mg
Middlebrook OADC enrichment	100.0mL
Glycerol	5.0mL

pH 6.6 ± 0.2 at 25°C

Source: This medium is available as a premixed powder from BBL Microbiology Systems and Difco Laboratories.

Middlebrook OADC Enrichment:

Composition per 100mL:

Bovine albumin fraction V	5.0g
Glucose	2.0g
NaCl	0.85g
Oleic acid	0.05g
Catalase	4.0mg

Source: Available as a prepared enrichment from Difco Laboratories.

Preparation of Middlebrook OADC Enrichment: Add components to distilled/deionized water and bring volume to 100.0mL. Mix thoroughly. Filter sterilize.

Preparation of Medium: Add glycerol to 900.0mL of distilled/deionized water and add remaining components, except Middlebrook OADC enrichment. Mix thoroughly. Gently heat and bring to boiling. Autoclave for 15 min at 15 psi pressure–121°C. Cool to 50°–55°C. Aseptically add 100.0mL of sterile Middlebrook OADC enrichment. Mix thoroughly. Pour into sterile Petri dishes or distribute into sterile tubes.

Use: For the isolation, cultivation and maintenance of *Mycobacterium* species, including *M. tuberculosis*. Also used for determining the antimicrobial susceptibility of mycobacteria.

Middlebrook 7H10 Agar with Middlebrook OADC Enrichment and Hemin (Hemin Medium for *Mycobacterium*)

Composition per liter:

Agar	15.0g
Na_2HPO_4	1.5g
KH_2PO_4	1.5g
$(NH_4)_2SO_4$	0.5g
L-Glutamic Acid	0.5g
Sodium citrate	0.4g
Ferric ammonium citrate	0.04g
$MgSO_4 \cdot 7H_2O$	0.025g
$ZnSO_4 \cdot 7H_2O$	1.0mg
$CuSO_4 \cdot 5H_2O$	1.0mg
Pyridoxine	1.0mg
Biotin	0.5mg
$CaCl_2 \cdot 2H_2O$	0.5mg
Malachite Green	.0.25mg
Middlebrook OADC enrichment	100.0mL
Glycerol	5.0mL
Hemin solution	3.9mL

pH 6.6 ± 0.2 at 25°C

Source: This medium is available as a premixed powder from BBL Microbiology Systems and Difco Laboratories.

Middlebrook OADC Enrichment:

Composition per 100mL:

Bovine albumin fraction V	5.0g
Glucose	2.0g
NaCl	0.85g
Oleic acid	0.05g
Catalase	4.0mg

Preparation of Middlebrook OADC Enrichment: Add components to distilled/deionized water and bring volume to 100.0mL. Mix thoroughly. Filter sterilize.

Hemin Solution:

Composition per 100mL:

Hemin	1.0g
NaOH (1*N* solution)	20.0mL

Preparation of Hemin Solution: Add hemin to 20.0mL of 1*N* NaOH solution. Mix thoroughly. Bring volume to 100.0mL with distilled/deionized water.

Preparation of Medium: Add glycerol to 891.1mL of distilled/deionized water and add remaining components, except Middlebrook OADC enrichment. Mix thoroughly. Gently heat and bring to boiling. Autoclave for 15 min at 15 psi pressure–121°C. Cool to 50°–55°C. Aseptically add 100.0mL of sterile Middlebrook OADC enrichment. Mix thoroughly. Pour into sterile Petri dishes or distribute into sterile tubes.

Use: For the isolation, cultivation and maintenance of *Mycobacterium* species, including *Mycobacterium tuberculosis*. For the cultivation and maintenance of *Mycobacterium haemophilum*. Also used for determining the antimicrobial susceptibility of mycobacteria.

Middlebrook 7H10 Agar with Middlebrook OADC Enrichment and Triton WR 1339

Composition per liter:

Agar	15.0g
Na_2HPO_4	1.5g
KH_2PO_4	1.5g
$(NH_4)_2SO_4$	0.5g
L-Glutamic Acid	0.5g
Sodium citrate	0.4g
Ferric ammonium citrate	0.04g
$MgSO_4 \cdot 7H_2O$	0.025g
$ZnSO_4 \cdot 7H_2O$	1.0mg
$CuSO_4 \cdot 5H_2O$	1.0mg
Pyridoxine	1.0mg
Biotin	0.5mg
$CaCl_2 \cdot 2H_2O$	0.5mg
Malachite Green	0.25mg
Middlebrook OADC enrichment with Triton WR 1339	100.0mL
Glycerol	5.0mL

pH 6.6 ± 0.2 at 25°C

Source: Available as a premixed powder from BBL Microbiology Systems and Difco Laboratories.

Middlebrook OADC Enrichment with Triton WR 1339:

Composition per 100mL:

Bovine albumin fraction V	5.0g
Glucose	2.0g
NaCl	0.85g
Triton WR-1339	0.25g
Oleic acid	0.05g
Catalase	4.0mg

Source: Available as a prepared enrichment from Difco Laboratories.

Preparation of Middlebrook OADC Enrichment with Triton WR 1339: Add components to distilled/deionized water and bring volume to 100.0mL. Mix thoroughly. Filter sterilize.

Preparation of Medium: Add glycerol to 900.0mL of distilled/deionized water and add remaining components, except Middlebrook OADC enrichment with Triton WR-1339. Mix thoroughly. Gently heat and bring to boiling. Autoclave for 15 min at 15 psi pressure–121°C. Cool to 50°–55°C. Aseptically add 100.0mL of sterile Middlebrook OADC enrichment with Triton WR-1339. Mix thoroughly. Pour into sterile Petri dishes or distribute into sterile tubes.

Use: For the isolation, cultivation and maintenance of *Mycobacterium* species, including *M. tuberculosis*. Also used for determining the antimicrobial susceptibility of mycobacteria.

Middlebrook 7H10 Agar with Streptomycin

Composition per liter:

Agar	15.0g
Na_2HPO_4	1.5g
KH_2PO_4	1.5g
$(NH_4)_2SO_4$	0.5g
L-Glutamic Acid	0.5g
Sodium citrate	0.4g
Ferric ammonium citrate	0.04g
$MgSO_4 \cdot 7H_2O$	0.025g
$ZnSO_4 \cdot 7H_2O$	1.0mg
$CuSO_4 \cdot 5H_2O$	1.0mg
Pyridoxine	1.0mg
Biotin	0.5mg
$CaCl_2 \cdot 2H_2O$	0.5mg
Malachite Green	0.25mg
Glycerol	5.0mL
Streptomycin	100.0mg

pH 6.6 ± 0.2 at 25°C

Source: This medium is available as a premixed powder from BBL Microbiology Systems and Difco Laboratories.

Preparation of Medium: Add glycerol to 1.0L of distilled/deionized water and add remaining components. Mix thoroughly. Gently heat and bring to boiling. Autoclave for 15 min at 15 psi pressure–121°C. Cool to 50°–55°C. Aseptically add streptomycin. Mix thoroughly. Pour into sterile Petri dishes or distribute into sterile tubes.

Use: For the isolation, cultivation and maintenance of *Mycobacterium kansasii*.

Middlebrook 7H11 Agar, Selective

Composition per liter:

Agar	15.0g
Na_2HPO_4	1.5g
KH_2PO_4	1.5g
Pancreatic digest of casein	1.0g
$(NH_4)_2SO_4$	0.5g
L-Glutamic acid	0.5g
Sodium citrate	0.4g
$MgSO_4 \cdot 7H_2O$	0.05g
Ferric ammonium citrate	0.04g
Pyridoxine	1.0mg
$ZnSO_4 \cdot 7H_2O$	1.0mg
$CuSO_4 \cdot 5H_2O$	1.0mg
$CaCl_2 \cdot 2H_2O$	0.5mg
Malachite Green	0.25mg
D-Biotin	0.5μg
Middlebrook OADC enrichment	100.0mL
Antibiotic solution	10.0mL
Glycerol	5.0mL

pH 6.6 ± 0.2 at 25°C

Middlebrook OADC Enrichment:

Composition per 100mL:

Bovine albumin fraction V	5.0g
Glucose	2.0g
NaCl	0.85g
Oleic acid	0.05g
Catalase	4.0mg

Source: Available as a prepared enrichment from Difco Laboratories.

Preparation of Middlebrook OADC Enrichment: Add components to distilled/deionized water and bring volume to 100.0mL. Mix thoroughly. Filter sterilize.

Antibiotic Solution:

Composition per 10mL:

Carbenicillin	0.050mg
Trimethoprim lactate	0.020mg
Amphotericin B	0.010mg
Polymyxin B	200,000U

Preparation of Antibiotic Solution: Add components to distilled/deionized water and bring volume to 10.0mL. Mix thoroughly. Filter sterilize.

Preparation of Medium: Add glycerol to 890.0mL of distilled/deionized water and add remaining components, except Middlebrook OADC enrichment and antibiotic solution. Mix thoroughly. Gently heat and bring to boiling. Autoclave for 15 min at 15 psi pressure–121°C. Cool to 50°–55°C. Aseptically add 100.0mL of sterile Middlebrook OADC enrichment and 10.0mL of sterile antibiotic solution. Mix thoroughly. Pour into sterile Petri dishes or distribute into sterile tubes.

Use: For the selective isolation and cultivation of pathogenic mycobacteria from specimens potentially contaminated with bacteria and fungi.

Middlebrook 7H11 Agar with Middlebrook ADC Enrichment (Mycobacteria 7H11 Agar with Middlebrook ADC Enrichment)

Composition per liter:

Agar	15.0g
Na_2HPO_4	1.5g
KH_2PO_4	1.5g
Pancreatic digest of casein	1.0g
$(NH_4)_2SO_4$	0.5g
L–Glutamic acid	0.5g
Sodium citrate	0.4g
$MgSO_4 \cdot 7H_2O$	0.05g
Ferric ammonium citrate	0.04g
Pyridoxine	1.0mg
Malachite Green	0.25mg
D–Biotin	0.5μg
Middlebrook ADC enrichment	100.0mL
Glycerol	5.0mL

pH 6.6 ± 0.2 at 25°C

Source: This medium is available as a premixed powder from BBL Microbiology Systems and Difco Laboratories.

Middlebrook ADC Enrichment:

Composition per 100mL:

Bovine albumin fraction V	5.0g
Glucose	2.0g
Catalase	0.003g
Distilled water	100.0mL

Source: Available as a prepared enrichment from Difco Laboratories.

Preparation of Middlebrook ADC Enrichment: Add components to distilled/deionized water and bring volume to 100.0mL. Mix thoroughly. Filter sterilize.

Preparation of Medium: Add glycerol to 900.0mL of distilled/deionized water and add remaining components, except Middlebrook ADC enrichment. Mix thoroughly. Gently heat and bring to boiling. Autoclave for 15 min at 15 psi pressure–121°C. Cool to 50°–55°C. Aseptically add 100.0mL of sterile Middlebrook ADC enrichment. Mix thoroughly. Pour into sterile Petri dishes or distribute into sterile tubes.

Use: For the cultivation of drug resistant (isoniazid [INH]) strains of *M. tuberculosis*. For the cultivation of particularly fastidious strains of tubercle bacilli which occur following treatment of tuberculosis patients with secondary anti-tubercular drugs. Generally these strains fail to grow on 7H10 medium.

Middlebrook 7H11 Agar with Middlebrook OADC Enrichment (Mycobacteria 7H11 Agar with Middlebrook OADC Enrichment)

Composition per liter:

Agar	15.0g
Na_2HPO_4	1.5g
KH_2PO_4	1.5g
Pancreatic digest of casein	1.0g
$(NH_4)_2SO_4$	0.5g
L–Glutamic acid	0.5g
Sodium citrate	0.4g
$MgSO_4 \cdot 7H_2O$	0.05g
Ferric ammonium citrate	0.04g
Pyridoxine	1.0mg
Malachite Green	0.25mg
D–Biotin	0.5μg
Middlebrook OADC enrichment	100.0mL
Glycerol	5.0mL

pH 6.6 ± 0.2 at 25°C

Source: This medium is available as a premixed powder from BBL Microbiology Systems and Difco Laboratories.

Middlebrook OADC Enrichment:

Composition per 100mL:

Bovine albumin fraction V	5.0g
Glucose	2.0g
NaCl	0.85g
Oleic acid	0.05g
Catalase	4.0mg

Source: Available as a prepared enrichment from Difco Laboratories.

Preparation of Middlebrook OADC Enrichment: Add components to distilled/deionized water and bring volume to 100.0mL. Mix thoroughly. Filter sterilize.

Preparation of Medium: Add glycerol to 900.0mL of distilled/deionized water and add remaining components, except Middlebrook OADC enrichment. Mix thoroughly. Gently heat and bring to boiling. Autoclave for 15 min at 15 psi pressure–121°C. Cool to 50°–55°C. Aseptically add 100.0mL of sterile Middlebrook OADC enrichment. Mix thoroughly. Pour into sterile Petri dishes or distribute into sterile tubes.

Use: For the cultivation of drug resistant (isoniazid [INH]) strains of *M. tuberculosis*. For the cultivation of particularly fastidious strains of tubercle bacilli which occur following treatment of tuberculosis patients with secondary anti-tubercular drugs. Generally these strains fail to grow on 7H10 medium.

Middlebrook 7H11 Agar with Middlebrook OADC Enrichment and Triton WR 1339 (Mycobacteria 7H11 Agar with Middlebrook OADC Enrichment and Triton WR 1339)

Composition per liter:

Agar	15.0g
Na_2HPO_4	1.5g
KH_2PO_4	1.5g
Pancreatic digest of casein	1.0g
$(NH_4)_2SO_4$	0.5g
L–Glutamic acid	0.5g
Sodium citrate	0.4g
$MgSO_4 \cdot 7H_2O$	0.05g
Ferric ammonium citrate	0.04g
Pyridoxine	1.0mg
Malachite Green	0.25mg
D–Biotin	0.5μg
Middlebrook OADC enrichment with Triton WR 1339	100.0mL
Glycerol	5.0mL

pH 6.6 ± 0.2 at 25°C

Source: This medium is available as a premixed powder from BBL Microbiology Systems and Difco Laboratories.

Middlebrook OADC Enrichment with Triton WR 1339:

Composition per 100mL:

Bovine albumin fraction V	5.0g
Glucose	2.0g
NaCl	0.85g
Triton WR-1339	0.25g
Oleic acid	0.05g
Catalase	4.0mg

Source: Available as a prepared enrichment from Difco Laboratories.

Preparation of Middlebrook OADC Enrichment with Triton WR 1339: Add components to distilled/deionized water and bring volume to 100.0mL. Mix thoroughly. Filter sterilize.

Preparation of Medium: Add glycerol to 900.0mL of distilled/deionized water and add remaining components, except Middlebrook OADC enrichment with Triton WR-1339. Mix thoroughly. Gently heat and bring to boiling. Autoclave for 15 min at 15 psi pressure–121°C. Cool to 50°–55°C. Aseptically add 100.0mL of sterile Middlebrook OADC enrichment with Triton WR-1339. Mix thoroughly. Pour into sterile Petri dishes or distribute into sterile tubes.

Use: For the cultivation of drug resistant (isoniazid [INH]) strains of *M. tuberculosis*. For the cultivation of particularly fastidious strains of tubercle bacilli which occur following treatment of tuberculosis patients with secondary anti-tubercular drugs. Generally these strains fail to grow on 7H10 medium.

Middlebrook 7H12 Medium

Composition per 102.5mL:

Bovine serum albumin	0.5g
Casein hydrolyslate	0.1g
Catalase	4,800U
^{14}C-Palmitic acid	100μCi
Middlebrook 7H9 broth	100.0mL
Antibiotic solution	2.5mL

pH 6.8 ± 0.1 at 25°C

Middlebrook 7H9 Broth:

Composition per liter:

Na_2HPO_4	2.5g
KH_2PO_4	1.0g
Monosodium glutamate	0.5g
$(NH_4)_2SO_4$	0.5g
Sodium citrate	0.1g
$MgSO_4·7H_2O$	0.05g
Ferric ammonium citrate	0.04g
$CuSO_4·5H_2O$	1.0mg
Pyridoxine	1.0mg
$ZnSO_4·7H_2O$	1.0mg
Biotin	0.5mg
$CaCl_2·2H_2O$	0.5mg
Glycerol	2.0mL

Preparation of Middlebrook 7H9 Broth: Add components to distilled/deionized water and bring volume to 1.0L. Mix thoroughly.

Antibiotic Solution:

Composition per 5mL:

Nalidixic acid	0.2g
Azlocillin	0.1g
Amphotericin B	0.050g
Trimethoprim	0.050g
Polymyxin B	500,000U

Preparation of Antibiotic Solution: Add components to distilled/deionized water and bring volume to 5.0mL. Mix thoroughly. Filter sterilize.

Preparation of Medium: To 100.0mL of Middlebrook 7H9 broth, add remaining components, except antibiotic solution. Mix thoroughly. Filter sterilize. Aseptically distribute into bottles in 4.0mL volumes. Prior to inoculation, aseptically add 0.1mL of antibiotic solution to each bottle. Mix thoroughly.

Use: For the cultivation of *Mycobacterium* species from the blood of patients suspected of having mycobacteremia.

Middlebrook 13A Medium

Composition per 112.5mL:

Casein hydrolysate	0.1g
Tween™ 80	0.02g
Sodium polyanetholesulfonate	0.025g
Catalase	36,000U
^{14}C-substrate	125μCi(185kBq)
Middlebrook 7H9 broth	100.0mL
Middlebrook 13A enrichment	12.5mL

pH 6.6 ± 0.2 at 25°C

Middlebrook 7H9 Broth:

Composition per liter:

Na_2HPO_4	2.5g
KH_2PO_4	1.0g
Monosodium glutamate	0.5g
$(NH_4)_2SO_4$	0.5g
Sodium citrate	0.1g
$MgSO_4·7H_2O$	0.05g
Ferric ammonium citrate	0.04g
$CuSO_4·5H_2O$	1.0mg
Pyridoxine	1.0mg
$ZnSO_4·7H_2O$	1.0mg
Biotin	0.5mg
$CaCl_2·2H_2O$	0.5mg
Glycerol	2.0mL

Preparation of Middlebrook 7H9 Broth: Add components to distilled/deionized water and bring volume to 1.0L. Mix thoroughly.

Middlebrook 13A Enrichment:

Composition per 20mL:

Bovine serum albumin	3.0g

Preparation of Middlebrook 13A Enrichment: Add bovine serum albumin to distilled/deionized water and bring volume to 20.0mL. Mix thoroughly. Filter sterilize.

Preparation of Medium: To 100.0mL of Middlebrook 7H9 broth, add remaining components, except Middlebrook 13A enrichment. Mix thoroughly. Filter sterilize. Aseptically distribute into bottles in 4.0mL volumes. Prior to inoculation, aseptically add 0.5mL of Middlebrook 13A enrichment to each bottle. Mix thoroughly.

Use: For the cultivation of *Mycobacterium* species from the blood of patients suspected of having mycobacteremia.

Middlebrook ADC Enrichment (Middlebrook Albumin Dextrose Catalase Enrichment)

Composition per 100mL:

Bovine albumin fraction V	5.0g
Glucose	2.0g
Catalase	0.003g

Source: Available as a prepared enrichment from Difco Laboratories.

Preparation of Enrichment: Add components to distilled/deionized water and bring volume to 100.0mL. Mix thoroughly. Filter sterilize.

Use: For use as a supplement to other Middlebrook media for the isolation, cultivation and maintenance of *Mycobacterium* species. Also used as a supplement to other Middlebrook media for determining the antimicrobial susceptibility of mycobacteria.

Middlebrook and Cohn 7H10 Agar *See:* Middlebrook 7H10 Agar with Middlebrook OADC Enrichment

Middlebrook OADC Enrichment (Middlebrook Oleic Albumin Dextrose Catalase Enrichment)

Composition per 100mL:

Bovine albumin fraction V	5.0g
Glucose	2.0g
NaCl	0.85g
Oleic acid	0.05g
Catalase	4.0mg

Source: Available as a prepared enrichment from Difco Laboratories.

Preparation of Enrichment: Add components to distilled/deionized water and bring volume to 100.0mL. Mix thoroughly. Filter sterilize.

Use: For use as a supplement to other Middlebrook media for the isolation, cultivation and maintenance of *Mycobacterium* species. Also used as a supplement to other Middlebrook media for determining the antimicrobial susceptibility of mycobacteria.

Middlebrook OADC Enrichment with Triton WR 1339 (Middlebrook Oleic Albumin Dextrose Catalase Enrichment with Triton WR 1339)

Composition per 100mL:

Bovine albumin fraction V	5.0g
Glucose	2.0g
NaCl	0.85g
Triton WR 1339	0.25g
Oleic acid	0.05g
Catalase	4.0mg

Source: Available as a prepared enrichment from Difco Laboratories.

Preparation of Enrichment: Add components to distilled/deionized water and bring volume to 100.0mL. Mix thoroughly. Filter sterilize.

Use: For use as a supplement to other Middlebrook media for the isolation, cultivation and maintenance of *Mycobacterium* species. Also used as a supplement to other Middlebrook media for determining the antimicrobial susceptibility of mycobacteria.

Middlebrook Oleic Albumin Dextrose Catalase Enrichment *See:* Middlebrook OADC Enrichment

Middlebrook Oleic Albumin Dextrose Catalase Enrichment with TritonWR 1339 *See:* Middlebrook OADC Enrichment with Triton WR 1339

MIL Medium (Motility Indole Lysine Medium)

Composition per liter:

Peptone	10.0g
Pancreatic digest of casein	10.0g
L-Lysine·HCl	10.0g
Yeast extract	3.0g
Agar	2.0g
Dextrose	1.0g

Ferric ammonium citrate....................................0.5g
Bromcresol Purple 0.02g

pH 6.6 ± 0.2 at 25°C

Source: Available as a premixed powder and prepared medium from Difco Laboratories.

Preparation of Medium: Add components to distilled/deionized water and bring volume to 1.0L. Mix thoroughly. Gently heat and bring to boiling. Distribute into tubes in 5.0mL volumes. Autoclave for 15 min at 15 psi pressure–121°C.

Use: For the cultivation and differentiation of members of the Enterobacteriaceae on the basis of motility, lysine decarboxylase activity, lysine deaminase activity and indole production.

Milk Agar
See: Skim Milk Agar

Milk Agar

Composition per liter:

Agar...15.0g
Peptone..5.0g
Yeast extract..3.0g
Milk solids or 1.0g
fresh milk) 10.0mL

pH 7.2 ± 0.2 at 25°C

Source: This medium is available as a premixed powder from Oxoid Unipath.

Preparation of Medium: Add components to distilled/deionized water and bring volume to 1.0L. Mix thoroughly. Gently heat and bring to boiling. Distribute into tubes or flasks. Autoclave for 15 min at 15 psi pressure–121°C.

Use: For the cultivation of microorganisms from dairy and water samples.

Milk Agar

Composition per liter:

Mixture A 500.0mL
Mixture B 500.0mL

Mixture A:

Composition per 500mL:

Instant nonfat milk 100.0g

Preparation of Mixture A: Add instant nonfat milk to distilled/deionized water and bring volume to 500.0mL. Mix thoroughly. Autoclave for 15 min at 15 psi pressure–121°C. Cool rapidly to 55°C.

Mixture B:

Composition per 500mL:

Agar...15.0g
Nutrient broth..12.5g
NaCl 2.5g

Preparation of Mixture B: Add components to distilled/deionized water and bring volume to 500.0mL. Mix thoroughly. Gently heat and bring to boiling. Autoclave for 15 min at 15 psi pressure–121°C. Cool rapidly to 55°C.

Preparation of Medium: Aseptically combine the cooled sterile mixture A with cooled sterile mixture B. Mix thoroughly. Pour into sterile Petri dishes in 20.0mL volumes.

Use: For the cultivation and estimation of numbers of *Pseudomonas aeruginosa* in water by the membrane filter method.

Milk Protein Hydrolysate Agar
See: MPH Agar

Mineral Base E for Autotrophic Growth

Composition per liter:

Noble agar..15.0g
K_2HPO_4 ....1.2g
KH_2PO_4 ....0.624g
$(NH_4)_2SO_4$ 0.5g
NaCl 0.1g
$CaCl_2 \cdot 6H_2O$ solution 10.0mL
$MgSO_4 \cdot 7H_2O$ solution.................................. 10.0mL
Mineral solution 1.0mL
p-Aminobenzoic acid solution 1.0mL

$CaCl_2 \cdot 6H_2O$ Solution:

Composition per liter:

$CaCl_2 \cdot 6H_2O$ 5.0g

Preparation of $CaCl_2 \cdot 6H_2O$ Solution: Add $CaCl_2 \cdot 6H_2O$ to distilled/deionized water and bring volume to 1.0L. Mix thoroughly. Autoclave for 15 min at 15 psi pressure–121°C.

$MgSO_4 \cdot 7H_2O$ Solution:

Composition per liter:

$MgSO_4 \cdot 7H_2O$ 20.0g

Preparation of $MgSO_4 \cdot 7H_2O$ Solution: Add $MgSO_4 \cdot 7H_2O$ to distilled/deionized water and bring volume to 1.0L. Mix thoroughly. Autoclave for 15 min at 15 psi pressure–121°C.

***p*-Aminobenzoic Acid Solution:**

Composition per 10mL:

p-Aminobenzoic acid 100.0mg

Preparation of *p*-Aminobenzoic Acid Solution: Add *p*-aminobenzoic acid to distilled/deionized water and bring volume to 10.0mL. Mix thoroughly. Autoclave for 15 min at 15 psi pressure–121°C.

Mineral Solution:
Composition per 100mL:

Disodium EDTA	1.58g
ZnSO4·7H2O	0.7g
$MnSO_4·4H_2O$	0.18g
$FeSO_4·7H_2O$	0.16g
$CoCl_2·6H_2O$	0.052g
$Na_2MoO_4·2H_2O$	0.047g
$CuSO_4·5H_2O$	0.047g

Preparation of Medium: Add components, except $CaCl_2·6H_2O$ solution, $MgSO_4·7H_2O$ solution, and *p*-aminobenzoic acid solution to distilled/deionized water and bring volume to 979.0mL. Mix thoroughly. Autoclave for 15 min at 15 psi pressure–121°C. Cool to 50°C. Aseptically add in the following order: 10.0mL of sterile $CaCl_2·6H_2O$ solution, 10.0mL of sterile $MgSO_4·7H_2O$ solution, and 1.0mL of sterile *p*-aminobenzoic acid solution. Mix thoroughly. Aseptically distribute into sterile tubes or flasks. Incubate inoculated tubes in 50% CO_2.

Use: For the autotrophic cultivation and maintenance of *Pseudomonas thermocarboxydovorans*.

Mineral Base E for Heterotrophic Growth

Composition per liter:

Noble agar	15.0g
K_2HPO_4	1.2g
KH_2PO_4	0.624g
$(NH_4)_2SO_4$	0.5g
NaCl	0.1g
$CaCl_2·6H_2O$ solution	10.0mL
$MgSO_4·7H_2O$ solution	10.0mL
Sodium pyruvate solution	10.0mL
Mineral solution	1.0mL
p-Aminobenzoic acid solution	1.0mL

$CaCl_2·6H_2O$ Solution:
Composition per liter:

$CaCl_2·6H_2O$	5.0g

Preparation of $CaCl_2·6H_2O$ Solution: Add $CaCl_2·6H_2O$ to distilled/deionized water and bring volume to 1.0L. Mix thoroughly. Autoclave for 15 min at 15 psi pressure–121°C.

$MgSO_4·7H_2O$ Solution:
Composition per liter:

$MgSO_4·7H_2O$	20.0g

Preparation of $MgSO_4·7H_2O$ Solution: Add $MgSO_4·7H_2O$ to distilled/deionized water and bring volume to 1.0L. Mix thoroughly. Autoclave for 15 min at 15 psi pressure–121°C.

Sodium Pyruvate Solution:
Composition per 10mL:

Sodium pyruvate	2.0g

Preparation of Sodium Pyruvate Solution: Add sodium pyruvate to distilled/deionized water and bring volume to 10.0mL. Mix thoroughly. Filter sterilize.

***p*-Aminobenzoic Acid Solution:**
Composition per 10mL:

p-Aminobenzoic acid	100.0mg

Preparation of *p*-Aminobenzoic Acid Solution: Add *p*-aminobenzoic acid to distilled/deionized water and bring volume to 10.0mL. Mix thoroughly. Autoclave for 15 min at 15 psi pressure–121°C.

Mineral Solution:
Composition per 100mL:

Disodium EDTA	1.58g
ZnSO4·7H2O	0.7g
$MnSO_4·4H_2O$	0.18g
$FeSO_4·7H_2O$	0.16g
$CoCl_2·6H_2O$	0.052g
$Na_2MoO_4·2H_2O$	0.047g
$CuSO_4·5H_2O$	0.047g

Preparation of Medium: Add components, except $CaCl_2·6H_2O$ solution, $MgSO_4·7H_2O$ solution, sodium pyruvate solution, and *p*-aminobenzoic acid solution to distilled/deionized water and bring volume to 969.0mL. Mix thoroughly. Autoclave for 15 min at 15 psi pressure–121°C. Cool to 45°–50°C. Aseptically add in the following order: 10.0mL of the sterile $CaCl_2·6H_2O$ solution, 10.0mL of the sterile $MgSO_4·7H_2O$ solution, 10.0mL of sterile sodium pyruvate solution and 1.0mL of sterile *p*-aminobenzoic acid solution. Mix thoroughly. Aseptically distribute into sterile tubes or flasks.

Use: For the heterotrophic cultivation and maintenance of *Pseudomonas thermocarboxydovorans*.

Mineral Base Medium with Acetate

See: **MBM Acetate Medium**

Mineral Medium

Composition per liter:

Yeast extract	2.0g
Mineral Base 5X	200.0mL

Trace element solution SL-6 .... 1.0mL
Thiamine·HCl 3.0μg
Biotin 0.2μg

pH 6.8 ± 0.2 at 25°C

Mineral Base 5X:
Composition per liter:

NaCl 5.0g
NH_4Cl 2.0g
KH_2PO_4 ....1.35g
$MgSO_4 \cdot 7H_2O$ 1.0g
K_2HPO_4 ....0.87g
$CaCl_2$ 0.05g
$FeCl_3 \cdot 6H_2O$ 1.25mg

Preparation of Mineral Base 5X: Add components to distilled/deionized water and bring volume to 1.0L. Mix thoroughly.

Trace Elements Solution SL-6:
Composition per liter:

H_3BO_3 ....0.3g
$CoCl_2 \cdot 6H_2O$ 0.2g
$ZnSO_4 \cdot 7H_2O$ 0.10g
$MnCl_2 \cdot 4H_2O$ 0.03g
$Na_2MoO_4 \cdot H_2O$ 0.03g
$NiCl_2 \cdot 6H_2O$ 0.02g
$CuCl_2 \cdot 2H_2O$ 0.01g

Preparation of Trace Elements Solution SL-6: Add components to distilled/deionized water and bring volume to 1.0L. Mix thoroughly. Adjust pH to 3.4.

Preparation of Medium: Add components to distilled/deionized water and bring volume to 1.0L. Mix thoroughly. Adjust pH to 6.8. Distribute into tubes or flasks. Autoclave for 15 min at 15 psi pressure–121°C.

Use: For the cultivation of the *Arthrobacter* species.

Mineral Medium A

Composition per liter:

$(NH_4)_2SO_4$ 1.0g
K_2HPO_4 ....1.0g

Preparation of Medium: Add components to tap water and bring volume to 1.0L. Mix thoroughly. Distribute into tubes or flasks. Autoclave for 15 min at 15 psi pressure–121°C.

Use: For the cultivation of *Saccharobacterium ovale*.

Mineral Medium for Hydrogen Bacteria

Composition per liter:

Agar 15.0g
$Na_2HPO_4 \cdot 2H_2O$ 2.9g
KH_2PO_4 ....2.3g
NH_4Cl 1.0g
$MgSO_4 \cdot 7H_2O$ 0.5g
$NaHCO_3$ 0.5g
$CaCl_2 \cdot 2H_2O$ 0.01g
Ferric ammonium citrate solution 20.0mL

Ferric Ammonium Citrate Solution:
Composition per 20mL:

Ferric ammonium citrate 0.05g

Preparation of Ferric Ammonium Citrate Solution: Add ferric ammonium citrate to 20.0mL of distilled/deionized water. Filter sterilize.

Preparation of Medium: Add components, except ferric ammonium citrate solution, to distilled/deionized water and bring volume to 980.0mL. Mix thoroughly. Gently heat and bring to boiling. Autoclave for 15 min at 15 psi pressure–121°C. Cool to 50°C. Aseptically add the sterile ferric ammonium citrate solution. Mix thoroughly. Pour into sterile Petri dishes or distribute into sterile tubes. Incubate inoculated medium at 30°C under 60% H_2 + 25% N_2 + 10% CO_2 + 5% O_2.

Use: For the cultivation and maintenance of *Alcaligenes eutrophus, Hydrogenophaga flava,* and *Hydrogenophaga pseudoflava.*

Mineral Medium S with 1% Sucrose

Composition per liter:

Sucrose 10.0g
$NH_4H_2PO_4$ 1.5g
$MgSO_4 \cdot 7H_2O$ 0.2g
$CaCl_2$ 0.1g
KCl 0.1g
$FeCl_3$ 0.005g
NaOH (1*N* solution) 10.0mL

Preparation of Medium: Add components to distilled/deionized water and bring volume to 1.0L. Mix thoroughly. Distribute into tubes or flasks. Autoclave for 15 min at 15 psi pressure–121°C.

Use: For the cultivation of *Saccharobacterium acuminatum*.

Mineral Medium with Antipyrin

Composition per liter:

Antipyrin 1.0g
$Na_2HPO_4 \cdot 12H_2O$ 0.7g
$(NH_4)_2HPO_4$ 0.7g
KH_2PO_4 ....0.3g

$(NH_4)H_2PO_4$	0.3g
$MgSO_4 \cdot 7H_2O$	0.25g
$(NH_4)_2SO_4$	0.1g
$CaCl_2 \cdot 6H_2O$	0.05g
H_3BO_3	0.5mg
$MnSO_4 \cdot 4H_2O$	0.4mg
$ZnSO_4 \cdot 7H_2O$	0.4mg
$FeCl_3 \cdot 6H_2O$	0.2mg
$(NH_4)_6Mo_7O_{24} \cdot 4H_2O$	0.2mg
KI	0.1mg
$CuSO_4 \cdot 5H_2O$	0.04mg
Vitamin solution	20.0mL

pH 6.8–7.0 at 25°C

Vitamin Solution:
Composition per 20mL:

Biotin	0.1mg
Vitamin B_{12}	0.03mg

Preparation of Vitamin Solution: Add biotin and vitamin B_{12} to 20.0mL of distilled/deionized water. Mix thoroughly. Filter sterilize.

Preparation of Medium: Add components, except vitamin solution, to distilled/deionized water and bring volume to 980.0mL. Mix thoroughly. Adjust pH to 6.8–7.0 with 1*N* NaOH. Autoclave for 20 min at 15 psi pressure–121°C. Cool to 45°–50°C. Aseptically add the sterile vitamin solution. Mix thoroughly. Distribute into sterile tubes or flasks.

Use: For the cultivation and maintenance of *Phenylobacterium immobile*.

Mineral Medium with Chloridazon

Composition per liter:

Chloridazon	1.0g
$Na_2HPO_4 \cdot 12H_2O$	0.7g
$(NH_4)_2HPO_4$	0.7g
KH_2PO_4	0.3g
$(NH_4)H_2PO_4$	0.3g
$MgSO_4 \cdot 7H_2O$	0.25g
$(NH_4)_2SO_4$	0.1g
$CaCl_2 \cdot 6H_2O$	0.05g
H_3BO_3	0.5mg
$MnSO_4 \cdot 4H_2O$	0.4mg
$ZnSO_4 \cdot 7H_2O$	0.4mg
$FeCl_3 \cdot 6H_2O$	0.2mg
$(NH_4)_6Mo_7O_{24} \cdot 4H_2O$	0.2mg
KI	0.1mg
$CuSO_4 \cdot 5H_2O$	0.04mg
Vitamin solution	20.0mL

pH 6.8–7.0 at 25°C

Vitamin Solution:
Composition per 20mL:

Biotin	0.1mg
Vitamin B_{12}	0.03mg

Preparation of Vitamin Solution: Add biotin and vitamin B_{12} to 20.0mL of distilled/deionized water. Mix thoroughly. Filter sterilize.

Preparation of Medium: Add components, except vitamin solution, to distilled/deionized water and bring volume to 980.0mL. Mix thoroughly. Adjust pH to 6.8–7.0 with 1*N* NaOH. Autoclave for 20 min at 15 psi pressure–121°C. Cool to 45°–50°C. Aseptically add the sterile vitamin solution. Mix thoroughly. Distribute into sterile tubes or flasks.

Use: For the cultivation and maintenance of *Phenylobacterium immobile*.

Mineral Medium with Dichlorobenzoate

Composition per liter:

Na_2HPO_4	2.78g
KH_2PO_4	2.78g
$(NH_4)_2SO_4$	1.0g
Hutner's mineral base	20.0mL
2,4-Dichlorobenzoate solution	10.0mL

pH 6.8 ± 0.2 at 25°C

Hutner's Mineral Base:
Composition per liter:

$MgSO_4 \cdot 7H_2O$	29.7g
Nitrilotriacetic acid	10.0g
$CaCl_2 \cdot 2H_2O$	3.34g
$FeSO_4 \cdot 7H_2O$	99.0mg
$(NH_4)_2MoO_4$	9.25mg
Metals "44"	50.0mL

Preparation of Hutner's Mineral Base: Add nitrilotriacetic acid to 500.0mL of distilled/deionized water. Dissolve by adjusting pH to 6.5 with KOH. Add remaining components. Readjust pH to 7.2 with H_2SO_4 or KOH. Add distilled/deionized water to 1.0L. Store at 5°C.

Metals "44":
Composition per 100mL:

$ZnSO_4 \cdot 7H_2O$	1.1g
$FeSO_4 \cdot 7H_2O$	0.5g
EDTA	0.25g
$MnSO_4 \cdot 7H_2O$	0.154g
$CuSO_4 \cdot 5H_2O$	0.04g
$Co(NO_3)_2 \cdot 6H_2O$	0.025g
$Na_2B_4O_7 \cdot 10H_2O$	0.018g

Preparation of Metals "44": Add a few drops of H_2SO_4 to distilled/deionized water to inhibit precipitate formation. Add components to acidified distilled/deionized water and bring volume to 100.0mL. Mix thoroughly.

2,4-Dichlorobenzoate Solution:
Composition per 10mL:

2,4-Dichlorobenzoate	5.0mg

Preparation of 2,4-Dichlorobenzoate Solution: Add 2,4-dichlorobenzoate to 10.0mL of distilled/deionized water. Mix thoroughly. Filter sterilize.

Preparation of Medium: Add components, except 2,4-dichlorobenzoate solution, to distilled/deionized water and bring volume to 990.0mL. Mix thoroughly. Adjust pH to 6.8 with 1*N* KOH. Autoclave for 15 min at 15 psi pressure–121°C. Cool to 45°–50°C. Aseptically add the sterile 2,4-dichlorobenzoate solution. Mix thoroughly. Distribute into sterile tubes or flasks.

Use: For the cultivation and maintenance of *Actinomyces viscosus.*

Mineral Medium with Glucose

Composition per liter:

Agar	20.0g
Na_2HPO_4	4.8g
KH_2PO_4	4.4g
NH_4Cl	1.0g
$MgSO_4 \cdot 7H_2O$	0.5g
Solution B	10.0mL
Solution A	5.0mL

pH 6.8 ± 0.2 at 25°C

Solution A:
Composition per 100mL:

Ferric ammonium citrate	1.0g
$CaCl_2$	0.1g

Preparation of Solution A: Add ferric ammonium citrate and $CaCl_2$ to distilled/deionized water and bring volume to 100.0mL. Mix thoroughly. Filter sterilize.

Solution B:
Composition per 100mL:

Glucose	10.0g

Preparation of Solution B: Add glucose to distilled/deionized water and bring volume to 100.0mL. Mix thoroughly. Filter sterilize.

Preparation of Medium: Add components, except solution A and solution B, to distilled/deionized water and bring volume to 985.0mL. Mix thoroughly. Gently heat and bring to boiling. Autoclave for 15 min at 15 psi pressure–121°C. Cool to 50°C. Aseptically add the sterile solution A and solution B. Mix thoroughly. Pour into sterile Petri dishes or distribute into sterile tubes.

Use: For the cultivation and maintenance of *Alcaligenes latus.*

Mineral Medium with Phenol

Composition per liter:

Phenol	1.0g
K_2HPO_4	1.0g
NH_4NO_3	1.0g
$(NH_4)_2SO_4$	0.5g
$MgSO_4$	0.5g
KH_2PO_4	0.5g
NaCl	0.5g
$CaCl_2$	0.02g
$FeSO_4$	0.02g
Wolfe's mineral solution	10.0mL

Wolfe's Mineral Solution:
Composition per liter

$MgSO_4 \cdot 7H_2O$	3.0g
Nitriloacetic acid	1.5g
NaCl	1.0g
$MnSO_4 \cdot H_2O$	0.5g
$FeSO_4 \cdot 7H_2O$	0.1g
$CoCl_2 \cdot 6H_2O$	0.1g
$CaCl_2$	0.1g
$ZnSO_4 \cdot 7H_2O$	0.1g
$CuSO_4 \cdot 5H_2O$	0.01g
$AlK(SO_4)_2 \cdot 12H_2O$	0.01g
H_3BO_3	0.01g
$Na_2MoO_4 \cdot 2H_2O$	0.01g

Preparation of Wolfe's Mineral Solution: Add nitrilotriacetic acid to 500.0mL of distilled/deionized water. Dissolve by adjusting pH to 6.5 with KOH. Add remaining components. Add distilled/deionized water to 1.0L.

Preparation of Medium: Add components, except phenol, to distilled/deionized water and bring volume to 1.0L. Mix thoroughly. Autoclave for 15 min at 15 psi pressure–121°C. Aseptically add the phenol. Mix thoroughly. Distribute into sterile tubes or flasks.

Use: For the cultivation and maintenance of *Pseudomonas putida.*

Mineral Medium with Santonin

Composition per liter:

K_2HPO_4	6.3g
α-Santonin	4.0g

KH_2PO_4	1.82g
NH_4NO_3	1.0g
$CaCl_2 \cdot 2H_2O$	0.75g
$MgSO_4 \cdot 7H_2O$	0.1g
$FeSO_4 \cdot 7H_2O$	0.06g
$MnSO_4$ (anhydrous)	600.0μg
$Na_2MoO_4 \cdot 2H_2O$	600.0μg

pH 7.0 ± 0.2 at 25°C

Preparation of Medium: Add components to distilled/deionized water and bring volume to 1.0L. Mix thoroughly. Autoclave for 15 min at 15 psi pressure–121°C. Distribute into sterile tubes or flasks.

Use: For the cultivation and maintenance of *Pseudomonas* species.

Mineral Pectin 5 Medium
See: **MP 5 Medium**

Mineral Pectin 7 Medium
See: **MP 7 Medium**

Mineral Salts Agar

Composition per liter:

Agar	15.0g
$NaNO_3$	2.0g
K_2HPO_4	1.2g
$MgSO_4$	0.5g
KCl	0.5g
KH_2PO_4	0.14g
Yeast extract	0.02g
$Fe_2(SO_4)_3 \cdot H_2O$	0.01g

pH 7.2 ± 0.2 at 25°C

Preparation of Medium: Add components to distilled/deionized water and bring volume to 1.0L. Mix thoroughly. Adjust pH to 7.2. Gently heat and bring to boiling. Distribute into tubes. Autoclave for 15 min at 15 psi pressure–121°C. Allow tubes to cool in a slanted position. Add a strip of sterile filter paper onto cooled slant. Inoculate organisms on filter paper.

Use: For the cultivation and maintenance of *Cytophaga aurantiaca* and *Sporocytophaga myxococcoides*.

Mineral Salts Enrichment Medium

Composition per liter:

KH_2PO_4	1.36g
$(NH_4)_2SO_4$	0.5g
$MgSO_4 \cdot 7H_2O$	0.2g
$CaCl_2 \cdot 2H_2O$	0.01g
$FeSO_4 \cdot 7H_2O$	5.0mg
$MnSO_4 \cdot 7H_2O$	2.5mg
$Na_2MoO_4 \cdot 2H_2O$	2.5mg
Na_2HPO_4	2.13mg

pH 7.2 ± 0.2 at 25°C

Preparation of Medium: Add components to distilled/deionized water and bring volume to 1.0L. Mix thoroughly. Distribute into tubes or flasks. Autoclave for 15 min at 15 psi pressure–121°C.

Use: For the enrichment and cultivation of *Caulobacter* species.

Mineral Salts for Thermophiles (L Salts for Thermophiles)

Composition per liter:

$NaNO_3$	0.25g
NH_4Cl	0.25g
Na_2HPO_4	0.21g
$MgSO_4 \cdot 7H_2O$	0.20g
NaH_2PO_4	0.09g
KCl	0.04g
$CaCl_2$	0.02g
$FeSO_4$	1.0mg
Trace mineral solution	10.0mL
n-Heptadecane	1.0mL

Trace Mineral Solution:

Composition per liter:

$ZnSO_4 \cdot 7H_2O$	7.0mg
H_3BO_4	1.0mg
MoO_3	1.0mg
$CuSO_4 \cdot 5H_2O$	500.0μg
$CoSO_4 \cdot 7H_2O$	18.0μg
$MnSO_4 \cdot 5H_2O$	7.0μg

Preparation of Trace Mineral Solution: Add components to distilled/deionized water and bring volume to 1.0L. Mix thoroughly.

Preparation of Medium: Add components, except *n*-heptadecane, to distilled/deionized water and bring volume to 1.0L. Mix thoroughly. Autoclave for 15 min at 15 psi pressure–121°C. Aseptically add the *n*-heptadecane. Mix thoroughly. Distribute into sterile tubes or flasks.

Use: For the cultivation and maintenance of *Bacillus thermoleovorans*.

Mineral Salts Medium

Composition per liter:

Na_2HPO_4	4.0g
KH_2PO_4	1.5g
NH_4Cl	1.0g
$MgSO_4 \cdot 7H_2O$	0.2g

Ferric ammonium citrate 5.0mg
Modified Hoagland trace element solution 1.0mL
pH 7.0 ± 0.2 at 25°C

Modified Hoagland Trace Element Solution:
Composition per 3.6 liters:

H_3BO_3 .... 11.0g
$MnCl_2 \cdot 4H_2O$ 7.0g
$AlCl_3$ 1.0g
$CoCl_2$ 1.0g
$CuCl_2$ 1.0g
KI 1.0g
$NiCl_2$ 1.0g
$ZnCl_2$ 1.0g
$BaCl_2$ 0.5g
KBr 0.5g
LiCl 0.5g
Na_2MoO_4 .... 0.5g
$SeCl_4$ 0.5g
$SnCl_2 \cdot 2H_2O$ 0.5g
$NaVO_3 \cdot H_2O$ 0.1g

Preparation of Modified Hoagland Trace Element Solution: Prepare each component as a separate solution. Dissolve each salt in approximately 100.0mL of distilled/deionized water. Adjust the pH of each solution to below 7.0. Combine all the salt solutions and bring the volume to 3.6L with distilled/deionized water. Adjust the pH to 3–4. A yellow precipitate may form after mixing. After a few days, it will turn into a fine white precipitate. Mix the solution thoroughly before using.

Preparation of Medium: Add components to distilled/deionized water and bring volume to 1.0L. Mix thoroughly. Distribute into tubes or flasks. Autoclave for 15 min at 15 psi pressure–121°C.

Use: For the cultivation and maintenance of *Rhodococcus rhodochrous.*

Mineral Salts Medium with Methanol

Composition per liter:

$NaNH_4HPO_4 \cdot 4H_2O$ 1.74g
$NaH_2PO_4 \cdot H_2O$ 0.54g
$MgSO_4 \cdot 7H_2O$ 0.2g
KCl 0.04g
$FeSO_4 \cdot 7H_2O$ 5.0mg
Methanol 5.0mL
Trace mineral solution 1.0mL
pH 7.2 ± 0.2 at 25°C

Trace Mineral Solution:
Composition per liter:

H_3BO_3 .... 2.86g
$MnCl_2 \cdot 4H_2O$ 1.81g
$ZnSO_4 \cdot 7H_2O$ 0.22g
$CuSO_4 \cdot 5H_2O$ 0.08g
$CoCl_2 \cdot 6H_2O$ 0.06g
$Na_2MoO_4 \cdot 2H_2O$ 25.0mg

Preparation of Trace Mineral Solution: Add components to distilled/deionized water and bring volume to 1.0L. Mix thoroughly.

Preparation of Medium: Add components, except methanol, to distilled/deionized water and bring volume to 1.0L. Mix thoroughly. Distribute into tubes or flasks. Autoclave for 15 min at 15 psi pressure–121°C. Cool to 50°C. Filter sterilize methanol. Aseptically add sterile methanol to cooled, sterile basal medium.

Use: For the cultivation and maintenance of *Rhodococcus rhodochrous.*

Mineral Salts Medium with Methanol and Yeast Extract

Composition per liter:

$NaNH_4HPO_4 \cdot 4H_2O$ 1.74g
$NaH_2PO_4 \cdot H_2O$ 0.54g
$MgSO_4 \cdot 7H_2O$ 0.2g
Yeast extract 0.2g
KCl 0.04g
$FeSO_4 \cdot 7H_2O$ 5.0mg
Methanol 5.0mL
Trace mineral solution 1.0mL
pH 7.2 ± 0.2 at 25°C

Trace Mineral Solution:
Composition per liter:

H_3BO_3 .... 2.86g
$MnCl_2 \cdot 4H_2O$ 1.81g
$ZnSO_4 \cdot 7H_2O$ 0.22g
$CuSO_4 \cdot 5H_2O$ 0.08g
$CoCl_2 \cdot 6H_2O$ 0.06g
$Na_2MoO_4 \cdot 2H_2O$ 25.0mg

Preparation of Trace Mineral Solution: Add components to distilled/deionized water and bring volume to 1.0L. Mix thoroughly.

Preparation of Medium: Add components, except methanol, to distilled/deionized water and bring volume to 1.0L. Mix thoroughly. Autoclave for 15 min at 15 psi pressure–121°C. Cool to 50°C. Filter sterilize methanol. Aseptically add sterile methanol to cooled, sterile basal medium. Aseptically distribute into sterile tubes or flasks.

Use: For the cultivation and maintenance of *Pseudomonas* species.

Mineral Salt Peptonized Milk Agar (SPMA)

Composition per liter:

Agar	15.0g
Milk, peptonized	1.0g
Mineral solution	100.0mL

Mineral Solution:

Composition per 100mL:

$MgSO_4 \cdot 7H_2O$	0.50g
$CaCl_2$	0.25g
K_2HPO_4	0.25g
$(NH_4)_2SO_4$	0.10g
$FeCl_3 \cdot 6H_2O$	0.01g
$MnCl_2$	0.1mg

Preparation of Mineral Solution: Add components to distilled/deionized water and bring volume to 100.0mL. Mix thoroughly. Filter sterilize.

Preparation of Medium: Add components, except mineral solution, to distilled/deionized water and bring volume to 900.0mL. Mix thoroughly. Gently heat and bring to boiling. Autoclave for 15 min at 15 psi pressure–121°C. Cool to 45°–50°C. Aseptically add 100.0mL of sterile mineral solution. Mix thoroughly. Pour into sterile Petri dishes or distribute into sterile tubes.

Use: For the cultivation of freshwater *Myxobacterium* species.

Mineral Salts with Butane

Composition per liter of tap water:

$(NH_4)_2HPO_4$	8.0g
$Na_2HPO_4 \cdot 12H_2O$	2.5g
KH_2PO_4	2.0g
$MgSO_4 \cdot 7H_2O$	0.5g
Yeast extract	100.0mg
$CaCl_2 \cdot 2H_2O$	60.0mg
$FeSO_4 \cdot 7H_2O$	30.0mg
$MnCl_2 \cdot 4H_2O$	60.0μg
$CuSO_4 \cdot 5H_2O$	15.0μg

pH 7.1 ± 0.2 at 25°C

Preparation of Medium: Add components to distilled/deionized water and bring volume to 1.0L. Mix thoroughly. Distribute into tubes or flasks. Autoclave for 15 min at 15 psi pressure–121°C. Incubate inoculated medium in 88% air + 7% *n*-butane + 5% CO_2.

Use: For the cultivation and maintenance of *Pseudomonas butanovora.*

Minerals Modified Glutamate Agar

Composition per liter:

Agar	15.0g
Lactose	10.0g
Sodium glutamate	6.35g
NH_4Cl	2.5g
K_2HPO_4	0.9g
Sodium formate	0.25g
$MgSO_4 \cdot 7H_2O$	0.1g
L-Aspartic acid	0.024g
L-Arginine	0.02g
L-Cystine	0.02g
Bromcresol Purple	0.01g
$CaCl_2 \cdot 2H_2O$	0.01g
Ferric ammonium citrate	0.01g
Nicotinic acid	1.0mg
Pantothenic acid	1.0mg
Thiamine	1.0mg

pH 6.7 ± 0.2 at 25°c

Preparation of Medium: Add components to distilled/deionized water and bring volume to 1.0L. Mix thoroughly. Gently heat and bring to boiling. Distribute into tubes or flasks. Autoclave for 10min at 11 psi pressure–116°C. Pour into sterile Petri dishes in 20.0mL volumes.

Use: For the cultivation of coliform bacteria from foods.

Minerals Modified Medium

Composition per liter:

Lactose	20.0g
Sodium glutamate	12.7g
NH_4Cl	5.0g
K_2HPO_4	1.80g
Sodium formate	0.5g
$MgSO_4 \cdot 7H_2O$	0.2g
L-Aspartic acid	0.048g
L-Cystine	0.04g
L-Arginine	0.04g
Ferric ammonium citrate	0.020g
$CaCl_2 \cdot 2H_2O$	0.020g
Bromcresol Purple	0.020g
Thiamine	2.0mg
Nicotinic acid	2.0mg
Pantothenic acid	2.0mg

pH 6.7 ± 0.2 at 25°C

Source: This medium is available as a premixed powder from Oxoid Unipath.

Preparation of Medium: Add NH_4Cl to distilled/deionized water and bring volume to 800.0mL. Add remaining components and bring volume to 1.0L.

Mix thoroughly. Adjust pH to 6.7. Distribute into tubes or flasks. Autoclave for 10 min at 10 psi pressure–116°C. Check pH after autoclaving. This medium is double-strength.

Use: For the enumeration of coliform bacteria in water.

Minimal Agar, Davis

Composition per liter:

Agar	15.0g
K_2HPO_4	7.0g
KH_2PO_4	2.0g
$(NH_4)_2SO_4$	1.0g
Glucose	1.0g
Sodium citrate	0.5g
$MgSO_4 \cdot 7H_2O$	0.1g

pH 7.0 ± 0.2 at 25°C

Source: This medium is available as a premixed powder from Difco Laboratories.

Preparation of Medium: Add components to cold distilled/deionized water and bring volume to 1.0L. Mix thoroughly. Gently heat and bring to boiling. Distribute into tubes or flasks. Autoclave for 15 min at 15 psi pressure–121°C. Pour into sterile Petri dishes or leave in tubes.

Use: For the isolation, cultivation and characterization of nutritional mutants of *Escherichia coli*.

Minimal Broth, Davis

Composition per liter:

K_2HPO_4	7.0g
KH_2PO_4	2.0g
$(NH_4)_2SO_4$	1.0g
Glucose	1.0g
Sodium citrate	0.5g
$MgSO_4 \cdot 7H_2O$	0.1g

pH 7.0 ± 0.2 at 25°C

Source: This medium is available as a premixed powder from Difco Laboratories.

Preparation of Medium: Add components to cold distilled/deionized water and bring volume to 1.0L. Mix thoroughly. Distribute into tubes or flasks. Autoclave for 15 min at 15 psi pressure–121°C.

Use: For the isolation, cultivation and characterization of nutritional mutants of *Escherichia coli*. Also recommended for the isolation and characterization of nutritional mutants from wild type strains of *Bacillus subtilis* when used in conjunction with Minimal Agar Davis and Antibiotic Medium 3.

Minimal F–Top Agar

Composition per liter:

NaCl	8.0g
Agar	4.5g
K_2HPO_4	2.1g
KH_2PO_4	0.6g
$(NH_4)_2SO_4$	0.3g
Glucose	0.3g
Sodium citrate	0.15g
$MgSO_4 \cdot 7H_2O$	0.03g

pH 7.0 ± 0.2 at 25°C

Preparation of Minimal Agar: Add components to cold distilled/deionized water and bring volume to 1.0L. Mix thoroughly. Gently heat and bring to boiling. Distribute into tubes or flasks. Autoclave for 15 min at 15 psi pressure–121°C. Pour into sterile Petri dishes or leave in tubes.

Use: For the distribution of bacteriophage or bacterial cells evenly in a thin layer over the surface of a plate.

Minimal Lactate Medium
See: **ML Medium**

Minimal Medium for Denitrifying Bacteria

Composition per liter:

Solution A	980.0mL
Solution B	10.0mL
Solution C	10.0mL

Solution A:

Composition per 980mL:

KNO_3	5.0g
Carbon source	4.0g
$(NH_4)_2SO_4$	1.0g
$K_2HPO_4 \cdot 3H_2O$	0.87g
KH_2PO_4	0.54g

Preparation of Solution A: Add components to distilled/deionized water and bring volume to 1.0L. Mix thoroughly. Autoclave for 15 min at 15 psi pressure–121°C. Cool to 25°C.

Solution B:

Composition per 100mL:

$MgSO_4 \cdot 7H_2O$	2.0g

Preparation of Solution B: Add $MgSO_4 \cdot 7H_2O$ to distilled/deionized water and bring volume to 100.0mL. Mix thoroughly. Autoclave for 15 min at 15 psi pressure–121°C. Cool to 25°C.

Solution C:

Composition per 100mL:

$CaCl_2 \cdot 2H_2O$	0.2g
$FeSO_4 \cdot 7H_2O$	0.1g
$MnSO_4 \cdot H_2O$	0.05g
$CuSO_4 \cdot 5H_2O$	0.01g
$Na_2MoO_4 \cdot 2H_2O$	0.01g
HCl (0.1*N* solution)	100.0mL

Preparation of Solution C: Combine components. Mix thoroughly. Autoclave for 15 min at 15 psi pressure–121°C. Cool to 25°C.

Preparation of Medium: Aseptically combine 980.0mL of cooled sterile solution A, 10.0mL of cooled sterile solution B, and 10.0mL of cooled sterile solution C. Mix thoroughly. Aseptically distribute into sterile tubes or flasks.

Use: For the isolation and cultivation of denitrifying bacteria.

MIO Medium
See: Motility Indole Ornithine Medium

Mist Agar

Composition per liter:

Cow-manure, dry	50.0g
Agar	15.0g

Preparation of Medium: Add cow manure to 1.0L of tap water. Boil for 1 hr. Filter through cheesecloth. Filter through paper. Add agar to filtrate and bring volume to 1.0L with tap water. Gently heat and bring to boiling. Distribute into tubes or flasks. Autoclave for 15 min at 15 psi pressure–121°C. Pour into sterile Petri dishes or leave in tubes.

Use: For the cultivation and maintenance of *Streptomyces fragmentosporus*.

Mitis–Salivarius Agar

Composition per liter:

Sucrose	50.0g
Agar	15.0g
Enzymatic digest of protein	10.0g
Proteose peptone	10.0g
K_2HPO_4	4.0g
Dextrose	1.0g
Trypan Blue	0.08g
Crystal Violet	0.8mg
Na_2TeO_3 solution	1.0mL

pH 7.0 ± 0.2 at 25°C

Source: This medium is available as a premixed powder from Difco Laboratories.

Na_2TeO_3 Solution:

Composition per 10mL:

Na_2TeO_3	0.1g

Preparation of Na_2TeO_3 Solution: Add Na_2TeO_3 to 10.0mL of distilled/deionized water. Mix thoroughly. Filter sterilize.

Caution: Potassium tellurite is toxic.

Preparation of Medium: Add components to distilled/deionized water and bring volume to 999.0mL. Mix thoroughly. Gently heat and bring to boiling. Autoclave for 15 min at 15 psi pressure–121°C. Cool medium to 50–55°C. Aseptically add 1.0mL of the sterile Na_2TeO_3 solution to the cooled basal medium. Mix thoroughly. Pour into sterile Petri dishes or distribute into sterile tubes.

Use: For the selective isolation of *Streptococcus mitis, Streptococcus salivarius,* other viridans streptococci and enterococci.

m-Kleb Agar
See: Kleb Agar

m–*Klebsiella* Medium
See: Klebsiella Medium

ML Medium
(Minimal Lactate Medium)

Composition per liter:

Sodium lactate	5.0g
$MgSO_4 \cdot 7H_2O$	2.0g
NH_4Cl	1.0g
Na_2SO_4	1.0g
Yeast extract	1.0g
K_2HPO_4	0.5g
Cysteine	0.5g
$CaCl_2 \cdot 6H_2O$	0.1g
Resazurin	1.0mg
$NaHCO_3$ solution	25.0mL
$FeSO_4 \cdot 7H_2O$ solution	25.0mL

pH 6.8 ± 0.2 at 25°C

$NaHCO_3$ Solution:

Composition per 25mL:

$NaHCO_3$	4.0g

Preparation of $NaHCO_3$ Solution: Add $NaHCO_3$ to distilled/deionized water and bring volume to 25.0mL. Mix thoroughly. Filter sterilize. Gas with O_2-free 97% N_2 + 3% H_2. Cap with a rubber stopper.

$FeSO_4 \cdot 7H_2O$ Solution:

Composition per 25mL:

$FeSO_4 \cdot 7H_2O$	4.0mg

Preparation of $FeSO_4·7H_2O$ Solution: Add $FeSO_4·7H_2O$ to distilled/deionized water and bring volume to 25.0mL. Mix thoroughly. Filter sterilize. Gas with O_2-free 97% N_2 + 3% H_2. Cap with a rubber stopper.

Preparation of Medium: Add components, except $NaHCO_3$ solution and $FeSO_4·7H_2O$ solution, to distilled/deionized water and bring volume to 1.0L. Gently heat and bring to boiling under O_2-free 97% N_2 + 3% H_2. Adjust pH to 6.8. Continue boiling until rezasurin becomes colorless, indicating reduction. Distribute anaerobically under O_2-free 97% N_2 + 3% H_2 into tubes in 10.0mL volumes. Cap with rubber stoppers. Place tubes in a press. Autoclave for 15 min at 15 psi pressure–121°C. Cool to room temperature. Prior to inoculation, add 0.25mL of sterile $NaHCO_3$ solution and 0.25mL of sterile $FeSO_4·7H_2O$ solution to each test tube containing 10.0mL of sterile basal medium.

Use: For the cultivation and maintenance of *Desulfovibrio* species.

m–Lauryl Sulfate Broth
See: **Lauryl Sulfate Broth**

MLCB Agar (Mannitol Lysine Crystal Violet Brilliant Green Agar)

Composition per liter:

Agar	15.0g
Peptone	10.0g
Yeast extract	5.0g
L-Lysine·HCl	5.0g
NaCl	4.0g
$Na_2S_2O_3$	4.0g
Mannitol	3.0g
Beef extract	2.0g
Ferric ammonium citrate	1.0g
Crystal Violet	0.01g
Brilliant Green	12.5mg

pH 6.8 ± 0.1 at 25°C

Source: This medium is available as a premixed powder from Oxoid Unipath.

Preparation of Medium: Add components to distilled/deionized water and bring volume to 1.0L. Mix thoroughly. Gently heat while stirring and bring to boiling. Do not autoclave. Cool to 50°C. Pour into sterile Petri dishes in 20.0mL volumes.

Use: For the selective isolation and cultivation of *Salmonella* species from fecal material and foods.

m–LES, Endo Agar
See: **Endo Agar, LES**

MM10 Agar (Modified Medium 10 Agar)

Composition per liter:

Base	954.0mL
Dithiothreitol solution	20.0mL
Sheep blood	20.0mL
Na_2CO_3 solution	5.0mL
Menadione solution	1.0mL

pH 7.2 ± 0.2 at 25°C

Base:

Composition per 954mL:

Agar	15.0g
Casein peptone	2.0g
Glucose	1.0g
Sodium formate	1.0g
KNO_3	0.5g
Yeast extract	0.5g
Hemin	0.01g
Mineral salt solution 1	38.0mL
Mineral salt solution 2	38.0mL
Sodium lactate solution	4.0mL

Preparation of Base: Add components to distilled/deionized water and bring volume to 954.0mL. Mix thoroughly. Gently heat and bring to boiling. Autoclave for 15 min at 15 psi pressure–121°C. Cool to 45°–50°C.

Mineral Salt Solution 1:

Composition per 100mL:

K_2HPO_4	0.6g

Preparation of Mineral Salt Solution 1: Add K_2HPO_4 to distilled/deionized water and bring volume to 100.0mL. Mix thoroughly.

Mineral Salt Solution 2:

Composition per 100mL:

NaCl	1.2g
$(NH_4)_2SO_4$	1.2g
KH_2PO_4	0.6g
$CaCl_2$	0.12g

Preparation of Mineral Salt Solution 2: Add components to distilled/deionized water and bring volume to 100.0mL. Mix thoroughly.

Sodium Lactate Solution:

Composition per 100mL:

Sodium lactate	60.0g

Preparation of Sodium Lactate Solution: Add sodium lactate to distilled/deionized water and bring volume to 100.0mL. Mix thoroughly.

Dithiothreitol Solution:
Composition per 100mL:

Dithiothreitol	1.0g

Preparation of Dithiothreitol Solution: Add dithiothreitol to distilled/deionized water and bring volume to 100.0mL. Mix thoroughly. Filter sterilize.

Menadione Solution:
Composition per 100mL:

Vitamin K_1 (phytomenadione)	0.05g
Ethanol (95% solution)	100.0mL

Preparation of Menadione Solution: Add vitamin K_1 to 100.0mL of ethanol. Mix thoroughly. Filter sterilize.

Na_2CO_3 Solution:
Composition per 100mL:

Na_2CO_3	8.0g

Preparation of Na_2CO_3 Solution: Add Na_2CO_3 to distilled/deionized water and bring volume to 100.0mL. Mix thoroughly. Filter sterilize.

Preparation of Medium: To 954.0mL of sterile cooled base aseptically add 20.0mL of the sterile dithiothreitol solution, 20.0mL of sterile, defibrinated sheep blood, 5.0mL of the sterile Na_2CO_3 solution and 1.0mL of the sterile menadione solution. Mix thoroughly. Pour into sterile Petri dishes or distribute into sterile screw-capped tubes.

Use: For the isolation and quantitation of plaque bacteria, especially *Streptococcus mutans, S. sanguis,* and *S. salivarius.*

MMS Medium for *Thermotoga neapolitana*

Composition per liter:

NaCl	6.93g
Starch	5.0g
$MgSO_4 \cdot 7H_2O$	1.75g
$MgCl_2 \cdot 6H_2O$	1.38g
KH_2PO_4	0.5g
$Na_2S \cdot 9H_2O$	0.5g
$CaCl_2$	0.38g
KCl	0.16g
NaBr	25.0mg
H_3BO_3	7.5mg
$SrCl_2 \cdot 6H_2O$	3.8mg
$(NH_4)_2Ni(SO_4)_2$	2.0mg
Resazurin	1.0mg
KI	0.025mg
Wolfe's mineral solution	15.0mL

pH 6.5 ± 0.2 at 25°C

Wolfe's Mineral Solution:
Composition per liter:

$MgSO_4 \cdot 7H_2O$	3.0g
Nitriloacetic acid	1.5g
NaCl	1.0g
$MnSO_4 \cdot H_2O$	0.5g
$FeSO_4 \cdot 7H_2O$	0.1g
$CoCl_2 \cdot 6H_2O$	0.1g
$CaCl_2$	0.1g
$ZnSO_4 \cdot 7H_2O$	0.1g
$CuSO_4 \cdot 5H_2O$	0.01g
$AlK(SO_4)_2 \cdot 12H_2O$	0.01g
H_3BO_3	0.01g
$Na_2MoO_4 \cdot 2H_2O$	0.01g

Preparation of Wolfe's Mineral Solution: Add nitrilotriacetic acid to 500.0mL of distilled/deionized water. Dissolve by adjusting pH to 6.5 with KOH. Add remaining components. Add distilled/deionized water to 1.0L.

Preparation of Medium: Prepare and dispense medium under 80% N_2 and 20% CO_2. Add components to distilled/deionized water and bring volume to 1.0L. Mix thoroughly. Adjust pH to 6.5 with H_2SO_4. Distribute into tubes or flasks. Autoclave for 15 min at 15 psi pressure–121°C.

Use: For the cultivation and maintenance of *Thermotoga neapolitana.*

Mn Agar No. 1
See: **Manganese Agar No. 1**

Mn Agar No. 2
See: **Manganese Agar No. 2**

MN Marine Medium

Composition per liter:

Noble agar	10.0g
$NaNO_3$	0.75g
$MgSO_4 \cdot 7H_2O$	0.04g
$CaCl_2 \cdot 2H_2O$	0.02g
$K_2HPO_4 \cdot 3H_2O$	0.02g
Na_2CO_3	0.02g
Citric acid	3.0mg
Ferric ammonium citrate	3.0mg
Disodium potassium EDTA	0.5mg
Trace metals A-5	1.0mL

pH 8.5 ± 0.2 at 25°C

A-5 Trace Metal Mix:
Composition per liter:

H_3BO_3	2.86g
$MnCl_2 \cdot 4H_2O$	1.81g
$ZnSO_4 \cdot 7H_2O$	0.222g

$CuSO_4 \cdot 5H_2O$	0.079g
$Na_2MoO_4 \cdot 2H_2O$	0.039g
$Co(NO_3)_2 \cdot 6H_2O$	0.049g

Preparation of A-5 Trace Metal Mix: Add components to distilled/deionized water and bring volume to 1.0L. Mix thoroughly.

Preparation of Medium: Add components to 750mL of sea water and bring volume to lL with glass-distilled water. Mix thoroughly. Gently heat and bring to boiling. Autoclave for 15 min at 15 psi pressure–121°C. After autoclaving, adjust pH to 8.5 with KOH.

Use: For cultivation and maintenance of marine cyanobacteria.

MN Marine Medium with Vitamin B_{12}

Composition per liter:

Noble agar	10.0g
$NaNO_3$	0.75g
$MgSO_4 \cdot 7H_2O$	0.04g
$CaCl_2 \cdot 2H_2O$	0.02g
$K_2HPO_4 \cdot 3H_2O$	0.02g
Na_2CO_3	0.02g
Citric acid	3.0mg
Ferric ammonium citrate	3.0mg
Disodium potassium EDTA	0.5mg
Vitamin B_{12}	20.0μg
Trace metals A-5	1.0mL

pH 8.5 ± 0.2 at 25°C

A-5 Trace Metal Mix:
Composition per liter:

H_3BO_3	2.86g
$MnCl_2 \cdot 4H_2O$	1.81g
$ZnSO_4 \cdot 7H_2O$	0.222g
$CuSO_4 \cdot 5H_2O$	0.079g
$Na_2MoO_4 \cdot 2H_2O$	0.039g
$Co(NO_3)_2 \cdot 6H_2O$	0.049g

Preparation of A-5 Trace Metal Mix: Add components to distilled/deionized water and bring volume to 1.0L. Mix thoroughly.

Preparation of Medium: Add components to 750.0mL of sea water and bring volume to lL with glass-distilled water. Mix thoroughly. Gently heat and bring to boiling. Autoclave for 15 min at 15 psi pressure–121°C. After autoclaving, adjust pH to 8.5 with KOH.

Use: For the cultivation and maintenance of *Dermocarpa* species, *Dermocarpella* species, *Myxosarcina* species, *Phormidium* species, *Pleurocapsa* species, *Synechococcus* species, *Synechocystis* species, and *Xenococcus* species.

Modified Salt Broth
See: **Salt Broth, Modified**

Modified Semi-Solid Rappaport Vassiliadis Medium (MSRV Medium)

Composition per liter:

$MgCl_2$, anhydrous	10.93g
NaCl	7.34g
Casein hydrolysate	4.59g
Tryptose	4.59g
Agar	2.7g
KH_2PO_4	1.47g
Malachite Green oxalate	0.037g
Novobiocin	10.0mL

pH 5.2 ± 0.2 at 25°C

Source: This medium is available as a premixed powder from Oxoid Unipath.

Novobiocin Solution:
Composition per 10mL:

Novobiocin	0.02g

Preparation of Novobiocin Solution: Add novobiocin to 10.0mL of distilled/deionized water. Mix thoroughly. Filter sterilize.

Preparation of Medium: Add components, except novobiocin solution, to distilled/deionized water and bring volume to 990.0mL. Mix thoroughly. Gently heat to boiling. Do not autoclave. Cool to 45°–50°C. Aseptically add 10.0mL of sterile novobiocin solution. Mix thoroughly. Pour into sterile Petri dishes. Air dry plates for at least 1 hr.

Use: For the isolation and cultivation of motile *Salmonella* species from food and environmental samples.

Modified Thayer–Martin Agar
See: **Thayer–Martin Agar, Modified**

Møller Decarboxylase Broth

Composition per liter:

Amino acid	10.0g
Peptic digest of animal tissue	5.0g
Beef extract	5.0g
Glucose	0.5g
Bromcresol Purple	0.01g
Cresol Red	5.0mg
Pyridoxal	5.0mg

pH 6.0 ± 0.2 at 25°C

Source: Available as a premixed powder from BBL Microbiology Systems and Difco Laboratories.

Preparation of Medium: Add components to distilled/deionized water and bring volume to 1.0L. Use L-lysine, L-arginine or L-ornithine. Mix thoroughly. Gently heat until dissolved. Distribute into screw-capped tubes in 5.0mL volumes. Autoclave for 15 min at 15 psi pressure–121°C. A slight precipitate may form in the ornithine broth.

Use: For the differentiation of Gram-negative enteric bacteria based on the production of arginine dihydrolase, lysine decarboxylase or ornithine decarboxylase.

Møller KCN Broth Base

Composition per liter:

Na_2HPO_4	5.64g
NaCl	5.0g
Pancreatic digest of casein	1.5g
Peptic digest of animal tissue	1.5g
KH_2PO_4	0.225g
KCN solution	0.15mL

pH 7.6 ± 0.2 at 25°C

Source: This medium is available as a premixed powder from BBL Microbiology Systems.

KCN Solution:

Composition per100mL:

KCN	0.5g

Preparation of KCN Solution: Add KCN to 100.0mL of cold distilled/deionized water. Mix thoroughly and cap. Do not mouth pipette.

Caution: Cyanide is toxic.

Preparation of Medium: Add components, except KCN solution, to distilled/deionized water and bring volume to 1.0L. Mix thoroughly. Autoclave for 15 min at 15 psi pressure–121°C. Cool to room temperature. Prior to use, add 0.15mL of KCN solution. Mix thoroughly. Aseptically distribute into sterile tubes.

Use: For the differentiation of Gram-negative enteric bacteria on the basis of their ability to grow in the presence of cyanide.

Molybdate Agar

Composition per101.5mL:

Base	100.0mL
Phosphomolybdic acid solution	1.5mL

pH 5.3 ± 0.2 at 25°C

Base:

Composition per liter:

Sucrose	40.0g
Agar	15.0g
Meat peptone	10.0g

Preparation of Base: Add components to distilled/deionized water and bring volume to 1.0L. Mix thoroughly. Adjust pH to 7.6. Gently heat and bring to boiling. Autoclave for 15 min at 15 psi pressure–121°C. Cool to 45°–50°C.

Phosphomolybdic Acid Solution:

Composition per 100mL:

$P_2O_5{\cdot}2OMoO_3$	12.5g

Preparation of Base: Add $P_2O_5{\cdot}2OMoO_3$ (phospho-12-molybdic acid, 12–molybdophosphoric acid, or PMA) to sterile distilled/deionized water. Mix thoroughly. Do not adjust pH.

Preparation of Medium: To 100.0mL of cooled sterile base add 1.5mL of phosphomolybdic acid solution. Mix thoroughly. Pour into sterile Petri dishes or distribute into sterile tubes.

Use: For the isolation and presumptive identification of yeast, especially *Candida* species. *Candida albicans* appears as smooth, medium olive colonies with medium olive bottoms. *Candida stellatoidea* appears as shiny, light gray colonies with light gray bottoms. *Candida tropicalis* appears as smooth, shiny, dark blue/gray colonies with dark blue/gray bottoms. *Candida krusei* appears as smooth, dull white colonies with white bottoms. *Saccaromyces cerevisiae* appears as smooth, shiny light blue/dark blue colonies with dark blue/green bottoms.

Monsur Agar (Taurocholate Tellurite Gelatin Agar)

Composition per liter:

Gelatin	30.0g
Agar	15.0g
Casein peptone	10.0g
NaCl	10.0g
Sodium taurocholate	5.0g
$Na_2CO_3{\cdot}H_2O$	1.0g
K_2TeO_3 solution	10.0mL

pH 8.5 ± 0.2 at 25°C

K_2TeO_3 Solution:

Composition per 10mL:

K_2TeO_3	0.02g

Preparation of K_2TeO_3 Solution: Add K_2TeO_3 to 10.0mL of distilled/deionized water. Mix thoroughly. Filter sterilize.

Caution: Potassium tellurite is toxic.

Preparation of Medium: Add components, except K_2TeO_3 solution, to distilled/deionized water and bring volume to 990.0mL. Mix thoroughly. Gently heat and bring to boiling. Autoclave for 15 min at 15 psi pressure–121°C. Cool to 45°–50°C. Add 10.0mL of sterile K_2TeO_3 solution. Mix thoroughly. Pour into sterile Petri dishes or distribute into sterile tubes.

Use: For the isolation of *Vibrio cholerae* from fecal specimens.

Motility GI Medium

Composition per liter:

Gelatin	53.4g
Heart infusion broth	25.0g
Agar	3.0g

pH 7.2 ± 0.2 at 25°C

Source: This medium is available as a premixed powder from Difco Laboratories.

Preparation of Medium: Add components to cold distilled/deionized water and bring to 1.0L. Mix thoroughly. Gently heat and bring to boiling. Distribute into tubes or flasks. Autoclave for 15 min at 15 psi pressure–121°C. Pour into sterile Petri dishes in 20.0mL volumes or leave in tubes.

Use: For demonstrating motility of microorganisms and for separating organisms in their motile phase.

Motility Indole Lysine Medium
See: **MIL Medium**

Motility Indole Ornithine Medium (MIO Medium)

Composition per liter:

Pancreatic digest of gelatin	10.0g
Pancreatic digest of casein	9.5g
L-Ornithine·HCl	5.0g
Yeast extract	3.0g
Agar	2.0g
Glucose	1.5g
Bromcresol Purple	0.02g

pH 6.6 ± 0.2 at 25°C

Source: This medium is available as a premixed powder from BBL Microbiology Systems and Difco Laboratories.

Preparation of Medium: Add components to distilled/deionized water and bring to 1.0L. Mix thoroughly. Gently heat and bring to boiling. Distribute into tubes or flasks. Autoclave for 15 min at 15 psi pressure–121°C.

Use: For the differentiation of Gram-negative enteric bacteria based on their motility, indole production and ornithine decarboxylase activity.

Motility Medium

Composition per liter:

Pancreatic digest of casein	10.0g
Glucose	5.0g
Agar	3.0g
Na_2HPO_4	2.5g
Yeast extract	2.5g

pH 7.4 ± 0.2 at 25°C

Preparation of Medium: Add components to distilled/deionized water and bring volume to 1.0L. Mix thoroughly. Gently heat and bring to boiling. Distribute into tubes in 2.0mL volumes. Autoclave for 15 min at 15 psi pressure–121°C. Allow tubes to stand at 25°C for 2 days prior to inoculation.

Use: For the cultivation and observation of motility of *Bacillus cereus*.

Motility Medium S

Composition per liter:

Beef heart solids from infusion	500.0g
Gelatin	30.0g
Enzymatic hydrolyzate of protein	10.0g
NaCl	5.0g
K_2HPO_4	2.0g
KNO_3	2.0g
Agar	1.0g
2,3,5-triphenyltetrazolium chloride solution	10.0mL

pH 7.2 ± 0.2 at 25°C

Source: This medium is available as a premixed powder from Difco Laboratories.

2,3,5-Triphenyltetrazolium Chloride Solution:
Composition per 10mL:

2,3,5-triphenyltetrazolium chloride	0.1g

Preparation of 2,3,5-Triphenyltetrazolium Chloride Solution: Add 2,3,5-triphenyltetrazolium chloride to distilled/deionized water and bring volume to 10.0mL. Mix thoroughly. Filter sterilize.

Preparation of Medium: Add components, except 2,3,5-triphenyltetrazolium chloride solution, to distilled/deionized water and bring volume to 990.0mL. Mix thoroughly. Gently heat while stirring and bring to boiling. Autoclave for 15 min at 15 psi pressure–121°C. Cool to 60°C. Aseptically add 10.0mL of the sterile 2,3,5-triphenyltetrazolium chloride solution. Mix thoroughly. Aseptically distribute into sterile tubes. Keep at 4°–8°C until used.

Use: For the determination of bacterial motility.

Motility Nitrate Agar

Composition per liter:

Beef heart, solids from infusion	100.0g
Tryptose	12.0g
Agar	3.0g
NaCl	1.0g
KNO_3	1.0g
Glucose	0.5g

pH 7.4 ± 0.2 at 25°C

Preparation of Medium: Add components to distilled/deionized water and bring volume to 1.0L. Mix thoroughly. Gently heat and bring to boiling. Distribute into tubes in 4.0mL volumes. Autoclave for 15 min at 15 psi pressure–121°C.

Use: For the cultivation and observation of motility and nitrate reduction in a variety of Gram-negative bacteria.

Motility Nitrate Medium (FDA M101)

Composition per liter:

Beef heart, solids from infusion	100.0g
Tryptose	12.0g
Agar	3.0g
NaCl	1.0g
KNO_3, nitrite-free	1.0g

pH 7.4 ± 0.2 at 25°C

Preparation of Medium: Add components to distilled/deionized water and bring volume to 1.0L. Mix thoroughly. Gently heat and bring to boiling. Distribute into screw-capped tubes in 4.0mL volumes. Autoclave for 15 min at 15 psi pressure–121°C.

Use: For the cultivation and differentiation of Gram-negative nonfermentative bacteria from cosmetics based on their motility and their ability to reduce nitrate to nitrite.

Motility Nitrate Medium

Composition per liter:

Beef heart, solids from infusion	100.0g
Tryptose	12.0g
Agar	3.0g
NaCl	1.0g
KNO_3	1.0g

Preparation of Medium: Add components to distilled/deionized water and bring volume to 1.0L. Mix thoroughly. Gently heat and bring to boiling. Distribute into tubes in 4.0mL volumes. Autoclave for 15 min at 15 psi pressure–121°C.

Use: For the cultivation and observation of motility and nitrate reduction in a variety of Gram-negative nonfermentative bacteria isolated from cosmetics.

Motility Nitrate Medium, Buffered

Composition per liter:

Peptone	5.0g
Galactose	5.0g
Agar	3.0g
Beef extract	3.0g
Na_2HPO_4	2.5g
KNO_3	1.0g
Glycerin	5.0mL

pH 7.3 ± 0.1 at 25°C

Preparation of Medium: Add components, except agar, to distilled/deionized water and bring volume to 1.0L. Mix thoroughly. Gently heat until dissolved. Add agar. Gently heat until boiling. Distribute into tubes in 11.0mL volumes. Autoclave for 15 min at 15 psi pressure–121°C. If not used within 4 hr, heat tubes to 100°C for 10 min.

Use: For the cultivation and observation of motility of *Clostridium perfringens.*

Motility Sulfide Medium

Composition per liter:

Gelatin	80.0g
Proteose peptone	10.0g
NaCl	5.0g
Agar	4.0g
Beef extract	3.0g
Sodium citrate	2.0g
L-Cystine	0.2g
Ferrous ammonium citrate	0.2g

pH 7.3 ± 0.2 at 25°C

Source: This medium is available as a premixed powder from Difco Laboratories.

Preparation of Medium: Add components to distilled/deionized water and bring volume to 1.0L. Mix thoroughly. Gently heat while stirring and bring to boiling. Distribute into tubes in 4–5.0mL volumes. Autoclave for 15 min at 10 psi pressure–116°C.

Use: For the determination of bacterial motility and the ability of bacteria to produce H_2S from L-cystine. Used for the differentiation of Gram-negative bacteria of the Enterobacteriaceae.

Motility Test and Maintenance Medium

Composition per liter:

Peptone	10.0g

NaCl 5.0g
Agar...4.0g
Beef extract 3.0g
2,3,5-triphenyltetrazolium chloride 0.05g

Preparation of Medium: Add components to distilled/deionized water and bring volume to 1.0L. Mix thoroughly. Distribute into screw-capped tubes in 8.0mL volumes. Autoclave for 15 min at 15 psi pressure–121°C.

Use: For the cultivation, maintenance and observation of motility of *Listeria monocytogenes*.

Motility Test and Maintenance Medium

Composition per liter:
Agar...9.0g
Tryptose 8.0g
NaCl 5.0g
Pancreatic digest of gelatin 2.5g
Beef extract 1.5g

pH 7.2 ± 0.1 at 25°C

Preparation of Medium: Add components to distilled/deionized water and bring volume to 1.0L. Mix thoroughly. Gently heat and bring to boiling. Distribute into tubes in 7.0mL volumes. Autoclave for 15 min at 15 psi pressure–121°C. Cool to 45°–50°C. Pass the cooled tubes into an anaerobic chamber containing 85% N_2 + 10% H_2 + 5% CO_2.

Use: For the cultivation, maintenance and observation of motility in a variety of anaerobic bacteria.

Motility Test and Maintenance Medium

Composition per liter:
Peptone..10.0g
NaCl 5.0g
Agar...4.0g
Beef extract 3.0g

pH 7.4 ± 0.1 at 25°C

Preparation of Medium: Add components to distilled/deionized water and bring volume to 1.0L. Mix thoroughly. Distribute into screw-capped tubes in 8.0mL volumes. Autoclave for 15 min at 15 psi pressure–121°C.

Use: For the cultivation, maintenance and observation of motility in members of the *Enterobacteriaceae*.

Motility Test and Maintenance Medium, Gilardi

Composition per liter:
Pancreatic digest of casein 10.0g
NaCl 5.0g
Agar...3.0g
Yeast extract 3.0g

pH 7.2 ± 0.1 at 25°C

Preparation of Medium: Add components to distilled/deionized water and bring volume to 1.0L. Mix thoroughly. Distribute into screw-capped tubes in 3.5mL volumes. Autoclave for 15 min at 15 psi pressure–121°C.

Use: For the cultivation, maintenance and observation of motility in nonfermenting Gram–negative bacteria.

Motility Test and Maintenance Medium, Tatum

Composition per liter:
Tryptose 8.0g
NaCl 5.0g
Agar...4.0g
Pancreatic digest of gelatin 2.5g
Beef extract 1.5g

pH 6.9 ± 0.2 at 25°C

Preparation of Medium: Add components to distilled/deionized water and bring volume to 1.0L. Mix thoroughly. Distribute into screw-capped tubes in 8.0mL volumes. Autoclave for 15 min at 15 psi pressure–121°C.

Use: For the cultivation, maintenance and observation of motility in nonfermenting Gram–negative bacteria.

Motility Test Medium

Composition per liter:
Pancreatic digest of gelatin 10.0g
NaCl 5.0g
Agar...4.0g
Beef extract 3.0g

pH 7.3 ± 0.2 at 25°C

Source: This medium is available as a premixed powder from BBL Microbiology Systems.

2,3,5-Triphenyltetrazolium Chloride Solution:
Composition per 10mL:
2,3,5-triphenyltetrazolium chloride 0.1g

Preparation of 2,3,5-Triphenyltetrazolium Chloride Solution: Add 2,3,5-triphenyltetrazolium chloride to distilled/deionized water and bring volume to 10.0mL. Mix thoroughly. Filter sterilize.

Preparation of Medium: Add components to distilled/deionized water and bring volume to 995.0mL. Mix thoroughly. Gently heat while stirring and bring to boiling. Autoclave for 15 min at 15 psi pressure–121°C. Cool to 45°–50°C. Aseptically add 5.0mL of sterile 2,3,5-triphenyltetrazolium chloride solution. Mix thoroughly. Aseptically distribute into sterile tubes.

Use: For the detection of motility of Gram-negative enteric bacteria.

Motility Test Medium

Composition per liter:

Tryptose	10.0g
NaCl	5.0g
Agar	5.0g

pH 7.2 ± 0.2 at 25°C

Source: This medium is available as a premixed powder from Difco Laboratories.

Preparation of Medium: Add components to distilled/deionized water and bring volume to 1.0L. Mix thoroughly. Gently heat while stirring and bring to boiling. Distribute into tubes in 4–5.0mL volumes. Autoclave for 15 min at 15 psi pressure–121°C. Cool tubes quickly in an upright position.

Use: For the determination of bacterial motility.

Motility Test Medium, Semisolid

Composition per liter:

Peptone	10.0g
NaCl	5.0g
Agar	4.0g
Beef extract	3.0g

pH 7.4 ± 0.2 at 25°C

Preparation of Medium: Add components to distilled/deionized water and bring volume to 1.0L. Mix thoroughly. Gently heat while stirring and bring to boiling. Distribute into screw-capped tubes in 8.0mL or 20.0mL volumes. Autoclave for 15 min at 15 psi pressure–121°C. Pour into sterile Petri dishes in 20.0mL volumes or leave in tubes.

Use: For the cultivation and observation of motility in a variety of bacteria, especially *Salmonella* species.

MOX Agar
See: **Magnesium Oxalate Agar**

MOX Agar
See: **Oxford Agar, Modified**

MP Agar

Composition per liter:

Agar	15.0g
Sodium acetate	0.1g
Basal medium	1.0L
Sodium sulfide solution	3.0mL

pH 7.0–7.5 at 25°C

Basal Medium:
Composition per liter:

$CaSO_4 \cdot 2H_2O$ (saturated solution)	20.0mL
$MgSO_4 \cdot 7H_2O$ (1% solution)	1.0mL
NH_4Cl (4% solution)	5.0mL
Trace elements solution	5.0mL
K_2HPO_4 (1% solution)	1.0mL

Preparation of Basal Medium: Add components to distilled/deionized water and bring volume to 1.0L. Mix thoroughly.

Trace Elements Solution:
Composition per liter:

$ZnSO_4 \cdot 7H_2O$ (0.1% solution)	10.0mL
$MnSO_4 \cdot 4H_2O$ (0.02% solution)	10.0mL
$CuSO_4 \cdot 5H_2O$ (0.00005% solution)	10.0mL
H_3BO_3 (0.1% solution)	10.0mL
$Co(NO_3)_2$ or $CoCl_2 \cdot 6H_2O$ (0.01% solution)	10.0mL
$Na_2MoO_4 \cdot 2H_2O$ (0.01% solution)	10.0mL
Ferrous EDTA solution	20.0mL

Preparation of Trace Elements Solution: Add components to distilled/deionized water and bring volume to 1.0L. Mix thoroughly.

Ferrous EDTA Solution:
Composition per 100mL:

$FeSO_4 \cdot 7H_2O$	7.0g
EDTA	2.0g
HCl, concentrated	1.0mL

Preparation of Ferrous EDTA Solution: Add components to distilled/deionized water and bring volume to 100.0mL. Mix thoroughly.

Sodium Sulfide Solution:
Composition per 10mL:

$Na_2S \cdot 9H_20$	1.0g

Preparation of Sodium Sulfide Solution: Add $Na_2S \cdot 9H_20$ to distilled/deionized water and bring volume to 10.0mL. Mix thoroughly. Autoclave for 15 min at 15 psi pressure–121°C. Prepare freshly.

Preparation of Medium: Add sodium acetate and agar to 1.0L of basal medium. Mix thoroughly. Adjust pH to 7.0–7.5. Gently heat and bring to boiling. Autoclave for 15 min at 15 psi pressure–121°C. Cool to 45°–50°C. Aseptically add 3.0mL of sterile sodium sulfide solution immediately prior to dispensing. Mix thoroughly. Pour into sterile Petri dishes or distribute into sterile screw-capped tubes.

Use: For the isolation and cultivation of *Beggiatoa* species and myxotrophic strains of *Thiothrix* species from water and environmental sources.

MP 5 Medium (Mineral Pectin 5 Medium)

Composition per liter:

Agar solution	500.0mL
Basal medium	250.0mL
Mineral solution	250.0mL

pH 5.0–6.0 at 25°C

Agar Solution:

Composition per 500mL:

Agar	15.0g

Preparation of Agar Solution: Add agar to distilled/deionized water and bring volume to 500.0mL. Mix thoroughly. Gently heat and bring to boiling. Autoclave for 15 min at 15 psi pressure–121°C. Cool to 45°–50°C.

Basal Medium:

Composition per 250mL:

Na_2HPO_4	6.0g
Pectin, citrus or apple	5.0g
KH_2PO_4	4.0g
NH_4SO_4	2.0g
Yeast extract	1.0g

Preparation of Basal Medium: Add components to distilled/deionized water and bring volume to 250.0mL. Mix thoroughly. Gently heat and bring to boiling.

Mineral Solution:

Composition per 250mL:

$FeSO_4$ (0.1% solution)	1.0mL
$MgSO_4 \cdot 7H_2O$ (20% solution)	1.0mL
$CaCl_2 \cdot 2H_2O$ (0.1% solution)	1.0mL
H_3BO_3 (0.001% solution)	1.0mL
$MnSO_4 \cdot H_2O$ (0.001% solution)	1.0mL
$ZnSO_4 \cdot 7H_2O$ (0.007% solution	1.0mL
$CuSO_4 \cdot 5H_2O$ (0.005% solution)	1.0mL
MoO_3 (0.001% solution)	1.0mL

Preparation of Mineral Solution: Add components to distilled/deionized water and bring volume to 250.0mL. Mix thoroughly.

Preparation of Medium: Combine 250.0mL of basal medium and 250.0mL of mineral solution. Mix thoroughly. Adjust pH to 5.0–6.0 with 1*N* HCl. Autoclave the basal medium-mineral solution and agar solution separately for 15 min at 15 psi pressure–121°C. Cool to 45°–50°C. Aseptically combine the two sterile solutions. Mix thoroughly. Pour immediately into sterile Petri dishes to prevent hydrolysis of the agar.

Use: For the cultivation of microorganisms that produce polygalactanase.

MP 7 Medium (Mineral Pectin 7 Medium)

Composition per liter:

Basal medium	500.0mL
Mineral solution	500.0mL

pH 7.2 ± 0.2 at 25°C

Basal Medium:

Composition per 500mL:

Agar	15.0g
Na_2HPO_4	6.0g
Pectin, citrus or apple	5.0g
KH_2PO_4	4.0g
NH_4SO_4	2.0g
Yeast extract	1.0g

Preparation of Basal Medium: Add components to distilled/deionized water and bring volume to 500.0mL. Mix thoroughly. Gently heat and bring to boiling.

Mineral Solution:

Composition per 500mL:

$FeSO_4$ (0.1% solution)	1.0mL
$MgSO_4 \cdot 7H_2O$ (20% solution)	1.0mL
$CaCl_2 \cdot 2H_2O$ (0.1% solution)	1.0mL
H_3BO_3 (0.001% solution)	1.0mL
$MnSO_4 \cdot H_2O$ (0.001% solution)	1.0mL
$ZnSO_4 \cdot 7H_2O$ (0.007% solution	1.0mL
$CuSO_4 \cdot 5H_2O$ (0.005% solution	1.0mL
MoO_3 (0.001% solution)	1.0mL

Preparation of Mineral Solution: Add components to distilled/deionized water and bring volume to 500.0mL. Mix thoroughly.

Preparation of Medium: Combine 500.0mL of basal medium and 500.0mL of mineral solution. Mix thoroughly. Adjust pH to 7.2. Autoclave for 15 min at 15 psi pressure–121°C. Cool to 50°C. Pour into sterile Petri dishes.

Use: For the cultivation of microorganisms that produce pectate lyase.

m–*PA* Agar

See: ***PA* Agar**

m–PA–C Agar

See: **PA-C Agar**

MPH Agar (Milk Protein Hydrolysate Agar)

Composition per liter:

Agar	15.0g
Casein hydrolysate	9.0g
Glucose	1.0g

pH 7.0 ± 0.2 at 25°C

Source: This medium is available as a premixed powder from BBL Microbiology Systems.

Preparation of Medium: Add components to distilled/deionized water and bring volume to 1.0L. Mix thoroughly. Gently heat while stirring and bring to boiling. Autoclave for 15 min at 15 psi pressure–121°C. Aseptically distibute into sterile tubes. Cool to 43°–45°C before using.

Use: For use in the enumeration of bacteria in water and dairy products.

m–Plate Count Broth
See: **Plate Count Broth**

m–*Pseudomonas aeruginosa* Agar
See: ***Pseudomonas aeruginosa* Agar**

MPSS Broth

Composition per liter:

Peptone	5.0g
$MgSO_4 \cdot 7H_2O$	1.0g
$(NH_4)_2SO_4$	1.0g
Succinic acid	1.0g
$FeCl_3 \cdot 6H_2O$ (0.2% solution)	1.0mL
$MnSO_4 \cdot H_2O$ (0.2% solution)	1.0mL

pH 6.8 ± 0.2 at 25°C

Preparation of Medium: Add components to distilled/deionized water and bring volume to 1.0L. Mix thoroughly. Distribute into tubes or flasks. Autoclave for 15 min at 15 psi pressure–121°C.

Use: For the cultivation of *Spirillum volutans*.

MPY Agar (Maltose Peptone Yeast Extract Medium) (ATCC Medium 518)

Composition per liter:

Agar	10.0g
Maltose	2.0g
Peptone	2.0g
Yeast extract	1.0g
Potassium phosphate buffer (1M, pH 7.5)	10.0mL

pH 7.5 ± 0.2 at 25°C

Preparation of Medium: Add components, except potassium phosphate buffer, to distilled/deionized water and bring volume to 990.0mL. Mix thoroughly. Gently heat and bring to boiling. Autoclave for 15 min at 15 psi pressure–121°C. Cool to 45°–50°C. Filter sterilize potassium phosphate bufffer. Aseptically add sterile potassium phosphate bufffer to sterile cooled basal medium. Distribute into sterile tubes or flasks.

Use: For the cultivation and maintenance of *Spirochaeta aurantia*.

MPY Agar (Malt Peptone Yeast Extract Agar) (ATCC Medium 582)

Composition per liter:

Agar	15.0g
Malt extract	5.0g
Xylose	2.0g
Fructose	2.0g
Lactose	2.0g
Peptone	1.0g
Yeast extract	1.0g

pH 7.0 ± 0.2 at 25°C

Preparation of Medium: Add components to distilled/deionized water and bring volume to 1.0L. Mix thoroughly. Gently heat and bring to boiling. Distribute into tubes or flasks. Autoclave for 15 min at 15 psi pressure–121°C. Pour into sterile Petri dishes or leave in tubes.

Use: For the cultivation and maintenance of *Streptomyces naniwaensis*.

MPY Broth (Maltose Peptone Yeast Extract Broth)

Composition per liter:

Maltose	2.0g
Peptone	2.0g
Yeast extract	1.0g
Potassium phosphate buffer (1M, pH 7.5)	10.0mL

pH 7.5 ± 0.2 at 25°C

Preparation of Medium: Add components, except potassium phosphate buffer, to distilled/deionized water and bring volume to 990.0mL. Mix thoroughly. Autoclave for 15 min at 15 psi pressure–

121°C. Filter sterilize potassium phosphate bufffer. Aseptically add sterile potassium phosphate bufffer to sterile cooled basal medium. Distribute into sterile tubes or flasks.

Use: For the cultivation and maintenance of *Spirochaeta aurantia*.

MPYG Medium
See: Peptone Yeast Extract Glucose Medium, Modified

MRS Agar (DeMan, Rogosa, Sharpe Agar)

Composition per liter:

Glucose	20.0g
Peptone	10.0g
Agar	10.0g
Beef extract	8.0g
Sodium acetate·$3H_2O$	5.0g
Yeast extract	4.0g
K_2HPO_4	2.0g
Triammonium citrate	2.0g
$MgSO_4·7H_2O$	0.2g
$MnSO_4·4H_2O$	0.05g
Sorbitan monooleate	1.0mL

pH 6.2 ± 0.2 at 25°C

Source: This medium is available as a premixed powder from Oxoid Unipath.

Preparation of Medium: Add components to distilled/deionized water and bring volume to 1.0L. Mix thoroughly. Gently heat and bring to boiling. Distribute into tubes or flasks. Autoclave for 15 min at 15 psi pressure–121°C. Pour into sterile Petri dishes or leave in tubes.

Use: For the cultivation of lactic acid bacteria.

MRS Agar (DeMan, Rogosa, Sharpe Agar)

Composition per liter:

Glucose	18.5g
Agar	13.5g
Pancreatic digest of gelatin	10.0g
Beef extract	8.0g
Yeast extract	4.0g
Sodium acetate	3.0g
K_2HPO_4	2.0g
Ammonium citrate	2.0g
Polysorbate 80	1.0g
$MgSO_4·7H_2O$	0.2g
$MnSO_4·4H_2O$	0.05g

pH 6.2 ± 0.2 at 25°C

Source: This medium is available as a premixed powder from BBL Microbiology Systems.

Preparation of Medium: Add components to distilled/deionized water and bring volume to 1.0L. Mix thoroughly. Gently heat while stirring and bring to boiling. Distribute into tubes or flasks. Autoclave for 15 min at 15 psi pressure–121°C. Pour into sterile Petri dishes or leave in tubes.

Use: For the isolation and cultivation of *Lactobacillus* species from clinical specimens, foods and dairy products.

MRS Broth (DeMan, Rogosa, Sharpe Broth)

Composition per liter:

Glucose	20.0g
Peptone	10.0g
Beef extract	8.0g
Sodium acetate·$3H_2O$	5.0g
Yeast extract	4.0g
K_2HPO_4	2.0g
Triammonium citrate	2.0g
$MgSO_4·7H_2O$	0.2g
$MnSO_4·4H_2O$	0.05g
Sorbitan monooleate	1.0mL

pH 6.2 ± 0.2 at 25°C

Source: This medium is available as a premixed powder from Oxoid Unipath.

Preparation of Medium: Add components to distilled/deionized water and bring volume to 1.0L. Mix thoroughly. Gently heat and bring to boiling. Distribute into tubes or flasks. Autoclave for 15 min at 15 psi pressure–121°C.

Use: For the cultivation of lactic acid bacteria.

MRS Broth (DeMan, Rogosa, Sharpe Broth)

Composition per liter:

Glucose	18.5g
Pancreatic digest of gelatin	10.0g
Beef extract	8.0g
Yeast extract	4.0g
Sodium acetate	3.0g
K_2HPO_4	2.0g
Ammonium citrate	2.0g
Polysorbate 80	1.0g
$MgSO_4·7H_2O$	0.2g
$MnSO_4·4H_2O$	0.05g

pH 6.2 ± 0.2 at 25°C

Source: This medium is available as a premixed powder from BBL Microbiology Systems.

Preparation of Medium: Add components to distilled/deionized water and bring volume to 1.0L. Mix thoroughly. Distribute into tubes or flasks. Autoclave for 15 min at 15 psi pressure–121°C.

Use: For the isolation and cultivation of *Lactobacillus* species from clinical specimens, foods and dairy products.

MRS, Modified
See: Lactobacillus Heteroferm Screen Broth

MRVP Broth (Methyl Red–Voges–Proskauer Broth)

Composition per liter:

Glucose	5.0g
KH_2PO_4	5.0g
Pancreatic digest of casein	3.5g
Peptic digest of animal tissue	3.5g

pH 6.9 ± 0.2 at 25°C

Source: Available as a premixed powder from BBL Microbiology Systems and as a prepared medium from Difco Laboratories.

Preparation of Medium: Add components to distilled/deionized water and bring volume to 1.0L. Mix thoroughly. Distribute into tubes or flasks. Autoclave for 15 min at 15 psi pressure–121°C.

Use: For the differentiation of bacteria based on acid production (methyl red test) and acetoin production (Voges-Proskauer reaction).

MRVP Medium (Methyl Red Voges–Proskauer Medium)

Composition per liter:

Glucose	5.0g
Peptone	5.0g
Phosphate buffer	5.0g

pH 7.5 ± 0.2 at 25°C

Source: This medium is available as a premixed powder from Oxoid Unipath.

Preparation of Medium: Add components to distilled/deionized water and bring volume to 1.0L. Mix thoroughly. Distribute into tubes or flasks. Autoclave for 15 min at 15 psi pressure–121°C.

Use: For the differentiation of bacteria based on acid production (methyl red test) and acetoin production (Voges-Proskauer reaction).

MS 1 Agar

Composition per liter:

Agar	15.0g

Preparation of Medium: Add agar to 1.0L of natural seawater. Mix thoroughly. Gently heat and bring to boiling. Distribute into tubes or flasks. Autoclave for 15 min at 15 psi pressure–121°C. Pour into sterile Petri dishes or leave in tubes.

Use: For the isolation and cultivation of *Cytophaga* species, *Herpetosiphon* species, *Saprospira* species, and *Flexithrix* species.

MS 3 Agar

Composition per liter:

Agar	15.0g
$(NH_4)_2SO_4$	1.0g

Preparation of Medium: Add agar to 500.0mL of natural seawater. Mix thoroughly. Gently heat and bring to boiling. In a separate flask, add $(NH_4)_2SO_4$ to 500.0mL of natural seawater. Mix thoroughly. Autoclave both solutions separately for 15 min at 15 psi pressure–121°C. Aseptically combine the two sterile solutions. Pour into sterile Petri dishes or distribute into sterile tubes.

Use: For the isolation and cultivation of *Cytophaga* species, *Herpetosiphon* species, *Saprospira* species, and *Flexithrix* species.

MS 4 Agar

Composition per liter:

Agar	15.0g
Glucose	2.0g
$(NH_4)_2SO_4$	1.0g

Preparation of Medium: Add agar to 500.0mL of natural seawater. Mix thoroughly. Gently heat and bring to boiling. In a separate flask, add $(NH_4)_2SO_4$ to 250.0mL of natural seawater. Mix thoroughly. In a third flask add glucose to 250.0mL of natural seawater. Mix thoroughly. Autoclave the three solutions separately for 15 min at 15 psi pressure–121°C. Aseptically combine the three sterile solutions. Pour into sterile Petri dishes or distribute into sterile tubes.

Use: For the isolation and cultivation of *Cytophaga* species, *Herpetosiphon* species, *Saprospira* species, and *Flexithrix* species.

MS Agar

Composition per liter:

Agar	15.0g
Peptone	1.0g

Yeast extract	1.0g
Glucose	1.0g

Preparation of Medium: Add components to distilled/deionized water and bring volume to 1.0L. Mix thoroughly. Gently heat and bring to boiling. Distribute into tubes or flasks. Autoclave for 15 min at 15 psi pressure–121°C. Pour into sterile Petri dishes or leave in tubes.

Use: For the cultivation and maintenance of *Runella slithyformis*.

MS Medium for Methanogens

Composition per 340mL:

Agar	8.0g
$NaHCO_3$	2.4g
Cysteine-sulfide reducing agent	16.0mL
Mineral solution 1	15.0mL
Mineral solution 2	15.0mL
Sodium formate (20% solution)	6.0mL
Yeast extract -soybean casein solution	4.0mL
Sodium acetate (25% solution)	4.0mL
Wolfe's vitamin solution	4.0mL
Wolfe's mineral solution	4.0mL
$FeSO_4 \cdot 7H_2O$ (0.2% solution)	0.4mL
Resazurin (0.1% solution)	0.4mL

pH 7.0 ± 0.2 at 25°C

Cysteine-Sulfide Reducing Agent:
Composition per 400mL:

L-Cysteine·HCl·H_2O	5.0g
Na_2S (12.5% solution)	40.0mL
NaOH (1*N* solution)	30.0mL

Preparation of Cysteine-Sulfide Reducing Agent: Add distilled/deionized water to a 500.0mL round-bottom flask. Add freshly prepared NaOH solution. Gently heat and bring to boiling under 100% N_2. Remove gassing probe. Add cysteine·HCl·H_2O. Add freshly prepared Na_2S solution. Renew gassing for several minutes. Cap with rubber stoppers. Distribute into 8.0mL/18 mm Hungate tubes.

Mineral Solution 1:
Composition per liter:

K_2HPO_4	6.0g

Preparation of Mineral Solution 1: Add K_2HPO_4 to distilled/deionized water and bring volume to 1.0L. Mix thoroughly.

Mineral Solution 2:
Composition per liter:

NaCl	12.0g
KH_2PO_4	6.0g
$(NH_4)_2SO_4$	6.0g
$MgSO_4 \cdot 7H_2O$	2.6g
$CaCl_2 \cdot 2H_2O$	0.16g

Preparation of Mineral Solution 2: Add components to distilled/deionized water and bring volume to 1.0L. Mix thoroughly.

Yeast Extract–Soybean Casein Solution:
Composition per 100mL:

Yeast extract	20.0g
Pancreatic digest of casein	20.0g

Preparation of Yeast Extract–Soybean Casein Solution: Add components to distilled/deionized water and bring volume to 100.0mL. Mix thoroughly.

Wolfe's Mineral Solution:
Composition per liter:

$MgSO_4 \cdot 7H_2O$	3.0g
Nitriloacetic acid	1.5g
NaCl	1.0g
$MnSO_4 \cdot H_2O$	0.5g
$FeSO_4 \cdot 7H_2O$	0.1g
$CoCl_2 \cdot 6H_2O$	0.1g
$CaCl_2$	0.1g
$ZnSO_4 \cdot 7H_2O$	0.1g
$CuSO_4 \cdot 5H_2O$	0.01g
$AlK(SO_4)_2 \cdot 12H_2O$	0.01g
H_3BO_3	0.01g
$Na_2MoO_4 \cdot 2H_2O$	0.01g

Preparation of Wolfe's Mineral Solution: Add nitrilotriacetic acid to 500.0mL of distilled/deionized water. Dissolve by adjusting pH to 6.5 with KOH. Add remaining components. Add distilled/deionized water to 1.0L.

Wolfe's Vitamin Solution:
Composition per liter:

Pyridoxine·HCl	10.0mg
Thiamine·HCl	5.0mg
Riboflavin	5.0mg
Nicotinic acid	5.0mg
Calcium pantothenate	5.0mg
p-Aminobenzoic acid	5.0mg
Thioctic acid	5.0mg
Biotin	2.0mg
Folic acid	2.0mg
Cyanocobalamin	100.0μg

Preparation of Wolfe's Vitamin Solution: Add components to distilled/deionized water and bring volume to 1.0L. Mix thoroughly.

Preparation of Medium: Add components to distilled/deionized water and bring volume to 408.0mL. Gently heat and bring to boiling under 80% N_2 + 20% CO_2. Continue boiling until rezasurin turns colorless, indicating reduction. Adjust pH to 7.0. Anaerobically distribute into into tubes under 80% N_2 + 20% CO_2. Cap with rubber stoppers and aluminum crimp closures. Autoclave for 15 min at 15 psi pressure–121°C.

Use: For the cultivation and maintenance of *Methanobacterium thermoautotrophicum, Methanobacterium wolfei, Methanobrevibacter smithii, Methanogenium bourgense,* and *Methanogenium* species.

m–Seven Hour Fecal Coliform Agar
See: Seven Hour Fecal Coliform Agar

MSGM, Revised
See: Magnetic *Spirillum* Growth Medium, Revised

m–Sporulation Agar
See: Sporulation Agar

MSRV Medium
See: Modified Semi-Solid Rappaport Vassiliadis Medium

m–ST Holding Medium,
See: ST Holding Medium

m–Standard Methods
See: Standard Methods Broth

m–*Staphylococcus* Broth
See: *Staphylococcus* Broth

MSV AcS Agar

Composition per liter:

$Na_2S \cdot 9H_2O$	0.187g
Sodium acetate	0.15g
MSV Agar	1.0L

pH 7.2–7.5 at 25°C

MSV Agar:
Composition per liter:

Agar	12.0g
$(NH_4)_2SO_4$	0.5g
K_2HPO_4	0.11g
KH_2PO_4	0.085g
$MgSO_4 \cdot 7H_2O$	0.05g
$CaCl_2 \cdot 2H_20$	0.05g
EDTA	3.0mg
$FeCl_3 \cdot H_2O$	2.0mg
Vitamin mix	1.0mL

Preparation of MSV Agar: Add components to distilled/deionized water and bring volume to 1.0L. Mix thoroughly.

Vitamin Mix:
Composition per 100mL:

Calcium pantothenate	0.01g
Niacin	0.01g
Pyridoxine	0.01g
p-Aminobenzoic acid	0.01g
Cocarboxylase	0.01g
Inositol	0.01g
Thiamine	0.01g
Riboflavin	0.01g
Biotin	0.5mg
Cyanocobalamin	0.5mg
Folic acid	0.5mg

Preparation of Vitamin Mix: Add components to distilled/deionized water and bring volume to 100.0mL. Mix thoroughly.

Preparation of Medium: To 1.0L of MSV Agar add sodium acetate and $Na_2S \cdot 9H_2O$. Adjust pH to 7.2–7.5. Gently heat to boiling. Distribute into tubes or flasks. Autoclave for 15 min at 15 psi pressure–121°C. Pour into sterile Petri dishes or leave in tubes.

Use: For the isolation, cultivation and enrichment of heterotrophic strains of *Thiothrix* species from water and environmental sources.

MSV Agar

Composition per liter:

Agar	12.0g
$(NH_4)_2SO_4$	0.5g
K_2HPO_4	0.11g
KH_2PO_4	0.085g
$MgSO_4 \cdot 7H_2O$	0.05g
$CaCl_2 \cdot 2H_20$	0.05g
EDTA	3.0mg
$FeCl_3 \cdot H_2O$	2.0mg
Vitamin mix	1.0mL

pH 7.2–7.5 at 25°C

Vitamin Mix:
Composition per 100mL:

Calcium pantothenate	0.01g
Niacin	0.01g
Pyridoxine	0.01g
p-Aminobenzoic acid	0.01g
Cocarboxylase	0.01g
Inositol	0.01g
Thiamine	0.01g

Riboflavin....0.01g
Biotin 0.5mg
Cyanocobalamin 0.5mg
Folic acid.... 0.5mg

Preparation of Vitamin Mix: Add components to distilled/deionized water and bring volume to 100.0mL. Mix thoroughly.

Preparation of Medium: Add components to distilled/deionized water and bring volume to 1.0L. Mix thoroughly. Adjust pH to 7.2–7.5. Gently heat to boiling. Distribute into tubes or flasks. Autoclave for 15 min at 15 psi pressure–121°C. Pour into sterile Petri dishes or leave in tubes.

Use: For the isolation, cultivation and enrichment of heterotrophic strains of *Thiothrix* species from water and environmental sources.

MSV Broth

Composition per liter:

$(NH_4)_2SO_4$....0.5g
K_2HPO_4....0.11g
$MgSO_4·7H_2O$....0.05g
$CaCl_2·2H_20$....0.05g
KH_2PO_4....0.085g
EDTA 3.0mg
$FeCl_3·H_2O$ 2.0mg
Vitamin mix 1.0mL

pH 7.2–7.5 at 25°C

Vitamin Mix:

Composition per 100mL:

Calcium pantothenate....0.01g
Niacin....0.01g
Pyridoxine....0.01g
p-Aminobenzoic acid....0.01g
Cocarboxylase....0.01g
Inositol 0.01g
Thiamine 0.01g
Riboflavin....0.01g
Biotin 0.5mg
Cyanocobalamin 0.5mg
Folic acid.... 0.5mg

Preparation of Vitamin Mix: Add components to distilled/deionized water and bring volume to 100.0mL. Mix thoroughly.

Preparation of Medium: Add components to distilled/deionized water and bring volume to 1.0L. Mix thoroughly. Adjust pH to 7.2–7.5. Distribute into tubes or flasks. Autoclave for 15 min at 15 psi pressure–121°C.

Use: For the isolation, cultivation and enrichment of heterotrophic strains of *Thiothrix* species from water and environmental sources.

MSV GS Agar

Composition per liter:

$Na_2S·9H_2O$....0.187g
Glucose 0.15g
MSV agar 1.0L

pH 7.2–7.5 at 25°C

MSV Agar:

Composition per liter:

Agar....12.0g
$(NH_4)_2SO_4$....0.5g
K_2HPO_4....0.11g
KH_2PO_4....0.085g
$MgSO_4·7H_2O$ 0.05g
$CaCl_2·2H_20$....0.05g
EDTA 3.0mg
$FeCl_3·H_2O$ 2.0mg
Vitamin mix 1.0mL

Preparation of MSV Agar: Add components to distilled/deionized water and bring volume to 1.0L. Mix thoroughly.

Vitamin Mix:

Composition per 100mL:

Calcium pantothenate....0.01g
Niacin....0.01g
Pyridoxine....0.01g
p-Aminobenzoic acid....0.01g
Cocarboxylase....0.01g
Inositol 0.01g
Thiamine 0.01g
Riboflavin....0.01g
Biotin 0.5mg
Cyanocobalamin 0.5mg
Folic acid.... 0.5mg

Preparation of Vitamin Mix: Add components to distilled/deionized water and bring volume to 100.0mL. Mix thoroughly.

Preparation of Medium: To 1.0L of MSV Agar add glucose and $Na_2S·9H_2O$. Adjust pH to 7.2–7.5. Gently heat to boiling. Distribute into tubes or flasks. Autoclave for 15 min at 15 psi pressure–121°C. Pour into sterile Petri dishes or leave in tubes.

Use: For the isolation, cultivation and enrichment of heterotrophic strains of *Thiothrix* species from water and environmental sources.

MSV I Agar

Composition per liter:

Glucose 0.15g
MSV agar.... 1.0L

pH 7.2–7.5 at 25°C

MSV Agar:

Composition per liter:

Agar	12.0g
$(NH_4)_2SO_4$	0.5g
K_2HPO_4	0.11g
KH_2PO_4	0.085g
$MgSO_4 \cdot 7H_2O$	0.05g
$CaCl_2 \cdot 2H_20$	0.05g
EDTA	3.0mg
$FeCl_3 \cdot H_2O$	2.0mg
Vitamin mix	1.0mL

Preparation of MSV Agar: Add components to distilled/deionized water and bring volume to 1.0L. Mix thoroughly.

Vitamin Mix:

Composition per 100mL:

Calcium pantothenate	0.01g
Niacin	0.01g
Pyridoxine	0.01g
p-Aminobenzoic acid	0.01g
Cocarboxylase	0.01g
Inositol	0.01g
Thiamine	0.01g
Riboflavin	0.01g
Biotin	0.5mg
Cyanocobalamin	0.5mg
Folic acid	0.5mg

Preparation of Vitamin Mix: Add components to distilled/deionized water and bring volume to 100.0mL. Mix thoroughly.

Preparation of Medium: To 1.0L of MSV Agar add glucose. Adjust pH to 7.2–7.5. Gently heat to boiling. Distribute into tubes or flasks. Autoclave for 15 min at 15 psi pressure–121°C. Pour into sterile Petri dishes or leave in tubes.

Use: For the isolation, cultivation and enrichment of heterotrophic strains of *Thiothrix* species from water and environmental sources.

MSV LT Agar

Composition per liter:

Sodium lactate	0.5g
$Na_2S_2O_3$	0.5g
MSV agar	1.0L

pH 7.2–7.5 at 25°C

MSV Agar:

Composition per liter:

Agar	12.0g
$(NH_4)_2SO_4$	0.5g
K_2HPO_4	0.11g
$MgSO_4 \cdot 7H_2O$	0.05g
$CaCl_2 \cdot 2H_20$	0.05g
KH_2PO_4	0.085g
EDTA	3.0mg
$FeCl_3 \cdot H_2O$	2.0mg
Vitamin mix	1.0mL

Preparation of MSV Agar: Add components to distilled/deionized water and bring volume to 1.0L. Mix thoroughly.

Vitamin Mix:

Composition per 100mL:

Calcium pantothenate	0.01g
Niacin	0.01g
Pyridoxine	0.01g
p-Aminobenzoic acid	0.01g
Cocarboxylase	0.01g
Inositol	0.01g
Thiamine	0.01g
Riboflavin	0.01g
Biotin	0.5mg
Cyanocobalamin	0.5mg
Folic acid	0.5mg

Preparation of Vitamin Mix: Add components to distilled/deionized water and bring volume to 100.0mL. Mix thoroughly.

Preparation of Medium: To 1.0L of MSV Agar add sodium lactate and $Na_2S_2O_3$. Adjust pH to 7.2–7.5. Gently heat to boiling. Distribute into tubes or flasks. Autoclave for 15 min at 15 psi pressure–121°C. Pour into sterile Petri dishes or leave in tubes.

Use: For the isolation, cultivation and enrichment of heterotrophic strains of *Thiothrix* species from water and environmental sources.

MSV S Agar

Composition per liter:

$Na_2S \cdot 9H_2O$	0.187g
MSV agar	1.0L

pH 7.2–7.5 at 25°C

MSV Agar:

Composition per liter:

Agar	12.0g
$(NH_4)_2SO_4$	0.5g
K_2HPO_4	0.11g
KH_2PO_4	0.085g
$MgSO_4 \cdot 7H_2O$	0.05g
$CaCl_2 \cdot 2H_20$	0.05g
EDTA	3.0mg
$FeCl_3 \cdot H_2O$	2.0mg
Vitamin mix	1.0mL

Preparation of MSV Agar: Add components to distilled/deionized water and bring volume to 1.0L. Mix thoroughly.

Vitamin Mix:

Composition per 100mL:

Calcium pantothenate	0.01g
Niacin	0.01g
Pyridoxine	0.01g
p-Aminobenzoic acid	0.01g
Cocarboxylase	0.01g
Inositol	0.01g
Thiamine	0.01g
Riboflavin	0.01g
Biotin	0.5mg
Cyanocobalamin	0.5mg
Folic acid	0.5mg

Preparation of Vitamin Mix: Add components to distilled/deionized water and bring volume to 100.0mL. Mix thoroughly.

Preparation of Medium: To 1.0L of MSV Agar add $Na_2S{\cdot}9H_2O$. Adjust pH to 7.2–7.5. Gently heat to boiling. Distribute into tubes or flasks. Autoclave for 15 min at 15 psi pressure–121°C. Pour into sterile Petri dishes or leave in tubes.

Use: For the isolation, cultivation and enrichment of heterotrophic strains of *Thiothrix* species from water and environmental sources.

MSV SS Agar

Composition per liter:

$Na_2S{\cdot}9H_2O$	0.187g
Sucrose	0.15g
MSV agar	1.0L

pH 7.2–7.5 at 25°C

MSV Agar:

Composition per liter:

Agar	12.0g
$(NH_4)_2SO_4$	0.5g
K_2HPO_4	0.11g
KH_2PO_4	0.085g
$MgSO_4{\cdot}7H_2O$	0.05g
$CaCl_2{\cdot}2H_20$	0.05g
EDTA	3.0mg
$FeCl_3{\cdot}H_2O$	2.0mg
Vitamin mix	1.0mL

Preparation of MSV Agar: Add components to distilled/deionized water and bring volume to 1.0L. Mix thoroughly.

Vitamin Mix:

Composition per 100mL:

Calcium pantothenate	0.01g
Niacin	0.01g
Pyridoxine	0.01g
p-Aminobenzoic acid	0.01g
Cocarboxylase	0.01g
Inositol	0.01g
Thiamine	0.01g
Riboflavin	0.01g
Biotin	0.5mg
Cyanocobalamin	0.5mg
Folic acid	0.5mg

Preparation of Vitamin Mix: Add components to distilled/deionized water and bring volume to 100.0mL. Mix thoroughly.

Preparation of Medium: To 1.0L of MSV Agar add $Na_2S{\cdot}9H_2O$ and sucrose. Adjust pH to 7.2–7.5. Gently heat to boiling. Distribute into tubes or flasks. Autoclave for 15 min at 15 psi pressure–121°C. Pour into sterile Petri dishes or leave in tubes.

Use: For the isolation, cultivation and enrichment of heterotrophic strains of *Thiothrix* species from water and environmental sources.

MSV SUC Agar

Composition per liter:

Sodium succinate	0.15g
MSV agar	1.0L

pH 7.2–7.5 at 25°C

MSV Agar:

Composition per liter:

Agar	12.0g
$(NH_4)_2SO_4$	0.5g
K_2HPO_4	0.11g
KH_2PO_4	0.085g
$MgSO_4{\cdot}7H_2O$	0.05g
$CaCl_2{\cdot}2H_20$	0.05g
EDTA	3.0mg
$FeCl_3{\cdot}H_2O$	2.0mg
Vitamin mix	1.0mL

Preparation of MSV Agar: Add components to distilled/deionized water and bring volume to 1.0L. Mix thoroughly.

Vitamin Mix:

Composition per 100mL:

Calcium pantothenate	0.01g
Niacin	0.01g
Pyridoxine	0.01g
p-Aminobenzoic acid	0.01g
Cocarboxylase	0.01g
Inositol	0.01g
Thiamine	0.01g
Riboflavin	0.01g
Biotin	0.5mg
Cyanocobalamin	0.5mg
Folic acid	0.5mg

Preparation of Vitamin Mix: Add components to distilled/deionized water and bring volume to 100.0mL. Mix thoroughly.

Preparation of Medium: To 1.0L of MSV Agar add sodium succinate. Adjust pH to 7.2–7.5. Gently heat to boiling. Distribute into tubes or flasks. Autoclave for 15 min at 15 psi pressure–121°C. Pour into sterile Petri dishes or leave in tubes.

Use: For the isolation, cultivation and enrichment of heterotrophic strains of *Thiothrix* species from water and environmental sources.

m–T7 Agar Base
See: T7 Agar Base

m–TEC Agar
See: TEC Agar

m–Teepol Broth, Enriched
See: Teepol Broth, Enriched

m–Tetrathionate Broth
See: Tetrathionate Broth

m–TGE Broth
See: TGE Broth

MTM II
See: Thayer–Martin Agar, Modified

m–TT Broth
See: Tetrathionate Broth

Mucate Broth

Composition per liter:

Mucic acid	10.0g
Peptone	10.0g
Bromthymol Blue	0.024g

pH 7.4 ± 0.1 at 25°C.

Preparation of Medium: Add components to distilled/deionized water and bring volume to 1.0L. Mix thoroughly. Add 5*N* NaOH while stirring until mucic acid dissolves. Distribute into screw-capped tubes in 5.0mL volumes. Autoclave for 10 min at 15 psi pressure–121°C.

Use: For the isolation and cultivation of enterovirulent *Escherichia coli* and *Shigella* species.

Mucate Control Broth

Composition per liter:

Peptone	10.0g
Bromthymol Blue	0.024g

pH 7.4 ± 0.1 at 25°C

Preparation of Medium: Add components to distilled/deionized water and bring volume to 1.0L. Mix thoroughly. Distribute into screw-capped tubes in 5.0mL volumes. Autoclave for 10 min at 15 psi pressure–121°C.

Use: For the isolation and cultivation of enterovirulent *Escherichia coli* and *Shigella* species.

Mueller–Hinton Agar

Composition per liter:

Beef infusion	300.0g
Acid hydrolysate of casein	17.5g
Agar	17.0g
Starch	1.5g

pH 7.4 ± 0.2 at 25°C

Source: This medium is available as a premixed powder from Difco Laboratories and Oxoid Unipath.

Preparation of Medium: Add components to distilled/deionized water and bring to 1.0L. Mix thoroughly. Gently heat and bring to boiling. Distribute into tubes or flasks. Autoclave for 15 min at 15 psi pressure–121°C. Pour into sterile Petri dishes or leave in tubes.

Use: For the isolation of pathogenic *Neisseria* species. For antimicrobial susceptibility testing of a variety of bacterial species. For the cultivation and maintenance of *Moraxella osloensis* and *Neisseria meningitidis*.

Mueller–Hinton Agar with IsoVitaleX® and Hemoglobin

Composition per liter:

Component A	490.0mL
Component B	490.0mL
IsoVitaleX® enrichment	20.0mL

pH 6.9 ± 0.2 at 25°C

Component A:

Composition per 490mL:

Beef infusion	300.0g
Acid hydrolysate of casein	17.5g
Agar	17.0g
Starch	1.5g

Preparation of Component A: Add components to distilled/deionized water and bring to 490.0mL. Mix thoroughly. Gently heat and bring to boiling.

Autoclave for 15 min at 15 psi pressure–121°C. Cool to 45°–50°C.

Component B:
Composition per 490mL:

Hemoglobin	10.0g

Preparation of Component B: Add hemoglobin to distilled/deionized water and bring to 490.0mL. Mix thoroughly. Gently heat and bring to boiling. Autoclave for 15 min at 15 psi pressure–121°C. Cool to 45°–50°C.

IsoVitaleX® Enrichment:
Composition per liter:

Glucose	100.0g
L-Cysteine·HCl	25.9g
L-Glutamine	10.0g
L-Cystine	1.1g
Adenine	1.0g
Nicotinamide adenine dinucleotide	0.25g
Vitamin B_{12}	0.1g
Thiamine pyrophosphate	0.1g
Guanine·HCl	0.03g
$Fe(NO_3)_3 \cdot 6H_2O$	0.02g
p-Aminobenzoic acid	0.013g
Thiamine·HCl	3.0mg

Source: The supplement solution IsoVitaleX® enrichment is available from BBL Microbiology Laboratories. This enrichment may be replaced by supplement VX from Difco Laboratories.

Preparation of IsoVitaleX® Enrichment: Add components to distilled/deionized water and bring volume to 1.0L. Mix thoroughly. Filter sterilize. Warm to 45°–50°C.

Preparation of Medium: Aseptically combine 490.0mL of component A, 490.0mL of component B, and 20.0mL of IsoVitaleX® enrichment. Mix thoroughly. Adjust pH to 6.9. Pour into sterile Petri dishes in 20.0mL volumes.

Use: For the isolation and cultivation of *Legionella pneumophila.*

Mueller–Hinton Broth

Composition per liter:

Acid hydrolysate of casein	17.5g
Beef extract	3.0g
Starch	1.5g

pH 7.3 ± 0.1 at 25°C

Source: This medium is available as a premixed powder from BBL Microbiology Systems, Difco Laboratories and Oxoid Unipath.

Preparation of Medium: Add components to distilled/deionized water and bring to 1.0L. Mix thoroughly. Gently heat and bring to boiling. Distribute into tubes or flasks. Autoclave for 10 min at 10 psi pressure–115°C. Do not overheat.

Use: For the cultivation of a wide variety of microorganisms. For antimicrobial susceptibility testing.

Mueller–Hinton Chocolate Agar

Beef infusion	300.0g
Acid hydrolysate of casein	17.5g
Agar	17.0g
Starch	1.5g
Sheep blood	50.0mL

pH 7.4 ± 0.2 at 25°C

Preparation of Medium: Add components, except sheep blood, to distilled/deionized water and bring volume to 950.0mL. Mix thoroughly. Gently heat and bring to boiling. Autoclave for 15 min at 15 psi pressure–121°C. Cool to 45°–50°C. Aseptically add sterile sheep blood. Mix thoroughly. Gently heat to 70°C for 10 min. Pour into sterile Petri dishes or distribute into sterile tubes.

Use: For the cultivation and maintenance of *Neisseria gonorrhoeae* and *Neisseria meningitidis*. For antimicrobial susceptibility testing offastidious microorganisms.

Mueller–Hinton II Agar

Composition per liter:

Acid hydrolysate of casein	17.5g
Agar	17.0g
Beef extract	2.0g
Starch	1.5g

pH 7.3 ± 0.1 at 25°C

Source: This medium is available as a premixed powder from BBL Microbiology Systems.

Preparation of Medium: Add components to distilled/deionized water and bring to 1.0L. Mix thoroughly. Gently heat and bring to boiling. Distribute into tubes or flasks. Autoclave for 15 min at 15 psi pressure–121°C. Pour into sterile Petri dishes or leave in tubes.

Use: For antimicrobial disc diffusion susceptibility testing by the Bauer-Kirby method of a variety of bacteria. This medium supplemented with 5% sheep blood is recommended for use in antimicrobial susceptibility testing of *Streptococcus pneumoniae* and *Haemophilus influenzae*.

Mueller–Hinton Medium with Garden Soil

Composition per liter:

Garden soil, sterile	300.0g
Beef infusion	300.0g

Acid hydrolysate of casein 17.5g
Agar 17.0g
Starch 1.5g

pH 7.4 ± 0.2 at 25°C

Preparation of Medium: Add components to distilled/deionized water and bring volume to 1.0L. Mix thoroughly. Gently heat and bring to boiling. Distribute into tubes or flasks. Autoclave for 15 min at 15 psi pressure–121°C. Pour into sterile Petri dishes or leave in tubes. Swirl flask while pouring to disperse soil.

Use: For the cultivation and maintenance of *Chromobacterium violaceum*.

Mueller–Hinton Medium with Rabbit Serum

Composition per liter:

Beef infusion 300.0g
Acid hydrolysate of casein 17.5g
Agar 17.0g
Starch 1.5g
Rabbit serum 100.0mL

pH 7.4 ± 0.2 at 25°C

Preparation of Medium: Add components, except rabbit serum, to distilled/deionized water and bring volume to 900.0mL. Mix thoroughly. Gently heat and bring to boiling. Autoclave for 15 min at 15 psi pressure–121°C. Cool to 45°–50°C. Aseptically add sterile rabbit serum. Pour into sterile Petri dishes or distribute into sterile tubes.

Use: For the cultivation and maintenance of *Corynebacterium* species.

Mueller Tellurite Medium

Composition per liter:

Casamino acids 20.0g
Agar 20.0g
Casein 5.0g
KH_2PO_4 .... 0.3g
$MgSO_4 \cdot 7H_2O$ 0.1g
L-Tryptophan 0.05g
Mueller tellurite serum 25.0mL

pH 7.4 ± 0.1 at 25°C

Mueller Tellurite Serum:

Composition per 100mL:

K_2TeO_3 solution 0.4g
Calcium pantothenate 0.2mg
Horse or beef serum, sterile 50.0mL
Sodium lactate solution 40.0mL
Ethyl alcohol 10.0mL

Preparation of Mueller Tellurite Serum: Add calcium pantothenate to 1.0mL of distilled/deionized water. Autoclave for 15 min at 15 psi pressure–121°C. Add K_2TeO_3 to 1.0mL of sterile distilled/deionized water. To 40.0mL of cooled, sterile sodium lactate solution add filter-sterilized ethanol, sterile calcium pantothenate solution, sterile serum and K_2TeO_3 solution. Mix thoroughly. Store at 4°–8°C.

Sodium Lactate Solution:

Composition per 100mL:

Lactic acid (85% solution) 50mL
Phenol Red solution (0.2g in 50% ethanol) 0.1mL

Preparation of Sodium Lactate Solution: Add lactic acid to distilled/deionized water and bring volume to 100.0mL. Add 0.1mL of Phenol Red solution. Add enough 40% NaOH solution to adjust pH to 7.0. Gently heat and bring to boiling for 5 min. Add more NaOH solution to retain red color, if necessary. Autoclave for 15 min at 15 psi pressure–121°C. Cool to 50°C.

Caution: Potassium tellurite is toxic.

Preparation of Medium: Add components to distilled/deionized water and bring volume to 975.0mL. Gently heat and bring to boiling. Autoclave for 15 min at 15 psi pressure–121°C. Cool quickly to 50°C. Aseptically add 25.0mL Mueller tellurite serum. Mix thoroughly. Distribute into sterile Petri dishes. Allow the surface of the plates to dry by partially removing the covers during solidification.

Use: For isolation, cultivation and differentiation of *Corynebacterium diphtheriae*.

Muller–Kauffmann Tetrathionate Broth

Composition per 1028mL:

$Na_2S_2O_3$ 40.7g
$CaCO_3$ 25.0g
Pancreatic digest of casein 7.0g
Ox bile 4.75g
Soya peptone 2.3g
NaCl 2.3g
Iodine solution 19.0mL
Brilliant Green solution 9.5mL

Iodine Solution:

Composition per 100mL:

Iodine 20.0g
KI 25.0g

Preparation of Iodine Solution: Add the KI to approximately 5.0mL of distilled/deionized water. Mix thoroughly. Add the iodine. Gently heat to dissolve. Bring volume to 100.0mL with distilled/deionized water. Filter sterilize.

Brilliant Green Solution:

Composition per 100mL:

Brilliant Green 0.1g

Preparation of Brilliant Green Solution: Add the Brilliant Green to distilled/deionized water and bring volume to 100.0mL. Mix thoroughly. Gently heat while stirring and bring to boiling. Continue boiling for 30 min while stirring until dye has dissolved. Filter sterilize. Store protected from light.

Preparation of Medium: Add components, except iodine solution and Brilliant Green solution, to distilled/deionized water and bring volume to 1.0L. Mix thoroughly. Gently heat and bring to boiling. Do not autoclave. Cool to 45°C. Prior to use, add 19.0mL of iodine solution and 9.5mL of Brilliant Green solution. Mix thoroughly. Aseptically distribute into sterile tubes or flasks.

Use: For the isolation and cultivation of *Salmonella* species from specimens with a mixed flora.

MVL Medium

Composition per liter:

Agar...15.0g
Pancreatic digest of casein..............................10.0g
Yeast extract..5.0g
Beef extract..2.0g
Glucose 2.0g
Mineral solution 1 ....75.0mL
Mineral solution 2..75.0mL
Na_2CO_3 (8.0% solution)...............................50.0mL
Hemin (0.07% solution)...............................10.0mL
L-Cysteine·HCl·H_2O (3.0% solution)...........10.0mL
Resazurin (0.1% solution).............................1.0mL

pH 7.0 ± 0.2 at 25°C

Mineral Solution 1:

Composition per 100mL:

K_2HPO_4..0.6g

Preparation of Mineral Solution 1: Add K_2HPO_4 to distilled/deionized water and bring volume to 100.0mL. Mix thoroughly.

Mineral Solution 2:

Composition per 100mL:

$(NH_4)_2SO_4$..1.2g
NaCl 1.2g
KH_2PO_4..0.6g
$MgSO_4·7H_2O$ 0.12g
$CaCl_2·2H_2O$...0.12g

Preparation of Mineral Solution 2: Add components to distilled/deionized water and bring volume to 100.0mL. Mix thoroughly.

Preparation of Medium: Prepare and dispense medium under 100% CO_2. Add components to distilled/deionized water and bring volume to 1.0L. Mix thoroughly. Adjust pH to 7.0. Anaerobically distribute into tubes. Autoclave for 15 min at 15 psi pressure–121°C.

Use: For the cultivation of *Tonsillophilus suis*.

MVTY Medium

Composition per liter:

Noble agar..7.0g
Yeast extract..2.0g
Pancreatic digest of casein................................2.0g
Cysteine·HCl·H_2O..1.0g
Salts A 200.0mL
Salts B 200.0mL
$NaHCO_3$ solution 100.0mL
Glucose (10% solution)................................20.0mL
Resazurin (0.1% solution)..............................1.0mL
n-Butyric acid..0.4mL
Isobutyric acid..0.2mL
DL-2-Methylbutyric acid 0.2mL
n-Valeric acid 0.2mL
Isovaleric acid 0.2mL

pH 6.9 ± 0.2 at 25°C

Salts A:

Composition per liter:

$CaCl_2$..0.45g
$MgSO_4$..0.45g

Preparation of Salts A: Add components to distilled/deionized water and bring volume to 1.0L. Mix thoroughly.

Salts B:

Composition per liter:

NaCl 4.5g
$(NH_4)_2SO_4$..4.5g
KH_2PO_4..2.25g
K_2HPO_4..2.25g

Preparation of Salts B: Add components to distilled/deionized water and bring volume to 1.0L. Mix thoroughly.

$NaHCO_3$ Solution:

Composition per 100mL:

$NaHCO_3$ 5.0g

Preparation of $NaHCO_3$ Solution: Add $NaHCO_3$ to distilled/deionized water and bring volume to 100.0mL. Mix thoroughly. Filter sterilize.

Preparation of Medium: Separately autoclave the glucose solution, *n*-butyric acid, isobutyric acid, DL-2-methylbutyric acid, *n*-valeric acid and isovaleric acid for 15 min at 15 psi pressure–121°C. Prepare and dispense the basal medium under 100% CO_2. Add the agar, yeast extract, pancreatic digest of casein , cysteine·HCl·H_2O, Salts A, Salts B and resazurin to distilled/deionized water and bring volume to 478.0mL. Mix thoroughly. Adjust pH to 7.0. Autoclave for 15 min at 15 psi pressure–121°C. Cool to 50°C. Add the sterile fatty acid solutions, the sterile $NaHCO_3$ solution and the sterile glucose solution. Mix thoroughly. Aseptically and anaerobically distribute into sterile tubes.

Use: For the cultivation and maintenance of *Treponema saccharophilum*.

MWY Medium (Wadowsky and Yee Medium, Modified)

Composition per liter:

Agar	13.0g
Yeast extract	10.0g
Glycine	3.0g
ACES buffer (2-[(2-Amino-2-oxoethyl)-amino]-ethane sulfonic acid)	2.0g
Charcoal, activated	2.0g
α-Ketoglutarate	0.2g
$Fe_4(P_2O_7)_3 \cdot 9H_2O$	0.05g
Bromcresol Purple	0.01g
Bromcresol Blue	0.01g
Antibiotic inhibitor	10.0mL
L-Cysteine·HCl·H_2O solution	10.0mL

pH 6.9 ± 0.2 at 25°C

Antibiotic Inhibitor:

Composition per 10mL:

Anisomycin	0.16g
Cefamandole	4.0mg
Vancomycin	1.0mg
Polymyxin B	130,000U

Preparation of Antibiotic Inhibitor: Add components to distilled/deionized water and bring volume to 10.0mL. Mix thoroughly. Filter sterilize.

L-Cysteine·HCl·H_2O Solution:

Composition per 10mL:

L-Cysteine·HCl·H_2O	0.08g

Preparation of L-Cysteine·HCl·H_2O Solution: Add L-Cysteine·HCl·H_2O to distilled/deionized water and bring volume to 10.0mL. Mix thoroughly. Filter sterilize.

Preparation of Medium: Add components, except cysteine and antibiotic inhibitor, to distilled/deionized water and bring volume to 980.0mL. Mix thoroughly. Adjust medium to pH 6.9 with 1*N* KOH. Heat gently and bring to boiling for 1 min. Autoclave for 15 min at 15 psi pressure–121°C. Cool to 50°–55°C. Add 10.0mL of the sterile L-cysteine·HCl·H_2O solution and 10.0mL of the sterile antibiotic solution. Mix thoroughly. Pour into sterile Petri dishes with constant agitation to keep charcoal in suspension.

Use: For the selective isolation and cultivation of *Legionella pneumophila* and other *Legionella* species.

MY40G

See: **Malt Extract Yeast Extract 40% Glucose Agar**

MY Agar

Composition per liter:

Agar	15.0g
Sodium acetate	0.1g
Yeast extract	0.1g
Pancreatic digest of gelatin	0.06g
Beef extract	0.04g
Basal medium	1.0L
Sodium sulfide solution	3.0mL

pH 7.0–7.5 at 25°C

Basal Medium:

Composition per liter:

$CaSO_4 \cdot 2H_2O$ (saturated solution)	20.0mL
NH_4Cl (4% solution)	5.0mL
Trace elements solution	5.0mL
K_2HPO_4 (1% solution)	1.0mL
$MgSO_4 \cdot 7H_2O$ (1% solution)	1.0mL

Preparation of Basal Medium: Add components to distilled/deionized water and bring volume to 1.0L. Mix thoroughly.

Trace Elements Solution:

Composition per liter:

Ferrous EDTA solution	20.0mL
$ZnSO_4 \cdot 7H_2O$ (0.1% solution)	10.0mL
$MnSO_4 \cdot 4H_2O$ (0.02% solution)	10.0mL
$CuSO_4 \cdot 5H_2O$ (0.00005% solution)	10.0mL
H_3BO_3 (0.1% solution)	10.0mL
$Co(NO_3)_2$ or $CoCl_2 \cdot 6H_2O$ (0.01% solution)	10.0mL
$Na_2MoO_4 \cdot 2H_2O$ (0.01% solution)	10.0mL

Preparation of Trace Elements Solution: Add components to distilled/deionized water and bring volume to 1.0L. Mix thoroughly.

Ferrous EDTA Solution:

Composition per 100mL:

$FeSO_4 \cdot 7H_2O$	7.0g
EDTA	2.0g
HCl, concentrated	1.0mL

Preparation of Ferrous EDTA Solution: Add components to distilled/deionized water and bring volume to 100.0mL. Mix thoroughly.

Sodium Sulfide Solution:

Composition per 10mL:

$Na_2S \cdot 9H_20$	1.0g

Preparation of Sodium Sulfide Solution: Add $Na_2S \cdot 9H_20$ to distilled/deionized water and bring volume to 10.0mL. Mix thoroughly. Autoclave for 15 min at 15 psi pressure–121°C. Prepare freshly.

Preparation of Medium: Add sodium acetate, nutrient broth powder, yeast extract and agar to 1.0L of basal medium. Mix thoroughly. Adjust pH to 7.0–7.5. Gently heat and bring to boiling. Autoclave for 15 min at 15 psi pressure–121°C. Cool to 45°–50°C. Aseptically add 3.0mL of sterile sodium sulfide solution immediately prior to dispensing. Mix thoroughly. Pour into sterile Petri dishes or distribute into sterile screw-capped tubes.

Use: For the isolation and cultivation of *Beggiatoa* species and myxotrophic strains of *Thiothrix* species from water and environmental sources.

Mycin Assay Agar

Composition per liter:

Dextrose	15.0g
Peptone	5.0g
Beef extract	3.0g

pH 7.9 ± 0.1 at 25°C

Preparation of Medium: Add components to distilled/deionized water and bring volume to 1.0L. Mix thoroughly. Distribute into tubes or flasks. Autoclave for 15 min at 15 psi pressure–121°C.

Use: For microbiological assay of antibiotics in pharmaceutical products, body fluids, feeds and other sample materials.

Mycobacteria 7H11 Agar with Middlebrook ADC Enrichment

See: **Middlebrook 7H11 Agar with Middlebrook ADC Enrichment**

Mycobacteria 7H11 Agar with Middlebrook OADC Enrichment

See: **Middlebrook 7H11 Agar with Middlebrook ADC Enrichment**

Mycobacteria 7H11 Agar with Middlebrook OADC Enrichment and Triton WR 1339

See: **Middlebrook 7H11 Agar with Middlebrook OADC Enrichment and Triton WR 1339**

Mycobacterium Medium

Composition per liter:

Noble agar	15.0g
$(NH_4)_2SO_4$	1.0g
Na_2HPO_4	0.5g
KH_2PO_4	0.5g
$MgSO_4$	0.2g
$FeSO_4 \cdot 7H_2O$	5.0mg
$MnSO_4$	2.0mg
Liquid paraffin	5.0mL

Preparation of Medium: Add components, except agar, to distilled/deionized water and bring volume to 1.0L. Homogenize in a blender. Add agar. Gently heat and bring to boiling. Distribute into tubes or flasks. Autoclave for 15 min at 15 psi pressure–121°C. Pour into sterile Petri dishes or leave in tubes.

Use: For the cultivation and maintenance of *Mycobacterium paraffinicum.*

Mycobacterium Yeast Extract Medium

Composition per liter:

Agar	15.0g
Pancreatic digest of casein	5.0g
Yeast extract	2.5g
Glucose	1.0g

Preparation of Medium: Add components to distilled/deionized water and bring volume to 1.0L. Mix thoroughly. Gently heat and bring to boiling. Distribute into tubes or flasks. Autoclave for 15 min at 15 psi pressure–121°C. Pour into sterile Petri dishes or leave in tubes.

Use: For the cultivation and maintenance of *Mycobacterium* species and *Rhodococcus* species.

Mycobactin Medium

Serum Agar Medium:

Composition per liter:

Noble agar	15.0g
Casamino acids	2.5g
Na_2HPO_4 (anhydrous)	2.5g
Sodium citrate	1.5g
KH_2PO_4	1.0g
$MgSO_4 \cdot 7H_2O$	0.6g
Asparagine	0.3g
Crude mycobactin	0.16g
Chloramphenicol	0.05g
Primaricine (myprozine)	0.05g
Penicillin	100,000U
Bovine serum , 56°C-inactivated	200.0mL
Tween™ 80 (1% solution)	50.0mL
Glycerol	25.0mL

pH 7.2 ± 0.2 at 25°C

Preparation of Crude Mycobactin: Grow *Mycobacterium phlei* in 600.0mL of Mycobactin Production Broth for 2 weeks at 37°C. Autoclave the culture for 15 min at 15 psi pressure–121°C. Filter the cells and wash with distilled/deionized water. Dry cells under $CaCl_2$. Treat 100.0g of dried culture with 3 successive acetone extractions—500.0mL of acetone for 30 min in a liter flask fitted with a reflux condenser. Evaporate the acetone to dryness. Extract the residue in a Soxhlet apparatus with petroleum ether for 18–20 hrs at 40°–60°C. A hard red residue will remain. Dissolve the residue in warm absolute ethanol. Centrifuge for 30 min at 2,250 rpm to remove debris. Evaporate the supernatant to dryness. Grind the residue to a powder of crude mycobactin.

Source: Purified mycobactin is available from Allied Labs, Inc., 2520 Hunt St., Ames, IA 50010.

Mycobactin Production Broth:

Composition per 600mL:

Solution B	500.0mL
Solution A	100.0mL

Preparation of Mycobactin Production Broth: Aseptically mix the cooled sterile solution A and solution B.

Solution A:

Composition per 100mL:

L-Asparagine	5.0g
Na_2HPO_4	2.0g
KH_2PO_4	1.0g
Glycerol	30.0mL

Preparation of Solution A: Add components to distilled/deionized water and bring volume to 100.0mL. Mix thoroughly. Autoclave for 15 min at 15 psi pressure–121°C. Cool to 45°–50°C.

Solution B:

Composition per 500mL:

Glucose	10.0g
$MgSO_4 \cdot 7H_2O$	0.2g

Preparation of Solution B: Add components to distilled/deionized water and bring volume to 500.0mL. Mix thoroughly. Autoclave for 15 min at 15 psi pressure–121°C. Cool to 45°–50°C.

Preparation of Medium: Add components, except penicillin, chloramphenicol, primaricine and bovine serum, to distilled/deionized water and bring volume to 800.0mL. Mix thoroughly. Gently heat with a minimum of heat. Autoclave for 15 min at 10 psi pressure–116°C. Cool to 50°C. Aseptically add penicillin, chloramphenicol, primaricine and sterile bovine serum. Mix thoroughly. Adjust pH to 7.2. Distribute into sterile tubes or flasks.

Use: For the cultivation and maintenance of *Mycobacterium avium* and *Mycobacterium paratuberculosis*.

Mycobactosel™ Agar

Composition per liter:

Agar	13.5g
Bovine albumin fraction V	5.0g
Glucose	2.0g
Na_2HPO_4	1.5g
KH_2PO_4	1.5g
Pancreatic digest of casein	1.0g
NaCl	0.85g
$(NH_4)_2SO_4$	0.5g
Monosodium glutamate	0.5g
Sodium citrate	0.4g
$MgSO_4 \cdot 7H_2O$	0.05g
Ferric ammonium citrate	0.04g
Pyridoxine	1.0mg
$ZnSO_4 \cdot 7H_2O$	1.0mg
$CuSO_4 \cdot 5H_2O$	1.0mg
Biotin	0.5mg
$CaCl_2 \cdot H_2O$	0.5mg
Malachite Green	0.25mg
Antibiotic solution	10.0mL
Catalase solution	10.0mL
Glycerol	5.0mL
Oleic Acid	0.06mL

pH 6.6 ± 0.2 at 25°C

Source: Available as a prepared medium from BBL Microbiology Systems.

Antibiotic Solution:

Composition per 10mL:

Cycloheximide	0.36g
Nalidixic acid	0.02g
Lincomycin	2.0mg

Preparation of Antibiotic Solution: Add components to distilled/deionized water and bring volume to 10.0mL. Mix thoroughly. Filter sterilize.

Catalase Solution:
Composition per 10mL:

Catalase	3.0mg

Preparation of Catalase Solution: Add catalase to distilled/deionized water and bring volume to 10.0mL. Mix thoroughly. Filter sterilize.

Preparation of Medium: Add components, except antibiotic solution and catalase solution, to distilled/deionized water and bring volume to 980.0mL. Mix thoroughly. Gently heat and bring to boiling. Autoclave for 15 min at 15 psi pressure–121°C. Cool to 45°–50°C. Aseptically add sterile antibiotic solution and sterile catalase solution. Mix thoroughly. Pour into sterile Petri dishes or distribute into sterile tubes.

Use: For the selective isolation of mycobacteria from specimens containing mixed flora.

Mycobactosel™ L–J Medium

Composition per liter:

Potato flour	30.0g
L-Asparagine	3.6g
KH_2PO_4, anhydrous	2.5g
Sodium citrate	0.6g
$MgSO_4 \cdot 7H_2O$	0.24g
Homogenized whole egg	1.0L
Malachite Green solution	20.0mL
Glycerol	12.0mL
Antibiotic solution	10.0mL

pH 7.0 ± 0.2 at 25°C

Source: Available as a prepared medium from BBL Microbiology Systems.

Homogenized Whole Egg:
Composition per liter:

Whole eggs	18–24

Preparation of Whole Egg: Use fresh eggs, less than 1 week old. Scrub the shells with soap. Let stand in a soap solution for 30 min. Rinse in running water. Soak eggs in 70% ethanol for 15 min. Break the eggs into a sterile container. Homogenize by shaking. Filter through four layers of sterile cheesecloth into a sterile graduated cylinder. Measure out 1.0L.

Malachite Green Solution:
Composition per 20mL:

Malachite Green	0.4g

Preparation of Malachite Green Solution: Add Malachite Green to sterile distilled/deionized water and bring volume to 20.0mL in a sterile container. Mix thoroughly.

Antibiotic Solution:
Composition per 10mL:

Cycloheximide	0.64g
Nalidixic acid	0.056g
Lincomycin	3.2mg

Preparation of Antibiotic Solution: Add components to distilled/deionized water and bring volume to 10.0mL. Mix thoroughly. Filter sterilize.

Preparation of Medium: Add components—except whole egg, Malachite Green solution, and antibiotic solution—to distilled/deionized water and bring volume to 600.0mL. Mix thoroughly. Autoclave for 30 min at 15 psi pressure–121°C. Cool to room temperature. Add the homogenized whole egg, Malachite Green solution, and antibiotic solution. Distribute into sterile tubes in 8.0mL volumes. Coagulate medium in a slanted position at 85°C (moist heat) for 50 min.

Use: For the isolation and cultivation of *Mycobacterium* species from clinical specimens.

Mycobiotic Agar (Cycloheximide Chloramphenicol Agar)

Composition per liter:

Agar	15.0g
Enzymatic hydrolysate of soybean meal	10.0g
Glucose	10.0g
Cycloheximide	0.5g
Chloramphenicol	0.05g

pH 6.5 ± 0.2 at 25°C

Source: This medium is available as a premixed powder from Difco Laboratories.

Preparation of Medium: Add components to distilled/deionized water and bring volume to 1.0L. Mix thoroughly. Gently heat and bring to boiling. Distribute into tubes or flasks. Autoclave for 15 min at 15 psi pressure–121°C. Cool tubes quickly in a slanted position.

Use: For the selective isolation and cultivation of pathogenic fungi.

Mycological Agar

Composition per liter:

Agar	15.0g
Enzymatic hydrolysate of soybean meal	10.0g
Glucose	10.0g

pH 7.0 ± 0.2 at 25°C

Source: This medium is available as a premixed powder from Difco Laboratories.

Preparation of Medium: Add components to distilled/deionized water and bring volume to 1.0L. Mix thoroughly. Gently heat and bring to boiling. Distribute into tubes or flasks. Autoclave for 15 min at 15 psi pressure–121°C.

Use: For the selective isolation, cultivation and maintenance of pathogenic fungi.

Mycological Agar with Low pH

Composition per liter:

Agar	15.0g
Enzymatic hydrolysate of soybean meal	10.0g
Glucose	10.0g

pH 4.8 ± 0.2 at 25°C

Source: This medium is available as a premixed powder from Difco Laboratories.

Preparation of Medium: Add components to distilled/deionized water and bring volume to 1.0L. Mix thoroughly. Gently heat and bring to boiling. Distribute into tubes or flasks. Autoclave for 15 min at 15 psi pressure–121°C.

Use: For the selective isolation, cultivation and maintenance of pathogenic fungi.

Mycological Broth

Composition per liter:

Glucose	40.0g
Enzymatic hydrolysate of soybean meal	10.0g

pH 7.0 ± 0.2 at 25°C

Source: This medium is available as a premixed powder from Difco Laboratories.

Preparation of Medium: Add components to distilled/deionized water and bring volume to 1.0L. Mix thoroughly. Gently heat and bring to boiling. Distribute into tubes or flasks. Autoclave for 15 min at 15 psi pressure–121°C.

Use: For the cultivation of fungi.

Mycological Broth with Low pH

Composition per liter:

Glucose	40.0g
Enzymatic hydrolysate of soybean meal	10.0g

pH 4.8 ± 0.2 at 25°C

Source: This medium is available as a premixed powder from Difco Laboratories.

Preparation of Medium: Add components to distilled/deionized water and bring volume to 1.0L. Mix thoroughly. Gently heat and bring to boiling. Adjust pH to 4.8. Distribute into tubes or flasks. Autoclave for 15 min at 15 psi pressure–121°C.

Use: For the cultivation of saprophytic species of yeasts and molds. It is also suitable for cultivation of aciduric bacteria.

Mycophil™ Agar

Composition per liter:

Agar	16.0g
Papaic digest of soybean meal	10.0g
Glucose	10.0g

pH 7.0 ± 0.2 at 25°C

Source: This medium is available as a premixed powder from BBL Microbiology Systems.

Preparation of Medium: Add components to distilled/deionized water and bring volume to 1.0L. Mix thoroughly. Gently heat and bring to boiling. Distribute into tubes or flasks. Autoclave for 15 min at 15 psi pressure–121°C. Pour into sterile Petri dishes or distribute into sterile tubes.

Use: For the cultivation, maintenance, and enumeration of fungi. For the demonstration of pigment production in fungal species. Also used for the cultivation and maintenance of *Bacillus* species.

Mycophil™ Agar with Low pH

Composition per liter:

Agar	18.0g
Papaic digest of soybean meal	10.0g
Glucose	10.0g

pH 4.7 ± 0.2 at 25°C

Source: This medium is available as a premixed powder from BBL Microbiology Systems.

Preparation of Medium: Add components to distilled/deionized water and bring volume to 1.0L. Mix thoroughly. Gently heat and bring to boiling. Distribute into tubes or flasks. Autoclave for 15 min at 15 psi pressure–121°C. Adjust pH to 4.7 by adding approximately 10.0mL of sterile 10% lactic acid. Mix thoroughly. Pour into sterile Petri dishes or distribute into sterile tubes.

Use: For the isolation and enumeration of yeasts and molds.

Mycophil™ Broth

Composition per liter:

Glucose	40.0g
Pancreatic digest of casein	5.0g
Peptic digest of animal tissue	5.0g

pH 7.0 ± 0.2 at 25°C

Source: This medium is available as a premixed powder from BBL Microbiology Systems.

Preparation of Medium: Add components to distilled/deionized water and bring volume to 1.0L. Mix thoroughly. Gently heat and bring to boiling. Distribute into tubes or flasks. Autoclave for 15 min at 15 psi pressure–118°C. Do not overheat.

Use: For the isolation and cultivation of a wide variety of fungi.

Mycoplasma Agar (ATCC Medium 1435)

Composition per 930mL:

Agar, noninhibitory to mycoplasmas	10.0g
Glucose	1.0g
Nicotinamide adenine dinucleotide	0.1g
PPLO broth without Crystal Violet	680.0mL
Swine serum (56°C, 30 min)	150.0mL
Yeast extract (25% w/v solution)	100.0mL

pH 7.8 ± 0.2 at 25°C

PPLO Broth without Crystal Violet:
Composition per 680mL:

Beef heart, infusion from	11.3g
Peptone	2.28g
NaCl	1.13g

Source: PPLO broth without Crystal Violet is available as a premixed powder from Difco Laboratories.

Preparation of PPLO Broth without Crystal Violet: Add components and agar to distilled/deionized water and bring volume to 680.0mL. Mix thoroughly. Gently heat and bring to boiling. Autoclave for 15 min at 15 psi pressure–121°C. Cool to 45°–50°C.

Preparation of Medium: Mix glucose, nicotinamide adenine dinucleotide, swine serum, and fresh yeast extract solution. Mix thoroughly. Heat to 56°C. Add to cooled, sterile PPLO broth without Crystal Violet. Mix thoroughly. Pour into sterile Petri dishes or distribute into sterile tubes.

Use: For the cultivation and maintenance of *Mycoplasma anseris* and *Mycoplasma lipofaciens*.

Mycoplasma Agar (ATCC Medium 555)

Composition per 103mL:

Noble agar	0.7g
Hartley's digest broth	30.0mL
Pig serum	20.0mL
Enzymatic hydrolysate of lactalbumin	10.0mL
Hanks' balanced salt solution, 10X	4.0mL
Fresh yeast extract solution	2.0mL
Phenol Red (0.25% solution)	1.0mL

pH 7.4 ± 0.2 at 25°C

Hartley's Digest Broth:
Composition per 10L:

Ox heart	3,000.0g
Pancreatin	50.0g
Na_2CO_3, anhydrous (0.8% solution)	5L
HCl, concentrated	80.0mL

pH 7.5 ± 0.2 at 25°C

Preparation of Hartley's Digest Broth: Finely mince the ox heart. Add the meat to 5.0L of distilled/deionized water. Gently heat and bring to 80°C. Add the 5.0L of Na_2CO_3 solution. Cool to 45°C. Add pancreatin and maintain at 45°C for 4 hr while stirring. Add the HCl and steam at 100°C for 30 min. Cool to room temperature. Adjust pH to 8.0 with 1*N* NaOH. Gently heat and bring to boiling. Continue boiling for 25 min. Filter while hot. Cool to room temperature. Adjust pH to 7.5. Autoclave for 15 min at 15 psi pressure–121°C.

Pig Serum:
Composition per 100mL:

Pig serum	100.0mL

Preparation of Pig Serum: Adjust pH of pig serum to 4.4 with sterile 1*N* HCl. Do not let pH go below 4.2. Let serum stand at 4°C for 18-20 hr. Adjust pH to 7.0 with sterile 1*N* NaOH. Centrifuge at 9,000 rpm for 20 min. Discard pellet. Filter supernatant solution through a 0.2 μm membrane. Store at –70°C.

Enzymatic Hydrolysate of Lactalbumin:
Composition per 100mL:

Enzymatic hydrolysate of lactalbumin	5.0g

Preparation of Enzymatic Hydrolysate of Lactalbumin: Add enzymatic hydrolysate of lactalbumin to 100.0mL of phosphate buffered saline, 1X, pH 7.0.

Phosphate Buffered Saline Solution, 1X:
Composition per liter:

NaCl	8.0g
$Na_2HPO_4 \cdot 7H_2O$	2.16g
KCl	0.2g
KH_2PO_4	0.2g
$MgCl_2 \cdot 6H_2O$	0.1g
$CaCl_2$	0.1g

Hanks' Balanced Salt Solution, 10X:
Composition per liter:

NaCl	80.0g
Glucose	10.0g
KCl	4.0g
$CaCl_2$	1.4g
$MgCl_2 \cdot 6H_2O$	1.0g

$MgSO_4 \cdot 7H_2O$ 1.0g
$Na_2HPO_4 \cdot 7H_2O$ 0.9g
KH_2PO_4 .... 0.6g

Preparation of Hanks' Balanced Salt Solution, 10X: Add components to distilled/deionized water and bring volume to 1.0L. Mix thoroughly.

Preparation of Medium: Combine components, except agar, in the following order: Hanks' balanced salt solution, 10X, Phenol Red, Hartley's Digest Broth, pig serum, enzymatic hydrolysate of lactalbumin, and fresh yeast extract solution. Mix thoroughly. Adjust pH to 7.4 with 1*N* NaOH. Filter sterilize through a 0.2μm membrane. Add 0.7g Noble agar to 36.0mL of distilled/deionized water. Autoclave for 15 min at 15 psi pressure–121°C. Cool to 56°C. Warm basal medium to 56°C. Aseptically combine the two solutions. Pour into sterile Petri dishes or distribute into sterile tubes.

Use: For the cultivation of *Mycoplasma* species.

Mycoplasma Agar

Composition per liter:

Basal medium 700.0mL
Horse serum 200.0mL
Fresh yeast extract solution 100.0mL

pH 7.5-7.8 at 25°C

Basal Medium:

Composition per 700mL:

Sorbitol 50.0g
Beef heart (solids from infusion) 16.2g
Noble agar 13.0g
Peptone 3.26g
NaCl 1.62g
Fructose 1.0g
Glucose 1.0g
Sucrose 1.0g
Pancreatic digest of casein 1.0g

Preparation of Basal Medium: Add components to distilled/deionized water and bring volume to 700.0mL. Mix thoroughly. Adjust pH to 7.5-7.8. Autoclave for 15 min at 15 psi pressure–121°C. Cool to 50°C.

Fresh Yeast Extract Solution:

Composition per 100mL:

Live, pressed, starch-free, Baker's yeast 25.0g

Preparation of Fresh Yeast Extract Solution: Add the live Baker's yeast to 100.0mL of distilled/deionized water. Autoclave for 90 min at 15 psi pressure–121°C. Allow to stand. Remove supernatant solution. Adjust pH to 6.6–6.8.

Preparation of Medium: Filter sterilize horse serum and fresh yeast extract solution. Aseptically add to cooled, sterile basal medium. Mix thoroughly. Aseptically distribute into sterile tubes or flasks.

Use: For the cultivation and maintenance of *Mycoplasma mycoides, Spiroplasma apis, Spiroplasma citri,* and *Spiroplasma melliferum.*

Mycoplasma Agar

Composition per1004mL:

Noble agar 8.0g
Arginine 1.0g
Glucose 1.0g
L-Cysteine·HCl·H_2O 1.0g
Mycoplasma broth base 850.0mL
Horse serum (not inactivated) 100.0mL
Fresh yeast extract (25% solution) 50.0mL
Phenol Red (1.0% solution) 2.0mL
DNA calf thymus solution 2.0mL

pH 7.8 ± 0.2 at 25°C

***Mycoplasma* Broth Base:**

Composition per 850mL:

Pancreatic digest of casein 7.0g
NaCl 5.0g
Beef extract 3.0g
Yeast extract 3.0g
Beef heart solids from infusion 2.0g

Preparation of *Mycoplasma* Broth Base: Add components to distilled/deionized water and bring volume to 850.0mL. Add the 8.0g of Noble agar. Mix thoroughly. Gently heat and bring to boiling. Autoclave for 15 min at 15 psi pressure–121°C. Cool to 50°C.

Fresh Yeast Extract Solution:

Composition per 100mL:

Live, pressed, starch-free, Baker's yeast 25.0g

Preparation of Fresh Yeast Extract Solution: Add the live Baker's yeast to 100.0mL of distilled/deionized water. Autoclave for 90 min at 15 psi pressure–121°C. Allow to stand. Remove supernatant solution. Adjust pH to 6.6–6.8.

DNA Calf Thymus Solution:

Composition per 10mL:

DNA calf thymus 1.0g

Preparation of DNA Calf Thymus Solution: Add DNA calf thymus to distilled/deionized water and bring volume to 10.0mL. Mix thoroughly. Filter sterilize.

Preparation of Medium: Combine components, except *Mycoplasma* Broth Base and DNA calf thymus solution, and mix thoroughly. Filter sterilize through a 0.2µm membrane. Add sterile solution to 850.0mL of cooled, sterile *Mycoplasma* Broth Base. Aseptically add 2.0mL of sterile DNA calf thymus solution. Mix thoroughly. Pour into sterile Petri dishes or distribute into sterile tubes.

Use: For the cultivation and maintenance of *Mycoplasma lipophilum* and *Mycoplasma* species.

Mycoplasma Agar Base (PPLO Agar Base)

Composition per 1300mL:

Agar	14.0g
Pancreatic digest of casein	7.0g
NaCl	5.0g
Beef extract	3.0g
Yeast extract	3.0g
Beef heart, infusion from (solids)	2.0g
Horse serum	260.0mL
Fresh yeast extract solution	65.0mL

pH 7.8 ± 0.2 at 25°C

Source: This medium is available as a premixed powder from BBL Microbiology Systems.

Fresh Yeast Extract Solution:
Composition per 100mL:

Baker's yeast, live, pressed, starch-free	25.0g

Preparation of Fresh Yeast Extract Solution: Add the live Baker's yeast to 100mL of distilled/deionized water. Autoclave for 90 min at 15 psi pressure–121°C. Allow to stand. Remove supernatant solution. Adjust pH to 6.6–6.8.

Preparation of Medium: Add components, except horse serum and special yeast extract, to distilled/deionized water and bring volume to 1.0L. Mix thoroughly. Gently heat and bring to boiling. Distribute into tubes or flasks. Autoclave for 15 min at 15 psi pressure–121°C. Cool to 50°C. To each 75.0mL of cooled, sterile basal medium add 20.0mL of sterile horse serum and 5.0mL of special yeast extract. Mix thoroughly. Pour into sterile Petri dishes or distribute into sterile tubes.

Use: For the preparation of media for cultivation of *Mycoplasma*.

Mycoplasma Agar with Increased Selectivity

Composition per 1300mL:

Agar	14.0g
Pancreatic digest of casein	7.0g
NaCl	5.0g
Beef extract	3.0g
Yeast extract	3.0g
Beef heart, solids from infusion	2.0g
Thallous acetate	0.7g
Penicillin	70,000U
Horse serum	260.0mL
Fresh yeast extract solution	65.0mL

pH 7.8 ± 0.2 at 25°C

Caution: Thallous acetate is a poison.

Fresh Yeast Extract Solution:
Composition per 100mL:

Baker's yeast, live, pressed, starch-free	25.0g

Preparation of Fresh Yeast Extract Solution: Add the live Baker's yeast to 100.0mL of distilled/deionized water. Autoclave for 90 min at 15 psi pressure–121°C. Allow to stand. Remove supernatant solution. Adjust pH to 6.6–6.8.

Preparation of Medium: Add components—except horse serum, special yeast extract, thallous acetate and penicillin—to distilled/deionized water and bring volume to 1.0L. Mix thoroughly. Gently heat and bring to boiling. Distribute into tubes or flasks. Autoclave for 15 min at 15 psi pressure–121°C. Cool to 50°C. To each of 70.0mL of cooled, sterile basal medium add 20.0mL of sterile horse serum, 10.0mL of special yeast extract, 50.0mg of thallous acetate and 5000U of penicillin. Mix thoroughly. Pour into sterile Petri dishes or distribute into sterile tubes.

Use: For the preparation of media for cultivation of *Mycoplasma* species.

Mycoplasma Agar with Supplement G

Composition per liter:

Agar	10.0g
Bacteriological peptone	10.0g
Beef extract	10.0g
NaCl	5.0g
Special mineral supplement, Oxoid Unipath	0.5g
Mycoplasma supplement G	250.0mL

pH 7.8 ± 0.2 at 25°C

Source: This medium is available as a premixed powder from Oxoid Unipath.

***Mycoplasma* Supplement G:**
Composition per 20mL:

Thallous acetate	25.0mg
Horse serum	20.0mL
Yeast extract (25% solution)	10.0mL
Penicillin	20,000U

Preparation of *Mycoplasma* Supplement G: Add components to distilled/deionized water and bring volume to 20.0mL. Mix thoroughly. Filter sterilize.

Caution: Thallous acetate is a poison.

Preparation of Medium: Add components, except *Mycoplasma* Supplement G, to distilled/deionized water and bring volume to 1.0L. Mix thoroughly. Gently heat and bring to boiling. Distribute into flasks in 80.0mL volumes. Autoclave for 15 min at 15 psi pressure–121°C. Cool to 50°C. Aseptically add 20.0mL of sterile *Mycoplasma* Supplement G to each 80.0mL of basal medium. Mix thoroughly.

Use: For the growth of *Mycoplasma* species.

Mycoplasma Agar with Supplement P

Composition per liter:

Agar	10.0g
Bacteriological peptone	10.0g
Beef extract	10.0g
NaCl	5.0g
Special mineral supplement, Oxoid Unipath	0.5g
Mycoplasma supplement P	250.0mL

pH 7.8 ± 0.2 at 25°C

Source: This medium is available as a premixed powder from Oxoid Unipath.

***Mycoplasma* Supplement P:**
Composition per 20mL:

Glucose	0.3g
Mycoplasma broth base	0.145g
Thallous acetate	8.0mg
Phenol Red	1.2mg
Methylene Blue chloride	0.3mg
Penicillin	12,000U
Horse serum	6.0mL
Yeast extract (25% solution)	3.0mL

Preparation of *Mycoplasma* Supplement P: Add components to distilled/deionized water and bring volume to 20.0mL. Mix thoroughly. Filter sterilize.

Caution: Thallous acetate is a poison.

Fresh Yeast Extract Solution:
Composition per 100mL:

Baker's yeast, live, pressed, starch-free	25.0g

Preparation of Fresh Yeast Extract Solution: Add the live Baker's yeast to 100.0mL of distilled/deionized water. Autoclave for 90 min at 15 psi pressure–121°C. Allow to stand. Remove supernatant solution. Adjust pH to 6.6–6.8.

Preparation of Medium: Add components, except *Mycoplasma* Supplement P, to distilled/deionized water and bring volume to 1.0L. Mix thoroughly. Gently heat and bring to boiling. Distribute into bottles in 1.0mL volumes. Autoclave for 15 min at 15 psi pressure–121°C. Cool to room temperature. Aseptically add 2.0mL of sterile *Mycoplasma* Supplement P to each bottle.

Use: For the growth of *Mycoplasma* species.

Mycoplasma Broth

Composition per liter:

Basal medium	700.0mL
Horse serum	200.0mL
Fresh yeast extract solution	100.0mL

pH 7.5–7.8 at 25°C

Basal Medium:
Composition per 700mL:

Sorbitol	50.0g
Beef heart, solids from infusion	16.2g
Peptone	3.26g
NaCl	1.62g
Fructose	1.0g
Glucose	1.0g
Sucrose	1.0g
Pancreatic digest of casein	1.0g

Preparation of Basal Medium: Add components to distilled/deionized water and bring volume to 700.0mL. Mix thoroughly. Adjust pH to 7.5–7.8. Autoclave for 15 min at 15 psi pressure–121°C. Cool to 50°C.

Fresh Yeast Extract Solution:
Composition per 100mL:

Baker's yeast, live, pressed, starch-free	25.0g

Preparation of Fresh Yeast Extract Solution: Add the live Baker's yeast to 100.0mL of distilled/deionized water. Autoclave for 90 min at 15 psi pressure–121°C. Allow to stand. Remove supernatant solution. Adjust pH to 6.6–6.8.

Preparation of Medium: Filter sterilize horse serum and fresh yeast extract solution. Aseptically add to cooled, sterile basal medium. Mix thoroughly. Aseptically distribute into sterile tubes or flasks.

Use: For the cultivation and maintenance of *Mycoplasma mycoides, Spiroplasma apis, Spiroplasma citri,* and *Spiroplasma melliferum.*

Mycoplasma Broth

Composition per 950mL:

Glucose	1.0g
Nicotinamide adenine dinucleotide	0.1g
PPLO broth without Crystal Violet	680.0mL
Swine serum (56°C, 30 min)	150.0mL
Fresh yeast extract solution	100.0mL
Phenol Red (0.1% w/v solution)	20.0mL

pH 7.8 ± 0.2 at 25°C

PPLO Broth without Crystal Violet:
Composition per 680mL:

Beef heart, solids from infusion	11.3g
Peptone	2.28g
NaCl	1.13g

Source: PPLO broth without crystal violet is available as a premixed powder from Difco Laboratories.

Preparation of PPLO Broth without Crystal Violet: Add components to distilled/deionized water and bring volume to 900.0mL. Autoclave for 15 min at 15 psi pressure–121°C. Cool to 56°C.

Fresh Yeast Extract Solution:
Composition per 100mL:

Baker's yeast, live, pressed, starch-free	25.0g

Preparation of Fresh Yeast Extract Solution: Add the live Baker's yeast to 100.0mL of distilled/deionized water. Autoclave for 90 min at 15 psi pressure–121°C. Allow to stand. Remove supernatant solution. Adjust pH to 6.6–6.8.

Preparation of Medium: Mix glucose, nicotinamide adenine dinucleotide, swine serum, fresh yeast extract solution, and Phenol Red. Mix thoroughly. Heat to 56°C. Add to cooled, sterile PPLO broth without Crystal Violet. Mix thoroughly. Aseptically distribute into sterile tubes or flasks.

Use: For the cultivation and maintenance of *Mycoplasma anseris* and *Mycoplasma lipofaciens*.

Mycoplasma Broth (ATCC Medium 555)

Composition per 103mL:

Hartley's digest broth	30.0mL
Pig serum	20.0mL
Enzymatic hydrolysate of lactalbumin	10.0mL
Hanks' balanced salt solution, 10X	4.0mL
Fresh yeast extract solution	2.0mL
Phenol Red (0.25% solution)	1.0mL

pH 7.4 ± 0.2 at 25°C

Hartley's Digest Broth:
Composition per 10L:

Ox heart	3,000.0g
Pancreatin	50.0g
Na_2CO_3, anhydrous (0.8% solution)	5L
HCl, concentrated	80.0mL

pH 7.5 ± 0.2 at 25°C

Preparation of Hartley's Digest Broth: Finely mince the ox heart. Add the meat to 5.0L of distilled/deionized water. Gently heat and bring to 80°C. Add the 5.0L of Na_2CO_3 solution. Cool to 45°C. Add pancreatin and maintain at 45°C for 4 hr while stirring. Add the HCl and steam at 100°C for 30 min. Cool to room temperature. Adjust pH to 8.0 with 1*N* NaOH. Gently heat and bring to boiling. Continue boiling for 25 min. Filter while hot. Cool to room temperature. Adjust pH to 7.5. Autoclave for 15 min at 15 psi pressure–121°C.

Pig Serum:
Composition per 100mL:

Pig serum	100.0mL

Preparation of Pig Serum: Adjust pH of pig serum to 4.4 with sterile 1*N* HCl. Do not let pH go below 4.2. Let serum stand at 4°C for 18-20 hr. Adjust pH to 7.0 with sterile 1*N* NaOH. Centrifuge at 9,000 rpm for 20 min. Discard pellet. Filter supernatant solution through a 0.2μm membrane. Store at −70°C.

Enzymatic Hydrolysate of Lactalbumin:
Composition per 100mL:

Enzymatic hydrolysate of lactalbumin	5.0g

Preparation of Enzymatic Hydrolysate of Lactalbumin: Add enzymatic hydrolysate of lactalbumin to 100.0mL of phosphate buffered saline, 1X, pH 7.0.

Phosphate Buffered Saline Solution, 1X:
Composition per liter:

NaCl	8.0g
$Na_2HPO_4 \cdot 7H_2O$	2.16g
KCl	0.2g
KH_2PO_4	0.2g
$MgCl_2 \cdot 6H_2O$	0.1g
$CaCl_2$	0.1g

Hanks' Balanced Salt Solution, 10X:
Composition per liter:

NaCl	80.0g
Glucose	10.0g
KCl	4.0g
$CaCl_2$	1.4g
$MgCl_2 \cdot 6H_2O$	1.0g
$MgSO_4 \cdot 7H_2O$	1.0g
$Na_2HPO_4 \cdot 7H_2O$	0.9g
KH_2PO_4	0.6g

Preparation of Hanks' Balanced Salt Solution, 10X: Add components to distilled/deionized water and bring volume to 1.0L. Mix thoroughly.

Preparation of Medium: Combine components in to following order: Hanks' balanced salt solution, 10X, Phenol Red, Hartley's Digest Broth, pig serum, enzymatic hydrolysate of lactalbumin, and fresh yeast extract solution. Mix thoroughly. Add 36.0mL of distilled/deionized water. Adjust pH to 7.4 with 1*N* NaOH. Filter sterilize through a 0.2μm membrane. Store at 4°C for up to 3 weeks.

Use: For the cultivation of *Mycoplasma* species.

Mycoplasma Broth Base (PPLO Broth Base without Crystal Violet)

Composition per liter:

Pancreatic digest of casein	7.0g
NaCl	5.0g
Beef extract	3.0g
Yeast extract	3.0g
Beef heart, solids from infusion	2.0g

pH 7.8 ± 0.2 at 25°C

Source: This medium is available as a premixed powder from BBL Microbiology Systems.

Preparation of Medium: Add components to distilled/deionized water and bring volume to 1.0L. Gently heat and bring to boiling. Mix thoroughly. Distribute into tubes or flasks. Autoclave for 15 min at 15 psi pressure–121°C.

Use: Use as a basal medium that should be enriched for the isolation and cultivation of *Mycoplasma* species.

Mycoplasma Broth Base, Frey

Composition per 1100mL:

Pancreatic digest	7.5g
Yeast extract	5.0g
NaCl	5.0g
Papaic digest of soybean meal	2.5g
Na_2HPO_4	1.6g
KCl	0.4g
$MgSO_4 \cdot 7H_2O$	0.2g
KH_2PO_4	0.1g
Horse serum	100.0mL

pH 7.7 ± 0.2 at 25°C

Source: This medium is available as a premixed powder from BBL Microbiology Systems.

Preparation of Medium: Add components, except horse serum, to distilled/deionized water and bring volume to 1.0L. Mix thoroughly. Gently heat and bring to boiling. Autoclave for 15 min at 15 psi pressure–121°C. Cool to 50°C. Add sterile, heat inactivated horse serum. Mix thoroughly. Aseptically distribute into sterile tubes.

Use: For the cultivation of avian mycoplasmas.

Mycoplasma Broth, Supplemented

Composition per liter:

Pancreatic digest of casein	7.0g
NaCl	5.0g
Beef extract	3.0g
Yeast extract	3.0g
Beef heart, solids from infusion	2.0g
Horse serum	260.0mL
Fresh yeast extract solution	65.0mL

pH 7.8 ± 0.2 at 25°C

Fresh Yeast Extract Solution:
Composition per 100mL:

Baker's yeast, live, pressed, starch-free	25.0g

Preparation of Fresh Yeast Extract Solution: Add the live Baker's yeast to 100.0mL of distilled/deionized water. Autoclave for 90 min at 15 psi pressure–121°C. Allow to stand. Remove supernatant solution. Adjust pH to 6.6–6.8.

Preparation of Medium: Add components, except horse serum and special yeast extract, to distilled/deionized water and bring volume to 1.0L. Mix thoroughly. Gently heat and bring to boiling. Distribute into tubes or flasks. Autoclave for 15 min at 15 psi pressure–121°C. Cool to 50°C. To each of 75.0mL of cooled, sterile basal medium add 20.0mL of sterile horse serum and 5.0mL of special yeast extract. Mix thoroughly. Aseptically distribute into sterile tubes.

Use: For the isolation and cultivation of *Mycoplasma* species.

Mycoplasma Broth with 10% Swine Serum

Composition per liter:

Pancreatic digest of casein	5.6g
NaCl	4.0g
Beef extract	2.4g
Yeast extract	2.6g
Beef heart, solids from infusion	1.6g
Swine serum (heat-inactivated)	100.0mL
Fresh yeast extract solution	100.0mL

pH 7.8 ± 0.2 at 25°C

Fresh Yeast Extract Solution:
Composition per 100mL:

Baker's yeast, live, pressed, starch-free	25.0g

Preparation of Fresh Yeast Extract Solution: Add the live Baker's yeast to 100.0mL of distilled/deionized water. Autoclave for 90 min at 15 psi pressure–121°C. Allow to stand. Remove supernatant solution. Adjust pH to 6.6–6.8.

Preparation of Medium: Add components, except swine serum and fresh yeast extract, to distilled/deionized water and bring volume to 800.0mL. Mix thoroughly. Gently heat and bring to boiling. Autoclave for 15 min at 15 psi pressure–121°C. Cool to 50°C. Add sterile swine serum and fresh yeast extract. Mix thoroughly. Aseptically distribute into sterile tubes.

Use: For the cultivation and maintenance of *Mycoplasma columbinum* and *Mycoplasma columborale*.

Mycoplasma Broth with Supplement G

Composition per liter:

Bacteriological peptone	10.0g
Beef extract	10.0g
NaCl	5.0g
Special mineral supplement, Oxoid Unipath	0.5g
Mycoplasma Supplement G	250.0mL

pH 7.8 ± 0.2 at 25°C

Source: This medium is available as a premixed powder from Oxoid Unipath.

***Mycoplasma* Supplement G:**
Composition per 20mL:

Thallous acetate	25.0mg
Horse serum	20.0mL
Yeast extract (25% solution)	10.0mL
Penicillin	20,000U

Preparation of *Mycoplasma* Supplement G: Add components to distilled/deionized water and bring volume to 20.0mL. Mix thoroughly. Filter sterilize.

Caution: Thallous acetate is a poison.

Preparation of Medium: Add components, except *Mycoplasma* Supplement G, to distilled/deionized water and bring volume to 1.0L. Mix thoroughly. Gently heat and bring to boiling. Distribute into flasks in 80.0mL volumes. Autoclave for 15 min at 15 psi pressure–121°C. Cool to 50°C. Aseptically add 20.0mL of sterile *Mycoplasma* Supplement G to each 80.0mL of basal medium. Mix thoroughly.

Use: For the growth of *Mycoplasma* species.

Mycoplasma Broth with Supplement P

Composition per liter:

Bacteriological peptone	10.0g
Beef extract	10.0g
NaCl	5.0g
Special mineral supplement, Oxoid Unipath	0.5g
Mycoplasma supplement P	250.0mL

pH 7.8 ± 0.2 at 25°C

Source: This medium is available as a premixed powder from Oxoid Unipath.

***Mycoplasma* Supplement P:**
Composition per 20mL:

Glucose	0.3g
Mycoplasma broth base	0.145g
Thallous acetate	8.0mg
Phenol Red	1.2mg
Methylene Blue chloride	0.3mg
Penicillin	12,000U
Horse serum	6.0mL
Yeast extract (25% solution)	3.0mL

Preparation of *Mycoplasma* Supplement P: Add components to distilled/deionized water and bring volume to 20.0mL. Mix thoroughly. Filter sterilize.

Caution: Thallous acetate is a poison.

Preparation of Medium: Add components, except *Mycoplasma* Supplement P, to distilled/deionized water and bring volume to 1.0L. Mix thoroughly. Gently heat and bring to boiling. Distribute into bottles in 1.0mL volumes. Autoclave for 15 min at 15 psi pressure–121°C. Cool to room temperature. Aseptically add 2.0mL of sterile *Mycoplasma* Supplement P to each bottle.

Use: For the cultivation of *Mycoplasma* species.

Mycoplasma Liquid Medium

Composition per1004mL:

Arginine	1.0g
Glucose	1.0g
L-Cysteine·HCl·H_2O	1.0g
Mycoplasma broth base	850.0mL
Horse serum (not inactivated)	100.0mL
Fresh yeast extract (25% solution)	50.0mL
Phenol Red (1.0% solution)	2.0mL
DNA calf thymus solution	2.0mL

pH 7.8 ± 0.2 at 25°C

***Mycoplasma* Broth Base:**
Composition per 850mL:

Pancreatic digest of casein	7.0g
NaCl	5.0g
Beef extract	3.0g
Yeast extract	3.0g
Beef heart, solids from infusion	2.0g

Source: *Mycoplasma* Broth Base is available as a premixed powder from BBL Microbiology Systems.

Preparation of *Mycoplasma* Broth Base: Add components to distilled/deionized water and bring volume to 850.0mL. Mix thoroughly. Gently heat and bring to boiling. Autoclave for 15 min at 15 psi pressure–121°C. Cool to 50°C.

Fresh Yeast Extract Solution:
Composition per 100mL:

Baker's yeast, live, pressed, starch-free	25.0g

Preparation of Fresh Yeast Extract Solution: Add the live Baker's yeast to 100.0mL of distilled/

deionized water. Autoclave for 90 min at 15 psi pressure–121°C. Allow to stand. Remove supernatant solution. Adjust pH to 6.6–6.8.

DNA Calf Thymus Solution:
Composition per 10mL:

DNA calf thymus	1.0g

Preparation of DNA Calf Thymus Solution: Add DNA calf thymus to distilled/deionized water and bring volume to 10.0mL. Mix thoroughly. Filter sterilize.

Preparation of Medium: Combine components, except *Mycoplasma* Broth Base and DNA calf thymus solution, and mix thoroughly. Filter sterilize through a 0.2μm membrane. Add sterile solution to 850.0mL of cooled, sterile *Mycoplasma* Broth Base. Aseptically add 2.0mL of sterile DNA calf thymus solution. Mix thoroughly. Aseptically distribute into sterile tubes or flasks.

Use: For the cultivation and maintenance of *Mycoplasma lipophilum* and *Mycoplasma* species.

Mycoplasma Medium

Composition per liter:

Heart infusion broth	25.0g
Mucin, bacteriological grade	5.0g
Hemoglobin	2.0g

pH 7.8 ± 0.2 at 25°C

Preparation of Medium: Adjust pH to 7.8. Heat mixture at 93°–95°C for 30 minutes in a water bath. Restore to original volume. Add 0.5% diatomaceous earth. Mix thoroughly. Filter through Whatman GFA (glass fiber paper) in Buchner filter. Clarify using 0.45μm Millipore filter. Add 15% inactivated turkey serum. Sterilize using S3 (0.1 microns) Seitz filter. Use positive pressure. For solid medium: Prepare 42.5 mL double strength broth and 42.5 mL distilled water containing 0.7 g purified agar. Sterilize the solutions separately and combine aseptically at 56°C with 15.0mL sterile inactivated turkey serum for a final volume of 100.0mL.

Use: For the cultivation and maintenance of *Mycoplasma hyosynoviae*.

Mycoplasma Medium, Revised

Composition per 1030mL:

Noble agar	10.0g
Distilled water	360.0mL
Heart infusion broth	300.0mL
Pig serum (heat-inactivated)	200.0mL
Enzymatic hydrolysate of lactalbumin	100.0mL
Hanks' balanced salt solution, 10X	40.0mL
Fresh yeast extract solution	20.0mL
Phenol Red (0.25% solution)	10.0mL

Heart Infusion Broth:
Composition per liter:

Beef heart, infusion from	500.0g
Tryptose	10.0g
NaCl	5.0g

Preparation of Heart Infusion Broth: Add components to distilled/deionized water and bring volume to 1.0L. Mix thoroughly. Gently heat and bring to boiling.

Enzymatic Hydrolysate of Lactalbumin:
Composition per 100mL:

Enzymatic hydrolysate of lactalbumin	5.0g

Preparation of Enzymatic Hydrolysate of Lactalbumin: Add enzymatic hydrolysate of lactalbumin to 100.0mL of phosphate buffered saline, 1X, pH 7.0.

Phosphate Buffered Saline Solution, 1X:
Composition per liter:

NaCl	8.0g
$Na_2HPO_4 \cdot 7H_2O$	2.16
KCl	0.2g
KH_2PO_4	0.2g
$MgCl_2 \cdot 6H_2O$	0.1g
$CaCl_2$	0.1g

Preparation of Phosphate Buffered Saline Solution, 1X: Add components to distilled/deionized water and bring volume to 1.0L. Mix thoroughly.

Hanks' Balanced Salt Solution, 10X:
Composition per liter:

Na_2Cl	80.0g
Glucose	10.0g
KCl	4.0g
$CaCl_2$	1.4g
$MgCl_2 \cdot 6H_2O$	1.0g
$MgSO_4 \cdot 7H_2O$	1.0g
$Na_2HPO_4 \cdot 7H_2O$	0.9g
KH_2PO_4	0.6g

Preparation of Hanks' Balanced Salt Solution, 10X: Add components to distilled/deionized water and bring volume to 1.0L. Mix thoroughly.

Fresh Yeast Extract Solution:
Composition per 100mL:

Baker's yeast, live, pressed, starch-free	25.0g

Preparation of Fresh Yeast Extract Solution: Add the live Baker's yeast to 100.0mL of distilled/deionized water. Autoclave for 90 min at 15 psi pressure–121°C. Allow to stand. Remove supernatant solution. Adjust pH to 6.6–6.8.

Preparation of Medium: Add agar and heart infusion broth to distilled/deionized water and bring volume to 660.0mL. Mix thoroughly. Adjust pH to 7.4. Autoclave for 15 min at 15 psi pressure–121°C. Aseptically add pig serum, enzymatic hydrolysate of lactalbumin, Hanks' balanced salt solution, 10X, fresh yeast extract solution, and Phenol Red solution. Mix thoroughly. Aseptically distribute into sterile tubes.

Use: For the cultivation and maintenance of *Mycoplasma dispar*, *Mycoplasma flocculare* and *Mycoplasma hyopneumoniae*.

Mycoplasma pneumoniae Isolation Medium

Composition per 1200mL:

Beef heart for infusion, Difco	50.0g
Peptone	10.0g
NaCl	5.0g
Water	900.0mL
Yeast extract solution	100.0mL
Agamma horse serum, unheated	200.0mL

pH 7.6–7.8 at 25°C

Yeast Extract Solution:
Composition per 10mL:

Yeast, active dry Baker's	250.0g

Preparation of Yeast Extract Solution: Add yeast to 1.0L of distilled/deionized water. Mix thoroughly. Gently heat and bring to boiling. Filter through Whatman #2 filter paper. Adjust the pH of the filtrate to 8.0 with NaOH. Distribute into tubes in 10.0mL volumes. Autoclave for 15 min at 15 psi pressure–121°C. Store at –20°C.

Preparation of Medium: Add components, except yeast extract solution and agamma horse serum, to distilled/deionized water and bring volume to 990.0mL. Mix thoroughly. Gently heat and bring to boiling. Autoclave for 15 min at 15 psi pressure–121°C. Cool to 45°–50°C. Aseptically add sterile yeast extract solution and agamma horse serum. Mix thoroughly. Aseptically distribute into sterile tubes.

Use: For the isolation and cultivation of *Mycoplasma pneumoniae*.

Mycoplasmal Agar

Composition per liter:

Papaic digest of soy meal	20.0g
Agarose	10.0g
NaCl	5.0g
Phenol Red (2% solution)	1.0mL

pH 7.3 ± 0.2 at 25°C

Preparation of Medium: Add components, except agarose, to distilled/deionized water and bring volume to 1.0L. Mix thoroughly. Adjust pH to 7.3 with 1*N* NaOH. Add agarose. Mix thoroughly. Gently heat and bring to boiling. Distribute into tubes or flasks. Autoclave for 15 min at 15 psi pressure–121°C. Pour into sterile Petri dishes or leave in tubes.

Use: For isolation and cultivation of human mycoplasmas and ureaplasmas.

Mycosel™ Agar (Cycloheximide Chloramphenicol Agar)

Composition per liter:

Agar	15.5g
Papaic digest of soybean meal	10.0g
Glucose	10.0g
Cycloheximide	0.4g
Chloramphenicol	0.05g

pH 6.9 ± 0.2 at 25°C

Source: This medium is available as a premixed powder from BBL Microbiology Systems.

Preparation of Medium: Add components to distilled/deionized water and bring volume to 1.0L. Mix thoroughly. Gently heat while stirring and bring to boiling. Autoclave for 15 min at 14 psi pressure–118°C. Avoid overheating. Pour into sterile Petri dishes or distribute into sterile tubes.

Use: For the selective isolation of pathogenic fungi from specimens having a large flora of other fungi and bacteria.

MYP Agar
See: **Mannitol Yolk Polymyxin Agar**

Mysorens Medium

Composition per liter:

Peptone	10.0g
Meat extract	10.0g
Yeast extract	5.0g
NaCl	3.0g

Preparation of Medium: Add components to distilled/deionized water and bring volume to 1.0L. Mix thoroughly. Distribute into tubes or flasks. Autoclave for 15 min at 15 psi pressure–121°C.

Use: For the cultivation and maintenance of *Arthrobacter mysorens*.

Myxobacteria Medium

Composition per liter:

Agar....................15.0g
Skim milk powder....................5.0g
Yeast extract....................0.5g

Preparation of Medium: Add components to distilled/deionized water and bring volume to 1.0L. Mix thoroughly. Do not adjust pH. Gently heat and bring to boiling. Distribute into tubes or flasks. Autoclave for 15 min at 15 psi pressure–121°C.

Use: For the cultivation and maintenance of *Archangium primigenium, Chondrococcus macrosporus* and *Myxococcus coralloides*.

Myxococcus Medium

Composition per liter:

Agar....................12.0g
Pancreatic digest of casein....................1.0g
Meat extract....................1.0g
Glucose solution....................50.0mL

pH 7.2 ± 0.2 at 25°C

Glucose Solution:
Composition per 50mL:

Glucose....................1.0g

Preparation of Glucose Solution: Add glucose to distilled/deionized water and bring volume to 50.0mL. Mix thoroughly. Autoclave for 15 min at 15 psi pressure–121°C. Cool to 25°C.

Preparation of Medium: Add components, except glucose solution, to distilled/deionized water and bring volume to 950.0mL. Mix thoroughly. Adjust pH to 7.2. Gently heat and bring to boiling. Autoclave for 15 min at 15 psi pressure–121°C. Cool to 45°–50°C. Aseptically add sterile glucose solution. Mix thoroughly. Pour into sterile Petri dishes or distribute into sterile tubes or bottles. Allow tubes or bottles to cool in a slanted position.

Use: For the cultivation of *Myxococcus* species.

Myxococcus xanthus Medium

Agar....................20.0g
Pancreatic digest of casein....................10.0g
$MgSO_4 \cdot 7H_2O$....................0.5g
K_2HPO_4....................0.148g
KH_2PO_4....................0.017g

pH 7.6 ± 0.2 at 25°C

Preparation of Medium: Add components to distilled/deionized water and bring volume to 1.0L. Mix thoroughly. Gently heat and bring to boiling. Distribute into tubes or flasks. Autoclave for 15 min at 15 psi pressure–121°C. Pour into sterile Petri dishes or leave in tubes.

Use: For the cultivation and maintenance of *Myxococcus xanthus*.

NAMn

See: **Nutrient Agar with Manganese**

NANAT Agar (Nalidixic Acid Novobiocin Actidione Tellurite Agar)

Composition per liter:

Pancreatic digest of casein	17.0g
Agar	15.0g
NaCl	5.0g
Tween™ 80	5.0g
Papaic digest of soybean meal	3.0g
K_2HPO_4	2.5g
Glucose	2.5g
Yeast extract	1.0g
Tellurite solution	10.0mL
Antibiotic solution	10.0mL

pH 7.2 ± 0.2 at 25°C

Tellurite Solution:

Composition per 100mL:

K_2TeO_3	0.05g

Preparation of Tellurite Solution: Add K_2TeO_3 to distilled/deionized water and bring volume to 100.0mL. Mix thoroughly. Filter sterilize.

Antibiotic Solution:

Composition per 10mL:

Actidione (cycloheximide)	0.04g
Novobiocin	0.025g
Nalidixic acid	0.020g
Polymyxin B (optional)	0.030g

Preparation of Antibiotic Solution: Add components to distilled/deionized water and bring volume to 10.0mL. Mix thoroughly. Filter sterilize.

Caution: Potassium tellurite is toxic.

Preparation of Medium: Add components, except tellurite solution and antibiotic solution, to distilled/deionized water and bring volume to 980.0mL. Mix thoroughly. Gently heat and bring to boiling. Autoclave for 15 min at 15 psi pressure–121°C. Cool to 45°–50°C. Aseptically add sterile tellurite solution and antibiotic solution. Mix thoroughly. Pour into sterile Petri dishes or distribute into sterile tubes.

Use: For the isolation and cultivation of *Rhodococcus (Corynebacterium) equi* from animal feces, especially from horses and swine. The addition of polymyxin B inhibits the growth of *Pseudomonas aeruginosa* which may interfere with the isolation of *R. equi.*

Nannocystis Agar

Composition per liter:

Agar	15.0g
$CaCl_2 \cdot 2H_2O$	1.0g

pH 7.2 ± 0.2 at 25°C

Preparation of Medium: Add components to distilled/deionized water and bring volume to 1.0L. Mix thoroughly. Gently heat and bring to boiling. Autoclave for 15 min at 15 psi pressure–121°C. Pour into sterile Petri dishes. After the agar has solidified, overlay the suface with 0.5mL of a suspension of dead (autoclaved) *Escherichia coli* cells.

Use: For the cultivation and maintenance of *Nannocystis* species.

Naphthalene Mineral Salts Medium

See: **Medium for Hydrocarbon–Degrading Bacteria**

Natronobacteria Medium

Composition per liter:

NaCl	200.0g
Agar	20.0g
Yeast extract	5.0g
Casamino acids	5.0g
KH_2PO_4	1.0g
KCl	1.0g
NH_4Cl	1.0g
Sodium glutamate	1.0g
$MgSO_4 \cdot 7H_2O$	0.24g
$CaSO_4 \cdot 2H_2O$	0.17g
Na_2CO_3 solution	100.0mL
Trace elements solution SL-6	1.0mL

pH 9.0 ± 0.2 at 25°C

Na_2CO_3 Solution:

Composition per 100mL:

Na_2CO_3	5.0g

Preparation of Na_2CO_3 Solution: Add Na_2CO_3 to distilled/deionized water and bring volume to 100.0mL. Mix thoroughly. Autoclave for 15 min at 15 psi pressure–121°C. Cool to 50°C.

Trace Elements Solution SL-6:

Composition per liter:

H_3BO_3	0.3g
$CoCl_2 \cdot 6H_2O$	0.2g
$ZnSO_4 \cdot 7H_2O$	0.10g
$MnCl_2 \cdot 4H_2O$	0.03g
$Na_2MoO_4 \cdot H_2O$	0.03g
$NiCl_2 \cdot 6H_2O$	0.02g
$CuCl_2 \cdot 2H_2O$	0.01g

Preparation of Trace Elements Solution SL-6: Add components to distilled/deionized water and bring volume to 1.0L. Mix thoroughly. Adjust pH to 3.4.

Preparation of Medium: Add components, except Na_2CO_3 solution, to distilled/deionized water and bring volume to 900.0mL. Mix thoroughly. Gently heat and bring to boiling. Autoclave for 15 min at 15 psi pressure–121°C. Cool to 45°–50°C. Aseptically add sterile Na_2CO_3 solution. Mix thoroughly. Adjust pH to 9.0, if necessary. Pour into sterile Petri dishes or distribute into sterile tubes.

Use: For the cultivation and maintenance of *Natronobacterium gregoryi*, *Natronobacterium magadii*, *Natronobacterium pharaonis* and *Natronococcus occultus*.

NBA Medium

Composition per liter:

Pancreatic digest of gelatin	5.0g
Casamino acids	5.0g
Beef extract	3.0g
Yeast extract	1.0g

pH 6.8 ± 0.2 at 25°C

Preparation of Medium: Add components to distilled/deionized water and bring volume to 1.0L. Mix thoroughly. Distribute into tubes or flasks. Autoclave for 15 min at 15 psi pressure–121°C.

Use: For the cultivation of *Bdellovibrio* species.

NBY Medium (Nutrient Broth Yeast Extract Medium) (ATCC Medium 763)

Composition per 940mL:

Agar	15.0g
Nutrient broth	8.0g
Yeast extract	2.0g
K_2HPO_4	2.0g
KH_2PO_4	0.5g
Glucose solution	50.0mL
$MgSO_4 \cdot 7H_2O$ solution	50.0mL

Glucose Solution:
Composition per 50mL:

D-Glucose	5.0g

Preparation of Glucose Solution: Add glucose to distilled/deionized water and bring volume to 50.0mL. Mix thoroughly. Filter sterilize.

$MgSO_4$ Solution:
Composition per 50mL:

$MgSO_4 \cdot 7H_2O$	0.25g

Preparation of $MgSO_4$ Solution: Add the solid $MgSO_4 \cdot 7H_2O$ to distilled/deionized water and bring volume to 50.0mL. Mix thoroughly. Autoclave for 15 min at 15 psi pressure–121°C. Cool to 50°C.

Preparation of Medium: Add components, except glucose solution and $MgSO_4 \cdot 7H_2O$ solution, to distilled/deionized water and bring volume to 900.0mL. Mix thoroughly. Gently heat and bring to boiling. Autoclave for 25 min at 15 psi pressure–121°C. Cool to 45°–50°C. Aseptically add sterile glucose solution and $MgSO_4 \cdot 7H_2O$ solution. Mix thoroughly. Pour into sterile Petri dishes or distribute into sterile tubes.

Use: For the cultivation and maintenance of *Bacillus sphaericus*.

NBY Medium (Nutrient Broth Yeast Extract Medium)

Composition per 950mL:

Nutrient broth, dehydrated	8.0g
Yeast extract	2.0g
K_2HPO_4	2.0g
KH_2PO_4	0.5g
Glucose solution	50.0mL
$MgSO_4 \cdot 7H_2O$ (1M solution)	1.0mL

Glucose Solution:
Composition per 50mL:

D-Glucose	5.0g

Preparation of Glucose Solution: Add glucose to distilled/deionized water and bring volume to 50.0mL. Mix thoroughly. Filter sterilize.

Preparation of Medium: Add components, except glucose solution, to distilled/deionized water and bring volume to 950.0mL. Mix thoroughly. Gently heat and bring to boiling. Autoclave for 15 min at 15 psi pressure–121°C. Cool to 45°–50°C. Aseptically add sterile glucose solution. Mix thoroughly. Pour into sterile Petri dishes or distribute into sterile tubes.

Use: For the cultivation and maintenance of *Curtobacterium flaccumfaciens* and *Pseudomonas syringae*.

Neisseria Medium

Composition per liter:

Biosate	10.0g
Polypeptone™	10.0g
NaCl	5.0g
Myosate	3.0g
Agar	1.5g

Phenol Red..0.017g
Carbohydrate solution.................................50.0mL

pH 7.4–7.6 at 25°C

Carbohydrate Solution:
Composition per 50mL:

Carbohydrate..10.0g

Preparation of Carbohydrate Solution: Add glucose, sucrose or maltose to distilled/deionized water and bring volume to 50.0mL. Mix thoroughly. Filter sterilize.

Preparation of Medium: Add components, except carbohydrate solution, to distilled/deionized water and bring volume to 950.0mL. Mix thoroughly. Gently heat and bring to boiling. Autoclave for 15 min at 15 psi pressure–121°C. Cool to 45°–50°C. Aseptically add sterile carbohydrate solution. Mix thoroughly. Pour into sterile Petri dishes or distribute into sterile tubes.

Use: For the cultivation of *Neisseria* species.

Neisseria meningitidis Medium

Composition per liter:

Beef infusion...300.0g
Acid hydrolysate of casein...............................17.5g
Agar...17.0g
Starch 1.5g
Antibiotic solution 10.0mL

pH 7.4 ± 0.2 at 25°C

Antibiotic Solution:
Composition per 10mL:

Vancomycin... 3.0mg
Colistin.. 7.5mg
Nystatin..12,500U

Preparation of Antibiotic Solution: Add components to distilled/deionized water and bring volume to 10.0mL. Mix thoroughly. Filter sterilize.

Preparation of Medium: Add components, except antibiotic solution, to distilled/deionized water and bring volume to 990.0mL. Mix thoroughly. Gently heat and bring to boiling. Autoclave for 15 min at 15 psi pressure–121°C. Cool to 45°–50°C. Aseptically add sterile antibiotic solution. Mix thoroughly. Pour into sterile Petri dishes or distribute into sterile tubes.

Use: For the selective isolation and cultivation of *Neisseria meningitidis*.

Nelson Medium for *Naegleria fowleri*

Composition per liter:

Glucose 1.0g
Ox liver digest...1.0g
Page's amoeba saline 1.0L
Fetal calf serum, inactivated 20.0mL

Page's Amoeba Saline:
Composition per liter:

Na_2HPO_4 ....0.142g
KH_2PO_4...0.136g
NaCl 0.12g
$MgSO_4 \cdot 7H_2O$ 4.0mg
$CaCl_2 \cdot 2H_2O$.. 4.0mg

Preparation of Page's Amoeba Saline: Add components to distilled/deionized water and bring volume to 10.0mL. Mix thoroughly.

Preparation of Medium: Add the glucose and ox liver digest to 1.0L of Page's amoeba saline. Mix thoroughly. Distribute into screw-capped tubes in 10.0mL volumes. Autoclave for 15 min at 15 psi pressure–121°C. Cool to 25°C. Aseptically add 0.2mL of sterile fetal calf serum to each tube. Mix thoroughly.

Use: For the cultivation of *Naegleria fowleri*.

Neomycin Agar

Composition per liter:

Agar...15.0g
Peptone..6.0g
Pancreatic digest of casein 4.0g
Yeast extract...3.0g
Beef extract 1.5g
Glucose 1.0g
Neomycin solution 10.0mL

pH 7.0 ± 0.2 at 25°C

Neomycin Solution:
Composition per 10mL:

Neomycin sulfate 1.0g

Preparation of Neomycin Solution: Add neomycin sulfate to distilled/deionized water and bring volume to 10.0mL. Mix thoroughly. Filter sterilize.

Preparation of Medium: Add components, except neomycin solution, to distilled/deionized water and bring volume to 990.0mL. Mix thoroughly. Gently heat and bring to boiling. Autoclave for 15 min at 15 psi pressure–121°C. Cool to 45°–50°C. Aseptically add sterile neomycin solution. Mix thoroughly. Pour into sterile Petri dishes or distribute into sterile tubes.

Use: For the cultivation and maintenance of *Micrococcus luteus*.

Neomycin Agar, Modified

Composition per liter:

Agar	15.0g
Peptone	6.0g
Pancreatic digest of casein	4.0g
Yeast extract	3.0g
Beef extract	1.5g
Glucose	1.0g
Methanol	20.0mL
Neomycin solution	10.0mL

pH 7.0 ± 0.2 at 25°C

Neomycin Solution:
Composition per 10mL:

Neomycin sulfate	1.0g

Preparation of Neomycin Solution: Add neomycin sulfate to distilled/deionized water and bring volume to 10.0mL. Mix thoroughly. Filter sterilize.

Preparation of Medium: Add components, except methanol and neomycin solution, to distilled/deionized water and bring volume to 970.0mL. Mix thoroughly. Gently heat and bring to boiling. Autoclave for 15 min at 15 psi pressure–121°C. Cool to 45°–50°C. Filter sterilize methanol. To cooled, sterile basal medium aseptically add sterile methanol and sterile neomycin solution. Mix thoroughly. Pour into sterile Petri dishes or distribute into sterile tubes.

Use: For the cultivation and maintenance of *Bordetella bronchiseptica.*

Neomycin Assay Agar
See: **Antibiotic Medium 11**

Neomycin Blood Agar

Composition per liter:

Pancreatic digest of casein	14.5g
Agar	14.0g
Papaic digest of soybean meal	5.0g
NaCl	5.0g
Growth factors	1.5g
Neomycin solution	10.0mL
Sheep blood, defibrinated	50.0mL

pH 7.3 ± 0.2 at 25°C

Source: This medium is available as a premixed powder from BBL Microbiology Systems.

Neomycin Solution:
Composition per 10mL:

Neomycin sulfate	0.030g

Preparation of Neomycin Solution: Add neomycin sulfate to distilled/deionized water and bring volume to 10.0mL. Mix thoroughly. Filter sterilize.

Preparation of Medium: Add components, except sheep blood and neomycin solution, to distilled/deionized water and bring volume to 940.0mL. Mix thoroughly. Gently heat and bring to boiling. Autoclave for 15 min at 15 psi pressure–121°C. Cool to 45°–50°C. Aseptically add sterile sheep blood and sterile neomycin solution. Mix thoroughly. Pour into sterile Petri dishes or distribute into sterile tubes.

Use: For the isolation and cultivation of Group A streptococci (*Streptococcus pyogenes)* and Group B streptococci (*S. agalactiae*) from throat cultures and other clinical specimens.

Neopeptone Glucose Agar

Composition per liter:

Agar	20.0g
Glucose	10.0g
Neopeptone	5.0g

pH 6.5 ± 0.2 at 25°C

Preparation of Medium: Add components to distilled/deionized water and bring volume to 1.0L. Mix thoroughly. Gently heat and bring to boiling. Distribute into tubes or flasks. Autoclave for 15 min at 15 psi pressure–121°C. Adjust pH to 6.5. Pour into sterile Petri dishes or leave in tubes.

Use: For the maintenance of stock cultures of a variety of microorganisms.

Neopeptone Glucose Rose Bengal Aureomycin® Agar

Composition per liter:

Agar	20.0g
Neopeptone	5.0g
Glucose	1.0g
Tetracycline solution	5.0mL
Rose Bengal solution	3.5mL

pH 6.5 ± 0.2 at 25°C

Tetracycline Solution:
Composition per 150mL:

Tetracycline	1.0g

Preparation of Tetracycline Solution: Add tetracycline to distilled/deionized water and bring volume to 150.0mL. Mix thoroughly. Filter sterilize.

Rose Bengal Solution:
Composition per 100mL:

Rose Bengal	1.0g

Preparation of Rose Bengal Solution: Add Rose Bengal to distilled/deionized water and bring volume to 100.0mL. Mix thoroughly. Filter sterilize.

Preparation of Medium: Add components, except tetracycline solution, to distilled/deionized water and bring volume to 995.0mL. Mix thoroughly. Gently heat and bring to boiling. Autoclave for 15 min at 15 psi pressure–121°C. Cool to 45°–50°C. Aseptically add 5.0mL of sterile tetracycline solution. Mix thoroughly. Pour into sterile Petri dishes or distribute into sterile tubes.

Use: For the isolation and cultivation of a wide variety of fungal species.

Neopeptone Infusion Agar

Composition per liter:

Beef heart, infusion from	500.0g
Neopeptone	20.0g
Agar	20.0g
NaCl	5.0g
Sheep blood, defibrinated	50.0mL

pH 7.4 ± 0.2 at 25°C

Preparation of Medium: Add components, except sheep blood, to distilled/deionized water and bring volume to 950.0mL. Mix thoroughly. Gently heat and bring to boiling. Autoclave for 15 min at 15 psi pressure–121°C. Cool to 45°–50°C. Aseptically add sterile sheep blood. Mix thoroughly. Pour into sterile Petri dishes or distribute into sterile tubes.

Use: For the cultivation of a wide variety of fastidious microorganisms.

Neurospora Culture Agar

Composition per liter:

Maltose	40.0g
Agar	15.0g
Proteose peptone No. 3	5.0g
Yeast extract	5.0g

pH 6.7 ± 0.2 at 25°C

Source: This medium is available as a premixed powder from Difco Laboratories.

Preparation of Medium: Add components to distilled/deionized water and bring volume to 1.0L. Mix thoroughly. Gently heat while stirring and bring to boiling. Distribute into tubes in 8.0mL volumes. Autoclave for 15 min at 15 psi pressure–121°C. Allow tubes to cool in a slanted position.

Use: For the cultivation of *Neurospora intermedia* used in the microbiological assay of pyridoxine. Also used for the cultivation of other fungi.

Neurospora Medium

Composition per liter:

Agar	15.0g
Glucose	5.0g
Malt syrup, spray-dried	5.0g
Sucrose	5.0g
Yeast extract	2.5g
Vitamin solution	10.0mL
Casein, hydrolyzed	5.0mL

Vitamin Solution:

Composition per liter:

Ribonucleic acid, alkali hydrolyzed	0.50g
Inositol	0.40g
Choline	0.20g
Nicotinamide	0.20g
Pantothenic acid	0.20g
Thiamine	0.10g
p-Aminobenzoic acid	0.05g
Pyridoxine	0.05g
Riboflavin	0.05g
Folic acid	4.0μg

Preparation of Vitamin Solution: Add components to distilled/deionized water and bring volume to 1.0L. Mix thoroughly. Filter sterilize.

Preparation of Medium: Add components, except vitamin solution, to distilled/deionized water and bring volume to 990.0mL. Mix thoroughly. Gently heat and bring to boiling. Autoclave for 15 min at 15 psi pressure–121°C. Cool to 45°–50°C. Aseptically add 10.0mL of sterile vitamin solution. Mix thoroughly. Pour into sterile Petri dishes or distribute into sterile tubes.

Use: For the cultivation of *Neurospora* species on complete medium.

Neurospora Medium

Composition per liter:

Sucrose	15.0g
Ammonium tartrate	5.0g
KH_2PO_4	1.0g
NH_4NO_3	1.0g
$MgSO_4 \cdot 7H_2O$	0.5g
$CaCl_2$	0.1g
NaCl	0.1g
$ZnCl_2$	2.0mg
$FeCl_3$	0.2mg
Cu Cl_2	0.1mg
$MnCl_2$	0.02mg
$Na_2MoO_4 \cdot 2H_2O$	0.02mg
H_3BO_3	0.01mg
Biotin	1.0μg

Preparation of Medium: Add components to distilled/deionized water and bring volume to 1.0L. Mix thoroughly. Distribute into tubes or flasks. Autoclave for 15 min at 15 psi pressure–121°C.

Use: For the cultivation of *Neurospora* species on minimal medium.

Neurospora Minimal Medium

Composition per liter:

Sucrose	20.0g
Ammonium tartrate	5.0g
KH_2PO_4	1.0g
$NaNO_3$	1.0g
$MgSO_4 \cdot 7H_2O$	0.5g
$CaCl_2 \cdot 2H_2O$	0.1g
NaCl	0.1g
$ZnSO_4 \cdot 7H_2O$	5.5mg
$FeSO_4 \cdot 7H_2O$	0.54mg
$CuSO_4 \cdot 5H_2O$	0.39mg
$MnSO_4 \cdot 4H_2O$	0.063mg
H_3BO_3	0.057mg
$Na_2MoO_4 \cdot 2H_2O$	0.050mg
Biotin	5.0μg

pH 7.3 ± 0.2 at 25°C

Preparation of Medium: Add components to distilled/deionized water and bring volume to 1.0L. Mix thoroughly. Gently heat and bring to boiling. Distribute into tubes or flasks. Autoclave for 15 min at 15 psi pressure–121°C.

Use: For the cultivation of *Neurospora* species and for the detection of *Neurospora* mutants.

Neutral Red Broth
See: LICNR Broth

New York City Medium

Composition per liter:

NYC basal medium	640.0mL
Horse blood cells	200.0mL
Horse plasma, citrated	120.0mL
Glucose solution	10.0mL
Yeast dialysate	25.0mL
Antibiotic VCNT solution	5.0mL

pH 7.4 ± 0.2 at 25°C

NYC Basal Medium:
Composition per 640mL:

Solution 1	400.0mL
Solution 3	200.0mL
Solution 2	40.0mL

Preparation of NYC Basal Medium: Combine solution 1, solution 2 and solution 3. Mix thoroughly. Autoclave for 15 min at 15 psi pressure–121°C. Cool to 45°–50°C.

Solution 1:
Composition per 400mL:

Agar	20.0g

Preparation of Solution 1: Add agar to distilled/deionized water and bring volume to 400.0mL. Mix thoroughly. Melt agar in autoclave for 10 min at 0 psi pressure–100°C. Cool to 45°–50°C.

Solution 2:
Composition per 40mL:

Cornstarch	1.0g

Preparation of Solution 2: Add cornstarch to distilled/deionized water and bring volume to 40.0mL. Mix thoroughly. Warm to 45°–50°C.

Solution 3:
Composition per 200mL:

Proteose peptone No. 3	15.0g
NaCl	5.0g
K_2HPO_4	4.0g
KH_2PO_4	1.0g

Preparation of Solution 3: Add components to distilled/deionized water and bring volume to 1.0L. Mix thoroughly. Gently heat and bring to boiling. Cool to 45°–50°C.

Horse Blood Cells:
Composition per 200mL:

Horse blood cells, sedimented	6.0mL

Preparation of Horse Blood Cells: Cow blood may be used instead of horse blood but do not use sheep blood. Use cells freshly packed by sedimentation. Do not pack by centrifugation. Aseptically add 6.0mL of sedimented blood cells to 200.0mL of sterile distilled/deionized water. Mix thoroughly.

Horse Plasma, Citrated:
Composition per 6 liters:

Horse blood	5400.0mL
Citrate solution	600.0mL

Preparation of Horse Plasma, Citrated: Place 600.0mL of sterile citrate solution into a receiving bottle. Draw horse blood to the 6.0L mark. Allow cells to sediment out. Aseptically remove plasma.

Citrate Solution:
Composition per liter:

Sodium citrate	150.0g
NaCl	81.13g

Preparation of Citrate Solution: Add components to distilled/deionized water and bring volume to 1.0L. Mix thoroughly. Filter sterilize.

Glucose Solution:
Composition per 10mL:

D-Glucose	5.0g

Preparation of Glucose Solution: Add D-glucose to distilled/deionized water and bring volume to 10.0mL. Mix thoroughly. Autoclave for 10 min at 10 psi pressure–115°C. Cool to 45°–50°C.

Yeast Dialysate:
Composition per 2500mL:

Baker's yeast, fresh	908.0g

Preparation of Yeast Dialysate: Add fresh baker's yeast to 2500.0mL of distilled/deionized water. Mix thoroughly. Autoclave for 10 min at 15 psi pressure–121°C. Cool to 25°C. Put into dialysis tubing. Dialyze against 2.0L of distilled/deionized water for 48 hr at 4°C.

Antibiotic VCNT Solution:
Composition per 5mL:

Colistin	7.5mg
Trimethorprim lactate	3.0mg
Vancomycin·HCl	2.0mg
Nystatin	12.5U

Preparation of Antibiotic Solution: Add components to distilled/deionized water and bring volume to 5.0mL. Mix thoroughly. Filter sterilize.

Preparation of Medium: Have all solutions prepared and at 45°–50°C. Aseptically combine components. Mix thoroughly. Pour into sterile Petri dishes.

Use: For the isolation and cultivation of pathogenic *Neisseria* species. Used as a transport medium for urogenital and other clinical specimens. For the isolation and presumptive identification of Mycoplasmatales, including large-colony species (*Mycoplasma pneumoniae*) and T–mycoplasmas from urogenital specimens.

New York City Medium, Modified

Composition per liter:

NYC basal medium	840.0mL
Agamma horse serum (Flow Labs)	120.0mL
Glucose solution	10.0mL
Yeast dialysate	25.0mL
Antibiotic LCNT solution	5.0mL

pH 7.4 ± 0.2 at 25°C

NYC Basal Medium:
Composition per 840mL:

Horse blood	5400.0mL
Solution 1	600.0mL
Solution 3	200.0mL
Solution 2	40.0mL

Preparation of NYC Basal Medium: Combine solution 1, solution 2 and solution 3. Mix thoroughly. Autoclave for 15 min at 15 psi pressure–121°C. Cool to 45°–50°C.

Solution 1:
Composition per 600mL:

Agar	20.0g

Preparation of Solution 1: Add agar to distilled/deionized water and bring volume to 600.0mL. Mix thoroughly. Melt agar in autoclave for 10 min at 0 psi pressure–100°C. Cool to 45°–50°C.

Solution 2:
Composition per 40mL:

Cornstarch	1.0g

Preparation of Solution 2: Add cornstarch to distilled/deionized water and bring volume to 40.0mL. Mix thoroughly. Warm to 45°–50°C.

Solution 3:
Composition per 200mL:

Proteose peptone No. 3	15.0g
NaCl	5.0g
K_2HPO_4	4.0g
KH_2PO_4	1.0g

Preparation of Solution 3: Add components to distilled/deionized water and bring volume to 200.0mL. Mix thoroughly. Gently heat and bring to boiling. Cool to 45°–50°C.

Glucose Solution:
Composition per 10mL:

D-Glucose	5.0g

Preparation of Glucose Solution: Add glucose to distilled/deionized water and bring volume to 10.0mL. Mix thoroughly. Autoclave for 10 min at 10 psi pressure–115°C. Cool to 45°–50°C.

Yeast Dialysate:
Composition per 2500mL:

Baker's yeast, fresh	908.0g

Preparation of Yeast Dialysate: Add fresh baker's yeast to 2500.0mL of distilled/deionized water. Mix thoroughly. Autoclave for 10 min at 15 psi pressure–121°C. Cool to 25°C. Put into dialysis tubing. Dialyze against 2.0L of distilled/deionized water for 48 hr at 4°C.

Antibiotic LCNT Solution:
Composition per 5mL:

Colistin	7.5mg
Lincomycin·HCl	4.0mg
Trimethorprim lactate	3.0mg
Nystatin	12.5U

Preparation of Antibiotic LCNT Solution: Add the components to distilled/deionized water and bring volume to 5.0mL. Mix thoroughly. Filter sterilize the solution.

Preparation of Medium: Have all solutions prepared and at 45°–50°C. Aseptically combine components. Mix thoroughly. Pour into sterile Petri dishes.

Use: For the isolation and cultivation of pathogenic *Neisseria* species. Used as a transport medium for urogenital and other clinical specimens. For the isolation and presumptive identification of Mycoplasmatales, including large-colony species (*Mycoplasma pneumoniae*) and T–mycoplasmas from urogenital specimens.

Niacin Assay Medium

Composition per liter:

Glucose	40.0g
Sodium acetate	20.0g
Vitamin assay casamino acids	12.0g
K_2HPO_4	1.0g
KH_2PO_4	1.0g
L-Cystine	0.4g
$MgSO_4 \cdot 7H_2O$	0.4g
DL-Tryptophan	0.2g
Adenine sulfate	0.020g
$FeSO_4 \cdot 5H_2O$	0.020g
Guanine·HCl	0.020g
$MnSO_4 \cdot H_2O$	0.020g
NaCl	0.020g
Uracil	0.020g
Pyridoxine·HCl	0.4mg
Riboflavin	0.4mg
Calcium pantothenate	0.2mg
Thiamine·HCl	0.2mg
p-Aminobenzoic acid	0.1mg
Biotin	0.8μg

pH 6.7 ± 0.2 at 25°C

Source: This medium is available as a premixed powder from Difco Laboratories.

Preparation of Medium: Add components to distilled/deionized water and bring volume to 1.0L. Mix thoroughly. Gently heat and bring to boiling. Continue boiling 2–3 min. Distribute into tubes in 5.0mL volumes. Add standard solution or test solutions to each tube. Adjust the volume of each tube to 10.0mL with distilled/deionized water. Autoclave for 10 min at 15 psi pressure–121°C.

Use: For the microbiological assay of nicotinic acid or nicotinamide (niacin) using *Lactobacillus plantarum* as the test organism.

Nickels and Leesment Agar, Modified

Composition per liter:

Part 1	750.0mL
Part 2	100.0mL
Part 3	100.0mL
Part 4	50.0mL

pH 6.65 ± 0.2 at 25°C

Part 1:

Composition per 750mL:

Pancreatic digest of casein	20.0g
Lactose	10.0g
Yeast extract	5.0g
NaCl	4.0g
Gelatin	2.5g
Sodium citrate	2.0g

Preparation of Part 1: Add components to distilled/deionized water and bring volume to 750.0mL. Mix thoroughly. Adjust pH to 6.65.

Part 2:

Composition per 500mL:

Nonfat dry milk	50.0g

Preparation of Part 2: Add nonfat dry milk to distilled/deionized water and bring volume to 500.0mL. Inoculate with *Lactobacillus bulgaricus*. Incubate at 20°C for 24 hr. Centrifuge at 5,000 rpm for 10 min to separate the curd. Collect the supernatant solution. Autoclave for 15 min at 15 psi pressure–121°C. Store at 4°C.

Part 3:

Composition per 100mL:

Calcium citrate	13.3g
Carboxymethylcellulose	0.8g

Preparation of Part 3: Combine the calcium citrate and carboxymethylcellulose in a mortar and grind until a fine powder. Add powder to 100.0mL of hot distilled/deionized water. Mix thoroughly. Filter through cheesecloth.

Part 4:

Composition per 50mL:

Calcium lactate	8.0g

Preparation of Part 4: Add calcium lactate to distilled/deionized water and bring volume to 50.0mL. Mix thoroughly. Gently heat until dissolved.

Preparation of Medium: Prepare each of the four parts separately. Autoclave each part for 15 min at 15 psi pressure–121°C. Aseptically combine part 1, part 2, part 3, and part 4. Mix thoroughly. Pour into sterile Petri dishes. Swirl flask while dispensing medium.

Use: For the isolation and cultivation of acid-producing microorganisms from foods.

Nickerson Medium
See: **BiGGY Agar**

Nigerseed Agar
See: **Birdseed Agar**

NIH Agar

Composition per liter:

Pancreatic digest of casein	15.0g
Agar	15.0g
Glucose	5.5g
Yeast extract	5.0g
NaCl	2.5g
L-Cystine	0.05g

pH 7.1 ± 0.2 at 25°C

Source: This medium is available as a premixed powder from Difco Laboratories.

Preparation of Medium: Add components to distilled/deionized water and bring volume to 1.0L. Mix thoroughly. Gently heat while stirring and bring to boiling. Distribute into tubes or flasks. Autoclave for 15 min at 15 psi pressure–121°C. Pour into sterile Petri dishes or leave in tubes.

Use: For the cultivation and maintenance of microorganisms isolated from sterility testing of biological products. Also used as a solid medium for sterility testing.

NIH Thioglycollate Broth

Composition per liter:

Pancreatic digest of casein	15.0g
Glucose	5.5g
Yeast extract	5.0g
NaCl	2.5g
L-Cystine	0.5g
Sodium thioglycollate	0.5g

pH 7.1 ± 0.2 at 25°C

Source: This medium is available as a premixed powder from Difco Laboratories.

Preparation of Medium: Add components to distilled/deionized water and bring volume to 1.0L. Mix thoroughly. Gently heat while stirring and bring to boiling. Distribute into tubes or flasks. Autoclave for 15 min at 15 psi pressure–121°C.

Use: For sterility testing of biological products that are turbid or otherwise cannot be cultivated in fluid thioglycollate broth because of its viscosity.

Nitrate Agar

Composition per liter:

Agar	12.0g
Peptone	5.0g
Beef extract	3.0g
KNO_3	1.0g

pH 6.8 ± 0.2 at 25°C

Preparation of Medium: Add components to distilled/deionized water and bring volume to 1.0L. Mix thoroughly. Gently heat and bring to boiling. Distribute into tubes. Autoclave for 15 min at 15 psi pressure–121°C. Allow tubes to cool in a slanted position.

Use: For the differentiation of aerobic and facultative Gram-negative microorganisms based on their ability to reduce nitrate. Test for nitrates with sulfanilic acid and α-naphthylamine reagents. Bacteria that reduce nitrate to nitrite turn the reagents red or pink.

Nitrate Assimilation Medium, Auxanographic Method for Yeast Identification

Composition per liter:

Noble agar	20.00g
Glucose	10.0g
KH_2PO_4	1.0g
$MgSO_4 \cdot 7H_2O$	0.5g
NaCl	0.1g
$CaCl_2 \cdot 2H_2O$	0.1g
DL-Methionine	0.02g
DL-Tryptophan	0.02g
L-Histidine·HCl	0.01g
Inositol	2.0mg
H_3BO_3	0.5mg
$ZnSO_4 \cdot 7H_2O$	0.4mg
$MnSO_4 \cdot 4H_2O$	0.4mg
Thiamine·HCl	0.4mg
Pyridoxine	0.4mg
Niacin	0.4mg
Calcium pantothenate	0.4mg
p-Aminobenzoic acid	0.2mg
Riboflavin	0.2mg
$FeCl_3$	0.2mg
$Na_2MoO_4 \cdot 4H_2O$	0.2mg
KI	0.1mg
$CuSO_4 \cdot 5H_2O$	0.04mg
Folic Acid	2.0μg
Biotin	2.0μg

pH 4.5 ± 0.2 at 25°C

Preparation of Medium: Add components to distilled/deionized water and bring volume to 1.0L. Mix thoroughly. Gently heat and bring to boiling. Distribute into screw-capped tubes in 20.0mL volumes. Autoclave for 15 min at 15 psi pressure–121°C.

Use: For nitrate assimilation tests by the auxanographic method.

Nitrate Broth (International Streptomyces Project Medium 8) (ISP Medium 8) (ATCC Medium 872)

Composition per liter:

Peptone	5.0g
Beef extract	3.0g
KNO_3	1.0g

pH 7.0 ± 0.2 at 25°C

Source: This medium is available as a premixed powder from Difco Laboratories.

Preparation of Medium: Add components to distilled/deionized water and bring volume to 1.0L. Mix thoroughly. Distribute into tubes or flasks. Autoclave for 15 min at 15 psi pressure–121°C.

Use: For the differentiation of aerobic and facultative Gram-negative microorganisms based on their ability to reduce nitrate. Test for nitrates with sulfanilic acid and α-naphthylamine reagents. Bacteria that reduce nitrate to nitrite turn the reagents red or pink.

Nitrate Broth

Composition per liter:

Pancreatic digest of gelatin	20.0g
KNO_3	2.0g

pH 7.2 ± 0.2 at 25°C

Source: This medium is available as a premixed powder from BBL Microbiology Systems.

Preparation of Medium: Add components to distilled/deionized water and bring volume to 1.0L. Mix thoroughly. Distribute into tubes or flasks. Autoclave for 15 min at 15 psi pressure–121°C.

Use: For the differentiation of aerobic and facultative Gram-negative microorganisms based on their ability to reduce nitrate. Test for nitrates with sulfanilic acid and α-naphthylamine reagents. Bacteria that reduce nitrate to nitrite turn the reagents red or pink.

Nitrate Broth, *Campylobacter*

Composition per liter:

Beef heart, solids from infusion	500.0g
Tryptose	10.0g
NaCl	5.0g
KNO_3	2.0g

pH 7.0 ± 0.2 at 25°C

Preparation of Medium: Add components to distilled/deionized water and bring volume to 1.0L. Mix thoroughly. Adjust pH to 7.0. Distribute 4.0mL volumes into test tubes that contain inverted Durham tubes. Autoclave for 15 min at 15 psi pressure–121°C.

Use: For the cultivation and differentiation of *Campylobacter* species based on their ability to reduce nitrate.

Nitrate Broth, Enriched

Composition per liter:

Pancreatic digest of casein	13.0g
NaCl	5.0g
Yeast extract	5.0g
Heart muscle, solids from infusion	2.0g
KNO_3	2.0g

pH 7.3 ± 0.2 at 25°C

Preparation of Medium: Add components to distilled/deionized water and bring volume to 1.0L. Mix thoroughly. Distribute into test tubes that contain an inverted Durham tube. Autoclave for 15 min at 15 psi pressure–121°C.

Use: For the differentiation of aerobic and facultative Gram-negative microorganisms based on their ability to reduce nitrate to nitrite or form N_2 gas. Test for nitrates with sulfanilic acid and α-naphthylamine reagents. Bacteria that reduce nitrate to nitrite turn the reagents red or pink.

Nitrate Liquid Medium

Composition per liter:

Solution A	500.0mL
Solution B	250.0mL
Solution C	250.0mL

Solution A:

Composition per 500mL:

Mannitol	10.0g
KNO_3	0.6g
$Na_2HPO_4 \cdot 12H_2O$	0.45g
Na_2SO_4	0.03g

Preparation of Solution A: Add components to distilled/deionized water and bring volume to 500.0mL. Mix thoroughly. Autoclave for 15 min at 15 psi pressure–121°C. Cool to 25°C.

Solution B:

Composition per 250mL:

$MgSO_4 \cdot 7H_2O$	0.12g
$CaCl_2 \cdot 6H_2O$	0.1g
$FeCl_3 \cdot 6H_2O$	0.01g

Preparation of Solution B: Add components to distilled/deionized water and bring volume to 250.0mL. Mix thoroughly. Autoclave for 15 min at 15 psi pressure–121°C. Cool to 25°C.

Solution C:

Composition per 250mL:

Calcium pantothenate	0.5mg
Thiamine·HCl	0.1mg
Biotin	0.5μg

Preparation of Solution C: Add components to distilled/deionized water and bring volume to 250.0mL. Mix thoroughly. Filter sterilize.

Preparation of Medium: Aseptically combine 500.0mL of cooled, sterile solution A, 250.0mL of cooled, sterile solution B, and 250.0mL of sterile solution C. Mix thoroughly. Aseptically distribute into sterile tubes or flasks.

Use: For the isolation and cultivation of *Rhizobium* species.

Nitrate Methanol Medium

Composition per liter:

$NaNO_3$	5.0g
K_2HPO_4	2.0g
NaCl	1.0g
$MgSO_4 \cdot 7H_2O$	0.02g
$Na_2MoO_4 \cdot H_2O$	1.0mg
Riboflavin	0.2mg
Calcium pantothenate	0.2mg
Pyridoxine·HCl	0.2mg
Nicotinic acid	0.2mg
Thiamine·HCl	0.1mg
p-Aminobenzoic acid	0.1mg
Biotin	0.01mg
Methanol	10.0mL

pH 7.0 ± 0.2 at 25°C

Preparation of Medium: Add components, except methanol, to distilled/deionized water and bring volume to 990.0mL. Mix thoroughly. Autoclave for 15 min at 15 psi pressure–121°C. Cool to 45°–50°C. Filter sterilize methanol. Aseptically add sterile methanol to cooled sterile medium. Mix thoroughly. Aseptically distribute into sterile tubes or flasks.

Use: For the cultivation and maintenance of *Methylobacterium rhodinum.*

Nitrate Mineral Salts Medium (NMS Medium)

Composition per liter:

Noble agar	12.5g
$MgSO_4 \cdot 7H_2O$	1.0g
KNO_3	1.0g
$Na_2HPO_4 \cdot 12H_2O$	0.717g
KH_2PO_4	0.272g
$CaCl_2 \cdot 6H_2O$	0.20g
Ferric ammonium EDTA	4.0mg
Trace element solution	0.5mL

pH 6.8 ± 0.2 at 25°C

Trace Element Solution:

Composition per liter:

Disodium EDTA	0.5g
$FeSO_4 \cdot 7H_2O$	0.2g
H_3BO_3	0.030g
$CoCl_2 \cdot 6H_2O$	0.020g
$ZnSO_4 \cdot 7H_2O$	0.010g
$MnCl_2 \cdot 4H_2O$	3.0mg
$Na_2MoO_4 \cdot 2H_2O$	3.0mg
$NiCl_2 \cdot 6H_2O$	2.0mg
$CaCl_2 \cdot 2H_2O$	1.0mg

Preparation of Trace Element Solution: Add components to distilled/deionized water and bring volume to 1.0L. Mix thoroughly.

Preparation of Medium: Add components to distilled/deionized water and bring volume to 1.0L. Mix thoroughly. Gently heat and bring to boiling. Adjust pH to 6.8. Distribute into tubes or flasks. Autoclave for 15 min at 15 psi pressure–121°C. Pour into sterile Petri dishes or leave in tubes.

Use: For the cultivation and maintenance of *Methylobacterium* species, *Methylococcus capsulatus*, *Methylomonas agile* and *Methylomonas methanica*.

Nitrate Mineral Salts Medium with Methanol (NMS Medium with Methanol)

Composition per liter:

Noble agar	12.5g
$MgSO_4 \cdot 7H_2O$	1.0g
KNO_3	1.0g
$Na_2HPO_4 \cdot 12H_2O$	0.717g
KH_2PO_4	0.272g
$CaCl_2 \cdot 6H_2O$	0.20g
Ferric ammonium EDTA	4.0mg
Trace element solution	0.5mL
Methanol	1.0mL

pH 6.8 ± 0.2 at 25°C

Trace Element Solution:

Composition per liter:

Disodium EDTA	0.5g
$FeSO_4 \cdot 7H_2O$	0.2g
H_3BO_3	0.030g
$CoCl_2 \cdot 6H_2O$	0.020g
$ZnSO_4 \cdot 7H_2O$	0.010g
$MnCl_2 \cdot 4H_2O$	3.0mg
$Na_2MoO_4 \cdot 2H_2O$	3.0mg
$NiCl_2 \cdot 6H_2O$	2.0mg
$CaCl_2 \cdot 2H_2O$	1.0mg

Preparation of Trace Element Solution: Add components to distilled/deionized water and bring volume to 1.0L. Mix thoroughly.

Preparation of Medium: Add components, except methanol, to distilled/deionized water and bring volume to 999.0mL. Mix thoroughly. Gently heat and bring to boiling. Adjust pH to 6.8. Distribute into tubes or flasks. Autoclave for 15 min at 15 psi pressure–121°C. Cool to 45°–50°C. Filter sterilize methanol. Aseptically add sterile methanol to cooled sterile medium. Mix thoroughly. Pour into sterile Petri dishes or leave in tubes.

Use: For the cultivation and maintenance of *Methylobacterium fujisawaense*, *Methylobacterium* species and *Methylomonas clara*.

Nitrate Reduction Broth

Composition per liter:

Pancreatic digest of gelatin	5.0g
Beef extract	3.0g
KNO_3	1.0g

pH 6.9 ± 0.2 at 25°C

Preparation of Medium: Add components to distilled/deionized water and bring volume to 1.0L. Mix thoroughly. Distribute into test tubes that contain an inverted Durham tube. Autoclave for 15 min at 15 psi pressure–121°C.

Use: For the differentiation of members of the Pseudomonadaceae based on their ability to reduce nitrate to nitrite or form N_2 gas. Test for nitrates with sulfanilic acid and α-naphthylamine reagents. Bacteria that reduce nitrate to nitrite turn the reagents red or pink.

Nitrate Reduction Broth

Composition per liter:

Pancreatic digest of casein	13.0g
NaCl	5.0g
Yeast extract	5.0g
Heart muscle, solids from infusion	2.0g
KNO_3 or $NaNO_3$	2.0g

pH 7.4 ± 0.2 at 25°C

Preparation of Medium: Add components to distilled/deionized water and bring volume to 1.0L. Mix thoroughly. Distribute into test tubes that contain an inverted Durham tube. Autoclave for 15 min at 15 psi pressure–121°C.

Use: For the differentiation of a variety of Gram-negative bacteria based on their ability to reduce nitrate to nitrite or form N_2 gas. Test for nitrates with sulfanilic acid and α-naphthylamine reagents. Bacteria that reduce nitrate to nitrite turn the reagents red or pink.

Nitrate Reduction Broth

Composition per liter:

Pancreatic digest of casein	13.0g
NaCl	5.0g
Yeast extract	5.0g
Heart muscle, solids from infusion	2.0g
KNO_3 or $NaNO_3$	2.0g

pH 7.4 ± 0.2 at 25°C

Preparation of Medium: Dispense in 3.0mL amounts into screw-cap tubes, add inverted vials and autoclave at 121°C for 15 min.

Use: For the differentiation of a variety of nonfermenting Gram-negative bacteria based on their ability to reduce nitrate to nitrite or form N_2 gas. Test for nitrates with sulfanilic acid and α-naphthylamine reagents. Bacteria that reduce nitrate to nitrite turn the reagents red or pink.

Nitrate Reduction Broth, Clark

Composition per liter:

Peptone	20.0g
KNO_3 or $NaNO_3$	2.0g

Preparation of Medium: Add components to distilled/deionized water and bring volume to 1.0L. Mix thoroughly. Distribute into test tubes that contain an inverted Durham tube. Autoclave for 15 min at 15 psi pressure–121°C.

Use: For the differentiation of a variety of Gram-negative bacteria based on their ability to reduce nitrate to nitrite or form N_2 gas. Test for nitrates with sulfanilic acid and α-naphthylamine reagents. Bacteria that reduce nitrate to nitrite turn the reagents red or pink.

Nitrate Reduction Broth for *Pseudomonas* and Related Genera

Composition per liter:

Peptone	5.0g
NaCl	5.0g
Yeast extract	2.0g
Beef extract	1.0g
$NaNO_3$	0.1g

pH 7.4 ± 0.2 at 25°C

Preparation of Medium: Add components to distilled/deionized water and bring volume to 1.0L. Mix thoroughly. Distribute into test tubes that contain an inverted Durham tube. Autoclave for 15 min at 15 psi pressure–121°C.

Use: For the differentiation of members of the Pseudomonadaceae based on their ability to reduce nitrate to nitrite or form N_2 gas. Test for nitrates with sulfanilic acid and α-naphthylamine reagents. Bacteria that reduce nitrate to nitrite turn the reagents red or pink.

Nitrobacter agilis Medium

Composition per liter:

$CaCO_3$	10.0g
NaCl	0.3g
Na_2CO_3	0.25g
KNO_2	0.17g
K_2HPO_4	0.14g
$MgSO_4 \cdot 7H_2O$	0.14g
$FeSO_4 \cdot 7H_2O$	0.03g
$MnSO_4 \cdot 4H_2O$	0.01g
Biotin solution	10.0mL

Biotin Solution:

Composition per 10mL:

Biotin	0.15g

Preparation of Biotin Solution: Add biotin to distilled/deionized water and bring volume to 10.0mL. Mix thoroughly. Filter sterilize.

Preparation of Medium: Add Na_2CO_3 to distilled/deionized water and bring volume to 200.0mL. Mix thoroughly. In a separate flask add the remaining components, except the biotin solution, to distilled/deionized water and bring volume to 790.0mL. Autoclave the Na_2CO_3 solution and salts solution separately for 15 min at 15 psi pressure–121°C. Cool to 25°C. Aseptically combine the sterile Na_2CO_3 solution, sterile salts solution, and sterile biotin solution. Mix thoroughly. Aseptically distribute into sterile tubes or flasks.

Use: For the cultivation of *Nitrobacter agilis*.

Nitrobacter Medium 203

Composition per liter:

Solution A	0.5mL
Solution B	0.5mL
Solution C	1.0mL
Solution D	0.5mL
Solution E	0.5mL
Solution F	0.2mL

Solution A:

Composition per 100mL:

$CaCl_2$	2.0g

Preparation of Solution A: Add $CaCl_2$ to distilled/deionized water and bring volume to 100.0mL. Mix thoroughly.

Solution B:

Composition per 100mL:

$MgSO_4 \cdot 7H_2O$	20.0g

Preparation of Solution B: Add $MgSO_4 \cdot 7H_2O$ to distilled/deionized water and bring volume to 100.0mL. Mix thoroughly.

Solution C:

Composition per 100mL:

Chelated iron (Sequestrene)	0.1g

Preparation of Solution C: Add chelated iron to distilled/deionized water and bring volume to 100.0mL. Mix thoroughly.

Solution D:

Composition per liter:

$MnCl_2 \cdot 4H_2O$	0.2g
$Na_2MoO_4 \cdot 2H_2O$	0.1g
$ZnSO_4 \cdot 7H_2O$	0.1g
$CuSO_4 \cdot 5H_2O$	0.02g
$CoCl_2 \cdot 6H_2O$	2.0mg

Preparation of Solution D: Add components to distilled/deionized water and bring volume to 1.0L. Mix thoroughly.

Solution E:

Composition per 100mL:

$NaNO_2$	41.4g

Preparation of Solution E: Add $NaNO_2$ to distilled/deionized water and bring volume to 100.0mL. Mix thoroughly.

Solution F:

Composition per 100mL:

K_2HPO_4	1.74g

Preparation of Solution F: Add K_2HPO_4 to distilled/deionized water and bring volume to 100.0mL. Mix thoroughly.

Preparation of Medium: Add the appropriate volumes of solutions A–F to distilled/deionized water and bring volume to 1.0L. Mix thoroughly. Distribute into tubes or flasks. Autoclave for 15 min at 15 psi pressure–121°C.

Use: For the cultivation and maintenance of *Nitrobacter* species and *Nitrobacter winogradskyi*.

Nitrobacter Medium 204

Composition per liter:

Sea water	700.0mL
Solution A	0.5mL
Solution B	0.5mL
Solution C	1.0mL
Solution D	0.5mL
Solution E	0.5mL
Solution F	0.2mL

Solution A:

Composition per 100mL:

$CaCl_2$	2.0g

Preparation of Solution A: Add $CaCl_2$ to distilled/deionized water and bring volume to 100.0mL. Mix thoroughly.

Solution B:

Composition per 100mL:

$MgSO_4·7H_2O$	20.0g

Preparation of Solution B: Add $MgSO_4·7H_2O$ to distilled/deionized water and bring volume to 100.0mL. Mix thoroughly.

Solution C:

Composition per 100mL:

Chelated iron (Sequestrene)	0.1g

Preparation of Solution C: Add chelated iron to distilled/deionized water and bring volume to 100.0mL. Mix thoroughly.

Solution D:

Composition per liter:

$MnCl_2·4H_2O$	0.2g
$Na_2MoO_4·2H_2O$	0.1g
$ZnSO_4·7H_2O$	0.1g
$CuSO_4·5H_2O$	0.02g
$CoCl_2·6H_2O$	2.0mg

Preparation of Solution D: Add components to distilled/deionized water and bring volume to 1.0L. Mix thoroughly.

Solution E:

Composition per 100mL:

$NaNO_2$	41.4g

Preparation of Solution E: Add $NaNO_2$ to distilled/deionized water and bring volume to 100.0mL. Mix thoroughly.

Solution F:

Composition per 100mL:

K_2HPO_4	1.74g

Preparation of Solution F: Add K_2HPO_4 to distilled/deionized water and bring volume to 100.0mL. Mix thoroughly.

Preparation of Medium: Add the appropriate volumes of solutions A–F and sea water to distilled/deionized water and bring volume to 1.0L. Mix thoroughly. Distribute into tubes or flasks. Autoclave for 15 min at 15 psi pressure–121°C.

Use: For the cultivation and maintenance of *Nitrococcus mobilis*.

Nitrobacter Medium B

Composition per liter:

$NaNO_2$	1.0g
K_2HPO_4	0.5g
$MgSO_4$	0.5g
NaCl	0.3g
$Fe_2(SO_4)_3$	5.0mg
$MnSO_4$	2.0mg
Marble chips	as needed

pH 7.5 ± 0.2 at 25°C

Preparation of Medium: Add components, except marble chips, to distilled/deionized water and bring volume to 1.0L. Mix thoroughly. Autoclave for 15 min at 15 psi pressure–121°C. Cool to 25°C. Wash marble chips in distilled/deionized water. Put a few chips into test tubes. Autoclave for 60 min at 15 psi pressure–121°C. Cool to 25°C. Aseptically distribute cooled sterile medium into test tubes to cover marble chips.

Use: For the cultivation of *Nitrobacter* species.

Nitrogen-Fixing Hydrocarbon Oxidizers Medium

Composition per liter:

Na_2HPO_4	0.3g
KH_2PO_4	0.2g
$MgSO_4·7H_2O$	0.1g
$FeSO_4·7H_2O$	5.0mg
$Na_2MoO_4·2H_2O$	2.0mg

Preparation of Medium: Add components to distilled/deionized water and bring volume to 1.0L. Mix thoroughly. Distribute into tubes or flasks. Autoclave for 15 min at 15 psi pressure–121°C.

Use: For the cultivation and enrichment of nitrogen-fixing hydrocarbon-oxidizing bacteria.

Nitrogen–Fixing Marine Medium

Composition per liter:

Noble agar	10.0g
$MgSO_4 \cdot 7H_2O$	0.04g
$CaCl_2 \cdot 2H_2O$	0.02g
$K_2HPO_4 \cdot 3H_2O$	0.02g
Na_2CO_3	0.02g
Citric acid	3.0mg
Ferric ammonium citrate	3.0mg
Disodium potassium EDTA	0.5mg
Sea water	750.0mL
Trace Metals A-5	1.0mL

pH 8.5 ± 0.2 at 25°C

Trace Metals A-5 Mix:

Composition per liter:

H_3BO_3	2.86g
$MnCl_2 \cdot 4H_2O$	1.81g
$ZnSO_4 \cdot 7H_2O$	0.222g
$CuSO_4 \cdot 5H_2O$	0.079g
$Na_2MoO_4 \cdot 2H_2O$	0.039g
$Co(NO_3)_2 \cdot 6H_2O$	0.050g

Preparation of Trace Metals A-5: Add components to distilled/deionized water and bring volume to 1.0L. Mix thoroughly.

Preparation of Medium: Add components to glass distilled water and bring volume to 1.0L. Mix thoroughly. Gently heat and bring to boiling. Autoclave for 15 min at 15 psi pressure–121°C. Adjust pH to 8.5 with KOH. Pour into sterile Petri dishes or distribute into sterile tubes.

Use: For the cultivation and maintenance of *Anabaena* species.

Nitrosococcus Medium

Composition per liter:

$(NH_4)_2SO_4$	1.32g
$MgSO_4 \cdot 7H_2O$	0.38g
$CaCl_2 \cdot 2H_2O$	0.020g
K_2HPO_4	8.7mg
Chelated iron	1.0mg
$MnCl_2 \cdot 4H_2O$	0.2mg
$Na_2MoO_4 \cdot 2H_2O$	0.1mg
$ZnSO_4 \cdot 7H_2O$	0.1mg
$CoCl_2 \cdot 6H_2O$	2.0μg
Phenol Red (0.04% solution)	3.25mL

pH 7.5–7.8 at 25°C

Preparation of Medium: Add components to filtered sea water and bring volume to 1.0L. Mix thoroughly. Adjust pH to 7.5–7.8 with 1*N* HCl. Distribute into tubes. Autoclave for 15 min at 15 psi pressure–121°C.

Use: For the cultivation of *Nitrosococcus oceanus*.

Nitrosolobus Medium (ATCC Medium 438)

Composition per liter:

$(NH_4)_2SO_4$	1.65g
$MgSO_4 \cdot 7H_2O$	0.2g
K_2HPO_4	0.087g
$CaCl_2 \cdot 2H_2O$	0.02g
Phenol Red	5.0mg
Disodium EDTA	1.0mg
$MnCl_2 \cdot 4H_2O$	0.2mg
$Na_2MoO_4 \cdot 2H_2O$	0.1mg
$ZnSO_4 \cdot 7H_2O$	0.1mg
$CuSO_4 \cdot 5H_2O$	0.02mg
$CoCl_2 \cdot 6H_2O$	2.0μg

pH 7.5 ± 0.2 at 25°C

Preparation of Medium: Add components to distilled/deionized water and bring volume to 1.0L. Mix thoroughly. Adjust pH to 7.5 with 0.1M K_2CO_3. Distribute into tubes or flasks. Autoclave for 15 min at 15 psi pressure–121°C.

Use: For the cultivation and maintenance of *Nitrosolobus multiformis*.

Nitrosolobus Medium (ATCC Medium 929)

Composition per liter:

$(NH_4)_2SO_4$	1.32g
$MgSO_4 \cdot 7H_2O$	0.38g
K_2HPO_4	0.087g
$CaCl_2 \cdot 2H_2O$	0.020g
Chelated iron	1.0mg
$MnCl_2 \cdot 4H_2O$	0.2mg
$Na_2MoO_4 \cdot 2H_2O$	0.1mg
$ZnSO_4 \cdot 7H_2O$	0.1mg
$CoCl_2 \cdot 6H_2O$	2.0μg
Phenol Red (0.5% solution)	0.25mL

pH 7.5 ± 0.2 at 25°C

Preparation of Medium: Add components to distilled/deionized water and bring volume to 1.0L. Mix thoroughly. Adjust pH to 7.5 with 0.1M K_2CO_3. Distribute into tubes or flasks. Autoclave for 15 min at 15 psi pressure–121°C.

Use: For the cultivation and maintenance of *Nitrosolobus multiformis*.

Nitrosomonas europaea Medium

Composition per liter:

$(NH_4)_2SO_4$	1.7g
$MgSO_4 \cdot 7H_2O$	0.2g
$CaCl_2 \cdot 2H_2O$	0.020g
K_2HPO_4	0.015g

Ferric EDTA 1.0mg
Trace elements solution 1.0mL
pH 7.5 ± 0.2 at 25°C

Trace Elements Solution:
Composition per 100mL:
$MnCl_2 \cdot 4H_2O$ 0.020g
$Na_2MoO_4 \cdot 2H_2O$ 0.010g
$ZnSO_4 \cdot 7H_2O$ 0.010g
$CuSO_4 \cdot 5H_2O$ 2.0mg
$CoCl_2 \cdot 6H_2O$ 0.2mg

Preparation of Trace Elements Solution: Add components to distilled/deionized water and bring volume to 1.0L. Mix thoroughly.

Preparation of Medium: Add components to distilled/deionized water and bring volume to 1.0L. Mix thoroughly. Adjust pH to 7.5 with K_2CO_3. Distribute into tubes or flasks. Autoclave for 15 min at 15 psi pressure–121°C. After inoculation, maintain pH at 7.5–7.8 with sterile 50% K_2CO_3 solution.

Use: For the cultivation and maintenance of *Nitrosomonas europaea.*

Nitrosomonas Medium

Composition per liter:
$(NH_4)_2SO_4$ 3.0g
K_2HPO_4 .... 0.5g
$MgSO_4 \cdot 7H_2O$ 0.05g
$CaCl_2 \cdot 2H_2O$ 4.0mg
Cresol Red (0.0005% solution) 25.0mL
Ferric EDTA solution 0.1mL
pH 8.2–8.4 at 25°C

Ferric EDTA Solution:
Composition per 100mL:
$FeSO_4 \cdot 7H_2O$ 0.50g
Disodium EDTA 0.14g
H_2SO_4, concentrated 0.05mL

Preparation of Ferric EDTA Solution: Add components to distilled/deionized water and bring volume to 100.0mL. Mix thoroughly.

Preparation of Medium: Add $CaCl_2 \cdot 2H_2O$ and $MgSO_4 \cdot 7H_2O$ to distilled/deionized water and bring volume to 500.0mL. Mix thoroughly. In a separate flask, add remaining components to distilled/deionized water and bring volume to 500.0mL. Mix thoroughly. Autoclave both solutions separately for 15 min at 15 psi pressure–121°C. Cool to 25°C. Aseptically combine the two sterile solutions. Mix thoroughly. Aseptically distribute into sterile tubes or flasks. After inoculation, maintain pH at 8.2–8.4 with sterile 50% K_2CO_3 solution.

Use: For the cultivation and maintenance of *Nitrosomonas europaea.*

NMS Medium
See: **Nitrate Mineral Salts Medium**

NMS Medium with Methanol
See: **Nitrate Mineral Salts Medium with Methanol**

NNN Medium (Novy, MacNeal and Nicole Medium)

Composition per liter:
Agar 7.0g
NaCl 3.0g
Rabbit blood, defibrinated 150.0mL

Preparation of Medium: Add components, except rabbit blood, to distilled/deionized water and bring volume to 850.0mL. Mix thoroughly. Gently heat and bring to boiling. Autoclave for 15 min at 15 psi pressure–121°C. Cool to 50°C. Aseptically add sterile rabbit blood. Mix thoroughly. Aseptically distribute into sterile tubes in 5.0mL volumes. Allow tubes to cool in a slanted position at 4°C.

Use: For the cultivation and maintenance of *Leishmania* species and *Trypanosoma cruzi.*

Nocardia histidans Medium

Composition per liter:
Agar 20.0g
Yeast extract 10.0g
Glucose 10.0g
Na_2HPO_4 .... 0.95g
KH_2PO_4 .... 0.91g
$MgSO_4 \cdot 7H_2O$ 0.5g
pH 7.0 ± 0.2 at 25°C

Preparation of Medium: Add components to distilled/deionized water and bring volume to 1.0L. Mix thoroughly. Gently heat and bring to boiling. Distribute into tubes or flasks. Autoclave for 15 min at 15 psi pressure–121°C. Pour into sterile Petri dishes or leave in tubes.

Use: For the cultivation and maintenance of *Nocardia histidans* and *Streptomyces* species.

Nocardia Medium

Composition per liter:
Agar 20.0g
Peptone 10.0g

Beef extract 5.0g
NaCl 2.5g

Preparation of Medium: Add components to distilled/deionized water and bring volume to 1.0L. Mix thoroughly. Gently heat and bring to boiling. Distribute into tubes or flasks. Autoclave for 15 min at 15 psi pressure–121°C. Pour into sterile Petri dishes or leave in tubes.

Use: For the cultivation and maintenance of *Rhodococcus globerulus* and *Nocardia* species.

Nocardia Medium 1

Composition per 1010mL:

Agar...12.0g
Proteose peptone 10.0g
Veal infusion solids 10.0g
NaCl 3.0g
Na_2HPO_4 ....2.0g
Glucose 2.0g
Sodium acetate 1.0g
Adenine sulfate 0.01g
Guanine·HCl 0.01g
Uracil..0.01g
Xanthine 0.01g
Thiamine 0.02mg
Additives solution 10.0mL

pH 7.4 ± 0.2 at 25°C

Additives Solution:

Composition per 10mL:

Actidione (cycloheximide)........................... 0.05mg
Mycostatin... 0.05mg
Dimethylchlortetracycline·HCl 5.0μg

Preparation of Additives Solution: Add components to distilled/deionized water and bring volume to 10.0mL. Mix thoroughly. Filter sterilize.

Preparation of Medium: Add components, except additives solution, to distilled/deionized water and bring volume to 990.0mL. Mix thoroughly. Gently heat and bring to boiling. Autoclave for 15 min at 15 psi pressure–121°C. Cool to 45°–50°C. Aseptically add sterile components. Mix thoroughly. Pour into sterile Petri dishes or distribute into sterile tubes.

Use: For the isolation and cultivation of *Nocardia*.

Nocardia Medium 2

Composition per 1010mL:

Agar...12.0g
Proteose peptone 10.0g
Veal infusion solids 10.0g
NaCl 3.0g
Na_2HPO_4 ....2.0g
Glucose 2.0g
Sodium acetate 1.0g
Adenine sulfate 0.01g
Guanine·HCl 0.01g
Uracil..0.01g
Xanthine 0.01g
Thiamine 0.02mg
Additives solution 10.0mL

pH 7.4 ± 0.2 at 25°C

Additives Solution:

Composition per 10mL:

Actidione (cycloheximide)........................... 0.05mg
Mycostatin... 0.05mg
Methacycline HCl 0.01mg

Preparation of Additives Solution: Add components to distilled/deionized water and bring volume to 10.0mL. Mix thoroughly. Filter sterilize.

Preparation of Medium: Add components, except additives solution, to distilled/deionized water and bring volume to 990.0mL. Mix thoroughly. Gently heat and bring to boiling. Autoclave for 15 min at 15 psi pressure–121°C. Cool to 45°–50°C. Aseptically add sterile components. Mix thoroughly. Pour into sterile Petri dishes or distribute into sterile tubes.

Use: For the isolation and cultivation of *Nocardia*.

Nocardia Medium 3

Composition per 1010mL:

Agar...12.0g
Proteose peptone 10.0g
Veal infusion solids 10.0g
NaCl 3.0g
Na_2HPO_4 ....2.0g
Glucose 2.0g
Sodium acetate 1.0g
Adenine sulfate 0.01g
Guanine·HCl 0.01g
Uracil..0.01g
Xanthine 0.01g
Thiamine 0.02mg
Actidione.. 0.05mg
Mycostatin... 0.05mg
Chlortetracycline HCl 0.045mg
Demethylchlortetracycline HCl 5.0μg
Additives solution 10.0mL

pH 7.4 ± 0.2 at 25°C

Additives Solution:

Composition per 10mL:

Actidione (cycloheximide)........................... 0.05mg
Mycostatin... 0.05mg
Dimethylchlortetracycline·HCl 5.0μg

Preparation of Additives Solution: Add components to distilled/deionized water and bring volume to 10.0mL. Mix thoroughly. Filter sterilize.

Preparation of Medium: Add components, except additives solution, to distilled/deionized water and bring volume to 990.0mL. Mix thoroughly. Gently heat and bring to boiling. Autoclave for 15 min at 15 psi pressure–121°C. Cool to 45°–50°C. Aseptically add sterile components. Mix thoroughly. Pour into sterile Petri dishes or distribute into sterile tubes.

Use: For the isolation and cultivation of *Nocardia*.

Nocardia Medium 4

Composition per 1010mL:

Agar	12.0g
Proteose peptone	10.0g
Veal infusion solids	10.0g
NaCl	3.0g
Na_2HPO_4	2.0g
Glucose	2.0g
Sodium acetate	1.0g
Adenine sulfate	0.01g
Guanine·HCl	0.01g
Uracil	0.01g
Xanthine	0.01g
Thiamine	0.02mg
Additives solution	10.0mL

pH 7.4 ± 0.2 at 25°C

Additives Solution:

Composition per 10mL:

Actidione (cycloheximide)	0.05mg
Mycostatin	0.05mg
Chlortetracycline·HCl	0.045mg
Methacycline·HCl	0.01mg

Preparation of Additives Solution: Add components to distilled/deionized water and bring volume to 10.0mL. Mix thoroughly. Filter sterilize.

Preparation of Medium: Add components, except additives solution, to distilled/deionized water and bring volume to 990.0mL. Mix thoroughly. Gently heat and bring to boiling. Autoclave for 15 min at 15 psi pressure–121°C. Cool to 45°–50°C. Aseptically add sterile components. Mix thoroughly. Pour into sterile Petri dishes or distribute into sterile tubes.

Use: For the isolation and cultivation of *Nocardia* species.

Nonfat Dry Milk, Reconstituted

Composition per liter:

Milk, nonfat dry	100.0g

pH 6.8 ± 0.2 at 25°C

Preparation of Medium: Add 100.0g nonfat dry milk to distilled/deionized water and bring volume to 1.0L. Mix thoroughly. Distribute into tubes or flasks. Autoclave for 15 min at 15 psi pressure–121°C.

Use: For the cultivation of *Salmonella* species and monkey kidney cells in tissue culture.

Nonnutrient Agar Plates

Composition per liter:

Agar	15.0g
Page's amoeba saline	1.0L

Page's Amoeba Saline:

Composition per liter:

Na_2HPO_4	0.142g
KH_2PO_4	0.136g
NaCl	0.12g
$MgSO_4·7H_2O$	4.0mg
$CaCl_2·2H_2O$	4.0mg

Preparation of Page's Amoeba Saline: Add components to distilled/deionized water and bring volume to 10.0mL. Mix thoroughly.

Preparation of Medium: Add agar to 1.0L of Page's amoeba saline. Mix thoroughly. Gently heat and bring to boiling. Autoclave for 15 min at 15 psi pressure–121°C. Cool to 60°C. Pour into sterile Petri dishes in 20.0mL volumes. Store at 4°C for up to three months.

Use: For isolation and cultivation of pathogenic free-living amoebae.

NOS Medium, Modified

Composition per 100.67mL:

Basal medium	94.0mL
$NaHCO_3$ solution	2.67mL
TPP/VFA mixture	2.0mL
Rabbit serum, heat inactivated	2.0mL

pH 7.4 ± 0.2 at 25°C

Basal Medium:

Composition per 94mL:

Pancreatic digest of casein	1.0g
Pancreatic digest of gelatin	0.48g
Yeast extract	0.25g
Brain heart, solids from infusion	0.2g
Peptic digest of animal tissue	0.2g
D-Glucose	0.2g
NaCl	0.17g
Glucose	0.1g
L-Cysteine·HCl·H_2O	0.1g
Na_2HPO_4	0.085g
Sodium thioglycollate	0.05g
L-Asparagine	0.025g
Resazurin (0.1% w/v solution)	0.1mL

Preparation of Basal Medium: Add components to distilled/deionized water and bring volume to 94.0mL. Mix thoroughly. Gently heat and bring to boiling. Gas under O_2-free 85% N_2 + 10% CO_2 + 5% H_2. Stopper and wire flask closed. Autoclave for 20 min at 15 psi pressure–121°C. Cool to 45°–50°C.

$NaHCO_3$ Solution:

Composition per 10mL:

$NaHCO_3$	0.75g

Preparation of $NaHCO_3$ Solution: Add the $NaHCO_3$ to distilled/deionized water and bring volume to 10.0mL. Mix thoroughly. Filter sterilize.

TPP/VFA Mixture:

Composition per 10.9mL:

Thiamine pyrophosphate (0.2% solution)	1.5mL
VFA Solution	1.0mL

Preparation of TPP/VFA Mixture: Add components to distilled/deionized water and bring volume to 10.9mL. Mix thoroughly. Filter sterilize. Store at –20°C.

VFA Solution:

Composition per 100mL:

NaOH (0.1*N* solution)	98.0mL
Isobutyric acid	0.5mL
2-Methylbutyric acid	0.5mL
Isovaleric acid	0.5mL
Valeric acid	0.5mL

Preparation of VFA Solution: Add volatile fatty acids to 98.0mL of NaOH solution. Mix thoroughly. Filter sterilize. Store at 4°C.

Preparation of Medium: Open the flask containing 94.0mL of cooled sterile basal medium while flushing with O_2-free 85% N_2 + 10% CO_2 + 5% H_2. Aseptically add sterile $NaHCO_3$ solution, sterile TPP/VFA mixture, and filter-sterilized rabbit serum. Mix thoroughly.

Use: For the cultivation and maintenance of *Treponema vincentii* and other *Treponema* species.

NOS Spirochete Medium

Composition per 1045mL:

Basal medium	1.0L
$NaHCO_3$ (10% solution)	20.0mL
Rabbit serum, heat inactivated	20.0mL
Thiamine pyrophosphate (0.2% solution)	3.0mL
VFA solution	2.0mL

pH 7.4 ± 0.2 at 25°C

Basal Medium:

Composition per liter:

Pancreatic digest of casein	10.0g
Pancreatic digest of gelatin	4.85g
Noble agar	3.0g
Yeast extract	2.5g
Brain heart, solids from infusion	2.0g
Peptic digest of animal tissue	2.0g
Glucose	2.0g
NaCl	1.65g
Glucose	1.0g
L-Cysteine·HCl·H_2O	1.0g
Na_2HPO_4	0.85g
Sodium thioglycollate	0.5g
L-Asparagine	0.25g

Preparation of Basal Medium: Add components to distilled/deionized water and bring volume to 1.0L. Mix thoroughly. Gently heat and bring to boiling. Gas under O_2-free 80% N_2 + 10% CO_2 + 10% H_2. Stopper and wire flask closed. Autoclave for 20 min at 15 psi pressure–121°C. Cool to 45°–50°C.

VFA Solution:

Composition per 100mL:

KOH (0.1*N* solution)	98.0mL
Isobutyric acid	0.5mL
2-Methylbutyric acid	0.5mL
Isovaleric acid	0.5mL
Valeric acid	0.5mL

Preparation of VFA Solution: Add volatile fatty acids to 98.0mL of KOH solution. Mix thoroughly. Filter sterilize. Store at 4°C.

Preparation of Medium: Combine 20.0mL of $NaHCO_3$ solution, 20.0mL of rabbit serum, 3.0mL of thiamine pyrophosphate solution, and 2.0mL of VFA solution. Mix thoroughly. Filter sterilize. Open the flask containing 1.0L of cooled sterile basal medium while flushing with O_2-free 85% N_2 + 10% CO_2 + 5% H_2. Aseptically add the filter-sterilized mixture. Mix thoroughly. Aseptically and anaerobically distribute into sterile tubes or flasks.

Use: For the cultivation and maintenance of *Treponema denticola* and *Treponema socranskii*.

NSMP, Modified

Composition per liter:

Casamino acids	5.0g
Glucose	2.0g
KH_2PO_4	0.86g
Sodium citrate	0.6g
K_2HPO_4	0.55g
$MgCl_2·6H_2O$	0.43g
$CaCl_2$	0.1g
$MnCl_2·4H_2O$	0.016g
$ZnCl_2$	7.0mg
$FeCl_3$	3.0mg

pH 6.5 ± 0.2 at 25°C

Preparation of Medium: Add components to distilled/deionized water and bring volume to 1.0L. Mix thoroughly. Distribute into tubes or flasks. Autoclave for 15 min at 15 psi pressure–121°C.

Use: For the cultivation and maintenance of *Bacillus thuringiensis.*

Nutrient Agar

Composition per liter:

Agar	15.0g
Peptone	5.0g
NaCl	5.0g
Yeast extract	2.0g
Beef extract	1.0g

pH 7.4 ± 0.2 at 25°C

Source: This medium is available as a premixed powder from Oxoid Unipath.

Preparation of Medium: Add components to distilled/deionized water and bring volume to 1.0L. Mix thoroughly. Gently heat and bring to boiling. Distribute into tubes or flasks. Autoclave for 15 min at 15 psi pressure–121°C. Pour into sterile Petri dishes or leave in tubes.

Use: For the cultivation and maintenance of a wide variety of microorganisms.

Nutrient Agar (ATCC Medium 3)

Composition per liter:

Agar	15.0g
Pancreatic digest of gelatin	5.0g
Beef extract	3.0g

pH 6.8 ± 0.2 at 25°C

Source: This medium is available as a premixed powder from BBL Microbiology Systems and Difco Laboratories.

Preparation of Medium: Add components to distilled/deionized water and bring volume to 1.0L. Mix thoroughly. Gently heat while stirring and bring to boiling. Distribute into tubes or flasks. Autoclave for 15 min at 15 psi pressure–121°C. Pour into sterile Petri dishes or leave in tubes.

Use: For the cultivation of a wide variety of bacteria and for the enumeration of organisms in water, sewage, feces and other materials.

Nutrient Agar, 1.5% (ATCC Medium 105)

Composition per liter:

Agar	15.0g
NaCl	8.0g
Pancreatic digest of gelatin	5.0g
Beef extract	3.0g

pH 7.3 ± 0.2 at 25°C

Source: This medium is available as a premixed powder from BBL Microbiology Systems and Difco Laboratories.

Preparation of Medium: Add components to distilled/deionized water and bring volume to 1.0L. Mix thoroughly. Gently heat while stirring and bring to boiling. Distribute into tubes or flasks. Autoclave for 15 min at 15 psi pressure–121°C. Pour into sterile Petri dishes or leave in tubes.

Use: For the cultivation and maintenance of a variety of nonfastidious bacteria.

Nutrient Agar, pH 5.0 (Oxoid Nutrient Agar No. 1, pH 5.0)

Composition per liter:

Agar	15.0g
Peptone	5.0g
NaCl	5.0g
Yeast extract	2.0g
Beef extract	1.0g

pH 5.0 ± 0.2 at 25°C

Source: This medium is available as a premixed powder from Oxoid Unipath.

Preparation of Medium: Add components to distilled/deionized water and bring volume to 1.0L. Mix thoroughly. Gently heat and bring to boiling. Adjust pH to 5.0. Distribute into tubes or flasks. Autoclave for 15 min at 15 psi pressure–121°C. Pour into sterile Petri dishes or leave in tubes.

Use: For the cultivation and maintenance of *Pseudomonas phenazinium.*

Nutrient Agar, pH 6.0

Composition per liter:

Agar	15.0g
Peptone	5.0g
Beef extract	3.0g

pH 6.0 ± 0.2 at 25°C

Source: This medium is available as a premixed powder from Difco Laboratories.

Preparation of Medium: Add components to distilled/deionized water and bring volume to 1.0L. Mix thoroughly. Gently heat and bring to boiling. Adjust pH to 6.0. Distribute into tubes or flasks. Autoclave for 15 min at 15 psi pressure–121°C. Pour into sterile Petri dishes or leave in tubes.

Use: For the cultivation of microorganisms that prefer a slightly acid nutrient agar.

Nutrient Agar, pH 8.0

Composition per liter:

Agar	15.0g
Pancreatic digest of gelatin	5.0g
Beef extract	3.0g

pH 8.0 ± 0.2 at 25°C.

Preparation of Medium: Add components to distilled/deionized water and bring volume to 1.0L. Mix thoroughly. Gently heat and bring to boiling. Adjust pH to 8.0. Distribute into tubes or flasks. Autoclave for 15 min at 15 psi pressure–121°C. Pour into sterile Petri dishes or leave in tubes. Allow tubes to cool in a slanted position.

Use: For the cultivation and maintenance of *Bacillus alcalophilus*.

Nutrient Agar with 0.5% NaCl

Composition per liter:

Agar	15.0g
NaCl	5.0g
Pancreatic digest of gelatin	5.0g
Beef extract	3.0g

pH 6.8 ± 0.2 at 25°C

Source: Nutrient Agar is available as a premixed powder from Difco Laboratories.

Preparation of Medium: Add components to distilled/deionized water and bring volume to 1.0L. Mix thoroughly. Gently heat and bring to boiling. Distribute into tubes or flasks. Autoclave for 15 min at 15 psi pressure–121°C. Pour into sterile Petri dishes or leave in tubes.

Use: For the cultivation and maintenance of *Agrobacterium tumefaciens*, *Escherichia coli*, *Pseudomonas aeruginosa*, *Salmonella choleraesuis*, *Shigella dysenteriae*, *Shigella flexneri*, *Vibrio* species, and *Yersinia* species.

Nutrient Agar with 0.5% NaCl and Sodium Citrate

Composition per liter:

Agar	15.0g
NaCl	5.0g
Pancreatic digest of gelatin	5.0g
Beef extract	3.0g
Sodium citrate	2.94g

pH 6.8 ± 0.2 at 25°C

Source: Nutrient Agar is available as a premixed powder from Difco Laboratories.

Preparation of Medium: Add components to distilled/deionized water and bring volume to 1.0L. Mix thoroughly. Gently heat and bring to boiling. Distribute into tubes or flasks. Autoclave for 15 min at 15 psi pressure–121°C. Pour into sterile Petri dishes or leave in tubes.

Use: For the cultivation and maintenance of *Escherichia coli*.

Nutrient Agar with 1.5% NaCl

Composition per liter:

NaCl	15.0g
Agar	15.0g
Pancreatic digest of gelatin	5.0g
Beef extract	3.0g

pH 6.8 ± 0.2 at 25°C

Source: Nutrient Agar is available as a premixed powder from Difco Laboratories.

Preparation of Medium: Add components to distilled/deionized water and bring volume to 1.0L. Mix thoroughly. Gently heat and bring to boiling. Distribute into tubes or flasks. Autoclave for 15 min at 15 psi pressure–121°C. Pour into sterile Petri dishes or leave in tubes.

Use: For the cultivation and maintenance of *Photobacterium leiognathi*, *Pseudomonas fluorescens*, and *Vibrio natriegens*.

Nutrient Agar with 3% NaCl

Composition per liter:

NaCl	30.0g
Agar	15.0g
Pancreatic digest of gelatin	5.0g
Beef extract	3.0g

pH 6.8 ± 0.2 at 25°C

Source: Nutrient Agar is available as a premixed powder from Difco Laboratories.

Preparation of Medium: Add components to distilled/deionized water and bring volume to 1.0L. Mix thoroughly. Gently heat and bring to boiling. Distribute into tubes or flasks. Autoclave for 15 min at 15 psi pressure–121°C. Pour into sterile Petri dishes or leave in tubes.

Use: For the cultivation and maintenance of *Bacillus* species, *Alteromonas nigrifaciens*, *Halococcus* species, *Planococcus citreus*, *Pseudomonas beijerinckii*, *Staphylococcus* species, *Streptococcus pyogenes* and *Vibrio* species.

Nutrient Agar with 10% NaCl

Composition per liter:

NaCl 100.0g
Agar 15.0g
Pancreatic digest of gelatin 5.0g
Beef extract 3.0g

pH 6.8 ± 0.2 at 25°C

Source: Nutrient Agar is available as a premixed powder from Difco Laboratories.

Preparation of Medium: Add components to distilled/deionized water and bring volume to 1.0L. Mix thoroughly. Gently heat and bring to boiling. Distribute into tubes or flasks. Autoclave for 15 min at 15 psi pressure–121°C. Pour into sterile Petri dishes or leave in tubes.

Use: For the cultivation and maintenance of *Paracoccus halodenitrificans* and *Micrococcus* species.

Nutrient Agar with 10% NaCl and Maltose

Composition per liter:

NaCl 100.0g
Agar 15.0g
Maltose 10.g
Pancreatic digest of gelatin 5.0g
Beef extract 3.0g

pH 6.8 ± 0.2 at 25°C

Source: Nutrient Agar is available as a premixed powder from Difco Laboratories.

Preparation of Medium: Add components to distilled/deionized water and bring volume to 1.0L. Mix thoroughly. Gently heat and bring to boiling. Distribute into tubes or flasks. Autoclave for 15 min at 15 psi pressure–121°C. Pour into sterile Petri dishes or leave in tubes.

Use: For the cultivation and maintenance of *Paracoccus halodenitrificans* and *Micrococcus* species.

Nutrient Agar with 1% Methanol (ATCC Medium 620)

Composition per liter:

Agar 15.0g
Pancreatic digest of gelatin 5.0g
Beef extract 3.0g
Methanol 10.0mL

pH 6.8 ± 0.2 at 25°C

Source: Nutrient Agar is available as a premixed powder from Difco Laboratories.

Preparation of Medium: Filter sterilize methanol. Add components, except methanol, to distilled/deionized water and bring volume to 990.0mL. Mix thoroughly. Gently heat and bring to boiling. Autoclave for 15 min at 15 psi pressure–121°C. Cool to 45°–50°C. Aseptically add sterile methanol. Mix thoroughly. Pour into sterile Petri dishes or distribute into sterile tubes.

Use: For the cultivation and maintenance of *Bacillus* species, *Methylomonas clara*, and *Pseudomonas methanolica*.

Nutrient Agar with 2% Methanol (ATCC Medium 628)

Composition per liter:

Agar 15.0g
Pancreatic digest of gelatin 5.0g
Beef extract 3.0g
Methanol 20.0mL

pH 6.8 ± 0.2 at 25°C

Source: Nutrient Agar is available as a premixed powder from Difco Laboratories.

Preparation of Medium: Filter sterilize methanol. Add components, except methanol, to distilled/deionized water and bring volume to 980.0mL. Mix thoroughly. Gently heat and bring to boiling. Autoclave for 15 min at 15 psi pressure–121°C. Cool to 45°–50°C. Aseptically add sterile methanol. Mix thoroughly. Pour into sterile Petri dishes or distribute into sterile tubes.

Use: For the cultivation and maintenance of *Pseudomonas* species.

Nutrient Agar with Cysteine

Composition per liter:

Agar 15.0g
Pancreatic digest of gelatin 5.0g
Beef extract 3.0g
L-Cysteine 0.1g

pH 6.8 ± 0.2 at 25°C

Source: Nutrient Agar is available as a premixed powder from Difco Laboratories.

Preparation of Medium: Add components to distilled/deionized water and bring volume to 1.0L. Mix thoroughly. Gently heat and bring to boiling. Distribute into tubes or flasks. Autoclave for 15 min at 15 psi pressure–121°C. Pour into sterile Petri dishes or leave in tubes.

Use: For the cultivation and maintenance of *Salmonella choleraesuis*.

Nutrient Agar with Dihydrostreptomycin

Composition per liter:

Agar	15.0g
Pancreatic digest of gelatin	5.0g
Beef extract	3.0g
Dihydrostreptomycin solution	10.0mL

pH 6.8 ± 0.2 at 25°C

Source: Nutrient Agar is available as a premixed powder from Difco Laboratories.

Dihydrostreptomycin Solution:
Composition per 10mL:

Dihydrostreptomycin	0.625g

Preparation of Dihydrostreptomycin Solution: Add dihydrostreptomycin to distilled/deionized water and bring volume to 10.0mL. Mix thoroughly. Filter sterilize.

Preparation of Medium: Add components, except dihydrostreptomycin solution, to distilled/deionized water and bring volume to 990.0mL. Mix thoroughly. Gently heat and bring to boiling. Autoclave for 15 min at 15 psi pressure–121°C. Cool to 45°–50°C. Aseptically add sterile dihydrostreptomycin solution. Mix thoroughly. Pour into sterile Petri dishes or distribute into sterile tubes.

Use: For the cultivation and maintenance of *Escherichia coli*, *Micrococcus luteus*, *Shigella* species and *Vibrio cholerae*.

Nutrient Agar with Erythromycin

Composition per liter:

Agar	15.0g
Peptone	5.0g
Beef extract	3.0g
Erythromycin solution	10.0mL

pH 6.8 ± 0.2 at 25°C.

Erythromycin Solution:
Composition per 10mL:

Erythromycin	0.6g

Preparation of Erythromycin Solution: Add erythromycin to distilled/deionized water and bring volume to 10.0mL. Mix thoroughly. Filter sterilize.

Preparation of Medium: Add components, except erythromycin solution, to distilled/deionized water and bring volume to 990.0mL. Mix thoroughly. Gently heat and bring to boiling. Autoclave for 15 min at 15 psi pressure–121°C. Cool to 45°–50°C. Aseptically add sterile erythromycin solution. Mix thoroughly. Pour into sterile Petri dishes or distribute into sterile tubes.

Use: For the cultivation and maintenance of *Micrococcus luteus*.

Nutrient Agar with Ethanolamine

Composition per liter:

Agar	15.0g
Pancreatic digest of gelatin	5.0g
Beef extract	3.0g
Ethanolamine	2.0mL

pH 6.8 ± 0.2 at 25°C

Source: Nutrient Agar is available as a premixed powder from Difco Laboratories.

Preparation of Medium: Add components to distilled/deionized water and bring volume to 1.0L. Mix thoroughly. Gently heat and bring to boiling. Distribute into tubes or flasks. Autoclave for 15 min at 15 psi pressure–121°C. Pour into sterile Petri dishes or leave in tubes.

Use: For the cultivation and maintenance of *Flavobacterium tirreniculum* and *Pseudomonas* species.

Nutrient Agar with Ethylene Glycol

Composition per liter:

Agar	15.0g
Pancreatic digest of gelatin	5.0g
Beef extract	3.0g
Ethylene glycol	2.0mL

pH 6.8 ± 0.2 at 25°C

Source: Nutrient Agar is available as a premixed powder from Difco Laboratories.

Preparation of Medium: Add components to distilled/deionized water and bring volume to 1.0L. Mix thoroughly. Gently heat and bring to boiling. Distribute into tubes or flasks. Autoclave for 15 min at 15 psi pressure–121°C. Pour into sterile Petri dishes or leave in tubes.

Use: For the cultivation and maintenance of *Pseudomonas putida*.

Nutrient Agar with Glucose

Composition per liter:

Agar	15.0g
Pancreatic digest of gelatin	5.0g
Beef extract	3.0g
Glucose	10.0g

pH 6.8 ± 0.2 at 25°C

Source: Nutrient Agar is available as a premixed powder from Difco Laboratories.

Preparation of Medium: Add components to distilled/deionized water and bring volume to 1.0L. Mix thoroughly. Gently heat and bring to boiling. Distribute into tubes or flasks. Autoclave for 15 min at 15 psi pressure–121°C. Pour into sterile Petri dishes or leave in tubes.

Use: For the cultivation and maintenance of *Amycolata saturnea, Arthrobacter* species, *Corynebacterium* species, *Curtobacterium flaccumfaciens, Deinococcus radiodurans, Escherichia coli, Hafnia alvei, Micrococcus aurantiacus, Myxomicrobium multiplex, Nocardia petroleophila, Nocardia* species, *Pseudomonas* species, *Rhodococcus rhodochrous, Streptomyces piedadensis* and *Xanthomonas* species.

Nutrient Agar with Horse Serum

Composition per liter:

Agar	15.0g
Pancreatic digest of gelatin	5.0g
Beef extract	3.0g
Horse serum	100.0mL

pH 6.8 ± 0.2 at 25°C

Source: Nutrient Agar is available as a premixed powder from Difco Laboratories.

Preparation of Medium: Add components, except horse serum, to distilled/deionized water and bring volume to 900.0mL. Mix thoroughly. Gently heat and bring to boiling. Autoclave for 15 min at 15 psi pressure–121°C. Cool to 45°–50°C. Aseptically add sterile horse serum. Mix thoroughly. Pour into sterile Petri dishes or distribute into sterile tubes.

Use: For the cultivation and maintenance of *Alysiella filiformis* and *Simonsiella crassa*.

Nutrient Agar with Manganese (NAMn)

Composition per liter:

Agar	15.0g
Pancreatic digest of gelatin	5.0g
Beef extract	3.0g
Manganese sulfate solution	1.0mL

pH 6.8 ± 0.2 at 25°C

Manganese Sulfate Solution:
Composition per 100mL:

$MnSO_4 \cdot H_2O$	3.08g

Preparation of Manganese Sulfate Solution: Add $MnSO_4 \cdot H_2O$ to distilled/deionized water and bring volume to 100.0mL. Mix thoroughly.

Preparation of Medium: Add components to distilled/deionized water and bring volume to 1.0L. Mix thoroughly. Gently heat and bring to boiling. Distribute into tubes or flasks. Autoclave for 15 min at 15 psi pressure–121°C. Pour into sterile Petri dishes or leave in tubes.

Use: For the cultivation of *Bacillus* species from canned foods. For enhanced spore production by *Bacillus* species.

Nutrient Agar with Phytone

Composition per liter:

Agar	15.0g
Phytone	10.0g
Pancreatic digest of gelatin	5.0g
Beef extract	3.0g

pH 6.8 ± 0.2 at 25°C

Preparation of Medium: Add components to distilled/deionized water and bring volume to 1.0L. Mix thoroughly. Gently heat while stirring and bring to boiling. Distribute into tubes or flasks. Autoclave for 15 min at 15 psi pressure–121°C. Pour into sterile Petri dishes or leave in tubes.

Use: For the cultivation of a wide variety of bacteria.

Nutrient Agar with Potato Starch

Composition per liter:

Agar	15.0g
Potato starch	10.0g
Pancreatic digest of gelatin	5.0g
Beef extract	3.0g

pH 6.8 ± 0.2 at 25°C

Source: Nutrient Agar is available as a premixed powder from Difco Laboratories.

Preparation of Medium: Add components to distilled/deionized water and bring volume to 1.0L. Mix thoroughly. Gently heat while stirring and bring to boiling. Distribute into tubes or flasks. Autoclave for 15 min at 15 psi pressure–121°C. Pour into sterile Petri dishes or leave in tubes.

Use: For the cultivation and maintenance of *Bacillus polymyxa* and *Bacillus subtilis*.

Nutrient Agar with Soil Extract

Composition per liter:

Agar	15.0g
Pancreatic digest of gelatin	5.0g
Beef extract	3.0g
Soil extract	250.0mL

pH 6.8 ± 0.2 at 25°C

Source: Nutrient Agar is available as a premixed powder from Difco Laboratories.

Soil Extract:

Composition per 300mL:

African Violet soil	115.5g
Na_2CO_3	0.3g

Preparation of Soil Extract: Add components to tap water and bring volume to 300.0mL. Autoclave for 60 min at 15 psi pressure–121°C. Filter through Whatman filter paper.

Preparation of Medium: Add components to distilled/deionized water and bring volume to 1.0L. Mix thoroughly. Gently heat and bring to boiling. Distribute into tubes or flasks. Autoclave for 15 min at 15 psi pressure–121°C. Pour into sterile Petri dishes.

Use: For the cultivation and maintenance of *Auerobacterium* species, *Bacillus* species, and *Saccharomonospora viridis*.

Nutrient Agar with Streptomycin

Composition per liter:

Agar	15.0g
Peptone	5.0g
Beef extract	3.0g
Streptomycin solution	10.0mL

pH 6.8 ± 0.2 at 25°C.

Streptomycin Solution:

Composition per 10mL:

Streptomycin	0.020g

Preparation of Streptomycin Solution: Add streptomycin to distilled/deionized water and bring volume to 10.0mL. Mix thoroughly. Filter sterilize.

Preparation of Medium: Add components, except streptomycin solution, to distilled/deionized water and bring volume to 990.0mL. Mix thoroughly. Gently heat and bring to boiling. Autoclave for 15 min at 15 psi pressure–121°C. Cool to 45°–50°C. Aseptically add sterile streptomycin solution. Mix thoroughly. Pour into sterile Petri dishes or distribute into sterile tubes.

Use: For the cultivation and maintenance of *Corynebacterium glutamicum* and *Corynebacterium herculis*.

Nutrient Agar with Sucrose

Composition per liter:

Sucrose	20.0g
Agar	15.0g
Pancreatic digest of gelatin	5.0g
Beef extract	3.0g

pH 6.8 ± 0.2 at 25°C

Source: Nutrient Agar is available as a premixed powder from Difco Laboratories.

Preparation of Medium: Add components to distilled/deionized water and bring volume to 1.0L. Mix thoroughly. Gently heat and bring to boiling. Distribute into tubes or flasks. Autoclave for 15 min at 15 psi pressure–121°C. Pour into sterile Petri dishes or leave in tubes.

Use: For the cultivation and maintenance of the *Pseudomonas* species.

Nutrient Agar with Tetracycline

Composition per liter:

Agar	15.0g
Pancreatic digest of gelatin	5.0g
Beef extract	3.0g
Tetracycline solution	10.0mL

pH 6.8 ± 0.2 at 25°C

Source: Nutrient Agar is available as a premixed powder from Difco Laboratories.

Tetracycline Solution:

Composition per 10mL:

Tetracycline	0.025g

Preparation of Tetracycline Solution: Add tetracycline to distilled/deionized water and bring volume to 10.0mL. Mix thoroughly. Filter sterilize.

Preparation of Medium: Add components, except tetracycline solution, to distilled/deionized water and bring volume to 990.0mL. Mix thoroughly. Gently heat and bring to boiling. Autoclave for 15 min at 15 psi pressure–121°C. Cool to 45°–50°C. Aseptically add sterile tetracycline solution. Mix thoroughly. Pour into sterile Petri dishes or distribute into sterile tubes.

Use: For the cultivation and maintenance of the *Salmonella choleraesuis*.

Nutrient Agar with Uracil

Composition per liter:

Agar...15.0g
Pancreatic digest of gelatin...5.0g
Beef extract...3.0g
Uracil...0.01g

pH 6.8 ± 0.2 at 25°C

Source: Nutrient Agar is available as a premixed powder from Difco Laboratories.

Preparation of Medium: Add components to distilled/deionized water and bring volume to 1.0L. Mix thoroughly. Gently heat and bring to boiling. Distribute into tubes or flasks. Autoclave for 15 min at 15 psi pressure–121°C. Pour into sterile Petri dishes or leave in tubes.

Use: For the cultivation and maintenance of *Escherichia coli*.

Nutrient Agar with V-8™ Juice

Composition per liter:

Agar...15.0g
Pancreatic digest of gelatin...5.0g
Beef extract...3.0g
V-8 Juice...200.0mL

pH 6.8 ± 0.2 at 25°C

Source: Nutrient Agar is available as a premixed powder from Difco Laboratories.

Preparation of Medium: Add components to distilled/deionized water and bring volume to 1.0L. Mix thoroughly. Gently heat and bring to boiling. Distribute into tubes or flasks. Autoclave for 15 min at 15 psi pressure–121°C. Pour into sterile Petri dishes or leave in tubes.

Use: For the cultivation and maintenance of *Pseudomonas tolaasii*.

Nutrient Agar with Yeast Extract

Composition per liter:

Yeast extract...20.0g
Agar...15.0g
Pancreatic digest of gelatin...5.0g
Beef extract...3.0g

pH 6.8 ± 0.2 at 25°C

Source: Nutrient Agar is available as a premixed powder from Difco Laboratories.

Preparation of Medium: Add components to distilled/deionized water and bring volume to 1.0L. Mix thoroughly. Gently heat and bring to boiling. Distribute into tubes or flasks. Autoclave for 15 min at 15 psi pressure–121°C. Pour into sterile Petri dishes or leave in tubes.

Use: For the cultivation and maintenance of *Bacillus anthracis* and *Comamonas testosteroni*.

Nutrient Broth

Composition per liter:

Peptone...5.0g
NaCl...5.0g
Yeast extract...2.0g
Beef extract...1.0g

pH 7.4 ± 0.2 at 25°C

Source: This medium is available as a premixed powder from Oxoid Unipath.

Preparation of Medium: Add components to distilled/deionized water and bring volume to 1.0L. Mix thoroughly. Distribute into tubes or flasks. Autoclave for 15 min at 15 psi pressure–121°C.

Use: For the cultivation of a wide variety of nonfastidious microorganisms.

Nutrient Broth

Composition per liter:

Pancreatic digest of gelatin...5.0g
Beef extract...3.0g

pH 6.9 ± 0.2 at 25°C

Source: This medium is available as a premixed powder from BBL Microbiology Systems and Difco Laboratories.

Preparation of Medium: Add components to distilled/deionized water and bring volume to 1.0L. Mix thoroughly. Distribute into tubes or flasks. Autoclave for 15 min at 15 psi pressure–121°C.

Use: For the cultivation of a wide variety of nonfastidious microorganisms.

Nutrient Broth Glycerol Medium

Composition per liter:

Glycerol...100.0g

Pancreatic digest of gelatin 5.0g
Beef extract 3.0g

pH 7.2–7.4 ± 0.2 at 25°C

Preparation of Medium: Add components to distilled/deionized water and bring volume to 1.0L. Mix thoroughly. Distribute into tubes or flasks. Autoclave for 15 min at 15 psi pressure–121°C.

Use: For the cultivation of *Pseudomonas* species.

Nutrient Broth, 1/2 Strength

Composition per liter:

Pancreatic digest of gelatin 2.5g
Beef extract 1.5g

pH 6.9 ± 0.2 at 25°C

Preparation of Medium: Add components to distilled/deionized water and bring volume to 1.0L. Mix thoroughly. Distribute into tubes or flasks. Autoclave for 15 min at 15 psi pressure–121°C.

Use: For the cultivation of *Cytophaga allerginae*.

Nutrient Broth, Diluted 1:100

Nutrient Broth:

Composition per liter:

Pancreatic digest of gelatin 5.0g
Beef extract 3.0g

pH 6.8 ± 0.2 at 25°C

Preparation of Nutrient Broth: Add components to distilled/deionized water and bring volume to 1.0L. Mix thoroughly.

Preparation of Medium: Add 10.0mL of nutrient broth to distilled/deionized water and bring volume to 1.0L. Mix thoroughly. Distribute into tubes or flasks. Autoclave for 15 min at 15 psi pressure–121°C.

Use: For the cultivation and maintenance of *Agromonas oligotrophica*.

Nutrient Broth NaCl Thymine Medium

Composition per liter:

Pancreatic digest of gelatin 5.0g
NaCl 5.0g
Beef extract 3.0g
Thymine 0.03g

pH 7.2 ± 0.2 at 25°C

Preparation of Medium: Add components to distilled/deionized water and bring volume to 1.0L. Mix thoroughly. Distribute into tubes or flasks. Autoclave for 15 min at 15 psi pressure–121°C.

Use: For the cultivation and maintenance of *Salmonella choleraesuis*.

Nutrient Broth No. 2

Composition per liter:

Beef extract 10.0g
Peptone 10.0g
NaCl 5.0g

pH 7.5 ± 0.2 at 25°C

Source: This medium is available as a premixed powder from Oxoid Unipath.

Preparation of Medium: Add components to distilled/deionized water and bring volume to 1.0L. Mix thoroughly. Distribute into tubes or flasks. Autoclave for 15 min at 15 psi pressure–121°C.

Use: For the cultivation of a variety of fastidious and nonfastidious microorganisms.

Nutrient Broth Salts Medium

Composition per liter:

Pancreatic digest of gelatin 5.0g
Beef extract 3.0g
NaCl 3.0g
$MgSO_4 \cdot 7H_2O$ 0.2g
$CaCl_2 \cdot 2H_2O$ 0.15g
$MnSO_4 \cdot H_2O$ 0.05g

pH 6.8 ± 0.2 at 25°C

Preparation of Medium: Add components to distilled/deionized water and bring volume to 1.0L. Mix thoroughly. Distribute into tubes or flasks. Autoclave for 15 min at 15 psi pressure–121°C.

Use: For the cultivation and maintenance of *Bacillus subtilis*.

Nutrient Broth, Standard II

Composition per liter:

Special peptone 8.6g
NaCl 6.4g

pH 7.5 ± 0.1 at 37°C

Preparation of Medium: Add components to distilled/deionized water and bring volume to 1.0L. Mix thoroughly. Distribute into tubes or flasks. Autoclave for 15 min at 15 psi pressure–121°C.

Use: For the cultivation of a variety of fastidious and nonfastidious microorganisms.

Nutrient Broth with Bovine Serum

Composition per liter:

NaCl	10.0g
Pancreatic digest of gelatin	5.0g
Beef extract	3.0g
Bovine serum	100.0mL

pH 6.9 ± 0.2 at 25°C

Preparation of Medium: Add components, except bovine serum, to distilled/deionized water and bring volume to 900.0mL. Mix thoroughly. Autoclave for 15 min at 15 psi pressure–121°C. Cool to 45°–50°C. Aseptically add sterile bovine serum. Mix thoroughly. Aseptically distribute into sterile tubes or flasks.

Use: For the cultivation and maintenance of *Pseudomonas anguilliseptica*.

Nutrient Broth with 6% NaCl

Composition per liter:

NaCl	60.0g
Pancreatic digest of gelatin	5.0g
Beef extract	3.0g

pH 6.8 ± 0.2 at 25°C

Preparation of Medium: Add components to distilled/deionized water and bring volume to 1.0L. Mix thoroughly. Distribute into tubes or flasks. Autoclave for 15 min at 15 psi pressure–121°C.

Use: For the cultivation of organisms in water, sewage, feces and other materials. For the cultivation and maintenance of *Paracoccus halodenitrificans*.

Nutrient Broth with Streptomycin

Composition per liter:

Peptone	5.0g
Beef extract	3.0g
Streptomycin solution	0.020g

pH 6.8 ± 0.2 at 25°C.

Streptomycin Solution:
Composition per 10mL:

Streptomycin	0.020g

Preparation of Streptomycin Solution: Add streptomycin to distilled/deionized water and bring volume to 10.0mL. Mix thoroughly. Filter sterilize.

Preparation of Medium: Add components, except streptomycin solution, to distilled/deionized water and bring volume to 990.0mL. Mix thoroughly. Gently heat and bring to boiling. Autoclave for 15 min at 15 psi pressure–121°C. Cool to 45°–50°C. Aseptically add sterile streptomycin solution. Mix thoroughly. Pour into sterile Petri dishes or distribute into sterile tubes.

Use: For the cultivation of *Corynebacterium glutamicum* and *Corynebacterium herculis*.

Nutrient Broth Yeast Extract Medium
See: **NY Medium**

Nutrient Broth Yeast Extract Medium
See: **NBY Medium**

Nutrient Gelatin

Composition per liter:

Gelatin	120.0g
Pancreatic digest of gelatin	5.0g
Beef extract	3.0g

pH 6.8 ± 0.2 at 25°C

Source: This medium is available as a premixed powder from BBL Microbiology Systems and Difco Laboratories and Oxoid Unipath.

Preparation of Medium: Add components to distilled/deionized water and bring volume to 1.0L. Mix thoroughly. Gently heat while stirring to 50°C. Distribute into tubes. Autoclave for 15 min at 15 psi pressure–121°C.

Use: For the cultivation and differentiation of bacteria based on their ability to liquefy gelatin.

Nutrient Gelatin

Composition per liter:

Beef heart, solids from infusion	500.0g
Gelatin	120.0g
Tryptose	10.0g
NaCl	5.0g

pH 7.4 ± 0.2 at 25°C

Preparation of Medium: Add components to distilled/deionized water and bring volume to 1.0L. Mix thoroughly. Gently heat and bring to boiling. Distribute into screw-capped tubes in 4.0mL volumes. Autoclave for 15 min at 15 psi pressure–121°C.

Use: For the cultivation of Gram-negative nonfermentative bacteria from foods.

Nutrient Yeast Glucose Medium

Composition per liter:

Glucose	10.0g
Pancreatic digest of gelatin	5.0g
Yeast extract	5.0g
Beef extract	3.0g

pH 6.8 ± 0.2 at 25°C

Preparation of Medium: Add components to distilled/deionized water and bring volume to 1.0L. Mix thoroughly. Distribute into tubes or flasks. Autoclave for 15 min at 15 psi pressure–121°C.

Use: For the cultivation and maintenance of *Erwinia amylovora*.

NWRI Agar (HPC Agar)

Composition per liter:

Agar	15.0g
Peptone	3.0g
Soluble casein	0.5g
K_2HPO_4	0.2g
$MgSO_4$	0.05g
$FeCl_3$	1.0mg

pH 7.2 ± 0.2 at 25°C.

Preparation of Medium: Add components to distilled/deionized water and bring volume to 1.0L. Mix thoroughly. Gently heat and bring to boiling. Adjust pH to 7.2. Distribute into tubes or flasks. Autoclave for 15 min at 15 psi pressure–121°C. Pour into sterile Petri dishes.

Use: For estimation of the number of live heterotrophic bacteria in water using the heterotrophic plate count technique.

NY Medium (Nutrient Broth Yeast Extract Medium)

Composition per liter:

NaCl	8.0g
Peptone	5.0g
Yeast extract	5.0g
Beef extract	3.0g

pH 7.2–7.4 at 25°C

Preparation of Medium: Add components to distilled/deionized water and bring volume to 1.0L. Mix thoroughly. Distribute into tubes or flasks. Autoclave for 15 min at 15 psi pressure–121°C.

Use: For the cultivation of *Pseudomonas* species.

Nystatin Assay Agar
See: **Antibiotic Medium 19**

N–Z Amine A™ Medium

Composition per liter:

N-Z-Amine A™	5.0g
Beef extract	1.0g
Glycerol	80.0mL

Preparation of Medium: Add components to tap water and bring volume to 1.0L. Mix thoroughly. Distribute into tubes or flasks. Autoclave for 15 min at 15 psi pressure–121°C.

Use: For the cultivation and maintenance of *Nocardia brevicatena*.

N–Z Amine A™ Medium with Soluble Starch and Glucose

Composition per liter:

Soluble starch	20.0g
Agar	15.0g
Glucose	10.0g
Yeast extract	5.0g
N-Z-Amine A™	5.0g
$CaCO_3$	1.0g

Preparation of Medium: Add components to distilled/deionized water and bring volume to 1.0L. Mix thoroughly. Gently heat and bring to boiling. Distribute into tubes or flasks. Autoclave for 15 min at 15 psi pressure–121°C. Pour into sterile Petri dishes or leave in tubes.

Use: For the cultivation and maintenance of *Actinomadura* species, *Actinoplanes* species, *Amycolatopsis fastidiosa*, *Catenuloplanes japonicus*, *Dactylosporangium* species, *Geodermatophilus obscurus*, *Glycomyces* species, *Kitasatosporia mediocidica*, *Micromonospora* species, *Saccharomonospora caesia*, *Saccharothrix aerocolonigenes*, *Streptomyces* species, and *Streptosporangium* species.

N–Z Amine A™ Glycerol Agar

Composition per liter:

Agar	15.0g
N-Z-Amine A™	5.0g
Beef extract	1.0g
Glycerol	70.0mL

pH 6.5–7.0 at 25°C

Preparation of Medium: Add components to distilled/deionized water and bring volume to 1.0L. Mix thoroughly. Gently heat and bring to boiling. Distrib-

ute into tubes or flasks. Autoclave for 15 min at 15 psi pressure–121°C. Pour into sterile Petri dishes or leave in tubes.

Use: For the isolation and cultivation of *Actinomadura* species, *Actinopolyspora* species, *Excellospora* species and *Microspora* species.

NZC Broth

Composition per liter:

N-Z-Amine A™	10.0g
NaCl	5.0g
$MgCl_2 \cdot 6H_2O$	2.0g
Casamino acids solution	5.0mL

Casamino Acids Solution:

Composition per 10mL:

Casamino acids	2.0g

Preparation of Casamino Acids Solution: Add casamino acids to distilled/deionized water and bring volume to 10.0mL. Mix thoroughly. Filter sterilize.

Preparation of Medium: Add components, except casamino acids solution, to distilled/deionized water and bring volume to 990.0mL. Mix thoroughly. Gently heat and bring to boiling. Autoclave for 15 min at 15 psi pressure–121°C. Cool to 45°–50°C. Aseptically add 5.0mL sterile casamino acids solution. Mix thoroughly. Aseptically distribute into sterile tubes or flasks.

Use: For the cultivation of *Escherichia coli.*

NZY Agar

Composition per liter:

Agar	20.0g
N-Z- Amine A™	10.0g
NaCl	5.0g
Yeast extract	5.0g
$MgCl_2 \cdot 6H_2O$	2.0g

Preparation of Medium: Add components to distilled/deionized water and bring volume to 1.0L. Mix thoroughly. Gently heat and bring to boiling. Distribute into tubes. Autoclave for 15 min at 15 psi pressure–121°C. Allow tubes to cool in a slanted position.

Use: For the cultivation and maintenance of a variety of microorganisms.

NZY Agar

Composition per liter:

Agar	11.0g
N-Z- Amine A™	10.0g
NaCl	5.0g
Yeast extract	5.0g
$MgCl_2 \cdot 6H_2O$	2.0g

Preparation of Medium: Add components to distilled/deionized water and bring volume to 1.0L. Mix thoroughly. Gently heat and bring to boiling. Distribute into tubes or flasks. Autoclave for 15 min at 15 psi pressure–121°C. Pour into sterile Petri dishes.

Use: For the cultivation and enumeration of a variety of microorganisms.

NZY Broth

Composition per liter:

N-Z- Amine A™	10.0g
NaCl	5.0g
Yeast extract	5.0g
$MgCl_2 \cdot 6H_2O$	2.0g

Preparation of Medium: Add components to distilled/deionized water and bring volume to 1.0L. Mix thoroughly. Gently heat and bring to boiling. Distribute into tubes. Autoclave for 15 min at 15 psi pressure–121°C.

Use: For the cultivation of a variety of microorganisms.

Oatmeal Agar (ATCC Medium 551)

Composition per liter:

Oatmeal	60.0g
Agar	12.5g

pH 6.0 ± 0.2 at 25°C

Source: This medium is available as a premixed powder from Difco Laboratories.

Preparation of Medium: Add components to distilled/deionized water and bring volume to 1.0L. Mix thoroughly. Gently heat and bring to boiling. Distribute into tubes or flasks. Autoclave for 15 min at 15 psi pressure–121°C. Pour into sterile Petri dishes or leave in tubes.

Use: Oatmeal Agar is used for cultivation of fungi and actinomycetes, particularly for macrospore formation.

Oatmeal Agar
See: **ISP Medium 3**

Oatmeal Nitrate Agar

Composition per liter:

Agar	15.0g
Oatmeal	3.0g
K_2HPO_4	0.5g
KNO_3	0.2g
$MgSO_4$	0.2g

pH 7.0 ± 0.2 at 25°C

Preparation of Medium: Add components to distilled/deionized water and bring volume to 1.0L. Mix thoroughly. Gently heat and bring to boiling. Distribute into tubes or flasks. Autoclave for 15 min at 15 psi pressure–121°C. Pour into sterile Petri dishes or leave in tubes.

Use: For the cultivation and maintenance of *Microtetraspora flexuosa.*

Oatmeal Soy Peptone Medium

Composition per liter:

Agar	20.0g
Oatmeal	20.0g
Glucose	2.0g
Soy peptone	2.0g

Preparation of Medium: Add components to distilled/deionized water and bring volume to 1.0L. Mix thoroughly. Gently heat and bring to boiling. Distribute into tubes or flasks. Autoclave for 15 min at 15 psi pressure–121°C. Pour into sterile Petri dishes or leave in tubes.

Use: For the cultivation and maintenance of *Streptomyces metachromogenes* and *Actinosporangium* species.

OF Glucose Medium, Semisolid
See: **Oxidative Fermentative Glucose Medium, Semisolid**

OF Glucose Medium, Semisolid with NaCl
See: **Oxidative Fermentative Glucose Medium, Semisolid, with NaCl**

OF Medium
See: **Oxidative Fermentative Medium**

Ogawa Egg Medium

Composition:

Chicken eggs, whole	200.0mL
Sodium glutamate-KH_2PO_4 solution	100.0mL
Glycerol	6.0mL
Malachite Green (2.0% solution)	6.0mL

pH 6.8 ± 0.2 at 25°C

Sodium Glutamate-KH_2PO_4 Solution:
Composition per 100mL:

Sodium glutamate	1.0g
KH_2PO_4	1.0g

Preparation of Sodium Glutamate-KH_2PO_4 Solution: Add components to distilled/deionized water and bring volume to 100.0mL. Mix thoroughly. Filter sterilize.

Preparation of Medium: Soak eggs with 1:100 dilution of saturated mercuric chloride solution for 1 min. Aseptically break eggs into a sterile graduated cylinder. Homogenize eggs. Add remaining components. Mix thoroughly. Aseptically distribute into sterile tubes in 10.0mL volumes. Inspissate at 90°C (moist heat) for 60 min.

Use: For the selective isolation and cultivation of *Nocardia* and *Rhodococci* species.

Ogawa TB Medium

Composition per 300mL:

KH_2PO_4	1.0g
Homogenized whole egg	200.0mL

Glycerol 6.0mL
Malachite Green (2% solution) 6.0mL

pH 6.5 ± 0.2 at 25°C

Homogenized Whole Egg:
Composition per liter:
Whole eggs 18–24

Preparation of Homogenized Whole Egg: Use fresh eggs, less than 1 week old. Scrub the shells with soap. Let stand in a soap solution for 30 min. Rinse in running water. Soak eggs in 70% ethanol for 15 min. Break the eggs into a sterile container. Homogenize by shaking. Filter through four layers of sterile cheesecloth into a sterile graduated cylinder. Measure out 1.0L.

Preparation of Medium: Add components, except homogenized whole egg, to distilled/deionized water and bring volume to 100.0mL. Mix thoroughly. Autoclave for 15 min at 15 psi pressure–121°C. Cool to 45°–50°C. Aseptically add 200.0mL of sterile homogenized whole egg. Mix thoroughly. Aseptically distribute into sterile screw-capped tubes in 7.0mL volumes. Inspissate at 85°–90°C (moist heat) for 60 min.

Use: For the isolation and cultivation of *Mycobacterium* species except for *M. leprae*.

OGYE Agar
See: Oxytetracycline Glucose Yeast Extract Agar

Oil Agar Medium

Composition per liter:
Agar, purified 20.0g
NaCl 10.0g
Oil powder 10.0g
NH_4NO_3 .... 1.0g
$MgSO_4$ 0.5g
Amphotericin B solution 10.0mL
K_2HPO_4 solution 7.0mL
KH_2PO_4 solution 3.0mL
$FeCl_3$ 0.1mL

Oil Powder:
Composition per 10g:
Hydrocarbon 10.0g
Silica gel 10.0g
Diethyl ether 30.0mL

Preparation of Oil Powder: Add 10.0g of hydrocarbon to 30.0mL of diethyl ether. Mix thoroughly. Add 10.0g of silica gel. Allow ether to evaporate.

Amphotericin B Solution:
Composition per 10mL:
Amphotericin B 0.01g

Preparation of Amphotericin B Solution: Add amphotericin B to distilled/deionized water and bring volume to 10.0mL. Mix thoroughly. Filter sterilize.

K_2HPO_4 Solution:
Composition per 100mL:
K_2HPO_4 .... 10.0g

Preparation of K_2HPO_4 Solution: Add K_2HPO_4 to distilled/deionized water and bring volume to 100.0mL. Mix thoroughly. Autoclave for 15 min at 15 psi pressure–121°C. Cool to 25°C.

KH_2PO_4 Solution:
Composition per 100mL:
KH_2PO_4 .... 10.0g

Preparation of KH_2PO_4 Solution: Add KH_2PO_4 to distilled/deionized water and bring volume to 100.0mL. Mix thoroughly. Autoclave for 15 min at 15 psi pressure–121°C. Cool to 25°C.

Preparation of Medium: Add components—except amphotericin B solution, K_2HPO_4 solution, and KH_2PO_4 solution—to distilled/deionized water and bring volume to 980.0mL. Mix thoroughly. Gently heat and bring to boiling. Autoclave for 15 min at 15 psi pressure–121°C. Cool to 45°–50°C. Aseptically add 10.0mL of sterile amphotericin B solution, 7.0mL of sterile K_2HPO_4 solution, and 3.0mL of sterile KH_2PO_4 solution. Mix thoroughly. Pour into sterile Petri dishes or distribute into sterile tubes.

Use: For the cultivation and enumeration of hydrocarbon-utilizing bacteria by direct plating of estuarine water and sediment samples.

Oleic Albumin Complex

Composition per liter:
Bovine albumin fraction V 50.0g
NaCl 8.5g
Oleic acid 0.6mL

Preparation of Medium: Add components to distilled/deionized water and bring volume to 1.0L. Mix thoroughly. Filter sterilize.

Use: For use in media employed for the cultivation of mycobacteria.

Önöz *Salmonella* Agar

Composition per liter:
Agar 15.0g
Sucrose 13.0g

Lactose	11.5g
Trisodium citrate–5, 5–hydrate	9.3g
Meat peptone	6.8g
Beef extract	6.0g
L-Phenylalanine	5.0g
$Na_2S_2O_3 \cdot 5H_2O$	4.25g
Bile salt mixture	3.825g
Yeast extract	3.0g
$Na_2HPO_4 \cdot 2H_2O$	1.0g
Ferric citrate	0.5g
Metachrome Yellow	0.47g
$MgSO_4 \cdot 7H_2O$	0.4g
Aniline Blue	0.25g
Neutral Red	0.022g
Brilliant Green	0.00166g

pH 7.1 ± 0.2 at 25°C

Preparation of Medium: Add components to distilled/deionized water and bring volume to 1.0L. Mix thoroughly. Gently heat and bring to boiling. Distribute into tubes or flasks. Autoclave for 15 min at 15 psi pressure–121°C. Pour into sterile Petri dishes or leave in tubes.

Use: For the isolation and cultivation of *Salmonella* from feces.

ONPG Broth

Composition per liter:

Peptone water	750.0mL
ONPG solution	250.0mL

pH 7.2–7.4 at 25°C

ONPG Solution:

Composition per 250mL:

ONPG (*o*-nitrophenyl-β-D-galactopyranoside)	1.5g
Sodium phosphate buffer (0.01M, pH 7.5)	250.0mL

Preparation of ONPG Solution: Add ONPG to 250.0mL of sodium phosphate buffer. Mix thoroughly. Filter sterilize.

Peptone Water:

Composition per 750mL:

Peptone	7.5g
NaCl	3.75g

Preparation of Peptone Water: Add components to distilled/deionized water and bring volume to 750.0mL. Mix thoroughly. Gently heat and bring to boiling. Adjust pH to 8.0–8.4. Continue boiling for 10 min. Filter through Whatman #1 filter paper. Readjust pH of filtrate to 7.2–7.4. Autoclave for 20 min at 10 psi pressure–115°C. Cool to 25°C.

Preparation of Medium: Aseptically combine the sterile ONPG solution with the cooled, sterile peptone water. Mix thoroughly. Aseptically distribute into tubes in 2.5–3.0.0mLvolumes. Store at 4°C for up to one month.

Use: For the differentiation of a variety of Gram-negative bacteria based on production of β-galactosidase. For the differentiation of lactose-delayed bacteria from lactose-negative bacteria. For the differentiation of *Pseudomonas cepacia* (positive) and *Pseudomonas maltophila* (positive) from other *Pseudomonas* species (negative). Bacteria that produce β-galactosidase turn the medium yellow.

OR Indicator Agar (Oxidation–Reduction Indicator Agar)

Composition per liter:

Agar	15.0g
Sodium glycerol phosphate	10.0g
Sodium thioglycollate	1.7g
$CaCl_2 \cdot 2H_2O$	0.1g
Methylene Blue	6.0mg

Preparation of Medium: Add components to distilled/deionized water and bring volume to 1.0L. Mix thoroughly. Gently heat and bring to boiling. Distribute into tubes or flasks. Autoclave for 15 min at 15 psi pressure–121°C. Pour into sterile Petri dishes or leave in tubes.

Use: For use as an indicator of oxygen free conditions in anaerobic culture chambers.

Oral *Fusobacterium* Medium

Composition per liter:

Agar	15.0g
Proteose peptone	10.0g
Na_2HPO_4	5.0g
Glucose	5.0g
Beef extract	3.0g
Soluble starch	2.0g
$NaNO_3$	1.0g
Yeast extract	1.0g
$Cysteine \cdot HCl \cdot H_2O$	0.5g
Ethyl Violet solution	10.0mL
Bacitracin solution	10.0mL

pH 7.6 ± 0.2 at 25°C

Ethyl Violet Solution:

Composition per 10mL:

Ethyl Violet	0.04g

Preparation of Ethyl Violet Solution: Add Ethyl Violet to distilled/deionized water and bring volume to 10.0mL. Mix thoroughly. Filter sterilize.

Bacitracin Solution:

Composition per 10mL:

Bacitracin	1.0mg

Preparation of Bacitracin Solution: Add bacitracin to distilled/deionized water and bring volume to 10.0mL. Mix thoroughly. Filter sterilize.

Preparation of Medium: Add components, except Ethyl Violet solution and bacitracin solution, to distilled/deionized water and bring volume to 980.0mL. Mix thoroughly. Gently heat and bring to boiling. Autoclave for 15 min at 15 psi pressure–121°C. Cool to 45°–50°C. Aseptically add sterile Ethyl Violet solution and bacitracin solution. Mix thoroughly. Pour into sterile Petri dishes or distribute into sterile tubes.

Use: For the selective isolation and cultivation of oral *Fusobacterium* species, especially *F. nucleatum*.

Oral *Treponema* Medium

Composition per 1250mL:

Veal heart, fresh ground	1.0Kg
Thiopeptone	20.0g
NaCl	10.0g
Ionagar No. 2	2.0g
Eggs, whole fresh	2
Glutathione (1% solution)	100.0mL
Rabbit serum or ascitic fluid	100.0mL

pH 6.8–7.0 at 25°C

Preparation of Medium: Add agar to 50.0mL of distilled/deionized water. Gently heat and bring to boiling. Autoclave for 15 min at 15 psi pressure–121°C. Cool to 45°–50°C. In a separate flask add finely ground veal heart to 1.0L of distilled/deionized water. Add remaining components—except glutathione, rabbit serum and agar. Gently heat and bring to 70°C. Adjust pH to 7.4. Gently heat and bring to 100°C. Continue heating at 100°C for 2 hr. Skim off fat. Filter through glass wool. Adjust pH to 7.6. Gently heat and bring to 100°C. Maintain at 100°C for 30 min. Store at 4°C for 18 hrs. Sterilize in an Arnold sterilizer for 30 min at 100°C on two consecutive days. Cool to 45°–50°C. Aseptically add rabbit serum, glutathione solution and sterile cooled agar solution. Mix thoroughly. Aseptically distribute into sterile tubes or flasks.

Use: For the isolation and cultivation of *Treponema denticola* and *Treponema oralis*.

Orange Serum Agar

Composition per liter:

Agar	17.0g
Pancreatic digest of casein	10.0g
Glucose	4.0g
Yeast extract	3.0g
K_2HPO_4	2.5g
Orange serum	200.0mL

pH 5.5 ± 0.2 at 25°C

Source: This medium is available as a premixed powder from Difco Laboratories.

Preparation of Medium: Add components to distilled/deionized water and bring volume to 1.0L. Mix thoroughly. Gently heat and bring to boiling. Distribute into tubes or flasks. Autoclave for 15 min at 15 psi pressure–121°C. Pour into sterile Petri dishes or leave in tubes.

Use: For the cultivation and enumeration of microorganisms associated with the spoilage of citrus products. For the cultivation of lactobacilli, other aciduric microorganisms and pathogenic fungi.

Orange Serum Agar

Composition per liter:

Agar	14.0g
Pancreatic digest of casein	10.0g
Glucose	4.0g
Orange serum, equivalent solids	3.5g
Yeast extract	3.0g
K_2HPO_4	2.5g

pH 5.5 ± 0.2 at 25°C

Source: This medium is available as a premixed powder from Oxoid Unipath.

Preparation of Medium: Add components to distilled/deionized water and bring volume to 1.0L. Mix thoroughly. Gently heat and bring to boiling. Distribute into tubes or flasks. Autoclave for 15 min at 15 psi pressure–121°C. Pour into sterile Petri dishes or leave in tubes.

Use: For the isolation and enumeration of spoilage organisms from citrus products.

Orange Serum Agar

Composition per liter:

Agar	15.5g
Orange serum	10.0g
Pancreatic digest of casein	10.0g
Glucose	4.0g
Yeast extract	3.0g
K_2HPO_4	2.5g

pH 5.5 ± 0.2 at 25°C

Source: This medium is available as a premixed powder from BBL Microbiology Systems.

Preparation of Medium: Add components to distilled/deionized water and bring volume to 1.0L. Mix thoroughly. Gently heat and bring to boiling. Distribute into tubes or flasks. Autoclave for 10 min at 15 psi pressure–121°C. Pour into sterile Petri dishes or leave in tubes.

Use: For the enumeration and cultivation of microorganisms from citrus juice and other products. For the cultivation of lactobacilli, pathogenic fungi and other aciduric microorganisms.

Orange Serum Broth Concentrate 10X

Composition per liter:

Pancreatic digest of casein	10.0g
Glucose	4.0g
Yeast extract	3.0g
K_2HPO_4	2.5g
Orange serum concentrate	100.0mL

pH 5.6 ± 0.2 at 25°C

Source: This medium is available as a premixed solution from Difco Laboratories.

Preparation of Medium: Add components to distilled/deionized water and bring volume to 1.0L. Mix thoroughly. Distribute into tubes or flasks. Autoclave for 15 min at 15 psi pressure–121°C.

Use: For the cultivation and enumeration of microorganisms associated with the spoilage of citrus products. For the cultivation of lactobacilli, other aciduric microorganisms and pathogenic fungi.

Organic Acid Medium KP (Organic Acid Medium, Kauffmann and Petersen)

Composition per liter:

Gelatin	10.0g
Bromthymol Blue	0.024g
Organic acid solution	100.0mL

pH 7.4 ± 0.2 at 25°C

Organic Acid Solution:
Composition per 100mL:

Organic acid	10.0g

Preparation of Medium: Add organic acid to distilled/deionized water and bring volume to 100.0mL. Sodium potassium D-tartrate, sodium citrate or mucic acid may be used. Mix thoroughly

Preparation of Medium: Add components, except organic acid solution, to distilled/deionized water and bring volume to 900.0mL. Mix thoroughly. Gently heat and bring to boiling. Autoclave for 15 min at 15 psi pressure–121°C. Cool to 45°–50°C. Aseptically add sterile organic acid solution. Mix thoroughly. Aseptically distribute into sterile tubes or flasks.

Use: For the cultivation and differentiation of members of the Enterobacteriaceae based on their ability to utilize different organic acids as carbon source. Bacteria that utilize tartrate, citrate or mucate turn the medium yellow.

OTI Medium, Modified

Composition per 1100mL:

Beef heart, solids from infusion	100.0g
NaCl	6.0g
Polypeptone™	5.0g
Yeast extract	5.0g
K_2HPO_4	2.0g
Tryptose	2.0g
Agar	1.6g
Pectin	0.8g
Glucose	0.8g
Starch	0.8g
Sucrose	0.8g
Maltose	0.8g
Sodium pyruvate	0.8g
Ribose	0.8g
Cysteine·HCl·H_2O	0.68g
$MgSO_4 \cdot 7H_2O$	0.1g
Rumen fluid, clarified	500.0 mL
Rabbit serum, inactivated	50.0mL
Thiamine pyrophosphate solution	50.0mL

pH 7.0 ± 0.2 at 25°C

Thiamine Pyrophosphate Solution:
Composition per 50mL:

Thiamine pyrophosphate	7.5mg

Preparation of Thiamine Pyrophosphate Solution: Add thiamine pyrophosphate to distilled/deionized water and bring volume to 50.0mL. Mix thoroughly. Filter sterilize.

Preparation of Medium: Prepare and dispense medium under O_2-free 100% N_2. Add components, except rabbit serum and thiamine pyrophosphate solution, to distilled/deionized water and bring volume to 1.0L. Mix thoroughly. Gently heat and bring to boiling. Adjust pH to 7.0. Anaerobically distribute into tubes in 10.0mL volumes. Autoclave for 15 min at 15 psi pressure–121°C. Cool to 45°–50°C. Aseptically add 0.5 mL of sterile rabbit solution and 0.5mL of sterile thiamine pyrophosphate solution to each tube. Mix thoroughly.

Use: For the cultivation and maintenance of *Treponema socranskii*.

Oxalate Maintenance Medium

Composition per liter:

Pancreatic digest of casein	10.0g
Sodium oxalate	5.0g
Na_2CO_3	4.0g
Yeast extract	1.0g
Sodium acetate	0.82g
$(NH_4)_2SO_4$	0.5g
Cysteine·HCl·H_2O	0.5g
K_2HPO_4	0.25g
KH_2PO_4	0.25g
$MgSO_4·7H_2O$	0.025g
Resazurin	0.001g

pH 6.8 ± 0.2 at 25°C

Preparation of Medium: Add components, except Na_2CO_3 and cysteine·HCl·H_2O, to distilled/deionized water and bring volume to 1.0L. Mix thoroughly. Adjust pH to 6.8. Gently heat and bring to boiling. Cool under O_2-free 100% CO_2. Add Na_2CO_3 and cysteine·HCl·H_2O. Mix thoroughly. Anaerobically distribute into tubes. Cap with rubber stoppers. Place tubes in a press. Autoclave for 15 min at 15 psi pressure–121°C with fast exhaust.

Use: For the cultivation and maintenance of *Oxalobacter formigenes*.

Oxalate Medium

Basal Medium:

Composition per liter:

K_2HPO_4	4.4g
KH_2PO_4	3.4g
Potasssium oxalate	2.0g
$(NH_4)_2SO_4$	0.5g
$MgSO_4·7H_2O$	0.2g
$FeCl_3$	0.015g
Phenol Red (0.4% solution)	5.0mL
Mineral stock solution	1.0mL

Mineral Stock Solution:

Composition per liter:

$ZnSO_4·7H_2O$	11.0g
$MnSO_4·H_2O$	5.0g
$Na_2MoO_4·2H_2O$	2.0g
$CoSO_4$	0.05g
H_3BO_3	0.05g
$CuSO_4·5H_2O$	7.0mg

Preparation of Mineral Stock Solution: Add components to distilled/deionized water and bring volume to 1.0L. Mix thoroughly.

Preparation of Medium: Add components to distilled/deionized water and bring volume to 1.0L. Mix thoroughly. Distribute into tubes or flasks. Autoclave for 15 min at 15 psi pressure–121°C.

Use: For the isolation and cultivation of oxalate–decomposing *Alcaligenes* species.

Oxalate Medium, Modified

Composition per liter:

Pancreatic digest of casein	10.0g
Sodium oxalate	5.0g
Na_2CO_3	4.0g
Yeast extract	1.0g
Sodium acetate	0.82g
$(NH_4)_2SO_4$	0.5g
Cysteine·HCl·H_2O	0.5g
K_2HPO_4	0.25g
KH_2PO_4	0.25g
$MgSO_4·7H_2O$	0.025g
Resazurin	0.001g
Trace element solution	20.0mL

pH 7.0 ± 0.2 at 25°C

Trace Element Solution:

Composition per liter:

H_3BO_3	0.03g
$CoCl_2·6H_2O$	0.02g
$ZnSO_4·7H_2O$	0.01g
$MnCl_2·4H_2O$	3.0mg
$Na_2MoO_4·2H_2O$	3.0mg
$NiCl_2·6H_2O$	2.0mg
$CuCl_2·6H_2O$	1.0mg

Preparation of Trace Element Solution: Add components to distilled/deionized water and bring volume to 1.0L. Mix thoroughly.

Preparation of Medium: Add components, except Na_2CO_3 and cysteine·HCl·H_2O, to distilled/deionized water and bring volume to 1.0L. Mix thoroughly. Adjust pH to 7.0. Gently heat and bring to boiling. Cool under O_2-free 100% CO_2. Add Na_2CO_3 and cysteine·HCl·H_2O. Mix thoroughly. Anaerobically distribute into tubes. Cap with rubber stoppers. Place tubes in a press. Autoclave for 15 min at 15 psi pressure–121°C with fast exhaust.

Use: For the cultivation and maintenance of *Oxalobacter formigenes*.

Oxalate Utilization Medium

Composition per liter:

Agar	12.0g
Potassium oxalate	1.0g
NaCl	1.0g
$(NH_4)_2HPO_4$	1.0g
KH_2PO_4	0.5g
$MgSO_4·7H_2O$	0.2g
$CaCl_2$ solution	80.0mL

pH 7.0 ± 0.2 at 25°C

$CaCl_2$ Solution:
Composition per 100mL:

$CaCl_2$	1.47g

Preparation of $CaCl_2$ Solution: Add $CaCl_2$ to distilled/deionized water and bring volume to 10.0mL. Mix thoroughly. Gently heat until dissolved. Filter sterilize.

Preparation of Medium: Add components, except $CaCl_2$ solution, to distilled/deionized water and bring volume to 920.0mL. Mix thoroughly. Gently heat and bring to boiling. Distribute into flasks in 92.0mL volumes. Autoclave for 15 min at 15 psi pressure–121°C. Cool to 45°–50°C. Aseptically add 8.0mL of sterile $CaCl_2$ solution to each flask. A fine precipitate of calcium oxalate will form. Mix thoroughly. Pour into sterile Petri dishes. Swirl flask while dispensing agar to disperse precipitate evenly.

Use: For the cultivation and differentiation of streptomycetes based on oxalate utilization. Bacteria that utilize oxalate turn the medium dark blue.

Oxford Agar (*Listeria* Selective Agar, Oxford)

Composition per liter:

Special peptone	23.0g
LiCl	15.0g
Agar	10.0g
NaCl	5.0g
Cornstarch	1.0g
Esculin	1.0g
Ferric ammonium citrate	0.5g
Antibiotic inhibitor	10.0mL

pH 7.0 ± 0.2 at 25°C

Source: This medium is available as a premixed powder from Oxoid Unipath.

Antibiotic Inhibitor:
Composition per 10mL:

Cycloheximide	0.4g
Colistin sulphate	0.02g
Fosfomycin	0.01g
Acriflavine	5.0mg
Cefotetan	2.0mg
Ethanol (50% solution)	10.0mL

Preparation of Antibiotic Inhibitor: Add antibiotics to 10.0mL of ethanol. Mix thoroughly. Filter sterilize.

Preparation of Medium: Add components, except antibiotic inhibitor, to distilled/deionized water and bring volume to 990.0mL. Mix thoroughly. Gently heat and bring to boiling. Autoclave for 15 min at 15 psi pressure–121°C. Cool to 45°–50°C. Aseptically add 10.0mL of sterile antibiotic inhibitor. Mix thoroughly. Pour into sterile Petri dishes or distribute into sterile tubes.

Use: For the isolation and cultivation of *Listeria monocytogenes* from specimens containing a mixed bacterial flora.

Oxford Agar, Modified (*Listeria* Selective Agar, Modified Oxford) (MOX Agar)

Composition per liter:

Special peptone	23.0g
LiCl	15.0g
Agar	12.0g
NaCl	5.0g
Cornstarch	1.0g
Esculin	1.0g
Ferric ammonium citrate	0.5g
Antibiotic inhibitor	10.0mL

pH 7.0 ± 0.2 at 25°C

Antibiotic Inhibitor:
Composition per 10mL:

Moxalactam	0.015g
Colistin sulfate	0.010g

Preparation of Antibiotic Inhibitor: Add components to distilled /deionized water and bring volume to 10.0mL. Mix thoroughly. Filter sterilize.

Preparation of Medium: Add components, except antibiotic inhibitor, to distilled/deionized water and bring volume to 990.0mL. Mix thoroughly. Gently heat and bring to boiling. Autoclave for 10 min at 15 psi pressure–121°C. Cool to 45°–50°C. Aseptically add 10.0mL of sterile antibiotic inhibitor. Mix thoroughly. Pour into sterile Petri dishes or distribute into sterile tubes.

Use: For the isolation and cultivation of *Listeria monocytogenes* from specimens containing a mixed bacterial flora.

Oxidative Fermentative Glucose Medium, Semisolid (OF Glucose Medium, Semisolid)

Composition per liter:

Glucose	10.0g
NaCl	5.0g
Agar	2.0g
Pancreatic digest of casein	2.0g
K_2HPO_4	0.3g
Bromthymol Blue dye	0.08g

pH 6.8 ± 0.2 at 25°C

Preparation of Medium: Add components to distilled/deionized water and bring volume to 1.0L. Mix thoroughly. Gently heat and bring to boiling. Distribute into tubes or flasks. Autoclave for 15 min at 15 psi pressure–121°C. Pour into sterile Petri dishes or leave in tubes.

Use: For differentiating Gram-negative bacteria based upon determining the oxidative and fermentative metabolism of glucose. Bacteria that ferment glucose turn the medium yellow.

Oxidation Fermentation Medium (OF Medium)

Composition per liter:

NaCl	5.0g
Agar	2.5g
Pancreatic digest of casein	2.0g
K_2HPO_4	0.3g
Bromthymol Blue	0.03g
Carbohydrate solution	100.0mL

pH 6.8 ± 0.1 at 25°C

Source: This medium is available as a premixed powder from BBL Microbiology Systems.

Carbohydrate Solution:
Composition per 100mL:

Carbohydrate	10.0g

Preparation of Carbohydrate Solution: Add carbohydrate to distilled/deionized water and bring volume to 100.0mL. Mix thoroughly. Filter sterilize.

Preparation of Medium: Add components, except carbohydrate solution, to distilled/deionized water and bring volume to 900.0mL. Mix thoroughly. Gently heat and bring to boiling. Autoclave for 15 min at 15 psi pressure–121°C. Cool to 45°–50°C. Aseptically add 100.0mL sterile carbohydrate solution. Mix thoroughly. Pour into sterile Petri dishes or distribute into sterile tubes.

Use: For differentiating Gram-negative bacteria based upon determining the oxidative and fermentative metabolism of carbohydrates.

Oxidation Fermentation Medium, Hugh–Leifson's (Hugh–Leifson's Oxidation Fermentation Medium)

Composition per liter:

NaCl	5.0g
Agar	3.0g
Peptone	2.0g
K_2HPO_4	0.3g
Carbohydrate solution	100.0mL
Bromthymol Blue solution (0.2%)	15.0mL

pH 7.1 ± 0.2 at 25°C

Carbohydrate Solution:
Composition per 100mL:

Carbohydrate	10.0g

Preparation of Carbohydrate Solution: Add carbohydrate to distilled/deionized water and bring volume to 100.0mL. Mix thoroughly. Filter sterilize.

Preparation of Medium: Add components, except carbohydrate solution, to distilled/deionized water and bring volume to 900.0mL. Mix thoroughly. Gently heat and bring to boiling. Autoclave for 15 min at 15 psi pressure–121°C. Cool to 45°–50°C. Aseptically add 100.0mL sterile carbohydrate solution. Mix thoroughly. Pour into sterile Petri dishes or distribute into sterile tubes.

Use: For differentiating Gram-negative bacteria, such as *Vibrio* species, based upon determining the oxidative and fermentative metabolism of carbohydrates. Bacteria that ferment the carbohydrate turn the medium yellow.

Oxidation Fermentation Medium, King's (King's OF Medium)

Composition per liter:

Base:

Agar	3.0g
Pancreatic digest of casein	2.0g
Carbohydrate solution	100.0mL
Phenol Red (1.5% solution)	2.0mL

pH to 7.3 ± 0.2

Carbohydrate Solution:
Composition per 100mL:

Carbohydrate	10.0g

Preparation of Carbohydrate Solution: Add carbohydrate to distilled/deionized water and bring volume to 100.0mL. Mix thoroughly. Filter sterilize.

Preparation of Medium: Add components, except carbohydrate solution, to distilled/deionized water and bring volume to 900.0mL. Mix thoroughly. Gently heat and bring to boiling. Autoclave for 15 min at 15 psi pressure–121°C. Cool to 45°–50°C. Aseptically add 100.0mL sterile carbohydrate solution. Mix thoroughly. Pour into sterile Petri dishes or distribute into sterile tubes.

Use: For differentiating bacteria based upon determining the oxidative and fermentative metabolism of carbohydrates. Bacteria that ferment the carbohydrate turn the medium yellow.

Oxidation Reduction Indicator Agar
See: **OR Indicator Agar**

Oxidative–Fermentative Glucose Medium, Semisolid, with NaCl (OF Glucose Medium, Semisolid with NaCl)

Composition per liter:

NaCl	20.0g
Glucose	10.0g
Agar	2.0g
Pancreatic digest of casein	2.0g
K_2HPO_4	0.3g
Bromthymol Blue dye	0.08g

pH 6.8 ± 0.2 at 25°C

Preparation of Medium: Add components to distilled/deionized water and bring volume to 1.0L. Mix thoroughly. Gently heat and bring to boiling. Distribute into tubes or flasks. Autoclave for 15 min at 15 psi pressure–121°C. Pour into sterile Petri dishes or leave in tubes.

Use: For differentiating halophilic *Vibrio* species based upon determining the oxidative and fermentative metabolism of glucose. Bacteria that ferment glucose turn the medium yellow.

Oxidative–Fermentative Medium (OF Medium)

Composition per liter:

NaCl	5.0g
Agar	2.0g
Pancreatic digest of casein	2.0g
K_2HPO_4	0.3g
Bromothymol Blue	0.08g
Carbohydrate solution	100.0mL

pH 6.8 ± 0.1 at 25°C

Source: This medium is available as a premixed powder from Difco Laboratories and Oxoid Unipath.

Carbohydrate Solution:
Composition per 100mL:

Carbohydrate	10.0g

Preparation of Carbohydrate Solution: Add carbohydrate to distilled/deionized water and bring volume to 100.0mL. Mix thoroughly. Filter sterilize.

Preparation of Medium: Add components, except carbohydrate solution, to distilled/deionized water and bring volume to 900.0mL. Mix thoroughly. Gently heat and bring to boiling. Autoclave for 15 min at 15 psi pressure–121°C. Cool to 45°–50°C. Aseptically add 100.0mL sterile carbohydrate solution. Mix thoroughly. Pour into sterile Petri dishes or distribute into sterile tubes.

Use: For differentiating bacteria based upon determining the oxidative and fermentative metabolism of carbohydrates. Bacteria that ferment the carbohydrate turn the medium yellow.

Oxidative–Fermentative Test Medium (OF Test Medium)

Composition per liter:

NaCl	5.0g
Agar	3.0g
Peptone	2.0g
K_2HPO_4	0.3g
Bromthymol Blue	0.03g
Carbohydrate solution	100.0mL

Carbohydrate Solution:
Composition per 100mL:

Carbohydrate	10.0g

Preparation of Carbohydrate Solution: Add carbohydrate to distilled/deionized water and bring volume to 100.0mL. Mix thoroughly. Filter sterilize.

Preparation of Medium: Add components, except carbohydrate solution, to distilled/deionized water and bring volume to 1.0L. Mix thoroughly. Gently heat and bring to boiling. Distribute into tubes in 3.0mL volumes. Autoclave for 15 min at 15 psi pressure–121°C. Cool to 45°–50°C. Aseptically add 0.3mL of sterile carbohydrate solution to each tube. Mix thoroughly.

Use: For the cultivation and differentiation of a variety of microorganisms based on their ability to ferment a specific carbohydrate. Bacteria that ferment the specific carbohydrate turn the medium yellow.

Oxoid Nutrient Agar No. 1, pH 5.0
See: **Nutrient Agar, pH 5.0**

Oxytetracycline Glucose Yeast Extract Agar (OGYE Agar)

Composition per liter:

Glucose	20.0g
Agar	12.0g

Yeast extract 5.0g
Biotin 0.1mg
Oxytetracycline solution 10.0mL
pH 7.0 ± 0.2 at 25°C

Source: This medium is available as a premixed powder from Oxoid Unipath.

Oxytetracycline Solution:
Composition per 10mL:
Oxytetracycline 0.1g
Tris(hydroxymethyl)
aminomethane buffer (0.1M, pH 7.0)..... 10.0mL

Preparation of Oxytetracycline Solution: Add oxytetracycline to 10.0mL of Tris buffer. Mix thoroughly. Filter sterilize.

Preparation of Medium: Add components, except oxytetracycline solution, to distilled/deionized water and bring volume to 990.0mL. Mix thoroughly. Gently heat and bring to boiling. Autoclave for 10 min at 10 psi pressure–115°C. Cool to 45°–50°C. Aseptically add sterile oxytetracycline solution. Mix thoroughly. Pour into sterile Petri dishes or distribute into sterile tubes.

Use: For the isolation, enumeration and cultivation of yeasts and other fungi from foods.

Oxytetracycline Glucose Yeast Extract Agar (OGYE Agar)

Composition per liter:
Glucose 20.0g
Agar...12.0g
Yeast extract...5.0g
Oxytetracycline solution 10.0mL
pH 7.0 ± 0.2 at 25°C

Source: This medium is available as a premixed powder from Difco Laboratories.

Oxytetracycline Solution:
Composition per 10mL:
Oxytetracycline 0.1g

Preparation of Oxytetracycline Solution: Add oxytetracycline to distilled/deionized water and bring volume to 10.0mL. Mix thoroughly. Filter sterilize.

Preparation of Medium: Add components, except oxytetracycline solution, to distilled/deionized water and bring volume to 990.0mL. Mix thoroughly. Gently heat and bring to boiling. Autoclave for 15 min at 15 psi pressure–121°C. Cool to 45°–50°C. Aseptically add sterile oxytetracycline solution. Mix thoroughly. Pour into sterile Petri dishes or distribute into sterile tubes.

Use: For the isolation, enumeration and cultivation of yeasts and other fungi from foods.

OZR Medium

Composition per liter:
Agar...15.0g
Pancreatic digest of casein 1.0g
Yeast extract...1.0g

Preparation of Medium: Add components to sea water and bring volume to 1.0L. Mix thoroughly. Gently heat and bring to boiling. Distribute into tubes or flasks. Autoclave for 15 min at 15 psi pressure–121°C. Pour into sterile Petri dishes or leave in tubes.

Use: For the cultivation and maintenance of *Leucothrix mucor*.

P Agar

Composition per liter:

Agar	15.0g
Peptone	10.0g
NaCl	5.0g
Yeast extract	5.0g
Glucose	1.0g

pH 7.5 ± 0.2 at 25°C

Preparation of Medium: Add components to distilled/deionized water and bring volume to 1.0L. Mix thoroughly. Gently heat and bring to boiling. Distribute into tubes or flasks. Autoclave for 15 min at 15 psi pressure–121°C. Pour into sterile Petri dishes or leave in tubes.

Use: For the cultivation of *Staphylococcus* species.

PA Agar
See: ***Pseudomonas aeruginosa*** **Agar**

PA Broth
See: **Presence Absence Broth**

PA C Agar (mPA C Agar)

Composition per liter:

Agar	12.0g
L-Lysine·HCl	5.0g
NaCl	5.0g
$Na_2S_2O_3$	5.0g
Yeast extract	2.0g
$MgSO_4 \cdot 7H_2O$	1.5g
Lactose	1.25g
Sucrose	1.25g
Xylose	1.25g
Ferric ammonium citrate	0.80g
Phenol Red	0.08g
Nalidixic acid	0.037g
Kanamycin	8.0mg

pH 7.2 ± 0.1 at 25°C

Source: This medium is available as a premixed powder BBL Microbiology Systems.

Preparation of Medium: Add components to distilled/deionized water and bring volume to 1.0L. Mix thoroughly. Gently heat and bring to boiling. Distribute into tubes or flasks. Autoclave for 15 min at 15 psi pressure–121°C. Pour into sterile Petri dishes or leave in tubes.

Use: For the selective recovery and enumeration of *Pseudomonas aeruginosa* from water samples.

Pablum Cereal Agar

Composition per liter:

Pablum cereal, precooked	100.0g
Agar	18.0g
Chloramphenicol	0.05g

Preparation of Medium: Add components to distilled/deionized water and bring volume to 1.0L. Mix thoroughly. Gently heat and bring to boiling. Distribute into tubes or flasks. Autoclave for 15 min at 15 psi pressure–121°C. Pour into sterile Petri dishes or leave in tubes.

Use: For the cultivation of dematiaceous fungi and stimulation of spore formation.

Packer's Agar
See: **Azide Blood Agar with Crystal Violet**

Pagano Levin Agar

Composition per liter:

Glucose	40.0g
Agar	15.0g
Peptone	10.0g
Yeast extract	1.0g
Neomycin	0.5g
2,3,5-Triphenyltetrazolium chloride	0.1g

pH 6.0 ± 0.1 at 25°C

Source: This medium is available as a premixed powder from Difco Laboratories.

Preparation of Medium: Add components, except neomycin and 2,3,5-triphenyltetrazolium chloride, to distilled/deionized water and bring volume to 1.0L. Mix thoroughly. Gently heat and bring to boiling. Autoclave for 15 min at 15 psi pressure–121°C. Cool to 45°–50°C. Aseptically add neomycin and 2,3,5-triphenyltetrazolium chloride. Mix thoroughly. Pour into sterile Petri dishes or distribute into sterile tubes. Allow tubes to cool in a slanted position.

Use: For the isolation, cultivation and differentiation of *Candida* species. *Candida albicans* appears as smooth, shiny cream-light pink colonies.

Pai Medium

Composition per liter:

Homogenized whole egg	666.0mL
NaCl (0.85% solution)	334.0mL

pH 6.75 ± 0.2 at 25°C

Homogenized Whole Egg:
Composition per liter:

Whole eggs	18-24

Preparation of Homogenized Whole Egg: Use fresh eggs, less than 1 week old. Scrub the shells with soap. Let stand in a soap solution for 30 min. Rinse in running water. Soak eggs in 70% ethanol for 15 min. Break the eggs into a sterile container. Homogenize by shaking. Filter through four layers of sterile cheesecloth into a sterile graduated cylinder. Measure out 1.0L.

Preparation of Medium: Combine components. Mix thoroughly. Aseptically distribute into sterile tubes. Inspissate tubes in a slanted position at 80°–90°C (moist heat) for 30 min.

Use: For the maintenance of stock cultures of *Salmonella typhi* and other *Salmonella* species.

Pai Medium

Composition per 1620mL:

Glucose	5.0g
Homogenized whole egg	1.0L
Glycerol	120.0mL

pH 6.75 ± 0.2 at 25°C

Homogenized Whole Egg:
Composition per liter:

Whole eggs	18–24

Preparation of Homogenized Whole Egg: Use fresh eggs, less than 1 week old. Scrub the shells with soap. Let stand in a soap solution for 30 min. Rinse in running water. Soak eggs in 70% ethanol for 15 min. Break the eggs into a sterile container. Homogenize by shaking. Filter through four layers of sterile cheesecloth into a sterile graduated cylinder. Measure out 1.0L.

Preparation of Medium: Combine components. Mix thoroughly. Aseptically distribute into sterile tubes. Inspissate tubes in a slanted position at 80°–90°C (moist heat) for 30 min.

Use: For the isolation and cultivation of *Corynebacterium* species.

PALCAM Agar (Polymyxin Acriflavine LiCl Ceftazidime Esculin Mannitol Agar)

Composition per liter:

Peptone	23.0g
LiCl	15.0g
Agar	10.0g
Mannitol	10.0g
NaCl	5.0g
Yeast extract	3.0g
Starch	1.0g
Esculin	0.8g
Ferric ammonium citrate	0.5g
Glucose	0.5g
Phenol Red	0.08g
PALCAM selective supplement	10.0mL

pH 7.2 ± 0.2 at 25°C

Source: This medium is available as a premixed powder from Oxoid Unipath.

PALCAM Selective Supplement:
Composition per 10mL:

Ceftazidime	20.0mg
Polymyxin B	10.0mg
Acriflavine·HCl	5.0mg

Preparation of PALCAM Selective Supplement: Add components to distilled/deionized water and bring volume to 10.0mL. Mix thoroughly. Filter sterilize.

Egg Yolk Emulsion:
Composition:

Chicken egg yolks	11
Whole chicken egg	1

Preparation of Egg Yolk Emulsion: Soak eggs with 1:100 dilution of saturated mercuric chloride solution for 1 min. Crack eggs and separate yolks from whites. Mix egg yolks with 1 chicken egg.

Preparation of Medium: Add components, except PALCAM selective supplement, to distilled/deionized water and bring volume to 990.0mL. Mix thoroughly. Gently heat and bring to boiling. Autoclave for 15 min at 15 psi pressure–121°C. Cool to 45°–50°C. Aseptically add sterile PALCAM selective supplement. Mix thoroughly. The addition of 25.0mL of egg yolk emulsion may aid in the recovery of damaged *Listeria*. Pour into sterile Petri dishes or distribute into sterile tubes.

Use: For the selective isolation cultivation and differentiation of *Listeria monocytogenes* and other *Listeria* species from foods.

Panthenol Assay Medium

Composition per 950mL:

Base medium	225.0mL
Panthenol supplement	475.0mL

pH 6.0 ± 0.2 at 25°C

Source: This medium is available as a premixed powder from Difco Laboratories.

Base Medium:
Composition per 900mL:

Glucose	15.0g
Pancreatic digest of casein, charcoal treated	10.0g

Sodium citrate	2.0g
Vitamin assay casamino acids	2.0g
K_2HPO_4	1.0g
KH_2PO_4	1.0g
$MgSO_4 \cdot 7H_2O$	0.8g
L-Tryptophan	0.2g
$MnSO_4 \cdot H_2O$	0.16g
L-Cystine	0.15g
$FeSO_4 \cdot 7H_2O$	0.04g
Liver concentrate	0.04g
NaCl	0.04g
Adenine sulfate	0.01g
Guanine·HCl	0.01g
Uracil	0.01g
p-Aminobenzoic acid	2.0mg
β-Alanine	2.0mg
Nicotinic acid	2.0mg
Pyridoxine·HCl	2.0mg
Riboflavin	2.0mg
Thiamine·HCl	2.0mg
Folic acid	0.02mg
Biotin	0.016mg

Preparation of Base Medium: Add components to distilled/deionized water and bring volume to 900.0mL. Mix thoroughly. Gently heat and bring to boiling.

Panthenol Supplement:
Composition per 100mL:

Glycerol	33.0g
Sorbitan monooleate complex	2.0g
Lactic acid	0.68g

Preparation of Panthenol Supplement: Add components to distilled/deionized water and bring volume to 100.0mL. Mix thoroughly. Filter sterilize.

Preparation of Medium: Distribute base medium into 50.0mL flasks in 4.5mL volumes. Add standard solution or test solution to each flask. Adjust the volume in each flask to 9.5mL with distilled/deionized water. Autoclave for 10 min at 15 psi pressure–121°C. Cool to 25°C. Aseptically add 0.5mL of panthenol supplement to each flask.

Use: For the microbiological assaying of panthenol using *Gluconobacter oxydans* subspecies *suboxydans* as the test organism.

Pantothenate Assay Medium

Composition per100mL:

Glucose	40.0g
Sodium acetate	20.0g
Vitamin assay casamino acids	10.0g
K_2HPO_4	1.0g
KH_2PO_4	1.0g
L-Cystine	0.4g
$MgSO_4 \cdot 7H_2O$	0.4g
DL-Tryptophan	0.2g
Adenine sulfate	0.02g
$FeSO_4 \cdot 7H_2O$	0.02g
Guanine·HCl	0.02g
$MnSO_4 \cdot H_2O$	0.02g
NaCl	0.02g
Uracil	0.02g
Niacin	1.0mg
Pyridoxine	0.8mg
Riboflavin	0.4mg
p-Aminobenzoic acid	0.2mg
Thiamine·HCl	0.2mg
Biotin	0.8μg

pH 6.7 ± 0.1 at 25°C

Source: This medium is available as a premixed powder from Difco Laboratories.

Preparation of Medium: Add components to distilled/deionized water and bring volume to 100.0mL. Mix thoroughly. Gently heat and bring to boiling. Continue boiling for 2–3 min. Distribute into tubes in 5.0mL volumes. Add standard solution or test solution to each tube. Adjust the volume of each tube to 10.0mL with distilled/deionized water. Autoclave for 15 min at 15 psi pressure–121°C.

Use: For the microbiological assaying of pantothenic acid and its salts using *Lactobacillus plantarum* as the test organism.

Pantothenate Assay Medium

Composition per liter:

Glucose	38.0g
Sodium acetate	20.0g
Vitamin-free casamino acids	10.0g
K_2HPO_4	3.0g
$(NH_4)_2SO_4$	2.0g
NaCl	1.0g
$MgSO_4 \cdot 7H_2O$	0.4g
L-Tryptophan	0.1g
$MnSO_4 \cdot H_2O$	0.026g
Xanthine	0.020g
Adenine	0.020g
Guanine	0.020g
Uracil	0.020g
$(NH_4)_2SO_4 \cdot FeSO_4 \cdot 6H_2O$	0.020g
Niacin	5.0mg
Pyridoxine·HCl	4.0mg
Riboflavin	2.0mg
Thiamine·HCl	1.0mg
p-Aminobenzoic acid	1.0mg
Biotin	0.05mg

Source: This medium is available as a premixed powder from BBL Microbiology Systems.

Preparation of Medium: Add components to distilled/deionized water and bring volume to 100.0mL. Mix thoroughly. Gently heat and bring to boiling. Continue boiling for 2–3 min. Distribute into tubes in 5.0mL volumes. Add standard solution or test solution to each tube. Adjust the volume of each tube to 10.0mL with distilled/deionized water. Autoclave for 15 min at 15 psi pressure–121°C.

Use: For the determination of pantothenate content of pharmaceutical products and other materials using *Lactobacillus plantarum* as the test organism.

Pantothenate Culture Agar, USP

Composition per liter:

Yeast extract	20.0g
Agar	15.0g
Glucose	5.0g
Sodium acetate	5.0g

pH 6.8 ± 0.2 at 25°C

Preparation of Medium: Add components to distilled/deionized water and bring volume to 1.0L. Mix thoroughly. Gently heat and bring to boiling. Distribute into tubes or flasks in 10.0mL volumes. Autoclave for 15 min at 15 psi pressure–121°C. Pour into sterile Petri dishes or leave in tubes.

Use: For the maintenance of *Lactobacillus plantarum* used in the microbiological assay of pantothenic acid or pantothenate. Also used for the cultivation of other *Lactobacillus* species.

Pantothenate Medium, AOAC USP

Composition per liter:

Glucose	40.0g
Sodium acetate	20.0g
Vitamin assay casamino acids	10.0g
K_2HPO_4	1.0g
KH_2PO_4	1.0g
L-Cystine	0.4g
$MgSO_4 \cdot 7H_2O$	0.4g
L-Tryptophan	0.1g
Sorbitan monooleate complex	0.1g
Adenine sulfate	0.020g
$FeSO_4 \cdot 7H_2O$	0.020g
Guanine·HCl	0.020g
$MnSO_4 \cdot H_2O$	0.020g
NaCl	0.020g
Uracil	0.020g
Nicotinic acid	1.0mg
Pyridoxine·HCl	0.8mg
Riboflavin	0.4mg
p-Aminobenzoic acid	0.2mg
Thiamine·HCl	0.2mg
Biotin	0.8μg

pH 6.7 ± 0.1 at 25°C

Source: This medium is available as a premixed powder from Difco Laboratories.

Preparation of Medium: Add components to distilled/deionized water and bring volume to 100.0mL. Mix thoroughly. Gently heat and bring to boiling. Continue boiling for 2–3 min. Distribute into tubes in 5.0mL volumes. Add standard solution or test solution to each tube. Adjust the volume of each tube to 10.0mL with distilled/deionized water. Autoclave for 15 min at 15 psi pressure–121°C.

Use: For the determination of pantothenic acid and its salts using *Lactobacillus plantarum* as the test organism.

Paraffin Agar

Composition per liter:

Agar	15.0g
K_2HPO_4	6.0g
NH_4NO_3	4.0g
KH_2PO_4	2.0g
Paraffin, liquid	1.0g
$ZnSO_4 \cdot 7H_2O$	0.049g
$MnCl_2 \cdot 4H_2O$	0.046g
$FeSO_4 \cdot 7H_2O$	5.4mg
$CuSO_4 \cdot 5H_2O$	2.5mg
$Na_2B_4O_7 \cdot 10H_2O$	0.94mg
$(NH_4)_6Mo_7O_{24} \cdot 4H_2O$	0.20mg

Preparation of Medium: Add components to distilled/deionized water and bring volume to 1.0L. Mix thoroughly. Gently heat and bring to boiling. Distribute into tubes or flasks. Autoclave for 15 min at 15 psi pressure–121°C. Pour into sterile Petri dishes or leave in tubes.

Use: For the selective isolation and cultivation of streptomycetes.

Paraffin Medium with McClung Carbon–Free Broth

Composition per liter:

Paraffin pellets	1lb
McClung carbon-free broth	1.0L

pH 7.2 ± 0.2 at 25°C

McClung Carbon–Free Broth:

Composition per liter:

$NaNO_3$	2.0g
K_2HPO_4	0.8g
$MgSO_4 \cdot 7H_2O$	0.5g
$FeCl_3$	0.010g

$MnCl_2 \cdot 4H_2O$	0.008g
$ZnSO_4$	0.002g

Preparation of McClung Carbon–Free Broth: Add components to distilled/deionized water and bring volume to 1.0L. Mix thoroughly. Gently heat (low heat for 15–30 min) until salts dissolve. Cool to 25°C. Adjust pH to 7.2, if necessary. Filter sterilize.

Preparation of Medium: Fill 16 by 125mm glass screw-cap tubes 60% full with paraffin pellets. Place on slanted rack and autoclave for 15 min at 121°C. Let tubes cool in a slanted position. Add 2.5mL of sterile McClung carbon-free broth to each paraffin slant. Tighten screw-caps. Sterility is tested by the addition of 2.5mL of sterile trypticase soy broth and sample to each slant.

Use: For use in the sterility testing of various specimens.

Paramecium Medium

Composition per liter:

Solution C	500.0mL
Solution A	10.0mL
Solution B	1.0mL

Solution A:

Composition per liter:

Thiamine·HCl	1.5g
Calcium pantothenate	1.0g
Nicotinamide	0.5g
Pyridoxal·HCl	0.5g
Riboflavin	0.5g
Folic acid	0.5g
α-Lipoic acid	0.1g
Biotin	0.1mg

Preparation of Solution A: Add components to distilled/deionized water and bring volume to 1.0L. Mix thoroughly. Distribute while stirring into screw-capped tubes in 10.0mL volumes. Store at –20°C. Thaw as needed.

Solution B:

Composition per 100mL:

TEM-4T (Hachmeister, Pittsburgh)	10.0g
Stigmasterol	0.5g
Ethanol, absolute	100.0mL

Preparation of Solution B: Add TEM-4T and stigmasterol to 100.0mL of hot ethanol. Mix thoroughly. Store at 4°C.

Solution C:

Composition per 500mL:

Proteose peptone	10.0g
Pancreatic digest of casein	5.0g
Ribonucleic acid	1.0g
$MgSO_4 \cdot 7H_2O$	0.5g

Preparation of Solution C: Add components to distilled/deionized water and bring volume to 500.0mL. Mix thoroughly.

Preparation of Medium: Combine 500.0mL of solution C, 10.0mL of solution A and 1.0mL of solution B. Mix thoroughly. Bring volume to 1.0L with distilled/deionized water. Adjust pH to 7.0–7.2 with 0.1*N* NaOH. Autoclave for 20 min at 15 psi pressure–121°C. Cool to 45°–50°C. Aseptically distribute into sterile tubes or flasks.

Use: For the cultivation of *Paramecium* species to be used as host cells by bacterial symbionts.

Park and Sanders Enrichment Broth

Composition per 1010mL:

Basal medium	1.0L
Supplement A	5.0mL
Supplement B	5.0mL

pH 7.0 ± 0.2 at 25°C

Basal Medium:

Composition per liter:

Pancreatic digest of casein	10.0g
Peptic digest of animal tissue	10.0g
NaCl	5.0g
Yeast extract	2.0g
Glucose	1.0g
Sodium pyruvate	0.25g
$NaHSO_3$	0.1g
Horse blood, lysed	50.0mL

Preparation of Basal Medium: Add components, except horse blood, to distilled/deionized water and bring volume to 950.0mL. Mix thoroughly. Gently heat and bring to boiling. Autoclave for 15 min at 15 psi pressure–121°C. Cool to 25°C. Aseptically add sterile horse blood. Mix thoroughly.

Supplement A:

Composition per 5mL:

Vancomycin	0.01g
Trimethoprim lactate	0.01g

Preparation of Supplement A: Add components to distilled/deionized water and bring volume to 5.0mL. Mix thoroughly. Filter sterilize.

Supplement B:

Composition per 5mL:

Cefoperazone	0.032g
Cycloheximide	0.1g

Preparation of Supplement B: Add components to distilled/deionized water and bring volume to 5.0mL. Mix thoroughly. Filter sterilize.

Preparation of Medium: To 1.0L of cooled, sterile basal medium, aseptically add 5.0mL of sterile supplement A. Mix thoroughly. Aseptically distribute into flasks in 100.0mL volumes. Inoculate medium with food samples. Incubate at 31°–32°C for 4 hr to recover and resuscitate injured cells. Aseptically add 0.5mL of supplement B to each 100.0mL of medium. Incubate cultures at 37°C for 2 hr.

Use: For the cultivation and enrichment of *Campylobacter* species from foods.

Paromomycin Vancomycin Blood Agar
See: PV Blood Agar

Pasteurella haemolytica Selective Medium

Composition per 1010mL:

Tryptose agar with peptic digest of blood	1.0L
Antibiotic solution	10.0mL

pH 7.2 ± 0.2 at 25°C

Tryptose Agar With Peptic Digest of Blood:
Composition per liter:

Agar	15.0g
Pancreatic digest of casein	10.0g
Peptic digest of animal tissue	10.0g
NaCl	5.0g
Glucose	1.0g
Peptic digest of blood	50.0mL

Preparation of Tryptose Agar With Peptic Digest of Blood: Add components to distilled/deionized water and bring volume to 950.0mL. Mix thoroughly. Gently heat and bring to boiling. Autoclave for 15 min at 15 psi pressure–121°C. Cool to 45°–50°C. Aseptically add peptic digest of blood. Mix thoroughly.

Antibiotic Solution:
Composition per 10mL:

Actidione (cycloheximide)	0.1g
Novobiocin	2.0mg
Neomycin	1.5mg

Preparation of Antibiotic Solution: Add components to distilled/deionized water and bring volume to 10.0mL. Mix thoroughly. Filter sterilize.

Preparation of Medium: To 1.0L of cooled, sterile tryptose agar with peptic digest of blood, aseptically add 10.0mL of sterile antibiotic solution. Mix thoroughly. Pour into sterile Petri dishes or distribute into sterile tubes.

Use: For the selective cultivation of *Pasteurella haemolytica*.

Pasteurella multocida Selective Medium

Composition per 1020mL:

Tryptose agar with peptic digest of blood	1.0L
Antibiotic solution	10.0mL
K_2TeO_3 solution	10.0mL

pH 7.2 ± 0.2 at 25°C

Tryptose Agar With Peptic Digest of Blood:
Composition per liter:

Agar	15.0g
Pancreatic digest of casein	10.0g
Peptic digest of animal tissue	10.0g
NaCl	5.0g
Glucose	1.0g
Peptic digest of blood	50.0mL

Preparation of Tryptose Agar With Peptic Digest of Blood: Add components to distilled/deionized water and bring volume to 950.0mL. Mix thoroughly. Gently heat and bring to boiling. Autoclave for 15 min at 15 psi pressure–121°C. Cool to 45°–50°C. Aseptically add peptic digest of blood. Mix thoroughly.

Antibiotic Solution:
Composition per 10mL:

Actidione (cycloheximide)	0.1g
Novobiocin	0.01g
Erythrocin	5.0mg

Preparation of Antibiotic Solution: Add components to distilled/deionized water and bring volume to 10.0mL. Mix thoroughly. Filter sterilize.

K_2TeO_3 Solution:
Composition per 10mL:

K_2TeO_3	5.0mg

Preparation of K_2TeO_3 Solution: Add K_2TeO_3 to distilled/deionized water and bring volume to 10.0mL. Mix thoroughly. Filter sterilize.

Caution: Potassium tellurite is toxic.

Preparation of Medium: To 1.0L of cooled, sterile tryptose agar with peptic digest of blood, aseptically add 10.0mL of sterile antibiotic solution and 10.0mL of sterile K_2TeO_3 solution. Mix thoroughly. Pour into sterile Petri dishes or distribute into sterile tubes.

Use: For the selective cultivation of *Pasteurella multocida*.

PD2 Medium

Composition per liter:

Agar	15.0g
Pancreatic digest of casein	4.0g

Papaic digest of soybean meal 2.0g
K_2HPO_4 ....1.5g
Disodium succinate 1.0g
KH_2PO_4 ....1.0g
$MgSO_4·7H_2O$ 1.0g
Trisodium citrate 1.0g
Bovine serum albumin solution 10.0mL
Hemin chloride solution 10.0mL

pH 7.0 ± 0.2 at 25°C

Bovine Serum Albumin Solution:
Composition per 10mL:
Bovine serum albumin 2.0g

Preparation of Bovine Serum Albumin Solution: Add bovine serum albumin to distilled/deionized water and bring volume to 10.0mL. Mix thoroughly. Filter sterilize.

Hemin Chloride Solution:
Composition per 100mL:
Hemin chloride 0.1g
NaOH (0.05*N* solution) 100.0mL

Preparation of Hemin Chloride Solution: Add hemin chloride to 100.0mL of NaOH solution. Mix thoroughly.

Preparation of Medium: Add components, except bovine serum albumin solution, to distilled/deionized water and bring volume to 990.0mL. Mix thoroughly. Gently heat and bring to boiling. Adjust pH to 7.0. Autoclave for 15 min at 15 psi pressure–121°C. Cool to 45°–50°C. Aseptically add sterile bovine serum albumin solution. Mix thoroughly. Pour into sterile Petri dishes or distribute into sterile tubes.

Use: For the isolation and cultivation of PD-ALS (Pierce's disease-almond leaf scorch) bacteria.

PDA Agar
See: **Potato Dextrose Agar**

PDA and Yeast Medium
See: **Potato Dextrose Agar and Yeast Medium**

PDY Agar
See: **Potato Dextrose Yeast Agar**

PE2 Medium

Composition per liter:
Peptone 20.0g
Yeast extract 3.0g
Alaska seed peas 416–520
Bromcresol Purple solution 2.0mL

Bromcresol Purple Solution:
Composition per 100mL:
Bromcresol Purple 2.0g
Ethanol 10.0mL

Preparation of Bromcresol Purple Solution: Add Bromcresol Purple to 10.0mL of ethanol. Mix thoroughly. Bring volume to 100.0mL with distilled/deionized water. Filter sterilize.

Preparation of Medium: Add components, except Alaska seed peas, to distilled/deionized water and bring volume to 1.0L. Mix thoroughly. Gently heat until dissolved. Add 8–10 Alaska seed peas to each of 18 × 150 mm screw-capped tubes. Distribute the broth into each tube in 19.0mL volumes. Allow tubes to stand for 1 hr. Autoclave for 15 min at 15 psi pressure–121°C.

Use: For the cultivation of *Clostridium botulinum* from foods.

Pectin Agar

Composition per plate:
Base agar 15.0mL
Pectin gel 7.0mL

Base Agar:
Composition per liter:
Agar 15.0g
K_2HPO_4 ....5.0g
$CaCl_2·2H_2O$ 3.0g
NH_4Cl 1.0g
$MgSO_4·7H_2O$ 0.2g
Tris(hydroxymethyl)amino–
methane buffer (1M, pH 8.0) 100.0mL
Trace element solution 1.0mL

Preparation of Base Agar: Add components to distilled/deionized water and bring volume to 1.0L. Mix thoroughly. Gently heat and bring to boiling. Autoclave for 15 min at 15 psi pressure–121°C. Cool to 45°–50°C.

Trace Element Solution:
Composition per liter:
Disodium EDTA 8.0g
$MnCl_2·4H_2O$ 0.1g
$CoCl_2·6H_2O$ 0.02g
$ZnCl_2$ 0.02g
KBr 0.02g
KI 0.02g
$CuSO_4$ 0.01g
$Na_2MoO_4·2H_2O$ 0.01g
H_3BO_3 ....0.01g
LiCl 5.0mg
$SnCl_2·2H_2O$ 5.0mg

Preparation of Trace Element Solution: Add components to distilled/deionized water and bring volume to 1.0L. Mix thoroughly.

Pectin Gel:
Composition per liter:

Pectin, low esterified	20.0g

Preparation of Pectin Gel: Add pectin to 70°C distilled/deionized water and bring volume to 1.0L. Autoclave for 10 min at 7 psi pressure–110°C. Cool to 45°–50°C.

Preparation of Medium: Pour cooled, sterile base agar into sterile Petri dishes in 15.0mL volumes. Allow agar to solidify. Overlay each plate with 7.0mL of cooled, sterile pectin gel. The pectin may take 5 hr to form a gel.

Use: For the isolation and cultivation of *Cytophaga* species, *Herpetosiphon* species, *Saprospira* species, and *Flexithrix* species.

Pectin Medium

Composition per liter:

Pectin	30.0g
Yeast extract	5.0g
Bromthymol Blue (0.1% solution)	1.0mL
$CaCl_2 \cdot 2H_2O$ (10.0% solution)	0.6mL

pH 7.3 ± 0.2 at 25°C

Preparation of Medium: Add 100.0mL of distilled/deionized water to a 2-liter flask and place on a magnetic stirrer with no heat. While stirring, slowly add $CaCl_2 \cdot 2H_2O$, Bromthymol Blue solution, yeast extract, and pectin. Add slowly to ensure each particle is wetted. Gently heat and bring to almost boiling. Adjust pH to 7.3 with 1N NaOH. Do not overshoot pH 7.3.

Use: For the isolation and presumptive identification of *Yersinia enterocolitica* and *Yersinia pseudotuberculosis* from other fermenting Gram–negative bacilli such as *Enterocolitica agglomerans* which is often confused with *Yersinia*. *E. agglomerans* does not produce pectinase. *Y. enterocolitica* is strongly positive for pectinase activity. *Y. pseudotuberculosis* is weakly positive for pectinase activity. *Yersinia pestis* is negative for pectinase activity. Also, use for the differentiation of *Klebsiella oxytoca* which is pectinase positive.

Pediocococcus cereviseae and *Aerococcus viridans* Medium

Composition per liter:

Pancreatic digest of casein	12.5g
Glucose	10.0g
Yeast extract	7.5g
K_2HPO_4	5.0g
NaCl	5.0g
Sodium citrate	5.0g
$MgSO_4$	0.8g
Tween™ 80	0.2g
$MnCl_2$	0.14g
$FeSO_4$	0.04g
Thiamine·HCl	0.1mg

pH 6.7 ± 0.2 at 25°C

Preparation of Medium: Add components to distilled/deionized water and bring volume to 1.0L. Mix thoroughly. Gently heat until dissolved. Distribute into tubes or flasks. Autoclave for 15 min at 15 psi pressure–121°C.

Use: For the cultivation of *Pediocococcus cereviseae* and *Aerococcus viridans*.

Pediococcus Medium (Lactobacilli MRS Broth)

Composition per 990mL:

Glucose	20.0g
Beef extract	10.0g
Peptone	10.0g
Sodium acetate	5.0g
Yeast extract	5.0g
Ammonium citrate	2.0g
Na_2HPO_4	2.0g
Tween™ 80	1.0g
$MgSO_4 \cdot 7H_2O$	0.1g
$MnSO_4 \cdot 5H_2O$	0.05g

pH 5.2 ± 0.2 at 25°C

Source: This medium is available from Difco Laboratories.

Preparation of Medium: Add components to distilled/deionized water and bring volume to 1.0L. Mix thoroughly. Gently heat and bring to boiling. Distribute into tubes or flasks. Autoclave for 15 min at 15 psi pressure–121°C.

Use: For the cultivation and maintenance of *Pediococcus damnosus*.

Pediococcus Medium with Mevalonic Acid (Lactobacilli MRS Broth with Melvalonic Acid)

Composition per 990mL:

Mevalonic acid	30.0g
Glucose	20.0g
Beef extract	10.0g
Peptone	10.0g
Sodium acetate	5.0g

Yeast extract 5.0g
Ammonium citrate 2.0g
Na_2HPO_4 .... 2.0g
Tween™ 80 .... 1.0g
$MgSO_4 \cdot 7H_2O$ 0.1g
$MnSO_4 \cdot 5H_2O$ 0.05g
pH 5.2 ± 0.2 at 25°C

Preparation of Medium: Add components to distilled/deionized water and bring volume to 1.0L. Mix thoroughly. Gently heat and bring to boiling. Distribute into tubes or flasks. Autoclave for 15 min at 15 psi pressure–121°C.

Use: For the cultivation and maintenance of *Pediococcus damnosus.*

Pedomicrobium PSM Medium

Composition per liter:

Yeast extract 0.5g
Sodium acetate solution 10.0mL
Vitamin solution 10.0mL
Trace elements solution 1.00mL
pH 7.1 ± 0.2 at 25°C

Sodium Acetate Solution:
Composition per 10mL:

Sodium acetate$\cdot 3H_2O$ 1.36g

Preparation of Sodium Acetate Solution: Add sodium acetate to distilled/deionized water and bring volume to 10.0mL. Mix thoroughly. Filter sterilize.

Vitamin Solution:
Composition per liter:

Pyridoxine·HCl 10.00mg
Thiamine·HCl 5.00mg
Ca lcium pantothenate 5.00mg
Riboflavin 5.00mg
Nicotinic acid 5.00mg
p-Aminobenzoic acid 5.00mg
D-Biotin 2.00mg
Folic acid 2.00mg
Cyanocobalamin 0.10mg

Preparation of Vitamin Solution: Add components to distilled/deionized water and bring volume to 1.0L. Mix thoroughly. Filter-sterilize.

Trace Elements Solution:
Composition per 50mL:

$ZnSO_4 \cdot 7H_2O$ 0.55g
$FeSO_4 \cdot 7H_2O$ 0.25g
EDTA 0.125g
$MnSO_4 \cdot H_2O$ 0.077g
$CuSO_4 \cdot 5H_2O$ 0.02g
$Co(NO_3)_2 \cdot 6H_2O$ 0.012g
$Na_2B_4O_7 \cdot 10H_2O$ 8.85mg

Preparation of Trace Elements Solution: Add components to distilled/deionized water and bring volume to 50.0mL. Mix thoroughly. Filter-sterilize.

Preparation of Medium: Add components, except sodium acetate solution and vitamin solution, to distilled/deionized water and bring volume to 980.0mL. Mix thoroughly. Adjust pH to 7.1 with KOH. Autoclave for 15 min at 15 psi pressure–121°C. Cool to 25°C. Aseptically add sterile sodium acetate solution and 10.0mL of sterile vitamin solution. Mix thoroughly. Aseptically distribute into sterile tubes or flasks.

Use: For the cultivation and maintenance of *Pedomicrobium americanum* and *Pedomicrobium ferrugineum.*

Pedomicrobium PSM Medium with Ribose

Composition per liter:

Peptone 0.25g
Yeast extract 0.25g
Ribose solution 10.0mL
Vitamin solution 10.0mL
Trace elements solution 1.00mL
pH 7.1 ± 0.2 at 25°C

Ribose Solution:
Composition per 100mL:

D-Ribose 10.0g

Preparation of Ribose Solution: Add ribose to distilled/deionized water and bring volume to 100.0mL. Mix thoroughly. Filter sterilize.

Vitamin Solution:
Composition per liter:

Pyridoxine·HCl 10.00mg
Thiamine·HCl 5.00mg
Calcium pantothenate 5.00mg
Riboflavin 5.00mg
Nicotinic acid 5.00mg
p-Aminobenzoic acid 5.00mg
D-Biotin 2.00mg
Folic acid 2.00mg
Cyanocobalamin 0.10mg

Preparation of Vitamin Solution: Add components to distilled/deionized water and bring volume to 1.0L. Mix thoroughly. Filter-sterilize.

Trace Elements Solution:
Composition per 50mL:

$ZnSO_4 \cdot 7H_2O$ 0.55g
$FeSO_4 \cdot 7H_2O$ 0.25g
EDTA 0.125g
$MnSO_4 \cdot H_2O$ 0.077g

$CuSO_4 \cdot 5H_2O$ 0.02g
$Co(NO_3)_2 \cdot 6H_2O$ 0.012g
$Na_2B_4O_7 \cdot 10H_2O$ 8.85mg

Preparation of Trace Elements Solution: Add components to distilled/deionized water and bring volume to 50.0mL. Mix thoroughly. Filter sterilize.

Preparation of Medium: Add components, except ribose solution and vitamin solution, to distilled/deionized water and bring volume to 980.0mL. Mix thoroughly. Adjust pH to 7.1 with KOH. Autoclave for 15 min at 15 psi pressure–121°C. Cool to 25°C. Aseptically add 10.0mL of sterile ribose solution and 10.0mL of sterile vitamin solution. Mix thoroughly. Aseptically distribute into sterile tubes or flasks.

Use: For the cultivation and maintenance of *Pedomicrobium americanum* and *Pedomicrobium ferrugineum.*

Pedomicrobium PYVM Medium

Composition per liter:

Peptone 0.25g
Yeast extract 0.25g
Malate solution 10.0mL
Vitamin solution 10.0mL
Trace elements solution 20.00mL
pH 7.1 ± 0.2 at 25°C

Malate Solution:

Composition per 100mL:

Malic acid 1.36g

Preparation of Malate Solution: Add malate to distilled/deionized water and bring volume to 100.0mL. Adjust pH to 7.0 with concentrated NaOH. Mix thoroughly. Filter sterilize.

Vitamin Solution:

Composition per liter:

Pyridoxine·HCl 10.00mg
Thiamine·HCl 5.00mg
Calcium pantothenate 5.00mg
Riboflavin 5.00mg
Nicotinic acid 5.00mg
p-Aminobenzoic acid 5.00mg
D-Biotin 2.00mg
Folic acid 2.00mg
Cyanocobalamin 0.10mg

Preparation of Vitamin Solution: Add components to distilled/deionized water and bring volume to 1.0L. Mix thoroughly. Filter sterilize.

Trace Elements Solution:

Composition per 50mL:

$ZnSO_4 \cdot 7H_2O$ 0.55g
$FeSO_4 \cdot 7H_2O$ 0.25g
EDTA 0.125g
$MnSO_4 \cdot H_2O$ 0.077g
$CuSO_4 \cdot 5H_2O$ 0.02g
$Co(NO_3)_2 \cdot 6H_2O$ 0.012g
$Na_2B_4O_7 \cdot 10H_2O$ 8.85mg

Preparation of Trace Elements Solution: Add components to distilled/deionized water and bring volume to 50.0mL. Mix thoroughly. Filter sterilize.

Preparation of Medium: Add components, except malate solution and vitamin solution, to distilled/deionized water and bring volume to 980.0mL. Mix thoroughly. Adjust pH to 7.1 with KOH. Autoclave for 15 min at 15 psi pressure–121°C. Cool to 25°C. Aseptically add 10.0mL of sterile malate solution and 10.0mL of sterile vitamin solution. Mix thoroughly. Aseptically distribute into sterile tubes or flasks.

Use: For the cultivation of *Pedomicrobium* species.

PEM
See: **Pre–Enrichment Medium**

PEMBA
See: **Polymyxin Pyruvate Egg Yolk Mannitol Bromthymol Blue Agar**

Penassay Base Agar
See: **Antibiotic Medium 2**

Penassay Broth
See: **Antibiotic Medium 3**

Penassay Broth with Chloramphenicol

Composition per liter:

Peptone 5.0g
K_2HPO_4 .... 3.68g
NaCl 3.5g
Beef extract 1.5g
Yeast extract 1.5g
KH_2PO_4 .... 1.32g
Glucose 1.0g
Chloramphenicol 10.0mg
pH 7.0 ± 0.05 at 25°C

Preparation of Medium: Add components to distilled/deionized water and bring volume to 1.0L. Mix thoroughly. Gently heat and bring to boiling. Distribute into tubes or flasks. Autoclave for 15 minutes at 15 psi–121°C.

Use: For the cultivation and maintenance of *Bacillus subtilis*.

Penassay Broth with Magnesium

Composition per liter:

Pancreatic digest of gelatin	5.0g
K_2HPO_4	3.68g
NaCl	3.5g
Beef extract	1.5g
Yeast extract	1.5g
KH_2PO_4	1.32g
Glucose	1.0g
$MgCl_2$	0.095g

pH 7.0 ± 0.05 at 25°C

Preparation of Medium: Add components to distilled/deionized water and bring volume to 1.0L. Mix thoroughly. Gently heat and bring to boiling. Distribute into tubes or flasks. Autoclave for 15 minutes at 15 psi–121°C.

Use: For the cultivation and maintenance of *Bacillus subtilis*.

Penassay G–THY Medium (Penassay Glucose Thymine Medium)

Composition per liter:

Glucose	21.0g
Pancreatic digest of gelatin	5.0g
K_2HPO_4	3.68g
NaCl	3.5g
Beef extract	1.5g
Yeast extract	1.5g
KH_2PO_4	1.32g
Thymine	0.05g

pH 7.0 ± 0.05 at 25°C

Preparation of Medium: Add components to distilled/deionized water and bring volume to 1.0L. Mix thoroughly. Gently heat and bring to boiling. Distribute into tubes or flasks. Autoclave for 15 minutes at 15 psi–121°C.

Use: For the cultivation and maintenance of *Bacillus subtilis*.

Penassay Seed Agar
See: Antibiotic Medium 1

Penicillinase-Producing *Neisseria gonorrhoeae* Medium
See: PPNG Selective Medium

Pentachloronitrobenzene Rose Bengal Yeast Extract Sucrose Agar (PRYES Agar)

Composition per liter:

Sucrose	150.0g
Agar	20.0g
Yeast extract	20.0g
Pentachloronitrobenzene (PCNB)	0.1g
Chloramphenicol	0.05g
Chlortetracycline·HCl	0.05g
Rose Bengal	0.025g

pH 5.6 ± 0.2 at 25°C

Preparation of Medium: Add components, except chloramphenicol and chlortetracycline, to distilled/deionized water and bring volume to 1.0L. Mix thoroughly. Adjust pH to 5.6 with tartaric acid. Autoclave for 15 min at 15 psi pressure–121°C. Cool to 45°–50°C. Aseptically add chloramphenicol and chlortetracycline. Mix thoroughly. Pour into sterile Petri dishes.

Use: For the cultivation and differentiation of nephrotoxin-producing strains of *Penicillium viridicatium* and related species isolated from foods. Colonies exhibiting a violet brown pigment on the reverse are counted as potential ochratoxin- and citrinin-producing strains of *P. viridicatium*. Colonies exhibiting a yellow reverse and obverse are counted as potential xanthomegnin- and viomellein-producing strains of *P. viridicatium* and *P. aurantiogriseum* (*P. cyclopium*).

PEP Medium

Composition per liter:

Agar	10.0g
Peptone	5.0g
Yeast extract	0.5g
K_2HPO_4	0.1g

Preparation of Medium: Add components to distilled/deionized water and bring volume to 1.0L. Mix thoroughly. Gently heat and bring to boiling. Distrib-

ute into tubes or flasks. Autoclave for 15 min at 15 psi pressure–121°C. Pour into sterile Petri dishes or leave in tubes.

Use: For the isolation and cultivation of *Spirochaeta aurantia*.

Pept Carb Soluble Starch Agar (Peptone Carbonate Starch Agar)

Composition per liter:

Solution A	900.0mL
Solution B	100.0mL

Solution A:

Composition per 900mL:

Soluble starch	20.0g
Agar	15.0g
Peptone	5.0g
Yeast extract	5.0g
K_2HPO_4	1.0g
$MgSO_4 \cdot 7H_2O$	0.2g

Preparation of Solution A: Add components to distilled/deionized water and bring volume to 900.0mL. Mix thoroughly. Gently heat and bring to boiling. Autoclave for 15 min at 10 psi pressure–115°C. Cool to 45°–50°C.

Solution B:

Composition per 100mL:

Na_2CO_3	10.0g

Preparation of Solution B: Add Na_2CO_3 to distilled/deionized water and bring volume to 100.0mL. Mix thoroughly. Autoclave for 15 min at 10 psi pressure–115°C. Cool to 45°–50°C.

Preparation of Medium: Aseptically combine cooled sterile solution A with cooled sterile solution B. Mix thoroughly. Aseptically distribute into sterile tubes or flasks.

Use: For the cultivation and maintenance of *Bacillus* species.

Peptococcus glycinophilus Medium

Composition per liter:

Pancreatic digest of casein	5.0g
Yeast extract	5.0g
Glycine	3.0g
Agar	2.0g
Reducing agent	20.0mL
Potassium phosphate buffer (1M, pH 7.1)	5.0mL
Salts B	1.0mL

Reducing Agent:

Composition per 100mL:

$NaHCO_3$	5.0g
$Na_2S_2O_4$	1.0g

Preparation of Reducing Agent: Add components to distilled/deionized water and bring volume to 100.0mL. Mix thoroughly.

Salts B:

Composition per 100mL:

$MgSO_4 \cdot 7H_2O$	20.0g
$FeSO_4 \cdot 7H_2O$	1.0g
$MnSO_4 \cdot H_2O$	0.5g

Preparation of Salts B: Add components to distilled/deionized water and bring volume to 100.0mL. Mix thoroughly.

Preparation of Medium: Add components to distilled/deionized water and bring volume to 1.0L. Mix thoroughly. Gently heat and bring to boiling. Distribute into tubes or flasks. Autoclave for 15 min at 15 psi pressure–121°C. Pour into sterile Petri dishes or leave in tubes.

Use: For the cultivation and maintenance of *Peptostreptococcus micros (Peptococcus glycinophilus)*.

Peptone Broth

Composition per liter:

Peptone	5.0g

pH 7.2 ± 0.2 at 25°C

Preparation of Medium: Add peptone to distilled/deionized water and bring volume to 1.0L. Mix thoroughly. Distribute into tubes which contain a 3 inch strip of Whatman #1 filter paper. Add enough broth to cover about two-thirds of the filter paper. Autoclave for 15 min at 15 psi pressure–121°C.

Use: For the cultivation and maintenance of *Cellvibrio gilvus* and *Pseudomonas* species.

Peptone Carbonate Starch Agar

See: **Pept Carb Soluble Starch Agar**

Peptone Cholic Acid Recovery

Composition per liter:

Meat extract	10.0g
Peptone	10.0g
Cholic acid	10.0g
NaCl	5.0g
NaOH	1.0g

pH 6.8 ± 0.2 at 25°C

Preparation of Medium: Add components to distilled/deionized water and bring volume to 1.0L. Mix thoroughly. Gently heat and bring to boiling. Adjust pH to 6.8. Distribute into tubes or flasks. Autoclave for 15 min at 15 psi pressure–121°C.

Use: For the cultivation and maintenance of *Arthrobacter* species and *Corynebacterium* species.

Peptone Fumarate Sulfate Medium
See: **PFS Medium**

Peptone Glucose Liver Extract Medium
See: **PGLE Medium**

Peptone Glucose Salt Agar
See: **PGS Agar**

Peptone Glucose Yeast Extract Agar
See: **PGY Agar**

Peptone Glycerol Phosphate Broth
See: **PGP Broth**

Peptone Iron Agar

Composition per liter:

Agar	15.0g
Peptone	15.0g
Proteose peptone	5.0g
Sodium glycerophosphate	1.0g
Ferric ammonium citrate	0.5g
$Na_2S_2O_3$	0.08g

pH 6.7 ± 0.2 at 25°C

Source: This medium is available as a premixed powder from Difco Laboratories.

Preparation of Medium: Add components to distilled/deionized water and bring volume to 1.0L. Mix thoroughly. Gently heat and bring to boiling. Distribute into tubes. Autoclave for 15 min at 15 psi pressure–121°C. Allow tubes to cool in an upright position.

Use: For the cultivation and differentiation of microorganisms based on their ability to produce H_2S. Microorganisms that produce H_2S turn the medium black.

Peptone Medium

Composition per liter:

Peptone	10.0g

Preparation of Medium: Add peptone to distilled/deionized water and bring volume to 1.0L. Mix thoroughly. Distribute into tubes or flasks. Autoclave for 15 min at 15 psi pressure–121°C.

Use: For the cultivation and maintenance of *Escherichia coli.*

Peptone Recovery Broth

Composition per liter:

Meat extract	10.0g
Peptone	10.0g
NaCl	5.0g

pH 6.8 ± 0.2 at 25°C

Preparation of Medium: Add components to distilled/deionized water and bring volume to 1.0L. Mix thoroughly. Adjust pH to 6.8. Distribute into tubes or flasks. Autoclave for 15 min at 15 psi pressure–121°C.

Use: For the cultivation of *Brevibacterium* species.

Peptone Sodium Cholate

Composition per liter:

Meat extract	10.0g
Peptone	10.0g
NaCl	5.0g
Sodium cholate	5.0g

pH 6.8 ± 0.2 at 25°C

Preparation of Medium: Add components to distilled/deionized water and bring volume to 1.0L. Mix thoroughly. Adjust pH to 6.8. Distribute into tubes or flasks. Autoclave for 15 min at 15 psi pressure–121°C.

Use: For the cultivation and maintenance of *Anthrobacter* species.

Peptone Sorbitol Bile Broth

Composition per liter:

Sorbitol	10.0g
Na_2HPO_4	8.23g
NaCl	5.0g
Peptone	5.0g
Bile salts No. 3	1.5g
NaH_2PO_4	1.2g

pH 7.6 ± 0.2 at 25°C

Preparation of Medium: Add components to distilled/deionized water and bring volume to 1.0L. Mix

thoroughly. Distribute into bottles in 100.0mL volumes. Autoclave for 15 min at 15 psi pressure–121°C.

Use: For the enrichment and cultivation of *Yersinia* species.

Peptone Starch Carbonate Medium

Composition per liter:

Agar	15.0g
Soluble starch	10.0g
Peptone	5.0g
Yeast extract	5.0g
K_2HPO_4	1.0g
$MgSO_4 \cdot 7H_2O$	0.2g
Na_2CO_3 solution	100.0mL

Na_2CO_3 Solution:
Composition per 100mL:

Na_2CO_3	10.0g

Preparation of Na_2CO_3 Solution: Add Na_2CO_3 to distilled/deionized water and bring volume to 100.0mL. Mix thoroughly. Autoclave for 15 min at 15 psi pressure–121°C. Cool to 45°–50°C.

Preparation of Medium: Add components, except Na_2CO_3 solution, to distilled/deionized water and bring volume to 990.0mL. Mix thoroughly. Gently heat and bring to boiling. Autoclave for 15 min at 15 psi pressure–121°C. Cool to 45°–50°C. Aseptically add sterile Na_2CO_3 solution. Mix thoroughly. Pour into sterile Petri dishes or distribute into sterile tubes.

Use: For the cultivation and maintenance of *Bacillus alcalophilus* and other *Bacillus* species.

Peptone Starch Dextrose Agar (PSD Agar) (Dunkelberg Agar)

Composition per liter:

Proteose peptone No. 3	20.0g
Agar	15.0g
Soluble starch	10.0g
Glucose	2.0g
Na_2HPO_4	1.0g
NaH_2PO_4	1.0g

pH 6.8 ± 0.2 at 25°C

Preparation of Medium: Add starch to approximately 100.0mL of cold distilled/deionized water. Mix thoroughly. Add starch solution to 400.0mL of boiling distilled/deionized water. Add remaining components. Mix thoroughly. Bring volume to 1.0L with distilled/deionized water. Autoclave for 12 min at 8 psi pressure–112°C. Pour into sterile Petri dishes or distribute into screw-capped tubes.

Use: For the selective isolation and cultivation of *Gardnerella vaginalis*.

Peptone Succinate Agar

Composition per liter:

Peptone	5.0g
Succinic acid	1.68g
Agar	1.5g
$MgSO_4 \cdot 7H_2O$	1.0g
$(NH_4)_2SO_4$	1.0g
$FeCl_3 \cdot 6H_2O$	2.0mg
$MnSO_4 \cdot H_2O$	2.0mg

pH 7.0 ± 0.2 at 25°C

Preparation of Medium: Add components to distilled/deionized water and bring volume to 1.0L. Mix thoroughly. Gently heat and bring to boiling. Distribute into tubes or flasks. Autoclave for 15 min at 15 psi pressure–121°C. Pour into sterile Petri dishes or leave in tubes.

Use: For the cultivation and maintenance of *Aquaspirillum bengal*, *Aquaspirillum dispar*, and *Spirillum volutans*.

Peptone Succinate Agar in Sea Water

Composition per liter:

Peptone	5.0g
Succinic acid	1.68g
Agar	1.5g
$MgSO_4 \cdot 7H_2O$	1.0g
$(NH_4)_2SO_4$	1.0g
$FeCl_3 \cdot 6H_2O$	2.0mg
$MnSO_4 \cdot H_2O$	2.0mg

pH 7.0 ± 0.2 at 25°C

Preparation of Medium: Add components to sea water and bring volume to 1.0L. Mix thoroughly. Gently heat and bring to boiling. Distribute into tubes or flasks. Autoclave for 15 min at 15 psi pressure–121°C. Pour into sterile Petri dishes or leave in tubes.

Use: For the cultivation and maintenance of *Oceanospirillum maris*.

Peptone Succinate Salts Broth (PSS Broth)

Composition per 100mL:

Peptone	1.0g
$MgSO_4 \cdot 7H_2O$	0.1g

$(NH_4)_2SO_4$ 0.1g
Succinic acid 0.1g
$FeCl_3 \cdot 6H_2O$ 0.2mg
$MnSO_4 \cdot H_2O$ 0.2mg
pH 6.8 ± 0.2 at 25°C

Preparation of Medium: Add components to distilled/deionized water and bring volume to 1.0L. Mix thoroughly. Adjust pH to 6.8 with KOH. Distribute into tubes or flasks. Autoclave for 15 min at 15 psi pressure–121°C.

Use: For the cultivation of *Spirillum* species.

Peptone Succinate Salts in Sea Water

Composition per liter:

Peptone 10.0g
$MgSO_4 \cdot 7H_2O$ 1.0g
$(NH_4)_2SO_4$ 1.0g
Succinic acid 1.0g
$FeCl_3 \cdot 6H_2O$ 2.0mg
$MnSO_4 \cdot H_2O$ 2.0mg
Synthetic sea water 1.0L
pH 6.8 ± 0.2 at 25°C

Synthetic Sea Water:
Composition per liter:

NaCl 27.5g
$MgCl_2$ 5.0g
$MgSO_4$ 2.0g
KCl 1.0g
$CaCl_2$ 0.5g
$FeSO_4$ 1.0μg

Preparation of Synthetic Sea Water: Add components to distilled/deionized water and bring volume to 1.0L. Mix thoroughly.

Preparation of Medium: Add solid components to synthetic sea water and bring volume to 1.0L. Mix thoroughly. Adjust pH to 6.8 with 2*N* KOH. Distribute into tubes or flasks. Autoclave for 15 min at 15 psi pressure–121°C.

Use: For the cultivation and maintenance of *Oceanospirillum maris*.

Peptone Succinate Salts Medium (PSS Medium)

Composition per liter:

Peptone 10.0g
$MgSO_4 \cdot 7H_2O$ 1.0g
$(NH_4)_2SO_4$ 1.0g
Succinic acid 1.0g
$FeCl_3 \cdot 6H_2O$ 2.0mg
$MnSO_4 \cdot H_2O$ 2.0mg
pH 6.8 ± 0.2 at 25°C

Preparation of Medium: Add solid components to synthetic sea water and bring volume to 1.0L. Mix thoroughly. Adjust pH to 6.8 with 2*N* KOH. Distribute into tubes or flasks. Autoclave for 15 min at 15 psi pressure–121°C.

Use: For the cultivation and maintenance of *Aquaspirillum anulus*.

Peptone Sucrose Broth

Composition per liter:

Sucrose 20.0g
Peptone 10.0g

Preparation of Medium: Add components to distilled/deionized water and bring volume to 1.0L. Mix thoroughly. Distribute into tubes or flasks. Autoclave for 15 min at 15 psi pressure–121°C.

Use: For the cultivation and maintenance of *Xanthomonas campestris*.

Peptone Water

Composition per liter:

Peptone 10.0g
NaCl 5.0g
pH 7.2 ± 0.2 at 25°C

Source: This medium is available as a premixed powder from Difco Laboratories and Oxoid Unipath.

Preparation of Medium: Add components to distilled/deionized water and bring volume to 1.0L. Mix thoroughly. Distribute into tubes or flasks. Autoclave for 15 min at 15 psi pressure–121°C.

Use: For the cultivation of nonfastidious microorganisms, for carbohydrate fermentation tests and for performing the indole test.

Peptone Water with Andrade's Indicator

Composition per liter:

Peptone 10.0g
NaCl 5.0g
Andrade's indicator 100.0mL
Carbohydrate solution 20.0mL
pH 7.4 ± 0.2 at 25°C

Source: This medium is available as a premixed powder from Oxoid Unipath.

Andrade's Indicator:

Composition per 100mL:

NaOH (1*N* solution)	16.0mL
Acid Fuchsin	0.1 g

Caution: Acid Fuchsin is a potential carcinogen and care must be taken to avoid inhalation of the powdered dye and contact with the skin.

Carbohydrate Solution:

Composition per 20mL:

Carbohydrate	5.0–10.0g

Preparation of Carbohydrate Solution: Add carbohydrate to distilled/deionized water and bring volume to 20.0mL. Mix thoroughly. Filter sterilize.

Preparation of Medium: Add components, except carbohydrate solution, to distilled/deionized water and bring volume to 980.0mL. Mix thoroughly. Adjust pH to 7.4 if necessary. Distribute into tubes containing an inverted Durham tube. Fill each tube with 9.8mL of medium. Autoclave for 15 min at 15 psi pressure–121°C. Aseptically add 0.2mL of sterile carbohydrate solution to each tube.

Use: For use in carbohydrate fermentation tests. Fermentation is determined by the production of acid—broth turns pink—and formation of gas—bubble trapped in Durham tube.

Peptone Yeast Extract Agar (ATCC Medium 1093)

Composition per liter:

Agar	15.0g
Peptone	0.5g
Yeast extract	0.5g

Preparation of Medium: Add components to distilled/deionized water and bring volume to 1.0L. Mix thoroughly. Gently heat and bring to boiling. Distribute into tubes or flasks. Autoclave for 15 min at 15 psi pressure–121°C. Pour into sterile Petri dishes or leave in tubes.

Use: For the cultivation and maintenance of *Angiococcus disciformis.*

Peptone Yeast Extract Agar (ATCC Medium 526)

Composition per liter:

Agar	15.0g
Peptone	10.0g
Yeast extract	3.0g

Preparation of Medium: Add components to distilled/deionized water and bring volume to 1.0L. Mix thoroughly. Gently heat and bring to boiling. Distribute into tubes or flasks. Autoclave for 15 min at 15 psi pressure–121°C. Pour into sterile Petri dishes or leave in tubes.

Use: For the cultivation and maintenance of *Bdellovibrio bacteriovorus* and *Bdellovibrio stolpii.*

Peptone Yeast Extract 1% Medium

See: **PY 1% Medium**

Peptone Yeast Extract Carboxymethyl Cellulose Medium

See: **PY CMC Medium**

Peptone Yeast Extract Glucose Broth

See: **PYG Broth**

Peptone Yeast Extract Glucose Maltose Medium

See: **PYGM Medium**

Peptone Yeast Extract Glucose Medium

See: **PYG Medium**

Peptone Yeast Extract Glucose Medium for *Spirillum*

See: **PYG Medium for *Spirillum***

Peptone Yeast Extract Glucose Medium, Modified (MPYG Medium)

Composition per 950mL:

Peptone	10.0g
Yeast extract	10.0g
Glucose	5.0g
Cysteine·HCl·H_2O	0.5g
$(NH_4)_2SO_4$	0.5g
Salt solution	40.0mL
Vitamin K-heme solution	10.0mL

Resazurin (0.025% solution)....4.0mL
Volatile fatty acid solution....3.1mL

pH 7.0 ± 0.2 at 25°C

Salt Solution:

Composition per liter:

$NaHCO_3$....10.0g
NaCl....2.0g
K_2HPO_4....1.0g
KH_2PO_4....1.0g
$CaCl_2$ (anhydrous)....0.2g
$MgSO_4$....0.2g

Preparation of Salt Solution: Add $CaCl_2$ and $MgSO_4$ to 300.0mL of distilled/deionized water. Mix thoroughly until dissolved. Bring volume to 800.0mL with distilled/deionized water. Add remaining components while stirring. Bring volume to 1.0L. Mix thoroughly. Store at 4°C.

Vitamin K-Heme Solution:

Composition per liter:

Part A....100.0mL
Part B....1.0mL

Preparation of Vitamin K-Heme Solution: Aseptically add 1.0mL of sterile part B to 100.0mL of cooled sterile part A. Mix thoroughly.

Part A:

Composition per 100mL:

Hemin....50.0mg
NaOH (1*N* solution)....1.0mL

Preparation of Part A: Add hemin to NaOH solution and bring volume to 100.0mL with distilled/deionized water. Mix thoroughly. Autoclave for 15 min at 15 psi pressure–121°C. Cool to 45°–50°C.

Part B:

Composition per 30mL:

Menadione (Vitamin K_3)....100.0mg
Ethanol (95% solution)....30.0mL

Preparation of Part B: Add menadione to ethanol. Mix thoroughly. Filter sterilize.

Volatile Fatty Acid Solution:

Composition per 31mL:

Acetic acid....17.0mL
Propionic acid....6.0mL
n-Butyric acid....4.0mL
n-Valeric acid....1.0mL
Isovaleric acid....1.0mL
Isobutyric acid....1.0mL
DL-α-Methyl butyric acid....1.0mL

Preparation of Volatile Fatty Acid Solution: Combine components. Mix thoroughly.

Preparation of Medium: Add components—except vitamin K-heme solution, cysteine·HCl·H_2O and volatile fatty acid solution—to distilled/deionized water and bring volume to 936.9mL. Gently heat and bring to boiling under 97% N_2 + 3% H_2. Continue boiling until resazurin turns colorless indicating reduction. Cool to 45°–50°C. Add vitamin K-heme solution, cysteine·HCl·H_2O and volatile fatty acid solution. Adjust pH to 7.0. Distribute into tubes under 97% N_2 + 3% H_2. Cap with rubber stoppers. Place tubes in a press. Autoclave for 15 min at 15 psi pressure–121°C with fast exhaust.

Use: For the cultivation and maintenance of *Acetivibrio ethanolgignens, Butyrivibrio fibrisolvens, Lachnospira multipara,* and *Succinivibrio dextrinosolvens.*

Peptone Yeast Extract Glucose Salt Medium

See: **PYEX Glucose Salt Medium**

Peptone Yeast Extract Glucose Vitamin Marine Medium

See: **PYGV Marine Medium**

Peptone Yeast Extract Glucose Vitamin Medium

See: **PYGV Medium**

Peptone Yeast Extract Inositol Medium

See: **PY Inositol Medium**

Peptone Yeast Extract Iron Agar

See: **ISP Medium 6**

Peptone Yeast Extract Medium (ATCC Medium 828)

Composition per liter:

Peptone....20.0g
Yeast extract....1.5g

pH 7.0 ± 0.2 at 25°C

Preparation of Medium: Add components to distilled/deionized water and bring volume to 1.0L. Mix thoroughly. Gently heat and bring to boiling. Distribute into tubes or flasks. Autoclave for 15 min at 15 psi pressure–121°C.

Use: For the cultivation and maintenance of *Acinetobacter lwoffii*.

Peptone Yeast Extract Medium (ATCC Medium 1366)

Composition per liter:

Peptone	10.0g
NaCl	5.0g
Yeast extract	5.0g

pH 7.2 ± 0.2 at 25°C

Preparation of Medium: Add components to distilled/deionized water and bring volume to 1.0L. Mix thoroughly. Gently heat and bring to boiling. Adjust pH to 7.2. Distribute into tubes or flasks. Autoclave for 15 min at 15 psi pressure–121°C.

Use: For the cultivation and maintenance of *Xenorhabdus nematophilus*.

Peptone Yeast Extract Medium (PY Medium) (ATCC Medium 1524)

Composition per 950mL:

Yeast extract	10.0g
Peptone	5.0g
Pancreatic digest of casein	5.0g
Cysteine·HCl·H_2O	0.5g
Salt solution	40.0mL
Hemin solution	10.0mL
Resazurin (0.025% solution)	4.0mL
Vitamin K_1 solution	0.2mL

pH 7.0 ± 0.2 at 25°C

Salt Solution:

Composition per liter:

$NaHCO_3$	10.0g
NaCl	2.0g
K_2HPO_4	1.0g
KH_2PO_4	1.0g
$CaCl_2$ (anhydrous)	0.2g
$MgSO_4$	0.2g

Preparation of Salt Solution: Add $CaCl_2$ and $MgSO_4$ to 300.0mL of distilled/deionized water. Mix thoroughly until dissolved. Bring volume to 800.0mL with distilled/deionized water. Add remaining components while stirring. Bring volume to 1.0L. Mix thoroughly. Store at 4°C.

Hemin Solution:

Composition per 100mL:

Hemin	0.05g
NaOH (1*N* solution)	1.0mL

Preparation of Hemin Solution: Add hemin to NaOH solution and bring volume to 100.0mL with distilled/deionized water. Mix thoroughly. Autoclave for 15 min at 15 psi pressure–121°C. Cool to 45°–50°C.

Vitamin K_1 Solution:

Composition per 30mL:

Vitamin K_1	0.15g
Ethanol (95% solution)	30.0mL

Preparation of Vitamin K_1 Solution: Add vitamin K_1 to ethanol. Mix thoroughly. Filter sterilize.

Preparation of Medium: Add components—except vitamin K_1 solution, hemin solution, and cysteine·HCl·H_2O—to distilled/deionized water and bring volume to 939.8mL. Gently heat and bring to boiling under 80% N_2 + 10% H_2 + 10% CO_2. Continue boiling until resazurin turns colorless indicating reduction. Cool to 45°–50°C. Add vitamin K_1 solution, hemin solution, and cysteine·HCl·H_2O. Adjust pH to 7.0. Distribute into tubes under 80% N_2 + 10% H_2 + 10% CO_2. Cap with rubber stoppers. Place tubes in a press. Autoclave for 15 min at 15 psi pressure–121°C with fast exhaust.

Use: For the cultivation and maintenance of *Megasphaera cerevisiae* and *Clostridium* species.

Peptone Yeast Extract Medium with Fructose

See: **PY Medium with Fructose**

Peptone Yeast Extract Medium with Glucose

See: **PY Medium with Glucose**

Peptone Yeast Extract Salt Medium

See: **PY Salt Medium**

Peptone Yeast Extract Salt Agar

See: **PYS Agar**

Peptone Yeast Glutamate Medium

Composition per liter:

Peptone	20.0g
Yeast extract	10.0g

Monosodium glutamate 4.0g
Sodium thioglycollate 1.0g
pH 7.2 ± 0.2 at 25°C

Preparation of Medium: Add components to distilled/deionized water and bring volume to 1.0L. Mix thoroughly. Distribute into tubes or flasks. Autoclave for 15 min at 15 psi pressure–121°C.

Use: For the cultivation of *Peptococcus aerogenes*.

Peptone Yeast Medium with $MgSO_4$

Composition per liter:

Peptone..10.0g
$MgSO_4 \cdot 7H_2O$ 2.0g
$(NH_4)_2SO_4$..2.0g
Yeast extract...1.0g
pH 7.0 ± 0.2 at 25°C

Preparation of Medium: Add components to distilled/deionized water and bring volume to 1.0L. Mix thoroughly. Distribute into tubes or flasks. Autoclave for 15 min at 15 psi pressure–121°C.

Use: For the cultivation and maintenance of *Aquaspirillum itersonii, Aquaspirillum peregrinum,* and *Aquaspirillum psychrophilum.*

Peptone, Czapek's

Composition per liter:

Agar...15.0g
Sucrose..15.0g
Peptone..5.0g
$NaNO_3$...3.0g
K_2HPO_4..1.0g
KCl...0.5g
$MgSO_4$..0.5g
$FeSO_4$..0.01g

Preparation of Medium: Add components to distilled/deionized water and bring volume to 1.0L. Mix thoroughly. Gently heat and bring to boiling. Distribute into tubes or flasks. Autoclave for 15 min at 15 psi pressure–121°C. Pour into sterile Petri dishes or leave in tubes.

Use: For the cultivation and maintenance of *Pilimelia anulata, Pilimelia terevasa* and other *Pilimelia* species.

Peptonized Milk Agar (PMA Medium)

Composition per liter:

Agar...15.0g
Milk, peptonized 1.0g

Preparation of Medium: Add components to distilled/deionized water and bring volume to 1.0L. Mix thoroughly. Gently heat and bring to boiling. Distribute into tubes or flasks. Autoclave for 15 min at 15 psi pressure–121°C. Pour into sterile Petri dishes or leave in tubes.

Use: For the cultivation of freshwater *Myxobacterium* species.

Perfringens Agar, OPSP
See: ***Clostridium perfringens*** **Agar, OPSP**

Petragnani Medium

Composition per 2398mL:

Skim milk..100.0g
Potato flour..36.4g
L-Asparagine 5.1g
Pancreatic digest of casein 5.1g
Malachite Green...1.2g
Whole egg 1277.0mL
Egg yolk 121.0mL
Glycerol..60.0mL
pH 7.0 ± 0.2 at 25°C

Source: Available as a prepared medium from BBL Microbiology Systems.

Preparation of Medium: Add components—except whole egg, egg yolk, and glycerol—to distilled/deionized water and bring volume to 940.0mL. Mix thoroughly. Add glycerol. Gently heat while stirring and bring to boiling. Autoclave for 15 min at 15 psi pressure–121°C. Cool to 45°–50°C. Scrub the eggshells with soap. Let stand in a soap solution for 30 min. Rinse in running water. Soak eggs in 70% ethanol for 15 min. Break the eggs into a sterile container. Homogenize by shaking. Filter through four layers of sterile cheesecloth into a sterile graduated cylinder. Measure out 1277.0mL. Add separated egg yolks to another sterile container. Measure out 121.0mL. Aseptically add homogenized whole egg and egg yolk to cooled sterile basal medium. Mix thoroughly. Aseptically distribute into sterile tubes. Inspissate at 85°–90°C (moist heat) for 45 min.

Use: For the isolation and cultivation of *Mycobacterium* species from clinical specimens. For the cultivation and maintenance of *Mycobacterium smegmatis.*

Petragnani Medium

Composition per 2285mL:

Potato 500.0g
Potato flour..36.0g

Malachite Green	1.2g
Whole egg	1200.0mL
Whole milk	900.0mL
Egg yolk	115.0mL
Glycerol	70.0mL

pH 7.2 ± 0.2 at 25°C

Source: Available as a prepared medium from Difco Laboratories.

Preparation of Medium: Peel and dice potato. Add potato to 500.0mL of distilled/deionized water. Gently heat and bring to boiling. Continue boiling for 30 min. Filter solids through two layers of cheesecloth. Combine potato solids with remaining components—except whole egg, egg yolk, and glycerol. Mix thoroughly. Add glycerol. Gently heat while stirring and bring to boiling. Autoclave for 15 min at 15 psi pressure–121°C. Cool to 45°–50°C. Scrub the eggshells with soap. Let stand in a soap solution for 30 min. Rinse in running water. Soak eggs in 70% ethanol for 15 min. Break the eggs into a sterile container. Homogenize by shaking. Filter through four layers of sterile cheesecloth into a sterile graduated cylinder. Measure out 1200.0mL. Add separated egg yolks to another sterile container. Measure out 115.0mL. Aseptically add homogenized whole egg and egg yolk to cooled sterile basal medium. Mix thoroughly. Aseptically distribute into sterile tubes. Inspissate at 85°–90°C (moist heat) for 45 min.

Use: For the isolation and cultivation of *Mycobacterium* species from clinical specimens. For the cultivation and maintenance of *Mycobacterium smegmatis.*

Pfizer Selective *Enterococcus* Agar (PSE Agar)

Peptone C	17.0g
Agar	15.0g
Bile	10.0g
NaCl	5.0g
Yeast extract	5.0g
Peptone B	3.0g
Esculin	1.0g
Sodium citrate	1.0g
Ferric ammonium citrate	0.5g
NaN_3	0.25g

pH 7.1 ± 0.2 at 25°C

Caution: Sodium azide is toxic. Azides also react with metals and disposal must be highly diluted.

Preparation of Medium: Add components to distilled/deionized water and bring volume to 1.0L. Mix thoroughly. Gently heat and bring to boiling. Distribute into tubes or flasks. Autoclave for 15 min at 15 psi pressure–121°C. Pour into sterile Petri dishes or leave in tubes.

Use: For the selective isolation, cultivation and enumeration of *Enterococcus* species by the multiple tube technique.

PFS Medium (Peptone Fumarate Sulfate Medium)

Composition per liter:

Peptone	10.0g
Fumaric acid	2.0g
$(NH_4)_2SO_4$	1.0g
$MgSO_4 \cdot 7H_2O$	0.5g
$FeCl_3 \cdot 6H_2O$	0.2mg
$MnSO_4 \cdot H_2O$	0.2mg

pH 7.0 ± 0.2 at 25°C

Preparation of Medium: Add components to distilled/deionized water and bring volume to 1.0L. Mix thoroughly. Adjust pH to 7.0 with KOH. Distribute into tubes or flasks. Autoclave for 15 min at 15 psi pressure–121°C.

Use: For the cultivation and maintenance of *Aquaspirillum fasciculus.*

PGLE Medium (Peptone Glucose Liver Extract Medium)

Composition per liter:

Agar	25.0g
Glucose	20.0g
Peptone	5.0g
Yeast extract	5.0g
K_2HPO_4	1.0g
Liver extract	0.5g

pH 9.0 ± 0.2 at 25°C

Preparation of Medium: Add components to distilled/deionized water and bring volume to 1.0L. Mix thoroughly. Gently heat and bring to boiling. Distribute into tubes or flasks. Autoclave for 15 min at 15 psi pressure–121°C. Pour into sterile Petri dishes or leave in tubes.

Use: For the cultivation and maintenance of *Enterobacter cloacae.*

PGP Broth (Peptone Glycerol Phosphate Broth)

Composition per liter:

Peptone	5.0g
K_2HPO_4	2.0g
Glycerol	10.0mL

Preparation of Medium: Add components to distilled/deionized water and bring volume to 1.0L. Mix thoroughly. Distribute into tubes or flasks. Autoclave for 15 min at 15 psi pressure–121°C.

Use: For the cultivation and maintenance of *Serratia marcescens*.

PGS Agar (Peptone Glucose Salt Agar)

Composition per liter:

Agar	15.0g
Glucose	10.0g
NaCl	10.0g
Peptone	10.0g

pH 7.2 ± 0.2 at 25°C

Preparation of Medium: Add components to distilled/deionized water and bring volume to 1.0L. Mix thoroughly. Gently heat and bring to boiling. Distribute into tubes or flasks. Autoclave for 15 min at 15 psi pressure–121°C. Pour into sterile Petri dishes or leave in tubes.

Use: For the cultivation and maintenance of *Rhodococcus australis*.

PGT Medium

Composition per liter:

Casamino acids	30.0g
L-Glutamic acid	0.5g
$MgSO_4 \cdot 7H_2O$	0.45g
Maltose	0.2g
L-Cystine	0.2g
DL-Tryptophan	0.1g
Solution 3	100.0mL
Solution 2	2.0mL
Calcium pantothenate (0.1% solution)	0.5mL

pH 6.8 ± 0.2 at 25°C

Solution 3:

Composition per 500mL:

Maltose	200.0g
$CaCl_2$	1.5g
Calcium pantothenate (0.1% solution)	3.0mL
$FeSO_4$ (1% in 1*N* HCl)	0.2mL

Preparation of Solution 3: Add components to distilled/deionized water and bring volume to 500.0mL. Mix thoroughly. Autoclave for 15 min at 7 psi pressure–111°C. Cool to 45°–50°C.

Solution 2:

Composition per 100mL:

β-Alanine	0.115g
Nicotinic acid	0.115g
$CuSO_4 \cdot 5H_2O$	0.05g
$ZnSO_4 \cdot 7H_2O$	0.045g
$MnCl_2 \cdot 4H_2O$	0.015g
Pimelic acid	7.5mg
HCl, concentrated	3.0mL

Preparation of Solution 2: Add components to distilled/deionized water and bring volume to 100.0mL. Mix thoroughly.

Preparation of Medium: Add components, except solution 3, to distilled/deionized water and bring volume to 900.0mL. Mix thoroughly. Adjust pH to 6.8 with 50% KOH. Gently heat and bring to boiling. Autoclave for 15 min at 15 psi pressure–121°C. Cool to 45°–50°C. Aseptically add sterile solution 3. Mix thoroughly. Aseptically distribute into sterile tubes or flasks.

Use: For the cultivation and maintenance of *Corynebacterium diphtheriae*.

PGY Agar (Peptone Glucose Yeast Extract Agar)

Composition per liter:

Agar	15.0g
Peptone	10.0g
Yeast extract	5.0g
Glucose	1.0g

Preparation of Medium: Add components to distilled/deionized water and bring volume to 1.0L. Mix thoroughly. Gently heat and bring to boiling. Distribute into tubes or flasks. Autoclave for 15 min at 15 psi pressure–121°C. Pour into sterile Petri dishes or leave in tubes.

Use: For the cultivation and maintenance of *Micrococcus luteus*.

PHB Medium

See: **Poly-β-Hydroxybutyrate Medium**

Phenethyl Alcohol Agar (Phenylethanol Agar) (Phenylethyl Alcohol Agar)

Composition per liter:

Agar	15.0g
Pancreatic digest of casein	15.0g
NaCl	5.0g
Papaic digest of soybean meal	5.0g
β-Phenethyl alcohol	2.5g
Blood	50.0mL

pH 7.3 ± 0.2 at 25°C

Source: This medium is available as a premixed powder from BBL Microbiology Systems.

Preparation of Medium: Add components, except blood, to distilled/deionized water and bring volume to 950.0mL. Mix thoroughly. Gently heat and bring to boiling. Autoclave for 15 min at 13 psi pressure–118°C. Cool to 45°–50°C. Aseptically add sterile defibrinated blood. Mix thoroughly. Pour into sterile Petri dishes or distribute into sterile tubes.

Use: For the selective isolation of Gram-positive bacteria, particularly Gram-positive cocci, from specimens with a mixed flora. Do not use for the observation of hemolytic reactions.

Phenol Nutrient–Supplemented Broth

Composition per liter:

$(NH_4)_2SO_4$	2.0g
Na_2HPO_4	1.0g
Nutrient broth	20.0mL
Phenol (90% solution)	1.0mL

pH 6.8 ± 0.2 at 25°C.

Nutrient Broth:
Composition per liter:

Peptone	5.0g
Beef extract	3.0g

Preparation of Nutrient Broth: Add components to distilled/deionized water and bring volume to 1.0L. Mix thoroughly. Gently heat and bring to boiling.

Preparation of Medium: Add components, except phenol, to distilled/deionized water and bring volume to 1.0L. Mix thoroughly. Autoclave for 15 min at 15 psi pressure–121°C. Cool to 45°–50°C. Aseptically add phenol. Mix thoroughly.

Use: For the cultivation of ATCC strain 1413.

Phenol Red Agar

Composition per liter:

Agar	15.0g
Pancreatic digest of casein	10.0g
NaCl	5.0g
Phenol Red	0.018g
Carbohydrate solution	20.0mL

pH 7.4 ± 0.2 at 25°C

Source: This medium is available as a premixed powder from BBL Microbiology Systems.

Carbohydrate Solution:
Composition per 20mL:

Carbohydrate	5.0–10.0g

Preparation of Carbohydrate Solution: Add carbohydrate to distilled/deionized water and bring volume to 20.0mL. Mix thoroughly. Filter sterilize.

Preparation of Medium: Add components, except carbohydrate solution, to distilled/deionized water and bring volume to 980.0mL. Mix thoroughly. Adjust pH to 7.4 if necessary. Autoclave for 15 min at 15 psi pressure–121°C. Cool to 45°–50°C. Aseptically add 20.0mL of sterile carbohydrate solution. Pour into sterile Petri dishes or distribute into sterile tubes. Allow tubes to cool in a slanted position.

Use: For the determination of fermentation reactions. Bacteria that can ferment the added carbohydrate turn the medium yellow.

Phenol Red Agar

Composition per liter:

Agar	15.0g
Proteose peptone No. 3	10.0g
NaCl	5.0g
Beef extract	1.0g
Phenol Red	0.025g
Carbohydrate solution	20.0mL

pH 7.4 ± 0.2 at 25°C

Source: This medium is available as a premixed powder from Difco Laboratories.

Carbohydrate Solution:
Composition per 20mL:

Carbohydrate	5.0–10.0g

Preparation of Carbohydrate Solution: Add carbohydrate to distilled/deionized water and bring volume to 20.0mL. Mix thoroughly. Filter sterilize.

Preparation of Medium: Add components, except carbohydrate solution, to distilled/deionized water and bring volume to 980.0mL. Mix thoroughly. Adjust pH to 7.4 if necessary. Autoclave for 15 min at 15 psi pressure–121°C. Cool to 45°–50°C. Aseptically add 20.0mL of sterile carbohydrate solution. Pour into sterile Petri dishes or distribute into sterile tubes. Allow tubes to cool in a slanted position.

Use: For the determination of fermentation reactions. Bacteria that can ferment the added carbohydrate turn the medium yellow.

Phenol Red Broth

Composition per liter:

Pancreatic digest of casein	10.0g
NaCl	5.0g
Phenol Red	0.018g
Carbohydrate solution	20.0mL

pH 7.4 ± 0.2 at 25°C

Source: This medium is available as a premixed powder from BBL Microbiology Systems.

Carbohydrate Solution:
Composition per 20mL:

Carbohydrate	5.0–10.0g

Preparation of Carbohydrate Solution: Add carbohydrate to distilled/deionized water and bring volume to 20.0mL. Mix thoroughly. Filter sterilize.

Preparation of Medium: Add components, except carbohydrate solution, to distilled/deionized water and bring volume to 980.0mL. Mix thoroughly. Adjust pH to 7.4 if necessary. Distribute into tubes containing an inverted Durham tube. Fill each tube with 9.8mL of medium. Autoclave for 15 min at 13 psi pressure–118°C. Cool to 45°–50°C. Aseptically add 0.2mL of sterile carbohydrate solution to each tube.

Use: For the determination of fermentation reactions in the differentiation of microorganisms. Fermentation is determined by the production of acid—broth turns yellow—and formation of gas—bubble trapped in Durham tube.

Phenol Red Glucose Broth

Composition per liter:

Pancreatic digest of casein	10.0g
Glucose	5.0g
NaCl	5.0g
Phenol Red	0.018g

pH 7.3 ± 0.2 at 25°C

Source: This medium is available as a premixed powder from BBL Microbiology Systems.

Preparation of Medium: Add components to distilled/deionized water and bring volume to 1.0L. Mix thoroughly. Adjust pH to 7.3 if necessary. Distribute into tubes containing an inverted Durham tube. Fill each tube with 10.0mL of medium. Autoclave for 15 min at 13 psi pressure–118°C.

Use: For the determination of the ability of a microorganism to ferment glucose. Fermentation is determined by the production of acid—broth turns yellow—and formation of gas—bubble trapped in Durham tube.

Phenol Red Lactose Agar

Composition per liter:

Agar	15.0g
Lactose	10.0g
Proteose peptone No. 3	10.0g
NaCl	5.0g
Beef extract	1.0g
Phenol Red	25.0mg

pH 7.4 ± 0.2 at 25°C

Source: This medium is available as a premixed powder from Difco Laboratories.

Preparation of Medium: Add components to distilled/deionized water and bring volume to 1.0L. Mix thoroughly. Gently heat and bring to boiling. Distribute into tubes or flasks. Autoclave for 15 min at 13 psi pressure–118°C. Pour into sterile Petri dishes or leave in tubes. Allow tubes to cool in a slanted position.

Use: For the determination of the ability of a microorganism to ferment lactose. Fermentation is determined by the production of acid—medium turns yellow.

Phenol Red Lactose Broth

Composition per liter:

Pancreatic digest of casein	10.0g
Lactose	5.0g
NaCl	5.0g
Phenol Red	0.018g

pH 7.3 ± 0.2 at 25°C

Source: This medium is available as a premixed powder from BBL Microbiology Systems.

Preparation of Medium: Add components to distilled/deionized water and bring volume to 1.0L. Mix thoroughly. Adjust pH to 7.3 if necessary. Distribute into tubes containing an inverted Durham tube. Fill each tube with 10.0mL of medium. Autoclave for 15 min at 13 psi pressure–118°C.

Use: For the determination of the ability of a microorganism to ferment lactose. Fermentation is determined by the production of acid—broth turns yellow—and formation of gas—bubble trapped in Durham tube.

Phenol Red Mannitol Agar

Composition per liter:

Agar	15.0g
Mannitol	10.0g
Proteose peptone No. 3	10.0g
NaCl	5.0g
Beef extract	1.0g
Phenol Red	25.0mg

pH 7.4 ± 0.2 at 25°C

Source: This medium is available as a premixed powder from Difco Laboratories.

Preparation of Medium: Add components to distilled/deionized water and bring volume to 1.0L. Mix thoroughly. Gently heat and bring to boiling. Distribute into tubes or flasks. Autoclave for 15 min at 13 psi pressure–118°C. Pour into sterile Petri dishes or leave in tubes. Allow tubes to cool in a slanted position.

Use: For the determination of the ability of a microorganism to ferment mannitol. Fermentation is determined by the production of acid—medium turns yellow.

Phenol Red Mannitol Broth

Composition per liter:

Pancreatic digest of casein	10.0g
D-Mannitol	5.0g
NaCl	5.0g
Phenol Red	0.018g

pH 7.3 ± 0.2 at 25°C

Source: This medium is available as a premixed powder from BBL Microbiology Systems.

Preparation of Medium: Add components to distilled/deionized water and bring volume to 1.0L. Mix thoroughly. Adjust pH to 7.3 if necessary. Distribute into tubes containing an inverted Durham tube. Fill each tube with 10.0mL of medium. Autoclave for 15 min at 13 psi pressure–118°C.

Use: For the determination of the ability of a microorganism to ferment mannitol. Fermentation is determined by the production of acid—broth turns yellow—and formation of gas—bubble trapped in Durham tube.

Phenol Red Sucrose Broth

Composition per liter:

Pancreatic digest of casein	10.0g
NaCl	5.0g
Sucrose	5.0g
Phenol Red	0.018g

pH 7.3 ± 0.2 at 25°C

Source: This medium is available as a premixed powder from BBL Microbiology Systems.

Preparation of Medium: Add components to distilled/deionized water and bring volume to 1.0L. Mix thoroughly. Adjust pH to 7.3 if necessary. Distribute into tubes containing an inverted Durham tube. Fill each tube with 10.0mL of medium. Autoclave for 15 min at 13 psi pressure–118°C.

Use: For the determination of the ability of a microorganism to ferment sucrose. Fermentation is determined by the production of acid—broth turns yellow—and formation of gas—bubble trapped in Durham tube.

Phenol Red Tartrate Agar

Composition per liter:

Agar	15.0g
Peptone	10.0g
Potassium tartrate	10.0g
NaCl	5.0g
Phenol Red	0.024g

pH 7.6 ± 0.2 at 25°C

Source: This medium is available as a premixed powder from Difco Laboratories.

Preparation of Medium: Add components to cold distilled/deionized water and bring volume to 1.0L. Mix thoroughly. Gently heat and bring to boiling. Distribute into tubes or flasks. Autoclave for 15 min at 13 psi pressure–118°C. Pour into sterile Petri dishes or leave in tubes. Allow tubes to cool in an upright position.

Use: For the differentiation of Gram-negative bacteria of the intestinal groups, particularly members of the *Salmonella* (paratyphoid) group based on their ability to ferment tartrate.

Phenol Red Tartrate Broth

Composition per liter:

Pancreatic digest of casein	10.0g
Potassium tartrate	10.0g
Agar	5.0g
NaCl	5.0g
Phenol Red	0.024g

pH 7.6 ± 0.2 at 25°C

Preparation of Medium: Add components to distilled/deionized water and bring volume to 1.0L. Mix thoroughly. Distribute into tubes or flasks. Autoclave for 15 min at 15 psi pressure–121°C.

Use: For the differentiation of Gram-negative bacteria of the intestinal groups, particularly members of the *Salmonella* (paratyphoid) group based on their ability to ferment tartrate.

Phenylalanine Agar (Phenylalanine Deaminase Medium)

Composition per liter:

Agar	12.0g
NaCl	5.0g
Yeast extract	3.0g
DL-Phenylalanine	2.0g
Na_2HPO_4	1.0g

pH 7.3 ± 0.2 at 25°C

Source: This medium is available as a premixed powder from BBL Microbiology Systems, and Difco Laboratories.

Preparation of Medium: Add components to distilled/deionized water and bring volume to 1.0L. Mix

thoroughly. Gently heat while stirring and bring to boiling. Distribute into tubes or flasks. Autoclave for 10 min at 15 psi pressure–121°C. Pour into sterile Petri dishes or leave in tubes.

Use: For the differentiation of enteric Gram-negative bacilli on the basis of their ability to produce phenylpyruvic acid from phenylalanine. After appropriate incubation of bacteria, ferric chloride reagent is added on the agar. Formation of a green color in 1–5 min indicates the production of phenylpyruvic acid.

Phenylalanine Malonate Broth

Composition per liter:

Sodium malonate	3.0g
DL-Phenylalanine	2.0g
NaCl	2.0g
$(NH_4)_2SO_4$	2.0g
Yeast extract	1.0g
K_2HPO_4	0.6g
KH_2PO_4	0.4g
Bromthymol Blue	0.025g

pH 7.3 ± 0.2 at 25°C

Source: This medium is available as a premixed powder from Difco Laboratories.

Preparation of Medium: Add components to distilled/deionized water and bring volume to 1.0L. Mix thoroughly. Distribute into tubes or flasks. Autoclave for 10 min at 10 psi pressure–115°C.

Use: For the differentiation of Gram-negative enteric bacilli on the basis of malonate utilization and formation of pyruvic acid from phenylalanine.

Phenylethanol Agar

Composition per liter:

Agar	15.0g
Tryptose	10.0g
NaCl	5.0g
Beef extract	3.0g
Phenylethanol	2.5g

pH 7.3 ± 0.2 at 25°C

Source: This medium is available as a premixed powder from Difco Laboratories.

Preparation of Medium: Add components to distilled/deionized water and bring volume to 1.0L. Mix thoroughly. Gently heat and bring to boiling. Distribute into tubes or flasks. Autoclave for 15 min at 15 psi pressure–121°C. Pour into sterile Petri dishes or leave in tubes.

Use: For the isolation of staphylococci and streptococci from specimens containing a mixed flora.

Phenylethanol Blood Agar

Composition per liter:

Agar	15.0g
Tryptose	10.0g
NaCl	5.0g
Beef extract	3.0g
Phenylethanol	2.5g
Blood, defibrinated	50.0mL

pH 7.3 ± 0.2 at 25°C

Preparation of Medium: Add components, except blood, to distilled/deionized water and bring volume to 950.0mL. Mix thoroughly. Gently heat and bring to boiling. Autoclave for 15 min at 13 psi pressure–118°C. Cool to 45°–50°C. Aseptically add sterile defibrinated blood. Mix thoroughly. Pour into sterile Petri dishes or distribute into sterile tubes.

Use: For the isolation of staphylococci and streptococci from specimens containing a mixed flora.

Phenylethyl Alcohol Agar
See: **Phenethyl Alcohol Agar)**

Phenylketouria Test Agar
See: **PKU Test Agar**

Phosphate Mineral Salts Medium with Octane

Composition per liter:

$(NH_4)_2HPO_4$	10.0g
K_2HPO_4	5.0g
Na_2SO_4	0.5g
Octane	10.0mL

Preparation of Medium: Add components, except octane, to tap water and bring volume to 990.0mL. Mix thoroughly. Autoclave for 15 min at 15 psi pressure–121°C. Prior to inoculation, filter sterilize octane. Aseptically add sterile octane to sterile medium. Aseptically distribute into sterile tubes or flasks.

Use: For the cultivation and maintenance of *Pseudomonas oleovorans*.

Photobacterium Broth

Composition per liter:

NaCl	30.0g
Sodium glycerol phosphate	23.5g
Pancreatic digest of casein	5.0g
KH_2PO_4	3.0g
Yeast extract	2.5g

$CaCO_3$ 1.0g
NH_4Cl 0.3g
$MgSO_4 \cdot 7H_2O$ 0.3g
$FeCl_3$ 0.01g

Source: This medium is available as a premixed powder from Difco Laboratories.

Preparation of Medium: Add components to distilled/deionized water and bring volume to 1.0L. Mix thoroughly. Distribute into tubes or flasks to form a shallow layer of medium. Autoclave for 15 min at 15 psi pressure–121°C.

Use: For the cultivation and demonstration of luminescence by photobacteria. For the cultivation and maintenance of *Alteromonas hanedai, Photobacterium phosphoreum, Shewanella hanedai, Vibrio fischeri, Vibrio harveyi,* and other *Vibrio* species.

Photobacterium MPY Medium

Composition per liter:
NaCl 28.2g
$MgSO_4 \cdot 7H_2O$ 6.9g
$MgCl_2 \cdot 6H_2O$ 5.5g
Peptone 5.0g
Yeast extract 3.0g
$CaCl_2 \cdot 2H_2O$ 1.5g
KCl 0.7g
pH 7.4 ± 0.2 at 25°C

Preparation of Medium: Add components to distilled/deionized water and bring volume to 1.0L. Mix thoroughly. Distribute into tubes or flasks. Autoclave for 15 min at 15 psi pressure–121°C.

Use: For the cultivation and maintenance of *Photobacterium leiognathi.*

Phthalic Acid Medium

Composition per liter:
Solution 1 .... 400.0mL
Solution 2 .... 400.0mL
Potassium hydrogen phthalate solution 200.0mL
pH 6.8 ± 0.2 at 25°C

Solution 1:
Composition per 400mL:
KH_2PO_4 .... 9.1g
$(NH_4)_2SO_4$ 1.2g

Preparation of Solution 1: Add components to distilled/deionized water and bring volume to 400.0mL. Mix thoroughly. Adjust pH to 6.8 with KOH. Autoclave for 15 min at 15 psi pressure–121°C. Cool to 25°C.

Solution 2:
Composition per 400mL:
$MgSO_4 \cdot 7H_2O$ 0.4g
$FeSO_4 \cdot 7H_2O$ 0.01g

Preparation of Solution 2: Add components to distilled/deionized water and bring volume to 400.0mL. Mix thoroughly. Adjust pH to 6.8 with KOH. Autoclave for 15 min at 15 psi pressure–121°C. Cool to 25°C.

Potassium Hydrogen Phthalate Solution:
Composition per 200mL:
Potassium hydrogen phthalate 1.0g

Preparation of Medium: Add component to distilled/deionized water and bring volume to 200.0mL. Mix thoroughly. Adjust pH to 6.8 with KOH. Autoclave for 15 min at 15 psi pressure–121°C. Cool to 25°C.

Preparation of Medium: Aseptically combine the three sterile solutions. Mix thoroughly. Aseptically distribute into sterile tubes or flasks.

Use: For the cultivation and maintenance of *Pseudomonas cepacia.*

PHYG Medium

Composition per 110.1mL:
Beef heart, solids from infusion 10.0g
Polypeptone™ 2.0g
Gelatin 1.0g
Glucose 1.0g
Yeast extract 1.0g
$NaHCO_3$ 0.5g
Tryptose 0.2g
Agar 0.16g
NaCl 0.1g
$Cysteine \cdot HCl \cdot H_2O$ 0.09g
$(NH_4)_2SO_4$ 0.05g
Resazurin 0.16mg
Salts solution 50.0mL
Rabbit serum, inactivated 10.0mL
Thiamine pyrophosphate solution 0.1mL
pH 7.2–7.5 at 25°C

Salts Solution:
Composition per 400mL:
K_2HPO_4 .... 0.9g
NaCl 0.8g
KH_2PO_4 .... 0.4g
$MnCl_2 \cdot 4H_2O$ 0.16g
$MgSO_4$ 0.08g

Preparation of Salts Solution: Add components to distilled/deionized water and bring volume to 400.0mL. Mix thoroughly.

Thiamine Pyrophosphate Solution:
Composition per 10mL:

Thiamine pyrophosphate	0.05g

Preparation of Thiamine Pyrophosphate Solution: Add thiamine pyrophosphate to distilled/deionized water and bring volume to 10.0mL. Mix thoroughly. Filter sterilize.

Preparation of Medium: Add components, except rabbit serum and thiamine pyrophosphate solution, to distilled/deionized water and bring volume to 100.0mL. Mix thoroughly. Gently heat and bring to boiling. Autoclave for 15 min at 15 psi pressure–121°C. Cool to 45°–50°C. Aseptically add 10.0mL of sterile rabbit serum and 0.1mL of sterile thiamine pyrophosphate solution. Mix thoroughly. Pour into sterile Petri dishes or distribute into sterile tubes.

Use: For the cultivation of treponemes.

Phytone™ Yeast Extract Agar

Composition per liter:

Glucose	40.0g
Agar	17.0g
Papaic digest of soybean meal	10.0g
Yeast extract	5.0g
Chloramphenicol	0.05g
Streptomycin	0.03g

pH 6.6 ± 0.2 at 25°C

Source: This medium is available as a premixed powder from BBL Microbiology Systems.

Preparation of Medium: Add components to distilled/deionized water and bring volume to 1.0L. Mix thoroughly. Gently heat and bring to boiling. Distribute into tubes or flasks. Autoclave for 15 min at 15 psi pressure–121°C. Pour into sterile Petri dishes or leave in tubes.

Use: For the selective isolation of dermatophytes, particularly *Trichophyton verrucosum* and other pathogenic fungi, from clinical specimens.

Pike Streptococcal Broth

Composition per liter:

Pancreatic digest of casein	10.0g
Tryptose	10.0g
Yeast extract	10.0g
Glucose	0.2g
NaN_3	0.065g
Crystal Violet	2.0mg
Rabbit blood, defibrinated	50.0mL

pH 7.4 ± 0.2 at 25°C

Caution: Sodium azide is toxic. Azides also react with metals and disposal must be highly diluted.

Preparation of Medium: Add components, except rabbit blood, to distilled/deionized water and bring volume to 950.0mL. Mix thoroughly. Gently heat and bring to boiling. Distribute into flasks in 100.0mL volumes. Autoclave for 15 min at 15 psi pressure–121°C. Cool to 45°–50°C. Aseptically add 5.0mL of sterile rabbit blood to each flask. Mix thoroughly.

Use: For the isolation and enrichment of hemolytic streptococci from throat swabs and other clinical specimens. After incubation of bacteria for 18–24 hr in this medium, they may be isolated by streaking the culture onto blood agar plates.

Pisu Medium

Composition per1512mL:

Agar base	960.0mL
Horse serum, sterile	360.0mL
Cystine solution	180.0mL
Lead acetate solution	12.0mL

pH 6.8 ± 0.2 at 25°C

Agar Base:
Composition per liter:

Proteose peptone No. 3	20.0g
Agar	7.0g
NaCl	5.0g
Meat extract	4.0g

Preparation for Agar Base: Add components to distilled/deionized water and bring volume to 1.0L. Mix thoroughly. Dissolve in steam for 15 min at 0 psi pressure–100°C. Cool to 45°–50°C. Adjust pH to 7.5. Filter sterilize. Distribute into 200.0mL Erlenmeyer flasks in 80.0mL volumes. Autoclave for 60 min at 0 psi pressure–100°C. Cool to 60°C.

Cystine Solution:
Composition per 200mL:

L-Cystine	2.0g

Preparation of Cystine Solution: Add cystine to distilled/deionized water and bring volume to 200.0mL. Mix thoroughly. Filter sterilize.

Lead Acetate Solution:
Composition per 100mL:

Lead acetate	10.0g

Preparation of Lead Acetate Solution: Add lead acetate to distilled/deionized water and bring volume to 100.0mL. Mix thoroughly. Filter sterilize.

Preparation for Medium: To each flask containing 80.0mL of cooled, sterile agar base, aseptically add 30.0mL of sterile horse serum, 15.0mL of sterile cystine solution, and 1.0mL of sterile lead acetate solution. Mix thoroughly. Aseptically distribute into small sterile tubes in 2.0–3.0mL volumes. Allow tubes to solidify in a vertical position.

Use: For the cultivation and differentiation of bacteria based on their ability to produce cystinase. Cystinase producing bacteria turn the medium black.

PKU Test Agar (Phenylketonuria Test Agar)

Composition per liter:

Agar	15.0g
K_2HPO_4	15.0g
Glucose	10.0g
KH_2PO_4	5.0g
$(NH_4)Cl$	2.5g
$(NH_4)NO_3$	0.5g
Asparagine	0.5g
DL-Alanine	0.5g
L-Glutamic acid	0.5g
Na_2SO_4	0.5g
$MgSO_4·7H_2O$	0.05g
$FeCl_3$	5.0mg
$MnCl_2·4H_2O$	5.0mg
$CaCl_2·2H_2O$	2.5mg
β-2-Thienylalanine	3.3mg
Bacillus subtilis spore suspension	10.0mL

pH 7.0 ± 0.2 at 25°C

Source: This medium is available as a premixed powder from Difco Laboratories.

Preparation of Medium: Add components to distilled/deionized water and bring volume to 1.0L. Mix thoroughly. Gently heat and bring to boiling. Continue boiling for 5 min. Do not autoclave. Cool to 50°C. Add 10.0mL of a suspension of *Bacillus subtilis* ATCC 6633 spores. Mix thoroughly. Pour into sterile Petri dishes or other containers.

Use: For the determination of phenylalanine concentrations in serum or urine. Use in the Guthrie-modified bacterial-inhibition assay procedure for screening newborn infants for phenylketonuria (PKU).

PKU Test Agar (Phenylketonuria Test Agar)

Composition per liter:

K_2HPO_4	15.0g
Agar	13.5g
Glucose	5.0g
KH_2PO_4	5.0g
$(NH_4)Cl$	2.5g
L-Asparagine	0.5g
L-Glutamic acid	0.5g
Na_2SO_4	0.5g
$(NH_4)NO_3$	0.5g
L-Alanine	0.25g
$MgSO_4·7H_2O$	0.05g
$FeCl_3$	0.005g
$MnSO_4$	0.005g
$CaCl_2·2H_2O$	2.5mg
β-2-Thienylalanine solution	1.0mL

pH 6.9 ± 0.2 at 25°C

Source: This medium is available as a premixed powder from BBL Microbiology Systems.

β-2-Thienylalanine Solution:
Composition per 100mL:

β-2-Thienylalanine	0.33g

Preparation of β-2-Thienylalanine Solution: Add β-2-thienylalanine to distilled/deionized water and bring volume to 100.0mL. Mix thoroughly. Filter sterilize.

Preparation of Medium: Add components, except β-2-thienylalanine solution, to distilled/deionized water and bring volume to 1.0L. Mix thoroughly. Gently heat and bring to boiling. Continue boiling for 5 min. Do not autoclave. Cool to 50°C. Add 10.0mL of a suspension of *Bacillus subtilis* ATCC 6633 spores and 1.0mL of sterile β-2-thienylalanine solution. Mix thoroughly. Pour into sterile Petri dishes or other containers.

Use: For the determination of phenylalanine concentrations in serum or urine. Used in the Guthrie-modified bacterial-inhibition assay procedure for screening newborn infants for phenylketonuria (PKU).

PL Agar

Composition per liter:

Agar	15.0g
Mannitol	7.5g
L-Arabinose	5.0g
Peptone	5.0g
NaCl	5.0g
Lysine	2.0g
Yeast extract	2.0g
Bile salts No. 2	1.0g
Inositol	1.0g
Phenol Red	0.08g

pH 7.4 ± 0.2 at 25°C

Preparation of Medium: Add components to distilled/deionized water and bring volume to 1.0L. Mix thoroughly. Gently heat and bring to boiling. Adjust pH to 7.4. Distribute into tubes or flasks. Autoclave for 15 min at 10 psi pressure–115°C. Pour into sterile Petri dishes or leave in tubes.

Use: For the isolation and cultivation of *Pleisomonas shigelloides* from foods.

Plant *Mycoplasma* Agar

Composition per 291.2mL:

Schneider's *Drosophila* medium	160.0mL
Solution 1	70.0mL
Fetal calf serum	50.0mL
Fresh yeast extract solution	10.0mL
Phenol Red (0.5% solution)	1.2mL

pH 7.4 ± 0.2 at 25°C

Schneider's *Drosophila* Medium:
Composition per liter:

$MgSO_4 \cdot 7H_2O$	3.7g
NaCl	2.1g
Yeast extract	2.0g
Trehalose	2.0g
D-Glucose	2.0g
L-Glutamine	1.8g
L-Lysine·HCl	1.7g
L-Proline	1.7g
KCl	1.6g
$Na_2HPO_4 \cdot 7H_2O$	1.3g
L-Glutamic acid	0.8g
L-Methionine	0.8g
$CaCl_2$, anhydrous	0.6g
KH_2PO_4	0.5g
β-Alanine	0.5g
L-Tyrosine	0.5g
L-Arginine	0.4g
L-Aspartic acid	0.4g
L-Histidine	0.4g
L-Threonine	0.4g
$NaHCO_3$	0.4g
Glycine	0.3g
L-Serine	0.3g
L-Valine	0.3g
L-Isoleucine	0.2g
L-Leucine	0.2g
L-Phenylalanine	0.2g
α-Ketoglutaric acid	0.2g
Fumaric acid	0.1g
Malic acid	0.1g
Succinic acid	0.1g
L-Cystine	0.1g
L-Tryptophan	0.1g
L-Cysteine	0.06g

Preparation of Schneider's *Drosophila* Medium: Add components to distilled/deionized water and bring volume to 1.0L. Mix thoroughly. Filter sterilize.

Solution 1:
Composition per 70mL:

Sorbitol	7.0g
Noble agar	5.0g
Beef heart, solids from infusion	5.0g
Peptone	1.8g
Pancreatic digest of casein	1.0g
Sucrose	1.0g
NaCl	0.5g
D-Fructose	0.1g
D-Glucose	0.1g

Preparation of Solution 1: Add components to distilled/deionized water and bring volume to 70.0mL. Mix thoroughly. Adjust pH to 7.8 with 1*N* NaOH. Autoclave for 15 min at 15 psi pressure–121°C. Cool to 50°C.

Fresh Yeast Extract Solution:
Composition per 100mL:

Baker's yeast live, pressed, starch-free	25.0g

Preparation of Fresh Yeast Extract Solution: Add the live Baker's yeast to 100.0mL of distilled/deionized water. Autoclave for 90 min at 15 psi pressure–121°C. Allow to stand. Remove supernatant solution. Adjust pH to 6.6–6.8.

Preparation of Medium: Bring fetal calf serum and Phenol Red solution to 56°C. Rapidly bring Schneider's *Drosophila* Medium to 37°C. Rapidly combine the components. Mix thoroughly. Pour into sterile Petri dishes or distribute into sterile tubes or flasks.

Use: For the cultivation and maintenance of *Spiroplasma floricola, Spiroplasma kunkelii, Spiroplasma melliferum,* and *Spiroplasma* species.

Plant *Mycoplasma* Broth

Composition per 291.2mL:

Schneider's *Drosophila* medium	160.0mL
Solution 1	70.0mL
Fetal calf serum	50.0mL
Fresh yeast extract solution	10.0mL
Phenol Red (0.5% solution)	1.2mL

pH 7.4 ± 0.2 at 25°C

Schneider's *Drosophila* Medium:
Composition per liter:

$MgSO_4 \cdot 7H_2O$	3.7g
NaCl	2.1g
Trehalose	2.0g
Yeast extract	2.0g
D-Glucose	2.0g
L-Glutamine	1.8g
L-Lysine·HCl	1.7g
L-Proline	1.7g
KCl	1.6g
$Na_2HPO_4 \cdot 7H_2O$	1.3g
L-Glutamic acid	0.8g

L-Methionine	0.8g
$CaCl_2$, anhydrous	0.6g
KH_2PO_4	0.5g
β-Alanine	0.5g
L-Tyrosine	0.5g
L-Arginine	0.4g
L-Aspartic acid	0.4g
L-Histidine	0.4g
L-Threonine	0.4g
$NaHCO_3$	0.4g
Glycine	0.3g
L-Serine	0.3g
L-Valine	0.3g
L-Isoleucine	0.2g
L-Leucine	0.2g
L-Phenylalanine	0.2g
α-Ketoglutaric acid	0.2g
Fumaric acid	0.1g
Malic acid	0.1g
Succinic acid	0.1g
L-Cystine	0.1g
L-Tryptophan	0.1g
L-Cysteine	0.06g

Preparation of Schneider's *Drosophila* Medium: Add components to distilled/deionized water and bring volume to 1.0L. Mix thoroughly. Filter sterilize.

Solution 1:
Composition per 70mL:

Sorbitol	7.0g
Beef heart, solids from infusion	5.0g
Peptone	1.8g
Pancreatic digest of casein	1.0g
Sucrose	1.0g
NaCl	0.5g
D-Fructose	0.1g
D-Glucose	0.1g

Preparation of Solution 1: Add components to distilled/deionized water and bring volume to 70.0mL. Mix thoroughly. Adjust pH to 7.8 with 1*N* NaOH. Autoclave for 15 min at 15 psi pressure–121°C. Cool to 25°C.

Fresh Yeast Extract Solution:
Composition per 100mL:

Baker's yeast live, pressed, starch-free	25.0g

Preparation of Fresh Yeast Extract Solution: Add the live Baker's yeast to 100.0mL of distilled/deionized water. Autoclave for 90 min at 15 psi pressure–121°C. Allow to stand. Remove supernatant solution. Adjust pH to 6.6–6.8.

Preparation of Medium: Bring fetal calf serum and Phenol Red solution to 56°C. Rapidly bring Schneider's *Drosophila* medium to 37°C. Rapidly combine the components. Mix thoroughly. Aseptically distribute into sterile tubes or flasks.

Use: For the cultivation and maintenance of *Spiroplasma floricola, Spiroplasma kunkelii, Spiroplasma melliferum,* and *Spiroplasma* species.

Plate Count Agar (Tryptone Glucose Yeast Agar)

Composition per liter:

Agar	9.0g
Pancreatic digest of casein	5.0g
Yeast extract	2.5g
Glucose	1.0g

pH 7.0 ± 0.2 at 25°C

Source: This medium is available as a premixed powder from Oxoid Unipath.

Preparation of Medium: Add components to distilled/deionized water and bring volume to 1.0L. Mix thoroughly. Gently heat while stirring and bring to boiling. Distribute into tubes or flasks. Autoclave for 15 min at 15 psi pressure–121°C. Pour into sterile Petri dishes or leave in tubes.

Use: For the enumeration of viable bacteria in milk and dairy products. Also used for the estimation of the number of live heterotrophic bacteria in water. For the cultivation and maintenance of *Brevibacterium casei, Brevibacterium epedermidis,* and *Methylobacterium mesophilicum.*

Plate Count Agar (ATCC Medium 1048)

Composition per liter:

Agar	15.0g
Pancreatic digest of casein	5.0g
Yeast extract	2.5g
Glucose	1.0g

pH 7.0 ± 0.2 at 25°C

Source: This medium is available as a premixed powder from Difco Laboratories and Oxoid Unipath.

Preparation of Medium: Add components to distilled/deionized water and bring volume to 1.0L. Mix thoroughly. Gently heat and bring to boiling. Distribute into tubes or flasks. Autoclave for 15 min at 15 psi pressure–121°C. Pour into sterile Petri dishes or leave in tubes.

Use: For the enumeration of bacteria in milk, water, food and dairy products.

Plate Count Agar, Modified

Composition per liter:

Pancreatic digest of casein 20.0g
Yeast extract 20.0g
Agar 10.0g
Glucose 4.0g
pH 7.0 ± 0.1 at 25°C

Preparation of Medium: Add components to distilled/deionized water and bring volume to 1.0L. Mix thoroughly. Gently heat and bring to boiling. Distribute into bottles. Autoclave for 15 min at 15 psi pressure–121°C.

Use: For the cultivation and enumeration of microorganisms from food by the plate count method.

Plate Count Agar, Special

Composition per liter:

Agar 30.1g
Pancreatic digest of casein 6.13g
Yeast extract 3.06g
Glucose 1.23g
pH 7.0 ± 0.2 at 25°C

Preparation of Medium: Add components to distilled/deionized water and bring volume to 1.0L. Mix thoroughly. Gently heat while stirring and bring to boiling. Distribute into tubes or flasks. Autoclave for 20 min at 15 psi pressure–121°C. Pour into sterile Petri dishes or leave in tubes.

Use: For the enumeration of viable bacteria in raw milk, milk and other dairy products.

Plate Count Agar with Antibiotic

Composition per liter:

Agar 9.0g
Pancreatic digest of casein 5.0g
Yeast extract 2.5g
Glucose 1.0g
Chloramphenicol 0.1g
pH 7.0 ± 0.2 at 25°C

Preparation of Medium: Add components to distilled/deionized water and bring volume to 1.0L. Mix thoroughly. Gently heat and bring to boiling. Distribute into tubes or flasks. Autoclave for 15 min at 15 psi pressure–121°C. Pour into sterile Petri dishes or leave in tubes.

Use: For the cultivation of yeasts from foods.

Plate Count Agar with Antibiotic-Free Skim Milk

Composition per liter:

Agar 10.0g
Pancreatic digest of casein 5.0g
Yeast extract 2.5g
Skim milk, antibiotic-free 1.0g
Glucose 1.0g
pH 6.9 ± 0.1 at 25°C

Source: This medium is available as a premixed powder from Oxoid Unipath.

Preparation of Medium: Add components to distilled/deionized water and bring volume to 1.0L. Mix thoroughly. Gently heat while stirring and bring to boiling. Distribute into tubes or flasks. Autoclave for 15 min at 15 psi pressure–121°C. Pour into sterile Petri dishes or leave in tubes.

Use: For the enumeration of viable bacteria in milk and dairy products.

Plate Count Broth (m–Plate Count Broth)

Composition per liter:

Yeast extract 5.0g
Glucose 2.0g
pH 7.0 ± 0.2 at 25°C

Source: This medium is available as a premixed powder from Difco Laboratories.

Preparation of Medium: Add components to distilled/deionized water and bring volume to 1.0L. Mix thoroughly. Distribute into tubes or flasks. Autoclave for 15 min at 15 psi pressure–121°C.

Use: For the determination of bacterial counts by the membrane filter method.

Plesiomonas Differential Agar

See: **Inositol Brilliant Green Bile Salts Agar**

PLET Agar

Composition per liter:

Beef heart, solids from infusion 500.0g
Agar 15.0g
Tryptose 10.0g
NaCl 5.0g
Ethylenediamine tetracetic acid,EDTA 0.3g
Thallous acetate 0.04g
Antibiotic inhibitor 10.0mL

Antibiotic Inhibitor:

Composition per 10mL:

Lysozyme 300,000U
Polymyxin 30,000U

Preparation of Antibiotic Inhibitor: Add components to distilled/deionized water and bring volume to 10.0mL. Mix thoroughly. Filter sterilize.

Preparation of Medium: Add components, except antibiotic inhibitor, to distilled/deionized water and bring volume to 990.0mL. Mix thoroughly. Gently heat and bring to boiling. Autoclave for 15 min at 15 psi pressure–121°C. Cool to 50°C. Aseptically add sterile antibiotic inhibitor. Mix thoroughly. Pour into sterile Petri dishes or distribute into sterile tubes.

Use: For the selective isolation and cultivation of *Bacillus anthracis*.

PM Indicator Agar

Composition per liter:

Agar	15.0g
Glucose	5.25g
Peptone	5.0g
Beef extract	3.0g
Pancreatic digest of casein	1.7g
Sorbitan monooleate complex	1.0g
NaCl	0.5g
Papaic digest of soybean meal	0.3g
K_2HPO_4	0.25g
Bromcresol Purple	0.06g

pH 7.8 ± 0.2 at 25°C

Source: This medium is available as a premixed powder from Difco Laboratories.

Preparation of Medium: Add components to distilled/deionized water and bring volume to 1.0L. Mix thoroughly. Gently heat and bring to boiling. Do not autoclave. Inoculate medium with a suspension of *Bacillus stearothermophilus* spores. Pour into 100mm flat-bottom Petri dishes in 6.0mL volumes. Use medium immediately.

Use: For the rapid detection of trace amounts of penicillin in milk using the AOAC *Bacillus stearothermophilus* Qualitative Disc Method II.

PMA Medium
See: **Peptonized Milk Agar**

PMP Broth

Composition per liter:

Na_2HPO_4	7.9g
D-Mannitol	2.5g
Peptone	2.5g
NaH_2PO_4	1.1g

pH 7.6 ± 0.2 at 25°C

Preparation of Medium: Add components to distilled/deionized water and bring volume to 1.0L. Mix thoroughly. Distribute into tubes or flasks. Autoclave for 15 min at 15 psi pressure–121°C.

Use: For the enrichment and cultivation of *Yersinia pseudotuberculosis* from food.

PMY Medium

Composition per liter:

Agar	15.0g
Glucose	10.0g
NaCl	5.0g
Polypeptone™	5.0g
Beef extract	2.0g
Yeast extract	1.0g
$MgSO_4 \cdot 7H_2O$	0.5g

pH 7.0 ± 0.2 at 25°C

Preparation of Medium: Add components to distilled/deionized water and bring volume to 1.0L. Mix thoroughly. Gently heat and bring to boiling. Distribute into tubes or flasks. Autoclave for 15 min at 15 psi pressure–121°C. Pour into sterile Petri dishes or leave in tubes.

Use: For the cultivation and maintenance of *Xanthomonas campestris*.

PMYA II Medium

Composition per liter:

Agar	15.0g
Milk, peptonized	1.0g
Yeast extract	0.2g
Sodium acetate	0.02g

Preparation of Medium: Add components to distilled/deionized water and bring volume to 1.0L. Mix thoroughly. Gently heat and bring to boiling. Distribute into tubes or flasks. Autoclave for 15 min at 15 psi pressure–121°C. Pour into sterile Petri dishes or leave in tubes.

Use: For the isolation and cultivation of *Flavobacterium* species.

Poly-β-Hydroxybutyrate Medium (PHB Medium)

Composition per liter:

Part A	900.0mL
Part B	100.0mL

pH 7.2 ± 0.2 at 25°C

Part A:

Composition per 900mL:

$K_2HPO_4 \cdot 3H_2O$	0.6g
KH_2PO_4	0.2g
$MgSO_4 \cdot 7H_2O$	0.2g
$(NH_4)_2SO_4$	0.2g

Preparation of Medium: Add components to distilled/deionized water and bring volume to 900.0mL.

Mix thoroughly. Adjust pH to 7.2. Autoclave for 15 min at 15 psi pressure–121°C. Cool to 25°C.

Part B:

Composition per 100mL:

Glucose 10.0g

Preparation of Medium: Add glucose to distilled/deionized water and bring volume to 100.0mL. Mix thoroughly. Autoclave for 15 min at 15 psi pressure–121°C. Cool to 25°C.

Preparation of Medium: Aseptically combine 900.0mL of cooled, sterile part A and 100.0mL of cooled, sterile part B. Mix thoroughly. Aseptically distribute into sterile tubes or flasks.

Use: For the cultivation and differentiation of *Pseudomonas* species based on their ability to produce intracellular poly-β-hydroxybutyrate. Production of poly-β-hydroxybutyrate is determined by staining cells with Sudan Black B.

Polymyxin Acriflavine LiCl Ceftazidime Esculin Mannitol Agar
See: **PALCAM Agar**

Polymyxin Base Agar
See: **Antibiotic Medium 9**

Polymyxin Pyruvate Egg Yolk Mannitol Bromthymol Blue Agar (PEMBA)

Composition per 110mL:

Agar 1.8g
D-Mannitol 1.0g
Sodium pyruvate 1.0g
Na_2HPO_4 .... 0.25g
NaCl 0.2g
Peptone 0.1g
KH_2PO_4 .... 0.025g
Bromthymol Blue 0.01g
$MgSO_4 \cdot 7H_2O$ 0.01g
Antibiotic inhibitor 5.0mL
Egg yolk emulsion (20% solution) 5.0mL

pH 7.4 ± 0.2 at 25°C

Antibiotic Inhibitor:

Composition per 5mL:

Polymyxin B 10,000U

Preparation of Antibiotic Inhibitor: Add components to distilled/deionized water and bring volume to 5.0mL. Mix thoroughly. Filter sterilize.

Egg Yolk Emulsion, 20% Solution:

Composition per 100mL:

Chicken egg yolks 11
Whole chicken egg 1
NaCl (0.9% solution) 80.0mL

Preparation of Egg Yolk Emulsion, 20% Solution: Soak eggs with 1:100 dilution of saturated mercuric chloride solution for 1 min. Crack eggs and separate yolks from whites. Mix egg yolks with 1 chicken egg. Measure 20.0mL of egg yolk emulsion and add to 80.0mL of 0.9% NaCl solution. Mix thoroughly. Filter sterilize. Warm to 45°–50°C.

Preparation of Medium: Add components, except antibiotic inhibitor and egg yolk emulsion, 20%, to distilled/deionized water and bring volume to 100.0mL. Mix thoroughly. Gently heat and bring to boiling. Autoclave for 15 min at 15 psi pressure–121°C. Cool to 45°–50°C. Aseptically add sterile antibiotic inhibitor and egg yolk emulsion, 20%. Mix thoroughly. Pour into sterile Petri dishes.

Use: For the cultivation of *Bacillus cereus*.

Polymyxin Seed Agar
See: **Antibiotic Medium 10**

Polymyxin *Staphylococcus* Medium

Composition per liter:

Agar 15.0g
Pancreatic digest of gelatin 5.0g
Beef extract 3.0g
Lecithin 0.7g
Polymyxin 0.075g
Tween™ 80 .... 10.2mL

pH 6.8 ± 0.2 at 25°C

Preparation of Medium: Add components to distilled/deionized water and bring volume to 1.0L. Mix thoroughly. Gently heat and bring to boiling. Distribute into tubes or flasks. Autoclave for 15 min at 15 psi pressure–121°C. Pour into sterile Petri dishes.

Use: For the selective isolation and cultivation of pathogenic, coagulase-positive *Staphylococcus aureus*. *Proteus* species will grow on this medium but appear as translucent colonies.

Polypectate Gel Medium

Composition per liter:

Sodium polypectate	70.0g
K_2HPO_4	5.0g
Peptone	5.0g
KH_2PO_4	1.0g
$CaCl_2{\cdot}2H_2O$	0.6g

pH 7.0 ± 0.2 at 25°C

Preparation of Medium: Add components, except sodium polypectate, to 500.0mL of boling water. Put into a blender and mix on low speed to dissolve. Add the sodium polypectate slowly with the blender on low speed to minimize air bubbles. Adjust pH to 7.0. Bring volume to 1.0L with distilled/deionized water. Autoclave for 15 min at 15 psi pressure–121°C. Cool to 45°–50°C. Pour into sterile Petri dishes.

Use: For the cultivation of microoganisms and detection of pectate degrading enzymes.

Polysorbate 80 Agar (Tween™ 80 Agar)

Composition per 1010mL:

Agar	15.0g
Peptone	10.0g
Tween™ 80	10.0mL

pH 7.2 ± 0.2 at 25°C

Preparation of Medium: Add components, except Tween™ 80, to distilled/deionized water and bring volume to 1.0L. Mix thoroughly. Gently heat and bring to boiling. Cool to 45°–50°C. Autoclave for 20 min at 15 psi pressure–121°C. Pour into sterile Petri dishes

Use: For the cultivation of a variety of microorganisms.

Porcine Heart Agar

Composition per liter:

Porcine heart, infusion from	375.0g
Agar	15.0g
Papaic digest of soybean meal	6.5g
Glucose	5.0g
NaCl	5.0g
Proteose peptone No. 3	5.0g
Yeast extract	3.5g
Sheep blood, defibrinated	50.0mL

pH 7.2 ± 0.2 at 25°C

Source: This medium is available as a premixed powder from Difco Laboratories.

Preparation of Medium: Add components, except sheep blood, to distilled/deionized water and bring volume to 950.0mL. Mix thoroughly. Gently heat and bring to boiling. Autoclave for 15 min at 15 psi pressure–121°C. Cool to 45°–50°C. Aseptically add sterile sheep blood. Mix thoroughly. Pour into sterile Petri dishes.

Use: For the determination of the sensitivity of microorganisms using in the disc plate technique.

Pork Plasma Fibrinogen Overlay Agar

Composition per plate:

Baird-Parker agar, modified	15.0mL
Pork plasma fibrinogen overlay agar	8.0mL

pH 7.0 ± 0.1 at 25°C

Baird-Parker Agar, Modified:

Composition per liter:

Agar	17.0g
Glycine	12.0g
Sodium pyruvate	10.0g
Pancreatic digest of casein	10.0g
Beef extract	5.0g
LiCl	5.0g
Yeast extract	1.0g
Chapman tellurite solution	1.0mL

pH 7.0 ± 0.2 at 25°C

Chapman Tellurite Solution:

Composition per 100mL:

K_2TeO_3	1.0g

Preparation of Chapman Tellurite Solution: Add K_2TeO_3 to distilled/deionized water and bring volume to 100.0mL. Mix thoroughly. Filter sterilize.

Preparation of Baird-Parker Agar, Modified: Add components, except Chapman tellurite solution, to distilled/deionized water and bring volume to 999.0mL. Mix thoroughly. Gently heat and bring to boiling. Autoclave for 15 min at 15 psi pressure–121°C. Cool to 45–50°C. Aseptically add 1.0mL of sterile Chapman tellurite solution. Mix thoroughly but gently.

Pork Plasma Fibrinogen Overlay Agar:

Composition per 100.5mL:

Agar solution	50.0mL
Bovine fibrinogen solution	47.5mL
Pork plasma	2.5mL
Trypsin inhibitor solution	0.5mL

Agar Solution:

Composition per 50mL:

Agar	0.7g

Preparation of Agar Solution: Add agar to distilled/deionized water and bring volume to 50.0mL. Mix thoroughly. Autoclave for 15 min at 15 psi pressure–121°C. Cool to 45°–50°C.

Bovine Fibrinogen Solution:

Composition per 50mL:

Bovine fibrinogen, fraction I 0.4g
Sodium phosphate buffer
(0.05M solution, pH 7.0) 50.0mL

Preparation of Bovine Fibrinogen Solution: Grind bovine fibrinogen in a mortar to a fine powder. Add bovine fibrinogen to 50.0mL of sodium phosphate buffer. Mix thoroughly on a magnetic stirrer for 30 min. Filter through Whatman #1 filter paper. Filter sterilize.

Pork Plasma:

Composition per 10mL:

Pork plasma-EDTA 10.0mL

Preparation of Pork Plasma: Filter sterilize fresh or rehydrated commercial pork plasma-EDTA.

Trypsin Inhibitor Solution:

Composition per 5mL:

Trypsin inhibitor 0.015g
Sodium phosphate buffer
(0.05M solution, pH 7.0) 5.0mL

Preparation of Trypsin Inhibitor Solution: Add trypsin inhibitor to 5.0mL of sodium phosphate buffer. Mix thoroughly. Filter sterilize.

Preparation of Pork Plasma Fibrinogen Overlay Agar: Aseptically combine 50.0mL of cooled, sterile agar solution, 47.5mL of sterile bovine fibrinogen solution, 2.5mL of sterile pork plasma, and 0.5mL of trypsin inhibitor solution. Mix thoroughly. Maintain at 45°–50°C but use within 1 hr.

Caution: Potassium tellurite is toxic.

Preparation of Medium: Pour cooled, sterile Baird-Parker agar, modified into sterile Petri dishes in 15.0mL volumes. Allow agar to solidify. Overlay each plate with 8.0mL of sterile pork plasma fibrinogen overlay agar.

Use: For the cultivation of *Staphylococcus aureus* from foods.

Postgate's Medium for Sulfate Reducers

See: **Medium for Sulfate Reducers**

Postgate's Medium B for Sulfate Reducers

See: **Medium B for Sulfate Reducers**

Postgate's Medium C for Sulfate Reducers

See: **Medium C for Sulfate Reducers**

Postgate's Medium D for Sulfate Reducers

See: **Medium D for Sulfate Reducers**

Postgate's Medium E for Sulfate Reducers

See: **Medium E for Sulfate Reducers**

Postgate's Medium F for Sulfate Reducers

See: **Medium F for Sulfate Reducers**

Postgate's Medium G for Sulfate Reducers

See: **Medium G for Sulfate Reducers**

Postgate's Medium N for Sulfate Reducers

See: **Medium N for Sulfate Reducers**

Potassium Cyanide Broth

Composition per liter:

Na_2HPO_4 ...... 5.64g
NaCl 5.0g
Proteose peptone No. 3 ...... 3.0g
KH_2PO_4 ...... 0.225g
KCN solution 15.0mL

pH 7.6 ± 0.2 at 25°C

KCN Solution:

Composition per 100mL:

KCN 0.5g

Preparation of KCN Solution: Add KCN to distilled/deionized water and bring volume to 100.0mL. Mix thoroughly.

Caution: Cyanide is toxic.

Preparation of Medium: Add components, except KCN solution, to distilled/deionized water and bring volume to 985.0mL. Mix thoroughly. Gently heat and bring to boiling. Autoclave for 15 min at 15 psi pressure–121°C. Cool to 25°C. Aseptically add 15.0mL of KCN solution. Mix thoroughly. Distribute into sterile screw-capped tubes or flasks in 1.0–1.5mL volumes. Close caps tightly.

Use: For the cultivation and differentiation of urease negative Gram-negative enteric bacteria. *Salmonella* species and *Shigella* species are nonmotile in this medium. *Proteus* species are motile in this medium.

Potassium Tellurite Agar

Composition per liter:

Beef heart, solids from infusion	500.0g
Agar	15.0g
Tryptose	10.0g
NaCl	5.0g
K_2TeO_3 solution	20.0mL
Blood, defibrinated	50.0mL

pH 6.0 ± 0.2 at 25°C

K_2TeO_3 Solution:

Composition per 20mL:

K_2TeO_3	0.5g

Preparation of K_2TeO_3 Solution: Add K_2TeO_3 to distilled/deionized water and bring volume to 10.0mL. Mix thoroughly. Filter sterilize.

Caution: Potassium tellurite is toxic.

Preparation of Medium: Add components, except K_2TeO_3 solution, to distilled/deionized water and bring volume to 930.0mL. Mix thoroughly. Gently heat and bring to boiling. Autoclave for 15 min at 15 psi pressure–121°C. Cool to 45°–50°C. Aseptically add sterile K_2TeO_3 solution and 50.0mL of blood. Rabbit or sheep blood may be used. Mix thoroughly. Pour into sterile Petri dishes or distribute into sterile tubes. Allow tubes to cool in a slanted position.

Use: For the cultivation and differentiation of *Enterococcus faecalis*. *E. faecalis* appears as black colonies.

Potato Carrot Medium

Composition per liter of tap water:

Potatoes (sliced with skin)	300.0g
Carrots (peeled and sliced)	25.0g
Agar	15.0g

Preparation of Medium: Slice potatoes with the skin on. Peel carrots and slice. Add potatoes and carrots to approximately 700.0mL of tap water. Gently heat and bring to boiling. Continue boiling for 20 min. Filter through cheesecloth. Bring volume of filtrate to 1.0L with distilled/deionized water. Add agar. Gently heat and bring to boiling. Distribute into tubes or flasks. Autoclave for 20 min at 15 psi pressure–121°C. Pour into sterile Petri dishes or leave in tubes.

Use: For the cultivation and maintenance of *Actinoplanes awajinensis, Actinoplanes nirasakinensis, Amorphosporangium auranticolor, Streptomyces flaveus*, and *Thermoactinomyces vulgaris*.

Potato Dextrose Agar

Composition per liter:

Glucose	20.0g
Agar	15.0g
Potato, infusion from	4.0g
Tartaric acid solution	14.0mL

pH 5.6 ± 0.2 at 25°C

Source: This medium is available as a premixed powder from BBL Microbiology Systems and Oxoid Unipath.

Tartaric Acid Solution:

Composition per 50mL:

Tartaric acid	5.0g

Preparation of Tartaric Acid Solution: Add tartaric acid to distilled/deionized water and bring volume to 50.0mL. Mix thoroughly. Filter sterilize.

Preparation of Medium: Add components to distilled/deionized water and bring volume to 986.0mL. Mix thoroughly. Gently heat and bring to boiling. Distribute into tubes or flasks. Autoclave for 15 min at 15 psi pressure–121°C. Cool to 45°–50°C. Aseptically add 14.0mL of sterile tartaric acid solution. Mix thoroughly. If medium is to be used for the enumeration of yeasts and molds in butter, adjust pH to 3.5. Pour into sterile Petri dishes or distribute into sterile tubes.

Use: For the cultivation and enumeration of yeasts and molds. For the enumeration of yeasts and molds in butter by the plate count method.

Potato Dextrose Agar (PDA Agar)

Composition per liter:

Glucose	20.0g
Agar	15.0g
Potatoes, infusion from	1.0L

Potatoes, Infusion From:

Composition per liter:

Potatoes	300.0g

Preparation of Potatoes, Infusion From: Peel and dice potatoes. Add 500.0mL of distilled/deionized water. Gently heat and bring to boiling. Continue boiling for 30 min. Filter through cheesecloth. Bring volume of filtrate to 1.0L.

Preparation of Medium: To 1.0L of potato infusion, add glucose and agar. Mix thoroughly. Gently heat and bring to boiling. Distribute into tubes or flasks. Autoclave for 15 min at 15 psi pressure–121°C. Pour into sterile Petri dishes or leave in tubes.

Use: For the cultivation and maintenance of *Bacillus megaterium, Bacillus subtilis, Pseudomonas lindbergii, Pseudomonas syringae, Streptomyces testaceus,* and *Xanthomonas campestris.*

Potato Dextrose Agar

Composition per liter:

Agar	20.0g
Glucose	20.0g
Potato infusion	200.0mL

pH 5.6 ± 0.2 at 25°C

Potato Infusion:

Composition per 10mL:

Potatoes, unpeeled and sliced	200.0g

Preparation of Potato Infusion: Add potato slices to 1.0L of distilled/deionized water. Gently heat and bring to boiling. Continue boiling for 30 min. Filter through cheesecloth.

Preparation of Medium: Add components to distilled/deionized water and bring volume to 1.0L. Mix thoroughly. Gently heat and bring to boiling. Distribute into tubes or flasks. Autoclave for 15 min at 15 psi pressure–121°C. Pour into sterile Petri dishes or leave in tubes.

Use: For the cultivation and enumeration of yeasts and filamentous fungi (molds) from foods.

Potato Dextrose Agar (PDA Agar) (ATCC Medium 336)

Composition per liter:

Potatoes, infusion from	200.0g
Glucose	20.0g
Agar	15.0g

pH 5.6 ± 0.2 at 25°C

Source: This medium is available as a premixed powder from Difco Laboratories.

Potatoes, Infusion From:

Composition per 500mL:

Potatoes	300.0g

Preparation of Potatoes, Infusion From: Peel and dice potatoes. Add 500.0mL of distilled/deionized water. Gently heat and bring to boiling. Continue boiling for 30 min. Filter through cheesecloth.

Preparation of Medium: Add components to distilled/deionized water and bring volume to 1.0L. Mix thoroughly. Gently heat and bring to boiling. Distribute into tubes or flasks. Autoclave for 15 min at 15 psi pressure–121°C. Pour into sterile Petri dishes or leave in tubes.

Use: For the cultivation yeasts and molds from dairy products and other foods. Also, used to induce sporulation in many fungi.

Potato Dextrose Agar and Yeast Medium (PDA and Yeast Medium) (ATCC Medium 104)

Composition per liter:

Potatoes, infusion from	200.0g
Glucose	20.0g
Agar	15.0g
KH_2PO_4	2.0g
Yeast extract	1.5g
$MgSO_4 \cdot 7H_2O$	0.5g

Potatoes, Infusion From:

Composition per 500mL:

Potatoes	300.0g

Preparation of Potatoes, Infusion From: Peel and dice potatoes. Add 500.0mL of distilled/deionized water. Gently heat and bring to boiling. Continue boiling for 30 min. Filter through cheesecloth.

Preparation of Medium: Add components to distilled/deionized water and bring volume to 1.0L. Mix thoroughly. Gently heat and bring to boiling. Distribute into tubes or flasks. Autoclave for 15 min at 15 psi pressure–121°C. Pour into sterile Petri dishes or leave in tubes.

Use: For the cultivation and maintenance of *Pseudomonas fluorescens, Streptoverticillium baldaccii,* and *Xanthomonas campestris.*

Potato Dextrose Agar with Antibiotics

Composition per liter:

Glucose	20.0g
Agar	15.0g
Potato, infusion from	4.0g
Antibiotic solution	20.0mL

Antibiotic Solution:

Composition per 100mL:

Chlortetracycline·HCl	0.5g
Chloramphenicol	0.5g

Preparation of Antibiotic Solution: Add components to distilled/deionized water and bring volume to 100.0mL. Mix thoroughly. Filter sterilize.

Preparation of Medium: Add components, except antibiotic solution, to distilled/deionized water and bring volume to 980.0mL. Mix thoroughly. Gently heat and bring to boiling. Autoclave for 15 min at 15 psi pressure–121°C. Cool to 45°–50°C. Aseptically add 20.0mL of sterile antibiotic solution. Mix thoroughly. Pour into sterile Petri dishes.

Use: For the cultivation of fungi from foods.

Potato Dextrose Agar with 2% Glucose and 60% Sucrose

Composition per liter:

Sucrose	600.0g
Glucose	20.0g
Agar	15.0g
Potato, solids from infusion	4.0g

Preparation of Medium: Add components to distilled/deionized water and bring volume to 1.0L. Mix thoroughly. Gently heat and bring to boiling. Distribute into tubes or flasks. Autoclave for 15 min at 15 psi pressure–121°C.

Use: For the isolation and cultivation of *Saccharomyces rouxii* from chocolate syrup.

Potato Dextrose Broth

Composition per liter:

Potatoes, infusion from	200.0g
Glucose	20.0g

pH 5.1 ± 0.2 at 25°C

Source: This medium is available as a premixed powder from Difco Laboratories.

Potatoes, Infusion From:

Composition per 500mL:

Potatoes	300.0g

Preparation of Potatoes, Infusion From: Peel and dice potatoes. Add 500.0mL of distilled/deionized water. Gently heat and bring to boiling. Continue boiling for 30 min. Filter through cheesecloth.

Preparation of Medium: Add components to distilled/deionized water and bring volume to 1.0L. Mix thoroughly. Distribute into tubes or flasks. Autoclave for 15 min at 15 psi pressure–121°C.

Use: For the cultivation of a wide variety of yeasts and molds.

Potato Dextrose Yeast Agar (PDY Agar)

Composition per liter:

Yeast extract	50.0g
Glucose	20.0g
Agar	15.0g
Potatoes, infusion from	1.0L

Potatoes, Infusion From:

Composition per liter:

Potatoes	300.0g

Preparation of Potatoes, Infusion From: Peel and dice potatoes. Add 500.0mL of distilled/deionized water. Gently heat and bring to boiling. Continue boiling for 30 min. Filter through cheesecloth. Bring volume of filtrate to 1.0L.

Preparation of Medium: To 1.0L of potato infusion, add glucose, yeast extract, and agar. Mix thoroughly. Gently heat and bring to boiling. Distribute into tubes or flasks. Autoclave for 15 min at 15 psi pressure–121°C. Pour into sterile Petri dishes or leave in tubes.

Use: For the cultivation and maintenance of *Bacillus* species.

Potato Extract Agar

Composition per liter:

Agar	15.0g
Peptone	5.0g
NaCl	5.0g
Yeast extract	2.0g
Beef extract powder	1.0g
Potato extract	20.0mL

pH 7.4 ± 0.2 at 25°C

Potato Extract:

Composition per liter:

Potatoes	300.0g

Preparation of Potato Extract: Peel and dice potatoes. Add 500.0mL of distilled/deionized water. Gently heat and bring to boiling. Continue boiling for 30 min. Filter through cheesecloth.

Use: For the cultivation of a wide variety of yeasts and molds.

Potato Flakes Agar

Composition per liter:

Potato flakes	20.0g
Agar	15.0g
Glucose	10.0g

Preparation of Medium: Add components to distilled/deionized water and bring volume to 1.0L. Mix thoroughly. Gently heat and bring to boiling. Distribute into tubes or flasks. Autoclave for 15 min at 15 psi pressure–121°C. Pour into sterile Petri dishes or leave in tubes.

Use: For the cultivation and induction of sporulation in all fungi.

Potato Glucose Agar

Composition per liter:

Potato, infusion from	500.0g
Glucose	20.0g
Agar	15.0g

Preparation of Medium: Peel and slice potatoes thinly. Add 800.0mL distilled/deionized water immediately to potatoes to prevent oxidation. Gently heat and bring to 60°C. Maintain at 60°C for 60 min. Filter through cheesecloth. Adjust volume of filtrate to 1.0L with distilled/deionized water. Add agar. Gently heat and bring to boiling. Add glucose. Mix thoroughly. Distribute into tubes or flasks. Autoclave for 20 min at 10 psi pressure–115°C. Pour into sterile Petri dishes or leave in tubes.

Use: For the cultivation and maintenance of *Nocardia asteroides, Pseudomonas caryophylli, Pseudomonas syringae, Rhodococcus* species, *Streptomyces nobilis, Streptomyces prasinosporus,* and *Streptomyces* species.

Potato Infusion Agar (ATCC Medium 421)

Composition per liter:

Potato	200.0g
Agar	15.0g

Preparation of Medium: Peel and finely dice potatoes. Add to 500.0mL of distilled/deionized water. Gently heat and bring to boiling. Continue boiling for 20 min. Filter through cheesecloth. Bring volume of filtrate to 1.0L with distilled/deionized water. Add agar. Gently heat and bring to boiling. Distribute into tubes or flasks. Autoclave for 15 min at 15 psi pressure–121°C. Pour into sterile Petri dishes or leave in tubes.

Use: For the cultivation and maintenance of *Streptomyces fradiae*.

Potato Infusion Agar

Composition per liter:

Potatoes, infusion from	200.0g
Agar	15.0g
Glucose	10.0g
Proteose peptone	10.0g
Beef extract	5.0g
NaCl	5.0g
Glycerol	20.0mL

pH 6.8 ± 0.2 at 25°C

Source: This medium is available as a premixed powder from Difco Laboratories.

Potatoes, Infusion From:
Composition per 500mL:

Potatoes	300.0g

Preparation of Potatoes, Infusion From: Peel and dice potatoes. Add 500.0mL of distilled/deionized water. Gently heat and bring to boiling. Continue boiling for 30 min. Filter through cheesecloth.

Preparation of Medium: Add glycerol to 500.0mL of distilled/deionized water. Add remaining components. Mix thoroughly. Gently heat and bring to boiling. Distribute into tubes or flasks. Autoclave for 15 min at 15 psi pressure–121°C. Pour into sterile Petri dishes or leave in tubes.

Use: For the isolation of *Brucella abortus*.

Potato Infusion with Inorganic Salts

Composition per liter:

Potato	200.0g
Agar	15.0g
K_2HPO_4	0.5g
$MgSO_4 \cdot 7H_2O$	0.4g
$CaCl_2 \cdot 2H_2O$	0.1g
$ZnSO_4 \cdot 7H_2O$	0.03g
$MnSO_4 \cdot 5H_2O$	0.02g
$CuSO_4 \cdot 5H_2O$	0.01g
$FeSO_4 \cdot 7H_2O$	0.01g

Preparation of Potato, Infusion From: Peel and dice potatoes. Add 500.0mL of distilled/deionized water. Gently heat and bring to boiling. Continue boiling for 30 min. Filter through cheesecloth. Bring volume of filtrate to 1.0L. Add agar. Mix thoroughly. Gently heat and bring to boiling. Add remaining components. Mix thoroughly. Distribute into tubes or flasks. Autoclave for 15 min at 15 psi pressure–121°C. Pour into sterile Petri dishes or leave in tubes.

Use: For the cultivation and maintenance of *Bacillus macquariensis*.

Potato Malt Agar

Composition per liter:

Potatoes, infusion from	200.0g
Sucrose	60.0g

Agar	20.0g
Malt extract	20.0g
Peptone	1.0g

pH 7.4 ± 0.2 at 25°C

Source: This medium is available as a premixed powder from Difco Laboratories.

Potatoes, Infusion From:
Composition per 500mL:

Potatoes	300.0g

Preparation of Potatoes, Infusion From: Peel and dice potatoes. Add 500.0mL of distilled/deionized water. Gently heat and bring to boiling. Continue boiling for 30 min. Filter through cheesecloth.

Preparation of Medium: Add components to distilled/deionized water and bring volume to 1.0L. Mix thoroughly. Gently heat and bring to boiling. Distribute into tubes or flasks. Autoclave for 15 min at 15 psi pressure–121°C. Pour into sterile Petri dishes or leave in tubes.

Use: For the cultivation and maintenance of fungi and other aciduric microorganisms.

Potato Medium

Composition per liter:

Potato	60.0g
Agar	15.0g
Glucose	10.0g
Peptone	10.0g
Yeast extract	5.0g
$CaCO_3$	1.0g

Preparation of Medium: Peel and dice potato. Homogenize in a blender. Add potato and remaining components to distilled/deionized water and bring volume to 1.0L. Mix thoroughly. Gently heat and bring to boiling. Distribute into tubes or flasks. Autoclave for 15 min at 15 psi pressure–121°C. Pour into sterile Petri dishes or leave in tubes.

Use: For the cultivation and maintenance of *Clostridium laniganii*.

Potato P–YE *Thermus* Medium

Composition per liter:

Agar	20.0g
Peptone	5.0g
Yeast extract	0.2g
Potatoes, infusion from	200.0mL

pH 7.8 ± 0.2 at 25°C

Potatoes, Infusion From:
Composition per 500mL:

Potatoes	300.0g

Preparation of Medium: Add components to distilled/deionized water and bring volume to 1.0L. Mix thoroughly. Gently heat and bring to boiling. Distribute into tubes or flasks. Autoclave for 15 min at 15 psi pressure–121°C. Pour into sterile Petri dishes or leave in tubes.

Use: For the cultivation and maintenance of *Thermus ruber*.

Powell and Errington's Medium

Composition per 1060mL:

Solution 1	50.0mL
Solution 3	50.0mL
Solution 4	10.0mL
Solution 2	5.0mL

pH 7.0 ± 0.2 at 25°C

Solution 1:
Composition per liter:

$(NH_4)_2HPO_4$	238.0g
K_2SO_4	70.0g
$NaH_2PO_4·2H_2O$	31.0g

Preparation of Solution 1: Add components to distilled/deionized water and bring volume to 1.0L. Mix thoroughly.

Solution 2:
Composition per liter:

MgO	10.0g
$FeCl_3·6H_2O$	5.4g
$CaCO_3$	2.0g
$ZnSO_4·7H_2O$	1.44g
$MnSO_4·4H_2O$	1.11g
$Na_2MoO_4·2H_2O$	0.49g
$CoSO_4·7H_2O$	0.28g
$CuSO_4·5H_2O$	0.25g
H_3BO_4	0.062g
HCl, concentrated	50.0mL

Preparation of Solution 2: Add components to distilled/deionized water and bring volume to 1.0L. Mix thoroughly.

Solution 3:
Composition per 50mL:

Citric acid	4.2g
Glucose	3.6g
L-Glutamic acid	2.94g
Succinic acid	1.18g

Preparation of Solution 3: Add components to distilled/deionized water and bring volume to 50.0mL. Mix thoroughly. Filter sterilize.

Solution 4:
Composition per 10mL:

$Na_2S_2O_3·5H_2O$	1.24g

Preparation of Solution 4: Add $Na_2S_2O_3 \cdot 5H_2O$ to distilled/deionized water and bring volume to 10.0mL. Mix thoroughly. Filter sterilize.

Preparation of Medium: Add 50.0mL of solution 1 and 5.0mL of solution 2. Mix thoroughly. Bring volume to 1.0L with distilled/deionized water. Autoclave for 15 min at 15 psi pressure–121°C. Cool to 25°C. Adjust pH to 7.0 with sterile NaOH. Aseptically add 50.0mL of solution 3 and 10.0mL of sterile solution 4. Mix thoroughly. Aseptically distribute into sterile tubes or flasks.

Use: For the cultivation of a variety of heterotrophic microorganisms.

PP Starch Medium

Composition per liter:

Polypeptone™	10.0g
Soluble starch	10.0g
K_2HPO_4	3.0g
$MgSO_4 \cdot 7H_2O$	1.0g

Preparation of Medium: Add components to distilled/deionized water and bring volume to 1.0L. Mix thoroughly. Gently heat while stirring and bring to boiling. Distribute into tubes or flasks. Autoclave for 15 min at 15 psi pressure–121°C.

Use: For the cultivation and maintenance of *Bacillus mycoides*.

PPLO Agar

Composition per liter:

Beef heart, infusion from	50.0g
Agar	14.0g
Peptone	10.0g
NaCl	5.0g
Bovine serum	100.0mL

pH 7.8 ± 0.2 at 25°C

Source: This medium is available as a premixed powder from Difco Laboratories.

Preparation of Medium: Add components, except bovine serum, to distilled/deionized water and bring volume to 900.0mL. Mix thoroughly. Gently heat and bring to boiling. Autoclave for 15 min at 15 psi pressure–121°C. Cool to 45°–50°C. Aseptically add sterile bovine serum. Mix thoroughly. Pour into sterile Petri dishes or distribute into sterile tubes.

Use: For the isolation and cultivation of *Mycoplasma* species (pleuro-pneumonia-like organisms).

PPLO Agar

Composition per liter:

Beef heart, infusion from	50.0g
Agar	14.0g
Peptone	10.0g
NaCl	5.0g
Ascitic fluid	250.0mL

pH 7.8 ± 0.2 at 25°C

Source: This medium is available as a premixed powder from Difco Laboratories.

Preparation of Medium: Add components, except ascitic fluid, to distilled/deionized water and bring volume to 750.0mL. Mix thoroughly. Gently heat and bring to boiling. Autoclave for 15 min at 15 psi pressure–121°C. Cool to 45°–50°C. Aseptically add sterile ascitic fluid. Mix thoroughly. Pour into sterile Petri dishes or distribute into sterile tubes.

Use: For the isolation and cultivation of *Mycoplasma* species (pleuro-pneumonia-like organisms).

PPLO Agar Base

See: ***Mycoplasma*** **Agar Base**

PPLO Agar with Additives for *Mycoplasma*

Composition per 1010mL:

Agar	15.0g
Arginine	1.74g
Glutamine	1.46g
Phenol Red	0.02g
PPLO broth	700.0mL
Horse serum, not inactivated	200.0mL
Yeast extract solution, fresh	100.0mL
Vitamins in Eagle's medium, 100X	10.0mL

pH 7.1 ± 0.2 at 25°C

PPLO Broth:

Composition per liter:

Beef heart, infusion from	50.0g
Peptone	10.0g
NaCl	5.0g

Preparation of PPLO Broth: Add components to distilled/deionized water and bring volume to 1.0L. Mix thoroughly. Gently heat and bring to boiling.

Yeast Extract Solution:

Composition per 300mL:

Baker's yeast, live, pressed, starch-free	75.0g

Preparation of Yeast Extract Solution: Add the live Baker's yeast to 300.0mL of distilled/deionized water. Autoclave for 90 min at 15 psi pressure–121°C. Allow to stand. Remove supernatant solution. Adjust pH to 6.6–6.8. Filter sterilize.

Vitamins in Eagle's Medium, 100X:
Composition per liter:

Inositol	2.0mg
Calcium pantothenate	1.0mg
Choline chloride	1.0mg
Folic acid	1.0mg
Nicotinamide	1.0mg
Pyridoxal	1.0mg
Thiamine·HCl	1.0mg
Riboflavin	0.1mg

Preparation of Vitamins in Eagle's Medium, 100X: Add components to distilled/deionized water and bring volume to 1.0L. Mix thoroughly. Filter sterilize.

Preparation of Medium: Add components—except fresh yeast extract solution, horse serum and vitamins in Eagle's medium, 100X—to distilled/deionized water and bring volume to 690.0mL. Mix thoroughly. Gently heat and bring to boiling. Autoclave for 15 min at 15 psi pressure–121°C. Cool to 45°–50°C. Aseptically add sterile fresh yeast extract solution, horse serum and vitamins in Eagle's medium, 100X. Mix thoroughly. Pour into sterile Petri dishes or distribute into sterile tubes.

Use: For the cultivation and maintenance of *Mycoplasma arginini* and *Spiroplasma apis*.

PPLO Agar, pH 7.6 with Additives for *Mycoplasma*

Composition per liter:

Agar	15.0g
L-Cysteine·HCl·H_2O	1.0g
PPLO broth	700.0mL
Horse serum, not inactivated	200.0mL
Yeast extract solution, fresh	100.0mL

pH 7.6 ± 0.2 at 25°C

PPLO Broth:
Composition per liter:

Beef heart, infusion from	50.0g
Peptone	10.0g
NaCl	5.0g

Preparation of PPLO Broth: Add components to distilled/deionized water and bring volume to 1.0L. Mix thoroughly. Gently heat and bring to boiling.

Yeast Extract Solution:
Composition per 300mL:

Baker's yeast, live, pressed, starch-free	75.0g

Preparation of Yeast Extract Solution: Add the live Baker's yeast to 300.0mL of distilled/deionized water. Autoclave for 90 min at 15 psi pressure–121°C. Allow to stand. Remove supernatant solution. Adjust pH to 6.6–6.8. Filter sterilize.

Preparation of Medium: Add components, except fresh yeast extract solution and horse serum, to distilled/deionized water and bring volume to 700.0mL. Mix thoroughly. Gently heat and bring to boiling. Autoclave for 15 min at 15 psi pressure–121°C. Cool to 45°–50°C. Aseptically add sterile fresh yeast extract solution and horse serum. Mix thoroughly. Pour into sterile Petri dishes or distribute into sterile tubes.

Use: For the cultivation and maintenance of *Mycoplasma faucium*.

PPLO Broth Base without Crystal Violet

See: Mycoplasma **Broth Base**

PPLO Broth with Additives for *Mycoplasma*

Composition per 1010mL:

Arginine	1.74g
Glutamine	1.46g
Phenol Red	0.02g
PPLO broth	700.0mL
Horse serum, not inactivated	200.0mL
Yeast extract solution, fresh	100.0mL
Vitamins in Eagle's medium, 100X	10.0mL

pH 7.1 ± 0.2 at 25°C

PPLO Broth:
Composition per liter:

Beef heart, infusion from	50.0g
Peptone	10.0g
NaCl	5.0g

Preparation of PPLO Broth: Add components to distilled/deionized water and bring volume to 1.0L. Mix thoroughly. Gently heat and bring to boiling.

Yeast Extract Solution:
Composition per 300mL:

Baker's yeast, live, pressed, starch-free	75.0g

Preparation of Yeast Extract Solution: Add the live Baker's yeast to 300.0mL of distilled/deionized water. Autoclave for 90 min at 15 psi pressure–121°C. Allow to stand. Remove supernatant solution. Adjust pH to 6.6–6.8. Filter sterilize.

Vitamins in Eagle's Medium, 100X:
Composition per liter:

Inositol	2.0mg
Calcium pantothenate	1.0mg
Choline chloride	1.0mg
Folic acid	1.0mg
Nicotinamide	1.0mg
Pyridoxal	1.0mg

Thiamine·HCl 1.0mg
Riboflavin 0.1mg

Preparation of Vitamins in Eagle's Medium, 100X: Add components to distilled/deionized water and bring volume to 1.0L. Mix thoroughly. Filter sterilize.

Preparation of Medium: Add components—except fresh yeast extract solution, horse serum and vitamins in Eagle's medium, 100X—to distilled/deionized water and bring volume to 690.0mL. Mix thoroughly. Gently heat and bring to boiling. Autoclave for 15 min at 15 psi pressure–121°C. Cool to 45°–50°C. Aseptically add sterile fresh yeast extract solution, horse serum and vitamins in Eagle's medium, 100X. Mix thoroughly. Aseptically distribute into sterile tubes or flasks.

Use: For the cultivation of *Mycoplasma arginini* and *Spiroplasma apis.*

PPLO Broth, pH 7.6 with Additives for *Mycoplasma*

Composition per liter:

L-Cysteine·HCl·H_2O	1.0g
PPLO broth	700.0mL
Horse serum, not inactivated	200.0mL
Yeast extract solution, fresh	100.0mL

pH 7.6 ± 0.2 at 25°C

PPLO Broth:

Composition per liter:

Beef heart, infusion from	50.0g
Peptone	10.0g
NaCl	5.0g

Preparation of PPLO Broth: Add components to distilled/deionized water and bring volume to 1.0L. Mix thoroughly. Gently heat and bring to boiling.

Yeast Extract Solution:

Composition per 300mL:

Baker's yeast, live, pressed, starch-free	75.0g

Preparation of Yeast Extract Solution: Add the live Baker's yeast to 300.0mL of distilled/deionized water. Autoclave for 90 min at 15 psi pressure–121°C. Allow to stand. Remove supernatant solution. Adjust pH to 6.6–6.8. Filter sterilize.

Preparation of Medium: Add components, except fresh yeast extract solution and horse serum, to distilled/deionized water and bring volume to 700.0mL. Mix thoroughly. Gently heat and bring to boiling. Autoclave for 15 min at 15 psi pressure–121°C. Cool to 45°–50°C. Aseptically add sterile fresh yeast extract solution and horse serum. Mix thoroughly. Aseptically distribute into sterile tubes or flasks.

Use: For the cultivation and maintenance of *Mycoplasma faucium.*

PPLO Broth with Bovine Serum

Composition per liter:

Beef heart, infusion from	50.0g
Peptone	10.0g
NaCl	5.0g
Phenol Red (1% solution)	2.0mL
Yeast extract solution, fresh	100.0mL
Glucose solution	25.0mL
Bovine serum, filter sterilized	10.0mL

pH 7.5 ± 0.2 at 25°C

Yeast Extract Solution:

Composition per 300mL:

Baker's yeast, live, pressed, starch-free	75.0g

Preparation of Yeast Extract Solution: Add the live Baker's yeast to 300.0mL of distilled/deionized water. Autoclave for 90 min at 15 psi pressure–121°C. Allow to stand. Remove supernatant solution. Adjust pH to 6.6–6.8. Filter sterilize.

Glucose Solution:

Composition per 100mL:

D-Glucose	20.0g

Preparation of Glucose Solution: Add glucose to distilled/deionized water and bring volume to 100.0mL. Mix thoroughly. Filter sterilize.

Preparation of Medium: Add components—except fresh yeast extract solution, glucose solution, and bovine serum—to distilled/deionized water and bring volume to 865.0mL. Mix thoroughly. Gently heat and bring to boiling. Adjust pH to 7.5. Autoclave for 15 min at 15 psi pressure–121°C. Cool to 45°–50°C. Aseptically add sterile fresh yeast extract solution, glucose solution, and bovine serum. Mix thoroughly. Aseptically distribute into sterile tubes or flasks.

Use: For the cultivation and maintenance of *Acholeplasma morum.*

PPLO Broth with Crystal Violet

Composition per liter:

Beef heart, infusion from	50.0g
Peptone	10.0g
NaCl	5.0g
Crystal Violet	0.01g
Ascitic fluid	250.0mL
Chapman tellurite solution	2.85mL

pH 7.8 ± 0.2 at 25°C

Source: This medium is available as a premixed powder from Difco Laboratories.

Chapman Tellurite Solution:
Composition per 100mL:
K_2TeO_3 .... 1.0g

Preparation of Chapman Tellurite Solution: Add K_2TeO_3 to distilled/deionized water and bring volume to 100.0mL. Mix thoroughly. Filter sterilize.

Caution: Potassium tellurite is toxic.

Preparation of Medium: Add components, except ascitic fluid and Chapman tellurite solution, to distilled/deionized water and bring volume to 747.15mL. Mix thoroughly. Autoclave for 15 min at 15 psi pressure–121°C. Cool to less than 37°C. Aseptically add sterile ascitic fluid and 2.85mL of Chapman tellurite solution. Mix thoroughly. Aseptically distribute into sterile tubes or flasks.

Use: For the isolation of *Mycoplasma* species from clinical specimens.

PPLO Broth without Crystal Violet

Composition per liter:
Beef heart, infusion from 50.0g
Peptone 10.0g
NaCl 5.0g
Ascitic fluid 250.0mL
pH 7.8 ± 0.2 at 25°C

Source: This medium is available as a premixed powder from Difco Laboratories.

Preparation of Medium: Add components, except ascitic fluid, to distilled/deionized water and bring volume to 750.0mL. Mix thoroughly. Autoclave for 15 min at 15 psi pressure–121°C. Cool to less than 37°C. Aseptically add sterile ascitic fluid. If desired, 0.5g of thallium acetate or 100,000U of penicillin may be added for a more selective medium. Mix thoroughly. Aseptically distribute into sterile tubes or flasks.

Use: For the enrichment of pleuro-pneumonia-like organisms (PPLOs) and *Mycoplasma* species from clinical specimens.

PPLO Broth without Crystal Violet with Calf Serum, Fresh Yeast Extract, and Sodium Acetate

Composition per liter:
Beef heart, infusion from 50.0g
Peptone 10.0g
Sodium acetate 9.0g
NaCl 5.0g
Yeast extract solution, fresh 250.0mL
Calf serum 100.0mL
pH 7.8 ± 0.2 at 25°C

Yeast Extract Solution:
Composition per 300mL:
Baker's yeast, live, pressed, starch-free 75.0g

Preparation of Yeast Extract Solution: Add the live Baker's yeast to 300.0mL of distilled/deionized water. Autoclave for 90 min at 15 psi pressure–121°C. Allow to stand. Remove supernatant solution. Adjust pH to 6.6–6.8. Filter sterilize.

Preparation of Medium: Add components, except fresh yeast extract solution and calf serum, to distilled/deionized water and bring volume to 550.0mL. Mix thoroughly. Autoclave for 15 min at 15 psi pressure–121°C. Cool to 45°–50°C. Aseptically add sterile fresh yeast extract solution and calf serum. Mix thoroughly. Aseptically distribute into sterile tubes or flasks.

Use: For the cultivation and maintenance of *Mycoplasma* species.

PPLO Broth without Crystal Violet with Horse Serum

Composition per liter:
Beef heart, infusion from 50.0g
Peptone 10.0g
NaCl 5.0g
Horse serum, inactivated 200.0mL
pH 7.8 ± 0.2 at 25°C

Preparation of Medium: Add components, except horse serum, to distilled/deionized water and bring volume to 800.0mL. Mix thoroughly. Autoclave for 15 min at 15 psi pressure–121°C. Cool to 45°–50°C. Aseptically add sterile horse serum. Mix thoroughly. Aseptically distribute into sterile tubes or flasks.

Use: For the cultivation and maintenance of *Acholeplasma* species and *Mycoplasma* species.

PPLO Broth without Crystal Violet with Horse Serum and Fresh Yeast Extract

Composition per liter:
Beef heart, infusion from, solids 50.0g
Peptone 10.0g

NaCl	5.0g
Yeast extract solution, fresh	250.0mL
Horse serum	200.0mL

pH 7.8 ± 0.2 at 25°C

Yeast Extract Solution:
Composition per 300mL:

Baker's yeast, live, pressed, starch-free	75.0g

Preparation of Yeast Extract Solution: Add the live Baker's yeast to 300.0mL of distilled/deionized water. Autoclave for 90 min at 15 psi pressure–121°C. Allow to stand. Remove supernatant solution. Adjust pH to 6.6–6.8. Filter sterilize.

Preparation of Medium: Add components, except fresh yeast extract solution and horse serum, to distilled/deionized water and bring volume to 550.0mL. Mix thoroughly. Autoclave for 15 min at 15 psi pressure–121°C. Cool to 45°–50°C. Aseptically add sterile fresh yeast extract solution and horse serum. Mix thoroughly. Aseptically distribute into sterile tubes or flasks.

Use: For the cultivation and maintenance of *Mycoplasma putrefaciens*.

PPLO Broth without Crystal Violet with Horse Serum, Glucose and Fresh Yeast Extract

Composition per liter:

Beef heart, infusion from	50.0g
Peptone	10.0g
Glucose	5.0g
NaCl	5.0g
Yeast extract solution, fresh	250.0mL
Horse serum	200.0mL

pH 7.8 ± 0.2 at 25°C

Yeast Extract Solution:
Composition per 300mL:

Baker's yeast, live, pressed, starch-free	75.0g

Preparation of Yeast Extract Solution: Add the live Baker's yeast to 300.0mL of distilled/deionized water. Autoclave for 90 min at 15 psi pressure–121°C. Allow to stand. Remove supernatant solution. Adjust pH to 6.6–6.8. Filter sterilize.

Preparation of Medium: Add components, except fresh yeast extract solution and horse serum, to distilled/deionized water and bring volume to 550.0mL. Mix thoroughly. Autoclave for 15 min at 15 psi pressure–121°C. Cool to 45°–50°C. Aseptically add sterile fresh yeast extract solution and horse serum. Mix thoroughly. Aseptically distribute into sterile tubes or flasks.

Use: For the cultivation and maintenance of *Mycoplasma putrefaciens*.

PPNG Selective Medium (Penicillinase-Producing *Neisseria gonorrhoeae* Medium)

Composition per plate:

Quadrant I	10.0mL
Quadrant II	10.0mL

Quadrant I:
Composition per 10mL:

Martin Lewis agar	10.0mL

Quadrant II:
Composition per 10mL:

Martin Lewis agar, enriched	10.0mL

Use: For the differentiation and presumptive identification of penicillinase-producing strains of *Neisseria gonorrhoeae*. The PPNG selective medium is a two-sectored plate each containing a different medium. See Martin-Lewis agars for additional information.

PPYG Medium

Composition per liter:

Agar	15.0g
Na_2CO_3	5.03g
Glucose	5.0g
Peptone	5.0g
NaCl	1.5g
$Na_2HPO_4 \cdot 12H_2O$	1.5g
Yeast extract	1.5g
$MgCl_2 \cdot 6H_2O$	0.1g
Na_2CO_3 solution	50.0mL
Glucose solution	50.0mL

pH 10.5–11.0 at 25°C

Na_2CO_3 Solution:
Composition per 50mL:

Na_2CO_3	5.03g

Preparation of Na_2CO_3 Solution: Add Na_2CO_3 to distilled/deionized water and bring volume to 50.0mL. Mix thoroughly. Filter sterilize.

Glucose Solution:
Composition per 50mL:

D-Glucose	5.0g

Preparation of Glucose Solution: Add glucose to distilled/deionized water and bring volume to 50.0mL. Mix thoroughly. Filter sterilize.

Preparation of Medium: Add components, except Na_2CO_3 solution and glucose solution, to distilled/deionized water and bring volume to 900.0mL. Mix thoroughly. Gently heat and bring to boiling. Autoclave for 15 min at 15 psi pressure–121°C. Cool to 45°–50°C. Aseptically add sterile Na_2CO_3 solution and glucose solution. Mix thoroughly. Pour into sterile Petri dishes or distribute into sterile tubes.

Use: For the cultivation and maintenance of *Exiguobacterium aurantiacum.*

Pre–Enrichment Medium (PEM)

Composition per liter:

Yeast extract 20.0g
Special peptone, Oxoid Unipath 10.0g
Na_2HPO_4 .... 7.1g
NaCl 1.0g
KCl 1.0g
$MgSO_4 \cdot 7H_2O$ solution 10.0mL
$CaCl_2 \cdot 2H_2O$ solution 10.0mL
pH 8.3 ± 0.2 at 25°C

$MgSO_4 \cdot 7H_2O$ Solution:
Composition per 10mL:

$MgSO_4 \cdot 7H_2O$ 0.010g

Preparation of $MgSO_4 \cdot 7H_2O$ Solution: Add $MgSO_4 \cdot 7H_2O$ to distilled/deionized water and bring volume to 10.0mL. Mix thoroughly. Filter sterilize.

$CaCl_2 \cdot 2H_2O$ Solution:

$CaCl_2 \cdot 2H_2O$ 0.010g

Preparation of $CaCl_2 \cdot 2H_2O$ Solution: Add the $CaCl_2 \cdot 2H_2O$ to distilled/deionized water and bring volume to 10.0mL. Mix thoroughly. Filter sterilize.

Preparation of Medium: Add components, except $MgSO_4 \cdot 7H_2O$ solution and $CaCl_2 \cdot 2H_2O$ solution, to distilled/deionized water and bring volume to 980.0mL. Mix thoroughly. Adjust pH to 8.3. Gently heat and bring to boiling. Autoclave for 15 min at 15 psi pressure–121°C. Cool to 45°–50°C. Aseptically add sterile $MgSO_4 \cdot 7H_2O$ solution and $CaCl_2 \cdot 2H_2O$ solution. Mix thoroughly. Aseptically distribute into sterile tubes.

Use: For the isolation and enrichment of *Yersinia enterocolitica* from foods.

Presence–Absence Broth (P–A Broth)

Composition per liter:

Pancreatic digest of casein 10.0g
Lactose 7.5g
Pancreatic digest of gelatin 5.0g
Beef extract 3.0g
NaCl 2.5g
K_2HPO_4 .... 1.375g
KH_2PO_4 .... 1.375g
Sodium lauryl sulfate 0.05g
Bromcresol Purple 8.5mg
pH 6.8 ± 0.2 at 25°C

Source: Available as a premixed powder from BBL Microbiology Systems and Difco Laboratories.

Preparation of Medium: Add components to distilled/deionized water and bring volume to 333.0mL. Mix thoroughly. Distribute into screw-capped 250.0mL milk dilution bottles in 50.0mL volumes. Autoclave for 15 min at 15 psi pressure–121°C.

Use: For the detection of coliform bacteria in water from treatment plants or distribution systems using the presence-absence coliform test.

Preston Blood–Free Medium
See: ***Campylobacter* Charcoal Differential Agar**

Preston Enrichment Broth

Composition per liter:

Beef extract 10.0g
Peptone 10.0g
NaCl 5.0g
Horse blood, lysed 50.0mL
Antibiotic solution 10.0mL
pH 7.5 ± 0.2 at 25°C

Antibiotic Solution:
Composition per 10mL:

Cycloheximide 0.1g
Rifampicin 0.01g
Trimethoprim lactate 0.01g
Polymyxin B 5,000U

Preparation of Antibiotic Solution: Add components to distilled/deionized water and bring volume to 10.0mL. Mix thoroughly. Filter sterilize.

Preparation of Medium: Add components, except horse blood and antibiotic solution, to distilled/deionized water and bring volume to 940.0mL. Mix thoroughly. Gently heat and bring to boiling. Autoclave for 15 min at 15 psi pressure–121°C. Cool to 45°–50°C. Aseptically add sterile horse blood and antibiotic solution. Mix thoroughly. Aseptically distribute into sterile tubes or flasks.

Use: For the isolation and enrichment of *Campylobacter* species from foods.

Preston's *Campylobacter* Medium
See: Campylobacter Selective Medium, Preston's

Presumpto Media

Composition per plate:

Quadrant I	5.0mL
Quadrant II	5.0mL
Quadrant III	5.0mL
Quadrant IV	5.0mL

Quadrant I:
Composition per 5mL:

Lombard-Dowell agar	5.0mL

Quadrant II:
Composition per 5mL:

Lombard-Dowell bile agar	5.0mL

Quadrant III:
Composition per 5mL:

Lombard-Dowell egg yolk agar	5.0mL

Quadrant IV:
Composition per 5mL:

Lombard-Dowell esculin agar	5.0mL

Use: For the differentiation and presumptive identification of anaerobic bacteria. The Presumpto media is a four-sectored plate each containing a different medium. See Lombard-Dowell agars for additional information.

Pril Xylose Ampicillin Agar (PXA Agar)

Composition per liter:

Agar	15.0g
Xylose	10.0g
Pancreatic digest of gelatin	5.0g
Beef extract	3.0g
Pril	0.2g
Ampicillin	0.03g
Phenol Red	0.025g

pH 6.8 ± 0.2 at 25°C

Note: Pril is a quaternary ammonium detergent composed of a mixture of primary alkyl sulfate, alkyl-benzyl sulfonate and salts. It is available from Böhme Fettchemie GmbH, Düsseldorf, Germany.

Preparation of Medium: Add components to distilled/deionized water and bring volume to 1.0L. Mix thoroughly. Gently heat and bring to boiling. Distribute into tubes or flasks. Autoclave for 15 min at 15 psi pressure–121°C. Pour into sterile Petri dishes or leave in tubes.

Use: For the selective isolation and cultivation of *Aeromonas hydrophila*.

Propionispira Medium

Composition per liter:

Sodium lactate	4.0g
Yeast extract	1.0g
Mineral solution 2	50.0mL
Sodium carbonate solution	50.0mL
Mineral solution 1	25.0mL
Cysteine-sulfide reducing agent	20.0mL
Wolfe's mineral solution	10.0mL
Vitamin solution	10.0mL
Resazurin (0.025% solution)	4.0mL

pH 7.2 ± 0.2 at 25°C

Mineral Solution 1:
Composition per liter:

K_2HPO_4	6.0g

Preparation of Medium: Add K_2HPO_4 to distilled/deionized water and bring volume to 1.0L. Mix thoroughly.

Mineral Solution 2:
Composition per liter:

NaCl	12.0g
KH_2PO_4	6.0g
$(NH_4)_2SO_4$	6.0g
$MgSO_4 \cdot 7H_2O$	2.4g
$CaCl_2 \cdot 2H_2O$	1.6g

Preparation of Mineral Solution 2: Add components to distilled/deionized water and bring volume to 1.0L. Mix thoroughly.

Sodium Carbonate Solution:
Composition per 100mL:

Na_2CO_3	8.0g

Preparation of Sodium Carbonate Solution: Add Na_2CO_3 to distilled/deionized water and bring volume to 100.0mL Mix thoroughly.

Cysteine-Sulfide Reducing Agent:
Composition per 20mL:

L-Cysteine·HCl·H_2O	300.0mg
$Na_2S \cdot 9H_2O$	300.0mg

Preparation of Cysteine-Sulfide Reducing Agent: Add L-Cysteine·HCl·H_2O to 10.0mL of distilled/deionized water. Mix thoroughly. In a separate tube, add $Na_2S \cdot 9H_2O$ to 10.0mL of distilled/deionized water. Mix thoroughly. Gas both solutions with 100% N_2 and cap tubes. Autoclave both solutions for 15 min at 15 psi pressure–121°C using fast exhaust. Cool to 50°C. Aseptically combine the two solutions under 100% N_2.

Wolfe's Mineral Solution:

Composition per liter

$MgSO_4 \cdot 7H_2O$	3.0g
Nitrilotriacetic acid	1.5g
NaCl	1.0g
$MnSO_4 \cdot H_2O$	0.5g
$FeSO_4 \cdot 7H_2O$	0.1g
$CoCl_2 \cdot 6H_2O$	0.1g
$CaCl_2$	0.1g
$ZnSO_4 \cdot 7H_2O$	0.1g
$CuSO_4 \cdot 5H_2O$	0.01g
$AlK(SO_4)_2 \cdot 12H_2O$	0.01g
H_3BO_3	0.01g
$Na_2MoO_4 \cdot 2H_2O$	0.01g

Preparation of Wolfe's Mineral Solution: Add nitrilotriacetic acid to 500.0mL of distilled/deionized water. Dissolve by adjusting pH to 6.5 with KOH. Add remaining components. Add distilled/deionized water to 1.0L.

Wolfe's Vitamin Solution:

Composition per liter:

Pyridoxine·HCl	10.0mg
Thiamine·HCl	5.0mg
Riboflavin	5.0mg
Nicotinic acid	5.0mg
Calcium pantothenate	5.0mg
p-Aminobenzoic acid	5.0mg
Thioctic acid	5.0mg
Biotin	2.0mg
Folic acid	2.0mg
Cyanocobalamin	100.0μg

Preparation of Wolfe's Vitamin Solution: Add components to distilled/deionized water and bring volume to 1.0L. Mix thoroughly. Filter sterilize.

Preparation of Medium: Add components, except vitamin solution and cysteine-sulfide reducing agent, to distilled/deionized water and bring volume to 970.0mL. Mix thoroughly. Autoclave for 15 min at 15 psi pressure–121°C. Cool under 80% N_2 + 20% CO_2. Aseptically add the sterile vitamin solution and then the sterile cysteine-sulfide reducing agent. Adjust the pH to 7.2. Distribute aseptically and anaerobically into sterile tubes.

Use: For the cultivation and maintenance of *Propionispira arboris*.

Proskauer–Beck Medium for *Mycobacterium*

Composition per liter:

Asparagine	5.0g
KH_2PO_4	5.0g
Magnesium citrate	2.5g
$MgSO_4 \cdot 7H_2O$	0.6g
Glycerol	20.0mL

pH 7.4 ± 0.2 at 25°C

Preparation of Medium: Add components one at a time to distilled/deionized water and bring volume to 1.0L. Mix thoroughly. Make sure one salt is totally dissolved before the next one is added. Adjust pH to 7.8 with 40% NaOH. Autoclave for 15 min at 15 psi pressure–121°C. The pH of the medium after autoclaving should be 7.4. Filter through Whatman #1 filter paper to remove any precipitate. Distribute into tubes or flasks. Autoclave again for 15 min at 15 psi pressure–121°C.

Use: For the cultivation and maintenance of *Mycobacterium tuberculosis*.

Prosthecobacter Medium

Composition per liter:

Glucose	0.25g
$(NH_4)_2SO_4$	0.25g
Na_2HPO_4	0.071g
Hutner's mineral base	20.0mL

Hutner's Mineral Base:

Composition per liter:

$MgSO_4 \cdot 7H_2O$	29.7g
Nitrilotriacetic acid	10.0g
$CaCl_2 \cdot 2H_2O$	3.34g
$FeSO_4 \cdot 7H_2O$	0.10g
$(NH_4)_2MoO_4$	9.25mg
Metals "44"	50.0mL

Preparation of Hutner's Mineral Base: Add nitrilotriacetic acid to 500.0mL of distilled/deionized water. Dissolve by adjusting pH to 6.5 with KOH. Add remaining components. Readjust pH to 7.2 with H_2SO_4 or KOH. Add distilled/deionized water to 1.0L. Store at 5°C.

Metals "44":

Composition per 100mL:

$ZnSO_4 \cdot 7H_2O$	1.1g
$FeSO_4 \cdot 7H_2O$	0.5g
EDTA	0.25g
$MnSO_4 \cdot 7H_2O$	0.154g
$CuSO_4 \cdot 5H_2O$	0.04g
$Co(NO_3)_2 \cdot 6H_2O$	0.025g
$Na_2B_4O_7 \cdot 10H_2O$	0.018g

Preparation of Metals "44": Add a few drops of H_2SO_4 to distilled/deionized water to inhibit precipitate formation. Add components to acidified distilled/deionized water and bring volume to 100.0mL. Mix thoroughly.

Preparation of Medium: Add components to distilled/deionized water and bring volume to 1.0L. Mix thoroughly. Distribute into tubes or flasks. Autoclave for 15 min at 15 psi pressure–121°C.

Use: For the cultivation of *Prosthecobacter fusiformis*.

Proteose Agar

Composition per liter:

Agar	15.0g
Proteose peptone No. 3	15.0g
Yeast extract	7.5g
Casamino acids	5.0g
K_2HPO_4	5.0g
$(NH_4)_2SO_4$	1.5g
Starch, soluble	1.0g

pH 9.0 ± 0.2 at 25°C

Preparation of Medium: Add components to distilled/deionized water and bring volume to 1.0L. Mix thoroughly. Gently heat and bring to boiling. Distribute into tubes in 10.0mL volumes. Autoclave for 15 min at 15 psi pressure–121°C. Allow tubes to cool in a slanted position.

Use: For the cultivation of *Vibrio* species from foods.

Proteose No. 3 Agar

Composition per 1010mL:

Proteose peptone No. 3	20.0g
Agar	15.0g
Na_2HPO_4	5.0g
NaCl	5.0g
Glucose	0.5g
Hemoglobin solution	500.0mL
Supplement A (Difco)	10.0mL

pH 7.3 ± 0.2 at 25°C

Source: This medium is available as a premixed powder from Difco Laboratories.

Hemoglobin Solution:

Composition per 500mL:

Hemoglobin	10.0g

Preparation of Hemoglobin Solution: Add hemoglobin to distilled/deionized water and bring volume to 500.0mL. Mix thoroughly. Autoclave for 15 min at 15 psi pressure–121°C. Cool to 45°–50°C.

Supplement A:

Composition per 10mL:

Supplement A contains yeast concentrate with Crystal Violet.

Preparation of Supplement A: Add components to distilled/deionized water and bring volume to 10.0mL. Mix thoroughly. Filter sterilize.

Preparation of Medium: Add components, except hemoglobin solution and supplement A, to distilled/deionized water and bring volume to 500.0mL. Mix thoroughly. Gently heat and bring to boiling. Autoclave for 15 min at 15 psi pressure–121°C. Cool to 50°–60°C. Aseptically add 500.0mL of sterile hemoglobin solution and 10.0mL of sterile supplement A. Mix thoroughly. Pour into sterile Petri dishes or distribute into sterile tubes.

Use: For the isolation and cultivation of *Neisseria* species, *Hemophilus* species, and other fastidious bacteria. For the cultivation and maintenance of *Escherichia coli*.

Proteose No. 3 Agar

Composition per 1010mL:

Proteose peptone No. 3	20.0g
Agar	15.0g
Na_2HPO_4	5.0g
NaCl	5.0g
Glucose	0.5g
Hemoglobin solution	500.0mL
Supplement B (Difco)	10.0mL

pH 7.3 ± 0.2 at 25°C

Source: This medium is available as a premixed powder from Difco Laboratories.

Hemoglobin Solution:

Composition per 500mL:

Hemoglobin	10.0g

Preparation of Hemoglobin Solution: Add hemoglobin to distilled/deionized water and bring volume to 500.0mL. Mix thoroughly. Autoclave for 15 min at 15 psi pressure–121°C. Cool to 45°–50°C.

Supplement B:

Composition per 10mL:

Supplement B contains yeast concentrate, glutamine, coenzyme, cocarboxylase, hematin and growth factors.

Preparation of Supplement B: Add components to distilled/deionized water and bring volume to 10.0mL. Mix thoroughly. Filter sterilize.

Preparation of Medium: Add components, except hemoglobin solution and supplement B, to distilled/deionized water and bring volume to 500.0mL. Mix thoroughly. Gently heat and bring to boiling. Autoclave for 15 min at 15 psi pressure–121°C. Cool to 50°–60°C. Aseptically add 500.0mL of sterile hemoglobin solution and 10.0mL of sterile supplement

B. Mix thoroughly. Pour into sterile Petri dishes or distribute into sterile tubes.

Use: For the isolation and cultivation of *Neisseria* species, *Hemophilus* species, and other fastidious bacteria. For the cultivation and maintenance of *Escherichia coli.*

Proteose No. 3 Agar

Composition per 1010mL:

Proteose peptone No. 3	20.0g
Agar	15.0g
Na_2HPO_4	5.0g
NaCl	5.0g
Glucose	0.5g
Hemoglobin solution	500.0mL
Supplement VX (Difco)	10.0mL

pH 7.3 ± 0.2 at 25°C

Source: This medium is available as a premixed powder from Difco Laboratories.

Hemoglobin Solution:

Composition per 500mL:

Hemoglobin	10.0g

Preparation of Hemoglobin Solution: Add hemoglobin to distilled/deionized water and bring volume to 500.0mL. Mix thoroughly. Autoclave for 15 min at 15 psi pressure–121°C. Cool to 45°–50°C.

Supplement VX:

Composition per 10mL:

Supplement B contains essential growth factors.

Preparation of Supplement VX: Add components to distilled/deionized water and bring volume to 10.0mL. Mix thoroughly. Filter sterilize.

Preparation of Medium: Add components, except hemoglobin solution and supplement VX, to distilled/deionized water and bring volume to 500.0mL. Mix thoroughly. Gently heat and bring to boiling. Autoclave for 15 min at 15 psi pressure–121°C. Cool to 50°–60°C. Aseptically add 500.0mL of sterile hemoglobin solution and 10.0mL of sterile supplement VX. Mix thoroughly. Pour into sterile Petri dishes or distribute into sterile tubes.

Use: For the isolation and cultivation of *Neisseria* species, *Hemophilus* species, and other fastidious bacteria. For the cultivation and maintenance of *Escherichia coli.*

Proteose Yeast Extract Medium

Composition per liter:

Proteose peptone	20.0g
Glucose	10.0g
Yeast extract	5.0g

Preparation of Medium: Add components to distilled/deionized water and bring volume to 1.0L. Mix thoroughly. Distribute into tubes or flasks. Autoclave for 15 min at 15 psi pressure–121°C.

Use: For the cultivation of a variety of bacteria.

Provasoli Medium

Composition per liter:

NaCl	11.75g
$MgCl_2·6H_2O$	5.35g
Na_2SO_4	2.0g
$CaCl_2·2H_2O$	0.75g
Tris(hydroxymethyl)aminomethane	0.5g
KCl	0.35g
Na_2HPO_4	0.05g

pH 7.6 ± 0.2 at 25°C

Preparation of Medium: Add components to distilled/deionized water and bring volume to 1.0L. Mix thoroughly. Distribute into tubes or flasks. Autoclave for 15 min at 15 psi pressure–121°C.

Use: For the isolation and cultivation of *Leucothrix* species from marine habitats.

PRYES Agar
See: **Pentachloronitrobenzene Rose Bengal Yeast Extract Sucrose Agar**

PSD Agar
See: **Peptone Starch Dextrose Agar**

PSE Agar
See: **Pfizer Selective *Enterococcus* Agar**

Pseudomonas aeruginosa Agar (*PA* Agar) (m–*PA* Agar) (m–*Pseudomonas aeruginosa* Agar)

Composition per liter:

Agar	15.0g
$Na_2S_2O_3$	6.8g
L-Lysine·HCl	5.0g
NaCl	5.0g
Xylose	2.5g

Yeast extract 2.0g
Lactose 1.25g
Sucrose 1.25g
Ferric ammonium citrate 0.8g
Sulfapyridine 0.176g
Cycloheximide 0.15g
Phenol Red 0.08g
Nalidixic acid 0.037g
Kanamycin 8.5mg
pH 7.1 ± 0.2 at 25°C

Preparation of Medium: Add components—except sulfapyridine, cycloheximide, nalidixic acid, and kanamycin—to distilled/deionized water and bring volume to 1.0L. Mix thoroughly. Adjust pH to 6.5. Autoclave for 15 min at 15 psi pressure–121°C. Cool to 55°–60°C. Readjust pH to 7.1. Aseptically add the sulfapyridine, cycloheximide, nalidixic acid, and kanamycin. Mix thoroughly. Pour into 50mm × 12mm Petri dishes in 3.0mL volumes.

Use: For the cultivation and estimation of numbers of *Pseudomonas aeruginosa* in water by the membrane filter method.

Pseudomonas Agar F

Composition per liter:
Proteose peptone No. 3 .... 20.0g
Agar 15.0g
Glycerol 10.0g
Pancreatic digest of casein 10.0g
K_2HPO_4 .... 1.5g
$MgSO_4 \cdot 7H_2O$ 0.73g
pH 7.0 ± 0.2 at 25°C

Preparation of Medium: Add components to distilled/deionized water and bring volume to 1.0L. Mix thoroughly. Gently heat and bring to boiling. Distribute into tubes or flasks. Autoclave for 15 min at 15 psi pressure–121°C. Pour into sterile Petri dishes or leave in tubes.

Use: For the cultivation and observation of fluorescein production in *Pseudomonas* species.

Pseudomonas Agar F

Composition per liter:
Agar 15.0g
Glycerol 10.0g
Proteose peptone No. 3 .... 10.0g
Pancreatic digest of casein 10.0g
K_2HPO_4 .... 1.5g
$MgSO_4 \cdot 7H_2O$ 1.5g
pH 7.0 ± 0.2 at 25°C

Source: This medium is available as a premixed powder from Difco Laboratories.

Preparation of Medium: Add components to distilled/deionized water and bring volume to 1.0L. Mix thoroughly. Gently heat and bring to boiling. Distribute into tubes or flasks. Autoclave for 15 min at 15 psi pressure–121°C. Pour into sterile Petri dishes or leave in tubes.

Use: For the isolation, cultivation and differentiation of *Pseudomonas aeruginosa* on the basis of pigment production.

Pseudomonas Agar P

Composition per liter:
Proteose peptone No. 3 .... 20.0g
Agar 15.0g
Glycerol 10.0g
K_2HPO_4 .... 10.0g
$MgCl_2 \cdot 6H_2O$ 1.4g
pH 7.0 ± 0.2 at 25°C

Source: This medium is available as a premixed powder from Difco Laboratories.

Preparation of Medium: Add components to distilled/deionized water and bring volume to 1.0L. Mix thoroughly. Gently heat and bring to boiling. Distribute into tubes or flasks. Autoclave for 15 min at 15 psi pressure–121°C. Pour into sterile Petri dishes or leave in tubes.

Use: For the isolation, cultivation and differentiation of *Pseudomonas aeruginosa* on the basis of pigment production.

Pseudomonas Basal Mineral Medium

Composition per liter:
K_2HPO_4 .... 12.5g
KH_2PO_4 .... 3.8g
$(NH_4)_2SO_4$ 1.0g
$MgSO_4 \cdot 7H_2O$ 0.1g
Carbon source (0.8M solution) 100.0mL
Trace element solution 5.0mL
pH 7.2 ± 0.2 at 25°C

Trace Element Solution:
Composition per liter:
H_3BO_3 .... 0.232g
$ZnSO_4 \cdot 7H_2O$ 0.174g
$FeSO_4(NH_4)_2SO_4 \cdot 6H_2O$ 0.116g
$CoSO_4 \cdot 7H_2O$ 0.096g
$(NH_4)_6Mo_7O_{24} \cdot 4H_2O$ 0.022g
$CuSO_4 \cdot 5H_2O$ 8.0mg
$MnSO_4 \cdot 4H_2O$ 8.0mg

Preparation of Trace Element Solution: Add components to distilled/deionized water and bring volume to 1.0L. Mix thoroughly.

Carbon Source:

Composition per 100mL:

Glucose 14.4g

Preparation of Carbon Source: Add glucose to distilled/deionized water and bring volume to 100.0mL. Mix thoroughly. Filter sterilize. Other carbon sources may replace glucose. Prepare 0.8M carbon source solution.

Preparation of Medium: Add components, except carbon source, to distilled/deionized water and bring volume to 900.0mL. Mix thoroughly. Gently heat and bring to boiling. Autoclave for 15 min at 15 psi pressure–121°C. Cool to 45°–50°C. Aseptically add 100.0mL of sterile carbon source. Mix thoroughly. Aseptically distribute into sterile tubes or flasks.

Use: For the cultivation and differentiation of *Pseudomonas* species based on their ability to grow on different carbon sources.

Pseudomonas bathycetes Medium

Composition per liter:

NaCl 24.0g
Proteose peptone 10.0g
$MgSO_4 \cdot 7H_2O$ 7.0g
$MgCl_2$ 5.3g
Yeast extract 3.0g
KCl 0.7g

pH 7.2–7.4 at 25°C

Preparation of Medium: Add components to distilled/deionized water and bring volume to 1.0L. Mix thoroughly. Distribute into tubes or flasks. Autoclave for 15 min at 15 psi pressure–121°C.

Use: For the cultivation and maintenance of *Alteromonas haloplanktis, Alteromonas nigrifaciens, Pseudomonas bathycetes,* and *Pseudomonas elongata.*

Pseudomonas CFC Agar

Composition per liter:

Pancreatic digest of gelatin 16.0g
Agar 11.0g
Pancreatic digest of casein 10.0g
K_2SO_4 .... 10.0g
$MgCl_2 \cdot 6H_2O$ 1.4g
CFC selective supplement 10.0mL
Glycerol 10.0mL

pH 7.1 ± 0.2 at 25°C

Source: This medium is available as a premixed powder from Oxoid Unipath.

CFC Selective Supplement:

Composition per 10mL:

Cephaloridine 0.05g
Fucidin 0.01g
Cetrimide 0.01g

Preparation of CFC Selective Supplement: Add components to distilled/deionized water and bring volume to 10.0mL. Mix thoroughly. Filter sterilize.

Preparation of Medium: Add components, except CFC selective supplement, to distilled/deionized water and bring volume to 990.0mL. Mix thoroughly. Gently heat and bring to boiling. Autoclave for 15 min at 15 psi pressure–121°C. Cool to 45°–50°C. Aseptically add sterile CFC selective supplement. Mix thoroughly. Pour into sterile Petri dishes or distribute into sterile tubes.

Use: For the selective isolation and cultivation of *Pseudomonas* species.

Pseudomonas CN Agar

Composition per liter:

Pancreatic digest of gelatin 16.0g
Agar 11.0g
Pancreatic digest of casein 10.0g
K_2SO_4 .... 10.0g
$MgCl_2 \cdot 6H_2O$ 1.4g
CN selective supplement 10.0mL
Glycerol 10.0mL

CN Selective Supplement:

Cetrimide 0.1g
Sodium nalidixate 7.5mg

pH 7.1 ± 0.2 at 25°C

Source: This medium is available as a premixed powder from Oxoid Unipath.

Preparation of Medium: Add components, except CN selective supplement, to distilled/deionized water and bring volume to 990.0mL. Mix thoroughly. Gently heat and bring to boiling. Autoclave for 15 min at 15 psi pressure–121°C. Cool to 45°–50°C. Aseptically add sterile CN selective supplement. Mix thoroughly. Pour into sterile Petri dishes or distribute into sterile tubes.

Use: For the selective isolation and cultivation of *Pseudomonas* species.

Pseudomonas denitrificans Medium

Composition per liter:

Agar 15.0g
Glucose 10.0g
Yeast extract 5.0g
$FeCl_3$ solution 20.0mL

$FeCl_3$ Solution:

Composition per 100mL:

$FeCl_3$	0.03g

Preparation of $FeCl_3$ Solution: Add $FeCl_3$ to distilled/deionized water and bring volume to 100.0mL. Mix thoroughly. Filter sterilize.

Preparation of Medium: Add components, except $FeCl_3$ solution, to distilled/deionized water and bring volume to 980.0mL. Mix thoroughly. Gently heat and bring to boiling. Autoclave for 15 min at 15 psi pressure–121°C. Cool to 45°–50°C. Aseptically add 20.0mL of sterile $FeCl_3$ solution. Mix thoroughly. Pour into sterile Petri dishes or distribute into sterile tubes.

Use: For the cultivation and maintenance of *Pseudomonas* species.

Pseudomonas Denitrification Medium

Composition per liter:

Glycerol	10.0g
KNO_3	10.0g
Yeast extract	3.0g
$(NH_4)_2SO_4$	1.5g
Agar	1.0g
$K_2HPO_4 \cdot 3H_2O$	0.8g
$MgSO_4 \cdot 7H_2O$	0.5g
KH_2PO_4	0.2g
$CaCl_2$	0.1g

pH 7.2 ± 0.2 at 25°C

Preparation of Medium: Add components to distilled/deionized water and bring volume to 1.0L. Mix thoroughly. Distribute into tubes in 10.0mL volumes. Autoclave for 15 min at 15 psi pressure–121°C.

Use: For the cultivation and differentiation of *Pseudomonas* species based on their ability to produce pyocin and other fluorescent pigments during denitrification.

Pseudomonas Isolation Agar

Composition per liter:

Peptone	20.0g
Agar	13.6g
K_2SO_4	10.0g
$MgCl_2 \cdot 6H_2O$	1.4g
Irgasan® (triclosan)	0.025g
Glycerol	20.0mL

pH 7.0 ± 0.2 at 25°C

Source: This medium is available as a premixed powder from BBL Microbiology Systems and Difco Laboratories.

Preparation of Medium: Add components to distilled/deionized water and bring volume to 1.0L. Mix thoroughly. Gently heat and bring to boiling. Distribute into tubes or flasks. Autoclave for 15 min at 15 psi pressure–121°C. Pour into sterile Petri dishes or leave in tubes.

Use: For the isolation and cultivation of *Pseudomonas* species.

Pseudomonas Medium (ATCC Medium 179)

Composition per 1020mL:

Solution 1	1.0L
Solution 2	5.0mL
Solution 3	15.0mL

pH 6.8 ± 0.2 at 25°C

Solution 1:

Composition per liter:

Agar	20.0g
K_2HPO_4	2.56g
KH_2PO_4	2.08g
NH_4Cl	1.0g
$MgSO_4 \cdot 7H_2O$	0.5g

Preparation of Solution 1: Add components to distilled/deionized water and bring volume to 1.0L. Mix thoroughly. Gently heat and bring to boiling. Adjust pH to 6.8. Distribute into tubes or flasks. Autoclave for 15 min at 15 psi pressure–121°C. Cool to 45°–50°C.

Solution 2:

Composition per 100mL:

Ferric ammonium citrate	1.0g
$CaCl_2$	0.1g

Preparation of Solution 2: Add components to distilled/deionized water and bring volume to 100.0mL. Mix thoroughly. Filter sterilize.

Solution 3:

Composition per 100mL:

Succinic acid	11.8g

Preparation of Solution 3: Add succinic acid to distilled/deionized water and bring volume to 100.0mL. Mix thoroughly. Adjust pH to 6.0 with NaOH. Filter sterilize.

Preparation of Medium: To 1.0L of cooled sterile solution 1, aseptically add 5.0mL of sterile solution 2 and 15.0mL of sterile solution 3. Mix thoroughly. Pour into sterile Petri dishes or distribute into sterile tubes.

Use: For the cultivation and maintenance of *Pseudomonas lemoignei* and *Pseudomonas putida*.

Pseudomonas Medium (ATCC Medium 775)

Composition per liter:

NH_4Cl	5.0g
K_2HPO_4	1.5g
L-Tryptophan	1.0g
KH_2PO_4	0.5g
Yeast extract	0.5g
$MgSO_4$	0.2g

Preparation of Medium: Add components to distilled/deionized water and bring volume to 1.0L. Mix thoroughly. Distribute into tubes or flasks. Autoclave for 15 min at 15 psi pressure–121°C.

Use: For the cultivation and maintenance of *Comamonas acidovorans.*

Pseudomonas Medium (ATCC Medium 186)

Composition per liter:

Marine salts mix	19.0g
Glucose	10.0g
Yeast extract	10.0g
K_2HPO_4	1.0g
Peptone	1.0g
KH_2PO_4	0.5g
NH_4Cl	0.5g
$CaCl_2$	0.1g
Na_2SO_3	0.1g
$NaHCO_3$	0.1g
$FeCl_3$	0.05g

Preparation of Medium: Add components to distilled/deionized water and bring volume to 1.0L. Mix thoroughly. Distribute into tubes or flasks. Autoclave for 15 min at 15 psi pressure–121°C.

Use: For the cultivation and maintenance of *Bacillus sphaericus.*

Pseudomonas Medium (ATCC Medium 226)

Composition per liter:

Agar	20.0g
Yeast extract	10.0g
Glucose	5.0g
Sodium acetate	0.5g

Preparation of Medium: Add components to distilled/deionized water and bring volume to 1.0L. Mix thoroughly. Gently heat and bring to boiling. Distribute into tubes or flasks. Autoclave for 15 min at 15 psi pressure–121°C. Pour into sterile Petri dishes or leave in tubes.

Use: For the cultivation and maintenance of *Pseudomonas indigofera.*

Pseudomonas Medium (ATCC Medium 59)

Composition per liter:

K_2HPO_4	1.15g
NH_4NO_3	1.0g
Yeast extract	1.0g
KH_2PO_4	0.625g
$MgSO_4 \cdot 7H_2O$	0.02g
Pyrrolidine	4.0mL

pH 7.0 ± 0.2 at 25°C

Preparation of Medium: Add pyrrolidine to approximately 500.0mL of distilled/deionized water. Mix thoroughly. Adjust pH to 7.0. Add remaining components. Bring volume to 1.0L with distilled/deionized water. Distribute into tubes or flasks. Autoclave for 15 min at 15 psi pressure–121°C.

Use: For the cultivation and maintenance of *Pseudomonas fluorescens.*

Pseudomonas Medium (ATCC Medium 609)

Composition per liter:

Agar	15.0g
K_2HPO_4	8.71g
Nitrilotriacetic acid	1.91g
Na_2SO_4	0.57g
$MgSO_4$	0.25g
$FeSO_4$	0.5mg
$Ca(NO_3)_2$	0.5mg

pH 6.5 ± 0.2 at 25°C

Preparation of Medium: Add nitrilotriacetic acid to approximately 500.0mL of distilled/deionized water. Mix thoroughly. Adjust pH to 6.5. Add remaining components. Bring volume to 1.0L with distilled/deionized water. Gently heat and bring to boiling. Distribute into tubes or flasks. Autoclave for 15 min at 15 psi pressure–121°C. Pour into sterile Petri dishes or leave in tubes.

Use: For the cultivation and maintenance of *Pseudomonas* species.

Pseudomonas Medium A

Composition per liter:

Peptone	20.0g
Agar	15.0g
Glycerol	10.0g

K_2SO_4 .... 10.0g
$MgCl_2$ 1.4g

pH 7.2 ± 0.2 at 25°C

Preparation of Medium: Add components to distilled/deionized water and bring volume to 1.0L. Mix thoroughly. Gently heat and bring to boiling. Distribute into tubes or flasks. Autoclave for 10 min at 10 psi pressure–115°C. Pour into sterile Petri dishes or leave in tubes.

Use: For the cultivation and production of pyocyanin by *Pseudomonas* species.

Pseudomonas Medium B

Composition per liter:

Peptone 20.0g
Agar 15.0g
Glycerol 10.0g
$MgSO_4·7H_2O$ 1.5g
K_2HPO_4 solution 100.0mL

pH 7.2 ± 0.2 at 25°C

K_2HPO_4 Solution:
Composition per 100mL:

K_2HPO_4 .... 1.5g

Preparation of K_2HPO_4 Solution: Add K_2HPO_4 to distilled/deionized water and bring volume to 100.0mL. Mix thoroughly. Autoclave for 15 min at 15 psi pressure–121°C. Cool to 45°–50°C.

Preparation of Medium: Add components, except K_2HPO_4 solution, to distilled/deionized water and bring volume to 900.0mL. Mix thoroughly. Gently heat and bring to boiling. Autoclave for 15 min at 15 psi pressure–121°C. Cool to 45°–50°C. Aseptically add 100.0mL of sterile K_2HPO_4 solution. Mix thoroughly. Pour into sterile Petri dishes or distribute into sterile tubes.

Use: For the cultivation and observation of fluorescin production by *Pseudomonas* species.

Pseudomonas Medium I

Composition:

Pancreatic digest of casein 10.0g
Yeast extract 10.0g
Glucose 5.0g
K_2HPO_4 .... 5.0g
Salts solution 5.0mL

Salts Solution:
Composition per 100mL:

$MgSO_4·4H_2O$ 4.0g
$FeSO_4$ 0.2g
$MnSO_4·4H_2O$ 0.2g
NaCl 0.2g

Preparation of Salts Solution: Add components to distilled/deionized water and bring volume to 100.0mL. Mix thoroughly.

Preparation of Medium: Add components to distilled/deionized water and bring volume to 1.0L. Mix thoroughly. Distribute into tubes or flasks. Autoclave for 15 min at 15 psi pressure–121°C.

Use: For the cultivation and maintenance of *Bacillus sphaericus*.

Pseudomonas Phage Medium

Composition per liter:

Agar 15.0g
Nutrient broth 10.0g
K_2HPO_4 .... 1.11g
Glucose 1.0g
KH_2PO_4 .... 0.49g

pH 7.0 ± 0.2 at 25°C

Preparation of Medium: Add components to distilled/deionized water and bring volume to 1.0L. Mix thoroughly. Gently heat and bring to boiling. Distribute into tubes or flasks. Autoclave for 15 min at 15 psi pressure–121°C. Pour into sterile Petri dishes or leave in tubes.

Use: For the cultivation and maintenance of *Pseudomonas fluorescens*.

Pseudomonas saccharophila Medium

Composition per 1015mL:

Agar 20.0g
Na_2HPO_4 .... 4.8g
KH_2PO_4 .... 4.4g
NH_4Cl 1.0g
$MgSO_4·7H_2O$ 0.5g
Solution A 5.0mL
Solution B 10.0mL

Solution A:
Composition per 100mL:

Ferric ammonium citrate 1.0g
$CaCl_2$ 0.1g

Preparation of Solution A: Add components to distilled/deionized water and bring volume to 100.0mL. Mix thoroughly. Filter sterilize.

Solution B:
Composition per 100mL:

Sucrose 10.0g

Preparation of Solution B: Add sucrose to distilled/deionized water and bring volume to 100.0mL. Mix thoroughly. Filter sterilize.

Preparation of Medium: Add components, except solution A and solution B, to distilled/deionized water and bring volume to 1.0L. Mix thoroughly. Gently heat and bring to boiling. Autoclave for 15 min at 15 psi pressure–121°C. Cool to 45°–50°C. Aseptically add sterile solution A and sterile solution B. Mix thoroughly. Pour into sterile Petri dishes or distribute into sterile tubes.

Use: For the cultivation and maintenance of *Pseudomonas saccharophila* and other *Pseudomonas* species.

Pseudomonas solanacearum Medium

Composition per liter:

Agar	17.0g
Peptone	10.0g
Glucose	5.0g
Pancreatic digest of casein	1.0g

Preparation of Medium: Add components to distilled/deionized water and bring volume to 1.0L. Mix thoroughly. Gently heat and bring to boiling. Distribute into tubes or flasks. Autoclave for 15 min at 15 psi pressure–121°C. Pour into sterile Petri dishes or leave in tubes.

Use: For the cultivation and maintenance of *Pseudomonas solanacearum*.

Pseudomonas syringae Selective Medium

Composition per liter:

Agar	15.0g
L-Proline	5.0g
$MgSO_4 \cdot 7H_2O$	0.2g
K_2HPO_4	0.08g
KH_2PO_4	0.02g
$MnSO_4 \cdot 4H_2O$ solution	10.0mL

pH 6.8 ± 0.2 at 25°C

$MnSO_4 \cdot 4H_2O$ Solution:
Composition per 10mL:

$MnSO_4 \cdot 4H_2O$	2.1g

Preparation of $MnSO_4 \cdot 4H_2O$ Solution: Add $MnSO_4 \cdot 4H_2O$ to distilled/deionized water and bring volume to 10.0mL. Mix thoroughly. Autoclave for 15 min at 15 psi pressure–121°C.

Preparation of Medium: Add components, except $MnSO_4 \cdot 4H_2O$ solution, to distilled/deionized water and bring volume to 990.0mL. Mix thoroughly. Gently heat and bring to boiling. Adjust pH to 6.8. Autoclave for 10 min at 10 psi pressure–115°C. Cool to 45°–50°C. Aseptically add sterile $MnSO_4 \cdot 4H_2O$ solution. Mix thoroughly. Pour into sterile Petri dishes.

Use: For the selective isolation and cultivation of *Pseudomonas syringae*.

Pseudosel™ Agar
See: **Cetrimide Agar, USP**

PSS Broth
See: **Peptone Succinate Salts Broth**

PSS Medium
See: **Peptone Succinate Salts Medium**

PT Agar

Composition per liter:

Agar	15.0g
Pancreatic digest of casein	4.0g
Yeast extract	4.0g
$MgSO_4 \cdot 7H_2O$	2.0g
$CaCl_2 \cdot 2H_2O$	1.0g

pH 7.2 ± 0.2 at 25°C

Preparation of Medium: Add components to distilled/deionized water and bring volume to 1.0L. Mix thoroughly. Gently heat and bring to boiling. Distribute into tubes or flasks. Autoclave for 15 min at 15 psi pressure–121°C. Pour into sterile Petri dishes or leave in tubes.

Use: For the cultivation of myxobacteria.

Purple Agar

Composition per liter:

Agar	15.0g
Proteose peptone No. 3	10.0g
NaCl	5.0g
Beef extract	1.0g
Bromcresol Purple	0.02g
Carbohydrate solution	20.0mL

pH 6.8 ± 0.2 at 25°C

Source: This medium is available as a premixed powder from Difco Laboratories.

Carbohydrate Solution:
Composition per 20mL:

Carbohydrate	10.0g

Preparation of Carbohydrate Solution: Add carbohydrate to distilled/deionized water and bring volume to 20.0mL. For expensive carbohydrates, 5.0g may be used instead of 10.0g. Mix thoroughly. Filter sterilize.

Preparation of Medium: Add components, except carbohydrate solution, to distilled/deionized water and bring volume to 980.0mL. Mix thoroughly. Gently heat and bring to boiling. Distribute into tubes in 9.8mL volumes. Autoclave for 15 min at 15 psi pressure–121°C. Cool to 45°–50°C. Aseptically add 0.2mL of sterile carbohydrate solution to each tube. Mix thoroughly. Allow tubes to cool in a slanted position.

Use: For the preparation of carbohydrate media used in fermentation studies for the identification of bacteria, especially members of the Enterobacteriaceae. Bacteria that can ferment the carbohydrate turn the medium yellow.

Purple Broth (Purple Carbohydrate Broth)

Composition per liter:

Proteose peptone No. 3	10.0g
NaCl	5.0g
Beef extract	1.0g
Bromcresol Purple	0.015g
Carbohydrate solution	20.0mL

pH 6.8 ± 0.2 at 25°C

Source: This medium is available as a premixed powder from Difco Laboratories.

Carbohydrate Solution:
Composition per 20mL:

Carbohydrate	10.0g

Preparation of Carbohydrate Solution: Add carbohydrate to distilled/deionized water and bring volume to 20.0mL. For expensive carbohydrates, 5.0g may be used instead of 10.0g. Mix thoroughly. Filter sterilize.

Preparation of Medium: Add components, except carbohydrate solution, to distilled/deionized water and bring volume to 980.0mL. Mix thoroughly. Gently heat and bring to boiling. Distribute into tubes in 9.8mL volumes. Autoclave for 15 min at 15 psi pressure–121°C. Cool to 25°C. Aseptically add 0.2mL of sterile carbohydrate solution to each tube. Mix thoroughly.

Use: For the preparation of carbohydrate media used in fermentation studies for the identification of bacteria, especially members of the Enterobacteriaceae. Bacteria that can ferment the carbohydrate turn the medium yellow.

Purple Broth

Composition per liter:

Pancreatic digest of gelatin	10.0g
NaCl	5.0g
Bromcresol Purple	0.02g
Carbohydrate solution	20.0mL

pH 6.8 ± 0.2 at 25°C

Source: This medium is available as a premixed powder from BBL Microbiology Systems.

Carbohydrate Solution:
Composition per 20mL:

Carbohydrate	10.0g

Preparation of Carbohydrate Solution: Add carbohydrate to distilled/deionized water and bring volume to 20.0mL. For expensive carbohydrates, 5.0g may be used instead of 10.0g. Mix thoroughly. Filter sterilize.

Preparation of Medium: Add components, except carbohydrate solution, to distilled/deionized water and bring volume to 980.0mL. Mix thoroughly. Adjust pH to 7.4 if necessary. Distribute into tubes containing an inverted Durham tube. Fill each tube with 9.8mL of medium. Autoclave for 15 min at 15 psi pressure–121°C. Aseptically add 0.2mL of sterile carbohydrate solution to each tube.

Use: For the preparation of liquid fermentation media. Bacteria that can ferment the carbohydrate turn the medium yellow.

Purple Carbohydrate Broth

See: **Purple Broth**

Purple Lactose Agar

Composition per liter:

Agar	10.0g
Lactose	10.0g
Peptone	5.0g
Beef extract	3.0g
Bromcresol Purple	0.025g

pH 6.8 ± 0.1 at 25°C

Source: This medium is available as a premixed powder from Difco Laboratories.

Preparation of Medium: Add components to distilled/deionized water and bring volume to 1.0L. Mix thoroughly. Gently heat and bring to boiling. Distribute into tubes or flasks. Autoclave for 15 min at 15 psi pressure–121°C. Pour into sterile Petri dishes or leave in tubes. Allow tubes to cool in a slanted position.

Use: For the detection and differentiation of members of the Enterobacteriaceae. Bacteria that can ferment lactose turn the medium yellow.

Purple Serum Agar Base

Composition per liter:

Agar	20.0g
Lactose	20.0g
Peptone	20.0g
NaCl	5.0g
Bromcresol Purple	0.030g
Phenol Eed	0.024g

pH 7.6 ± 0.2 at 25°C

Preparation of Medium: Add components to distilled/deionized water and bring volume to 1.0L. Mix thoroughly. Gently heat and bring to boiling. Distribute into tubes or flasks. Autoclave for 15 min at 15 psi pressure–121°C. Pour into sterile Petri dishes or leave in tubes.

Use: For the cultivation and differentiation of Gram-negative bacteria isolated from the urinary tract. Bacteria that can ferment lactose turn the medium yellow.

PV Blood Agar (Paromomycin Vancomycin Blood Agar)

Composition per liter:

Agar	20.0g
Pancreatic digest of casein	15.0g
NaCl	5.0g
Papaic digest of soybean meal	5.0g
Yeast extract	5.0g
L-Cystine	0.4g
Paromomycin	0.1g
Vancomycin	7.5mg
Hemin	5.0mg
Sheep blood, defibrinated	50.0mL
Vitamin K_1 solution	10.0mL

pH 7.5 ± 0.2 at 25°C

Vitamin K_1 Solution:
Composition per 10mL:

Vitamin K_1	0.01g
Ethanol	10.0mL

Preparation of Vitamin K_1 Solution: Add vitamin K_1 to 10.0mL of absolute ethanol. Mix thoroughly. Filter sterilize.

Preparation of Medium: Add components—except vitamin K_1 solution, sheep blood, paromomycin and vancomycin—to distilled/deionized water and bring volume to 940.0mL. Mix thoroughly. Gently heat and bring to boiling for 1 min. Autoclave for 15 min at 15 psi pressure–121°C. Cool to 50–55°C. Aseptically add the sterile Vitamin K_1 solution, sheep blood, vancomycin and paromomycin. Mix thoroughly. Pour into sterile Petri dishes.

Use: For the selective cultivation of fastidious anaerobic bacteria.

PXA Agar
See: **Pril Xylose Ampicillin Agar**

PY 1% Medium (Peptone Yeast Extract 1% Medium)

Composition per liter:

Peptone	10.0g
Yeast extract	10.0g
NaCl	5.0g

Preparation of Medium: Add components to distilled/deionized water and bring volume to 1.0L. Mix thoroughly. Distribute into tubes or flasks. Autoclave for 15 min at 15 psi pressure–121°C.

Use: For the cultivation and maintenance of *Brevibacterium lactofermentum* and *Corynebacterium glutamicum*.

PY CMC Medium (Peptone Yeast Extract Carboxymethyl Cellulose Medium)

Composition per liter:

Agar	15.0g
Carboxymethyl cellulose	10.0g
NaCl	5.0g
Polypeptone™	5.0g
Yeast extract	5.0g
$MgSO_4 \cdot 7H_2O$	2.0g
KH_2PO_4	1.0g
Na_2CO_3 solution	100.0mL

pH 9.5 ± 0.2 at 25°C

Na_2CO_3 Solution:
Composition per 100mL:

Na_2CO_3	10.0g

Preparation of Na_2CO_3 Solution: Add Na_2CO_3 to distilled/deionized water and bring volume to 100.0mL. Mix thoroughly. Autoclave for 15 min at 15 psi pressure–121°C. Cool to 45°–50°C.

Preparation of Medium: Add components, except Na_2CO_3 solution, to distilled/deionized water and bring volume to 900.0mL. Mix thoroughly. Gently heat and bring to boiling. Autoclave for 15 min at 15 psi pressure–121°C. Cool to 45°–50°C. Aseptically add sterile Na_2CO_3 solution. Mix thoroughly. Adjust pH to 9.5 if necessary. Pour into sterile Petri dishes or distribute into sterile tubes.

Use: For the cultivation and maintenance of alkalophilic *Bacillus* species.

PYGV Medium

Composition per liter:

Agar	15.0g
Peptone	0.25g
Yeast extract	0.25g
Mineral solution	20.0mL
Glucose solution	10.0mL
Vitamin solution	5.0mL

pH 7.5 ± 0.2 at 25°C

Mineral Solution:

Composition per liter:

$MgSO_4 \cdot 7H_20$	29.7g
$NaMoO_4 \cdot 2H_2O$	12.67g
Nitrilotriacetic acid	10.0g
$CaCl_2 \cdot 2H_2O$	3.34g
$FeSO_4 \cdot 7H_2O$	0.1g
Metals “44” solution	50.0mL

Preparation of Mineral Solution: Add nitrilotriacetic acid to 500.0mL of distilled/deionized water. Dissolve by adjusting pH to 6.5 with KOH. Add remaining components. Readjust pH to 7.2 with H_2SO_4 or KOH. Add distilled/deionized water to 1.0L. Store at 5°C.

Metals “44”:

Composition per 100mL:

$ZnSO_4 \cdot 7H_2O$	1.1g
$FeSO_4 \cdot 7H_2O$	0.5g
EDTA	0.25g
$MnSO_4 \cdot 7H_2O$	0.154g
$CuSO_4 \cdot 5H_2O$	0.04g
$Co(NO_3)_2 \cdot 6H_2O$	0.025g
$Na_2B_4O_7 \cdot 10H_2O$	0.018g

Preparation of Metals “44”: Add components to distilled/deionized water and bring volume to 100.0mL. Mix thoroughly.

Glucose Solution:

Composition per 100mL:

D-Glucose	2.5g

Preparation of Glucose Solution: Add D-glucose to distilled/deionized water and bring volume to 100.0mL. Mix thoroughly. Filter sterilize.

Vitamin Solution:

Composition per liter:

Pyridoxin·HCl	0.02g
p-Aminobenzoic acid	0.01g
Ca-panthothenate	0.01g
Nicotinamide	0.01g
Riboflavin	0.01g
Thiamine·HCl	0.01g
Biotin	4.0mg
Folic acid	4.0mg
Vitamin B_{12}	0.2mg

Preparation of Vitamin Solution: Add components to distilled/deionized water and bring volume to 1.0L.

Preparation of Medium: Add components, except glucose solution and vitamin solution, to distilled/deionized water and bring volume to 985.0mL. Mix thoroughly. Gently heat and bring to boiling. Autoclave for 20 min at 15 psi pressure–121°C. Cool to 60°C. Aseptically add 10.0mL of sterile glucose solution and 5.0mL of sterile vitamin solution. Mix thoroughly. Pour into sterile Petri dishes or distribute into sterile tubes.

Use: For the enrichment of *Stella* species from polluted waters.

PY Basal Medium

Composition per 104mL:

Yeast extract	1.0g
$Cysteine \cdot HCl \cdot H_2O$	0.5g
Peptone	0.5g
Pancreatic digest of casein	0.5g
Resazurin	0.16mg
Salts solution	4.0mL

Salts Solution:

Composition per liter:

$NaHCO_3$	10.0g
NaCl	2.0g
K_2HPO_4	1.0g
KH_2PO_4	1.0g
$CaCl_2$	0.2g
$MgSO_4$	0.2g

Preparation of Salts Solution: Add components to distilled/deionized water and bring volume to 1.0L. Mix thoroughly.

Preparation of Medium: Add components to distilled/deionized water and bring volume to 104.0mL. Mix thoroughly. Distribute into tubes or flasks. Autoclave for 15 min at 15 psi pressure–121°C.

Use: For the identification of treponemes.

PY Carbohydrate Medium

Composition per 110mL:

Polypeptone™	1.0g
Yeast extract	1.0g
Glucose	0.4g
Maltose	0.4g
Ribose	0.4g
Starch, soluble	0.4g
Cysteine·HCl·H_2O	0.09g
Resazurin	0.16mg
Serum VFAH supplement	10.0mL
$NaHCO_3$ solution	10.0mL
Salts solution	4.0mL

pH 7.2–7.5 at 25°C

Serum VFAH Supplement:

Composition per 107.5mL:

Rabbit serum	100.0mL
Thiamine pyrophosphate solution	1.0mL
Heme solution	5.0mL
VFA solution	1.5mL

Preparation of Serum VFAH: Combine the four solutions. Mix thoroughly. Filter sterilize.

Thiamine Pyrophosphate Solution:

Composition per 10mL:

Thiamine pyrophosphate	0.05g

Preparation of Thiamine Pyrophosphate Solution: Add thiamine pyrophosphate to distilled/deionized water and bring volume to 10.0mL. Mix thoroughly. Filter sterilize.

Heme Solution:

Composition per 100mL:

Hemin	0.5g
NaOH (1*N* solution)	1.0mL

Preparation of Heme Solution: Add heme to NaOH solution. Mix thoroughly. Bring volume to 100.0mL with distilled/deionized water.

VFA Solution:

Composition per 100mL:

Acetic acid, glacial	5.0mL
n-Butyric acid	4.0mL
n-Valeric acid	1.0mL
Isobutyric acid	1.0mL
Isovaleric acid	1.0mL
n-Butyric acid	1.0mL

Preparation of VFA Solution: Add components to distilled/deionized water and bring volume to 100.0mL. Mix thoroughly. Adjust pH to 7.0.

$NaHCO_3$ Solution:

Composition per 10mL:

$NaHCO_3$	0.5g

Preparation of $NaHCO_3$ Solution: Add the $NaHCO_3$ to distilled/deionized water and bring volume to 10.0mL. Mix thoroughly. Filter sterilize.

Salts Solution:

Composition per 400mL:

K_2HPO_4	0.9g
NaCl	0.8g
KH_2PO_4	0.4g
$MnCl_2·4H_2O$	0.16g
$MgSO_4$	0.08g

Preparation of Salts Solution: Add components to distilled/deionized water and bring volume to 400.0mL. Mix thoroughly.

Preparation of Medium: Add components, except serum VFAH supplement and $NaHCO_3$ solution, to distilled/deionized water and bring volume to 90.0mL. Mix thoroughly. Gently heat and bring to boiling. Continue boiling until resazurin turns colorless indicating reduction. Autoclave for 15 min at 15 psi pressure–121°C. Cool to 25°C. Aseptically and anaerobically add 10.0mL of sterile serum VFAH supplement and 10.0mL of sterile $NaHCO_3$ solution. Anaerobically distribute into sterile tubes or flasks under 100% N_2.

Use: For the cultivation of oral treponemes.

PY Inositol Medium (Peptone Yeast Extract Inositol Medium)

Composition per liter:

i-Inositol	10.0g
Yeast extract	10.0g
Peptone	5.0g
Pancreatic digest of casein	5.0g
Cysteine·HCl·H_2O	0.5g
Salts solution	40.0mL
Hemin solution	10.0mL
Resazurin solution	4.0mL
Vitamin K_1 solution	0.2mL

pH 6.9 ± 0.2 at 25°C

Salts Solution:

Composition per liter:

$NaHCO_3$	10.0g
NaCl	2.0g
K_2HPO_4	1.0g
KH_2PO_4	1.0g
$CaCl_2$, anhydrous	0.2g
$MgSO_4$	0.2g

Preparation of Salts Solution: Add $CaCl_2$ and $MgSO_4$ to 300.0mL of distilled/deionized water. Mix thoroughly until dissolved. Bring volume to

800.0mL with distilled/deionized water. Add remaining components while stirring. Bring volume to 1.0L. Mix thoroughly. Store at 4°C.

Hemin Solution:

Composition per 100mL:

Hemin	0.050g
NaOH (1*N* solution)	1.0mL

Preparation of Hemin Solution: Add hemin to NaOH solution. Mix thoroughly. Adjust volume to 100.0mL with distilled/deionized water. Autoclave for 15 min at 15 psi pressure–121°C. Cool to 45°–50°C.

Resazurin Solution:

Composition per 44mL:

Resazurin	0.044g

Preparation of Resazurin Solution: Add resazurin to distilled/deionized water and bring volume to 44.0mL. Mix thoroughly.

Vitamin K_1 Solution:

Composition per 30mL:

Vitamin K_1	0.15g
Ethanol (95% solution)	30.0mL

Preparation of Vitamin K_1 Solution: Add vitamin K_1 to ethanol. Mix thoroughly. Store in brown bottle and keep under refrigeration. Discard after one month.

Preparation of Medium: Add components—except cysteine·HCl·H_2O, hemin solution, and vitamin K_1 solution—to distilled/deionized water and bring volume to 989.8mL. Mix thoroughly. Gently heat and bring to boiling under 80% N_2 + 10% CO_2 + 10% H_2. Continue boiling until resazurin turns colorless indicating reduction. Cool to 45°–50°C. Add the cysteine·HCl·H_2O, hemin solution, and vitamin K_1 solution. Adjust pH to 6.9, if necessary. Anaerobically distribute into tubes under 80% N_2 + 10% CO_2 + 10% H_2. Cap the tubes with rubber stoppers. Place tubes in a press. Autoclave for 15 min at 15 psi pressure–121°C with fast exhaust.

Use: For the cultivation and maintenance of *Eubacterium desmolans*.

PY Medium
See: Peptone Yeast Extract Medium

PY Medium with Fructose (Peptone Yeast Extract Medium with Fructose)

Composition per liter:

Fructose	10.0g
Yeast extract	10.0g
Peptone	5.0g
Pancreatic digest of casein	5.0g
Cysteine·HCl·H_2O	0.5g
Salts solution	40.0mL
Hemin solution	10.0mL
Resazurin solution	4.0mL
Vitamin K_1 solution	0.2mL

pH 6.9 ± 0.2 at 25°C

Salts Solution:

Composition per liter:

$NaHCO_3$	10.0g
NaCl	2.0g
K_2HPO_4	1.0g
KH_2PO_4	1.0g
$CaCl_2$, anhydrous	0.2g
$MgSO_4$	0.2g

Preparation of Salts Solution: Add $CaCl_2$ and $MgSO_4$ to 300.0mL of distilled/deionized water. Mix thoroughly until dissolved. Bring volume to 800.0mL with distilled/deionized water. Add remaining components while stirring. Bring volume to 1.0L. Mix thoroughly. Store at 4°C.

Hemin Solution:

Composition per 100mL:

Hemin	0.050g
NaOH (1*N* solution)	1.0mL

Preparation of Hemin Solution: Add hemin to NaOH solution. Mix thoroughly. Adjust volume to 100.0mL with distilled/deionized water. Autoclave for 15 min at 15 psi pressure–121°C. Cool to 45°–50°C.

Resazurin Solution:

Composition per 44mL:

Resazurin	0.044g

Preparation of Resazurin Solution: Add resazurin to distilled/deionized water and bring volume to 44.0mL. Mix thoroughly.

Vitamin K_1 Solution:

Composition per 30mL:

Vitamin K_1	0.15g
Ethanol (95% solution)	30.0mL

Preparation of Vitamin K_1 Solution: Add vitamin K_1 to ethanol. Mix thoroughly. Store in brown bottle and keep under refrigeration. Discard after one month.

Preparation of Medium: Add components—except cysteine·HCl·H_2O, hemin solution, and vitamin K_1 solution—to distilled/deionized water and bring volume to 989.8mL. Mix thoroughly. Gently heat and bring to boiling under 80% N_2 + 10% CO_2 + 10% H_2. Continue boiling until resazurin turns colorless indicating reduction. Cool to 45°–50°C. Add the cysteine·HCl·H_2O, hemin solution, and vitamin K_1 solution. Adjust pH to 6.9, if necessary. Anaerobically distribute into tubes under 80% N_2 + 10% CO_2 + 10% H_2. Cap the tubes with rubber stoppers. Place tubes in a press. Autoclave for 15 min at 15 psi pressure–121°C with fast exhaust.

Use: For the cultivation and maintenance of *Megasphaera cerevisiae.*

PY Medium with Glucose (Peptone Yeast Extract Medium with Glucose) (PYG Medium)

Composition per liter:

Glucose	10.0g
Yeast extract	10.0g
Peptone	5.0g
Pancreatic digest of casein	5.0g
Cysteine·HCl·H_2O	0.5g
Salts solution	40.0mL
Hemin solution	10.0mL
Resazurin solution	4.0mL
Vitamin K_1 solution	0.2mL

pH 6.9 ± 0.2 at 25°C

Salts Solution:

Composition per liter:

$NaHCO_3$	10.0g
NaCl	2.0g
K_2HPO_4	1.0g
KH_2PO_4	1.0g
$CaCl_2$, anhydrous	0.2g
$MgSO_4$	0.2g

Preparation of Salts Solution: Add $CaCl_2$ and $MgSO_4$ to 300.0mL of distilled/deionized water. Mix thoroughly until dissolved. Bring volume to 800.0mL with distilled/deionized water. Add remaining components while stirring. Bring volume to 1.0L. Mix thoroughly. Store at 4°C.

Hemin Solution:

Composition per 100mL:

Hemin	0.050g
NaOH (1*N* solution)	1.0mL

Preparation of Hemin Solution: Add hemin to NaOH solution. Mix thoroughly. Adjust volume to 100.0mL with distilled/deionized water. Autoclave for 15 min at 15 psi pressure–121°C. Cool to 45°–50°C.

Resazurin Solution:

Composition per 44mL:

Resazurin	0.044g

Preparation of Resazurin Solution: Add resazurin to distilled/deionized water and bring volume to 44.0mL. Mix thoroughly.

Vitamin K_1 Solution:

Composition per 30mL:

Vitamin K_1	0.15g
Ethanol (95% solution)	30.0mL

Preparation of Vitamin K_1 Solution: Add vitamin K_1 to ethanol. Mix thoroughly. Store in brown bottle and keep under refrigeration. Discard after one month.

Preparation of Medium: Add components—except cysteine·HCl·H_2O, hemin solution, and vitamin K_1 solution—to distilled/deionized water and bring volume to 989.8mL. Mix thoroughly. Gently heat and bring to boiling under 80% N_2 + 10% CO_2 + 10% H_2. Continue boiling until resazurin turns colorless indicating reduction. Cool to 45°–50°C. Add the cysteine·HCl·H_2O, hemin solution, and vitamin K_1 solution. Adjust pH to 6.9, if necessary. Anaerobically distribute into tubes under 80% N_2 + 10% CO_2 + 10% H_2. Cap the tubes with rubber stoppers. Place tubes in a press. Autoclave for 15 min at 15 psi pressure–121°C with fast exhaust.

Use: For the cultivation and maintenance of *Clostridium* species.

PY Salt Medium (Peptone Yeast Extract Salt Medium)

Composition per liter:

Peptone	9.0g
NaCl	5.0g
Yeast extract	5.0g

pH 7.2 ± 0.2 at 25°C

Preparation of Medium: Add components to distilled/deionized water and bring volume to 1.0L. Mix thoroughly. Distribute into tubes or flasks. Autoclave for 15 min at 15 psi pressure–121°C.

Use: For the cultivation and maintenance of *Bacillus brevis.*

PYEX Glucose Salt Medium (Peptone Yeast Extract Glucose Salt Medium)

Composition per liter:

Peptone	10.0g
NaCl	5.0g
Yeast extract	5.0g
Glucose	1.0g

Preparation of Medium: Add components to distilled/deionized water and bring volume to 1.0L. Mix thoroughly. Distribute into tubes or flasks. Autoclave for 15 min at 15 psi pressure–121°C.

Use: For the cultivation and maintenance of *Micrococcus luteus.*

PYG Broth (Peptone Yeast Extract Glucose Broth)

Composition per liter:

Peptone	20.0g
D-Glucose	10.0g
Yeast extract	10.0g
Cysteine·HCl·H_2O	0.5g
VPI salt solution	40.0mL
Resazurin solution	4.0mL

pH 7.2 ± 0.2 at 25°C

Resazurin Solution:

Composition per 44mL:

Resazurin	0.044g

Preparation of Resazurin Solution: Add resazurin to distilled/deionized water and bring volume to 44.0mL. Mix thoroughly.

VPI Salt Solution:

Composition per 40mL:

$CaCl_2$	0.2g
$MgSO_4$	0.2g
K_2HPO_4	1.0g
KH_2PO_4	1.0g

Preparation of VPI Salt Solution: Add $CaCl_2$ and $MgSO_4$ to 300.0mL of distilled/deionized water. Mix thoroughly until dissolved. Bring volume to 800.0mL with distilled/deionized water. Add remaining components while stirring. Bring volume to 1.0L. Mix thoroughly. Store at 4°C.

Preparation of Medium: Add components to distilled/deionized water and bring volume to 1.0L. Mix thoroughly. Distribute into screw-capped tubes in 7.0mL volumes. Autoclave for 15 min at 15 psi pressure–121°C. Cool to 45°–50°C under 100% N_2.

Use: For the cultivation of a wide variety of anaerobic bacteria.

PYG Medium
See: **PY Medium with Glucose**

PYG Medium (Peptone Yeast Extract Glucose Medium)

Composition per liter:

Proteose peptone	20.0g
Glucose	18.0g
Yeast extract	2.0g
Sodium citrate·$2H_2O$	1.0g
$MgSO_4·7H_2O$	0.98g
$Na_2HPO_4·7H_2O$	0.355g
KH_2PO_4	0.34g
$CaCl_2$	0.059g
$Fe(NH_4)_2(SO_4)_2·6H_2O$	0.02g

pH 6.5 ± 0.2 at 25°C

Preparation of Medium: Add components, except $CaCl_2$, to distilled/deionized water and bring volume to 900.0mL. Mix thoroughly until dissolved. Add $CaCl_2$. Mix thoroughly. Bring volume to 1.0L with distilled/deionized water. Distribute into screw-capped tubes in 5.0mL volumes. Autoclave for 15 min at 15 psi pressure–121°C.

Use: For the cultivation of *Acanthamoeba* species.

PYG Medium (Peptone Yeast Extract Glucose Medium)

Composition per liter:

Agar	15.0g
Glucose	0.25g
Peptone	0.25g
Yeast extract	0.25g
Hutner's modified salt solution	20.0mL
Vitamin solution	10.0mL

Hutner's Modified Salts Solution:

Composition per liter:

$MgSO_4·7H_2O$	29.7g
Nitrilotriacetic acid	10.0g
$CaCl_2·2H_2O$	3.34g
$FeSO_4·7H_2O$	0.10g
Metals "44"	50.0mL

Preparation of Hutner's Modified Salts Solution: Add nitrilotriacetic acid to 500.0mL of distilled/deionized water. Dissolve by adjusting pH to 6.5 with KOH. Add remaining components. Add distilled/deionized water to 1.0L.

Metals "44":

Composition per 100mL:

$ZnSO_4·7H_2O$	1.1g
$FeSO_4·7H_2O$	0.5g
EDTA	0.25g

$MnSO_4 \cdot 7H_2O$ 0.154g
$CuSO_4 \cdot 5H_2O$ 0.04g
$Co(NO_3)_2 \cdot 6H_2O$ 0.025g
$Na_2B_4O_7 \cdot 10H_2O$ 0.018g

Preparation of Metals "44": Add components to distilled/deionized water and bring volume to 100.0mL. Mix thoroughly.

Vitamin Solution:

Composition per liter:

Pyridoxine HCl 0.01g
Calcium pantothenate 5.0mg
Nicotinamide 5.0mg
Riboflavin 5.0mg
Thiamine HCl 5.0mg
Biotin 2.0mg
Folic acid 2.0mg
Vitamin B_{12} 0.1mg

Preparation of Vitamin Solution: Add components to distilled/deionized water and bring volume to 1.0L. Mix thoroughly. Filter sterilize.

Preparation of Medium: Add components, except vitamin solution, to distilled/deionized water and bring volume to 990.0mL. Mix thoroughly. Gently heat and bring to boiling. Autoclave for 15 min at 15 psi pressure–121°C. Cool to 45°–50°C. Aseptically add 10.0mL of sterile vitamin solution. Mix thoroughly. Pour into sterile Petri dishes or distribute into sterile tubes.

Use: For the isolation and cultivation of *Pasteuria ramosa*.

PYG Medium (Peptone Yeast Extract Glucose Medium) (ATCC Medium 663)

Composition per liter:

Agar 20.0g
Glucose 3.0g
Peptone 1.25g
Yeast extract 1.25g

Preparation of Medium: Add components to distilled/deionized water and bring volume to 1.0L. Mix thoroughly. Gently heat and bring to boiling. Distribute into tubes or flasks. Autoclave for 15 min at 15 psi pressure–121°C. Pour into sterile Petri dishes or leave in tubes.

Use: For the cultivation of *Eikenella corrodens*.

PYG Medium for *Spirillum* (Peptone Yeast Extract Glucose Medium for *Spirillum*)

Composition per liter:

Agar 15.0g
Peptone 10.0g
Yeast extract 5.0g
Glucose 3.0g

pH 7.2 ± 0.2 at 25°C

Glucose Solution:

Composition per 10mL:

D-Glucose 3.0g

Preparation of Glucose Solution: Add glucose to distilled/deionized water and bring volume to 10.0mL. Mix thoroughly. Filter sterilize.

Preparation of Medium: Add components, except glucose solution, to distilled/deionized water and bring volume to 990.0mL. Mix thoroughly. Gently heat and bring to boiling. Autoclave for 15 min at 15 psi pressure–121°C. Cool to 45°–50°C. Aseptically add sterile glucose solution. Mix thoroughly. Pour into sterile Petri dishes or distribute into sterile tubes.

Use: For the cultivation and maintenance of *Spirillum pleomorphum*.

PYGHS Medium

Composition per 60mL:

Beef heart, solids from infusion 10.0g
Glucose 1.0g
Polypeptone™ 1.0g
Yeast extract 1.0g
$Cysteine \cdot HCl \cdot H_2O$ 0.9g
Ionagar No. 2 .... 0.72g
$NaHCO_3$ 0.5g
Tryptose 0.2g
NaCl 0.1g
Resazurin 0.16mg
Rabbit serum, inactivated 5.0mL
Salts solution 5.0mL

pH 7.2–7.5 at 25°C

Salts Solution:

Composition per 400mL:

K_2HPO_4 .... 0.9g
NaCl 0.8g
KH_2PO_4 .... 0.4g
$MnCl_2 \cdot 4H_2O$ 0.16g
$MgSO_4$ 0.08g

Preparation of Salts Solution: Add components to distilled/deionized water and bring volume to 1.0L. Mix thoroughly.

Preparation of Medium: Add components, except rabbit serum, to distilled/deionized water and bring volume to 55.0mL. Mix thoroughly. Gently heat and bring to boiling. Autoclave for 15 min at 15 psi pressure–121°C. Cool to 45°–50°C. Aseptically add 5.0mL of sterile rabbit serum. Mix thoroughly.

Use: For the cultivation of treponemes.

PYGM Medium (Peptone Yeast Extract Glucose Maltose Medium)

Composition per 261mL:

Peptone	5.0g
Yeast extract	2.5g
Glucose	1.25g
Maltose	1.25g
Cysteine·HCl·H_2O	0.125g
Salts solution	10.0mL
Resazurin (0.025% solution)	1.0mL

Salts Solution:

Composition per 100mL:

$NaHCO_3$	1.0g
NaCl	0.2g
K_2HPO_4	0.1g
KH_2PO_4	0.1g
$CaCl_2$, anhydrous	0.02g
$MgSO_4$	0.02g
H_2SO_4 (50% solution)	0.3mL
$Na_2MoO_4 \cdot 2H_2O$	1.0μg
$CoCl_2 \cdot 6H_2O$	1.0μg

Preparation of Salts Solution: Add components to distilled/deionized water and bring volume to 100.0mL. Mix thoroughly.

Preparation of Medium: Add components to distilled/deionized water and bring volume to 261.0mL. Mix thoroughly. Distribute into tubes or flasks. Autoclave for 15 min at 15 psi pressure–121°C.

Use: For the cultivation of *Bacteroides praeacutus, Eubacterium nitritogenes, Lactobacillus ruminis,* and *Tissierella praeacuta.*

PYGV Marine Medium (Peptone Yeast Extract Glucose Vitamin Marine Medium)

Composition per liter:

Agar	15.0g
Peptone	0.25g
Yeast extract	0.25g
Mineral salt solution	20.0mL
Glucose solution	10.0mL
Vitamin solution	5.0mL

pH 7.5 ± 0.2 at 25°C

Mineral Salt Solution:

Composition per liter:

$MgSO_4 \cdot 7H_2O$	29.7g
Nitrilotriacetic acid	10.0g
$CaCl_2 \cdot 2H_2O$	3.34g
$FeSO_4 \cdot 7H_2O$	0.099g
$Na_2MoO_4 \cdot 2H_2O$	0.013g
Metals "44"	50.0mL

Preparation of Mineral Salt Solution: Add nitrilotriacetic acid to 500.0mL of distilled/deionized water. Dissolve by adjusting pH to 6.5 with KOH. Add remaining components. Add distilled/deionized water to 1.0L. Readjust pH to 7.2.

Metals "44":

Composition per 100mL:

$ZnSO_4 \cdot 7H_2O$	1.1g
$FeSO_4 \cdot 7H_2O$	0.5g
EDTA	0.25g
$MnSO_4 \cdot 7H_2O$	0.154g
$CuSO_4 \cdot 5H_2O$	0.04g
$Co(NO_3)_2 \cdot 6H_2O$	0.025g
$Na_2B_4O_7 \cdot 10H_2O$	0.018g

Preparation of Metals "44": Add a few drops of H_2SO_4 to distilled/deionized water to inhibit precipitate formation. Add components to acidified distilled/deionized water and bring volume to 100.0mL. Mix thoroughly.

Glucose Solution:

Composition per 100mL:

D-Glucose	2.5g

Preparation of Glucose Solution: Add glucose to distilled/deionized water and bring volume to 100.0mL. Mix thoroughly. Filter sterilize.

Vitamin Solution:

Composition per liter:

Pyridoxine·HCl	0.02g
p-Aminobenzoic acid	0.01g
Calcium D-pantothenate	0.01g
Nicotinamide	0.01g
Riboflavin	0.01g
Thiamine·HCl	0.01g
Biotin	4.0mg
Folic acid	4.0mg
Cyanocobalamin	0.2mg

Preparation of Vitamin Solution: Add components to distilled/deionized water and bring volume to 1.0L. Mix thoroughly. Filter sterilize.

Preparation of Medium: Add components, except glucose solution and vitamin solution, to sea water and bring volume to 985.0mL. Mix thoroughly. Gently heat and bring to boiling. Autoclave for 15 min at 15 psi pressure–121°C. Cool to 45°–50°C. Aseptically add 10.0mL of sterile glucose solution and 5.0mL of sterile vitamin solution. Mix thoroughly. Adjust pH to 7.5 with sterile KOH, if necessary. Pour into sterile Petri dishes or distribute into sterile tubes.

Use: For the cultivation and maintenance of *Planctomyces brasiliensis.*

PYGV Medium (Peptone Yeast Extract Glucose Vitamin Medium)

Composition per liter:

Agar	15.0g
Peptone	0.25g
Yeast extract	0.25g
Mineral salt solution	20.0mL
Glucose solution	10.0mL
Vitamin solution	5.0mL

pH 7.5 ± 0.2 at 25°C

Mineral Salt Solution:

Composition per liter:

$MgSO_4 \cdot 7H_2O$	29.7g
Nitrilotriacetic acid	10.0g
$CaCl_2 \cdot 2H_2O$	3.34g
$FeSO_4 \cdot 7H_2O$	99.0mg
$Na_2MoO_4 \cdot 2H_2O$	12.67mg
Metals "44"	50.0mL

Preparation of Mineral Salt Solution: Add nitrilotriacetic acid to 500.0mL of distilled/deionized water. Dissolve by adjusting pH to 6.5 with KOH. Add remaining components. Add distilled/deionized water to 1.0L. Readjust pH to 7.2.

Metals "44":

Composition per 100mL:

$ZnSO_4 \cdot 7H_2O$	1.1g
$FeSO_4 \cdot 7H_2O$	0.5g
EDTA	0.25g
$MnSO_4 \cdot 7H_2O$	0.154g
$CuSO_4 \cdot 5H_2O$	0.04g
$Co(NO_3)_2 \cdot 6H_2O$	0.025g
$Na_2B_4O_7 \cdot 10H_2O$	0.018g

Preparation of Metals "44": Add a few drops of H_2SO_4 to distilled/deionized water to inhibit precipitate formation. Add components to acidified distilled/deionized water and bring volume to 100.0mL. Mix thoroughly.

Glucose Solution:

Composition per 100mL:

D-Glucose	2.5g

Preparation of Glucose Solution: Add glucose to distilled/deionized water and bring volume to 100.0mL. Mix thoroughly. Filter sterilize.

Vitamin Solution:

Composition per liter:

Pyridoxine·HCl	0.02g
p-Aminobenzoic acid	0.01g
Calcium D-pantothenate	0.01g
Nicotinamide	0.01g
Riboflavin	0.01g
Thiamine·HCl	0.01g
Biotin	4.0mg
Folic acid	4.0mg
Cyanocobalamin	0.2mg

Preparation of Vitamin Solution: Add components to distilled/deionized water and bring volume to 1.0L. Mix thoroughly. Filter sterilize.

Preparation of Medium: Add components, except glucose solution and vitamin solution, to distilled/deionized water and bring volume to 985.0mL. Mix thoroughly. Gently heat and bring to boiling. Autoclave for 15 min at 15 psi pressure–121°C. Cool to 45°–50°C. Aseptically add 10.0mL of sterile glucose solution and 5.0mL of sterile vitamin solution. Mix thoroughly. Adjust pH to 7.5 with sterile KOH, if necessary. Pour into sterile Petri dishes or distribute into sterile tubes.

Use: For the cultivation and maintenance of *Blastobacter aggregatus, Blastobacter capsulatus, Blastobacter denitrificans,* and *Planctomyces limnophilus.*

Pyrazinamidase Agar (Pyrazinamide Medium)

Composition per liter:

Pancreatic digest of casein	15.0g
Agar	15.0g
Papaic digest of soybean meal	5.0g
NaCl	5.0g
Yeast extract	3.0g
Pyrazinecarboxamide	1.0g
Tris(hydroxymethyl)aminomethane maleate buffer (0.2M, pH 6.0)	1.0L

pH 6.0 ± 0.2 at 25°C

Preparation of Medium: Combine components. Mix thoroughly. Gently heat and bring to boiling. Distribute into tubes in 5.0mL volumes. Autoclave for 15 min at 15 psi pressure–121°C. Allow tubes to cool in a slanted position.

Use: For the cultivation, differentiation and maintenance of pathogenic *Yersinia* species. Bacteria that produce pyrazinamidase turn the medium pink.

Pyrazinamide Medium

Composition per liter:

Agar	15.0g
Na_2HPO_4	2.5g
Sodium pyruvate	2.0g
L-Asparagine	2.0g
KH_2PO_4	1.0g
Pancreatic digest of casein	0.5g
Tween™ 80	0.2g
Pyrazinamide	0.1g
$CaCl_2 \cdot 2H_2O$	0.5mg
$CuSO_4 \cdot 5H_2O$	0.1mg
$ZnSO_4 \cdot 7H_2O$	0.1mg
Ferric ammonium citrate	0.05g
$MgSO_4 \cdot 7H_2O$	0.01g

pH 6.6 ± 0.2 at 25°C

Preparation of Medium: Combine components. Mix thoroughly. Gently heat and bring to boiling. Distribute into tubes in 5.0mL volumes. Autoclave for 15 min at 15 psi pressure–121°C. Allow tubes to cool in a slanted position.

Use: For cultivation and differentiation of *Corynebacterium* species and related organisms. Bacteria that produce pyrazinamidase turn the medium pink.

Pyridoxine Assay Medium

Composition per liter:

Sucrose	30.0g
Ammonium tartrate	10.0g
KH_2PO_4	5.0g
Sodium dihydrogen citrate	4.0g
$MgSO_4 \cdot 7H_2O$	1.0g
$CaCl_2 \cdot 2H_2O$	0.2g
NaCl	0.2g
Choline chloride	0.01g
$FeCl_3$	0.01g
Thiamine·HCl	0.01g
$ZnSO_4 \cdot 7H_2O$	4.0mg
Nicotinic acid	2.0mg
Calcium pantothenate	1.0mg
Riboflavin	1.0mg
p-Aminobenzoic acid	200μg
Biotin	8μg

pH 4.5 ± 0.2 at 25°C

Source: Available as a prepared medium from Difco Laboratories.

Preparation of Medium: Add components to distilled/deionized water and bring volume to 1.0L. Mix thoroughly. Gently heat and bring to boiling. Continue boiling for 2–3 min. Distribute into tubes in 5.0mL volumes. Add standard solutions and test solutions to each tube. Bring volume of each tube to 10.0mL with distilled/deionized water. Autoclave for 15 min at 15 psi pressure–121°C.

Use: For the microbiological assay of pyridoxine using *Neurospora sitophila* as the test organism.

Pyridoxine Y Medium

Composition per liter:

Glucose	40.0g
$(NH_4)_2SO_4$	4.0g
L-Asparagine	4.0g
KH_2PO_4	3.0g
$MgSO_4 \cdot 7H_2O$	1.0g
$CaCl_2 \cdot 2H_2O$	0.49g
DL-Isoleucine	0.04g
DL-Methionine	0.04g
DL-Tryptophan	0.04g
DL-Valine	0.04g
L-Histidine·HCl	0.02g
Riboflavin	0.02g
Biotin salt	8.0mg
Inositol	5.0mg
$FeSO_4 \cdot 7H_2O$	0.5mg
Calciun pantothenate	0.4mg
Nicotinic acid	0.4mg
Thiamine·HCl	0.4mg
H_3BO_3	0.2mg
KI	0.2mg
$CuSO_4 \cdot 5H_2O$	0.09mg
$MnSO_4 \cdot H_2O$	0.08mg
$ZnSO_4 \cdot 7H_2O$	0.08mg
$(NH_4)_2MoO_4$	0.04mg

pH 4.4 ± 0.2 at 25°C

Source: Available as a prepared medium from Difco Laboratories.

Preparation of Medium: Add components to distilled/deionized water and bring volume to 1.0L. Mix thoroughly. Gently heat and bring to boiling. Continue boiling for 2–3 min. Distribute into tubes in 5.0mL volumes. Add standard solutions and test solutions to each tube. Bring volume of each tube to 10.0mL with distilled/deionized water. Autoclave for 15 min at 15 psi pressure–121°C.

Use: For the microbiological assay of pyridoxine using *Saccharomyces uvarum* as the test organism.

Pyrrolidone Agar

Composition per liter:

Noble agar	21.0g
K_2HPO_4	5.65g

KH_2PO_4 .... 2.95g
$MgSO_4 \cdot 7H_2O$ 1.0g
Pyrrolidone carboxylic acid solution 30.0mL
NaOH solution 30.0mL
Trace metals 6.3mL

Pyrrolidone Carboxylic Acid Solution:
Composition per 300mL:
Pyrrolidone carboxylic acid 50.0g

Preparation of Pyrrolidone Carboxylic Acid Solution: Add pyrrolidone carboxylic acid to distilled/deionized water and bring volume to 300.0mL. Mix thoroughly. Filter sterilize.

NaOH Solution:
Composition per 100mL:
NaOH 5.0g

Preparation of Resazurin Solution: Add NaOH to distilled/deionized water and bring volume to 100.0mL. Mix thoroughly. Filter sterilize.

Trace Metals:
Composition per 100mL:
$FeSO_4 \cdot 7H_2O$ 0.18g
$MnCl_2 \cdot 2H_2O$ 0.13g
$CuSO_4 \cdot 5H_2O$ 0.10g
$ZnSO_4 . 7H_2O$ 0.02g

Preparation of Trace Metals: Add a few drops of H_2SO_4 to distilled/deionized water to inhibit precipitate formation. Add components to acidified distilled/deionized water and bring volume to 100.0mL. Mix thoroughly.

Preparation of Medium: Add components, except pyrrolidone carboxylic acid solution and NaOH solution, to distilled/deionized water and bring volume to 940.0mL. Mix thoroughly. Gently heat and bring to boiling. Autoclave for 15 min at 15 psi pressure–121°C. Cool to 45°–50°C. Aseptically add 30.0mL of sterile pyrrolidone carboxylic acid solution and 30.0mL of sterile NaOH solution. Mix thoroughly. Pour into sterile Petri dishes or distribute into sterile tubes.

Use: For the cultivation and maintenance of *Pseudomonas fluorescens*.

Pyruvate Utilization Medium

Composition per liter:
Sodium pyruvate 10.0g
Pancreatic digest of casein 10.0g
K_2HPO_4 .... 5.0g
NaCl 5.0g
Yeast extract 5.0g
Bromthymol Blue 0.1g
pH 7.1–7.4 at 25°C

Preparation of Medium: Add components to distilled/deionized water and bring volume to 1.0L. Mix thoroughly. Gently heat and bring to boiling. Adjust pH to 7.1–7.4. Distribute into tubes in 5.0mL volumes. Autoclave for 15 min at 15 psi pressure–121°C.

Use: For the cultivation of bacteria that can metabolize pyruvate. Bacteria that can utilize pyruvate turn the medium yellow.

Pyruvic Acid Egg Medium

Composition per 1640mL:
KH_2PO_4 .... 11.4g
Na_2HPO_4 .... 6.0g
Pyruvic acid 3.0g
$MgSO_4 \cdot 7H_2O$ 0.3g
Malachite Green 0.125g
D-Glucose 10.0g
Egg, homogenized whole 1.0L
Penicillin solution 10.0mL

Source: This medium is available as a prepared medium from Oxoid Unipath.

Penicillin Solution:
Composition per 10mL:
Penicillin G 100,000U

Preparation of Penicillin Solution: Add penicillin G to distilled/deionized water and bring volume to 10.0mL. Mix thoroughly. Filter sterilize.

Homogenized Whole Egg:
Composition per liter:
Whole eggs 18-24

Preparation of Homogenized Whole Egg: Use fresh eggs, less than 1 week old. Scrub the shells with soap. Let stand in a soap solution for 30 min. Rinse in running water. Soak eggs in 70% ethanol for 15 min. Break the eggs into a sterile container. Homogenize by shaking. Filter through four layers of sterile cheesecloth into a sterile graduated cylinder. Measure out 1.0L.

Preparation of Medium: Add components, except homogenized whole egg and penicillin solution, to distilled/deionized water and bring volume to 630.0mL. Mix thoroughly. Autoclave for 15 min at 15 psi pressure–121°C. Cool to 45°–50°C. Aseptically add homogenized whole egg and penicillin solution to cooled sterile basal medium. Mix thoroughly. Aseptically distribute into sterile tubes. Inspissate at 85°–90°C (moist heat) for 45 min.

Use: For the isolation and cultivation of *Mycobacterium* species, especially ones that are drug-resistant and difficult to grow.

PYS Agar (Peptone Yeast Extract Salt Agar)

Composition per liter:

Agar	15.0g
Peptone	15.0g
NaCl	5.0g
Yeast extract	5.0g

pH 7.2–7.4 at 25°C

Preparation of Medium: Add components to tap water and bring volume to 1.0L. Mix thoroughly. Gently heat and bring to boiling. Distribute into tubes or flasks. Autoclave for 15 min at 15 psi pressure–121°C. Pour into sterile Petri dishes or leave in tubes.

Use: For the cultivation and maintenance of *Actinomadura madurae*.

Quinolinic Acid Medium

Composition per liter:

Quinolinic acid	1.5g
K_2HPO_4	1.1g
NH_4NO_3	1.0g
KH_2PO_4	0.5g
$MgSO_4 \cdot 7H_2O$	0.25g

Preparation: Add quinolinic acid to distilled/deionized water and bring volume to 900.0mL. Mix thoroughly. Bring pH to 7.0 with NaOH. Add other components. Bring volume to 1.0L. Mix thoroughly. Distribute into tubes or flasks. Autoclave for 15 min at 15 psi pressure–121°C.

Use: For the cultivation of microorganisms that can utilize quinolinic acid as sole carbon source.

R Agar for Phage Lysates

Composition per liter:

Agar	12.0g
Pancreatic digest of casein	10.0g
NaCl	8.0g
Yeast extract	1.0g
Glucose solution	5.0mL
$CaCl_2 \cdot 2H_2O$ solution	2.0mL

pH 6.8 ± 0.2 at 25°C

Glucose Solution:

Composition per 10mL:

D-Glucose	2.0g

Preparation of Glucose Solution: Add glucose to distilled/deionized water and bring volume to 10.0mL. Mix thoroughly. Filter sterilize.

$CaCl_2 \cdot 2H_2O$ Solution:

Composition per 10mL:

$CaCl_2 \cdot 2H_2O$	1.47g

Preparation of $CaCl_2 \cdot 2H_2O$ Solution: Add the $CaCl_2 \cdot 2H_2O$ to distilled/deionized water and bring volume to 10.0mL. Mix thoroughly. Filter sterilize.

Preparation of Medium: Add components, except glucose solution and $CaCl_2 \cdot 2H_2O$ solution, to distilled/deionized water and bring volume to 993.0mL. Mix thoroughly. Gently heat and bring to boiling. Autoclave for 15 min at 15 psi pressure–121°C. Cool to 45°–50°C. Aseptically add 5.0mL of glucose solution and 2.0mL of $CaCl_2 \cdot 2H_2O$ solution. Mix thoroughly. Pour into sterile Petri dishes.

Use: For the cultivation of bacterial host cells in the production of bacteriophage lysates.

R Broth for Phage Lysates

Composition per liter:

Pancreatic digest of casein	10.0g
NaCl	8.0g
Yeast extract	1.0g
Glucose solution	5.0mL
$CaCl_2 \cdot 2H_2O$ solution	2.0mL

pH 6.8 ± 0.2 at 25°C

Glucose Solution:

Composition per 10mL:

D-Glucose	2.0g

Preparation of Glucose Solution: Add glucose to distilled/deionized water and bring volume to 10.0mL. Mix thoroughly. Filter sterilize.

$CaCl_2 \cdot 2H_2O$ Solution:

Composition per 10mL:

$CaCl_2 \cdot 2H_2O$	1.47g

Preparation of $CaCl_2 \cdot 2H_2O$ Solution: Add the $CaCl_2 \cdot 2H_2O$ to distilled/deionized water and bring volume to 10.0mL. Mix thoroughly. Filter sterilize.

Preparation of Medium: Add components, except glucose solution and $CaCl_2 \cdot 2H_2O$ solution, to distilled/deionized water and bring volume to 993.0mL. Mix thoroughly. Gently heat and bring to boiling. Autoclave for 15 min at 15 psi pressure–121°C. Cool to 45°–50°C. Aseptically add 5.0mL of glucose solution and 2.0mL of $CaCl_2 \cdot 2H_2O$ solution. Mix thoroughly. Aseptically distribute into sterile tubes or flasks.

Use: For the cultivation of bacterial host cells in the production of bacteriophage lysates.

R Medium

Composition per liter:

Agar	30.0g
$NaHCO_3$	8.0g
K_2HPO_4	3.0g
Glucose	2.5g
Glutamine	0.61g
Serine	0.24g
Leucine	0.23g
Lysine	0.23g
Asparagine	0.18g
Valine	0.17g
Isoleucine	0.17g
Tyrosine	0.14g
Arginine·HCl	0.125g
Phenylalanine	0.125g
Threonine	0.12g
Methionine	0.073g
Glycine	0.065g
Histidine·HCl	0.055g
Proline	0.043g
Tryptophan	0.035g
Cystine	0.025g
$MgSO_4 \cdot H_2O$	9.9mg
$CaCl_2 \cdot 2H_2O$	7.4mg
Adenine sulfate	2.1mg
Uracil	1.4mg
Thiamine·HCl	1.0mg
$MnSO_4 \cdot H_2O$	0.9mg

pH 8.0 ± 0.2 at 25°C

Preparation of Medium: Add components, except agar, to distilled/deionized water and bring volume to 500.0mL. Mix thoroughly. Filter sterilize. Warm to 45°–50°C. Add agar to distilled/deionized water and bring volume to 500.0mL. Mix thoroughly. Autoclave for 15 min at 15 psi pressure–121°C. Cool to 45°–50°C. Aseptically combine both solutions. Mix thoroughly. Pour into sterile Petri dishes or distribute into sterile tubes.

Use: For the cultivation of *Bacillus anthracis* especially for the production of toxins.

R2A Agar

Composition per liter:

Agar	15.0g
Yeast extract	0.5g
Acid hydrolysate of casein	0.5g
Glucose	0.5g
Soluble starch	0.5g
K_2HPO_4	0.3g
Sodium pyruvate	0.3g
Pancreatic digest of casein	0.25g
Peptic digest of animal tissue	0.25g
$MgSO_4$, anhydrous	0.024g

pH 7.2 ± 0.2 at 25°C

Source: Available as a premixed powder from BBL Microbiology Systems and Difco Laboratories.

Preparation of Medium: Add components to distilled/deionized water and bring volume to 1.0L. Mix thoroughly. Gently heat with mixing and bring to boiling. Distribute into tubes or flasks. Autoclave for 15 min at 15 psi pressure–121°C. Do not overheat. Pour into sterile Petri dishes or leave in tubes.

Use: For use in standard methods for pour plate, spread plate and membrane filter analysis to enumerate heterotrophic bacteria from potable waters.

R2YE Medium

Composition per 1062.2mL:

Thiostrepton	50.0mg
Basal solution	800.0mL
TES (N-tris[hydroxymethyl] methyl-2-amino–ethanesulfonic acid) buffer	100.0mL
$CaCl_2 \cdot 2H_2O$ solution	80.2mL
Yeast extract solution	50.0mL
L-Proline solution	15.0mL
KH_2PO_4 solution	10.0mL
NaOH solution	5.0mL
Trace element solution	2.0mL

pH 7.2 ± 0.2 at 25°C

Basal Solution:

Composition per 800mL:

Sucrose	103.0g
$MgCl_2 \cdot 6H_2O$	10.12g
D-Glucose	10.0g
K_2SO_4	0.25g
Casamino acids	0.1g

Preparation of Basal Solution: Add components to distilled/deionized water and bring volume to 800.0mL. Mix thoroughly. Autoclave for 15 min at 15 psi pressure–121°C. Cool to 25°C.

TES Buffer:

Composition per liter:

TES (N-tris[hydroxymethyl] methyl-2-amino–ethanesulfonic acid) buffer	57.3g

Preparation of TES Buffer: Add TES to distilled/deionized water and bring volume to 1.0L. Mix thoroughly. Adjust pH to 7.2. Filter sterilize.

$CaCl_2 \cdot 2H_2O$ Solution:

Composition per 100mL:

$CaCl_2 \cdot 2H_2O$	3.68g

Preparation of $CaCl_2 \cdot 2H_2O$ Solution: Add the $CaCl_2 \cdot 2H_2O$ to distilled/deionized water and bring volume to 100.0mL. Mix thoroughly. Filter sterilize.

Yeast Extract Solution:

Composition per 100mL:

Yeast extract	10.0g

Preparation of Yeast Extract Solution: Add yeast extract to distilled/deionized water and bring volume to 100.0mL. Mix thoroughly. Filter sterilize.

L-Proline Solution:

Composition per 100mL:

L-Proline	20.0g

Preparation of L-Proline Solution: Add the proline to distilled/deionized water and bring volume to 100.0mL. Mix thoroughly. Filter sterilize.

KH_2PO_4 Solution:

Composition per 100mL:

KH_2PO_4	0.5g

Preparation of KH_2PO_4 Solution Solution: Add the KH_2PO_4 to distilled/deionized water and bring volume to 100.0mL. Mix thoroughly. Filter sterilize.

NaOH Solution:

Composition per 100mL:

NaOH	40.0g

Preparation of NaOH Solution: Add the NaOH to distilled/deionized water and bring volume to 100.0mL. Mix thoroughly. Filter sterilize.

Trace Element Solution:

Composition per liter:

$FeCl_3 \cdot 6H_2O$	0.2g
$ZnCl_2$	0.04g
$CuCl_2 \cdot 2H_2O$	0.01g
$MnCl_2 \cdot 4H_2O$	0.01g
$Na_2B_4O_7 \cdot 10H_2O$	0.01g
$(NH_4)_6MoO_7O_{24} \cdot 4H_2O$	0.01g

Preparation of Trace Element Solution: Add components to distilled/deionized water and bring volume to 1.0L. Mix thoroughly. Filter sterilize.

Preparation of Medium: To 800.0mL of cooled, sterile basal solution aseptically add the remaining components. Mix thoroughly. Aseptically distribute into sterile tubes or flasks.

Use: For the cultivation and maintenance of *Streptomyces lividans*.

R3A Agar

Composition per liter:

Agar	15.0g
Yeast extract	1.0g
Acid hydrolysate of casein	1.0g
Glucose	1.0g
Soluble starch	1.0g
K_2HPO_4	0.6g
Sodium pyruvate	0.6g
Pancreatic digest of casein	0.5g
Peptic digest of animal tissue	0.5g
$MgSO_4$, anhydrous	0.048g

pH 7.2 ± 0.2 at 25°C

Source: This medium is available as a premixed powder from BBL Microbiology Systems.

Preparation of Medium: Add components to distilled/deionized water and bring volume to 1.0L. Mix thoroughly. Gently heat with mixing and bring to boiling. Distribute into tubes or flasks. Autoclave for 15 min at 15 psi pressure–121°C. Do not overheat. Pour into sterile Petri dishes or leave in tubes.

Use: For the cultivation and maintenance of heterotrophic bacteria from potable waters.

R8AH Medium

Composition per liter:

Malic acid	2.5g
$(NH_4)_2SO_4$	1.25g
Yeast extract	1.0g
K_2HPO_4	0.9g
KH_2PO_4	0.6g
$MgSO_4 \cdot 7H_2O$	0.2g
$CaCl_2 \cdot 2H_2O$	0.07g
EDTA	0.02g
Ferric citrate	0.01g
Vitamin solution	7.5mL
Trace element solution	1.0mL

pH 6.9 ± 0.2 at 25°C

Trace Element Solution:

Composition per 100mL:

Ferric citrate	0.3g
EDTA	0.05g
$CaCl_2 \cdot 2H_2O$	0.02g
$MnSO_4 \cdot H_2O$	2.0mg
$(NH_4)_6Mo_7O_{24} \cdot 4H_2O$	2.0mg
H_3BO_3	1.0mg
$CuSO_4 \cdot 5H_2O$	1.0mg
$ZnSO_4$	1.0mg

Preparation of Trace Element Solution: Add components to distilled/deionized water and bring volume to 100.0mL. Mix thoroughly.

Vitamin Solution:

Composition per liter:

Thiamine·HCl	0.4g
Nicotinic acid	0.2g
Nicotinamide	0.2g
Biotin	8.0mg

Preparation of Vitamin Solution: Add components to distilled/deionized water and bring volume to 1.0L. Mix thoroughly.

Preparation of Medium: Add malic acid to approximately 500.0mL of distilled/deionized water. Adjust pH to 6.9 with NaOH. Add remaining components. Bring volume to 1.0L with distilled/deionized water. Mix thoroughly. Adjust pH to 6.9. Distribute into tubes or flasks. Autoclave for 15 min at 15 psi pressure–121°C.

Use: For the cultivation and maintenance of *Rhodobacter sphaeroides, Rhodocyclus tenuis, Rhodopseudomonas rutila, Rhodospirillum photometricum,* and *Rhodospirillum rubrum.*

R70-2 Agar, Modified with Fructose

Composition per liter:

Fructose	20.0g
Agar	15.0g
Yeast extract	5.0g
Dimethyl glutaric acid	4.01g
$(NH_4)_2SO_4$	3.30g
Trisodium citrate·$2H_2O$	1.18g
KH_2PO_4	1.00g
$MgSO_4 \cdot 7H_2O$	0.25g
Wolfe's vitamin solution	10.0mL
100X modified salts	10.0mL

pH 5.0 ± 0.2 at 25°C

Wolfe's Vitamin Solution:

Composition per liter:

Pyridoxine·HCl	0.01g
Thiamine·HCl	5.0mg
Riboflavin	5.0mg
Nicotinic acid	5.0mg
Calcium pantothenate	5.0mg

p-Aminobenzoic acid 5.0mg
Thioctic acid 5.0mg
Biotin 2.0mg
Folic acid 2.0mg
Cyanocobalamin 0.1mg

Preparation of Wolfe's Vitamin Solution: Add components to distilled/deionized water and bring volume to 1.0L. Mix thoroughly. Filter sterilize.

100X Modified Salts:

Composition per liter:

$CaCl_2 \cdot 2H_2O$ 1.47g
$FeCl_3 \cdot 6H_2O$ 0.27g
$ZnSO_4 \cdot 7H_2O$ 0.144g
$MnSO_4 \cdot H_2O$ 0.085g
$CoCl_2 \cdot 6H_2O$ 0.024g
$NiCl_2 \cdot 6H_2O$ 0.024g
$Na_2MoO_4 \cdot 2H_2O$ 0.024g
$CuSO_4 \cdot 5H_2O$ 0.016g
HCl, concentrated 4.1mL

Preparation of 100X Modified Salts: Add components to distilled/deionized water and bring volume to 1.0L. Mix thoroughly. Filter sterilize.

Preparation of Medium: Add components, except Wolfe's vitamin solution and 100X modified salts, to distilled/deionized water and bring volume to 980.0mL. Mix thoroughly. Adjust pH to 5.0. Gently heat and bring to boiling. Autoclave for 15 min at 15 psi pressure–121°C. Cool to 45°–50°C. Aseptically add sterile Wolfe's vitamin solution and 100X modified salts. Mix thoroughly. Pour into sterile Petri dishes or distribute into sterile tubes.

Use: For the cultivation of *Acetobacter xylinum.*

R70-2 Agar, Modified with Glucose

Composition per liter:

Glucose 20.0g
Agar 15.0g
Yeast extract 5.0g
Dimethyl glutaric acid 4.01g
$(NH_4)_2SO_4$ 3.30g
Trisodium citrate·$2H_2O$ 1.18g
KH_2PO_4 .... 1.00g
$MgSO_4 \cdot 7H_2O$ 0.25g
Wolfe's vitamin solution 10.0mL
100X modified salts 10.0mL

pH 5.0 ± 0.2 at 25°C

Wolfe's Vitamin Solution:

Composition per liter:

Pyridoxine·HCl 0.01g
Thiamine·HCl 5.0mg
Riboflavin 5.0mg
Nicotinic acid 5.0mg
Calcium pantothenate 5.0mg
p-Aminobenzoic acid 5.0mg
Thioctic acid 5.0mg
Biotin 2.0mg
Folic acid 2.0mg
Cyanocobalamin 0.1mg

Preparation of Wolfe's Vitamin Solution: Add components to distilled/deionized water and bring volume to 1.0L. Mix thoroughly. Filter sterilize.

100X Modified Salts:

Composition per liter:

$CaCl_2 \cdot 2H_2O$ 1.47g
$FeCl_3 \cdot 6H_2O$ 0.27g
$ZnSO_4 \cdot 7H_2O$ 0.144g
$MnSO_4 \cdot H_2O$ 0.085g
$CoCl_2 \cdot 6H_2O$ 0.024g
$NiCl_2 \cdot 6H_2O$ 0.024g
$Na_2MoO_4 \cdot 2H_2O$ 0.024g
$CuSO_4 \cdot 5H_2O$ 0.016g
HCl, concentrated 4.1mL

Preparation of 100X Modified Salts: Add components to distilled/deionized water and bring volume to 1.0L. Mix thoroughly. Filter sterilize.

Preparation of Medium: Add components, except Wolfe's vitamin solution and 100X modified salts, to distilled/deionized water and bring volume to 980.0mL. Mix thoroughly. Adjust pH to 5.0. Gently heat and bring to boiling. Autoclave for 15 min at 15 psi pressure–121°C. Cool to 45°–50°C. Aseptically add sterile Wolfe's vitamin solution and 100X modified salts. Mix thoroughly. Pour into sterile Petri dishes or distribute into sterile tubes.

Use: For the cultivation of *Acetobacter xylinum.*

R70-2 Broth, Modified with Fructose

Composition per liter:

Fructose 30.0g
Yeast extract 5.0g
Dimethyl glutaric acid 4.01g
$(NH_4)_2SO_4$ 3.30g
Trisodium citrate·$2H_2O$ 1.18g
KH_2PO_4 .... 1.00g
$MgSO_4 \cdot 7H_2O$ 0.25g
Wolfe's vitamin solution 10.0mL
100X modified salts 10.0mL

pH 5.0 ± 0.2 at 25°C

Wolfe's Vitamin Solution:

Composition per liter:

Pyridoxine·HCl 0.01g

Thiamine·HCl 5.0mg
Riboflavin 5.0mg
Nicotinic acid 5.0mg
Calcium pantothenate 5.0mg
p-Aminobenzoic acid 5.0mg
Thioctic acid 5.0mg
Biotin 2.0mg
Folic acid 2.0mg
Cyanocobalamin 0.1mg

Preparation of Wolfe's Vitamin Solution: Add components to distilled/deionized water and bring volume to 1.0L. Mix thoroughly. Filter sterilize.

100X Modified Salts:

Composition per liter:

$CaCl_2 \cdot 2H_2O$ 1.47g
$FeCl_3 \cdot 6H_2O$ 0.27g
$ZnSO_4 \cdot 7H_2O$ 0.144g
$MnSO_4 \cdot H_2O$ 0.085g
$CoCl_2 \cdot 6H_2O$ 0.024g
$NiCl_2 \cdot 6H_2O$ 0.024g
$Na_2MoO_4 \cdot 2H_2O$ 0.024g
$CuSO_4 \cdot 5H_2O$ 0.016g
HCl, concentrated 4.1mL

Preparation of 100X Modified Salts: Add components to distilled/deionized water and bring volume to 1.0L. Mix thoroughly. Filter sterilize.

Preparation of Medium: Add components, except Wolfe's vitamin solution and 100X modified salts, to distilled/deionized water and bring volume to 980.0mL. Mix thoroughly. Autoclave for 15 min at 15 psi pressure–121°C. Cool to 45°–50°C. Aseptically add sterile Wolfe's vitamin solution and 100X modified salts. Mix thoroughly. Aseptically distribute into sterile tubes or flasks.

Use: For the cultivation of *Acetobacter xylinum.*

R70-2 Broth, Modified with Glucose

Composition per liter:

Fructose 30.0g
Yeast extract 5.0g
Dimethyl glutaric acid 4.01g
$(NH_4)_2SO_4$ 3.30g
Trisodium citrate·$2H_2O$ 1.18g
KH_2PO_4 .... 1.00g
$MgSO_4 \cdot 7H_2O$ 0.25g
Wolfe's vitamin solution 10.0mL
100X modified salts 10.0mL

pH 5.0 ± 0.2 at 25°C

Wolfe's Vitamin Solution:

Composition per liter:

Pyridoxine·HCl 0.01g

Thiamine·HCl 5.0mg
Riboflavin 5.0mg
Nicotinic acid 5.0mg
Calcium pantothenate 5.0mg
p-Aminobenzoic acid 5.0mg
Thioctic acid 5.0mg
Biotin 2.0mg
Folic acid 2.0mg
Cyanocobalamin 0.1mg

Preparation of Wolfe's Vitamin Solution: Add components to distilled/deionized water and bring volume to 1.0L. Mix thoroughly. Filter sterilize.

100X Modified Salts:

Composition per liter:

$CaCl_2 \cdot 2H_2O$ 1.47g
$FeCl_3 \cdot 6H_2O$ 0.27g
$ZnSO_4 \cdot 7H_2O$ 0.144g
$MnSO_4 \cdot H_2O$ 0.085g
$CoCl_2 \cdot 6H_2O$ 0.024g
$NiCl_2 \cdot 6H_2O$ 0.024g
$Na_2MoO_4 \cdot 2H_2O$ 0.024g
$CuSO_4 \cdot 5H_2O$ 0.016g
HCl, concentrated 4.1mL

Preparation of 100X Modified Salts: Add components to distilled/deionized water and bring volume to 1.0L. Mix thoroughly. Filter sterilize.

Preparation of Medium: Add components, except Wolfe's vitamin solution and 100X modified salts, to distilled/deionized water and bring volume to 980.0mL. Mix thoroughly. Autoclave for 15 min at 15 psi pressure–121°C. Cool to 45°–50°C. Aseptically add sterile Wolfe's vitamin solution and 100X modified salts. Mix thoroughly. Aseptically distribute into sterile tubes or flasks.

Use: For the cultivation of *Acetobacter xylinum.*

Rabbit Blood Agar

Composition per 1250mL:

Pancreatic digest of casein 16.0g
Agar 13.5g
Brain heart, solids from infusion 8.0g
Peptic digest of animal tissue 5.0g
NaCl 5.0g
Glucose 2.0g
Na_2HPO_4 .... 2.5g
Rabbit blood, defibrinated 250.0mL

pH 7.4 ± 0.2 at 25°C

Preparation of Medium: Add components, except rabbit blood, to distilled/deionized water and bring volume to 1.0L. Mix thoroughly. Autoclave for 15 min at 15 psi–121°C. Aseptically add sterile rabbit blood. Pour into sterile Petri dishes or aseptically distribute into sterile tubes or flasks while shaking to disperse precipitate.

Use: For the cultivation and maintenance of *Haemophilus ducreyi* and *Actinobacillus lignieresii*.

Rabbit Dung Agar

Composition per liter:

Rabbit dung	20.0g
Agar	15.0g

pH 7.2 ± 0.2 at 25°C

Preparation of Medium: Add rabbit dung to 1.0L of distilled/deionized water. Gently heat and bring to boiling. Continue boiling for 20 min. Filter through Whatman #1 filter paper. Bring volume of filtrate to 1.0L with distilled/deionized water. Add agar. Adjust pH to 7.2. Distribute into tubes or flasks. Autoclave for 15 min at 15 psi pressure–121°C. Pour into sterile Petri dishes or leave in tubes.

Use: For the cultivation of myxobacteria.

Rabbit Laked Blood Agar

Composition per liter:

Agar	15.0g
Pancreatic digest of casein	10.0g
Peptic digest of animal tissue	10.0g
NaCl	5.0g
Yeast extract	2.0g
Glucose	1.0g
$NaHSO_3$	0.1g
Hemin solution	1.0mL
Vitamin K_1 solution	1.0mL
Rabbit blood, laked	50.0mL

pH 7.0 ± 0.2 at 25°C

Hemin Solution:

Composition per 100mL:

Hemin	0.5g
NaOH (1*N* solution)	10.0mL

Preparation of Hemin Solution: Add hemin to 10.0mL of NaOH solution. Mix thoroughly.

Vitamin K_1 Solution:

Composition per 20mL:

Vitamin K_1 (phytomenadione)	0.2g
Ethanol (95% solution)	20.0mL

Preparation of Vitamin K_1 Solution: Add vitamin K_1 to 20.0mL of ethanol. Mix thoroughly.

Preparation of Medium: Add components, except vitamin K_1 solution and laked rabbit blood, to distilled/deionized water and bring volume to 849.0mL. Mix thoroughly. Gently heat and bring to boiling. Autoclave for 15 min at 15 psi pressure–121°C. Cool to 45°–50°C. Aseptically add 1.0mL of sterile vitamin K_1 solution and 50.0mL of sterile laked rabbit blood. Laked blood is prepared by freezing whole blood overnight and thawing to room temperature. Mix thoroughly. Pour into sterile Petri dishes or distribute into sterile tubes.

Use: For the cultivation and enhancement of pigment production of a variety of anaerobic bacteria.

Rabbit Serum Medium (Rabbit Serum Bovine Serum Albumin Tween™ 80 Medium) (Rabbit Serum BSA Tween™ 80 Medium)

Composition per liter:

Basal medium	900.0mL
Rabbit serum with supplements	100.0mL

pH 7.4 ± 0.2 at 25°C

Basal Medium:

Composition per 900mL:

Na_2HPO_4	1.0g
NaCl	1.0g
KH_2PO_4	0.3g
Glycerol (10% solution)	1.0mL
NH_4Cl (25% solution)	1.0mL
Sodium pyruvate (10% solution)	1.0mL
Thiamine (0.5% solution	1.0mL

Preparation of Basal Medium: Add components to distilled/deionized water and bring volume to 900.0mL. Mix thoroughly. Gently heat and bring to boiling. Autoclave for 20 min at 15 psi pressure–121°C. Cool to 25°C.

Rabbit Serum with Supplements :

Composition per 106mL:

Rabbit serum	100.0mL
L-Asparagine (3% solution)	5.0mL
$MgCl_2$-$CaCl_2$ solution	1.0mL

Preparation of Rabbit Serum with Supplements: Combine the three solutions. Mix thoroughly. Filter sterilize.

$MgCl_2$–$CaCl_2$ Solution:

Composition per 100mL:

$CaCl_2 \cdot 2H_2O$	1.5g
$MgCl_2 \cdot 6H_2O$	1.5g

Preparation of $MgCl_2$–$CaCl_2$ Solution: Add components to distilled/deionized water and bring volume to 100.0mL. Mix thoroughly.

Preparation of Medium: Aseptically combine 900.0mL of cooled sterile basal medium and

100.0mL of sterile rabbit serum with supplements. Mix thoroughly. Aseptically distribute into sterile tubes or flasks.

Use: For the cultivation of *Leptospira* species.

Raka–Ray Agar

Composition per liter:

Pancreatic digest of casein	20.0g
Agar	17.0g
Maltose	10.0g
Fructose	5.0g
Glucose	5.0g
Yeast extract	5.0g
2-Phenylethanol	3.0g
Potassium aspartate	2.5g
Potassium glutamate	2.5g
Betaine·HCL	2.0g
Diammonium hydrogen citrate	2.0g
$MgSO_4 \cdot 7H_2O$	2.0g
KH2PO4	2.0g
Liver concentrate	1.0g
$MnSO_4 \cdot H_2O$	0.66g
N-Acetylglucosamine	0.5g
Cycloheximide	7.0mg
Sorbitan monooleate	10.0mL

pH 5.4 ± 0.2 at 25°C

Source: This medium is available as a premixed powder from Oxoid Unipath.

Preparation of Medium: Add components, except phenylethanol, to distilled/deionized water and bring volume to 1.0L. Mix thoroughly. Gently heat and bring to boiling. Autoclave for 15 min at 15 psi pressure–121°C. Cool to 45°–50°C. Aseptically add 3.0g of 2-phenylethanol. Mix thoroughly. Pour into sterile Petri dishes or distribute into sterile tubes.

Use: For the isolation of lactic acid bacteria in beer and brewing processes.

Rap Broth, Modified
See: **Rappaport Broth, Modified**

Raper *Achyla* Medium No. 1

Composition per liter:

Agar	20.0g
Lentil (hot water extract)	10.0g
Starch, soluble	3.0g
Peptone	1.0g
$CaCl_2$	1.0μg
$FeCl_3$	1.0μg
KH_2PO_4	1.0μg
$MgSO_4$	1.0μg
$ZnSO_4$	1.0μg

Preparation of Medium: Add components to distilled/deionized water and bring volume to 1.0L. Mix thoroughly. Gently heat and bring to boiling. Distribute into tubes or flasks. Autoclave for 15 min at 15 psi pressure–121°C. Pour into sterile Petri dishes or leave in tubes.

Use: For the cultivation of *Achyla* species.

Raper *Achyla* Medium No. 2

Composition per liter:

Agar	20.0g
Starch, soluble	3.0g
Inositol	1.0g
Peptone	1.0g

Preparation of Medium: Add components to distilled/deionized water and bring volume to 1.0L. Mix thoroughly. Gently heat and bring to boiling. Distribute into tubes or flasks. Autoclave for 15 min at 15 psi pressure–121°C. Pour into sterile Petri dishes or leave in tubes.

Use: For the cultivation of *Achyla* species.

Rapid Fermentation Medium

Composition per liter:

Pancreatic digest of casein	20.0g
NaCl	5.0g
Agar	3.5g
Cystine	0.5g
Na_2SO_3	0.5g
Phenol Red	0.017g

pH 7.3 ± 0.2 at 25°C

Source: This medium is available as a premixed powder from BBL Microbiology Systems.

Preparation of Medium: Add components to distilled/deionized water and bring volume to 1.0L. Mix thoroughly. Distribute into tubes or flasks. Autoclave for 15 min at 15 psi pressure–121°C.

Use: For the differentiation of *Neisseria* species isolated from clinical specimens.

Rappaport Broth, Modified (Rap Broth, Modified)

Composition per 250.2mL:

Solution A	155.0mL
Solution C	53.0mL
Solution B	40.0mL
Solution D	1.6mL
Solution E	0.6mL

Solution A:

Composition per liter:

Pancreatic digest of casein	10.0g

Preparation of Solution A: Add pancreatic digest of casein to distilled/deionized water and bring volume to 1.0L. Mix thoroughly.

Solution B:

Composition per liter:

Na_2HPO_4	9.5g

Preparation of Solution B: Add Na_2HPO_4 to distilled/deionized water and bring volume to 1.0L. Mix thoroughly.

Solution C:

Composition per 100mL:

$MgCl_2 \cdot 6H_2O$	40.0g

Preparation of Solution C: Add $MgCl_2 \cdot 6H_2O$ to distilled/deionized water and bring volume to 100.0mL. Mix thoroughly. Autoclave for 15 min at 15 psi pressure–121°C. Cool to 25°C.

Solution D:

Composition per 100mL:

Malachite Green	0.2g

Preparation of Solution D: Add Malachite Green to sterile distilled/deionized water and bring volume to 100.0mL. Mix thoroughly. Do not sterilize.

Solution E:

Composition per 10mL:

Carbenicillin	0.010g

Preparation of Solution E: Add carbenicillin to distilled/deionized water and bring volume to 10.0mL. Mix thoroughly. Filter sterilize.

Preparation of Medium: Combine 155.0mL of solution A and 40.0mL of solution B. Mix thoroughly. Autoclave for 15 min at 15 psi pressure–121°C. Cool to 45°–50°C. Aseptically add 53.0mL of sterile solution C, 1.6mL of solution D and 0.6mL of sterile solution E. Mix thoroughly. Aseptically distribute into sterile tubes or flasks.

Use: For the isolation and cultivation of *Yersinia enterocolitica* from foods.

Rappaport–Vassiliadis Enrichment Broth (RV Enrichment Broth)

Composition per 1110mL:

NaCl	8.0g
Papaic digest of soybean meal	5.0g
KH_2PO_4	1.6g
Magnesium chloride solution	100.0mL
Malachite Green solution	10.0mL

pH 5.2 ± 0.2 at 25°C

Source: This medium is available as a premixed powder from Oxoid Unipath.

Magnesium Chloride Solution:

Composition per 100mL:

$MgCl_2 \cdot 6H_2O$	40.0g

Preparation of Magnesium Chloride Solution: Add $MgCl_2 \cdot 6H_2O$ to distilled/deionized water and bring volume to 100.0mL. Mix thoroughly. Autoclave for 15 min at 15 psi pressure–121°C. Cool to 45°–50°C.

Malachite Green Solution:

Composition per 10.0mL:

Malachite Green oxalate	0.04g

Preparation of Malachite Green Solution: Add Malachite Green to distilled/deionized water and bring volume to 10.0mL. Mix thoroughly. Autoclave for 15 min at 15 psi pressure–121°C. Cool to 45°–50°C.

Preparation of Medium: Add components to distilled/deionized water and bring volume to 1.0L. Mix thoroughly. Distribute into tubes in 10.0mL volumes. Autoclave for 15 min at 10 psi pressure–115°C.

Use: For the isolation and cultivation of *Salmonella* species from food and environmental specimens.

Rappaport–Vassiliadis R10 Broth

Composition per liter:

$MgCl_2$, anhydrous	13.4g
NaCl	7.2g
Papaic digest of soybean meal	4.54g
KH_2PO_4	1.45g
Malachite Green oxalate	0.036g

pH 5.1 ± 0.2 at 25°C

Source: This medium is available as a premixed powder from Difco Laboratories.

Preparation of Medium: Add components to distilled/deionized water and bring volume to 1.0L. Mix thoroughly. Distribute into screw-capped tubes in 10.0mL volumes. Autoclave for 15 min at 10 psi pressure–116°C.

Use: For the isolation and cultivation of *Salmonella* species from food and environmental specimens.

Rappaport–Vassiliadis Soya Peptone Broth (RVS Broth)

Composition per liter:

$MgCl_2$, anhydrous	13.58g
NaCl	7.2g
Papaic digest of soybean meal	4.5g

KH_2PO_4 .... 1.26g
K_2HPO_4 .... 0.18g
Malachite Green 0.036g
pH 5.2 ± 0.2 at 25°C

Source: Available as a premixed powder from Oxoid Unipath.

Preparation of Medium: Add components to distilled/deionized water and bring volume to 1.0L. Mix thoroughly. Distribute into screw-capped tubes in 10.0mL volumes. Autoclave for 15 min at 10 psi pressure–115°C.

Use: For the isolation and cultivation of *Salmonella* species from food and environmental specimens.

RC Agar
See: Rippey–Cabelli Agar

Reduced Salt Solution Medium (RSS Medium)

Composition per liter:
$CaCl_2 \cdot H_2O$ 20.0g
$NaHCO_3$ 10.0g
Dithiothreitol 2.0g
$MgSO_4 \cdot 7H_2O$ 2.0g
K_2HPO_4 .... 1.0g
KH_2PO_4 .... 1.0g
NaCl 0.2g
pH 9.2 ± 0.2 at 25°C

Preparation of Medium: Add components to distilled/deionized water and bring volume to 1.0L. Mix thoroughly. Distribute into screw-capped tubes or flasks. Autoclave for 15 min at 15 psi pressure–121°C.

Use: For the transport and isolation of bacteria from dental plaque, especially, *Streptococcus mutans, Streptococcus sanguis*, and *Lactobacillus* species.

Reduced Transport Fluid

Composition per liter:
$(NH_4)_2SO_4$ 9.0g
NaCl 9.0g
K_2HPO_4 .... 4.5g
KH_2PO_4 .... 4.5g
Na_2CO_3 .... 4.0g
EDTA (ethylenediamine tetraacetic acid) 3.8g
Dithiothreitol 2.0g
$MgSO_4 \cdot 7H_2O$ 1.8g
pH 8.0 ± 0.2 at 25°C

Preparation of Medium: Add components to distilled/deionized water and bring volume to 1.0L. Mix thoroughly. Filter sterilize. Aseptically distribute into sterile tubes with rubber stoppers.

Use: For the transport and isolation of bacteria from dental plaque, especially *Streptococcus mutans* and *Streptococcus sanguis*. Also, used for the cultivation of a variety of Gram-positive bacteria from the oral cavity, especially streptococci, actinomycetes, lactobacilli, clostridia, *Bacteroides* species, *Fusobacterium* species, and *Veillonella* species.

Reduced Transport Fluid

Composition per liter:
Stock mineral salt solution No. 1 .... 75.0mL
Stock mineral salt solution No. 2 .... 75.0mL
Ethylenediamine tetraacetic acid
(1M solution) 10.0mL
Na_2CO_3 (8% solution) 5.0mL
Dithiothreitol (1% solution) 20.0mL
Resazurin (0.1% solution) 1.0mL
pH 8.0 ± 0.2 at 25°C

Stock Mineral Salt Solution No. 1:
Composition per 100mL:
K_2HPO_4 .... 0.6g

Preparation of Stock Mineral Salt Solution No. 1: Add K_2HPO_4 to distilled/deionized water and bring volume to 100.0mL. Mix thoroughly.

Stock Mineral Salt Solution No. 2:
Composition per 100mL:
NaCl 1.2g
$(NH_4)_2SO_4$ 1.2g
K_2HPO_4 .... 0.6g
$MgSO_4 \cdot 7H_2O$ 0.25g

Preparation of Stock Mineral Salt Solution No. 2: Add components to distilled/deionized water and bring volume to 100.0mL. Mix thoroughly.

Preparation of Medium: Add components to distilled/deionized water and bring volume to 1.0L. Mix thoroughly. Filter sterilize. Aseptically distribute into sterile tubes with rubber stoppers.

Use: For the transport and isolation of bacteria from dental plaque, especially *Streptococcus mutans* and *Streptococcus sanguis*. Also, used for the cultivation of a variety of Gram-positive bacteria from the oral cavity, especially streptococci, actinomycetes, lactobacilli, clostrida, *Bacteroides*, Fusobacteria, and *Veillonella*.

Regan–Lowe Charcoal Agar (Regan–Lowe Medium)

Composition per liter:
Agar 12.0g
Beef extract 10.0g
Pancreatic digest of gelatin 10.0g

Soluble starch..10.0g
NaCl...5.0g
Charcoal 4.0g
Niacin...0.01g
Horse blood, defibrinated..........................100.0mL
Cephalexin solution 10.0mL
pH 7.4 ± 0.2 at 25°C

Source: This medium is available as a premixed powder from BBL Microbiology Systems.

Cephalexin Solution:
Composition per 10mL:
Cephalexin 0.04g

Preparation of Cephalexin Solution: Add cephalexin to distilled/deionized water and bring volume to 10.0mL. Mix thoroughly. Filter sterilize.

Preparation of Medium: Add components, except horse blood and cephalexin solution, to distilled/deionized water and bring volume to 890.0mL. Mix thoroughly. Gently heat and bring to boiling. Autoclave for 15 min at 15 psi pressure–121°C. Cool to 45°–50°C. Aseptically add sterile horse blood and sterile cephalexin solution. Mix thoroughly. Pour into sterile Petri dishes or distribute into sterile tubes. Swirl medium while dispensing to keep charcoal in suspension.

Use: For the selective isolation and cultivation of *Bordetella pertussis* and *Bordetella parapertussis* from clinical specimens.

Regan–Lowe Semisolid Transport Medium

Composition per liter:
Agar...6.0g
Beef extract..5.0g
Pancreatic digest of gelatin..................................5.0g
Soluble starch...5.0g
NaCl...2.5g
Charcoal 2.0g
Niacin...0.01g
Horse blood, defibrinated..........................100.0mL
Cephalexin solution 10.0mL
pH 7.4 ± 0.2 at 25°C

Cephalexin Solution:
Composition per 10mL:
Cephalexin 0.04g

Preparation of Cephalexin Solution: Add cephalexin to distilled/deionized water and bring volume to 10.0mL. Mix thoroughly. Filter sterilize.

Preparation of Medium: Add components, except horse blood and cephalexin solution, to distilled/deionized water and bring volume to 890.0mL. Mix thoroughly. Gently heat and bring to boiling. Autoclave for 15 min at 15 psi pressure–121°C. Cool to 45°–50°C. Aseptically add sterile horse blood and sterile cephalexin solution. Mix thoroughly. Aseptically distribute into small, sterile, screw-capped tubes. Fill tubes half-full. Swirl medium while dispensing to keep charcoal in suspension.

Use: For the transport of *Bordetella pertussis* and *Bordetella parapertussis* isolated from clinical specimens.

Reinforced Clostridial Agar

Composition per liter:
Agar...13.5g
Beef extract..10.0g
Pancreatic digest of casein..................................10.0g
NaCl...5.0g
Glucose 5.0g
Yeast extract...3.0g
Sodium acetate 3.0g
Soluble starch..1.0g
L-Cysteine·HCl·H_2O..0.5g
pH 6.8 ± 0.2 at 25°C

Source: This medium is available as a premixed powder from BBL Microbiology Systems, Difco Laboratories and Oxoid Unipath.

Preparation of Medium: Add components to distilled/deionized water and bring volume to 1.0L. Mix thoroughly. Gently heat and bring to boiling. Distribute into tubes or flasks. Autoclave for 15 min at 10 psi pressure–115°C. Pour into sterile Petri dishes or leave in tubes.

Use: For the cultivation and enumeration of *Clostridium* species, *Bifidobacterium* species, other anaerobes (e.g., lactobacilli) and facultative organisms from clinical specimens and foods.

Reinforced Clostridial Medium

Composition per liter:
Tryptose 10.0g
Beef extract...10.0g
Glucose 5.0g
NaCl...5.0g
Yeast extract...3.0g
Sodium acetate 3.0g
Soluble starch..1.0g
Cysteine·HCl·H_2O...0.5g
Agar...0.5g
pH 6.8 ± 0.2 at 25°C

Source: This medium is available as a premixed powder from Difco Laboratories and Oxoid Unipath.

Preparation of Medium: Add components to distilled/deionized water and bring volume to 1.0L. Mix thoroughly. Gently heat and bring to boiling. Distribute into tubes or flasks. Autoclave for 15 min at 10 psi pressure–115°C. Pour into sterile Petri dishes or leave in tubes.

Use: For the non-selective cultivation and enumeration of *Clostridium* species, other anaerobes such as lactobacilli, and facultative organisms from clinical specimens and foods.

Reinforced Clostridial Medium with Sodium Lactate

Composition per liter:

Tryptose	10.0g
Beef extract	10.0g
Glucose	5.0g
NaCl	5.0g
Yeast extract	3.0g
Sodium acetate	3.0g
Soluble starch	1.0g
Cysteine·HCl·H_2O	0.5g
Agar	0.5g
Sodium lactate (60% solution)	15.0mL

pH 6.8 ± 0.2 at 25°C

Preparation of Medium: Add components to distilled/deionized water and bring volume to 1.0L. Mix thoroughly. Gently heat and bring to boiling. Distribute into tubes or flasks. Autoclave for 15 min at 10 psi pressure–115°C. Pour into sterile Petri dishes or leave in tubes.

Use: For the non-selective cultivation and enumeration of *Clostridium* species, other anaerobes such as lactobacilli, and facultative organisms from clinical specimens and foods.

Renibacterium KDM2 Medium

Composition per liter:

Agar	15.0g
Peptone	10.0g
L-Cysteine·HCl·H_2O	1.0g
Yeast extract	0.5g
Fetal calf serum	200.0mL

pH 6.5 ± 0.2 at 25°C

Preparation of Medium: Add components, except fetal calf serum and agar, to distilled/deionized water and bring volume to 800.0mL. Mix thoroughly. Adjust pH to 6.5 with NaOH. Add agar. Gently heat while stirring and bring to boiling. Autoclave for 15 min at 15 psi pressure–121°C. Cool to 45°–50°C. Aseptically add fetal calf serum. Mix thoroughly. Pour into sterile Petri dishes or distribute into sterile tubes.

Use: For the cultivation and maintenance of *Renibacterium salmoninarum*.

RF Medium

Composition per liter:

Yeast extract	0.05g
Peptone	0.05g
$(NH_4)_2SO_4$	0.05g
L-Cysteine·HCl·H_2O	0.05g
Salt solution	50.0mL
Rumen fluid, clarified	30.0mL
Resazurin (1% solution)	0.1mL

pH 7.4 ± 0.2 at 25°C

Salt Solution:

Composition per liter:

$NaHCO_3$	10.0g
NaCl	2.0g
K_2HPO_4	1.0g
KH_2PO_4	1.0g
$CaCl_2$, anhydrous	0.2g
$MgSO_4$	0.2g

Preparation of Salts Solution: Add $CaCl_2$ and $MgSO_4$ to 300.0mL of distilled/deionized water. Mix thoroughly until dissolved. Bring volume to 800.0mL with distilled/deionized water. Add remaining components while stirring. Bring volume to 1.0L. Mix thoroughly. Store at 4°C.

Preparation of Medium: Add components to distilled/deionized water and bring volume to 1.0L. Mix thoroughly. Adjust pH to 6.2–6.3 with 4*N* HCl. Gently heat and bring to boiling under 100% N_2. Anaerobically distribute into tubes in 7.0mL volumes. Cap with rubber stoppers. Place tubes in a press. Autoclave for 20 min at 15 psi pressure–121°C with fast exhaust. The pH of the medium should be 7.4 after autoclaving.

Use: For the cultivation and maintenance of *Treponema bryantii*.

RFC Agar
See: **Rumen Fluid Cellobiose Agar**

RGCA Medium

Composition per 300.3mL:

Rumen fluid	120.0mL
Solution IV	65.0mL
Mineral solution I	45.0mL

Mineral solution II 45.0mL
Na_2CO_3 solution...20.0mL
Cysteine·HCl·H_2O solution............................5.0mL
Solution III 0.3mL

pH 6.6 ± 0.2 at 25°C

Mineral Solution I:

Composition per 100mL:

K_2HPO_4 ....0.3g

Preparation of Mineral Solution I: Add K_2HPO_4 to distilled/deionized water and bring volume to 100.0mL. Mix thoroughly.

Mineral Solution II:

Composition per 100mL:

$(NH_4)_2SO_4$ 0.6g
NaCl 0.6g
K_2HPO_4 ....0.3g
$MgSO_4$...0.06g
$CaCl_2$ 0.06g

Preparation of Mineral Solution II: Add K_2HPO_4 to distilled/deionized water and bring volume to 100.0mL. Mix thoroughly.

Solution III:

Composition per 10mL:

Resazurin...0.01g

Preparation of Solution III: Add resazurin to 10.0mL of distilled/deionized water. Mix thoroughly.

Solution IV:

Composition per 65mL:

Agar...4.5g
Glucose 0.6g
Cellobiose 0.6g

Preparation of Solution IV: Add components to distilled/deionized water and bring volume to 65.0mL. Mix thoroughly.

Cysteine·HCl·H_2O Solution:

Composition per 100mL:

Cysteine·HCl·H_2O...3.0g

Preparation of Cysteine·HCl·H_2O Solution: Add cysteine·HCl·H_2O to distilled/deionized water and bring volume to 100.0mL. Mix thoroughly. Filter sterilize.

Na_2CO_3 Solution:

Composition per 100mL:

Na_2CO_3...6.0g

Preparation of Na_2CO_3 Solution: Add Na_2CO_3 to distilled/deionized water and bring volume to 100.0mL. Mix thoroughly. Filter sterilize.

Rumen Fluid:

Composition per 120mL:

Rumen fluid...120.0mL

Preparation of Rumen Fluid: Filter rumen contents, obtained from a cow on an alfalfa-hay concentrate ration, through two layers of cheesecloth to remove larger particles. Store under CO_2 in quart milk bottles in the refrigerator. Much of the particulate matter settles out. Use the supernatant fluid.

Preparation of Medium: Combine 45.0mL of mineral solution I, 45.0mL of mineral solution II, 0.3mL of solution III and 65.0mL of solution IV in a 500mL flask. Gently heat and bring to boiling. Add 120.0mL of rumen fluid. Gently heat and bring to boiling under 100% CO_2. Cap with a rubber stopper and wire the stopper secure. Autoclave for 20 min at 15 psi pressure–121°C. Cool to 45°–50°C. Remove stopper and gas with 100% CO_2 to eliminate O_2. Aseptically add 5.0mL of sterile cysteine·HCl·H_2O solution and 20.0mL of sterile Na_2CO_3 solution. Mix thoroughly. Aseptically and anaerobically distribute into tubes under 100% CO_2 in 6.0mL volumes. Cap with rubber stoppers.

Use: For the cultivation and maintenance of *Ruminococcus albus, Ruminococcus flavefaciens,* and *Succinimonas amylolytica.*

Rhizobium BIII Defined Agar

Composition per liter:

Agar...13.0g
Mannitol 10.0g
Sodium glutamate 1.10g
K_2HPO_4 ....0.23g
$MgSO_4 \cdot 7H_2O$ 0.10g
Trace element stock 1.0mL
Vitamin stock 1.0mL

pH 7.0 ± 0.2 at 25°C

Trace Element Stock:

Composition per liter:

Nitrilotriacetic acid 7.0g
$CaCl_2 \cdot 2H_2O$...6.62g
H_3BO_3 ....0.145g
$FeSO_4 \cdot 7H_2O$ 0.125g
Na_2MoO_4..0.125g
$ZnSO_4 \cdot 7H_2O$ 0.108g
$CoSO_4 \cdot 7H_2O$ 0.070g
$CuSO_4 \cdot 5H_2O$ 5.0mg
$MnCl_2 \cdot 4H_2O$.. 4.3mg

Preparation of Trace Element Stock: Add components to 500.0mL of distilled/deionized water in the following order: $CaCl_2 \cdot 2H_2O$, H_3BO_3, $FeSO_4 \cdot 7H_2O$, $CoSO_4 \cdot 7H_2O$, $CuSO_4 \cdot 5H_2O$, $MnCl_2 \cdot 4H_2O$, $ZnSO_4 \cdot 7H_2O$, and Na_2MoO_4. Adjust pH to 5.0. Add nitrilotriacetic acid. Bring volume to 1.0L with distilled/deionized water.

Vitamin Stock:

Composition per liter:

Inositol	0.12g
p-Aminobenzoic acid	0.02g
Biotin	0.02g
Calcium pantothenate	0.02g
Nicotinic acid	0.02g
Pyridoxine·HCl	0.02g
Riboflavin	0.02g
Thiamine·HCl	0.02g
Sodium phosphate buffer (50.0mM solution, pH 7.0)	1.0L

Preparation of Vitamin Stock: Combine components. Mix thoroughly. Filter sterilize. Store at 4°C in the dark.

Preparation of Medium: Add components, except vitamin stock, to distilled/deionized water and bring volume to 999.0mL. Mix thoroughly. Gently heat and bring to boiling. Autoclave for 15 min at 15 psi pressure–121°C. Cool to 45°–50°C. Aseptically add 1.0mL of sterile vitamin stock. Mix thoroughly. Pour into sterile Petri dishes or distribute into sterile tubes.

Use: For the isolation and cultivation of *Rhizobium* species from root nodules.

Rhizobium BIII Defined Broth

Composition per liter:

Mannitol	10.0g
Sodium glutamate	1.10g
K_2HPO_4	0.23g
$MgSO_4 \cdot 7H_2O$	0.10g
Trace element stock	1.0mL
Vitamin stock	1.0mL

pH 7.0 ± 0.2 at 25°C

Trace Element Stock:

Composition per liter:

Nitrilotriacetic acid	7.0g
$CaCl_2 \cdot 2H_2O$	6.62g
H_3BO_3	0.145g
$FeSO_4 \cdot 7H_2O$	0.125g
Na_2MoO_4	0.125g
$ZnSO_4 \cdot 7H_2O$	0.108g
$CoSO_4 \cdot 7H_2O$	0.070g
$CuSO_4 \cdot 5H_2O$	5.0mg
$MnCl_2 \cdot 4H_2O$	4.3mg

Preparation of Trace Element Stock: Add components to 500.0mL of distilled/deionized water in the following order: $CaCl_2 \cdot 2H_2O$, H_3BO_3, $FeSO_4 \cdot 7H_2O$, $CoSO_4 \cdot 7H_2O$, $CuSO_4 \cdot 5H_2O$, $MnCl_2 \cdot 4H_2O$, $ZnSO_4 \cdot 7H_2O$, and Na_2MoO_4. Adjust pH to 5.0. Add nitrilotriacetic acid. Bring volume to 1.0L with distilled/deionized water.

Vitamin Stock:

Composition per liter:

Inositol	0.12g
p-Aminobenzoic acid	0.02g
Biotin	0.02g
Calcium pantothenate	0.02g
Nicotinic acid	0.02g
Pyridoxine·HCl	0.02g
Riboflavin	0.02g
Thiamine·HCl	0.02g
Sodium phosphate buffer (50.0mM solution, pH 7.0)	1.0L

Preparation of Vitamin Stock: Combine components. Mix thoroughly. Filter sterilize. Store at 4°C in the dark.

Preparation of Medium: Add components, except vitamin stock, to distilled/deionized water and bring volume to 999.0mL. Mix thoroughly. Autoclave for 15 min at 15 psi pressure–121°C. Cool to 25°C. Aseptically add 1.0mL of sterile vitamin stock. Mix thoroughly. Aseptically distribute into sterile tubes or flasks.

Use: For the isolation and cultivation of *Rhizobium* species.

Rhizobium Medium 1

Composition per liter:

Agar	15.0g
Yeast extract	10.0g
K_2HPO_4	0.5g
$MgSO_4 \cdot 7H_2O$	0.2g
NaCl	0.2g
$FeCl_3 \cdot 6H_2O$	0.002g

pH 7.2 ± 0.2 at 25°C

Preparation of Medium: Add components, except agar, to distilled/deionized water and bring volume to 1.0L. Mix thoroughly. Adjust pH to 7.2. Add agar. Gently heat and bring to boiling. Distribute into tubes or flasks. Autoclave for 15 min at 15 psi pressure–121°C. Pour into sterile Petri dishes or leave in tubes.

Use: For the cultivation of members of the Rhizobiaceae.

Rhizobium Medium 2

Composition per liter:

Agar	15.0g
Glycerol	4.6g
$CaSO_4$	1.3g
K_2HPO_4	1.0g
L-Arabinose	1.0g

Yeast extract 1,0g
KNO_3 .... 0.7g
$MgSO_4 \cdot 7H_2O$ 0.36g
$FeCl_3 \cdot 6H_2O$ 4.0mg
pH 7.2 ± 0.2 at 25°C

Preparation of Medium: Add components, except agar, to distilled/deionized water and bring volume to 1.0L. Mix thoroughly. Adjust pH to 7.2. Add agar. Gently heat and bring to boiling. Distribute into tubes or flasks. Autoclave for 15 min at 15 psi pressure–121°C. Pour into sterile Petri dishes or leave in tubes.

Use: For the cultivation of members of the Rhizobiaceae.

Rhizobium X Medium

Composition per liter:
Agar 15.0g
Mannitol 10.0g
Yeast extract 1.0g
Soil extract 200.0mL
pH 7.2 ± 0.2 at 25°C

Soil Extract:
Composition per 200mL:
African Violet soil 77.0g
Na_2CO_3 .... 0.2g

Preparation of Soil Extract: Add components to tap water and bring volume to 200.0mL. Autoclave for 60 min at 15 psi pressure–121°C. Filter through Whatman #1 filter paper.

Preparation of Medium: Add components to distilled/deionized water and bring volume to 1.0L. Mix thoroughly. Gently heat and bring to boiling. Distribute into tubes or flasks. Autoclave for 15 min at 15 psi pressure–121°C. Pour into sterile Petri dishes or leave in tubes.

Use: For the cultivation and maintenance of *Bradyrhizobium japonicum, Rhizobium* species, and *Sinorhizobium xinjiangensis.*

Rhizobium X Medium with Thiram

Composition per liter:
Agar 15.0g
Mannitol 10.0g
Yeast extract 1.0g
Soil extract 200.0mL
Thiram solution 10.0mL
pH 7.2 ± 0.2 at 25°C

Thiram Solution:
Composition per 10mL:
Thiram 1.0mg
Ethanol, absolute 10.0mL

Preparation of Thiram Solution: Add thiram to 10.0mL of absolute ethanol. Mix thoroughly. Filter sterilize.

Soil Extract:
Composition per 200mL:
African Violet soil 77.0g
Na_2CO_3 .... 0.2g

Preparation of Soil Extract: Add components to tap water and bring volume to 200.0mL.

Preparation of Medium: Add components, except thiram solution, to distilled/deionized water and bring volume to 990.0mL. Mix thoroughly. Gently heat and bring to boiling. Autoclave for 15 min at 15 psi pressure–121°C. Cool to 50°C. Aseptically add 10.0mL of sterile thiram solution. Pour into sterile Petri dishes or distribute into sterile tubes.

Use: For the cultivation and maintenance of *Bradyrhizobium japonicum, Rhizobium* species, and *Sinorhizobium xinjiangensis.*

Rhizoctonia Isolation Medium

Composition per liter:
Agar 20.0g
K_2HPO_4 .... 1.0g
KCl 0.5g
$MgSO_4 \cdot 7H_2O$ 0.5g
$NaNO_2$ 0.2g
$FeSO_4 \cdot 7H_2O$ 0.01g
Dexon® solution 10.0mL
Antibiotic solution 10.0mL
Gallic acid solution 10.0mL

Antibiotic Solution:
Composition per 10mL:
Chloramphenicol 0.05g
Streptomycin 0.05g

Preparation of Antibiotic Solution: Add components to distilled/deionized water and bring volume to 10.0mL. Mix thoroughly. Filter sterilize.

Dexon® Solution:
Composition per 10mL:
Dexon® (Chemagro®) wettable powder 0.09g

Preparation of Dexon® Solution: Add Dexon to distilled/deionized water and bring volume to 10.0mL. Mix thoroughly. Filter sterilize.

Gallic Acid Solution:
Composition per 10mL:
Gallic acid 0.4g

Preparation of Gallic Acid Solution: Add gallic acid to distilled/deionized water and bring volume to 10.0mL. Mix thoroughly. Filter sterilize.

Preparation of Medium: Add components—except Dexon® solution, antibiotic solution, and gallic acid solution—to distilled/deionized water and bring volume to 970.0mL. Mix thoroughly. Gently heat and bring to boiling. Autoclave for 15 min at 15 psi pressure–121°C. Cool to 45°–50°C. Aseptically add sterile Dexon® solution, sterile antibiotic solution, and sterile gallic acid solution. Mix thoroughly. Pour into sterile Petri dishes or distribute into sterile tubes.

Use: For the isolation and cultivation of *Rhizoctonia* species.

Rhizomonas Medium

Composition per liter:

Noble agar	11.0g
Pancreatic digest of casein	5.0g
Glucose	2.5g
K_2HPO_4	1.0g
$MgSO_4 \cdot 7H_2O$	0.5g
KNO_3	0.5g
$Ca(NO_3)_2 \cdot 4H_2O$	0.06g

pH 7.2 ± 0.2 at 25°C

Preparation of Medium: Add components to distilled/deionized water and bring volume to 1.0L. Mix thoroughly. Gently heat and bring to boiling. Adjust pH to 7.2. Distribute into tubes or flasks. Autoclave for 15 min at 15 psi pressure–121°C. Pour into sterile Petri dishes or leave in tubes.

Use: For the cultivation and maintenance of *Rhizomonas suberifaciens*.

Rhodobacter veldkampii Medium

Composition per 127mL:

Solution 1	76.2mL
Solution 2 + Solution 3	44.8mL
Solution 4	6.0mL

Solution 1:

Composition per 2.5 liters:

$CaCl_2$	2.0g

Preparation of Solution 1: Add $CaCl_2$ to distilled/deionized water and bring volume to 2.5L. Distribute in 80.0mL volumes into 127.0mL screw-capped bottles. Autoclave for 15 min at 15 psi pressure–121°C.

Solution 2:

Composition per 100mL:

Sodium ascorbate	2.4g
Sodium acetate	1.0g
KC1	1.0g
KH_2PO_4	1.0g
$MgCl_2 \cdot 6H_2O$	0.8g
NH_4Cl	0.8g
Heavy metal solution	50.0mL
Vitamin solution	15.0mL
Vitamin B_{12} solution	3.0mL

Preparation of Solution 2: Add components to distilled/deionized water and bring volume to 100.0mL. Mix thoroughly.

Heavy Metal Solution:

Composition per liter:

Ethylenediamine tetraacetate (EDTA)	1.5g
$FeSO_4 \cdot 7H_2O$	0.2g
$ZnSO_4 \cdot 7H_2O$	0.1g
$MnCl_2 \cdot 4H_2O$	0.02g
Modified Hoagland trace element solution	6.0mL

Preparation of Heavy Metal Solution: Dissolve EDTA in 800.0mL of distilled/deionized water. Add remaining components. Bring volume to 1.0L with distilled/deionized water. Mix thoroughly.

Modified Hoagland Trace Element Solution:

Composition per 3.6L:

H_3BO_3	11.0g
$MnCl_2 \cdot 4H_2O$	7.0g
$AlCl_3$	1.0g
$CoCl_2$	1.0g
$CuCl_2$	1.0g
KI	1.0g
$NiCl_2$	1.0g
$ZnCl_2$	1.0g
$BaCl_2$	0.5g
KBr	0.5g
LiCl	0.5g
Na_2MoO_4	0.5g
$SeCl_4$	0.5g
$SnCl_2 \cdot 2H_2O$	0.5g
$NaVO_3 \cdot H_2O$	0.1g

Preparation of Modified Hoagland Trace Element Solution: Prepare each component as a separate solution. Dissolve each salt in approximately 100.0mL of distilled/deionized water. Adjust the pH of each solution to below 7.0. Combine all the salt solutions and bring the volume to 3.6L with distilled/deionized water. Adjust the pH to 3–4. A yellow precipitate may form after mixing. After a few days, it will turn into a fine white precipitate. Mix the solution thoroughly before using.

Vitamin Solution:

Composition per 100mL:

Pyridoxamine·2HCl	5.0mg
Nicotinic acid	2.0mg

Thiamine 1.0mg
Pantothenic acid 0.5mg
Biotin 0.2mg
p-Aminobenzoic acid 0.1mg

Preparation of Vitamin Solution: Add components to distilled/deionized water and bring volume to 100.0mL. Mix thoroughly.

Vitamin B_{12} Solution:
Composition per 100mL:
Vitamin B_{12} (cyanocobalamin)....................... 2.0mg

Preparation of Vitamin B_{12} Solution: Add vitamin B_{12} to distilled/deionized water and bring volume to 100.0mL. Mix thoroughly.

Solution 3:
Composition per 900mL:
$NaHCO_3$ 4.5g

Preparation of Solution 3: Add $NaHCO_3$ to distilled/deionized water and bring volume to 900.0mL. Mix thoroughly. Bubble 100% CO_2 through the solution for 30 min. After CO_2 saturation of Solution 3, add Solution 2 and immediately filter the mixture through a Seitz filter (or a Millipore) using positive CO_2 pressure to push the liquid through.

Solution 4:
Composition per 200mL:
$Na_2S{\cdot}9H_2O$ 3.0g

Preparation of Solution 4: Add $Na_2S{\cdot}9H_2O$ to distilled/deionized water and bring volume to 200.0mL. Add a magnetic stir bar to the flask. Autoclave for 15 min at 15 psi pressure–121°C. On a magnetic stirrer, slowly add 2.0mL of sterile 2M H_2SO_4. This partially neutralizes the solution. The solution should turn yellow. H_2S gas will be liberated—neutralization and distribution of the solution should be done as rapidly as possible under adequate ventilation.

Preparation of Medium: To the 80.0mL of sterile solution 1 in screw-capped bottles, add combined solutions 2 and 3 immediately after filtration and fill bottles to capacity. Mix thoroughly. Aseptically remove 6.0mL of the medium from the bottles and replace it with 6.0mL of neutralized solution 4. Let stand for 24 hr. The medium should form a fine white precipitate before using. To inoculate, remove 6.0mL of the completed medium from the bottles and replace it with 6.0mL of inoculum.

Use: For the cultivation and maintenance of *Rhodobacter veldkampii*.

Rhodopila globiformis Medium

Composition per liter:
Mannitol..1.5g
Sodium gluconate...0.56g
KH_2PO_4 ....0.4g
NaCl 0.4g
$MgCl_2{\cdot}6H_2O$..0.4g
NH_4Cl..0.4g
$Na_2S_2O_3{\cdot}5H_2O$..0.2g
$CaCl_2{\cdot}2H_2O$..0.05g
Ferric citrate 5.0mg
VA vitamins.. 1.0mL
Trace elements solution SL-6 .... 1.0mL
pH 4.9 ± 0.2 at 25°C

VA Vitamins:
Composition per 500mL:
Nicotinamide...0.175g
Thiamine·HCl...0.15g
p-Aminobenzoic acid..0.1g
Biotin 0.05g
Pyridoxine·2HCl 0.05g
Calcium pantothenate..0.05g
Cyanocobalamin 0.025g

Preparation of VA Vitamins: Add components to distilled/deionized water and bring volume to 500.0mL. Mix thoroughly.

Trace Elements Solution SL-6:
Composition per liter:
H_3BO_3 ....0.3g
$CoCl_2{\cdot}6H_2O$...0.2g
$ZnSO_4{\cdot}7H_2O$...0.10g
$MnCl_2{\cdot}4H_2O$...0.03g
$Na_2MoO_4{\cdot}H_2O$..0.03g
$NiCl_2{\cdot}6H_2O$ 0.02g
$CuCl_2{\cdot}2H_2O$..0.01g

Preparation of Trace Elements Solution SL-6: Add components to distilled/deionized water and bring volume to 1.0L. Mix thoroughly. Adjust pH to 3.4.

Preparation of Medium: Add components to distilled/deionized water and bring volume to 1.0L. Mix thoroughly. Adjust pH to 4.9. Distribute into tubes or flasks. Autoclave for 15 min at 15 psi pressure–121°C.

Use: For the cultivation and maintenance of *Rhodopila globiformis*.

Rhodopseudomonas blastica Medium

Composition per liter:
Sodium pyruvate 1.5g
Sodium hydrogen malate 1.5g
Yeast extract 1.0g
NH_4Cl...0.5g

$MgSO_4 \cdot 7H_2O$	0.4g
NaCl	0.4g
$CaCl_2 \cdot 2H_2O$	0.05g
Sodium phosphate buffer (0.1M, pH 6.8)	50.0mL

pH 6.8 ± 0.2 at 25°C

Preparation of Medium: Add components—except sodium pyruvate solution, sodium hydrogen malate solution and sodium phosphate buffer—to distilled/deionized water and bring volume to 950.0mL. Mix thoroughly. Gently heat and bring to boiling. Adjust pH to 6.8 with KOH. Autoclave for 15 min at 15 psi pressure–121°C. Cool to 45°–50°C. Filter sterilize the sodium pyruvate solution, sodium hydrogen malate solution and sodium phosphate buffer solution. Aseptically add 1.5g of sterile sodium pyruvate solution, 1.5g of sodium hydrogen malate solution and 50.0mL of sodium phosphate buffer solution to cooled basal medium. Mix thoroughly. Pour into sterile Petri dishes or distribute into sterile tubes.

Use: For the cultivation and maintenance of *Rhodopseudomonas blastica* and other *Rhodopseudomonas* species.

Rhodopseudomonas Medium (ATCC Medium 543)

Composition per liter:

Sodium succinate	2.5g
$(NH_4)_2SO_4$	1.25g
K_2HPO_4	0.9g
KH_2PO_4	0.6g
Yeast extract	0.5g
$MgSO_4 \cdot 7H_2O$	0.2g
$CaCl_2$	0.07g
Ferric citrate	3.0mg
Ethylenediamine tetraacetate (EDTA)	2.0mg

pH 7.0 ± 0.2 at 25°C

Preparation of Medium: Add components to distilled/deionized water and bring volume to 1.0L. Mix thoroughly. Distribute into tubes or flasks. Autoclave for 15 min at 15 psi pressure–121°C.

Use: For the cultivation and maintenance of *Rhodopseudomonas* species.

Rhodopseudomonas Medium (ATCC Medium 650)

Composition per liter:

Sodium succinate	1.5g
KH_2PO_4	1.0g
NH_4Cl	0.5g
$MgSO_4 \cdot 7H_2O$	0.4g
NaCl	0.4g
$CaCl_2 \cdot 2H_2O$	0.05g
Trace metals solution	10.0mL

pH 5.6–6.0 at 25°C

Trace Metals Solution:
Composition per 100mL:

Ferric citrate	0.3g
Ethylenediamine tetraacetic acid (EDTA)	0.05g
$CaCl_2 \cdot 2H_2O$	0.02g
$MnSO_4 \cdot H_2O$	0.002g
$(NH_4)_6Mo_7O_{24} \cdot 4H_2O$	0.002g
H_3BO_3	0.001g
$CuSO_4 \cdot 5H_2O$	0.001g
$ZnSO_4$	0.001g

Preparation of Trace Metals Solution: Add components to distilled/deionized water and bring volume to 100.0mL. Mix thoroughly. Filter sterilize.

Preparation of Medium: Add components, except trace metals solution, to distilled/deionized water and bring volume to 990.0mL. Mix thoroughly. Autoclave for 15 min at 15 psi pressure–121°C. Cool to 25°C. Aseptically add 10.0mL of trace metals solution. Mix thoroughly. Aseptically distribute into sterile tubes or flasks.

Use: For the cultivation and maintenance of *Rhodopseudomonas viridis, Rhodopseudomonas acidophila,* and other *Rhodopseudomonas* species.

Rhodospirillaceae Enrichment Medium

Composition per liter:

Dicarboxylic acid substrate	1.0g
KH_2PO_4	0.5g
NaCl	0.4g
NH_4Cl	0.4g
$MgSO_4 \cdot 7H_2O$	0.2g
Yeast extract	0.2g
$CaCl_2 \cdot 2H_2O$	0.05g
Ferric citrate solution	5.0mL
Trace element solution SL-7	1.0mL
Vitamin B_{12} solution	1.0mL

pH 6.8 ± 0.2 at 25°C

Ferric Citrate Solution:
Composition per 100mL:

Ferric citrate	0.1g

Preparation of Ferric Citrate Solution: Add ferric citrate to distilled/deionized water and bring volume to 100.0mL. Mix thoroughly.

Trace Element Solution SL-7:
Composition per liter:

$CoCl_2 \cdot 6H_2O$	0.2g

$MnCl_2 \cdot 4H_2O$	0.1g
$ZnCl_2$	0.07g
H_3BO_3	0.06g
$NaMoO_4 \cdot 2H_2O$	0.04g
$CuCl_2 \cdot 2H_2O$	0.02g
$NiCl_2 \cdot 6H_2O$	0.02g
HCl (25% solution)	1.0mL

Preparation of Trace Element Solution SL-7: Add components to distilled/deionized water and bring volume to 1.0L. Mix thoroughly.

Vitamin B_{12} Solution:
Composition per 100mL:

Vitamin B_{12}	1.0mg

Preparation of Vitamin B_{12} Solution: Add vitamin B_{12} to distilled/deionized water and bring volume to 100.0mL. Mix thoroughly.

Preparation of Medium: Add components to distilled/deionized water and bring volume to 1.0L. Succinic acid or glutaric acid may be used for the dicarboxylic acid substrate. Mix thoroughly. Adjust pH to 6.8. Distribute into tubes or flasks. Autoclave for 15 min at 15 psi pressure–121°C.

Use: For the enrichment and isolation of members of the Rhodospirillaceae.

Rhodospirillaceae Enrichment Medium

Composition per liter:

Fatty acid substrate	1.0g
KH_2PO_4	0.5g
NaCl	0.4g
NH_4Cl	0.4g
$MgSO_4 \cdot 7H_2O$	0.2g
Yeast extract	0.2g
$CaCl_2 \cdot 2H_2O$	0.05g
$NaHCO_3$ solution	40.0mL
Ferric citrate solution	5.0mL
Trace element solution SL-7	1.0mL
Vitamin B_{12} solution	1.0mL

pH 7.3 ± 0.2 at 25°C

Ferric Citrate Solution:
Composition per 100mL:

Ferric citrate	0.1g

Preparation of Ferric Citrate Solution: Add ferric citrate to distilled/deionized water and bring volume to 100.0mL. Mix thoroughly.

Trace Element Solution SL-7:
Composition per liter:

$CoCl_2 \cdot 6H_2O$	0.2g
$MnCl_2 \cdot 4H_2O$	0.1g
$ZnCl_2$	0.07g
H_3BO_3	0.06g
$NaMoO_4 \cdot 2H_2O$	0.04g
$CuCl_2 \cdot 2H_2O$	0.02g
$NiCl_2 \cdot 6H_2O$	0.02g
HCl (25% solution)	1.0mL

Preparation of Trace Element Solution SL-7: Add components to distilled/deionized water and bring volume to 1.0L. Mix thoroughly.

Vitamin B_{12} Solution:
Composition per 100mL:

Vitamin B_{12}	1.0mg

Preparation of Vitamin B_{12} Solution: Add vitamin B_{12} to distilled/deionized water and bring volume to 100.0mL. Mix thoroughly.

$NaHCO_3$ Solution:
Composition per 100mL:

$NaHCO_3$	5.0g

Preparation of $NaHCO_3$ Solution: Add $NaHCO_3$ to distilled/deionized water and bring volume to 100.0mL. Mix thoroughly. Filter sterilize.

Preparation of Medium: Add components, except $NaHCO_3$ solution, to distilled/deionized water and bring volume to 1.0L. Acetate, propionate, or butyrate salts may be used for the fatty acid substrate. Mix thoroughly. Adjust pH to 7.3. Distribute into flasks in 50.0mL volumes. Autoclave for 15 min at 15 psi pressure–121°C. Cool to 25°C. Immediately prior to inoculation, aseptically add 2.0mL of sterile $NaHCO_3$ solution to each flask containing 50.0mL of medium.

Use: For the enrichment and isolation of members of the Rhodospirillaceae.

Rhodospirillaceae Enrichment Medium

Composition per liter:

Fatty acid or dicarboxylic acid substrate	1.0g
KH_2PO_4	0.5g
NaCl	0.4g
NH_4Cl	0.4g
$MgSO_4 \cdot 7H_2O$	0.2g
Yeast extract	0.2g
$CaCl_2 \cdot 2H_2O$	0.05g
Ferric citrate solution	5.0mL
Trace element solution SL-7	1.0mL
Vitamin B_{12} solution	1.0mL

pH 5.2–5.5 at 25°C

Ferric Citrate Solution:
Composition per 100mL:

Ferric citrate	0.1g

Preparation of Ferric Citrate Solution: Add ferric citrate to distilled/deionized water and bring volume to 100.0mL. Mix thoroughly.

Trace Element Solution SL-7:

Composition per liter:

$CoCl_2 \cdot 6H_2O$	0.2g
$MnCl_2 \cdot 4H_2O$	0.1g
$ZnCl_2$	0.07g
H_3BO_3	0.06g
$NaMoO_4 \cdot 2H_2O$	0.04g
$CuCl_2 \cdot 2H_2O$	0.02g
$NiCl_2 \cdot 6H_2O$	0.02g
HCl (25% solution)	1.0mL

Preparation of Trace Element Solution SL-7: Add components to distilled/deionized water and bring volume to 1.0L. Mix thoroughly.

Vitamin B_{12} Solution:

Composition per 100mL:

Vitamin B_{12}	1.0mg

Preparation of Vitamin B_{12} Solution: Add vitamin B_{12} to distilled/deionized water and bring volume to 100.0mL. Mix thoroughly.

Preparation of Medium: Add components to distilled/deionized water and bring volume to 1.0L. Acetate, propionate, or butyrate salts may be used for the fatty acid substrate. Succinic acid or glutaric acid may be used for the dicarboxylic acid substrate. Lactate or ethanol may be used as an alternate substrate. Mix thoroughly. Adjust pH to 5.2–5.5. Distribute into tubes or flasks. Autoclave for 15 min at 15 psi pressure–121°C.

Use: For the enrichment and isolation of *Rhodopseudomonas acidophila* and *Rhodomicrobium vannielii.*

Rhodospirillum Medium (ATCC Medium 1308)

Composition per liter:

Yeast extract	1.0g
Disodium succinate	1.0g
KH_2PO_4	0.5g
Sodium ascorbate	0.5g
$MgSO_4 \cdot 7H_2O$	0.4g
NaCl	0.4g
NH_4Cl	0.4g
$CaCl_2 \cdot 2H_2O$	0.05g
Ferric citrate (0.1% solution)	5.0mL
Trace elements solution SL-6	1.0mL
Ethanol	0.5mL

pH 6.0 ± 0.2 at 25°C

Trace Elements Solution SL-6:

Composition per liter:

H_3BO_3	0.3g
$CoCl_2 \cdot 6H_2O$	0.2g
$ZnSO_4 \cdot 7H_2O$	0.10g
$MnCl_2 \cdot 4H_2O$	0.03g
$Na_2MoO_4 \cdot H_2O$	0.03g
$NiCl_2 \cdot 6H_2O$	0.02g
$CuCl_2 \cdot 2H_2O$	0.01g

Preparation of Trace Elements Solution SL-6: Add components to distilled/deionized water and bring volume to 1.0L. Mix thoroughly. Adjust pH to 3.4.

Preparation of Medium: Add components to distilled/deionized water and bring volume to 1.0L. Mix thoroughly. Adjust pH to 6.0. Distribute into tubes or flasks. Autoclave for 15 min at 15 psi pressure–121°C.

Use: For the cultivation and maintenance of *Rhodospirillum* species.

Rhodospirillum Medium (ATCC Medium 1408)

Composition per liter:

NaCl	100.0g
$MgCl_2 \cdot 6H_2O$	3.5g
Yeast extract	1.5g
Peptone	1.5g
Sodium malate	1.4g
KH_2PO_4	0.3g
SLA trace elements	1.0mL

pH 7.0 ± 0.2 at 25°C

SLA Trace Elements:

Composition per liter:

$FeCl_2 \cdot 4H_2O$	1.8g
H_3BO_3	0.5g
$CoCl_2 \cdot 6H_2O$	0.25g
$ZnCl_2$	0.1g
$MnCl_2 \cdot 4H_2O$	0.07g
$Na_2MoO_4 \cdot 2H_2O$	0.03g
$NiCl_2 \cdot 6H_2O$	0.01g
$CuCl_2 \cdot 2H_2O$	0.01g
$Na_2SeO_3 \cdot 5H_2O$	0.01g

Preparation of SLA Trace Elements: Add components to distilled/deionized water and bring volume to 1.0L. Mix thoroughly. Adjust pH to 2–3.

Preparation of Medium: Adjust medium for final pH 7.0. Sterilize by autoclaving at 121°C, 15 minutes.

Use: For the cultivation of *Rhodospirillum* species.

Riboflavin Assay Medium

Composition per liter:

Peptone, photolyzed	22.0g
Glucose	20.0g
Yeast supplement	2.0g
Sodium acetate	1.8g
K_2HPO_4	1.0g
KH_2PO_4	1.0g
$MgSO_4 \cdot 7H_2O$	0.4g
L-Cystine	0.2g
$FeSO_4 \cdot 7H_2O$	20.0mg
$MnSO_4 \cdot H_2O$	20.0mg
NaCl	20.0mg

pH 6.8 ± 0.2 at 25°C

Source: This medium is available as a premixed powder from Difco Laboratories.

Preparation of Medium: Add components to distilled/deionized water and bring volume to 1.0L. Mix thoroughly. Distribute into tubes in 5.0mL volumes. Add standard solutions and test solutions to each tube. Bring volume of each tube to 10.0mL. Autoclave for 10 min at 15 psi pressure–121°C.

Use: For the microbiological assaying of riboflavin using *Lactobacillus casei* as the test organism.

Ribose Production Medium

Composition per liter:

D-Glucose	150.0g
$CaCO_3$	20.0g
Dried yeast	10.0g
$(NH_4)_2SO_4$	5.0g
Tryptophan	0.05g
Tyrosine	0.05g
Phenylalanine	0.05g

Preparation of Medium: Add components to distilled/deionized water and bring volume to 1.0L. Mix thoroughly. Distribute into tubes or flasks. Autoclave for 15 min at 15 psi pressure–121°C.

Use: For the cultivation and maintenance of *Bacillus subtilis*.

Rice Extract Agar

Composition per liter:

Agar	20.0g
White rice, solids from extract	5.0g
Polysorbate 80	10.0mL

pH 6.6 ± 0.2 at 25°C

Source: This medium is available as a premixed powder from BBL Microbiology Systems.

Preparation of Medium: Add components, except polysorbate 80, to distilled/deionized water and bring volume to 990.0mL. Mix thoroughly. Gently heat and bring to boiling. Add polysorbate 80. Mix thoroughly. Distribute into tubes or flasks. Autoclave for 15 min at 15 psi pressure–121°C. Pour into sterile Petri dishes.

Use: For the cultivation and differentiation of *Candida albicans* and *Candida stellatoidea* from other *Candida* species based on chlamydospore formation.

Rice Extract Agar

Composition per liter:

Agar	20.0g
White rice, solids from extract	20.0g

pH 7.1 ± 0.2 at 25°C

Source: This medium is available as a premixed powder from Difco Laboratories.

Preparation of Medium: Add components, except polysorbate 80, to distilled/deionized water and bring volume to 990.0mL. Mix thoroughly. Gently heat and bring to boiling. Add polysorbate 80. Mix thoroughly. Distribute into tubes or flasks. Autoclave for 15 min at 15 psi pressure–121°C. Pour into sterile Petri dishes.

Use: For the cultivation and differentiation of *Candida albicans* and *Candida stellatoidea* from other *Candida* species based on chlamydospore formation.

Rice Grain Medium

Composition per 25mL:

White rice, polished and without added vitamins	8.0g

Preparation of Medium: Add 8.0g white rice to 25.0mL of distilled/deionized water. Mix thoroughly. Autoclave for 15 min at 15 psi pressure–121°C. Pour into sterile Petri dishes.

Use: For the identification of *Microsporum audovini* (no growth) from other *Microsporum* species (growth and sporulation).

Rice Infusion Oxgall Tween™ 80 Agar

See: **RIOT Agar**

Rila Marine Medium

Composition per liter:

Agar	15.0g
Peptone	0.5g

Yeast extract....0.5g
Pancreatic digest of casein....0.5g
Marine salts mixture....800.0mL
pH 7.6-8.0 at 25°C

Preparation of Medium: Add components to distilled/deionized water and bring volume to 1.0L. Mix thoroughly. Gently heat and bring to boiling. Autoclave for 15 min at 15 psi pressure–121°C. Adjust pH to 7.6–8.0. Pour into sterile Petri dishes or distribute into sterile tubes.

Use: For the cultivation and maintenance of *Alteromonas denitrificans*.

Rimler–Shotts Medium (RS Medium)

Composition per liter:
Agar....13.5g
$Na_2S_2O_3 \cdot 5H_2O$....6.8g
L-Ornithine·HCl....6.5g
NaCl....5.0g
L-Lysine·HCl....5.0g
Maltose....3.5g
Yeast extract....3.0g
Sodium deoxycholate....1.0g
Ferric ammonium citrate....0.8g
L-Cysteine·HCl....0.3g
Bromthymol Blue....0.03g
Novobiocin solution....10.0mL
pH 7.0 ± 0.2 at 25°C

Novobiocin Solution:
Composition per 10mL:
Novobiocin....5.0mg

Preparation of Novobiocin Solution: Add novobiocin to distilled/deionized water and bring volume to 10.0mL. Mix thoroughly. Filter sterilize.

Preparation of Medium: Add components, except novobiocin solution, to distilled/deionized water and bring volume to 990.0mL. Mix thoroughly. Gently heat and bring to boiling. Autoclave for 15 min at 15 psi pressure–121°C. Cool to 45°–50°C. Aseptically add sterile components. Mix thoroughly. Pour into sterile Petri dishes or distribute into sterile tubes.

Use: For the selective isolation, cultivation and presumptive identification of *Aeromonas hydrophila* and other Gram-negative bacteria based on their ability to decarboxylate lysine and ornithine, ferment maltose and produce H_2S. Maltose-fermenting bacteria appear as yellow colonies. Bacteria that produce lysine or ornithine decarboxylase turn the medium greenish-yellow to yellow. Bacteria that produce H_2S appear as colonies with black centers.

RIOT Agar (Rice Infusion Oxgall Tween™ 80 Agar)

Composition per 1010mL:
Agar....10.0g
Oxgall....10.0g
Rice extract....1.0L
Tween™ 80....10.0mL
pH 7.3 ± 0.2 at 25°C

Rice Extract:
Composition per liter:
Cream of rice cereal....10.0g

Preparation of Rice Extract: Add cream of rice cereal to 1.0L of boiling tap water. Mix thoroughly. Filter quickly through cheesecloth. Bring volume of filtrate to 1.0L with tap water.

Preparation of Medium: Combine components. Mix thoroughly. Gently heat and bring to boiling. Distribute into tubes or flasks. Autoclave for 15 min at 15 psi pressure–121°C. Pour into sterile Petri dishes or leave in tubes.

Use: For the cultivation and differentiation of *Candida albicans* and *Candida stellatoidea* from other *Candida* species based on chlamydospore formation.

Rippey–Cabelli Agar (RC Agar)

Composition per liter:
Agar....15.0g
Meat peptone....5.0g
Trehalose....5.0g
NaCl....3.0g
KCl....2.0g
Yeast extract....2.0g
Bromthymol Blue....0.44g
$MgSO_4 \cdot 7H_2O$....0.2g
$FeCl_3 \cdot 6H_2O$....0.1g
Sodium deoxycholate....0.1g
Ampicillin solution....10.0mL
Ethanol....10.0mL
pH 8.0 ± 0.2 at 25°C

Ampicillin Solution:
Composition per 10mL:
Ampicillin....0.02g

Preparation of Ampicillin Solution: Add ampicillin to distilled/deionized water and bring volume to 10.0mL. Mix thoroughly. Filter sterilize.

Preparation of Medium: Add components—except sodium deoxycholate, ampicillin solution and ethanol—to distilled/deionized water and bring volume to 990.0mL. Mix thoroughly. Gently heat and bring to boiling. Autoclave for 15 min at 15 psi pressure–121°C. Cool to 45°–50°C. Aseptically add sodium deoxycholate, 10.0mL of sterile ampicillin solution and 10.0mL of ethanol. Mix thoroughly. Pour into sterile Petri dishes or distribute into sterile tubes.

Use: For the isolation, cultivation and differentiation of *Aeromonas* species and *Plesiomonas* species from water samples using the membrane filter method. This medium differentiates bacteria on the basis of trehalose fermentation. Bacteria that ferment trehalose turn the medium yellow.

RM Medium

Composition per liter:

Glucose	20.0g
Agar	15.0g
Yeast extract	10.0g
KH_2PO_4	2.0g
Solution 1	250.0mL
Solution 2	250.0mL
Solution 3	250.0mL
Solution 4	250.0mL

pH 6.0 ± 0.2 at 25°C

Solution 1:

Composition per 250mL:

Glucose	20.0g

Preparation of Solution 1: Add glucose to distilled/deionized water and bring volume to 250.0mL. Mix thoroughly. Autoclave for 15 min at 15 psi pressure–121°C. Cool to 45°–50°C.

Solution 2:

Composition per 250mL:

Agar	15.0g

Preparation of Solution 2: Add agar to distilled/deionized water and bring volume to 250.0mL. Mix thoroughly. Autoclave for 15 min at 15 psi pressure–121°C. Cool to 45°–50°C.

Solution 3:

Composition per 250mL:

Yeast extract	10.0g

Preparation of Solution 3: Add yeast extract to distilled/deionized water and bring volume to 250.0mL. Mix thoroughly. Autoclave for 15 min at 15 psi pressure–121°C. Cool to 45°–50°C.

Solution 4:

Composition per 250mL:

KH_2PO_4	2.0g

Preparation of Solution 4: Add KH_2PO_4 to distilled/deionized water and bring volume to 250.0mL. Mix thoroughly. Autoclave for 15 min at 15 psi pressure–121°C. Cool to 45°–50°C.

Preparation of Medium: Aseptically combine the four sterile solutions. Mix thoroughly. Adjust pH to 6.0. Pour into sterile Petri dishes or distribute into sterile tubes.

Use: For the cultivation and maintenance of *Zymomonas mobilis*.

Rogosa Agar

Composition per liter:

Sodium acetate	25.0g
Agar	20.0g
Glucose	20.0g
Pancreatic digest of casein	10.0g
KH_2PO_4	6.0g
Yeast extract	5.0g
Ammonium citrate	2.0g
Sorbitan monooleate	1.0g
$MgSO_4 \cdot 7H_2O$	0.575g
$MnSO_4 \cdot H_2O$	0.12g
$FeSO_4 \cdot 7H_2O$	0.4mg
Acetic acid, glacial	1.32mL

pH 5.4 ± 0.2 at 25°C

Source: This medium is available as a premixed powder from Oxoid Unipath.

Preparation of Medium: Add components, except acetic acid, to distilled/deionized water and bring volume to 998.7mL. Mix thoroughly. Gently heat and bring to boiling. Add glacial acetic acid. Mix thoroughly. Gently heat while stirring and bring to 90°–100°C for 2–3 min. Do not autoclave. Pour into sterile Petri dishes or distribute into sterile tubes.

Use: For the isolation, cultivation and enumeration of lactobacilli, especially from feces, saliva, vaginal specimens and dairy products.

Rogosa SL Agar (Rogosa Selective *Lactobacillus* Agar)

Composition per liter:

Agar	15.0g
Sodium acetate	15.0g
Glucose	10.0g
Pancreatic digest of casein	10.0g
K_2HPO_4	6.0g
Yeast extract	5.0g
Arabinose	5.0g
Sucrose	5.0g
Ammonium citrate	2.0g
Sorbitan monooleate	1.0g

$MgSO_4 \cdot 7H_2O$ 0.57g
$MnSO_4 \cdot 7H_2O$ 0.12g
$FeSO_4 \cdot H_2O$ 0.03g
Acetic acid, glacial 1.32mL
pH 5.4 ± 0.2 at 25°C

Source: This medium is available as a premixed powder from Difco Laboratories.

Preparation of Medium: Add components, except glacial acetic acid, to distilled/deionized water and bring volume to 998.7mL. Mix thoroughly. Gently heat and bring to boiling. Add glacial acetic acid. Mix thoroughly. Gently heat while stirring and bring to 90°–100°C for 2–3 min. Do not autoclave. Pour into sterile Petri dishes or distribute into sterile tubes.

Use: For the isolation, cultivation and enumeration of lactobacilli, especially from feces, saliva, vaginal specimens and dairy products.

Rogosa SL Broth (Rogosa Selective *Lactobacillus* Broth)

Composition per liter:
Sodium acetate 15.0g
Glucose 10.0g
Pancreatic digest of casein 10.0g
K_2HPO_4 .... 6.0g
Yeast extract 5.0g
Arabinose 5.0g
Sucrose 5.0g
Ammonium citrate 2.0g
Sorbitan monooleate 1.0g
$MgSO_4 \cdot 7H_2O$ 0.57g
$MnSO_4 \cdot 7H_2O$ 0.12g
$FeSO_4 \cdot H_2O$ 0.03g
Acetic acid, glacial 1.32mL
pH 5.4 ± 0.2 at 25°C

Source: This medium is available as a premixed powder from Difco Laboratories.

Preparation of Medium: Add components, except glacial acetic acid, to distilled/deionized water and bring volume to 998.7mL. Mix thoroughly. Gently heat and bring to boiling. Add glacial acetic acid. Mix thoroughly. Gently heat while stirring and bring to 90°–100°C for 2–3 min. Do not autoclave. Aseptically distribute into sterile tubes.

Use: For the isolation, cultivation and enumeration of lactobacilli, especially from feces, saliva, vaginal specimens and dairy products.

Rose Bengal Chloramphenicol Agar

Composition per liter:
Agar 15.0g
Glucose 10.0g
Papaic digest of soybean meal 5.0g
KH_2PO_4 .... 1.0g
$MgSO_4 \cdot 7H_2O$ 0.5g
Rose Bengal 0.05g
Chloramphenicol solution 10.0mL
pH 7.0 ± 0.2 at 25°C

Source: This medium is available as a premixed powder from Difco Laboratories and Oxoid Unipath.

Chloramphenicol Solution:
Composition per 10mL:
Chloramphenicol 0.10g

Preparation of Chloramphenicol Solution: Add chloramphenicol to distilled/deionized water and bring volume to 10.0mL. Mix thoroughly. Filter sterilize.

Preparation of Medium: Add components, except chloramphenicol solution, to distilled/deionized water and bring volume to 990.0mL. Mix thoroughly. Gently heat and bring to boiling. Autoclave for 15 min at 15 psi pressure–121°C. Cool to 45°C. Aseptically add sterile chloramphenicol solution. Mix thoroughly. Pour into sterile Petri dishes or distribute into sterile tubes.

Use: For the selective isolation, cultivation and enumeration of yeasts and molds from environmental specimens and foods.

RPMI 1640 Medium with L-Glutamine

Composition per liter:
NaCl 6.0g
$NaHCO_3$ 2.0g
D-Glucose 2.0g
$Na_2HPO_4 \cdot 7H_2O$ 1.5g
KCl 0.4g
L-Glutamine 0.3g
L-Arginine 0.2g
$Ca(NO_3)_2 \cdot 4H_2O$ 0.1g
$MgSO_4 \cdot 7H_2O$ 0.1g
L-Asparagine 0.05g
L-Cystine 0.05g
L-Isoleucine, allo free 0.05g
L-Leucine, methionine free 0.05g
L-Lysine·HCl 0.04g
i-Inositol 0.035g
L-Serine 0.03g

L-Aspartic acid 0.02g
L-Glutamic acid 0.02g
L-Hydroxyproline 0.02g
L-Proline, hydroxy-L-proline free 0.02g
L-Threonine, allo free 0.02g
L-Tyrosine 0.02g
L-Valine 0.02g
L-Histidine, free base 0.015g
L-Methionine 0.015g
L-Phenylalanine 0.015g
Glycine 0.01g
L-Tryptophan 5.0mg
Phenol Red 5.0mg
Choline chloride 3.0mg
Glutathione, reduced 1.0mg
p-Aminobenzoic acid 1.0mg
Folic acid 1.0mg
Nicotinamide 1.0mg
Pyridoxine·HCl 1.0mg
Thiamine·HCl 1.0mg
D-Calcium pantothenate 0.25mg
Biotin 0.20mg
Riboflavin 0.20mg
Vitamin B_{12} 5.0μg
pH 7.3 ± 0.2 at 25°C

Preparation of Medium: Add components to distilled/deionized water and bring volume to 1.0L. Adjust pH to 7.3 with 1N HCl or 1N NaOH. Filter sterilize. Aseptically distribute into sterile tubes or flasks.

Use: For the cultivation of mammalian cells in tissue culture. Culture media for human immunodeficiency viruses.

RS Medium
See: **Rimler–Shotts Medium**

RSS Medium
See: **Reduced Salt Solution Medium**

R–Top Agar

Composition per liter:

Pancreatic digest of casein 10.0g
NaCl 8.0g
Agar 5.0g
K_2HPO_4 ....2.3g
Yeast extract 1.0g
KH_2PO_4 ....0.67g
$(NH_4)_2SO_4$ 0.33g
Glucose 0.33g
Sodium citrate 0.17g
$MgSO_4·7H_2O$ 0.03g
Glucose solution 5.0mL
$CaCl_2·2H_2O$ solution 2.0mL
pH 7.0 ± 0.2 at 25°C

Glucose Solution:
Composition per 10mL:

D-Glucose 2.0g

Preparation of Glucose Solution: Add glucose to distilled/deionized water and bring volume to 10.0mL. Mix thoroughly. Filter sterilize.

$CaCl_2·2H_2O$ Solution:
Composition per 10mL:

$CaCl_2·2H_2O$ 1.47g

Preparation of $CaCl_2·2H_2O$ Solution: Add the $CaCl_2·2H_2O$ to distilled/deionized water and bring volume to 10.0mL. Mix thoroughly. Filter sterilize.

Preparation of Medium: Add components, except glucose solution and $CaCl_2·2H_2O$ solution, to distilled/deionized water and bring volume to 993.0mL. Mix thoroughly. Gently heat and bring to boiling. Autoclave for 15 min at 15 psi pressure–121°C. Cool to 45°–50°C. Aseptically add 5.0mL of glucose solution and 2.0mL of $CaCl_2·2H_2O$ solution. Mix thoroughly. Pour into sterile Petri dishes.

Use: For use as a top agar in the cultivation of bacterial host cells for the production of bacteriophage lysates.

Rumen Fluid Cellobiose Agar (RFC Agar)

Composition per 10mL:

Rumen fluid cellobiose base medium 8.9mL
$NaHCO_3$–rifampin solution 1.0mL
Cellobiose solution 0.1mL

Rumen Fluid Cellobiose Base Medium:
Composition per 89mL:

Noble agar 0.7g
Cysteine·HCl·H_2O 0.1g
Clarified rumen fluid 30.0mL
Salts solution A 20.0mL
Salts solution B 20.0mL
Resazurin (0.1% solution) 0.1mL
pH 6.7–7.0 at 25°C

Preparation of Rumen Fluid Cellobiose Base Medium: Add components to distilled/deionized water and bring volume to 89.0mL. Mix thoroughly. Gently heat and bring to boiling. Continue boiling until resazurin turns colorless indicating reduction. Anaerobically distribute into tubes in 8.9mL volumes under 100% CO_2. Cap tubes with rubber stoppers. Autoclave for 15 min at 15 psi pressure–121°C. Cool to 25°C.

Salts Solution A:

Composition per liter:

$CaCl_2$	0.45g
$MgSO_4$	0.45g

Preparation of Salts Solution A: Add components to distilled/deionized water and bring volume to 1.0L. Mix thoroughly.

Salts Solution B:

Composition per liter:

NaCl	4.5g
$(NH_4)_2SO_4$	4.5g
KH_2PO_4	2.25g
K_2HPO_4	2.25g

Preparation of Salts Solution B: Add components to distilled/deionized water and bring volume to 1.0L. Mix thoroughly.

$NaHCO_3$–Rifampin Solution:

Composition per 10mL:

$NaHCO_3$	0.5g
Rifampin	0.1mg

Preparation of $NaHCO_3$–Rifampin Solution: Add components to distilled/deionized water and bring volume to 10.0mL. Mix thoroughly. Filter sterilize.

Cellobiose Solution:

Composition per 10mL:

Cellobiose	1.0g

Preparation of Cellobiose Solution: Add cellobiose to distilled/deionized water and bring volume to 10.0mL. Mix thoroughly. Filter sterilize.

Preparation of Medium: To each tube containing 8.9mL of sterile rumen fluid cellobiose base medium, aseptically add 1.0mL of sterile $NaHCO_3$–rifampin solution and 0.1mL of sterile cellobiose solution. Mix thoroughly.

Use: For the selective isolation of rumen treponemes.

Ruminococcus pasteuri Medium

Composition per liter:

$NaHCO_3$	2.5g
Sodium tartrate	2.0g
NaCl	1.0g
KCl	0.5g
$MgCl_2 \cdot 6H_2O$	0.4g
$Na_2S \cdot 9H_2O$	0.36g
NH_4Cl	0.25g
KH_2PO_4	0.2g
$CaCl_2 \cdot 2H_2O$	0.15g
Resazurin	1.0mg
Trace elements solution SL-7	1.0mL

pH 7.2 ± 0.2 at 25°C

Trace Elements Solution SL-7:

Composition per liter:

$FeCl_2 \cdot 4H_2O$	1.5g
$CoCl_2 \cdot 6H_2O$	0.19g
$MnCl_2 \cdot 4H_2O$	0.1g
$ZnCl_2$	0.07g
H_3BO_3	0.062g
$Na_2MoO_4 \cdot 2H_2O$	0.036g
$NiCl_2 \cdot 6H_2O$	0.024g
$CuCl_2 \cdot 2H_2O$	0.017g
HCl (25% solution)	10.0mL

Preparation of Trace Elements Solution SL-7: Add the $FeCl_2 \cdot 4H_2O$ to the HCl. Add distilled/deionized water and bring volume to 1.0L. Add remaining components. Mix thoroughly. Autoclave for 15 min at 15 psi pressure–121°C under 100% N_2. Cool to room temperature.

$NaHCO_3$ Solution:

Composition per 10mL:

$NaHCO_3$	2.5g

Preparation of $NaHCO_3$ Solution: Add the $NaHCO_3$ to distilled/deionized water and bring volume to 10.0mL. Mix thoroughly. Filter sterilize.

$Na_2S \cdot 9H_2O$ Solution:

Composition per 10mL:

$Na_2S \cdot 9H_2O$	0.36g

Preparation of $Na_2S \cdot 9H_2O$ Solution: Add $Na_2S \cdot 9H_2O$ to distilled/deionized water and bring volume to 10.0mL. Mix thoroughly. Autoclave for 15 min at 15 psi pressure–121°C under 100% N_2.

Preparation of Medium: Add components—except $NaHCO_3$ solution, $Na_2S \cdot 9H_2O$ solution, and trace elements solution SL-7—to distilled/deionized water and bring volume to 999.0mL. Mix thoroughly. Adjust pH to 7.2. Gently heat and bring to boiling under 80% N_2 + 20% CO_2. Distribute into tubes in 9.8mL volumes under 80% N_2 + 20% CO_2. Cool to 25°C. Aseptically add 0.1mL of sterile $NaHCO_3$ solution and 0.01mL of sterile trace elements solution SL-7 to each tube. Mix thoroughly. Immediately prior to inoculation aseptically add 0.1mL of sterile $Na_2S \cdot 9H_2O$ solution to each tube.

Use: For the cultivation and maintenance of *Ruminococcus pasteuri*.

Russell Double Sugar Agar

Composition per liter:

Agar	15.0g
Proteose peptone No. 3	12.0g
Lactose	10.0g
NaCl	5.0g
Beef extract	1.0g
Glucose	1.0g
Phenol Red	0.025g

pH 7.5 ± 0.2 at 25°C

Preparation of Medium: Add components to distilled/deionized water and bring volume to 1.0L. Mix thoroughly. Gently heat and bring to boiling. Distribute into tubes. Autoclave for 15 min at 15 psi pressure–121°C. Allow tubes to cool in a slanted position.

Use: For the identification of Gram-negative enteric bacilli based on their fermentation of glucose and lactose. Bacteria that ferment both glucose and lactose produce a yellow slant and yellow butt. Bacteria that ferment glucose but do not ferment lactose produce a red slant and a yellow butt. Bacteria that ferment neither glucose nor lactose produce an unchanged pink-orange color.

RV Enrichment Broth
See: **Rappaport–Vassiliadis Enrichment Broth**

RVS Broth
See: **Rappaport–Vassiliadis Soya Peptone Broth**

Ryan's *Aeromonas* Medium
See: ***Aeromonas* Medium**

S Salts

Composition per liter:

Dibenzothiophene	5.00g
NH_4Cl	0.50g
KH_2PO_4	0.25g
$MgCl_2 \cdot 6H_2O$	0.25g

pH 6.5–7.0 at 25°C

Preparation of Medium: Add components to distilled/deionized water and bring volume to 1.0L. Mix thoroughly. Adjust pH to 6.5–7.0 with KOH. Distribute into tubes or flasks. Autoclave for 15 min at 15 psi pressure–121°C.

Use: For the cultivation of *Bacillus sulfasportare*.

S6 Medium for Thiobacilli

Composition per liter:

Agar	15.0g
$Na_2S_2O_3$	10.0g
KH_2PO_4	11.8g
Na_2HPO_4	1.2g
$MgSO_4 \cdot 7H_2O$	0.1g
$(NH_4)_2SO_4$	0.1g
$CaCl_2$	0.03g
$FeCl_3$	0.02g
$MnSO_4$	0.02g

Preparation of Medium: Add components to distilled/deionized water and bring volume to 1.0L. Mix thoroughly. Gently heat and bring to boiling. Distribute into tubes or flasks. Autoclave for 15 min at 15 psi pressure–121°C. Pour into sterile Petri dishes or leave in tubes.

Use: For the cultivation and maintenance of *Thiobacillus denitrificans* and *Thiobacillus thioparus*.

S8 Medium for Thiobacilli

Composition per liter:

Agar	15.0g
KH_2PO_4	11.8g
$Na_2S_2O_3$	10.0g
KNO_3	5.0g
Na_2HPO_4	1.2g
$NaHCO_3$	0.5g
$MgSO_4 \cdot 7H_2O$	0.1g
$(NH_4)_2SO_4$	0.1g
$CaCl_2$	0.03g
$FeCl_3$	0.02g
$MnSO_4$	0.02g

Preparation of Medium: Add components to distilled/deionized water and bring volume to 1.0L. Mix thoroughly. Gently heat and bring to boiling. Distribute into tubes or flasks. Autoclave for 15 min at 15 psi pressure–121°C. Pour into sterile Petri dishes or leave in tubes.

Use: For the cultivation and maintenance of *Thiobacillus neapolitanus*.

S Medium

Composition per liter:

Glycogen	3.0g
$MgSO_4 \cdot 7H_2O$	2.0g
L-Leucine	1.0g
L-Tyrosine	0.60g
L-Asparagine	0.50g
L-Isoleucine	0.50g
L-Proline	0.50g
L-Lysine	0.25g
KH_2PO_4	0.13g
Djenkolic acid	0.10g
L-Arginine	0.10g
L-Serine	0.10g
L-Threonine	0.10g
L-Valine	0.10g
L-Alanine	0.05g
L-Glycine	0.05g
L-Histidine	0.05g
L-Methionine	0.05g
L-Tryptophan	0.05g

pH 7.6 ± 0.2 at 25°C

Preparation of Medium: Add components to distilled/deionized water and bring volume to 1.0L. Mix thoroughly. Filter sterilize. Aseptically distribute into tubes or flasks.

Use: For the cultivation of *Myxococcus xanthus*.

SA Agar

Composition per liter:

Agar	15.0g
Pancreatic digest of casein	10.0g
NaCl	5.0g
Starch, soluble	1.0g
Ampicillin	0.01g
Phenol Red	0.018g

pH 7.4 ± 0.2 at 25°C

Preparation of Medium: Add components, except ampicillin, to distilled/deionized water and bring volume to 1.0L. Mix thoroughly. Gently heat and bring to boiling. Autoclave for 15 min at 15 psi pressure–121°C. Cool to 45°–50°C. Aseptically add ampicillin. Mix thoroughly. Pour into sterile Petri dishes.

Use: For the isolation, cultivation, and differentiation, based on starch hydrolysis, of *Aeromonas hy-*

drophila from foods. After inoculation of plates and growth of cultures, starch hydrolysis is determined by flooding each plate with 5.0mL of Lugol's iodine solution.

SA Agar, Modified (Lachica's Medium)

Composition per liter:

Beef heart, solids from infusion	500.0g
Agar	15.0g
Tryptose	10.0g
NaCl	5.0g
Amylose azure	3.0g
Ampicillin	0.01mg

pH 7.4 ± 0.2 at 25°C

Preparation of Medium: Add components to distilled/deionized water and bring volume to 1.0L. Mix thoroughly. Gently heat and bring to boiling. Distribute into tubes or flasks. Autoclave for 15 min at 15 psi pressure–121°C. Pour into sterile Petri dishes.

Use: For the isolation and cultivation of *Aeromonas hydrophila* from foods. *Aeromonas hydrophila* appear as colonies surrounded by a light halo on a light blue background.

SABHI Agar (Sabouraud Glucose and Brain Heart Infusion Agar)

Composition per liter:

Glucose	21.0g
Agar	15.0g
Pancreatic digest of casein	10.5g
Peptic digest of animal tissue	5.0g
Brain heart, solids from infusion	4.0g
NaCl	2.5g
Na_2HPO_4	1.25g

pH 6.8 ± 0.2 at 25°C

Source: This medium is available as a premixed powder from BBL Microbiology Systems.

Preparation of Medium: Add components to distilled/deionized water and bring volume to 1.0L. Mix thoroughly. Gently heat and bring to boiling. Distribute into tubes or flasks. Autoclave for 15 min at 15 psi pressure–121°C. Pour into sterile Petri dishes in 20.0mL volumes or leave in tubes.

Use: For the cultivation of dermatophytes and other pathogenic and nonpathogenic fungi from clinical and nonclinical specimens.

SABHI Agar

Composition per liter:

Beef heart, infusion from	125.0g
Calf brains, infusion from	100.0g
Glucose	21.0g
Agar	15.0g
Neopeptone	5.0g
Proteose peptone	5.0g
NaCl	2.5g
Na_2HPO_4	1.25g
Chloromycetin solution	1.0mL

pH 7.0 ± 0.2 at 25°C

Source: This medium is available as a premixed powder from Difco Laboratories.

Chloromycetin Solution:
Composition per 10mL:

Chloromycetin	1.0g

Preparation of Chloromycetin Solution: Add chloromycetin to distilled/deionized water and bring volume to 10.0mL. Mix thoroughly. Filter sterilize.

Preparation of Medium: Add components, except chloromycetin solution, to distilled/deionized water and bring volume to 999.0mL. Mix thoroughly. Gently heat and bring to boiling. Autoclave for 15 min at 15 psi pressure–121°C. Cool to 45°–50°C. Aseptically add 1.0mL of sterile chloromycetin solution. Mix thoroughly. Aseptically distribute into sterile tubes in 5.0mL volumes.

Use: For the cultivation of dermatophytes and other pathogenic and nonpathogenic fungi from clinical and nonclinical specimens.

SABHI Agar, Modified

Composition per liter:

Beef heart, infusion from	62.5g
Calf brain, infusion from	50.0g
Glucose	20.5g
Brain heart infusion broth	18.6g
Agar	7.5g
Neopeptone	5.0g
Pancreatic digest of gelatin	2.50g
NaCl	1.25g
Na_2HPO_4	0.625g

pH 6.8 ± 0.2 at 25°C

Preparation of Medium: Dissolve, autoclave at 121°C for 15 min. Cool to 50°C and add 1.0mL of sterile chloramphenicol solution (100.0mg/mL). Mix well and dispense into sterile tubes. Slant and allow to harden. Refrigerate until needed.

Use: For the cultivation of dermatophytes and other pathogenic and nonpathogenic fungi from clinical and nonclinical specimens.

SABHI Blood Agar

Composition per liter:

Beef heart, infusion from	125.0g
Calf brains, infusion from	100.0g
Glucose	21.0g
Agar	15.0g
Neopeptone	5.0g
Proteose peptone	5.0g
NaCl	2.5g
Na_2HPO_4	1.25g
Blood	100.0mL
Chloromycetin solution	1.0mL

pH 7.0 ± 0.2 at 25°C

Source: This medium is available as a premixed powder from Difco Laboratories.

Chloromycetin Solution:
Composition per 10mL:

Chloromycetin	1.0g

Preparation of Chloromycetin Solution: Add chloromycetin to distilled/deionized water and bring volume to 10.0mL. Mix thoroughly. Filter sterilize.

Preparation of Medium: Add components, except blood and chloromycetin solution, to distilled/deionized water and bring volume to 899.0mL. Mix thoroughly. Gently heat and bring to boiling. Autoclave for 15 min at 15 psi pressure–121°C. Cool to 45°–50°C. Aseptically add 100.0mL of sterile blood and 1.0mL of sterile chloromycetin solution. Sheep blood or human blood may be used. Mix thoroughly. Aseptically distribute into sterile tubes in 5.0mL volumes.

Use: For the cultivation of dermatophytes and other pathogenic and nonpathogenic fungi from clinical and nonclinical specimens. Blood enhances the recovery of *Blastomyces dermatitidis* and *Histoplasma capsulatum* and their conversion to the yeast phase.

Sabouraud Agar with CCG and 3% NaCl

Composition per 3031.5mL:

Glucose	120.0g
NaCl	90.0g
Agar	45.0g
Peptone	30.0g
Chloramphenicol solution	15.0mL
Cycloheximide solution	15.0mL
Gentamicin solution	1.5mL

Chloramphenicol Solution:
Composition per 15mL:

Chloramphenicol	0.15g

Preparation of Chloramphenicol Solution: Add chloramphenicol to distilled/deionized water and bring volume to 15.0mL. Mix thoroughly. Filter sterilize.

Cycloheximide Solution:
Composition per 15mL:

Cycloheximide	0.3g

Preparation of Cycloheximide Solution: Add cycloheximide to distilled/deionized water and bring volume to 15.0mL. Mix thoroughly. Filter sterilize.

Gentamicin Solution:
Composition per 10mL:

Gentamicin	0.4g

Preparation of Gentamicin Solution: Add gentamicin to distilled/deionized water and bring volume to 10.0mL. Mix thoroughly. Filter sterilize.

Preparation of Medium: Add components—except chloramphenicol solution, cycloheximide solution, and gentamicin solution—to distilled/deionized water and bring volume to 3.0L. Mix thoroughly. Gently heat and bring to boiling. Autoclave for 15 min at 15 psi pressure–121°C. Cool to 45°–50°C. Aseptically add 15.0mL of sterile chloramphenicol solution, 15.0mL of sterile cycloheximide solution, and 1.5mL of sterile gentamicin solution. Mix thoroughly. Aseptically distribute into sterile tubes. Allow tubes to cool in a slanted position.

Use: For the selective isolation and cultivation of fungi from specimens with a mixed flora.

Sabouraud Agar with CCG and 5% NaCl

Composition per 3031.5mL:

NaCl	150.0g
Glucose	120.0g
Agar	45.0g
Peptone	30.0g
Chloramphenicol solution	15.0mL
Cycloheximide solution	15.0mL
Gentamicin solution	1.5mL

Chloramphenicol Solution:
Composition per 15mL:

Chloramphenicol	0.15g

Preparation of Chloramphenicol Solution: Add chloramphenicol to distilled/deionized water and bring volume to 15.0mL. Mix thoroughly. Filter sterilize.

Cycloheximide Solution:
Composition per 15mL:

Cycloheximide	0.3g

Preparation of Cycloheximide Solution: Add cycloheximide to distilled/deionized water and bring volume to 15.0mL. Mix thoroughly. Filter sterilize.

Gentamicin Solution:
Composition per 10mL:

Gentamicin	0.4g

Preparation of Gentamicin Solution: Add gentamicin to distilled/deionized water and bring volume to 10.0mL. Mix thoroughly. Filter sterilize.

Preparation of Medium: Add components—except chloramphenicol solution, cycloheximide solution, and gentamicin solution—to distilled/deionized water and bring volume to 3.0L. Mix thoroughly. Gently heat and bring to boiling. Autoclave for 15 min at 15 psi pressure–121°C. Cool to 45°–50°C. Aseptically add 15.0mL of sterile chloramphenicol solution, 15.0mL of sterile cycloheximide solution, and 1.5mL of sterile gentamicin solution. Mix thoroughly. Aseptically distribute into sterile tubes. Allow tubes to cool in a slanted position.

Use: For the selective isolation and cultivation of fungi from specimens with a mixed flora.

Sabouraud Agar, Modified

Composition per liter:

Agar	20.0g
Glucose	20.0g
Neopeptone	10.0g

pH 7.0 ± 0.2 at 25°C

Source: This medium is available as a premixed powder from Difco Laboratories.

Preparation of Medium: Add components to distilled/deionized water and bring volume to 1.0L. Mix thoroughly. Gently heat and bring to boiling. Distribute into tubes or flasks. Autoclave for 15 min at 15 psi pressure–121°C. Pour into sterile Petri dishes or leave in tubes.

Use: For the cultivation of yeasts, molds and aciduric bacteria.

Sabouraud Glucose Agar

Composition per liter:

Glucose	40.0g
Agar	15.0g
Pancreatic digest of casein	5.0g
Peptic digest of animal tissue	5.0g

pH 5.6 ± 0.2 at 25°C

Source: This medium is available as a premixed powder from BBL Microbiology Systems, Difco Laboratories and Oxoid Unipath.

Preparation of Medium: Add components to distilled/deionized water and bring volume to 1.0L. Mix thoroughly. Gently heat and bring to boiling. Distribute into tubes or flasks. Autoclave for 15 min at 15 psi pressure–121°C. Pour into sterile Petri dishes or leave in tubes.

Use: For the cultivation of pathogenic and nonpathogenic fungi, especially dermatophytes. The medium may be made more selective for fungi by the addition of chloramphenicol. For the cultivation of yeast and filamentous fungi.

Sabouraud Glucose Agar with Chloramphenicol and Cycloheximide

Composition per liter:

Glucose	40.0g
Agar	15.0g
Pancreatic digest of casein	5.0g
Peptic digest of animal tissue	5.0g
Cycloheximide solution	10.0mL
Chloramphenicol solution	10.0mL

pH 5.6 ± 0.2 at 25°C

Cycloheximide Solution:
Composition per 10mL:

Cycloheximide	0.5g
Acetone	10.0mL

Preparation of Cycloheximide Solution: Add cycloheximide to acetone. Mix thoroughly.

Chloramphenicol Solution:
Composition per 10mL:

Chloramphenicol	0.05g
Ethanol (95% solution)	10.0mL

Preparation of Chloramphenicol Solution: Add chloramphenicol to ethanol. Mix thoroughly.

Preparation of Medium: Add components, except cycloheximide solution and chloramphenicol solution, to distilled/deionized water and bring volume to 980.0mL. Mix thoroughly. Gently heat and bring to boiling. Add the cycloheximide solution and chloramphenicol solution. Mix thoroughly. Distribute into tubes or flasks. Autoclave for 15 min at 15 psi pressure–121°C. Pour into sterile Petri dishes or leave in tubes.

Use: For the cultivation and identification of yeasts.

Sabouraud Glucose Agar with Olive Oil

Composition per liter:

Glucose	40.0g

Agar	15.0g
Pancreatic digest of casein	5.0g
Peptic digest of animal tissue	5.0g
Olive oil	20.0mL
Tween™ 80	2.0mL

pH 5.6 ± 0.2 at 25°C

Preparation of Medium: Add components to distilled/deionized water and bring volume to 1.0L. Mix thoroughly. Gently heat and bring to boiling. Distribute into tubes or flasks. Autoclave for 15 min at 15 psi pressure–121°C. Allow tubes to cool in a slanted position.

Use: For the cultivation and maintenance of *Malassezia* species.

Sabouraud Glucose Agar, Emmons

Composition per liter:

Glucose	20.0g
Agar	17.0g
Pancreatic digest of casein	5.0g
Peptic digest of animal tissue	5.0g

pH 6.9 ± 0.2 at 25°C

Source: This medium is available as a premixed powder from BBL Microbiology Systems.

Preparation of Medium: Add components to distilled/deionized water and bring volume to 1.0L. Mix thoroughly. Gently heat and bring to boiling. Distribute into tubes or flasks. Autoclave for 15 min at 13 psi pressure–118°C. Pour into sterile Petri dishes or leave in tubes.

Use: For the cultivation of dermatophytes and other pathogenic and nonpathogenic fungi from clinical and nonclinical specimens. For the cultivation of yeast and filamentous fungi.

Sabouraud Glucose and Brain Heart Infusion Agar
See: **SABHI Agar**

Sabouraud Glucose Broth

Composition per liter:

Glucose	20.0g
Neopeptone	10.0g

pH 5.6 ± 0.2 at 25°C

Source: This medium is available as a premixed powder from Difco Laboratories.

Preparation of Medium: Add components to distilled/deionized water and bring volume to 1.0L. Mix thoroughly. Distribute into tubes or flasks. Autoclave for 15 min at 15 psi pressure–121°C. Avoid overheating.

Use: For the cultivation of pathogenic and nonpathogenic fungi, especially dermatophytes. The medium may be made more selective for fungi by the addition of chloramphenicol.

Sabouraud Liquid Broth, Modified
See: **Antibiotic Medium 13**

Sabouraud Maltose Agar

Composition per liter:

Maltose	40.0g
Agar	15.0g
Pancreatic digest of casein	5.0g
Peptic digest of animal tissue	5.0g

pH 5.6 ± 0.2 at 25°C

Source: This medium is available as a premixed powder from BBL Microbiology Systems, Difco Laboratories and Oxoid Unipath.

Preparation of Medium: Add components to distilled/deionized water and bring volume to 1.0L. Mix thoroughly. Gently heat and bring to boiling. Distribute into tubes or flasks. Autoclave for 15 min at 15 psi pressure–121°C. Avoid overheating. Pour into sterile Petri dishes or leave in tubes.

Use: For the cultivation and maintenance of a variety of fungi.

Sabouraud Maltose Broth

Composition per liter:

Maltose	40.0g
Neopeptone	10.0g

pH 5.6 ± 0.2 at 25°C

Source: This medium is available as a premixed powder from Difco Laboratories.

Preparation of Medium: Add components to distilled/deionized water and bring volume to 1.0L. Mix thoroughly. Distribute into tubes or flasks. Autoclave for 15 min at 15 psi pressure–121°C. Avoid overheating.

Use: For the cultivation of a variety of fungi.

Sabouraud Medium, Fluid

Composition per liter:

Glucose	20.0g
Pancreatic digest of casein	5.0g
Peptamin	5.0g

pH 5.7 ± 0.2at 25°C

Source: This medium is available as a premixed powder from Difco Laboratories and Oxoid Unipath.

Preparation of Medium: Add components to distilled/deionized water and bring volume to 1.0L. Mix thoroughly. Distribute into tubes or flasks. Autoclave for 15 min at 15 psi pressure–121°C. Avoid overheating.

Use: For isolation and cultivation of yeasts and molds.

Saccharolytic Clostridia Medium

Composition per liter:

Pancreatic digest of casein	10.0g
Yeast extract	6.0g
Sodium thioglycollate	0.5g
Carbohydrate solution	100.0mL
Potassium phosphate (1M solution, pH 7.5)	30.0mL
$MgSO_4$ (1M solution)	1.0mL
Solution M	0.5mL
$FeSO_4$ solution	0.2mL

pH 7.0–7.2 at 25°C

Carbohydrate Solution:

Composition per 100mL:

Glucose or sucrose	20.0g

Preparation of Carbohydrate Solution: Add carbohydrate to distilled/deionized water and bring volume to 100.0mL. Mix thoroughly. Filter sterilize.

Solution M:

Composition per liter:

$CaCl_2$	3.33g
$MnCl_2 \cdot 4H_2O$	1.98g
$Na_2MoO_4 \cdot 2H_2O$	1.21g
$CoCl_2 \cdot 6H_2O$	1.19g

Preparation of Solution M: Add components to distilled/deionized water and bring volume to 1.0L. Mix thoroughly.

$FeSO_4 \cdot 7H_2O$ Solution:

Composition per 10mL:

$FeSO_4 \cdot 7H_2O$	55.6g
H_2SO_4 (0.1M solution)	10.0mL

Preparation of $FeSO_4 \cdot 7H_2O$ Solution: Add the $FeSO_4 \cdot 7H_2O$ to 10.0mL of H_2SO_4 solution. Mix thoroughly. Filter sterilize.

Preparation of Medium: Add components, except carbohydrate solution, to distilled/deionized water and bring volume to 900.0mL. Mix thoroughly. Gently heat and bring to boiling. Autoclave for 20 min at 15 psi pressure–121°C. Cool to 45°–50°C. Aseptically add sterile carbohydrate solution. Mix thoroughly. Aseptically distribute into sterile tubes.

Use: For the cultivation of saccharolytic *Clostridium* species.

Saccharolytic Clostridia Medium

Composition per liter:

Sodium thioglycollate	1.0g
K_2HPO_4	0.8g
KH_2PO_4	0.2g
$MgSO_4 \cdot 7H_2O$	0.2g
NaCl	0.2g
$Na_2MoO_4 \cdot 2H_2O$	0.025g
Yeast extract	0.010g
$FeSO_4 \cdot 7H_2O$	0.010g
$MnSO_4 \cdot 4H_2O$	0.010g
CaCl2	0.010g
Carbohydrate solution	100.0mL
Soil extract	10.0mL
Trace element solution	1.0mL

pH 7.2 ± 0.2 at 25°C

Carbohydrate Solution:

Composition per 100mL:

Glucose or sucrose	10.0g

Preparation of Carbohydrate Solution: Add glucose or sucrose to distilled/deionized water and bring volume to 100.0mL. Mix thoroughly. Filter sterilize.

Soil Extract:

Composition per 200mL:

Garden soil, neutral	100.0g

Preparation of Soil Extract: Add garden soil to 100.0mL of tap water. Gently heat and bring to 130°C for 60 min. Cool to 45°C. Filter through Whatman #1 filter paper. Autoclave for 15 min at 15 psi pressure–121°C. Cool to 45°–50°C.

Trace Element Solution:

Composition per liter:

$Na_2B_4O_7 \cdot 10H_2O$	0.05g
$CoNO_3 \cdot 6H_2O$	0.05g
$CdSO_4 \cdot 2H_2O$	0.05g
$CuSO_4 \cdot 5H_2O$	0.05g
$ZnSO_4 \cdot 7H_2O$	0.05g
$MnSO_4 \cdot H_2O$	0.05g

Preparation of Trace Element Solution: Add components to distilled/deionized water and bring volume to 1.0L. Mix thoroughly.

Preparation of Medium: Add components, except sodium thioglycollate and carbohydrate solution, to distilled/deionized water and bring volume to 900.0mL. Mix thoroughly. Gently heat and bring to boiling. Add sodium thioglycollate. Mix thoroughly. Distribute 9.5mL into test tubes that contain inverted Durham tubes. Autoclave for 15 min at 15 psi pressure–121°C. Cool to 45°–50°C. Aseptically add 0.5mL of sterile carbohydrate solution to each tube. Mix thoroughly.

Use: For the isolation of N_2-fixing, saccharolytic *Clostridium* species.

Salmonella Medium

Composition per liter:

Pancreatic digest of casein	10.0g
NaCl	5.0g

pH 7.4 ± 0.2 at 25°C

Preparation of Medium: Add components to distilled/deionized water and bring volume to 1.0L. Mix thoroughly. Distribute into tubes or flasks. Autoclave for 15 min at 15 psi pressure–121°C.

Use: For the cultivation and maintenance of *Escherichia coli* and *Salmonella choleraesuis.*

Salmonella Shigella Agar (SS Agar)

Composition per liter:

Agar	13.5g
Lactose	10.0g
Bile salts	8.5g
$Na_2S_2O_3$	8.5g
Sodium citrate	8.5g
Beef extract	5.0g
Pancreatic digest of casein	2.5g
Peptic digest of animal tissue	2.5g
Ferric citrate	1.0g
Neutral Red	0.025g
Brilliant Green	0.330mg

pH 7.0 ± 0.2 at 25°C

Source: This medium is available as a premixed powder from BBL Microbiology Systems, Difco Laboratories and Oxoid Unipath.

Preparation of Medium: Add components to distilled/deionized water and bring volume to 1.0L. Mix thoroughly. Gently heat while stirring and bring to boiling. Do not autoclave. Cool to 45°–50°C. Pour into sterile Petri dishes in 20.0mL volumes. Allow the surface of the plates to dry before inoculation.

Use: For the selective isolation and differentiation of pathogenic enteric bacilli, especially those belonging to the genus *Salmonella*. This medium is not recommended for the primary isolation of *Shigella* species. Lactose-fermenting bacteria such as *Escherichia coli* or *Klebsiella pneumoniae* appear as small pink or red colonies. Lactose-nonfermenting bacteria—such as *Salmonella* species, *Proteus* species and *Shigella* species—appear as colorless colonies. Production of H_2S by *Salmonella* species turns the center of the colonies black.

Salmonella Shigella Agar, Modified (SS Agar, Modified)

Composition per liter:

Agar	12.0g
Lactose	10.0g
Sodium citrate	10.0g
$Na_2S_2O_3$	8.5g
Bile salts	5.5g
Beef extract	5.0g
Peptone	5.0g
Ferric citrate	1.0g
Neutral Red	0.025g
Brilliant Green	0.330mg

pH 7.3 ± 0.2 at 25°C

Source: This medium is available as a premixed powder from Oxoid Unipath.

Preparation of Medium: Add components to distilled/deionized water and bring volume to 1.0L. Mix thoroughly. Gently heat while stirring and bring to boiling. Do not autoclave. Cool to 45°–50°C. Pour into sterile Petri dishes in 20.0mL volumes. Allow the surface of the plates to dry before inoculation.

Use: For the selective isolation and differentiation of pathogenic enteric bacilli, especially those belonging to the genus *Salmonella*. This medium provides better growth of *Shigella* species. Lactose-fermenting bacteria such as *Escherichia coli* or *Klebsiella pneumoniae* appear as small pink or red colonies. Lactose-nonfermenting bacteria—such as *Salmonella* species, *Proteus* species and *Shigella* species—appear as colorless colonies. Production of H_2S by *Salmonella* species turns the center of the colonies black.

Salmonella Shigella Deoxycholate Agar

See: **SS Deoxycholate Agar**

Salt Broth, Modified

Composition per liter:

NaCl	65.0g
Enzymatic digest of animal tissue	10.0g
Heart digest	10.0g
Glucose	1.0g
Bromcresol Purple	0.016g

pH 7.2 ± 0.2 at 25°C

Source: Available as a prepared medium from BBL Microbiology Systems.

Preparation of Medium: Add components to distilled/deionized water and bring volume to 1.0L. Mix thoroughly. Distribute into tubes or flasks. Autoclave for 15 min at 15 psi pressure–121°C.

Use: For the cultivation and differentiation of the enterococcal group D streptococci from nonenterococcal group D streptococci based on salt tolerance.

Salt Colistin Broth

Composition per liter:

NaCl	20.0g
Peptone	10.0g
Yeast extract	3.0g
Colistin solution	10.0mL

pH 7.4 ± 0.2 at 25°C

Colistin Solution:

Composition per 10mL:

Colistin methane sulfonate	500,000U

Preparation of Colistin Solution: Add colistin methane sulfonate to distilled/deionized water and bring volume to 10.0mL. Mix thoroughly. Filter sterilize.

Preparation of Medium: Add components, except colistin solution, to distilled/deionized water and bring volume to 990.0mL. Mix thoroughly. Gently heat until dissolved. Autoclave for 15 min at 15 psi pressure–121°C. Cool to 25°C. Aseptically add sterile colistin solution. Mix thoroughly. Aseptically distribute into sterile tubes or flasks.

Use: For the cultivation of halophilic *Vibrio* species.

Salt Meat Broth

Composition per liter:

NaCl	100.0g
Neutral ox-heart tissue	30.0g
Beef extract	10.0g
Peptone	10.0g

pH 7.6 ± 0.2 at 25°C

Source: This medium is available as tablets from Oxoid Unipath.

Preparation of Medium: Add components to distilled/deionized water and bring volume to 1.0L. Mix thoroughly. Distribute into tubes or flasks. Autoclave for 15 min at 15 psi pressure–121°C.

Use: For the isolation and cultivation of staphylococci from specimens with a mixed flora such as fecal specimens, especially during the investigation of staphylococcal food poisoning.

Salt Medium

Composition per liter:

NaCl	58.4g
Agar	15.0g
Proteose peptone	5.0g
Pancreatic digest of casein	5.0g

pH 6.9 ± 0.2 at 25°C

Preparation of Medium: Add components to distilled/deionized water and bring volume to 1.0L. Mix thoroughly. Gently heat and bring to boiling. Distribute into tubes or flasks. Autoclave for 15 min at 15 psi pressure–121°C. Pour into sterile Petri dishes or leave in tubes.

Use: For the cultivation and maintenance of *Marinococcus halophilus.*

Salt Polymyxin Broth (SPB)

Composition per liter:

NaCl	20.0g
Pancreatic digest of casein	10.0g
Yeast extract	3.0g
Polymyxin B	250,000U

pH 8.8 ± 0.2 at 25°C

Preparation of Medium: Add components to distilled/deionized water and bring volume to 1.0L. Mix thoroughly. Adjust pH to 8.8. Distribute into tubes or flasks. Autoclave for 10 min at 10 psi pressure–115°C.

Use: For the isolation and cultivation of *Vibrio* species from foods.

Salt Tolerance Medium

Composition per liter:

Beef heart, infusion from	500.0g
NaCl	65.0g
Tryptose	10.0g
Glucose	1.0g
Indicator solution	1.0mL

pH 7.4 ± 0.2 at 25°C

Indicator Solution:

Composition per 100mL:

Bromcresol Purple	1.6g
Ethanol (95% solution)	100.0mL

Preparation of Indicator Solution: Add Bromcresol Purple to ethanol. Mix thoroughly.

Preparation of Medium: Add components to distilled/deionized water and bring volume to 1.0L. Mix thoroughly. Distribute into tubes or flasks. Autoclave for 15 min at 15 psi pressure–121°C.

Use: For the cultivation of salt-tolerant *Streptococcus* species and other salt-tolerant Gram-positive cocci. For the differentiation of Gram-positive cocci based on salt tolerance.

Salt Tolerance Medium

Composition per liter:

NaCl	60.0g
Peptone	5.0g
Yeast extract	2.0g
Beef extract	1.0g

pH 7.4 ± 0.2 at 25°C

Preparation of Medium: Add components to distilled/deionized water and bring volume to 1.0L. Mix thoroughly. Distribute into tubes or flasks. Autoclave for 15 min at 15 psi pressure–121°C.

Use: For the cultivation and differentiation of *Aeromonas* and *Plesiomonas* species based on salt tolerance.

Salt Tolerance Medium

Composition per liter:

Beef heart, solids from infusion	500.0g
NaCl	65.0g
Tryptose	10.0g

pH 7.4 ± 0.2 at 25°C

Preparation of Medium: Add components to distilled/deionized water and bring volume to 1.0L. Mix thoroughly. Distribute into tubes or flasks. Autoclave for 15 min at 15 psi pressure–121°C.

Use: For testing the salt tolerance of a variety of microorganisms.

Salt Tolerance Medium, Gilardi

Composition per liter:

NaCl	65.0g
Pancreatic digest of casein	15.0g
Agar	15.0g
Papaic digest of soybean meal	5.0g

pH 7.3 ± 0.2 at 25°C

Preparation of Medium: Add components to distilled/deionized water and bring volume to 1.0L. Mix thoroughly. Gently heat and bring to boiling. Distribute into tubes or flasks. Autoclave for 15 min at 15 psi pressure–121°C. Do not overheat. Pour into sterile Petri dishes or leave in tubes.

Use: For the cultivation and maintenance of salt-tolerant, nonfermenting Gram-negative bacteria. For the differentiation of nonfermenting Gram-negative bacteria based on salt tolerance.

Salt Tolerance Medium, Tatum

Composition per liter:

NaCl	65.0g
Peptone	5.0g
Yeast extract	2.0g
Beef extract	1.0g

pH 7.4 ± 0.2 at 25°C

Preparation of Medium: Add components to distilled/deionized water and bring volume to 1.0L. Mix thoroughly. Distribute into tubes or flasks. Autoclave for 15 min at 15 psi pressure–121°C.

Use: For the cultivation of salt-tolerant, nonfermenting Gram-negative bacteria. For the differentiation of nonfermenting Gram-negative bacteria based on salt tolerance.

SAP 1 Agar

Composition per liter:

Agar	15.0g
Pancreatic digest of casein	5.0g
Yeast extract	5.0g
Artificial seawater	1.0L

pH 7.2 ± 0.2 at 25°C

Artificial Seawater:
Composition per liter:

NaCl	24.7g
$MgSO_4 \cdot 7H_2O$	6.3g
$MgCl_2 \cdot 6H_2O$	4.6g
$CaCl_2$	1.0g
KCl	0.7g
$NaHCO_3$	0.2g

Preparation of Artificial Seawater: Add components to distilled/deionized water and bring volume to 1.0L. Mix thoroughly.

Preparation of Medium: Add solid components to 1.0L of artificial seawater. Mix thoroughly. Gently heat and bring to boiling. Distribute into tubes or flasks. Autoclave for 15 min at 15 psi pressure–121°C. Pour into sterile Petri dishes or leave in tubes.

Use: For the isolation and cultivation of *Cytophaga* species, *Herpetosiphon* species, *Saprospira* species, and *Flexithrix* species.

SAP 2 Agar

Composition per liter:

Agar	15.0g
Pancreatic digest of casein	1.0g
Yeast extract	1.0g
Artificial seawater	1.0L

pH 7.2 ± 0.2 at 25°C

Artificial Seawater:

Composition per liter:

NaCl	24.7g
$MgSO_4 \cdot 7H_2O$	6.3g
$MgCl_2 \cdot 6H_2O$	4.6g
$CaCl_2$	1.0g
KCl	0.7g
$NaHCO_3$	0.2g

Preparation of Artificial Seawater: Add components to distilled/deionized water and bring volume to 1.0L. Mix thoroughly.

Preparation of Medium: Add solid components to 1.0L of artificial seawater. Mix thoroughly. Gently heat and bring to boiling. Distribute into tubes or flasks. Autoclave for 15 min at 15 psi pressure–121°C. Pour into sterile Petri dishes or leave in tubes.

Use: For the isolation and cultivation of *Cytophaga* species, *Herpetosiphon* species, *Saprospira* species, and *Flexithrix* species.

Saprospira grandis Medium

Composition per 1010mL:

Pancreatic digest of casein	5.0g
Yeast extract	5.0g
$Ca(NO_3)_2 \cdot 4H_2O$	0.1g
K_2HPO_4	0.02g
Sea water, filtered	1.0L
Trace elements	10.0mL

pH 7.0 ± 0.2 at 25°C

Trace Elements:

Composition per liter:

$FeSO_4 \cdot 7H_2O$	0.5mg
$ZnSO_4 \cdot 7H_2O$	0.3mg
H_3BO_3	0.1mg
$CoCl_2 \cdot 6H_2O$	0.1mg
$CuSO_4 \cdot 5H_2O$	0.1mg
$MnSO_4 \cdot 4H_2O$	0.1mg
$Na_2MoO_4 \cdot 2H_2O$	0.1mg

Preparation of Trace Elements: Add components to distilled/deionized water and bring volume to 1.0L. Mix thoroughly.

Preparation of Medium: Combine components. Mix thoroughly. Adjust pH to 7.0. Filter sterilize.

Use: For the cultivation of *Saprospira grandis*.

Sarcina maxima Medium

Composition per liter:

Glucose	10.0g
Peptone	10.0g
Yeast extract	5.0g
Cysteine·HCl solution	10.0mL

pH 6.0 ± 0.2 at 25°C

Cysteine·HCl Solution:

Composition per 10mL:

Cysteine·HCl	0.5g

Preparation of Cysteine·HCl Solution: Add cysteine·HCl to distilled/deionized water and bring volume to 10.0mL. Mix thoroughly. Filter sterilize.

Preparation of Medium: Add components, except cysteine·HCl solution, to distilled/deionized water and bring volume to 990.0mL. Mix thoroughly. Gently heat and bring to boiling. Autoclave for 15 min at 15 psi pressure–121°C. Cool to 25°C. Aseptically add sterile cysteine·HCl solution. Mix thoroughly. Aseptically distribute into sterile tubes or flasks.

Use: For the cultivation of *Sarcina maxima*.

Sarcina ventriculi Growth Medium

Composition per liter:

Glucose	30.0g
Peptone	5.0g
Yeast extract	5.0g

pH 6.0 ± 0.2 at 25°C

Preparation of Medium: Add components to distilled/deionized water and bring volume to 1.0L. Mix thoroughly. Distribute into tubes in 10.0mL volumes. Autoclave for 20 min at 15 psi pressure–121°C.

Use: For the cultivation and maintenance of *Sarcina maxima* and *Sarcina ventriculi*.

Sauton's Medium

Composition per liter:

L-Asparagine	4.0g
Citric acid	2.0g
K_2HPO_4	0.5g
$MgSO_4$	0.5g
Triton® WR 1339	0.25g
Ferric ammonium citrate	0.05g
Glycerol	40.0mL

Preparation of Medium: Add components to distilled/deionized water and bring volume to 1.0L. Mix thoroughly. Distribute into tubes or flasks. Autoclave for 15 min at 15 psi pressure–121°C.

Use: For the cultivation of *Mycobacterium tuberculosis* Bacille Calmette-Guèrin (BCG) for vaccine production.

SBG Enrichment Broth (Selenite Brilliant Green Enrichment Broth)

Composition per liter:

D-Mannitol	5.0g
Peptone	5.0g
Yeast extract	5.0g
$Na_2SeO_3 \cdot 5H_2O$	4.0g
K_2HPO_4	2.65g
KH_2PO_4	1.02g
Sodium taurocholate	1.0g
Brilliant Green	5.0mg

pH 7.2 ± 0.2 at 25°C

Source: This medium is available as a premixed powder from Difco Laboratories.

Preparation of Medium: Add components to distilled/deionized water and bring volume to 1.0L. Mix thoroughly. Gently heat and bring to boiling. Continue boiling for 5–10 min. Do not autoclave. Distribute into sterile tubes or flasks.

Use: For the selective isolation of *Salmonella* species, especially from eggs and egg products.

SBG Sulfa Enrichment

Composition per liter:

D-Mannitol	5.0g
Peptone	5.0g
Yeast extract	5.0g
$Na_2SeO_3 \cdot 5H_2O$	4.0g
K_2HPO_4	2.65g
KH_2PO_4	1.02g
Sodium taurocholate	1.0g
Sodium sulfapyridine	0.5g
Brilliant Green	5.0mg

pH 7.2 ± 0.2 at 25°C

Source: This medium is available as a premixed powder from Difco Laboratories.

Preparation of Medium: Add components to distilled/deionized water and bring volume to 1.0L. Mix thoroughly. Gently heat and bring to boiling. Continue boiling for 5–10 min. Do not autoclave. Distribute into sterile tubes or flasks.

Use: For the selective isolation of *Salmonella* species, especially from eggs and egg products.

SC Agar

Composition per liter:

Agar	15.0g
Papaic digest of soybean meal	8.0g
Corn meal (solids from infusion)	2.0g
K_2HPO_4	1.0g
KH_2PO_4	1.0g
$MgSO_4 \cdot 7H_2O$	0.2g
Hemin solution	15.0mL
Bovine serum albumin, fraction V	10.0mL
$Cysteine \cdot H_2O$ solution	10.0mL
Glucose solution	1.0mL

pH 6.6 ± 0.2 at 25°C

Hemin Solution:
Composition per 100mL:

Hemin	0.1g
NaOH (0.05*N* solution)	100.0mL

Preparation of Hemin Solution: Add hemin to NaOH solution. Mix thoroughly.

Bovine Serum Albumin, Fraction V Solution:
Composition per 10mL:

Bovine serum albumin, fraction V	2.0g

Preparation of Bovine Serum Albumin, Fraction V Solution: Add bovine serum albumin to distilled/deionized water and bring volume to 10.0mL. Mix thoroughly. Filter sterilize.

Cysteine·H_2O Solution:
Composition per 10mL:

$Cysteine \cdot H_2O$	1.0g

Preparation of Cysteine·H_2O Solution: Add cysteine·H_2O to distilled/deionized water and bring volume to 10.0mL. Mix thoroughly. Filter sterilize.

Glucose Solution:
Composition per 10mL:

D-Glucose	5.0g

Preparation of Glucose Solution: Add glucose to distilled/deionized water and bring volume to 10.0mL. Mix thoroughly. Filter sterilize.

Preparation of Medium: Add components—except bovine serum albumin, cysteine·H_2O solution, and glucose solution—to distilled/deionized water and bring volume to 979.0mL. Mix thoroughly. Adjust pH to 6.6 with NaOH. Gently heat and bring to boiling. Autoclave for 15 min at 15 psi pressure–121°C. Cool to 45°–50°C. Aseptically add 10.0mL of sterile bovine serum albumin, 10.0mL of sterile cysteine·H_2O solution, and 1.0mL of sterile glucose solution. Mix thoroughly. Pour into sterile Petri dishes or distribute into sterile tubes.

Use: For the cultivation and maintenance of *Clavibacter xyli*.

SC Broth

Composition per liter:

Papaic digest of soybean meal	8.0g

KH_2PO_4 .... 1.5g
K_2HPO_4 .... 0.5g
$MgSO_4 \cdot 7H_2O$ 0.2g
Hemin solution 15.0mL
Bovine serum albumin, fraction V 10.0mL
Cysteine·H_2O solution 10.0mL
Glucose solution 1.0mL

pH 6.6 ± 0.2 at 25°C

Hemin Solution:
Composition per 100mL:

Hemin 0.1g
NaOH (0.05*N* solution) 100.0mL

Preparation of Hemin Solution: Add hemin to NaOH solution. Mix thoroughly.

Bovine Serum Albumin, Fraction V Solution:
Composition per 10mL:

Bovine serum albumin, fraction V 2.0g

Preparation of Bovine Serum Albumin, Fraction V Solution: Add bovine serum albumin to distilled/deionized water and bring volume to 10.0mL. Mix thoroughly. Filter sterilize.

Cysteine·H_2O Solution:
Composition per 10mL:

Cysteine·H_2O 1.0g

Preparation of Cysteine·H_2O Solution: Add cysteine·H_2O to distilled/deionized water and bring volume to 10.0mL. Mix thoroughly. Filter sterilize.

Glucose Solution:
Composition per 10mL:

D-Glucose 5.0g

Preparation of Glucose Solution: Add glucose to distilled/deionized water and bring volume to 10.0mL. Mix thoroughly. Filter sterilize.

Preparation of Medium: Add components—except bovine serum albumin, cysteine·H_2O solution, and glucose solution—to distilled/deionized water and bring volume to 979.0mL. Mix thoroughly. Adjust pH to 6.6 with NaOH. Gently heat and bring to boiling. Autoclave for 15 min at 15 psi pressure–121°C. Cool to 45°–50°C. Aseptically add 10.0mL of sterile bovine serum albumin, 10.0mL of sterile cysteine·H_2O solution, and 1.0mL of sterile glucose solution. Mix thoroughly. Aseptically distribute into sterile tubes or flasks.

Use: For the cultivation of *Clavibacter xyli*.

SC Medium

Composition per 1021mL:

Agar 15.0g
Papaic digest of soybean meal 8.0g
Corn meal, solids from infusion 2.0g
Tween™ 80 .... 1.0g
K_2HPO_4 .... 1.0g
KH_2PO_4 .... 1.0g
$MgSO_4 \cdot 7H_2O$ 0.2g
Hemin chloride solution 15.0mL
Bovine serum albumin solution 10.0mL
Cysteine solution 10.0mL
Glucose solution 1.0mL

pH 6.6 at 25°C

Hemin Chloride Solution:
Composition per 100mL:

Hemin chloride 0.1g
NaOH (0.05*N* solution) 100.0mL

Preparation of Hemin Chloride Solution: Add hemin chloride to 100.0mL of NaOH solution. Mix thoroughly.

Bovine Serum Albumin Solution:
Composition per 10mL:

Bovine serum albumin 2.0g

Preparation of Bovine Serum Albumin Solution: Add bovine serum albumin to distilled/deionized water and bring volume to 10.0mL. Mix thoroughly. Filter sterilize.

Cysteine Solution:
Composition per 10mL:

Cysteine, free base 1.0g

Preparation of Cysteine Solution: Add cysteine to distilled/deionized water and bring volume to 10.0mL. Mix thoroughly. Filter sterilize.

Glucose Solution:
Composition per 10mL:

Glucose 5.0g

Preparation of Glucose Solution: Add glucose to distilled/deionized water and bring volume to 10.0mL. Mix thoroughly. Autoclave for 15 min at 15 psi pressure–121°C. Cool to 25°C.

Preparation of Medium: Add components—except bovine serum albumin solution, cysteine solution, and glucose solution—to distilled/deionized water and bring volume to 1.0L. Mix thoroughly. Gently heat and bring to boiling. Autoclave for 15 min at 15 psi pressure–121°C. Cool to 45°–50°C. Aseptically add 10.0mL of sterile bovine serum albumin solution, 10.0mL of sterile cysteine solution, and 1.0mL of sterile glucose solution. Mix thoroughly. Pour into sterile Petri dishes or distribute into sterile tubes.

Use: For the isolation and cultivation of coryneform bacteria that cause ratoon stunting disease of sugarcane.

Schaedler Agar (Schaedler Anaerobic Agar)

Composition per liter:

Agar	13.5g
Glucose	5.83g
Pancreatic digest of casein	5.7g
Proteose peptone No. 3	5.0g
Yeast extract	5.0g
Tris(hydroxymethyl)aminomethane buffer	3.0g
NaCl	1.65g
Papaic digest of soybean meal	1.0g
K_2HPO_4	0.83g
L-Cystine	0.4g
Hemin	0.01g

pH 7.6 ± 0.2 at 25°C

Source: This medium is available as a premixed powder from Difco Laboratories and Oxoid Unipath.

Preparation of Medium: Add components to distilled/deionized water and bring volume to 1.0L. Mix thoroughly. Gently heat and bring to boiling. Distribute into tubes or flasks. Autoclave for 15 min at 15 psi pressure–121°C. Pour into sterile Petri dishes or leave in tubes.

Use: For the isolation, cultivation and enumeration of anaerobic and aerobic microorganisms.

Schaedler Agar

Composition per liter:

Agar	13.5g
Pancreatic digest of casein	8.2g
Glucose	5.8g
Yeast extract	5.0g
Tris(hydroxymethyl)aminomethane buffer	3.0g
Peptic digest of animal tissue	2.5g
NaCl	1.7g
Papaic digest of soybean meal	1.0g
K_2HPO_4	0.8g
L-Cystine	0.4g
Hemin	0.01g

pH 7.6 ± 0.2 at 25°C

Source: This medium is available as a premixed powder from BBL Microbiology Systems.

Preparation of Medium: Add components to distilled/deionized water and bring volume to 1.0L. Mix thoroughly. Gently heat and bring to boiling. Distribute into tubes or flasks. Autoclave for 15 min at 15 psi pressure–121°C. Pour into sterile Petri dishes or leave in tubes.

Use: For the isolation, cultivation and enumeration of anaerobic and aerobic microorganisms.

Schaedler Agar with Vitamin K_1 and Sheep Blood

Composition per liter:

Agar	13.5g
Pancreatic digest of casein	8.2g
Glucose	5.8g
Yeast extract	5.0g
Tris(hydroxymethyl)aminomethane buffer	3.0g
Peptic digest of animal tissue	2.5g
Papaic digest of soybean meal	1.0g
NaCl	1.7g
K_2HPO_4	0.8g
L-Cystine	0.4g
Hemin	0.01g
Sheep blood, defibrinated	50.0mL
Vitamin K_1 solution	1.0mL

pH 7.6 ± 0.2 at 25°C

Vitamin K_1 Solution:
Composition per 10mL:

Vitamin K_1	5.0g
Ethanol, absolute	10.0mL

Preparation of Vitamin K_1 Solution: Add vitamin K_1 to ethanol. Mix thoroughly.

Preparation of Medium: Add components, except sheep blood, to distilled/deionized water and bring volume to 950.0mL. Mix thoroughly. Gently heat and bring to boiling. Autoclave for 15 min at 15 psi pressure–121°C. Cool to 45°–50°C. Aseptically add sterile sheep blood. Mix thoroughly. Pour into sterile Petri dishes or distribute into sterile tubes.

Use: For the recovery of fastidious anaerobic bacteria such as *Bacteroides* species.

Schaedler Anaerobic Agar
See: **Schaedler Agar**

Schaedler Anaerobic Broth
See: **Schaedler Broth**

Schaedler Broth (Schaedler Anaerobic Broth)

Composition per liter:

Pancreatic digest of casein	8.2g
Glucose	5.8g
Yeast extract	5.0g
Tris(hydroxymethyl)aminomethane buffer	3.0g
Peptic digest of animal tissue	2.5g
NaCl	1.7g
Papaic digest of soybean meal	1.0g

K_2HPO_4 .... 0.8g
L-Cystine 0.4g
Hemin 0.01g

pH 7.6 ± 0.2 at 25°C

Source: This medium is available as a premixed powder from BBL Microbiology Systems, Difco Laboratories and Oxoid Unipath.

Preparation of Medium: Add components to distilled/deionized water and bring volume to 1.0L. Mix thoroughly. Distribute into tubes or flasks. Autoclave for 15 min at 15 psi pressure–121°C.

Use: For the cultivation and maintenance of *Eubacterium combesii, Eubacterium contortum* and a variety of other anaerobic bacteria.

Schaedler CNA Agar with Vitamin K_1 and Sheep Blood

Composition per liter:

Agar 13.5g
Pancreatic digest of casein 8.2g
Glucose 5.8g
Yeast extract 5.0g
Tris(hydroxymethyl)aminomethane buffer 3.0g
Peptic digest of animal tissue 2.5g
Papaic digest of soybean meal 1.0g
NaCl 1.7g
K_2HPO_4 .... 0.8g
L-Cystine 0.4g
Hemin 0.01g
Colistin 0.01g
Nalidixic acid 0.01g
Sheep blood, defibrinated 50.0mL
Vitamin K_1 solution 1.0mL

pH 7.6 ± 0.2 at 25°C

Vitamin K_1 Solution:
Composition per 10mL:

Vitamin K_1 .... 5.0g
Ethanol, absolute 10.0mL

Preparation of Vitamin K_1 Solution: Add vitamin K_1 to ethanol. Mix thoroughly.

Preparation of Medium: Add components, except sheep blood, to distilled/deionized water and bring volume to 950.0mL. Mix thoroughly. Gently heat and bring to boiling. Autoclave for 15 min at 15 psi pressure–121°C. Cool to 45°–50°C. Aseptically add sterile sheep blood. Mix thoroughly. Pour into sterile Petri dishes or distribute into sterile tubes.

Use: For the selective isolation of anaerobic, Gram-positive cocci, especially *Peptococcus* species and *Peptostreptococcus* species.

Schaedler KV Agar with Vitamin K_1 and Sheep Blood

Composition per liter:

Agar 13.5g
Pancreatic digest of casein 8.2g
Glucose 5.8g
Yeast extract 5.0g
Tris(hydroxymethyl)aminomethane buffer 3.0g
Peptic digest of animal tissue 2.5g
Papaic digest of soybean meal 1.0g
NaCl 1.7g
K_2HPO_4 .... 0.8g
L-Cystine 0.4g
Hemin 0.01g
Kanamycin 0.01g
Vancomycin 7.5mg
Sheep blood, defibrinated .... 50.0 mL
Vitamin K_1 solution 1.0mL

pH 7.6 ± 0.2 at 25°C

Vitamin K_1 Solution:
Composition per 10mL:

Vitamin K_1 .... 5.0g
Ethanol, absolute 10.0mL

Preparation of Vitamin K_1 Solution: Add vitamin K_1 to ethanol. Mix thoroughly.

Preparation of Medium: Add components, except sheep blood, to distilled/deionized water and bring volume to 950.0mL. Mix thoroughly. Gently heat and bring to boiling. Autoclave for 15 min at 15 psi pressure–121°C. Cool to 45°–50°C. Aseptically add sterile sheep blood. Mix thoroughly. Pour into sterile Petri dishes or distribute into sterile tubes.

Use: For the selective isolation of Gram-negative anaerobic bacteria.

Schleifer–Krämer Agar (SK Agar)

Composition per liter:

Agar 13.0g
Glycerol 10.0g
Sodium pyruvate 10.0g
Pancreatic digest of casein 10.0g
Beef extract 5.0g
Yeast extract 3.0g
Potassium isothiocyanate 2.25g
LiCl 2.0g
$Na_2HPO_4 \cdot 2H_2O$ 0.9g
$NaH_2PO_4 \cdot H_2O$ 0.6g
Glycine 0.5g
NaN_3 solution 10.0mL

pH 7.2 ± 0.2 at 25°C

NaN_3 Solution:
Composition per 10mL:
NaN_3 .... 0.045g

Preparation of NaN_3 Solution: Add NaN_3 to distilled/deionized water and bring volume to 10.0mL. Mix thoroughly. Filter sterilize.

Preparation of Medium: Add components, except NaN_3 solution, to distilled/deionized water and bring volume to 990.0mL. Mix thoroughly. Adjust pH to 7.2. Gently heat and bring to boiling. Autoclave for 15 min at 15 psi pressure–121°C. Cool to 45°–50°C. Aseptically add sterile NaN_3 solution. Mix thoroughly. Pour into sterile Petri dishes or distribute into sterile tubes.

Use: For the isolation and cultivation of *Staphylococcus* species.

Schneider's *Drosophila* Medium

Composition per liter:

$MgSO_4 \cdot 7H_2O$	3.7g
NaCl	2.1g
Yeast extract	2.0g
Trehalose	2.0g
D-Glucose	2.0g
L-Glutamine	1.8g
L-Lysine·HCl	1.7g
L-Proline	1.7g
KCl	1.6g
$Na_2HPO_4 \cdot 7H_2O$	1.3g
L-Glutamic acid	0.8g
L-Methionine	0.8g
$CaCl_2$, anhydrous	0.6g
KH_2PO_4	0.5g
β-Alanine	0.5g
L-Tyrosine	0.5g
L-Arginine	0.4g
L-Aspartic acid	0.4g
L-Histidine	0.4g
L-Threonine	0.4g
$NaHCO_3$	0.4g
Glycine	0.3g
L-Serine	0.3g
L-Valine	0.3g
L-Isoleucine	0.2g
L-Leucine	0.2g
L-Phenylalanine	0.2g
α-Ketoglutaric acid	0.2g
Fumaric acid	0.1g
Malic acid	0.1g
Succinic acid	0.1g
L-Cystine	0.1g
L-Tryptophan	0.1g
L-Cysteine	0.06g

Preparation of Medium: Add components to distilled/deionized water and bring volume to 1.0L. Mix thoroughly. Filter sterilize.

Use: For the cultivation of *Drosophila* and other insect species. Also use as a chemically-defined supplement for the cultivation of fastidious microorganisms.

SCY Medium (Maintenance SCY Medium)

Composition per liter:

Agar	10.0g
Sucrose	1.0g
Pancreatic digest of casein	0.92g
Yeast extract	0.25g
NaCl	0.05g
Papaic digest of soybean meal	0.03g
K_2HPO_4	0.025g
Thiamine	0.4mg
Cyanocobalamin	0.01mg

pH 7.3 ± 0.2 at 25°C

Preparation of Medium: Add components to distilled/deionized water and bring volume to 1.0L. Mix thoroughly. Filter sterilize.

Use: For the cultivation and maintenance of iron and sulfur bacteria.

SCY Medium (Maintenance SCY Medium)

Composition per liter:

Solution A	1.0L
Solution B	200.0mL

Solution A:
Composition per liter:

Agar	10.0g
Pancreatic digest of casein	0.92g
NaCl	0.05g
Papaic digest of soybean meal	0.03g
K_2HPO_4	0.025g

pH 7.0 ± 0.2 at 25°C

Preparation of Solution A: Add components to distilled/deionized water and bring volume to 1.0L. Mix thoroughly. Gently heat and bring to boiling. Distribute into tubes in 10.0mL volumes. Autoclave for 15 min at 15 psi pressure–121°C. Allow tubes to cool in a slanted position.

Solution B:
Composition per 200mL:

Sucrose	2.0g
Yeast extract	0.5g

Thiamine 0.8mg
Vitamin B_{12} 0.02mg
pH 8.5 ± 0.2 at 25°C

Preparation of Solution B: Add components to slightly alkaline tap water, pH 8.5, and bring volume to 200.0mL. Mix thoroughly. Filter sterilize.

Preparation of Medium: Inoculate bacteria onto prepared slants of solution A. After inoculation of tubes, aseptically add 2.0mL of sterile solution B on top of each slant.

Use: For the cultivation and maintenance of iron bacteria. For the cultivation and maintenance of *Haliscomenobacter hydrossis*.

SD Medium
See: Serratia **Differential Medium**

Sea Water Agar (SWA)

Composition per liter:

Agar 15.0g
Peptone 5.0g
Yeast extract 5.0g
Beef extract 3.0g
Sea water, synthetic 1.0L
pH 7.5 ± 0.2 at 25°C

Sea Water, Synthetic:

Composition per liter:

NaCl 27.0g
$MgSO_4 \cdot 7H_2O$ 7.0g
Tris(hydroxymethyl)aminomethane buffer 2.0g
KCl 0.6g
$CaCl_2$ 0.3g

Preparation of Sea Water, Synthetic: Add components to distilled/deionized water and bring volume to 1.0L. Mix thoroughly.

Preparation of Medium: Combine components. Mix thoroughly. Gently heat and bring to boiling. Distribute into tubes or flasks. Autoclave for 15 min at 15 psi pressure–121°C. Pour into sterile Petri dishes or leave in tubes.

Use: For the isolation and cultivation of halophilic microorganisms from foods, such as *Pseudomonas* species and *Vibrio* species from fish.

Sea Water Agar

Composition per 250mL:

Agar 20.0g
Beef extract 10.0g
Peptone 10.0g
Sea water 750.0mL
pH 7.2 ± 0.2 at 25°C

Preparation of Medium: Add components to tap water and bring volume to 1.0L. Mix thoroughly. Gently heat and bring to boiling. Distribute into tubes or flasks. Autoclave for 15 min at 15 psi pressure–121°C. Pour into sterile Petri dishes or leave in tubes.

Use: For the selective isolation and cultivation of *Planococcus* species.

Sea Water Agar (SWA)

Composition per liter:

Agar 15.0g
Peptone 5.0g
Yeast extract 5.0g
Beef extract 3.0g
Sea water, synthetic 1.0L
pH 7.5 ± 0.2 at 25°C

Sea Water, Synthetic:

Composition per liter:

NaCl 24.0g
$MgSO_4 \cdot 7H_2O$ 7.0g
$MgCl_2 \cdot 6H_2O$ 5.3g
KCl 0.7g
$CaCl_2$ 0.1g

Preparation of Sea Water, Synthetic: Add components to distilled/deionized water and bring volume to 1.0L. Mix thoroughly. Adjust pH to 7.5.

Preparation of Medium: Combine components. Mix thoroughly. Gently heat and bring to boiling. Distribute into tubes or flasks. Autoclave for 15 min at 15 psi pressure–121°C. Pour into sterile Petri dishes or leave in tubes.

Use: For the isolation and cultivation of halophilic microorganisms from foods, such as *Pseudomonas* species and *Vibrio* species from fish.

Sea Water Agar Medium

Composition per liter:

Agar 15.0g
Beef extract 10.0g
Peptone 10.0g
Sea water, aged 750.0mL
pH 7.2–7.3 at 25°C

Preparation of Medium: Add components to distilled/deionized water and bring volume to 1.0L. Mix thoroughly. Gently heat and bring to boiling. Distribute into tubes or flasks. Autoclave for 15 min at 15 psi

pressure–121°C. Pour into sterile Petri dishes or leave in tubes.

Use: For the isolation and cultivation of marine *Flavobacterium* species.

Sea Water Complete Medium

Composition per liter:

Pancreatic digest of casein	5.0g
Yeast extract	3.0g
Sea water	750.0mL
Glycerol	3.0mL

Preparation of Medium: Add components to distilled/deionized water and bring volume to 1.0L. Mix thoroughly. Distribute into tubes or flasks. Autoclave for 15 min at 15 psi pressure–121°C.

Use: For the cultivation and maintenance of *Vibrio fischeri*.

Sea Water Medium

Composition per liter:

Agar	15.0g
Peptone	5.0g
Beef extract	2.0g
KNO_3	0.5g
Sea water, aged	1.0L

pH 7.8 ± 0.2 at 25°C

Preparation of Medium: Combine components. Mix thoroughly. Gently heat and bring to boiling. Distribute into tubes or flasks. Autoclave for 15 min at 15 psi pressure–121°C. Pour into sterile Petri dishes or leave in tubes.

Use: For the cultivation of halophilic bacteria.

Sea Water Medium

Composition per liter:

Agar	15.0g
Peptone	5.0g
Yeast extract	1.0g
$FeSO_4$	0.2g

Preparation of Medium: Add components to 1.0L of sea water. Mix thoroughly. Gently heat and bring to boiling. Distribute into tubes or flasks. Autoclave for 15 min at 15 psi pressure–121°C. Pour into sterile Petri dishes or leave in tubes.

Use: For the cultivation and maintenance of *Cyclobacterium marinus*.

Sea Water Yeast Extract Agar

Composition per liter:

Marine salts mix	37.9g
Agar	15.0g
Proteose peptone	10.0g
Yeast extract	3.0g

pH 7.2–7.4 at 25°C

Preparation of Medium: Add components to distilled/deionized water and bring volume to 1.0L. Mix thoroughly. Gently heat and bring to boiling. Distribute into tubes or flasks. Autoclave for 15 min at 15 psi pressure–121°C. Pour into sterile Petri dishes or leave in tubes.

Use: For the cultivation and maintenance of *Alteromonas* species, *Caulobacter halobacteroides*, *Caulobacter maris*, *Cytophaga marinoflava* and *Cytophaga salmonicolor*.

Sea Water Yeast Extract Broth, Modified

Composition per liter:

NaCl	23.4g
$MgSO_4 \cdot 7H_2O$	6.9g
Peptone	1.0g
Yeast extract	1.0g
KCl	0.75g

Preparation of Medium: Add components to distilled/deionized water and bring volume to 1.0L. Mix thoroughly. Distribute into tubes or flasks. Autoclave for 15 min at 15 psi pressure–121°C.

Use: For the cultivation and maintenance of the *Proteus* species and the *Vibrio* species.

Sea Water Yeast Extract Peptone Medium

Composition per liter:

Agar	15.0g
Peptone	5.0g
Yeast extract	3.0g
Sea water, aged and filtered	750.0mL

pH 7.3 ± 0.2 at 25°C

Preparation of Medium: Add components, except agar, to distilled/deionized water and bring volume to 1.0L. Mix thoroughly. Adjust pH to 7.8. Gently heat and bring to boiling. Continue boiling for 3–5 min. Filter through Whatman filter paper. Adjust pH to 7.3. Add agar. Gently heat and bring to boiling. Distribute into tubes or flasks. Autoclave for 15 min at 15 psi pressure–121°C. Pour into sterile Petri dishes or leave in tubes.

Use: For the cultivation of *Planococcus kocurii*.

Seed Agar
See: **Antibiotic Medium 1**

Selective 7H11 Agar
See: **Seven H11 Agar**

Selenite Brilliant Green Enrichment Broth
See: **SBG Enrichment Broth**

Selenite Broth (Selenite Broth, Lactose) (Selenite F Enrichment Medium) (Sodium Biselenite Medium) (Sodium Hydrogen Selenite Medium)

Composition per liter:

Na_2HPO_4	10.0g
Pancreatic digest of casein	5.0g
Lactose	4.0g
$NaHSeO_3 \cdot 5H_2O$	4.0g

pH 7.0 ± 0.2 at 25°C

Source: This medium is available as a premixed powder from Difco Laboratories and a prepared media from Oxoid Unipath.

Caution: Sodium biselenite is toxic and a potential teratogen and care must be taken to avoid inhalation of the powdered dye, contact with the skin, or ingestion, especially in pregnant laboratory workers.

Preparation of Medium: Add components to distilled/deionized water and bring volume to 1.0L. Mix thoroughly. Gently heat and bring to boiling. Do not autoclave. Distribute into sterile tubes in 10.0mL volumes.

Use: For the isolation and enrichment of *Salmonella* species from clinical specimens and food products.

Selenite Broth Base, Mannitol

Composition per liter:

Na_2HPO_4	10.0g
Peptone	5.0g
Mannitol	4.0g
$NaHSeO_3 \cdot 5H_2O$	4.0g

pH 7.1 ± 0.2 at 25°C

Source: This medium is available as a premixed powder from Oxoid Unipath.

Caution: Sodium selenite is toxic and a potential teratogen and care must be taken to avoid inhalation of the powdered dye, contact with the skin, or ingestion, especially in pregnant laboratory workers.

Preparation of Medium: Add components to distilled/deionized water and bring volume to 1.0L. Mix thoroughly. Gently heat. Do not autoclave. Distribute into sterile tubes in 10.0mL volumes. Sterilize for 10 min at 0 psi pressure–100°C.

Use: For the isolation and cultivation of *Salmonella typhi* and *Salmonella paratyphi B*.

Selenite Cystine Broth

Composition per liter:

Na_2HPO_4	10.0g
Pancreatic digest of casein	5.0g
Lactose	4.0g
$Na_2SeO_3 \cdot 5H_2O$	4.0g
L-Cystine	0.02g

pH 7.0 ± 0.2 at 25°C

Source: This medium is available as a premixed powder from BBL Microbiology Systems, Difco Laboratories and Oxoid Unipath.

Caution: Sodium selenite is toxic and a potential teratogen and care must be taken to avoid inhalation of the powdered dye, contact with the skin, or ingestion, especially in pregnant laboratory workers.

Preparation of Medium: Add components to distilled/deionized water and bring volume to 1.0L. Mix thoroughly. Gently heat. Do not autoclave. Distribute into sterile tubes in 10.0mL volumes. Sterilize for 15 min at 0 psi pressure–100°C.

Use: For the isolation and cultivation of *Salmonella* species from feces, dairy products and other specimens.

Selenite F Broth

Composition per liter:

KH_2PO_4	7.0g
Pancreatic digest of casein	5.0g
Lactose	4.0g
$Na_2SeO_3 \cdot 5H_2O$	4.0g
Na_2HPO_4	3.0g

pH 7.0 ± 0.2 at 25°C

Source: This medium is available as a premixed powder from BBL Microbiology Systems.

Caution: Sodium selenite is toxic and a potential teratogen and care must be taken to avoid inhalation of the powdered dye, contact with the skin, or ingestion, especially in pregnant laboratory workers.

Preparation of Medium: Add components to distilled/deionized water and bring volume to 1.0L. Mix thoroughly. Gently heat. Do not autoclave. Distribute into sterile tubes in 10.0mL volumes. Sterilize for 30 min at 0 psi pressure–100°C.

Use: For the isolation and cultivation of *Salmonella* species from feces, dairy products and other specimens.

Selenite F Enrichment Medium
See: **Selenite Broth**

Selenomonas acidaminophila Medium

Composition per liter:

Disodium β-glycerophosphate	19.0g
Beef extract	5.0g
Lactose	5.0g
Papaic digest of soybean meal	5.0g
Sodium glutamate	3.4g
Pancreatic digest of casein	2.5g
Peptic digest of animal tissue	2.5g
Yeast extract	2.5g
Ascorbic acid	0.5g
$MgSO_4 \cdot 7H_2O$	0.25g

pH 7.15 ± 0.05 at 25°C

Preparation of Medium: Add components to distilled/deionized water and bring volume to 1.0L. Mix thoroughly. Distribute into tubes or flasks. Autoclave for 15 min at 15 psi pressure–121°C.

Use: For the cultivation and maintenance of *Selenomonas acidaminophila*.

Selenomonas Selective Medium (SS Medium)

Composition per 100mL:

Pancreatic digest of casein	0.5g
Mannitol	0.2g
$FeSO_4 \cdot 7H_2O$	0.1g
Sodium acetate	0.1g
Yeast extract	0.1g
L-Cysteine·HCl	0.08g
Mineral solution S	4.0mL
Sodium carbonate (8% solution)	2.5mL
n-Valeric acid	0.05mL

pH 5.9–6.1 at 25°C

Mineral Solution S:

Composition per liter:

KH_2PO_4	12.0g
NaCl	12.0g
$(NH_4)_2SO_4$	6.0g
$MgSO_4 \cdot 7H_2O$	2.5g
$CaCl_2 \cdot 2H_2O$	1.6g

Preparation of Mineral Solution S: Add components to distilled/deionized water and bring volume to 1.0L. Mix thoroughly.

Preparation of Medium: Add components to distilled/deionized water and bring volume to 100.0mL. Mix thoroughly. Filter sterilize. Aseptically distribute into sterile tubes or flasks.

Use: For the isolation and cultivation of *Selenomonas* species.

Sellers Agar (Sellers Differential Agar)

Composition per 1015mL:

Pancreatic digest of gelatin	20.0g
Agar	13.5g
D-Mannitol	2.0g
NaCl	2.0g
$MgSO_4 \cdot 7H_2O$	1.5g
K_2HPO_4	1.0g
L-Arginine	1.0g
$NaNO_3$	1.0g
Yeast extract	1.0g
$NaNO_3$	0.35g
Bromthymol Blue	0.04g
Phenol Red	8.0mg
Glucose solution	15.0mL

pH 6.7 ± 0.2 at 25°C

Source: This medium is available as a premixed powder from BBL Microbiology Systems.

Glucose Solution:

Composition per 10mL:

D-Glucose	5.0g

Preparation of Glucose Solution: Add glucose to distilled/deionized water and bring volume to 10.0mL. Mix thoroughly. Filter sterilize.

Preparation of Medium: Add components, except glucose solution, to distilled/deionized water and bring volume to 1.0L. Mix thoroughly. Gently heat and bring to boiling. Distribute into tubes in 10.0mL volumes. Autoclave for 15 min at 15 psi pressure–121°C. Allow tubes to cool in a slanted position to form a 3 inch slant with a 1.5 inch butt. Immediately prior to inoculation aseptically add 0.15mL of sterile glucose solution to each tube. Let the glucose solution run down the side of the tube opposite the slant.

Use: For the cultivation and differentiation of nonfermentative Gram-negative bacilli, especially *Pseudomonas aeruginosa, Herellea vaginicola (Acinetobacter calcoaceticus) Mima polymorpha (Acinetobacter lwoffii), Alcaligenes faecalis* and *Bacterium anitratum (Acinetobacter calcoaceticus).*

Semiselective Medium for *Legionella pneumophila* *See:* BMPA–α Medium

Semisolid BSA Tween™ 80 Medium *See:* Bovine Serum Albumin Tween™ 80 Soft Agar

Semisolid Medium for Motility

Composition per liter:

Biosate	5.0g
Polypeptone™	5.0g
NaCl	5.0g
Agar	4.0g
Myosate	1.5g
Triphenyltetrazolium chloride solution	2.5mL

pH 6.9–7.1at 25°C

Triphenyltetrazolium Chloride Solution:

Composition per 10mL:

Triphenyltetrazolium chloride	0.1g
Ethanol (95% solution)	10.0mL

Preparation of Triphenyltetrazolium Chloride Solution: Add triphenyltetrazolium chloride to 10.0mL of ethanol. Mix thoroughly.

Preparation of Medium: Add components, except triphenyltetrazolium chloride solution, to distilled/deionized water and bring volume to 997.5mL. Mix thoroughly. Gently heat and bring to boiling. Add 2.5mL of triphenyltetrazolium chloride solution. Mix thoroughly. Distribute into tubes in 10.0mL volumes. Autoclave for 15 min at 15 psi pressure–121°C.

Use: For the differentiation of bacteria based on motility.

Semisolid Pectin Agar

Composition per liter:

Pectin	30.0g
Yeast extract	5.0g
Agar	3.0g
Bromthymol Blue (0.1% solution)	1.0mL
$CaCl_2 \cdot 2H_2O$ (10% solution	0.6mL

pH 7.3 ± 0.2 at 25°C

Preparation of Medium: Add approximately 100.0mL of distilled/deionized water to a 2-L flask. Place on a magnetic stirrer without heat. Slowly add the $CaCl_2 \cdot 2H_2O$ solution, Bromthymol Blue solution, yeast extract, and pectin while stirring. Mix thoroughly to ensure uniform wetting of the particles. Add agar. Gently heat and bring to almost boiling. Adjust pH to 7.3 with 1*N* NaOH if necessary. Do not overshoot pH above 7.3. Distribute into tubes or flasks. Autoclave for 15 min at 15 psi pressure–121°C.

Use: For the isolation and cultivation of bacteria such as *Erwinia* species and some *Klebsiella* species based on their ability to degrade pectin.

Sensitest Agar

Composition per liter:

Pancreatic digest of casein	11.0g
Agar	8.0g
Buffer salts	3.3g
Peptone	3.0g
NaCl	3.0g
Glucose	2.0g
Starch	1.0g
Nucleoside bases	0.02g
Thiamine	0.02mg

pH 7.4 ± 0.2 at 25°C

Source: This medium is available as a premixed powder from Oxoid Unipath.

Preparation of Medium: Add components to distilled/deionized water and bring volume to 1.0L. Mix thoroughly. Gently heat and bring to boiling. Distribute into tubes or flasks. Autoclave for 15 min at 15 psi pressure–121°C. Pour into sterile Petri dishes.

Use: For performance of antibiotic sensitivity assays.

Serratia Differential Medium (SD Medium)

Composition per 102mL:

Solution A	92.0mL
Solution B	10.0mL

pH 6.7 ± 0.2 at 25°C

Solution A:

Composition per 92mL:

Yeast extract	1.0g
L-Ornithine	1.0g
NaCl	0.5g

Agar..0.4g
Irgasan inhibitor...1.0mL
Indicator solution...1.0mL

Preparation of Solution A: Add components to distilled/deionized water and bring volume to 92.0mL. Mix thoroughly. Adjust pH to 6.7 with 1*N* NaOH.

Irgasan Inhibitor:
Composition per 100mL:
Irgasan-DP-300 (4,2´, 4´-trichloro-
2-hydroxydiphenylether)..............................0.1g
NaOH (1*N* solution).....................................10.0mL

Preparation of Irgasan Inhibitor: Add irgasan to 10.0mL of NaOH solution. Mix thoroughly. Gently heat to dissolve. Bring volume to 100.0mL with distilled/deionized water.

Indicator Solution:
Composition per 100mL:
Bromthymol Blue...0.2g
Phenol Red...0.1g

Preparation of Indicator Solution: Add components to 50.0mL of distilled/deionized water. Mix thoroughly for 1 hr. Bring volume to 100.0mL with distilled/deionized water.

Solution B:
Composition per 10mL:
L-Arabinose..1.0g

Preparation of Solution B: Add arabinose to distilled/deionized water and bring volume to 10.0mL. Mix thoroughly.

Preparation of Medium: Combine 92.0mL of solution A with 10.0mL of solution B. Mix thoroughly. Distribute into tubes. Autoclave for 15 min at 15 psi pressure–121°C. Allow tubes to cool in an upright position.

Use: For the cultivation and differentiation of *Serratia* species based on fermentation of arabinose and production of ornithine decarboxylase. *S. marcescens* changes the medium to purple throughout the tube. *S. liquefaciens* changes the medium to a band of purple at the top of the tube with a green/yellow butt. *S. rubidaea* changes the medium to yellow throughout the tube.

Serratia Hd–MHr

Composition per liter:
Agar..15.0g
K_2HPO_4..7.0g
Glucose...5.0g
KH_2PO_4..3.0g
2-Methyl-DL-histidine·2HCl...............................1.0g
$(NH_4)_2SO_4$...1.0g
$MgSO_4 \cdot 7H_2O$..0.5g

Preparation of Medium: Add components to distilled/deionized water and bring volume to 1.0L. Mix thoroughly. Gently heat and bring to boiling. Distribute into tubes or flasks. Autoclave for 15 min at 15 psi pressure–121°C. Pour into sterile Petri dishes or leave in tubes.

Use: For the cultivation and maintenance of *Serratia marcescens*.

Serratia Medium (ATCC Medium 181)

Composition per liter:
Agar..20.0g
Pancreatic digest of casein...............................5.0g
Yeast extract...5.0g
Glucose...1.0g
K_2HPO_4..1.0g
pH 7.0 ± 0.2 at 25°C

Preparation of Medium: Add components to distilled/deionized water and bring volume to 1.0L. Mix thoroughly. Gently heat and bring to boiling. Distribute into tubes or flasks. Autoclave for 15 min at 15 psi pressure–121°C. Pour into sterile Petri dishes or leave in tubes.

Use: For the cultivation and maintenance of *Serratia marcescens*.

Serratia Medium (ATCC Medium 1399)

Composition per liter:
Agar..15.0g
K_2HPO_4..7.0g
Glucose...5.0g
KH_2PO_4..3.0g
Casein hydrolysate...1.0g
$(NH_4)_2SO_4$...1.0g
Yeast extract...1.0g
$MgSO_4 \cdot 7H_2O$..0.1g
pH 7.0 ± 0.2 at 25°C

Preparation of Medium: Add components to distilled/deionized water and bring volume to 1.0L. Mix thoroughly. Gently heat and bring to boiling. Distribute into tubes or flasks. Autoclave for 15 min at 15 psi pressure–121°C. Pour into sterile Petri dishes or leave in tubes.

Use: For the cultivation and maintenance of *Serratia marcescens*.

Serum Glucose Agar

Composition per 1060mL:

Agar	15.0g
Peptone	10.0g
Beef extract	5.0g
NaCl	5.0g
Serum-glucose solution	60.0mL

pH 7.3 ± 0.2 at 25°C

Serum-Glucose Solution:

Composition per 60mL:

D-Glucose	10.0g
Serum (inactivated at 56°C, 30 min)	50.0mL

Preparation of Serum-Glucose Solution: Add glucose to 50.0mL of heat-inactivated serum. Horse serum or ox serum may be used. Mix thoroughly. Filter sterilize.

Preparation of Medium: Add components, except serum-glucose solution, to distilled/deionized water and bring volume to 1.0L. Mix thoroughly. Gently heat and bring to boiling. Autoclave for 15 min at 10 psi pressure–115°C. Cool to 50°C. Aseptically add 60.0mL of sterile serum-glucose solution. Mix thoroughly. Pour into sterile Petri dishes or distribute into sterile tubes. Allow tubes to cool in a slanted position.

Use: For the cultivation and maintenance of *Brucella* species.

Serum Glucose Agar, Farrell Modified

Composition per 1086.9mL:

Agar	15.0g
Peptone	10.0g
Beef extract	5.0g
NaCl	5.0g
Serum-glucose solution	60.0mL
Bacitracin solution	12.5mL
Cycloheximide solution	10.0mL
Nystatin solution	2.0mL
Polymyxin B solution	1.0mL
Nalidixic acid solution	1.0mL
Vancomycin solution	0.4mL

pH 7.3 ± 0.2 at 25°C

Serum-Glucose Solution:

Composition per 60mL:

D-Glucose	10.0g
Serum (inactivated at 56°C, 30 min)	50.0mL

Preparation of Serum-Glucose Solution: Add glucose to 50.0mL of heat-inactivated serum. Horse serum or ox serum may be used. Mix thoroughly. Filter sterilize.

Bacitracin Solution:

Composition per 12.5mL:

Bacitracin	25,000U

Preparation of Bacitracin Solution: Add Bacitracin to distilled/deionized water and bring volume to 12.5mL. Mix thoroughly. Filter sterilize.

Cycloheximide Solution:

Composition per 100mL:

Cycloheximide	1.0g
Acetone	5.0mL

Preparation of Cycloheximide Solution: Add cycloheximide to 5.0mL of acetone. Mix thoroughly. Bring volume to 100.0mL with distilled/deionized water. Mix thoroughly. Filter sterilize.

Nystatin Solution:

Composition per 5mL:

Nystatin	250,000U

Preparation of Nystatin Solution: Add nystatin to distilled/deionized water and bring volume to 5.0mL. Mix thoroughly. Filter sterilize.

Polymyxin B Solution:

Composition per 2mL:

Polymyxin B	10,000U

Preparation of Polymyxin B Solution: Add polymyxin B to distilled/deionized water and bring volume to 2.0mL. Mix thoroughly. Filter sterilize.

Nalidixic Acid Solution:

Composition per 2mL:

Nalidixic acid	0.1g
NaOH (0.5*N* solution)	2.0mL

Preparation of Nalidixic Acid Solution: Add nalidixic acid to 2.0mL of NaOH solution. Mix thoroughly. Immediately before use add 1.0mL of this stock solution to 9.0mL of distilled/deionized water. Mix thoroughly. Filter sterilize.

Vancomycin Solution:

Composition per 1mL:

Vancomycin	0.05g

Preparation of Vancomycin Solution: Add vancomycin to distilled/deionized water and bring volume to 1.0mL. Mix thoroughly. Filter sterilize.

Preparation of Medium: Add components—except serum-glucose solution, bacitracin solution, cycloheximide solution, nystatin solution, polymyxin B solution, nalidixic acid solution, and vancomycin solution—to distilled/deionized water and bring volume to 1.0L. Mix thoroughly. Gently heat and bring to boiling. Autoclave for 15 min at 10 psi pressure–115°C. Cool to 50°C. Aseptically add 60.0mL of sterile serum-glucose solution, 12.5mL of sterile bacitracin solution, 10.0mL of sterile cycloheximide solution, 2.0mL of sterile nystatin solution, 1.0mL of sterile polymyxin B solution, 1.0mL of sterile nalidixic acid solution, and 0.4mL of sterile vancomycin solution. Mix thoroughly. Pour into sterile Petri dishes or distribute into sterile tubes. Allow tubes to cool in a slanted position.

Use: For the selective isolation and cultivation of *Brucella* species.

Serum Potato Infusion Agar

Composition per 1120mL:

Agar	15.0g
Peptone	10.0g
Meat extract	5.0g
NaCl	5.0g
Potato infusion	1.0L
Horse serum, heat inactivated	100.0mL
Glycerol	20.0mL

pH 6.8 ± 0.2 at 25°C

Potato Infusion:

Composition per 10mL:

Potatoes	250.0g

Preparation of Potato Infusion: Add peeled, thinly sliced potatoes to 1.0L of distilled/deionized water. Infuse overnight at 60°C. Filter through Whatman #1 filter paper. Bring volume to 1.0L with distilled/deionized water.

Preparation of Medium: Combine components, except horse serum. Mix thoroughly. Gently heat and bring to boiling. Autoclave for 15 min at 15 psi pressure–121°C. Cool to 45°–50°C. Aseptically add 100.0mL of sterile horse serum. Mix thoroughly. Pour into sterile Petri dishes or distribute into sterile tubes.

Use: For the cultivation of *Brucella* species.

Serum Tellurite Agar

Composition per liter:

Agar	20.0g
Pancreatic digest of casein	10.0g
Peptic digest of animal tissue	10.0g
NaCl	5.0g
Glucose	2.0g
Lamb serum	50.0mL
Chapman tellurite solution	10.0mL

pH 7.5 ± 0.2 at 25°C

Source: This medium is available as a premixed powder from BBL Microbiology Systems.

Chapman Tellurite Solution:

Composition per 100mL:

K_2TeO_3	1.0g

Preparation of Chapman Tellurite Solution: Add K_2TeO_3 to distilled/deionized water and bring volume to 100.0mL. Mix thoroughly. Filter sterilize.

Preparation of Medium: Add components, except lamb serum and Chapman tellurite solution, to distilled/deionized water and bring volume to 940.0mL. Mix thoroughly. Gently heat and bring to boiling. Autoclave for 15 min at 15 psi pressure–121°C. Cool to 45°–50°C. Aseptically add sterile lamb serum and 10.0mL of sterile Chapman tellurite solution. Mix thoroughly. Pour into sterile Petri dishes or distribute into sterile tubes.

Use: For the isolation and cultivation of *Corynebacterium* species, especially in the laboratory diagnosis of diphtheria.

Seven H11 Agar (Selective 7H11 Agar)

Composition per 1010mL:

Agar	13.5g
KH_2PO_4	1.5g
Na_2HPO_4	1.5g
Pancreatic digest of casein	1.0g
NaCl	0.85g
Monosodium glutamate	0.5g
$(NH_4)_2SO_4$	0.5g
Sodium citrate	0.4g
$MgSO_4·7H_2O$	0.05g
Ferric ammonium citrate	0.04g
$CuSO_4·5H_2O$	1.0mg
Pyridoxine	1.0mg
$ZnSO_4·7H_2O$	1.0mg
Biotin	0.5mg
$CaCl_2·2H_2O$	0.5mg
Malachite Green	0.25mg
Glycerol	5.0mL
Middlebrook OADC enrichment	100.0mL
Antibiotic inhibitor	10.0mL

pH 6.6 ± 0.2 at 25°C

Source: Available as a prepared medium from BBL Microbiology Systems.

Middlebrook OADC Enrichment:

Bovine albumin, fraction V	5.0g
Glucose	2.0g
NaCl	0.85g
Catalase	3.0mg
Oleic acid	0.06mL

Preparation of Middlebrook OADC Enrichment: Add components to distilled/deionized water and bring volume to 100.0mL. Mix thoroughly. Filter sterilize.

Antibiotic Inhibitor:

Composition per 10mL:

Carbenicillin	0.05g
Trimethoprim lactate	0.02g
Amphotericin B	0.01g
Polymyxin B	200,000U

Preparation of Antibiotic Inhibitor: Add components to distilled/deionized water and bring volume to 10.0mL. Mix thoroughly. Filter sterilize.

Preparation of Medium: Add glycerol to 900.0mL of distilled/deionized water. Mix thoroughly. Add remaining components, except Middlebrook OADC enrichment and antibiotic inhibitor. Mix thoroughly. Gently heat. Do not boil. Autoclave for 10 min at 15 psi pressure–121°C. Cool to 50°–55°C. Aseptically add 100.0mL of sterile Middlebrook OADC enrichment and 10.0mL of sterile antibiotic solution. Mix thoroughly. Pour into sterile Petri dishes or distribute into sterile tubes.

Use: For the isolation and cultivation of *Mycobacterium* species from specimens with a mixed flora.

Seven Hour Fecal Coliform Agar (Seven Hour FC Agar) (m–Seven Hour Fecal Coliform Agar)

Composition per liter:

Agar	15.0g
Lactose	10.0g
NaCl	7.5g
D-Mannitol	5.0g
Proteose peptone No. 3	5.0g
Yeast extract	3.0g
Bromcresol Purple	0.35g
Phenol Red	0.3g
Sodium lauryl sulfate	0.2g
Sodium deoxycholate	0.1g

pH 7.3 ± 0.1 at 25°C

Preparation of Medium: Add components to distilled/deionized water and bring volume to 1.0L. Mix thoroughly. Gently heat and bring to boiling. Continue boiling for 5 min. Cool to 55°–60°C. Adjust pH to 7.3 with 0.1*N* NaOH. Cool to 45°–50°C. Pour into sterile Petri dishes with tight-fitting lids in 5.0mL volumes. Store at 2°–10°C.

Use: For the rapid estimation of the bacteriological quality of water using the membrane filter method.

SF Broth (*Streptococcus faecalis* Broth)

Composition per liter:

Pancreatic digest of casein	20.0g
Glucose	5.0g
NaCl	5.0g
K_2HPO_4	4.0g
KH_2PO_4	1.5g
NaN_3	0.5g
Bromcresol Purple	0.032g

pH 6.9 ± 0.2 at 25°C

Source: This medium is available as a premixed powder from BBL Microbiology Systems and Difco Laboratories.

Preparation of Medium: Add components to distilled/deionized water and bring volume to 1.0L. Mix thoroughly. Distribute into tubes or flasks. Autoclave for 15 min at 15 psi pressure–121°C.

Use: For the cultivation and differentiation of Group D enterococci (*Streptococcus faecalis* and *Streptococcus faecium*) from Group D nonenterococci and from other *Streptococcus* species. Group D enterococci turn the medium turbid and yellow-brown.

SFP Agar (Shahidi–Ferguson Perfringens Agar)

Composition per 2020mL:

Basal layer	1010.0mL
Cover layer	1010.0mL

Source: This medium is available as a premixed powder from Difco Laboratories and Oxoid Unipath.

Basal Layer:

Composition per 1010mL:

Agar	20.0g
Tryptose	15.0g
Papaic digest of soybean meal	5.0g
Yeast extract	5.0g
Ferric ammonium citrate	1.0g
$NaHSO_3$	1.0g
Egg yolk emulsion, 50%	100.0mL
Antibiotic inhibitor	10.0mL

pH 7.6 ± 0.2 at 25°C

Egg Yolk Emulsion, 50%:

Composition per 100mL:

Chicken egg yolks	11
Whole chicken egg	1
NaCl (0.9% solution)	50.0mL

Preparation of Egg Yolk Emulsion, 50%: Soak eggs with 1:100 dilution of saturated mercuric chloride solution for 1 min. Crack eggs and separate yolks from whites. Mix egg yolks with 1 chicken egg. Measure 50.0mL of egg yolk emulsion and add to 50.0mL of 0.9% NaCl solution. Mix thoroughly. Filter sterilize. Warm to 45°–50°C.

Antibiotic Inhibitor:
Composition per 10mL:

Kanamycin	0.012g
Polymyxin B sulfate	30,000U

Preparation of Antibiotic Inhibitor: Add components to distilled/deionized water and bring volume to 10.0mL. Mix thoroughly. Filter sterilize.

Preparation of Basal Layer: Add components—except egg yolk emulsion, 50% and antibiotic inhibitor—to distilled/deionized water and bring volume to 990.0mL. Mix thoroughly. Gently heat and bring to boiling. Autoclave for 15 min at 15 psi pressure–121°C. Cool to 45°–50°C. Aseptically add sterile egg yolk emulsion, 50% and antibiotic inhibitor. Mix thoroughly. Pour into sterile Petri dishes in 10.0mL volumes.

Cover Layer:
Composition per 1010mL:

Agar	20.0g
Tryptose	15.0g
Papaic digest of soybean meal	5.0g
Yeast extract	5.0g
Ferric ammonium citrate	1.0g
$NaHSO_3$	1.0g
Antibiotic inhibitor	10.0mL

pH 7.6 ± 0.2 at 25°C

Preparation of Cover Layer: Add components—except antibiotic inhibitor—to distilled/deionized water and bring volume to 1.0L. Mix thoroughly. Gently heat and bring to boiling. Autoclave for 15 min at 15 psi pressure–121°C. Cool to 45°–50°C. Aseptically add sterile antibiotic inhibitor. Mix thoroughly.

Preparation of Medium: Prepare and dispense basal layer into sterile Petri dishes in 10.0mL volumes. Incubate overnight to dry plates and test for sterility. Inoculate plates using 0.1mL volume. Spread inoculum over suface of agar. Aseptically add 10.0mL of cover layer to each plate. Incubate at 37°C under 90% N_2 + 10% CO_2.

Use: For the isolation and enumeration of *Clostridium perfringens* from foods. *Clostridium perfringens* appears as black colonies surrounded by a precipitate.

SG Agar

Composition per liter:

Agar	15.0g
Pancreatic digest of casein	15.0g
$CaCl_2 \cdot 2H_2O$	2.0g
$MgSO_4 \cdot 7H_2O$	1.0g

pH 7.0 ± 0.2 at 25°C

Preparation of Medium: Add components to distilled/deionized water and bring volume to 1.0L. Mix thoroughly. Gently heat and bring to boiling. Distribute into tubes or flasks. Autoclave for 15 min at 15 psi pressure–121°C. Pour into sterile Petri dishes or leave in tubes.

Use: For the cultivation of myxobacteria.

Shepard's Differential Agar
See: **A7 Agar**

Shepard's M10 Medium
See: **Standard Fluid Medium 10B**

Shigella Broth

Composition per liter:

Pancreatic digest of casein	20.0g
NaCl	5.0g
K_2HPO_4	2.0g
KH_2PO_4	2.0g
Glucose	1.0g
Novobiocin solution	11.1mL
Tween™ 80	1.5mL

pH 7.0 ± 0.2 at 25°C

Novobiocin Solution:
Composition per liter:

Novobiocin	0.050g

Preparation of Novobiocin Solution: Add novobiocin to distilled/deionized water and bring volume to 1.0L. Mix thoroughly. Filter sterilize.

Preparation of Medium: Add components, except novobiocin solution, to distilled/deionized water and bring volume to 988.9mL. Mix thoroughly. Gently heat and bring to boiling. Autoclave for 15 min at 15 psi pressure–121°C. Cool to 45°–50°C. Aseptically add sterile novobiocin solution. Mix thoroughly. Aseptically distribute into sterile tubes.

Use: For the isolation and cultivation of *Shigella* species from food.

SI Agar

Composition per liter:

Peptone	15.6g
Agar	12.0g
NaCl	5.6g

Yeast extract......2.8g
D-Glucose......1.0g

pH 7.5 ± 0.2 at 25°C

Preparation of Medium: Add components to distilled/deionized water and bring volume to 1.0L. Mix thoroughly. Gently heat and bring to boiling. Distribute into tubes or flasks. Autoclave for 15 min at 15 psi pressure–121°C. Pour into sterile Petri dishes or leave in tubes.

Use: For the cultivation and maintenance of *Escherichia coli.*

Siderophore Mineral Medium

Composition per liter:

KH_2PO_4......8.2g
NaOH......1.6g
NH_4Cl......1.0g
KCl......0.5g
$CaSO_4 \cdot 2H_2O$......0.5mg
$CuSO_4 \cdot 5H_2O$......0.5mg
$FeCl_3 \cdot 6H_2O$......0.5mg
$ZnSO_4 \cdot 7H_2O$......0.5mg
Deferrioxamine B solution......10.0mL
$MgSO_4 \cdot 7H_2O$ solution......10.0mL
Wolfe's vitamin solution......5.0mL

Deferrioxamine B Solution:

Composition per 10mL:

Deferrioxamine B......1.0g

Preparation of Antibiotic Inhibitor: Add deferrioxamine B to distilled/deionized water and bring volume to 10.0mL. Mix thoroughly. Filter sterilize.

$MgSO_4 \cdot 7H_2O$ Solution:

Composition per 10mL:

$MgSO_4 \cdot 7H_2O$......0.5g

Preparation of $MgSO_4 \cdot 7H_2O$ Solution: Add $MgSO_4 \cdot 7H_2O$ to distilled/deionized water and bring volume to 10.0mL. Mix thoroughly. Filter sterilize.

Wolfe's Vitamin Solution:

Composition per liter:

Pyridoxine·HCl......0.01g
Thiamine·HCl......5.0mg
Riboflavin......5.0mg
Nicotinic acid......5.0mg
Calcium pantothenate......5.0mg
p-Aminobenzoic acid......5.0mg
Thioctic acid......5.0mg
Biotin......2.0mg
Folic acid......2.0mg
Cyanocobalamin......0.1mg

Preparation of Wolfe's Vitamin Solution: Add components to distilled/deionized water and bring volume to 1.0L. Mix thoroughly. Filter sterilize.

Preparation of Medium: Add components—except deferrioxamine B solution, $MgSO_4 \cdot 7H_2O$ solution, and Wolfe's vitamin solution—to distilled/deionized water and bring volume to 975.0mL. Mix thoroughly. Gently heat and bring to boiling. Autoclave for 15 min at 15 psi pressure–121°C. Cool to 45°–50°C. Aseptically add 10.0mL of sterile deferrioxamine B solution, 10.0mL of sterile $MgSO_4 \cdot 7H_2O$ solution, and 5.0mL of sterile Wolfe's vitamin solution. Mix thoroughly. Pour into sterile Petri dishes or distribute into sterile tubes.

Use: For the cultivation of ATCC strain 49538.

Sierra Medium

Composition per liter:

Agar......15.0g
Peptone......10.0g
NaCl......5.0g
$CaCl_2 \cdot H_2O$......0.1g
Tween™ 80......10.0mL

pH 7.4 ± 0.2 at 25°C

Preparation of Medium: Add components, except Tween™ 80, to distilled/deionized water and bring volume to 990.0mL. Mix thoroughly. Gently heat and bring to boiling. Autoclave for 15 min at 15 psi pressure–121°C. Cool to 45°–50°C. Separately autoclave Tween™ 80 for 15 min at 15 psi pressure–121°C. Cool to 45°–50°C. Aseptically add 10.0mL of sterile Tween™ 80. Mix thoroughly. Pour into sterile Petri dishes.

Use: For the differentiation of bacteria based on lipase activity. Bacteria with lipase activity form colonies surrounded by a white precipitate.

SIM Medium

Composition per liter:

Peptone......30.0g
Agar......3.0g
Beef extract......3.0g
Peptonized iron (Difco)......0.2g
$Na_2S_2O_3 \cdot 5H_2O$......0.025g

pH 7.3 ± 0.2 at 25°C

Source: This medium is available as a premixed powder from Difco Laboratories.

Preparation of Medium: Add components to distilled/deionized water and bring volume to 1.0L. Mix thoroughly. Gently heat and bring to boiling. Distribute into tubes in 15.0mL volumes. Autoclave for 15 min at 15 psi pressure–121°C. Allow tubes to cool in an upright position.

Use: For the differentiation of members of the Enterobacteriaceae, based on H_2S production, indole production and motility.

SIM Medium

Composition per liter:

Pancreatic digest of casein	20.0g
Peptic digest of animal tissue	6.1g
Agar	3.5g
$Fe(NH_4)_2(SO_4)_2 \cdot 6H_2O$	0.2g
$Na_2S_2O_3 \cdot 5H_2O$	0.2g

pH 7.3 ± 0.2 at 25°C

Source: This medium is available as a premixed powder from BBL Microbiology Systems and Oxoid Unipath.

Preparation of Medium: Add components to distilled/deionized water and bring volume to 1.0L. Mix thoroughly. Gently heat and bring to boiling. Distribute into tubes in 15.0mL volumes. Autoclave for 15 min at 15 psi pressure–121°C. Allow tubes to cool in an upright position.

Use: For the differentiation of members of the Enterobacteriaceae, based on H_2S production, indole production and motility.

Simmons' Citrate Agar (Citrate Agar)

Composition per liter:

Agar	15.0g
NaCl	5.0g
Sodium citrate	2.0g
K_2HPO_4	1.0g
$(NH_4)H_2PO_4$	1.0g
$MgSO_4 \cdot 7H_2O$	0.2g
Bromthymol Blue	0.08g

pH 6.9 ± 0.2 at 25°C

Source: This medium is available as a premixed powder from BBL Microbiology Systems, Difco Laboratories and Oxoid Unipath.

Preparation of Medium: Add components to distilled/deionized water and bring volume to 1.0L. Mix thoroughly. Gently heat while stirring and bring to boiling. Distribute into tubes or flasks. Autoclave for 15 min at 15 psi pressure–121°C. Pour into sterile Petri dishes or leave in tubes.

Use: For the differentiation of Gram-negative bacteria on the basis of citrate utilization. Bacteria which can utilize citrate as sole carbon source turn the medium blue.

Simmons' Citrate Agar, Modified
See: Acetate Differential Agar

Singh's Medium, Modified

Composition per liter:

NaCl	8.75g
Lactalbumin hydrolysate	8.13g
Yeast extract	6.25g
D-Glucose	5.00g
$CaCl_2 \cdot 2H_2O$	0.25g
KCl	0.25g
$NaH_2PO_4 \cdot H_2O$	0.25g
$NaHCO_3$	0.15g
$MgCl_2 \cdot 6H_2O$	0.13g
Phenol Red	0.01g
Fetal bovine serum (heat-inactivated at 56°C, 30 min)	200.0mL

pH 7.0 ± 0.2 at 25°C

Preparation of Medium: Add components to distilled/deionized water and bring volume to 1.0L. Mix thoroughly. Adjust pH to 7.0 with NaOH, if necessary. Filter sterilize. Aseptically distribute into sterile tubes or flasks.

Use: For the cultivation and maintenance of *Spiroplasma* species.

Single Layer Agar

Composition per 1050mL:

Basal Medium	1.0L
Tributyrin substrate	50.0g

pH 6.8 ± 0.2 at 25°C

Basal Medium:
Composition per liter:

Agar	15.0g
Pancreatic digest of gelatin	5.0g
Beef extract	3.0g

Preparation of Basal Medium: Add agar to 1.0L of distilled/deionized water. Autoclave for 15 min at 15 psi pressure–121°C. Cool to 50°C.Victoria Blue B Solution:

Tributyrin Substrate:
Composition:

Tributyrin substrate	50.0g

Preparation of Fat Substrate: Remove free fatty acids in the tributyrin by dissolving 50.0g in 500.0mL of petroleum ether. Pass the solution through an activated alumina column. Remove the petroleum ether by evaporation on a steam table under 100% N_2. Autoclave for 30 min at 15 psi pressure–121°C. Cool to 50°C.

Preparation of Medium: Aseptically combine 1.0L of sterile basal medium with 50.0g of sterile tributyrin substrate in a warm sterile blender container. Blend for 1 min until homogenized. Rapidly pour into sterile Petri dishes in 7.0mL volumes. Dry the surface of the plates by partially opening the lids in a laminar flow hood for 15 min.

Use: For the isolation, cultivation and identification of lipolytic microorganisms from food.

Single Layer Agar

Composition per 1050mL:

Basal Medium	1.0L
Fat substrate	50.0g

pH 6.8 ± 0.2 at 25°C

Basal Medium:

Composition per liter:

Agar	15.0g
Pancreatic digest of gelatin	5.0g
Beef extract	3.0g
Victoria Blue B solution	200.0mL

Preparation of Basal Medium: Add agar to 800.0mL of distilled/deionized water. Autoclave for 15 min at 15 psi pressure–121°C. Cool to 50°C. Aseptically add 200.0mL of Victoria Blue B solution. Mix thoroughly.

Victoria Blue B Solution:

Composition per 200mL:

Victoria Blue B	0.12g

Preparation of Victoria Blue B Solution: Add the Victoria Blue B to 200.0mL of distilled/deionized water. Mix thoroughly. Filter sterilize. Warm to 50°C.

Fat Substrate:

Composition:

Fat substrate	50.0g

Preparation of Fat Substrate: Corn oil, soybean oil, any cooking oil, lard, tallow, or triglycerides that do not contain antioxidants or other inhibitory substances may be used. Remove free fatty acids in the fat substrate by dissolving 50.0g of fat substrate in 500.0mL of petroleum ether. Pass the solution through an activated alumina column. Remove the petroleum ether by evaporation on a steam table under 100% N_2. Autoclave for 30 min at 15 psi pressure–121°C. Cool to 50°C.

Preparation of Medium: Aseptically combine 1.0L of sterile basal medium with 50.0g of sterile fat substrate in a warm sterile blender container. Blend for 1 min until homogenized. Rapidly pour into sterile Petri dishes in 7.0mL volumes. Dry the surface of the plates by partially opening the lids in a laminar flow hood for 15 min.

Use: For the isolation, cultivation and identification of lipolytic microorganisms from food.

SJ Agar

Composition per liter:

Agar	15.0g
K_2HPO_4	1.0g
KCl	0.5g
$MgSO_4 \cdot 7H_2O$	0.5g
$NaNO_3$	0.5g
$FeSO_4 \cdot 7H_2O$	0.01g
Glucose solution	100.0mL

pH 7.2 ± 0.2 at 25°C

Glucose Solution:

Composition per 100mL:

D-Glucose	1.0g

Preparation of Glucose Solution: Add D-glucose to distilled/deionized water and bring volume to 100.0mL. Mix thoroughly. Autoclave for 15 min at 15 psi pressure–121°C. Cool to 25°C.

Preparation of Medium: Add components, except glucose solution, to distilled/deionized water and bring volume to 900.0mL. Mix thoroughly. Gently heat and bring to boiling. Autoclave for 15 min at 15 psi pressure–121°C. Cool to 45°–50°C. Aseptically add sterile glucose solution. Mix thoroughly. Pour into sterile Petri dishes or distribute into sterile tubes.

Use: For the isolation and cultivation of *Cytophaga* species, *Herpetosiphon* species, *Saprospira* species, and *Flexithrix* species.

SK Agar

See: **Schleifer–Krämer Agar**

Skim Milk Acetate Medium

Composition per liter:

Agar	15.0g
Skim milk powder	5.0g
Yeast extract	0.5g
Sodium acetate	0.2g

Preparation of Medium: Add components to distilled/deionized water and bring volume to 1.0L. Mix thoroughly. Gently heat and bring to boiling. Distribute into tubes or flasks. Autoclave for 15 min at 15 psi pressure–121°C. Pour into sterile Petri dishes or leave in tubes.

Use: For the cultivation and maintenance of *Cytophaga johnsonae*.

Skim Milk Agar (Milk Agar) (ATCC Medium 377)

Composition per liter:

Agar	15.0g
Skim milk	8.0g

Preparation of Medium: Add components to distilled/deionized water and bring volume to 1.0L. Mix thoroughly. Gently heat and bring to boiling. Distribute into tubes or flasks. Autoclave for 15 min at 15 psi pressure–121°C. Pour into sterile Petri dishes or leave in tubes.

Use: For the cultivation and maintenance of *Herpetosiphon aurantiacus*.

Skim Milk Agar

Composition per 1100mL:

Agar	15.0g
Pancretic digest of casein	5.0g
Yeast extract	2.5g
Glucose	1.0g
Skim milk solution	100.0mL

pH 7.0 ± 0.1 at 25°C

Preparation of Skim Milk Solution: Add skim milk solids to distilled/deionized water and bring volume to 100.0mL. Mix thoroughly. Autoclave for 15 min at 15 psi pressure–121°C. Cool to 45°–50°C.

Preparation of Medium: Add components, except skim milk solution, to distilled/deionized water and bring volume to 1.0L. Mix thoroughly. Gently heat and bring to boiling. Distribute into tubes or flasks. Autoclave for 15 min at 15 psi pressure–121°C. Cool to 45°–50°C. Aseptically add 100.0mL of cooled, sterile skim milk solution. Mix thoroughly. Pour into sterile Petri dishes or aseptically distribute into sterile tubes.

Use: For the cultivation and differentiation of bacteria based on proteolytic activity.

Skirrow *Brucella* Medium

Composition per liter:

Blood agar base No. 2	940.0mL
Horse blood, lysed defibrinated	50.0mL
Antibiotic solution	10.0mL

pH 7.4 ± 0.2 at 25°C

Blood Agar Base No. 2

Composition per 940mL:

Proteose peptone	15.0g
Agar	12.0g
NaCl	5.0g
Yeast extract	5.0g
Liver digest	2.5g

pH 7.4 ± 0.2 at 25°C

Preparation of Blood Agar Base No. 2: Add components to distilled/deionized water and bring volume to 940.0mL. Mix thoroughly. Gently heat while stirring and bring to boiling. Autoclave for 15 min at 15 psi–121°C. Cool to 45°–50°C.

Antibiotic Solution:

Composition per 10mL:

Vancomycin	0.01g
Trimethoprim	5.0mg
Polymyxin B	2,500U

Preparation of Antibiotic Solution: Add components to distilled/deionized water and bring volume to 10.0mL. Mix thoroughly. Filter sterilize.

Preparation of Medium: To 940.0mL of sterile cooled blood agar base No. 2, aseptically add 50.0mL of sterile, lysed defibrinated horse blood and 10.0mL of sterile antibiotic solution. Pour into sterile Petri dishes or distribute into sterile tubes.

Use: For the selective isolation and cultivation of *Campylobacter* species.

Skirrow's *Campylobacter* Agar

See: ***Campylobacter*** **Agar, Skirrow's**

SL Medium

Composition per liter:

Agar	15.0g
Pancreatic digest of casein	10.0g
Glucose	10.0g
KH_2PO_4	6.0g
Arabinose	5.0g
Sucrose	5.0g
Yeast extract	5.0g
Sodium acetate·$3H_2O$	2.5g
Diammonium citrate	2.0g
Tween™ 80	1.0g
$MgSO_4·7H_2O$	0.58g
$MnSO_4·4H_2O$	0.28g

pH 5.4 ± 0.2 at 25°C

Preparation of Medium: Add agar to 500.0mL of distilled/deionized water. Gently heat and bring to boiling. In a separate flask, add the remaining components, except sodium acetate, to distilled/deionized water and bring volume to 300.0mL. Mix thoroughly. Combine the two solutions. In a separate flask add sodium acetate to distilled/deionized water and bring volume to100.0mL. Mix thoroughly. Adjust pH to 5.4 with glacial acetic acid. Add this solution to the agar solution. Mix thoroughly. Adjust pH to 5.4 with additional glacial acetic acid. Bring volume to 1.0L with distilled/deionized water. Do not autoclave. Aseptically distribute into sterile screw-capped tubes or flasks.

Use: For the selective isolation and cultivation of *Lactobacillus* species.

Slanetz and Bartley Medium

Composition per liter:

Tryptose	20.0g
Agar	10.0g
Yeast extract	5.0g
Na_2HPO_4 $2H_2O$	4.0g
Glucose	2.0g
NaN_3	0.4g
Tetrazolium chloride	0.1g

pH 7.2 ± 0.2 at 25°C

Source: This medium is available as a premixed powder from Oxoid Unipath.

Preparation of Medium: Add components to distilled/deionized water and bring volume to 1.0L. Mix thoroughly. Gently heat and bring to boiling. Distribute into tubes or flasks. Autoclave for 15 min at 15 psi pressure–121°C. Pour into sterile Petri dishes.

Use: For the detection and enumeration of enterococci by the membrane filter method.

Sludge Medium for Methanobacteria

Composition per liter:

$NaHCO_3$	4.0g
Sodium formate	2.0g
Sodium acetate	1.0g
Yeast extract	1.0g
Cysteine·HCl·H_2O	0.5g
KH_2PO_4	0.5g
$Na_2S \cdot 9H_2O$	0.5g
$MgSO_4 \cdot 7H_2O$	0.4g
NaCl	0.4g
NH_4Cl	0.4g
$CaCl_2 \cdot 2H_2O$	0.05g
$FeSO_4 \cdot 7H_2O$	2.0mg
Resazurin	1.0mg
Sludge fluid	50.0mL
Fatty acid mixture	20.0mL
Trace elements SL-6	1.0mL

pH 6.7-7.0 at 25°C

Sludge Fluid:

Composition per 100mL:

Sludge	100.0mL
Yeast extract	0.4g

Preparation of Sludge Fluid: Add yeast extract to a concentration of 0.4% to sludge taken from an anaerobic digester. Gas with 100% N_2 for 5 min. Incubate at 37°C for 24 hr. Centrifuge at 13,000 × *g*. Remove the supernatant fluid. Gas with 100% N_2 for 5 min. Autoclave for 15 min at 15 psi pressure–121°C. Store at 25°C protected from light.

Fatty Acid Mixture:

Composition per 20mL:

α-Methylbutyric acid	0.5g
Isobutyric acid	0.5g
Isovaleric acid	0.5g
Valeric acid	0.5g

Preparation of Fatty Acid Mixture: Add components to distilled/deionized water and bring volume to 20.0mL. Mix thoroughly. Adjust pH to 7.5 with concentrated NaOH.

Trace Elements Solution SL-6:

Composition per liter:

H_3BO_3	0.3g
$CoCl_2 \cdot 6H_2O$	0.2g
$ZnSO_4 \cdot 7H_2O$	0.10g
$MnCl_2 \cdot 4H_2O$	0.03g
$Na_2MoO_4 \cdot H_2O$	0.03g
$NiCl_2 \cdot 6H_2O$	0.02g
$CuCl_2 . 2H_2O$	0.01g

Preparation of Trace Elements Solution SL-6: Add components to distilled/deionized water and bring volume to 1.0L. Mix thoroughly. Adjust pH to 3.4.

Preparation of Medium: Prepare and dispense medium under 80% N_2 + 20% CO_2. Add components to distilled/deionized water and bring volume to 1.0L. Mix thoroughly. Adjust pH to 6.7–7.0. Distribute anaerobically into tubes or bottles with aluminum seals. Autoclave for 15 min at 15 psi pressure–121°C with fast exhaust.

Use: For the cultivation and maintenance of *Methanobacterium uliginosum* and *Methanobrevibacter ruminantium*.

Sludge Medium for Methanobacteria, pH 7.9

Composition per liter:

$NaHCO_3$	4.0g

Sodium formate	2.0g
Sodium acetate	1.0g
Yeast extract	1.0g
Cysteine·HCl·H_2O	0.5g
KH_2PO_4	0.5g
$Na_2S·9H_2O$	0.5g
$MgSO_4·7H_2O$	0.4g
NaCl	0.4g
NH_4Cl	0.4g
$CaCl_2·2H_2O$	0.05g
$FeSO_4·7H_2O$	2.0mg
Resazurin	1.0mg
Sludge fluid	50.0mL
Fatty acid mixture	20.0mL
Trace elements SL-6	1.0mL

pH 7.9 ± 0.2 at 25°C

Sludge Fluid:

Composition per 100mL:

Sludge	100.0mL
Yeast extract	0.4g

Preparation of Sludge Fluid: Add yeast extract to a concentration of 0.4% to sludge taken from an anaerobic digester. Gas with 100% N_2 for 5 min. Incubate at 37°C for 24 hr. Centrifuge at 13,000 × *g*. Remove the supernatant fluid. Gas with 100% N_2 for 5 min. Autoclave for 15 min at 15 psi pressure–121°C. Store at 25°C protected from light.

Fatty Acid Mixture:

Composition per 20mL:

α-Methylbutyric acid	0.5g
Isobutyric acid	0.5g
Isovaleric acid	0.5g
Valeric acid	0.5g

Preparation of Fatty Acid Mixture: Add components to distilled/deionized water and bring volume to 20.0mL. Mix thoroughly. Adjust pH to 7.5 with concentrated NaOH.

Trace Elements Solution SL-6:

Composition per liter:

H_3BO_3	0.3g
$CoCl_2·6H_2O$	0.2g
$ZnSO_4·7H_2O$	0.10g
$MnCl_2·4H_2O$	0.03g
$Na_2MoO_4·H_2O$	0.03g
$NiCl_2·6H_2O$	0.02g
$CuCl_2.2H_2O$	0.01g

Preparation of Trace Elements Solution SL-6: Add components to distilled/deionized water and bring volume to 1.0L. Mix thoroughly. Adjust pH to 3.4.

Preparation of Medium: Prepare and dispense medium under 80% N_2 + 20% CO_2. Add components to distilled/deionized water and bring volume to 1.0L. Mix thoroughly. Adjust pH to 7.9. Distribute anaerobically into tubes or bottles with aluminum seals. Autoclave for 15 min at 15 psi pressure–121°C with fast exhaust.

Use: For the cultivation and maintenance of *Methanobacterium alcaliphilum* and *Methanobacterium thermoalcalip.*

SM Selective Medium

Composition per liter:

Agar	15.0g
Mannitol	2.5g
L-Glutamic acid	1.0g
$MgSO_4·7H_2O$	0.16g
2,3,5-Triphenyltetrazolium·HCl solution	10.0mL
Antibiotic solution	10.0mL
KH_2PO_4 (0.2mM solution)	1.0mL
Metals solution	0.05mL

pH 7.2 ± 0.2 at 25°C

Antibiotic Solution:

Composition per 10mL:

Bacitracin	0.05g
Cycloheximide	0.05g
Tyrothricin	0.02g
Captan	0.01g
Vancomycin	0.01g
Chloromycetin	5.0μg
Penicillin G	1.0μg

Preparation of Antibiotic Solution: Add components to distilled/deionized water and bring volume to 10.0mL. Mix thoroughly. Filter sterilize.

Metals Solution:

Composition per 100mL:

$ZnSO_4·7H_2O$	1.1g
$MnSO_4·H_2O$	0.62g
$Fe(NH_4)_2(SO_4)_2·6H_2O$	0.18g
$CuSO_4·5H_2O$	0.029g
$CaSO_4·5H_2O$	0.029g
H_3PO_3	0.011g
KI	0.013mg

Preparation of Metals Solution: Add components to distilled/deionized water and bring volume to 100.0mL. Mix thoroughly.

Preparation of Medium: Add components, except antibiotic solution, to distilled/deionized water and bring volume to 990.0mL. Mix thoroughly. Gently heat and bring to boiling. Adjust pH to 7.2 with KOH. Autoclave for 15 min at 15 psi pressure–121°C. Cool to 45°–50°C. Aseptically add sterile antibiotic solution. Mix thoroughly. Pour into sterile Petri dishes or distribute into sterile tubes.

Use: For the isolation and cultivation of *Pseudomonas solanacearum.*

SMC Medium

Composition per liter:

Sorbitol	70.0g
Pancreatic digest of casein	17.0g
NaCl	5.0g
Beef extract	3.0g
Yeast extract	3.0g
Beef heart, solids from infusion	2.0g
Horse serum	200.0mL
Yeast extract solution	100.0mL
Phenol Red solution	20.0mL
Sucrose solution	20.0mL
L-Arginine·HCl solution	10.0mL
Fructose solution	2.0mL
Glucose solution	2.0mL

pH 7.5 ± 0.2 at 25°C

Yeast Extract Solution:

Composition per 100mL:

Yeast extract	25.0g

Preparation of Yeast Extract Solution: Add yeast extract to distilled/deionized water and bring volume to 100.0mL. Mix thoroughly. Autoclave for 15 min at 15 psi pressure–121°C. Cool to 45°–50°C.

Phenol Red Solution:

Composition per 100mL:

Phenol Red	0.01g

Preparation of Phenol Red Solution: Add Phenol Red to distilled/deionized water and bring volume to 100.0mL. Mix thoroughly. Autoclave for 15 min at 15 psi pressure–121°C. Cool to 45°–50°C.

Sucrose Solution:

Composition per 20mL:

Sucrose	10.0g

Preparation of Sucrose Solution: Add sucrose to distilled/deionized water and bring volume to 20.0mL. Mix thoroughly. Autoclave for 15 min at 15 psi pressure–121°C. Cool to 45°–50°C.

L-Arginine·HCl Solution:

Composition per 10mL:

L-Arginine·HCl	4.2g

Preparation of L-Arginine·HCl Solution: Add arginine·HCl to distilled/deionized water and bring volume to 10.0mL. Mix thoroughly. Autoclave for 15 min at 15 psi pressure–121°C. Cool to 45°–50°C.

Fructose Solution:

Composition per 10mL:

Fructose	5.0g

Preparation of Fructose Solution: Add fructose to distilled/deionized water and bring volume to 10.0mL. Mix thoroughly. Autoclave for 15 min at 15 psi pressure–121°C. Cool to 45°–50°C.

Glucose Solution:

Composition per 10mL:

Glucose	5.0g

Preparation of Glucose Solution: Add glucose to distilled/deionized water and bring volume to 10.0mL. Mix thoroughly. Autoclave for 15 min at 15 psi pressure–121°C. Cool to 45°–50°C.

Preparation of Medium: Add components—except horse serum, yeast extract solution, Phenol Red solution, sucrose solution, L-arginine·HCl solution, fructose solution, and glucose solution—to distilled/deionized water and bring volume to 646.0mL. Mix thoroughly. Gently heat and bring to boiling. Autoclave for 15 min at 15 psi pressure–121°C. Cool to 45°–50°C. Aseptically add 200.0mL of sterile horse serum, 100.0mL of sterile yeast extract solution, 20.0mL of sterile Phenol Red solution, 20.0mL of sterile sucrose solution, 10.0mL of sterile L-arginine·HCl solution, 2.0mL of sterile fructose solution, and 2.0mL of sterile glucose solution. Mix thoroughly. Aseptically distribute into sterile tubes or flasks.

Use: For the cultivation and maintenance of *Spiroplasma citri*.

SMC, Modified

Composition per liter:

Sorbitol	70.0g
Pancreatic digest of casein	17.0g
NaCl	5.0g
Beef extract	3.0g
Yeast extract	3.0g
Beef heart, solids from infusion	2.0g
Solution 1	100.0mL
Solution 3	20.0mL
Solution 2	10.0mL
NaOH (1*N* solution)	6.0mL

pH 7.7–7.8 at 25°C

Solution 1:

Composition per 100mL:

Sucrose	10.0g
Yeast extract	2.0g
Fructose	1.0g
Glucose	1.0g
Phenol Red	0.02g

Preparation of Solution 1: Add components to distilled/deionized water and bring volume to 100.0mL. Mix thoroughly. Filter sterilize.

Solution 2:

Composition per 10mL:

Bovine serum albumin, fraction V 0.1g

Preparation of Solution 2: Add bovine serum albumin to distilled/deionized water and bring volume to 10.0mL. Mix thoroughly. Filter sterilize.

Solution 3:

Composition per 20mL:

Horse serum 20.0mL

Preparation of Solution 3: Inactivate horse serum at 56°C for 30 min. Filter sterilize.

Preparation of Medium: Add components—except solution 1, solution 2, and solution 3—to distilled/deionized water and bring volume to 870.0mL. Autoclave for 15 min at 15 psi pressure–121°C. Cool to 45°–50°C. Aseptically add 100.0mL of sterile solution 1, 20.0mL of sterile solution 3, and 10.0mL of sterile solution 1. Mix thoroughly. Adjust pH to 7.7–7.8. Aseptically distribute into sterile tubes or flasks.

Use: For the cultivation and maintenance of *Spiroplasma citri.*

Snyder Agar

Composition per liter:

Glucose 20.0g
Agar 16.0g
Pancreatic digest of casein 13.5g
Yeast extract 6.5g
NaCl 5.0g
Bromcresol Green 0.02g

pH 4.8 ± 0.2 at 25°C

Source: This medium is available as a premixed powder from BBL Microbiology Systems.

Preparation of Medium: Add components to distilled/deionized water and bring volume to 1.0L. Mix thoroughly. Gently heat and bring to boiling. Distribute into tubes in 10.0mL volumes. Autoclave for 15 min at 13 psi pressure–118°C. Do not overheat. Pour into sterile Petri dishes or leave in tubes.

Use: For the cultivation and enumeration of lactobacilli in saliva and indication of dental caries activity.

Snyder Test Agar

Composition per liter:

Agar 20.0g
Glucose 20.0g
Tryptose 20.0g
NaCl 5.0g
Bromcresol Green 0.02g

pH 4.8 ± 0.2 at 25°C

Source: This medium is available as a premixed powder from Difco Laboratories.

Preparation of Medium: Add components to distilled/deionized water and bring volume to 1.0L. Mix thoroughly. Gently heat and bring to boiling. Distribute into tubes in 10.0mL volumes. Autoclave for 15 min at 13 psi pressure–118°C. Do not overheat. Pour into sterile Petri dishes or leave in tubes.

Use: For the cultivation and enumeration of lactobacilli in saliva and indication of dental caries activity.

Sodium Acetate Agar
See: Acetate Differential Agar

Sodium Acetate Medium I

Composition per liter:

Sodium acetate 33.0g
Agar 15.0g
Glucose 10.0g
Peptone 10.0g
K_2HPO_4 .... 5.0g
NaCl 5.0g
Yeast extract 3.0g

pH 7.0 ± 0.2 at 25°C

Preparation of Medium: Add components to distilled/deionized water and bring volume to 1.0L. Mix thoroughly. Gently heat and bring to boiling. Adjust pH to 7.0. Distribute into tubes or flasks. Autoclave for 10 min at 15 psi pressure–121°C. Pour into sterile Petri dishes or leave in tubes.

Use: For the cultivation and maintenance of *Pediococcus halophilus* and *Staphylococcus simulans.*

Sodium Biselenite Medium
See: Selenite Broth

Sodium Chloride Broth, 6.5

Composition per liter:

Beef heart, solids from infusion 500.0g
NaCl 65.0g
Tryptose 10.0g

pH 7.4 ± 0.2 at 25°C

Preparation of Medium: Add components to distilled/deionized water and bring volume to 1.0L. Mix thoroughly. Distribute into tubes or flasks. Autoclave for 15 min at 15 psi pressure–121°C.

Use: For the cultivation of enterococci and other salt-tolerant organisms. Use for the differentiation of microorganisms based on salt tolerance.

Sodium Chloride Sucrose Medium 900 (Sodium Chloride SUC Medium 900)

Composition per liter:

Sucrose	97.3g
Pancreatic digest of gelatin	14.5g
NaCl	14.3g
Agar	13.3g
Brain heart, solids from infusion	6.0g
Peptic digest of animal tissue	6.0g
Yeast extract	5.0g
Glucose	3.0g
Na_2HPO_4	2.5g
$MgSO_4$	0.25g
Horse serum (γ-globulin-free, inactivated 30 min at 56°C)	100.0mL
Carbenicillin solution	10.0mL

pH 7.4 ± 0.2 at 25°C

Carbenicillin Solution:

Composition per 10mL:

Carbenicillin	5.0g

Preparation of Carbenicillin Solution: Add carbenicillin to distilled/deionized water and bring volume to 10.0mL. Mix thoroughly. Filter sterilize.

Preparation of Medium: Add components, except carbenicillin solution and horse serum, to distilled/deionized water and bring volume to 890.0mL. Mix thoroughly. Gently heat and bring to boiling. Autoclave for 15 min at 15 psi pressure–121°C. Cool to 45°–50°C. Aseptically add carbenicillin solution and horse serum. Mix thoroughly. Pour into sterile Petri dishes or distribute into sterile tubes.

Use: For the cultivation of *Pseudomonas aeruginosa.*

Sodium Chloride Sucrose Medium 900 with Penicillin G

Composition per liter:

Sucrose	97.3g
Pancreatic digest of gelatin	14.5g
NaCl	14.3g
Agar	13.3g
Brain heart, solids from infusion	6.0g
Peptic digest of animal tissue	6.0g
Yeast extract	5.0g
Glucose	3.0g
Na_2HPO_4	2.5g
$MgSO_4$	0.25g
Horse serum (γ-globulin-free, inactivated 30 min at 56°C)	100.0mL
Penicillin G	10.0mL

pH 7.4 ± 0.2 at 25°C

Penicillin Solution:

Composition per 10mL:

Penicillin G	500,000U

Preparation of Penicillin Solution: Add penicillin to distilled/deionized water and bring volume to 10.0mL. Mix thoroughly. Filter sterilize.

Preparation of Medium: Add components, except penicillin G and horse serum, to distilled/deionized water and bring volume to 900.0mL. Mix thoroughly. Gently heat and bring to boiling. Autoclave for 15 min at 15 psi pressure–121°C. Cool to 45°–50°C. Aseptically add penicillin G and horse serum. Mix thoroughly. Pour into sterile Petri dishes or distribute into sterile tubes.

Use: For the cultivation of *Pseudomonas aeruginosa.*

Sodium Dodecyl Sulfate Polymyxin Sucrose Agar

Composition per liter:

NaCl	20.0g
Agar	15.0g
Sucrose	15.0g
Proteose peptone	10.0g
Beef extract	5.0g
Sodium lauryl sulfate	1.0g
Bromthymol Blue	0.04g
Cresol Red	0.04g
Polymyxin B solution	10.0mL

pH 7.6 ± 0.2 at 25°C

Polymyxin B Solution:

Composition per 10mL:

Polymyxin B sulfate	100,000U

Preparation of Polymyxin B Solution: Add polymyxin B sulfate to distilled/deionized water and bring volume to 10.0mL. Mix thoroughly. Filter sterilize.

Preparation of Medium: Add components, except polymyxin B solution, to distilled/deionized water and bring volume to 990.0mL. Mix thoroughly. Gently heat and bring to boiling. Autoclave for 15 min at 15 psi pressure–121°C. Cool to 45°–50°C. Aseptically add sterile polymyxin B solution. Mix thoroughly. Pour immediately into sterile Petri dishes or distribute into sterile tubes.

Use: For the isolation and cultivation of *Vibrio* species from foods.

Sodium Hippurate Broth (Hippurate Broth)

Composition per liter:

Beef heart, solids from infusion	500.0g
Tryptose	10.0g
Sodium hippurate	10.0g
NaCl	5.0g

pH 7.4 ± 0.2 at 25°C

Source: Heart infusion broth is available as a premixed powder from Difco Laboratories and BBL Microbiology Systems.

Preparation of Medium: Add components to distilled/deionized water and bring volume to 1.0L. Mix thoroughly. Gently heat and bring to boiling. Distribute into screw-capped tubes or flasks. Autoclave for 15 min at 15 psi pressure–121°C. Tighten caps to prevent drying.

Use: For the identification and differentiation of β-hemolytic streptococci based on hippurate hydrolysis. After inoculation and incubation, tubes are treated with $FeCl_3$ reagent. A heavy precipitate remaining after 10–15 min indicates that hippurate has been hydrolyzed.

Sodium Hydrogen Selenite Medium

See: **Selenite Broth**

Sodium Lactate Agar

Composition per liter:

Agar	15.0g
Pancreatic digest of casein	10.0g
Sodium lactate	10.0g
Yeast extract	10.0g
K_2HPO_4	0.25g

pH 7.0 ± 0.2 at 25°C

Preparation of Medium: Add components to distilled/deionized water and bring volume to 1.0L. Mix thoroughly. Gently heat and bring to boiling. Distribute into tubes or flasks. Adjust pH to 7.0. Autoclave for 15 min at 15 psi pressure–121°C. Pour into sterile Petri dishes or leave in tubes.

Use: For the cultivation and isolation of *Propionibacterium* species from foods. For the isolation and cultivation of propionic acid producing bacteria from cheese.

Sodium Lactate Agar, Modified

Composition per liter:

Agar	15.0g
Yeast extract	10.0g
Pancreatic digest of casein	5.1g
NaCl	1.5g
Papaic digest of soybean meal	0.9g
K_2HPO_4	0.75g
Glucose	0.75g
Sodium lactate, 60% syrup	20.0mL

pH 7.0 ± 0.2 at 25°C

Preparation of Medium: Add components to distilled/deionized water and bring volume to 1.0L. Mix thoroughly. Gently heat and bring to boiling. Adjust pH to 7.0. Distribute into tubes or flasks. Autoclave for 15 min at 15 psi pressure–121°C. Pour into sterile Petri dishes or leave in tubes.

Use: For the cultivation and isolation of *Propionibacterium* species from foods. For the isolation and cultivation of propionic acid producing bacteria from cheese.

Soft Agar Gelatin Overlay

Composition per plate:

Base agar	15.0mL
Soft agar gelatin overlay	2.5mL

pH 7.0 ± 0.2 at 25°C

Base Agar:
Composition per liter:

Agar	15.0g
Peptone	5.0g
NaCl	5.0g
Beef extract	3.0g
$MnSO_4 \cdot H_2O$	0.05g

Preparation of Base Agar: Add components to distilled/deionized water and bring volume to 1.0L. Mix thoroughly. Gently heat and bring to boiling. Autoclave for 15 min at 15 psi pressure–121°C. Cool to 45°–50°C.

Soft Agar Gelatin Overlay:
Composition per liter:

Gelatin	15.0g
Agar	8.0g
Peptone	5.0g
NaCl	5.0g
Beef extract	3.0g
$MnSO_4 \cdot H_2O$	0.05g

Preparation of Soft Agar Gelatin Overlay: Add components to distilled/deionized water and bring volume to 1.0L. Mix thoroughly. Gently heat and bring to boiling. Autoclave for 15 min at 15 psi pressure–121°C. Cool to 45°–50°C.

Preparation of Medium: Aseptically pour cooled, sterile base agar into sterile Petri dishes in 15.0mL volumes. Allow agar to solidify. Inoculate plates with samples. Overlay each plate with 2.5mL of soft agar gelatin overlay.

Use: For the cultivation and differentiation of microorganisms based on proteolytic activity.

Soil Extract Agar

Composition per liter:

Agar	20.0g
Glucose	1.0g
K_2HPO_4	1.0g
Peptone	1.0g
Yeast extract	1.0g
Soil extract	400.0mL
Cycloheximide solution	10.0mL

pH 6.6 ± 0.2 at 25°C

Soil Extract:

Composition per liter:

Garden soil, neutral	1.0Kg

Preparation of Soil Extract: Add garden soil to 1.0L of tap water. Autoclave for 20 min at 15 psi pressure–121°C. Filter through Whatman filter paper. Bring volume to 1.0L with tap water.

Cycloheximide Solution:

Composition per 10mL:

Cycloheximide	0.04g

Preparation of Cycloheximide Solution: Add cycloheximide to distilled/deionized water and bring volume to 10.0mL. Mix thoroughly. Filter sterilize.

Preparation of Medium: Add components, except cycloheximide solution, to distilled/deionized water and bring volume to 990.0mL. Mix thoroughly. Gently heat and bring to boiling. Autoclave for 15 min at 15 psi pressure–121°C. Cool to 45°–50°C. Aseptically add sterile cycloheximide solution. Mix thoroughly. Pour into sterile Petri dishes or distribute into sterile tubes.

Use: For the isolation and cultivation of *Arthrobacter* species.

Soil Extract Agar

Composition per liter:

Soil	500.0g
Agar	15.0g
Glucose	2.0g
Yeast extract	1.0g
KH_2PO_4	0.5g

Preparation of Medium: Add 500.0g of garden soil to 1.0L of tap water. Autoclave for 3 hours at 15 psi pressure–121°C. Filter through Whatman No. 2 filter paper. Add remaining components to filtrate. Bring volume to 1.0L with tap water. Gently heat and bring to boiling. Distribute into tubes in 7.0mL volumes. Autoclave for 15 min at 15 psi pressure–121°C. Allow tubes to cool in a slanted position.

Use: For the cultivation and identification of *Histoplasma capsulatum, Blastomyces dermatitidis*, and *Bacillus* species based on formation of typical conidia.

Soil Extract Glucose Yeast Extract Agar

Composition per liter:

Agar	15.0g
Glucose	2.0g
Yeast extract	1.0g
Soil extract	250.0mL

pH 6.8 ± 0.2 at 25°C

Soil Extract:

Composition per liter:

Garden soil	500.0g

Preparation of Soil Extract: Add 500.0g of garden soil to 1.0L of tap water. Autoclave for 1 hour at 15 psi pressure–121°C. Filter through Whatman No. 2 filter paper.

Preparation of Medium: Add components to distilled/deionized water and bring volume to 1.0L. Mix thoroughly. Gently heat and bring to boiling. Distribute into tubes or flasks. Autoclave for 15 min at 15 psi pressure–121°C. Pour into sterile Petri dishes or leave in tubes.

Use: For the cultivation and maintenance of *Streptomyces rectus*.

Soil Extract Medium

Composition per liter:

Agar	15.0g
Pancreatic digest of gelatin	5.0g
Beef extract	3.0g
Soil extract	250.0mL

pH 6.8 ± 0.2 at 25°C

Soil Extract:

Composition per liter:

Garden soil	500.0g

Preparation of Soil Extract: Add 500.0g of garden soil to 1.0L of tap water. Autoclave for 1 hour at 15 psi pressure–121°C. Filter through Whatman No. 2 filter paper.

Preparation of Medium: Add components to distilled/deionized water and bring volume to 1.0L. Mix thoroughly. Gently heat and bring to boiling. Distribute into tubes or flasks. Autoclave for 15 min at 15 psi

pressure–121°C. Pour into sterile Petri dishes or leave in tubes.

Use: For the cultivation and maintenance of *Streptomyces rectus*.

Soil Extract Peptone Beef Extract Medium

Composition per liter:

Agar	15.0g
Peptone	5.0g
Beef extract	3.0g
Soil extract	1.0L

pH 7.0 ± 0.2 at 25°C

Soil Extract:

Composition per liter:

Garden soil	400.0g

Preparation of Soil Extract: Add garden soil to 1.0L of tap water. Autoclave for 1 hour at 15 psi pressure–121°C. Filter through cheesecloth and Whatman No.2 filter paper. Autoclave filtrate again for 20 min at 15 psi pressure–121°C. Filter through Whatman No.2 filter paper.

Preparation of Medium: Add agar, peptone, and beef extract to 1.0L of soil extract. Mix thoroughly. Gently heat and bring to boiling. Distribute into tubes or flasks. Autoclave for 15 min at 15 psi pressure–121°C. Pour into sterile Petri dishes or leave in tubes.

Use: For the cultivation and maintenance of *Oerskovia turbata* and *Oerskovia xanthineolytica*.

Soil Extract Potato Extract Medium

Composition per 510mL:

Malt extract	10.0g
Yeast extract	4.0g
Potato extract	250.0mL
Soil extract	250.0mL

pH 7.0 ± 0.2 at 25°C

Soil Extract:

Composition per liter:

Garden soil	400.0g

Preparation of Soil Extract: Add garden soil to 1.0L of tap water. Autoclave for 45 min at 15 psi pressure–121°C. Filter through cheesecloth.

Potato Extract:

Composition per liter:

Potatoes	400.0g

Preparation of Potato Extract: Peel and dice potatoes. Add 500.0mL of distilled/deionized water. Gently heat and bring to boiling. Continue boiling for 15 min. Filter through cheesecloth. Bring volume to 1.0L with distilled/deionized water.

Preparation of Medium: Combine components. Mix thoroughly. Distribute into tubes or flasks. Autoclave for 15 min at 15 psi pressure–121°C.

Use: For the cultivation and maintenance of *Saccharopolyspora rectivirgula*.

Sorangium Medium

Composition per liter:

Agar	10.0g
KNO_3	1.0g
K_2HPO_4	1.0g
$MgSO_4$	0.2g
$CaCl_2$	0.1g
$FeCl_3$	0.02g

Preparation of Medium: Add components to tap water and bring volume to 1.0L. Mix thoroughly. Gently heat and bring to boiling. Distribute into tubes or flasks. Autoclave for 15 min at 15 psi pressure–121°C. Pour into sterile Petri dishes or leave in tubes. Allow tubes to cool in a slanted position. Aseptically add a sterile strip (4.5cm × 1.0cm) of Whatman No. 1 filter paper to the surface of each slant or 4–6 sterile strips of filter paper to the surface of each agar plate.

Use: For the cultivation and maintenance of *Polyangium cellulosum*.

Sorbitol Agar

Composition per liter:

Agar	20.0g
Peptone	10.0g
NaCl	2.0g
Yeast extract	2.0g
Sorbitol solution	50.0mL

pH 7.0 ± 0.2 at 25°C

Sorbitol Solution:

Composition per 50mL:

Sorbitol	5.0g

Preparation of Sorbitol Solution: Add sorbitol to distilled/deionized water and bring volume to 50.0mL. Mix thoroughly. Filter sterilize.

Preparation of Medium: Add components, except sorbitol solution, to distilled/deionized water and bring volume to 950.0mL. Mix thoroughly. Adjust pH to 7.0. Gently heat and bring to boiling. Autoclave for 15 min at 15 psi pressure–121°C. Cool to 45°–50°C. Aseptically add sterile sorbitol solution. Mix thoroughly. Pour into sterile Petri dishes or distribute into sterile tubes.

Use: For the cultivation and maintenance of *Pseudomonas* species.

Sorbitol MacConkey Agar (MacConkey Agar with Sorbitol)

Composition per liter:

Peptone	20.0g
Agar	15.0g
Sorbitol	10.0g
NaCl	5.0g
Bile salts No.3	1.5g
Neutral Red	0.03g
Crystal Violet	1.0mg

pH 7.1 ± 0.2 at 25°C

Source: This medium is available as a premixed powder from Difco Laboratories and Oxoid Unipath.

Preparation of Medium: Add components to distilled/deionized water and bring volume to 1.0L. Mix thoroughly. Gently heat and bring to boiling. Distribute into tubes or flasks. Autoclave for 15 min at 15 psi pressure–121°C. Pour into sterile Petri dishes or leave in tubes.

Use: For the isolation and cultivation of pathogenic *Escherichia coli*.

Sorbitol Medium

Composition per liter:

Sorbitol	50.0g
Agar	15.0g
Peptone	10.0g
Yeast extract	10.0g

pH 6.0 ± 0.2 at 25°C

Preparation of Medium: Add components to distilled/deionized water and bring volume to 1.0L. Mix thoroughly. Adjust pH to 6.0 with HCl. Gently heat and bring to boiling. Distribute into tubes or flasks. Autoclave for 15 min at 15 psi pressure–121°C. Pour into sterile Petri dishes or leave in tubes.

Use: For the cultivation and maintenance of *Gluconobacter asaii* and *Gluconobacter frateurii*.

SOT Medium

Composition per liter:

$NaHCO_3$	16.8g
$NaNO_3$	2.5g
K_2SO_4	1.0g
NaCl	1.0g
K_2HPO_4	0.5g
$MgSO_4 \cdot 7H_2O$	0.2g
Disodium EDTA·$2H_2O$	0.08g
$CaCl_2 \cdot 2H_2O$	0.04g
$FeSO_4 \cdot 7H_2O$	0.01g
Trace metal mix A5	1.0mL
Trace metal mix B6, modified	1.0mL

pH 9.0 ± 0.2 at 25°C

Trace Metal Mix A5:
Composition per liter:

H_3BO_3	2.86g
$MnCl_2 \cdot 4H_2O$	1.81g
$Na_2MoO_4 \cdot 2H_2O$	0.39g
$ZnSO_4 \cdot 7H_2O$	0.222g
$CuSO_4 \cdot 5H_2O$	0.079g
$Co(NO_3)_2 \cdot 6H_2O$	0.049g

Preparation of Trace Metal Mix A5: Add components to distilled/deionized water and bring volume to 1.0L. Mix thoroughly.

Trace Metal Mix B6, Modified:
Composition per liter:

NH_4NO_3	0.23g
$K_2Cr_2(SO_4)_4 \cdot 24H_2O$	0.096g
$NiSO_4 \cdot 7H_2O$	0.048g
$Ti_2(SO_4)_3$	0.040g
$Na_2WO_4 \cdot 2H_2O$	0.018g

Preparation of Trace Metal Mix B6, Modified: Add components to distilled/deionized water and bring volume to 1.0L. Mix thoroughly.

Preparation of Medium: Add components to distilled/deionized water and bring volume to 1.0L. Mix thoroughly. Adjust pH to 9.0. Distribute into tubes or flasks. Autoclave for 15 min at 15 psi pressure–121°C.

Use: For the cultivation and maintenance of *Spirulina maxima* and *Spirulina platensis*.

Sour Dough Medium

Composition per liter:

Maltose	20.0g
Pancreatic digest of casein	6.0g
Yeast extract	3.0g
Tween™ 80	0.3g
Fresh yeast extract solution	15.0mL

pH 5.6 ± 0.2 at 25°C

Fresh Yeast Extract Solution:
Composition per 100mL:

Baker's yeast, live, pressed, starch-free,	25.0g

Preparation of Fresh Yeast Extract Solution: Add the live Baker's yeast to 100.0mL of distilled/deionized water. Autoclave for 90 min at 15 psi pressure–121°C. Allow to stand. Remove supernatant solution. Adjust pH to 6.6–6.8.

Preparation of Medium: Add components to distilled/deionized water and bring volume to 1.0L. Mix thoroughly. Adjust pH to 5.6 with 20% lactic acid or 6*N* HCl. Distribute into tubes or flasks. Autoclave for 15 min at 15 psi pressure–121°C.

Use: For the cultivation and maintenance of *Lactobacillus sanfrancisco*.

Soy Peptone Broth

Composition per liter:

Papaic digest of soybean meal	20.0g
NaCl	5.0g
Phenol Red (2% solution)	1.0mL

pH 7.3 ± 0.2 at 25°C

Preparation of Medium: Add components to distilled/deionized water and bring volume to 1.0L. Mix thoroughly. Adjust pH to 7.3. Distribute into tubes or flasks. Autoclave for 15 min at 15 psi pressure–121°C.

Use: For the isolation and cultivation of *Mycoplasma* species and *Ureaplasma* species.

Soybean Agar

Composition per liter:

White soybeans	100.0g
Agar	15.0g

Preparation of Medium: Add soybeans to 1.0L of distilled/deionized water. Soak overnight. Autoclave for 60 min at 15 psi pressure–121°C. Filter through cheesecloth. Measure volume of filtrate. Add agar to a concentration of 1.5%. Gently heat and bring to boiling. Distribute into tubes or flasks. Autoclave for 15 min at 15 psi pressure–121°C. Pour into sterile Petri dishes or leave in tubes.

Use: For the cultivation and maintenance of *Bacillus subtilis* and *Pseudomonas syringae*.

Soybean Casein Digest Agar
See: **Trypticase Soy Agar**

Soybean Casein Digest Broth, USP
See: **Trypticase Soy Broth**

Soybean Extract, M-1

Composition per liter:

Soybeans	50.0g
Soluble starch	15.0g
$(NH_4)_2HPO_3$	10.0g
KCl	0.2g
$MgSO_4 \cdot 7H_2O$	0.2g

pH 7.0 ± 0.2 at 25°C

Preparation of Medium: Add soybeans to 1.0L of distilled/deionized water. Soak overnight. Add 2.0g NaOH. Adjust pH to 7.0 with HCl. Autoclave for 60 min at 0 psi pressure–100°C. Filter through cheesecloth. Bring volume of filtrate to 1.0L with distilled/deionized water. Add remaining components. Mix thoroughly. Distribute into tubes or flasks. Autoclave for 15 min at 15 psi pressure–121°C.

Use: For the cultivation and maintenance of *Bacillus subtilis*.

SP Agar

Composition per liter:

Agar	15.0g
Pancreatic digest of casein	2.5g
Galactose	1.0g
Raffinose	1.0g
Sucrose	1.0g
$MgSO_4 \cdot 7H_2O$	0.50g
K_2HPO_4	0.25g
Vitamin solution	2.5mL

Vitamin Solution:

Composition per liter:

Inositol	1.0g
Calcium pantothenate	0.2g
Choline hydrochloride	0.2g
Thiamine	0.1g
Nicotinamide	0.75g
Pyridoxin	0.75g
Riboflavin	0.75g
p-Aminobenzoic acid	5.0mg
Folic acid	1.0mg
Biotin	0.05mg
Vitamin B_{12}	0.05mg
Ethanol	1.0L

Preparation of Vitamin Solution: Add solid components to 1.0L of ethanol. Mix thoroughly.

Preparation of Medium: Add components to distilled/deionized water and bring volume to 1.0L. Mix thoroughly. Gently heat and bring to boiling. Distribute into tubes or flasks. Autoclave for 15 min at 15 psi pressure–121°C. Pour into sterile Petri dishes or leave in tubes.

Use: For the cultivation of myxobacteria.

SP 2 Agar

Composition per liter:

Agar	15.0g

Pancreatic digest of casein	3.0g
Yeast extract	1.0g
Sodium acetate	0.02g
Artificial seawater	1.0L

pH 7.2 ± 0.2 at 25°C

Artificial Seawater:

Composition per liter:

NaCl	24.7g
$MgSO_4 \cdot 7H_2O$	6.3g
$MgCl_2 \cdot 6H_2O$	4.6g
$CaCl_2$	1.0g
KCl	0.7g
$NaHCO_3$	0.2g

Preparation of Artificial Seawater: Add components to distilled/deionized water and bring volume to 1.0L. Mix thoroughly.

Preparation of Medium: Add solid components to 1.0L of artificial seawater. Mix thoroughly. Gently heat and bring to boiling. Distribute into tubes or flasks. Autoclave for 15 min at 15 psi pressure–121°C. Pour into sterile Petri dishes or leave in tubes.

Use: For the isolation and cultivation of *Cytophaga* species, *Herpetosiphon* species, *Saprospira* species, and *Flexithrix* species.

SP 6 Agar

Composition per liter:

Agar	15.0g
Pancreatic digest of casein	3.0g
Yeast extract	1.0g
Artificial seawater	1.0L

pH 7.2 ± 0.2 at 25°C

Artificial Seawater:

Composition per liter:

NaCl	24.7g
$MgSO_4 \cdot 7H_2O$	6.3g
$MgCl_2 \cdot 6H_2O$	4.6g
$CaCl_2$	1.0g
KCl	0.7g
$NaHCO_3$	0.2g

Preparation of Artificial Seawater: Add components to distilled/deionized water and bring volume to 1.0L. Mix thoroughly.

Preparation of Medium: Add solid components to 1.0L of artificial seawater. Mix thoroughly. Gently heat and bring to boiling. Distribute into tubes or flasks. Autoclave for 15 min at 15 psi pressure–121°C. Pour into sterile Petri dishes or leave in tubes.

Use: For the isolation and cultivation of *Cytophaga* species, *Herpetosiphon* species, *Saprospira* species, and *Flexithrix* species.

SP5 Broth

Composition per liter:

Pancreatic digest of casein	9.0g
Yeast extract	1.0g
Artificial seawater	1.0L

pH 7.2 ± 0.2 at 25°C

Artificial Seawater:

Composition per liter:

NaCl	24.7g
$MgSO_4 \cdot 7H_2O$	6.3g
$MgCl_2 \cdot 6H_2O$	4.6g
$CaCl_2$	1.0g
KCl	0.7g
$NaHCO_3$	0.2g

Preparation of Artificial Seawater: Add components to distilled/deionized water and bring volume to 1.0L. Mix thoroughly.

Preparation of Medium: Add solid components to 1.0L of artificial seawater. Mix thoroughly. Gently heat and bring to boiling. Distribute into tubes or flasks. Autoclave for 15 min at 15 psi pressure–121°C.

Use: For the isolation and cultivation of *Cytophaga* species, *Herpetosiphon* species, *Saprospira* species, and *Flexithrix* species.

SP Medium

Composition per liter:

Agar	15.0g
Soluble starch	5.0g
Pancreatic digest of casein	2.5g
Galactose	1.0g
Raffinose	1.0g
Sucrose	1.0g
$MgSO_4 \cdot 7H_2O$	0.5g
K_2HPO_4	0.25g

Preparation of Medium: Add components to distilled/deionized water and bring volume to 1.0L. Mix thoroughly. Gently heat and bring to boiling. Distribute into tubes or flasks. Autoclave for 15 min at 15 psi pressure–121°C. Pour into sterile Petri dishes or leave in tubes.

Use: For the cultivation and maintenance of *Archangium gephyra*, *Cystobacter fuscus*, *Melittangium lichenicola*, *Myxococcus* species, *Polyangium brachysporum*, *Stigmatella aurantiaca* and *Stigmatella erecta*.

SP4 Medium

Composition per liter:

Base solution	615.0mL

Fetal calf serum
(inactivated at 56°C, 1 hr) 170.0mL
Yeast extract (2% solution) 100.0mL
CMRL 1066, 10X with glutamine 50.0mL
Fresh yeast extract solution 35.0mL
Phenol Red (0.1% solution) 20.0mL
Penicillin solution 10.0mL
pH 7.0–7.4 ± 0.2 at 25°C

Base Solution:
Composition per 615mL:
Pancreatic digest of casein 11.2g
Noble agar 8.0g
Pancreatic digest of gelatin 5.3g
Glucose 5.0g
NaCl 0.875g
Beef extract 0.525g
Yeast extract 0.525g
Beef heart, solids from infusion 0.35g

Preparation of Base Solution: Add components to distilled/deionized water and bring volume to 615.0mL. Mix thoroughly. Adjust pH to 7.5. Gently heat and bring to boiling. Autoclave for 15 min at 15 psi pressure–121°C. Cool to 45°–50°C.

CMRL 1066, 10X With Glutamine:
Composition per liter:
NaCl 6.8g
$NaHCO_3$ 2.2g
D-Glucose 1.0g
KCl 0.4g
L-Cysteine·HCl·H_2O 0.26g
$CaCl_2$, anhydrous 0.2g
$MgSO_4 \cdot 7H_2O$ 0.2g
$NaH_2PO_4 \cdot H_2O$ 0.14g
L-Glutamine 0.1g
Sodium acetate·$3H_2O$ 0.083g
L-Glutamic acid 0.075g
L-Arginine·HCl 0.070g
L-Lysine·HCl 0.070g
L-Leucine 0.060g
Glycine 0.050g
Ascorbic acid 0.050g
L-Proline 0.040g
L-Tyrosine 0.040g
L-Aspartic acid 0.030g
L-Threonine 0.030g
L-Alanine 0.025g
L-Phenylalanine 0.025g
L-Serine 0.025g
L-Valine 0.025g
L-Cystine 0.020g
L-Histidine·HCl·H_2O 0.020g
L-Isoleucine 0.020g
Phenol Red 0.020g
L-Methionine 0.015g
Deoxyadenosine 0.010g
Deoxycytidine 0.010g
Deoxyguanosine 0.010g
Glutathione, reduced 0.010g
Thymidine 0.010g
Hydroxy-L-proline 0.010g
L-Tryptophan 0.010g
Nicotinamide adenine dinucleotide 7.0mg
Tween™ 80 .... 5.0mg
Sodium glucoronate·H_2O 4.2mg
Coenzyme A 2.5mg
Cocarboxylase 1.0mg
Flavin adenine dinucleotide 1.0mg
Nicotinamide adenine
dinucleotide phosphate 1.0mg
Uridine triphosphate 1.0mg
Choline chloride 0.50mg
Cholesterol 0.20mg
5-Methyldeoxycytidine 0.10mg
Inositol 0.05mg
p-Aminobenzoic acid 0.05mg
Niacin 0.025mg
Niacinamide 0.025mg
Pyridoxine 0.025mg
Pyridoxal·HCl 0.025mg
Biotin 0.01mg
D-Calcium pantothenate 0.01mg
Folic acid 0.01mg
Riboflavin 0.01mg
Thiamine·HCl 0.01mg

Preparation of CMRL 1066, 10X With Glutamine: Add components to distilled/deionized water and bring volume to 1.0L. Mix thoroughly. Adjust pH to 7.2. Filter sterilize.

Fresh Yeast Extract Solution:
Composition per 100mL:
Baker's yeast, live, pressed, starch-free, ... 25.0g

Preparation of Fresh Yeast Extract Solution: Add the live Baker's yeast to 100.0mL of distilled/deionized water. Autoclave for 90 min at 15 psi pressure–121°C. Allow to stand. Remove supernatant solution. Adjust pH to 6.6–6.8.

Penicillin Solution:
Composition per 10mL:
Penicillin G 1,000,000U

Preparation of Penicillin Solution: Add penicillin to distilled/deionized water and bring volume to 10.0mL. Mix thoroughly. Filter sterilize.

Preparation of Medium: To 615.0mL of cooled sterile base solution, aseptically add 170.0mL of sterile inactivated fetal calf serum, 100.0mL of sterile yeast extract, 50.0mL of sterile CMRL 1066, 10X with glutamine, 35.0mL of sterile fresh yeast extract solution, 20.0mL of Phenol Red solution, and 10.0mL of sterile penicillin solution. Mix thoroughly. Aseptically distribute into sterile tubes. Allow tubes to cool in a slanted position.

Use: For the cultivation of tick-derived *Mycoplasma* (*Spiroplasma*). Used for the enhanced recovery of *Mycoplasma pneumoniae*, *M. alvi*, and *M. hyopneumoniae*.

SPB
See: **Salt Polymyxin Broth**

Specimen Preservative Medium

Composition per liter:

NaCl	5.0g
Sodium citrate ·2 H_2O	5.0g
$(NH_4)_2HPO_4$	4.0g
KH_2PO_4	2.0g
Yeast extract	1.0g
Sodium deoxycholate	0.5g
$MgSO_4 \cdot 7H_2O$	0.4g
Glycerol	300.0mL

pH 7.0 ± 0.2 at 25°C

Preparation of Medium: Add components, except glycerol, to distilled/deionized water and bring volume to 700.0mL. Mix thoroughly. Gently heat and bring to boiling. Add 300.0mL glycerol. Mix thoroughly. Distribute into tubes or flasks. Autoclave for 10 min at 11 psi pressure–116°C.

Use: For the preservation of viable microorganisms in stool specimens. For transport of fecal material.

Sphaericus Spore Medium

Composition per liter:

Agar	15.0g
Pancreatic digest of gelatin	5.0g
Beef extract	3.0g
Yeast extract	0.5g
$MgCl_2$	0.095g
$CaCl_2$	0.078g
$MnCl_2$	6.0mg

Preparation of Medium: Add components to distilled/deionized water and bring volume to 1.0L. Mix thoroughly. Gently heat and bring to boiling. Distribute into tubes or flasks. Autoclave for 15 min at 15 psi pressure–121°C. Pour into sterile Petri dishes or leave in tubes.

Use: For the cultivation and maintenance of *Bacillus sphaericus*.

Sphaerotilus and *Leptothrix* Enrichment Medium

Composition per liter:

Glucose	1.0g
Peptone	1.0g
$MgSO_4 \cdot 7H_2O$	0.2g
$FeCl_3 \cdot 6H_2O$	0.1g
$CaCl_2 \cdot 2H_2O$	0.05g

pH 7.0 ± 0.2 at 25°C

Preparation of Medium: Add components to distilled/deionized water and bring volume to 1.0L. Mix thoroughly. Distribute into tubes or flasks. Autoclave for 15 min at 15 psi pressure–121°C.

Use: For the enrichment and cultivation of *Sphaerotilus* species and *Leptothrix* species.

Sphaerotilus CGYA Medium

Composition per liter:

Glycerol	10.0g
Pancreatic digest of casein	5.0g
Yeast extract	1.0g

Preparation of Medium: Add components to distilled/deionized water and bring volume to 1.0L. Mix thoroughly. Distribute into tubes or flasks. Autoclave for 15 min at 15 psi pressure–121°C.

Use: For the cultivation and maintenance of *Sphaerotilus natans* and *Sphaerotilus* species.

Sphaerotilus Defined Medium

Composition per liter:

Agar	15.0g
Glycerol	5.0g
Glutamic acid	0.9g
$FeSO_4 \cdot 7H_2O$	0.5g
$MgSO_4 \cdot 7H_2O$	0.1g
$CaCl_2 \cdot 2H_2O$	0.03g
$ZnSO_4 \cdot 7H_2O$	0.03g
Phosphate solution	100.0mL

pH 7.0 ± 0.2 at 25°C

Phosphate Solution:

Composition per 500mL:

K_2HPO_4	5.7g
KH_2PO_4	2.3g

Preparation of Phosphate Solution: Add components to distilled/deionized water and bring volume to 500.0mL. Mix thoroughly. Gently heat until dissolved. Autoclave for 15 min at 15 psi pressure–121°C.

Preparation of Medium: Add components, except phosphate solution, to distilled/deionized water and bring volume to 900.0mL. Mix thoroughly. Gently heat and bring to boiling. Autoclave for 10 min at 15 psi pressure–121°C. Cool to 45°–50°C. Aseptically add 100.0mL of sterile phosphate solution. Mix thoroughly. Pour into sterile Petri dishes or distribute into sterile tubes.

Use: For the cultivation of *Sphaerotilus* species.

Sphaerotilus discophorus Medium

Composition per liter:

Agar	12.0g
Peptone	5.0g
$MgSO_4 \cdot 7H_2O$	0.2g
$CaCl_2$	0.05g
$MnSO_4 \cdot H_2O$	0.05g
Ferric solution	100.0mL

pH 7.0 ± 0.2 at 25°C

Ferric Solution:

Composition per 100mL:

Ferric ammonium citrate	0.5g
$FeCl_3 \cdot 6H_2O$	0.01g

Preparation of Ferric Solution: Add components to distilled/deionized water and bring volume to 100.0mL. Mix thoroughly. Filter sterilize.

Preparation of Medium: Add components, except ferric solution, to tap water and bring volume to 900.0mL. Mix thoroughly. Gently heat and bring to boiling. Autoclave for 15 min at 15 psi pressure–121°C. Cool to 45°–50°C. Aseptically add sterile ferric solution. Mix thoroughly. Pour into sterile Petri dishes or distribute into sterile tubes.

Use: For the cultivation of *Sphaerotilus discophorus*.

Sphaerotilus Isolation Medium

Composition per liter:

Agar	15.0g
Glycerol	10.0g
Pancreatic digest of casein	5.0g
Yeast extract	1.0g

pH 7.0 ± 0.2 at 25°C

Preparation of Medium: Add components to distilled/deionized water and bring volume to 1.0L. Mix thoroughly. Gently heat and bring to boiling. Distribute into tubes or flasks. Autoclave for 15 min at 15 psi pressure–121°C. Pour into sterile Petri dishes or leave in tubes.

Use: For the isolation and cultivation of *Sphaerotilus* species.

Sphaerotilus natans Enrichment Medium

Composition per liter:

Sodium lactate	0.1g
$Na_2HPO_4 \cdot 7H_2O$	0.034g
$CaCl_2$	0.027g
$MgSO_4 \cdot 7H_2O$	0.023g
K_2HPO_4	0.022g
KH_2PO_4	8.5mg
NH_4Cl	1.7mg
$FeCl_3 \cdot 6H_2O$	0.25mg

pH 7.1–7.2 at 25°C

Preparation of Medium: Add components to distilled/deionized water and bring volume to 1.0L. Mix thoroughly. Distribute into tubes or flasks. Autoclave for 15 min at 15 psi pressure–121°C.

Use: For the enrichment and cultivation of *Sphaerotilus natans*.

Sphaerotilus natans Isolation Agar

Composition per liter:

Agar	15.0g
Meat extract	0.5g

Preparation of Medium: Add components to tap water and bring volume to 1.0L. Mix thoroughly. Gently heat and bring to boiling. Distribute into tubes or flasks. Autoclave for 15 min at 15 psi pressure–121°C. Pour into sterile Petri dishes or leave in tubes.

Use: For the isolation and cultivation of *Sphaerotilus natans*.

Sphaerotilus natans Isolation Agar

Composition per liter:

Agar	15.0g
Casein hydrolysate	1.5g

Preparation of Medium: Add components to tap water and bring volume to 1.0L. Mix thoroughly. Gently heat and bring to boiling. Distribute into tubes or flasks. Autoclave for 15 min at 15 psi pressure–121°C. Pour into sterile Petri dishes or leave in tubes.

Use: For the isolation and cultivation of *Sphaerotilus natans*.

Sphingobacterium Medium

Composition per liter:

Pancreatic digest of casein	10.0g
NaCl	5.0g
Yeast extract	3.0g

Preparation of Medium: Add components to distilled/deionized water and bring volume to 1.0L. Mix thoroughly. Distribute into tubes or flasks. Autoclave for 15 min at 15 psi pressure–121°C.

Use: For the cultivation and maintenance of *Sphingobacterium mizutae*, *S. multivorum* and *S. spiritivorum.*

Spirillum gracile Medium

Composition per liter:

Agar	15.0g
Peptone	5.0g
Yeast extract	0.5g
K_2HPO_4	0.1g
Tween™ 80	0.02g

pH 7.2 ± 0.2 at 25°C

Preparation of Medium: Add components to tap water and bring volume to 1.0L. Mix thoroughly. Gently heat and bring to boiling. Distribute into tubes or flasks. Autoclave for 15 min at 15 psi pressure–121°C. Pour into sterile Petri dishes or leave in tubes.

Use: For the cultivation and maintenance of *Aquaspirillum gracile.*

Spirillum lipoferum Medium

Composition per liter:

Sodium malate	5.0g
Agar	3.5g
KH_2PO_4	0.4g
$MgSO_4 \cdot 7H_2O$	0.2g
K_2HPO_4	0.1g
NaCl	0.1g
$CaCl_2$	0.02g
$FeCl_3$	0.01g
$NaMoO_4 \cdot 2H_2O$	2.0mg
Bromthymol Blue solution	5.0mL

pH 6.8 ± 0.2 at 25°C

Bromthymol Blue Solution:

Composition per 10mL:

Bromthymol Blue	0.5g
Ethanol	10.0mL

Preparation of Bromthymol Blue Solution: Add Bromthymol Blue to 10.0mL of ethanol. Mix thoroughly.

Preparation of Medium: Add components to distilled/deionized water and bring volume to 1.0L. Mix thoroughly. Gently heat and bring to boiling. Distribute into tubes or flasks. Autoclave for 15 min at 15 psi pressure–121°C.

Use: For the isolation and cultivation of *Spirillum leptoferum.*

Spirillum lipoferum Medium

Composition per liter:

Malic acid	5.0g
NaOH	4.7g
Agar	1.75g
KH_2PO_4	0.4g
$MgSO_4 \cdot 7H_2O$	0.2g
K_2HPO_4	0.1g
NaCl	0.1g
$CaCl_2$	0.02g
$FeCl_3$	0.01g
$NaMoO_4 \cdot 2H_2O$	2.0mg
Bromthymol Blue solution	5.0mL

pH 6.8 ± 0.2 at 25°C

Bromthymol Blue Solution:

Composition per 10mL:

Bromthymol Blue	0.5g
Ethanol	10.0mL

Preparation of Bromthymol Blue Solution: Add Bromthymol Blue to 10.0mL of ethanol. Mix thoroughly.

Preparation of Medium: Add components to distilled/deionized water and bring volume to 1.0L. Mix thoroughly. Gently heat and bring to boiling. Distribute into tubes or flasks. Autoclave for 15 min at 15 psi pressure–121°C.

Use: For the isolation and cultivation of *Spirillum leptoferum.*

Spirillum lipoferum Medium

Composition per liter:

Malic acid	5.0g
KOH	4.0g
Agar	1.75g
$FeSO_4 \cdot 7H_2O$	0.5g
K_2HPO_4	0.5g
$MgSO_4 \cdot 7H_2O$	0.2g
NaCl	0.1g
$CaCl_2$	0.02g
$MnSO_4 \cdot H_2O$	0.01g
$NaMoO_4 \cdot 2H_2O$	2.0mg
Bromthymol Blue solution	5.0mL

pH 6.8 ± 0.2 at 25°C

Bromthymol Blue Solution:

Composition per 10mL:

Bromthymol Blue	0.5g
Ethanol	10.0mL

Preparation of Bromthymol Blue Solution: Add Bromthymol Blue to 10.0mL of ethanol. Mix thoroughly.

Preparation of Medium: Add components to distilled/deionized water and bring volume to 1.0L. Mix thoroughly. Gently heat and bring to boiling. Distribute into tubes or flasks. Autoclave for 15 min at 15 psi pressure–121°C.

Use: For the isolation and cultivation of *Spirillum leptoferum*.

Spirillum Medium

Composition per liter:

Calcium lactate	10.0g
Peptone	5.0g
Beef extract	3.0g
Yeast extract	3.0g

pH 7.0 ± 0.2 at 25°C

Preparation of Medium: Add components to distilled/deionized water and bring volume to 1.0L. Mix thoroughly. Adjust pH to 7.0. Distribute into tubes or flasks. Autoclave for 20 min at 11 psi pressure–116°C. A precipitate will form during autoclaving.

Use: For the cultivation of *Spirillum* species.

Spirillum Nitrogen–Fixing Medium

Composition per liter:

Sodium malate	5.0g
KH_2PO_4	0.4g
$MgSO_4 \cdot 7H_2O$	0.2g
K_2HPO_4	0.1g
NaCl	0.1g
Yeast extract	0.05g
$CaCl_2$	0.02g
$FeCl_3$	0.01g
$NaMoO_4 \cdot 2H_2O$	2.0mg

pH 7.2-7.4 ± 0.2 at 25°C

Preparation of Medium: Add components to distilled/deionized water and bring volume to 1.0L. Mix thoroughly. Distribute into tubes or flasks. Autoclave for 15 min at 15 psi pressure–121°C.

Use: For the cultivation and maintenance of *Azospirillum brasilense*, *Azospirillum lipoferum* and *Herbaspirillum seropedicae*.

Spirillum volutans Defined Medium

Composition per liter:

BES (*N,N*-bis[2-hydroxyethyl]-2-aminoethane sulfonic acid) buffer	1.07g
$MgSO_4 \cdot 7H_2O$	1.0g
$(NH_4)_2SO_4$	1.0g
Succinic acid	1.0g
L-Histidine	0.2g
L-Isoleucine	0.2g
L-Methionine	0.2g
L-Threonine	0.2g
NaCl	0.085g
L-Cystine	0.025g
K_2HPO_4	0.02g
$FeCl_3 \cdot 6H_2O$	3.0mg
DL-Norepinephrine	2.0mg
$MnSO_4 \cdot H_2O$	2.0mg
$CaCO_3$	1.0mg
$ZnSO_4 \cdot 7H_2O$	0.72mg
$Na_2MoO_4 \cdot 2H_2O$	0.245mg
$CoSO_4 \cdot 7H_2O$	0.14mg
$CuSO_4 \cdot 5H_2O$	0.13mg
H_2BO_3	0.031mg

pH 6.8 ± 0.2 at 25°C

Preparation of Medium: Add components to distilled/deionized water and bring volume to 1.0L. Mix thoroughly. Adjust pH to 6.8. Distribute into tubes or flasks. Autoclave for 15 min at 15 psi pressure–121°C.

Use: For the cultivation of *Spirillum volutans*.

Spirit Blue Agar

Composition per liter:

Agar	20.0g
Pancreatic digest of casein	10.0g
Yeast extract	5.0g
Spirit Blue	0.15g
Lipoidal emulsion	30.0mL

pH 6.8 ± 0.2 at 25°C

Source: This medium is available as a premixed powder from Difco Laboratories.

Lipoidal Emulsion:

Composition per 500mL:

Tween™ 80	1.0mL
Cottonseed oil or olive oil	100.0mL

Preparation of Lipoidal Emulsion: Add Tween™ 80 to 400.0mL of warm distilled/deionized water. Mix thoroughly. Add 100.0mL of cottonseed or olive oil. Emulsify in a blender. Autoclave for 15 min at 15 psi pressure–121°C. Cool to 45°–50°C.

Preparation of Medium: Add components, except lipoidal emulsion, to distilled/deionized water and bring volume to 970.0mL. Mix thoroughly. Gently heat and bring to boiling. Autoclave for 15 min at 15 psi pressure–121°C. Cool to 45°–50°C. Aseptically add 30.0mL sterile lipoidal emulsion. Mix thoroughly. Pour into sterile Petri dishes while shaking flask to keep emulsion dispersed.

Use: For the detection, enumeration and study of lipolytic microorganisms.

Spirochaeta aurantia Growth Medium

Composition per liter:

Yeast extract	4.0g
Maltose	2.0g
Peptone	2.0g
Potassium phosphate buffer (0.1M solution, pH 7.0)	100.0mL

pH 7.2 ± 0.2 at 25°C

Preparation of Medium: Filter sterilize potassium phosphate buffer. Add components, except potassium phosphate buffer, to distilled/deionized water and bring volume to 900.0mL. Mix thoroughly. Gently heat and bring to boiling. Autoclave for 15 min at 15 psi pressure–121°C. Cool to 45°–50°C. Aseptically add sterile potassium phosphate buffer. Mix thoroughly. Adjust pH to 7.2. Aseptically distribute into sterile tubes or flasks.

Use: For the cultivation of *Spirochetes aurantia*.

Spirochaeta aurantia Isolation Medium

Composition per liter:

Peptone	1.0g
Yeast extract	1.0g
Hay extract	500.0mL

pH 6.5 ± 0.2 at 25°C

Hay Extract:

Composition per liter:

Hay, dried	5.0g

Preparation of Hay Extract: Add hay to distilled/deionized water and bring volume to 1.0L. Mix thoroughly. Gently heat and bring to boiling. Continue boiling for 10 min. Filter through Whatman #1 filter paper.

Preparation of Medium: Add components to distilled/deionized water and bring volume to 1.0L. Mix thoroughly. Distribute into tubes or flasks. Autoclave for 15 min at 15 psi pressure–121°C.

Use: For the isolation and cultivation of *Spirochaeta aurantia*.

Spirochaeta halophila Medium

Composition per liter:

Glucose salts solution	970.0mL
Yeast extract peptone solution	30.0mL

Glucose Salts Solution:

Composition per liter:

NaCl	49.3g
$MgSO_4 \cdot 7H_2O$	49.2g
$CaCl_2 \cdot 2H_2O$	5.9g
Glucose solution	100.0mL
Sulfide solution	100.0mL

Preparation of Glucose Salts Solution: Add components, except glucose solution and sulfide solution, to distilled/deionized water and bring volume to 800.0mL. Mix thoroughly. Autoclave for 15 min at 15 psi pressure–121°C. Cool to 25°C. Aseptically add sterile glucose solution and sulfide solution. Mix thoroughly.

Sulfide Solution:

Composition per 100mL:

$Na_2S \cdot 9H_2O$	0.5g

Preparation of Sulfide Solution: Add $Na_2S \cdot 9H_2O$ to distilled/deionized water and bring volume to 100.0mL. Mix thoroughly. Autoclave for 15 min at 15 psi pressure–121°C. Cool to 25°C.

Glucose Solution:

Composition per 100mL:

Glucose	5.0g

Preparation of Glucose Solution: Add glucose to distilled/deionized water and bring volume to 100.0mL. Mix thoroughly. Filter sterilize.

Yeast Extract Peptone Solution:

Composition per 30mL:

Yeast exract	4.0g
Peptone	2.0g

Preparation of Yeast Extract Peptone Solution: Add components to distilled/deionized water and bring volume to 30.0mL. Mix thoroughly. Autoclave for 15 min at 15 psi pressure–121°C. Cool to 25°C.

Preparation of Medium: Aseptically combine 30.0mL yeast extract peptone solution with 970.0mL glucose slats solution. Mix thoroughly. Aseptically distribute into sterile tubes or flasks.

Use: For the isolation and cultivation of *Spirochaeta halophila*.

Spirochaeta litoralis Medium

Composition per liter:

Pancreatic digest of casein	3.0g
NaCl	2.0g
Yeast extract	0.5g
Glucose solution	2.0mL

Potasssium phosphate
buffer (1M, pH 7.4) 2.0mL
Sulfide solution 0.5mL
Salts solution 0.2mL
pH 7.3 ± 0.2 at 25°C

Glucose Solution:

Composition per 100mL:

D-Glucose 25.0g

Preparation of Glucose Solution: Add D-glucose to distilled/deionized water and bring volume to 100.0mL. Mix thoroughly. Filter sterilize.

Sulfide Solution:

Composition per 100mL:

$Na_2S \cdot 9H_2O$ 10.0g

Preparation of Sulfide Solution: Add $Na_2S \cdot 9H_2O$ to distilled/deionized water and bring volume to 100.0mL. Autoclave for 15 min at 15 psi pressure–121°C. Cool to 25°C.

Salts Solution:

Composition per 100mL:

$MgSO_4 \cdot 7H_2O$ 12.5g
$CaCl_2 \cdot 2H_2O$ 3.75g
EDTA 1.0g
$FeSO_4 \cdot 7H_2O$ 0.5g
Trace elements solution 25.0mL

Preparation of Salts Solution: Add components to distilled/deionized water and bring volume to 100.0mL. Mix thoroughly.

Trace Elements Solution:

Composition per 1800mL:

H_3BO_3 .... 5.5g
$MnCl_2 \cdot 4H_2O$ 3.5g
$AlCl_3 \cdot 6H_2O$ 0.50g
$CoCl_2 \cdot 6H_2O$ 0.50g
$CuCl_2 \cdot 2H_2O$ 0.50g
$NiCl_2 \cdot 6H_2O$ 0.50g
$ZnCl_2$ 0.50g
KI 0.25g
LiCl 0.25g
$Na_2MoO_4 \cdot 2H_2O$ 0.25g
$BaCl_2 \cdot 2H_2O$ 0.15g
$SnCl_2 \cdot 2H_2O$ 0.15g
$NaVO_3$ 0.05g

Preparation of Trace Elements Solution: Add components to distilled/deionized water and bring volume to 1800.0mL. Mix thoroughly. Adjust pH to 3–4 with HCl.

Preparation of Medium: Add components, except glucose solution and sulfide solution, to distilled/deionized water and bring volume to 997.5mL. Mix thoroughly. Gently heat and bring to boiling. Autoclave for 15 min at 15 psi pressure–121°C. Cool to 45°–50°C. Aseptically add sterile glucose solution and sulfide solution. Mix thoroughly. Aseptically distribute into sterile tubes or flasks.

Use: For the isolation of *Spirochaeta litoralis* from marine habitats.

Spirochaeta stenostrepta Medium

Composition per liter:

Glucose 5.0g
Peptone 2.0g
Yeast extract 0.3g
Vitamin B_{12} 0.01mg
Salts solution 100.0mL
Phosphate solution 15.0mL
Sulfide solution 10.0mL
pH 7.0 ± 0.2 at 25°C

Phosphate Solution:

Composition per liter:

KH_2PO_4 .... 30.0g
K_2HPO_4 .... 70.0g

Preparation of Phosphate Solution: Add components to distilled/deionized water and bring volume to 1.0L. Mix thoroughly. Filter sterilize.

Salts Solution:

Composition per liter:

$MgSO_4 \cdot 7H_2O$ 2.0g
$CaCl_2 \cdot 2H_2O$ 0.75g
EDTA 0.2g
$FeSO_4 \cdot 7H_2O$ 0.1g
Trace elements solution 5.0mL

Preparation of Salts Solution: Add EDTA to approximately 800.0mL of distilled/deionized water. Gently heat until dissolved. Adjust pH to 7.0 with 2.5% KOH. Add the remaining components. Mix thoroughly. Bring volume to 1.0L with distilled/deionized water.

Trace Elements Solution:

Composition per 1800mL:

H_3BO_3 .... 5.5g
$MnCl_2 \cdot 4H_2O$ 3.5g
$AlCl_3 \cdot 6H_2O$ 0.50g
$CoCl_2 \cdot 6H_2O$ 0.50g
$CuCl_2 \cdot 2H_2O$ 0.50g
$NiCl_2 \cdot 6H_2O$ 0.50g
$ZnCl_2$ 0.50g
KI 0.25g
LiCl 0.25g
$Na_2MoO_4 \cdot 2H_2O$ 0.25g
$BaCl_2 \cdot 2H_2O$ 0.15g
$SnCl_2 \cdot 2H_2O$ 0.15g
$NaVO_3$ 0.05g

Preparation of Trace Elements Solution: Add components to distilled/deionized water and bring volume to 1800.0mL. Mix thoroughly. Adjust pH to 3–4 with HCl.

Sulfide Solution:

Composition per 100mL:

$Na_2S \cdot 9H_2O$	2.0g

Preparation of Sulfide Solution: Add $Na_2S \cdot 9H_2O$ to distilled/deionized water and bring volume to 100.0mL. Autoclave for 15 min at 15 psi pressure–121°C. Cool to 25°C. Prepare solution freshly.

Preparation of Medium: Add components, except sulfide solution, to distilled/deionized water and bring volume to 990.0mL. Mix thoroughly. Gently heat and bring to boiling. Autoclave for 15 min at 15 psi pressure–121°C. Cool to 45°–50°C. Aseptically add 10.0mL of sterile sulfide solution. Mix thoroughly. Aseptically distribute into sterile tubes or flasks.

Use: For the isolation of *Spirochaeta stenostrepta*.

Spirochaeta zuelzerae Medium

Composition per liter:

Solution 1	480.0mL
Solution 2	480.0mL
Solution 3	20.0mL
Solution 4	20.0mL

pH 7.2 ± 0.2 at 25°C

Solution 1:

Composition per 480mL:

KH_2PO_4	0.75g
Cysteine·HCl·H_2O	0.5g
$NaH_2PO_4 \cdot H_2O$	0.25g

Preparation of Solution 1: Add components to distilled/deionized water and bring volume to 480.0mL. Mix thoroughly. Adjust pH to 7.2 with 5.0% KOH. Autoclave for 15 min at 15 psi pressure–121°C. Cool to 45°–50°C.

Solution 2:

Composition per 480mL:

NH_4Cl	1.0g
$MgSO_4 \cdot 7H_2O$	0.5g
Yeast extract	0.2g
$CaCl_2$	0.02g
Resazurin	1.0mg
$FeCl_3 \cdot 6H_2O$ solution (0.25g/L)	10.0mL
Trace element solution	2.0mL

Preparation of Solution 2: Add components to distilled/deionized water and bring volume to 480.0mL. Mix thoroughly. Autoclave for 15 min at 15 psi pressure–121°C. Cool to 45°–50°C.

Trace Element Solution:

Composition per 100mL:

$Na_2MoO_4 \cdot 2H_2O$	0.075g
H_3BO_3	0.056g
$ZnSO_4 \cdot 7H_2O$	0.044g
$CoCl_2 \cdot 6H_2O$	0.020g
$CuSO_4 \cdot 5H_2O$	2.0mg
$MnCl_2$	2.0mg

Preparation of Trace Element Solution: Add components to distilled/deionized water and bring volume to 100.0mL. Mix thoroughly.

Solution 3:

Composition per 20mL:

$NaHCO_3$	1.0g

Preparation of Solution 3: Add $NaHCO_3$ to distilled/deionized water and bring volume to 20.0mL. Mix thoroughly. Filter sterilize under pressure.

Solution 4:

Composition per 20mL:

Glucose	2.0g

Preparation of Solution 4: Add glucose to distilled/deionized water and bring volume to 20.0mL. Mix thoroughly. Filter sterilize.

Preparation of Medium: Aseptically add 480.0 mL of sterile Solution 1 to 480.0mL of sterile Solution 2 under 80% N_2 + 20% CO_2. While gassing, add 20.0mL of sterile solution 3 and 20.0mL of sterile solution 4. Mix thoroughly. Adjust pH to 7.2. Aseptically and anaerobically distribute into tubes. Cap with rubber stoppers.

Use: For the cultivation and maintenance of *Spirochaeta zuelzerae*.

Spirochete Enrichment Medium

Composition per liter:

Agar	10.0g
Beef extract	1.0g
Peptone	1.0g
Yeast extract	1.0g
Seawater	500.0mL

Preparation of Medium: Add components to distilled/deionized water and bring volume to 1.0L. Mix thoroughly. Gently heat and bring to boiling. Distribute into tubes or flasks. Autoclave for 15 min at 15 psi pressure–121°C. Pour into sterile Petri dishes.

Use: For the isolation of spirochetes from muds. A well is cut into the agar plate and filled with mud samples. Spirochetes migrate out of the mud into the agar surrounding the well.

Spirochete Medium (ATCC Medium 1712)

Composition per liter:

Tris(hydroxymethyl)aminomethane buffer	7.52g
Pancreatic digest of casein	1.0g
Yeast extract	1.0g
Cysteine HCl·$2H_2O$	0.5g
Resazurin	1.0mg
Sea water	750.0mL
Glucose solution	20.0mL

pH 7.2 ± 0.2 at 25°C

Glucose Solution:

Composition per 20mL:

Glucose	2.0g

Preparation of Glucose Solution: Add glucose to distilled/deionized water and bring volume to 20.0mL. Mix thoroughly. Filter sterilize.

Preparation of Medium: Prepare and dispense medium under 100% N_2. Add components, except glucose solution, to distilled/deionized water and bring volume to 980.0mL. Mix thoroughly. Adjust pH to 7.5. Autoclave for 15 min at 15 psi pressure–121°C. Cool to 50°C. Aseptically add sterile glucose solution. Mix thoroughly. Aseptically distribute into sterile tubes or flasks.

Use: For the cultivation and maintenance of *Spirochaeta litoralis*.

Spirochete Medium (ATCC Medium 164)

Composition per liter:

Agar	15.0g
KH_2PO_4	1.0g
NH_4Cl	1.0g
Yeast extract	1.0g
$MgSO_4$	0.5g
$CaCl_2$	0.04g
$FeCl_3·6H_2O$	1.25mg
$NaHCO_3$ solution	20.0mL
Glucose solution	10.0mL
$Na_2S·9H_2O$ solution	5.0mL

Glucose Solution:

Composition per 100mL:

Glucose	10.0g

Preparation of Glucose Solution: Add glucose to distilled/deionized water and bring volume to 100.0mL. Mix thoroughly. Filter sterilize.

$NaHCO_3$ Solution:

Composition per 100mL:

$NaHCO_3$	5.0g

Preparation of $NaHCO_3$ Solution: Add the $NaHCO_3$ to distilled/deionized water and bring volume to 100.0mL. Mix thoroughly. Filter sterilize.

$Na_2S·9H_2O$ Solution:

Composition per 100mL:

$Na_2S·9H_2O$	10.0g

Preparation of $Na_2S·9H_2O$ Solution: Add $Na_2S·9H_2O$ to distilled/deionized water and bring volume to 100.0mL. Mix thoroughly. Autoclave for 15 min at 15 psi pressure–121°C.

Preparation of Medium: Add components—except $NaHCO_3$ solution, glucose solution and $Na_2S·9H_2O$ solution—to distilled/deionized water and bring volume to 965.0mL. Autoclave for 15 min at 15 psi pressure–121°C. Cool to 50°C. Aseptically add 20.0mL of sterile $NaHCO_3$ solution, 10.0mL of sterile glucose solution and 5.0mL of sterile $Na_2S·9H_2O$ solution. Mix thoroughly. Pour into sterile Petri dishes or distribute into sterile tubes.

Use: For the cultivation of spirochetes.

Spirolate Broth

Composition per liter:

Pancreatic digest of casein	15.0g
Glucose	5.0g
Yeast extract	5.0g
NaCl	2.5g
L-Cysteine·HCl·H_2O	1.0g
Sodium thioglycollate	0.5g
Palmitic acid	0.05g
Stearic acid	0.05g
Oleic acid	0.05g
Linoleic acid	0.05g
Serum	100.0mL

pH 7.1 ± 0.2 at 25°C

Source: This medium is available as a premixed powder from BBL Microbiology Systems.

Preparation of Medium: Add components, except serum, to distilled/deionized water and bring volume to 900.0mL. Mix thoroughly. Distribute into screw-capped tubes in 20.0mL volumes. Autoclave for 15 min at 15 psi pressure–121°C. Cool to 25°C. Aseptically add 2.0mL of serum to each tube. Heat-inactivated sheep, rabbit or bovine serum may be used. Tighten caps. Mix thoroughly.

Use: For the cultivation of *Treponema phagedenis* and other spirochetes.

Spiroplasma Agar MID

Composition per 291.2mL:

Schneider's *Drosophila* medium	160.0mL

Solution 1 .... 80.0mL
Fetal calf serum 50.0mL
Phenol Red (0.5% solution) 1.2mL
pH 7.4 ± 0.2 at 25°C

Schneider's *Drosophila* Medium:
Composition per liter:

$MgSO_4 \cdot 7H_2O$	3.7g
NaCl	2.1g
Yeast extract	2.0g
Trehalose	2.0g
D-Glucose	2.0g
L-Glutamine	1.8g
L-Lysine·HCl	1.7g
L-Proline	1.7g
KCl	1.6g
$Na_2HPO_4 \cdot 7H_2O$	1.3g
L-Glutamic acid	0.8g
L-Methionine	0.8g
$CaCl_2$, anhydrous	0.6g
KH_2PO_4	0.5g
β-Alanine	0.5g
L-Tyrosine	0.5g
L-Arginine	0.4g
L-Aspartic acid	0.4g
L-Histidine	0.4g
L-Threonine	0.4g
$NaHCO_3$	0.4g
Glycine	0.3g
L-Serine	0.3g
L-Valine	0.3g
L-Isoleucine	0.2g
L-Leucine	0.2g
L-Phenylalanine	0.2g
α-Ketoglutaric acid	0.2g
Fumaric acid	0.1g
Malic acid	0.1g
Succinic acid	0.1g
L-Cystine	0.1g
L-Tryptophan	0.1g
L-Cysteine	0.06g

Preparation of Schneider's *Drosophila* Medium: Add components to 1.0L of distilled/deionized water. Mix thoroughly. Filter sterilize.

Solution 1:
Composition per 80mL:

Sorbitol	7.0g
Noble agar	5.0g
Beef heart (solids from infusion)	5.0g
Peptone	1.8g
Sucrose	1.0g
Pancreatic digest of casein	1.0g
NaCl	0.5g
D-Fructose	0.1g
D-Glucose	0.1g

Preparation of Solution 1: Add components to distilled/deionized water and bring volume to 80.0mL. Mix thoroughly. Adjust pH to 7.8 with 1*N* NaOH. Autoclave for 15 min at 15 psi pressure–121°C. Cool to 50°C.

Preparation of Medium: Bring fetal calf serum and Phenol Red solution to 56°C. Rapidly bring Schneider's *Drosophila* Medium to 37°C. Rapidly combine the components. Mix thoroughly. Pour into sterile Petri dishes or distribute into sterile tubes or flasks.

Use: For the cultivation and maintenance of *Spiroplasma kunkelii* and *Spiroplasma* species.

Spiroplasma Broth MID

Composition per 291.2mL:

Schneider's *Drosophila* Medium	160.0mL
Solution 1	80.0mL
Fetal calf serum	50.0mL
Phenol Red (0.5% solution)	1.2mL

pH 7.4 ± 0.2 at 25°C

Schneider's Drosophila Medium:
Composition per liter:

$MgSO_4 \cdot 7H_2O$	3.7g
NaCl	2.1g
Yeast extract	2.0g
Trehalose	2.0g
D-Glucose	2.0g
L-Glutamine	1.8g
L-Lysine·HCl	1.7g
L-Proline	1.7g
KCl	1.6g
$Na_2HPO_4 \cdot 7H_2O$	1.3g
L-Glutamic acid	0.8g
L-Methionine	0.8g
$CaCl_2$, anhydrous	0.6g
KH_2PO_4	0.5g
β-Alanine	0.5g
L-Tyrosine	0.5g
L-Arginine	0.4g
L-Aspartic acid	0.4g
L-Histidine	0.4g
L-Threonine	0.4g
$NaHCO_3$	0.4g
Glycine	0.3g
L-Serine	0.3g
L-Valine	0.3g
L-Isoleucine	0.2g
L-Leucine	0.2g
L-Phenylalanine	0.2g
α-Ketoglutaric acid	0.2g
Fumaric acid	0.1g
Malic acid	0.1g

Succinic acid 0.1g
L-Cystine 0.1g
L-Tryptophan 0.1g
L-Cysteine 0.06g

Preparation of Schneider's *Drosophila* Medium: Add components to distilled/deionized water and bring volume to 1.0L. Mix thoroughly. Filter sterilize.

Solution 1:

Composition per 80mL:

Sorbitol 7.0g
Beef heart (solids from infusion) 5.0g
Peptone 1.8g
Sucrose 1.0g
Pancreatic digest of casein 1.0g
NaCl 0.5g
D-Fructose 0.1g
D-Glucose 0.1g

Preparation of Solution 1: Add components to distilled/deionized water and bring volume to 80.0mL. Mix thoroughly. Adjust pH to 7.8 with 1*N* NaOH. Autoclave for 15 min at 15 psi pressure–121°C. Cool to 25°C.

Preparation of Medium: Bring fetal calf serum and Phenol Red solution to 56°C. Rapidly bring Schneider's Drosophila Medium to 37°C. Rapidly combine the components. Mix thoroughly. Aseptically distribute into sterile tubes or flasks.

Use: For the cultivation and maintenance of *Spiroplasma kunkelii* and *Spiroplasma* species.

Spiroplasma Medium

Composition per liter:

Sucrose 80.0g
Beef heart (solids from infusion) 34.7g
Peptone 6.9g
NaCl 3.5g
Horse serum, heat-inactivated 100.0mL

pH 7.2 ± 0.2 at 25°C

Preparation of Medium: Add components, except horse serum, to distilled/deionized water and bring volume to 900.0mL. Mix thoroughly. Autoclave for 15 min at 15 psi pressure–121°C. Cool to 25°C. Aseptically add horse serum. Mix thoroughly. Aseptically distribute into sterile tubes or flasks.

Use: For the cultivation and maintenance of *Spiroplasma* species.

Spiroplasma Medium with 25 mg/L Phenol Red

Composition per liter:

Sucrose 80.0g
Beef heart (solids from infusion) 34.7g
Peptone 6.9g
NaCl 3.5g
Phenol Red 25.0mg
Horse serum, heat-inactivated 100.0mL

pH 7.2 ± 0.2 at 25°C

Preparation of Medium: Add components, except horse serum, to distilled/deionized water and bring volume to 900.0mL. Mix thoroughly. Gently heat and bring to boiling. Autoclave for 15 min at 15 psi pressure–121°C. Cool to 45°–50°C. Aseptically add heat-inactivated horse serum. Mix thoroughly. Aseptically distribute into sterile tubes or flasks.

Use: For the cultivation and maintenance of *Spiroplasma floricola*.

Spizizen Potato Agar

Composition per liter:

Potatoes 200.0g
Agar 15.0g
$MnSO_4$ 5.0mg

pH 6.8 ± 0.2 at 25°C

Preparation of Medium: Peel and dice potatoes. Add potatoes to 1.0L of tap water. Gently heat and bring to boiling. Continue boiling for 30 min. Filter through cheesecloth. Add $MnSO_4$ to filtrate and bring volume to 1.0L with tap water. Mix thoroughly. Adjust pH to 6.8. Add agar. Gently heat and bring to boiling. Distribute into tubes or flasks. Autoclave for 15 min at 15 psi pressure–121°C. Pour into sterile Petri dishes or leave in tubes.

Use: For the cultivation and maintenance of *Bacillus amyloliquefaciens*.

SPMA
See: **Mineral Salt Peptonized Milk Agar**

Spore Strip Broth

Composition per liter:

Spore strip broth 9.0g

Preparation of Medium: Add 9.0g of spore strip broth powder (a mixture of glucose, buffer salts, growth factors and Bromthymol Blue) to distilled/deionized water and bring volume to 1.0L. Mix thoroughly. Distribute into tubes or flasks. Autoclave for 15 min at 15 psi pressure–121°C.

Use: For the recovery of spores of *Bacillus stearothermophilus* on spore strips used to determine the sterilization efficiency of autoclaves.

Sporomusa Medium

Composition per 877mL:

NaCl	2.25g
Pancreatic digest of casein	2.0g
Yeast extract	2.0g
$MgSO_4 \cdot 7H_2O$	0.5g
NH_4Cl	0.5g
K_2HPO_4	0.35g
KH_2PO_4	0.23g
$CaCl_2 \cdot 2H_2O$	0.025g
$FeSO_4 \cdot 7H_2O$	2.0mg
Resazurin	1.0mg
$NaHSeO_3$	15.0μg
$NaHCO_3$ solution	50.0mL
Glycine betaine solution	50.0mL
Wolfe's vitamin solution	10.0mL
Cysteine·HCl·H_2O solution	10.0mL
Trace elements solution SL-6	3.0mL

pH 7.0–7.2 at 25°C

$NaHCO_3$ Solution:

Composition per 50mL:

$NaHCO_3$	4.0g

Preparation of $NaHCO_3$ Solution: Add the $NaHCO_3$ to distilled/deionized water and bring volume to 50.0mL. Mix thoroughly. Gas under 80% N_2 + 20% CO_2 for 20 min.

Glycine Betaine Solution:

Composition per 50mL:

Glycine betaine	5.0g

Preparation of Glycine Betaine Solution: Add the glycine betaine to distilled/deionized water and bring volume to 50.0mL. Mix thoroughly. Filter sterilize. Aseptically gas under 80% N_2 + 20% CO_2.

Wolfe's Vitamin Solution:

Composition per liter:

Pyridoxine·HCl	0.01g
Thiamine·HCl	5.0mg
Riboflavin	5.0mg
Nicotinic acid	5.0mg
Calcium pantothenate	5.0mg
p-Aminobenzoic acid	5.0mg
Thioctic acid	5.0mg
Biotin	2.0mg
Folic acid	2.0mg
Cyanocobalamin	0.1mg

Preparation of Wolfe's Vitamin Solution: Add components to distilled/deionized water and bring volume to 1.0L. Mix thoroughly.

Cysteine·HCl·H_2O Solution:

Composition per 10mL:

Cysteine·HCl·H_2O	0.3g

Preparation of Cysteine·HCl·H_2O Solution: Add cysteine·HCl·H_2O to distilled/deionized water and bring volume to 10.0mL. Mix thoroughly. Filter sterilize. Aseptically gas under 100% N_2.

Trace Elements Solution SL-6:

Composition per liter:

H_3BO_3	0.3g
$CoCl_2 \cdot 6H_2O$	0.2g
$ZnSO_4 \cdot 7H_2O$	0.10g
$MnCl_2 \cdot 4H_2O$	0.03g
$Na_2MoO_4 \cdot H_2O$	0.03g
$NiCl_2 \cdot 6H_2O$	0.02g
$CuCl_2 . 2H_2O$	0.01g

Preparation of Trace Elements Solution SL-6: Add components to distilled/deionized water and bring volume to 1.0L. Mix thoroughly. Adjust pH to 3.4.

Preparation of Medium: Add components—except $NaHCO_3$ solution, glycine betaine solution, and cysteine·HCl·H_2O solution—to distilled/deionized water and bring volume to 890.0mL. Mix thoroughly. Gently heat and bring to boiling. Continue boiling for 5 min. Cool rapidly to 25°C under 80% N_2 + 20% CO_2. Add 50.0mL of $NaHCO_3$ solution. Mix thoroughly. Autoclave anaerobically for 15 min at 15 psi pressure–121°C. Cool to 25°C. Aseptically add sterile glycine betaine solution. Mix thoroughly. Immediately prior to inoculation aseptically and anaerobically add cysteine·HCl·H_2O solution.

Use: For the cultivation and maintenance of *Sporomusa ovata* and *Sporomusa sphaeroides*.

Sporosarcina ureae

Composition per liter:

L-Asparagine·H_2O or L-glutamine	30.0g
KCl	3.4g
NaCl	2.92g
K_2HPO_4	0.25g
$(NH_4)_2SO_4$	0.2g
$MgSO_4 \cdot 7H_2O$	0.05g
$FeSO_4 \cdot 7H_2O$	2.5mg
$MnCl_2 \cdot 4H_2O$	0.25mg
Biotin solution	10.0mL
Cysteine solution	10.0mL
$(NH_4)_2SO_4$ solution	10.0mL

pH 8.7 ± 0.2 at 25°C

Biotin Solution:
Composition per 10mL:
D-Biotin 1.0mg

Preparation of Biotin Solution: Add biotin to distilled/deionized water and bring volume to 10.0mL. Mix thoroughly. Filter sterilize.

Cysteine Solution:
Composition per 10mL:
Cysteine.......... 0.04g

Preparation of Cysteine Solution: Add cysteine to distilled/deionized water and bring volume to 10.0mL. Mix thoroughly. Filter sterilize.

$(NH_4)_2SO_4$ Solution:
Composition per 10mL:
$(NH_4)_2SO_4$.......... 0.2g

Preparation of $(NH_4)_2SO_4$ Solution: Add $(NH_4)_2SO_4$ to distilled/deionized water and bring volume to 10.0mL. Mix thoroughly. Filter sterilize.

Preparation of Medium: Add components—except biotin solution, cysteine solution, and $(NH_4)_2SO_4$ solution—to distilled/deionized water and bring volume to 970.0mL. Mix thoroughly. Adjust pH to 8.7 with 1*N* NaOH. Autoclave for 15 min at 15 psi pressure–121°C. Cool to 45°–50°C. Aseptically add sterile biotin solution, cysteine solution, and $(NH_4)_2SO_4$ solution. Mix thoroughly. Aseptically distribute into sterile tubes.

Use: For the cultivation of *Sporosarcina ureae*.

Sporosarcina ureae Medium

Composition per liter:
Agar.......... 30.0g
Glucose 4.0g
$(NH_4)_2SO_4$.......... 4.0g
Malt extract 3.0g
Peptone.......... 3.0g
Yeast extract.......... 2.0g
K_2HPO_4.......... 1.0g
$MgSO_4$.......... 0.8g
$CaCl_2$.......... 0.1g
$MnSO_4·H_2O$ 0.1g
$CuSO_4·5H_2O$ 0.01g
$ZnSO_4$.......... 0.01g
$FeSO_4·7H_2O$.......... 1.0mg

Preparation of Medium: Add components to 1.0L of distilled/deionized water. Mix thoroughly. Gently heat and bring to boiling. Distribute into tubes or flasks. Autoclave for 15 min at 15 psi pressure–121°C. Pour into sterile Petri dishes or leave in tubes.

Use: For the cultivation and induction of sporulation of *Sporosarcina ureae*.

Sporulating Agar
See: **AK Agar No. 2**

Sporulation Agar (m–Sporulation Agar)

Composition per liter:
Agar.......... 15.0g
Glucose 10.0g
Tryptose 2.0g
Beef extract.......... 1.0g
Yeast extract.......... 1.0g
$FeSO_4$.......... 1.0μg
pH 7.2 ± 0.2 at 25°C

Preparation of Medium: Add components to distilled/deionized water and bring volume to 1.0L. Mix thoroughly. Gently heat and bring to boiling. Distribute into tubes or flasks. Autoclave for 15 min at 15 psi pressure–121°C. Pour into sterile Petri dishes or leave in tubes.

Use: For the cultivation and sporulation of *Streptomyces*, *Streptoverticillium*, and *Thermoactinomyces* species. For the identification of sporulating bacteria by the membrane filter method.

Sporulation Broth

Composition per liter:
Polypeptone™.......... 15.0g
Na_2HPO_4.......... 11.0g
Starch, soluble.......... 3.0g
Yeast extract.......... 3.0g
Sodium thioglycollate.......... 1.0g
$MgSO_4$, anhydrous.......... 0.1g
pH 7.8 ± 0.1 at 25°C

Preparation of Medium: Add components to distilled/deionized water and bring volume to 1.0L. Mix thoroughly. Distribute into tubes in 15mL volumes. Autoclave for 15 min at 15 psi pressure–121°C.

Use: For the cultivation and observation of sporulation of *Clostridium perfringens*.

Sporulation Broth

Composition per liter:
Glucose 3.3g
Tryptose 0.66g
Beef extract.......... 0.33g
Yeast extract.......... 0.33g
$FeSO_4$.......... 0.33μg
pH 7.2 ± 0.2 at 25°C

Preparation of Medium: Add components to distilled/deionized water and bring volume to 1.0L. Mix thoroughly. Distribute into tubes or flasks. Autoclave for 15 min at 15 psi pressure–121°C.

Use: For the cultivation and sporulation of *Streptomyces*, *Streptoverticillium*, and *Thermoactinomyces* species.

Sporulation Medium, Modified *See*: Duncan–Strong Sporulation Medium, Modified

Spray's Fermentation Medium

Composition per 1100mL:

Neopeptone	10.0g
Pancreatic digest of casein	10.0g
Agar	2.0g
Sodium thioglycollate	0.025g
Carbohydrate solution	110.0mL

pH 7.4 ± 0.1 at 25°C

Carbohydrate Solution:
Composition per 200mL:

Carbohydrate	20.0g

Preparation of Carbohydrate Solution: Add carbohydrate to distilled/deionized water and bring volume to 200.0mL. Glucose or glycerol may be used. Mix thoroughly. Filter sterilize.

Preparation of Medium: Add components, except agar and carbohydrate solution, to distilled/deionized water and bring volume to 990.0mL. Mix thoroughly. Adjust pH to 7.4. Add agar. Gently heat and bring to boiling. Distribute into tubes in 9.0mL volumes. Autoclave for 15 min at 15 psi pressure–121°C. Cool to 25°C. Immediately prior to use heat tubes in a boiling water bath for 10 min. Cool to 45°C. Aseptically add 1.0mL of sterile carbohydrate solution. Mix thoroughly.

Use: For the cultivation and differentiation of *Clostridium perfringens* based on carbohydrate fermentation patterns.

SPS Agar (Sulfite Polymyxin Sulfadiazine Agar

Composition per liter:

Pancreatic digest of casein	15.0g
Agar	13.9g
Yeast extract	10.0g
Ferric citrate	0.5g
Na_2SO_3	0.5g
Polymyxin sulfate	0.01g
Sulfadiazine	0.12g

pH 7.0 ± 0.2 at 25°C

Source: This medium is available as a premixed powder from BBL Microbiology Systems and Difco Laboratories.

Preparation of Medium: Add components to distilled/deionized water and bring volume to 1.0L. Mix thoroughly. Gently heat while stirring and bring to boiling. Distribute into tubes or flasks. Autoclave for 15 min at 13 psi pressure–118°C. Pour into sterile Petri dishes or leave in tubes.

Use: For the isolation and detection of *Clostridium perfringens* and *Clostridium botulinum* in foods and other materials.

SS Agar *See: Salmonella–Shigella* Agar

SS Agar, Modified *See: Salmonella–Shigella* Agar, Modified

SS Deoxycholate Agar (*Salmonella–Shigella* Deoxycholate Agar) (SSDC)

Composition per liter:

Agar	13.5g
Lactose	10.0g
Sodium desoxycholate	10.0g
Bile salts	8.5g
$Na_2S_2O_3$	8.5g
Sodium citrate	8.5g
Beef extract	5.0g
Pancreatic digest of casein	2.5g
Peptic digest of animal tissue	2.5g
$CaCl_2 \cdot 2H_2O$	1.0g
Ferric citrate	1.0g
Neutral Red	0.025g
Brilliant Green	0.330mg

pH 7.0 ± 0.2 at 25°C

Preparation of Medium: Add components to distilled/deionized water and bring volume to 1.0L. Mix thoroughly. Gently heat while stirring and bring to boiling. Do not autoclave. Cool to 45°–50°C. Pour into sterile Petri dishes in 20.0mL volumes. Allow the surface of the plates to dry before inoculation.

Use: For the isolation and cultivation of *Yersinia enterocolitica* from foods.

SSL Agar

Composition per liter:

Agar	2.5g
$CaCl_2 \cdot 2H_2O$	1.0g
Gelatin	1.0g
KNO_3	1.0g
$MgSO_4 \cdot 7H_2O$	1.0g
NaCl	1.0g
Pancreatic digest of casein	1.0g
Yeast extract	1.0g
Sodium glycerophosphate	0.1g
Cyanocobalamin	1.0μg
Trace element solution	1.0mL

pH 7.5 ± 0.2 at 25°C

Trace Element Solution:
Composition per liter:

Disodium EDTA	8.0g
$MnCl_2 \cdot 4H_2O$	0.1g
$CoCl_2 \cdot 6H_2O$	0.02g
KBr	0.02g
KI	0.02g
$ZnCl_2$	0.02g
$CuSO_4$	0.01g
H_3BO_3	0.01g
$Na_2MoO_4 \cdot 2H_2O$	0.01g
LiCl	5.0mg
$SnCl_2 \cdot 2H_2O$	5.0mg

Preparation of Trace Element Solution: Add components to distilled/deionized water and bring volume to 1.0L. Mix thoroughly.

Preparation of Medium: Add components to distilled/deionized water and bring volume to 1.0L. Mix thoroughly. Gently heat and bring to boiling. Distribute into tubes or flasks. Autoclave for 15 min at 15 psi pressure–121°C. Pour into sterile Petri dishes or leave in tubes.

Use: For the isolation and cultivation of *Cytophaga* species, *Herpetosiphon* species, *Saprospira* species, and *Flexithrix* species.

ST Agar

See: ***Streptococcus thermophilus*** **Agar**

ST Holding Medium (m–ST Holding Medium)

Composition per liter:

KH_2PO_4	3.0g
Tris(hydroxymethyl)aminomethane buffer	3.0g
Sulfanilamide	1.5g
$NaH_2PO_4 \cdot H_2O$	0.1g
Ethanol (95% solution)	10.0mL

pH 8.6 ± 0.2 at 25°C

Preparation of Medium: Dissolve the sulfanilamide in the ethanol. Add all components to distilled/deionized water and bring volume to 1.0L. Mix thoroughly. Autoclave for 15 min at 15 psi pressure–121°C. Distribute in 1.8mL volumes to sterile Petri dishes with tight-fitting lids and an absorbent filter.

Use: For the cultivation and enumeration of coliform bacteria by the delayed-incubation total coliform procedure. For use as a holding or transport medium to keep coliform bacteria viable between sampling and laboratory culture.

STAA Agar Base

Composition per liter:

Peptone	20.0g
Agar	13.0g
Glycerol	7.5g
Yeast extract	2.0g
K_2HPO_4	1.0g
$MgSO_4 \cdot 7H_2O$	1.0g
STAA selective supplement	10.0mL

pH 7.0 ± 0.1 at 25°C

Source: This medium is available as a premixed powder from Oxoid Unipath.

STAA Selective Supplement:
Composition per 10mL:

Streptomycin sulfate	0.5g
Cycloheximide	0.05g
Thallous acetate	0.05g

Preparation of STAA Selective Supplement: Add components to distilled/deionized water and bring volume to 10.0mL. Mix thoroughly. Filter sterilize.

Preparation of Medium: Add components, except glycerol and STAA selective supplement, to distilled/deionized water and bring volume to 985.0mL. Mix thoroughly. Gently heat and bring to boiling. Add glycerol. Mix thoroughly. Autoclave for 15 min at 15 psi pressure–121°C. Cool to 45°–50°C. Aseptically add sterile STAA selective supplement. Mix thoroughly. Pour into sterile Petri dishes or distribute into sterile tubes.

Use: For the isolation of *Brochothrix thermosphacta* from meat products.

Stab Agar

Composition per liter:

Nutrient broth	10.0g
Agar	6.0g
NaCl	5.0g
L-Cysteine·HCl·H_2O	0.01g
Thymine	0.01g

pH 7.0 ± 0.2 at 25°C

Preparation of Medium: Add components to distilled/deionized water and bring volume to 1.0L. Mix thoroughly. Gently heat and bring to boiling. Adjust pH to 7.0. Distribute into tubes. Autoclave for 15 min at 15 psi pressure–121°C.

Use: For the maintenance of bacterial strains, especially *Escherichia coli.*

Staib Agar
See: **Birdseed Agar**

Stan 4 Agar

Composition per liter:

Solution B	650.0mL
Solution A	350.0mL

Solution A:

Composition per 350mL:

$CaCl_2·2H_2O$	1.0g
KNO_3	1.0g
$MgSO_4·7H_2O$	1.0g
Trace element solution	1.0mL

Preparation of Solution A: Add components to distilled/deionized water and bring volume to 350.0mL. Mix thoroughly. Gently heat and bring to boiling. Autoclave for 15 min at 15 psi pressure–121°C. Cool to 45°–50°C.

Trace Element Solution:

Composition per liter:

EDTA	8.0g
$MnCl_2·4H_2O$	0.1g
$CoCl_2$	0.02g
KBr	0.02g
$ZnCl_2$	0.02g
$CuSO_4$	0.01g
H_3BO_3	0.01g
$NaMoO_4·2H_2O$	0.01g
$BaCl_2$	5.0mg
LiCl	5.0mg
$SnCl_2·2H_2O$	5.0mg

Preparation of Trace Element Solution: Add components to distilled/deionized water and bring volume to 1.0L. Mix thoroughly.

Solution B:

Composition per 650mL:

Agar	10.0g
K_2HPO_4	1.0g

Preparation of Solution B: Add components to distilled/deionized water and bring volume to 650.0mL. Mix thoroughly. Gently heat and bring to boiling. Autoclave for 15 min at 15 psi pressure–121°C. Cool to 45°–50°C.

Preparation of Medium: Aseptically combine 350.0mL of cooled, sterile solution A and 650.0mL of cooled sterile solution B. Mix thoroughly. Pour into sterile Petri dishes or distribute into sterile tubes.

Use: For the cultivation of myxobacteria.

Stan 5 Agar

Composition per liter:

Solution B	650.0mL
Solution A	350.0mL

Solution A:

Composition per 350mL:

$CaCl_2·2H_2O$	1.0g
$(NH_4)_2SO_4$	1.0g
$MgSO_4·7H_2O$	1.0g
Trace element solution	1.0mL

Preparation of Solution A: Add components to distilled/deionized water and bring volume to 350.0mL. Mix thoroughly. Gently heat and bring to boiling. Autoclave for 15 min at 15 psi pressure–121°C. Cool to 45°–50°C.

Trace Element Solution:

Composition per liter:

EDTA	8.0g
$MnCl_2·4H_2O$	0.1g
$CoCl_2$	0.02g
KBr	0.02g
$ZnCl_2$	0.02g
$CuSO_4$	0.01g
H_3BO_3	0.01g
$NaMoO_4·2H_2O$	0.01g
$BaCl_2$	5.0mg
LiCl	5.0mg
$SnCl_2·2H_2O$	5.0mg

Preparation of Trace Element Solution: Add components to distilled/deionized water and bring volume to 1.0L. Mix thoroughly.

Solution B:

Composition per 650mL:

Agar	10.0g
K_2HPO_4	1.0g

Preparation of Solution B: Add components to distilled/deionized water and bring volume to 650.0mL. Mix thoroughly. Gently heat and bring to boiling. Autoclave for 15 min at 15 psi pressure–121°C. Cool to 45°–50°C.

Preparation of Medium: Aseptically combine 350.0mL of cooled, sterile solution A and 650.0mL of cooled sterile solution B. Mix thoroughly. Pour into sterile Petri dishes or distribute into sterile tubes.

Use: For the cultivation of myxobacteria.

Stan 6 Agar

Composition per liter:

Agar	10.0g
$CaCl_2 \cdot 2H_2O$	1.0g
K_2HPO_4	1.0g
$MgSO_4 \cdot 7H_2O$	1.0g
$(NH_4)_2SO_4$	1.0g
$FeCl_3$	0.2g
$MnSO_4 \cdot 7H_2O$	0.1g
Yeast extract	0.02g
Trace element solution	1.0mL

Trace Element Solution:

Composition per liter:

Disodium EDTA	8.0g
$MnCl_2 \cdot 4H_2O$	0.1g
$CoCl_2 \cdot 6H_2O$	0.02g
KBr	0.02g
KI	0.02g
$ZnCl_2$	0.02g
$CuSO_4$	0.01g
H_3BO_3	0.01g
$Na_2MoO_4 \cdot 2H_2O$	0.01g
LiCl	5.0mg
$SnCl_2 \cdot 2H_2O$	5.0mg

Preparation of Trace Element Solution: Add components to distilled/deionized water and bring volume to 1.0L. Mix thoroughly.

Preparation of Medium: Add components to distilled/deionized water and bring volume to 1.0L. Mix thoroughly. Gently heat and bring to boiling. Distribute into tubes or flasks. Autoclave for 15 min at 15 psi pressure–121°C. Pour into sterile Petri dishes or leave in tubes.

Use: For the isolation and cultivation of *Cytophaga* species, *Herpetosiphon* species, *Saprospira* species, and *Flexithrix* species.

Stan 5 Mineral Medium

Composition per liter:

K_2HPO_4	1.0g
$(NH_4)_2SO_4$	1.0g
$MgSO_4 \cdot 7H_2O$	0.2g
$CaCl_2 \cdot 2H_2O$	0.1g
$FeCl_3$	0.02g

pH 7.0–7.5 at 25°C

Preparation of Medium: Add components to distilled/deionized water and bring volume to 1.0L. Mix thoroughly. Distribute into tubes or flasks. Autoclave for 15 min at 15 psi pressure–121°C.

Use: For the isolation and cultivation of *Cytophaga* species, *Herpetosiphon* species, *Saprospira* species, and *Flexithrix* species.

Standard Agar with Methanol and Yeast Extract

Composition per liter:

Base solution	982.0mL
Methanol	10.0mL
Solution B	5.0mL
Solution A	1.0mL
Solution C	1.0mL
Solution D	1.0mL

pH 4.0–4.5 at 25°C

Base Solution:

Composition per 982mL:

Agar	20.0g
Yeast extract	0.5g

Preparation of Base Solution: Add components to distilled/deionized water and bring volume to 982.0mL. Mix thoroughly. Gently heat and bring to boiling. Autoclave for 15 min at 15 psi pressure–121°C. Cool to 45°–50°C.

Methanol:

Composition per 10mL:

Methanol	10.0mL

Preparation of Methanol: Filter sterilize.

Solution A:

Composition per liter:

K_2HPO_4	87.09g
KH_2PO_4	68.05g

Preparation of Solution A: Add components to distilled/deionized water and bring volume to 1.0L. Mix thoroughly. Filter sterilize.

Solution B:

Composition per liter:

NH_4Cl	152.28g

Preparation of Solution B: Add NH_4Cl to distilled/deionized water and bring volume to 1.0L. Mix thoroughly. Filter sterilize.

Solution C:
Composition per liter:
$CaCl_2 \cdot 6H_2O$ 5.47g

Preparation of Solution C: Add $CaCl_2 \cdot 6H_2O$ to distilled/deionized water and bring volume to 1.0L. Mix thoroughly. Filter sterilize.

Solution D:
Composition per liter:
$MgSO_4 \cdot 7H_2O$ 71.2g
$FeSO_4 \cdot 7H_2O$ 5.0g
$MnSO_4 \cdot 4H_2O$ 0.81g
$CuSO_4 \cdot 5H_2O$ 0.79g
$ZnSO_4 \cdot 7H_2O$ 0.44g
$Na_2MoO_4 \cdot 2H_2O$ 0.25g

Preparation of Solution D: Add components to distilled/deionized water and bring volume to 1.0L. Mix thoroughly. Filter sterilize.

Preparation of Medium: To 982.0mL of cooled sterile Base solution, aseptically add 10.0mL of sterile methanol, 1.0mL of sterile solution A, 5.0mL of sterile solution B, 1.0mL of sterile solution C, and 1.0mL of sterile solution D. Mix thoroughly. Adjust pH to 4.0–4.5 if necessary. Pour into sterile Petri dishes or distribute into sterile tubes.

Use: For the cultivation and maintenance of *Acetobacter methanolicus*.

Standard Fluid Medium 10B (Shepard's M10 Medium)

Composition per 102.5mL:
Base solution 70.0mL
Horse serum, unheated 20.0mL
Fresh yeast extract solution 10.0mL
Penicillin solution 1.0mL
CVA enrichment 0.5mL
L-Cysteine·HCl·H_2O solution 0.5mL
Urea solution 0.4mL
Phenol Red 0.1mL
pH 6.0 ± 0.2 at 25°C

Base Solution:
Composition per 70mL:
Beef heart (solids from infusion) 5.0g
Peptone 1.0g
NaCl 0.5g

Preparation of Base Solution: Add components to distilled/deionized water and bring volume to 70.0mL. Mix thoroughly. Adjust pH to 5.5 with 2*N* HCl. Autoclave for 15 min at 15 psi pressure–121°C. Cool to 45°–50°C.

Fresh Yeast Extract Solution:
Composition per 100mL:
Baker's yeast, live, pressed, starch-free, ... 25.0g

Preparation of Fresh Yeast Extract Solution: Add the live Baker's yeast to 100.0mL of distilled/deionized water. Autoclave for 90 min at 15 psi pressure–121°C. Allow to stand. Remove supernatant solution. Adjust pH to 6.6–6.8.

Penicillin Solution:
Composition per 10mL:
Penicillin G 1,000,000U

Preparation of Penicillin Solution: Add penicillin to distilled/deionized water and bring volume to 10.0mL. Mix thoroughly. Filter sterilize.

CVA Enrichment:
Composition per liter:
Glucose 100.0g
L-Cysteine·HCl·H_2O 25.9g
L-Glutamine 10.0g
L-Cystine·2HCl 1.0g
Adenine 1.0g
Nicotinamide adenine dinucleotide 0.25g
Cocarboxylase 0.1g
Guanine·HCl 0.03g
$Fe(NO_3)_3$ 0.02g
p-Aminobenzoic acid 0.013g
Vitamin B_{12} 0.01g
Thiamine·HCl 3.0mg

Preparation of CVA Enrichment: Add components to distilled/deionized water and bring volume to 1.0L. Mix thoroughly. Filter sterilize.

Cysteine·HCl·H_2O Solution:
Composition per 10mL:
Cysteine·HCl·H_2O 0.2g

Preparation of Cysteine·HCl·H_2O Solution: Add cysteine·HCl·H_2O to distilled/deionized water and bring volume to 10.0mL. Mix thoroughly. Filter sterilize.

Urea Solution:
Composition per 10mL:
Urea 1.0g

Preparation of Urea Solution: Add urea to distilled/deionized water and bring volume to 10.0mL. Mix thoroughly. Filter sterilize.

Phenol Red Solution:
Composition per 10mL:
Phenol Red 0.1g

Preparation of Phenol Red Solution: Add Phenol Red to distilled/deionized water and bring volume to 10.0mL. Mix thoroughly. Autoclave for 15 min at 15 psi pressure–121°C.

Preparation of Medium: To 70.0mL of cooled sterile base solution, aseptically add 20.0mL of sterile horse serum, 10.0mL of sterile fresh yeast extract solution, 1.0mL of sterile penicillin solution, 0.5mL of sterile CVA enrichment, 0.5mL of sterile L-cysteine·HCl·H_2O solution, 0.4mL of sterile urea solution and 0.1mL of sterile Phenol Red solution. Mix thoroughly. Aseptically distribute into sterile tubes or flasks.

Use: For the isolation and cultivation of *Ureaplasma urealyticum* from clinical specimens.

Standard Methods Agar (Tryptone Glucose Yeast Agar) (Plate Count Agar)

Composition per liter:

Agar	15.0g
Pancretic digest of casein	5.0g
Yeast extract	2.5g
Glucose	1.0g

pH 7.0 ± 0.1 at 25°C

Source: Available as a premixed powder from BBL Microbiology Systems and Difco Laboratories.

Preparation of Medium: Add components to distilled/deionized water and bring volume to 1.0L. Mix thoroughly. Gently heat and bring to boiling. Distribute into tubes or flasks. Autoclave for 15 min at 15 psi pressure–121°C. Pour into sterile Petri dishes or leave in tubes.

Use: For the cultivation and enumeration by microbial plate counts of microorganisms isolated from milk and dairy products, foods, water and other specimens.

Standard Methods Agar with Lecithin and Polysorbate 80

Composition per liter:

Agar	15.0g
Pancreatic digest of casein	5.0g
Polysorbate 80	5.0g
Yeast extract	2.5g
Glucose	1.0g
Lecithin	0.7g

pH 7.0 ± 0.2 at 25°C

Source: This medium is available as a premixed powder from BBL Microbiology Systems.

Preparation of Medium: Add components to distilled/deionized water and bring volume to 1.0L. Mix thoroughly. Gently heat and bring to boiling. Distribute into tubes or flasks. Autoclave for 15 min at 15 psi pressure–121°C. Pour into sterile Petri dishes or leave in tubes.

Use: For determination of the sterility of surfaces.

Standard Methods Broth (m–Standard Methods Broth) (Tryptone Glucose Yeast Broth) (m–Plate Count Broth)

Composition per liter:

Pancreatic digest of casein	10.0g
Yeast extract	5.0g
Glucose	2.0g

pH 7.0 ± 0.2 at 25°C

Source: Available as a premixed powder from BBL Microbiology Systems and Difco Laboratories.

Preparation of Medium: Add components to distilled/deionized water and bring volume to 1.0L. Mix thoroughly. Distribute into tubes or flasks. Autoclave for 15 min at 15 psi pressure–121°C.

Use: For the enumeration of total number of microorganisms by the membrane filter method.

Standard Methods Caseinate Agar

Composition per liter:

Agar	15.0g
Sodium caseinate	10.0g
Pancreatic digest of casein	5.0g
Yeast extract	2.5g
Glucose	1.0g
Trisodium citrate solution	1.0L
$CaCl_2 \cdot 2H_2O$ (1M solution)	20.0mL

Trisodium Citrate Solution:

Composition per liter:

Trisodium citrate·$2H_2O$	4.41g

Preparation of Trisodium Citrate Solution: Add trisodium citrate·$2H_2O$ to distilled/deionized water and bring volume to 1.0L. Mix thoroughly.

Preparation of Medium: Add yeast extract, pancreatic digest of casein, glucose and agar to 500.0mL of the trisodium citrate solution. Mix thoroughly. Gently heat until boiling. Add the remaining components. Bring volume to 1.0L with trisodium citrate solution. Autoclave for 15 min at 15 psi pressure–121°C. Cool to 45°–50°C. Pour into sterile Petri dishes.

Use: For thecultivation of proteolytic bacteria from foods.

Standard II Nutrient Agar

Composition per liter:

Agar	13.0g
Tryptose	7.0g
NaCl	5.0g

pH 7.5 ± 0.2 at 25°C

Source: This medium is available as a premixed powder from Difco Laboratories.

Preparation of Medium: Add components to distilled/deionized water and bring volume to 1.0L. Mix thoroughly. Gently heat and bring to boiling. Distribute into tubes or flasks. Autoclave for 15 min at 15 psi pressure–121°C. Pour into sterile Petri dishes or leave in tubes.

Use: For the cultivation of nonfastidious microorganisms. For the maintenance of cultures of a wide variety of nonfastidious bacteria. May also be used as a base for blood and other enrichments for the cultivation of fastidious microorganisms. May be used to determine indole production.

Stanier's Basal Medium with Pyridoxine and Yeast Extract

Composition per liter:

KH_2PO_4	2.78g
Na_2HPO_4	2.78g
$(NH_4)_2SO_4$	1.0g
Yeast extract	0.2g
Hutner's mineral base	20.0mL
Pyridoxine solution	10.0mL

pH 6.8 ± 0.2 at 25°C

Pyridoxine Solution:

Composition per 10mL:

Pyridoxine	2.0g

Preparation of Pyridoxine Solution: Add pyridoxine to distille/deionized water and bring volume to 10.0mL. Mix thoroughly. Filter sterilize.

Hutner's Mineral Base:

Composition per liter:

$MgSO_4 \cdot 7H_2O$	29.7g
Nitrilotriacetic acid	10.0g
$CaCl_2 \cdot 2H_2O$	3.34g
$FeSO_4 \cdot 7H_2O$	99.0mg
$(NH_4)_2MoO_4$	9.25mg
Metals "44"	50.0mL

Preparation of Hutner's Mineral Base: Add nitrilotriacetic acid to 500.0mL of distilled/deionized water. Dissolve by adjusting pH to 6.5 with KOH. Add remaining components. Readjust pH to 7.2 with H_2SO_4 or KOH. Add distilled/deionized water to 1.0L. Store at 5°C.

Metals "44":

Composition per100mL:

$ZnSO_4 \cdot 7H_2O$	1.1g
$FeSO_4 \cdot 7H_2O$	0.5g
EDTA	0.25g
$MnSO_4 \cdot 7H_2O$	0.154g
$CuSO_4 \cdot 5H_2O$	0.04g
$Co(NO_3)_2 \cdot 6H_2O$	0.025g
$Na_2B_4O_7 \cdot 10H_2O$	0.018g

Preparation of Metals "44": Add a few drops of H_2SO_4 to distilled/deionized water to inhibit precipitate formation. Add components to acidified distilled/deionized water and bring volume to 100.0mL. Mix thoroughly.

Preparation of Medium: Add components, except pyridoxine solution, to distilled/deionized water and bring volume to 990.0mL. Mix thoroughly. Adjust pH to 6.8 with 1*N* KOH. Autoclave for 15 min at 15 psi pressure–121°C. Cool to 45°–50°C. Aseptically add sterile pyridoxine solution. Mix thoroughly. Aseptically distribute into sterile tubes or flasks.

Use: For the cultivation and maintenance of *Pseudomonas* species.

Stanier's Basal Medium with Trichlorophenoxyacetate

Composition per liter:

KH_2PO_4	2.78g
Na_2HPO_4	2.78g
$(NH_4)_2SO_4$	1.0g
2,4,5-trichlorophenoxyacetate	1.0g
Hutner's mineral base	20.0mL

Hutner's Mineral Base:

Composition per liter:

$MgSO_4 \cdot 7H_2O$	29.7g
Nitrilotriacetic acid	10.0g
$CaCl_2 \cdot 2H_2O$	3.34g
$FeSO_4 \cdot 7H_2O$	0.10g
$(NH_4)_2MoO_4$	9.25mg
Metals "44"	50.0mL

Preparation of Hutner's Mineral Base: Add nitrilotriacetic acid to 500.0mL of distilled/deionized water. Dissolve by adjusting pH to 6.5 with KOH. Add remaining components. Readjust pH to 7.2 with H_2SO_4 or KOH. Add distilled/deionized water to 1.0L. Store at 5°C.

Metals "44":

Composition per 100mL:

$ZnSO_4 \cdot 7H_2O$	1.1g
$FeSO_4 \cdot 7H_2O$	0.5g
EDTA	0.25g
$MnSO_4 \cdot 7H_2O$	0.154g
$CuSO_4 \cdot 5H_2O$	0.04g
$Co(NO_3)_2 \cdot 6H_2O$	0.025g
$Na_2B_4O_7 \cdot 10H_2O$	0.018g

Preparation of Metals "44": Add a few drops of H_2SO_4 to distilled/deionized water to inhibit precipitate formation. Add components to acidified distilled/deionized water and bring volume to 100.0mL. Mix thoroughly.

Preparation of Medium: Add components to distilled/deionized water and bring volume to 1.0L. Mix thoroughly. Distribute into tubes or flasks. Autoclave for 15 min at 15 psi pressure–121°C.

Use: For the cultivation and maintenance of *Pseudomonas cepacia*.

Staphylococcus Agar No. 110

Composition per liter:

NaCl	75.0g
Gelatin	30.0g
Agar	15.0g
D-Mannitol	10.0g
Pancreatic digest of casein	10.0g
K_2HPO_4	5.0g
Yeast extract	2.5g
Lactose	2.0g

pH 7.0 ± 0.2 at 25°C

Source: This medium is available as a premixed powder from BBL Microbiology Systems and Oxoid Unipath.

Preparation of Medium: Add components to distilled/deionized water and bring volume to 1.0L. Mix thoroughly. Gently heat and bring to boiling. Distribute into tubes or flasks. Autoclave for 15 min at 15 psi pressure–121°C. Pour into sterile Petri dishes or leave in tubes. Swirl flask while pouring plates to disperse precipitate.

Use: For the isolation and enumeration of staphylococci from clinical and nonclinical specimens.

Staphylococcus Broth (m–*Staphylococcus* Broth)

Composition per liter:

NaCl	75.0g
Mannitol	10.0g
Pancreatic digest of casein	10.0g
K_2HPO_4	5.0g
Yeast extract	2.5g
Lactose	2.0g

pH 7.0 ± 0.2 at 25°C

Source: This medium is available as a premixed powder from Difco Laboratories.

Preparation of Medium: Add components to distilled/deionized water and bring volume to 1.0L. Mix thoroughly. Distribute into tubes or flasks. Autoclave for 15 min at 15 psi pressure–121°C.

Use: For the cultivation and enumeration of pathogenic and enterotoxigenic staphylococci by the membrane filter method. Also, when used in conjunction with Lipovitellin-salt-mannitol agar, for the detection of *Staphylococcus aureus* in swimming pool water.

Staphylococcus Medium

Composition per liter:

Agar	15.0g
Peptone	6.0g
Pancreatic digest of casein	4.0g
Yeast extract	3.0g
Beef extract	1.5g
Glucose	1.0g

pH 6.6 ± 0.2 at 25°C

Preparation of Medium: Add components to distilled/deionized water and bring volume to 1.0L. Mix thoroughly. Gently heat and bring to boiling. Distribute into tubes or flasks. Autoclave for 15 min at 15 psi pressure–121°C. Pour into sterile Petri dishes or leave in tubes.

Use: For the cultivation and maintenance of *Staphylococcus aureus*. For the enumeration of pathogenic and enterotoxigenic staphylococci by the membrane filter method.

Staphylococcus Medium No. 110

Composition per liter:

NaCl	75.0g
Gelatin	30.0g
Agar	15.0g
Mannitol	10.0g
Pancreatic digest of casein	10.0g
K_2HPO_4	5.0g
Yeast extract	2.5g
Lactose	2.0g

pH 7.1 ± 0.2 at 25°C

Source: This medium is available as a premixed powder from Difco Laboratories.

Preparation of Medium: Add components to distilled/deionized water and bring volume to 1.0L. Mix thoroughly. Gently heat and bring to boiling. Distribute into tubes or flasks. Autoclave for 15 min at 15 psi pressure–121°C. Mix thoroughly. Pour into sterile Petri dishes or leave in tubes.

Use: For the selective isolation, cultivation and maintenance of *Staphylococcus* species.

Staphylococcus–Streptococcus Selective Medium

Composition per 1060mL:

Columbia blood agar base	1.0L
Horse blood, defibrinated	50.0mL
Antibiotic inhibitor	10.0mL

pH 7.3 ± 0.2 at 25°C

Columbia Blood Agar Base:

Composition per liter:

Special peptone	23.0g
Agar	10.0g
NaCl	5.0g
Starch	1.0g

Source: Columbia blood agar base is available as a premixed powder from Oxoid Unipath.

Preparation of Columbia Blood Agar Base: Add components to distilled/deionized water and bring volume to 1.0L. Mix thoroughly. Gently heat and bring to boiling. Autoclave for 15 min at 15 psi pressure–121°C. Cool to 45°–50°C.

Antibiotic Inhibitor:

Composition per 10mL:

Nalidixic acid	0.015g
Colistin sulfate	0.010g
Ethanol (95% solution)	10.0mL

Preparation of Antibiotic Inhibitor: Add components to 10.0mL ethanol. Mix thoroughly. Filter sterilize.

Preparation of Medium: To 1.0L of cooled sterile Columbia blood agar base aseptically add sterile horse blood and sterile antibiotic inhibitor. Mix thoroughly. Pour into sterile Petri dishes or distribute into sterile tubes.

Use: For the selective isolation of *Staphylococcus aureus* and streptococci from clinical specimens or foods.

Starch Agar

Composition per liter:

Starch, soluble	20.0g
Agar	10.0g
$NaNO_3$	2.5g
K_2HPO_4	1.0g
$MgSO_4 \cdot 7H_2O$	0.6g
$CaCl_2 \cdot 2H_2O$	0.1g
NaCl	0.1g
$FeCl_3$	1mg

pH 7.2 ± 0.2 at 25°C

Preparation of Medium: Add components to distilled/deionized water and bring volume to 1.0L. Mix thoroughly. Gently heat and bring to boiling. Distribute into tubes or flasks. Autoclave for 15 min at 15 psi pressure–121°C. Pour into sterile Petri dishes or leave in tubes.

Use: For the cultivation of myxobacteria.

Starch Agar

Composition per liter:

Agar	15.0g
Potato starch	10.0g
Pancreatic digest of gelatin	5.0g
Beef extract	3.0g

pH 6.8 ± 0.2 at 25°C

Preparation of Medium: Add components, except potato starch, to distilled/deionized water and bring volume to 500.0mL. Mix thoroughly. Gently heat and bring to boiling. Add potato starch to distilled/deionized water and bring volume to 250.0mL. Gently heat and bring to boiling. Combine the two solutions and bring the volume to 1.0L with distilled/deionized water. Autoclave for 15 min at 15 psi pressure–121°C. Pour into sterile Petri dishes.

Use: For the cultivation and differentiation of aerobic *Actinomycetes* species based on amylase production. After incubation, starch hydrolysis is determined by the addition of Gram's or Lugol's iodine solution. Organisms which produce amylase appear as colonies surrounded by a clear zone.

Starch Agar

Composition per liter:

Agar	12.0g
Soluble starch	10.0g
Beef extract	3.0g

pH 7.5 ± 0.2 at 25°C

Source: This medium is available as a premixed powder from Difco Laboratories.

Preparation of Medium: Add components to distilled/deionized water and bring volume to 1.0L. Mix thoroughly. Gently heat and bring to boiling. Distribute into tubes or flasks. Autoclave for 15 min at 15 psi pressure–121°C. Pour into sterile Petri dishes.

Use: For the cultivation and differentiation of a variety of microorganisms based on amylase production. After incubation, starch hydrolysis is determined by the addition of Gram's or Lugol's iodine solution. Organisms which produce amylase appear as colonies surrounded by a clear zone.

Starch Agar Medium for *Pseudomonas*

Composition per liter:

Agar	15.0g
Peptone	5.0g
Yeast extract	5.0g
Soluble starch	3.0g

pH 7.0 ± 0.2 at 25°C

Preparation of Medium: Add components to distilled/deionized water and bring volume to 1.0L. Mix thoroughly. Gently heat and bring to boiling. Distribute into tubes or flasks. Autoclave for 15 min at 15 psi pressure–121°C. Pour into sterile Petri dishes or leave in tubes.

Use: For the cultivation and maintenance of *Pseudomonas* species and *Erwinia herbicola*.

Starch Agar with Bromcresol Purple

Composition per liter:

Agar	15.0g
Cornstarch	10.0g
Meat peptone	10.0g
Bromcresol Purple solution	1.2mL

pH 6.8 ± 0.2 at 25°C

Bromcresol Purple Solution:

Composition per 10mL:

Bromcresol Purple	0.16g
Ethanol (95% solution)	10.0mL

Preparation of Bromcresol Purple Solution: Add Bromcresol Purple to 10.0mL of 95% ethanol. Mix thoroughly.

Preparation of Medium: Add components to distilled/deionized water and bring volume to 1.0L. Mix thoroughly. Gently heat and bring to boiling. Distribute into tubes or flasks. Autoclave for 15 min at 15 psi pressure–121°C. Pour into sterile Petri dishes or leave in tubes.

Use: For the differentiation of *Gardnerella vaginalis (Haemophilus vaginalis, Corynebacterium vaginale)* from other microorganisms found in the genitourinary tract with the exception of some strains of *Streptococcus* and *Loctobacillus*. Differentiation is based on starch hydrolysis. Bacteria that can hydrolyze starch appear as colonies surrounded by a yellow zone.

Starch Agar with Bromcresol Purple

Composition per liter:

Solution 1	200.0mL
Solution 2	20.0mL

pH 7.8 ± 0.2 at 25°C

Solution 1:

Composition per 200mL:

Heart infusion agar	5.0mL
Bromcresol Purple solution	0.2mL

Preparation of Solution 1: Add components to distilled/deionized water and bring volume to 200.0mL. Mix thoroughly. Gently heat while stirring and bring to boiling.

Heart Infusion Agar:

Composition per liter:

Beef heart, solids from infusion	500.0g
Agar	15.0g
Tryptose	10.0g
NaCl	5.0g

Preparation of Heart Infusion Agar: Add components to distilled/deionized water and bring volume to 1.0L. Mix thoroughly. Gently heat and bring to boiling.

Bromcresol Purple Solution:

Composition per 10mL:

Bromcresol Purple	0.16g
Ethanol (95% solution)	10.0mL

Preparation of Bromcresol Purple Solution: Add Bromcresol Purple to 10.0mL of ethanol. Mix thoroughly.

Solution 2:

Composition per 20mL:

Starch	0.4g

Preparation of Solution 2: Add starch to distilled/deionized water and bring volume to 20.0mL. Mix thoroughly. Gently heat while stirring and bring to boiling.

Preparation of Medium: Combine solution 1 and solution 2. Mix thoroughly. Autoclave for 15 min at 15 psi pressure–121°C. Pour into sterile Petri dishes or distribute into sterile tubes.

Use: For the differentiation of *Gardnerella vaginalis (Haemophilus vaginals, Corynebacterium vaginale)* from other microorganisms found in the genitourinary tract with the exception of some strains of *Streptococcus* and *Lactobacillus*. Differentiation is based on starch hydrolysis. Bacteria that can hydrolyze starch appear as colonies surrounded by a yellow zone.

Starch Casein Agar

Composition per liter:

Agar	15.0g
Soluble starch	10.0g
K_2HPO_4	2.0g
KNO_3	2.0g
NaCl	2.0g
Casein	0.3g
$MgSO_4 \cdot 7H_2O$	0.05g
$CaCO_3$	0.02g
$FeSO_4 \cdot 7H_2O$	0.01g

Preparation of Medium: Add components to distilled/deionized water and bring volume to 1.0L. Mix thoroughly. Gently heat and bring to boiling. Distribute into tubes or flasks. For bottom layers, distribute into tubes in 15.0mL volumes. For top layers, distribute into tubes in 17.0mL volumes. Autoclave for 15 min at 15 psi pressure–121°C. Pour into sterile Petri dishes or leave in tubes.

Use: For the cultivation and enumeration of *Actinomycetes* species from water and soil samples by the double-layer agar technique.

Starch Casein KNO_3 Agar

Composition per liter:

Agar	18.0g
Starch	10.0g
KNO_3	2.0g
K_2HPO_4	2.0g
NaCl	2.0g
Casein	0.3g
$MgSO_4 \cdot 7H_2O$	0.05g
$CaCO_3$	0.02g
$FeSO_4 \cdot 7H_2O$	0.01g

Preparation of Medium: Add components to distilled/deionized water and bring volume to 1.0L. Mix thoroughly. Gently heat and bring to boiling. Distribute into tubes or flasks. Autoclave for 15 min at 15 psi pressure–121°C. Pour into sterile Petri dishes or leave in tubes.

Use: For the selective isolation and cultivation of streptomycetes.

Starch Fermentation Broth

Composition per 225.2mL:

Heart infusion broth	5.0mL
Bromcresol Purple solution	0.2mL
Starch solution	20.0mL

pH 7.8 ± 0.2 at 25°C

Heart Infusion Broth:
Composition per liter:

Beef heart, infusion from	500.0g
Tryptose	10.0g
NaCl	5.0g

pH 7.4 ± 0.2 at 25°C

Source: Heart infusionbroth is available as a premixed powder from Difco Laboratories.

Preparation of Heart Infusion Broth: Add components to distilled/deionized water and bring volume to 1.0L. Mix thoroughly. Distribute into tubes or flasks. Autoclave for 15 min at 15 psi pressure–121°C.

Bromcresol Purple Solution:
Composition per 10mL:

Bromcresol Purple	0.1g
Ethanol (95% solution)	10.0mL

Preparation of Bromcresol Purple Solution: Add Bromcresol Purple to 10.0mL ethanol. Mix thoroughly.

Starch Solution:
Composition per 20mL:

Starch	0.4g

Preparation of Starch Solution: Add starch to distilled/deionized water and bring volume to 20.0mL. Mix thoroughly. Gently heat while stirring and bring to boiling.

Preparation of Medium: Combine 5.0mL heart infusion, 0.2mL of Bromcresol Purple solution, 200.0mL of distilled/deionized water and 20.0mL of starch solution. Mix thoroughly. Distribute into tubes or flasks. Autoclave for 15 min at 15 psi pressure–121°C. Pour into sterile Petri dishes or leave in tubes.

Use: For the cultivation of *Corynebacterium* species.

Starch Hydrolysis Agar

Composition per liter:

Beef heart, infusion from	500.0g
Soluble starch	20.0g
Agar	15.0g
Tryptose	10.0g
NaCl	5.0g

pH 7.4 ± 0.2 at 25°C

Preparation of Medium: Add components to distilled/deionized water and bring volume to 1.0L. Mix thoroughly. Gently heat and bring to boiling. Distribute into tubes or flasks. Autoclave for 15 min at 15 psi pressure–121°C. Pour into sterile Petri dishes or leave in tubes.

Use: For the cultivation and differentiation of a variety of microorganisms based on amylase production. After incubation, starch hydrolysis is determined by the addition of Gram's or Lugol's iodine solution.

Organisms that produce amylase appear as colonies surrounded by a clear zone.

Starkey's Medium C, Modified

Composition per liter:

Sodium lactate	3.5g
$MgSO_4 \cdot 7H_2O$	2.0g
Na_2SO_4	1.0g
NH_4Cl	1.0g
Yeast extract	1.0g
KH_2PO_4	0.5g
$CaCl_2 \cdot 2H_2O$	0.1g
Ferrous ammonium sulfate solution	50.0mL
Cysteine·HCl·H_2O solution	10.0mL

pH 7.5 ± 0.2 at 25°C

Ferrous Ammonium Sulfate Solution:
Composition per 100mL:

$Fe(NH_4)_2(SO_4)_2 \cdot 6H_2O$	1.0g

Preparation of Ferrous Ammonium Sulfate Solution: Add $Fe(NH_4)_2(SO_4)_2 \cdot 6H_2O$ to distilled/deionized water and bring volume to 100.0mL. Mix thoroughly. Filter sterilize.

Cysteine·HCl·H_2O Solution:
Composition per 10mL:

Cysteine·HCl·H_2O	0.75g

Preparation of Cysteine·HCl·H_2O Solution: Add cysteine·HCl·H_2O to distilled/deionized water and bring volume to 10.0mL. Mix thoroughly. Filter sterilize.

Preparation of Medium: Add components, except ferrous ammonium sulfate solution and cysteine·HCl·H_2O solution, to tap water and bring volume to 940.0mL. Mix thoroughly. Gently heat and bring to boiling. Autoclave for 15 min at 15 psi pressure–121°C. Cool to 45°–50°C. Aseptically add 50.0mL of sterile ferrous ammonium sulfate solution and 10.0mL of sterile cysteine·HCl·H_2O solution. Mix thoroughly. Adjust pH to 7.5 with filter-sterilized 2*N* NaOH. Pour into sterile Petri dishes or distribute into sterile tubes.

Use: For the cultivation and maintenance of *Desulfotomaculum* species and *Desulfovibrio* species.

Starkey's Medium C, Modified with Salt

Composition per liter:

NaCl	25.0g
Sodium lactate	3.5g
$MgSO_4 \cdot 7H_2O$	2.0g
Na_2SO_4	1.0g
NH_4Cl	1.0g
Yeast extract	1.0g
KH_2PO_4	0.5g
$CaCl_2 \cdot 2H_2O$	0.1g
Ferrous ammonium sulfate solution	50.0mL
Cysteine·HCl·H_2O solution	10.0mL

pH 7.5 ± 0.2 at 25°C

Ferrous Ammonium Sulfate Solution:
Composition per 100mL:

$Fe(NH_4)_2(SO_4)_2 \cdot 6H_2O$	1.0g

Preparation of Ferrous Ammonium Sulfate Solution: Add $Fe(NH_4)_2(SO_4)_2 \cdot 6H_2O$ to distilled/deionized water and bring volume to 100.0mL. Mix thoroughly. Filter sterilize.

Cysteine·HCl·H_2O Solution:
Composition per 10mL:

Cysteine·HCl·H_2O	0.75g

Preparation of Cysteine·HCl·H_2O Solution: Add cysteine·HCl·H_2O to distilled/deionized water and bring volume to 10.0mL. Mix thoroughly. Filter sterilize.

Preparation of Medium: Add components, except ferrous ammonium sulfate solution and cysteine·HCl·H_2O solution, to tap water and bring volume to 940.0mL. Mix thoroughly. Gently heat and bring to boiling. Autoclave for 15 min at 15 psi pressure–121°C. Cool to 45°–50°C. Aseptically add 50.0mL of sterile ferrous ammonium sulfate solution and 10.0mL of sterile cysteine·HCl·H_2O solution. Mix thoroughly. Adjust pH to 7.5 with filter-sterilized 2*N* NaOH. Pour into sterile Petri dishes or distribute into sterile tubes.

Use: For the cultivation and maintenance of halophilic *Desulfovibrio* species.

Steenken and Smith Agar (Hohn's Medium, Modified)

Composition per 2065mL:

Homogenized whole egg	1500.0mL
Stock salts solution	500.0mL
Lacmoid solution	25.0mL
HCl (1*N* solution)	40.0mL

pH 6.6 ± 0.2 at 25°C

Stock Salts Solution:
Composition per 500mL:

KH_2PO_4	2.0g
Asparagine	1.5g
Magnesium citrate	1.25 g
Na_2HPO_4, anhydrous	1.2 g
$MgSO_4$	0.3g
Glycerol	60.0mL

Preparation of Stock Salts Solution: Add components, except glycerol, to distilled/deionized water which has been warmed to 80°C. Bring volume to 440.0mL. Mix thoroughly. Add 60.0mL of glycerol. Mix thoroughly. Autoclave for 20 min at 10 psi pressure–115°C. Cool to 25°C. Aseptically divide the solution into two 250.0mL parts.

Lacmoid Solution:

Composition per 100mL:

Lacmoid	1.0g
Ethanol (50% solution)	100.0mL

Preparation of Lacmoid Solution: Add lacmoid to 100.0mL of ethanol solution. Mix thoroughly.

Preparation of Medium: To one 250.0mL part of sterile stock salts solution, add 25.0mL of lacmoid solution. Mix thoroughly. To the other 250.0mL part of sterile stock salts solution, add 40.0mL of HCl solution. Mix thoroughly. Soak eggs in 70% ethanol for 10 min. Dry between sterile towels. Break eggs into a sterile container. Aseptically homogenize the whole eggs with a sterile glass rod. Add both stock salt solutions to the homogenized whole egg mixture. Mix thoroughly. Filter through sterile cheesecloth. Aseptically distribute into sterile tubes. Inspissate medium at 85°C (moist heat) for 90 min on two consecutive days.

Use: For the cultivation and maintenance of *Mycobacterium microti*.

Sterility Test Broth (USP Alternative Thioglycollate Medium)

Composition per liter:

Pancreatic digest of casein	15.0g
Glucose	5.0g
Yeast extract	5.0g
NaCl	2.5g
L-Cystine	0.5g
Sodium thioglycollate	0.5g

pH 7.1 ± 0.2 at 25°C

Source: This medium is available as a premixed powder from BBL Microbiology Systems.

Preparation of Medium: Add components to distilled/deionized water and bring volume to 1.0L. Mix thoroughly. Gently heat and bring to boiling. Distribute into tubes or flasks. Autoclave for 15 min at 15 psi pressure–121°C. Cool to 25°C. If not used immediately, prior to inoculation heat tubes in a boiling water bath for 5–10 min. Cool to 25°C.

Use: As an alternate medium, instead of Fluid Thioglycollate Broth, for testing the sterility of a variety of specimens.

STL Broth

Composition per liter:

Casamino acids	1.0g
Glucose	1.0g
Sodium glutamate	1.0g
$CaCl_2 \cdot 2H_2O$	0.1g
KNO_3	0.1g
$MgSO_4 \cdot 7H_2O$	0.1g
Sodium glycerophosphate	0.1g
Thiamine	1.0mg
Vitamin B_{12}	1.0μg
Trace element solution	1.0mL

pH 7.5 ± 0.2 at 25°C

Trace Element Solution:

Composition per liter:

Disodium EDTA	8.0g
$MnCl_2 \cdot 4H_2O$	0.1g
$CoCl_2 \cdot 6H_2O$	0.02g
KBr	0.02g
KI	0.02g
$ZnCl_2$	0.02g
$CuSO_4$	0.01g
H_3BO_3	0.01g
$Na_2MoO_4 \cdot 2H_2O$	0.01g
LiCl	5.0mg
$SnCl_2 \cdot 2H_2O$	5.0mg

Preparation of Trace Element Solution: Add components to distilled/deionized water and bring volume to 1.0L. Mix thoroughly.

Preparation of Medium: Add components to distilled/deionized water and bring volume to 1.0L. Mix thoroughly. Gently heat and bring to boiling. Distribute into tubes or flasks. Autoclave for 15 min at 15 psi pressure–121°C. Pour into sterile Petri dishes or leave in tubes.

Use: For the isolation and cultivation of *Cytophaga* species, *Herpetosiphon* species, *Saprospira* species, and *Flexithrix* species.

Stock Culture Agar

Composition per liter:

Beef heart infusion	500.0g
Gelatin	10.0g
Proteose peptone	10.0g
Agar	7.5g
Casein	5.0g
Na_2HPO_4	4.0g
Sodium citrate	3.0g
Glucose	0.5g

pH 7.5 ± 0.2 at 25°C

Source: This medium is available as a premixed powder from Difco Laboratories.

Preparation of Medium: Add components to cold distilled/deionized water and bring volume to 1.0L. Mix thoroughly. Gently heat while stirring and bring to boiling. Distribute into tubes or flasks. Autoclave for 15 min at 15 psi pressure–121°C. Pour into sterile Petri dishes or leave in tubes.

Use: For the maintenance of pathogenic and nonpathogenic bacteria, especially streptococci.

Stock Culture Agar with L-Asparagine

Composition per liter:

Beef heart infusion	500.0g
Gelatin	10.0g
Proteose peptone	10.0g
Agar	7.5g
Casein	5.0g
Na_2HPO_4	4.0g
Sodium citrate	3.0g
L-Asparagine	1.0g
Glucose	0.5g

pH 7.5 ± 0.2 at 25°C

Preparation of Medium: Add components to cold distilled/deionized water and bring volume to 1.0L. Mix thoroughly. Gently heat while stirring and bring to boiling. Distribute into tubes or flasks. Autoclave for 15 min at 15 psi pressure–121°C. Pour into sterile Petri dishes or leave in tubes.

Use: For the maintenance of pathogenic and nonpathogenic bacteria, especiallly streptococci.

Stokes Agar

Composition per liter:

Agar	12.5g
Glucose	1.0g
Peptone	1.0g
$MgSO_4 \cdot 7H_2O$	0.2g
$CaCl_2$	0.05g
$FeCl_3 \cdot 6H_2O$	0.01g

Preparation of Medium: Add components to tap water and bring volume to 1.0L. Mix thoroughly. Gently heat and bring to boiling. Distribute into tubes or flasks. Autoclave for 15 min at 15 psi pressure–121°C. Pour into sterile Petri dishes or leave in tubes.

Use: For the isolation and cultivation of *Sphaerotilus natans*.

Strep ID Quad Plate

Composition per liter:

Quadrant I	5.0mL
Quadrant II	5.0mL
Quadrant III	5.0mL
Quadrant IV	5.0mL

Source: Available as a prepared medium from BBL Microbiology Systems.

Quadrant I:
Composition per 5.0mL:

Bacitracin	0.5mg
TSA II agar	5.0mL

Quadrant II:
Composition per 5.0mL:

TSA II agar	5.0mL
Sheep blood, defibrinated	0.25mL

Quadrant III:
Composition per 5.0mL:

Bile esculin agar	5.0mL

Quadrant IV:
Composition per 5.0mL:

Blood agar base with 6.5% NaCl	5.0mL

Use: For the differentiation and presumptive identification of streptococci. The Strep (*Streptococcus)* ID (Identification) Quad Plate is a four-sectored plate each containing a different medium.

Streptococcal Growth Medium

Composition per liter:

Beef heart, solids from infusion	500.0g
Tryptose	10.0g
NaCl	5.0g
Glucose	1.0g
Bromcresol Purple solution	1.0mL

pH 7.4 ± 0.2 at 25°C

Bromcresol Purple Solution:
Composition per 10mL:

Bromcresol Purple	0.16g
Ethanol (95% solution)	10.0mL

Preparation of Bromcresol Purple Solution: Add Bromcresol Purple to 10.0mL of ethanol. Mix thoroughly.

Preparation of Medium: Add components to distilled/deionized water and bring volume to 1.0L. Mix thoroughly. Distribute into tubes in 5.0mL volumes. Autoclave for 15 min at 15 psi pressure–121°C.

Use: For the cultivation of *Streptococcus* species and other Gram-positive cocci. Growth in this medium turns the indicator yellow and the solution turbid.

Streptococcus Agar

Composition per liter:

Glucose	20.0g

Pancreatic digest of casein 20.0g
Agar....................15.0g
K_2HPO_4....................2.0g
$MgSO_4 \cdot 7H_2O$....................0.1g
pH 6.8 ± 0.2 at 25°C

Preparation of Medium: Add components to distilled/deionized water and bring volume to 1.0L. Mix thoroughly. Gently heat and bring to boiling. Distribute into tubes or flasks. Autoclave for 15 min at 15 psi pressure–121°C. Pour into sterile Petri dishes or leave in tubes.

Use: For the cultivation and maintenance of *Streptococcus* species.

Streptococcus Blood Agar, Selective

Composition per liter:
Agar....................15.0g
Pancreatic digest of casein....................10.0g
Beef extract....................6.7g
Nucleic acid....................6.0g
NaCl....................5.0g
Sheep blood, defibrinated....................50.0mL
Maltose solution....................10.0mL
Antibiotic inhibitor....................10.0mL
pH 7.3 ± 0.2 at 25°C

Maltose Solution:
Composition per 10mL:
Maltose....................0.25–5.0g

Preparation of Maltose Solution: Add maltose to distilled/deionized water and bring volume to 10.0mL. Mix thoroughly. Filter sterilize.

Antibiotic Inhibitor:
Composition per 10mL:
Polymyxin B sulfate....................0.020g
Neomycin sulfate....................0.010g

Preparation of Antibiotic Inhibitor: Add components to distilled/deionized water and bring volume to 10.0mL. Mix thoroughly. Filter sterilize.

Preparation of Medium: Add components—except sheep blood, maltose solution, and antibiotic inhibitor—to distilled/deionized water and bring volume to 930.0mL. Mix thoroughly. Gently heat and bring to boiling. Autoclave for 15 min at 15 psi pressure–121°C. Cool to 45°–50°C. Aseptically add sterile sheep blood, sterile maltose solution, and sterile antibiotic inhibitor. Mix thoroughly. Pour into sterile Petri dishes or distribute into sterile tubes.

Use: For the isolation and cultivation of Group A hemolytic *Streptococcus* species from the human respiratory tract.

Streptococcus faecalis Broth
See: ***SF*** **Broth**

Streptococcus Medium

Composition per liter:
Agar....................15.0g
Glucose....................4.0g
K_2HPO_4....................3.8g
Pancreatic digest of casein....................2.5g
Yeast extract....................2.5g
pH 7.6 ± 0.2 at 25°C

Preparation of Medium: Add components to distilled/deionized water and bring volume to 1.0L. Mix thoroughly. Gently heat and bring to boiling. Distribute into tubes or flasks. Autoclave for 15 min at 15 psi pressure–121°C. Pour into sterile Petri dishes or leave in tubes.

Use: For the cultivation and maintenance of *Enterococcus faecalis*.

Streptococcus mutans Medium

Composition per 100mL:
Pancreatic digest of casein....................2.0g
Mannitol....................0.5g
NaCl....................0.25g
Lactoalbumin....................0.25g
Agar....................0.075g
L-Cystine....................0.05g
Sodium thioglycollate....................0.05g
Thallium acetate....................0.025g
Crystal Violet....................0.1mg
Bromcresol Purple (0.04% solution)....................15.0mL
pH 7.1 ± 0.2 at 25°C

Caution: Thallium salts are toxic.

Preparation of Medium: Add components to distilled/deionized water and bring volume to 1.0L. Mix thoroughly. Gently heat and bring to boiling. Distribute into tubes in 5.0mL volumes. Autoclave for 15 min at 15 psi pressure–121°C.

Use: For the selective isolation and cultivation of *Streptococcus mutans*. Bacteria that turn the medium yellow are presuptive for *S. mutans*.

Streptococcus Selective Medium

Composition per liter:
Special peptone....................23.0g
Agar....................10.0g
NaCl....................5.0g
Starch....................1.0g

Horse blood, defibrinated	50.0mL
Antibiotic inhibitor	10.0mL

pH 7.3± 0.2 at 25°C

Source: This medium is available as a premixed powder from Oxoid Unipath.

Antibiotic Inhibitor:
Composition per 10mL:

Colistin sulfate	10.0mg
Oxolinic acid	5.0mg

Preparation of Antibiotic Inhibitor: Add components to distilled/deionized water and bring volume to 10.0mL. Mix thoroughly. Filter sterilize.

Preparation of Medium: Add components, except horse blood and antibiotic inhibitor, to distilled/deionized water and bring volume to 940.0mL. Mix thoroughly. Gently heat and bring to boiling. Autoclave for 15 min at 15 psi pressure–121°C. Cool to 45°–50°C. Aseptically add sterile horse blood, and sterile antibiotic inhibitor. Mix thoroughly. Pour into sterile Petri dishes or distribute into sterile tubes.

Use: For the selective isolation of streptococci from clinical specimens or foodstuffs.

Streptococcus thermophilus Agar (ST Agar)

Composition per liter:

Agar	15.0g
Sucrose	10.0g
Pancreatic digest of casein	10.0g
Yeast extract	5.0g
K_2HPO_4	2.0g

pH 6.8 ± 0.2 at 25°C

Preparation of Medium: Add components to distilled/deionized water and bring volume to 1.0L. Mix thoroughly. Gently heat and bring to boiling. Distribute into tubes or flasks. Autoclave for 15 min at 15 psi pressure–121°C. Pour into sterile Petri dishes or leave in tubes.

Use: For the isolation and cultivation of *Streptococcus thermophilus* from dairy products.

Streptomyces Medium

Composition per liter:

Agar	25.0g
Glucose	5.0g
L-Glutamic acid	4.0g
KH_2PO_4	1.0g
NaCl	1.0g
$MgSO_4 \cdot 7H_2O$	0.7g
$FeSO_4 \cdot 7H_2O$	3.0mg

pH 7.0 ± 0.2 at 25°C

Preparation of Medium: Add components to distilled/deionized water and bring volume to 1.0L. Mix thoroughly. Gently heat and bring to boiling. Distribute into tubes or flasks. Autoclave for 15 min at 15 psi pressure–121°C. Pour into sterile Petri dishes or leave in tubes.

Use: For the cultivation and maintenance of *Streptomyces kanamyceticus.*

Streptomycete Antibiotic Activity Inoculum Medium

Composition per liter:

Pancreatic digest of casein	10.0g
D-Glucose	5.0g
Yeast extract	5.0g
K_2HPO_4	1.0g
Liver extract	100.0mL

pH 6.9 ± 0.2 at 25°C

Preparation of Medium: Add components to distilled/deionized water and bring volume to 1.0L. Mix thoroughly. Distribute into tubes or flasks. Autoclave for 15 min at 15 psi pressure–121°C.

Use: For the cultivation of *Streptomyces* species to be used in the antibiotic activity assay.

Streptomycete Antibiotic Activity Medium

Composition per liter:

Agar	15.0g
D-Glucose	15.0g
Soybean meal	15.0g
NaCl	5.0g
Yeast extract	1.0g
$CaCO_3$	1.0g
Glycerol	2.5mL

pH 6.8 ± 0.2 at 25°C

Preparation of Medium: Add components to distilled/deionized water and bring volume to 1.0L. Mix thoroughly. Gently heat and bring to boiling. Distribute into tubes or flasks. Autoclave for 15 min at 15 psi pressure–121°C. Pour into sterile Petri dishes.

Use: For the the cultivation and determination of antibiotic production of *Streptomyces* species by the streak method. *Bacillus subtilis* NRRL B-765, *Sarcina lutea* NRRL B-1018, *Escherichia coli* NRRL B-766, *Saccharomyces pasteurianus* NRRL Y-139, *Candida albicans* NRRL Y-477, and *Mucor ramannianus* NRRL 1839 are used as test organisms.

Streptomycete Medium

Composition per liter:

Solution B 500.0mL
Solution A 400.0mL
Solution C 100.0mL

Solution A:
Composition per 400mL:

Glucose 20.0g
Agar 4.0g
Yeast extract 1.2g
$MgSO_4 \cdot 7H_2O$ 0.25g
Bromcresol Purple 0.012g

Preparation of Solution A: Add components to distilled/deionized water and bring volume to 400.0mL. Mix thoroughly. Autoclave for 15 min at 15 psi pressure–121°C. Cool to 45°–50°C.

Solution B:
Composition per 500mL:

$Na_2HPO_4 \cdot 2H_2O$ 534.0mg
KH_2PO_4 .... 272.0mg

Preparation of Solution B: Add components to distilled/deionized water and bring volume to 500.0mL. Mix thoroughly. Autoclave for 15 min at 15 psi pressure–121°C. Cool to 45°–50°C.

Solution C:
Composition per 100mL:

$CaCO_3$ 1.0g

Preparation of Solution C: Add $CaCO_3$ to distilled/deionized water and bring volume to 100.0mL. Mix thoroughly.

Preparation of Medium: Distribute solution C into test tubes in 0.2mL volumes. Autoclave for 15 min at 15 psi pressure–121°C. Cool to 45°–50°C. Combine cooled, sterile solution A and cooled, sterile solution B. Mix thoroughly. Add 1.8mL of solution A-B to each test tube containing sterile solution C. Mix thoroughly to distribute the $CaCO_3$. Cool tubes rapidly in an ice water bath.

Use: For the cultivation and differentiation of streptomycetes based on their formation of organic acids. Bacteria that form organic acids turn the medium yellow and dissolve the $CaCO_3$.

Streptomycete Medium

Composition per liter:

Glycerol 5.0g
Agar 4.0g
NaCl 2.0g
KNO_3 .... 1.0g
$Na_2HPO_4 \cdot 2H_2O$ 0.534g
$MgSO_4 \cdot 7H_2O$ 0.5g
KH_2PO_4 .... 0.272g
Trace elements solution 1.0mL

pH 6.8 ± 0.2 at 25°C

Trace Elements Solution:
Composition per 100mL:

$FeSO_4 \cdot 7H_2O$ 0.1g
$MnCl_2 \cdot 4H_2O$ 0.1g
$ZnSO_4 \cdot 7H_2O$ 0.1g

Preparation of Trace Elements Solution: Add components to distilled/deionized water and bring volume to 100.0mL. Mix thoroughly.

Preparation of Medium: Add components to distilled/deionized water and bring volume to 1.0L. Mix thoroughly. Gently heat and bring to boiling. Distribute into tubes in 1.0mL volumes. Autoclave for 15 min at 15 psi pressure–121°C.

Use: For the cultivation and differentiation of streptomycetes based on their reduction of nitrate to nitrite. Bacteria that reduce nitrate to nitrite for a red color after the addition of Griess-Ilosvay reagent.

Streptomycete Medium

Composition per liter:

Agar 12.0g
NaCl 5.0g
$Na_2HPO_4 \cdot 2H_2O$ 1.98g
KH_2PO_4 .... 1.51g
Glucose 1.0g
Pancreatic digest of casein 1.0g
$MgSO_4 \cdot 7H_2O$ 0.5g
Phenol Red 0.012g
Urea solution 100.0mL

pH 6.8 ± 0.2 at 25°C

Urea Solution:
Composition per 100mL:

Urea 20.0g

Preparation of Urea Solution: Add urea to distilled/deionized water and bring volume to 100.0mL. Mix thoroughly. Filter sterilize.

Preparation of Medium: Add components, except urea solution, to distilled/deionized water and bring volume to 900.0mL. Mix thoroughly. Gently heat and bring to boiling. Autoclave for 15 min at 15 psi pressure–121°C. Cool to 45°–50°C. Aseptically add 100.0mL of sterile urea solution. Mix thoroughly. Aseptically distribute into sterile tubes. Allow tubes to cool in a slanted position.

Use: For the cultivation and differentiation of streptomycetes based on their ability to produce urease.

Streptomycete Medium

Composition per liter:

Sodium hippurate	10.0g
Na_2HPO_4	5.0g
Glucose	2.0g
Meat extract	2.0g
Peptone	2.0g
Yeast extract	2.0g

pH 7.0 ± 0.2 at 25°C

Preparation of Medium: Add components to distilled/deionized water and bring volume to 1.0L. Mix thoroughly. Distribute into tubes in 3.0mL volumes. Autoclave for 15 min at 15 psi pressure–121°C.

Use: For the cultivation and differentiation of streptomycetes based on their ability to hydrolyze hippurate.

Streptomycin Assay Agar with Yeast Extract *See:* Antibiotic Medium 5

Streptomycin Terramycin® Malt Extract Agar

Composition per liter:

Malt extract	30.0g
Agar	15.0g
Peptone	5.0g
Streptomycin solution	100.0mL
Terramycin solution	100.0mL

pH 5.4 ± 0.2 at 25°C

Streptomycin Solution:
Composition per 100mL:

Streptomycin	0.07g

Preparation of Streptomycin Solution: Add streptomycin to distilled/deionized water and bring volume to 100.0mL. Mix thoroughly. Filter sterilize.

Terramycin Solution:
Composition per 10mL:

Terramycin	0.07g

Preparation of Terramycin Solution: Add terramycin to distilled/deionized water and bring volume to 100.0mL. Mix thoroughly. Filter sterilize.

Preparation of Medium: Add components, except streptomycin solution and terramycin solution, to distilled/deionized water and bring volume to 800.0mL. Mix thoroughly. Gently heat and bring to boiling. Autoclave for 15 min at 15 psi pressure–121°C. Cool to 45°–50°C. Aseptically add 100.0mL of sterile streptomycin solution and 100.0mL of sterile terramycin solution. Mix thoroughly. Pour into sterile Petri dishes in 20.0mL volumes.

Use: For the cultivation and enumeration of fungi isolated from sewage and polluted waters.

Streptosel™ Agar

Composition per liter:

Pancreatic digest of casein	15.0g
Agar	12.0g
Glucose	5.0g
Papaic digest of soybean meal	5.0g
NaCl	4.0g
Sodium citrate	1.0g
L-Cystine	0.2g
NaN_3	0.2g
Na_2SO_3	0.2g
Crystal Violet	0.2mg

pH 7.4 ± 0.2 at 25°C

Source: This medium is available as a premixed powder from BBL Microbiology Systems.

Preparation of Medium: Add components to distilled/deionized water and bring volume to 1.0L. Mix thoroughly. Gently heat and bring to boiling. Distribute into tubes or flasks. If medium is used the same day, do not autoclve. Pour into sterile Petri dishes or leave in tubes. If medium is to be stored, autoclave for 15 min at 13 psi pressure–118°C. Pour into sterile Petri dishes or leave in tubes.

Use: For the selective isolation, cultivation and enumeration of streptococci from specimens containing a mixed flora.

Streptosel™ Broth

Composition per liter:

Pancreatic digest of casein	15.0g
Glucose	5.0g
Papaic digest of soybean meal	5.0g
NaCl	4.0g
Sodium citrate	1.0g
L-Cystine	0.2g
Na_2SO_3	0.2g
NaN_3	0.2g
Crystal Violet	0.2mg

pH 7.4 ± 0.2 at 25°C

Source: This medium is available as a premixed powder from BBL Microbiology Systems.

Preparation of Medium: Add components to distilled/deionized water and bring volume to 1.0L. Mix thoroughly. Distribute into tubes or flasks. Autoclave for 15 min at 13 psi pressure–118°C.

Use: For the selective isolation and cultivation of streptococci from specimens containing a mixed flora.

STT Agar
See: **Sucrose Teepol Tellurite Agar**

STTA Medium

Composition per liter:

Peptone	20.0g
Glycerol	15.0g
Agar	13.0g
Yeast extract	2.0g
K_2HPO_4	1.0g
$MgSO_4{\cdot}4H_2O$	1.0g
Antibiotic solution	10.0mL
Thallous acetate solution	10.0mL

pH 7.0 ± 0.2 at 25°C

Antibiotic Solution:
Composition per 10mL:

Streptomycin sulfate	0.5g
Cycloheximide	0.05g

Preparation of Antibiotic Solution: Add cephalexin to distilled/deionized water and bring volume to 10.0mL. Mix thoroughly. Filter sterilize.

Thallous Acetate Solution:
Composition per 10mL:

Thallous acetate	0.05g

Preparation of Thallous Acetate Solution: Add thallous acetate to distilled/deionized water and bring volume to 10.0mL. Mix thoroughly. Filter sterilize.

Preparation of Medium: Add components, except antibiotic solution and thallous acetate solution, to distilled/deionized water and bring volume to 980.0mL. Mix thoroughly. Gently heat and bring to boiling. Autoclave for 15 min at 15 psi pressure–121°C. Cool to 45°–50°C. Aseptically add sterile antibiotic solution and thallous acetate solution. Mix thoroughly. Pour into sterile Petri dishes or distribute into sterile tubes.

Use: For the selective isolation and cultivation of *Brochothrix thermosphacta*.

Stuart *Leptospira* Broth, Modified

Composition per liter:

NaCl	1.93g
Na_2HPO_4	0.66g
NH_4Cl	0.34g
$MgCl_2{\cdot}6H_2O$	0.19g
L-Asparagine	0.13g
KH_2PO_4	0.08g
Glycerol	5.0mL
Rabbit serum, inactivated at 56°C, 30 min	100.0mL

pH 7.4 ± 0.2 at 25°C

Preparation of Medium: Add each component, except rabbit serum, to distilled/deionized water in separate flasks and bring each volume to 100.0mL. Mix thoroughly. Combine the seven solutions, except the rabbit serum. Mix thoroughly. Gently heat and bring to boiling. Autoclave for 15 min at 15 psi pressure–121°C. Cool to 45°–50°C. Aseptically add sterile rabbit serum. Mix thoroughly. Aseptically distribute into sterile tubes or flasks.

Use: For the cultivation of *Leptospira* species.

Stuart Medium Base

Composition per 1100mL:

NaCl	1.8g
Na_2HPO_4	0.67g
$MgCl_2{\cdot}6H_2O$	0.41g
NH_4Cl	0.27g
Asparagine	0.13g
KH_2PO_4	0.09g
Phenol Red	0.01g
Glycerol	5.0mL
Leptospira enrichment (Difco)	100.0mL

pH 7.6 ± 0.2 at 25°C

Source: This medium is available as a premixed powder from Difco Laboratories. *Leptospira* enrichment contains rabbit serum and hemoglobin and is available from Difco Laboratories.

Preparation of Medium: Add components, except glycerol and *Leptospira* enrichment, to distilled/deionized water and bring volume to 995.0mL. Mix thoroughly. Add glycerol. Mix thoroughly. Autoclave for 15 min at 15 psi pressure–121°C. Cool to 45°–50°C. Aseptically add *Leptospira* enrichment. Mix thoroughly. Aseptically distribute into sterile screw-capped tubes in 10.0mL volumes.

Use: For the cultivation of *Leptospira* species.

Stuart Transport Medium

Composition per liter:

Sodium glycerophosphate	10.0g
Sodium thioglycollate	1.0g
$CaCl_2{\cdot}2H_2O$	0.1g
Methylene Blue	2.0mg

pH 7.4 ± 0.2 at 25°C

Preparation of Medium: Add components to distilled/deionized water and bring volume to 1.0L. Mix thoroughly. Gently heat and bring to boiling. Distribute into 7.0mL screw-capped tubes. Fill tubes to capacity. Autoclave for 15 min at 15 psi pressure–121°C.

Use: For the preservation of *Neisseria* species and other fastidious organisms during their transport from clinic to laboratory.

Stuart Transport Medium, Modified

Composition per liter:

Sodium glycerophosphate	10.0g
Agar	5.0g
Cysteine·HCl·H_2O	0.5g
Sodium thioglycollate	0.5g
$CaCl_2 \cdot 2H_2O$	0.1g
Methylene Blue	1.0mg

pH 7.4 ± 0.2 at 25°C

Source: This medium is available as a premixed powder from Oxoid Unipath.

Preparation of Medium: Add components to distilled/deionized water and bring volume to 1.0L. Mix thoroughly. Gently heat and bring to boiling. Distribute into 7.0mL screw-capped tubes. Fill tubes to capacity. Autoclave for 15 min at 15 psi pressure–121°C.

Use: For the preservation of *Neisseria* species and other fastidious organisms during their transport from clinic to laboratory.

Sucrose Agar

Composition per liter:

Sucrose	50.0g
Agar	20.0g
Peptone	10.0g
NaCl	5.0g
Yeast extract	5.0g
$CaCO_3$	3.0g
Phenylethyl alcohol	3.0g
$MgSO_4 \cdot 7H_2O$	0.5g
$MnSO_4 \cdot 4H_2O$	0.5g
Tween™ 80	0.1g
Bromcresol Green	20.0mg
Cycloheximide solution	10.0mL

pH 6.2 ± 0.2 at 25°C

Cycloheximide Solution:
Composition per 10mL:

Cycloheximide	0.01g

Preparation of Cycloheximide Solution: Add cycloheximide to distilled/deionized water and bring volume to 10.0mL. Mix thoroughly. Filter sterilize.

Preparation of Medium: Add agar, to distilled/deionized water and bring volume to 500.0mL. Mix thoroughly. Gently heat and bring to boiling. Add remaining components, except cycloheximide solution and phenylethyl alcohol, and bring volume to 990.0mL with distilled/deionized water. Adjust pH to 6.2. Autoclave for 15 min at 15 psi pressure–121°C. Cool to 45°–50°C. Aseptically add 10.0mL of sterile cycloheximide solution and 3.0g of phenylethyl alcohol. Mix thoroughly. Pour into sterile Petri dishes or distribute into sterile tubes.

Use: For the isolation and cultivation of *Lactobacillus* species from brewery isolates.

Sucrose Agar

Composition per liter:

Beef heart (solids from infusion)	500.0g
Sucrose	50.0g
Agar	15.0g
Tryptose	10.0g
NaCl	5.0g

pH 7.4 ± 0.2 at 25°C

Preparation of Medium: Add components to distilled/deionized water and bring volume to 1.0L. Mix thoroughly. Gently heat and bring to boiling. Distribute into tubes or flasks. Autoclave for 15 min at 15 psi pressure–121°C. Pour into sterile Petri dishes or leave in tubes.

Use: For the differentiation of bacteria based on their ability to produce glucan. Dextran production, typical of *Streptococcus sanguis* and *Streptococcus mutans* results in highly refractile-adherent or dry-adherent colonies. Levan production, typical of *Streptococcus salivarius*, results in opaque, gummy, nonadherent colonies. Colonies of *Streptococcus bovis* and *Leuconostoc mesenteroides* are similar to those of *S. salivarius* but are somewhat less gummy and rarely adhere to the medium. Large or small colonies that are mucoidal and nonadherent are considered negative or have no extracellular polysaccharide production.

Sucrose Broth

Composition per liter:

Solution A	500.0mL
Solution B	500.0mL

pH 7.1 ± 0.2 at 25°C

Solution A:
Composition per 500mL:

Pancreatic digest of casein	15.0g
Sodium acetate	12.0g
K_2HPO_4	10.0g
Glucose	5.5g
Yeast extract	5.0g
NaCl	2.5g

L-Cystine 0.5g
Sodium thioglycollate 0.5g

Preparation of Solution A: Add components to distilled/deionized water and bring volume to 500.0mL. Mix thoroughly. Autoclave for 15 min at 15 psi pressure–121°C. Cool to 45°–50°C.

Solution B:
Composition per 500mL:
Sucrose 50.0g

Preparation of Solution B: Add sucrose to distilled/deionized water and bring volume to 500.0mL. Mix thoroughly. Autoclave for 15 min at 15 psi pressure–121°C. Cool to 45°–50°C.

Preparation of Medium: Combine sterile solution A with sterile solution B. Mix thoroughly. Aseptically distribute into sterile tubes or flasks.

Use: For the differentiation of bacteria based on their ability to produce glucan. Production of glucan is indicated when the broth is partially or completely gelled—typical of *Streptococcus sanguis*, or when gelatinous, adherent deposits form on the bottom and walls of the tube—typical of *Streptococcus mutans*. An increase in the viscosity indicates the production of slime (unknown polysaccharide)—typical of *Streptococcus bovis*.

Sucrose Peptone Agar

Composition per liter:
Sucrose 20.0g
Agar 12.0g
Peptone 5.0g
K_2HPO_4 ....0.5g
$MgSO_4 \cdot 7H_2O$ 0.25g
pH 7.2–7.4 at 25°C

Preparation of Medium: Add components to distilled/deionized water and bring volume to 1.0L. Mix thoroughly. Gently heat and bring to boiling. Adjust pH to 7.2–7.4. Distribute into tubes or flasks. Autoclave for 15 min at 15 psi pressure–121°C. Pour into sterile Petri dishes or leave in tubes.

Use: For the cultivation and maintenance of *Pseudomonas solanacearum* and *Xanthomonas albilineans*.

Sucrose Phosphate Glutamate Transport Medium

Composition per liter:
Sucrose 75.0g
K_2HPO_4 ....0.52g
Na_2HPO_4 ....1.22g
Glutamic acid 0.72g
Bovine serum 50.0mL
Antibiotic inhibitor 10.0mL
pH 7.4–7.6 at 25°C

Antibiotic Inhibitor:
Composition per 10mL:
Vancomycin 0.1g
Streptomycin 0.050g
Nystatin 25,000U

Preparation of Antibiotic Inhibitor: Add components to distilled/deionized water and bring volume to 10.0mL. Mix thoroughly. Filter sterilize.

Preparation of Medium: Add components, except bovine serum and antibiotic inhibitor, to distilled/deionized water and bring volume to 940.0mL. Mix thoroughly. Gently heat and bring to boiling. Adjust pH to 7.4–7.6. Autoclave for 15 min at 15 psi pressure–121°C. Cool to 45°–50°C. Aseptically add sterile bovine serum and sterile antibiotic inhibitor. Mix thoroughly. Aseptically distribute into sterile tubes or flasks.

Use: For the maintenance of *Chlamydia* species during transport.

Sucrose Phosphate Transport Medium

Composition per liter:
Sucrose 68.5g
K_2HPO_4 ....2.1g
KH_2PO_4 ....1.1g
Bovine serum 50.0mL
Antibiotic inhibitor 10.0mL
pH 7.0 ± 0.2 at 25°C

Antibiotic Inhibitor:
Composition per 10mL:
Vancomycin 0.1g
Streptomycin 0.050g
Nystatin 25,000U

Preparation of Antibiotic Inhibitor: Add components to distilled/deionized water and bring volume to 10.0mL. Mix thoroughly. Filter sterilize.

Preparation of Medium: Add components, except bovine serum and antibiotic inhibitor, to distilled/deionized water and bring volume to 940.0mL. Mix thoroughly. Gently heat and bring to boiling. Adjust pH to 7.0. Autoclave for 15 min at 15 psi pressure–121°C. Cool to 45°–50°C. Aseptically add sterile bovine serum and sterile antibiotic inhibitor. Mix thoroughly. Aseptically distribute into sterile tubes or flasks.

Use: For the maintenance of *Chlamydia* species during transport.

Sucrose Teepol Tellurite Agar (STT Agar)

Composition per liter:

Agar	20.0g
Beef extract	1.0g
Peptone	1.0g
Sucrose	1.0g
NaCl	0.5g
Bromthymol Blue (0.2% solution)	2.5mL
Tellurite solution	2.5mL
Sodium lauryl sulfate (Teepol—0.1% solution)	0.2mL

pH 8.0 ± 0.2 at 25°C

Tellurite Solution:

Composition per 100mL:

K_2TeO_3	0.05g

Preparation of Tellurite Solution: Add the K_2TeO_3 to distilled/deionized water and bring the volume to 100.0mL. Mix thoroughly. Filter sterilize. Use freshly prepared solution.

Caution: Potassium tellurite is toxic.

Preparation of Medium: Add components to distilled/deionized water and bring volume to 1.0L. Mix thoroughly. Gently heat and bring to boiling. Do not autoclave. Pour into sterile Petri dishes.

Use: For the selective isolation, cultivation and differentiation of *Vibrio* species based on their ability to ferment sucrose. *Vibrio cholerae* appears as flat yellow colonies. *Vibrio parahaemolyticus* appears as levated green-yellow mucoid colonies.

Sulfate API Broth

Composition per liter:

NaCl	10.0g
Sodium lactate	5.2g
Yeast extract	1.0g
$MgSO_4 \cdot 7H_2O$	0.2g
Ascorbic acid	0.1g
$Fe(NH_4)_2(SO_4)_2 \cdot 6H_2O$	0.1g
K_2HPO_4	0.01g

pH 7.5± 0.2 at 25°C

Source: This medium is available as a premixed powder from Difco Laboratories.

Preparation of Medium: Add the components to distilled/deionized water and bring volume to 1.0L. Mix thoroughly until dissolved. Distribute into tubes in 9.0mL volumes. Autoclave for 10 min at 15 psi pressure–121°C.

Use: For the detection, differentiation and estimation of sulfate-reducing bacteria.

Sulfate-Reducing Bacteria Enrichment Medium

Composition per 1018mL:

Solution 1	970.0mL
Solution 4	30.0mL
Solution 6A, 6B, 6C, 6D or 6E	10.0mL
Solution 5	3.0mL
Solution 2	1.0mL
Solution 3	1.0mL
Solution 7	1.0mL
Solution 8	1.0mL
Solution 9	1.0mL

pH 7.2 ± 0.2 at 25°C

Solution 1:

Composition per 970mL:

Na_2SO_4	3.0g
NaCl	1.2g
$MgCl_2 \cdot 6H_2O$	0.4g
KCl	0.3g
NH_4Cl	0.3g
KH_2PO_4	0.2g
$CaCl_2 \cdot 2H_2O$	0.15g

Preparation of Solution 1: Add components to distilled/deionized water and bring volume to 970.0mL. Mix thoroughly. Autoclave for 30 min at 15 psi pressure–121°C. Cool to 25°C under 90% N_2 + 10% CO_2.

Solution 2:

Composition per liter:

$FeCl_2 \cdot 4H_2O$	1.5g
$CoCl_2 \cdot 6H_2O$	0.12g
$MnCl_2 \cdot 4H_2O$	0.1g
$ZnCl_2$	0.07g
H_3BO_3	0.06g
$Na_2MoO_4 \cdot 2H_2O$	0.025g
$NiCl_2 \cdot 6H_2O$	0.025g
$CuCl_2 \cdot 2H_2O$	0.015g
HCl (25% solution)	6.5mL

Preparation of Solution 2: Add components to distilled/deionized water and bring volume to 1.0L. Mix thoroughly. Autoclave for 15 min at 15 psi pressure–121°C. Cool to 25°C.

Solution 3:

Composition per liter:

NaOH	0.5g
Na_2SeO_3	3.0mg

Preparation of Solution 3: Add components to distilled/deionized water and bring volume to 1.0L. Mix thoroughly. Autoclave for 15 min at 15 psi pressure–121°C. Cool to 25°C.

Solution 4:

Composition per 100mL:

$NaHCO_3$ 8.5g

Preparation of Solution 4: Add $NaHCO_3$ to distilled/deionized water and bring volume to 100.0mL. Mix thoroughly. Saturate with 100% CO_2. Filter sterilize. Aseptically add solution to sterile, gas-tight, screw-capped bottles.

Solution 5:

Composition per 100mL:

$Na_2S{\cdot}9H_2O$ 12.0g

Preparation of Solution 5: Add $Na_2S{\cdot}9H_2O$ to distilled/deionized water and bring volume to 100.0mL. Mix thoroughly. Add solution to gas-tight, screw-capped bottles. Gas under 100% N_2 for 20 min. Close caps tightly. Autoclave for 15 min at 15 psi pressure–121°C. Cool to 25°C.

Solution 6A:

Composition per 100mL:

Sodium acetate·$3H_2O$ 20.0g

Preparation of Solution 6A: Add sodium acetate·$3H_2O$ to distilled/deionized water and bring volume to 100.0mL. Autoclave for 15 min at 15 psi pressure–121°C. Cool to 25°C.

Solution 6B:

Composition per 100mL:

n-Butyric acid 8.0g

Preparation of Solution 6B: Add *n*-butyric acid to distilled/deionized water and bring volume to 100.0mL. Adjust pH to 9.0 with NaOH. Autoclave for 15 min at 15 psi pressure–121°C. Cool to 25°C.

Solution 6C:

Composition per 100mL:

Propionic acid 7.0g

Preparation of Solution 6C: Add propionic acid to 100.0mL of distilled/deionized water. Adjust pH to 9.0 with NaOH. Autoclave for 15 min at 15 psi pressure–121°C. Cool to 25°C.

Solution 6D:

Composition per 100mL:

Benzoic acid 5.0g

Preparation of Solution 6D: Add benzoic acid to distilled/deionized water and bring volume to 100.0mL. Adjust pH to 9.0 with NaOH. Autoclave for 15 min at 15 psi pressure–121°C. Cool to 25°C.

Solution 6E:

Composition per 100mL:

n-Palmitic acid 5.0g

NaOH 0.78g

Preparation of Solution 6E: Add *n*-palmitic acid and NaOH to distilled/deionized water and bring volume to 100.0mL. Heat in a water bath until clear. Autoclave for 15 min at 15 psi pressure–121°C. Cool to 25°C.

Solution 7:

Composition per 100mL:

Thiamine 0.01g

p-Aminobenzoic acid 5.0mg

Vitamin B_{12} 5.0mg

Biotin 1.0mg

Preparation of Solution 7: Add components to distilled/deionized water and bring volume to 100.0mL. Mix thoroughly. Filter sterilize.

Solution 8:

Composition per liter:

Succinic acid 0.6g

Isobutyric acid 0.5g

2-Methylbutyric acid 0.5g

3-Methylbutyric acid 0.5g

Valeric acid 0.5g

Caproic acid 0.2g

Preparation of Solution 8: Add components to distilled/deionized water and bring volume to 100.0mL. Mix thoroughly. Adjust pH to 9.0 with NaOH. Autoclave for 15 min at 15 psi pressure–121°C. Cool to 25°C.

Solution 9:

Composition per 100mL:

$Na_2S_2O_4$ 3.0g

Preparation of Solution 9: Add $Na_2S_2O_4$ to 100.0mL of O_2-free distilled/deionized water. Mix thoroughly. Anaerobically filter sterilize.

Preparation Medium: To 970.0mL of cooled, sterile solution 1, aseptically and anaerobically add 1.0mL of sterile solution 2, 1.0mL of sterile solution 3, 30.0mL of sterile solution 4, and 3.0mL of sterile solution 5. Mix thoroughly. Adjust pH to 7.2 with sterile HCl solution or sterile Na_2CO_3 solution. Aseptically and anaerobically distribute into sterile screw-capped bottles in 100.0mL volumes. Add 1.0mL of solution 6A, 6B, 6C, 6D, or 6E to each bottle containing 100.0mL of basal medium. Add 0.1mL of solution 7, 0.1mL of solution 8, and 0.1mL of solution 9 to each bottle containing 100.0mL of basal medium. Mix thoroughly.

Use: For the isolation, cultivation and enrichment of sulfate-reducing bacteria.

Sulfate-Reducing Bacteria Medium with Lactate

Composition per liter:

Solution 1 .... 980.0mL

Solution 2 .... 10.0mL
Solution 3 .... 10.0mL

pH 7.4 ± 0.2 at 25°C

Solution 1:
Composition per 980mL:

Sodium lactate (70% solution) 3.5g
$MgSO_4 \cdot 7H_2O$ 2.0g
NH_4Cl 1.0g
Na_2SO_4 .... 1.0g
Yeast extract 1.0g
K_2HPO_4 .... 0.5g
$CaCl_2 \cdot 2H_2O$ 0.1g

Preparation of Solution 1: Add components to distilled/deionized water and bring volume to 980.0mL. Mix thoroughly. Autoclave for 15 min at 15 psi pressure–121°C. Cool to 50°C.

Solution 2:
Composition per 10mL:

$FeSO_4 \cdot 7H_2O$ 0.5g

Preparation of Solution 2: Add $FeSO_4 \cdot 7H_2O$ to distilled/deionized water and bring volume to 10.0mL. Mix thoroughly. Autoclave for 15 min at 15 psi pressure–121°C. Cool to 50°C.

Solution 3:
Composition per 10mL:

Ascorbic acid 0.1g
Sodium thioglycollate 0.1g

Preparation of Solution 2: Add components to distilled/deionized water and bring volume to 10.0mL. Mix thoroughly. Autoclave for 15 min at 15 psi pressure–121°C. Cool to 50°C.

Preparation of Medium: Aseptically combine 980.0mL of cooled sterile solution 1, 10.0mL of cooled, sterile solution 2, and 10.0mL of cooled, sterile solution 3. Mix thoroughly. Aseptically distribute into sterile tubes or flasks.

Use: For the enrichment and isolation of sulfate-reducing bacteria.

Sulfate–Reducing Medium

Composition per liter:

Sodium lactate 3.5g
$MgSO_4 \cdot 7H_2O$ 2.0g
Peptone 2.0g
Na_2SO_4 .... 1.5g
Beef extract 1.0g
K_2HPO_4 .... 0.5g
$CaCl_2$ 0.10g
$Fe(NH_4)_2(SO_4)_2 \cdot 6H_2O$ solution 10.0mL
Sodium ascorbate solution 10.0mL

pH 7.5 ± 0.3 at 25°C

$Fe(NH_4)_2(SO_4)_2 \cdot 6H_2O$ Solution:
Composition per 100mL:

$Fe(NH_4)_2(SO_4)_2 \cdot 6H_2O$ 3.92g

Preparation of $Fe(NH_4)_2(SO_4)_2 \cdot 6H_2O$ Solution: Add $Fe(NH_4)_2(SO_4)_2 \cdot 6H_2O$ to distilled/deionized water and bring volume to 100.0mL. Mix thoroughly. Filter sterilize. Use freshly prepared medium.

Sodium Ascorbate Solution:
Composition per 100mL:

Sodium ascorbate 0.050g

Preparation of Sodium Ascorbate Solution: Add sodium ascorbate to distilled/deionized water and bring volume to 100.0mL. Mix thoroughly. Filter sterilize. Use freshly prepared medium.

Preparation of Medium: Add components, except $Fe(NH_4)_2(SO_4)_2 \cdot 6H_2O$ solution and sodium ascorbate solution, to distilled/deionized water and bring volume to 980.0mL. Mix thoroughly. Distribute into screw-capped tubes in 10.0mL volumes. Autoclave for 15 min at 15 psi pressure–121°C. Tubes must be filled to capacity after inoculation so prepare extra medium and sterilize in a screw-capped flask or bottle. Prior to inoculation, aseptically add 0.1mL of freshly prepared sterile $Fe(NH_4)_2(SO_4)_2 \cdot 6H_2O$ solution for each 10.0mL of medium in the tubes. Also aseptically add 0.1mL of freshly prepared sterile sodium ascorbate solution for each 10.0mL of medium in the tubes. Inoculate tubes. Fill tubes to capacity with extra sterile medium. Screw caps tight.

Use: For the isolation, cultivation, and enumeration of iron and sulfur bacteria.

Sulfite Agar

Composition per liter:

Agar 20.0g
Pancreatic digest of casein 10.0g
Na_2SO_3 .... 1.0g
Iron nails 66

pH 7.6 ± 0.2 at 25°C

Source: This medium is available as a premixed powder from Difco Laboratories.

Preparation of Medium: Add components to distilled/deionized water and bring volume to 1.0L. Mix thoroughly. Gently heat and bring to boiling. Distribute into screw-capped tubes in 15.0mL volumes. Add a clean iron nail to each tube. Autoclave for 15 min at 15 psi pressure–121°C. Cool to 45°–50°C until ready to inoculate.

Use: For the detection and cultivation of thermophilic anaerobes that can produce H_2S from sulfite. Sulfite reduction appears as a blackening of the medium.

Sulfite Polymyxin Sulfadiazine Agar
See: **SPS Agar**

Sulfolobus Medium

Composition per liter:

$(NH_4)_2SO_4$	1.3g
Yeast extract	1.0g
KH_2PO_4	0.28g
$MgSO_4 \cdot 7H_2O$	0.25g
$CaCl_2 \cdot 2H_2O$	0.07g
$FeCl_3 \cdot 6H_2O$	0.02g
$Na_2B_4O_7 \cdot 10H_2O$	4.5mg
$MnCl_2 \cdot 4H_2O$	1.8mg
$ZnSO_4 \cdot 7H_2O$	0.22mg
$CuCl_2 \cdot 2H_2O$	0.05mg
$Na_2MoO_4 \cdot 2H_2O$	0.03mg
$VOSO_4 \cdot 2H_2O$	0.03mg
$CoSO_4$	0.01mg

pH 2.0 ± 0.2 at 25°C

Preparation of Medium: Add components to distilled/deionized water and bring volume to 1.0L. Mix thoroughly. Adjust pH at 25°C to 2.0 with 10*N* H_2SO_4. Filter sterilize. Aseptically distribute into tubes or flasks.

Use: For the cultivation and maintenance of *Sulfolobus acidocaldarius.*

Sulfolobus Medium, Revised

Composition per liter:

$(NH_4)_2SO_4$	1.3g
Tryptone	1.0g
KH_2PO_4	0.28g
$MgSO_4 \cdot 7H_2O$	0.25g
$CaCl_2 \cdot 2H_2O$	0.07g
Yeast Extract	0.05g
$FeCl_3 \cdot 6H_2O$	0.02g
$Na_2B_4O_7$	4.5mg
$MnCl_2 \cdot 4H_2O$	1.8mg
$ZnSO_4 \cdot 7H_2O$	0.22mg
$CuCl_2 \cdot H_2O$	0.05mg
$Na_2MoO_4 \cdot H_2O$	0.03mg
$VOSO_4 \cdot 2H_2O$	0.03mg
$CoSO_4$	0.01mg

pH 3.0 ± 0.2 at 25°C

Preparation of Medium: Add components to distilled/deionized water and bring volume to 1.0L. Mix thoroughly. Adjust pH at 25°C to 3.0 with 10*N* H_2SO_4. Filter sterilize. Aseptically distribute into tubes or flasks.

Use: For the cultivation and maintenance of *Sulfolobus* species.

Sulfolobus solfataricus Medium

Composition per liter:

KH_2PO_4	3.1g
$(NH_4)_2SO_4$	2.5g
Casamino acids	1.0g
Yeast extract	1.0g
$CaCl_2 \cdot 2H_2O$	0.25g
$MgSO_4 \cdot 7H_2O$	0.2g
$Na_2B_4O_7 \cdot 10H_2O$	4.5mg
$MnCl_2 \cdot 4H_2O$	1.8mg
$ZnSO_4 \cdot 7H_2O$	0.22mg
$CuCl_2 \cdot 2H_2O$	0.05mg
$Na_2MoO_4 \cdot 2H_2O$	0.03mg
$VOSO_4 \cdot 2H_2O$	0.03mg
$CoSO_4 \cdot 7H_2O$	0.01mg

pH 4.0–4.2 at 25°C

Preparation of Medium: Add components to distilled/deionized water and bring volume to 1.0L. Mix thoroughly. Adjust pH at 25°C to 4.0–4.2 with 10*N* H_2SO_4. Filter sterilize. Aseptically distribute into tubes or flasks.

Use: For the cultivation and maintenance of *Sulfolobus solfataricus*.

Sulfur Medium

Composition per liter:

Sulfur, elemental	10.0g
KH_2PO_4	3.0g
$MgSO_4 \cdot 7H_2O$	0.5g
$(NH_4)_2SO_4$	0.3g
$CaCl_2 \cdot 2H_2O$	0.25g
$FeCl_3 \cdot 6H_2O$	0.02g

pH 4.8± 0.2 at 25°C

Preparation of Medium: Add components, except sulfur, to distilled/deionized water and bring volume to 1.0L. Mix thoroughly. Add 1.0g sulfur to each of ten 250.0mL flasks. Add 100.0mL of medium to each flask. Autoclave for 30 min at 0 psi pressure–100°C on 3 consecutive days.

Use: For isolation, cultivation and enumeration of iron and sulfur bacteria.

Superbroth

Composition per liter:

Pancreatic digest of casein	32.0g

Yeast extract 20.0g
NaCl 5.0g
NaOH (1*N* solution) 5.0mL

Preparation of Medium: Add components to distilled/deionized water and bring volume to 1.0L. Mix thoroughly. Distribute into tubes or flasks. Autoclave for 15 min at 15 psi pressure–121°C.

Use: For the cultivation of *Escherichia coli*.

SW 2 Agar

Composition per liter:

Agar 15.0g
NH_4Cl 1.0g
Sodium acetate 0.02g
Artificial seawater 1.0L

Artificial Seawater:

Composition per liter:

NaCl 24.7g
$MgSO_4 \cdot 7H_2O$ 6.3g
$MgCl_2 \cdot 6H_2O$ 4.6g
$CaCl_2$ 1.0g
KCl 0.7g
$NaHCO_3$ 0.2g

Preparation of Artificial Seawater: Add components to distilled/deionized water and bring volume to 1.0L. Mix thoroughly.

Preparation of Medium: Add solid components to 1.0L of artificial seawater. Mix thoroughly. Gently heat and bring to boiling. Distribute into tubes or flasks. Autoclave for 15 min at 15 psi pressure–121°C. Pour into sterile Petri dishes or leave in tubes.

Use: For the isolation and cultivation of *Cytophaga* species, *Herpetosiphon* species, *Saprospira* species, and *Flexithrix* species.

SWA
See: Sea Water Agar

Swampy Medium

Composition per liter:

Agar 10.0g
$CaCO_3$ 10.0g
Peptone 0.5g
Yeast extract 0.5g

Preparation of Medium: Add components to sea water and bring volume to 1.0L. Mix thoroughly. Gently heat and bring to boiling. Distribute into tubes or flasks. Autoclave for 15 min at 15 psi pressure–121°C. Pour into sterile Petri dishes or leave in tubes.

Use: For the cultivation and maintenance of *Vibrio liquefaciens*.

Sweet E Broth for Anaerobes

Composition per 100mL:

Gelatin 0.3g
Cellobiose 0.1g
Fructose 0.1g
Glucose 0.1g
L-Arabinose 0.1g
Maltose 0.1g
Starch 0.1g
Agar 0.075g
Peptone 0.05g
L-Cysteine·HCl·H_2O 0.05g
$(NH_4)_2SO_4$ 0.05g
Yeast extract 0.05g
Salts solution 50.0mL
Rumen fluid 30.0mL
Resazurin solution 0.4mL
Pyruvic acid 0.01mL

pH 6.5 ± 0.2 at 25°C

Salts Solution:

Composition per liter:

$NaHCO_3$ 10.0g
NaCl 2.0g
K_2HPO_4 ....1.0g
KH_2PO_4 ....1.0g
$CaCl_2$, anhydrous 0.2g
$MgSO_4 \cdot 7H_2O$ 0.2g

Preparation of Salts Solution: Add $CaCl_2$ and $MgSO_4 \cdot 7H_2O$ to distilled/deionized water and bring volume to 300.0mL. Mix thoroughly. Bring volume to 800.0mL with distilled/deionized water. Add remaining components while stirring. Bring volume to 1.0L with distilled/deionized water. Mix thoroughly. Store at 4°C.

Resazurin Solution:

Composition per 44mL:

Resazurin 0.011g

Preparation of Resazurin Solution: Add resazurin to distilled/deionized water and bring volume to 44.0mL. Mix thoroughly.

Preparation of Medium: Add components to distilled/deionized water and bring volume to 100.0mL. Mix thoroughly. Gently heat and bring to boiling under O_2-free 97% N_2 + 3% H_2. Adjust the pH to 6.5 if necessary. Continue boiling until the medium turns yellow. Distribute into tubes or flasks under O_2-free 97% N_2 + 3% H_2. Cap tubes with rubber stoppers. Place tubes in a press. Autoclave for 15 min at 15 psi pressure–121°C with fast exhaust.

Use: For the cultivation and maintenance of *Clostridium cocleatum* and *Clostridium spiroforme*.

SWMTY Marine Medium

Composition per liter:

Marine salts mix	38.0g
Agar	15.0g
Pancreatic digest of casein	2.0g
Yeast extract	2.0g
Tris(hydroxymethyl)aminomethane buffer	1.0g
KNO_3	0.5g
Sodium glycerophosphate	0.1g
Trace element solution HO-LE	1.0mL

pH 7.0 ± 0.2 at 25°C

Trace Element Solution HO-LE:

Composition per liter:

H_3BO_3	2.85g
$MnCl_2 \cdot 4H_2O$	1.8g
Sodium tartrate	1.77g
$FeSO_4 \cdot 7H_2O$	1.36g
$CoCl_2 \cdot 6H_2O$	0.04g
$CuCl_2 . 2H_2O$	0.027g
$Na_2MoO_4 \cdot 2H_2O$	0.025g
$ZnCl_2$	0.020g

Preparation of Trace Element Solution HO-LE: Add components to distilled/deionized water and bring volume to 1.0L. Mix thoroughly. Filter sterilize.

Preparation of Medium: Add components to distilled/deionized water and bring volume to 1.0L. Mix thoroughly. Gently heat and bring to boiling. Distribute into tubes or flasks. Autoclave for 15 min at 15 psi pressure–121°C. Pour into sterile Petri dishes or leave in tubes.

Use: For the cultivation and maintenance of a variety heterotrophic marine bacterial species.

SXT Blood Agar

Composition per liter:

Pancreatic digest of casein	14.5g
Agar	14.0g
NaCl	5.0g
Papaic digest of soybean meal	5.0g
Growth factor, BBL	1.5g
Sulfamethoxazole	0.024g
Trimethoprim	1.25mg
Sheep blood, defibrinated	50.0mL

pH 7.3 ± 0.2 at 25°C

Source: This medium is available as a premixed powder from BBL Microbiology Systems.

Preparation of Medium: Add components, except defibrinated sheep blood, to distilled/deionized water and bring volume to 950.0mL. Mix thoroughly. Gently heat and bring to boiling. Autoclave for 15 min at 15 psi pressure–121°C. Cool to 45°–50°C. Aseptically add 50.0mL defibrinated sheep blood. Mix thoroughly. Pour into sterile Petri dishes or distribute into sterile tubes.

Use: For the selective isolation of Lancefield Group A and Group B streptococci from throat cultures and other clinical specimens.

SY Broth

Composition per liter:

$(NH_4)_2SO_4$	2.0g
$Na_2HPO_4 \cdot 2H_2O$	1.4g
KH_2PO_4	0.7g
$MgSO_4 \cdot 7H_2O$	0.2g
$FeSO_4$	5.0mg
$MnSO_4$	5.0mg
Glucose solution	100.0mL

Glucose Solution:

Composition per 100mL:

D-Glucose	10.0g

Preparation of Glucose Solution: Add D-glucose to distilled/deionized water and bring volume to 100.0mL. Mix thoroughly. Autoclave for 15 min at 15 psi pressure–121°C. Cool to 25°C.

Preparation of Medium: Add components, except glucose solution, to distilled/deionized water and bring volume to 900.0mL. Mix thoroughly. Gently heat and bring to boiling. Autoclave for 15 min at 15 psi pressure–121°C. Cool to 45°–50°C. Aseptically add sterile glucose solution. Mix thoroughly. Aseptically distribute into sterile tubes or flasks.

Use: For the isolation and cultivation of *Cytophaga* species, *Herpetosiphon* species, *Saprospira* species, and *Flexithrix* species.

SYA Medium

Composition per liter:

Agar	20.0g
Soluble starch	10.0g
Yeast extract	2.0g

Preparation of Medium: Add components to tap water and bring volume to 1.0L. Mix thoroughly. Gently heat and bring to boiling. Distribute into tubes or flasks. Autoclave for 15 min at 15 psi pressure–121°C. Pour into sterile Petri dishes or leave in tubes.

Use: For the cultivation and maintenance of *Streptomyces chartreusis*.

SYC Medium

Composition per liter:

Sucrose	10.0g
Pancreatic digest of casein	8.0g
Yeast extract	4.0g
K_2HPO_4	3.0g
$MgSO_4 \cdot 7H_2O$	0.3g

pH 7.0 ± 0.2 at 25°C

Preparation of Medium: Add components to distilled/deionized water and bring volume to 1.0L. Mix thoroughly. Distribute into tubes or flasks. Autoclave for 15 min at 15 psi pressure–121°C.

Use: For the cultivation and maintenance of *Agrobacterium tumefaciens*.

Syncase Broth

Composition per liter:

Casamino acids	20.0g
K_2HPO_4	8.71g
Yeast extract	6.0g
NaCl	2.5g

pH 8.5 ± 0.2 at 25°C

Preparation of Medium: Add components to distilled/deionized water and bring volume to 1.0L. Mix thoroughly. Adjust pH to 8.5. Distribute into tubes or flasks. Autoclave for 15 min at 15 psi pressure–121°C.

Use: For the cultivation of heat-labile toxin-producing *Escherichia coli* from foods.

Synthetic Broth, AOAC (Synthetic Broth, Association of Official Analytical Chemists)

Composition per liter:

Na_2HPO_4	4.0g
NaCl	3.0g
K_2HPO_4	1.5g
L-Glutamic acid	1.3g
DL-Valine	1.0g
L-Lysine	0.85g
L-Leucine	0.8g
DL-Serine	0.61g
DL-Threonine	0.5g
L-Aspartic acid	0.45g
DL-Isoleucine	0.44g
DL-Alanine	0.43g
L-Arginine	0.4g
DL-Methionine	0.37g
DL-Histidine	0.3g
DL-Phenylalanine	0.26g
L-Tyrosine	0.21g
KCl	0.2g
Aminoacetic acid	0.06g
L-Cystine	0.05g
$MgSO_4$	0.05g
L-Proline	0.05g
DL-Tryptophan	0.05g
Nicotinamide	0.01g
Thiamine·HCl	0.01g

pH 7.1 ± 0.1 at 25°C

Source: This medium is available in a premixed powder from Difco Laboratories.

Preparation of Medium: Add components to distilled/deionized water and bring volume to 1.0L. Mix thoroughly. Gently heat and bring to boiling. Distribute into tubes or flasks. Autoclave for 20 min at 15 psi pressure–121°C.

Use: For the determination of phenol coefficients of disinfectants.

Synthetic Sea Water Medium

Composition per liter:

NaCl	27.0g
$MgSO_4 \cdot 7H_2O$	7.0g
Monosodium glutamate	5.0g
Tris(hydroxymethyl)aminomethane buffer	2.0g
Glucose	1.0g
KCl	0.6g
$CaCl_2$	0.3g
Sodium glycerophosphate	0.2g
Vitamin B_{12}	1.0μg

pH 7.5 ± 0.2 at 25°C

Preparation of Medium: Add components to distilled/deionized water and bring volume to 1.0L. Mix thoroughly. Adjust pH to 7.5. Distribute into tubes or flasks. Autoclave for 15 min at 15 psi pressure–121°C.

Use: For the cultivation and maintenance of *Leucothrix mucor*.

T_1N_0 Broth (Tryptone Broth)

Composition per liter:

Pancreatic digest of casein 10.0g

pH 7.1 ± 0.2 at 25°C

Preparation of Medium: Add components to distilled/deionized water and bring volume to 1.0L. Mix thoroughly. Gently heat and bring to boiling. Distribute into tubes or flasks. Autoclave for 15 min at 15 psi pressure–121°C.

Use: For the cultivation of *Vibrio cholerae* and other *Vibrio* species.

T_1N_1 Agar (Tryptone Salt Agar)

Composition per liter:

Agar 20.0g
NaCl 10.0g
Pancreatic digest of casein 10.0g

pH 7.1 ± 0.2 at 25°C

Preparation of Medium: Add components to distilled/deionized water and bring volume to 1.0L. Mix thoroughly. Gently heat and bring to boiling. Distribute into tubes or flasks. Autoclave for 15 min at 15 psi pressure–121°C. Pour into sterile Petri dishes or leave in tubes. Allow tubes to cool in a slanted position.

Use: For the cultivation of *Vibrio cholerae* and other *Vibrio* species.

T_1N_2 Agar (Tryptone Salt Agar)

Composition per liter:

Agar 20.0g
NaCl 20.0g
Pancreatic digest of casein 10.0g

pH 7.1 ± 0.2 at 25°C

Preparation of Medium: Add components to distilled/deionized water and bring volume to 1.0L. Mix thoroughly. Gently heat and bring to boiling. Distribute into tubes or flasks. Autoclave for 15 min at 15 psi pressure–121°C. Pour into sterile Petri dishes or leave in tubes. Allow tubes to cool in a slanted position.

Use: For the cultivation of *Vibrio cholerae* and other *Vibrio* species.

T_1N_1 Broth (Tryptone Salt Broth)

Composition per liter:

NaCl 10.0g
Pancreatic digest of casein 10.0g

pH 7.1 ± 0.2 at 25°C

Preparation of Medium: Add components to distilled/deionized water and bring volume to 1.0L. Mix thoroughly. Gently heat and bring to boiling. Distribute into tubes or flasks. Autoclave for 15 min at 15 psi pressure–121°C.

Use: For the cultivation of *Vibrio cholerae* and other *Vibrio* species.

T_1N_3 Broth (Tryptone Salt Broth)

Composition per liter:

NaCl 30.0g
Pancreatic digest of casein 10.0g

pH 7.1 ± 0.2 at 25°C

Preparation of Medium: Add components to distilled/deionized water and bring volume to 1.0L. Mix thoroughly. Gently heat and bring to boiling. Distribute into tubes or flasks. Autoclave for 15 min at 15 psi pressure–121°C.

Use: For the cultivation of *Vibrio cholerae* and other *Vibrio* species.

T_1N_6 Broth (Tryptone Salt Broth)

Composition per liter:

NaCl 60.0g
Pancreatic digest of casein 10.0g

pH 7.1 ± 0.2 at 25°C

Preparation of Medium: Add components to distilled/deionized water and bring volume to 1.0L. Mix thoroughly. Gently heat and bring to boiling. Distribute into tubes or flasks. Autoclave for 15 min at 15 psi pressure–121°C.

Use: For the cultivation of *Vibrio cholerae* and other *Vibrio* species.

T_1N_8 Broth (Tryptone Salt Broth)

Composition per liter:

NaCl 80.0g
Pancreatic digest of casein 10.0g

pH 7.1 ± 0.2 at 25°C

Preparation of Medium: Add components to distilled/deionized water and bring volume to 1.0L. Mix thoroughly. Gently heat and bring to boiling. Distribute into tubes or flasks. Autoclave for 15 min at 15 psi pressure–121°C.

Use: For the cultivation of *Vibrio cholerae* and other *Vibrio* species.

T_1N_{10} Broth (Tryptone Salt Broth)

Composition per liter:

NaCl	100.0g
Pancreatic digest of casein	10.0g

pH 7.1 ± 0.2 at 25°C

Preparation of Medium: Add components to distilled/deionized water and bring volume to 1.0L. Mix thoroughly. Gently heat and bring to boiling. Distribute into tubes or flasks. Autoclave for 15 min at 15 psi pressure–121°C.

Use: For the cultivation of *Vibrio cholerae* and other *Vibrio* species.

T2 Medium for *Thiobacillus*

Composition per liter:

Solution A	250.0mL
Solution B	250.0mL
Solution C	250.0mL
Solution D	250.0mL

pH 7.0 ± 0.2 at 25°C

Solution A:

Composition per 250mL:

$Na_2S_2O_3 \cdot 5H_2O$	5.0g
KNO_3	2.0g
NH_4Cl	1.0g

Preparation of Solution A: Add components to distilled/deionized water and bring volume to 250.0mL. Mix thoroughly. Filter sterilize.

Solution B:

Composition per 250mL

KH_2PO_4	2.0g

Preparation of Solution B: Add KH_2PO_4 to distilled/deionized water and bring volume to 250.0mL. Mix thoroughly. Filter sterilize.

Solution C:

Composition per 250mL

$NaHCO_3$	2.0g

Preparation of Solution C: Add $NaHCO_3$ to distilled/deionized water and bring volume to 250.0mL. Mix thoroughly. Filter sterilize.

Solution D:

Composition per 250mL

$MgSO_4 \cdot 7H_2O$	0.8g
$FeSO_4 \cdot 7H_2O$ (2%, w/v, in *N* HCl)	1.0mL
Trace metal solution	1.0mL

Preparation of Solution D: Add components to distilled/deionized water and bring volume to 250.0mL. Mix thoroughly. Filter sterilize.

$FeSO_4 \cdot 7H_2O$ Solution:

Composition per 100mL

$FeSO_4 \cdot 7H_2O$	2.0g
HCl (1*N* solution)	100.0mL

Preparation of $FeSO_4 \cdot 7H_2O$ Solution: Add the $FeSO_4 \cdot 7H_2O$ to the HCl solution. Mix thoroughly.

Trace Metal Solution:

Composition per liter:

EDTA	50.0g
$ZnSO_4$	22.0g
$CaCl_2$	5.54g
$MnCl_2$	5.06g
$FeSO_4 \cdot 7H_2O$	4.99g
$CoCl_2$	1.61g
$CuSO_4$	1.57g
$(NH_4)_2MoO_4$	1.10g

Preparation of Trace Metal Solution: Add components to distilled/deionized water and bring volume to 1.0L. Mix thoroughly. Adjust pH to 6.0 with KOH.

Preparation of Medium: Aseptically combine the four sterile solutions: solution A, solution B, solution C and solution D. Adjust the pH to 7.0. Aseptically distribute into sterile tubes or flasks.

Use: For the cultivation and maintenance of *Thiobacillus denitrificans* and other thiobacilli.

T3 Agar

See: ***Thiobacillus* A2 Agar**

T3 Broth

See: ***Thiobacillus* A2 Broth**

T7 Agar Base (m–T7 Agar Base)

Composition per liter:

Lactose	20.0g
Agar	15.0g
Polyoxyethylene ether W-1	5.0g
Yeast extract	3.0g
Pancreatic digest of casein	2.5g

Peptic digest of animal tissue............................2.5g
Sodium heptadecyl sulfate 0.1g
Bromthymol Blue...0.1g
Bromcresol Purple 0.1g
pH 7.4 ± 0.2 at 25°C

Source: Available as a premixed powder from BBL Microbiology Systems and Difco Laboratories.

Preparation of Medium: Add components to distilled/deionized water and bring volume to 1.0L. Mix thoroughly. Gently heat while stirring and bring to boiling. Distribute into tubes or flasks. Autoclave for 15 min at 15 psi pressure–121°C. Cool to 45°–50°C. The medium may be made more selective by adding 1.0mg penicillin G per liter. Pour into sterile Petri dishes or leave in tubes.

Use: For the selective recovery and differential identification of injured coliform microorganisms from chlorinated water by the membrane filter method. Also, for rapid estimation of the bacteriological quality of water using the membrane filter method.

Tap Water Agar

Composition per liter:

Agar...15.0g
Tap water.. 1.0L

Preparation of Medium: Add agar to 1.0L of tap water. Mix thoroughly. Gently heat and bring to boiling. Autoclave for 15 min at 15 psi pressure–121°C. Pour into sterile Petri dishes.

Use: For the cultivation and differentiation of fungi and aerobic actinomycetes based on filament and aerial hyphae morphology.

Tarshis Blood Agar

Composition per 1050mL:

Beef heart infusion..500.0g
Agar...15.0g
Meat peptone..10.0g
NaCl 5.0g
Penicillin G, sterile...................................100,000U
Sheep blood, sterile..................................300.0mL
Glycerol..10.0mL
pH 6.6 ± 0.2 at 25°C

Preparation of Medium: Add components, except sheep blood and penicillin G, to distilled/deionized water and bring volume to 750.0mL. Mix thoroughly. Gently heat and bring to boiling. Autoclave for 15 min at 15 psi pressure–121°C. Cool to 45°–50°C. Aseptically add sterile sheep blood and sterile penicillin G. Mix thoroughly. Pour into sterile Petri dishes or distribute into sterile tubes.

Use: For the isolation and cultivation of *Mycobacterium tuberculosis*.

TAT Broth Base (Trypticase Azolectin Tween™ Broth Base)

Composition per liter:

Pancreatic digest of casein..............................20.0g
Lecithin 5.0g
Polysorbate 20 (Tween™ 20) 40.0mL
pH 7.2 ± 0.2 at 25°C

Source: This medium is available as a premixed powder from BBL Microbiology Systems and Difco Laboratories.

Preparation of Medium: Add pancreatic digest of casein and lecithin to distilled/deionized water and bring volume to 960.0mL. Add the Tween™ 20. Mix thoroughly. Gently heat and bring to 48°–50°C for 30 min. Distribute into tubes or flasks. Autoclave for 15 min at 15 psi pressure–121°C.

Use: For the isolation of Gram-negative organisms from topical drugs and cosmetics.

Tatum Motility Test and Maintenance Medium

See: **Motility Test and Maintenance Medium, Tatum**

Taurocholate Tellurite Gelatin Agar

See: **Monsur Agar**

TB Nitrate Reduction Broth

Composition per 100mL:

$Na_2HPO_4 \cdot 12H_2O$ 0.485g
KH_2PO_4 ....0.117g
$NaNO_3$ 0.085g
pH 7.0 ± 0.2 at 25°C

Preparation of Medium: Add components to distilled/deionized water and bring volume to 100.0mL. Mix thoroughly. Distribute into tubes or flasks. Autoclave for 15 min at 15 psi pressure–121°C.

Use: For the differentiation of *Mycobacterium* species based on nitrate reduction. After growth of cells in appropriate medium, nitrate reduction is determined by making a suspension of cells in TB nitrate reduction broth and adding hydrochloric acid, sulfanilamide and *N*-naphylenendiamine. Nitrate reduction turns the medium pink. *M. tuberculosis* reduces nitrate and turns the medium deep pink within 1 min. *M. bovis* does not reduce nitrate and does not change the medium.

TBAB 298 Medium (Tryptose 298 Blood Agar Base Medium)

Tryptose Blood Agar Base:
Composition per 409.6mL:

Agar	15.0g
Tryptose	10.0g
NaCl	5.0g
Beef extract	3.0g
Glucose solution	4.0mL
Thymine solution	2.0mL
D-Alanine solution	2.0mL
Streptomycin sulfate solution	1.6mL

pH 7.2 ± 0.2 at 25°C

Glucose Solution:
Composition per 100mL

Glucose	50.0g

Preparation of Glucose Solution: Add KH_2PO_4 to distilled/deionized water and bring volume to 100.0mL. Mix thoroughly. Filter sterilize.

Thymine Solution:
Composition per 10mL

Thymine	0.1g

Preparation of Thymine Solution: Add Thymine to distilled/deionized water and bring volume to 10.0mL. Mix thoroughly. Filter sterilize.

D-Alanine Solution:
Composition per 10mL

D-Alanine	0.1g

Preparation of Solution B: Add D-Alanine to distilled/deionized water and bring volume to 10.0mL. Mix thoroughly. Filter sterilize.

Streptomycin Sulfate Solution:
Composition per 10mL

Streptomycin sulfate	2.5g

Preparation of Streptomycin Sulfate Solution: Add streptomycin sulfate to distilled/deionized water and bring volume to 10.0mL. Mix thoroughly. Filter sterilize.

Preparation of Medium: Add agar, tryptose, NaCl and beef extract to distilled/deionized water and bring volume to 400.0mL. Mix thoroughly. Autoclave for 15 min at 15 psi pressure–121°C. Cool to 50°C. Aseptically add 4.0mL of the sterile glucose solution, 2.0mL of the sterile thymine solution, 2.0mL of the sterile alanine solution and 1.6mL of the sterile streptomycin sulfate solution. Mix thoroughly. Aseptically distribute into sterile tubes or flasks.

Use: For the cultivation of *Bacillus subtilis*.

TC Amino Acids, HeLa 100X
See: Tissue Culture Amino Acids, HeLa 100X

TC Amino Acids, Minimal Eagle 50X
See: Tissue Culture Amino Acids, Minimal Eagle 50X

TC Dulbecco Solution
See: Tissue Culture Dulbecco Solution

TC Earle Solution
See: Tissue Culture Earle Solution

TC Hanks Solution
See: Tissue Culture Hanks Solution

TC Medium 199
See: Tissue Culture Medium 199

TC Medium Eagle, HeLa
See: Tissue Culture Medium Eagle, HeLa

TC Medium Eagle with Earle BSS
See: Tissue Culture Medium Eagle with Earle Balanced Salt Solution

TC Medium Eagle with Hanks BSS
See: Tissue Culture Medium Eagle with Hanks Balanced Salt Solution

TC Medium Ham F10
See: Tissue Culture Medium Ham F10

TC Medium NCTC 109
See: Tissue Culture Medium NCTC 109

TC Medium RPMI #1640
See: Tissue Culture Medium RPMI #1640

TC Minimal Medium Eagle Spinner Modified MEM–S
See: Tissue Culture Minimal Medium Eagle Spinner Modified

TC Minimal Medium Eagle with Earle BSS
See: Tissue Culture Minimal Medium Eagle with Earle Balanced Salts Solution

TC Tyrode Solution
See: Tissue Culture Tyrode Solution

TC Vitamins Minimal Eagle, 100X
See: Tissue Culture Vitamins Minimal Eagle, 100X

TCBS Agar (Thiosulfate Citrate Bile Salt Sucrose Agar)

Composition per liter:

Sucrose	20.0g
Agar	14.0g
NaCl	10.0g
Sodium citrate	10.0g
$Na_2S_2O_3$	10.0g
Yeast extract	5.0g
Pancreatic digest of casein	5.0g
Peptic digest of animal tissue	5.0g
Oxgall	5.0g
Sodium cholate	3.0g
Ferric citrate	1.0g
Thymol Blue	0.04g
Bromthymol Blue	0.04g

pH 8.6 ± 0.2 at 25°C

Source: This medium is available as a premixed powder from BBL Microbiology Systems and Difco Laboratories.

Preparation of Medium: Add components to distilled/deionized water and bring volume to 1.0L. Mix thoroughly. Gently heat while stirring and bring to boiling. Do not autoclave. Cool to 45°–50°C. Pour into sterile Petri dishes or distribute into sterile tubes.

Use: For the selective isolation of *Vibrio cholerae* and *Vibrio parahaemolyticus* from a variety of clinical and nonclinical specimens.

TCH Medium (Thiophene 2 Carboxylic Acid Hydrazide Medium)

Composition per 1105mL:

Thiophene-2-carboxylic acid hydrazide	1.1mg
Middlebrook 7H10 Agar Base	1.0L
OADC Enrichment	100.0mL
Glycerol	5.0mL

pH 6.6 ± 0.2 at 25°C

Middlebrook 7H10 Agar Base:
Composition per liter:

Agar	15.0g
Na_2HPO_4	1.5g
KH_2PO_4	1.5g
$(NH_4)_2SO_4$	0.5g
L-Glutamic Acid	0.5g
Sodium citrate	0.4g
Ferric ammonium citrate	0.04g
$MgSO_4 \cdot 7H_2O$	0.025g
$ZnSO_4 \cdot 7H_2O$	1.0mg
$CuSO_4 \cdot 5H_2O$	1.0mg
Pyridoxine	1.0mg
Biotin	0.5mg
$CaCl_2 \cdot 2H_2O$	0.5mg
Malachite Green	.0.25mg

Preparation of Middlebrook 7H10 Agar Base: Add glycerol to 900.0mL of distilled/deionized water and add remaining components. Mix thoroughly. Gently heat and bring to boiling.

Middlebrook OADC Enrichment:
Composition per 100mL:

Bovine albumin fraction V	5.0g
Glucose	2.0g

NaCl 0.85g
Oleic acid 0.05g
Catalase 4.0mg

Source: Available as a prepared enrichment from Difco Laboratories.

Preparation of Middlebrook OADC Enrichment: Add components to distilled/deionized water and bring volume to 100.0mL. Mix thoroughly. Filter sterilize.

Preparation for Medium: Combine components. Mix thoroughly. Distribute into tubes or flasks. Autoclave for 15 min at 15 psi pressure–121°C. Pour into sterile Petri dishes or leave in tubes.

Use: For the differentiation of *Mycobacterium* species based on sensitivity to TCH. *Mycobacterium bovis* is inhibited by TCH. *M. tuberculosis* and other mycobacteria are generally resistant to low concentrations of TCH. This distinguishes *M. bovis* from other non-chromogenic, slow-growing mycobacteria.

TCN Medium
See: **Tyrosine Casein Nitrate Medium**

TDC Medium

Composition per liter:

Agar 20.0g
$CaCO_3$ 10.0g
Glucose 5.0g
K_2HPO_4 ..... 1.0g
$MgSO_4$ 1.0g

Preparation of Medium: Add components to tap water and bring volume to 1.0L. Mix thoroughly. Gently heat and bring to boiling. Distribute into tubes or flasks. Autoclave for 15 min at 15 psi pressure–121°C. Pour into sterile Petri dishes or leave in tubes.

Use: For the cultivation and maintenance of *Azotobacter beijerinckii* and other *Azotobacter* species.

TEC Agar (m–TEC Agar)

Composition per liter:

Agar 15.0g
Lactose 10.0g
NaCl 7.5g
Proteose peptone 5.0g
K_2HPO_4 ..... 3.3g
Yeast extract 3.0g
KH_2PO_4 ..... 1.0g
Sodium lauryl sulfate 0.2g
Sodium deoxycholate 0.1g
Bromcresol Purple 0.08g
Bromphenol Red 0.08g
pH 5.0 ± 0.2 at 25°C

Source: Available as a premixed powder from Difco Laboratories.

Preparation of Medium: Add components to distilled/deionized water and bring volume to 1.0L. Mix thoroughly. Gently heat and bring to boiling. Adjust pH to 5.0. Sterilization is unnecessary. Pour into sterile Petri dishes or distribute into sterile tubes or flasks. Store at 2°–8°C. Use within 1 week.

Use: For detection of *Escherichia coli* in recreational waters by the membrane filter method. This agar is used in conjunction with a urea substrate to detect urease production. After addition of the urea substrate, *E. coli* appears as yellow-yellow/brown colonies when viewed under a fluorescent lamp.

Tech Agar

Composition per liter:

Pancreatic digest of gelatin 20.0g
Agar 13.6g
$K_2SO_4 \cdot 7H_2O$ 10.0g
$MgCl_2 \cdot 6H_2O$ 1.4g
Glycerol 10.0mL
pH 7.2 ± 0.2 at 25°C

Source: This medium is available as a premixed powder from BBL Microbiology Systems.

Preparation of Medium: Add components, except glycerol, to distilled/deionized water and bring volume to 990.0mL. Mix thoroughly. Add glycerol. Gently heat and bring to boiling. Distribute into tubes or flasks. Autoclave for 15 min at 15 psi pressure–121°C. Pour into sterile Petri dishes or leave in tubes.

Use: For the production of pyocyanin pigment by *Pseudomonas* species.

Teepol Broth, Enriched (m–Teepol Broth, Enriched)

Composition per liter:

Peptone 40.0g
Lactose 30.0g
Yeast extract 6.0g
Phenol Red 0.2g
Sodium lauryl sulfate
(Teepol—0.1% solution) 4.0mL
pH 7.4 ± 0.2 at 25°C

Preparation of Medium: Add components to distilled/deionized water and bring volume to 1.0L. Mix thoroughly. Distribute into tubes or flasks. Autoclave for 15 min at 15 psi pressure–121°C.

Use: For the enumeration of coliform organisms and *Escherichia coli* in water by the membrane filter method.

Tellurite Blood Agar
See: Chocolate Tellurite Agar

Tellurite Glycine Agar

Composition per liter:

Agar	17.5g
Pancreatic digest of casein	10.0g
Glycine	10.0g
Yeast extract	6.5g
D-Mannitol	5.0g
K_2HPO_4	5.0g
LiCl	5.0g
Enzymatic hydrolysate of soybean meal	3.5g
Chapman tellurite solution	10.0mL

pH 7.2 ± 0.2 at 25°C

Source: This medium is available as a premixed powder from Difco Laboratories.

Chapman Tellurite Solution:
Composition per 100mL:

K_2TeO_3	1.0g

Preparation of Chapman Tellurite Solution: Add K_2TeO_3 to distilled/deionized water and bring volume to 100.0mL. Mix thoroughly. Filter sterilize.

Caution: Potassium tellurite is toxic.

Preparation of Medium: Add components, except Chapman tellurite solution, to distilled/deionized water and bring volume to 990.0mL. Mix thoroughly. Gently heat and bring to boiling. Autoclave for 15 min at 15 psi pressure–121°C. Cool to 50°–55°C. Aseptically add 10.0mL of sterile Chapman tellurite solution. Mix thoroughly. Pour into sterile Petri dishes or distribute into sterile tubes. Allow the surface of the plates to dry before inoculating.

Use: For the isolation and cultivation of of coagulase-positive staphylococci.

Tellurite Glycine Agar

Composition per liter:

Agar	16.0g
Pancreatic digest of casein	10.0g
Glycine	10.0g
Yeast extract	5.0g
D-Mannitol	5.0g
K_2HPO_4	5.0g
LiCl	5.0g
Chapman tellurite solution	20.0mL

pH 7.2 ± 0.2 at 25°C

Source: This medium is available as a premixed powder from BBL Microbiology Systems.

Chapman Tellurite Solution:
Composition per 100mL:

K_2TeO_3	1.0g

Preparation of Chapman Tellurite Solution: Add K_2TeO_3 to distilled/deionized water and bring volume to 100.0mL. Mix thoroughly. Filter sterilize.

Caution: Potassium tellurite is toxic.

Preparation of Medium: Add components, except Chapman tellurite solution, to distilled/deionized water and bring volume to 980.0mL. Mix thoroughly. Gently heat and bring to boiling. Autoclave for 15 min at 15 psi pressure–121°C. Cool to 50°–55°C. Aseptically add 20.0mL of sterile Chapman tellurite solution. Mix thoroughly. Pour into sterile Petri dishes or distribute into sterile tubes. Allow the surface of the plates to dry before inoculating.

Use: For the quantitative detection of coagulase-positive staphylococci from foods and other sources.

Tellurite Polymyxin Egg Yolk Agar
See: TPEY Agar

TEP Uric Acid Medium

Composition per liter:

Agar	20.0g
$Na_2HPO_4 \cdot 12H_2O$	9.0g
Uric acid	4.0g
Pancreatic digest of casein	1.7g
KH_2PO_4	1.5g
NaCl	0.5g
Papaic digest of soybean meal	0.3g
K_2HPO_4	0.25g
Glucose	0.25g
$MgSO_4 \cdot 7H_2O$	0.2g
$CaCl_2$	0.02g
Ferric ammonium citrate	1.2mg
$MnCl_2 \cdot 4H_2O$	1.0mg

pH 7.2 ± 0.2 at 25°C

Preparation of Medium: Add components to distilled/deionized water and bring volume to 1.0L. Mix thoroughly. Gently heat and bring to boiling. Distribute into tubes or flasks. Autoclave for 15 min at 15 psi pressure–121°C. Pour into sterile Petri dishes or leave in tubes.

Use: For the cultivation and maintenance of *Bacillus fastidiosus* and other microorganisms that can utilize uric acid as a carbon source.

Tergitol 7 Agar

Composition per liter:

Lactose	20.0g
Agar	13.0g
Peptone	10.0g
Yeast extract	6.0g
Meat extract	5.0g
Tergitol-7	0.1g
Bromothymol Blue	0.05g
TTC solution	5.0mL

pH 7.2 ± 0.2 at 25°C

Source: This medium is available as a premixed powder from Oxoid Unipath.

TTC Solution:

Composition per 100mL:

Triphenyltetrazolium chloride	0.05g

Preparation of TTC Solution: Add triphenyltetrazolium chloride to distilled/deionized water and bring volume to 100.0mL. Mix thoroughly. Filter sterilize.

Preparation of Medium: Add components to distilled/deionized water and bring volume to 995.0mL. Mix thoroughly. Gently heat and bring to boiling. Autoclave for 15 min at 15 psi pressure–121°C. Cool to 50°C. Aseptically add 5.0mL of sterile TTC solution. Mix thoroughly. Pour into sterile Petri dishes or distribute into sterile tubes.

Use: For the detection and enumeration of coliforms. Lactose-fermenting bacteria appear as yellow colonies. Lactose-nonfermenting bacteria appear as blue colonies.

Tergitol 7 Agar

Composition per liter:

Agar	15.0g
Lactose	10.0g
Yeast extract	3.0g
Pancreatic digest of casein	2.5g
Peptic digest of animal tissue	2.5g
Tergitol 7	0.1g
Bromthymol Blue	25.0mg
TTC solution	3.0mL

pH 6.9 ± 0.2 at 25°C

Source: This medium is available as a premixed powder from BBL Microbiology Systems and Difco Laboratories.

TTC Solution:

Composition per 100mL:

Triphenyl tetrazolium chloride	1.0g

Preparation of TTC Solution: Add triphenyltetrazolium chloride to distilled/deionized water and bring volume to 100.0mL. Mix thoroughly. Filter sterilize.

Preparation of Medium: Add components to distilled/deionized water and bring volume to 997.0mL. Mix thoroughly. Gently heat and bring to boiling. Autoclave for 15 min at 15 psi pressure–121°C. Cool to 50°C. Aseptically add 3.0mL of sterile TTC solution. Mix thoroughly. Pour into sterile Petri dishes or distribute into sterile tubes.

Use: For the selective isolation and differentiation of of coliform bacteria based on lactose fermentation. Lactose-fermenting bacteria appear as yellow colonies. Lactose-nonfermenting bacteria appear as blue colonies.

Tergitol 7 Agar H

Composition per liter:

Agar	15.0g
Lactose	10.0g
Yeast extract	3.0g
Pancreatic digest of casein	2.5g
Peptic digest of animal tissue	2.5g
Ferric ammonium citrate	0.5g
$Na_2S_2O_3$	0.5g
Tergitol 7	0.1g
Bromthymol Blue	0.025g

pH 7.2 ± 0.2

Preparation of Medium: Add components to distilled/deionized water and bring volume to 1.0L. Mix thoroughly. Gently heat and bring to boiling. Distribute into tubes or flasks. Autoclave for 15 min at 15 psi pressure–121°C. Pour into sterile Petri dishes or leave in tubes.

Use: For the selective isolation and differentiation of enteric bacteria from urine.

Tergitol 7 Broth

Composition per liter:

Lactose	10.0g
Yeast extract	3.0g
Pancreatic digest of casein	2.5g
Peptic digest of animal tissue	2.5g
Tergitol 7	0.1g
Bromthymol Blue	25.0mg
TTC solution	3.0mL

pH 6.9 ± 0.2 at 25°C

Source: This medium is available as a premixed powder from BBL Microbiology Systems and Difco Laboratories.

TTC Solution:

Composition per 100mL:

Triphenyl tetrazolium chloride	1.0g

Preparation of TTC Solution: Add tri-phenyltetrazolium chloride to distilled/deionized water and bring volume to 100.0mL. Mix thoroughly. Filter sterilize.

Preparation of Medium: Add components to distilled/deionized water and bring volume to 997.0mL. Mix thoroughly. Gently heat while stirring and bring to boiling. Autoclave for 15 min at 15 psi pressure–121°C. Cool to 25°C. Aseptically add 3.0mL of sterile TTC solution. Mix thoroughly.

Use: For the isolation and cultivation of coliforms, *Salmonella* and other enteric bacteria.

Tetrathionate Broth (FDA M145)

Composition per 1030mL:

Tetrathionate broth base	1.0L
Iodine-potassium iodide solution	20.0mL
Brilliant Green solution	10.0mL

pH 8.4 ± 0.2 at 25°C

Tetrathionate Broth Base:

Composition per liter:

$Na_2S_2O_3 \cdot 5H_2O$	30.0g
$CaCO_3$	10.0g
Polypeptone™	5.0g
Bile salts	1.0g

Preparation of Tetrathionate Broth Base: Add components to distilled/deionized water and bring volume to 1.0L. Mix thoroughly. Gently heat and bring to boiling. A slight precipitate will remain. Do not autoclave. Cool to 25°C. Store at 4°C.

Iodine–Potassium Iodide Solution:

Composition per 20mL:

Iodine, resublimed	6.0g
KI	5.0g

Preparation of Iodine–Potassium Iodide Solution: Add KI to 5.0mL of sterile distilled/deionized water. Mix thoroughly. Add iodine. Mix thoroughly. Bring volume to 20.0mL with sterile distilled/deionized water.

Brilliant Green Solution:

Composition per 100mL:

Brilliant Green	0.1g

Preparation of Brilliant Green Solution: Add Brilliant Green to sterile distilled/deionized water. Mix thoroughly.

Preparation of Medium: Combine 1.0L of tetrathionate broth base, 20.0mL of iodine-potassium iodide solution, and 10.0mL of Brilliant Green solution. Mix thoroughly. Aseptically distribute into tubes in 10.0mL volumes. Do not heat medium after it has been mixed.

Use: For the selective isolation and cultivation of *Salmonella* species from foods.

Tetrathionate Broth

Composition per liter:

$Na_2S_2O_3$	40.7g
$CaCO_3$	25.0g
NaCl	4.5g
Peptone	4.5g
Yeast extract	1.8g
Beef extract	0.9g
Iodine solution	20.0mL

Source: This medium is available as a premixed powder from Oxoid Unipath.

Iodine Solution:

Composition per 20mL:

Iodine	6.0g
KI	5.0g

Preparation of Iodine Solution: Add iodine and KI to distilled/deionized water and bring volume to 20.0mL. Mix thoroughly.

Preparation of Medium: Add components, except iodine solution, to distilled/deionized water and bring volume to 980.0mL. Mix thoroughly. Gently heat and bring to boiling. Do not autoclave. Cool to 40°C. Add 20.0mL of iodine solution. Mix thoroughly. Distribute into tubes in 10.0mL volumes. Use medium the same day it is prepared.

Use: For the selective isolation and enrichment of *Salmonella typhi* and other salmonellae from fecal specimens, sewage and other specimens.

Tetrathionate Broth (TT Broth)

Composition per liter:

$Na_2S_2O_3$	30.0g
$CaCO_3$	10.0g
Proteose peptone	5.0g
Bile salts	1.0g
Iodine solution	20.0mL

pH 8.4± 0.2 at 25°C

Source: This medium is available as a premixed powder from Difco Laboratories and BBL Microbiology Systems.

Iodine Solution:
Composition per 20mL:
Iodine 6.0g
KI 5.0g

Preparation of Iodine Solution: Add iodine and KI to distilled/deionized water and bring volume to 20.0mL. Mix thoroughly.

Preparation of Medium: Add components, except iodine solution, to distilled/deionized water and bring volume to 980.0mL. Mix thoroughly. Gently heat and bring to boiling. Do not autoclave. Cool to 40°C. Add 20.0mL of iodine solution. Mix thoroughly. Distribute into tubes in 10.0mL volumes. Use medium the same day it is prepared.

Use: For the selective isolation and enrichment of *Salmonella typhi* and other salmonellae from infectious material.

Tetrathionate Broth (m–Tetrathionate Broth) (m–TT Broth)

Composition per liter:
$Na_2S_2O_3$ 30.0g
$CaCO_3$ 10.0g
Pancreatic digest of casein 2.5g
Peptic digest of animal tissue 2.5g
Iodine-iodide solution 20.0mL
pH 8.0 ± 0.2 at 25°C

Iodine-Iodide Solution:
Composition per 20mL:
Iodine 6.0g
KI 5.0g

Preparation of Iodine-Iodide Solution: Add iodine and KI to distilled/deionized water and bring volume to 20.0mL. Mix thoroughly.

Preparation of Medium: Add components, except iodine-iodide solution, to distilled/deionized water and bring volume to 980.0mL. Mix thoroughly. Gently heat and bring to boiling. Do not autoclave. Cool to 40°C. Add 20.0mL of iodine-iodide solution. Mix thoroughly. Distribute into tubes in 10.0mL volumes. Use medium the same day it is prepared.

Use: For the selective isolation in the membrane filter method of *Salmonella* species from feces, urine, foods and other specimens of sanitary importance.

Tetrathionate Broth (m–Tetrathionate Broth)

Composition per liter:
$Na_2S_2O_3$ 30.0g
Proteose peptone 5.0g
Bile salts 1.0g
Iodine solution 20.0mL
pH 8.0 ± 0.2 at 25°C

Source: This medium is available as a premixed powder from Difco Laboratories.

Iodine Solution:
Composition per 20mL:
Iodine 6.0g
KI 5.0g

Preparation of Iodine Solution: Add iodine and KI to distilled/deionized water and bring volume to 20.0mL. Mix thoroughly.

Preparation of Medium: Add components, except iodine solution, to distilled/deionized water and bring volume to 980.0mL. Mix thoroughly. Gently heat and bring to boiling. Do not autoclave. Cool to 40°C. Add 20.0mL of iodine solution. Mix thoroughly. Use medium the same day it is prepared.

Use: For the enrichment of *Salmonella* species in the membrane filter method prior to placing the filter on selective media such as Brilliant Green broth.

Tetrathionate Broth, Hajna (TT Broth, Hajna)

Composition per liter:
$Na_2S_2O_3$ 38.0g
$CaCO_3$ 25.0g
Casein/meat peptone (50/50) 18.0g
NaCl 5.0g
D–Mannitol 2.5g
Yeast extract 2.0g
Glucose 0.5g
Sodium deoxycholate 0.5g
Brilliant Green 0.01g
Iodine solution 40.0mL
pH 7.5–7.8 at 25°C

Source: This medium is available as a premixed powder from Difco Laboratories.

Iodine Solution:
Composition per 40mL:
KI 8.0g
Iodine 5.0g

Preparation of Iodine Solution: Add iodine and KI to distilled/deionized water and bring volume to 40.0mL. Mix thoroughly.

Preparation of Medium: Add components, except iodine solution, to distilled/deionized water and bring volume to 960.0mL. Mix thoroughly. Gently heat and bring to boiling. Do not autoclave. Cool to 40°C. Add 40.0mL of iodine solution. Mix thoroughly. Distribute into tubes in 10.0mL volumes. Use medium the same day it is prepared.

Use: For the isolation of *Salmonella* species, except *Salmonella typhi*, and *Arizona* species from fecal specimens, urine, food samples and other specimens of sanitary significance.

Tetrathionate Broth, USA (TT Broth, USA)

$Na_2S_2O_3$	30.0g
$CaCO_3$	10.0g
Casein peptone	2.5g
Meat peptone	2.5g
Bile salts	1.0g

Source: This medium is available as a premixed powder from Oxoid Unipath.

Iodine-Iodide Solution:
Composition per 20mL:

Iodine	6.0g
KI	5.0g

Preparation of Iodine-Iodide Solution: Add iodine and KI to distilled/deionized water and bring volume to 20.0mL. Mix thoroughly.

Preparation of Medium: Add components, except iodine solution, to distilled/deionized water and bring volume to 980.0mL. Mix thoroughly. Gently heat and bring to boiling. Do not autoclave. Cool to 40°C. Add 20.0mL of iodine solution. Mix thoroughly. Distribute into tubes in 10.0mL volumes. Use medium the same day it is prepared.

Use: For the selective enrichment of *Salmonella* species from feces, urine, foods and other specimens of sanitary importance.

Tetrathionate Broth with Novobiocin

Composition per liter:

$Na_2S_2O_3$	38.0g
$CaCO_3$	25.0g
Casein/meat peptone (50/50)	18.0g
NaCl	5.0g
Yeast extract	2.0g
D–Mannitol	0.5g
Glucose	0.5g
Sodium deoxycholate	0.5g
Brilliant Green	0.01g
Novobiocin	4.0mg
Iodine solution	40.0mL

pH 7.5–7.8 at 25°C

Iodine Solution:
Composition per 40mL:

KI	8.0g
Iodine	5.0g

Preparation of Iodine Solution: Add iodine and KI to distilled/deionized water and bring volume to 20.0mL. Mix thoroughly.

Preparation of Medium: Add components, except iodine solution, to distilled/deionized water and bring volume to 960.0mL. Mix thoroughly. Gently heat and bring to boiling. Do not autoclave. Cool to 40°C. Add 40.0mL of iodine solution. Mix thoroughly. Distribute into tubes in 10.0mL volumes. Use medium the same day it is prepared.

Use: For the isolation of *Salmonella* species, except *Salmonella typhi*, and *Arizona* species from fecal specimens, urine, food and other specimens of sanitary importance. Novobiocin suppresses the growth of *Proteus* species.

Tetrathionate Crystal Violet Enhancement Broth

Composition per liter:

Potassium tetrathionate	20.0g
Casein/meat peptone (50/50)	8.6g
NaCl	6.4g
Crystal Violet	0.005g

pH 6.5 ± 0.2 at 25°C

Preparation of Medium: Add components to distilled/deionized water and bring volume to 1.0L. Mix thoroughly. Distribute into tubes or flasks. Autoclave for 15 min at 15 psi pressure–121°C.

Use:For the isolation of *Salmonella* (except *Salmonella typhi*) and *Arizona* from fecal specimens, urine, food samples and other specimens of sanitary significance.

Tetrathionate Reductase Medium

Composition per tube:

Solution I	10.0mL
Solution III	0.2mL
Solution II	0.1mL
Solution IV	0.1mL

Solution I:
Composition per liter:

$Na_2HPO_4 \cdot 12H_2O$	3.6g
KH_2PO_4	1.0g
NH_4Cl	0.5g
Peptone	0.25g
Yeast Extract	0.25g
$MgSO_4 \cdot 7H_2O$	0.03g

Preparation of Solution I: Add components to distilled/deionized water and bring volume to 1.0L. Mix thoroughly. Gently heat and bring to boiling. Distribute into tubes in 10.0mL volumes. Autoclave for 15 min at 15 psi pressure–121°C. Cool to 25°C.

Solution II:

Composition per 100mL:

$CaCl_2 \cdot 2H_2O$	0.1g
Ferric ammonium citrate	0.05g

Preparation of Solution II: Add components to distilled/deionized water and bring volume to 100.0mL. Mix thoroughly. Gently heat and bring to boiling. Autoclave for 15 min at 15 psi pressure–121°C. Cool to 25°C.

Solution III:

Composition per 100mL:

Sodium succinate	15.0g

Preparation of Solution III: Add sodium succinate to distilled/deionized water and bring volume to 100.0mL. Mix thoroughly. Gently heat until dissolved. Autoclave for 15 min at 15 psi pressure–121°C. Cool to 25°C.

Solution IV

Composition per 100mL:

$Na_2S_4O_6 \cdot 2H_2O$	10.0g

Preparation of Solution IV: Add $Na_2S_4O_6 \cdot 2H_2O$ to distilled/deionized water and bring volume to 100.0mL. Mix thoroughly. Sterilize by filtration. Store at 4°C.

Preparation of Medium: To each tube containing 10.0mL of sterile solution I, aseptically add 0.1mL of sterile solution II, 0.2mL of sterile solution III, and 0.1mL of sterile solution IV. Mix thoroughly. Use immediately.

Use: For the cultivation and differentiation of hydrogen-oxidizing bacteria based on their production of tetrathionate reductase.

Tetrathionate Reductase Test Medium

Composition per 1025mL:

$K_2S_4O_6$	5.0g
Peptone water	1.0L
Bromthymol Blue (0.2% solution)	25.0mL

pH 7.4 ± 0.2 at 25°C

Peptone Water:

Composition per liter:

Peptone	10.0g
NaCl	5.0g

Preparation of Peptone Water: Add components to distilled/deionized water and bring volume to 1.0L. Mix thoroughly.

Preparation of Medium: Combine components. Mix thoroughly. Adjust pH to 7.4. Filter sterilize. Dispense into tubes in 1.0mL volumes or into wells of sterile microculture plates for replica inoculation.

Use: For the cultivation and identification of *Serratia* species based on their ability to reduce tetrathionate. Bacteria that reduce tetrathionate turn the medium yellow.

Tetrazolium Thallium Glucose Agar

Composition per liter:

Agar	14.0g
Beef extract	10.0g
Peptone	10.0g
Glucose solution	100.0mL
2,3,5-Triphenyltetrazolium·HCl solution	10.0mL
Thallous acetate solution	10.0mL

2,3,5-Triphenyltetrazolium·HCl Solution:

Composition per 10mL:

2,3,5-Triphenyltetrazolium·HCl	0.1g

Preparation of 2,3,5-Triphenyltetrazolium-HCl Solution: Add 2,3,5-triphenyltetrazolium·HCl to distilled/deionized water and bring volume to 10.0mL. Mix thoroughly. Autoclave for 7 min at 15 psi pressure–121°C.

Glucose Solution:

Composition per 100mL:

Glucose	10.0g

Preparation of Glucose Solution: Add glucose to distilled/deionized water and bring volume to 100.0mL. Mix thoroughly. Filter sterilize.

Thallous Acetate Solution:

Composition per 10mL:

Thallous acetate	1.0g

Preparation of Thallous Acetate Solution: Add thallous acetate to distilled/deionized water and bring volume to 10.0mL. Mix thoroughly. Filter sterilize.

Preparation of Medium: Add components—except glucose solution, 2,3,5-triphenyltetrazolium·HCl solution, and thallous acetate solution—to distilled/deionized water and bring volume to 880.0mL. Mix thoroughly. Gently heat and bring to boiling. Autoclave for 15 min at 15 psi pressure–121°C. Cool to 45°–50°C. Aseptically add sterile glucose solution, 2,3,5-triphenyltetrazolium·HCl solution, and thallous acetate solution. Mix thoroughly. Pour into sterile Petri dishes or distribute into sterile tubes.

Use: For the cultivation of *Streptococcus* species.

Tetrazolium Tolerance Agar (TTC Agar)

Composition per liter:

Pancreatic digest of casein	15.0g
Agar	15.0g
Triphenyltetrazolium chloride	10.0g
Papaic digest of soybean meal	5.0g
NaCl	5.0g

pH 7.3 ± 0.2 at 25°C

Preparation of Medium: Add components to distilled/deionized water and bring volume to 1.0L. Mix thoroughly. Gently heat and bring to boiling. Distribute into tubes or flasks. Autoclave for 15 min at 15 psi pressure–121°C. Do not overheat. Pour into sterile Petri dishes or leave in tubes.

Use: For the differentiation of bacteria based upon the ability to tolerate and grow in the presence of tetrazolium. *Streptococcus faecalis* (enterococci) rapidly reduce tetrazolium.

TGE Broth

Composition per liter:

Pancreatic digest of casein	10.0g
Beef extract	6.0g
Glucose	2.0g

pH 7.0 ± 0.2 at 25°C

Source: This medium is available as a premixed powder from Difco Laboratories.

Preparation of Medium: Add components to distilled/deionized water and bring volume to 1.0L. Mix thoroughly. Distribute into tubes or flasks. Autoclave for 15 min at 15 psi pressure–121°C.

Use: For the enumeration of bacteria by the membrane filter method.

TGY Medium (Tryptone Glucose Yeast Extract Medium)

Composition per liter of tap water:

Agar	20.0g
Pancreatic digest of casein	5.0g
Yeast extract	5.0g
Glucose	1.0g
K_2HPO_4	1.0g

pH 7.0 ± 0.2 at 25°C

Preparation of Medium: Add components to distilled/deionized water and bring volume to 1.0L. Mix thoroughly. Gently heat and bring to boiling. Distribute into tubes or flasks. Autoclave for 15 min at 15 psi pressure–121°C. Pour into sterile Petri dishes or leave in tubes.

Use: For the cultivation and maintenance of a variety of bacteria including *Bacillus* species, *Corynebacterium* species, *Enterococcus* species and *Pseudomonas* species.

TGYM Medium (Tryptone Glucose Yeast Extract Methionine Medium)

Composition per liter:

Pancreatic digest of casein	5.0g
Yeast extract	3.0g
Glucose	1.0g
DL-Methionine	0.5g

Preparation of Medium: Add components to distilled/deionized water and bring volume to 1.0L. Mix thoroughly. Distribute into tubes or flasks. Autoclave for 15 min at 15 psi pressure–121°C.

Use: For the cultivation and maintenance of *Deinococcus* species.

Thayer–Martin Agar, Modified (MTM II) (Modified Thayer–Martin Agar)

Composition per liter:

Agar	12.0g
Hemoglobin	10.0g
Pancreatic digest of casein	7.5g
Selected meat peptone	7.5g
NaCl	5.0g
K_2HPO_4	4.0g
Cornstarch	1.0g
KH_2PO_4	1.0g
C-N-V-T inhibitor	10.0mL
Supplement solution	10.0mL

pH 7.2 ± 0.2 at 25°C

CNVT Inhibitor:

Composition per 10mL:

Colistin sulfate	7.5mg
Trimethoprim lactate	5.0mg
Vancomycin	3.0mg
Nystatin	12,500U

Preparation of CNVT Inhibitor: Add components to distilled/deionized water and bring volume to 10.0mL. Mix thoroughly. Filter sterilize.

Supplement Solution:

Composition per liter:

Glucose	100.0g
L-Cysteine·HCl	25.9g
L-Glutamine	10.0g
L-Cystine	1.1g
Adenine	1.0g
Nicotinamide adenine dinucleotide	0.25g
Vitamin B_{12}	0.1g
Thiamine pyrophosphate	0.1g
Guanine·HCl	0.03g
$Fe(NO_3)_3 \cdot 6H_2O$	0.02g
p-Aminobenzoic acid	0.013g
Thiamine·HCl	3.0mg

Source: The supplement solution IsoVitaleX® enrichment is available from BBL Microbiology Laboratories. This enrichment may be replaced by supplement VX from Difco Laboratories.

Preparation of Supplement Solution: Add components to distilled/deionized water and bring volume to 1.0L. Mix thoroughly. Filter sterilize.

Preparation of Medium: Add components, except C-N-V-T inhibitor and supplement solution, to distilled/deionized water and bring volume to 990.0mL. Mix thoroughly. Gently heat and bring to boiling. Distribute into tubes or flasks. Autoclave for 15 min at 15 psi pressure–121°C. Cool to 45°–50°C. Aseptically add 10.0mL of sterile C-N-V-T inhibitor and 10.0mL of sterile Isosupplement solution. Mix thoroughly. Pour into sterile Petri dishes or distribute into sterile tubes.

Use: For the isolation of *Neisseria* species from specimens containing mixed flora of bacteria and fungi.

Thayer–Martin Medium

Composition per liter:

GC agar base	740.0mL
Hemoglobin solution	250.0mL
Vitox supplement	10.0mL

pH 7.3 ± 0.2 at 25°C

GC Agar Base:

Composition per 740mL:

Special peptone	15.0g
Agar	10.0g
NaCl	5.0g
K_2HPO_4	4.0g
Cornstarch	1.0g
KH_2PO_4	1.0g

pH 7.2 ± 0.2 at 25°C

Preparation of GC Agar Base: Add components of GC medium base and the hemoglobin to distilled/deionized water and bring volume to 740.0mL. Mix thoroughly. Gently heat until boiling. Autoclave for 15 min at 15 psi pressure–121°C. Cool to 45°–50°C.

Hemoglobin Solution:

Composition per 250mL:

Hemoglobin	5.0g

Preparation of Hemoglobin Solution: Add hemoglobin to distilled/deionized water and bring volume to 250.0mL. Mix thoroughly. Autoclave for 15 min at 15 psi pressure–121°C. Cool to 45°–50°C.

Vitox Supplement:

Composition per 10.0mL:

Glucose	2.0g
L-Cysteine·HCl	0.518g
L-Glutamine	0.2g
L-Cystine	0.022g
Adenine sulfate	0.010g
Nicotinamide adenine dinucleotide	5.0mg
Cocarboxylase	2.0mg
Guanine·HCl	0.6mg
$Fe(NO_3)_3 \cdot 6H_2O$	0.4mg
p-Aminobenzoic acid	0.26mg
Vitamin B_{12}	0.2mg
Thiamine·HCl	0.06mg

Preparation of Vitox Supplement: Add components to distilled/deionized water and bring volume to 10.0mL. Mix thoroughly. Filter sterilize.

Preparation of Medium: To 740.0mL of cooled sterile GC agar base, aseptically add 250.0mL of sterile hemoglobin solution and 10.0mL of sterile Vitox supplement. Mix thoroughly. Pour into sterile Petri dishes or distribute into sterile tubes.

Use: For the isolation and cultivation of fastidious microorganisms, especially *Neisseria* species.

Thayer–Martin Medium

Composition per liter:

Hemoglobin	10.0g
GC medium base	980.0mL
CNVT inhibitor	10.0mL
Supplement B (Difco)	10.0mL

pH 7.3 ± 0.2 at 25°C

Source: Available as a prepared medium in tubes from Difco Laboratories.

GC Medium Base:

Composition per 980mL:

Proteose Peptone No. 3	15.0g
Agar	10.0g
NaCl	5.0g
K_2HPO_4	4.0g

Cornstarch 1.0g
KH_2PO_4 ..... 1.0g
pH 7.2 ± 0.2 at 25°C

Preparation of GC Medium Base: Add components of GC medium base and the hemoglobin to distilled/deionized water and bring volume to 1.0L. Mix thoroughly. Gently heat until boiling. Autoclave for 15 min at 15 psi pressure–121°C. Cool to 45°–50°C.

CNVT Inhibitor:
Composition per 10mL:

Colistin sulfate	7.5mg
Trimethoprim lactate	5.0mg
Vancomycin	3.0mg
Nystatin	12,500U

Preparation of CNVT Inhibitor: Add components to distilled/deionized water and bring volume to 10.0mL. Mix thoroughly. Filter sterilize.

Preparation of Medium: To 980.0mL of cooled sterile GC medium base, aseptically add 10.0mL of sterile CNVT inhibitor and 10.0mL of sterile supplement B. Mix thoroughly. Pour into sterile Petri dishes or distribute into sterile tubes.

Use: For the isolation and cultivation of fastidious microorganisms, especially *Neisseria* species.

Thayer–Martin Medium, Modified (Modified Thayer–Martin Agar)

Composition per liter:

GC agar base	720.0mL
Hemoglobin solution	250.0mL
GC supplement	30.0mL

pH 7.3 ± 0.2 at 25°C

GC Agar Base:
Composition per 720mL:

Special peptone	15.0g
Agar	10.0g
NaCl	5.0g
K_2HPO_4	4.0g
Cornstarch	1.0g
KH_2PO_4	1.0g

pH 7.2 ± 0.2 at 25°C

Preparation of GC Agar Base: Add components of GC medium base and the hemoglobin to distilled/deionized water and bring volume to 720.0mL. Mix thoroughly. Gently heat until boiling. Autoclave for 15 min at 15 psi pressure–121°C. Cool to 45°–50°C.

Hemoglobin Solution:
Composition per 250mL:
Hemoglobin 5.0g

Preparation of Hemoglobin Solution: Add hemoglobin to distilled/deionized water and bring volume to 250.0mL. Mix thoroughly. Autoclave for 15 min at 15 psi pressure–121°C. Cool to 45°–50°C.

GC Supplement:
Composition per 30.0mL:

Yeast autolysate	10.0g
Glucose	1.5g
$NaHCO_3$	0.15g
Colistin sulfate	7.5mg
Trimethoprim lactate	5.0mg
Vancomycin	3.0mg
Nystatin	12,500U

Preparation of GC Supplement: Add components to distilled/deionized water and bring volume to 30.0mL. Mix thoroughly. Filter sterilize.

Preparation of Medium: To 720.0mL of cooled sterile GC agar base, aseptically add 250.0mL of sterile hemoglobin solution, and 30.0mL of sterile GC supplement. Mix thoroughly. Pour into sterile Petri dishes or distribute into sterile tubes.

Use: For the selective isolation and cultivation of fastidious microorganisms, especially *Neisseria* species.

Thayer–Martin Medium, Modified (Modified Thayer–Martin Agar)

Composition per liter:

GC agar base	730.0mL
Hemoglobin solution	250.0mL
Vitox supplement	10.0mL
VCNT antibiotic solution	10.0mL

pH 7.3 ± 0.2 at 25°C

GC Agar Base:
Composition per 730mL:

Special peptone	15.0g
Agar	10.0g
NaCl	5.0g
K_2HPO_4	4.0g
Cornstarch	1.0g
KH_2PO_4	1.0g

pH 7.2 ± 0.2 at 25°C

Preparation of GC Agar Base: Add components of GC medium base and the hemoglobin to distilled/deionized water and bring volume to 730.0mL. Mix thoroughly. Gently heat until boiling. Autoclave for 15 min at 15 psi pressure–121°C. Cool to 45°–50°C.

Hemoglobin Solution:
Composition per 250mL:
Hemoglobin 5.0g

Preparation of Hemoglobin Solution: Add hemoglobin to distilled/deionized water and bring volume to 250.0mL. Mix thoroughly. Autoclave for 15 min at 15 psi pressure–121°C. Cool to 45°–50°C.

Vitox Supplement:

Composition per 10.0mL:

Glucose	2.0g
L-Cysteine·HCl	0.518g
L-Glutamine	0.2g
L-Cystine	0.022g
Adenine sulfate	0.010g
Nicotinamide adenine dinucleotide	5.0mg
Cocarboxylase	2.0mg
Guanine·HCl	0.6mg
$Fe(NO_3)_3 \cdot 6H_2O$	0.4mg
p-Aminobenzoic acid	0.26mg
Vitamin B_{12}	0.2mg
Thiamine·HCl	0.06mg

Preparation of Vitox Supplement: Add components to distilled/deionized water and bring volume to 10.0mL. Mix thoroughly. Filter sterilize.

VCNT Antibiotic Solution:

Composition per 10mL:

Colistin methane sulfonate	7.5mg
Trimethoprim lactate	5.0mg
Vancomycin	3.0mg
Nystatin	12,500U

Preparation of VCNT Antibiotic Solution: Add components to distilled/deionized water and bring volume to 10.0mL. Mix thoroughly. Filter sterilize.

Preparation of Medium: To 730.0mL of cooled sterile GC agar base, aseptically add 250.0mL of sterile hemoglobin solution , 10.0mL of sterile Vitox supplement, and 10.0mL of VCNT antibiotic solution. Mix thoroughly. Pour into sterile Petri dishes or distribute into sterile tubes.

Use: For the selective isolation and cultivation of fastidious microorganisms, especially *Neisseria* species.

Thayer–Martin Medium, Selective

Composition per liter:

GC agar base	730.0mL
Hemoglobin solution	250.0mL
Vitox supplement	10.0mL
VCN antibiotic solution	10.0mL

pH 7.3 ± 0.2 at 25°C

GC Agar Base:

Composition per 730mL:

Special peptone	15.0g
Agar	10.0g
NaCl	5.0g
K_2HPO_4	4.0g
Cornstarch	1.0g
KH_2PO_4	1.0g

pH 7.2 ± 0.2 at 25°C

Preparation of GC Agar Base: Add components of GC medium base and the hemoglobin to distilled/deionized water and bring volume to 730.0mL. Mix thoroughly. Gently heat until boiling. Autoclave for 15 min at 15 psi pressure–121°C. Cool to 45°–50°C.

Hemoglobin Solution:

Composition per 250mL:

Hemoglobin	5.0g

Preparation of Hemoglobin Solution: Add hemoglobin to distilled/deionized water and bring volume to 250.0mL. Mix thoroughly. Autoclave for 15 min at 15 psi pressure–121°C. Cool to 45°–50°C.

Vitox Supplement:

Composition per 10.0mL:

Glucose	2.0g
L-Cysteine·HCl	0.518g
L-Glutamine	0.2g
L-Cystine	0.022g
Adenine sulfate	0.010g
Nicotinamide adenine dinucleotide	5.0mg
Cocarboxylase	2.0mg
Guanine·HCl	0.6mg
$Fe(NO_3)_3 \cdot 6H_2O$	0.4mg
p-Aminobenzoic acid	0.26mg
Vitamin B_{12}	0.2mg
Thiamine·HCl	0.06mg

Preparation of Vitox Supplement: Add components to distilled/deionized water and bring volume to 10.0mL. Mix thoroughly. Filter sterilize.

VCN Antibiotic Solution:

Composition per 10mL:

Colistin methane sulfonate	7.5mg
Vancomycin	3.0mg
Nystatin	12,500U

Preparation of VCN Antibiotic Solution: Add components to distilled/deionized water and bring volume to 10.0mL. Mix thoroughly. Filter sterilize.

Preparation of Medium: To 730.0mL of cooled sterile GC agar base, aseptically add 250.0mL of sterile hemoglobin solution , 10.0mL of sterile Vitox supplement, and 10.0mL of VCN antibiotic solution. Mix thoroughly. Pour into sterile Petri dishes or distribute into sterile tubes.

Use: For the selective isolation and cultivation of fastidious microorganisms, especially *Neisseria* species.

Thayer–Martin Selective Agar

Composition per liter:

Agar	12.0g
Hemoglobin	10.0g
Pancreatic digest of casein	7.5g
Selected meat peptone	7.5g
NaCl	5.0g
K_2HPO_4	4.0g
Cornstarch	1.0g
KH_2PO_4	1.0g
Supplement solution	10.0mL
V-C-N inhibitor	10.0mL

pH 7.2 ± 0.2 at 25°C

Source: This medium is available as a premixed powder from BBL Microbiology Systems.

Supplement Solution:
Composition per liter:

Glucose	100.0g
L-Cysteine·HCl	25.9g
L-Glutamine	10.0g
L-Cystine	1.1g
Adenine	1.0g
Nicotinamide adenine dinucleotide	0.25g
Vitamin B_{12}	0.1g
Thiamine pyrophosphate	0.1g
Guanine·HCl	0.03g
$Fe(NO_3)_3 \cdot 6H_2O$	0.02g
p-Aminobenzoic acid	0.013g
Thiamine·HCl	3.0mg

Source: The supplement solution IsoVitaleX® enrichment is available from BBL Microbiology Laboratories. This enrichment may be replaced by supplement VX from Difco Laboratories.

Preparation of Supplement Solution: Add components to distilled/deionized water and bring volume to 1.0L. Mix thoroughly. Filter sterilize.

VCN Inhibitor:
Composition per 10mL:

Colistin	7.5mg
Vancomycin	3.0mg
Nystatin	12,500U

Preparation of VCN Inhibitor: Add components to distilled/deionized water and bring volume to 10.0mL. Mix thoroughly. Filter sterilize.

Preparation of Medium: Add components, except supplement solution and V-C-N inhibitor, to distilled/deionized water and bring volume to 980.0mL. Mix thoroughly. Gently heat and bring to boiling. Autoclave for 15 min at 15 psi pressure–121°C. Cool to 45°–50°C. Aseptically add sterile V-C-N inhibitor and sterile supplement solution. Mix thoroughly. Pour into sterile Petri dishes or distribute into sterile tubes.

Use: For the selective isolation of *Neisseria gonorrhoeae* and *Neisseria meningitidis* from specimens containing mixed flora of bacteria and fungi.

Thermoacidurans Agar

Composition per liter:

Agar	20.0g
Yeast extract	5.0g
Proteose peptone	5.0g
Glucose	5.0g
K_2HPO_4	4.0g

pH 5.0 ± 0.2 at 25°C

Source: This medium is available as a premixed powder from Difco Laboratories.

Preparation of Medium: Add components to distilled/deionized water and bring volume to 1.0L. Mix thoroughly. Gently heat and bring to boiling. Distribute into tubes or flasks. Autoclave for 15 min at 15 psi pressure–121°C. Do not overheat. Pour into sterile Petri dishes or leave in tubes.

Use: For the isolation and cultivation of *Bacillus thermoacidurans* from food products.

Thermoactinomyces dichotomicus Medium

Composition per liter:

Maize, split	50.0g
Agar	20.0g
Starch	10.0g
NaCl	5.0g
Peptone	5.0g
$CaCl_2$	0.5g

pH 7.2 ± 0.2 at 25°C

Preparation of Medium: Add split maize (crushed corn) to 1.0L of boiling water. Steam for 30 min. Filter through Whatman #1 filter paper. Add remaining components to maize filtrate. Gently heat and bring to boiling. Distribute into tubes or flasks. Autoclave for 15 min at 15 psi pressure–121°C. Pour into sterile Petri dishes or leave in tubes.

Use: For the cultivation of *Thermoactinomyces dichotomicus*.

Thermoactinomyces Medium

Composition per liter:

Agar	20.0g
Malt extract	10.0g
Yeast extract	4.0g
Glucose	4.0g

pH 7.3 ± 0.2 at 25°C

Preparation of Medium: Add components to distilled/deionized water and bring volume to 1.0L. Mix thoroughly. Gently heat and bring to boiling. Distribute into tubes or flasks. Autoclave for 15 min at 15 psi pressure–121°C. Pour into sterile Petri dishes or leave in tubes.

Use: For the cultivation and maintenance of *Thermoactinomyces sacchari.*

Thermoactinopolyspora Medium

Composition per liter:

Maltose	20.0g
Agar	15.0g
Papaic digest of soybean meal	15.0g
Yeast extract	2.0g

pH 7.2 ± 0.2 at 25°C

Preparation of Medium: Add components to tap water and bring volume to 1.0L. Mix thoroughly. Gently heat and bring to boiling. Distribute into tubes or flasks. Autoclave for 15 min at 15 psi pressure–121°C. Pour into sterile Petri dishes or leave in tubes.

Use: For the cultivation and maintenance of *Thermoactinomyces* and *Thermoactinopolyspora* species.

Thermoanaerobacter ethanolicus Medium

Composition per liter:

Glucose	8.0g
$Na_2HPO_4 \cdot 12H_2O$	4.2g
Yeast extract	2.0g
KH_2PO_4	1.5g
NH_4Cl	0.5g
$MgCl_2 \cdot 6H_2O$	0.18g
Reducing solution	40.0mL
Wolfe's modified mineral solution	5.0mL
Resazurin (0.1% solution)	1.0mL
Vitamin solution	0.5mL

Caution: This medium contains Na_2S, and H_2S production will occur, especially upon prolonged boiling. H_2S is hazardous and preparation of this medium should be done in a chemical fume hood.

Reducing Solution:

Composition per 200mL:

Cysteine·HCl·H_2O	2.5g
$Na_2S \cdot 9H_2O$	2.5g
NaOH (0.2*N* solution)	200.0mL

Preparation of Reducing Solution: Gently heat the NaOH solution and bring to boiling. Gas with 95% N_2 + 5% H_2. Cool to room temperature. Add the cysteine·HCl·H_2O and $Na_2S \cdot 9H_2O$. Anaerobically distibute into tubes. Cap with rubber stoppers. Autoclave for 15 min at 15 psi pressure–121°C.

Vitamin Solution:

Composition per 500 mL:

Pyridoxine HCl	0.1g
p-Aminobenzoic acid	0.05g
Calcium pantothenate	0.05g
Nicotinic acid	0.05g
Thioctic acid	0.05g
Biotin	0.02g
Folic acid	0.02g
Riboflavin	5.0mg
Thiamine·HCl	5.0mg
Vitamin B_{12}	1.0mg

Preparation of Vitamin Solution: Add components to distilled/deionized water and bring volume to 500.0mL. Mix thoroughly. Store solution in the dark at −10°C.

Wolfe's Modified Mineral Solution:

Composition per liter:

$MgSO_4 \cdot 7H_2O$	3.0g
Nitrilotriacetic acid	1.5g
NaCl	1.0g
$MnSO_4 \cdot H_2O$	0.5g
$CaCl_2$ (anhydrous)	0.1g
$Co(NO_3)_2 \cdot 6H_2O$	0.1g
$FeSO_4 \cdot 7H_2O$	0.1g
$ZnSO_4 \cdot 7H_2O$	0.1g
$AlK(SO_4)_2$ (anhydrous)	0.01g
$CuSO_4 \cdot 5H_2O$	0.01g
H_3BO_3	0.01g
$Na_2MoO_4 \cdot 2H_2O$	0.01g
Na_2SeO_3 (anhydrous)	1.0mg

Preparation of Wolfe's Modified Mineral Solution: Add nitrilotriacetic acid to 500.0mL of distilled/deionized water. Dissolve by adjusting pH to 6.5 with KOH. Add remaining components. Add distilled/deionized water to 1.0L.

Preparation of Medium: Add components, except reducing solution, to distilled/deionized water and bring volume to 1.0L. Gently heat and bring to boiling under 95% N_2 + 5% H_2. Continue boiling until color changes from blue to pink. Add the reducing solution. The pink color will disappear, indicating that the solution has been reduced. Distribute into tubes or flasks under 95% N_2 + 5% H_2 using anerobic techniques. Cap tubes with rubber stoppers. Autoclave for 15 min at 15 psi pressure–121°C.

Use: For the cultivation and maintenance of thermophilic anaerobes such as *Thermoanaerobacter* species and some *Clostridium* species.

Thermoanaerobium brockii Medium

Composition per liter:

Pancreatic digest of casein	10.0g
Yeast extract	3.0g
K_2HPO_4	1.5g
NH_4Cl	0.9g
NaCl	0.9g
KH_2PO_4	0.75g
$MgCl_2 \cdot 6H_2O$	0.2g
Glucose solution	25.0mL
$Na_2S \cdot 9H_2O$ (10% solution)	10.0mL
Trace element solution	9.0mL
Vitamin solution	5.0mL
Resazurin (0.025% solution)	4.0mL
$FeSO_4 \cdot 7H_2O$ (10% solution)	0.03mL

pH 7.3 ± 0.2 at 25°C

Glucose Solution:

Composition per 100mL:

Glucose	20.0g

Preparation of Glucose Solution: Add glucose to distilled/deionized water and bring volume to 100.0mL. Mix thoroughly. Filter sterilize.

Trace Element Solution:

Composition per liter:

Nitrilotriacetic acid	12.5g
NaCl	1.0g
$FeCl_3 \cdot 4H_2O$	0.2g
$MnCl_2 \cdot 4H_2O$	0.1g
$CaCl_2 \cdot 2H_2O$	0.1g
$ZnCl_2$	0.1g
$CuCl_2$	0.02g
Na_2SeO_3	0.02g
$CoCl_2 \cdot 6H_2O$	0.017g
H_3BO_3	0.01g
$Na_2MoO_4 \cdot 2H_2O$	0.01g

Preparation of Trace Element Solution: Add nitrilotriacetic acid to 500.0mL of distilled/deionized water. Adjusting pH to 6.5 with KOH. Add remaining components. Add distilled/deionized water to 1.0L.

Wolfe's Vitamin Solution:

Composition per liter:

Pyridoxine·HCl	10.0mg
Thiamine·HCl	5.0mg
Riboflavin	5.0mg
Nicotinic acid	5.0mg
Calcium pantothenate	5.0mg
p-Aminobenzoic acid	5.0mg
Thioctic acid	5.0mg
Biotin	2.0mg
Folic acid	2.0mg
Cyanocobalamin	100.0μg

Preparation of Wolfe's Vitamin Solution: Add components to distilled/deionized water and bring volume to 1.0L. Mix thoroughly.

Preparation of Medium: Add components, except glucose solution, to distilled/deionized water and bring volume to 975.0mL. Mix thoroughly. Autoclave for 15 min at 15 psi pressure–121°C. While still hot, aseptically add 25.0mL of the sterile glucose solution under 97% N_2 + 3% H_2. Adjust pH to 7.3 if necessary. Aseptically and anaerobically distribute into tubes. Cap with rubber stoppers.

Use: For the cultivation and maintenance of *Thermoanaerobium brockii.*

Thermobacterium Medium

Composition per liter:

Agar	20.0g
$(NH_4)_2SO_4$	1.3g
Yeast extract	1.0g
Pancreatic digest of casein	1.0g
KH_2PO_4	0.28g
$MgSO_4 \cdot 7H_2O$	0.247g
$CaCl_2 \cdot 2H_2O$	0.074g
$FeCl_3.6H_2O$	0.019g
Salt solution	1.0mL

pH 8.5 + 0.2 at 25°C

Salt Solution:

Composition per liter:

$Na_2B_4O_7 \cdot 10H_2O$	4.4g
$MnCl_2 \cdot 4H_2O$	1.8g
$ZnSO_4 \cdot 7H_2O$	0.22g
$CuCl_2 \cdot H_2O$	0.05g
$Na_2MoO_4.2H_2O$	0.03g
$VOSO_4 \cdot 2H_2O$	0.03g

Preparation of Salt Solution: Add components to distilled/deionized water and bring volume to 1.0L. Mix thoroughly. Adjust pH to 2.0 with H_2SO_4.

Preparation of Medium: Add components to distilled/deionized water and bring volume to 1.0L. Mix thoroughly. Gently heat and bring to boiling. Distribute into tubes in 11.0–12.0mL volumes. Autoclave for 15 min at 15 psi pressure–121°C. Allow tubes to solidify in a slanted position.

Use: For the cultivation and maintenance of *Thermomicrobium roseum.*

Thermodesulfotobacterium Agar

Composition per liter:

Na_2SO_4	30.0g
Agar	20.0g

Sodium lactate	4.0g
Yeast extract	1.0g
Mineral solution 2	50.0mL
Na_2CO_3 solution	50.0mL
Mineral solution 1	25.0mL
Cysteine-sulfide reducing agent	20.0mL
Wolfe's Mineral solution	10.0mL
Wolfe's Vitamin solution	10.0mL
Resazurin (0.025% solution)	4.0mL

pH 7.2 ± 0.2 at 25°C

Mineral Solution 1:

Composition per liter:

K_2HPO_4	6.0g

Preparation of Mineral Solution 1: Add K_2HPO_4 to distilled/deionized water and bring volume to 1.0L. Mix thoroughly.

Mineral Solution 2:

Composition per liter:

NaCl	12.0g
KH_2PO_4	6.0g
$(NH_4)_2SO_4$	6.0g
$MgSO_4 \cdot 7H_2O$	2.4g
$CaCl_2 \cdot 2H_2O$	1.6g

Preparation of Mineral Solution 2: Add components to distilled/deionized water and bring volume to 1.0L. Mix thoroughly.

Na_2CO_3 Solution:

Composition per 100mL:

Na_2CO_3	8.0g

Preparation of Na_2CO_3 Solution: Add Na_2CO_3 to distilled/deionized water and bring volume to 100.0mL. Mix thoroughly.

Cysteine-Sulfide Reducing Agent:

Composition per 20mL:

L-Cysteine·HCl·H_2O	300.0mg
$Na_2S \cdot 9H_2O$	300.0mg

Preparation of Cysteine-Sulfide Reducing Agent: Add L-Cysteine·HCl·H_2O to 10.0mL of distilled/deionized water. Mix thoroughly. In a separate tube, add $Na_2S \cdot 9H_2O$ to 10.0mL of distilled/deionized water. Mix thoroughly. Gas both solutions with 100% N_2 and cap tubes. Autoclave both solutions for 15 min at 15 psi pressure–121°C using fast exhaust. Cool to 50°C. Aseptically combine the two solutions under 100% N_2.

Wolfe's Mineral Solution:

Composition per liter

$MgSO_4 \cdot 7H_2O$	3.0g
Nitriloacetic acid	1.5g
NaCl	1.0g
$MnSO_4 \cdot H_2O$	0.5g
$FeSO_4 \cdot 7H_2O$	0.1g
$CoCl_2 \cdot 6H_2O$	0.1g
$CaCl_2$	0.1g
$ZnSO_4 \cdot 7H_2O$	0.1g
$CuSO_4 \cdot 5H_2O$	0.01g
$AlK(SO_4)_2 \cdot 12H_2O$	0.01g
H_3BO_3	0.01g
$Na_2MoO_4 \cdot 2H_2O$	0.01g

Preparation of Wolfe's Mineral Solution: Add nitrilotriacetic acid to 500.0mL of distilled/deionized water. Dissolve by adjusting pH to 6.5 with KOH. Add remaining components. Add distilled/deionized water to 1.0L.

Wolfe's Vitamin Solution:

Composition per liter:

Pyridoxine·HCl	10.0mg
Thiamine·HCl	5.0mg
Riboflavin	5.0mg
Nicotinic acid	5.0mg
Calcium pantothenate	5.0mg
p-Aminobenzoic acid	5.0mg
Thioctic acid	5.0mg
Biotin	2.0mg
Folic acid	2.0mg
Cyanocobalamin	100.0μg

Preparation of Wolfe's Vitamin Solution: Add components to distilled/deionized water and bring volume to 1.0L. Mix thoroughly. Filter sterilize.

Preparation of Medium: Add components, except vitamin solution and cysteine-sulfide reducing agent, to distilled/deionized water and bring volume to 970.0mL. Mix thoroughly. Gently heat and bring to boiling. Autoclave for 15 min at 15 psi pressure–121°C. Cool to 50°–55°C under 80% N_2 + 20% CO_2. Aseptically add the sterile vitamin solution and then the sterile cysteine-sulfide reducing agent. Adjust the pH to 7.2. Distribute aseptically and anaerobically into sterile tubes.

Use: For the cultivation and maintenance of *Thermodesulfobacterium commune* and other *Thermodesulfobacterium* species.

Thermodesulfotobacterium Broth

Composition per liter:

Na_2SO_4	30.0g
Sodium lactate	4.0g
Yeast extract	1.0g
Mineral solution 2	50.0mL
Na_2CO_3 solution	50.0mL
Mineral solution 1	25.0mL

Cysteine-sulfide reducing agent	20.0mL
Wolfe's Mineral solution	10.0mL
Wolfe's Vitamin solution	10.0mL
Resazurin (0.025% solution)	4.0mL

pH 7.2 ± 0.2 at 25°C

Mineral Solution 1:

Composition per liter:

K_2HPO_4	6.0g

Preparation of Mineral Solution 1: Add K_2HPO_4 to distilled/deionized water and bring volume to 1.0L. Mix thoroughly.

Mineral Solution 2:

Composition per liter:

NaCl	12.0g
KH_2PO_4	6.0g
$(NH_4)_2SO_4$	6.0g
$MgSO_4 \cdot 7H_2O$	2.4g
$CaCl_2 \cdot 2H_2O$	1.6g

Preparation of Mineral Solution 2: Add components to distilled/deionized water and bring volume to 1.0L. Mix thoroughly.

Na_2CO_3 Solution:

Composition per 100mL:

Na_2CO_3	8.0g

Preparation of Na_2CO_3 Solution: Add Na_2CO_3 to distilled/deionized water and bring volume to 100.0mL Mix thoroughly.

Cysteine-Sulfide Reducing Agent:

Composition per 20mL:

L-Cysteine·HCl·H_2O	300.0mg
$Na_2S \cdot 9H_2O$	300.0mg

Preparation of Cysteine-Sulfide Reducing Agent: Add L-Cysteine·HCl·H_2O to 10.0mL of distilled/deionized water. Mix thoroughly. In a separate tube, add $Na_2S \cdot 9H_2O$ to 10.0mL of distilled/deionized water. Mix thoroughly. Gas both solutions with 100% N_2 and cap tubes. Autoclave both solutions for 15 min at 15 psi pressure–121°C using fast exhaust. Cool to 50°C. Aseptically combine the two solutions under 100% N_2.

Wolfe's Mineral Solution:

Composition per liter

$MgSO_4 \cdot 7H_2O$	3.0g
Nitriloacetic acid	1.5g
NaCl	1.0g
$MnSO_4 \cdot H_2O$	0.5g
$FeSO_4 \cdot 7H_2O$	0.1g
$CoCl_2 \cdot 6H_2O$	0.1g
$CaCl_2$	0.1g
$ZnSO_4 \cdot 7H_2O$	0.1g
$CuSO_4 \cdot 5H_2O$	0.01g
$AlK(SO_4)_2 \cdot 12H_2O$	0.01g
H_3BO_3	0.01g
$Na_2MoO_4 \cdot 2H_2O$	0.01g

Preparation of Wolfe's Mineral Solution: Add nitrilotriacetic acid to 500.0mL of distilled/deionized water. Dissolve by adjusting pH to 6.5 with KOH. Add remaining components. Add distilled/deionized water to 1.0L.

Wolfe's Vitamin Solution:

Composition per liter:

Pyridoxine·HCl	10.0mg
Thiamine·HCl	5.0mg
Riboflavin	5.0mg
Nicotinic acid	5.0mg
Calcium pantothenate	5.0mg
p-Aminobenzoic acid	5.0mg
Thioctic acid	5.0mg
Biotin	2.0mg
Folic acid	2.0mg
Cyanocobalamin	100.0μg

Preparation of Wolfe's Vitamin Solution: Add components to distilled/deionized water and bring volume to 1.0L. Mix thoroughly. Filter sterilize.

Preparation of Medium: Add components, except vitamin solution and cysteine-sulfide reducing agent, to distilled/deionized water and bring volume to 970.0mL. Mix thoroughly. Autoclave for 15 min at 15 psi pressure–121°C. Cool under 80% N_2 + 20% CO_2. Aseptically add the sterile vitamin solution and then the sterile cysteine-sulfide reducing agent. Adjust the pH to 7.2. Distribute aseptically and anaerobically into sterile tubes.

Use: For the cultivation and maintenance of *Thermodesulfobacterium commune* and other *Thermodesulfobacterium* species.

Thermophilic *Bacillus* Medium

Composition per liter:

Peptone	8.0g
Yeast extract	4.0g
NaCl	3.0g

pH 7.5 ± 0.2 at 25°C

Preparation of Medium: Add components to distilled/deionized water and bring volume to 1.0L. Mix thoroughly. Distribute into tubes or flasks. Autoclave for 15 min at 15 psi pressure–121°C.

Use: For the cultivation and maintenance of a variety of thermophilic *Bacillus* species.

Thermophilic Maintenance Medium

Composition per liter:

$NaHCO_3$	3.0g
Yeast extract	1.0g
NH_4Cl	1.0g
KH_2PO_4	0.4g
K_2HPO_4	0.4g
$MgSO_4 \cdot 7H_2O$	0.1g
Cysteine-sulfide reducing solution	40.0mL
Fructose solution	25.0mL
Wolfe's vitamin solution	10.0mL
Wolfe's mineral solution	10.0mL
Resazurin (0.01% solution)	1.0mL

pH 5.6 ± 0.2 at 25°C

Cysteine-Sulfide Reducing Solution:
Composition per 100mL:

$Cysteine \cdot HCl \cdot H_2O$	1.25g
$Na_2S \cdot 9H_2O$	1.25g

Preparation of Cysteine-Sulfide Reducing Solution: Add $Cysteine \cdot HCl \cdot H_2O$ and $Na_2S \cdot 9H_2O$ to distilled/deionized water and bring volume to 100.0mL. Mix thoroughly.

Fructose Solution:
Composition per 100mL:

Fructose	20.0g

Preparation of Fructose Solution: Add fructose to distilled/deionized water and bring volume to 100.0mL. Mix thoroughly. Filter sterilize.

Wolfe's Vitamin Solution:
Composition per liter:

Pyridoxine·HCl	0.01g
Thiamine·HCl	5.0mg
Riboflavin	5.0mg
Nicotinic acid	5.0mg
Calcium pantothenate	5.0mg
p-Aminobenzoic acid	5.0mg
Thioctic acid	5.0mg
Biotin	2.0mg
Folic acid	2.0mg
Cyanocobalamin	100.0μg

Preparation of Wolfe's Vitamin Solution: Add components to distilled/deionized water and bring volume to 1.0L. Mix thoroughly.

Wolfe's Mineral Solution:
Composition per liter

$MgSO_4 \cdot 7H_2O$	3.0g
Nitriloacetic acid	1.5g
NaCl	1.0g
$MnSO_4 \cdot H_2O$	0.5g
$FeSO_4 \cdot 7H_2O$	0.1g
$CoCl_2 \cdot 6H_2O$	0.1g
$CaCl_2$	0.1g
$ZnSO_4 \cdot 7H_2O$	0.1g
$CuSO_4 \cdot 5H_2O$	0.01g
$AlK(SO_4)_2 \cdot 12H_2O$	0.01g
H_3BO_3	0.01g
$Na_2MoO_4 \cdot 2H_2O$	0.01g

Preparation of Wolfe's Mineral Solution: Add nitrilotriacetic acid to 500mL of distilled/deionized water. Dissolve by adjusting pH to 6.5 with KOH. Add remaining components. Add distilled/deionized water to 1.0L.

Preparation of Medium: Add components, except fructose solution, to distilled/deionized water and bring volume to 935.0mL. Mix thoroughly. Gently heat and bring to boiling. Continue boiling until resazurin turns colorless indicating reduction. Add 40.0mL of the cysteine-sulfide reducing solution. Autoclave for 15 min at 15 psi pressure–121°C. Cool to 50°C under O_2-free 90% N_2 + 10% CO_2. Add 25.0mL of the sterile fructose solution. Adjust the pH to 5.6 if necessary. Aseptically and anaerobically distribute into sterile tubes. Cap with rubber stoppers.

Use: For the cultivation and maintenance of a variety of thermophilic anaerobes including *Clostridium thermoautotrophicum*.

Thermophilic Streptomycete Medium

Composition per liter:

Agar	20.0g
Maltose	20.0g
Soybean meal	5.0g
Yeast extract	2.0g

pH 6.5 ± 0.2 at 25°C

Preparation of Medium: Add components to tap water and bring volume to 1.0L. Mix thoroughly. Gently heat and bring to boiling. Distribute into tubes or flasks. Autoclave for 15 min at 15 psi pressure–121°C. Pour into sterile Petri dishes or leave in tubes.

Use: For the isolation and cultivation of thermophilic streptomycetes.

Thermophilic Streptomycete Medium

Composition per liter:

Soybean oil meal	20.0g
Glucose	10.0g
NaCl	10.0g
Pancreatic digest of casein	10.0g
Silica solution (Ludox)	500.0mL

Preparation of Medium: Add components, except silica solution, to distilled/deionized water and bring volume to 500.0mL. Mix thoroughly. Gently heat until dissolved. Autoclave this solution and the 500.0mL of silica solution separately for 15 min at 15 psi pressure–121°C. Cool to 25°C. Adjust the pH of both solutions to 7.0. Aseptically combine the two sterile solutions. Mix thoroughly. Pour into sterile Petri dishes in 40.0mL volumes.

Use: For the isolation and cultivation of thermophilic streptomycetes.

Thermophilic Streptomycete Medium Ia

Composition per liter:

Agar	20.0g
Sucrose	5.0g
Pancreatic digest of casein	5.0g
Yeast extract	3.0g
$MgSO_4 \cdot 7H_2O$	0.5g
$FeSO_4 \cdot 7H_2O$	0.01g
Dung extract	5.0mL
Molasses	5.0mL
Trace element solution	1.0mL

pH 7.2 ± 0.2 at 25°C

Dung Extract:

Composition per 100mL:

Sheep manure, dried	25.0g

Preparation of Dung Extract: Add dried sheep manure to 100.0mL of tap water. Mix thoroughly. Autoclave for 30 min at 15 psi pressure–121°C. Filter through Whatman #1 filter paper. Store at 4°C under toluene.

Trace Element Solution:

Composition per 100mL:

$Fe(NH_4)_2SO_4$	0.1g
$ZnSO_4$	0.1g
$MnSO_4$	0.05g
$CoSO_4$	0.01g
H_3BO_3	0.01g
$CuSO_4$	8.0mg

Preparation of Trace Element Solution: Add components to distilled/deionized water and bring volume to 100.0mL. Mix thoroughly.

Preparation of Medium: Add components to distilled/deionized water and bring volume to 1.0L. Mix thoroughly. Gently heat and bring to boiling. Distribute into tubes or flasks. Autoclave for 15 min at 15 psi pressure–121°C. Pour into sterile Petri dishes or leave in tubes.

Use: For the isolation and cultivation of thermophilic streptomycetes.

Thermoplasma Agar

Composition per liter:

Basal solution	450.0mL
Solution B	450.0mL
Solution C	100.0mL

pH 2.0 ± 0.2 at 25°C

Basal Solution:

Composition per 500mL:

KH_2PO_4	3.0g
Yeast extract	1.0g
$MgSO_4 \cdot 7H_2O$	0.5g
$CaCl_2 \cdot 2H_2O$	0.25g
$(NH_4)_2SO_4$	0.2g

Preparation of Basal Solution: Add components to distilled/deionized water and bring volume to 500.0mL. Mix thoroughly. Adjust pH to 2.0 with 10*N* H_2SO_4. Autoclave for 15 min at 15 psi pressure–121°C. Cool to 55°C.

Solution B:

Composition per 450mL:

Noble agar	12.0g

Preparation of Solution B: Add agar to distilled/deionized water and bring volume to 450.0mL. Mix thoroughly. Gently heat and bring to boiling. Autoclave for 15 min at 15 psi pressure–121°C. Cool to 55°C.

Solution C:

Composition per 100mL:

Glucose	10.0g

Preparation of Solution C: Add glucose to distilled/deionized water and bring volume to 100.0mL. Mix thoroughly. Filter sterilize.

Preparation of Medium: Aseptically combine the cooled sterile basal medium with the sterile solution B and sterile solution C. Mix thoroughly. Pour into sterile Petri dishes or distribute into sterile tubes.

Use: For the cultivation and maintenance of *Thermoplasma acidophilum* and other *Thermoplasma* species.

Thermoplasma Broth

Composition per liter:

Basal solution	500.0mL
Solution C	100.0mL

pH 2.0 ± 0.2 at 25°C

Basal Solution:

Composition per 500mL:

KH_2PO_4	3.0g
Yeast extract	1.0g
$MgSO_4 \cdot 7H_2O$	0.5g

$CaCl_2 \cdot 2H_2O$	0.25g
$(NH_4)_2SO_4$	0.2g

Preparation of Basal Solution: Add components to distilled/deionized water and bring volume to 500.0mL. Mix thoroughly. Adjust pH to 2.0 with 10*N* H_2SO_4.

Solution C:

Composition per 100mL:

Glucose	10.0g

Preparation of Solution C: Add glucose to distilled/deionized water and bring volume to 100.0mL. Mix thoroughly. Filter sterilize.

Preparation of Medium: Add 500.0mL of basal solution to 400.0mL of distilled/deionized water. Autoclave for 15 min at 15 psi pressure–121°C. Cool to 55°C. Aseptically add 100.0mL of sterile glucose solution. Mix thoroughly. Aseptically distribute into sterile tubes.

Use: For the cultivation and maintenance of *Thermoplasma acidophilum* and other *Thermoplasma* species.

Thermoproteus Medium

Composition per liter:

Solution A	500.0mL
Solution B	450.0mL
Solution C	50.0mL

pH 4.8–5.6 at 25°C

Solution A:

Composition per 500mL:

Glucose	10.0g
$FeSO_4 \cdot 7H_2O$	0.556g
$MgSO_4 \cdot 7H_2O$	0.492g
$CaSO_4 \cdot 2H_2O$	0.344g
$(NH_4)_2SO_4$	0.264g
Yeast extract	0.2g
KH_2PO_4	0.014g
Resazurin	1.0mg
Trace elements	10.0mL

Preparation of Solution A: Add components to distilled/deionized water and bring volume to 500.0mL. Mix thoroughly. Immediately filter sterilize.

Trace Elements:

Composition per liter:

$Na_2B_4O_7 \cdot 10H_2O$	0.45g
$MnCl_2 \cdot 4H_2O$	0.18g
$ZnSO_4 \cdot 7H_2O$	0.022g
$CuCl_2 \cdot 2H_2O$	5.0mg
$Na_2MoO_4 \cdot 2H_2O$	3.6mg
$VOSO_4 \cdot 5H_2O$	3.6mg
$CoSO_4 \cdot 7H_2O$	1.2mg

Preparation of Trace Elements: Add components to distilled/deionized water and bring volume to 1.0L. Mix thoroughly. Adjust pH to 3.0 with H_2SO_4 to retard precipitation.

Solution B:

Composition per 450mL:

Sulfur	10.0g

Preparation of Solution B: Add sulfur to 450.0mL of distilled/deionized water. Autoclave for 30 min at 0 psi pressure–100°C on three consecutive days.

Solution C:

Composition per 50mL:

$Na_2S \cdot 9H_2O$	0.85g

Preparation of Solution C: Add $Na_2S \cdot 9H_2O$ to distilled/deionized water and bring volume to 50.0mL. Mix thoroughly. Autoclave for 15 min at 15 psi pressure–121°C.

Preparation of Medium: Aseptically combine solutions A, B and C under 97% N_2 + 3% H_2. Adjust pH to 4.8–5.6 with H_2SO_4. Aseptically and anaerobically distribute into sterile tubes or flasks under 97% N_2 + 3% H_2.

Use: For the cultivation and maintenance of *Thermoproteus tenax* and other *Thermoproteus* species.

Thermus Beef Extract Polypeptone™ Medium

See: Thermus BP Medium

Thermus BP Medium (*Thermus* Beef Extract Polypeptone™ Medium)

Composition per liter:

Agar	25.0g
Beef extract	4.0g
Polypeptone™	4.0g
K_2HPO_4	3.0g
KH_2PO_4	1.0g

pH 7.0 ± 0.2 at 25°C

Preparation of Medium: Add components to distilled/deionized water and bring volume to 1.0L. Mix thoroughly. Gently heat and bring to boiling. Distribute into tubes or flasks. Autoclave for 15 min at 15 psi pressure–121°C. Pour into sterile Petri dishes or leave in tubes.

Use: For the cultivation and maintenance of *Thermus aquaticus* and other *Thermus* species.

Thermus Medium

Composition per liter:

Agar	30.0g
Polypeptone™	8.0g
Yeast extract	4.0g
NaCl	2.0g

pH 7.5 ± 0.2 at 25°C

Preparation of Medium: Add components to distilled/deionized water and bring volume to 1.0L. Mix thoroughly. Gently heat and bring to boiling. Distribute into tubes or flasks. Autoclave for 15 min at 15 psi pressure–121°C. Pour into sterile Petri dishes or leave in tubes.

Use: For the cultivation and maintenance of *Thermus aquaticus* and other *Thermus* species.

Thermus Peptone Meat Extract Yeast Extract Agar

See: Thermus **PMY Agar**

Thermus Peptone Meat Extract Yeast Extract Broth

See: Thermus **PMY Broth**

Thermus PMY Agar (*Thermus* Peptone Meat Extract Yeast Extract Agar)

Composition per liter:

Agar	15.0g
Peptone	5.0g
Meat extract	3.5g
Yeast extract	1.5g
NaCl	1.5g

pH 7.0 ± 0.2 at 25°C

Preparation of Medium: Add components to distilled/deionized water and bring volume to 1.0L. Mix thoroughly. Gently heat and bring to boiling. Distribute into tubes or flasks. Autoclave for 15 min at 15 psi pressure–121°C. Pour into sterile Petri dishes or leave in tubes.

Use: For the cultivation and maintenance of *Thermus aquaticus* and other *Thermus* species.

Thermus PMY Broth (*Thermus* Peptone Meat Extract Yeast Extract Broth)

Composition per liter:

Peptone	5.0g
Meat extract	3.5g
Agar	3.0g
Yeast extract	1.5g
NaCl	1.5g

pH 7.0 ± 0.2 at 25°C

Preparation of Medium: Add components to distilled/deionized water and bring volume to 1.0L. Mix thoroughly. Distribute into tubes or flasks. Autoclave for 15 min at 15 psi pressure–121°C.

Use: For the cultivation and maintenance of *Thermus aquaticus* and other *Thermus* species.

Thiamine Assay Medium

Composition per liter:

Glucose	40.0g
Peptone	22.0g
Sodium acetate	15.0g
Vitamin assay casamino acids	5.0g
K_2HPO_4	1.0g
KH_2PO_4	1.0g
$MgSO_4 \cdot 7H_2O$	0.4g
Adenine sulfate	0.02g
$FeSO_4 \cdot 7H_2O$	0.02g
Guanine·HCl	0.02g
$MnSO_4 \cdot 5H_2O$	0.02g
NaCl	0.02g
Uracil	0.02g
L-Cystine	0.2mg
p-Aminobenzoic acid	0.2mg
Calcium pantothenate	0.2mg
Niacin	0.2mg
Pyridoxine·HCl	0.2mg
Riboflavin	0.2mg
Folic acid	0.5μg
Biotin	0.8μg

pH 6.5 ± 0.2 at 25°C

Source: This medium is available as a premixed powder from Difco Laboratories.

Preparation of Medium: Add components to distilled/deionized water and bring volume to 1.0L. Mix thoroughly. Gently heat and bring to boiling. Continue boiling for 2–3 min. Distribute into tubes or flasks in 5.0mL volumes while swirling the flask to disperse the precipitate. Add standard solutions or test solutions and bring volume of each tube to 10.0mL with distilled/deionized water. Autoclave for 15 min at 15 psi pressure–121°C.

Use: For the microbiological assaying of thiamine using *Lactobacillus fermentum* as the test organism.

Thiamine Assay Medium LV

Composition per liter:

Glucose	20.0g
Pancreatic digest of casein	20.0g

K_2HPO_4	10.0g
NaCl	10.0g
Sodium citrate	10.0g
Yeast extract, thiamine-free	10.0g
Sorbitan monooleate complex	2.0g
$MgSO_4 \cdot 7H_2O$	1.6g
$MnCl_2 \cdot 4H_2O$	0.28g
$FeSO_4 \cdot 7H_2O$	0.08g

pH 6.0 ± 0.2 at 25°C

Source: This medium is available as a premixed powder from Difco Laboratories.

Preparation of Medium: Add components to distilled/deionized water and bring volume to 1.0L. Mix thoroughly. Gently heat and bring to boiling. Continue boiling for 2–3 min. Distribute into tubes or flasks in 5.0mL volumes while swirling the flask to disperse the precipitate. Add standard solutions or test solutions and bring volume of each tube to 10.0mL with distilled/deionized water. Autoclave for 15 min at 15 psi pressure–121°C.

Use: For the microbiological assaying of thiamine using *Lactobacillus viridescens* as the test organism.

Thiamine Salts Medium

Composition per liter:

KH_2PO_4	1.0g
$FeSO_4 \cdot 7H_2O$	0.05g
$MgSO_4 \cdot 7H_2O$	0.02g
$CaCl_2$	0.02g
$MnCl_2 \cdot 4H_2O$	1.0mg
$Na_2MoO_4 \cdot 2H_2O$	1.0mg
Thiamine·HCl solution	10.0mL

pH 7.0 ± 0.2 at 25°C

Thiamine·HCl solution:

Composition per 10mL:

Thiamine·HCl	3.0g

Preparation of Solution C: Add thiamine·HCl to distilled/deionized water and bring volume to 10.0mL. Mix thoroughly. Filter sterilize.

Preparation of Medium: Add components, except thiamine·HCl solution, to distilled/deionized water and bring volume to 990.0mL. Mix thoroughly. Adjust pH to 7.0 with KOH. Autoclave for 20 min at 15 psi pressure–121°C. Cool to 45°–50°C. Aseptically add the sterile thiamine·HCl solution. Mix thoroughly. Distribute into sterile tubes or flasks.

Use: For the cultivation of ATCC strain 25589.

THIO Medium
See: Thioglycollate Medium, Enriched

THIO + Bile Medium
See: Thioglycollate Medium with 20% Bile

Thiobacillus A2 Agar (T3 Agar)

Composition per 1100mL:

Solution A	100.0mL
Solution B	1.0L

pH 8.5 ± 0.2 at 25°C

Solution A:

Composition per 100mL:

$Na_2S_2O_3 \cdot 5H_2O$	5.0g
Na_2HPO_4	4.2g
KH_2PO_4	1.5g
NH_4Cl	1.0g
Phenol Red (0.2% solution)	1.0mL

Preparation of Solution A: Add components to distilled/deionized water and bring volume to 100.0mL. Mix thoroughly. Adjust pH to 9.0. Autoclave for 15 min at 15 psi pressure–121°C. Cool to 45°–50°C.

Solution B:

Composition per liter:

Agar	15.0g
$MgSO_4 \cdot 7H_2O$	0.1g
Trace metal solution	5.0mL

Preparation of Solution B: Add components to distilled/deionized water and bring volume to 1.0mL. Mix thoroughly. Autoclave for 15 min at 15 psi pressure–121°C. Cool to 45°–50°C.

Trace Metal Solution:

Composition per liter:

EDTA	50.0g
$ZnSO_4$	22.0g
$CaCl_2$	5.54g
$MnCl_2$	5.06g
$FeSO_4 \cdot 7H_2O$	4.99g
$CoCl_2$	1.61g
$CuSO_4$	1.57g
$(NH_4)_2MoO_4 \cdot 4H_2O$	1.10g

Preparation of Trace Metal Solution: Add components to distilled/deionized water and bring volume to 1.0L. Mix thoroughly. Adjust pH to 6.0 with KOH.

Preparation of Medium: Aseptically add 100.0mL of sterile solution A to 1.0L of sterile solution B. Mix thoroughly. Adjust pH to 8.5 if necessary. Pour into sterile Petri dishes or distribute into sterile tubes.

Use: For the cultivation and maintenance of *Thiobacillus versutus* and other *Thiobacillus* species.

Thiobacillus A2 Broth (T3 Broth)

Composition per 1100mL:

Solution B	1.0L
Solution A	100.0mL

pH 8.5 ± 0.2 at 25°C

Solution A:

Composition per 100mL:

$Na_2S_2O_3 \cdot 5H_2O$	5.0g
Na_2HPO_4	4.2g
KH_2PO_4	1.5g
NH_4Cl	1.0g
Phenol Red (0.2% solution)	1.0mL

Preparation of Solution A: Add components to distilled/deionized water and bring volume to 100.0mL. Mix thoroughly. Adjust pH to 9.0. Autoclave for 15 min at 15 psi pressure–121°C. Cool to 45°–50°C.

Solution B:

Composition per liter:

$MgSO_4 \cdot 7H_2O$	0.1g
Trace metal solution	5.0mL

Preparation of Solution B: Add components to distilled/deionized water and bring volume to 1.0mL. Mix thoroughly. Autoclave for 15 min at 15 psi pressure–121°C. Cool to 45°–50°C.

Trace Metal Solution:

Composition per liter:

EDTA	50.0g
$ZnSO_4$	22.0g
$CaCl_2$	5.54g
$MnCl_2$	5.06g
$FeSO_4 \cdot 7H_2O$	4.99g
$CoCl_2$	1.61g
$CuSO_4$	1.57g
$(NH_4)_2MoO_4 \cdot 4H_2O$	1.10g

Preparation of Trace Metal Solution: Add components to distilled/deionized water and bring volume to 1.0mL. Mix thoroughly. Adjust pH to 6.0 with KOH.

Preparation of Medium: Aseptically add 100.0mL of sterile solution A to 1.0L of sterile solution B. Mix thoroughly. Adjust pH to 8.5 if necessary. Distribute into sterile tubes or flasks.

Use: For the cultivation and maintenance of *Thiobacillus versutus* and other *Thiobacillus* species.

Thiobacillus denitrificans Medium

Composition per liter:

KNO_3	5.0g
$Na_2S_2O_3 \cdot 5H_2O$	5.0g
$NaHCO_3$	1.0g
K_2HPO_4	0.2g
$MgCl_2$	0.1g

pH 7.0 ± 0.2 at 25°C

Preparation of Medium: Add components to distilled/deionized water and bring volume to 1.0L. Mix thoroughly. Distribute into tubes or flasks. Autoclave for 15 min at 15 psi pressure–121°C.

Use: For the cultivation of *Thiobacillus denitrificans.*

Thiobacillus denitrificans Medium

Composition per liter:

$Na_2S_2O_3 \cdot 5H_2O$	5.0g
KNO_3	2.0g
KH_2PO_4	2.0g
$NaHCO_3$	2.0g
NH_4Cl	1.0g
$MgSO_4 \cdot 7H_2O$	0.8g
Trace metals solution	1.0mL

pH 6.8–7.0 at 25°C

Trace Metals Solution:

Composition per liter:

Disodium EDTA	50.0g
NaOH	11.0g
$CaCl_2 \cdot 2H_2O$	7.34g
$FeSO_4 \cdot 7H_2O$	5.0g
$MnCl_2 \cdot 2H_2O$	2.5g
$ZnSO_4 \cdot 7H_2O$	2.2g
$CoCl_2 \cdot 6H_2O$	0.5g
$(NH_4)_6Mo_7O_{24} \cdot 4H_2O$	0.5g
$CuSO_4 \cdot 5H_2O$	0.2g

Preparation of Trace Metals Solution: Add EDTA to distilled/deionized water and bring volume to 500.0mL. Mix thoroughly. Adjust pH to 6.0 with NaOH. Add remaining components, one by one. Maintain the pH at 6.0. After dissolution of all the salts, adjust the pH to 4.0 with HCl. Store at 4°C.

Preparation of Medium: Add components to distilled/deionized water and bring volume to 1.0L. Mix thoroughly. Distribute into tubes or flasks. Autoclave for 15 min at 15 psi pressure–121°C.

Use: For the isolation and cultivation of *Thiobacillus denitrificans.*

Thiobacillus ferrooxidans Medium

Composition per liter:

$Al_2(SO_4)_3 \cdot 12H_2O$	1.4g
NaCl	1.0g

KH_2PO_4 .... 0.4g
$MgSO_4 \cdot 7H_2O$ 0.1g
$(NH_4)_2SO_4$ 0.1g
$CaCl_2$ 0.03g
$MnSO_4 \cdot 4H_2O$ 0.02g
$FeSO_4 \cdot 7H_2O$ solution 100.0mL

$FeSO_4 \cdot 7H_2O$ Solution:
Composition per 100mL:
$FeSO_4 \cdot 7H_2O$ 10.0g
H_2SO_4, concentrated 0.09mL

Preparation of $FeSO_4 \cdot 7H_2O$ Solution: Add $FeSO_4 \cdot 7H_2O$ and H_2SO_4 to distilled/deionized water and bring volume to 100.0mL. Mix thoroughly. Autoclave for 15 min at 15 psi pressure–121°C.

Preparation of Medium: Add components, except $FeSO_4 \cdot 7H_2O$ solution, to distilled/deionized water and bring volume to 900.0mL. Mix thoroughly. Gently heat and bring to boiling. Distribute into flasks in 90.0mL volumes. Autoclave for 15 min at 15 psi pressure–121°C. Cool to 25°C. Aseptically add 10.0mL of sterile $FeSO_4 \cdot 7H_2O$ solution to each flask. Mix thoroughly.

Use: For the cultivation of *Thiobacillus ferrooxidans*.

Thiobacillus ferrooxidans Medium

Composition per liter:
Solution I 400.0mL
Solution III 400.0mL
Solution II 200.0mL

Solution I:
Composition per 500mL:
K_2HPO_4 .... 0.5g
$MgSO_4 \cdot 7H_2O$ 0.5g
$(NH_4)_2SO_4$ 0.5g
1*N* H_2SO_4 .... 5.0mL

Preparation of Solution I: Add components to distilled/deionized water and bring volume to 500.0mL. Mix thoroughly. Autoclave for 15 min at 15 psi pressure–121°C. Cool to 45°–50°C.

Solution II:
Composition per liter:
$FeSO_4 \cdot 7H_2O$ 167.0g
1*N* H_2SO_4 .... 50.0mL

Preparation of Solution II: Add components to distilled/deionized water and bring volume to 1.0L. Mix thoroughly. Filter sterilize. Warm to 45°–50°C.

Solution III:
Composition per liter:
Agar 10.0g

Preparation of Solution III: Add agar to distilled/deionized water and bring volume to 1.0L. Mix thoroughly.

Preparation of Medium: Aseptically combine 400.0mL of sterile solution I, 200.0mL of sterile solution II and 400.0mL of sterile solution III. Mix thoroughly. Aseptically distribute into sterile tubes or flasks.

Use: For the isolation and cultivation of *Thiobacillus ferrooxidans*.

Thiobacillus Heterotrophic Medium

Composition per liter:
Glucose 5.0g
$MgSO_4 \cdot 7H_2O$ 0.5g
$(NH_4)_2SO_4$ 0.15g
KH_2PO_4 .... 0.1g
KCl 0.05g
$Ca(NO_3)_2$ 0.01g
pH 3.0 ± 0.2 at 25°C

Preparation of Medium: Add components to distilled/deionized water and bring volume to 1.0L. Mix thoroughly. Filter sterilize.

Use: For the cultivation and maintenance of *Thiobacillus organoparus* and other heterotrophic *Thiobacillus* species.

Thiobacillus intermedius Medium

Composition per 1010mL:
$Na_2S_2O_3 \cdot 5H_2O$ 10.0g
Solution I 1.0L
Solution II 10.0mL

Solution I:
Composition per liter:
NH_4Cl 1.0g
K_2HPO_4 .... 0.6g
$MgCl_2 \cdot 6H_2O$ 0.5g
KH_2PO_4 .... 0.4g
$MgSO_4$ 0.3g
$CaCl_2 \cdot 2H_2O$ 0.2g
$FeCl_3 \cdot 6H_2O$ 0.02g

Preparation of Solution I: Add components to distilled/deionized water and bring volume to 1.0L. Mix thoroughly.

Solution II:
Composition per liter:
$CaCl_2 \cdot 2H_2O$ 0.1g
$ZnSO_4 \cdot 7H_2O$ 0.09g
$CuSO_4 \cdot 5H_2O$ 0.04g

$MnSO_4$ 0.02g
$Na_2B_4O_7$ 0.01g
$(NH_4)_6Mo_7O_{24}\cdot 4H_2O$ 5.0mg

Preparation of Solution II: Add components to distilled/deionized water and bring volume to 1.0L. Mix thoroughly.

Preparation of Medium: Combine solution I, 10.0mL of solution II and 10.0g of $Na_2S_2O_3\cdot 5H_2O$. Mix thoroughly. Filter sterilize. Aseptically distribute into sterile tubes or flasks.

Use: For the isolation and autotrophic cultivation of *Thiobacillus intermedius*.

Thiobacillus intermedius Medium

Composition per 1010mL:

Glucose 10.0g
$Na_2S_2O_3\cdot 5H_2O$ 10.0g
Solution I 1.0L
Solution II 10.0mL

Solution I:

Composition per liter:

NH_4Cl 1.0g
K_2HPO_4 .... 0.6g
$MgCl_2\cdot 6H_2O$ 0.5g
KH_2PO_4 .... 0.4g
$MgSO_4$ 0.3g
$CaCl_2\cdot 2H_2O$ 0.2g
$FeCl_3\cdot 6H_2O$ 0.02g

Preparation of Solution I: Add components to distilled/deionized water and bring volume to 1.0L. Mix thoroughly.

Solution II:

Composition per liter:

$CaCl_2\cdot 2H_2O$ 0.1g
$ZnSO_4\cdot 7H_2O$ 0.09g
$CuSO_4\cdot 5H_2O$ 0.04g
$MnSO_4$ 0.02g
$Na_2B_4O_7$ 0.01g
$(NH_4)_6Mo_7O_{24}\cdot 4H_2O$ 5.0mg

Preparation of Solution II: Add components to distilled/deionized water and bring volume to 1.0L. Mix thoroughly.

Preparation of Medium: Combine 1.0L of solution I, 10.0mL of solution II, 10.0g of glucose and 10.0g of $Na_2S_2O_3\cdot 5H_2O$. Mix thoroughly. Filter sterilize. Aseptically distribute into sterile tubes or flasks.

Use: For the isolation and mixotrophic cultivation of *Thiobacillus intermedius*.

Thiobacillus intermedius Medium

Composition per 1010mL:

Glucose 10.0g
Yeast extract 0.3g
Solution I 1.0L
Solution II 10.0mL

Solution I:

Composition per liter:

NH_4Cl 1.0g
K_2HPO_4 .... 0.6g
$MgCl_2\cdot 6H_2O$ 0.5g
KH_2PO_4 .... 0.4g
$MgSO_4$ 0.3g
$CaCl_2\cdot 2H_2O$ 0.2g
$FeCl_3\cdot 6H_2O$ 0.02g

Preparation of Solution I: Add components to distilled/deionized water and bring volume to 1.0L. Mix thoroughly.

Solution II:

Composition per liter:

$CaCl_2\cdot 2H_2O$ 0.1g
$ZnSO_4\cdot 7H_2O$ 0.09g
$CuSO_4\cdot 5H_2O$ 0.04g
$MnSO_4$ 0.02g
$Na_2B_4O_7$ 0.01g
$(NH_4)_6Mo_7O_{24}\cdot 4H_2O$ 5.0mg

Preparation of Solution II: Add components to distilled/deionized water and bring volume to 1.0L. Mix thoroughly.

Preparation of Medium: Combine 1.0L of solution I, 10.0mL of solution II, 10.0g of glucose, and 0.3g of yeast extract. Mix thoroughly. Filter sterilize. Aseptically distribute into sterile tubes or flasks.

Use: For the isolation and heterotrophic cultivation of *Thiobacillus intermedius*.

Thiobacillus Medium

Composition per 100mL:

$Na_2S_2O_3\cdot 5H_2O$ 1.0g
KH_2PO_4 .... 0.1g
NH_4Cl 0.1g
$MgCl_2\cdot 7H_2O$ 0.05g

pH 6.8 ± 0.2 at 25°C

Preparation of Medium: Add components to distilled/deionized water and bring volume to 1.0L. Mix thoroughly. Distribute into tubes or flasks. Autoclave for 15 min at 15 psi pressure–121°C.

Use: For the cultivation of *Thiobacillus thioparus* and *Thiobacillus thiooxidans*.

Thiobacillus Medium

Composition per liter:

$Na_2S_2O_3 \cdot 5H_2O$	10.0g
K_2HPO_4	4.0g
KH_2PO_4	4.0g
$CaCl_2$	0.1g
$MgSO_4 \cdot 7H_2O$	0.1g
$(NH_4)_2SO_4$	0.1g
$FeCl_3 \cdot 6H_2O$	0.02g
$MnSO_4 \cdot 4H_2O$	0.02g

pH 6.6 ± 0.2 at 25°C

Preparation of Medium: Add components to distilled/deionized water and bring volume to 1.0L. Mix thoroughly. Distribute into flasks in 100.0mL volumes. Autoclave for 60 min at 0 psi pressure–100°C on three consecutive days.

Use: For the cultivation of nonaciduric *Thiobacillus* species.

Thiobacillus Medium (ATCC Medium 64)

Composition per 500mL:

Solution A	400.0mL
Solution B	100.0mL

pH 2.8 ± 0.2 at 25°C

Solution A:

Composition per 400mL:

$(NH_4)_2SO_4$	0.4g
KH_2PO_4	0.2g
$MgSO_4 \cdot 7H_2O$	0.08g

Preparation of Solution A: Add components to distilled/deionized water and bring volume to 400.0mL. Mix thoroughly. Autoclave for 15 min at 15 psi pressure–121°C. Cool to 45°–50°C.

Solution B:

Composition per 100mL:

$FeSO_4 \cdot 7H_2O$	10.0g
H_2SO_4 (1*N* solution)	1.0mL

Preparation of Solution B: Add components to distilled/deionized water and bring volume to 100.0mL. Mix thoroughly. Autoclave for 15 min at 15 psi pressure–121°C. Cool to 45°–50°C.

Preparation of Medium: Aseptically add 100.0mL of cooled sterile solution B to 400.0mL of cooled sterile solution A. Mix thoroughly. Adjust pH to 2.8. Aseptically distribute into sterile tubes or flasks.

Use: For the cultivation and maintenance of a variety of *Thiobacillus* species.

Thiobacillus Medium

Composition per liter:

$Na_2S_2O_3 \cdot 5H_2O$	10.0g
$Na_2HPO_4 \cdot 7H_2O$	7.9g
Sodium formate	6.8g
Glucose	3.6g
KNO_3	2.0g
KH_2PO_4	1.5g
NH_4Cl	0.3g
$MgSO_4 \cdot 7H_2O$	0.1g
Trace metals solution	5.0mL

pH 7.6–8.5 at 25°C

Trace Metals Solution:

Composition per liter:

Disodium EDTA	50.0g
NaOH	11.0g
$CaCl_2 \cdot 2H_2O$	7.34g
$FeSO_4 \cdot 7H_2O$	5.0g
$MnCl_2 \cdot 2H_2O$	2.5g
$ZnSO_4 \cdot 7H_2O$	2.2g
$CoCl_2 \cdot 6H_2O$	0.5g
$(NH_4)_6Mo_7O_{24} \cdot 4H_2O$	0.5g
$CuSO_4 \cdot 5H_2O$	0.2g

Preparation of Trace Metals Solution: Add EDTA to distilled/deionized water and bring volume to 500.0mL. Mix thoroughly. Adjust pH to 6.0 with NaOH. Add remaining components, one by one. Maintain the pH at 6.0. After dissolution of all the salts, adjust the pH to 4.0 with HCl. Store at 4°C.

Preparation of Medium: Add components to distilled/deionized water and bring volume to 1.0L. Mix thoroughly. Adjust pH to 7.6–8.5. Filter sterilize. Aseptically distribute into sterile tubes or flasks.

Use: For the isolation and anaerobic cultivation of *Thiobacillus* species.

Thiobacillus Medium

Composition per liter:

$Na_2S_2O_3 \cdot 5H_2O$	10.0g
$Na_2HPO_4 \cdot 7H_2O$	7.9g
Sodium formate	6.8g
Glucose	3.6g
KH_2PO_4	1.5g
NH_4Cl	0.3g
$MgSO_4 \cdot 7H_2O$	0.1g
Trace metals solution	5.0mL

pH 7.6–8.5 at 25°C

Trace Metals Solution:

Composition per liter:

Disodium EDTA	50.0g
NaOH	11.0g
$CaCl_2 \cdot 2H_2O$	7.34g

$FeSO_4 \cdot 7H_2O$ 5.0g
$MnCl_2 \cdot 2H_2O$ 2.5g
$ZnSO_4 \cdot 7H_2O$ 2.2g
$CoCl_2 \cdot 6H_2O$ 0.5g
$(NH_4)_6Mo_7O_{24} \cdot 4H_2O$ 0.5g
$CuSO_4 \cdot 5H_2O$ 0.2g

Preparation of Trace Metals Solution: Add EDTA to distilled/deionized water and bring volume to 500.0mL. Mix thoroughly. Adjust pH to 6.0 with NaOH. Add remaining components, one by one. Maintain the pH at 6.0. After dissolution of all the salts, adjust the pH to 4.0 with HCl. Store at 4°C.

Preparation of Medium: Add components to distilled/deionized water and bring volume to 1.0L. Mix thoroughly. Adjust pH to 7.6–8.5. Filter sterilize. Aseptically distribute into sterile tubes or flasks.

Use: For the isolation and aerobic cultivation of *Thiobacillus* species.

Thiobacillus Medium (ATCC Medium 125)

Composition per liter:

Sulfur 10.0g
KH_2PO_4 .... 3.0g
$MgSO_4 \cdot 7H_2O$ 0.5g
$CaCl_2$ 0.25g
$(NH_4)_2SO_4$ 0.2g
$FeSO_4 \cdot 7H_2O$ 5.0mg

Preparation of Medium: Add components, except sulfur, to tap water and bring volume to 1.0L. Mix thoroughly. Add 1.0g of sulfur to each of 10 flasks. Distribute the broth in 100.0mL volumes into the flasks. Pour the broth down the side of the flask so that the sulfur is not wetted. Autoclave for 30 min at 0 psi pressure–100°C on three consecutive days. Be sure that sulfur remains on the surface of the broth during the sterilization.

Use: For the cultivation and maintenance of a variety of *Thiobacillus* species.

Thiobacillus Medium (ATCC Medium 152)

Composition per liter:

Agar 15.0g
$Na_2S_2O_3 \cdot 5H_2O$ 10.0g
NH_4Cl 1.0g
Yeast extract 1.0g
K_2HPO_4 .... 0.6g
$MgCl_2$ 0.5g
KH_2PO_4 .... 0.4g
Chlorophenol Red 0.08g
$FeCl_3$ 0.02g

Preparation of Medium: Add components to distilled/deionized water and bring volume to 1.0L. Mix thoroughly. Gently heat and bring to boiling. Distribute into tubes or flasks. Autoclave for 15 min at 15 psi pressure–121°C. Pour into sterile Petri dishes or leave in tubes.

Use: For the cultivation and maintenance of a variety of *Thiobacillus* species.

Thiobacillus Medium (ATCC Medium 426)

Composition per liter:

$Na_2S_2O_3 \cdot 5H_2O$ 10.0g
$Na_2HPO_4 \cdot 7H_2O$ 7.9g
KH_2PO_4 .... 1.5g
NH_4Cl 0.3g
$MgSO_4 \cdot 7H_2O$ 0.1g
Phenol Red 2.0mg
Trace metal solution 5.0mL

pH 8.5 ± 0.2 at 25°C

Trace Metal Solution:

Composition per liter:

EDTA 50.0g
$ZnSO_4$ 22.0g
$CaCl_2$ 5.54g
$MnCl_2$ 5.06g
$FeSO_4 \cdot 7H_2O$ 4.99g
$CoCl_2$ 1.61g
$CuSO_4$ 1.57g
$(NH_4)_2MoO_4 \cdot 4H_2O$ 1.10g

Preparation of Trace Metal Solution: Add components to distilled/deionized water and bring volume to 1.0mL. Mix thoroughly. Adjust pH to 6.0 with KOH.

Preparation of Medium: Add components to distilled/deionized water and bring volume to 1.0L. Mix thoroughly. Adjust pH to 8.5 with 10% Na_2CO_3. Distribute into tubes or flasks. Autoclave for 15 min at 15 psi pressure–121°C. Adjust pH to 8.5 with sterile 10% Na_2CO_3 if necessary. The broth should be pink.

Use: For the cultivation and maintenance of a variety of *Thiobacillus* species.

Thiobacillus Medium (ATCC Medium 528)

Composition per liter:

$Na_2S_2O_3$ 10.0g
Yeast extract 5.0g
NH_4Cl 1.0g

K_2HPO_4	0.6g
$MgCl_2$	0.5g
KH_2PO_4	0.4g
$MgSO_4$	0.3g
Bromthymol Blue	0.03g
$FeCl_3$	0.02g
Heavy metal solution	30.0mL

pH 6.8 ± 0.2 at 25°C

Heavy Metal Solution:

Composition per liter:

Ethylenediamine tetraacetate	1.5g
$FeSO_4 \cdot 7H_2O$	0.2g
$ZnSO_4 \cdot 7H_2O$	0.1g
$MnCl_2 \cdot 4H_2O$	0.02g
Modified Hoagland trace element solution	6.0mL

Preparation of Heavy Metal Solution: Add EDTA to approximately 900.0mL of distilled/deionized water. Dissolve by adjusting pH to 7.0 with NaOH. Bring volume to 1.0L with distilled/deionized water.

Modified Hoagland Trace Element Solution:

Composition per 3.6 liters:

H_3BO_3	11.0g
$MnCl_2 \cdot 4H_2O$	7.0g
$AlCl_3$	1.0g
$CoCl_2$	1.0g
$CuCl_2$	1.0g
KI	1.0g
$NiCl_2$	1.0g
$ZnCl_2$	1.0g
$BaCl_2$	0.5g
KBr	0.5g
LiCl	0.5g
Na_2MoO_4	0.5g
$SeCl_4$	0.5g
$SnCl_2 \cdot 2H_2O$	0.5g
$NaVO_3 \cdot H_2O$	0.1g

Preparation of Modified Hoagland Trace Element Solution: Prepare each component as a separate solution. Dissolve each salt in approximately 100.0mL of distilled /deionized water. Adjust the pH of each solution to below 7.0. Combine all the salt solutions and bring the volume to 3.6L with distilled/deionized water. Adjust the pH to 3–4. A yellow precipitate may form after mixing. After a few days, it will turn into a fine white precipitate. Mix the solution thoroughly before using.

Preparation of Medium: Add components to distilled/deionized water and bring volume to 1.0L. Mix thoroughly. Distribute into tubes or flasks. Autoclave for 15 min at 15 psi pressure–121°C.

Use: For the cultivation and maintenance of a variety of *Thiobacillus* species.

Thiobacillus Medium B

Composition per liter:

Noble agar	15.0g
$Na_2S_2O_3 \cdot 5H_2O$	5.0g
KH_2PO_4	3.0g
NH_4Cl	0.1g
$MgCl_2$	0.1g
$CaCl_2$	0.1g

pH 4.2 ± 0.2 at 25°C

Preparation of Medium: Add components to distilled/deionized water and bring volume to 1.0L. Mix thoroughly. Gently heat and bring to boiling. Distribute into tubes or flasks. Autoclave for 15 min at 15 psi pressure–121°C. Pour into sterile Petri dishes or leave in tubes.

Use: For the cultivation and maintenance of *Thiobacillus thiooxidans* and *Streptomyces scabies*.

Thiobacillus neapolitanus Medium

Composition per 1002mL:

Solution I	1.0L
Solution II	2.0mL

pH 6.2–7.0 at 25°C

Solution I:

Composition per liter:

$Na_2S_2O_3 \cdot 5H_2O$	10.0g
KH_2PO_4	4.0g
K_2HPO_4	4.0g
$MgSO_4 \cdot 7H_2O$	0.8g
$KHCO_3$	0.7g
NH_4Cl	0.4g

Preparation of Solution I: Add components to distilled/deionized water and bring volume to 1.0L. Mix thoroughly.

Solution II:

Composition per liter:

Disodium EDTA	50.0g
NaOH	11.0g
$CaCl_2 \cdot 2H_2O$	7.34g
$FeSO_4 \cdot 7H_2O$	5.0g
$MnCl_2 \cdot 2H_2O$	2.5g
$ZnSO_4 \cdot 7H_2O$	2.2g
$CoCl_2 \cdot 6H_2O$	0.5g
$(NH_4)_6Mo_7O_{24} \cdot 4H_2O$	0.5g
$CuSO_4 \cdot 5H_2O$	0.2g

Preparation of Solution II: Add EDTA to distilled/deionized water and bring volume to 500.0mL. Mix thoroughly. Adjust pH to 6.0 with NaOH. Add remaining components, one by one. Maintain the pH at 6.0. After dissolution of all the salts, adjust the pH to 4.0 with HCl. Store at 4°C.

Preparation of Medium: Aseptically combine 1.0L of solution I and 2.0mL of solution II. Mix thoroughly. Adjust pH to 6.2–7.0. Distribute into tubes or flasks. Autoclave for 15 min at 15 psi pressure–121°C.

Use: For the isolation and cultivation of *Thiobacillus neapolitanus.*

Thiobacillus novellus Medium

Composition per liter:

$Na_2S_2O_3 \cdot 5H_2O$	10.0g
K_2HPO	4.0g
KH_2PO_4	1.5g
$MgSO_4 \cdot 7H_2O$	0.5g
$(NH_4)_2SO_4$	0.3g
Yeast extract	0.3g
Trace metals solution	10.0mL

pH 6.8–7.2 at 25°C

Trace Metals Solution:

Composition per liter:

Disodium EDTA	50.0g
NaOH	11.0g
$CaCl_2 \cdot 2H_2O$	7.34g
$FeSO_4 \cdot 7H_2O$	5.0g
$MnCl_2 \cdot 2H_2O$	2.5g
$ZnSO_4 \cdot 7H_2O$	2.2g
$CoCl_2 \cdot 6H_2O$	0.5g
$(NH_4)_6Mo_7O_{24} \cdot 4H_2O$	0.5g
$CuSO_4 \cdot 5H_2O$	0.2g

Preparation of Trace Metals Solution: Add EDTA to distilled/deionized water and bring volume to 500.0mL. Mix thoroughly. Adjust pH to 6.0 with NaOH. Add remaining components, one by one. Maintain the pH at 6.0. After dissolution of all the salts, adjust the pH to 4.0 with HCl. Store at 4°C.

Preparation of Medium: Add components to distilled/deionized water and bring volume to 1.0L. Mix thoroughly. Distribute into tubes or flasks. Autoclave for 15 min at 15 psi pressure–121°C.

Use: For the isolation and cultivation of *Thiobacillus novellus.*

Thiobacillus tepidarius Medium

Composition per liter:

Agar	10.0g
$Na_2S_2O_3 \cdot 5H_2O$	4.96g
$MgSO_4 \cdot 7H_2O$	0.8g
NH_4Cl	0.4g
Phosphate solution	100.0mL
Bromcresol Purple, saturated solution	2.0mL
Trace metals A-5	1.0mL

Phosphate Solution:

Composition per 100mL:

KH_2PO_4	4.0g
K_2HPO_4	4.0g

Preparation of Phosphate Solution: Add components to distilled/deionized water and bring volume to 100.0mL. Mix thoroughly. Autoclave for 15 min at 15 psi pressure–121°C.

Trace Metals A-5:

Composition per liter:

H_3BO_3	2.86g
$MnCl_2 \cdot 4H_2O$	1.81g
$Na_2MoO_4 \cdot 2H_2O$	0.39g
$ZnSO_4 \cdot 7H_2O$	0.222g
$CuSO_4 \cdot 5H_2O$	0.079g
$Co(NO_3)_2 \cdot 6H_2O$	49.4mg

Preparation of Trace Metals A-5: Add components to distilled/deionized water and bring volume to 1.0L. Mix thoroughly.

Preparation of Medium: Add components, except phosphate solution, to distilled/deionized water and bring volume to 900.0mL. Autoclave for 15 min at 15 psi pressure–121°C. Aseptically add 100.0mL of the sterile phosphate solution. Mix thoroughly. Aseptically distribute into sterile tubes or flasks.

Use: For the cultivation and maintenance of *Thiobacillus tepidarius.*

Thiobacillus thiooxidans Medium

Composition per liter:

Sulfur, powdered	10.0g
KH_2PO_4	5.0g
$MgSO_4 \cdot 7H_2O$	0.5g
$CaCl_2$	0.25g
$(NH_4)_2SO_4$	0.2g
$FeSO_4$	0.01g

pH 7.0 ± 0.2 at 25°C

Preparation of Medium: Add components, except sulfur, to distilled/deionized water and bring volume to 1.0L. Mix thoroughly. Distribute into flasks in 100.0mL volumes. Add 1.0g of sulfur to each flask. Autoclave for 30 min at 0 psi pressure–100°C on three consecutive days.

Use: For the cultivation of *Thiobacillus thiooxidans.*

Thiobacillus thiooxidans Medium

Composition per liter:

Flowers of sulfur	5.0g
K_2HPO	3.5g
$MgSO_4 \cdot 7H_2O$	0.5g
$(NH_4)_2SO_4$	0.3g

$CaCl_2$ 0.25g
$FeSO_4 \cdot 7H_2O$ 0.02g

pH 4.5 ± 0.2 at 25°C

Preparation of Medium: Add components, except flowers of sulfur, to distilled/deionized water and bring volume to 1.0L. Mix thoroughly. Gently heat and bring to boiling. Distribute into flasks or bottles in 100.0mL volumes. Add 0.5g of flowers of sulfur to each flask or bottle. Autoclave for 15 min at 15 psi pressure–121°C.

Use: For the isolation and cultivation of *Thiobacillus thiooxidans*.

Thiobacillus thioparus Medium

Composition per liter:

$Na_2S_2O_3 \cdot 5H_2O$ 5.0g
K_2HPO_4 .... 4.0g
$MgSO_4 \cdot 7H_2O$ 0.5g
$(NH_4)_2SO_4$ 0.4g
$CaCl_2$ 0.25g
$FeSO_4$ 0.01g

pH 7.0 ± 0.2 at 25°C

Preparation of Medium: Add components to distilled/deionized water and bring volume to 1.0L. Mix thoroughly. Distribute into tubes or flasks. Autoclave for 15 min at 15 psi pressure–121°C.

Use: For the cultivation of *Thiobacillus thioparus*.

Thiocapsa Medium

Composition per 127mL:

Solution 1 .... 76.2mL
Solution 2 + Solution 3 .... 44.8mL
Solution 4 .... 6.0mL

Solution 1:

Composition per 2.5L:

NaCl 39.68g
$CaCl_2$ 2.0g

Preparation of Solution 1: Add components to distilled/deionized water and bring volume to 2.5L. Distribute in 80.0mL volumes into 127mL screw-capped bottles. Autoclave for 15 min at 15 psi pressure–121°C.

Solution 2:

Composition per 100mL:

Sodium ascorbate 2.4g
KCl 1.0g
KH_2PO_4 .... 1.0g
$MgCl_2 \cdot 6H_2O$ 0.8g
NH_4Cl 0.8g
Heavy metal solution 50.0mL
Vitamin solution 15.0mL
Vitamin B_{12} solution 3.0mL

Preparation of Solution 2: Add components to distilled/deionized water and bring volume to 100.0mL. Mix thoroughly.

Heavy Metal Solution:

Composition per liter:

Ethylenediamine tetraacetate (EDTA) 1.5g
$FeSO_4 \cdot 7H_2O$ 0.2g
$ZnSO_4 \cdot 7H_2O$ 0.1g
$MnCl_2 \cdot 4H_2O$ 0.02g
Modified Hoagland trace element solution 6.0mL

Preparation of Heavy Metal Solution: Dissolve EDTA in approximately 800.0mL of distilled/deionized water. Add remaining components. Bring volume to 1.0L with distilled/deionized water. Mix thoroughly.

Modified Hoagland Trace Element Solution:

Composition per 3.6L:

H_3BO_3 .... 11.0g
$MnCl_2 \cdot 4H_2O$ 7.0g
$AlCl_3$ 1.0g
$CoCl_2$ 1.0g
$CuCl_2$ 1.0g
KI 1.0g
$NiCl_2$ 1.0g
$ZnCl_2$ 1.0g
$BaCl_2$ 0.5g
KBr 0.5g
LiCl 0.5g
Na_2MoO_4 .... 0.5g
$SeCl_4$ 0.5g
$SnCl_2 \cdot 2H_2O$ 0.5g
$NaVO_3 \cdot H_2O$ 0.1g

Preparation of Modified Hoagland Trace Element Solution: Prepare each component as a separate solution. Dissolve each salt in approximately 100.0mL of distilled/deionized water. Adjust the pH of each solution to below 7.0. Combine all the salt solutions and bring the volume to 3.6L with distilled/deionized water. Adjust the pH to 3–4. A yellow precipitate may form after mixing. After a few days, it will turn into a fine white precipitate. Mix the solution thoroughly before using.

Vitamin Solution:

Composition per 100mL:

Pyridoxamine·2HCl 5.0mg
Nicotinic acid 2.0mg
Thiamine 1.0mg
Pantothenic acid 0.5mg
Biotin 0.2mg
p-Aminobenzoic acid 0.1mg

Preparation of Vitamin Solution: Add components to distilled/deionized water and bring volume to 100.0mL. Mix thoroughly.

Vitamin B_{12} Solution:

Composition per 100mL:

Vitamin B_{12} (cyanocobalamin) 2.0mg

Preparation of Vitamin B_{12} Solution: Add vitamin B_{12} to distilled/deionized water and bring volume to 100.0mL. Mix thoroughly.

Solution 3:

Composition per 900mL:

$NaHCO_3$ 4.5g

Preparation of Solution 3: Add $NaHCO_3$ to distilled/deionized water and bring volume to 900.0mL. Mix thoroughly. Bubble 100% CO_2 through the solution for 30 min. After CO_2 saturation of Solution 3, add Solution 2 and immediately filter the mixture through a Seitz filter (or a Millipore) using positive CO_2 pressure to push the liquid through.

Solution 4:

Composition per 200mL:

$Na_2S \cdot 9H_2O$ 3.0g

Preparation of Solution 4: Add $Na_2S \cdot 9H_2O$ to distilled/deionized water and bring volume to 200.0mL. Add a magnetic stir bar to the flask. Autoclave for 15 min at 15 psi pressure–121°C. On a magnetic stirrer, slowly add 2.0mL of sterile 2M H_2SO_4. This partially neutralizes the solution. The solution should turn yellow. H_2S gas will be liberated—neutralization and distribution of the solution should be done as rapidly as possible under adequate ventilation.

Preparation of Medium: To the 80.0mL of sterile solution 1 in screw-capped bottles, add combined solutions 2 and 3 immediately after filtration and fill bottles to capacity. Mix thoroughly. Aseptically remove 6.0mL of the medium from the bottles and replace it with 6.0mL of neutralized solution 4. Let stand for 24 hr. The medium should form a fine white precipitate before using. To inoculate, remove 6.0mL of the completed medium from the bottles and replace it with 6.0mL of inoculum.

Use: For the cultivation and maintenance of a variety of *Thiocapsa* species.

Thiocyanate Utilization Medium

Composition per 1225mL:

Basal solution.. 1.0L
Solution C 200.0mL
Solution B 20.0mL
Solution A 5.0mL

Basal Solution:

Composition per liter:

Na_2HPO_4 ....4.8g
KH_2PO_4 ....4.4g
$MgSO_4 \cdot 7H_2O$ 0.5g

Preparation of Basal Solution: Add components to distilled/deionized water and bring volume to 1.0L. Mix thoroughly. Autoclave for 15 min at 15 psi pressure–121°C. Cool to 45°–50°C.

Solution A:

Composition per100mL:

$FeCl_3 \cdot 6H_2O$ 1.0g
$CaCl_2$ 0.1g

Preparation of Solution A: Add components to distilled/deionized water and bring volume to 100.0mL. Mix thoroughly. Filter sterilize.

Solution B:

Composition per 100mL:

D-Glucose 10.0g

Preparation of Solution B: Add glucose to distilled/deionized water and bring volume to 100.0mL. Mix thoroughly. Filter sterilize.

Solution C:

Composition per 200mL:

NaSCN 1.0g

Preparation of Solution C: Add NaSCN to distilled/deionized water and bring volume to 200.0mL. Mix thoroughly. Filter sterilize.

Preparation of Medium: To 1.0L of cooled, sterile basal solution, aseptically add 5.0mL of sterile solution A, 20.0mL of sterile solution B, and 200.0mL of sterile solution C. Mix thoroughly. Aseptically distribute into sterile tubes or flasks.

Use: For the cultivation and maintenance of a variety of microorganisms which can utilize thiocyanate as sole source of nitrogen and sulfur.

Thiogel® Medium

Composition per liter:

Gelatin..50.0g
Pancreatic digest of casein..............................17.0g
Glucose 6.0g
Papaic digest of soybean meal 3.0g
NaCl 2.5g
Sodium thioglycollate 0.5g
Agar..0.7g
Na_2SO_3 ....0.1g
L-Cystine 0.25g

pH 7.0 ± 0.2 at 25°C

Source: This medium is available as a premixed powder from BBL Microbiology Systems.

Preparation of Medium: Add components to distilled/deionized water preheated to 50°C and bring volume to 1.0L. Mix thoroughly. Let stand for 5 min. Gently heat while stirring and bring to boiling. Distribute into tubes filling them half full. Autoclave for 15 min at 13 psi pressure–118°C. Pour into sterile Petri dishes or leave in tubes.

Use: For the differentiation of microorganisms based on their ability to liquefy gelatin.

Thioglycollate Bile Broth

Composition per 1050mL:

Pancreatic digest of casein	15.0g
Glucose	5.5g
Yeast extract	5.0g
NaCl	2.5g
Agar	0.75g
L-Cystine	0.5g
Sodium thioglycollate	0.5g
Bile solution	50.0mL

pH 7.1 ± 0.2 at 25°C

Bile Solution:

Composition per 100mL:

Oxgall	40.0g
Sodium deoxycholate	2.0g

Preparation of Bile Solution: Add components to distilled/deionized water and bring volume to 100.0mL. Mix thoroughly. Filter sterilize.

Preparation of Medium: Add components, except bile solution, to distilled/deionized water and bring volume to 1.0L. Mix thoroughly. Gently heat and bring to boiling. Distribute into tubes in 10.0mL volumes. Autoclave for 15 min at 15 psi pressure–121°C. Cool to 45°–50°C. Aseptically add 0.5mL of sterile bile solution to each tube. Mix thoroughly.

Use: For the cultivation of *Bacteroides fragilis* and *Clostridium perfringens* from clinical specimens.

Thioglycollate Broth USP, Alternative

Composition per liter:

Pancreatic digest of casein	15.0g
Glucose	5.5g
Yeast extract	5.0g
NaCl	2.5g
L-Cystine	0.5g
Sodium thioglycollate	0.5g

pH 7.1 ± 0.2 at 25°C

Source: This medium is available as a premixed powder from Oxoid Unipath.

Preparation of Medium: Add components to distilled/deionized water and bring volume to 1.0L. Mix thoroughly. Distribute into tubes or flasks. Autoclave for 15 min at 15 psi pressure–121°C. Prepare freshly or boil and cool the medium just before use.

Use: For the cultivation of both aerobic and anaerobic organisms in the performance of sterility tests of turbid or viscous specimens.

Thioglycollate Gelatin Medium

Composition per liter:

Gelatin	50.0g
Pancreatic digest of casein	15.0g
Yeast extract	5.0g
NaCl	2.5g
Glucose	2.0g
Agar	0.75g
L-Cystine	0.25g
Na_2SO_3	0.1g
Thioglycollic acid	0.3mL

pH 7.0 ± 0.2 at 25°C

Source: This medium is available as a premixed powder from Difco Laboratories.

Preparation of Medium: Add components to distilled/deionized water and bring volume to 1.0L. Mix thoroughly. Gently heat and bring to 50°C. Let stand 5 min. Gently heat and bring to boiling. Distribute into tubes or flasks. Autoclave for 15 min at 15 psi pressure–121°C.

Use: For the determination of gelatin liquefaction by aerobes, microaerophiles and anaerobes without special incubation.

Thioglycollate Medium, Brewer

Composition per liter:

Glucose	5.0g
Peptone	5.0g
NaCl	5.0g
Yeast extract	2.0g
Sodium thioglycollate	1.1g
Agar	1.0g
Beef extract	1.0g
Methylene Blue	2.0mg

pH 7.2 ± 0.2 at 25°C

Source: This medium is available as a premixed powder from Oxoid Unipath.

Preparation of Medium: Add components to distilled/deionized water and bring volume to 1.0L. Mix thoroughly. Gently heat and bring to boiling. Distribute into tubes or flasks. Autoclave for 15 min at 15 psi pressure–121°C.

Use: For the determination of sterility of solutions containing mercurial preservatives.

Thioglycollate Medium, Brewer Modified

Composition per liter:

Pancreatic digest of casein	17.5g
Glucose	10.0g
NaCl	5.0g
Papaic digest of soybean meal	2.5g
K_2HPO_4	2.0g
Sodium thioglycollate	1.0g
Agar	0.5g
Methylene Blue	0.002g

pH 7.2 ± 0.2 at 25°C

Source: This medium is available as a premixed powder from BBL Microbiology Systems.

Preparation of Medium: Add components to distilled/deionized water and bring volume to 1.0L. Mix thoroughly. Gently heat while stirring and bring to boiling. Distribute into tubes or flasks, filling them half full. Autoclave for 15 min at 15 psi pressure–121°C.

Use: For the cultivation of obligate anaerobes, microaerophiles and facultative organisms.

Thioglycollate Medium, Enriched (THIO Medium) (Thioglycollate Medium with Vitamin K_1 and Hemin)

Composition per liter:

Thioglycollate medium without indicator	1.0L
Hemin solution	0.5mL
Vitamin K_1 solution	0.1mL

pH 7.0 ± 0.2 at 25°C

Thioglycollate Medium without Indicator:
Composition per liter:

Pancreatic digest of casein	17.0g
Glucose	6.0g
Papaic digest of soybean meal	3.0g
NaCl	2.5g
Agar	0.7g
Sodium thioglycollate	0.5g
L-Cystine	0.25g
Na_2SO_3	0.1g

Source: Thioglycollate Medium without Indicator is available as a premixed powder from Oxoid Unipath and BBL Microbiology Systems.

Preparation of Thioglycollate Medium without Indicator: Add components to distilled/deionized water and bring volume to 1.0L. Mix thoroughly.

Vitamin K_1 Solution:
Composition per 100mL:

Vitamin K_1	1.0g

Preparation of Vitamin K_1 Solution: Add vitamin K_1 to 99.0mL of absolute ethanol. Mix thoroughly.

Hemin Solution:
Composition per 100mL:

Hemin	1.0g
NaOH (1*N* solution)	20.0mL

Preparation of Hemin Solution: Add hemin to 20.0mL of 1*N* NaOH solution. Mix thoroughly. Bring volume to 100.0mL with distilled/deionized water.

Preparation of Medium: Add 0.5mL of hemin solution and 0.1mL of vitamin K_1 solution to 1.0L of thioglycollate medium without indicator. Mix thoroughly. Distribute into screw-capped tubes or flasks. Autoclave for 15 min at 15 psi pressure–121°C. Cool tubes or flasks under 85% N_2 + 10% H_2 + 5% CO_2. Tighten caps.

Use: For the isolation, cultivation and identification of a wide variety of obligate anaerobic bacteria.

Thioglycollate Medium, Supplemented

See: **Thioglycollate Medium without Indicator with Hemin**

Thioglycollate Medium, USP

Composition per liter:

Pancreatic digest of casein	15.0g
Glucose	5.5g
Yeast extract	5.0g
NaCl	2.5g
Agar	0.5g
L-cystine	0.5g
Sodium thioglycollate	0.5g
Resazurin	1.0mg

pH 7.1 ± 0.2 at 25°C

Source: This medium is available as a premixed powder from Oxoid Unipath.

Preparation of Medium: Add components to distilled/deionized water and bring volume to 1.0L. Mix thoroughly. Gently heat and bring to boiling. Distribute into tubes or flasks. Autoclave for 15 min at 15 psi pressure–121°C.

Use: For the cultivation of both aerobic and anaerobic organisms in the performance of sterility tests.

Thioglycollate Medium with 20% Bile (THIO + Bile Medium)

Composition per liter:

Oxgall	20.0g
Thioglycollate medium without indicator	1.0L
Hemin solution	0.5mL
Vitamin K_1 solution	0.1mL

pH 7.0 ± 0.2 at 25°C

Thioglycollate Medium without Indicator:
Composition per liter:

Pancreatic digest of casein	17.0g
Glucose	6.0g
Papaic digest of soybean meal	3.0g
NaCl	2.5g
Agar	0.7g
Sodium thioglycollate	0.5g
L-Cystine	0.25g
Na_2SO_3	0.1g

Preparation of Thioglycollate Medium Without Indicator: Add components to distilled/deionized water and bring volume to 1.0L. Mix thoroughly.

Vitamin K_1 Solution:
Composition per 100mL:

Vitamin K_1	1.0g

Preparation of Vitamin K_1 Solution: Add vitamin K_1 to 99.0mL of absolute ethanol. Mix thoroughly.

Hemin Solution:
Composition per 100mL:

Hemin	1.0g
NaOH (1*N* solution)	20.0mL

Preparation of Hemin Solution: Add hemin to 20.0mL of 1*N* NaOH solution. Mix thoroughly. Bring volume to 100.0mL with distilled/deionized water.

Preparation of Medium: Add 0.5mL of hemin solution, 0.1mL of vitamin K_1 solution, and oxgall to 1.0L of Thioglycolate Medium without Indicator. Mix thoroughly. Distribute into screw-capped tubes or flasks. Autoclave for 15 min at 15 psi pressure–121°C. Cool tubes or flasks under 85% N_2 + 10% H_2 + 5% CO_2. Tighten caps.

Use: For the isolation, cultivation and identification of a variety of obligate anaerobic bile-tolerant bacteria.

Thioglycollate Medium with Vitamin K_1 and Hemin

See: **Thioglycollate Medium, Enriched**

Thioglycollate Medium without Glucose

Composition per liter:

Pancreatic digest of casein	15.0g
Yeast extract	5.0g
NaCl	2.5g
Agar	0.75g
L-Cystine	0.25g
Methylene Blue	2.0mg
Thioglycollic acid	0.3mL

pH 7.2 ± 0.2 at 25°C

Source: This medium is available as a premixed powder from Difco Laboratories.

Preparation of Medium: Add components to distilled/deionized water and bring volume to 1.0L. Mix thoroughly. Gently heat and bring to boiling. Distribute into tubes or flasks. Autoclave for 15 min at 15 psi pressure–121°C. If medium becomes oxidized before use (Methylene Blue turns blue) heat in a boiling water bath to expel absorbed O_2. Cool to 25°C.

Use: For the cultivation of anaerobic, microaerophilic and aerobic microorganisms. For use in sterility testing of a variety of specimens.

Thioglycollate Medium without Glucose

Composition per liter:

Pancreatic digest of casein	20.0g
NaCl	2.5g
K_2HPO_4	1.5g
Sodium thioglycollate	0.6g
Agar	0.5g
L-Cystine	0.4g
Na_2SO_3	0.2g
Methylene Blue	2.0mg

pH 7.2 ± 0.2 at 25°C

Source: This medium is available as a premixed powder from BBL Microbiology Systems.

Preparation of Medium: Add components to distilled/deionized water and bring volume to 1.0L. Mix thoroughly. Gently heat while stirring and bring to boiling. Distribute into tubes or flasks, filling them half full. Autoclave for 15 min at 15 psi pressure–121°C.

Use: Use as a base for fermentation studies of anaerobic bacteria and for the promotion of endospore formation.

Thioglycollate Medium without Glucose and Indicator

Composition per liter:

Pancreatic digest of casein	15.0g
Yeast extract	5.0g
NaCl	2.5g
Agar	0.75g
L-Cystine	0.25g
Thioglycollic acid	0.3mL

pH 7.2 ± 0.2 at 25°C

Source: This medium is available as a premixed powder from Difco Laboratories.

Preparation of Medium: Add components to distilled/deionized water and bring volume to 1.0L. Mix thoroughly. Gently heat and bring to boiling. Distribute into tubes or flasks. Autoclave for 15 min at 15 psi pressure–121°C. If medium becomes oxidized before use, heat in a boiling water bath to expel absorbed O_2. Cool to 25°C.

Use: For the cultivation of anaerobic, microaerophilic and aerobic microorganisms. For use in sterility testing of a variety of specimens.

Thioglycollate Medium without Indicator

Composition per liter:

Pancreatic digest of casein	15.0g
Yeast extract	5.0g
Glucose	5.0g
NaCl	2.5g
Agar	0.75g
Sodium thioglycollate	0.5g
L-Cystine	0.25g

pH 7.2 ± 0.2 at 25°C

Source: This medium is available as a premixed powder from Difco Laboratories.

Preparation of Medium: Add components to distilled/deionized water and bring volume to 1.0L. Mix thoroughly. Gently heat and bring to boiling. Distribute into tubes or flasks. Autoclave for 15 min at 15 psi pressure–121°C. If medium becomes oxidized before use, heat in a boiling water bath to expel absorbed O_2. Cool to 25°C.

Use: For the cultivation of anaerobic, microaerophilic and aerobic microorganisms. For use in sterility testing of a variety of specimens.

Thioglycollate Medium without Indicator

Composition per liter:

Pancreatic digest of casein	17.0g
Glucose	6.0g
Papaic digest of soybean meal	3.0g
NaCl	2.5g
Agar	0.7g
Sodium thioglycollate	0.5g
L-Cystine	0.25g
Na_2SO_3	0.1g

pH 7.0 ± 0.2 at 25°C

Source: This medium is available as a premixed powder from Oxoid Unipath.

Preparation of Medium: Add components to distilled/deionized water and bring volume to 1.0L. Mix thoroughly. Distribute into tubes or flasks. Autoclave for 15 min at 15 psi pressure–121°C. Prepare freshly or boil and cool the medium just before use.

Use: For the growth of aerobic and anaerobic microorganisms in diagnostic bacteriology.

Thioglycollate Medium without Indicator–135C

Composition per liter:

Pancreatic digest of casein	17.0g
Glucose	6.0g
Papaic digest of soybean meal	3.0g
NaCl	2.5g
Agar	0.7g
Sodium thioglycollate	0.5g
Na_2SO_3	0.1g
L-Cystine	0.25g

pH 7.0 ± 0.2 at 25°C

Source: This medium is available as a premixed powder from BBL Microbiology Systems.

Preparation of Medium: Add components to distilled/deionized water and bring volume to 1.0L. Mix thoroughly. Gently heat while stirring and bring to boiling. Distribute into tubes or flasks filling them half full. For maintenance of cultures, a small quantity of $CaCO_3$ may be added to tubes before adding medium. Autoclave for 15 min at 13 psi pressure–118°C. Prepare freshly or boil and cool the medium just before use. Store prepared medium at 2°–8°C in the dark.

Use: For the isolation and cultivation of a wide variety of microorganisms, particularly obligate anaerobes, from clinical specimens and other materials.

Thioglycollate Medium without Indicator with Hemin (Thioglycollate Medium, Supplemented)

Composition per liter:

Pancreatic digest of casein	17.0g
$CaCO_3$, chips or powder	10.0g
Glucose	6.0g
Papaic digest of soybean meal	3.0g
NaCl	2.50g
Agar	0.70g
Sodium thioglycollate	0.50g
L-Cystine	0.25g
Na_2SO_3	0.10g
Hemin	5.0mg
Na_2CO_3 solution	10.0mL
Vitamin K_1 solution	10.0mL

pH 7.2 ± 0.2 at 25°C

Na_2CO_3 Solution:

Composition per 10mL:

Na_2CO_3	1.0g

Preparation of Na_2CO_3 Solution: Add Na_2CO_3 to distilled/deionized water and bring volume to 10.0mL. Mix thoroughly. Filter sterilize.

Vitamin K_1 Solution:

Composition per 100mL:

Vitamin K_1	1.0g
Ethanol, absolute	99.0mL

Preparation of Vitamin K_1 Solution: Add vitamin K_1 to 99.0mL of absolute ethanol. Mix thoroughly.

Preparation of Medium: Add components, except $CaCO_3$, Na_2CO_3 solution and vitamin K_1 solution, to distilled/deionized water and bring volume to 990.0mL. Mix thoroughly. Gently heat and bring to boiling. Add 0.1g of $CaCO_3$ chips or powder to each of 100 test tubes. Distribute broth into the same tubes in 10.0mL volumes. Autoclave for 15 min at 15 psi pressure–121°C. Cool to 45°–50°C. Aseptically add 0.1mL of sterile Na_2CO_3 solution and 0.1mL of sterile vitamin K_1 solution to each tube. Mix thoroughly.

Use: For the cultivation of a wide variety of obligate anaerobes.

Thioglycollate Peptone Glucose Yeast Extract Medium *See:* TPGY Medium

Thioglycollate Potato Liver Medium *See:* TPL Medium

Thiol Broth

Composition per liter:

Proteose peptone, No. 3	10.0g
Thiol complex (Difco)	8.0g
Yeast extract	5.0g
NaCl	5.0g
Glucose	1.0g
p-Aminobenzoic acid	0.05g

pH 7.1 ± 0.2 at 25°C

Source: This medium is available as a premixed powder from Difco Laboratories.

Preparation of Medium: Add components to distilled/deionized water and bring volume to 1.0L. Mix thoroughly. Gently heat and bring to boiling. Distribute into tubes or flasks. For neutralization of penicillin, distribute medium into tubes to a depth of 60 mm. For neutralization of streptomycin, distribute medium into tubes in shallow layers. Autoclave for 15 min at 15 psi pressure–121°C.

Use: For the cultivation of bacteria from body fluids and other materials containing penicillin, streptomycin or sulfonamides. Also used for the cultivation and maintenance of *Bifidobacterium* species.

Thiol Medium

Composition per liter:

Proteose peptone, No. 3	10.0g
Thiol complex	8.0g
Yeast extract	5.0g
NaCl	5.0g
Glucose	1.0g
Agar	1.0g
p-Aminobenzoic acid	0.05g

pH 7.1 ± 0.2 at 25°C

Source: This medium is available as a premixed powder from Difco Laboratories.

Preparation of Medium: Add components to distilled/deionized water and bring volume to 1.0L. Mix thoroughly. Gently heat and bring to boiling. Distribute into tubes or flasks. For neutralization of penicillin, distribute medium into tubes to a depth of 60 mm. For neutralization of streptomycin, distribute medium into tubes in shallow layers. Autoclave for 15 min at 15 psi pressure–121°C.

Use: For the cultivation of bacteria from body fluids and other materials containing penicillin, streptomycin or sulfonamides. Also used for the cultivation and maintenance of *Bifidobacterium* species.

Thiomicrospira denitrificans Agar

Composition per 1001mL:

Solution A 940.0mL
Solution B 40.0mL
Solution C 20.0mL
Solution D 1.0mL

pH 7.0 ± 0.2 at 25°C

Solution A:

Composition per 940mL:

Agar 15.0g
KH_2PO_4 .... 2.0g
KNO_3 .... 2.0g
NH_4Cl 1.0g
$MgSO_4 \cdot 7H_2O$ 0.8g
Trace element solution SL-4 .... 2.0mL

Preparation of Solution A: Add components to distilled/deionized water and bring volume to 940.0mL. Mix thoroughly. Gently heat and bring to boiling. Adjust pH to 7.0 with NaOH. Autoclave for 15 min at 15 psi pressure–121°C. Cool to 45°–50°C.

Trace Element Solution Sl-4:

Composition per liter:

EDTA 0.5g
$FeSO_4 \cdot 7H_2O$ 0.2g
Trace element solution SL-6 .... 100.0

Preparation of Trace Elements Solution SL-4: Add components to distilled/deionized water and bring volume to 1.0L. Mix thoroughly.

Trace Elements Solution SL-6:

Composition per liter:

H_3BO_3 .... 0.3g
$CoCl_2 \cdot 6H_2O$ 0.2g
$ZnSO_4 \cdot 7H_2O$ 0.10g
$MnCl_2 \cdot 4H_2O$ 0.03g
$Na_2MoO_4 \cdot H_2O$ 0.03g
$NiCl_2 \cdot 6H_2O$ 0.02g
$CuCl_2.2H_2O$ 0.01g

Preparation of Trace Elements Solution SL-6: Add components to distilled/deionized water and bring volume to 1.0L. Mix thoroughly. Adjust pH to 3.4.

Solution B:

Composition per 40mL:

$Na_2S_2O_3 \cdot 5H_2O$ 5.0g

Preparation of Solution B: Add $Na_2S_2O_3 \cdot 5H_2O$ to distilled/deionized water and bring volume to 40.0mL. Mix thoroughly. Autoclave for 15 min at 15 psi pressure–121°C. Cool to 45°–50°C.

Solution C:

Composition per 20mL:

$NaHCO_3$ 1.0g

Preparation of Solution C: Add $NaHCO_3$ to distilled/deionized water and bring volume to 20.0mL. Mix thoroughly. Filter sterilize.

Solution D:

Composition per liter:

$FeSO_4 \cdot 7H_2O$ 2.0mg
H_2SO_4 (0.1*N* solution) 1.0mL

Preparation of Solution D: Add $FeSO_4 \cdot 7H_2O$ to 1.0mL of 0.1*N* H_2SO_4 solution. Mix thoroughly. Autoclave for 15 min at 15 psi pressure–121°C. Cool to 45°–50°C.

Preparation of Medium: Aseptically add 40.0mL of sterile solution B, 20.0mL of sterile solution C and 1.0mL of sterile solution D to 940.0mL of sterile solution A. Mix thoroughly. Aseptically and anaerobically distribute into sterile tubes under 100% N_2.

Use: For the cultivation and maintenance of *Thiomicrospira denitrificans.*

Thiomicrospira denitrificans Broth

Composition per 1001mL:

Solution A 940.0mL
Solution B 40.0mL
Solution C 20.0mL
Solution D 1.0mL

pH 7.0 ± 0.2 at 25°C

Solution A:

Composition per 940mL:

KH_2PO_4 .... 2.0g
KNO_3 .... 2.0g
NH_4Cl 1.0g
$MgSO_4 \cdot 7H_2O$ 0.8g
Trace element solution SL-4 .... 2.0mL

Preparation of Solution A: Add components to distilled/deionized water and bring volume to 940.0mL. Mix thoroughly. Adjust pH to 7.0 with NaOH. Autoclave for 15 min at 15 psi pressure–121°C. Cool to 45°–50°C.

Solution B:

Composition per 40mL:

$Na_2S_2O_3 \cdot 5H_2O$ 5.0g

Preparation of Solution B: Add $Na_2S_2O_3 \cdot 5H_2O$ to distilled/deionized water and bring volume to 40.0mL. Mix thoroughly. Autoclave for 15 min at 15 psi pressure–121°C. Cool to 45°–50°C.

Solution C:

Composition per 20mL:

$NaHCO_3$ 1.0g

Preparation of Solution C: Add $NaHCO_3$ to distilled/deionized water and bring volume to 20.0mL. Mix thoroughly. Filter sterilize.

Solution D:

Composition per liter:

$FeSO_4 \cdot 7H_2O$ 2.0mg
H_2SO_4 (0.1*N* solution) 1.0mL

Preparation of Solution D: Add $FeSO_4 \cdot 7H_2O$ to 1.0mL of 0.1*N* H_2SO_4 solution. Mix thoroughly. Autoclave for 15 min at 15 psi pressure–121°C. Cool to 45°–50°C.

Trace Element Solution Sl-4:

Composition per liter:

EDTA 0.5g
$FeSO_4 \cdot 7H_2O$ 0.2g
Trace element solution SL-6 .... 100.0mL

Preparation of Trace Element Solution SL-4: Add components to distilled/deionized water and bring volume to 1.0L. Mix thoroughly.

Trace Element Solution SL-6:

Composition per liter:

H_3BO_3 .... 0.3g
$CoCl_2 \cdot 6H_2O$ 0.2g
$ZnSO_4 \cdot 7H_2O$ 0.10g
$MnCl_2 \cdot 4H_2O$ 0.03g
$Na_2MoO_4 \cdot H_2O$ 0.03g
$NiCl_2 \cdot 6H_2O$ 0.02g
$CuCl_2 \cdot 2H_2O$ 0.01g

Preparation of Trace Element Solution SL-6: Add components to distilled/deionized water and bring volume to 1.0L. Mix thoroughly. Adjust pH to 3.4.

Preparation of Medium: Aseptically add 40.0mL of sterile solution B, 20.0mL of sterile solution C and 1.0mL of sterile solution D to 940.0mL of sterile solution A. Mix thoroughly. Aseptically and anaerobically distribute into sterile tubes under 100% N_2.

Use: For the cultivation and maintenance of *Thiomicrospira denitrificans*.

Thiomicrospira denitrificans Medium

Composition per liter:

Part I 500.0mL
Part II 500.0mL

pH 7.0–8.0 at 25°C

Part I:

Composition per liter:

NaCl 20.0g
KNO_3 .... 4.0g
$(NH_4)_2SO_4$ 2.0g
$MgSO_4 \cdot 7H_2O$ 1.5g
K_2HPO_4 .... 0.6g
KH_2PO_4 .... 0.4g
$FeSO_4$ solution 2.0mL
Trace metals solution 2.0mL
HCl, concentrated 1.0mL

Preparation of Part I: Add components to distilled/deionized water and bring volume to 1.0L. Mix thoroughly. Autoclave for 15 min at 15 psi pressure–121°C. Cool to 25°C.

$FeSO_4$ Solution:

Composition per 100mL:

$FeSO_4 \cdot 7H_2O$ 0.5g
HCl (1*N* solution) 100.0mL

Preparation of $FeSO_4$ Solution: Combine the $FeSO_4 \cdot 7H_2O$ and 100.0mL of HCl solution. Mix thoroughly.

Trace Metals Solution:

Composition per liter:

Disodium EDTA 50.0g
NaOH 11.0g
$CaCl_2 \cdot 2H_2O$ 7.34g
$MnCl_2 \cdot 2H_2O$ 2.5g
$ZnSO_4 \cdot 7H_2O$ 2.2g
$CoCl_2 \cdot 6H_2O$ 0.5g
$(NH_4)_6Mo_7O_{24} \cdot 4H_2O$ 0.5g
$CuSO_4 \cdot 5H_2O$ 0.2g

Preparation of Trace Metals Solution: Add EDTA to distilled/deionized water and bring volume to 500.0mL. Mix thoroughly. Adjust pH to 6.0 with NaOH. Add remaining components, one by one. Maintain the pH at 6.0. After dissolution of all the salts, adjust the pH to 4.0 with HCl. Store at 4°C.

Part II:

Composition per liter:

$Na_2S_2O_3 \cdot 5H_2O$ 10.0g
$NaHCO_3$ 3.0g
NaOH 0.05g

Preparation of Part II: Add components to distilled/deionized water and bring volume to 1.0L. Mix thoroughly. Autoclave for 15 min at 15 psi pressure–121°C. Cool to 25°C.

Preparation of Medium: Aseptically combine 500.0mL of sterile part I and 500.0mL of sterile part II. Mix thoroughly. Aseptically distribute into sterile tubes or flasks.

Use: For the isolation and cultivation of *Thiomicrospira denitrificans.*

Thiomicrospira Medium (ATCC Medium 1422)

Composition per liter:

NaCl	25.1g
Tris·HCl	3.07g
$Na_2S_2O_3 \cdot 5H_2O$	2.48g
$MgSO_4 \cdot 7H_2O$	1.5g
$(NH_4)_2SO_4$	1.0g
KH_2PO_4	0.42g
$CaCl_2 \cdot 2H_2O$	0.29g
$NaHCO_3$	0.20g
Phenol Red (0.5% solution)	1.0mL
Vishniac and Santer trace metals	0.2mL

pH 7.5 ± 0.2 at 25°C

Vishniac and Santer Trace Metals:

Composition per liter:

Ethylenediamine tetraacetic acid (EDTA)	50.0g
$ZnSO_4 \cdot 7H_2O$	22.0g
$CaCl_2$	5.54g
$MnCl_2 \cdot 4H_2O$	5.06g
$FeSO_4 \cdot 7H_2O$	4.99g
$CoCl_2 \cdot 6H_2O$	1.61g
$CuSO_4 \cdot 5H_2O$	1.57g
$(NH_4)_6Mo_7O_{24} \cdot 4H_2O$	1.10g

Preparation of Vishniac and Santer Trace Metals: Add components to distilled/deionized water and bring volume to 1.0L. Adjust pH to 6.0 with KOH. Mix thoroughly.

Preparation of Medium: Add components to distilled/deionized water and bring volume to 1.0L. Mix thoroughly. Adjust pH to 7.5. Filter sterilize. Aseptically distribute into sterile tubes or flasks.

Use: For the cultivation and maintenance of *Thiomicrospira* species.

Thiomicrospira Medium (ATCC Medium 1036)

Composition per liter:

NaCl	25.0g
$Na_2S_2O_3 \cdot 5H_2O$	8.0g
$MgSO_4 \cdot 7H_2O$	1.5g
$(NH_4)_2SO_4$	1.0g
K_2HPO_4	0.5g
$CaCl_2$	0.3g
Vitamin B_{12}	15.0μg
Vishniac and Santer trace metals	0.2mL
Bromcresol Purple (0.05% solution)	0.1mL

pH 7.2 ± 0.2 at 25°C

Vishniac and Santer Trace Metals:

Composition per liter:

Ethylenediamine tetraacetic acid (EDTA)	50.0g
$ZnSO_4 \cdot 7H_2O$	22.0g
$CaCl_2$	5.54g
$MnCl_2 \cdot 4H_2O$	5.06g
$FeSO_4 \cdot 7H_2O$	4.99g
$CoCl_2 \cdot 6H_2O$	1.61g
$CuSO_4 \cdot 5H_2O$	1.57g
$(NH_4)_6Mo_7O_{24} \cdot 4H_2O$	1.10g

Preparation of Vishniac and Santer Trace Metals: Add components to distilled/deionized water and bring volume to 1.0L. Adjust pH to 6.0 with KOH. Mix thoroughly.

Preparation of Medium: Add components to distilled/deionized water and bring volume to 1.0L. Mix thoroughly. Adjust pH to 7.2. Filter sterilize. Aseptically distribute into sterile tubes or flasks.

Use: For the cultivation and maintenance of *Thiomicrospira* species.

Thiomicrospira pelophila Medium

Composition per liter:

NaCl	25.0g
Agar	10.0g
$Na_2S_2O_3 \cdot 5H_2O$	5.0-8.0g
$MgSO_4 \cdot 7H_2O$	1.5g
$(NH_4)_2SO_4$	1.0g
K_2HPO_4	0.5g
$CaCl_2$	0.3g
Vitamin B_{12}	0.15mg
Trace metals solution	0.2mL

Trace Metals Solution:

Composition per liter:

Disodium EDTA	50.0g
NaOH	11.0g
$CaCl_2 \cdot 2H_2O$	7.34g
$FeSO_4 \cdot 7H_2O$	5.0g
$MnCl_2 \cdot 2H_2O$	2.5g
$ZnSO_4 \cdot 7H_2O$	2.2g
$CoCl_2 \cdot 6H_2O$	0.5g
$(NH_4)_6Mo_7O_{24} \cdot 4H_2O$	0.5g
$CuSO_4 \cdot 5H_2O$	0.2g

Preparation of Trace Metals Solution: Add EDTA to distilled/deionized water and bring volume to 500.0mL. Mix thoroughly. Adjust pH to 6.0 with NaOH. Add remaining components, one by one. Maintain the pH at 6.0. After dissolution of all the salts, adjust the pH to 4.0 with HCl. Store at 4°C.

Preparation of Medium: Add components to distilled/deionized water and bring volume to 1.0L. Mix thoroughly. Gently heat and bring to boiling. Distribute into tubes or flasks. Autoclave for 15 min at 15 psi pressure–121°C. Pour into sterile Petri dishes or leave in tubes.

Use: For the cultivation of *Thiomicrospira pelophila*.

Thiophene–2–Carboxylic Acid Hydrazide

See: **TCH Medium**

Thiosphaera Agar

Composition per liter:

Agar	15.0g
Na_2HPO_4	4.2g
KH_2PO_4	1.5g
NH_4Cl	0.3g
$MgSO_4 \cdot 7H_2O$	0.1g
KNO_3	0.1g
Vishniac and Santer trace metals	2.0mL

pH 8.0–8.2 at 25°C

Vishniac and Santer Trace Metals:

Composition per liter:

Ethylenediamine tetraacetic acid (EDTA)	50.0g
$ZnSO_4 \cdot 7H_2O$	22.0g
$CaCl_2$	5.54g
$MnCl_2 \cdot 4H_2O$	5.06g
$FeSO_4 \cdot 7H_2O$	4.99g
$CoCl_2 \cdot 6H_2O$	1.61g
$CuSO_4 \cdot 5H_2O$	1.57g
$(NH_4)_6Mo_7O_{24} \cdot 4H_2O$	1.10g

Preparation of Vishniac and Santer Trace Metals: Add components to distilled/deionized water and bring volume to 1.0L. Adjust pH to 6.0 with KOH. Mix thoroughly.

Preparation of Medium: Add components, except agar, to distilled/deionized water and bring volume to 500.0mL. Mix thoroughly. Adjust pH to 8.0–8.2. Filter sterilize. Warm to 45°–50°C. Add agar to distilled/deionized water and bring volume to 500.0mL. Mix thoroughly. Gently heat and bring to boiling. Autoclave for 15 min at 15 psi pressure–121°C. Cool to 45°–50°C. Aseptically combine the two sterile solutions. Mix thoroughly. Pour into sterile Petri dishes or distribute into sterile tubes.

Use: For the cultivation and maintenance of *Thiosphaera pantotropha*.

Thiosphaera Broth

Composition per liter:

Na_2HPO_4	4.2g
KH_2PO_4	1.5g
NH_4Cl	0.3g
$MgSO_4 \cdot 7H_2O$	0.1g
KNO_3	0.1g
Vishniac and Santer trace metals	2.0mL

pH 8.0–8.2 at 25°C

Vishniac and Santer Trace Metals:

Composition per liter:

Ethylenediamine tetraacetic acid (EDTA)	50.0g
$ZnSO_4 \cdot 7H_2O$	22.0g
$CaCl_2$	5.54g
$MnCl_2 \cdot 4H_2O$	5.06g
$FeSO_4 \cdot 7H_2O$	4.99g
$CoCl_2 \cdot 6H_2O$	1.61g
$CuSO_4 \cdot 5H_2O$	1.57g
$(NH_4)_6Mo_7O_{24} \cdot 4H_2O$	1.10g

Preparation of Vishniac and Santer Trace Metals: Add components to distilled/deionized water and bring volume to 1.0L. Adjust pH to 6.0 with KOH. Mix thoroughly.

Preparation of Medium: Add components to distilled/deionized water and bring volume to 1.0L. Mix thoroughly. Adjust pH to 8.0–8.2. Filter sterilize. Aseptically distribute into sterile tubes or flasks.

Use: For the cultivation and maintenance of *Thiosphaera pantotropha*.

Thiosulfate Citrate Bile Salts Sucrose Agar

See: **TCBS Agar**

Thiosulfate–Oxidizing Medium

Composition per liter:

K_2HPO_4	2.0g
$MgSO_4 \cdot 7H_2O$	0.1g
$CaCl_2 \cdot 2H_2O$	0.1g
$FeCl_3 \cdot 6H_2O$	0.02g
$(NH_4)_2SO_4$ solution	100.0mL
Thiosulfate solution	100.0mL

pH 7.8 ± 0.2 at 25°C

$(NH_4)_2SO_4$ Solution:

Composition per 100mL:

$(NH_4)_2SO_4$	0.1g

Preparation of $(NH_4)_2SO_4$ Solution: Add the $(NH_4)_2SO_4$ to distilled/deionized water and bring volume to 100.0mL. Mix thoroughly. Autoclave for 15 min at 15 psi pressure–121°C. Cool to 45°–50°C

Thiosulfate Solution:

Composition per 100mL:

$Na_2S_2O_3 \cdot 5H_2O$	10.0g

Preparation of Thiosulfate Solution: Add the $Na_2S_2O_3 \cdot 5H_2O$ to distilled/deionized water and bring volume to 100.0mL. Mix thoroughly. Autoclave for 15 min at 15 psi pressure–121°C. Cool to 45°–50°C.

Preparation of Medium: Add components, except $(NH_4)_2SO_4$ solution and thiosulfate solution, to distilled/deionized water and bring volume to 800.0mL. Mix thoroughly. Autoclave for 15 min at 15 psi pressure–121°C. Cool to 45°–50°C. Aseptically add the sterile $(NH_4)_2SO_4$ solution and the sterile thiosulfate solution. Mix thoroughly. Adjust the pH to 7.8 if necessary. Aseptically distribute into sterile tubes or flasks.

Use: For the isolation and cultivation of iron and sulfur bacteria.

Tinsdale Agar

Composition per 1100mL:

Proteose peptone	20.0g
Agar	15.0g
NaCl	5.0g
Yeast extract	5.0g
L-Cystine	0.24g
Tinsdale supplement	150.0mL

pH 7.4 ± 0.2 at 25°C

Source: This medium is available as a premixed powder from Difco Laboratories and Oxoid Unipath.

Tinsdale Supplement:

Composition per100mL:

$Na_2S_2O_3$	0.43g
K_2TeO_3	0.35g
Serum	100.0mL

Preparation of Tinsdale Supplement: Add $Na_2S_2O_3$ and K_2TeO_3 to serum. Mix thoroughly. Filter sterilize.

Caution: Potassium tellurite is toxic.

Preparation of Medium: Add components, except Tinsdale supplement, to distilled/deionized water and bring volume to 1.0L. Mix thoroughly. Gently heat and bring to boiling. Autoclave for 15 min at 15 psi pressure–121°C. Cool to 50°–55°C. Aseptically add 100.0mL of sterile Tinsdale supplement. Mix thoroughly. Pour into sterile Petri dishes or distribute into sterile tubes.

Use: For the primary isolation and identification of *Corynebacterium diphtheriae*.

Tissue Culture Amino Acids, HeLa 100X (TC Amino Acids, HeLa 100X)

Composition per liter:

L-Lysine	0.029g
L-Isoleucine	0.026g
L-Leucine	0.026g
L-Threonine	0.023g
L-Valine	0.023g
L-Tyrosine	0.018g
L-Arginine	0.017g
L-Phenylalanine	0.016g
L-Cystine	0.012g
L-Histidine	7.8mg
L-Methionine	7.5mg
L-Tryptophan	4.1mg

pH 7.2–7.4 at 25°C

Preparation of Tissue Culture Amino Acids, HeLa 100X: Add components to distilled/deionized water and bring volume to 1.0L. Mix thoroughly. Adjust pH to 7.2–7.4. Filter sterilize.

Use: For the preparation of Eagle HeLa Medium for tissue culture procedures and virus studies.

Tissue Culture Amino Acids, Minimal Eagle 50X (TC Amino Acids, Minimal Eagle 50X)

Composition per liter:

L-Arginine	0.1g
L-Lysine	0.058g
L-Isoleucine	0.052g
L-Leucine	0.052g
L-Threonine	0.048g
L-Valine	0.046g
L-Tyrosine	0.036g
L-Phenylalanine	0.032g
L-Histidine	0.031g
L-Cystine	0.024g
L-Methionine	0.015g
L-Tryptophan	0.010g

pH 7.2–7.4 at 25°C

Preparation of Tissue Culture Amino Acids, Minimal Eagle 50X: Add components to distilled/deionized water and bring volume to 1.0L. Mix thoroughly. Adjust pH to 7.2–7.4. Filter sterilize.

Use: For the preparation of TC Minimal Medium Eagle for tissue culture procedures and virus studies.

Tissue Culture Dulbecco Solution (TC Dulbecco Solution)

Composition per liter:

NaCl	8.0g
Na_2HPO_4	1.15g
KH_2PO_4	0.2g
KCl	0.2g
$CaCl_2 \cdot 2H_2O$	0.1g
$MgCl_2 \cdot 6H_2O$	0.1g

pH 7.2–7.4 at 25°C

Preparation of Tissue Culture Dulbecco Solution: Add components to distilled/deionized water and bring volume to 1.0L. Mix thoroughly. Adjust pH to 7.2–7.4. Filter sterilize.

Use: For use in tissue culture and virus preparations.

Tissue Culture Earle Solution (TC Earle Solution)

Composition per 1002mL:

NaCl	6.8g
$NaHCO_3$	2.2g
Glucose	1.0g
KCl	0.4g
$CaCl_2 \cdot 2H_2O$	0.2g
NaH_2PO_4	0.125g
$MgSO_4 \cdot 7H_2O$	0.1g
Phenol Red (1% solution)	2.0mL

pH 7.2–7.4 at 25°C

Preparation of Tissue Culture Earle Solution: Add components, except phenol red, to distilled/deionized water and bring volume to 1.0L. Mix thoroughly. Add 2.0mL of phenol red solution. Adjust pH to 7.2–7.4. Filter sterilize.

Use: For use in tissue culture and virus preparations.

Tissue Culture Hanks Solution (TC Hanks Solution)

Composition per liter:

NaCl	8.0g
Glucose	1.0g
KCl	0.4g
$NaHCO_3$	0.35g
$CaCl_2 \cdot 2H_2O$	0.14g
$MgCl_2 \cdot 6H_2O$	0.1g
$MgSO_4 \cdot 7H_2O$	0.1g
Na_2HPO_4	0.06g
KH_2PO_4	0.06g
Phenol Red	0.02g

pH 7.2–7.4 at 25°C

Source: This medium is available as a premixed solution from Difco Laboratories.

Preparation of Tissue Culture Hanks Solution: Add components to distilled/deionized water and bring volume to 1.0L. Mix thoroughly. Adjust pH to 7.2–7.4. Filter sterilize.

Use: For use in tissue culture procedures.

Tissue Culture Medium 199 (TC Medium 199)

Composition per 1050mL:

NaCl	8.0g
Glucose	1.0g
KCl	0.4g
$NaHCO_3$	0.35g
DL-Glutamic acid	0.15g
$CaCl_2 \cdot 2H_2O$	0.14g
DL-Leucine	0.12g
L-Glutamine	0.1g
$MgSO_4 \cdot 7H_2O$	0.1g
L-Arginine	0.07g
L-Lysine	0.07g
DL-Aspartic acid	0.06g
Na_2HPO_4	0.06g
KH_2PO_4	0.06g
DL-Threonine	0.06g
DL-Alanine	0.05g
Glycine	0.05g
DL-Phenylalanine	0.05g
DL-Serine	0.05g
Sodium acetate	0.05g
DL-Valine	0.05g
DL-Isoleucine	0.04g
L-Proline	0.04g
L-Tyrosine	0.04g
DL-Methionine	0.03g
L-Cystine	0.02g
L-Histidine	0.02g
Phenol Red	0.02g
DL-Tryptophan	0.02g
Adenine	0.01g
L-Hydroxyproline	0.01g
Tween™ 80	5.0mg
Adenosine triphosphate	1.0mg
Choline	0.5mg
Deoxyribose	0.5mg
Ribose	0.5mg
Guanine	0.3mg
Hypoxanthine	0.3mg
Thymine	0.3mg
Uracil	0.3mg
Xanthine	0.3mg
Adenylic acid	0.2mg
Cholesterol	0.2mg
Calciferol	0.1mg

$Fe(NO_3)_3 \cdot 9H_2O$	0.1mg
L-Cysteine	0.1mg
Vitamin A	0.1mg
α-Tocopherol phosphate	0.01mg
Biotin	0.01mg
Calcium pantothenate	0.01mg
Folic acid	0.01mg
Menadione	0.01mg
Riboflavin	0.01mg
Thiamine·HCl	0.01mg
p-Aminobenzoic acid	0.05mg
Ascorbic acid	0.05mg
Glutathione	0.05mg
Inositol	0.05mg
Niacin	0.025mg
Niacinamide	0.025mg
Pyridoxine·HCl	0.025mg
Pyridoxal·HCl	0.025mg
Serum	50.0–100.0mL

pH 7.2–7.4 at 25°C

Preparation of Medium: Add components, except serum, to distilled/deionized water and bring volume to 1.0L. Mix thoroughly. Adjust pH to 7.2–7.4 with 10% Na_2CO_3 solution. Filter sterilize. Aseptically add 50.0–100.0mL of sterile serum. Human serum, bovine serum, horse serum or fetal calf serum may be used. Mix thoroughly. If desired antibacterial inhibitors may be added. Aseptically add 500,000U penicillin and 0.5g streptomycin to 1050.0mL of the complete medium.

Use: For cultivation of a wide variety of cell lines in tissue culture. It is especially useful for detection, titering and identification of viruses in tissue culture cells.

Tissue Culture Medium Eagle, HeLa (TC Medium Eagle, HeLa)

Composition per 1056mL:

NaCl	5.85g
$NaHCO_3$	1.68g
Glucose	0.9g
KCl	0.373g
NaH_2PO_4	0.138g
$MgCl_2 \cdot 6H_2O$	0.120g
$CaCl_2 \cdot 2H_2O$	0.11g
L-Lysine	0.0269g
L-Isoleucine	0.0262g
L-Leucine	0.0262g
L-Threonine	0.0238g
L-Valine	0.0234g
L-Tyrosine	0.0181g
L-Arginine	0.0174g
L-Phenylalanine	0.0165g
L-Cystine	0.012g
L-Histidine	7.8mg
L-Methionine	7.5mg
Phenol Red	5.0mg
L-Tryptophan	4.1mg
Folic acid	0.44mg
Thiamine·HCl	0.34mg
Biotin	0.24mg
Pantothenic acid	0.22mg
Pyridoxal·HCl	0.2mg
Choline chloride	0.14mg
Nicotinamide	0.12mg
Riboflavin	0.04mg
Serum	50.0mL–100.0mL
Glutamine solution	6.0mL

pH 7.2–7.4 at 25°C

Glutamine Solution:

Composition per 100mL:

L-Glutamine	5.0g
NaCl (0.85% solution)	100.0mL

Preparation of Glutamine Solution: Add the glutamine to the 0.85% NaCl solution. Mix thoroughly. Filter sterilize.

Preparation of Medium: Add components, except glutamine and serum, to distilled/deionized water and bring volume to 1.0L. Mix thoroughly. Adjust pH to 7.2–7.4. Filter sterilize. Aseptically add 6.0mL of sterile glutamine solution and 50.0–100.0mL of sterile serum. Human serum, bovine serum, horse serum or fetal calf serum may be used. Mix thoroughly.

Use: For the cultivation and maintenance of HeLa and other cell lines in tissue culture, and for studying the cytopathogenicity of viral agents.

Tissue Culture Medium Eagle with Earle Balanced Salt Solution (TC Medium Eagle with Earle BSS)

Composition per 1056mL:

NaCl	6.8g
$NaHCO_3$	2.2g
Glucose	1.0g
KCl	0.4g
$CaCl_2 \cdot 2H_2O$	0.2g
NaH_2PO_4	0.125g
$MgSO_4 \cdot 7H_2O$	0.1g
L-Isoleucine	0.026g
L-Leucine	0.026g
L-Lysine	0.026g
L-Threonine	0.024g
L-Valine	0.0235g

L-Tyrosine	0.018g
L-Arginine	0.0174g
L-Phenylalanine	0.0165g
L-Cystine	0.012g
L-Histidine	8.0mg
L-Methionine	7.5mg
Phenol Red	5.0mg
L-Tryptophan	4.0mg
Inositol	1.8mg
Biotin	1.0mg
Calcium pantothenate	1.0mg
Choline chloride	1.0mg
Folic acid	1.0mg
Nicotinamide	1.0mg
Pyridoxal·HCl	1.0mg
Thiamine·HCl	1.0mg
Riboflavin	0.1mg
Serum	50.0mL–100.0mL
Glutamine solution	6.0mL

pH 7.2–7.4 at 25°C

Glutamine Solution:

Composition per 100mL:

L-Glutamine	5.0g
NaCl (0.85% solution)	100.0mL

Preparation of Glutamine Solution: Add the glutamine to the 0.85% NaCl solution. Mix thoroughly. Filter sterilize.

Preparation of Medium: Add components, except glutamine and serum, to distilled/deionized water and bring volume to 1.0L. Mix thoroughly. Adjust pH to 7.2–7.4. Filter sterilize. Aseptically add 6.0mL of sterile glutamine solution and 50.0–100.0mL of sterile serum. Human serum, bovine serum, horse serum or fetal calf serum may be used. Mix thoroughly.

Use: For the cultivation of HeLa, KB and other tissue culture cell lines.

Tissue Culture Medium Eagle with Hanks Balanced Salt Solution (TC Medium Eagle with Hanks BSS)

Composition per 1056mL:

NaCl	8.0g
Glucose	1.0g
KCl	0.4g
$CaCl_2·2H_2O$	0.14g
$MgSO_4·7H_2O$	0.1g
KH_2PO_4	0.06g
Na_2HPO_4	0.05g
L-Isoleucine	0.026g
L-Leucine	0.026g
L-Lysine	0.026g
L-Threonine	0.024g
L-Valine	0.0235g
L-Tyrosine	0.018g
L-Arginine	0.0174g
L-Phenylalanine	0.0165g
L-Cystine	0.012g
L-Histidine	8.0mg
L-Methionine	7.5mg
Phenol Red	5.0mg
L-Tryptophan	4.0mg
Inositol	1.8mg
Biotin	1.0mg
Folic acid	1.0mg
Calcium pantothenate	1.0mg
Choline chloride	1.0mg
Nicotinamide	1.0mg
Pyridoxal·HCl	1.0mg
Thiamine·HCl	1.0mg
Riboflavin	0.1mg
Serum	50.0mL–100.0mL
Glutamine solution	6.0mL

pH 7.2–7.4 at 25°C

Glutamine Solution:

Composition per 100mL:

L-Glutamine	5.0g
NaCl (0.85% solution)	100.0mL

Preparation of Glutamine Solution: Add the glutamine to the 0.85% NaCl solution. Mix thoroughly. Filter sterilize.

Preparation of Medium: Add components, except glutamine and serum, to distilled/deionized water and bring volume to 1.0L. Mix thoroughly. Adjust pH to 7.2–7.4. Filter sterilize. Aseptically add 6.0mL of sterile glutamine solution and 50.0–100.0mL of sterile serum. Human serum, bovine serum, horse serum or fetal calf serum may be used. Mix thoroughly.

Use: For use as a base in the preparation of liquid media used for the cultivation of tissue culture cell lines.

Tissue Culture Medium Ham F10 (TC Medium Ham F10)

Composition per 1050mL:

NaCl	7.4g
Glucose	1.1g
Na_2HPO_4	0.290g
KCl	0.285g
L-Arginine	0.211g
$MgSO_4·7H_2O$	0.153g
L-Glutamine	0.1462g
Sodium pyruvate	0.110g

KH_2PO_4	0.083g
$CaCl_2 \cdot 2H_2O$	0.044g
L-Cystine	0.0315g
L-Lysine	0.0293g
L-Histidine	0.021g
L-Asparagine	0.015g
L-Glutamic acid	0.0147g
L-Aspartic acid	0.0133g
L-Leucine	0.0131g
L-Proline	0.0115g
L-Serine	0.0105g
L-Alanine	8.91mg
Glycine	7.51mg
L-Phenylalanine	4.96mg
L-Methionine	4.48mg
Hypoxanthine	4.0mg
L-Threonine	3.57mg
L-Valine	3.5mg
L-Isoleucine	2.6mg
L-Tyrosine	1.81mg
Cyanocobalamin	1.3mg
Folic acid	1.3mg
Phenol Red	1.2mg
Thiamine·HCl	1.0mg
$FeSO_4 \cdot 7H_2O$	0.83mg
Calcium pantothenate	0.7mg
Thymidine	0.7mg
Choline chloride	0.69mg
Niacinamide	0.6mg
L-Tryptophan	0.6mg
i-Inositol	0.54mg
Riboflavin	0.37mg
Lipoic acid	0.2mg
Pyridoxine·HCl	0.2mg
$ZnSO_4 \cdot 7H_2O$	0.028mg
Biotin	0.024mg
$CuSO_4 \cdot 5H_2O$	2.5μg
Fetal calf serum	50.0–100.0mL

pH 7.2–7.4 at 25°C

Preparation of Medium: Add components, except fetal calf serum, to distilled/deionized water and bring volume to 1.0L. Mix thoroughly. Adjust pH to 7.2–7.4 with 10% Na_2CO_3 solution. Filter sterilize. Aseptically add 50.0–100.0mL of sterile fetal calf serum. Mix thoroughly.

Use: For cultivation of a wide variety of cell lines in tissue culture.

Tissue Culture Medium NCTC 109 (TC Medium NCTC 109)

Composition per 1050mL:

NaCl	6.8g
$NaHCO_3$	2.2g
Glucose	1.0g
KCl	0.4g
L-Cysteine	0.26g
$CaCl_2 \cdot 2H_2O$	0.20g
NaH_2PO_4	0.14g
L-Glutamine	0.14g
$MgSO_4 \cdot 7H_2O$	0.10g
Sodium acetate	0.05g
Ascorbic acid	0.05g
L-Alanine	0.03g
L-Lysine	0.03g
L-Arginine	0.026g
L-Valine	0.025g
L-Leucine	0.020g
Phenol Red	0.020g
L-Histidine	0.019g
L-Threonine	0.019g
L-Isoleucine	0.018g
L-Tryptophan	0.017.g
L-Phenylalanine	0.017g
L-Tyrosine	0.016g
Glycine	0.014g
Tween™ 80	0.012g
L-Serine	0.011g
L-Cystine	0.010g
Glutathione	0.010g
Cyanocobalamin	0.010g
Deoxycytidine	0.010g
Deoxyguanosine	0.010g
Deoxyadenosine	0.010g
Thymidine	0.010g
L-Aspartic acid	9.91mg
L-Glutamic acid	8.26mg
L-Arginine	8.09mg
L-Ornithine	7.38mg
Nicotinamide adenine dinucleotide	7.0mg
L-Proline	6.13mg
L-α-N-Butyric acid	5.51mg
L-Methionine	4.44mg
L-Taurine	4.18mg
L-Hydroxyproline	4.09mg
D-Glucosamine	3.2mg
Coenzyme A	2.5mg
Glucuronolactone	1.8mg
Sodium glucuronate	1.8mg
Choline chloride	1.25mg
Cocarboxylase	1.0mg
Flavin adenine dinucleotide	1.0mg
Uridine triphosphate	1.0mg
Nicotinamide adenine dinucleotide phosphate	1.0mg
Vitamin A	0.25mg
Calciferol	0.25mg
i-Inositol	0.125mg
p-Aminobenzoic acid	0.125mg

5-Methylcytosine	0.1mg
Pyridoxine·HCl	0.0625mg
Pyridoxal·HCl	0.0625mg
Niacin	0.0625mg
Niacinamide	0.0625mg
Biotin	0.025mg
Folic acid	0.025mg
Menadione	0.025mg
Pantothenate	0.025mg
Riboflavin	0.025mg
Thiamine·HCl	0.025mg
α-Tocopherol phosphate	0.025mg
Serum	50.0–100.0mL

pH 7.2–7.4 at 25°C

Preparation of Medium: Add components, except serum, to distilled/deionized water and bring volume to 1.0L. Mix thoroughly. Adjust pH to 7.2–7.4 with 10% Na_2CO_3 solution. Filter sterilize. Aseptically add 50.0–100.0mL of sterile serum. Human serum, bovine serum, horse serum or fetal calf serum may be used. Mix thoroughly.

Use: For cultivation of a wide variety of cell lines in tissue culture.

Tissue Culture Medium RPMI #1640 (TC Medium RPMI #1640)

Composition per liter:

NaCl	6.46g
Glucose	2.0g
$NaHCO_3$	2.0g
NaH_2PO_4	1.512g
KCl	0.4g
L-Glutamine	0.3g
L-Arginine	0.2g
Calcium nitrate	0.1g
$MgSO_4 \cdot 7H_2O$	0.1g
L-Asparagine	0.05g
L-Cystine	0.05g
L-Isoleucine	0.05g
L-Leucine	0.05g
L-Lysine HCl	0.04g
Inositol	0.035g
L-Serine	0.030g
Hydroxy-L-Proline	0.020g
L-Aspartic acid	0.020g
L-Glutamic acid	0.020g
L-Proline	0.020g
L-Threonine	0.020g
L-Tyrosine	0.020g
L-Valine	0.020g
L-Histidine	0.015g
L-Methionine	0.015g
L-Phenylalanine	0.015g
Glycine	0.01g
L-Tryptophan	5.0mg
Phenol Red	5.0mg
Choline chloride	3.0mg
p-Aminobenzoic acid	1.0mg
Folic acid	1.0mg
Glutathione	1.0mg
Nicotinamide	1.0mg
Pyridoxine HCl	1.0mg
Thiamine HCl	1.0mg
Calcium pantothenate	0.25mg
Biotin	0.20mg
Riboflavin	0.20mg
Vitamin B_{12}	5.0mg
Serum	50.0–100.0mL

pH 7.2–7.4 at 25°C

Source: This medium is available as a premixed powder and solution from Difco Laboratories.

Preparation of Medium: Add components, except serum, to distilled/deionized water and bring volume to 1.0L. Mix thoroughly. Adjust pH to 7.2–7.4 with 10% Na_2CO_3 solution. Filter sterilize. Aseptically add 50.0–100.0mL of sterile serum. Human serum, bovine serum, horse serum or fetal calf serum may be used. Mix thoroughly.

Use: For cultivation of a wide variety of cell lines in tissue culture.

Tissue Culture Minimal Medium Eagle

Composition per liter:

Sterile salt solution	944.0mL
TC amino acids, minimal Eagle 50X	20.0mL
TC $NaHCO_3$, 10%	20.0mL
TC vitamins minimal Eagle 100X	10.0mL
TC glutamine, 5%	6.0mL

pH 7.2–7.4 at 25°C

Sterile Salt Solution:

Composition per 944mL:

NaCl	6.8g
Glucose	1.0g
KCl	0.4g
$CaCl_2$	0.2g
$MgCl_2$	0.2g
NaH_2PO_4	0.15g

Preparation of Sterile Salt Solution: Add components to distilled/deionized water and bring volume to 944.0mL. Mix thoroughly. Filter sterilize.

Tissue Culture Amino Acids, Minimal Eagle 50X:

Composition per liter:

L-Arginine	0.1g
L-Lysine	0.06g
L-Isoleucine	0.05g
L-Leucine	0.05g
L-Threonine	0.05g
L-Valine	0.05g
L-Tyrosine	0.04g
L-Phenylalanine	0.03g
L-Histidine	0.03g
L-Cystine	0.02g
L-Methionine	0.02g
L-Tryptophan	0.01g

Preparation of Tissue Culture Amino Acids, Minimal Eagle 50X: Add components to distilled/deionized water and bring volume to 1.0L. Mix thoroughly. Adjust pH to 7.2–7.4. Filter sterilize.

TC $NaHCO_3$, 10%:

Composition per 100mL:

$NaHCO_3$	10.0g

Preparation of TC $NaHCO_3$, 10%: Add $NaHCO_3$ to distilled/deionized water and bring volume to 100.0mL. Mix thoroughly. Filter sterilize.

TC Vitamins, Minimal Eagle 100X:

Composition per liter:

Inositol	2.0mg
Calcium pantothenate	1.0mg
Choline chloride	1.0mg
Folic acid	1.0mg
Nicotinamide	1.0mg
Pyridoxal	1.0mg
Thiamine·HCl	1.0mg
Riboflavin	0.1mg

Preparation of TC Vitamins, Minimal Eagle 100X: Add components to distilled/deionized water and bring volume to 1.0L. Mix thoroughly. Filter sterilize.

TC Glutamine, 5%:

Composition per 100mL:

L-Glutamine	5.0g
NaCl (0.85% solution)	100.0mL

Preparation of TC Glutamine, 5%: Add the glutamine to the 0.85% NaCl solution. Mix thoroughly. Filter sterilize.

Preparation of Medium: Aseptically combine 944.0mL of sterile salt solution, 20.0mL of sterile TC amino acids, minimal Eagle 50X, 20.0mL of sterile TC $NaHCO_3$, 10%, 10.0mL of sterile TC vitamins minimal Eagle 100X, and 6.0mL of sterile TC glutamine, 5%. Mix thoroughly. Adjust pH to 7.2–7.4 if necessary.

Use: For the cultivation of mammalian cells in monolayer or suspension for tissue culture procedures and virus preparation.

Tissue Culture Minimal Medium Eagle Spinner Modified (TC Minimal Medium Eagle Spinner Modified MEM–S)

Composition per 1056mL:

NaH_2PO_4	1.35g
NaCl	6.8g
$NaHCO_3$	2.2g
Glucose	1.0g
KCl	0.4g
$CaCl_2·2H_2O$	0.2g
NaH_2PO_4	0.125g
$MgSO_4·7H_2O$	0.1g
L-Isoleucine	0.026g
L-Leucine	0.026g
L-Lysine	0.026g
L-Threonine	0.024g
L-Valine	0.0235g
L-Tyrosine	0.018g
L-Arginine	0.0174g
L-Phenylalanine	0.0165g
L-Cystine	0.012g
L-Histidine	8.0mg
L-Methionine	7.5mg
Phenol Red	5.0mg
L-Tryptophan	4.0mg
Inositol	1.8mg
Biotin	1.0mg
Calcium pantothenate	1.0mg
Choline chloride	1.0mg
Folic acid	1.0mg
Nicotinamide	1.0mg
Pyridoxal·HCl	1.0mg
Thiamine·HCl	1.0mg
Riboflavin	0.1mg
Serum	50.0mL–100.0mL
Glutamine solution	6.0mL

pH 7.2–7.4 at 25°C

Glutamine Solution:

Composition per 100mL:

L-Glutamine	5.0g
NaCl (0.85% solution)	100.0mL

Preparation of Glutamine Solution: Add the glutamine to the 0.85% NaCl solution. Mix thoroughly. Filter sterilize.

Preparation of Medium: Add components, except glutamine and serum, to distilled/deionized water and bring volume to 1.0L. Mix thoroughly. Adjust pH to 7.2–7.4 with 10% Na_2CO_3 solution. Filter sterilize. Aseptically add 6.0mL of sterile glutamine solution and 50.0–100.0mL of sterile serum. Human serum, bovine serum, horse serum or fetal calf serum may be used. Mix thoroughly.

Use: For the cultivation of mammalian cells in suspension.

Tissue Culture Minimal Medium Eagle with Earle Balanced Salts Solution (TC Minimal Medium Eagle with Earle BSS)

Composition per 1056mL:

NaCl	6.8g
Glucose	1.0g
KCl	0.4g
$CaCl_2 \cdot 2H_2O$	0.2g
$MgCl_2 \cdot 6H_2O$	0.2g
NaH_2PO_4	0.15g
L-Arginine	0.10g
L-Lysine	0.06g
L-Isoleucine	0.05g
L-Leucine	0.05g
L-Threonine	0.05g
L-Valine	0.05g
L-Tyrosine	0.04g
L-Phenylalanine	0.03g
L-Histidine	0.03g
L-Cystine	0.02g
L-Methionine	0.02g
L-Tryptophan	0.01g
i-Inositol	2.0mg
Calcium pantothenate	1.0mg
Choline chloride	1.0mg
Folic acid	1.0mg
Nicotinamide	1.0mg
Pyridoxal	1.0mg
Thiamine·HCl	1.0mg
Riboflavin	0.1mg
Serum	50.0mL–100.0mL
Glutamine solution	6.0mL

pH 7.2–7.4 at 25°C

Glutamine Solution:

Composition per 100mL:

L-Glutamine	5.0g
NaCl (0.85% solution)	100.0mL

Preparation of Glutamine Solution: Add the glutamine to the 0.85% NaCl solution. Mix thoroughly. Filter sterilize.

Preparation of Medium: Add components, except glutamine and serum, to distilled/deionized water and bring volume to 1.0L. Mix thoroughly. Adjust pH to 7.2–7.4 with 10% Na_2CO_3 solution. Filter sterilize. Aseptically add 6.0mL of sterile glutamine solution and 50.0–100.0mL of sterile serum. Human serum, bovine serum, horse serum or fetal calf serum may be used. Mix thoroughly. To grow cells in a monolayer, aseptically add 2.0mL of a sterile 10% $CaCl_2 \cdot 2H_2O$ solution. To grow cells in suspension, omit the $CaCl_2 \cdot 2H_2O$ solution.

Use: For preparation of Eagle's Minimal Medium for the cultivation of cells in monolayer or suspension in tissue culture.

Tissue Culture Tyrode Solution (TC Tyrode Solution)

Composition per 1002mL:

NaCl	8.0g
Glucose	1.0g
$NaHCO_3$	1.0g
$CaCl_2 \cdot 2H_2O$	0.2g
KCl	0.2g
$MgCl_2 \cdot 6H_2O$	0.1g
NaH_2PO_4	0.05g
Phenol Red (1% solution)	2.0mL

pH 7.2–7.4 at 25°C

Preparation of Tissue Culture Tyrode Solution: Add components, except Phenol Red, to distilled/deionized water and bring volume to 1.0L. Mix thoroughly. Add 2.0mL of Phenol Red solution. Adjust pH to 7.2–7.4. Filter sterilize.

Use: For use in tissue culture procedures.

Tissue Culture Vitamins Minimal Eagle, 100X (TC Vitamins Minimal Eagle, 100X)

Composition per liter:

Inositol	2.0mg
Calcium pantothenate	1.0mg
Choline chloride	1.0mg
Folic acid	1.0mg
Nicotinamide	1.0mg
Pyridoxal	1.0mg
Thiamine·HCl	1.0mg
Riboflavin	0.1mg

pH 7.2–7.4 at 25°C

Preparation of TC Vitamins, Minimal Eagle 100X: Add components to distilled/deionized water and bring volume to 1.0L. Mix thoroughly. Filter sterilize.

Use: For the preparation of TC Minimal Medium Eagle used in tissue culture procedures.

TMAO Medium
See: **Trimethylamine N–Oxide Medium**

TN Broth
See: **Trypticase Novobiocin Broth**

TNSA Agar
See: **Trypaflavin Nalidixic Acid Serum Agar**

TNT Medium (Tryptone NaCl Thiamine Medium)

Composition per liter:

Pancreatic digest of casein	10.0g
NaCl	5.0g
Thiamine·HCl	1.0mg

pH 7.3 ± 0.2 at 25°C

Preparation of Medium: Add components to distilled/deionized water and bring volume to 1.0L. Mix thoroughly. Distribute into tubes or flasks. Autoclave for 15 min at 15 psi pressure–121°C.

Use: For the cultivation and maintenance of *Escherichia coli.*

TOC Agar (Tween™ 80 Oxgall Caffeic acid Agar)

Composition per liter:

Agar	20.0g
Oxgall	10.0g
Caffeic acid	0.3g
Tween™ 80	10.0mL

Source: Available as a prepared medium from BBL Microbiology Systems.

Preparation of Medium: Add components to distilled/deionized water and bring volume to 1.0L. Mix thoroughly. Gently heat and bring to boiling. Autoclave for 15 min at 15 psi pressure–121°C. Pour into sterile Petri dishes.

Use: For the differentiation and identification of *Candida albicans* and *Cryptococcus neoformans. C. albicans* produces germ tubes and chlamydospores when grown on this medium. *C. neoformans* appears as tan to brown colonies.

Todd–Hewitt Broth

Composition per liter:

Beef heart, infusion from	500.0g
Neopeptone	20.0g
Na_2CO_3	2.5g
Glucose	2.0g
NaCl	2.0g
Na_2HPO_4	0.4g

pH 7.8 ± 0.2 at 25°C

Source: This medium is available as a premixed powder from Difco Laboratories.

Preparation of Medium: Add components to distilled/deionized water and bring volume to 1.0L. Mix thoroughly. Distribute into tubes or flasks. Autoclave for 15 min at 15 psi pressure–121°C.

Use: For the cultivation of Group A streptococci used in serological typing, and for the cultivation of a variety of pathogenic microorganisms.

Todd–Hewitt Broth

Composition per liter:

Pancreatic digest of casein	20.0g
Infusion from 450g fat-free minced meat	10.0g
Glucose	2.0g
$NaHCO_3$	2.0g
NaCl	2.0g
Na_2HPO_4	0.4g

pH 7.8 ± 0.2 at 25°C

Source: This medium is available as a premixed powder from Oxoid Unipath.

Preparation of Medium: Add components to distilled/deionized water and bring volume to 1.0L. Mix thoroughly. Distribute into tubes or flasks. Autoclave for 10 min at 10 psi pressure–115°C.

Use: For the cultivation of Group A streptococci used in serological typing, and for the cultivation of a variety of pathogenic microorganisms.

Todd–Hewitt Broth (ATCC Medium 235)

Composition per liter:

Peptone	20.0g
Beef heart, solids from infusion	3.1g
Na_2CO_3	2.5g
Glucose	2.0g
NaCl	2.0g
Na_2HPO_4	0.4g

pH 7.8 ± 0.2 at 25°C

Source: This medium is available as a premixed powder from BBL Microbiology Systems.

Preparation of Medium: Add components to distilled/deionized water and bring volume to 1.0L. Mix thoroughly. Distribute into tubes or flasks. Autoclave for 15 min at 15 psi pressure–121°C.

Use: For the cultivation of Group A streptococci used in serological typing, and for the cultivation of a variety of pathogenic microorganisms.

Todd–Hewitt Broth, Modified

Composition per liter:

Neopeptone	20.0g
Glucose	2.0g
$NaHCO_3$	2.0g
NaCl	2.0g
Na_2HPO_4	0.4g
Beef heart infusion	1.0L

pH 7.8 ± 0.2 at 25°C

Preparation of Medium: Add components to distilled/deionized water and bring volume to 1.0L. Mix thoroughly. Distribute into tubes or flasks. Autoclave for 10 min at 10 psi pressure–115°C.

Use: For cultivation of streptococci for serological identification.

Toluidine Blue DNA Agar

Composition per liter:

Agar	10.0g
NaCl	10.0g
Tris(hydroxymethyl)aminomethane buffer	6.1g
Deoxyribonucleic acid	0.3g
Toluidine Blue O	0.083g
$CaCl_2$, anhydrous	1.1mg

pH 9.0 ± 0.2 at 25°C

Preparation of Medium: Add tris(hydroxymethyl)aminomethane buffer to distilled/deionized water and bring volume to 1.0L. Mix thoroughly. Adjust pH to 9.0. Add the remaining components, except Toluidine Blue O. Mix thoroughly. Gently heat and bring to boiling. Add Toluidine Blue O. Mix thoroughly. If used the same day, sterilization is not necessary. Cool to 50°C. Pour into sterile Petri dishes or distribute into sterile tubes.

Use: For the cultivation and differentiation of *Staphylococcus aureus* from foods.

Toluidine Blue DNA Agar

Composition per liter:

Agar	10.0g
NaCl	10.0g
Tris(hydroxymethyl)aminomethane buffer	6.1g
Deoxyribonucleic acid (DNA)	0.3g
Toluidine Blue O	0.083g
$CaCl_2$, anhydrous	1.1mg

pH 7.3 ± 0.2 at 25°C

Preparation of Medium: Add components, except Toluidine Blue O, to distilled/deionized water and bring volume to 1.0L. Mix thoroughly. Gently heat and bring to boiling. Add Toluidine Blue O. Mix thoroughly. Medium does not have to be sterilized if used immediately. Pour into sterile Petri dishes or distibute into sterile tubes. Allow tubes to cool in a slanted position.

Use: For the cultivation and differentiation of bacteria based on their production of deoxyribonuclease (DNase). Bacteria that produce DNase turn the medium pink.

Tomato Dextrin Yeast Medium

Composition per liter:

Tomato paste	20.0g
Dextrin	20.0g
Agar	20.0g
Baker's yeast	10.0g
$CoCl_2 \cdot 6H_2O$	5.0mg

pH 7.2-7.4 at 25°C

Preparation of Medium: Add components, except agar, to distilled/deionized water and bring volume to 1.0L. Mix thoroughly. Adjust pH to 7.2-7.4. Add agar. Gently heat and bring to boiling. Distribute into tubes or flasks. Autoclave for 15 min at 15 psi pressure–121°C. Pour into sterile Petri dishes or leave in tubes.

Use: For the cultivation of *Streptomyces avermitilis*.

Tomato Juice Agar

Composition per liter:

Agar	12.0g
Pancreatic digest of casein	10.0g
Peptonized milk	10.0g
Tomato juice	400.0mL

pH 6.1 ± 0.2 at 25°C

Source: This medium is available as a premixed powder from BBL Microbiology Systems, Difco Laboratories and Oxoid Unipath.

Preparation of Medium: Add components to distilled/deionized water and bring volume to 1.0L. Mix thoroughly. Gently heat and bring to boiling. Distribute into tubes or flasks. Autoclave for 15 min at 15 psi pressure–121°C. Pour into sterile Petri dishes or leave in tubes.

Use: For the cultivation of lactobacilli, especially *Lactobacillus acidophilus*.

Tomato Juice Agar (ATCC Medium 33)

Composition per liter:

Agar	11.0g
Pancreatic digest of casein	10.0g
Yeast extract	10.0g
Tomato juice, filtered	200.0mL

pH 7.2 ± 0.2 at 25°C

Preparation of Medium: Add components to distilled/deionized water and bring volume to 1.0L. Mix thoroughly. Adjust pH to 7.2. Gently heat and bring to boiling. Distribute into tubes or flasks. Autoclave for 15 min at 15 psi pressure–121°C. Pour into sterile Petri dishes or leave in tubes.

Use: For the cultivation and maintenance of a variety of bacteria including *Lactobacillus, Leuconostoc, Pediococcus* and *Propionibacterium* species.

Tomato Juice Agar Special

Composition per liter:

Agar	20.0g
Pancreatic digest of casein	10.0g
Peptonized milk	10.0g
Tomato juice	400.0mL

pH 5.0 ± 0.2 at 25°C

Source: This medium is available as a premixed powder from Difco Laboratories.

Preparation of Medium: Add components to distilled/deionized water and bring volume to 1.0L. Mix thoroughly. Gently heat and bring to boiling. Distribute into tubes or flasks. Autoclave for 15 min at 15 psi pressure–121°C. Avoid overheating—it results in a soft agar. Pour into sterile Petri dishes or leave in tubes.

Use: For the cultivation and enumeration of lactobacilli.

Tomato Juice Broth (ATCC Medium 433)

Composition per liter:

Pancreatic digest of casein	10.0g
Yeast extract	10.0g
Tomato juice, filtered	200.0mL

pH 7.2 ± 0.2 at 25°C

Preparation of Medium: Add components to distilled/deionized water and bring volume to 1.0L. Mix thoroughly. Adjust pH to 7.2. Gently heat and bring to boiling. Distribute into tubes or flasks. Autoclave for 15 min at 15 psi pressure–121°C.

Use: For the cultivation and maintenance of a variety of fastidious bacteria which require complex growth factors including *Lactobacillus, Aerococcus, Bifidobacterium* and *Pediococcus* species.

Tomato Juice Broth

Composition per liter:

Tomato juice, dessicated	20.0g
Glucose	10.0g
Yeast extract	10.0g
K_2HPO_4	0.5g
KH_2PO_4	0.5g
$MgSO_4·7H_2O$	0.2g
$FeSO_4·7H_2O$	0.01g
$MnSO_4·7H_2O$	0.01g
NaCl	0.01g

pH 6.7 ± 0.2 at 25°C

Source: This medium is available as a premixed powder from Difco Laboratories.

Preparation of Medium: Add components to distilled/deionized water and bring volume to 1.0L. Mix thoroughly. Distribute into tubes or flasks. Autoclave for 15 min at 15 psi pressure–121°C.

Use: For the cultivation of yeast and other aciduric microorganisms.

Tomato Juice Medium

Composition per liter:

Agar	15.0g
Glucose	10.0g
Polypeptone™	5.0g
Yeast extract	5.0g
KH_2PO_4	0.5g
$CaCl_2·2H_2O$	0.125g
KCl	0.125g
$MgSO_4·7H_2O$	0.125g
NaCl	0.125g
Bromcresol Green	0.03g
$MnSO_4·4H_2O$	3.0mg
Tomato juice, canned	150.0mL
Cycloheximide solution	10.0mL

pH 5.0 ± 0.2 at 25°C

Cycloheximide Solution:

Composition per 10mL:

Cycloheximide	0.1g

Preparation of Cycloheximide Solution: Add cycloheximide to distilled/deionized water and bring volume to 10.0mL. Mix thoroughly. Filter sterilize.

Preparation of Medium: Add components, except cycloheximide solution, to distilled/deionized water and bring volume to 990.0mL. Mix thoroughly. Gently heat and bring to boiling. Autoclave for 15 min at 15 psi pressure–121°C. Cool to 45°–50°C. Aseptically add sterile cycloheximide solution. Mix thoroughly. Pour into sterile Petri dishes or distribute into sterile tubes.

Use: For the isolation and cultivation of *Lactobacilli* from wine.

Tomato Juice Yeast Extract Milk Medium

Composition per liter:

Skim milk	100.0g
Yeast extract	5.0g
Tomato juice, filtered	100.0mL

pH 7.0 ± 0.2 at 25°C

Preparation of Medium: Filter canned tomato juice through paper. Let stand overnight at 10°C. Add remaining components and bring to 1.0L with distilled/deionized water. Mix thoroughly. Distribute into tubes or flasks. Autoclave for 15 min at 15 psi pressure–121°C.

Use: For the cultivation and maintenance of a variety of fastidious bacteria which require complex growth factors including *Lactobacillus, Streptococcus* and *Enterococcus* species.

Tomato Paste Oatmeal Agar

Composition per liter:

Oatmeal (dried baby food)	20.0g
Tomato paste	20.0g
Agar	15.0g

pH 7.0 ± 0.2 at 25°C

Preparation of Medium: Add components to distilled/deionized water and bring volume to 1.0L. Mix thoroughly. Melt agar by steaming for 20–30 min at 0 psi pressure–100°C. Distribute into tubes or flasks. Autoclave for 15 min at 15 psi pressure–121°C. Pour into sterile Petri dishes or leave in tubes.

Use: For the cultivation and maintenance of *Flexibacter* species.

Top Agarose

Composition per liter:

Pancreatic digest of casein	10.0g
NaCl	8.0g
Agarose	6.0g

pH 7.0 ± 0.2 at 25°C

Preparation of Medium: Add components to distilled/deionized water and bring volume to 1.0L. Mix thoroughly. Distribute into tubes or flasks. Autoclave for 25 min at 15 psi pressure–121°C.

Use: For the distribution of bacteriophage or bacterial cells, especially *Escherichia coli*, evenly in a thin layer over the surface of a plate.

TPBY

See: **Tryptone Phosphate Brain Heart Infusion Yeast Extract Agar**

TPEY Agar (Tellurite Polymyxin Egg Yolk Agar)

Composition per liter:

NaCl	20.0g
Agar	15.5g
Pancreatic digest of casein	10.0g
Yeast extract	5.0g
D-Mannitol	5.0g
LiCl	2.0g
Egg yolk emulsion (30% solution)	100.0mL
Chapman tellurite solution	10.0mL
Polymyxin B solution	0.4mL

pH 7.1 ± 0.2 at 25°C

Source: This medium is available as a premixed powder from BBL Microbiology Systems and Difco Laboratories.

Egg Yolk Emulsion (30% Solution):

Composition per 100mL:

NaCl	0.6g
Egg yolk	30.0mL

Preparation of Egg Yolk Emulsion (30% Solution): Add NaCl and egg yolk to distilled/deionized water and bring volume to 100.0mL. Mix thoroughly. Filter sterilize.

Chapman Tellurite Solution:

Composition per 100mL:

K_2TeO_3	1.0g

Preparation of Chapman Tellurite Solution: Add K_2TeO_3 to distilled/deionized water and bring volume to 100.0mL. Mix thoroughly. Filter sterilize.

Polymyxin B Solution:

Composition per 100mL:

Polymyxin B	1.0g

Preparation of Polymyxin B Solution: Add polymyxin B to distilled/deionized water and bring volume to 100.0mL. Mix thoroughly. Filter sterilize.

Caution: Potassium tellurite is toxic.

Preparation of Medium: Add components—except 30% egg yolk emulsion, Chapman tellurite solution, and Polymyxin B solution—to distilled/deionized water and bring volume to 890.0mL. Mix thoroughly. Gently heat and bring to boiling. Autoclave for 15 min at 15 psi pressure–121°C. Cool to 45°–50°C. Aseptically add 100.0mL of sterile 30% egg yolk emulsion, 10.0mL of sterile Chapman tellurite solution, and 0.4mL of sterile polymyxin B solution. Mix thoroughly. Pour into sterile Petri dishes or distribute into sterile tubes.

Use: For the recovery of staphylococci from foods and other materials.

TPGY Broth
See: Trypticase Peptone Glucose Yeast Extract Broth

TPGY Medium (Thioglycollate Peptone Glucose Yeast Extract Medium)

Composition per liter:

Pancreatic digest of casein	50.0g
Peptone	5.0g
Yeast extract	5.0g
Glucose	1.0g
Sodium thioglycollate	1.0g

pH 7.1 ± 0.2 at 25°C

Preparation of Medium: Add components to distilled/deionized water and bring volume to 1.0L. Mix thoroughly. Distribute into tubes or flasks. Autoclave for 15 min at 15 psi pressure–121°C.

Use: For the cultivation of a variety of anaerobic bacteria.

TPGYT Broth
See: Trypticase Peptone Glucose Yeast Extract Broth with Trypsin

TPL Medium (Thioglycollate Potato Liver Medium)

Composition per liter:

Potato	200.0g
Yeast extract	31.0g
Liver	25.0g
Glycerol	15.0g
Agar	15.0g
Meat extract	5.5g
Glucose	7.5g
Peptone	2.5g
NaCl	2.5g
Sodium thioglycollate	0.5g
Methylene Blue	1.0mg

pH 7.0 ± 0.2 at 25°C

Preparation of Medium: Add peeled, sliced potato to approximately 500.0mL of distilled/deionized water. Gently heat and bring to boiling. Continue boiling for 30 min. Filter through cheesecloth. Cut up liver into small pieces and to approximately 150.0mL of distilled/deionized water. Gently heat and bring to boiling. Continue boiling for 30 min. Filter through cheesecloth. Add boiled potato solids, boiled liver solids and remaining components to distilled/deionized water and bring volume to 1.0L. Mix thoroughly. Gently heat and bring to boiling. Distribute into tubes or flasks. Make sure each of the tubes receive a few pieces of liver. Autoclave for 15 min at 15 psi pressure–121°C.

Use: For the cultivation and maintenance of *Pseudomonas* species.

TPY Medium

Composition per liter:

Pancreatic digest of casein	10.0g
Glucose	5.0g
Pancreatic digest of soybean meal	5.0g
Yeast extract	2.5g
K_2HPO_4	2.0g
Agar	1.5g
Cysteine·HCl	0.5g
$MgCl_2 \cdot 6H_2O$	0.5g
$ZnSO_4 \cdot 7H_2O$	0.25g
$CaCl_2$	0.15g
$FeCl_3$	1.0μg
Tween™ 80	1.0mL

pH 6.5 ± 0.2 at 25°C

Preparation of Medium: Add components to distilled/deionized water and bring volume to 1.0L. Mix thoroughly. Gently heat until dissolved. Distribute into tubes or flasks. Autoclave for 15 min at 15 psi pressure–121°C.

Use: For the isolation and cultivation of *Bifidobacterium* species.

Trace Element Solution HO–LE

Composition per liter:

H_3BO_3	2.85g
$MnCl_2 \cdot 4H_2O$	1.8g
Sodium tartrate	1.77g
$FeSO_4 \cdot 7H_2O$	1.36g
$CoCl_2 \cdot 6H_2O$	0.04g
$CuCl_2 \cdot 2H_2O$	0.027g
$Na_2MoO_4 \cdot 2H_2O$	0.025g
$ZnCl_2$	0.020g

Preparation of Trace Element Solution HO-LE: Add components to distilled/deionized water and bring volume to 1.0L. Mix thoroughly. Filter sterilize.

Use: For the enrichment of other media requiring added trace metals.

Transgrow Medium

Composition per liter:

GC agar base	730.0mL
Hemoglobin solution	250.0mL
Vitox supplement	10.0mL
VCN antibiotic solution	10.0mL

pH 7.3 ± 0.2 at 25°C

GC Agar Base:
Composition per 730mL:

Special peptone	15.0g
Agar	20.0g
NaCl	5.0g
K_2HPO_4	4.0g
Cornstarch	1.0g
KH_2PO_4	1.0g

pH 7.2 ± 0.2 at 25°C

Preparation of GC Agar Base: Add components of GC medium base and the hemoglobin to distilled/deionized water and bring volume to 730.0mL. Mix thoroughly. Gently heat until boiling. Autoclave for 15 min at 15 psi pressure–121°C. Cool to 45°–50°C.

Hemoglobin Solution:
Composition per 250mL:

Hemoglobin	5.0g

Preparation of Hemoglobin Solution: Add hemoglobin to distilled/deionized water and bring volume to 250.0mL. Mix thoroughly. Autoclave for 15 min at 15 psi pressure–121°C. Cool to 45°–50°C.

Vitox Supplement:
Composition per 10.0mL:

Glucose	2.0g
L-Cysteine·HCl	0.518g
L-Glutamine	0.2g
L-Cystine	0.022g
Adenine sulfate	0.010g
Nicotinamide adenine dinucleotide	5.0mg
Cocarboxylase	2.0mg
Guanine·HCl	0.6mg
$Fe(NO_3)_3 \cdot 6H_2O$	0.4mg
p-Aminobenzoic acid	0.26mg
Vitamin B_{12}	0.2mg
Thiamine·HCl	0.06mg

Preparation of Vitox Supplement: Add components to distilled/deionized water and bring volume to 10.0mL. Mix thoroughly. Filter sterilize.

VCN Antibiotic Solution:
Composition per 10mL:

Colistin methane sulfonate	7.5mg
Vancomycin	3.0mg
Nystatin	12,500U

Preparation of VCN Antibiotic Solution: Add components to distilled/deionized water and bring volume to 10.0mL. Mix thoroughly. Filter sterilize.

Preparation of Medium: To 730.0mL of cooled sterile GC agar base, aseptically add 250.0mL of sterile hemoglobin solution , 10.0mL of sterile Vitox supplement, and 10.0mL of VCN antibiotic solution. Mix thoroughly. Pour into sterile Petri dishes or distribute into sterile tubes.

Use: For the cultivation and transport of fastidious microorganisms, especially *Neisseria* species.

Transgrow Medium

Composition per liter:

GC medium base	730.0mL
Hemoglobin solution	250.0mL
Supplement B (Difco)	10.0mL
VCNT antibiotic solution	10.0mL

pH 7.3 ± 0.2 at 25°C

GC Medium Base:
Composition per 730mL:

Proteose peptone No. 3	15.0g
Agar	20.0g
NaCl	5.0g
K_2HPO_4	4.0g
Glucose	1.5g
Cornstarch	1.0g
KH_2PO_4	1.0g

pH 7.2 ± 0.2 at 25°C

Preparation of GC Medium Base: Add components to distilled/deionized water and bring volume to 730.0mL. Mix thoroughly. Gently heat until boiling. Autoclave for 15 min at 15 psi pressure–121°C. Cool to 45°–50°C.

Hemoglobin Solution:
Composition per 250mL:

Hemoglobin	10.0g

Preparation of Hemoglobin Solution: Add hemoglobin to distilled/deionized water and bring volume to 250.0mL. Mix thoroughly. Autoclave for 15 min at 15 psi pressure–121°C. Cool to 45°–50°C.

Supplement B:
Composition per 10mL:

Supplement B contains yeast concentrate, glutamine, coenzyme, cocarboxylase, hematin and growth factors.

Preparation of Supplement B: Add components to distilled/deionized water and bring volume to 10.0mL. Mix thoroughly. Filter sterilize.

VCNT Antibiotic Solution:
Composition per 10mL:

Colistin methane sulfonate	7.5mg
Trimethoprim lactate	5.0mg
Vancomycin	3.0mg
Nystatin	12,500U

Preparation of VCNT Antibiotic Solution: Add components to distilled/deionized water and bring volume to 10.0mL. Mix thoroughly. Filter sterilize.

Preparation of Medium: To 730.0mL of cooled sterile GC medium base, aseptically add 250.0mL of sterile hemoglobin solution , 10.0mL of sterile supplement B, and 10.0mL of VCNT antibiotic solution. Mix thoroughly. Pour into sterile Petri dishes or distribute into sterile tubes.

Use: For the cultivation and transport of fastidious microorganisms, especially *Neisseria* species.

Transgrow Medium with Trimethoprim

Composition per liter:

Agar	20.0g
Hemoglobin	10.0g
Pancreatic digest of casein	7.5g
Selected meat peptone	7.5g
NaCl	5.0g
K_2HPO_4	4.0g
Glucose	1.5g
Cornstarch	1.0g
KH_2PO_4	1.0g
Supplement solution	10.0mL
V-C-N-T inhibitor	10.0mL

pH 6.7 ± 0.2 at 25°C

Source: Available as a prepared medium from BBL Microbiology Systems.

Supplement Solution:
Composition per liter:

Glucose	100.0g
L-Cysteine·HCl	25.9g
L-Glutamine	10.0g
L-Cystine	1.1g
Adenine	1.0g
Nicotinamide adenine dinucleotide	0.25g
Vitamin B_{12}	0.1g
Thiamine pyrophosphate	0.1g
Guanine·HCl	0.03g
$Fe(NO_3)_3 \cdot 6H_2O$	0.02g
p-Aminobenzoic acid	0.013g
Thiamine·HCl	3.0mg

Source: The supplement solution IsoVitaleX® enrichment is available from BBL Microbiology Laboratories. This enrichment may be replaced by supplement VX from Difco Laboratories.

Preparation of Supplement Solution: Add components to distilled/deionized water and bring volume to 1.0L. Mix thoroughly. Filter sterilize.

VCNT Inhibitor:
Composition per 10mL:

Colistin	7.5mg
Trimethoprim lactate	5.0mg
Vancomycin	3.0mg
Nystatin	12,500U

Preparation of VCNT Inhibitor: Add components to distilled/deionized water and bring volume to 10.0mL. Mix thoroughly. Filter sterilize.

Preparation of Medium: Add components, except supplement solution and V-C-N-T inhibitor, to distilled/deionized water and bring volume to 980.0mL. Mix thoroughly. Gently heat and bring to boiling. Autoclave for 15 min at 15 psi pressure–121°C. Cool to 45°–50°C under 5–30% CO_2. Aseptically add 10.0mL of sterile supplement solution and 10.0mL of sterile V-C-N-T inhibitor. Mix thoroughly. Aseptically distribute under 5–30% CO_2 into sterile screw-capped tubes.

Use: For the transportation and recovery of pathogenic *Neisseria* species.

Transgrow Medium without Trimethoprim

Composition per liter:

Agar	20.0g
Hemoglobin	10.0g
Pancreatic digest of casein	7.5g
Selected meat peptone	7.5g
NaCl	5.0g
K_2HPO_4	4.0g
Glucose	1.5g
Cornstarch	1.0g
KH_2PO_4	1.0g

Supplement solution.................................... 10.0mL
V-C-N inhibitor.. 10.0mL
pH 6.7 ± 0.2 at 25°C

Source: Available as a prepared medium from BBL Microbiology Systems.

Supplemement Solution:
Composition per liter:
Glucose 100.0g
L-Cysteine·HCl 25.9g
L-Glutamine 10.0g
L-Cystine 1.1g
Adenine 1.0g
Nicotinamide adenine dinucleotide.................. 0.25g
Vitamin B_{12} 0.1g
Thiamine pyrophosphate.................................... 0.1g
Guanine·HCl 0.03g
$Fe(NO_3)_3 \cdot 6H_2O$ 0.02g
p-Aminobenzoic acid 0.013g
Thiamine·HCl.. 3.0mg

Source: The supplement solution IsoVitaleX® enrichment is available from BBL Microbiology Laboratories. This enrichment may be replaced by supplement VX from Difco Laboratories.

Preparation of Supplement Solution: Add components to distilled/deionized water and bring volume to 1.0L. Mix thoroughly. Filter sterilize.

VCN Inhibitor:
Composition per 10mL:
Colistin.. 7.5mg
Vancomycin.. 3.0mg
Nystatin 12,500U

Preparation of VCN Inhibitor: Add components to distilled/deionized water and bring volume to 10.0mL. Mix thoroughly. Filter sterilize.

Preparation of Medium: Add components, except supplement solution and V-C-N inhibitor, to distilled/deionized water and bring volume to 980.0mL. Mix thoroughly. Gently heat and bring to boiling. Autoclave for 15 min at 15 psi pressure–121°C. Cool to 45°–50°C under 5–30% CO_2. Aseptically add 10.0mL of sterile supplement solution and 10.0mL of sterile V-C-N inhibitor. Mix thoroughly. Aseptically distribute under 5–30% CO_2 into sterile screw-capped tubes.

Use: For the transportation and recovery of pathogenic *Neisseria* species.

Transport Medium

Composition per liter:
Sodium glycerophosphate................................ 10.0g
Agar.. 3.0g
Sodium thioglycollate 1.0g
$CaCl_2 \cdot 2H_2O$ 0.1g
Methylene Blue 2.0mg
pH 7.3 ± 0.2 at 25°C

Source: This medium is available as a premixed powder from BBL Microbiology Systems.

Preparation of Medium: Add components to distilled/deionized water and bring volume to 1.0L. Mix thoroughly. Gently heat while stirring and bring to boiling. Distribute into screw-capped tubes or vials. Fill tubes nearly to capacity. Leave only enough space so that when a small swab is introduced the tube does not overflow. Autoclave for 10 min at 15 psi pressure–121°C. Tighten caps on tubes.

Use: For the transportation of swab specimens for the recovery of a wide variety of microorganisms, including *Neisseria gonorrhoeae*.

Transport Medium Stuart

Composition per liter:
Sodium glycerophosphate................................ 10.0g
Agar.. 3.0g
Sodium thioglycollate 0.9g
$CaCl_2 \cdot 2H_2O$ 0.1g
Methylene Blue 2.0mg
pH 7.3 ± 0.2 at 25°C

Source: This medium is available as a premixed powder from Difco Laboratories.

Preparation of Medium: Add components to distilled/deionized water and bring volume to 1.0L. Mix thoroughly. Gently heat while stirring and bring to boiling. Distribute into screw-capped tubes or vials. Fill tubes nearly to capacity. Leave only enough space so that when a small swab is introduced the tube does not overflow. Autoclave for 10 min at 15 psi pressure–121°C. Tighten caps on tubes.

Use: For the transportation of swab specimens for the recovery of a wide variety of microorganisms, including *Neisseria gonorrhoeae*.

Treponema Isolation Medium

Composition per liter:
Solution A 450.0mL
Spirolate broth... 450.0mL
Rabbit serum,
incativated at 56°C for 30min............... 100.0mL
pH 7.4 ± 0.2 at 25°C

Solution A:
Composition per 450mL:
Agar.. 8.0g
Asparagine 0.25g

Sodium thioglycollate	0.25g
Pancreatic digest of casein	0.25g
Brain heart infusion broth	450.0mL

Preparation of Solution A: Combine components. Mix thoroughly. Gently heat and bring to boiling. Autoclave for 15 min at 15 psi pressure–121°C. Cool to 45°–50°C.

Brain Heart Infusion Broth:
Composition per liter:

Pancreatic digest of gelatin	14.5g
Brain heart, solids from infusion	6.0g
Peptic digest of animal tissue	6.0g
NaCl	5.0g
Casein	5.0g
Glucose	3.0g
Na_2HPO_4	2.5g

Preparation of Brain Heart Infusion Broth: Add components to distilled/deionized water and bring volume to 1.0L. Mix thoroughly.

Spirolate Broth:
Composition per liter:

Pancreatic digest of casein	15.0g
Glucose	5.0g
Yeast extract	5.0g
NaCl	2.5g
L-Cysteine·HCl·H_2O	1.0g
Sodium thioglycollate	0.5g
Palmitic acid	0.05g
Stearic acid	0.05g
Oleic acid	0.05g
Linoleic acid	0.05g

Preparation of Spirolate Broth: Add components to distilled/deionized water and bring volume to 1.0L. Mix thoroughly. Autoclave for 15 min at 15 psi pressure–121°C. Cool to 25°.

Preparation of Medium: Combine 450.0mL of sterile solution A, 450.0mL of sterile spirolate broth and 100.0mL of rabbit serum. Mix thoroughly. Aseptically distribute into sterile tubes or flasks.

Use: For the isolation and cultivaton of oral, genital, and fecal treponemes.

Treponema Isolation Medium

Composition per liter:

Beef heart, solids from infusion	20.0g
Ionagar No. 2	7.2g
K_2HPO_4	2.0g
Arabinose	0.8g
Glucose	0.8g
Maltose	0.8g
Polypeptone™	0.8g
Pyruvate	0.8g
Starch, soluble	0.8g
Sucrose	0.8g
Cysteine·HCl	0.68g
$(NH_4)_2SO_4$	0.6g
Serine	0.4g
Tryptose	0.4g
Yeast extract	0.4g
NaCl	0.2g
Rumen fluid	500.0mL
Rabbit serum–cocarboxylase solution	100.0mL

pH 7.2 ± 0.2 at 25°C

Rabbit Serum–Cocarboxylase Solution:
Composition per liter:

Rabbit serum, heat inactivated	100.0mL
Cocarboxylase solution	1.0mL

Preparation of Rabbit Serum–Cocarboxylase Solution: Heat rabbit serum at 56°C for 1hr. Add 1.0mL cocarboxylase solution. Mix thoroughly.

Cocarboxylase Solution:
Composition per 1mL:

Cocarboxylase	0.5g

Preparation of Cocarboxylase Solution: Add cocarboxylase to 1.0mL of distilled/deionized water. Mix thoroughly. Filter sterilize.

Preparation of Medium: Add components, except rumen fluid and rabbit serum–cocarboxylase solution, to distilled/deionized water and bring volume to 400.0mL. Mix thoroughly. Gently heat and bring to boiling. Autoclave for 15 min at 15 psi pressure–121°C. Cool to 45°–50°C. Aseptically add 500.0mL of sterile rumen fluid and 100.0mL of sterile rabbit serum–cocarboxylase solution. Mix thoroughly. Pour into sterile Petri dishes or distribute into sterile tubes.

Use: For the isolation of oral treponemes.

Treponema macrodentium Medium

Composition per liter:

Glucose	1.0g
Nicotinamide	0.4g
Spermine·4HCl	0.15g
Sodium isobutyrate	0.02g
Carboxylase	5.0mg
PPLO agar	900.0mL
Bovine serum	100.0mL

pH 7.0 ± 0.2 at 25°C

PPLO Agar:
Composition per 900mL:

Beef heart, infusion from	50.0g
Agar	14.0g

Peptone 10.0g
NaCl 5.0g
pH 7.8 ± 0.2 at 25°C

Preparation of PPLO Agar: Add components to distilled/deionized water and bring volume to 900.0mL. Mix thoroughly.

Preparation of Medium: Combine components, except bovine serum. Mix thoroughly. Autoclave for 15 min at 15 psi pressure–121°C. Cool to 45°–50°C. Aseptically add sterile bovine serum. Mix thoroughly. Aseptically distribute into sterile tubes or flasks.

Use: For the isolation and cultivation of *Treponema macrodentium.*

Treponema Medium

Composition per liter:

Pancreatic digest of casein 30.0g
Ionagar No. 2 .... 8.0g
Glucose 5.0g
Yeast extract 5.0g
NaCl 2.5g
Cysteine·HCl 0.75g
Horse serum, inactivated 100.0mL
pH 7.4 ± 0.2 at 25°C

Preparation of Medium: Add components, except horse serum, to distilled/deionized water and bring volume to 900.0mL. Mix thoroughly. Gently heat and bring to boiling. Autoclave for 15 min at 15 psi pressure–121°C. Cool to 45°–50°C. Aseptically add 100.0mL sterile horse serum. Mix thoroughly. Pour into sterile Petri dishes or distribute into sterile tubes.

Use: For the isolation and cultivation of oral treponemes.

Treponema Medium

Composition per liter:

Spirolate agar 900.0mL
Rabbit serum,
incativated at 56°C for 30min 100.0mL

Spirolate Agar:

Composition per liter:

Pancreatic digest of casein 15.0g
Agar 14.0g
Glucose 5.0g
Yeast extract 5.0g
NaCl 2.5g
L-Cysteine·HCl·H_2O 1.0g
Sodium thioglycollate 0.5g
Palmitic acid 0.05g
Stearic acid 0.05g
Oleic acid 0.05g
Linoleic acid 0.05g

Preparation of Spirolate Agar: Add components to distilled/deionized water and bring volume to 1.0L. Mix thoroughly. Gently heat and bring to boiling Autoclave for 15 min at 15 psi pressure–121°C. Cool to 45°–50°C.

Preparation of Medium: To 900.0mL of cooled sterile spirolate agar, aseptically add 100.0mL of rabbit serum. Mix thoroughly. Aseptically distribute into sterile tubes or flasks.

Use: For the isolation of oral treponemes.

Treponema Medium

Composition per liter:

Spirolate agar 675.0mL
Brain heart infusion broth 225.0mL
Rabbit serum,
incativated at 56°C for 30min 100.0mL
pH 7.0–7.2 ± 0.2 at 25°C

Spirolate Agar:

Composition per 675mL:

Pancreatic digest of casein 15.0g
Ionagar No. 2 .... 8.0g
Glucose 5.0g
Yeast extract 5.0g
NaCl 2.5g
L-Cysteine·HCl·H_2O 1.0g
Sodium thioglycollate 0.5g
Palmitic acid 0.05g
Stearic acid 0.05g
Oleic acid 0.05g
Linoleic acid 0.05g

Preparation of Spirolate Agar: Add components to distilled/deionized water and bring volume to 675.0mL. Mix thoroughly. Autoclave for 15 min at 15 psi pressure–121°C. Cool to 45°–50°C.

Brain Heart Infusion Broth:

Composition per liter:

Pancreatic digest of gelatin 14.5g
Brain heart, solids from infusion 6.0g
Peptic digest of animal tissue 6.0g
NaCl 5.0g
Casein 5.0g
Glucose 3.0g
Na_2HPO_4 .... 2.5g

Preparation of Brain Heart Infusion Broth: Add components to distilled/deionized water and bring volume to 1.0L. Mix thoroughly.

Preparation of Medium: Aseptically combine 675.0mL of cooled sterile spirolate agar, 225.0mL of cooled sterile brain heart infusion broth, and

100.0mL of rabbit serum. Mix thoroughly. Pour into sterile Petri dishes or distribute into sterile tubes.

Use: For the isolation of oral treponemes.

Treponema Medium

Composition per liter:

Solution A	440.0mL
Spirolate broth	440.0mL
Rabbit serum, incactivated at 56°C for 30min	100.0mL
Mucin solution	20.0mL

pH 7.8 ± 0.2 at 25°C

Solution A:

Composition per 440mL:

Ionagar No. 2	8.0g
Brain heart infusion broth	440.0mL

Brain Heart Infusion Broth:

Composition per liter:

Pancreatic digest of gelatin	14.5g
Brain heart, solids from infusion	6.0g
Peptic digest of animal tissue	6.0g
NaCl	5.0g
Casein	5.0g
Glucose	3.0g
Na_2HPO_4	2.5g

Preparation of Brain Heart Infusion Broth: Add components to distilled/deionized water and bring volume to 1.0L. Mix thoroughly.

Preparation of Solution A: Add 8.0g of ionagar to 440.0mL of brain heart infusion broth, mix thoroughly. Gently heat and bring to boiling. Autoclave for 15 min at 15 psi pressure–121°C. Cool to 45°–50°C.

Spirolate Broth:

Composition per liter:

Pancreatic digest of casein	15.0g
Glucose	5.0g
Yeast extract	5.0g
NaCl	2.5g
L-Cysteine·HCl·H_2O	1.0g
Sodium thioglycollate	0.5g
Palmitic acid	0.05g
Stearic acid	0.05g
Oleic acid	0.05g
Linoleic acid	0.05g

Preparation of Spirolate Broth: Add components to distilled/deionized water and bring volume to 1.0L. Mix thoroughly. Autoclave for 15 min at 15 psi pressure–121°C. Cool to 25°.

Mucin Solution:

Composition per 20mL:

Mucin	0.2g

Preparation of Mucin Solution: Add mucin to distilled/deionized water and bring volume to 20.0mL. Mix thoroughly. Filter sterilize.

Preparation of Medium: Aseptically combine 440.0mL of solution A, 440.0mL of spirolate broth, 100.0mL of rabbit serum, and 20.0mL of mucin solution. Mix thoroughly. Aseptically distribute into sterile tubes or flasks.

Use: For the isolation of intestinal treponemes.

Treponema Medium

Composition per liter:

Agar	13.0g
Glucose	1.4g
Cysteine·HCl	0.64g
$(NH_4)_2SO_4$	0.5g
Polypeptone™	0.5g
Starch, soluble	0.5g
Yeast extract	0.5g
Resazurin	1.6mg
Salts solution	500.0mL
Bovine rumen fluid	280.0mL

pH 7.2–7.5 at 25°C

Salts Solution:

Composition per liter:

$NaHCO_3$	10.0g
NaCl	2.0g
K_2HPO_4	1.0g
KH_2PO_4	1.0g
$CaCl_2$	0.2g
$MgSO_4$	0.2g
CoCl	3.4mg
$MnSO_4$	3.4mg
$NaMoO_4$	3.4mg

Preparation of Salts Solution: Add components to distilled/deionized water and bring volume to 1.0L. Mix thoroughly.

Preparation of Medium: Add components, except bovine rumen fluid, to distilled/deionized water and bring volume to 720.0mL. Mix thoroughly. Gently heat and bring to boiling. Autoclave for 15 min at 15 psi pressure–121°C. Cool to 45°–50°C. Aseptically add bovine rumen fluid. Mix thoroughly. Pour into sterile Petri dishes or distribute into sterile tubes.

Use: For the isolation of intestinal treponemes.

Treponema Medium

Composition per liter:

Cysteine·HCl·H_2O	1.0g
Glucose	1.0g

Nicotinamide....0.4g
Spermidine·4HCl....0.15g
Sodium isobutyrate....0.02g
Thiamine pyrophosphate....5.0mg
PPLO Broth....900.0mL
Rabbit serum, inactivated....100.0mL

pH 7.8 ± 0.2 at 25°C

PPLO Broth:

Composition per 900mL:

Beef heart, infusion from solids....50.0g
Peptone....10.0g
NaCl....5.0g

Preparation of PPLO Broth: Add components to distilled/deionized water and bring volume to 900.0mL. Mix thoroughly.

Preparation of Medium: Combine components, except rabbit serum. Mix thoroughly. Filter sterilize. Aseptically add sterile rabbit serum. Mix thoroughly. Aseptically distribute into sterile tubes or flasks.

Use: For the cultivation of oral treponemes. For cultivation of *Treponema denticola*, *T. macrodentium*, and *T. oralis*.

Treponema Medium

Composition per liter:

Spirolate broth....675.0mL
Brain heart infusion broth....225.0mL
Rabbit serum....100.0mL

Spirolate Broth:

Composition per liter:

Pancreatic digest of casein....15.0g
Glucose....5.0g
Yeast extract....5.0g
NaCl....2.5g
L-Cysteine·HCl·H_2O....1.0g
Sodium thioglycollate....0.5g
Palmitic acid....0.05g
Stearic acid....0.05g
Oleic acid....0.05g
Linoleic acid....0.05g

Preparation of Spirolate Broth: Add components to distilled/deionized water and bring volume to 1.0L. Mix thoroughly. Autoclave for 15 min at 15 psi pressure–121°C. Cool to 25°.

Brain Heart Infusion Broth:

Composition per liter:

Pancreatic digest of gelatin....14.5g
Brain heart, solids from infusion....6.0g
Peptic digest of animal tissue....6.0g
NaCl....5.0g
Casein....5.0g
Glucose....3.0g
Na_2HPO_4....2.5g

Preparation of Brain Heart Infusion Broth: Add components to distilled/deionized water and bring volume to 1.0L. Mix thoroughly. Autoclave for 15 min at 15 psi pressure–121°C. Cool to 25°.

Preparation of Medium: Aseptically combine 675.0mL of cooled, sterile spirolate broth, 225.0mL of cooled, sterile brain heart infusion broth, and 100.0mL of rabbit serum. Mix thoroughly.

Use: For the cultivation of oral treponemes.

Treponema Medium

Composition per liter:

Heart infusion broth, modified....450.0mL
Spirolate broth....450.0mL
Rabbit serum, inactivated....100.0mL

pH 7.4 ± 0.2 at 25°C

Heart Infusion Broth, Modified:

Composition per liter:

Beef heart, solids from infusion....500.0g
Tryptose....10.0g
NaCl....5.0g
Asparagine....2.5g
Sodium thioglycollate....2.5g
Pancreatic digest of casein....2.5g

Preparation of Heart Infusion Broth, Modified: Add components to distilled/deionized water and bring volume to 1.0L. Mix thoroughly. Gently heat and bring to boiling. Autoclave for 15 min at 15 psi pressure–121°C. Cool to 25°.

Spirolate Broth:

Composition per liter:

Pancreatic digest of casein....15.0g
Glucose....5.0g
Yeast extract....5.0g
NaCl....2.5g
L-Cysteine·HCl·H_2O....1.0g
Sodium thioglycollate....0.5g
Palmitic acid....0.05g
Stearic acid....0.05g
Oleic acid....0.05g
Linoleic acid....0.05g

Preparation of Spirolate Broth: Add components to distilled/deionized water and bring volume to 1.0L. Mix thoroughly. Autoclave for 15 min at 15 psi pressure–121°C. Cool to 25°.

Preparation of Medium: Aseptically combine 450.0mL of cooled, sterile spirolate broth, 450.0mL of cooled, sterile heart infusion broth, modified and

100.0mL of rabbit serum. Mix thoroughly. Aseptically distribute into sterile tubes or flasks.

Use: For the cultivation of treponemes.

Treponema Medium

Composition per 500mL :

Beef heart, solids from infusion	250.0g
Sucrose	50.0g
Tryptose	5.0g
NaCl	2.5g
Yeast extract	2.5g
Agar	0.5g
Sodium thioglycollate	0.38g
$MgSO_4$	0.05g
Horse serum, inactivated	100.0mL

pH 7.4 ± 0.2 at 25°C

Preparation of Medium: Add components, except horse serum, to distilled/deionized water and bring volume to 400.0mL. Mix thoroughly. Adjust pH to 7.4. Gently heat and bring to boiling. Distribute into tubes in 4.0mL volumes. Autoclave for 15 min at 15 psi pressure–121°C. Cool to 25°C. Prior to inoculation, add 1.0mL sterile horse serum to each tube.

Use: For the cultivation and maintenance of *Treponema pallidum* and other *Treponema* species.

Treponema Medium 1

Composition per liter:

Thioglycollate agar USP, alternate	900.0mL
Normal calf serum	100.0mL

pH 7.1 ± 0.2 at 25°C

Thioglycollate Agar USP, Alternate:

Composition per 900mL:

Pancreatic digest of casein	15.0g
Ionagar No. 2	7.0g
Glucose	5.5g
Yeast extract	5.0g
NaCl	2.5g
L-Cystine	0.5g
Sodium thioglycollate	0.5g

Preparation of Thioglycollate Agar USP, Alternate: Add components to distilled/deionized water and bring volume to 1.0L. Mix thoroughly. Autoclave for 15 min at 15 psi pressure–121°C. Cool to 45°–50°C.

Preparation of Medium: Aseptically combine 900.0mL of cooled sterile thioglycollate agar USP, alternate and 100.0mL of calf serum. Mix thoroughly. Pour into sterile Petri dishes or distribute into sterile tubes.

Use: For the cultivation of treponemes.

Treponema Medium 2

Composition per liter:

Pancreatic digest of casein	30.0g
Ionagar No. 2	7.0g
Glucose	5.0g
Yeast extract	5.0g
NaCl	2.5g
L-Cysteine·HCl·H_2O	2.0g
Rabbit serum	100.0mL

pH 7.2 ± 0.2 at 25°C

Preparation of Medium: Add components, except rabbit serum, to distilled/deionized water and bring volume to 900.0mL. Mix thoroughly. Gently heat and bring to boiling. Autoclave for 15 min at 15 psi pressure–121°C. Cool to 45°–50°C. Aseptically add sterile rabbit serum. Mix thoroughly. Pour into sterile Petri dishes or distribute into sterile tubes.

Use: For the cultivation of treponemes.

Treponema Medium 3

Composition per liter:

Spirolate agar	675.0mL
Brain heart infusion broth	225.0mL
Rabbit serum	100.0mL

Spirolate Agar:

Composition per liter:

Pancreatic digest of casein	15.0g
Ionagar No. 2	7.0g
Glucose	5.0g
Yeast extract	5.0g
NaCl	2.5g
L-Cysteine·HCl·H_2O	1.0g
Sodium thioglycollate	0.5g
Palmitic acid	0.05g
Stearic acid	0.05g
Oleic acid	0.05g
Linoleic acid	0.05g

Preparation of Spirolate Agar: Add components to distilled/deionized water and bring volume to 1.0L. Mix thoroughly. Gently heat and bring to boiling. Autoclave for 15 min at 15 psi pressure–121°C. Cool to 25°.

Brain Heart Infusion Broth:

Composition per liter:

Pancreatic digest of gelatin	14.5g
Brain heart, solids from infusion	6.0g
Peptic digest of animal tissue	6.0g
NaCl	5.0g
Casein	5.0g
Glucose	3.0g
Na_2HPO_4	2.5g

Preparation of Brain Heart Infusion Broth: Add components to distilled/deionized water and bring volume to 1.0L. Mix thoroughly. Gently heat and bring to boiling. Autoclave for 15 min at 15 psi pressure–121°C. Cool to 25°.

Preparation of Medium: Aseptically combine 675.0mL of cooled, sterile spirolate broth, 225.0mL of cooled, sterile brain heart infusion broth, and 100.0mL of rabbit serum. Mix thoroughly.

Use: For the cultivation of treponemes.

Treponema Medium, Prereduced

Composition per liter:

Agar	1.6g
Glucose	1.4g
Cysteine·HCl·H_2O	0.64g
$(NH_4)_2SO_4$	0.5g
Polypeptone™	0.5g
Starch, soluble	0.5g
Yeast extract	0.5g
Resazurin	1.6mg
Salts solution	500.0mL
Bovine rumen fluid	280.0mL

pH 7.2–7.5 at 25°C

Salts Solution:

Composition per liter:

$NaHCO_3$	10.0g
NaCl	2.0g
K_2HPO_4	1.0g
KH_2PO_4	1.0g
$CaCl_2$	0.2g
$MgSO_4$	0.2g
CoCl	3.4mg
$MnSO_4$	3.4mg
$NaMoO_4$	3.4mg

Preparation of Salts Solution: Add components to distilled/deionized water and bring volume to 1.0L. Mix thoroughly.

Preparation of Medium: Add components, except bovine rumen fluid, to distilled/deionized water and bring volume to 720.0mL. Mix thoroughly. Gently heat and bring to boiling. Autoclave for 15 min at 15 psi pressure–121°C. Cool to 45°–50°C. Aseptically add 280.0mL of sterile bovine rumen fluid. Mix thoroughly. Aseptically and anaerobically distribute into sterile tubes or flasks under 100% N_2.

Use: For the cultivation of fecal and intestinal treponemes.

Tributyrin Agar

Composition per liter:

Agar	15.0g
Tributyrin (glyceryl tributyrate)	10.0g
Peptone	5.0g
Yeast extract	3.0g

pH 7.5 ± 0.2 at 25°C

Source: Available as a prepared medium from Oxoid Unipath.

Preparation of Medium: Add components to distilled/deionized water and bring volume to 1.0L. Mix thoroughly. Gently heat and bring to boiling. Distribute into tubes or flasks. Autoclave for 15 min at 15 psi pressure–121°C. Pour into sterile Petri dishes.

Use: For the cultivation and enumeration of lipolytic fungi and bacteria, especially *Staphylococcus* species, *Flavobacterium* species, *Clostridium* species and *Pseudomonas* species from butter. Lipolytic bacteria appear as colonies surrounded by a clear zone.

Trichlorophenol Medium

Composition per liter:

Pancreatic digest of casein	8.5g
NaCl	2.5g
Papaic digest of soybean meal	1.5g
K_2HPO_4	1.25g
Glucose	1.25g
2,4,6-Trichlorophenol	1.25g

pH 7.3 ± 0.2 at 25°C

Preparation of Medium: Add components to distilled/deionized water and bring volume to 1.0L. Mix thoroughly. Gently heat until dissolved. Distribute into tubes or flasks. Autoclave for 15 min at 15 psi pressure–121°C.

Use: For the cultivation and maintenance of *Arthrobacter* species and other microorganisms which can degrade chlorinated phenols.

Trichomonas Medium

Composition per liter:

Liver digest	25.0g
NaCl	6.5g
Glucose	5.0g
Agar	1.0g
Horse serum	80.0mL

pH 6.4 ± 0.2 at 25°C

Source: This medium is available as a premixed powder from Oxoid Unipath.

Horse Serum:
Composition per 80mL:

Horse serum	80.0mL

Preparation of Horse Serum: Gently heat sterile horse serum to 56°C for 30 min. Aseptically adjust pH to 6.0 with 0.1*N* HCl. Use immediately.

Preparation of Medium: Add components, except horse serum, to distilled/deionized water and bring volume to 920.0mL. Mix thoroughly. Gently heat and bring to boiling. Autoclave for 15 min at 15 psi pressure–121°C. Cool to 45°–50°C. Aseptically add 80.0mL of freshly prepared sterile horse serum. Mix thoroughly. Aseptically distribute into sterile tubes or flasks.

Use: For the cultivation of *Trichomonas vaginalis*.

Trichomonas Medium No. 2

Composition per liter:

Glucose	22.5g
Liver digest	18.0g
Pancreatic digest of casein	17.0g
NaCl	5.0g
Pancreatic digest of soybean meal	3.0g
K_2HPO_4	2.5g
Chloramphenicol	0.125g
Horse serum	250.0mL
Calcium pantothenate (0.5% solution)	1.0mL

pH 6.2 ± 0.2 at 25°C

Source: Available as a prepared medium from Oxoid Unipath.

Preparation of Medium: Add components, except horse serum, to distilled/deionized water and bring volume to 750.0mL. Mix thoroughly. Autoclave for 15 min at 5 psi pressure–108°C. Cool to 45°–50°C. Aseptically add 250.0mL of sterile horse serum. Mix thoroughly. Aseptically distribute into sterile tubes or flasks.

Use: For the isolation of *Trichomonas vaginalis*.

Trichomonas Selective Medium

Composition per liter:

Liver digest	25.0g
NaCl	6.5g
Glucose	5.0g
Agar	1.0g
Horse serum	80.0mL
Antibiotic inhibitor	10.0mL

pH 6.4 ± 0.2 at 25°C

Source: This medium is available as a premixed powder from Oxoid Unipath.

Horse Serum:
Composition per 80mL:

Horse serum	80.0mL

Preparation of Horse Serum: Gently heat sterile horse serum to 56°C for 30 min. Aseptically adjust pH to 6.0 with 0.1*N* HCl. Use immediately.

Antibiotic Inhibitor:
Composition per 10mL:

Penicillin G	1,000,000U
Streptomycin	500.0mg

Preparation of Antibiotic Inhibitor: Add components to distilled/deionized water and bring volume to 10.0mL. Mix thoroughly. Filter sterilize.

Preparation of Medium: Add components, except horse serum, to distilled/deionized water and bring volume to 910.0mL. Mix thoroughly. Gently heat and bring to boiling. Autoclave for 15 min at 15 psi pressure–121°C. Cool to 45°–50°C. Aseptically add 80.0mL of freshly prepared sterile horse serum and 10.0mL of sterile antibiotic inhibitor. Mix thoroughly. Aseptically distribute into sterile tubes or flasks.

Use: For the cultivation of *Trichomonas vaginalis* from specimens with a mixed bacterial flora.

Trichomonas Selective Medium

Composition per liter:

Liver digest	25.0g
NaCl	6.5g
Glucose	5.0g
Agar	1.0g
Horse serum	80.0mL
Antibiotic inhibitor	10.0mL

pH 6.4 ± 0.2 at 25°C

Horse Serum:
Composition per 80mL:

Horse serum	80.0mL

Preparation of Horse Serum: Gently heat sterile horse serum to 56°C for 30 min. Aseptically adjust pH to 6.0 with 0.1*N* HCl. Use immediately.

Antibiotic Inhibitor:
Composition per 10mL:

Chloramphenicol	100.0mg

Preparation of Antibiotic Inhibitor: Add components to distilled/deionized water and bring volume to 10.0mL. Mix thoroughly. Filter sterilize.

Preparation of Medium: Add components, except horse serum, to distilled/deionized water and bring volume to 910.0mL. Mix thoroughly. Gently heat and bring to boiling. Autoclave for 15 min at 15 psi pressure–121°C. Cool to 45°–50°C. Aseptically add

80.0mL of freshly prepared sterile horse serum and 10.0mL of sterile antibiotic inhibitor. Mix thoroughly. Aseptically distribute into sterile tubes or flasks.

Use: For the cultivation of *Trichomonas vaginalis* from specimens with a mixed bacterial flora.

Trichophyton Agar 1

Composition per liter:

Glucose 40.0g
Agar 15.0g
Vitamin assay casamino acids 2.5g
KH_2PO_4 .... 1.8g
$MgSO_4{\cdot}7H_2O$ 0.1g
pH 6.8 ± 0.2 at 25°C

Source: This medium is available as a premixed powder from Difco Laboratories.

Preparation of Medium: Add components to distilled/deionized water and bring volume to 1.0L. Mix thoroughly. Gently heat and bring to boiling. Distribute into tubes. Autoclave for 15 min at 15 psi pressure–121°C. Allow tubes to cool in a slanted position.

Use: For the differentiation of the *Trichophyton* species.

Trichophyton Agar 2

Composition per liter:

Glucose 40.0g
Agar 15.0g
Vitamin assay casamino acids 2.5g
KH_2PO_4 .... 1.8g
$MgSO_4{\cdot}7H_2O$ 0.1g
Inositol 50.0mg
pH 6.8 ± 0.2 at 25°C

Source: This medium is available as a premixed powder from Difco Laboratories.

Preparation of Medium: Add components to distilled/deionized water and bring volume to 1.0L. Mix thoroughly. Gently heat and bring to boiling. Distribute into tubes. Autoclave for 15 min at 15 psi pressure–121°C. Allow tubes to cool in a slanted position.

Use: For the differentiation of the *Trichophyton* species.

Trichophyton Agar 3

Composition per liter:

Glucose 40.0g
Agar 15.0g
Vitamin assay casamino acids 2.5g
KH_2PO_4 .... 1.8g
$MgSO_4{\cdot}7H_2O$ 0.1g
Inositol 0.05g
Thiamine·HCl 0.2mg
pH 6.8 ± 0.2 at 25°C

Source: This medium is available as a premixed powder from Difco Laboratories.

Preparation of Medium: Add components to distilled/deionized water and bring volume to 1.0L. Mix thoroughly. Gently heat and bring to boiling. Distribute into tubes. Autoclave for 15 min at 15 psi pressure–121°C. Allow tubes to cool in a slanted position.

Use: For the differentiation of the *Trichophyton* species.

Trichophyton Agar 4

Composition per liter:

Glucose 40.0g
Agar 15.0g
Vitaminassay casamino acids 2.5g
KH_2PO_4 .... 1.8g
$MgSO_4{\cdot}7H_2O$ 0.1g
Thiamine·HCl USP 200.0μg
pH 6.8 ± 0.2 at 25°C

Source: This medium is available as a premixed powder from Difco Laboratories.

Preparation of Medium: Add components to distilled/deionized water and bring volume to 1.0L. Mix thoroughly. Gently heat and bring to boiling. Distribute into tubes. Autoclave for 15 min at 15 psi pressure–121°C. Allow tubes to cool in a slanted position.

Use: For the differentiation of the *Trichophyton* species.

Trichophyton Agar 5

Composition per liter:

Glucose 40.0g
Agar 15.0g
Vitamin Assay Casamino Acids 2.5g
KH_2PO_4 .... 1.8g
$MgSO_4{\cdot}7H_2O$ 0.1g
Nicotinic acid 2.0mg
pH 6.8 ± 0.2 at 25°C

Source: This medium is available as a premixed powder from Difco Laboratories.

Preparation of Medium: Add components to distilled/deionized water and bring volume to 1.0L. Mix thoroughly. Gently heat and bring to boiling. Distribute into tubes. Autoclave for 15 min at 15 psi pressure–121°C. Allow tubes to cool in a slanted position.

Use: For the differentiation of the *Trichophyton* species.

Trichophyton Agar 6

Composition per liter:

Glucose	40.0g
Agar	15.0g
KH_2PO_4	1.8g
Ammonium nitrate	1.5g
$MgSO_4 \cdot 7H_2O$	0.1g

pH 6.8 ± 0.2 at 25°C

Source: This medium is available as a premixed powder from Difco Laboratories.

Preparation of Medium: Add components to distilled/deionized water and bring volume to 1.0L. Mix thoroughly. Gently heat and bring to boiling. Distribute into tubes. Autoclave for 15 min at 15 psi pressure–121°C. Allow tubes to cool in a slanted position.

Use: For the differentiation of the *Trichophyton* species.

Trichophyton Agar 7

Composition per liter:

Glucose	40.0g
Agar	15.0g
KH_2PO_4	1.8g
Ammonium nitrate	1.5g
$MgSO_4 \cdot 7H_2O$	0.1g
Histidine·HCl	0.03g

pH 6.8 ± 0.2 at 25°C

Source: This medium is available as a premixed powder from Difco Laboratories.

Preparation of Medium: Add components to distilled/deionized water and bring volume to 1.0L. Mix thoroughly. Gently heat and bring to boiling. Distribute into tubes. Autoclave for 15 min at 15 psi pressure–121°C. Allow tubes to cool in a slanted position.

Use: For the differentiation of the *Trichophyton* species.

Trichosel™ Broth, Modified

Composition per liter:

Pancreatic digest of casein	12.0g
Yeast extract	5.0g
Liver extract	2.0g
Maltose	2.0g
L-Cysteine·HCl	1.0g
Agar	1.0g
Chloramphenicol	0.1g
Methylene Blue	3.0mg
Horse serum	50.0mL

pH 6.0 ± 0.2 at 25°C

Source: This medium is available as a premixed powder from BBL Microbiology Systems.

Preparation of Medium: Add components to distilled/deionized water and bring volume to 950.0mL. Mix thoroughly. Gently heat while stirring and bring to boiling. Autoclave for 15 min at 13 psi pressure–118°C. Cool to 45°–50°C. Aseptically add 50.0mL of sterile horse serum. Mix thoroughly. Aseptically distribute into sterile tubes or flasks.

Use: For the isolation and cultivation of *Trichomonas* species.

Trimethylamine N–Oxide Medium (TMAO Medium)

Composition per liter:

Beef extract	10.0g
Peptone	10.0g
NaCl	5.0g
Agar	2.0g
Trimethylamine N-oxide	1.0g
Yeast extract	1.0g

pH 7.5 ± 0.2 at 25°C

Source: This medium is available as a premixed powder from Oxoid Unipath.

Preparation of Medium: Add components to distilled/deionized water and bring volume to 1.0L. Mix thoroughly. Gently heat and bring to boiling. Distribute into screw-capped tubes in 4.0mL volumes. Autoclave for 15 min at 15 psi pressure–121°C. Allow tubes to cool in an upright position. Store at 4°C.

Use: For the cultivation and differentiation of *Campylobacter* species from foods. *C. jejuni* and *C. coli* will not grow on this medium.

Triple Sugar Iron Agar (TSI Agar)

Composition per liter:

Peptone	20.0g
Agar	12.0g
Lactose	10.0g
Sucrose	10.0g
NaCl	5.0g
Beef extract	3.0g
Yeast extract	3.0g
Glucose	1.0g
Ferric citrate	0.3g
$Na_2S_2O_3$	0.3g
Phenol Red	0.025g

pH 7.4 ± 0.2 at 25°C

Source: This medium is available as a premixed powder from Difco Laboratories and Oxoid Unipath.

Preparation of Medium: Add components to distilled/deionized water and bring volume to 1.0L. Mix thoroughly. Gently heat and bring to boiling. Distribute into tubes or flasks. Autoclave for 15 min at 15 psi pressure–121°C. Allow tubes to cool in a slanted position to form a 1.0 inch butt.

Use: For the differentiation of members of the Enterobacteriaceae based on their fermentation of lactose, sucrose and glucose and the production of H_2S.

Triple Sugar Iron Agar (TSI Agar)

Composition per liter:

Agar	13.0g
Pancreatic digest of casein	10.0g
Peptic digest of animal tissue	10.0g
Lactose	10.0g
Sucrose	10.0g
NaCl	5.0g
Glucose	1.0g
$Fe(NH_4)_2(SO_4)_2 \cdot 6H_2O$	0.2g
$Na_2S_2O_3$	0.2g
Phenol Red	0.025g

pH 7.3 ± 0.2 at 25°C

Source: This medium is available as a premixed powder from BBL Microbiology Systems.

Preparation of Medium: Add components to distilled/deionized water and bring volume to 1.0L. Mix thoroughly. Gently heat and bring to boiling. Distribute into tubes or flasks. Autoclave for 15 min at 15 psi pressure–121°C. Allow tubes to cool in a slanted position to form a 1.0 inch butt.

Use: For the differentiation of members of the Enterobacteriaceae based on their fermentation of lactose, sucrose and glucose and the production of H_2S.

Tris YP Agar (Tris Yeast Extract Peptone Agar)

Composition per liter:

Agar	19.0g
Yeast extract	3.0g
Glucose	1.0g
Peptone	0.6g
Tris-buffer (0.05 M, pH 7.5)	1.0L

pH 7.5 ± 0.2 at 25°C

Preparation of Medium: Add components to distilled/deionized water and bring volume to 1.0L. For top layer agar add 6.0g of agar instead of 19.0g. Mix thoroughly. Gently heat and bring to boiling. Distribute into tubes or flasks. Autoclave for 15 min at 15 psi pressure–121°C. Pour into sterile Petri dishes.

Use: For the cultivation and maintenance of *Bdello–vibrio* species.

Tris YP Broth (Tris Yeast Extract Peptone Broth)

Composition per liter:

Yeast extract	3.0g
Glucose	1.0g
Peptone	0.6g
Tris-buffer (0.05 M, pH 7.5)	1.0L

pH 7.5 ± 0.2 at 25°C

Preparation of Medium: Add components to distilled/deionized water and bring volume to 1.0L. Mix thoroughly. Distribute into tubes or flasks. Autoclave for 15 min at 15 psi pressure–121°C.

Use: For the cultivation and maintenance of *Bdello–vibrio* species.

Trypaflavin Nalidixic Acid Serum Agar (TNSA Agar)

Composition per liter:

Ionagar No. 2	12.0g
Peptone	10.0g
Beef extract	3.0g
H_2O	926.5mL
Bovine serum, heat inactivated	50.0mL
Nalidixic acid solution	20.0mL
Trypaflavin solution	3.5mL

pH 7.2–7.4 at 25°C

Nalidixic Acid Solution:
Composition per 10mL:

Nalidixic acid	0.02g

Preparation of Nalidixic Acid Solution: Add nalidixic acid to distilled/deionized water and bring volume to 10.0mL. Mix thoroughly. Filter sterilize.

Trypaflavin Solution:
Composition per 10mL:

Trypaflavin	0.1g

Preparation of Trypaflavin Solution: Add trypaflavin to distilled/deionized water and bring volume to 10.0mL. Mix thoroughly. Filter sterilize.

Preparation of Medium: Add components—except bovine serum, nalidixic acid solution, and trypaflavin solution—to distilled/deionized water and bring volume to 926.5mL. Mix thoroughly. Gently heat and bring to boiling. Autoclave for 15 min at 15 psi pressure–121°C. Cool to 45°–50°C. Aseptically add 50.0mL of sterile bovine serum, 20.0mL of sterile nalidixic acid solution, and 3.5mL of trypaflavin solution. Mix thoroughly. Pour into sterile Petri dishes or distribute into sterile tubes.

Use: For the isolation and cultivation of *Listeria* species from pre-enriched specimens.

Tryptic Digest Broth

Composition per liter:

Tryptic digest of beef heart	10.0g
NaCl	5.0g
Glucose	1.0g

pH 7.6 ± 0.2 at 25°C

Source: This medium is available as a premixed powder from Difco Laboratories.

Preparation of Medium: Add components to distilled/deionized water and bring volume to 1.0L. Mix thoroughly. Distribute into tubes or flasks. Autoclave for 15 min at 15 psi pressure–121°C.

Use: For use as a base medium to which enrichments are added for cultivation of fastidious microorganisms.

Tryptic Nitrate Medium

Composition per liter:

Tryptose	20.0g
Na_2HPO_4	2.0g
Agar	1.0g
Glucose	1.0g
KNO_3	1.0g

pH 7.6 ± 0.2 at 25°C

Source: This medium is available as a premixed powder from Difco Laboratories.

Preparation of Medium: Add components to distilled/deionized water and bring volume to 1.0L. Mix thoroughly. Gently heat and bring to boiling. Distribute into tubes in 10.0mL volumes. Autoclave for 15 min at 15 psi pressure–121°C.

Use: For the cultivation and differentiation of *Pseudomonas* and related genera. For the differentiation of bacteria based on their reduction of nitrate to nitrite. After incubation of the bacterium in tryptic nitrate medium for 18–24 hr, sulfanillic acid and α-naphthol reagents are added. Nitrate reduction is indicated by the development of a red to violet color.

Tryptic Soy Agar
See: **Trypticase™ Soy Agar**

Tryptic Soy Agar Blood Agar Base
See: **Trypticase Soy Agar with Sheep Blood**

Tryptic Soy Agar with Magnesium Sulfate

Composition per liter:

Agar	15.0g
Pancreatic digest of casein	15.0g
NaCl	5.0g
Pancreatic digest of soybean meal	5.0g
$MgSO_4 \cdot 7H_2O$	1.5g

pH 7.3 ± 0.2 at 25°C

Preparation of Medium: Add components to distilled/deionized water and bring volume to 1.0L. Mix thoroughly. Gently heat and bring to boiling. Autoclave for 15 min at 15 psi pressure–121°C. Pour into sterile Petri dishes in 20.0mL volumes.

Use: For the cultivation of *Escherichia coli* from foods.

Tryptic Soy Agar with Magnesium Sulfate and Sodium Chloride

Composition per liter:

Pancreatic digest of casein	50.0g
NaCl	30.0g
Agar	15.0g
Pancreatic digest of soybean meal	5.0g
$MgSO_4 \cdot 7H_2O$	1.5g

pH 7.3 ± 0.2 at 25°C

Preparation of Medium: Add components to distilled/deionized water and bring volume to 1.0L. Mix thoroughly. Gently heat and bring to boiling. Autoclave for 15 min at 15 psi pressure–121°C. Pour into sterile Petri dishes in 20.0mL volumes.

Use: For the cultivation of *Vibrio* species from foods.

Tryptic Soy Agar with 0.6% Yeast Extract

Composition per liter:

Agar	15.0g
Pancreatic digest of casein	15.0g

Yeast extract 6.0g
Pancreatic digest of soybean meal 5.0g
NaCl 5.0g

pH 7.0–7.5 at 25°C

Source: This medium is available as a premixed powder from Difco Laboratories.

Preparation of Medium: Add components to distilled/deionized water and bring volume to 1.0L. Mix thoroughly. Gently heat and bring to boiling. Distribute into tubes or flasks. Autoclave for 15 min at 15 psi pressure–121°C. Pour into sterile Petri dishes or leave in tubes.

Use: For the isolation and cultivation of *Listeria monocytogenes* from foods.

Tryptic Soy Blood Agar
See: Trypticase™ Soy Agar with Sheep Blood

Tryptic Soy Fast Green Agar (TSFA)

Composition per liter:

Pancreatic digest of casein 17.0g
Agar 15.0g
NaCl 5.0g
Papaic digest of soybean meal 3.0g
K_2HPO_4 ....2.5g
Glucose 2.5g
Fast Green FCF 0.25g

pH 7.3 ± 0.2 at 25°C

Preparation of Medium: Add components to distilled/deionized water and bring volume to 1.0L. Mix thoroughly. Gently heat and bring to boiling. Distribute into tubes or flasks. Autoclave for 15 min at 15 psi pressure–121°C. Cool to 45°–50°C. Aseptically adjust pH to 7.3. Pour into sterile Petri dishes.

Use: For the isolation and cultivation of *Salmonella* species form foods.

Trypticase™ 1% Solution
See: Tryptophan 1% Solution

Trypticase™ Agar Base

Composition per liter:

Pancreatic digest of casein 20.0g
Agar 3.5g
Phenol Red 0.02g

pH 7.4 ± 0.2 at 25°C

Source: This medium is available as a premixed powder from BBL Microbiology Systems, Difco Laboratories and Oxoid Unipath.

Preparation of Medium: Add components to distilled/deionized water and bring volume to 1.0L. Mix thoroughly. Gently heat and bring to boiling. Distribute into tubes or flasks. Autoclave for 15 min at 15 psi pressure–121°C. Pour into sterile Petri dishes or leave in tubes.

Use: For the differentiation of microorganisms based on their motility.

Trypticase™ Agar Base with Carbohydrate

Composition per liter:

Pancreatic digest of casein 20.0g
Carbohydrate 5.0g
Agar 3.5g
Phenol Red 0.02g

pH 7.4 ± 0.2 at 25°C

Preparation of Medium: Add components to distilled/deionized water and bring volume to 1.0L. Mix thoroughly. Gently heat and bring to boiling. Distribute into tubes. Autoclave for 15 min at 13 psi pressure–118°C. Do not overheat. Pour into sterile Petri dishes or leave in tubes.

Use: For differentiation of microorganisms based on their motility and fermentation reactions. Fermentation of carbohydrate changes the medium yellow.

Trypticase™ Azolectin Tween™ Broth Base
See: TAT Broth Base

Trypticase™ Broth, Supplemented

Composition per liter:

Pancreatic digest of casein 20.0g
$MgSO_4 \cdot 7H_2O$ 0.015g
$FeCl_3$ 7.0mg

pH 7.2 ± 0.2 at 25°C

Preparation of Medium: Add components to distilled/deionized water and bring volume to 1.0L. Mix thoroughly. Distribute into tubes or flasks. Autoclave for 15 min at 15 psi pressure–121°C.

Use: For the cultivation of *Bacillus stearothermophilus.*

Trypticase™ Glucose Extract Agar

Composition per liter:

Agar	15.0g
Pancreatic digest of casein	5.0g
Beef extract	3.0g
Glucose	1.0g

pH 7.0 ± 0.2 at 25°C

Source: This medium is available as a premixed powder from BBL Microbiology Systems.

Preparation of Medium: Add components to distilled/deionized water and bring volume to 1.0L. Mix thoroughly. Gently heat and bring to boiling. Distribute into tubes or flasks. Autoclave for 15 min at 15 psi pressure–121°C. Pour into sterile Petri dishes or leave in tubes.

Use: For enumeration of bacteria in water, milk and other specimens.

Trypticase™ Nitrate Broth

See: **Tryptic Nitrate Medium**

Trypticase™ Novobiocin Broth (TN Broth)

Composition per liter:

Pancreatic digest of casein	17.0g
NaCl	5.0g
Papaic digest of soybean meal	3.0g
K_2HPO_4	2.5g
Glucose	2.5g
Bile salts No. 3	1.5g
K_2HPO_4	1.5g
Novobiocin solution	10.0mL

pH 7.3 ± 0.2 at 25°C

Novobiocin Solution:

Composition per 10mL:

Novobiocin	0.02g

Preparation of Novobiocin Solution: Add novobiocin to distilled/deionized water and bring volume to 10.0mL. Mix thoroughly. Filter sterilize.

Preparation of Medium: Add components, except novobiocin solution, to distilled/deionized water and bring volume to 990.0mL. Mix thoroughly. Gently heat and bring to boiling. Autoclave for 15 min at 15 psi pressure–121°C. Cool to 45°–50°C. Aseptically add sterile novobiocin solution. Mix thoroughly. Pour into sterile Petri dishes or distribute into sterile tubes.

Use: For the cultivation of verotoxin-producing *Escherichia coli.*

Trypticase™ Peptone Glucose Yeast Extract Broth (TPGY Broth)

Composition per liter:

Pancreatic digest of casein	50.0g
Yeast extract	20.0g
Peptone	5.0g
Glucose	4.0g
Sodium thioglycollate	1.0g

pH 7.0 ± 0.2 at 25°C

Preparation of Medium: Add components to distilled/deionized water and bring volume to 1.0L. Mix thoroughly. Distribute into tubes in 15mL volumes. Autoclave for 10 min at 15 psi pressure–121°C.

Use: For the cultivation of *Clostridium botulinum.*

Trypticase™ Peptone Glucose Yeast Extract Broth with Trypsin (TPGYT Broth)

Composition per 1067mL:

Pancreatic digest of casein	50.0g
Yeast extract	20.0g
Peptone	5.0g
Glucose	4.0g
Sodium thioglycollate	1.0g
Trypsin solution	67.0mL

pH 7.0 ± 0.2 at 25°C

Trypsin Solution:

Composition per 100mL:

Trypsin	1.5g

Preparation of Trypsin Solution: Add trypsin to distilled/deionized water and bring volume to 100.0mL. Mix thoroughly. Filter sterilize.

Preparation of Medium: Add components, except trypsin solution, to distilled/deionized water and bring volume to 1.0L. Mix thoroughly. Gently heat and bring to boiling. Distribute into tubes in 15.0mL volumes. Autoclave for 10 min at 15 psi pressure–121°C. Immediately prior to use, aseptically add 1.0mL sterile trypsin solution to each tube. Mix thoroughly.

Use: For the cultivation of *Clostridium botulinum.*

Trypticase™ Peptone Glucose Yeast Extract Broth, Buffered

Composition per liter:

Pancreatic digest of casein	50.0g
Yeast extract	20.0g
Na_2HPO_4	5.0g
Peptone	5.0g

Glucose 4.0g
Sodium thioglycollate 1.0g

pH 7.3 ± 0.2 at 25°C

Preparation of Medium: Add components to distilled/deionized water and bring volume to 1.0L. Mix thoroughly.Gently heat until dissolved. Adjust pH to 7.3. Distribute into tubes in 15.0mL volumes. Autoclave for 8 min at 15 psi pressure–121°C.

Use: For the isolation and cultivation of *Clostridium perfringens* from foods.

Trypticase™ Phytone Glucose Medium

Composition per liter:

Glucose 15.0g
Pancreatic digest of casein 10.0g
Agar...8.0g
Papaic digest of soybean meal 5.0g
Yeast extract...2.5g
K_2HPO_4 ....2.0g
L-Cysteine·HCl·H_2O 0.5g
$MgCl_2$ 0.5g
$ZnSO_4·7H_2O$ 0.25g
$FeCl_3$ 1.0mg

Preparation of Medium: Add $ZnSO_4$ to approximately 100mL of distilled/deionized water and dissolve. Add remaining components and bring volume to 1.0L with distilled/deionized water. Mix thoroughly. Distribute into tubes or flasks. Autoclave for 15 min at 15 psi pressure–121°C. Pour into sterile Petri dishes or leave in tubes.

Use: For the cultivation and maintenance of *Bifidobacterium* species.

Trypticase™ Phytone Glucose Medium with Tween™ 80

Composition per liter:

Glucose 15.0g
Pancreatic digest of casein 10.0g
Agar...8.0g
Papaic digest of soybean meal 5.0g
Yeast extract...2.5g
K_2HPO_4 ....2.0g
L-Cysteine·HCl·H_2O 0.5g
$MgCl_2$ 0.5g
$ZnSO_4·7H_2O$ 0.25g
$FeCl_3$ 1.0mg
Tween™ 80 ....2.0mL

Preparation of Medium: Add $ZnSO_4$ to approximately 100.0mL of distilled/deionized water and dissolve. Add remaining components and bring volume to 1.0L with distilled/deionized water. Mix thoroughly. Distribute into tubes or flasks. Autoclave for 15 min at 15 psi pressure–121°C. Pour into sterile Petri dishes or leave in tubes.

Use: For cultivation and maintenance of *Bifidobacterium* species.

Trypticase™ Soy Agar

Composition per liter:

Pancreatic digest of casein 17.0g
Agar...15.0g
NaCl 5.0g
Papaic digest of soybean meal 3.0g
K_2HPO_4 ....2.5g
Glucose 2.5g

pH 7.3 ± 0.2 at 25°C

Preparation of Medium: Add components to distilled/deionized water and bring volume to 1.0L. Mix thoroughly. Gently heat and bring to boiling. Distribute into tubes or flasks. Autoclave for 15 min at 15 psi pressure–121°C. Pour into sterile Petri dishes or leave in tubes.

Use: For the cultivation and maintenance of a wide variety of heterotrophic microorganisms.

Trypticase™ Soy Agar (ATCC Medium 18)

Composition per liter:

Pancreatic digest of casein 17.0g
Agar...15.0g
NaCl 5.0g
Papaic digest of soybean meal 3.0g
K_2HPO_4 ....2.5g
Glucose 2.5g

pH 7.3 ± 0.2 at 25°C

Preparation of Medium: Add components to distilled/deionized water and bring volume to 1.0L. Mix thoroughly. Distribute into tubes or flasks. Autoclave for 15 min at 15 psi pressure–121°C.

Use: For the cultivation of a wide variety of fastidious and nonfastidious microorganisms from clinical and nonclinical specimens. Also used for the rapid estimation of the bacteriological quality of water.

Trypticase™ Soy Agar (Tryptic Soy Agar) (Soybean Casein Digest Agar) (ATCC Medium 77)

Composition per liter:

Pancreatic digest of casein 15.0g
Agar...15.0g

Papaic digest of soybean meal	5.0g
NaCl	5.0g

pH 7.3 ± 0.2 at 25°C

Source: This medium is available as a premixed powder from BBL Microbiology Systems and Difco Laboratories.

Preparation of Medium: Add components to distilled/deionized water and bring volume to 1.0L. Mix thoroughly. Gently heat and bring to boiling. Distribute into tubes or flasks. Autoclave for 15 min at 15 psi pressure–121°C. Do not overheat. Pour into sterile Petri dishes or leave in tubes.

Use: For the isolation and cultivation of a wide variety of fastidious as well as nonfastidious microorganisms.

Trypticase™ Soy Agar, Modified (ATCC Medium 1386)

Composition per liter:

Agar	18.0g
Pancreatic digest of casein	17.0g
NaCl	5.0g
Papaic digest of soybean meal	3.0g
K_2HPO_4	2.5g
Glucose	2.5g
Yeast extract	0.4g
NH_4OH, concentrated	0.035mL

pH 7.5 ± 0.2 at 25°C

Preparation of Medium: Add components, except NH_4OH, to distilled/deionized water and bring volume to 1.0L. Mix thoroughly. Gently heat and bring to boiling. Autoclave for 15 min at 15 psi pressure–121°C. Cool to 45°–50°C. Aseptically add NH_4OH. Mix thoroughly. Adjust pH to 7.5 if necessary. Pour into sterile Petri dishes or distribute into sterile tubes.

Use: For the cultivation of ATCC strain 31205.

Trypticase™ Soy Agar, Modified (ATCC Medium 1481)

Composition per liter:

Pancreatic digest of casein	17.0g
Agar	15.0g
NaCl	5.0g
Papaic digest of soybean meal	3.0g
K_2HPO_4	2.5g
Glucose	2.5g
L-Glutamine	10.0mL

pH 6.5 ± 0.2 at 25°C

Preparation of Medium: Add components, except glutamine, to distilled/deionized water and bring volume to 990.0mL. Mix thoroughly. Gently heat and bring to boiling. Autoclave for 15 min at 15 psi pressure–121°C. Cool to 45°–50°C. Aseptically add 10.0mL of sterile glutamine. Mix thoroughly. Adjust pH to 6.5. Pour into sterile Petri dishes or distribute into sterile tubes.

Use: Use as a base that is supplemented for cultivation of fastidious microorganisms. When supplemented with sheep blood, this medium is useful for the observation of hemolytic reactions of a variety of bacteria.

Trypticase™ Soy Agar, Modified

Composition per liter:

Pancreatic digest of casein	17.0g
Agar	15.0g
NaCl	5.0g
Yeast extract	4.0g
Papaic digest of soybean meal	3.0g
K_2HPO_4	2.5g

Preparation of Medium: Add components to distilled/deionized water and bring volume to 1.0L. Mix thoroughly. Gently heat and bring to boiling. Distribute into tubes or flasks. Autoclave for 15 min at 15 psi pressure–121°C. Pour into sterile Petri dishes or leave in tubes.

Use: For the cultivation and maintenance of the *Simonsiella* species.

Trypticase™ Soy Agar, Modified (TSA II™)

Composition per liter:

Pancreatic digest of casein	14.5g
Agar	14.0g
Papaic digest of soybean meal	5.0g
NaCl	5.0g
Growth factors (BBL)	1.5g

pH 7.3 ± 0.2 at 25°C

Source: This medium is available as a premixed powder from BBL Microbiology Systems.

Preparation of Medium: Add components to distilled/deionized water and bring volume to 1.0L. Mix thoroughly. Gently heat while stirring and bring to boiling. Distribute into tubes or flasks. Autoclave for 15 min at 15 psi pressure–121°C. Do not overheat. Pour into sterile Petri dishes or leave in tubes. For blood plates, 50.0–100.0mL of sterile defibrinated sheep blood may be added to sterile medium that has been melted and cooled to 45°–50°C.

Use: Use as a base that is supplemented for cultivation of fastidious microorganisms. When supplemented with sheep blood, this medium is useful for the observation of hemolytic reactions of a variety of bacteria. It may be used to perform the CAMP test for the presumptive identification of group B streptococci (*Streptococcus agalactiae*).

Trypticase™ Soy Agar, Modified with Horse Serum

Composition per liter:

Pancreatic digest of casein	17.0g
Agar	15.0g
NaCl	5.0g
Yeast extract	4.0g
Papaic digest of soybean meal	3.0g
K_2HPO_4	2.5g
Horse serum	100.0mL

pH 7.3 ± 0.2 at 25°C

Preparation of Medium: Add components, except horse serum, to distilled/deionized water and bring volume to 900.0mL. Mix thoroughly. Gently heat and bring to boiling. Autoclave for 15 min at 15 psi pressure–121°C. Cool to 45°–50°C. Aseptically add sterile horse serum. Mix thoroughly. Pour into sterile Petri dishes or distribute into sterile tubes.

Use: For the cultivation and maintenance of the *Simonsiella* species, the *Alysiella* species, and the *Moraxella* species.

Trypticase™ Soy Agar with Glycerol

Composition per liter:

Pancreatic digest of casein	15.0g
Agar	15.0g
Papaic digest of soybean meal	5.0g
NaCl	5.0g
Glycerol	50.0mL

pH 7.3 ± 0.2 at 25°C

Preparation of Medium: Add components to distilled/deionized water and bring volume to 1.0L. Mix thoroughly. Gently heat while stirring and bring to boiling. Distribute into tubes or flasks. Autoclave for 15 min at 15 psi pressure–121°C. Do not overheat. Pour into sterile Petri dishes or leave in tubes. For blood plates, 50.0–100.0mL of sterile defibrinated sheep blood may be added to sterile medium that has been melted and cooled to 45°–50°C.

Use: For the cultivation and maintenance of *Acinetobacter calcoaceticus*.

Trypticase™ Soy Agar with Lecithin and Polysorbate 80 (Microbial Content Test Agar)

Composition per liter:

Pancreatic digest of casein	15.0g
Agar	15.0g
Papaic digest of soybean meal	5.0g
NaCl	5.0g
Polysorbate 80 (Tween™ 80)	5.0g
Lecithin	0.7g

pH 7.3 ± 0.2 at 25°C

Source: This medium is available as a premixed powder from BBL Microbiology Systems.

Preparation of Medium: Add components to distilled/deionized water and bring volume to 1.0L. Mix thoroughly. Gently heat and bring to boiling. Distribute into tubes or flasks. Autoclave for 15 min at 13 psi pressure–118°C. Cool to 45°–50°C. Pour into sterile Petri dishes in 17.0mL volumes or leave in tubes.

Use: For the detection and enumeration of microorganisms in replicate plating techniques. Also used for the detection and enumeration of microorganisms present on surfaces of sanitary importance.

Trypticase™ Soy Agar with 3% NaCl (TSA NaCl)

Composition per liter:

NaCl	30.0g
Agar	15.0g
Pancreatic digest of casein	15.0g
Papaic digest of soybean meal	5.0g

pH 7.3 ± 0.2 at 25°C

Preparation of Medium: Add components to distilled/deionized water and bring volume to 1.0L. Mix thoroughly. Gently heat and bring to boiling. Distribute into tubes or flasks. Autoclave for 15 min at 15 psi pressure–121°C. Pour into sterile Petri dishes or leave in tubes.

Use: For the cultivation of halophilic microorganisms isolated from foods.

Trypticase™ Soy Agar with NaCl

Composition per liter:

NaCl	30.0g
Pancreatic digest of casein	15.0g
Agar	15.0g

Papaic digest of soybean meal 5.0g
Bile salts No. 3 ....1.0g

pH 7.3 ± 0.2 at 25°C

Preparation of Medium: Add components to distilled/deionized water and bring volume to 1.0L. Mix thoroughly. Gently heat and bring to boiling. Distribute into tubes or flasks. Autoclave for 15 min at 15 psi pressure–121°C. Pour into sterile Petri dishes or leave in tubes.

Use: For the cultivation and maintenance of *Vibrio alginolyticus*.

Trypticase™ Soy Agar with NaCl, Horse Serum and Penicillin

Composition per liter:

Pancreatic digest of casein 15.0g
Agar..15.0g
Papaic digest of soybean meal 5.0g
NaCl 35.0g
Horse serum, inactivated............................ 100.0mL
Penicillin solution 10.0mL

pH 7.3 ± 0.2 at 25°C

Penicillin Solution:

Composition per 10mL:

Penicillin 1,000,000U

Preparation of Penicillin Solution: Add penicillin to distilled/deionized water and bring volume to 10.0mL. Mix thoroughly. Filter sterilize.

Preparation of Medium: Add components, except horse serum and penicillin solution, to distilled/deionized water and bring volume to 890.0mL. Mix thoroughly. Gently heat and bring to boiling. Autoclave for 15 min at 15 psi pressure–121°C. Do not overheat. Cool to 50°C. Aseptically add 100.0mL of sterile horse serum and 10.0mL of sterile penicillin solution. Mix thoroughly. Pour into sterile Petri dishes or distribute into sterile tubes.

Use: For the isolation and cultivation of fungi.

Trypticase™ Soy Agar with Human Blood

Composition per liter:

Pancreatic digest of casein 15.0g
Agar..15.0g
Papaic digest of soybean meal 5.0g
NaCl 5.0g
Human blood, defibrinated........................... 50.0mL

pH 7.3 ± 0.2 at 25°C

Preparation of Medium: Add components, except human blood, to distilled/deionized water and bring volume to 950.0mL. Mix thoroughly. Gently heat and bring to boiling. Autoclave for 15 min at 15 psi pressure–121°C. Cool to 45°–50°C. Aseptically add sterile human blood. Mix thoroughly. Pour into sterile Petri dishes in 17.0mL volumes or distribute into sterile tubes.

Use: For the cultivation of a wide variety of fastidious microorganisms. For the observation of hemolytic reactions of a variety of bacteria. It may be used to perform the CAMP test for the presumptive identification of group B streptococci (*Streptococcus agalactiae*).

Trypticase™ Soy Agar with Sheep Blood (Tryptic Soy Blood Agar) (TSA Blood Agar)

Composition per liter:

Pancreatic digest of casein 15.0g
Agar..15.0g
Papaic digest of soybean meal 5.0g
NaCl 5.0g
Sheep blood, defibrinated............................. 50.0mL

pH 7.3 ± 0.2 at 25°C

Preparation of Medium: Add components, except sheep blood, to distilled/deionized water and bring volume to 950.0mL. Mix thoroughly. Gently heat and bring to boiling. Autoclave for 15 min at 15 psi pressure–121°C. Cool to 45°–50°C. Aseptically add sterile sheep blood. Mix thoroughly. Pour into sterile Petri dishes in 17.0mL volumes or distribute into sterile tubes.

Use: For the cultivation of a wide variety of fastidious microorganisms. For the observation of hemolytic reactions of a variety of bacteria. It may be used to perform the CAMP test for the presumptive identification of group B streptococci (*Streptococcus agalactiae*).

Trypticase™ Soy Agar with Sheep Blood and Gentamicin (TSA II™ with Sheep Blood and Gentamicin)

Composition per liter:

Pancreatic digest of casein 14.5g
Agar..14.0g
Papaic digest of soybean meal 5.0g

NaCl 5.0g
Growth factors (BBL) 1.5g
Sheep blood, defibrinated 50.0mL
Gentamicin solution 10.0mL
pH 7.3 ± 0.2 at 25°C

Source: This medium is available as a premixed powder from BBL Microbiology Systems.

Gentamicin Solution:
Composition per 10mL:
Gentamicin 2.5mg

Preparation of Gentamicin Solution: Add gentamicin to distilled/deionized water and bring volume to 10.0mL. Mix thoroughly. Filter sterilize.

Preparation of Medium: Add components, except sheep blood and gentamicin solution, to distilled/deionized water and bring volume to 940.0mL. Mix thoroughly. Gently heat and bring to boiling. Autoclave for 15 min at 15 psi pressure–121°C. Cool to 45°–50°C. Aseptically add sterile sheep blood and sterile gentamicin solution. Mix thoroughly. Pour into sterile Petri dishes or distribute into sterile tubes.

Use: For the isolation of *Streptococcus pneumoniae* from a variety of clinical specimens.

Trypticase™ Soy Agar with Sheep Blood, Formate and Fumarate

Composition per liter:
Pancreatic digest of casein 14.5g
Agar 14.0g
Papaic digest of soybean meal 5.0g
NaCl 5.0g
Sucrose 2.0g
Growth factors (BBL) 1.5g
Sheep blood, defibrinated 50.0mL
Formate-fumarate solution 13.0mL
pH 7.3 ± 0.2 at 25°C

Formate-Fumarate Solution:
Composition per 100.0mL:
Sodium formate 6.0g
Fumaric acid 6.0g

Preparation of Formate-Fumarate Solution: Add components to distilled/deionized water and bring volume to 100.0mL. Mix thoroughly.Adjust pH to 7.0. Filter sterilize.

Preparation of Medium: Add components, except sheep blood, to distilled/deionized water and bring volume to 950.0mL. Mix thoroughly. Gently heat and bring to boiling. Autoclave for 15 min at 15 psi pressure–121°C. Cool to 45°–50°C. Aseptically add sterile sheep blood. Mix thoroughly. Pour into sterile Petri dishes. Prior to inoculation aseptically spread 0.2mL of sterile formate-fumarate solution on each plate.

Use: For the isolation of *Streptococcus pneumoniae* from a variety of clinical specimens.

Trypticase™ Soy Agar with Sheep Blood, Sucrose and Tetracycline

Composition per liter:
Pancreatic digest of casein 14.5g
Agar 14.0g
Papaic digest of soybean meal 5.0g
NaCl 5.0g
Sucrose 2.0g
Growth factors (BBL) 1.5g
Sheep blood, defibrinated 50.0mL
Tetracycline solution 10.0mL
pH 7.3 ± 0.2 at 25°C

Tetracycline Solution:
Composition per 10mL:
Tetracycline 0.5mg

Preparation of Tetracycline Solution: Add tetracycline to distilled/deionized water and bring volume to 10.0mL. Mix thoroughly. Filter sterilize.

Preparation of Medium: Add components, except sheep blood and tetracycline solution, to distilled/deionized water and bring volume to 940.0mL. Mix thoroughly. Gently heat and bring to boiling. Autoclave for 15 min at 15 psi pressure–121°C. Cool to 45°–50°C. Aseptically add sterile sheep blood and sterile tetracycline solution. Mix thoroughly. Pour into sterile Petri dishes or distribute into sterile tubes.

Use: For the isolation of *Streptococcus pneumoniae* from a variety of clinical specimens.

Trypticase™ Soy Agar with Tobramycin

Composition per liter:
Pancreatic digest of casein 17.0g
Agar 15.0g
NaCl 5.0g
Papaic digest of soybean meal 3.0g
K_2HPO_4 .... 2.5g
Glucose 2.5g
Tobramycin solution 10.0mL
pH 7.3 ± 0.2 at 25°C

Tobromycin Solution:
Composition per 10mL:
Tobramycin 8.0mg

Preparation of Tobromycin Solution: Add tobramycin to distilled/deionized water and bring volume to 10.0mL. Mix thoroughly. Filter sterilize.

Preparation of Medium: Add components, except tobramycin solution, to distilled/deionized water and bring volume to 990.0mL. Mix thoroughly. Gently heat and bring to boiling. Autoclave for 15 min at 15 psi pressure–121°C. Cool to 45°–50°C. Aseptically add sterile tobramycin solution. Mix thoroughly. Pour into sterile Petri dishes or distribute into sterile tubes.

Use: For cultivation and maintenance of *Serratia marcescens*.

Trypticase™ Soy Agar with Yeast Extract and Glucose

Composition per liter:

Pancreatic digest of casein	17.0g
Agar	15.0g
Glucose	7.5g
NaCl	5.0g
Yeast extract	3.0g
Papaic digest of soybean meal	3.0g
K_2HPO_4	2.5g

pH 7.0–7.2 at 25°C

Preparation of Medium: Add components to distilled/deionized water and bring volume to 1.0L. Mix thoroughly. Gently heat and bring to boiling. Distribute into tubes or flasks. Autoclave for 15 min at 15 psi pressure–121°C. Pour into sterile Petri dishes or leave in tubes.

Use: For cultivation and maintenance of *Pediococcus urinaeequi*.

Trypticase™ Soy Agar Yeast Extract (TSAYE)

Composition per liter:

Pancreatic digest of casein	17.0g
Agar	15.0g
Yeast extract	6.0g
NaCl	5.0g
Papaic digest of soybean meal	3.0g
K_2HPO_4	2.5g
Glucose	2.5g

pH 7.3 ± 0.2 at 25°C

Preparation of Medium: Add components to distilled/deionized water and bring volume to 1.0L. Mix thoroughly. Gently heat and bring to boiling. Distribute into tubes or flasks. Autoclave for 15 min at 15 psi pressure–121°C. Pour into sterile Petri dishes or leave in tubes.

Use: For the cultivation and maintenance of a wide variety of heterotrophic microorganisms. For the isolation and cultivation of *Listeria monocytogenes* from foods.

Trypticase™ Soy Broth (Soybean Casein Digest Broth, USP)

Composition per liter:

Pancreatic digest of casein	17.0g
NaCl	5.0g
Papaic digest of soybean meal	3.0g
K_2HPO_4	2.5g
Glucose	2.5g

pH 7.3 ± 0.2 at 25°C

Source: This medium is available as a premixed powder from BBL Microbiology Systems and Difco Laboratories.

Preparation of Medium: Add components to distilled/deionized water and bring volume to 1.0L. Mix thoroughly. Distribute into tubes or flasks. Autoclave for 15 min at 15 psi pressure–121°C.

Use: For the cultivation of a wide variety of fastidious and nonfastidious microorganisms from clinical and nonclinical specimens. Also used for the rapid estimation of the bacteriological quality of water.

Trypticase™ Soy Broth, Modified

Composition per 1000.2mL:

Pancreatic digest of casein	17.0g
NaCl	15.0g
K_2HPO_4	4.0g
Papaic digest of soybean meal	3.0g
Glucose	2.5g
Bile salts No. 3	1.5g
Novobiocin solution	0.2mL

pH 7.3 ± 0.2 at 25°C

Novobiocin Solution:

Composition per liter:

Novobiocin	0.050g

Preparation of Novobiocin Solution: Add novobiocin to distilled/deionized water and bring volume to 1.0L. Mix thoroughly. Filter sterilize.

Preparation of Medium: Add components, except novobiocin solution, to distilled/deionized water and bring volume to 1.0L. Mix thoroughly. Gently heat and bring to boiling. Autoclave for 15 min at 15 psi pressure–121°C. Cool to 45°–50°C. Aseptically add sterile novobiocin solution. Mix thoroughly. Aseptically distribute into sterile tubes.

Use: For the isolation and cultivation of *Shigella* species from food.

Trypticase™ Soy Broth with 0.1% Agar

Composition per liter:

Pancreatic digest of casein	17.0g
NaCl	5.0g
Papaic digest of soybean meal	3.0g
K_2HPO_4	2.5g
Glucose	2.5g
Agar	1.0g

pH 7.3 ± 0.2 at 25°C

Source: This medium is available as a premixed powder from BBL Microbiology Systems.

Preparation of Medium: Add components to distilled/deionized water and bring volume to 1.0L. Mix thoroughly. Distribute into tubes or flasks. Autoclave for 15 min at 15 psi pressure–121°C.

Use: For the cultivation of anaerobic microorganisms. For the cultivation of microorganisms isolated of root canals and other clinical specimens.

Trypticase™ Soy Broth with 10mM Glucose

Composition per liter:

Pancreatic digest of casein	17.0g
NaCl	5.0g
Papaic digest of soybean meal	3.0g
K_2HPO_4	2.5g
Glucose	1.8g

pH 7.3 ± 0.2 at 25°C

Preparation of Medium: Add components to distilled/deionized water and bring volume to 1.0L. Mix thoroughly. Distribute into tubes or flasks. Autoclave for 15 min at 15 psi pressure–121°C.

Use: For the cultivation of a variety of fastidious and nonfastidious microorganisms from clinical and nonclinical specimens.

Trypticase™ Soy Broth with 1.5% NaCl

Composition per liter:

Pancreatic digest of casein	17.0g
NaCl	15.0g
Papaic digest of soybean meal	3.0g
K_2HPO_4	2.5g
Glucose	2.5g

pH 7.3 ± 0.2 at 25°C

Preparation of Medium: Add components to distilled/deionized water and bring volume to 1.0L. Mix thoroughly. Distribute into tubes or flasks. Autoclave for 15 min at 15 psi pressure–121°C.

Use: For cultivation and maintenance of *Pasteurella* species and *Listonella anguillarum*.

Trypticase™ Soy Broth with Calcium Chloride

Composition per liter:

Pancreatic digest of casein	17.0g
NaCl	5.0g
Papaic digest of soybean meal	3.0g
K_2HPO_4	2.5g
Glucose	2.5g
$CaCl_2 \cdot 2H_2O$	0.15g

pH 7.3 ± 0.2 at 25°C

Preparation of Medium: Add components to distilled/deionized water and bring volume to 1.0L. Mix thoroughly. Distribute into tubes or flasks. Autoclave for 15 min at 15 psi pressure–121°C.

Use: For cultivation and maintenance of *Brochothrix thermosphacta*.

Trypticase™ Soy Broth with Fetal Calf Serum

Composition per liter:

Pancreatic digest of casein	17.0g
NaCl	5.0g
Papaic digest of soybean meal	3.0g
K_2HPO_4	2.5g
Glucose	2.5g
Fetal calf serum	100.0mL

pH 7.3 ± 0.2 at 25°C

Preparation of Medium: Add components, except fetal calf serum, to distilled/deionized water and bring volume to 900.0mL. Mix thoroughly. Autoclave for 15 min at 15 psi pressure–121°C. Cool to 25°C. Aseptically add sterile fetal calf serum. Mix thoroughly. Aseptically distribute into sterile tubes.

Use: For cultivation and maintenance of *Serpula innocens*.

Trypticase™ Soy Broth with Glycerol

Composition per liter:

Pancreatic digest of casein	17.0g
NaCl	15.0g
Papaic digest of soybean meal	3.0g

K_2HPO_4 .... 2.5g
Glycerol 240.0mL

pH 7.3 ± 0.2 at 25°C

Preparation of Medium: Add components to distilled/deionized water and bring volume to 1.0L. Mix thoroughly. Gently heat until dissolved. Adjust pH to 7.3. Distribute into tubes in 10.0mL volumes. Autoclave for 15 min at 15 psi pressure–121°C.

Use: For the cultivation and maintenance of a wide variety of microorganisms from foods.

Trypticase™ Soy Broth with Horse Serum

Composition per liter:

Pancreatic digest of casein 17.0g
NaCl 5.0g
Papaic digest of soybean meal 3.0g
K_2HPO_4 .... 2.5g
Glucose 2.5g
Horse serum, inactivated 100.0mL

pH 7.3 ± 0.2 at 25°C

Preparation of Medium: Add components, except horse serum, to distilled/deionized water and bring volume to 900.0mL. Mix thoroughly. Autoclave for 15 min at 15 psi pressure–121°C. Cool to 25°C. Aseptically add 100.0mL of sterile horse serum. Mix thoroughly. Aseptically distribute into sterile tubes.

Use: For cultivation and maintenance of *Serpula innocens.*

Trypticase™ Soy Broth with Human Blood

Composition per liter:

Pancreatic digest of casein 17.0g
NaCl 5.0g
Papaic digest of soybean meal 3.0g
K_2HPO_4 .... 2.5g
Glucose 2.5g
Human blood, defibrinated 50.0mL

pH 7.3 ± 0.2 at 25°C

Preparation of Medium: Add components, except human blood, to distilled/deionized water and bring volume to 950.0mL. Mix thoroughly. Autoclave for 15 min at 15 psi pressure–121°C. Cool to 25°C. Aseptically add 50.0mL of sterile human blood. Mix thoroughly. Aseptically distribute into sterile tubes.

Use: For cultivation and maintenance of *Serpula innocens.*

Trypticase™ Soy Broth with Neomycin

Composition per liter:

Pancreatic digest of casein 17.0g
NaCl 5.0g
Papaic digest of soybean meal 3.0g
K_2HPO_4 .... 2.5g
Glucose 2.5g
Agar 1.0g
Neomycin solution 10.0mL

pH 7.3 ± 0.2 at 25°C

Neomycin Solution:
Composition per 10mL:

Neomycin 5.0mg

Preparation of Neomycin Solution: Add neomycin to distilled/deionized water and bring volume to 10.0mL. Mix thoroughly. Filter sterilize.

Preparation of Medium: Add components, except neomycin solution, to distilled/deionized water and bring volume to 990.0mL. Mix thoroughly. Gently heat and bring to boiling. Autoclave for 15 min at 15 psi pressure–121°C. Cool to 45°–50°C. Aseptically add sterile neomycin solution. Mix thoroughly. Pour into sterile Petri dishes or distribute into sterile tubes.

Use: For cultivation and maintenance of *Bacillus megaterium.*

Trypticase™ Soy Broth with Sodium Chloride and Sodium Pyruvate

Composition per liter:

NaCl 100.0g
Pancreatic digest of casein 17.0g
Sodium pyruvate 10.0g
Papaic digest of soybean meal 3.0g
Glucose 2.5g
K_2HPO_4 .... 2.5g

pH 7.3 ± 0.2 at 25°C

Preparation of Medium: Add components to distilled/deionized water and bring volume to 1.0L. Mix thoroughly. Gently heat until dissolved. Adjust pH to 7.3. Distribute into tubes in 10.0mL volumes. Autoclave for 15 min at 15 psi pressure–121°C.

Use: For the isolation and cultivation of *Staphylococcus aureus* from foods.

Trypticase™ Soy Broth with Tobramycin

Composition per liter:

Pancreatic digest of casein 17.0g

NaCl 5.0g
Papaic digest of soybean meal 3.0g
K_2HPO_4 .... 2.5g
Glucose 2.5g
Agar 1.0g
Tobramycin solution 10.0mL
pH 7.3 ± 0.2 at 25°C

Tobromycin Solution:
Composition per 10mL:
Tobramycin 8.0mg

Preparation of Tobromycin Solution: Add tobramycin to distilled/deionized water and bring volume to 10.0mL. Mix thoroughly. Filter sterilize.

Preparation of Medium: Add components, except tobramycin solution, to distilled/deionized water and bring volume to 990.0mL. Mix thoroughly. Autoclave for 15 min at 15 psi pressure–121°C. Cool to 45°–50°C. Aseptically add sterile tobramycin solution. Mix thoroughly. Aseptically distribute into sterile tubes.

Use: For cultivation and maintenance of *Serratia marcescens.*

Trypticase™ Soy Broth with Tween™ 80

Composition per liter:
Pancreatic digest of casein 17.0g
NaCl 5.0g
Papaic digest of soybean meal 3.0g
K_2HPO_4 .... 2.5g
Glucose 2.5g
Agar 1.0g
Tween™ 80 .... 1.0mL
pH 7.3 ± 0.2 at 25°C

Preparation of Medium: Add components to distilled/deionized water and bring volume to 1.0L. Mix thoroughly. Distribute into tubes or flasks. Autoclave for 15 min at 15 psi pressure–121°C.

Use: For cultivation and maintenance of *Corynebacterium genitalium.*

Trypticase™ Soy Broth with Yeast Extract

Composition per liter:
Pancreatic digest of casein 17.0g
Yeast extract 6.0g
NaCl 5.0g
Papaic digest of soybean meal 3.0g
K_2HPO_4 .... 2.5g
Glucose 2.5g
pH 7.3 ± 0.2 at 25°C

Preparation of Medium: Add components to distilled/deionized water and bring volume to 1.0L. Mix thoroughly. Gently heat and bring to boiling. Distribute into tubes or flasks. Autoclave for 15 min at 15 psi pressure–121°C.

Use: For the cultivation and maintenance of a wide variety of heterotrophic microorganisms.

Trypticase™ Soy Broth without Glucose

Composition per liter:
Pancreatic digest of casein 17.0g
NaCl 5.0g
Papaic digest of soybean meal 3.0g
K_2HPO_4 .... 2.5g
pH 7.3 ± 0.2 at 25°C

Source: This medium is available as a premixed powder from BBL Microbiology Systems and Difco Laboratories.

Preparation of Medium: Add components to distilled/deionized water and bring volume to 1.0L. Mix thoroughly. Distribute into tubes or flasks. Autoclave for 15 min at 15 psi pressure–121°C.

Use: For the cultivation of a wide variety of microorganisms when the presence of carbohydrate is undesirable.

Trypticase™ Soy Broth Yeast Extract (TSBYE)

Composition per liter:
Pancreatic digest of casein 17.0g
Yeast extract 6.0g
NaCl 5.0g
Papaic digest of soybean meal 3.0g
K_2HPO_4 .... 2.5g
Glucose 2.5g
pH 7.3 ± 0.2 at 25°C

Preparation of Medium: Add components to distilled/deionized water and bring volume to 1.0L. Mix thoroughly. Gently heat and bring to boiling. Distribute into tubes or flasks. Autoclave for 15 min at 15 psi pressure–121°C.

Use: For the cultivation of *Listeria monocytogenes* from foods.

Trypticase™ Soy Glucose Medium

Composition per liter:
Glucose 50.0g
Pancreatic digest of casein 7.5g

Agar..7.5g
Papaic digest of soybean meal 2.5g
NaCl 2.5g

pH 7.3 ± 0.2 at 25°C

Preparation of Medium: Add components to distilled/deionized water and bring volume to 1.0L. Mix thoroughly. Gently heat and bring to boiling. Distribute into tubes or flasks. Autoclave for 15 min at 15 psi pressure–121°C.

Use: For cultivation and maintenance of *Pseudomonas cepacia*.

Trypticase™ Soy Polymyxin Broth

Composition per 1006.67mL:

Pancreatic digest of casein 17.0g
NaCl 5.0g
Papaic digest of soybean meal 3.0g
K_2HPO_4 ....2.5g
Glucose 2.5g
Polymyxin B solution 6.67mL

pH 7.3 ± 0.2 at 25°C

Polymyxin B Solution:

Composition per 10mL:

Polymyxin B 0.015g

Preparation of Polymyxin B Solution: Add polymyxin B to distilled/deionized water and bring volume to 10.0mL. Mix thoroughly. Filter sterilize.

Preparation of Medium: Add components, except polymyxin B solution, to distilled/deionized water and bring volume to 1.0L. Mix thoroughly. Gently heat and bring to boiling. Distribute into tubes in 15.0mL volumes. Autoclave for 15 min at 15 psi pressure–121°C. Cool to 45°–50°C. Aseptically add 0.1mL of sterile polymyxin B solution to each tube. Mix thoroughly.

Use: For the isolation and cultivation of *Bacillus cereus* from foods.

Trypticase™ Soy Soil Extract
See: **TS Soil Extract**

Trypticase™ Soy Tryptose Broth

Composition per liter:

Pancreatic digest of casein 13.5g
Peptic digest of animal tissue.............................5.0g
NaCl 5.0g
Yeast extract 3.0g
Glucose 1.75g
Papaic digest of soybean meal 1.5g
K_2HPO_4 ....1.25g

pH 7.2 ± 0.2 at 25°C

Preparation of Medium: Add components to distilled/deionized water and bring volume to 1.0L. Mix thoroughly. Distribute into tubes in 5.0mL volumes. Autoclave for 15 min at 15 psi pressure–121°C.

Use: For the enrichment of *Salmonella* species from foods.

Trypticase™ Soy Yeast Extract Medium
See: **TSY Medium**

Trypticase™ Soy Yeast Extract Starch Medium
See: **TSYES Medium**

Trypticase™ Sulfite Neomycin Agar
See: **TSN Agar**

Trypticase™ Tellurite Agar Base

Composition per liter:

Agar..20.0g
Pancreatic digest of casein 10.0g
Peptic digest of animal tissue...........................10.0g
NaCl 5.0g
Glucose 2.0g
Serum 50.0mL
Chapman tellurite solution 10.0mL

pH 7.5 ± 0.2 at 25°C

Source: This medium is available as a premixed powder from BBL Microbiology Systems.

Chapman Tellurite Solution:

Composition per 100mL:

K_2TeO_3 ....1.0g

Preparation of Chapman Tellurite Solution: Add K_2TeO_3 to distilled/deionized water and bring volume to 100.0mL. Mix thoroughly. Filter sterilize.

Caution: Potassium tellurite is toxic.

Preparation of Medium: Add components, except serum and Chapman tellurite solution, to distilled/deionized water and bring volume to 940.0mL. Mix thoroughly. Gently heat and bring to boiling. Autoclave for 15 min at 15 psi pressure–121°C. Cool to 45°–50°C. Aseptically add sterile serum and sterile Chapman tellurite solution. Sheep serum, rabbit serum or human serum may be used. Mix thoroughly. Pour into sterile Petri dishes.

Use: For the selective isolation of microorganisms from clinical specimens, especially from the nose, throat and vagina.

Trypticase™ Yeast Extract Glucose Medium

Composition per liter:

Pancreatic digest of casein	10.0g
Glucose	10.0g
KH_2PO_4	6.8g
Yeast extract	5.0g
$NaHCO_3$	1.0g
Tween™ 80	0.5g
Sodium formaldehyde sulfoxalate	0.5g
$CaCl_2$	0.02g
$MgSO_4$	0.02g
NaCl	0.02g

pH 7.0 ± 0.2 at 25°C

Preparation of Medium: Add components to distilled/deionized water and bring volume to 1.0L. Mix thoroughly. Adjust pH to 7.0. Distribute into tubes or flasks. Autoclave for 15 min at 15 psi pressure–121°C.

Use: For the cultivation of *Propionibacterium* species.

Trypticase™ Yeast Extract Glucose Medium *See:* TYEG Medium

Trypticase™ Yeast Extract Glucose Medium *See:* TYG Medium

Tryptone Agar

Composition per liter:

Agar	15.0g
Pancreatic digest of casein	8.0g
NaCl	8.0g

pH 7.2 ± 0.2 at 25°C

Preparation of Medium: Add components to distilled/deionized water and bring volume to 1.0L. Mix thoroughly. Gently heat and bring to boiling. Distribute into tubes or flasks. Autoclave for 15 min at 15 psi pressure–121°C. Pour into sterile Petri dishes or leave in tubes.

Use: For the cultivation and maintenance of fastidious aerobic and facultative microorganisms such as *Escherichia coli* and *Pseudomonas* species.

Tryptone Bile Agar

Composition per liter:

Pancreatic digest of casein	20.0g
Agar	15.0g
Bile salts No. 3	1.5g

pH 7.2 ± 0.2 at 25°C

Source: This medium is available as a premixed powder from Difco Laboratories and Oxoid Unipath.

Preparation of Medium: Add components to distilled/deionized water and bring volume to 1.0L. Mix thoroughly. Gently heat and bring to boiling. Distribute into tubes or flasks. Autoclave for 15 min at 15 psi pressure–121°C. Pour into sterile Petri dishes.

Use: For the isolation and enumeration of *Escherichia coli* from foods.

Tryptone Broth (ATCC Medium 274)

Composition per liter:

Pancreatic digest of casein	10.0g

pH 7.2 ± 0.2 at 25°C

Preparation of Medium: Add components to distilled/deionized water and bring volume to 1.0L. Mix thoroughly. Distribute into tubes or flasks. Autoclave for 15 min at 15 psi pressure–121°C.

Use: For the cultivation and maintenance of fastidious aerobic and facultative microorganisms such as *Escherichia coli* and *Pseudomonas* species.

Tryptone Broth

Composition per liter:

Pancreatic digest of casein	10.0g
Glucose	5.0g
K_2HPO_4	1.25g
Yeast extract	1.0g
Bromcresol Purple solution	2.0mL

Bromcresol Purple Solution:
Composition per 100mL:

Bromcresol Purple	2.0g
Ethanol	10.0mL

Preparation of Bromcresol Purple Solution: Add Bromcresol Purple to 10.0mL of ethanol. Mix thoroughly. Bring volume to 100.0mL with distilled/deionized water.

Preparation of Medium: Add components to distilled/deionized water and bring volume to 1.0L. Mix thoroughly. Distribute into screw-capped tubes in 10.0mL volumes. Autoclave for 20 min at 15 psi pressure–121°C.

Use: For the cultivation of *Salmonella* species from foods.

Tryptone Broth
See: T_1N_0 Broth

Tryptone Broth
See: Tryptone Water Broth

Tryptone Broth with $CaCl_2$

Composition per liter:

Pancreatic digest of casein	10.0g
$CaCl_2$	5.5g

pH 7.2 ± 0.2 at 25°C

Preparation of Medium: Add components to distilled/deionized water and bring volume to 1.0L. Mix thoroughly. Distribute into tubes or flasks. Autoclave for 15 min at 15 psi pressure–121°C.

Use: For the cultivation and maintenance of fastidious aerobic and facultative microorganisms such as *Escherichia coli.*

Tryptone Glucose Beef Extract Agar (Tryptone Glucose Extract Agar)

Composition per liter:

Agar	15.0g
Pancreatic digest of casein	5.0g
Beef extract	3.0g
Glucose	1.0g

pH 7.0 ± 0.2 at 25°C

Source: This medium is available as a premixed powder from Difco Laboratories and Oxoid Unipath.

Preparation of Medium: Add components to distilled/deionized water and bring volume to 1.0L. Mix thoroughly. Gently heat and bring to boiling. Distribute into tubes or flasks. Autoclave for 15 min at 15 psi pressure–121°C. Cool to 45°–50°C. If the dilution of the specimen is greater than 1:10, add 10.0mL of sterile 10% skim milk solution. Mix thoroughly. Pour into sterile Petri dishes or leave in tubes.

Use: For the enumeration of bacteria by the Standard Plate Count procedure. Use for the cultivation and enumeration of bacteria from milk and dairy products. Also, for the detection of thermophilic organisms.

Tryptone Glucose Beef Extract Agar with Sucrose

Composition per liter:

Agar	15.0g
Pancreatic digest of casein	5.0g
Beef extract	3.0g
Glucose	1.0g
Sucrose	5.0g

pH 7.0 ± 0.2 at 25°C

Preparation of Medium: Add components to distilled/deionized water and bring volume to 1.0L. Mix thoroughly. Gently heat and bring to boiling. Distribute into tubes or flasks. Autoclave for 15 min at 15 psi pressure–121°C. Pour into sterile Petri dishes or leave in tubes.

Use: For the cultivation and maintenance of *Saccharococcus thermophilus.*

Tryptone Glucose Beef Extract Agar with Yeast Extract

Composition per liter:

Agar	15.0g
Pancreatic digest of casein	5.0g
Beef extract	3.0g
Glucose	1.0g
Yeast extract	1.0g

pH 7.0 ± 0.2 at 25°C

Preparation of Medium: Add components to distilled/deionized water and bring volume to 1.0L. Mix thoroughly. Gently heat and bring to boiling. Distribute into tubes or flasks. Autoclave for 15 min at 15 psi pressure–121°C. Pour into sterile Petri dishes or leave in tubes.

Use: For the cultivation and maintenance of *Ancylobacter aquaticus* and *Spirosoma linguale.*

Tryptone Glucose Extract Agar
See: Tryptone Glucose Beef Extract Agar

Tryptone Glucose Yeast Agar
See: Plate Count Agar

Tryptone Glucose Yeast Agar
See: Standard Methods Agar

Tryptone Glucose Yeast Broth
See: **Standard Methods Broth**

Tryptone Glucose Yeast Extract Medium
See: **TGY Medium**

Tryptone Glucose Yeast Extract Methionine Medium
See: **TGYM Medium**

Tryptone in Sea Water Agar

Composition per liter:

Agar	15.0g
Pancreatic digest of casein	1.0g

pH 7.0 ± 0.2 at 25°C

Preparation of Medium: Add components to sea water and bring volume to 1.0L. Mix thoroughly. Gently heat and bring to boiling. Distribute into tubes or flasks. Autoclave for 15 min at 15 psi pressure–121°C. Pour into sterile Petri dishes or leave in tubes.

Use: For the cultivation and maintenance of *Bacillus pacificus*.

Tryptone NaCl Thiamine Medium
See: **TNT Medium**

Tryptone Phosphate Brain Heart Infusion Yeast Extract Agar (TPBY)

Composition per liter:

Pancreatic digest of casein	20.0g
Agar	15.0g
NaCl	5.14g
K_2HPO_4	2.0g
KH_2PO_4	2.0g
Yeast extract	1.0g
Oxgall	0.5g
Pancreatic digest of gelatin	0.4g
Brain heart, solids from infusion	0.16g
Peptic digest of animal tissue	0.16g
Glucose	0.08g
Na_2HPO_4	0.06g
Tween™ 80	1.5mL

pH 7.0 ± 0.2 at 25°C

Preparation of Medium: Add components to distilled/deionized water and bring volume to 1.0L. Mix thoroughly. Gently heat and bring to boiling. Adjust pH to 7.0. Distribute into tubes or flasks. Autoclave for 15 min at 15 psi pressure–121°C. Pour into sterile Petri dishes or leave in tubes.

Use: For the cultivation of coliform bacteria, such as *Escherichia coli*, from foods.

Tryptone Phosphate Broth

Composition per liter:

Pancreatic digest of casein	20.0g
NaCl	5.0g
K_2HPO_4	2.0g
KH_2PO_4	2.0g
Tween™ 80	15.0mL

pH 7.0 ± 0.2 at 25°C

Preparation of Medium: Add components to distilled/deionized water and bring volume to 1.0L. Mix thoroughly. Distribute into tubes or flasks. Autoclave for 15 min at 15 psi pressure–121°C.

Use: For the cultivation of enteropathogenic *Escherichia coli*.

Tryptone Phosphate Broth

Composition per liter:

Pancreatic digest of casein	20.0g
NaCl	5.0g
K_2HPO_4	2.0g
KH_2PO_4	2.0g
Tween™ 80	1.5mL

pH 7.0 ± 0.2 at 25°C

Preparation of Medium: Add components to distilled/deionized water and bring volume to 1.0L. Mix thoroughly. Distribute into tubes or flasks. Autoclave for 15 min at 15 psi pressure–121°C.

Use: For the cultivation of coliform bacteria, such as *Escherichia coli*, from foods.

Tryptone Salt Agar
See: **T_1N_1 Agar**

Tryptone Salt Agar
See: **T_1N_2 Agar**

Tryptone Salt Broth
See: **T_1N_1 Broth**

Tryptone Salt Broth
See: T_1N_3 Broth

Tryptone Salt Broth
See: T_1N_6 Broth

Tryptone Salt Broth
See: T_1N_8 Broth

Tryptone Salt Broth
See: T_1N_{10} Broth

Tryptone Soya Agar

Composition per liter:

Agar	15.0g
Pancreatic digest of casein	15.0g
NaCl	5.0g
Pancreatic digest of soybean meal	5.0g

pH 7.3 ± 0.2 at 25°C

Source: This medium is available as a premixed powder from Oxoid Unipath.

Preparation of Medium: Add components to distilled/deionized water and bring volume to 1.0L. Mix thoroughly. Gently heat and bring to boiling. Distribute into tubes or flasks. Autoclave for 15 min at 15 psi pressure–121°C. Pour into sterile Petri dishes or leave in tubes.

Use: For cultivation and maintenance of a wide variety of microorganisms.

Tryptone Soya Agar with 4.5% NaCl

Composition per liter:

NaCl	45.0g
Pancreatic digest of casein	17.0g
Agar	12.0g
Papaic digest of soybean meal	3.0g
Glucose	2.5g
K_2HPO_4	2.5g

pH 7.0 ± 0.2 at 25°C

Preparation of Medium: Add components to distilled/deionized water and bring volume to 1.0L. Mix thoroughly. Gently heat and bring to boiling. Distribute into tubes or flasks. Autoclave for 15 min at 15 psi pressure–121°C. Pour into sterile Petri dishes or leave in tubes.

Use: For the isolation and cultivation of *Brevibacterium linens*.

Tryptone Soya Broth

Composition per liter:

Pancreatic digest of casein	17.0g
NaCl	5.0g
Pancreatic digest of soybean meal	3.0g
K_2HPO_4	2.5g
Glucose	2.5g

pH 7.3 ± 0.2 at 25°C

Source: This medium is available as a premixed powder from Oxoid Unipath.

Preparation of Medium: Add components to distilled/deionized water and bring volume to 1.0L. Mix thoroughly. Distribute into tubes or flasks. Autoclave for 15 min at 15 psi pressure–121°C.

Use: For the cultivation of a wide variety of microorganisms.

Tryptone Water Broth (Tryptone Broth)

Composition per liter:

Pancreatic digest of casein	10.0g
NaCl	5.0g

pH 7.5 ± 0.2 at 25°C

Source: This medium is available as a premixed powder from Oxoid Unipath.

Preparation of Medium: Dissolve 15.0g in 1 liter of distilled water and distribute into final containers. Sterilize by autoclaving at 121°C for 15 minutes.

Use: For the cultivation of production of indole by microorganisms.

Tryptone with NaCl Broth

Composition per liter:

Pancreatic digest of casein	8.0g
NaCl	0.5g

Preparation of Medium: Add components to distilled/deionized water and bring volume to 1.0L. Mix thoroughly. Distribute into tubes or flasks. Autoclave for 15 min at 15 psi pressure–121°C.

Use: For the cultivation and maintenance of fastidious aerobic and facultative microorganisms such as *Escherichia coli* and *Pseudomonas* species.

Tryptone Yeast Extract Agar

Composition per liter:

Pancreatic digest of casein	10.0g
Agar	2.0g
Yeast extract	1.0g

Bromcresol Purple 0.04g
Carbohydrate solution 100.0mL

pH 7.0 ± 0.2 at 25°C

Carbohydrate Solution:
Composition per 100mL:

Carbohydrate 10.0g

Preparation of Carbohydrate Solution: Add carbohydrate to distilled/deionized water and bring volume to 100.0mL. Glucose or mannitol may be used. Mix thoroughly. Filter sterilize.

Preparation of Medium: Add components, except carbohydrate solution, to distilled/deionized water and bring volume to 900.0mL. Mix thoroughly. Adjust pH to 7.0. Gently heat and bring to boiling. Distribute into tubes in 13.5mL volumes. Autoclave for 20 min at 10 psi pressure–115°C. Cool to 45°–50°C. Aseptically add 1.5mL of carbohydrate solution to each tube. Mix thoroughly. Solidify agar quickly by placing tubes in ice water.

Use: For the cultivation and differentiation of *Staphylococcus aureus* based on glucose and mannitol fermentation. Bacteria that ferment the added carbohydrate turn the medium yellow.

Tryptone Yeast Extract Broth
See: ISP Medium 1

Tryptone Yeast Extract Glucose Medium
See: TYG Medium

Tryptone Yeast Extract Glucose Salt Medium
See: TYGS Medium

Tryptone Yeast Extract Salt Medium
See: TYES Medium

Tryptophan 1% Solution (Trypticase™ 1% Solution) (Tryptone 1% Solution)

Composition per liter:

Pancreatic digest of casein 10.0g

pH 7.0 ± 0.2 at 25°C

Source: Available as a premixed powder from BBL Microbiology Systems and Difco Laboratories.

Preparation of Medium: Add components to distilled/deionized water and bring volume to 1.0L. Mix thoroughly. Distribute into tubes or flasks. Autoclave for 15 min at 15 psi pressure–121°C.

Use: For the differentiation of bacteria, especially members of the Enterobacteaceae, based on their production of indole.

Tryptophan Assay Medium

Composition per liter:

Glucose 40.0g
Sodium acetate 20.0g
Casamino acids 12.0g
K_2HPO_4 ....1.0g
KH_2PO_4 ....1.0g
$MgSO_4 \cdot 7H_2O$ 0.4g
L-Cystine 0.2g
Adenine sulfate 0.02g
$FeSO_4 \cdot 7H_2O$ 0.02g
Guanine·HCl 0.02g
$MnSO_4 \cdot 7H_2O$ 0.02g
NaCl 0.02g
Uracil...0.02g
Pyridoxine·HCl 0.4mg
Riboflavin... 0.4mg
p-Aminobenzoic acid 0.2mg
Calcium pantothenate...................................... 0.2mg
Niacin 0.2mg
Thiamine·HCl.. 0.2mg
Biotin 0.8μg

pH 6.7 ± 0.2 at 25°C

Source: This medium is available as a premixed powder from Difco Laboratories.

Preparation of Medium: Add components to distilled/deionized water and bring volume to 1.0L. Mix thoroughly. Distribute into tubes in 5.0mL volumes. Add standard solutions and test solutions to each tube. Bring volume of each tube to 10.0mL. Autoclave for 15 min at 15 psi pressure–121°C.

Use: For the assay of tryptophan using *Lactobacillus plantarum* as an indicator organism.

Tryptophan Broth

Composition per 100mL:

L-Tryptophan 0.5g
NaCl 0.5g
KH_2PO_4 ....0.25g

pH 7.4 ± 0.2 at 25°C

Preparation of Medium: Add components to distilled/deionized water and bring volume to 100.0mL. Mix thoroughly. Adjust pH to 7.4. Filter sterilize.

Aseptically distribute in 1.0mL volumes into sterile screw-capped tubes.

Use: For the cultivation of *Flavobacterium* species and a variety of other bacteria. Also used to differentiate bacteria based on indole production. Indole is determined by the addition of modified Kovacs reagent to cultures which have incubated for 18–24 hr. Formation of a red color in the upper layer indicates indole formation.

Tryptose Agar

Composition per liter:

Agar	15.0g
Pancreatic digest of casein	10.0g
Peptic digest of animal tissue	10.0g
NaCl	5.0g
Glucose	1.0g

pH 7.2 ± 0.2 at 25°C

Source: This medium is available as a premixed powder from Difco Laboratories.

Preparation of Medium: Add components to distilled/deionized water and bring volume to 1.0L. Mix thoroughly. Gently heat and bring to boiling. Distribute into tubes or flasks. Autoclave for 15 min at 15 psi pressure–121°C. Pour into sterile Petri dishes or leave in tubes.

Use: For the cultivation and maintenance of fastidious aerobic and facultative microorganisms.

Tryptose Agar with Citrate

Composition per liter:

Agar	15.0g
Pancreatic digest of casein	10.0g
Peptic digest of animal tissue	10.0g
Sodium citrate	10.0g
NaCl	5.0g
Glucose	1.0g

pH 7.2 ± 0.2 at 25°C

Preparation of Medium: Add components to distilled/deionized water and bring volume to 1.0L. Mix thoroughly. Gently heat and bring to boiling. Distribute into tubes or flasks. Autoclave for 15 min at 15 psi pressure–121°C. Pour into sterile Petri dishes or leave in tubes.

Use: For the cultivation and maintenance of fastidious aerobic and facultative microorganisms including *Brucella* species and streptococci.

Tryptose Agar with Thiamine

Composition per liter:

Agar	15.0g
Pancreatic digest of casein	10.0g
Peptic digest of animal tissue	10.0g
NaCl	5.0g
Glucose	1.0g
Thiamine·HCl	5.0mg

pH 7.2 ± 0.2 at 25°C

Preparation of Medium: Add components to distilled/deionized water and bring volume to 1.0L. Mix thoroughly. Gently heat and bring to boiling. Distribute into tubes or flasks. Autoclave for 15 min at 15 psi pressure–121°C. Pour into sterile Petri dishes or leave in tubes.

Use: For the cultivation and maintenance of fastidious aerobic and facultative microorganisms including *Brucella* species and streptococci.

Tryptose 298 Blood Agar Base Medium
See: **TBAB 298 Medium**

Tryptose Blood Agar

Composition per liter:

Agar	12.0g
Tryptose	10.0g
NaCl	5.0g
Beef extract	3.0g
Sheep blood, defibrinated	70.0mL

pH 7.2 ± 0.2 at 25°C

Source: This medium is available as a premixed powder from Oxoid Unipath.

Preparation of Medium: Add components, except sheep blood, to distilled/deionized water and bring volume to 930.0mL. Mix thoroughly. Gently heat and bring to boiling. Autoclave for 15 min at 15 psi pressure–121°C. Cool to 45°–50°C. Aseptically add sterile sheep blood. Mix thoroughly. Pour into sterile Petri dishes in 17.0mL volumes or distribute into sterile tubes.

Use: For the cultivation and maintenance of a wide variety of fastidious microorganisms.

Tryptose Blood Agar Base

Composition per liter:

Agar	15.0g
Tryptose	10.0g
NaCl	5.0g
Beef extract	3.0g

Preparation of Medium: Add components to distilled/deionized water and bring volume to 1.0L. Mix thoroughly. Gently heat and bring to boiling. Distribute into tubes. Autoclave for 15 min at 15 psi pressure–121°C. Allow tubes to cool in a slanted position to obtain a 4–5 cm slant and a 2–3 cm butt.

Use: For the cultivation and enumeration of *Salmonella* species from foods.

Tryptose Blood Agar Base with Yeast Extract

Composition per liter:

Agar	15.0g
Tryptose	10.0g
NaCl	5.0g
Beef extract	3.0g
Yeast extract	1.0g
Sheep blood, defibrinated	50.0mL

pH 7.3 ± 0.2 at 25°C

Source: This medium is available as a premixed powder from Difco Laboratories.

Preparation of Medium: Add components, except sheep blood, to distilled/deionized water and bring volume to 950.0mL. Mix thoroughly. Gently heat and bring to boiling. Autoclave for 15 min at 15 psi pressure–121°C. Cool to 45°–50°C. Aseptically add sterile sheep blood. Mix thoroughly. Pour into sterile Petri dishes in 17.0mL volumes or distribute into sterile tubes.

Use: For the cultivation and maintenance of a wide variety of fastidious microorganisms.

Tryptose Broth

Composition per liter:

Pancreatic digest of casein	10.0g
Peptic digest of animal tissue	10.0g
NaCl	5.0g
Glucose	1.0g

pH 7.2 ± 0.2 at 25°C

Source: This medium is available as a premixed powder from Difco Laboratories.

Preparation of Medium: Add components to distilled/deionized water and bring volume to 1.0L. Mix thoroughly. Distribute into tubes or flasks. Autoclave for 15 min at 15 psi pressure–121°C.

Use: For the cultivation of fastidious aerobic and facultative microorganisms including streptococci.

Tryptose Broth

Composition per liter:

Pancreatic digest of casein	10.0g
Peptic digest of animal tissue	10.0g
NaCl	5.0g
Glucose	1.0g
Thiamine·HCl	5.0mg

pH 7.2 ± 0.2 at 25°C

Source: This medium is available as a premixed powder from BBL Microbiology Systems.

Preparation of Medium: Add components to distilled/deionized water and bring volume to 1.0L. Mix thoroughly. Distribute into tubes or flasks. Autoclave for 15 min at 15 psi pressure–121°C.

Use: For the cultivation of fastidious aerobic and facultative microorganisms.

Tryptose Broth with Citrate

Composition per liter:

Pancreatic digest of casein	10.0g
Peptic digest of animal tissue	10.0g
Sodium citrate	10.0g
NaCl	5.0g
Glucose	1.0g
Thiamine·HCl	5.0mg

pH 7.2 ± 0.2 at 25°C

Preparation of Medium: Add components to distilled/deionized water and bring volume to 1.0L. Mix thoroughly. Distribute into tubes or flasks. Autoclave for 15 min at 15 psi pressure–121°C.

Use: For the isolation and cultivation of a variety of fastidious aerobic microorganisms especially *Brucella* species from clinical sources and dairy products.

Tryptose Cycloserine Dextrose Agar

Composition per liter:

Agar	20.0g
Tryptose	15.0g
Pancreatic digest of soybean meal	5.0g
Yeast extract	5.0g
Ferric ammonium citrate	1.0g
Cycloserine solution	10.0mL

pH 7.6 ± 0.2 at 25°C

Cycloserine Solution:
Composition per 10mL:

D-Cycloserine	0.4g

Preparation of Cycloserine Solution: Add cycloserine to distilled/deionized water and bring volume to 10.0mL. Mix thoroughly. Filter sterilize.

Preparation of Medium: Add components, except cycloserine solution, to distilled/deionized water and bring volume to 990.0mL. Mix thoroughly. Gently heat and bring to boiling. Autoclave for 15 min at 15 psi pressure–121°C. Cool to 45°–50°C. Aseptically add sterile cycloserine solution. Mix thoroughly. Pour into sterile Petri dishes or distribute into sterile tubes.

Use: For the isolation and cultivation of *Clostridium* species, especially *C. botulinum*, from foods.

Tryptose Phosphate Broth

Composition per liter:

Tryptose	20.0g
NaCl	5.0g
Na_2HPO_4	2.5g
Glucose	2.0g

pH 7.3 ± 0.2 at 25°C

Source: This medium is available as a premixed powder from Difco Laboratories and Oxoid Unipath.

Preparation of Medium: Add components to distilled/deionized water and bring volume to 1.0L. Mix thoroughly. Distribute into tubes or flasks. Autoclave for 15 min at 15 psi pressure–121°C. Prior to the inoculation of anaerobic microorganisms, place tubes of sterile medium in a 100°C bath for 15 min and cool undisturbed.

Use: For the cultivation of a variety of fastidious bacteria.

Tryptose Phosphate Broth, Modified

Composition per liter:

Enzymatic digest of casein	20.0g
NaCl	5.0g
Na_2HPO_4	2.5g
Glucose	2.0g

pH 7.3 ± 0.2 at 25°C

Source: This medium is available as a premixed powder from BBL Microbiology Systems.

Preparation of Medium: Add components to distilled/deionized water and bring volume to 1.0L. Mix thoroughly. Distribute into tubes or flasks. Autoclave for 15 min at 15 psi pressure–121°C. Prior to the inoculation of anaerobic microorganisms, place tubes of sterile medium in a 100°C bath for 15 min and cool undisturbed.

Use: For the cultivation of a variety of fastidious microorganisms including pneumococci, streptococci and meningococci.

Tryptose Sulfite Cycloserine Agar (TSC Agar)

Composition per liter:

Tryptose	15.0g
Agar	14.0g
Pancreatic digest of soybean meal	5.0g
Yeast extract	5.0g
Ferric ammonium citrate	1.0g
$Na_2S_2O_5$	1.0g
Cycloserine solution	10.0mL

pH 7.6 ± 0.2 at 25°C

Cycloserine Solution:
Composition per 10mL:

D-Cycloserine	0.4g

Preparation of Cycloserine Solution: Add cycloserine to distilled/deionized water and bring volume to 10.0mL. Mix thoroughly. Filter sterilize.

Preparation of Medium: Add components, except cycloserine solution, to distilled/deionized water and bring volume to 990.0mL. Mix thoroughly. Gently heat and bring to boiling. Autoclave for 15 min at 15 psi pressure–121°C. Cool to 45°–50°C. Aseptically add sterile cycloserine solution. Mix thoroughly. Pour into sterile Petri dishes.

Use: For the presumptive identification and enumeration of *Clostridium perfringens*.

Tryptose Sulfite Cycloserine Agar (TSC Agar)

Composition per liter:

Tryptose	15.0g
Agar	14.0g
Beef extract	5.0g
Pancreatic digest of soybean meal	5.0g
Yeast extract	5.0g
Ferric ammonium citrate	1.0g
$Na_2S_2O_5$	1.0g
Egg yolk emulsion	50.0mL
Cycloserine solution	10.0mL

pH 7.6 ± 0.2 at 25°C

Source: This medium is available as a premixed powder from Oxoid Unipath.

Egg Yolk Emulsion:
Composition:

Chicken egg yolks	11
Whole chicken egg	1

Preparation of Egg Yolk Emulsion: Soak eggs with 1:100 dilution of saturated mercuric chloride solution for 1 min. Crack eggs and separate yolks from whites. Mix egg yolks with 1 chicken egg.

Cycloserine Solution:
Composition per 10mL:

D-Cycloserine	0.4g

Preparation of Cycloserine Solution: Add cycloserine to distilled/deionized water and bring volume to 10.0mL. Mix thoroughly. Filter sterilize.

Preparation of Medium: Add components, except cycloserine solution and egg yolk emulsion, to distilled/deionized water and bring volume to 940.0mL. Mix thoroughly. Gently heat and bring to boiling. Autoclave for 15 min at 15 psi pressure–121°C. Cool to 45°–50°C. Aseptically add sterile cycloserine solution and egg yolk emulsion. Mix thoroughly. Pour into sterile Petri dishes.

Use: For the presumptive identification and enumeration of *Clostridium perfringens*.

Tryptose Sulfite Cycloserine Agar with Polymyxin and Kanamycin

Composition per liter:

Tryptose	15.0g
Agar	14.0g
Beef extract	5.0g
Pancreatic digest of soybean meal	5.0g
Yeast extract	5.0g
Ferric ammonium citrate	1.0g
$Na_2S_2O_5$	1.0g
Antibiotic solution	10.0mL

pH 7.6 ± 0.2 at 25°C

Antibiotic Solution:
Composition per 10.0mL:

D-Cycloserine	0.4g
Polymyxin B sulfate	0.03g
Kanamycin sulfate	0.012g

Preparation of Antibiotic Solution: Add components to distilled/deionized water and bring volume to 10.0mL. Mix thoroughly. Filter sterilize.

Preparation of Medium: Add components, except antibiotic solution, to distilled/deionized water and bring volume to 990.0mL. Mix thoroughly. Gently heat and bring to boiling. Autoclave for 15 min at 15 psi pressure–121°C. Cool to 45°–50°C. Aseptically add sterile antibiotic solution. Mix thoroughly. Pour into sterile Petri dishes.

Use: For the isolation and enumeration of *Clostridium perfringens* from foods and a variety of clinical specimens.

Tryptose Sulfite Cycloserine Agar without Egg Yolk (TSC Agar without Egg Yolk)

Composition per liter:

Tryptose	15.0g
Agar	14.0g
Beef extract	5.0g
Pancreatic digest of soybean meal	5.0g
Yeast extract	5.0g
Ferric ammonium citrate	1.0g
$Na_2S_2O_5$	1.0g
Cycloserine solution	10.0mL

pH 7.6 ± 0.2 at 25°C

Cycloserine Solution:
Composition per 10mL:

D-Cycloserine	0.4g

Preparation of Cycloserine Solution: Add cycloserine to distilled/deionized water and bring volume to 10.0mL. Mix thoroughly. Filter sterilize.

Preparation of Medium: Add components, except cycloserine solution, to distilled/deionized water and bring volume to 990.0mL. Mix thoroughly. Gently heat and bring to boiling. Autoclave for 15 min at 15 psi pressure–121°C. Cool to 45°–50°C. Aseptically add sterile cycloserine solution. Mix thoroughly. Pour into sterile Petri dishes.

Use: For the presumptive identification and enumeration of *Clostridium perfringens*.

TSFA

See: **Tryptic Soy Fast Green Agar**

TS Soil Extract (Trypticase™ Soy Soil Extract)

Composition per liter:

Pancreatic digest of casein	17.0g
Agar	15.0g
NaCl	5.0g
Papaic digest of soybean meal	3.0g
K_2HPO_4	2.5g
Glucose	2.5g
Soil extract	250.0mL

Soil Extract:
Composition per 400mL:

African violet soil	154.0g
Na_2CO_3	0.4g

Preparation of Soil Extract: Add components to tap water and bring volume to 400.0mL. Autoclave for 60 min at 15 psi pressure–121°C. Filter through Whatman filter paper.

Preparation of Medium: Add components to tap water and bring volume to 1.0L. Mix thoroughly. Gently heat and bring to boiling. Distribute into tubes or flasks. Autoclave for 15 min at 15 psi pressure–121°C. Pour into sterile Petri dishes or leave in tubes.

Use: For cultivation and maintenance of *Bacillus xerothermodurans*.

TSA 5400 Selective Isolation Medium

Composition per liter:

Pancreatic digest of casein	15.0g
Agar	15.0g
Papaic digest of soybean meal	5.0g
NaCl	5.0g
Bovine blood, citrated	100.0mL
Spectinomycin solution	10.0mL

pH 7.3 ± 0.2 at 25°C

Spectinomycin Solution:
Composition per 10mL:

Spectinomycin	0.4g

Preparation of Spectinomycin Solution: Add spectinomycin to distilled/deionized water and bring volume to 10.0mL. Mix thoroughly. Filter sterilize.

Preparation of Medium: Add components, except bovine blood and spectinomycin solution, to distilled/deionized water and bring volume to 890.0mL. Mix thoroughly. Gently heat and bring to boiling. Autoclave for 15 min at 15 psi pressure–121°C. Cool to 45°–50°C. Aseptically add sterile bovine blood and 10.0mL of sterile spectinomycin solution. Mix thoroughly. Pour into sterile Petri dishes or distribute into sterile tubes.

Use: For the isolation of *Treponema hyodysenteriae*.

TSA Blood Agar
See: **Trypticase™ Soy Agar with Sheep Blood**

TSA NaCl
See: **Trypticase™ Soy Agar with 3% NaCl**

TSA II™
See: **Trypticase™ Soy Agar, Modified**

TSA II™ with Sheep Blood and Gentamicin
See: **Trypticase™ Soy Agar with Sheep Blood and Gentamicin**

TSC Agar
See: **Tryptose Sulfite Cycloserine Agar**

TSC Agar without Egg Yolk
See: **Tryptose Sulfite Cycloserine Agar without Egg Yolk**

TSI Agar
See: **Triple Sugar Iron Agar**

TSN Agar (Trypticase Sulfite Neomycin Agar)

Composition per liter:

Pancreatic digest of casein	15.0g
Agar	13.5g
Yeast extract	10.0g
Na_2SO_3	1.0g
Ferric citrate	0.5g
Neomycin sulfate	0.05g
Polymyxin sulfate	0.02g
Buffered thioglycollate solution	50.0mL

pH 7.2 ± 0.2 at 25°C

Source: This medium is available as a premixed powder from BBL Microbiology Systems.

Buffered Thioglycollate Solution:
Composition per 50mL:

Buffer solution	35.0mL
Sodium thioglycollate solution	15.0mL

Preparation of Buffered Thioglycollate Solution: Combine components. Mix thoroughly. Autoclave for 15 min at 15 psi pressure–121°C. Cool to 45°–50°C.

Buffer Solution:
Composition per 100mL:

Na_2CO_3	28.0g
K_2HPO_4	5.7g

Preparation of Buffer Solution: Add components to distilled/deionized water and bring volume to 100.0mL. Mix thoroughly.

Thioglycollate Solution:
Composition per 100mL:

Sodium thioglycollate	13.3g

Preparation of Thioglycollate Solution: Add sodium thioglycollate to distilled/deionized water and bring volume to 100.0mL. Mix thoroughly.

Preparation of Medium: Add components, except buffered thioglycollate solution, to distilled/deionized water and bring volume to 950.0mL. Mix thoroughly. Gently heat and bring to boiling. Autoclave for 12 min at 13 psi pressure–118°C. Do not overheat. Cool to 45°–50°C. Aseptically add buffered thioglycollate solution. Mix thoroughly. Pour into sterile Petri dishes or distribute into sterile tubes.

Use: For the selective isolation of *Clostridium perfringens.*

TSY Medium (Trypticase™ Soy Yeast Extract Medium)

Agar	20.0g
Pancreatic digest of casein	17.0g
Yeast extract	5.0g
NaCl	5.0g
Papaic digest of soybean meal	3.0g
K_2HPO_4	2.5g
Glucose	2.5g

pH 7.3 ± 0.2 at 25°C

Preparation of Medium: Add components to distilled/deionized water and bring volume to 1.0L. Mix thoroughly. Gently heat and bring to boiling. Distribute into tubes or flasks. Autoclave for 15 min at 15 psi pressure–121°C. Pour into sterile Petri dishes or leave in tubes.

Use: For cultivation and maintenance of *Escherichia coli.*

TSYES Medium (Trypticase™ Soy Yeast Extract Starch Medium)

Composition per liter:

Pancreatic digest of casein	17.0g
Agar	15.0g
NaCl	5.0g
Papaic digest of soybean meal	3.0g
K_2HPO_4	2.5g
Glucose	2.5g
Yeast extract	2.0g
Soluble starch	1.0g

pH 7.3 ± 0.2 at 25°C

Preparation of Medium: Add components to distilled/deionized water and bring volume to 1.0L. Mix thoroughly. Gently heat while stirring and bring to boiling. Distribute into tubes or flasks. Autoclave for 15 min at 15 psi pressure–121°C. Pour into sterile Petri dishes or leave in tubes.

Use: For cultivation and maintenance of *Bacillus* species.

TT Broth
See: **Tetrathionate Broth**

TT Broth , Hajna
See: **Tetrathionate Broth , Hajna**

TT Broth, USA
See: **Tetrathionate Broth, USA**

TTC Agar
See: **Tetrazolium Tolerance Agar**

Tween™ 80 Agar
See: **Polysorbate 80 Agar**

Tween™ 80 Hydrolysis Broth

Composition per liter:

Na_2HPO_4	5.79g
NaH_2PO_4	3.53g
Neutral Red	0.02g
Tween™ 80	5.00mL

pH 7.0 ± 0.2 at 25°C

Preparation of Medium: Add components to distilled/deionized water and bring volume to 1.0L. Mix thoroughly. Distribute into tubes or flasks. Autoclave for 15 min at 15 psi pressure–121°C.

Use: For the differentiation of *Mycobacterium* species. Strains that hydrolyze Tween™ 80 within 5 days turn the medium pink to red.

Tween™ 80 Hydrolysis Broth

Composition per 125mL:

Neutral Red	0.1g
Solution 1	38.9mL
Solution 2	61.1mL
Tween™ 80	25.0mL

pH 7.0 ± 0.2 at 25°C

Solution 1:
Composition per 400mL:

KH_2PO_4	22.7g

Preparation of Solution 1: Add KH_2PO_4 to distilled/deionized water and bring volume to 400.0mL. Mix thoroughly.

Solution 2:
Composition per 400mL:

Na_2HPO_4	23.8g

Preparation of Solution 2: Add Na_2HPO_4 to distilled/deionized water and bring volume to 400.0mL. Mix thoroughly.

Preparation of Medium: Combine components. Mix thoroughly. Distribute into tubes or flasks. Autoclave for 15 min at 15 psi pressure–121°C.

Use: For the differentiation of *Mycobacterium* species. Strains that hydrolyze Tween™ 80 within 5 days turn the medium pink to red.

Tween™ 80 Hydrolysis Broth

Composition per 102.5mL:

$NaHPO_4$ (0.066M solution)	61.1mL
KH_2PO_4 (0.066M solution)	38.9mL
Neutral Red (0.1% solution)	2.0mL
Tween™ 80	0.5mL

pH 7.0 ± 0.2 at 25°C

Preparation of Medium: Combine components. Mix thoroughly. Distribute into tubes or flasks. Autoclave for 15 min at 15 psi pressure–121°C.

Use: For the differentiation of *Mycobacterium* species. Strains that hydrolyze Tween™ 80 within 5 days turn the medium pink to red.

Tween™ 80 Hydrolysis Medium

Composition per liter:

Agar	12.0g
Peptone	10.0g
NaCl	5.0g
$CaCl_2$	0.1g
Tween™ 80	10.0mL

pH 7.2–7.4 at 25°C

Preparation of Medium: Add components to distilled/deionized water and bring volume to 1.0L. Mix thoroughly. Gently heat and bring to boiling. Distribute into tubes or flasks. Autoclave for 15 min at 15 psi pressure–121°C. Pour into sterile Petri dishes.

Use: For the cultivation and differentiation of *Pseudomonas* species based on their ability to hydrolyze Tween™ 80. Bacteria that hydrolyze Tween™ 80 appear as colonies surrounded by an opaque zone.

Tween™ 80 Oxgall Caffeic Acid Agar
See: **TOC Agar**

TY Medium, 2X

Composition per liter:

Pancreatic digest of casein	16.0g
Yeast extract	10.0g
NaCl	5.0g

pH 7.0 ± 0.2 at 25°C

Preparation of Medium: Add components to distilled/deionized water and bring volume to 1.0L. Mix thoroughly. Distribute into tubes or flasks. Autoclave for 25 min at 15 psi pressure–121°C.

Use: For the cultivation of *Escherichia coli.*

TYE HES Medium

Composition per 950mL:

NaCl	49.7g
$MgSO_4.7H_2O$	49.3g
Noble agar	10.0g
Yeast extract	0.5g
Pancreatic digest of casein	0.5g
$CaCl_2·2H_2O$ solution	50.0mL

pH 7.2 ± 0.2 at 25°C

$CaCl_2·2H_2O$ Solution:
Composition per 100mL:

$CaCl_2·2H_2O$	0.3g

Preparation of $CaCl_2·2H_2O$ Solution: Add the $CaCl_2·2H_2O$ to distilled/deionized water and bring volume to 100.0mL. Mix thoroughly. Autoclave for 15 min at 15 psi pressure–121°C. Cool to 45°–50°C.

Preparation of Medium: Add components, except $CaCl_2·2H_2O$ solution, to distilled/deionized water and bring volume to 950.0mL. Mix thoroughly. Gently heat and bring to boiling. Autoclave for 15 min at 15 psi pressure–121°C. Cool to 45°–50°C. Aseptically add 50.0mL of sterile $CaCl_2·2H_2O$ solution. Mix thoroughly. Adjust pH to 7.2. Pour into sterile Petri dishes or distribute into sterile tubes.

Use: For the cultivation of *Planococcus* species.

TYEG Medium (Trypticase™ Yeast Extract Glucose Medium)

Composition per 1050mL:

NaCl	100.0g
Pancreatic digest of casein	10.0g
$Na_2HPO_4 \cdot 7H_2O$	2.1g
NH_4Cl	1.0g
KH_2PO_4	0.3g
$MgCl_2 \cdot 6H_2O$	0.2g
Glucose solution	50.0mL
$Na_2S \cdot 7H_2O$ solution	25.0mL
Trace mineral solution II	10.0mL
Wolfe's vitamin solution	10.0mL
Yeast extract solution	5.0mL
Resazurin (0.2% solution)	1.0mL
$FeSO_4 \cdot 9H_2O$ (2.5% solution)	25.0µl

pH 7.3 ± 0.1 at 25°C

Glucose Solution:

Composition per 100mL:

D-Glucose	10.0g

Preparation of Glucose Solution: Add glucose to distilled/deionized water and bring volume to 100.0mL. Mix thoroughly. Filter sterilize. Aseptically bubble with 90% N_2 + 10% CO_2 to reduce.

$Na_2S \cdot 7H_2O$ Solution:

Composition per 100mL:

$Na_2S \cdot 7H_2O$	2.5g

Preparation of $Na_2S \cdot 7H_2O$ Solution: Add $Na_2S \cdot 7H_2O$ to distilled/deionized water and bring volume to 100.0mL. Mix thoroughly. Autoclave for 15 min at 15 psi pressure–121°C. Use freshly prepared solution.

Trace Mineral Solution II:

Composition per liter:

Nitrilotriacetic acid	12.8g
$CoCl_2 \cdot 6H_2O$	0.17g
$CaCl_2 \cdot 2H_2O$	0.1g
FeSO4·7H2O	0.1g
$MnCl_2 \cdot 4H_2O$	0.1g
NaCl	0.1g
$ZnCl_2$	0.1g
$NiSO_4 \cdot 6H_2O$	0.026g
$CuCl_2 \cdot 2H_2O$	0.02g
Na_2SeO_3	0.017g
H_3BO_3	0.01g
$Na_2MoO_4 \cdot 2H_2O$	0.01g

Preparation of Trace Mineral Solution II: Add nitrilotriacetic acid to 500.0mL of distilled/deionized water. Dissolve by adjusting pH to 6.5 with KOH. Add remaining components. Add distilled/deionized water to 1.0L. Filter through Whatman filter paper. Store under N_2.

Wolfe's Vitamin Solution:

Composition per liter:

Pyridoxine·HCl	10.0mg
Thiamine·HCl	5.0mg
Riboflavin	5.0mg
Nicotinic acid	5.0mg
Calcium pantothenate	5.0mg
p-Aminobenzoic acid	5.0mg
Thioctic acid	5.0mg
Biotin	2.0mg
Folic acid	2.0mg
Cyanocobalamin	100.0µg

Preparation of Wolfe's Vitamin Solution: Add components to distilled/deionized water and bring volume to 1.0L. Mix thoroughly. Filter sterilize. Aseptically bubble with 90% N_2 + 10% CO_2 to reduce.

Yeast Extract Solution:

Composition per 100mL:

Yeast extract	10.0g

Preparation of Yeast Extract Solution: Add yeast extract to distilled/deionized water and bring volume to 100.0mL. Mix thoroughly. Filter sterilize. Aseptically bubble with 90% N_2 + 10% CO_2 to reduce.

Preparation of Medium: Add components—except glucose solution, yeast extract solution, and Wolfe's vitamin solution—to distilled/deionized water and bring volume to 960.0mL. Mix thoroughly. Adjust pH to 7.3. Gently heat and bring to boiling under 90% N_2 + 10% CO_2. Autoclave for 15 min at 15 psi pressure–121°C. Cool to 45°–50°C. Aseptically and anaerobically add 50.0mL of sterile glucose solution, 10.0mL of sterile Wolfe's vitamin solution, and 5.0mL of sterile yeast extract solution. Aseptically and anaerobically distribute into sterile tubes in 5.0mL volumes. Immediately prior to inoculation aseptically add 0.125mL of sterile $Na_2S \cdot 9H_2O$ solution per tube.

Use: For the cultivation and maintenenace of *Halobacteroides acetoethylicus*.

TYES Medium (Tryptone Yeast Extract Salt Medium)

Composition per liter:

Agar	15.0g
Pancreatic digest of casein	10.0g

NaCl 8.0g
Yeast extract 1.0g
$CaCl_2$ 0.3g

pH 7.2 ± 0.2 at 25°C

Preparation of Medium: Add components to distilled/deionized water and bring volume to 1.0L. Mix thoroughly. Gently heat and bring to boiling. Distribute into tubes or flasks. Autoclave for 15 min at 15 psi pressure–121°C. Pour into sterile Petri dishes or leave in tubes.

Use: For the cultivation and maintenenace of *Escherichia coli.*

TYG Medium (Tryptone Yeast Extract Glucose Medium) (ATCC Medium 741)

Composition per liter:

Agar 20.0g
Pancreatic digest of casein 3.0g
Yeast extract 3.0g
Glucose 3.0g
K_2HPO_4 .... 1.0g

pH 7.4 ± 0.2 at 25°C

Preparation of Medium: Add components to distilled/deionized water and bring volume to 1.0L. Mix thoroughly. Gently heat and bring to boiling. Distribute into tubes or flasks. Autoclave for 15 min at 15 psi pressure–121°C. Pour into sterile Petri dishes or leave in tubes.

Use: For the cultivation and maintenenace of *Thermomonospora fusca.*

TYG Medium (Trypticase Yeast Extract Glucose Medium) (ATCC Medium 603)

Composition per liter:

Pancreatic digest of casein 10.0g
NaCl 8.0g
Yeast extract 1.0g
Glucose 1.0g
$CaCl_2 \cdot 2H_2O$ 0.3g

Preparation of Medium: Add components to distilled/deionized water and bring volume to 1.0L. Mix thoroughly. Distribute into tubes or flasks. Autoclave for 15 min at 15 psi pressure–121°C.

Use: For the cultivation and maintenenace of *Escherichia coli.*

TYGPN Medium

Composition per liter:

Pancreatic digest of casein 20.0g
KNO_3 .... 10.0g
Yeast extract 10.0g
Na_2HPO_4 .... 5.0g
Glycerol (80% solution) 10.0mL

pH 7.0 ± 0.2 at 25°C

Preparation of Medium: Add components to distilled/deionized water and bring volume to 1.0L. Mix thoroughly. Distribute into tubes or flasks. Autoclave for 25 min at 15 psi pressure–121°C.

Use: For the cultivation of *Escherichia coli.*

TYGS Medium (Tryptone Yeast Extract Glucose Salt Medium)

Composition per liter:

Agar 15.0g
Pancreatic digest of casein 10.0g
NaCl 8.0g
Yeast extract 1.0g
$CaCl_2 \cdot 2H_2O$ solution 100.0mL
Glucose solution 100.0mL

$CaCl_2 \cdot 2H_2O$ Solution:

Composition per 100mL:

$CaCl_2 \cdot 2H_2O$ 0.3g

Preparation of $CaCl_2 \cdot 2H_2O$ Solution: Add the $CaCl_2 \cdot 2H_2O$ to distilled/deionized water and bring volume to 100.0mL. Mix thoroughly. Filter sterilize.

Glucose Solution:

Composition per 100mL:

D-Glucose 1.0g

Preparation of Glucose Solution: Add glucose to distilled/deionized water and bring volume to 100.0mL. Mix thoroughly. Filter sterilize.

Preparation of Medium: Add components, except $CaCl_2 \cdot 2H_2O$ solution and glucose solution, to distilled/deionized water and bring volume to 800.0mL. Mix thoroughly. Gently heat and bring to boiling. Autoclave for 15 min at 15 psi pressure–121°C. Cool to 45°–50°C. Aseptically add the sterile $CaCl_2 \cdot 2H_2O$ solution and sterile glucose solution. Mix thoroughly. Pour into sterile Petri dishes or distribute into sterile tubes.

Use: For the cultivation and maintenance of a variety of bacteria.

TYN Medium

Composition per liter:

$Na_2S_2O_3 \cdot 5H_2O$	10.0g
Pancreatic digest of casein	1.0g
Yeast extract	1.0g
Na_2SO_4	1.0g

Preparation of Medium: Add components to distilled/deionized water and bring volume to 1.0L. Mix thoroughly. Distribute into tubes or flasks. Autoclave for 15 min at 15 psi pressure–121°C.

Use: For cultivation and maintenance of *Thiobacillus* species.

Tyrosine Agar (International Streptomyces Project Medium 7) (ISP Medium 7) (ATCC Medium 1776)

Composition per liter:

Agar	20.0g
Glycerol	15.0g
L-Tyrosine	0.5g
L-Asparagine	1.0g
K_2HPO_4	0.5g
$MgSO_4 \cdot 7H_2O$	0.5g
NaCl	0.5g
$FeSO_4 \cdot 7H_2O$	0.01g
Trace elements solution HO-LE	1.0mL

pH 7.3 ± 0.1 at 25°C

Trace Element Solution HO-LE:
Composition per liter:

H_3BO_3	2.85g
$MnCl_2 \cdot 4H_2O$	1.8g
Sodium tartrate	1.77g
$FeSO_4 \cdot 7H_2O$	1.36g
$CoCl_2 \cdot 6H_2O$	0.04g
$CuCl_2 . 2H_2O$	0.027g
$Na_2MoO_4 \cdot 2H_2O$	0.025g
$ZnCl_2$	0.020g

Preparation of Trace Element Solution HO-LE: Add components to distilled/deionized water and bring volume to 1.0L. Mix thoroughly. Filter sterilize.

Preparation of Medium: Add components to distilled/deionized water and bring volume to 1.0L. Mix thoroughly. Adjust pH to 7.3. Gently heat and bring to boiling. Distribute into tubes or flasks. Autoclave for 15 min at 15 psi pressure–121°C. Pour into sterile Petri dishes or leave in tubes.

Use: For cultivation and maintenance of *Streptoalloteichus* species. For the differentiation of *Streptomyces* species based on melanine production.

Tyrosine Agar

Composition per liter:

Solution 1	900.0mL
Solution 2	100.0mL

pH 7.0 ± 0.2 at 25°C

Solution 1:
Composition per 900mL:

Agar	15.0g
Pancreatic digest of gelatin	5.0g
Beef extract	3.0g

Preparation of Solution 1: Add components to distilled/deionized water and bring volume to 900.0mL. Mix thoroughly. Gently heat and bring to boiling.

Solution 2:
Composition per 100mL:

Tyrosine	5.0g

Preparation of Solution 2: Add tyrosine to distilled/deionized water and bring volume to 100.0mL. Mix thoroughly. Gently heat and bring to boiling.

Preparation of Medium: Combine solutions 1 and 2. Mix thoroughly. Distribute into tubes or flasks. Autoclave for 15 min at 15 psi pressure–121°C. Pour into sterile Petri dishes or leave in tubes.

Use: For the differentiation of aerobic *Actinomycete* species. Clearing around a colony indicates utilization of tyrosine. *Streptomyces* and *Actinomadura* species utilize tyrosine. *Nocaridia asteroides, N. caviae,* and *Mycobacterium fortuitum* do not utilize tyrosine.

Tyrosine Casein Nitrate Medium (TCN Medium)

Composition per liter:

Sodium caseinate	25.0g
Agar	15.0g
$NaNO_3$	10.0g
L-Tyrosine	1.0g

Preparation of Medium: Add components to tap water and bring volume to 1.0L. Mix thoroughly. Gently heat and bring to boiling. Distribute into tubes or flasks. Autoclave for 15 min at 15 psi pressure–121°C. Pour into sterile Petri dishes or leave in tubes.

Use: For the isolation and cultivation of streptomycetes from infected plants.

TZC Selective Medium

Composition per liter:

Agar	17.0g
Peptone	1.0g
Glucose	0.5g

Pancreatic digest of casein 0.1g
2,3,5-Triphenyltetrazolium·HCl solution...... 10.0mL

2,3,5-Triphenyltetrazolium·HCl Solution:
Composition per 10mL:
2,3,5-Triphenyltetrazolium·HCl........................0.05g

Preparation of 2,3,5-Triphenyltetrazolium-HCl Solution: Add 2,3,5-triphenyltetrazolium·HCl to distilled/deionized water and bring volume to 10.0mL. Mix thoroughly. Autoclave for 15 min at 15 psi pressure–121°C.

Preparation of Medium: Add components, except 2,3,5-triphenyltetrazolium·HCl solution, to distilled/deionized water and bring volume to 990.0mL. Mix thoroughly. Gently heat and bring to boiling. Autoclave for 15 min at 15 psi pressure–121°C. Cool to 45°–50°C. Aseptically add 10.0mL of sterile 2,3,5-triphenyltetrazolium·HCl solution. Mix thoroughly. Pour into sterile Petri dishes.

Use: For the isolation, cultivation, and differentiation of *Pseudomonas solanacearum.* The virulent, wild type strains appear as irregular to round white colonies with a pink center. Avirulent mutants, which readily occur in nature, appear as round deep red colonies with a narrow blue border.

U Agar Plates
(*Ureaplasma* Agar Plates)
(MES Agar)

Composition per 100.2mL:

Base agar	65.0mL
Horse serum	20.0mL
Yeast dialysate	10.0mL
MES (2-N-Morpholinoethane sulfonic acid) buffer solution	3.0mL
Penicillin solution	2.0mL
Urea solution	0.2mL

pH 5.5 ± 0.2 at 25°C

Base Agar:

Composition per liter:

Papaic digest of soybean meal	20.0g
Agarose	10.0g
NaCl	5.0g
Phenol Red (2% solution)	1.0mL

Preparation of Base Agar: Add components to distilled/deionized water and bring volume to 1.0L. Mix thoroughly. Gently heat and bring to boiling. Adjust pH to 7.3. Autoclave for 15 min at 15 psi pressure–121°C. Cool to 45°–50°C.

Yeast Dialysate:

Composition per 10mL:

Yeast, active dried	450.0g

Preparation of Yeast Dialysate: Add active, dried yeast to distilled/deionized water and bring volume to 1250.0mL. Gently heat and bring to 40°C. Autoclave for 15 min at 15 psi pressure–121°C. Put into dialysis tubing. Dialyze against 1.0L of distilled/deionized water for 2 days at 4°C. Discard tubing and its contents. Autoclave dialysate for 15 min at 15 psi pressure–121°C. Store at –20°C.

MES Buffer Solution:

Composition per 100mL:

MES (2-N-Morpholinoethane sulfonic acid) buffer	19.52g

Preparation of MES Buffer Solution: Add MES buffer to distilled/deionized water and bring volume to 100.0mL. Mix thoroughly. Adjust pH to 5.5. Filter sterilize.

Penicillin Solution:

Composition per 10mL:

Penicillin	100,000U

Preparation of Penicillin Solution: Add penicillin to distilled/deionized water and bring volume to 10.0mL. Mix thoroughly. Filter sterilize.

Urea Solution:

Composition per 100mL:

Urea	6.0g

Preparation of Urea Solution: Add urea to distilled/deionized water and bring volume to 100.0mL. Mix thoroughly. Filter sterilize.

Preparation of Medium: To 65.0mL of cooled, sterile base agar, aseptically add 10.0mL of sterile yeast dialysate, 20.0mL of horse serum, 2.0mL of sterile penicillin solution, 3.0mL of sterile MES buffer solution, and 0.2mL of sterile urea solution. Mix thoroughly. Pour into 10mm × 35mm Petri dishes in 5.0mL volumes. Allow plates to stand overnight at 25°C to remove excess surface moisture.

Use: For the isolation and cultivation of *Ureaplasma* species.

U Broth
(*Ureaplasma* Broth)

Composition per 99.5mL:

Base agar	65.0mL
Horse serum	20.0mL
Yeast dialysate	10.0mL
Penicillin solution	2.0mL
MES (2-N-Morpholinoethane sulfonic acid) buffer solution	1.0mL
Na_2SO_3 solution	1.0mL
Urea solution	0.5mL

pH 5.5 ± 0.2 at 25°C

Base Agar:

Composition per liter:

Papaic digest of soybean meal	20.0g
Agarose	10.0g
NaCl	5.0g
Phenol Red (2% solution)	1.0mL

Preparation of Base Agar: Add components to distilled/deionized water and bring volume to 1.0L. Mix thoroughly. Gently heat and bring to boiling. Adjust pH to 7.3. Autoclave for 15 min at 15 psi pressure–121°C. Cool to 45°–50°C.

Yeast Dialysate:

Composition per 10mL:

Yeast, active dried	450.0g

Preparation of Yeast Dialysate: Add active, dried yeast to distilled/deionized water and bring volume to 1250.0mL. Gently heat and bring to 40°C. Autoclave for 15 min at 15 psi pressure–121°C. Put into dialysis tubing. Dialyze against 1.0L of distilled/deionized water for 2 days at 4°C. Discard tubing and its contents. Autoclave dialysate for 15 min at 15 psi pressure–121°C. Store at –20°C.

Penicillin Solution:

Composition per 10mL:

Penicillin	100,000U

Preparation of Penicillin Solution: Add penicillin to distilled/deionized water and bring volume to 10.0mL. Mix thoroughly. Filter sterilize.

MES Buffer Solution:
Composition per 100mL:

MES (2-N-Morpholinoethane sulfonic acid) buffer	19.52g

Preparation of MES Buffer Solution: Add MES buffer to distilled/deionized water and bring volume to 100.0mL. Mix thoroughly. Adjust pH to 5.5. Filter sterilize.

Na_2SO_3 Solution:
Composition per 10mL:

Na_2SO_3	0.126g

Preparation of Na_2SO_3 Solution: Add Na_2SO_3 to distilled/deionized water and bring volume to 10.0mL. Mix thoroughly. Filter sterilize.

Urea Solution:
Composition per 100mL:

Urea	6.0g

Preparation of Urea Solution: Add urea to distilled/deionized water and bring volume to 100.0mL. Mix thoroughly. Filter sterilize.

Preparation of Medium: To 65.0mL of cooled, sterile base agar, aseptically add 10.0mL of sterile yeast dialysate, 20.0mL of horse serum, 2.0mL of sterile penicillin solution, 1.0mL of sterile MES buffer solution, 1.0mL of sterile Na_2SO_3 solution, and 0.5mL of sterile urea solution. Mix thoroughly. Pour into 10mm × 35mm Petri dishes in 5.0mL volumes. Allow plates to stand overnight at 25°C to remove excess surface moisture. Use within 48 hr.

Use: For the isolation and cultivation of *Ureaplasma urealyticum.*

U4 Medium

Composition per 100mL:

Hartley's digest broth	20.0mL
Fetal calf serum	15.0mL
Fresh yeast extract solution	10.0mL
Hanks balanced salt solution, 10X	4.0mL
$MgSO_4 \cdot 5H_2O$ (0.025% solution)	1.0mL
Urea (20% solution)	0.25mL
Phenol Red (1% solution)	0.2mL

pH 6.0–6.2 at 25°C

Hartley's Digest Broth:
Composition per 10L:

Ox heart	3000.0g
Pancreatin	50.0g
Na_2CO_3, anhydrous (0.8% solution)	5.0L
HCl, concentrated	80.0mL

pH 7.5 ± 0.2 at 25°C

Preparation of Hartley's Digest Broth: Finely mince the ox heart. Add the meat to 5.0L of distilled/deionized water. Gently heat and bring to 80°C. Add the 5.0L of Na_2CO_3 solution. Cool to 45°C. Add pancreatin and maintain at 45°C for 4 hr while stirring. Add the HCl and steam at 100°C for 30 min. Cool to room temperature. Adjust pH to 8.0 with 1*N* NaOH. Gently heat and bring to boiling. Continue boiling for 25 min. Filter while hot. Cool to room temperature. Adjust pH to 7.5. Autoclave for 15 min at 15 psi pressure–121°C.

Fresh Yeast Extract Solution:
Composition per 100mL:

Baker's yeast, live, pressed, starch-free	25.0g

Preparation of Fresh Yeast Extract Solution: Add the live Baker's yeast to 100 .0mL of distilled/deionized water. Autoclave for 90 min at 15 psi pressure–121°C. Allow to stand. Remove supernatant solution. Adjust pH to 6.6–6.8.

Hanks' Balanced Salt Solution, 10X:
Composition per liter:

Na_2Cl	80.0g
Glucose	10.0g
KCl	4.0g
$CaCl_2$	1.4g
$MgCl_2 \cdot 6H_2O$	1.0g
$MgSO_4 \cdot 7H_2O$	1.0g
$Na_2HPO_4 \cdot 7H_2O$	0.9g
KH_2PO_4	0.6g

Preparation of Hanks' Balanced Salt Solution, 10X: Add components to distilled/deionized water and bring volume to 100.0mL. Mix thoroughly.

Preparation of Medium: Add components to distilled/deionized water and bring volume to 1.0L. Mix thoroughly. Adjust pH to 6.0–6.2 with HCl. Filter-sterilize medium. Aseptically distribute into sterile tubes or flasks.

Use: For the cultivation and maintenance of *Ureaplasma diversum.*

U9 Broth (Urease Color Test Medium)

Composition per 101.6mL:

U9 base	95.0mL
Horse serum, unheated	5.0mL
Penicillin G solution	1.0mL

Urea solution 0.5mL
Phenol Red solution 0.1mL

pH 6.0 ± 0.2 at 25°C

U9 Base:

Composition per 100mL:

NaCl 0.63g
Pancreatic digest of casein 0.425g
Papaic digest of soybean meal 0.075g
K_2HPO_4 .... 0.063g
Glucose 0.063g
KH_2PO_4 .... 0.02g

Preparation of U9 Base: Add components to distilled/deionized water and bring volume to 100.0mL. Mix thoroughly. Adjust pH to 5.5 with 1*N* HCl. Autoclave for 15 min at 15 psi pressure–121°C. Cool to 45°–50°C.

Penicillin G Solution:

Composition per 10mL:

Penicillin G .... 0.63 g

Preparation of Penicillin G Solution: Add Penicillin G to distilled/deionized water and bring volume to 10.0mL. Mix thoroughly. Filter sterilize.

Urea Solution:

Composition per 30mL:

Urea 3.0g

Preparation of Urea Solution: Add urea to distilled/deionized water and bring volume to 30.0mL. Mix thoroughly. Filter sterilize.

Phenol Red Solution:

Composition per 10mL:

Phenol Red 0.1g

Preparation of Phenol Red Solution: Add Phenol Red to distilled/deionized water and bring volume to 10.0mL. Mix thoroughly. Filter sterilize.

Preparation of Medium: To 95.0mL of cooled, sterile U9 base, aseptically add 5.0mL of sterile horse serum, 1.0mL of sterile penicillin G solution, 0.5mL of sterile urea solution, and 0.1mL of sterile Phenol Red solution. Mix thoroughly. Aseptically distribute into sterile tubes or flasks.

Use: For the isolation and identification of T-strain mycoplasmas from clinical specimens, especially *Ureaplasma urealyticum*. T-mycoplasmas are the only members of the *Mycoplasma* group known to contain urease. Bacteria with urease activity turn the medium dark pink.

U9 Broth with Amphotericin B

Composition per 101.6mL:

U9 base 95.0mL
Horse serum, unheated 5.0mL
Antibiotic solution 1.0mL
Urea solution 0.5mL
Phenol Red solution 0.1mL

pH 6.0 ± 0.2 at 25°C

U9 Base:

Composition per 100mL:

NaCl 0.63g
Pancreatic digest of casein 0.425g
Papaic digest of soybean meal 0.075g
K_2HPO_4 .... 0.063g
Glucose 0.063g
KH_2PO_4 .... 0.02g

Preparation of U9 Base: Add components to distilled/deionized water and bring volume to 100.0mL. Mix thoroughly. Adjust pH to 5.5 with 1*N* HCl. Autoclave for 15 min at 15 psi pressure–121°C. Cool to 45°–50°C.

Antibiotic Solution:

Composition per 10mL:

Penicillin G 0.63g
Amphotericin B 2.5mg

Preparation of Antibiotic Solution: Add Penicillin G and Amphotericin B to distilled/deionized water and bring volume to 10.0mL. Mix thoroughly. Filter sterilize.

Urea Solution:

Composition per30mL:

Urea 3.0g

Preparation of Urea Solution: Add urea to distilled/deionized water and bring volume to 30.0mL. Mix thoroughly. Filter sterilize.

Phenol Red Solution:

Composition per 10mL:

Phenol Red 0.1g

Preparation of Phenol Red Solution: Add Phenol Red to distilled/deionized water and bring volume to 10.0mL. Mix thoroughly. Filter sterilize.

Preparation of Medium: To 95.0mL of cooled, sterile U9 base, aseptically add 5.0mL of sterile horse serum, 1.0mL of sterile antibiotic solution, 0.5mL of sterile urea solution, and 0.1mL of sterile Phenol Red solution. Mix thoroughly. Aseptically distribute into sterile tubes or flasks.

Use: For the isolation and identification of T-strain mycoplasmas from clinical specimens, especially *Ureaplasma urealyticum*. T-mycoplasmas are the only members of the *Mycoplasma* group known to contain urease. Bacteria with urease activity turn the medium dark pink.

U9B Broth

Composition per 102.1mL:

U9 base	95.0mL
Horse serum, unheated	5.0mL
Penicillin G solution	1.0mL
Urea solution	0.5mL
Cysteine·HCl·H_2O solution	0.5mL
Phenol Red solution	0.1mL

pH 6.0 ± 0.2 at 25°C

U9 Base:

Composition per 100mL:

NaCl	0.63g
Pancreatic digest of casein	0.425g
Papaic digest of soybean meal	0.075g
K_2HPO_4	0.063g
Glucose	0.063g
KH_2PO_4	0.02g

Preparation of U9 Base: Add components to distilled/deionized water and bring volume to 100.0mL. Mix thoroughly. Adjust pH to 5.5 with 1*N* HCl. Autoclave for 15 min at 15 psi pressure–121°C. Cool to 45°–50°C.

Penicillin G Solution:

Composition per 10mL:

Penicillin G	0.63 g

Preparation of Penicillin G Solution: Add Penicillin G to distilled/deionized water and bring volume to 10.0mL. Mix thoroughly. Filter sterilize.

Urea Solution:

Composition per 30mL:

Urea	3.0g

Preparation of Urea Solution: Add urea to distilled/deionized water and bring volume to 30.0mL. Mix thoroughly. Filter sterilize.

Cysteine·HCl·H_2O Solution:

Composition per50mL:

Cysteine·HCl·H_2O	1.0g

Preparation of Cysteine·HCl·H_2O Solution: Add urea to distilled/deionized water and bring volume to 50.0mL. Mix thoroughly. Filter sterilize.

Phenol Red Solution:

Composition per 10mL:

Phenol Red	0.1g

Preparation of Phenol Red Solution: Add Phenol Red to distilled/deionized water and bring volume to 10.0mL. Mix thoroughly. Filter sterilize.

Preparation of Medium: To 95.0mL of cooled, sterile U9 base, aseptically add 5.0mL of sterile horse serum, 1.0mL of sterile penicillin G solution, 0.5mL of sterile urea solution, 0.5mL of sterile cysteine·HCl·H_2O solution and 0.1mL of sterile Phenol Red solution. Mix thoroughly. Aseptically distribute into sterile tubes or flasks.

Use: For the isolation and identification of T-strain mycoplasmas from clinical specimens, especially *Ureaplasma urealyticum*. T-mycoplasmas are the only members of the *Mycoplasma* group known to contain urease. Bacteria with urease activity turn the medium dark pink.

U9C Broth

Composition per 102mL:

U9C base	90.0mL
Horse serum, unheated	10.0mL
Penicillin G solution	1.0mL
Urea solution	0.3mL
Cysteine·HCl·H_2O solution	0.5mL
GHL tripeptide solution	0.1mL
Phenol Red solution	0.1mL

pH 6.0 ± 0.2 at 25°C

U9C Base:

Composition per 100mL:

NaCl	0.85g
Pancreatic digest of casein	0.25g
Papaic digest of soybean meal	0.15g
K_2HPO_4	0.12g
Glucose	0.12g
$MgCl_2·6H_2O$	0.2g
Yeast extract	0.1g
KH_2PO_4	0.02g

Preparation of U9C Base: Add components to distilled/deionized water and bring volume to 100.0mL. Mix thoroughly. Adjust pH to 5.5 with 2*N* HCl. Autoclave for 15 min at 15 psi pressure–121°C. Cool to 45°–50°C.

Penicillin G Solution:

Composition per 10mL:

Penicillin G	0.63 g

Preparation of Penicillin G Solution: Add Penicillin G to distilled/deionized water and bring volume to 10.0mL. Mix thoroughly. Filter sterilize.

Urea Solution:

Composition per 30mL:

Urea	3.0g

Preparation of Urea Solution: Add urea to distilled/deionized water and bring volume to 30.0mL. Mix thoroughly. Filter sterilize.

Cysteine·HCl·H_2O Solution:

Composition per 50mL:

Cysteine·HCl·H_2O	1.0g

Preparation of Cysteine·HCl·H_2O Solution: Add urea to distilled/deionized water and bring volume to 50.0mL. Mix thoroughly. Filter sterilize.

GHL Tripeptide Solution:

Composition per 10mL:

GHL tripeptide	0.2mg

Preparation of GHL Tripeptide Solution: Add GHL tripeptide (glycyl-L-histidyl-L-lysine acetate) to distilled/deionized water and bring volume to 10.0mL. Mix thoroughly. Filter sterilize.

Phenol Red Solution:

Composition per 10mL:

Phenol Red	0.1g

Preparation of Phenol Red Solution: Add Phenol Red to distilled/deionized water and bring volume to 10.0mL. Mix thoroughly. Filter sterilize.

Preparation of Medium: To 90.0mL of cooled, sterile U9C base, aseptically add 10.0mL of sterile horse serum, 1.0mL of sterile penicillin G solution, 0.3mL of sterile urea solution, 0.5mL of sterile cysteine·HCl·H_2O solution, 0.1mL of sterile GHL tripeptide solution and 0.1mL of sterile Phenol Red solution. Mix thoroughly. Aseptically distribute into sterile tubes or flasks.

Use: For the isolation and identification of T-strain mycoplasmas from clinical specimens, especially *Ureaplasma urealyticum*. T-mycoplasmas are the only members of the *Mycoplasma* group known to contain urease. Bacteria with urease activity turn the medium dark pink.

UBA Medium (Universal Beer Agar)

Composition per liter:

Glucose	16.1g
Peptonized milk	15.0g
Agar	12.0g
Tomato juice, dessicated	12.2g
Yeast extract	6.1g
K_2HPO_4	0.31g
KH_2PO_4	0.31g
$MgSO_4·7H_2O$	0.12g
$FeSO_4$	6.0mg
$MnSO_4·5H_2O$	6.0mg
NaCl	6.0mg
Beer	250.0mL

pH 6.3 ± 0.2 at 25°C

Source: This medium is available as a premixed powder from Difco Laboratories and Oxoid Unipath.

Preparation of Medium: Add components, except beer, to distilled/deionized water and bring volume to 750.0mL. Mix thoroughly. Gently heat and bring to boiling. Add beer. Mix thoroughly. Distribute into tubes or flasks. Autoclave for 10 min at 15 psi pressure–121°C. Pour into sterile Petri dishes or leave in tubes.

Use: For cultivation of microorganisms of significance in the brewing industry.

Universal Beer Agar

Composition per liter:

Peptonized milk	15.0g
Agar	12.0g
Yeast extract	10.0g
Glucose	10.0g
Tomato juice solids	7.0g
K_2HPO_4	0.5g
KH_2PO_4	0.5g
$MgSO_4·5H_2O$	0.2g
NaCl	0.01g
$FeSO_4·7H_2O$	0.01g
$MnSO_4·H_2O$	0.01g
Beer	250.0mL

pH 6.3 ± 0.2 at 25°C

Source: This medium is available as a premixed powder from BBL Microbiology Systems.

Preparation of Medium: Add components, except beer, to distilled/deionized water and bring volume to 750.0mL. Mix thoroughly. Gently heat and bring to boiling. Add beer. Mix thoroughly. Distribute into tubes or flasks. Autoclave for 10 min at 15 psi pressure–121°C. Cool to 50°C. Aseptically add 250.0mL beer without degassing. Pour into sterile Petri dishes or leave in tubes

Use: For the enumeration of contaminating bacteria and yeasts encountered in wort and beer.

Universal Beer Agar
See: UBA Medium

University of Vermont *Listeria* Enrichment Broth
See: UVM *Listeria* Enrichment Broth

University of Vermont I *Listeria* Primary Selective Enrichment Broth
See: *Listeria* Enrichment Broth I, USDA FSIS

University of Vermont II *Listeria* Primary Selective enrichment Broth
See: *Listeria* Enrichment Broth II, USDA FSIS

University of Vermont Modified *Listeria* Enrichment Broth *See:* UVM Modified *Listeria* Enrichment Broth

Urea Agar (Urease Test Agar) (Urea Agar Base, Christensen)

Composition per liter:

Urea	20.0g
Agar	15.0g
NaCl	5.0g
KH_2PO_4	2.0g
Peptone	1.0g
Glucose	1.0g
Phenol Red	0.012g

pH 6.8 ± 0.2 at 25°C

Source: This medium is available as a premixed powder from BBL Microbiology Systems and Difco Laboratories.

Preparation of Medium: Add components, except agar, to distilled/deionized water and bring volume to 100.0mL. Mix thoroughly. Filter sterilize. Add agar to distilled/deionized water and bring volume to 900.0mL. Mix thoroughly. Gently heat and bring to boiling. Autoclave for 15 min at 15 psi pressure–121°C. Cool to 50°C. Aseptically add the 100.0mL of sterile basal medium. Mix thoroughly. Distribute into sterile tubes. Allow tubes to solidify in a slanted position.

Use: For the differentiation of a variety of microorganisms, especially members of the Enterobacteriaceae, aerobic actinomycetes, streptococci and nonfermenting Gram-negative bacteria, on the basis of urease production.

Urea Agar Base

Composition per liter:

Agar	15.0g
NaCl	5.0g
Na_2HPO_4	1.2g
Peptone	1.0g
Glucose	1.0g
KH_2PO_4	0.8g
Phenol Red	0.012g
Urea solution	50.0mL

pH 6.8 ± 0.2 at 25°C

Source: This medium is available as a premixed powder from Oxoid Unipath.

Urea Solution:
Composition per 100mL:

Urea	40.0g

Preparation of Urea Solution: Add urea to distilled/deionized water and bring volume to 100.0mL. Mix thoroughly. Filter sterilize.

Preparation of Medium: Add components, except urea solution, to distilled/deionized water and bring volume to 950.0mL. Mix thoroughly. Gently heat and bring to boiling. Autoclave for 20 min at 10 psi pressure–115°C. Cool to 50°C. Aseptically add 50.0mL of sterile urea solution. Mix thoroughly. Pour into sterile Petri dishes or distribute into sterile tubes. Allow tubes to solidify in a slanted position.

Use: For the detection of *Proteus* species based on rapid urease activity and the identification of other members of the Enterobacteriaceae based on urease activity. Urease positive bacteria turn the medium pink.

Urea Agar Base, Christensen *See:* Urea Agar

Urea Broth *See*: Urease Test Broth

Urea Broth 10B for *Ureaplasma urealyticum*

Composition per 100.5mL:

PPLO broth without Crystal Violet	70.0mL
Horse serum, unheated	20.0mL
Fresh yeast extract solution	10.0mL
L-Cysteine·HCl·H_2O	0.5mL
CVA enrichment	0.5mL
Urea solution	0.4mL
Phenol Red	0.1mL

PPLO Broth without Crystal Violet:
Composition per 900mL:

Beef heart, solids from infusion	16.1g
Peptone	3.25g
NaCl	1.61g

Preparation of PPLO Broth without Crystal Violet: Add components to distilled/deionized water and bring volume to 900mL. Adjust pH to 5.5 with 2*N* HCl. Autoclave for 15 min at 15 psi pressure–121°C. Cool to 37°C.

Fresh Yeast Extract Solution:
Composition per 100mL:

Baker's yeast live, pressed, starch-free	25.0g

Preparation of Fresh Yeast Extract Solution: Add the live Baker's yeast to 100.0mL of distilled/deionized water. Autoclave for 90 min at 15 psi pressure–121°C. Allow to stand. Remove supernatant solution. Adjust pH to 6.6–6.8.

Cysteine·HCl·H_2O Solution:
Composition per 50mL:
Cysteine·HCl·H_2O 1.0g

Preparation of Cysteine·HCl·H_2O Solution: Add cysteine·HCl·H_2O to distilled/deionized water and bring volume to 50.0mL. Mix thoroughly. Filter sterilize.

CVA Enrichment:
Composition per liter:

Glucose	100.0g
L-Cysteine·HCl·H_2O	25.9g
L-Glutamine	10.0g
Adenine	1.0g
L-Cystine·2HCl	1.0g
Nicotinamide adenine dinucleotide	0.25g
Cocarboxylase	0.1g
Guanine·HCl	0.03g
$Fe(NO_3)_3$	0.02g
Vitamin B_{12}	0.01g
p-Aminobenzoic acid	0.013g
Thiamine·HCl	3.0mg

Preparation of CVA Enrichment: Add components to distilled/deionized water and bring volume to 1.0L. Mix thoroughly. Filter sterilize.

Urea Solution:
Composition per 30mL:
Urea 3.0g

Preparation of Urea Solution: Add urea to distilled/deionized water and bring volume to 30.0mL. Mix thoroughly. Filter sterilize.

Preparation of Medium: Aseptically combine the components, except the PPLO broth without Crystal Violet. Aseptically add this mixture to the cooled, sterile PPLO broth without Crystal Violet. Mix thoroughly. Aseptically distribute into sterile tubes or flasks.

Use: For the cultivation and maintenance of *Ureaplasma urealyticum* and other *Ureaplasma* species. Urease positive bacteria turn the medium to peach orange.

Urea Broth Base

Composition per liter:

NaCl	5.0g
Na_2HPO_4	1.2g
Peptone	1.0g
Glucose	1.0g
KH_2PO_4	0.8g
Phenol Red	0.012g
Urea solution	50.0mL

pH 6.8 ± 0.2 at 25°C

Source: This medium is available as a premixed powder from Oxoid Unipath.

Urea Solution:
Composition per 100mL:
Urea 40.0g

Preparation of Urea Solution: Add urea to distilled/deionized water and bring volume to 100.0mL. Mix thoroughly. Filter sterilize.

Preparation of Medium: Add components, except urea solution, to distilled/deionized water and bring volume to 950.0mL. Mix thoroughly. Autoclave for 20 min at 10 psi pressure–115°C. Cool to 50°C. Aseptically add 50.0mL of sterile urea solution. Mix thoroughly. Aseptically distribute into sterile tubes or flasks.

Use: For the differentiation of members of the Enterobacteriaceae based on urease production. Urease positive bacteria turn the medium pink.

Urea R Broth (Urea Rapid Broth)

Composition per liter:

Urea	20.0g
Yeast extract	0.1g
Na_2HPO_4	0.095g
KH_2PO_4	0.091g
Phenol Red	0.01g

pH 6.9 ± 0.2 at 25°C

Source: Available as a prepared medium from Difco Laboratories.

Preparation of Medium: Add components to distilled/deionized water and bring volume to 1.0L. Mix thoroughly. Filter sterilize. Aseptically distribute into sterile tubes or flasks.

Use: For the differentiation of members of the Enterobacteriaceae based on rapid detection of urease activity. Urease positive bacteria turn the medium cerise.

Urea Semisolid Medium

Composition per liter:

Solution A	400.0mL
Solution B	50.0mL

Solution A:

Composition per 400mL:

Pancreatic digest of casein	6.0g
Yeast extract	2.0g
NaCl	1.0g
Yeast extract	0.8g
Agar	0.3g
L-Cystine	0.1g
Thioglycollic acid	0.12mL

pH 7.2 ± 0.2 at 25°C

Preparation of Solution A: Add components to distilled/deionized water and bring volume to 400.0mL. Mix thoroughly. Gently heat and bring to boiling. Autoclave for 15 min at 15 psi pressure–121°C. Cool to 60°C.

Solution B:

Composition per 50mL:

Urea	8.0g
Na_2HPO_4	3.8g
KH_2PO_4	3.64g
Yeast extract	0.04g
Phenol Red	4.0mg

Preparation of Solution B: Add components to distilled/deionized water and bring volume to 50.0mL. Mix thoroughly. Filter sterilize.

Preparation of Medium: Aseptically combine 400.0mL of sterile solution A and 50.0mL of sterile solution B. Mix thoroughly. Aseptically distribute into sterile screw-capped tubes in 7.0mL volumes. Pass the tubes into an anaerobic chamber containing 85% N_2 + 10% H_2 + 5% CO_2 for 60 min. Close screw caps tightly.

Use: For the cultivation and differentiation of anaerobic bacteria based on their production of urease. Bacteria that produce urease turn the medium bright red.

Urea Test Broth

Composition per liter:

Urea	20.0g
Na_2HPO_4	9.5g
KH_2PO_4	9.1g
Yeast extract	0.1g
Phenol Red	0.01g
Urea solution	100.0mL

Urea Solution:

Composition per 100mL:

Urea	20.0g

Preparation of Urea Solution: Add urea to distilled/deionized water and bring volume to 100.0mL. Mix thoroughly. Filter sterilize.

Preparation of Medium: Add components, except urea solution, to distilled/deionized water and bring volume to 900.0mL. Mix thoroughly. Autoclave for 15 min at 15 psi pressure–121°C. Cool to 45°–50°C. Aseptically add sterile urea solution. Mix thoroughly. Aseptically distribute into sterile tubes in 3.0mL volumes.

Use: For the cultivation and differentiation of members of the Enterobacteriaceae and aerobic actinomycetes based on their production of urease. Bacteria that produce urease turn the medium bright red.

Urea Test Broth

Composition per 99.6mL:

H broth base	85.0mL
Horse serum	10.0mL
Penicillin solution	2.0mL
MES (2-N-Morpholinoethane sulfonic acid) buffer solution	1.0mL
Na_2SO_3 solution	1.0mL
Urea solution	0.5mL
Phenol Red solution	0.1mL

pH 7.2 ± 0.2 at 25°C

H Broth Base:

Composition per liter:

NaCl	5.0g
Pancreatic digest of casein	5.0g
Peptone	5.0g
Beef extract	3.0g
K_2HPO_4	2.5g
Glucose	1.0g

Preparation of H Broth Base: Add components to distilled/deionized water and bring volume to 1.0L. Mix thoroughly. Gently heat and bring to boiling. Distribute into tubes in 4.0mL volumes. Autoclave for 15 min at 10 psi pressure–115°C. Cool to 45°–50°C.

Penicillin Solution:

Composition per 10mL:

Penicillin	100,000U

Preparation of Penicillin Solution: Add penicillin to distilled/deionized water and bring volume to 10.0mL. Mix thoroughly. Filter sterilize.

MES Buffer Solution:

Composition per 100mL:

MES (2-N-Morpholinoethane sulfonic acid) buffer	19.52g

Preparation of MES Buffer Solution: Add MES buffer to distilled/deionized water and bring volume to 100.0mL. Mix thoroughly. Adjust pH to 5.5. Filter sterilize.

Na_2SO_3 Solution:
Composition per 10mL:

Na_2SO_3	0.126g

Preparation of Na_2SO_3 Solution: Add Na_2SO_3 to distilled/deionized water and bring volume to 10.0mL. Mix thoroughly. Filter sterilize.

Urea Solution:
Composition per 100mL:

Urea	6.0g

Preparation of Urea Solution: Add urea to distilled/deionized water and bring volume to 100.0mL. Mix thoroughly. Filter sterilize.

Phenol Red Solution:
Composition per 10mL:

Phenol Red	0.1g

Preparation of Phenol Red Solution: Add Phenol Red to distilled/deionized water and bring volume to 10.0mL. Mix thoroughly. Filter sterilize.

Preparation of Medium: To 85.0mL of cooled sterile H broth base, aseptically add 10.0mL of sterile horse serum, 2.0mL of sterile penicillin solution, 1.0mL of MES buffer solution, 1.0mL of Na_2SO_3 solution, 0.5mL of urea solution, and 0.1 mL of sterile Phenol Red solution. Mix thoroughly. Aseptically distribute into test tubes in 3.0mL volumes.

Use: For the cultivation and differentiation of *Ureaplasma* species based on their production of urease.

Urease Color Test Medium
See: **U9 Broth**

Urease Indole Test Broth
See: **F35M Hajna Broth**

Urease Test Agar
See: **Urea Agar**

Urease Test Broth (Urea Broth)

Composition per liter:

Urea	20.0g
Na_2HPO_4	9.5g
KH_2PO_4	9.1g
Yeast extract	0.1g
Phenol Red	0.01g

pH 6.8 ± 0.2 at 25°C

Source: This medium is available as a premixed powder from BBL Microbiology Systems and Difco Laboratories.

Preparation of Medium: Add components to distilled/deionized water and bring volume to 1.0L. Mix thoroughly. Filter sterilize. Aseptically distribute into sterile tubes or flasks.

Use: For the differentiation of organisms, especially the Enterobacteriaceae, on the basis of urease production. Urease positive bacteria turn the medium pink.

Uric Acid Agar

Composition per liter:

Agar	20.0g
Uric acid	10.0g
KH_2PO_4	0.5g

pH 7.0 ± 0.2 at 25°C

Preparation of Medium: Add components to distilled/deionized water and bring volume to 1.0L. Mix thoroughly. Adjust pH to 7.0. Gently heat and bring to boiling. Distribute into tubes or flasks. Autoclave for 15 min at 15 psi pressure–121°C. Pour into sterile Petri dishes or leave in tubes. Swirl flask while pouring medium.

Use: For the cultivation and maintenance of microorganisms, such as *Bacillus fastidiosus*, that can utilize uric acid as sole source of carbon, nitrogen and energy.

Uric Acid Agar for Clostridia

Composition per liter:

Agar	20.0g
K_2HPO_4	4.0g
Uric acid	3.0g
Yeast extract	1.0g
Sodium thioglycollate	0.5g
$MgSO_4 \cdot 7H_2O$	0.1g
$FeSO_4 \cdot 7H_2O$	5.0mg
Phenol Red (0.04% solution)	1.0mL

pH 7.6-8.0 at 25°C

Preparation of Medium: Add components, except uric acid, to approximately 900.0mL of distilled/deionized water. For a semi-solid agar use 2.0g agar instead of 20.0g. Mix thoroughly. Gently heat and bring to boiling. Adjust pH to 7.6 with 1*N* NaOH. Add the uric acid. Mix thoroughly. Adjust pH to 7.6. Distribute into tubes or flasks. Autoclave for 15 min at 15 psi pressure–121°C. Pour into sterile Petri dishes or leave in tubes.

Use: For the cultivation and maintenance of anaerobic bacteria, such as *Clostridium acidiurici* and *Clostridium cylindrosporum*, which can utilize uric acid as sole source of carbon and energy.

Uric Acid Broth for Clostridia

Composition per liter:

K_2HPO_4	4.0g
Uric acid	3.0g
Yeast extract	1.0g
Sodium thioglycollate	0.5g
$MgSO_4 \cdot 7H_2O$	0.1g
$FeSO_4 \cdot 7H_2O$	5.0mg
Phenol Red (0.04% solution)	1.0mL

pH 7.6-8.0 at 25°C

Preparation of Medium: Add components, except uric acid, to approximately 900.0mL of distilled/deionized water. Mix thoroughly. Gently heat and bring to boiling. Adjust pH to 7.6 with 1*N* NaOH. Add the uric acid. Mix thoroughly. Adjust pH to 7.6. Distribute into tubes or flasks. Autoclave for 15 min at 15 psi pressure–121°C.

Use: For the cultivation and maintenance of anaerobic bacteria, such as *Clostridium acidiurici* and *Clostridium cylindrosporum*, which can utilize uric acid as sole source of carbon and energy.

Uric Acid Utilization Agar

Composition per liter:

Agar	15.0g
Uric acid	10.0g
K_2HPO_4	0.5g

pH 7.2 ± 0.2 at 25°C

Preparation of Medium: Add components to distilled/deionized water and bring volume to 1.0L. Mix thoroughly. Gently heat and bring to boiling. Distribute into tubes or flasks. Autoclave for 15 min at 15 psi pressure–121°C. Pour into sterile Petri dishes or leave in tubes.

Use: For the cultivation and maintenance of anaerobic bacteria, such as *Bacillus fastidiosus*, which can utilize uric acid as sole source of carbon and energy.

USP Alternative Thioglycollate Medium
See: **Sterility Test Broth**

Ustilago Medium

Composition per liter:

Yeast extract	11.0g
Glucose	10.0g
NH_4NO_3	1.5g
Salt solution	62.5mL
Vitamin solution	10.0mL

Salt Solution:

Composition per liter:

KH_2PO_4	16.0g
KCl	8.0g
Na_2SO_4	4.0g
$MgSO_4 \cdot 7H_2O$	2.0g
$CaCl_2$	1.0g
Trace elements solution	8.0mL

Preparation of Salt Solution: Add components to distilled/deionized water and bring volume to 1.0L. Mix thoroughly.

Trace Elements Solution:

Composition per 500mL:

$CuSO_4 \cdot 5H_2O$	0.2g
$ZnCl_2$	0.2g
$MnCl_2 \cdot 4H_2O$	0.07g
$FeCl_3 \cdot 6H_2O$	0.05g
H_3BO_3	0.03g
$Na_2MoO_4 \cdot 2H_2O$	0.02g

Preparation of Trace Elements Solution: Add components to distilled/deionized water and bring volume to 500.0mL. Mix thoroughly.

Vitamin Solution:

Composition per liter:

Inositol	0.4g
Calcium pantothenate	0.2g
Choline chloride	0.2g
Nicotinic acid	0.2g
Thiamine	0.1g
Pyridoxine	0.05g
Riboflavin	0.05g

Preparation of Vitamin Solution: Add components to distilled/deionized water and bring volume to 1.0L. Mix thoroughly. Filter sterilize.

Preparation of Medium: Add components, except vitamin solution, to distilled/deionized water and bring volume to 990.0mL. Mix thoroughly. Gently heat and bring to boiling. Autoclave for 15 min at 15 psi pressure–121°C. Cool to 45°–50°C. Aseptically add 10.0mL of sterile vitamin solution. Mix thoroughly. Aseptically distribute into sterile tubes or flasks.

Use: For the cultivation of *Ustilago* species.

Ustilago Minimal Medium

Composition per liter:

Glucose 10.0g
KNO_3 .... 3.0g
Salt solution 62.5mL

Salt Solution:

Composition per liter:

KH_2PO_4 .... 16.0g
KCl 8.0g
Na_2SO_4 .... 4.0g
$MgSO_4 \cdot 7H_2O$ 2.0g
$CaCl_2$ 1.0g
Trace elements solution 8.0mL

Preparation of Salt Solution: Add components to distilled/deionized water and bring volume to 1.0L. Mix thoroughly.

Trace Elements Solution:

Composition per 500mL:

$CuSO_4 \cdot 5H_2O$ 0.2g
$ZnCl_2$ 0.2g
$MnCl_2 \cdot 4H_2O$ 0.07g
$FeCl_3 \cdot 6H_2O$ 0.05g
H_3BO_3 .... 0.03g
$Na_2MoO_4 \cdot 2H_2O$ 0.02g

Preparation of Trace Elements Solution: Add components to distilled/deionized water and bring volume to 500.0mL. Mix thoroughly.

Preparation of Medium: Add components to distilled/deionized water and bring volume to 1.0L. Mix thoroughly. Distribute into tubes or flasks. Autoclave for 15 min at 15 psi pressure–121°C.

Use: For the cultivation of *Ustilago* species.

UVM *Listeria* Enrichment Broth (University of Vermont *Listeria* Enrichment Broth)

Composition per liter:

NaCl 20.0g
Na_2HPO_4 .... 9.6g
Pancreatic digest of casein 5.0g
Peptic digest of animal tissue 5.0g
Beef extract 5.0g
Yeast extract 5.0g
KH_2PO_4 .... 1.35g
Esculin 1.0g
Nalidixic acid 40.0mg
Acriflavine·HCl 12.0mg

pH 7.2 ± 0.2 at 25°C

Preparation of Medium: Add components to distilled/deionized water and bring volume to 1.0L. Mix thoroughly. Distribute into tubes or flasks. Autoclave for 15 min at 15 psi pressure–121°C.

Use: For the selective isolation of *Listeria monocytogenes.*

UVM Modified *Listeria* Enrichment Broth (University of Vermont Modified *Listeria* Enrichment Broth)

Composition per liter:

NaCl 20.0g
Na_2HPO_4 .... 9.6g
Pancreatic digest of casein 5.0g
Peptic digest of animal tissue 5.0g
Beef extract 5.0g
Yeast extract 5.0g
KH_2PO_4 .... 1.35g
Esculin 1.0g
Nalidixic acid 20.0mg
Acriflavine·HCl 12.0mg

pH 7.2 ± 0.2 at 25°C

Source: This medium is available as a premixed powder from BBL Microbiology Systems.

Preparation of Medium: Add components to distilled/deionized water and bring volume to 1.0L. Mix thoroughly. Distribute into tubes or flasks. Autoclave for 15 min at 15 psi pressure–121°C.

Use: For the selective isolation of *Listeria monocytogenes.*

V–8™ Agar

Composition per liter:

Agar	20.0g
$CaCO_3$	4.0g
V-8 canned vegetable juice	200.0mL

pH 7.3 ± 0.2 at 25°C

Preparation of Medium: Add components to distilled/deionized water and bring volume to 1.0L. Mix thoroughly. Gently heat and bring to boiling. Distribute into tubes or flasks. Autoclave for 15 min at 15 psi pressure–121°C. Pour into sterile Petri dishes or leave in tubes.

Use: For the isolation and cultivation of *Actinomadura* species, *Actinopolyspora* species, *Excellospora* species and *Microspora* species.

V Agar

Composition per liter:

Agar	13.5g
Pancreatic digest of casein	12.0g
Peptone	10.0g
Peptic digest of animal tissue	5.0g
NaCl	5.0g
Beef extract	3.0g
Yeast extract	3.0g
Cornstarch	1.0g
Human blood, anticoagulated	50.0mL

pH 7.4 ± 0.2 at 25°C

Source: This medium is available as a prepared medium from BBL Microbiology Systems.

Preparation of Medium: Add components, except human blood, to distilled/deionized water and bring volume to 950.0mL. Mix thoroughly. Gently heat and bring to boiling. Distribute into tubes or flasks. Autoclave for 15 min at 15 psi pressure–121°C. Cool to 50°C. Aseptically add 50.0mL of human blood. Mix thoroughly. Pour into sterile Petri dishes or leave in tubes.

Use: For the isolation and differentiation of *Gardnerella vaginalis* from clinical specimens. Plates are incubated under an atmosphere with 3-10% CO_2. *G. vaginalis* appears as small white colonies with diffuse β-hemolysis.

Van Niel's Medium, Modified

Composition per liter:

Yeast extract	10.0g
$MgSO_4$	0.1g
EDTA	2.0mg
Trace elements solution	10.0mL
K_2HPO_4 solution	2.5mL

pH 7.1 ± 0.2 at 25°C

Trace Elements Solution:
Composition per 100mL:

$CaCl_2 \cdot 2H_2O$	0.3g
Ferric ammonium citrate	0.2g

Preparation of Trace Elements Solution: Add components to distilled/deionized water and bring volume to 100.0mL. Mix thoroughly. Filter sterilize.

K_2HPO_4 Solution:
Composition per 100mL:

K_2HPO_4	4.0g

Preparation of K_2HPO_4 Solution: Add K_2HPO_4 to distilled/deionized water and bring volume to 100.0mL. Mix thoroughly. Filter sterilize.

Preparation of Medium: Add components, except K_2HPO_4 and trace elements solution, to distilled/deionized water and bring volume to 987.5mL. Mix thoroughly. Autoclave for 15 min at 15 psi pressure–121°C. Cool to 50°C. Aseptically add trace elements and K_2HPO_4 solutions. Mix thoroughly. Distribute into sterile tubes or flasks.

Use: For the cultivation and maintenance of *Rhodobacter sphaeroides*.

Van Niel's Yeast Medium with Pyruvate, Modified

Composition per liter:

Yeast extract	10.0g
$MgSO_4$	0.1g
EDTA	2.0mg
Sodium pyruvate solution	100.0mL
Trace elements solution	10.0mL
K_2HPO_4 solution	5.0mL
Trace metal A-5 solution	1.0mL

pH 7.1± 0.1 at 25°C

Sodium Pyruvate Solution:
Composition per 100mL:

Sodium pyruvate	1.1g

Preparation of K_2HPO_4 Solution: Add sodium pyruvate to distilled/deionized water and bring volume to 100.0mL. Mix thoroughly. Filter sterilize.

Trace Elements Solution:
Composition per 100 mL:

$CaCl_2 \cdot 2H_2O$	0.3g
Ferric ammonium citrate	0.2g

Preparation of Trace Elements Solution: Add components to distilled/deionized water and bring volume to 100.0mL. Mix thoroughly. Filter sterilize.

K_2HPO_4 Solution:
Composition per 100mL:

K_2HPO_4	4.0g

Preparation of K_2HPO_4 Solution: Add K_2HPO_4 to distilled/deionized water and bring volume to 100.0mL. Mix thoroughly. Filter sterilize.

Trace Metal A-5 Solution:

Composition per liter:

H_3BO_3	2.86g
$MnCl_2 \cdot 4H_2O$	1.81g
$Na_2MoO_4 \cdot 2H_2O$	0.39g
$ZnSO_4 \cdot 7H_2O$	0.222g
$CuSO_4 \cdot 5H_2O$	0.079g
$Co(NO_3)_2 \cdot 6H_2O$	0.049.g

Preparation of Trace Metal A-5 Solution: Add components to distilled/deionized water and bring volume to 1.0L. Mix thoroughly. Filter sterilize.

Preparation of Medium: Add components, except sodium pyruvate, trace elements, K_2HPO_4 and trace metal A-5 solutions, to distilled/deionized water and bring volume to 884.0mL. Mix thoroughly. Autoclave for 15 min at 15 psi pressure–121°C. Cool to 25°C. Aseptically add 100.0mL of sterile sodium pyruvate solution, 10.0mL of sterile trace elements soultion, 5.0mL of sterile K_2HPO_4 solution, and 1.0mL of sterile trace metal A-5 solution. Mix thoroughly. Adjust pH to 7.1± 0.1. Aseptically distribute into sterile tubes or flasks.

Use: For the cultivation and maintenance of photosynthetic bacteria, such as *Heliobacillus mobilis* and *Rhodopseudomonas palustris*.

Veal Infusion Agar

Composition per liter:

Agar	15.0g
Veal, infusion from	10.0g
Pancreatic digest of casein	5.0g
Peptic digest of animal tissue	5.0g
NaCl	5.0g

pH 7.4 ± 0.2 at 25°C

Source: This medium is available as a premixed powder from BBL Microbiology Systems.

Preparation of Medium: Add components to distilled/deionized water and bring volume to 1.0L. Mix thoroughly. Gently heat and bring to boiling. Distribute into tubes or flasks. Autoclave for 15 min at 15 psi pressure–121°C. Pour into sterile Petri dishes or leave in tubes.

Use: For the cultivation and maintenance of a variety of microorganisms. Can be used for cultivation of fastidious microorganisms when enriched with blood or serum.

Veal Infusion Agar (ATCC Medium 521)

Composition per liter:

Veal, infusion from	500.0g
Agar	15.0g
Pancreatic digest of casein	5.0g
Peptic digest of animal tissue	5.0g
NaCl	5.0g

pH 7.4 ± 0.2 at 25°C

Source: This medium is available as a premixed powder from Difco Laboratories.

Preparation of Medium: Add components to distilled/deionized water and bring volume to 1.0L. Mix thoroughly. Gently heat and bring to boiling. Distribute into tubes or flasks. Autoclave for 15 min at 15 psi pressure–121°C. Pour into sterile Petri dishes or leave in tubes.

Use: For the cultivation and maintenance of a variety of microorganisms. Can be used for cultivation of fastidious microorganisms when enriched with blood or serum.

Veal Infusion Broth (ATCC Medium 521)

Composition per liter:

Veal, infusion from	500.0g
Pancreatic digest of casein	5.0g
Peptic digest of animal tissue	5.0g
NaCl	5.0g

pH 7.4 ± 0.2 at 25°C

Source: This medium is available as a premixed powder from Difco Laboratories.

Preparation of Medium: Add components to distilled/deionized water and bring volume to 1.0L. Mix thoroughly. Distribute into tubes or flasks. Autoclave for 15 min at 15 psi pressure–121°C. Use freshly prepared solution.

Use: For the cultivation and maintenance of *Arthrobacter* species, streptococci and other microorganisms.

Veal Infusion Broth

Composition per liter:

Veal, infusion from	10.0g
Pancreatic digest of casein	5.0g
Peptic digest of animal tissue	5.0g
NaCl	5.0g

pH 7.4 ± 0.2 at 25°C

Source: This medium is available as a premixed powder from BBL Microbiology Systems.

Preparation of Medium: Add components to distilled/deionized water and bring volume to 1.0L. Mix thoroughly. Gently heat and bring to boiling. Distribute into tubes or flasks. Autoclave for 15 min at 15 psi pressure–121°C. Use freshly prepared solution.

Use: For the cultivation of streptococci and other microorganisms.

Veal Infusion Broth with Horse Serum

Composition per liter:

Veal, infusion from	500.0g
Pancreatic digest of casein	5.0g
Peptic digest of animal tissue	5.0g
NaCl	5.0g
Horse serum, heat-inactivated	100.0mL

pH 7.4 ± 0.2 at 25°C

Preparation of Medium: Add components, except horse serum, to distilled/deionized water and bring volume to 900.0mL. Mix thoroughly. Gently heat and bring to boiling. Autoclave for 15 min at 15 psi pressure–121°C.Cool to 50°C. Aseptically add 100.0mL horse serum. Mix thoroughly. Aseptically distribute into tubes or flasks. Use freshly prepared solution or boil without mixing prior to use.

Use: For the cultivation and maintenance of *Streptococcus pyogenes*.

Veal Infusion Broth with Rabbit Serum

Composition per liter:

Veal, infusion from	500.0g
Pancreatic digest of casein	5.0g
Peptic digest of animal tissue	5.0g
NaCl	5.0g
Rabbit serum, heat-inactivated	150.0mL

pH 7.4 ± 0.2 at 25°C

Preparation of Medium: Add components, except rabbit serum, to distilled/deionized water and bring volume to 850.0mL. Mix thoroughly. Gently heat and bring to boiling. Autoclave for 15 min at 15 psi pressure–121°C. Cool to 50°C. Aseptically add 150.0mL rabbit serum. Mix thoroughly. Aseptically distribute into sterile tubes or flasks. Boil without agitation prior to use. Use freshly prepared solution or boil without mixing prior to use.

Use: For the cultivation and maintenance of *Proteus mirabilis*.

Veal Yeast Extract Medium
See: **VY Medium**

Veillonella Agar

Composition per liter:

Agar	15.0g
Pancreatic digest of casein	5.0g
Yeast extract	3.0g
Sodium thioglycollate	0.75g
Vancomycin	7.5mg
Basic Fuchsin	2.0mg
Sodium lactate (60% solution)	21.0mL

pH 7.5± 0.2 at 25°C

Source: This medium is available as a premixed powder from Difco Laboratories.

Caution: Basic Fuchsin is a potential carcinogen and care must be taken to avoid inhalation of the powdered dye and contact with the skin.

Preparation of Medium: Add components, except vancomycin, to distilled/deionized water and bring volume to 1.0L. Mix thoroughly. Gently heat and bring to boiling. Distribute into tubes or flasks. Autoclave for 15 min at 15 psi pressure–121°C. Cool to 50°C. Aseptically add vancomycin. Mix thoroughly. Pour into sterile Petri dishes or leave in tubes.

Use: For the selective isolation and cultivation of *Veillonella* species.

Veillonella Medium

Composition per liter:

Pancreatic digest of casein	5.0g
Yeast extract	3.0g
Tween™ 80	1.0g
Glucose	1.0g
Sodium thioglycollate	0.75g
Sodium lactate (60% solution)	21.0mL

pH 7.5 ± 0.2 at 25°C

Preparation of Medium: Add components to distilled/deionized water and bring volume to 1.0L. Mix thoroughly. Adjust pH to 7.5 with K_2CO_3. Distribute into tubes or flasks. Autoclave for 15 min at 15 psi pressure–121°C.

Use: For the cultivation and maintenance of *Veillonella* species.

Veillonella Medium, DSM

Composition per liter:

Sodium lactate (60% solution)	7.5g
Pancreatic digest of casein	5.0g

Yeast extract	3.0g
Tween™ 80	1.0g
Glucose	1.0g
Sodium thioglycollate	0.75g
Putrescine	3.0mg
Resazurin	1.0mg

pH 7.5 ± 0.2 at 25°C

Preparation of Medium: Prepare and dispense medium anaerobically under 100% N_2. Add components to distilled/deionized water and bring volume to 1.0L. Mix thoroughly. Adjust pH to 7.5 with K_2CO_3. Anaerobically distribute into tubes or flasks. Autoclave for 15 min at 15 psi pressure–121°C.

Use: For the cultivation and maintenance of *Veillonella parvula* and other *Veillonella* species.

Veillonella Selective Medium

Composition per liter:

Pancreatic digest of casein	5.0g
Yeast extract	3.0g
Tween™ 80	1.0g
Sodium thioglycollate	0.75g
Sodium lactate (50% solution)	25.0mL
Streptomycin solution	10.0mL

pH 6.6 ± 0.2 at 25°C

Streptomycin Solution:
Composition per 10mL:

Streptomycin	5.0mg

Preparation of Streptomycin Solution: Add streptomycin to distilled/deionized water and bring volume to 10.0mL. Mix thoroughly. Filter sterilize.

Preparation of Medium: Add components, except streptomycin solution, to distilled/deionized water and bring volume to 990.0mL. Mix thoroughly. Gently heat and bring to boiling. Adjust pH to 6.6 with K_2CO_3. Autoclave for 15 min at 15 psi pressure–121°C. Cool to 45°–50°C. Aseptically add sterile streptomycin solution. Mix thoroughly. Aseptically distribute into sterile tubes or flasks.

Use: For the cultivation of *Veillonella* species.

Viability–Preserving Microbiostatic Medium
See: VMGII Medium

Vibrio Agar

Composition per liter:

Sucrose	20.0g
Agar	15.0g
NaCl	10.0g
Sodium citrate·$2H_2O$	10.0g
$Na_2S_2O_3·5H_2O$	6.5g
Ox-gall	5.0g
Yeast extract	5.0g
Pancreatic digest of casein	4.0g
Proteose peptone	3.0g
Sodium deoxycholate	1.0g
Sodium lauryl sulfate	0.2g
Water Blue	0.2g
Cresol Red	0.02g

pH 8.5 ± 0.2 at 25°C

Preparation of Medium: Add components to distilled/deionized water and bring volume to 1.0L. Mix thoroughly. Adjust pH to 8.5. Gently heat and bring to boiling. Do not autoclave. Pour into sterile Petri dishes or distribute into sterile tubes.

Use: For the isolation and cultivation of the *Vibrio cholerae*.

Vibrio parahaemolyticus Agar (VP Agar)

Composition per liter:

Agar	20.0g
NaCl	20.0g
Sucrose	20.0g
Sodium citrate	10.0g
$Na_2S_2O_3·5H_2O$	10.0g
Peptone	10.0g
Sodium taurocholate	5.0g
Yeast extract	5.0g
Sodium lauryl sulfate	0.2g
Bromthymol Blue	0.04g
Thymol Blue	0.04g

pH 8.6 ± 0.2 at 25°C

Preparation of Medium: Add components to distilled/deionized water and bring volume to 1.0L. Mix thoroughly. Gently heat and bring to boiling. Do not autoclave. Pour into sterile Petri dishes.

Use: For the isolation, cultivation, enumeration and presumptive identification of coliforms in milk, food, and other specimens of sanitary significance. For enumeration of bacteria in cheese, especially *Pseudomonas fragi, Pseudomonas viscosa,* and *Alcaligenes metalcaligenes*. Sucrose-fermenting bacteria appear as yellow colonies with pale yellow peripheries. Sucrose-nonfermenting bacteria appear as mucoid, green colonies with a dark green center.

Vibrio parahaemolyticus Sucrose Agar (VPSA)

Composition per liter:

NaCl	30.0g
Agar	15.0g
Sucrose	10.0g
Yeast extract	7.0g
Tryptose	5.0g
Pancreatic digest of casein	5.0g
Bile salts No. 3	1.5g
Bromthymol Blue	0.025g

pH 8.6 ± 0.2 at 25°C

Preparation of Medium: Add components to distilled/deionized water and bring volume to 1.0L. Mix thoroughly. Gently heat and bring to boiling. Do not autoclave. Cool to 50°C. Pour into sterile Petri dishes in 20.0mL volumes. Allow plates to dry before using.

Use: For the isolation, cultivation, and differentiation of *Vibrio parahaemolyticus* from seafood. *V. parahaemolyticus* and *V. vulnificus* appear as blue to green colonies. Other *Vibrio* species appear as yellow colonies.

Violet Peptone Bile Lactose Broth

Composition per liter:

Lactose	10.0g
Peptone	10.0g
Bile salts	5.0g
Gentian Violet	0.04g

pH 7.6 ± 0.2 at 25°C

Preparation of Medium: Add components to distilled/deionized water and bring volume to 1.0L. Mix thoroughly. Gently heat and bring to boiling. Distribute into tubes or flasks. Autoclave for 15 min at 15 psi pressure–121°C. Pour into sterile Petri dishes or leave in tubes.

Use: For the selective cultivation of members of the Enterobacteriaceae.

Violet Red Bile Agar

Composition per liter:

Agar	15.0g
Lactose	10.0g
Glucose	10.0g
Pancreatic digest of gelatin	7.0g
NaCl	5.0g
Yeast extract	3.0g
Bile salts	1.5g
Neutral Red	0.03g
Crystal Violet	2.0mg

pH 7.4 ± 0.2 at 25°C

Preparation of Medium: Add components to distilled/deionized water and bring volume to 1.0L. Mix thoroughly. Gently heat while stirring and bring to boiling. Distribute into tubes or flasks. Autoclave for 15 min at 15 psi pressure–121°C. Pour immediately into sterile Petri dishes or leave in tubes.

Use: For the isolation and cultivation of members of the Enterobacteriaceae from brined vegetables. For the enumeration of members of the Enterobacteriaceae from brined vegetables by the pour plate technique.

Violet Red Bile Agar (VRB Agar)

Composition per liter:

Agar	15.0g
Lactose	10.0g
Pancreatic digest of gelatin	7.0g
NaCl	5.0g
Yeast extract	3.0g
Bile salts	1.5g
Neutral Red	0.03g
Crystal Violet	2.0mg

pH 7.4 ± 0.2 at 25°C

Source: This medium is available as a premixed powder from BBL Microbiology Systems, Difco Laboratores and Oxoid Unipath.

Preparation of Medium: Add components to distilled/deionized water and bring volume to 1.0L. Mix thoroughly. Gently heat while stirring and bring to boiling. Distribute into tubes or flasks. Autoclave for 15 min at 15 psi pressure–121°C. Pour immediately into sterile Petri dishes or leave in tubes.

Use: For the detection of coliform bacteria in water and food.

Violet Red Bile Agar with MUG

Composition per liter:

Agar	15.0g
Lactose	10.0g
Pancreatic digest of gelatin	7.0g
NaCl	5.0g
Yeast extract	3.0g
Bile salts	1.5g
MUG (4-methylumbeliferyl-β-D-glucuronide)	0.1g
Neutral Red	0.03g
Crystal Violet	2.0mg

pH 7.4 ± 0.2 at 25°C

Source: This medium is available as a premixed powder from BBL Microbiology Systems and Difco Laboratories.

Preparation of Medium: Add components to distilled/deionized water and bring volume to 1.0L. Mix thoroughly. Gently heat while stirring and bring to boiling. Distribute into tubes or flasks. Autoclave for 15 min at 15 psi pressure–121°C. Pour immediately into sterile Petri dishes or leave in tubes.

Use: For the differentiation of *Escherichia coli* from dairy products and other foods based on their ability to produce β-glucuronidase.

Violet Red Bile Glucose Agar

Composition per liter:

Agar	12.0g
Glucose	10.0g
Peptone	7.0g
NaCl	5.0g
Yeast extract	3.0g
Bile salts No. 3	1.5g
Neutral Red	0.03g
Crystal Violet	2.0mg

pH 7.4 ± 0.2 at 25°C

Source: This medium is available as a premixed powder from Oxoid Unipath.

Preparation of Medium: Add components to distilled/deionized water and bring volume to 1.0L. Mix thoroughly. Gently heat and bring to boiling. Do not autoclave. Pour into sterile Petri dishes or distribute into sterile tubes.

Use: For the detection and enumeration of Enterobacteriaceae from foods.

Viral Transport Medium (VTM)

Composition per 104.1mL:

Bovine serum albumin	0.5g
Veal infusion broth	100.0mL
Phenol Red	0.4mL
Amphotericin B solution	2.0mL
Gentamicin	1.0mL
Vancomycin	0.2mL

pH 7.4 ± 0.2 at 25°C

Veal Infusion Broth:

Composition per liter:

Veal, infusion from	500.0g
NaCl	5.0g
Pancreatic digest of casein	5.0g
Peptic digest of animal tissue	5.0g

Preparation of Veal Infusion Broth: Add components to distilled/deionized water and bring volume to 1.0L. Mix thoroughly. Distribute into tubes or flasks. Autoclave for 15 min at 15 psi pressure–121°C. Use freshly prepared solution.

Amphotericin B Solution:

Composition per 10mL:

Amphotericin B	2.5g

Preparation of Amphotericin B Solution: Add amphotericin B to distilled/deionized water and bring volume to 10.0mL. Mix thoroughly. Filter sterilize.

Gentamicin Solution:

Composition per 10mL:

Gentamicin	0.5g

Preparation of Gentamicin Solution: Add gentamicin to distilled/deionized water and bring volume to 10.0mL. Mix thoroughly. Filter sterilize.

Vancomycin Solution:

Composition per 10mL:

Vancomycin	0.5g

Preparation of Vancomycin Solution: Add vancomycin to distilled/deionized water and bring volume to 10.0mL. Mix thoroughly. Filter sterilize.

Preparation of Medium: To 100.0mL of sterile veal infusion broth, aseptically add bovine serum albumin, Phenol Red, amphotericin B solution, gentamicin solution and vancomycin solution. Mix thoroughly. Dispense 2.0mL of medium into serum vials. Store at 4°C and use for up to 2 months.

Use: For the maintenance and transport of specimens suspected of being virally infected.

Vitamin B_{12} Assay Medium

Composition per liter:

Glucose	40.0g
Sodium acetate	20.0g
Vitamin assay casamino acids	12.0g
Sorbitan monooleate complex	2.0g
K_2HPO_4	1.0g
KH_2PO_4	1.0g
$MgSO_4 \cdot 7H_2O$	0.4g
DL-Tryptophan	0.2g
L-Cystine	0.2g
Adenine	0.02g
$FeSO_4$	0.02g
Guanine	0.02g
$MnSO_4 \cdot 5H_2O$	0.02g
NaCl	0.02g
Uracil	0.02g
Pyridoxine·HCl	4.0mg

Niacin	2.0mg
Riboflavin	2.0mg
Thiamine·HCl	2.0mg
Xanthine	1.0mg
Calcium pantothenate	200μg
p-Aminobenzoic acid	200μg
Folic acid	100μg
Biotin	10μg

pH 6.3 ± 0.2 at 25°C

Source: This medium is available as a premixed powder from Difco Laboratories.

Preparation of Medium: Add components to distilled/deionized water and bring volume to 1.0L. Mix thoroughly. Gently heat and bring to boiling. Continue boiling 2–3 min. Distribute into tubes in 5.0mL volumes. Add standard solution or test solutions to each tube. Adjust the volume of each tube to 10.0mL with distilled/deionized water. Autoclave for 15 min at 15 psi pressure–121°C.

Use: For the microbiological assaying of vitamin B_{12} using *Lactobacillus leichmannii* as the test organism.

Vitamin B_6 Blood Agar

Composition per liter:

Agar	15.0g
Pancreatic digest of casein	15.0g
Papaic digest of soybean meal	5.0g
NaCl	5.0g
Pyridoxal·HCl	0.01g
Sheep blood, defibrinated	50.0mL

pH 7.3 ± 0.2 at 25°C

Preparation of Medium: Add components, except sheep blood, to distilled/deionized water and bring volume to 1.0L. Mix thoroughly. Gently heat and bring to boiling. Autoclave for 15 min at 15 psi pressure–121°C. Cool to 45°–50°C. Aseptically add sterile, defibrinated sheep blood. Pour into sterile Petri dishes or distribute into sterile tubes.

Use: For the cultivation and maintenance of fastidious microorganisms, especially *Streptococcus* species.

Vitamin Medium for *Microbacterium*

Composition per liter:

Casamino acids	10.0g
Glucose	10.0g
$(NH_4)_2SO_4$	5.0g
KH_2PO_4	5.0g
K_2HPO_4	5.0g
$MgSO_4·7H_2O$	0.5g
Vitamin solution	4.0mL

pH 7.0 ± 0.2 at 25°C

Vitamin Solution:

Composition per 100 mL:

Thiamine	0.05g
Riboflavin	0.05g
Pyridoxine·HCl	0.05g
Calcium pantothenate	0.05g
Nicotinic acid	0.01g
Biotin	0.01g
Folic acid	0.01g
p-Aminobenzoic acid	0.01g

Preparation of Vitamin Solution: Add components to distilled/deionized water and bring volume to 100.0mL. Mix thoroughly. Filter-sterilize.

Preparation of Medium: Add components, except vitamin solution, to distilled/deionized water and bring volume to 996.0mL. Mix thoroughly. Autoclave for 10 min at 15 psi pressure–121°C. Cool to 45°–50°C. Aseptically add 4.0mL of sterile vitamin solution. Mix thoroughly. Distribute into sterile tubes or flasks.

Use: For the cultivation and maintenance of *Microbacterium* species.

VL Medium

Composition per liter:

Pancreatic digest of casein	10.0g
Agar	6.0g
NaCl	5.0g
Yeast extract	5.0g
Meat extract	2.0g
Glucose	2.0g
Cysteine·HCl·H_2O	0.3g
Antibiotic solution	10.0mL

pH 7.4 ± 0.2 at 25°C

Antibiotic Solution:

Composition per 10mL:

Kanamycin	0.1g
Vancomycin	7.5mg

Preparation of Antibiotic Solution: Add components to distilled/deionized water and bring volume to 10.0mL. Mix thoroughly. Filter sterilize.

Preparation of Medium: Add components, except antibiotic solution, to distilled/deionized water and bring volume to 990.0mL. Mix thoroughly. Gently heat and bring to boiling. Autoclave for 15 min at 15 psi pressure–121°C. Cool to 45°–50°C. Aseptically add sterile antibiotic solution. Mix thoroughly. Aseptically distribute into sterile tubes or flasks.

Use: For the isolation and cultivation of *Bacteroides* species.

VL Medium

Composition per liter:

Pancreatic digest of casein	10.0g
Agar	6.0g
NaCl	5.0g
Yeast extract	5.0g
Meat extract	2.0g
Glucose	2.0g
Cysteine·HCl·H_2O	0.3g
NaN_3	0.05g
Ethyl Violet	0.05g

pH 7.4 ± 0.2 at 25°C

Caution: Sodium azide is toxic. Azides also react with metals and disposal must be highly diluted.

Preparation of Medium: Add components to distilled/deionized water and bring volume to 1.0L. Mix thoroughly. Gently heat and bring to boiling. Distribute into tubes or flasks. Autoclave for 15 min at 15 psi pressure–121°C.

Use: For the isolation and cultivation of *Fusobacterium* species.

VMGII Medium (Viability–Preserving Microbiostatic Medium)

Composition per 1100mL:

Solution 1	900.0mL
Solution 2	100.0mL
Salt stock solution	100.0mL

Solution 1:

Composition per 900mL:

Noble agar	0.1g

Preparation of Solution 1: Add agar to distilled/deionized water and bring volume to 900.0mL. Mix thoroughly. Gently heat and bring to boiling. Cool to 45°–50°C.

Solution 2:

Composition per 100mL:

Charcoal, bacteriological	10.0g
Gelatin peptone	10.0g
Meat peptone	1.0g
Cysteine·HCl	0.5g
Thioglycollic acid	0.5mL

Preparation of Solution 2: Add components to distilled/deionized water and bring volume to 100.0mL. Mix thoroughly.

Stock Salt Solution:

Composition per liter:

Sodium glycerophosphate	100.0g
NaCl	10.0g
KCl	4.2g
$CaCl_2 \cdot 6H_2O$	2.4g
$MgSO_4 \cdot 7H_2O$	1.0g
Phenylmercuric acetate	0.03g

Preparation of Stock Salt Solution: Add phenylmercuric acetate to approximately 800.0mL of distilled/deionized water. Gently heat. Add remaining components. Bring volume to 1.0L with distilled/deionized water.

Preparation of Medium: To 900.0mL of cooled solution 1, add 100.0mL of solution 2 and 100.0mL of stock salt solution. Mix thoroughly. Distribute into screw-capped tubes. Autoclave for 15 min at 15 psi pressure–121°C.

Use: For the isolation and cultivation of oral streptococci, including *Streptococcus mutans* and *Streptococcus sanguis*, and nonspore-forming bacteria, including *Lactobacillus* species from human dental plaque.

Vogel and Johnson Agar

Composition per liter:

Agar	16.0g
Pancreatic digest of casein	10.0g
D-Mannitol	10.0g
Glycine	10.0g
Yeast extract	5.0g
K_2HPO_4	5.0g
LiCl	5.0g
Phenol Red	0.025g
K_2TeO_3 solution	20.0mL

pH 7.2 ± 0.2 at 25°C

Source: This medium is available as a premixed powder from BBL Microbiology Systems, Difco Laboratories and Oxoid Unipath.

K_2TeO_3 Solution:

Composition per 100mL:

K_2TeO_3	1.0g

Preparation of K_2TeO_3 Solution: Add K_2TeO_3 to distilled/deionized water and bring volume to 100.0mL. Mix thoroughly. Filter sterilize.

Caution: Potassium tellurite is toxic.

Preparation of Medium: Add components, except K_2TeO_3 solution, to distilled/deionized water and bring volume to 980.0mL. Mix thoroughly. Gently heat and bring to boiling. Autoclave for 15 min at 15 psi pressure–121°C. Cool to 45°–50°C. Aseptically add 20.0mL of sterile K_2TeO_3 solution. Mix thoroughly. Pour into sterile Petri dishes or distribute into sterile tubes.

Use: For the detection of coagulase-positive *Staphylococcus aureus*.

Voges-Proskauer Medium
See: **VP Medium**

VP Agar
See: ***Vibrio parahaemolyticus* Agar**

VP Broth, Modified, Smith, Gordon and Clark

Composition per liter:

Proteose peptone	7.0g
Glucose	5.0g
NaCl	5.0g

Preparation of Medium: Add components to distilled/deionized water and bring volume to 1.0L. Mix thoroughly. Distribute into tubes in 5.0mL volumes. Autoclave for 15 min at 15 psi pressure–121°C.

Use: For the isolation and cultivation of *Bacillus cereus* from foods.

VP Medium (Voges-Proskauer Medium)

Composition per liter:

Peptone	7.0g
K_2HPO_4	5.0g
Glucose	5.0g

pH 6.9 ± 0.2 at 25°C

Preparation of Medium: Add components to distilled/deionized water and bring volume to 1.0L. Mix thoroughly. Adjust pH to 6.9. Distribute into tubes in 3.0mL volumes. Autoclave for 15 min at 15 psi pressure–121°C.

Use: For the cultivation and differentiation of bacteria based on their ability to produce acetoin.

VPSA
See: ***Vibrio parahaemolyticus* Sucrose Agar**

VRB Agar
See: **Violet Red Bile Agar**

VTM
See: **Viral Transport Medium**

VY Agar

Composition per liter:

Agar	15.0g
Baker's yeast	10.0g
$CaCl_2 \cdot 2H_2O$	1.0g
Cyanocobalamin	5.0mg

pH 7.2 ± 0.2 at 25°C

Preparation of Medium: Add components to distilled/deionized water and bring volume to 1.0L. Mix thoroughly. Gently heat and bring to boiling. Distribute into tubes or flasks. Autoclave for 15 min at 15 psi pressure–121°C. Pour into sterile Petri dishes or leave in tubes.

Use: For the cultivation and maintenance of myxobacteria.

VY2 Agar

Composition per liter:

Agar	15.0g
Baker's yeast	5.0g
$CaCl_2 \cdot 2H_2O$	1.0g
Cyanocobalamin	5.0mg

pH 7.2 ± 0.2 at 25°C

Preparation of Medium: Add components to distilled/deionized water and bring volume to 1.0L. Mix thoroughly. Gently heat and bring to boiling. Distribute into tubes or flasks. Autoclave for 15 min at 15 psi pressure–121°C. Pour into sterile Petri dishes or leave in tubes.

Use: For the cultivation and maintenance of *Myxococcus amylovorans*.

VY5 Agar

Composition per liter:

Agar	15.0g
Baker's yeast	2.0g
$CaCl_2 \cdot 2H_2O$	1.0g
Cyanocobalamin	5.0mg

pH 7.2 ± 0.2 at 25°C

Preparation of Medium: Add components to distilled/deionized water and bring volume to 1.0L. Mix thoroughly. Gently heat and bring to boiling. Distribute into tubes or flasks. Autoclave for 15 min at 15 psi pressure–121°C. Pour into sterile Petri dishes or leave in tubes.

Use: For the cultivation and maintenance of myxobacteria.

VY Medium
(Veal Yeast Extract Medium)

Composition per liter:

Veal, solids from infusion	10.0g
Pancreatic digest of casein	5.0g
Peptic digest of animal tissue	5.0g
NaCl	5.0g
Yeast extract	5.0g

Preparation of Medium: Add components to distilled/deionized water and bring volume to 1.0L. Mix thoroughly. Distribute into tubes or flasks. Autoclave for 15 min at 15 psi pressure–121°C.

Use: For the cultivation and maintenance of *Bacillus subtilis*.

Wadowsky–Yee Medium
See: **BCYE Selective Agar with PAV**

Wadowsky and Yee Medium, Modified
See: **MWY Medium**

Wagatsuma Agar

Composition per 1050mL:

NaCl	70.0g
Agar	15.0g
Mannitol	10.0g
Peptone	10.0g
K_2HPO_4	5.0g
Yeast extract	3.0g
Crystal Violet	1.0mg
Red blood cells	50.0mL

pH 8.0 ± 0.2 at 25°C

Red Blood Cells:

Composition per 100mL:

Blood, human or rabbit	100.0mL

Preparation of Red Blood Cells: Mix freshly drawn human or rabbit blood with anticoagulant and an equal volume of sterile 0.85% saline solution. Centrifuge cells at 4000 × *g* at 4°C for 15 min. Pour off saline and wash two more times with sterile saline. After last wash, pour off saline and resuspend cells to their original volume.

Preparation of Medium: Add components, except blood, to distilled/deionized water and bring volume to 1.0L. Mix thoroughly. Adjust pH to 8.0. Place in a steam bath for 30 min. Do not autoclave. Cool to 45°–50°C. Add 50.0mL of washed red blood cells. Mix thoroughly. Pour into sterile Petri dishes. Dry plates before using.

Use: For the cultivation and detection of the thermostable hemolysin of *Vibrio parahaemolyticus* by the Kanagawa reaction.

Wakimoto Medium, Modified

Composition per liter:

Agar	15.0g
Sucrose	15.0g
Peptone	5.0g
$Na_2HPO_4 \cdot 12\,H_2O$	2.0g
$Ca(NO_3)_2 \cdot 4H_2O$	0.5g
$FeSO_4 \cdot 7H_2O$	0.5g

Preparation of Medium: Add components to distilled/deionized water and bring volume to 1.0L. Mix thoroughly. Gently heat and bring to boiling. Distribute into tubes or flasks. Autoclave for 15 min at 15 psi pressure–121°C. Pour into sterile Petri dishes or leave in tubes.

Use: For the cultivation and maintenance of *Corynebacterium* species and *Pseudomonas* species.

Waksman's Glucose Agar

Composition per liter:

Agar	12.5g
Glucose	10.0g
Peptone	5.0g
Beef extract	5.0g
NaCl	5.0g

pH 7.4-7.6 at 25°C

Preparation of Medium: Add components to distilled/deionized water and bring volume to 1.0L. Mix thoroughly. Gently heat and bring to boiling. Distribute into tubes or flasks. Autoclave for 15 min at 15 psi pressure–121°C. Pour into sterile Petri dishes or leave in tubes.

Use: For the cultivation and maintenance of *Streptomyces* species.

Waksman's Sulfur Medium

Composition per liter:

KH_2PO_4	3.0g
$MgSO_4 \cdot 7H_2O$	0.5g
$(NH_4)_2SO_4$	0.2g
$CaCl_2 \cdot 2H_2O$	0.2g
$Fe_2(SO_4)_3$	0.1mg

Preparation of Medium: Add components to distilled/deionized water and bring volume to 1.0L. Mix thoroughly. It is not necessary to sterilize this medium. Distribute into sterile tubes or flasks.

Use: For the cultivation of sulfate-reducing microorganisms from soil.

Wall Defective Bacterial Medium

Composition per liter:

Sucrose	100.0g
Papaic digest of soybean meal	20.0g
Agarose	10.0g
Yeast extract	10.0g
NaCl	5.0g
$MgSO_4 \cdot 7H_2O$	2.5g
Horse serum	200.0mL
Cholesterol solution	10.0mL

pH 7.8 ± 0.2 at 25°C

Cholesterol Solution:

Composition per 10mL:

Cholesterol	0.04g
Ethanol (95% solution)	10.0mL

Preparation of Cholesterol Solution: Add cholesterol to 10.0mL of 95% ethanol. Mix thoroughly. Filter sterilize.

Preparation of Medium: Add components, except cholesterol solution and horse serum, to distilled/deionized water and bring volume to 790.0mL. Mix thoroughly. Adjust pH to 7.8. Add 10.0mL cholesterol solution. Mix thoroughly. Gently heat and bring to boiling. Autoclave for 15 min at 15 psi pressure–121°C. Cool to 50°–55°C. Aseptically add 200.0mL sterile horse serum. Mix thoroughly. Pour into sterile Petri dishes or distribute into sterile tubes.

Use: For the cultivation of cell wall deficient bacteria, such as L forms, that depend on osmotic stabilization.

Wallenstein Medium

Composition per 4.225L:

Malachite Green	0.75g
Egg yolk emulsion	3.125L
Glycerol	100.0mL

pH 6.75 ± 0.2 at 25°C

Egg Yolk Emulsion:

Composition:

Chicken egg yolks	66
Whole chicken egg	6

Preparation of Egg Yolk Emulsion: Soak eggs with 1:100 dilution of saturated mercuric chloride solution for 1 min. Crack eggs and separate yolks from whites. Mix egg yolks with 6 chicken eggs.

Preparation of Medium: Add components to distilled/deionized water and bring volume to 1.0L. Mix thoroughly. Distribute into tubes. Autoclave for 15 min at 15 psi pressure–121°C.

Use: For the isolation of *Mycobacterium* species other than *M. leprae*.

Wallerstein Laboratory Differential Agar
See: **WL Differential Agar**

Wallerstein Laboratory Differential Medium
See: **WL Differential Medium**

Wallerstein Laboratory Medium with Tomato Juice
See: **WL Medium with Tomato Juice**

Wallerstein Laboratory Nutrient Agar
See: **WL Nutrient Agar**

Wallerstein Laboratory Nutrient Broth
See: **WL Nutrient Broth**

Walsby Medium

Composition per liter:

$MgSO_4 \cdot 7H_2O$	0.075g
K_2HPO_4	0.039g
Na_2CO_3	0.02g
$CaCl_2 \cdot 2H_2O$	0.018g
H_3BO_3	2.8mg
$MnSO_4 \cdot 4H_2O$	2.0mg
$ZnSO_4$	0.22mg
MoO_3	0.18mg
$CuSO_4 \cdot 5H_2O$	0.08mg
$Co(NO_3)_2 \cdot 6H_2O$	0.05mg
Iron-EDTA solution	1.0mL

pH 8.5 ± 0.2 at 25°C

Iron-EDTA Solution:

Composition per liter:

EDTA	12.7g
$FeSO_4 \cdot 7H_2O$	4.98g

Preparation of Iron-EDTA Solution: Add components to distilled/deionized water and bring volume to 1.0L. Mix thoroughly.

Preparation of Medium: Add components to distilled/deionized water and bring volume to 1.0L. Mix thoroughly. Distribute into tubes or flasks. Autoclave for 15 min at 15 psi pressure–121°C.

Use: For the isolation and cultivation of planktonic gas-vacuolate cyanobacteria.

Wang's Transport Storage Medium

Composition per liter:

Pancreatic digest of casein	10.0g
Peptic digest of animal tissue	10.0g
NaCl	5.0g
Agar	4.0g
Yeast extract	2.0g

Glucose 1.0g
$NaHSO_3$ 0.1g
Sheep blood, defibrinated 100.0mL
pH 7.0 ± 0.2 at 25°C

Preparation of Medium: Add components, except sheep blood, to distilled/deionized water and bring volume to 900.0mL. Mix thoroughly. Gently heat and bring to boiling. Autoclave for 15 min at 15 psi pressure–121°C. Cool to 45°–50°C. Aseptically add sterile sheep blood. Sheep blood may be replaced by 50.0mL of horse blood and 50.0mL of sterile distilled/deionized water. Mix thoroughly. Aseptically distribute into sterile screw-capped tubes in 4.0mL volumes. Allow tubes to cool in an upright position.

Use: For the cultivation, transport, and maintenance of *Campylobacter* species from foods.

Water Agar

Composition per liter:
Agar 20.0g

Preparation of Medium: Add agar to distilled/deionized water and bring volume to 1.0L. Mix thoroughly. Gently heat and bring to boiling. Distribute into tubes or flasks. Autoclave for 15 min at 15 psi pressure–121°C. Pour into sterile Petri dishes or leave in tubes.

Use: For the cultivation and observation of sporulation of some fungi.

Water Agar

Composition per liter:
Agar 15.0g
$CaCl_2 \cdot 2H_2O$ 1.0g
pH 7.2 ± 0.2 at 25°C

Preparation of Medium: Add components to distilled/deionized water and bring volume to 1.0L. Mix thoroughly. Gently heat and bring to boiling. Distribute into tubes or flasks. Autoclave for 15 min at 15 psi pressure–121°C. Pour into sterile Petri dishes or leave in tubes.

Use: For the cultivation of myxobacteria.

Waxy Maize Starch Medium

Composition per liter:
Agar 20.0g
Waxy maize starch 5.0g
Pancreatic digest of casein 5.0g
Yeast extract 5.0g
$CoCl_2 \cdot 6H_2O$ 0.1g
$CaCl_2 \cdot 2H_2O$ 0.1g
Maltose solution 100.0mL
pH 6.7 ± 0.2 at 25°C

Maltose Solution:
Composition per 100mL:
Maltose 10.0g

Preparation of Maltose Solution: Add maltose to distilled/deionized water and bring volume to 100.0mL. Mix thoroughly. Filter sterilize.

Preparation of Medium: Add components, except maltose solution, to distilled/deionized water and bring volume to 900.0mL. Mix thoroughly. Gently heat and bring to boiling. Adjust pH to 6.7. Distribute into tubes or flasks. Autoclave for 15 min at 15 psi pressure–121°C. Aseptically add maltose solution. Pour into sterile Petri dishes or leave in tubes.

Use: For the cultivation and maintenance of *Bacillus* species.

Wayne Sulfatase Agar
See: **Arylsulfatase Agar**

WCX Agar

Composition per liter:
Agar 15.0g
$CaCl_2 \cdot 2H_2O$ 1.0g
Cycloheximide solution 100.0mL
pH 7.2 ± 0.2 at 25°C

Cycloheximide Solution
Composition per 100mL:
Cycloheximide 2.5mg

Preparation of Cycloheximide Solution: Add components to distilled/deionized water and bring volume to 100.0mL. Mix thoroughly. Filter sterilize.

Preparation of Medium: Add components, except cycloheximide solution, to distilled/deionized water and bring volume to 900.0mL. Mix thoroughly. Gently heat and bring to boiling. Autoclave for 15 min at 15 psi pressure–121°C. Cool to 45°–50°C. Aseptically add sterile cycloheximide solution. Mix thoroughly. Pour into sterile Petri dishes or distribute into sterile tubes.

Use: For the cultivation of myxobacteria.

Wesley Broth

Composition per liter:
Tryptose 20.0g
Bicine (N,N-bis-2-[hydroxyethyl]glycine) buffer 10.0g

NaCl	5.0g
Yeast extract	2.5g
Agar	1.0g
$FeSO_4$	0.25g
$Na_2S_2O_5$	0.25g
Sodium pyruvate	0.25g
Antibiotic solution	10.0mL
Alkaline hematin solution	6.25mL

Antibiotic Solution:
Composition per 10mL:

Rifampin	0.025g
Cefsulodin	6.25mg
Polymyxin B sulfate	20,000U

Preparation of Antibiotic Solution: Add components to distilled/deionized water and bring volume to 10.0mL. Mix thoroughly. Filter sterilize.

Alkaline Hematin Solution:
Composition per 10mL:

Hemin	0.032g
NaOH (0.15*N* solution)	10.0mL

Preparation of Hemin Solution: Add hemin to 10.0mL of NaOH solution. Mix thoroughly. Autoclave for 30 min at 5 psi pressure–108°C. Cool to 25°C.

Preparation of Medium: Add components, except antibiotic solution and alkaline hematin solution, to distilled/deionized water and bring volume to 983.75mL. Mix thoroughly. Gently heat and bring to boiling. Autoclave for 15 min at 15 psi pressure–121°C. Cool to 45°–50°C. Aseptically add 10.0mL of sterile antibiotic solution and 6.25mL of sterile alkaline hematin solution. Mix thoroughly. Aseptically distribute into sterile tubes. Use medium immediately or store overnight at 4°C.

Use: For the enrichment of *Campylobacter* species from foods.

Wickerham Broth

Composition per 100mL:

Carbohydrate	10.0g
Yeast nitrogen base	100.0mL

Yeast Nitrogen Base, 10X:
Composition per liter:

Glucose	10.0g
KH_2PO_4	1.0g
$MgSO_4 \cdot 7H_2O$	0.5g
NaCl	0.1g
$CaCl_2 \cdot 2H_2O$	0.1g
DL-Methionine	0.02g
DL-Tryptophan	0.02g
L-Histidine·HCl	0.01g
Inositol	2.0mg
H_3BO_3	0.5mg
$ZnSO_4 \cdot 7H_2O$	0.4mg
$MnSO_4 \cdot 4H_2O$	0.4mg
Thiamine·HCl	0.4mg
Pyridoxine	0.4mg
Niacin	0.4mg
Calcium pantothenate	0.4mg
p-Aminobenzoic acid	0.2mg
Riboflavin	0.2mg
$FeCl_3$	0.2mg
$Na_2MoO_4 \cdot 4H_2O$	0.2mg
KI	0.1mg
$CuSO_4 \cdot 5H_2O$	0.04mg
Folic Acid	2.0μg
Biotin	2.0μg

pH 4.5 ± 0.2 at 25°C

Preparation of Yeast Nitrogen Base: Add components to distilled/deionized water and bring volume to 1.0L. Mix thoroughly.

Preparation of Medium: To 100.0mL of yeast nitrogen base, add 10.0g of carbohydrate. Mix thoroughly. Filter sterilize. Aseptically distribute 0.5mL into tubes containing 4.5mL of sterile, distilled/deionized water.

Use: For the cultivation and differentiation of bacteria based on carbohydrate assimilation.

Wickerham Broth

Composition per 100mL:

KNO_3	0.78g
Yeast carbon base	100.0mL

pH 4.5 ± 0.2 at 25°C

Yeast Carbon Base:
Composition per liter:

Glucose	10.0g
KH_2PO_4	1.0g
$MgSO_4 \cdot 7H_2O$	0.5g
NaCl	0.1g
$CaCl_2 \cdot 2H_2O$	0.1g
DL-Methionine	0.02g
DL-Tryptophan	0.02g
L-Histidine·HCl	0.01g
Inositol	2.0mg
H_3BO_3	0.5mg
$ZnSO_4 \cdot 7H_2O$	0.4mg
$MnSO_4 \cdot 4H_2O$	0.4mg
Thiamine·HCl	0.4mg
Pyridoxine	0.4mg
Niacin	0.4mg
Calcium pantothenate	0.4mg
p-Aminobenzoic acid	0.2mg
Riboflavin	0.2mg
$FeCl_3$	0.2mg

$Na_2MoO_4 \cdot 4H_2O$ 0.2mg
KI 0.1mg
$CuSO_4 \cdot 5H_2O$ 0.04mg
Folic Acid 2.0μg
Biotin 2.0μg

Preparation of Yeast Carbon Base: Add components to distilled/deionized water and bring volume to 1.0L. Mix thoroughly.

Preparation of Medium: To 100.0mL of yeast carbon base, add 0.78g of KNO_3 (or peptone). Mix thoroughly. Filter sterilize. Aseptically distribute 0.5mL into tubes containing 4.5mL of sterile, distilled/deionized water.

Use: For the cultivation and differentiation of bacteria based on nitrate assimilation.

Wickerham Broth with Raffinose

Composition per 100mL:

Raffinose 20.0g
Yeast nitrogen base 100.0mL

Yeast Nitrogen Base, 10 X:
Composition per liter:

Glucose 10.0g
KH_2PO_4 .... 1.0g
$MgSO_4 \cdot 7H_2O$ 0.5g
NaCl 0.1g
$CaCl_2 \cdot 2H_2O$ 0.1g
DL-Methionine 0.02g
DL-Tryptophan 0.02g
L-Histidine·HCl 0.01g
Inositol 2.0mg
H_3BO_3 .... 0.5mg
$ZnSO_4 \cdot 7H_2O$ 0.4mg
$MnSO_4 \cdot 4H_2O$ 0.4mg
Thiamine·HCl 0.4mg
Pyridoxine 0.4mg
Niacin 0.4mg
Calcium pantothenate 0.4mg
p-Aminobenzoic acid 0.2mg
Riboflavin 0.2mg
$FeCl_3$ 0.2mg
$Na_2MoO_4 \cdot 4H_2O$ 0.2mg
KI 0.1mg
$CuSO_4 \cdot 5H_2O$ 0.04mg
Folic Acid 2.0μg
Biotin 2.0μg

pH 4.5 ± 0.2 at 25°C

Preparation of Yeast Nitrogen Base: Add components to distilled/deionized water and bring volume to 1.0L. Mix thoroughly.

Preparation of Medium: To 100.0mL of yeast nitrogen base, add 20.0g of raffinose. Mix thoroughly. Filter sterilize. Aseptically distribute 0.5mL into tubes containing 4.5mL of sterile, distilled/deionized water.

Use: For the cultivation and differentiation of bacteria based on carbohydrate assimilation.

Wickerham Carbon Base Broth
See: **Yeast Carbon Base, 10X**

Wilbrinck's Agar for *Xanthomonas albilineans*

Composition per liter:

Agar 20.0g
Sucrose 10.0g
Peptone 5.0g
K_2HPO_4 .... 0.5g
$MgSO_4 \cdot 7H_2O$ 0.25g
Na_2SO_3 (anhydrous) 0.05g

pH 7.2 ± 0.2 at 25°C

Preparation of Medium: Add components to distilled/deionized water and bring volume to 1.0L. Mix thoroughly. Gently heat and bring to boiling. Distribute into tubes or flasks. Autoclave for 15 min at 15 psi pressure–121°C. Pour into sterile Petri dishes or leave in tubes.

Use: For the cultivation and maintenance of *Xanthomonas albilineans* and other *Xanthomonas* species.

Wilkins–Chalgren Agar

Composition per liter:

Agar 15.0g
Gelatin peptone 10.0g
Pancreatic digest of casein 10.0g
NaCl 5.0g
Yeast extract 5.0g
Glucose 1.0g
L-Arginine 1.0g
Sodium pyruvate 1.0g
Hemin 5.0mg
Vitamin K_1 (Menadione) 0.5mg

pH 7.1 ± 0.2 at 25°C

Source: This medium is available as a premixed powder from Difco Laboratories.

Preparation of Medium: Add components to distilled/deionized water and bring volume to 1.0L. Mix thoroughly. Gently heat and bring to boiling. Distribute into tubes or flasks. Autoclave for 15 min at 15 psi pressure–121°C. Cool to 50–55°C. Add antibiotic to be assayed; varying concentrations of antibiotics are used. Mix thoroughly. Pour into sterile Petri dishes or leave in tubes.

Use: For the cultivation and maintenance of anaerobic bacteria. For standardized antimicrobic susceptibility testing to determine the minimum inhibibitory concentrations of antimicrobics for anaerobic bacteria.

Wilkins–Chalgren Anaerobe Agar

Composition per liter:

Agar	10.0g
Pancreatic digest of casein	10.0g
Gelatin peptone	10.0g
NaCl	5.0g
Yeast extract	5.0g
Glucose	1.0g
L-Arginine	1.0g
Sodium pyruvate	1.0g
Haemin	5.0mg
Menadione	0.5mg
Defibrinated blood	50.0mL
Tween™ 80	1.0mL

pH 7.1 ± 0.2 at 25°C

Source: This medium is available as a premixed powder from Oxoid Unipath.

Preparation of Medium: Add components, except defibrinated blood, to distilled/deionized water and bring volume to 950.0mL. Mix thoroughly. Gently heat and bring to boiling. Distribute into tubes or flasks. Autoclave for 15 min at 15 psi pressure–121°C. Cool to 50–55°C. Aseptically add 50.0mL of defibrinated blood. Mix thoroughly. Pour into sterile Petri dishes or leave in tubes.

Use: For the cultivation of nonsporulating anaerobes.

Wilkins–Chalgren Anaerobe Agar with GN Supplement

Composition per liter:

Agar	10.0g
Pancreatic digest of casein	10.0g
Gelatin peptone	10.0g
NaCl	5.0g
Yeast extract	5.0g
Glucose	1.0g
L-Arginine	1.0g
Sodium pyruvate	1.0g
Hemin	5.0mg
Menadione	0.5mg
Defibrinated blood	50.0mL
G-N anaerobe selective supplement	20.0 mL

pH 7.1 ± 0.2 at 25°C

Source: This medium is available as a premixed powder from Oxoid Unipath.

GN Anaerobe Selective Supplement

Composition per 20mL:

Nalidixic acid	10.0mg
Hemin	5.0mg
Sodium succinate	2.5mg
Vancomycin	2.5mg
Menadione	0.5mg

Preparation of GN Anaerobe Selective Supplement: Add components to distilled/deionized water and bring volume to 20.0mL. Mix thoroughly. Filter sterilize.

Preparation of Medium: Add components, except defibrinated blood and GN anaerobe selective supplement, to distilled/deionized water and bring volume to 900.0mL. Mix thoroughly. Gently heat and bring to boiling. Distribute into tubes or flasks. Autoclave for 15 min at 15 psi pressure–121°C. Cool to 50–55°C. Aseptically add 20.0mL of G-N anaerobe selective supplement and 50.0mL of defibrinated blood. Bring volume to 1.0L with distilled/deionized water. Mix thoroughly. Pour into sterile Petri dishes or leave in tubes.

Use: For the selective isolation of Gram-negative anaerobes.

Wilkins–Chalgren Anaerobe Agar with NS Supplement

Composition per liter:

Agar	10.0g
Pancreatic digest of casein	10.0g
Gelatin peptone	10.0g
NaCl	5.0g
Yeast extract	5.0g
Glucose	1.0g
L-Arginine	1.0g
Sodium pyruvate	1.0g
Hemin	5.0mg
Menadione	0.5mg
Defibrinated blood	50.0mL
N-S anaerobe selective supplement	20.0 mL
Tween™ 80	1.0mL

pH 7.1 ± 0.2 at 25°C

Source: This medium is available as a premixed powder from Oxoid Unipath.

NS Anaerobe Selective Supplement:

Composition per 20mL:

Sodium pyruvate	1.0g
Nalidixic acid	0.01g
Hemin	5.0mg
Menadione	0.5mg

Preparation of NS Anaerobe Selective Supplement: Add components to distilled/deionized water and bring volume to 20.0mL. Mix thoroughly. Filter sterilize.

Preparation of Medium: Add components, except defibrinated blood and N-S anaerobe selective supplement, to distilled/deionized water and bring volume to 900.0mL. Mix thoroughly. Gently heat and bring to boiling. Distribute into tubes or flasks. Autoclave for 15 min at 15 psi pressure–121°C. Cool to 50–55°C. Aseptically add 20.0mL of N-S anaerobe selective supplement and 50.0mL of defibrinated blood. Bring volume to 1.0L with distilled/deionized water. Mix thoroughly. Pour into sterile Petri dishes or leave in tubes.

Use: For the selective isolation of nonsporulating anaerobes.

Wilkins–Chalgren Anaerobe Broth (Anaerobe Broth, MIC)

Composition per liter:

Pancreatic digest of casein	10.0g
Gelatin peptone	10.0g
NaCl	5.0g
Yeast extract	5.0g
Glucose	1.0g
L-Arginine	1.0g
Sodium pyruvate	1.0g
Hemin	5.0mg
Menadione	0.5mg

pH 7.1 ± 0.2 at 25°C

Source: This medium is available as a premixed powder from Difco Laboratories and Oxoid Unipath.

Preparation of Medium: Add components to distilled/deionized water and bring volume to 1.0L. Mix thoroughly. Distribute into tubes or flasks. Autoclave for 15 min at 15 psi pressure–121°C.

Use: For the cultivation and antimicrobial susceptibility (MIC) testing of anaerobic bacteria.

Wilson and Blair's Medium

Composition per 165mL:

Nutrient agar solution	100.0mL
Solution B	45.0mL
Solution A	20.0mL

pH 6.8 ± 0.2 at 25°C

Nutrient Agar Solution:

Composition per 100mL:

Agar	1.95g
Pancreatic digest of gelatin	0.65g
Beef extract	0.39g

Preparation of Nutrient Agar Solution: Aseptically add components to distilled/deionized water and bring volume to 100.0mL. Mix thoroughly. Autoclave for 15 min at 15 psi pressure–121°C. Cool to 45°–50°C.

Solution A:

Composition per liter:

Na_2HPO_4	100.0g
Na_2SO_3	100.0g
Glucose	50.0g
Bismuth ammonium citrate	30.0g

Preparation of Solution A: Aseptically add 30.0g of bismuth ammonium citrate to 250.0mL of boiling distilled/deionized water. Aseptically add Na_2SO_3 to 500.0mL of boiling distilled/deionized water. Mix the two solutions. Add the Na_2HPO_4 to the boiling mixture. Cool to 45°C. In a separate flask, aseptically add glucose to 250.0mL of sterile distilled/deionized water. Mix thoroughly. Aseptically add the glucose solution to the other cooled solution.

Solution B:

Composition per 225mL:

Ferric citrate	2.0g
Brilliant Green	0.55g

Preparation of Solution B: Aseptically add 2.0g of ferric citrate to 200.0mL of sterile distilled/deionized water. Mix thoroughly. Aseptically add Brilliant Green to 25.0mL of sterile distilled/deionized water. Mix thoroughly. Aseptically combine the two solutions.

Preparation of Medium: Aseptically combine 100.0mL of nutrient agar solution, 20.0mL of solution A and 45.0mL of solution B. Mix thoroughly. Pour into sterile Petri dishes or distribute into sterile tubes.

Use: For the cultivation of *Clostridium welchii.*

Wilson Blair Base

Composition per liter:

Agar	30.0g
Proteose peptone No. 3	10.0g
Glucose	10.0g
Beef extract	5.0g
NaCl	5.0g
Selective reagent	70.0mL
Brilliant Green (1% solution)	4.0mL

pH 7.3 ± 0.2 at 25°C

Selective Reagent:

Composition per 100mL:

Solution 1	100.0mL
Solution 2	100.0mL
Solution 3	100.0mL
Solution 4	20.2mL

Preparation of Selective Reagent: Combine 100.0mL of solution 1, 100.0mL of solution 2, 100.0mL of solution 3, and 20.2mL of solution 4. Mix thoroughly. Gently heat to boiling until a slate grey color develops. Cool to 50°C.

Solution 1:

Composition per 100mL:

$NaHSO_3$	40.0g

Preparation of Solution 1: Add $NaHSO_3$ to 100.0mL distilled/deionized water. Mix thoroughly.

Solution 2:

Composition per 100mL:

NaH_2PO_4	21.0g

Preparation of Solution 2: Add NaH_2PO_4 to 100.0mL distilled/deionized water. Mix thoroughly.

Solution 3:

Composition per 100mL:

Bismuth ammonium citrate	12.5g

Preparation of Solution 3: Add bismuth ammonium citrate to 100.0mL distilled/deionized water. Mix thoroughly.

Solution 4:

Composition per 20.2mL:

$FeSO_4$	0.96g

Preparation of Solution 4: Add $FeSO_4$ to 20.0mL distilled/deionized water. Add 0.2mL concentrated HCl. Mix thoroughly.

Preparation of Medium: Add components, except selective reagent and brilliant green solution, to distilled/deionized water and bring volume to 976.0mL. Mix thoroughly. Gently heat and bring to boiling. Distribute into tubes or flasks. Autoclave for 15 min at 15 psi pressure–121°C. Cool to 50°C. Aseptically add selective reagent and brilliant green solution. Mix thoroughly. Pour into sterile Petri dishes or leave in tubes.

Use: For isolation and cultivation of *Salmonella*, especially *Salmonella typhi*.

Winogradsky's Medium, Modified

Composition per liter:

$CaCO_3$	5.0g
$(NH_4)_2SO_4$	1.0g
K_2HPO_4	1.0g
NaCl	1.0g
$MgSO_4 \cdot 7H_2O$	0.5g
$FeSO_4$	0.4g

Preparation of Medium: Add components to distilled/deionized water and bring volume to 1.0L. Mix thoroughly. Gently heat until dissolved. Do not autoclave. Distribute into tubes or flasks. Swirl flask while dispensing to suspend precipitate.

Use: For the cultivation of nitrifying bacteria.

Winogradsky's N–Free Medium

Composition per liter:

Agar	20.0g
$CaCO_3$	5.0mg
Sugar solution	100.0mL
Concentrated salt solution	5.0mL

pH 7.2 ± 0.2 at 25°C

Sugar Solution:

Composition per 100mL:

Sucrose or glucose	10.0g

Preparation of Sugar Solution: Add sugar to 100.0mL distilled/deionized water. Mix thoroughly. Autoclave for 10 min at 10 psi pressure–115°C. Cool to 50°C.

Concentrated Salt Solution:

Composition per liter:

KH_2PO_4	50.0g
$MgSO_4 \cdot 7H_2O$	25.0g
NaCl	25.0g
$FeSO_4 \cdot 7H_2O$	1.0g
$MnSO_4 \cdot 4H_2O$	1.0g
$Na_2MoO_4 \cdot 4H_2O$	1.0g

Preparation of Concentrated Salt Solution: Add components to tap water and bring volume to 1.0L. Mix thoroughly. Filter sterilize.

Preparation of Medium: Add components, except sugar solution, to distilled/deionized water and bring volume to 900.0mL. Mix thoroughly. Gently heat and bring to boiling. Distribute into tubes or flasks. Autoclave for 15 min at 15 psi pressure–121°C. Cool to 50°C. Aseptically add sugar solution. Adjust pH to 7.2. Mix thoroughly. Pour into sterile Petri dishes or leave in tubes.

Use: For the cultivation and maintenance of *Azomonas insignis*.

Winogradsky's Nitrite Medium

Composition per liter:

Agar	15.0g
$NaNO_2$	2.0g

Na_2CO_3, anhydrous 1.0g
K_2HPO_4 .... 0.5g

Preparation of Medium: Add components to distilled/deionized water and bring volume to 1.0L. Mix thoroughly. Gently heat and bring to boiling. Distribute into tubes. Autoclave for 15 min at 15 psi pressure–121°C.

Use: For the selective isolation and cultivation of *Nocardia* species and *Rhodococcus* species.

WL Differential Agar (Wallerstein Laboratory Differential Agar)

Composition per liter:

Glucose	50.0g
Agar	20.0g
Pancreatic digest of casein	5.0g
Yeast extract	4.0g
KH_2PO_4	0.55g
KCl	0.425g
$CaCl_2 \cdot 2H_2O$	0.125g
$MgSO_4 \cdot 7H_2O$	0.125g
Bromcresol Green	0.022g
Actidione® (Cycloheximide)	4.0mg
$FeCl_3$	2.5mg
$MnSO_4 \cdot 4H_2O$	2.5mg

pH 5.5 ± 0.2 at 25°C

Source: This medium is available as a premixed powder from BBL Microbiology Systems.

Preparation of Medium: Add components to distilled/deionized water and bring volume to 1.0L. Mix thoroughly. Gently heat with mixing and bring to boiling. Distribute into tubes or flasks. Autoclave for 15 min at 15 psi pressure–121°C. Pour into sterile Petri dishes or leave in tubes.

Use: For the differential cultivation of bacteria from industrial fermentation processes. Growth of yeasts and molds is inhibited.

WL Differential Medium (Wallerstein Laboratory Differential Medium)

Composition per liter:

Glucose	50.0g
Agar	20.0g
Pancreatic digest of casein	5.0g
Yeast extract	4.0g
KH_2PO_4	0.55g
KCl	0.425g
$CaCl_2 \cdot 2H_2O$	0.125g
$MgSO_4 \cdot 7H_2O$	0.125g
Bromcresol Green	0.022g
Actidione® (cycloheximide)	4.0mg
$FeCl_3$	2.5mg
$MnSO_4 \cdot 4H_2O$	2.5mg

pH 5.5 ± 0.2 at 25°C

Source: This medium is available as a premixed powder from Difco Laboratories.

Preparation of Medium: Add components to distilled/deionized water and bring volume to 1.0L. Mix thoroughly. Gently heat with mixing and bring to boiling. Distribute into tubes or flasks. Autoclave for 15 min at 15 psi pressure–121°C. Pour into sterile Petri dishes or leave in tubes.

Use: For the differential cultivation of bacteria from industrial fermentation processes. Growth of yeasts and molds is inhibited.

WL Differential Medium (Wallerstein Laboratory Differential Medium)

Composition per liter:

Glucose	50.0g
Agar	20.0g
Pancreatic digest of casein	5.0g
Yeast extract	4.0g
KH_2PO_4	0.55g
KCl	0.425g
$CaCl_2 \cdot 2H_2O$	0.125g
$MgSO_4 \cdot 7H_2O$	0.125g
Bromcresol Green	0.022g
Actidione® (cycloheximide)	4.0mg
$FeCl_3$	2.5mg
$MnSO_4 \cdot 4H_2O$	2.5mg
Na_2CO_3 (1% solution)	30.0mL

pH 6.5 ± 0.2 at 25°C

Source: This medium is available as a premixed powder from Difco Laboratories.

Preparation of Medium: Add components to distilled/deionized water and bring volume to 1.0L. Mix thoroughly. Gently heat with mixing and bring to boiling. Distribute into tubes or flasks. Autoclave for 15 min at 15 psi pressure–121°C. Pour into sterile Petri dishes or leave in tubes.

Use: For the differential cultivation of bacteria from industrial fermentation processes. Growth of yeasts and molds is inhibited.

WL Medium with Tomato Juice (Wallerstein Laboratory Medium with Tomato Juice)

Composition per liter:

Glucose	50.0g
Pancreatic digest of casein	5.0g
Yeast extract	4.0g
KH_2PO_4	0.55g
KCl	0.425g
$CaCl_2$	0.125g
$MgSO_4 \cdot 7H_2O$	0.125g
Bromcresol Green	0.022g
$FeCl_3$	2.5mg
$MnSO_4 \cdot 4H_2O$	2.5mg
Tomato juice, canned, clarified	400.0mL

pH 5.5 ± 0.2 at 25°C

Preparation of Medium: Add components to distilled/deionized water and bring volume to 1.0L. Mix thoroughly. Distribute into tubes or flasks. Autoclave for 15 min at 15 psi pressure–121°C.

Use: For the cultivation of microorganisms from alcoholic mash.

WL Nutrient Agar (Wallerstein Laboratory Nutrient Agar)

Composition per liter:

Glucose	50.0g
Agar	20.0g
Pancreatic digest of casein	5.0g
Yeast extract	4.0g
KH_2PO_4	0.55g
KCl	0.425g
$CaCl_2 \cdot 2H_2O$	0.125g
$MgSO_4 \cdot 7H_2O$	0.125g
Bromcresol Green	0.022g
$FeCl_3$	2.5mg
$MnSO_4 \cdot 4H_2O$	2.5mg

pH 5.5 ± 0.2 at 25°C

Source: This medium is available as a premixed powder from BBL Microbiology Systems, Difco Laboratories and Oxoid Unipath.

Preparation of Medium: Add components to distilled/deionized water and bring volume to 1.0L. Mix thoroughly. Gently heat with mixing and bring to boiling. Distribute into tubes or flasks. Autoclave for 15 min at 15 psi pressure–121°C. Pour into sterile Petri dishes or leave in tubes.

Use: For detection of bacteria and yeasts in industrial fermentation processes, particularly from beer processing.

WL Nutrient Agar (Wallerstein Laboratory Nutrient Agar)

Composition per liter:

Glucose	50.0g
Agar	20.0g
Pancreatic digest of casein	5.0g
Yeast extract	4.0g
KH_2PO_4	0.55g
KCl	0.425g
$CaCl_2 \cdot 2H_2O$	0.125g
$MgSO_4 \cdot 7H_2O$	0.125g
Bromcresol Green	0.022g
$FeCl_3$	2.5mg
$MnSO_4 \cdot 4H_2O$	2.5mg
Na_2CO_3 (1% solution)	30.0mL

pH 6.5 ± 0.2 at 25°C

Source: This medium is available as a premixed powder from BBL Microbiology Systems and Oxoid Unipath.

Preparation of Medium: Add components to distilled/deionized water and bring volume to 1.0L. Mix thoroughly. Gently heat with mixing and bring to boiling. Distribute into tubes or flasks. Autoclave for 15 min at 15 psi pressure–121°C. Pour into sterile Petri dishes or leave in tubes.

Use: For the detection of bacteria and yeasts from industrial fermentation processes, particularly from beer processing.

WL Nutrient Broth (Wallerstein Laboratory Nutrient Broth)

Composition per liter:

Glucose	50.0g
Pancreatic digest of casein	5.0g
Yeast extract	4.0g
KH_2PO_4	0.55g
KCl	0.425g
$CaCl_2$	0.125g
$MgSO_4 \cdot 7H_2O$	0.125g
Bromcresol Green	0.022g
$FeCl_3$	2.5mg
$MnSO_4 \cdot 4H_2O$	2.5mg

pH 5.5 ± 0.2 at 25°C

Source: This medium is available as a premixed powder from Difco Laboratories and Oxoid Unipath.

Preparation of Medium: Add components to distilled/deionized water and bring volume to 1.0L. Mix thoroughly. Distribute into tubes or flasks. Autoclave for 15 min at 15 psi pressure–121°C.

Use: For the cultivation of yeasts, molds and bacteria found in brewing and other industrial fermentation processes.

WL Nutrient Broth (Wallerstein Laboratory Nutrient Broth)

Composition per liter:

Glucose	50.0g
Pancreatic digest of casein	5.0g
Yeast extract	4.0g
KH_2PO_4	0.55g
KCl	0.425g
$CaCl_2$	0.125g
$MgSO_4 \cdot 7H_2O$	0.125g
Bromcresol Green	0.022g
$FeCl_3$	2.5mg
$MnSO_4 \cdot 4H_2O$	2.5mg
Na_2CO_3 (1% solution)	30.0mL

pH 6.5 ± 0.2 at 25°C

Source: This medium is available as a premixed powder from Difco Laboratories and Oxoid Unipath.

Preparation of Medium: Add components to distilled/deionized water and bring volume to 1.0L. Mix thoroughly. Distribute into tubes or flasks. Autoclave for 15 min at 15 psi pressure–121°C.

Use: For the cultivation of yeasts, molds and bacteria found in brewing and other industrial fermentation processes.

WL Nutrient Broth (Wallerstein Laboratory Nutrient Broth)

Composition per liter:

Glucose	50.0g
Pancreatic digest of casein	5.0g
Yeast extract	4.0g
$FeCl_3$	2.5g
$MnSO_4 \cdot 4H_2O$	2.5g
KH_2PO_4	0.55g
KCl	0.425g
$CaCl_2 \cdot 2H_2O$	0.125g
$MgSO_4 \cdot 7H_2O$	0.125g
Bromcresol Green	0.022g

pH 5.5 ± 0.2 at 25°C

Preparation of Medium: Add components to distilled/deionized water and bring volume to 1.0L. Mix thoroughly. Distribute into tubes or flasks. Autoclave for 15 min at 15 psi pressure–121°C.

Use: For the control of brewing and other fermentation processes.

Wolin–Bevis Medium

Composition per liter:

Agar	20.0g
$(NH_4)_2SO_4$	1.0g
KH_2HPO_4	1.0g
Glucose	0.25g
L-Histidine·HCl	0.25g
Tween™ 80 (polysorbate 80)	3.0mL

pH 5.4 ± 0.2 at 25°C

Preparation of Medium: Add components to distilled/deionized water and bring volume to 1.0L. Mix thoroughly. Gently heat and bring to boiling. Distribute into tubes or flasks. Autoclave for 15 min at 15 psi pressure–121°C. Pour into sterile Petri dishes or leave in tubes.

Use: For the enhanced production of chlamydospores by *Candida albicans*.

Worfel–Ferguson Agar

Composition per liter:

Sucrose	20.0g
Agar	15.0g
NaCl	2.0g
Yeast extract	2.0g
K_2SO_4	1.0g
$MgSO_4 \cdot 7H_2O$	0.25g

pH 6.5 ± 0.2 at 25°C

Preparation of Medium: Add components to distilled/deionized water and bring volume to 1.0L. Mix thoroughly. Gently heat and bring to boiling. Distribute into tubes or flasks. Autoclave for 15 min at 15 psi pressure–121°C. Pour into sterile Petri dishes or leave in tubes.

Use: For the detection of capsule production by *Klebsiella*. For serological detection of the Neufeld (Quellung) reaction.

Wort Agar

Composition per liter:

Agar	15.0g
Malt extract	15.0g
Maltose	12.75g
Dextrin	2.75g
Glycerol	2.35g
K_2HPO_4	1.0g
NH_4Cl	1.0g
Pancreatic digest of gelatin	0.78g

pH 4.8 ± 0.2 at 25°C

Source: This medium is available as a premixed powder from BBL Microbiology Systems, Difco Laboratories and Oxoid Unipath.

Preparation of Medium: Add components to distilled/deionized water and bring volume to 1.0L. Mix thoroughly. Gently heat and bring to boiling. Boil for 1 min with mixing. Distribute into tubes or flasks. Autoclave for 15 min at 15 psi pressure–121°C. Do not overheat as this will result in hydrolysis of the agar. An additional 5.0g of agar can be used to make a firmer agar. Pour into sterile Petri dishes or leave in tubes.

Use: For the cultivation and enumeration of yeasts. The low pH of the agar selectively inhibits bacterial growth.

Xanthine Agar

Composition per liter:

Solution 1 .... 900.0mL
Solution 2 .... 100.0mL

pH 7.0 ± 0.2 at 25°C

Solution 1:

Composition per 900mL:

Agar...15.0g
Pancreatic digest of gelatin 5.0g
Beef extract 3.0g

Preparation of Solution 1: Add components to distilled/deionized water and bring volume to 900.0mL. Mix thoroughly. Gently heat and bring to boiling.

Solution 2:

Composition per 100mL:

Xanthine 4.0g

Preparation of Solution 2: Add xanthine to distilled/deionized water and bring volume to 100.0mL. Mix thoroughly. Gently heat and bring to boiling.

Preparation of Medium: Combine solutions 1 and 2. Mix thoroughly. Distribute into tubes or flasks. Autoclave for 15 min at 15 psi pressure–121°C. Pour into sterile Petri dishes or leave in tubes.

Use: For the differentiation of aerobic *Actinomycete* species. Clearing around a colony indicates utilization of xanthine. *Streptomyces* species utilize xanthine; most *Nocardia* and *Actinomadura* species do not utilize xanthine.

Xanthomonas Agar

Composition per liter:

Agar...15.0g
Pancreatic digest of gelatin 10.0g
Sucrose..10.0g
Beef extract 6.0g

pH 6.8 ± 0.2 at 25°C

Preparation of Medium: Add components to distilled/deionized water and bring volume to 1.0L. Mix thoroughly. Gently heat and bring to boiling. Distribute into tubes or flasks. Autoclave for 15 min at 15 psi pressure–121°C. Pour into sterile Petri dishes or leave in tubes.

Use: For cultivation and maintenance of *Xanthomonas* species.

Xanthomonas Medium

Composition per liter:

Pancreatic digest of gelatin 10.0g
Sucrose..10.0g
Beef extract 6.0g

pH 6.8 ± 0.2 at 25°C

Preparation of Medium: Add components to distilled/deionized water and bring volume to 1.0L. Mix thoroughly. Gently heat with mixing. Distribute into screw cap test tubes. Autoclave for 15 min at 15 psi pressure–121°C.

Use: For cultivation and maintenance of *Xanthomonas* species.

Xanthomonas TYG Agar (*Xanthomonas* Tryptone Yeast Extract Glucose Agar)

Composition per liter:

Agar...20.0g
Pancreatic digest of casein.................................5.0g
Glucose 5.0g
Yeast extract...3.0g
K_2HPO_4..0.7g
$MgSO_4 \cdot 7H_2O$ 0.25g

Preparation of Medium: Add components to distilled/deionized water and bring volume to 1.0L. Mix thoroughly. Gently heat and bring to boiling. Distribute into tubes or flasks. Autoclave for 15 min at 15 psi pressure–121°C. Pour into sterile Petri dishes or leave in tubes.

Use: For the cultivation and maintenance of *Xanthomonas* species.

XL Agar Base (Xylose Lysine Agar Base)

Composition per liter:

Agar...13.5g
Lactose 7.5g
Sucrose..7.5g
L-Lysine...5.0g
NaCl 5.0g
Xylose 3.5g
Yeast extract...3.0g
Phenol Red...0.08g
Thiosulfate-citrate solution 20.0mL

pH 7.5 ± 0.2 at 25°C

Source: This medium is available as a premixed powder from BBL Microbiology Systems.

Thiosulfate-Citrate Solution:

Composition per 100mL:

$Na_2S_2O_3$..34.0g
Ferric ammonium citrate....................................4.0g

Preparation of Thiosulfate-Citrate Solution: Add components to distilled/deionized water and bring volume to 100.0mL. Mix thoroughly.

Preparation of Medium: Add components, except thiosulfate-citrate solution, to distilled/deionized water and bring volume to 980.0mL. Mix thoroughly. Gently heat while stirring and bring to boiling. Distribute into tubes or flasks. Autoclave for 10 min at 14 psi pressure–118°C. Cool to 55°C. Aseptically add 20.0 mL of the sterile thiosulfate-citrate solution. Mix thoroughly. Pour into sterile Petri dishes or leave in tubes.

Use: For the isolation, cultivation, and differentiation of enteric pathogens. Nonfermenting xylose/lactose/sucrose bacteria appear as red colonies. Xylose-fermenting lysine-decarboxylating bacteria appear as red colonies. Xylose-fermenting lysine-non-decarboxylating bacteria appear as opaque yellow colonies. Lactose or sucrose-fermenting bacteria appear as yellow colonies.

XL Agar Base

Composition per liter:

Agar	15 g
Lactose	7.5g
Sucrose	7.5g
L-Lysine	5.0g
NaCl	5.0g
Xylose	3.75g
Yeast extract	3.0g
Phenol Red	0.08g
Thiosulfate-citrate solution	20.0mL

pH 7.4 ± 0.2 at 25°C

Source: This medium is available as a premixed powder from Difco Laboratories.

Thiosulfate-Citrate Solution:
Composition per 100mL:

$Na_2S_2O_3$	34.0g
Ferric ammonium citrate	4.0g

Preparation of Thiosulfate-Citrate Solution: Add components to distilled/deionized water and bring volume to 100.0mL. Mix thoroughly.

Preparation of Medium: Add components, except thiosulfate-citrate solution, to distilled/deionized water and bring volume to 980.0mL. Mix thoroughly. Gently heat while stirring and bring to boiling. Distribute into tubes or flasks. Autoclave for 10 min at 14 psi pressure–118°C. Cool to 55°C. Aseptically add 20.0 mL of the sterile thiosulfate-citrate solution. Mix thoroughly. Pour into sterile Petri dishes or leave in tubes.

Use: For the isolation, cultivation, and differentiation of enteric pathogens. Nonfermenting xylose/lactose/sucrose bacteria appear as red colonies. Xylose-fermenting lysine-decarboxylating bacteria appear as red colonies. Xylose-fermenting lysine-non-decarboxylating bacteria appear as opaque yellow colonies. Lactose or sucrose-fermenting bacteria appear as yellow colonies.

XLD Agar (Xylose Lysine Deoxycholate Agar)

Composition per liter:

Agar	13.5g
Lactose	7.5g
Sucrose	7.5g
$Na_2S_2O_3$	6.8g
L-Lysine	5.0g
NaCl	5.0g
Xylose	3.5g
Yeast extract	3.0g
Sodium desoxycholate	2.5g
Ferric ammonium citrate	0.8g
Phenol Red	0.08g

pH 7.5 ± 0.2 at 25°C

Source: This medium is available as a premixed powder from BBL Microbiology Systems, Difco Laboratories and Oxoid Unipath.

Preparation of Medium: Add components to distilled/deionized water and bring volume to 1.0L. Mix thoroughly. Gently heat and bring to boiling. Do not overheat. Distribute into tubes or flasks. Autoclave for 15 min at 15 psi pressure–121°C. Pour into sterile Petri dishes or leave in tubes. Plates should be poured as soon as possible to avoid precipitation.

Use: For isolation and differentiation of enteric pathogens, especially *Shigella* and *Providencia* species. Nonfermenting xylose/lactose/sucrose bacteria appear as red colonies. Xylose-fermenting lysine-decarboxylating bacteria appear as red colonies. Xylose-fermenting lysine-non-decarboxylating bacteria appear as opaque yellow colonies. Lactose or sucrose-fermenting bacteria appear as yellow colonies.

XSM Agar

Composition per liter:

Agar	15.0g
Glucose	5.0g
Sucrose	2.0g
Malt extract	1.0g

Yeast extract 1.0g
Liver extract concentrate 1.0g
Corn steep liquor 1.0g
pH 7.0 ± 0.2 at 25°C

Preparation of Medium: Add components to tap water and bring volume to 1.0L. Mix thoroughly. Gently heat and bring to boiling. Distribute into screw cap test tubes. Autoclave for 15 min at 15 psi pressure–121°C. Pour into sterile Petri dishes or leave in tubes.

Use: For cultivation and maintenance of *Streptomyces cinereus* and *Streptomyces flaveus*.

Xylan Medium

Composition per liter:

Xylan 30.0g
Agar 12.0g
Peptone 2.0g
Yeast extract 0.5g
Cysteine·HCl·H_2O 0.25g
$Na_2S \cdot 9H_2O$ 0.25g
Rumen fluid 400.0mL
$NaHCO_3$ solution 40.0mL
Mineral solution I 25.0mL
Mineral solution II 25.0mL
Wolfe's vitamin solution 10.0mL
VFA solution 10.0mL
Hemin solution 10.0mL
Trace elements solution SL-6 .... 1.0mL
pH 7.0 ± 0.2 at 25°C

$NaHCO_3$ Solution:
Composition per 100mL:

$NaHCO_3$ 3.96g

Preparation of $NaHCO_3$ Solution: Add $NaHCO_3$ to distilled/deionized water and bring volume to 100.0mL. Mix thoroughly. Gas with 100% CO_2.

Mineral Solution I:
Composition per liter:

K_2HPO_4 .... 3.0g

Preparation of Mineral Solution I: Add K_2HPO_4 to distilled/deionized water and bring volume to 1.0L. Mix thoroughly.

Mineral Solution II:
Composition per liter:

Sodium citrate 20.0g
NaCl 12.0g
KH_2PO_4 .... 6.0g
$MgCl_2 \cdot 6H_2O$ 2.0g
$CaCl_2$ 1.2g

Preparation of Mineral Solution II: Add components to distilled/deionized water and bring volume to 1.0L. Mix thoroughly.

Wolfe's Vitamin Solution:
Composition per liter:

Pyridoxine·HCl 10.0mg
Thiamine·HCl 5.0mg
Riboflavin 5.0mg
Nicotinic acid 5.0mg
Calcium pantothenate 5.0mg
p-Aminobenzoic acid 5.0mg
Thioctic acid 5.0mg
Biotin 2.0mg
Folic acid 2.0mg
Cyanocobalamin 100.0μg

Preparation of Wolfe's Vitamin Solution: Add components to distilled/deionized water and bring volume to 1.0L. Mix thoroughly.

VFA Solution:
Composition per liter:

Acetic acid 178.3mL
Propionic acid 59.6mL
n-Butyric acid 38.4mL
Isobutyric acid 9.5mL
n-Valeric acid 9.4mL
Isovaleric acid 9.3mL
DL α-Methylbutyric acid 4.4mL

Preparation of VFA Solution: Add components to distilled/deionized water and bring volume to approximately 500.0mL. Adjust pH to 7.5 with NaOH. Mix thoroughly. Bring volume to 1.0L with distilled/deionized water.

Hemin Solution:
Composition per 100mL:

Hemin 0.01g

Preparation of Hemin Solution: Add hemin to 100.0mL of 0.01*N* NaOH. Mix thoroughly.

Trace Elements Solution SL-6:
Composition per liter:

H_3BO_3 .... 0.30g
$CoCl_2 \cdot 6H_2O$ 0.20g
$ZnSO_4 \cdot 7H_2O$ 0.10g
$MnCl_2 \cdot 4H_2O$ 0.03g
$Na_2MoO_4 \cdot H_2O$ 0.03g
$NiCl_2 \cdot 6H_2O$ 0.02g
$CuCl_2 \cdot 2H_2O$ 0.01g

Preparation of Trace Elements Solution SL-6: Add components to distilled/deionized water and bring volume to 1.0L. Mix thoroughly. Adjust pH to 3.4.

Preparation of Medium: Add components, except cysteine·HCl·H_2O, $Na_2S \cdot 9H_2O$, and sodium bicarbonate solution, to distilled/deionized water and bring volume to 1.0L. Mix thoroughly. Gently heat and bring to boiling. Cool under 80% N_2 + 20% CO_2. Add cysteine·HCl·H_2O and $Na_2S \cdot 9H_2O$. Add sufficient sodium bicarbonate solution to bring pH to 7.2 under 80% N_2 + 20% CO_2. Anaerobically distribute into tubes under 80% N_2 + 20% CO_2. Autoclave for 15 min at 15 psi pressure–121°C.

Use: For cultivation and maintenance of *Clostridium xylanolyticum* and other microorganisms that can utilize xylan as a carbon source.

Xylose Lysine Agar Base
See: XL Agar Base

Xylose Lysine Desoxycholate Agar
See: XLD Agar

Xylose Sodium Deoxycholate Citrate Agar

Composition per liter:

Agar	12.0g
Xylose	10.0g
Sodium citrate	5.0g
$Na_2S_2O_3 \cdot 5H_2O$	5.0g
Beef extract	5.0g
Peptone	5.0g
NaCl	2.5g
Sodium deoxycholate	2.5g
Ferric ammonium citrate	1.0g
Neutral Red (1% solution)	2.5mL

pH 7.5 ± 0.2 at 25°C

Preparation of Medium: Add components to distilled/deionized water and bring volume to 1.0L. Mix thoroughly. Gently heat and bring to boiling for 20 sec. Do not autoclave. Cool to 45°–50°C. Pour into sterile Petri dishes.

Use: For the cultivation of *Salmonella* species and some *Shigella* species.

Xylose YP Agar (Xylose Yeast Extract Peptone Agar)

Composition per liter:

$CaCO_3$	20.0g
Agar	15.0g
Xylose	10.0g
Yeast extract	10.0g
Peptone	10.0g
$MgSO_4 \cdot 7H_2O$	0.20g
$MnSO_4 \cdot 4H_2O$	0.01g
$FeSO_4 \cdot 7H_2O$	0.01g
NaCl	0.01g

pH 6.8 ± 0.2 at 25°C

Preparation of Medium: Add components to distilled/deionized water and bring volume to 1.0L. Mix thoroughly. Gently heat and bring to boiling. Distribute into screw cap test tubes. Autoclave for 15 min at 15 psi pressure–121°C. Adjust pH to 6.8. Mix thoroughly. Pour into sterile Petri dishes or leave in tubes.

Use: For cultivation and maintenance of *Lactobacillus vaccinostercus* and other microorganisms that utilize xylose as a carbon source.

Xylose YP Broth (Xylose Yeast Extract Peptone Broth)

Composition per liter:

Xylose	10.0g
Yeast extract	10.0g
Peptone	10.0g
$MgSO_4 \cdot 7H_2O$	0.20g
$MnSO_4 \cdot 4H_2O$	0.01g
$FeSO_4 \cdot 7H_2O$	0.01g
NaCl	0.01g

pH 6.8 ± 0.2 at 25°C

Preparation of Medium: Add components to distilled/deionized water and bring volume to 1.0L. Mix thoroughly. Distribute into screw cap test tubes. Autoclave for 15 min at 15 psi pressure–121°C.

Use: For cultivation of *Lactobacillus vaccinostercus* and other microorganisms that utilize xylose as a carbon source.

Y 1 Adrenal Cell Growth Medium

Composition per 101mL:

Ham's F-10 Medium 90.0mL
Fetal bovine serum 10.0mL
Penicillin-streptomycin solution 1.0mL
pH 7.0 ± 0.2 at 25°C

Ham's F-10 Medium:
Composition per liter:

NaCl 7.4g
$NaHCO_3$ 1.2g
Glucose 1.1g
$NaH_2PO_4 \cdot H_2O$ 0.29g
KCl 0.28g
L-Arginine·HCl 0.21g
L-Glutamine 0.15g
$MgSO_4 \cdot 7H_2O$ 0.15g
Sodium pyruvate 0.11g
KH_2PO_4 .... 0.08g
$CaCl_2 \cdot 2H_2O$ 0.04g
L-Cystine·2HCl 0.04g
L-Histidine·HCl·H_2O 0.02g
L-Lysine·HCl 0.02g
L-Asparagine-H_2O 0.01g
L-Aspartic Acid 0.01g
L-Glutamic acid 0.01g
L-Leucine 0.01g
L-Proline 0.01g
L-Serine 0.01g
L-Alanine 8.9mg
Glycine 7.5mg
D-Phenylalanine 5.0mg
L-Methionine 4.5mg
Hypoxanthine 4.1mg
L-Threonine 3.6mg
L-Valine 3.5mg
L-Isoleucine 2.6mg
L-Tyrosine 1.8mg
Vitamin B_{12} 1.4mg
Folic acid 1.3mg
Phenol Red 1.2mg
Thiamine·HCl 1.0mg
$FeSO_4 \cdot 7H_2O$ 0.8mg
Choline chloride 0.7mg
D-Calcium pantothenate 0.7mg
Thymidine 0.7mg
Niacinamide 0.6mg
L-Tryptophan 0.6mg
Isoinositol 0.5mg
Riboflavin 0.4mg
Lipoic acid 0.2mg
Pyridoxine·HCl 0.2mg
$ZnSO_4 \cdot 7H_2O$ 0.03mg
Biotin 0.02mg
$CuSO_4 \cdot 5H_2O$ 3.0μg

Preparation of Ham's F-10 Medium: Add components to distilled/deionized water and bring volume to 1.0L. Mix thoroughly.

Penicillin-Streptomycin Solution:
Composition per 100mL:

Streptomycin 0.5g
Penicillin G 500,000U

Preparation of Penicillin-Streptomycin Solution: Add components to distilled/deionized water and bring volume to 100.0mL. Mix thoroughly. Filter sterilize.

Preparation of Medium: Aseptically combine components. Filter sterilize. Store at 4-5°C.

Use: For the cultivation of Y-1 mouse adrenal tissue culture cells used for the detection of heat-labile toxin (LT) produced by enterotoxigenic strains of *Escherichia coli*. LT causes the conversion of elongated fibroblast-like cells into round, refractile cells.

Y 1 Adrenal Cell Growth Medium

Composition per 580mL:

Ham's F-10 Medium 500.0mL
Fetal bovine serum 75.0mL
Penicillin-streptomycin solution 5.0mL
pH 7.0 ± 0.2 at 25°C

Ham's F-10 Medium:
Composition per liter:

NaCl 7.4g
$NaHCO_3$ 1.2g
Glucose 1.1g
$NaH_2PO_4 \cdot H_2O$ 0.29g
KCl 0.28g
L-Arginine·HCl 0.21g
L-Glutamine 0.15g
$MgSO_4 \cdot 7H_2O$ 0.15g
Sodium pyruvate 0.11g
KH_2PO_4 .... 0.08g
$CaCl_2 \cdot 2H_2O$ 0.04g
L-Cystine·2HCl 0.04g
L-Histidine·HCl·H_2O 0.02g
L-Lysine·HCl 0.02g
L-Asparagine-H_2O 0.01g
L-Aspartic Acid 0.01g
L-Glutamic acid 0.01g
L-Leucine 0.01g
L-Proline 0.01g
L-Serine 0.01g
L-Alanine 8.9mg
Glycine 7.5mg

D-Phenylalanine	5.0mg
L-Methionine	4.5mg
Hypoxanthine	4.1mg
L-Threonine	3.6mg
L-Valine	3.5mg
L-Isoleucine	2.6mg
L-Tyrosine	1.8mg
Vitamin B_{12}	1.4mg
Folic acid	1.3mg
Phenol Red	1.2mg
Thiamine·HCl	1.0mg
$FeSO_4 \cdot 7H_2O$	0.8mg
Choline chloride	0.7mg
D-Calcium pantothenate	0.7mg
Thymidine	0.7mg
Niacinamide	0.6mg
L-Tryptophan	0.6mg
Isoinositol	0.5mg
Riboflavin	0.4mg
Lipoic acid	0.2mg
Pyridoxine·HCl	0.2mg
$ZnSO_4 \cdot 7H_2O$	0.03mg
Biotin	0.02mg
$CuSO_4 \cdot 5H_2O$	3.0μg

Preparation of Ham's F-10 Medium: Add components to distilled/deionized water and bring volume to 1.0L. Mix thoroughly.

Penicillin-Streptomycin Solution:
Composition per 100mL:

Streptomycin	0.5g
Penicillin G	500,000U

Preparation of Penicillin-Streptomycin Solution: Add components to distilled/deionized water and bring volume to 100.0mL. Mix thoroughly. Filter sterilize.

Preparation of Medium: Aseptically combine components. Filter sterilize. Store at 4°–5°C.

Use: For the cultivation of Y-1 mouse adrenal tissue culture cells used for the detection of cholera enterotoxin (CT) produced by enterotoxigenic strains of *Vibrio cholerae* or *Vibrio mimicus*. CT causes the conversion of elongated fibroblast-like cells into round, refractile cells.

YA Halophile Medium

Composition per liter:

NaCl	100.0g
Agar	15.0g
Sodium acetate·$3H_2O$	10.0g
Na_2HPO_4	3.8g
KH_2PO_4	1.3g
$Mg(NO_3)_2 \cdot 6H_2O$	1.0g
$(NH_4)_2SO_4$	1.0g
Yeast extract	1.0g

pH 7.2 ± 0.2 at 25°C

Preparation of Medium: Add components except magnesium nitrate to tap water and bring volume to 1.0L. Mix thoroughly. Distribute into tubes or flasks. Autoclave for 15 min at 15 psi pressure–121°C. Aseptically add magnesium nitrate. Adjust pH 7.2 with sterile KOH. Pour into sterile Petri dishes or leave in tubes.

Use: For the cultivation and maintenance of halophilic microorganisms, including *Bacillus halodenitrificans.*

YB Medium (Yeast Extract Beef Extract Medium)

Composition per liter:

Agar	20.0g
Peptone	10.0g
Beef extract	7.0g
Yeast extract	5.0g
NaCl	3.0g
Thiourea	0.1g
Methanol	20.0mL

pH 7.2 ± 0.2 at 25°C

Preparation of Medium: Add components except methanol to distilled/deionized water and bring volume to 1.0L. Mix thoroughly. Distribute into tubes or flasks. Autoclave for 15 min at 15 psi pressure–121°C. Aseptically add filter sterilized methanol. Pour into sterile Petri dishes or leave in tubes.

Use: For the cultivation and maintenance of bacteria that can utilize methanol as a carbon source, including *Achromobacter methanolophila, Methanomonas methylovora, Methylobacterium* species and *Pseudomonas methanolica.*

YDC Agar (Yeast Extract Dextrose Calcium Carbonate Agar)

Composition per liter:

$CaCO_3$, finely divided	20.0g
Glucose	20.0g
Agar	15.0g
Yeast extract	10.0g

pH 7. 0 ± 0.2 at 25°C

Preparation of Medium: Add components to distilled/deionized water and bring volume to 1.0L. Mix thoroughly. Gently heat and bring to boiling. Distrib-

ute into tubes in 10.0mL volumes. Autoclave for 15 min at 15 psi pressure–121°C. Allow tubes to cool in a slanted position.

Use: For the cultivation and maintenance of *Pseudomonas* species on agar slants.

YDC Medium

Composition per liter:

$CaCO_3$	20.0g
Glucose	20.0g
Agar	15.0g
Yeast extract	10.0g

pH 7.2 ± 0.2 at 25°C

Preparation of Medium: Add components to distilled/deionized water and bring volume to 1.0L. Mix thoroughly. Gently heat and bring to boiling. Distribute into tubes or flasks. Autoclave for 15 min at 15 psi pressure–121°C. Pour into sterile Petri dishes or leave in tubes.

Use: For the cultivation of *Bdellovibrio* species.

Yeast Agar, Van Niel's

Composition per liter:

Agar	20.0g
Yeast extract	10.0g
K_2HPO_4	1.0g
$MgSO_4 \cdot 7H_2O$	0.5g

pH 7.0–7.2 at 25°C

Preparation of Medium: Add components to tap water and bring volume to 1.0L. Mix thoroughly. Gently heat and bring to boiling. Distribute into tubes or flasks. Autoclave for 15 min at 15 psi pressure–121°C. Pour into sterile Petri dishes or leave in tubes.

Use: For the cultivation and maintenance of a variety of microorganisms including *Cytophaga* species, *Heliobacterium chlorum, Lysobacter enzymogenes, Rhodobacter* species, *Rhodocyclus gelatinosus, Rhodomicrobium vannielii, Rhodopseudomonas palustris* and *Rhodospirillum rubrum.*

Yeast Agar, Van Niel's with 2.5% NaCl (ATCC Medium 1370)

Composition per liter:

NaCl	25.0g
Agar	20.0g
Yeast extract	10.0g
K_2HPO_4	1.0g
$MgSO_4$	0.5g

pH 7.0–7.2 at 25°C

Preparation of Medium: Add components to tap water and bring volume to 1.0L. Mix thoroughly. Gently heat and bring to boiling. Distribute into tubes or flasks. Autoclave for 15 min at 15 psi pressure–121°C. Pour into sterile Petri dishes or leave in tubes.

Use: For the cultivation and maintenance of *Chromatium vinosum* and *Rhodopseudomonas* species.

Yeast Agar, Van Niel's with 25% NaCl

Composition per liter:

NaCl	250.0g
Agar	20.0g
Yeast extract	10.0g
K_2HPO_4	1.0g
$MgSO_4$	0.5g

pH 7.0–7.2 at 25°C

Preparation of Medium: Add components to tap water and bring volume to 1.0L. Mix thoroughly. Gently heat and bring to boiling. Distribute into tubes or flasks. Autoclave for 15 min at 15 psi pressure–121°C. Pour into sterile Petri dishes or leave in tubes.

Use: For the cultivation and maintenance of halophilic bacteria including *Haloarcula vallismortis, Halococcus morrhuae* and *Halobacterium* species.

Yeast Agar, Van Niel's with Glutamate

Composition per liter:

Agar	20.0g
Yeast extract	10.0g
K_2HPO_4	1.0g
$MgSO_4$	0.5g
Monosodium glutamate	0.85g

pH 7.0–7.2 at 25°C

Preparation of Medium: Add components to tap water and bring volume to 1.0L. Mix thoroughly. Gently heat and bring to boiling. Distribute into tubes or flasks. Autoclave for 15 min at 15 psi pressure–121°C. Pour into sterile Petri dishes or leave in tubes.

Use: For the cultivation and maintenance of a variety of bacteria, including *Bacillus firmus, Cytophaga johnsonae, Heliobacterium chlorum, Lysobacter enzymogenes, Rhodobacter capsulatus, Rhodobacter sphaeroides, Rhodocyclus gelatinosus, Rhodocyclus gelatinosus, Rhodomicrobium vannielii, Rhodopseudomonas palustris,* and *Rhodospirillum rubrum.*

Yeast Agar, Van Niel's with Succinate

Composition per liter:

Agar	20.0g
Yeast extract	10.0g
Sodium succinate	1.35
K_2HPO_4	1.0g
$MgSO_4·7H_2O$	0.5g

pH 7.0–7.2 at 25°C

Preparation of Medium: Add components to tap water and bring volume to 1.0L. Mix thoroughly. Gently heat and bring to boiling. Distribute into tubes or flasks. Autoclave for 15 min at 15 psi pressure–121°C. Pour into sterile Petri dishes or leave in tubes.

Use: For the cultivation and maintenance of *Rhodobacter capsulatus*.

Yeast Ascospore Agar

Composition per liter:

Agar	30.0g
Potassium acetate	10.0g
Yeast extract	2.5g
Glucose	1.0g

Preparation of Medium: Add components to distilled/deionized water and bring volume to 1.0L. Mix thoroughly. Gently heat and bring to boiling. Distribute into tubes or flasks. Autoclave for 15 min at 15 psi pressure–121°C.

Use: For the cultivation and observation of ascospore formation of yeast.

Yeast Beef Agar
See: **Antibiotic Medium 4**

Yeast Carbon Base, 10X (Wickerham Carbon Base Broth)

Composition per liter:

Glucose	10.0g
KH_2PO_4	1.0g
$MgSO_4·7H_2O$	0.5g
NaCl	0.1g
$CaCl_2·2H_2O$	0.1g
DL-Methionine	0.02g
DL-Tryptophan	0.02g
L-Histidine·HCl	0.01g
Inositol	2.0mg
H_3BO_3	0.5mg
$ZnSO_4·7H_2O$	0.4mg
$MnSO_4·4H_2O$	0.4mg
Thiamine·HCl	0.4mg
Pyridoxine	0.4mg
Niacin	0.4mg
Calcium pantothenate	0.4mg
p-Aminobenzoic acid	0.2mg
Riboflavin	0.2mg
$FeCl_3$	0.2mg
$Na_2MoO_4·4H_2O$	0.2mg
KI	0.1mg
$CuSO_4·5H_2O$	0.04mg
Folic Acid	2.0μg
Biotin	2.0μg

pH 5.5 ± 0.2 at 25°C

Source: This medium is available as a premixed powder from Difco Laboratories.

Preparation of Medium: Add components to distilled/deionized water and bring volume to 1.0L. Mix thoroughly. Filter sterilize.

Use: Used as a base to which different nitrogen sources may be added. For the cultivation and differentiation of bacteria based on their ability to utilize diverse added nitrogen sources.

Yeast Dextrose Agar

Composition per liter:

Agar	15.0g
Glucose	10.0g
Yeast extract	10.0g

pH 7.0 ± 0.2 at 25°C

Preparation of Medium: Add components to distilled/deionized water and bring volume to 1.0L. Mix thoroughly. Gently heat and bring to boiling. Adjust pH to 7.0. Distribute into tubes or flasks. Autoclave for 15 min at 15 psi pressure–121°C.

Use: For the cultivation of a variety of heterotrophic microorganisms.

Yeast Extract Agar

Composition per liter:

Agar	15.0g
Malt extract	10.0g
Glucose	4.0g
Yeast extract	4.0g

pH 7.0 ± 0.2 at 25°C

Preparation of Medium: Add components to distilled/deionized water and bring volume to 1.0L. Mix thoroughly. Gently heat and bring to boiling. Distribute into tubes or flasks. Autoclave for 15 min at 15 psi pressure–121°C. Pour into sterile Petri dishes or leave in tubes.

Use: For the isolation and cultivation of *Actinomadura* species, *Actinopolyspora* species, *Excellospora* species and *Microspora* species.

Yeast Extract Agar

Composition per liter:

Agar	15.0g
Peptone	5.0g
Yeast extract	3.0g

pH 7.2 ± 0.2 at 25°C

Source: This medium is available as a premixed powder from Oxoid Unipath.

Preparation of Medium: Add components to distilled/deionized water and bring volume to 1.0L. Mix thoroughly. Gently heat and bring to boiling. Distribute into tubes or flasks. Autoclave for 15 min at 15 psi pressure–121°C. Pour into sterile Petri dishes or leave in tubes.

Use: For the enumeration of microorganisms in potable and freshwater samples.

Yeast Extract Agar

Composition per liter:

Agar	15.0g
Peptone	9.5g
Yeast extract	7.0g
Beef extract	5.0g
NaCl	5.0g

pH 7.0 ± 0.2 at 25°C

Preparation of Medium: Add components to distilled/deionized water and bring volume to 1.0L. Mix thoroughly. Gently heat and bring to boiling. Distribute into tubes or flasks. Autoclave for 15 min at 15 psi pressure–121°C. Pour into sterile Petri dishes or leave in tubes.

Use: For the cultivation of *Aeromonas salmonicida*.

Yeast Extract Agar

Composition per liter:

Agar	20.0g
Yeast extract	1.0g
Buffer solution	2.0mL

pH 6.0 ± 0.2 at 25°C

Buffer Solution:

Composition per 400mL:

KH_2PO_4	60.0g
Na_2HPO_4	40.0g

Preparation of Buffer Solution: Add 40.0g Na_2HPO_4 to 300.0mL of distilled/deionized water. Mix thoroughly. Add 60.0g of KH_2PO_4. Mix thoroughly. Adjust pH to 6.0.

Preparation of Medium: Add components to distilled/deionized water and bring volume to 1.0L. Mix thoroughly. Autoclave for 15 min at 15 psi pressure–121°C. Pour into sterile Petri dishes.

Use: For identification of *Histoplasma capsulatum*, *Blastomyces dermatitidis* and *Coccidioides immitis*.

Yeast Extract Agar

Composition per liter:

Agar	15.0g
Proteose peptone	10.0g
NaCl	5.0g
Yeast extract	3.0g

Preparation of Medium: Add components to distilled/deionized water and bring volume to 1.0L. Mix thoroughly. Gently heat and bring to boiling. Distribute into tubes or flasks. Autoclave for 15 min at 15 psi pressure–121°C. Pour into sterile Petri dishes or leave in tubes.

Use: For the cultivation of a variety of heterotrophic microorganisms.

Yeast Extract Beef Extract Medium
See: **YB Medium**

Yeast Extract Dextrose Calcium Carbonate Agar
See: **YDC Agar**

Yeast Extract Glucose Calcium Carbonate Agar

Composition per liter:

$CaCO_3$	20.0g
Glucose	20.0g
Agar	15.0g
Yeast extract	10.0g

Preparation of Medium: Add components to distilled/deionized water and bring volume to 1.0L. Mix thoroughly. Gently heat and bring to boiling. Distribute into tubes or flasks. Autoclave for 15 min at 15 psi pressure–121°C. Pour into sterile Petri dishes or leave in tubes.

Use: For the isolation and cultivation of *Erwinia* species.

Yeast Extract Glucose Carbonate Medium
See: **YGC Medium**

Yeast Extract Glucose Carbonate Peptone Medium
See: **YGCP Medium**

Yeast Extract Glucose Citrate Medium
See: **YGC Medium**

Yeast Extract Glucose Citrate Medium with Cysteine
See: **YGC Medium with Cysteine**

Yeast Extract Glucose Citrate Medium with Glutamic Acid
See: **YGC Medium with Glutamic Acid**

Yeast Extract Glucose Medium

Composition per liter:

Agar	15.0g
Yeast extract	10.0g
Glucose	10.0g

Preparation of Medium: Add components to tap water and bring volume to 1.0L. Mix thoroughly. Gently heat and bring to boiling. Distribute into tubes or flasks. Autoclave for 15 min at 15 psi pressure–121°C. Pour into sterile Petri dishes or leave in tubes.

Use: For the cultivation of a variety of bacteria, including *Streptomyces* species, *Rhodococcus* species and others.

Yeast Extract Glycerol Medium

Composition per liter:

Agar	15.0g
Yeast extract	5.0g
Glycerol	50.0mL

Preparation of Medium: Add components to distilled/deionized water and bring volume to 1.0L. Mix thoroughly. Gently heat and bring to boiling. Distribute into tubes or flasks. Autoclave for 15 min at 15 psi pressure–121°C. Pour into sterile Petri dishes or leave in tubes.

Use: For the cultivation and maintenance of *Geodermatophilus obscurus* subspecies *utahensis*.

Yeast Extract Malt Extract Agar
See: **ISP Medium 2**

Yeast Extract Malt Extract Glucose Agar

Composition per liter:

Agar	20.0g
Glucose	10.0g
Neopeptone	5.0g
Malt extract	3.0g
Yeast extract	3.0g

Preparation of Medium: Add components to distilled/deionized water and bring volume to 1.0L. Mix thoroughly. Gently heat and bring to boiling. Distribute into tubes or flasks. Autoclave for 15 min at 15 psi pressure–121°C. Pour into sterile Petri dishes or leave in tubes.

Use: For the isolation and cultivation of yeasts.

Yeast Extract Mannitol Agar

Composition per liter:

Agar	15.0g
Mannitol	10.0g
$CaCO_3$	4.0g
K_2HPO_4	0.5g
Yeast extract	0.4g
$MgSO_4 \cdot 7H_2O$	0.2g
NaCl	0.1g

pH 6.8–7.0 ± 0.2 at 25°C

Preparation of Medium: Add components to distilled/deionized water and bring volume to 1.0L. Omit $CaCO_3$ if a clear solution is needed. Mix thoroughly. Gently heat and bring to boiling. Distribute into tubes or flasks. Autoclave for 15 min at 15 psi pressure–121°C. Pour into sterile Petri dishes or leave in tubes.

Use: For the cultivation of members of the Rhizobiaceae.

Yeast Extract Medium

Composition per liter:

Yeast extract	10.0g

Preparation of Medium: Add yeast extract to distilled/deionized water and bring volume to 1.0L.

Mix thoroughly. Distribute into tubes or flasks. Autoclave for 15 min at 15 psi pressure–121°C.

Use: For the cultivation of *Pseudomonas cepacia*.

Yeast Extract Nutrient Agar Medium
See: **YNA Medium**

Yeast Extract Nutrient Gelatin Medium
See: **YNG Medium**

Yeast Extract Peptone Beef Extract Medium
See: **YEPB Medium**

Yeast Extract Peptone Starch Agar

Composition per liter:

Agar	18.0g
Soluble starch	10.0g
Peptone	10.0g
$CaCO_3$	5.0g
Sodium acetate	5.0g
Yeast extract	3.0g
KH_2PO_4	0.5g
K_2HPO_4	0.5g
$MgSO_4 \cdot 7H_2O$	0.3g
Sodium citrate	0.027g
NaCl	0.01g
$MnSO_4 \cdot 5H_2O$	0.01g
$CuSO_4 \cdot 5H_2O$	1.0mg
$CoCl_2 \cdot 6H_2O$	1.0mg
$FeSO_4 \cdot 7H_2O$	1.0mg

Preparation of Medium: Add components to tap water and bring volume to 1.0L. Mix thoroughly. Gently heat and bring to boiling. Distribute into tubes or flasks. Autoclave for 15 min at 15 psi pressure–121°C. Pour into sterile Petri dishes or leave in tubes.

Use: For the cultivation and maintenance of *Bacillus* species that utilize starch as a carbon source.

Yeast Extract Peptone Sulfate Cysteine Medium
See: **YPSC Medium**

Yeast Extract Peptone Sulfate Cysteine Soft Agar
See: **YPSC Soft Agar**

Yeast Extract Phosphate Agar (YEP Agar)

Composition per liter:

Agar	20.0g
Yeast extract	1.0g
KH_2PO_4	0.3g
Na_2HPO_4	0.2g
Phenol Red	1.0mg

Source: This medium is available as a premixed powder from BBL Microbiology Systems.

Preparation of Medium: Add components to distilled/deionized water and bring volume to 1.0L. Mix thoroughly. Gently heat and bring to boiling. Distribute into tubes or flasks. Autoclave for 15 min at 15 psi pressure–121°C. Pour into sterile Petri dishes or leave in tubes.

Use: For the isolation of dimorphic pathogenic fungi from clinical specimens.

Yeast Extract Proteose Peptone Medium
See: **YEPP Medium**

Yeast Extract Sodium Lactate Medium

Composition per liter:

Agar	15.0g
Pancreatic digest of casein	10.0g
Yeast extract	10.0g
Sodium lactate	10.0g
KH_2PO_4	2.5g
$MnSO_4$	5.0mg

pH 7.0 ± 0.2 at 25°C

Preparation of Medium: Add components to distilled/deionized water and bring volume to 1.0L. Mix thoroughly. Gently heat and bring to boiling. Distribute into tubes or flasks. Autoclave for 15 min at 15 psi pressure–121°C. Pour into sterile Petri dishes or leave in tubes.

Use: For the isolation, cultivation and maintenance of *Propionibacterium* species.

Yeast Extract Tryptone Medium
See: **YT Medium**

Yeast Extract Tryptone NaCl Medium
See: **YTN Medium**

Yeast Fermentation Broth

Composition per liter:

Carbohydrate	10g
Pancreatic digest of gelatin	7.5g
Yeast extract	5.5g
Bromcresol Purple	16.0mg

Source: This medium is available as a premixed powder from BBL Microbiology Systems.

Preparation of Medium: Add components to distilled/deionized water and bring volume to 1.0L. Mix thoroughly. Distribute into test tubes, each containing an inverted Durham tube. Autoclave for 15 min at 15 psi pressure–121°C.

Use: For fermentation tests of specific carbohydrates used in the characterization and identification of yeasts. Gas accumulation in the Durham tube and a color change of the medium to yellow indicates carbohydrate fermentation.

Yeast Fermentation Medium

Composition per liter:

Peptone	7.5g
Yeast extract	4.5g
Bromthymol Blue (1.6% solution)	1.0mL
Carbohydrate solution	1.0mL

Carbohydrate Solution:
Composition per 10mL:

Carbohydrate	0.6g

Preparation of Carbohydrate Solution: Add carbohydrate to distilled/deionized water and bring volume to 10.0mL. Glucose, maltose, lactose, galactose or trehalose may be used. If raffinose is used, prepare a 12% solution. Mix thoroughly. Filter sterilize.

Preparation of Medium: Add components, except carbohydrate solution, to distilled/deionized water and bring volume to 1.0L. Mix thoroughly. Gently heat and bring to boiling. Distribute in 2.0mL volumes into test tubes that contain an inverted Durham tube. Autoclave for 15 min at 15 psi pressure–121°C. Cool to 45°–50°C. Aseptically add 1.0mL of sterile carbohydrate solution. Mix thoroughly.

Use: For the cultivation and differentiation of yeast based on carbohydrate fermentation patterns. Yeast that can ferment a specific carbohydrate turn the medium yellow.

Yeast Glucose Broth

Composition per liter:

Pancreatic digest of gelatin	7.75g
Beef extract	4.75g
Yeast extract	2.5g
K_2HPO_4	2.5g
Glucose	1.0g

pH 7.0 ± 0.2 at 25°C

Preparation of Medium: Add components to distilled/deionized water and bring volume to 1.0L. Mix thoroughly. Distribute into tubes or flasks. Autoclave for 15 min at 15 psi pressure–121°C.

Use: For the cultivation of *Staphylococcus caseolyticus*.

Yeast Malate Medium

Composition per liter:

Yeast extract	5.0g
Sodium malate	1.0g

pH 7.0 ± 0.2 at 25°C

Preparation of Medium: Add components to distilled/deionized water and bring volume to 1.0L. Mix thoroughly. Distribute into tubes or flasks. Autoclave for 15 min at 15 psi pressure–121°C.

Use: For the cultivation of *Rhodopseudomonas viridis*.

Yeast Malt Extract Agar (YM Agar)

Composition per liter:

Agar	20.0g
Glucose	10.0g
Peptone	5.0g
Yeast extract	3.0g
Malt extract	3.0g

pH 6.2 ± 0.2 at 25°C

Source: This medium is available as a premixed powder from Difco Laboratories.

Preparation of Medium: Add components to distilled/deionized water and bring volume to 1.0L. Mix thoroughly. Gently heat and bring to boiling. Distribute into tubes or flasks. Autoclave for 15 min at 15 psi pressure–121°C. The medium may be rendered selective by adjusting the pH to 3.0–4.0 at 45–55°C or by the addition of antibiotics at 45–50°C or below. Pour into sterile Petri dishes or leave in tubes.

Use: For the cultivation of fungi, including yeasts, and other aciduric microorganisms such as *Actino-*

planes species, *Streptomyces* species, *Streptoverticillium* species, and *Nocardia* species.

Yeast Malt Extract Broth (YM Broth)

Composition per liter:

Glucose	10.0g
Peptone	5.0g
Yeast extract	3.0g
Malt extract	3.0g

pH 6.2 ± 0.2 at 25°C

Source: This medium is available as a premixed powder from Difco Laboratories.

Preparation of Medium: Add components to distilled/deionized water and bring volume to 1.0L. Mix thoroughly. Distribute into tubes or flasks. Autoclave for 15 min at 15 psi pressure–121°C. The medium may be rendered selective by adjusting the pH to 3.0–4.0 at 45–55°C or by the addition of antibiotics at 45–50°C or below.

Use: For the cultivation of yeasts, molds and other aciduric microorganisms such as *Actinoplanes* species, *Streptomyces* species, *Streptoverticillium* species, and *Nocardia* species.

Yeast Malt Extract Catalase Agar (YM Catalase Agar)

Composition per liter:

Agar	15.0g
K_2HPO_4	5.74g
Malt extract	5.0g
Yeast extract	5.0g
$NH_4H_2PO_4$	1.15g
$MgSO_4·7H_2O$ solution	10.0mL
Catalase solution	10.0mL

Catalase Solution:

Composition per 10mL:

Catalase	60.0mg

Preparation of Catalase Solution: Add catalase to distilled/deionized water and bring volume to 10.0mL. Mix thoroughly. Filter sterilize.

Magnesium Sulfate Solution:

Composition per 10mL:

$MgSO_4·7H_2O$	205.0mg

Preparation of Magnesium Sulfate Solution: Add $MgSO_4·7H_2O$ to distilled/deionized water and bring volume to 10.0mL. Mix thoroughly. Filter sterilize.

Preparation of Medium: Add components, except catalase and magnesium sulfate, to distilled/deionized water and bring volume to 980.0mL. Mix thoroughly. Gently heat and bring to boiling. Distribute into tubes or flasks. Autoclave for 15 min at 15 psi pressure–121°C. Cool to 50°C. Aseptically add filter-sterilized catalase and $MgSO_4·7H_2O$ solutions.

Use: For the cultivation and maintenance of *Rarobacter faecitabidus*.

Yeast Mannitol Agar

Composition per liter:

Agar	15.0g
Mannitol	10.0g
K_2HPO_4	0.5g
Yeast extract	0.4g
$MgSO_4·7H_2O$	0.2g
NaCl	0.1g

Preparation of Medium: Add components to distilled/deionized water and bring volume to 1.0L. Mix thoroughly. Gently heat and bring to boiling. Distribute into tubes or flasks. Autoclave for 15 min at 15 psi pressure–121°C. Pour into sterile Petri dishes or leave in tubes.

Use: For the cultivation of *Rhizobium* and *Azorhizobium* species.

Yeast Milk Medium

Composition per liter:

Agar	15.0g
Skim milk	10.0g
Yeast extract	1.0g

Preparation of Medium: Add components to distilled/deionized water and bring volume to 1.0L. Mix thoroughly. Gently heat and bring to boiling. Distribute into tubes or flasks. Autoclave for 15 min at 15 psi pressure–121°C. Pour into sterile Petri dishes or leave in tubes.

Use: For the cultivation and maintenance of *Lysobacter enzymogenes*.

Yeast Morphology Agar

Composition per liter:

$(NH_4)_2SO_4$	32.5g
Agar	18.0g
Glucose	10.0g
Asparagine	1.5g
KH_2PO_4	1.0g
$MgSO_4·7H_2O$	0.5g
NaCl	0.5g
$CaCl_2·2H_2O$	0.1g
DL-Methionine	0.02g
DL-Tryptophan	0.02g

L-Histidine·HCl	0.01g
Inositol	2.0mg
H_3BO_3	0.5mg
Calcium pantothenate	0.4mg
$MgSO_4 \cdot 7H_2O$	0.4mg
Niacin	0.4mg
Pyridoxine·HCl	0.4mg
Thiamine·HCl	0.4mg
$ZnSO_4 \cdot 7H_2O$	0.4mg
p-Aminobenzoic acid	0.2mg
$FeCl_3$	0.2mg
Riboflavin	0.2mg
$Na_2MoO_4 \cdot 4H_2O$	0.2mg
KI	0.1mg
$CuSO_4 \cdot 5H_2O$	0.04mg
Biotin	2.0μg
Folic acid	2.0μg

pH 5.6 ± 0.2 at 25°C

Source: This medium is available as a premixed powder from Difco Laboratories.

Preparation of Medium: Add components to distilled/deionized water and bring volume to 1.0L. Mix thoroughly. Gently heat and bring to boiling. Distribute into tubes or flasks. Autoclave for 15 min at 15 psi pressure–121°C.

Use: For agar dilution and diffusion disk testing with 5-flucytosine. For the microbiological assay of flucytosine using *Candida kefyr* ATCC 28838 or *Saccharomyces cerevisiae* ATCC 36375 as the indicator microorganisms.

Yeast Nitrogen Base

Composition per liter:

$(NH_4)_2SO_4$	5.0g
KH_2PO_4	1.0g
$MgSO_4 \cdot 7H_2O$	0.5g
NaCl	0.1g
$CaCl_2 \cdot 2H_2O$	0.1g
DL-Methionine	0.02g
DL-Tryptophan	0.02g
L-Histidine·HCl	0.01g
Inositol	2.0mg
KI	0.1mg
H_3BO_3	0.5mg
$ZnSO_4 \cdot 7H_2O$	0.4mg
$MnSO_4 \cdot 4H_2O$	0.4mg
Thiamine·HCl	0.4mg
Pyroxidine·HCl	0.4mg
Niacin	0.4mg
Calcium pantothenate	0.4mg
p-Aminobenzoic acid	0.2mg
Riboflavin	0.2mg
$FeCl_3$	0.2mg
$Na_2MoO_4 \cdot 4H_2O$	0.2mg
$CuSO_4 \cdot 5H_2O$	0.04mg
Folic acid	2.0μg
Biotin	2.0μg

pH 5.5 ± 0.2 at 25°C

Source: This medium is available as a premixed powder from BBL Microbiology Systems and Difco Laboratories.

Preparation of Medium: Add components to distilled/deionized water and bring volume to 1.0L. Mix thoroughly. Distribute into tubes or flasks. Autoclave for 15 min at 15 psi pressure–121°C. Alternately for carbon assimilation tests, prepare a 10× concentrated solution by adding components to distilled/deionized water and bring volume to 100.0mL. Mix thoroughly. Distribute into tubes or flasks. Autoclave for 15 min at 15 psi pressure–121°C. Prepare a carbohydrate solution by adding 0.5g of carbohydrate to 90.0mL of distilled/deionized water. Mix thoroughly. Filter sterilize. Aseptically add 0.5mL of the 10× concentrated solution to 4.5mL of the filter sterilized carbohydrate solution. Mix thoroughly.

Use: For carbohydrate assimilation tests in the characterization and identification of yeasts.

Yeast Nitrogen Base Glucose Broth

Composition per liter:

Yeast nitrogen base	25.0mL
Glucose solution	25.0mL

pH 5.6 ± 0.2 at 25°C

Yeast Nitrogen Base:

Composition per 500mL:

$(NH_4)_2SO_4$	5.0g
KH_2PO_4	1.0g
$MgSO_4 \cdot 7H_2O$	0.5g
NaCl	0.1g
$CaCl_2 \cdot 2H_2O$	0.1g
DL-Methionine	0.02g
DL-Tryptophan	0.02g
L-Histidine·HCl	0.01g
Inositol	2.0mg
H_3BO_3	0.5mg
$ZnSO_4 \cdot 7H_2O$	0.4mg
$MnSO_4 \cdot 4H_2O$	0.4mg
Thiamine·HCl	0.4mg
Pyroxidine·HCl	0.4mg
Niacin	0.4mg
Calcium pantothenate	0.4mg
p-Aminobenzoic acid	0.2mg
Riboflavin	0.2mg
$FeCl_3$	0.2mg

$Na_2MoO_4·4H_2O$	0.2mg
KI	0.1mg
$CuSO_4·5H_2O$	0.04mg
Folic acid	2.0μg
Biotin	2.0μg

Source: Yeast nitrogen base is available as a premixed powder from BBL Microbiology Systems.

Preparation of Yeast Nitrogen Base: Add components to distilled/deionized water and bring volume to 500.0mL. Mix thoroughly. Filter sterilize.

Glucose Solution:
Composition per 500mL:

Glucose	10.0g

Preparation of Glucose Solution: Add glucose to distilled/deionized water and bring volume to 500.0mL. Mix thoroughly. Filter sterilize.

Preparation of Medium: Aseptically combine 25.0mL of sterile yeast nitrogen base and 25.0mL of sterile glucose solution. Mix thoroughly.

Use: For the cultivation and enrichment of yeast from sewage and polluted waters.

Yeast Nitrogen Base with Carbohydrate

Composition per liter:

Carbohydrate	5.0g
$(NH_4)_2SO_4$	5.0g
KH_2PO_4	1.0g
$MgSO_4·7H_2O$	0.5g
NaCl	0.1g
$CaCl_2·2H_2O$	0.1g
DL-Methionine	0.02g
DL-Tryptophan	0.02g
L-Histidine·HCl	0.01g
Inositol	2.0mg
KI	0.1mg
H_3BO_3	0.5mg
$ZnSO_4·7H_2O$	0.4mg
$MnSO_4·4H_2O$	0.4mg
Thiamine·HCl	0.4mg
Pyroxidine·HCl	0.4mg
Niacin	0.4mg
Calcium pantothenate	0.4mg
p-Aminobenzoic acid	0.2mg
Riboflavin	0.2mg
$FeCl_3$	0.2mg
$Na_2MoO_4·4H_2O$	0.2mg
$CuSO_4·5H_2O$	0.04mg
Folic acid	2.0μg
Biotin	2.0μg

pH 5.6 ± 0.2 at 25°C

Preparation of Medium: Add components to distilled/deionized water and bring volume to 1.0L. Mix thoroughly. Filter sterilize. Aseptically distribute into tubes or flasks.

Use: For carbohydrate assimilation tests in the characterization and identification of yeasts.

Yeast Nitrogen Base, 10X with Asparagine and Glucose

Composition per liter:

Glucose	10.0g
$(NH_4)_2SO_4$	5.0g
L-Asparagine	1.5g
KH_2PO_4	1.0g
$MgSO_4·7H_2O$	0.5g
NaCl	0.1g
$CaCl_2·2H_2O$	0.1g
DL-Methionine	0.02g
DL-Tryptophan	0.02g
L-Histidine·HCl	0.01g
Inositol	2.0mg
H_3BO_3	0.5mg
$ZnSO_4·7H_2O$	0.4mg
$MnSO_4·4H_2O$	0.4mg
Thiamine·HCl	0.4mg
Pyridoxine·HCl	0.4mg
Niacin	0.4mg
Calcium pantothenate	0.4mg
p-Aminobenzoic acid	0.2mg
Riboflavin	0.2mg
$FeCl_3$	0.2mg
$Na_2MoO_4·4H_2O$	0.2mg
KI	0.1mg
$CuSO_4·5H_2O$	0.04mg
Folic Acid	2.0μg
Biotin	2.0μg

pH 5.6 ± 0.2 at 25°C

Preparation of Medium: Add components to distilled/deionized water and bring volume to 1.0L. Dilute 100.0mL of 10× medium with 900.0mL of distilled/deionized water. Mix thoroughly. Filter sterilize. Aseptically distribute into sterile tubes or flasks.

Use: For susceptibility tests with yeasts and fungi.

Yeast Peptone Broth

Composition per liter:

Yeast extract	2.5g
Peptone	2.5g

pH 7.0 ± 0.2 at 25°C

Preparation of Medium: Add components to distilled/deionized water and bring volume to 1.0L. Mix

thoroughly. Distribute into tubes or flasks. Autoclave for 15 min at 15 psi pressure–121°C.

Use: For the cultivation of *Rhodopseudomonas* species.

Yeast Synthetic Minimal Medium

Composition per liter:

D-Glucose	20.0g
Agar	15.0g
$(NH_4)_2SO_4$	5.0g
KH_2PO_4	1.0g
$MgSO_4 \cdot 7H_2O$	0.5g
NaCl	0.1g
$CaCl_2 \cdot 2H_2O$	0.1g
Inositol	2.0mg
H_3BO_3	0.5mg
$ZnSO_4 \cdot 7H_2O$	0.4mg
$MnSO_4 \cdot 4H_2O$	0.4mg
Thiamine·HCl	0.4mg
Pyridoxine·HCl	0.4mg
Niacin	0.4mg
Calcium pantothenate	0.4mg
p-Aminobenzoic acid	0.2mg
Riboflavin	0.2mg
$FeCl_3$	0.2mg
$Na_2MoO_4 \cdot 4H_2O$	0.2mg
KI	0.1mg
$CuSO_4 \cdot 5H_2O$	0.04mg
Folic acid	2.0μg
Biotin	2.0μg

pH 5.6 ± 0.2 at 25°C

Preparation of Medium: Add agar to 900.0mL of distilled/deionized water. Mix thoroughly. Gently heat and bring to boiling. Distribute into tubes or flasks. Autoclave for 15 min at 15 psi pressure–121°C. Cool to 45°–50°C. In a separate flask, add remaining components to 100.0mL of distilled/deionized water. Mix thoroughly. Filter sterilize. Aseptically combine the two sterile solutions. Mix thoroughly. Pour into sterile Petri dishes.

Use: For the cultivation of a wide variety of heterotrophic microorganisms.

Yeast Tryptone Medium

Composition per liter:

Pancreatic digest of casein	10.0g
NaCl	10.0g
Yeast extract	5.0g

pH 7.0 ± 0.2 at 25°C

Preparation of Medium: Add components to distilled/deionized water and bring volume to 1.0L. Mix thoroughly. Distribute into tubes or flasks. Autoclave for 15 min at 15 psi pressure–121°C.

Use: For the cultivation of *Escherichia coli.*

Yeast Tryptone Medium with Streptomycin

Composition per liter:

Pancreatic digest of casein	10.0g
NaCl	10.0g
Yeast extract	5.0g
Streptomycin	0.2g

pH 7.0 ± 0.2 at 25°C

Preparation of Medium: Add components to distilled/deionized water and bring volume to 1.0L. Mix thoroughly. Distribute into tubes or flasks. Autoclave for 15 min at 15 psi pressure–121°C.

Use: For the cultivation of *Escherichia coli.*

Yeast Tryptone Starch Medium

Composition per liter:

Agar	15.0g
Soluble starch	10.0g
Yeast extract	5.0g
Pancreatic digest of casein	5.0g
KH_2PO_4	2.0g
$CaCl_2 \cdot 2H_2O$	0.5g
$MnCl_2 \cdot 4H_2O$	0.5g

Preparation of Medium: Add components to distilled/deionized water and bring volume to 1.0L. Mix thoroughly. Gently heat and bring to boiling. Distribute into tubes or flasks. Autoclave for 15 min at 15 psi pressure–121°C. Pour into sterile Petri dishes or leave in tubes.

Use: For the cultivation and maintenance of *Bacillus circulans.*

Yeast Water Agar

Composition per liter:

Glucose	20.0g
Agar	15.0g
Casein hydrolysate	5.0g
Yeast extract	4.0g
KH_2PO_4	0.55g
KCl	0.40g
$CaCl_2$	0.13g
$MgCl_2 \cdot 7H_2O$	0.13g
$FeCl_3 \cdot 6H_2O$	2.5mg
$MnSO_4 \cdot 4H_2O$	2.5mg
Bromcresol Green solution	1.0mL

Bromcresol Green Solution:

Composition per 10mL:

Bromcresol Green	0.22g
Ethanol	10.0mL

Preparation of Bromcresol Green Solution: Add Bromcresol Green to 10.0mL of ethanol. Mix thoroughly. Filter sterilize.

Preparation of Medium: Add components to distilled/deionized water and bring volume to 1.0L. Mix thoroughly. Gently heat and bring to boiling. Distribute into tubes or flasks. Autoclave for 15 min at 15 psi pressure–121°C. Pour into sterile Petri dishes or leave in tubes.

Use: For the cultivation of *Zymomonas* species.

YEP Agar
See: Yeast Extract Phosphate Agar

YEP Galactose Agar

Composition per liter:

Agar	20.0g
Galactose	20.0g
Peptone	20.0g
Yeast extract	10.0g

Preparation of Medium: Add components to distilled/deionized water and bring volume to 1.0L. Mix thoroughly. Gently heat and bring to boiling. Distribute into tubes or flasks. Autoclave for 15 min at 15 psi pressure–121°C. Pour into sterile Petri dishes or leave in tubes.

Use: For the cultivation of a variety of heterotrophic microorganisms.

YEPB Medium (Yeast Extract Peptone Beef Extract Medium)

Composition per liter:

Beef extract	10.0g
Polypeptone™	10.0g
Yeast extract	5.0g
NaCl	3.0g
$MnCl_2 \cdot 4H_2O$	0.1g

pH 7.0 ± 0.2 at 25°C

Preparation of Medium: Add components to distilled/deionized water and bring volume to 1.0L. Mix thoroughly. Adjust pH to 7.0 with KOH. Distribute into tubes or flasks. Autoclave for 15 min at 15 psi pressure–121°C. Adjust pH to 7.0 with KOH.

Use: For the cultivation of *Microbacterium* species.

YEPD Medium

Composition per liter:

Agar	20.0g
Glucose	20.0g
Peptone	20.0g
Yeast extract	10.0g

Preparation of Medium: Add components to distilled/deionized water and bring volume to 1.0L. Mix thoroughly. Gently heat and bring to boiling. Distribute into tubes or flasks. Autoclave for 15 min at 15 psi pressure–121°C. Pour into sterile Petri dishes or leave in tubes.

Use: For the cultivation of a variety of heterotrophic microorganisms.

YEPP Medium (Yeast Extract Proteose Peptone Medium)

Composition per liter:

Agar	15.0g
Proteose peptone	10.0g
NaCl	5.0g
Yeast extract	3.0g

pH 7.2-7.4 ± 0.2 at 25°C

Preparation of Medium: Add components to distilled/deionized water and bring volume to 1.0L. Mix thoroughly. Gently heat and bring to boiling. Distribute into tubes or flasks. Autoclave for 15 min at 15 psi pressure–121°C. Pour into sterile Petri dishes or leave in tubes.

Use: For the cultivation and maintenance of *Pseudomonas* species.

Yersinia Isolation Agar
See: CAL Agar

Yersinia Selective Agar Base

Composition per liter:

Mannitol	20.0g
Peptone	17.0g
Agar	12.5g
Proteose peptone	3.0g
Yeast extract	2.0g
Sodium pyruvate	2.0g
NaCl	1.0g
Sodium desoxycholate	0.5g
$MgSO_4 \cdot 7H_2O$	0.01g
Neutral Red	0.03g
Crystal Violet	1.0mg
Selective supplement	6.0mL

pH 7.4 ± 0.2 at 25°C

Source: This medium is available as a premixed powder from Difco Laboratories and Oxoid Unipath.

Selective Supplement:
Composition per 6mL:

Cefsulodin	15.0mg
Irgasan	4.0mg
Novobiocin	2.5mg
Ethanol	2.0mL

Preparation of Selective Supplement: Aseptically add components to 4.0mL distilled/deionized water and 2.0mL ethanol. Mix thoroughly.

Preparation of Medium: Add components to distilled/deionized water and bring volume to 1.0L. Mix thoroughly. Gently heat and bring to boiling. Distribute into tubes or flasks. Autoclave for 15 min at 15 psi pressure–121°C. Cool to 50°C. Aseptically add selective supplement. Mix thoroughly. Pour into sterile Petri dishes or leave in tubes.

Use: For the isolation and enumeration of *Yersinia enterocolitica* from food and clinical specimens.

Yersinia Selective Agar
See: **CIN Agar**

YGC Medium (Yeast Extract Glucose Carbonate Medium) (ATCC Medium 73)

Composition per liter:

Agar	20.0g
Glucose	20.0g
$CaCO_3$	20.0g
Yeast extract	10.0g

Preparation of Medium: Add components to distilled/deionized water and bring volume to 1.0L. Mix thoroughly. Gently heat and bring to boiling. Distribute into tubes or flasks. Autoclave for 30 min at 10 psi pressure–115°C. Cool to 48°C. Mix thoroughly. Pour into sterile Petri dishes or leave in tubes.

Use: For the cultivation of *Xanthomonas* species, *Erwinia* species, *Kluyvera* species, *Rhodococcus* species, *Streptomyces* species, *Pseudomonas psueudoalcaligenes* and *Xylophilus ampelinus*.

YGC Medium (Yeast Extract Glucose Carbonate Medium) (ATCC Medium 459)

Composition per liter:

Glucose	50.0g
Agar	15.0g
$CaCO_3$	12.5g
Yeast extract	5.0g

Preparation of Medium: Add components to distilled/deionized water and bring volume to 1.0L. Mix thoroughly. Gently heat and bring to boiling. Distribute into tubes or flasks. Autoclave for 15 min at 15 psi pressure–121°C. Cool to 50°C. Pour into sterile Petri dishes or leave in tubes.

Use: For the cultivation and maintenance of *Acetobacter* species.

YGC Medium (Yeast Extract Glucose Citrate Medium) (ATCC Medium 216)

Composition per liter:

Beef extract	10.0g
Glucose	10.0g
Peptone	10.0g
Ammonium citrate	5.0g
Yeast extract	5.0g
Sodium acetate	2.0g
Tween™ 80	1.0g
$MgSO_4 \cdot 7H_2O$	0.2g
$MnSO_4 \cdot 4H_2O$	0.05g

pH 6.5 ± 0.2 at 25°C

Preparation of Medium: Add components to distilled/deionized water and bring volume to 1.0L. Mix thoroughly. Distribute into tubes or flasks. Autoclave for 15 min at 15 psi pressure–121°C.

Use: For the isolation and cultivation of *Leuconostoc* species.

YGC Medium with Cysteine (Yeast Extract Glucose Citrate Medium with Cysteine)

Composition per liter:

Glucose	10.0g
Peptone	10.0g
Beef extract	10.0g
Yeast extract	5.0g
Ammonium citrate	5.0g
Sodium acetate	2.0g
Tween™ 80	1.0g
$MgSO_4 \cdot 7H_2O$	0.2g
$MnSO_4 \cdot 4H_2O$	0.05g
Cysteine·HCl·H_2O solution	10.0mL

pH 6.5 ± 0.2 at 25°C

Cysteine Hydrochloride Solution:
Composition per 10mL:

Cysteine·HCl·H_2O	0.5g

Preparation of Cysteine Hydrochloride Solution: Add Cysteine·HCl·H_2O to distilled/deionized water and bring volume to 10.0mL. Mix thoroughly. Filter sterilize.

Preparation of Medium: Add components, except cysteine·HCl·H_2O, to distilled/deionized water and bring volume to 990.0mL. Mix thoroughly. Distribute into tubes or flasks. Autoclave for 15 min at 15 psi pressure–121°C. Aseptically add cysteine hydrochloride solution.

Use: For the cultivation and maintenance of *Leuconostoc mesenteroides*.

YGC Medium with Glutamic Acid (Yeast Extract Glucose Carbonate Medium with Glutamic Acid)

Composition per liter:

Agar	20.0g
Glucose	20.0g
$CaCO_3$	20.0g
Agar	20.0g
Yeast extract	10.0g
Glutamic acid	0.1g

Preparation of Medium: Add components to distilled/deionized water and bring volume to 1.0L. Mix thoroughly. Gently heat and bring to boiling. Distribute into tubes or flasks. Autoclave for 30 min at 10 psi pressure–115°C. Cool to 48°C. Mix thoroughly. Pour into sterile Petri dishes or leave in tubes.

Use: For the cultivation and maintenance of *Xanthomonas campestris*.

YGCP Medium (Yeast Extract Glucose Carbonate Peptone Medium)

Composition per liter:

Glucose	20.0g
Agar	17.5g
$CaCO_3$	10.0g
Yeast extract	2.5g
Peptone	2.5g
NaCl	1.0g
K_2HPO_4	1.0g
$MgSO_4$	0.5g

Preparation of Medium: Add components, except calcium carbonate, to distilled/deionized water and bring volume to 1.0L. Mix thoroughly. Gently heat and bring to boiling. Adjust pH to 7.2. Add calcium carbonate. Mix thoroughly. Distribute into tubes or flasks. Autoclave for 15 min at 15 psi pressure–121°C. Pour into sterile Petri dishes or leave in tubes.

Use: For the cultivation and maintenance of*Xanthomonas campestris* and *Xanthomonas oryzae*.

YM Agar
See: **Yeast Malt Extract Agar**

YM Broth
See: **Yeast Malt Extract Broth**

YM Catalase Agar
See: **Yeast Malt Extract Catalase Agar**

YNA Medium (Yeast Extract Nutrient Agar Medium)

Composition per liter:

Agar	15.0g
NaCl	5.0g
Peptone	5.0g
Meat extract	4.0g
Yeast extract	2.5g

pH 7.0 ± 0.2 at 25°C

Preparation of Medium: Add components to distilled/deionized water and bring volume to 1.0L. Mix thoroughly. Gently heat and bring to boiling. Distribute into tubes or flasks. Autoclave for 15 min at 15 psi pressure–121°C. Pour into sterile Petri dishes or leave in tubes.

Use: For the isolation and cultivation of *Kurthia* species according to the agar streak method.

YNG Medium (Yeast Extract Nutrient Gelatin Medium)

Composition per liter:

Gelatin	100.0g
NaCl	5.0g
Peptone	5.0g

Meat extract 4.0g
Yeast extract...2.5g
pH 7.0 ± 0.2 at 25°C

Preparation of Medium: Add components to distilled/deionized water and bring volume to 1.0L. Mix thoroughly. Gently heat until dissolved. Distribute into tubes or flasks. Autoclave for 30 min at 10 psi pressure–115°C.

Use: For the isolation and cultivation of *Kurthia* species using the gelatin streak method.

Yopp's Medium

Composition per liter:

NaCl...116.88g
$MgCl_2 \cdot 6H_2O$..10.68g
$MgSO_4 \cdot 7H_2O$ 10.0g
KCl..2.0g
$CaNO_3 \cdot 4H_2O$...1.0g
Glycyl-glycine buffer..0.5g
$K_2HPO_4 \cdot 3H_2O$..0.065g
Ferric EDTA.. 5.0mg
Trace metal solution....................................... 1.0mL
pH 7.8 ± 0.2 at 25°C

Trace Metal Solution :

Composition per liter:

$MnCl_2 \cdot 4H_2O$..2.0g
H_3BO_3 ....0.5g
$ZnNO_3 \cdot 6H_2O$...0.5g
$Co(NO_3)_2 \cdot 6H_2O$..0.025g
$CuCl_2 \cdot 2H_2O$..0.025g
$Na_2MoO_4 \cdot 2H_2O$..0.025g
$VOSO_4 \cdot 6H_2O$ 0.025g
HCl.. 3.0mL

Preparation of Trace Metal Solution: Add components to distilled/deionized water and bring volume to 1.0L. Mix thoroughly.

Preparation of Medium: Add components to distilled/deionized water and bring volume to 1.0L. Mix thoroughly. Gently heat and bring to boiling. Distribute into tubes or flasks. Autoclave for 15 min at 15 psi pressure–121°C.

Use: For the isolation and cultivation of halophilic cyanobacteria.

YPC Medium

Composition per liter:

Agar...15.0g
Proteose peptone 15.0g
Yeast extract...5.0g
KH_2PO_4...4.0g
Sucrose..2.5g
Glucose 2.0g
L-Cystine 0.5g
Na_2SO_3 ....0.2g
pH 7.2 ± 0.2 at 25°C

Preparation of Medium: Add components to distilled/deionized water and bring volume to 1.0L. Mix thoroughly. Gently heat and bring to boiling. Distribute into tubes or flasks. Autoclave for 15 min at 15 psi pressure–121°C. Pour into sterile Petri dishes or leave in tubes.

Use: For the cultivation of *Pasteurella multocida*.

YPSC Agar (Yeast Extract Peptone Sulfate Cysteine Agar)

Composition per liter:

Agar...15.0g
Yeast extract...1.0g
Peptone...1.0g
Sodium acetate·$3H_2O$...0.5g
$MgSO_4 \cdot 7H_2O$ 0.25g
$CaCl_2 \cdot 2H_2O$...0.25g
L-Cysteine·HCl·H_2O..0.05g
pH 7.5 ± 0.2 at 25°C

Preparation of Medium: Add components to distilled/deionized water and bring volume to 1.0L. Mix thoroughly. Gently heat and bring to boiling. Distribute into tubes or flasks. Autoclave for 15 min at 15 psi pressure–121°C. Adjust pH to 7.5 with sterile 10M NaOH. Pour into sterile Petri dishes or leave in tubes.

Use: For the cultivation and maintenance of *Bdellovibrio* species.

YPSC Agar, Cation–Supplemented

Composition per liter:

Sodium acetate·$3H_2O$...50.0g
Agar...15.0g
Peptone...10.0g
Yeast extract...10.0g
$MgSO_4 \cdot 7H_2O$ 0.74g
$CaCl_2 \cdot 2H_2O$...0.29g
L-Cysteine·HCl·H_2O..0.05g
Bacitracin solution 10.0mL

Bacitracin Solution:

Composition per 10mL:

Bacitracin 6,000U

Preparation of Bacitracin Solution: Add bacitracin to distilled/deionized water and bring volume to 10.0mL. Mix thoroughly. Filter sterilize.

Preparation of Medium: Add components, except bacitracin solution, to distilled/deionized water and bring volume to 990.0mL. Mix thoroughly. Gently heat and bring to boiling. Autoclave for 15 min at 15 psi pressure–121°C. Cool to 45°–50°C. Aseptically add sterile bacitracin solution. Mix thoroughly. Pour into sterile Petri dishes or distribute into sterile tubes.

Use: For the cultivation and enumeration of *Bdellovibrio* species.

YPSC Medium (Yeast Extract Peptone Sulfate Cysteine Medium)

Composition per liter:

Yeast extract	1.0g
Peptone	1.0g
Sodium acetate·$3H_2O$	0.5g
$MgSO_4 \cdot 7H_2O$	0.25g
$CaCl_2 \cdot 2H_2O$	0.25g
L-Cysteine·HCl·H_2O	0.05g

pH 7.5 ± 0.2 at 25°C

Preparation of Medium: Add components to distilled/deionized water and bring volume to 1.0L. Mix thoroughly. Distribute into tubes or flasks. Autoclave for 15 min at 15 psi pressure–121°C. Adjust pH to 7.5 with sterile 10M NaOH.

Use: For the cultivation and maintenance of *Bdellovibrio* species.

YPSC Soft Agar (Yeast Extract Peptone Sulfate Cysteine Soft Agar)

Composition per liter:

Agar	6.0g
Yeast extract	1.0g
Peptone	1.0g
Sodium acetate·$3H_2O$	0.5g
$MgSO_4 \cdot 7H_2O$	0.25g
$CaCl_2 \cdot 2H_2O$	0.25g
L-Cysteine·HCl·H_2O	0.05g

pH 7.5 ± 0.2 at 25°C

Preparation of Medium: Add components to distilled/deionized water and bring volume to 1.0L. Mix thoroughly. Gently heat and bring to boiling. Distribute into tubes or flasks. Autoclave for 15 min at 15 psi pressure–121°C. Adjust pH to 7.5 with sterile 10M NaOH. Pour into sterile Petri dishes or leave in tubes.

Use: For the cultivation and maintenance of *Bdellovibrio* species.

YT Medium (Yeast Extract Tryptone Medium)

Composition per liter:

Pancreatic digest of casein	8.0g
Yeast extract	5.0g
NaCl	5.0g

Preparation of Medium: Add components to distilled/deionized water and bring volume to 1.0L. Mix thoroughly. Distribute into tubes or flasks. Autoclave for 15 min at 15 psi pressure–121°C.

Use: For the cultivation of *Escherichia coli.*

YTN Medium (Yeast Extract Tryptone NaCl Medium)

Composition per liter:

NaCl	30.0g
Agar	15.0g
Yeast extract	10.0g
Pancreatic digest of casein	10.0g
Glucose	1.0g
Trace elements solution	1.0mL

Trace Elements Solution:

Composition per liter:

H_3BO_3	2.85g
$MnCl_2 \cdot 4H_2O$	1.8g
Sodium tartrate	1.77g
$FeSO_4$	1.36g
$CoCl_2 \cdot 6H_2O$	0.04g
$CuCl_2.2H_2O$	0.027g
$Na_2MoO_4 \cdot 2H_2O$	0.025g
$ZnCl_2$	0.021g

pH 7.2 ± 0.2 at 25°C

Preparation of Trace Elements Solution: Add components to distilled/deionized water and bring volume to 1.0L. Mix thoroughly.

Preparation of Medium: Add components to tap water and bring volume to 1.0L. Mix thoroughly. Gently heat and bring to boiling. Distribute into tubes or flasks. Autoclave for 15 min at 15 psi pressure–121°C. Pour into sterile Petri dishes or leave in tubes.

Use: For the cultivation of ATCC strain 21588.

Zoogloea Medium

Composition per liter:

Agar	15.0g
Pancreatic digest of casein	5.0g
Glycerol	5.0g
Yeast autolysate	1.0g
Sodium lactate	0.5g

Preparation of Medium: Add components to distilled/deionized water and bring volume to 1.0L. Mix thoroughly. Gently heat and bring to boiling. Distribute into screw cap test tubes. Autoclave for 15 min at 15 psi pressure–121°C. Pour into sterile Petri dishes or leave in tubes.

Use: For cultivation and maintenance of *Zoogloea ramigera* and other *Zoogloea* species.

Zymobacterium Agar

Composition per liter:

Pancreatic digest of casein	20.0g
Agar	15.0g
Orotic acid	2.0g
$Na_3PO_4 \cdot 12H_2O$	1.5g
Sodium thioglycollate	1.0g
NaOH	0.5g
Riboflavin	0.015g

pH 7.9 ± 0.2 at 25°C

Preparation of Medium: Add components to distilled/deionized water in the following order: tryptone, orotic acid, sodium hydroxide, sodium, phosphate, sodium thioglycollate, riboflavin, agar and bring volume to 1.0L . Mix thoroughly. Some orotic acid will remain undissolved. Adjust pH to 7.9 using NaH_2PO_4. Distribute into tubes or flasks. Autoclave for 15 min at 15 psi pressure–121°C. Mix thoroughly to dissolve orotic acid. Pour into sterile Petri dishes or leave in tubes.

Use: For cultivation and maintenance of *Clostridium (Zymobacterium) oroticum.*

Zymobacterium Broth

Composition per liter:

Pancreatic digest of casein	20.0g
Orotic acid	2.0g
$Na_3PO_4 \cdot 12H_2O$	1.5g
Sodium thioglycollate	1.0g
NaOH	0.5g
Riboflavin	0.015g

pH 7.9 ± 0.2 at 25°C

Preparation of Medium: Add components to distilled/deionized water in the following order: tryptone, orotic acid, sodium hydroxide, sodium, phosphate, sodium thioglycollate, riboflavin and bring volume to 1.0L. Mix thoroughly. Some orotic acid will remain undissolved. Adjust pH to 7.9 using NaH_2PO_4. Distribute into tubes or flasks. Autoclave for 15 min at 15 psi pressure–121°C. Mix again to dissolve orotic acid.

Use: For cultivation and maintenance of *Clostridium (Zymobacterium) oroticu*m.

Zymomonas Medium

Composition per liter:

Glucose	20.0g
Agar	15.0g
Peptone	10.0g
Yeast extract	10.0g

pH 6.8 ± 0.2 at 25°C

Preparation of Medium: Add components to distilled/deionized water and bring volume to 1.0L. Mix thoroughly. Gently heat and bring to boiling. Distribute into tubes or flasks. Autoclave for 15 min at 15 psi pressure–121°C. Pour into sterile Petri dishes or leave in tubes.

Use: For the cultivation of *Zymomonas anaerobia*.

Zymomonas Medium (ATCC Medium 845) (ATCC Medium 948)

Composition per liter:

Glucose	20.0g
Yeast extract	5.0g

Preparation of Medium: Add components to distilled/deionized water and bring volume to 1.0L. Mix thoroughly. Autoclave for 15 min at 15 psi pressure–121°C. Pour into sterile Petri dishes or distribute into sterile tubes.

Use: For cultivation of *Zymomonas mobilis* and other *Zymomonas* species.

Zymomonas Sucrose Medium

Composition per liter:

Sucrose	150.0g
Yeast extract	2.0g
Peptone	2.0g
KH_2PO_4	2.0g
$(NH_4)_2SO_4$	2.0g
$MgSO_4 \cdot 7H_2O$	2.0g

pH 7.0 ± 0.2 at 25°C

Preparation of Medium: Add components to distilled/deionized water and bring volume to 1.0L. Mix thoroughly. Adjust pH to 7.0. Distribute into tubes or flasks. Autoclave for 15 min at 15 psi pressure–121°C.

Use: For cultivation of *Zymomonas mobilis*.